12th International Conference on Experimental Fluid Mechanics (EFM 2017)

EPJ Web of Conferences Volume 180 (2018)

Mikulov, Czech Republic
21 - 24 November 2017

Editor:

Petra Dancova

ISBN: 978-1-5108-6391-0

Printed from e-media with permission by:

Curran Associates, Inc.
57 Morehouse Lane
Red Hook, NY 12571

proceedings.com

Some format issues inherent in the e-media version may also appear in this print version.

Printed by Curran Associates, Inc. (2018)

For additional information, please contact EDP Sciences – Web of Conferences at the address below.

EDP Sciences – Web of Conferences
17, Avenue du Hoggar
Parc d'Activité de Courtabœuf
BP 112
F-91944 Les Ulis Cedex A
France

Phone: +33 (0) 1 69 18 75 75
Fax: +33 (0) 1 69 28 84 91

contact-edps@webofconferences.org

Additional copies of this publication are available from:

Curran Associates, Inc.
57 Morehouse Lane
Red Hook, NY 12571 USA
Phone: 845-758-0400
Fax: 845-758-2633
Email: curran@proceedings.com
Web: www.proceedings.com

TABLE OF CONTENTS

https://doi.org/10.1051/epjconf/201818000001

Experimental Fluid Mechanics 2017

Mikulov, Czech Republic, 21st–24th November 2017

efm.kez.tul.cz

12th International Conference „Experimental Fluid Mechanics 2017" was held at 21st–24th November 2017 at a Kastle in Mikulov, Czech Republic.

The EFM conference is focused on obtaining overviews of projects under investigation and experimental methods used in the field of fluid mechanic and thermodynamic. Its aim is to organize a network of research workers who are interested in experimental, and also theoretical, work in the field of fluid mechanic and thermodynamic. The conference should contribute the creation of closer contacts and sharing experience in application of various experimental methods, preparation and implementation of experiments, processing of results and numerical simulations of experiments.

The main topics of the conference are:

- Experimental methods used in fluid mechanics and thermodynamics.
- Preparation and implementation of experiments.
- Application of numerical simulation for preparation of experiments, comparison of the results obtained by numerical or analytical methods with experiments.
- Processing the results.
- Experiments in the field of compressible and uncompressible fluid flow, stability of flow, heat and mass transfer, cavitation etc.
- Experiments and measuring carried out on power and regulating equipment.
- Cryogenic fluid dynamics.

Members of Sciences Committee are:

- Tomáš Vít – Head of the committee (TU Liberec, Czech Republic)
- Sergey Alekseenko (Novosibirsk State University, Russia)
- Petra Dančová (TU Liberec, Czech Republic)
- Simone Ferrari (University of Cagliari, Italy)
- Matěj Forman (ESI – OpenCFD)
- Jan Hrubý (Academy of Science of Czech Republic)
- Hendric de Lange (Eindhoven Univ. of Technology, Netherlands)
- Václav Kopecký (TU Liberec, Czech Republic)
- Alexander Kusyumov (A. Tupolev KNRTU-KAI, Russia)
- Robert Pastuszko (Kielce University of Technology, Poland)
- Kazimierz Peszynski (UTP University of Science and Technology, Poland)
- Ludvík Prášil (TU Liberec, Czech Republic)
- Jaromír Příhoda (Academy of Science of Czech Republic)
- Rudi Santbergen (Delft University of Technology, Netherlands)
- Ladislav Skrbek (Charles University, Czech Republic)
- Zdeněk Trávníček (Academy of Science of Czech Republic)
- Václav Uruba (Academy of Science of Czech Republic)
- Magda Vestfálová (TU Liberec, Czech Republic)
- An-Bang Wang (National Taiwan University, Taiwan)
- Yoshifumi Yokoi (National Defense Academy of Japan, Japan)

Steering Committee:

- Petra Dančová – Head of the committee (TU Liberec, Czech Republic)
- Jan Novosád (TU Liberec, Czech Republic)

Experimental Fluid Mechanics 2017

Mikulov, Czech Republic, 21st–24th November 2017

efm.kez.tul.cz

Keynote lectures for 2017:

- Assoc. Prof. Michel Speetjens (Eindhoven University of Technology, Netherlands)
  Experiments on mixing and heat transfer in industrial fluids processing
- Assoc. Prof. Marcin Zych (AGH University Science and Technology, Poland)
  Prospects for the application of radiometric methods in the measurement of two-phase flows

List of Sponsors (in alphabetical order):

- DANTEC DYNAMICS GmbH (www.dantecdynamics.com)
- LAVISION (http://www.lavision.de/en/)
- MIT s.r.o. (www.mit-laser.cz)
- TSI GmbH (www.tsi.com)

Conference venue:

Every year, the organizers want to prepare the conference at nice and interesting place of the Czech Republic. In 2017, the conference venue was Mikulov.

The town of Mikulov offers perception with literally all senses. For many the reasons for a visit include history embodied in architectural monuments, and for nature lovers and people preferring active rest it is the unique natural conditions of the region. Those who appreciate good wine Mikulov offers rich viticulture tradition.

The Mikulov castle is in the process of admissions procedure for UNESCO world heritage list.

The organizers would like to invite you for the next year of the EFM conference, which will be held in Prague, November 13th – 16th, 2018 (see our web pages efm.kez.tul.cz).

Mikulov

Beginning of the EFM 2017

Conference venue

We prepared a cipher game

sciences

Web of Conferences

Statement of Peer review

In submitting conference proceedings to *Web of Conferences*, the editors of the proceedings certify to the Publisher that

1. They adhere to its **Policy on Publishing Integrity** in order to safeguard good scientific practice in publishing.
2. All articles have been subjected to peer review administered by the proceedings editors.
3. Reviews have been conducted by expert referees, who have been requested to provide unbiased and constructive comments aimed, whenever possible, at improving the work.
4. Proceedings editors have taken all reasonable steps to ensure the quality of the materials they publish and their decision to accept or reject a paper for publication has been based only on the merits of the work and the relevance to the journal.

Title, date and place of the conference

EXPERIMENTAL FLUID MECHANICS
21-24/NOVEMBER/2017
MIKULOV, CZECH REPUBLIC

Proceedings editor(s):

PETRA DANCOVA

Date and editor's signature 14/5/2018

Prospects for the application of radiometric methods in the measurement of two-phase flows

Marcin Zych

AGH University Science and Technology, Faculty of Geology, Geophysics and Environmental Protection, 30-059 Kraków, Poland

Abstract. The article constitutes an overview of the application of radiometric methods in the research of two-phase flows: liquid-solid particles and liquid-gas flows. The methods which were used were described on the basis of the experiments which were conducted in the Water Laboratory of the Wrocław University of Environmental and Life Sciences and in the Sedimentological Laboratory of the Faculty of Geology, Geophysics and Environmental Protection, AGH-UST in Kraków. The advanced mathematical methods for the analysis of signals from scintillation probes that were applied enable the acquisition of a number of parameters associated with the flowing two-phase mixture, such as: average velocities of the particular phases, concentration of the solid phase, and void fraction for a liquid-gas mixture. Despite the fact that the application of radioactive sources requires considerable carefulness and a number of state permits, in many cases these sources become useful in the experiments which are presented.

1 Introduction

Radiometric methods were applied as early as in the 1930s in the research of the kinetics of chemical reactions by György von Hevesy. A considerable contribution to the description of the application of radiometric markers in industrial processes was provided by P. V. Danckwerst in the 1950s [1].

Further development of radiometric methods is also associated with the application of the measurements of the absorption of gamma and X-rays in the flowing mixtures [2-6]. The application of mathematical analyses of signals derived from nuclear probes enables the determination of many parameters of multi-phase mixtures transported in open channels [7] and pipelines, the research and control of hydrometallurgical processes and the extraction of crude oil. What is also important is that the algorithms of the analysis of signals which are used may also be applied in the analysis of data, e.g. in optical and electrical methods [8-14].

By applying radiometric methods, or, to be more precise, synchrotronic radiation, one may also analyse flows through nano- and microstructures in rocks [15, 16].

The article will present the measurement techniques featuring the application of radiometric tracers and the absorption of gamma rays in the measurements of two-phase flows on the basis of research work in which the author participated.

2 The radiotracer method

Measurements which employed radiometric tracers were conducted during the research work in the experimental research facility located in the Water Laboratory of the Wrocław University of Environmental and Life Sciences, Poland [17-19]. The model of the research installation was presented in Fig. 1. Due to the fact that measurements which employed closed sources of gamma radiation were conducted at the same time, also absorption sets were indicated as well. The method of the absorption of gamma radiation is described in Chapter 3.

The radiotracer method is widely used in various types of measurements, although it requires specific precautionary measures due to the application of open radioactive isotopes, e.g. in the examining of the tightness of tanks, diffusion processes and multi-phase flow [18-22]. In measurements of this type a chemical substance containing radioactive isotopes is introduced into the selected flowing phase and then its movement or accumulation is traced with the use of suitable detectors.

The principle of the radiotracer method used to monitor the flow of marked grains in a vertical pipeline is presented in Fig. 2. Isotope marked grain (3) is added to the two-phase mixture flowing in the pipeline (1) and then while it flows through cross-sections "*v*" and "*z*", it generates signals $I_v(t)$, $I_z(t)$ at the outlets of scintillation detectors (4). The scintillation probes (with 2'' NaI(Tl) crystals), distributed with the spatial distance L_2, were placed in lead collimators with rectangular 4 mm × 40 mm slots. The applied tracer was an isotope of technetium, Tc-99m, with gamma radiation energy of 141 keV. Due to its short half-life of 6 h, technetium was obtained from a Mo-Tc medical generator (courtesy of POLATOM), and its absolute activity was 30 GBq. The chemical reaction of precipitation [19]:

Corresponding author: zych@geol.agh.edu.pl

$$NaTcO_4 + 3FeCl_2 + 5NaOH + 2H_2O = TcO_2\downarrow + 3Fe(OH)_3 + 6NaCl, \quad (1)$$

produced technetium oxide TcO_2 which subsequently was inserted into bores drilled in grains of selected dimensions. This method of marking was performed in grains of three groups constituting ceramic models of nodules of various sizes: large, with maximum dimensions of 50-70 mm, medium (30-50 mm) and small (15-30 mm). The activity of the isotope inserted into the grains was selected in such a way as to not interfere with the absorption measurement. The marked grains are shown in Fig. 3.

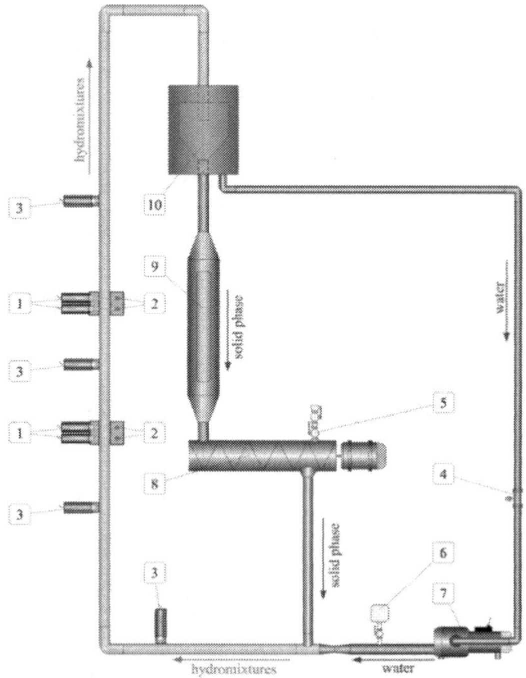

Fig. 1. Diagram of experimental set-up: 1 – absorption probes, 2 – sealed sources, 3 – probes recording radiotracers, 4 – electromagnetic flow meter, 5 – marked particles feeder, 6 – feeder of marked water, 7 – pump, 8 – conveyor, 9 – container for solid particles, 10 – separating container [18].

The radiotracer method also enables measurement of the instantaneous velocity of water. In such a case, in accordance with Fig. 1 a $NaTcO_4$ solution is injected into the installation with injector 6 by means of compressed air. The solution undergoes gradual mixing (Fig. 2) as it moves along the pipeline, which causes the emergence of a peculiar signal, indicated in Fig. 4 with the letter W.

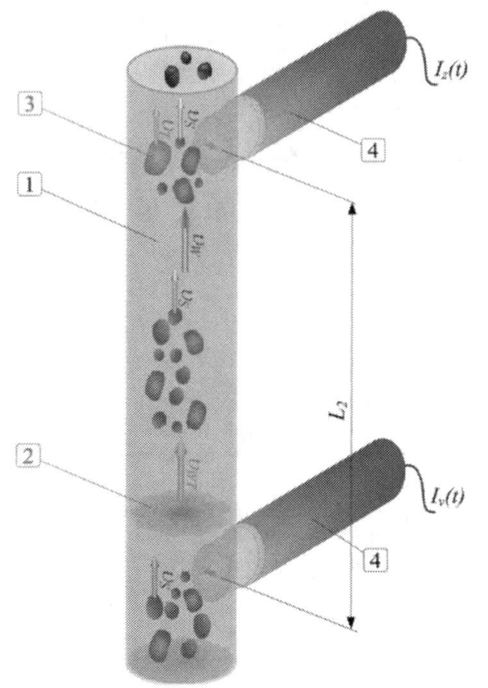

Fig. 2. The radiotracer measuring equipment: 1 – pipeline with two-phase flow, 2 – tracing water, 3 – marked particle, 4 – scintillation probe for radiotracer measurements, υ_S, υ_W – velocities of solid particles and water, respectively, υ_T – velocity of marked particles, υ_{WT} – velocity of traced water.

Fig. 3. The image of the grains (the right-hand side: the particle with a bore for the isotope Tc-99m) [19].

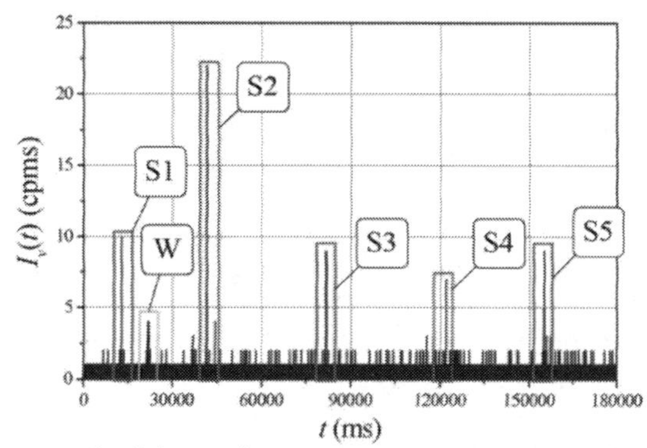

Fig. 4. Signal from radiotracer: S1, S2, S3, S4, S5 – medium grain run, W – tracer water run.

3 Absorption method

Measurements of the absorption of gamma radiation associated with an analysis of two-phase liquid-gas flow were conducted at the research installation built in AGH – University of Science and Technology in Kraków. A simplified model of the stand is shown in Figure 5 [23, 24].

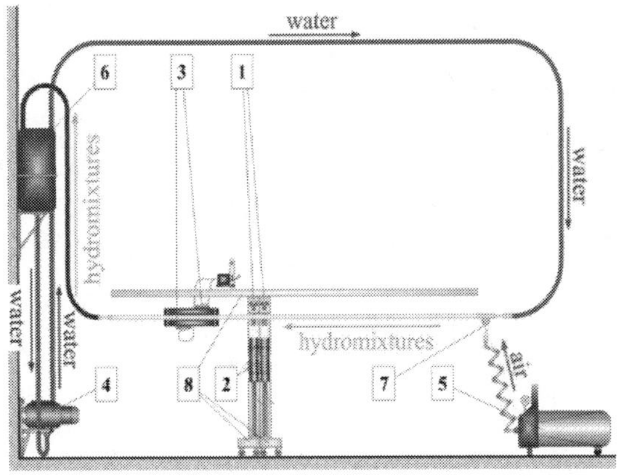

Fig. 5. A model of the experimental hydraulic installation: 1 – γ-ray sealed source, 2 – scintillation probe, 3 – ultrasound flow meter, 4 – pump, 5 – air compressor, 6 – air-removing vessel, 7 – air nozzle, 8 – shifting system [23].

The installation consists of a transparent acrylic glass tube with an internal diameter of 30 mm and a length of 4.5 m, to which water is pumped by a rotary pump (4) and air from the compressor (5) by an injector nozzle (7). Due to that, the controlled mixture of water and air supplies the measuring pipe and flows up to the air-removing tank (6). The measurement of the velocity of air bubbles and the recognition of the flow regime was arranged by the measuring system consisting of two linear γ-radioactive sources, ^{241}Am (1) with an activity of 100 mCi and two probes with a NaI(Tl) (2) scintillation detector. The measuring set is mounted on a special trolley (8), which enables the set to move along the pipeline. The water flow rate was continuously measured by a Uniflow 990 ultrasonic flowmeter (3). The rotary pump (4) controlled by a waver enables the adjustment of the revolution in the range between 1000 and 2800 rpm, which in turn enables the selection of water velocity in the measuring pipe between 0.5-2.5 m/s [23].

The data acquisition system was connected to a PC by a dedicated counter card and a USB port.

Moreover, the transparent measuring section of the pipeline enabled photographic documentation of the observed phenomena.

The idea and geometry of the measurement are shown in Figure 6 [24]. Two Am-241 sealed radioactive sources (2) emit γ radiation beams (6, 7), shaped by the collimator (1). Photons pass through the pipeline with the analyzed mixture (5) and collimator (3) before they reach a detector (4).

The research procedure utilized linear radiation sources – Am-241, QSA Global – whose structure and dimensions are shown in Figure 7.

The data acquisition equipment enables the recording of $I_x(t)$ and $I_y(t)$ intensity representing the composition of the flowing mixture. Results recorded in the BUB006 experiment are shown in Figure 8. In contradistinction to the radiotracer method, the signal in absorption measurements is a record of the changes in the intensity of radiation which passes through the pipeline, i.e. the changes in the density of the moving two-phase mixture.

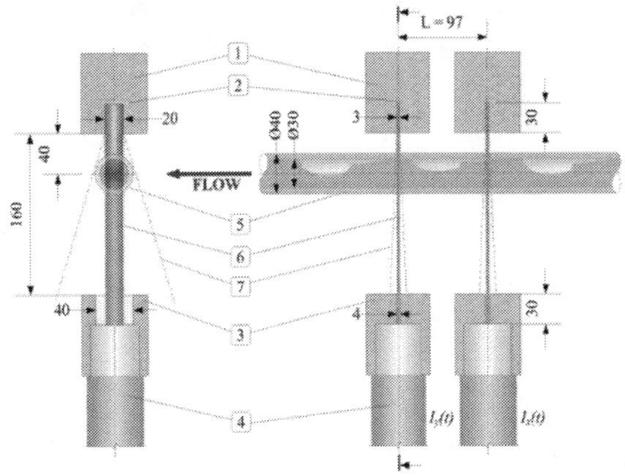

Fig. 6. The principal geometry of the γ-ray absorption measurement: 1 – collimator, 2 – Am-241 source, 3 – collimator of the detector 4 – scintillation probe, 5 – pipeline, 6 – main γ-ray beam, 7 – external γradiation. All dimensions are given in mm [24].

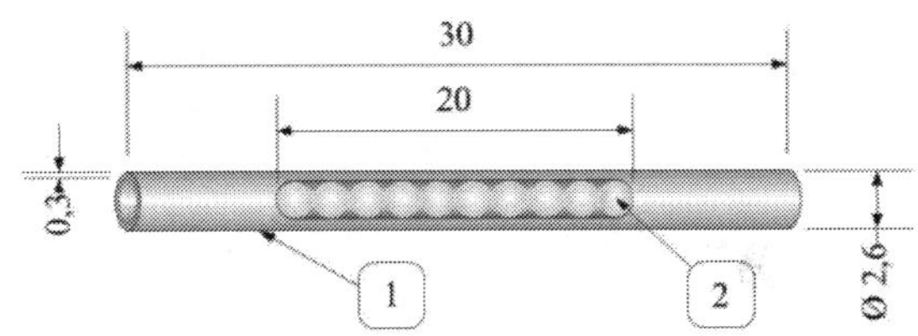

Fig. 7. A model of γ source Am-241, QSA Global: 1 – stainless steel, 2 – radioactive material. All dimensions are given in mm [23].

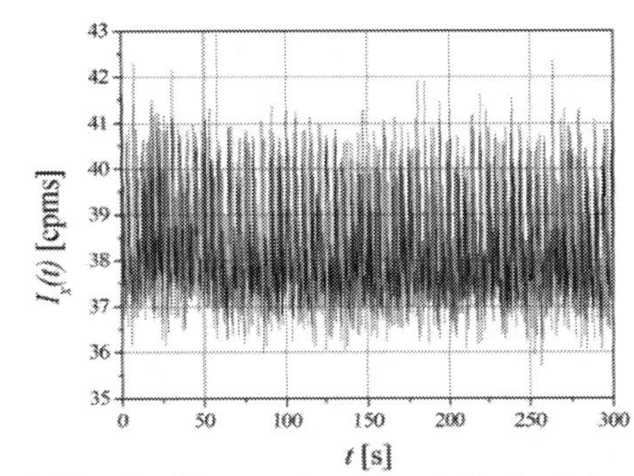

Fig. 8. Intensity of the γ-ray beam recorded by the first probe in the BUB006 run [23].

4 Methods of signal analysis

4.1 Time delay estimation

When we compare the signals indicated in Figure 4 and 8, we may discern a fundamental difference between radiotracer measurements and absorption measurements. In spite of this, in both cases one may apply similar procedures in order to calculate the transport time delay, and then the velocities of the particular phases [17, 18]. A universal method has to do with the function of mutual correlation, which may be efficiently used even in reference to signals which feature a weak signal-to-noise ratio [17-21, 23-27]:

$$R_{xy}(\tau) = \lim_{T\to\infty}\frac{1}{T}\int_0^T x(t)\cdot y(t+\tau)dt\,, \qquad (2)$$

where T is the averaging time, τ – time delay.
It is assumed that the most probable transportation time delay of the solid phase is expressed by the argument of the main CCF maximum:

$$\hat{\tau}_0 = \arg\{\max R_{ij}(\tau)\} = \arg\{R_{ij}(\tau_0)\}\ . \qquad (3)$$

Hence the average velocity of the movement of the particular phases may be calculated on the basis of the following relationship:

$$\upsilon = \frac{L}{\hat{\tau}_0}\,, \qquad (4)$$

where: L – distance between probes.

Articles [17, 25, 28, 29] present other methods of the establishment of the transportation time delay, including ones based on the modifications of the mutual correlation functions, among which one may classify the following:

- Deconvolution,
- Phase of the cross-spectral density function,
- Hilbert Transformation,
- Average magnitude difference function,
- Average square difference function.

If the uncertainty of transport delay τ_0 is known, the uncertainty which results from the precision of the measurement and calculation methodology which was applied, one may estimate the uncertainty of velocity on the basis of the law of the propagation of uncertainty:

$$u_C(\upsilon) = \sqrt{\left[\frac{\partial\upsilon}{\partial L}\cdot u_B(L)\right]^2 + \left[\frac{\partial\upsilon}{\partial\tau_0}\cdot u_A(\tau_0)\right]^2}\ . \qquad (5)$$

In the majority of the cases which were analyzed, the precision of the establishment of the velocity by means of the aforementioned methods does not exceed 1% of the value.

4.2 Concentration and void fraction

There is a much greater number of the possibilities of the application of the absorption of gamma radiation in the flowing mixture than merely the establishment of the average velocity of the minority phase. Provided that the measurement system is appropriately calibrated beforehand, one may establish the following by means of the Lambert-Beer law [30]:

- for the liquid-solid particles mixture – the cross-section volume concentration C_{VP} [31],
- for the liquid-gas mixture – the void fraction of gas α [32].

The graphs in Fig. 9 and 10 present calibration curves for the establishment of the concentration of ceramic grains for the flow studied in the Water Laboratory of the Wrocław University of Environmental and Life Sciences and the void fraction for the air in the water-air flow simulated in the research facility in AGH – University of Science and Technology, respectively. In the case of the establishment of C_{VP} or α, precision ranges from 10% to 15% of the value [31, 32]. Such a great uncertainty of the measurement is associated with the precision of the performance of the calibration and the establishment of the magnitude of the proper background of the detectors. In the case when radiometric markers are utilized in the measurements it is also important to take into consideration the interference of the absorption measurement.

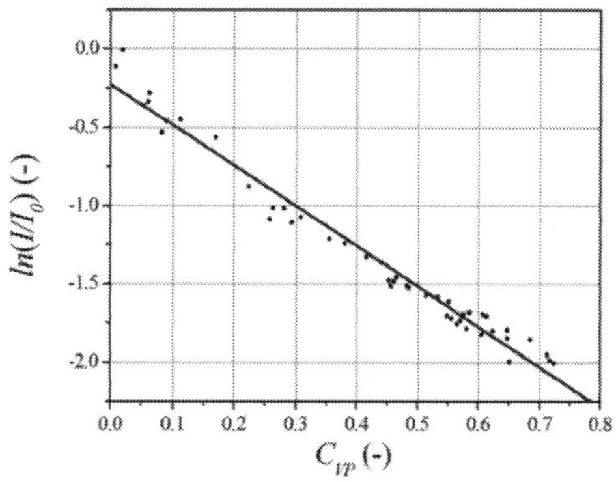

Fig. 9. The relation between I/I_0 and the volumetric concentration C_{VP} [31].

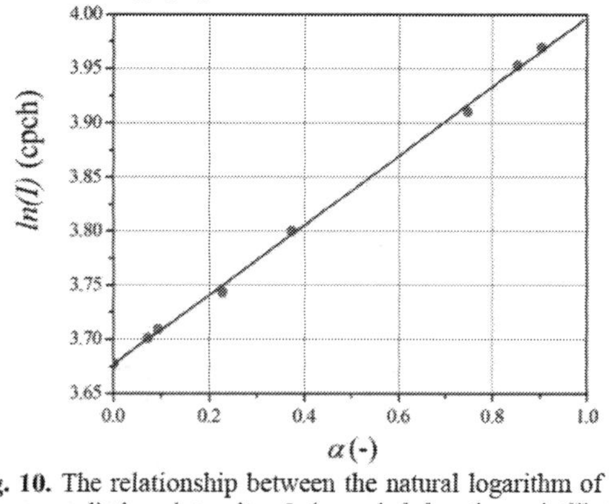

Fig. 10. The relationship between the natural logarithm of the gamma radiation intensity I (recorded by the scintillation probe) and the fixed void fraction α [32].

The enhancement of the precision of the results which are obtained is possible due to the application of additional calculations conducted in the MCNP (Monte

Carlo N-Particle) environment [33, 34]. This procedure enables the obtaining of more precise calibration curves and – in many cases – correction of the measurement geometry.

4.3 Other possibilities of gamma absorption methods

An in-depth statistical analysis of the signals, including the application of advanced mathematical methods, facilitates the obtaining of more information about the flow.

For dual-detector systems, in the case of intermittent liquid-gas flows, when one uses the Fourier transform or chirp analysis, one may distinguish peculiar frequency groups associated with the dimensions of the structures which emerge during the flow of a mixture [35].

The distinguishing of statistical parameters of the signal, such as: the average value, standard deviation, kurtosis, etc. enables the establishment of the type of flow. In order to perform automatic pattern flow recognition one may utilize artificial neuron networks or other methods of the so-called calculation intelligence. The particular applications were described in publications [36-39].

5 Conclusions

Radiometric methods applied to the research of two-phase flows offer a great deal of possibilities in the field of the determination of many parameters of flows. The basic quantities which may be determined include the following: transport delay time, velocity of the average minority phase or of the particular phases, concentration of solid particles or the void fraction of gas.

The application of advanced calculation methods enables the enhancement of the precision of the parameters which are determined and the information which is obtained. Moreover, one may also determine the dimensions (the length) of the structures which arise in intermittent flows and establish the types of flows.

However, such measurements require particular carefulness and appropriately developed technology in order to provide radiological security to the personnel. This is also associated with the necessity of obtaining the relevant permits.

Acknowledgment

The author would like to thank dr Leszek Petryka, dr Robert Hanus, dr Marek Jaszczur, dr Włodzimierz Mosorov, prof. Pavel Vlasak, dr Dariusz Świsulski and prof. Jerzy Sobota for their cooperation during the measurements.

This publication is funded by AGH University of Science and Technology (No 11.11.140.645).

References

1. P.V. Danckwerst, *Insights into chemical engineering* (Pergamon Press, Oxford, 1981)
2. G.A. Johansen, P. Jackson, *Radioisotope gauges for industrial process measurements* (Wiley, New York, 2004)
3. P. Vlasak et al., J. Hydrol. Hydrom. **62**, 241–247 (2014)
4. P.R. Tortora et al., Int. J. Multiphas. Flow **32**, 972–995 (2006)
5. H.J. Pant, et al., Appl. Radiat. Isotop. **67**, 1609–1615 (2009)
6. G.H. Roshani et al., Flow Meas. Instrum. **54**, 39–45 (2017)
7. M. Jaszczur, L. Portela, Quality and Reliability of Large-Eddy Simulations, Ercoftac Series, **12**, 343-354, (Springer, Dordrecht, 2008)
8. V. Mosorov, Meas. Sci. Technol. **17,** 753-760 (2006)
9. M. Bottin, et al, Int. J. Multiphase Flow **60**, 161-179 (2014)
10. T. Xue, et al., Flow Meas. Instrum. **27**, 29–36 (2012)
11. P. Psota, et al., EPJ WoC **143**, 02097 (2017)
12. E. J. Mohamad, et al., Flow Meas. Instrum. 47 62–70, (2016)
13. X. Zeng, et al., Sens. **16**, 587 (2016)
14. M. Arkani, et al., Metrol. Meas. Syst. **24**, 441-461 (2017)
15. P. Krakowska, et al., Nat. Gas Sci. Engin. **31**, 67-75 (2016)
16. J. Jarzyna, et al., Comput. Geosci. **20**, 1155–1167 (2016)
17. R. Hanus, et al., Flow Meas. Instrum. **40**, 58–63 (2014)
18. M. Zych, et al., EPJ WoC **114**, 02145 (2016)
19. M. Zych, et al., Powder Technol. **318**, 491–500 (2017)
20. L. Petryka, et al., Nukleonika **50**, 43–46 (2005)
21. H.J. Pant, et al., Appl. Radiat. Isot. **13**, 54–60 (2015)
22. N. Singh (Ed.), *Radioisotopes – Applications in physical sciences* (InTech, Croatia, 2011)
23. M. Zych, et al., MATEC WoC **18**, 02004 (2014)
24. M. Zych, et al., EPJ WoC **92**, 02122 (2015)
25. R. Hanus, Int. J. Multiphas. Flow **72**, 210-217 (2015)
26. M.S. Beck, A. Plaskowski, *Cross correlation flowmeters - their design and application.* (Adam Hilger, Bristol, 1987)
27. J.S. Bendat, A.G. Piersol, *Random data - analysis and measurement procedures* (Wiley, New York, 2010)
28. M. Zych, et al., EPJ WoC **92**, 02122 (2016)
29. R. Hanus, et al., Math. Probl. Eng. 2014, 475735 (2014)
30. V. Mosorov, Appl. Radiat. Isotop. **128**, 1-5 (2017)
31. M. Zych, et al., EPJ WoC **114**, 02146 (2016)
32. M. Zych, et al., JPCS **745**, 032124 (2016)
33. V. Mosorov, et al., Appl. Radiat. Isotop. **112**, 136-140 (2016)
34. E. Nazemi, et al., Nucl. Eng. Technol. **51**, 64-71 (2016)
35. M. Zych, et al., *Proc. EFM16: Experimental Fluid Mechanics 2016*, 996-999 (2016)
36. G.H. Roshani, E. Nazemi, Flow Meas. Instrum. **58**, 6–11 (2017)
37. R. Hanus, et al., EPJ WoC **143**, 02033 (2017)
38. R. Hanus, et al., EPJ WoC **114**, 02035 (2016)
39. R. Hanus, et al., EPJ WoC **114**, 02036 (2016)

Vortex valve

Karel Adámek[1,*], *Jan* Kolář[1], and *Pavel* Peukert[1]

[1]VÚTS Liberec, a.s., Department of Flow Simulations, Svárovská 619, 46001 Liberec, Czech Republic

Abstract. The paper deals with a flow field inside the so-called vortex valve, used as an outlet device on retention reservoirs for retention of rainstorms and later slow outflow into sewerage etc. The system is very simple, without moving parts. Using the method of numerical flow simulation, the unusual flow characteristic $\Delta p = f(Q)$, containing two branches, is explained. Further, there it is studied influence of both inlet/outlet opening sizes on the form of the characteristic. Results can be used for designing the new model series of valves for various flows.

Keywords: Flow numerical simulation, Vortex valve, Flow characteristic.

1 Introduction

The aim of the presented study is the statement of the resistance–flow characteristic $\Delta p[Pa] = f(Q[m^3/s])$ for the so-called vortex valve, mounted usually on the outlet from the retention reservoir. Such a reservoir is used for instance as the retention of the rainstorm or similarly and for the consecutive slow outflow into drainage etc. The influence of main design parameters, first of all inlet/outlet diameters, is discussed, too, as guidelines for practical commercial design.

From a simple search, some essential information, only, is going out. The principle is described in [1], but the presented geometry is very complicated. Next studies [2], [3] describe theory and contain next 50 references, approx. No more references are added, because the presented results were made by own simulations and measuring and using standard knowledge of fluid mechanics, well known for technical community. It is a pity that this knowledge is not used in daily practice.

Some actual designs are presented in many internet references, but as commercial messages, only and the principle is described curiously by Mr. Google and many dealers and producers, for instance as "the effect is given by flow phenomenon" or "it works on the principle of hydraulic phenomenon". Reality is not so mysterious, the flow inside of such a valve is possible to describe by numerical flow simulation, completed by some essential formulas of fluid mechanics.

2 Initial model

2.1. Geometry

The model contains six liquid bodies, see Fig. 2-1. The surrounding retention reservoir is not modelled here – it is assumed that the velocity range in such surrounding can be neglected in comparison with the flow inside the studied valve.

1. Inlet – cylindrical tangential, situated from below.
2. Chamber – cylindrical, with tangential inlet and axial outlet.
3. Outlet – axial hollow on the forefront.
4. Outlet flange – outlet prolongation.
5. Vent opening – small orifice in the highest point of the chamber.

6. Outer surroundings (air) – is minimalized here, a smaller volume of the model means shorter time of the solution. In this model, the flow resistance of the valve is solved, only, the shape of the outer flow (or its spraying quality) is not important here.

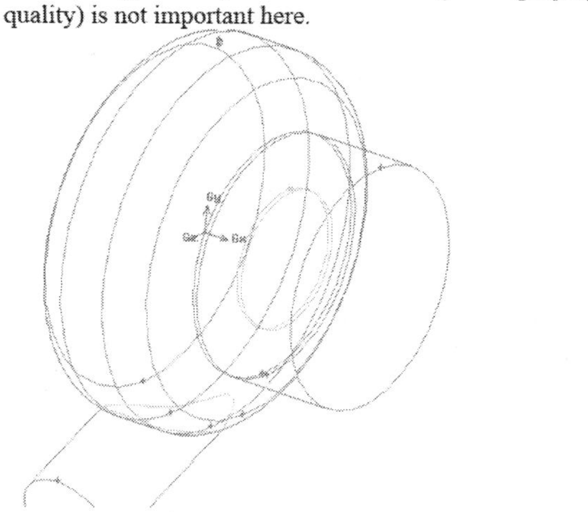

Fig. 2-1. Model composed from six liquid bodies.

2.2 Mesh

Regarding the complicated two-phase flow it is necessary to use a fine calculating mesh after Fig. 2-2, first of all, in the vicinity of the outlet, where the flow is disintegrating. The initial mesh can be modified yet during calculation.

* Corresponding author: k.adamek@volny.cz

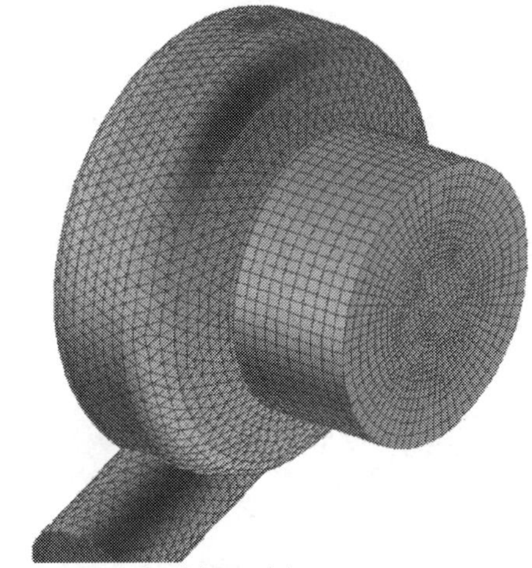

Fig. 2-2. Mesh of the model.

2.3 Boundary and initial conditions

In the inlet cross-section, there it is simply defined constant (hydrostatic) pressure, given by the height of the surrounding water level. Zero pressure is defined here in the middle of this inlet cross-section. For the used dimensions of the valve, the water level on the outlet axis reaches of 0.2 m approx. (pressure 2 kPa) and at the upper contour of the chamber 0.36 m of water approx. (pressure 3.6 kPa).

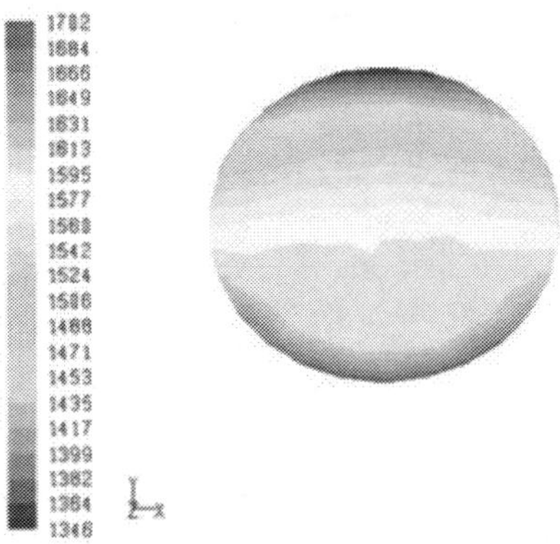

Fig. 2-3. Uneven inlet pressure field (water level of 0.16 m).

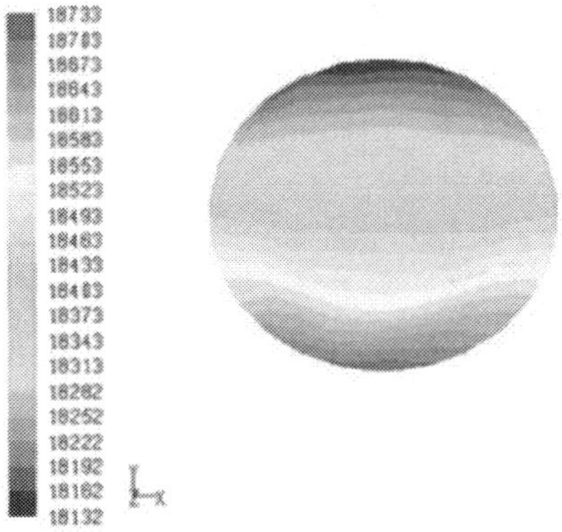

Fig. 2-4. Uneven inlet pressure field (water level of 2.0 m).

Really, the pressure distribution is not absolutely constant here, this influence is neglected hereafter. Fig. 2-3 and Fig. 2-4 present real pressure fields in the inlet cross-section for min. and max. water levels of 0.16 m and 2.0 m – some part of the defined static pressure is consumed yet for kinetic energy of the flow.

In the surroundings, there is zero pressure (atmosphere).

Operational condition – gravity acceleration of 9.81 m/s^2 in the down direction - axis (-y).

Initial conditions: the mixture pressure in the inlet max. 20 kPa and the value of the phase f2 is equal to zero (water only), in the outlet, the mixture pressure 0 Pa, with defined backflow of the phase 2 equal to 1 (air only).

2.4 Solver

Isothermal model (if speculating about friction losses, the water warming would be of some 0.01 K, only). Turbulence model k-ω SST.

Two models of two-phase (water-air) flow were tested: the unsteady model „VOF“ (volume of fluid) with a very short time step and the steady model „mixture“.

2.4.1 Model VOF (volume of fluid)

As an illustration, only, the testing images of both phases after several first time steps of the solution are presented in Fig. 2-5 (low pressure difference) and in Fig. 2-6 (high pressure difference). The solution with a time step of 10 ms is diverging so that the time step shortening to 0.1 ms approx. should be necessary.

The field of both phases contains more details, captured in the actual time step of the unsteady solution. Using small pressure (Fig. 2-5), the water (blue, phase f2 = 0) is flowing mainly down, owing to gravitation – inside, there is not any important rotating flow yet. In the surrounding, there is air (red, phase f2 = 1). Using a higher pressure (Fig. 2-6), the axial flow becomes more dominant.

The large surroundings volume (right) is deleted later, due to a shortening of the solution time.

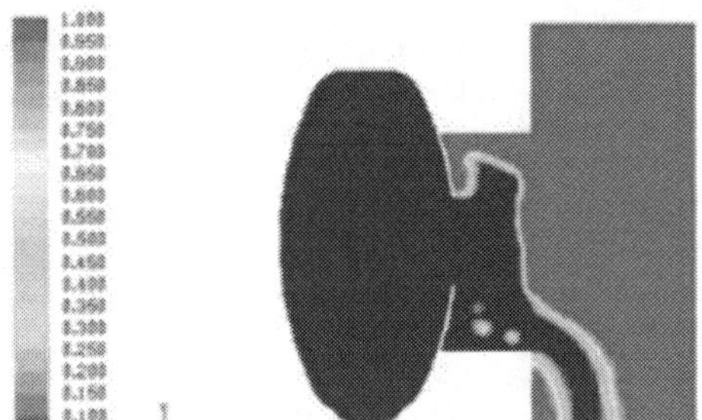

Fig. 2-5. Phase interface, low pressure difference, 0.7 s after start (model COUPLED).

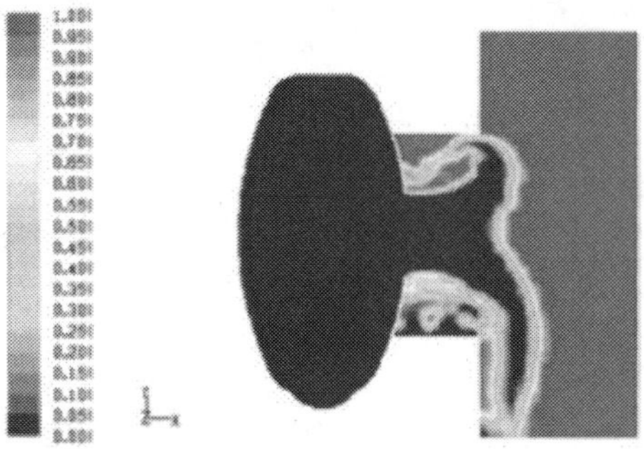

Fig. 2-6. Phase interface, high pressure difference, 0.35 s after start (model SIMPLE).

2.4.1 Model "mixture"

The results of the first simulation are not presented here; they were used for verification of the proper progress of the solution and are used in the models below. From the results, it follows that the interface water-air is changing in the time so that the presented result is any fictive steady flow field, captured in any actual time of the solution. The relatively large boundary water-air is given by this unsteady time-dependent water level (surface).

2.5 Conclusions for the next models

Pressure difference (instead of water level height) is defined in the middle of the inlet cross-section. Difference between lowest and highest point of this cross-section can be neglected toward other dimensions.

The steady two-phase model „mixture", is used in preference. Above all, because the calculation of each point of the operational characteristics is converging faster and the balance error (difference between inflow and outflow) is small enough. The flow value of the converged solution is oscillating a little; therefore, the recorded values can be time-different in the range of this oscillation.

Two-phases model VOF – if the method is used from the initial state, it is necessary a very short time step of the solution, 0.1 ms and less, and the time of the solution is very long. So it is better to use this method as a finishing of the previous steady model „mixture". Then, the time step of 0.01 s can be used and the solution finishing is real. The values of the flow, received with the time step of 10 ms, are of 6-7% smaller in comparison with the method „mixture" and with the time step of 0.1 ms of 3% smaller. Probably, it is some numerical mistake between both used models – without next investigation.

Vent hole – the solution is unstable when water reaches the level of this hole. Really, it makes no problem, but during the simulation, only. Maybe, using a larger diameter of the orifice, the simulation could be more stable.

The primary aim is the creation of the resistance characteristics of the valve $\Delta p = f(Q)$, the shape of the water splashing in the surroundings is not important. A prospective error in consequence of an unsuitable two-phase model will be the same for all solved cases; therefore, the mutual comparison of all cases remains in the same rate.

As a piece of information, only, the converged solution by the method „mixture" needs many time steps, using the method VOF. An ideal procedure is the beginning with „mixture" and to continue with VOF. Then, for higher water levels, the time step of 10 ms is sufficient, but when the chamber is not full, it is necessary to shorten the time step at least to 0.1 ms. Thereby, the solution time is considerably longer.

The definition of the 2-branches characteristic of the observed valve needs for each geometry to simulate as minimum 4+4 cases for various pressures.

Flow resistance of outlet flange: The check simulation of one case confirms that for the model without flange, the flow is of 3% higher approx. It means that the main resistance of the valve is creating inside the valve, the flange influence is not important.

For the shortening of the solution time for the next models, the outer area behind the flange (air area) is removed and bigger mesh elements are used. The aim of the solution is to observe water flow inside the valve, not outside.

3 Basic model

Following the results of the initial model above, the steady model of two-phases flow „mixture" is used, the steady shape of the water level is displayed.
The thick interface water/air represents the real level waviness; therefore, the time-averaged positions of the water level are displayed. The tangential bottom inlet creates rotational flow, but at a partially filled chamber, the water level is very corrugated. This action is unsteady and the used steady model „mixture" records any steady state in any actual time.

3.1 Flow fields for various water levels

For various water levels, there are presented the interfaces water/air (blue/red). For the low water levels, the water is flowing from the bottom tangential inlet (crosswise to the paper plane) directly into the axial outlet – a typical vessel outflow, for higher water levels, the rotating flow in the chamber is creating. All Figures are presented as lengthwise vertical cross-sections through the outlet axis.

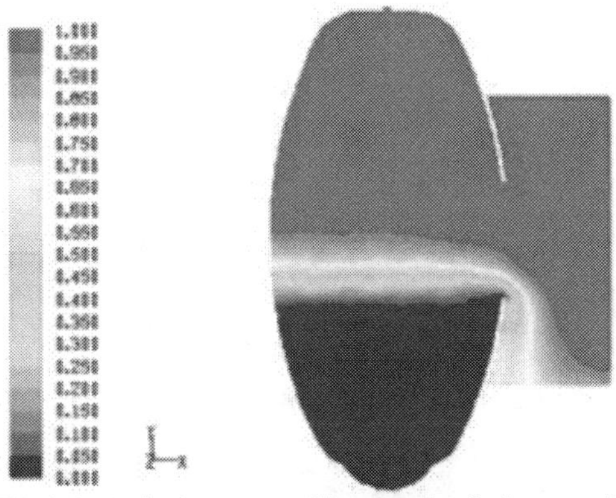

Fig. 3-1. At the lowest possible water level, the water hardly flows over the bottom edge of the outlet directly down on the bottom surface of the flange; the surrounding air is coming in from the right surroundings.

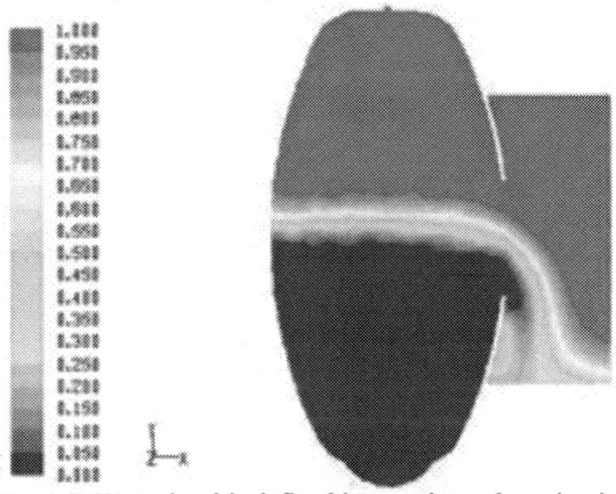

Fig. 3-2. Water level is defined just on the outlet axis, situation is similar, and the outflow is more intensive. The level (theoretically horizontal) is inclined to the outlet.

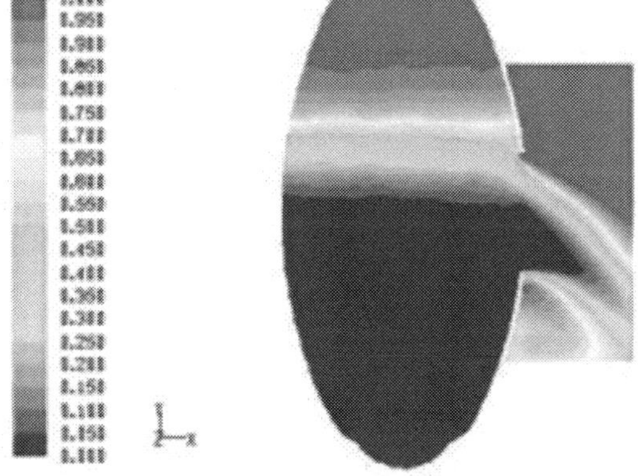

Fig. 3-3. Water level defined at the upper edge of the outlet, the flow is separated from the chamber wall, it is visible the beginning rotating flow, the air flow in the vent hole practically does not flow out.

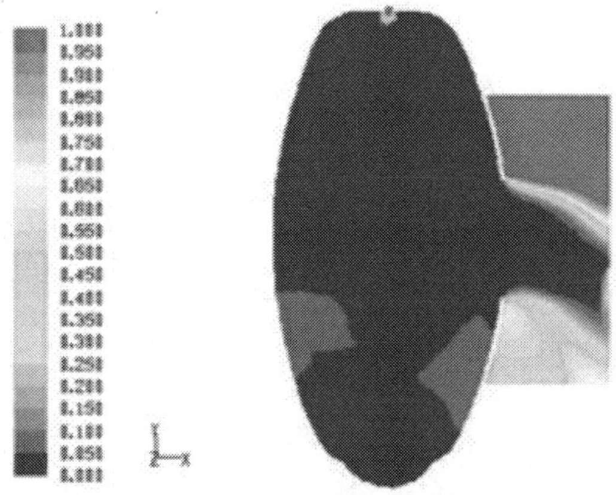

Fig. 3-4. Water level just under the upper edge of the chamber, the rotating flow inside is transformed into axial outer flow; the gravity influence is visible yet.

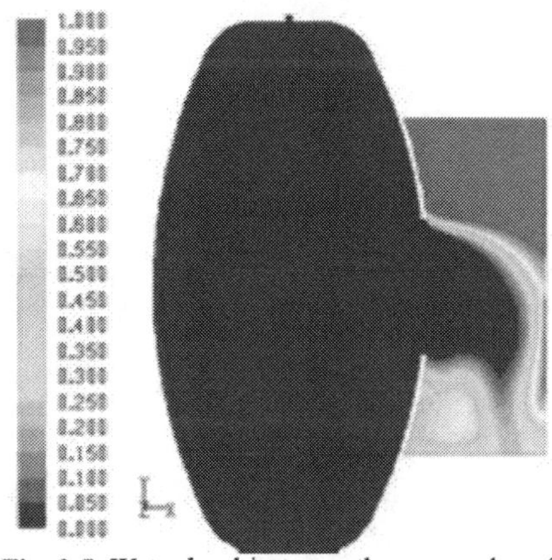

Fig. 3-5. Water level just over the upper edge of the chamber, the expressive rotating flow inside; outside it begins the transformation of gravitational flow into water splashing (spraying).

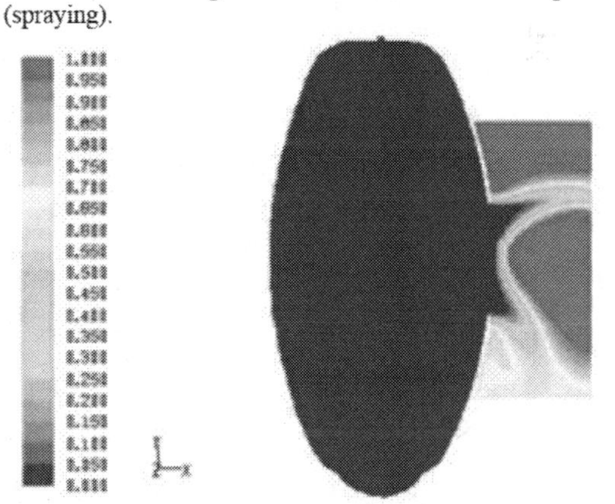

Fig. 3-6. The expressive inner rotating flow causes the spraying of the axial water flow into a conical shape, in the middle filled by back flowed air, deformation due to the gravity. For the chamber of a larger outlet diameter (see below), the air is coming up to the chamber interior.

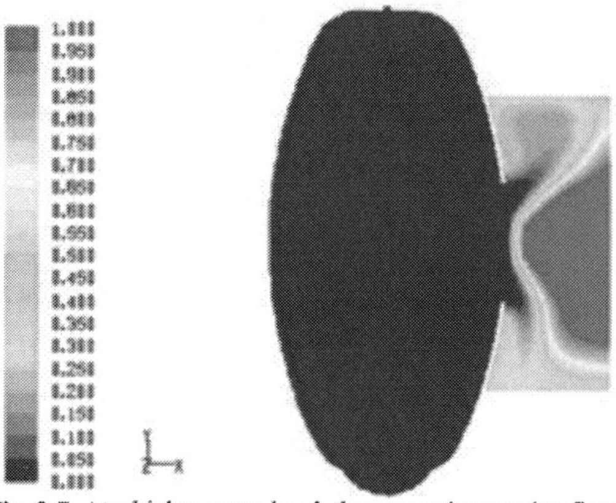

Fig. 3-7. At a higher water level, the expressive rotating flow is persisting, the conical water spray partially hits on the inner wall of the flange and is coming back. The gravity influence is less, but still evident it is the some flow along the bottom wall of the flange.

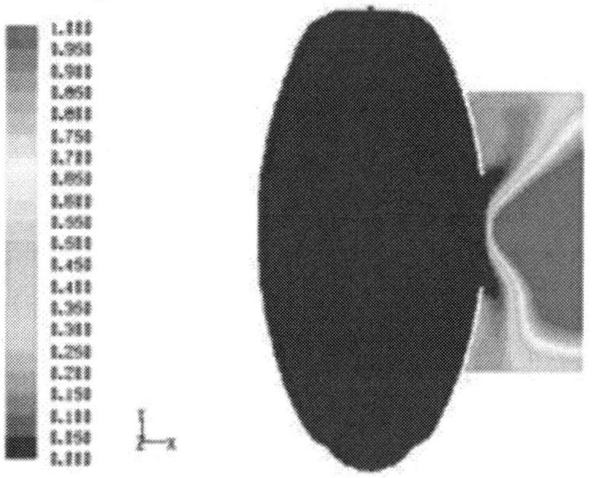

Fig. 3-8. Continuation of the previous for the max. defined water level (pressure). The higher part of the sprayed flow hits the inner walls of the flange and is rotating there.

3.2 Resistance characteristic

For the several water levels (pressure differences), relevant cases were solved, see above Fig. 3-1 to Fig. 3-8. From the received results, the flow characteristic

$$Q[m^3/s] = f(\Delta p[Pa])$$

was evaluated and the inverse resistance characteristic respectively

$$\Delta p[Pa] = f(m[kg/s]) \text{ resp. } \Delta p[Pa] = f(Q[m^3/s]),$$

see the graph in Fig. 3-9.

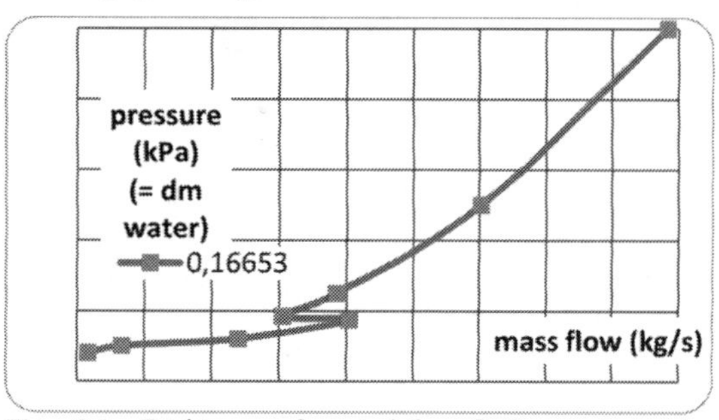

Fig. 3-9. Resistance characteristic of the vortex valve (numerical simulations).

The characteristic has two parabolic branches of different gradient. The „switching-over" is coming when the water level is just around the upper outline of the chamber. Flow through the system without rotation (or with partial rotation, only) is passing into the flow of fully created rotating flow. The numerical solution in this „switching-over" area is not stable, the stabilization needs a longer calculating time. Probably, the instability is given by an alternating air flow in and out through the small vent hole, which is given by a wavy water level. Such a small orifice acts as a large flow resistance and the response on the chamber filling is delayed. Maybe, more stable solution could be possible with a large vent diameter (not tested). Another influence is the transition into the fully rotating flow.

For a better description of this area it should be created more calculations. For the smallest flows, the curve gradient is higher – at the lowest water level the flow width is much less than the outlet diameter.

There are visible two different modes of the flow:

Flow with free water level inside the chamber (smaller water level, smaller gradient of characteristic) – neglecting the flow resistance in the chamber, the theoretical flow velocity w [m/s] and volume flow V [m^3/s] at water level H[m] can be solved from simple formulas

$$w = \sqrt{(2.g.H)}$$
$$V = \pi/4 \,.\, D^2 \,.\, w \,.$$

Flow without free water level inside the chamber (higher water level, higher characteristic gradient) – water from the tangential inlet is rotating inside the chamber and on the outlet edge, the compact flow is sprayed into individual drops.

3.3 Detailed analysis of the results

For an absolutely free flow, the theoretical outlet velocity is logically higher, comparing with the real flow velocity in a real valve with flow resistances. The comparison of such a theoretical value with a simulated value is in Fig. 3-10.

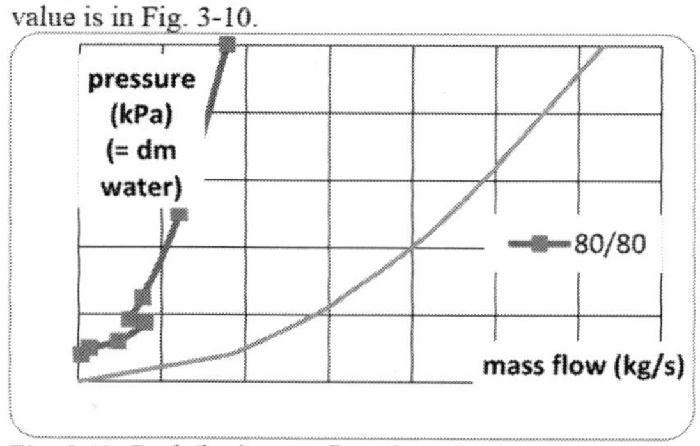

Fig. 3-10. Real (low) mass flow through the simulated valve and ideal (high) mass flow through the simple opening.

Further, there is derived the so-called flow coefficient

$$\varphi = w_{real} / w,$$

containing all losses of the valve (inlet, friction in the inlet tube, inlet in the chamber, rotation and friction in the chamber, passing into the outlet, flow in the outlet, flow separation, all as three-dimensional flow). The received values of flow coefficient for the observed case are presented in Fig. 3-11.

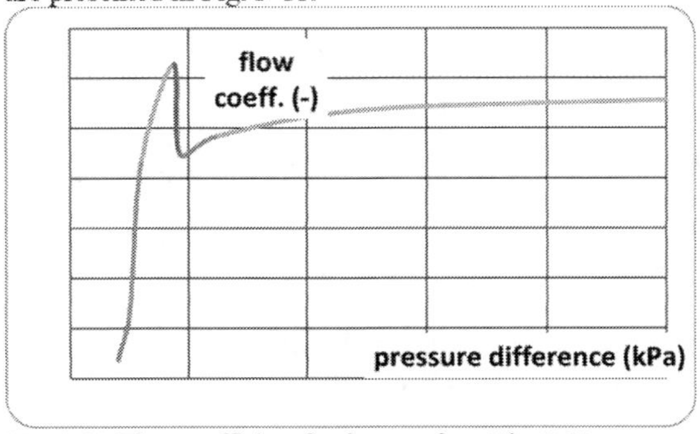

Fig. 3-11. Flow coefficient for the case from Fig. 3-10.

For high flows, the value is approaching to 0.28 and with decreased flow the value is decreased to the value of 0.2. For small flows, the maximum value is 0.31 and with flow decreasing, the value is decreasing quickly, first of all, for a water level under 0.2 m, where the cross-section of the outlet is decreasing in the transverse direction, too. Extrapolating the simulations up to a very high value of 200 kPa, we can state that the flow coefficient does not exceed the value of 0.35 approx.

3.4 Remark to the effect of the vortex valve

Fig. 3-12 presents the profile of dynamic pressure in the vertical cross section through the outlet plane for the previous case. In the large central part, the pressure is equal to zero (atmospheric). At the periphery, only, the values of 1.2 kPa (negative x-axis, down, here left) and 0.6 kPa (positive x-axis, up, here right) respectively are reached - due to the gravitation, the flow is slightly bent down. Mass flow of 8.8 kg/s.

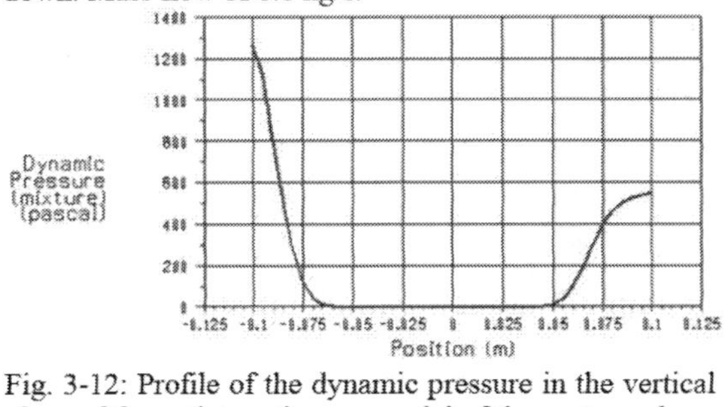

Fig. 3-12: Profile of the dynamic pressure in the vertical plane of the outlet section – a model of the vortex valve

The next Fig. 3-13 presents for the comparison, only, the similar graph of dynamic pressure for a simple cylindrical outlet. In this model, the outlet dimensions and the pressure difference are the same, but the rotational effect is none. The result is typical for sharp outlet flow – the maximum of 18 kPa for mass flow of 23.8 kg/s, i.e. 2.7 times higher, relative to the previous Fig. 3-12. Decreasing the outlet at 50 mm approx., the mass flow could be decreased at the above mentioned value of 8.8 kg/s and the dynamic pressure could be theoretically 2.5 kPa approx., so still higher, comparing with the vortex valve in Fig. 3-12. The value down (here left) is slightly higher then up (here right) again, due to the gravitation effect.

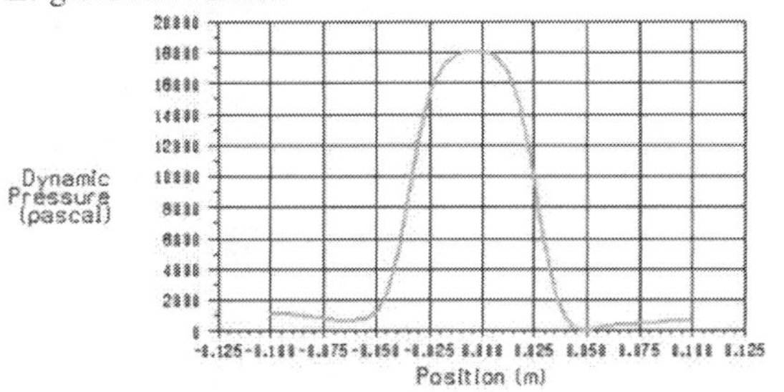

Fig. 3-13: Profile of the dynamic pressure in the vertical plane of the outlet section – a model of the straight outlet

4 Outlet changes

After the detailed analysis in Par. 3 above, the next similar solutions were made for valves of other outlets. Using suitable outlet inserts for the same valve body, it is possible to get the range of flow up to 6.25:1. Similar sets of flow fields as above are not presented here again, due to their large extent.

Reduced outlet is operating as a classical nozzle, used as the flow acceleration. The characteristic shape is fluent, at higher water levels, the outlet flow remains compact and rotating, with a slight bending due to the gravity, but without spraying, see Fig. 4-1 and Fig. 4-2 for pressure gradients of 2 kPa and 20 kPa.

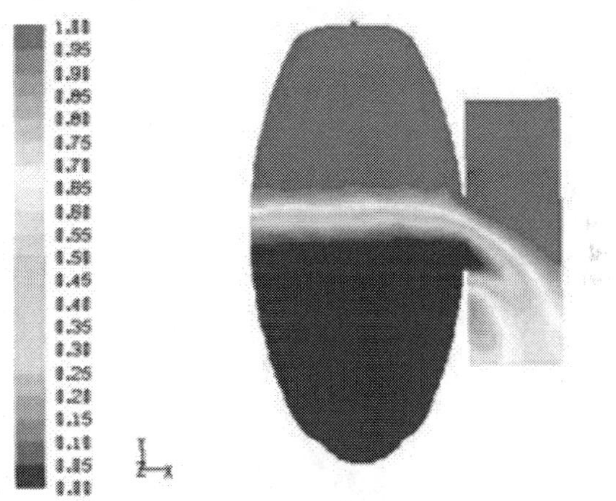

Fig. 4-1. Phase interface, $\Delta p = 2$ kPa, reduced outlet.

An unsuitable strong dynamic effect of the outgoing flow on the walls in the surroundings.

 https://doi.org/10.1051/epjconf/201818002001

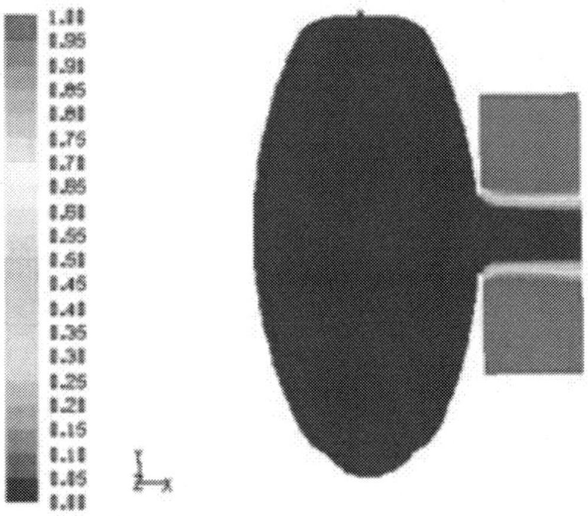

Fig. 4-2. Phase interface, Δp = 20 kPa, reduced outlet.

Using the <u>enlarged outlet cross section</u>, inside the inner volume, there it is created an expressive rotating water ring. The outside air is coming inside along the rotation axis, the air amount is greater than that for original outlet diameter. Fig. 4-3 and Fig. 4-4 present the situation for the outlet area 2x larger in comparison with the basic case from Par. 3.

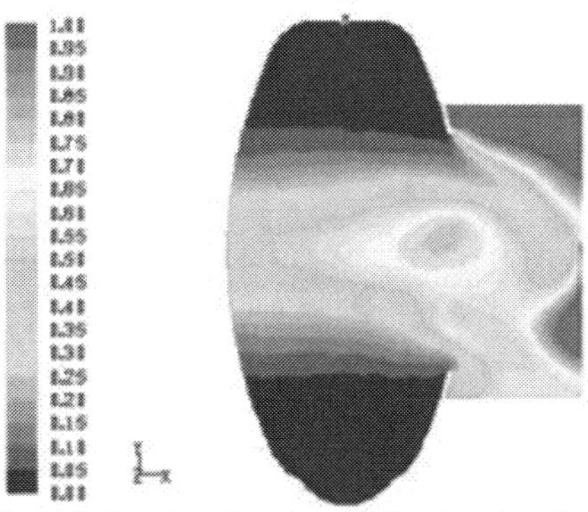

Fig. 4-3. Phase interface, Δp = 2 kPa, 2x enlarged outlet area.

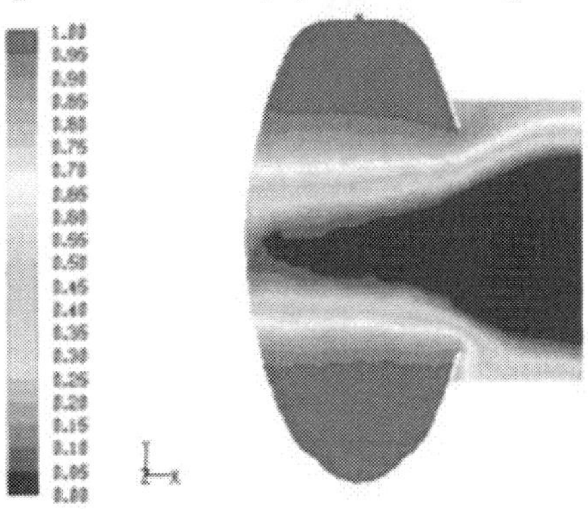

Fig. 4-4. Phase interface, Δp = 20 kPa, 2x enlarged outlet area. (Attention, error: here, only, the water is red and the air is blue!)

Using the 4x larger outlet cross section (Fig. 4-5 and Fig. 4-6), the simulation is unstable for pressure under 3.5 kPa, i.e. for water level under the upper outline of the chamber. The pressure field is not uniform, too, due to the very large outlet cross-section, the rotating water ring is not being created. Along the axis, there is a visible large area filled by outside air.

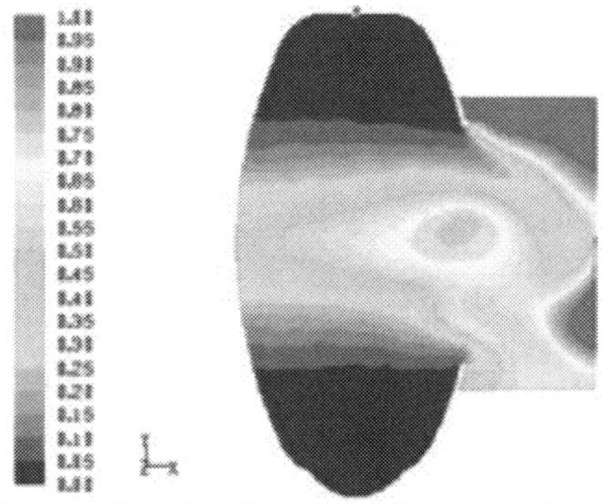

Fig. 4-5. Phase interface, Δp = 3.5 kPa, 4x enlarged outlet area.

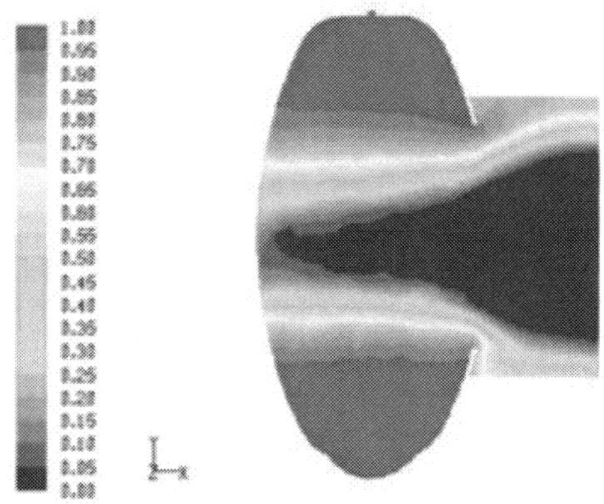

Fig. 4-6. Phase interface, Δp = 20 kPa, 4x enlarged outlet area. (Attention, error: here, only, the water is red and the air is blue!)

Remark: The usable effect of a dynamic pressure suppressing remains, the spare quality is not important here.

4.1 Summary characteristics

The summary of all long-lasting simulations with the resistive characteristics of all solved cases is the sole graph in Fig. 4-7.

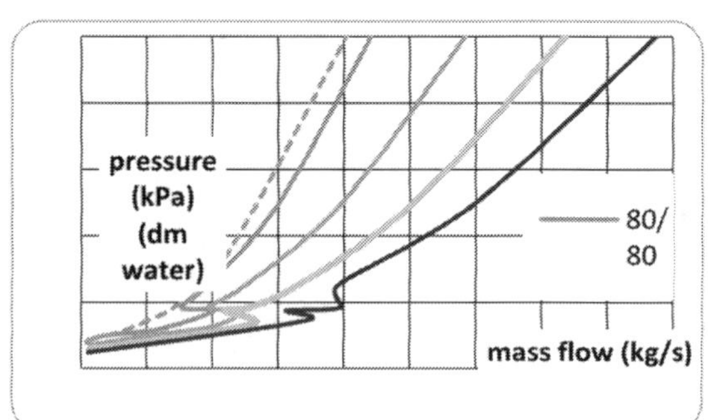

Fig. 4-7. Summary characteristics – changes of the outlet cross-section.

The geometry remains the same; different are the outlet diameters, only. In the unstable transitional area between two branches of each characteristic, the numerical errors are higher than in the rest of each characteristic, containing two branches.

The Graph in Fig. 4-8, derived from the previous simulations, can be used for a prediction of an outlet diameter for the needed mass flow. In linear correlation it means y = outlet diameter, x = needed mass flow.

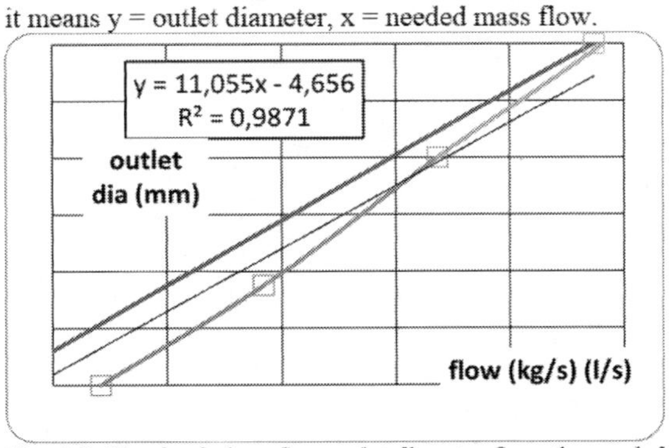

Fig. 4-8. Simple choice of an outlet diameter from the needed mass flow – from the simulations above.

5 Testing measurement

The results of the presented numerical flow simulations could be useful for the designing and for a subsequent production of vortex valves. Therefore, the results of the simulations should be verified by real measurements.

The tested valve created after Par. 3 with one inlet cross section and with several outlet cross sections, realized as exchangeable inserts as 1-2-4x larger inlet cross sections. The water level maintained at several certain values, max. 2.0 m. At the outlet, there it is a mounted measuring device.

For the simulations, there were used water levels of 0.5 – 1.0 – 2.0 m, for a measurement with larger outlets, but the used pump allows the level 1.6 or 1.2 m, only, so the results of the simulations are extrapolated for those levels.

The results of the measurements were confronted with the results of the simulations (Fig. 4-7). Comparison of both procedures, simulation and measuring, is in Tab. 5-1.

Table 5-1. Simulations compared with measurements (inlet 80 mm).

out/in	level	flow		diff.
-	m	l/s		measur.-simul.
		measur.	simul.	%
1	2.0	9.2	8.8	**+4.5**
1	1.6	7.6	8.0	**-5.0**
2	1.6	10.6	10.6	**0.0**
4	1.2	12.0	11.55	**+3.9**

The coincidence of both procedures is very good, the error does not exceed the error of standard operating measuring devices (±4%). The found differences can be caused by several influences:

- The water level before the valve was not absolutely constant, as well as the water level after the outlet, where the flow meter was installed.
- A next error is given by the definition of zero pressure – constant for simulation, slightly uneven for real measuring, see Fig. 2-3 and Fig. 2-4 above.
- Some error it is brought in the used method of numerical simulation, for instance the coarse mesh elements.

Following such a positive comparison, it is possible to suppose that the application of a numerical flow simulation in the vortex valve is the suitable method for determining the valve main dimensions, operating at the defined flow. Further, it can be supposed that for the geometrically similar shapes of the valves, their operational characteristics are similar, too.

6 Conclusion

For a determination of resistance characteristic $\Delta p = f(m)$ resp. $\Delta p = f(V)$ of the observed model, it should be to test at first suitable procedures of the solution. Because of a very high number of necessary simulations, the procedure should be the shortest, but with warranted and verified results. Main conclusions are mentioned in Par. 2.

The prepared method was used at first for the so-called basic model – Par. 3 and later for various shape variations of it – Par. 4.

From the many received results, the summary characteristics are evaluated and finally simple graphs, too, for a prediction of the new model series of valves.

At the end, it should be noted that any simulating model is a better or worse approach to the reality, only, which should be tested by measuring on the real piece and so to make sure that the results of a simulation are in the conformity, either very good, or acceptable, or approximate only.

Acknowledgment

Our acknowledgment is given to VUTS Liberec – Centre for Development in Machinery Research for the support in the framework of the grant NPU-LO1213, "National program of sustainability", granted by the Czech ministry for education, youth and sport.

References

1. Vortex valve, patent US 3219048 from 23.11.1965
2. Jarman, D., Butler, D., Tabor, G., Andoh, R.: Modelling of vortex flow control at high drainage flow rates, ICE proceedings - Engineering and computational mechanics, Vol. 168 issue EM1
3. Andoh, R.Y.G., Faram, M.G., Osey, K.: Vortex flow controls in integrated stormwater management for urban environments, Proc. of the 6th int. water sensitive urban design conference and hydropolis, Perth Australia, 2009

Experimental study on the formation of subaqueous barchan dunes in closed conduits

Carlos A. Alvarez[1*], and *Erick Franklin*[1]

[1]School of Mechanical Engineering - University of Campinas - UNICAMP, Rua Mendeleyev, 200 - Campinas - SP - CEP: 13083-860, Brazil

Abstract. The present paper reports the formation of subaqueous barchan dunes by analyzing the temporal evolution of their main geometrical characteristics (width W, length L and horn lengths Lh). After certain time, the dunes reach an equilibrium state and it is possible to study the relation between W versus L, and the dependence of the dune velocity on L. The barchan dunes were formed from spherical glass and zirconium beads. An initial conical heap of beads was placed on the bottom wall of a rectangular channel and it was entrained by a water turbulent flow. The evolution of the dunes was filmed with a CCD camera placed above the channel and mounted on a traveling system. Our results show that after a characteristic time the dune shape does not change and it travels with a roughly constant velocity. Once the equilibrium state is reach, W and L are measured, showing linear dependence. Furthermore, we show that the dune velocity V_d scales with the inverse of the dune length.

1 Introduction

The transport of solid particles entrained by a fluid flow is frequently found in both nature and industry. When shear stresses exerted by the fluid on the bed of particles are bounded to some limits, a mobile layer of particles known as bed-load takes place in which the particles stay in contact with the fixed bed. If it takes place over a non-erodible ground or with limited sediment supply from the bed, they form dunes showing a remarkable crescentic shape with horns pointing downstream [1]. These dunes, known as barchan dunes, are commonly observed in deserts. However, field measurements suffer from the lack of control on meteorology conditions as well as the large length and time scales (100 m and 1 year) involved in the physics of dunes [2]. The shape and size selection of dunes, their dynamics, and even their stability remain as open issues [3]. Creating dunes in a controlled way and using shorter scales could help to improve our knowledge on those questions. Similar barchan dunes have also been observed under water flows (with much shorter lengths than in air). Barchan dunes under water have received much less attention than aeolian dunes, and important issues remain unanswered. Recently, some numerical studies have been conducted to investigate the evolution of barchan dunes [4, 5].

Nowadays, the availability of high-performance computational tools allows the use of numerical simulations to study complex phenomena like the formation of barchan dunes. Therefore, experimental studies are vital to generate data and validate those numerical simulations. This work presents some experiments with glass and zirconium beads entrained by a turbulent water flow. Its objective is to investigate the formation and velocity of barchan dunes.

2 Experimental device

The experimental device consisted of a water reservoir, two centrifugal pumps, a flow straightener, a 5 m long channel, a settling tank, and a return line. The flow straightener was a divergent-convergent nozzle filled with d = 3 mm glass spheres, the function of which was to homogenize the flow profile at the channel inlet. The channel had a rectangular cross-section (width = 160 mm and height 2δ = 50 mm) and was made of transparent material. Figure 1 shows the scheme of the experimental loop. The channel test section was 1 m long and started 40 hydraulic diameters (3 m) downstream of the channel inlet (the turbulent flow was completely developed in the test section). The particles were placed in the channel test section, which was previously filled with water. The grains settled in the water at rest and formed a conical heap. Next, a water flow was imposed in the channel, and the heap deformed into a barchan dune.

The evolution of the dune shape was recorded with a CCD camera placed above the channel and mounted on a traveling system. With this procedure, each experiment concerns one single isolated dune. In our experiments, we used water as the fluid media and the granular material was spherical beads: glass beads with density ρ_s = 2500 kg/m^3 and mean diameter d = 0.37 mm, and heavier zirconium particles with density ρ_s = 4000 kg/m^3 and mean diameter d = 0.50 mm.

* Corresponding author: calvarez@fem.unicamp.br

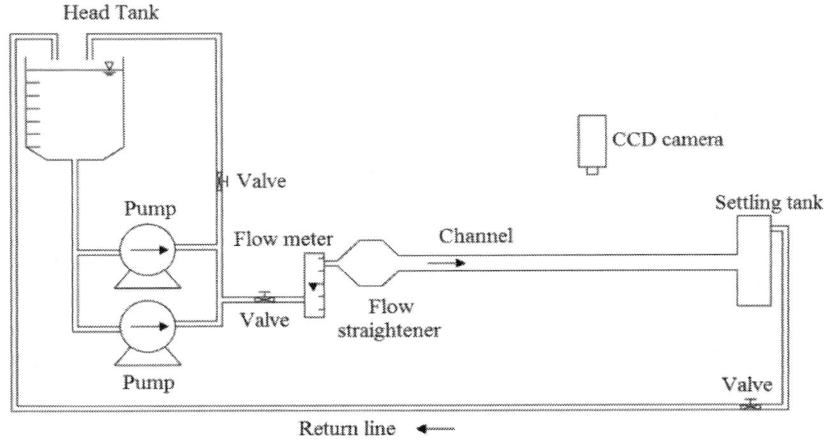

Fig. 1. Schematic of the experimental loop. (Figure extracted from the Supplemental Material of [6]).

The water flow rate was varied between 6.5 and 9.5 m^3/h. The flow velocity $\overline{U}$, defined as the ratio of the measured volumetric flow rate and the channel cross-section, was varied between 0.22 and 0.33 m/s. The corresponding Reynolds number, with υ as the kinematic viscosity, was calculated using the Eq. 1 and it was in the range 11280 – 16700.

$$Re = \frac{\overline{U}2\delta}{\upsilon} \quad (1)$$

3 Dune morphology

Once a conical heap of grains is formed in the channel, the flow is started up at a constant rate. Before being displaced over a measurable distance, the pile is deformed and adopts a crescentic shape as shown in Fig. 2.

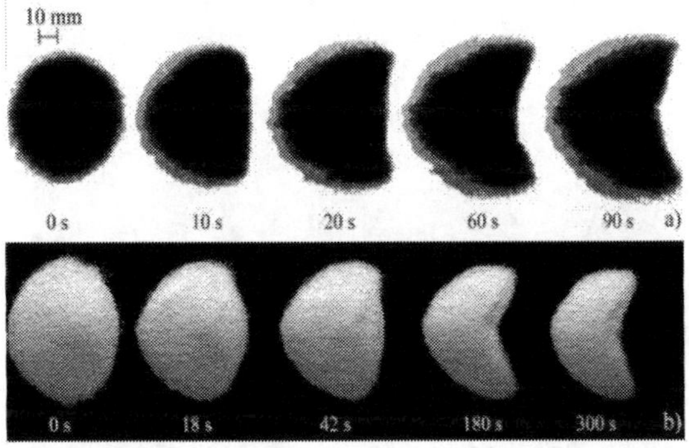

Fig. 2. Top views of initially conical heaps deposited on the channel test section and deformed by the water flow at different times, the water flow is from left to right. a) Black glass beads with mean diameter of 0.37 mm at Re = 14700. b) Zirconium beads with mean diameter of 0.50 mm at Re = 16700.

A schematic view of a barchan dune is shown in Fig. 3, which defines its dimensions. L is the distance, in the symmetry plane, from the front to the rear of the dune, not taking into consideration the horns. W is the dune width, L_h is the horns length and H is the dune height. An image processing code, written as a Matlab script, was developed during this work to automatically treat the acquired images. The code tracks the dunes along the images and computes its L, W, L_h, and the centroid position; from the latter the dune velocity V_d is computed.

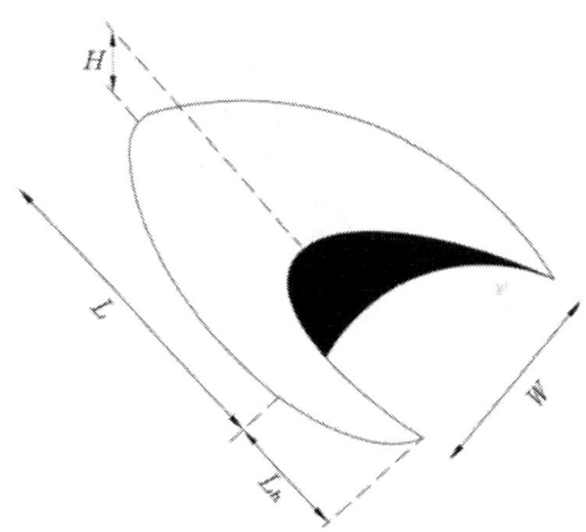

Fig. 3. Scheme of a barchan dune showing its dimensions.

4 Results and discussion

To analyze the formation of barchan dunes, we started by evaluating their temporal evolution. It is considered as initial state a conical heap of grains. As soon as the water flow begins, the heap deforms and to adopt a "croissant" shape.

Our results show the evolution of some geometrical parameters of dunes along the time. Figure 4 shows the formation of barchan dunes, through W and W/L, considering the initial state at time equals to zero. Figure 5 shows the centroid positions along the time, the dune velocity V_d can be obtained by computing the slopes in this figure. V_d can be calculated once the dunes had reached their equilibrium shape, i.e., after they had traveled a distance of the order of the size of the initial heap [7]. For the tests displayed in Fig. 5, the equilibrium state is approximately reached at time equals to 8 s. That time also corresponds to when the dune width W is roughly the same as the length L (see Fig. 4). Each continuous line in the Figs 4 and 5 corresponds to a barchan dune analyzed during a test run with Re = 16500. Asterisks represent a mean value for the observations analyzed.

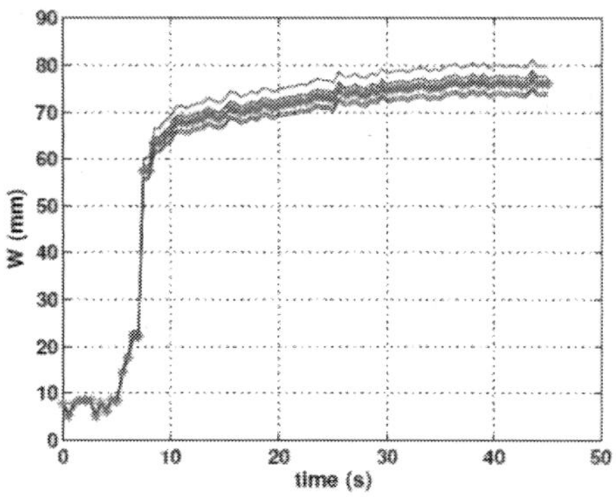

(a)

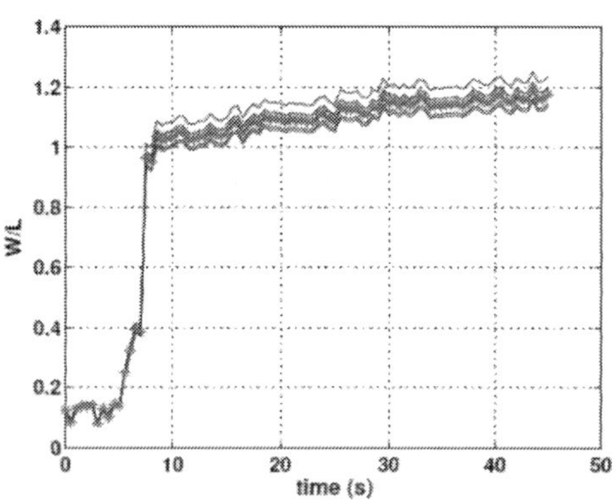

(b)

Fig. 4. (a) Evolution of dune width W along the time. (b) Evolution of W normalized by the dune length L. The particles employed were spherical glass beads with mean diameter of 0.37 mm, Re = 16700. Each line corresponds to a different barchan dune. (*) mean value of W along the time

Once the equilibrium state was reached we calculated the length L and the width W of barchan dunes. This procedure was performed for a large number of test runs. Figure 6a and 6b show the width W versus the length L, respectively, for black glass beads (d = 0.37 mm) and zirconium beads (d = 0.50 mm), for four different Reynolds numbers. Each one of the symbols in these figures represents one test run. W is approximately related to L with a linear dependence, and this relationship appears to be independent of the particle diameter and density.

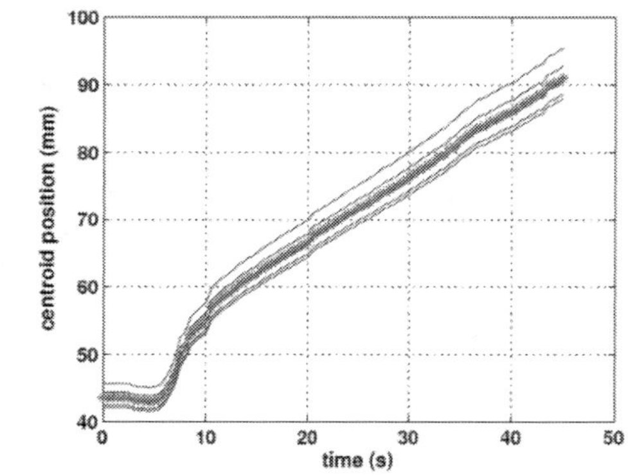

Fig. 5. Centroid position versus the time. The inclination gives the dune velocity V_d, Re = 16700. Each line corresponds to a barchan dune observed in a test run. (*) mean value of observations.

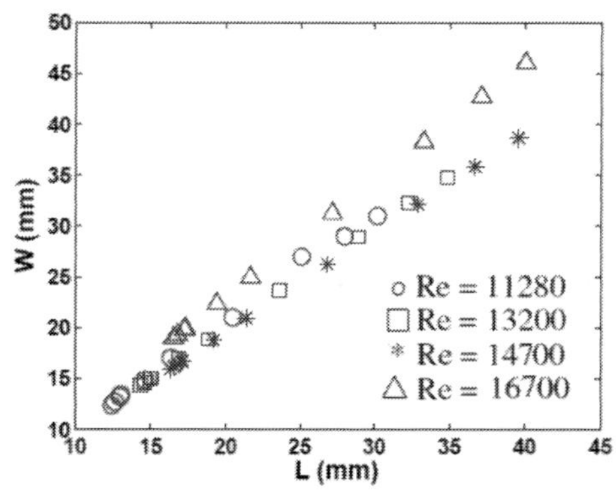

(a)

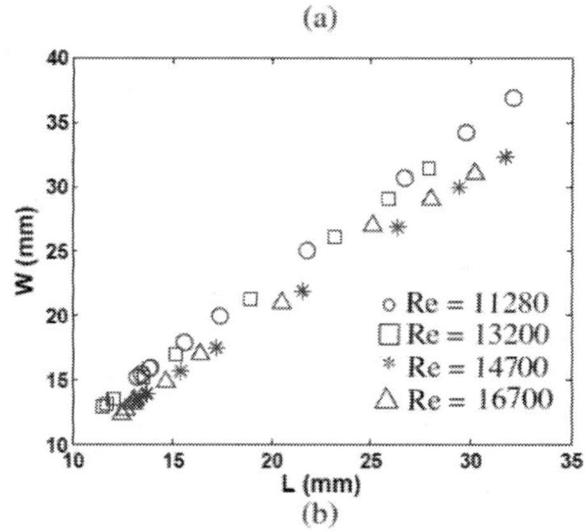

(b)

Fig. 6. (a) Width W versus the length L for four different Reynolds number. The symbols used are explained in the key. (a) Black glass beads with mean diameter d = 0.37 mm. (b) Zirconium beads with d = 0.50 mm.

Figure 7 displays the velocity V_d as a function of the inverse dune length (1/L), for four different Reynold numbers, and for the two types of beads employed in this study. It is possible to identify that for a given Reynolds number, the dune velocity scales with the inverse length, $V_d \propto 1/L$. Furthermore, in Fig. 7 it is clearly visible that, given a dune length, the dune velocity increases with the Reynolds number. These results are in good agreement with previously published works [2, 8].

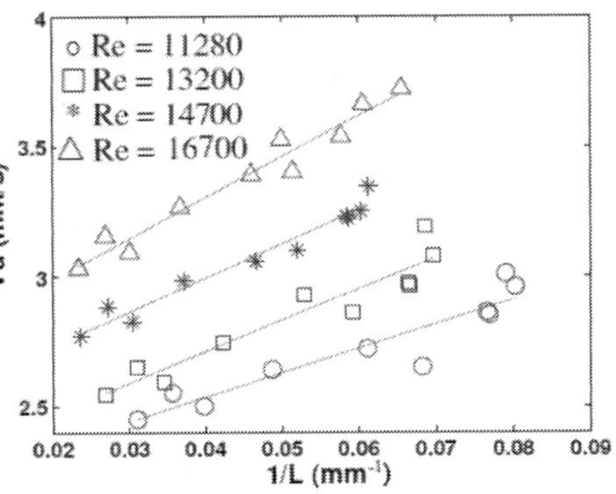

(a)

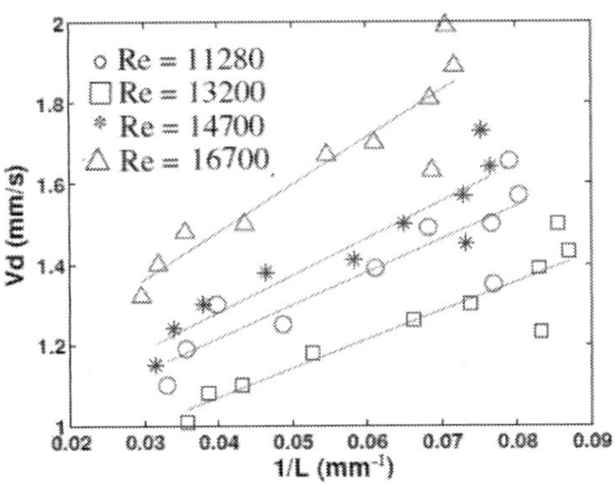

(b)

Fig. 7. Dune velocity V_d versus the inverse length 1/L for four different Reynolds number. The symbols used are explained in the key. (a) Black glass beads with mean diameter d = 0.37 mm. (b) Zirconium beads with d = 0.50 mm. The continuous lines emphasize the linear dependence $V_d \propto 1/L$ for the given flow conditions.

5 Conclusions

The present paper reported the evolution of a conical heap of grains entrained by a turbulent flow that evolved to a barchan dune. The grains were formed from glass and zirconium beads. The formation of the dunes was filmed with a CCD camera. An image processing code was developed to identify the dunes, follow them along the images, and to calculate some geometrical parameters that define their morphology. Our results showed that after a characteristic time the dune shape does not change, and it travels with a roughly constant velocity.

The linear relationship between the width and length appears to be independent of the particle diameter and density. The velocity scales with the inverse of the dune length, and for the two grain types grain used the dune velocity increases with the Reynolds number for a given dune length. An important issue to future is to study the formation of barchan dunes employing grains with different characteristics, for instance, using beads with irregular shapes. Other important task is to test the evolution of barchan consisting of grains of two different specific masses.

Carlos A. Alvarez is grateful to the Ecuadorian government foundation SENESCYT for the scholarship grant (no. 2013-AR2Q2850) and to CNPq (grant no. 140773/2016-9). Erick Franklin is grateful to FAPESP (grant nos. 2012/19562-6 and 2016/13474-9), to CNPq (grant no. 400284/2016-2) and to FAEPEX/UNICAMP (grant no. 2210/17)) for the financial support provided.

References

1. H.J. Herrmann, G. Sauermann, Physica A **283**, 24–30 (2000)
2. P. Hersen, S. Douady, B. Andreotti, Phys. Rev. Lett. **89**, No. 264301 (2002)
3. P. Hersen, K.H. Andersen, H. Elbelrhiti, B. Andreotti, P. Claudin, S. Douady, Phys. Rev. E **69**, No. 011304 (2004)
4. A. Khosronejad, F. Sotiropoulos, J. Fluid Mech. **815**, 117–148 (2017)
5. A.G. Kidanemariam, M. Uhlmann, J. Fluid Mech. **818**, 716–743 (2017)
6. C.A. Alvarez, E. Franklin, Phys. Rev. E **96**, No. 062906 (2017)
7. E.M. Franklin, F. Charru, Powder Technology **190**, 247–251 (2009)
8. E.M. Franklin, F. Charru, J. Fluid Mech. **675**, 199–222 (2011)

EPJ Web of Conferences **180**, 02003 (2018) https://doi.org/10.1051/epjconf/201818002003
EFM 2017

Temperature measurement and performance assessment of the experimental composting bioreactor

Jaroslav Bajko[1,★], *Jan* Fišer[1], and *Miroslav* Jícha[1]

[1] *Faculty of Mechanical Engineering, Brno University of Technology, Technická 2896/2, 616 69 Brno, Czech Republic*

Abstract. Considerable amount of heat produced during composting of organic matter is usually lost to the surrounding environment. In order to utilize this potential heat source, biomass is composted in mounds or vessels and excess thermal energy is captured via heat exchangers and transported to the site for space heating, preparation of utility water and other applications. The aim of this experimental research is to measure the temperature profiles in a pilot-scale composting bioreactor and assess its performance.

1 Introduction

Composting is widely known as aerobic decomposition of organic materials by microorganisms, [1], producing valuable soil amendment and conditioner - compost - as well as heat, since the bio-chemical reactions presented in compost piles are mostly exothermic. The decomposition progresses slowly and steadily, thus the compost pile may be seen as a bioreactor or low grade source of combustion-free thermal energy.

Various studies have been devoted to this heat source ranging from laboratory scale experiments [2–4], to pilot-scale reactors [5–8] as well as commercial scale enterprises [9, 10] with the aim of characterizing the amount of heat that can be captured and utilized. Recently, all systems using heat from composting processes have been united under term "Compost Heat Recovery Systems" (CHRSs), cf. [11, 12].

Subject of this paper is to present a pilot-scale composting bioreactor with the long-term within-pile temperature measurement. The CHRS based on hydronic system, cf. [5], is studied with the analysis of internal compost thermodynamics when heat is extracted from the pile. This experiment was conducted with the intent to create the simplest form of pilot-scale CHRS using buried hydronic heat exchanger and to measure the heat recovery rate of the system.

1.1 General composting practise

Design of CHRS include many variables that need to be taken into account for succesfull composting operation and thus the production of heat, [1, 13]. These include external factors: climatic (weather, seasons of the year), exposure to surrounding environment (open or in-vessel facility), insulation, duration of composting as well as internal factors: energy and nutrient content of the composting material, its degradability, porosity. All of these factors create conditions for microbial activity that is responsible for the composting process as a whole.

Considering evidence based on [14, 15], basic measurable variables (and their optimal range) involved in composting read as follows:

- Carbon-to-Nitrogen (C:N) ratio of the whole mix around 30 : 1,
- Moisture between 50 – 60 %,
- Oxygen content between 10 – 18 % (active phase) and 1 – 5 % (maturation phase),
- Temperature under 65 °C (150 °F),
- Porosity, i.e. free air space 35 – 50 %,
- pH between 5.5 – 8.0.

Composted organic material should be thoroughly mixed, should not contain particles larger than 8 cm and toxic substances. Moreover, drainage should be included and compost leachate reused in order to avoid the development of anaerobic conditions, cf. [16].

1.2 Thermodynamic model of compost piles

Thermal balance of a composting system, cf. [17–19], can be written as the heat accumulation term equals to the input $Q_{\rm in}$, output $Q_{\rm out}$ and transformation $Q_{\rm tran}$ energy terms as follows:

$$\frac{\mathrm{d}(mcT)}{\mathrm{d}t} = Q_{\rm in} - Q_{\rm out} + Q_{\rm tran}, \qquad (1)$$

where the left-hand side refers to sensible heat of the composting mass, input term includes heat gain due to solar radiation and enthalpy of incoming air. Output term consists of heat losses through system boundaries (including

★e-mail: bajko@fme.vutbr.cz

conductive, convective and radiative losses), enthalpy of exiting gas and heat taken away via heat recovery system. Finally, the transformation term refers to bio-chemical reactions occuring during composting process that can be both exo- or endo-thermic.

Among the terms involved in the heat balance (1), the most significant contributors are the biologically generated heat and enthalpy of exit water vapour, cf. [11, 18].

2 Materials and Methods

2.1 Experimental composting bioreactor

The composting bioreactor was initially designed as a standalone cylindrical pile with diameter of 3 m and height of 2 m. Volume of this composting reactor is 14.137 m^3 resulting in the surface area-to-volume ratio $1,83 : 1\,[m^2/m^3]$, assuming the pile stands on the ground in direct contact with soil, i.e. base contact area is excluded.

Compost feedstock consists of the following mixture of materials:

- Wood chips - softwood, 3140 kg,
- Horse manure - fresh, 2403 kg,
- Vegetable wastes and leaves, 260 kg,

where the moisture content of wood chips, wet weight of all materials were obtained by measurement. Other material properties were taken as average values from [1]. Resulting moisture of the initial compost recipe is estimated to $M_{\mathrm{mix}} = 47,5\ \%$.

Estimation of the Carbon-to-Nitrogen ratio of the whole mixture then follows from

$$(\mathrm{C{:}N})_{\mathrm{mix}} = \frac{\sum_{i=1}^{n} C_i W_i (1 - M_i)}{\sum_{i=1}^{n} N_i W_i (1 - M_i)} = 113,2 : 1, \qquad (2)$$

where C_i is the carbon content, N_i is the nitrogen content (dry weight), W_i is the wet weight and M_i is moisture content of the i-th material. Note, that wood chips as carbon-rich material were used intentionally in abundance (accordingly to J. Pain method, [5, 8]) in order to prolong the thermophilic phase due to higher C:N ratio.

Fig. 1 shows the reactor design with dimensions and initial position of the within-pile hydronic heat exchanger as well as polypropylene (PP) pipes containing thermal sensors.

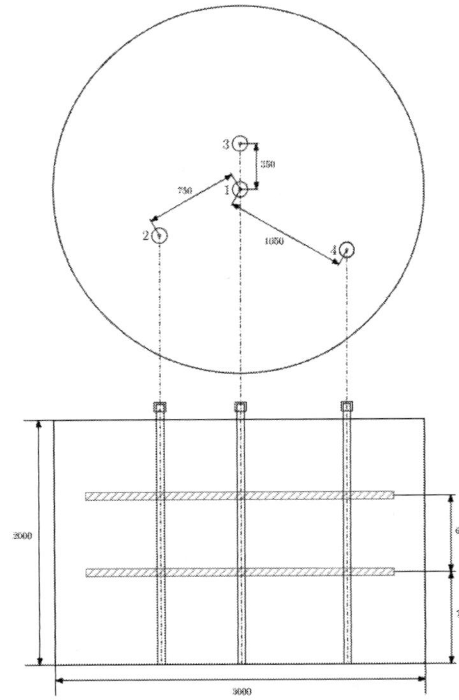

Fig. 1: Composting bioreactor shape, dimensions and positions of PP pipes containing thermal sensors. Location of the hydronic heat exchanger coils.

2.2 Within-pile temperature measurement

Conditions created inside the composting reactor represent a challenge for long-term continual measurements. These involve high humidity, pH variations, presence of chemical reactions and microorganisms that could adversely affect any part of the measuring electronics. For this reason, simple replacement and access to each part of the system was a crucial condition when designing the experimental measurement.

Therefore, for continuous within-pile temperature measurement, we have designed a simple, inexpensive solution consisting of 2.1 m long PP pipe and a wooden rod that can be inserted inside the pipe and pulled up again. At the beginning of the reactor construction, 4 PP pipes were installed, i.e. a central pipe and another three pipes were placed radially in 120° angle with distance 350, 750 and 1050 mm from the center in accordance with Fig. 1. This configuration of sensors was chosen with the assumtion of radial symmetry of temperature distribution, i.e. that the organic matter is evenly distributed in each horizontal layer of the bioreactor.

Into each PP pipe was inserted a wooden rod (Fig. 2) equiped with 2 measuring sections, each containing a thermal digital thermometer (DS18S20, accuracy ±0.5° C from −10°C to +85°C). The wooden rod was used in order to minimize vertical conductive heat transfer that could affect neighbouring thermal sensor. Moreover, convection inside the pipe was eliminated via polyuretane (PU) foam plugs.

For faster registration of temperature variatons the thermometers were tightly attached to the aluminium plates (Fig. 3) due to their high thermal conductivity and direct contact with the internal pipe surface. Data from within-pile and ambient sensors were collected by 1-wire line (Bus A) and processed with PLC device.

2.3 The heating system

Fig. 1 depicts also the position of two horizontal layers of the heat exchanger which were built in the composting matter 700 and 1350 mm above the ground (initially at the

https://doi.org/10.1051/epjconf/201818002003

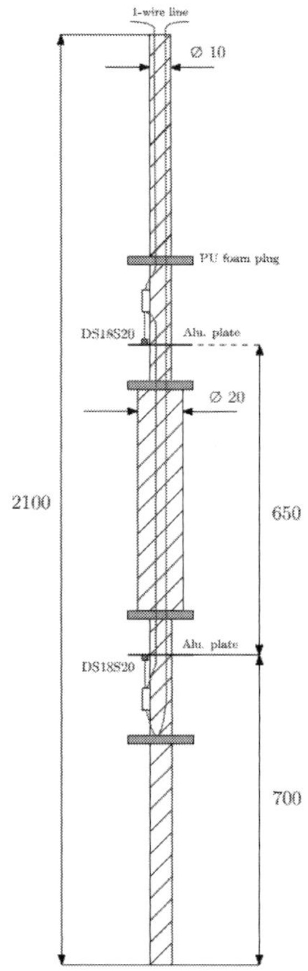

Fig. 2: Internal wooden rod (inserted into the PP pipe) with two separated measuring sections.

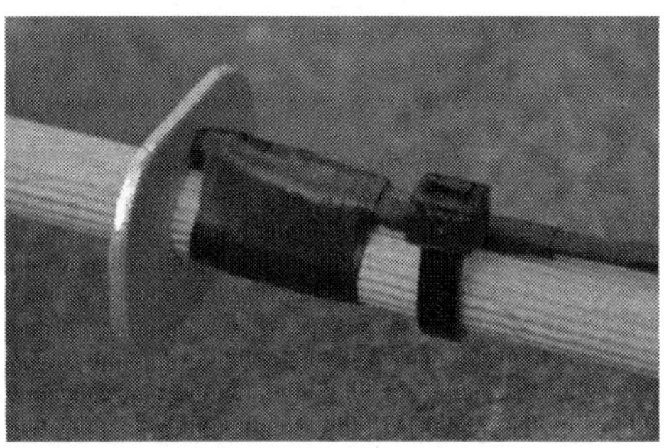

Fig. 3: Detail of the thermometer DS18S20 attached to the aluminium plate.

same height as the thermometers in the PP pipes). Each layer of the heat exchanger formed a spiral (as shown in Fig. 4 and 5) from low density polyethylene (LDPE) pipe (diameter 25 mm, wall thickness 3.5 mm) of length 26.5 m, which was attached to the concrete wire mesh for the sake of its from stability. The overall heat exchange surface of the within-pile heat exchanger was therefore 4.16 m^2.

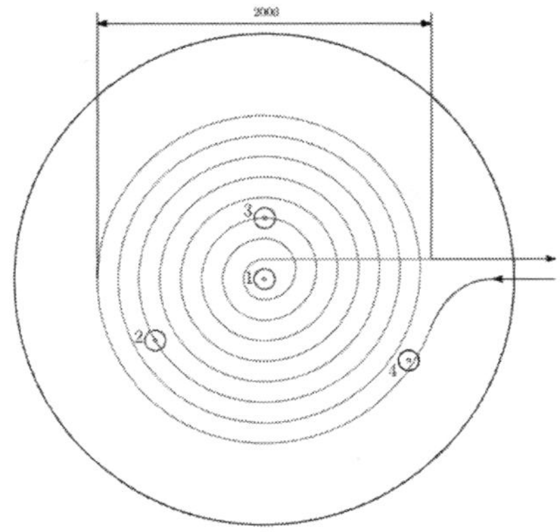

Fig. 4: A layer of the spiral heat exchanger, directions of fluid flow and positions of PP pipes with thermometers.

Both layers of the within-pile heat exchanger were connected with 12 m long insulated line to the radiator (heat sink), expansion vessel, air release valve and the flow meter allowing the measurement of the volumetric flow rate. The working fluid (water) circulated in a closed-loop circuit by electically powered circulating pump (UPS 25-60) with adjustable input power.

Fig. 5: A part of the heat exchanger made of LDPE pipe (25 × 3.5 mm) attached to the concrete wire mesh.

Second 1-wire line (Bus B) comprised of sensors measuring approximate temperature of water inside the heating system. Thermometers attached to the brass fittings (Fig. 6) provided estimation of water temperature entering and leaving the bioreactor allowing us to compute direct heat recovery rate. Insulation by PE foam was tightly wrapped around the fitting.

3 Results

Experimental raw data were collected in a period of 5 s from each thermometer starting on Nov 5. at 16:14. For graphical output the data were averaged for each minute. Graph in Fig. 7 shows first 9 days of within-pile temperature measurements.

Fig. 6: Thermometer attached to the brass fitting.

Central pole P1 (sensors 3C and F8, cf. legend of Fig. 7 and 8) measures core temperature of the mound, which represents the overall composting dynamics. Self-heating of the pile is well documented as the core temperature raises exponentially up to 78 °C followed by similar pattern for both layers in poles P2 (sensors 6C and 3F) and P3 (sensors FD and EF).

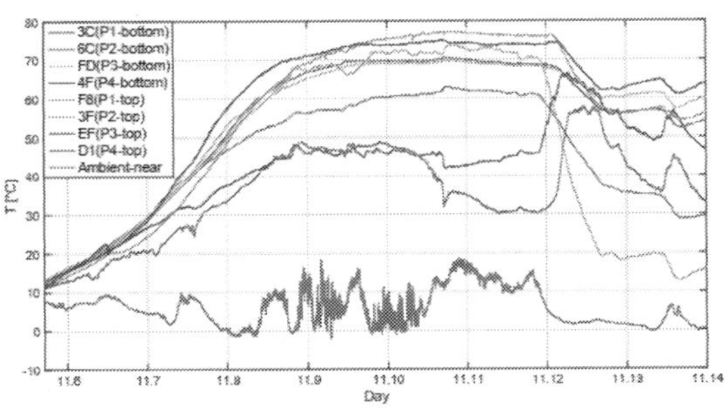

Fig. 7: Temperature profiles at 8 points (bottom layer 700 mm and top layer 1350 mm above the ground) within the bioreactor and ambient (Nov 5. to Nov 14.). P1 - center, P2 - 750 mm, P3 - 350 mm, P4 - 1050 mm from the center.

Measurement in pole P4 (sensors 4F and D1) shows slower temperature growth due to their shortest radial distance (0.45 m) from the surface. Note, the average ambient temperature during this early period was 5.9 °C.

After rapid temperature growth into the thermophilic stage and exceeding the optimal range of 45 – 65 °C, sudden fall accompanied with lower ambient temperatures was observed on Nov 11. This behaviour is well known from practise, when microorganisms start to perish due to the excess heat they produce.

Although the temperature drop occured, a stable thermophilic phase has recovered and lasted in average in the range 45 – 65 °C till Dec 1. (27 days, the end of this period is shown in Fig. 8.) For the computation of the power output of the bioreactor, the goal is to find an amount of heat that could be recovered without adversely affecting the composting process. To reach this goal, we first run the heating circuit for 15, 30 and 45 minutes on Nov 25, 26 and 27, respectively, with the volumetric flow rate $\dot{V} = 4.6$ l/min $= 7.66 \times 10^{-5}$ m^3/s.

All three runs and heating system dynamics (transient and stable state) are depicted in Fig. 8. Among 8 within-pile sensors, another two are showing temperature profiles of water entering (1A - Compost in) and leaving (1ACD - Compost out) the pile.

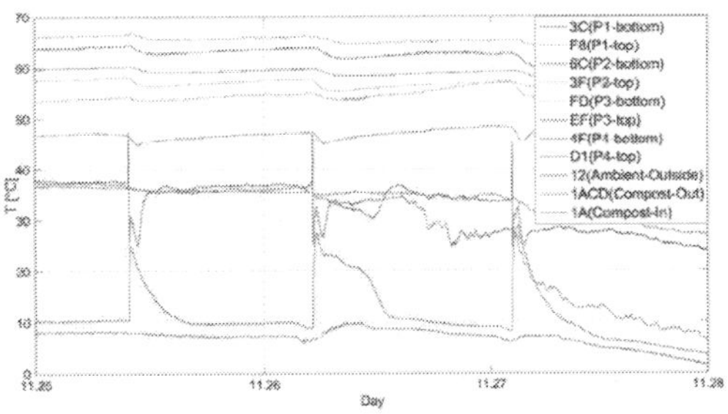

Fig. 8: Within-pile temperature profiles, ambient and temperature of water circulating at points Compost In/Out in the heating circuit (Nov 25. to Nov 28.).

From stable states for each run, we obtained the following average temperature difference $\Delta T = 6.01; 4.88$ and 4.33 °C between water entering and leaving the bioreactor. The average power output for each run than follows from

$$P = \dot{m} c_p \Delta T, \quad [\mathrm{W}] \tag{3}$$

where $\dot{m} = \rho \dot{V}$ [kg/s] is the mass flow rate and c_p is the heat capacity at constant pressure. Taking properties of water at 30 °C ($\rho = 995.6$ kg/m^3 and $c_p = 4176$ J/kg.K) we get thermal power output of the bioreactor $P = 1914; 1554$ and 1379 W for each run, respectively.

4 Discussion

The pilot-scale standalone composting reactor performed well in early days as shown in Fig. 7, i.e. it reached thermophilic stage in three days after construction and even exceeded optimal range which led to sharp fall in average temperature within 24 hours. However, stable thermophilic period has established for almost a month and after that it remained in lower mesophilic range 15 – 44 °C for the whole winter even if the temperatures dropped deeply bellow zero. Average within-pile temperatures have been continuously recovering during spring reaching middle and higher mesophilic range in summer.

Fig. 7 also shows how within-pile temperatures change over the distance from the pile center, especially lower average temperatures measured in pole P4 indicate the rate of thermal loses through mound boundaries. In order to reduce heat loses and improve the whole performance of the composting reactor, it was confirmed that the insulation plays a crucial role. Simple layer of straw or hay covering the surface would satisfy insulation need as well as aeration from sides, cf. [8].

The heat exchanger built from two spirals of LDPE pipe (total heat exchange suraface of 4.16 m^2) proved to be well designed for the reactor with volume 14 m^3. The whole circuit also worked good without any significant problems transfering enough heat from the pile.

Graph in Fig. 8 clearly shows, how each run of the heating circuit led to visible temperature drop at all within-pile sensors. Moreover, regeneration of temperatures as well as the slight descending tendency of the average reactor temperature can be observed. This behaviour suggests that achieved heat recovery rate was too high in each run, thus the long-term performance of the bioreactor at these power outputs is questionable. Therefore, real long-term overall performance is limited by 1379 W from top for this pilot-scale bioreactor.

Comparing results published in [11], where lab-scale reactors reached heat recovery rate 1895 kJ/h = 0.526 kW and pilot-scale reactors reached 20035 kJ/h = 5.565 kW, our composting system with top output of 4964 kJ/h = 1.379 kW belongs to weakly performing pilot-scale experimental bioreactors.

Acknowledgement

Authors greatly acknowledge support from the Project Fond vědy 2017 (No. RV90700003).

References

1. R. Rynk, M. van de Kamp, G. Willson, M. Singley, T. Richard, J. Kolega, F. Couin, L. Laliberty, D. Kay, D. Murphy et al., *On-Farm Composting Handbook* (Ithaca, NY: Northeast Regional Agricultural Engineering Service (NRAES-54), 1992)
2. D. Chambers, Master's thesis, Galway-Mayo Institute of Technology (2009)
3. R.A. Gnanaraj, Asian Journal of Scientific Research **5**, 185 (2012)
4. H. Seki, S. Kiyose, S. Sakida, Journal of Agricultural Meteorology **70**, 1 (2014)
5. I. Pain, J. Pain, *The methods of Jean Pain: Another kind of garden* (Draguignan: Ancienne Imprimerie NEGRO, 1972)
6. Z. Adams, Master's thesis, University of Vermont, Burlington (2005)
7. C. Gilson, Master's thesis, University of Waterloo (2009)
8. G. Brown, *The compost-powered water heater* (Woodstock: The Countryman Press, 2014)
9. F. Di Maria, M. Benavoli, M. Zoppitelli, Waste Management **28**, 805 (2008)
10. G. Irvine, E.R. Lamont, B. Antizar-Ladislao, International Journal of Chemical Engineering **2010**, 10 (2010)
11. M.M. Smith, J.D. Aber, R. Rynk, Compost Science & Utilization (2016)
12. R.F. Zhao, W. Gao, H. Guo, Int J Agric & Biol Eng **10**, 1 (2017)
13. R.T. Haug, *The Practical Handbook of Compost Engineering* (CRC Press, 1993)
14. E. Epstein, *Industrial composting: Environmental engineering and facilities management* (CRC Press, 2011)
15. A. Chiumenti, R. Chiumenti, L. Diaz, G. Savage, L. Eggerth, N. Goldstein, *Modern composting technologies* (Emmaus, PA: JG Press, 2005)
16. M.M. Smith, J.D. Aber, Tech. rep., Durham, NH: University of New Hampshire Cooperative Extension (2017), 64p
17. E.G. Khater, A.H. Bahnasawy, S.A. Ali, Journal of Environmental & Analytical Toxicology **4** (2014)
18. I.G. Mason, Waste Management **26**, 3 (2006)
19. I.G. Mason, Ph.D. thesis, University of Canterbury (2007)

Experimental Investigation of Vortex Structures in Wake of Hyperboloid-Shaped Model by Means of 2D Particle Image Velocimetry Measurement

V. Barraclough *, J. Novotný **, and P. Šafařík**
* *Aerospace Research and Test Establishment, PLC. (ARTE), Prague, Czech Republic*
** *Czech Technical University in Prague, Czech Republic*

Abstract. This paper deals with flow around a bluff body of hyperboloid shape. It consists of results gathered in the course of research by means of Particle Image Velocimetry (PIV). The experiments were carried out by means of low-frequency 2D PIV in a range of Reynolds numbers from 40000 to 50000. A hyperboloid-shaped model was measured in a wind tunnel with a modelled atmospheric boundary layer (and additionally, in a low-speed wind tunnel with low turbulence). The model was tested in a subcritical range of Reynolds numbers and various planes in a wake of the model were captured with the intention of getting an estimation of 3D flow structures. The tunnel with the modelled atmospheric boundary layer has a high rate of turbulence, so the influence of the turbulence of incoming flow on the wake could be outlined. The ratio of the height of the model to a thickness of the modelled boundary layer in the tunnel was 1/3, meaning the turbulence in the boundary layer strongly influenced the flow around the model; it suppresses the wake which leads to a lot shorter area of recirculation than low turbulence incoming flow would cause.

1 Introduction

In this contribution, research in the mechanics of fluids, that developed from the experimental testing of a model for civil engineering will be presented. Originally one of the often-investigated problems of civil engineering, wind loading on cooling towers, had been investigated. However, this paper is not going to deal with the cooling towers, it only originated there.

It was explained in [1] and [2], that the model of the cooling tower later started to be called a hyperboloid-shaped model because an original civil-engineering problem solved in a classical way (wind loadings on cooling towers were measured and evaluated) later turned into a pure mechanics of fluids problem. After all the pressures on the model had been measured, Particle Image Velocimetry got involved and the velocity fields around the model were caught.

As e.g. Zdravkovich in [3] illustrates, the cooling towers problem can be remotely compared to a circular cylinder problem. To get a comparable picture of the flow, it is possible to come out of the mean flow topology of the cylinder, as depicted in [4] for instance. Therefore, in the flow around the hyperboloid object we could expect a horseshoe vortex and then, depending on the model geometry, a tip vortex, Kármán street vortex formation or arch vortex respectively and a very influential trailing vortex developed at the free end from the upwash in front of the object and on its sides, and the downwash at the back of the object. A detailed and thoroughly processed summary of these structures and under which conditions they appear, with many references to the relevant papers and literature was given in [5].

It might seem that this problem is already well described, however, there are many difficulties, when it comes to the model slightly differing from the simple geometry of regular and symmetric cylinder. The main difference is that we have a cylinder with variable diameter, thus the Reynolds number of the experiment varies from 40000 to 50000. To further complicate the problem, we are talking about the model submerged in the boundary layer, which is to be explained later, so the velocity of incoming flow changes as the altitude of the model changes. This means that the boundary layer thickness developed on the model will vary as well. Additionally, the model remained covered by small grains from the civil engineering experiments, hence our last piece of known theory fails. Such rough surface forces the transition in the boundary layer on the model to occur at higher Reynolds numbers. On the other hand, the high rate of turbulence of incoming flow has opposite effect.

This phenomenon together with quite large high wall blockage and the low aspect ratio of the model, ensures the measurement with such model involved is filled with results different from the classic circular cylinder knowledge. Some background of the aspect ratios of the cylinders can be found in [6].

There is another separate object of discussion. According to [7], the shedding of vortices changes from a Kármán asymmetric type to a symmetric arch type at a geometrical aspect ratio of an object (cylinder) of 2.5. As the aspect ratio of our measured hyperboloid object varies

*Corresponding author: barraclough@vzlu.cz

from 1.3 to 2.54, the narrowest area of it could be measured with respect to this fact. Unfortunately, the equipment used for the experiment limited us, and this paper could not go deeper into this particular detail.

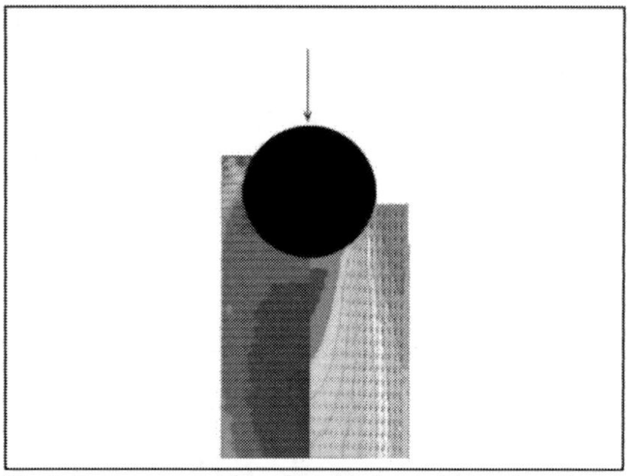

Fig. 1. A comparison of a wake behind a model in a Low Speed Wind Tunnel with a low rate of turbulence (left side of the wake) to the wake in a Boundary Layer Wind Tunnel with a high rate of turbulence (right side). The result is a field of time-averaged velocity field. The scalar map shows velocity u/U from -0,24 to 1.

2 Experimental Facility

The experimental part was provided by means of 2D Particle Image Velocimetry (PIV). The measurement was provided at Reynolds numbers Re = 40 000 ÷ 50000 with free stream velocity U = 5 m/s measured independently on PIV by Prandtl probe. The tunnel used for the measurement was the Boundary Layer Wind Tunnel (BLWT) for simulating atmospheric boundary layer and built mainly for civil engineering measurements. It is equipped with a special surface helping the development of the boundary layer. This special surface is a foil with regularly placed bumps and, according to the standard for civil engineers [8], this equipment simulates a terrain of the 3rd category: suburbs and unbroken forests. Detailed characteristics of this particular turbulent boundary layer, the tunnel itself, and the atmospheric boundary layer modelling can be found in report [9]. It has to be mentioned here, that the turbulence intensity falls from 30% close to the ground to 5% at 900 mm altitude. To complete the description of BLWT, the dimensions of the rectangular measurement area were 1.8 x 1.5 meters. For completeness, the second tunnel used for comparison was Low Speed Wind Tunnel of Eiffel Type with the low turbulence level (up to 1%).

The model was printed using a 3D printer as a 0,32 m tall plastic object. The bottom part of the model consists of pillars and is penetrable for the flow. The model is hollow as well, so the suction effect of this bottom part will somehow influence the free end and the flow around it. The surface of the model is covered by grains of 0,6-0,7 mm diameter. The roughness of the surface is one of the influencing parameters [10]. Effects of this parameter can be studied in [10], [11] and [12],. The ratio of the height of the model to the thickness of the modelled boundary layer in the BLWT is 1/3. The velocity profile of the incoming flow in BLWT tunnel is depicted in Fig. 2.

The PIV experiment was provided in a classical way - the measurement area was seeded with olive oil droplets produced by Laskin nozzles and a green laser light was used to illuminate the area. The 2D experimental part was provided with a laser with Nd:YAG crystals and low frequency (15 Hz) of the pulses. A double pulse mode was used, which means, that two consecutive pictures were captured within few tens of microseconds (usually 100) and the whole measurement lasted at least 60 seconds, so at least 900 of these double pictures were captured. For evaluation, a method called adaptive correlation was used with these parameters: final interrogation area was 64 x 64 pixels, which represented 7 - 8.9 x 7 - 8.9 square millimetres of real area. The capturing camera contained a chip with 2560 x 2160 pixels.

The main advantage of this method of getting experimental data is that the large area can be captured at one single moment. The main disadvantage is, apart from the amount of data and large storage media needed, that the strong illuminating light also strongly reflects from the surface. Meaning that the boundary layer around the model, with its thickness of few millimetres, can´t be measured at the same moment as the large field of wake. The area close to the model is usually overexposed, and as the light reflects in all possible angles from the grains, the laser sheet there is useless.

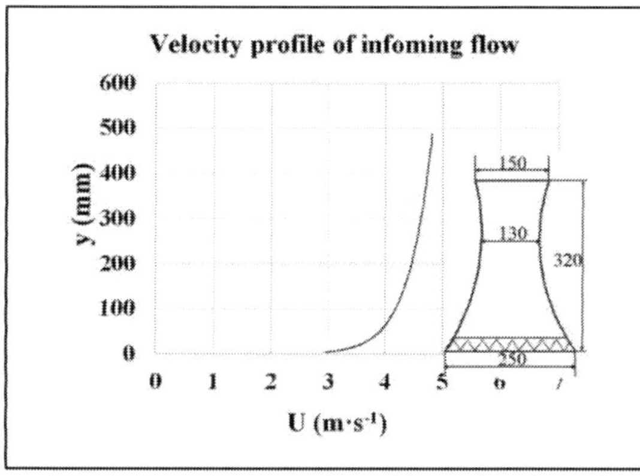

Fig. 2. The hyperboloid-shaped model with a vented base and its dimensions submerged in the boundary layer of depicted velocity profile.

3 Measurements and results

Some of the results were already presented in [1] and [2]. For this paper, special attention was paid to the measurement in the wake. The aim was to compare the influence of the turbulence intensity on a wake character.

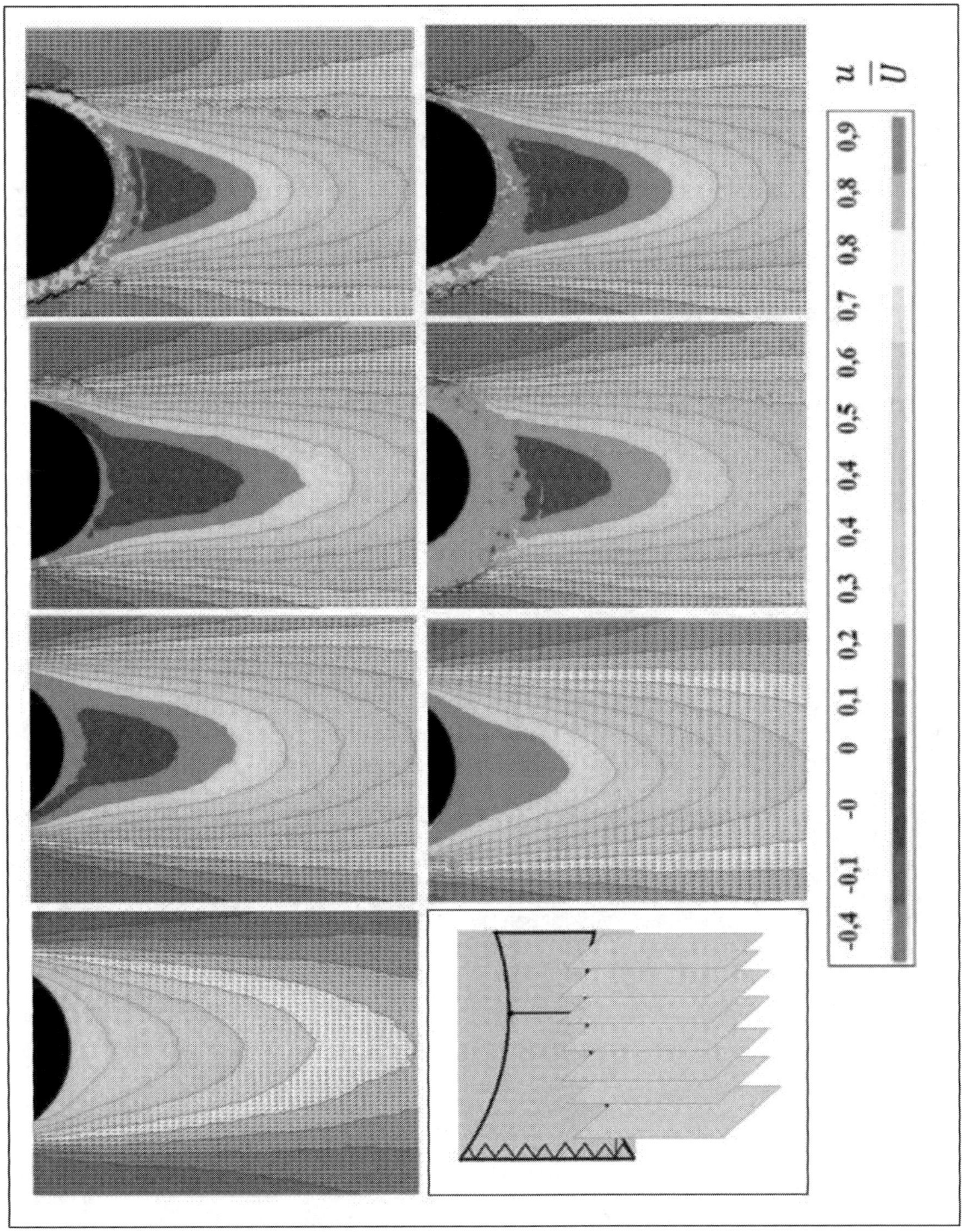

Fig. 3. The averaged velocity fields calculated in the planes parallel to the ground, varying altitudes; from top: y/L = 0,2; 0,34; 0,47; 0,59; 0,72; 0,84; 0,97.

The areas x/D x z/D of each field are as follows:

y/L=0,2: 2,73 x 2,31; 0,34: 2,65 x 2,23; 0,47: 2,53 x 2,13; 0,59: 2,44 x 2,05; 0,72: 2,34 x 1,98; 0,84: 2,24 x 1,89; 0,97: 2,15 x 1,81.

Dimension D represents minimum diameter of the model, L altitude of the model and velocity U represents velocity of incoming flow. Bottom pictogram shows the measured areas.

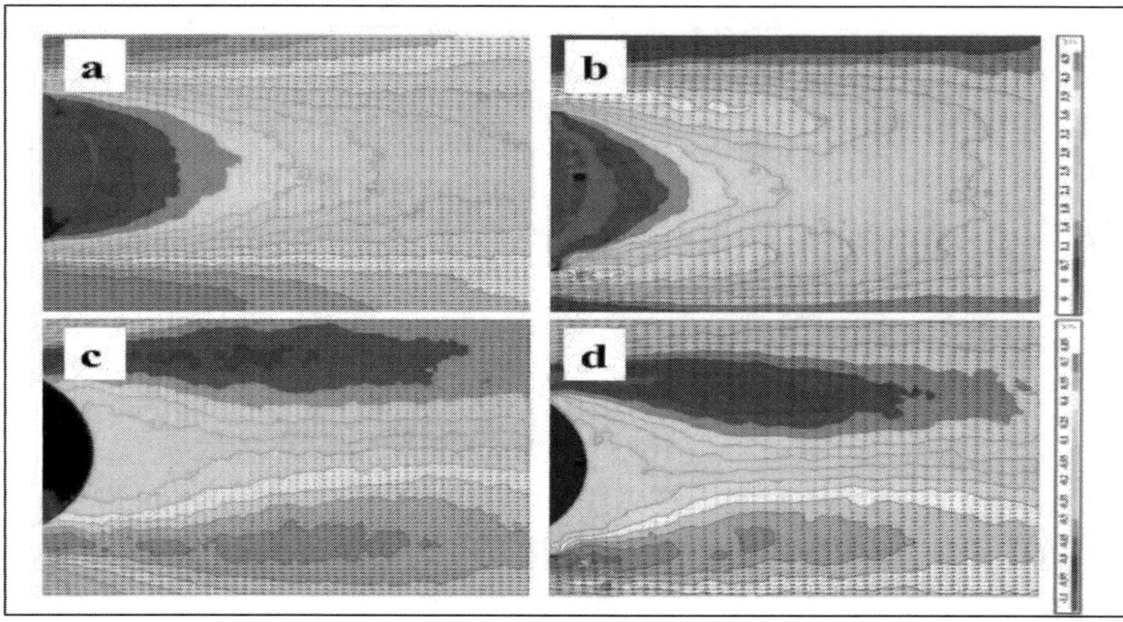

Fig. 4. Upper left and right: sum of variances (proportional to turbulent kinetic energy) at y/L=0,97 (a) and 0,84 (b) altitude and bottom left and right: covariances (proportional to Reynolds stresses) at 0,97 (a) and 0,84 (b) altitude. Dimension L represents the altitude of the model.

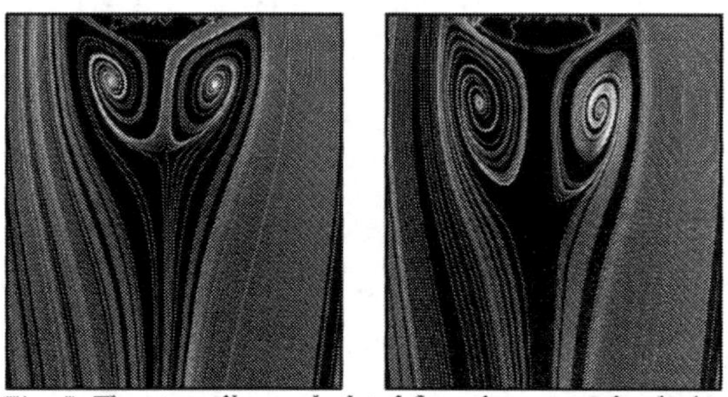

Fig. 5. The streamlines calculated from the averaged velocity field revealed two symmetric eddies (altitudes y/L 0,84 a 0,97).

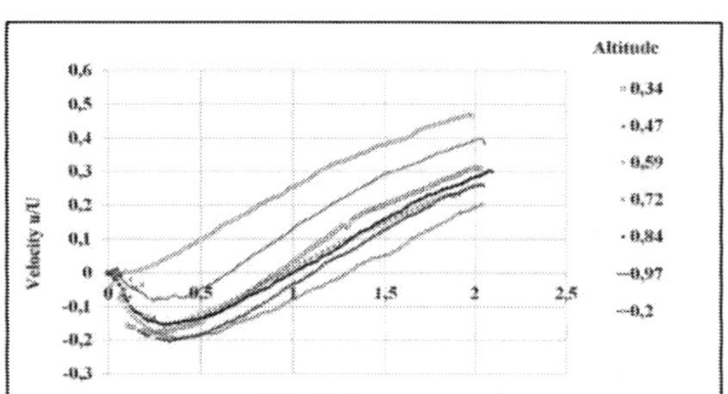

Fig. 6. u/U velocity x-component profiles throughout wake at varying altitudes y/L. Dimension D represents minimum diameter of the model, L altitude of the model and velocity U represents velocity of incoming flow.

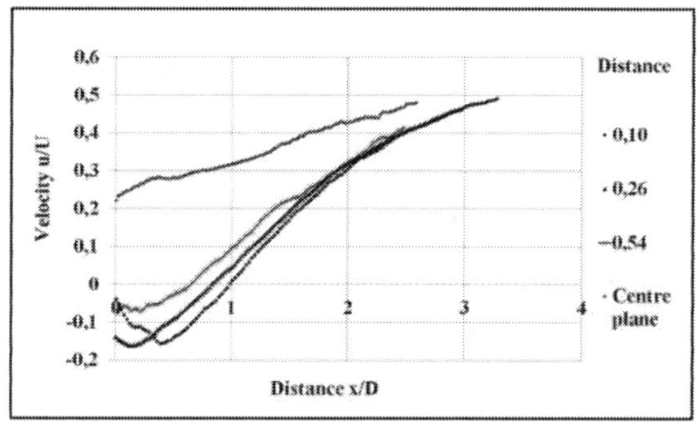

Fig. 7. u/U velocity x-component profiles throughout the wake at various distances z/D off the object axis plane as extracted at altitude y/L=0,84. Dimension D represents minimum diameter of the model, velocity U represents velocity of incoming flow.

The result of this phenomenon – influence of different turbulence intensity on the wake behind the model - can be seen in Fig. 1. The measuring plane for this particular test was set as parallel to the ground plane, at an altitude of y/L=0,63 from it. The influence of the turbulent intensity of the inlet flow on the wake is considerable, as it was already noticed in the abstract.

Afterwards, other measuring planes for more tests were parallel to the main one described above, in varying distances (see a pictogram in Fig. 3).

The statistically averaged velocity fields were calculated for each measurement, and then streamlines showing recirculating areas and possible location of most expected 3D structure, arch vortex, were evaluated. To confirm or disprove the existence of the arch vortex, other measurement planes for the wake were gained - parallel to the flow direction in the object axis and perpendicular to the ground this time (see a pictogram in Fig. 8).

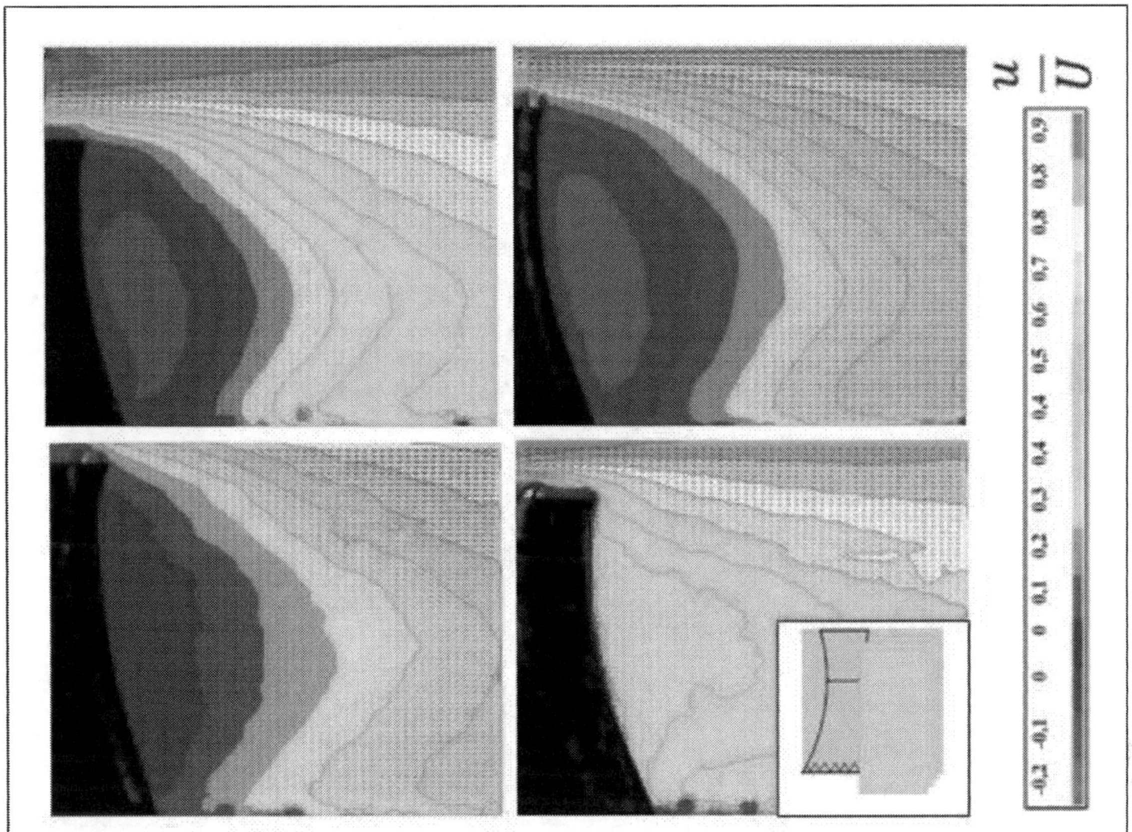

Fig. 8. The averaged velocity fields calculated from the measurements in the planes perpendicular to ground, parallel to the object axis plane, varying distances z/D from a centre from top: 0; 0,10; 0,26; 0,54. Dimension L represents altitude of the model and velocity U represents velocity of incoming flow. The scalar map shows x-component of velocity, a range u/U is -0,2 to 0,9. Bottom pictogram shows measured areas. The areas of fields of view x/D x y/L are as follows: z/D=0: 1,3 x 4,5; z/D=0,10: 1,2 x 3,3; z/D=0,26: 0,93 x 2,7; z/D=0,54: 0,9 x 2,85.

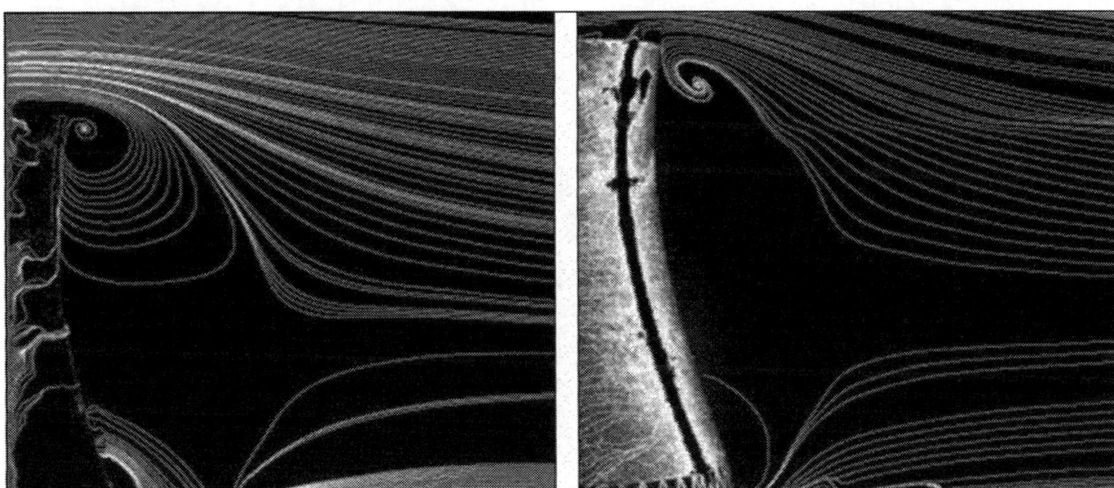

Fig. 9. Streamlines calculated from the averaged velocity field show the structure known as arch vortex (model-axis plane and z/D=0,26 off the axis plane).

4 Conclusion

The wake of a hyperboloid-shaped model was measured in the wind tunnel with a modelled atmospheric boundary layer. The measurement was provided by means of 2D Particle Image Velocimetry. The ratio of the height of the model to the thickness of the modelled boundary layer in the BLWT was 1/3, meaning that turbulence in the boundary layer strongly influences the flow around the model. Presented in Fig. 1, this parameter supresses the wake, leading to a much shorter area of recirculation than low turbulence incoming flow would cause.

The wake influenced by the high level of turbulence in the incoming flow was much shorter and followed turbulent separation; while the wake not disturbed by the high level of turbulence was longer and followed laminar separation. The turbulent separation is present in the subcritical range of the Reynolds number.

From averaged velocity fields (Fig. 3), one can see the size of the recirculating area in different planes. In addition, if the sum of variances and covariance pictures, which are proportional to turbulent kinetic energy and Reynolds stresses respectively, are displayed (Fig. 4), one can surmise a shear layer border and an area of boundary layer separation.

The single illustrations in Fig. 5 indicate the run of the vortex filament in a perpendicular plane to the ground. It actually copies the shape of the recirculating area. The idea obtained from the measurement is that the arch vortex is also bent in the x direction. This could be the contribution of the three dimensions of the model, or the boundary layer, or both equally.

An interesting, but not surprising, result was given by extracting the x-component velocity plots from the central part of the wake (Fig. 6 and 7). The zero on the x-axis was taken in the surface of the model. A closer view to the single graphs detects, that the running of the velocity at various altitude levels differs and there is no recirculating area close to the top. In the near wake, up to the $x/D = 0{,}3$ behind the cylinder, the results seem to be chaotic. There is also no regularity. We cannot just say the closer to the ground the slower the flow is.

Looking at the results in Fig. 9, the streamlines calculated from the averaged velocity field revealed supposed 3D structure - the desired arch-shaped vortex. Its upper part appears at the dimensionless altitude between y/L 0,84 and 0,97. Other planes measured parallel to this one were in the distances of $z/D = 0{,}1$ and 0,26, and still showed possible vortex filament. It is obvious that at a distance of $z/D = 0{,}26$ the arch already veers towards the ground.

Acknowledgements

The support from the Project No. CZ.2.16/3.1.00/21569 Centre 3D Volumetric Anemometry is gratefully acknowledged.

This work was supported by the Grant Agency of the Czech Technical University in Prague, grant No. SGS17/074/OHK2/1T/12.

The research was performed with financial institutional support from the government budget through the Ministry of Industry and Trade of the Czech Republic.

References

1. V. Barraclough, J. Novotný, P. Šafařík: Study of Flow around Model of Cooling Tower by Means of 2D Particle Image Velocimetry Measurement. Experimental Fluid Mechanics, *EPJ Web of Conferences,* 01/2017 (2016).
2. V. Barraclough, J. Novotný, P. Šafařík: Experimental Investigation of 3D Vortex Structures in Wake of Hyperboloid-Shaped Model, Topical Problems of Fluid Mechanics, *Conference Proceedings* (2017).
3. M.M. Zdravkovich: Flow around Circular Cylinders. Vol.2: Applications, *Oxford University Press*, (2003).
4. R.J. Pattenden, S.R. Turnock, X. Zhang.: Measurements of the Flow over a Low-Aspect-Ratio Cylinder Mounted on a Ground Plane, *Experiments in Fluids* 39: 10–21. (2005).
5. J.L. Heseltine: Flow around a Circular Cylinder with a Free End. *A Thesis for the Degree of Master of Science*, The Department of Mechanical Engineering, University of Saskatchewan, Canada (2003).
6. D. Sumner, J.L. Heseltine: Tip Vortex Structure for a Circular Cylinder with a Free End, *Journal of Wind Engineering and Industrial Aerodynamics 96,* (2008).
7. H. Sakamoto, M. Arie: Vortex Shedding from a Rectangular Prism and a Circular Cylinder Placed Vertically in a Turbulent Boundary Layer. *Journal of Fluid Mechanics*, pp. 147–165, vol. 126, (1983).
8. European Standard EN 1991-1-4:2005 Eurocode 1: Actions on Structures.
9. M. Jirsák: Physical Principles of Work with Boundary Layer Wind Tunnel. *An Internal Report (in Czech)*, Aeronautical Research and Test Establishment, Plc., Prague, Czech Republic (2012).
10. M.M. Zdravkovich: Flow around Circular Cylinders, *Oxford University Press* (1997).
11. M.M. Zdravkovich: Conceptual Overview of Laminar and Turbulent Flows Past Smooth and Rough Circular Cylinders. *Journal of Wind Engineering and Industrial Aerodynamics*, 33, 53-62 (1990).
12. R. Merrick, G.T. Bitsuamlak: Control of Flow around a Circular Cylinder by the Use of Surface Roughness: A Computational and Experimental Approach, *www.researchgate.net* (2014).

Wet stack – Analysis of condensation

Ondřej Bartoš[1,*], *Jan Havlík*[1], and *František Hrdlička*[1]

[1] CTU in Prague, Department of Energy Engineering, Technická 4, Prague 6, 166 07, Czechia

Abstract. The aim of the paper is to introduce the study of condensation processes in the wet stack. Because of the high cost of reheating (due to the loss of useful heat of flue gases), wet stacks are being considered for new or retrofit applications of wet flue gas desulfurization (FGD) systems around the word. Wet stacks, in contrast to classical chimneys where flue gases are heated up to avoid any condensation, works with wet flue gases and condensation is welcome. The study provides a quantitative analysis of the condensation.

1 Introduction

At present, we distinguish between two types of industrial stacks according to the water saturation of flue gases. In the "classical" stack, the flue gas temperature is significantly above the dew point of the flue gases. This can be achieved either by exhausting the flue gases without the use of cleaning devices or by the additional heating of the flue gases after desulfurization. In both cases, no condensate under steady operating conditions will be generated in the flue gas liner, although condensation may occur in the start-up and non-design conditions.

In "wet stacks" the exhaust flue gases are saturated with moisture. A typical wet stack is located downstream of a wet Flue Gas Desulfurization (FGD) system. FGD systems reduce the sulphur dioxide (SO_2) content by spraying slurry of limestone and lime into the flue gas stream. While SO_2 is scrubbed from the gas stream, water from the slurry saturates the flue gas. In addition, the relatively cool slurry typically reduces the temperature of the flue gas to the range of 45°C to 55°C. Since the exhaust gas flow is saturated with water, condensation will occur in these stacks during steady-state and start-up conditions.

The construction of the typical stack has a reinforced concrete windshield with one or more flue gas liners inside. Depending on the operating conditions of the chimney and the surrounding conditions, there are three basic sources of the liquid in the flue gas liner.

We consider these resources as::

- Condensation on the liner wall due to temperature and concentration gradients
- Adiabatic condensation
- Droplets which passed through the eliminator after FGD

The condensation layer especially on the wall of the stack may be pulled back into the flue gas stream. This phenomenon may result in large drops drifting out of the chimney where they can fall on the ground nearby. The amount of falling droplets depends on the rate of combustion and the properties of the condensate layer, heat transfer and mass transfer.

Fine droplets usually evaporate before they impact the ground. However, large droplets can fall on the ground like an acid rain and cause nearby objects, equipment and vehicles to corrode. Large droplets mainly come from the condensate, which is accumulated on the wall of the liner and is pulled back from the wall and is blown out of the chimney by the flue gas. This phenomenon is sometimes called "spitting" from the upper part of the chimneys. An important aspect of the liner head arrangement is therefore to reduce the droplet size emitted from the chimney by flue gases to prevent droplets from falling to the ground.

In addition to the potential corrosion caused by the dispersion of droplets into surrounding areas, wet and acid flue gases from FGD may chemically interact with the reinforced concrete structure of its own chimney, especially on the head of the stack. Properly selected materials and well-designed condensate removal equipment can minimize the corrosion. Commonly used lining materials are acid-resistant bricks, closed-cell ceramic foam, FRP and nickel alloys. Although corrosion resistance is an important feature of the liner, thermomechanical properties such as thermal conductivity, specific heat capacity and density can also significantly affect heat transfer and mass transfer in the liner. Heat and mass transfer can then affect the amount of condensation created on the walls.

2 Stack heat transfer

As an example of wet stack condensation analysis the typical industrial chimney would be expected to have one liner in the construction of its reinforced concrete windshield.

* Corresponding author: ondrej.bartos@fs.cvut.cz

The total heat transfer through the heat transferring area is given by the equation (1).

$$Q = U \cdot A \cdot \Delta T_{log}, \quad (1)$$

where U is the overall Heat transfer coefficient (HTC), A is the heat transfer surface and is the logarithmic mean temperature difference. For the case of the wet stack it is necessary to divide the heat transfer into two parts. The overall HTC for the liner is calculated from the term

$$U_d = \frac{1/D_{d-o}}{\frac{1}{d_{d-i} \cdot h_{d-i}} + \frac{1}{2k} ln\left(\frac{D_{d-o}}{d_{d-i}}\right) + \frac{1}{D_{d-o} \cdot h_{d-o}}}, \quad (2)$$

where is the inside diameter, k is the thermal conductivity, and is the outside diameter. The overall HTC for the concrete construction is calculated from the term

$$U_{st} = \frac{1/D_{st-o}}{\frac{1}{d_{ts-i} \cdot h_{st-i}} + \frac{1}{2k} ln\left(\frac{D_{st-o}}{d_{st-i}}\right) + \frac{1}{D_{st-o} \cdot h_{st-o}}}, \quad (3)$$

The heat transfer through the liner is equal to the heat transferred through the stack

$$Q_d = Q_{st} \quad (4)$$

The HTCs are determined on the basis of EN 13084-1 [1], except for the HTC on the inner side of the duct where the water vapor condenses. For the through-air gap between the duct and the stack, the HTC on the outer side of the duct is 8 $Wm^{-2}K^{-1}$, the HTC on the inner side of the stack is 8 $Wm^{-2}K^{-1}$ and the HTC on the outer side of the stack is 24 $Wm^{-2}K^{-1}$ Since the condensation of water vapor in a mixture with flue gases is a more complex process compared to the condensation of pure water vapor, this issue is dealt with in a separate part 2.1. The physical properties of flue gas mixture are calculated according to the composition of the specified flue gases with a consideration of the temperature.

2.1 Condensation of water vapour in mixture with flue gas in a flue gas liner

Condensation of water vapor in the mixture of flue gas is different from condensation of pure water vapor. In a mixture, the molecules of steam diffuse through flue gas towards the vapor-liquid interface. There is a difference in the concentrations of steam in the bulk gas mixture C_∞ and at the condensation phase interface C_i and mass transfer occurs [2], [3]. Steam concentrations decrease along the length of the tube at the same rate as corresponding steam partial pressures p_∞ resp. p_i and temperatures T_∞ resp. T_i (see Figure 1). On the contrary in the case of pure steam condensation, the temperatures T_∞ and T_i are identical and constant during the process.

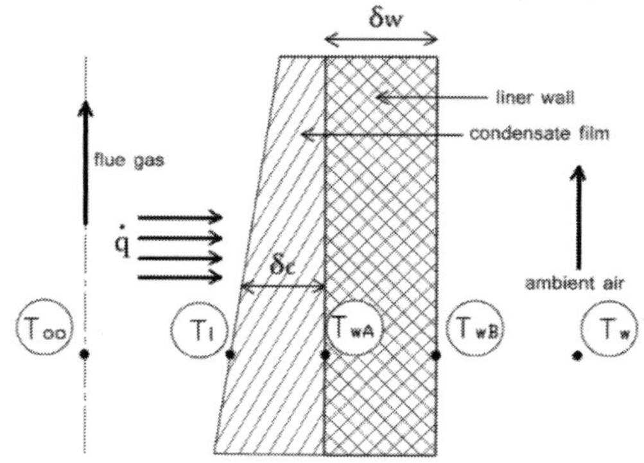

Fig. 1. Heat transfer in a flue gas liner

The water vapor concentration in the mixture decreases during the process due to water vapor condensation. In the case of a wet stack the decrease of concentration is very small; hence the dew point temperature decreases very slowly. Conversely, the concentration of flue gas, the absolute amount of which will not change, will increase. This also reduces the condensation temperature of the water vapor. The effect of the presence of non-condensing gas is twofold, apart from reducing the heat transfer rate, it is also decreases the condensing temperature during the process.

2.2 Heat balance calculation

The basis of the calculation model is the heat balance of the process. The heat flux from water vapor condensation is transferred through the condensate by conduction, thought the wall by conduction and between the wall and the ambient air. For the calculation it is necessary to know the temperatures of both media, the ambient air T_a and the flue gas T_∞. The other temperatures between these temperature (condensation temperature Ti and temperature on both pipe surfaces T_s and T_{s0}) are determined by the equality of the heat flux.

In the total heat transfer process, the value of the heat flux $\dot{q}$ consists of the following processes:

- heat flux from the condensing steam
$$\dot{q} = \dot{m}_c \cdot h'_{fg}, \quad (5)$$
where $\dot{m}_c\ [kg/sm^2]$ is the amount of condensing steam, and h'_{fg} is the latent heat of condensation.
- heat flux through condensate film
$$\dot{q} = h_{ci} \cdot (T_i - T_{wA}), \quad (6)$$
where h_{ci} is the condensation HTC, where T_i is the steam condensation temperature and T_{wA} is the wall temperature.
- heat flux through the tube wall
$$\dot{q} = \frac{k_w}{\delta_w} \cdot (T_{wA} - T_{wB}), \quad (7)$$

where k_w is the thermal conductivity of the wall, δ_w is the thickness of the wall, T_{wB} is the wall temperature.

- heat flux to the ambient air

$$\dot{q} = h_w \cdot (T_{wB} - T_a), \quad (8)$$

where T_a is the ambient air temperature.

To determine the HTC value T_i, it is necessary to calculate the condensation temperature at the phase interface T_i taking into account the mass transfer generated by the different partial pressures at the interface between the gas and the liquid phases p_i and bulk gas mixture p_∞ [1], [3], [4]. In accordance with the analogy between the mass transfer and the heat transfer, the condensation HTC is evaluated according to the Nusselt model combined with the application of the suitable criterion equation for the mass transfer.
Condensing molar flux $\dot{n}_v$ describes the balance of the process

$$\dot{n}_v = \beta C_T ln \left[\frac{1 - \bar{y}_{v,\infty}}{1 - \bar{y}_{v,i}}\right] \quad (1)$$

where β is the mass transfer coefficient, C_T is the total molar concentration, $\bar{y}_{v,\infty}$ is the steam mole fraction in the bulk gas mixture, respective at the liquid-vapor interface $\bar{y}_{v,i}$. The mass transfer coefficient β is determined according to a suitable criterial equation analogically with the heat transfer. The partial pressure $p_{v,i}$ at the liquid-vapor interface is given as

$$p_{v,i} = \bar{y}_{v,i} p \quad (2)$$

and the temperature at the liquid-vapor interface T_i as

$$T_i = T_{sat}(p_{v,i}) \quad (3)$$

The HTC on the inner side of the duct h_{d-i} is given after calculating the heat balance as

$$h_{d-i} = \frac{\dot{q}}{(T_\infty - T_{wA})} \quad (9)$$

2.3 Results

A common geometry of stacks used for steam power plants was chosen for the calculation.

Stack geometry:
Stack height 250 m
Stack inner diameter 21.771 m to 15 m in level of 194 m and up constant
Stack outer diameter 23.271 m to 15,4 m in level of 194 m and up constant
Liner inner diameter 5.3 m
Fireclay thickness 0.1 m
Foam glass thickness 0.05 m

Calculation parameters were chosen in respect to the operating and climate conditions in the Czech Republic.
Operating conditions:
Flue gas temperature 50 °C
Ambient air temperature -10 °C, + 10 °C; resp. +30 °C
Flue gas volume 250Nm^3/s resp. 166.7 Nm^3/s a 83.3 Nm^3/s

The state with a flow rate of 250 Nm^{3}s^{-1} and an ambient temperature of 10 °C is calculated as the basic option. Subsequently, the results were calculated for other modes of operation, taking into account changes in the ambient air temperature, in the flow rate of the flue gases and the stack construction with and without foam glass insulation.

2.3.1 The case with foam glass insulation

The results for the case using foam glass insulation with changes in the ambient air temperature, respective to the flue gas volume, are shown in Tab. 1 and Tab. 2.

Tab. 1 Effect of changing the ambient air temperature

Ambient air temperature	°C	-10	10	30
Flue gas volume	Nm3/s	250	250	250
Transferred heat	kW	175	117	58
Amount of condensate	kg/h	279	186	93

Tab. 2 Effect of changing the flue gas volume flow

Ambient air temperature	°C	10	10	10
Flue gas volume	Nm3/s	250	166,7	83,3
Transferred heat	kW	117	116	116
Amount of condensate	kg/h	186	186	185

2.3.2 The case without foam glass insulation

The results for the case without foam gas insulation with changes in the ambient air temperature, respective to the flue gas volume, are shown in Tab. 3 and Tab. 4.

Tab. 3 Effect of changing the ambient air temperature

Ambient air temperature	°C	-10	10	30
Flue gas volume	Nm3/s	250	250	250
Transferred heat	kW	621	414	207
Amount of condensate	kg/h	991	661	330

Tab. 4 Effect of changing the flue gas flow

Ambient air temperature	°C	10	10	10
Flue gas volume	Nm3/s	250	166.7	88.9
Transferred heat	kW	414	412	408
Amount of condensate	kg/h	661	657	651

https://doi.org/10.1051/epjconf/201818002005

3 Adiabatic condensation

This mechanism is named after the basic thermodynamic changes in which no heat is shared, the so-called adiabatic change. The change can be described by a classical relationship

$$pv^{\kappa} = konst. \tag{10}$$

or it can be expressed from the first law of thermodynamics

$$dq = 0 = di - vdp = du + pdv \tag{11}$$

However, in the case of flue gas flow in the chimney, this change cannot be considered as an isentropic change because there are additional processes (condensation, mixing) that are associated with entropy growth. The idea of adiabatic expansion in a chimney is to be considered as an approximate description, because the heat transfer is continuously going to the environment as flue gas passes through the liner of the chimney and the chimney's own concrete shaft. On the other hand, the diameter of the liner is usually in the order of meters, and therefore we can assume that a process without heat transfer occurs in the core of the flow due to the speed of transport processes such as diffusion and turbulence.
The condensation of water vapor in the flue gas stream is mainly due to the decrease of the aerostatic pressure of the flue gas with the height of the liner. This decrease can be described by a barometric equation.

$$p = p_0 e^{\frac{\rho_0 g dh}{p_0}} \tag{12}$$

When considering the decrease of the flue gas pressure with the height of the liner and assuming that the flue flows in the chimney without the heat release to the environment, we obtain from (12) a relation (13) for determining the flue gas temperature during flow in the chimney.

$$T_h = T_0 \left(\frac{p_h}{p_0}\right)^{\frac{\kappa-1}{\kappa}} \tag{13}$$

The condensation occurs due to the faster drop in water vapour pressure than the decrease in total water partial pressure drop.
The flue gas density and adiabatic exponent κ should be determined on the basis of the actual flue gas properties. Flue gas can be considered as a mixture of different gases in which each has its own specific properties. For the purposes of determining the properties in this paper, IAWPS 97 water/steam tables were used to determine the water properties (H_2O) and other gases are determined using the ideal gas model for a specific molecule according to the state equation:

$$pv = rT \tag{14}$$

Where r is the specific gas constant and our case of gas mixtures according to Table 5. has a value of 285 J / kgK and κ is 1.37. This approach can be considered to be accurate enough for most engineering applications.

Tab. 5 Volumetric composition of the reference flue gases

Gas	Volume ratio [%]
H_2O	cca 10-12
CO_2	cca 11,5
O_2	cca 7-9
N_2	cca 70

From the equations 12 to 14 we can obtain the theoretical maximum drop in temperature given by the pressure drop in the vent and the associated temperature drop in Figure 2.

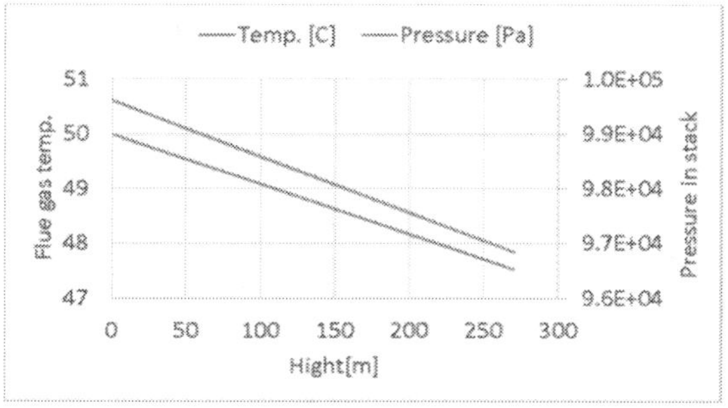

Fig. 2. Pressure and temperature profile along the stack height without condensation influence.

Figure 2. shows that the theoretical drop of the temperature is approximately 2.5 °C. In this way, we could model the temperature drop in the case of unsaturated flue gas, till the saturation. In our case, we already have saturated fumes at the stack inlet. The latent heat released by the condensation is going against the temperature drop by the expansion of the flue gas in the stack.
To determine the output temperature at the liner outlet one can use a simple iteration process where the amount of heat released from the condensation will be equal to the heat required to heat the flue gas flow to the outlet temperature. The resulting output temperature drop from the iterative calculation is 0.6°C. The actual output temperature may differ depending on the intensity of the flue gas cooling from the wall of the stack, against which various dissipative effects occur in the flue gas stream. The maximum amount of condensed liquid is about 850 Kg / hour in our case. This value is high, but it should be remembered that this condensation takes place in the flue gas stream. Primarily, there are already existing droplets in the flue gas stream and the surface these droplets the takes place on the surface. The transport of these droplets to the surface of the liner where they are trapped is mainly due to turbulence in the chimney. Other transport phenomena do not have a significant influence on the transfer of droplets from the main stream to the walls of the liner.

3.1 Droplet growth

In the following paragraph, we will focus on how much the diameter in drops present in the flue gas stream beyond the absorber will increase. The formation of new droplets resulting from homogeneous nucleation is less probable due to low supersaturation of the exhaust gas.

https://doi.org/10.1051/epjconf/201818002005

The growth of drops in the flue gas environment can be written according to equation 15.

$$r = \left(r_*^2 + \frac{2(S-1)}{C_1+C_2}(t-t_0)\right)^{0.5} \tag{15}$$

Where S is the supersaturation and can be expressed as the ratio of the partial pressure of water vapor and the saturation pressure (16)

$$S = \frac{p_v}{p_s} \tag{16}$$

r* is the original droplet diameter at the beginning of the process. The C1 and C2 values respect the water vapor diffusivity in the carrier gas and the thermal conductivity.
To solve this project, there is no assigned water drop distribution function at the chimney inlet, so we calculate the droplet growth for 4 different diameters which are interesting for the input.

Tab. 6 Dependence of the outlet diameter of the droplet on the inlet diameter.

D_{inlet} [μm]	5	50	150	300
D_{outlet} [μm]	42	65	156	303

Table 6 shows that smaller droplets are growing faster. For larger droplets, growth is very slow, although the transport of matter is essentially the same. The driving force of condensation here is the overflow, which moves in very small quantities, so condensation cannot be expected to involve other condensing particles apart from the existing droplets. There may be such a situation when the input number of condensation seeds will not be able to satisfy the condensation flow of H_2O molecules to the droplet surface, and thus the flue gas in the non-equilibrium state will leave the chimney. This would again mean a smaller amount of condensed moisture. However, this problem is very complex and requires a detailed study for each stuck situation, including the acquisition of experimental data that would determine the dimensional structure of the liquid phase at the chimney inlet as well as its output.

3.2 Deposition of droplets on the wall

When flue gas flows through the chimney, we assume a turbulent flow. The deposition of droplets on the wall of the liner in this type of flow is difficult to accurately model. In the literature [16], the particle concentration is assumed to be constant with respect to the cross section of the channel, and the wall is the boundary layer area where a linear decrease in concentration is considered. However, determining the thickness of the boundary layer is difficult, so semi-empirical solutions are commonly used. The main capture dominates two main mechanisms - turbulent diffusion and inertial capture. In the case of chimney breaks, we use an empirical relationship that respects both of these mechanisms.

$$P = exp\left(\frac{-4V_{dep}L}{d_t U}\right) \tag{17}$$

Where V_{dep} is the deposition velocity, which could be written as

$$V_{dep} = \frac{0{,}04U}{Re^{0{,}25}}\left(\frac{\rho_g D}{\mu}\right)^{2/3} \tag{18}$$

L is the length of the vent, U is the flue gas velocity, D the diffusion coefficient, Re the Reynolds flow rate and dt is the diameter of the liner.
The results are shown in Table 7 below

Tab. 7 Results of the deposition model

$D_{droplet}$ [m]	5 x10^{-6}	5 x10^{-5}	1 x10^{-4}	3 x10^{-4}
D [m^2/s]	5x10^{-12}	5 x10^{-13}	2 x10^{-13}	8 x10^{-14}
V_{dep} [m/s]	5 x10^{-7}	1 x10^{-7}	6 x10^{-8}	3 x10^{-8}
P [-]	0.99	0.99	0.99	1

P indicates the probability of passing through the stack liner. From the values in the table, it is clear that if it were smooth without any obstructions, the capture of droplets on the wall from the mainstream would be negligible. Here, however, it is important to emphasize that obstacles in cross-section can increase the amount of drops caught, especially for larger droplet diameters. On the other hand, these obstacles need to be riddled with sharp edges from which the condensate could be pulled back into the flue gas stream.

4 Discussion

The analysis of the liquid formation in the wet stack provides the result that the liquid is coming mostly from the heat transfer to the environment. The amount of the condensate in the presented case is in the range of 500-600 kilograms of condensate. The liquid from the adiabatic condensation is generally at the same order of magnitude as in the case of the heat transfer but most of the condensate moisture remains in the flue gas flow and is spread to the environment. For the designers of the wet stack the presented paper can be used as a guide to determine the condensate collecting system from the stack. The condensate could be usually returned to the FGD.

Together with the study of the condensation in the wet stack. The possibility [5,6] of reducing the pollutants collected in the condensate liquid is clear. It is known that the condensate contains a higher concentration of pollutants in comparison with the bulk flue gases. Due to the lower temperature of the flue gas, the wet stack has slightly worse dispersion characteristic in the vicinity of the stack. The use of the wet stack can provide positive effects on the efficiency of the cycle and reduction of pollutants which are captured in the condensation process. At the moment the pollutants are measured before they enter the stack and the operators of energy facilities are not motivated to reduce the amount in the wet stack. The wider use of the wet stack with respect to lower outlet velocities and the temperature of the flue gases can provide additional decreases of pollutants.

Acknowledgements: We gratefully acknowledge the support by the employees of Omega – Teplotechna, a.s. who provided their experience from the real-life use of wet stacks.

References

1. G. F. Hewitt, G. L. Shires, T. R. Bott, *Process Heat Transfer* (Begell House, New York, 2000).
2. J. Sestak, F. Rieger, *Prenos hybnosti tepla a hmoty* (CTU, Prague, 2004) (in Czech)
3. ČSN EN 13084-1 Volně stojící komíny – Část 1: Všeobecné požadavky, ICS 91.060.40, Praha: ÚNMZ, září 2007
4. W.C. Hinds, Aerosol technology (Second Ed.). Wiley, New York, 1999.
5. de Kreij, A.; Anderson, D.K.; Increasing power plant efficiency for safe wet stack operation, Cicind report, Vol. 32, No. 1
6. B. Jezek at al,: Industrial Chimneys vs Cooling Towers: International Symposium on Industrial Chimneys and Cooling Towers, Prague, Oct 8-11, 2014

https://doi.org/10.1051/epjconf/201818002006

New methodology for the heat flow prediction for different construction materials

Elhadj Benachour[1,2,a], *Belkacem* Draoui[1], *Bachir* Imine[2], *Khadidja* Asnoune[3], *Allaoua* Boumediene [3], *Brahim* Mebarki[1]

[1]Laboratory of Energy in Arid Regions, University of Tahri Mohamed - Bechar, P.O. P 417. 08000 , Road Kenedza , Bechar, Algeria
[2]Department of Mechanical Engineering, University of Sciences and Technology of Oran -Mohamed Boudiaf, P.O.1505 El-Mnaouar, Oran, Algeria
[3]Laboratory of Smart Grids and Renewable Energies, TAHRI Mohammed , University of Bechar, BP. 417 Bechar (08000), Algeria

Abstract. Among the ways of energy transfer by conduction is that of molecular interaction, in which the greater motion of a molecule at a higher energy level (temperature) imparts energy to adjacent molecules at lower energy levels. This type of transfer is present, to some degree, in all systems in which a temperature gradient exists and in which molecules of a solid, liquid, or gas are present. So, in heat transfer, the thermal conductivity of a substance is an intensive property that indicates its ability to conduct heat In particular in the building sector. The thermal flux is often measured with a mathematical analysis but for the same material, on the other hand the estimate will be disruptive and sometimes very difficult when the material changes. In this paper, a single equation for predicting heat flux of different materials is given.

1 Introduction

According to the Second Law of Thermodynamics, heat transfer is only possible in the direction from higher temperatures to a lower one. It becomes zero if temperatures are equal. The heat loss through an envelope should therefore be proportional to the difference Temp (inside) – Temp (outside), or to a positive power of it for small differences. For a simple formula, a linear dependence on temperature difference is sufficient. Accepting further that heat loss grows linearly with the surface. Knowing how heat "flows" inside a structure is significant for designers to make sure that the structure will perform its intended heat conduction function properly [1]. Several works are interested in the conduction heat transfer in several materials and several domains [2-11], and other works are interested in convective heat transfer and conduction convection coupling [12-13]. Irrespective of the mode of heat transfer, the researcher is interested in the energy flow through which is diffused by this building material among the most important physical and thermal properties there is thermal conductivity. In this work, we will vary the conductivity thermal analysis of four different materials: wood, concrete, brick and granite, and mathematical calculation of the thermal flux to find a point cloud that helps us to use Lagrange's polynomial interpolation method to develop a only correlation to estimate the heat flux diffused by one of these four materials proposed at all times.

2 Geometric configuration

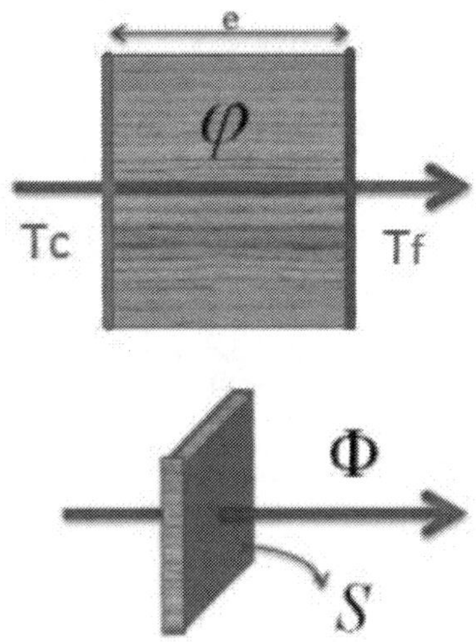

Fig. 1. Schematic of the studied configuration

To simplify the problem, assume that:

- Heat transfer is in the direction of decreasing temperature (basis for minus sign).

[a] Corresponding author: benachour_elhadj@yahoo.fr

- Direction of heat transfer is perpendicular to lines of constant temperature (isotherms).
- Heat flux vector may be resolved into orthogonal components
- For transient conduction, heat equation is first order in time, requiring specification of an initial temperature distribution

3 Objectives

We can express our aim for this study in the following points:

- study of heat transfer by conduction in different types of material
- Studied the effect of thermal conductivity on thermal flux and conduction heat transfer in different types of material
- Prediction of the Heat flow for different construction materials.

4 Mathematical model

The differential form of the Fourier's Law of heat conduction is

$$\bar{q} = -\lambda \ \nabla T \tag{1}$$

where $\bar{q}$ is the local heat flux, λ is the thermal conductivity, and T is the local temperature. The Fourier's law is very simple in mathematics and has been widely used in engineering applications and in most scientific research at life scale, even though it is only an empirical relationship.

Experience leads us to believe that as e increases Φ should get smaller. All of these lead to the generalization (made by Fourier in 1807).

A more useful quantity to work with is the heat transfer per unit area, defined as:

$$\Phi = q \ . \ S \tag{2}$$

where:

$$q = -\lambda \ \frac{dT}{dx} \tag{3}$$

the quantity q is called the heat flux and its units are W/m^2 . The expression in (2) can be written in terms of heat flux as:

$$\Phi = -\lambda \ \frac{dT}{dx} \ . \ S \tag{4}$$

for one-dimensional heat conduction (temperature depending on one variable only), we can devise a basic description of the process.

The heat transfer rate in at the left (at x) is:

$$\Phi(x) = -\lambda \ (S \ \frac{dT}{dx})_x \tag{5}$$

the heat transfer rate on the right is

$$\Phi(x + dx) = (\Phi(x) + \left. \frac{d\Phi}{dx} \right|_x dx + ... \tag{6}$$

using the conditions on the overall heat flow and taking the limit as dx approaches zero we obtain

$$\frac{d\Phi(x)}{dx} = 0 \tag{7}$$

or

$$\frac{d}{dx}(\lambda \ \frac{dT}{dx} \ . \ S) = 0 \tag{8}$$

if λ is constant (i.e. if the properties independent of temperature), this reduces to

$$\frac{d}{dx}(\frac{dT}{dx} \ . \ S) = 0 \tag{9}$$

or (using the chain rule)

$$\frac{d^2T}{dx^2} + (\frac{1}{S}\frac{dS}{dx})\frac{dT}{dx} = 0 \tag{10}$$

equations (9) or (10) describe the temperature field for quasi-one-dimensional steady state (no time dependence) heat transfer. We now apply this to our example.

For this configuration (Figure1), the area is not a function of x, i.e. S = constant. Equation (10) thus became

$$\frac{d^2T}{dx^2} = 0 \tag{11}$$

$$\frac{dT}{dx} = A \tag{12}$$

$$T(x) = A(x) + B \tag{13}$$

equation (13) is an expression for the temperature field where A and B are constants of integration. For a second order equation, such as (11), we need two boundary conditions to determine A and B. One such set of boundary conditions can be the specification of the temperatures at both sides of the slab as shown in Figure, say T (0) = Tc; T (e)=Tf.

The condition T (0) = Tc implies that b = Tc. The condition Tf = T (e) implies that Tf = A e + Tc, or

$$A = \frac{Tf - Tc}{e} \tag{14}$$

$$T(x) = A(x) + B \tag{15}$$

with these expressions for a and b the temperature distribution can be written as

$$T(x) = Tc + \left(\frac{Tf - Tc}{e}\right)x \quad (16)$$

this linear variation in temperature is shown in Figure 2 for a situation in which Tc > Tf.

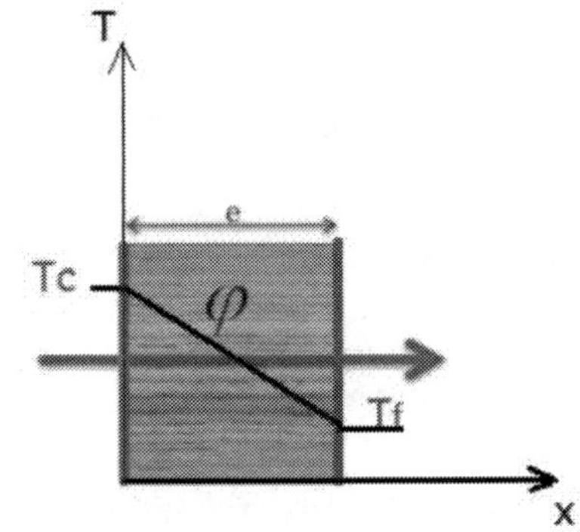

Fig. 2. Temperature distribution through a slab

The heat flux q is also of interest. This is given by

$$q = -\lambda \frac{dT}{dx} = -\lambda \frac{(Tf - Tc)}{e} = \text{constant} \quad (17)$$

There is an electrical analogy with conduction heat transfer that can be exploited in problem solving. The analog of Q is current, and the analog of the temperature difference, Tc - Tf, is voltage difference. From this perspective the slab is a pure resistance to heat transfer and we can define

$$\Phi = \frac{(Tf - Tc)}{R} = \frac{\Delta T}{R} \quad (18)$$

where the thermal resistance is given by:

$$R = \frac{e}{\lambda S} \quad (19)$$

5 Procedure of Simulation

Of course the only mode of heat transfer is conduction but the resistance of the different materials change despite us being interested at the moment when the temperature gradients are equal, due to change in thermal conductivity.

Then, firstly, the thermal flux of each material is calculated mathematically to obtain a cloud of points. then the polynomial interpolation technique will be used to develop a correlation that aggregates all thermal transfer diffusion data in all materials regardless of thermal conductivity.

6 Mathematical model for The Lagrange polynomial Interpolation Method

The Lagrange interpolating polynomial is the polynomial of degree ≤ (n-1) that passes through the n points (x1, y1=f(x1)) , (x2, y2=f(x2)) , ..., (xn, yn=f(xn)), the interpolation polynomial in the Lagrange form is a linear combination [14] and is given by :

$$P_n(x) = \sum_{i=0}^{n} y_i \; L_i(x) \quad (20)$$

for Lagrange basis polynomials

$$L_i(x) = \frac{\prod_{j=0, j\neq i}^{n} (x - x_j)}{\prod_{j=0, j\neq i}^{n} (x_i - x_j)} = \prod_{j=0, j\neq i}^{n} \frac{(x - x_j)}{(x_i - x_j)} \quad (21)$$

where

$$\begin{cases} L_i(x_j) = 0 & \text{if } i \neq j \\ L_i(x_j) = 1 & \text{if } i = j \end{cases} \quad (22)$$

7 Results and Discussion

7.1. The polynomial interpolation points

We wish to find the polynomial interpolating the points

Table 1. The sets of polynomial interpolation points.

	Wood	Concrete	Brick	Granite
λ	0.17	1.75	1.31	0.12
Φ	19.55	201.25	150.65	13.8

where, the points of Φ (λ) are obtained by the mathematical method.

If (0.12, 13.8), (0.17, 19.55), (1.31, 150.65), (1.75, 201.25), are given data points, then the cubic polynomial passing through these points can be expressed as,

$$P_3(x) = y_0 l_0 + y_1 l_1 + y_2 l_2 + y_3 l_3$$

We would have the four basis polynomials

$$l_0 = \frac{(x - 0.17)(x - 1.31)(x - 1.75)}{(0.12 - 0.17)(0.12 - 1.31)(0.12 - 1.75)} \quad (23)$$

$$l_1 = \frac{(x - 0.12)(x - 1.31)(x - 1.75)}{(0.17 - 0.12)(0.17 - 1.31)(0.17 - 1.75)} \quad (24)$$

$$l_2 = \frac{(x - 0.12)(x - 1.17)(x - 1.75)}{(1.31 - 0.12)(1.31 - 1.17)(1.31 - 1.75)} \quad (25)$$

$$l_3 = \frac{(x - 0.12)(x - 0.17)(x - 1.31)}{(1.75 - 0.12)(1.75 - 0.17)(1.75 - 1.31)} \quad (26)$$

We can see that each term is of degree three, so the entire sum has degree at most three.

The polynomial P(x) given by the above formula is called Lagrange's interpolating polynomial and the functions Eq. (23), Eq. (24), Eq (25), Eq.(26) are called Lagrange's interpolating basis functions.

7.2 Heat flow correlation

It should be noted that the numerical result given by the equation Eq.(20). Remarkably similar to the estimate of the Heat flow by the Lagrange interpolation method for each material which her thermal conductivity between [0.12, 1.75].

$$P_3(x) = y_0 l_0 + y_1\ l_1 + y_2\ l_2 + y_3\ l_3 \quad (27)$$

$$P_3(x) = 0.12\ l_0 + 0.17\ l_1 + 1.31\ l_2 + 1.75\ l_3 \quad (28)$$

Therefore, we can write,

$$\Phi(x) = -0.0126\ x^3 - 3.565\ .10^{-3}\ x^2 + 114.28\ x - 0.127 \quad (29)$$

So, since the variable is the thermal conductivity, finally we can write

$$\Phi(\lambda) = -0.0126\ \lambda^3 - 3.565\ .10^{-3}\ \lambda^2 + 114.28\ \lambda - 0.127 \quad (30)$$

8 Comparison and validation of equation

Table 2. The errors estimate.

The material	λ	Φ (λ) The mathematically obtained result (Eq 18) (exact solution)	Φ(λ) The results obtained by our equation (Eq 30) (approximate solution)	Relative error
Granite	0.12	13.80	13.5865269	1.5 %
Wood	0.17	19.55	19.3004341	1.2 %
Brick	1.31	150.65	149.545349	0.73 %
Concrete	1.75	201.25	199.784546	0.73 %

From this comparison, it is very clear that the error is very small for any proposed material that it grows and shows the effectiveness of this developed formula.

9 Methodology testing

Our model comparison has demonstrated that different structures material can be useful in evaluating heat flow transmission In any chosen material with this equation (see fig.3)

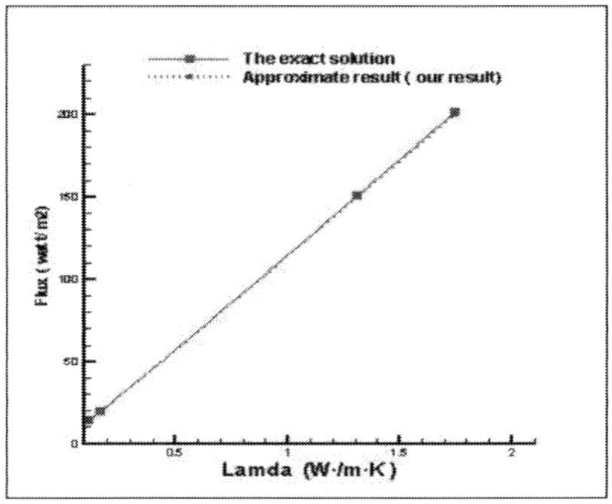

Fig. 3. Comparative study between the exact and approximate solution to estimate the heat flux for different materials

10 Conclusion

Knowing how heat "flows" inside a structure is significant for designers to make sure that the structure will perform its intended heat conduction function properly. This paper describes the principle and the process of defining an optimal topology for heat conduction.

Predictive simple correlation (Eq 30) was developed to estimate the Heat flow by the Lagrange interpolation method for each material: wood, concrete, brick and granite And even for all materials that have their thermal conductivity which varies between [0.12, 1.75] W/m·K. Due to their simplicity and reliability it is very straightforward to use in practice.

Acknowledgment

This work supported by the Research program sponsored by ENERGARID laboratory, the Faculty of Science and Technology,Tahri Mohamed University, Bechar, Algeria.

I would like to thank all reviewers for taking the time and energy to review our work.

https://doi.org/10.1051/epjconf/201818002006

References

1. Baotong Li, Jun Hong, Xiangyang Tian, (2016) Generating optimal topologies for heat conduction by heat flow paths identification, International Communications in Heat and Mass Transfer, Volume 75, July 2016, Pages 177-182.
2. Marcin Zych, Robert Hanus, Pavel Vlasak ,Leszek Petryka and Marek Jaszczur, (2016). Application of nuclear techniques in two-phase liquid-solid particles hydrotransport investigations , EPJ Web of Conferences 114- 0 2145 (2016).
3. Arkadi Berezovski,(2016) On the influence of microstructure on heat conduction in solids, International Journal of Heat and Mass Transfer,Vo lume 103, Pages 516-520.
4. G. Medvedev,(2016) High-temperature heating of liquid in a round tube in terms of axial heat conduction, EPJ Web of Conferences 110- 01039 (2016).
5. Vyacheslav I. Maksimov, Tatiana A. Nagornova and Viktor P. Glazyrin, (2016) Conjugate Heat Transfer in a Closed Volume with the Local Heat Sources and Non-Uniform Heat Dissipation on the Boundaries of Heat Conducting Walls, EPJ Web of Conferences 110- 01038 (2016).
6. Mustafa Turkyilmazoglu,(2015) Exact Multiple Solutions for the Slip Flow and Heat Transfer in a Converging Channel, Journal of Heat Transfer October 01, 2015 | Volume 137, Issue 10
7. Syed Ashraf Ali and Sandip Mazumder,(2015) Phonon Heat Conduction in Multidimensional Heterostructures: Predictions Using the Boltzmann Transport Equation, J. Heat Transfer 137(10), 102401.
8. J. Ordóñez-Miranda and J. J. Alvarado-Gil, (2011) Effective Thermal Properties of Layered Systems Under the Parabolic and Hyperbolic Heat Conduction Models Using Pulsed Heat Sources, Journal of Heat Transfer, September, 2011 | Volume 133 | Issue 9
9. N. A. Roberts and D. G. Walker, (2011) Computational Study of Thermal Rectification From Nanostructured Interfaces, , Journal of Heat Transfer, September, 2011 | Volume 133 | Issue 9
10. Jing Fan and Liqiu Wang,(2011) Erratum: "Review of Heat Conduction in Nanofluids"[Journal of Heat Transfer, 2011, 133(4), p. 040801].
11. Moran wang, bin-yang cao , zeng-yuan guo,(2010) general heat conduction equations based on the thermomass theory, frontiers in heat and mass transfer (FHMT), 1, 013004 (2010), global digital central issn: 2151-8629.
12. Elhadj Benachour, Belkacem Draoui, Bachir Imine, Mohammed Hasnat and Lakhdar Rahmani(2016), Effect of the insulation by the mud on the convection in building in the Sahara of Algeria - case of Bechar City, EPJ Web of Conferences 114-02003(2016).
13. Amina Hammadou, Abdelkrim Missoum, Elmir Mohamed and Belkacem Draoui
The effect of the Saharan climate on the indoor environment of a station of electric energy transformation MV/LV
EPJ Web of Conferences 143, 02032 (2017).
14. Elhadj Benachour, Belkacem Draoui, Bachir Imine, Khadidja Asnoune and Elmir Mohamed (2017), New methodology for the walls design in buildings by numerical simulation of the thermal convection, EPJ Web of Conferences 143, 02007 (2017)

Optimization of the cleaning process on a pilot filtration setup for waste water treatment accompanied by flow visualization

Petr Bílek[1*], *Jakub* Hrůza[1]

Technical University of Liberec, Studentská 1402/2, 461 17, Liberec 1, Czech Republic[1]

Abstract. This paper deals with an optimization of the cleaning process on a liquid flat-sheet filter accompanied by visualization of the inlet side of a filter. The cleaning process has a crucial impact on the hydrodynamic properties of flat-sheet filters. Cleaning methods avoid depositing of particles on the filter surface and forming a filtration cake. Visualization significantly helps to optimize the cleaning methods, because it brings new overall view on the filtration process in time. The optical method, described in the article, enables to see flow behaviour in a thin laser sheet on the inlet side of a tested filter during the cleaning process. Visualization is a strong tool for investigation of the processes on filters in details and it is also possible to determine concentration of particles after an image analysis. The impact of air flow rate, inverse pressure drop and duration on the cleaning mechanism is investigated in the article. Images of the cleaning process are compared to the hydrodynamic data. The tests are carried out on a pilot filtration setup for waste water treatment.

1 Introduction

Water became a strategic commodity as an outcome of world population growth. One of the most promising way, how to safe drinking water sources, is waste water treatment and recycling. The goal is to get rid of solid particles and bacteria as well as soluble materials as hormones, minerals, acids, etc. Filtration is a key mechanism for separation most of the solids and making solid particle free water. Unfortunately filtration membranes suffer flux declension because of formation of a filtration cake and pore blocking. Cleaning process is a very important part of the filtration process which limits the growth of the filtration cake, improves hydrodynamic properties and extends lifetime of the filter.

Optimization of the cleaning process is discussed in [1-4]. Effect of ultrasound on membrane filtration and cleaning process is investigated in [1]. Chemical cleaning optimization on milk filtration is carried out in [2]. It was found out that the chemical cleaning is a very effective way to regenerate membrane flux. Selected chemicals have to be safe, cheap and water washable. Optimization was carried out using multi-regresion in software Modde 8. Operation conditions were optimized such as cleaning time, temperature, cross-flow velocity and concentration of chemicals. The effect of wastewater particle size, concentration, operation conditions and backwashing on filtration of waste water is shown in [3]. The membrane is cleaned by continuous tangential flow and in some time intervals by backwashing. Permeate is mixed with sodium hypochlorite during the backwashing and the retentate with lifted solids leads to the drain. After the backwashing period, the fast flush period is applied, when cross-flow velocity is 10-20 times higher than in the usual filtration cycle. The retentate is mixed with feed water again during the usual filtration cycle at the end. It was found out that small particles cause more resistance to flow because of pore blocking. In the article were also summarized operational costs while using the filtration system. Physical and chemical cleaning is optimized during waste water treatment by a statistical factorial design in [4].

A real-time non-invasive monitoring provides more information about the filtration and the cleaning process next to hydrodynamic data [5-9]. An extensive review of non-invasive and mainly optical methods is written in [5]. Optical methods provide real time video or images which can be stored for later image analysis. These methods include for instance well known direct visual observation method (DVO) and particle image velocimetry (PIV). DVO provides images of filter surface and the resolution is given by visible light – 1 μm. PIV is a strong tool mainly used for velocity field measurement, but it provides also images of illuminated particles. The minimum size of the particles is 10 μm, they have to be spherical, scatter sufficient amount of light and have the same density as water. DVO method was also applied during tangential flow filtration of yeast cells and latex particles accompanied with a surface transport model in [6]. In [7, 8] are discussed non-invasive methods for a filtration cake thickness measurement and observation of a bio-fouling. Removing ultrafine $Mg(OH)_2$ particles from ceramic membrane accompanied by DVO method is shown in [9].

* petr.bilek@tul.cz

 https://doi.org/10.1051/epjconf/201818002007

In the article the optimization of the cleaning process is performed supplemented by visualization of flow. The tests are carried out on a pilot filtration setup for waste water treatment. The optical method enables to see flow behaviour in a thin laser sheet. The impact of air flow rate, inverse pressure drop and duration on the cleaning mechanism is investigated. Images of the cleaning process are compared to the hydrodynamic data.

2 Experimental setup

The scheme of the filtration setup is seen on Figure 1. Three filter modules are mounted in the filtration chamber. The filter module is composed from a plastic frame, where a membrane is deposited from each side. Contaminated water passes through the membrane into the frame, gets rid of the particles and then it flows from the membrane module out into the measuring cylinder.

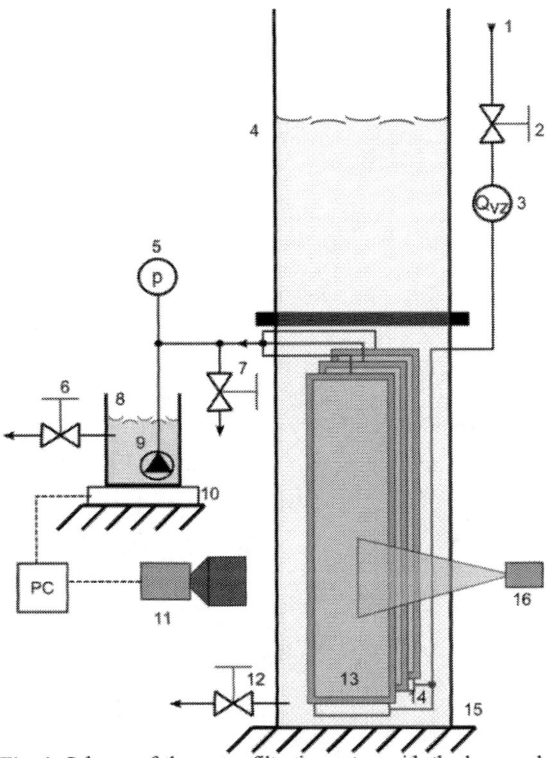

Fig. 1. Scheme of the water filtration setup with the laser and the camera (1: Compressed air supply, 2: pressure reducer, 3: Digital air flowmeter, 4: Constant water level, 5: Pressure sensor, 6: Outlet valve for cleaned water, 7: Outlet valve for samples of filtered water, 8: Measuring cylinder, 9: Submerged pump for backflow, 10: digital scale, 11: Digital camera, 12: Outlet valve, 13: Nanofibrous filter modules, 14: Two perforated tubes for air bubbles, 15: Filtration chamber, 16: Laser module.

Pressure drop is caused by a height of water level, which is controlled by the unit HRH-1 and it is constant (500 mm) during the tests. The pressure drop Δp [kPa] is also constant (5 kPa). Flow Q_v [m^3/s] through the tested filters is measured by the digital scale PCB 10000-1. Permeability

$$k_2 = \frac{Q_v}{A_f \cdot \Delta p} \left[\frac{\mathrm{m}}{\mathrm{s} \cdot \mathrm{Pa}}\right] \quad (1)$$

is calculated from flow-rate, filter area A_f [m^2] and pressure drop. Concentration of particles in inlet C_1 [ug/l] and outlet C_2 is measured by a laboratory nephelometer and the filtration efficiency is calculated according to

$$EF = \left(1 - \frac{C_2}{C_1}\right) \cdot 100\ [\%]. \quad (2)$$

Continual cleaning of the filters is performed by air bubbles and also by reverse flow, which is generated by a small submerged pump supplied by 6 V. Inverse pressure on the filter is measured by GMSD 350 MRE, Greisinger. Compressed air leads to two perforated tubes mounted under filter frames. Air bubbling avoids growing of a thick filtration cake on the inlet side of tested filters. Air flow measured by Testo 6441 is stopped only during the visualisation process because of better visibility.

The cleaning process is investigated in the filter chamber made from transparent plexiglass seen on Figure 2. The inlet side of a tested filter is illuminated by the laser sheet generated by Powell lens (30°) and LED laser module (532 nm, 121 mW). Power of the laser unit was measured by the device LabMaster Ultima v. 2.35 with an attenuator 1:1000. The illuminated flow was taken by the digital camera (Pike F-210B/C, 50 fps, 1920 × 1080 px) equipped by lens (Samyang 35 mm, F1.4).

Fig. 2. Picture of the filtration chamber with a laser unit (1) and a digital camera (2).

3 Experimental and results

3.1 Materials and methods

The tested nanofibrous filter is made of polyamide and the support layer is made of polyester. The nanofibrous layer is made by the electrospinning technique and it was pressed on the support layer via a polymer mesh. The mean pore size is 0.5 μm and the maximum pore size is 0.9 μm. An active area of the one filter module is 0.2 m^2 then the used three modules have the total active area 0.6 m^2. Air permeability is 7 – 10 m/s·Pa.

Distilled water was used with added artificial particles (Bang Laboratories Inc.), which are easier to detect and to visualize compared to the bacteria in real waste water. Size of the particles was chosen considering scattering sufficient amount of light in a laser sheet as well as being small enough to simulate most of bacteria in water. According to the Mie theory [10] the 0.29, 0.54, 0.75 and 1 μm monodisperse polystyrene microspheres are the most suitable, because they occur in the local maximums of the scattering diagram Figure 3. The scattering curve was made in software MiePlot for the scattering angle 90° [11]. Refractive index for water is 1.334 and for polystyrene (testing particles) is 1.595. Size of the most bacteria in wastewater is over one micrometer, the 0.96 μm polystyrene microspheres were chosen as model testing particles [12].

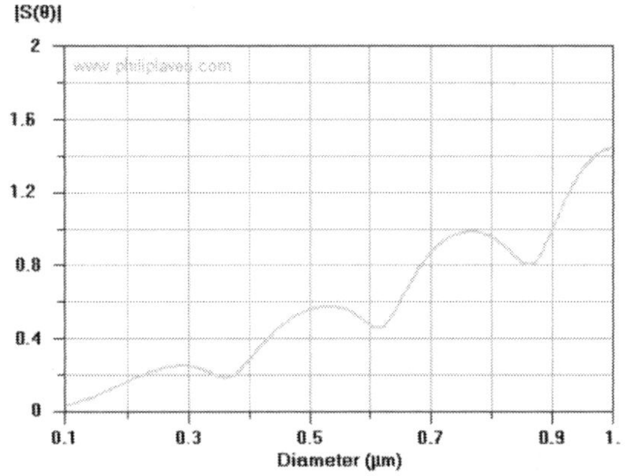

Fig. 3. Scattering diagram for polystyrene spherical particles in water generated by MiePlot software [11].

Flow-rate and filtration efficiency was measured during the tests. Permeability was calculated according to (1) from the flow-rate, which was measured by digital scales. Filtration efficiency was calculated according to (2) from mass concentration measured by a nephelometer. Filtration efficiency was measured for the confidence, that the filtration samples are not damaged and the flow-rate is not wrong. Mass concentration of the particles in the feed water was 200 μg/liter. Measured data are compared to the pictures of the cleaning process.

3.2 Air flow-rate

The filter is cleaned by two methods: continuously by air bubbles and every 50 minutes by reverse flow. To find the optimum air flow rate to the bubble maker (perforated tube under the membrane module), the air flow rate was changed from 0 to 135 litres per minute (LPM). The reverse flow was not activated. The graph of forward water flowrate (green dots) versus air flowrate is seen on Figure 4. Data in the graph were measured every hour. The measured flowrate was corrected because of clogging of the membrane in time (10 % per hour) and influence of air bubbling to water level, to pressure drop and to forward water flow rate (0.56 % per 1 LPM of air flow rate) in our pilot setup. From the graph is seen that the membrane permeability is not sensitive on the air flowrate (bubbling) in a short term of filtration process. The minimum air flowrate was chosen on 30 LPM, it is 50 LPM of air per 1 m^2 of active filter area. According to the direct observing (Figure 5), the all filter modules were in touch with air bubbles to avoid of growing of a dense filtration cake. The filtration efficiency was 100 %.

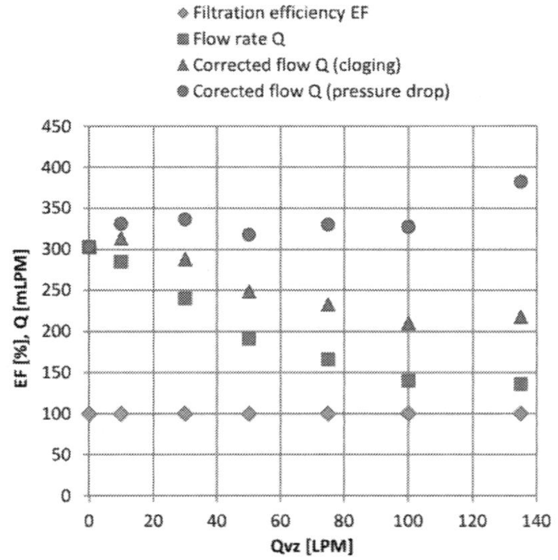

Fig. 4. Graph of the water flowrate versus flowrate of air.

Fig. 5. Image of a filtration module during air flow rate 30 LPM.

3.3 Inverse flow

We can see from the graph on Figure 6, that it is not possible to use higher inverse pressure drop than 15 kPa. Otherwise the membrane is damaged (lower filtration efficiency indicates it). From the graph it is also seen that

https://doi.org/10.1051/epjconf/201818002007

the flowrate after the cleaning process is not changed that much depending on applied inverse pressure drop.

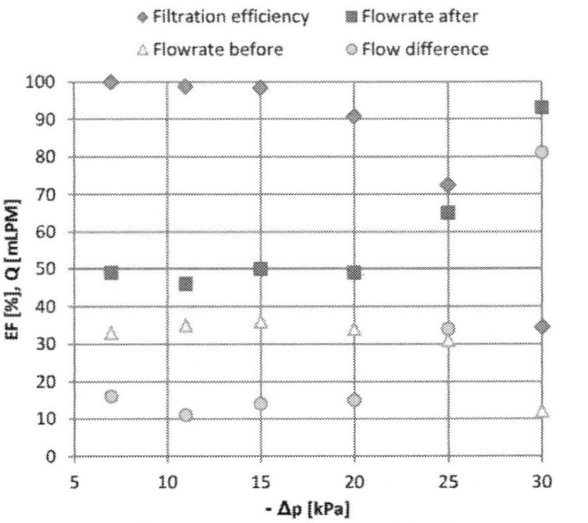

Fig. 6. Graph of the filtration efficiency and the flowrate versus inverse pressure drop on filter (particles: 0.5 μm, duration: 24 hours, active filtration area: 0.2 m^2).

In the second graph on Figure 7 we focused on the inverse pressure drop in range from 0 to 15 kPa and we measured data again in more detail. The measurement was accompanied by visualization helping to find the optimum conditions. We can see from the graph a flow difference (flow after cleaning minus flow before cleaning) versus applied inverse pressure drop. In case of zero inverse pressure drop the membrane was not cleaned and the flow difference was 0.

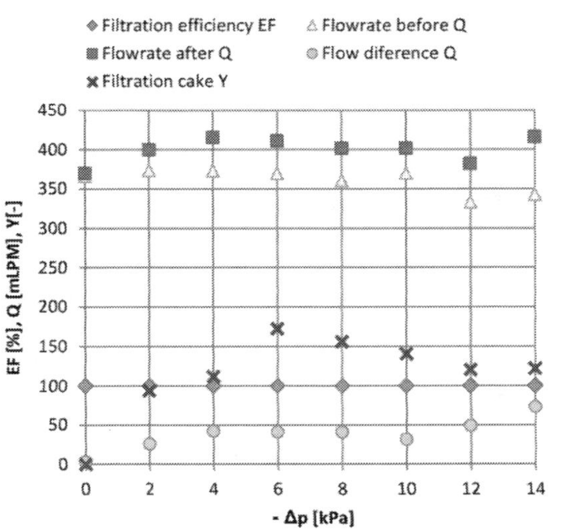

Fig. 7. Graph of the filtration efficiency, flowrate and digital grey value versus inverse pressure drop on the filter (particles: 0.96 μm, duration: 50 minutes, active filtration area: 0.6 m^2).

The first image from the Figure 8 confirms the fact of insufficient cleaning. There is no evidence of removing of the filtration cake from the inlet side of the tested filter. The higher applied inverse pressure drop is the higher flow difference is and the more removed particles are observed. The local maximum is seen between 4 and 8 kPa of the inverse pressure drop. The optimum inverse pressure drop has to be high enough to remove filtration cake properly. On the other hand for ensuring safety of the membrane (concerning of the mechanical robustness), the optimum inverse pressure drop was chosen as a minimum value from the range: 4 kPa.

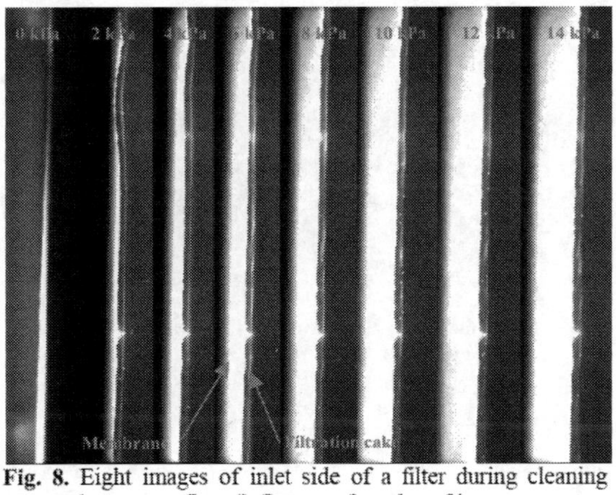

Fig. 8. Eight images of inlet side of a filter during cleaning process by reverse flow (influence of a value of inverse pressure drop).

3.3 Duration of the inverse flow

Influence of the duration of the applied inverse pressure drop is seen in graph on Figure 9. The flow difference grows with higher duration with the local maximum around 1 hour.

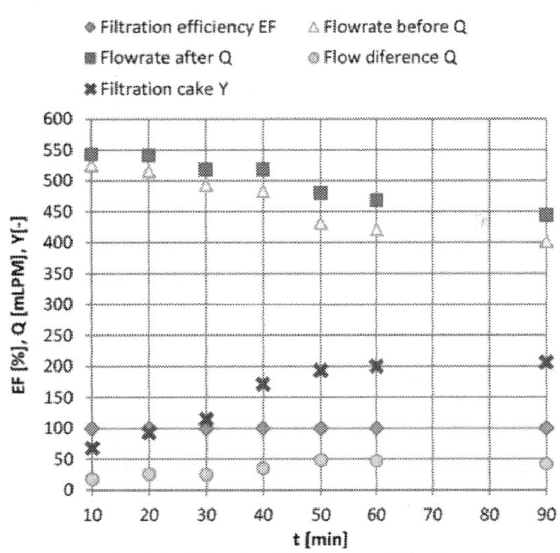

Fig. 9. Graph of the filtration efficiency and the flowrate versus duration of applied inverse pressure drop (inverse pressure drop: 4 kPa).

It is seen from the images on Figure 10 that the longer time the filter is not cleaned, the more particles are removed (deposited). The limitation of the process is in irreversible clogging. If the filter is not cleaned sufficiently often, the filtration cake becomes compacted and it is more complicated to remove it completely. The irreversible clogging indicates the declination of the flowrate (red dots). The optimum duration shifts because of irreversible clogging to the lower values and it is in rage from 30 to 50 minutes.

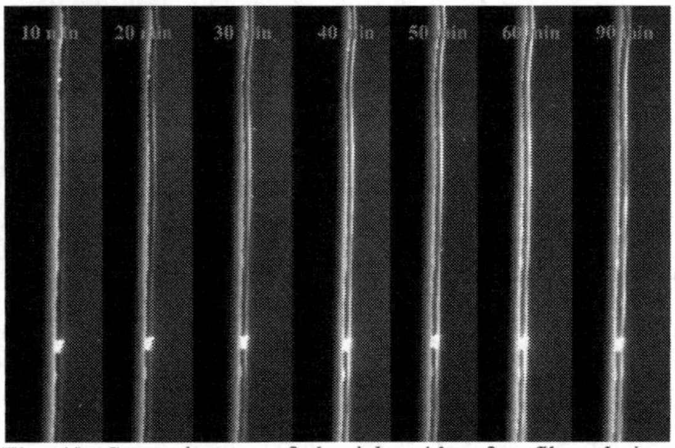

Fig. 10. Seven images of the inlet side of a filter during cleaning process by reverse flow (influence of a duration of applied inverse pressure drop).

4 Conclusion

The cleaning process is a very important part of the filtration process which limits the growth of the filtration cake, improves hydrodynamic properties and extends lifetime of the filter. In the article the optimization of the cleaning process was carried out supplemented by visualization of flow. The tests were carried out on a pilot filtration setup for waste water treatment. Model water was used instead of real waste water because of visibility and easier analysis of filtration efficiency. The model water was composed from distilled water with polystyrene particles, size was 0.96 μm, concentration was 500 μg per litre. The impact of air flow rate, inverse pressure drop and duration on the pressure drop was investigated. Images of the cleaning process were obtained and compared to the hydrodynamic data.

Air bubbles clean the filter surface all the time during filtration. From the hydrodynamic data is seen (Figure 4) no significant influence of the bubbling intensity to the forward flow in short term perspective. The optimum air flow rate was chosen with reference to covering whole the surface by air bubbles. The optimum air flow rate related to the active filtration area is 50 LPM per m^2.

The second cleaning method applied during filtration is reverse flow caused by inverse pressure drop. It was found out that the higher applied inverse pressure drop was the higher flow difference was indicated and the more removed particles were observed (Figure 7 and 8). The local maximum was found between 4 and 8 kPa for the inverse pressure drop. With reference to the mechanical robustness of the nanolayer the minimum value from the range: 4 kPa was chosen. Another parameter: duration of the inverse pressure drop was optimized. The flow difference grew with higher duration about the local maximum around 1 hour (Figure 9). Considering the irreversible clogging the optimum duration was chosen on 30 minutes.

The optical method enables to see flow behaviour in a thin laser sheet. Amount of removing particles is possible to determine on the basis of image analysis. Visualization of the cleaning process completes information about the flow rate and the filtration efficiency and proofs rightness of the measured hydrodynamic data.

Acknowledgments

The results of this project LO1201 were obtained through the financial support of the Ministry of Education, Youth and Sports in the framework of the targeted support of the "National Programme for Sustainability I".

References

1. Ng, K.-K., Wu, C.-J., Yang, H.-L., Panchangam, S.C., Lin, Y.-C., Hong, P.-K.A., Wu, C.-H., Lin, C.-F. Effect of Ultrasound on Membrane Filtration and Cleaning Operations (2012) Separation Science and Technology (Philadelphia), 48 (2), pp. 215-222
2. Madaeni, S.S., Tavakolian, H.R., Rahimpour, F. Cleaning optimization of microfiltration membrane employed for milk sterilization (2011) Separation Science and Technology, 46 (4), pp. 571-580
3. Bourgeous, K.N., Darby, J.L., Tchobanoglous, G. Ultrafiltration of wastewater: Effects of particles, mode of operation, and backwash effectiveness (2001) Water Research, 35 (1), pp. 77-90
4. Chen, J.P., Kim, S.L., Ting, Y.P. Optimization of membrane physical and chemical cleaning by a statistically designed approach (2003) Journal of Membrane Science, 219 (1-2), pp. 27-45
5. Chen, V., Li, H., Fane, A.G. Non-invasive observation of synthetic membrane processes - A review of methods (2004) Journal of Membrane Science, 241 (1), pp. 23-44
6. Knutsen, J.S., Davis, R.H. Deposition of foulant particles during tangential flow filtration (2006) Journal of Membrane Science, 271 (1-2), pp. 101-113
7. Mendret, J., Guigui, C., Schmitz, P., Cabassud, C., Duru, P. An optical method for in situ characterization of fouling during filtration (2007) AIChE Journal, 53 (9), pp. 2265-2274
8. Vanysacker, L., Declerck, P., Vankelecom, I. Development of a high throughput cross-flow filtration system for detailed investigation of fouling processes (2013) Journal of Membrane Science, 442, pp. 168-176
9. Zhong, Z., Li, F., Xing, W. Fouling formation and removal in the microfiltration of Mg(OH)2 suspension with ceramic membrane (2013) Desalination, 325, pp. 132-137
10. Hulst H G van de 1957 Light Scattering by Small Particles Willey New York p 470 ISBN: 0-486-64228-3
11. Laven P MiePlot software [online: 30.1.2017] URL: http://www.philiplaven.com/mieplot.htm
12. Spellman F R, Handbook of water and wastewater treatment plant operations, CRC Press 2009, p 826, ISBN: 13-978-1-4200-7530-4

Time-Fractional Fourier Law in a finite hollow cylinder under Gaussian-distributed heat flux

Slawomir Blasiak [1,*]

[1] Kielce University of Technology, Faculty of Mechatronics and Mechanical Engineering, Department of Mechanical Technology and Metrology, Aleja Tysiaclecia Panstwa Polskiego 7, 25-314 Kielce, Poland

Abstract. This paper presents the solution of the theoretical model of heat conduction based on time-fractional Fourier equation for a finite hollow cylinder treated with heat flux on one of the front surfaces. A derivative of fractional order in the Caputo sense was applied to record the temperature derivative in time. The distributions of temperature fields in the hollow cylinder were determined with the use of Fourier-Bessel series, as surface functions of two variables (r, θ). The distributions of temperature fields were determined using analytical methods and applying integral transformation methods. The Laplace transform with reference to time, the Fourier finite cosine transform with reference to axial coordinate z and Marchi-Zgrablich transform for radial coordinate r. The fractional heat conduction equation was analysed for $0 < \alpha \leq 2$.

1 Introduction

The research papers concerning heat conduction in structural elements in machines and devices are based on the classical Fourier's law. They might be generally classified e.g. in terms of the shape of the medium where the heat conduction is considered. Most research papers describe heat conduction in axial-symmetrical elements: rods, cylinders, plates of finite or infinite length or thickness [1–3]. There are as well studies which include discussions concerning heat conduction in relation to thermo-elastic deformations in different types of mechatronic devices [4–7]. The classical theory of heat conduction is based on the Fourier law, which might be recorded in the following form:

$$\mathbf{q} = -\lambda \, grad\,(\vartheta) \tag{1}$$

where $\mathbf{q}$ is the heat flux vector, $\vartheta = T - T_0$ denotes change in temperature, and λ is the thermal conductivity. In combination with the law of conservation of energy is expressed locally by:

$$-\nabla \mathbf{q}(t) = \rho C_p \frac{\partial \vartheta(t)}{\partial t} \tag{2}$$

where ρ is density and C_p is the specific heat capacity at constant pressure.

Another aspect of the research papers which has gained in popularity over the last thirty years is differential and integral calculus conducted on derivatives of fractional order [8–16]. The differential calculus (and integral calculus) of fractional order is a natural extension of the following concepts: integral and differential included in traditional differential and integral calculus. The interest of a large number of scientists is caused by numerous observations which demonstrate that a lot of phenomena might be explained in a more detailed way and modelled using the above mentioned differential calculus of fractional order [17]. The most crucial examples include the application of the differential calculus of fractional order to build controllers, such as $PI^{\lambda}D^{\mu}$, model the processes of heating and heat conduction, thermo-elasticity, thermal shock [18, 19], complex nonlinear systems, super capacitors, electrical systems, electrical filters, phenomena of dielectric relaxation, phenomena of diffusion and viscoelastic phenomena. The time fractional heat conduction equation might be recorded in the following form:

$$\frac{\partial^{\alpha} \vartheta}{\partial t^{\alpha}} = \kappa \Delta \vartheta \qquad 0 < \alpha \leq 2, \tag{3}$$

where $\kappa = \dfrac{K}{\rho C_p}$

The Caputo fractional derivative is expressed with the dependency:

$$\frac{\partial^{\alpha} f(t)}{\partial t^{\alpha}} \equiv D_C^{\alpha} f(t) = I^{n-\alpha} D^n f(t)$$
$$= \frac{1}{\Gamma(n-\alpha)} \int_0^t (t-\tau)^{n-\alpha-1} \frac{d^n f(\tau)}{d\tau^n} d\tau \tag{4}$$
$$t > 0, \quad n-1 < \alpha < n$$

With reference to the subject of this paper, the results of heat conduction in the hollow cylinder under Gaussian-distributed heat flux were presented.

* Corresponding author: sblasiak@tu.kielce.pl

2 The boundary and initial conditions for heat conduction

In order to solve the equation describing the temperature distribution in the hollow cylinder, it is necessary to introduce initial and boundary conditions. On the surface of the hollow cylinder, the initial conditions were applied which were presented graphically in Fig. 1.

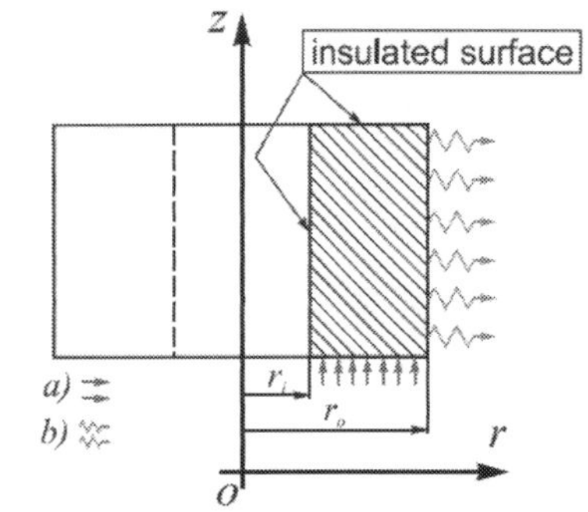

Fig. 1. Boundary conditions for heat transfer in the hollow cylinder. a) heat flux, b) convection.

It was assumed that the cylindrical surface on the inner radius r_i and the top surface of the cyliner are completely isolated from the surrounding which takes the general form:

$$\frac{\partial \vartheta}{\partial n} = 0 . \tag{5}$$

For the cylinder surface (on the outer radius r_o) which has a direct contact with the surrounding medium, the heat is transferred via convection, which was recorded as:

$$-\lambda \left.\frac{\partial \vartheta}{\partial r}\right|_{r=r_o} = \alpha^f \left.\vartheta\right|_{r=r_o} . \tag{6}$$

where α^f is the heat transfer coefficient. It was assumed that the lower surface of the hollow cylinder is exposed to heat flux expressed with the dependency:

$$\varphi_q = \frac{1200}{\sigma\sqrt{2\pi}} \exp\left(-\frac{(r-\bar{r})^2}{2\sigma^2}\right) \tag{7}$$

where $\sigma = 0.0005$, whereas $\bar{r} = r_i + 0.5(r_o - r_i)$. The heat flux distribution on the lower surface of the hollow cylinder was presented in Fig. 2.Next chapter presents the analytical solution of the above presented issue.

3 Analytical solution

By adopting a cylindrical coordinate system consistent with the geometry of the physical model in question, the equation (3) might be recorded in the following form:

$$\frac{\partial^\alpha \vartheta}{\partial \tau^\alpha} = \frac{1}{r}\frac{\partial \vartheta}{\partial r} + \frac{\partial^2 \vartheta}{\partial r^2} + \frac{\partial^2 \vartheta}{\partial z^2} \tag{8}$$

It was adopted that $\tau = \kappa t$. To solve the equation (8), Marchi–Zgrablich transform, Fourier finite cosine transform and the Laplace transform for time-fractional differential equations were applied.

The general form of Marchi–Zgrablich integral transform might be defined in the following way:

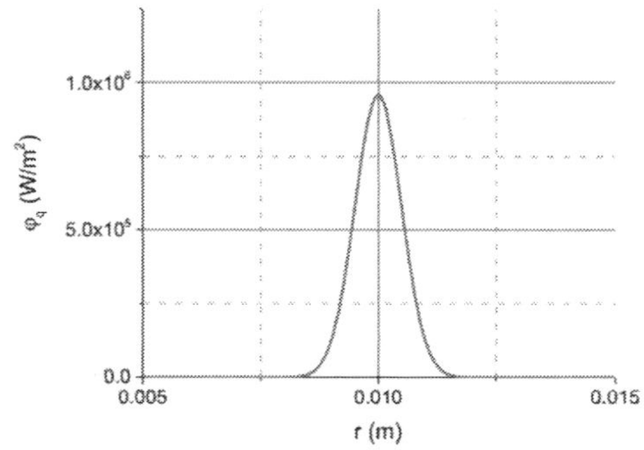

Fig. 2. The heat flux distribution on the lower surface of the hollow cylinder.

$$H(f(x)) = \bar{f}(n) = \int_{r_i}^{r_o} r f(x) S_p(\lambda, \alpha, k_n r) dr \tag{9}$$

The reverse integral transform takes the following form:

$$H^{-1}(\bar{f}(x)) = f(x) = \sum_n a_n S_p(\lambda, \alpha, k_n r) \tag{10}$$

where:

$$a_n = \frac{\bar{f}_p(n)}{C_n} \tag{11}$$

$$C_n = \int_{r_i}^{r_o} r\left[S_p(\lambda, \alpha, k_n r)\right]^2 dr \tag{12}$$

$$H^{-1}(\bar{f}(x)) = f(x) = = \sum_n \frac{\bar{f}_p(n)}{\int_{r_i}^{r_o} r\left[S_p(\lambda, \alpha, k_n \cdot r)\right]^2 dr} S_p(\lambda, \alpha, k_n r) \tag{13}$$

The equation (8) takes then the following form:

$$\frac{\partial^\alpha \bar{\vartheta}}{\partial \tau^\alpha} = \left(\frac{\partial^2 \bar{\vartheta}}{\partial z^2} - k_n^2 \bar{\vartheta}\right) \tag{14}$$

Fourier finite cosine transform is defined as follows:

$$J_c\left[g(x)\right] = g^*(n) = \int_0^L g(x)\cos\left(\frac{n\pi x}{L}\right)dx \tag{15}$$

Where $n = 1, 2, 3, \ldots$

$$J_c\left[\frac{d^2 g(x)}{dx^2}\right] = (-1)^n \left.\frac{dg(x)}{dx}\right|_{x=L} - \left.\frac{dg(x)}{dx}\right|_{x=0} + - \frac{n^2\pi^2}{L^2} \left. g^*(n) \right|_{g^*(n)=J_c[g(x)]} \tag{16}$$

The reverse transform is described by:

$$J_c^{-1}\left[g^*(n)\right] = g(x) = \frac{g^*(n=0)}{L} + + \frac{2}{L}\sum_{n=1}^{\infty} g^*(n)\cos\left(\frac{n\pi x}{L}\right) \tag{17}$$

By applying Fourier finite cosine transform to particular elements of the equation (14) with regard to the dependency:

$$J_c\left[\frac{\partial^2 \bar{\vartheta}}{\partial z^2}\right] = (-1)^n \left.\frac{d\bar{\vartheta}}{dz}\right|_{z=h} - \left.\frac{d\bar{\vartheta}}{dz}\right|_{z=0} - \frac{n^2\pi^2}{L^2}\bar{\vartheta}^*(n) = = \frac{\varphi_q}{\lambda} - \frac{n^2\pi^2}{L^2}\bar{\vartheta}^*(n) \tag{18}$$

$$J_c\left[\bar{\vartheta}\right] = \bar{\vartheta}^*(n) = \int_0^L \bar{\vartheta}(x)\cos\left(\frac{n\pi x}{L}\right)dx \tag{19}$$

The following was achieved:

$$\frac{\partial^{\alpha}\bar{\vartheta}^{*}}{\partial\tau^{\alpha}}=\left(\frac{\bar{\varphi}_{q}}{\lambda}-\frac{n^{2}\pi^{2}}{L^{2}}\bar{\vartheta}^{*}(n)-k_{n}^{2}\bar{\vartheta}^{*}(n)\right)\tag{20}$$

The dependence (8) describes the time-fractional heat conduction equation, for which the following initial conditions were adopted.

$$\begin{aligned} &t=0 \quad \vartheta=0, \quad 0<\alpha\leq 2,\\ &t=0 \quad \frac{\partial\vartheta}{\partial t}=0, \quad 1<\alpha\leq 2. \end{aligned}\tag{21}$$

as well as the boundary conditions for the area of the ring cross-section.

The left side in the equation (8) is the Caputo frictional derivative. Using the Laplace transform rule in the general form:

$$\mathrm{L}\left\{D_{C}^{\alpha}f(t)\right\}=s^{\alpha}\hat{f}(s)-\sum_{k=0}^{n-1}f^{(k)}(0^{+})s^{\alpha-1-k},\; n-1<\alpha<n\tag{22}$$

the equation (8) took the following form:

$$s^{\alpha}\hat{\vartheta}^{*}(s)=\frac{\bar{\varphi}_{q}}{\lambda s}-\frac{m^{2}\pi^{2}}{L^{2}}\hat{\vartheta}^{*}(s)-k_{n}^{2}\hat{\vartheta}^{*}(s)\tag{23}$$

By applying subsequent algebraic transformations, the following was achieved:

$$\hat{\vartheta}^{*}(s)=\frac{\bar{\varphi}_{q}}{\lambda}\frac{1}{\omega^{2}}\left(\frac{1}{s}-\frac{s^{\alpha-1}}{(s^{\alpha}+\omega^{2})}\right)\tag{24}$$

Where $\left(\frac{m^{2}\pi^{2}+L^{2}k_{n}^{2}}{L^{2}}\right)=\omega^{2}$

After using the reverse Laplace transform, the dependence (24) takes the following form:

$$\begin{aligned}\bar{\vartheta}^{*}(s)&=\frac{\bar{\varphi}_{q}}{\lambda\omega^{2}}\left(1-E_{\alpha}(-\kappa\omega^{2}t^{\alpha})\right)=\\ &=\frac{\bar{q}}{\lambda}\left(\frac{1}{\omega^{2}}-\frac{1}{\omega^{2}}E_{\alpha}(-\kappa\omega^{2}t^{\alpha})\right)\end{aligned}\tag{25}$$

Where $E_{\alpha,\beta}(z)$ is Mittag-Leffler type function defined by the series [20, 21]:

$$E_{\alpha,\beta}(z)=\sum_{\nu=1}^{\infty}\frac{z^{\nu}}{\Gamma(\alpha\nu+\beta)} \qquad \alpha>0, \quad \beta>0\tag{26}$$

After applying two subsequent integral transforms, the equation describing the temperature distribution in the hollow cylinder takes its final form:

$$\begin{aligned}\vartheta=&\frac{1}{L}\sum_{n=1}^{\infty}\frac{\int_{r_i}^{r_o}r\bar{\varphi}_{q}S_{p}(\lambda^{s},\alpha^{f},k_{n}r)dr}{\lambda\int_{r_i}^{r_o}r\left[S_{p}(\lambda,\alpha,k_{n}r)\right]^{2}dr}\cdot\\ &S_{p}(\lambda,\alpha,k_{n}r)\underbrace{\sum_{m=1}^{\infty}\left(\frac{1}{\omega^{2}}-\frac{1}{\omega^{2}}E_{\alpha}(-\kappa\omega^{2}t^{\alpha})\right)}_{\omega_{0}}+\\ &+\frac{2}{L}\sum_{n=1}^{\infty}\frac{\int_{r_i}^{r_o}r\bar{\varphi}_{q}S_{p}(\lambda^{s},\alpha^{f},k_{n}r)dr}{\lambda\int_{r_i}^{r_o}r\left[S_{p}(\lambda,\alpha,k_{n}r)\right]^{2}dr}\cdot\\ &S_{p}(\lambda,\alpha,k_{n}r)\sum_{m=1}^{\infty}\cos\left(\frac{m\pi x}{L}\right)\left(\frac{1}{\omega^{2}}-\frac{1}{\omega^{2}}E_{\alpha}(-\kappa\omega^{2}t^{\alpha})\right)\end{aligned}\tag{27}$$

When $\alpha=1$, the equation (27) might be recorded as:

$$\begin{aligned}\vartheta=&\frac{1}{L}\sum_{n=1}^{\infty}\frac{\int_{r_i}^{r_o}r\bar{\varphi}_{q}S_{p}(\lambda^{s},\alpha^{f},k_{n}r)dr}{\lambda\int_{r_i}^{r_o}r\left[S_{p}(\lambda,\alpha,k_{n}r)\right]^{2}dr}\cdot\\ &S_{p}(\lambda,\alpha,k_{n}r)\underbrace{\sum_{m=1}^{\infty}\left(\frac{1}{\omega^{2}}-\frac{1}{\omega^{2}}\exp(-\kappa\omega^{2}t^{\alpha})\right)}_{\omega_{0}}+\\ &+\frac{2}{L}\sum_{n=1}^{\infty}\frac{\int_{r_i}^{r_o}r\bar{\varphi}_{q}S_{p}(\lambda^{s},\alpha^{f},k_{n}r)dr}{\lambda\int_{r_i}^{r_o}r\left[S_{p}(\lambda,\alpha,k_{n}r)\right]^{2}dr}\cdot\\ &S_{p}(\lambda,\alpha,k_{n}r)\sum_{m=1}^{\infty}\cos\left(\frac{m\pi x}{L}\right)\left(\frac{1}{\omega^{2}}-\frac{1}{\omega^{2}}\exp(-\kappa\omega^{2}t^{\alpha})\right)\end{aligned}\tag{28}$$

4 RESULTS AND DISCUSSION

The parameters characterising the material used for the hollow cylinder are presented in the table below.

Table 1. Properties of the Carbon Steels material

Density ρ (10^3 kg/m^3)	7.85	Thermal Expansion τ (10^{-6}/K)	11
Young's modulus E (GPa)	190	Thermal conductivity λ (W/mK)	24.3
Poisson ratio ν	0.27	Specific Heat C_p (J/kgK)	450

The characteristic dimensions of the hollow cylinder as well as the reference operating parameters were summarized in Table 2. It was assumed that the studied cylinder was surrounded with fluid T_o of $20\,^{\circ}C$.

The presented solution of the mathematical model makes it possible to demonstrate the obtained results for two cases:

- The first case which might be called the classical one, when $\alpha=1$.
- The second case when α in the equation (8) takes the value of incomplete order $0<\alpha\leq 2$.

In this part of this paper, the results of the simulation studies concerning the influence of geometric and operating parameters on the temperature distribution in the ring were collected.

Table 2. Geometry- and performance-related parameters

Geometry-related parameters	
Inner radius r_i (m)	0.005
Outer radius r_o (m)	0.015
Thickness of the hollow cylinder h (m)	0.015
Performance-related parameters	
Heat transfer coefficient (water) α^f (W/m^2 K)	1200
Outer Fluid temperature T_o (°C)	20

 https://doi.org/10.1051/epjconf/201818002008

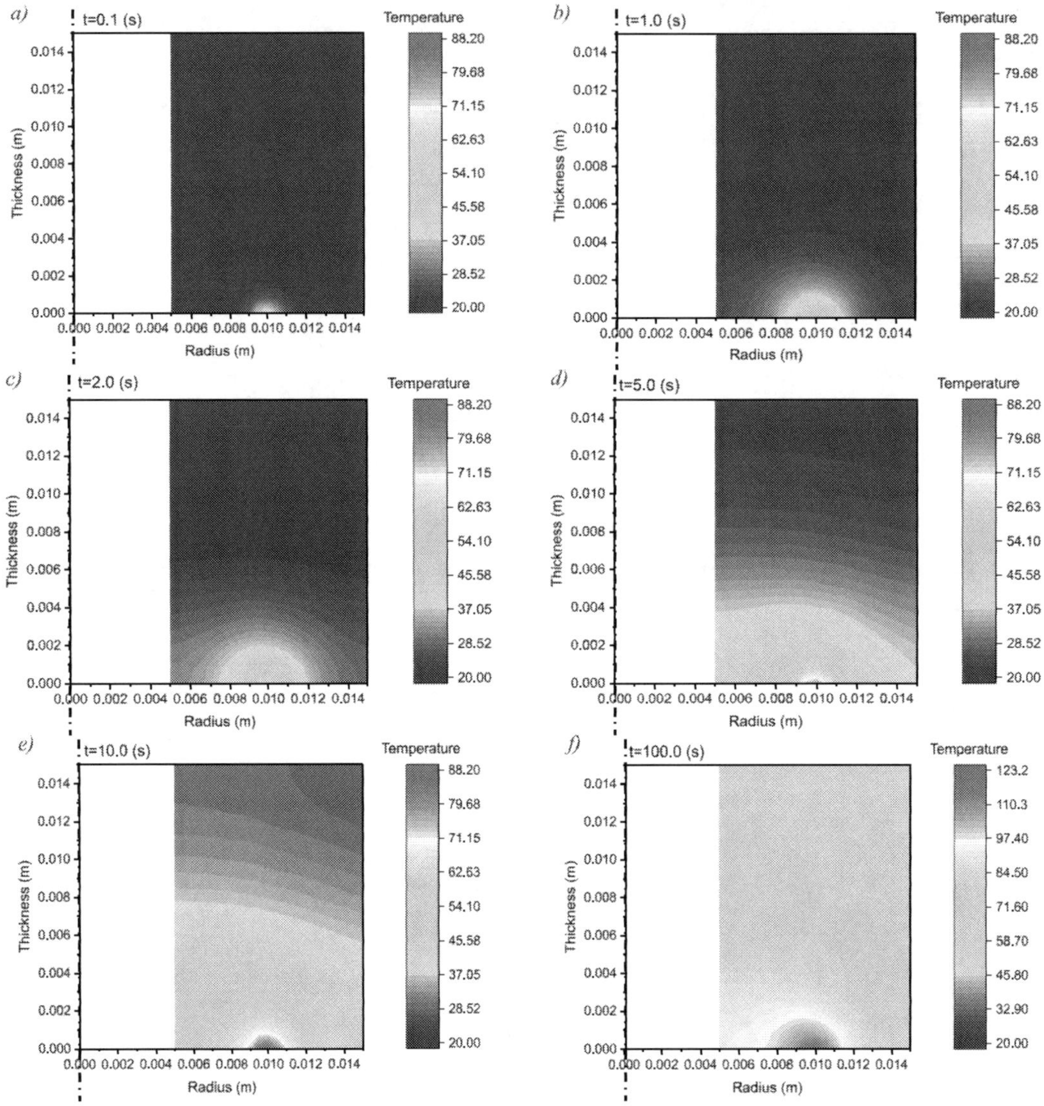

Fig. 3. The temperature distribution in the cross-section of the hollow cylinder for different t.

The temperature changes were presented graphically as temperature distributions for the cross section of the hollow cylinder. and Fig. 4 present the temperature changes for the simulation time within the range $0.1-100(s)$. The results presented in the graphs below were developed for $\alpha = 1$, i.e. for the classical solution of the Fourier equation (see eq. (28)).shows the process of heating the studied element with heat flux. The maximum temperature value is located on the radius $\bar{r}$, i.e. in the place with the highest density of the heat flux.

When $\alpha = 1$ and when $t \to \infty$, then the temperature ϑ tends to achieve steady state temperature. Such a state in the physical model in question takes place after approximately 50 (s).

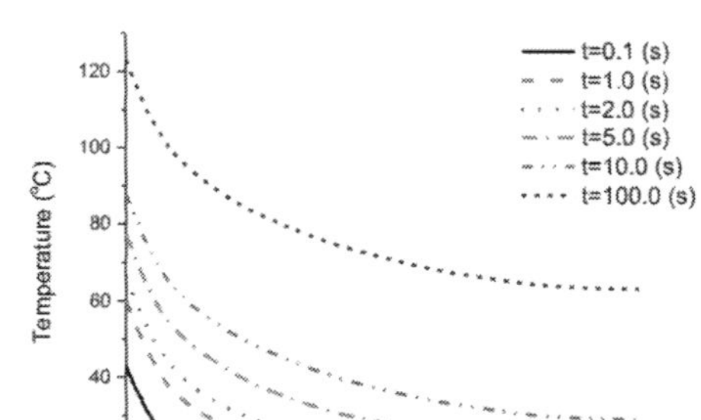

Fig. 4. The temperature distribution along the ring thickness on the radius $\bar{r}$ for different time values t.

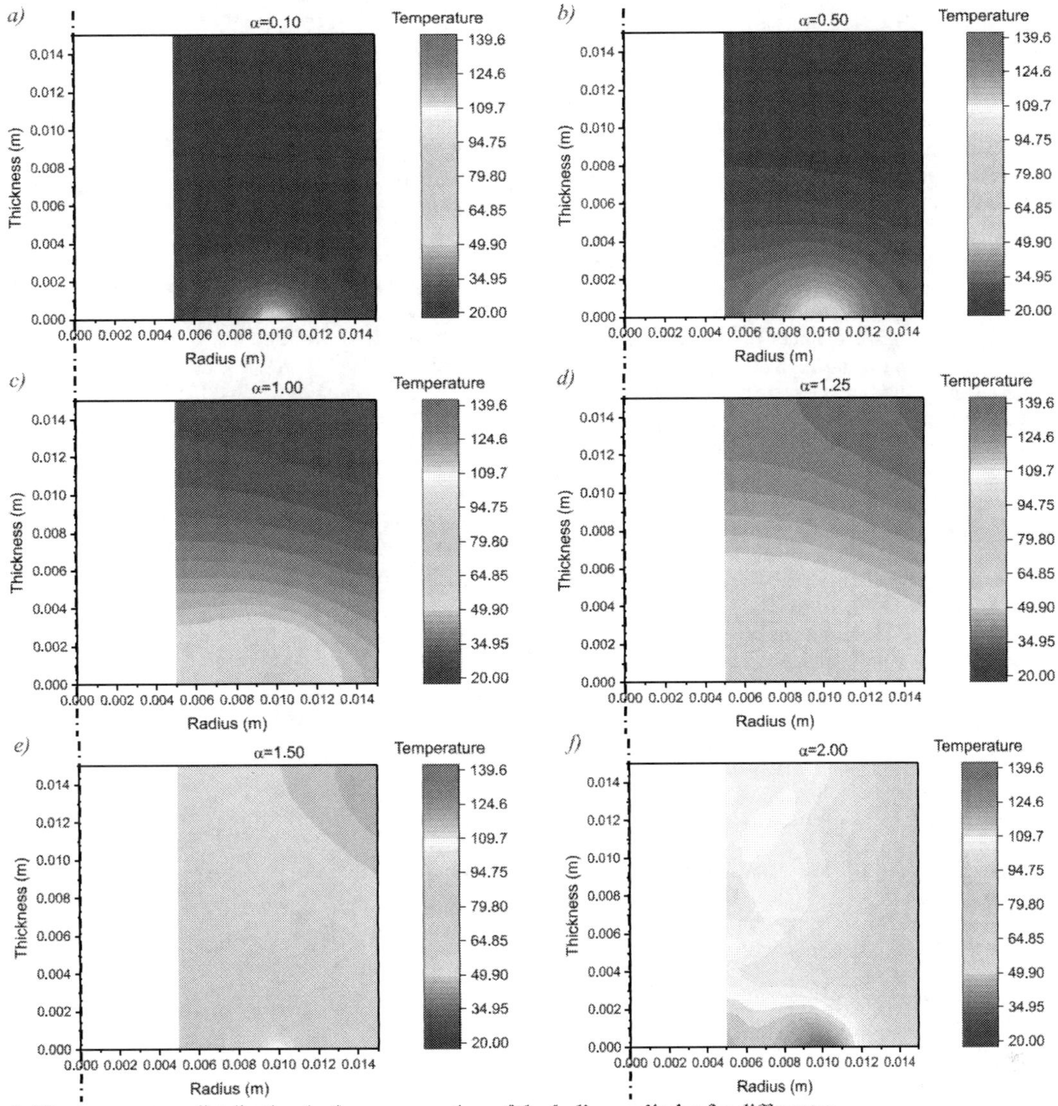

Fig. 5. The temperature distribution in the cross-section of the hollow cylinder for different α.

Fig. 5 presents the temperature distributions for $t = 10(s)$ and the order of fractional derivative α within $\alpha = 0.1 - 2.0$. By analysing the collected results, it might be noticed that for α from the range of $0 < \alpha < 1.0$ the velocity of the overtemperature changes over time is low and does not exceed the values which are to be found for $\alpha = 1$.

In the case of the order of derivative of the overtemperature over time with the value α from the range of $1.0 < \alpha < 2.0$, the temperature is approximately 1.5 times higher than for $\alpha = 1$. In the case of $\alpha = 2$ the so called ballistic heat conduction equation is to be found.

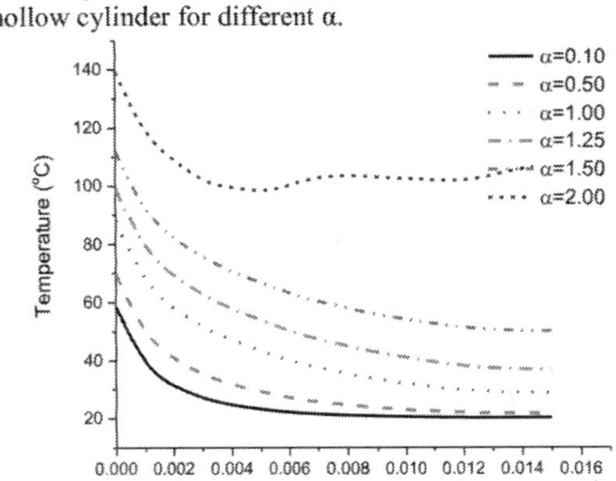

Fig. 6. The temperature distribution along the ring thickness on the radius $\bar{r}$ for different values of the equation order α.

5 Conclusion

The paper presents the mathematical model of heat conduction for the hollow cylinder, where one front surface was treated with Gaussian-distributed heat flux. The results of the analytical solution of the mathematical model were presented in two aspects. The first one referred to the classical Fourier law (for $\alpha = 1$), while the second one regarded applying the derivative of fractional order with respect to time (for $0 < \alpha \le 2$). The use of differential equations of fractional order expanded the possibilities of analysing the phenomenon of heat conduction in the hollow cylinder to a great extent. It is particularly visible in Fig. 6 which depicts the temperature distributions for t=10 (s) in the form of a bundle of curves created for different values of the coefficient $0 < \alpha \le 2$, which is the order of fractional derivative. The change of the coefficient α from 1 to 2 illustrates the transition from the heat equation to the wave differential equation. The character of the latter is clearly visible in Fig. 5f in the form of "chaotic" temperature changes in the ring cross-section. It should be emphasized that the mathematical analysis using the differential equation of fractional order is useful to construct more detailed mathematical models describing the physical phenomena.

References

1. S. Blasiak, *Int. J. Heat Mass Transfer* 81 (2015), 90–102
2. S. D. Warbhe, J. J. Tripathi, K. C. Deshmukh et al., *J. Heat Transfer* 139(4) (2017), 44502
3. M. A. Ezzat, A. A. El-Bary, *Int. J. Therm. Sci.* 108 (2016), 62–69
4. P. A. Laski. In: Zolotarev I, Radolf V (eds) ENGINEERING MECHANICS 2016. ACAD SCI CZECH REPUBLIC, INST THERMOMECHANICS, DOLEJSKOVA 5, PRAGUE 8, 182 00, CZECH REPUBLIC (2016), pp 350–353
5. S. Blasiak, J. E. Takosoglu, P. A. Laski, *J THERM SCI TECH-JPN* 9(2) (2014), JTST0011-JTST0011
6. J. Bochnia. In: Martsynkovskyy V, Zahorulko A (eds) XIIITH INTERNATIONAL SCIENTIFIC AND ENGINEERING CONFERENCE HERMETIC SEALING, VIBRATION RELIABILITY AND ECOLOGICAL SAFETY OF PUMP AND COMPRESSOR MACHINERY-HERVICON-2011, vol 39. ELSEVIER SCIENCE BV, SARA BURGERHARTSTRAAT 25, PO BOX 211, 1000 AE AMSTERDAM, NETHERLANDS (2012), pp 98–110
7. J. E. Takosoglu. In: Zolotarev I, Radolf V (eds) ENGINEERING MECHANICS 2016. ACAD SCI CZECH REPUBLIC, INST THERMOMECHANICS, DOLEJSKOVA 5, PRAGUE 8, 182 00, CZECH REPUBLIC (2016), pp 546–549
8. S. Blasiak, *Int. J. Heat Mass Transfer* 100 (2016), 79–88
9. C. Marugán-Cruz, O. Flores, D. Santana et al., *Int. J. Heat Mass Transfer* 96 (2016), 256–266
10. Y. Z. Povstenko, *Int. J. Diff. Eq.* 2012(7) (2012), 1–13
11. Y. Z. Povstenko, *Math. Prob. in Eng.* 2014(3) (2014), 1–7
12. Y. Povstenko, *Arch Appl Mech* 82(3) (2012), 345–362
13. Y. Povstenko, *Entropy* 15(10) (2013), 4122–4133
14. Y. Povstenko, *Comm in App Ind. Math.* 6(1) (2014)
15. H.-T. Qi, H.-Y. Xu, X.-W. Guo, *Comput. Math. Appl.* 66(5) (2013), 824–831
16. X. Jiang, M. Xu, *Physica A* 389(17) (2010), 3368–3374
17. W. E. Raslan, *jtam* (2016), 295
18. H. M. Youssef, *Acta Mech* 223(6) (2012), 1219–1231
19. X.-Y. Zhang, X.-F. Li, *Eng. Fract. Mech.* 171 (2017), 22–34
20. S. Rogosin, *Mathematics* 3(2) (2015), 368–381
21. H. J. Haubold, A. M. Mathai, R. K. Saxena, *J. Appl. Math.* 2011(3) (2011), 1–51

 https://doi.org/10.1051/epjconf/201818002009

High Temperature Modification of SNCR Technology and its Impact on NOx Removal Process

Tomáš Blejchař[1,*], *Jaroslav* Konvička[2], *Bernd* von der Heide[3], *Rostislav* Malý[1], and *Miloš* Maier[1]

[1]ORGREZ, a.s., Hudcova 321/76, 612 00 Brno-Medlánky, Czech Republic
[2]VŠB Technical university of Ostrava, 17. listopadu 15/2172, 708 33, Ostrava Poruba, Czech Republic
[3]Mehldau & Steinfath Umwelttechnik GmbH, Alfredstraße 279, D-45133 Essen, Deutschland

Abstract. SNCR *(Selective non-catalytic reduction)* Technology is currently being used to reach the emission limit for nitrogen oxides at fossil fuel fired power plant and/or heating plant and optimum temperature for SNCR process is in range 850 - 1050°C. Modified SNCR technology is able to reach reduction 60% of nitrogen oxides at temperature up to 1250°C. So the technology can also be installed where the flue gas temperature is too high in combustion chamber. Modified SNCR was tested using generally known SNCR chemistry implemented in CFD *(Computation fluid dynamics)* code. CFD model was focused on detail simulation of reagent injection and influence of flue gas temperature. Than CFD simulation was compared with operating data of boiler where the modified SNCR technology is installed. By comparing the experiment results with the model, the effect on nitrous oxides removal process and temperature of flue gas at the injection region.

1 Introduction

The retrofitting of large coal fired boilers with NO_X control systems has been concluded in Western Europe many years ago and in Central Europe has been finishing.

Starting in the nineties a number of Central and Eastern European countries has joined the EU and as a consequence they also had to accept the emission limits which are in force. This means for most of the power plants that the existing boilers either have to be shut down or measures have to reach limits specified by EU regulations like NO_X.

For the SCR technology, which might be the first choice and it was considered as the best available technology (BAT) in the eighties and nineties, many reliable results and experiences are available to estimate the feasibility as well as investment and operating costs with high accuracy. However, beside this fact, that the investment costs for a SCR *(Selective catalytic reduction)* are about ten times as high as for SNCR system, it is necessary to take into consideration also other disadvantages.

Installation of the catalyst is often very problematic especially when the boiler has a large economiser instead of an air pre-heater so that heat exchangers would have to be re-placed to install the catalyst. Design of SCR layers would generally cause static problems due to the height and weight of the catalyst and additional steel structure. The downtime of the boiler for retrofitting with SCR could cause a considerable loss of profit.

Pressure drop and problematic installation of SCR catalyst could result in higher operating costs for SCR and the investment volume increase to the multiple of the cost of a SCNR system.

The SNCR process has been enhanced for smaller boilers like waste incineration plants in last years, and it is widely considered now as the 'Best Available Technology' (BAT) for this size of boilers. Based on this information an increasing number of owners of power plants are seriously investigating today whether the SNCR process is feasible for their large boilers. Special attention is typically being paid to the overall cost compared to SCR.

2 Elementary description of the SNCR technology

The basic principle of NOx reduction by the NH_2^- radical has been known since the 1970s. A dominant chemical reaction is very simple and is defined by the following chemical formula [1], [2], [3], [4].

$$NH_2^- + NO \rightarrow N_2 + H_2O \qquad (1)$$

This reaction occurs dominantly without a catalyser in the temperature window between 850-1050°C. The main problem of SNCR method is a relatively narrow temperature window (850-1050°C). If the reagent is injected into colder temperature zone in a furnace, nitrogen oxides NO_X do not react with NH_2^- radical, because the reaction rate is slow and high concentrations of Ammonia NH_3 occur in flue gas. If the reagent is

* Corresponding author: tomas.blejchar@orgrez.cz

https://doi.org/10.1051/epjconf/201818002009

injected into warmer temperature zone of the furnace, NH_2^- reacts with oxygen O_2. Therefore, nitrogen oxides are not reduced but they are produced by combustion of ammonia. However the reduction of NO_X is only 50-65% at stochiometric ratio ca 2,8 in typical boiler and the SNCR processes achieve reasonable NO_X reduction only in a narrow range of temperature (extremity 850-1050°C, optimum 950°C). [1], [4]

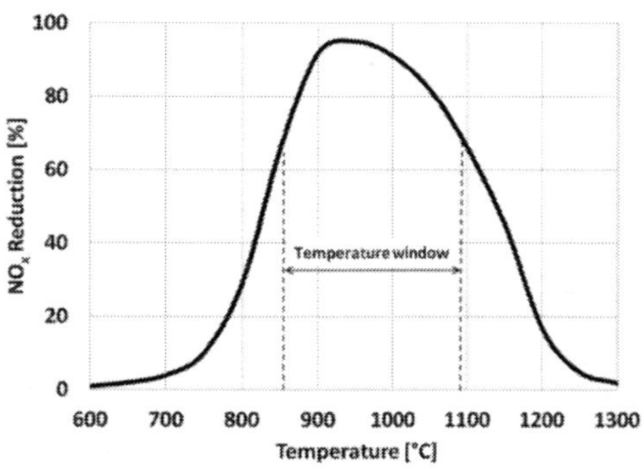

Fig. 1. NO_X reduction as a function of temperature and illustration of temperature window [1], [4]

2.1. Elementary chemistry of SNCR process

Three different commercial processes are usually used at the present time: Thermal DeNOx, NOxOUT and RAPRENOx These methods are based on injection of Ammonia NH_3, Urea $CO(NH_2)_2$ and Cyanuric acid $(HCNO)_3$ into combustion chamber of boiler.

The Thermal DeNOx process was originally proposed by Lyon in the USA in 1975, with ammonia as a reduction reagent. This process was developed as a substitution of SCR, which eliminates disadvantage of catalytic reduction such a high cost, poisoning of catalyst, required space, etc. This process involves addition of ammonia NH_3 or ammonia water NH_4OH as a reducing reagent. The DeNOx process can be described by the following chemical net reaction [1], [4]

$$4NH_3 + 4NO + O_2 \rightarrow 4N_2 + 6H_2O \qquad (2)$$

First test of injecting urea solution downstream to the combustion zone to reduce NO_X in a non-catalytic way was reported in 1977. This process was later patented in USA with a market name NOxOUT. The use of a urea as a reducing reagent is mainly interesting for reason of storage and handling. The NOxOUT process can be described by the following chemical net reaction [1], [4]

$$2CO(NH_2)_2 + 4NO + O_2 \rightarrow 4N_2 + 2CO_2 + 4H_2O \qquad (3)$$

Furthermore, the paper deals only with a method based on the injection of an aqueous solution of urea.

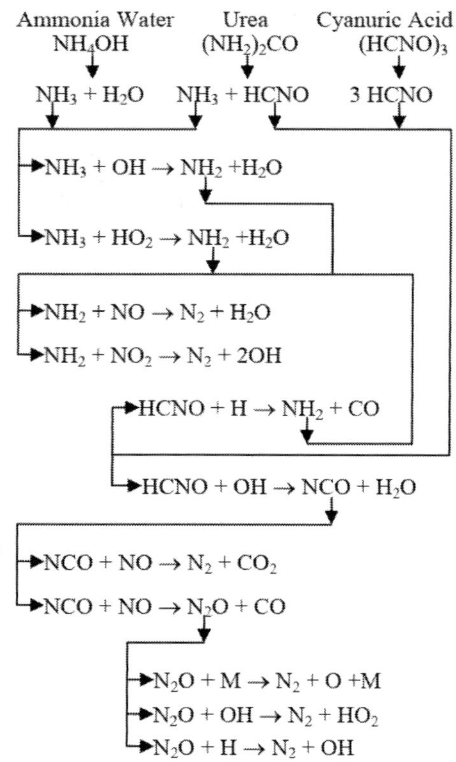

Fig. 2. Simplified chemical mechanism of SNCR method

2.2 Kinetics data of SNCR process

Chemical reactions included in CFD are based on the Arrhenius equation (4), which describes the reaction rate [11].

$$k = A \cdot e^{\frac{-E}{RT}} \cdot T^{\beta} \left[\left(\frac{mol}{m^3} \right)^{1-n} \cdot \frac{K^{-\beta}}{s} \right] \qquad (4)$$

where: k is reaction rate, A is pre-exponential factor, E is activation energy, R is universal gas constant, T is temperature, β is temperature exponent, and n is the sum of the number of reaction orders.

The Arrhenius equation describes the rate of reaction only. The chemical mechanism of every reaction must be specified with parent and children materials of reaction and with stoichiometric coefficients [11], [12], [6], [7].

The SNCR model described in this paper is based on seven-step kinetic mechanism of NO defined by Brouwer at al [3] and two-step mechanism for primary breakdown of Urea into NH_3 and HCNO defined by Rota at al. [10]

$$(NH_2)_2CO \rightarrow NH_3 + HCNO \qquad (5)$$

$$(NH_2)_2CO + H_2O \rightarrow 2NH_3 + CO_2 \qquad (6)$$

$$NH_3 + NO \rightarrow N_2 + H_2O \qquad (7)$$

$$NH_3 + NO_2 \rightarrow N_2 + H_2O \qquad (8)$$

$$NH_3 + O_2 \rightarrow NO + H_2O \quad (9)$$

$$HCNO + ¼O_2 \rightarrow NCO + ½H_2O \quad (10)$$

$$NCO + NO \rightarrow N_2O + CO \quad (11)$$

$$N_2O \rightarrow N_2 + O_2 \quad (12)$$

$$CO + ½O_2 \rightarrow CO_2 \quad (13)$$

Another modification of SNCR kinetics mechanism can be found in many studies and papers e.g.[8],[9], and [10].

The simplified SNCR reaction mechanism is described by previous reactions. Whereas the reaction mechanism with NH_2^- radical is complicated that is why the proposed mechanism does not include Ammonia radical NH_2^-, that means the problematic non-stable transitional material, is not necessary to be specified. The proposed chemical mechanism for SNCR with Urea solution is relatively precise and simple. The complexity and computational intensity of CFD simulation depend on many materials and reactions. The reaction mechanism was proposed with respect to simulation speed and usability.

Table 1. Kinetics parameters of SNCR Model.

Reaction	A $\left[\left(\frac{mol}{m^3}\right)^{1-n} \cdot \frac{K^{-\beta}}{s}\right]$	β [1]	E $\left[\frac{J}{mol}\right]$	n [1]
Eqv. (5)	1,27E+04	0	65048	1 $[(NH_2)_2CO]^1$
Eqv. (6)	6,13E+04	0	87819	2 $[(NH_2)_2CO]^1$ $[H_2O]^1$
Eqv. (7)	4,20 E+02	5,3	315937	2 $[NH_3]^1$ $[NO]^1$
Eqv. (8)	4,20E+02	5,3	349937	2 $[NH_3]^1$ $[NO_2]^1$
Eqv. (9)	3,50E-03	7,65	544487	2 $[NH_3]^1$ $[O_2]^1$
Eqv. (10)	6,24E+09	0,85	284637	1,25 $[HNCO]^1$ $[O_2]^1$
Eqv. (11)	1,00E+07	0	-1632	2 $[NCO]^1$ $[NO]^1$
Eqv. (12)	6,90E+12	-2,5	151075	1 $[N_2O]^1$
Eqv. (13)	1,26E+06	0	167742	1,75 $[CO]^1$ $[O_2]^{0,25}$ $[H_2O]^{0,5}$

2.3 Influence of CO concentration on SNCR process

The presence of CO, where reagent is injected, affects kinetics of NO_X reduction with urea. Effect of CO oxidation leads to production of H, O, and OH due to reactions (14)-(16).

$$CO + OH \leftrightarrow CO_2 + H \quad (14)$$

$$H + O_2 \leftrightarrow O + OH \quad (15)$$

$$O + H_2O \leftrightarrow OH + OH \quad (16)$$

The influence of CO on the SNCR process indicating that the presence of CO shifts the selectivity by increasing the rate of NH_2^- formation and the rate of NH_3 oxidation to NO. Higher CO level shifts the optimal temperature for NO reduction to lower temperatures due to enhanced chemical rates in presence of CO. The shift in selectivity is accomplished by two mechanisms:

- Increasing local temperature
- Increasing net production of OH radical.

The firs effect is due to exothermicity of the main CO burnout reaction (14), while the second effects is due to chain branching that results from the H produced in this reaction. The contribution of CO oxidation to the local temperature is accounted for in the model. Empirical adjustment to the rates is used to account for the effect of CO without including details radical chemistry. The empirical adjustment is applied in the form of shift in effective temperature to account for the radical chemistry effects of CO as follows and the rate of reactions (7), (8), (9), (10) is modified as [3]

$$k = A \cdot e^{\frac{-E}{R(T+S_{CO})}} \cdot (T + S_{CO})^{\beta} \quad (15)$$

S_{CO} is the shift in the temperature required to achieve the same NO reduction and reagent decomposition compared to the case without CO [3].

$$S_{CO} = 17{,}5 \cdot \ln(X_{CO}) + 173 \quad (16)$$

where X_{CO} is volume fraction of CO. Application of modified rate of reaction is not possible to set in CFD code ANSYS CFX. Analysis of equation (15) shown that this equation can be simplified as follows and term X_{CO}^{γ} describes CO shift in simplified form.

$$k = A \cdot X_{CO}^{\chi} \cdot e^{\frac{-E}{RT}} \cdot T^{\beta} \quad (17)$$

Table 2. Kinetics parameters of SNCR Model with CO shift.

Reaction	A $\left[\left(\frac{mol}{m^3}\right)^{1-n} \cdot \frac{K^{-\beta}}{s}\right]$	β [1]	γ [1]	E $\left[\frac{J}{mol}\right]$	n [1]
Eqv. (7)	8,63E+04	5,27	0,58	311822	2; $[NH_3]^1$ $[NO]^1$
Eqv. (8)	1,39E+05	5,27	0,63	345403	2; $[NH_3]^1$ $[NO_2]^1$

Eqv. (9)	2,63E+01	7,61	0,97	537456	2; $[NH_3]^1$ $[O_2]^1$
Eqv. (10)	3,88E+11	0,84	0,46	281090	1,25 $[HNCO]^1$ $[O_2]^1$

Previous equation was used to analysis of kinetics data of reactions (7), (8), (9), (10) calculated by equation (15) and (16). The influence of CO on the SNCR process can be described by modifying the kinetic parameters of the equations (7), (8), (9), (10).

2.4 Influence of O_2 concentration

The influence of oxygen concentration in flue gas in the SNCR process was presented in several studies. Higher concentration of oxygen shifts the temperature window towards lower temperatures. The higher concentration of oxygen did not involve reduction of NO_X.

2.5 Influence of H_2O concentration

Presence of H_2O in flue gas also affects the efficiency of NO_X reduction. Concentration of H_2O in reacting zone effects due to radical O. This radical inhibits the formation of $NH_2^{\cdot}$. The process of inhibition is described by following reactions [1]:

$$O + H_2O \leftrightarrow OH + OH \quad (18)$$

$$NH_2 + OH \leftrightarrow NH_3 + O \quad (19)$$

Higher concentration of water in flue gas shifts the temperature window to higher temperatures. But this effect is not cross-validated.

2.6 Influence of H_2 concentration

Addition of hydrogen shifts the temperature to lower values. Process of dissociation and/or oxidation of hydrogen produce radicals OH, H and O. These radicals influence the concentration of $NH_2^{\cdot}$ due to the $NH_3 \rightarrow NH_2^{\cdot}$ conversion, see following equations

$$NH_3 + OH \leftrightarrow NH_2 + H_2O \quad (20)$$

$$NH_3 + O \leftrightarrow NH_2 + OH \quad (21)$$

$$NH_3 + H \leftrightarrow NH_2 + H_2 \quad (22)$$

3 High temperature modification of SNCR - Selective cooling of the flue gas

SNCR technology can operate in a narrow range of temperature (extremity 850-1050°C, optimum 950°C). This temperature window occurs usually upstream heat exchangers like super heater, over heater etc. Sometimes the temperature window reaches up to domain where the heat exchangers are located. If the flue gas reaches heat exchangers than the temperature of flue gas fall very fast. It means if the reagent (urea) is injected just upstream heat exchangers than the reagent reaches cold walls of exchanger very quick and the process of NO_X reduction is stopped due to low temperature of flue gas. Another necessary condition is residence time of reagent at temperature window which cannot be met in the case of injection of reagent just upstream heat exchangers. Sometimes SNCR technology cannot be designed for the above-mentioned reasons because temperature of flue gas is too high in combustion chamber and upstream heat exchangers respectively.

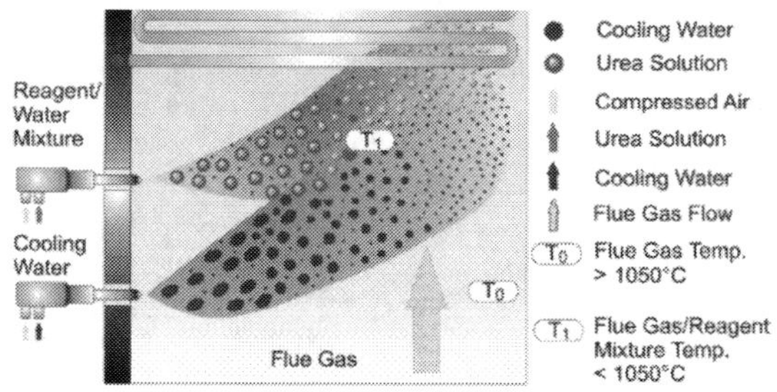

Fig. 3. Basic principle of selective cooling of flue gas

As has been said, the presence of H_2O, where reagent is injected, affects kinetics of NO_X reduction and higher concentration of H_2O in flue gas shifts the temperature window to higher temperatures. This leads to the idea that if additional water can be transported to the injection region, SNCR technology will be able to operate at higher temperatures.

In cases when the flue gas temperatures are too hot, as is mostly the case at load peaks and/or in limited areas, cooling of the flue gas might be a viable alternative. This simply could be accomplished by increasing the quantity of process water. Since the changing of the fluid flow in the injectors would change the spraying pattern as well, a better solution is to install additional injectors for cooling water close to the reagent injectors.

With this concept, cooling water can be switched on or off for each injector independently when needed depending on the flue gas temperature in this location, while the distribution of the reagent would not be affected.

4 CFD model of selective cooling of the flue gas

The problem of the jet penetration and mixing in turbulent cross flow is relevant in many industrial applications, and it has been studied by several researchers. The characteristic mixing time is not exactly the time required for molecular mixing of primary cross flow with the lateral jet stream. However, the mixing time can be an important tool when compared with the reaction time scales, in order to predict the operating conditions where the mixing can play an important role.

The SNCR technology is relatively a simple technological unit, but it includes some important parts which affect dominantly the efficiency of NO_X reduction. One of important parts is a nozzle. Three types of nozzles are generally used at present. The following nozzles are used: 1) cone spraying nozzle, 2) plane spraying nozzle, 3) cone spraying nozzle with an angle of 45°. The main parameters of the nozzle and injection lance are affected by the sprayed reagent and the depth of penetration of flue gas by reagent respectively. It is not possible to measure the spraying process in situ, because the nozzles are situated in the combustion chamber. Due to the insufficient information about the process of spraying and cooling of the flue gas, it was decided to analyse the flow in the nozzle in detail. The spraying process is theoretically demonstrated by mixing of two perpendicular flows of fluids. The main flow is flue gas and the second flow is the air with droplets of urea solution.

The penetration of flue gas is ca 2 m. This is a low value according to the dimension of the combustion furnace. The NO_X molecules are reduced near the wall of combustion chamber and the core of flue gas flow is not affected by reagent. Therefore, there is a significant problem with balanced penetration of flue gas and this is the main reason of low efficiency of SNCR reaching ca 65% at high stoichiometric ratio ca 2.8.

Table 3. Penetration of flue gas by reagent.

Nozzle type	Penetration [m]
Cone	ca 1,2 - 2,4
Plane	ca 0,6 - 2,1
Cone 45°	ca 0,9 - 2,4

The process of selective cooling of flue gas was simulated on the base of real design of SNCR technology. Simulation domain and boundary condition was focused on detail of couple of injection lances (regent and cooling water) with cone nozzle.

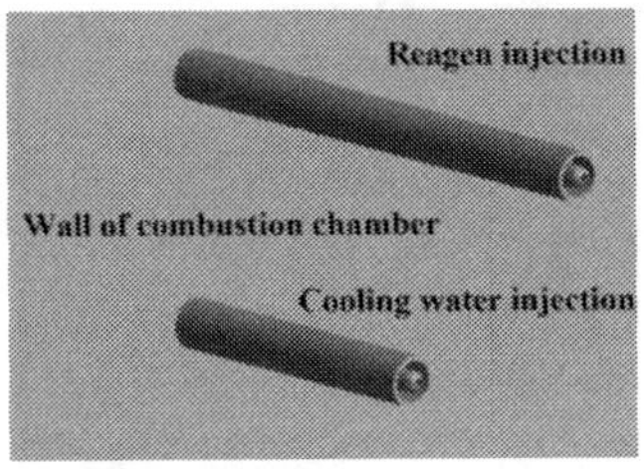

Fig. 4. Detail of simulation domain

So the domain was basically a large duct with two injection lances, which was perpendicular to main flow of flue gas in duct. Grid was strictly built form hexahedral elements and total number of elements was approximately 850 000.

The basic equation set, which describes both regimes of flow, represents the application of the conservation laws. The law of mass, momentum and energy conservation are used for numerical modelling of flow [12]. The law of momentum conservation represents the Navier-Stokes equation [12]. The law of mass conservation represents the continuity equation [12], and the law of energy equation represents the energy equation [12]. The CFD code ANSYS CFX was used for simulations. Turbulent flow is stochastic, but it is statistic stable. Arbitrary physical quantity can be decompiled on sum of average value and fluctuation value $u = \bar{u} + u'$. Commonly used turbulent models are based on description of local turbulent intensity by length and velocity scale. Flow regime was analysed by means of Reynolds number.

$$Re = \frac{\rho \cdot v \cdot D}{\eta} \approx 100000 \qquad (22)$$

Where: v is velocity [m/s], D is characteristic dimension [m], ρ is density [kg/m^3] and η is dynamic viscosity [Pa*s] of the fluid.

As is evident, this is a turbulent flow and the jet flow produces shear stress. That is why Shear Stress Transport k-ω turbulence model is used for CFD simulation [12]. The turbulent viscosity is specified by two transport equations. These equations solve turbulent kinetic energy and vorticity ω [12]. The production of turbulence by expansion is included in the simulation too

As already mentioned, the CFD simulation was focused on effect of water injection upstream injection area of reagent at higher temperature of flue gas than optimal. The practical experience with the selective cooling system already exists and this system is installed and operated. But the process of cooling of flue gas is not fully understood yet.

Boundary condition was specified on the base of the process data that was recorded during the commissioning and optimizing of the SNCR technology with selective cooling system. Test of selective cooling was realised at maximum load of the boiler if the temperature of the flue gas is maximum. Effect of cooling water was evaluated on the base of time record of boiler process data.

The first step of the test was stabilisation of maximum load of the boiler and then was started SNCR technology in automatic mode. After the suppression of the effects of unsteady running (PID regulator reach the set value of NO_X) were installed suction pyrometers 1,3 m upstream couple of cooling and injection lances. The suction pyrometers were used do measure of flue gas temperature. Flue gas sucked up through the pyrometer were analysed by FTIR [8] equipment and the analysis was mainly focused on concentration of O_2, CO and NO_x.

Once the boiler was stabilised and all important values were approximately constant so technology was blocked. PID regulator was deactivated so all valves and other regulation equipment were "frozen" in current status. Next SNCR technology was ready for selective cooling test.

The test was very simple and consisted of a manual closing of the cooling water supply. This test thus

https://doi.org/10.1051/epjconf/201818002009

showed the response of the SNCR system to the presence of cooling water in the injection area and 100 mm upstream respectively.

In the picture we can see that if the cooling water was not in the injection area than the concentration of NOx increased from ca 185 to ca 275 mg/m$_N^3$ in the stack of power plant.

As already mentioned, the CFD model and its boundary conditions were defined on the base of cooling water test. Computation domain represented approximately one sixth of the real combustion chamber. CFD simulation the simulation was calculated for two basic conditions, 1) the cooling nozzle switched on and 2) the cooling nozzle switched off. This corresponded to the selective cooling test.

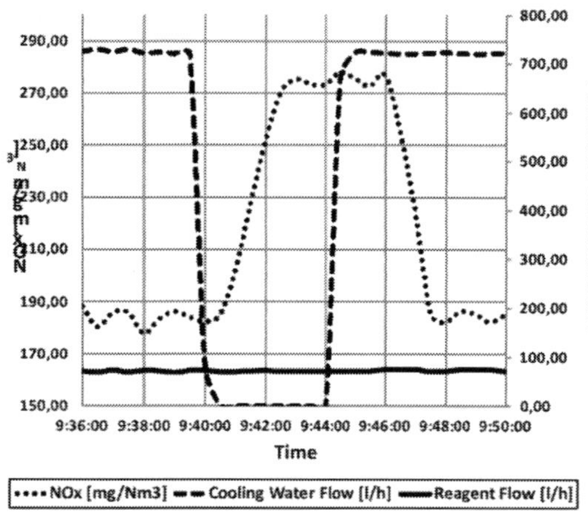

Fig. 5. Results of cooling system test

3 Results of CFD simulation

Results of CFD simulation were compared with time record data of selective cooling test. The most important is the effect of cooling water on NO_X reduction that is why the effect of cooling water was evaluated by reduction of NOx concentration, which means reduced amount of reduced NO_X and NO respectively, see next table.

Table 4. Penetration of flue gas by reagent.

Reduction of NO_X [%]	Test	CFD
Cooling lance ON	43	22
Cooling lance OFF	16	6
Effect of Cooling water	+27	+16

As it is evident from the previous table CFD model is not reach the same reduction. Compared to the test, the reduction is approximately 50% lower. This means, that the model is pessimistic and understates the effect of cooling water in injection area.

Next the effect of cooling water was investigated in injection area. Following figures shown the concentration field of NO in plane parallel to flue gas flow and the plane is located in the centre of lances. Cooling water in injection area creates cold cover where the temperature is lower than temperature of surrounding gas which is too high. Inside this area the temperature is ideal for reduction of NO (7) and NO_2 (8).

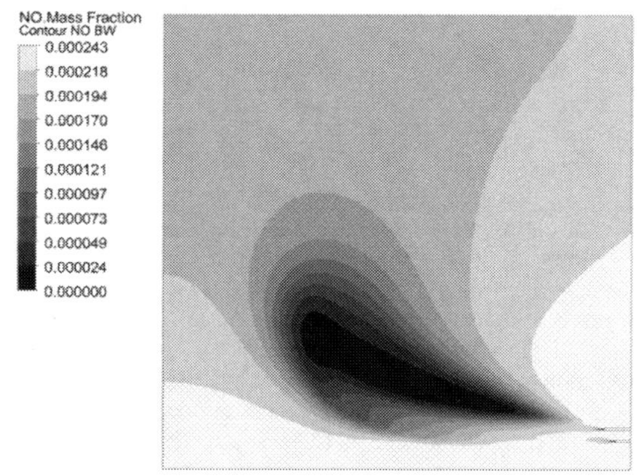

Fig. 6. Concentration field of NO, horizontal plane parallel to flue gas flow, cooling lance ON

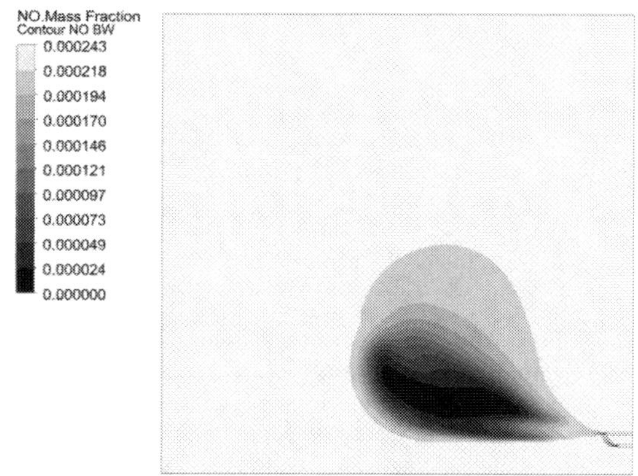

Fig. 7. Concentration field of NO, horizontal plane parallel to flue gas flow, cooling lance OFF

When the cooling water is not in injection area than temperature is too high. NO and NO_2 are reduced but only in area where reagent is evaporated. Temperature exceeds the upper limit of the temperature window outside and reagent produces NO (9).

5 Conclusion

The paper is focused on the simulation of high temperature modification of SNCR technology which is based on injection of cooling water upstream injection area of reagent. Chemical reactions used to describing of SNCR process are able to predict quite accurately the reduction of NO, but chemistry of SNCR process is strongly simplified due to practical usability and limitation of number of chemical equations included in the CFD model. Described CFD model of SNCR process can be used for prediction of the efficiency of SNCR technology, but the model in not able to describe complexity of the phenomena.

Effect of cooling water injection upstream injection area of reagent was analysed on the base of real test. Recorded data of the test was used to definition of CFD model of couple of injection and cooling lance. Whereas the proposed SNCR model does not include detail chemistry of OH radical, than the CFD model is not able to describe full effect of cooling water in injection area. This is evident from a comparison of the test results and CFD model. Effect of cooling water is reduced to local cooling due to evaporation of water.

The project will continue and future goal will be modification of kinetics of SNCR process, which describes effect which fully describes the effect of water on the rate of chemical reactions and shift of temperature window to higher temperatures.

Acknowledgment

The paper was prepared within the Project of MPO ČR TIP No. FI-IM4/193 "Research and Development of Nozzles for SNCR Technology in Power Engineering Focused on Elimination of Negative Effects of SNCR

References

1. A. Kohl, R. Nielsen, *Gas Purification, Fifth Edition*, Gulf Publishing Company, Houston, Texas, (1997), ISBN 0-88415-220-0
2. X. Hou, H. Zhang, M. Pilawska, J. Lu, G. Yue, The formation of N_2O during the reduction of NO by NH_3, Fuel, p. 3271-3277, **87** (2008)
3. J. Brouwer, M. P. Heap, D. W. Pershing, and P. J. Smith, A Model for Prediction of Selective Non-Catalytic Reduction of Nitrogen Oxides by Ammonia, Urea, and Cyanuric Acid with Mixing Limitations in the Presence of CO. *26th Symposium (Int'l) on Combustion*. The Combustion Institute. 1996.
4. B. von der Heide, Ist das SNCR-Verfahren noch Stand der Technik, Thomé-Kosmiensky, Michael Beckmann (Hrsg.): *Energie aus Abfall*, TK Verlag Karl Thomé-Kozmiensky, 275 – 293, **4**, ISBN 978-3-935317-32-0
5. Ch. R. Choi, Ch. N. Kim, Numerical investigation on the flow, combustion and NO_X emission characteristics in a 500 MWe tangentially fired pulverized-coal boiler, Fuel, 1720-1731, **88** (2009)
6. Y. Lu, I. Hippinen, A. Jahkola Control of NOx and N2O in pressurized fluidized-bed combustion Fuel, 317-322, **74** (1995)
7. R. Rota, E. F. Zanoelo, Influence of oxygenated additives on the NOxOUT process efficiency Fuel, 765-770, **82** (2003)
8. D. R. Tree, A. W Clark Advanced reburning measurements of temperature and species in a pulverized coal flame Fuel, 1687-1695, **79** (2000)
9. J. A. Miller, G. T, Bowman, Mechanism and modeling of nitrogen chemistry, Combustion, Progress in Energy and Combustion Science , 287-338 , **15** (1989)
10. R. Rota, D. Antos, E. F. Zanoelo, M. Morbidelli, Experimental and Modeling Analysis of the NOxOUT Process, Chemical Engineering Science, 27–38, **57** (2002)
11. M. Čarnogurská, T. Brestovič, M. Příhoda, Modelling of Nitrogen Oxides Formation Applying Dimensional Analysis, Chemical and Process Engineering, 175-184, **32** (2011), ISSN 0208-6425
12. Ansys, Inc., *ANSYS CFX-Solver Theory Guide*, (2016)

Mathematical Simulation of Drying Process of Fibrous Material

Tomáš Blejchař[1,*], *Jiří* Raška[1], and *Jana* Jablonská[1]

[1]VŠB-Technical University of Ostrava, Faculty of Mechanical Engineering, Department of Hydromechanics and Hydraulic Equipment, tř. 17 listopadu 15/2172, Ostrava-Poruba, Czech Republic

Abstract. The article describes mathematical simulation of flowing air through porous zone and water vaporisation from mentioned porous area which actually represents dried fibrous material - cotton towel. Simulation is based on finite volume method. Wet towel is placed in pipe and hot air flow through the towel. Water from towel is evaporated. Simulation of airflow through porous element is described first. Eulerian multiphase model is then used for simulation of water vaporisation from porous medium. Results of simulation are compared with experiment. Ansys Fluent 13.0 was used for calculation.

1 Introduction

Drying process is used in wide range of industrial application. Drying of textile is mainly used in commercial laundries, hotels etc. Textile is hydroscopic material which has the ability to attract water molecules. To remove water from linen is usually done by evaporation of water into hot air. Mass flow of hot air through textile can be interpreted as flow through porous zone. The mathematical modelling is a very good way how to simulate airflow through the porous medium. The porous area with defined parameters can be represented as linen - cotton towel. Porous parameters and next inputs for simulation are derived from the real experiment [2], [4]. The mathematical model of wet linen is achieved by patching liquid phase into porous zone. The simulation based on finite volume method can be complemented by evaporation model in Ansys Fluent so liquid phase is evaporated from porous zone and drying process is simulated.

The group of authors refers about simulation of drying process realized in industry. The example of simulation process of drying is the drying of soybean [5], spraying and evaporation of particles into hot airflow [6], vacuum drying of wood [7] or drying of the iron ore pellets [8]. Mathematical modelling of multiphase flow with phase change in energy facilities is refered [9].

2 Experiment Data - Base for Ansys Fluent Data

The drying process and a detailed description of the measurement procedure, measuring devices and measurement results have been described in [4]. The towel was placed in the long pipe (diameter of pipe is 200 mm) and fixed (length of the pipe with towel is 110 mm (Figure 2). The towel contained certain amount of water. The airflow through wet towel caused evaporation of water. Temperature of air was regulated by thermostat. The drying rate was measured from wet towel with using the relative humidity sensor and temperature sensor placed behind towel. The pressure sensors were placed in the same sampling points. The schema of the test shows Figure 1. Hot air flows in arrows direction.

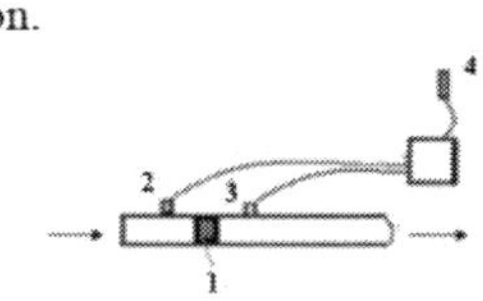

1 – wet towel
2 – temperature, flow rate and pressure sensors
3 – relative humidity, temperature and pressure sensors
4 – ambient humidity and temperature sensor

Fig. 1. Experimental equipment.

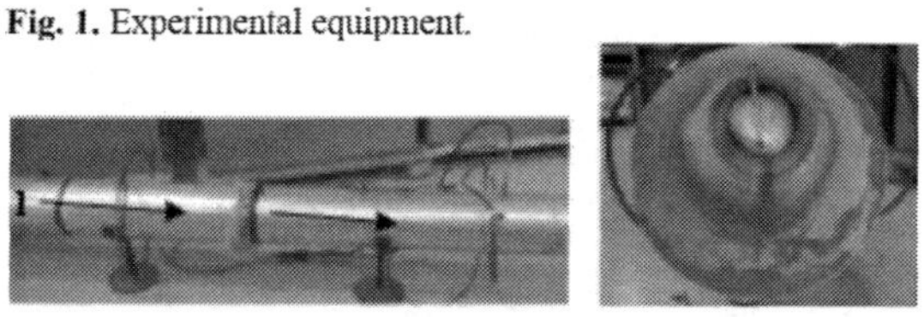

Fig. 2. Experiment layout with fixed towel.

Diameter of the pipe was 0,2 m, distance between sampling points 2 and 3 was 2,14 m. The towel was placed between sampling points 2 and 3 and length of the pipe with towel was 0,11 m.

Water content before and after drying was defined as weight. Evaporation process removed 0,25 kg of water from towel (Table 1) and after the test the towel becomes totally dry. Weight of the dry towel was 0,2 kg. Measurement of evaporation taken 16 minutes.

* Corresponding author: tomas.blejchar@vsb.cz

Table 1. Weight of towel and water.

Weight - wet towel	Weight after drying	Weight difference
0,45 kg	0,2 kg	0,25 kg.

3 Theory of Eulerian Model

The main purpose of using this type of simulation model is its ability to solve each phases by multiple separate way [1]. The phases of air, water-vapour and water - liquid are set in the model. To simulate drying of the wet fibrous material the phase water - liquid can be patched into porous zone and the velocity of water - liquid phase can be fixed to 0 m/s in this zone. Air phase (with included vapour phase to simulate ambient wet air) flows through porous zone. Eulerian model also allows use the vaporisation model and water-liquid is evaporated from the porous area.

3.1 Continuity equation

The continuity equation for phase q is described by equation (1), where $\vec{v}_q$ is velocity of phase q, γ is porosity of the porous area, ρ_q is density of the phase q, α_q is phase volume fraction, m_{pq} is mass transfer from phase p to phase q, S_q is source term considered to be zero in our case [1].3.1.

$$\frac{\partial(\gamma\alpha_q\rho_q)}{\partial t}+\nabla\cdot(\gamma\ \alpha_q\rho_q\vec{v}_q)=\gamma\sum_{p=1}^{n}(\dot{m}_{pq}-\dot{m}_{qp})+\gamma\ S_q \qquad (1)$$

3.2 Momentum equation

Momentum balance for the phase q shows (2), where $\vec{R}_{pq}$ is interaction force between phases, $\vec{F}_q$ is external body force, $\vec{F}_{lift,q}$ is lift force, $\vec{F}_{vm,q}$ is virtual mass force, $\vec{v}_{pq}$ is interphase velocity, τ is stress tensor [1].The continuity equation for phase is

$$\frac{\partial(\gamma\alpha_q\rho_q\vec{v}_q)}{\partial t}+\nabla\cdot(\gamma\ \alpha_q\rho_q\vec{v}_q\vec{v}_q)=-\gamma\alpha_q\nabla p+\nabla\cdot(\gamma\ \tau_q)+\gamma\alpha_q\rho_q\vec{g}+ \qquad (2)$$
$$+\gamma\sum_{p=1}(\vec{R}_{pq}+\dot{m}_{pq}\vec{v}_{pq}-\dot{m}_{qp}\vec{v}_{qp})+\gamma(\vec{F}_q+\vec{F}_{lift,q}+\vec{F}_{vm,q})+\alpha_q\left(\frac{\mu}{K}+\frac{C_2\rho}{2}|\vec{v}_q|\right)\vec{v}_q$$

Coefficients C2 and 1/α for porous area were determined by manner described in reference [2].

3.3 Energy equation

Energy equation for multiphase Eulerian model has form (3), where h_q is enthalpy of phase q, is heat flux, Q_{pq} is intensity of heat exchange between phases p and q [1].

$$\frac{\partial(\gamma\alpha_q\rho_q h_q)}{\partial t}+\nabla\cdot(\gamma\ \alpha_q\rho_q\vec{v}_q h_q)=-\gamma\alpha_q\frac{\partial(p_q)}{\partial t}+\gamma\tau_q:\nabla\vec{v}_q-\nabla(\gamma\ \vec{q}_q)+\gamma\ S_q+ \qquad (3)$$
$$+\gamma\sum_{p=1}(Q_{pq}+\dot{m}_{pq}h_{pq}-\dot{m}_{qp}h_{qp})$$

Ranz - Marshall correlation (5) [1] was used for solving the heat transfer coefficient h_{pq} (4) between phases air and water. Here κ_q is thermal conductivity, Nu is Nusselt number, d_p is particle diameter, Re is Reynolds number, Pr is Prandtl number.

$$h_{pq}=\frac{6\kappa_q\alpha_p\alpha_q Nu_p}{d_p^{\,2}} \qquad (4)$$

$$N_u=2.0+0.6R_e^{\,1/2}\,\mathrm{Pr}^{1/3} \qquad (5)$$

3.4 Ansys Fluent Evaporation model

Momentum equation for evaporation model (6) [1] contents mass transfer expression from liquid state to gas state (evaporation). Mass transfer from gas to liquid state (condensation) is zero for our case. The symbols in equation (6) means: ***v*** is vapour index; ***l*** is liquid index; α is volume fraction of vapour; ρ_v is vapour density; $\vec{V}_v$ is velocity of gas state; $\dot{m}_{l\to v}-\dot{m}_{v\to l}$ is evaporation rate [kg/(s·m^3)].

$$\frac{\partial}{\partial t}(\alpha\cdot\rho_v)+\nabla\cdot(\alpha\cdot\rho_v\vec{V}_v)=\dot{m}_{l\to v}-\dot{m}_{v\to l} \qquad (6)$$

Mass transfer for evaporation is started when saturation temperature T_{sat} (input for Fluent) is exceeded by ambient temperature in certain part of domain. Mass transfer is defined by equation [1]:

$$\dot{m}_{l\to v}=coeff\cdot\alpha_l\rho_l\frac{(T-T_{sat})}{T_{sat}} \qquad (7)$$

4 Simulation in Ansys Fluent 13

Mathematical model consists from pipe zone and from porous zone. Model is simplified and defined as 2D axisymmetric case, see Figure 4. Diameter of the pipe is 0,2 m and length of the porous zone is 0,11 m.

Fig. 4. Mathematical model

4.1 Boundary conditions

4.1.1 Inlet of airflow and its humidity

At the inlet the mass flow of hot wet air (0,084 kg/s) containing the water vapour (mass fraction 0,0087) is defined. Ambient air temperature is 26 °C and relative humidity 35%. Specific humidity is then 0,0075 kg/kg. Basically specie defined in the calculation as mass fraction 0,0087 represents humidity available in the air.

4.1.2 Inlet of air temperature

Temperature values at inlet were measured and regulated by the thermostat, see Figure 5. Boundary condition of air temperature is driven by values from file.

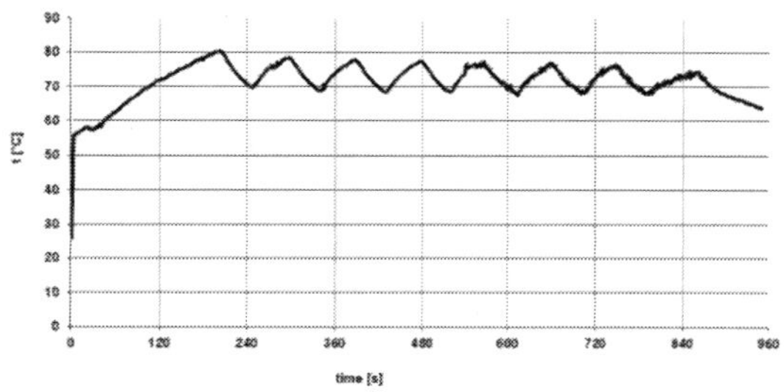

Fig. 5. Inlet temperature

4.2 Initial conditions

4.2.1 Pressure and content of moisture in porous zone

Pressure initialization is specified by value 101300 Pa. Other important initial value is volume fraction of water in the porous zone. Practically it means moisture content in fibrous material (cotton towel). Defined volume fraction 0,058 of phase water reflects 0,25 kg of water in the porous area. Water droplet diameter is set to value 0,0025 mm. It reflects good results of vaporisation process in simulation process in Fluent.

Porosity of the porous area is set to 0,95. This value is determined on basis of experimental and numerical testing. Water phase has fixed value of velocity in porous zone (0 m/s) so it does not move in the zone.

4.3 Saturation temperature and pressure

The evaporation model described in chapter 3.4 needs to define saturation temperature in terms of pressure. For saturation temperature the value of dew point is used. It corresponds to relative humidity 8% (mainly measured value in experiment) and operating temperature [4]. Table 2 shows set of values.

Table 2. Saturation temperature and pressure

t [°C]	t_{dew} [°C]	p_p [Pa]
50	6,7	986
60	13,9	1592
70	21	2491
80	28	3789

5 Results of Calculation and Comparison with Experiment

The results of the calculation are compared with data measured during experiment [4]. It is mainly pressure drop across the pipe, temperature, specific humidity, relative humidity.

5.1 Pressure drop in the pipe

Result of pressure drop from experiment [4] is 83 Pa at airflow velocity 2,37 m/s and corresponds with result from simulation. For calculated velocity of airflow 2,66 m/s (right part of Figure 6) the simulation shows the pressure drop 86 Pa (left part of Figure 6). Velocity through porous zone is modelled as a model of physical velocity [2] it leads to higher values of velocity in porous area (Figure 6).

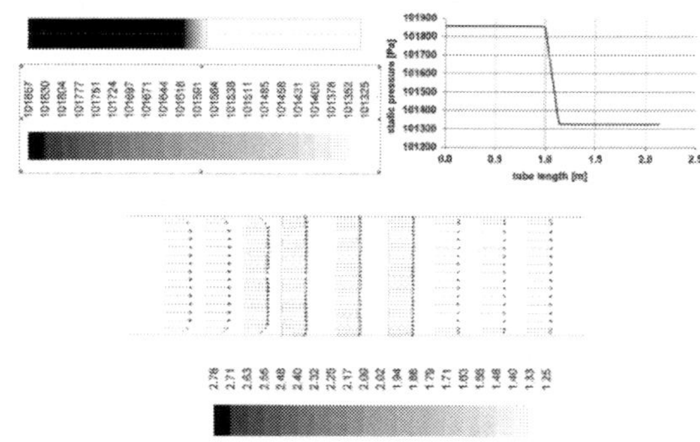

Fig. 6. Static pressure and velocity profile in porous zone

5.2 Relative humidity

Relative humidity indicated at the outlet varies and dependents on temperature and intensity of water evaporation. Ambient relative humidity is 35,5% at the temperature 26 °C. Calculated relative humidity at outlet - Fluent shows higher values till flow time 360 s then it continues with values placed under measured data (Figure 7). Relative error of simulation results for certain time is shown in Table 3.

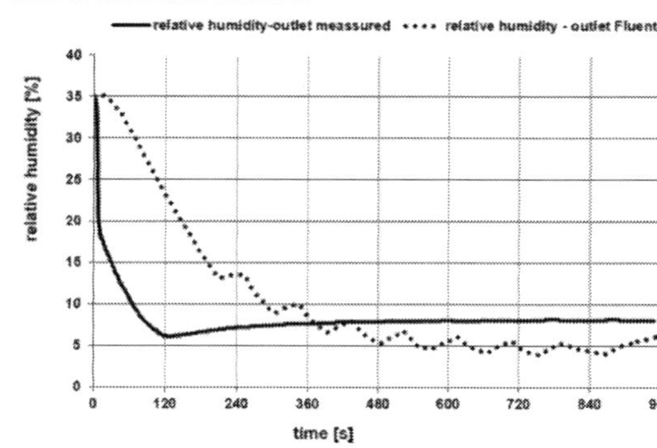

Fig. 7. Relative humidity outlet Fluent

Table 3. Relative error

Flow time [s]	Relative error [%]
205	49,2
520	17,4
920	30

5.3 Temperature

Inlet temperature reported by Ansys Fluent consists from same values which was measured Figure 5. Temperature t - outlet is gained from experiment and is lower than at the inlet due to water evaporation. Temperature t - outlet Fluent is calculated from simulation. It is lower at the beginning and with flow time rises to measured values at the outlet. This behaviour is logical and reflects results from chapter 5.1. Dew point temperature in front and behind of the towel is lower than temperature t - inlet and t - outlet (Figure 8). It means air is not saturated and no condensation occurs, just evaporation. Relative error of calculated temperature t - outlet Fluent in comparison with measured temperature at the outlet for selected time is shown in Table 4.

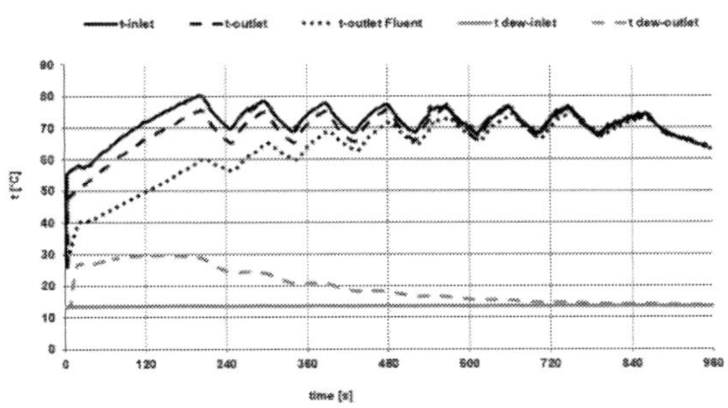

Fig. 7. Measured and calculated temperature

Table 4. Relative error

Flow time [s]	Relative error [%]
205	20,2
520	1,2
920	0,8

5.4 Specific humidity

Specific humidity is derived from measured values of temperature and relative humidity at the outlet. Graph (Figure 8) show ambient specific humidity which is constant (0,0075 kg/kg of dry air) because amount of moisture does not change. The same situation will be at the inlet where just temperature varies but humidity does not. More interesting is state behind porous zone and at the outlet. Water is evaporated from porous zone. Measured specific humidity at the outlet is higher than calculated from time 250 s. Results from simulation shows more intensive evaporation at the beginning and then decreasing trend. Relative error of calculated specific humidity - outlet Fluent related to measured specific humidity for selected time is shown Table 5.

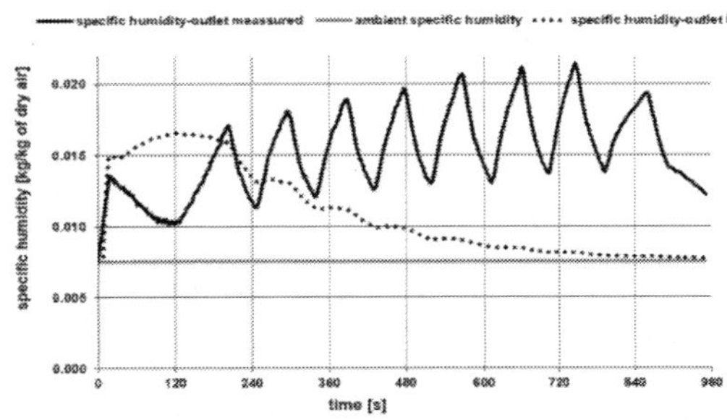

Fig. 8. Calculated and measured specific humidity

Table 5. Relative error

Flow time [s]	Relative error [%]
205	1,8
520	23
920	34

5.5 Volume fraction of water in porous zone

Calculated evaporation process removes volume fraction of water from porous zone. Volume fraction patched into porous zone is set to 0,058 and it is derived from water weight in the towel (0,25 kg) at the beginning of experiment. At the end of calculation volume fraction is 0,0007 and it corresponds to 0,003 kg of water. Almost all the water is evaporated to vapour during calculation and the flow time of vaporisation is specified. The results correspond with Table 1.

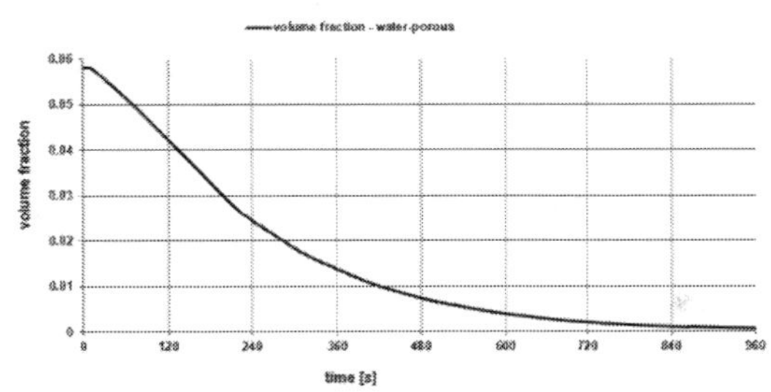

Fig. 9. Water volume fraction in porous zone vs time

5.4.1 Contours of evaporation

Example of evaporation (Figure 10) gives an idea how the simulation principle of drying works and that water liquid is transferred to water vapour state.

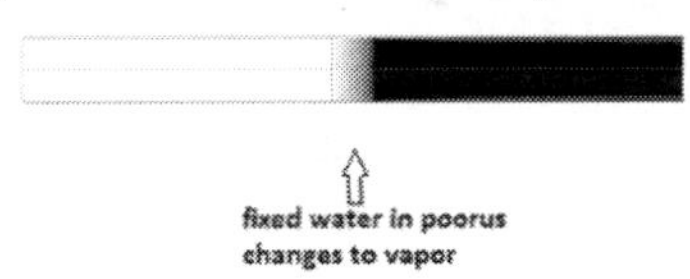

Fig. 10. Evaporation from the porous zone

6. Conclusions

Mathematical simulation of drying process is continuation of the physical experiment of drying [4]. Computational case can be set in Ansys Fluent 13 by a very meaningful way. Due to ability to fix the phase velocity (water-liquid) to zero in the porous zone then the wet fibrous material can be simulated in the mathematical model. The multiphase Eulerian model, thanks of its complexity, is more than suitable for this kind of settings. Final results of the temperature and relative humidity at the outlet are close to measured values. The specific humidity which is calculated [3] has higher relative error in comparison with mathematical simulation. The simulated and measured drying process removes almost all water from linen (porous area).

REFERENCES

1. ANSYS FLUENT INC. *Fluent 13.0, Help –Theory guide*. Fluent Inc. (2010).
2. ANSYS FLUENT INC. *Fluent 13.0, Help -User's guide*. Fluent Inc. (2010).
3. MUJUMDAR, A. S.: *Handbook of industrial drying, third edition,* 1280 pages, (2007)
4. RASZKA, J.: *The experiment of drying process of the fibrous material,* Transactions of the VŠB – Technical University of Ostrava, Mechanical Series No. 1, **vol. LXI** article No. 1996 (2015)
5. RAFIEE, S., KEYHANI, A., MOHAMMADI, A.: *Soybean seeds mass transfer simulation during drying using finite element method,* IDOSI publication 2008, p. 284-288, (2008).
6. HUANG, L., KUMAR, K., MUJUMDAR, A. S.: *Computational fluid dynamic simulation of droplet drying in a spray dryer*, Sau Paulo IDS 2004, 25.-26. August, p. 326-332, (2004)
7. ERRIGUIBLE, A., BERNADA, P., COUTURE, F., ROQUES, M. A.: *Modeling and numerical simulation of heat and mass transfer during vacuum drying of porous medium*, Sau Paulo IDS 2004, 25.-26. August, p. 726-732., (2004)
8. LJUNG, A. L..: *Drying of iron ore pellets-Analysis with CFD,* Lulea University of technology, (2008)
9. BOJKO, M. *Metody matematického modelování vícefázového proudění s uvažováním fázových změn v energetických zařízenich*: teze habilitační práce k habilitačnímu řízení v oboru Aplikovaná mechanika. Ostrava: VŠB - Technická univerzita Ostrava, ISBN 978-80-248-3842-7 (2015) .

The influence of flow modification on air and PCM temperatures in an accumulative heat exchanger

Marcin Borcuch[1*], *Michał* Musiał[1], *Karol* Sztekler[1], *Wojciech* Kalawa[1], *Stanisław* Gumuła[1] and *Sebastian* Stefański[1]

[1]AGH University of Science and Technology, A. Mickiewicza Av. 30, 30-059 Krakow, Poland

Abstract. The paper presents the influence of flow modification on the operation of an accumulative heat exchanger. This device can be used as a regenerator in ventilation and air supply systems. A heat exchanger uses ceresine (a mixture of paraffins) as a phase change material (PCM). The aim of this research was to determine the effect of flow modification on temperature distribution and pressure drops in the device. The introduction contains a short description of the test stand used, including the accumulative heat exchanger, the guide vanes, and the locations of measurement and control equipment. We found that additional objects limited vortex structures, increased the inside temperature, and dropped the pressure along the heat exchanger. Guidelines for further research are proposed and briefly discussed.

1 Introduction

The increasing use of energy today has necessitated a search for alternatives in the energy sector. Harsher UE energy politics are commissioning research on new technologies that can lower energy costs, decrease environmental impact and improve energy efficiency. Renewable technologies that have rapidly developed within the last decade, especially those for harnessing wind and solar energy, are proof of their growing popularity. Unfortunately, the application of wind turbines and PV panels strongly depends on weather conditions. It is difficult to operate with high efficiency, and seasonal changes limit the effectiveness and profitability of these technologies. One solution is to store energy produced in off-peak energy consumption periods, and use it later to meet demands. The accumulative heat exchanger constructed at the Department of Thermal and Fluid Flow Machines is a device that allows storage of heat in a PCM. Due to its high latent heat values [1], the device is a suitable solution for heat storage. The application of PCMs in heat exchangers has been studied since the late 1980s [2]. However, the design and development of such devices intended for ventilation and air conditioning systems is relatively new [3,4]. Much research is being carried out in an effort to improve the heat transfer coefficient, minimize flow losses, and improve the insulation of whole construction. The geometry proposed in [5,6] has been investigated, but the results obtained showed that the flow arrangement should be modified. The PCM used ceresine as a heat storage material, which is a neutral product formed from the distillation of crude oil. The energy accumulated ranged from 240 kJ to 880 kJ.

2 Developing simulation models

The experimental test stand allows measurement and control of all operating parameters of the accumulative heat exchanger (Fig. 1). An electric air heater is placed at the beginning of the test stand to control temperature and monitor the relative humidity of air entering the system. The accumulative heat exchanger is equipped with 6 thermocouples: two for measuring the temperature of ceresin in the selected cassette; and four for measuring the temperature of air between cassettes. Additionally, two thermocouples measure the temperature of air entering and leaving the heat exchanger. The throttling valve fitted at the heat exchanger exhaust is used to regulate the rate at which air flows into the system. The flowmeter with guide vanes determines the volumetric airflow rate in the heat exchanger. The entire system is driven by a centrifugal fan, located at the end of a duct. The rotational speed of the fan is set precisely by an inverter. Pressure losses Δp in the heat exchanger are estimated by measuring static pressure at the inlet and outlet.

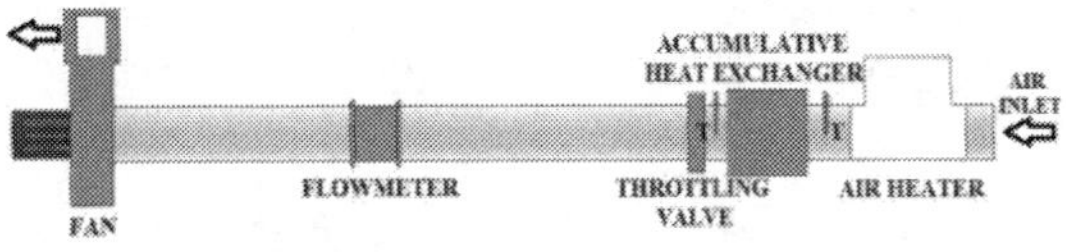

Fig. 1. Experimental stand (T - temperature measurement points in the center of the duct).

Two configurations of the accumulative heat exchanger have been examined. At the beginning, measurements were carried out on four cassettes. Each of them was filled with 1.25 kg of ceresine. In the

* Corresponding author: borcuch@agh.edu.pl

second step, steel guide vanes were prepared and fitted on the edges of the cassettes. These two configurations are shown in Fig. 2 and Fig. 3. The ceresine parameters are presented in Table 1.

Fig. 2. Two configurations of the accumulative heat exchanger and the location of the guide vanes.

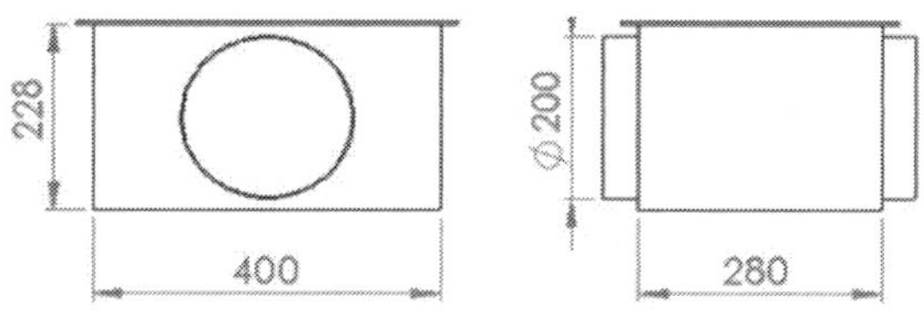

Fig. 3. Dimensions of the device.

Table 1. Ceresine parameters.

Parameter	Value
melting temperature range	61°C to 78°C
density	0.91 kg/m^3
specific heat capacity	2.9 kJ/(kg-K)
thermal conductivity	0.25 W/(m-K)

The dimensions of the device are presented in Fig. 3, while the experimental test stand prepared for measurements is presented in Fig. 4. After placing all the thermocouples, the exchanger was covered by a plexiglas sheet and wool-insulated top cover to minimize heat losses.

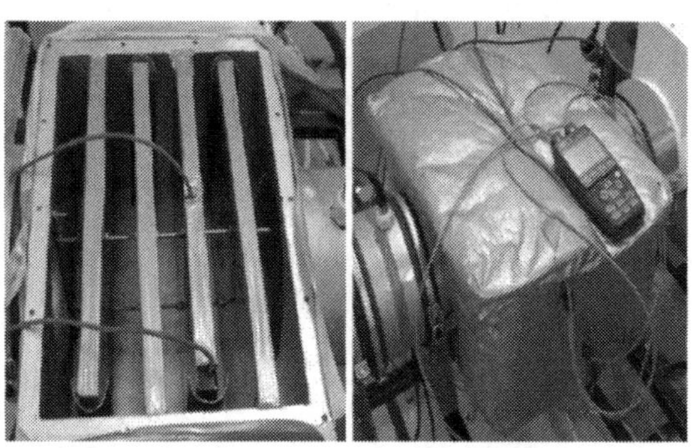

Fig. 4. The accumulative heat exchanger: left – internal view of the exchanger, placement of cassettes, guide vanes, thermocouples to measure the temperature of ceresin; right – fully insulated heat exchanger fitted on the experimental stand.

Figure 5 shows a scheme of the analyzed region of the heat exchanger. The arrows indicate the flow directions inside the device. Measuring the temperature at selected points enables estimation of the extent to which the flow modification influences the conditions of heat transfer between the air and the ceresine.

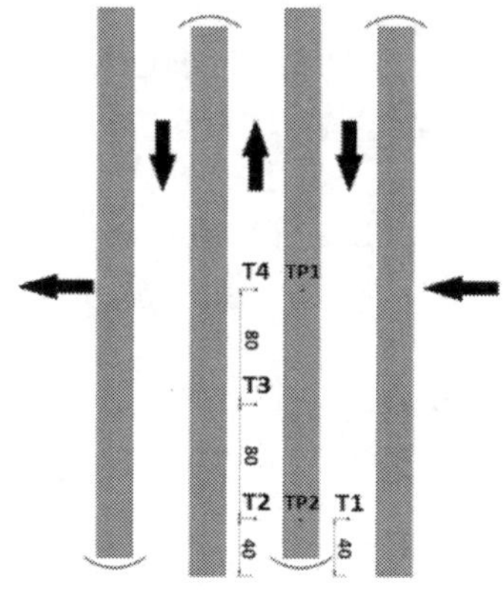

Fig. 5. The air flow direction, guide vanes, the location of thermocouples inside the accumulative heat exchanger.

3 Experiment and results

3.1 Pressure losses

Experiments were done with and without guide vanes. The influence of the guide vanes can be observed when comparing the thermal and flow parameters of the heat exchanger. Flow modifications introduced additional Δp in the heat exchanger, so the entire system required a higher volume flow rate $\dot{V}_{\text{air}}$, maintained by an increased fan rotational speed. This was done to compare the operational parameters of heat exchangers with and without guide vanes under the same flow conditions. The values for static Δp in the exchanger are presented in Fig. 6.

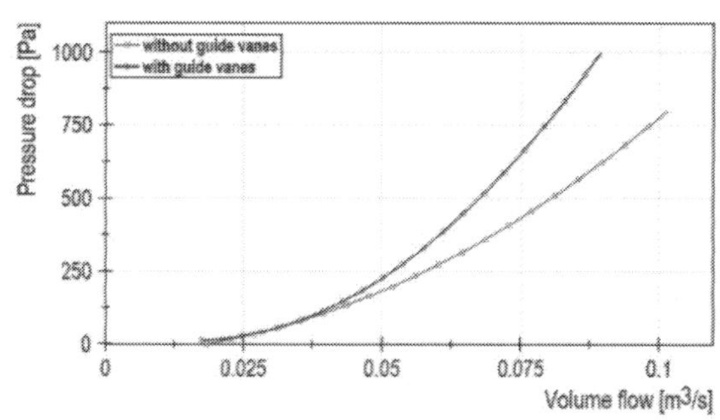

Fig. 6. Measured Δp in the accumulative heat exchanger with and without guide vanes.

Higher temperatures of air and ceresin due to the guide vanes was observed, but this came at the price of additional Δp (which reached up to 50% compared to when vanes were not used).

https://doi.org/10.1051/epjconf/201818002011

3.2 Ceresine and air temperature

The points of measurement of the air temperature inside the exchanger were located after the second cassette due to stable conditions, and the vortices and disturbances were the same as those after the third cassette. Before the second cassette, there was a meaningful influence of inlet stream, so stable conditions were impossible because of the turbulent nature of the inlet flow. Experiments were done with a $\dot{V}_{air}$ range of 0.038–0.070 m^3/s, but results are presented only for the two most interesting cases. An increase in air temperature due to the presence of the guide vanes was observed. In particular, temperatures in the vicinity of the guide vanes tended to increase. The increment reached about 7°C (Fig. 7 and 8). The temperatures in the modified accumulative heat exchanger appeared to be more greatly varied.

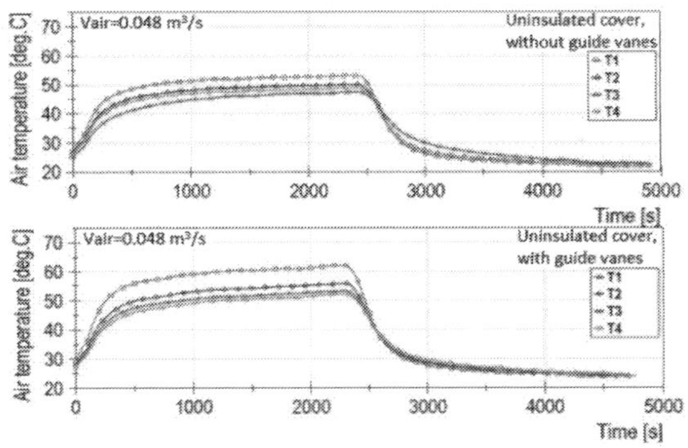

Fig. 7. Air temperature inside the heat exchanger vs time.

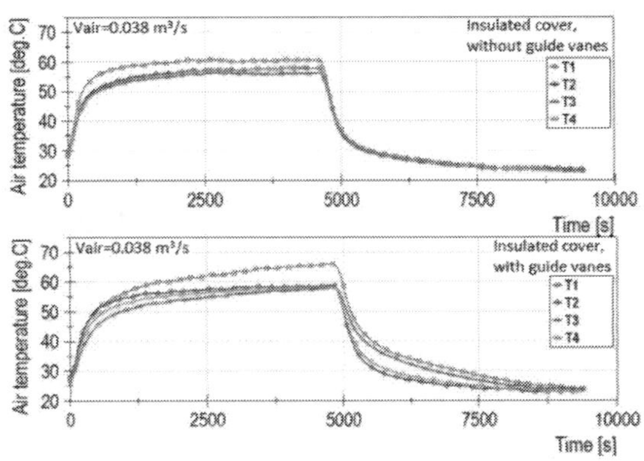

Fig. 8. Air temperature inside the heat exchanger vs time.

Clear differences were observed between T1, T2, T3, and T4, especially during the discharge stage (e.g. Fig. 8 – from 5000 to 7500 s). This phenomenon occurred for all flow rates. A noticeable influence caused by the guide vanes was observed in ceresine temperatures as well. The positions of the thermocouples in the cassette allowed measurement of the temperature at two main points: in the middle and on the edge of the cassette. The guide vanes induced faster air flow around the cassettes, so a higher heat transfer coefficient could be expected. This assumption has been confirmed by experiments. The application of guide vanes caused higher temperatures on the edge of the cassette (Fig. 9). Furthermore, the temperature in the middle of the cassette appeared to be a little lower (Fig. 10), which may have been caused by the absence of vortical structures between modules, and which lead to a smaller flux of heat from the air to the ceresine.

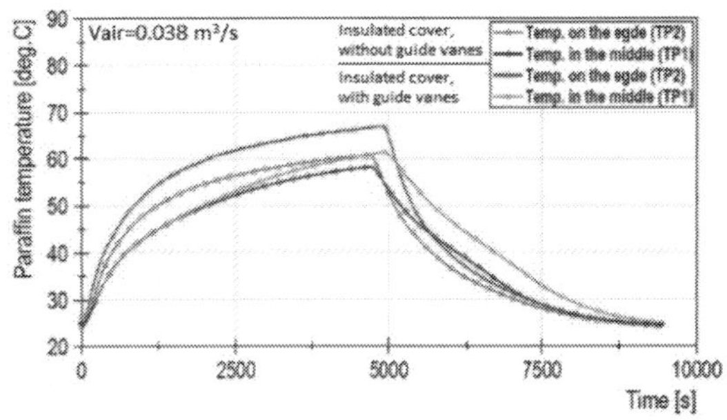

Fig. 9. Heat storage material temperatures inside the second cassette ($\dot{V}_{air}$ = 0.038 m^3/s) vs time.

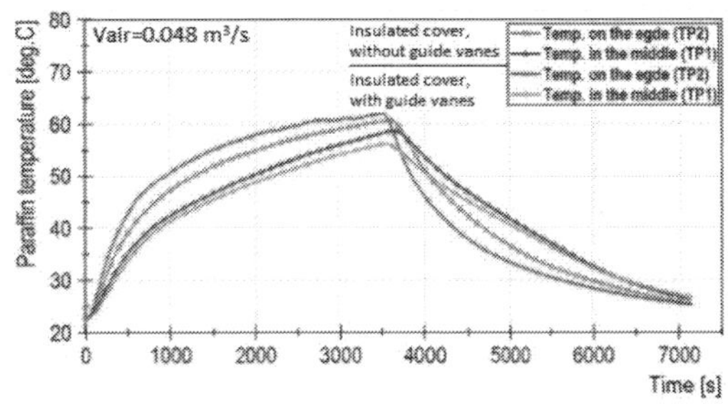

Fig. 10. Heat storage material temperatures inside the second cassette ($\dot{V}_{air}$ = 0.048 m^3/s) vs time.

The temperature measured in the center of the duct is higher than average. Temperature profiles are presented in Fig. 11. The temperature profile at the inlet and outlet pipes of the uninsulated exchanger was measured (Fig. 12).

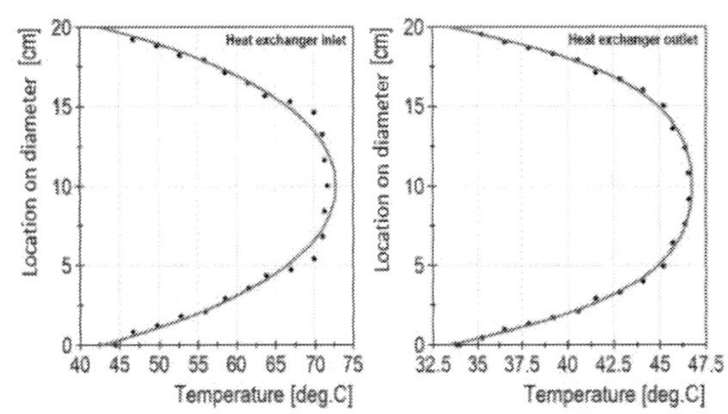

Fig. 11. Temperature profiles at the inlet and outlet of the accumulative heat exchanger.

It can be assumed that average temperatures are: a measured value multiplied by 0.88 at the inlet; and a measured value multiplied by 0.93 at the outlet (Fig. 11). For one case, the values have been recalculated and presented in Fig. 12. One cycle of the exchanger operation was divided into two separate parts: the charging stage on the left side; and the discharging stage on the right side. Different temperatures correlate to various air densities:

- density of air at 25°C: ρ_{25} = 1.184 kg/m^3;
- density of air at 40°C: ρ_{40} = 1.127 kg/m^3.

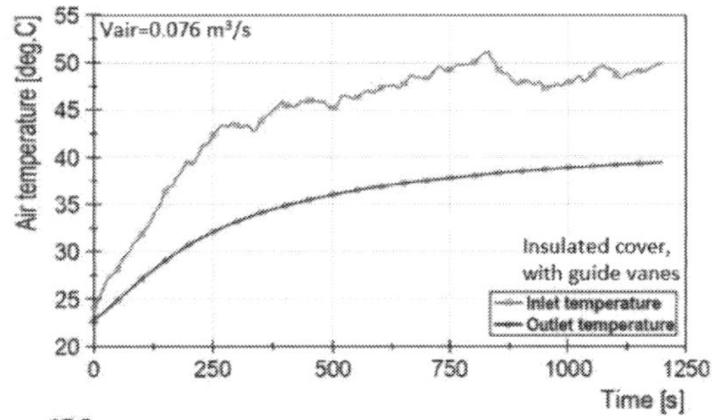

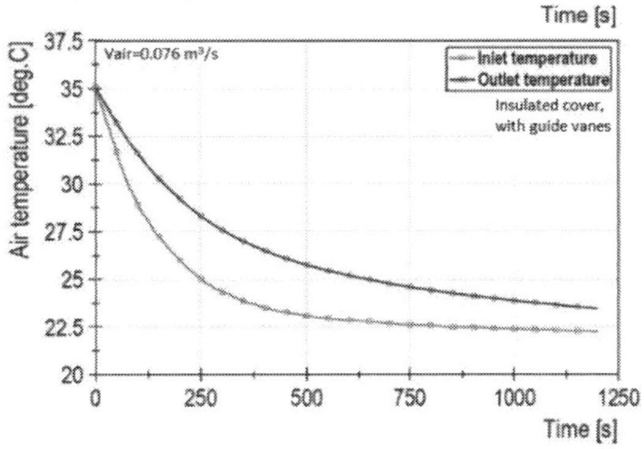

Fig. 12. Average temperatures at inlet and outlet of the exchanger vs time; upper fig. – charging, lower – discharging.

3 Conclusions

The application of guide vanes limited the influence of the vortex structures on temperature distribution inside the exchanger. The vanes also increased pressure losses, which induced larger temperatures in flow channels. The compromise between the heat transfer efficiency and flow losses can be balanced by an increase of the rotational speed of the fan. Further research on minimizing the negative impact of guide vanes on flow conditions by modification of their shape and location must be carried out to improve heat exchanger efficiency. For an exchanger with a smaller amount of heat storage modules, the use of cassettes should be considered. The influence of the number of cassettes, their arrangement, and their thickness on accumulated heat are the goals of further research.

Acknowledgment

The paper was funded from government money Faculty of Energy and Fuels number 11.11.210.216.

References

1. A. Sharma, V.V. Tyag, C.R. Chen, D. Buddhi, *Renew. Sust. Energy Rev.* **13** 318 (2009)
2. M. Okada, *J.of Enh. Heat Trasfer* **2** 273 (1995)
3. M. Labat, J. Virgone, D. David, F. Kuznik, *Appl. Thermal Engineering* **66** 375 (2014)
4. P. Dolado, A. Lazaro, J. M. Marin, B. Zalba, *Renewable Enerrgy* **36** 2906 (2011)
5. M. Borcuch, M. Musiał, T. M. Wójcik, *Transactions of the Inst. Of Fluid-Flow Mach.* **128** 19 (2015)
6. M. Borcuch, M. Musiał, S. Gumuła, K. Sztekler, K. Wojciechowski, *Appl. Thermal Engineering* **127** 1335 (2017)

Mathematical model of compact type evaporator

Martin Borovička [1,*], *Tomáš* Hyhlík [1]

[1]CTU in Prague, Department of Fluid Mechanics and Thermodynamics, Czech Republic

Abstract. In this paper, development of the mathematical model for evaporator used in heat pump circuits is covered, with focus on air dehumidification application. Main target of this ad-hoc numerical model is to simulate heat and mass transfer in evaporator for prescribed inlet conditions and different geometrical parameters. Simplified 2D mathematical model is developed in MATLAB SW. Solvers for multiple heat and mass transfer problems - plate surface temperature, condensate film temperature, local heat and mass transfer coefficients, refrigerant temperature distribution, humid air enthalpy change are included as subprocedures of this model. An automatic procedure of data transfer is developed in order to use results of MATLAB model in more complex simulation within commercial CFD code. In the end, Proper Orthogonal Decomposition (POD) method is introduced and implemented into MATLAB model.

1 Introduction

An increasing effort to decrease the cost of any device requires more advanced methods for precise simulation and modeling of physical phenomena. Heat and mass transfer in heat exchangers/evaporator with complex geometry is still a complicated problem, because an analytical solution is not often available, so numerical simulation along with measurement is the only option that can be used for heat duty estimation and geometry optimization. In this paper, focus is on modeling heat and mass transfer inside evaporators for air dehumidification. Selected evaporator for this application is classified as a cross flow compact type heat exchanger (or finned tube heat exchanger) - tube bundle (single tube with multiple passes) with common fins (every rectangular fin is in contact with all passes of the tube), using R134a as a refrigerant. An evaporator ("Inside Heat exchanger" in **Fig. 1.**) is placed in heat pump circuit that can be often found in many air dehumidification devices. Other parts of heat pump circuits are: condenser, compressor and capillary tube/reduction valve.

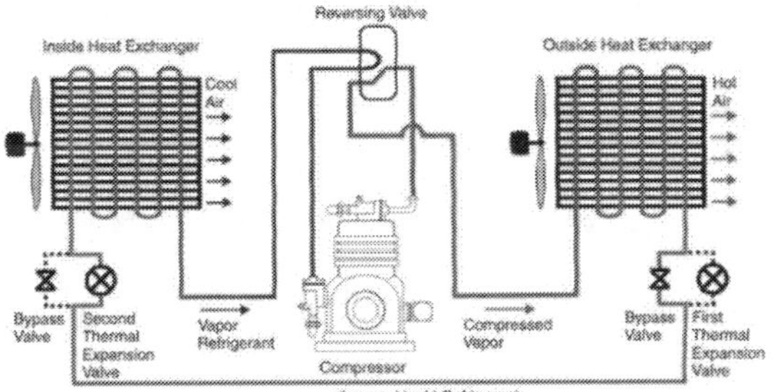

Fig. 1. Heat pump circuit [1]

2 Mathematical model

The mathematical model is developed in MATLAB SW, because of user-friendly environment, simple implementation of matrix operations and ability to use advanced optimization tools for model validation and development. The model contains simple menu interface for user convenience and tool for data transfer into commercial CFD code. Optimization of code has been performed so model can be used for optimization of heat exchanger geometry.

2.1 Model properties

The mathematical model simulates heat transfer between humid air and coolant/refrigerant flowing through tubes. Next, simulates mass transfer on outer side - water vapor condensation from air into water film on fins and tubes, and refrigerant evaporation on internal (tube) side. The model predicts locations, where condensation occurs, based on local flow parameters and local fin/tube temperature. Model is parametric in inlet conditions (air/refrigerant temperature and mass flow, humidity, pressure) and also in heat exchanger geometrical parameters. A semi-automatic procedure, that will manage data transfer of results from the MATLAB model into the simulation in commercial CFD package, using scripts and macros is also developed and implemented. A very important attribute of the model is its dimension – in how many axis parameter/property distribution is solved. Choice of model dimension and its complexity affects precision, computation time and computational resources. To obtain a balance between model precision, computation time and computation requirements, 2D model is used. 2 dimensions (coordinate x and y) where flow parameters are solved correspond to 2 edges of a fin as shown in **Fig. 2**.

* Corresponding author: Martin.Borovicka@fs.cvut.cz

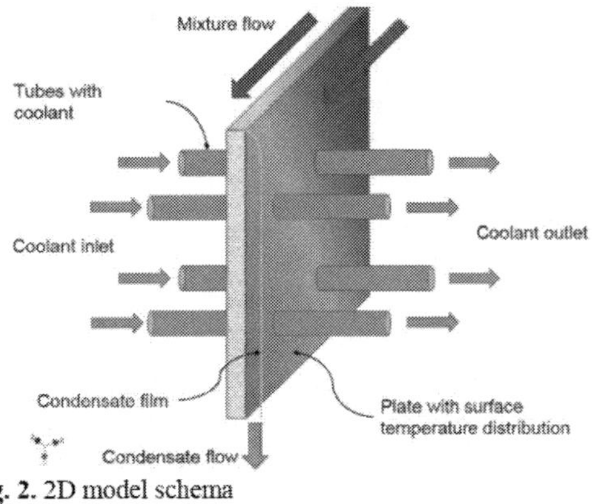

Fig. 2. 2D model schema

From point of view of data transfer into commercial CFD, 2D model is not the best, because in remaining, not simulated coordinate, all parameters are constant. This fact might cause imprecise results if big non-uniformities in this coordinate are present – for example non-uniform velocity in inlet caused by inlet duct.

2.2 Coordinate system and flow orientation

Heat exchanger with arbitrary dimensions and its coordinate system is in **Fig. 3.** (x – length, y – height, z - width)

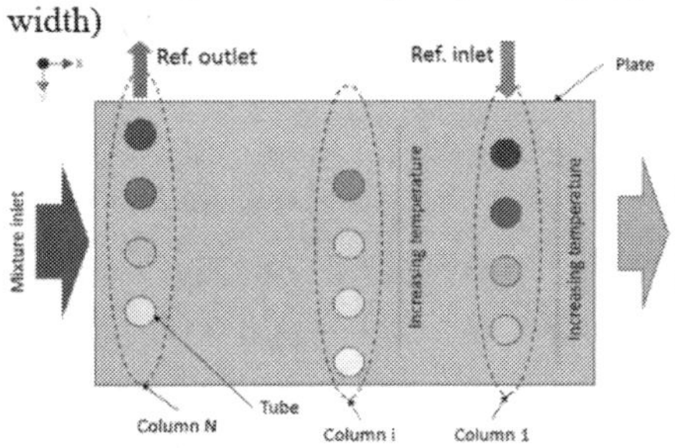

Fig. 3. Coordinate systemorientation

Air flow is always orientated in **x** direction (from left to right), and refrigerant flow is oriented in **z** direction – "inlet tube section" is placed at upper right corner of fin and "outlet tube section" at upper left corner (only even number of tube columns with staggered pattern is supported in this version).

2.3 Simplifying assumptions

Modeling heat and mass transfer is complicated geometry is a complex problem, so couple simplifications are assumed: [2]

1) The fin is flat. (corrugated fins are usually present, the effect of corrugation in included in *AreaFx*, *AreaFy* parameters – area increase factors)
2) For the first guess, refrigerant temperature is constant (equal to boiling temperature) through whole evaporator. During simulation, refrigerant temperature distribution is developing.
3) Condensate from tubes is dropping down, not flowing down the fin.
4) Temperature of fin is constant through its thickness.
5) No heat flow from fin edges.
6) All fields of flow parameters (temperature, velocity, humidity …) are constant in z direction in every slit.
7) Steady-state flow.
8) Perfect contact between outer tube surface and fin.

2.4 Governing equations

Differential equations describing heat and mass transfer problems are listed below. Equations (1) and (2) describe the general balance of mass for air and film flow. Ad hoc equation (3), (4) and (5) describe conservation of enthalpy in 1 slot (space between 2 neighboring fins). Discretization of the equation (6) is used for a surface temperature analysis.

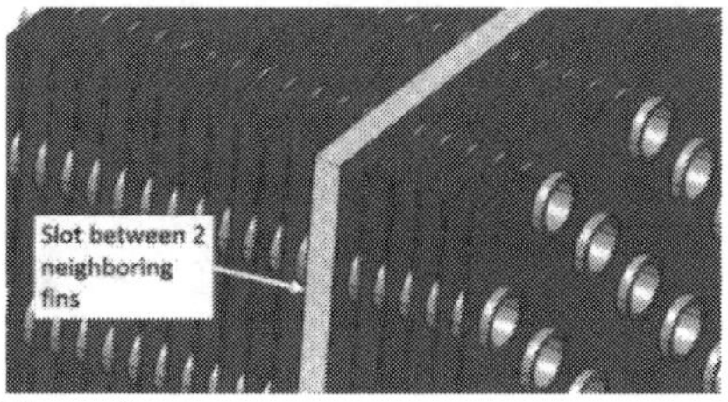

Fig. 4. Flow volume region

- Continuity equation for air can be written as

$$\frac{\partial \rho_a}{\partial t} + \frac{\partial(\rho u_{a,i})}{\partial x_i} = 0. \tag{1}$$

- Continuity equation for water film is

$$\frac{\partial \rho_f}{\partial t} + \frac{\partial(\rho_f u_{f,i})}{\partial y} = 0. \tag{2}$$

- Enthalpy conservation equation for humid air is

$$dQ_{a\to f} = 2q_{a\to f}dxdy = dH_a - V_a dp_a. \tag{3}$$

- Enthalpy conservation equation for water film is

$$-dQ_{f\to w} + dQ_{a\to f} = \left(-q_{f\to w} + q_{a\to f}\right)dxdy = dH_f. \tag{4}$$

- Enthalpy conservation equation for refrigerant/coolant can be written as

$$dQ_{w\to c} = dH_c - V_c dp_c. \tag{5}$$

- Fourier's law for heat convection is

$$\vec{q} = -\lambda \cdot \nabla T. \tag{6}$$

- Equation of state for ideal gas is

$$p = \rho RT \tag{7}$$

2.5 Meshing

Automatic mesh generation (and tube bundle generation) on fin is included in the model. (only parameter of the mesh is *Telem* - #of elements per tube cross-section). Number of elements in x direction - N and y direction - M is solved automatically for given *Telem* parameter by sub-optimization (minimization) of tube cross section area error - $ErrSurf$.

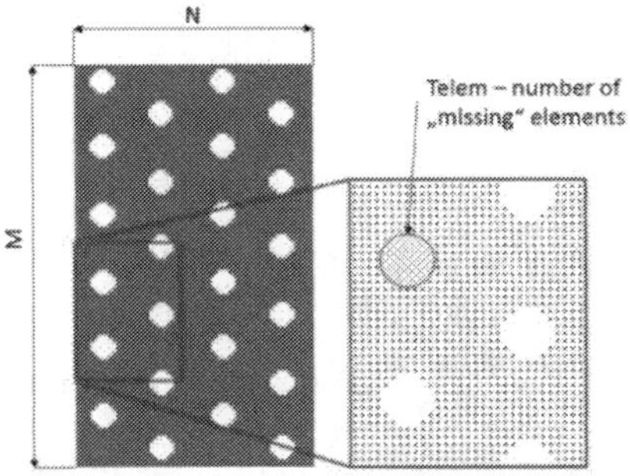

Fig. 5. Volume mesh

For numerical simulation, it is necessary to divide tube into elements/segments in which temperature/enthalpy increments are evaluated. In this model, tube element is defined by tube segment between 2 neighboring fins.

2.6 Solver sub-procedures

MATLAB model itself is composed of multiple sub-procedures, which solve various heat and mass transfer problems. [2]

- Surface temperature solver
- Refrigerant temperature solver
- "i + j" cycles to solve air and film local parameters (Air/film enthalpy solver)
- Film thickness solver
- Film temperature solver

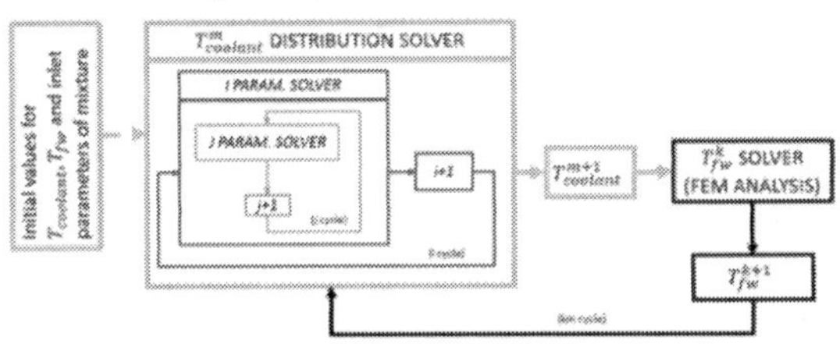

Fig. 6. Solver schema

The iterative solution procedure is obvious from the diagram above. First, the initial distribution of air temperature and condensate flow field is solved from guessed refrigerant temperature distribution. Based on new air temperature, a new refrigerant temperature distribution is solved, from which new fin surface temperature is evaluated. New refrigerant and surface temperature are then used to solve for new air temperature and condensate flow and iteration loop is completed. A number of iterations required to obtain stable solution depends on inlet conditions, but varies from 50 to 150.

2.6.1 Film thickness solver

Because the focus is on evaporators for air dehumidification, low condensate flow is expected, thus, small thickness of a film is assumed. Velocity gradient for a simple 1D case of fluid film on vertical wall can be then derived as [3] [4]

$$\frac{du_f}{dz} = \frac{\rho_f - \rho_g}{\mu_l} g(\delta - z) + \frac{\tau_g}{\mu_l}. \qquad (8)$$

Ordinary differential equation above is integrated and boundary conditions applied, velocity profile is obtained.

$$u_f(z) = \frac{\rho_f - \rho_g}{\mu_l} g \left[\delta z - \frac{1}{2} z^2\right] + \frac{\tau_g}{\mu_l} z. \qquad (9)$$

Where range for z coordinate is $0 \leq z \leq \delta_{j,i}$, ρ_f – condensate film density, ρ_g – mixture density and τ_g is shear stress at interface between mixture and film. Local mean velocity in condensate film is obtained by integration over local film thickness

$$\bar{u}_f(x, y) = \bar{u}_{f,j,i} = \frac{1}{\delta_{j,i}} \int_0^{\delta_{j,i}} u_f(x, y, z)\, dz. \qquad (10)$$

Local mass flow of condensate film is given by

$$\dot{m}_{f,j,i} = \rho_f \bar{u}_{f,j,i} \delta_{j,i} \Delta x. \qquad (11)$$

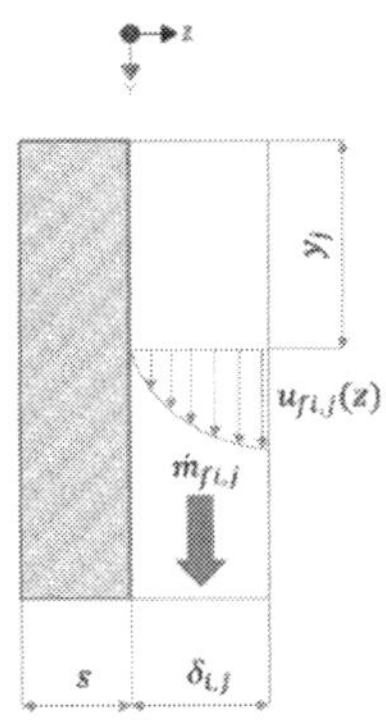

Fig. 7. Viscosity effect in condensate film

These equations are used to solve local film thickness - $\delta_{i,j}$, that is necessary for solving heat transfer through film.

2.6.2 Film Temperature solver

Local temperature of film on tubes and fins is very important parameter for heat transfer and condensation and it is determined from heat balance between humid air, film and fin/tube surface. [3] [6]

Heat balance between air, film and surface can be written as

$$k'(T_F - T_{fw}) = \dot{m}_f'' \Delta h_{23} + \alpha_{fg} E_T (T_G - T_F). \qquad (12)$$

Increment of film mass flow

$$\dot{m}_f'' = \rho_g \beta_g ln\left(\frac{1 - \tilde{y}_v\{T_F\}}{1 - y_v}\right), \qquad (13)$$

$$\phi_T = \frac{\rho_g \beta_g c_{pv}}{\alpha_{fg}} ln\left(\frac{1 - \tilde{y}_v\{T_F\}}{1 - y_v}\right). \qquad (14)$$

Combining equations above we obtain

$$k'(T_F - T_{fw}) = \alpha_{fg}\phi_T\left(\frac{\Delta h_{23}}{c_{pv}} + \frac{T_G - T_F}{1 - exp(-\phi_T)}\right). \qquad (15)$$

Equation (15) is solved iteratively

$$\rightarrow T_F^{i+1} = T_c + \frac{\alpha_{fg}\phi_T}{k'}\left(\frac{\Delta h_{23}}{c_{pv}} + \frac{T_G - T_F^i}{1 - exp(-\phi_T)}\right). \qquad (16)$$

From equation (16), mean film temperature is solved after couple iterations. This equation is also used as a part of solution stability monitoring – the first indication of solution divergence is unstable film temperature, if this fact is detected, simulation is terminated.

2.6.3 Fin surface temperature solver

After considering simplifications, problem of fin surface temperature can be then solved separately (based on informations from air and refrigerant side) as 2D heat conduction problem in rectangular domain. To ensure unique solution, boundary conditions are required – refrigerant temperatures at fin/tube intersections are used. As a volumetric heat loads, local heat flow from gas (or film) into fin is used. [7] [2]
Heat flow (flux) vector is

$$\boldsymbol{q} = \begin{bmatrix} q_x \\ q_y \end{bmatrix}. \tag{17}$$

Fourier's law in matrix form can be written as

$$\boldsymbol{q} = -\boldsymbol{D}\nabla T. \tag{18}$$

Linear shape functions (*N*) are chosen for interpolation, weak formulation is established as

$$\oiint_{\Omega} \boldsymbol{B}^T s\boldsymbol{DB} d\Omega \boldsymbol{a} = - \oint_{\partial\Omega_q} \boldsymbol{N}^T t q_b d\partial\Omega - \oint_{\partial\Omega_T} \boldsymbol{N}^T s q_n d\partial\Omega + \oiint_{\Omega} \boldsymbol{N} s Q d\Omega. \tag{19}$$

Where

$$\boldsymbol{B} = \nabla \boldsymbol{N}, \tag{20}$$

$$\boldsymbol{a} = [T_i]. \tag{21}$$

Equations from weak formulation can be then rewritten as a simple matrix equation

$$\boldsymbol{Ka} = \boldsymbol{f}_b + \boldsymbol{f}_l = \boldsymbol{f}. \tag{22}$$

This method is then applied on heat exchanger geometry for arbitrary inlet conditions, resulting surface temperature is shown in **Fig. 8.**

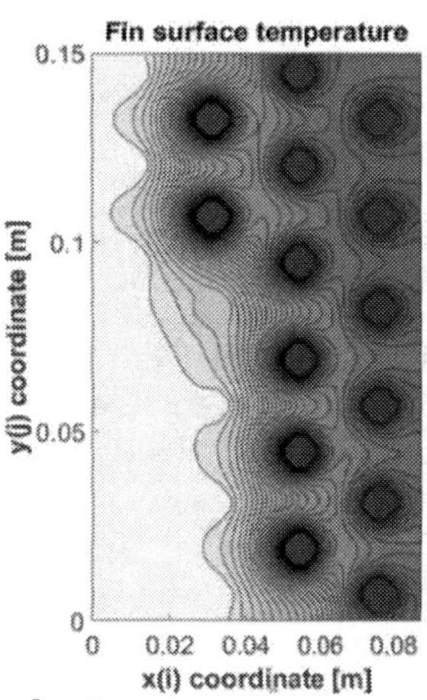

Fig. 8. Fin surface temperature

Because plate/fin surface temperature is solved every iteration, code optimization is required to decrease overall computation time even for fine meshes. Temperature in mesh nodes can solved from equation (22) by

$$\boldsymbol{a} = \boldsymbol{K}^{-1}\boldsymbol{f} \tag{23}$$

Matrix inverse operation (equation (23)) is very demanding and imprecise for large matrices. The dimension of stiffness matrix for a coarse mesh is typically 7300x7300, which takes 22 seconds on average (4 core CPU, 3,3GHz). This is not suitable for simulations that usually take 150 iterations to converge. Matrix inverse can be replaced by *mldivide* operation implemented in MATLAB. Another improvement is using algorithms for sparse matrices (CRS form – compressed row storage) and LU factorization.

$$[\boldsymbol{L_K}, \boldsymbol{U_K}] = lu(\boldsymbol{K}_{sparse}) \tag{24}$$

Equation (23) is then replaced by 2 equations

$$\widetilde{\boldsymbol{a}}_{temp} = \boldsymbol{L_K} \backslash \boldsymbol{f} \tag{25}$$

$$\boldsymbol{a} = \boldsymbol{U_K} \backslash \widetilde{\boldsymbol{a}}_{temp}. \tag{26}$$

Plate/fin surface also needs to be divided into sub-regions (areas), from where all the heat flow coming from the air is assigned to specific tube cross section. This allows evaluating enthalpy/temperature increment of tube pass. In general, these sub-regions can be traced by finding locations, where local heat flux inside the fin is 0 or very small. A boundary of these sub-regions are very complicated (**Fig. 9**); so an additional simplification is required.

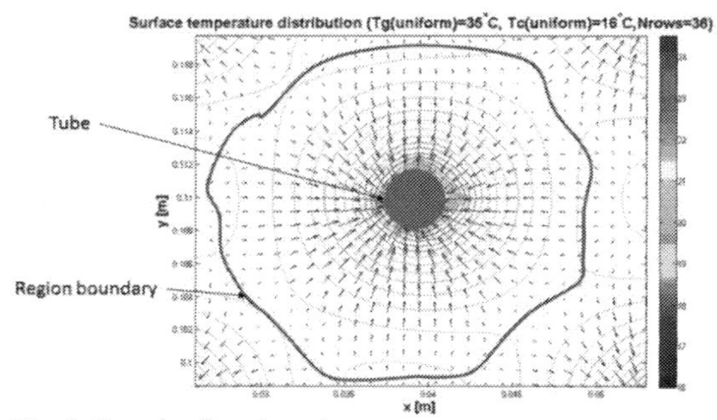

Fig. 9. Complex fin sub-regions

Simplified sub-regions of fin are shown in **Fig. 10.**

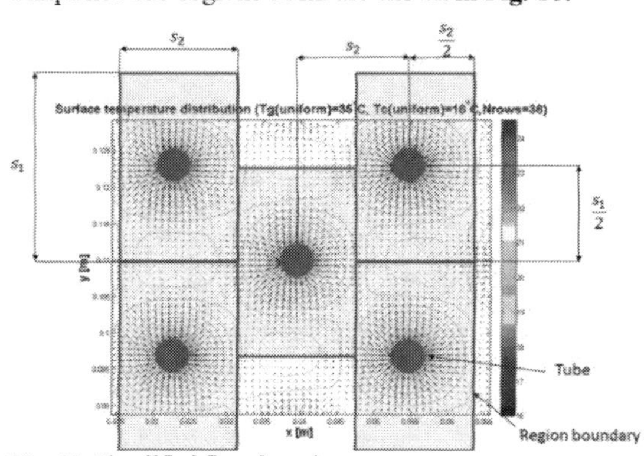

Fig. 10. Simplified fin sub-regions

This simplification also brings new problem – for a case with big temperature gradients between 2 neighboring tubes, some of the heat that in reality "flows" into tube 1 is within this simplification associated with tube 2. This causes an error in heat balance (difference of enthalpy on air side is not equal to enthalpy on the refrigerant side). To prevent this situation, *Redistribution Correction Solver* is added. RCS eliminates this error by transferring additional heat from 1 sub-region to the other, based on estimation from heat flux integral on its boundary.

2.6.4 Superposition of heat and mass transfer

Theoretical heat transfer coefficient on fin and tube is known, but effect of tube presence on heat transfer coefficient on fins (and fin presence to heat transfer coefficient on tubes), needs to be included. If we assume that evaporator is only composed of tubes, total heat duty without condensation (mass transfer) can be written symbolically as [2] [8]

$$Q_{tubes} = \tilde{\alpha}_{tube} \cdot A \cdot \overline{(T_g - T_f)}, \tag{27}$$

where $\tilde{\alpha}_{\text{tube}}$ is heat transfer coefficient on tubes. Similarly, if evaporator (heat exchanger) is made only from fins, total heat duty is

$$Q_{fins} = \tilde{\alpha}_{plate} \cdot A \cdot \overline{(T_g - T_f)}, \tag{28}$$

where $\tilde{\alpha}_{plate}$ is heat transfer coefficient on fins/plates. Superposing fins and tubes, total heat duty without condensation (mass transfer) is symbolically:

$$Q_{tot} = \left(R_p \cdot \tilde{\alpha}_{plate} + R_t \cdot \tilde{\alpha}_{tube}\right) \cdot A \cdot \overline{(T_g - T_f)}. \tag{29}$$

Where superposition (effectiveness) coefficients R_p, R_t are derived from experimental data by optimization – fitting the model results to results from experiment. These coefficients describe influence of temperature and velocity field of one case to the other. Similar hypothesis can be made with condensate mass flow, so another 2 effectiveness coefficients are introduced. Local heat and mass transfer coefficients are obtained from correlations for the case of flow in small slot (as a limit of concentric annular duct) and flow over staggered tube bundle. These coefficients are then used to solve overall heat transfer coefficient according to **Fig. 11.**

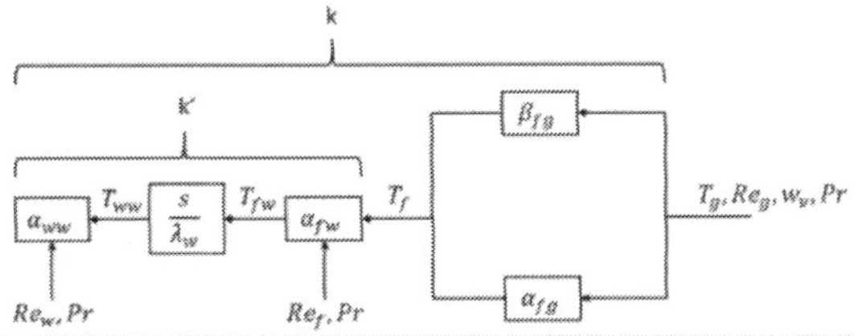

Fig. 11. Overall HTC

2.6.5 POD and RBF method

Another improvement of data transfer process is done by incorporating the discrete version of POD (Proper Orthogonal Decomposition) method - Method of Snapshots. This method allows to simplify a large amount of data into a smaller package and also to reconstruct solution for given (directly unsolved) inlet parameters very quickly. Input for POD method is set of direct MATLAB solutions for different parameters – inlet conditions.

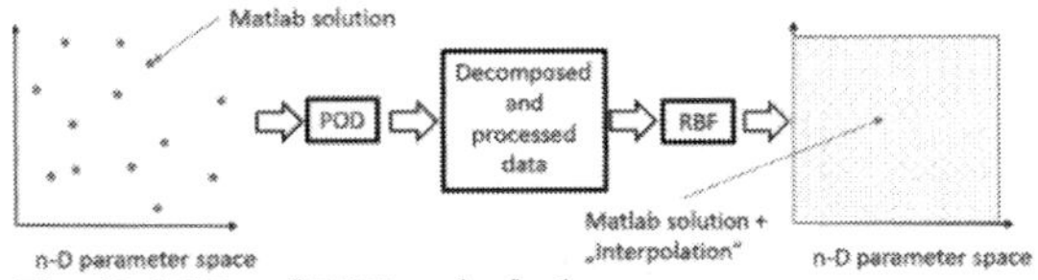

Fig. 12. POD and RBF method schema

All fields from solution are then decomposed into average values and fluctuation, which is then decomposed into modes and amplitudes. Reconstructed solution for given inlet conditions can be written symbolically as [9]

$$\boldsymbol{OUTPUT}_{param_k} = \boldsymbol{OUTPUT}_{mean} + \sum_{j=1}^{n} \boldsymbol{MOD}_j \cdot \boldsymbol{AMPLITUDE}_{param_k,j} \tag{30}$$

where n is number of modes. Fields $\boldsymbol{OUTPUT}_{mean}$ and $\boldsymbol{MOD}$ are constant, but $\boldsymbol{AMPLITUDE}$ is a function of inlet conditions that needs to be constructed based on direct solutions with *Radial Basis Functions* (RBF) in Gaussian form. Overall approximation function can be interpreted as RBF network.

$$h(x) = b + \sum_{k=1}^{N} h_k(x) = b + \sum_{k=1}^{N} w_k e^{-\gamma\|x-x_k\|^2}. \tag{31}$$

Where x is arbitrary points in n-D parameter space and x_k is a point for which direct simulation is performed. Coefficients b and w_k are trained with data from direct solutions by MATLAB optimization method *fminsearch*. Target (fitness) function is defined as

$$\Delta_{tot} = \sum_{k=1}^{N} \Delta(x_k), \tag{32}$$

where sample error function is evaluated as

$$\Delta(x_k) = (h(x_k) - y_k)^2 \tag{33}$$

Total approximation error function can be then written as

$$\Delta_{tot} = \sum_{k=1}^{N} \Delta(x_k) = \sum_{k=1}^{N} \left(b + \sum_{l=1}^{N} w_l e^{-\gamma\|x_k-x_l\|^2} - y_k\right)^2. \tag{34}$$

Necessary number of modes required to "precise" approximation is determined from quantity called level of correlation and it is defined as

$$L_n = \frac{\sum_{j=1}^{n} \lambda_j}{\sum_{j=1}^{N} \lambda_j};\ n = 1,2,\dots,N \tag{35}$$

Where N is a number of direct simulations and n is a number of modes, choose so $L_n > 0.99$. Implementation of POD and RBF method into the mathematical model is covered more deeply in [2].

Conclusion

The main subject of this paper is the heat and mass transfer modeling inside heat exchangers (evaporator) used in air dehumidification applications. Theoretical and experimental study of evaporator performance have been carried out. Requirements and properties of the mathematical model are introduced. The MATLAB software is selected as a model development environment, because of its user-friendly environment, the relatively simple implementation of matrix operations and the available optimization tools for model validation and correction. The model itself is composed of multiple sub-procedures that solve various heat and mass transfer problems – fin surface temperature, condensate film temperature, local humid air parameters and conditions for condensation as well as the refrigerant temperature distribution along the tube bundle. The differential equations for heat and mass balance used in the model are solved with explicit first order Euler method. As a possible improvement of MATLAB model, additional code optimization and generalization can be mentioned: the current computation time is approximately 2 second per iteration for reasonable mesh density. Instead of using a finite difference method for the air temperature distribution and a finite element method for surface temperature distribution, it is preferable to use a 2D finite volume method, which may lead to further decrease in the computation time. The next step is validation of the model. It will be divided into multiple stages, to test all necessary components of MATLAB model and maximize the precision of results. As a continuation of this work, the model of the condenser should be created. It can be simply derived from evaporator model, by excluding condensation procedures and other simplifications, so further time reduction of computation time is expected. To complete the whole heat pump circuit, simplified model of a compressor and reduction valve should be also developed, to be able to simulate circuit as a unit, without measurement of particular parameters, like refrigerant temperature in condenser outlet and evaporator inlet, etc.

We gratefully acknowledge the support from Centre of 3D volumetric velocimetry (COLA), reg. no. CZ.2.16/3.1.00/21569, supported by operational programme Prague Competitiveness.

Nomenclature

Used symbols

Symbol	*Unit*	*Name - description*
$\boldsymbol{q}$	$W \cdot m^{-2}$	heat flux vector
$\boldsymbol{D}$	$W \cdot m^{-1} \cdot K^{-1}$	thermal conductivity matrix
$\boldsymbol{\nabla T}$	$K \cdot m^{-1}$	temperature gradient
k	$W \cdot m^{-2} \cdot K^{-1}$	overall heat transfer coefficient
p	Pa	pressure
p_0	Pa	barometric pressure
p_v	Pa	water vapor saturation pressure
T	°C, K	temperature
β	$m \cdot s^{-1}$	mass transfer coefficient
s	mm, m	fin/plate thickness
δ	mm, m	condensate film thickness
h	$J \cdot kg^{-1}$	enthalpy
q_{gw}	$W \cdot m^{-2}$	heat flux from air to fin/film
q_b, q_n	$W \cdot m^{-2}$	boundary heat flows
Q	$W \cdot m^{-3}$	internal heat supply
Ω	-	area of elements
$\Omega_{q,T}$	-	elements boundary with heat or temperature condition
μ	$Pa \cdot s$	dynamic viscosity
ν	$m^2 \cdot s^{-1}$	kinematic viscosity
τ_g	Pa	shear stress at air-film interface
R	$J \cdot mol^{-1} \cdot K^{-1}$	gas constant
u	$m \cdot s^{-1}$	velocity
$\dot{m}$	$kg \cdot s^{-1}$	mass flow rate
$\dot{m}''$	$kg \cdot m^{-2} \cdot s^{-1}$	mass flux
l_{230}	$J \cdot kg^{-1}$	latent heat of evaporation
Δh_{23}	$J \cdot kg^{-1}$	enthalpy of evaporation
α	$W \cdot m^{-2} \cdot s^{-1}$	heat transfer coefficient
N	1	number of elements in x direction
M	1	number of elements in y direction
ErrSurf	%	error of tube section area
q_b, q_n	$W \cdot m^{-2}$	boundary heat flows
Telem	1	number of elements in tube cross-section
λ	$W \cdot m^{-1} \cdot K^{-1}$	heat/thermal conductivity
y	1	mole fraction
$\tilde{y}$	1	mole fraction at saturation
c_p	$J \cdot kg^{-1} \cdot K^{-1}$	specific heat/heat capacity at constant pressure

Subscripts

a	air
w	wall (fin or tube surface)
v	water vapor
$f/film$	condensate film
c	coolant
a	air

References

1. Reversing valve cooling. Ecorenovator. © EcoRenovator (2012)
2. M. Borovička, Student's conf. STC, 7 (2016)
3. Li Jun-De, S. Mohammad, G. Thorpe. IJHMT. **54**, 4078 (2011).
4. J. Šesták, F. Rieger, *Přenos hybnosti, tepla a hmoty*. (CTU, Prague, 2005)
5. Verein Deutscher Ingenieure. *VDI-Wärmeatlas*. (7th ed., VDI-Verlag, 1994)
6. F. P. Incropera, *Fundamentals of heat and mass transfer*. (7th ed. John Wiley& Sons, Inc , USA, 2011)
7. AALBORG UNIVERSITET. Finite Element Method – 2D heat conduction (2015)
8. J. Pavlů, *Vývoj výpočetního modelu a metodiky pro výpočet kondenzátorů s minikanálky* (Czech technical university in Brno, Faculty of Mechanical Engineering, Brno, 2012)
9. J. Adamec, *Modelování přestupu tepla a hmoty ve filmové výplni* (CTU, Faculty of Mechanical Engineering, Department of Fluid Dynamics and Thermodynamics, Prague, 2014)

Comparison of turbulence models and CFD solution options for a plain pipe

Eyub Canli[1,*], *Ali* Ates[1], and *Sefik* Bilir[1]

[1]Selcuk University, Mechanical Engineering Department, P.K. 42003 Selcuklu Konya, Turkey

Abstract. Present paper is partly a declaration of state of a currently ongoing PhD work about turbulent flow in a thick walled pipe in order to analyze conjugate heat transfer. An ongoing effort on CFD investigation of this problem using cylindrical coordinates and dimensionless governing equations is identified alongside a literature review. The mentioned PhD work will be conducted using an in-house developed code. However it needs preliminary evaluation by means of commercial codes available in the field. Accordingly ANSYS CFD was utilized in order to evaluate mesh structure needs and asses the turbulence models and solution options in terms of computational power versus difference signification. Present work contains a literature survey, an arrangement of governing equations of the PhD work, CFD essentials of the preliminary analysis and findings about the mesh structure and solution options. Mesh element number was changed between 5,000 and 320,000. k-ϵ, k-ω, Spalart-Allmaras and Viscous-Laminar models were compared. Reynolds number was changed between 1,000 and 50,000. As it may be expected due to the literature, k-ϵ yields more favorable results near the pipe axis and k-ω yields more convenient results near the wall. However k-ϵ is found sufficient to give turbulent structures for a conjugate heat transfer problem in a thick walled plain pipe.

1 Introduction

Flow inside a thick walled pipe and related heat transfer is of interest. Numerical and computational means can be used for investing a wide range of mentioned interest and compact results can be reported in a way that the results can be converted and applied to a wide application variety and field. The compact results can be derived by scaling and using dimensionless governing equations.

A dedicated team in Turkey has been performing numerical studies on unsteady and steady conjugate heat transfer related to thick walled pipes under various conditions such as periodic boundary conditions, simultaneously developing flow, rarefaction effects and slippery flow, heat transfer in tangential direction and etc by using cylindrical coordinates and their own developed codes. As mentioned earlier, the results are dimensionless and can be expanded for certain magnitudes. Remarkable developments and major events are summarized below according to the publishing timeline.

After a detailed and informative PhD dissertation was published in 1988 [1] about heat transfer in laminar pipe flow in the thermal development region at low Peclet numbers including axial conduction, Bilir (1995) reported a numerical laminar heat transfer analysis including two dimensional axial conduction for the wall and for the fluid [2]. The pipe in the problem was constituted by two regions and second region causes a step change in outside surface temperature. This causes heat transfer through the wall and the fluid. Peclet number, thickness ratio and wall-to-fluid conductivity ratio were taken as the investigated parameters while Peclet number was found as the most effective parameter among them. It should also worth that the numerical code of [1] is also given in it. Bilir then reported a solution for transient regime of the previously mentioned problem [3]. This time parameters were increased as wall-to-fluid thermal diffusivity ratio was added into the analysis. As a main conclusion of the work, author reported that the time passed in order to reach the steady state conditions does not change significantly with the parameters although the early and intermediate periods exhibit differences. It was also stressed that flow conditions determine the thermal inertia rather than the wall specifications.

In 2003, a different boundary condition was reported by Bilir and Ates [4]. Convective boundary condition was applied to downstream region of the numerical domain which was consist of two regions. Also Biot number was taken as an additional parameter which makes the total parameter number as five. Considerable amount of axial conduction towards upstream was reported. The effect of parameters on the time to reach steady state condition was found to be significant. Also it is given that there are certain values of parameter after which results are not changing significantly. The details also can be found in the dissertation [5]. Problem then modified in respect of the parameters and results are

* Corresponding author: ecanli@selcuk.edu.tr

reported [6]. Their main findings can be summarized as the preheating phenomenon due to the axial heat transfer towards the upstream; reverse heat flow (from fluid to wall) at the early transient stages in upstream; peak values at the interface of upstream and downstream due to the surpassing radial wall conduction; and axial heat conduction towards downstream region through the wall in steady state conditions making the heat flux values at fluid wall interface greater than the outer surface where heat flux boundary condition exists.

First time numerical solution of the flow of the team was in 2004 where laminar flow development was obtained for simultaneously developing flow [7]. The effect of the velocity profile other than the parabolic velocity profile is investigated and results are given comparatively in terms of the outer and inner wall temperatures, bulk temperatures, radial distributions of the fluid temperatures and Nusselt numbers [8]. It is said that velocity profile has no effect on interface and outer surface temperatures. Bulk temperature values on the other hand are higher for developing flow comparing to the developed flow. Slug flow for the thick walled pipe is also examined numerically by the team [9]. For this flow type, temperature profiles develop faster in the order of outer surface, interface and bulk flow. They validated their results by considering the Nusselt number for developed flow with constant temperature condition. Since thick walled pipes can be seen in applications involving micro-channels and mini-pipes frequently, the parameters were arranged by the authors and reported [10]. There are six parameters which are wall thickness ratio, wall-to-fluid thermal conductivity ratio, wall-to-fluid thermal diffusivity ratio, the Peclet number, the Biot number and the Prandtl number. It is stated that the interfacial heat flux rises rapidly at initiation of the transient regime due to fast radial conduction in the wall at the entrance of the pipe. Then it reaches a constant value in a short axial distance. Again it is noticed that heat flux values decrease in the mean and uniformity of the curves are deformed as time elapses because of the increasing effect of convection in the fluid side and after a maximum they decrease in the flow direction. As an analogy to electrical current, conjugation increases with the diameter. Remaining parameters reduce it. The critic values of the parameters are also given and after these critic values towards a certain direction, results don't change significantly. Only Pr number doesn't affect time to reach steady state conditions significantly.

Periodic boundary conditions for the thick walled pipe in which heat conjugation takes place for laminar flow was also dealt numerically by Altun [11]. A paper was also published on periodic outer wall temperature changing with time [12]. It can be inferred from the results that the periodicity can be observed in the plots identically for all parameter values while parameters change the peak values. The additional parameter of the work is the angular frequency. Demirpolat [13] repeated the analysis for periodic convective boundary conditions.

Heat transfer in tangential direction is taken into consideration in the dissertation of Atmaca [14]. Transient conjugated heat transfer in circumferentially half heated thick walled pipes for thermally developing laminar flows is investigated in the study. As expected, an intensive heat transfer occurs towards the insulated part of the pipe by tangential conduction.

Finally Sen completed a PhD thesis on slippery flow which can be attributed to gas flow in micro channels [15]. In its paper [16], it is stated that momentum and energy equations are solved with the boundary conditions of the first degree velocity slip and temperature jump. The independent parameters are Peclet number, the Knudsen number, the Brinkman number and the wall thickness ratio. Their major conclusion can be summarized as; fluid axial conduction is effective in the entrance region while it can be neglected in the remaining part; Interfacial heat flux values decrease with increasing Knudsen numbers; reverse heat flow occurs in upstream region with increase in Brinkman number; and thermal development length increases with increasing diameter and decreasing Peclet number.

Currently a PhD thesis (of the first author) is going on by the team in order to investigate the heat transfer for a turbulent flow in a thick walled pipe numerically by using finite elements. This should be considered a further step of the accumulated knowledge. Literature is also surveyed for similar cases.

The deterministic near-wall turbulence model was used by Kasagi et al. for calculation and simulation of near wall turbulent heat transfer involving the effect of unsteady heat conduction in the solid wall [17]. They introduced three fluctuating velocity components into the governing equations not by using the Reynolds decomposition method but using algebraic expressions along with their approach. Temperature variance, turbulent heat flux, turbulent Prandtl number and some more statistical data are reported for fluids with different Prandtl numbers. They stated that the wall thickness and thermal properties affect the near wall values.

Cao and Faghri studied turbulent forced convection of a transport fluid by using k-ϵ model and coupling it with the phase change solution because the application of interest is a phase changing energy storage application [18]. They validated their code with different experimental data and then used the code for other conjugated heat transfer scenarios. A correlation for the energy storage capacity was derived and its empirical constants were determined by the numerical data and presented.

Jaeger investigated the heat transfer to liquid metals for forced convection in various geometries by using empirical models [19]. Since liquid metals are mentioned here, one can think that laminar flow should be considered here. The author also emphasized that this fact. The empirical formulas in order to represent thermal-hydraulic behavior of several liquid metals in the numerical codes were compiled and used. This work can be regarded as a validation reference where liquid metals or specific conditions that can be attributed to the liquid metals are of interest.

A direct numerical solution of turbulent heat transfer in a channel while wall conduction exists was reported by Tiselj et al. [20]. They investigated unsteady conduction in the heated walls for fully developed flow conditions. The flow field is two dimensional. They proposed that it can be decided, which behavior has to be expected in a real fluid-solid system and which one of the limiting boundary conditions is valid for calculation, or whether more expensive conjugate heat transfer calculation is required. They also stated the weakness of the previous studies as underestimating the wall temperature fluctuations for the limiting cases of the ideal-isoflux boundary conditions.

DNS of turbulent channel flow with conjugate heat transfer at Prandtl number 0.01 is performed by Tiselj and Cizelj [21]. This Pr number practically corresponds to the liquid sodium. They used pseudo-spectral channel-flow code. As a summary of their findings, it can be said that the log law region cannot be observed by the temperature profile of the fluid, even for the highest implemented Reynolds number; the near-wall RMS temperature fluctuations show Reynolds number dependence; an intensive turbulent temperature fluctuation penetrates towards the wall according to the conjugate heat transfer simulations.

In the subsequent sections, governing equations of the quoted study and supplementary information are given. Also the design of the ANSYS CFD preliminary analysis is presented. Turbulent flow is calculated at the entrance region of a pipe by means of four viscous flow model using ANSYS CFD. Results are given comparatively in terms of u velocity profiles versus coordinate axes, wall y+, turbulence viscosity and intensity values.

2 Method

Governing equations of a case involving numerical investigation of a turbulent flow and heat transfer for incompressible constant viscosity flow can be given in vector form are listed below between (1-3) [22]:

$$\frac{\partial \rho}{\partial t} + \nabla \cdot (\rho \mathbf{V}) = 0 \tag{1}$$

$$\rho \frac{D\mathbf{V}}{Dt} = \mathbf{f} - \nabla p + \mu \nabla^2 \mathbf{V} \tag{2}$$

$$\rho \frac{\partial e}{\partial t} + p(\nabla \mathbf{V}) = \nabla(k \nabla T) + \Phi \tag{3}$$

(1) is continuity, (2) is the momentum equation and (3) is the energy equation. The extended forms of (1-3) for two dimensional steady flow in cylindrical coordinate system after Reynolds decomposition and neglecting viscous dissipation are given below [9,22]:

$$\frac{\partial \bar{u}}{\partial x} + \frac{\bar{v}}{r} + \frac{\partial \bar{v}}{\partial r} = 0 \tag{4}$$

$$\bar{u}\frac{\partial \bar{u}}{\partial x} + \bar{v}\frac{\partial \bar{u}}{\partial r} = -\frac{1}{\rho}\frac{\partial p}{\partial x} + \nu\left[\frac{1}{r}\frac{\partial}{\partial r}\left(r\frac{\partial \bar{u}}{\partial r}\right) + \frac{\partial^2 \bar{u}}{\partial x^2}\right] + \frac{1}{r}\frac{\partial}{\partial r}(-r\overline{v'u'}) + \frac{\partial}{\partial x}(-\overline{u'u'}) \tag{5}$$

$$\bar{u}\frac{\partial \bar{v}}{\partial x} + \bar{v}\frac{\partial \bar{v}}{\partial r} = -\frac{1}{\rho}\frac{\partial p}{\partial r} + \nu\left[\frac{\partial}{\partial r}\left(\frac{1}{r}\frac{\partial (r\bar{v})}{\partial r}\right) + \frac{\partial^2 \bar{v}}{\partial x^2}\right] + \frac{1}{r}\frac{\partial}{\partial r}(-r\overline{v'v'}) + \frac{\partial}{\partial x}(-\overline{u'v'}) \tag{6}$$

$$\bar{u}\frac{\partial \bar{T}}{\partial x} + \bar{v}\frac{\partial \bar{T}}{\partial r} = \alpha\left[\frac{\partial}{\partial x}\left(\frac{\partial \bar{T}}{\partial x}\right) + \frac{1}{r}\frac{\partial}{\partial r}\left(r\frac{\partial \bar{T}}{\partial r}\right)\right] + \frac{\partial}{\partial x}(-\overline{u'T'}) + \frac{1}{r}\frac{\partial}{\partial r}(-r\overline{v'T'}) \tag{7}$$

Fluctuating velocity component and temperature values take place in the equations as it can be seen from (4-7). These averaged products will be modeled by k-ϵ while they can be given as;

$$-\overline{u'_i u'_j} = \nu_T\left(\frac{\partial \bar{u}_i}{\partial x_j} + \frac{\partial \bar{u}_j}{\partial x_i}\right) - \frac{2}{3}k\,\delta_{ij} \tag{8}$$

$$-\overline{v'u'} = \overline{u'v'} = \nu_T\left(\frac{1}{r}\frac{\partial r\bar{u}}{\partial r} + \frac{\partial \bar{v}}{\partial x}\right) \tag{9}$$

$$-\overline{u'u'} = \nu_T\left(2\frac{\partial \bar{u}}{\partial x}\right) - k \tag{10}$$

$$-\overline{v'v'} = \nu_T\left(\frac{2}{r}\frac{\partial r\bar{v}}{\partial r}\right) - k \tag{11}$$

$$-\overline{u'T'} = \alpha_T\frac{\partial \bar{T}}{\partial x} \tag{12}$$

$$-\overline{v'T'} = \alpha_T\frac{1}{r}\frac{\partial \bar{T}}{\partial r} \tag{13}$$

$$k = \frac{1}{2}\left(\overline{u'^2} + \overline{v'^2}\right) \tag{14}$$

Here ν_T is the turbulence or eddy viscosity and it is not a property of the fluid. Instead it is a property of the flow and it changes according to the location and can be related to the average flow properties. Kronecker delta δ_{ij} enables to consider situations only for $i=j$. When $i=j$ the first parenthesis of right hand side of the (8) would yield a normal stress and the addition of the two normal stresses would be zero due to the continuity. However all normal stresses are positive by their definition. Also it is known that the sum of normal stresses gives the kinetic energy of the turbulent flow which is multiplied by two. The second part of (8) ensures that the sum of the normal stresses is not zero. k then can be absorbed into the pressure when inserting turbulence viscosity into the governing equations and therefore it is not necessary to calculate k. Instead, ν_T distribution is tried to be obtained. In the k-ϵ, ν_T is modeled by a velocity scale ($\sqrt{k}$) and a length scale ($k^{3/2}/\varepsilon$). (15) is given as transport equation of k and (16) is given as the transport equation of ϵ [23,24].

$$\frac{\partial k}{\partial t}+\bar{u}_i\frac{\partial k}{\partial x_i}=\frac{\partial}{\partial x_i}\left(\frac{\nu_T}{\sigma_k}\frac{\partial k}{\partial x_i}\right)+G-\epsilon \quad (15)$$

$$\frac{\partial \epsilon}{\partial t}+\bar{u}_i\frac{\partial \epsilon}{\partial x_i}=\frac{\partial}{\partial x_i}\left(\frac{\nu_T}{\sigma_\epsilon}\frac{\partial \epsilon}{\partial x_i}\right)+C_{1\epsilon}\frac{\epsilon}{k}G-C_{2\epsilon}\frac{\epsilon^2}{k} \quad (16)$$

Here G (17) is the generation term and represent the generation of k. k and ϵ can be related to each other as in (18). The coefficients appear in the equations can be viewed in (19).

$$G=\nu_T\left(\frac{\partial \bar{u}_i}{\partial x_j}+\frac{\partial \bar{u}_j}{\partial x_i}\right)\frac{\partial \bar{u}_i}{\partial x_j} \quad (17)$$

$$\nu_T=C_\mu\frac{k^2}{\epsilon} \quad (18)$$

$$C_\mu=0.09, C_{1\epsilon}=1.44, C_{2\epsilon}=1.92, \sigma_k=1.0, \sigma_\epsilon=1.3 \quad (19)$$

When the equations are put in an order and reorganized for 2D steady flow in cylindrical coordinates; a similar approach to above can be applied for the energy equation. The turbulence viscosity can be related to the turbulence diffusion coefficient as in (23).

$$G=\nu_t\left[2\left(\frac{\partial \bar{u}}{\partial x}\right)^2+2\left(\frac{1}{r}\frac{\partial r\bar{v}}{\partial r}\right)^2+\left(\frac{1}{r}\frac{\partial r\bar{u}}{\partial r}+\frac{\partial \bar{v}}{\partial x}\right)^2\right] \quad (20)$$

$$\bar{u}\frac{\partial k}{\partial x}+\bar{v}\frac{\partial k}{\partial r}=\frac{\partial}{\partial x}\left(\frac{\nu_T}{\sigma_k}\frac{\partial k}{\partial x}\right)+\frac{1}{r}\frac{\partial}{\partial r}\left(\frac{\nu_T}{r\sigma_k}\frac{\partial rk}{\partial r}\right)+G-\epsilon \quad (21)$$

$$\bar{u}\frac{\partial \epsilon}{\partial x}+\bar{v}\frac{\partial \epsilon}{\partial r}=\frac{\partial}{\partial x}\left(\frac{\nu_T}{\sigma_\epsilon}\frac{\partial \epsilon}{\partial x}\right)+\frac{1}{r}\frac{\partial}{\partial r}\left(\frac{\nu_T}{r\sigma_\epsilon}\frac{\partial r\epsilon}{\partial r}\right)$$
$$+C_{1\epsilon}\frac{\epsilon}{k}G-C_{2\epsilon}\frac{\epsilon^2}{k} \quad (22)$$

$$\alpha_T=\frac{\nu_T}{\sigma_T} \quad (23)$$

σ_T is named as turbulence Prandtl number or Schmidt number and it was shown that its value is nearly constant by means of experiments. 0.9 is reported as an appropriate value. However it is also stated that the coefficient is affected by buoyancy and streamline curvature. If the RANS equations are re-written by using turbulence viscosity and turbulence diffusion coefficient for k-ϵ model;

$$\bar{u}\frac{\partial \bar{u}}{\partial x}+\bar{v}\frac{\partial \bar{u}}{\partial r}=-\frac{1}{\rho}\frac{\partial p}{\partial x}+\nu\left[\frac{1}{r}\frac{\partial}{\partial r}\left(r\frac{\partial \bar{u}}{\partial r}\right)+\frac{\partial^2 \bar{u}}{\partial x^2}\right]\cdot$$
$$+\frac{1}{r}\frac{\partial}{\partial r}\left(r\nu_T\left(\frac{1}{r}\frac{\partial r\bar{u}}{\partial r}+\frac{\partial \bar{v}}{\partial x}\right)\right)+\frac{\partial}{\partial x}\left(\nu_T\left(2\frac{\partial \bar{u}}{\partial x}\right)-k\right) \quad (24)$$

$$\bar{u}\frac{\partial \bar{v}}{\partial x}+\bar{v}\frac{\partial \bar{v}}{\partial r}=-\frac{1}{\rho}\frac{\partial p}{\partial r}+\nu\left[\frac{\partial}{\partial r}\left(\frac{1}{r}\frac{\partial (r\bar{v})}{\partial r}\right)+\frac{\partial^2 \bar{v}}{\partial x^2}\right]$$
$$+\frac{1}{r}\frac{\partial}{\partial r}\left(r\nu_T\left(\frac{2}{r}\frac{\partial r\bar{v}}{\partial r}\right)-k\right)+\frac{\partial}{\partial x}\left(\nu_T\left(\frac{1}{r}\frac{\partial r\bar{u}}{\partial r}+\frac{\partial \bar{v}}{\partial x}\right)\right) \quad (25)$$

$$\bar{u}\frac{\partial \bar{T}}{\partial x}+\bar{v}\frac{\partial \bar{T}}{\partial r}=\frac{\partial}{\partial x}\left[(\alpha+\alpha_T)\frac{\partial \bar{T}}{\partial x}\right]+\frac{1}{r}\frac{\partial}{\partial r}\left[(\alpha r+\alpha_T)\frac{\partial \bar{T}}{\partial r}\right] \quad (26)$$

As summarized above, this work will be conducted by an in-house developed code. In order to narrow the starting range of investigation and validate results comparatively, a series of preliminary analyses is targeted by utilizing commercial codes and/along with open sourced codes. As a first step, present paper contains CFD results of a case using Ansys CFD. Details are given below.

The Ansys CFD analysis is two dimensional and axisymmetric. So the flow domain was drawn as a rectangle after the settings of Ansys Geometry was set to "2D". A schematic drawing of the geometry is given in Fig. 1.

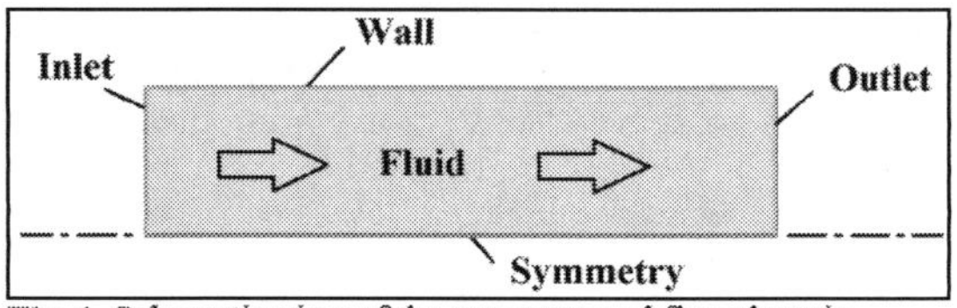

Fig. 1. Schematic view of the geometry and flow domain

The grid (mesh) was constituted by 20,000 (50x400) quadrilateral cells. 50 cells were located at the inlet along the radius and 400 cells were located at the symmetry along the symmetry axis. A bias was applied to all edges of the geometry by "edge sizing" option in a way that the cell number for an edge stays constant but the edge length of the cell on the edge of the geometry stretches towards symmetry axis and outlet by the bias factor of 10. A "face mesh" option was also applied to have a structured mesh. The node number of the mesh is 20,451. Maximum mesh skeweness is $1.31\text{x}10^{-10}$ which can be viewed a success of the structural mesh. The view of the inlet and outlet parts of the mesh is given in Fig.3. In order to asses if the mesh is adequate for the solution, the element number at x and y axes are doubled and halved without touching the bias factors and face meshing of the structural mesh. At this point, 5,000, 80,000 and 320,000 elements were tried. Table 1 shows five different results at the outlet of the pipe for four different element number values. Figure 2 also given for a comparison of different element numbers. According to the results, 80,000 element number was chosen for the comparison of the viscous flow models since tradeoff between time cost and change in turbulent values is satisfactory.

Table 1. Grid comparison - k-ϵ Re=10,000 at pipe outlet

	5,000	20,000	80,000	320,000
Vel. (m/s)	0.1219	0.1091 (10%)	0.1041 (4.5%)	0.0942 (9.5%)
Tot. Pres. (Pa)	7.45	5.95 (20%)	5.42 (9%)	4.44 (18%)
Turb. Int.	0.1362	0.3494 (156%)	0.443 (26%)	0.4385 (1%)
Turb. Vis. (Pa*s)	0.0008	0.019 (2387%)	0.0265 (33%)	0.0215 (18%)
y+	2.049	1.058 (48%)	0.587 (44%)	0.349 (40%)
Av. % Dif.	-	524.2%	23.3%	17.3%

https://doi.org/10.1051/epjconf/201818002013

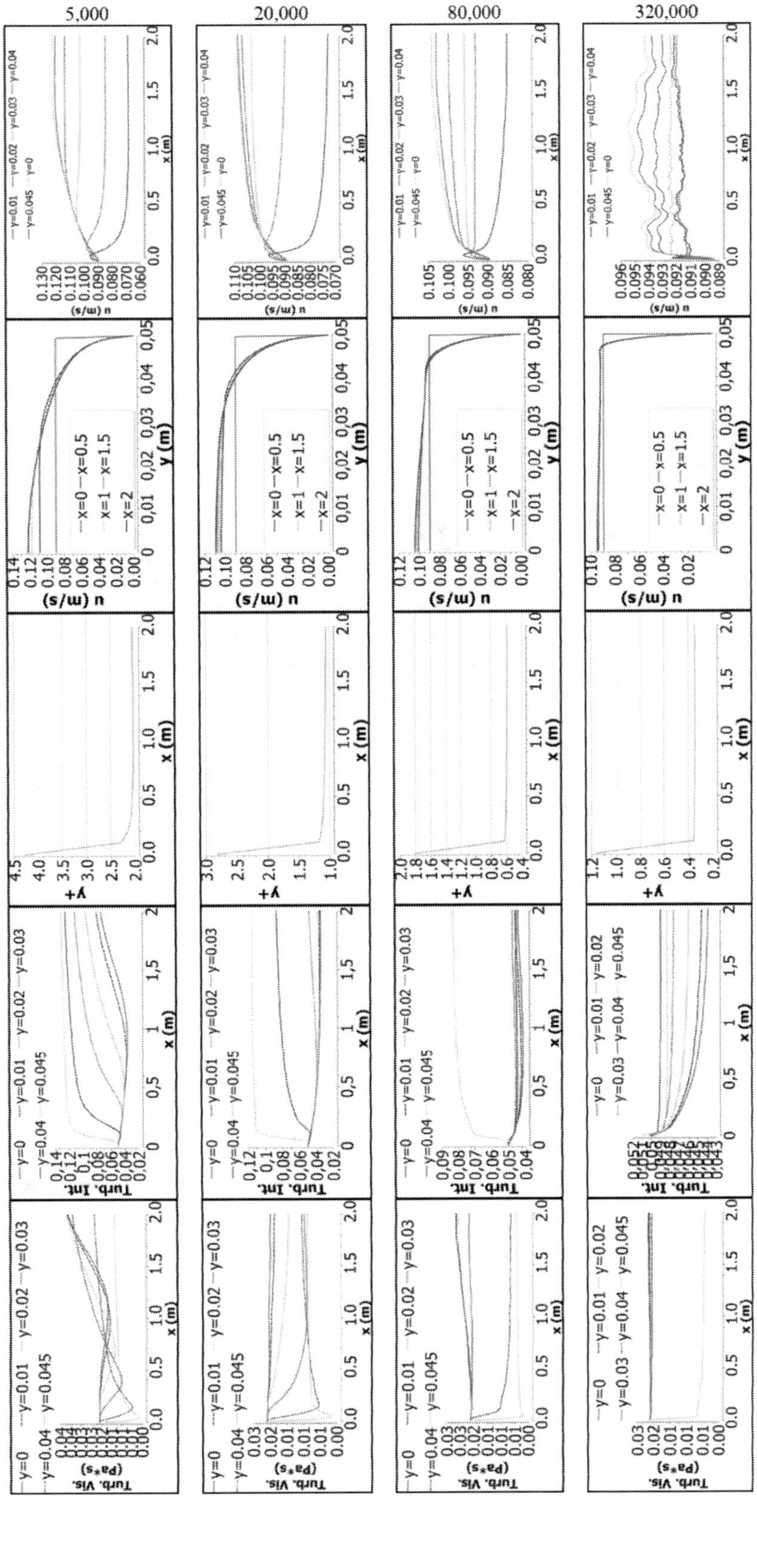

Fig. 2. Comparison of different mesh element number results for *k-ε* at Re=10,000

Although element number changes for the grid independency, Figure 3 gives an idea about the structural mesh and the bias factor.

Four viscous flow models were used; Viscous Laminar, Spalart-Allmaras, k-ϵ and k-ω. Then the software was run for four different Re numbers; 50,000, 10,000, 5,000 and 1,000 for the selected turbulence model. Five different two dimensional plot were drawn after the computations where it was possible; u velocity versus x axis (or pipe length), u velocity versus y axis (or pipe radius), wall y+ versus x axis, turbulent viscosity versus x axis and turbulent intensity versus x axis. These two dimensional plots were drawn for different line positions; for instance, lines starting at different "y" axis points at the inlet lie towards the outlet parallel to the wall and the symmetry axis were used for the profiles.

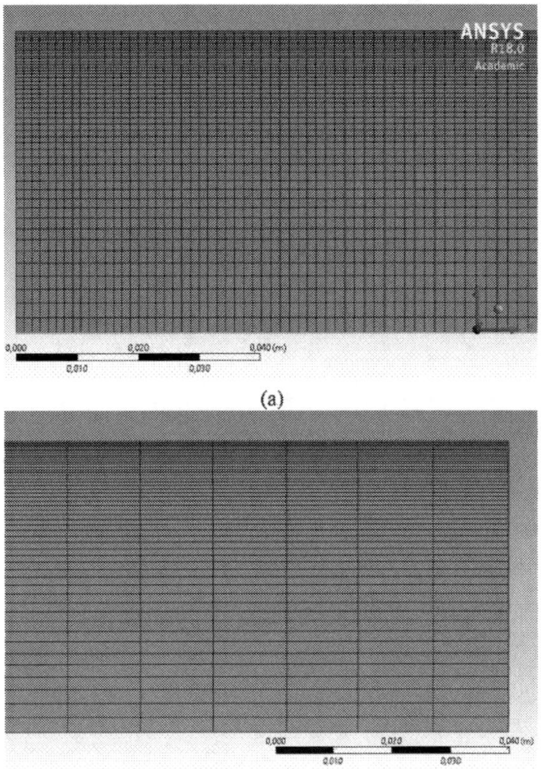

(a)

(b)

Fig.3. (a) mesh view at inlet (b) mesh view at outlet

Fluent solver of the Ansys CFD ran for viscous flow models. General setup can be summarized as; 2D, single precision, serial processing, pressure based, axisymmetric and without gravity. Inlet boundary condition was selected as constant velocity in a way that is ensuring the desired Re number according to the pipe diameter. Outlet boundary condition was set to pressure outlet and its value is fixed as 0 Pa gauge pressure. Hybrid initialization was used. Iteration numbers change between 29 and 145. Solution converges in seconds (between 5 and 30 seconds) since 320,000 elements only used once for the grid examination and mainly 80,000 elements were used. Transition Re number (i.e. 5,000 in this case) caused highest iteration numbers except for k-ϵ. The highest iteration number was 145 for k-ϵ and Re=1,000. The least iteration number was 88 for k-ϵ and Re=50,000. 0.001 was selected as the convergence criteria for all residuals. ϵ was the last converged magnitude among k, continuity and velocity components for k-ϵ. ω was the first to converge for k-ω. "y" velocity was converge in the order of first two for all models as expected. Most of the time for the analysis was spent for structural mesh and reporting.

CFD-Post was used for the post processing. The 3D view of the 2D flow domain in CFD-Post is given in Figure 4.

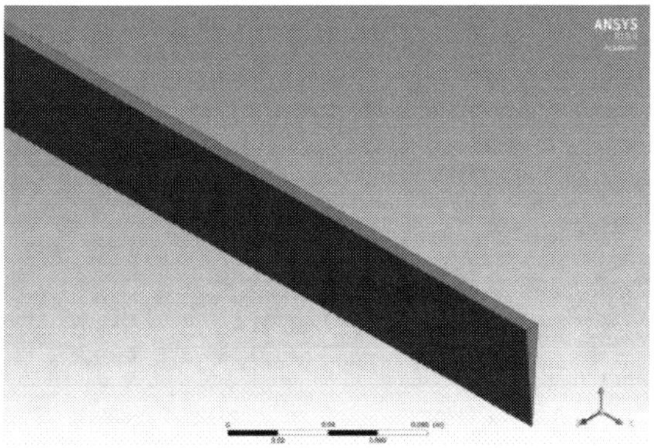

Fig.4. 3D view of the 2D flow domain in CFD-Post

3 Results

First of all, the four different viscous flow models are compared in respect of x-velocity, or in other words, streamwise velocity component, wall y+ value, turbulence intensity and turbulence viscosity at Re=10,000 in Figure 5. The velocity component in the graphics is denoted with "u" and its unit is "m/s". The pipe length or pipe axis is denoted with "x" axis and pipe radius is indicated with "y" axis. Velocity profiles are important since there are information in the literature about velocity profiles versus Re number, characteristic length and distance from the pipe axis. Therefore it is relatively easier to evaluate and assess the obtained results from the calculations. The change of "u" along the "x" axis gives an idea about the hydrodynamic development of the flow. These profiles have steeper changes starting from "x"=0 and then their change becomes smoother after some distance. It is expected that the profiles have a zero change at the far end of the pipe after the flow is developed. The profile at the pipe axis is important here because it indicates the velocity magnitude that starts in the inviscid core flow region and continues towards to the developed flow region. So it is easier to evaluate the development of the flow by examining this profile. On the other hand, the profile very close to the wall gives an idea of the development of the boundary layer.

https://doi.org/10.1051/epjconf/201818002013

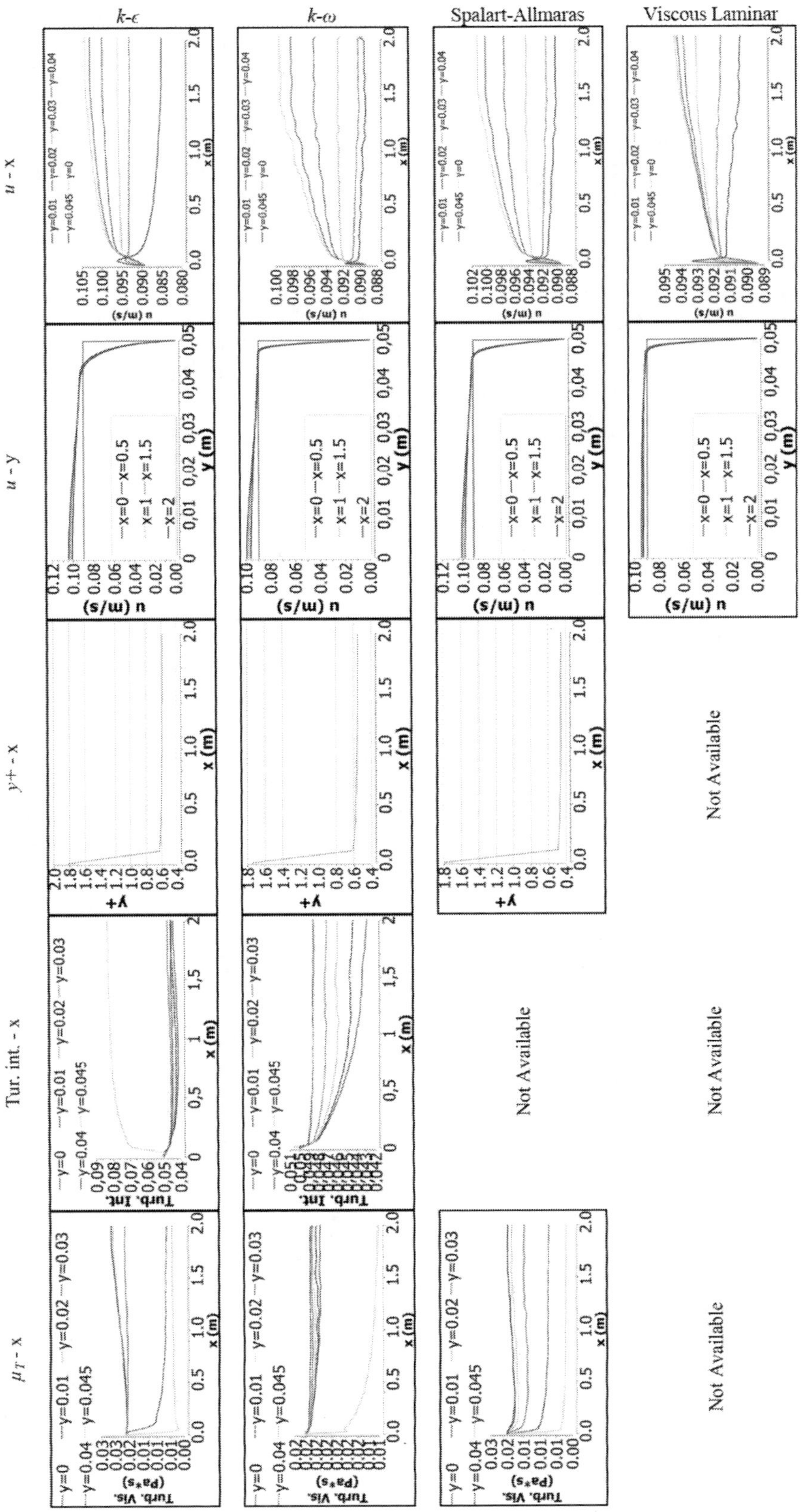

Fig. 5. Comparison of four viscous flow models at Re=10,000

At the first glance to Fig. 5, k-ϵ does not yield too distinguished profiles comparing to the remaining models. The range in velocity magnitude considering all profiles is the least for viscous laminar (VL) model while k-ω and Spalart-Allmaras (S-A) model follow it respectively. k-ϵ profiles here have the widest range indicating that this model distinguish more from the VL for Re=10,000. Actually the models other than k-ϵ imply a higher Re number comparing to profiles of k-ϵ or in other words vice versa. Turbulent flow develops faster than laminar flow due to the presence of turbulent structures that have much more impact on the flow than the molecular viscosity. The boundary layer rises and is restricted because of the turbulent momentum transfer. In the absence of such momentum transfer, the boundary layer of the laminar flow rises slowly with length towards the center. Since more length is needed for the laminar flow development, the initial part of the developing laminar flow looks like a turbulent one. This is the case here because the length of the pipe is 2 m and it is well known that practically development length is 10 D_h for turbulent flows, meaning that 1 m in the present case. 50 m is needed for the laminar developed flow. It is interesting to see that the velocity profiles for "u" is similar that of VL model and current turbulence models for the turbulence entrance region. One can propose that physical phenomena that are related to "u" component can benefit from the ease of the VL model for the turbulence entrance region.

The k-ϵ developed a thicker boundary layer before remaining models, yet its velocity profile changes smoothly after a distance more close to the 1 m length. This statement is justified by looking at the "y"=0.045 profile (purple colored). It can be inferred that some of the flow energy is converted into turbulence and hence velocity profile develops faster for k-ϵ than remaining models. u-y profiles show identical inlet velocity profiles for four models. At the outlet, however, VL velocity profile shows that only a big portion has the same velocity magnitude and the maximum magnitude deviates only about 0.003 m/s from the initial velocity magnitude while this value is between 0.003-0.005 for S-A, 0.004-0.006 for k-ω and 0.007-0.01 for k-ϵ.

Wall treatment in numerical investigation considers y+ dimensionless length. Since VL doesn't give this information, only wall y+ values of remaining turbulence models along the pipe length are given in Fig.5. This value is calculated for the cell next to the wall. y+ is desired about 1 for k-ϵ and below 1 for k-ω according to the literature. Fig. 5 shows that wall y+ is about 0.6 for k-ϵ and k-ω. For lower mesh element numbers, y+ gets higher values. Although the turbulent flow can be modeled, there is a region close to the wall where viscous effects become stronger and flow behaves like laminar. The molecular viscosity here plays the major role. This region is characterized by y+ value which is a dimensionless distance. It is defined as "y" times the friction velocity over kinematic viscosity. y+ between 0-6 indicates viscous sub-layer where u+ is equal to y+. 6-30 indicate buffer zone and 30-500 is the log-law zone. In the viscous sub-layer, the fluctuating velocity components diminish and shear stress is only due to the viscosity. The wall functions to be used are determined by looking to y+. Also development of y+ along the perpendicular axis gives information about the development of the boundary layer. Considering above mentioned facts, wall y+ is expected between 0-6.

More turbulent viscosity values are calculated with k-ϵ than the remaining models which means fluctuating velocity components have higher magnitudes over the flow domain. This should be due to the approach in which turbulent viscosity is calculated. Again k-ϵ is found favorable here because it reveals more information about the turbulence comparing to other models.

Turbulence intensity was estimated as 5% at the inlet for the initial conditions of the simulation. k-ϵ yields turbulence intensity values greater than 5% along the pipe length while k-ω yields results implying that turbulence diminishes with length and becomes weaker. Still, it is understood that a big portion of the flow field has turbulence intensity value about 5% in k-ϵ and 4% in k-ω. Near the wall, turbulence intensity has values as high as 15% for k-ϵ and 6% for k-ω.

Figure 6 is given for comparison of different Reynolds numbers considering that k-ϵ is the turbulence model. Looking to the Fig. 6, only Re=50,000 shows a different u-x profile near the wall, indicating that boundary layer becomes narrower as expected. u-y velocity profiles are similar for Re=50,000 and 10,000 while they become more parabolic for transition region of Re and laminar Re. The maximum u velocity values for Re=50,000, 10,000, 5,000 and 1,000 respectively are 0.475 m/s, 0.1 m/s, 0.05 m/s and 0.011 m/s. The maximum u velocity values also indicate that the maximum values increase more as it goes further into the laminar region since we can compare 10 fold decreased Re number maximum u velocity value with the u velocity value of the Re number by omitting one "0" after the decimal. The wall y+ become closer to 6 value for Re=50,000. y+ values develop after 0.1 m x length. Turbulence intensity is about 3.5 fold of Re=10,000 for Re=50,000 which implies that the increase of intensity is not linear comparing to the increase in Re. Turbulence intensity changes between 27% and 20% for Re=50,000, 8% and 5% for Re=10,000, 5% and 2% for Re=5,000 and 2% and 0.5 % for Re=1,000. Turbulent viscosity tends to get a common value for further pipe length and Re=10,000 and below. However Re=50,000 is different in this respect since pipe radius values away from the wall (starting from y=0.03 and goes towards y=0.01) yield turbulent viscosity values diverge from turbulent viscosity values of pipe radiuses close to wall. These results also imply that turbulence magnitude through the flow domain is nearly homogenous or in other words doesn't get too much different values. Looking to the ranges of the results in the 2D plots, turbulence is calculated for Re=1,000 however the calculated results have relatively small magnitudes. Therefore the model can also be thought for the solution of the laminar flows.

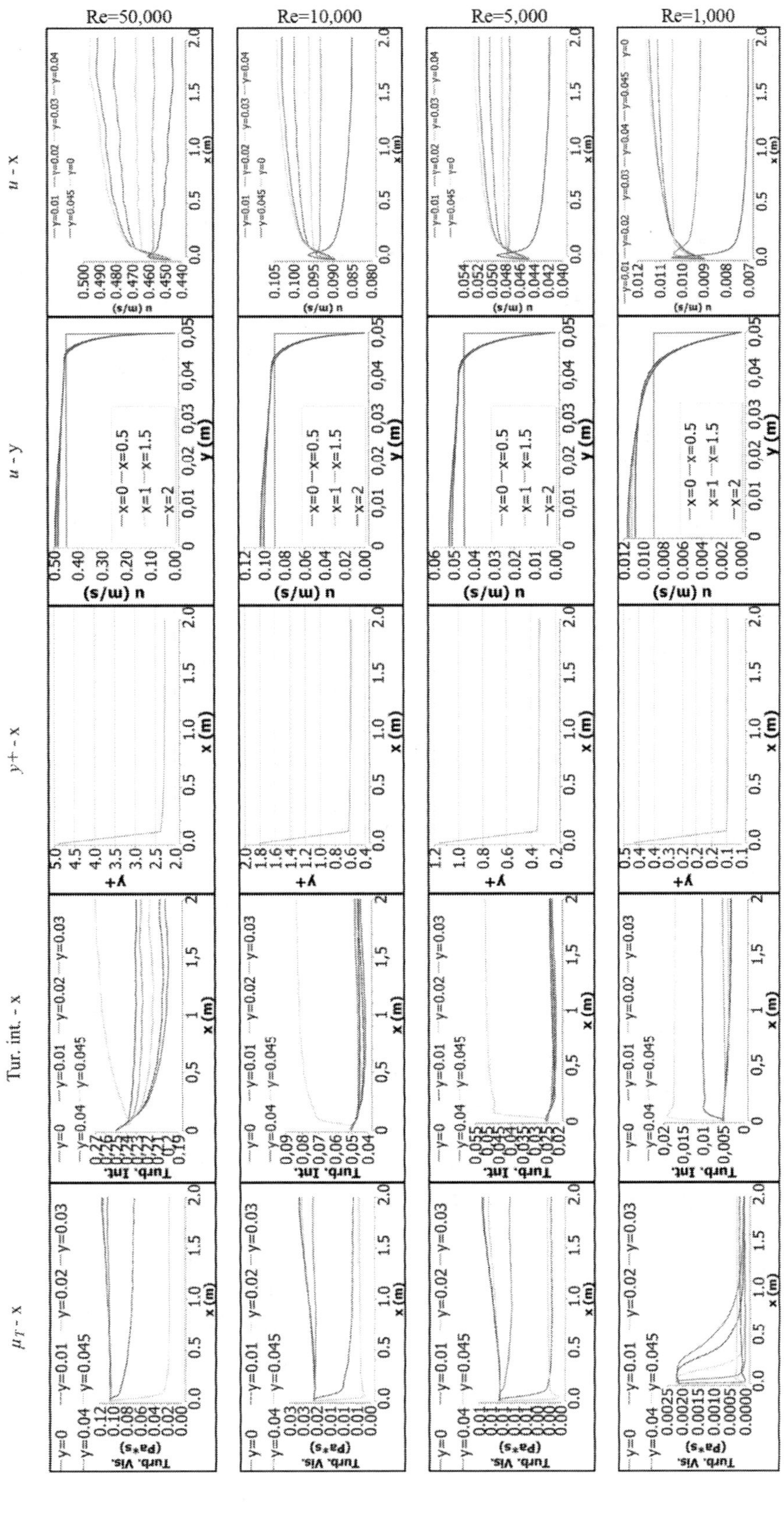

Fig. 6. Comparison of four different Re numbers for k-ϵ

4 Conclusions

This work aims to declare an ongoing PhD work and its foundation. The motivation of the work is tried to be explained by a literature timeline. The governing equations in 2D cylindrical coordinates are given. Then a preliminary computational fluid dynamics analysis is reported in this paper for a similar turbulent pipe flow case. Ansys CFD commercial software is utilized and four different viscous flow models were compared. Finally k-ϵ turbulence model results for four different Reynolds numbers are presented. Following remarks can be drawn from the text.

- The u-y velocity profiles show that the turbulent flow develops after 10 hydraulic diameters as an experimental justification of the results.
- The increase in mesh element number yield unfavorable results after a certain level and it is thought that this is due to the very small lengths leading to numerical errors. The information of free stream turbulence is also decreasing with increasing mesh element number.
- k-ϵ yields more information about the turbulence considering the whole of the flow domain.
- k-ω can be preferred if near wall region is specifically important.
- The models don't make a significant difference in time cost of the simulation.
- Mesh (grid) elements between 20,000 and 80,000 elements are found favorable for a similar case for further analyses.

Although the method and the findings are related with CFD, experimental results and facts both establish a basis for justification and provide data for developing modeling. For instance, u-y velocity profiles of the viscous flow models of the present study is compared to the practical flow development length of turbulent flows in reality which was experimentally obtained in the past. Also it is known that the CFD is used for phenomena that are hard to be measured but in the light of experimentally obtained experimental boundary conditions.

Deeper parametrical analyses of the computational events related to the results will be done in the future for a deeper understanding of the turbulence modeling. Also it is planned to establish a laboratory setup that can enable justification of CFD results by hotwire measurements, led type PIV and thermal camera applications.

Acknowledgment

The academic license of the software used in this work is belonged to Selcuk Univertsity. Authors also would like to acknowledge the financial support of Selcuk University Scientific Research Projects Coordination Unit for the conference participation (project no: 17701387).

References

1. S. Bilir, Selcuk University, Natural and Applied Sciences Ins., PhD Thesis, Konya, TURKEY (1988)
2. S. Bilir, Int. J. Heat Mass Transfer., **38**, 9 (1995)
3. S. Bilir, International Journal of Heat and Mass Transfer, **45** (2002)
4. S. Bilir, A. Ates, International Journal of Heat and Mass Transfer, **46** (2003)
5. A. Ates, Selcuk University, Natural and Applied Sciences Ins., PhD Thesis, Konya, TURKEY (1998)
6. A. Ates, S. Darici, S. Bilir, International Journal of Heat and Mass Transfer, **53** (2010)
7. S. Darici, Selcuk University, Natural and Applied Sciences Ins., PhD Thesis, Konya, TURKEY (2004)
8. S. Darici, A. Ates, S. Bilir, J. Fac.Eng.Arch. Selcuk Univ., **20**, 3 (2005)
9. S. Darici, A. Ates, S. Bilir, Journal of Selcuk-Technic, **7**, 2 (2008)
10. S. Darici, S. Bilir, A. Ates, International Journal of Heat and Mass Transfer, **84** (2015)
11. A.H. Altun, Selcuk University, Natural and Applied Sciences Ins., PhD Thesis, Konya, TURKEY (2013)
12. A.H. Altun, S. Bilir, A. Ates, International Journal of Heat and Mass Transfer, **92** (2016)
13. H. Demirpolat, Selcuk University, Natural and Applied Sciences Ins., PhD Thesis, Konya, TURKEY (2015)
14. S.U. Atmaca, Selcuk University, Natural and Applied Sciences Ins., PhD Thesis, Konya, TURKEY (2013)
15. S. Sen, Selcuk University, Natural and Applied Sciences Ins., PhD Thesis, Konya, TURKEY (2016)
16. S. Sen, S. Darici, Applied Thermal Engineering, **111** (2017)
17. N. Kasagi, A. Kuroda, M. Hirata, J. Heat Transfer, **111**, 2 (1989)
18. Y. Cao, A. Faghri, J. Heat Transfer, **114**, 4 (1992)
19. W. Jaeger, Nuclear Engineering and Design, **319** (2017)
20. I. Tiselj, R. Bergant, B. Mavko, I. Bajsic, G. Hetsroni, J. Heat Transfer, **123** (2001)
21. I. Tiselj, L. Cizelj, Nuclear Engineering and Design, (2012) DOI: 10.1016/j.nucengdes.2012.08.008
22. D.A. Anderson, J.C. Tannehill, R.H. Pletcher, Computational Fluid Mechanics and Heat Transfer, *McGraw-Hill* (1984)
23. W.P. Jones, B.E. Launder, Int. J. Heat Mass Transfer, **15** (1972)
24. B.E. Launder, D.B. Spalding, Comput. Meth. Appl. Mech. Engg, **3** (1974)

Conjugate heat transfer for turbulent flow in a thick walled plain pipe

Eyub Canli[1,*], *Ali* Ates[1], and *Sefik* Bilir[1]

[1]Selcuk University, Mechanical Engineering Department, P.K. 42003 Selcuklu Konya, Turkey

Abstract. Laminar and turbulent flow have their own characteristics in respect of heat transfer in pipes. While conjugate heat transfer is a major concern for a thick walled pipe with laminar flow inside it, there are limited studies about a turbulent flow in a thick walled plain pipe considering the conjugate heat transfer. In order to conduct such a work by means of in-house developed code, it was desired to make a preliminary investigation with commercially available CFD codes. ANSYS CFD was selected as the tool since it has a positive reputation in the literature for reliability. Defined heat transfer problem was solved with SIMPLE and Coupled Schemes for pressure velocity coupling and results are presented accordingly.

1 Introduction

Certain cases of heat transfer involve thick walled pipes and flow inside them. For such a case, not only heat convection transfers heat but also effect of conduction can be important for solid side and may be for the fluid side. This event is called conjugated heat transfer and has been subject of numerical and computational analysis efforts for several decades. Although there are vast amount of literature, some recent works are summarized below.

Iida et al. utilized Direct Numerical Simulation (DNS) for a turbulent channel flow under stable density difference [1]. Authors compared DNS results with experimental results since the study evaluates DNS as a tool and reported that the Nusselt Number (Nu) results agree with each other. Authors identified Internal Gravity Waves as being formed in the core region of the flow. These waves are followed by steep turbulent heat flux suppression and gradients of velocity and temperatures. As a novel output, they reported that the density stratification makes the flow laminar on one wall while the flow on the outher wall remains turbulent [1].

Varol et al. investigated conjugate heat transfer for different vertical walls and natural convection in them [2]. Authors specifically focused on the entropy generation due to the conjugate heat transfer. Entropy generation due to heat transfer is reported to be higher than that of fluid flow irreversibility [2].

Conjugate heat transfer for natural convection seems more popular recently than the forced convection. Kuznetsov and Sheremet [3] conducted a numerical study in order to analyze turbulent natural convection in terms of conjugate heat transfer due to thick wall enclosure. They deliberately defined a relatively complex thermal case. Standard k-ϵ was used with wall functions. As a different parameter, Grashof number between 10^8 and 10^{10} was taken as one of the parameters. They presented very similar results to some examples of the literature in order to validate their work. Then investigated the effects of parameters they choose. Finally they present a correlation of Nu against Grashof Number [3].

Similar to the natural convection for conjugate heat transfer, another trending research field is on DNS of the conjugate heat transfer. DNS of turbulent channel flow with conjugate heat transfer at Prandtl Number (Pr) 0.01 is reported by Tiselj and Cizelj [4]. Additionally fluctuating temperature boundary condition was investigated alongside with the non-fluctuating temperature boundary condition. The selected Pr corresponds to the liquid sodium-steel contact. It is indicated that the turbulent temperature fluctuations can affect temperature distribution of the wall [4].

The most relevant work to the target investigation of the authors of the present paper is reported by Yahya et al. [5]. They utilized Thermal Large Eddy Simulation (T-LES). Since this model only models sub grid eddies, more realistic flow structures can be attained and authors visualized their results in a successful manner in this respect. They considered effect of changing viscosity. As a main result they found that the turbulent kinetic energy decreases near the hot wall as the viscosity of the fluid increases and cold wall enhances the turbulence [5]. A similar work is presented by Satish and Venkatasubbaiah [6]. The flow is enclosed between two flat plates but with a significant difference; one of the plates is moving. They reported that heat conduction through the wall changes the heat transfer characteristics significantly. Actually their major findings are expected phenomena however the addition of heat transfer enhancement by means of increasing moving plate speed is important. They also used k-ϵ turbulence model [6].

* Corresponding author: ecanli@selcuk.edu.tr

Analytical solution of conjugate turbulent forced convection boundary layer flow over plates is presented by Shariatzadeh [7]. A semi-analytical solution based on differential transform method is presented in the paper. The method is for solving the non-linear differential equation in the problem. The author indicates that a slight temperature difference occurs in conjugate case comparing to the non-conjugate case.

Conjugate heat transfer along a heated flat plate is analyzed by using wall resolved LES [8]. The authors indicate that a locally dynamic Smagorinsky SGS model was implemented into the LES solver to enable the backscatter phenomenon intrinsic to transitioning boundary layer flows for the subgrid scale closure. This kind of LES is reported to be convenient considering the response capability and experimental results [8].

Laminar-turbulent transition in case of a conjugate heat transfer scenario is reported for assessing RANS [9]. Performance of totally 8 different turbulence model are given. They also add some other parameters to the study such as inclination angle and Richardson number and investigate the location of the transition point.

This work presents temperature distribution results of ANSYS CFD for a thick walled plain circular pipe and turbulent flow inside it. The results will also be used as a comparison tool for an ongoing PhD work.

2 Method

A 2 m plain pipe having a thick wall and turbulent flow inside it are considered for a conjugated heat transfer case. Computational Fluid Dynamics (CFD) was chosen as the tool of investigation. Since there is an ongoing work to develop an in-house code, it is desired to obtain preliminary results from commercial codes for comparison and also mention about the ongoing work. Therefore ANSYS software was used for all the CFD steps. The two dimensional, axis-symmetrical flow domain was created for two different dimensions in ANSYS Space Claim module. Figure 1 shows the two drawings. The only difference between flow domains are the wall thickness. The first flow domain is 0.01 m thick and the second one is 0.025 m thick. Flow length is 2 m and the pipe radius is 0.05 m. Although materials can be selected in the solver, wall material was selected as 405 stainless steel and fluid was selected as water at room temperatures. Naming edges and domains and relating edges with each other were left to mesh generator.

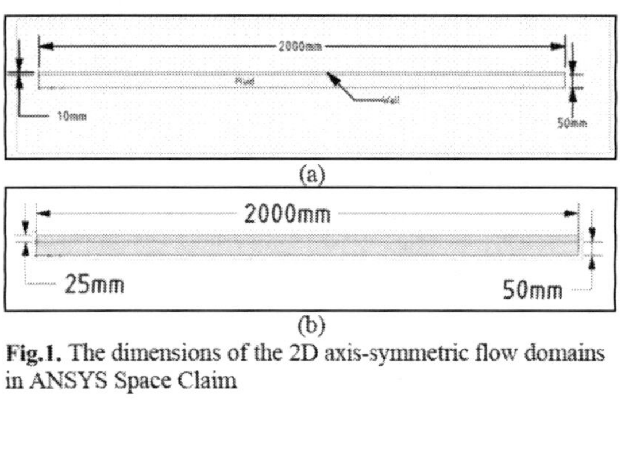

Fig.1. The dimensions of the 2D axis-symmetric flow domains in ANSYS Space Claim

Mesh was generated in the ANSYS Mesh module. Two different mesh files were generated due to the two different thickness of the wall. However the mesh element number of the flow domain remained same. The proper mesh number for the flow domain was found to be 80,000 elements and reported in "Comparison of turbulence models and solution options for a thick walled plain pipe" titled conference paper [10] in the present conference. Therefore same structured mesh was arranged for the flow domain. 0.05 m pipe radius was divided into 100 units with a bias factor of 10. The thinnest cell is next to the wall and the widest cell is next to the symmetry axis. 2 m pipe length was divided 800 units with again a bias factor of 10. Therefore the flow domain mesh exhibits a view that it is stretched towards the outlet of the pipe and the symmetry axis. These arrangements were done by using 4 edge sizing and 1 face mesh options. All the sizing arrangements were set to "hard" behavior meaning that they forced the sizes by setting the edge sizes at the beginning of mesh generation incident. Wall edges parallel to the symmetry axis were divided similar to the pipe due to assure that the cells correspond to each other perfectly. The two side edges of the wall were divided into 10 uniform pieces for 0.01 m wall thickness and 25 uniform pieces for 0.025 m wall thickness. All edges and faces are named in the mesh generator. The south edge of the wall and the north edge of the fluid according to view plain were set as contact region. Figure 2 shows the inlet and outlet views of the meshes. Figure 3 is for the edge names and Figure 4 shows the contact region.

According to the mesh report, the geometry is fully defined. Although the drawing and the solution is two dimensional, the mesh generator always puts a unit length to the third dimension. Therefore a volume value is seen equal to the surface area of the analysis domain. For the 0.01 m wall thickness case, there is 0.12 m^2 surface area therefore 0.12 m^3 volume can be seen in the mesh report. 85,707 nodes and 84,000 elements exist for 0.01 m wall thickness and 101,727 nodes and 100,000 elements exist for 0.025 m wall thickness. Element quality values are min. 0.04, max. 0.99, av. 0.46. However these element quality values are better for the wall side because of the uniform distribution of the cells. The wall side element quality values are min. 0.48, max. 0.99, av. 0.82. Preferred physics option was set to CFD and solver preference was set to Fluent. Remaining settings were leaved as default and no inflation was arranged since the near wall region has sufficient mesh elements in it. This is ensured by checking wall y+ values.

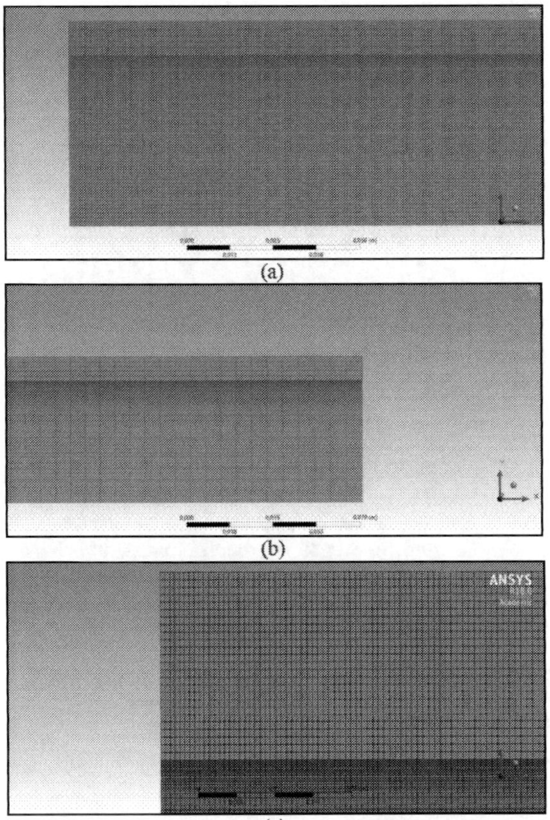

(a)
(b)
(c)

Fig.2. Mesh views (a) inlet of the 0.01 m wall pipe (b) outlet of the 0.01 m wall pipe (c) inlet of the 0.025 m wall pipe

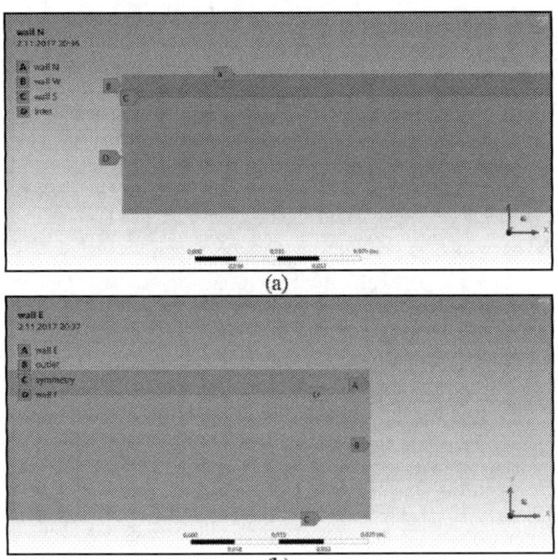

(a)
(b)

Fig.3. Names of the boundaries (a) inlet side (b) outlet side

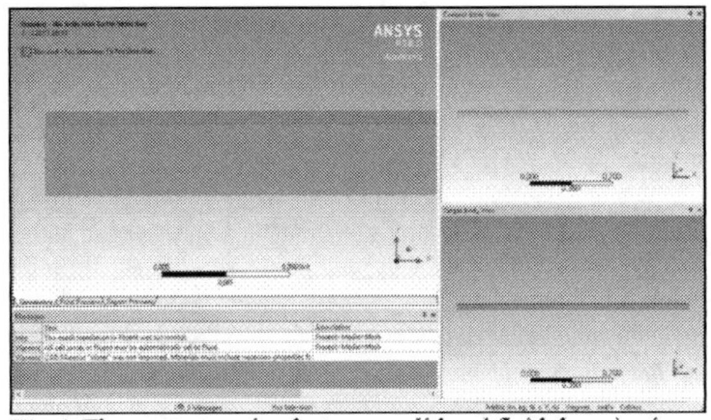

Fig.4. The contact region between solid and fluid domains in mesh generator

Fluent Solver initial setup settings are: single precision, serial processing, 2D. The general settings for the solver are pressure based, absolute velocity formulation, steady and 2D axis-symmetric. Effect of the gravity was ignored. Standard k-ϵ with standard wall function was selected as the turbulence model alongside with the energy equation. Fluid material was selected as liquid water and wall material was selected as steel. Thermodynamic properties were selected as constant. Density of the water was 998.2 kg/m^3 and steel was 8030 kg/m^3. Specific heat values were 4182 and 502.48 j/kg·K for water and steel respectfully. Water had 0.6 W/m·K thermal conductivity while steel had 16.27 W/m·K. Water viscosity was 0.001003 kg/m·s. Water inlet was set to 0.08937 m/s absolute velocity magnitude that corresponds to 10,000 Reynolds Number (Re) while initial turbulence intensity was set to 5% and hydraulic diameter was set to 0.1 m. The inlet water temperature was 300 K. Outlet boundary condition was selected as 0 Pa gauge pressure. All walls were set as wall boundary condition with 0 W/m^2 constant heat flux except for the north edge of the wall. North edge of the wall set to 400 K and 500 K for separate calculations. Symmetry axis was set to symmetry boundary condition and connection edges were similarly set as connection boundary condition. Reference values were taken from the water inlet. Two different solution schemes were used; namely SIMPLE and Coupled. Therefore it can be said that three parameters were changed during analysis; wall thickness, pipe outside surface temperature and solution scheme. First of all, the flow dynamics obtained by SIMPLE and Coupled schemes are compared and then SIMPLE scheme was used for different temperature and thickness values. The point values comparison of the SIMPLE and Coupled schemes are given in Table 1.

Table 1. The point values comparison of the SIMPLE and Coupled schemes

	SIMPLE	Coupled
Vel. (m/s)) at symmetry outlet	0.105	0.119 (13%)
Tot. Pres. (Pa)) at symmetry outlet	5.48	7.07 (29%)
Turb. Int.) at symmetry outlet	0.045	0.054 (20%)
Turb. Vis. (Pa*s)) at symmetry outlet	0.0275	0.03154 (14%)
y+ at connection outlet	0.587	0.56 (4.5%)
Temperature (K) at symmetry outlet	305.557	309.258 (1.2%)
Temperature (K) at connection outlet	368.929	366.253 (0.7%)
Av. % Dif.	-	11.7%

The spatial discretization options were left as default and they were Least Squares Cell Based Gradient; Second Order Pressure; Second Order Upwind Momentum; First Order Upwind Turbulent Kinetic Energy; First Order Upwind Turbulent Dissipation Rate; Second Order Upwind Energy. Solution control under relaxation factors were selected as 0.3 for pressure, 1 for density, 1 for body forces, 0.7 for momentum, 0.8 for turbulent kinetic energy, 0.8 for turbulent dissipation rate, 1 for turbulent viscosity, 1 for energy. Hybrid initialization was used. Solution was converged about 100 iterations in minutes.

CFD Post was used for post processing. The computational domain is seen three dimensional though the design is two dimensional. This is thought to be due to the infrastructure of the ANSYS CFD software when axis-symmetrical solution is used. The 3D view of the computational domain is given in Figure 5.

Since the aspect ratio of the domain is very small (height/length=0.03), x-y plots are utilized for the results mostly. Only close inlet and outlet contour views of the temperature are given as graphical results.

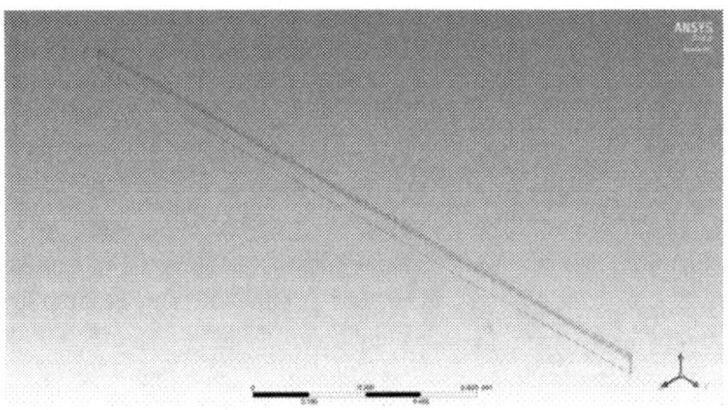

(a)

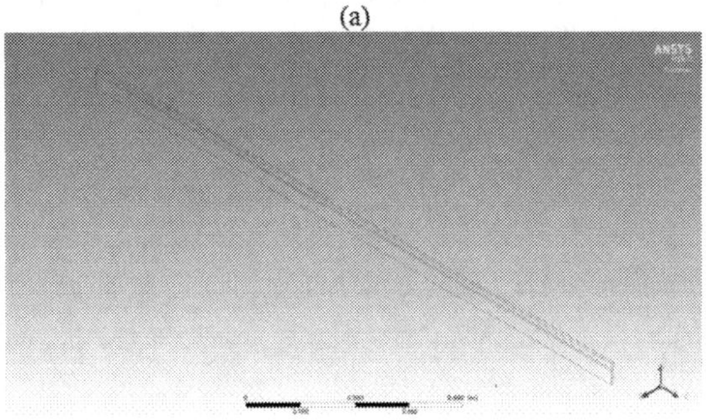

(b)

Fig.5. 3D CFD Post views (a) 0.01 m wall thickness (b) 0.025 wall thickness

3 Results

Initially comparison of SIMPLE scheme with Coupled scheme for pressure velocity coupling is given in terms of velocity, wall y+, turbulence intensity and viscosity in Figure 6.

The inlet fluid temperature contours of two different wall thicknesses and two different wall outside temperature boundary conditions are given in Figure 7. Similarly inlet wall temperature contours are given in Figure 8. The outlet fluid temperature contours of two different wall thicknesses and two different wall outside temperature boundary conditions are given in Figure 9. And finally outlet wall temperature contours are presented in Figure 10.

The pressure velocity coupling by Coupled Scheme shows smoother flow values comparing to SIMPLE scheme, however the u-y velocity profiles of the Coupled Scheme show that the flow develops later. Additionally wall y+ values are lower for Coupled scheme. The turbulent viscosity and intensity values of the Coupled Scheme indicate that this scheme produces additional turbulence information however the temperature profiles of the two schemes are very similar. Only T-y profiles of the Coupler Scheme show greater distribution of the temperature. According to the present results, it is decided that the less work burden scheme can be chosen for the present case.

Temperature profiles show that the thermal development is not acquired with the current pipe length. This can be evaluated with a dimensionless analysis with a semi-infinite pipe length through downstream. It can also be inferred that a transient analysis depending on time can yield different temperature profiles for a period of time and then the temperature profiles change with location.

Temperature contours tended identical for different wall outside temperatures since the thermodynamic properties were taken constant. The change of wall thickness has an effect only on magnitudes similar to the wall outside thickness.

4 Conclusions

This work is a preliminary study surveying the results of a commercial CFD code for comparison with an in-house developed code while assessing some fundamental settings and methods in CFD for the investigated case. A thick walled plain pipe is analyzed with 2D axis-symmetrical design for turbulent flow of Re=10,000 and heat transfer at constant outside wall temperature. The interface condition is unknown and obtained as a result. Results of SIMPLE and Coupled pressure velocity coupling schemes are compared. Following remarks can be noted.

- Pressure velocity coupling avoids velocity anomalies of the k-ϵ with the present structured mesh. The low element quality at the medium and far downstream region of the fluid domain may cause these anomalies with SIMPLE Scheme. Therefore Coupled Scheme can give more information about the turbulence for the present case and setup.
- When thermodynamic properties are taken constant, changing value of the constant outside surface temperature and wall thickness only changes the magnitudes but the behavior or the tendency remain same for the steady calculation. Transient solution may yield different results according to time.

https://doi.org/10.1051/epjconf/201818002014

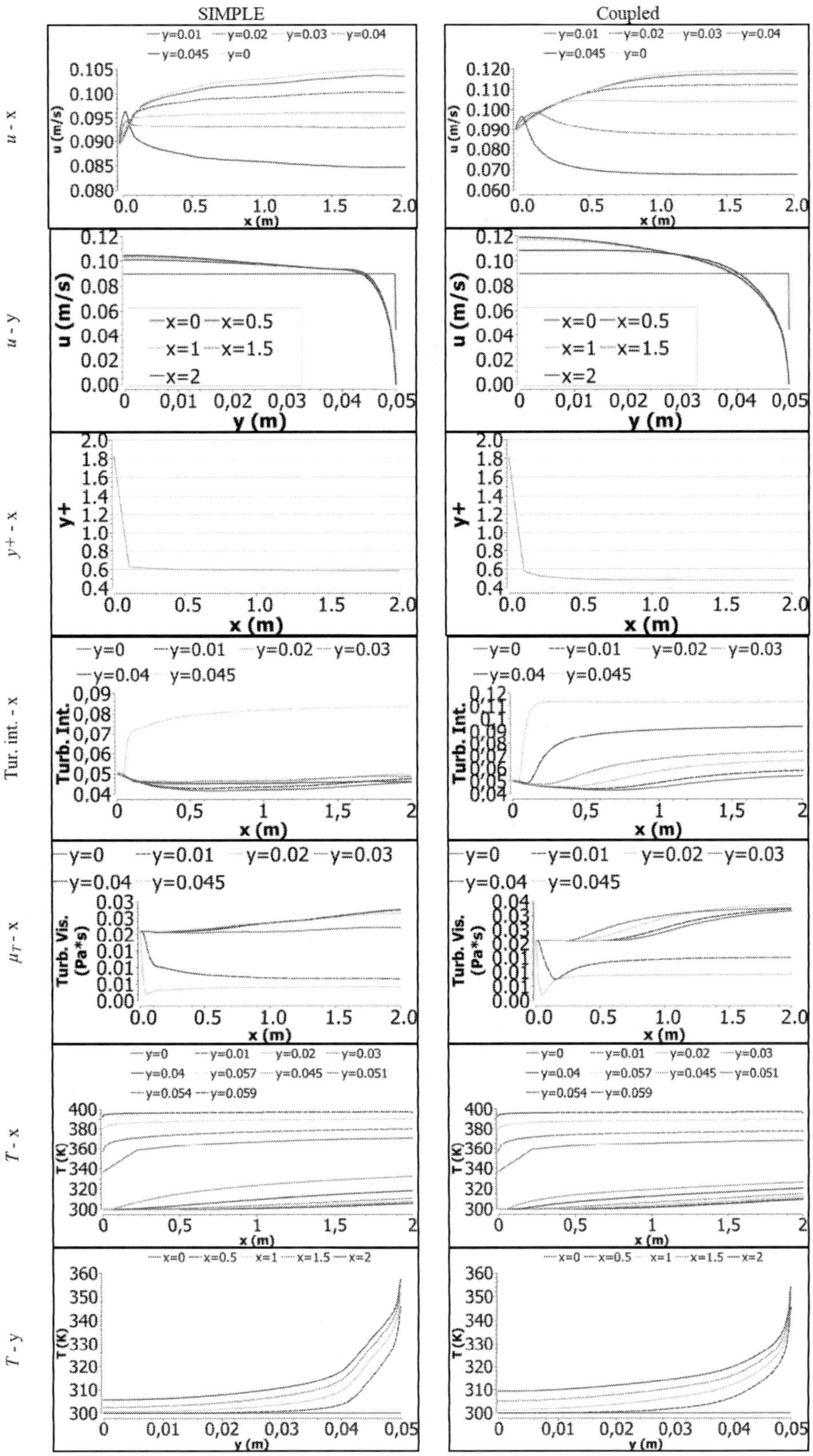

Fig. 6. Comparison of SIMPLE and Coupled schemes

- Conjugate heat transfer cannot be identified in the present turbulent flow with characteristics similar to the laminar flow since the thermal boundary condition is applied at the beginning of the pipe and lasts till the end. The steady simulation and relatively fast flow also avoided effects of axial heat conduction of the fluid.

In the future, authors are planning to make a parametrical study for conjugated heat transfer in a circular thick walled pipe while turbulent flow exists inside it by constituting and using an in-house developed code. Not only physics will be investigated but also numerical and computational approach will be tried to be improved.

Acknowledgment

The academic license of the software used in this work is belonged to Selcuk Univertsity. Authors also would like to acknowledge the financial support of Selcuk University Scientific Research Projects Coordination Unit for the conference participation (project no: 17701387).

References

1. O. Iida, N. Kasagi, Y. Nagano, International Journal of Heat and Mass Transfer **45**, (2002)
2. Y. Varol, H. F. Oztop, A. Koca, International Communications in Heat and Mass Transfer **35**, (2008)
3. G. V. Kuznetsov, M. A. Sheremet, International Journal of Heat and Mass Transfer **53**, (2010)
4. I. Tiselj, L. Cizelj, Nuclear Engineering and Design **253**, (2012)
5. S. M. Yahya, S. F. Anwer, S. Sanghi, International Journal of Thermal Sciences **88**, (2015)
6. N. Satish, K. Venkatasubbaiah, Applied Thermal Engineering **100**, (2016)
7. O. J. Shariatzadeh, Thermal Science **20**, 5, (2016)
8. A. Ojofeitimi, Y. Hattori, International Journal of Heat and Mass Transfer **109**, (2017)
9. M. Abdollahzadeh, M. Esmaeilpour, R. Vizinho, A. Younesi, J.C. Pàscoa, International Journal of Heat and Mass Transfer **115**, (2017)
10. E. Canli, A. Ates, S. Bilir, Experimental Fluid Mechanics 2017 (EFM 2017) (to be published in the proceedings)

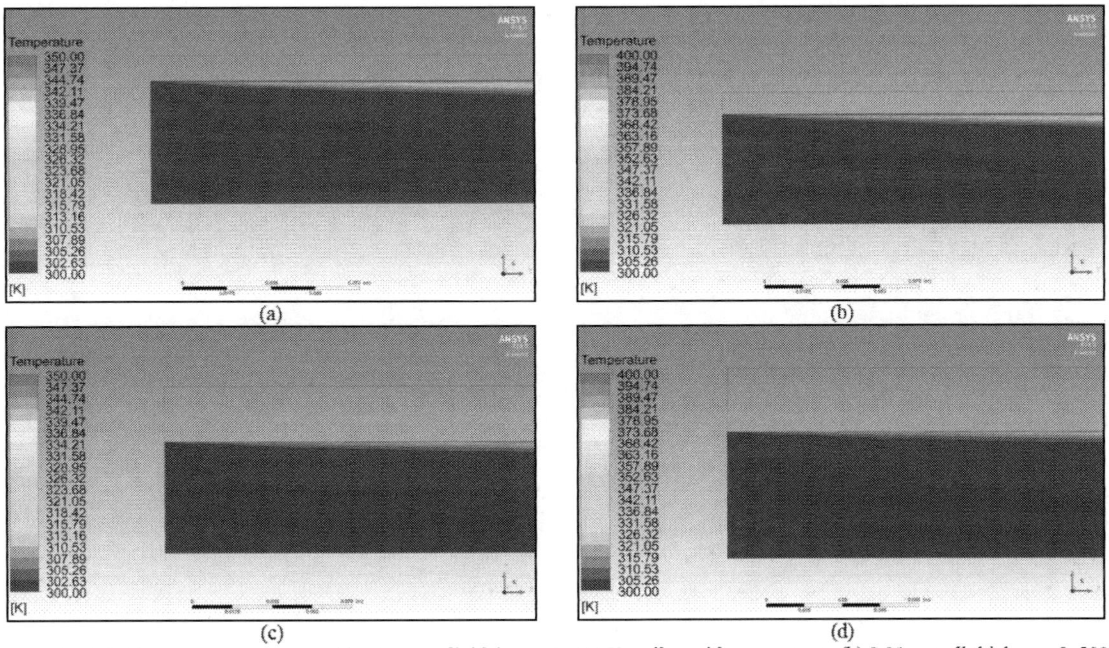

Fig.7. Inlet fluid temperature contours (a) 0.01 m wall thickness & 400 K wall outside temperature (b) 0.01 m wall thickness & 500 K wall outside temperature (c) 0.025 m wall thickness & 400 K wall outside temperature (d) 0.025 m wall thickness & 500 K wall outside temperature

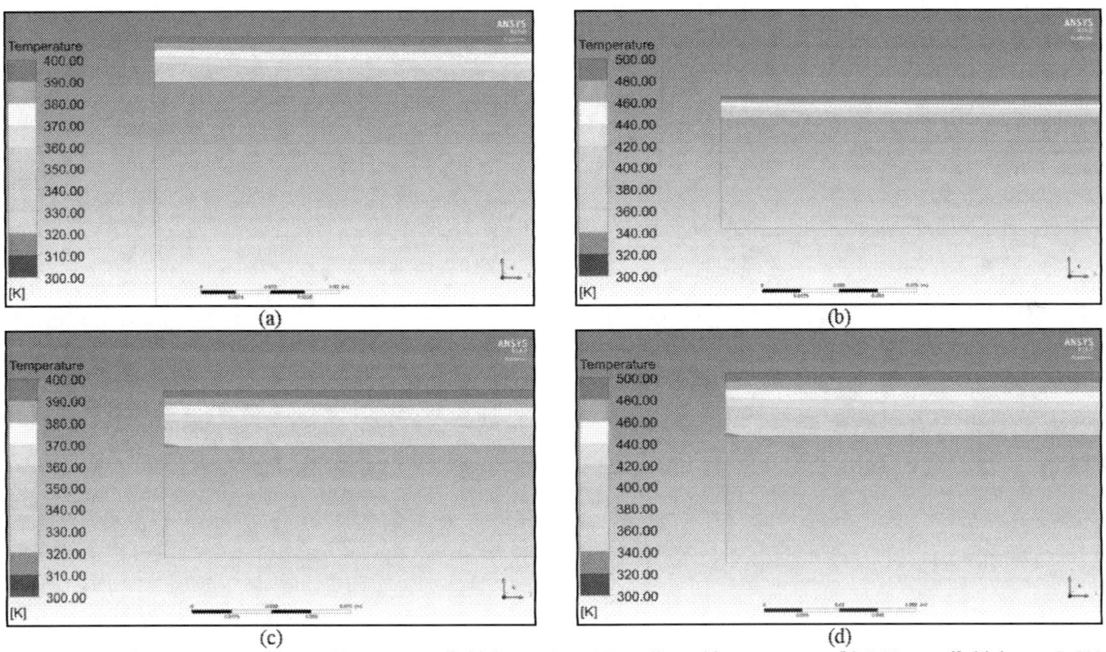

Fig.8. Inlet wall temperature contours (a) 0.01 m wall thickness & 400 K wall outside temperature (b) 0.01 m wall thickness & 500 K wall outside temperature (c) 0.025 m wall thickness & 400 K wall outside temperature (d) 0.025 m wall thickness & 500 K wall outside temperature

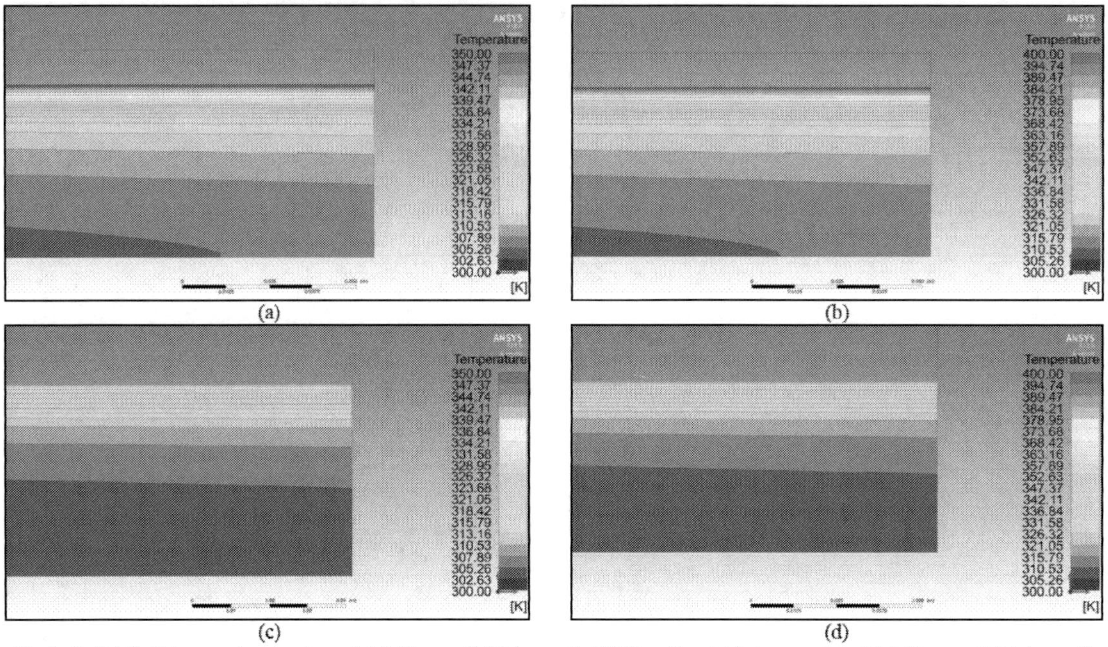

Fig.9. Outlet fluid temperature contours (a) 0.01 m wall thickness & 400 K wall outside temperature (b) 0.01 m wall thickness & 500 K wall outside temperature (c) 0.025 m wall thickness & 400 K wall outside temperature (d) 0.025 m wall thickness & 500 K wall outside temperature

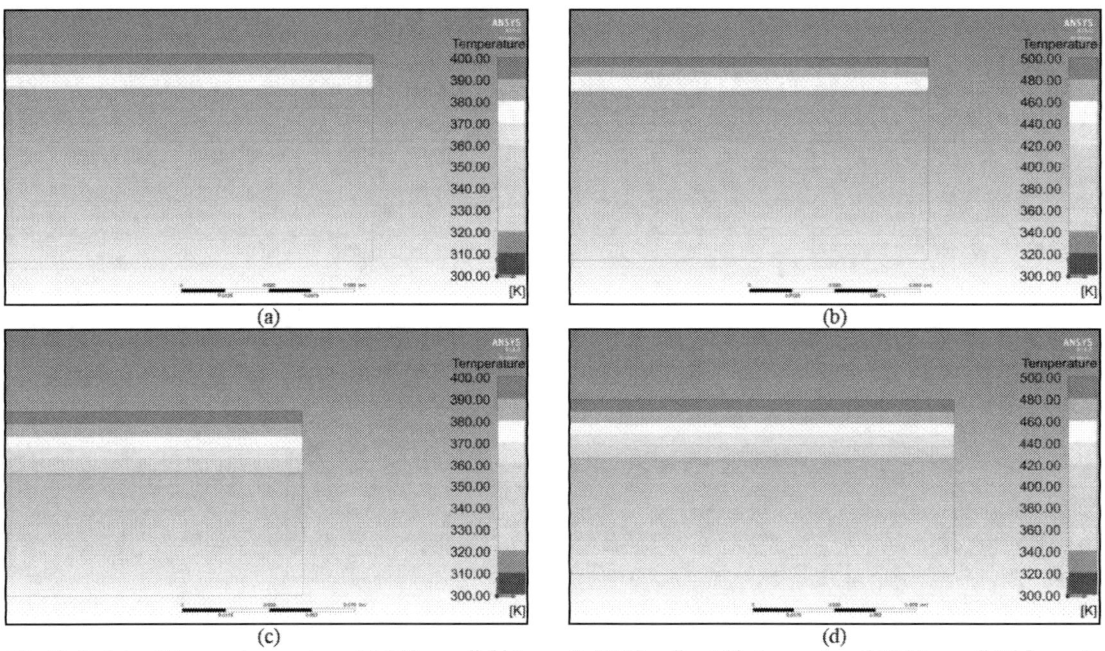

Fig.10. Outlet wall temperature contours (a) 0.01 m wall thickness & 400 K wall outside temperature (b) 0.01 m wall thickness & 500 K wall outside temperature (c) 0.025 m wall thickness & 400 K wall outside temperature (d) 0.025 m wall thickness & 500 K wall outside temperature

Investigation of droplet nucleation in CCS relevant systems – design and testing of the expansion chamber

Miroslav Čenský[1,*], Jan Hrubý[1], Václav Vinš[1], Jiří Hykl[1] and Bohuslav Šmíd[1]

[1]Institute of Thermomechanics of the CAS, Dolejškova 1402/5, Prague 8, 182 00, Czech Republic

Abstract. A unique in-house designed experimental apparatus for investigation of nucleation of droplets in CCS relevant systems is being developed by the present team. The apparatus allows simulating various processes relevant to CCS technologies. Gaseous mixtures with CO_2 are prepared in a Mixture Preparation Device (MPD) based on accurate adjustment of flow rates of individual components [EPJ Web of Conferences 143, 02140 (2017)]. The mixture then flows into an expansion chamber, where it undergoes a rapid adiabatic expansion. As a consequence of adiabatic cooling, the mixture becomes supersaturated and nucleation and simultaneous growth of droplets occurs. In this study, we describe the design and testing of the expansion part of the experimental setup. The rapid expansion was realized using two valve systems, one for low pressures (up to 0.7 MPa) and the other for high pressures (up to 10 MPa). A challenge for a proper design of the expansion system is avoiding acoustic oscillations. These can occur either in the mode of Helmholtz resonator, where the compressible gas in the chamber acts as a spring and the rapidly moving gas in the valve system as a mass, or in the "flute" mode, where acoustic waves are generated in a long outlet tubing.

1 Introduction

The team of the Laboratory of Phase Transition Kinetics at the Institute of Thermomechanics of the Czech Academy of Sciences (CAS) is developing a unique experimental apparatus for investing the formation of droplets in CO_2-rich systems.

To obtain new valuable data for nucleation of droplets in binary and multicomponent mixtures important in the design of new Carbon Capture and Storage (CCS) technologies [1,2] is the main motivation for this research. The apparatus allows for simulating droplet nucleation relevant for large variety of processes of technologies.

Gaseous mixtures with CO_2 are prepared in a Mixture Preparation Device (MPD) based on an accurate adjustment of flow rates of individual components [1]. The mixture then flows into an expansion chamber, where it undergoes a rapid adiabatic expansion. As a consequence of adiabatic cooling, the mixture becomes supersaturated and nucleation and simultaneous growth of droplets occurs. The droplets are indicated optically based on laser light scattering and the nucleation and growth rates can be evaluated.

Expansion cloud chambers have been used for the research of droplet formation for more than a century [3] and many high-quality data sets for homogeneous nucleation rates have been obtained using the so-called nucleation pulse method implemented in expansion chambers [4, 5]. In this method, nucleation occurs in a short period of time (typically 1 ms) and the supersaturation is reduced afterwards such that the existing droplets can grow to observable sizes (diameter of several hundred nanometers), but practically no new droplets appear. This approach is, however, limited to systems, where the growth is slow enough. In case of nucleation in CO_2 rich mixtures, the growth is very rapid. Therefore, rather than the nucleation pulse method [6], we apply the continuous expansion method [7]. Here, the main control parameter is the expansion rate. The expansion rate, together with the nucleation and growth kinetics, determines the maximum nucleation rate achieved in the process. The higher is the expansion rate, the higher is the maximum nucleation rate. In order to have a broad experimental window, it is necessary to have the possibility of adjusting the nucleation rate in a broad range.

New apparatus for investigation of nucleation in CO_2-rich systems was described in our previous paper [1], which was focused on the CO_2 branch of the mixture preparation device. In this study, we describe the design and testing of the expansion part of the experimental setup.

2 Experimental Apparatus

The rapid expansion was realized using two valve systems, one for low pressures (up to 0.7 MPa) and the other for high pressures (up to 10 MPa). A challenge for a proper design of the expansion system is avoiding acoustic oscillations. These can occur either in the mode of Helmholtz resonator, where the compressible gas in

* Corresponding author: censky@it.cas.cz

https://doi.org/10.1051/epjconf/201818002015

the chamber acts as a spring and the rapidly moving gas in the valve system as a mass, or in the "flute" mode, where acoustic waves are generated in a long outlet tubing. Small acoustic oscillations lower the accuracy of the experiments; large oscillations can completely distort the expansion and even lead to the malfunction of the piston (or diaphragm) valve, which repeatedly closes and opens.

The setup of low-pressure expansion system is shown in Figure 1 and of high-pressure expansion system in Figure 2. Both low and high-pressure apparatuses consist of high pressure expansion chamber (1) pressurized from gas vessel (11) by nitrogen which was used as a model gas in the tests performed in this study. The initial pressure in the expansion chamber was set with a precision back pressure regulator from Equilibar, LLC[a] manufacturer, (12) equipped with manometer (13). Precise measurement of the initial pressure inside the expansion chamber and a rapid pressure drop during the expansion in both low and high pressure expansion systems are measured with fast-response pressure transducers HEM-375 from Kulite Semiconductor Products[b] (3) available for pressure range up to 100, 500 and 2500 PSI. Electric signal from pressure transducer is amplified and connected to terminal block BNC-2110 used as an interface to PC with multifunction I/O device PCI-6115 from National Instruments[c] manufacturer.
The low-pressure expansion system (Fig. 1) consists of a pilot-operated diaphragm valve from ASCO Valve, Inc.[d] (2) followed by a throttling valve (5). For very slow expansions, the low pressure system uses a metering needle valve from HOKE[e] (6) followed by a manually controlled solenoid valve ASCO[d] (7).

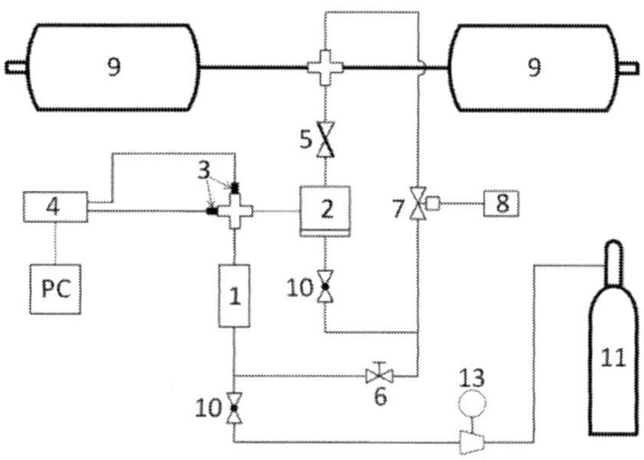

Fig. 1. Experimental apparatus for rapid expansion with throttle and needle valve. 1 – high pressure expansion chamber, 2 – pilot-operated diaphragm valve, 3 – rapid high pressure transducers, 4 – BNC terminal block connected with PC, 5 – throttling valve, 6 – metering needle valve, 7 – solenoid valve, 8 – solenoid valve controller, 9 – expansion vessels, 10 – ball valves, 11 – gas vessel with nitrogen, 12 – pressure regulator, 13 – manometer.

[a] www.equilibar.com
[b] www.kulite.com
[c] www.ni.com
[d] www.asco.com
[e] www.hoke.com

The high-pressure system shown in Figure 2 is intended for investigation of most important CO_2-rich mixtures relevant for the CCS technologies. The high-pressure setup was based on a commercially available pilot-operated piston valve from ASCO Valve, Inc.[d] manufacturer which was modified such that its saddle was directly followed by a throttling needle in order to minimize acoustic oscillations. The throttling needle has a special shape engineered such that the flow area can be reproducibly adjusted from fully open down to a fraction of percent. The adjustable position of the stem is measured with a digital distance indicator ID-H0530 from Mitutoyo[f] manufacturer with micrometer resolution. Expanded gas is exhausted via two expansion vessels (9) opened to atmospheric pressure in the tests reported.

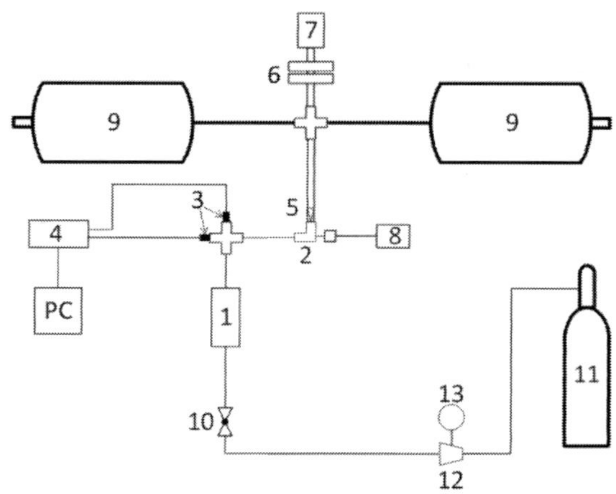

Fig. 2. Experimental apparatus for slow expansion special throttle valve. 1 – high pressure expansion chamber, 2 – pilot-operated piston valve, 3 – rapid high pressure transducers, 4 – BNC terminal block connected with PC, 5 – throttling needle valve with a special shape, 6 – manual throttle valve controller, 7 – precise distance indicator, 8 – electromagnetic valve controller, 9 – expansion vessels, 10 – ball valves, 11 – gas vessel with nitrogen, 12 – pressure regulator, 13 – manometer.

3 Results and Discussion

Several series of expansions were done with nitrogen in both low and high pressure systems and corresponding expansion rates were evaluated based on following equation:

$$\text{expansion rate} = \frac{d\ln p}{dt} = \frac{1}{p}\cdot\frac{dp}{dt}$$

where p is pressure in expansion chamber and t is time. Figure 3 shows time dependence of pressure during the low-pressure slow expansion from approximately 700 kPa via the metering needle valve (6 in Figure 1). Higher

[f] www.mitutoyo.com

number of revolutions, given as rev. in the figure legend, is directly linked to wider opening of the needle valve. Expansion rate is evaluated from a slope of the pressure on time between pressures 400 and 500 kPa, where the pressure dependence is nearly linear. Dependence of the evaluated expansion rate on opening of the metering needle valve is shown in Figure 4.

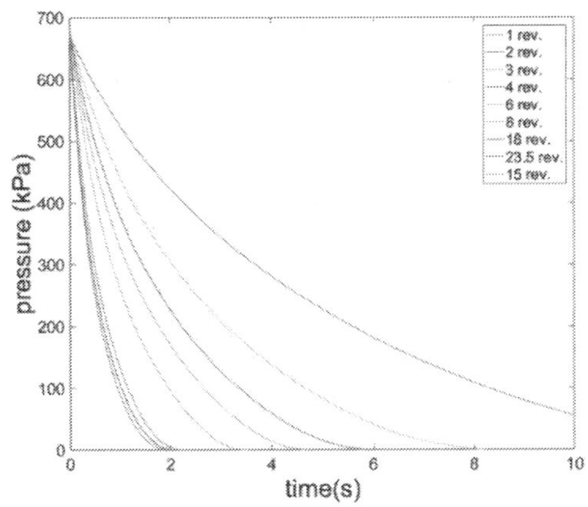

Fig. 3. Nitrogen pressure dependence on time during expansion from 700 kPa through various openings of metering needle valve.

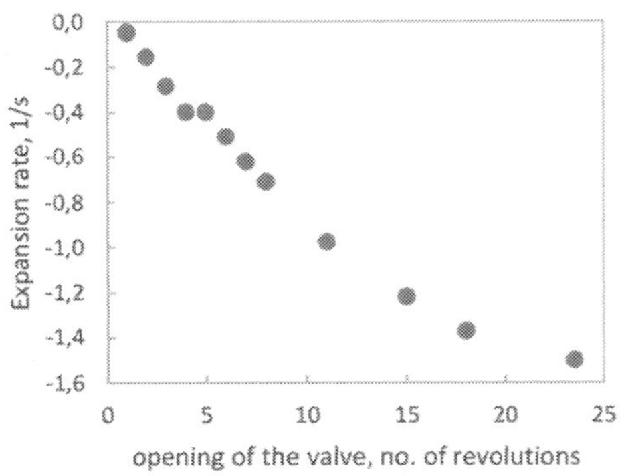

Fig. 4. Expansion rate dependence on opening of metering needle valve for expansion from 700 kPa. Slope of the expansion rate evaluated from linear interpolation between 400 and 500 kPa.

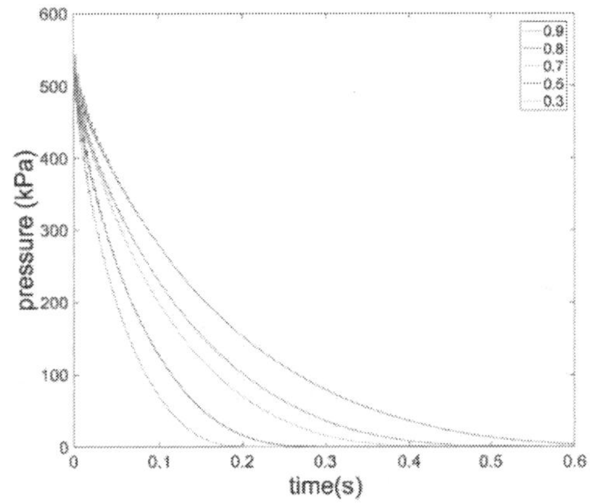

Fig. 5. Nitrogen pressure dependence on time during expansion from 600 kPa through various small openings of throttling valve.

Time dependence of pressure during the low-pressure expansion from approximately 600 kPa via the throttling valve (5 in Figure 1) is shown in Figures 5 and 6. Higher value of a relative number corresponds to wider opening of the valve and consequently to higher expansion rates. Expansion rates were evaluated from a slope of the pressure depending on time between pressures 400 and 500 kPa where this dependence can be considered nearly linear. Dependence of the evaluated expansion rates on opening of the throttling valve is shown in Figure 7.

In the system with throttling valve, behaviour in the mode of Helmholtz resonator can be observed particularly in the beginning of the expansion, where generated acoustic waves caused oscillations of the pressure.

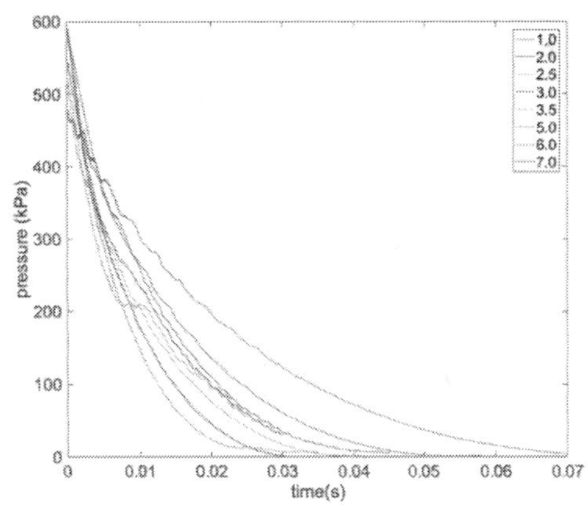

Fig. 6. Nitrogen pressure dependence on time during expansion from 600 kPa through various wide openings of throttling valve.

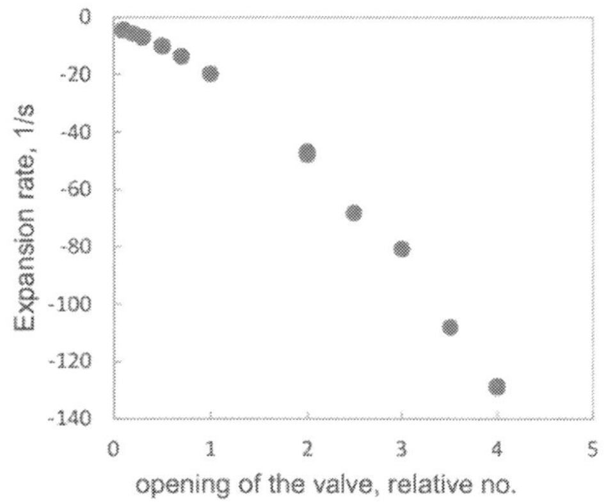

Fig. 7. Expansion rate dependence on opening of throttling valve for expansion from 600 kPa. Slope of the expansion rate evaluated from linear interpolation between 400 and 500 kPa.

The high-pressure system with the throttling needle valve with a special shape was tested for two initial pressures of 5 and 10 MPa.

Time dependences of pressure during the expansion from 5 MPa and 10 MPa via throttling needle valve with a special shape (5 in Figure 2) is shown in Figures 8 and 9, respectively. Opening of the valve was measured precisely with digital distance indicator. Distance around 6.5 mm corresponds to the maximum opening of the valve. Expansion rates were evaluated from a nearly linear slope of the pressure on time in the pressure ranges of 4.5 to 4 MPa and 9 to 8 MPa, respectively. Dependence of the evaluated expansion rates on opening of the throttling needle valve is shown in Figure 10.

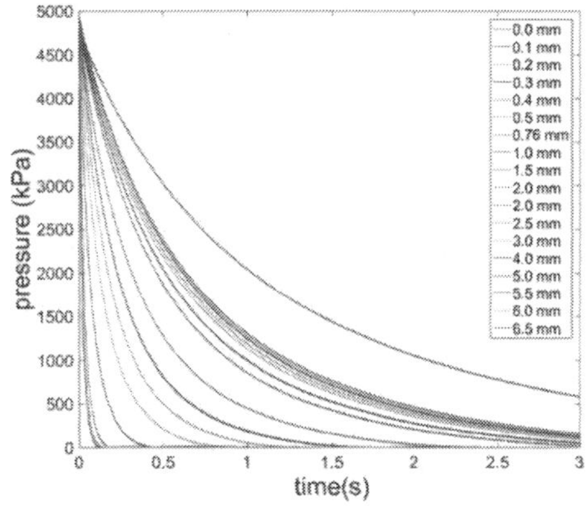

Fig. 8. Nitrogen pressure dependence on time during expansion from 5 MPa through various openings of throttling needle valve with a special shape.

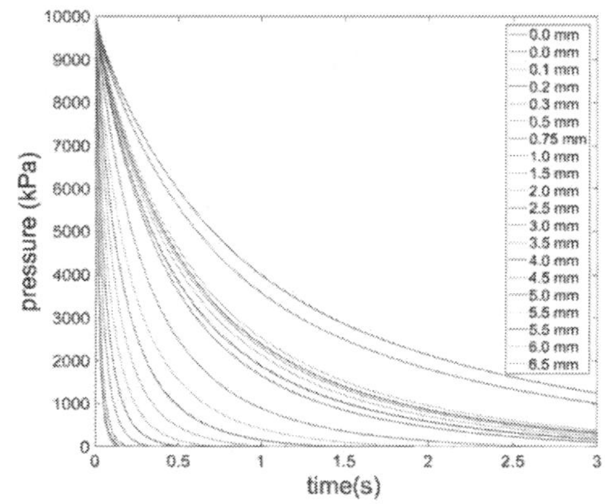

Fig. 9. Nitrogen pressure dependence on time during expansion from 10 MPa through various openings of throttling needle valve with a special shape.

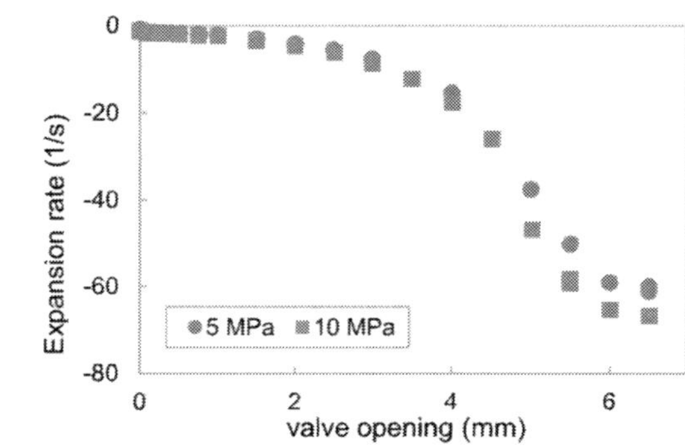

Fig. 10. Expansion rate dependence on opening of throttling needle valve with a special shape for expansion (item 5 in Figure 2) from 5 and 10 MPa. Slope of the expansion rate evaluated from linear interpolation between 9 and 8 MPa.

4 Conclusions

A droplet nucleation apparatus for CCS relevant systems is under construction in our laboratory. A newly designed expansion part of this apparatus was tested in this work. We used two arrangements for expansions in low and high pressure system using nitrogen as an expansion gas. The experiments showed that for the low-pressure system the expansion rate could be adjusted between 0.05 and 120 s^{-1} and for the high-pressure system the expansion rate could be adjusted between 0.87 and 67 s^{-1}. It has been shown that these ranges of expansion rates can be achieved in the whole range of initial pressures considered, i.e. from 0.2 to 10 MPa.

Acknowledgement

The research leading to these results has received funding from the Norwegian Financial Mechanism 2009–2014 under Project Contract No. 7F14466. Additional experiments have been performed in frame of project No. 16-02647S by the Czech Science

Foundation. Institutional support RVO:61388998 is also gratefully acknowledged.

References

1. V. Vinš, M. Čenský, J. Hykl, J. Hrubý: EPJ Web of Conferences **143**, 02140 (2017)
2. IPCC Special Report: Carbon Dioxide Capture and Storage – Summary for Policymakers. Intergovernmental Panel on Climate Change, Eighth Session of IPCC Working Group III, Montreal, (Canada) 22-24 September 2005
3. C.R.T. Wilson: Phil. Trans. R. Soc. A., London **189**, 265-307 (1897)
4. R.C. Miller, R.J. Anderson, J.L. Kassner, D.E. Hagen: J. Chem. Phys. **78**, 3204-3211 (1983)
5. J. Wölk, R. Strey: J. Phys. Chem. B, **105**, 11683-11701 (2001)
6. A. Fladerer, R. Strey: J. Chem. Phys. **124**, 164710 (2006)
7. H.J.S. Smolders: Ph.D Thesis, Eindhoven University of Technology, (1992).

Space-time discontinuous Galerkin method for the numerical simulation of viscous compressible gas flow with the k-omega turbulence model

Jan Česenek[1,a]

[1]VZLU - Czech Aerospace Research Center, Beranových 130, 199 05 Praha - Letňany, Czech Republic

Abstract. In this article we deal with the numerical simulation of the non-stationary compressible turbulent flow described by the Reynolds-Averaged Navier-Stokes (RANS) equations. This RANS system is equipped with two-equation k-omega turbulence model. The discretization of these two systems is carried out separately by the space-time discontinuous Galerkin method. This method is based on the piecewise polynomial discontinuous approximation of the sought solution in space and in time. We use the numerical experiments to demonstrate the applicability of the shown approach. All presented results were computed with the own-developed code.

1 Introduction

During the last decade the space-time discontinuous Galerkin finite element method (ST-DG), which is based on piecewise polynomial discontinuous approximations of the sought solution, became very popular in the field of numerical simulation of the fluid flow. This method of higher order was successfully used for the simulation of the Navier-Stokes equations [1–6]. In the case of compressible turbulent flow the finite volume - space-time discontinuous Galerkin method was used [7–9], where the equations of the turbulence model were discretized by the implicit finite volume method. This article is devoted to the discretization of viscous compressible turbulent gas flow using ST-DG applied also for the equations of the turbulence model. The flow is described by the system of the RANS equations equipped with the system of $k-\omega$ equations. These two systems are solved separately and discretized by ST-DG. In this article Wilcox $k-\omega$ turbulence model was chosen [10]. The flow around airfoil RAE2822 will be used for the numerical simulations. These results are compared with experiments. Computational results show that the ST-DG method is good approach for the solution of these types of problems.

2 Formulation of the $k-\omega$ turbulence model

We consider compressible turbulent flow in a bounded domain $\Omega \subset R^2$. We assume that the boundary Ω consists of three disjoint parts $\partial\Omega = \Gamma_I \cup \Gamma_O \cup \Gamma_W$, where Γ_I is the inlet, Γ_O is the outlet and Γ_W is impermeable wall.

The system of the RANS equations describing the viscous compressible turbulent flow can be written in the form

$$\frac{\partial \mathbf{w}}{\partial t} + \sum_{s=1}^{2} \frac{\partial \mathbf{f}_s(\mathbf{w})}{\partial x_s} + \sum_{s=1}^{2} \frac{\partial \mathbf{f}_s^p(\mathbf{w})}{\partial x_s}$$

$$= \sum_{s=1}^{2} \frac{\partial \mathbf{k}_s(\mathbf{w})}{\partial x_s} + \sum_{s=1}^{2} \frac{\partial \mathbf{R}_s(\mathbf{w}, \nabla \mathbf{w})}{\partial x_s}, \qquad (1)$$

where for $s = 1, 2$ we have

$$\begin{aligned}
\mathbf{w} &= (w_1, ..., w_4)^T = (\rho, \rho v_1, \rho v_2, E)^T \in R^4,\\
\mathbf{f}_s(\mathbf{w}) &= (f_{s,1}, ..., f_{s,4})^T\\
&= (\rho v_s, \rho v_1 v_s + \delta_{1s} p,\\
&\quad \rho v_2 v_s + \delta_{2s} p, (E+p) v_s)^T,\\
\mathbf{f}_s^p(\mathbf{w}) &= (f_{s,1}^p, ..., f_{s,4}^p)^T\\
&= (0, \tfrac{2}{3}\rho k \delta_{1s}, \tfrac{2}{3}\rho k \delta_{2s}, \tfrac{2}{3}\rho k v_s)^T,\\
\mathbf{R}_s(\mathbf{w}, \nabla \mathbf{w}) &= (R_{s,1}, ..., R_{s,4})^T\\
&= \left(0, \tau_{s1}^V, \tau_{s2}^V, \sum_{r=1}^{2} \tau_{sr}^V v_r + \left(\frac{c_p \mu_L}{Pr} + \frac{c_p \mu_T}{Pr_T}\right) \frac{\partial \theta}{\partial x_s}\right)^T,\\
\mathbf{k}_s(\mathbf{w}) &= (k_{s,1}, ..., k_{s,4})^T\\
&= \left(0, ..., 0, \left(\mu_L + \sigma_k \frac{\rho k}{e^{\tilde{\omega}}}\right) \frac{\partial k}{\partial x_s}\right)^T,
\end{aligned}$$

where

$$\begin{aligned}
\tau_{sr}^V &= -\frac{2}{3}(\mu_L + \mu_T)\, \mathrm{div} \boldsymbol{v}\, \delta_{sr} + 2(\mu_L + \mu_T)\, d_{sr}(\boldsymbol{v}),\\
d_{sr}(\boldsymbol{v}) &= \frac{1}{2}\left(\frac{\partial v_s}{\partial x_r} + \frac{\partial v_r}{\partial x_s}\right).
\end{aligned}$$

We use the following notation: $\boldsymbol{v} = (v_1, v_2)$ - velocity, ρ - density, p - pressure, θ - absolute temperature, E - total energy, γ - Poisson adiabatic constant, κ - heat conduction coefficient, c_v - specific heat at constant volume, c_p - specific heat at constant pressure, where $c_p = \gamma c_v$, Pr is the laminar Prandtl number, which can be express in the form $Pr = \frac{c_p \mu_L}{\kappa}$, Pr_T is the turbulent Prandtl constant number, μ_T is the eddy-viscosity coefficient, μ_L is the dynamic viscosity coefficient dependent on temperature via Sutherland's

[a] e-mail: jan.cessa@seznam.cz, cesenek@vzlu.cz

formula. The above system is completed by the thermodynamical relations

$$p = (\gamma - 1)\left(E - \frac{1}{2}\rho\,|\boldsymbol{v}|^2 - \rho k\right), \quad \theta = \frac{1}{c_v}\left(\frac{E}{\rho} - \frac{1}{2}\,|\boldsymbol{v}|^2 - k\right)$$

and is equipped with the initial condition

$$\mathbf{w}(x, 0) = \mathbf{w}^0(x), \quad x \in \Omega$$

and the following boundary conditions

a) $\rho|_{\Gamma_I} = \rho_D$,
b) $\boldsymbol{v}|_{\Gamma_I} = \boldsymbol{v}_D = (v_{D1}, v_{D2})^T$,
c) $\displaystyle\sum_{s,r=1}^{2} \tau^V_{sr} n_s v_r + \left(\frac{c_p\mu_L}{Pr} + \frac{c_p\mu_T}{Pr_T}\right)\frac{\partial\theta}{\partial n} = 0 \quad \text{on } \Gamma_I,$
a) $\boldsymbol{v}|_{\Gamma_W} = 0$,
b) $\displaystyle\left.\frac{\partial\theta}{\partial n}\right|_{\Gamma_W} = 0$,
a) $\displaystyle\sum_{s=1}^{2} \tau^V_{sr} n_s = 0,\ r = 1, 2 \quad \text{on } \Gamma_O,$ b) $\displaystyle\left.\frac{\partial\theta}{\partial n}\right|_{\Gamma_O} = 0,$

with given data $\mathbf{w}^0, \rho_D, \boldsymbol{v}_D$. It is possible to show that $\mathbf{f}_s(\alpha\mathbf{w}) = \alpha\,\mathbf{f}_s(\mathbf{w})$ for $\alpha > 0$. This property implies that

$$\mathbf{f}_s(\mathbf{w}) = \mathbf{A}_s(\mathbf{w})\mathbf{w}, \quad s = 1, 2, \tag{2}$$

where

$$\mathbf{A}_s(\mathbf{w}) = \frac{D\mathbf{f}_s(\mathbf{w})}{D\mathbf{w}}, \quad s = 1, 2,$$

are the Jacobi matrices of the mappings $\mathbf{f}_s$. Similary we can express

$$\mathbf{f}^P_s(\mathbf{w}) = \mathbf{A}^P_s(\mathbf{w})\mathbf{w}, \quad s = 1, 2.$$

The viscous terms $\mathbf{R}_s(\mathbf{w}, \nabla\mathbf{w})$ can be expressed in the form

$$\mathbf{R}_s(\mathbf{w}, \nabla\mathbf{w}) = \sum_{k=1}^{2} \mathbf{K}_{s,k}(\mathbf{w})\frac{\partial\mathbf{w}}{\partial x_k}, \quad s = 1, 2, \tag{3}$$

where $\mathbf{K}_{s,k}(\mathbf{w}) \in R^{4\times4}$ are matrices depending on $\mathbf{w}$.

The aforesaid system of the RANS equations is completed by the Wilcox's turbulence model (see [10]) for μ_T. For the sake of the stability we introduce new variable $\tilde{\omega} = \ln\omega$. (For details see [11]). Then the turbulence model reads

$$\frac{\partial\rho\tilde{\mathbf{w}}}{\partial t} + \sum_{s=1}^{2}\frac{\partial\tilde{\mathbf{f}}_s(\tilde{\mathbf{w}})}{\partial x_s} = \sum_{s=1}^{2}\frac{\partial\tilde{\mathbf{R}}_s(\tilde{\mathbf{w}}, \nabla\tilde{\mathbf{w}})}{\partial x_s} + \tilde{\mathbf{s}}(\tilde{\mathbf{w}}), \tag{4}$$

where for $s = 1, 2$ we have

$$\begin{aligned}
\tilde{\mathbf{w}} &= (\tilde{w}_1, \tilde{w}_2)^T = (\tilde{\omega}, k)^T \in R^2,\\
\tilde{\mathbf{f}}_s(\tilde{\mathbf{w}}) &= (\tilde{f}_{s,1}, \tilde{f}_{s,2})^T = (\rho\tilde{\omega}v_s, \rho k v_s)^T,\\
\tilde{\mathbf{R}}_s(\tilde{\mathbf{w}}, \nabla\tilde{\mathbf{w}}) &= (\tilde{R}_{s,1}, \tilde{R}_{s,2})^T =\\
&\left((\mu_L + \sigma_\omega\mu_T)\frac{\partial\tilde{\omega}}{\partial x_s}, (\mu_L + \sigma_k\mu_T)\frac{\partial k}{\partial x_s}\right)^T,\\
\tilde{\mathbf{s}}(\tilde{\mathbf{w}}) &= (\alpha_\omega\frac{P_k}{k} - \beta\rho e^{\tilde{\omega}} + \sum_{s=1}^{2}(\mu_L + \sigma_\omega\mu_T)\frac{\partial\tilde{\omega}}{\partial x_s}\frac{\partial\tilde{\omega}}{\partial x_s} + \tilde{C}_D,\\
&P_k - \beta^*\rho e^{\tilde{\omega}}k)^T.
\end{aligned}$$

Here ω is the turbulence dissipation, k is the turbulence kinetic energy and μ_T is the eddy viscosity. We can write the production term as

$$P_k = \sum_{s=1}^{2}\sum_{r=1}^{2}\tau^T_{sr}\frac{\partial v_s}{\partial x_r},$$

where

$$\begin{aligned}
\tau^T_{sr} &= -\frac{2}{3}\mu_T\,\mathrm{div}\boldsymbol{v}\,\delta_{sr} - \frac{2}{3}\rho k\,\delta_{sr} + 2\mu_T\,d_{sr}(\boldsymbol{v}),\\
d_{sr}(\boldsymbol{v}) &= \frac{1}{2}\left(\frac{\partial v_s}{\partial x_r} + \frac{\partial v_r}{\partial x_s}\right),\\
\mu_T &= \frac{\rho k}{\omega_{lim}}.
\end{aligned}$$

Limited eddy viscosity ω_{lim} is given by the term

$$\begin{aligned}
\omega_{lim} &= \max\left\{e^{\tilde{\omega}}, C_{lim}\sqrt{\frac{1}{2\beta^*}\sum_{r,s=1}^{2}\tilde{\tau}_{rs}\tilde{\tau}_{rs}}\right\},\\
\tilde{\tau}_{rs} &= -\frac{2}{3}\mathrm{div}\boldsymbol{v}\,\delta_{rs} + 2d_{rs}(\boldsymbol{v}).
\end{aligned}$$

The cross-diffusion term C_D is defined as

$$C_D = \sigma_D\frac{\rho}{e^{\tilde{\omega}}}\max\left\{\sum_{s=1}^{2}\frac{\partial k}{\partial x_s}\frac{\partial\tilde{\omega}}{\partial x_s}, 0\right\}.$$

The coefficients $\beta, \beta^*, \sigma_k, \sigma_\omega, \sigma_D, \alpha_\omega, C_{lim}, Pr_T$ are chosen by [10]:

$$\begin{aligned}
&\alpha_\omega = \frac{13}{25}, \quad \beta = 0.0708, \quad \beta^* = 0.09, \quad \sigma_k = 0.6, \quad \sigma_\omega = 0.5,\\
&C_{lim} = \frac{7}{8}, \quad \sigma_D = \frac{1}{8}, \quad Pr_T = \frac{8}{9}.
\end{aligned}$$

This system is also equipped with the initial condition

$$\tilde{\mathbf{w}}(x, 0) = \tilde{\mathbf{w}}^0(x), \quad x \in \Omega$$

and the following boundary conditions

a) $\tilde{\omega}|_{\Gamma_I} = \tilde{\omega}_D$, b) $k|_{\Gamma_I} = k_D$,
a) $\tilde{\omega}|_{\Gamma_W} = \tilde{\omega}_{wall}$, b) $k|_{\Gamma_W} = 0$,
a) $\displaystyle\left.\frac{\partial\tilde{\omega}}{\partial n}\right|_{\Gamma_O} = 0$, b) $\displaystyle\left.\frac{\partial k}{\partial n}\right|_{\Gamma_O} = 0$,

with given data $\tilde{\mathbf{w}}^0, \tilde{\omega}_D, k_D, \tilde{\omega}_{wall}$. Similary like in the RANS case we can express convect terms

$$\tilde{\mathbf{f}}^P_s(\mathbf{w}) = \tilde{\mathbf{A}}^P_s(\tilde{\mathbf{w}})\tilde{\mathbf{w}}, \quad s = 1, 2 \tag{5}$$

and diffusion terms

$$\tilde{\mathbf{R}}_s(\tilde{\mathbf{w}}, \nabla\tilde{\mathbf{w}}) = \tilde{\mathbf{K}}_s(\tilde{\mathbf{w}})\frac{\partial\tilde{\mathbf{w}}}{\partial x_s}, \quad s = 1, 2, \tag{6}$$

where $\tilde{\mathbf{K}}_s(\mathbf{w}) \in R^{2\times2}$ are matrices depending on $\tilde{\mathbf{w}}$.

 https://doi.org/10.1051/epjconf/201818002016

3 Discretization

3.1 Space discretization of the problem

By Ω_h we denote polygonal approximation of the domain Ω. Let $\mathcal{T}_h$ be a partition of the domain Ω_h into finite number of closed elements with mutually disjoint interiors such that $\overline{\Omega_h} = \bigcup_{K\in\mathcal{T}_h} K$. In 2D problems, we usually choose $K \in \mathcal{T}_h$ as triangles or quadrilaterals. By $\mathcal{F}_h$ we denote the system of all faces of all elements $K \in \mathcal{T}_h$. Further, we introduce the set of boundary faces $\mathcal{F}_h^B = \{\Gamma \in \mathcal{F}_h; \Gamma \subset \partial\Omega_h\}$ and the set of inner faces $\mathcal{F}_h^I = \mathcal{F}_h \backslash \mathcal{F}_h^B$. Each $\Gamma \in \mathcal{F}_h$ is associated with a unit normal vector $\boldsymbol{n}_\Gamma$. For $\Gamma \in \mathcal{F}_h^B$ the normal $\boldsymbol{n}_\Gamma$ has the same orientation as the outer normal to $\partial\Omega_h$. For each $\Gamma \in \mathcal{F}_h^I$ there exist two neighbouring elements $K_\Gamma^L, K_\Gamma^R \in \mathcal{T}_h$ such that $\Gamma \subset \partial K_\Gamma^L \cap \partial K_\Gamma^R$. We use the convention that K_Γ^R lies in the direction of $\boldsymbol{n}_\Gamma$ and K_Γ^L lies in the opposite direction to $\boldsymbol{n}_\Gamma$. If $\Gamma \in \mathcal{F}_h^B$, then the element adjacent to Γ will be denoted by K_Γ^L.

We shall look for an approximate solution of the problem in the space of piecewise polynomial functions

$$\mathbf{S}_h^p = (S_h^p)^4, \qquad S_h^p = \{v;\, v|_K \in P^p(K), \forall K \in \mathcal{T}_h\},$$

where $p > 0$ is an integer and $P^p(K)$ denotes the space of all polynomials on K of degree $\leq p$. A function $\Phi \in \mathbf{S}_h^p$ is, in general, discontinuous on interfaces $\Gamma \in \mathcal{F}_h^I$. By Φ_Γ^L and Φ_Γ^R we denote the values of Φ on Γ considered from the interior and the exterior of K_Γ^L, respectively, and set

$$\langle \Phi \rangle_\Gamma = \frac{1}{2}\left(\Phi_\Gamma^L + \Phi_\Gamma^R\right),$$
$$[\Phi]_\Gamma = \Phi_\Gamma^L - \Phi_\Gamma^R,$$

which denotes the average and jump of Φ on Γ.

The discrete problem is derived in the following way: For arbitrary $t \in [0, T]$ we can multiply the system by a test function $\mathbf{S}_h^p$, integrate over $K \in \mathcal{T}_h$, apply Green's theorem, sum over all elements $K \in \mathcal{T}_h$, use the concept of the numerical flux and introduce suitable terms mutually vanishing for a regular exact solution. Moreover, we carry out a suitable partial linearization of nonlinear terms and then we can define the following forms.

In order to evaluate the integrals over $\Gamma \in \mathcal{F}_h$ in inviscid term we use the approximation

$$\mathbf{H}(\mathbf{w}_\Gamma^L, \mathbf{w}_\Gamma^R, \boldsymbol{n}_\Gamma) \approx \sum_{s=1}^{2} \mathbf{f}_s(\mathbf{w})(\boldsymbol{n}_\Gamma)_s,$$

where $\mathbf{H}$ is a numerical flux. For the construction of the numerical flux we use the properties (2) of $\mathbf{f}_s$. Let us define the matrix

$$\mathbf{P}(\mathbf{w}, \boldsymbol{n}) := \sum_{s=1}^{2} \mathbf{A}_s(\mathbf{w})n_s,$$

where $\boldsymbol{n} = (n_1, n_2)$, $n_1^2 + n_2^2 = 1$. Then we have

$$\mathbf{P}(\mathbf{w}, \boldsymbol{n})\mathbf{w} := \sum_{s=1}^{2} \mathbf{f}_s(\mathbf{w})n_s.$$

It is possible to show that the matrix $\mathbf{P}$ is diagonalizable. It means that there exists a nonsingular matrix $\mathbf{T} = \mathbf{T}(\mathbf{w}, \boldsymbol{n})$ and a diagonal matrix $\Lambda = \Lambda(\mathbf{w}, \boldsymbol{n})$ such that

$$\mathbf{P} = \mathbf{T}\Lambda\mathbf{T}^{-1}, \quad \Lambda = \operatorname{diag}(\lambda_1, ..., \lambda_4),$$

where $\lambda_i = \lambda_i(\mathbf{w}, \boldsymbol{n})$ are eigenvalues of the matrix $\mathbf{P}$. Then we can define the "positive" and "negative" parts of the matrix $\mathbf{P}$ by

$$\mathbf{P}^\pm = \mathbf{T}\Lambda^\pm\mathbf{T}^{-1}, \quad \Lambda^\pm = \operatorname{diag}(\lambda_1^\pm, ..., \lambda_4^\pm),$$

where $\lambda^+ = \max(\lambda, 0)$, $\lambda^- = \min(\lambda, 0)$. Using this concept, we introduce the so-called Vijayasundaram numerical flux

$$\mathbf{H}(\mathbf{w}^L, \mathbf{w}^R, \boldsymbol{n}) = \mathbf{P}^+\left(\frac{\mathbf{w}^L + \mathbf{w}^R}{2}, \boldsymbol{n}\right)\mathbf{w}^L + \mathbf{P}^-\left(\frac{\mathbf{w}^L + \mathbf{w}^R}{2}, \boldsymbol{n}\right)\mathbf{w}^R.$$

This numerical flux has suitable form for a linearization. Now we can define inviscid form in the following way

$$\begin{aligned} b_h(\bar{\mathbf{w}}_h, \mathbf{w}_h, \Phi_h) :=& \\ -\sum_{K\in\mathcal{T}_h} \int_K &\sum_{s=1}^{2} \mathbf{A}_s(\bar{\mathbf{w}}_h)\mathbf{w}_h \cdot \frac{\partial \Phi_h}{\partial x_s}\, \mathrm{d}x \\ +\sum_{\Gamma\in\mathcal{F}_h^I} \int_\Gamma &\left(\mathbf{P}^+\left(\langle\bar{\mathbf{w}}_h\rangle, \boldsymbol{n}_\Gamma\right)\mathbf{w}_h^L + \mathbf{P}^-\left(\langle\bar{\mathbf{w}}_h\rangle, \boldsymbol{n}_\Gamma\right)\mathbf{w}_h^R\right) \cdot [\Phi_h]\, \mathrm{d}S \\ +\sum_{\Gamma\in\mathcal{F}_h^B} \int_\Gamma &\left(\mathbf{P}^+\left(\langle\bar{\mathbf{w}}_h\rangle, \boldsymbol{n}_\Gamma\right)\mathbf{w}_h^L + \mathbf{P}^-\left(\langle\bar{\mathbf{w}}_h\rangle, \boldsymbol{n}_\Gamma\right)\bar{\mathbf{w}}_h^R\right) \cdot \Phi_h\, \mathrm{d}S, \end{aligned}$$

where the boundary state $\bar{\mathbf{w}}_h^R$ is evaluated with the aid of the local linearized Riemann problem described in [1]. For the discretization of the viscous terms we use the property (3) and get the viscous form

$$\begin{aligned} a_h(\bar{\mathbf{w}}_h, \mathbf{w}_h, \Phi_h) :=& \\ +\sum_{K\in\mathcal{T}_h} \int_K &\sum_{s=1}^{2}\sum_{k=1}^{2} \mathbf{K}_{s,k}(\bar{\mathbf{w}}_h)\frac{\partial \mathbf{w}_h}{\partial x_k} \cdot \frac{\partial \Phi_h}{\partial x_s}\, \mathrm{d}x \\ -\sum_{\Gamma\in\mathcal{F}_h^I} \int_\Gamma &\sum_{s=1}^{2}\left\langle\sum_{k=1}^{2} \mathbf{K}_{s,k}(\bar{\mathbf{w}}_h)\frac{\partial \mathbf{w}_h}{\partial x_k}\right\rangle(\boldsymbol{n}_\Gamma)_s \cdot [\Phi_h]\, \mathrm{d}S \\ -\sum_{\Gamma\in\mathcal{F}_h^B} \int_\Gamma &\sum_{s=1}^{2}\sum_{k=1}^{2} \mathbf{K}_{s,k}(\bar{\mathbf{w}}_h)\frac{\partial \mathbf{w}_h}{\partial x_k}(\boldsymbol{n}_\Gamma)_s \cdot \Phi_h\, \mathrm{d}S \\ -\Theta\sum_{\Gamma\in\mathcal{F}_h^I} \int_\Gamma &\sum_{s=1}^{2}\left\langle\sum_{k=1}^{2} \mathbf{K}_{k,s}^T(\bar{\mathbf{w}}_h)\frac{\partial \Phi_h}{\partial x_k}\right\rangle(\boldsymbol{n}_\Gamma)_s \cdot [\mathbf{w}_h]\, \mathrm{d}S \\ -\Theta\sum_{\Gamma\in\mathcal{F}_h^B} \int_\Gamma &\sum_{s=1}^{2}\sum_{k=1}^{2} \mathbf{K}_{k,s}^T(\bar{\mathbf{w}}_h)\frac{\partial \Phi_h}{\partial x_k}(\boldsymbol{n}_\Gamma)_s \cdot \mathbf{w}_h\, \mathrm{d}S \end{aligned}$$

We set $\Theta = 1$ or $\Theta = 0$ or $\Theta = -1$ and get the so-called symmetric version (SIPG) or incomplete version (IIPG) or nonsymetric version (NIPG), respectively, of the discretization of the viscous terms.

Further, we define the turbulent forms p_h, k_h, the interior and boundary penalty form J_h^σ and the right-hand side form l_h in the following way:

$$p_h(\bar{\mathbf{w}}_h, \mathbf{w}_h, \Phi_h) := \sum_{\Gamma\in\mathcal{F}_h^I} \int_\Gamma \sum_{s=1}^{2} \mathbf{A}_s^P(\bar{\mathbf{w}}_h^L)\mathbf{w}_h^L \cdot \Phi_h^L(\boldsymbol{n}_\Gamma)_s\, \mathrm{d}S$$

$$- \sum_{\Gamma \in \mathcal{F}_h^I} \int_\Gamma \sum_{s=1}^2 \mathbf{A}_s^P(\bar{\mathbf{w}}_h^R)\mathbf{w}_h^R \cdot \Phi_h^R \, (\boldsymbol{n}_\Gamma)_s \, \mathrm{d}S$$
$$+ \sum_{\Gamma \in \mathcal{F}_h^B} \int_\Gamma \sum_{s=1}^2 \mathbf{A}_s^P(\bar{\mathbf{w}}_h)\mathbf{w}_h \cdot \Phi_h \, (\boldsymbol{n}_\Gamma)_s \, \mathrm{d}S$$
$$- \sum_{K \in \mathcal{T}_h} \int_K \sum_{s=1}^2 \mathbf{A}_s^P(\bar{\mathbf{w}}_h)\mathbf{w}_h \cdot \frac{\partial \Phi_h}{\partial x_s} \, \mathrm{d}\boldsymbol{x},$$

$$k_h(\bar{\mathbf{w}}_h, \Phi_h) := \sum_{\Gamma \in \mathcal{F}_h^I} \int_\Gamma \sum_{s=1}^2 \mathbf{k}_s(\bar{\mathbf{w}}_h^L) \cdot \Phi_h^L \, (\boldsymbol{n}_\Gamma)_s \, \mathrm{d}S$$
$$- \sum_{\Gamma \in \mathcal{F}_h^I} \int_\Gamma \sum_{s=1}^2 \mathbf{k}_s(\bar{\mathbf{w}}_h^R) \cdot \Phi_h^R \, (\boldsymbol{n}_\Gamma)_s \, \mathrm{d}S$$
$$+ \sum_{\Gamma \in \mathcal{F}_h^B} \int_\Gamma \sum_{s=1}^2 \mathbf{k}_s(\bar{\mathbf{w}}_h) \cdot \Phi_h \, (\boldsymbol{n}_\Gamma)_s \, \mathrm{d}S$$
$$- \sum_{K \in \mathcal{T}_h} \int_K \sum_{s=1}^2 \mathbf{k}_s(\bar{\mathbf{w}}_h) \cdot \frac{\partial \Phi_h}{\partial x_s} \, \mathrm{d}\boldsymbol{x},$$

$$J_h^\sigma(\mathbf{w}_h, \Phi_h) := \sum_{\Gamma \in \mathcal{F}_h^I} \int_\Gamma \sigma[\mathbf{w}_h] \cdot [\Phi_h] \, \mathrm{d}S$$
$$+ \sum_{\Gamma \in \mathcal{F}_h^B} \int_\Gamma \sigma \mathbf{w}_h \cdot \Phi_h \, \mathrm{d}S,$$
$$l_h(\bar{\mathbf{w}}_h, \Phi_h) := \sum_{\Gamma \in \mathcal{F}_h^B} \int_\Gamma \sigma \mathbf{w}_B \cdot \Phi_h \, \mathrm{d}S$$
$$- \Theta \sum_{\Gamma \in \mathcal{F}_h^B} \int_\Gamma \sum_{s=1}^2 \sum_{k=1}^2 \mathbf{K}_{k,s}^T(\bar{\mathbf{w}}_h) \frac{\partial \Phi_h}{\partial x_k} (\boldsymbol{n}_\Gamma)_s \cdot \mathbf{w}_B \, \mathrm{d}S,$$

where σ is a parameter of the method and boundary state $\mathbf{w}_B$ is defined on the basis of the Dirichlet boundary conditions and extrapolation.

In the vicinity of discontinuities or steep gradients non-physical oscillations can appear in the approximate solution. In order to overcome this difficulty we employ artificial viscosity forms, see [3]. They are based on the discontinuity indicator

$$g(K) := \frac{1}{d(K)\,|K|^{3/4}} \int_{\partial K} [\bar{\rho}_h]^2 dS, \quad K \in \mathcal{T}_h,$$

where $[\bar{\rho}_h]$ is the jump of the function $\bar{\rho}_h$ (= the first component of the vector function $\bar{\mathbf{w}}_h$) on the boundary ∂K, $d(K)$ denotes the diameter of K and $|K|$ denotes the area of the element K. Then we define the discrete discontinuity indicator

$$G(K) := 0 \quad \text{if } g(K) < 1,$$
$$G(K) := 1 \quad \text{if } g(K) \geq 1, \; K \in \mathcal{T}_h,$$

and the artificial viscosity forms

$$\tilde{\beta}_h(\bar{\mathbf{w}}_h, \mathbf{w}_h, \Phi_h) :=$$
$$\nu_1 \sum_{K \in \mathcal{T}_h} d(K)G(K) \int_K \nabla \mathbf{w}_h \cdot \nabla \Phi_h \, \mathrm{d}\boldsymbol{x},$$

$$\tilde{J}_h(\bar{\mathbf{w}}_h, \mathbf{w}_h, \Phi_h) :=$$
$$\nu_2 \sum_{\Gamma \in \mathcal{F}_h^I} \frac{1}{2}\left(G(K_\Gamma^L) + G(K_\Gamma^R)\right) \int_\Gamma [\mathbf{w}_h] \cdot [\Phi_h] \, \mathrm{d}S,$$

with constants ν_1 and ν_2.

All these forms are linear with respect to $\mathbf{w}_h$ and nonlinear with respect to $\bar{\mathbf{w}}_h$.

Finally, we set

$$(\varphi, \psi) = \int_{\Omega_h} \varphi \psi \, \mathrm{d}x.$$

3.2 Full space-time DG discretization

Let $0 = t_0 < t_1 < ... < t_M = T$ be a partition of the interval $[0, T]$ and let us denote $I_m = (t_{m-1}, t_m)$, $\tau_m = t_m - t_{m-1}$ for $m = 1, ..., M$. We define the space $\mathbf{S}_{h,\tau}^{p,q} = (S_{h,\tau}^{p,q})^4$, where

$$S_{h,\tau}^{p,q} := \left\{ \phi \, ; \, \phi|_{I_m} = \sum_{i=0}^q \zeta_i \phi_i, \text{ where } \phi_i \in S_h^p, \; \zeta_i \in P^q(I_m) \right\}.$$

with integers $p, q \geq 1$. $P^q(I_m)$ denotes the space of all polynomials in t on I_m of degree $\leq q$. Moreover for $\Phi \in \mathbf{S}_{h,\tau}^{p,q}$ we introduce the following notation:

$$\Phi_m^\pm = \Phi(t_m^\pm) = \lim_{t \to t_{m\pm}} \Phi(t),$$
$$\{\Phi\}_m = \Phi_m^+ - \Phi_m^-.$$

Approximate solution $\mathbf{w}_{h\tau}$ of the problem will be sought in the space $\mathbf{S}_{h,\tau}^{p,q}$. Since the functions of this space are in general discontinuous in time, we ensure the connection between I_{m-1} and I_m by the penalty term in time

$$\left(\{\mathbf{w}_{h\tau}\}_{m-1}, \Phi_{h\tau}(t_{m-1}^+)\right).$$

The initial state $\mathbf{w}_{h\tau}$ is included by the $L^2(\Omega_h(t_0))$-projection of $\mathbf{w}^0$ on $\mathbf{S}_h^p(t_0)$:

$$\left(\mathbf{w}_{h\tau}(t_0^+), \Phi_{h\tau}(t_0^+)\right) = \left(\mathbf{w}^0, \Phi_{h\tau}(t_0^+)\right) \quad \forall \Phi_{h\tau} \in \mathbf{S}_{h,\tau}^{p,q}.$$

Now we introduce a suitable linearization. We can use two possibilities.

1) We put $\bar{\mathbf{w}}_{h\tau}(t) := \mathbf{w}_h(t_{m-1}^-)$ for $t \in I_m$.

2) We prolong the solution from the time interval I_{m-1} to the time interval I_m.

We say that a function $\mathbf{w}_{h\tau} \in \mathbf{S}_{h,\tau}^{p,q}$ is the approximate solution of the problem (1) obtained by the ST-DG method, if it satisfies the following conditions

$$\sum_{m=1}^M \int_{I_m} \left(\frac{\partial \mathbf{w}_{h\tau}}{\partial t}, \Phi_{h\tau} \right) \mathrm{d}t$$
$$+ \sum_{m=1}^M \int_{I_m} (a_h(\bar{\mathbf{w}}_{h\tau}, \mathbf{w}_{h\tau}, \Phi_{h\tau}) + b_h(\bar{\mathbf{w}}_{h\tau}, \mathbf{w}_{h\tau}, \Phi_{h\tau})) \, \mathrm{d}t$$
$$+ \sum_{m=1}^M \int_{I_m} \left(\tilde{\beta}_h(\bar{\mathbf{w}}_{h\tau}, \mathbf{w}_{h\tau}, \Phi_{h\tau}) + \tilde{J}_h(\bar{\mathbf{w}}_{h\tau}, \mathbf{w}_{h\tau}, \Phi_{h\tau})\right) \mathrm{d}t$$
$$+ \sum_{m=1}^M \int_{I_m} (J_h^\sigma(\mathbf{w}_{h\tau}, \Phi_{h\tau}) + p_h(\bar{\mathbf{w}}_{h\tau}, \mathbf{w}_{h\tau}, \Phi_{h\tau})) \, \mathrm{d}t$$

$$+\sum_{m=2}^{M}\Big(\{\mathbf{w}_{h\tau}\}_{m-1},\Phi_{h\tau}(t_{m-1}^{+})\Big)+\Big(\mathbf{w}_{h\tau}(t_0^{+}),\Phi_{h\tau}(t_0^{+})\Big)$$

$$=\sum_{m=1}^{M}\int_{I_m} l_h(\bar{\mathbf{w}}_{h\tau},\Phi_{h\tau})\,\mathrm{d}t+\Big(\mathbf{w}^0,\Phi_{h\tau}(t_0^{+})\Big)$$

$$+\sum_{m=1}^{M}\int_{I_m} k_h(\bar{\mathbf{w}}_{h\tau},\Phi_{h\tau})\,\mathrm{d}t\quad\forall\Phi_{h\tau}\in\mathbf{S}_{h,\tau}^{p,q}.$$

4 Discretization of the $k-\omega$ turbulence model

4.1 Space discretization of the problem

We apply discontinuous Galerkin method in the similar way as in the previous section. Discretization is carried out on the same mesh for simplicity. An approximate solution of the problem is looked for in the space $\tilde{\mathbf{S}}_h^{\tilde{p}}=(S_h^{\tilde{p}})^2$ and a function $\tilde{\Phi}\in\tilde{\mathbf{S}}_h^{\tilde{p}}$ is, in general, discontinuous on interfaces $\Gamma\in\mathcal{F}_h^I$.

If we define $\tilde{\mathbf{P}}^{\pm}=(\langle\rho\mathbf{v}\rangle\cdot\mathbf{n}_\Gamma)^{\pm}\mathbf{I}$, where $\mathbf{I}=\mathrm{diag}(1,1)$, and use the property (5) then we can define convect form

$$\tilde{b}_h(\bar{\mathbf{w}}_h,\tilde{\mathbf{w}}_h,\tilde{\Phi}_h):=$$

$$-\sum_{K\in\mathcal{T}_h}\int_K\sum_{s=1}^{2}\tilde{\mathbf{A}}_s\tilde{\mathbf{w}}_h\cdot\frac{\partial\tilde{\Phi}_h}{\partial x_s}\,\mathrm{d}\boldsymbol{x}$$

$$+\sum_{\Gamma\in\mathcal{F}_h^I}\int_\Gamma\Big(\tilde{\mathbf{P}}^{+}\,\tilde{\mathbf{w}}_h^L+\tilde{\mathbf{P}}^{-}\,\tilde{\mathbf{w}}_h^R\Big)\cdot[\tilde{\Phi}_h]\,\mathrm{d}S$$

$$+\sum_{\Gamma\in\mathcal{F}_h^B}\int_\Gamma\Big(\tilde{\mathbf{P}}^{+}\,\tilde{\mathbf{w}}_h^L+\tilde{\mathbf{P}}^{-}\,\bar{\tilde{\mathbf{w}}}_h^R\Big)\cdot\tilde{\Phi}_h\,\mathrm{d}S,$$

where the state $\bar{\tilde{\mathbf{w}}}_h^R$ is based on the boundary conditions.

For the discretization of the diffusion term we use the property (6) and then we have

$$\tilde{a}_h(\bar{\tilde{\mathbf{w}}}_h,\tilde{\mathbf{w}}_h,\tilde{\Phi}_h):=$$

$$+\sum_{K\in\mathcal{T}_h}\int_K\sum_{s=1}^{2}\tilde{\mathbf{K}}_s(\bar{\tilde{\mathbf{w}}}_h)\,\frac{\partial\tilde{\mathbf{w}}_h}{\partial x_s}\cdot\frac{\partial\tilde{\Phi}_h}{\partial x_s}\,\mathrm{d}\boldsymbol{x}$$

$$-\sum_{\Gamma\in\mathcal{F}_h^I}\int_\Gamma\sum_{s=1}^{2}\left\langle\tilde{\mathbf{K}}_s(\bar{\tilde{\mathbf{w}}}_h)\frac{\partial\tilde{\mathbf{w}}_h}{\partial x_s}\right\rangle(\boldsymbol{n}_\Gamma)_s\cdot[\tilde{\Phi}_h]\,\mathrm{d}S$$

$$-\sum_{\Gamma\in\mathcal{F}_h^B}\int_\Gamma\sum_{s=1}^{2}\tilde{\mathbf{K}}_s(\bar{\tilde{\mathbf{w}}}_h)\frac{\partial\tilde{\mathbf{w}}_h}{\partial x_s}(\boldsymbol{n}_\Gamma)_s\cdot\tilde{\Phi}_h\,\mathrm{d}S$$

$$-\Theta\sum_{\Gamma\in\mathcal{F}_h^I}\int_\Gamma\sum_{s=1}^{2}\left\langle\tilde{\mathbf{K}}_s(\bar{\tilde{\mathbf{w}}}_h)\frac{\partial\tilde{\Phi}_h}{\partial x_s}\right\rangle(\boldsymbol{n}_\Gamma)_s\cdot[\tilde{\mathbf{w}}_h]\,\mathrm{d}S$$

$$-\Theta\sum_{\Gamma\in\mathcal{F}_h^B}\int_\Gamma\sum_{s=1}^{2}\tilde{\mathbf{K}}_s(\bar{\tilde{\mathbf{w}}}_h)\frac{\partial\tilde{\Phi}_h}{\partial x_s}(\boldsymbol{n}_\Gamma)_s\cdot\tilde{\mathbf{w}}_h\,\mathrm{d}S.$$

Further we define the turbulent form $\tilde{s}_h$, the interior and boundary penalty form $\tilde{J}_h^{\tilde{\sigma}}$ and the right-hand side form $\tilde{l}_h$ in the following way:

$$\tilde{J}_h^{\tilde{\sigma}}(\tilde{\mathbf{w}}_h,\tilde{\Phi}_h):=\sum_{\Gamma\in\mathcal{F}_h^I}\int_\Gamma\tilde{\sigma}[\tilde{\mathbf{w}}_h]\cdot[\tilde{\Phi}_h]\,\mathrm{d}S$$

$$+\sum_{\Gamma\in\mathcal{F}_h^B}\int_\Gamma\tilde{\sigma}\tilde{\mathbf{w}}_h\cdot\tilde{\Phi}_h\,\mathrm{d}S,$$

$$\tilde{l}_h(\bar{\tilde{\mathbf{w}}}_h,\tilde{\Phi}_h):=\sum_{\Gamma\in\mathcal{F}_h^B}\int_\Gamma\tilde{\sigma}\tilde{\mathbf{w}}_B\cdot\tilde{\Phi}_h\,\mathrm{d}S$$

$$-\Theta\sum_{\Gamma\in\mathcal{F}_h^B}\int_\Gamma\sum_{s=1}^{2}\tilde{\mathbf{K}}_s(\bar{\tilde{\mathbf{w}}}_h)\frac{\partial\tilde{\Phi}_h}{\partial x_s}(\boldsymbol{n}_\Gamma)_s\cdot\tilde{\mathbf{w}}_B\,\mathrm{d}S.$$

$$\tilde{s}_h(\bar{\tilde{\mathbf{w}}}_h,\tilde{\Phi}_h):=\sum_{K\in\mathcal{T}_h}\int_K\tilde{s}(\bar{\tilde{\mathbf{w}}}_h)\cdot\tilde{\Phi}_h\,\mathrm{d}\boldsymbol{x}$$

Boundary state $\tilde{\mathbf{w}}_B$ is defined on the basis of the Dirichlet boundary conditions and extrapolation.

4.2 Full space-time DG discretization

For the full space-time discontinuous Galerkin discretization we use the same partition of the interval $[0,T]$ as in the previous section. We say that a function $\tilde{\mathbf{w}}_{h\tau}\in\tilde{\mathbf{S}}_{h,\tau}^{\tilde{p},\tilde{q}}$ is the approximate solution of the problem (4) obtained by the ST-DG method, if it satisfies the following conditions

$$\sum_{m=0}^{M-1}\int_{I_m}\left(\frac{\partial\rho\tilde{\mathbf{w}}_{h\tau}}{\partial t},\tilde{\Phi}_{h\tau}\right)\mathrm{d}t$$

$$+\sum_{m=0}^{M-1}\int_{I_m}\Big(\tilde{a}_h(\bar{\tilde{\mathbf{w}}}_{h\tau},\tilde{\mathbf{w}}_{h\tau},\tilde{\Phi}_{h\tau})+\tilde{b}_h(\bar{\tilde{\mathbf{w}}}_{h\tau},\tilde{\mathbf{w}}_{h\tau},\tilde{\Phi}_{h\tau})\Big)\,\mathrm{d}t$$

$$+\sum_{m=0}^{M-1}\int_{I_m}\tilde{J}_h^{\tilde{\sigma}}(\tilde{\mathbf{w}}_{h\tau},\tilde{\Phi}_{h\tau})\,\mathrm{d}t$$

$$+\sum_{m=1}^{M-1}\Big(\{\tilde{\mathbf{w}}_{h\tau}\}_m,\tilde{\Phi}_{h\tau}(t_m^{+})\Big)+\Big(\tilde{\mathbf{w}}_{h\tau}(t_0^{+}),\tilde{\Phi}_{h\tau}(t_0^{+})\Big)$$

$$=\sum_{m=0}^{M-1}\int_{I_m}(\tilde{l}_h(\bar{\tilde{\mathbf{w}}}_{h\tau},\tilde{\Phi}_{h\tau})+\tilde{s}_h(\bar{\tilde{\mathbf{w}}}_{h\tau},\tilde{\Phi}_{h\tau}))\,\mathrm{d}t$$

$$+\Big(\tilde{\mathbf{w}}^0,\tilde{\Phi}_{h\tau}(t_0^{+})\Big)\quad\forall\tilde{\Phi}_{h\tau}\in\tilde{\mathbf{S}}_{h,\tau}^{\tilde{p},\tilde{q}}.$$

5 Numerical experiments

In order to demonstrate the applicability of the developed method we shall present two numerical simulations of compressible turbulent flow in 2D. For this purpose we chose two regimes for the profile RAE2822 and compared it with experimental data, which can be found in [12]. First regime is characterized by the far-field Mach number $M_\infty=0.6$ and the angle of the attack $\alpha=2.57^o$ ('case 3' in [12]) and second case is characterized by the Mach number $M_\infty=0.73$ and the angle of the attack $\alpha=3.19^o$ ('case 9' in [12]). For both cases we set free-stream turbulence intensity $Tu=1\%$. Figures 1-3 and 5-7 show the pressure distribution, Mach number distribution and turbulent kinetic energy distribution. Comparison of the computed pressure coefficient c_p with the experimental data is shown in Figures 4 and 8. Figure 9 shows detail of the mesh which consist of 64150 elements in total.

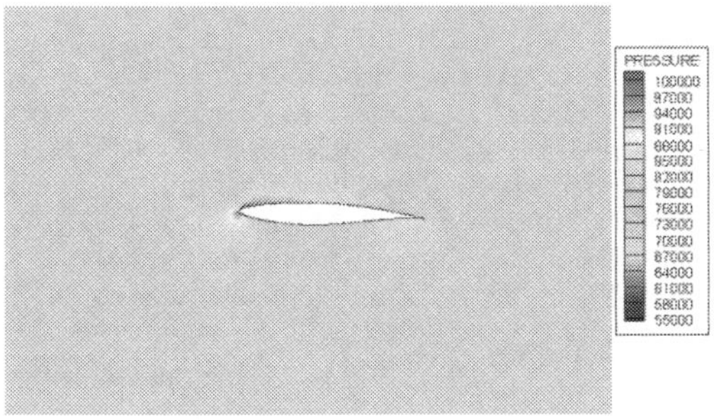

Fig. 1. The distribution of the pressure for the case 3.

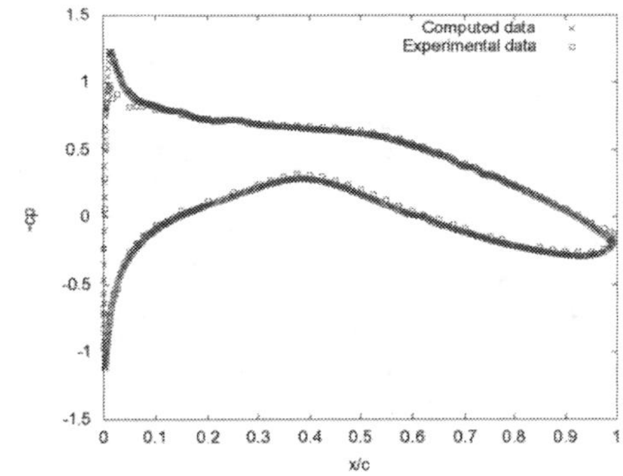

Fig. 4. Comparison of the computed pressure coefficient c_p with the experimental data for the case 3.

Fig. 2. The distribution of the Mach number for the case 3.

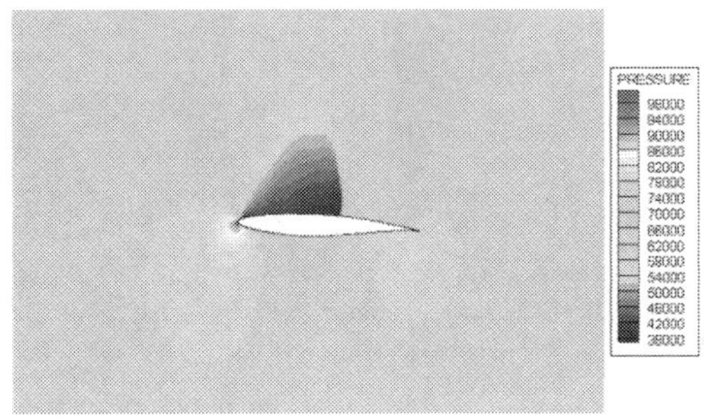

Fig. 5. The distribution of the pressure for the case 9.

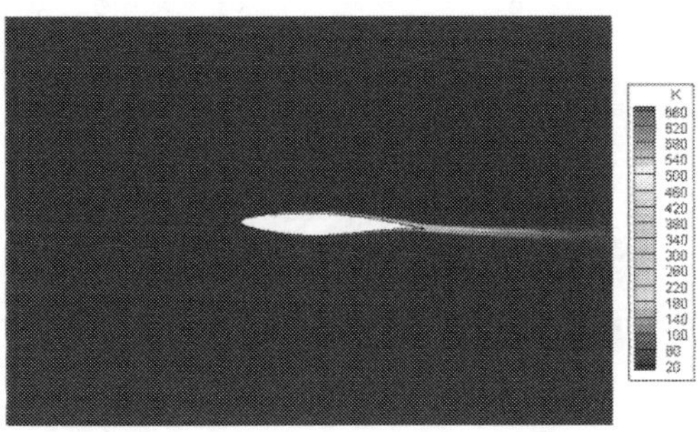

Fig. 3. The distribution of the turbulent kinetic energy for the case 3.

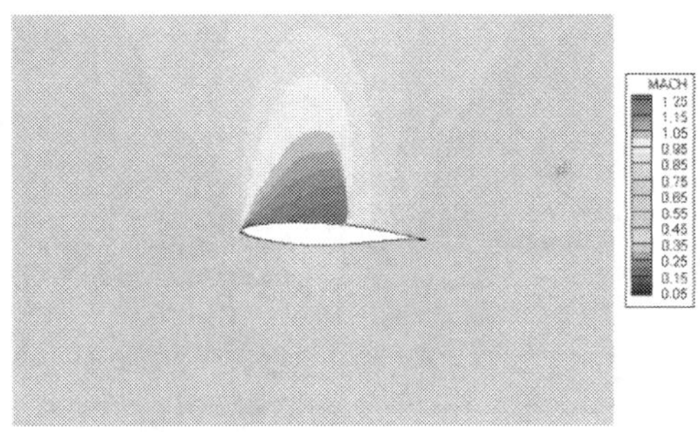

Fig. 6. The distribution of the Mach number for the case 9.

6 Conclusion

In this paper we dealt with the space-time discontinuous Galerkin method for the numerical solution of the viscous compressible turbulent flow. The applicability of the proposed method was demonstrated on the examples which show that the presented method is an efficient numerical scheme for the solution of the viscous compressible turbulent flow.

Acknowledgment

This result originated with the support of Ministry of Industry and Trade of the Czech Republic for the long-term strategic development of the research organization. The author acknowledge this support.

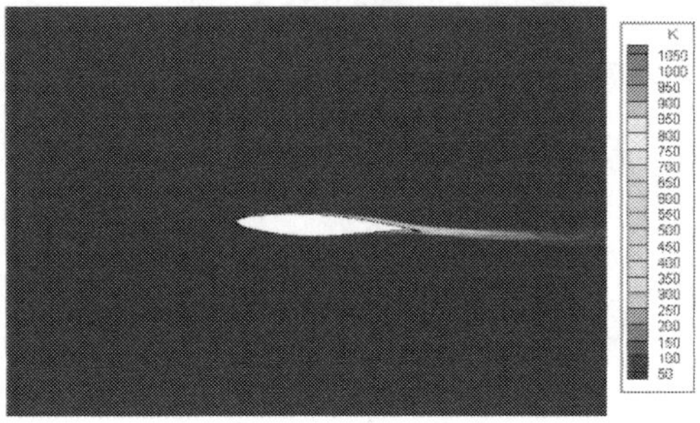

Fig. 7. The distribution of the turbulent kinetic energy for the case 9.

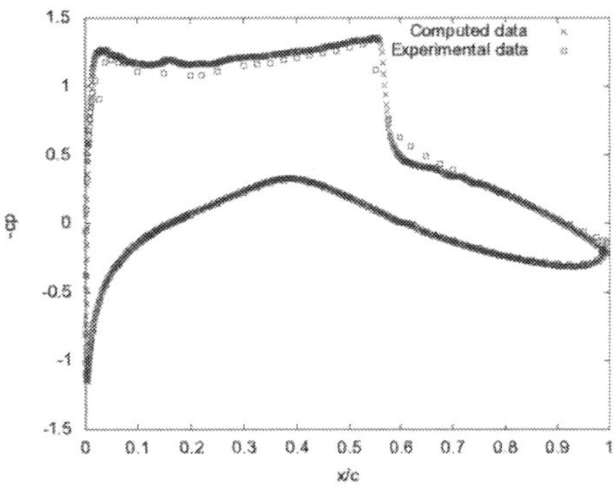

Fig. 8. Comparison of the computed pressure coefficient c_p with the experimental data for the case 9.

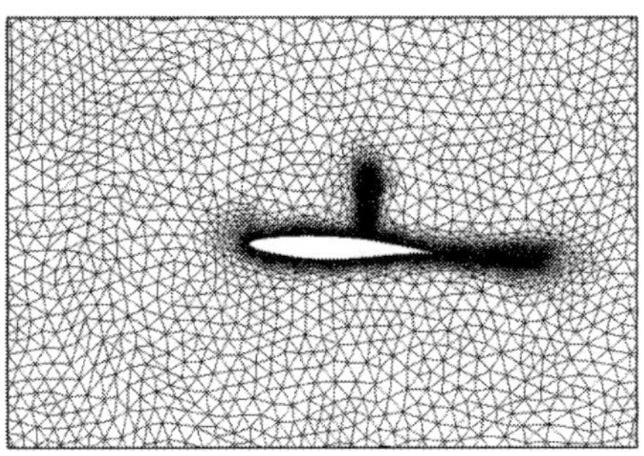

Fig. 9. Detail of the mesh.

References

1. M. Feistauer, J. Felcman, and I. Straškraba *Mathematical and Computational Methods for Compressible Flow*, Clarendon Press, Oxford (2003)
2. J. Česenek, *Discontinuous Galerkin method for solving compressible viscous flow*, Doctoral Thesis, (2011)
3. V. Dolejší, M. Feistauer, *Discontinuous Galerkin Method*, Springer (2015)
4. J. Česenek, et al., *Appl. Math. Comput.*, doi:10.1016/1.amc.2011.08.077 (2011)
5. J. Česenek, M. Feistauer, A. Kosik, *ZAMM – Z. Angew. Math. Mech.* **93**, 6 – 7, 387 – 402 (2013)
6. J. Česenek, *EPJ WoC* **67**, 02015 (2014)
7. J. Česenek, *Combined finite volume - space-time discontinuous Galerkin method for the 2D compressible turbulent flow*. Technical report R-6154, VZLÚ (2014)
8. J. Česenek, *EPJ WoC* **114**, 02012 (2016)
9. J. Česenek, *EPJ WoC* **143**, 02014 (2017)
10. D.C. Wilcox, *AIAA Journal* **46,11**, (2008)
11. Bassi, et al., *Computers and Fluids* **34**, 507-540 (2005)
12. P.H. Cook, M.A. McDonald and M.C.P. Firmin, *AGARD Advisory Report* **138**, (1979)

The Influence of Surface Roughness on the Flow Fields Generated by an Oscillating Cantilever

Ciaran Conway[1,⋆] and *Jeff* Punch[1]

[1] *Stokes Laboratories, University of Limerick, Ireland*

Abstract. With the current trend of miniaturisation of electronic devices, piezoelectric fans have attracted increasing interest as a means of inducing forced convection cooling, instead of traditional rotary solutions. Although there exists an abundance of research on various piezo-actuated flapping fans in the literature, the geometries of these fans all consist of a smooth rectangular cross section with thicknesses typically of the order of 100 μm. The focus of these studies has primarily been on variables such as frequency, amplitude and, in some cases, resonance mode. It is generally noted that the induced flow dynamics are a direct consequence of the pressure differential at the fan tip as well as the pressure driven 'over the top' vortices generated at the upper and lower edges of the fan. Rough surfaces such as golf ball dimples or vortex generators on an aircraft wing have proven to be beneficial by tripping the boundary layer and energising the adjacent airflow. This paper aims to examine the influence of surface roughness on the airflow generation of a flapping fan, and to determine if the induced wake can be manipulated or enhanced by energising the airflow around the fan tip. Particle-Image Velocimetry (PIV) is carried out on mechanically oscillated rigid fans with surfaces consisting of protruding pillars and dimples. A smooth rigid fan surface is also investigated as a control. No significant difference was noted between the smooth and roughened fans through observation of the induced flow fields. Both fans produced results that were largely consistent with the existing literature on oscillating cantilevers. The results of this paper may be used to inform the design of piezoelectric fans and to aid in understanding the complex aerodynamics inherent in flapping wing flight.

1 Introduction

Advances in electronics, coupled with modern micro-fabrication methods, have accelerated the trend of miniaturisation for many compact electrical devices. This has resulted in increasing power densities and significantly high heat fluxes that can be generated when in use. Existing rotary fan solutions may struggle to provide adequate cooling due to strict size and space requirements within these devices. As an alternative, oscillating cantilevers, or piezoelectric fans, are seen as a promising thermal solution due to their low power, frictionless air moving capabilities [1].

Research on piezoelectric fans was first carried out in the 1970's by Toda [2] who presented both a theoretical model and experimental analysis of the airflow induced by a resonating piezoelectric fan. Later work by Toda and Osaka [3] first proposed the use of piezoelectric fans for thermal management in electronic devices. Acikalin *et al.* [4] investigated the feasibility of using piezoelectric fans to cool portable electronic devices. In this study, a heat source was placed into an enclosure of similar dimensions to a mobile phone and subsequently cooled by a piezoelectric fan operating in various orientations. A heat transfer enhancement of over 100% was recorded when operating in the horizontal orientation relative to passive cooling. Kimber and Garimella [5] investigated the influence of variables such as amplitude, frequency and fan height on the thermal performance of piezoelectric fans oscillating normal to a heated surface. It was found that the performance of a given fan can be described in terms of oscillation frequency and amplitude only, with frequency being more influential on maximum heat transfer rates.

Many studies have also focused entirely on the fluid mechanics of oscillating cantilevers. Kim *et al.* [6] conducted flow visualisation and velocity field measurements on a piezo-actuated oscillating cantilever confined between two end plates. They noted the development of two counter-rotating vortices generated by the pressure differentials across the tip of the fan as it oscillated over a period of one cycle. The velocity between these vortices reached over four times the maximum fan tip velocity and resulted in a jet-like flow originating at the fan tip. The three-dimensional nature of oscillating cantilever aerodynamics was investigated by Agarwal *et al.* [1]. In this study, three-dimensional vorticity plots were interpolated from a series of two-dimensional velocity field measurements. These plots showed the development of a horseshoe vortex around the fan that evolves into hairpin vortices following shedding at the fan tip.

⋆e-mail: ciaran.conway@ul.ie

The majority of the literature on the fluid mechanics of oscillating cantilevers is mainly conducted on smooth surfaced blades where the primary variables under investigation are vibrating frequency and amplitude. There is relatively little focus on the influence of surface roughness on the induced flow dynamics of oscillating cantilevers. In many fluid mechanics applications, surface structures are used to change the local flow dynamics. A well known example of this is the dimpled surface of a golf ball. The dimples trip the boundary layer around the ball and transition it to a turbulent regime. The energised flow adjacent to the surface delays the separation of airflow from the ball and thus reduces pressure drag, allowing it to travel a further distance [7]. Similarly, the presence of protruding vortex generators on aircraft wings can also promote separation and delay the onset of aerodyamic stall. This study considers the influence of surface roughness on the aerodynamics of rigid fans that are mechanically oscillated at a fixed frequency and amplitude. Two surface roughnesses are considered. The first is a dimpled surface similar to that of a golf ball. The second is a reverse of the dimpled surface with protruding pillars. The induced flow fields are recorded and visualised using Particle-Image Velocimetry (PIV). The setup and procedure for the PIV experimentation is described in the next section and, following this, the PIV results are presented. Vorticity, $\omega = \nabla \times \mathbf{v}$, is used to highlight and quantify the flow structures generated by the fans as they oscillate, whilst wake regions are visualised through the use of velocity magnitude plots. The results of this work may inform the design of oscillating fans for use in thermal management applications, as well as contribute to the literature on the complex phenomenon of flapping wing flight.

2 Experimental Method

In this section, the experimental apparatus used to mechanically oscillate the fans is described. Following this, the geometries of the two fans under investigation in this study are defined, and the final part of this section outlines the experimental methodology.

2.1 Mechanical Oscillator

Mechanically oscillated rigid fans were used in this study in order to achieve desired amplitudes, frequencies and surface roughness patterns. A custom designed mechanical oscillator was used which converts rotary motion into oscillatory motion of the fan tips. The rotary motion – provided by a Maxon A-max32 brushed motor – was initially converted into reciprocating linear motion of a 3D-printed driveshaft. The driveshaft linearly displaces the "dead" side of the oscillating fan relative to a fixed pivot point thus inducing an oscillation at the fan tip; this is illustrated in Figure 1. Oscillating frequency was controlled by the motor RPM where f = RPM/60. Tip amplitude depended both on the length of the fan and the distance between the 3D-printed driveshaft and fixed pivot point.

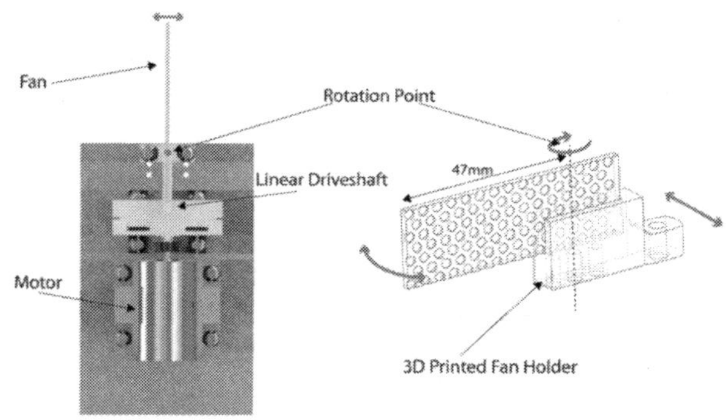

Fig. 1. Mechanical oscillator and method of oscillation

2.2 Fan Characterisation

The fans in this work were cut from 2.8 mm thick acrylic sheeting. The fans were rectangular in shape with cross-sectional dimensions of 60 mm x 20 mm. A surface roughness was imparted on two of the fans by etching the material with a GCC LaserPro X380 laser cutter. A diagonal grid of circular dimples was etched on to both sides of one fan as depicted in Figure 2. A reverse of this pattern was etched onto another fan which created a series of circular protrusions, or pillars. The depth of the etchings on the roughened fans was approximately 1 mm. No surface roughness was imparted on the final fan so the smooth surface of the acrylic could be used as a control. Each fan was placed in a custom designed 3D-printed holder during experimentation.

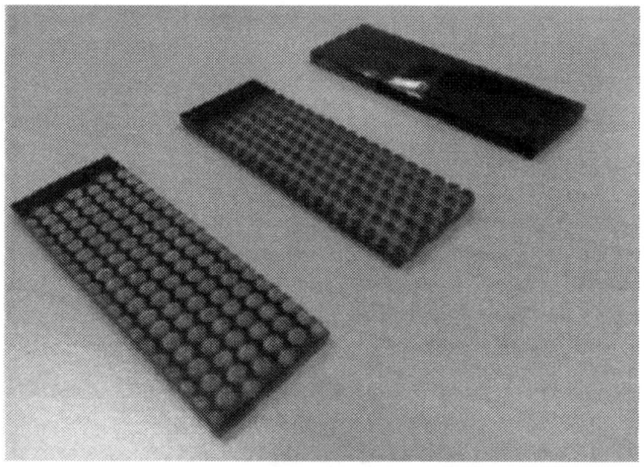

Fig. 2. Tested fans with dimpled surface (nearest), protruding pillars (middle) and smooth surface (furthest)

The point about which the fans were rotated by the oscillating mechanism was 47 mm behind the fan tip. The fans were oscillated at a frequency of 40 Hz with a tip-to-tip amplitude of 14 mm. This resulted in a tip Reynolds number of 4,480 according to Equation 1, where f, a, c, and ν are oscillating frequency, oscillating amplitude, fan width and kinematic viscosity respectively.

$$Re = \frac{2\pi f a c}{\upsilon} \tag{1}$$

2.3 Flow Field Measurement

The velocity fields induced by the oscillating fans were recorded using PIV. A Quantel Big Sky Laser was used to generated a double pulsed, 532 nm wavelength laser sheet which illuminated the seeded flow field. Seeding was provided by water based tracer particles issued from a Martin Magnum 2500 Hz hazer. A TSI PowerView Plus 2MP CCD camera recorded a series of image pairs of the illuminated flowfield. The camera and laser were synchronised using a TSI Laser Pulse Synchroniser (Model no: 610035). The mechanism was placed within an acrylic enclosure with dimensions of 880 x 540 x 600 mm to protect against external air currents affecting the induced flow fields. The distance between the fan and the nearest wall was 20 cm therefore the fan was considered to be unconfined. The enclosure was perforated to allow air to enter and exit the system, and it had an open slot at one side to allow the laser sheet to enter unobstructed.

Phase-locked image pairs were obtained by triggering the PIV system to capture images at various points of the fan oscillating cycle. A 5V TTL pulse emitted from the motor corresponding to each revolution was sent to the TSI synchroniser as a triggering signal.

Desired phase angles were achieved by setting appropriate time delays between the trigger signal and the beginning of the resulting image capture sequence. Image pairs were recorded at incremental time delays of 1 ms from 0 up to 24 ms. This corresponded to one period of the fan motion when oscillating at 40 Hz.

PIV images were recorded for two different configurations of the oscillating fan. In the first configuration, the camera was placed normal to the oscillating plane (Figure 3(a)), resulting in a plan view of the induced flow fields in the x-y plane. The laser sheet intersected the fan at mid-chord in this configuration. In the second configuration, the camera was positioned such that it captured flow dynamics in the y-z plane as shown in Figure 3(b). Here, the laser sheet was positioned 5 mm behind the fan tip. The recorded image pairs were processed with a multi-pass, image deformation processing algorithm using TSI Insight4G software.

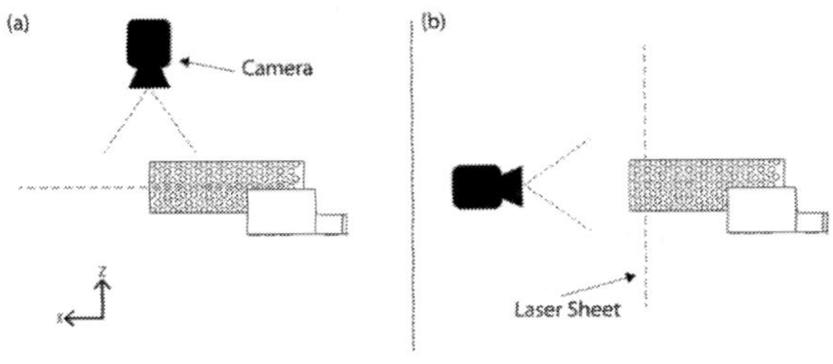

Fig. 3. PIV configurations: (a) laser sheet, oscillation plane, and camera image plane are all parallel to capture data in the x-y plane; (b) laser sheet and camera image plane are normal to the oscillation plane to capture data in the y-z plane

3 Results and Discussion

The primary objective of this study is investigate the influence of surface roughness on the flow fields induced by rigid, oscillating fans. The PIV results are initially presented in the x-y plane though vorticity and velocity magnitude plots. The three-dimensional nature of the flow is subsequently displayed through measurements in the y-z plane.

3.1 x-y Plane Measurements

Vorticity plots for the smooth surfaced fan are largely consistent with existing literature on oscillating fan aerodynamics [1, 6, 8]. Vortex formation is driven by the pressure differential across the fan tip. This pressure differential arises due to a pressure increase on the leading surface of the fan as it accelerates from points of zero velocity (ϕ = 90°/270°) to points of maximum velocity (ϕ = 0°/180°). Figure 4(a) shows the smooth fan at a phase angle of approximately 300°. Here, the fan is accelerating towards a phase angle of 360°/0°. Airflow from the high pressure leading surface of the fan is seen to flow around the fan tip to the low pressure surface resulting in the generation of a counter-clockwise tip vortex. The clockwise vortex shed from the previous half period (90° → 270°) is visible adjacent to the fan tip.

There is constructive interference in the region between the shed vortex and the developing vortex at the fan tip whereby a high velocity jet flow is induced. As the fan accelerates through a phase angle of ϕ = 0°, the counter-clockwise tip vortex develops as it remains attached to the fan structure. The shed clockwise vortex from the previous half period propagates away from the fan and its energy dissipates due to shear with the stationary ambient air. At ϕ = 90° in Figure 4(c), similar flow behaviours to those of ϕ = 300° are observed. The counter-clockwise tip vortex has been shed and a new clockwise vortex has begun to develop as the fan accelerates away from the point of zero velocity at ϕ = 270°. Again, a high velocty jet region is visible between the two vortices that helps to generate jet-like wakes emanating from the fan tip as seen in the time-averaged results.

The time-averaged velocity flow field generated by the smooth oscillating fan is displayed in Figure 5. There is a clear symmetry in the induced flow field, consistent with the findings of Stafford and Jeffers [9] and Bidakhvidi *et al.* [10]. Steamlines reveal a large inflow region in front of the fan tip from which air is drawn before being expelled by the oscillating fan into two high velocity lateral jet regions. The direction of these jet regions is approximately 90° to the initial inflow direction with no apparent airflow generated by the fan in the positive x direction. This is in contrast to much of the literature on oscillating piezoelectric fans where airflow emanates in a forward direction from the fan tip. The cooling capabilities of these airflows as they impinge on a heated surface have been studied extensively ([5], [11], [12], [13]). The lateral wake regions in the current work

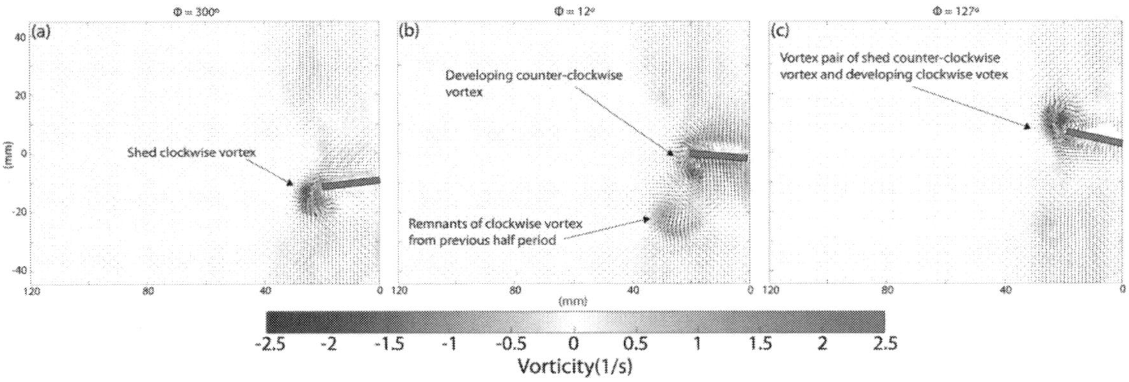

Fig. 4. Vorticity plots for the smooth fan at phase angles of (a) 300°, (b) 12° and (c) 127° in the x-y plane

are a consequence of the fan thickness (2.8 mm) which is significantly larger than the thicknesses of piezofans investigated in literature ($O \approx 0.1$ mm). The effect of thickness on the flowfields induced by oscillating fans is investigated further by Conway *et al.* [14]. Regions of maximum average velocity in Figure 5 occur immediately adjacent to the fan tip at points of maximum deflection (zero tip velocity). This is consistent with the high velocity regions noted in the previous section between clockwise and counter-clockwise rotating vortices.

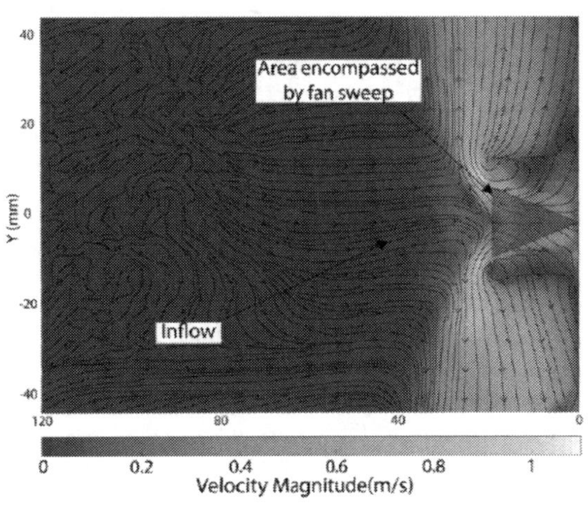

Fig. 5. Time-averaged velocity magnitude plot for the smooth fan in the x-y plane

The rough surfaced fans shown in Figure 6 give similar results to those of the smooth fan. Figure 7 shows the development of tip vortices for the two roughened fans in this study. A counter-clockwise, pressure driven tip vortex is again developed as the fans move from $\phi = 270°$ to $\phi = 90°$. Vorticity values are largely consistent with the smooth fan. Maximum vorticity magnitudes of approximately 2.5/s were recorded for all three fans indicating that the induced flow structures are entirely pressure driven, with surface roughness having little influence at this Reynolds number. The time-averaged velocity magnitude plots for the roughened fans displayed in Figure 6 again bear similarities to the smooth fan.

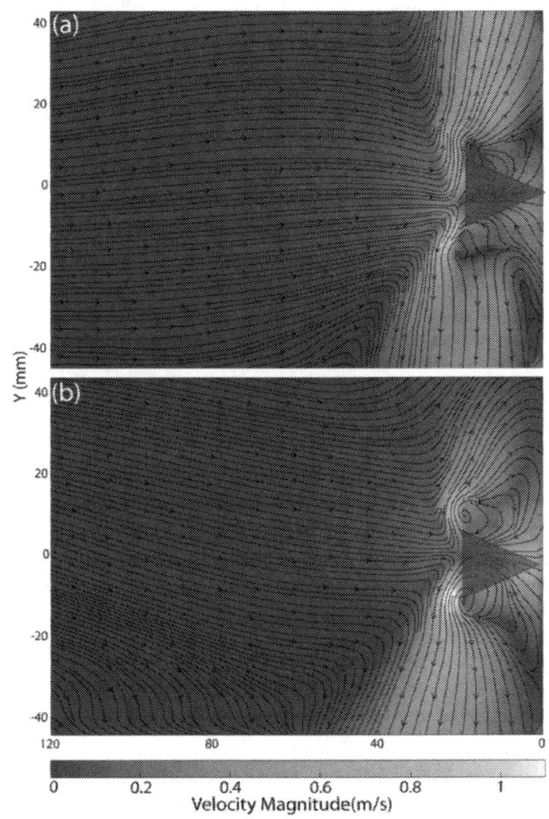

Fig. 6. Time-averaged velocity magnitude plots for (a) dimpled surface and (b) protruding pillar surface in the x-y plane

3.2 *y-z* Plane Measurements

The time-averaged velocity magnitude plots in the y-z plane for the smooth and rough surfaced fans are displayed in Figure 8 and Figure 9 respectively. The shadow on the right hand side of the plots is a consequence of the opaque material from which the fans were fabricated. Symmetry is therefore assumed and the detail on the left side of the fan is considered to be representative of the time-averaged fluid mechanics on both sides of the fan as it oscillates.

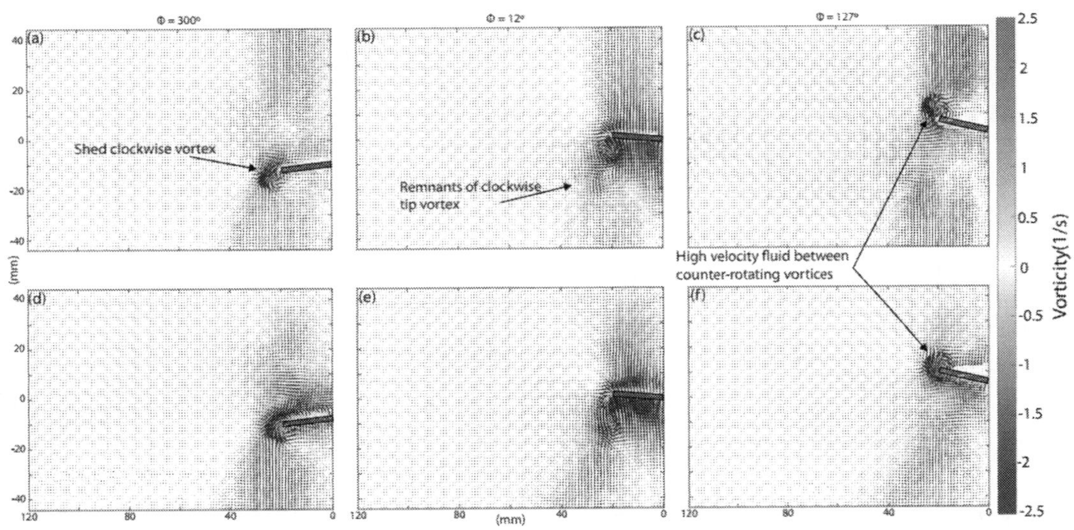

Fig. 7. Vorticity plots for the roughened surface fans in the *x-y* plane: (a-c) dimpled surface and (d-f) protruding pillars

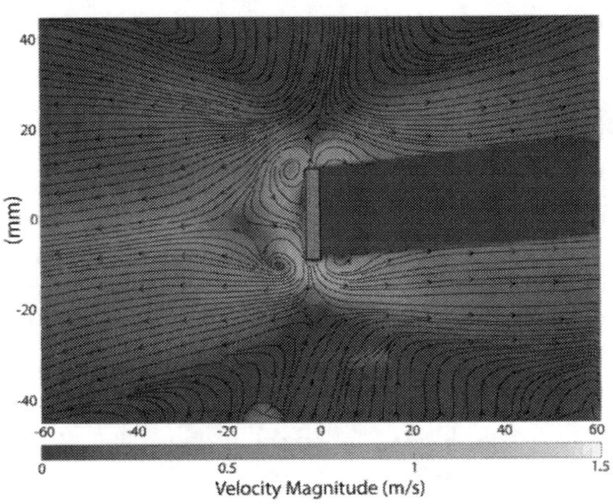

Fig. 8. Time-averaged velocity magnitude profiles in the *y-z* plane for smooth surfaced fan

As shown in Figure 8 and Figure 9, each fan generates strong lateral flow components that are symmetrical in nature – consistent with the time-averaged velocity plots in the *x-y* plane. The roughened suface fans of Figure 9 appear to generate a slightly angled flow field. This may be an effect of the unaligned surface structures on each side of the fan, as the smooth surfaced fan of Figure 8 produces a relatively unbiased lateral wake. The fan with the protruding surface structures appears to generate the smallest wake (Figure 9(b)). The average velocity reaches a maximum of approximately 1.5 m/s in the immediate vicinity of the fan structure. This is slightly larger than the maximum velocites of approximately 1 m/s surrounding the smooth and dimpled fans. Downstream of the fan structures, the jet velocities are largely consistent for all three of the tested fans at 0.5 – 0.7 m/s.

Streamlines within the plots in Figure 9 illustrate the development of the flow fields that are induced by the fans. The pressure-driven counter rotating vortices generated at

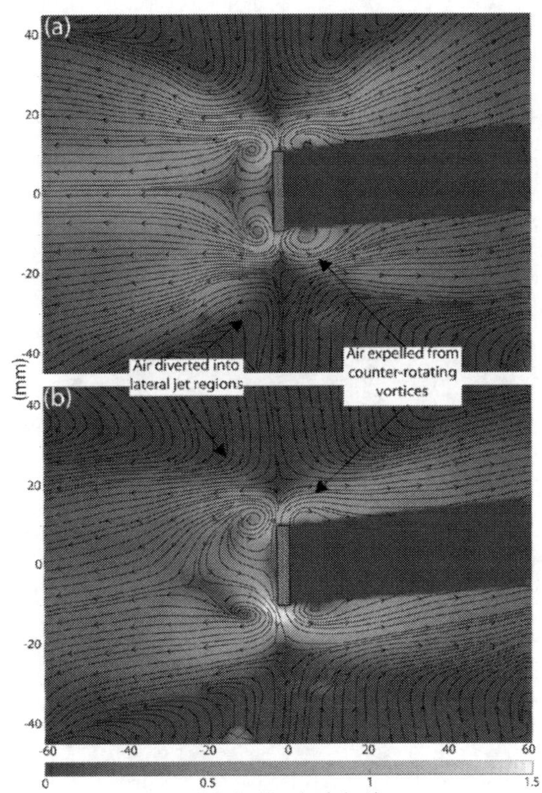

Fig. 9. Time-averaged velocity magnitude profiles in the *y-z* plane for the fans with (a) dimpled surface and (b) protrusions

the upper and lower edges of the fan structures are clearly visible in the time averaged plots. Air is drawn in from above and below the fans by the counter rotating vortices and is then expelled into the lateral jet regions. Local airflow not affected by the vortices is instead diverted

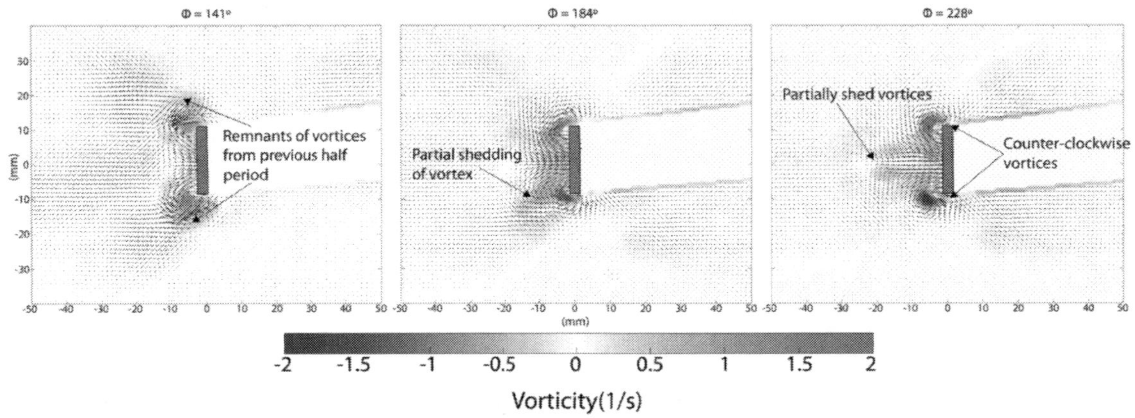

Fig. 10. Vorticity plots in the *y-z* plane for the smooth fan as it moves from ϕ = 141° to ϕ = 228°

into the lateral jets through the influence of viscous shear. The streamlines also highlight the presence of saddle points at the lower velocity regions to the left of the fan structures in Figures 8 and 9. The low velocity regions and diverging streamlines indicate out-of-plane flow components parallel to the span of the fan blades.

Spanwise vorticity results for the smooth fan as it moves from ϕ = 90° to ϕ = 270° are presented in Figure 10. Pressure driven vortices are generated at the upper and lower edges of the fan and develop as the fan accelerates. The vortices appear to partly separate from the fan at the midpoint of its half period sweep however the cores remain attached, adjacent to the fan surfaces with decreasing vorticity as the fan decelerates. The roughened fans produced identical results.

4 Conclusions

PIV experimentation was carried out on three oscillating fans with different surface roughnesses. Flow fields were recorded in two planes to capture the three-dimensional nature of oscillating fan aerodynamics. The development and behaviour of the vortices around the fan structure was similar for each of the three fans tested, and these results were also consistent with the literature on piezoelectric fan aerodynamics. The development and propagation of the pressure-driven vortices surrounding the fan blades did not appear to be inhibited or aided by the presence of the dimpled and protruding surface structures. Time-averaged velocity plots revealed similar wake profiles generated by the fans in both the *x-y* plane and *y-z* planes. The time-averaged wakes are lateral in direction with little disturbance ahead of the fan tip. This is a consequence of the relatively thick fan structures used in this study in contrast to the significantly thinner blades found in the literature. There appears to be no noticeable difference in the flow fields induced by the surface roughened fans in comparison to the smooth surfaced fan at the tested Reynolds number. This would suggest that flow structures and flow fields generated by the fans at this Reynolds number are purely a result of pressure differentials across the surfaces of the fans as they oscillate.

Acknowledgements

The author acknowledges the financial support of the Irish Research Council (IRC) for funding under the Government of Ireland Postgraduate Scholarship Scheme: Project ID GOIP/2014/1480.

References

1. A. Agarwal, K.P. Nolan, J. Stafford, N. Jeffers, Journal of Fluids and Structures **70**, 450 (2017)
2. M. Toda, Ferroelectrics **22**, 911 (1978)
3. M. Toda, S. Osaka, **6**, 1171 (1979)
4. T. Açikalin, S.M. Wait, S.V. Garimella, A. Raman, Heat Transfer Engineering **25**, 4 (2004)
5. M. Kimber, S.V. Garimella, International Journal of Heat and Mass Transfer **52**, 4470 (2009)
6. Y.H. Kim, S.T. Wereley, C.H. Chun, Physics of Fluids **16**, 145 (2004)
7. J.A. Scobie, C.M. Sangan, G.D. Lock, Procedia Engineering **72**, 738 (2014), the Engineering of Sport 10
8. M. Choi, C. Cierpka, Y.H. Kim, Journal of Fluids and Structures **31**, 67 (2012)
9. J. Stafford, N. Jeffers, IEEE Transactions on Components, Packaging and Manufacturing Technology **7**, 751 (2017)
10. M.A. Bidakhvidi, R. Shirzadeh, G. Steenackers, S. Vanlanduit, Experiments in Fluids **54** (2013)
11. T. Açikalin, S.V. Garimella, A. Raman, J. Petroski, International Journal of Heat and Fluid Flow **28**, 806 (2007)
12. C.N. Lin, International Journal of Heat and Mass Transfer **55**, 3043 (2012)
13. M.K. Abdullah, B.H. Murni, M.Z. Abdullah, M.a. Mujeebu, F. Hussin, H. Yusoff, N.C. Ismail, K.a. Ahmad, Z. Mohd Ripin, Isi Bilimi Ve Teknigi Dergisi/ Journal of Thermal Science and Technology **32**, 41 (2012)
14. C. Conway, N. Jeffers, J. Punch, *'The Influence of Blade Thickness on the Flow Field Generated by an Oscillating Fan'*, under review for publication (2017)

Experimental investigation of five parallel plane jets with variation of Reynolds number and outlet conditions

Tomas Daubner[1,*], *Jens* Kizhofer[1], and *Mircea* Dinulescu[1]
[1]Apex Research B.V., Apex Research Laboratory, Westeinde 10 2275AD Voorburg, The Netherlands

Abstract. This article describes an experimental investigation in the near field of five parallel plane jets. The study applies 2D Particle Image Velocimetry (PIV) for ventilated and unventilated jets, where ventilated means exiting into a duct with expansion ratio 3.5 and unventilated means exiting to the free atmosphere. Results are presented for Reynolds numbers 1408, 5857 and 10510. The Reynolds number is calculated for the middle channel and is based on the height of the nozzle (channel) equivalent diameter 2h. All characteristic regions of the methodology to describe multiple interacting jets are observed by the PIV measurements - converging, merging and combined. Each of the five parallel channels has an aspect ratio of 25 defined as nozzle width (w) to height (h). The channels have a length of 185 times the channel height guaranteeing a fully developed velocity profile at the exit from the channel. Spacing between the single plane jets is 3 times the channel height. The near field of multiple mixing jets is depended on outlet nozzle geometry. Blunt geometry of the nozzle was chosen (sudden contraction).

1 Introduction

Main focus of the Apex Research Laboratory is the improvement of plate type heat exchangers, the reduction of the pressure drop through the ductwork and the mixing effectiveness in various industrial processes, for example mixing of ammonia in DENOx applications. In the sake of the mixing effectiveness Apex Research Laboratory have developed an experimental setup for understanding the physical phenomena like the mixing of multiple jets. Apex Research Laboratory has published its first paper in the WASET journal in 2017 on the topic "Experimental investigation of plane jets exiting five parallel channels with large aspect ratio" [1], describing the difference between 2 outlet nozzle geometries. Further paper continues in the discussion about multiple jet and its main focus is on the difference in the outlet conditions for multiple jets and its effect on mixing performance.

Multiple parallel plane jets can be found in Heating Ventilating and Air Conditioning systems (HVAC), drying systems and in the combustion chambers. The parameters based on which is necessary to judge the mixing effectiveness are pressure drop and certain mixing when the jets generate a desired velocity distribution.

Single plane jets have been studied extensively since the works of Schlichting [2]. Following studies mainly focussed on characteristics of jet propagation and pointed out the importance and influence of initial boundary conditions [3-4]. L. Moruz [1] studied the mixing effectiveness and Reynolds stress distribution concluding that the outlet geometry is important for this phenomena. We assume that outlet conditions will have also influence on mixing effectiveness express as magnitude – Reynolds stress or as turbulence intensity. This implies that for this paper we have identified following parameters for jet formation: boundary conditions influencing the jet formation are Reynolds number based on channel height (nozzle), nozzle aspect ratio, velocity profiles at exit including mean velocity profile as well as turbulence characteristics and nozzle geometry and spacing between the jets. A Reynolds number dependency will be examined.

Dual parallel jets have been studied by Miller [5] and Nasr [6] beside others. Here, two cases are defined, a ventilated and an unventilated case. The ventilated case has duct downstream from the outlet from the channels, whereas in the unventilated case the jets exit to the free atmosphere. Three main regions have been identified in literature, the converging region, the merging region, and the combined region. When the jets exit the nozzle, a recirculation zone with lower pressure is generated between the jets in the converging region. This low pressure region deflects the jets towards each other. The beginning of the merging region is defined by the merging point, where the recirculation zone vanishes, thus all streamwise velocities are positive. In the merging region the two jets merge together up to the combined point, where only a single combined jet can be identified. For definition of these regions for 2 parallel jets see figure 1 from Nasr [6].

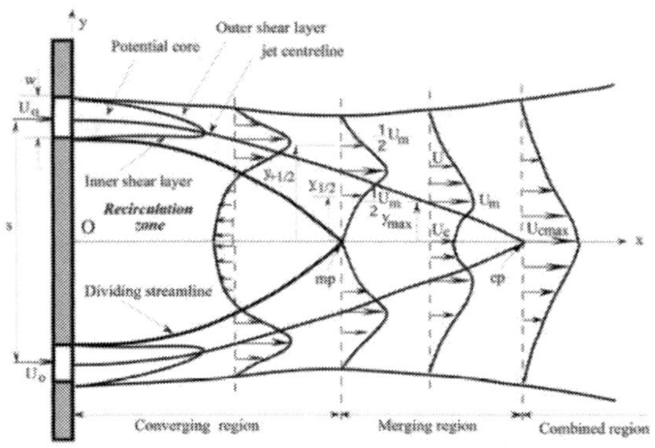

Fig. 1. The definition of the regions and points for two parallel jets [6]

Ghahremanian [7] applied latest measurement equipment like Laser Doppler Anemometry and Particle Image Velocimetry and identified the importance of the consideration of coherent structures in the jet shear layers for the analysis of multiple parallel jets. Furthermore, he discussed the unstable behaviour of the merging points and discussed statistical tools for the prediction of the jet behaviour.

This paper is meant to set up experimental data including the definition of boundary conditions for multiple parallel plane jets exiting to different ambient conditions. For that purpose, chapter 2 describes the experimental set up, the applied measurement techniques like Particle Image Velocimetry and Laser Doppler Anemometry and the measured boundary conditions upstream of the inlet into the five channel geometry as well as the boundary conditions exiting the channels at the nozzle outlet. The third chapter describes and discusses the experimental data with respect to mean velocity field and fluctuating velocity field in terms of Turbulence intensity.

2 Experimental setup

Figure 2 shows the outlet from the 5 parallel channels. Figure 3 shows overall view on the facility.

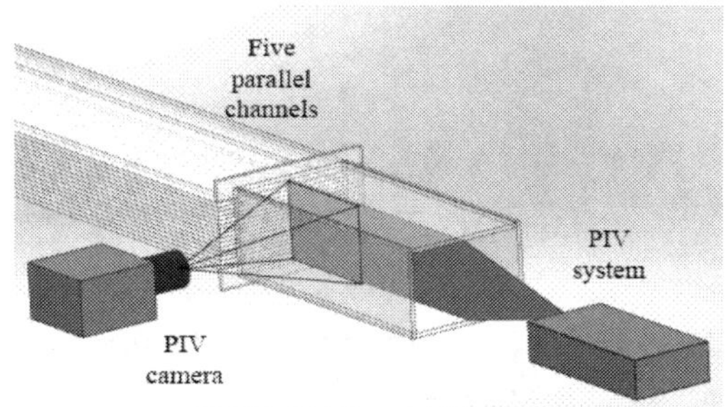

Fig. 2. Closer isometric view for ventilated configuration.

2.1. Overview of flow facility

A transparent wind tunnel have been designed to facilitate the air flow distribution in laminar, transient and turbulent regime upstream of the 5 plane parallel channels (see figure 3). A centrifugal fan with maximum volume flow rate of 534 m^3/h and 1.1 kW electric power drives the air through all the test facility. The volume flow rate is measured with a flow velocity probe from Testo. To control the air flow rate a digital variable power supply is used.

The air is transferred to the measurements position by a round duct, small angle diffuser, square bend, settling chamber, contraction duct, 30° inclined duct, Apex Delfino Flow Conditioner, five plane parallel channels, nozzle geometries and outlet duct.

The diffuser has a total angle of 12° for reducing the flow sidewall separation and total pressure drop, the settling chamber has a squared section of 0.3 m x 0.3 m followed by a contraction duct with an aspect ratio of 1:3. The contraction duct is followed by flow straightener of 50 mm length. The ratio between the honeycomb cell diameter and length guarantees a removal of secondary flow like swirl. The 30° inclined duct is used to facilitate a non-uniform flow profile just upstream of Apex Delfino Flow Conditioner, which consist of rectangular honeycomb structure and gradually changing fine mesh downstream of the honeycomb, to remove above mentioned influence of 30° bend.

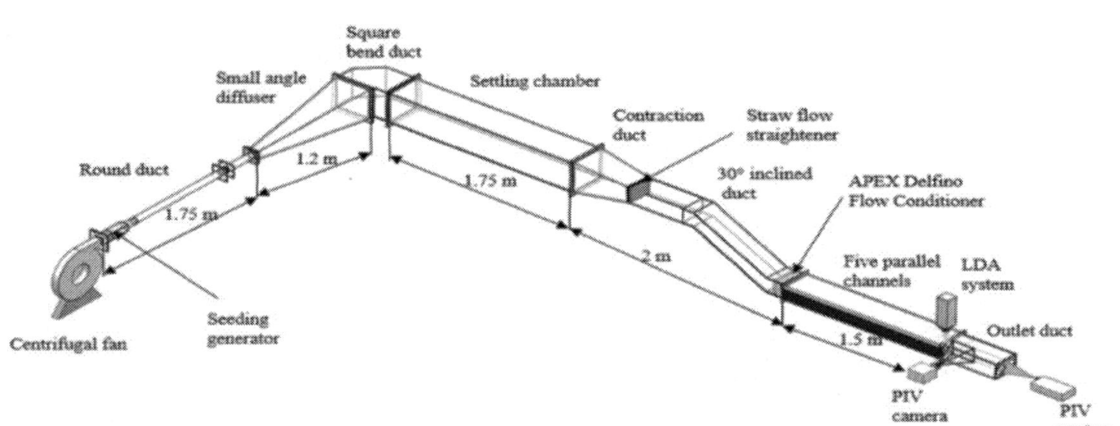

Fig. 3. Transparent flow facility and installed measurement equipment, isometric view

The five channel geometry consist of transparent plexiglass plates allowing optical measurement techniques. The channel height is h = 8 mm, width is w = 200 mm and the length is L = 1500 mm (and hence a large aspect ratio of 25 between channel width / channel height). The non-dimensional spacing ratio between the channels is 3 and the contraction/expansion ratio at inlet/outlet is 3.5, velocity profile inside the channel is found to be turbulent for lower and higher Reynolds number (see table 1). The Reynolds number is defined as: $Re=(ro \times u \times d_h)/\mu$ where u is the mean velocity based on the volume flow rate, d_h is equivalent diameter of the channel which equals with 2h and μ is dynamic viscosity. The outlet conditions are varied between ventilated (figure 2) and unventilated case. To be very clear the unventilated in this report means outlet to the free atmosphere and ventilated means outlet to the ducting as show in figure 2. The duct is 0.5 meter long, the width is equal with width of the channels and the height is shown in figure 4 and is 136 mm.

2.2. Overview of measurement equipment

The measurement techniques range from static pressure measurements to laser optical measurement techniques like Particle Image Velocimetry (PIV) or Laser Doppler Anemometry (LDA). For that purpose, a seeding generator introduces DEHS particles in the range of 1 μm into the flow immediately downstream of the fan, see figure 3.

Laser Doppler Anemometry is applied to measure the upstream velocity profile in the center channel. The system is a 2D Flow Explorer from Dantec Dynamics. The LDA system is equipped with a 150 mm focal length lens generating a measurement volume of 40 μm in diameter and 200 μm in length. The system is traversed on an ISEL traverse with a positioning accuracy of ≈ 7 μm. The analysis is performed in the BSA Flow Software. The stop criteria are defined with a measurement length of 30 s for each point resulting in about 30000 samples or a data rate of 1000 bursts/s

2D Particle Image Velocimetry is applied at the outlet duct to identify the topological behaviour of the five parallel plane jets. The seeding particles are illuminated in the xz-plane at y=0.5 w by a 145 mJ dual cavity laser equipped with a Dantec light sheet optics. The pulse length of the laser is 9 ns to avoid displacement of particles during the illumination. The scattered light of the particles is recorded by a FlowSense EO 4M double framing camera running at 10 Hz. The separation time between the images is adapted to the jets velocity to yield a maximum displacement of 10 Pix. The number of double frames is 200 resulting in a measurement time of 20 s per geometry and Reynolds Number. The analysis of the particle images is performed in Dynamic Studio with the "Adaptive PIV" routine. The Cross Correlation algorithm automatically chooses an interrogation area size with respect to number of particles. The minimum number of particles are chosen to be 6/interrogation area. For increased accuracy in shear flows, the algorithm deforms the interrogation area to cover rotation, strain and shear. Erroneous vectors are detected by the N-Sigma outlier detection method with N=6. N-Sigma method is based on computation of mean, RMS and σ, from series of data and subsequently reject all samples lying more than N times σ from the mean. The scaling is 12 Pix/mm. The resulting vector field has a vector pitch of 16 Pix (1.3vectors/mm). Instantaneous picture is shown in figure 4 and figure 5.

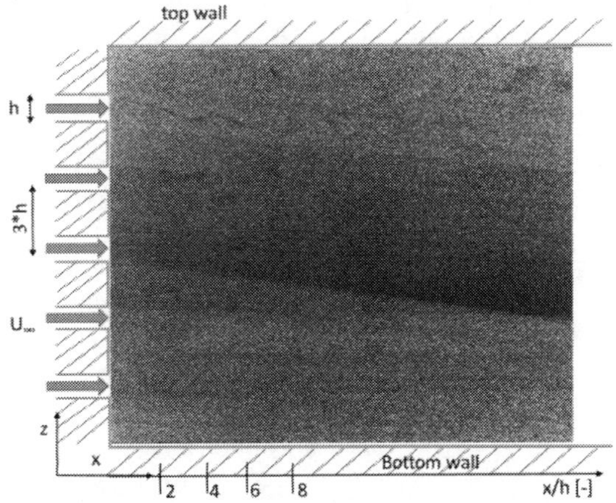

Fig. 4. Example of raw instantaneous PIV image for ventilated configuration for Re=1408

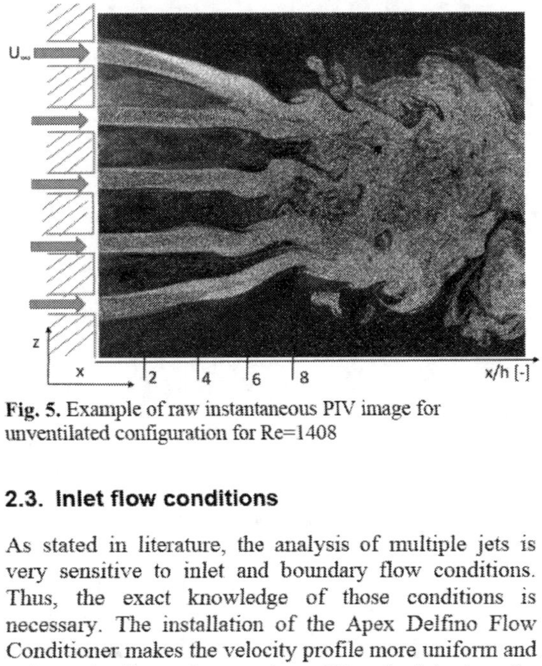

Fig. 5. Example of raw instantaneous PIV image for unventilated configuration for Re=1408

2.3. Inlet flow conditions

As stated in literature, the analysis of multiple jets is very sensitive to inlet and boundary flow conditions. Thus, the exact knowledge of those conditions is necessary. The installation of the Apex Delfino Flow Conditioner makes the velocity profile more uniform and reduces the fluctuations to about 5% at the inlet into the channel model. The streamwise velocity profile at the outlet from the middle channel is shown in figure 6, normalized streamwise velocity profile is shown figure 7 as well as turbulence intensity in figure 8. Both Figures were acquired by applying LDA measurement technique. Table 1 shows the cases for which the measurement was taken. At Re = 1408, the turbulence intensity is very low, indication a laminar flow field. At Re = 5857 and 10510,

https://doi.org/10.1051/epjconf/201818002018

the turbulence intensity in the centre of the channel reach similar amplitudes (5%).

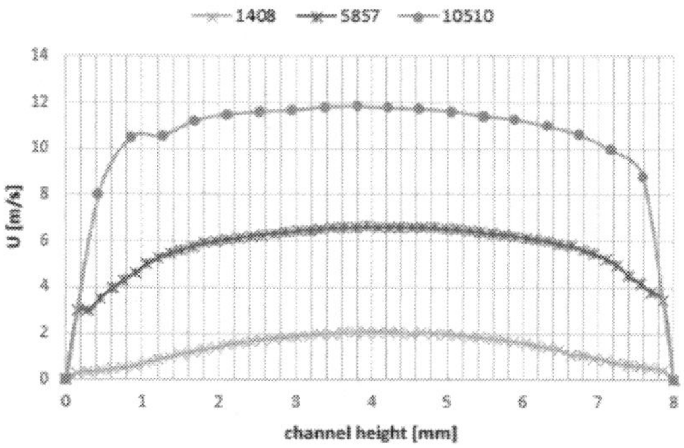

Fig. 6. Streamwise velocity profile for the middle channel for different Re

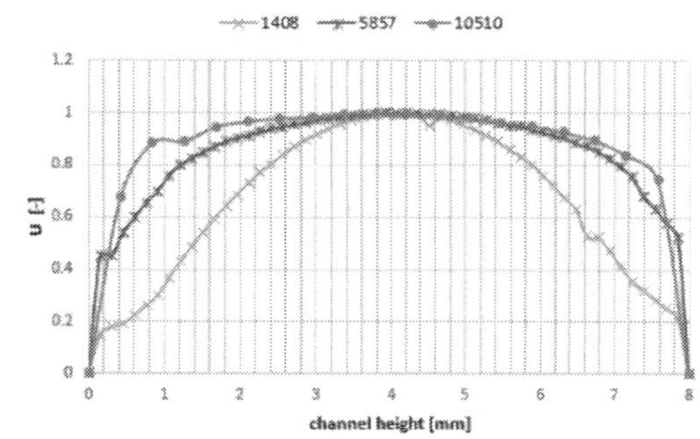

Fig. 7. Normalized streamwise velocity profile for the middle channel for different Re

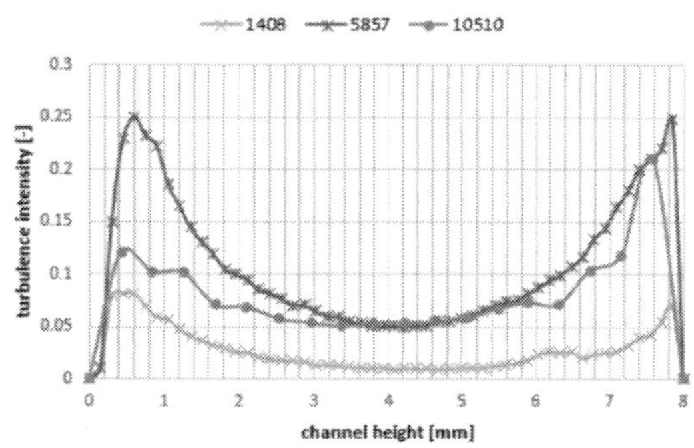

Fig. 8. Turbulence intensity for the middle channel for different Re

Table 1. Measurement characteristics at position 10 times h upstream from the nozzle outlet

v mean [m/s]	Re	Tu_x min [%]	Tu_x max [%]
1.32	1408	1	8
5.49	5857	5	21
9.85	10510	5	25

3 Results and discussion

3.1. Streamwise velocity

Figures (9-13) show the normalized positive streamwise mean velocity field (contour plot) and absolute velocity vectors (pitch is 2 times h) for different Reynolds numbers. Main coordinate system can be found in these figures as well. The scale bar for these figures ranges from u=0 up to $u=U_\infty$. Where U_∞ is the exiting velocity of the center jet. Scale is for of dimensionless streamwise velocity.

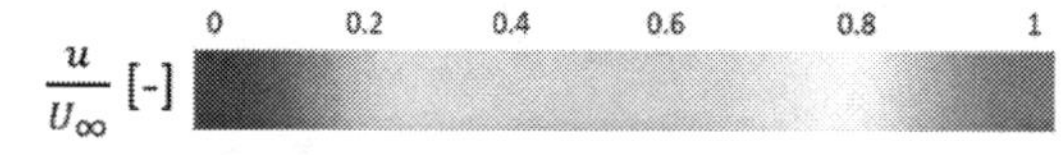

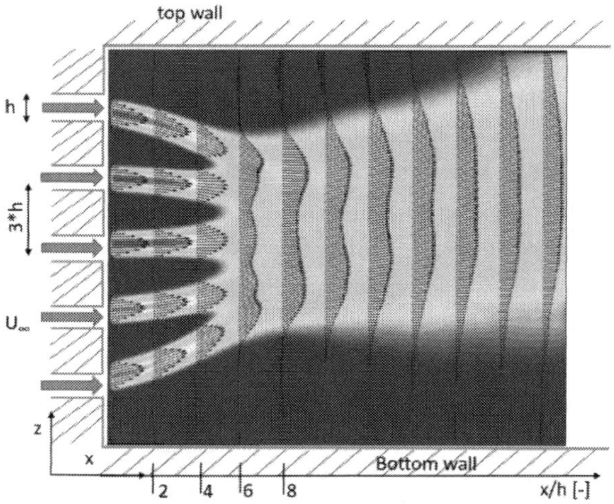

Fig. 9. Ventilated configuration Re=1408

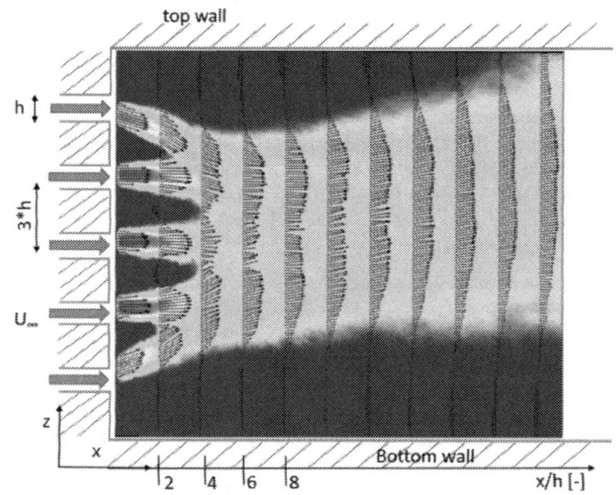

Fig. 10. Ventilated configuration Re=5857

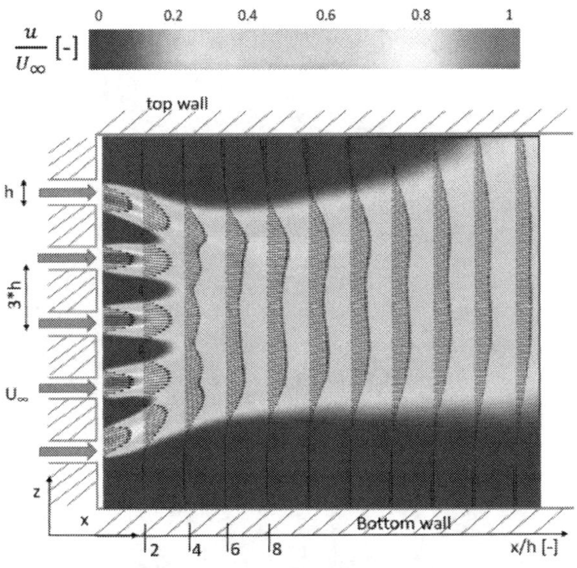

Fig. 11. Ventilated configuration Re=10510

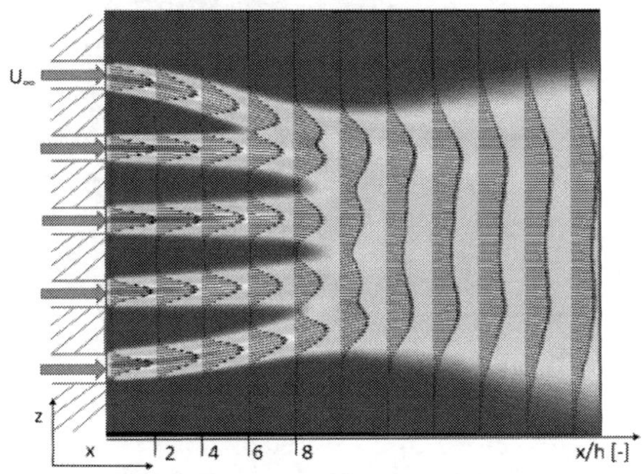

Fig. 12. Unventilated configuration Re=1408

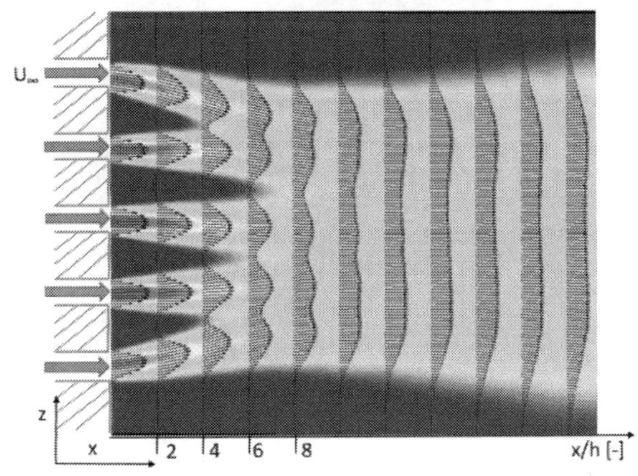

Fig. 13. Unventilated configuration Re=5857

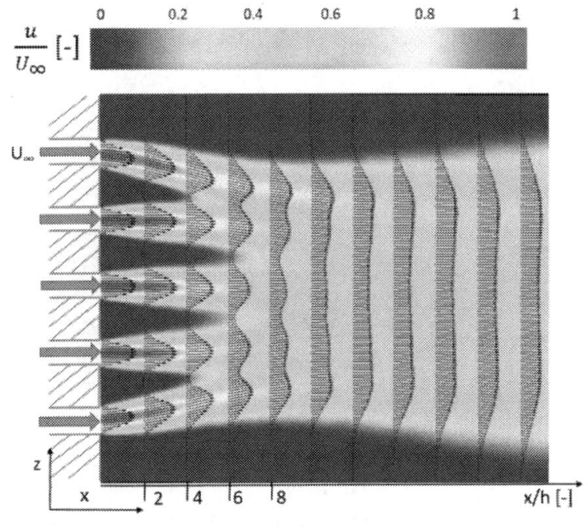

Fig. 14. Unventilated configuration Re=10510

Distance from the nozzle outlet to the merging points also called converging region tends to decrease with increase of the Reynolds number for both configurations. This is valid for laminar and transient flow. After transition to turbulent flow, the flow patterns stabilize and does not change anymore. This distance is decreased for outer jets for both configurations. We have observed that for the ventilated configuration this distance is shorter in comparison with unventilated configuration, where this finding is independent of the Reynolds number.

Big difference is in x/h=4 for unventilated configuration the multiple jets are more pronounced which is confirming that the converging region is shortened. The mean merging point for ventilated case is at x/h=5 for Re=1408 and the mean merging point for unventilated case is at x/h=7 for Re=1408. This can also be observed by looking at the velocity profiles. In other words, the ventilated case tends to have larger contraction angle, defined as inclination angle of the outer jets. This is caused by the pressure gradient downstream from nozzle outlet. Outer jets seem to have longer penetration length.

3.2. Turbulence intensity

Figures (16-21) show the turbulence intensity fields in streamwise direction. The scale ranges from 0 up to 5% for all figures. Moreover, the figures contain the magnitude of velocity in vector form of presentation.

 https://doi.org/10.1051/epjconf/201818002018

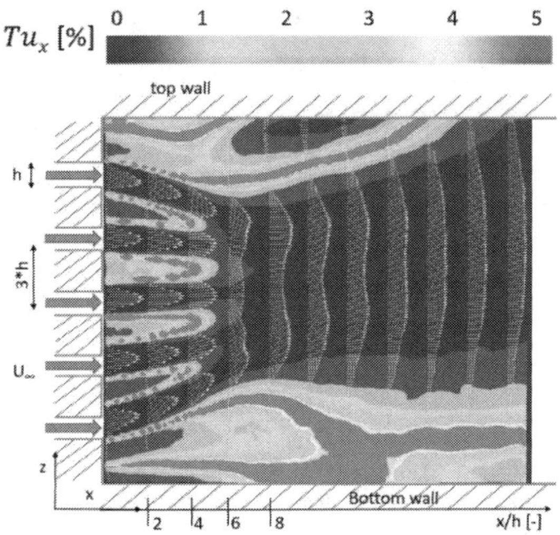

Fig. 15. Turbulence intensity for ventilated case for Re=1408

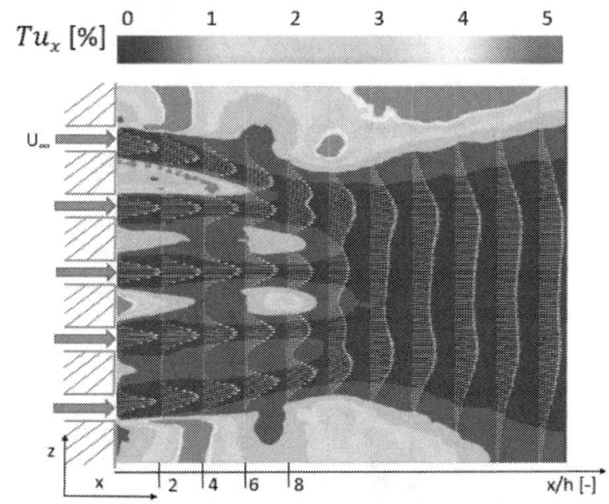

Fig. 18. Turbulence intensity for unventilated case for Re=1408

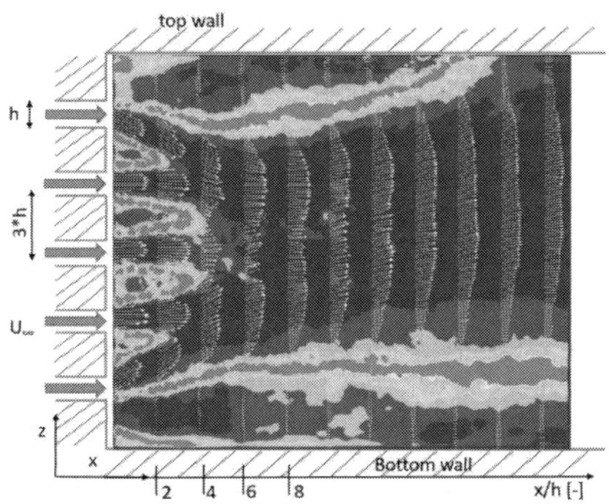

Fig. 16. Turbulence intensity for ventilated case for Re=5857

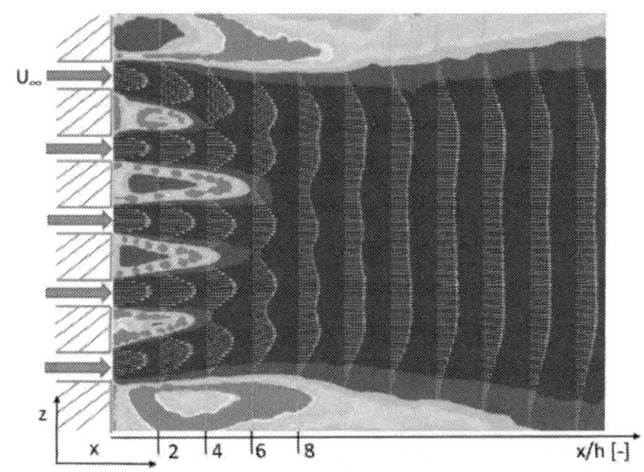

Fig. 19. Turbulence intensity for unventilated case for Re=5857

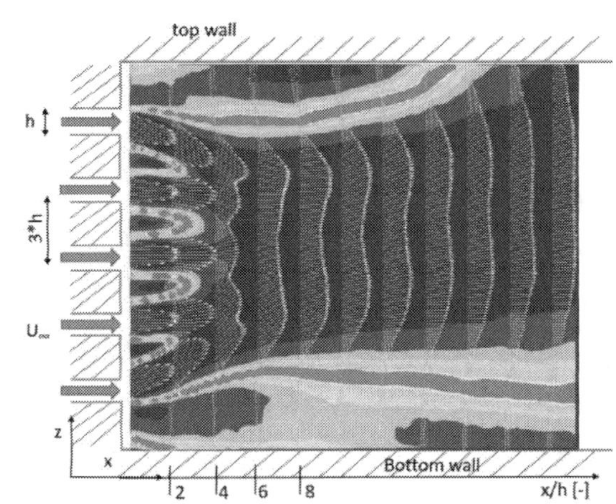

Fig. 17. Turbulence intensity for ventilated case for Re=10510

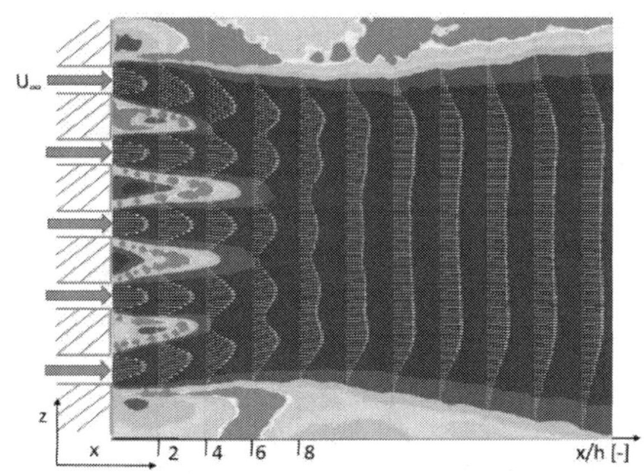

Fig. 20. Turbulence intensity for unventilated case for Re=10510

In the ventilated case, recirculation zones on top and bottom wall are generated resulting in a larger tilt angle of the outer jets compared with the unventilated case. These larger angles cause shorter converging regions, which can be observed by the shorter recirculation zones between the jets. Another difference between the configurations clear from the following turbulence intensity figures is that the border between recirculation zone and the jet flow is clearly more visible for ventilated configuration while in ventilated configuration the border is more disperse or missing. In other words, in the unventilated case the jet flow attracts the ambient atmospheric air while in the ventilated configuration the jet flow recirculates the air inside of the duct. The backward phasing step on top and bottom impact the flow differently in un/ventilated case.

3.3. Turbulence intensity in section and in point

Figure 21 and figure 22 show z component of the velocity as a function of time at position x/h=4 from outlet of the middle jet for Re=5857. The fluctuations are larger for ventilated configuration. Figure 23 shows turbulence intensity at position x/h=4 from the outlet for Re=5857. The middle jets have larger turbulence intensity than outer jets. For unventilated case the turbulence intensity is more pronounced.

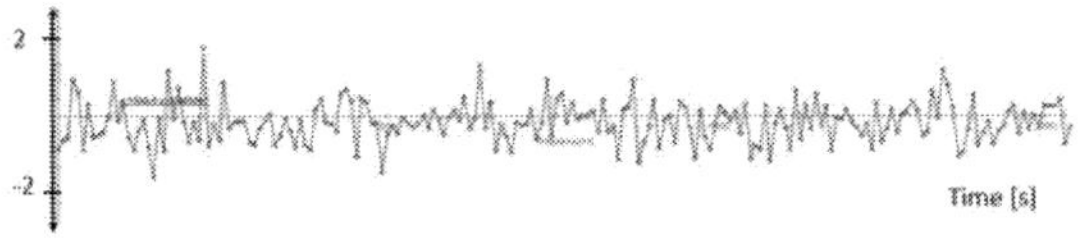

Fig. 21. – Unventilated – Outlet from middle jet at x/h=3 for Re=5857

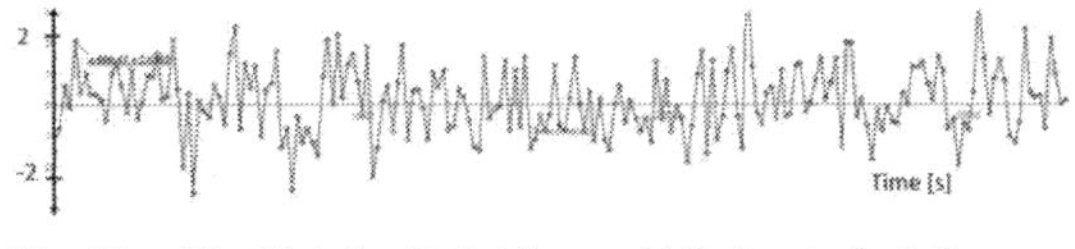

Fig. 22. – Ventilated – Outlet from middle jet at x/h=3 for Re=5857

4 Conclusion

In the ventilated case, recirculation zones on top and bottom wall are generated resulting in a larger tilt angle of the outer jets compared with the unventilated case. This effect is independent of Reynolds number. These larger angles cause shorter converging regions, which can be observed by the shorter recirculation zones between the jets. The mean merging point for ventilated case is at x/h=5 and the mean merging point for unventilated case is at x/h=7 for Re=1408. After turbulence flow is reached the merging point stays constant. The merging point is not changing from Re=5857 to Re=10510.

The large turbulence intensity of the ventilated configuration indicates strong mixing in comparison with lower turbulence intensity of the unventilated configuration. Laser Optical Measurement equipment is well suited for the analysis of such complicated velocity fields as multiple interacting jets.

Based on turbulence intensity figures the further research can be done on examining outer shear layer of multiple parallel jets.

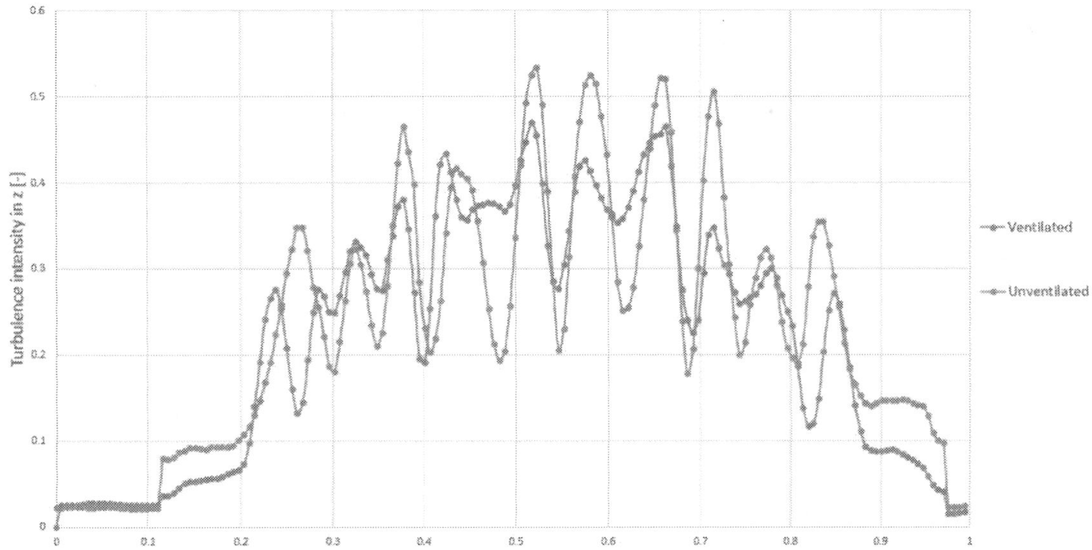

Fig. 23. Turbulence intensity in section for Re=5857

References

1. L. Moruz, J. Kitzhofer, M. Dinulescu, International Journal of Mechanical, Aerospace, Industrial, Mechatronic and Manufacturing Engineering, *Experimental investigation of plane jets exiting five parallel channels with large aspect ratio*, **11(4)**, 801 - 808, (2017)
2. H. Schlichting, *Laminare Strahlausbreitungs* (ZAMM 13, 260, 1933)
3. R.C. Deo, PhD. thesis, *Experimental investigations of the influence of Reynolds number and boundary conditions on a plane air jet*, University of Adelaide, Australia, (2005)
4. R. C. Deo, Physics of Fluids, *The influence of Reynolds number on a plane jet*, **20**, 75-108, (2008)
5. D.R. Miller, E.W. Comings, Journal of Fluid Mechanics, *Force-momentum fields in a dual-jet flow*, **7 (2)**, 237-256, (1960)
6. A. Nasr, J.C.S Lai, International Journal of Transport Phenomena, *The effects of nozzle spacing on the development of two parallel plane jets*, in International Journal of Transport Phenomena 2 (1), pp. 57-70, 2000
7. S. Ghahremanian, PhD. Thesis, *A near-field study of multiple interacting jets: Confluent jets*, Linköpimg University, Institute of Technology, Sweden, (2014)

The effect of mixed convection on the thermal field of horizontal channel flow

Jakub Devera[1,*], *Tomáš* Hyhlík[2]

[1]CTU in Prague, FME, Department of Fluid Dynamics and Thermodynamics, Technická 4, 166 07 Prague

Abstract. An experimental study was conducted to investigate the thermal field of low-speed flow of moist-air in a horizontal rectangular duct, uniformly heated from below (Poiseuille–Benard flow). Averaged temperature fields were measured in vertical mid-plane using thermocouples. The buoyancy-induced secondary flow is interacting with forced Poiseuille flow which leads to complex flow behaviour with flow reversal along the top surface. Two types of bottom boundary conditions were examined: aluminium plate (only heat transfer) and water film (simultaneous heat and mass transfer) heated to the same temperature. Presence of mass transfer causes stronger transient behaviour above the bottom surface and enhancing of heat transfer, but the overall character of the thermal field remains the same as in case of heat transfer only.

1 Introduction

Combined free and forced convection can be found in many industrial applications: heat exchangers, chemical vapour deposition reactors and in cooling of electronic components. Mixed convection flow in the horizontal channel is common for these applications. To flow in horizontal channel heated from below is referred as Poiseuille–Benard flow [1].

Bottom heating induces buoyancy driven secondary transverse flow, which alters the velocity field and temperature field. As consequence, heat transfer from the bottom wall is enhanced. Earlier studies were focused on qualifying the heat transfer enhancement and evaluating Nusselt number and its variations along the heated wall [2,3]. To understand the nature of the flow and identifying its patterns, flow visualisation studies [4,5] were conducted. Recently, PIV studies were realized to measure the velocity profiles, asymmetric due to the convection from the bottom surface [6,7].

Mixed convection can be characterized by Richardson number [8]

$$Ri = \frac{Gr}{Re^2}, \quad (1)$$

where Gr is Grashof number and Re is Reynolds number. Richardson number characterizes the ratio of buoyancy and inertia forces. If $Ri \gg 1$, natural convection is the main convection mode and contribution from forced convection is negligible and vice versa for $Ri \ll 1$. If $Ri \approx 1$, contribution from buoyancy and inertia forces are in the same relative order of magnitude.

The aim of the study is a visualisation of the thermal field in the horizontal channel subjected to the bottom heating with thermocouples and obtain average temperature field. Two types of bottom isothermal boundary conditions are inspected, aluminium plate and water film, to see whether the simultaneous heat and mass transfer influences the structure of temperature field. Obtained results will be used in upcoming CFD simulation.

2 Experimental apparatus

The experimental setup used in the present investigation is shown schematically in fig. 1. At the inlet section of the test rig is a nozzle, which straightens the flow and minimalizes its disturbances. Experiments were conducted in 300mm x 300mm square channel. The test section is 1m long and the top and side walls are made from 8mm thick plexiglass and insulated by 25mm mirelon plates. At the bottom of the test section is water tank (at the bottom and side walls are power regulated heating foils) containing heated water for maintaining stable conditions. An aluminium plate, equipped with 18 temperature sensors (digital thermometers Dallas Ds18b20) to monitor plate's temperature and its uniformity, is dipped in the tank. The maximal deviation of the temperature from the mean value is less than 1°C. The vertical position of the plate is adjustable, thus it is possible to create water film by lowering the plate. Such arrangement enables investigation of two boundary conditions at the bottom of the test section: heated aluminium plate (heat transfer) and water film (simultaneous heat and mass transfer).

The air mass flow rate through the test section is measured by orifice plate. Differential pressure transducer Setra 265 with range ±125Pa is used for measuring the orifice pressure difference. The mass flow rate is evaluated according to standard CSN EN ISO 5176-2.

Inlet and outlet temperature and specific humidity is measured by psychrometers. Each psychrometer consists

* Corresponding author: jakub.devera@fs.cvut.cz

of two sheathed RTD probes (PT1000). One of the probes is wrapped by a moisten sock (wet-bulb thermometer). One psychrometer is located at the inlet of the nozzle and two psychrometers are in outlet pipe. Measured outlet specific humidity is used for the calculation of properties of moist air (dynamic viscosity, density) necessary for mass flow rate evaluation. Mass flow rate of dry air is evaluated as

$$\dot{m}_{dry-air} = \frac{\dot{m}_{orifice}}{1 + x_{outlet}}, \quad (2)$$

where $\dot{m}$ is mass flow rate and x is specific humidity.

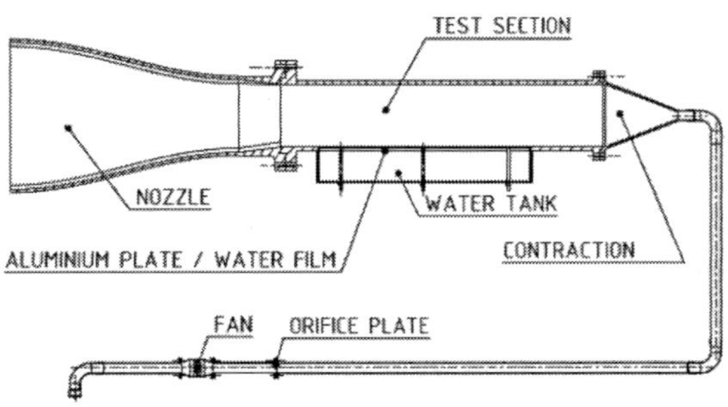

Fig. 1. Schematic of the experimental rig

2.1. Thermal field measurement

The thermal field is measured with 10 thermocouples, type K. Data are acquired via NI cDAQ-9174 chassis with four NI 9211 modules which include cold junction compensation. The data are obtained by Matlab (via Instrument control toolbox) and further processed. Thermocouples are mounted to a horizontal traverser unit, vertical distance between probes is 31mm, the bottom thermocouple is at height of 10mm (see fig. 2). Whole 1m long test section is traversed by the probe stand, also 0.3m in front of and 0.1m behind the test section is included, in total 15 horizontal positions (horizontal distance between positions is 100mm). Measurements are conducted in streamwise central vertical plane. In each position is waited for thermal stability, measurement itself in one horizontal position takes 5 minutes. Processing of the data is done in Matlab. Data from each position are averaged

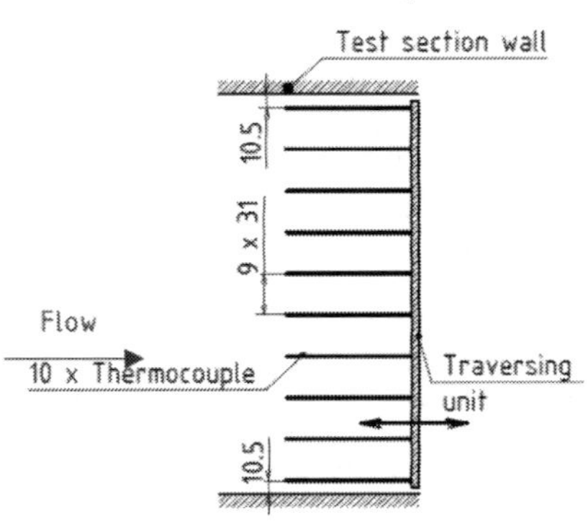

Fig. 2. Horizontal traversing stand with 10 thermocouples used for the thermal filed measurement, horizontal distance between positions was 100mm

and then interpolated on the measurement grid. Due to the traversing, only averaged temperature field is obtained, any instantaneous behaviour is not captured. The measurement was repeated 3 times in the following days to test the repeatability, the results were comparable.

2.2. Evaluating dimensionless numbers

To identify the regime of the flow, it is necessary to evaluate dimensionless criteria from the inlet and boundary conditions, the Reynolds number is calculated according to

$$Re = \frac{u\, D_h}{\nu}, \quad (3)$$

where u is the mean velocity in the test section based on the mass flow rate, ν is kinematic viscosity evaluated based on inlet conditions, D_h is hydraulic diameter. Grashof number is calculated as

$$Gr = \frac{g\, \Delta T\, \beta\, D_h^3}{\nu^2}, \quad (4)$$

where $\Delta T = T_{boundary} - T_{in}$ and β is the coefficient of thermal expansion, calculated as $\beta = \left(\frac{T_{in}+T_{boundary}}{2}\right)^{-1}$. $T_{boundary}$ is the temperature of the aluminium plate, water film resp., and T_{in} is the temperature at the inlet of the experimental rig, g is acceleration due to Earth's gravity.

3 Results and discussion

The parameter, which qualifies the relative contribution of forced and mixed convection is Richardson number, $Ri = Gr/Re^2$. Both, Reynolds and Grashof number is evaluated based on hydraulic diameter D_h. In the present study, the value of Reynolds number is 2045, Grashof number $1.33 \cdot 10^8$ and Richardson number is 32, which implies, that the natural convection is the dominant heat transfer mode. The bottom heating causes buoyancy driven secondary flow, that has a large impact on the thermal field in the channel. Measured thermal field for the vertical mid-plane is shown in fig. 3a for the aluminium plate case (heat transfer) and in 3b for the water film case (simultaneous heat and mass transfer). The heated plate, resp. water film, starts at $z = 0mm$ and ends at $z = 1000mm$. The temperature of the plate, water film resp., is $52.5°C$. The y axis is normalized by hydraulic diameter

$$y^* = \frac{y}{D_h}, \quad (5)$$

the thermal field is normalized as

$$T^* = \frac{T - T_{in}}{T_{boundary} - T_{in}}. \quad (6)$$

Both thermal fields have the same complex structure. Forced Poiseuille flow interacts with buoyancy driven secondary flow. The peak of the cold stream's core is shifted toward the heated plate ($y^* \approx 0.35$), which caused very steep temperature gradient above the heated plate /

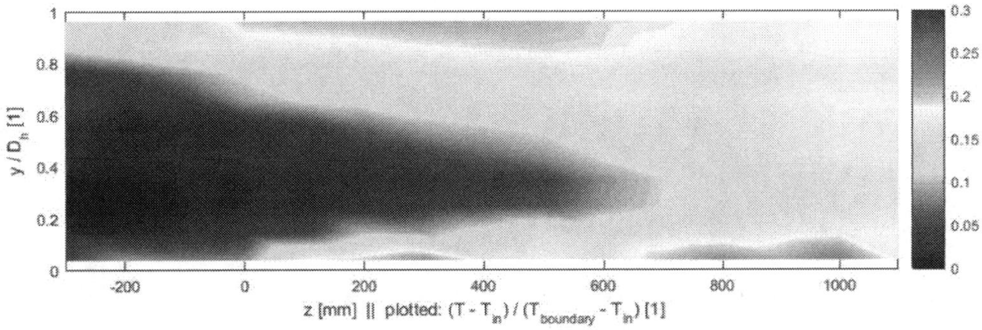

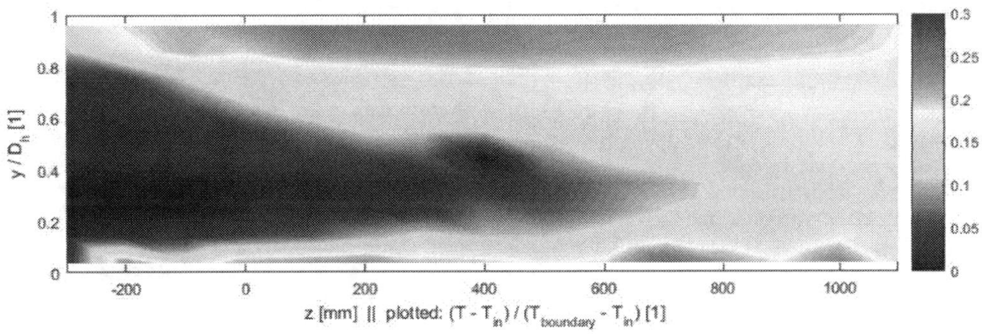

Fig. 3a) (up): mid-plane temperature field above aluminum plate (heat transfer); **3b)** (down): mid-plane temperature field above water film (simultaneous heat and mass transfer)

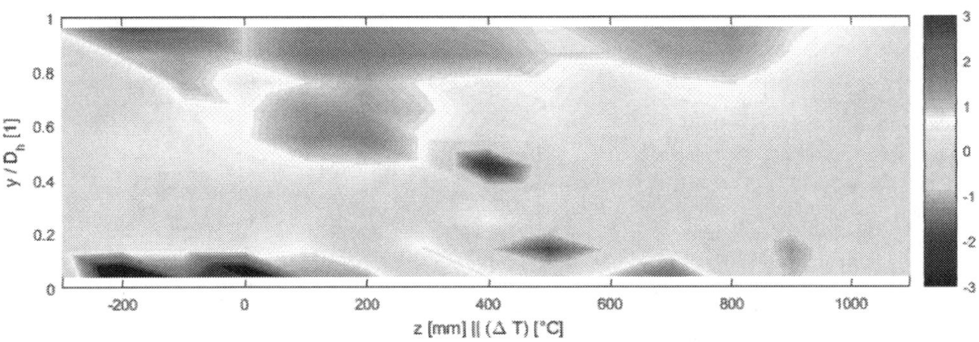

Fig. 4. Difference of temperature fields (nonnormalized); the temperature field above aluminum plate is subtracted from temperature field above water film

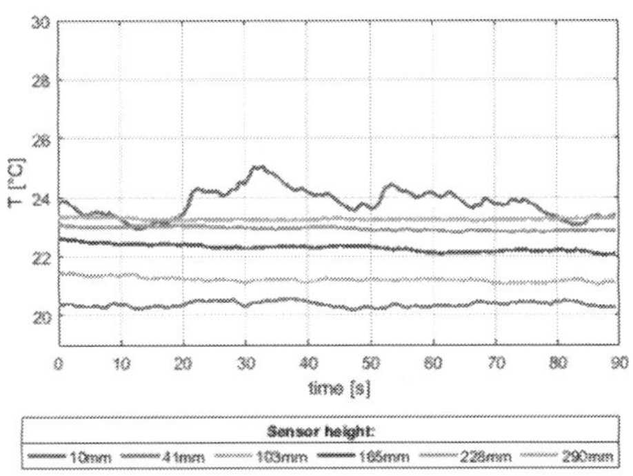

Fig. 5a. Temperature fluctuations vs. time from thermocouples, horizontal position $z = 500mm$ for aluminum plate case

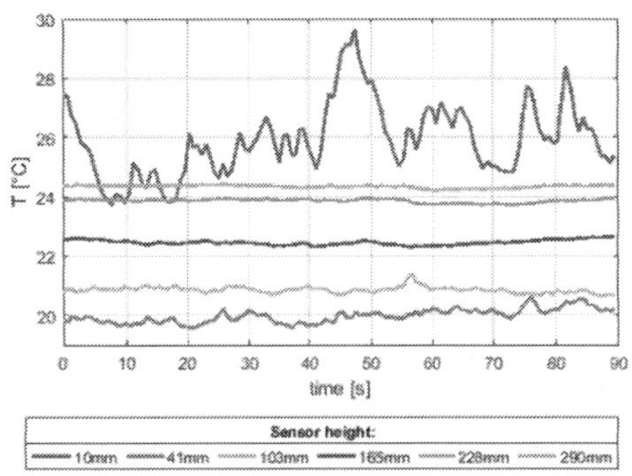

Fig. 5b. Temperature fluctuations vs. time from thermocouples, horizontal position $z = 500mm$, water film case

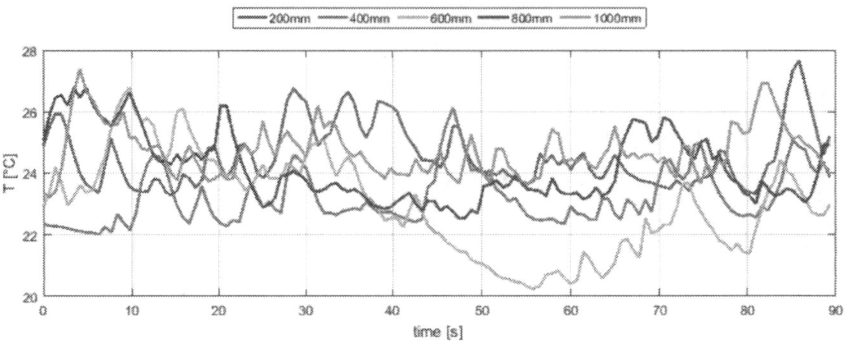

Fig. 6. Temperature fluctuations vs. time from thermocouples, y=10mm, different horizontal positions, water film case measurement

hot fluid film. The heated region below the top surface steps into area $x \in (-300, 0)mm$, which is in upwind direction. That is caused by reversed flow along the top surface. The backflow was observed by other researchers studying horizontal mixed convection flow with $Re > 10$ [5-7, 9]. The flow reversal makes the fluid in the upper region hotter than that below due to the convective heat transfer from the downstream fluid [9]. The heated region along the top surfaces covers the entire section and continues to nozzle, thus it not any local behaviour. A large convective cell is created along the entire heated section. The reversed flow a product of complicated 3D flow behaviour because of buoyancy forces, which creates crossflow and positive streamwise pressure gradient [9]. In experiments with water as a working fluid was discovered that reversed flow is induced when $Gr/Re^2 > 55$ [6]. During this measurement with air, the reverse occurs already when $Gr/Re^2 = 32$.

As was mentioned, two boundary conditions were investigated: aluminium plate and water film. Mass transfer doesn't have any effect on the structure of the thermal field, it is the same in both cases, but the thermal field in the second case is hotter in the upper region. The fluid is enhanced by the hot water vapour, which leads to the increase of the temperature. The absolute temperature difference of the both fields is in fig. 4. The airflow is hotter in the upper region ($y^* > 0.6$) by 2°C. Differences are also at height of 10mm, but there is problem with steep temperature gradient, where even a small displacement of the probe can cause an error in order of degrees Celsius.

The temperature variations versus time are also inspected. The character of natural convection is unsteady, but the signal from the thermocouples is almost stable in both cases, see fig. 5a and fig. 5b except the signal of the lowest thermocouple, from 10mm. The temperature at this height is fluctuating, the changes are almost 5°C for the water film case, 1.5°C for the aluminium plate. At height of 41mm above the surface, the signal was much less transient, fluctuations were maximally 0.5°C for both cases. The transient behaviour is the same in other horizontal positions, in fig. 6 is plotted dependency of temperature on time from the lowest thermocouple from another 5 horizontal positions.

The temperature signal from 10mm height from the water film case was subjected to spectral analysis using Fast-Fourier transformation, but any significant frequency wasn't discovered.

4 Conclusion

The thermal field of the Poiseuille–Benard ($Ri = 32$) flow was measured in the vertical mid-plane using thermocouples. Natural convection is dominant heat transfer mode and has significant impact on the thermal field. Along the top surface of the test section, reversed flow was observed. In case of water film, as a bottom boundary, the airflow in the upper region was hotter by 2°C. For this case was also observed strong transient behaviour at height of 10mm above the water film, the temperature was fluctuating in range of 5°C opposite to the aluminium plate case, where the fluctuations was in range of 1.5°C.

Acknowledgement

We gratefully acknowledge the support from Centre of 3D volumetric velocimetry (COLA), reg. no. CZ.2.16/3.1.00/21569, supported by operational programme Prague Competitiveness.

References

1. Bonnefoi, C., C. Abid, M. Medale and F. Papini , *Int. J. Thermal Sciences,* **43,** 791 (2004)
2. Mahaney, H. V., F. P. Incropera and S. Ramadhyani, *Numerical heat transfer,* **12,** 137 (1987)
3. Maughan, J. R., and F. P. Incropera, *Int. J. Heat Mass Transfer,* **30,** 1307 (1987)
4. Wang L. W., K. H. Hou, I. G. Lu and C. F. Hsu, *J. Thermal Energy Generation, Transport, Storage, and Conversion*, **9**, 257 (1996)
5. K. Toriyama and K. Ichimiya, *J. Flow Visualization and Image Processing*, **17**, 69 (2010)
6. Elatar A. and K. Siddiqui, *Int. J. Heat and Fluid Flow*, **46**, 29 (2014)
7. Elatar A. and K. Siddiqui, *Int. J. Thermal Sciences,* **90,** 351 (2015)
8. F. P. Incropera and T. L. Bergman, *Fundamentals of heat and mass transfer*, 7th ed, Wiley (2011)
9. Choi D.K. and D.H. Choi, *Int. J. Heat Mass Transfer,* **37,** 1899 (1994)

Temperature profiles measurements in turbulent Rayleigh-Bénard convection by optical fibre system at the Barrel of Ilmenau

Jakub Drahotský[1,*], *Pavel* Hanzelka[1], *Věra* Musilová[1], *Michal* Macek[1], *Ronald* du Puits[2], and *Pavel* Urban[1]

[1] *Institute of Scientific Instruments of the CAS, v.v.i., Královopolská 147, Brno, Czech Republic*
[2] *Technische Universität Ilmenau, Institute of Thermodynamics and Fluid Mechanics, 98684 Ilmenau, Germany*

Abstract. Modelling of large-scale natural (thermally-generated) turbulent flows (such as the turbulent convection in Earth's atmosphere, oceans, or Sun) is approached in laboratory experiments in the simplified model system called the Rayleigh-Bénard convection (RBC). We present preliminary measurements of vertical temperature profiles in the cell with the height of 4.7 m, 7.15 m in diameter, obtained at the Barrel of Ilmenau (BOI), the worldwide largest experimental setup to study highly turbulent RBC, newly equipped with the Luna ODiSI-B optical fibre system. In our configuration, the system permits to measure the temperature with a high spatial resolution of 5 mm along a very thin glass optical fibre with the length of 5 m and seems to be perfectly suited for measurement of time series of instantaneous vertical temperature profiles. The system was supplemented with the two Pt100 vertically movable probes specially designed by us for reference temperature profiles measurements.

1 Introduction

Natural thermally-generated turbulent flows play a vital role in heat transfers in nature, be it the circulation in the atmosphere or the oceans, the flows under the surface of stars or in diverse branches of industry. The Rayleigh-Bénard convection (RBC) is a simplified physical model serving for fundamental studies of thermally-generated flows occurring in an Oberbeck-Boussinesq (OB) fluid layer in gravitational field confined between two horizontal, laterally infinite and perfectly conducting plates heated from below and cooled from above [1]. RBC is fully characterized by two dimensionless parameters, the Rayleigh numbers (Ra) and the Prandtl number (Pr), which are defined as follows: $Ra = g(\alpha/\nu\kappa)\Delta T L^3$, $Pr = \nu/\kappa$. Here g is acceleration due to gravity, ΔT is the temperature difference between the temperature T_b of the bottom plate and temperature T_t of the top plate. The plates are mutually separated by the vertical distance L. When the experiment is conducted in a vertical cylindrical cell with the diameter D, the geometry of the cell is characterised by aspect ratio $\Gamma = D/L$. The physical properties of the fluid in the OB approximation are represented by temperature independent material properties: the isobaric thermal expansivity α, the kinematic viscosity ν, the thermal diffusivity κ. The temperature variation of density ρ is supposed to be less than a few percent. Breaking these conditions leads to Non-Oberbeck-Boussinesq (NOB) effects.

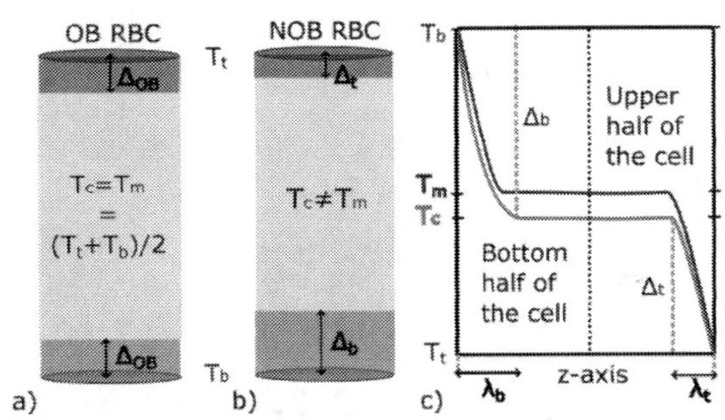

Fig. 1. Schematic illustration of the thermal boundary layers and vertical temperature profiles in RBC: a) Thermal boundary layers in OB RBC where the temperature drops (Δ_{OB}) on the top and bottom boundary layers are equal. The turbulent core temperature T_c equals the mean temperature T_m. b) Thermal boundary layers in NOB RBC, where the temperature drops on the top ($\Delta_t = T_c - T_t$) and bottom ($\Delta_b = T_b - T_c$) layers are asymmetric. The turbulent core temperature T_c differs from the mean temperature T_m. c) The vertical temperature profile in the OB (blue) and NOB (red) RBC. λ_t and λ_b represents the thickness of the top and bottom boundary layers.

The principal motivation of our long-term work is to shed light on the discrepancies among published results on heat transfer efficiency by RBC for Ra values above 10^{11} [2–4] and influence of NOB effects on RBC, especially on a possible transition of RBC to the ultimate regime [5, 6]. It was observed experimentally that NOB effects do lead to asymmetry of the thermal boundary layers adjacent to the

*e-mail: drahotsky@isibrno.cz

top and bottom plates (see schematic illustration in Fig. 1), which subsequently alters the heat transfer efficiency at high *Ra* [5, 7].

In our previous work, we studied NOB effects in cryogenic RBC system using Czech Cryogenic Turbulence Facility (CCTF-3) [8, 9] with the $\Gamma = 1$ cell ($D = L = 0.3$ m) using cryogenic helium as a working fluid. The current work aims to extend our knowledge about NOB effects in RBC via measurement of vertical temperature profiles in the cell of $\Gamma \sim 1$ for varying *Ra* numbers in the range of 10^{10} to 10^{12}. The experiments presented here took place in the Barrel of Ilmenau (BOI) [10–12] using dry air as a working fluid. The BOI is newly equipped with the optical fibre system Luna ODISI-B (Optical Distributed Sensor Interrogator) [13] for distributed measurement of temperature along the whole fibre length. In this paper we report the first experiment verifying capabilities of Luna ODISI-B for measurement of temperature profiles in RBC systems.

2 Experimental arrangement

2.1 Barrel of Ilmenau

The Barrel of Ilmenau (BOI) is an experimental facility (see Fig. 2) for the study of highly turbulent convection using dry air as a working fluid ($Pr = 0.7$ at atmospheric pressure and working temperatures). The cylindrical cell ($D = 7.15$ m) with movable top plate allows to adjust the cell height L in the range of 0.2 m to 6.3 m. The corresponding *Ra* number and the aspect ratio Γ vary in the range of 10^5 to 10^{12} and 1.13 to 36, respectively [10]. The cooling (heating) system of the top (bottom) plate allows to set the temperatures of the plates within the range from 10 °C up to 80 °C.

Fig. 2. Experimental facility Barrel of Ilmenau (BOI) is a world-wide largest experiment (diameter $D = 7.15$ m, adjustable height L up to 6.3 m) for the study of highly turbulent convection in dry air.

2.2 Optical fibre measurement system Luna ODISI-B

Luna ODISI-B is a system that uses a thin glass optical fibre as a sensor for distributed temperature or strain measurements based on Rayleigh scattering, evaluated by Optical Frequency Domain Reflectometry (OFDR) combined with a Mach Zehnder interferometer [13]. The resulting quantity indicated by the system is called the "frequency shift" that is converted into a temperature information by a suitable calibration (see Sec. 3.2). In general, the system permits to measure the temperature profiles along the fibre with the length of up to 20 m with acquisition rate of 250 Hz and spatial resolution down to 1.25 mm in dependence on type of the system configuration. Detailed parameters of the system used in our experiments are given in Table 1.

Table 1. Specifications of the Luna ODiSI-B device installed in the BOI.

Length of the fibre	5 m
Temperature repeatability	±0.4 °C
Temperature range	−268 to 900 °C
Spatial resolution	5 mm
Acquisition rate	up to 100 Hz

The installation of Luna ODISI-B together with Pt100 probes in the BOI is schematically shown in Fig. 3. The fibre is fastened to a special holder above the top plate, passes through a hole in the plate down to the bottom plate where it is fixed. The last ∼ 10 cm of the fibre cannot be used for the measurements (because of the low refraction termination). To obtain also temperature profile within the region of the bottom boundary layer, we bent this part of the fibre and fixed it to the bottom plate.

The temperature profiles were measured at two verticals: along the cell axis and a vertical line displaced 1 m from the sidewall (side position). The arrangement of the Luna ODISI-B system above the top plate for measurements at the central axis is shown in Fig. 4.

The thermal response of the fibre is dominated mainly by dn/dT effect, where n is the index of refraction and T is the temperature. To fix the fibre in the defined position, we glued it to a supporting fishing line with little bends by epoxide every 30 cm. The fishing line was vertically tightened between the plates. We suppose that in this configuration, the deformation of fibre and its effect on measured profiles is minimized.

2.3 Pt100 temperature probes for reference temperature profiles measurements

The Luna ODISI-B system was supplemented with two vertically movable Pt100 probes specially designed at the Institute of Scientific Instruments of the CAS (ISI) for reference measurements of the temperature profiles (Fig. 5). The lower Pt100 probe was used for the measurement in the region of the bottom boundary layer and turbulent core, the upper Pt100 probe was used only for measurement in the region of the top boundary layer (Fig. 3).

Both Pt100 probes were suspended on the feed cables passing through the hole in the top plate. We could move the probes by hand via cables up and down.

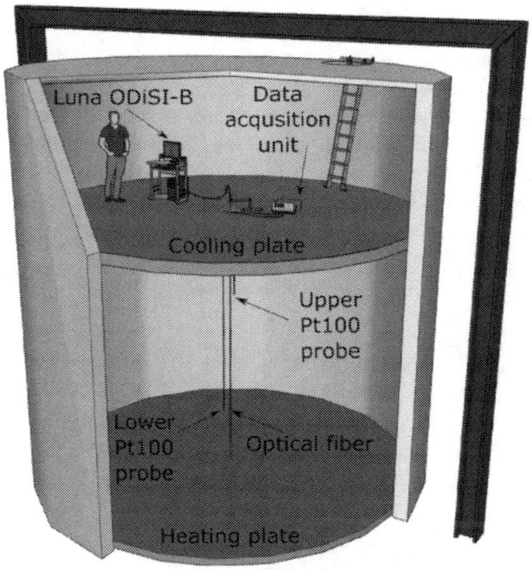

Fig. 3. Schematic illustration of Luna ODISI-B and Pt100 probes installation in the axis of BOI. Luna ODiSi-B system and data acquisition unit for Pt100 probes were placed on an upper side of the top plate. The optical fibre passes through a hole in the plate and is fixed to the bottom plate. Both Pt100 probes were suspended on the feed cables passing through the hole in the top plate together with the fibre. The upper Pt100 probe serves only for measurement of temperatures profiles across the top boundary layer, whereas the lower Pt100 probe serves for measurement both across the bottom layer and in the turbulent core.

The Pt100 probes consist of a 0.3 mm thin platinum resistor film (type 03 002, tolerance class A with 0.25 °C uncertainty at 45 °C) produced by ZPA EKOREG [14]. The resistors with the nominal resistance value of 100 Ω (measurement range of −40 °C to 400 °C) and nominal size of 2.3 mm × 5.0 mm is fastened to a small brass plate with a gold tip. The brass plate (0.7 mm in thickness) is linked to the 1.5 mm thick rectangular holder (6 mm × 250 mm) made of fibreglass. The sensor is connected to the copper contacts of a cable connector via 4-wire "ribbon cable" Lake Shore WQL 36 composed of thin phosphor bronze wires (Fig. 5). The gold tip works as an electrical indicator of a mechanical contact of the probe with the bottom plate and it is also electrically connected to the cable connector. To prevent swinging of the lower Pt100 probe on the 4.7 m long feed cable, we equipped the holder with two guide handles sliding along a fishing line vertically tightened between the plates. Moreover, an additional weight in the form of a brass ring was placed on the lower probe. This weight keeps the feed cable tight. The materials chosen for manufacturing of the probes have low thermal conductivity with the aim to minimize the influence of the probes on the measured temperature profiles.

3 Results and discussions

We performed the test measurements of vertical temperature profiles by Luna ODISI-B and reference Pt100 probes

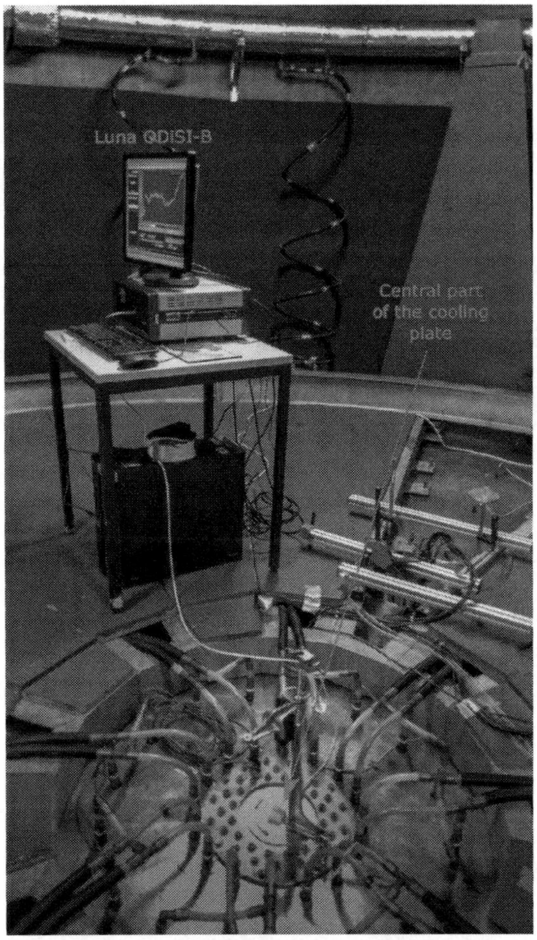

Fig. 4. The arrangement of Luna ODISI-B in the space above the top plate. Optical fibre here enters the cell in the centre of the plate. The pipelines belong to the water cooling system of the plate.

along (i) the cell axis and (ii) along a vertical at a radial position of 1 m from the cell sidewall. The measurements were done for various temperature differences ΔT between the plates within the range of Ra numbers of 1×10^{10} to 4×10^{11}. In all measurements, the cell was filled with dry air at atmospheric pressure. The mean temperature T_m of air was kept at 30 °C ± 0.1 °C. In the following sections (3.1 - 3.3) we present example test measurements obtained at two Ra numbers.

3.1 Temperature profiles measured by Pt100 probes

Time-averaged temperature profiles $\langle T_{Pt}(z)\rangle$ measured by Pt100 probes in both locations along the central axis and out of the centre are presented in Fig. 6. The $\langle T_{Pt}(z)\rangle$ profiles represent time-averaged temperatures measured in each distance z within 1 minute interval with the acquisition rate of 10 Hz. The profiles were measured at $Ra = 2.18 \times 10^{11}$ with the temperature difference of $\Delta T =$

20 °C. In the region of the boundary layers, the measurements were done with steps of 5 mm. Due to the dimensions of the probes, it was not possible to obtain temperatures closer than 5 mm from the plates. Although only few points could be measured directly in the boundary layers, the transitions of temperature profiles from the boundary layers to the bulk are clearly visible. The shift of the side position profile to higher temperatures is expected as a result of upward convection flow near the side wall. The differences between the time-averaged temperatures $\langle T_{Pt}(z)\rangle$ and the mean temperature T_m in the region of turbulent core can be ascribed to the NOB effects [3, 5]. The situation corresponds to the schematic illustration in Fig. 1b, however with the BL asymmetry inverted (the measured bottom BL is in fact narrower than the top BL).

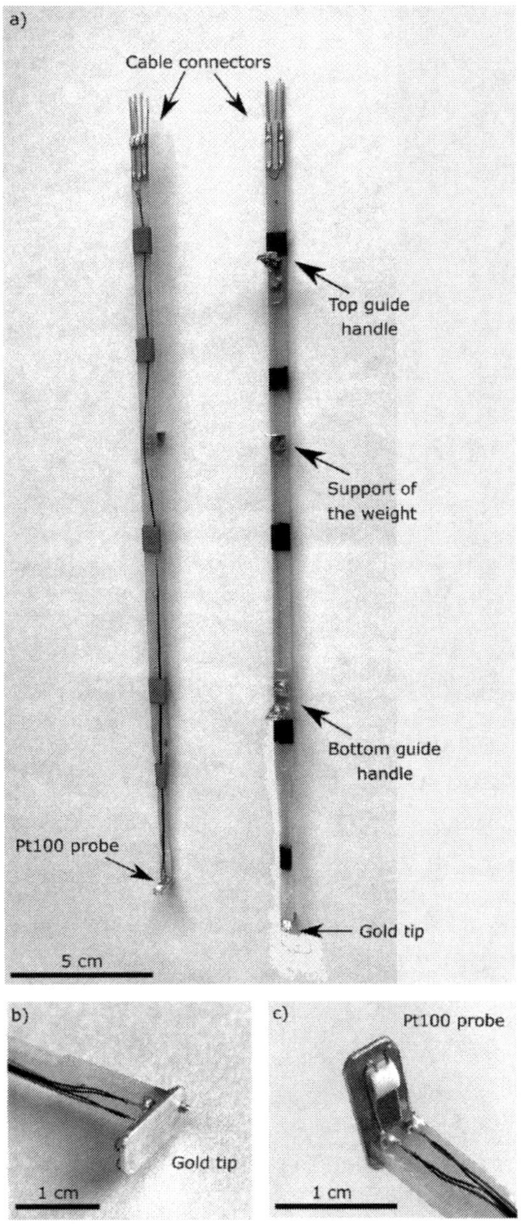

Fig. 5. Lower Pt100 probe for the reference temperature profiles measurements designed at ISI. a) The probe is shown from the front and back side. b) Detail view on of the gold tip which serves as the electrical indicator of mechanical contact of the probe with the bottom plate. Opposite to the tip, the small electrode made of the copper sheet is installed on the surface of the bottom plate to secure the electrical connection. c) Detail view of the Pt100 sensor fastened to the brass plate.

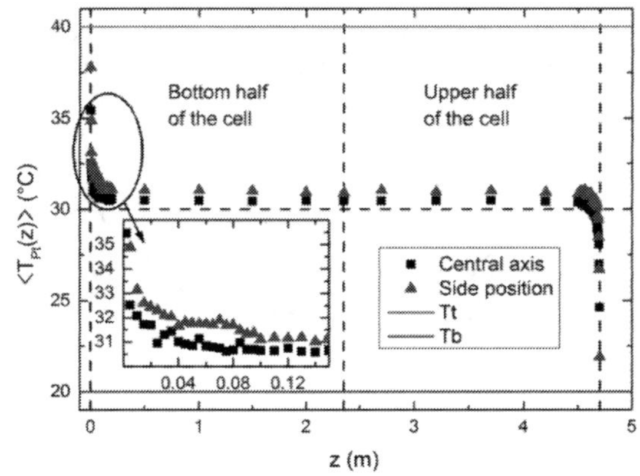

Fig. 6. Time-averaged temperature profiles as $\langle T_{Pt}(z)\rangle$ a function of distance z from the bottom conducting plate measured by Pt100 probes ($\Delta T = 20\,°C, Ra = 2.18 \times 10^{11}$). The vertical dashed black lines indicate the location of the cell centre and the plates position. The blue filled triangles represent the measurements along the central axis of the cell, the black filled squares represent the measurements along the axis 1 m away from the cell sidewall. The inset shows detail view on the profiles in the region of the bottom boundary layer. The solid horizontal lines represent the mean temperatures of the top (T_t - blue line) and bottom (T_b - red line) plates.

3.2 Calibration of Luna ODISI-B

Fig. 7 shows the measured time-averaged frequency shift as a function of distance z away from the bottom plate at $\Delta T = 30\,°C$ ($Ra = 3.27 \times 10^{11}$). The data correspond to an average obtained from a one and half hour long measurement with the acquisition rate of 10 Hz. The inset shows a calibration curve used for the transformation of the time-averaged frequency shift onto the time-averaged temperature $\langle T(z)\rangle$. The red solid curve is a second order polynomial fit of the three selected data points at the plates and in the centre of the cell. Temperatures used for calibration curve were measured by the lower Pt100 probe moved to the centre of the cell and by temperature sensors embedded in the plates. Minimal (maximal) frequency shift was assigned to the temperature of the bottom heating (top cooling) plate.

3.3 Temperature profiles measured by Luna ODISI-B

Resulting time-averaged temperature profile $\langle T(z)\rangle$ is shown in Fig. 8 (green solid line and filled green circles).

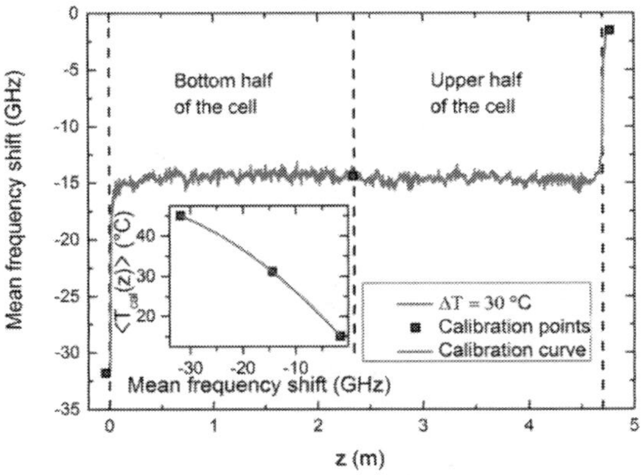

Fig. 7. The time-averaged frequency shift as a function of the vertical distance z from the bottom conducting plate (ΔT = 30 °C, $Ra = 3.27 \times 10^{11}$) measured in a vertical 1 m from the cell sidewall. The vertical dashed black lines indicate the location of the cell centre and the positions of the plates. The inset represents a calibration curve used for transformation of the time-averaged frequency shift to the time-averaged temperature $\langle T(z)\rangle$. The red curve is second order polynomial fit of the three selected data points shown as blue squares. The temperature $\langle T_{cal}(z)\rangle$ of the middle point was measured by the lower Pt100 probe, the other two points were measured by sensors placed in plates.

Here $\langle T(z)\rangle$ profile is compared with the profile measured by Pt100 probes (blue filled triangles). The detailed view on the profiles in the region of the boundary layer at the bottom plate shows that the both profiles do not fit one to each other ideally (see inset). It is probably caused by imperfect calibration curve which must be refined in the future. In addition, the correction of the background spatial noise pattern ~ ± 1 °C, due to fibre imperfection (seen now in Figs. 7 and 8) needs to be done.

4 Conclusions

We applied a new optical system LUNA ODiSI-B and presented here preliminary results on vertical temperature profiles measurement in turbulent RBC across the whole height of the experimental cell, involving the turbulent core as well as the thermal boundary layers. This system, using a glass optical fibre, has not yet been used in any RBC experiments. The LUNA ODiSI-B should enable us to get instantaneous profiles with a high spatial resolution of about 5 mm along the fibre with the length of 4.7 m. We compared the time averaged profiles obtained by this method and by movable Pt100 temperatures probes. The temperatures obtained by Luna ODISI-B with calibration curve delivered by the manufacturer failed in our measurements. Therefore we performed our own simple three point calibration before each of the measurement for various Ra numbers. The evaluated temperature profiles within boundary layers still differed markedly from the profiles obtained by Pt100 probes. The calibration will be refined in the future data processing, and corrections extracting the background spatial noise due to fibre imperfection will be done. Reference temperature profiles were

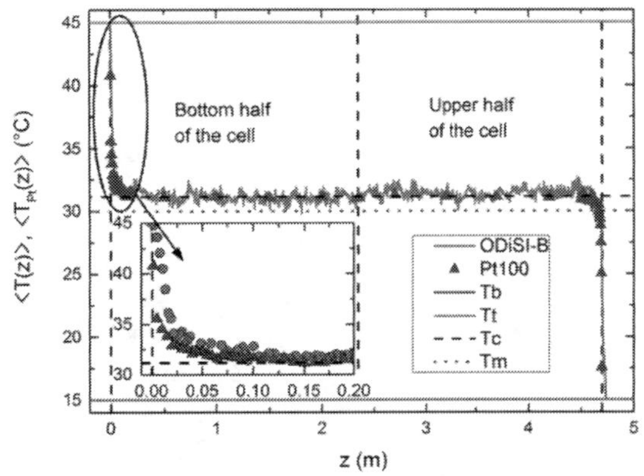

Fig. 8. Comparison of the time-averaged temperatures measured by Luna ODISI-B (green solid curve, filled green circles) and Pt100 probes (blue filled triangles) as a function of distance z from the bottom plate (ΔT = 30 °C, $Ra = 3.27 \times 10^{11}$). Both profiles were measured along the vertical in the radial position 1 m from the cell sidewall. The vertical dashed black lines indicate the location of the cell centre and the plates. The solid horizontal lines represent the time-averaged temperatures of the top (T_t - blue line) and bottom (T_b - red line) plates. The horizontal dotted black line corresponds to the time-averaged mean plate temperature T_m. The dashed black line represents the time-averaged turbulent core temperature T_c measured in the cell centre by Pt100 probe.

obtained by Pt100 probe measurement. In these, asymmetry between top and bottom thermal boundary layers (NOB effect) was observed. Opposite to Pt100 probes, the Luna ODISI-B is promising for study of the time-dependence [15] of temperature profiles and space-time temperature correlations [16]. This motivates us to continue the study of applicability of this method in RBC experiments.

Acknowledgements

We thank L. Skrbek, T. Králík, and A. Srnka for stimulating discussions, and further especially to P. Krejčí for technical support with the Pt100 probes manufacturing and S. Abawi and V. Mitschunas for help with performing the measurements at BOI. We acknowledge the support of this research by the Czech Science Foundation under the project GA17-03572S and by the European Community Framework Programme 7, EuHIT (European High-performance Infrastructures in Turbulence) Grant Agreement No. 312778. M. Macek acknowledges the support by the MEYS CR project LO1212.

References

1. G. Ahlers, S. Grossmann, D. Lohse, Rev. Mod. Phys., **81**, 503 (2009)
2. F. Chillà and J. Schumacher, Eur. Phys. J. E, **35**, 58 (2012)
3. P. Urban, P. Hanzelka, V. Musilová, T. Králík, M. La Mantia, A. Srnka, L. Skrbek, New J. Phys., **16**, 053042 (2014)

4. P. Urban, V. Musilová, L. Skrbek, Phys. Rev. Lett., **107**, 014302 (2011)
5. L. Skrbek and P. Urban, J. Fluid Mech., **785**, 270 (2015)
6. X. He, E. Bodenschatz, G. Ahlers, J. Fluid Mech., **791**, R3 (2016)
7. P. Urban, P. Hanzelka, T. Králík, V. Musilová, A. Srnka and L. Skrbek, Phys. Rev. Lett., **109**, 154301 (2012)
8. P. Urban, P. Hanzelka, T. Králík, V. Musilová, L. Skrbek, and A. Srnka, Rev. Sci. Instrum., **81**, 085103 (2010)
9. https://www.euhit.org/infras/czech-cryogenic-turbulence-facility/facilities
10. R du Puits, Ch. Resagk and A. Thess, New J. Phys., **15**, 013040 (2013)
11. R du Puits, Ch. Resagk, A. Tilgner, F. H. Busse, A. Thess, J. Fluid Mech., **572**, 231 (2007)
12. https://www.euhit.org/infras/barrel-of-ilmenau
13. http://lunainc.com/odisi
14. http://www.zpaul.cz/
15. Y. Wang, X. He, P. Tong, Phys. Rev. Fluids, **1**, 082301(R) (2016)
16. V. Musilová, T. Králík, M. La Mantia, M. Macek, P. Urban, L. Skrbek, J. Fluid. Mech. **832**, 721 (2017)

Complete Description of Forces Acting on a Flying Beach Volleyball

Jan Dumek[1*], *Pavel* Šafařík[1],

[1] Department of Fluid Dynamics and Thermodynamics, Faculty of Mechanical Engeneering, Czech Technical University in Prague, Technická 4, 190 00, Czech Republic

Abstract. Complete description of all forces acting on a flying Beach Volleyball was made based on measurements in the wind tunnel. Forces (drag, lift and side force) were measured for different angle of attack β which varies from 0° to 47°. Velocity region was from 10 to 25 m/s and revolution region was from 0 to 12.5 rps. Moments (Roll, Yaw, Pitch) were detected. Results are described by means of non-dimensional numbers, such as Reynolds number Re, spin s, drag C_D, lift C_L and side force C_S coefficients. Differences in results of C_D, C_L and C_S were detected for various angle β and are further described in the article. Conclusions of the investigation can be utilized 1st by ball producers for practical use in development, 2nd for sport Methodist to build more exact methodology for Beach Volleyball, 3rd in basic and applied aerodynamic research.

1 Introduction

Forces acting on rotating sport ball (as well as any other body in flow) which is in the wind tunnel test section are three, drag force F_D, lift force F_L and side force F_S. Three moments are acting on balls in the flow in a non-symmetric case of flow "Roll" M_X, "Yaw" M_Y, "Pitch" M_Z.

Experimental data of forces acting on the flying rotating sphere were measured in different studies and were published in articles [1], [2], [3], [4], [5] and others. Different approaches were used. Free observation of rotating spheres, which are thrown into water is one approach [1], second describes trajectory of flight of a volleyball in real conditions [4] and others [3] measured forces acting on the volleyball fixed in a special stand in an aerodynamic wind tunnel test section. Wind tunnel measurements can describe with forces acting (forces and moments) on a rotating sphere, therefore a right set-up is required.

The measurement set-up, which can measure all three components of force and all three components of moments, is described in next chapter, Description of Experiment. The results of the realized measurements are presented in the article.

2 Description of Experiment

The experiment was conducted in a 1.8 m-diameter low-speed wind tunnel in the laboratories of Aerospace Research and Test Establishment in Prague. The set-up principal is visible in Figure 2. Forces were measured by 6 dynamometers, two in direction of the lift force F_L, three in direction of the drag F_D and one in direction of side force. Air-flowing velocities of v = <10 – 25> m/s were observed. Reynolds number of measurement is in the range of 1.5 – 3.8 x 10^5. The ball was rotating in these revolutions: n = <0, 12.5> rps.

Non dimensional parameters such as Reynolds number Re and the spin s are calculated according to the equations (1) and (2):

$$Re = \frac{vd}{\nu}, \tag{1}$$

$$s = \frac{\pi n d}{v}, \tag{2}$$

where v is velocity[m/s], d is diameter of the ball [m], ν is kinematic viscosity [m^2/s], n is revolution [rps]. Spin is a nondimesional expression of movement in the boundary condition and in the fluid. Meanwhile Strouhal number is criterium of impulsive and inertial forces.Values of Reynolds number are shown tab. 1.

Table 1. - Reynolds number values

v [m/s]	10	15	17	19	21	25
Re [-]	152 420	228 630	259 120	289 600	320 090	381 060

The spin number is in the range s = (0 – 0.85).

Revolutions were conducted and measured by a "stepping motor" and in this article are conducted as a

* Coresponding author: jandumek@gmail.com

independent value and recalculations of the spin number are not performed. All the values were recorded by the Labview program and evaluated afterwards by using a Matlab based program.

An experimental set-up is depicted in figures 1. and 2.

Fig. 1. Special stand for measurement of all 3 forces without volleyball

Fig. 2. Special stand for measurement of all 3 forces with rotating volleyball

Influence of moments and influence of side forces must be taken in consideration for a complete description of the situation. This experimental set up enable an observation by measurement of forces and moments acting on a rotating sphere in different angle of the airstream – β: axis of a rotating ball is not just normal to the airstream, but has also a defined angle (β = 0 – 45°).

Forces and moments are calculated according to the equations (3) - (8):

- Drag:

$$F_D = F_3 + F_4 + F_5 \tag{3}$$

- Lift:

$$F_L = F_1 + F_2 \tag{4}$$

- Side force:

$$F_S = F_6 \tag{5}$$

- Roll:

$$M_X = F_1 \cdot a - F_2 \cdot a \tag{6}$$

- Yaw:

$$M_Y = F_4 \cdot f - F_3 \cdot f \tag{7}$$

- Pitch:

$$M_Z = (F_3 + F_4) \cdot b - F_5 \cdot c \tag{8}$$

From kinematic scheme, which can be seen in Figure 3, is defined in a complete picture of forces (F_D, F_L, F_S) and torques (M_X, M_Y, M_Z) acting on the rotating volleyball. Based on the kinematic scheme, the real stand and frame were constructed and installed in a wind tunnel test section, depicted in Figure 2. The measurement is prepared.

Forces acting on the special stand without volleyball were measured firstly and then charcteristics of the volleyball in the stand were measured in the same conditions (v and n) as in the experiment – in purpose to deduct influence of the stand. Volleyball characteristics: I. must be spherical, II. circumference is in the range of 650 – 670 mm, III. weight of 260 – 280 g, IV. internal pressure 0,300 – 0,325 kg/cm^2. The volleyball MIKASA VLS 300 was measured.

3 Results

Results of coefficients C_D, C_L, C_S and moments M_X, M_Y, M_Z are depicted at figures 4. – 21.

4 Discussion

Basic principles are observed in the figures of results:

1. Drag coefficient for β = 0° (Fig. 4.). Dependence of Drag coefficient for n = 0 rps shows that results include region of critical Reynolds number (according to theory critical Reynolds number smooth sphere Re_{CR} = 300 000, [6]). Drag coefficient firstly decreses to critical region and increases in turbulent region.
2. With increasing revolutions area of critical Re moves (Fig. 4.). Revolutions apparently have impact on drag force – increasing revolutions decrease the drag. This phenomena will be studied and described more detialed in further articles.
3. Lift coefficient increases with increasing revolutions, as in [7].
4. Influence of the rotation n and angle of attack β is evident on each depandences.
5. Strict concept of symetry is not confirmed in the observation.
   Theoreticaly lift force (coefficient C_L) should be equal to zero for n = 0 rps.

 https://doi.org/10.1051/epjconf/201818002021

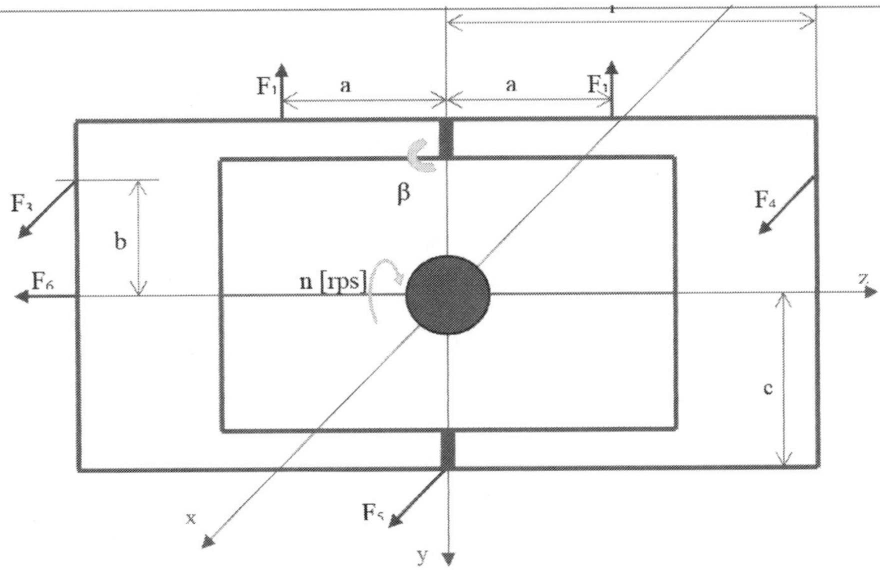

Fig. 3. Force scheme for the set-up

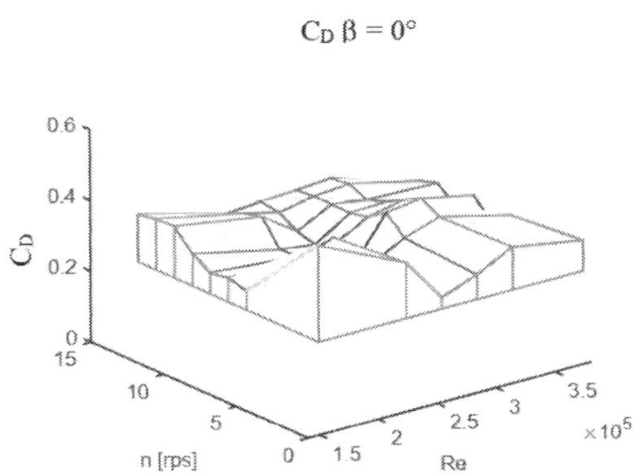

Fig. 4. Drag coefficient C_D $\beta = 0°$

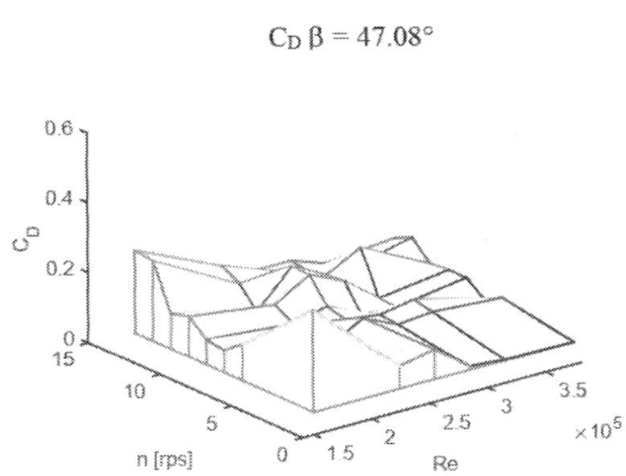

Fig. 5. Drag coefficient C_D $\beta = 47.08°$

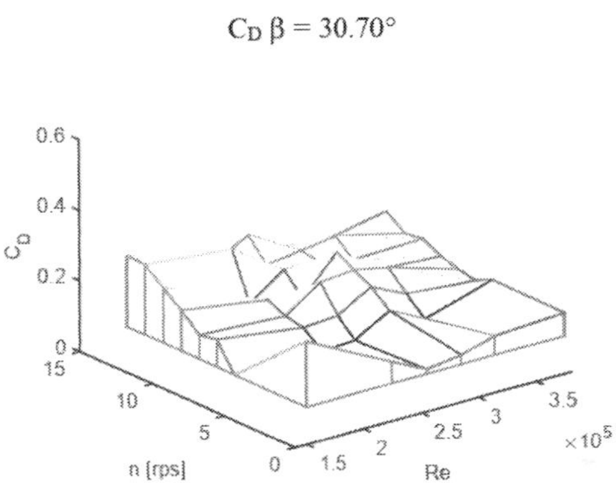

Fig. 6. Drag coefficient C_D $\beta = 30.70°$

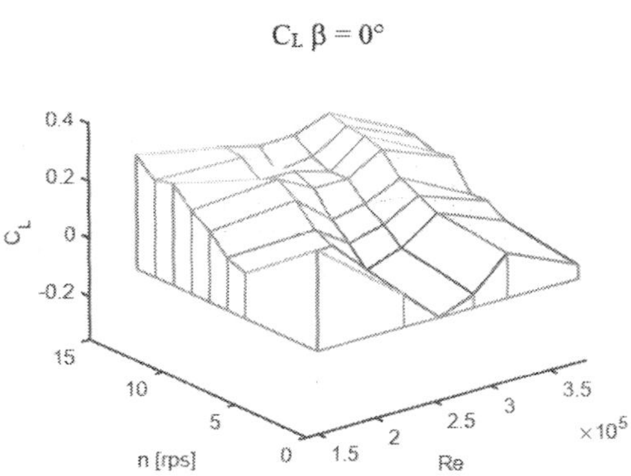

Fig. 7. Lift coefficient C_L $\beta = 0°$

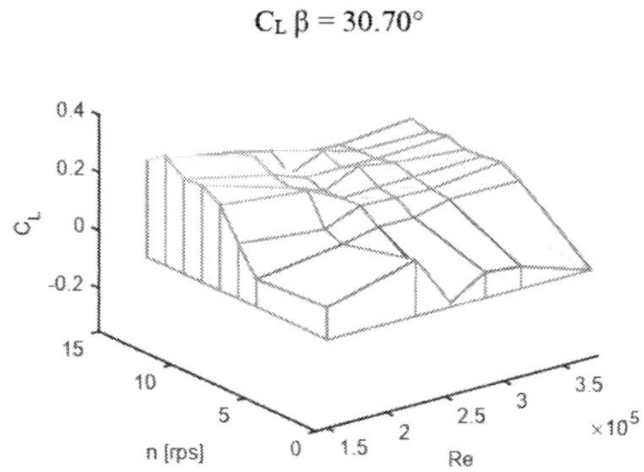

Fig. 8. Lift coefficient C_L β = 30.70°

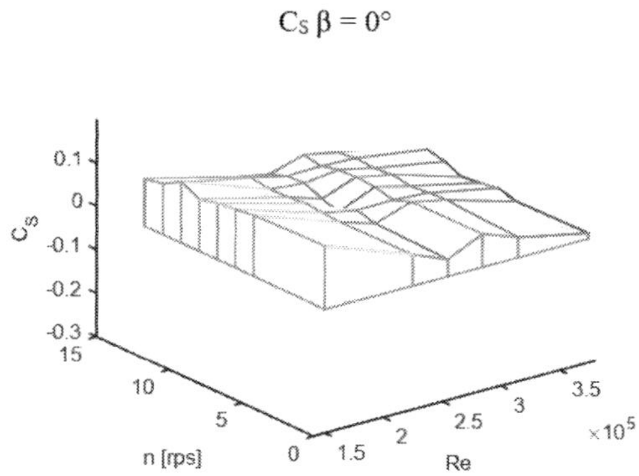

Fig. 9. Side force coefficient C_S β = 0°

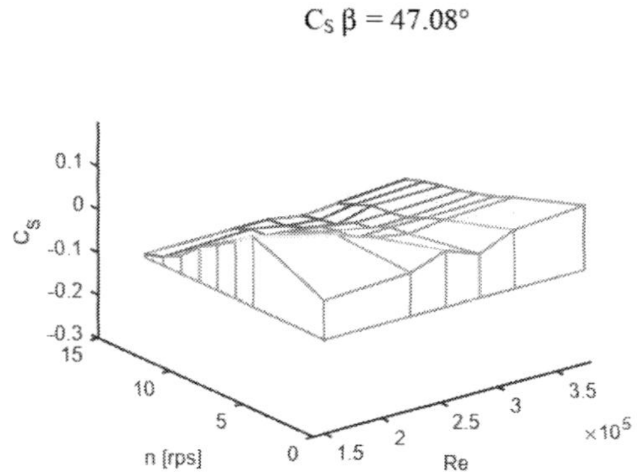

Fig. 10. Side force coefficient C_S β = 47.08°

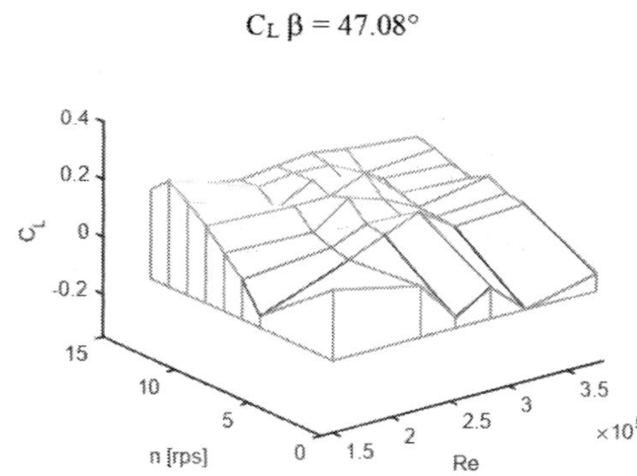

Fig. 11. Lift coefficient C_L β = 47.08°

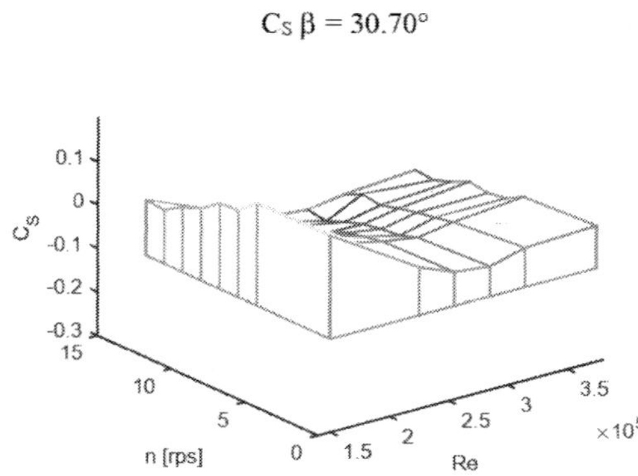

Fig. 12. Side force coefficient C_S β = 30.70°

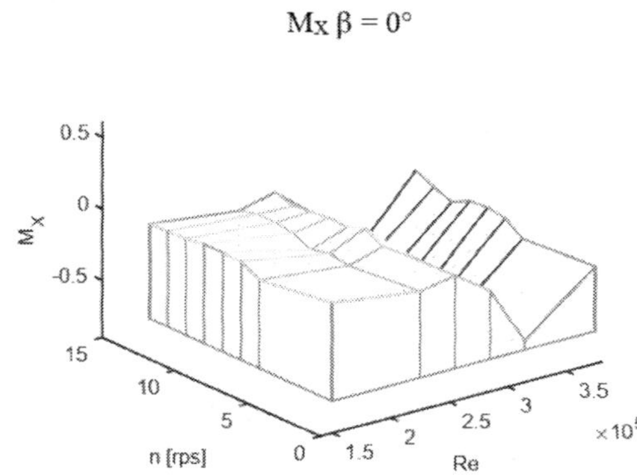

Fig. 13. Moment M_X β = 0°

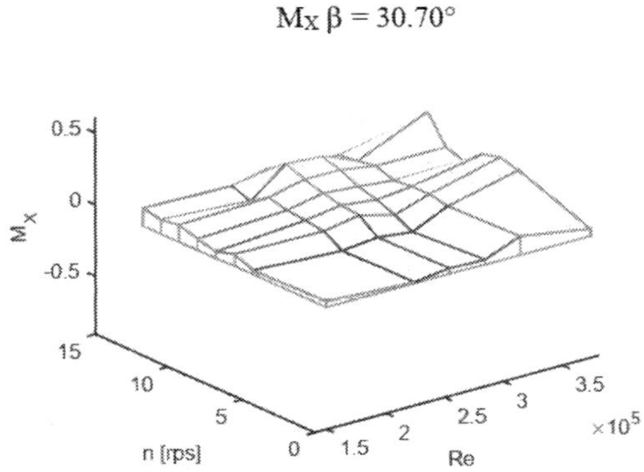

Fig. 14. Moment M_X $\beta = 30.70°$

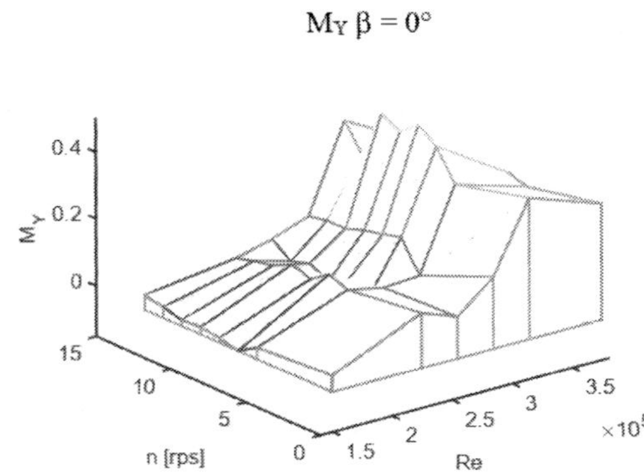

Fig. 15. Moment M_Y $\beta = 0°$

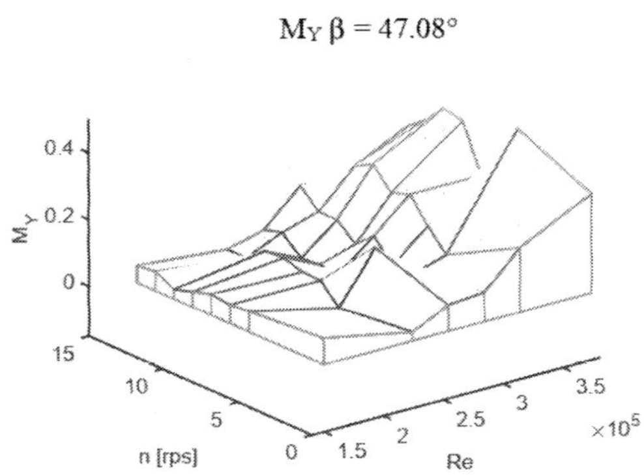

Fig. 16. Moment M_Y $\beta = 47.08°$

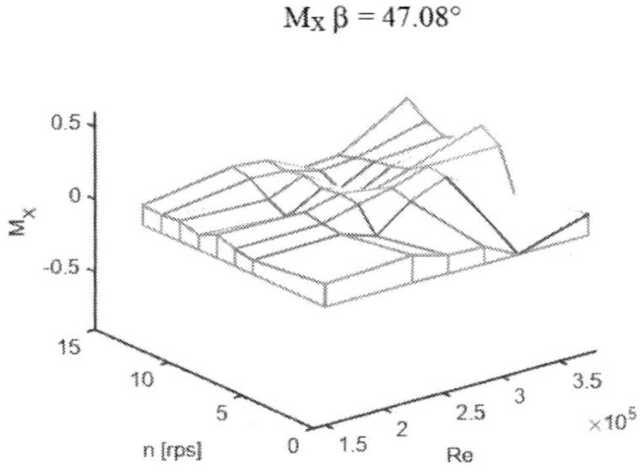

Fig. 17. Moment M_X $\beta = 47.08°$

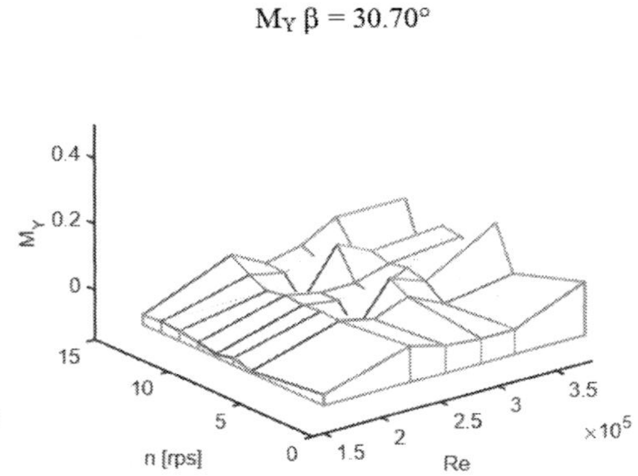

Fig. 18. Moment M_Y $\beta = 30.70°$

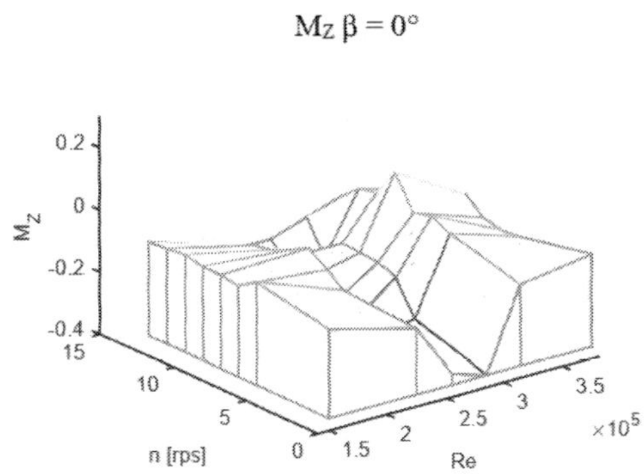

Fig. 19. Moment M_Z $\beta = 0°$

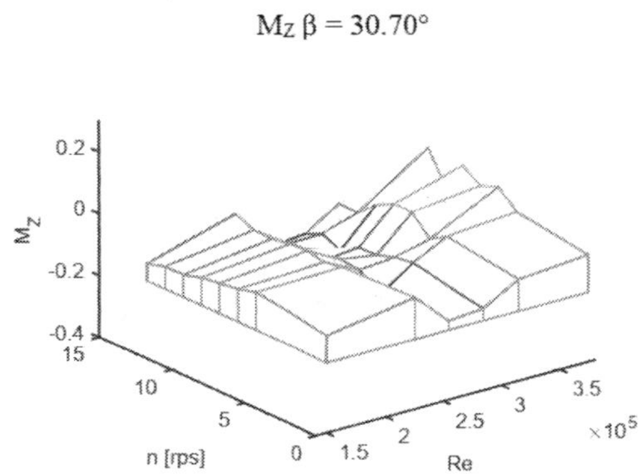

Fig. 20. Moment Mz β = 30.70°

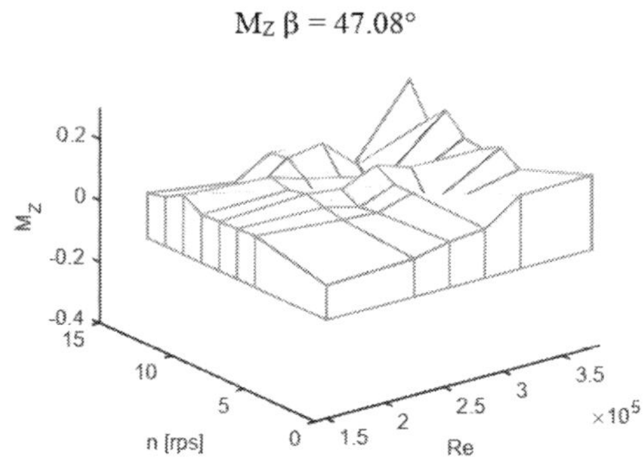

Fig. 21. Moment Mz β = 47.08°

5 Conclusions

Experimental set up of volleyball, which rotates in two planes in the wind tunnel was made and experiment was performed. Results are shown in the Section 3 Results. Basic principles, which are evident from results were described in the Section 4 Discussion. Results will be studied more thoroughly. Based on the principels described in Discussion, main conclusion is made:

1. detailed results study will be made,
2. measurement uncertainties must be counted and described,
3. moments coefficients will be counted and results will be described,
4. aerodynamic characteristic C_L vs. C_D will be described.

Acknowledgments

Support from from the Project No. CZ.2.16/3.1.00/21569 Centre 3D Volumetric Anemometry is gratefully acknowledged.

References

1. A. Kharlamov, Z. Chára, P. Vlasák: Experimental investigation of Magnus force acting on smooth sphere at high Reynolds numbers, *Colloquium Fluid Dynamics 2008*, Prague, (2008)
2. R. G. Watts, R. Ferrer: The lateral force on a spinning sphere: *Aerodynamics of a Curvebal. Am. J. Phys.*, **55**, 40-44, (1987)
3. T. Asaia, S. Itob, K. Seoc, A. Hitotsubashi: Aerodynamics of a new volleyball, *Procedia Engineering 2 2493–2498*, Elsevier, (2010)
4. R. D. Mehta: Aerodynamics of sport balls, Annual Rewiew of Fluid Mechanics, page 151 – 189, In: *M. van Dyke, etal.: Annual Rewiews* , Palo Alto, Vol. **17**, (1985)
5. Wei Qing-Ding, Lin Rong-Sheng, Liu Zhi-Jin : Vortex-induced dynamic loads on a non-spinning volleyball, Fluid dynamics research, North Holland, 2000
6. R. W. Fox, A. T. McDonald, A.T.: *Introduction to fluid mechanics*, John Wiley & Sons, New York, (1994)
7. Dumek, J., Šafařík, P.: Aerodynamic experimental tests of forces and torques acting on the volleyball, page 113 – 116 In: *Šimurda, D., Bodnár, T.: Topical Problems of Fluid Mechanics 2017*, Conference, Prague, (2017)

Compressed Air Savings During Cooling and Cleaning of Product

Lukáš Dvořák[1,*], *Kamil* Fojtášek[2], *Petr* Holiš[3]

[1,2,3] VŠB-Technical University of Ostrava, Faculty of Mechanical Engineering, Department of Hydrodynamics and Hydraulic Equipment, 17. listopadu 15/2172, Ostrava 70833, Czech Republic

Abstract. The paper deals with possibilities of reduction of compressed air consumption during cooling of drill bit and pressure cleaning of products. First part is focused on the experimental verification of the air consumption of the production machine. On the basis of the experiment in second part of paper the modifications of machine which are aimed to reduction of air consumption are proposed.

1 Introduction

Compressed air is used not only to drive of pneumatic mechanisms but also in some applications for cooling or cleaning products or tools. In these cases, often simple solution is used - the air flow from tubes or hoses. The amount of flowing air can be controlled by a throttle valve. This solution, see Figure 1, is structurally simple, the production costs are low but in operation it shows low efficiency and unnecessary waste of the compressed air.

Fig. 1. Photo of drilling part of the product machine

Based on the industry demand, we have dealt with the possibilities of air savings on production machines [1]. As a part of our research, we first measured air consumption when drilling holes on two machines - the production (assembly) machine and the CNC drill. In a second phase, we proposed the device modifications and on the base of it we quantified possible savings of the compressed air. Consumption measurement was performed such that the flow meter has been included in the inlet tube for cooling and blowing of chips. At the same time, the pressure at the inlet to the tube was measured. Based on the measurements, the operating costs were calculated. In the calculation we used the price of the compressed air $0{,}5\ CZK/m^3\ (ANR)$ (Atmospheric Normal Reference). The results are in Table 1.

Table 1. Consumption of compressed air and its cost

quantity	assembly machine		CNC drill
	cooling	chips blowing	chips blowing
flow rate $[dm^3 \cdot min^{-1}\ (ANR)]$	52.7	130	71.5
inlet pressure $[MPa]$	0.65	0.63	0.68
total consumption per 1 hour $[m^3\ (ANR)]$	3.162	7.800	4.29
compressed air cost per 1 hour $[CZK]$	1.58	3.9	2.15
compressed air cost per 1 year* $[CZK]$	6570	16220	8945

*260 working days, two-shift operation (16 hours)

At this point, it should be noted that the air has been flowing out all the time since the machine was switched on until it was switched off. Air was wasted outside of production time, for example during breaks, when the tools was replacing, etc. This, of course, greatly increases total costs. Modifications and possible savings will be further described on the example of the drill bit cooling.

* Corresponding author: lukas.dvorak@vsb.cz

In this case, the year-round operation cost is about 6570 CZK.

2 Proposed modifications of the machine

In this case, there are two ways to reduce air consumption. The first is to use the amplifier (high efficiency nozzle) instead of the tube. The second possibility is to switch on cooling only when it is needed. Of course, it would be possible to combine both ways. These options are compared below.

2.1 High efficiency nozzle

Producers of pneumatic components offer a number of components for blowing, cleaning and cooling products or working tools. These elements include various types of nozzles and so-called air knives.

Due to the optimised shape, the air knife provide slim output of air flow with high power and speed, Figure 2.

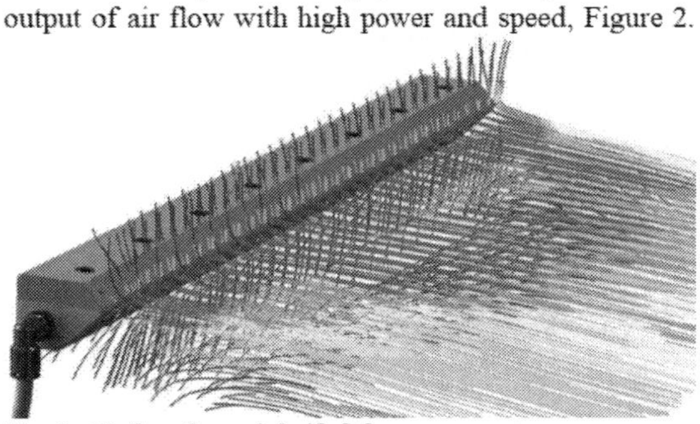

Fig. 2. Air flow from air knife [2]

Unique knife shape causes suction of surrounding air which increase the total capacity of air flow.

Another possibility is to use some kind of nozzle. For cooling applications, an amplifier nozzle or high efficiency nozzle is suitable. Like the air knives, these nozzles also suck surrounding air as shown in Figure 3 and Figure 4.

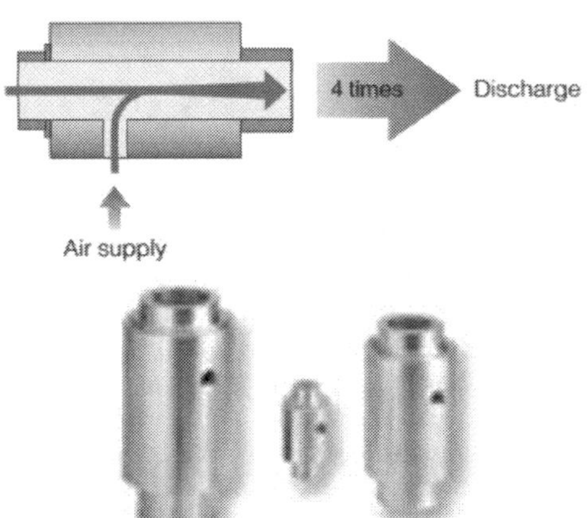

Fig. 3. Air amplifier [3]

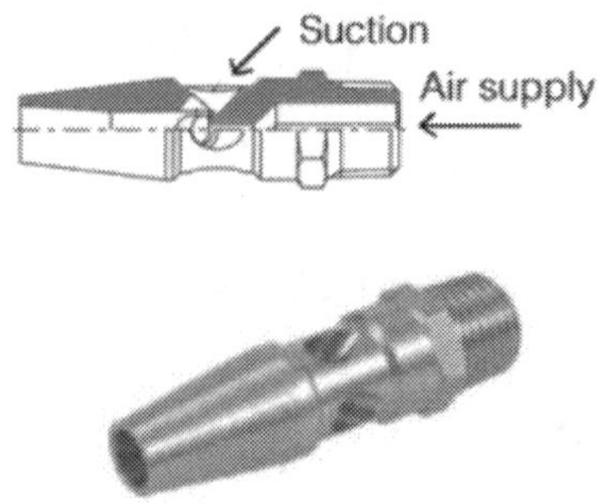

Fig. 4. High efficiency nozzle [4]

Due to the minimal requirements for modifications the machine design, the highly efficient nozzle type KNH-R02-150 from SMC was chosen. The nozzle has a diameter of $1.5\,mm$. At a maximum inlet pressure of 1 MPa the air consumption is $200\,dm^3 \cdot min^{-1}\,(ANR)$.

To compare cooling efficiency with a 3.8 mm diameter tube and a nozzle, a simple experiment was performed. Using a Prandtl tube, the velocity profile was measured from the axis of the air flow to its edge, see Figure 5. The velocity profile was measured at a distance of $35\,mm$ from the end of the tube. This distance corresponds to the actual conditions on the machine. The same pressure and flow rate as on the machine were set using the pressure regulator and throttle valve. Measurements were first carried out at an inlet flow rate (air consumption) of $52\,dm^3 \cdot min^{-1}\,(ANR)$.

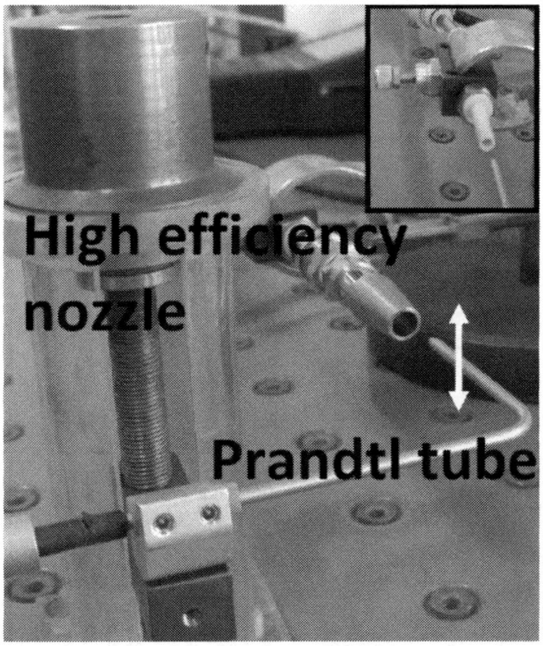

Fig. 5. Measurement device

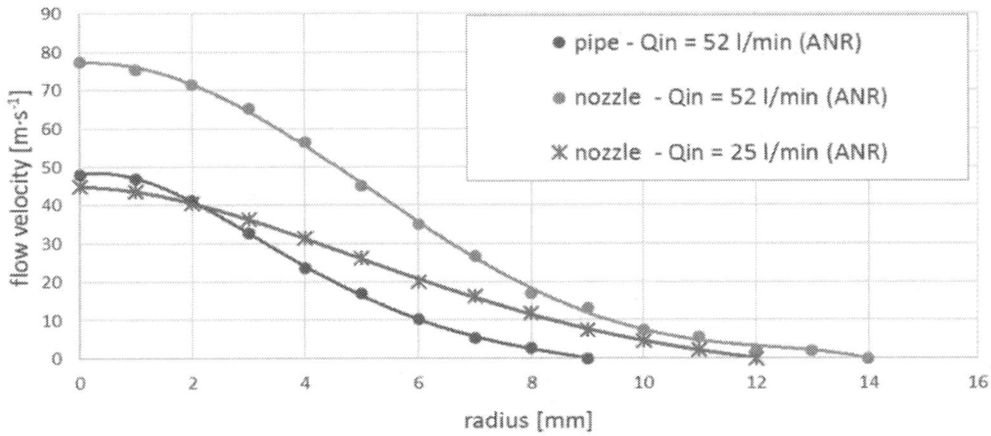

Fig. 6. Velocity profile

The measurement results are shown in the graph in Figure 6. It can be seen from the graph that the velocity in the axis of flow at the outlet of the tube is $48\ m \cdot s^{-1}$. Upon discharge from the nozzle, the maximum speed is $77.5\ m \cdot s^{-1}$. There is also a difference in the stream dimension. In the case of the outflow from the tube, the radius of the stream is $9\ mm$, with the nozzle it is $14\ mm$. From the comparison follows that, a smaller input flow rate is needed to achieve the same cooling effect when using the nozzle. Input flow was therefore reduced to approximately half - $25\ dm^3 \cdot min^{-1}\ (ANR)$. In this case, the maximum speed is $44.8\ m \cdot s^{-1}$ and the radius of the stream is $12\ mm$. This means that using a high efficiency nozzle can reduce air consumption by half. In addition, the cooling effect will increase as the drill will be cooled over a length $24\ mm$ instead of the original $18\ mm$.

Table 2. Air flow from the tube

tube, inlet flow (air consumption) $52\ dm^3 \cdot min^{-1}\ (ANR),\ L = 35\ mm$		
radius - r	velocity - v	Output flow – Q_N
$[mm]$	$[m \cdot s^{-1}]$	$[\ dm^3 \cdot min^{-1}\ (ANR)]$
0	48	-
1	46.9	8.9
2	41.3	24.9
3	32.7	34.9
4	23.7	37.2
5	17	34.5
6	10.2	28.2
7	5.3	19.0
8	2.8	11.5
9	0	4.5
Total flow rate		203.6

From the results of measurement of the velocity profile was also calculated a total flow rate at a distance of $35\ mm$ from the tube or nozzle. The total flow was determined as the sum of the flows through the individual annuluses calculated according to the formula

$$Q = \frac{v_2 + v_3}{2} \cdot \pi \cdot (r_3^2 - r_2^2)$$

It has already been mentioned that in the outflow from the nozzle there is a sucking of surrounding air. This is also the case when the air flows out of the tube. At an inlet flow of $52\ dm^3 \cdot min^{-1}\ (ANR)$ the flow is increased to approximately $200\ dm^3 \cdot min^{-1}\ (ANR)$. In the case of a nozzle at the same inlet flow it is up to $650\ dm^3 \cdot min^{-1}\ (ANR)$. At half-consumption, the total flow rate is approximately $365\ dm^3 \cdot min^{-1}\ (ANR)$.

Table 3. Air flow from the nozzle

Nozzle, inlet flow (air consumption) $25\ dm^3 \cdot min^{-1}\ (ANR),\ L = 35\ mm$		
radius - r	velocity - v	Output flow – Q_N
$[mm]$	$[m \cdot s^{-1}]$	$[\ dm^3 \cdot min^{-1}\ (ANR)]$
0	44.8	
1	43.5	8.3
2	40.5	23.8
3	36.3	36.2
4	31.5	44.7
5	26.2	48.9
6	20	47.9
7	16.3	44.5
8	11.9	39.9
9	7.3	30.8
10	4.5	21.1
11	2.3	13.5
12	0	5.0
Total flow rate		364.5

From the above results, it is clear that even at half the consumption it is possible to achieve better cooling efficiency.

Therefore, the first recommendation was to equip the machine with high efficiency nozzles. Due to its size, the nozzles cannot be mounted on all applications. In these cases, there can make other arrangements.

2.2 Turn on and off the flow

At the beginning of the article, it has already been mentioned that the air flows from the tubes for cooling and blowing constantly. By switching on and off the flow, consumption can be limited only to the time it is justified. First, let us introduce the cycle. It lasts 6 seconds. The cycle begins with the drill bit moving to the workpiece - 0.5 s, then drilling for 4 s, moving back 0.5 s and replacing the workpiece 1 s. Throughout the drilling time, the drill bit is completely out of the cooling air flow and this leads to unnecessary consumption. By switching off the airflow to this time, consumption can be reduced to only a third. For flow switching, a 2/2 valve must be included in the system and the control program adjusted.

A similar situation occurs when the chips are blown out of the hole. In this case, there is a need for a stream of air at the end of the drilling and above all after it. Even in this case, it would be enough to turn the air on for about 2 seconds.

3 Conclusion

Now we can evaluate the possible savings of air and thus of the financial losses. The above mentioned modifications are described primarily for cooling. If we make the same adjustments for chip blowing, we can express the cost of the year-round operation of the whole machine.

The Table 4 shows the price of the year-round operation of the machine in its original state and the cost of operation with nozzles, on / off variant and a combination of both.

Table 4. Year-round operation cost

	cooling	chips blowing	all machine
original state - per 1 year* [CZK]	6570	16220	22790
with nozzle - per 1 year* [CZK]	3285	8110	11395
on/off - per 1 year* [CZK]	2190	5407	7597
Combination - per 1 year* [CZK]	1095	2703	3798

*260 working days, two-shift operation (16 hours)

The table shows that by using both adjustments can be the total costs of compressed air reduced up to 83%. But it is also necessary to look at the whole matter in terms of return on investment. Only the price of the elements is counted in the next comparison. Costs for installation and adjustment time are not included. The price of one high efficiency nozzle is 1300 CZK including VAT. The price of one 2/2 valve, connector, cable and fittings is approximately CZK 1500 including VAT.

The Table 5 shows the financial savings achieved by the individual ways and the return on investment. It can be stated that in all cases the return on investment is within four months.

Table 5. Return on investment

	original state	with nozzle	on/off	combina-tion
operation cost - all machine [CZK]	22790	11395	7597	3798
savings [CZK]	-	11395	15193	18992
savings per day [CZK]	-	43.8	58.4	73.0
investment [CZK]	-	2600	3000	5600
return on investment [$days$]	-	59	51	77

From the above, the relatively significant savings of compressed air and thus the finance through simple changes are evident. At first glance it may seem that a saving of 11 000 up to 19 000 may not play a significant role in the operation of the company. On the other hand, if more than one plant is in operation, year-round savings can range from hundreds of thousands to millions of crowns.

Acknowledgements

The work presented in this paper was supported by a grant SGS „Research in field of fluid system dynamics" SP2017/103.

References

1. P. Holiš: *Possibilities of Reduction of Compressed Air Consumption at Pressure Cleaning of Products: Bachelor Thesis* (2016)
2. MOS technik. *Air Knives*. www.mos-pneumatics.com/ (2017)
3. SMC. *Nozzles – catalogue*. www.smc.cz. (2017)
4. SMC. *Air amplifier – catalogue*. www.smc.cz. (2017)

EPJ Web of Conferences **180**, 02023 (2018) https://doi.org/10.1051/epjconf/201818002023
EFM 2017

Experimental investigation of centrifugal fans for personal protection equipment – effect of used 3D printing technologies

Václav Dvořák[1,*], *Radek* Votrubec[1], *Jiří* Šafka[1] and *Jan* Kracík[1]

[1]Technical University of Liberec, Studentska 2, 46117 Liberec, Czech Republic

Abstract. The aim of the research is experimental investigation of centrifugal fans for a personal protection equipment. The aim of the fan is to drive the contaminated air containing harmful or irritating particles through the filters and then into the mask of workers, such as a fireman, a labourer or a lab worker. The fan is measured on the test stand, the characteristics and performances are evaluated, i.e. the dependencies of the working pressure on the flow rate. The characteristics are measured for three constant speed settings. The characteristics of the wheels produced by the different 3D printing technology are compared. It is found that the production technology has only a minimal effect, the performance of the wheels is more influenced by the position of the impeller on the motor shaft and hence by the mutual position of the impeller and the diffuser.

1 Introduction

The aim of research is development of new generation of powered air purifying respirators designed to filtrate contaminants in the form of gases, vapours, particles and their combinations. These high performance units guarantee sufficient protection of the wearer even in heavy industrial environments, the chemical industry, laboratories and the pharmaceutical industry [1].

To help people while breathing through filter, a fan is used to propel air flowing in personal respiratory protection systems and units. Because, the reliability and efficiency are crucial for operation of these systems, our work is focused on improvement and optimization of a centrifugal fan used.

Centrifugal fans are used in many applications and therefore, a lot of works can be found. Flow in centrifugal fan is very complex and large number of researchers investigated it both numerically and experimentally.

As Gesner showed, the flow in centrifugal fans is very complicated and complex [2]. Gholamian et al. [3] tested the effect of axial gap between inlet nozzle and impeller on efficiency and flow pattern in centrifugal fans and showed that the minimal axial gap between inlet nozzle and the wheel is not the most favorable.

Even more researchers use CFD [4-6] to calculate flow in centrifugal fans. CFD is also used for optimization of both – backward curved airfoils [7] or the whole centrifugal fan [8].

The aim of the investigation is to verify designed test stand and experimentally investigate high pressure centrifugal fan wheels produced by different technologies.

2 Experimental investigation

The section of geometry of the fan wheel is visible in Fig. 1 and Fig. 2. The wheel has a classic design of high pressure fan, i.e. it has twelve long backward curved blades and the width of the wheel decreases, i.e. $b_2 < b_1$, while $d_1/d_2 \approx 0.5$.

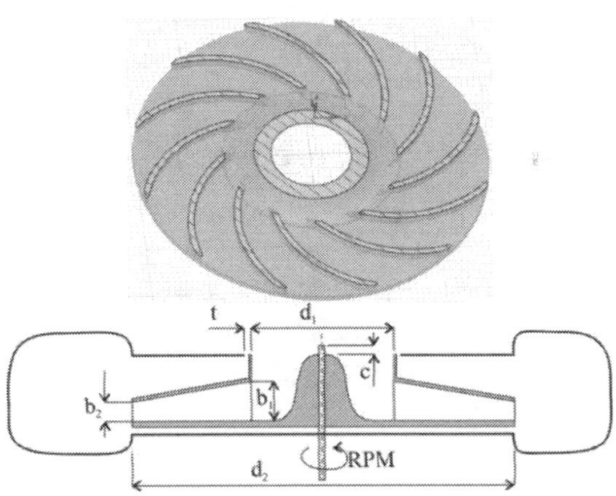

Fig. 1 Design of fan wheel under investigation.

All wheels were mounted in diffuser and engine section, see Fig. 2. It was ensured to keep the axial positon of the wheel on the shaft constant for all variants during investigation, i.e. $c \cong 1.8$ mm.

* Corresponding author: vaclav.dvorak@tul.cz

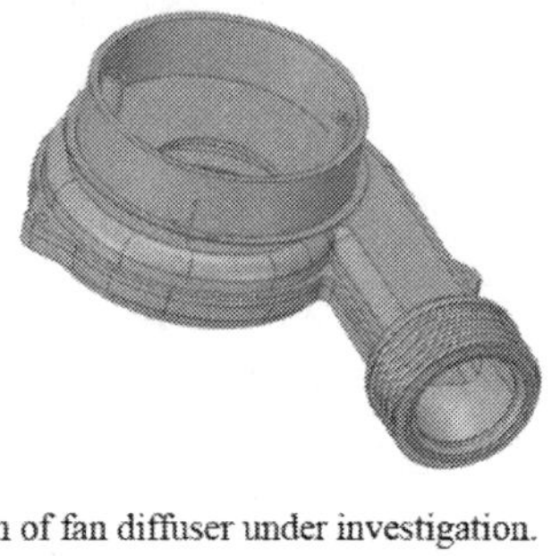

Fig. 2 Design of fan diffuser under investigation.

Eight different wheels manufactured by different technologies were measured. The first wheel (variant 0) was plastic, the rest of wheels (variants 1 – 7) were made by Rapid prototyping using 3D printers while using several materials and printing technologies. All wheels are listed in Table 1.

The first technology which we used for 3D printing was PolyJet. It is a photopolymers technology that uses a UV-sensitive liquid photopolymer resin. PolyJet technology works similar like inkjet printing, but instead of jetting drops of ink onto paper, the PolyJet jets layers of sensitive liquid onto a build platform.

This material is cured by thin layers of ultraviolet radiation, see Fig. 3. The size of one layer of the model built in this way is 0.016 mm.

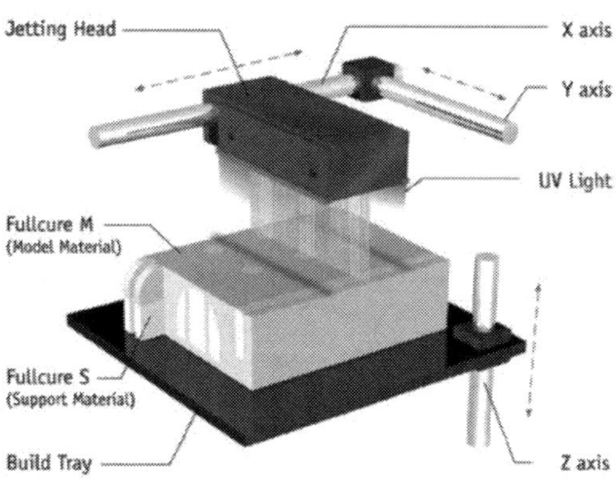

Fig. 3 The principle of PolyJet Matrix method [9]

The second technology of 3D printing was Fused Deposition Modelling (FDM). For this technology material is supplied to a 3D printer in the form of a fibre on a spool. The fibre is extruded through the print head. The print head moves in the xy plane. When the layer is completed, the head moves in the vertical direction z. This technology uses one material to create the geometry of the model and the second material for building the support structures, see Fig. 4. The FDM technology is suitable for solid models without surface quality requirements. It is possible used different thermoplastic polymers: ABS, PLA, PC-ABS, etc.

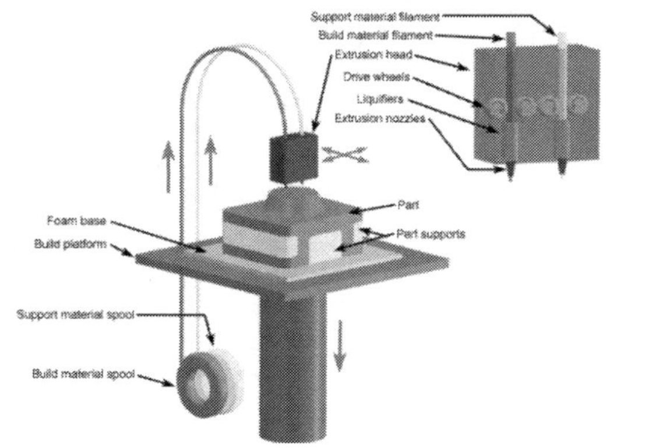

Fig. 4 The principle of FDM technology [10]

As we can see in Table 1, dimensions of printed wheels vary and differ from original plastic wheel. While outer diameters d_2 of the wheels differ within 0.05 mm, i.e. within 0.1%, the outlet width b_2 differ within 0.3 mm, i.e. within 8%. Also the inlet width b_1 of printed wheels was approximately bigger by 0.2 mm, i.e. by 3%.

These differences and also different surface roughness will affect the flow and thus also performance of the fan.

Table 1. Tested wheels, technology, material and dimensions.

No.	Technology / material	3D print layer thickness	d_2 x b_2
		(mm)	(mm)
0	Injection / plastic	-	60 x 3.86
1	PolyJet / VeroClear	0.014	60 x 4.17
2	PolyJet / VeroBlack	0.014	60.05 x 4.14
3	PolyJet / VeroGrey	0.016	60.1 x 4.12
4	PolyJet / ABS-like	0.03	59.95 x 4.1
5	FDM / ABS-M30	0.127	59.95 x 4.21
6	FDM / PC-ABS	0.127	60.05 x 4.18
7	FDM / ABS- P400	0.178	60.05 x 4.02

The testing stand is visible in Fig. 5. The fan wheel (2) is propelled by electronically controlled brushless DC motor (3). Two different ways were used to control engine output: First, electric current was set to obtain required rotations for low flow resistance. However, it showed that the rotation increased during increasing working pressure. Therefore, the second way was used after that – rotations were specified directly and kept constant during measurements, this way is shown in Fig. 5.

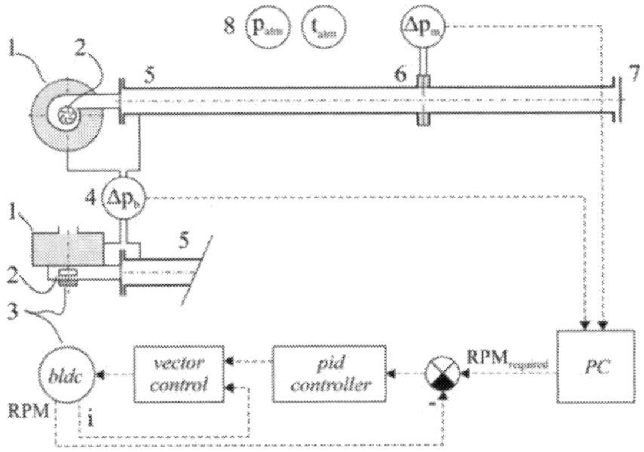

Fig. 5 Testing stand: 1 – suction chamber with filter assembling, 2 – fan wheel, 3 – brushless DC motor, 4 – working pressure measuring (Δp_b), 5 – outflow tube, 6 – differential pressure Δp_m measuring for mass flow specification, 7 – chocking, 8 – measuring of atmospheric pressure and temperature.

The fan was investigated for three values of rotation speed: 6000, 9000 and 11000 (min^{-1}).

The air is sucked through suction chamber (1), which will later allow assembling of filters, and is compressed and transported into the outflow tube (5). The back pressure and thus the working pressure Δp_b of the fan is controlled manually by chocking (7) at the tube exit and measured by differential pressure transducer. The mass flow rate is measured by orifice (6), where differential pressure Δp_m is measured.

The mass flow rate is than specified by relation

$$\dot{m} = \frac{1}{\sqrt{1-\beta^4}} \frac{\pi}{4} d^2 \varepsilon C \sqrt{2 \Delta p_m \rho}, \qquad (1)$$

where ρ is air density specified from atmospheric pressure p_{atm} and temperature t_{atm} outside the tube, d is orifice diameter, β is ratio between orifice and tube diameters, $\varepsilon \cong 1$ is expansion coefficient and C is flow coefficient according standards [11].

3 Results

The results of preliminary measurements are carried out in diagram in Fig. 6. There are results only for rotation speed 9000 /min and for two regimes in the diagram: curve A is for constant current and curve B for constant rotation. As we can see, curves differ for low flow rates and higher working pressure. In further works, we used only constant rotation regime, because, if non-constant rotation regime is chosen, final engine rotation can be affected by many factors: flow resistance, the mutual position of the wheel and the shaft, surface roughness of the wheel, etc.

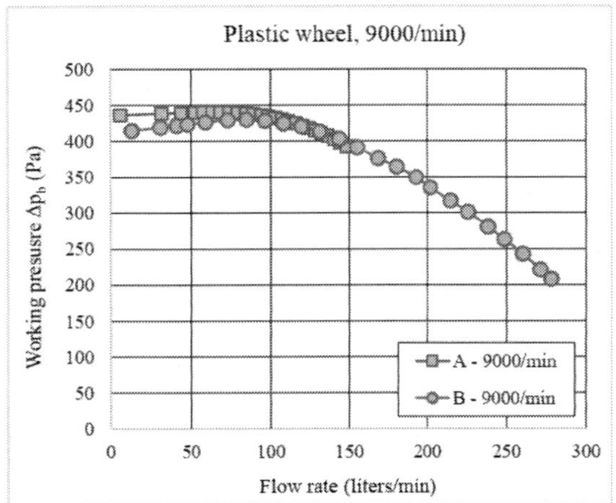

Fig. 6 Fan output for different way of engine control.

The performance curves for plastic wheel, i.e. dependencies of working pressure on flow rate and rotation speed are curried in logarithmic scale in Fig. 7.

It is clear that the fan works according our expectations and that the testing stand is well designed, has good repeatability and reliability. Although, the flow resistant is acceptable for this measuring, it does not allow to measure with filters. Thus, before further investigation, the flow resistance of the test stand should be lowered.

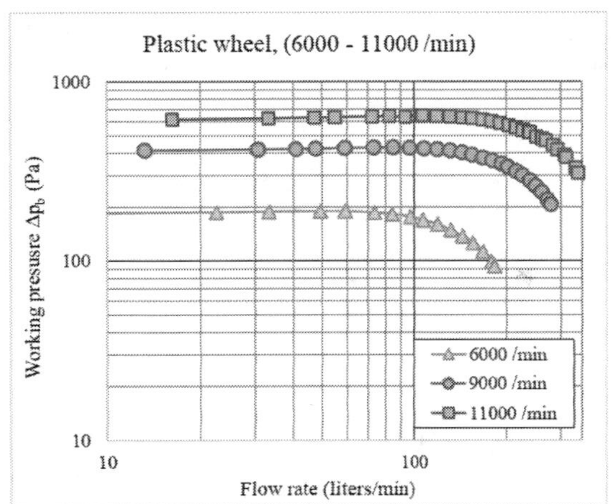

Fig. 7 Fan output for different motor speeds.

The performance curves for all wheels and motor speed 9000 /min are curried in diagram in Fig. 8. Again, we can see that the measuring is very precise with very good repeatability, all curves are quite close to each other. Further, we can obtain that all printed wheels yielded higher power, i.e. higher maximal flow rate and higher maximal working pressure than plastic wheel. We can assume that this difference is connected with bigger outlet width b_2 of printed wheels, which influences the flow rate proportionally.

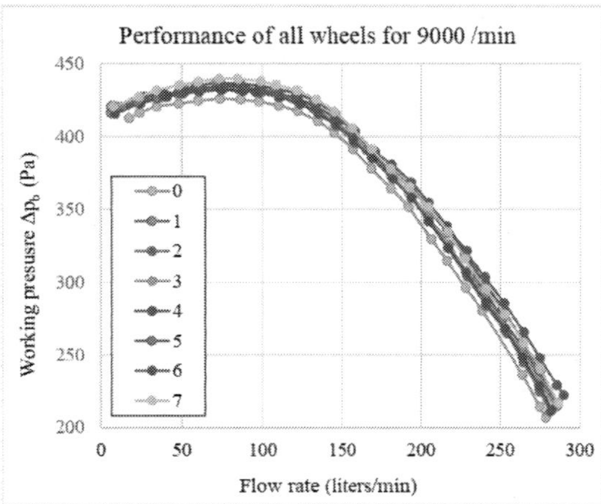

Fig. 8 Fan output for plastic and seven printed wheels for engine speed 9000 /min.

Therefore we corrected obtained data according average measured value b_2, the resulting diagram is in Fig. 9. In this diagram, the maximum flow rate is provided by wheel 7, the second one is plastic. On the other hand, the plastic wheel still provides the lowest pressure difference for low flow rates.

The highest working pressure provides wheel 7. According Table 1, this wheel has extremely high roughness which results from high thickness of layers of 3D printing. Also wheels 5 and 6 belong to variants with high performance, while wheels with significantly smoother surfaces, i.e. 1, 2, 3 and 4, generally provide lower flow rates.

We can assume that high roughness of wheel surfaces influences positively power curves by two effects: Firstly, it enables better power transfer from wheel into the fluid, and secondly, it reduces air leak through gap between wheel and fan case around inlet. It seems that reduction of air leak has significant affect for low flow rates.

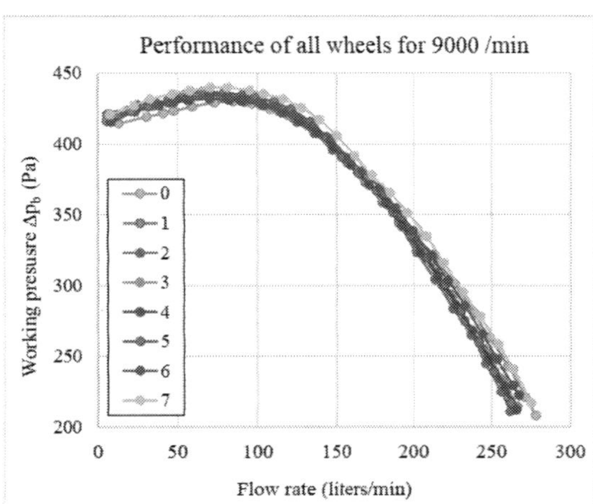

Fig. 9 Fan output for all wheels for motor speed 9000 /min corrected according outlet width b_2.

Mutual position of wheel and the shaft c (mm) was another parameter that was investigated. This research, which was performed for wheel no. 7, was done to answer the question of how inaccurate assembling of a wheel on the motor shaft affects its characteristics.

As it is visible in Fig. 10, where low flow rates are presented, the lowest maxima of working pressure is obtained for lowest values of c. If $c \geq 1.3$ mm, the maximum is approximately the same. We can observe similar situation for high flow rates in Fig. 11. The lowest value of c brings the lowest working pressure.

On the other hand, the situation is opposite for middle values of flow rates: The lower c is, the higher working pressure is achieved. How this can be explained?

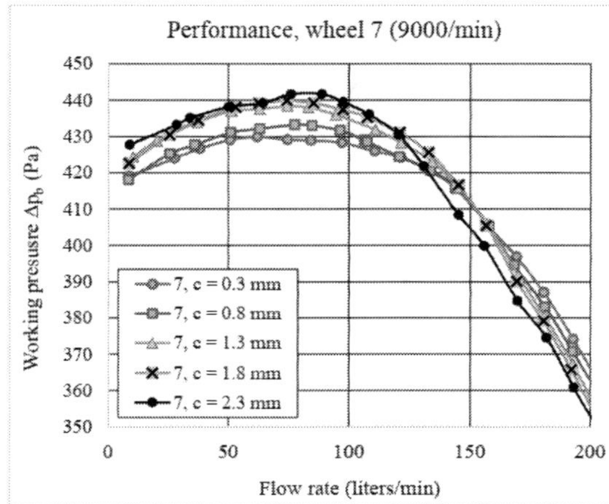

Fig. 10 Fan output for wheel 7 and various mutual position of wheel and the shaft c, maximum working pressure.

The answer is following: For low working pressures, there is additional suction into the fan through gab t between the wheel and the diffuser, see Fig. 1. Therefore, low c means low additional suction and lower flow rates for the same pressure. If $c \geq 1.3$ mm, the additional suction is restricted by t and these variant shows almost the same data.

For $\Delta p_b > 270$ Pa, see Fig. 11, the additional suction turns into leak and low c becomes an advantage. However, for $\Delta p_b > 400$ Pa and flow rates < 150 litres /min, see Fig. 10, another effect becomes significant: The working pressure of a fan is a results of flow inside the wheel. As flow rate decreases, the pressure rise inside the wheel becomes more and more problematic and obtained working pressure decreases. Now, the leak between the rotor and stator allows additional flow through the wheel and thus plays positive role for resulting pressure rise. As a result, the maximal working pressure is obtained for maximal c.

Because the standard value of c is 1.8 mm, we can conclude: Until c lies between 1.3 mm and 1.8 mm, the inaccurate assembling of a wheel on the motor shaft has low or even negligible effect on fan performance.

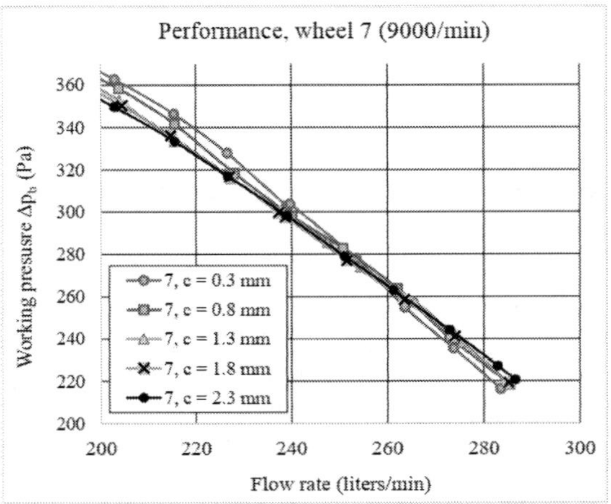

Fig. 11 Fan output for wheel 7 and various mutual position of wheel and the shaft c, maximum flow rate.

4 Conclusions

One plastic and seven fan wheels printed from various materials and by different technologies were experimentally tested. It was proofed that designed test stand has high accuracy and very good repeatability, even small differences between investigated wheels can be revealed and tested.

The wheels had various dimensions, therefore obtained flow rate was corrected. Obtained data proofed, how inaccurate assembling, i.e. mutual position of a fan wheel and the shaft, can affect fan performance. However, for certain range of position, the influence is negligible.

It follows from measured data that higher surface roughness could be advantageous and bring higher fan performance.

For further research, it would be beneficial to decrease flow resistant of the test stand to allow measuring with assembled filters. Shape and dimension inaccuracies of printed prototypes will be the main problem for further testing and development of new fans. Therefore, it will be necessary and beneficial to engage numerical research too.

5 Acknowledgement

This paper was created in the framework of project "Applied research in the field of the new generation of personal protective equipment for the demands of joint rescue service" no. VI20172020052, supported by Ministry of Interior of the Czech Republic.

References

1. https://www.clean-air.cz
2. F.B. Gessner, ASME Paper no. 67 (1967)
3. M. Gholamian, G. K. M. Rao, B. Panitapu, Case Studies in Therm. Eng. **1,** (2013) 26-37
4. J. Zhang, W. Chu, H. Zhang, Y. Wu, X. Dong, Applied Acoustics, 110, (2016) 256-267
5. T. Wang, F. Kong, B. Xia, Y. Bai, C. Wang, Renewable Energy 109 (2017) 518-528
6. S.-C. Lin, M.-L. Tsai, Computers & Fluids **56** (2012) 24–38
7. M. M. Mohaideen, Procedia Eng. 38 (2012) 1592-1598
8. B. D. Baloni, Y. Pathak, S.A. Channiwala, Comp. & Fluids **112** (2015) 72-78
9. http://www.3daddfab.com
10. http://www.custompartnet.com
11. ISO 5167-1:2003

An investigation of the near-wake flow structure of a cylinder with guiding plates

Fırat Ekinci[1], *Erhan* Fırat[2,*], *Göktürk M.* Özkan[3,4], and *Hüseyin* Akıllı[4]

[1]Adana Science and Technology University, Department of Energy System Engineering, 01180, Adana, Turkey
[2]Munzur University, Department of Mechanical Engineering, 62000, Tunceli, Turkey
[3]Delft University of Technology, Process and Energy Department, 2628 CA, Delft, Netherlands
[4]Çukurova University, Department of Mechanical Engineering, 01330, Adana, Turkey

Abstract. In this study, the flow behind a circular cylinder with a pair of outer identical guiding plates was investigated using particle image velocimetry (PIV) for various angular positions of the plates (i.e. $\alpha=\pm70°$, $\pm100°$, and $\pm130°$). The gaps between these plates and cylinder are equal and are $0.3D$. Experiments were carried out at a subcritical Reynolds ($Re=\rho \cdot U_\infty \cdot D/\mu$) number of 7500, based on the cylinder diameter (D) and the flow velocity (U_∞). The features of the near-wake with and without the guiding plates were interpreted in terms of patterns of time-averaged vorticity and streamlines, time-averaged and fluctuating velocity components. The spectral analysis was also carried out to determine the time-dependent variation of the transverse velocity at given locations in the near-wake. Two-dimensional computations of flow around circular cylinders with and without guiding plates have also been performed to predict the time-averaged and root-mean-square of force coefficients of the various models. It was seen that the guiding plates at an appropriate angular position can lead to substantial attenuation, or retardation, of the process of large-scale vortex formation in the near-wake, thus can lead to vortex-induced vibration (VIV) suppression without any increase in drag.

1 Introduction

When two-dimensional bodies that are not streamlined with respect to the free-stream direction (high-rise structural systems, bridge piers, some parts of onshore/ offshore systems, etc) subjected to a fluid flow at adequately large Reynolds numbers (*Re*), each of the shear layers from this body roll up in an alternating, periodic fashion, into large-scale vortices (Kármán vortices) behind these bodies (near-wake). Relatively high aerodynamic/hydrodynamic drag and fluctuating forces on the body are several natural consequences of the Kármán vortex formation. These forces are generally undesired. To decrease them on the body, shear layers must be controlled before (boundary layer) and/or after (free shear layers) the separation [1-10]. The aim of this study is firstly to alter the near-wake structure of the cylinder using guiding plates, and then, to interpret the foregoing physics of the near-wake in terms of patterns of time-averaged vorticity and streamlines, time-averaged and fluctuating velocity components, and spectral analysis; secondly to predict the trend of time-averaged drag coefficient and root-mean-square of the fluctuating lift coefficient numerically in the absence of the direct force measurements.

2 Experimental setup

The experiments were carried out at Çukurova University in a closed circuit water channel with a test section of 8000 mm long, 1000 mm wide (W) and 750 mm high. The test section has 15 mm thick Plexiglas walls to allow optical access. The water level was maintained at 500±2 mm (H). The water was pumped

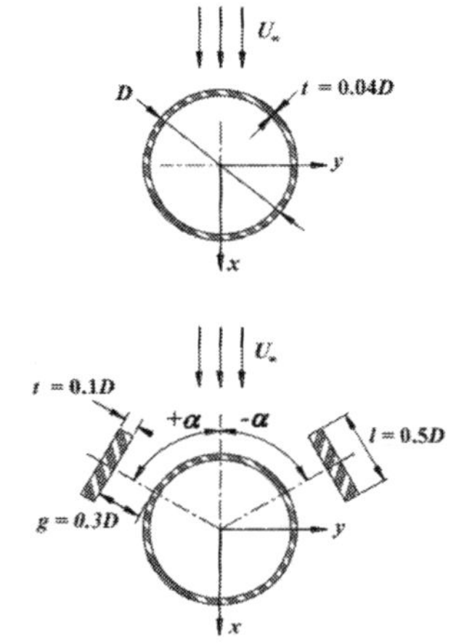

Fig. 1. Cross-sectional view of the circular cylinder without (upper one) and with guiding plates.

* Corresponding author: efirat@munzur.edu.tr

from the downstream reservoir through external piping into the upstream reservoir. It then passed through flow straightener, 2:1 contraction and test section of the channel. The water flow was driven by a 15 kW centrifugal pump. The rotation speed of the pump was controlled by a variable frequency drive controller unit and set to a constant value during tests. A water channel velocity of 0.140 m/s gives a Reynolds number of (Re) approximately 7500, based on the cylinder diameter (D). The free-stream turbulence intensity (Tu_x) was less than 1.45%.

One cylinder model without any modification (BC) and three cylinder models with guiding plates at various angular plate positions ($\alpha=\pm70°$, $\pm100°$ and $\pm130°$) were investigated (Fig. 1). The models made of Plexiglas were mounted vertically in the test section of the water channel. Cylinder models were producing a maximum blockage ratio ($BR=D/W$) of approximately 8.5% and have an aspect ratio ($AR=l/D$) of 8. Guiding plates may undergo large deflections due to the hydrodynamic load. To prevent this, relatively thick plates were employed.

It was stated that a reasonable representation of two-dimensional flow can be achieved for models with an aspect ratio greater than about 6 between walls or end plates [11]. Our models fulfill this requirement; therefore, two-dimensional, two-component digital particle image velocimetry (Dantec Dynamics A/S) measurements were obtained only in horizontal plane (x-y) which is parallel to the floor of the channel test section and 250 mm higher than it (mid-depth). A double pulsed Nd:YAG laser with a wavelength of 532 nm (New Wave Research, Solo 120XT) was used to illuminate the above-mentioned plane. The laser sheet was perpendicular to the axis of the cylinder. The flow was seeded by silver-coated hollow glass spheres with a nominal particle diameter of 10 μm. One CCD camera with a spatial resolution of 1200 × 1600 pixels (Dantec Dynamics, FlowSense 2M) was used to detect the movement of seeding particles. The camera was equipped with 60 mm lens (Nikon AF Micro-Nikkor 60 mm f/2.8D). Each frame covered an area of approximately 158 × 210.7 mm^2 (3.16D × 4.21D). Each vector in an instantaneous vector map was obtained by using frame to frame adaptive correlation technique (double-frame mode). The size of the dimension of interrogation area in pixels was 32 × 32. The percentage amount of overlap was 50. Each vector map contained 7326 (74 × 99) instantaneous velocity vectors. It was seen that 1000 instantaneous vector maps can sufficiently represent the time-averaged vector map. Therefore, the image pairs were recorded at 10 Hz (at least 17 times higher than the vortex shedding frequency of the bare cylinder) for 100 s to produce 1000 independent vector map samples.

3 Numerical methods

It is known that today's CFD codes are generally unable to predict hydrodynamic coefficients with an acceptable degree of reliability but yield reasonable results for many practical engineering problems. Therefore, our aim here is to predict the trend of hydrodynamic statistics (i.e. time-averaged drag coefficient, root-mean-square of the fluctuating lift coefficient) in the absence of the direct force measurements.

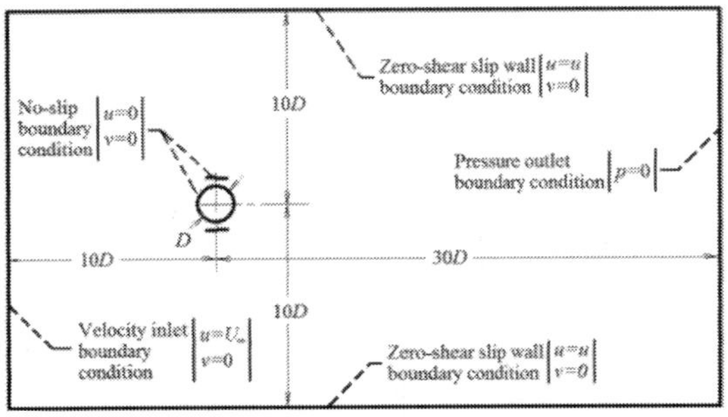

Fig. 2. Geometry of computational domain where length is in diameter of the cylinder (D) and boundary conditions. For the present computations, the inflow boundary condition is taken from the experimental results.

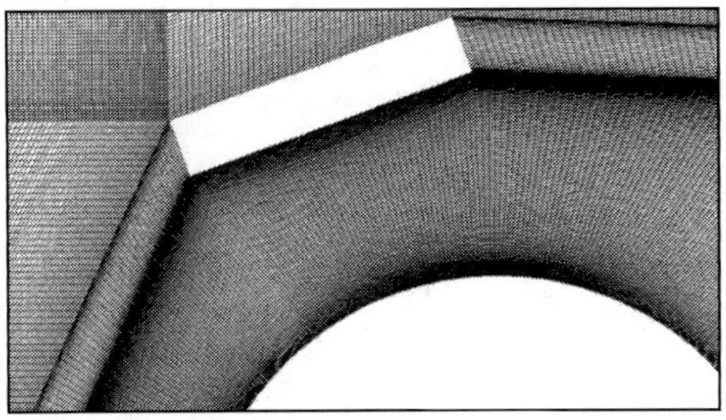

Fig. 3. Image showing near field meshes from the cylinder for A70 case before the mesh adaption process. Flow is from left to right. Computational domain was divided into various blocks with respect to the case, but each block was meshed with structured grid.

Computations were performed using the commercial CFD code ANSYS Fluent 6.3. A Reynolds-averaged Navier-Stokes (RANS) approach was taken due to the lack of the computational effort. The shear stress transport (SST) k-ω turbulence model which is developed by Menter and provided in FLUENT was chosen since it shows a better performance in complex flows with boundary layer separation when compared to RANS based one- and two-equations turbulence models [13, 14]. Second-order accuracy is used for the viscous terms. The diffusion terms in the discretized governing equations are central-differenced and second-order accurate. The second-order implicit scheme is adopted for the temporal discretization. SIMPLE (Semi-Implicit Method for the Pressure-Linked Equation) was adopted as pressure-velocity coupling algorithm. The time step was obtained from the quotient of approximate vortex shedding period of the single square prism at Re=7500 by 500 (Δt=0.0034 s) in order to avoid any unphysical oscillation and obtain a smooth time-independent solution. Although the initial mesh around the cylinder and guiding plates are fine, refinements close to the cylinder surface, plate surfaces and in the wake of the cylinder using boundary adaption and velocity gradient adaption were also made. The number of cells was varied from 700000 to 1000000 after adaption processes. The height of the first mesh on the cylinder surface was decreased to about $1\times10^{-4}D$ after adaption process. For bare cylinder, the predictions are compared with experimental and numerical results (Table 1). It is clear from the table 1 that the SST *k-ω* turbulence model results in over-prediction of hydrodynamic force and vortex shedding values compared with experiment and LES.

Table 1. Comparison of the current results with those of other authors for bare cylinder.

Reference/Technique	*Re*	C_D	$C_{L\,RMS}$	*St*
[12] / Experimental	7500	-	0.350	0.203
[15] / Experimental	7233	1.12	0.177	0.235
[16] / Experimental	7500	1.14	0.411	0.203
[16] / Numerical (LES)	7500	1.23	0.359	0.200
[17] / Numerical (SST)	10000	1.57	1.20	0.235
Present / Numerical (SST)	7500	1.59	1.19	0.258
Present / Experimental	7500	-	-	0.200

4 Results and discussion

All positions and lengths were normalized by diameter of the circular cylinder (*D*) and all time-averaged velocity components and root-mean-square (RMS) values of the fluctuating velocity components were normalized by freestream velocity (U_∞). Figure 4 shows the averaged streamline topology on the measurement plane. Four cases are shown:

1) Bare cylinder, BC.
2) Cylinder with guiding plates located at $\alpha=\pm70°$, A70.
3) Cylinder with guiding plates located at $\alpha=\pm100°$, A100.
4) Cylinder with guiding plates located at $\alpha=\pm130°$, A130.

It is evident that there is a pair of recirculation cells in the near-wake of BC. It shows that vortex shedding from the BC occurs. For the case of A70, the size of these recirculation cells increased, i.e., vortex formation region is expanded. The size of the vortex formation region almost remained unchanged as angular positions of the plates from the front stagnation point of the cylinder increased from ±70° to ±110°. In addition, a well-defined recirculation cell in the wake of the left plate as well as its counterpart on the opposite side of the wake centerline, are clearly apparent. It shows that vortex shedding from the guiding plates also begins to occur.

For the case of A100, two portion of the oncoming flow that entered the gaps between the cylinder and plates were directed slightly into the vortex formation region, whereas for the case of A130, they were penetrated the vortex formation region and substantially reduced the size of it. Guiding plates increased the streamwise velocity in the very near-wake (Fig. 5), i.e., reduced the pressure difference between the front and rear surfaces of the cylinder and presumably lead to a considerable reduction in the pressure (form) drag acting on the cylinder. On the other hand, at $\alpha=\pm130°$, one small recirculating cell in the wakes of the plates increased to two large recirculating cells, that is, pressure drag of the plates were increased.

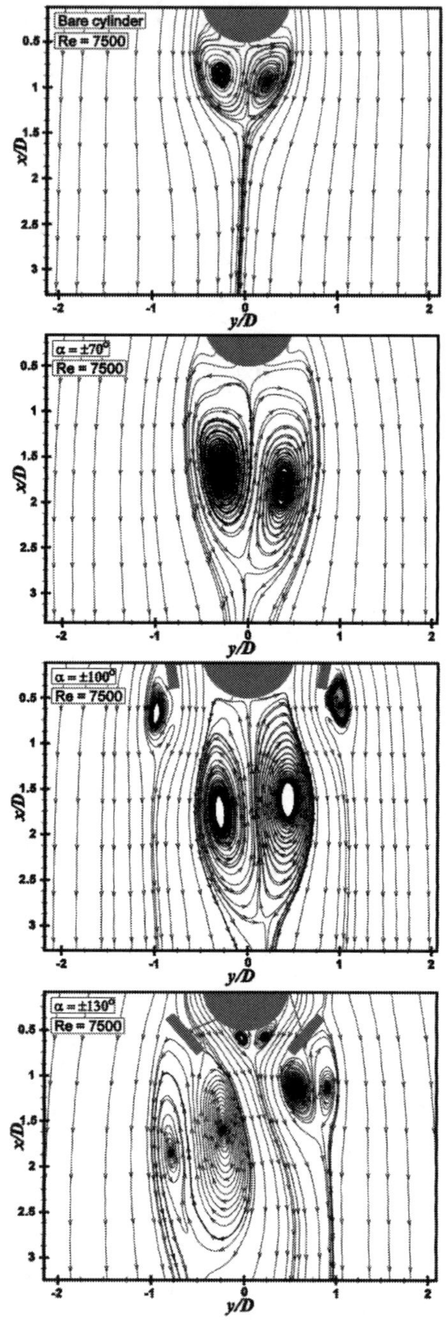

Fig. 4. Patterns of time-averaged streamline topology in the near-wake of the cylinder with and without the guiding plates.

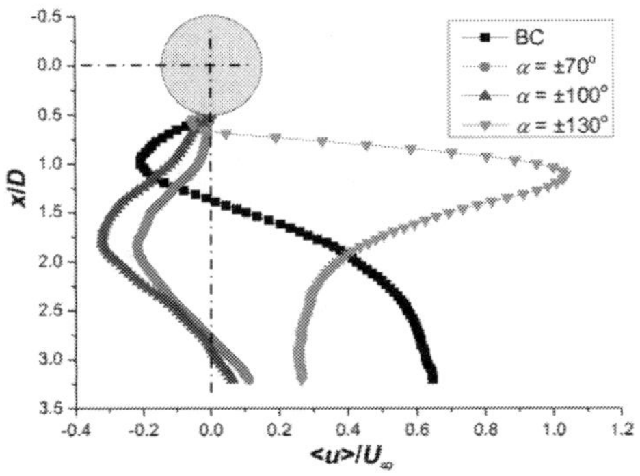

Fig. 5. Variation of $<u>/U_\infty$ along the wake centerline for the BC, A70, A100 and A130 cases.

Figure 6 shows the patterns of averaged vorticity on the measurement plane. Low pressure region in the near-wake of the bare cylinder bent the shear layers from either sides of the cylinder towards the wake centerline. For the case of A70, the aforementioned shear layers were significantly extended in the streamwise direction. That is, vortex formation was displaced significantly downstream, relative to the case of BC. This is also supported by the Fig. 7. Figure 7 shows that the vortex formation length (i.e. vertical distance between the center of the body and a point downstream of the body where the transverse velocity fluctuation level, v_{rms}, has grown to a maximum) reaches a maximum in the A70 case. On the other hand, it reduced sharply with increased angle of the plate position simply because guiding plates started to obstruct the paths of the shear layers from the cylinder in the A100 case and greatly obstructed them in the A130 case. Hence, in A100 and A130 cases, shear layers from the cylinder bent towards the very near-wake of the cylinder. That is, in these cases, guiding plates are making the near-wake flow behaviour close to potential flow ($Re<1$) patterns. Galvao et al. [7] stated that to make flow behavior closer to potential-flow patterns is a good guiding principle to eliminate vortex-induced vibration (VIV) while also reducing drag. Figure 7 also shows that the level of interaction between the two free shear layers from the cylinder is reduced by the existence of the guiding plates. That is, their strengths were reduced. Thus, the energy transferred from the fluid to the cylinder is reduced. As a result of this, VIV is reduced.

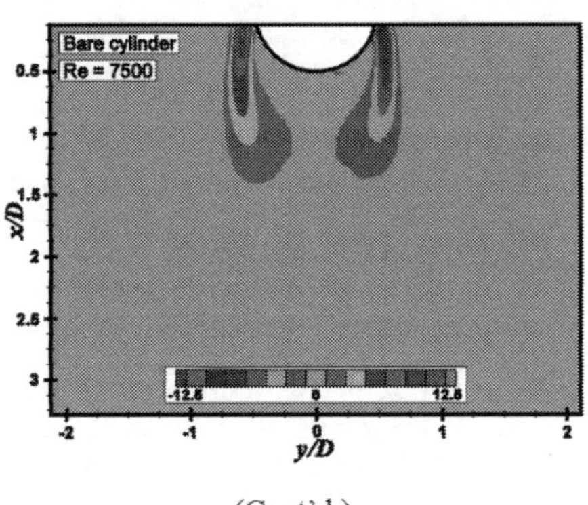

(Cont'd.)

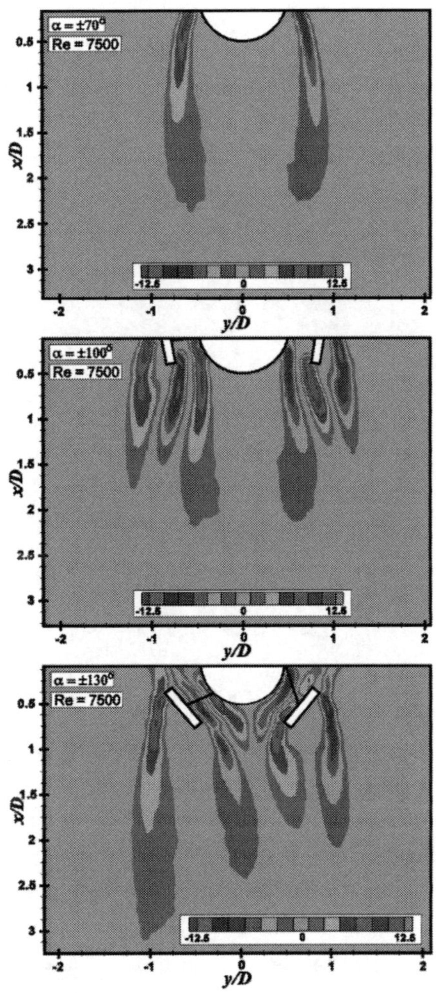

Fig. 6. Time-averaged patterns of spanwise vorticity in the near-wake of the cylinder with and without the guiding plates.

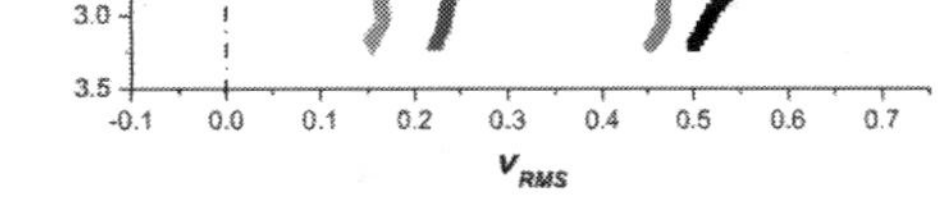

Fig. 7. Variation of v_{rms} along the wake centreline for the BC, A70, A100 and A130 cases.

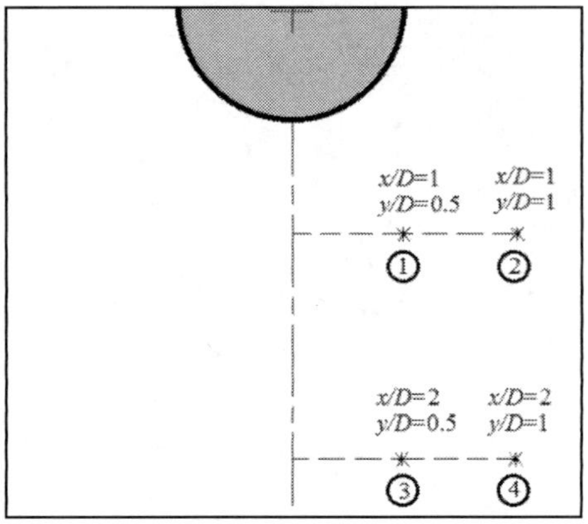

Fig. 8. The location and number of virtual transverse velocity probes behind the cylinder with and without the plates.

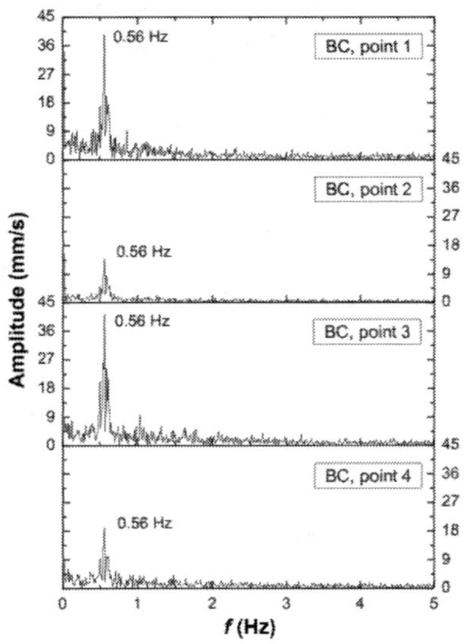

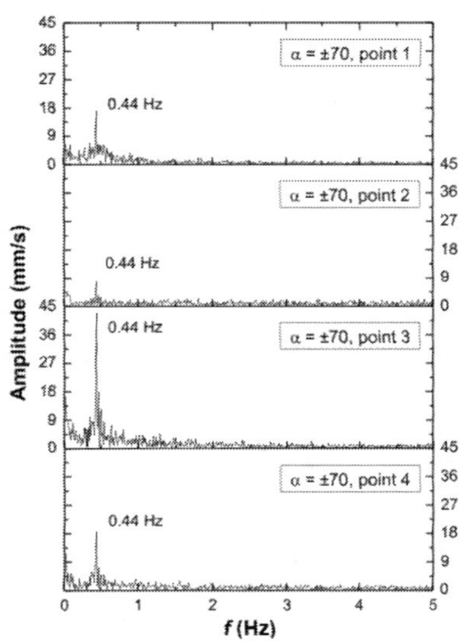

(Cont'd.)

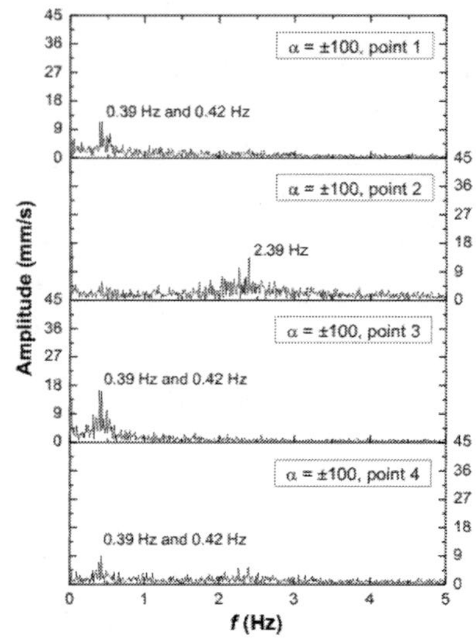

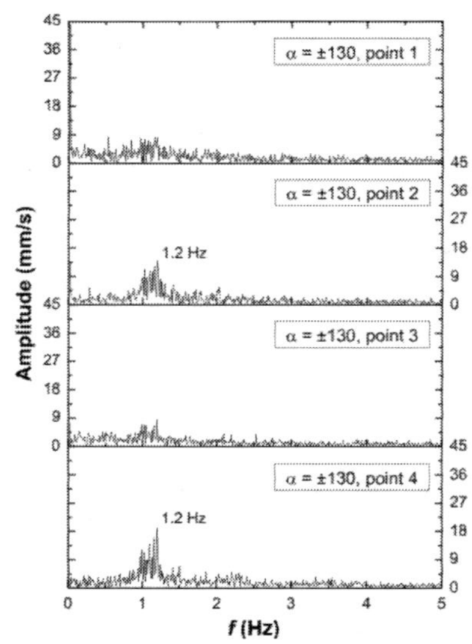

Fig. 9. Power spectra of the transverse velocity – time history at four locations.

A typical set of amplitude spectra of wake instantaneous transverse velocities at various positions through the near-wake (Fig. 8) is shown in Fig. 9. For the case of BC, irrespective of the location of virtual probe, a well-defined spectral peak is observed at the vortex shedding frequency, f=0.56 Hz, corresponded to a Strouhal number ($St=f{\cdot}D/U_{\infty}$) of St=0.200, and this is in agreement with value obtained by Norberg [12], 0.203.

The amplitude values of the spectral peaks at locations close to the wake centerline (i.e. points 1 and 3) are greater than those at locations far to the wake centerline (i.e. points 2 and 4) because their coordinates are inside the approximate boundary of vortex formation region where strong interaction occurs between the two shear layers from the cylinder. For the case of A70, the region of vortex formation was displaced noticeably downstream relative to the case of BC. Therefore, the amplitude value of the spectral peak at point 1 (Fig. 8) was reduced. Moreover, guiding plates at $\alpha=\pm70°$ noticeably retarded the onset and development of large-scale vortical structures in the near-wake, relative to those occurring from a bare cylinder; thus, reduced the vortex shedding frequency from 0.56 Hz to 0.44 Hz. For the cases of A100 and A130, irrespective of the location of the probe, there are no high peaks in the spectra. This means that guiding plates stabilized the near-wake by delaying the separation from the cylinder (Fig. 6) hence remove the tendency of the shear layers to react strongly with each other across the wake.

The force coefficients for the various cases considered are listed in table 2. Meanwhile, the frontal and planform area of the models are kept constant at $D\cdot1$ m^2 to compare directly the forces acting on models. The SST predictions suggest that guiding plates are not effective at drag reduction, but quite effective at vortex shedding suppression.

Table 2. Predicted time-averaged and rms force coefficients for cylinders with and without guiding plates.

Cases simulated	C_D	$C_{L\,RMS}$
Bare cylinder (BC)	1.59	1.19
A70 case ($\alpha=\pm70°$)	2.01	0.759
A100 case ($\alpha=\pm100°$)	1.58	0.538
A130 case ($\alpha=\pm130°$)	2.20	0.651

5 Conclusions

The effects of guiding plates and their angular positions on the near-wake flow structure have been investigated experimentally and on the hydrodynamic coefficients have been predicted numerically. The results showed that guiding plates made a significant contribution to the reduction of the energy transfer from fluid to cylinder, that is, they are effective in suppression of vortex-induced vibrations (VIV) by extending the shear layers further downstream of the wake (A70) or by delaying the separation (A100 and A130). However, hydrodynamic coefficients predicted by two-dimensional unsteady RANS with SST k-ω model indicate that guiding plates are not effective at drag reduction.

References

1. J.C. Owen, P.W. Bearman, J. Fluid Struct. **15**, 597-605 (2001)
2. H.-C. Lim, S.-J. Lee, AIAA J. **40,** 2027-2036 (2002)
3. C.P. Shao, Q.D. Wei, AIAA J. **46,** 397-407 (2008)
4. B. Zhou, X. Wang, W. Guo, J. Zheng, S.K. Tan, J Wind Eng. Ind. Aerodyn. **145,** 30-41 (2015)
5. M. Sarioglu, Flow Meas. Instrum. **53,** 221-229 (2017)
6. A.M. Hamed, J. Vega, B. Liu, L.P. Chamorro, Exp Fluids **58,** 22 (2017)
7. R. Galvao, E. Lee, D. Farrell, F. Hover, M. Triantafyllou, N. Kitney, P. Beynet, J. Fluid Struct. **24**, 1216-1226 (2008)
8. Y.E. Akansu, E. Firat, M. Hacıalioğulları, Exp. Therm. Fluid Sci. **86**, 11-22 (2017)
9. E. Firat, Y.E. Akansu, H. Akilli, Ocean Eng. **108**, 504-518 (2015)
10. G.M. Ozkan, E. Firat, H. Akilli, Ocean Eng. **134**, 35-49 (2017)
11. Engineering Science Data Unit, Data Item: ESDU 80025
12. C. Norberg, J. Fluid Struct. **17**, 57-96 (2003)
13. F.R. Menter, AIAA J. **32**, 1598-1605 (1994)
14. *Fluent 6.3 User's Guide*, 853-858 (2006)
15. A. Khalak, C.H.K. Williamson, J. Fluid Struct. **10**, 455-472 (1996)
16. K. Lam, Y.F. Lin, L. Zou, Y. Liu, Int. J. Heat Fluid Flow **31**, 32-44 (2010)
17. R.M. Stringer, J. Zang, A.J. Hills, Ocean Eng. **87**, 1-9 (2014)

An Investigation on the Effects of Different Stratifications on Negatively Buoyant Jets

Simone Ferrari[1,*], *Maria Grazia* Badas[1], and *Giorgio* Querzoli[1]

[1] DICAAR – Dip. Ingegneria Civile, Ambientale e Architettura, University of Cagliari, via Marengo 2, 09123 Cagliari, Italy

Abstract. Negatively buoyant jets develop when fluids are released upwards into a lighter fluid or, vice versa, downwards into a heavier fluid. There are many engineering applications, such as the discharge, via submerged outfalls, of brine from desalination plants into the sea. Some concerns are raised about the potential negative environmental impacts of this discharge. The increase in salinity is the major cause for environmental impact, as it is very harmful to many marine species. The diffusers for brine discharge are typically inclined upwards, to increase the path before the brine reaches the sea bottom, as it tends to fall downwards driven by negative buoyancy. The negatively buoyant jet that develops conserves axisymmetry only when released vertically, so that it is not possible to use the well-known equations for axisymmetric jets. The main target of this paper is to investigate on a laboratory model the effects of different stratifications on the features of negatively buoyant jets. This has been done via a LIF (Light Induced Fluorescence) technique, testing various release angles on the horizontal and densimetric Froude numbers. Except for the initial stage, a different widening rate for the upper boundary and the lower boundary has been highlighted.

1 Introduction

Negatively buoyant jets (NBJs) develop when fluids are released upwards into a lighter fluid or, vice versa, downwards into a heavier fluid.

NBJs are implied in many fields, from the passive containment cooling system of nuclear power plants (Wang et al., 2017 [1]) to the volcanic eruption columns (Kaminski et al. 2005 [2]), see Ferrari and Querzoli, 2010 [3], for a list. In particular, the discharge of brine from desalination plants is one of the most investigated (see, for instance, Cipollina et al., 2004 [4]; Kikkert et al., 2007 [5], Besalduch et al., 2013 [6], Ferrari and Querzoli, 2015 [7], Crowe et al., 2016 [8], Nikiforakis et al., 2017 [9]), because of the concerns linked to the potential negative environmental impacts of the brine discharge (Campbell et al., 2005 [10]): this is mainly due to the increase in salinity which brine discharges impose on the receiving body, harmful to a lot of marine species, in particular to the Posidonia Oceanica (see Latorre, 2005 [11]; Sànchez-Lizaso et al., 2008 [12]).

The typical configuration for marine brine discharge is a submerged outfall equipped with upward-inclined diffusers: in this way, the path followed by the brine is increased, as brine is driven downwards by negative buoyancy and the path is delimited by the sea bottom (e.g. Roberts and Toms, 1988 [13], and Jirka, 2008 [14]). Even if, as above showed, there have been several investigations about NBJs, the high complexity of the physical phenomenon is not yet fully understood. As a matter of fact, NBJs conserve axisymmetry only as long as they are vertically released, so it is not possible to use the well-known equations for axisymmetric jets. As this lack in axisymmetry is mainly due to the different conditions of stratifications in the upper and lower boundaries of NBJs, the main target of this paper is to investigate the different effects of these stratifications on the features of NBJs.

2 Dimensional Analysis

When a NBJ is released from an upward-inclined diffuser, the flow is driven from two sources: one of momentum and one of negative buoyancy. As a consequence (see Figure 1), if the initial part is driven mostly by the momentum (more "simple-jet-like"), as far as the fluid moves away from the outlet, the buoyancy prevails and the axis bends down (more "plume-like"). As stated by List, 1979 [15], when the receiving fluid is at rest, the only parameters characterizing the boundary conditions are the initial flux of volume Q, the initial flux of momentum M and the initial flux of buoyancy per unit mass B: performing a dimensional analysis, two length scales l_Q and l_M can be obtained. The former, i.e. l_Q, gives the magnitude of the distance from the orifice where the jet becomes independent from the initial conditions:

$$l_Q = \frac{Q}{\sqrt{M}} \qquad (1)$$

* Corresponding author: ferraris@unica.it

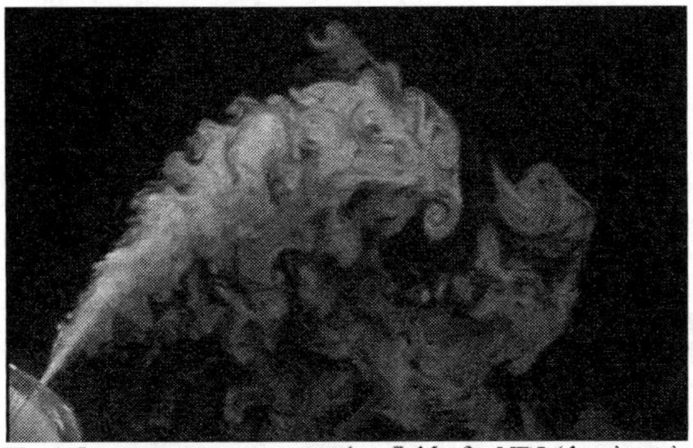

Fig 1. Instantaneous concentration field of a NBJ (densimetric Froude number Fr: 14.8; release angle to the horizontal θ: 55°).

In the case of a circular outlet, such as the one employed in the experiments presented here, l_Q becomes:

$$l_Q = \sqrt{\pi/4}D \cong D \quad (2)$$

The latter, i.e. l_M, gives the magnitude of the distance from the orifice where the plume-like behaviour prevails:

$$l_M = \frac{M^{3/4}}{\sqrt{B}} \quad (3)$$

From the ratio of l_Q to l_M, a non-dimensional number, the Richardson number Ri, measuring the relative importance of buoyancy and momentum, can be obtained:

$$Ri = \frac{l_Q}{l_M} \quad (4)$$

The most important non-dimensional parameter for NBJ is usually considered the densimetric Froude number Fr, i.e. the ratio of inertia to buoyancy:

$$Fr = \frac{U}{\sqrt{g\frac{\rho_{DISC} - \rho_{REC}}{\rho_{REC}}D}} \quad (5)$$

where U is the mean initial jet velocity, ρ_{DISC} the discharged fluid density, ρ_{REC} the receiving fluid density and D the outlet diameter. Fr an Ri are inversely proportional:

$$Fr = \sqrt[4]{\frac{\pi}{4}}Ri^{-1} \quad (6)$$

In addition to Fr, the non-dimensional parameters influencing NBJs are the Reynolds number

$$\mathrm{Re} = \frac{UD}{\nu} \quad (7)$$

where υ is the kinematic viscosity of the brine, and the angle to the horizontal θ. As θ measures the misalignment between M and B, a NBJ will be axisymmetric only if θ is ±90°.

3 Experimental set-up and data elaboration

Experiments were carried out in the Hydraulics Laboratory of the DICAAR, University of Cagliari. In particular, a 30 cm wide, 21 m long flume was employed. The flume was filled with tap water to simulate the sea at rest and a solution of water, sodium-sulphate (a low corrosive salt employed to increase the density of the solution) and fluoresceine (a fluorescent salt) was used to simulate the brine. The solution flew (through a pipe) from a constant head tank (supplied by a closed hydraulic circuit) to a cylindrical vessel, simulating a portion of a pipe laid down the sea bottom, with a sharp-edged orifice on its lateral wall. This was done to simulate a widely employed configuration for the marine outfall design (see, for instance, Wright et al., 1982 [16]; Avanzini, 2006 [17]).

To allow measurements via a Digital Image Analysis (DIA) technique (see below), the NBJ middle vertical section was lighted by a 4 mm thick light sheet and a video-camera, orthogonal to the light sheet, filmed the experiments at a frequency of 25 Hz and a resolution of 720x576 pixels. The flow was investigated by means of LIF (Light Induced Fluorescence), a DIA technique: in Fig. 1, the reader can see an example of an instantaneous image, where a light NBJ is above a dark background. With the employed concentrations, the concentration field is proportional to the light intensity, so a quantitative measure of the concentration is allowed. Digital Image Analysis (DIA) techniques have many advantages, if compared to the traditional techniques employing physical point probes: DIA are non-intrusive and quasi-continuous in space, as every pixel on the camera sensor works as a single probe (see Ferrari, 2017 [18] for a review). The LIF codes were designed in-house by the Hydraulics Group of the University of Cagliari, which has a considerable expertise in the development of DIA techniques to measure concentrations (Ferrari and Querzoli, 2010 [3], Ferrari et al., 2016 [19]), velocity (Besalduch et al., 2013 [6] and 2014 [20], Ferrari et al, 2017 [21]) and accelerations (Ferrari et al, 2007 [22], Ferrari and Rossi, 2008 [23], Ferrari et al, 2008 [24], Lardeau et al., 2008 [25], Rossi et al, 2009 [26]) in the environmental (such as in the present paper), biomedical (Badas et al., 2015 [27], Espa et al, 2012 [28]) or civil and industrial fields (Badas et al., 2017 [29]).

Experiments we were performed with θ = 45° ÷ 90° and Fr = 4.9 ÷ 40.2: by this way, the whole field of practical interest was investigated. Moreover, Q was kept constant and high enough to have a *Re* larger than its critical value for the apparatus.

Under the hypothesis of ergodicity, the statistics of mean and variance of the concentration field were obtained by time averaging. On Fig. 2 and 3, the mean and the variance of the concentration field are shown. The mean concentration is normalized by the initial concentration, the variance by the squared maximum concentration. The jet axis was assumed to be coincident with the locus of the concentration maxima on jet cross-sections. Detailed information on these parameters, as well as on the experimental set-up and data elaboration, can be found on Ferrari and Querzoli, 2010 [3] and on Ferrari et al., 2016 [19].

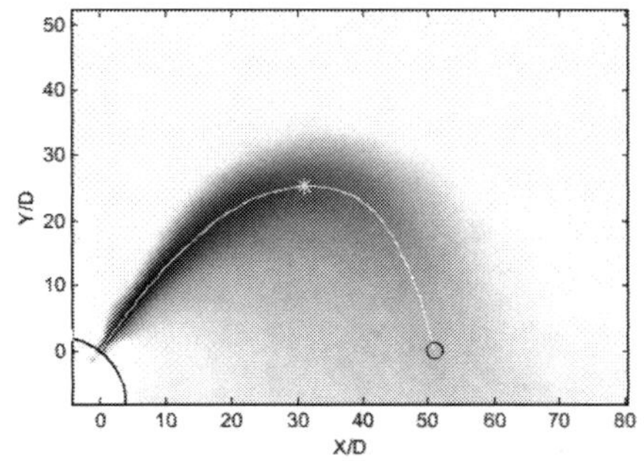

Fig. 2. Map of the non-dimensional mean concentration for the same NBJ of Fig. 1; the white line is the jet axis, the white asterisk is the point of maximum height and the black circle is the impact point.

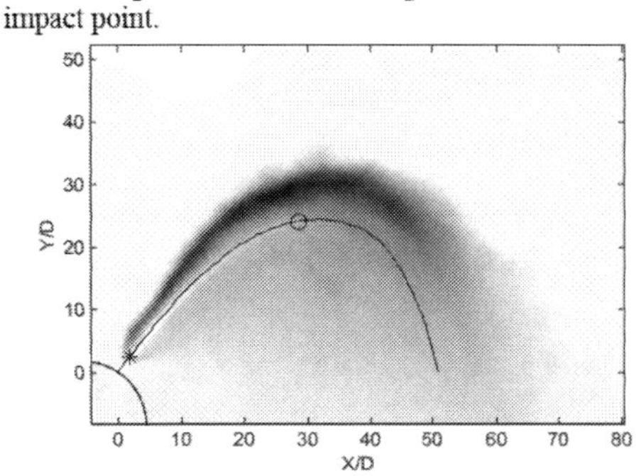

Fig. 3. Map of the variance of the concentration, non-dimensionalised by the square of the maximum concentration, for the same NBJ of Fig. 1; the black line is the jet axis, the black asterisk is the point of onset of the Kelvin-Helmholtz instability and the black circle is the point of transition to turbulence of the whole NBJ.

4 Results

This paragraph is organised as follows: first of all, a discussion about the different stratifications, stable on the upper region and unstable in the lower region of the NBJ, will be given; then, the concentration profiles, orthogonal to the NBJ axis, at various distances from the outlet will be shown, in order to discuss their shape, with particular regard to the Gaussian shape hypothesis often employed in the study of these phenomena; lastly, the focus will be on the widening features of the NBJ upper and lower region.

4.1. The different stratifications on the NBJ boundaries

From Figures 1, 2 and 3, we can see that NBJs released upwards are characterised by an initial region with a width similar to the outlet diameter *D*: afterwards, the onset of Kelvin-Helmholtz (KH) instabilities (in the region pointed out by a black asterisk on Figure 3) causes a sudden widening, as the consequent billows drag external fluid into the NBJ, resulting in mixing and dilution. This dilution, as demonstrated by the reduction in the grey intensity along the axis on Figure 2, grows along the trajectory followed by the jet.

On Figure 2, we can see that the mean concentration field of a NBJ is symmetrical in the first part of the jet, where the flow is driven essentially by *M*; then, as *B* becomes relevant, the span-wise sections become neither symmetrical nor self-similar (see chapter 4.2). As a consequence, above the axis there is a narrow region characterized by high upwards concentration gradients, whilst below the axis there is a wide region of smooth variations. This is due to the fact that, on the upper boundary, the local stable stratification (*M* and *B* have different directions) allows a full development of compact KH billows, followed by their breaking; on the contrary, the local unstable stratification on the lower boundary (*M* and *B* have the same direction) tends to transform the growing waves into downward plumes.

On Figure 3, the variance field shows the lowest values in the core region of the NBJ, close to the axis, values that increase towards the upper and lower boundaries. The stable stratification at the upper boundary of the NBJ has the consequence that the maximum variance values are located at the upper, external region, where the KH waves develop upwards but rapidly break, because of the downwards *B* contrasting their growth. On the opposite, the unstable stratification at the lower boundary of the NBJ has the consequence that the downwards-developing, plume-like structures are driven away from the NBJ core by the buoyancy, with more distributed but less intense concentration fluctuations.

4.2 Shape and Gaussianity of the concentration cross-sectional profiles

As previously stated, when the receiving fluid is at rest, a NBJ conserves axisymmetry only if it is released vertically (*M* and *B* have the same direction) but, if the discharge is done with a different θ, the NBJ rapidly loses its symmetry. This is clearly visible in Figure 4 and 5, where cross-sectional profiles of C/C_0 (the non-dimensional mean concentration *C*, normalized by the initial mean concentration C_0), are shown; *R/D* is the non-dimensional coordinate orthogonal to the jet axis (positive values of *R/D* identify the lower portion of the jet, while negative ones identify the upper portion). The legend shows the non-dimensional distance *S/D*, measured from the outlet along the axis, where the profiles are measured. The different shape between the upper and lower region increases with increasing *S/D*.

The measured points on the concentration profiles (circles, asterisks, crosses and triangles) are shown together with the best-fit Gaussian curves (in a least-squares sense), are shown on Figure 6; the NBJ upper region is on the left-hand side (negative *R/D*), the lower one on the right-hand side (positive *R/D*).

This graph highlights another difference between the upper and lower regions: the concentration profiles have a Gaussian shape, both in the lower and upper part, in the initial region of the NBJ, close to the outlet (solid lines in Figure 6), even if they are not identical. As *S/D* grows, the upper region conserves a Gaussian shape (dashed and dotted lines on the left), whilst the lower region loses its Gaussian shape (as shown by the dashed and dotted lines on the right that do not correctly fit the experimental data). When the NBJ becomes to an end (last section, *S/D* = 50), both the profiles have a Gaussian shape (dash-dot lines), but again with different coefficients.

We can consequently state that an inclined NBJ cannot be considered as a Gaussian phenomenon.

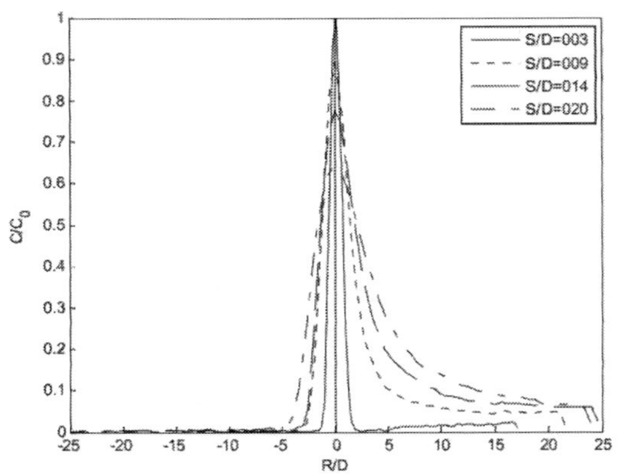

Fig. 4. cross-sectional non-dimensional concentration profiles for a NBJ with θ = 45°, *Fr* = 8.0; positive values of *R/D* identify the lower portion of the NBJ, negative ones identify the upper portion; *S/D* is the non-dimensional distance from the outlet measured along the axis; *S/D* = 3÷20.

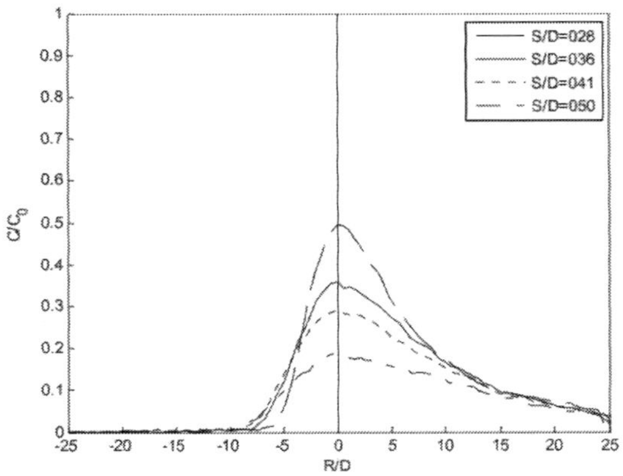

Fig. 5. cross-sectional non-dimensional concentration profiles for the same NBJ of Fig. 4; *S/D* = 28÷50.

4.3 NBJ widening features

To study the widening features of inclined NBJs, Figures 7, 8, 9 and 10 show the experimental data for the widening of the upper boundary (red asterisks) and lower boundary (blue circles) of NBJs, taking into account all the θ in the range 45° ÷ 70° and *Fr* of 30.8 (Fig.7), 23.8 (8), 14.8 (9) and 8.0 (10). The data for the highest θ have been not taken into account because at those θs NBJs experiences the so called re-entrainment. Re-entrainment is the mixing of the NBJ fluid with itself instead of with external fluid; under these conditions, there is not a sharp separation between the uprising and descending branches of the NBJs, as shown by Ferrari and Querzoli, 2010 [3], and consequently it is not possible to define the lower boundaries of the NBJs. The widening is computed as the *R/D* where the concentration decreases to a fraction 1/e of the maximum concentration on that particular profile.

The NBJs with the highest *Fr* (30.8 on Figure 7) experience the largest values of *S/D*, because they are the lightest and, as a consequence, they cover the longest distance before falling down to the sea bottom; of course, exactly the opposite happens for the heaviest NBJs (*Fr* = 8.0 on Figure 10), that cover the shortest path. For *S/D* lower than the length scale l_M (in the range between 29, for *Fr* = 30.8 and 7.6, for *Fr* = 8.0), the widening is similar for the upper and lower regions, as the NBJ is still behaving in a simple-jet-like matter, that with a similar linear growth for both the upper and lower region. When *S/D* becomes much larger than l_M, the lower region, because of the unstable stratification, begins to widen with a faster rate than the upper region, whose widening is slowed down by the stable stratification (see paragraph 4.1).

To study the law of widening rate of the upper and lower region of these NBJs, the upper region data have been fitted, on a least-square sense, with the linear law of eq. 8, the lower boundary with both the eq. 8 and the 5/3 power law of eq. 9:

$$\frac{R}{D} = c\frac{S}{D} + \frac{1}{e} \tag{8}$$

$$\frac{R}{D} = \kappa\left(\frac{S}{D}\right)^{5/3} + \frac{1}{e} \tag{9}$$

Obviously, the constant 1/e has been added because, as a consequence of the chosen definition for the widening, the NBJ width at the outlet has this value for all the *Fr* and θ.

Looking at Figures from 7 to 10, equation 8 seems to correctly fit the data relative to the upper region. Focusing on the lower region, both equations 8 and 9 are able to catch some of the data features, with some lacks in precision for the lowest and highest *S/D*. More in details, equation 8 strongly overestimates the initial widening of the lower region and underestimates the final growth.

https://doi.org/10.1051/epjconf/201818002025

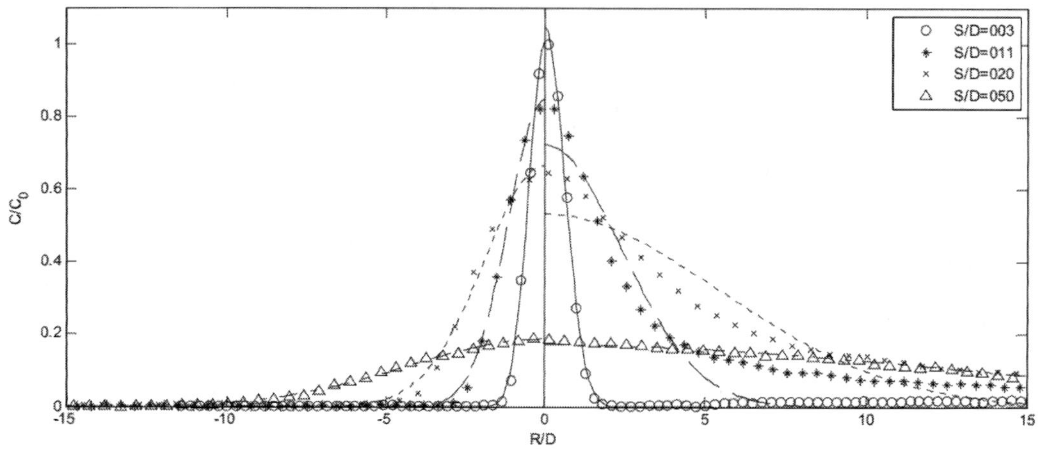

Fig. 6. Interpolation with best-fit (least mean squares) Gaussian curves of cross-sectional non-dimensional concentration profiles: on the left the upper region of the NBJ, on the right the lower region of the NBJ (same NBJ as in Figures 4 and 5).

Equation 9, instead, slightly underestimates the initial widening (and it better follows the trend of the data for low values of *S/D*) but it is able to better catch the trend for increasing *S/D*, particularly for the lowest *Fr*s. As a consequence, we can state that equation 9 seems to better fit the points with the highest *S/D*, which could mean that, if the nearest region to the origin is not taken into account, the lower region tends to a faster than linear evolution or, in other words, doesn't behave like a simple jet of a plume (both linear).

It seems, thereby, that the upper boundary of the inclined NBJs tends to develop following a linear law, whilst the lower boundary tends to evolve in a more complex way, nor jet- neither plume-like, in particular when *S/D* increases and becomes much larger than l_M.

On Table 1, the coefficient of eq. 8 and 9 for the curves of Figures 7, 8, 9 and 10 have been reported; the subscript *U* refers to the upper region, the subscript *L* to the lower region. The values of the coefficient c_U (representing the rate of growth of the upper part of the inclined NBJs) decrease with decreasing *Fr*, suggesting that, as the NBJs become heavier, the stable stratification on the upper boundary prevents the Kelvin-Helmholtz billows to develop vertically and, consequently, the NBJs to widen more. On the opposite, the coefficients κ increase with decreasing *Fr*, suggesting that, as the NBJs become heavier, the unstable stratification on the lower region helps the descending plumes to detach and, as a consequence, the NBJs to widen more. Given a fixed *S/D*, as *Fr* decreases (i.e., *B* increases), the lower region of an inclined NBJ tends to widen more whilst, vice-versa, the upper boundary tends to widen less.

Finally, considering all the showed results, we can state that an incline NBJ cannot be considered as an axisymmetric or Gaussian phenomenon, as some features of simple jets, plumes and more complex ones coexist: the upper region tends to evolve in a simple-jet-like way, following a linear law widening rate; the lower boundary in a more complex manner, following a power law widening rate. The investigation on inclined NBJs can be considered, consequently, a complicated one, because we are not allowed to employ the well-known integral equations usually adopted to forecast the behaviour of axisymmetric jets or plumes.

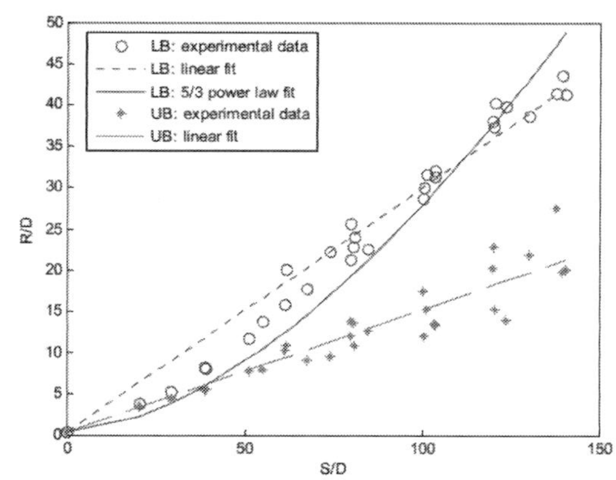

Fig. 7. Experimental data for the widening of the upper boundary (UB) and lower boundary (LB) of NBJs with Fr = 30.8 and $\theta = 45° \div 70°$.

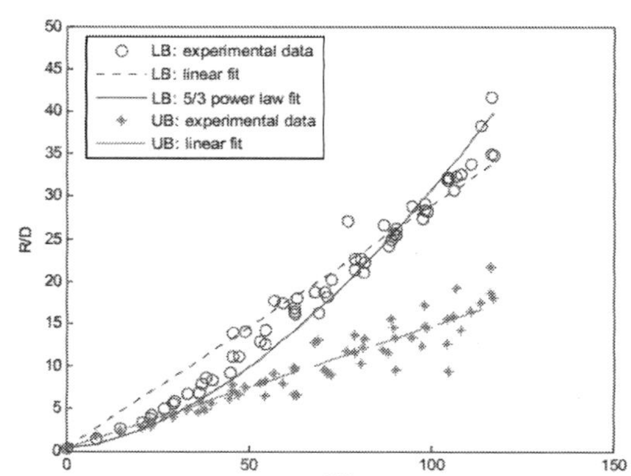

Fig. 8. Experimental data for the widening of the upper boundary (UB) and lower boundary (LB) of NBJs with Fr = 23.8 and $\theta = 45° \div 70°$.

* Corresponding author: ferraris@unica.it

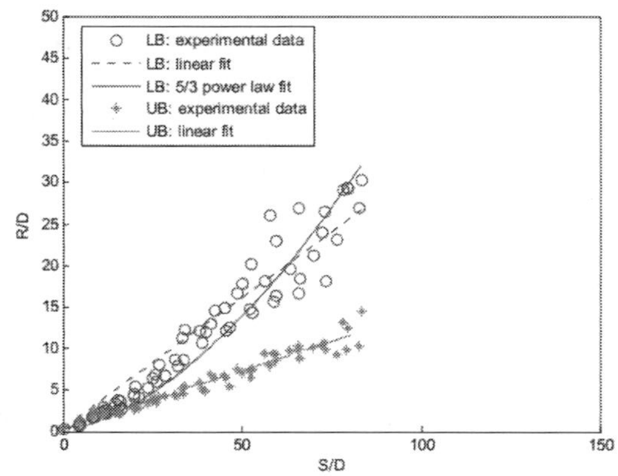

Fig. 9. Experimental data for the widening of the upper boundary (UB) and lower boundary (LB) of NBJs with Fr = 14.8 and $\theta = 45°\div70°$.

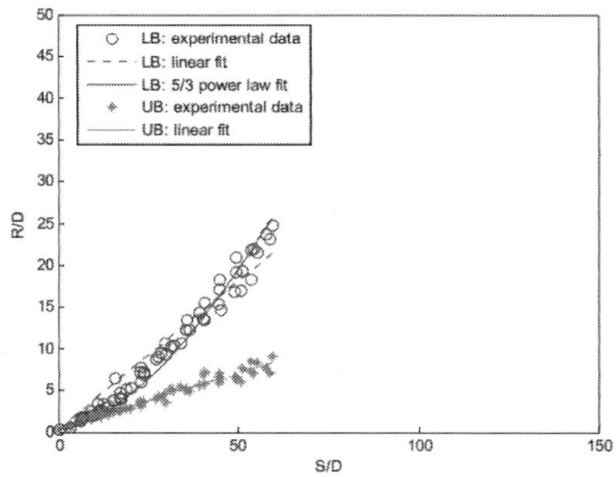

Fig. 10. Experimental data for the widening of the upper boundary (UB) and lower boundary (LB) of NBJs with Fr = 8.0 and $\theta = 45°\div70°$.

Table 1. coefficients of the eq. 6 and 7 for the data of Figure 5; U = upper boundary, L = lower boundary.

Fr	c_L	κ	c_U
30.8	0.297	0.013	0.149
23.8	0.286	0.014	0.143
14.8	0.317	0.020	0.140
8.0	0.355	0.028	0.136

5 Conclusions

In order to have a better insight into the effects of stable and unstable stratification on NBJs released with a certain angle to the horizontal θ, an experimental investigation have been carried out in the Hydraulic Laboratory of the DICAAR, University of Cagliari. A non-intrusive Digital Image Analysis technique, namely LIF (Light Induced Fluorescence), was in-house developed and employed. The experimental parameters, in particular θ and the densimetric Froude number Fr, were chosen in order to investigate the whole field of practical interest for the marine brine discharge.

The results show that the different conditions of stratifications, in particular a stable stratification on the upper region and an unstable stratification on the lower region, force the lower region of the NBJ to widens with a faster rate than the upper region, leading to a strongly asymmetric and non-Gaussian phenomenon. As a consequence, a great care should then be employed when applying the integral equations for axisymmetric jets or plume to inclined NBJs. Moreover, the experimental data have shown that the upper region has a linear widening rate, whilst the lower region tends to widen with a faster 5/3 power law rate.

Acknowledgement

The Authors would like to acknowledge Dr. Luigi Antonio Besalduch, Dr. Michela Garau, Mr. Antonio Mascia, Dr. Lionel Rossi and Prof. John Christos Vassilicos.

References

1. Wang, S., Wang, Y., Zhuo, W., Niu, F., Yu, Y., Guo, Z., Shen, Y., Chen, W., Jiang, X., Experimental analysis of steam mixing and thermal stratification phenomena related to small steel containment studies, Ann. Nucl. En. 109, pp 103–112 (2017)
2. Kaminski, E., Tait, S., Carazzo, G., Turbulent entrainment in jets with arbitrary buoyancy, J. Fluid Mech., 526, pp. 361-376 (2005)
3. Ferrari, S., Querzoli, G., Mixing and re-entrainment in a negatively buoyant jet, Journal of Hydraulic Research, 48 (5), pp. 632-640. (2010)
4. Cipollina, A., Bonfiglio, A., Micale, G. Brucato A., Dense jet modeling applied to the design of dense effluent diffusers, Desalination, 167, pp. 459-468 (2004)
5. Kikkert, G.A., Davidson, M.J., Nokes, R.I. Inclined negatively buoyant discharges. J.Hydr.Eng 133(5), pp. 545-554 (2007)
6. Besalduch, L.A., Badas, M.G., Ferrari, S., Querzoli, G., Experimental Studies for the characterization of the mixing processes in negative buoyant jets, EPJ Web of Conferences, 45, art. no. 01012 (2013)
7. Ferrari, S., Querzoli, G., Laboratory experiments on the interaction between inclined negatively buoyant jets and regular waves, EPJ Web of Conferences, 92, art. no. 02018 (2015)
8. Crowe, A.T., Davidson, M.J., Nokes, R.I., Velocity measurements in inclined negatively buoyant jets, Environ Fluid Mech, 16, pp. 503-520 (2016)
9. Nikiforakis, I.K., Stamou, A.I., Christodoulou, G.C., Integrated modeling of single port brine discharges into unstratified stagnant ambient, Environ Fluid Mech, 17, pp. 247-275 (2017)
10. Campbell, R.L., Jones, A.T., Appropriate disposal of effluent from coastal desalination facilities, Desalination, 182, pp. 365–372 (2005)

11. Latorre, M., Environmental impact of brine disposal on Posidonia seagrasses, Desalination, 182, pp. 517-524 (2005)
12. Sànchez-Lizaso, J.L., Romero, J., Ruiz, J., Gacia, E., Buceta, J.L., Invers, O., Fernàndez Torquemada, Y., Mas, J., Ruiz-Mateo, A., Manzanera, M., Salinity tolerance of the Mediterranean seagrass Posidonia Oceanica: recommendations to minimize the impact of brine discharges from desalination plants, Desalination, 221, pp. 602-607 (2008)
13. Roberts, P.J.W., Toms, G., Ocean outfall system for dense and buoyant effluents, J.Environ.Eng., 114(5), pp. 1175-1191 (1988)
14. Jirka, G.H., Improved Discharge Configurations for Brine Effluents from Desalination Plants, J.Hydr.Eng., 134, pp. 116-120 (2008)
15. List, E.J., Turbulent jets and plumes, in: Fisher, H.B., List, E.J., Koh, R.C.Y., Imberger, J., Brooks, N.H., (eds), Mixing in inland and coastal water. New York, USA, Academic Press, pp. 315-389 (1979)
16. Wright, S.J., Wong, D.R., Zimmerman, K.E., Wallace, R.B., Outfall diffuser behaviour in stratified ambient fluid, J.Hyd.Div. ASCE, 108(4), pp. 483-501 (1982)
17. Avanzini, C., An outline of common practical problems and solutions in the diffusers design, in: Avanzini, C. (ed.), Proceedings of MWWD & IEMES 2006, Antalya (Turkey), 6-10 November 2006, ISBN: 9944-5566-0-2 (2006)
18. Ferrari, S., Image analysis techniques for the study of turbulent flows, EPJ Web of Conferences, 143, art. no. 01001 (2017)
19. Ferrari, S., Badas, M.G., Querzoli, G., A non-intrusive and continuous-in-space technique to investigate the wave transformation and breaking over a breakwater, EPJ Web of Conferences, 114, art. no. 02022 (2016)
20. Besalduch, L.A., Badas, M.G., Ferrari, S., Querzoli, G., On the near field behavior of inclined negatively buoyant jets, EPJ Web of Conferences, 67, art. no. 02007 (2014)
21. Ferrari, S., Badas, M.G., Garau, M., Seoni, A. Querzoli, G., The air quality in narrow two-dimensional urban canyons with pitched and flat roof buildings, Int. J.Environment and Pollution, in press (2017)
22. Ferrari, S., Rossi, L., Vassilicos, J.C., Acceleration measurements in turbulent-like flows, Advances in Turbulence XI - Proceedings of the 11th EUROMECH European Turbulence Conference, pp. 485-487 (2007)
23. Ferrari, S., Rossi, L., Particle tracking velocimetry and accelerometry (PTVA) measurements applied to quasi-two-dimensional multi-scale flows, Experiments in Fluids, 44 (6), pp. 873-886 (2008)
24. Ferrari, S., Kewcharoenwong, P., Rossi, L., Vassilicos, J.C., Multi-scale flow control for efficient mixing: Laboratory generation of unsteady multi-scale flows controlled by multi-scale electromagnetic forces, Solid Mechanics and its Applications, 7, pp. 267-272 (2008)
25. Lardeau, S., Ferrari, S., Rossi, L., Three-dimensional direct numerical simulation of electromagnetically driven multiscale shallow layer flows: Numerical modeling and physical properties, Physics of Fluids, 20 (12), art. no. 127101 (2008)
26. Rossi, L., Bocquet, S., Ferrari, S., Garcia de la Cruz, J.M., Lardeau, S., Control of flow geometry using electromagnetic body forcing, International Journal of Heat and Fluid Flow, 30 (3), pp. 505-513 (2009)
27. Badas, M.G., Espa, S., Fortini, S., Querzoli, G., 3D Finite Time Lyapunov Exponents in a left ventricle laboratory model, EPJ Web of Conferences, 92, art. no. 02004 (2015)
28. Espa, S., Badas, M.G., Fortini, S., Querzoli, G., Cenedese, A., A Lagrangian investigation of the flow inside the left ventricle, European Journal of Mechanics, B/Fluids, 35, pp. 9-19, (2012)
29. Badas, M.G., Ferrari, S., Garau, M., Querzoli, G., On the effect of gable roof on natural ventilation in two-dimensional urban canyons, J Wind Eng Ind Aerodyn 162, pp. 24–34 (2017)

Experimental investigation of the Multipoint Ultrasonic Flowmeter

Filipský Jakub[1]

[1] *CTU in Prague, Faculty of Mechanical Engineering, Department of Fluid Mechanics and Thermodynamics, Technická 4, 166 07 Praha 6, Czech Republic*

Abstract. The Multipoint Ultrasonic Flowmeter is a vector tomographic device capable of reconstructing all three components of velocity field based solely on boundary ultrasonic measurements. Computer simulations have shown the feasibility of such a device and have been published previously. This paper describes an experimental investigation of achievable accuracy of such a method. Doubled acoustic tripoles used to obtain information of the solenoidal part of vector field show extremely short differences between the Time Of Flights (TOFs) of individual sensors and are therefore sensitive to parasitic effects of TOF measurements. Sampling at 40MHz and correlation method is used to measure the TOF.

1 Introduction

The Multipoint Ultrasonic Flowmeter introduced previously in [1] is an implementation of a vector tomographic method. Tomographic methods have been widely used in several fields such as medical imaging, non-destructive testing, or oceanography. The general concept resides in the reconstruction of interior of media based on boundary measurements of some signal, which passes through that media and is influenced by its properties. This means that tomographic methods are highly non-invasive (that is the reason why their driving field is medical imaging). Also we generally receive information about properties of medium at a much higher spatial resolution compared to conventional single-point measurement methods, although point-wise precision of tomographic methods is somewhat lower.

Vector tomography specifically aims at the reconstruction of a vector quantity inside a volume, Norton [2] was the first one to introduce the necessary mathematical apparatus needed to reconstruct the vector field. When the vector quantity is fluid velocity, then ultrasound is the signal of choice.

The analytical solution to reconstruction problem given by [2] uses the Radon transform and needs a complete data set of parallel measurements taken from all possible angles. This is rather impractical for general implementation and soon many alternative reconstruction solutions began to emerge. A nice overview and division of possible approaches into groups is given by Jovanovic [3]. The solution in [1] is a l_2 norm based regularized solution that is additionally regularized by the equation of continuity.

Norton [2] realized, as well as some others later on, that only the irrotational vector field can be reconstructed from the longitudinal interaction with the measurement signal (i.e. time-of-flight), while both longitudinal and transversal interactions are needed to reconstruct the solenoidal component of the vector field [4]. When talking about fluid velocity fields and ultrasonic waves, then the transversal interaction influences the angle-of-departure and angle-of-arrival at the transmitter/receiver side respectively [3]. While the longitudinal interaction is well understood and easily described (and generally used in sonic anemometers, e.g. [5]) the transversal interaction has not been given much interest by the researchers and it is of interest to experimentally investigate the accuracy that can be achieved. Two different approaches to a practical realization of such a measurement were suggested by Jovanovic [3] – the Blumlein microphone and the acoustic tripole. The acoustic tripole method is verified experimentally in this paper as it is not very well examined in literature and still remains one of the uncertainties of the Multipoint Ultrasonic Flowmeter.

2 Transversal flow interaction measurement

Each vector field can be decomposed into its solenoidal (divergence free) and irrotational (curl free) components, we call this the Helmholtz decomposition. It was first shown by Norton [2] and later by others that only irrotational part of vector fields can be reconstructed from longitudal interaction (time-of-flight) measurement, while both longitudinal and transversal interaction are needed to reconstruct the solenoidal part [4]. The transversal interaction can be obtained from ultrasonic measurement also but besides simple time-of-flight between transceiver pairs there also needs to be an information of the angle-of-arrival (AoA) or angle-of-departure from the transmitter or receiver respectively.

The more straight forward method of AoA measurement is the use of the Blumlein pair which is widely used in stereo recording. It consists of two matched receivers (microphones, ultrasonic receivers) with a known directional sensitivity, positioned at different angles. Jovanovic [3], [6] proposes a different technique called the acoustic tripole. An acoustic tripole consists of three nodes – either three transmitters for angle-of-departure measurement, or three receivers for AoA measurement. Such an acoustic tripole consisting of receivers A, B, and C is sketched in Figure 1. Let us suppose that the receiver is sufficiently far away from the transmitter and we can neglect the curvature of the ray path. The ray arrives at the tripole traveling at group velocity $\boldsymbol{u_g}$ which is the vector sum of local flow velocity and the wavefront velocity. The fact that the wavefront is not traveling at its normal direction is important since the difference between TOF sensed by receivers A, B, and C is not dependent on the wavefront normal but rather on the group velocity.

The original acoustic tripole introduced by Jovanovic [3] is different in the sense that the receiver C on Figure 1 would be to the left of receivers A and B. The choice of different layout is solely for the purpose of easier practical implementation. Also Jovanovic uses the absolute TOF sensed by the receivers for AoA computation. The author of this paper finds that rather impractical since the precise detection of ultrasonic signal arrival is complicated with physical receivers implementation. A differential TOF measurement is implemented for reasons stated later in this paper, resulting in the calculation of TOF difference with respect to some reference measurement. Let us suppose that the reference measurement will be sampled with surrounding fluid at rest. Further let τ_A, τ_B, τ_C be the TOF sensed by receivers A, B and C respectively; $\tau_A^0, \tau_B^0, \tau_C^0$ are TOF sensed at rest, $\Delta s_A, \Delta s_B$ are differences of path lenghts that the ray needs to reach the receivers; rest of the notation is apparent from Figure 1. It is only a matter of geometrical constraints resulting from the sketch in Figure 1 that the difference between absolute TOF values sensed by receivers A and C is:

$$\tau_C - \tau_A = \frac{d}{u_g}\sin(\delta - \Theta) \quad (1)$$

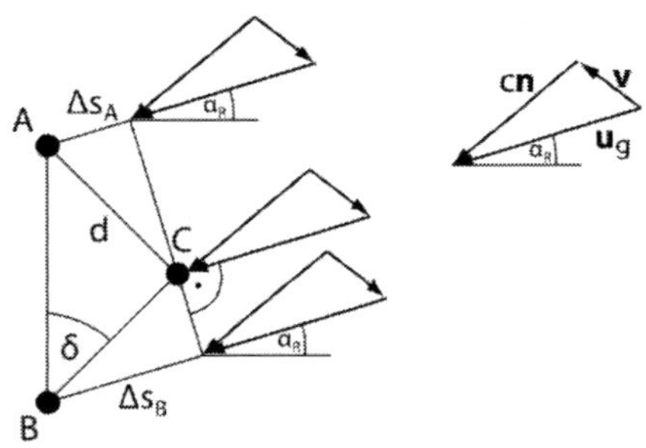

Fig. 1. Acoustic tripole

For the case of reference measurement the group velocity in equation 1 will be replaced by speed of sound when the fluid is at rest :

$$\tau_C^0 - \tau_A^0 = \frac{d}{c_0} sin(\delta - \theta) \quad (2)$$

Combining equations 1 and 2 and expressing the same for receiver pair B and C we obtain a set of nonlinear equations for two unknowns Θ and u_g:

$$sin(\delta - \theta) - v\left(\frac{\Delta\tau_A}{d} + sin\frac{\delta}{c_0}\right) = 0 \quad (3)$$

$$sin(\delta + \theta) - v\left(\frac{\Delta\tau_B}{d} + sin\frac{\delta}{c_0}\right) = 0 \quad (4)$$

The system needs to be solved numerically. Once we know the group velocity $\boldsymbol{u_g}$ we further need the knowledge of local fluid velocity to be subtracted from $\boldsymbol{u_g}$ in order to arrive at the angle Θ_R. Jovanovic suggests to use the acoustic tripole as a standard sonic anemometer to measure the local fluid velocity. It is important to note that the work of Jovanovic concerns tomographic reconstruction of wind speeds in the atmosphere and so the scale of the acoustic tripole may be as large as practical constrains allow it to be. This means that fluid can freely flow between the receivers or transmitters. However in our case the scale of the acoustic tripole is of concern since the size of the experiment is a lot smaller. It is also important to note that in the case when the receivers or transmitters of the acoustic tripole are not submerged in the fluid flow (e.g. local fluid velocity is zero), then we only need receivers A and B as seen in Figure 1 (acoustic dipole). This can be advantageous when receivers/transmitters can be positioned outside of the fluid flow (e.g. further in the wind tunnel test section).

3 Ray tracing

The general task of ray tracing is to find such a path of ultrasonic waves propagating through media which satisfies the shortest time-of-arrival at its destination. This topic of acoustics geometry is well covered by several authors in [7], [8], [9], [10]. The original solution by Uginčius [8], [11] is implemented in this paper, only a brief description is presented here and the reader is referred to available literature for full mathematical derivation. The solution is based on the eikonal equation well known from optics geometry, for the case of acoustic rays in a moving, inhomogenous medium the eikonal equation takes the form

$$|\boldsymbol{\nabla\Phi}| = \frac{c_0}{c+\boldsymbol{v}\cdot\boldsymbol{\hat{n}}} \equiv \mu, \quad (5)$$

where $\boldsymbol{\hat{n}} = \boldsymbol{\nabla\Phi}/\mu$ is the unit vector in the direction normal to the wavefront. Further quantities are introduced to simplify the relations

$$N \equiv (\mu/c)u_g, \boldsymbol{V} \equiv (\mu/c)\boldsymbol{v}, \quad (6)$$

where $\boldsymbol{v}$ represents local fluid velocity and u_g is the magnitude of local group velocity. The eikonal equation then takes the form

$$\nabla\Phi = N\boldsymbol{r}' - \boldsymbol{V}, \quad (7)$$

where $\boldsymbol{r}$ is the position vector of ray path expressed in global coordinate system and $\boldsymbol{r}'$ is its corresponding derivative with respect to the arc length. Equation (7) is then differentiated with respect to the arc length:

$$(N\boldsymbol{r}')' + \boldsymbol{r}' \times (\nabla \times \boldsymbol{V}) = \nabla N \quad (8)$$

Equation (8) represents a set of three ordinary second-order differential equations in $\mathbb{R}^3$ space. The system needs to be integrated numerically with initial conditions for $\boldsymbol{r}$ (position of the transmitter) and $\boldsymbol{r}'$ (initial ray angle). If ray angle measurements are taken at the receiver side, then initial ray angle at the transmitter side needs to be iterated in order for the ray path to hit the receiver. It is of interest that equation (8) does not contain the term $\boldsymbol{V}$ in its zeroth derivative and so it is only the curl of local velocity $\boldsymbol{v}$ that matters. There are two important practical impacts on AoA measurement and fluid transversal interaction in general:

- The ray path will always be a straight line in a homogenous velocity field.
- Let us consider a case of homogenous velocity field with both transmitting and receiving site being submerged in the fluid in such a layout that the homogenous flow velocity is perpendicular to the line defined by transmitter and receiver. The ultrasonic ray travelling as a straight line will arrive at the receiver with zero AOA and thus will carry no information about the velocity field.

4 Ultrasonic TOF measurement

Ultrasonic TOF measurement has become a standard part of multiple industries ranging from sonic anemometers, medical imaging, non-destructive testing, to car parking sensors. The driving industry in this field has been medical imaging, much useful information is found in [12] and [13].

The requirement for precision of TOF measurement is directly connected with the AoA sensing principle. If we suppose two ultrasonic receivers spaced by 36mm (that is the case of the experimental verification introduced later), then for $5°$ of incidence angle the difference between TOF sensed by the two receivers will be in the order of $10\mu s$ when operating in air at normal temperatures. There are several ways of detecting the beginning of the ultrasonic signal, ranging from simple thresholding, through envelope fitting, to the brute-force of correlation. Comparison of four selected methods is given in [14] with the expected result of correlation being the most precise method with the highest cost and computational burden. This will be the method of choice for the experimental investigation introduced later in this paper.

Operational frequency of ultrasonic apparatus is often chosen high since higher frequencies offer higher bandwidth and higher spatial (and temporal) resolution of the signal. As shown by Cheeke [12], the directivity of the transceivers lowers down (e.g. narrower beam) with higher frequencies for constant dimensions of the transceiver. The ultrasonic vector tomograph, as suggested in the introduction to this paper, requires the transceivers to be as wide-angle as possible to get a valid signal between as many ultrasonic pairs as possible. This requirement comes handy since there is a wide range of inexpensive 40 kHz transceivers available.

5 Experimental verification

A simple experiment was carried out in order to obtain validation data for the above apparatus. A setup consisting of a single transmitter and 3 receivers forming an acoustic tripole was prepared. Inexpensive off-the-shelf 40 kHz transmitter/receivers were used, signals at the electrical port of the receivers were sampled by an 8-channel (only 3 channels were connected) synchronous 40 MHz A/D converter. The ultrasonic setup was positioned at the exit of a wind tunnel nozzle with a homogenous (piston) flow. There were three configurations of transmitter/receiver positions with respect to the flow: both ends well outside of the flow (Figure 3.A); transmitter at the border of the flow (Figure 3.B) and both ends at the border (Figure 3.C). No experiment with ultrasonic transmitter/receivers submerged in the flow was done since the ray path would take a straight line connecting the transmitter and receiver as noted previously.

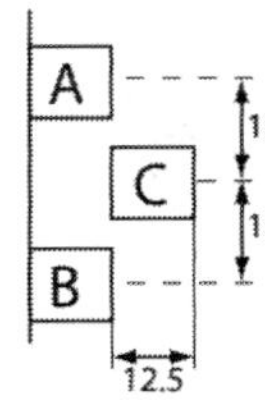

Fig. 2. Dimensions of the acoustic tripole

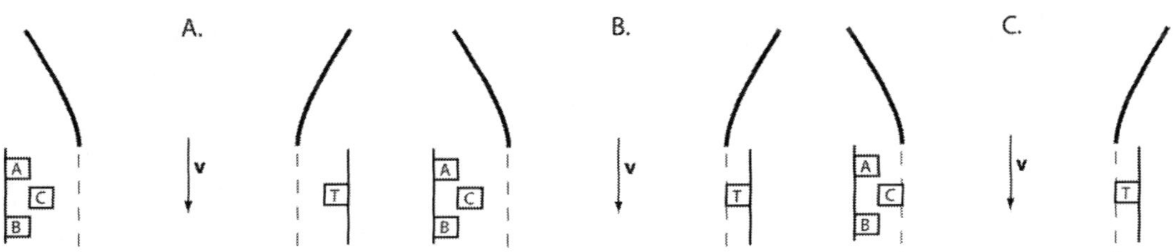

Fig. 3. Three geometrical configurations of the experimental setup

The dimensions of the acoustic tripole are sketched in Figure 2. The width of the flow (ray path length intersecting the flow) was 950mm, when the nodes were positioned outside of the flow their distance to the flow stream was 120mm. Each measurement was done for 0, 5, 10, 20, and 30m/s, 3 samples per each prescribed velocity.

AOA dependence on velocity is plotted for each measurement setup A, B, and C in Figure 4. It is apparent that highest sensitivity was achieved with setup C where the ray path intersecting fluid with zero velocity was shortest. Lowest sensitivity is observed in setup A.

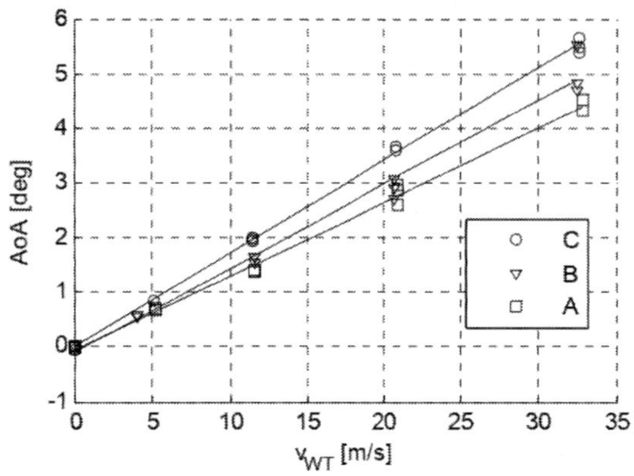

Fig. 4. Measurdd AoA dependence on velocity

Let us now convert the AoA to velocity domain. Of course there is not enough information to reconstruct the velocity distribution along the ray path, not to mention that only the rotational part of the flow field is being seen by the AoA measurement alone. Instead a normalized velocity field was prescribed with the value of one where the ray path crosses the fluid stream and zero elsewhere. Ray tracing algorithm was then run on this etalon flow field, iteratively scaling it so that the ray arrives at the acoustic tripole at the same AoA as evaluated from the experiment. The whole iterative scheme was a cascade of two iteration loops:

- Inner loop iterates the angle of departure at the transmitter side so that the ray hits the receiver
- Outer loop iterates the scaling of the etalon flow field until the desired AoA was reached

The result of the iteration is a simplified velocity profile along the ray path, which, to the author's best knowledge, is a valid assumption of the flow conditions near the exit of the wind tunnel nozzle. The velocity of the simplified profile can be directly compared to the velocity indicated by the wind tunnel. Percentage of error is plotted in Figure 5 such that positive error indicates that velocity computed from AoA is higher than velocity measured by the wind tunnel.

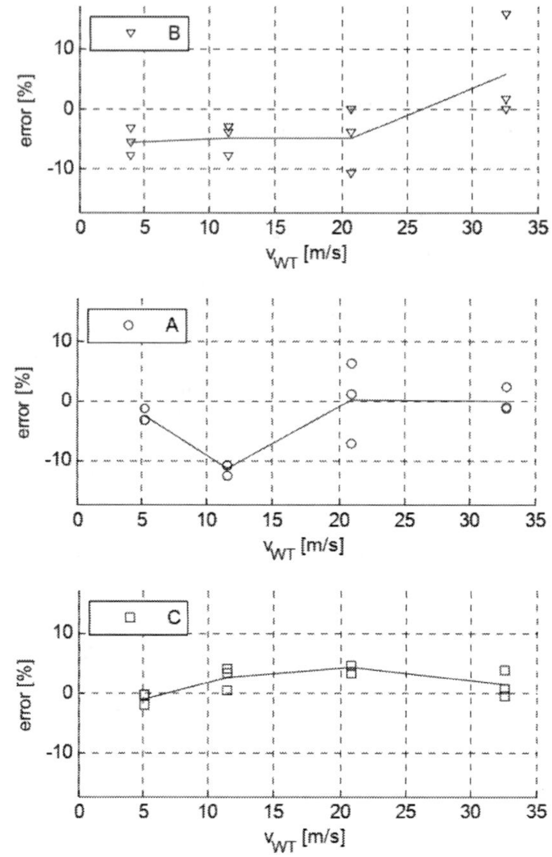

Fig. 5. Velocity error of simulation compared to the experiment

6 Conclusion

A simple yet robust experimental setup was proposed and carried out to test the accuracy achieved by the acoustic tripole at three geometric configurations with respect to the flow. The receiving side was positioned outside of the flow, thus surrounded by fluid at rest – similar test results might have been obtained by a simpler acoustic dipole although that was not the scope of this paper. Sensitivity of AoA to velocity change was lower as there was more fluid along the ray path with zero velocity; this is in good agreement with the ray tracing simulation. The error of evaluated velocity is higher for the cases with lower sensitivity (also the noise seems a lot higher) but the large increase in performance for geometrical case C can hardly be explained simply by sensitivity rise. The fluid outside the flow was probably not at full rest as there are vortices forming near the stream exiting the nozzle. These vortices give a good reason for the worsened performance of cases A and B. If we consider case C as the only valid experiment then the resulting accuracy is well within the range of ±5%. The method chosen for TOF detection – 40MHz sampling and subsequent correlation – leaves very little doubts on the side of signal processing and most of the error will be caused by geometrical uncertainties on the side of the acoustic tripole and also on the side of the velocity field exiting wind tunnel nozzle. Any small deviations of AoA measurement caused by misalignment of the acoustic tripole were mostly eliminated by the use of reference reading at zero velocity. This would be valid exactly if the AoA on TOF difference (sensed by the receivers of the acoustic tripole) dependence was linear, which is true for very small misalignment angles. Further improvements in accuracy could be achieved by identification of coordinates of the receivers with respect to each other, rather than measuring them physically. This could be done by several reference readings from different angles with fluid at rest.

References

1. Filipský, J.; Nožička, J., (2016): *Simulation of Multipoint Ultrasonic Flowmeter*, In: STČ 2016. Praha: ČVUT, FS Ú-12242, VCSVTT, 2016. pp. 81-86. ISBN 978-80-01-05929-6.
2. Norton, S.J., (1988): *Tomographic reconstruction of 2-D vector fields: application for flow imaging,* In: Geophysical Journal, 97:161-168
3. Jovanović, I., L. Sbaiz, and M. Vetterli, (2009): *Acoustic Tomography for Scalar and Vector Fields: Theory and Application to Temperature and Wind Estimation*, In: J. Atmos. Oceanic Technol., 26, 1475–1492, https://doi.org/10.1175/2009JTECHA1266.1
4. Braun, H., Axel Hauck, (1991): *Tomographic reconstruction of vector fields*. IEEE Trans. Signal Processing 39: 464-471.
5. A. Coppin, P & J. Taylor, K., (1983): *A three component sonic anemometer/thermometer system for general micromeleorology research. Boundary-Layer* Meteorology. 27. 27-42. 10.1007/BF00119970.
6. Jovanovic, I., Sbaiz, L., Vetterli, M., (2006): *Acoustic Tomography Method for Measuring Temperature and Wind Velocity*. IV - IV. 10.1109/ICASSP.2006.1661175.
7. Landau, L. D. ; Lifshitz, E. M., (1959): Sykes, J. B.; Reid, W. H.: *Fluid Mechanics* : Pergamon Press Oxford, England
8. Ugincius, P., (1971): *Acoustic Rays in a Moving Inhomogeneous Medium*. Journal of The Acoustical Society of America - J ACOUST SOC AMER. 49. . 10.1121/1.1975587
9. J. Thompson, Robert. (1972). *Ray Theory for an Inhomogeneous Moving Medium*. The Journal of the Acoustical Society of America. 51. 1675. 10.1121/1.1913014.
10. Ostashev, V. E., (1997): *Acoustics in Moving Inhomogeneous Media*. E&FN SPON, 259 pp.
11. Munk, W, Worcester P, Wunsch C. (1995): *Ocean acoustic tomography.* , Cambridge, England: Cambridge University Press
12. J. David N. Cheeke, (2017): *Fundamentals and Applications of Ultrasonic Waves*, Second Edition, March 29, 2017 by CRC Press Reference - 504 Pages - 161 B/W Illustrations ISBN 9781138077201 - CAT# K34283
13. Szabo, Thomas L.: *Diagnostic Ultrasound Imaging Inside Out* A volume in Biomedical Engineering, ISBN: 978-0-12-680145-3
14. Billur, B. (2000): *Fast processing techniques for accurate ultrasonic range.* 0957-023

Experimental verification of the properties of pneumatic elements

K. Fojtášek[1,a], L. Dvořák[2] and Š. Chmura[3]

[1,2,3]VŠB-Technical University of Ostrava, Faculty of Mechanical Engineering, Department of Hydrodynamics and Hydraulic Equipment, 17. listopadu 15/2172, Ostrava 70833, Czech Republic

Abstract. The subject of this article is experimental verification of properties of pneumatic elements. These elements have been a part of equipment for measuring of rotary air motors parameters. When comparing the measured data with the mathematical model, there were slight variations. The main objective is to verify the correct functioning of these pneumatic elements, comparing the measured results with theoretical data from the catalogs of manufacturers and determine their potential impact on the entire measuring system.

1 Introduction

Last year they were published two papers on the topic of measuring characteristics [1] and mathematical modeling of pneumatic rotary motors [2]. These publications included a description of the measurement and also the fact that one of the approaches to modeling did not show good results. This could be caused not only by the model of the rotary motor itself, but also influenced by elements of the measuring device used before the motor. For this reason, the characteristics of the basic elements used in the measuring circuit were measured to verify their correct functionality.

The subject of this article is the experimental verification of properties of pneumatic elements. When measuring it is necessary to comply with the general requirements for testing of pneumatic components. It is necessary to ensure a certain accuracy and credibility of the measured results as well as reproducibility and repeatability of each measurement. These general requirements are described in the ČSN 10 9006 standard for pneumatic elements. First of all, it is important to inspect the tested element properly, to ensure the inner and outer cleanliness of the element and its integrity. Check the functionality of the element and, if necessary, adjust the element according to the instructions or as specified in the technical specifications. The whole test circuit must then be checked for proper arrangement, functionality, leakage and used measuring equipment. The measuring equipment must have the correct measuring ranges and the correct accuracy class. The whole circuit must be built according to a predetermined scheme. The elements are mounted in the prescribed position and emphasis must be placed on safety. All measuring devices and sensors must undergo inspection or calibration prior to the measurement. If they change their parameters as a result of use during the test, they must be checked again after the test. If a deviation is greater than permitted class of precision the measurements must be repeated. It is also necessary that each measurement be carried out in a steady state unless there are special requirements for the dynamics of the measured element and the measurements must be made at least three times. All measured values are recorded during the measurement, test conditions (ambient temperature, air pollution class, atmospheric pressure, vibration, impact, etc.), date, place of measurement, data on measurement methodology and measuring devices and also the test system scheme. These data are based on test report, which contains the main technical parameters of the test element and their comparison with the measurement results. The measurement results are displayed either in the table or in the graph as functional dependencies [3].

[a] Corresponding author: kamil.fojtasek@vsb.cz

2 Theory of measurement

Measured elements are a throttling valve, a pressure reducing valve and a directional control valve whose measurement methodology is described in ISO 6358 standard. This is sonic conductance measurements, critical pressure ratio and normal nominal flow rate. Another alternative is measurement according to the Japanese standard JIS B 8390, where the effective area (equivalent cross-section) is measured.

According to the International Standard, the flow characteristics are given as a result of the comparison between the sonic conductance $C\,[dm^3 \cdot s^{-1} \cdot bar^{-1}]$ and the critical pressure ratio $b\,[-]$. Sonic conductance C is the maximum flow capacity of any pneumatic element. Maximum capacity is reached when the passing gas reaches the choked flow. The critical pressure ratio b

defines the pressure ratio at which the flow in the element is changing Fig. 1.

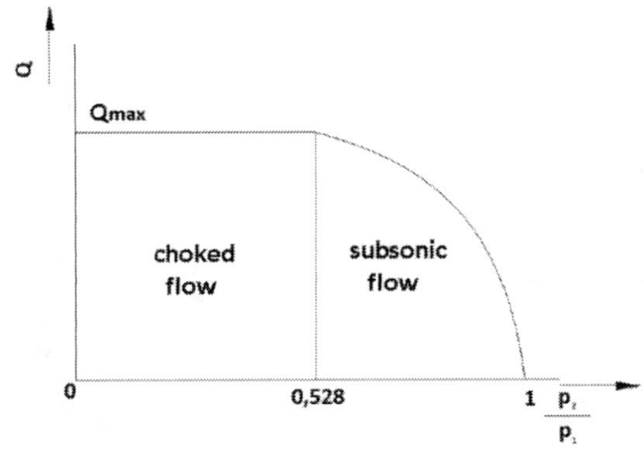

Fig. 1. Types of flow

At throttled flow, the input (upstream) pressure is higher than the output (downstream) pressure on the measured element, and at some part of the element the sound speed is reached. The standard defines the basic conditions, air temperature $20°C$, absolute pressure 1 bar and relative humidity 65%. To calculate the flow rate through the measured element, if $\frac{p_2+1}{p_1+1} \leq b$ (choked flow)

$$Q = 60 \cdot C(p_1 + 1)\sqrt{\frac{293}{273 + t}} \tag{1}$$

For subsonic flow, if $\frac{p_2+1}{p_1+1} > b$ is valid

$$Q = 60 \cdot C(p_1 + 1)\sqrt{\left[\frac{\frac{p_2 + 1}{p_1 + 1} \cdot b}{b}\right]^2}\sqrt{\frac{293}{273 + t}} \tag{2}$$

Figure 2 shows a measuring circuit according to ISO 6358 standard. By attaching a test equipment with the test circuit, the process of measurement and determination of parameters are follows. Upstream pressure p_1 have to be minimally 3 bar and all the time of measurement have to be constant. Throttling valve is fully open. Now maximal (sonic) flow rate Q_m^* can be measured. Sonic conductance C can be determine by formula

$$C = \frac{Q_m^*}{\rho_0 \cdot p_1}\sqrt{\frac{T_1^*}{T_0}} \tag{3}$$

Calculation of pressure ratio b requires measured pressure p_1, p_2, temperature T_1 and flow rate by the four other settings of flow rate Q_m yet (80%, 60%, 40% and 20% of sonic flow Q_m^*). Flow can be changed by throttling valve. Critical pressure ratio b is determined as a mean value of pressure ratios b for each of flow rate Q_m. Single pressure ratios can be calculated by formula

$$b = 1 - \frac{\frac{\Delta p}{p_1}}{1 - \sqrt{1 - \left(\frac{Q_m}{Q_m^*}\right)^2}} \tag{4}$$

On the same track, the normal nominal flow rate is measured at the inlet pressure of 6 bar and 5 bar at the outlet of the element.

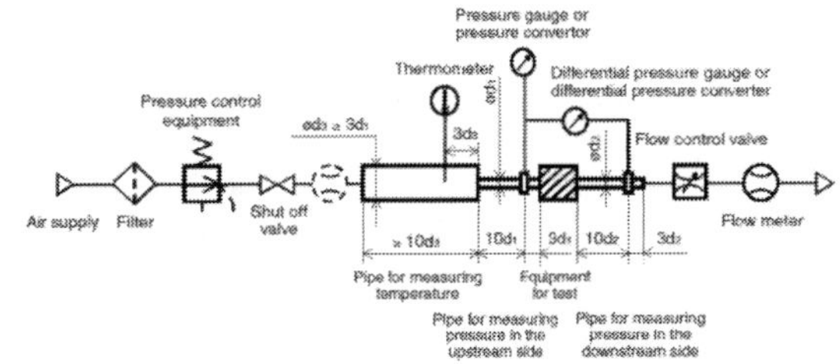

Fig. 2. Measuring circuit according to ISO 6358 standard [4]

JIS B 8390 standard defines another measurement method. Describe the effective area S, which is the cross-sectional area having an ideal throttle without friction deduced from the calculation of the pressure changes inside an air tank or without reduced flow when discharging the compressed air in a choked flow, from an equipment attached to the air tank. Again according to the type of the flow, two basic equations can be determined.

For choked flow $\frac{p_2+1}{p_1+1} \leq 0.5$ is flow rate

$$Q = 12 \cdot S(p_1 + 1)\sqrt{\frac{293}{273 + t}} \tag{5}$$

For subsonic flow $\frac{p_2+1}{p_1+1} > 0.5$

$$Q = 24 \cdot S\sqrt{(p_2 + 1)(p_1 - p_2)}\sqrt{\frac{293}{273 + t}} \tag{6}$$

Measuring circuit according to JIS B 8390 standard is in Fig. 3. The air tank must be pressurized to a certain pressure which must not fall below 6 bar. When compressed air is discharged, the time at which the pressure in the air tank drops to 2.5 bar is measured.

Now it is also possible to calculate the effective area S according to the equation (7). The volume of the air tank shall be selected within a given range corresponding to the relative cross-section of the test element.

$$S = 1.21\frac{V}{t}\log_{10}\left(\frac{p_s + 1}{p + 1}\right)\sqrt{\frac{293}{T}} \tag{7}$$

S is effective area $[mm^2]$, V is air tank capacity $[dm^3]$, t is discharging time $[s]$, p_s is pressure inside air tank before discharging $[bar]$, p is residual pressure inside air tank after discharging $[bar]$ and T is temperature inside air tank before discharging $[T]$.

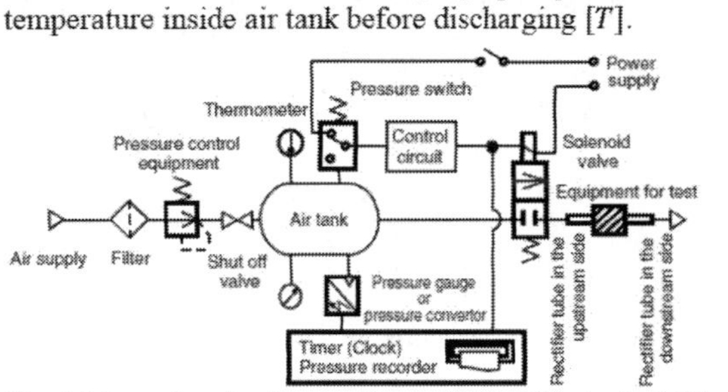

Fig. 3. Measuring circuit according to JIS B 8390 standard [4]

3 Measurement methodology

The design of the track for measuring the flow characteristics of the throttling and pressure reducing valve was based on ISO 6358 standard. Measured values was inlet pressure p_1, outlet pressure p_2 (before and behind the measured element), temperature T and flow rate Q. The temperature was measured at the beginning of the measuring track, the flow meter was positioned at the end of the track with respect to its type. The scheme of the measuring track is in Fig. 4.

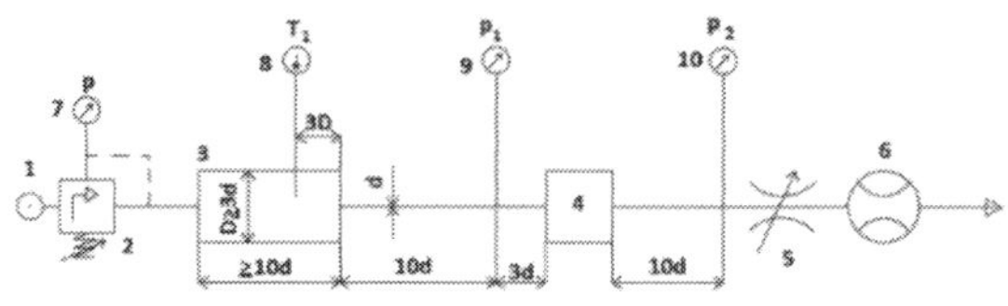

Fig. 4. Scheme of the measuring track for throttling and pressure reducing valves
1 - source of compressed air, 2 – pressure reducing valve, 3 – air tank , 4 – measured element, 5 – throttling valve, 6 – flow meter, 7 – control manometer, 8 - thermometer, 9 – manometer for inlet pressure measurement, 10 - manometer for outlet pressure measurement, d - inner diameter of the pipe, D – diameter of air tank

In the case of the measurement of the throttling valve, after the source of compressed air 1, a reducing valve 2 is located which adjusts and regulates the constant pressure p_1 at the inlet of the measured element 4. The throttling valve 5 is completely closed at the beginning of the measurement. Turn the measured throttling valve 4 to a certain number of revolutions and start measuring. By gradual opening of the throttling valve 5 by certain revolutions, the outlet pressure p_2 and the flow rate Q (measured by the flow meter 6) are also increasing. Measurement is shown in Fig. 5.

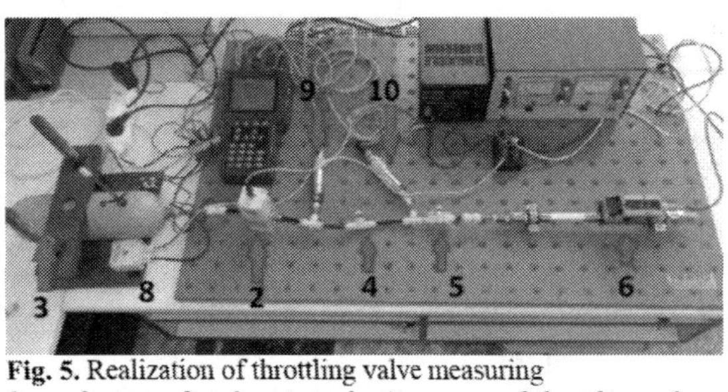

Fig. 5. Realization of throttling valve measuring
2 – reducing valve, 3 – air tank , 4 – measured throttling valve, 5 – throttling valve, 6 – flow meter, 8 - thermometer, 9 – inlet pressure sensor, 10 – outlet pressure sensor

The measured element was a throttling valve AS 3001F. The working range of the valve is $1 \div 10\,bar$. The operating temperature is $-5 \div 60°C$ and nominal size is $10\,mm$. The throttle valve was measured for 4 different revolutions settings and set to $5.5\,rev$, $6.5\,rev$, 8 and 10 revolutions.

The measuring track for the reduction valve is the same as in Fig. 4. Behind the source of compressed air 1 is a reduction valve 2, which sets the constant pressure p_1 at the inlet of the masured valve 4. On the pressure valve 4 we set the same pressure like at the inlet or the less pressure. The measurement is carried out by gradually opening the throttle valve 5, thereby changing the pressure and flow, at the same time the inlet pressure must be controlled by the pressure regulator 2. The flow rate and temperature is measured as in the previous measurement. Realization of measurement is shown in Fig. 6.

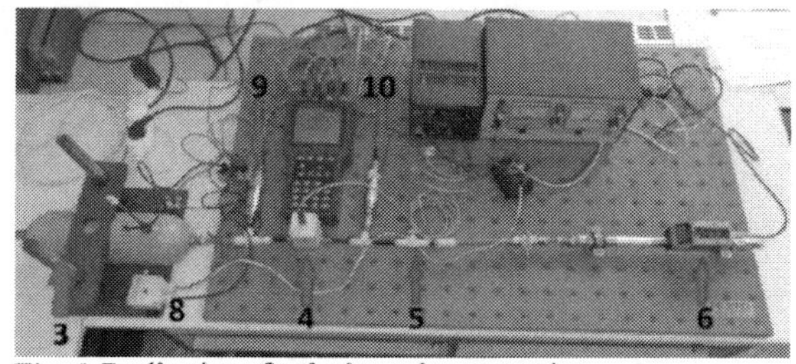

Fig. 6. Realization of reducing valve measuring
3 – air tank , 4 – measured reducing valve, 5 – throttling valve, 6 – flow meter, 8 - thermometer, 9 - imput pressure sensor, 10 - output pressure sensor

The measured element was a pressure reducing valve VPPE-3-1-1/8-10-010-E1 Festo. The control range of this valve is $0.1 \div 10\,bar$ and is capable of operating at ambient temperature $0 \div 60°C$ and the temperature of the medium$10 \div 50°C$. The reducing valve was measured for 4 different settings 3, 4, 5 and 6 bar.

The scheme of the measuring track for measuring the 5/2 directional valve is shown in Fig. 7. Behind the source of compressed air 1, is a reducing valve 2 to adjust the inlet pressure to the directional valve 6. Shut off valve 5 open to full flow and measure the discharge time of the air tanks 3 and 4. Connection is made only in one direction of the valve, return strand is blinded. During the measurement, the directional valve is not switching. The layout of the elements on the scheme is based on the JIS B 8390 standard.

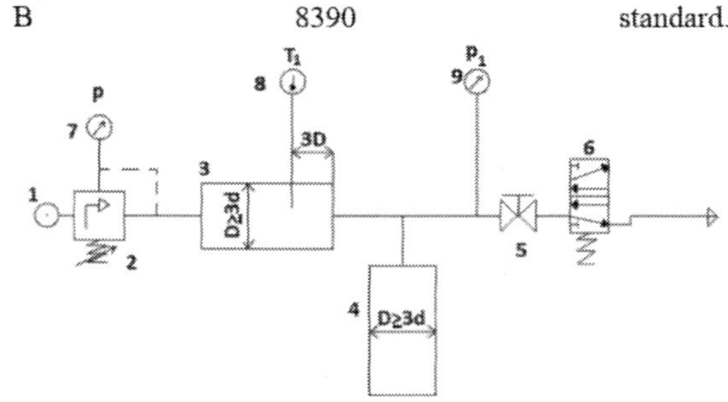

Fig. 7. Scheme of the measuring track for directional valve
1 - source of compressed air, 2 – pressure reducing valve, 3 – air tank , 4 – air tank, 5 – shut off valve, 6 – measured directional valve, 7 - control manometer, 8 - thermometer, 9 - manometer for inlet pressure measurement

Measured directional valve was 5/2 SMC SYA3120 – M5. The working range of the valve is $1.5 \div 7\,bar$ and is capable of operating at ambient and medium temperature $-10 \div 60°C$. Nominal flow rate is $171\,dm^3 \cdot min^{-1}$. Realization of measuring is shown in Fig. 8.

Overall, there were three measurements. The first measurement can only be considered as a test. The remaining two measurements can be considered as final. During the measuring were recorded time values of discharging of air tank. These values were measured until to pressure 2.5 bar. Then, according to equation (7) was calculate effective area S.

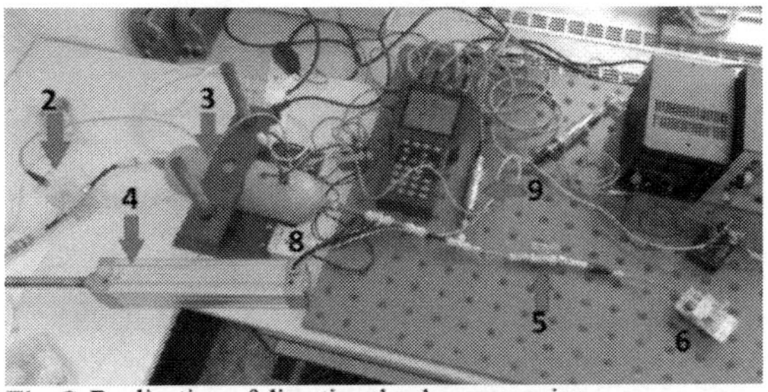

Fig. 8. Realization of directional valve measuring
2 – reduction valve, 3,4 air tank, 5- shut off valve, 6 - measured directional valve, 8 - thermometer, 9 – inlet pressure sensor

4 Conclusion

This chapter lists the measurement results and the conclusions that follow. In the case of a throttling valve, the manufacturer does not specify the characteristics of this valve in the catalog, so it cannot be compared with it. However, according to the visual comparison with the theoretical characteristics of the throttle valves, it can be assumed that the throttle valve has the correct flow throughput at the given setting. The resulting measured flow characteristic is shown in Fig. 9.

In the case of a reduction valve, the measured characteristics are consistent with the characteristics stated by the manufacturer. Due to the conditions in the pneumatic mechanism laboratory, it was not possible to verify the properties of the reduction valve over the entire flow range. However, this was not necessary with respect to flow rates when measuring rotary motors. Measurements at a pressure of $p_1 = 6\ bar$ were influenced by the system before the measuring line and lack of permeability. The resulting characteristic with its course coincides with the characteristics of the manufacturer and their comparison is shown in Fig. 10.

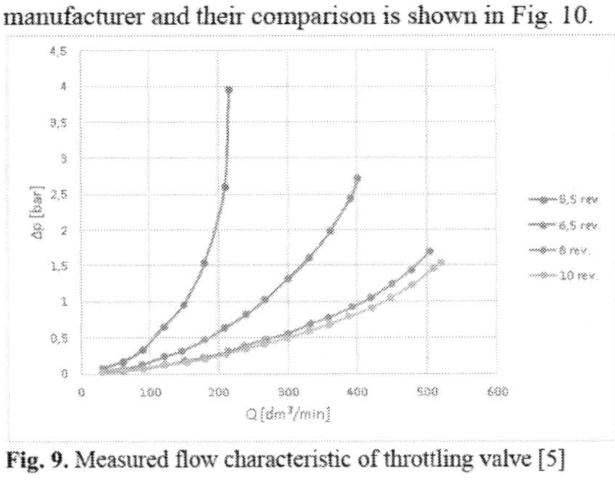

Fig. 9. Measured flow characteristic of throttling valve [5]

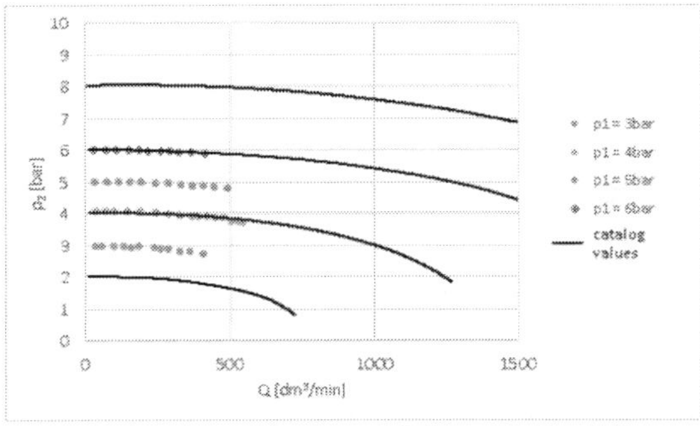

Fig. 10. Comparison of the measured characteristics of the reduction valve with the information given in the manufacturer's catalog [5]

For directional valves was determined effective area S by the measured values of discharging air tank. Its average value for all measurements was $S = 2.619\ mm^2$. Subsequently, for comparison with catalog data, the effective area was recalculated to the C_v factor. Resulting values of this factor was 0.144, which is lower than the value stated in the manufacturer's catalog (0.16). This deviation could be due to the wear of the measured element or insufficient volume of the air tank. To confirm this statement would have to be carried out another series of measurements, with a longer discharging time of air tank.

The measured flow characteristics of the pneumatic elements will be further used to compare and modify the mathematical model of the pneumatic system with the rotary pneumatic motor. Based on the measurements, it was confirmed correct operation of the throttling and a pressure reducing valve. For subsequent control measurement of rotary air motors, will be replaced directional valve by a new element.

Acknowledgements

The work presented in this paper was supported by a grant SGS „Research in field of fluid system dynamics" SP2017/103.

References

1. L. Dvořák, K. Fojtášek, V. Řeháček, Experimental Equipment for Measuring of Rotary Air Motors Parameters, *EPJ Web of Conferences*, Volume 143 (2017)
2. L. Dvořák, K. Fojtášek, V. Řeháček, Calculations of Parameters and Mathematical Model of Rotary air Motor, *EPJ Web of Conferences*, Volume 143 (2017)
3. J. Kopáček, *Pneumatické mechanismy* (1998)
4. SMC, *Solenoid Valves Flow Characteristics* (1998)
5. Š. Chmura, Measurement of Characteristics of Pneumatic Components, *Bachelor Thesis* (2017)

CFD analysis of a twin scroll radial turbine

Jiří Fürst[1,★] and *Zdeněk* Žák[2,★★]

[1] *Faculty of Mechanical Engineering, Czech Technical University in Prague, Karlovo nám. 13, 121 35 Praha 2, Czech Republic*
[2] *Centre of Vehicles for Sustainable Mobility, VTP Roztoky, Přílepská 1920 252 63 Roztoky u Prahy, Czech Republic*

Abstract. The contribution deals with the application of coupled implicit solver for compressible flows to CFD analysis of a twin scroll radial turbine. The solver is based on the finite volume method, convective terms are approximated using AUSM+up scheme, viscous terms use central approximation and the time evolution is achieved with lower-upper symmetric Gauss-Seidel (LU-SGS) method. The solver allows steady simulation with the so called frozen rotor approach as well as the fully unsteady solution. Both approaches are at first validated for the case of ERCOFTAC pump [1]. Then the CFD analysis of the flow through a twin scroll radial turbine and the predictions of the efficiency and turbine power is performed and the results are compared to experimental data obtained in the framework of Josef Božek – Competence Centre for Automotive Industry.

1 Introduction

CFD simulations play an important role in turbomachinery design and analysis. Although there are a number of ready-made software packages (both commercial and in-house), the reliable simulations remains still a challenging task especially in the case of high rotational speed radial turbines or low-pressure stages of axial turbines. The flow passing through such types of turbines achieves high Mach number and has to therefore considered as compressible.

Most software packages for the simulation of compressible flows are based on finite volume approach and use either a segregated or a coupled approach. When using the segregated solver the equations expressing conservation of momentum, density, and total energy are solved independently in a loop in order to obtain steady state or unsteady solution. The algorithm is usually based on the so called SIMPLE algorithm proposed originally Patankar and Spalding in [2] for stationary case, or on the PISO algorithm of Issa [3] for unsteady case. Although the segregated solvers have been extended also to compressible flow problems including transonic or supersonic flows (see e.g. [4] or [5]), the efficiency of these method is in the case of high speed flows usually worse than the efficiency of implicit fully coupled solvers [6].

So far several attempts have been done in order to develop coupled solvers for OpenFOAM platform. Let us mention the solver based on the central-upwind scheme of Kurganov and Tadmor [7] or the density based solvers on OpenFOAM-extend package based on AUSM+up [8] or HLLC [9] fluxes. Unfortunately all these solvers use explicit time-marching method which puts a strong limit to the time step and therefore the efficiency is very weak especially for steady state flows at high Reynolds number. Therefore we developed in [6] an implicit coupled solver based on the symmetric lower-upper Gauss-Seidel procedure, see e.g. [10] or [11] and we further extend this solver for the case of turbomachinery flows using either the frozen rotor or unsteady approaches.

2 LU-SGS solver for turbomachinery flows

2.1 Description of the solver

The system of time averaged Navier-Stokes equations in conservative form are given as follows

$$\frac{\partial \rho}{\partial t} + \nabla \cdot (\rho \vec{V}) = 0, \qquad (1)$$

$$\frac{\partial (\rho \vec{U})}{\partial t} + \nabla \cdot (\rho \vec{V} \otimes \vec{U} - \vec{\tau}) + \nabla p + \rho \vec{\Omega} \times \vec{U} = 0, \qquad (2)$$

$$\frac{\partial (\rho E)}{\partial t} + \nabla \cdot \left(\rho E \vec{V} + p \vec{U} - \vec{\tau} \cdot \vec{U} - k \nabla T\right) = 0, \qquad (3)$$

where ρ is the density, $\vec{U}$ is the absolute velocity vector, $\vec{V} = \vec{U} - \vec{\Omega} \times \vec{r}$ is the velocity vector in the reference frame rotating with constant angular velocity $\vec{\Omega}$, p is the pressure, $\vec{\tau}$ is the (effective) stress tensor, E is the specific total energy, k is the (effective) thermal conductivity, and T is the temperature. The system is closed by the equation of state for ideal gas $p/\rho = rT$.

The discretization in space is achieved using finite volume method where convective fluxes are replaced by an approximate Riemann solver (in our case with the HLLC flux, see [9]) and viscous fluxes are discretized using central differences. The temporal discretization is achieved by using the implicit Euler method of first or second order, thus for the first order method

$$\vec{W}_i^{n+1} = \vec{W}_i^n - \frac{\Delta t}{|\Omega_i|} \vec{R}(\vec{W}^{n+1})_i. \qquad (4)$$

★e-mail: Jiri.Furst@fs.cvut.cz
★★e-mail: Zdenek.Zak@fs.cvut.cz

Here $\vec{W} = [\rho, \rho\vec{u}, \rho E]$ and $\vec{R}(\vec{W})$ is the discretization of spatial terms.

The resulting system of non-linear equations is then solved by a matrix-free lower-upper symmetric Gauss-Seidel method using the following two step procedure:

$$D_i \Delta \vec{W}_i^{(1)} = -\vec{R}_i - \frac{1}{2}\sum_{j<i}\left[\Delta F_j^{(1)} \cdot \vec{S}_{ij} + \lambda_{ij}^* \Delta \vec{W}_j^{(1)}\right],$$

$$D_i \Delta \vec{W}_i = D_i \Delta \vec{W}_i^{(1)} - \frac{1}{2}\sum_{j>i}\left[\Delta F_j \cdot \vec{S}_{ij} + \lambda_{ij}^* \Delta \vec{W}_j\right].$$

Here

$$\Delta \vec{W}_i^{(1)} = \vec{W}_i^{(1)} - \vec{W}_i^n, \quad (5)$$

$$\Delta F_i^{(1)} = F(\vec{W}_i^{(1)}) - F(\vec{W}_i^n), \quad (6)$$

$$\Delta F_i = F(\vec{W}_i^{n+1}) - F(\vec{W}_i^n), \quad (7)$$

λ^* being the spectral radius of jacobains of fluxes and

$$D_i = \left(\frac{|\Omega_i|}{\Delta t_i} + \frac{1}{2}\sum_{j\in N_i}\lambda_{ij}^*\right) I. \quad (8)$$

See [10] or [6] for details.

In order to solve turbomachinery flows the solver has been extended to support calculations both in multiple reference frames allowing the so called frozen-rotor method and using the so called sliding mesh method [12].

2.2 Validation for the case of ERCOFTAC pump

In order to validate the solver and to assess suitability of frozen rotor approach a simulation of flow through a two-dimensional model of the ERCOFTAC radial pump has been performed. The results are compared to experimental data obtained by Ubaldi et al. [1]. The case was previously studied by [13] assuming incompressible flow and detailed data together with numerical results were published in the form of validation test case for evaluation of rotor-stator interaction.

The simplified model of the pump has 7 impeller blades, 12 diffuser vanes and 6% vaneless radial gap. The parameters of the model are given bellow:

Impeller :

- Inlet blade diameter $D_1 = 240$ mm,
- Outlet blade diameter $D_2 = 420$ mm,
- Number of blades $Z_i = 7$,
- Blade span $z = 40.4$ mm,

Diffuser :

- Inlet vane diameter $D_3 = 444$ mm,
- Outlet vane diameter $D_2 = 664$ mm,
- Number of vanes $Z_d = 12$,

Operating conditions :

- Rotational speed $n = 2000$ rpm,
- Impeller tip speed $U_2 = 43.98$ m/s,
- Flow rate coefficient $\phi = 4Q/(U_2 \pi D_2^2) = 0.048$,

Fig. 1. Structured 2D multiblock mesh for ERCOFTAC centrifugal pump, rotating zone in red.

- Flow rate $Q = 0.292\,\mathrm{m^3/s}$,
- Reynolds number $Re = \frac{U_2 D_2}{2\nu} = 6.5 \times 10^5$.

The numerical solution has been calculated using a domain with inlet diameter $D_0 = 200$ mm and outlet diameter $D_4 = 730$ mm. The multi-block structured mesh provided by Petit at al. [13] contains 93 886 quadrilateral cells with average near wall cell size about $y_1^+ \approx 20$, see fig. 1.

2.2.1 *Steady state frozen rotor simulation*

In order to assess the accuracy and suitability of frozen rotor approach two simulations have been run. First one using standard segregated solver based on SIMPLE loop (namely `simpleFoam`) assuming incompressible flow and the second one using coupled LU-SGS scheme for compressible flow.

Left half of figure 2 shows the absolute velocity magnitude obtained with segregated solver and the right half shows results obtained with coupled LU-SGS solver. One can see that both models give very similar flow field.

Figures 3 and 4 show a comparison of the radial and tangential components of the velocity at impeller outlet $D_m/D_2 = 1.02$ obtained with segregated and LU-SGS solver with experimental data of [1]. Note that the velocity is normalized by the impeller tip speed and the arc length y_i is normalized by the impeller pitch $G_i = \pi D_2/Z_i$. These results correspond to data obtained with another numerical method based on segregated approach by Petit et al. [14].

One can see that the frozen approach is capable to resolve some basic features of the flow. Namely the radial component of the velocity corresponds well to experimental data. The tangential velocity is slightly over-estimated by both calculations. On the other hand the shape and position of wakes is completely wrong due to frozen rotor approach.

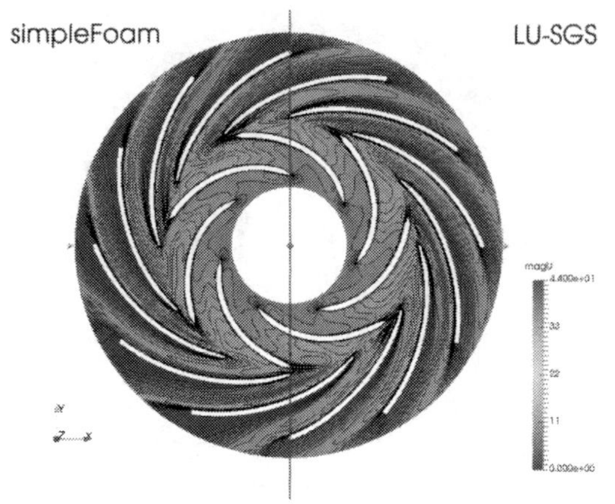

Fig. 2. Comparison of calculated velocity magnitude using frozen rotor approach and segregated solver (left part) or coupled LU-SGS solver (right part).

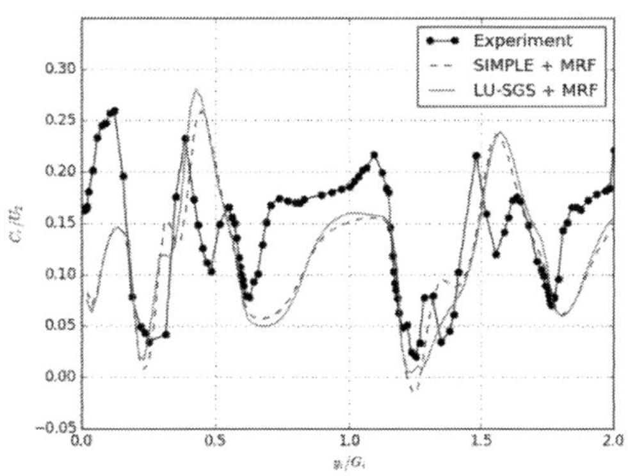

Fig. 3. Radial component of absolute velocity at the ERCOFTAC impeller outlet, frozen rotor approach.

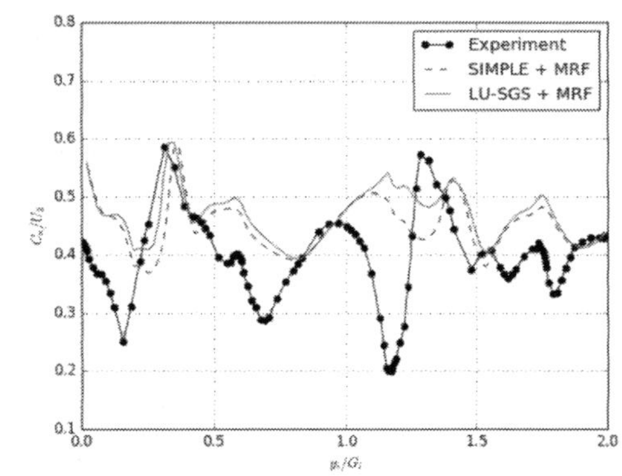

Fig. 4. Tangential component of absolute velocity at the ERCOFTAC impeller outlet, frozen rotor approach.

Table 1. Efficiency evaluation of the ERCOFTAC pump.

	Q (m^3/s)	Δp (Pa)	M (Nm)	η (%)
Steady state simulation				
LU-SGS	0.296	931.12	1.48	88.9
simpleFoam	0.292	921.21	1.43	89.6
Unsteady simulation				
LU-SGS	0.296	940.90	1.43	93.1
pimpleFoam	0.292	931.09	1.39	93.6

2.2.2 Unsteady sliding mesh simulation

In order to resolve unsteady rotor-stator interactions the unsteady simulation with moving rotor zone had to be carried out. The calculation has been run with the same mesh as in the previous case using second order backward differencing in time combined with dual time stepping technique in the case of LU-SGS solver. The physical time step was set to $\Delta t = 3 \times 10^{-5}$ s which corresponds to $1/1000$ of revolution period.

Figure 5 shows comparison of phase locked ensemble averaged velocity components calculated with coupled LU-SGS solver with results obtained with segregated solver (`pimpleDyMFoam`) and with experimental data of Ubaldi et al. [1]. The figures correspond to dimensionless time $t/T_i = 0.126, 0.226, 0.326$, and 0.426 where $T_i = 2\pi/(\omega Zi)$ is the impeller blade passing time.

One can see that both solvers predict very similar velocity at the impeller outlet. The most pronounced difference is in the tangential component of the velocity for $t/Ti = 0.126$, see fig. 5b. The coupled solver predicts higher peak in the velocity component at $y_i/G_i \approx 0.3$ whereas the results obtained with segregated solver correspond very well to experimental data.

The table 1 sumarizes results obtained with coupled LU-SGS solver for compressible flow and with segregated solver for incompressible flows. One can see that the LU-SGS solver predicts both in steady and unsteady cases slightly higher pressure rise Δp and torque M. On the other hand both solvers predict similar pump efficiency $\eta = \Delta p Q/(M\omega)$. The efficiency predicted by steady-state frozen rotor simulations is slightly lower than the one calculated with unsteady method, nevertheless the difference isn't excessively large and therefore the frozen rotor approach is a viable method for estimating efficiency of the pump.

3 Analysis of twin scroll turbine

Twin scroll radial centripetal turbocharger turbines are used in the automotive industry. The twin scroll design is beneficial in conjunction with the internal combustion

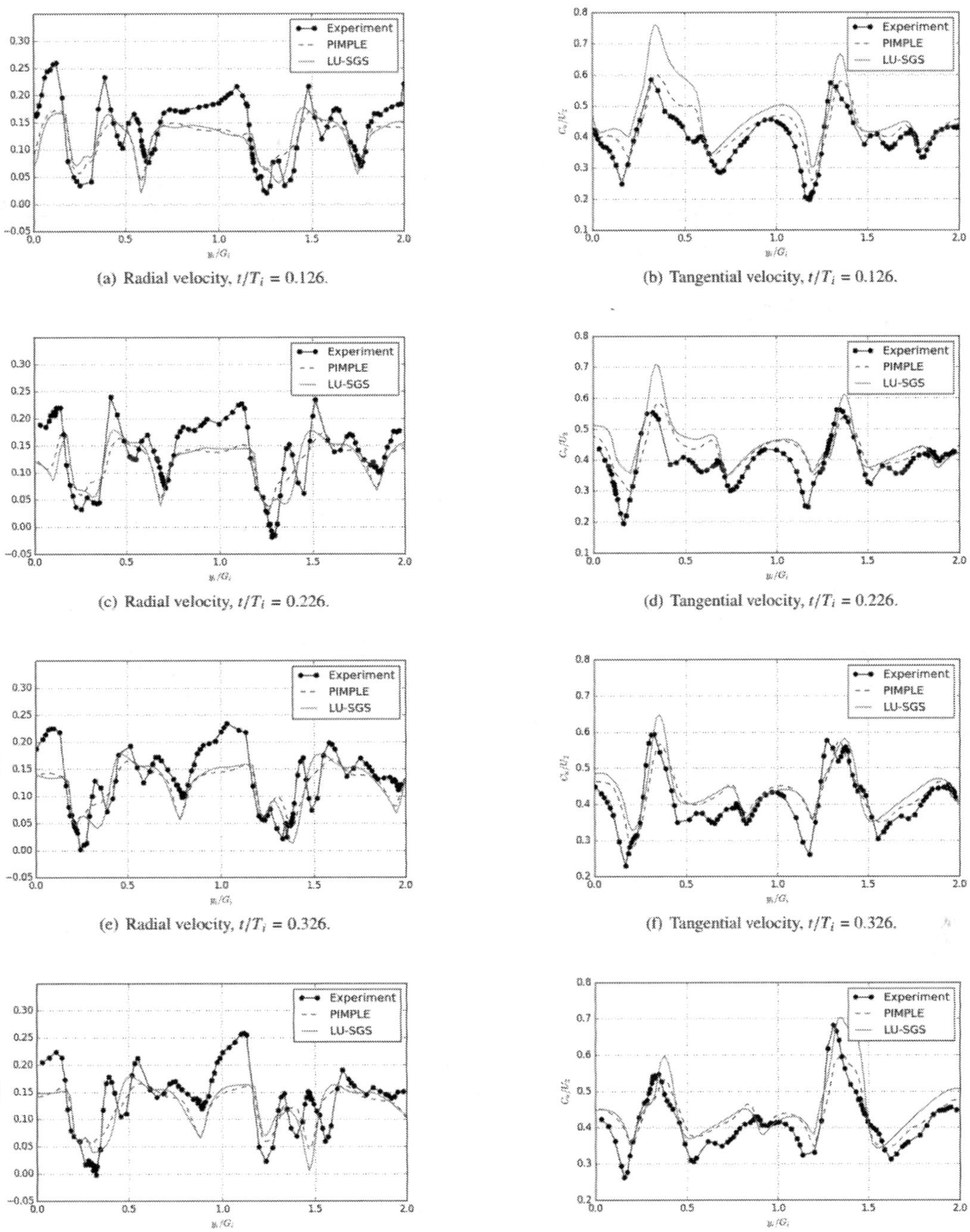

(a) Radial velocity, $t/T_i = 0.126$.

(b) Tangential velocity, $t/T_i = 0.126$.

(c) Radial velocity, $t/T_i = 0.226$.

(d) Tangential velocity, $t/T_i = 0.226$.

(e) Radial velocity, $t/T_i = 0.326$.

(f) Tangential velocity, $t/T_i = 0.326$.

(g) Radial velocity, $t/T_i = 0.426$.

(h) Tangential velocity, $t/T_i = 0.426$.

Fig. 5. Ensemble averaged velocity components at the ERCOFTAC impeller outlet.

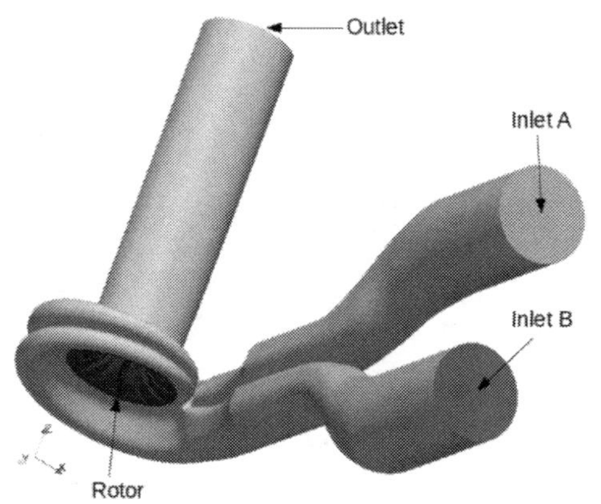

Fig. 6. Computational domain for twin scroll turbine simulations.

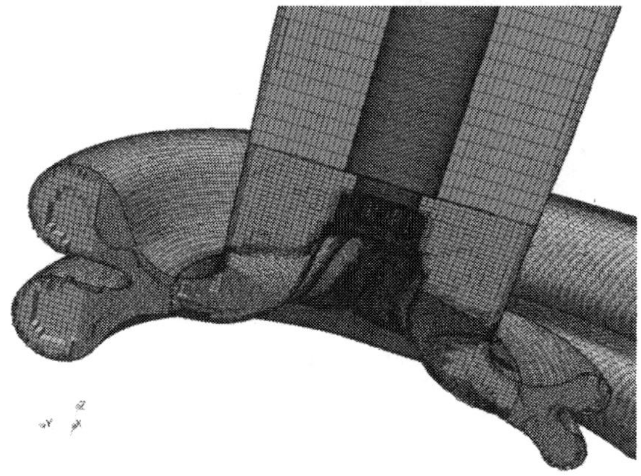

Fig. 7. Computational mesh for twin scroll turbine simulations.

Fig. 8. Surface mesh at the rotor for twin scroll turbine simulations.

engine with firing interval between cylinders around 240 degrees in exhaust manifold branches, i.e. six cylinder engines. The advantages consist in the higher efficiency of a turbine and better engine response during the transients due to better utilization of exhaust gas energy. The twin scroll turbines are also used for the four cylinder engines to improve the engine behaviour during transients. The firing interval between cylinders in mentioned case is 360° in exhaust manifold branches, thus out of optimum and the high pulsating flow upstream of a turbine cannot be utilized as effective as in the case of six cylinder engine.

The tested twin scroll turbine with symmetric design of sections is used for turbocharging of a six cylinder Diesel engine. The impeller has 12 vanes. The impeller diameter at inlet equals to 76.6 mm and averaged outlet diameter equals to 46.7 mm.

3.1 Simulation setup

Due to high rotational speed together with high expansion ratio, the maximal Mach number exceeds one in some cases and the flow has to be definitely considered as a compressible gas. Simulations have been run with LU-SGS solver which is developed especially for dealing with transonic flows. The domain (see figure 6) was divided to three zones: the stationary inlet zone containing both inlet tubes and volutes, the rotating zone containing turbine rotor, and the stationary zone including outlet tube. Each zone was meshed independently using `snappyHexMesh` software and the three parts were then glued into single mesh with interfaces between zones. The final unstructured hexa-dominant mesh contained 2.2×10^6 cells with near-wall mesh size $y_1^+ < 160$, (60 at the rotor) with average $y_1^+ \approx 10$, see fig. 7 and 8..

The spatial discretization was performed using second order scheme with Barth-Jaspersen limiter [15] and HLLC flux [9] for convective terms and central scheme for diffusive terms. The implicit Euler method with local time stepping was used for advancing in pseudo-time in the case of steady frozen rotor simulation whereas second order backward Euler method combined with local pseudo-time stepping was used in the transient case.

The flow regime was characterized by the following boundary conditions:

- Inlets - the value of total pressures p_{0A}, p_{0B} and total temperatures T_{0A}, T_{0B} was prescribed at both inlets. The velocity direction was normal to the inlet planes and the inlet turbulence intensity $Tu = 5\%$ together with mixing length scale $l = 2.8$ mm was assumed in all cases.
- Outlet - the value of static pressure p_{out} was prescribed according to given expansion ratio.
- Shroud - the no-slip condition was assumed together with $\partial p/\partial n$ and given wall temperature T_{wall} at the shroud. Due to relatively coarse mesh the wall functions were used for turbulence model.
- Rotor - the same set of boundary conditions as for the shroud was assumed here with the exception of velocity at the rotor wall which was set to the velocity of the rotor.

The gas was governed by the ideal equation of state with constant heat capacity c_p corresponding to air at the

Fig. 9. Twin scroll turbocharger test bed with separated sections upstream of a turbine

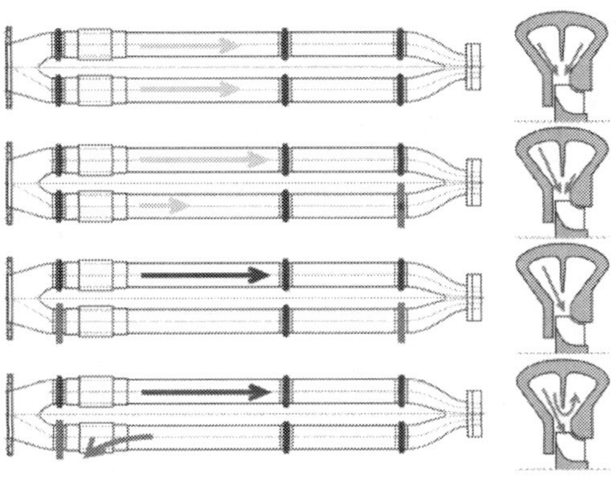

Fig. 10. Turbocharger test bed (down from top) - uniform admission; partial admission (throttling in one turbine section); extreme partial admission with closed section; backflow

temperature T_{wall}. The effects of turbulence were modeled by Menter's SST turbulence model [16].

The table 2 sumarizes boundary conditions and gas properties used for the simulations.

3.2 Experimental data

The goal of the experiments on the steady flow test bed (see fig. 9), with separated sections upstream of a turbine, was to obtain a sufficient set of data, which describes the twin scroll turbine behaviour under different level of the impeller admission. The specific hot gas stand with open loop is suitable for achievement of the arbitrary level of turbine impeller admission. The unequal partial level of admission is reached via the throttling in a turbine section and extreme partial admission by closure of one section, see the sketch at fig. 10.

The selected twin scroll turbine was tested at different pressure ratios, relatively broad range of load and under different levels of the impeller admission. The turbine load is described by the blade speed ratio - BSR. Low blade speed ratios, i.e. high load of the turbine, were achieved by the combination of the compressor with larger wheel as a load and relatively high temperature upstream of the turbine about 1100 K. With the temperature of exhaust gases about 900 K upstream of the turbine, the turbine was able to run with isentropic efficiency close to optimum. The turbine, driven by cold air only, was unloaded and achieved high blade speed ratios. Three levels of impeller admission were measured in our case - full, partial (throttling in one section) and extreme partial with closed section.

3.3 Results

Although the experimental data covers different partial levels of admission, the CFD study was limited to cases with uniform admission only. The results of simulations were postprocessed in order to make comparison with experimental data. First, the reduced mass flow rate defined by

$$\dot{m}_{red} = (\dot{m}_A + \dot{m}_B)\frac{\sqrt{T_{0AB}}}{p_{0AB}}10^5, \tag{9}$$

was evaluated. Here $\dot{m}_A$ and $\dot{m}_B$ are the mass flow rates through inlet A and B, $T_{0AB} = (T_{0A} + T_{0B})/2$, and $p_{0AB} = (p_{0A} + p_{0B})/2$. Next, the torque M has been evaluated and used for the rotor power calculation

$$P_{rotor} = M\omega_{rotor}. \tag{10}$$

The ventilation losses were estimated using empirical formula

$$P_{vent} = K\rho_2 D_2^2 U_2^3 Re_2^{-0.2} \tag{11}$$

where $K = 0.07$ is an empirical constant, D_2 is the rotor diameter, $U_2 = \omega_{rotor}D_2/2$ is the rotor inlet edge speed and $Re_2 = \rho_2 D_2 U_2/$ is the Reynolds number and finally the turbine efficiency was evaluated as

$$\eta = \frac{P_{rotor} - P_{vent}}{(\dot{m}_A + \dot{m}_B)c_p T_{0AB}\left(1 - \left(\frac{p_{out}}{p_{0AB}}\right)^{\frac{\kappa-1}{\kappa}}\right)} \cdot 100\%. \tag{12}$$

Note that in the case of transient simulations calculations have been started from steady the initial condition corresponding to steady state frozen rotor simulation and the ten revolutions of the rotor have been calculated. The resulting mass flow rate and torque were then calculated by averaging over the last revolution. Figure 11 shows a time dependent torque obtained by transient simulation for TP1 regime. From the amplitude spectrum displayed at the figure 12 one can see that time dependent signal contains only one important harmonics corresponding to the blade passage frequency. Similar results were obtained also for other regimes.

Figures 13, 14, and 15 display the distribution of the pressure at solid and rotating walls, the absolute Mach number and velocity vectors in the longitudinal cut through the domain, and the relative Mach number in the rotor zone for TP1 regime. One can see that both absolute and relative Mach numbers exceed unity and therefore the compressibility plays an important role in the simulations.

The table 4 shows the comparison of experimental data with the results of steady and unsteady simulations.

Table 2. Boundary conditions and gas properties for twin scroll turbine analysis

Run	p_{0A} (Pa)	p_{0B} (Pa)	p_{out} (Pa)	T_{0A} (K)	T_{0B} (K)	T_{wall} (K)	RPM (rpm)	c_p (J/kg/K)	κ (-)
TP 1	2.070×10^5	2.077×10^5	9.232×10^4	877.030	891.656	800	104 850	1097.997	1.354
TP 2	2.066×10^5	2.075×10^5	9.236×10^4	874.636	889.859	800	109 011	1097.997	1.354
TP 3	2.084×10^5	2.092×10^5	9.146×10^4	1077.040	1080.259	1000	88 776	1142.580	1.336
TP 4	1.544×10^5	1.550×10^5	9.512×10^4	873.127	891.125	800	85 227	1097.997	1.354
TP 5	2.063×10^5	2.067×10^5	9.333×10^4	288.478	288.191	288	73 521	1005.051	1.400
TP 6	1.545×10^5	1.550×10^5	9.296×10^4	1075.105	1080.552	1000	72 814	1142.580	1.336
TP 7	1.550×10^5	1.554×10^5	9.310×10^4	1079.464	1084.990	1000	67 489	1142.580	1.336
TP 8	1.216×10^5	1.219×10^5	9.654×10^4	881.779	888.423	800	63 430	1097.997	1.354
TP 9	1.534×10^5	1.535×10^5	9.623×10^4	289.580	289.341	288	57 048	1005.051	1.400
TP 10	1.219×10^5	1.223×10^5	9.629×10^4	874.801	888.021	800	54 418	1097.997	1.354

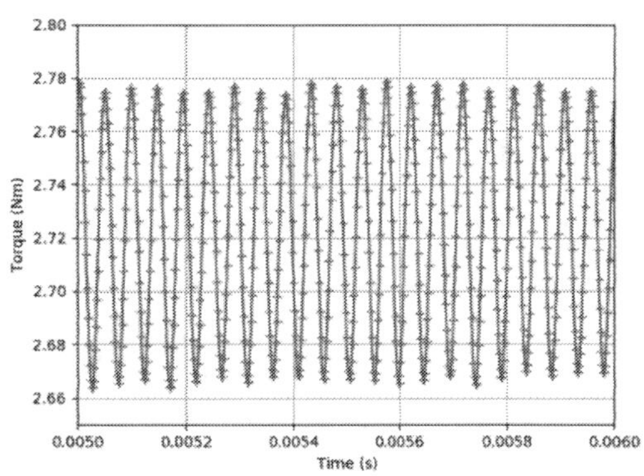

Fig. 11. Time dependent torque as a result of transient simulation for TP1 regime.

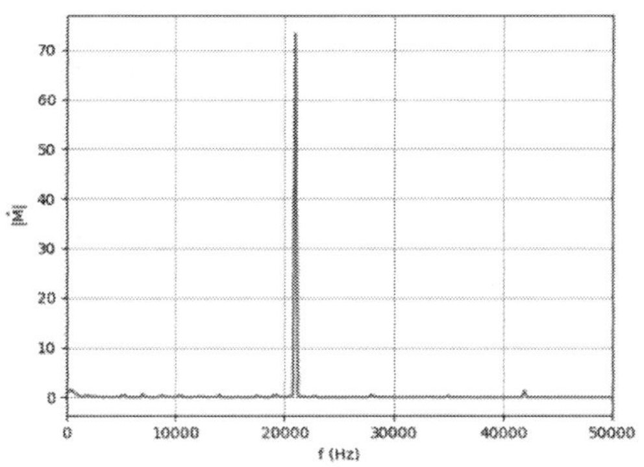

Fig. 12. Amplitudes of FFT of time dependent torque for TP1 regime.

Figures 16 and 17 show graphically the reduced mass flow rates and turbine efficiency as a function of turbine expansion ratio $\Pi = p_{0AB}/p_{out}$. Each number represents corresponding regime (see table 2) and its color corresponds to a method for obtaining the result: black color means experimental data, red color means steady state CFD, and blue color represents transient CFD. Both figures contains also lines displaying results of steady state calculation with three fixed rotation speeds 50 000 rpm, 75 000 rpm and 100 000 rpm with $T_{0AB} = 850$ K.

A model for reduced mass flow rate and turbine efficiency has been built from the frozen rotor results at three fixed rotation speeds. The reduced mass flow rate is approximated as

$$\dot{m}_{red} \approx a_0 + a_1\Pi + a_2 n_{red} + a_{11}\Pi^2 + a_{12}\Pi n_{red} + a_{22} n_{red}^2, \quad (13)$$

where the coefficients are $a_0 = -3.902$, $a_1 = 8.192$, $a_2 = -0.621 \times 10^{-3}$, $a_{11} = -2.213$, $a_{12} = 0.473 \times 10^{-3}$, and $a_{22} = -0.105 \times 10^{-6}$. The figure 18 compares the reduced mass flow rate predictions given by the model (13) with experimental data and with the results of CFD simulations. One can see that the differences between the model predictions and the results are less than 2 % with the exception of TP5 and TP9 regimes. In TP5 regime the unsteady simulation gives lower value of the mass flow rate whereas experimental data corresponds well to the model. Similarly in the TP9 the unsteady CFD predicts

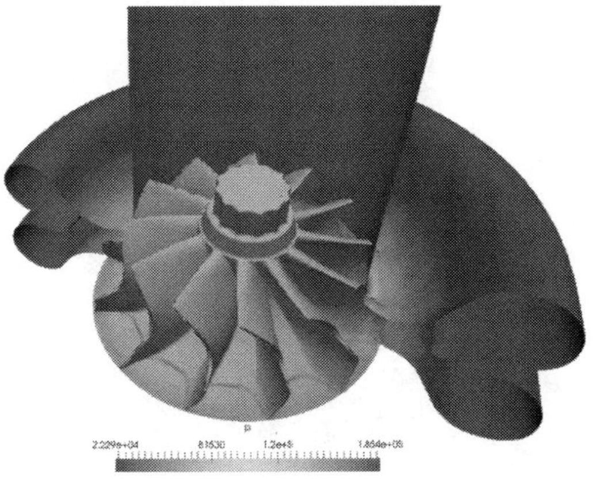

Fig. 13. Pressure distribution at solid walls in the twin scroll turbine, regime TP 1.

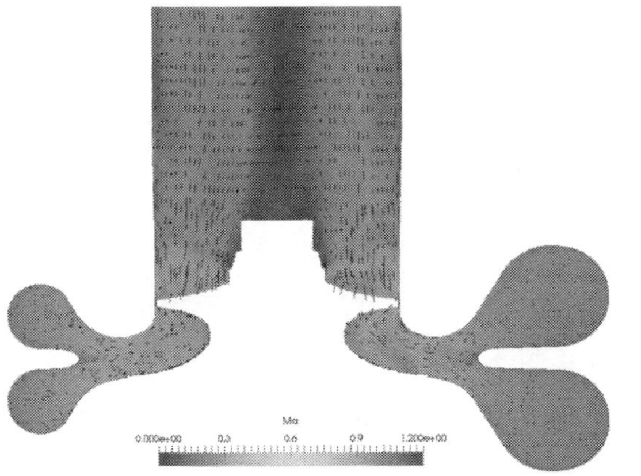

Fig. 14. Mach number in the cut through twin scroll turbine, regime TP 1.

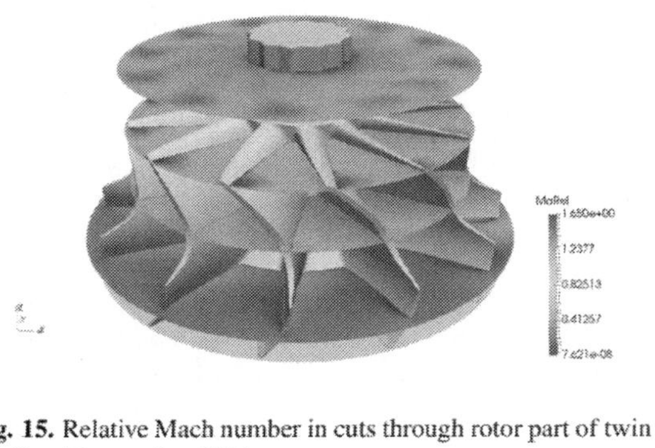

Fig. 15. Relative Mach number in cuts through rotor part of twin scroll turbine, regime TP 1.

Table 3. Expansion ratio $\Pi = p_{0AB}/p_{out}$ and reduced rotation speed $n_{red} = RPM/\sqrt{T_{0AB}}$ for twin scroll turbine analysis.

Run	Π (−)	n_{red} (rpm/$\sqrt{K}$)
TP 1	2.246	3526
TP 2	2.242	3670
TP 3	2.283	2703
TP 4	1.627	2870
TP 5	2.213	4330
TP 6	1.665	2218
TP 7	1.667	2052
TP 8	1.261	2132
TP 9	1.595	3353
TP 10	1.268	1833

Table 4. Comparison of reduced mass flow rate and turbine efficiency obtained with the experiment (columns "Exp"), by the steady state simulation (columns "Steady") and transient simulation (columns "Trans").

Run	$\dot{m}_{red}$ (kg s^{-1} $\sqrt{K}$/bar)			η (%)		
	Exp	Steady	Trans	Exp	Steady	Trans
TP 1	3.370	3.411	3.365	64.6	65.3	65.3
TP 2	3.303	3.371	3.309	64.4	64.2	63.9
TP 3	3.613	3.570	3.540	60.8	61.4	61.7
TP 4	3.030	3.066	3.021	68.1	62.1	62.4
TP 5	3.092	3.112	2.839	52.4	54.6	45.7
TP 6	3.321	3.329	3.300	64.8	63.4	62.9
TP 7	3.373	3.371	3.352	61.0	61.8	61.7
TP 8	2.332	2.242	2.219	70.5	58.2	57.7
TP 9	2.815	2.665	2.487	61.1	47.4	45.0
TP 10	2.457	2.444	2.424	70.9	64.4	64.2

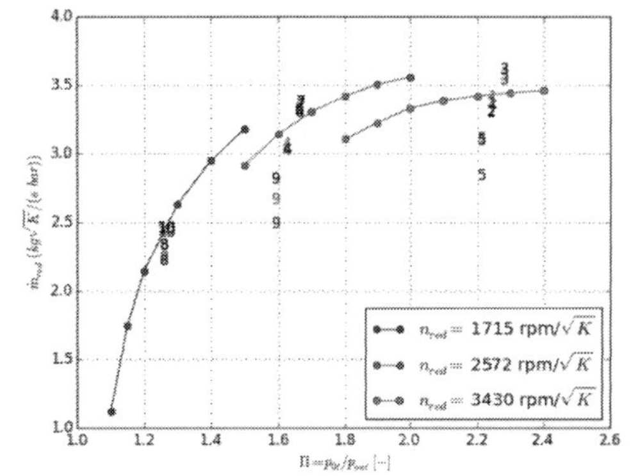

Fig. 16. Reduced mass flow rate for twin scroll turbine (experimental data depicted as black numbers, steady state simulation as red numbers, and transient simulation as blue numbers).

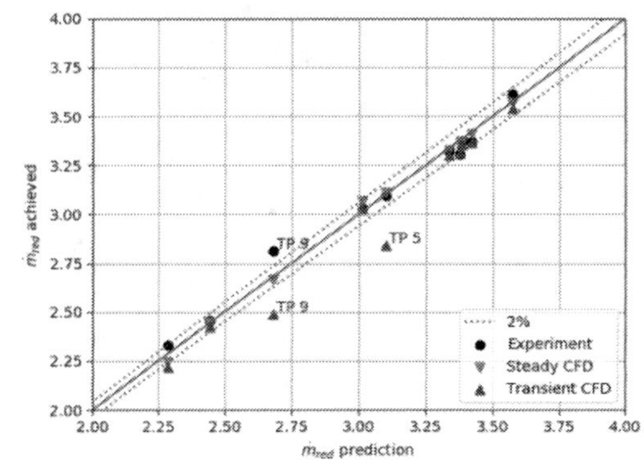

Fig. 18. Comparison of the mass flow rate model (13) with experimental and CFD results.

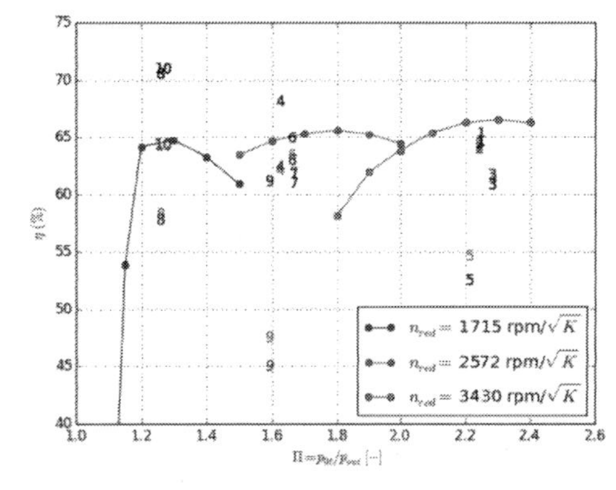

Fig. 17. Efficiency of twin scroll turbine (experimental data depicted as black numbers, steady state simulation as red numbers, and transient simulation as blue numbers).

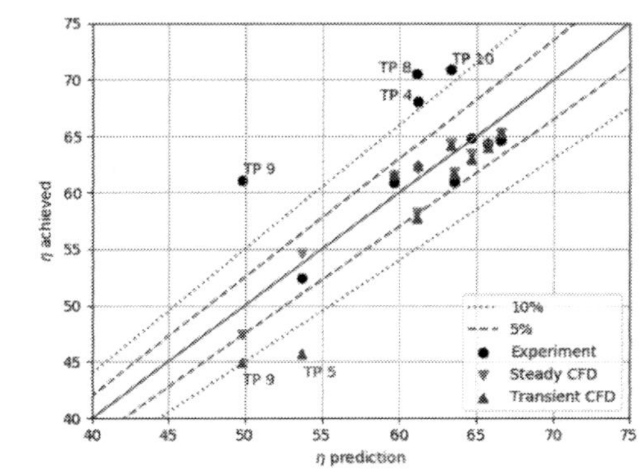

Fig. 19. Comparison of the efficiency model (14) with experimental and CFD results.

lower mass flow rate whereas the experimental data gives a slightly higher value.

Similarly the efficiency can be approximated as

$$\dot\eta \approx b_0 + b_1\Pi + b_2 n_{red} + b_{11}\Pi^2 + b_{12}\Pi n_{red} + b_{22} n_{red}^2, \quad (14)$$

with $b_0 = 53.333$, $b_1 = 26.649$, $b_2 = -11.255 \times 10^{-3}$, $b_{11} = -38.737$, $b_{12} = 44.195 \times 10^{-3}$, and $b_{22} = -12.994 \times 10^{-6}$.

The figure 19 compares the predicted efficiency given by the model (14) with the efficiency evaluated from the experimental data and from the CFD simulations. One can see that most of the data falls into 5 % region, however the experimental data give slightly higher efficiencies in regimes TP4, TP8, TP9, and TP10 whereas the unsteady simulations predict lower efficiency in regimes TP5 and TP9.

Note that the regimes TP5 and TP9 where the models for $\dot m_{red}$ and η differ from the experimental data and unsteady simulations are characterized by lower inlet temperature resulting in higher Mach numbers in the turbine when comparing to other regimes.

4 Conclusion

The LU-SGS solver has been extended for turbomachinery flows using stationary frozen-rotor and unsteady sliding mesh approaches. The solver has been validated using the ERCOFTAC centrifugal pump comparing the results both with experimental data and with the pressure based solver from the OpenFOAM package. Finally the solver has been used for the analysis of the flows through a twin-scroll turbine. The comparison of simulation results using both approaches with the experimental data shows that the mass flow rate is predicted within 2 % error by both computational approaches and that a reasonable agreement between predicted and measured efficiency can be achieved. Although the frozen rotor approach completely disregards the unsteady character of turbomachinery flow, the current study shows that it can be viable alternative to much more computationally demanding sliding mesh method for flows in radial turbines.

Acknowledgements. Authors acknowledge support from the EU Operational Programme Research, Development and Education, and from the Center of Advanced Aerospace Technology (CZ.02.1.01/0.0/0.0/16_019/0000826), Faculty of Mechanical Engineering, Czech Technical University in Prague and support by TACR, project no. TE01020020.

References

1. M. Ubaldi, P. Zunino, G. Barigozzi, A. Cattanei, J. Turbomach. **118**, 41 (1996)
2. S.V. Patankar, D.B. Spalding, Int. J. Heat Mass Transf. **15**, 1787 (1972)
3. R. Issa, J. Comput. Phys. **62**, 40 (1986)
4. J. Benajes, J. Galindo, P. Fajardo, R. Navarro, J. Appl. Fluid Mech. **7**, 673 (2014)
5. L. Mangani, M. Darwish, F. Moukalled, in *Proceedings of ASME TURBO EXPO 2014* (ASME, Düsseldorf, 2014), p. 13
6. J. Fürst, in *EPJ Web of Conferences* (2017), Vol. 143, p. 02027
7. A. Kurganov, E. Tadmor, J. Comput. Phys. **160**, 241 (2000)
8. M.S. Liou, J. Comput. Phys. **214**, 137 (2006)
9. E.F. Toro, in *Riemann Solvers and Numerical Methods for Fluid Dynamics* (Springer Berlin Heidelberg, Berlin, Heidelberg, 1997), pp. 293–311
10. J. Blazek, *Computational fluid dynamics : principles and applications* (Butterworth-Heinemann, 2015)
11. C. Shen, X.l. Xia, Y.z. Wang, F. Yu, Z.w. Jiao, Adv. Eng. Softw. **91**, 80 (2016)
12. J. Fürst, in *12th OpenFOAM workshop* (Exeter, 2017)
13. O. Petit, M. Page, M. Beaudoin, H. Nilsson, in *IAHR International Meeting of the Workgroup on Cavitation and Dynamic Problems in Hydraulic Machinery and Systems* (2009), pp. 523–532
14. O. Petit, H. Nilsson, Int. J. Rot. Mach. **2013**, 1 (2013)
15. T. Barth, D. Jespersen, in *27th Aerospace Sciences Meeting* (AIAA, Reston, Virigina, 1989)
16. F.R. Menter, AIAA Journal **32**, 1598 (1994)

Flow Dynamics in the Vicinity of Tandem Buildings

Renata Gnatowska[1,*], *Radka* Kellnerová[2], and *Václav* Uruba[2,3]

[1]Czestochowa University of Technology, Faculty of Mechanical Engineering and Computer Science, 21 Armii Krajowej Av, 42-200 Czestochowa, Poland
[2]Institute of Thermomechanics, ASCR, v.v.i., Dolejškova 5, Praha 8, CR
[3]UWB, Faculty of Mechanical Engineering, Department of Power System Engineering, Universitní 8, Plzeň, CR

Abstract. The flow-field in the vicinity of tandem building model in a wind tunnel will be subjected to analysis of dynamics. The model is 3D consisting of the two blocks of different sizes arranged in a streamwise direction. Experiments were performed using time-resolved PIV technique in several measuring planes to capture both spatial and dynamical features.

1 Introduction

With increasing urbanization and density of population, the understanding of the problem of flow fields in urban environments and microclimate becomes more important. Increased interest in has been observed over the past few decades. It is governed by potential applications of the results in the design process of new buildings as well as optimization of the existing urban areas and city centres. Nowadays, this issue plays an important role in environmental protection, it concerns in particular the so called wind comfort around and between buildings, the life quality in urban areas as well as specific economic aspects of utilization of the defined zones. The studies of the wind environment around buildings are generally performed as wind tunnel experiments since it is the most well-established way to simulate a natural wind. Wind-tunnel measurements are performed with low-cost techniques such as hot-wire or hot-film anemometry (HWA or HFA) (e.g. Refs. [1-6]. Particle Image Velocimetry (PIV) is less often used because it is more elaborate and more expensive [7-8].

The objective of this paper is to provide a detailed experimental analysis of the flow interactions of buildings models situated one after another in a so-called tandem arrangement. The main aim is to clarify the flow field, found in the experiment performed using time-resolved PIV technique in several measuring planes to capture both spatial and dynamical features.

2 Experimental setup

The experiment with Particle Image Velocimetry (PIV) technics was carried out in the open-circuit wind tunnel at the Institute of Thermomechanics of the Academy of Sciences of the Czech Republic.

The analysed configuration of building models was located within ground boundary layer generated by the facility output from the contraction nozzle of a cross section of 250 x 250 mm. The test section of the wind tunnel is 3 m long, the measurements were performed in the central part of the tunnel. The SAFEX fog generator has been used to seed particles.

The laser beam from a dual-head Nd:YLF laser with a pulse energy of 10mJ at 1kHz illuminated the particles. Dantec CCD camera (SpeedSense 611) was used to capture pairs of particle images at a frame rate of 1 kHz, 4000 snapshots representing 4 s of real-time. The PIV measurements provided the velocity vector field with two velocity components (U and V) in the measuring plane of symmetry (XY) – see Fig. 1.

The research object represents the arrangement of two buildings modelled as rectangular of square base (0.025 m) and different heights settled in the flow domain. As depicted in Fig. 1, the height ratio of the buildings models is 0.6 and the distance between them is set to 1.5 of the length of the cuboid base side. The inflow boundary layer has been prescribed according to the equation (1).

$$U(y) = U(\delta)(y/\delta)^{\alpha} \qquad (1)$$

where:
α – terrain type factor,
δ – thickness of the ground layer.
The parameters chosen for the research were as follows: $\alpha = 0.15$ (suburban boundary layer), $\delta = 0.05$ m and $U(\delta) = 11$ m/s.

The same geometry has been studied in [8], however only the time-mean structure has been studied. In the presented study the space between the buildings has been subjected to analysis of the flow dynamics.

3 Results

The acquired vector fields were analysed using standard methods. First, classical statistical methods have been applied, then, to study dynamics of the flow-field, the POD method has been used as shown in [9].

* Corresponding author: gnatowska@imc.pcz.pl

3.1 Statistical analysis

The distributions of various statistical quantities will be represented in non-dimensional form, the reference velocity U_{ref} was the velocity outside the boundary layer in the region just upstream the models at X = 0. The spatial coordinates X and Y were related to the buildings dimension D. In the measuring plane Z = 0.

In all figures to be presented in this chapter the vector-lines derived from the time-mean velocity field are added. Flow comes from left.

In Fig. 2 distribution of the streamwise mean velocity component U_{mean}, which is the component in X direction, is shown.

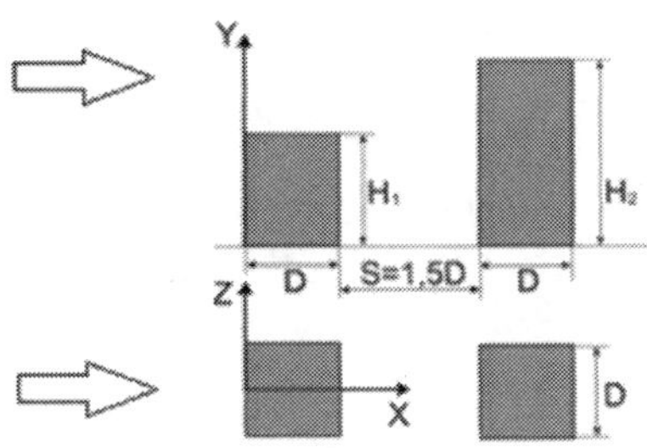

Fig. 1. The research object (H_1 = 0.03 m, H_2 = 0.05 m, D = 0.025 m).

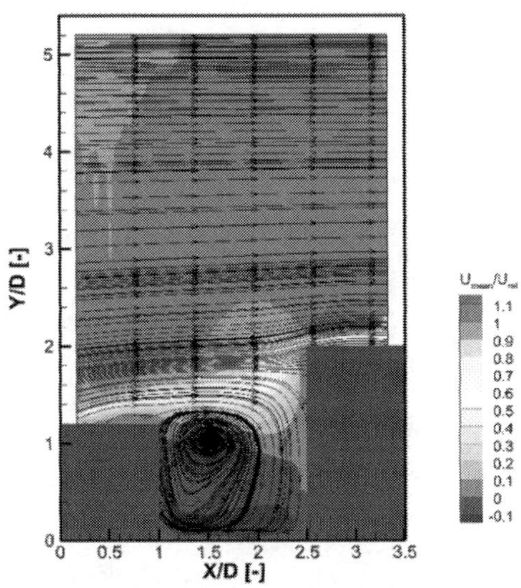

Fig. 2. Distribution of the streamwise velocity component U_{mean}.

The result shows not perturbed flow above the buildings, while in the space between the buildings a very strong vortex takes place, however the mean streamwise velocity component is quite small, positive or negative. The negative value means local backflow located close to the space between buildings bottom.

The vertical (direction of Y-axis) mean velocity component W_{mean} distribution in Fig. 3 reveals strong downward flow region along the higher building windward wall, while the ascending leg behind the smaller building is considerably weaker. This fact indicates the strong 3D structure of the flow in this region with the important mean velocity component in the out-of-plane direction Z. However, of course not precisely in the plane of symmetry, but close to it. This result is supported by data presented in [8].

The dynamical activity could be quantified by means of the turbulent kinetic energy TKE, however in this case, only two velocity components are taken into account (X and Y directions). The distribution of the TKE is shown in Fig. 4.

Turbulence activity in the main flow above the buildings is relatively very low, as expected. Maximum of the TKE is located within the free shear layer above the smaller building roof and in the descending flow in front of the higher leeward building. However reaching the floor between the buildings calms the turbulence. Note also that very close to the windward wall of the higher building very week turbulent activity is detected, seems to be damped by interaction with the wall.

Evaluation of turbulence production is demonstrated using the velocity components covariance. In Fig. 5 distribution of the covariance of the in-plane X and Y velocity components is shown, normalized again using the U_{ref}.

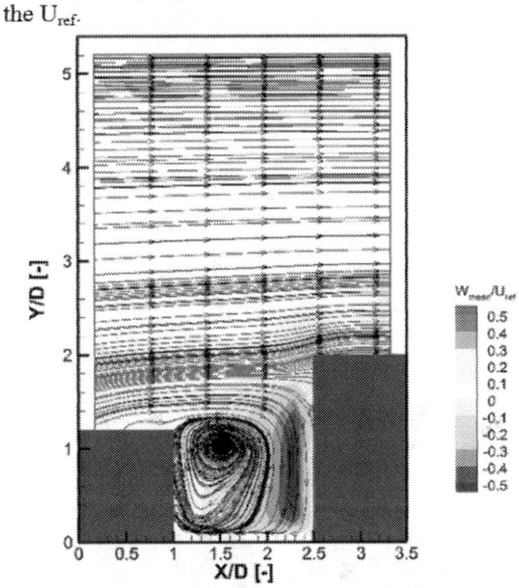

Fig. 3. Distribution of the vertical velocity component W_{mean}.

The physical interpretation of the velocity components covariance is the production of turbulence in regions of negative value, while positive value has no straightforward interpretation. The strong covariance negative region has been detected within the free shear layer above the smaller building roof, little upstream the location where the maximum of TKE is indicated (compare with Fig. 4).

3.2 Dynamical analysis

The spatial aspects of the flow dynamics have been studied using the Proper Orthogonal Decomposition (POD).

Multi-point velocity data from PIV allows us to use the POD for detection of large organised structures in the flow. The POD is actually interpreted as a re-expression of original data into new basis. The basis is chosen to emphasize turbulent kinetic energy (TKE). We applied a 'snapshot' modification, published by Sirovich (1987) [10], which is not so computationally demanding compared to the classic POD.

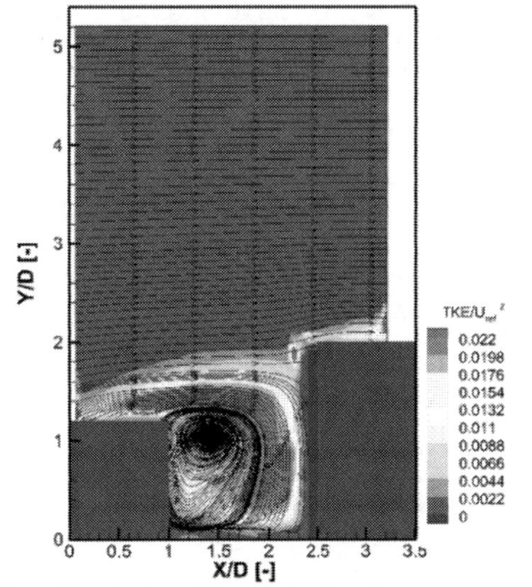

Fig. 4. Distribution of the turbulent kinetic energy.

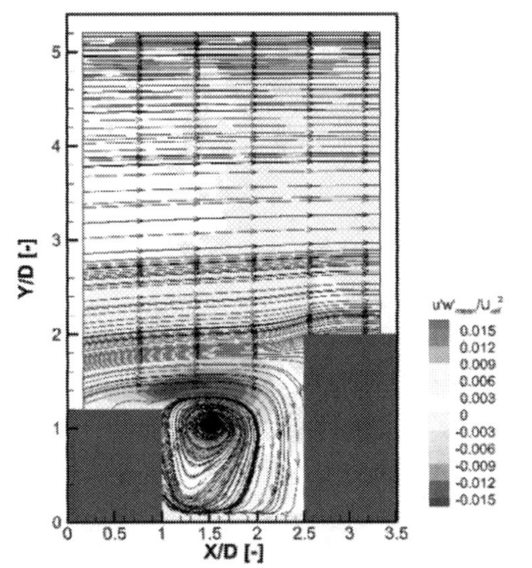

Fig. 5. Distribution of the covariance of in-plane velocity components.

Based on TKE, we are able to obtain spatial information about the most dominant TKE modes in the flow. Further, the speed of a convergence of the cumulative contribution from particular modes to the total TKE reveals how organised the flow is.

The calculation involves two velocity components, longitudinal U in the streamwise X direction and vertical W in wall-normal (Y) direction.

POD yields an eigenvalue for each mode. We can arrange numerically the eigenvalues from the largest one downwards. If we calculate contribution of particular eigenvalue to the total sum, we obtain its fractional the contribution to the entire TKE contained in the flow. The plot of relative and cumulative contributions from each mode can be seen in Fig. 6.

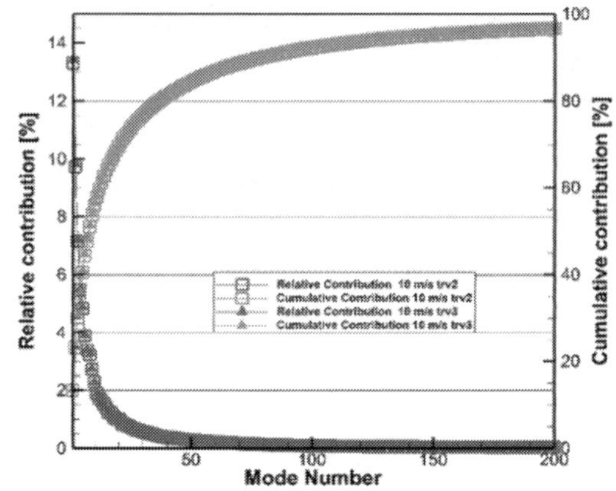

Fig. 6. Relative and cumulative contribution to the TKE from each mode.

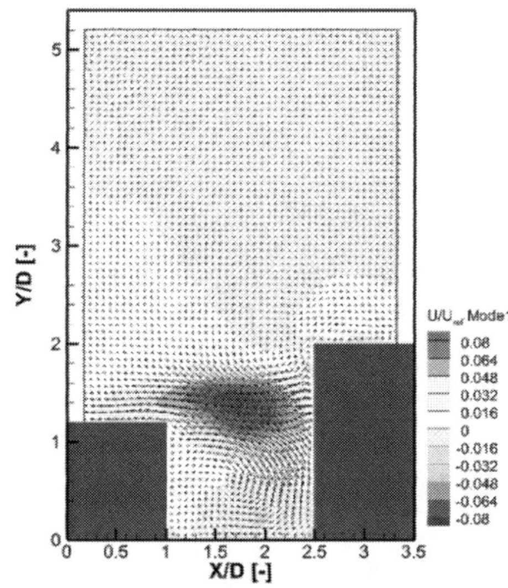

Fig. 7. The first POD mode topology, vector field with streamwise velocity component distribution.

The afore-mentioned first three modes contain 13%, 10% and 7% of TKE respectively.

To show the POD modes the vector fields could be used. As an example, the first POD mode is shown in Fig. 7, together with the distribution of the streamwise velocity component.

However in the vector field in Fig. 7 the vortical structures are not very clear. To represent the POD modes better the vector-lines were added. In Fig. 8 the first three modes most energetic for longitudinal velocity are depicted by contour-lines from left to right, respectively. The second row represents the three modes for wall-normal velocity.

Figure 8 reveals that the most dominant structure is the large vortex between the buildings and small vortex with opposite orientation upstream of the higher building present in the first POD mode. Below the large vortex, there is a jet-flow penetrating the space between the buildings. The second mode shows similar large vortex between buildings accompanied by weaker one below, within the gap between the buildings. In this mode, the other couple of small whirls was detected sitting below and above the leeward bigger building higher corner. The third mode has the similar structure as the first mode except for the structure of the orientation of the flow above the buildings.

The flow dynamics could be characterized by the system of vortices located between the buildings and close to the upper part of the downstream bigger building. This vortical unsteady behaviour will influence possible forced diffusion in the space between the building considerably.

4 Summaries

The flow structure between the tandem buildings has been studied using time-resolved PIV method in the plane of symmetry. Various vortical structures have been detected in the flow in the vicinity of the tandem buildings.

The time-mean flow-field is characterized by dominant clock-wise vortex filling the entire space between the buildings with the clock-wise orientation.

The flow dynamics in the region between the buildings is governed by smaller whirls appearing in particular close to upper part of the leeward bigger building.

This work was supported by the Grant Agency of the Czech Republic, projects No. 17-01088S and 15-18964S.

References

1. Y. Uematsu, M. Yamada, H. Higashiyama, T. Orimo, *J. Wind Eng. Ind. Aerodyn.* **44** (1-3) 2289-2300 (1992)
2. N.J. Jamieson, P. Carpenter, P.D. Cenek, *J. Wind Eng. Ind. Aerodyn.* **44** (1-3) 2301-2312 (1992)
3. B. Blocken, T. Stathopoulos, J. Carmeliet, *J. Aerosp. Eng. - ASCE* **21** (4) 280-287 (2008)
4. A. Jarża, R. Gnatowska, *Chem Process Eng-Inz.* **27** (3/1), 721-735 (2006)
5. R. Gnatowska, *Int. J Turbo Jet. Eng.* **28** (1) 21-29 (2011)
6. R. Gnatowska, Pol J Environ Stud, 24(4), 1571-1577 (2015)
7. J.P.A.J. Van Beeck, G. Dezso, P. Planquart, *Microclimate Assessment by Sand € Erosion and Irwin Probes for Atmospheric Boundary Layer Wind Tunnels*, PHYSMOD 2009, Von Karman Institute (VKI), Sint-Genesius-Rode, Belgium, (2009)
8. R. Gnatowska, M. Sosnowski, V. Uruba, CFD modelling and PIV experimental validation of flow fields in urban environments, *E3S Web of Conferences*, **14**, 01034 (2017)
9. V. Uruba, Decomposition methods in turbulent research, EFM11, *EPJ Web of Conferences*, **25** 01095 (2012)
10. L. Sirovich, Turbulence and the dynamics of coherent structures. Part I: coherent structures. *Q. Appl. Math.* , **45** (3), 561-571 (1987)

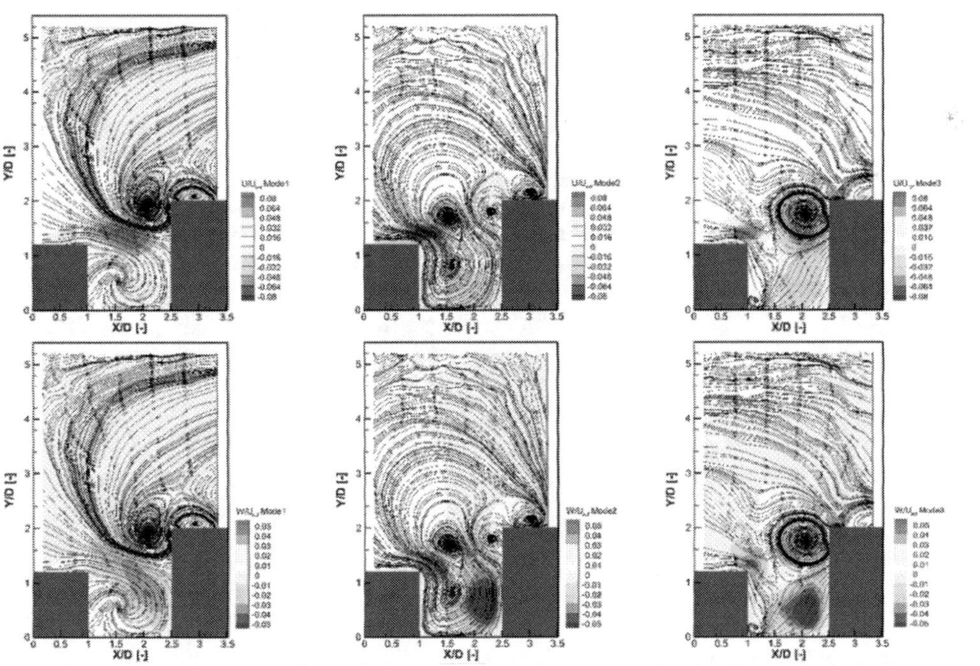

Fig. 8. Three POD modes between the buildings for longitudinal velocity (upper row) and wall-normal velocity (lower row).

The influence of distance between vehicles in platoon on aerodynamic parameters

Renata Gnatowska[1*], *Marcin* Sosnowski[2]

[1]Czestochowa University of Technology, Faculty of Mechanical Engineering and Computer Science, 21 Armii Krajowej Av, 42-200 Czestochowa, Poland
[2]Jan Dlugosz University in Czestochowa, Faculty of Mathematics and Natural Sciences, 13/15 Armii Krajowej Av, 42-200 Czestochowa, Poland

Abstract. The paper presents the results of experimental and numerical research focused on the reduction of fuel consumption of vehicles driving one after another in a so-called platoon arrangement. The aerodynamic parameters and safety issues were analyzed in order to determine the optimal distance between the vehicles in traffic conditions. The experimental research delivered the results concerning the drag and was performed for simplified model of two vehicles positioned in wind tunnel equipped with aerodynamic balance. The additional numerical analysis allowed investigating the pressure and velocity fields as well as other aerodynamics parameters of the test case.

1 Introduction

The basic role of platooning in driving at a short intermediate distance between the vehicles in tandem arrangement results in a considerable reduction in drag coefficient and finally fuel consumption for the cars. Results of other researches showed also that 20% of emission reduction, 10% reduction in death rate and a smoother traffic flow with potential increase in traffic flow can be achieved through vehicle platooning. On the other hand, by growing the traffic complexity, the action of one vehicle may affect all the others vehicles in a group. Therefore, safety limits of distance of one car to another vehicle become a challenge. Aerodynamic parameters of two objects (cubes, cylinders, other shapes) in platooning arrangement were investigated previous by many researchers [1-5]. From these research the drag coe□cient of the follower body was bigger than that of the leading one at near inter-vehicle distances. Research on the effect of vehicles spacing on the drag coefficient with vehicles in platooning was published by Zabat et al. [1]. For all cases, the drag coefficient of the follower object was bigger than that of the leader one at close inter-vehicle distances ($x/L < 0.35$). Browand, Hammache et al. [4, 5] have investigated the interaction of two truck vehicles with various combinations of bodies geometry. They observed, that for the cases with sharp corners in front of the leading vehicle, the drag coefficient of the leading body is higher than that of the one in isolation. Watkins and Vino [6] have studied the e□ect of distances between two vehicles on the lift and drag coe□cients. Their experiment showed that the drag force of the follower body increases for spacing of 0.1-1.0, compared to the corresponding drag value of a single body in isolation. Mirzaei et. al. [7, 8] numerically analysed the e□ect of inter-vehicle distance on the flow structure of two vehicles in platoon. Large Eddy Simulations were made for two Ahmed bodies for three inter-vehicle distances, $x/L = 0.3$, 0.5 and 1 at $Re = 30000$. A sudden increase of the drag coe□cient was observed for the follower with $x/L = 0.5$, compared to the drag value of a single body in isolation.

The objective of this paper is to provide a detailed experimental and numerical analysis of the flow interactions of vehicles driving one after another in a so-called platoon arrangement. The main aim is to clarify the increase of the drag force of the follower vehicle at some particular close distances, found in the experiment part of the present investigation. Accurate numerical simulations at three different distances are performed to describe the flow structures around the body and to hopefully obtain more information about the reasons causing the change of the drag coefficient.

2 Description of the work

The research concerns numerical and experimental investigations of the pressure and velocity fields as well as other aerodynamics parameters between two vehicles in platoon arrangement. The vehicles are represented by Ahmed bodies [9]. Two identical bodies are used as the test models. The Ahmed body is one of the most popular models employed as a simplified vehicle shape to predict the fundamental flow physics associated with vehicles' wake. The original shape of the body and flow structure around it can be found in [9]. The dimension of the analysed bodies used in the research are 261×97,25×72 mm (LxWxH) and the frontal area is $0.007 m^2$. The slant angle is set to 25 degrees. The distance between the

* Corresponding author: gnatowska@imc.pcz.pl

bodies S is varied in the range of S=0÷1.2L (L is the body length). The investigation of both of the bodies (leading or trailing) are compared to the drag coefficient of a single body in isolation.

2.1 Experimental setup

The experimental measurements are made in the open-circuit wind tunnel facility of the Institute of Thermal Machinery, Czestochowa University of Technology. The test section of the wind tunnel is characterized by cross-section of $1m^2$ and 10m length. The maximum inlet velocity of air in the wind tunnel is 8m/s resulting in the Reynolds number of $1,4 \times 10^5$, based on the length of the body. Air pressure and temperature are measured continuously during wind tunnel operation. Schematic view of the test section is presented in Figure 1. The vehicle models (Ahmed bodies) are made of wood and are aligned with the direction of the wind in the tunnel. The loads on the bodies are measured with an aerodynamic balance, which is located under the tunnel floor and is attached to the body through a holder.

One car is attached to the balance and the other one can move longitudinally (through traversing system) along the wind tunnel axis to achieve different vehicle spacing configurations.

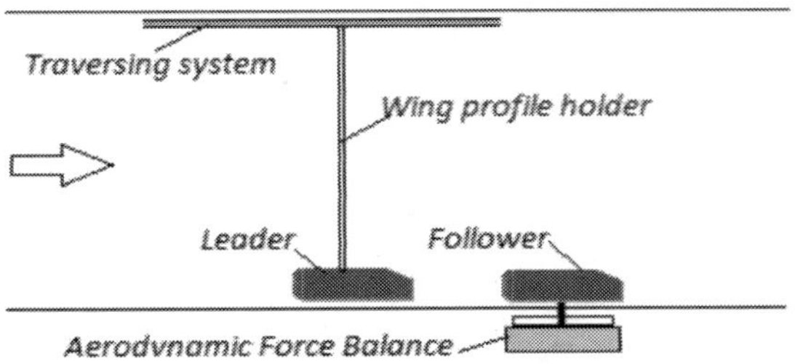

Fig. 1. Schematic side view of the test section

2.2 Numerical simulations

The simulations were performed with the commercial CFD code ANSYS Fluent 18.1. Two identical computational models (except for the distance between the vehicles and the inlet velocity) were used for all analysed cases. The size of the computation domain is 2.0m long, 0.75m wide and 0.75m high. The distance of the inlet plane to the single or leading car was set to 0.5m.

The bottom surface of the Ahmed bodies are located at 0.02m above the ground. The solver was configured as pressure-based and the analysis were performed for steady state. The 3D steady RANS equations were solved with the standard k– ε model. The choice of the standard k– ε model was made based on a previous extensive validation study for the aerodynamics of vehicles [10 - 12].

Pressure–velocity coupling was taken care of by the SIMPLEC algorithm based on the relation between velocity and pressure to enforce mass conservation and to obtain the pressure field. Pressure interpolation was second order and second-order discretisation schemes were used for both the convection terms and the viscous terms of the governing equations. Convergence was monitored carefully and the iterations were terminated when all residuals showed no further reduction with increasing number of iterations.

Air at normal conditions was the fluid medium. Velocity-inlet boundary condition type was assigned to the inlet with velocity magnitude normal to the boundary equal 8 m/s (corresponding to experiment) and 88 m/s (representing the real traffic situation for $Re_L=1,4\times10^6$). The outlet of the computational domain was defined as pressure-outlet with ambient static pressure. Wall with no slip shear condition and no roughness was assigned to the remaining surfaces (car models and wind tunnel walls).

The meshing process was performed using ANSYS Meshing 18.1. Tetrahedral (TET) elements were used to discretize the computational domain. Figure 2 shows part of the grid distribution in the vicinity and on the Ahmed bodies. The size function was set to UNIFORM in order to obtain the most regular mesh. Moreover the maximal face size as well as maximal TET size constrains were set to the value of 50mm. Additional face sizing on surfaces representing car models was defined as 5mm with 1.1 growth rate in order to refine the mesh in the vicinity of the analysed Ahmed bodies. Such meshing procedure was previously validated in [13]. The above mentioned meshing settings resulted in the final mesh consisting of approx. 600 thousand elements. The mesh resolution was based on the mesh sensitivity analysis and generation guidelines [14, 15]. The mesh quality was also taken into consideration. The quality analysis provides a metric that ranges between 0 and 1. A value of 1 indicates a perfect element while a value of 0 indicates that the element has a zero or negative volume. The element of the worst quality generated throughout the entire research (ten meshes) was characterized by the quality index of 0.16 and therefore it was assumed that the mesh did not affect the numerical stability of the computational model. The near-wall region in a turbulent boundary layer was considered by y+ >30.

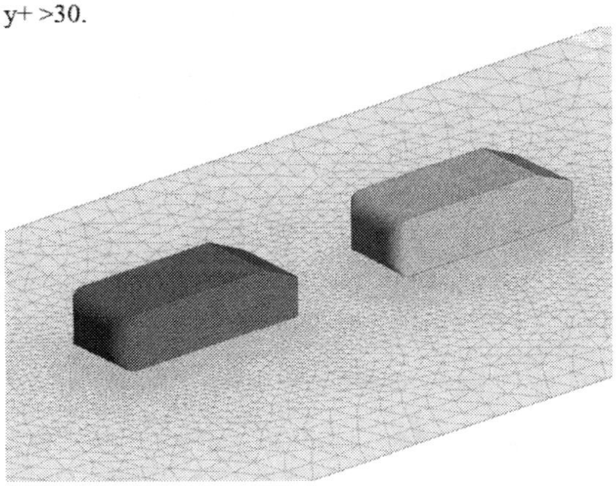

Fig. 2. Computational domain with Ahmed bodies.

3 Results and discussions

CFD simulations are performed based on the steady-state Reynolds-Averaged Navier–Stokes equations with the standard k– ε model and standard wall function. The simulations are validated based on a series of wind tunnel measurements. Both CFD simulations and wind tunnel measurements allowed ascertaining and quantifying the upstream effect by a downstream car on the drag of the first one.

Figure 3 shows the averaged drag coe□cient (C_D) of two identical Ahmed bodies measured for various distances of the bodies in the range of $0 < x/L < 1.2$, where L is the body length. The drag coefficient values are normalized with the drag of a single body in isolation and are presented for the leader and the downstream vehicle. The experiment is conducted at inlet velocity of 8m/s. CFD simulations are conducted at two inlet velocity of 8m/s and 88m/s (Figure 3b). The comparison of the results (Figure 3a) shows a good agreement between the data and the maximum di□erence is approx. 10%. Only for the configuration when bodies are at the closest distance (x/L = 0.2) differences of results are more significant. As inferred from the figure 3b, the drag coefficient variations show similar trends when the velocity increases from 8m/s to 88m/s. The drag value of the leader decreases about 10% by increasing the distance in the range of x/L = 0.2 – 0.5. Then it starts to increase and approach towards a single body drag. More interesting results can be noticed for the downstream vehicle. In spite of locating the first vehicle in the wake region of the leader in which, the drag force of the downstream body is expected to be reduced compared to the value of the single body in isolation (the concept of the platooning), the drag increases to almost 20% higher value than the drag of a single body. Significant amount of drag coefficient increase is observed in the range of x/L = 0.2 – 0.5, where the drag value of the downstream vehicle increases from 5% less than for a single body at x/L = 0.2 to 20% more than that at x/L = 0.5. The numerical simulations of the present study performed for three di□erent inter-body distances of S = 0.5W, 1.5W and 2.5W are described. The distances are selected within the regions where the increase or decrease of the drag coefficient starts and reaches the maximum value. The flow characteristics are compared and thoroughly discussed for all three di□erent gaps on the base the static pressure and velocity fields. The similar trend can be found in both numerical and experimental predictions. Figure 4 compares the averaged static pressure distribution on the symmetric plane for single Ahmed body and three di□erent inter-body distances for two inlet velocities 8m/s and 88m/s. The contours present the wake regions in the gap between the bodies, behind the downstream vehicle and also region in front of the objects. The vortical structures, one over the other one, can be noticed in the gap of the bodies for three distances. For S = 0.5W, these vortical regions are small and attached to the downstream body. Increasing the distance to S=1,5W, two vortices are still attached to the downstream vehicle. The vortices are stretched out towards the leader, move to the leader and become closer to the middle of the gap.

a)

b)

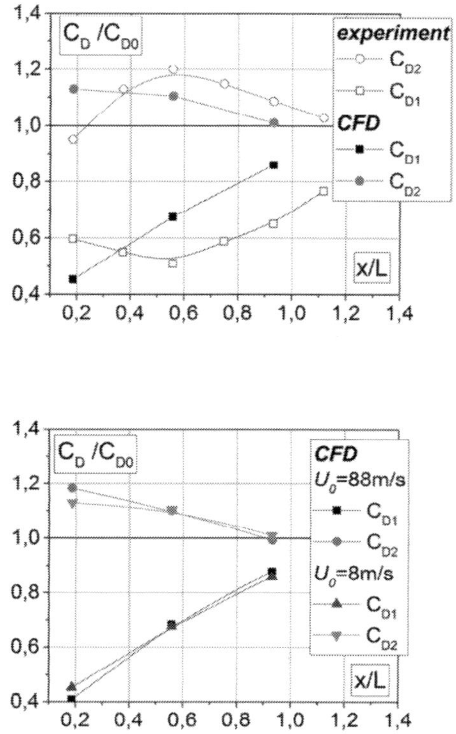

Fig. 3. Averaged normalized drag coefficient C_D/C_{D0} (C_{D0} drag value of a single body in isolation) of two Ahmed bodies for various inter-vehicle distances. Comparison of numerical simulation results and experimental data (a) and comparison of numerical results for velocities 8m/s and 88m/s (b).

Increasing the distance of the bodies and therefore, growing the size of the wake behind the leader, reduces the pressure in the center of the gap and produces a low-pressure suction area which biases the shear layers towards the center. On the other side, the size of the follower's wake does not change by increasing the distance of the bodies. Increasing the distance of the bodies would eventually increase the pressure over the follower's body. The streamlines are distributed more uniformly around the leader and the high-pressure impingement zone moves towards the middle of the downstream vehicle frontal face. According to the Figure 4, the pressure variations show similar trends when the velocity increases from 8m/s to 88m/s. The time-averaged velocity for three di□erent gaps of the bodies is plotted in Figure 5. It is also displays the flow field for the single isolated body. These pictures show how increasing the distance of the bodies stretches out the vortical region behind the leader. With increasing the distance, the recirculation zone is biased to the ground and the low-pressure region and the flow impingement point on the follower thus move to the center of the frontal surface.

https://doi.org/10.1051/epjconf/201818002030

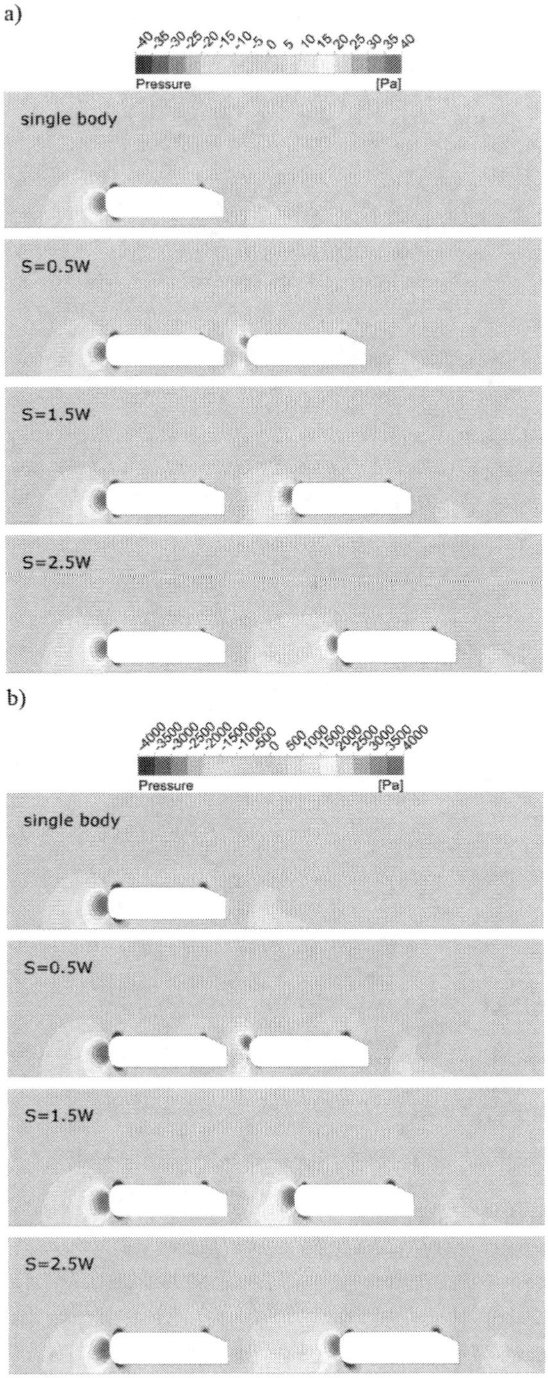

Fig. 4. Averaged static pressure distribution around two Ahmed bodies for various inter-vehicle distances on the symmetric plane. The numerical results at different velocities 8m/s (a) and 88m/s (b).

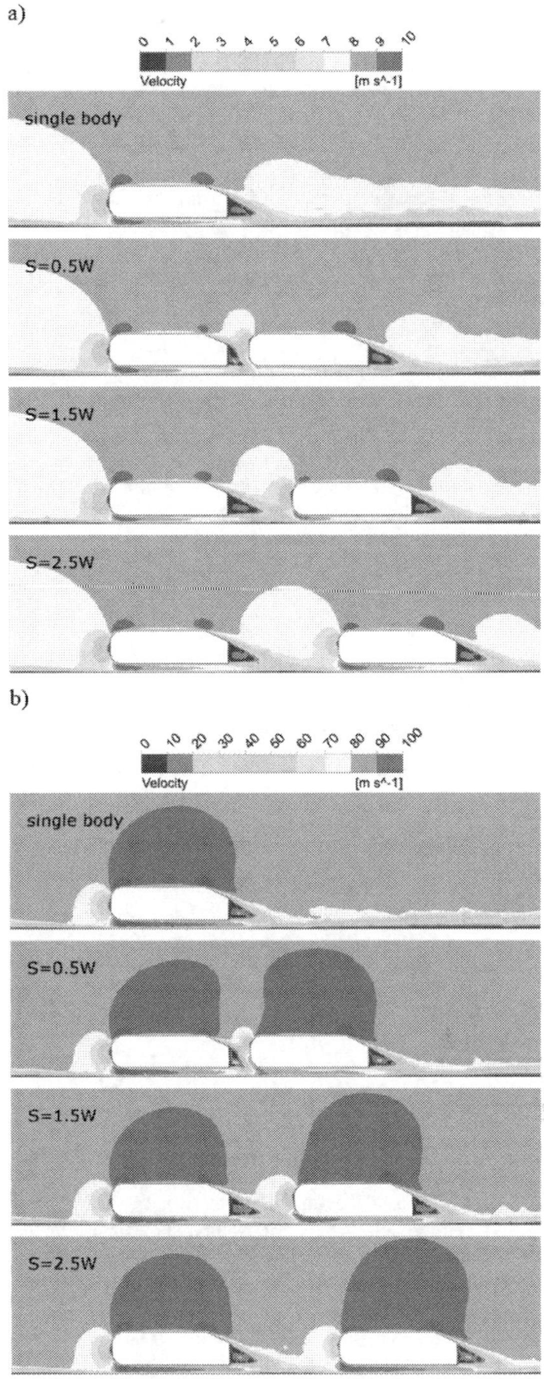

Fig. 5. Mean velocity field around two Ahmed bodies for various inter-vehicle distances. The numerical results at different velocities 8m/s (a) and 88m/s (b).

4 Conclusions

This paper describes the detailed experimental and numerical analysis of the effect of inter-vehicle distance in the platoon arrangement on the mutual flow interactions. The main attention was concerned with the significant increase of the drag coefficient of the follower at some particular close distances, found in the experiment part of the present investigation. Accurate numerical simulations at three di□erent inter-body distances are performed to describe the flow structures around the body and to obtain more information about the reasons causing the increase of the drag coefficient.

The main findings from this study are:

• Complex flow fields have been numerically visualized showing various velocity and pressure fields formed at different regions.

• The flow structures in the wake region behind the downstream Ahmed vehicle is more or less the same for all the analysed cases (x/L = 0.2-1.2).

• The recirculation zone is formed in the gap between two Ahmed vehicles and the interaction between the flow on the top of vehicles and the flow inside the gap region is very weak especially for x/L = 0.2. When the gap size increases further the strong interaction is observed between the flow on top of the two vehicles and the flow inside the gap region.

• Comparison of various flow characteristics revealed that a significant change in flow structure between the vehicles in the platoon with 0.3<x/L<1.2 is the underlying reason for the increase in drag coefficient.

References

1. M. Zabat, N. Stabile, S. Farascaroli, F. Browand, *The aerodynamic performance of platoons: A final report.* California Partners for Advanced Transit and Highways (PATH, 1995
2. R. Gnatowska, Arch. Mech. **60** (6), 475-490 (2008)
3. R. Gnatowska, Int. J Turbo Jet. Eng. **28** (1) 21-29 (2011)
4. F. Browand, M. Hammache, M. The limits of drag behavior for two bluff bodies in tandem (No. 2004-01-1145). SAE Technical Paper. (2004)
5. M. Hammache, M. Michaelian, F. Browand, F. Aerodynamic forces on truck models, including two trucks in tandem (No. 2002-01-0530). SAE Technical Paper (2002)
6. S. Watkin, G. Vino, J. Wind Eng. Ind. Aerodyn. **96**, 1232-1239 (2008)
7. M. Mirzaei, S. Krajnović,. Large Eddy Simulations of Flow Around Two Generic Vehicles in a Platoon. In *Proceedings of the 5th International Conference on Jets, Wakes and Separated Flows (ICJWSF2015)* (pp. 283-288). Springer International Publishing, 2016)
8. M. Mirzaei, A. Pavlenko, S. Krajnovic, V.G. Chernoray, *Experimental and numerical study of flow structures between two Ahmed bodies with various inter-body distances.* In *ICHMT DIGITAL LIBRARY ONLINE.* Begel House Inc.. (2015)
9. S.R. Ahmed ATJFE **105**, 429-434 (1984)
10. T. Defraeye, B. Blocken, E. Koninckx, P. Hespel, J. Carmeliet, J. Biomech. **43** (12), 2281–2287 (2010)
11. B. Blocken, Build. Environ. **91**, 219–245 (2015)
12. R. Gnatowska, M. Sosnowski, V. Uruba, *CFD modelling and PIV experimental validation of flow fields in urban environments*, E3S Web of Conferences, 14, 01034 (2017)
13. M. Sosnowski, J. Krzywanski, R. Gnatowska, *Polyhedral meshing as an innovative approach to computational domain discretization of a cyclone in a fluidized bed CLC unit.* E3S Web of Conferences, 14, 01027 (2017)
14. J. Franke, A. Hellsten, H. Schlünzen, B. Carissimo, *Best Practice Guideline for the CFD Simulation of Flows in the Urban Environment*, COST Action 732: Quality Assurance and Improvement of Microscale Meteorological Models. (Hamburg, Germany. 2007)
15. Y. Tominaga, A. Mochida, R. Yoshie, H. Kataoka, T. Nozu, M. Yoshikawa, T. Shirasawa, J. Wind Eng. Ind. Aerodyn. **96** (10–11), 1749–1761. (2008)

 https://doi.org/10.1051/epjconf/201818002031

Cable vibration caused by wind

Irena Golebiowska[1,a], *Kazimierz* Peszynski[2]

[1]UTP University of Science and Technology, Faculty of Civil Eng., Al. prof. S. Kaliskiego 7, 85-796 Bydgoszcz, Poland
[2]UTP University of Science and Technology, Faculty of Mech. Eng., Al. prof. S. Kaliskiego 7, 85-796 Bydgoszcz, Poland

Abstract. The paper briefly presents selected basic kinds of excitation of cable vibration caused by dynamic effect of wind. It describes the aerodynamic phenomena such as vortex excitation, wind-rain excitation, galloping and buffeting. Cables are structures which are characterised by low internal damping, low rigidity and low weight, so they are not capable of total excitation energy dissipation, hence they can reach large amplitudes of vibration. Large amplitude of vibration causes excessive stress, thereby lowering the safety of the structure.

1 Introduction

In the twentieth century, along with the evolution of new materials, technologies and advanced calculation methods, modern bridges with long spans are being built. Jiashao Bridge (2800 m, 2013, China) is currently the longest and the widest six-pylon cable-stayed bridge in the world, while Messina bridge (3300 m, Italy), designed in the 90s of the twentieth century, will be the longest suspension bridge.

Increase of span length of such bridges requires the installation of increasingly longer cables. The cables are extremely sensitive to dynamic excitation due to low internal damping, high flexibility, light mass. Therefore, they can reach high vibration amplitudes. These high oscillations can cause fatigue destruction of the cables and/or damages at the mounting points.

The measurements of internal cable damping of cable-stayed bridges in Japan show a low level of structural damping, approximately 0.1% of critical damping (Fig. 1) [1].

Even though many various tests (observation, tunnel tests, numerical simulations etc.) regarding the understanding of cable vibration excitation mechanisms (mainly of inclined cables), have been carried out, some of phenomena still need to be explained.

Types and level of vibration depend on cable structural properties (damping, stiffness, mass), wind force conditions and interaction between wind and cables.

There are several types of cable vibration induced by wind [2]:

- Vortex excitation of an isolated cable or groups of cables;
- Wake galloping of groups of cables;
- Galloping of a single inclined cable to the direction of wind;
- Rain- and wind-induced vibration of cables;
- Galloping of ice-covered cables;
- Galloping of cables in a wake of other structural elements;
- Motion due to turbulent buffeting;
- Motion due to fluctuating cable tensions;
- Cable vibration induced by aerodynamic excitation of a bridge.

In order to assume an effective and efficient method of lowering the level of cable vibration, we need to have deep understanding of physical phenomena that take place under the influence of wind.

The paper describes basic types of cable vibration caused by wind, observable only in the cables of long-span bridges (suspension and cables-stay): vortex excitation, vortex excitation at high velocity, rain-wind induced excitation, galloping, galloping of dry inclined cables, wake galloping for groups of cables and buffeting.

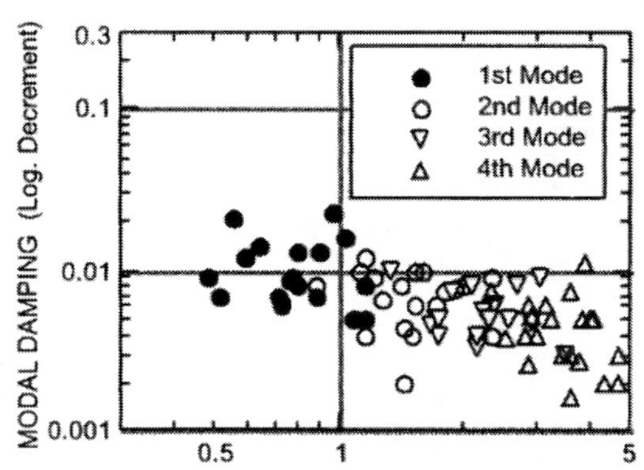

Fig. 1. Relation between modal damping and natural frequency in stay cables [1].

[a] Corresponding author: irena_golebiowska@wp.pl

2 Vortex excitation

Vortex excitation is characterised by vibration with limited amplitude occurring at relatively low velocities of wind. Vortex-induced cable vibrations are common and it is known that amplitudes of these vibrations are small and rarely reach the size of the cable cross-section [3].

Excitation of a single cable induced by vortexes occurs as the vortexes alternately detach from either side of the cable (vortex shedding) when the wind direction makes an approximately right angle with the cable axis. Occurrence and detachment of vortexes create oscillatory transverse force to the wind direction. Vortex shedding generates a periodic asymmetric flow, which is known as Karman vortex street. The frequency of vortex detaching is proportional to the wind velocity and inversely proportional to the diameter of cable cross-section. This type of vibration is particularly dangerous when the vortex shedding frequency f_s is similar or equal to the any engine-frequency of cable f_{ci}; resonance will occur and the cable will be excited (Fig. 1).

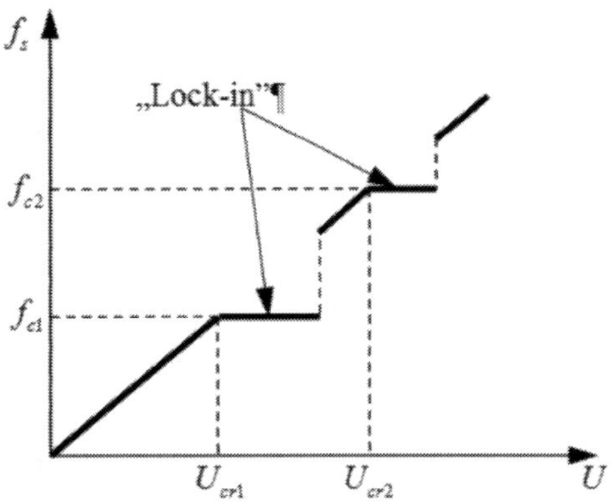

Fig. 2. Lock-in phenomenon during vortex excitation.

The vortex resonance leads to strong interaction of cables with the surrounding flow when the critical velocities of wind U_{cri} is reached. An increase of the flow velocity by a few percent does not change the shedding frequency which coincides with the natural frequency of cable. This phenomenon of synchronisation is referred to as the lock-in state.

Williamson and Roshko [4] studied the effect of the amplitude and frequency of oscillations on a wake of an oscillating cylinder, forced to translate in a sinusoidal trajectory. They gave a map showing when the various vortex patterns would occur using a non-dimensional wavelength - amplitude plane. The Reynolds number was not defined as an independent parameter in the study. The Reynolds number was kept within a certain range 300<Re<1000, but was never held fixed.

Each of periodic vortex wake pattern consisted of single vorticed (S) and vortex pairs (P), leading to combinations such as 2S (similar to the classical Karman vortices), 2P and P+S modes, which are the principal modes near the fundamental lock-in region. At the critical wavelength only two vortices are formed in each cycle, resulting in more concentrated shed vorticity than at other wavelengths. Larger forces are induced by the shedding of more concentrated vorticity. This state is called a resonant synchronisation. Figure 3 shows the mapping of different patterns, or modes identified as functions of non-dimensional amplitude and wavelength (frequency).

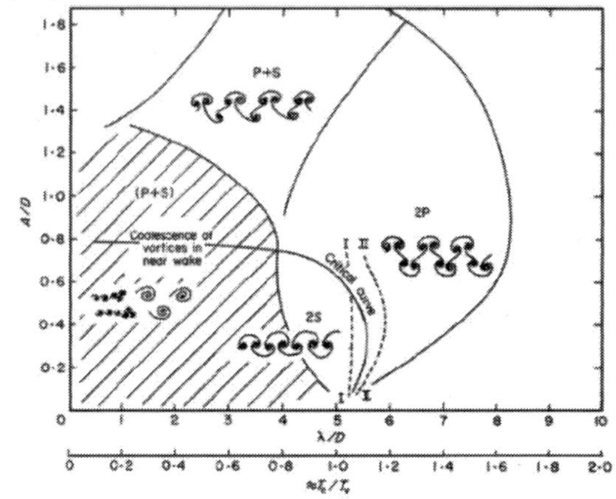

Fig. 3. Map of vortex synchronisation patterns near the fundamental lock-in region, A/D is the amplitude ratio and λ/D is the wavelength ratio. The critical curve represents the transformation from one mode of vortex formation to another. Curves I and II represent locations of a sharp jump of the forces (acting) on the body [4].

3 High-speed vortex excitation

High-speed vortex excitation is characterised by limited amplitude of vibration, occurring for a restricted interval of wind velocities in the critical Reynolds number range.

Vortex excitation of dry inclined cables at high wind velocities has been investigated, among others, by Matsumoto [5]. Their research demonstrates that this type of excitation is related to the interaction of Karman's conventional two-dimensional vortexes with the axial vortexes which shed in a wake of the inclined cable, generating an amplified vortex once after every three Karman vortexes (Fig. 4). Frequency of the axial vortex shedding is approximately one-third of the conventional Karman frequency.

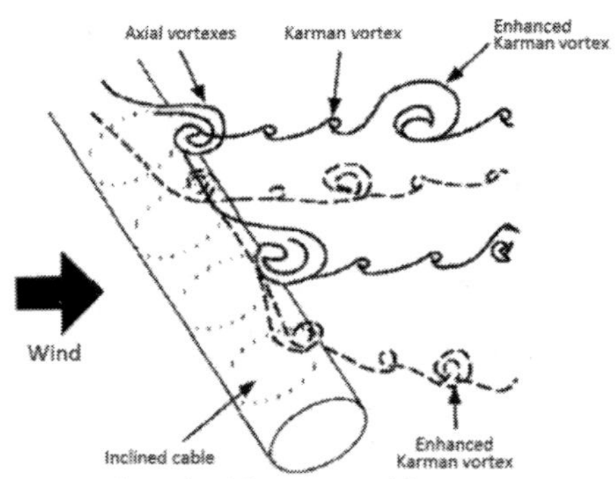

Fig. 4. Interaction of axial vortexes with Karman vortexes in a wake of inclined cable [6].

Axial vortex is proven by a wake of yawed cylinder of flow visualisation tests (Fig. 5)

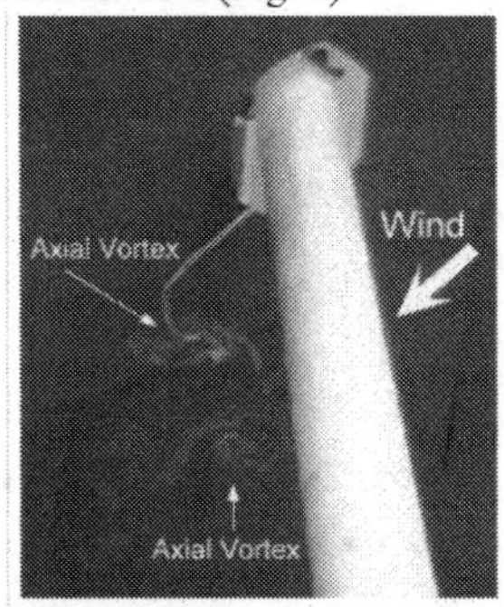

Fig. 5. Visualised axial vortex in a wake of inclined cable (without end plate, β=45°, V=0,5 m/s in smooth flow) [6].

Axial flow interrupts the interaction between two separated shear layers in a wake of the cable. This effect leads to the mitigation of Karman vortexes. The Karman vortex shedding suppression increases the sensitivity of separated flow to external excitation like body motion, resulting in enhancement of the instability [6, 7].

3 Rain-wind excitation

The rain - wind vibration induced instability are limited amplitude - type vibration.

Rain and wind induced vibration occurs under certain wind velocity range. However, only during rains and only on inclined cables. Vibration can reach high amplitudes at low frequencies.

During windy and rainy weather, on the surface of an inclined cable, water trickles as a single drop at low wind velocities, but these drops are knocked from the cable at high velocities. Only in a certain range of medium velocity, it is possible that along the surface of a cable, a rivulet of water will occur. Hikami et al. [8] noticed that when the wind velocity is low during the rain, raindrops gather on a cable forming a rivulet at the bottom part of a cable. The rivulet oscillates circumferentially with the same frequency as the vibrating cable. When wind velocity gets between 8 and 15 m/s, the second rivulet forms at the top of a cable. The rivulets oscillate with the same frequency of the cable movement.

Results of research in a wind tunnel carried out by Hikami, Y., Shiraisi N. demonstrate that the second rivulet occurred on the top windward surface of the cable and the occurrence of this second rivulet was related to the wind velocity. When the wind velocity was increased to a certain value, rain drops overcame the forces of friction and gravitation, forming the top rivulet [8]. The dependence between the angle of forming rivulets, their aerodynamic forces, and wind velocity is presented in Fig. 6.

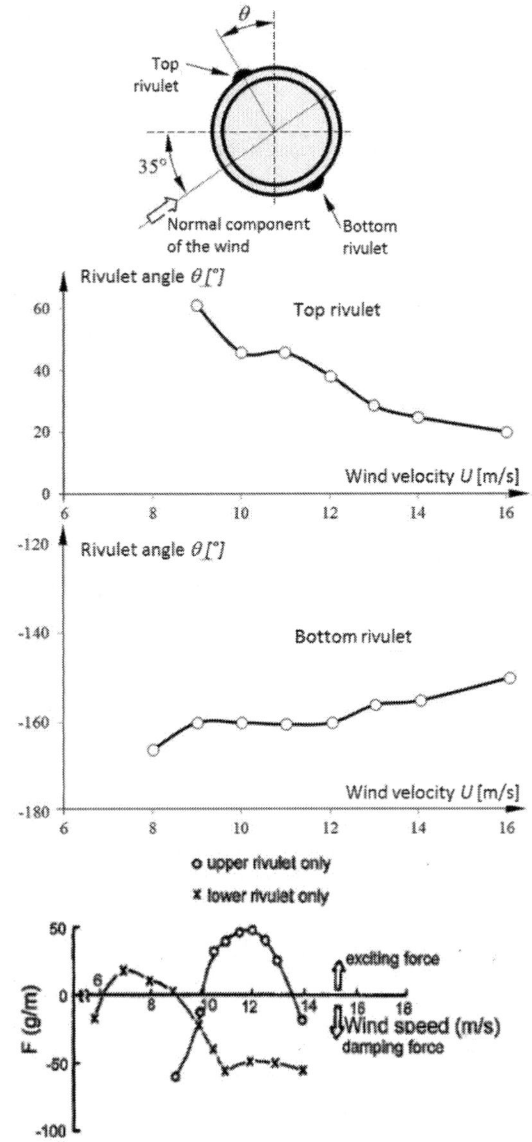

Fig. 6. Variations of the angle of water rivulets and their aerodynamic forces with wind velocity [8].

Verwiebe has described three different mechanisms of rain-wind excitation: parallel and transverse to the direction of wind, depending on the rivulet flow on the cable surface – Fig. 7. The first one occurred when the wind direction is the same as the direction of the cable vibration. Symmetrical oscillations of two symmetrical water rivulets at the leeward side intensified the cable vibration and the maximum amplitude of vibration occured at the wind velocity of 25 m/s. The second mechanism took place when the wind velocity was 18 m/s and cable vibration was transverse to the wind direction. Vibration induced asymmetrical oscillations of the bottom water rivulets. Third mechanism was observed when the cables were transverse oscillated to the wind direction at the wind velocity less than 19 m/s.

Then, the oscillation of a single bottom water rivulet took place on the lee side of the cable. However, when the wind velocity exceeded 19 m/s, oscillations of both water rivulets, top and bottom, were observed [9]. According to Verwiebe, water rivulets circumferentially oscillate on the cable. The rivulet flow is affected by interaction, structure - liquid and forces acting on them. Changing the location of rivulets causes a continuous variation of cable cross-section, which leads to change of aerodynamic forces acting on the cable.

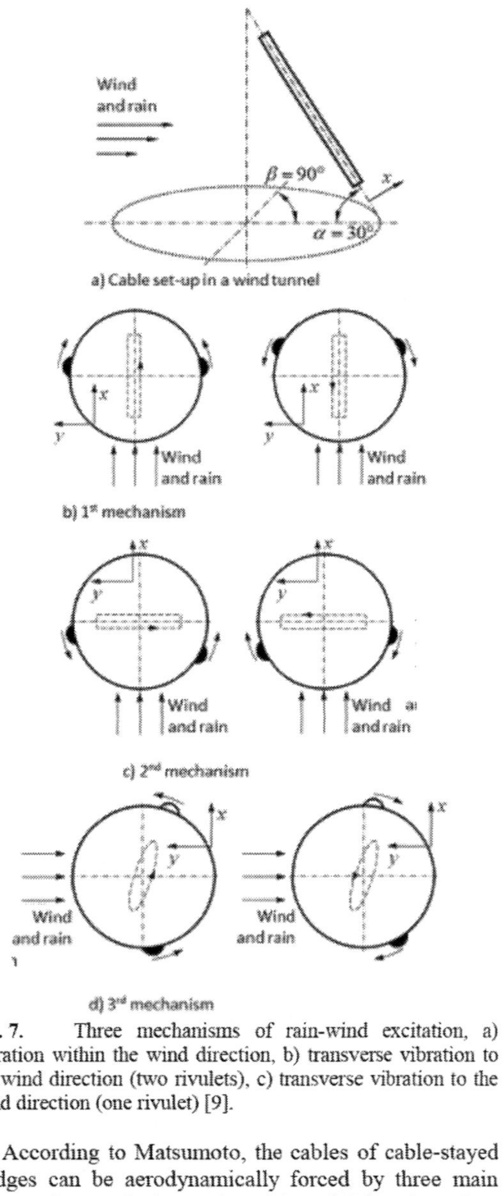

Fig. 7. Three mechanisms of rain-wind excitation, a) vibration within the wind direction, b) transverse vibration to the wind direction (two rivulets), c) transverse vibration to the wind direction (one rivulet) [9].

According to Matsumoto, the cables of cable-stayed bridges can be aerodynamically forced by three main factors: the top rivulet on the cable, axial flow in a close wake of inclined / yawed cable and Reynolds number (in the critical Reynolds number range) [10]. These factors can cause mainly unsteady galloping. However, the formation of a rivulet in a specific location can generate significant slope of the lift force (ie. $\mathrm{d}C_L/\mathrm{d}\alpha < 0$ and may consequently cause quasi – steady galloping [10]

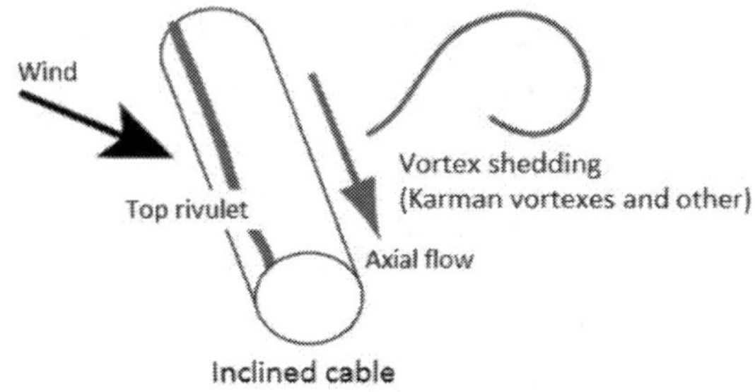

Fig. 8. Factors considered in the aerodynamics of inclined cables [11].

Fig. 8 illustrates factors considered in the aero-dynamics of inclined cable during rain-wind excitation.

According to experimental studies carried out by Saito, vibration induced by rain and wind may be reduced to an acceptable level, assuming that the Scruton number ($S = m\xi/\rho D^2$, where m – cable mass per unit of length, ξ – damping as ratio of critical damping, ρ – air density [kg/m^3], D – cable diameter [m]), is higher than 10. Based on this criterion, we may determine the damping level which must be added to the cable in order to reduce vibration caused by this excitation [12], [13].

Although there are many causes of excessive cable vibration, around 95% of all reported problems with cable-stay bridges were caused by rain-wind excitation, according to literature.

Despite numerous analytical and experimental studies carried out by many scientific institutes, the basic physical mechanisms of this phenomenon are still unclear.

4 Galloping

Galloping is an aeroelastic self-excited phenomenon, characterized by low frequencies and high amplitudes. It was observed for the first time on ice-covered power lines subject to strong wind. Galloping is a typical instability induced by the coupling of aerodynamic forces, which affect the structure along with its vibration. Vibration of the structure periodically changes the angle of attack of wind. Changes in the angle of attack induce change in aerodynamic forces affecting the structure, changing the structure's response. The first simplified criterion (assuming the model of a single-degree-of-freedom system) with regard to the instability connected with galloping was introduced by Glauert-Den Hartog and is as follows:

$$\left(\frac{dC_L}{d\alpha} + C_D\right)_{\alpha=0} < 0 \qquad (1)$$

where: α is the angle of attack, C_L and C_D are aerodynamic coefficients of the lift force and drag force, respectively. A necessary condition for galloping to take place (as part of quasi-steady theory) is the occurrence of negative aeroelastic damping in the system. Cable with circular cross-section cannot gallop, because of its geometrical symmetry $\left(dC_{L}/d\alpha=0\right)$ unless the cross-section is changed. Ice accumulation on a cable causes changes in its cross-section and, as a result, it leads to the cable's aerodynamic instability.

According to Matsumoto [10], Den Hartog criterion is not always an essential condition for unsteady galloping.

4.1. Dry inclined cables galloping

Galloping of dry inclined cables is a new term within wind engineering, invented in the 80s. Single cables with circular cross-sections do not gallop if they are situated perpendicularly to the wind direction. However, if the wind velocity has a component which is not perpendicular to cable axis, we can observe an instability with the same features as galloping. Cables then make an elliptical motion. In the case of a single inclined cable, wind affects its elliptical cross-section [2]. Nakamura demonstrates that the mechanism which induces galloping is a broken connection between the upper and lower flow. Connection of these two separated flows may cause zero pressure difference at the top and bottom surface of a cylinder [14]. According to Matsumoto, "dry" galloping may be boiled down to two different mechanisms: quasi - steady galloping (divergent type) and unsteady galloping. The first mechanism corresponds to the conventional galloping and its responses may be explained with the quasi-steady theory. Unsteady galloping takes place when the mitigation of the Karman vortexes is insufficient and non-stationary and the amplitude of the response is changeable due to the fluctuation of Karman vortexes intensity [7, 15]. Unsteady cross-flow response of yawed / inclined cables at lower, reduced velocity than reduced velocity of divergent, according to Matsumoto, should be a kind of unsteady galloping which is generated by unsteady intensity of Karman vortex due to unsteady axial flow [10]. The possibility of forcing the divergent-type galloping of inclined cables was confirmed by numerous experimental studies carried out by Saito, with the assumption that the angle between a cable axis and wind direction ranged from 30° to 60° and the level of damping was low [13].

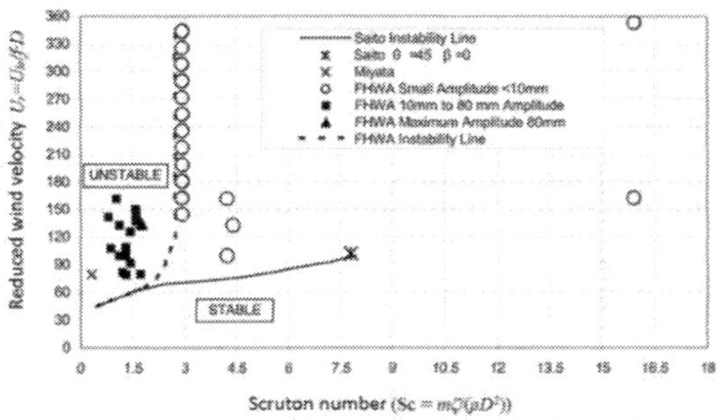

Fig. 9. Dependence between wind velocity and Scruton number [2].

For the engineering purposes, FHWA (Federal Highway Administration) proposes an original diagram of dependencies between wind velocity and Scruton number, used for specifying required level of damping [2].

Figure 9 illustrates the results of experimental studies conducted by FHWA and compares them with the instability line determined by Saito. The research suggests that even at low levels of damping ($\xi > 0.003$), vibration caused by vortex shedding and galloping of inclined cables are insignificant. Such level of damping corresponds to the Scruton number amounting to approximately 3, which is lower than the minimum required to reduce the level of vibration caused by rain and wind. As illustrated in Figure 9, there is a big difference between both criteria. According to Matsumoto, FHWA criterion and Saito criterion correspond respectively to steady galloping and unsteady galloping. Parameters regulating the intensity of axial flow, as well as the mechanism of its occurrence have not been explained yet. However, it is known that cable inclination causes axial flow, which is the reason for the instability of aerodynamic phenomena.

4.2 Wake galloping for groups of cables

Wake galloping is an elliptical motion caused by the variability of forces that affects cables which are in the wake of other structural elements (e.g. pylons or other cables). This phenomenon occurs at high wind velocities and is associated with high vibration amplitudes. Cooper [16] has proposed an approximate global stability criterion in the form of minimum critical wind velocity above which wake galloping can be expected. Minimum critical velocity is proportional to a square root of the Scruton number, frequency of cable's own vibration and to its diameter. At small distances between cables (from 2 to 6 x cable diameters), leeward cables move along an almost circular orbit; for larger distances, the orbit becomes elliptical, with the main axis of the ellipsis situated approximately in the same direction as the wind direction. These types of vibration have been observed on overhead power lines, with the amplitudes around 20 x cable diameters and with cables arranged with a span between them of 10 to 20 x diameters [2].

5 Buffeting

Buffeting is a kind of vibration forced randomly, caused by the turbulent component of wind. Characteristics of the responses depend on the turbulent nature of wind; vibration amplitudes are relatively small.

Random vibration induced by gust of wind concerns mainly power lines. For bridge cables, buffeting does not constitute the problem. High tension of bridge cables helps to reduce the amplitudes of that kind of vibration.

Conclusions

Semi-streamlined structures, such as cables in cable-stay bridges, hanger cables in suspension bridges, overhead power lines, are susceptible to wind-induced vibration, because of their low rigidity, low weight, and very low internal damping. Frequent excessive amplitudes of cable vibration cause their fatigue destruction at anchoring points. Understanding the excitations of cable vibration induced by wind and basic instabilities related to them are key to effective vibration control. Cable vibration caused by wind are a very complex phenomenon. Despite numerous experimental and theoretical studies, the problems of the instability of galloping or rain-wind excitation still remain unsolved.

References

1. Fujino Y., Vibration, control and monitoring of long-span bridges – recent, research, developments and practice in Japan, Journal od Constructional Steel Research 58, 2002, pp. 71-97
2. Kumarasena S., Jense N.P., Irvin P., Taylor P., Wind-Induced Vibration of Stay Cables. Report No. FHWA-HRT-05-083. US Department of Transportation, Turner-Fairbank Highway Research Center, 2007
3. Davenport, A. A Simple Representation of the Dynamics of a Massive Stay Cable in Wind, Proceedings of the IABSE/FIP International Conference on Cable-Stayed and Suspension Bridges, 1994
4. Williamson C.H,K, Roshko A., Vortex formation in the wake of an oscillating cylinder. Journal of Fluids and Structures 2, 1988, pp. 355-381
5. Matsumoto, M., Yamagishi, M., Aoki, J., Shiraishi, N., Various Mechanism of Inclined Cable Aerodynamics, Ninth International Conference on Wind Engineering, 1995
6. Matsumoto, M., Yagi, T., Shigemura, Y., Tsushima, D., Vortex-Induced Cable Vibration of Cable-Stayed Bridges at High Reduced Wind Velocity, Journal of Wind Engineering and Industrial Aero-dynamics 89(7-8), 2001, pp. 633-647
7. Matsumoto M., Yagi T., Hatsuda H., Shima T. and Tanaka M., Sensitivity of Dry-State Galloping of Cable Stayed Bridges to Scruton Number. Proceedings of the 7th International Symposium on Cable Dynamics, Vienna, Austria, December 10-13, 2007, 331-338
8. Hikami, Y., Shiraisi, N., Rain-Wind Induced Vibrations of Cables in Cable-Stayed Bridges, Journal of Wind Engineering and Industrial Aerodynamics, 1988
9. Verwiebe, C., Rain-Wind-Induced Vibrations of Cables and Bars, Bridge Aerodynamics, 1998.
10. Matsumoto. M., On galloping instability of stay cables of cable – stayed bridges, ACEM' 12, Seul, Korea, 2012, pp. 25-44
11. Fujino Y., Kimura K, Tanaka H., Wind Resistant Design of Bridges in Japan, Springer, 2012
12. Irwin P.A., Nedim A., Telang N., Wind induced stay cable vibration – A case study, in proceedings of the 3rd ISCD, Trondheim pp. 171-176, 1999
13. Saito T., Matsumoto M., Kitazawa M., Rain-Wind Excitation of Cables on Cable-Stayed Higashi-Kobe Bridge and Cable Vibration Control, Proceedings of the International Conference on Cable- Stayed and Suspension Bridges, Vol. 2, Deauville, October 1994
14. Nakamura K., Hirata K. (1994) The Aerodynamic Mechanism of Galloping. Transaction of the Japan Society for Aeronautical and Space Science 36 (114), pp. 257-269
15. Matsumoto M., Yagi T., Hatsuda H. Shima T., Tanaka M. and Naito H., Dry Galloping Characteristics and its Mechanism of Inclined/Yawed Cables, 2010
16. Cooper, K. R., A Note on the Wind-Induced Vibrations of Bundled Bridge Stay Cables. National Research Council of Canada, Note provided to RWDI, 1985

Radioisotope measurements of the liquid-gas flow in the horizontal pipeline using phase method

Robert Hanus[1,*], *Marcin* Zych[2], *Marek* Jaszczur[3], *Leszek* Petryka[4], and *Dariusz* Świsulski[5]

[1]Rzeszów University of Technology, Faculty of Electrical and Computer Engineering, 35-959 Rzeszów, Poland
[2]AGH University of Science and Technology, Faculty of Geology, Geophysics and Environmental Protection, 30-059 Kraków, Poland
[3]AGH University Science and Technology, Faculty of Energy and Fuels, 30-059 Kraków, Poland
[4]AGH University of Science and Technology, Faculty of Physics and Applied Computer Science, 30-059 Kraków, Poland
[5]Gdańsk University of Technology, Faculty of Electrical and Control Engineering, 80-233 Gdańsk, Poland

Abstract. The paper presents application of the gamma-absorption method to a two-phase liquid-gas flow investigation in a horizontal pipeline. The water-air mixture was examined by a set of two Am-241 radioactive sources and two NaI(Tl) scintillation probes. For analysis of the electrical signals obtained from detectors the cross-spectral density function (CSDF) was applied. Results of the gas phase average velocity measurements for CSDF were compared with results obtained by application of the classical cross-correlation function (CCF). It was found that the combined uncertainties of the gas-phase velocity in the presented experiments did not exceed 1.6% for CSDF method and 5.5% for CCF.

1 Introduction

Two-phase liquid-gas flows - when gas is transported by liquid - commonly occurs in industry, especially in mining, nuclear, pharmaceutical, petrochemical, energy as well as in environmental engineering. This type of flows are of particular importance in describing heat transfer and mass processes in a number of devices such as, reactors, distillation columns, heat exchangers, absorption and rectification columns.

Two-phase flows are very difficult for detailed the mathematical description. Therefore, experimental research and the development of the new methods for measuring such flows are of major importance. Two-phase flows can be studied using several methods, such as computer tomography, particle image velocimetry (PIV), magnetic resonance imaging (MRI), laser Doppler anemometry (LDA) or using Coriolis flowmeters, optical and nuclear techniques [1-11]. The nuclear methods use radiotracers or sealed gamma-ray sources and scintillation detectors for the flow analysis. The gamma-absorption method is particularly convenient for continuous investigation of two-phase flows. The absorption measurement is relatively safe, non-invasive, and allows determining the velocity as well as concentration of a dispersed phase using the same set-up. Most important in such measurements is the analysis of the electrical signals received from the probes. Due to the stochastic nature of obtained signals, statistical methods have to be used for this purpose. The measured time delay of signals is applied to determine the averaged velocity of the dispersed phase and other flow parameters.

The most known methods of time delay estimation for stationary random signals is the cross-correlation function (CCF) [8, 12]. Other methods (in time domain) include differential and combined methods [13, 14], Hilbert Transform-based CCF [15, 16] or methods based on the conditional averaging of signals [17, 18]. In the frequency domain the phase of cross-spectral density function (CSDF) can be used [19, 20]. This paper presents an application of the phase method to evaluate the mean velocity of gas phase in the two phase gas-water mixture in a horizontal pipe.

The paper consists of five sections. In Section 2 the basis of gamma-absorption method for liquid-gas flow evaluation and the laboratory set-up are described. Section 3 presents principles of time delay estimation applying phase of CSDF. In section 4 the comparison between results of average velocity measurements obtained using phase method and CCF are presented. The last section contains the final conclusions.

2 Gamma-absorption method and laboratory set-up

The gamma absorption technique is based on exponential decay of a gamma radiation beam in function of composition and geometry of the absorbent [9]. The changes of the intensity of radiation are detected by the scintillation probes and converted into electrical impulses [7].

The typical gamma-absorption set for two-phase liquid-gas flow measurement is presented in Fig. 1. Two sealed radioactive sources emit gamma radiation beams shaped by collimators (1). Photons pass through the pipe

* Corresponding author: rohan@prz.edu.pl

and two-phase flows mixture (4). Two scintillation detectors (2) with collimators (3) are placed on the other in referent to the suorces side of the pipe at $L = 97$ mm distance between them. In the presented experiment two sets consisting of linear Am-241 sources with an activity of 100 mCi and probes with NaI(Tl) scintillation detectors were used. At the outputs of two scintillation probes, count signals $I_x(t)$ and $I_y(t)$ are registered for the first and second probe respectively [21].

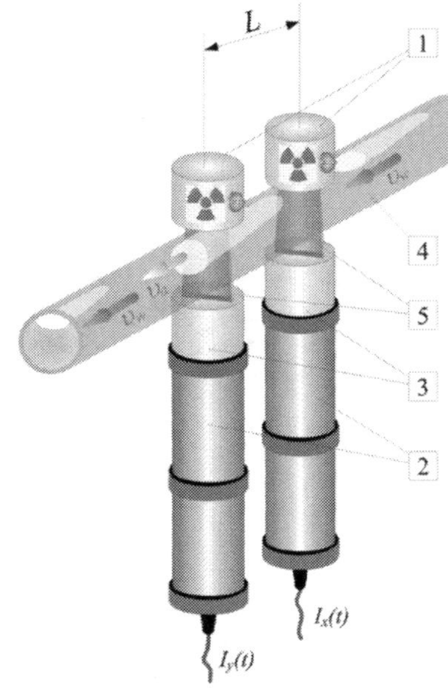

Fig. 1. The gamma-absorption measuring set: 1 – gamma ray sources with collimators, 2 – scintillation probes, 3 – detectors' collimators, 4 – pipeline, 5 – the main beam of gamma rays, υ_a – velocity of air, υ_W – velocity of water. [21]

The gamma-absorption equipment described above was applied in the experimental measurement set-up, performed in Industrial Radiometry Laboratory of the AGH University of Science and Technology in Krakow, Poland. Measurement data waere collected during experiments carried out on a laboratory hydraulic installation with a horizontal plexiglas pipe of 4.5 m length and inner diameter of 30 mm. A detailed description of experimental set-up can be found in [16, 18]. Figure 2 shows the general view of the measurement section.

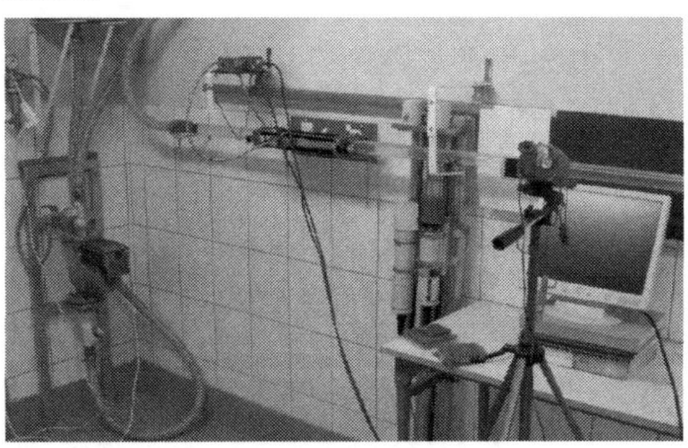

Fig. 2. General view of the experimental set-up.

The data acquisition system include a dedicated 8-channel counter - HSC 8000 (EC Electronics) connected to work station using an USB port. Voltage pulses I_x and I_y are counted within the sampling time $\Delta t = 1$ ms and create mutually delayed discrete stochastic signals x and y. An exemplary time records of signals obtained in the LIT0003 experiments are presented in Fig. 3.

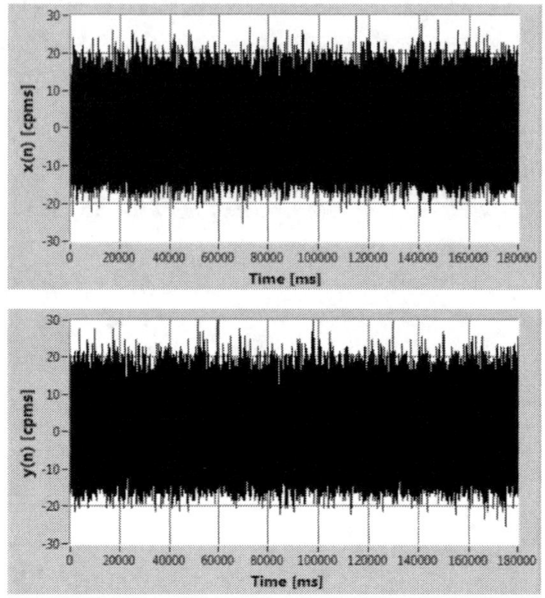

Fig. 3. Time records of signal $x(t)$ and $y(t)$ (after centering) obtained in the LIT0003 experiment.

3 Analysis of signals

The analyses of signals from scintillation probes in the time or frequency domain allow determining of the transportation time delay τ_0 and then the mean velocity υ_a of gas phase transportation:

$$\upsilon_a = L / \tau_0 \qquad (1)$$

The most known classical method of time delay estimation of random signals is based on the CCF and defined as follow [12]:

$$R_{xy}(\tau) = \lim_{T \to \infty} \frac{1}{T} \int_0^T x(t) y(t+\tau) dt \qquad (2)$$

where T is the averaging time and τ is the time delay.

The τ_0 transportation time delay is determined by finding the location for the CCF maximum.

Time delay estimation in the frequency domain is possible by use of the phase of CSDF. In practice only one-sided CSDF for the frequency range $0 < f < \infty$ can be designated. This is a complex quantity defined as follows [19]:

$$\underline{G}_{xy}(f) = 2 \int_{-\infty}^{\infty} R_{xy}(\tau) \cdot e^{-j2\pi f \tau} d\tau \qquad (3)$$

Normalized CSDF is referred to as coherence function $\gamma_{xy}(f)$. In practice, the real-valued squared coherence function is used:

$$\gamma_{xy}^2(f) = \frac{\left|\underline{G}_{xy}(f)\right|^2}{G_{xx}(f) \cdot G_{yy}(f)} \quad (4)$$

where $G_{xx}(f)$ and $G_{yy}(f)$ are autospectral density functions of signals $x(t)$ and $y(t)$ respectively.
The squared coherence $\gamma^2_{xy}(f)$ satisfies the condition: $0 \le \gamma^2_{xy}(f) \le 1$.

The $\Phi_{xy}(f)$ phase of the CSDF is a linear function of frequency f:

$$\Phi_{xy}(f) = arctg\left\{\mathrm{Im}\left[\underline{G}_{xy}(f)\right] \Big/ \mathrm{Re}\left[\underline{G}_{xy}(f)\right]\right\} = 2\pi \cdot f \cdot \tau_0 \quad (5)$$

Currently most often the discrete Fourier transform (DFT) is applied to calculate the spectral densities directly for the discrete signals $x(n)$ and $y(n)$, where $n = t / \Delta t$. In analysis of stochastic signals an averaging of the obtained spectral density estimators is necessary. The most popular Welch procedure requires the sets of signal samples to be divided into N_d windowing segments, each of the N samples. The N_d and N values should be big enough to limit the statistical fluctuations of CSDF estimation [19]. DFT gives the k useful frequencies f_k $(k = 0\ldots(N - 1) / 2)$, with the spectral resolution $\Delta f = 1 / N\Delta t$. Consequently the transportation time delay τ_0 is determined from the smoothed $\widetilde{\Phi}_{xy}(f)$ estimator after applying the procedure of the phase unwrapping [22]. Figure 4 presents the plot of the $\widetilde{\Phi}_{xy}(f) = f(f)$ function determined in the LIT0003 experiment ($N_d = 533$, $N = 500$, overlapping 35% and Hanning window).

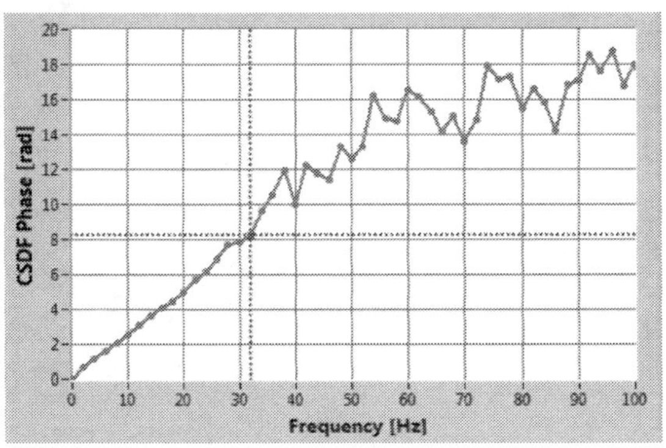

Fig. 4. Plot of the smoothed CSDF phase.

The function $\widetilde{\Phi}_{xy}(f)$ is nominally linear and passes through the coordinate origin, so the $\hat{\tau}_0$ time delay estimator can be determined by use slope a of the regression line fitted to the CSDF phase:

$$\hat{\tau}_0 = \frac{1}{2\pi} \cdot \hat{a} = \frac{1}{2\pi}\left[\sum_{k=1}^{m} f_k \cdot \widetilde{\Phi}_{xy}(f_k) \Big/ \sum_{k=1}^{m} f_k^2\right] \quad (6)$$

where m is the number of the pairs of both phase and frequency values used for the regression analysis [20, 23].

The selection of the appropriate frequency linearization range is vital due to fluctuations of the CSDF phase above the cut-off frequency band of the recorded signals. In this case the cut-off frequency can be determined based on the significant value of coherence function [20]. Figure 5 presents the graph of the smoothed coherence $\widetilde{\gamma}_{xy}^2(f)$ obtained in the LIT0003 experiment.

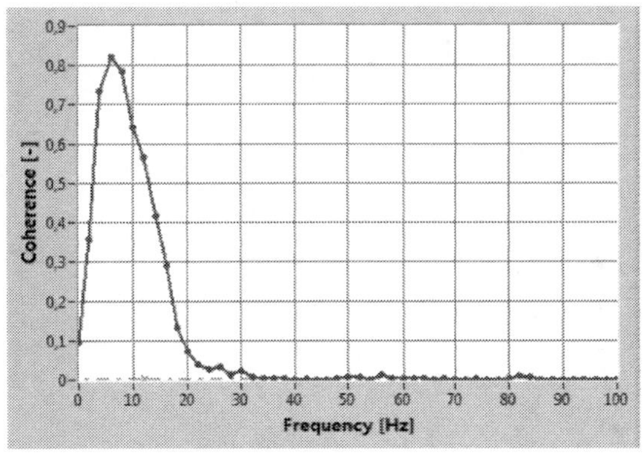

Fig. 5. Smoothed coherence function.

The dotted red line in the Fig. 5 marks the border of one-sided confidence interval E_α fixed for the significance level $\alpha = 0.05$ from the relation [24]:

$$E_\alpha = (1-\alpha)^{1/(N_d - 1)} \quad (7)$$

Based on the E_α value, the cut-off frequency and m value for $\hat{\tau}_0$ time delay estimation are determined.

4 Exemplary results

The plot of the CSDF phase obtained in the LIT0003 experiment for the selected frequency range from 0 to 32 Hz is shown in Fig. 6. The solid line represents the fitted regression line determined to find out the transportation time delay estimator from eq. (6).

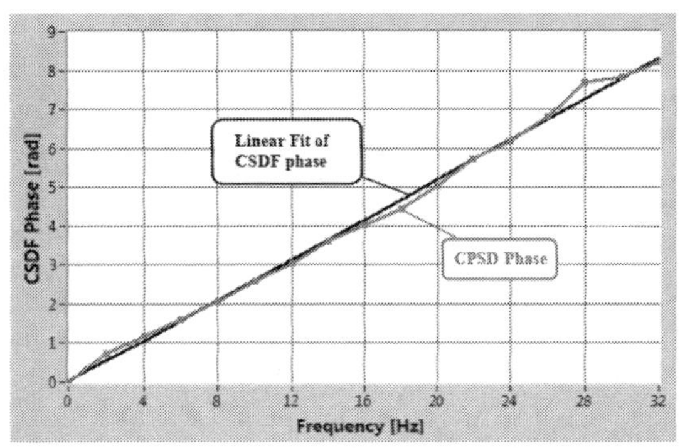

Fig. 6. A linear fit of the smoothed CSDF phase determined in the LIT0003 experiment.

The standard uncertainty $u(\hat{\tau}_0)$ calculated using CSDF phase is given by [20, 23]:

$$u(\hat{\tau}_0)=\left\{\frac{1}{4\pi^2(m-1)\sum_{k=1}^{m}f_k^2}\left[\sum_{k=1}^{m}\widetilde{\Phi}_{xyk}^2(f_k)-\frac{\left[\sum_{k=1}^{m}f_k\cdot\widetilde{\Phi}_{xyk}(f_k)\right]^2}{\sum_{k=1}^{m}f_k^2}\right]\right\}^{1/2} \quad (8)$$

Assuming negligible small uncertainties of the data acquisition system, the combined standard uncertainty $u_c(\upsilon_a)$ of the gas phase average velocity depends on an inaccuracy of uncorrelated L and $\hat{\tau}_0$ determination [25]:

$$u_c(\upsilon_a)=\left[\left(\frac{\partial\upsilon_a}{\partial L}\right)^2\cdot u^2(L)+\left(\frac{\partial\upsilon_a}{\partial\hat{\tau}_0}\right)^2\cdot u^2(\hat{\tau}_0)\right]^{1/2} \quad (9)$$

where $u(L)$ is the standard uncertainty of the distance between detectors.

In table 1 the results of the average velocity υ_a and its combined uncertainties $u_c(\upsilon_a)$ obtained in the selected LIT experiments by the use of CSDF phase method are presented. For comparison, the results obtained under analysis of the CCF distribution are also given.

Table 1. Exemplary results of experiments.

Method	Average velocity υ_a [m/s]	Combined uncertainty $u_c(\upsilon_a)$ [m/s]	Experiment
CSDF phase	1,781	0,016	LIT0002
CCF	1,798	0,093	
CSDF phase	2,314	0,015	LIT0003
CCF	2,295	0,125	
CSDF phase	2,679	0,044	LIT0004
CCF	2,712	0,144	

5 Conclusions

This paper presents an application of the CSDF phase method to radioisotope measurement of the velocity of gas phase in water-air mixture transportation by horizontal pipeline. It was found that the combined uncertainties of the average velocity of air in the exemplary LIT experiments did not exceed 1.6% for the CSDF phase method and 5.5% for the CCF method respectively. Moreover, in the cross-correlation method, the filtration of the recorded signals should be applied in order to reduce noises [16]. On contrary to CCF method, the filtration of signals in the phase method is not necessary.

Based on the obtained results, it can be stated that the CSDF phase method may be successfully applied in investigation of liquid-gas flow in horizontal pipeline in connection with gamma-ray absorption method.

References

1. M. Bottin, et al, Int. J. Multiphas. Flow **60**, 161-179 (2014)
2. T. Xue, et al, Flow Meas. Instrum. **27**, 29–36 (2012)
3. G. Falcone, G.F. Hewitt, C. Alimonti, *Multiphase flow metering: principles and applications* (Elsevier, Amsterdam 2009)
4. R.L. Powell, Phys. Fluids. **20**, 040605 (2008)
5. I. Ismail, et al, Flow Meas. Instrum. **16**, 145–155 (2005)
6. G.A. Johansen, P. Jackson, *Radioisotope gauges for industrial process measurements* (John Wiley, New York 2004)
7. Z. Yu, B. Qincheng, H. Richa, Appl. Therm. Eng. **60,** 398-410 (2013)
8. S.L. Soo (ed.), *Instrumentation for fluid-particle flow* (Noyes Publications, New Jersey, 1999)
9. W.A.S. Kumara, B.M. Halvorsen, M.C. Melaaen, Int. J. Multiphas. Flow **36,** 467-480 (2010)
10. V. Mosorov, Meas. Sci. Technol. **17,** 753-760 (2006)
11. P. Vlasak, et al, J. Hydrol. Hydrom. **62,** 241-247 (2014)
12. M.S. Beck, A. Plaskowski, *Cross correlation flowmeters - their design and application.* (Adam Hilger, Bristol, 1987)
13. G. Jacovitti, G. Scarano, IEEE Trans. Sign. Proc **41**, 525-533 (1993)
14. R. Hanus, et al., Math. Probl. Eng. **2014,** 475735 (2014)
15. J.S. Bendat, *The Hilbert Transform and applications to correlation measurements* Brüel&Kjær BT0008-11 (Naerum, Denmark, 1985)
16. R. Hanus, Int. J. Multiphas. Flow **72,** 210–217 (2015)
17. A. Kowalczyk, *Measurement applications of conditional signal averaging* (O.W. Politechniki Rzeszowskiej, Rzeszów, 2015) (in Polish)
18. R. Hanus, et al, *EPJ WoC* **92,** 02021 (2015)
19. J.S. Bendat, A.G. Piersol, *Random data - analysis and measurement procedures* (John Wiley, New York, 2010)
20. R. Hanus, et al, *EPJ WoC* **92,** 02020 (2015)
21. R. Hanus, et al, *EPJ WoC* **143**, 02033 (2017)
22. R.G. Lyons, *Understanding digital signal processing* (Prentice Hall, New York, 2011)
23. A.G. Piersol, IEEE Trans. Acoust. Speech Sign Proc. **29**, 471-477 (1981)
24. D.M. Halliday, et al, Prog. Biophys. Mol. Bio. **64**, 237-278 (1995)
25. *Guide to the expression of uncertainly in measurement* (International Organisation for Standardisation, 1995)

Effect of flow velocity on the process of air-steam condensation in a vertical tube condenser

Jan Havlík[1,*], *Tomáš* Dlouhý[1]

[1]Czech Technical University in Prague, Faculty of Mechanical Engineering, Department of Energy Engineering, Technicka 4, 16607 Prague 6, Czech Republic

Abstract. This article describes the influence of flow velocity on the condensation process in a vertical tube. For the case of condensation in a vertical tube condenser, both the pure steam condensation process and the air-steam mixture condensation process were theoretically and experimentally analyzed. The influence of steam flow velocity on the value of the heat transfer coefficient during the condensation process was evaluated. For the condensation of pure steam, the influence of flow velocity on the value of the heat transfer coefficient begins to be seen at higher speeds, conversely, this effect is negligible at low values of steam velocity. On the other hand, for the air-steam mixture condensation, the influence of flow velocity must always be taken into account. The flow velocity affects the water vapor diffusion process through non-condensing air. The presence of air significantly reduces the value of the heat transfer coefficient. This drop in the heat transfer coefficient is significant at low velocities; on the contrary, the decrease is relatively small at high values of the velocity.

1 Introduction

Condensers are used in a variety of chemical, petroleum, processing and power facilities for distillation, refrigeration and power generation. Most condensers used in the chemical process industries are water cooled shell-and-tube exchangers and air cooled tube or platen exchangers. Shell-and-tube condensers, which are used for the condensation of process vapors, are classified according to orientation (horizontal and vertical) and placement of condensing vapor (shell-side and tube-side) [1]. For industrial waste vapor, which may contain small solid particles (e.g. vapors from drying processes and other material processes [2]), a vertical tube-side configuration is used. These solid particles, which can stick to the tube wall, are spontaneously carried away from the tubes by the condensate blowing out. This design is very flexible and is suitable where a particularly low pressure drop is specified for the condensing fluid.

A vertical tube-side condenser consists of a shell with fixed tubesheets. The lower head is over-sized to accumulate the condensate and must be equipped with a vent for removing non-condensable gases. The condensate flows down the tubes in the form of an annular film of liquid, thereby maintaining good contact with both the cooling surface and the remaining vapor. The disadvantages are that the coolant, which is often more prone to fouling, is on the shell side, and the use of finned tubes is precluded.

Calculation of the overall heat transfer coefficient (HTC) necessary for the design of the condenser heat transfer area is well described in the literature, however for limited operating conditions only. The Nusselt condensation model commonly recommended for the calculation of condensing side HTC is derived for conditions which need not be satisfied in real operation [3], [4]. In many process industry applications, the vapor to be condensed is multicomponent in nature. Such cases include those where the single condensing vapor is mixed with a non-condensable gas. The presence of such gases can have profound effects on condensation greatly reducing the effective HTC.

2 Theory of condensation in a vertical tube

The basic heat-transfer model for film condensation was first derived by Nusselt to describe how a pure-component saturated vapor condenses on a vertical wall, forming a thin film of condensate that flows downward due to gravity [3], [4]. Then the following equation for the calculation of the HTC h_c was introduced

$$h_c = \frac{2\sqrt{2}}{3}\left[\frac{\rho_L g(\rho_L - \rho_p)h'_{fg}k_L^3}{\mu_L \Delta T_{sat} L}\right]^{1/4} \quad (1)$$

where ρ_L is the density of the condensate, ρ_p is the density of the water vapor, h'_{fg} is the latent heat of condensation, k_L is the thermal conductivity of the condensate, μ_L is the dynamic viscosity of the condensate, ΔT_{sat} is the difference between the saturation

* Corresponding author: jan.havlik@fs.cvut.cz

temperature and the wall temperature, and L is the wall length.

The model was derived with the assumptions of a laminar flow in the condensate film, a linear temperature profile across the condensate film (this assumption is reasonable for a very thin film), a negligible shear stress at the vapor-liquid interface and a low fluid velocity in the film. The operating conditions of condensers may be different from the assumptions adopted in the basic Nusselt theory. Deviations in the operation of real devices may be caused for example by the influence of condensing vapor velocity or the presence of non-condensable air in the vapor on the condensation process.

2.1 Effect of flow velocity

During condensation inside vertical tubes, steam works on the film surface by shear force. When the steam flows downward, film flows accelerate and the HTC increases slightly. On the contrary, moving upward slows down the film flow [4], [5].

The velocity in the liquid film for gravity controlled condensation (the basic Nusselt model) can be written as

$$u(y) = \frac{g(\rho_L - \rho_v)}{\mu_L}\left[y\delta - \frac{1}{2}y^2\right] \quad (2)$$

where u is the liquid velocity within the film at a distance y from the wall. At higher steam velocities, the interfacial shear becomes significant compared to the gravitational force on the liquid phase. For a partial velocity of the film given by shear stress, we can write

$$u(y) = \frac{\tau_g}{\mu_L}y \quad (3)$$

where τ_g is the interfacial shear stress. The resulting equation for the velocity combines Eq. (2) and Eq. (3) in the form

$$u(y) = \frac{g(\rho_L - \rho_v)}{\mu_L}\left[y\delta - \frac{1}{2}y^2\right] + \frac{\tau_g}{\mu_L}y \quad (4)$$

τ_g can be determined according to the following relation, where u_g is the velocity of the flowing gas and f is the coefficient of friction losses, its determination is described e.g. in [1], [2], [5].

$$\tau_g = \frac{1}{2}\rho {u_g}^2 f \quad (5)$$

A change in the velocity profile of the condensate film influences the calculation of the condensation HTC. The calculation process is described in detail in the works of Kageyama [7], Rohsenow [8] and Blangetti [9].

2.2 Condensation of steam mixture

The problem of steam mixture condensation is described in detail in [9]. If steam condenses in a mixture with inert gases, there is a difference in the concentrations of the steam in the bulk gas mixture and at the condensation phase interface. Steam concentrations decrease along the length of the tube equally with corresponding steam partial pressures p_∞ resp. p_i and temperatures T_∞ resp. T_i (see Figure 1). The molecules of steam diffuse through an inert gas towards the vapor-liquid interface and mass transfer occurs [1], [4]. On the contrary in the case of pure steam condensation, the temperatures T_∞ and T_i are identical and constant during the process [10].

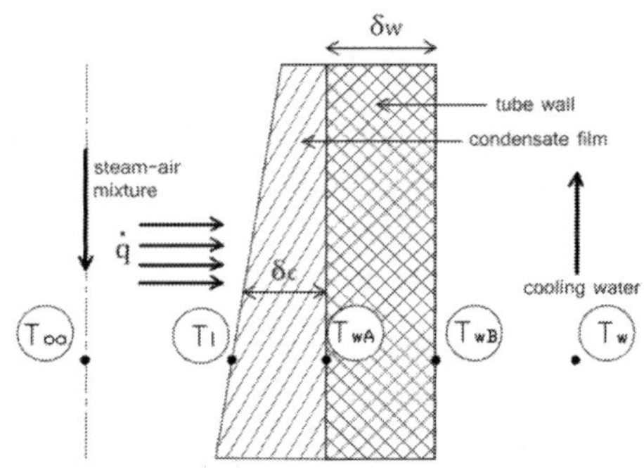

Fig. 1. Scheme of the condensation process [10]

To determine the HTC value, it is necessary to calculate the condensation temperature at the phase interface T_i taking into account the mass transfer generated by the different partial pressures at the interface between the gas and the liquid phases p_i and the bulk gas mixture p_∞ [1], [4], [5]. In accordance with the analogy between the mass transfer and the heat transfer, the condensation HTC is evaluated according to the Nusselt model combined with the application of the suitable criterion equation for the mass transfer [10].

Condensing molar flux $\dot{n}_v$ describes the balance of the process

$$\dot{n}_v = \beta C_T ln\left[\frac{1 - \bar{y}_{v,\infty}}{1 - \bar{y}_{v,i}}\right] \quad (6)$$

where β is the mass transfer coefficient, C_T is the total molar concentration, $\bar{y}_{v,\infty}$ is the steam mole fraction in the bulk gas mixture, respective at the liquid-vapor interface $\bar{y}_{v,i}$. The mass transfer coefficient β is determined according to the suitable criterial equation analogically with the heat transfer. The partial pressure $p_{v,i}$ at the liquid-vapor interface is given as

$$p_{v,i} = \bar{y}_{v,i}p \quad (7)$$

and the temperature at the liquid-vapor interface T_i as

$$T_i = T_{sat}(p_{v,i}) \quad (8)$$

3 Experiment

In order to verify of the described theory experiments were carried out on a vertical shell-and-tube heat exchanger in which the condensing water vapor flows downwards inside vertical tubes and the cooling water flows in a counter current in the outer shell.

https://doi.org/10.1051/epjconf/201818002033

3.1 Experimental set-up

The condenser is shown in Fig. 1. The tube bundle is formed by 49 tubes 865 mm in length, 28 mm in external diameter and 24 mm in internal diameter. The tubes are arranged in staggered arrays with a triangular tube pitch of 35 mm. The cross-section of the shell is rectangular in shape, 223 mm by 270 mm in size. Seven segmental baffles (223 x 230 mm) are used in the shell section. The tubes are made from stainless steel 1.4301 (AISI 304).

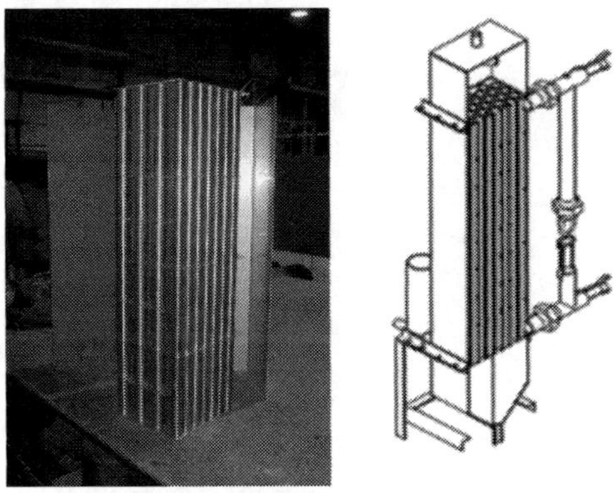

Fig. 1. Vertical tube condenser

The testing loop is shown in Fig. 2. Steam is produced in a steam generator. Before the steam enters the vertical tube condenser, its parameters are reduced to the required values. Hot water recirculation makes it possible to set the cooling water temperature and the cooling water flow rate in the condenser to the required values.

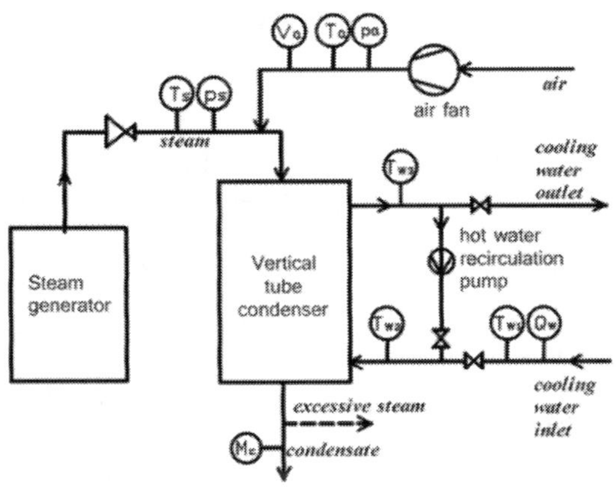

Fig. 2. Scheme of the testing loop

The measured parameters are the inlet cooling water temperature T_{w1}, the temperature of the water after recirculation T_{w2}, the outlet cooling water temperature T_{w3}, the cooling water flow rate Q_w, the inlet steam pressure p_s , the inlet steam temperature T_s and the amount of steam condensate M_c.

3.2 Determination of the condensation HTC

Calculation of the HTC is based on the heat balance of the condenser [1]. The total heat power Q is given by the equation

$$Q = M_w \cdot c_w \cdot \left(T_{w,out} - T_{w,in}\right), \tag{9}$$

where M_w is the cooling water flow, c_w is the specific heat capacity of water, $T_{w,out}$ is the outlet cooling water temperature, and $T_{w,in}$ is the inlet cooling water temperature. The overall HTC U is given by the equation

$$U = \frac{Q}{A \cdot \Delta T_{log}}, \tag{10}$$

where A is the heat transfer surface and ΔT_{log} is the logarithmic mean temperature difference. It is possible to calculate the tube-side HTC (vapor condensation) h_c from the term for determining the overall HTC

$$U = \frac{{}^{1}/{}_{D_o}}{\frac{1}{d_i \cdot h_c} + \frac{1}{2k} ln\left(\frac{D_o}{d_i}\right) + \frac{1}{D_o \cdot h_w}} \tag{11}$$

where d_i is the inside diameter of the tubes, k is the thermal conductivity of the tubes, h_w is the tube-side HTC (cooling water side), and D_o is the outside diameter of the tubes.

The value of the cooling water HTC h_w was experimentally evaluated [11], [12]. The criterion equation for calculation of the HTC in the cooling water side of the experimental condenser was determined by the Wilson plot method [13], [14]. The experimentally derived equation has the following form [11] as

$$Nu = 0.270 Re^{0.56} Pr^{0.33} \left(\frac{\mu}{\mu_s}\right)^{0.14} \tag{12}$$

afterwards h_w is given by the equation

$$h_w = \frac{Nu \cdot k_w}{D} \tag{13}$$

The process for the experimental determination of the heat transfer coefficient and the proposal of the criterion equation of shell-and-tube condensers using the Wilson plot method is described in detail in [11].

4 Results

Experiments were carried for the case of pure steam condensation and the case of air-steam mixture condensation. During the condensation process the gas velocity decreases together with a decrease of the volume flow of the condensing gas. Therefore, the range of small values of the velocity was analyzed.

4.1 Pure steam

After determination of h_w from Eq. (13), the value of h_c for various gas velocities is provided from Eq. (11). It is necessary to maintain constant conditions on the cooling water side. The results for 3 different levels of cooling conditions are shown in Fig. 4.

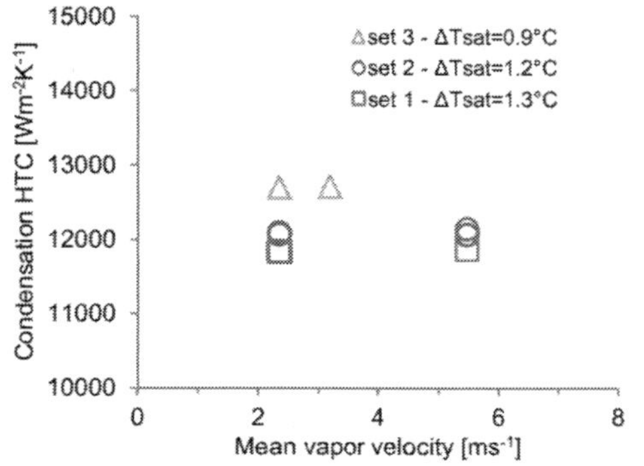

Fig. 4. Experimentally given dependencies of the HTC on pure steam velocity

The values of condensation HTC reach similar values for various velocities in all 3 analyzed cases. The effect of steam velocity is negligible at low velocity values. Condensation HTC is only sensitive to the difference ΔT_{sat} (Eq. 1) between the saturation temperature T_i and the wall temperature T_{wA} (see Fig. 1).

Based on the Blangetti model, the dependence of the HTC on the condensing steam velocity was created. In Fig. 5, an increase of the condensation HTC is shown to be caused by an increase of the steam velocity.

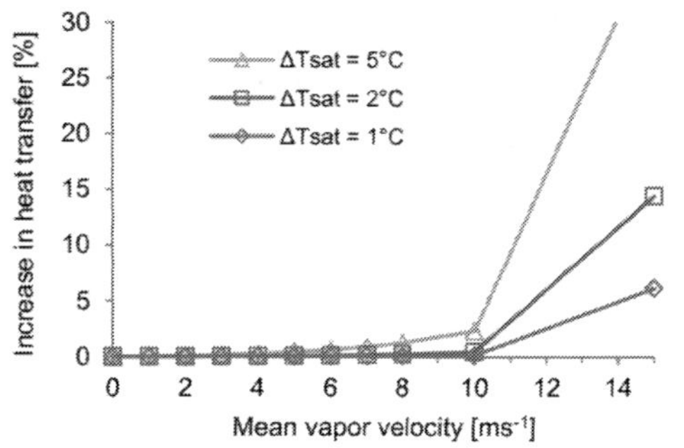

Fig. 5. Theoretically calculated effect of pure steam velocity on the condensation HTC

It can be seen that the influence of velocity is also dependent on the difference between the temperature of the condensation on the film surface and the wall temperature. In general, the rate for a temperature difference of less than 5 °C begins to influence the magnitude of the heat transfer coefficient at steam velocities above 5 ms^{-1}. At lower speeds it is not necessary to take this influence into account.

4.2 Air-steam mixture

The condensation of the air-steam mixture is a complex process where the change of any parameter (e.g., concentration, cooling water temperature, intensity of diffusion of steam particles in noncondensable air) influences its heat equilibrium. Therefore, each measured state has to be individually balanced. Experiments were carried out for two air concentrations in a mixture with steam – 1.1% (see Fig. 6) and 2.3% (see Fig. 7).

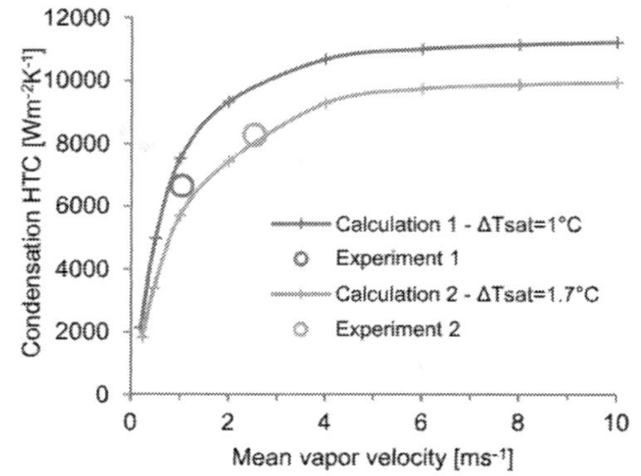

Fig. 6. Effect of flow velocity on the HTC for vapor with 1.1% of air

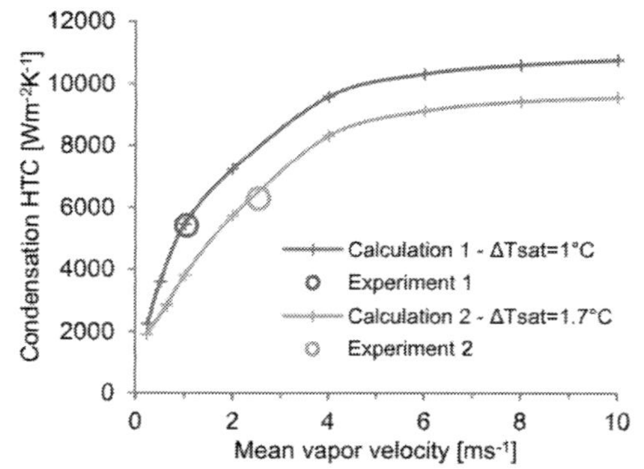

Fig.7. Effect of flow velocity on the HTC for vapor with 2.3% of air

For each analyzed state, the heat equilibrium of the tested condenser is different. It is evident that the heat transfer affects the thickness of the condensate film and the corresponding temperature difference ΔT_{sat} between the condensation and wall side of the film.

Experimental and theoretical results are in good agreement. For air-steam mixture condensation, a significant dependence of the condensation HTC value on the velocity begins at very low velocities.

5 Discussion

Comparison of the results for all analyzed cases is shown in Fig. 8. For the range of low velocities, the values of the HTC for pure steam condensation are not sensitive to steam velocity while for air-steam mixture condensation the values of the HTC increase with an increase of the velocity. The presence of the air significantly reduces the value of the HTC for low velocities; on the contrary, the decrease is relatively small at high values of the velocity. However in both cases, the HTC is influenced by the temperature difference ΔT_{sat} in the condensate film. The difference

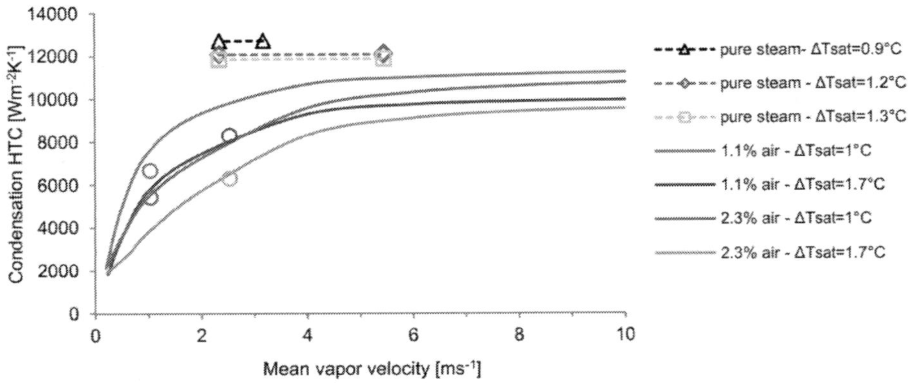

Fig. 8. Total comparison of the effect of gas velocity on the HTC

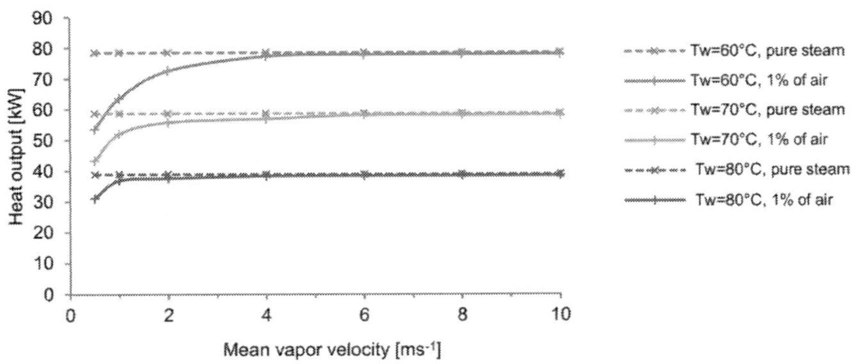

Fig. 9. Comparison of the heat output for various coolling water temperatures

mainly arises from the cooling intensity, respective to the thickness of the condensate film.

The effect of vapor velocity on the heat output of the tested condenser is shown in Fig. 9. For the experimental condenser (Fig. 1), the heat output is theoretically analyzed for various cooling water temperatures (60, 70 and 80 °C) for the case of the condensation of pure steam and 1% air-steam mixture. The heat output increases with a lower temperature of the cooling water. The presence of the air significantly reduces the value of the heat output as well as the HTC for low velocities; on the contrary, the decrease is relatively small at high velocity values.

6 Conclusion

The influence of steam flow velocity on the value of the heat transfer coefficient during the condensation process was theoretically and experimentally investigated.

For the condensation of pure steam, the influence of the flow velocity on the value of the HTC begins to be observed at higher velocity values, conversely, this effect is negligible at low velocity values. On the other hand, when the steam-air mixture is condensed, the influence of flow velocity must always be taken into account. The flow velocity affects the water vapor diffusion process through non-condensing air. The presence of the air significantly reduces the value of the heat transfer coefficient. This drop in the heat transfer coefficient is significant at low velocities; on the contrary, the decrease is relatively small at high velocity values.

This work has been supported by Technology Agency of the Czech Republic, project No. TJ01000192.

Nomenclature

A	heat transfer area (m^2)
c	specific heat capacity ($Jkg^{-1}K^{-1}$)
d	diameter (m)
D	characteristic diameter (m)
g	gravitational acceleration (ms^{-2})
h	heat transfer coefficient ($Wm^{-2}K^{-1}$)
h_{fg}	latent heat of condensation ($kJkg^{-1}$)
k	thermal conductivity ($Wm^{-1}K^{-1}$)
L	tube length (m)
M	mass flow rate (kgs^{-1})
$\dot{n}$	molar flux ($kmolm^{-2}s^{-1}$)
Nu	Nusselt number (-)
Q	Rate of heat transfer (W)
T	temperature (K)
ΔT_{log}	logarithmic mean temperature difference (K)
U	overall heat transfer coefficient ($Wm^{-2}K^{-1}$)
u	velocity (ms^{-1})

y mole fraction (-)
β mass transfer coefficient (ms^{-1})
μ dynamic viscosity (Pas)
ν kinematic viscosity (m^2s^{-1})
τ shear stress (Pa)

Subscripts
f fluid
g gas
i inner
i interfacial
in inlet
L liquid
o outer
out outlet
ov overall
s wall
t tube
w water
∞ bulk gas mixture

References

1. G. F. Hewitt, G. L. Shires, T. R. Bott, *Process Heat Transfer* (Begell House, New York, 2000)
2. J. Havlik, T. Dlouhy, JCEJ **50**, 792-798 (2017)
3. F. P. Incropera et al., *Principles of Heat and Mass Transfer* (John Whiley, Singapore, 2013).
4. J. Sestak, F. Rieger, *Přenos hybnosti tepla a hmoty* (CTU, Prague, 2004) (in Czech)
5. H. D. Baehr, K. Stephan. *Heat and Mass Transfer* (Springer, Berlin, 1998)
6. F. P. Incropera, F. P. et al., *Fundamentals of heat and mass transfer* (J. Wiley, Hoboken, 2007)
7. T. Kageyama, P. F. Peterson, V. E. Schrock. Nuclear Engineering and Design **141**, 289-302 (1993)
8. W. M. Rosenhow, J. H. Webber, A. T. Ling, Trans. ASME **78**, 1637-1643 (1956)
9. F. Blangetti, R. Krebs, E. U. Schunder, Chemical Engineering Fundamnetals **1**, 20-63 (1982)
10. J. Havlik, T. Dlouhy, EPJ web of conferences **114**, 02037 (2016)
11. J. Havlik, T. Dlouhy, EPJ web of conferences **143**, 02035 (2017)
12. J. Havlik, T. Dlouhy, *Engineering Mechanics 2014*, 216-219 (2014)
13. J. Opatril, J. Havlik, O. Bartos, T. Dlouhy, Acta Polytechnica **56**, 367-372 (2016)
14. J. Fernández-Seara, F. J. Uhía, J. Sieres, A. Campo, Applied Thermal Engineering **27**, 2745 (2007)

Study on the impulsive pressure of tank oscillating by force towards multiple degrees of freedom

Shigeyuki Hibi[1,*]

[1] The National Defense Academy, Department of Mechanical Systems Engineering, 1-10-20 Hashirimizu, Yokosuka, Kanagawa 239-8686, Japan

Abstract. Impulsive loads should be excited under nonlinear phenomena with free surface fluctuating severely such as sloshing and slamming. Estimating impulsive loads properly are important to recent numerical simulations. But it is still difficult to rely on the results of simulations perfectly because of the nonlinearity of the phenomena. In order to develop the algorithm of numerical simulations experimental results of nonlinear phenomena are needed. In this study an apparatus which can oscillate a tank by force was introduced in order to investigate impulsive pressure on the wall of the tank. This apparatus can oscillate it simultaneously towards 3 degrees of freedom with each phase differences. The impulsive pressure under the various combinations of oscillation direction was examined and the specific phase differences to appear the largest peak values of pressure were identified. Experimental results were verified through FFT analysis and statistical methods.

1 Introduction

Impulsive loads[1,2] should be excited under nonlinear phenomena with free surface fluctuating severely such as sloshing and slamming in the field of naval architecture.

Many studies about not only the estimation of the impulsive loads but also elastic behavior of structures have been carried out along with the recent developments of the numerical simulation methods. However the number of experiments to verify those numerical techniques is not enough because of the nonlinearity of the phenomena.

In this study an apparatus which can oscillate a tank by force was introduced in order to investigate impulsive pressure on the wall of the tank. This apparatus can oscillate it simultaneously towards 3 degrees of freedom (up-down, left-right and rotation) with each phase differences. It could happen that larger impulsive pressure is excited under oscillating the tank simultaneously towards multiple degrees of freedom than under single direction oscilation. The author examined the impulsive pressure under the various combinations of oscillation direction and identified the specific phase differences to appear the largest peak values of pressure. Therefor the author showed larger impulsive pressure could appear under the oscillation towards multiple directions in comparison with the oscillation towards a single direction.

In the meantime the uncertainty of the result of experiments has been settled by the GUM[3] (Guide to the Expression of Uncertainty in Measurement) since 1993. Based on this a FFT analysis about measured pressure values was carried out and the confidence interval of them based on t-distribution was also determined. At last uncertainty analysis based on the GUM was executed and the reliability of the experiment was assured.

2 Experimental apparatus and measuring instruments

2.1 Specification of oscillation apparatus

The apparatus used in this study can oscillate a tank attached on the bracket harmonically and simultaneously towards 3 degrees of freedom with each phase differences (See Fig. 1 and Table 1).

* Corresponding author: hibiteru@nda.ac.jp

The size of tank is $400 \times 400 \times 100$ (mm) (Height $\times$ Width $\times$ Depth).

We define the symbols for the oscillation directions of 3 degrees of freedom as follows in Fig.2. Heave oscillation (Up – down) stands for Z and Sway oscillation (Left – right) stands for X and Roll oscillation (Rotation) stands for θ.

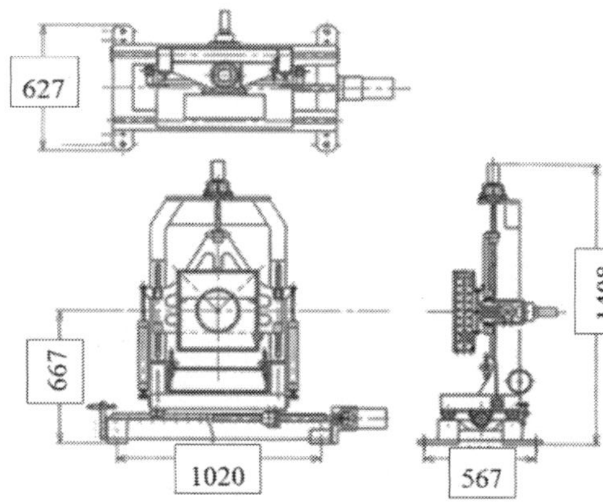

Fig. 1. Oscillation apparatus with 3 degrees of freedom.

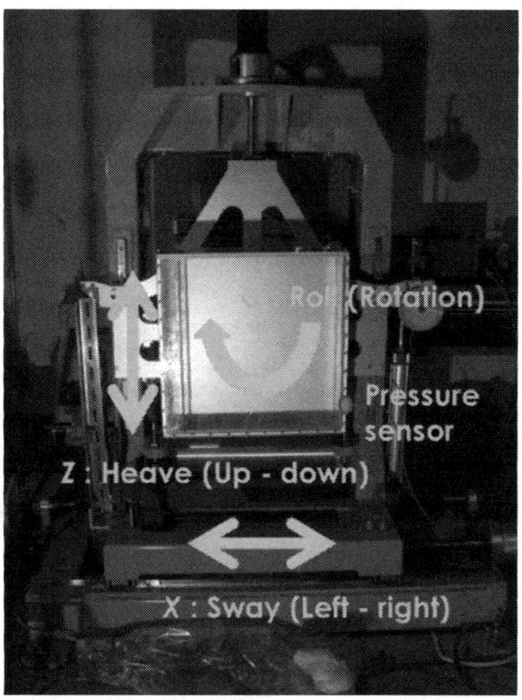

Fig. 2. Oscillation direction and pressure sensor position.

Table 1. Specification of the apparatus about oscillation.

Direction of oscillation	Maximum amplitude	Minimum period
Heave (Z) (Up-down)	± 50 mm	1.0 sec.
Sway (X) (Left-right)	± 50 mm	1.0 sec.
Roll (θ) (rotation)	± 45 deg.	1.0 sec.

2.2 Measuring instruments

(1) Potentiometer

In order to confirm the acculacy of the roll oscillation control of a servo motor a potentiometer is attached to measure a rotation angle at every moment.

(2) Laser displacement gauge

In order to confirm the acculacy of both heave and sway oscillation control of servo motors laser displacement gauges are placed to measure displacements at every moment.

(3) Pressure sensor

A pressure sensor is attached on the right side of the tank mounted on the apparatus. The position of the sensor is 40mm from the bottom. The pressure receiver is a round with a diameter of 8mm.

3 Analytical methods for measured data

3.1. FFT (Fast Fourier Transform) analysis

A FFT analysis are applied to the measured pressure data of time history in this study. Frequency components including in each data wave profile are discussed.

3.2 Determining the confidence intervals based on the *t*-distribution

Independent randam variables subject to the normal distribution $X_1, X_2, \dots X_N$ are assumed that mean values are μ and variances are σ^2 (usually unkonwn). Then sample mean $\overline{X}$ and unbiased variance u^2 are defined by eq. (1).

$$\overline{X} = \frac{1}{N}\sum_{i=1}^{N} X_i, \quad u^2 = \frac{1}{N-1}\sum_{i=1}^{N}(X_i - \overline{X})^2 \qquad (1)$$

Then randam variable T defined by eq. (2) is subject to Student's t – distribution of $N-1$ degree of freedom.

$$T = \left(\overline{X} - \mu\right) / \left(u / \sqrt{N}\right) \quad (2)$$

Probability density function of the t – distribution $f_{N-1}(t)$ is given by eq. (3).

$$f_{N-1}(t) = \frac{\Gamma\left(\frac{N}{2}\right)}{\sqrt{(N-1)\pi}\,\Gamma\left(\frac{N-1}{2}\right)}\left(1 + \frac{t^2}{N-1}\right)^{-N/2} \quad (3)$$

Here Γ is the gamma function.
Therefore in case of a sample of size N the confidence interval at a confidence level z % is given by eq. (4).

$$\bar{x} - t_{N-1}(z\%)\frac{u}{\sqrt{N}} \le \mu \le \bar{x} + t_{N-1}(z\%)\frac{u}{\sqrt{N}} \quad (4)$$

3.3 Uncertainty analysis

Uncertainty about measuring according to the GUM guidance is evaluated as follows. First of all we determine factors of dispersion in the measuring procedure.

There are 2 types of methods for the estimation of uncertainty in the GUM guidance.

One is called 'A' type of uncertainty. Its type is based on a statistical method. We estimate uncertainty of each factor using mean values and standard deviations in this type of uncertainty.

The other is called 'B' type of uncertainty. Its type is based on another method (e.g. technical information or some experience) without a statistical approach.

After we select the dispersion factors into 2 types of uncertainty, we combine the result to estimate the total uncertainty. the expanded uncertainty U is obtained by multiplying the combined standard uncertainty $u_c(y)$ by a coverage factor k (k = 2. This means the 95% confidence level).

4 Experiment and discussion

4.1 Investigating phase differences during simultaneous oscillation towards 2 degrees of freedom

The reliability of oscillation amplitude towards each degree of freedom is assured even with a period close to the natural period of the tank by previous study.

In this study we chose 3 pairs of 2 degrees of freedom from 3 one. Those will be subscribed as follows ($\theta - X$, $\theta - Z$, $X - Z$). The oscillation period is uniformly 1.08 (sec.) which is close to the natural period of the tank in case of the water depth is 60 mm (fixed). The oscillation amplitudes are 15 mm for X and Z and 20 degrees for θ.

We measured time historys of pressure on the wall of the tank repeatedly while changing the phase difference from 0 degree to 360 degrees. We identified the phase difference which cause the largest peak value of pressure for each oscillation pair ($\theta - X$, $\theta - Z$, $X - Z$). The results are shown in Table 2.

Table 2. Phase difference which cause the largest peak value.

Oscillation pair	Phase difference
$\theta - X$	15 degrees
$\theta - Z$	335 degrees
$X - Z$	220 degrees

Example results of time history of pressure for each oscillation pair are shown in Figs. 3(a), (b) and (c). In any cases it can be deduced that there is periodicity in the pressure fluctuation, but it is found that the peak values have variations.

4.2 FFT analysis for measured time history of pressure

FFT analysis was carried out with 1 ,12 and 19 period intervals from the measured data of time history of pressure for each oscillation pair. The results are shown in Figs. 4(a), (b) and (c). The waveform of the spectrum for 1 period

https://doi.org/10.1051/epjconf/201818002034

interval contains the ones for 12 and 19 period intervals. We can also deduce that although the obtained
peak values varies, the waveform itself seems periodic.

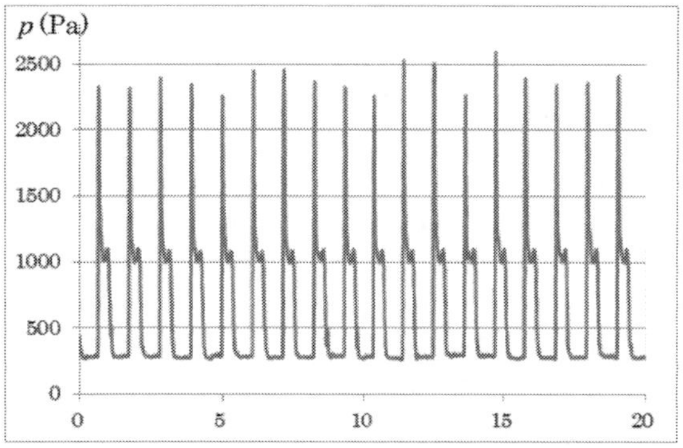

Fig. 3(a). Time history of pressure about $\theta - X$ oscillation.

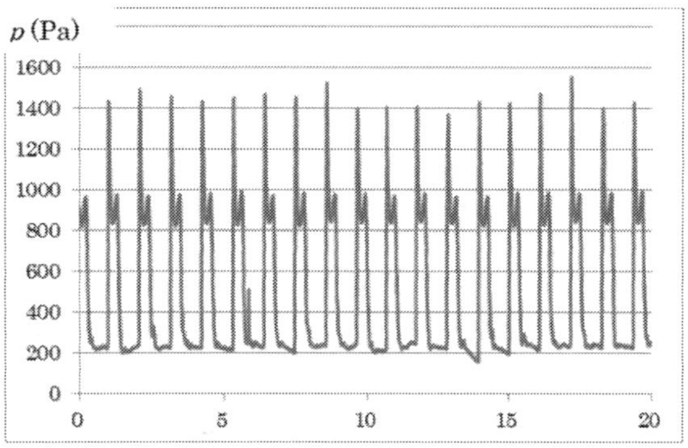

Fig. 3(b). Time history of pressure about $\theta - Z$ oscillation.

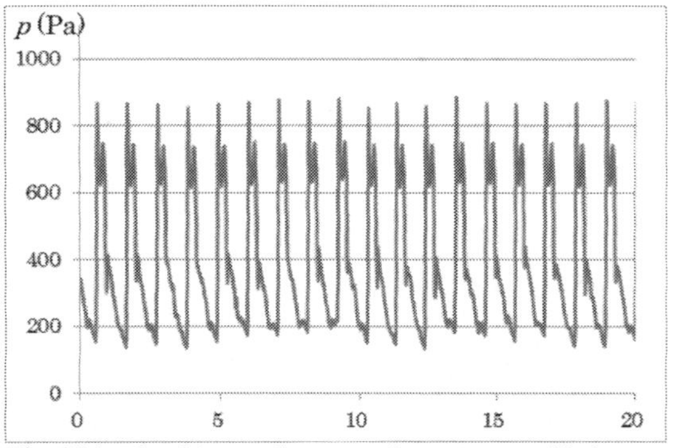

Fig. 3(c). Time history of pressure about $X - Z$ oscillation.

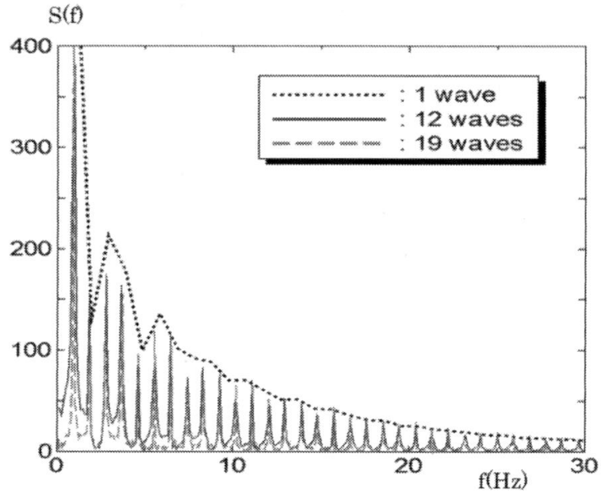

Fig. 4(a). FFT analysis about $\theta - X$ oscillation.

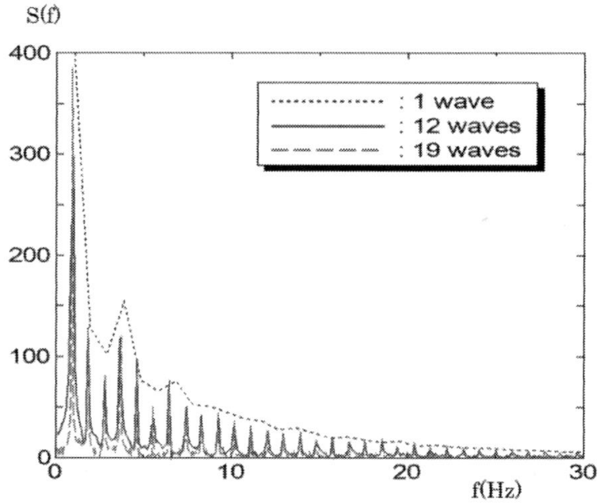

Fig. 4(b). FFT analysis about $\theta - Z$ oscillation.

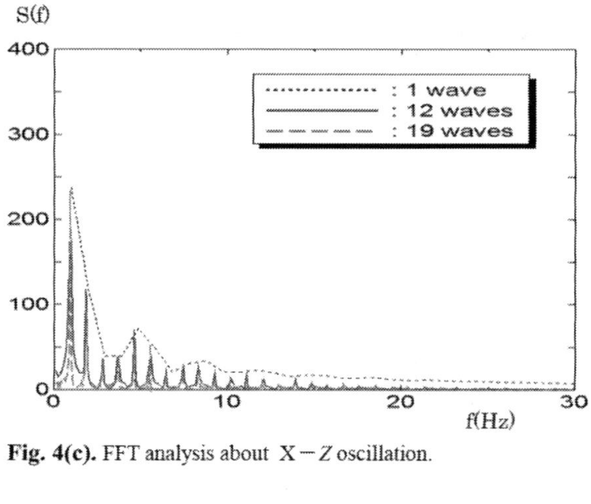

Fig. 4(c). FFT analysis about $X - Z$ oscillation.

4.3 Estimation of confidence intervals for the confidence level 95%

Experiments were repeated as one measurement for 20 seconds. Peak values were extracted from the time history of pressure for each oscillation pair. Next confidence intervals for the confidence level 95 % were determined. The results are shown in the Fig. 4.

Although the width of the confidence interval tends to decrease when the number of samples increases, it is considered that there is a significant width, which may indicate the nonlinearity of the present experimental data.

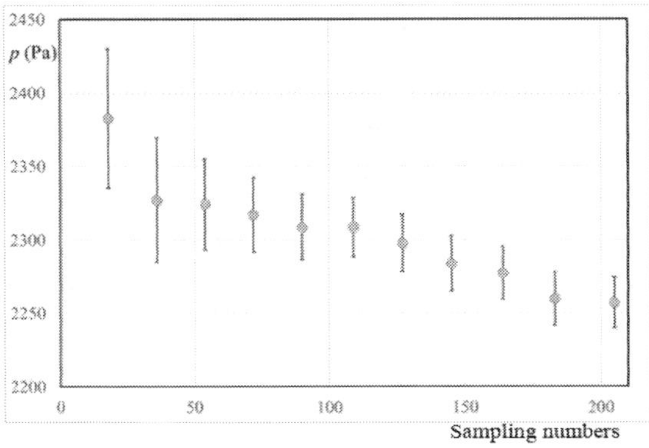

Fig. 5(a). Estimation of peak values at confidence level 95% about $\theta - X$ oscillation.

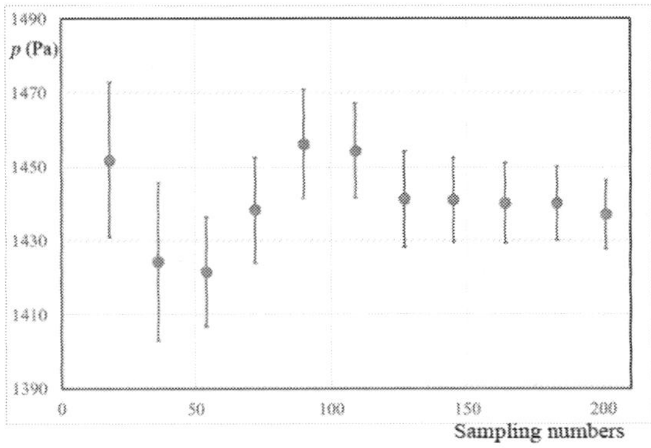

Fig. 5(b). Estimation of peak values at confidence level 95% about $\theta - Z$ oscillation.

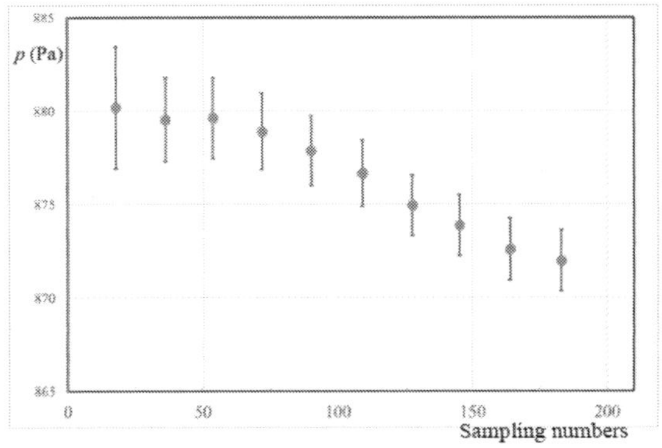

Fig. 5(c). Estimation of peak values at confidence level 95% about $X - Z$ oscillation.

4.4 Determining the expanded uncertainty

Expanded uncertainty was determined according to the budget table shown in Table 3. The function f used to estimate the uncertainty when measuring pressure values is in eq. (5).

$$f = \rho g a (V(d) - V(20)) \tag{5}$$

Where ρ is the density of water and g is the gravitational acceleration. 'a' is the coefficient for conversion from voltage to pressure. V(d) is the voltage value of an amplifier at the d mm water depth and V(20) is the voltage value of an amplifier at the 20 mm water depth.

The expanded uncertainty calculated from Table 3 is 101.9 (Pa). The number of iterations used here is 11 times. In most cases it can be seen that the peak value varies around a mean value at this interval.

5 Conclusion

In this study impulsive pressure on the tank wall were measured and some information were found about it as below.

- The fluctuation measured had some periodism but the peak values had variation.
- From the estimation of confidence intervals for the measured values at the confidence level 95% significant intervals were found to exist.
- Since characteristics of impulsive pressure for the oscillation directions of 2 degrees of freedom is known, confirmation of that for the oscillation directions of 3 degrees of freedom is a future task.

References

1. Hakan Akyildiza, Erdem U¨nal, Ocean Engineering **32**, 1503 (2005)
2. P. Temarel, W. Bai, A. Bruns, Q. Derbanne, D. Dessi,S. Dhavalikar, N. Fonseca, T. Fukasawa, X. Gu, A. Nestegård,A. Papanikolaou, J. Parunov, K.H. Song m, S. Wang, Ocean Engineering **119**, 274 (2016)
3. JCGM2008, Evaluation of measurement data - Guide to the expression of uncertainty in measurement, (2008)

Table 3. Budget table for the estimation of uncertainty

Factors of uncertainty : χ_i		Cause of uncertainty	Type of uncertainty	Shapes of probabilistic distribution	Mean value	Sensitivity factor $\frac{\partial f}{\partial x_i}$	Standard uncertainty		
a	Coefficient for conversion from voltage to pressure	Repeated experiment	A	Normal	159.0	3.888	0.8828		
$V(20)$	Voltage value at the 20mm water depth	Repeated experiment	A	Normal	2.815	1558	0.03199		
Vp	Calibration of the pressure sensor	Calibration error of the pressure sensor	B	Uniform	2.418	1	4.35		
$Vamp$	Calibration of the amplifier	Calibration error of the amplifier	B	Uniform	5.00	1	9.00		
u_c	Combined standart uncertainty							50.96	(Pa)
u	Expanded uncertainty							101.9	(Pa)

https://doi.org/10.1051/epjconf/201818002035

Additional effects of fluid on the piston

Daniel Himr[a], Vladimír Habán, Jan Kůrečka, and František Pochylý

Brno University of Technology, Faculty of Mechanical Engineering, Energy Institute, Victor Kaplan Department of Fluid Engineering, Technická 2, 616 69, Brno, Czech Republic

Abstract. Additional effect of fluid on the solid body is often ignored phenomenon. Although it is possible to do an analytic derivation of equations describing additional effects for many various conditions, it can be difficult to solve them. The paper shows pressure pulsations in the pipe with piston solved by transfer matrix and harmonic analysis with ANSYS Acoustics, where effect of second viscosity is included.

1 Introduction

The classic mechanics is based on the law, that all forces acting on the body bring about the acceleration according to (1):

$$\sum F_i = m \cdot \frac{\mathrm{d}^2 x_i}{\mathrm{d}t^2}. \tag{1}$$

The forces can act throughout the volume of a body (e.g. gravity force), on the surface (e.g. pressure force) or at a point. The forces can depend on the acceleration, velocity or position. When only one degree of freedom is supposed (see figure 1), the equation of motion for a rigid body can be written in the following form:

$$m \cdot \frac{\mathrm{d}^2 x_i}{\mathrm{d}t^2} + b \cdot \frac{\mathrm{d}x_i}{\mathrm{d}t} + k \cdot x_i = F_i, \tag{2}$$

where F_i is an exciting force, damping and stiffness originate from body mounting. The mass represents an effect of innertia, the damping represents an energy loss and the stiffness represents a returnable part of the energy, which is connected with a deformation.

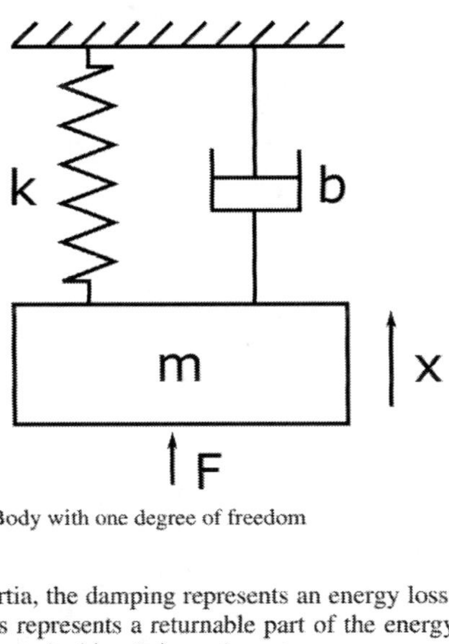

Fig. 1. Body with one degree of freedom

[a] e-mail: himr@fme.vutbr.cz

Fluids also behave according to (1). Thus, it is possible to apply appropriate assumptions and to derive an equation of motion for fluids (3) [1].

$$\frac{\partial v_i}{\partial t} + \frac{\partial v_i}{\partial x_j} \cdot v_j + \frac{1}{\rho} \cdot \frac{\partial p}{\partial x_i} - \frac{1}{\rho} \cdot \frac{\partial \Pi_{ij}}{\partial x_j} = g_i. \tag{3}$$

Here, an energy loss is represented by the stress tensor Π_{ij}, which is defined as follows:

$$\Pi_{ij} = \eta \cdot \left(\frac{\partial v_i}{\partial x_j} + \frac{\partial v_j}{\partial x_i} \right) + \delta_{ij} \cdot \eta_2 \cdot \frac{\partial v_k}{\partial x_k}, \tag{4}$$

where condition $\eta_2 = -\frac{2}{3} \cdot \eta$ is often used (e.g. [2]), but this condition originate from Stokes' work [3] and is valid only for monoatomic gases. It means that the gas is always in the thermodynamic equilibrium. This assumption fails in case of polyatomic gases and liquids [4].

When the fluid structure interaction is solved, equations (2) and (3) can be used together with conditions that velocity and stress of the body surface is the same as velocity and stress of the fluid on this surface. Thus, it is possible to derive an equation of motion for a body in the fluid:

$$(m + m_a) \cdot \frac{\mathrm{d}^2 x_i}{\mathrm{d}t^2} + (b + b_a) \cdot \frac{\mathrm{d}x_i}{\mathrm{d}t} + (k + k_a) \cdot x_i = F_i, \tag{5}$$

where added mass, added damping and added stiffness originate from the simple fact; when the body is to move, also surrounding fluid has to move. The additional effects can be neglected in some cases. E.g. motion in the air, because the air has low density, but water is almost thousand times thicker. Thus, additional effects start being important.

From practical point of view, eigenfrequency of the body is changed when it is submerged in the water. When this effect is overlooked it can cause problems with vibrations (e.g. runner blades) [5].

Equation (5) is not universal. Its shape can differ and depends on the body geometry, fluid properties, whether the fluid flows or not. More information can be found e.g. in [6]. Thus, equation (5) is simplified version of more complex equation, where non-linear terms are neglected. It causes that the range of validity is limited.

Table 1. System properties

Entry	Value	Unit
Length	1	m
Cross-section	1	m^2
Diameter	$\approx$1.128	m
Fluid density	1000	kg m^{-3}
Speed of sound	1000	m s^{-1}
Shear viscosity	0	m^2 s^{-1}

2 Model task

The problem of fluid additional effects is shown in this paper on a simple example. It is a piston, which is moving in a pipe as shown in figure 2. The pipe has dead ends and

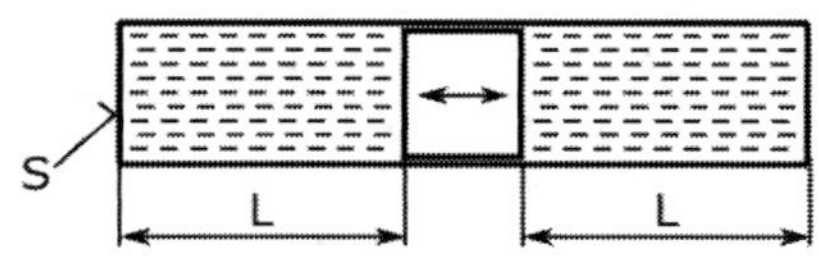

Fig. 2. Solved task; a piston oscillating in the pipe.

rigid walls. The piston oscillates in the axial direction and there is no gap between the piston and the pipe wall. Thus, the fluid cannot flow from the one side of the piston to the other and vice versa. No friction between piston and pipe is assumed and the piston has no mass. The system properties are listed in table 1.

The task is symetrical. Thus, it is possible to solve only one half. The piston velocity is described as follows:

$$v = i \cdot \frac{\omega}{1000} \cdot e^{i \cdot \omega \cdot t}. \tag{6}$$

The corresponding displacement of the piston is:

$$x = \frac{v}{i \cdot \omega}. \tag{7}$$

Now, Laplace transform is applied to (5). When mass, damping and stiffness are assumed to be zero, it is possible to write:

$$\left(m_a \cdot s^2 + b_a \cdot s + k_a\right) \cdot \hat{x} = \hat{F}. \tag{8}$$

When $i \cdot \omega$ is substituted for the parameter of Laplace transformation, it is possible to write:

$$\left(k_a - m_a \cdot \omega^2\right) + i \cdot \omega \cdot b_a = \frac{\hat{F}}{\hat{x}}. \tag{9}$$

When frequency amplitude characteristic is known (e.g. from a simulation or an experiment), it is possible to determine added mass and stiffness from real part, which is a parabolic function of the angular speed, and added damping from imaginary part, which is a linerar function of the angular speed.

2.1 Flow simulation

Two approaches to solve the fluid flow were used. The first one is by transfer matrix, the other one is with a commercial software ANSYS Acoustics. The solution by transfer matrix of the system originates from the momentum equation (3) and the continuity equation (10):

$$\rho \cdot a^2 \cdot \frac{\partial v}{\partial x} + \frac{\partial p}{\partial t} = 0. \tag{10}$$

When the laminar flow is supposed, it is possible to write the result as follows:

$$\hat{u}(x, s) = P(x, s) \cdot \hat{u}(0, s), \tag{11}$$

where u represents a state vector:

$$\hat{u}(x, s) = \begin{pmatrix} \hat{v}(x, s) \\ \hat{p}(x, s) \end{pmatrix} \tag{12}$$

and P represents the transfer matrix:

$$P = \begin{pmatrix} \cosh\left(\sqrt{\frac{s \cdot B}{A \cdot C}} \cdot x\right) & -\sqrt{\frac{s \cdot A}{B \cdot C}} \cdot \sinh\left(\sqrt{\frac{s \cdot B}{A \cdot C}} \cdot x\right) \\ -\sqrt{\frac{B \cdot C}{s \cdot A}} \cdot \sinh\left(\sqrt{\frac{s \cdot B}{A \cdot C}} \cdot x\right) & \cosh\left(\sqrt{\frac{s \cdot B}{A \cdot C}} \cdot x\right) \end{pmatrix}, \tag{13}$$

where:

$$A = \frac{1}{\rho} + \frac{2 \cdot \eta + \eta_2}{\rho^2} \cdot \frac{s}{a^2}, \tag{14}$$

$$B = s - \frac{s \cdot J_1(y)}{s \cdot J_1(y) - \frac{y}{2} \cdot J_0(y)}, \tag{15}$$

$$C = a^2 \cdot \rho, \tag{16}$$

$$y = \frac{D}{2} \cdot i \cdot \sqrt{\frac{s \cdot \rho}{\eta}}. \tag{17}$$

Equation (15) represents an influence of velocity profile, which does not have any effect because zero shear viscosity is supposed.

Thus, equation (11) has a solution, when two boundary conditions are given; zero velocity at dead end of the pipe and excitation (6) at the other end. The computed pressure on the piston allows determining the force as a function of the angular velocity. The function has the real and imaginary part and it is possible to determine additional effects, when (9) is used.

The simulation was performed with constant second viscosity and with the second viscosity as a function of angular velocity:

$$\eta_2 = \frac{h}{\omega}. \tag{18}$$

Simulation in ANSYS Acoustics was performed using Harmonic response analysis. Geometry was represented by one-meter-long cylinder with cross section one square meter and this volume was meshed with appropriate element density. Boundary conditions of rigid wall (zero particle velocity) was used on cylindrical surface and on one end, on the opposite end acoustic surface velocity in normal direction with amplitude of 0.001 m s^{-1} was applied. Physical constants were identical to the ones in Table 1 except for shear viscosity which was 10^{-6} m^2 s^{-1}. Unfortunately, when zero shear viscosity is used, the software does not compute effect of second viscosity either. It is unpleasant property of this software, but, in this case, the effect of shear viscosity is small in comparison with the effect of second viscosity.

The ANSYS Acoustics was used because it has the second viscosity as an optional parameter. This software is an extension of ANSYS Mechanical and is designated for simulations of noise propagation, sound acoustics, vibroacoustics etc. The solver is based on the wave equation.

https://doi.org/10.1051/epjconf/201818002035

3 Results

This section is divided into three subsections. First one is devoted to comparison of result by the transfer matrix with ANSYS Acoustic result. Then, an influence of the second viscosity on the result was investigated and finally, computation with the frequency depended second viscosity was performed. The range of computed frequencies is from 0 Hz to 1273 Hz (0 rad s^{-1} to 8000 rad s^{-1}).

3.1 Transfer matrix vs commercial software

Computation by transfer matrix gives the same results as computation with ANSYS Acoustics. The difference is given only by the step on the solved frequency; see figures 3 and 4.

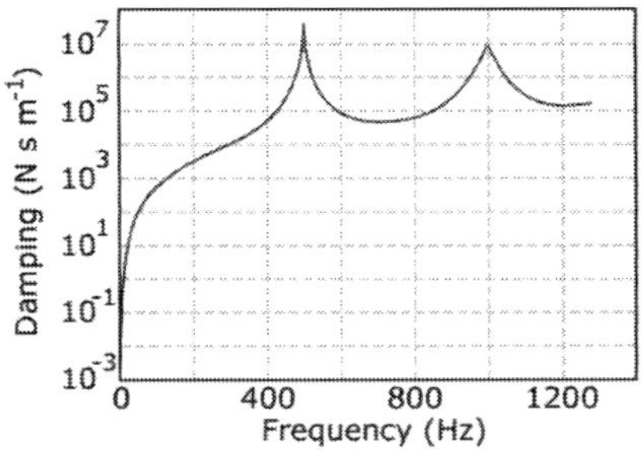

Fig. 3. Comparison of computed damping by transfer matrix (solid black curve) and ANSYS (grey dashed curve)

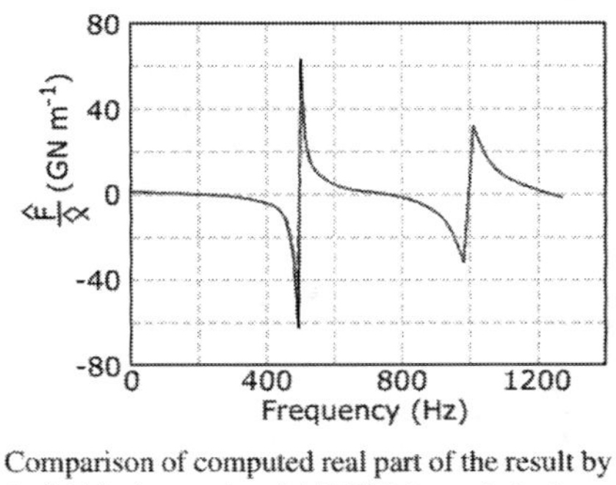

Fig. 4. Comparison of computed real part of the result by transfer matrix (solid black curve) and ANSYS (grey dashed curve)

3.2 Influence of second viscosity

Pressure pulsation with variable value of second viscosity gives different damping. The second viscosity is independent on the frequency in this case. The results are compared in the figure 5. Generally, the higher value of second viscosity the higher value of damping, but the damping is more intense for lower values of the second viscosity at the eigenfrequency of the system.

The real part of the result for higher values of the second viscosity is shown in figure 6. The result for lower values of the second viscosity is shown in figure 7. Unlike the

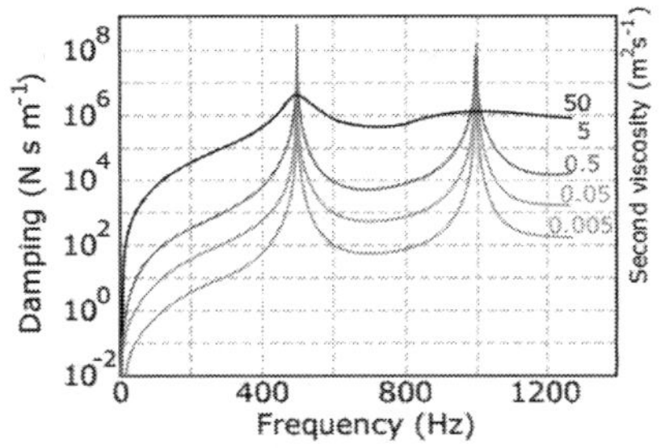

Fig. 5. Evaluated damping for various values of the second viscosity

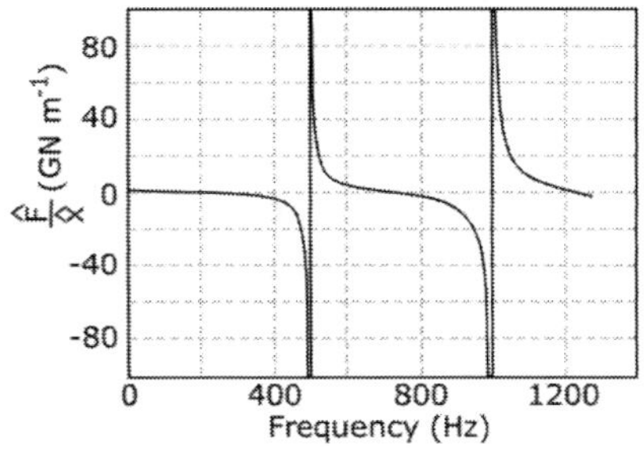

Fig. 6. The real part of the result for values of the second viscosity (50 $m^2 s^{-1}$, 5 $m^2 s^{-1}$ and 0.5 $m^2 s^{-1}$)

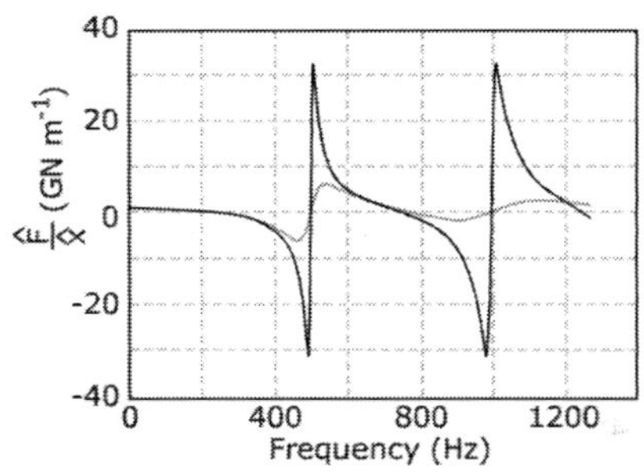

Fig. 7. The real part of the result for values of the second viscosity 0.05 $m^2 s^{-1}$ (black curve) and 0.005 $m^2 s^{-1}$ (grey curve)

imaginary part of the result, the real part is almost the same for second viscosities 0.5 $m^2 s^{-1}$, 5 $m^2 s^{-1}$ and 50 $m^2 s^{-1}$.

Figures 5 to 7 are drawn according to transfer matrix (or ANSYS, the results are the same). When model according to (5) is used it is possible to get valid result up to frequency 160 Hz. Figure 8 compares result by transfer matrix and (5). Added stiffness is 10 GN m^{-1} and added mass is 350 kg. The error of description by (5) is drawn in figure 9. One can see that the relative error is small for low frequencies. The error is 1% for 160 Hz, which is approximately 1/3 of the first eigenfrequency. From this point the error is rapidly growing.

3.3 Frequency depended second viscosity

When the second viscosity is assumed to follow (18), the damping is higher for low frequencies and decreases with increasing frequency. One computation was performed with hysteresis damping coefficient $h = 5000\,m^2 s^{-2}$.

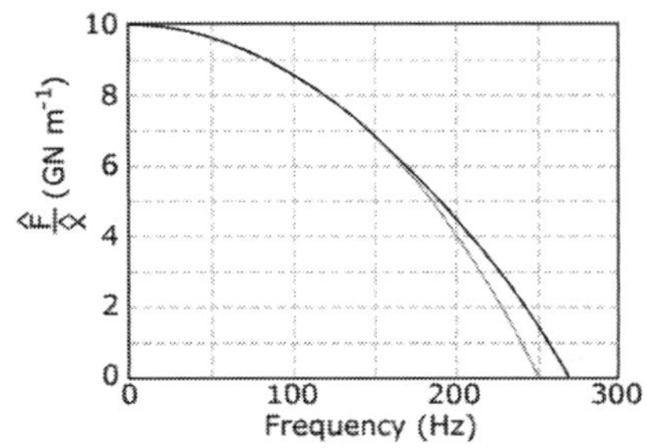

Fig. 8. Real part of the result. Grey curve is obtained by transfer matrix, the black curve is drawn according to (5).

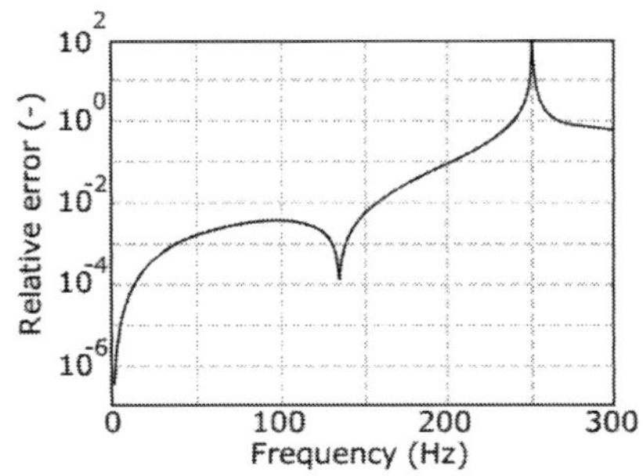

Fig. 9. The error of model with additional effects (5)

Figure 10 compares the result with simulation where constant second viscosity was used. The damping is stronger than case with second viscosity 50 $m^2\,s^{-1}$ but converges to case 5 $m^2\,s^{-1}$ with increasing frequency. The result is the same for angular velocity 1000 rad s^{-1}.

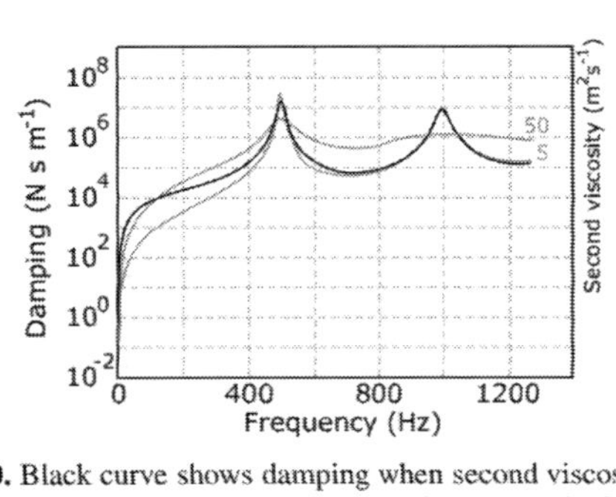

Fig. 10. Black curve shows damping when second viscosity follows hyperbolic function(18). The result is compared with cases where constant second viscosity is used (grey curves).

The real part of the result does not exhibit such a big deviations from the case with constant second viscosity; see figure 11.

4 Conclusion

The paper shows possibilities how to determine additional effects of a fluid on a piston. The solution was performed by transfer matrix and with ANSYS Acoustics software. The comparison showed that both approaches give the same results. The complicated models of the fluid flow can be replaced with the momentum equation of rigid body with

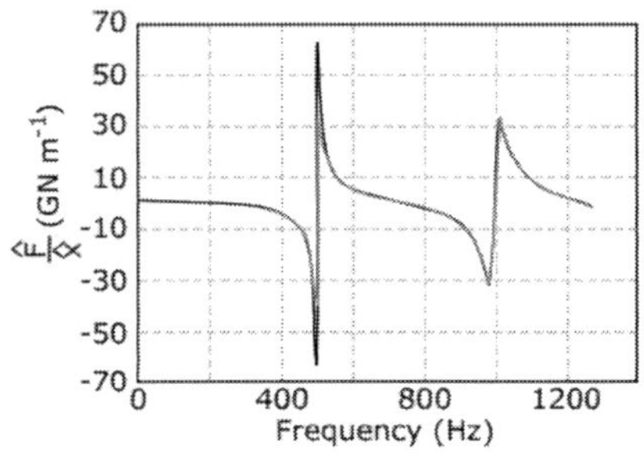

Fig. 11. Real part of the result when second viscosity follows the hyperbolic function (black curve) is compared with the case with second viscosity is 5 $m^2\,s^{-1}$ (grey curve)

added mass, added damping and added stiffness to describe the force acting on the piston. The added mass and added stiffness are constant at low frequencies (up to 1/3 the of first eigenfrequency). The added damping is a linear function of the angular velocity at low frequencies, when the second viscosity follows (18). When the second viscosity is constant the added damping is a parabolic function.

The attention was also paid to influence of the second viscosity, which has dominant effect on the energy loss in the solved task.

Acknowledgement

This work is an output of cooperation between projects of Technology Agency of the Czech Republic No. TE02000232 "Special rotary machine engineering centre" and NETME Centre, regional R&D centre built with the financial support from the Operational Programme Research and Development for Innovations within the project NETME Centre (New Technologies for Mechanical Engineering), Reg. no. CZ.1.05/2.1.00/ 01.0002 and, in the follow-up sustainability stage, supported through NETME CENTRE PLUS (LO1202) by financial means from the Ministry of Education, Youth and Sports under the "National Sustainability Programme I."

Nomenclature

a	($m\,s^{-1}$)	speed of sound,
b	($N\,s\,m^{-1}$)	damping,
b_a	($N\,s\,m^{-1}$)	added damping,
D	(m)	diameter,
F	(N)	force,
g	($m\,s^{-2}$)	gravity acceleration,
h	($m^2\,s^{-2}$)	hysteresis damping coefficient,
k	($N\,m^{-1}$)	stiffness,
k_a	($N\,m^{-1}$)	added stiffness,
L	(m)	length,
m	(kg)	mass,
m_a	(kg)	added mass,
P	(-)	transfermatrix,
p	(Pa)	pressure,
S	(m^2)	cross-section,
s	(s^{-1})	parameter of the Laplace transform,
t	(s)	time,

u		state vector,
v	(m s^{-1})	velocity,
x	(m)	space coordinates,
δ_{ij}	(-)	Kronecker delta,
η	(Pa s)	shear viscosity,
η_2	(Pa s)	second viscosity,
Π	(Pa)	stress tensor,
ρ	(kg m^{-3})	density,
ω	(rad s^{-1})	angular velocity,
$\hat{}$		variable after the Laplace transform.

References

1. V. L. Streeter, E. B. Wylie, K. W. Bedford, *Fluid Mechanics, Ninth Edition* (WCB/McGraw Hill, 1998).
2. L. Hružík, M. Vašina, A. Bureček, Advances in Mechanism Design II **44**, 137 – 143, (2017).
3. G. G. Stokes, Trans. of the Cambridge phil. soc. **8**, 287 – 305, (1845).
4. L. Tisza, Physical Review **61**, (1942).
5. F. Pochylý, S. Fialová, M. Kozubková, M. Bojko, IOP Conf. Series-Earth and Environmental Science **15**, 072027, (2013).
6. C. Conca, A. Osses, J. Planchard, Comp. methods in appl. mech. and eng. **146** (1997) 387 – 405.

Calibration of PVDF Film Transducers for the Cavitation Impact Measurement

Jan Hujer[1,*], *Miloš* Müller[1]

[1]Technical University of Liberec, Dep. of Power Engineering Equipment, Studentska 2, 461 17 Liberec, Czech Republic

Abstract. This paper describes investigation of the influence of the protective layer thickness on the calibration sensitivity of PVDF films sensors for the cavitation impacts measurements. The PVDF film sensor is casted into an aluminium block. The drop ball method is used for the measurement of the relation between impact force and the voltage detected on the PVDF film sensor. The calibration constants are measured for three different protective layers thicknesses. Five different ball weights for 400 mm drop height are used to reach the required impact force range. The ball positions for the evaluation of the impact force are measured with a high speed camera. The voltage signal detected on the PVDF film clamps was measured with a high speed digitizer. The measured signals are analysed in LabVIEW Signal Express.

1 Introduction

The cavitating flow appearing in hydraulic machines is usually associated with a noise, vibration, loss in the efficiency and cavitation erosion. These undesirable phenomena are connected with activities of cavitation bubbles or rather cavitation bubble structures – the bubble clouds. In the case of cavitation erosion, microjets and shock waves are formed during the final stage of the cavitation bubble cloud collapse. These effects interacts with near walls and causes the material damage.

The research focusing on the cavitation erosion covers mainly mass loss tests and pitting tests. During the test the material is undergoing cavitation load for a certain time period [1-2]. Another way how to study the cavitation erosion is to measure and count pulses due to impacts of microjets and shock waves by sensors mounted at exposed positions on the material. The piezoelectric transducers and especially thin piezoelectric polyvinylidene fluoride (PVDF) polymer transducers are applicable for these measurement. The very fast response of the PVDF transducers enables the measurement of impact forces of a high intensity and a short duration, which is just the case of impacts due to the cavitation bubble cloud collapse. During the cavitation erosion the mounted sensor is highly loaded by the cavitating flow. A protective layers of a resin or a tape are used to avoid the transducer damage. The presence of the protective layers can however influence the sensor sensitivity. Methods used to protect the PVDF films transducers are presented in the following text.

Momma et al. [3] counted pulses and measured their height by the 110 μm thick PVDF film transducer with protection in cavitating jet apparatus. Authors tested effect of two, four and six protective layers of polyimide tapes on pulse height distribution. The output in all cases was indistinguishable.

Soyama et al. [4] used 110 μm thick PVDF film transducer to count and measure pulses in cavitating jet apparatus. The sensor in this work was protected by 60 μm thick polyimide tape.

Wang et al. [5] used 28 μm thick PVDF film transducer to measure impact forces due to a single bubble collapses near a solid boundary. Authors protected used PVDF film transducer by a one layer of scotch tape. The measured sensitivity constant was 0,013 V/N.

Grinspan et al. [6] used 52 μm thick PVDF film transducer protected by the cellophane tape. The sensor sensitivity of used design was 0,116 V/N.

In a case of very aggressive flow regimes or testing conditions, transducers need to be protected by higher number of protective layers. These could be made of glue. The aim of this study is to calibrate a 28 μm thick piezoelectric PVDF film transducer for the cavitation impact load measurement in a vibratory cavitation apparatus. Another task is to evaluate the influence of a protective layer thickness on the sensor sensitivity.

2 Calibration – ball drop method

Three methods – a pendulum-typed ball impact method, ball drop method and lead breaking method are commonly used for the calibration of the PVDF film transducers [7]. The ball drop method was chosen for the calibrations in presented work. The scheme of the method is presented in Fig. 1.

A stainless steel ball of certain weight m drops vertically from position 1 of height h_1 above the calibrated transducer, hits the calibrated transducer and rebounds to position 2 of height h_2 above the calibrated transducer. A plastic tube is used for the ball guiding. A camera and

* Corresponding author: jan.hujer@tul.cz

a rule are used to measure the heights h_1 and h_2. During the impact, a part of the initial ball energy is transformed in a deformation of transducer which generates an electric charge. Electric charge is measured as an electric voltage output from which the maximum voltage U_{max} and the time duration of the impact τ are determined.

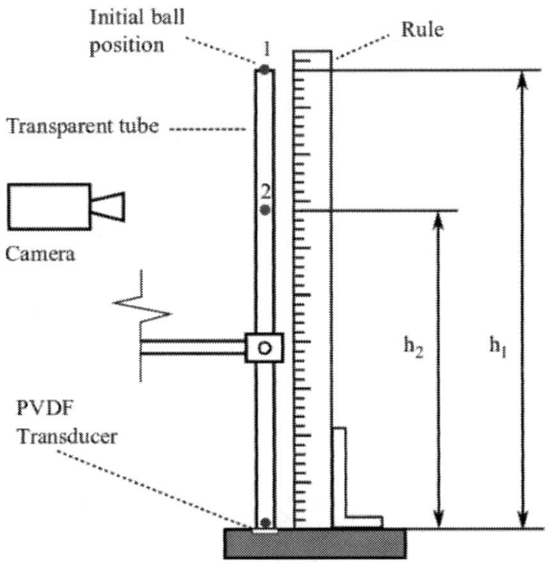

Fig. 1. Calibration - Ball drop method.

Mean impact force is calculated from the following formula based on the impulse-momentum theorem

$$F_{mean} = \frac{1}{\tau} \int_{t_1}^{t_2} F(t)dt = \frac{m}{\tau}(v_1 + v_2). \quad (1)$$

Velocities before and after the impact are calculated from the following formula

$$v_{1,2} = \sqrt{2gh_{1,2}} \quad (2)$$

where $h_{1,2}$ are heights and g is gravity acceleration. Maximum force is double of the mean force, which results from the shape of the measured signal

$$F_{max} = 2 \cdot F_{mean} . \quad (3)$$

3 Experimental setup

In this study a 28 µm thick PVDF film transducer (DT1-028K/L) manufactured by TE Connectivity was chosen. The active area of a 12 mm x 30 mm of the transducer was reduced to 12 mm x 1,5 mm (dimensions b x c in Fig. 2.). Transducers were placed in a machined holes of aluminium blocks. The protective layers thicknesses were 1 mm, 1,5 mm and 2 mm (dimension a in Fig. 2.). The mounting procedure included a gluing of the active transducer area by a cyanoacrylate glue (Loctite Super Attak Power Flex Gel), the casting of the hole by a transparent two component epoxy glue (Bison 5 min) and a grinding and polishing the upper surface.

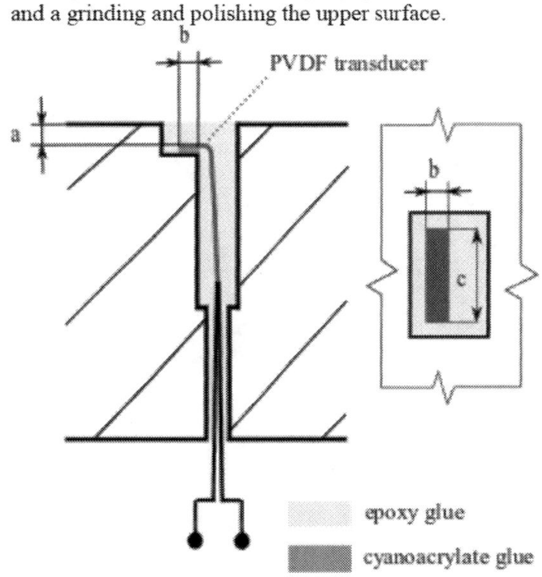

Fig. 2. Mounting of the PVDF transducer.

The transducer was connected directly to the NI USB 5132 digitizer input, which was used for the signal acquisition (Fig. 3.). Signals were analysed in LabVIEW Signal Express software, where the impact duration τ and maximum voltage U_{max} were detected. The digitizer, computer, aluminium blocks and negative pole of PVDF transducer were grounded to avoid a signal noise (Fig. 3.).

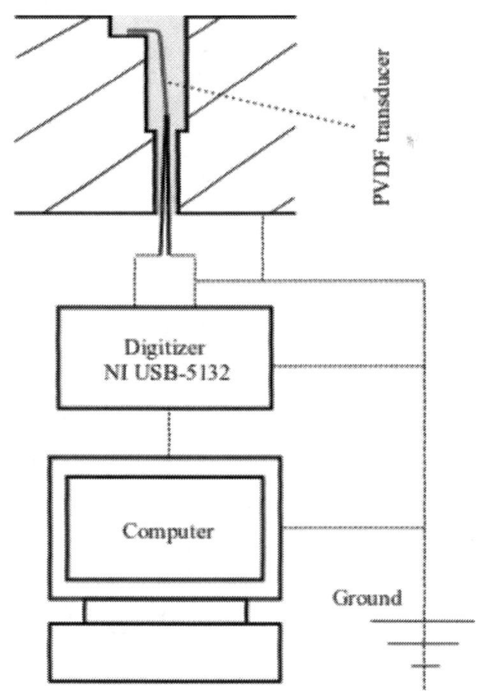

Fig. 3. Experimental setup.

Five different stainless steel balls (mass: 0,44 g; 0,7 g; 1,046 g; 1,398 g; 2,04 g) were used for the measurement. Total number of tests for each protective layer thickness were 25. Initial position of the ball was defined by the length of guiding tube of 400 mm for all tests.

In Fig. 4. an example of the calibration signal for 1,5 mm protective layer thickness with ball of 0.7 g is presented. Maximum measured voltage in this case was 1,91 V and pulse duration was 53 μs.

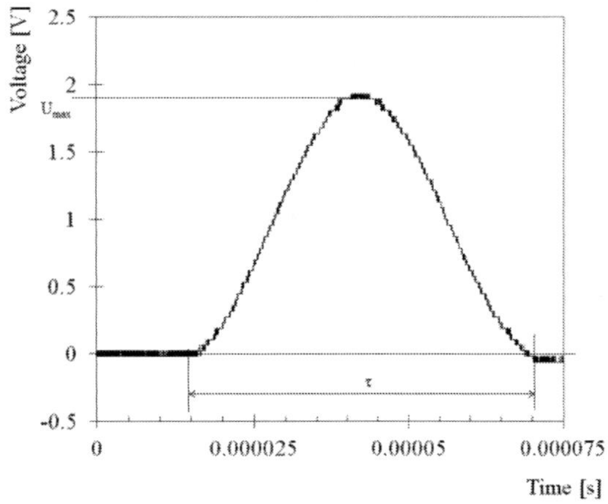

Fig. 4. An example of a calibration signal (ball 0,7g; tube 400 mm).

4 Results

The results of the measurements are presented in Fig. 5. – Fig. 7. The linear least square method was used to fit the measured data. The sensitivity constants for 1 mm, 1,5 mm and 2 mm thick protective layer are 0,0156 V/N, 0,0136 V/N and 0,0115 V/N respectively. Fig. 8. shows the sensitivity decrease with the increasing the protective layer thickness. The approximation method usually results in a nonzero voltage at the zero force, which is physically not possible. For this reason the zero voltage value was artificially added to the zero force for the fitting procedure.

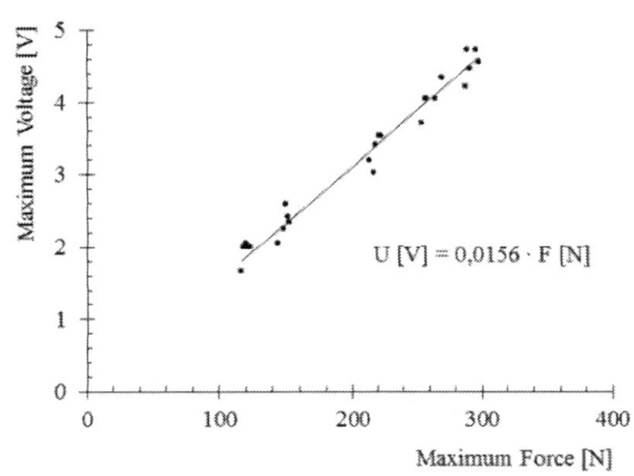

Fig. 5. The calibration - 1 mm thick protective layer.

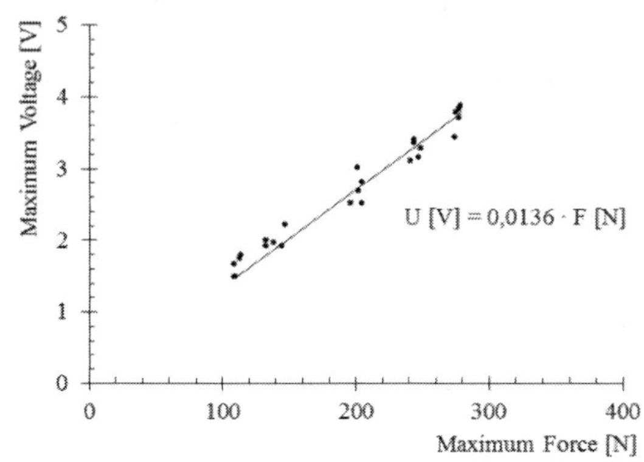

Fig. 6. The calibration - 1,5 mm thick protective layer.

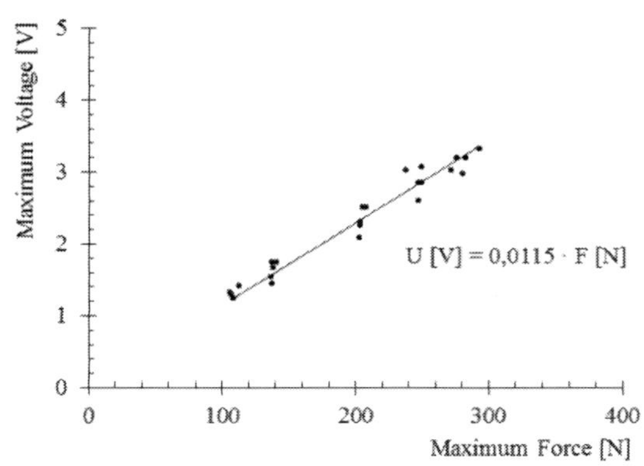

Fig. 7. The calibration - 2 mm thick protective layer.

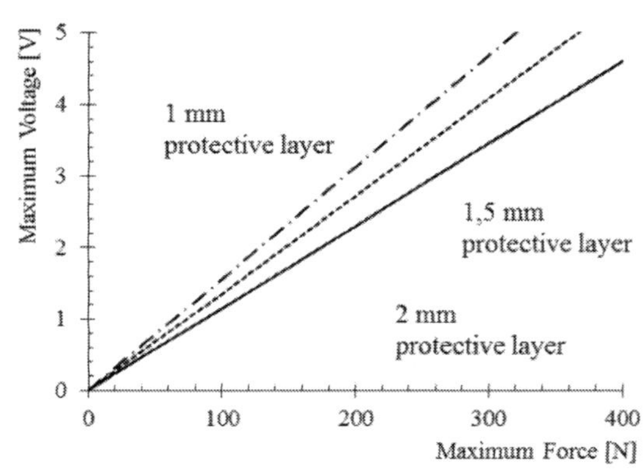

Fig. 8. The calibration curves recalculated according to measured calibration constants.

5 Conclusion

This work presents influence of the thickness of the protective layer for PVDF film transducers. It was shown that the presence of protective layer decreases the PVDF film sensitivity. From the practical point of view this means, that a new calibration should be required for each sensor mounting. The thickness of the protective layer also defines the possible duration of the measurement.

The results shows, that the maximum force dependence is almost linear and the calibration constants can be extended outside the calibration range.

Acknowledgements

This publication was written at the Technical University of Liberec as part of the project SGS no. 21182 "The cavitation phenomenon and its erosion potential" with the support of the Specific University Research Grant, as provided by the Ministry of Education, Youth and Sports of the Czech Republic in the year 2017.

References

1. K.H. Kim, G. Chahine, J.-P. Franc, A.Karimi, *Advanced experimental and numerical techniques for* cavitation *erosion prediction* (Springer, Dordrecht, 2014)
2. Ch.E. Brennen, *Cavitation and Bubble Dynamics* (Oxford University Press, Oxford, 1995)
3. T. Momma, A. Lichtarowicz, Wear **186-187**, 425-436 (1995)
4. H. Soyama, Y. Sekine, K. Saito, Measurement **44**, 1279-1283 (2011)
5. Y.Ch. Wang, Y.W. Chen, Experimental Thermal and Fluid Science **32** 2, 403-414 (2007)
6. A.S. Grinspan, R. Gnanamoorthy, Colloids and Surfaces A: Physicochem. Eng. Aspects **356**, 162-168 (2010)
7. H. Soyama, A. Lichtarowicz, T. Momma, E.J. Williams, J. Fluids Eng. **120** (4), 712-718 (1998)

https://doi.org/10.1051/epjconf/201818002037

Development of Modified Incompressible Ideal Gas Model for Natural Draft Cooling Tower Flow Simulation

Tomáš Hyhlík[1,*]

[1]CTU in Prague, FME, Department of Fluid Dynamics and Thermodynamics, Technická 4, 166 07 Prague, Czechia

Abstract. The article deals with the development of incompressible ideal gas like model, which can be used as a part of mathematical model describing natural draft wet-cooling tower flow, heat and mass transfer. It is shown, based on the results of a complex mathematical model of natural draft wet-cooling tower flow, that behaviour of pressure, temperature and density is very similar to the case of hydrostatics of moist air, where heat and mass transfer in the fill zone must be taken into account. The behaviour inside the cooling tower is documented using density, pressure and temperature distributions. The proposed equation for the density is based on the same idea like the incompressible ideal gas model, which is only dependent on temperature, specific humidity and in this case on elevation. It is shown that normalized density difference of the density based on proposed model and density based on the nonsimplified model is in the order of 10^{-4}. The classical incompressible ideal gas model, Boussinesq model and generalised Boussinesq model are also tested. These models show deviation in percentages.

1 Introduction

The working principle of natural draft wet-cooling tower is connected with releasing of latent heat during water evaporation. This effect is connected with a decrease of moist air density. There is a density difference between surrounding and inside of cooling tower which can create a natural draft. The density change in the natural draft cooling tower is not related only to heat and mass transfer between water and moist air, but there is also the change due to gravity.

The aim of this article is to identify the influences affecting the density change inside the natural draft wet-cooling tower. There are several simplified density models like Boussinesq model or so called incompressible ideal gas model. The article deals with the testing and modifying of known simplified density models using data from the non-simplified model of flow, heat and mass transfer.

This work is motivated by an effort to get numerical method which provides results quickly. The numerical solutions based on previously published model [1] of flow, heat and mass transfer with general equation of state is very slow. Due to the fact, that the value of velocity in the cooling tower is in the order of meters per second, it seems natural that changes in pressure, temperature and density are almost independent of velocity except for the fill zone. The flow in the natural draft wet-cooling tower is one of the problems where Mach mumber $Ma \to 0$.

*e-mail: tomas.hyhlik@fs.cvut.cz

2 Governing equations of quasi one-dimensional model

System of governing equations for quasi one-dimensional flow is

$$\frac{\partial(\mathbf{W}A)}{\partial t} + \frac{\partial(\mathbf{F}A)}{\partial z} = \mathbf{Q}, \tag{1}$$

where vector of conservative properties **W** and vector of fluxes **F** are

$$\mathbf{W} = \begin{bmatrix} \rho \\ \rho v \\ \rho w_v \\ \rho e \end{bmatrix}, \quad \mathbf{F} = \begin{bmatrix} \rho v \\ \rho v^2 + p \\ \rho v w_v \\ v(\rho e + p) \end{bmatrix}. \tag{2}$$

Vector of sources is

$$\mathbf{Q} = \begin{bmatrix} A(z)\sigma_v(z) \\ p\frac{\mathrm{d}A}{\mathrm{d}z} - A(z)(\rho g + \zeta\rho\frac{v^2}{2}) \\ A(z)\sigma_v(z) \\ A(z)(\sigma_{qmodel}(z) + \rho g v) \end{bmatrix}. \tag{3}$$

First equation is overall continuity equation, second equation is momentum equation, third equation is water vapour

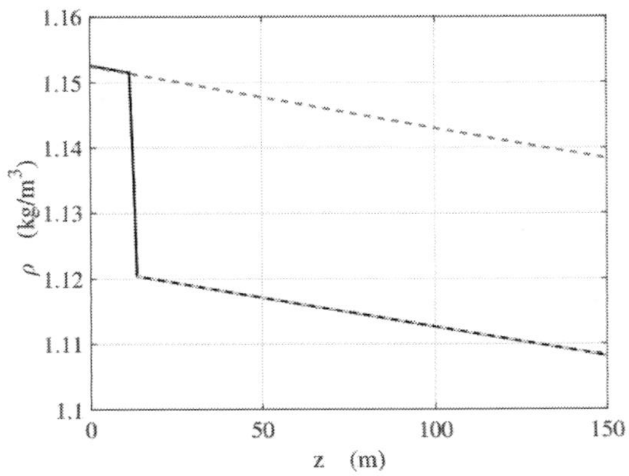

Fig. 1. Distribution of moist air density based on the numerical simulation - full black line; density distribution outside the cooling tower - dashed red line; density distribution based on hydrostatics of moist air with initial moist air parameters as calculated above the fill zone - dashed green line

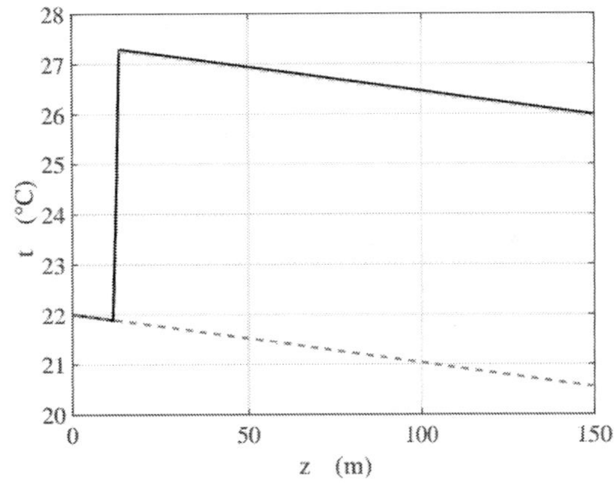

Fig. 2. Distribution of moist air temperature based on the numerical simulation - full black line; temperature distribution outside the cooling tower - dashed red line; temperature distribution based on equation (16) with initial temperature and specific humidity as calculated above the fill zone - dashed green line

continuity equation and the last equation is equation of energy. We define ad hoc internal energy of unsaturated humid air as

$$u = (w_a c_{v_a} + w_v c_{v_v})T = c_v T. \quad (4)$$

Water vapour mass fraction w_v is connected with specific humidity x as

$$w_v = \frac{x}{1+x} \quad (5)$$

and dry air mass fraction is

$$w_a = 1 - w_v. \quad (6)$$

The total energy of unsaturated moist air is expressed in the form

$$\rho e = \rho\left(c_v T + \frac{v_k v_k}{2}\right). \quad (7)$$

Pressure can be calculated using equation of state as

$$p = \frac{\rho R T\left(1+\frac{M_a}{M_v}x\right)}{M_a(1+x)} = \rho R T\left(\frac{M_v - w_v(M_v - M_a)}{M_a M_v}\right), \quad (8)$$

where M_v is molecular weight of water vapour and M_a is molecular weight of dry air. This model is described in detail together with the method of its numerical solution in [1].

3 Hydrostatics of moist air

There is an equilibrium between gravity and pressure force. The pressure gradient is

$$\frac{\mathrm{d}p}{\mathrm{d}z} = -\rho g. \quad (9)$$

The moist air density at given specific humidity x, pressure p and temperature T is

$$\rho = \frac{M_a(1+x)}{\left(1+\frac{M_a}{M_v}x\right)}\frac{p}{RT}. \quad (10)$$

After substituting equation (10) into equation (9) is

$$\frac{\mathrm{d}p}{p} = -\frac{M_a(1+x)}{\left(1+\frac{M_a}{M_v}x\right)}\frac{g}{R}\frac{\mathrm{d}z}{T}. \quad (11)$$

For an isoentropic proces is

$$\frac{p}{\rho^\kappa} = constant, \quad (12)$$

where κ is Poisson constant of moist air

$$\kappa = \frac{c_p}{c_v}. \quad (13)$$

It is possible to substitute equation (10) into equation (12) and differentiate with respect z direction similarly like in the reference [2]

$$\frac{(1-\kappa)}{\kappa}\frac{1}{p}\frac{\mathrm{d}p}{\mathrm{d}z} + \frac{1}{T}\frac{\mathrm{d}T}{\mathrm{d}z} = 0. \quad (14)$$

This relation is based on the assumption of constant specific humidity x which is not function of z direction. By using equations (9), (10) and (14) can be the temperature gradient expressed as

$$\frac{\mathrm{d}T}{\mathrm{d}z} = -\frac{(\kappa-1)}{\kappa}\frac{M_a(1+x)}{\left(1+\frac{M_a}{M_v}x\right)}\frac{g}{R}. \quad (15)$$

Temperature distribution is after integration of previous equation

$$T(z) = T(z_1) - (z - z_1)\frac{(\kappa-1)}{\kappa}\frac{M_a(1+x)}{\left(1+\frac{M_a}{M_v}x\right)}\frac{g}{R}. \quad (16)$$

It is possible to substitute equation (16) into equation (11) and integrate with z coordinate to obtain pressure distribution

$$p(z) = p(z_1)\left[1 - (z - z_1)\frac{g(\kappa-1)}{\kappa R T(z_1)}\frac{M_a(1+x)}{\left(1+\frac{M_a}{M_v}x\right)}\right]^{\frac{\kappa}{\kappa-1}}. \quad (17)$$

https://doi.org/10.1051/epjconf/201818002037

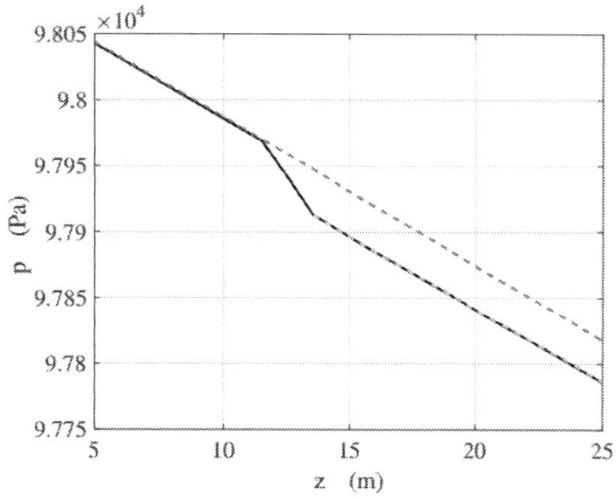

Fig. 3. Detail of the distribution of moist air pressure based on the numerical simulation - full black line; pressure distribution outside the cooling tower - dashed red line; pressure distribution based on equation (17) with initial moist air parameters as calculated above the fill zone - dashed green line

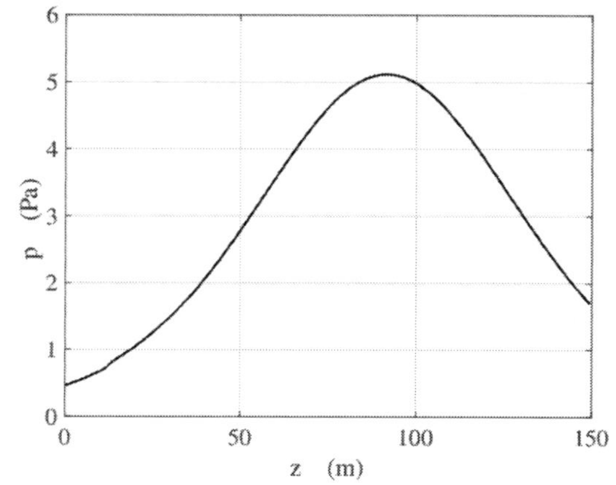

Fig. 4. Distribution of kinetic pressure based on the numerical simulation of the flow inside cooling tower

4 Simplified density models

One of the models used in the calculation of natural convection is Boussinesq model which is based on the definition of volumetric thermal expansion coefficient [3]

$$\beta_T = -\frac{1}{\rho}\left(\frac{\partial \rho}{\partial T}\right)_{p,\, w_v} . \quad (18)$$

Based on the aproximation of previous definition is

$$\rho = \rho_0 \left[1 - \beta_T (T - T_0)\right], \quad (19)$$

where ρ_0 and T_0 are constants and thermal expansion coefficiet is

$$\beta_T = \frac{1}{T_0}. \quad (20)$$

It is possible to take into account change in humidity of moist air and create generalised Boussinesq model based on thermal expansion coefficient and vapour expansion coefficient

$$\beta_w = -\frac{1}{\rho}\left(\frac{\partial \rho}{\partial w_v}\right)_{p,\, T} . \quad (21)$$

Generalised Boussinesq model [4] is

$$\rho = \rho_0 \left[1 - \beta_T (T - T_0) - \beta_w (w_v - w_{v_0})\right], \quad (22)$$

where w_{v_0} is contant and vapour expansion coefficient is

$$\beta_w = \frac{M_a - M_v}{M_v - w_v(M_v - M_a)} \approx \frac{M_a}{M_v} - 1, \quad (23)$$

where M_a is molecular weight of dry air and M_v is water vapour molecular weight.

Incompressible ideal gas model, which is taking into account humidity change is

$$\rho = \frac{M_a (1 + x)}{\left(1 + \frac{M_a}{M_v} x\right)} \frac{p_0}{RT}. \quad (24)$$

Let's take into account that dominant pressure change is connected with gravity. Incompressible ideal gas like model which is taking into account pressure change with gravity is

$$\rho(z) = \frac{M_a (1 + x)}{\left(1 + \frac{M_a}{M_v} x\right)} \frac{p_0}{RT(z)} \left[1 - z \frac{g(\kappa - 1)}{\kappa R T_0} \frac{M_a (1 + x)}{\left(1 + \frac{M_a}{M_v} x\right)}\right]^{\frac{\kappa}{\kappa - 1}}, \quad (25)$$

where p_0 is the pressure at ground level, T is predicted temperature, T_0 is the temperature at ground level, x is predicted specific humidity and κ is the Poisson constant of moist air.

5 Results

There are comparisons of the distribution of density, temperature and pressure calculated using general model of flow, heat and mass transfer, where non-simplified equation of state is used, with distributions based on hydrostatics of moist air in figures 1, 2 and 3. It is possible to observe that behaviour in the input section corresponds very well with the behaviour outside the cooling tower. It is also visible that the behaviour above the fill zone practically corresponds with the behaviour connected with hydrostatics of moist air, where the parameters above the fill are taken into account. This behaviour can be explained if figure 4 is taking into account. It is clear that kinetic pressure inside cooling tower is very low and pressure changes due to gravity field are much higher. It is possible to observe that the slope of the pressure distribution outside the cooling tower is different than the slope of pressure distribution above the fill zone which is based on the hydrostatics or on the numerical solution, but both lines intersect at the cooling tower outflow. Pressure decreased above the fill zone due to losses, temperature increased due to heat transfer, humidity increased due to mass transfer and these effects leads to the density decrease.

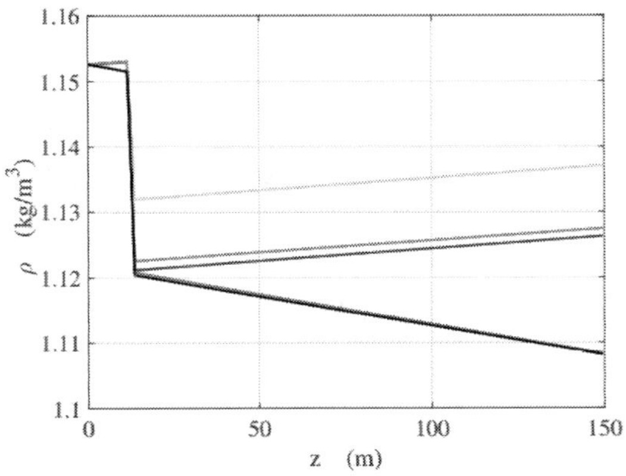

Fig. 5. Density distribution based on the numerical simulation - black line; density distribution based on modified incompressible ideal gas model represented by equation (25) - red line; density distribution based on incompressible ideal gas represented by equation (24) - magenta line; density distribution based on Boussinesq model represented by equation (19) - green line; density distribution based on generalised Boussinesq model represented by equation (22) - blue line

Figure 5 shows the comparison of the density prediction based on several simplified density models. It is depicted that the only modified incompressible ideal gas model which is taking into account hydrostatics of moist air shows a decrease above the fill similarly like the non-simplified equation of state. It is visible that the second best model is the generalised Boussinesq model, the third best model is the classical incompressible ideal gas model and the worst is the Boussinesq model. The normalized deviations of calculated densities are shown in figure 6. It is seen that the normalized deviations connected with modified incompressible ideal gas which is taking into account hydrostatics of moist air are in the order 10^{-4}. The other simplified density models show normalized deviations in percentages. The biggest deviation is connected with Boussinesq model and is more than 2.5%. It is possible to observe that the biggest increase of normalized deviation in the fill zone is also connected with Boussinesq model. This fact is possible to explain by ignoring changes in specific humidity in the Boussinesq model.

6 Conclusions

It has been shown that the behaviour of pressure, temperature and density in the natural draft wet-cooling tower corresponds very well with the behaviour of stagnant moist air in the gravity field excluding fill zone, where the heat and mass transfer occurs and minor losses have to be taken into account. Four simplified density models have been compared with the solution based on the general model. It is shown that the best match is achieved using the modified incompressible ideal gas model which is taking into account hydrostatics of moist air. Presented simplified density models are useful not only in the case of the solution of the natural draft wet-cooling towers but also in the case of the modelling of natural convection in the atmosphere. It is possible to imagine that the model which is taking into account supersaturated moist air can be created on the same basis.

Acknowledgement This work has been supported by *Technology Agency of the Czech Republic* under the project *Advanced Technologies for Heat and Electricity Production - TE01020036.*

References

[1] T. Hyhlík, EPJ Web of Conferences, **92**, (2015), 02027
[2] D. G. Kröger, *Air-Cooled Heat Exchangers and Cooling Towers.* (Penn Well Corporation, Tulsa, 2004)
[3] M.J. Moran, H.N. Shapiro, *Fundamentals of engineering thermodynamics* (John Wiley & Sons, Inc., 1992)
[4] P. Sosnowski, A. Petronio, V. Armenio, International Journal of Heat and Mass Transfer, **66**, (2013), 382-395

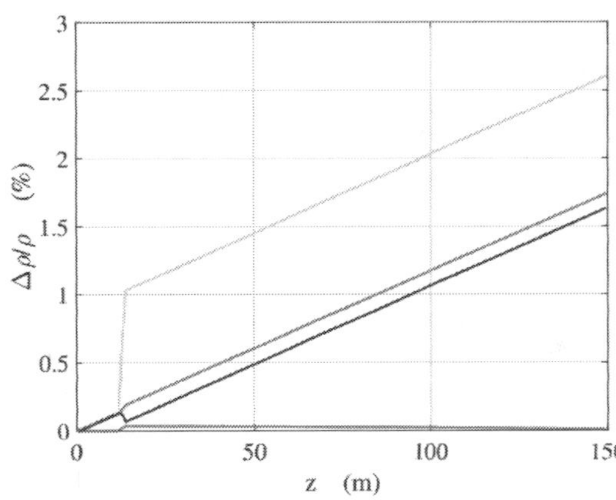

Fig. 6. Normalized deviation of density based on modified incompressible ideal gas model represented by equation (25) and density based on the numerical simulation - red line; normalized deviation of density based on incompressible ideal gas model (24) and density based on the numerical simulation - magenta line; normalized deviation of density based on Boussinesq model (19) and density based on the numerical simulation - green line; normalized deviation of density based on generalised Boussinesq model (22) and density based on the numerical simulation - blue line

Comparison of cavitation bubbles evolution in viscous media

Darina Jasikova[*], *Petr* Schovanec, *Michal* Kotek, *Vaclav* Kopecky

Department of Physical Measurement, Technical University of Liberec, The Institute for Nanomaterials, Advanced Technology and Innovation, Studentska 1402/2 Liberec 1 461 17, Czech Republic

Abstract. There have been tried many types of liquids with different ranges of viscosity values that have been tested to form a single cavitation bubble. The purpose of these experiments was to observe the behaviour of cavitation bubbles in media with different ranges of absorbance. The most of the method was based on spark to induced superheat limit of liquid. Here we used arrangement of the laser-induced breakdown (LIB) method. There were described the set cavitation setting that affects the size bubble in media with different absorbance. We visualized the cavitation bubble with a 60 kHz high speed camera. We used here shadowgraphy setup for the bubble visualization. There were observed time development and bubble extinction in various media, where the size of the bubble in the silicone oil was extremely small, due to the absorbance size of silicon oil.

1 Introduction

The cavitation phenomena is mainly known for its undesirable effects in turbo machinery, however there is a great potential for its utilization in the industry, medicine, biology, pharmacy, or tissue engineering.

The research in the field of cavitation was previously focused mainly on the investigation of the bubble behavior in the vicinity of rigid or flexible boundaries; [1-4] however the current investigations require including the description of response of the impacted material itself. [1-2] The definition of the material response can help in development of new, more resistant structures or layers which better withstand the action of cavitation bubbles and prolong the service life of products. The key in the understanding of the cavitation interaction with various materials is the investigation of the impact of individual bubbles and its structures [1].

Cavitation can be defined as a collection of effects connected to the origin, activities, and collapse of macroscopic bubbles in liquid. Cavitation bubbles are usually not separated in real applications. The bubbles create structures, which acts collectively, however essential elements of these structures are individual bubbles [2].

Although the current state of technique is at very high level, we are still not able to produce one controllable individual bubble by pressure decrease in the liquid volume, following the cavitation definition. All other methods as the spark or laser generated bubbles are closer to the boiling as these are based on the evaporation of small volume of liquid. The bubbles generated by the ultrasonic field satisfy the cavitation definition, however to produce one single bubble is almost impossible. Anyway, we dealing with great technique LIB, there still exists a lack of information in experimental part of cavitation bubble investigation [1].

The one of the method for bubble generation is spark in liquid, or a heated top of a wire. Once the bubble is stable, and of certain volume, it is either over heated, or expose to force impact. The convenient method for generating single cavitation bubble that can be very precisely geometrically placed in the volume of the liquid, and close to the sample, is Laser Induced Breakdown (LIB). [5-9] The LIB method enables to generate cavitation bubble using ultrashort pulses of milijoules energies. This method is thermal breakdown based on natural plasma generation. This plasma is in form of optical breakdown, if the pulse exposure is from microseconds to femtoseconds' time. There occurs direct, multiphoton, and cascade ionization during the LIB. The significant role plays impurities of the medium, spot size, light wavelength, and pulse width during the breakdown. The whole mechanism of the ionization is very well explained by Kennedy [6].

The research of bubble behavior focuses primarily on the movement of bubbles in viscous liquids. The motion of bubbles in fluids is of great importance in various gas–liquid reactors and processes, as well as numerous natural phenomena. As a result, extensive studies have been conducted in the past (see reviews by Clift et al. (1978), Magnaudet and Eames (2000), and Kulkarni and Joshi (2005)), although various aspects still remain indeterminate, particularly in relation to the dynamic behaviour of the bubbles in liquids [10].

Bubble generated in a higher viscosity liquid is being formation in two differs stages. For the evolution of the volume of the bubble in the breakup stage is always greater than that in the expansion stage, due to the different role of the liquid phase in the evolution of the bubble for the two stages [11].

[*] Corresponding author: darina.jasikova@tul.cz

Furthermore, the effect of absorbance and absorption of light on bubble generation in the liquid is important, where absorption of light in the sample is described in the Lambert-Beer Law. The measured values are expressed as absorbance or transmittance. Absorption is usually measure in a linear arrangement. Absorption spectrometers: single-beam or two-beam. The intensity of the beam passing through the sample also weakens scattering. The measurement gives reasonable results for the absorbance in the interval (0.005 - 1) [12].

Absorbance takes values from 0 to infinity. If the absorbance has a zero value, the substance does not absorb any radiation and the permeability is equal to one. If the absorbance is infinite, the substance under investigation absorbs all the radiation and the throughput is zero. It should also be noted that less than 10% of the light passes through the absorbance of more than 1, and a small error in passing radiation can cause a large error in the determination of the absorbance. Most determinations are therefore performed so that the maximum absorbance value lies around the value of 1 [13].

2 Experimental

The bubbles can be generated by several physically different mechanisms. The most obvious in nature is the hydrodynamics cavitation, where bubbles are produced due to local pressure decrease caused by the flow acceleration in vicinity of obstacles. Acoustic cavitation is produced by imposing an intensive acoustic field into the bulk of liquid [5].

Here, we used laser induced breakdown for cavitation bubble generating in liquid. Experiments run in optical glass cuvette.

2.1 Laser Induced Breakdown

Optical breakdown in liquid is usually produced by focusing of the laser light trough suitably designed optics. Laser induced breakdown in aqueous media and its collateral effects are described in detail by Kennedy in [6]. The energy distribution during the growth and collapse of laser induced bubble was described e.g. by Vogel in [14] and [15]. Authors investigated the influence of laser pulse duration and input laser energy on the bubble dynamics and shock waves emission.

Here we chosen green light as exciting light pulse. The absorption coefficient of the 532 nm laser light is only 0.02 m^{-1} in distillated pure water. This wave length requires higher energy input to the system to induce cavitation process. For the single bubble generation we used here the setup for the LIB. The 10 ns short laser pulse was generated using Q-switched Nd: YAG New Wave Gemini pulse laser. This laser worked with one cavity for single shot generation on the wavelength 532 nm. The Q-switch signal synchronized the high speed camera running in triggering mode.

LIB was set as an optical direct way. The outlet diameter of laser beam was 5mm with Gaussian characteristics of the intensity. This setup was followed with concave lens f_s = 200 mm, and convex lens f_s=25 mm of 1inch diameter. The focused laser beam created the laser point - probe (diameter < 0.1 mm). Due the losses in the optical path on each of optical elements, comparable to this, we had to increase the energy level that enters the whole system. The set output energy of the laser is taken in account in the relation to the bubble diameter. The energies that are required for the bubble generation are seen in Fig. 1.

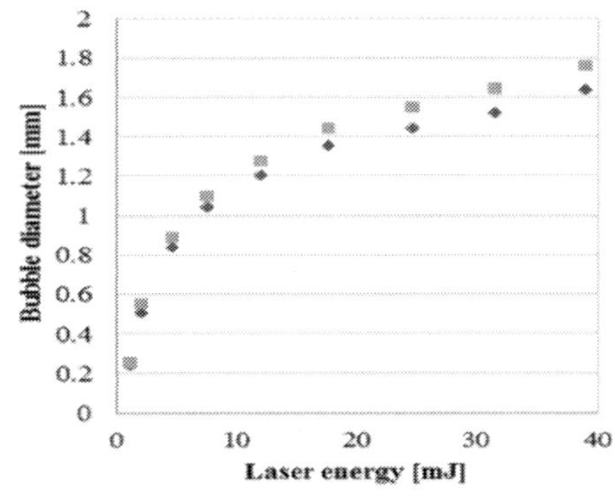

Fig. 1. The graph of maximal bubble diameter on laser energy.

For the relation between input laser energy and the bubble size that was generated with our optical setup, we performed the set of calibration and size measurement. The energy of the laser beam was adjusted and measured with Ophir pyroelectric energy sensor. We measured the laser energy in three positions. There is seen the characteristic of laser light energy calibrated to the position of attenuator. NewWave Gemini laser enables setting of light intensity adjusting flash lamp power and fine adjustment using attenuator. The attenuator enables controllable backward loop, as an actual value is seen on the display. We set the laser on full flash lamp power and adjust the output energy using attenuator. It is seen the energy disorder in Fig. 2, when the attenuator is 0, the output laser energy is higher than for attenuator 50. We neglect this effect as we are working on the attenuator position starting on 120 to generate cavitation bubble, anyway. We suppose this is laser attenuator characteristic.

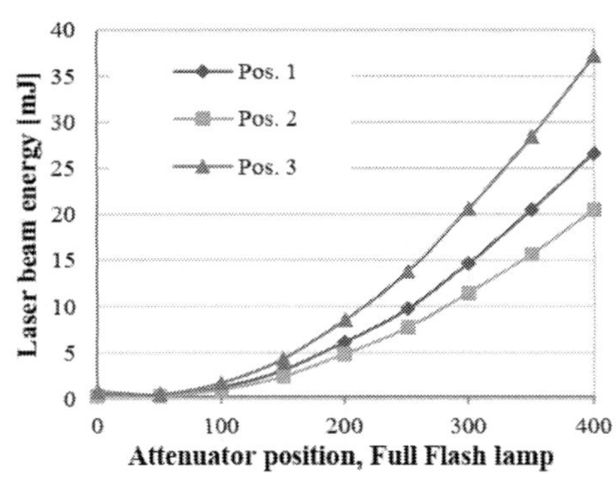

Fig. 2. The calibration curve of output energy on laser attenuator; Position 1 was at the outcome of the laser head,

position 2 was after the collimating lens, and position 3 was after the focusing mirror

According to Kennedy [16], plasma temperature shows asymptotic dependence on laser pulse energy as well. With higher laser energy we recognized negative influence of impurities and presence of segmentations on the bubble surface.

2.2 Optical setup

Here we used two kind of setup for LIB: the direct way with 50 mm Plano-Convex Lens and 30 mm Plano-Convex Lens. The direct optical setup enables easy handling and replacement of single optical component. Here we work with the lowest possible number of optical components, so we expecting lowest light intensity lost in system. The setup and used optical components is seen in Fig. 3.

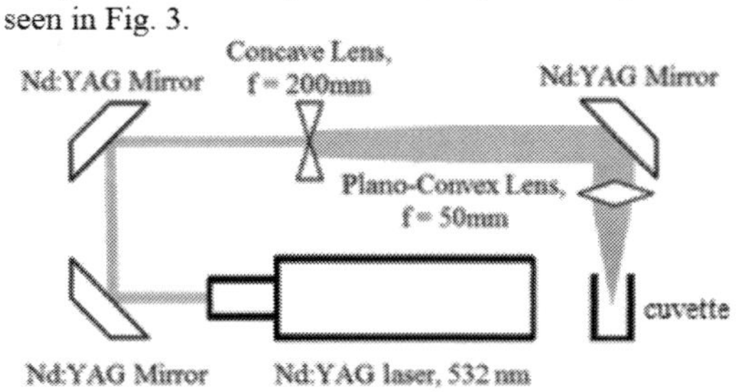

Fig. 3. The optical setup of LIB technique for generating cavitation bubble.

2.3 Visualization setup

We used here shadowgraphy setup for the bubble visualization. This setup consists of continuous LED matrix daylight lamp Veritas, MiniConstellation 120 – 5000 K of illuminance 92 klux in 0.5 m, set with optical diffuse filter. Opposite to the light source was placed high speed CMOS camera SpeedSense. This camera is working on frequency 60 kHz with resolution of (128 x 128) px or lower frequency with higher resolution up to (1280 x 800) px, and the dynamic range 12 bit.

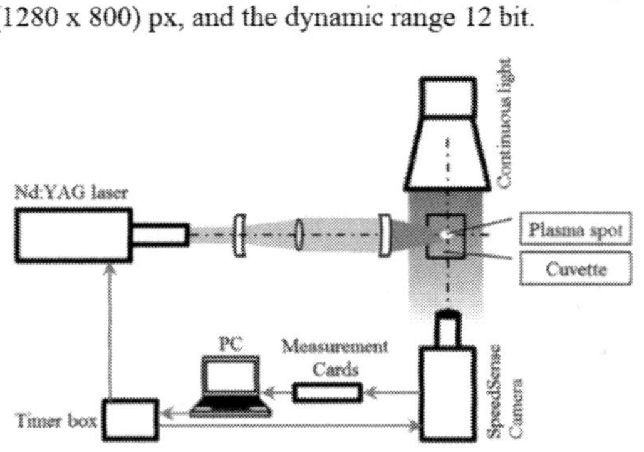

Fig. 4. The experimental setup of the laser, and the visualization system.

The camera exposure time was 1μs. The sub-pixel resolution was 20μm. The camera was mounted with optical lens system INFINIPROBE™ TS-160 universal macro/micro imaging system that enables 4x, and 16x magnification. The camera was mounted with edge pass and long pass filter cut-on wavelength 550 nm low pass optical filter to reduce the backward laser flashes to the camera and also to eliminate the flash generated while plasmatic breakdown.

We also used a light reflecting setup for visualization of the outer bubble structures. We placed the light under 45° from the camera direction.

2.4 Absorbance

The Lambert-Beer law expresses the relationship between the absorbance and the concentration of the investigated substance.

$$A = \varepsilon . c . l \qquad [1]$$

where A is the absorbance, c molar concentration of the test sample (mol/l), l cell thickness (cm) and ε molar absorption coefficient ($l.mol^{-1}.cm^{-1}$). The molecular absorption coefficient is a constant for a given substance at a given wavelength [13].

Liquid	Ethyl Alcohol	Ethylene Glycol	Glycerine
Surface tension [mN/m], temp.	22,39 [20°C]	47,30 [25°C]	63,4 [20°C]
Kinematic viscosity [mm^2/s]	1,52 [20°C]	17,80 [21,1°C]	648 [20,3°C]
Density [kg/m^3]	789 [25°C]	1097 [25°C]	1259 [25°C]
Absorption coeff. [cm^{-1}], wavelength	0,0182 [535nm]	0,0113 [535nm]	0,9837 [1200nm]

Liquid	Water distilled	Water fresh	Silicone oil
Surface tension [mN/m], temp.	72,80 [20°C]	72,86 [20°C]	20,10 [25°C]
Kinematic viscosity [mm^2/s]	1,0038 [20°C]	1,13 [15,6°C]	16,50 [40°C]
Density [kg/m^3]	0,9980 [20°C]	0,9982 [20°C]	980 [25°C]
Absorption coeff. [cm^{-1}], wavelength	0,0002 [535nm]	0,0006 [535nm]	2211 [535nm]

3 Results

There is generated the plasma in the spot due the concentration of the laser energy. The plasma is visible due the emission. The temperature in liquid increase to 10^3 K, and pressure to 10^3 bar in the spot volume. This leads to plasma expansion at supersonic velocities, producing an acoustic shock wave following with cavitation bubble effect. It is important to mark for the following figures 5 - 7 that the laser light comes from the up side on the pictures.

There have been discovered that at the same energy of the laser the bubble has a different lifetime depending on the type of liquid. Here we expected the bubble generated in liquid with the lower absorption coefficient to have the longest lifetime.

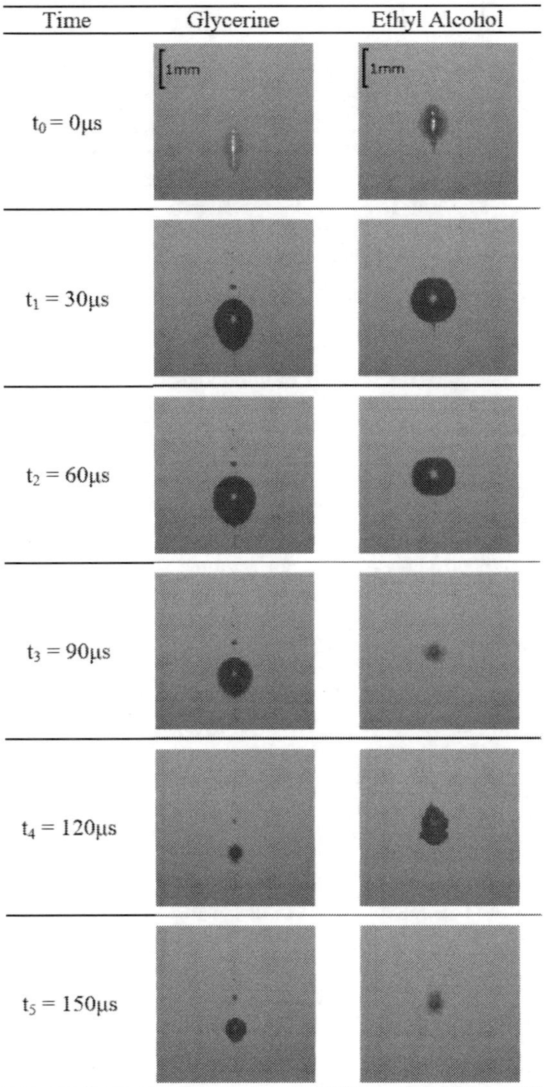

Fig. 5. Comparison of LIB generated bubble in Glycerine and Ethyl Alcohol for laser input energy 5,6mJ.

Figure 5 shows the development of the bubble in Glycerine and Ethyl Alcohol, where the size of the bubble is identical but its lifetime is different. Figure 6 shows the development of the bubble in Ethylene Glycol and water, where the size of the bubble and its lifetime is different.

It can be seen that the input laser beam differs in Figures 5 and 6 at time t_0 at the same laser energy. Figure 7 shows the bubble generated in Ethylene Glycol at double the energy of the laser and an interesting chain effect of bubbles orientation is seen. The plasma is avalanched and formed to almost line geometry with increasing energy from the source. This corresponds to temporal and spatial plasma evolution. [6] The energy in the spot is not distributed continuously that is also shown in bubble formation – from the small to the bigger one.

The parts of bubbles are joined together. The part of small bubbles perishes independently on the rest. The multiple bubbles develop also in reverse focusing setup, but more compact.

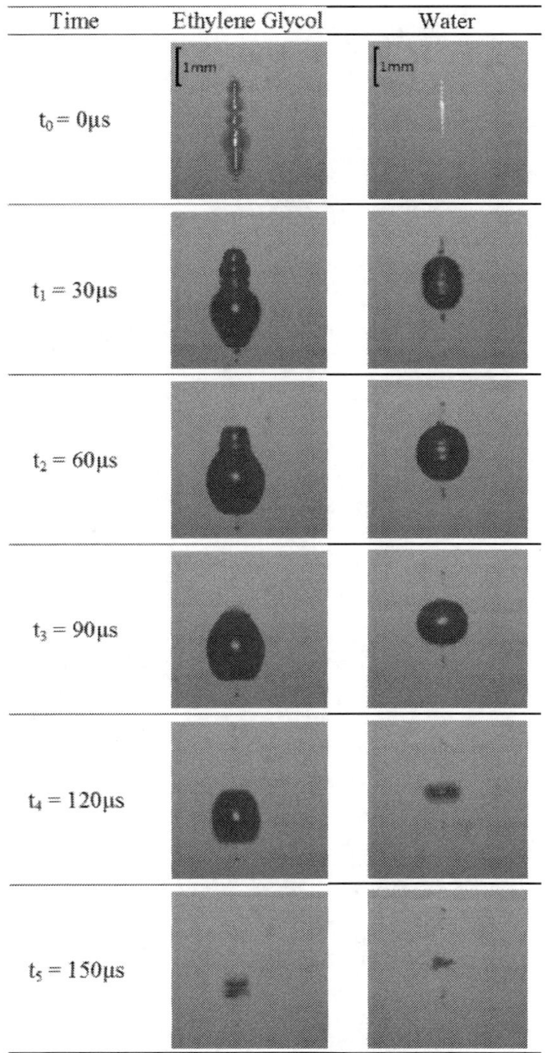

Fig. 6. Comparison of LIB generated bubble in Ethylene Glycol and water for laser input energy 5,6mJ.

They are further joined into big bubble that behaves as single one. As the bubbles were distributed at the beginning of cavitation process it is seen separated during the phase of implosion. Anyway the bubbles joined into single one, they are segregated again. In Figure 7, a primary show micro-jet creation through collapsing cavitation bubbles. It was well caught a specifically at time $t_8 - t_{14}$.

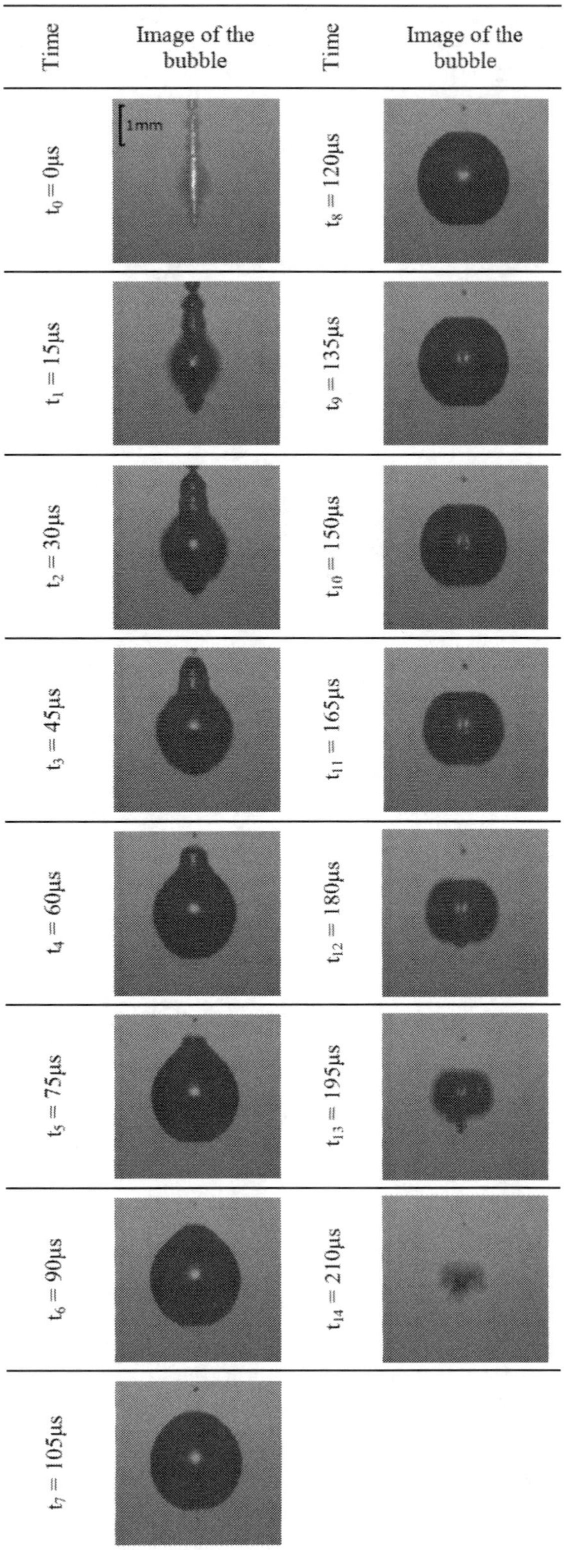

Fig. 7. LIB generated bubble in Ethylene Glycol for laser input energy 10,8mJ.

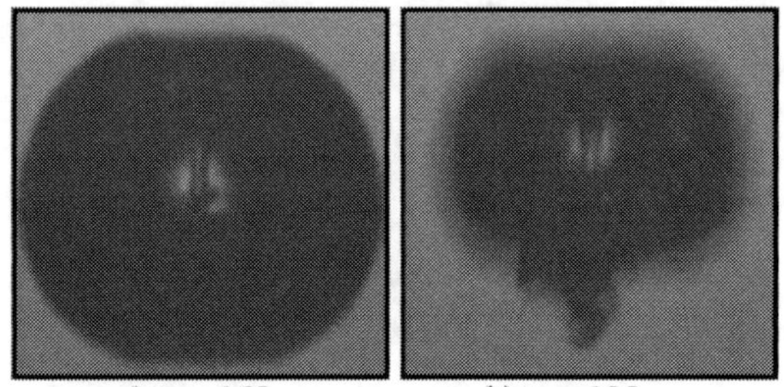

Fig. 8. Detail on micro-jet creation through collapsing cavitation bubble. LIB generated bubble in Ethylene Glycol for laser input energy 10,8mJ.

4 Conclusions

The visualization of cavitation bubbles is very important and useful tool in study of cavitation process. Here we use several type of liquid to study their advantages. We visualized the limits of the bubbles in their growth and size. This study enables us to get to know the importance of the suitable liquid selection and behaviour of bubble.

The volume in Figure 9 was obtained by calculating the pixel area of each captured image by the approximation of the sphere.

The curves in Figure 9 show the first phase of bubble life, with the exception of the Ethyl Alcohol curve. The Ethyl Alcohol Curve gets to zero at time of 90μs and then begins to grow again as a result of the second phase of the bubble lifespan. This fully corresponds to the development of the bubble cavitation.

The plasma expansion is followed by the shock wave and during this sequence the liquid is vaporized and a cavitation bubble is growing. This bubble is filled with water vapour. The bubble collapse occurs during the second phase of the bubble lifespan. The collapse is caused by the reduction of interior pressure and cooling that is influenced by the external environment.

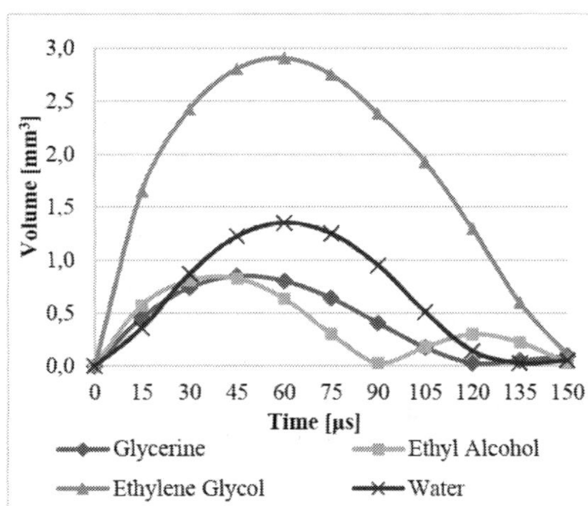

Fig. 9. Comparison of bubble size in different liquids.

The bubble has been generated with the same energy of the laser 5,6mJ, where depending on the type of liquid was measured different gas volumes of bubbles. The largest bubble was formed in Ethylene Glycol as seen in Figure 9. It is further shown in Figure 9 that the bubble formed in Ethyl Alcohol had the collapse of the bubble most rapidly to 90μs. The higher the viscosity of the liquid, the slower the bubble collapsed. Furthermore, in a liquid with higher absorbance was formed a bubble with a smaller volume of gas.

Here we managed to acquire a technique and procedure for evaluating bubble behavior over time that can be quantified and compared. This technique and procedure is based on image analysis. This first step is important for further investigation, but requires further optimization, such as approximating the bubble in a suitable shape.

Acknowledgment

The results of this project LO1201 were obtained through the financial support of the Ministry of Education, Youth and Sports in the framework of the targeted support of the "National Programme for Sustainability I", and the SGS project no. 21176/115. The authors gratefully thank to the support of Technology Agency of the Czech Republic (TA ČR) TH01020982 – Efficiency improvement of energy storage and ensuring grid stability by extending the operating range of pumped storage. Authors also acknowledge the institutional support of the Faculty of Mechatronics, Informatics and Interdisciplinary Studies of the Technical University of Liberec.

References

1. D. Jasikova, P. Schovanec, M. Kotek, M. Muller, V. Kopecky, Comparison of ultrasound and LIB generated cavitation bubble, *EPJ Web of Conferences*, 143, 2017.
2. D. Jasikova, P. Schovanec, M. Kotek, M. Muller, V. Kopecky, Experimental setup for laser-induced breakdown in aqueous media, *Proceedings of SPIE*, 10151, 2016.
3. D. Jasikova, M. Kotek, M. Muller, V. Kopecky, Single cavitation bubble interaction close to hydrophobic surface, *International Journal of Mechanics*, 11, 2017, 73-81.
4. D. Jasikova, M. Muller, M. Kotek, V. Kopecky, The synchronized force impact measurement and visualization of single cavitation bubble generated with LIB, *International Journal of Mechanics*, 9, 2015, 138-144.
5. Ch. E. Brennen, *Cavitation and bubble dynamics*, (Cambridge University Press, 2015)
6. P. K. Kennedy et.al, Prog.QuantElectr. **21**, 3, (1997)
7. E. A. Brujan, et.al, J.Fluid Mech. **433**, (2001)
8. W. Lauterborn, et.al, Archives of Acoustics **33**, (2008)
9. A. Shima, Y. Tomita, Ingenieur-Archiv **51**, (1981)
10. M. Gumulya, et. al, Bubbles in viscous liquids: Time dependent behaviour and wake characteristics, *In Chemical Engineering Science*, 144, 2016.
11. Chong Zhang, et. al, Dynamics of bubble formation in highly viscous liquids in a flow-focusing device, *In Chemical Engineering Science*, 172, 2017.
12. M. Kubala: Optical spectroscopy 1: Absorption spectroscopy. Olomouc 2011.
13. P. Šmejkal, I. Trejbalová, E. Urválková: Experiments of general and physical chemistry practicum. Prague 2007.
14. A. Vogel, S. Busch, U. Parlitz, J. Acoust. Soc. Am. **100**, (1996)
15. A. Vogel, et. al, Applied Physics B: Lasers and Optics **68**, (1999)
16. P. K. Kennedy et al., Prog. Quant. Electr. **21**, 3, (1997)

Compiled visualization with IPI method for analysing of liquid-liquid mixing process

Darina Jasikova[1,*], *Michal* Kotek[1], *Bohus* Kysela[2], *Radek* Sulc[3], and *Vaclav* Kopecky[1]
[1]Department of physical measurements, Institute for Nanomaterials, Advanced Technology and Innovation, Studentska 1402/2 460 01 Liberec 1, Czech Republic
[2] Institute of Hydrodynamics AS CR, v. v. i., Pod Patankou 30/5, 166 12 Prague 6, Czech Republic
[3] Czech Technical University in Prague, Faculty of Mechanical Engineering, Department of Process Engineering, Technicka 4, 166 07 Prague, Czech Republic

Abstract. The article deals with the research of mixing process using visualization techniques and IPI method. Characteristics of the size distribution and the evolution of two liquid-liquid phase's disintegration were studied. A methodology has been proposed for visualization and image analysis of data acquired during the initial phase of the mixing process. IPI method was used for subsequent detailed study of the disintegrated droplets. The article describes advantages of usage of appropriate method, presents the limits of each method, and compares them.

1 Introduction

Mixing is a very important operation in chemical industry and process engineering. It represents more than sixty percent of all industrial processes. A huge amount of mass is mixed in vessels stirred by an impeller. Large agitated tanks with impellers are used in food industry, pharmacy, and also in mining industry, waste water treatment etc. The global trend in process optimization involves new technologies providing higher quality products with lower energy demands. Trend is to develop more efficient mixing equipment. The optimization of stirred vessel and impeller design is done with help of Computational Fluid Dynamics simulations. The importance of these simulations validation was in recent past overlooked. Today, we are returning again to verification and the value of experimental validation is increasing.

Modern CFD methods are dealing with prediction of drop size distributions and their dynamics. This parameter determines mixing process evolution. The drop breakage, coalescence rates and final droplet size distribution depends on geometry of vessel and impeller, agitation rate and physical properties of mixed phases. Kolmogorov [1] and Hinze [2] were the first, who derived a relation for the maximum drop diameter. The maximum drop size d_{max} is determined by external deforming forces and restoration forces. Kolmogorov found the relation between d_{max} and local rate of energy dissipation ε. The time evolution of drop size distribution is based on theory of breakage and coalescence functions. Though, physical background is well described the basic relations and boundary conditions have to be improved depending on actual setup. [3]

Separate branch of research is becoming a development of methods enabling experimental validation of CFD simulations. Basically, measurement methods can be divided in two groups: the first one is based on sampling procedure; the second one is based on ex-situ measurements. [4] The ex-situ measurements can be done on emulsion before it coalesces, or on stabilized emulsions. The first group of measurements can be run directly with the mixing process and gives information about its evolution. These methods are mostly based on optical principles. The advantages of optical observations are in their non-invasive character. The disadvantage is limitation in the measurable range of droplet sizes. Expected droplet size distribution is above units of micrometres in mixing process. It can also reach sub-millimetre sizes at very beginning. This range can be covered with automated imaging, light back-scattering, and ballistic imaging methods. The first investigation of two-phase breakups of liquid in air is dating to before 1904. [2] These researchers used short-flash photographs showing the breakup of drops. This simple method upgraded with industrial sensitive camera and powerful light source, and helps reveal initial instabilities, shape of bulks and disintegration of one phase in these days as well in past. Even, we refill the camera with magnifying lens system, for example long distance microscopy, the limits of imaging are in micrometre size of droplets. Especially, it is essential if we look at a larger area for statistical evaluation. [5]

We searched for method that enables investigation of micrometre droplet sizes reaching limits of simple imaging methods. Suitable options are optical methods based on interference structure, which is created by light scattering on particle in collimated light. Expansion of

* Corresponding author: darina.jasikova@tul.cz

lasers in 60's brought more sophisticated methods such as PDA, and ILIDS, or IPI. [6, 7]

Phase Doppler anemometry (PDA) is an extension of laser Doppler anemometry (LDA). It is single-point measurement method. We can provide size and velocity measurement using this technique. The extremely difficulty of this methods is based on its principle. It is a local measuring method, so if the mixing process develops rapidly, it requires fast traversing over investigated area. On the other hand PDA enables measurement of large scale of particle sizes (0 -300) um and their velocities. It is useful method for study of stabilized dispersions.

Interferometric Particle Imaging (ILIDS or IPI) was developed as the alternative to the PDA. ILIDS can provide instantaneous spatial distribution of particle diameter. The spatial pattern is generated by the interference of external reflection and direct refraction of the droplet that is placed in the laser sheet. The limits of this method are in the size of defocused image, when in denser systems particles interfere together. Deploying this method is appropriate when primary disintegration of liquid bulks has already occurred and disintegration of liquids leads to separated droplets in volume. This method gives results of long-term measurements for evaluation and validation final droplet size. It works very well on ranges of particle size (10 - 300) um.

The motivation of our research was finding way to describe evolution of the mixing process and droplets disintegration as well as the stable dispersion that reach units from millimetres to micrometres. We tried to follow droplet size and shape. We have compiled a visualization setup with knowledge of the limitations of above mentioned methods that enables record initial part of rough disintegration during mixing process and also further settling phase. We combined shadowgraph method using backlight and industrial camera with IPI camera setup and laser light. The connection of both methods and the comparison of the results can give us tool that can be applied on scaled model of stirrer vessel.

2 Experimental Methods

2.1 Stirred Vessel Setup

We designed the scaled model of stirred vessel with inner diameter of cylindrical tank D = 150 mm and height H=D, position C = 1/4H.

We also used scaled model of Rushton turbine impeller of size D_1 = 37 mm, w= 10 mm, l = 12 mm and t= 1 mm. The vessel was set with four baffles of width B/D=1/10, so was B = 15 mm fixed evenly around circumference of the vessel. The height of the liquid level was covered.

There was used degassed water as a working liquid (density ρ = 1000 kg.m^3, dynamic viscosity μ = 1 mPa.s) under constant impeller speed 540 rpm. The volume of water was 2.65l.

The second phase was Rhodosil Oil 47V50 (density ρ = 959 kg.m^3, dynamic viscosity μ = 50 mPa.s, and surface tension σ = 20. 7 mN/m) injected directly into the vessel. The volume of oil was set 1% of liquid volume, so we used 26 ml.

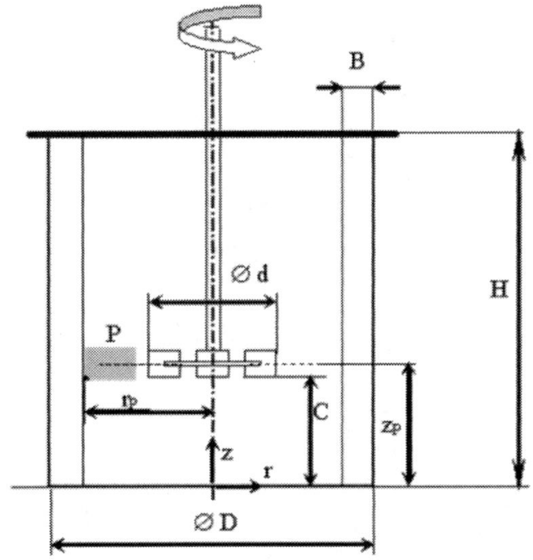

Fig. 1. Geometry of stirred vessel.

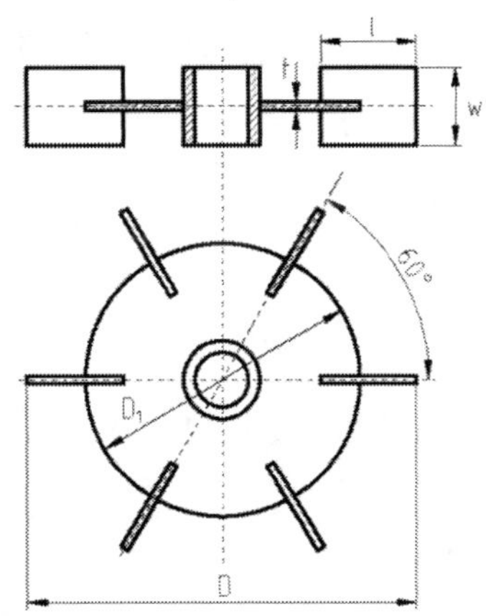

Fig. 2. Sketch of Rushton turbine impeller.

2.2 Shadowgraph Method

Shadowgraph method is well known in experimental fluid dynamics, especially for simple flow visualizations. The method requires clear optical transparent medium with local changes of refractive index. These disturbances are called shadows. We studied two-phase flow, shadows comes from disintegrated drops of oil. The setup was based on continuous Light Emitting Diode (LED). The LED matrix daylight lamp Veritas from Edmund Optics co., MiniConstellation 120 – light temperature 5000 K of illuminance 92 klux in 0.5 m, was equipped with optical diffuse filter. The industrial camera Basler acA2500-14gm with resolution 2592 px x 1944 px was used for visualization. The camera was fitted with optical lens system Sigma Macro. The camera exposure time was set 26 usec. That was the shortest exposure time controlled by the Pylon Camera Software

Suite. The intensity of light impacting the camera chip was manually set with the aperture. The back-lighting allowed the visualization of droplet shadows that appear as a black bounded object on white or light-grey background. Weakness in this conventional setup is in picture of drops that are out of camera focusing plane. These unfocused droplets move the statistics to higher values. The inverse mode of this setup enables capture light reflected on the surface of droplet. The droplets are seen as closed rings on the dark background. These rings appear only in focusing plane of the lens system. The rest of particles that are out of focus don't have sufficient contrast and cannot be edged detected. The inversed setup turned out to be more advantageous for further image processing and analysis.

The capturing frequency was adapted to the evolution of emulsification process. It was set 0.5 Hz to obtain the initial phase and rough disintegration into separated droplets.

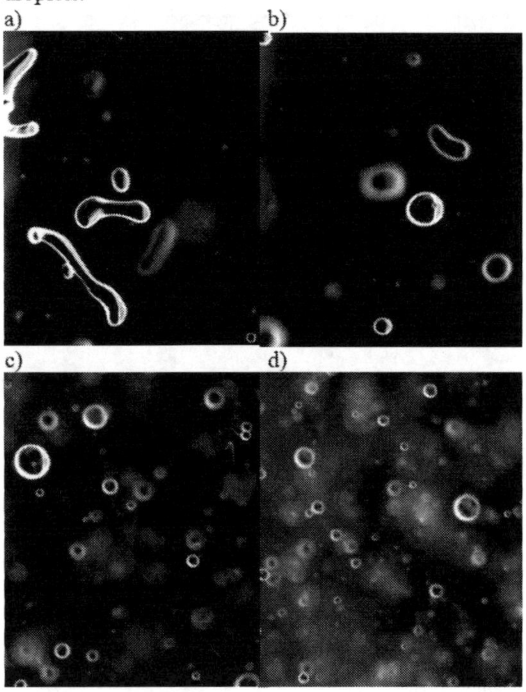

Fig. 3. Examples of records captured with shadowgraph method. a), and b) bulks of liquid at very beginning of mixing process; c) middle part with disintegration, d) stabilized disintegrated droplets of oil.

2.2.1 Image Processing

The image sequence was captured with frequency 0.5 Hz. The time of the record was set according to evolution of the process and last 140 sec. The image processing run in ImageJ open source software. The images were adjusted on brightness and contrast to eliminate unfocused and blurred droplets. The threshold function was used to remove dark background, highlight droplets and transform image to binary level. Next step of image preparation contained set of binary operation. There was applied mathematical morphology closing operator. Closing operator enlarged foreground region and kept original boundary shape. We managed improve the closure of rings (drops) and repair it, where it was missing due uneven light reflection. The prepared rings could be filled with another morphology operator – filling holes.

Software ImageJ provide special function for particle detection and analysis. This function works by scanning the image until it finds the edge of an object. The object can be specified by the size of particle, and circularity. The circularity set in range 0.8-1.0 enables to remove non spherical particles, such as bulks of liquid or particles joined together due overlap in the focusing plane. This variable set is based on researchers' experience, and an unhealthy setting may change the evaluation of statistical interpretation. We removed very small particles (noise due image processing) with setting the minimum area of particle to $10um^2$.

It was quite a problem at beginning of the mixing; when the liquid was formed into bulks of non-spherical shape, and diameter above 1mm (see Fig. 3). Result of analysis was presented as Sauter Mean Diameter D[3,2].

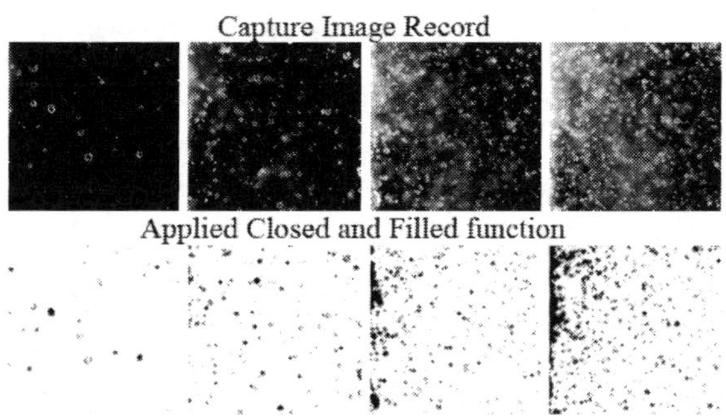

Fig. 4. Results after application of sequence image processing method.

Fig. 5. Processed image after Particle Analysis function with marked number of single particle.

2.3 Interferometric Particle Imaging Method (IPI)

The interferometric method of droplet sizing or interferometric particle imaging (IPI or ILIDS) can be attributed to the class of field optical investigation methods for two phase flows. One phase is liquid and second phase can be gas or liquid with different density, viscosity or immiscible liquid. It gives information about droplets is the whole chosen cross-section of the flow. That is great advantage of this method. The IPI can be apply to flow with lower densities of second phase and also requires optically transparent medium that enables light reflection and refraction.

The IPI method uses the distance between the fringes of interference pattern which is formed by light reflected and once refracted by a droplet to measure the droplet diameter. The fringe pattern is created by defocusing of camera. The level of defocused particle image is between 50 px and 120 px. There are mathematical suggestions, but proper correction depends on actual scene, and light conditions. The measurement system is based on laser, formed in the light sheet, and digital camera with defocused lens. The camera registers interference patterns from all droplets illuminated by the light sheet. Here comes the restriction by droplet concentration.

Other limitations come with the physical background of IPI method. There are set lower and upper limits of measured particle diameter for each experimental setup. The first limit relates to geometrical optics. The second limitation depends on actual setup, i.e. camera distance, objective aperture, collecting angle. These limits are reflected in the number of interference fringes. The minimum number of fringes and therefore measurable diameter is N=2. Such a pattern can be easily created by secondary reflection. Evaluation of such noise as a valid particle is incorrect. This reading is reflected by shifting the peak to the smallest particles in the resulting histogram. Limits for maximal size of particles are related to Nyquist theorem. There is set the minimum distance between fringes in interference pattern Δx=2px. [8, 9, 10]

The experimental setup for IPI technique is based on laser source and high-resolution camera. The simplest version doesn't require two cameras, when velocity field calculation is not need. Here we used Nd:YAG pulse laser NewGemini, working on 532 nm with output energy of 80 mJ per pulse. The laser illuminated region of size (56 x 68) mm that was also set as investigated area. The laser sheet thickness was 3mm. The camera HiSense Neo with resolution (2560x2160) px and 12-bit dynamic range. This camera was fitted with an objective with a focal length 60 mm (f/2.8). The distance between laser sheet and camera was 800 mm. This experimental setup with scattering angle 60°, aperture diameter of 21.430 enables measurement of droplet diameter minimum 21.58 um and maximum 323.63 um.

We run IPI measurements every 5 minutes and there were captured 256 records in each run. Sampling frequency was set 10 Hz. The first run was set immediately after the injection of Rhodosil oil. Immediate creation of large bulks of oil drops disabled measurement due huge light scattering. Emulsion started to be measurable after 8minutes of mixing, this correspond to shadowgraph method. Droplets are significantly separate and starting to have uniform distribution.

Analysis of results also contained image post processing procedures such as subtract of mean image to reduce scattered light, masking of objects and dark areas, and filters (edge detection, threshold).

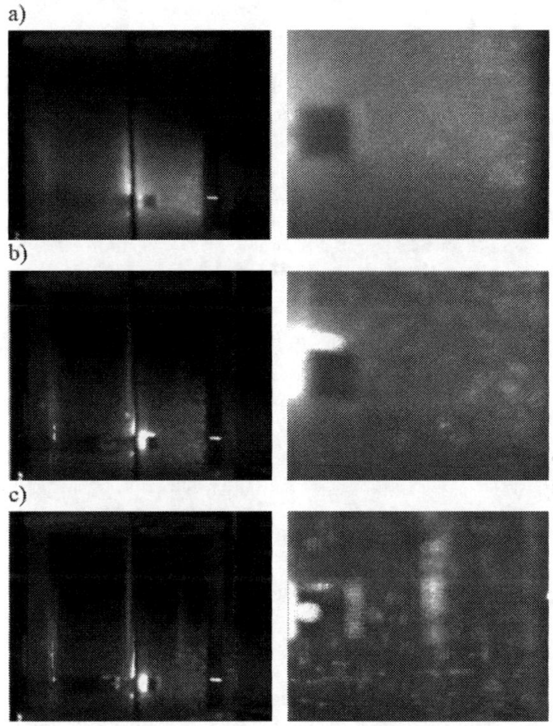

Fig. 6. Camera view on mixing process. The focused picture of interference structure. a) blurred investigated area due scattered light from bulks of oil, b) phase of forming droplets, c) measurable medium with well separated droplets of uniform size.

3 Results and Discussion

The theoretical considerations of stirring on liquid-liquid dispersion include definition of Reynolds and Weber numbers. Reynolds number classifies the flow regime. We were at the stage of looking for a methodology for evaluation, so we could not afford the Re > 10^4. We needed to monitor the relatively slow start of the process so we adjusted the speed of rotation. The Reynold number calculated:

$$Re = \frac{\rho . N . D^2}{\mu} \quad (1)$$

where ρ is liquid density [kg.m^3], μ is dynamic viscosity [Pa.s], N is stirrer rotational velocity (s^{-1}), and D is diameter of the stirrer [m].
The Reynolds number was 20x10^3 in our case, so we followed condition for turbulent flow regime, when inertial forces are higher than the viscous effects.

The balance between break-up and coalescence of the droplets can be expressed by Weber number:

$$We = \frac{\rho.N^2.D^3}{\sigma} \quad (2)$$

where σ is interfacial tension [N/m].

Using Weber number we are able estimated d_{max}:

$$d_{max} = 0.125.We^{-0.6}.D \quad (3)$$

Predicted size of particle was calculated as 160 um. The calculation is based on the Weber number with exponent 0.6. The exponent itself is subject of optimization. It seems to be dependent on the uniqueness of the stirred mixture of liquids. We used this value to set the experiment, especially the IPI method's range.

We measured the experiment with Rhodosil Oil in two separate runs. The mixing process was captured with shadowgraph method during its evaluation. The IPI method was applied after first 2 minutes of immeasurable stadium. Fig. 7 and 8 shows stable state 10 minutes after injection. Fig. 7 and 8 are comparable. We did this experiment to validate dataset measured with IPI technique using different method. There is seen difference on the level (20 - 40) um. The IPI histogram of oil droplet size contains characteristic values. This is due to shadowgraph method image processing. The lower limits were set during the Particle Analyse function. These limits reduced the population of droplets with this diameter. There is seen missing column in fig. 7. The visualization method can be improved with regard to the estimated process development. We were not sure about the development of the disintegration at this initial experimental setup. The optics of the visualization system has been chosen to hit large particles (above 1mm in diameter). We were interested in a larger evaluation area. We already know that the smallest droplets reach about 30 um. We choose optics that will allow greater magnification in the next step of research.

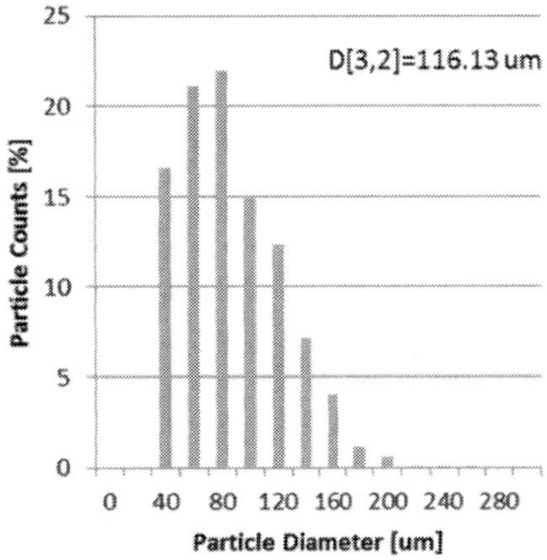

Fig. 7. Histogram of particle diameters of oil droplets evaluated from shadowgraph records in stabile part of dispersion.

We calculated Sauter Mean Diameter SMD for both measurements. This value is presented as volume diameter D[3,2]. There is agreement for both measurements with different technique. The D[3,2] for shadowgraph method was 116,13um and D[3,2] for IPI method was 119,9um. The values differ in 3%. This error is award to methods. The second influence is uneven adhesion of larger oil bulk to vessel walls, baffles, and lid in unmeasurable amounts, which affects the overall viscosity of the liquid.

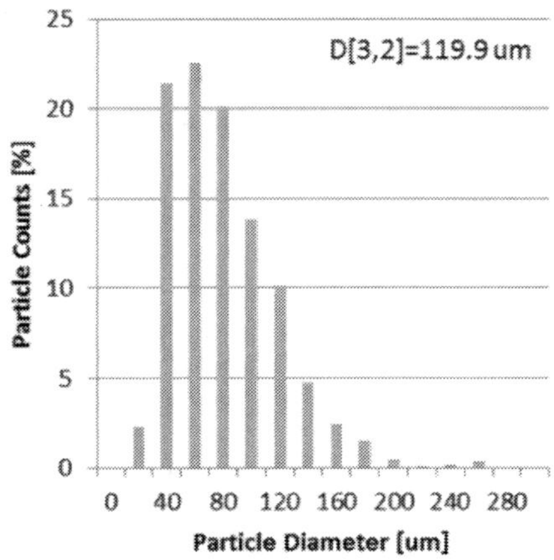

Fig. 8. Histogram of particle diameters of oil droplets evaluated from IPI method in stabile part of dispersion.

The finer histogram, see Fig. 9, represents detailed IPI graph. Previous interpretations of measured data by the IPI method were tailored to the occurrence of classes as set by the visualization method. The distribution in Fig. 9 is separated into finer steps. The IPI is more sensitive to the circular nature of the particles, and eliminates large particles outside the measurable range.

Evidence that the IPI method was set correctly is that it does not arise from the analysis of secondary interfering patterns. There is no false peak (phantom pattern) due to noise.

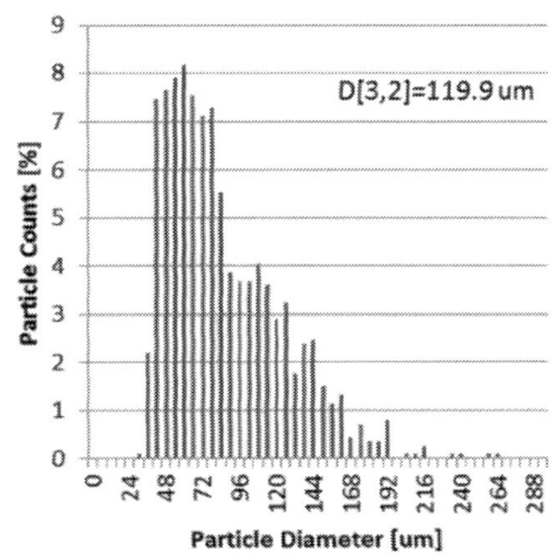

Fig. 9. Detailed view on histogram of particle diameters of oil droplets evaluated from IPI method.

These evaluations show that in the stabile part of mixing process we can compare both methods and the results are not significantly different.

Comparison of the time course in figure 10 is combining results of both methods. The a) represents dataset results captured by shadowgraphy method. This part of chart fluently continues in part b). Part b) of Fig. 10 represents dataset measured by IPI method. Each dot in the figure 10 a) indicates the evaluation and average value AVG dimeter of the detected drops in one record. Dots in figure 10 b) means IPI statistic evaluation within a few minutes of 256 records collected with 10Hz frequencies during 26 sec. The shadowgraphy scanning frequency was 0.5 Hz. Therefore, scattering of points is higher, because they are not statically averaged.

Smaller time sampling was more advantageous due to the rapid dispersion of the oil phase. This is a unit of seconds. While the staging phase of the process takes a few minutes. Once, we have tool for automated analysis the sampling frequency can be set higher. The upper limit for sampling frequency is the velocity of moving droplets.

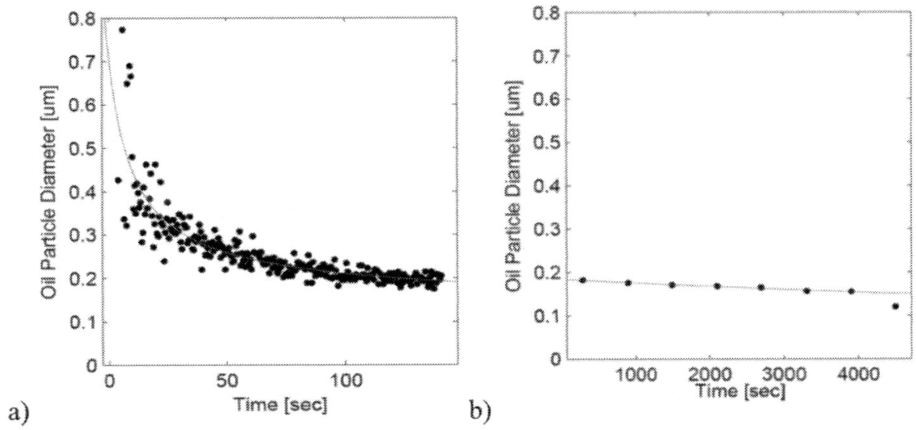

Fig. 10. Evaluation of time evolution of mixing process – comparison of shadowgraph images and IPI method. a) Results of analysis of shadowgraph images – AVG diameters, b) IPI – AVG diameter statistics

It is required that the particle population in each image changes and the same particles are not measured several times.

The question is whether the gradual trend of decreasing of average particle diameter is due to progressive disintegration, or rather by settling of the bulk of liquids on the walls of the vessel, or separating them on the cover.

Each of these methods is suitable for study of different part of the dispersion process. The chosen method is complementary and directly related to the size distribution of the particles. To verify the accuracy of the methods, an overlay time window can be found in which it was possible to evaluate the size distribution of both methods (Figure 7, and 8)

It appears that the size stops significantly decreasing after 2 minutes of mixing, depending on evolution in time. It is possible, in this particular case (ratio 1% of volume), to mix two liquids in a given ratio to be stabilized in this time.

It has been shown that combining these two methods (visualization and IPI method) can include a comprehensive overview of the development and stabilization of the mixing process of two liquids.

Acknowledgment

Authors gratefully thank to the support of Grant Agency of the Czech Republic (GA ČR) 16-20175S Local turbulent energy dissipation rate in dispersion systems, and LO1201 co-funding from the Ministry of Education, Youth and Sports as part of targeted support from the „National Programme for Sustainability I".

References

1. A.N. Kolmogorov, Dokl.Akad.Nauk SSSR, **6** (1949)
2. J. O. Hinze, A.I.Ch.E. Journal, **1**,3 (1955)
3. P. Ditl, F. Rieger, Chemical Engineering Progress, **102** (2006)
4. A. Khalil, Chem.Eng.J., **165** (2010)
5. T. Kawaguchi, Meas. Sci. Technol., **13** (2002)
6. F. Pereira, Meas. Sci. Technol., **15** (2004)
7. Ch. Dunker, Meas. Sci. Technol., **27** (2016)
8. N. Semidetnov, Meas. Sci. Technol., **15** (2004)
9. M. Itoh, Optical Review, **5** (1998)
10. G. Lacagnina, Exp Fluids, **50** (2011)

Experimental Investigation of the Boundary Layer Transition on a Laminar Airfoil Using Infrared Thermography

Tomáš Jelínek[1,*]

[1]VZLU – Czech Aerospace Research Centre, Beranových 130, 199 05 Prague, Czechia

Abstract: The transition in the boundary layer is investigated using infrared thermography (IRT). The study is carried out on a laminar airfoil in the transonic intermittent in-draft wind tunnel. The transition in the boundary layer is evocated using transition-generator strips of different thicknesses at two Mach numbers: 0.4 and 0.8. The tested transition-generators thickness to boundary layer displacement thickness ratio was from 0.42 to 1.25. The Reynolds number respect to the airfoil chord is: Re = 1.0 – 1.4·10^6. The six cases for different transition-generators thickness ratios were compared. The behaviours of laminar and turbulent boundary layers are discussed. The use of IRT has been proven to be an appropriate tool for detecting the transition of the boundary layer in high-speed wind tunnel testing.

1 Introduction

To study the development of the boundary layer and the transition from the laminar to the turbulent flow in the boundary layer is very important because it fundamentally changes the character of the boundary layer; it means the parameters of the boundary layer and its behaviour. How the boundary layer behaves especially in critical flow regimes depends mainly on its nature and how the boundary layer is developed.This is important both in aerospace applications where the airplane's lift depends on the separation of the boundary layer from the wing airfoil and in the applications in the drive units and other turbomachinery applications of various power aggregates where the performance and efficiency of these machines depends on the nature of the boundary layer and its possible separation.

For these reasons, it is necessary to address the boundary layer's nature, to examine under what conditions and how soon the transition of the laminar boundary layer into turbulence occurs.

This phenomenon can be monitored, for example, by measuring the surface tension by means of hot wire anemometry. It requires equipping the surface of the object under investigation with hot films and to measure their response to cooling caused by fluctuations of the turbulence flow.

However, the formation of the transition of the boundary layer into turbulence also depends on the surface of the model under investigation, where small inequalities or changes on the surface may cause or bypass the transition of the boundary layer into turbulence. Thus, if the model is of small dimensions, it is problematic to equip the surface of such a model with sensors that always have the final dimensions, and so the nature of the boundary layer can be significantly affected. In these cases, it is appropriate to use a method that would detect the transition in the boundary layer without having to change the surface of the examined model.

One of the methods commonly used to visualize the transition of the boundary layer without any need to change the model´s surface is visualization using high resolution infrared thermography. Its use is widespread in both aerodynamic wind tunnel testing and real-conditions experiments.

In aerodynamic wind tunnel testing, it is sometimes necessary to artificially trigger the acceleration of the turbulence in those cases where the measurement results are significantly affected by the existing laminar boundary layer. Very often, the strips with roughness elements are used at varying distances at the leading edge of testing airfoil or wings when there is a need to fix the boundary layer transition.

This can also be the case of prismatic blades testing when inlet flow conditions (low inlet turbulence intensity along with a low Reynolds number) can cause the boundary layer to remain laminar up to the blade´s trailing edge. This may not be the case with the real operation conditions of the blades, where the transition may take place much closer to the leading edge due to different inlet flow conditions.

The transition in the boundary layer can also be simulated using various mathematical models in Computational Fluid Dynamics (CFD). Today mathematical solvers are available that use different transition models and predict the transition in the boundary layer based on the boundary conditions. The

* Corresponding author: jelinek@vzlu.cz

obtained characteristics of boundary layers experimentally are therefore important also for validation of these mathematical models [1], [2].

To gain a better idea of how to use the transition-generator strips and how they can affect the transition in the boundary layer, an experiment was performed on the laminar airfoil in the transonic testing section at two Mach numbers and three different strips thickness.

2 Principle of IR thermography

There are many options how to detect the transition in the boundary layer experimentally. These methods are clearly described in [3]. For qualitative transition evaluation, the hot-wire anemometry (hot-films) is best to use. This method gives a very good conception of the behaviour of the boundary layer, but the installation of the surface sensors may, in particular in the case of small dimensions of the model, significantly affects the transition in the boundary layer.

Therefore, if there is need to determine the transition in the boundary layer, for example, for comparing (validating) with a CFD calculation, which uses transition models in the boundary layer, using IR thermography appears to be very good method.

The principle of detecting the character of boundary layer using high resolution IR thermography is a common technique in many research areas that has been used since the 1960s.

Scanning the infrared radiation one can detect both the transition to turbulence and the separation of the boundary layer. Increased heat transfer in turbulent flow compared to laminar leads to a change in the surface temperature of the model. It works providing that the fluid and the surface of the model have different temperatures. This technique is applicable to stationary measurements where the conditions of the boundary layer are stable in the long term, but it is also possible to measure unsteady phenomena with sufficient equipment.

The basic principle of IR thermography for detecting the transition of laminar boundary layer into turbulence is the difference in coefficient of convection of heat transfer between the laminar and turbulent boundary layers.

The laminar boundary layer acts as an insulator in the heat transfer between the ambient air and the surface of the model. On the other hand, in the turbulent boundary layer the vorticity of the flow increase and thus the heat transfer between the surrounding flow and the surface of the model increase as well. This will cause that the model surface temperature change. The surface temperature gets closer to the temperature of the flow.

Thanks to this phenomenon, the transition region can be detected as an area of steep temperature change between two areas with different temperatures. (The direction of temperature change is depending on whether the model is warmer than the ambient air or vice versa). The greater the difference between the temperature of the model and the flow, the more readable is the area of the transition.

IR thermography of turbulence boundary layer detection is therefore a temperature measuring technique related to forced thermal convection on the surface of the model under investigation. Heat transfer q can be calculated using the heat transfer coefficient h, the stagnation temperature of the flow T_0 and the model surface temperature T_w:

$$q = h \cdot (T_w - T_0), \quad (1)$$

where h is closely related to the friction coefficient c_f, the velocity of the flow v, thermal conductivity of fluid λ_c and kinematic viscosity ν. For the heat transfer coefficient in the distance x the dependence is given by:

$$h(x) = \tfrac{1}{2} \cdot c_f(x) \cdot v(x) \cdot \lambda_c / \nu. \quad (2)$$

Equation (1) is often called the Newton's cooling law, equation (2) is determined from the Reynolds's analogy, which refers to the heat (energy) transfer in the boundary layer when the value of Prandtl number is close to one.

Usually, c_f is one order higher for a turbulent case than a laminar one. If the heat transfer from the model to the flow q (in the case of a warmer model´s surface and a cooler flow) is kept constant (eg. by constant electrical heating), then if the c_f increases (transition from laminar flow to turbulent), the surface temperature of the model T_w will decrease. The resulting change in surface temperature of the model can then be measured, for example, by IR camera.

3 Configuration of experiment

3.1 Wind tunnel

The experiments were carried out in the transonic wind tunnel with testing section dimensions 0.4 m (width) and 0.8 m (height). It is an intermittent in-draft wind tunnel, which employs a vacuum chamber with a capacity of about 4100 m^3, in which a vacuum is produced using a set of water-ring vacuum pumps. The maximum level of vacuum is about 40 kPa of absolute pressure. Once the necessary pressure has been reached, the vacuum chamber is connected to atmospheric pressure and atmospheric air flows through the test section and through the adjustable throat. The humidity of intake air can be reduced in the wind tunnel inlet by a dryer plant. The air in the test section had a temperature of 284 - 287 °K (11-14 °C); the surface temperature of the model was between 15 - 18 °C during the performed experiments.

The advantage of using an intermittent aerodynamic tunnel in this case is that the model does not need to be permanently heated to a different temperature because it still keeps the ambient temperature of the aerodynamic tunnel test section area. Exposure the model to the flow of lower temperature (since it was measured in winter time with a lower air temperature without switching on the air dryer) was only about 5 seconds. There was no change for the model to keep the ambient flow temperature during the time of wind tunnel run.

3.2 Model

For the aerodynamic wind tunnel tests, the airfoil NACA 16-012 was selected. Geometry is visible in figure 1. The model chord was 100 mm; the span of the model was 200 mm.

This is a symmetric laminar airfoil with a maximum thickness in 60% of the chord. The static pressure distribution on the surface of the model was available from the previous measurement.

3.3 Transition-generator strips

To accelerate the boundary layer transition on the laminar airfoil, the transition-generator strips were attached to the observed side of the model at a distance $x/c = 0.35$ from the leading edge. Strips were made using the scotch tape of different thicknesses. The width of the strips was always 2 mm. The three different thicknesses of the transition-generators, K = 0.04, 0.08, 0.11 mm were tested in total.

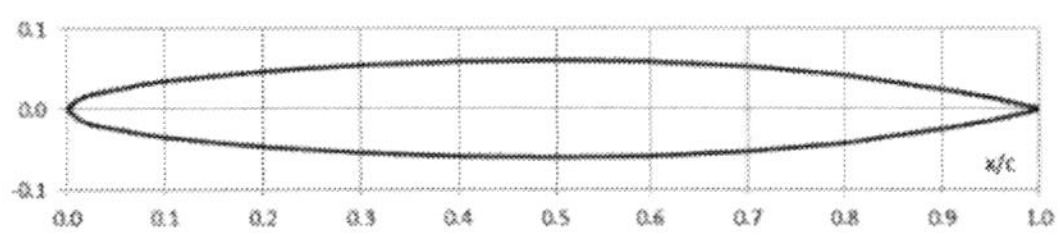

Fig. 1. Geometry of tested airfoil NACA 16-012.

3.4 China clay visualization

In addition to monitoring the boundary layer transition by means of IR camera, the surface visualization method was applied in the experiments using an oil mixture and visualization particles. Air kerosene (kinematic viscosity at 20 °C is 1.3 mm^2/s) was used as oil.

Magnesium oxide powder was used as the visualization particles.

The mixing ratio was in all cases the same in a volume ratio of 1:4 (one unit of particle to four units of oil).

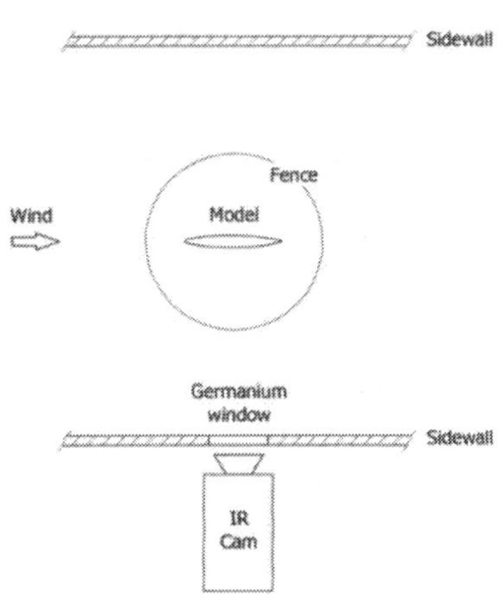

Fig. 2. Schema of the xperiment layout in the wind tunnel test section (0.8 m height, 0.4 m width)

3.5 IR Thermocamera

The high-resolution FLIR A655sc thermocamera was used for the experiment. The resolution of the video taken is 640 x 480 pixels collected at 50 Hz. The camera's thermal resolution is less than 50 mK. The temperature measurement accuracy is ± 2 °C.

For the data post-processing the ResearchIR© software was used.

3.6 Experiment layout

Since it was necessary to provide optical access to the surface of the model, the model was placed in the test section in a non-standard manner, parallel to the side walls of the wind tunnel (see figure 2). It was necessary to attach the model to special brackets that were attached to the upper and lower walls of the test section. Since the model was not bounded by walls in the tunnel, the airfoil was equipped with circular fences of radius r = 170 mm made of 1 mm thick metal sheet.

From the sidewall of the wind tunnel, it was possible to capture the surface temperature of the model with a thermocamera through a germanium-glass window.

4 Numerical simulations

To imagine the thickness of the boundary layer, numerical 2D flow calculations were performed for two inlet Mach numbers (0.48 and 0.81). By calculation, the boundary layer parameters were found in 35% of the profile chord, where the transition-generators were placed on the surface of the model.

The calculation was done using CFD code Edge [4].

The numerical simulation was performed as a laminar flow with fixed position of transition to turbulence at 60% of the chord length.

For the numerical simulation, the inlet conditions parameters were set as follows: p_0 = 100,000 Pa, T_0 = 288.15 K, $\alpha = 0°$

The calculation results are summarized in table 1.

Table 1. Boundary layer parameters in $x/c = 0.35$

	$M_\infty = 0.48$	$M_\infty = 0.81$
δ [mm]	0.340140	0.277734
δ^* [mm]	0.094918	0.087988
δ^{**} [mm]	0.044597	0.038229
H [1]	2.128346	2.301966
τ_w [kg·m^{-1}·s^{-2}]	22.798541	52.932635
u_τ [m·s^{-1}]	4.796121	9.187901
c_f [1]	0.001655	0.001774

5 Results analysis

The transition to turbulence in the boundary layer on the planar plate is a process that is very well described by Morkovin [4] and, of course, by Schlichting [5]. On the emerging, initially stable laminar boundary layer, unstable (Tollmien-Schlichting) waves gradually begin to occur after reaching the Re_{kr}, which gradually passes

 https://doi.org/10.1051/epjconf/201818002040

into three-dimensional vortices (the so-called lambda-structures). These vortices begin to disintegrate, turbulent areas are formed, and turbulent flow begins to develop.

On a bent plate (airfoil, blade cascades), this mechanism is additionally subjected to a different pressure gradient that can accelerate or delay the process of transition of the boundary layer.

If any irregularity, impurity, etc. occurs on the surface, this may cause a local wedge of a turbulent boundary layer (figure 3), if the impurity is sufficiently large.

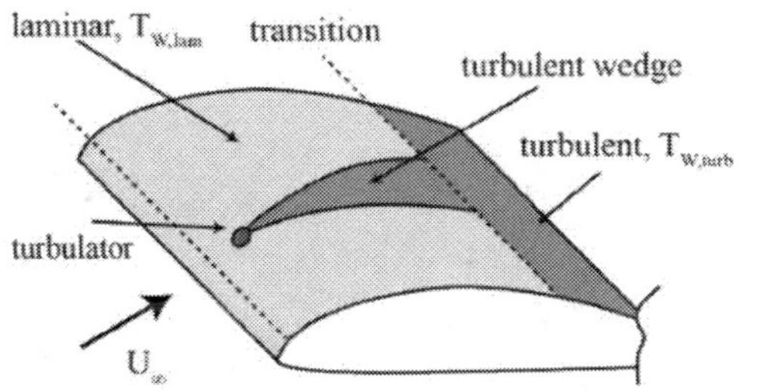

Fig. 3. Scheme of transition of the boundary layer on the airfoil with the representation of the turbulization wedge.

In the images obtained using IR thermography, it is possible to distinguish different color shades that represent the temperature that the surface of the examined model emits at each point. The darkest shades represent the coldest surface area and the lightest shade the warmest surface area. In some cases where surface visualizations using oil solution were also used, it is also possible to identify areas with higher concentrations of solid particles on the surface.

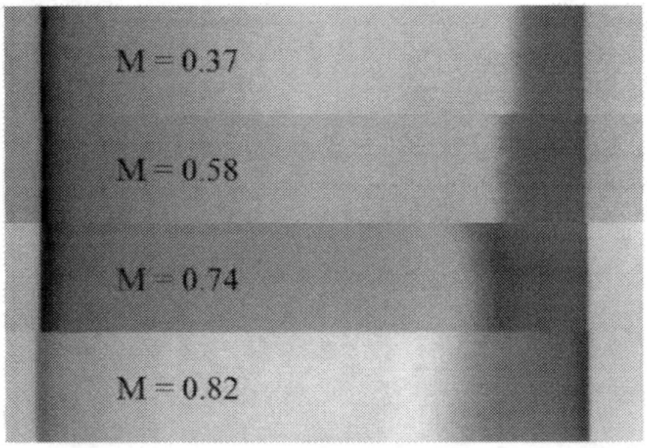

Fig. 4. The comparison of IRT images of the Airfoil without the transition-generator for different Mach numbers.

5.1 Airfoil without the transition-generator

Without a transition-generator strip, a total of five regimes were measured for the progressively increasing Mach number: 0.37; 0.58; 0.74; 0.82 without the use of surface oil visualization. From the results (figure 4), it is obvious that as the Mach number rises, the transition area of the boundary layer moves towards the leading edge.

In the case of transonic regimes, the shock wave oscillates due to the non-stationary tearing of the laminar boundary layer and its position is unclear (figure 5). For the cases of the turbulent boundary layer, the shock wave is considerably more stable (figure 6).

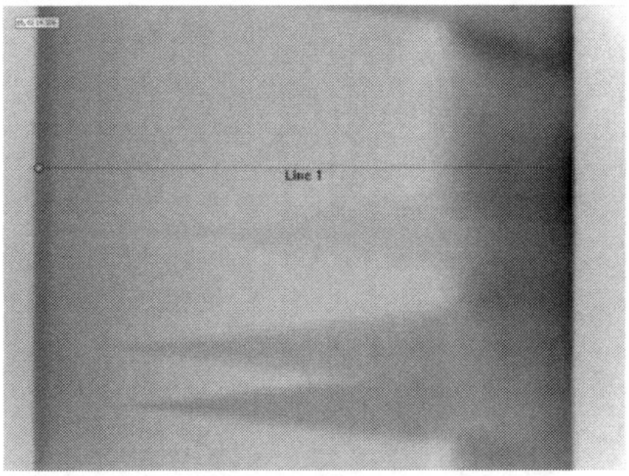

Fig. 5. The IRT image of the Airfoil without the transition-generator strip (M = 0.8)

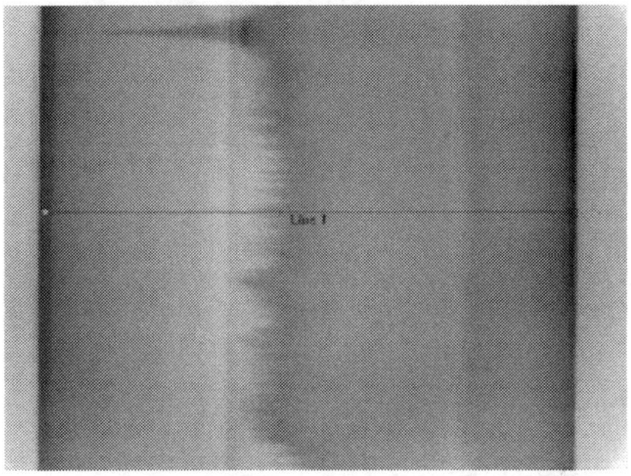

Fig. 6. The IRT image of the Airfoil with the transition-generator strip in x/c = 0,35 (M = 0.8).

5.2 Comparison of the results for different thicknesses of transition-generator strips

Figure 7 compares measured data obtained by IR thermography for two Mach number levels and three transition-generator strips thicknesses.

Values of K′ and L′ are displayed in table 2:

Table 2. Values of K′ and L′ for tested cases.

	M = 0.5	M = 0.8	
	δ^* = 0.095	δ^* = 0.088	L'
K = 0.04 mm	K' = 0.42	K' = 0.45	50
K = 0.08 mm	K' = 0.84	K' = 0.91	25
K = 0.11 mm	K' = 1.16	K' = 1.25	18

From the results obtained by the IR thermographs, it is evident that even the lowest value of the thickness (K' = 0.42) causes a transition in the boundary layer. If the thickness further increased (K' = 0.45, 0.84, 0.91), the transition area is gradually shortened. The same can be observed when increasing Mach number what means reducing the thickness of the boundary layer under the same conditions, thus increasing the value of K'.

For values of $K' > 1$ then we can observe the local boundary layer separation (bubble) behind the transition-generator strip. Behind this bubble, there is an almost immediate transition to turbulence (figure 7, e,f).

In All figures depicted in figure 7, there is a line, indicated as a "Line 1". The temperature profile along the airfoil chord in this line is plotted in figure 8.

By comparing the temperature profiles along the airfoil chord for the individual investigated cases (figure 8), we can observe a gradual increase in temperature from the leading edge to the position of the transition-generator strip (in 35% of the chord).

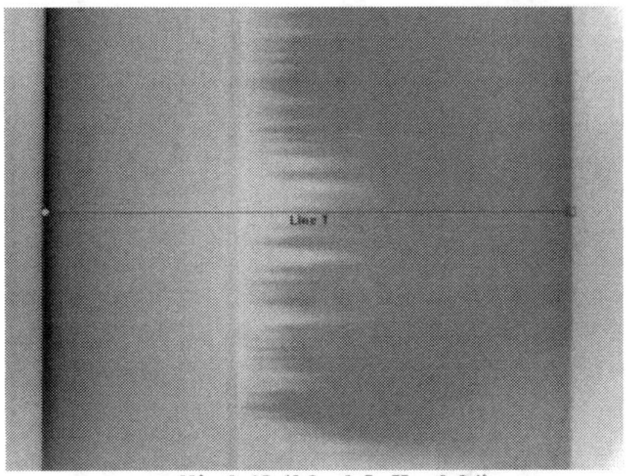

a: K'= 0.42 (M = 0.5, K = 0.04)

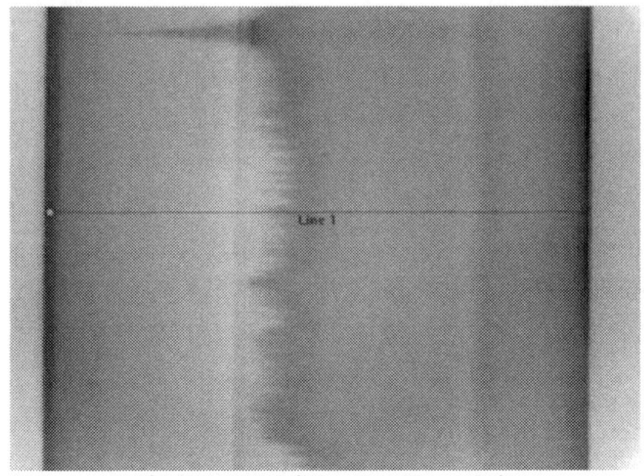

b: K´ = 0.45 (M = 0.8, K = 0.04)

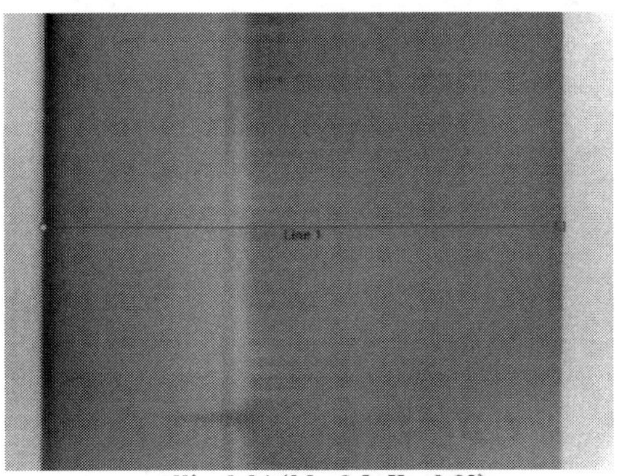

c: K' = 0.84 (M = 0.5, K = 0.08)

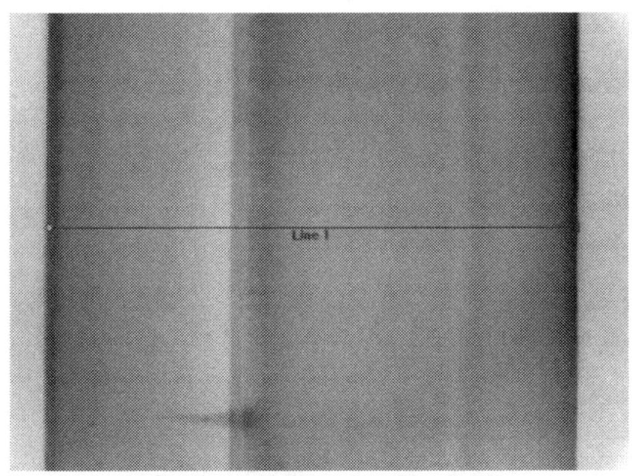

d: K' = 0.91 (M = 0.8, K = 0.08)

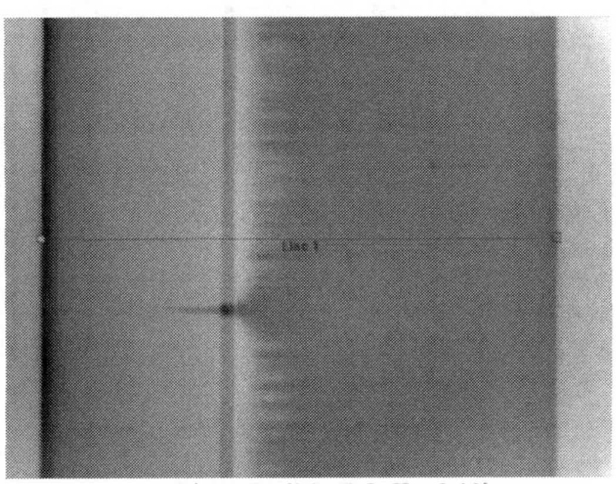

e: K' = 1.16 (M = 0.5, K = 0.11)

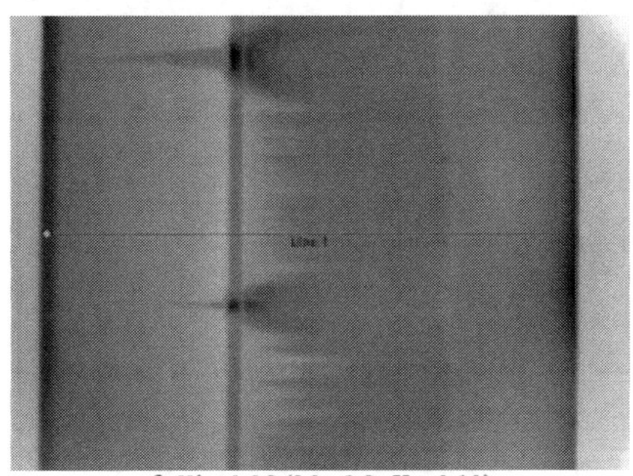

f: K' = 1.25 (M = 0.8, K = 0.11)

Fig. 7 (a, c, e): The IRT images of the airfoil with the transition-generator strip in x/c = 0.35 with various K.

Fig. 7 (b, d, f). The IRT images of the airfoil with the transition-generator strip in x/c = 0.35 with various K.

This corresponds to increase the laminar boundary layer thickness, which acts as an insulator and the surface maintains its original temperature.

The subsequent drop in the temperature means the transition area in the boundary layer (turbulization of the boundary layer). Here it can be observed that for the case where $K' = 0.42$, the transition area is the longest and the most gradual compared to the case where $K' = 1.25$, when there is a relatively rapid drop in the surface temperature of the model.

A subsequent region of more or less constant temperature means the area of the turbulent boundary layer. If there is a sharp decrease in temperature at the end of the airfoil, it means there is a boundary layer separation.

Schlichting [6] and Tani et al. [7], show the dependence of the surface disparity height K for the turbulence transition condition (relation of Re^*_t over K/δ^*). This dependency is determined for a single roughness element.

In figure 9 the comparison of the results obtained on the laminar airfoil with the transition-generator strips of different heights with the dependence obtained for the single roughness element is depicted.

It should be noted here that the experiments that led to the determination of dependence on figure 9 and the generalized line according to H.L. Drydena [8], were made assuming the development of a boundary layer on a flat plate, the use of a circular single roughness element and an incompressible flow.

As can be seen in figure 9, the points from the experiment performed in this work are divided by the line of the completed transition in the boundary layer into two areas: For $K' = 0.42$, 0.45, 0.81 lies in the unfinished transition area, while values for $K' = 0.91$, 1.16, 1.25 lies in the transition area.

If we follow the temperature curve in figure 8, it is evident that for the values of $K' = 1.25$ and $K' = 1.16$ there is a clear transition to the boundary layer and the lengths of transition for both cases is roughly $x/c = 0.1$. For the case of $K' = 0.91$, however, the temperature progression indicates a partial relaminarization of the boundary layer after its transition to turbulence.

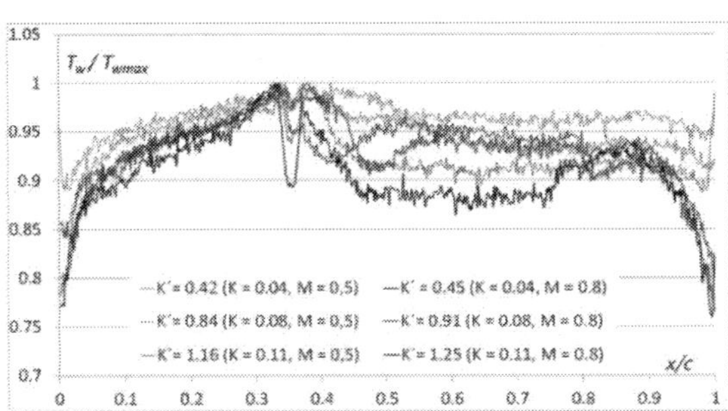

Fig. 8. Temperature profile along the airfoil chord acording to "Line 1" depicted in figure 7.

What can be further observed from the visualization of the transition using IR thermography (in figure 7), is the way of transition itself. It is evident the formation of longitudinal vortices that gradually merge at some distance. In the area of transition, laminar flow lasts in some parts (lighter shades - a warmer surface), in some parts turbulization occurs earlier (darker shades - cooled surface).

This proved the theory of instability in the boundary layer transition. Therefore, if we want to properly predict the transition in the boundary layer using numerical models, it is not appropriate to use simplistic models for 2D numerical calculations since the transition always develops three-dimensionally.

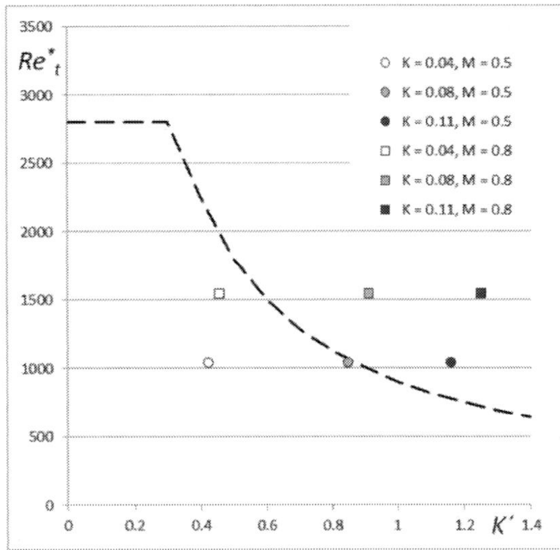

Fig. 9. Effect of roughness element on transition: the dashed line is a transition trashold defined by Tani et al. [7], the points from experiment are inserted.

6 Conclusions

The results in this paper demonstrate the violence transition of the boundary layer on a laminar airfoil using the flat roughness element. There is shown how the transition-generator height can influence the boundary layer transition.

The results are compared with well-known dependence of single roughness element and some differences were observed.

The IR thermography was verified for such experiment in high subsonic and transonic Mach numbers application. It nicely visualized the transition region without any surface influence.

The continuation of this work is to use this turbulence generator in a prismatic blade cascade to force the transition of boundary layer in front of the channel throat.

Acknowledgements

The work was performed with a great acknowledge to TACR for a financial support on the project

TA04020129 "Investigation of flows in a multi-stage axial turbine stages arrangement".

The work was supported by the Long-term Framework Advancement Plan provided by the Ministry of Industry And Trade of the Czech Republic.

Nomenclature

c	airfoil chord
c_f	friction coefficient
h	heat transfer coefficient
H	shape factor of boundary layer
K	transition-generator strip thickness
K'	transition-generator strip thickness related to δ*
L	transition-generator strip length
L'	transition-generator strip length related to δ*
M	Mach number
p	pressure
Pr	Prandtl number
q	heat flow
Re	Reynolds number
Re_{kr}	Critical Reynolds number: $Re_{kr} = v_1 \cdot x/\nu$ Reynolds number at transition point: $Re^*_t = v_1 \cdot \delta^*/\nu$
T	temperature
u_τ	friction velocity
v	velocity
x	coordinate
y	coordinate
α	angel of attack
δ	boundary layer thickness
δ^*	displacement thickness
δ^{**}	momentum thickness
λ_c	thermal conductivity
ν	kinematic viscosity
τ_w	shear stress

Indexes:

0	total condition
∞	value in far distance in front of the model
max	maximum value
t	at transition point
w	wall condition

References

1. P. Straka, J. Příhoda, Proc. Conf. Experimental Fluid Mechanics, TU Liberec, pp. 636-641, (2010).
2. P. Straka, J. Příhoda, P. Šafařík, Proc. Conf. Application of Experimental and Numerical Methods in Fluid Mechanics and Energy, University of Žilina, pp. 261-268, (2012).
3. S. Zuccher, W. S. Saric, Experiments in Fluids, Vol. **44**, pp.145–157, (2008)
4. P. Eliasson, Proceedings to Finite Volume for Complex Applications III., ISTE Ltd., London, pp. 527–534, (2002)
5. M. V. Morkovin, Graham R (ed) Transition in turbines. NASA Conf Publ 2386, **161**, (1985)
6. Schlichting H., Gersten K., *Boundary Layer Theory* (8th edd.), Springer-Verlag (2004)
7. I. Tani, at al., Rep. Inst. Sci. Technol. Tokyo University, Vol. **8**., (1954)
8. H.L. Drydena, Journal of the Aeronautical Sciences, Vol. **20**, 477-482, (1953)

Comparison of heat transfer coefficients of open micro-channels and plain micro-fins

Robert Kaniowski[1], *Robert* Pastuszko[1,*]
[1]Kielce University of Technology, Chair of Mechanics, al. 1000-lecia Państwa Polskiego 7, PL-25-314 Kielce, Poland

Abstract. The paper describes results of analysis of pool boiling heat transfer on enhanced surfaces. Two types of structural surfaces were used: open microchannel surfaces consisting of a system of parallel micro-channels 0.3 mm wide, from 0.2 to 0.5 mm deep and with a pitch of 0.6 mm, and plain micro-fins 0.5 mm in height, uniformly spaced on the base surface with a spacing from 0.6 to1.5 mm. Pool boiling data at atmospheric pressure were obtained for saturated water, ethanol and FC-72. The effects of micro-channel/micro-fin dimensions on heat transfer coefficient in nucleate pool boiling were examined. Substantial enhancement of heat transfer coefficient was observed.

1 Introduction

Miniaturization and significant increase in the power of electronic chips requires mastering cooling techniques. High heat fluxes are typically generated by compact heat exchangers, heat pipes, nuclear reactors, internal combustion engines, and gas turbines.

Passive cooling techniques such as like fins, micro-channels, etc. are widely used for boiling heat transfer enhancement for their compatibility and potential for cooling miniaturized devices [1]. Manufacture of surfaces with micro-fins and micro-channels is relatively simple and does not generate excessive costs.

Passive methods encompass a number of enhancement techniques. Surfaces with plain micro-fins and open micro-channels fall into three categories:

- Surfaces with uniformly spaced micro-fins. Rainey and You [2] used surfaces with 0 to 8 mm long fins. Heat transfer enhancement was significant only for the fins with maximum length of 5 mm. Pastuszko [3] used surfaces with plain micro-fins for boiling of FC-72. Heat transfer coefficients were higher than from micro-fins with a wire mesh. Yu and Lu [4] applied a system of 1 mm thick micro-fins with heights from 0.5 to 4 mm for FC-72. Chan et al. [5] used finned surfaces for boiling water at near vacuum pressures and fins 0.75 – 15 mm high and 0.5 – 2 mm wide. Honda et al. [6], Wei and Honda [7] and Wei et al. [8] investigated pool boiling of FC-72 on micro-fins (0.05 mm thick and up to 0.270 mm high) formed by dry etching directly on silicon semiconductor wafers.
- Surfaces with micro-fins spaced uniformly or non-uniformly. Gugliemini et al. [9] used 3 or 6 mm long uniformly spaced fins; the heat transfer rate increased with decrease in fin width and spacing.
- Micro-channels, e.g. with width of 60 – 200 μm and depths from 180 to 275 μm, etched in silicon plates [10] and with widths of 0.2 – 0.4 mm and depths from 0.100 to 0.400 mm [11]. Tang et al. [12] used interconnected micro-channels (0.4 – 0.7 mm wide and 1.1 mm deep), formed in the copper porous matrix.

The objective of the present work was to compare heat transfer coefficients (HTCs) and heat fluxes of plain micro-fins and open micro-channels for flow boiling of water, ethanol and FC-72.

2 Experimental investigations

2.1 Experimental apparatus

The test station allows determination of the boiling curves [3,13,14]. The main section consisted of a cylindrical vessel made of glass filled with working fluid, i.e., ethanol, water, and FC-72, placed above the specimen. The specimen was soldered with a tin layer to a copper tubular rod. The thickness of the tin layer did not exceed 0.1 mm. In the rod, a 1 kW electric heater was fitted and K type thermocouples were installed to determine the heat flux from the temperature gradient. The entire heating module was thermally insulated (Fig. 1). Experimental heat transfer experiments on the micro-fins and open micro-channels were performed under pool boiling conditions. Prior to the measurements, fluids were degassed in the test station. Measurement data were recorded using a Hydra 2635A data acquisition station (FLUKE) after temperature stabilization. The heat flux was increased until the boiling crisis was reached. The experiment was carried out at atmospheric pressure.

The estimated uncertainties were as follows [3]: at low heat flux (2 kW/m^2): wall superheat at the micro-fin

* Corresponding author: tmprp@tu.kielce.pl

and micro-channel base ±0.2 K, heat flux ±35%, heat transfer coefficient ±40%; at high heat flux (400 kW/m^2): wall superheat at the micro-fin and micro-channel base ±0.5 K, heat flux ±2.4%, heat transfer coefficient ±2.7%.

The base heat transfer coefficient was expressed as:

$$\alpha = \frac{q}{\Delta T_{sat}} \tag{1}$$

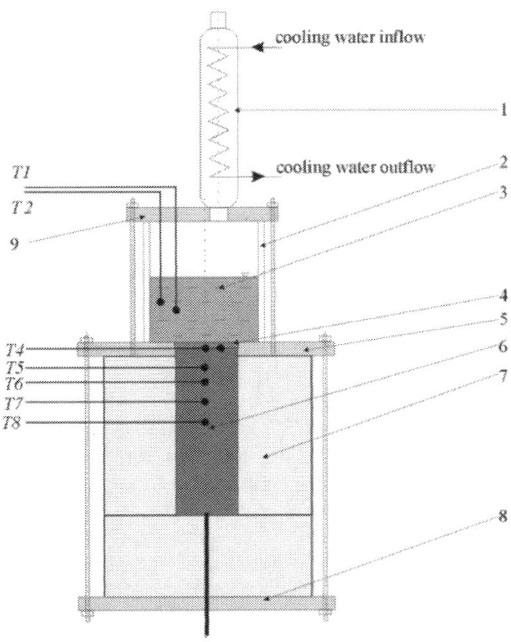

Fig. 1. Main section of pool boiling experimental station: 1 – condenser, 2 – glass vessel, 3 – test fluid, 4 – test surface, 5 – teflon lid, 6 – copper bar with cartridge heater, 7 – insulation, 8 – teflon base, 9 – top flange.

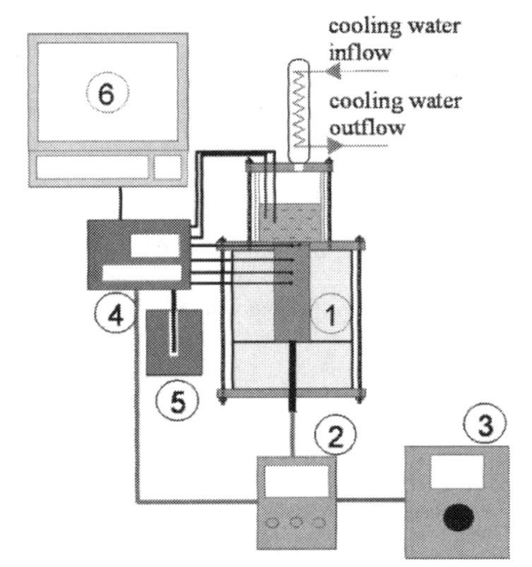

Fig. 2. Pool boiling measurement stand: 1 – main module, 2 – wattmeter, 3 – autotransformer, 4 – data logger, 5 – dry-well calibrator, 6 – PC.

2.2 Test surfaces

The HTC values were compared on two types of structured surfaces: micro-fins and micro-channels. The test section consisted of a 32.5 x 32.5 mm^2 square copper specimen with a 26.8 x 26.8 mm^2 boiling region. The surfaces were produced with CNC milling cutters. The specimen surface had a mean roughness of 0.127 m. Figure 3 shows photographs of the specimens with micro-fins and micro-channels, as well as typical dimensions in the pictorial diagrams. Table 1 compiles the surface codes and specifications. Surface extension coefficient was defined as:

$$\varphi = \frac{A_{ext}}{A_b} \tag{2}$$

a)

b)

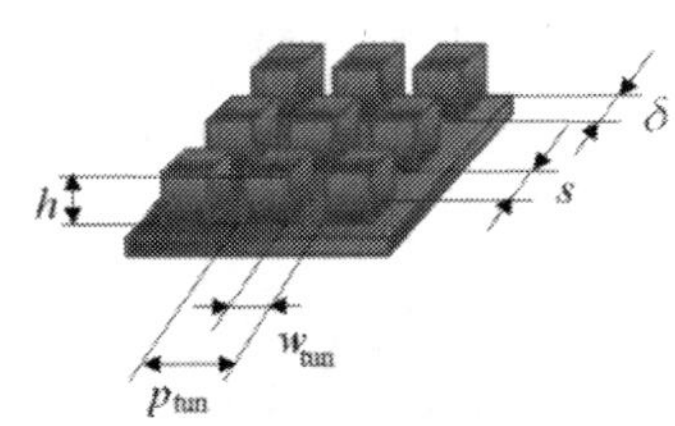

c)

d)

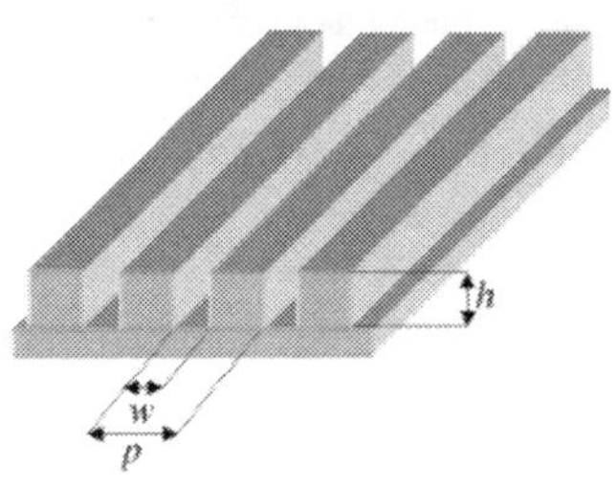

Fig. 3. a) View of the MF specimen, b) MF dimension symbols, c) view of the MC specimen, d) MC dimension symbols.

Table 1. Surface codes and specifications.

Specimen code	w mm	h mm	p mm	φ
MC-0.3-0.5-0.6	0.3	0.5	0.6	2,67
MC-0.3-0.4-0.6	0.3	0.4	0.6	2.33
MC-0.3-0.3-0.6	0.3	0.3	0.6	2.00
MC-0.3-0.2-0.6	0.3	0.2	0.6	1.67

Specimen code	w_{fun} mm	h mm	δ mm	s mm	p_{fun} mm	φ
MF-0.5-0.6	0.6	0.5	2	1.5	2	1.49
MF-0.5-1.0	1.0	0.5	2	1.5	2	1.43
MF-0.5-1.5	1.5	0.5	2	1.5	2	1.36

3 Results

The Bo number, defined as $Bo = w_{fun}/L$ for micro-fins and $Bo = w/L$ for micro-channel, was introduced to enable the comparison of the results for boiling of water, ethanol and FC-72. The capillary length is expressed by:

$$L = \sqrt{\frac{\sigma}{g(\rho_l - \rho_v)}} \quad (3)$$

The $L = 2.5$ mm for boiling water, $L = 1.55$ mm for boiling ethanol and $L = 0.72$ mm for boiling of FC-72. Capillary length is about 3.5 times greater for water than for FC-72 and 1.6 times greater for water than for ethanol. Table 2 compiles the calculated values of the *Bo* number for different specimens and fluids.

Table 2. Bond number values for MF and MC specimens

Specimen code	*Bo*		
	water	ethanol	FC-72
MF-0.5-0.6	0.24	0.38	0.83
MF-0.5-1.0	0.40	0.65	1.39
MF-0.5-1.5	0.60	0.97	2.10
MC-0.3-0.5-0.6	0.12	0.19	0.42
MC-0.3-0.4-0.6	0.12	0.19	0.42
MC-0.3-0.3-0.6	0.12	0.19	0.42
MC-0.3-0.2-0.6	0.12	0.19	0.42

Boiling of water

Figures 4a and 4b show HTC/heat flux and heat flux/ temperature superheat relationships at increasing heat flux on MF and MC surfaces for the boiling of water. Some of the data were taken from [3,15,16].

Surfaces with micro-channels provide higher coefficients, as shown by the noticeable increase in heat transfer relative to surfaces with micro-fins. The coefficient depends on the micro-channel depth. The best results were shown by MC-0.3-0.3-0.6 specimen, where the maximum heat transfer coefficient $\alpha = 64.4$ kW/m^{2}K at $\Delta T \approx 10.5$ K. Maximum HTCs of the MF surfaces were three times lower (max. 22.5 kW/m^{2}K at about 320 kW/m^2).

Boiling of ethanol

Figure 5 shows ethanol boiling curves on the surfaces with micro-fins and micro-channels.

Slight variations in heat transfer intensities were observed between the MF and MC surfaces tested. Compared to the micro-fin surfaces, the only advantage of the micro-channel surfaces was the higher heat flux (up to 400 kW/m^2). The heat transfer coefficient was independent of the micro-channel depth. Of all the specimens tested, the best results were obtained for MC-0.3-0.3-0.6, where the heat transfer coefficient was $\alpha = 20.3$ W/m^{2}K at $\Delta T \approx 15.7$ K. For medium flux values (20 – 120 kW/m^2), the MF-0.5-1.0 surface allowed obtaining the highest HTC.

Boiling of FC-72

The FC-72 boiling curves on the structured surfaces are shown in Fig. 6.

As with the boiling water, the highest HTC was obtained for micro-channels 0.3 and 0.5 mm deep – about twice as high as those on the MF surface. The highest heat transfer enhancement was for MC-0.3-0.5-0.6, where the coefficient was $\alpha = 9.8$ W/m^{2}K for superheat $\Delta T \approx 9.3$ K and heat flux of 90 kW/m^2. The highest maximum heat flux, however, were provided by the surfaces with 0.3 and 0.4 mm deep micro-channels.

4 Conclusions

The experimental studies discussed in this paper lead to the following conclusions:

- Lower *Bo* number values obtained for 0.3 mm wide micro-channels and higher surface extension coefficients compared to micro-fin surfaces facilitate obtaining higher HTCs for MC surfaces.
- With boiling of water and FC-72, micro-channel surfaces have more than two times higher HTCs than those obtained by micro-fin surfaces.
- The effect of the structure type is the least during the boiling of ethanol – the differences between HTCs of MC and MF surfaces are negligible.
- The highest heat transfer coefficient – 63 kW/m^{2}K at $q \approx 700$ kW/m^2 for the boiling of wáter and 20 kW/m^{2}K at $q \approx 390$ kW/m^2 for the boiling of ethanol – was obtained for the surface with 0.3 mm deep open micro-channels.
- The performance of a 0.2 mm deep micro-channel is similar to that of 0.5 mm high micro-fins.

Nomenclature

A	– area, m^2,
Bo	– Bond number,
h	– micro-channel depth, m,
HTC	– heat transfer coefficient,
L	– capillary length, mm,
MC	– micro-channel,
MF	– micro-fin,
p	– pitch, m,
q	– heat flux, kW/m^2,
w	– width, m.

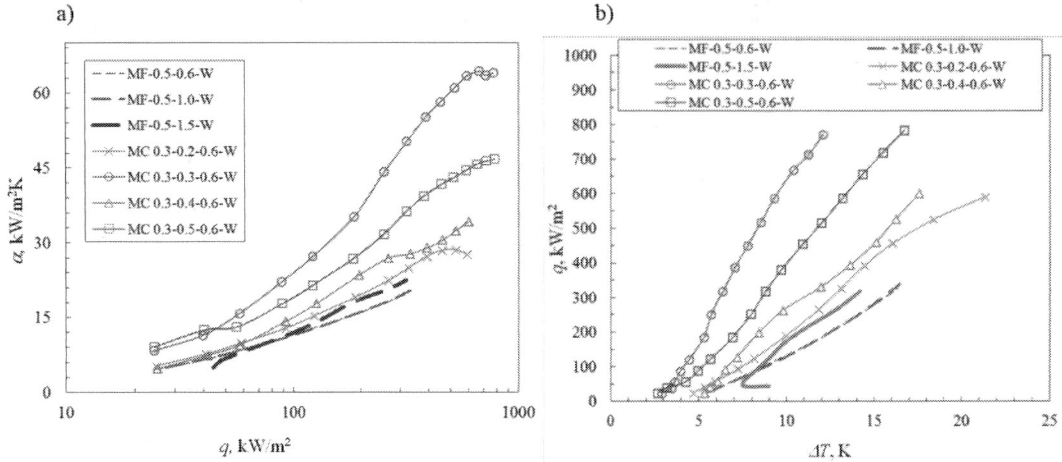

Fig. 4. Boiling curves for water, a) heat transfer coefficient vs. heat flux, b) heat flux vs. temperature difference.

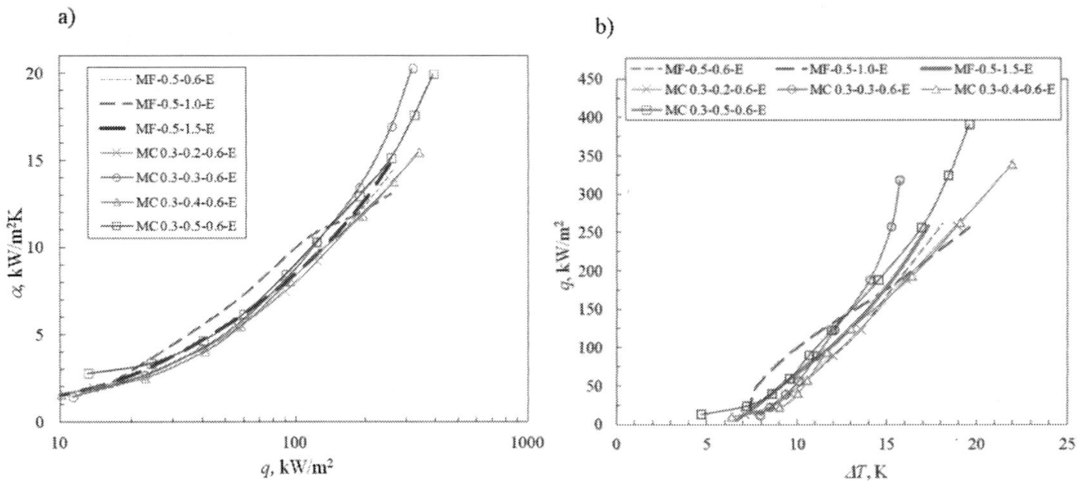

Fig. 5. Boiling curves for ethanol, a) heat transfer coefficient vs. heat flux, b) heat flux vs. temperature difference.

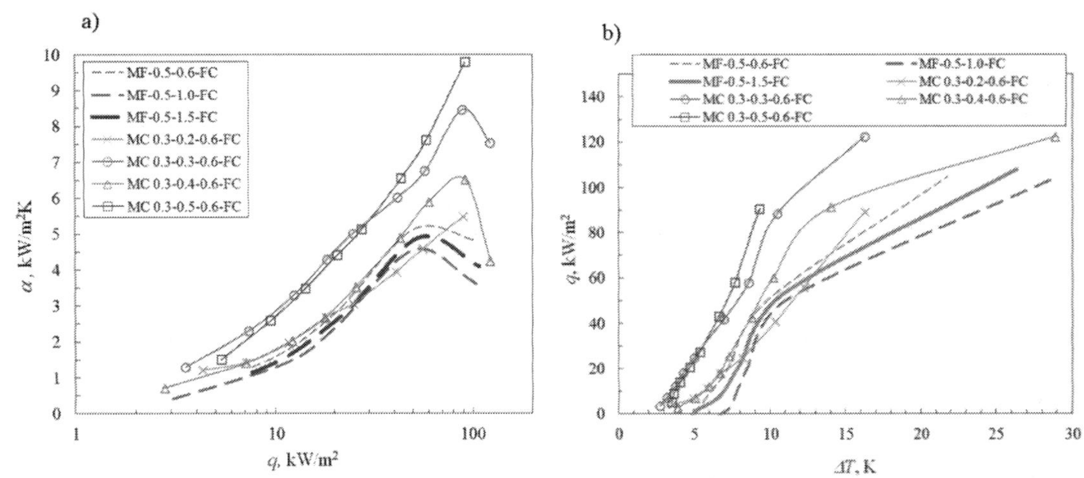

Fig. 6. Boiling curves for FC-72, a) heat transfer coefficient vs. heat flux, b) heat flux vs. temperature difference.

Greek symbols

α – heat transfer coefficient, $W/(m^2K)$,
ΔT – difference of temperature, K,
φ – surface extension coefficient,
ρ – density, kg/m^3,
σ – surface tension, N/m.

Subscripts

b – base,
ext – extended,
l – liquid,
sat – saturation,
tun – tunnel,
v – vapor.

References

1. A. Walunj, A. Sathyabhama, Appl. Therm. Eng. **128** (2018)
2. K. N. Rainey, S. M. You, J. Heat Transfer **122** (2000)
3. R. Pastuszko, Int. J. Therm. Sci. **49** (2010)
4. C. K. Yu, D. C. Lu, Int. J. Heat Mass Transfer **50** (2007)
5. M. A. Chan, C. R. Yap, K. C. Ng, J. Heat Transfer, **132** (2010)
6. H. Honda, H. Takamastu, J. J. Wei, J. Heat Transfer, **124** (2002)
7. J. J. Wei, H. Honda, Int. J. Heat Mass Transfer **46** (2003)
8. J. J. Wei, L. J. Guo, H. Honda, Heat Mass Transfer, **41** (2005)
9. G. Guglielmini, M. Misale, C. Schenone, Int. J. Therm. Sci. **41** (2002)
10. D. Cooke, S. G. Kandlikar, J. Heat Transfer **133** (2011)
11. D. Cooke, S. G. Kandlikar, Int. J. Heat Mass Transfer **55** (2012)
12. Y. Tang, J. Zeng, S. Zhang, C. Chen, J. Chen, Int. J. Heat and Mass Transfer **93** (2016)
13. R. Pastuszko, EPJ Web of Conferences **45**, 01020 (2013)
14. M. Piasecka, Ann. Nucl. Energy **73** (2014)
15. R. Kaniowski, R. Pastuszko, L. Nowakowski, EPJ Web of Conferences **143**, 02049 (2017)
16. R. Kaniowski, R. Pastuszko, EPJ Web of Conferences **143**, 02050 (2017)

Pool boiling of ethanol and FC-72 on open microchannel surfaces

Robert Kaniowski[1], *Robert* Pastuszko[1,*]
[1]Kielce University of Technology, Chair of Mechanics, al. 1000-lecia Państwa Polskiego 7, PL-25-314 Kielce, Poland

Abstract. The paper presents experimental investigations into pool boiling heat transfer for open microchannel surfaces. Parallel microchannels fabricated by machining were about 0.3 mm wide, and 0.2 to 0.5 mm deep and spaced every 0.1 mm. The experiments were carried out for ethanol, and FC-72 at atmospheric pressure. The image acquisition speed was 493 fps (at resolution 400 x 300 pixels with Photonfocus PHOT MV-D1024-160-CL camera). Visualization investigations aimed to identify nucleation sites and flow patterns and to determine the bubble departure diameter and frequency at various superheats. The primary factor in the increase of heat transfer coefficient at increasing heat flux was a growing number of active pores and increased departure frequency. Heat transfer coefficients obtained in this study were noticeably higher than those from a smooth surface.

1 Introduction

Miniaturization and increasing power of electronic systems requires dissipation of a considerable amount of heat. Modern technology miniaturizes devices and increases their demand for energy. Equipment that generates large heat fluxes is, for example, digital systems, nuclear reactors, internal combustion engines, gas turbines, etc. In the construction of heat exchangers, the aim is to reduce their volume and mass. Pool boiling is a passive cooling technique that allows dissipating large amounts of heat flux at low superheats.

The boiling heat transfer coefficient (HTC) can be further increased by surface modification that involves changing roughness parameters, adding a porous layer, changing the structure of the surface by making small holes in it, and forming microfins or microchannels.

As the practical options of surface structure modification are diverse, the potential for boiling intensification is considerable. Structured surfaces provide higher heat transfer coefficients compared to plain surfaces.

Kalani and Kandlikar [1] used surfaces with open microchannels to investigate pool boiling performance at various channel depths (245 – 470 μm) and widths (194 – 406 μm) with ethanol as the working fluid. Microchannels were machined over an area of 10 x 10 mm^2 in the middle of the 20 x 20 mm^2 test section. The authors concluded that the highest HTC values were obtained for microchannels that were 400 μm deep and 200 μm wide.

Gheitaghy et al. [2] reported increased HTC on surfaces with inclined minichannels. The experiments were performed with water at atmospheric pressure. Visualization studies carried out on these surfaces indicated increased growing bubble dynamics.

Jaikumar and Kandlikar [3] investigated surfaces with three different interfin channel widths, 300 μm, 500 μm and 762 μm, with water at atmospheric pressure. Investigations into the effect of interfin channel width on the performance of the surface and heat transfer mechanisms were conducted on open microchannel surfaces with three coating configurations: sintered-throughout, sintered-fin-tops, and sintered-channels. The heat transfer coefficients obtained, up to 2900 kW/m^{2}K, were high relative to the structured 10 x 10 mm^2 surface providing about 1000 kW/m^{2}K relative to the total surface of heat transfer (17 x 17 mm^2).

A. Walunj and A. Sathyabhama [4] studied the effect of a channel shape and geometrical parameters on boiling heat transfer in rectangular, parabolic and stepped microchannels. Microchannels were fabricated on the 10 mm diameter circular copper test piece. At the heat flux of 100 kW/m^2, the HTC for water increased up to 35 kW/m^{2}K.

Research on pool boiling heat transfer with water, ethanol, FC-72 and Novec-649 from tunnel structures, microfins, microchannels, microcavities [5—10] and boiling heat transfer with FC-72 flowing in narrow channels [11—13] has been conducted at the Kielce University of Technology for more a decade.

The objectives of the present work are to determine the effect of microchannel width on the HTC for ethanol and FC-72 and to identify diameters and frequency of departing bubbles.

2 Experimental investigations

The measurements were performed on the setup [6,14] shown in Figs 1 and 2. The setup was designed to allow plotting boiling curves and video recording of the boiling

* Corresponding author: tmprp@tu.kielce.pl

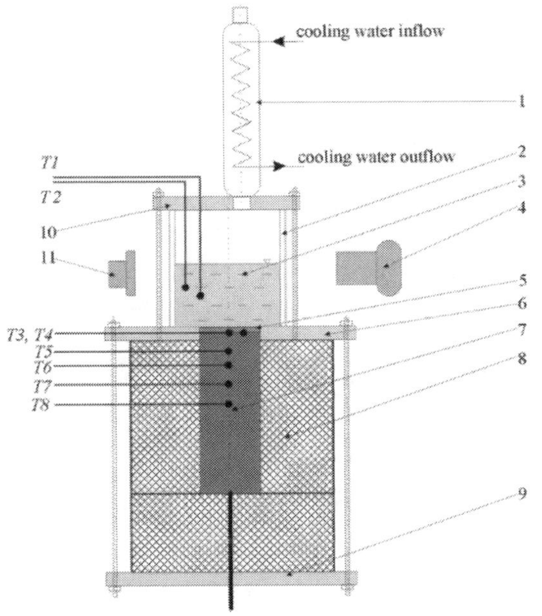

Fig. 1. Main section of pool boiling experimental setup: 1 – condenser, 2 – glass vessel, 3 – test fluid, 4 – lights, 5 – test surface, 6 – teflon lid, 7 – copper bar with cartridge heater, 8 – insulation, 9 – teflon base, 10 – top flange, 11 – high speed camera.

process. From the visualization of boiling, it is possible to determine the diameter of departing vapor bubbles and the frequency of departure from the heat transfer surface. The heat transfer experiments on microchannels were performed under pool boiling conditions, thus the liquid level was more than 50 mm above the specimen surface. Prior to the measurements, ethanol and FC-72 were degassed by boiling for about 15 minutes. Measurement data were recorded after the stabilization of all temperature values, duration of about 14 min. The heat flux was increased until the boiling crisis was reached. The experiment was performed at atmospheric pressure.

Data reduction was achieved using [6]. Seven K type thermocouples (NiCr-NiAl) were placed in the boiling liquid, under the specimen and in the cylinder. The heat flux, temperature superheat and boiling heat transfer coefficient were related to the level of the micro-fins base. The heat flux was determined from the temperature gradient in the upper part of the heating cylinder, assuming one-dimensional heat conduction.

The base heat transfer coefficient was expressed as:

$$\alpha = \frac{q}{T_{mb} - T_{sat}} = \frac{q}{\Delta T} \quad (1)$$

The estimated uncertainties were as follows: low heat flux (2 kW/m^2): heat flux ±35%, heat transfer coefficient ±40%, high heat flux (400 kW/m^2): heat flux ±2.4%, heat transfer coefficient ±2.7%.

The images of the bubbles at nucleation, growth and departure were obtained using a PHOT MV-D1024-160-CL (*Photonfocus*) digital monochromatic camera with a resolution of 1024x758 pixels.

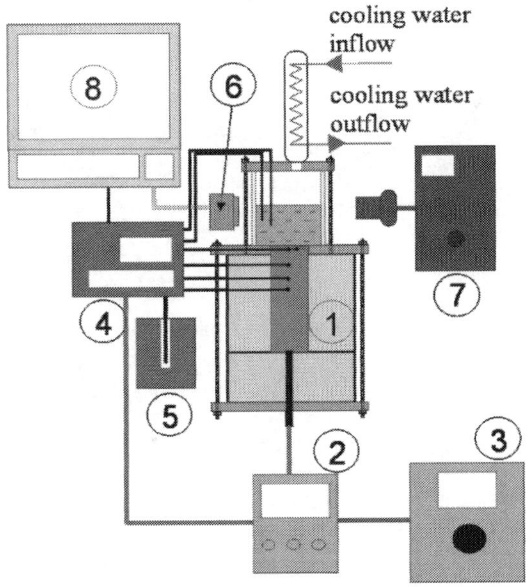

Fig. 2. Pool boiling measurement setup: 1 – main section, 2 – wattmeter, 3 – autotransformer, 4 – data logger, 5 – dry-well calibrator, 6 – high speed camera, 7 –lights, 8 – PC.

The camera recorded the photos at a speed of 428 fps at a resolution of 500x250 pixels and the mean error of 2 fps. An EX-FH20 (*Casio*) camera was used to record the images of the entire surface of the specimen. Lighting was provided by a halogen lamp with light pipes (front light and back light).

The images of the bubbles at nucleation, growth and departure were obtained using a PHOT MV-D1024-160-CL (*Photonfocus*) digital monochromatic camera with a resolution of 1024x758 pixels. The camera recorded the photos at a speed of 428 fps at a resolution of 500x250 pixels and the mean error of 2 fps. An EX-FH20 (*Casio*) camera was used to record the images of the entire surface of the specimen. Lighting was provided by a halogen lamp with light pipes (front light and back light).

The diameters and departure frequencies were determined as the average value from 10 nucleation sites for 30 consecutive vapor bubbles. Bubble diameters were measured in two directions and the arithmetic mean was determined at a constant heat flux. The absolute error of bubble diameter measurement was 2 pixels, i.e., 0.2 mm [7].

The specimens with test surfaces were made of copper and had parallel grooves with a constant pitch, made with a end mill of 0.3 mm in diameter (CNC machining process), Fig.3. The test section consists of a 32.5 x 32.5 mm^2 square copper specimen with a 26.8 x 26.8 mm^2 boiling region. Figure 3a shows a magnification of the microchannels. The mean roughness of the microchannel surface was 0.127 µm. Table 1 compiles the surface codes and specifications according to Fig. 3b.

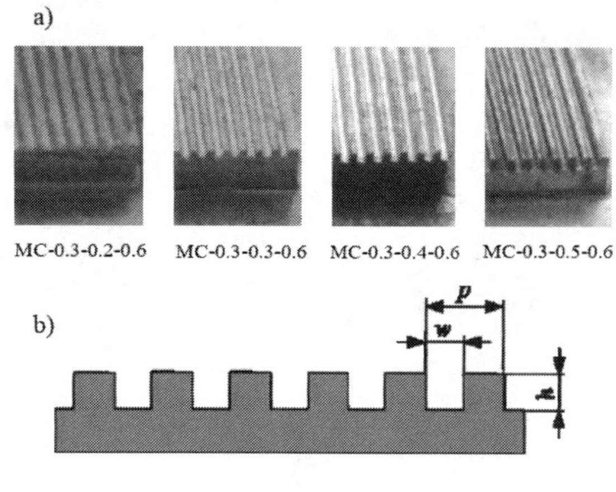

Fig. 3. a) View of the specimen, b) dimension symbols.

Table 1. Surface codes and specifications.

Specimen code	*w* mm	*h* mm	*p* mm
MC-0.3-0.5-0.6	0.3	0.5	0.6
MC-0.3-0.4-0.6	0.3	0.4	0.6
MC-0.3-0.3-0.6	0.3	0.3	0.6
MC-0.3-0.2-0.6	0.3	0.2	0.6

3 Results

Figures 4a and 4b show the relationship between the heat flux and temperature superheat on MC surfaces for boiling ethanol and FC-72. The data for ethanol was taken from [14].

Table 2 summarizes calculation results of constant C and exponent n at $q = C\Delta T^n$ for ethanol and FC-72 on four MC surfaces and one smooth surface.

A slight heat transfer enhancement was observed on the MC surfaces compared to smooth surface (ethanol). The heat transfer coefficient was independent of the microchannel depth. Of the samples tested, the best results were recorded for MC-0.3-0.3-0.6, where the heat transfer coefficient was 20.3 W/m^{2}K at a superheat of 15.7 K. With FC-72, a significant heat transfer increase was attained by MC-0.3-0.5-0.6 (α = 9.8 W/m^{2}K at $\Delta T \approx 9.3$ K).

Figures 5 and 6 illustrate nucleate boiling of ethanol and FC-72 respectively on the surface with microchannels at increasing atmospheric pressure. With ethanol, small vapor bubbles formed, all of approximately 0.5 mm in size, at a low heat flux of about 10 kW/m^2. The nucleation sites located at the bottom of the microchannels, spread over their entire lengths, were active uninterruptedly, and their influence areas did not overlap. Similar phenomena were observed for FC-72. For higher heat flux (q > 20 kW/m^2 ethanol) and (q > 7 kW/m^2 FC-72), the bubble diameters and frequency stabilized taking the values d_b = 2.2 mm (dash line), f = 70 Hz for ethanol and d_b = 1.4 mm (dash line), f = 105 Hz for FC-72, Figs 7, 8. Most centers were active during the entire cycle of increasing heat flux, producing bubbles continuously. The bubbles started to coalesce both before departure (horizontal coalescence) and after departure (vertical coalescence). The bubbles developed funnel-like shapes, extending in the lower part. For q > 58 kW/m^2 (ethanol) and q > 18 kW/m^2 (FC-72), quantitative observations became impossible due to the bubbles obscuring the image by coalescing and forming vapor tunnels.

Heat flux/diameter and heat flux/frequency relationships for ethanol and FC-72 are illustrated in Figs 7 and 8 respectively.

The authors proposed calculating the departing bubble diameter for the open microchannel on the basis of buoyancy force and surface tension balance referred to the incircle diameter of the microchannel space. It was also assumed that the contact angle was 90^0, and inertia forces and lift force were neglected. The departing bubble diameter dependence can be calculated from the formula:

$$d_b = \left(\frac{6\sigma w}{g(\rho_l - \rho_v)} \right)^{1/3} \tag{2}$$

For the microchannel width w = 0.3 mm, the calculated values d_b = 1.6 mm and d_b = 1.0 mm for ethanol and FC-72, respectively, are lower than the described earlier stabilized diameters of about 28 % (Fig. 5).

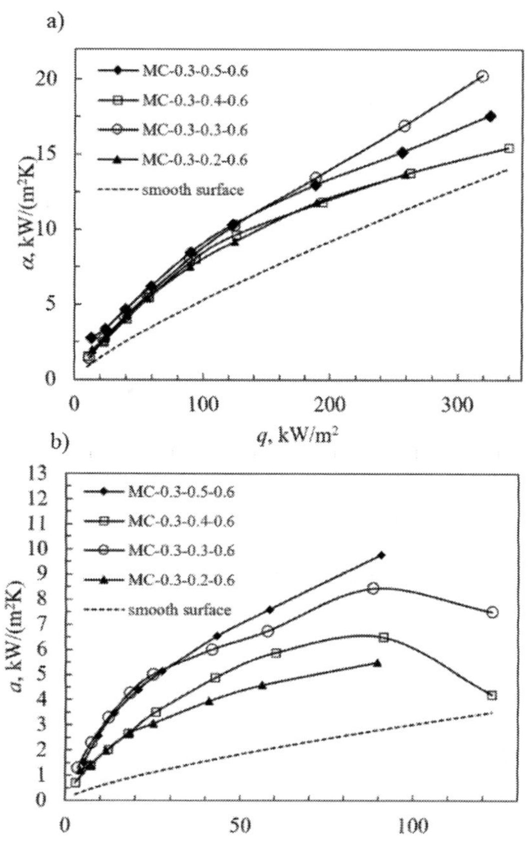

Fig. 4. Heat transfer coefficient vs. heat flux, a) ethanol, b) FC-72.

Table 2. Constant C and exponent n as a function of $q(\Delta T)$

Specimen code	fluid	C	n
MC-0.3-0.2-0.6	ethanol	0.036	3.09
MC-0.3-0.3-0.6	ethanol	0.047	2.96
MC-0.3-0.4-0.6	ethanol	0.002	4.45
MC-0.3-0.5-0.6	ethanol	0.220	2.50
Smooth surface	ethanol	0.001	5.00
MC-0.3-0.2-0.6	FC-72	0.302	2.08
MC-0.3-0.3-0.6	FC-72	0.574	2.20
MC-0.3-0.4-0.6	FC-72	0.078	2.81
MC-0.3-0.5-0.6	FC-72	0.305	2.61
Smooth surface	FC-72	0.001	3.57

(a) q = 10 kW/m^2 (b) q = 22.6 kW/m^2

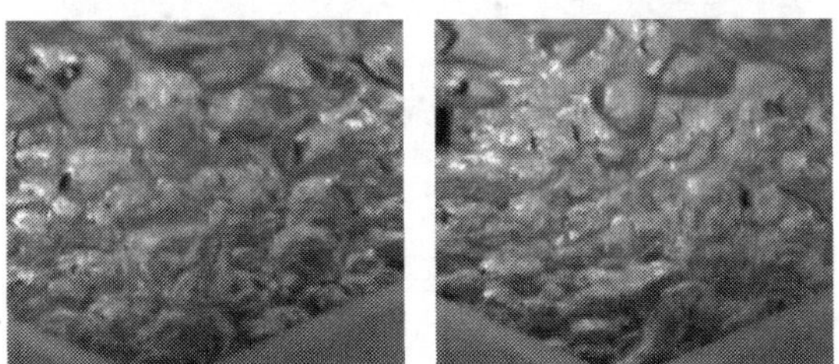

(c) q = 40.4 kW/m^2 (d) q = 57,8 kW/m^2

Fig. 5. Visualization observations of pool boiling of ethanol on the microchannel surface MC-0.3-0.4-0.6.

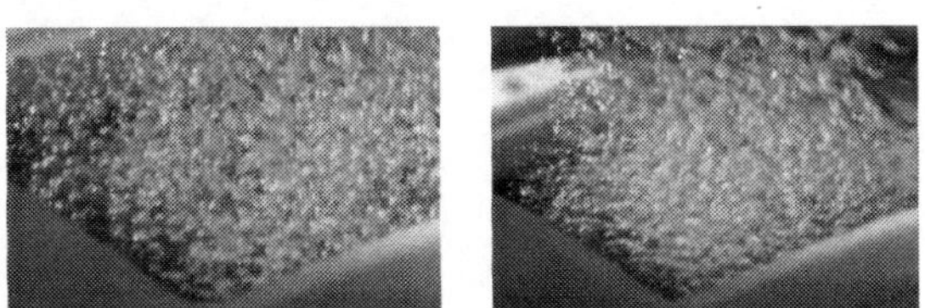

(a) q = 2.8 kW/m^2 (b) q = 7.1 kW/m^2

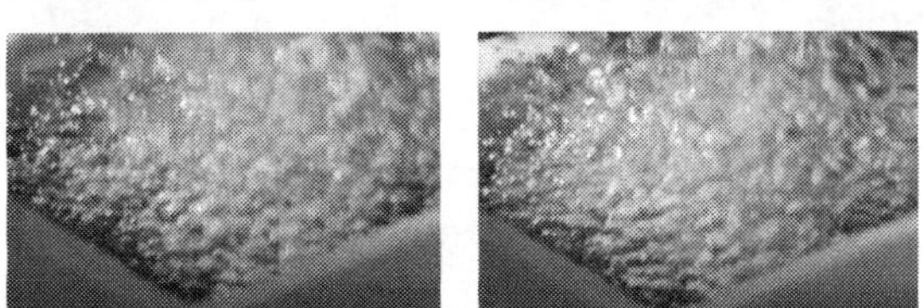

(c) q = 12.1 kW/m^2 (d) q = 17.9 kW/m^2

Fig. 6. Visualization observations of pool boiling of FC-72 on the microchannel surface MC-0.3-0.4-0.6.

a)

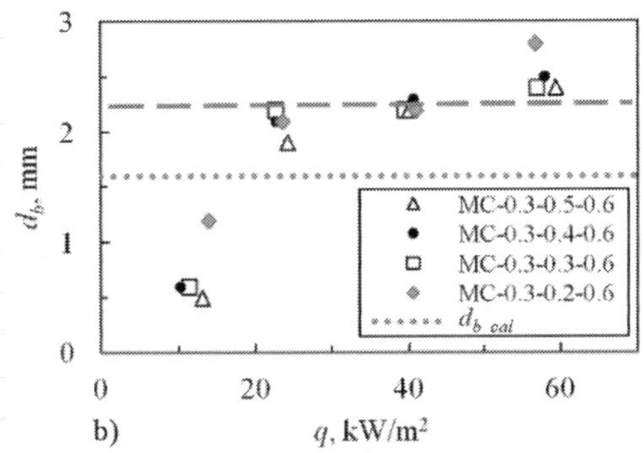

b)

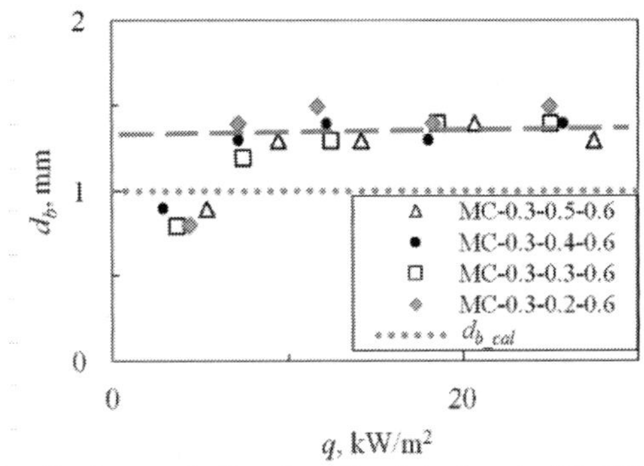

Fig. 7. Averaged diameters for (a) ethanol, (b) FC-72, theoretical value d_{b_cal} was calculated from eq. 2.

a)

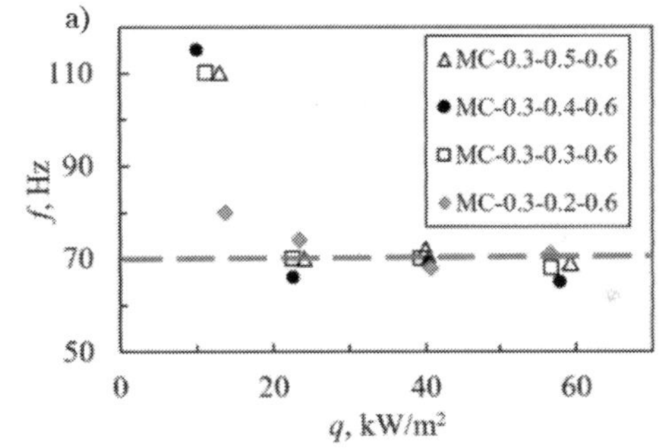

b)

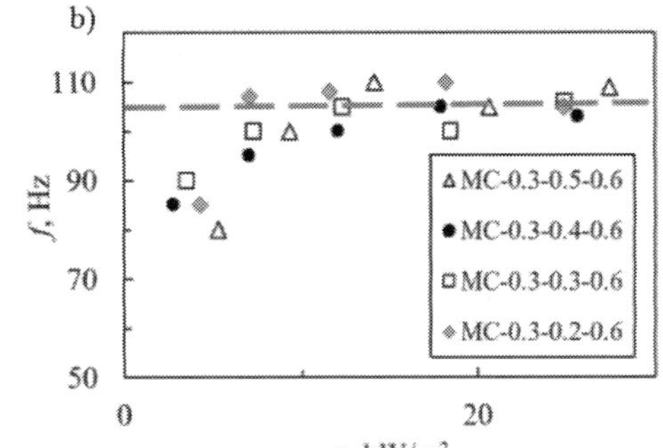

Fig. 8. Averaged frequencies for (a) ethanol, (b) FC-72.

4 Conclusions

The experimental studies discussed in this paper led to the following conclusions:

- HTC on surfaces with best performance were two times higher for ethanol and more than three times higher for FC-72 than HTC for plain surfaces.
- For small superheats between the heating surface and the liquid, vapor bubbles take spherical shapes and do not coalesce. At larger superheats, the bubbles interact strongly forming vapor tunnels.
- Bubble diameters and departure frequencies became stable after reaching the limit of 20 kW/m^2 with ethanol and 7 kW/m^2 with FC-72.
- Experimental studies for other liquids, channel widths and pitches are necessary to check the effect of at least three geometrical parameters on the heat transfer process.

Nomenclature

C – constant,
d – diameter, m,
f – frequency, Hz,
h – microchannel depth, m,
HTC – heat transfer coefficient,
n – constant,
p – pitch, m,
q – heat flux, kW/m^2,
w – width, m.

Greek symbols

α – heat transfer coefficient (HTC), $W/(m^2K)$,
ΔT – difference of temperature, K,
ρ – density, kg/m^3,
σ – surface tension, N/m,

Subscripts

b – bubble,
cal – calculated,
l – liquid,
mb – microchannel base (bottom),
sat – saturation,
v – vapor.

References

1. Kalani A., Kandlikar S.G., proc. ASME 10th Int. Conf. Nanochannels, Microchannels and Minichannels, Rio Grande, Puerto Rico (2012)
2. A. M. Gheitaghy, A. Samimi, H. Saffari, Appl. Therm. Eng. **126** (2017)
3. A. Jaikumar, S. G. Kandlikar, Int. J. Heat and Mass Transfer **95** (2016)
4. A. Walunj, A. Sathyabhama, Appl. Therm. Eng. 128 (2018)
5. R. Pastuszko, R. Kaniowski, EPJ Web of Conferences, **25**, 02019 (2012)
6. R. Pastuszko, EPJ Web of Conferences 114, 01002 (2016)
7. R. Kaniowski, R. Pastuszko, L. Nowakowski, EPJ Web of Conferences **143**, 02049 (2017)
8. R. Pastuszko, EPJ Web of Conferences **45**, 01020 (2013)
9. R. Pastuszko, M. Piasecka, Journal of Physics: Conference Series **395,** 012137 (2012)
10. R. Pastuszko, M.E. Poniewski, M. Koziol, Heat Mass Transfer, **48** (2012)
11. S. Hozejowska, R. Kaniowski, M. Poniewski, Exp. Therm. Fluid Sci.**78** (2016)
12. M. Piasecka, K. Strak, B. Grabas, Arch. Metall. Mater. **62** (2017)
13. M. Piasecka, K. Strak, B. Maciejewska, Heat Transf. Eng. **38** (2017)
14. R. Kaniowski, R. Pastuszko, Proc. XXIII Convention of Thermodynamics Researchers, Beskid Śląski, Poland, 2017

Dynamics of the nozzle valve with regard to the properties of the piping system

Roman Klas[1,*], *Vladimír* Habán[1], and *Pavel* Rudolf[1]

[1]Brno University of Technology, Faculty of Mechanical Engineering, Victor Kaplan Department of Fluids Engineering, Technická 2, 61669 Brno, Czech Republic

Abstract. It is obvious that the main function of the nozzle valve is to shut off the stream of fluid in the piping system. The response rate of the valve to the decreasing or reversing flow in the system will then depend on the valve properties and equally on the properties of the piping system. The interaction of these two elements is also important for the origin of pressure pulsations in the system. While the pressure pulsations were the cause for design of this particular valve it should be noted that the general design of the valve for any pipeline system is not possible. The valve cannot properly work under all circumstances and operating conditions. With respect to this, the dynamic properties of the valve will be assessed on the basis of the valve equation of motion and the pipeline model. An adequate response of the whole system can be obtained by combining both approaches. The valve equations of motion are also complemented by CFD simulations, which enable to capture the movement of the valve disc with respect to flow rate.

1 Introduction

The presented study follows the main simulation of the in-line check valve [1]. An attempt to predict the in-line valve poppet lift in dependence on the flow rate occurred in these simulations based on a one-dimensional model. Verification of the one-dimensional model was realized only partially through the 1DOF CFD simulations. This paper will complements the CFD simulations of the in-line check valve, but it will preferably focus on the nozzle check valve. The 1DOF CFD in-line check valve simulations will also serve to compare with 1DOF CFD by simulating the nozzle check valve. The fact that the designing of the in-line check valve and the nozzle check valve was underway within the same development of the check valves is the reason. Each of these valves was one of the concepts that had been part of a prototype search for a real device. Therefore, the comparison is actually expected. Another important factor is that experimental data are already available in the case of an in-line valve.

The other circumstance that will be considered is the fact that the dynamic testing of the valve can be realized in two directions. The first is the dynamic behavior of the check valve itself, i.e. the valve without the connected pipe system. Of course, it is necessary to note at this point that the obtained data obtained can hardly match the operational properties of the valve in the real pipeline. Such an approach is somewhat more favorable for CFD simulation, because CFD simulations in generally characterize considerable computational requirements. It would be inconvenient in common conditions to simulate the whole piping system through CFD and still consider a dynamic network, even in a small computational domain. Of course, it is also possible to combine CFD simulations and one-dimensional models representing the piping system with all its elements. However, if there was a one-dimensional model in the form of the valve equation of motion, it would lose that justification and the comparison could be made directly. The use of CFD simulations is particularly useful in the determination of those valve parameters that are elusive by the experimental testing. These data can then serve to accuracy a one-dimensional model. It is also necessary to say that the valve disc movement prediction through CFD brings problems of greater or lesser loss of control over the quality of the computational mesh, which varies depending on the remeshing or smoothing mode in the individual time steps. This has an impact on the choice of wall functions and the turbulent model in CFD simulations [2, 3].

The second option is to create a one-dimensional model of the pipeline system supplemented with a check valve model. Verification of the obtained results is possible from experimental data.

2 CFD simulations and description of the check valves and the piping system

All CFD simulations were carried out by using the commercial code ANSYS ANSYS Fluent 17.2. The component that is part of this software was used to realize the disc movement in the valve axis (1DOF) by considering the dynamic computational mesh. The valve geometry and computational mesh were created in ANSYS Design Modeler and Meshing. The computational mesh was conceived as mapped except the

* Corresponding author: klas@fme.vutbr.cz

nearness of the disc and the valve seat, i.e. the area of the dynamic mesh.

The computational mesh was also improved with respect to determine the exact magnitudes of the forces acting on the disc. Therefore, in incremental steps, the number of mesh cells was increased until the force magnitudes in the corresponding flow modes remained unchanged. However, it is necessary to mention at this point a certain complication in the selection of wall functions and in their correct setting with respect to the wall y + parameter. The right size of the wall y + parameter was the most conflicting with reaching the relevant force magnitudes. 2D CFD simulations were used as the basic mode of calculation, which at least in case of the in-line check valve proved to be interchangeable with 3D simulations [1]. In addition, it is to be noted that due to the number of computational cells of the three-dimensional computational mesh, the simulation of the valve's disc movement while maintaining the correct conditions is time-consuming. This is also why the aim of the paper is to minimize computational requirements on prediction of valve disc motion. Besides the above mentioned, the flow field parameters necessary to create a one-dimensional model were obtained by CFD simulations. The parameters of CFD simulations are presented in Table 1.

Table 1. Description of CFD numerical model.

	2D
Pipeline diameter	DN100
Number of computational cells	~ 400000
Turbulence model and near wall modeling	realizable k – ε non equilibrium wall function
Boundary conditions	Inlet: velocity inlet Outlet: pressure outlet
Calculation mode	unsteady, incompressible flow, 1DOF

The nozzle check valve is shown in Fig. 1 and the in-line valve on Fig. 2.

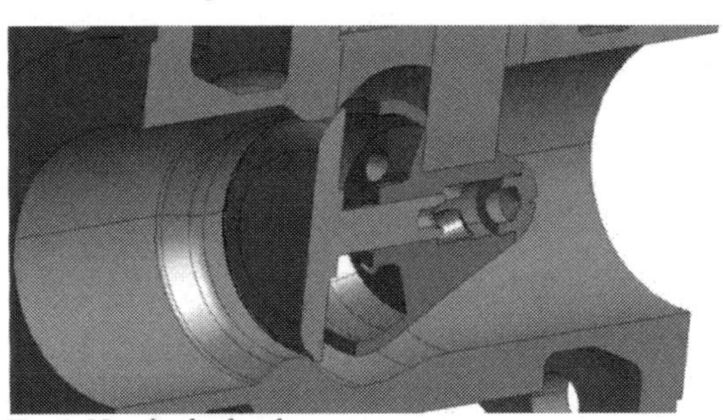

Fig. 1. Nozzle check valve.

The piping system (Fig. 3) used in the one-dimensional model should be as close as possible to the real valve testing circuit. However, due to the character of the circuit, its model is in principle universal.

Circuit consists of the connecting pipes, pressurized vessel, which is connected to vacuum pump (for cavitation testing), the valve itself and hydrodynamic pump for the case of turbine and valve testing. The hydrodynamic pump is driven by dynamometer, which also measures pump torque and speed. Flow rate is measured using induction flowmeter, pressure difference between valve inlet and outlet is assessed using pressure sensors. Measurement procedure and data evaluation is according to valid standard ČSN EN ISO 9906.

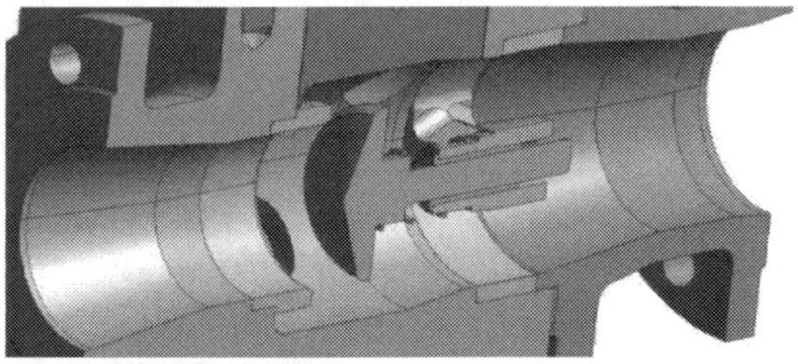

Fig. 2. In-line check valve.

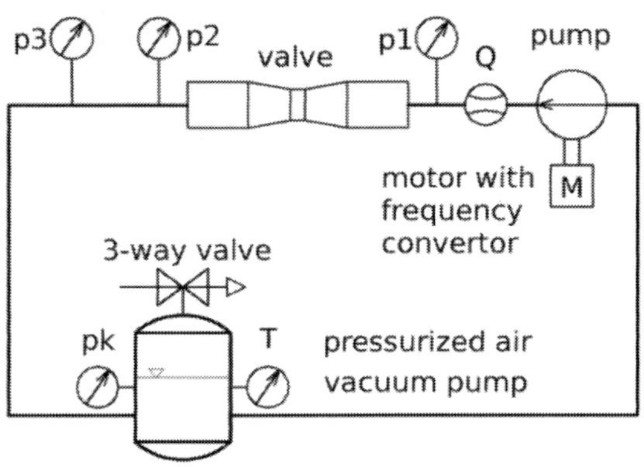

Fig. 3. The pipeline and the testing circuit diagram

The Table 2 lists the quantities and their markings used in the paper and in the following equations.

Table 2. List of symbols.

A_Y, B_Y, C_Y	coefficients of specific energy
b_t	damping
J	moment of inertia
k, k_S	gradient of the straight line, stiffness of the spring
K	fin and guide bushing
l_a	Sa surface center of gravity coordinates
l_b	Sb surface center of gravity coordinates
m_j, n_i, n_j	normal vectors
M_s	friction moment
n	pump speed
p, pa, pb	static pressure and static pressure on the inlet and outlet surfaces
P	solid wall of the valve housing

P_D, P_o	dissipated power, dissipated power at the shut-off point
Q, Q_m	volumetric and mass flow rate
R_{ab}	hydraulic resistance of the valve between the surfaces Sa and Sb
Sa, Sb, S	areas of the inlet and outlet surfaces of the valve
v	absolute fluid velocity
V_Γ	volume of the valve disc
x_0, x_3	minimum spring preload, disc position
Y	specific energy of the pump
Γ	moving wall of the disc
Π	strain rate tensor
η	pump efficiency
ρ_F, ρ_Γ	density of the liquid and the valve disc
ω	angular speed
1DOF, 1AM	one degree of freedom, one dimensional numerical model

From the Navier-Stokes equation (1), from the continuity equation (2) and from boundary and initial conditions we can obtain the fluid force acting on the valve disc.

$$\rho_F\left(\frac{\partial v_i}{\partial t}+\frac{\partial v_i}{\partial x_j}v_j\right)=\rho_F g_i-\frac{\partial p}{\partial x_i}+\frac{\partial \Pi_{ji}}{\partial x_j} \quad (1)$$

$$\frac{\partial v_j}{\partial x_j}=0 \quad (2)$$

The equation (3) defining the force magnitude is, of course, dependent on the choice of boundaries and boundary conditions. It seems logical to identify the position of the boundary conditions with the valve boundaries in particular as regards the location of its inlet and outlet parts. However, the problem of determining some of the members in the equation (3) increases. It is possible to accept the assumption of rotational symmetry and, above all, to simplify the integration of the static pressure on the surface **P**, but problems with the expression of strain rate tensor ***Π****ij* persist. Any possible neglect of these members in the equation, of course, will reduce the accuracy of the obtained results.

The second option is to move the boundaries of the area to the limit positions of the valve disc. The above mentioned difficulties would persist, though the inaccuracy of the one-dimensional model would be less, for example, by neglecting shear stresses. Of course, we plan to use CFD methods to determine them in both cases. However, the simplest description of the physical phenomenon should be the objective of the analytical model. Therefore, it would be more acceptable to keep the boundaries on the boundary surfaces of the valve and to evaluate the valve losses as if they were only generated by the valve disc. Thus, it is necessary to determine only the forces acting on the disc, not on the remaining part of the valve, although it is expedient to evaluate these forces from equation (3). This gives us an idea of the synergy of the valve disc and the housing with regard to the designing of the valve interior.

$$\begin{aligned}F_F &= \rho_F \ddot{x}_i V_\Gamma - \rho_F(l_a+l_b)\frac{\partial Q}{\partial t}+p_a S_a - p_b S_b + \\ &\iint_{S_a\cup S_b\cup P\cup K}\Pi_{ij}n_j dS + \iint_{S_a}\rho_F v_i v_j m_j dS - \iint_{S_b}\rho_F v_i v_j n_j dS \\ &- \iint_{P\cup K} p\, n_i dS\end{aligned} \quad (3)$$

The fluid force acting on the disc can be simplified in equation (4) by defining the boundary conditions according to Fig. 4.

$$F_F = \rho_F \ddot{x}_i V_\Gamma - \rho_F(l_a+l_b)\frac{\partial Q}{\partial t}+R_{ab}Q^2 S \quad (4)$$

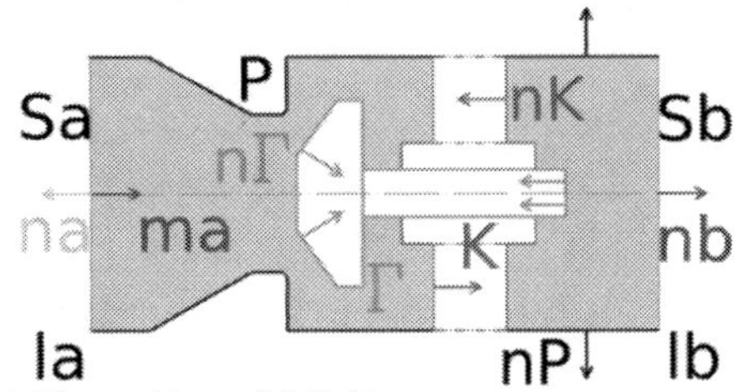

Fig. 4. The markings of definition areas.

In most common cases, the hydrodynamic pump is a feeding unit in a real pipeline system, except the gravity water system. The design of this pump may be axial, semi-axial or radial depending on the impeller type. We will focus on further considerations regarding the different properties of these pumps to systems with hydrodynamic pumps radial and peripherally also semi-axial [4].

Pump and fluid in piping are a source of inertial effects in the pipeline, which has a significant impact on the check valve dynamics and its closing times. It can be expected that the hydrodynamic pump will have favorably effects in terms of the danger of water hammer, and vice versa, it may have a negative effect on the long closing times of the valve. It is also important differentiate low and high pressure systems with regard to simulation of compressible media [5].

It is necessary to consider the operation of the system in order to assess the effect of the pump. We can expect a piping system failure and a subsequent decelerating of impeller rotation, which is the most frequent consequence of a power outage. In addition, we can consider a pump failure consisting of an immediate stop of the impeller, or pump speed control and its slow shutdown [6]. The last possibility will not be further discussed as it offers full control over pipeline dynamics.

Of course, it is necessary to complement the previous considerations with the relevant equations that include the effect of the hydrodynamic pump, the fluid in the pipeline, the elements of the piping system and the check valve itself. The described system will be simulated until a possible water hammer can occur and the incompressible liquid will be further considered. Fig. 3 in combination with Tab. 2. includes the parameters of the fluid and the piping system.

$$\omega J\frac{d\omega}{dt}+P_D=0 \quad (5)$$

The equation (5) is the basic relationship describing the balance between the power consumption of the machine and its inertial effects. Determining the power consumption P_D, which in terms of dissipated power in the pump interior except the spiral case, is the main difficulty of this equation. There are basically two main possibilities in this respect. The first is a detailed analysis of a centrifugal pump and a subsequent determination of the losses in the main parts of the pump interior. The literature offers several basic resources in this respect, among which the most well-known are [7-13]. In [13] is possible find a description of the hydraulic losses in the spiral case.

Operation of centrifugal pumps is usually assumed at constant speed. However, even at constant impeller speeds, parameters such as specific energy and flow rate may change. The pump power is, of course, changing as a result of these parameters. Changing pump speeds, such as machine downtime until it stops completely, complicates this situation. The power of the pumping unit is derived from the inertia effects of the engine and pump assembly. This power consumption then decides the time at which the impeller stops. It is necessary to consider that the impeller and fluid flow stopping time will most likely not match. It is obvious that the remaining parts of the piping system, including the check valve, will affect the fluid flow. In the event of an undesired backflow through the valve, which is a negative factor for the possible high peak pressure peak in the pipeline, consideration should also be given to possible backflow of the fluid to the impeller. The impeller can rotate in the pump mode or even operate at a low speed like a turbine for a short time. However, the turbine modes of operation will no longer be the focus of our research. The pump is therefore analysed primarily in terms of its impact on the check valve dynamics.

Affinity laws (6) will be used to overcome the problem of determining change of power with a change in speed.

$$\frac{P}{\overline{P}}=\frac{n(t)^3}{\overline{n}^3},\ \frac{Q(t)}{\overline{Q(t)}}=\frac{n(t)}{\overline{n}},\ \frac{Y(Q)}{\overline{Y(Q)}}=\frac{n(t)^2}{\overline{n}^2} \quad (6)$$

Further, it is assumed that the hydraulic and volumetric efficiency is constant, depending on pump speed (7). The mechanical efficiency is then considered for the corresponding pump speed at the mode of its operation.

$$\eta_{h,o}=\overline{\eta_{h,o}}=\frac{\overline{\rho Q(t)Y(Q)}}{\overline{P_o+kQ(t)-M_s\omega}}=\frac{\rho Q(t)Y(Q)}{P_o+kQ(t)-M_s\omega} \quad (7)$$

It is also possible to consider approximately linear load dependence on pump flow rate (8) for radial flow pumps.

$$P_D=P_o+kQ(t) \quad (8)$$

The equation (8) becomes the equation (9), which can be used to determine the speed of the impeller. Speed will represent basic information about the mode of pump operation.

$$(2\pi)^2 J\,n(t)\frac{d\,n(t)}{dt}+\frac{n(t)^3}{\overline{n}^3}\left(\overline{P_o}+\overline{k\,n}\frac{\overline{Q(t)}}{n(t)}-\overline{M_s}\,2\pi\,\overline{n}\right)=-2\pi M_s\,n(t) \quad (9)$$

Equation (8) can also be used when power consumption is not a linear dependence of pump flow rate. Just consider the variable coefficient k. The ability to determine pump consumption and efficiency $\eta_{h,o}$ is also offered. In this way, it would be difficult to determine the pump consumption at the shut-off point, see (8).

In order to be relatively simple to descrive the pump's specific energy, it is possible to approximate it by the corresponding curve (10). It would be possible to use a more accurate substitution, but this would not mean a greater contribution to the problem analysis. The dependence of the specific energy on the speed of the impeller corresponds again to the affine laws (6). A typical dependence of specific energy can be described at a constant pump speed according to (10). More can be found in [6].

$$\overline{Y(Q)}=A_Y-B_Y\overline{Q(t)}^2-C_Y\overline{Q(t)} \quad (10)$$

Similarly, the specific energy Y(Q) corresponds to the equation (11) at variable speeds.

$$Y(Q)=A_Y\left(\frac{n(t)}{\overline{n}}\right)^2-B_YQ(t)^2-C_Y\frac{n(t)}{\overline{n}}Q(t) \quad (11)$$

The equation (12) describes the movement of the valve disc.

$$\rho_\Gamma V_\Gamma \ddot{x}_t+b_t\dot{x}_t+k_S(x_0+x_t)=\rho_F\ddot{x}_tV_\Gamma-\rho_F(l_a+l_b)\frac{\partial Q}{\partial t}+R_{ab}Q^2S \quad (12)$$

The following equation (13) symbolically describes the pipeline in which the valve was connected.

$$\frac{1}{S}\sum_j l_j\frac{dQ}{dt}+\frac{Q|Q|}{2S^2}\left(\sum_i\xi_i+\sum_j\lambda_j\frac{l_j}{d_j}\right)=\pm Y(Q) \quad (13)$$

It is advisable to consider improvement the one-dimensional circuit model by simulating unsteady friction as the flow is unsteady. This issue is more clearly explained e.g. in [14]. However, unsteady friction in the pipeline model had almost no effect on the obtained

 https://doi.org/10.1051/epjconf/201818002043

results, and therefore unsteady friction was further neglected.

3 Numerical simulations

As already mentioned, valve testing involves two approaches and combines CFD methods and one-dimensional models of the pipeline system and the valve itself. However, it would not be possible to create a one-dimensional model without CFD simulations. The meaningfulness of a one-dimensional model, of course, lies in smaller computational demands and in the analysis of parameters influencing the movement of the valve plug. Of course, the meaningfulness of a one-dimensional model is based on lesser computational requirements, as well as in the analysis of parameters influencing the movement of the valve disc.

Several results will be presented. 1DOF CFD simulation of the axial valve itself without the pipe system Fig. 5, which is not finished in [1], will be completed first. The change of flow rate Q is considered linear. CFD simulation is also complemented by experimental data.

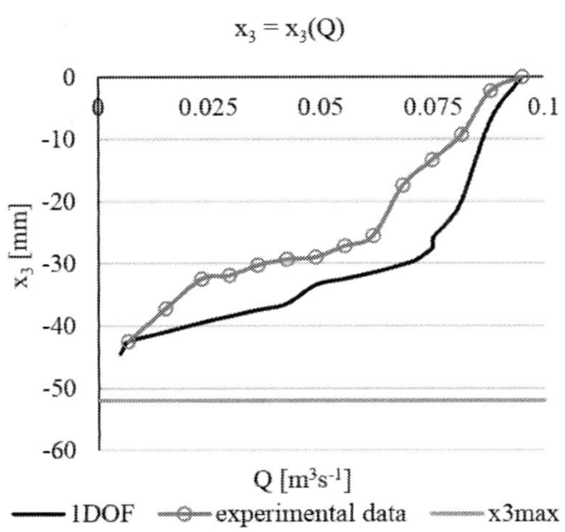

Fig. 5. 1DOF CFD In-line check valve simulation and comparison with experimental data, Ts = 1.5s.

It was possible to simulate about 90% of the maximum opening of the valve, considering the real closing time of the valve Ts = 1.5s. However, the valve is usually only open from 20 to 30% in real operation. The 1DOF CFD simulation was therefore realized for this opening, Fig. 6. The opening corresponds to 30% of its maximum opening. This simulation is also complemented by a one-dimensional model (1AM) following the equation of motion (12).

The agreement of both simulations is basically good, but their comparison does not exist near the valve seat. The computational stability of 1DOF CFD was week, as in the previous case. Similarly, data 1DOF CFD and 1AM simulation of the nozzle check valve itself are shown in Fig. 7. The closing time Ts = 1.5s was set for the second check valve so that both valves could be compared.

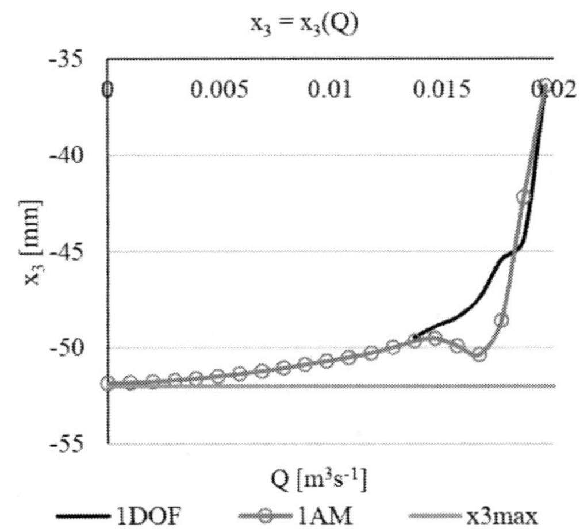

Fig. 6. 1DOF CFD and 1AM in-line valve simulations, Ts = 0.5s and initial opening 30%.

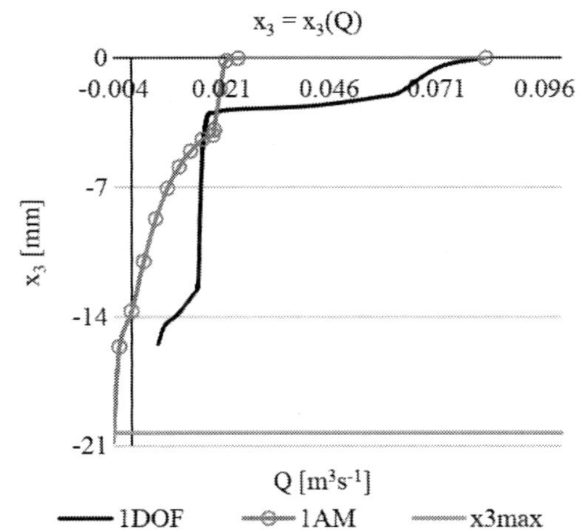

Fig. 7. 1DOF CFD and 1AM simulations of the disc position of the nozzle check valve, Ts = 1.5s.

The biggest difference in Fig. 7 is visible at maximum valve opening. The length of time the disc does not move differs slightly. Rapid closing is characteristic in both cases. The fact that full valve closure occurs with a small backflow through the valve could be unfavorable. The backflow may be undesirable in relation to the pressure peak size if a water hammer occurs. It is shown with respect to the time delay at the beginning of the valve closing that the exact determination of the hydraulic resistance of $\boldsymbol{R_{ab}}$ is decisive in the times approaching t = 0s. Even a very small change in $\boldsymbol{R_{ab}}$ can be very significant in the length of time the valve disc remains in its initial position.

The other simulations already concern the movement of the disc in the check valve that is inserted into the piping systém. 1AM in-line and nozzle check valve simulations in the piping system are shown in the following Figs. 8 and 9.

Note: The literature [1] contains only 1AM simulation of the in-line check valve itself.

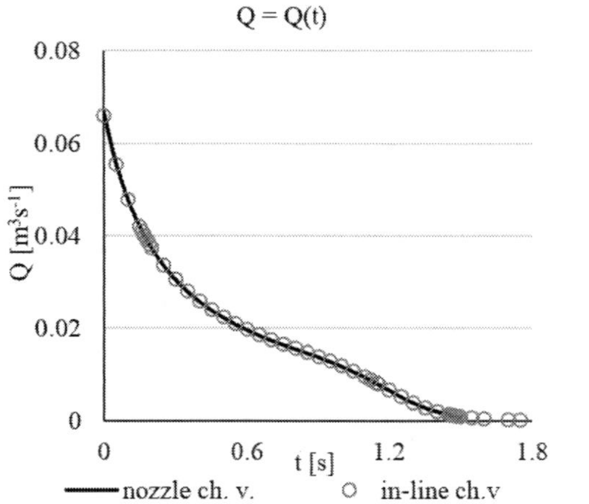

Fig. 8. 1AM simulations of the in-line and nozzle check valve in the piping system, the flow rate.

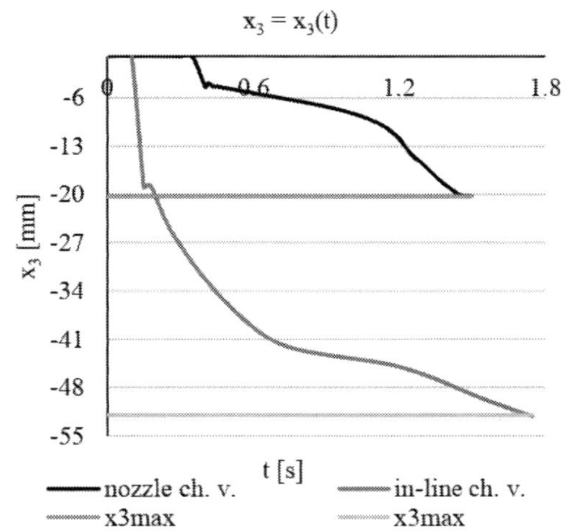

Fig. 9. 1AM simulations of the in-line and nozzle check valve in the piping system, the disc and the poppet positons.

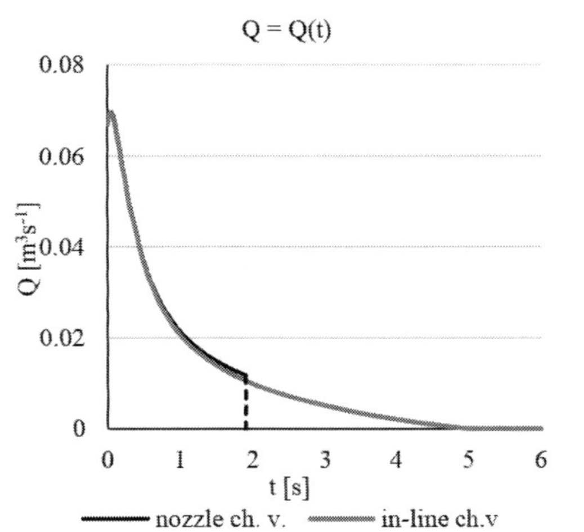

Fig. 10. 1AM in-line valve and nozzle check valve simulations in the piping system in case of a sudden failure of centrifugal pump during a power outage, the flow rate.

The effect of inertia in the system and, primarily, other hydraulic resistances in the circuit appears in the case of both valves (Figs. 8 and 9). Also, the closure curves of both valves will be plotted in addition to the above mentioned characteristics in case of a sudden failure of centrifugal pump during a power outage, Figs. 10 and 11.

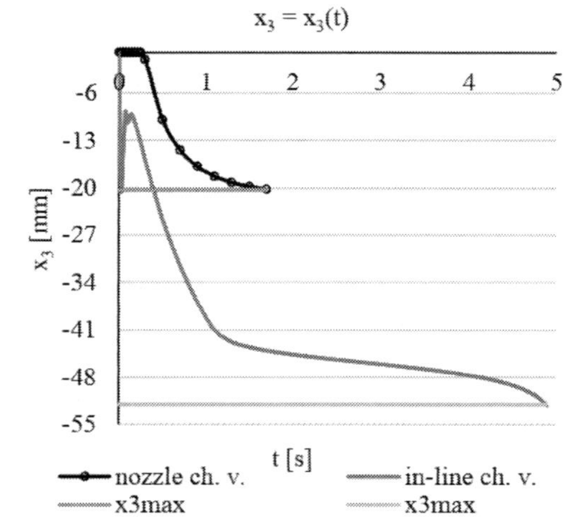

Fig. 11. 1AM in-line valve and nozzle check valve simulations in the piping system in case of a sudden failure of centrifugal pump during a power outage, the disc and the poppet positons.

It is evident from Figs. 10 and 11 that due to the inertia moment of the pump impeller and the inertia effects of the fluid in the pipeline, the closing times of both valves are relatively long. The influence of the hydrodynamic pump or its characteristics also appears at times close to t = 0s. There may be a slight increase in the flow rate for a very short time period and thus the movement of the poppet in the direction of flow. This movement is eliminated in the case of the nozzle check valve due to its maximum operational opening. But it is more important that the in-line valve closes in all cases without any backflow. Also, there is no backflow in the case of a nozzle check valve, unlike the simulation of the nozzle check valve itself. On the contrary, the nozzle check valve closes suddenly in low flows. The amount of this flow is determined by the moment of inertia of a pump impeller, the friction torque on the pump shaft and the inertia effects of the fluid. Also, the magnitude of the spring preload, which is of course given by the opening pressure, has a similar effect. The moment of inertia of the impeller is related to the appropriate choice of pump parameters and optimal operation. This can affect the size of the moment of inertia, which may be small or unnecessarily large depending on the pump. Not considering a flywheel. Also, the efficiency of the pump will affect the minimum flow rate at which the valve closes. The dependence of efficiency on flow rate is, of course, related to the charakteristic of power consumption but also to the stability of the head – flow curve. The disk friction losses in the pump have a stabilizing effect with rexpect to the head – flow curve, and just the magnitude of the disk friction and mechanical and spiral case losses at the shut-off point is determined by the gradient of efficiency curve at the shut-off point and thus the stability of the head-flow curve.

The time dependencies of the flow rate and the valve position for different sizes of the moment of inertia ***J*** and for different efficiency curves of the pump (Fig. 12) are shown in Figs. 13 - 16. The pump parameters at the operating point remain the same. The difference of the efficiency curves is based only on the different magnitude of the dissipated power ***Po*** at the shut-off point. This consideration is related to the upcoming detailed analysis of losses in the centrifugal pump and to the effect of these losses on the dynamic behavior of the valve. The moment of inertia and the dissipated power in the figures are related to their nominal values.

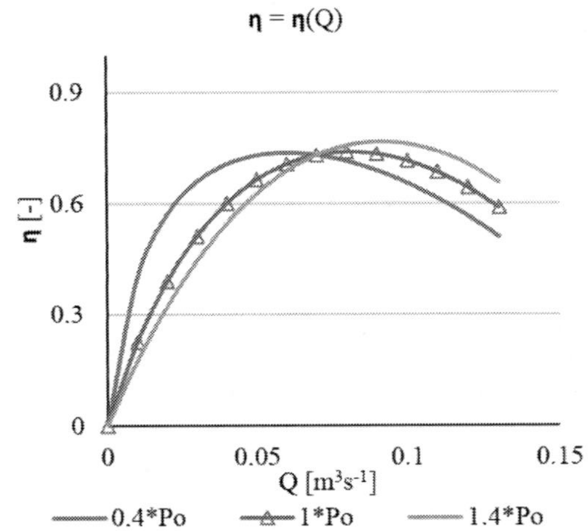

Fig. 12. The pump efficiency curves.

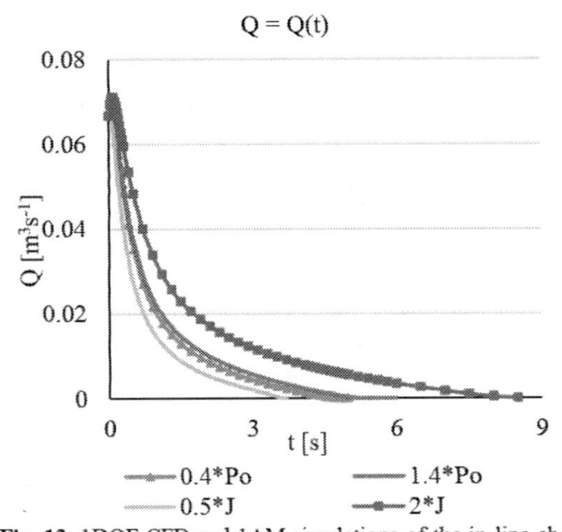

Fig. 13. 1DOF CFD and 1AM simulations of the in-line check valve, the flow rate.

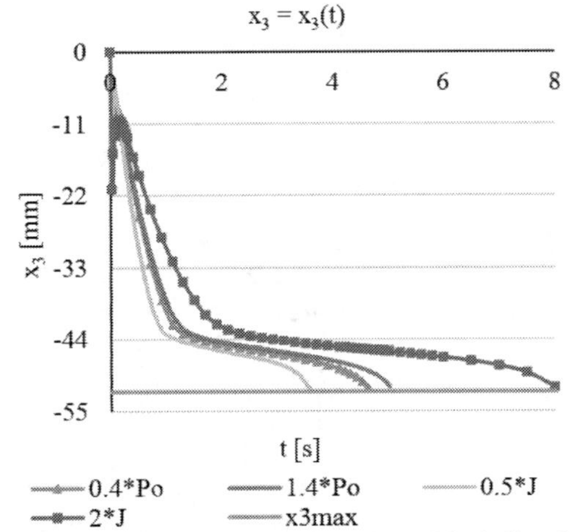

Fig. 14. 1DOF CFD and 1AM simulations of the in-line check valve, the positions of the poppet.

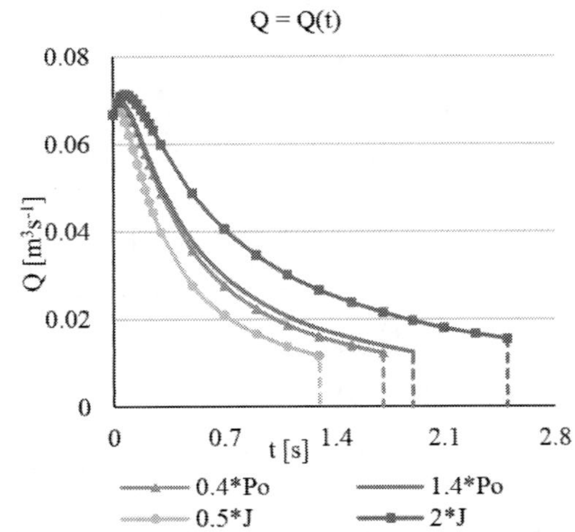

Fig. 15. 1DOF CFD and 1AM simulations of the nozzle check valve, the flow rate.

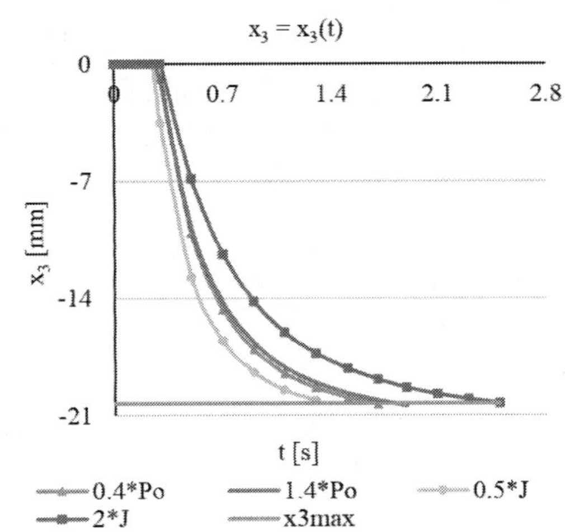

Fig. 16. 1DOF CFD and 1AM simulations of the nozzle check valve, the positions of the disc.

4 Conclusions

Finding a one-dimensional check valve model in the piping system should provide a basic analysis of the dynamic properties of the nozzle check valve. Since this is a follow-up study, the previous design (i.e. the in-line valve) was also reassessed in this respect.

The possibilities of positioning the boundary surfaces defining the valve were dealt with in the first phase. This step is important with respect to the definition of the magnitude of force acting on the disc by a one-dimensional model. However, the boundary surfaces of the valve can not be moved considerably due to the attempt to create a one-dimensional model of the check valve itself supplemented by a piping system. At the same time, it is necessary to use CFD methods to describe the variable velocity and pressure fields at the boundary surfaces of the valve.

The centrifugal pump as a typical part of the hydraulic system was simulated cumulatively by the power consumption, head - flow curve approximation, and the pump affinity laws. Also, the volume efficiency of the pump, even with the constant hydraulic resistance of the sealing rings, has been considered. In the future, the pump will be divided according to its basic parts and the losses determined in individual parts of the pump. However, this procedure does not have to increase the accuracy of the one-dimensional model because the pump power consumption used in the 1AM model can be determined quite accurately, leaving only defining spiral case losses The advantage will be to specify the spontaneous rotation of the impeller depending on the flow, which is essential with respect to the prediction of the water hammer in the pipeline. However, this study did not examine the water hammer.

A comparison of experimental data of in-line check valve and CFD simulation results in a relatively good agreement. The prediction of the poppet movement near the valve seat differs most. The practically non-existent backflow through the valve is important in assessing valve behavior. The nearness of the valve seat appeared problematic for 1DOF CFD simulations for both valves. An experimental testing in the form of the disc's position dependence on the flow was not realised for the nozzle check valve. The study of the valve was therefore realised with 1AM and 1DOF CFD simulations. The steeper gradient of closure curve of the valve depending on time, and the existence of backflow results from these simulations. However, the backflow was detected only in the simulation of the nozzle check valve itself without the connected pipeline. During the creation of the one-dimensional model, a considerable sensitivity of the hydraulic resistance was shown in extremely short times after the simulation started, when the disc was moved from the fully open position to the closed position. The magnitude of this hydraulic resistance has a significant effect on the time delay in which the valve remains in its fully open position. The dynamic behaviour of the pump with respect to full valve closure is shown in the pipeline system model with centrifugal pump. This may result in a rapid valve closure depending on inertia and mechanical friction in the pump.

It is also important to mention the necessity of experimental testing, which will allow the validation of the 1AM model and its further improvement.

This work has been supported by Technology Agency of the Czech Republic under the project Innovative research of check valves for extreme operating conditions in energetics TH01011352.

References

1. R. Klas, V. Habán, P. Rudolf, EPJ Web of Conferences, **143** 02051 (2017)
2. D. C. Wilcox, *Turbulence Modeling for CFD* (DCW Industries, Incorporated, 1994)
3. V. C. Patel, W. Rodi, G. Scheuerer, *Turbulence models for near-wall and low Reynolds numbers flows: A review*, AIAA J. **23(9)**, pp. 1308-1319 (1985)
4. J. F. Gülich, *Centrifugal Pumps* (Heidelberg Springer-Verlag, 2010)
5. P. Beierer, *Experimental and Numerical Analysis of the Hydraulic Cirquit of a High Pressure Common Rail Diesel Fuel Injection System* (Tampere, Tampere University of Technology, Publication 715, 2007)
6. J. Záruba, *Water Hammer in Pipe – Line Systems* (Academia Praha, 1993)
7. J. F. Gülich, World Pumps **1999(395)**, pp. 30 – 34 (1999)
8. J. F. Gülich, World Pumps **1999(396)**, pp. 39 – 42 (1999)
9. J. F. Gülich, Forsch Ingenieurwes **68**, pp. 87 – 95 (2003)
10. A. Nemdili, D. H. Hellmann, Forsch Ingenieurwes **71** pp. 59 -67 (2007)
11. A. Ladouani, A. Nemdili, Forsch Ingenieurwes **77 (3-4)** pp. 59 – 69 (2013)
12. M. H. Shojaeefard, B. Salimian Rizi, A.Khalkhali, M. Tahani, JAFM **8(4)**, pp. 673- 681 (2015)
13. A. K. Omar, A. Khaldi, A. Ladouani, Australian J. of Mech. Eng. pp. 1 – 12 (2016)
14. B. Brunone, U. M. Golia, M. Greco, J. of Hydr. Eng. **121(12)** pp. 906 – 912 (1995)

Study of the nozzle check valve with respect to its operating characteristics

Roman Klas[1,*], *Vladimír* Habán[1], and *Pavel* Rudolf[1]

[1]Brno University of Technology, Faculty of Mechanical Engineering, Victor Kaplan Department of Fluid Engineering, Technická 2, 61669 Brno, Czech Republic

Abstract. Several modifications were developed when designing the nozzle valve. This study offers an assessment of the properties of new modifications of the nozzle valve design. The main operating characteristics, such as loss and flow coefficients, were determined using a CFD methods. Besides mentioned coefficients, the forces acting on the valve disc are also decisive for the behavior of the valve, both in its steady and unsteady operation. It is important to examine the possible simplification and matching of CFD simulation results from 2D and 3D geometries in terms of subsequent dynamic analysis of the valve. This will be taken into consideration by comparing the above-mentioned operating characteristics, by analyzing the forces acting on the valve disc and comparing the velocity and pressure fields.

1 Introduction

The presented work proceeds on the development and study of check valves and swing check valves. The construction of these valves, as well as the ways of their application, differs considerably. Of course, this is due, among other things, to the different parameters of the pipe system, in which these fittings are used. From a basic division to low and high pressure hydraulics, we will deal with a low pressure hydraulics. The currently examined check valve is described as the nozzle check valve. As it has already been mentioned, it is a newly developed product based on a well-known principle. In a broader context, the goal is an optimal hydraulic design of check valves. A similar reasearch is also discussed in [1, 2]. Presented data obtained mainly from CFD simulations served as a support for follow-up experiments on real device. This approach has also been chosen in order to explore some of the phenomena in more detail, e.g. fully visualize flow in the valve, and examine physical quantities that would be difficult to obtain from the experiment.

We are usually interested in the operation of the check valve in terms of the dynamic effect of the whole pipeline. Therefore, it is difficult to optimally design a valve to perform as well as possible under all conditions and without generation of other adverse phenomena. Of course, the problematic operation of the valve is its closure, which depending on the other parameters of the pipeline, leads to a water hammer. The water hammer is pressure wave, which, depending on its frequency, can be transmitted undesirably to the sensitive parts of the pipeline system. Connected sensors or controls for valves are usually affected.

2 CFD methods and description of the valve

The static and some dynamic valve characteristics were evaluated using CFD simulations. Velocity and pressure fields were also monitored and the ANSYS 17.2 ANSYS Fluent was used as a basic tool for the analysis. The valve geometry was considered both as two and three dimensional and the computational mesh was created in ANSYS Meshing. More information can be found in Table 1.

Table 1. Description of numerical model.

	2D	3D
Pipeline diameter	DN100	
Number of computational cells	400000	9 - 16 millions
Turbulence model and near wall modeling	realizable k – ε non equilibrium wall function	
	realizable k – ε scalable wall function	
Boundary conditions	Inlet: velocity inlet Outlet: pressure outlet	
Calculation mode	unsteady, incompressible flow	

Two methods of near-wall treatment were chosen in the test tasks, since the size of the Reynolds number values ranged from 0 to 10^6 and the computational mesh was built in terms of its quality and the way of modeling

* Corresponding author: klas@fme.vutbr.cz

the boundary layer regions. More can be found, for example, in [3-7]. Of course, wall functions have been chosen with respect to the smallest number of computational mesh cells and the lowest computational requirements. The RNG k - ε turbulence model is recommended in the literature with regard to the higher accuracy of CFD simulations with respect to experimental data [8, 9]. However, using the RNG k - ε turbulence model settings in this case provided a worse agreement between CFD simulations results and the experimental data than the realizable k – ε model. Hydraulic valve losses in the case of RNG settings have been underestimated.

The nozzle check valve excels by relatively small installation dimensions. The valve can also be described as approximately axisymmetric. This, of course, can help to improve the accuracy of a simpler two-dimensional analysis. Of course, this may help to improve the accuracy of a simpler two-dimensional analysis. However, the properties of the devices that are adapted to the experimental testing and are different in some aspects from the production version of the valve, are problematic, see Fig.1.

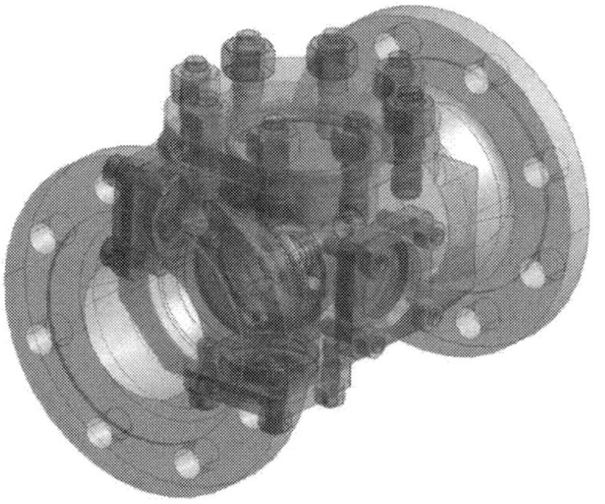

Fig. 1. The nozzle check valve, 3D P simulation.

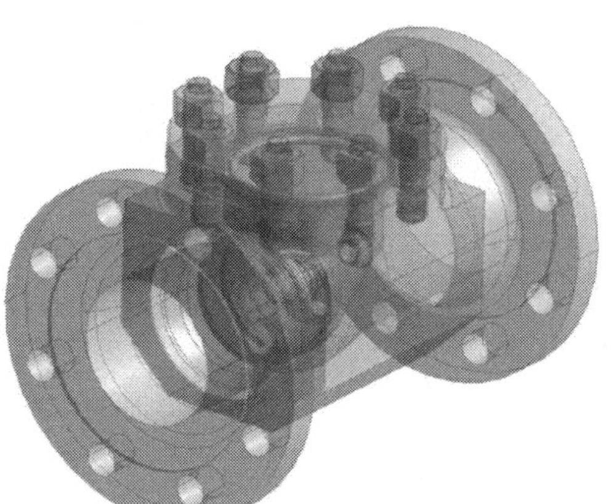

Fig. 2. The nozzle check valve without windows, 3D simulation.

As can be seen from the Fig. 1, the device was completed with several windows to ensure the visualization of the velocity field. The mounting hole at the top of the valve is fitted with an insert. The hole may also serve as an observation window if the insert is removed. Two 3D models were created due to observation windows, see Fig.2. The windows are omitted in the second 3D model the. Therefore, the number of computational cells in the three-dimensional model in Tab. 1 is quite different.

The CFD simulation task will, among other things, decide the difference in the results of the experimental device (Fig.1) and the nozzle check valve, which is very close to the current production version. A view of the interior of the valve is provided in Fig.3.

Fig. 3. Interior of the nozzle check valve, 3D simulation.

Fig. 4. Interior of the nozzle check valve, 2D simulation.

Fig.4 shows the model used in 2D simulation. This simulation was solved as axisymmetric. Of course, the pipeline parts of the 2D and 3D simulations were also considered to eliminate the effect of the boundary conditions on the interior of the valve. The influence of boundary conditions is discussed in more detail e.g. in [10, 11]. The disc itself was also tested in the valve chamber in addition to the above-mentioned simulations (Fig.5). The reason is the future study of the dynamic behavior of the valve and decsription of the force magnitude from the fluid acting on the disc.

Fig. 5. The disc in the valve interior, 2D simulation.

3 Static characteristics

The valve was tested in operating modes that were defined by mass flow. Tab. 2 provides informations about the mean values of absolute velocities in the inlet section of the valve and the sizes of its opening. Values in the columns do not necessarily match each other.

Table 2. Range of mean absolute velocities in the valve and the valve opening sizes.

v[m.s^{-1}]	-3	-2	-1	1	2	4	6	8	10	10.5
z[mm]					2	5	7	10	15	20.5

The additional symbols used in the following figures for the pressure and velocity fields in the valve are shown in Table 3.

Table 3. The additional list of symbols.

N	non equilibrium wall function
ON	disc only
P	windows in the valve walls
S	scalable wall function
SM	non-friction valve walls, disc with friction
Sa - Sb	the difference between the inlet and outlet sections of the valve
Inlet - Outlet	the difference between surfaces with boundary conditions

Since the static characteristics of the valve will be presented, the flow in the valve was simulated for several basic positions of the disc according to the degree of its opening. Convergence of CFD simultions, especially 3D ones, was relatively weak due to high valve velocities with small valve openings and high flow rates. For this reason, it was not possible to realize all the simulations in all modes corresponding to Tab. 2. Of course, many of them do not match any of the operating modes.

The dependence of static pressure difference on flow rate is one of the most important characteristics that describe the static behavior of the valve. For the time being, the dependencies corresponding to the valve opening $z = 7$ mm will only be shown, see Fig. 6 and Fig. 7. The results of 2D and 3D simulations as well as the choice of different wall functions for three-dimensional problems are compared. Preliminary, we can say that non equilibrium wall functions will be preferred. Just 2D simulations should bring simplification of further valve analysis, as already mentioned.

The evaluations of the flow parametres on the different boundary surfaces are also presented, see Tab. 3. It is not possible to trust the accuracy of static pressure evaluation on Inlet - Outlet surfaces. At the same time, the differencies in the areas of Sa - Sb due to the boundary conditions are shown. Of course, it is paramount that the turbulent or laminar velocity profile was fully developed before the computational domain of the valve interior itself. The above mentioned is also partly related to non-matching positions of pressure sensors in CFD and experiment. More details on agreement of numerical and experimental approach can be found in [12] for example. Of course, the location of the sensors does not completely match the input and output surfaces to the valve. The hydraulic losses produced by the valve are relatively small, and the inclusion of a part of the pipeline into the evaluated section may produce an undesirable error or misrepresentation.

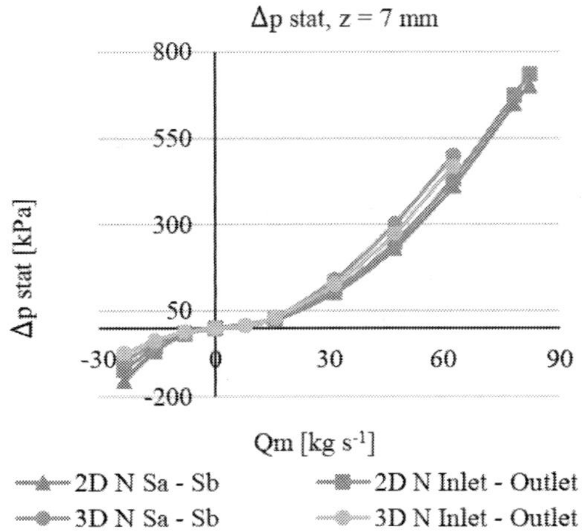

Fig. 6. Downstream and upstream static pressure differencies.

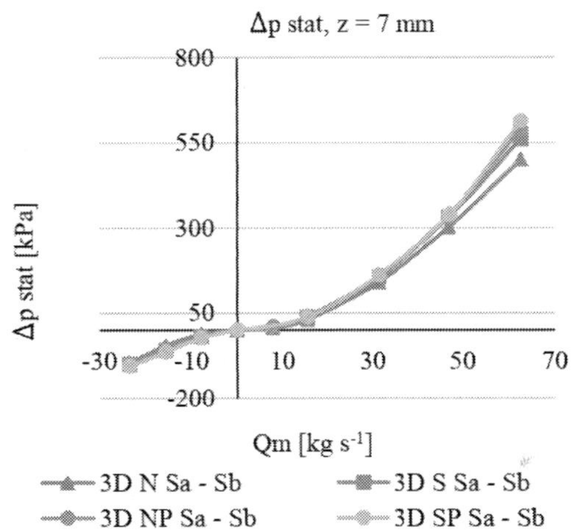

Fig. 7. Downstream and upstream static pressure differencies.

It is evident from Fig. 6 that there are some differences between 2D and 3D simulations. These differences mainly correspond to the areas on which the boundary conditions are located. On the other hand, the agreement of static pressure differencies is very good on the surfaces corresponding to the inlet and outlet of the valve. The following igure (Fig. 7) offers a comparison of 3D valves with and without windows (Fig.1 and Fig.2) as well as comparison of the use of two different wall functions, see Tab.1 and Tab.3. It seems that scalable wall functions predict a more pronounced pressure loss. The greater static pressure differences of the valve with the windows was foreseeable.

It is also evident that the pressure difference was also evaluated into the backflow region, considering that the backflow through the valve may theoretically occur. This condition is undesirable from the water hammer pressure extreme point of view. The flow coefficients $\boldsymbol{K_v}$ as defined in (1) are plotted in the following Figures 8 and 9.

The flow coefficients Kv standardly defined by (1) are captured in the following Figs. 8 and 9.

$$K_V = Q\sqrt{\frac{\Delta p_{ref}\ \rho}{\Delta p\ \rho_{ref}}} \tag{1}$$

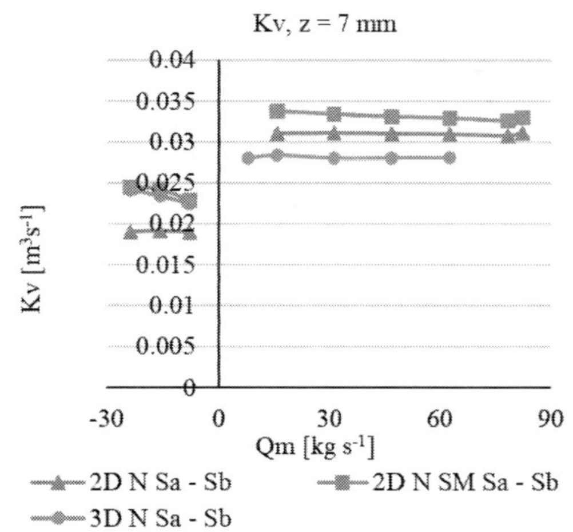

Fig. 8. Downstream and upstream flow coefficients.

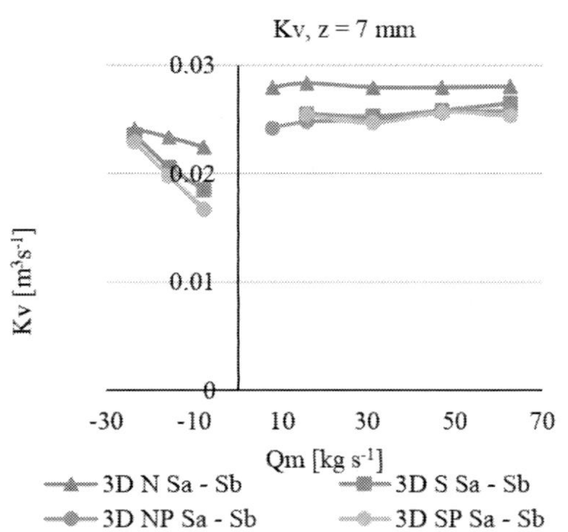

Fig. 9. Downstream and upstream flow coefficients.

It results from Fig. 8 that the highest values of the flow coefficient are achieved by a valve whose walls are defined as friction-free walls. The reason for including this simulation in the attached statistics is to analyze the force acting on the valve disc. Furthermore, it is clear from Fig. 9 that the flow coefficient Kv is slightly higher even for the 3D simulation of the valve without windows. However, the differences are not large. The choice of different wall functions was most evident in the model without windows (Fig.2).

Another pair of graphs are Fig.10 and Fig.11, which capture the magnitude of the loss coefficient according to the mass flow through the valve (2).

$$\xi = \frac{\Delta p}{\rho \frac{v^2}{2}} \tag{2}$$

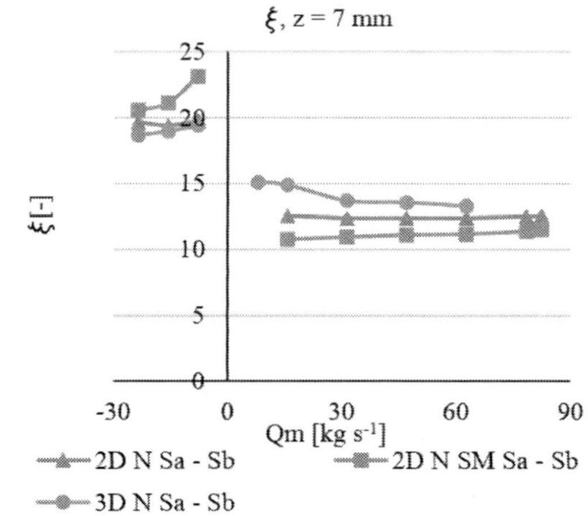

Fig. 10. Loss coefficient.

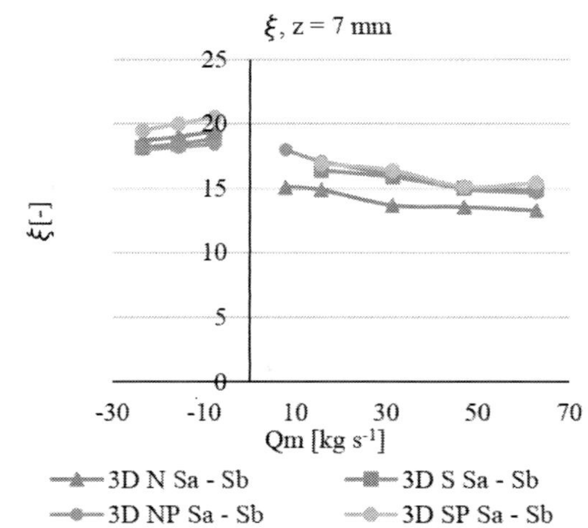

Fig. 11. Loss coefficient.

The loss coefficient increases with a decreasing flow rate in accordance with the assumptions. Slightly greater difference between 2D N and 3D N simulations for $z = 7$ mm takes into consideration the effect of the fin on which the valve disc bushing is located, see Fig. Fig.3. The windows in the valve wall increase the deviation from the axisymmetric flow, Fig.11. There is a very small difference between 3D S and 3D SP simulations for downstream flow rates. Such situation is inconsistent with expectations. Therefor the N model will be preferred, Tab.3.

It will be important for the dynamic behavior of the valve to determine the magnitude of the force by which fluid acts on the valve disc. The forces Fx3 acting on the axis of the valve, depending on the mean value of the static pressure, are shown in Fig. 12 and Fig. 13.

https://doi.org/10.1051/epjconf/201818002044

Note: No force from the spring is taken into consideration in any of the graphs. Positive effect is assigned to the force acting in the direction of downstream flow through the valve.

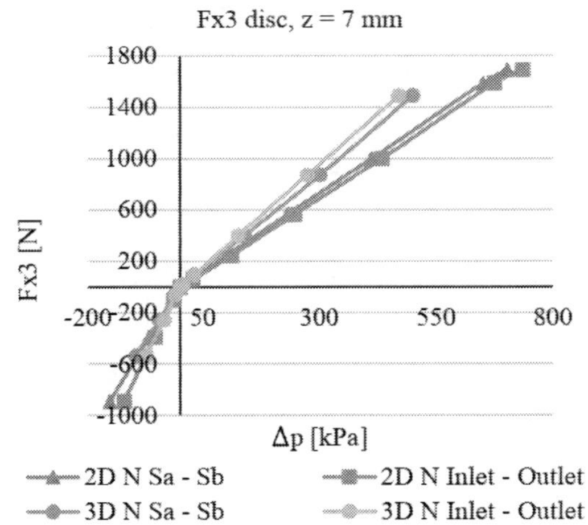

Fig. 12. The forces acting on the valve disc.

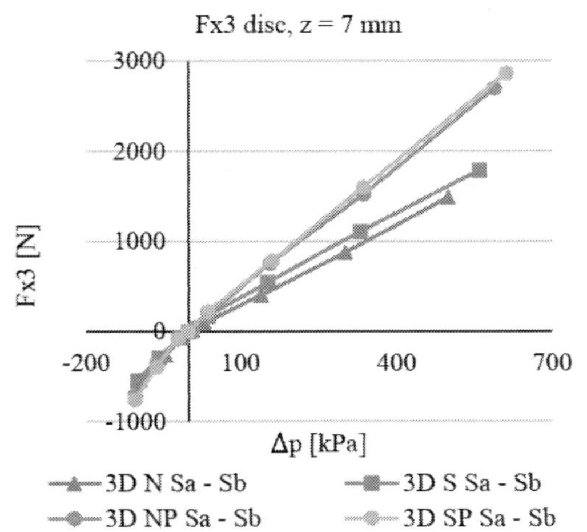

Fig. 13. The forces acting on the valve disc.

All forces are almost linear. Conformity of 2D and 3D models is generally better negative mass floe rate values or static pressure differences. The magnitude of the valve forces with or without windows do not differ for both cases of selected wall functions, which is convenient. This allows more usability of the already obtained data and also presents a lesser risk of choosing unsuitable wall functions. It can also be said that the force on the disc is higher than in case of the smooth shape of the valve with the windows. This may have an impact on a visualization experiment that verifies the dynamic behavior of the valve, which should be realized just on the valve with the windows (Fig.1).

Valve force acting on the immobile solid wall of the valve was also evaluated in addition to the force acting on the plug (excluding the inlet and outlet pipes). The ratio of these forces will be interesting from this point of view.

Force analysis can be done by integrating the Navier-Stokes equation (3) within the valve interior.

$$\rho_K\left(\frac{\partial v_i}{\partial t}+\frac{\partial v_i}{\partial x_j}v_j\right)=\rho_K g_i-\frac{\partial p}{\partial x_i}+\frac{\partial \Pi_{ji}}{\partial x_j} \qquad (3)$$

Obviously, it would be problematic to integrate all the terms in equation (3) without CFD simulations. This fact makes it more difficult to build a corresponding one-dimensional model that predicts the behavior of the valve in the piping system. Even with knowledge of CFD simulation data, it is not easy to simply substitute a term in (3) in which includes the irreversible stress tensor $\boldsymbol{\Pi}$. Of course, the force from the fluid affects the fixed and moving walls of the valve. We can assume at least approximately an axisymmetric static pressure $\boldsymbol{p}$ distribution when determining the resultant force effect. However, such simplification will not be useful for shear stresses. The adhesion of liquid to the walls (no slip boundary condition) can only be considered. The fluid force on the valve disc $\boldsymbol{F_K}$ can then be expressed, for example, by equation (4).

$$F_K=\rho_K x_3 V_\Gamma-\rho_K\left(x_{3Sa}+x_{3Sb}\right)\frac{\partial Q}{\partial t}+R_{ab}Q^2S \qquad (4)$$

$$R_{ab}=\frac{\xi}{2S^2} \qquad (5)$$

The hydraulic resistance $\boldsymbol{R_{ab}}$ (5) then represents the term in (3), which describes the magnitudes and distribution of static pressures and shear stresses on the solid walls. The normal and tangential forces acting on the moving valve disc are represented by $\boldsymbol{F_K}$. However, it follows from this consideration that $\boldsymbol{R_{ab}}$ can not be determined from the total loss coefficient in the valve (eg experimental data), because it includes the hydraulic losses caused by the disc itself.

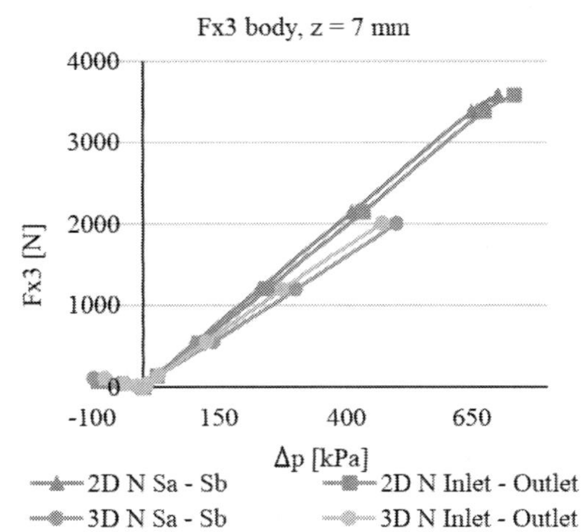

Fig. 14. The forces acting on the valve body.

For these reasons, it is important to get an idea of the ratio of forces only acting on the valve disc and on its fixed solid walls, respectively, of the hydraulic losses caused by the disc and the non-moving solid walls of the valve. Although it is necessary to consider mutual interaction of that the pressure and velocity fields around the disc and around the solid walls. Equation (4) also indicates that it is necessary to focus attention on the unsteady flow rate diferencies that could significantly affect $\boldsymbol{F_K}$. However, the magnitudes of forces on the whole body of the valve will be presented first, Fig.14 and Fig.15.

Fig. 14 shows that the similarity of 2D and 3D simulations is sufficient. Also, the forces determined by the negative size of the static pressure difference are important, i.e., during backflow of the fluid through the valve. In some cases, the magnitude of these forces may be positive and not negative, as in Fig. 14. Obviously, the orientation of the force vector is given by the shape of the valve, the fin and the guide bushing. Some body of guide bushings are also designed as through-flow channel. Even in the case of evaluation of the axial force on the valve wall, there is a good consistency between two different wall functions (Fig.15). The forces also maintain a linear dependence on the pressure difference in the valve.

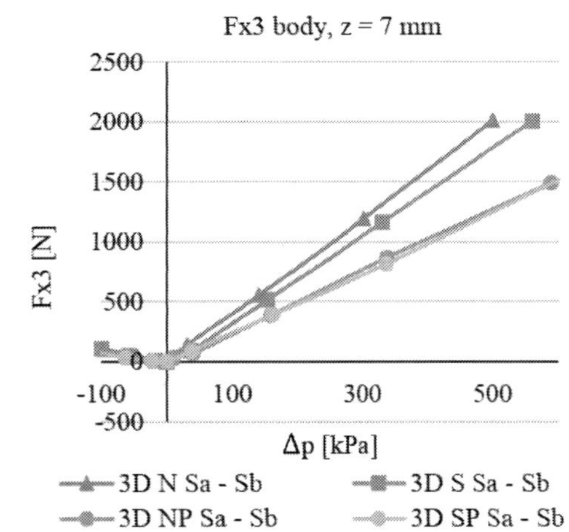

Fig. 15. The forces acting on the valve body.

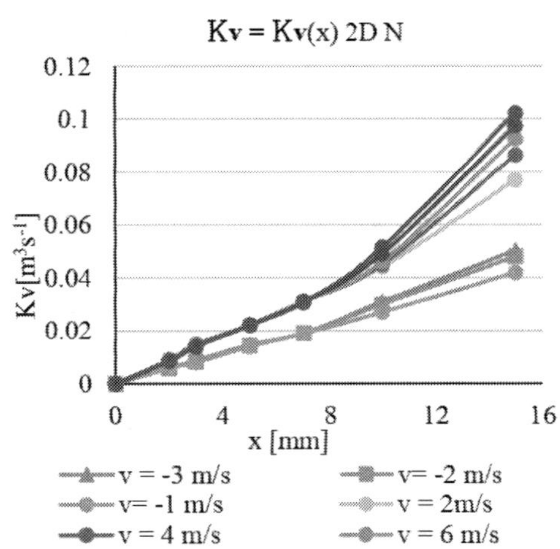

Fig. 16. Flow coefficient.

All previously presented characteristics corresponded to one position of the valve disc. The aim was to specify the typical characters of the physical quantities. In addition, the static characteristics will be shown for the different positions of the disc z. Fig. 16 shows the flow coefficient $\boldsymbol{K_v}$ in case of simulation 2D N (Tab.3).

It is clear that the curves are divided into two basic parts according to the direction of flow. The obtained data are not easily separable due to the number of curves corresponding to the velocities in Tab. 2. The lower curves correspond to backflow.

Similarly, loss coefficients ξ for 2D N simulation are shown, see Fig. 17. It is clear from this figure that there is a clearer separation of downstream flow and backflow. And it is also obvious that the backflow velocities almost correspond to the highest loss coefficients ξ in range of disc position z.

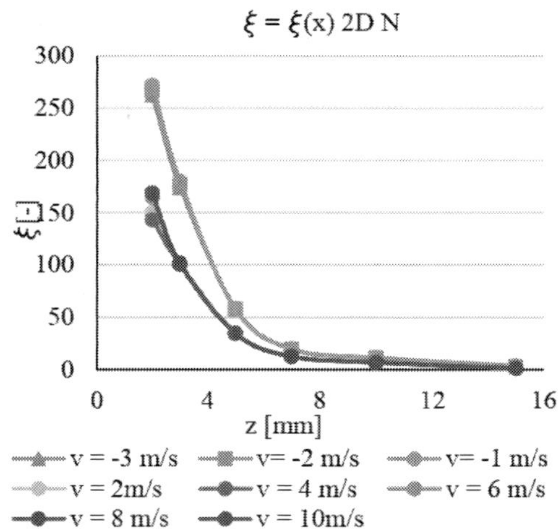

Fig. 17. Loss coefficient.

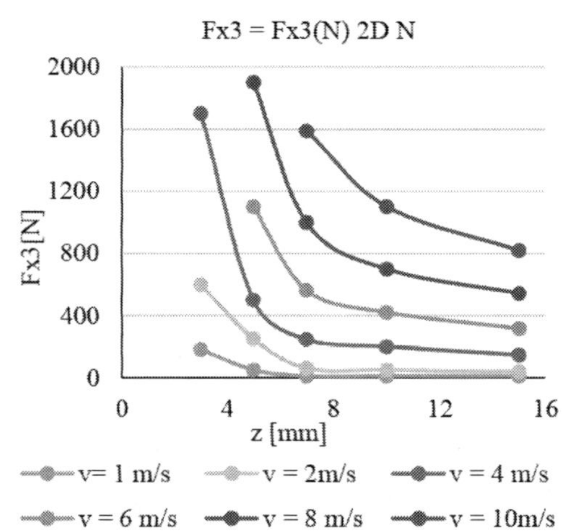

Fig. 18. The forces acting on the valve disc.

The force magnitude for the 2D N simulation are shown in Fig.18. Each curve corresponds to approximately hyperboles and should increase theoretically for $z = 0$ to infinity. However in the case of a real valve the magnitude of the force acting on the disc corresponds to the spring preload force, for $z = 0$.

Undoubtedly, a crucial point of view for assessing static characteristics will be their comparison with the experimental data. Experimental data were available for valve without windows (Fig. 2) at the time of writing this paper. Of course, the valve disc is moved when the flow is changed, so its position, which can be associated with the flow rate, needs to be known. Flow and loss coefficients were evaluated for constant flow during the experiment. The problem with the unknown valve disc position has been resolved through CFD simulations. Software ANSYS 17.2 allows 1DOF (degree of freedom) simulation of valve disc movement. Therefore, the dynamic simulation has been realized for the constant flow rate, for which the valve disc has found its operating position. The dependence of the loss and flow coefficients was then obtained by interpolation of the static characteristics in dependence on the flow rate (Fig. 19 and Fig. 20).

The valve was tested on the closed test circuit. Circuit consists of the connecting pipes, pressurized vessel, which is connected to vacuum pump (for cavitation testing), the tested valve itself and feeding pump. Flow rate was measured using induction flowmeter, pressure difference between valve inlet and outlet was assessed using pressure sensors. Measurement procedure and data evaluation is according to valid standard ČSN EN ISO 9906.

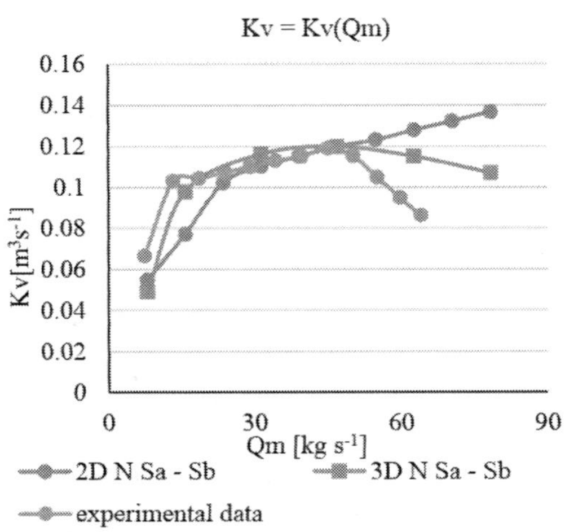

Fig. 19. Flow coefficient.

Agreement of CFD simulation with experiment is good for medium operating flow rates in both cases. The difference in results is most noticeable in 2D simulations in the high flow rate area. The loss coefficient usually decreases with increasing flow rate and the flow coefficient increases. However, there are coefficient break points for this valve and both coefficients manifest non-standard characteristics. The reason is the embedded fin in 3D, the cavitation and the separated and swirling flow around the inner parts of the valve.

Since the magnitude of the loss and flow coefficient in 3D simulations shows differences in comparison with the experiment, the dependencies of these coefficients for the maximum opening of the valve are also shown, namely $z = 20.5$ mm, Fig. 21 and Fig. 22. However, it should be noted that the valve may not be fully open in normal operating mode. Trend of coefficients from 3D simulations and experiment occurs with increasing valve opening approximately from $z = 10$ mm but first very slightly (Fig. 21 and Fig. 22). This disc position corresponds to half of the maximum lift of the disc.

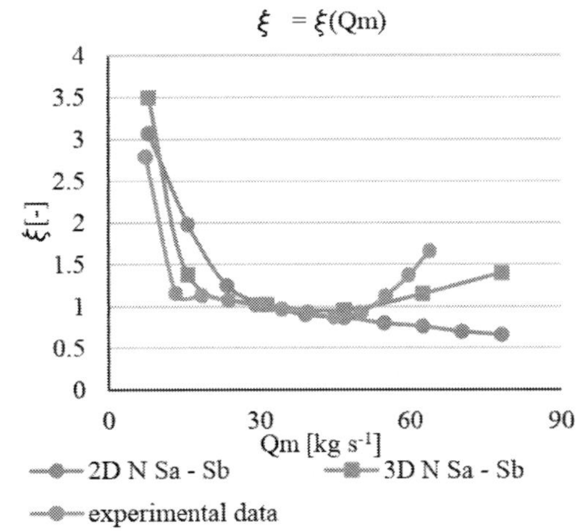

Fig. 20. Loss coefficient.

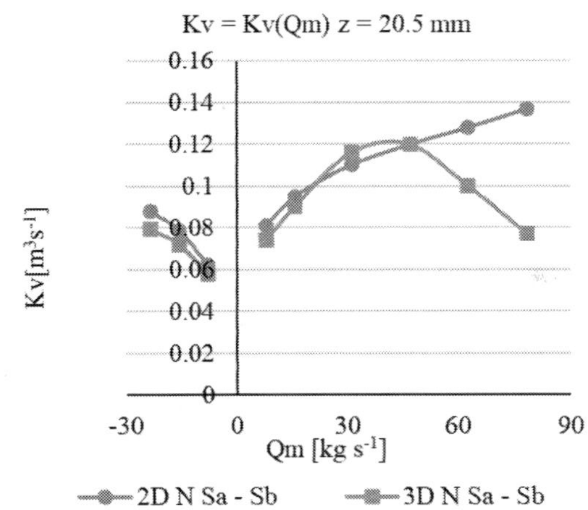

Fig. 21. Flow coefficient

Valve characteristics with constant disc position and variable flow rate are the last dependencies to be provided. A closer study of the mathematical model of force acting on the valve disc (4) is the reason for their presentation. An idea of the actual magnitude of the unsteady term in (4) will be built upon them. The magnitude of the unsteady term in the force equation was determined first. This magnitude was then compared as a force increment with the increase in force in the unsteady change of flow in CFD simulation. The change in test flow was linear and ranged from positive to negative for the time corresponding to the closing time $\boldsymbol{Ts}$ of the valve.

The Ts values are described in the corresponding curve legend in the graph. The dependencies of disc forces on the static pressure difference were again linear. It seems that, due to different sizes Δp, the magnitude of these forces differs considerably. For this reason, the forces were plotted according to the constant and variable flow through the valve, see Fig.23.

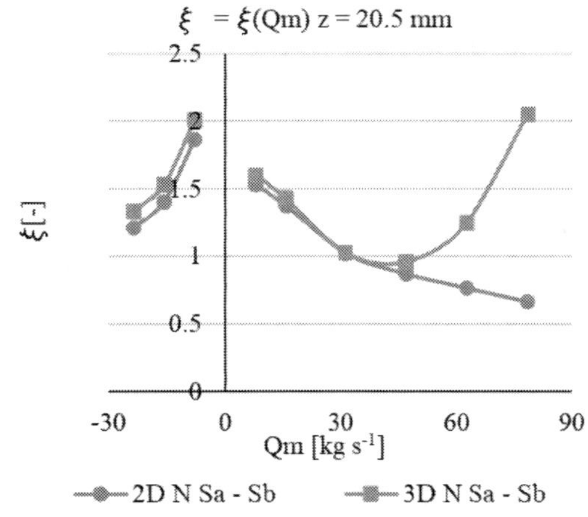

Fig. 22. Loss coefficient.

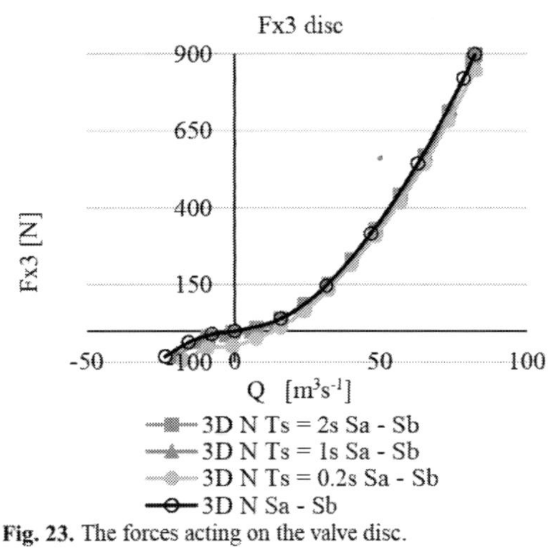

Fig. 23. The forces acting on the valve disc.

Valve shut-off time Ts = 0.2 s is extreme and can only be realized in case of valve bypass or pipe failure. Fig. 24 shows the ratio of the force gains on the disc versus the stationary state in the CFD simulations and in prediction from the analytical model of force (4). The ratio of the force gains is more balanced when the valve closes faster. The model of the unsteady force component in (4) assumes a constant distribution of local acceleration on the Sa and Sb surfaces, respectively, considers its mean value according to the flow rate. A shorter valve closing time achieves a more balanced change of the absolute velocity vectors on the surfaces. In the case of closing time Ts = 0.2s, the ratio of the analytical model of the unsteady force component F_{x3} is 40-80% of the value obtained from the CFD simulation. This ratio increases with increasing time and becomes an extreme whose value has a growing character. On the other hand, the flow rate at which the extreme occurs decreases. The whole problem can also be described as a prediction of an unsteady velocity profile. Therefore, it is more advantageous to extend the valve area to the inlet and outlet pipe sections in the mathematical model so that the velocity profile is not significantly affected by the flow in the valve on the Sa and Sb surfaces. However, those parts of the pipeline would be included in the simulation, which do not actually belong to the valve.

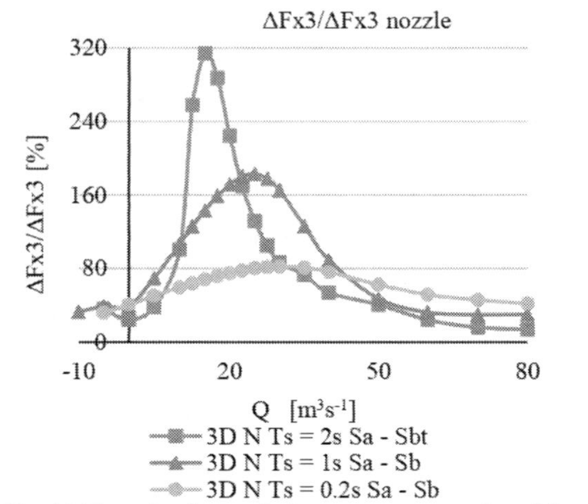

Fig. 24. The ratio of unsteady force components evaluated from analytical approach and from CFD simulations.

4 Comparison of the velocity and pressure field

It will be important to compare the flow field in the valve interior to complement the study of the valve geometries. Of course, the main objective is to determine the differences and to assess the possibilities of reducing the computational requirements of CFD simulations. The velocity and pressure fields in two directions of flow will only be monitored. Flow rates correspond to the mean velocity magnitude of the absolute velocity, $v = 2\mathrm{m.s^{-1}}$ for Q1 and $v = -1\ \mathrm{m.s^{-1}}$ for Q2. The flow rate Q1 represents the normal flow through the valve. The other labels used in the figures correspond to the symbols in Tab. 3. The valve opening z = 7 mm is identical to the position of the disc in the previous static characteristics.

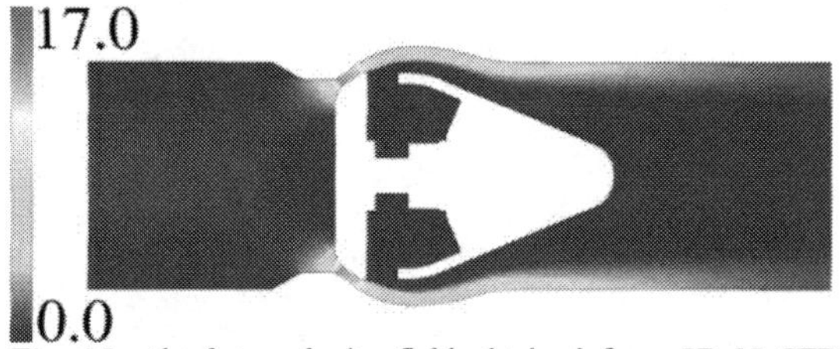

Fig. 25. Absolute velocity field obtained from 2D N CFD simulation, flow rate Q1.

First, Fig. 25 - 34 present a velocity field. Significant differences are not observable in Fig. 25-28. The embedded fin on which the disc and guide bushing are attached affects the flow in the 3D geometry on Fig.28. However, the flow is axisymmetric in the remaining parts of the valve, and the magnitudes of the absolute velocities are almost identical in all cases.

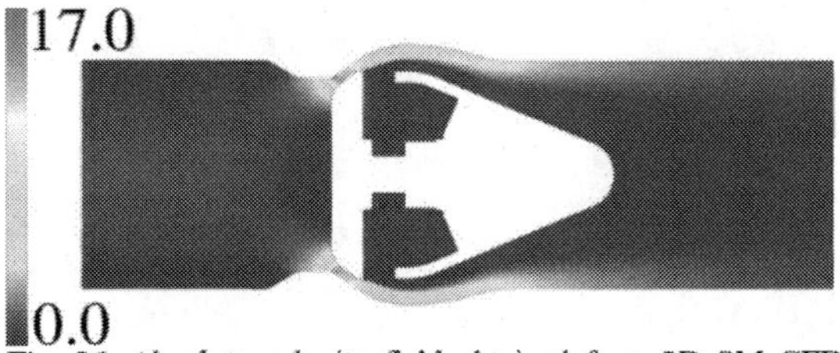

Fig. 26. Absolute velocity field obtained from 2D SM CFD simulation, flow rate Q1.

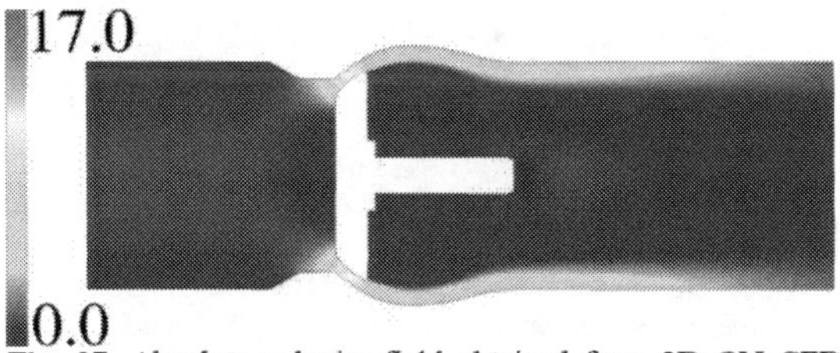

Fig. 27. Absolute velocity field obtained from 2D ON CFD simulation, flow rate Q1.

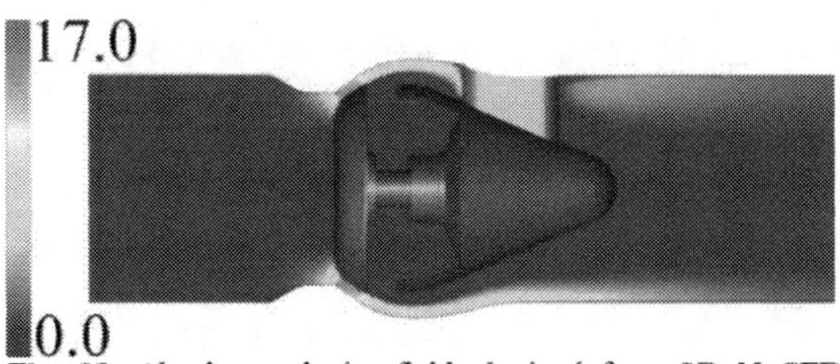

Fig. 28. Absolute velocity field obtained from 3D N CFD simulation, flow rate Q1.

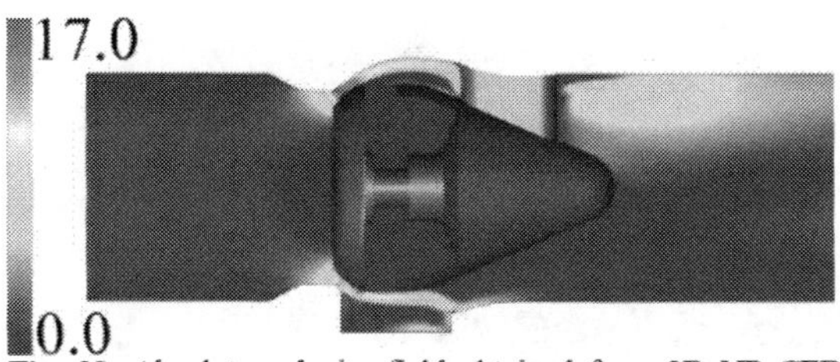

Fig. 29. Absolute velocity field obtained from 3D NP CFD simulation, flow rate Q1.

The valve windows (Fig. 1), which significantly interfere with the flow field in Fig. 29, appear to be problematic. The windows are an element that deforms significantly the flow around the valve disc. The window frames intersect approximately one-third of the disc circumference. However, the situation in the case of backflow through the valve represented by the flow Q2 is significantly different. The flow around the disc and inner parts of the valve in Fig. 30 and Fig. 31 is noticeably different from the flow around the disc itself in Fig. 32 and the 3D geometry in Fig. 33. The computational meshes were very similar in 2D geometries, and the feature that could only have affected the resulting flow character is the axisymmetric condition. Also, the k - ω SST turbulence model has been tested for 2D and 3D geometry to be sure in this particular case with respect to the possible flow separation of the 3D model. However, the velocity field did not change its character in any of the cases. The disc itself appears to be the most identical in terms of flow behind the disc in backflow for 2D geometry (Fig.32) and complete 3D geometry for the valve without windows (Fig.33). It should be noted that the high velocity magnitude regions in the figures are relatively short and the fluid on the walls does not slip. It is therefore important for the resulting flow field to reflect the shape of the streamlines between the disc and the guide disc as well as the ratio between the shear and the pressure forces. The whole body of the valve is a slip wall in Fig. 31, and the flow field has the character of an intermediate step between the results of 2D and 3D simulations, i.e., Fig. 30 and Fig. 33. Even simulation of the flow around the disc (Fig.32), in which the shear force part due to the absence of the guide disc and bushing is smaller, indicates this.

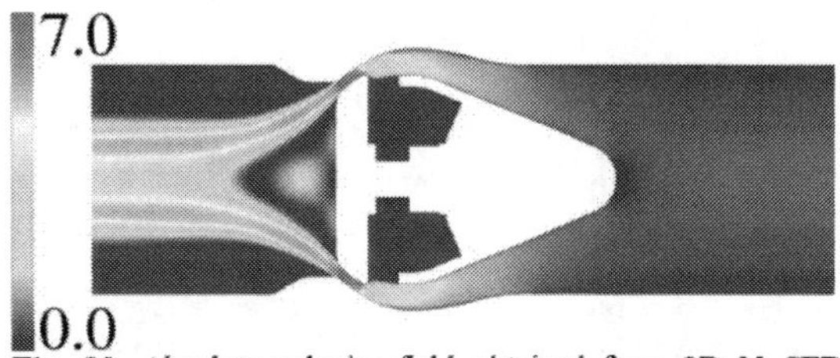

Fig. 30. Absolute velocity field obtained from 2D N CFD simulation, flow rate Q2.

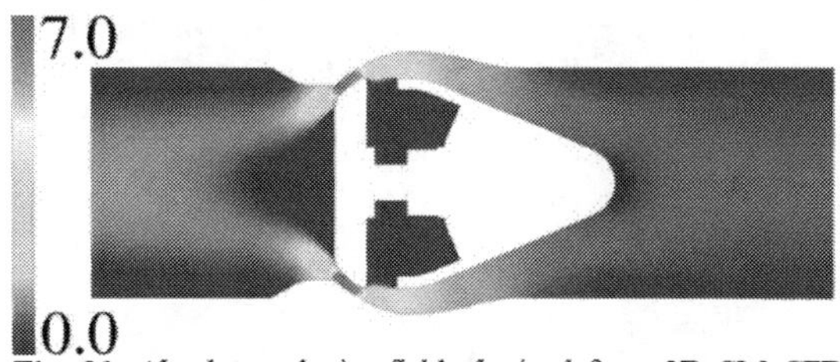

Fig. 31. Absolute velocity field obtained from 2D SM CFD simulation, flow rate Q2.

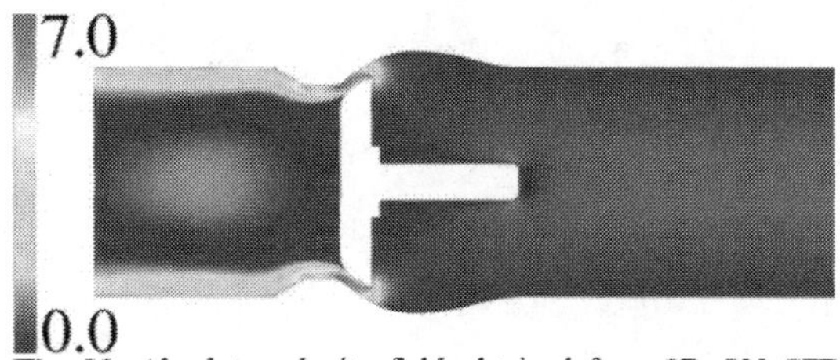

Fig. 32. Absolute velocity field obtained from 2D ON CFD simulation, flow rate Q2.

As could be expected, the most disordered velocity field is represented by geometry with windows, see Fig. 34. The window frame itself also splits the stream and causes the flow between the disc and the guide disc. These facts may affect the subsequent assessment of the dynamic properties of the valve during the experimental testing. Note: The velocity magnitude scale has to be changed for both 3D simulations because the maximum values are slightly higher than for the 2D geometry.

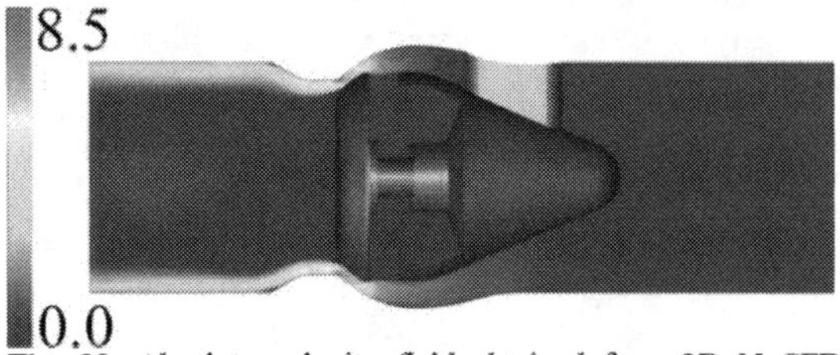

Fig. 33. Absolute velocity field obtained from 3D N CFD simulation, flow rate Q2.

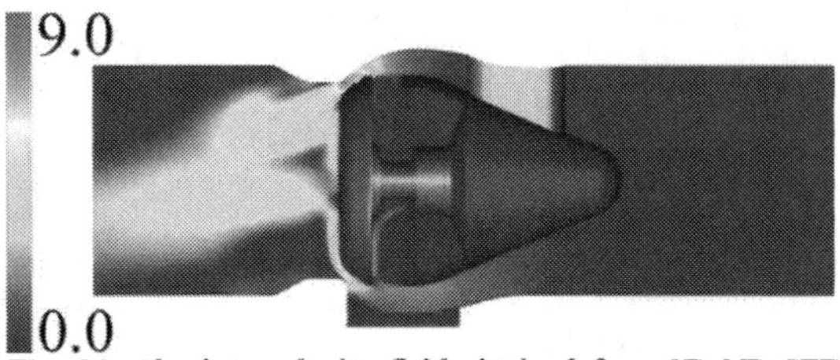

Fig. 34. Absolute velocity field obtained from 3D NP CFD simulation, flow rate Q2.

In addition to the velocity field, the field of static pressures will be plotted (Fig. 35 - 44). The same scale was retained in the figures wherever it was possible, or at least the comparative value was marked. The first is shown the pressure field for the flow direction Q1. Present pressures are relative and their real value would be obtained by adding the absolute static pressure in the pipeline system. Significant differences are not observable in addition to the pressure magnitudes before the disc in Fig. 35 and 36. The extension of the lower pressure zone behind the plug is shown in Fig. 37. The guide disc in combination with the valve body partly behaves as a diffuser and causes a faster pressure increase in the guide disc area (Figs. 35 and 36).

Fig. 35. Static pressure field obtained from 2D N CFD simulation, flow rate Q1.

Fig. 36. Static pressure field obtained from 2D SM CFD simulation, flow rate Q1.

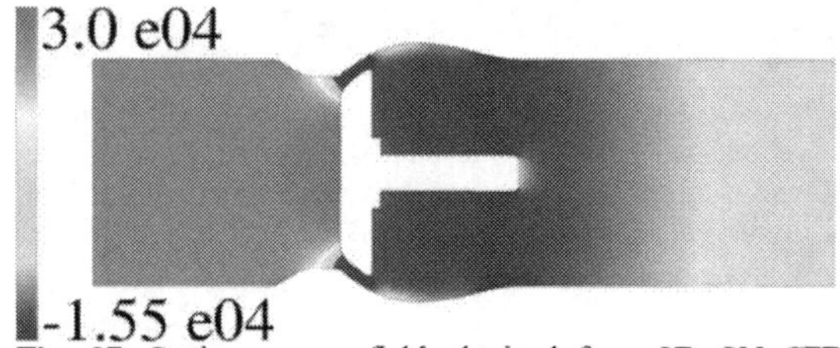

Fig. 37. Static pressure field obtained from 2D ON CFD simulation, flow rate Q1.

The lowest pressure value from the previous 2D simulations is marked on the scale map in Fig.38. Static pressure magnitudes do not disagree from the previous simulations significantly, but the pressure drop in the region between the disc seat and the disc is noticeable. The pressure extreme region is observable in the vicinity of the valve seat also in Fig. 39. The pressure values in Fig.38 and Fig. 39 are almost identical in the area between the disc and the guide disc. The value of -1.55 E-04 in Fig. 39 is marked down on the scale map. The window at the bottom of the valve is again the most significant difference that appreciably changes the pressure distribution around the seat of the disc. The increase in static pressure in front of the valve disc compared to the valve without the windows is also noticeable. The expected increase in static pressure around the valve fin is evident.

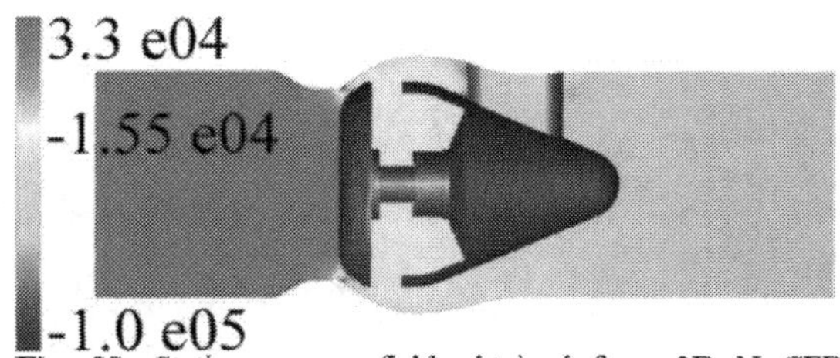

Fig. 38. Static pressure field obtained from 3D N CFD simulation, flow rate Q1.

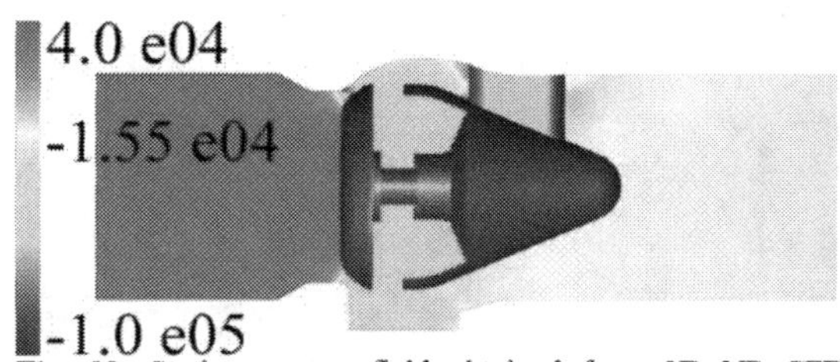

Fig. 39. Static pressure field obtained from 3D NP CFD simulation, flow rate Q1.

The remaining figures of the pressure fields (Figures 40 - 44) represent the upstream flow rate Q2 through the valve. Again, it was only possible to keep the same scale map on 2D simulations. In comparison with 2D and 3D simulations (Figures 40, 43 and 44), the difference between the highest and lowest values of the static pressure is approximately equivalent to the variants 2D N and 3D N. However, the static pressure distribution is different. This situation has already been mentioned in the description of the velocity field behind the valve plug in the direction of flow in Fig. 30. It is again possible to see the essential differences in the 3D NP geometry in Fig.

44. Extreme pressure values vary, and the viewports create remarkable asymmetry of the pressure field.

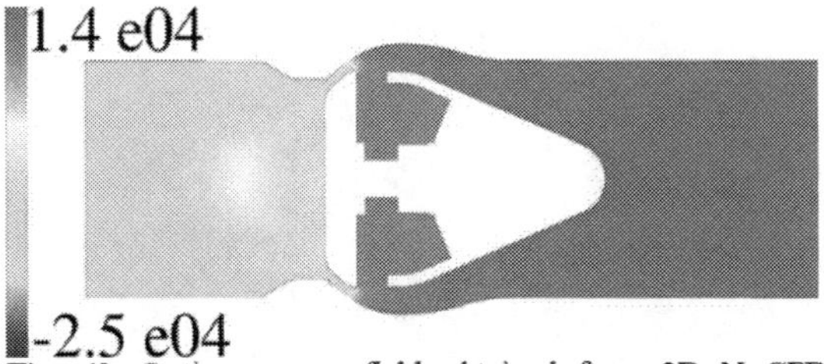

Fig. 40. Static pressure field obtained from 2D N CFD simulation, flow rate Q2.

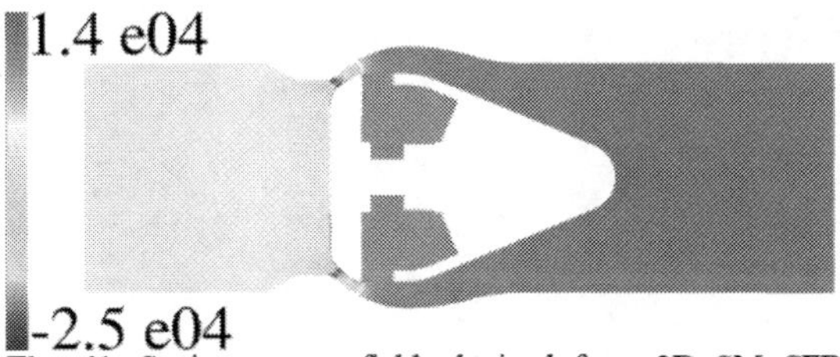

Fig. 41. Static pressure field obtained from 2D SM CFD simulation, flow rate Q2.

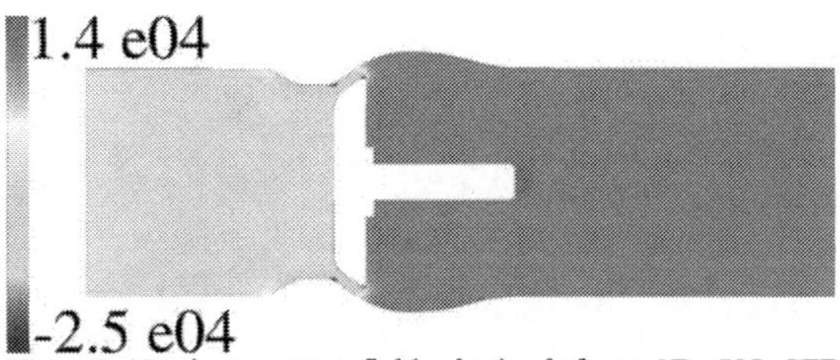

Fig. 42. Static pressure field obtained from 2D ON CFD simulation, flow rate Q2.

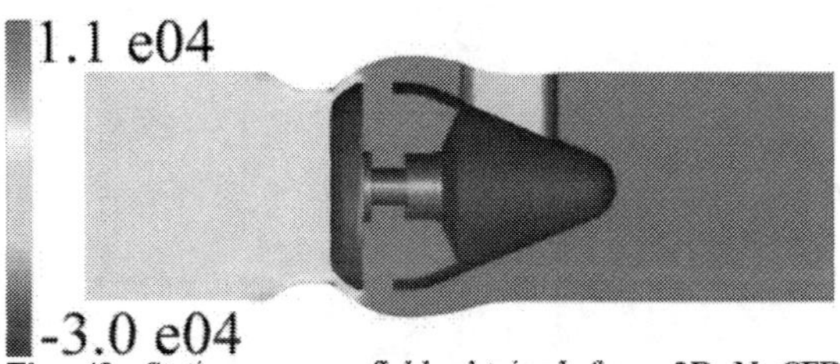

Fig. 43. Static pressure field obtained from 3D N CFD simulation, flow rate Q2.

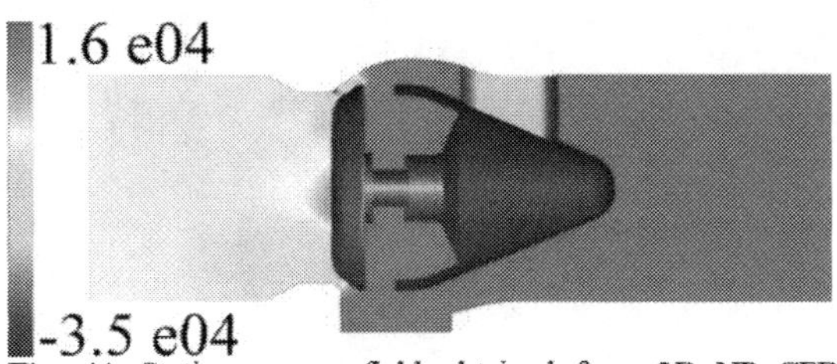

Fig. 44. Static pressure field obtained from 3D NP CFD simulation, flow rate Q2.

5 Conclusions

The presented study was intended as an information supplement that could not be directly obtained from the experimental testing and which should to serve as a basis for further simulation of dynamic valve behavior. An experiment was realized a little later for verification purposes. Another task was to monitor and assess possible simplifications with regard to 2D CFD simulation, one-dimensional valve analysis and computational requirements.

The first question that dealt with was the general setting of CFD simulations and the choice of appropriate wall functions. Non-equilibrium wall function and scalable wall function were selected for comparison. Non-equilibrium wall function was evaluated as more accurate for 3D geometry after confrontation with experimental data,. The RNG k - ε turbulence model recommended in the literature did not achieve better results.

One of the experimental devices was equipped with windows to better understand the operation of the valve. The windows should be important especially for real-time analysis of the velocity field and also for basic evaluation of the disc movement depending on flow rate through the valve. Two 3D geometries with and without windows were created for this purpose. It was possible in advance to anticipate that the windows would be reflected in the static characteristics of the valve. The deformation of the velocity and pressure fields around the window frames, which is manifested in the magnitude of forces acting on the disc, is more important in this respect. Therefore, it is shown that the analysis of dynamic properties on the experimental valve with visors is not simply transferable to real device. Of course, the analysis is rather complicated as it is difficult to build a one-dimensional mathematical model and predict the valve motion for different flow rates.

There is a very good agreement visual qualitative comparison of the velocity and pressure fields in the operating flow direction. Of course, results from 2D and 3D axisymmetric geometries are compared. The 2D geometry fails in the backflow, and especially the velocity field varies considerably. The applicability of 2D simulations is therefore problematic with this valve. A more important comparison of 2D and 3D is apparent in static valve characteristics. It could be said that both geometries differ only by the embedded fin, yet this fin contributes to the deformation of the loss and flow coefficient characteristics. It is usual that the loss coefficient with a higher flow rate decreases and the flow coefficient increases on the contrary. However, the trend of both coefficients in 3D geometry changes to higher flow rates and the loss coefficient increases and the flow coefficient decreases. The described effect was not apparent in 2D geometry, despite the revision of the computational mesh. It can be said that the 2D and 3D simulations correspond to velocities of up to 6 $m.s^{-1}$.

The possibilities of building a one-dimensional mathematical model describing the valve disc movement depending on the flow rate in the case of two-dimensional simulations were also tested. The force effects on the disc can be described by force acting on the remaining surfaces of the valve definition volume, as can be seen from the definition of force on the disc or directly from the integration of the Navier-Stokes equation. However, it is necessary to know the differences in static pressure within the valve. For this reason, CFD simulation of the slip valve body and CFD simulation of the valve body equipped with the plug itself was created. The ratio of the loss coefficients of the valve to the non-friction body valves is quite different. This ratio is approximately 90-95% for full valve opening, and drops to 65-70% when the disc is evidently closer to the valve seat. This fact should be taken into consideration in a one-dimensional model if only the aggregate data from the experimental testing is available. The most unfavorable situation occurs in the case of the analytical model of the unsteady force component of the fluid acting on the valve disc. The CFD simulation shows that it significantly differs from the prediction of the analytical model. This difference increases with the shut-off time of the valve and has its extreme corresponding to a specific flow rate. The best agreement of the analytical model with numerical arrives at the normal piping system unrealistic shut-off time of the valve $Ts = 0.2$ s. An analytical model of unsteady force at this Ts value achieves 40-80% predictions from CFD simulations.

This work has been supported by Technology Agency of the Czech Republic under the project Innovative research of check valves for extreme operating conditions in energetics TH01011352.

References

1. V. J. Sonawane, T. J. Rane, A. D. Monde, R. V. Vajarinkar, P. C. Gawade, *Design and Analysis of Globe Valve as Control Valve Using CFD Software,* IOSR J. of Mech. and Civil Eng. pp. 63 – 71 (2013)
2. P. Kurian, C. R. Krishnamurthy, R. Rajesh, Int. J. of Emerging Tech. and Adv. Eng. **4**, 10 pp. 391-397 (2014)
3. D. C. Wilcox, *Turbulence Modeling for CFD* (DCW Industries, Incorporated, 1994)
4. V. C. Patel, W. Rodi, G. Scheuerer, *Turbulence models for near-wall and low Reynolds numbers flows: A review*, AIAA J. **23(9)**, pp.1308-1319 (1985)
5. J. Bredberg, Internal Report **00**, 4, (Department of Thermo and Fuid Dynamics, Göteborg, Sweden, 2000)
6. M. Chmielewski, CMST **19(2)**, pp. 107-114 (2013)
7. K. Hanjali´c, B.E. Lauder, J. Fluid Mech. **74**, pp. 593-610 (1976)
8. S. Sibilla, M. Gallati, J. Fluids Eng. **130**, 121101 (2008)
9. Q. Yang, Z. Zhang, M. Liu, J. Hu, Procedia Eng. **23**, p. 547 (2011)
10. A. Montorfano, F. Piscaglia, G. Ferrari, Mathematical and Computer Modelling. **57**, 7-8, pp. 1640-1647 (2013)
11. J. Nordström, *Well Posed Problems and Boundary Conditions in Computational Fluid Dynamics* (22nd AIAA Computational Fluid Dynamics Conference, Dallas, 2015)
12. J. A. Valverde, S. H. Frankel, G. P. Salvador, *Three-dimensional control valve with complex geometry: CFD modelling and experimental validation* (34th AIAA Fluid Dynamics Conference and Exhibit, Portland, Oregon, 2004)

CFD analysis of optical probe interaction with wet steam flow field

Michal Kolovratník[a], Gukchol Jun, Ondřej Bartoš

CTU in Prague, Department of Energy Engineering, Technická 4, Prague 6, 166 07, Czech Republic.

Abstract. In the frame of the measurement feasibility study of the liquid phase size distribution structure in steam turbines at intermediate and high pressures, on CTU the interaction of optical probes with the wet steam flow field is investigated. In order to validate and refine the existing knowledge, a new series of CFD simulations were performed, considering turbine flow geometry, water steam characteristics according to IAPWS97 formulation, and improved boundary conditions and quality of the computing mesh. This paper briefly presents the newly obtained results

1 Introduction

The current worldwide energy consumption of primary sources is about 19 TW. More than one third of this consumption is used in the technologies converting the primary sources to electricity. There is no doubt that for the next few decades the dominant conversion technology will be steam turbines in power plants using fossil fuels and nuclear energy as well as renewable energy sources. This is tightly linked with the emphasis on sustainability, which requires an increase in the primary energy utilization efficiency. One of the options and challenges is the further development and improvement of current steam turbines, specifically in handling the issue of two-phase flow of wet steam.

The presence of wet steam in the turbine flow part negatively affects their reliability and efficiency. Erosion and corrosion-erosion damage are known and the additional energy losses due to wetness are also significant. In practice, however, it is not advantageous or practical to exclude wet steam from steam turbines. It is therefore necessary to address the problems of two-phase flow both theoretically and experimentally; to actively model and modify the geometry of the flow part of the turbines based on the acquired knowledge; to develop a methodology of their operation so that the negative effects of wet steam flow are suppressed and to keep these effects within acceptable limits.

Experimental studies of the liquid phase structure of wet steam expanding in a steam turbine are integral to these activities. Wet steam research has been going on for decades with varying intensity. However, it has consistently focused on low-pressure (LP) turbines and mainly on the last stages of these turbines operating below atmospheric pressure. For a number of reasons, attention has also recently been focused to phase change and two-phase flow at high-pressure (HP) [1, 2]. The starting point for these theoretical analyses and the demanding laboratory experiments are several remarkable works of the last century, especially [3, 4, 5, 6]. The focus of these activities is primarily connected with the HP part of nuclear power plant technology with PWR, more recently with the use of renewable energy sources as well as with conversion technologies in which heat transferring water is replaced by other liquids which are more suitable for the given conditions.

Similarly, at Czech Technical University in Prague (CTU), research was launched on wet steam at high pressure in power systems. Attention has mainly been focused on the modelling of the formation and development of two-phase flow in multi-stage HP steam turbines and on the modelling of energy losses.

CTU traditionally focuses on the development of optical probes for wet steam measurement in LP steam turbines in conventional and nuclear power plants and on the application of these probes for measurement in the turbines in cooperation with Doosan Škoda Power s.r.o. (DSP). Therefore, this work is also focused on considering various possibilities for transferring the experimental research of HP wet steam from the laboratory to the turbines or, in the case of nuclear power plants, to the separators. In addition to the many technical problems associated with the higher pressure and temperature, such as inputs for the instrumentation and seals for its insertion into the operating systems, the question arises as to the choice of a suitable method of measurement in an optically dense environment.

One partial task is to solve the interaction of the probe with the flow field in the steam turbine, which, due to limited experimental data, is still limited to regions of higher pressures before the last stage of the LP part of the steam turbine in the nuclear power plant Temelin (ETE) [7]. This paper focuses on improvements in CFD analysis

[a] Corresponding author: michal.kolovratnik@fs.cvut.cz

of this interaction. The results will also subsequently be applied to the conditions of the stages of the HP steam turbine.

The presence of the probe always affects the wet steam flow field. In the first step, the character and intensity of this influence must be solved for the specific flow-through section and the geometry of the selected probe. The second step is to assess the impacts on the measured values and subsequently evaluated parameters. Therefore the following is a description of the geometrical setup and the dependence of the parameters for the given optical probe.

2 Optical measurement in the steam turbine

The shape design of the CTU probe head for measuring the wet steam characteristics, the positioning and placement of the probe in the turbine, and the principles of the optical measurement methods used have been described in more detail in the previous work [7]. Therefore, the following is only a brief overview of this issue, clearly documented by Fig. 1

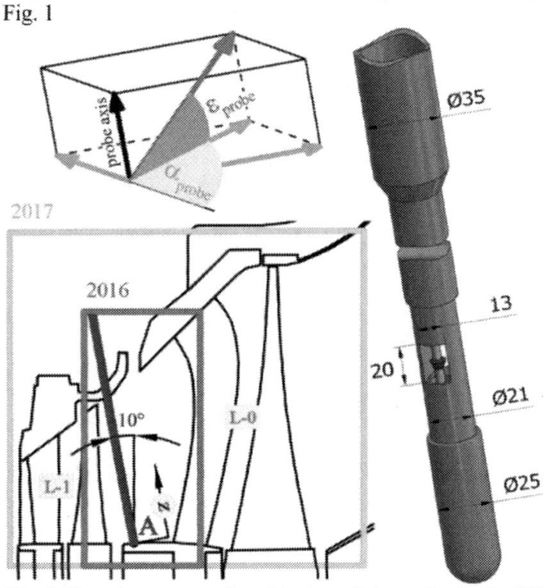

Fig. 1 The placement and positioning of the probe in the LP turbine

The end part of the cylindrical probe consisting of its head (21/25 mm diameters) and a positioning carrier (35 mm in diameter, 4.5 m long) can be seen on the right side of Fig. 1. A measuring slot of 20 x 13 mm is cut out in the cylindrical head, its borders on the sides are defined by ribs carrying the tip of the probe. Parallel to the probe axis, a 4 mm diameter light beam passes between glass windows through the measurement slot. These windows are embedded 0.7 mm below the plane of the wall of the measuring area. Geometrically the active measurement volume in the measuring slot, which the wet steam passes through, is a cylinder with a diameter of 4 mm and a length of $\ell = 21.4$ mm (shown in Fig. 1 in yellow). From the decrease of the light intensity ($I_0 \rightarrow I$) due to light scattering on the droplets passing through this measurement volume in the wet steam, it is possible to evaluate the desired characteristics of the liquid phase, i.e. droplets size distribution and wetness at the measuring point. The dependence of the measured I/I_0 relative light attenuation on the measured quantities is described by the Lambert-Beer law. It can be written in a simple illustrative form valid for a monodisperse droplet size distribution:

$$\ln\left(\frac{I_0}{I}\right) = 1.5\frac{y}{D}\ell\frac{\rho_{WS}(\rho',\rho'',y)}{\rho'}Q\left(\frac{\pi D}{\lambda}\right) \tag{1}$$

where I_0 is the intensity of light on the inlet of the measuring volume; I the intensity of light on the outlet of the measuring volume; y is steam wetness; D is droplets diameter; ℓ is the length of the measuring beam of the probe; ρ', $\rho''$and ρ_{WS} is the density of the liquid phase, gaseous phase and wet steam respectively; $Q(\pi D/\lambda)$ is the coefficient of light extinction according to Mie theory and λ is the wavelength of the light. The measurement is performed with a wave length between $\lambda \sim 200$-1000 nm. On the bottom left side of Fig. 1 is schematically shown the location and measurement position of the probe in the gap between the rotor blade plane of the second last stage *L-1* and the last *L-0* stage of the LP part of the turbine as the measuring plane *A* with a slope of ~ 10° from the radial plane.

The measured light intensity attenuation is used to characterize local conditions of the wet steam in the measurement volume. To obtain more comprehensive information the measurement is repeated in a number of positions z along the blade height. The traversing of the probe between these positions and its rotation for alignment with the flow field (according to the angle of the probe α_{probe}) is enabled by the carrier. Only the probe angle ($\varepsilon_{\text{probe}}$) is fixed by the design and cannot be changed.

3 Flow field properties influenced by the probe

Ideally, without affecting the flow field, the wet steam parameters in the measuring volume would match the parameters at the same location without the presence of the probe.

The density ρ_{WS} in the probe would correspond to the pressure measured in the same position by the calibrated DSP air probes *(Note: the effect of these probes on the flow field is not within the framework of this paper).*

Under real conditions, the flow field around the probe is influenced by the measuring probe head and varies in intensity depending on the length of the probe carrier inserted into the turbine. At the same time, in the measuring section of the probe due to acceleration, breaking and flow separation, the wet steam may change its density and other parameters. If we accept the simplistic assumption that the droplet dimensions and steam wetness would not change, the deviation of the measured attenuation of the light would arise from the change in density as derived from equation (1).

The results of the initial analyses [7] confirmed that from the measured results, the influence to the flow field cannot be neglected. Due to the many simplifications used in the initial CFD analyses (hereafter OLD), the focus of the CFD approach to real-life conditions was taken into account in the new analyses (hereafter NEW). The primary objective of these works was to verify the validity of the existing results and to enhance them further.

4 Previous versus new results

In the new analyses some enhancements were made. The most significant changes included: the extension of the modelling region, an improved computational mesh, the enhancement of boundary conditions, and the originally used ideal gas was replaced by real steam.
Two coloured rectangles in Fig. 1 characterize the extended modelling region of the flow field used for the CFD analyses. The original area used in 2016 is bound by the red line; the new one is shown by the green rectangle.
The change in boundary conditions is related to the extension of the computational model to the entire turbine stages. This allows for more detailed and relevant specification of the overall pressure and temperature profiles and flow directions into the subdomain between the stages comprising the probe. The same applies to output parameters from this subdomain, especially for the static pressure distribution and the output flow direction profile. A more detailed specification of the boundary conditions should provide a numerical simulation closer to the real steam flow in the probe and its vicinity.
Another change concerns the working fluid media. In the new calculations the real steam is considered according to IAPWS IF 97. Compared to the old calculations with the ideal gas, this has a number of advantages including the potential for refining the results. One example of such an advantage is the possibility for direct monitoring of the influence of the probe on the equilibrium wetness as documented in Fig. 2. Of course, the results of the original calculations with the ideal gas could not provide any information about the wetness at all.

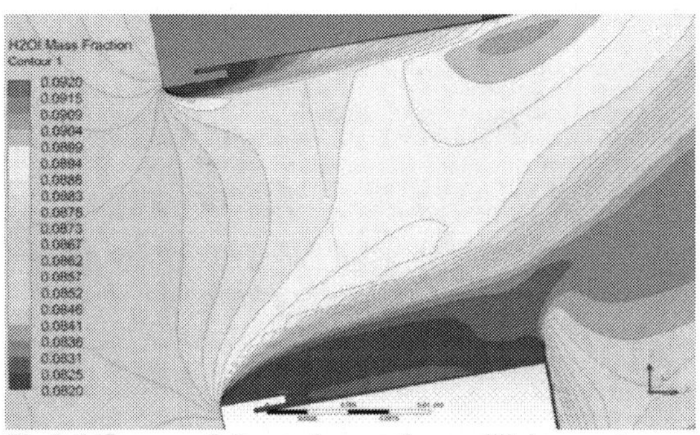

Fig.2 Influence of the probe on the equilibrium wetness - longitudinal section of the computing domain

A significant change is also achieved in the quality of the computing mesh. The computational mesh of the old calculations was prepared in ANSYS Meshing and contained mainly tetrahedral and wedge cells. The mesh for the new computations is prepared using ANSYS ICEM with the use of the appropriate topology technique. As a result, the new mesh only contains hexahedral cells and a significantly improved convergence is achieved in the new calculations. At the same time, significant improvements have been achieved in the boundary layer behaviour modelling around the probe surfaces. In particular, the newly prepared mesh provides values of the $y+ < 2.3$ as documented in Fig. 3.

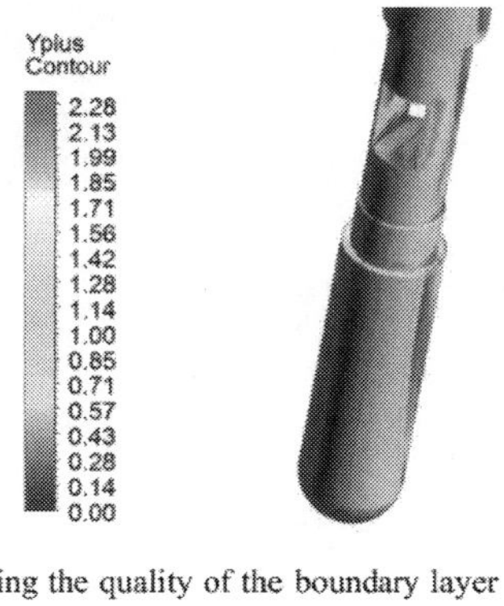

Fig.3 Improving the quality of the boundary layer modelling - $y+$ distribution

For comparison, in the old computing mesh, the maximum $y+$ value was approximately 50.

5 CFD modelling of the probe influence

Computational model and the mesh
The domain of the new computational model is shown in Fig. 4. The model contains both the second last (L-1) and the last stage (L-0) of the ETE turbine as well as a part of the axial diffusor. The optical probe is located in the space between the rotor blades of the second last stage and the stator blades of the last stage of the turbine. The probe is "traversed" for each measured point along an axis which is inclined from the radial by 10°. The mesh for the blades of both stages is prepared in ANSYS TurboGrid, and the part of the model where the probe is and the part of the axial diffuser are meshed in ANSYS ICEM. Thanks to the suitably chosen topology technique, the model is completely meshed in hexahedral cells. The total number of computing cells is approximately 30 mil.

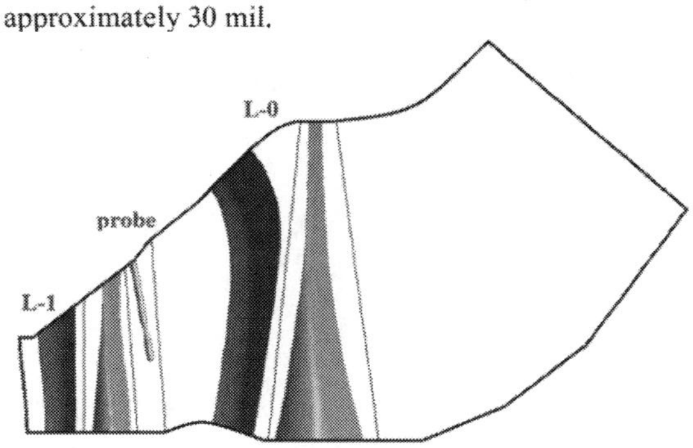

Fig. 4 A longitudinal view of the axial section of the new computational model

A detailed view of the computational mesh can be seen in Fig. 5.

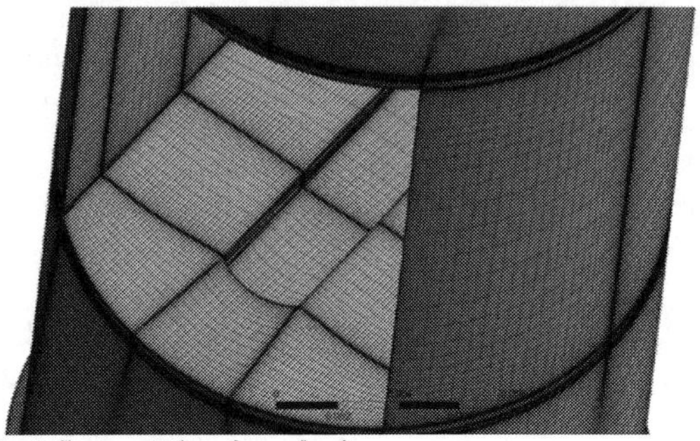

Fig.5 Computational mesh view

Boundary condition setting

At the inlet of the domain a "pressure - inlet" condition type with total pressure and enthalpy profiles was set together with the corresponding flow directions. Instead of the total temperature profile, the total enthalpy profile was used, because the steam is already wet at the inlet to the second last stage. The total pressure and the total enthalpy profiles of the steam set for the inlet are shown in Fig. 6.

The outflow of the model is a "pressure - outlet" condition type. Here the corresponding static pressure is set with respect to a "Radial Equilibrium" condition. The boundary conditions are summarized in Table 1.

Tab. 1 Summary of the boundary conditions

			Inlet	Outlet
Boundary condition			Pressure inlet	Pressure outlet
Set values	Total pressure	[Pa]	Profile	-
	Static pressure	[Pa]	-	28800
	Total enthalpy	[K]	Profile	-
	Turb. intensity	[%]	5	-

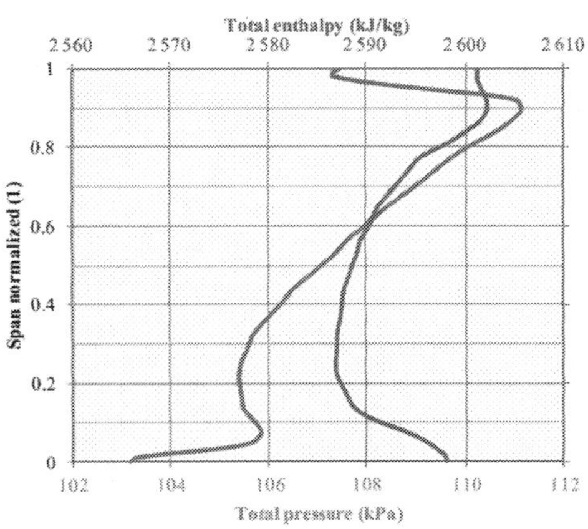

Fig.6 Total pressure and enthalpy distribution along the radial coordinate of the blade

For all the calculations, the two-equation SST k-ω turbulent model was used. The real steam properties according to IAPWS IF 97 were considered.

6 Result – examples

The examples of the results are presented first for longitudinal effects and later for the angular effects of the probe position on the measured values. Starting with the longitudinal effects, in Fig. 7a, the Mach number contours of the steam flow are shown in the longitudinal section of the probe measuring the volume for 3 selected traverse positions z of the probe along the blade height (z = 140, 385 and 585 mm). It is obvious that the flow rate within the measuring slot is heavily dependent on the radial component of the steam velocity from the rotor of the second last stage.

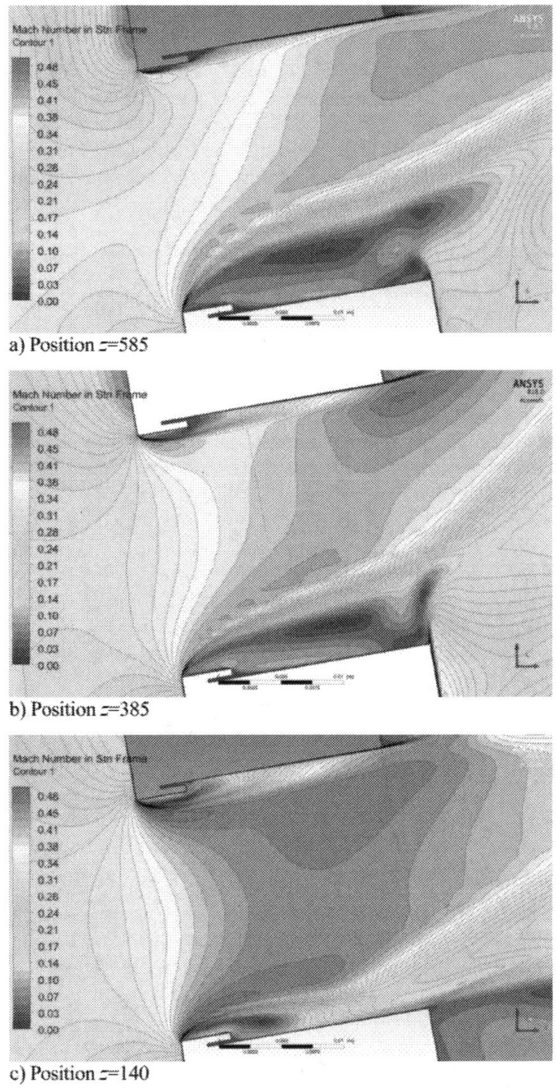

Fig. 7 Longitudinal contours of the Mach number in the probe's measurement volume at three positions z along the blade

In order to apply CFD analyses to correct the optical data measured by the probe in the affected flow field, the most significant data needed is the change of density [7]. Therefore, only normalised densities of the flow passing the measurement volume are used in the results comparison. Normalization is related to the maximum steam density $\rho_{WP\ MAX}$ in the same volume as determined by the reference calculations in the model domain without the probe (denoted as WP).

Fig. 8 compares the normalized density distributions along the light beam ρ_{IB} (red lines) for 3 probe positions z along the blade height. These distributions are complemented by normalized distribution of density ρ_{WP} (blue lines) and normalized density averages in the light beam ρ_{IBA} (green lines).

For the blade stage mean diameter (position z = 385 mm), NEW results on Fig. 8b are compared with similar values evaluated from the results of OLD CFD analyses (dashed lines and circles of corresponding colours). The clear difference between the results of the original and the new calculations, which is only partially documented in Fig. 8b, confirms the reason for the work to improve the CFD model described in Chapter 4. (*Note: For the original profile values, a secondary horizontal axis is used at the top of the chart with a different scale.*)

It is evident from Fig. 8 that the character and intensity of the influence of the steam density by the probe changes significantly with the position along the blade. At the bottom and centre of the blade, the density of the flow in the probe is lower than that seen in the uninfluenced flow along the length of the measuring beam. Only in the region of the blade tip is the deformation of the flow field different due to the significant radial component of the velocity interacting with the outer limiting wall of the probe. At the top of the beam, the density of the steam is higher and on the opposite side it is lower than in the flow without the probe. The blue and green curves in Fig. 8 illustrate the more general nature of the influence of the probe on the steam density along the blade radius. By comparing them it follows that the largest change is the relative drop in density at the bottom of the blade. At the tip of the blade, however, there is a less pronounced increase in the average density.

A more detailed evaluation and discussion of the results obtained from the CFD calculations goes beyond the scope of this paper. It can be said that the present knowledge makes it possible to quantify a simple Probe Influence Factor (PIF) to correct the density influence by the probe at the specific analysed positions.

The PIF was defined as the ratio of the average density in the place of measurement with the probe influence and without the probe [7].

$$\mathrm{PIF} = \rho_{\mathrm{IBA}}/\rho_{\mathrm{WP}} \tag{2}$$

A generalization of the PIF correction for any position z of the probe or for probe use in other turbines is not yet possible. This practically requires modelling the impact of the probe on the flow field in all measurement cases.

The complexity of the problem regarded from other possible off-design positions of the probe is shown in the second example of the results as presented in Figures 9 and 10. This example represents the variation of the steam parameter in the measuring space when the probe angle of attack α_{probe} does not match the flow direction $\alpha_{\mathrm{flowCFD}}$. Results are presented for z=385 mm and several angular deviations:

$$\Delta\alpha = \alpha_{\mathrm{flowCFD}} - \alpha_{\mathrm{probe}} \tag{3}$$

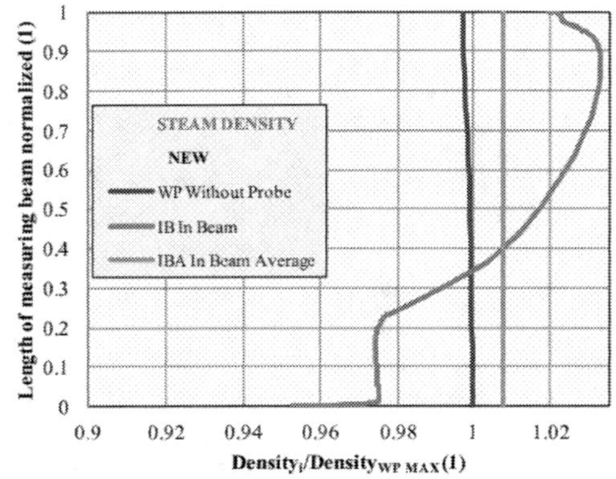

a) Position z=585

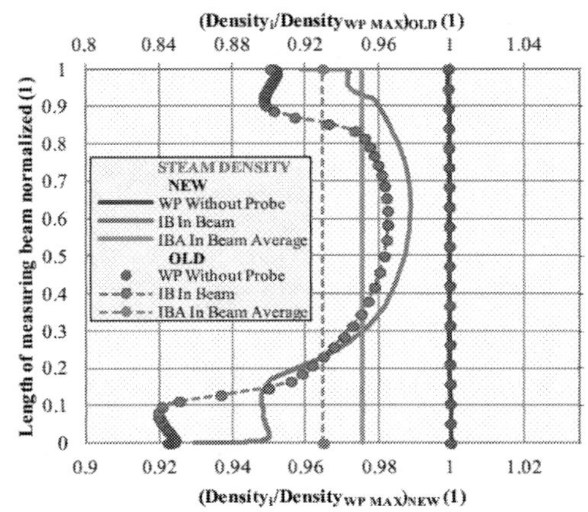

b) Position z=385

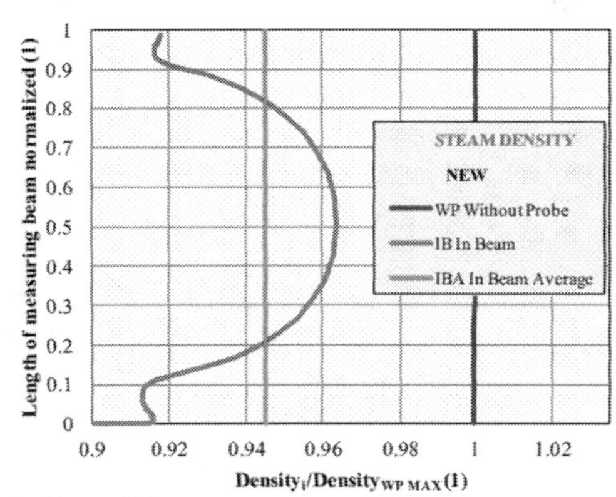

c) Position z=140

Fig. 8 Density distribution at three positions z along the blade hight

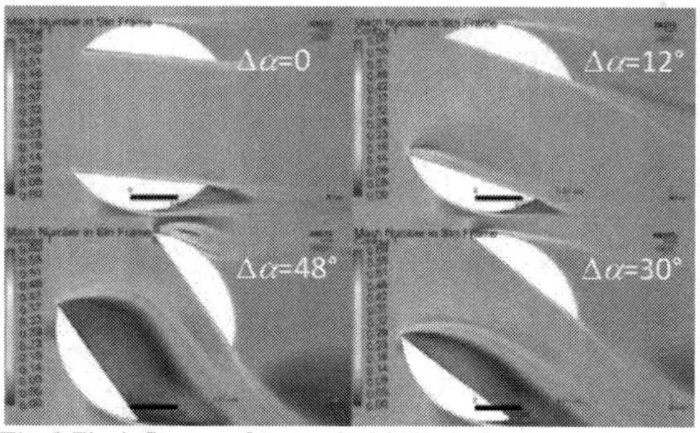

Fig. 9 The influence of $\Delta\alpha$ on the Mach number contours

Fig. 9 shows the Mach number contours on the cross-section for 4 different deviation values of the probe angle of attack settings $\Delta\alpha$=0°, 12°, 30° and 48°. It is obvious that the probe rotation also affects the flow rate within the measuring space. However, the intensity of this effect does not increase by increasing $\Delta\alpha$ linearly. A strong disturbance of the flow within the measuring area is evident in the cases of $\Delta\alpha$ = 30° and 48°.

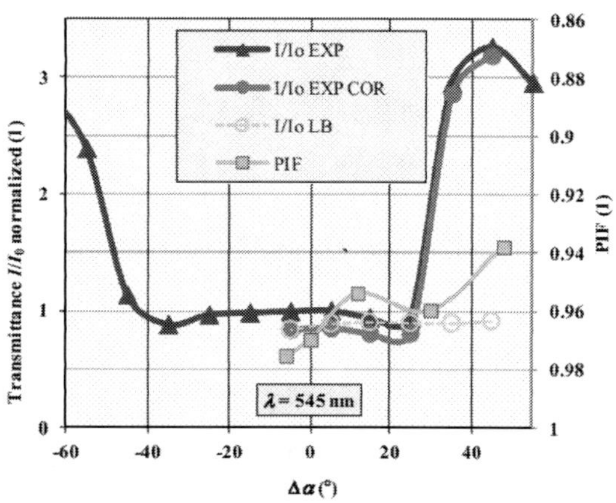

Fig. 10 Dependence of $I/I_0(\Delta\alpha)$ and PIF($\Delta\alpha$)

Fig. 10 shows the PIF dependence on angle deviation ($\Delta\alpha$) evaluated from CFD flow analyses in the probe measurement volume when it is rotated from the direction of the flow field. The flow direction in the non-disturbed flow field corresponds to the angle α_{flowCFD} = 85°. As a demonstration of the PIF application, Fig. 10 also compares the normalized profile of three dependencies I/I_0 on the angular deviation of the probe setting in the flow field at z = 385 mm. The full blue line shows the dependence $I/I_{0\,\text{EXP}}$ obtained experimentally in the steam turbine at ETE (particularly for the wavelength λ = 545 nm). The red line represents the profile of the dependence corrected using PIF (for respective $\Delta\alpha$) according to the CFD analysis results - $I/I_{0\,\text{EXP COR}}$. The green line then shows the dependence of $I/I_{0\,\text{LB}}$ calculated according to the Lambert-Beer law with the application of the hypothesis of Chapter 3 limiting the impact of the probe in the flow field only to the change in the density. This means that identical steam wetness values and droplet dimensions were considered for all rotation positions as was obtained in the probe position $\Delta\alpha$ = 0 by the evaluation from the $I/I_{0\,\text{EXP COR}}$ values. Therefore the green and red lines start from the same point.

Fig. 10 illustrates several significant findings for the analysed position z and the deviation $\Delta\alpha$:

- the extent of deviation of the flow field in the probe's measurement section depends on its rotation (see the orange curve)
- this influence clearly changes the measured light attenuation (differences between the blue and red curves)
- the impact of the probe cannot be simplified only to a function of the wet steam density (differences in the red and green curves, especially for$\Delta\alpha$ > 30°).

These points can most likely by applied generally to other probes and their systems. The last point shows that the current simple correction factor PIF (see equation 3) should be supplemented by the influence of the probe on the measured parameters of the liquid phase D and y. Our current CFD simulations do not allow for a direct assessment of this impact, but the next stage will focus on this issue.

8 Conclusions

In the research of high pressure wet steam in energy systems, CTU is focusing on the analysis of the self-influence of optical probes on the flow field in steam turbines using CFD.

The first objective is to develop a methodology for modelling the influence of probes in turbines and to verify it in experimentally available places with increased pressure in LP parts. Subsequently, the methodology will be used for the modelling of probe conditions in high pressure parts of turbines.

The presented work follows on the initial analysis from the year 2016. Special attention was paid to improving the CFD model. This process involved 3 major changes. The computational model was extended to include the entire second last and last turbine stages, which allowed for more precise specification of the boundary conditions. Furthermore, the quality of the computing mesh has been improved. This has resulted in better convergence for all the calculations made. At the same time the new calculations were considered for real steam according to IAWPS IF 97. Real steam calculations have more advantages than the computations with the ideal gas. An interesting example is the direct visualization of changes in the equilibrium steam wetness in the probe measuring volume.

This improved CFD model was used to calculate the flow field characteristics in the area before the last stage of the steam turbine ETE without the probe and with the probe. The results of the calculations were analysed and compared in part with the experimental data.

This allowed us to fulfil the secondary objective of our research, i.e. to obtain data for the correction of data measured by CTU optical probes considering the influence of the probe on the flow field, and to help in optimising the shape of the probes.

The obtained results of the calculations and their analyses are only presented quite briefly in this paper. In general, both of these objectives have been met. Yet, there remains a wide scope for further work on this issue. The basic challenges include: generalizing the correction methodology, adding characteristics of the liquid phase to the PIF correction factor, modelling the influence of purging the optical windows with a stream of air.

Acknowledgements

We gratefully acknowledge the support by Grant No. 16-20123S of the Grant Agency of the Czech Republic. The authors would also like to thank all DSP and ETE employs who have made it possible to take measurements in the low pressure part of the steam turbine and who have been involved in their implementation.

References

1. Guo, T., Burnett, M., Turnquist, N. and Moraga, F.: *High pressure subsonic nucleation in 1d nozzles - an experimental study*, Proceedings of ASME Turbo Expo 2014: Turbine Technical Conference and Exposition, GT2014-27241, pp. V01BT27A051
2. Azzini, L., Pini, M.: *Numerical investigation of high pressure condensing flows in supersonic nozzles*, J. Phys.: Conf. Ser.821 (2017) 012008
3. Gyarmathy, G.: *Nucleation of Steam in High-Pressure Nozzle Experiments*, Proc. IMechE 2005, 219 Part A: J. Power and Energy , 6, 511-521
4. Bakhtar, F., Zidi K.: *Nucleation Phenomena in Flowing High-Pressure Steam: Experimental results*, Proc. IMechE 1989, 203 Part A: J. Power and Energy , 3, 195–200
5. Bakhtar, F., Zidi K.: *Nucleation Phenomena in Flowing High-Pressure Steam Part 2: Theoretical Analysis*, Proc. IMechE 1990, 204 Part A: J. Power and Energy , 4, 233–242
6. Valha J., Ryley D. J.: *Optical Studies of Nucleation in High-Pressure Steam*. In ASME Conference on Condensation in High Speed Flows, Yale University, Connecticut, 1977, 27-42
7. Kolovratník, M., Jun, G., Bartoš, O.: *Control of the probe influence on the flow field in LP steam turbine*. EPJ Web of Conferences 143, 02052, (2017)

Volumetric flow rate in simulations of microfluidic devices +

Kristína Kovalčíková[1,*], *Martin* Slavík[1,*], *Katarína* Bachratá[1], *Hynek* Bachratý[1], and *Alžbeta* Bohiniková[1]

[1] Cell-in-fluid Research Group, cell-in-fluid.fri.uniza.sk; Department of Software Technologies; Faculty of Management Science and Informatics, Žilina, Slovakia

Abstract. In this work, we examine the volumetric flow rate of microfluidic devices. The volumetric flow rate is a parameter which is necessary to correctly set up a simulation of a real device and to check the conformity of a simulation and a laboratory experiments [1]. Instead of defining the volumetric rate at the beginning as a simulation parameter, a parameter of external force is set. The proposed hypothesis is that for a fixed set of other parameters (topology, viscosity of the liquid, ...) the volumetric flow rate is linearly dependent on external force in typical ranges of fluid velocity used in our simulations. To confirm this linearity hypothesis and to find numerical limits of this approach, we test several values of the external force parameter. The tests are designed for three different topologies of simulation box and for various haematocrits. The topologies of the microfluidic devices are inspired by existing laboratory experiments [3 - 6]. The linear relationship between the external force and the volumetric flow rate is verified in orders of magnitudes similar to the values obtained from laboratory experiments.

1 Introduction

The research group Cell in fluid is interested in modelling of elastic objects in fluid flow in solid channels, in order to simulate the flow of the plasma and blood cells in microfluidic devices. The program is developed within open-source software ESPResSo. The model is described in [2]. The numerical simulation of such a device is faster and cheaper to establish, compare with the fabrication of the real microchip.

The long-time goal of the group is to make models of the microfluid devices capturing Circulating Tumour Cells (CTC). At today's stage of the software's development, some calibration calculations are needed, to compare the results of the numeric simulations with results of laboratory experiments. In order to do so, the input parameters influencing the movement of the cells in the blood plasma have to be the same, such as the viscosity of the liquid, the hematocrit, or the volumetric flow rate.

The volumetric flow rate is closely related with the difference of the pressures on the two ends of the device. In the numerical models, this dependence is present as well, and the difference of the pressures is somehow represented by an external fluid force parameter.

We have very often an information about a volumetric flow rate from a laboratory experiment. So we have to find a way how to introduce a known volumetric flow rate into the simulation. That's why we would like to find a way how to define it in the simulation at the beginning, as an input parameter.

It is possible to define the volumetric flow rate directly as an input parameter. The disadvantage of this approach is that we must define the course of the volumetric flow rate across the device, as well. However, it is quite difficult to obtain such an information from laboratory experiment. That's why we look for another option of defining the input flow rate, through the parameter of external fluid force.

The aim of this article is to verify, whether the volumetric flow rate in the simulations of the microfluidic devices is linearly dependent on the parameter of external force, set at the beginning as the simulation parameter. This approach demands only the knowledge of the external force parameter. If it is the case, the wished volumetric flow rate could be obtained by only one iterative simulation.

Authors in [5], p 3296, in sequence "*Using a simple syringe to control flow rate*" considered different volumetric flow rates in their experiments, and they noticed that "*The elastic PDMS device becoming deformed, and inflated, at high pressures. Consequently, the flow rate is not linearly proportional to the applied driving pressure.*" We do not ignore this feature of the microfluidic devices in our work – our goal is not to show that the volumetric flow rate can be increased infinitely in the laboratory experiments. The aim is to find a way how to define it correctly and rapidly from few calibration simulations.

+ Supported by the Slovak Research and Development Agency under the contract No. APVV-15-0751 and by the Ministry of Education, Science, Research and Sport of the Slovak Republic under the contract No. VEGA 1/0643/17.

* Corresponding authors: kristina.kovalcikova@fri.uniza.sk, martin.slavik@fri.uniza.sk

2 Methodology

To confirm the hypotheses of the linear relationship between the volumetric flow rate and the parameter of external force, we tested three different topologies. All of them are inspired by existing laboratory microfluidic devices. Each tested topology consisted of cylindrical or spherical obstacles placed in a rectangular box. The simulation parameter of the external force was set to different values, to verify the linear relationship between the volumetric flow rate and the simulation parameter. In the beginning, we tested the volumetric flow rate in the topologies without the Red Blood Cells (RBC). Afterwards, we ran the simulations without RBC, in order to include more realistic haematocrit in the laboratory experiments.

2.1 Cylindrical adhesive topology – capturing of CTCs by antibodies-coated obstacles in micro channel

First tested topology is inspired by works [3] and [4]. Both articles are concerned with a separation of the CTCs. For this purpose, they use a microfluidic device which can capture common CTCs from the patients' blood. The CTC are bigger than ordinary blood cells. While the size of ordinary blood cells ranges from approximately 4 to 18 μm, CTCs, in contrast, are larger and range from 15 to 25 μm in diameter [7]. However, those devices are not designed to isolate the cells by a size-sorting mechanism. Instead, they are using antibodies, spread over the obstacles in microchips. These antibodies capture the CTCs that flow through the channel. The dimensions of the obstacles are designed to stimulate the highest possibility of collisions between the cells and the obstacles.

In [3], the microfluidic devices contain cylindrical obstacles with radius of 50 μm placed in equilateral triangular arrangement, with a 50 μm distance between obstacles and a 50 μm shift after every 3 rows. The device is 100 μm deep. The active capture area of the laboratory device was 19 mm x 51 mm. The volumetric flow rate in the device was about 1-2 ml/h.

In [4], the experiments were made on a GEDI device. It contains cylindrical obstacles with radius of 80 μm. The gaps between the obstacles are 100 μm large, and each subsequent row is shifted by 7 μm. The device is 100 μm deep, 8 mm wide and 25 mm long. The volumetric flow rate maintained in the device was of 1 ml/h.

The chosen topology of the simulation channel is presented in the **Fig. 1**. The radius of the obstacles is 50 μm, the dimension of the gaps between the obstacles is 50 μm, and the obstacles are placed in an equilateral triangular array. The simulation channel is 100 μm deep, 150 μm wide and 260 μm long. The volumetric flow rate which should correspond to a 1ml/h for a 8 mm wide device, corresponds to 4,47 μm3/μs.

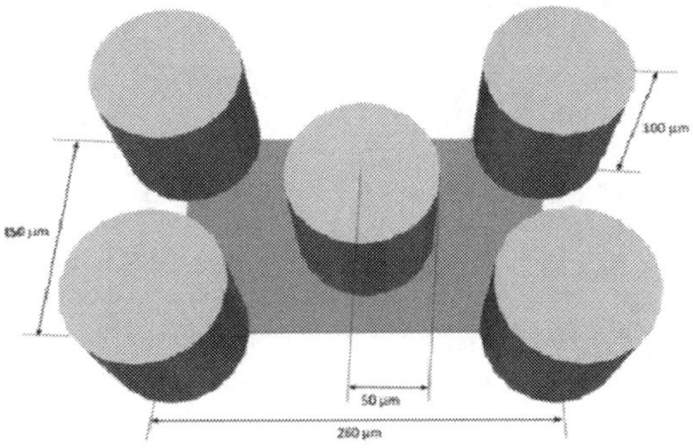

Fig. 1. Description of the simulation channel topology of cylindrical adhesive microfluidic device.

2.2 Cylindrical sorting topology – sorting of the parasites in the blood according to their size

The second tested topology is inspired by a device used in the field detection of blood parasites [5]. The aim is to sort the T. cyclops parasites from erythrocytes and leucocytes in human blood, using a pattern of obstacles which helps to sort the objects in blood according to their size. The microfluidic device is composed from 3 parts. First part separates the big leucocytes from the smaller erythrocytes and T. cyclops, which have similar dimensions. The second part of the device places the erythrocytes and the parasites close to the boundaries of the device. And the last part of the device keeps the erythrocytes close to the boundaries, while the parasites are placed in the middle stream of the device. The numerical model is inspired by the second part of the microchip.

This second part this microfluidic device consists of cylindrical obstacles, with radius of approximately 10 μm. The gap between the obstacles within one row is approximately 20 μm, and the gap between the rows is approximately 20 μm as well. Each row is shifted by one fifth of the distance between the central axes of the obstacles, which means approximately 8 μm. The depth of the device is 3,5 μm. The total width of the device is about 2 mm. The volumetric flow rates used in the experiment were in range of 2 to 4 μl/min. The device is dimensioned to be used in field, so we considered a haematocrit of a non-diluted blood, which means approximately 40%.

The chosen topology for this device is presented in the **Fig.**. The radius of the obstacles is 10 μm. The gap between the obstacles in the row and the distance between the rows is 20 μm as well. Each row is shifted by 8 μm. The depth of the device is 4 μm. The simulation box has dimensions 200x40x4 um. The volumetric flow rate which correspond to 2-4 μl/min, is about 0.7-1.3 μm3/μs.

https://doi.org/10.1051/epjconf/201818002046

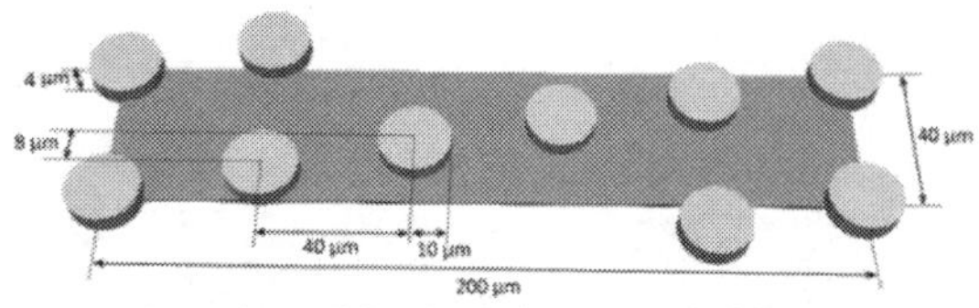

Fig. 2. Dimensions of the simulation channel of the cylindrical sorting topology.

2.2 Spherical adhesive topology – capturing of the CTCs using antibodies-coated magnetic beads

The last topology is inspired by work [6] and [8]. Both of them use magnetic beads to construct the microfluidic device. The idea, as well as in [3] and [4], is not to sort the cells according to their size, but to use antibodies which could capture the cells and keep them glued to the obstacle in the device. The advantage of the magnetic beads is that they do not need to be treated with the antibodies inside the channel during or after its construction, as it is a case for devices with cylindrical obstacles. The magnetic beads can be prepared separately, and after the treatment, they are poured into the device, and kept in the right position by an external magnetic field.

The magnetic beads used in [8] have diameter of 4.5 µm. In the bottom side of the channel, there is a magnetic pattern which defines the topology of the whole device. Once the beads are poured into the microfluidic channel, the magnetic field is activated, and the beads form chains. Those chains are starting at the specific points defined by the pattern at the bottom of the device. This pattern has a form of an equilateral triangular field. Distance between the chains is 40 µm. The depth of the device is 50 µm. It is 0.5 mm large, and 10 mm long. The volumetric flow rate in the channel in experiments is few µL/min.

The simulation box inspired by this geometry (**Fig. 3**) has dimensions of 40x70 µm. It contains five chains of 12 beads with radius of 2.25 µm. Four chains are in the corners, and one is in the middle of the box. This way, they form an array of regular equilateral triangular field. The depth of the device is 50 µm. The volumetric flow rate through this box corresponding to the laboratory experiment is fixed to be 4 $\mu m^3/\mu s$.

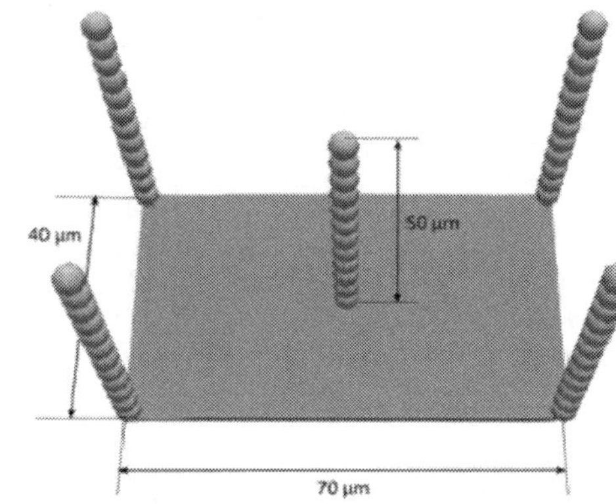

Fig. 3. Dimensions of the simulation channel of the spherical adhesive topology.

3 Measurement of the volumetric flow rate

For each topology, the volumetric flow rate was measured in three different positions. In theory, it should have the same value over any cross section of the channel. The possible differences could have origin in numerical imprecision of the simulation.

The three different positions of volumetric flow rate measurement for the cylindrical adhesive topology are shown in the **Fig. 4**. The evolution of the volumetric flow rate in the three different positions is presented in the **Fig. 5** for maximale and minimal examined value of external force parameter (with multiplier of 1 and 500).

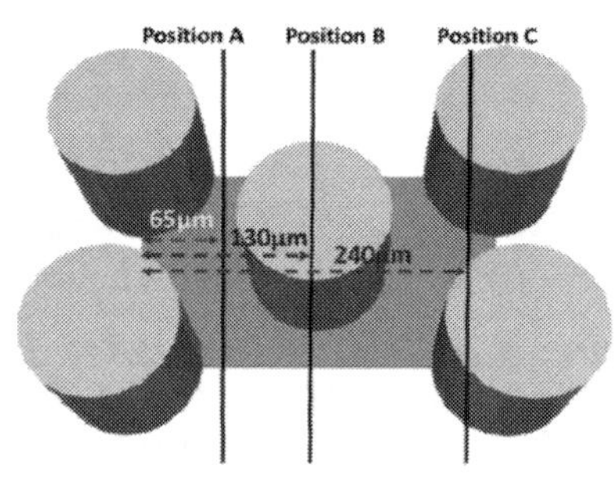

Fig. 4. Dimensions of the simulation channel of the cylindrical sorting topology.

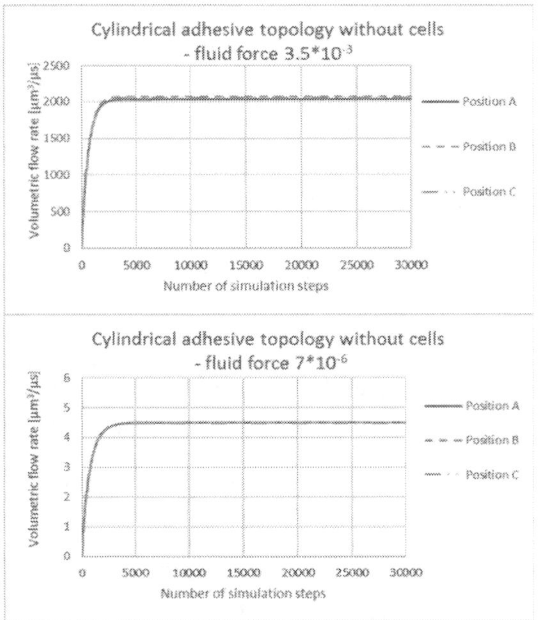

Fig. 5. Cylindrical adhesive topology without cells: comparison between the evolution of the volumetric flow rate measured in three different positions A, B and C. The presented graphs came from the simulation with the smallest and with the highest parameter of external force.

For the cylindrical sorting topology, the volumetric flow rate was measured in three different positions, as well (**Fig. 6**). The evolution of the volumetric flow rate in the three different positions is presented in the **Fig. 7**, for the maximal and the minimal examined value of external force parameter – for multipliers 1 and 100. We can observe that for the multiplier of 100, the volumetric flow rate is not the same for all positions, but its values differ by 8%. For the multiplier of 50, its values differ by 3%, and for multiplier of 10, the measured volumetric flow rate is the same for all the three positions.

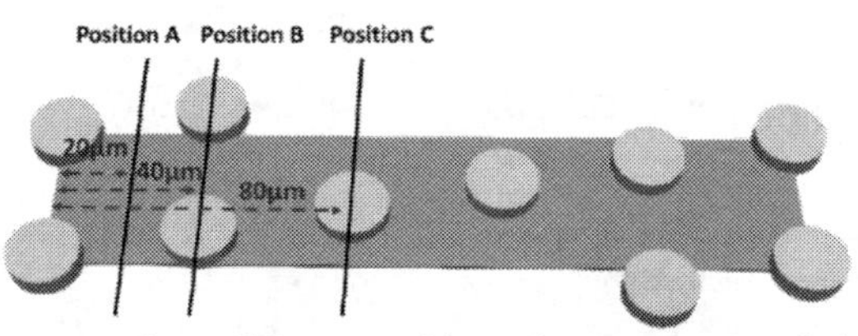

Fig. 6. Three different position of measurement of the volumetric flow rate in the channel with cylindrical sorting topology

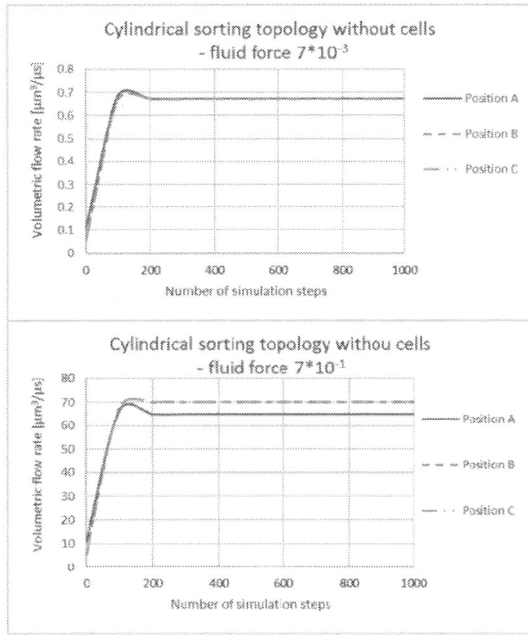

Fig. 7. Cylindrical sorting topology without cells: comparison between the evolution of the volumetric flow rate measured in three different positions A, B and C. The presented graphs came from the simulation with the smallest and with the highest parameter of external force.

In the case of spherical adhesive topology, similarly, the volumetric flow rate was measured in three different sections, presented in the **Fig. 8**. The volumetric flow rates measured in three different positions are presented in **Fig. 9**. The curves were obtained from simulations with the multiplier 1 and 500. For the multiplier 500, the difference between steady values is 2%.

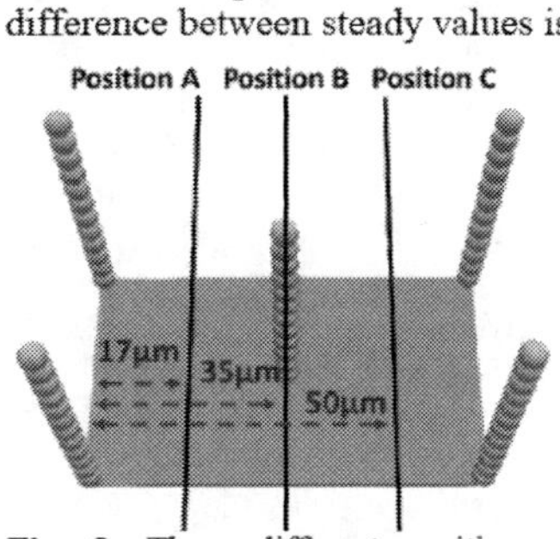

Fig. 8. Three different position of measurement of the volumetric flow rate in the channel with spherical adhesive topology

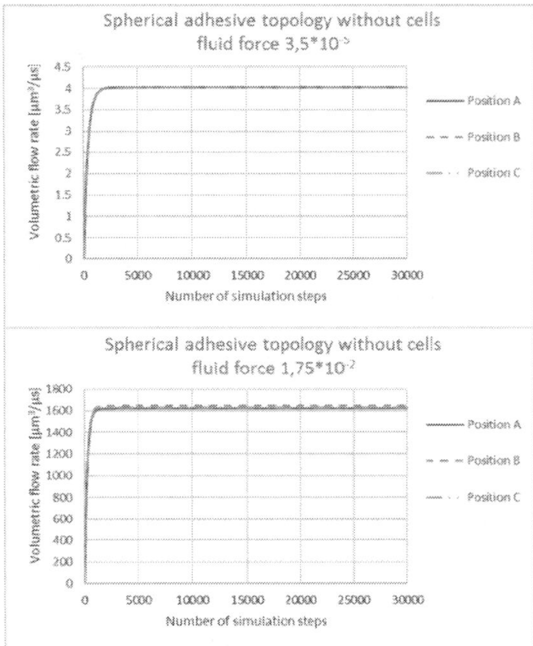

Fig. 9. Spherical adhesive topology without cells: comparison between the evolution of the volumetric flow rate measured in three different positions A, B and C. The presented graphs came from the simulation with the smallest and with the highest parameter of external force.

4 Results

4.1. Cylindrical adhesive topology

The initial value of the external force parameter for the simulation without RBCs was set to 0.000007, which results in volumetric flow rate of 4,48 4 $\mu m^3/\mu s$. The value of external force parameter was multiplied afterwards. The results of simulations run with different multipliers are presented in **Fig. 10** and **Fig. 11**.

The initial value of the external force parameter for the simulation without RBCs was set to 0.000007, which results in volumetric flow rate of 4,48 4 $\mu m^3/\mu s$. The value of external force parameter was multiplied afterwards. The results of simulations run with different multipliers are presented in Figure 10 and Figure 11.

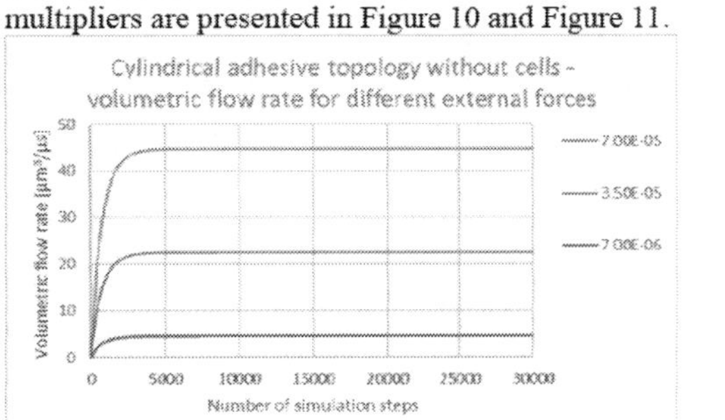

Fig. 10. Example of evolution of the volumetric flow rate during the simulation for three different values of external force for simulations without cells.

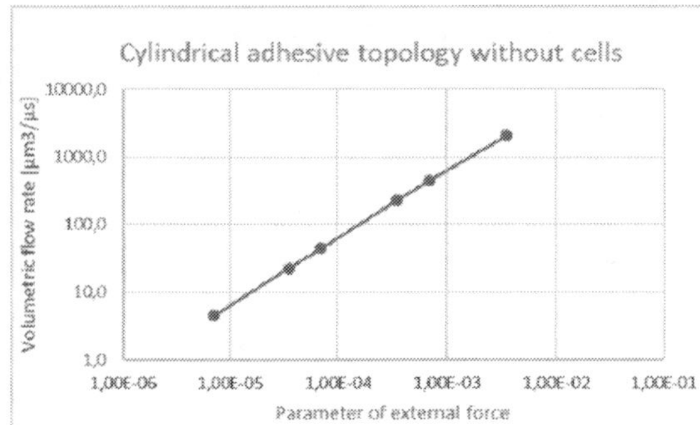

Fig. 11. Relationship between the value of external force parameter and the volumetric flow rate for simulations without cells

For the simulation with cells, we used a hematocrit of approximately 0,5 %. Three simulations were run with this level of hematocrit, with three different values of the external force parameter. The simulation with multiplier of 100 did not reached the end, it finished prematurely because of instabilities in the simulation. The obtained values of volumetric flow rates are presented in the **Fig. 12.**

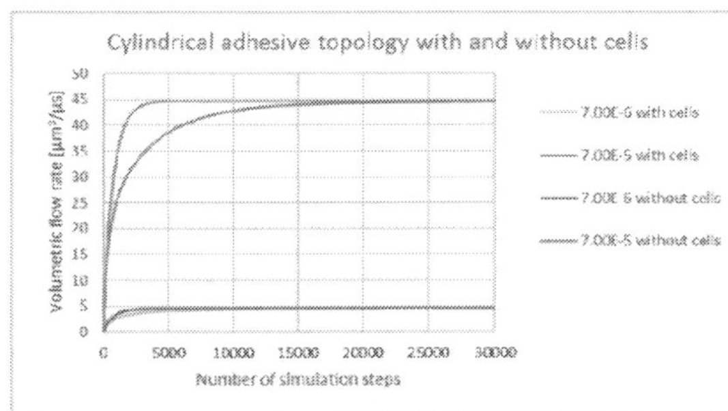

Fig. 12. Comparison between the evolution of the volumetric flow rate for simulations with and without cells.

As we can see in the **Table 1**, the volumetric flow rate is linearly dependent on the parameter of the external force for the small multipliers. The linearity is more significantly disturbed when the multiplier reaches value of 500. The percental difference between the obtained and the expected value is about 7.5%.

Another observation we can make is the similarity between values of volumetric flow rates obtained from simulations without RBCs, and the values obtained from simulations with the RBCs. The models with cells get obviously more time to get a stabilized volumetric flow rate, but the final value is similar to the one without cells. More detailed analysis of the similarity between flow rates obtained from simulations with and without cells is presented in the end of this chapter.

Table 1. Values of volumetric flow rate appropriated to different values of external force parameter.

external force	multiplier applied to the external force	volumetric flow rate [µm³/µs]	obtained multiplier for volumetric flow rate	percentual difference between the two multipliers
		without cells		
7,00E-06	1	4,5	1,0	0,0%
3,50E-05	5	22,4	5,0	0,0%
7,00E-05	10	44,8	10,0	0,0%
3,50E-04	50	223,6	49,9	-0,1%
7,00E-04	100	445,7	99,6	-0,4%
3,50E-03	500	2071,4	462,7	-7,5%
		with cells		
external force	multiplier applied to the external force	volumetric flow rate [µm³/µs]	obtained multiplier for volumetric flow rate	percentual difference between the two multipliers
7,00E-06	1	4,4	1,0	0,0%
1,40E-05	2	8,9	2,0	0,0%
7,00E-05	10	44,5	10,0	0,3%

4.2 Cylindrical sorting topology

The value of external force was set initially to 0,007, and the obtained volumetric flow rate is 0,67 µm3/µs. The results for different values of the external force parameter settings are presented in the **Fig. 13** and **Fig. 14**. The results from the simulations with cells with hematocrit 20 % are presented in the **Fig. 15**.

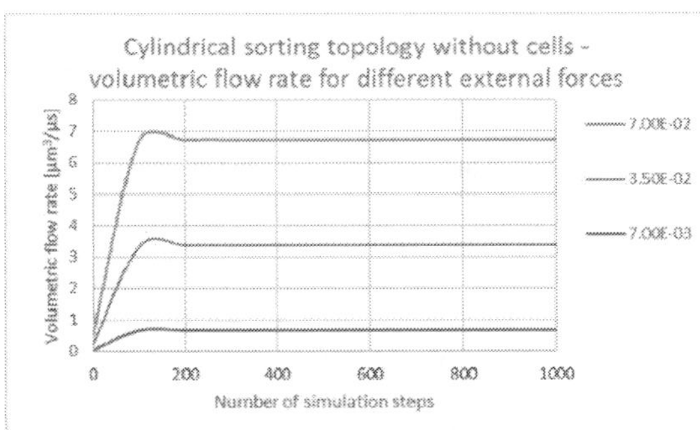

Fig. 13. Example of evolution of the volumetric flow rate during the simulation for three different values of external force for simulations without cells

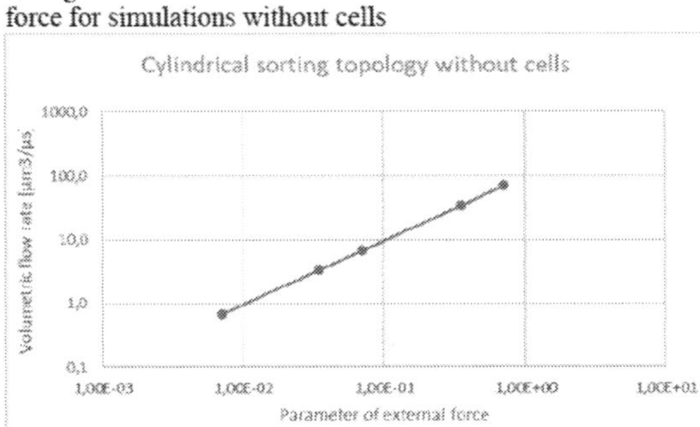

Fig. 14. Relationship between the value of external force parameter and the volumetric flow rate for simulations without cells

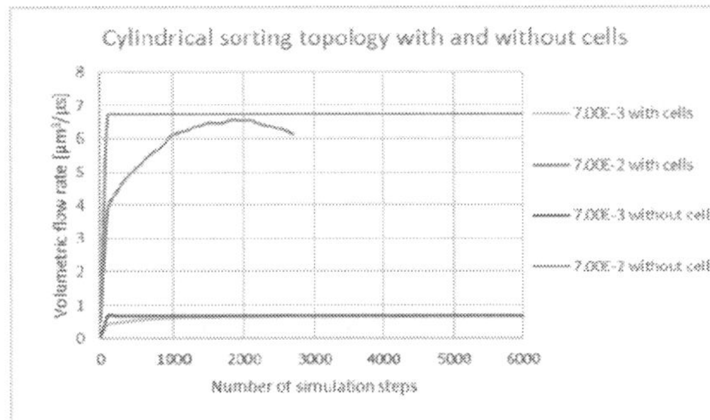

Fig. 15. Comparison between the evolution of the volumetric flow rate for simulations with and without cells.

The observation we make is the very same as for the previous topologies – the volumetric flow rate is linearly dependent on the parameter of external force for the small multipliers (**Table 2**). The simulations without cells and with a multiplier of 500 finished earlier than expected, due to the numerical instability. However, the expected value of the volumetric flow rate in this simulation corresponds to 1000 µl/min, a value 500 times bigger than the value of the real volumetric flow rate in the laboratory experiment. This means that the linearity is preserved for the common range of the volumetric flow rate used in laboratory experiments.

Table 2. Values of volumetric flow rate appropriated to different values of external force parameter. The linearity is slightly perturbed for the big multipliers.

external force	multiplier applied to the external force	volumetric flow rate [µm³/µs]	obtained multiplier for volumetric flow rate	percentual difference between the two multipliers
		without cells		
7,00E-03	1	0,7	1,0	0,0%
3,50E-02	5	3,4	5,0	0,1%
7,00E-02	10	6,7	10,0	0,2%
3,50E-01	50	34,1	50,7	1,4%
7,00E-01	100	70,1	104,3	4,3%
		with cells		
external force	multiplier applied to the external force	volumetric flow rate [µm³/µs]	obtained multiplier for volumetric flow rate	percentual difference between the two multipliers
7,00E-03	1	0,6	1	0%
1,40E-02	2	1,3	2,1	3%

4.3 Spherical adhesive topology

The initial value of the external force parameter was set to 0.000035, which results in volumetric flow rate of 4.02 µm³/µs. The results obtained by modifying this parameter are presented in the **Fig. 16** and **Fig. 17** for without cells simulations, and in the **Fig. 18** for simulations with cells. The hematocrit used for simulations with cells is 15 %.

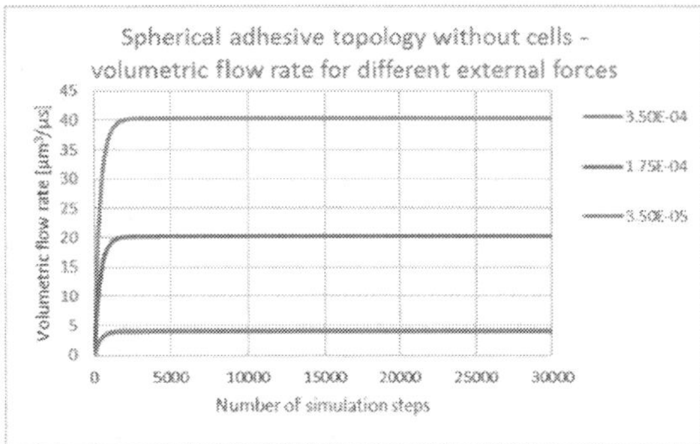

Fig. 16. Example of evolution of the volumetric flow rate during the simulation for three different values of external force for simulations without cells

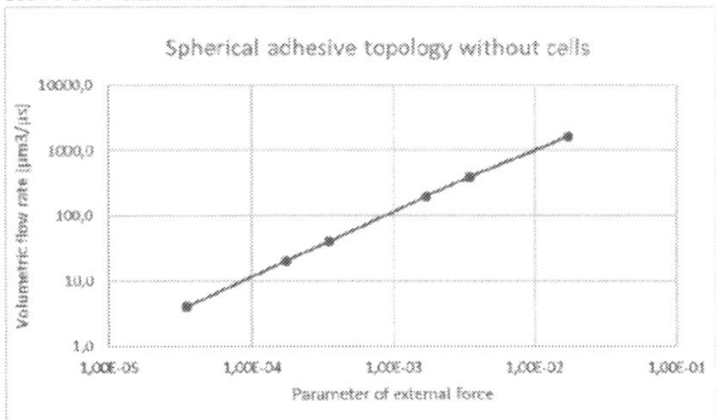

Fig. 17. Relationship between the value of external force parameter and the volumetric flow rate for simulations without cells.

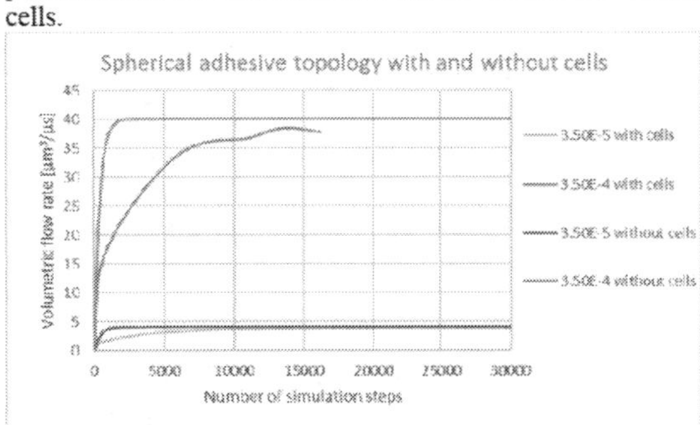

Fig. 18. Comparison between the evolution of the volumetric flow rate for simulations with and without cells.

Table 3. Values of volumetric flow rate appropriated to different values of external force parameter. The linearity is slightly perturbed for the big multipliers.

		without cells		
external force	multiplier applied to the external force	volumetric flow rate [µm³/µs]	obtained multiplier for volumetric flow rate	percentual difference between the two multipliers
3,50E-05	1	4,0	1,0	0,0%
1,75E-04	5	20,1	5,0	0,0%
3,50E-04	10	40,2	10,0	0,0%
1,75E-03	50	199,6	49,6	-0,7%
3,50E-03	100	391,8	97,4	-2,6%
1,75E-02	500	1644,6	409,1	-18,2%
		with cells		
external force	multiplier applied to the external force	volumetric flow rate [µm³/µs]	obtained multiplier for volumetric flow rate	percentual difference between the two multipliers
3,50E-05	1	4,0	1,0	0,0%
7,00E-05	2	7,83	2,0	-1,1%

The linear relationship between the volumetric flow rate and the external force parameter is partially verified (**Table 3**). The linearity is valid for multipliers smaller than 100. For the multiplier of 500, the difference between the obtained and the expected volumetric flow rate is about 18%.

The value of external force was set initially to 0,007, and the obtained

4.4 Comparison of volumetric flow rate for simulations with and without cells

The values are comparable for simulations with and without the cells, especially for simulations with a low hematocrit. The simulation with a high hematocrit shows that it influences the volumetric flow rate in a significant way, even for values close to the laboratory conditions. The comparison between the volumetric flow rates obtained from simulations with and without the cells are presented in the **Table 4**.

Table 4. Volumetric flow rate variation due to presence of the cells in the simulation.

Type of topology	External force	Volumetric flow rate [µm3/µs]		Percentuel difference for vomuletric flow rate with and without cells
		without cells	with cells	
Cylindrical adhesive topology	7,00E-06	4,48	4,44	-1%
	7,00E-05	44,76	44,52	-1%
Cylindrical sorting topology	7,00E-03	0,67	0,63	-6%
Spherical adhesive topology	3,50E-05	4,02	3,96	-2%

4 Conclusions

In our study, we have tested the linear dependence of the volumetric flow rate on the parameter of the external fluid force. For this purpose, we have tested the hypothesis on three different topologies inspired by existing laboratory experiments. The results show that for the range of values commonly used in the laboratory, the hypothesis is valid. For volumetric flow rates which exceed the laboratory conditions by a factor of hundreds, the linearity is not certain.

Another observation is that for high values of the external force parameter, simulations are less stable, particularly the simulations with cells.

An interesting observation can be done about the time necessary for the simulation to get a stabilized volumetric flow rate. For simulations without cells, this time is 2 to 3 times shorter than for the simulations with the same topology and with cells inside the liquid.

References

[1] I. Cimrák, M. Gusenbauer, T. Schrefl, Computers and Mathematics with Applications, **64(3)**, 278-288 (2012)

[2] I. Cimrák, M. Gusenbauer, I. Jančigová, Computer Physics Communications, **185**, 900-907 (2014)

https://doi.org/10.1051/epjconf/201818002046

[3] S. Nagrath, L.V. Sequist et al., Nature, **450**, 1235–1239 (2007)
[4] J.P. Gleghorn et al., Lab Chip, **10**, 27–29 (2010)
[5] S.H. Holm, J.P. Beech, M.P. Barret, J.O. Tegenfeldt, Anal. Methods, **8**, 3291-3300 (2016)
[6] D. Horák, Biomed Mater Res Part A, **101A**, 23–32 (2013)
[7] S. Zheng, H. Lin, J. Liu, M. Balic, R. Datar, R. Cote, Y. Tai, Journal of Chromatography A, **1162**, 154–161 (2007)
[8] A. E. Saliba et al., PNAS, **107**, 14524–14529 (2010)

Experimental and Numerical Study on Supersonic Ejectors Working with R-1234ze(E)

Jan Kracik[1,*], *Vaclav* Dvorak[1], *Vu* Nguyen Van[1] and *Kamil* Smierciew[2]

[1]Department of Power Engineering Equipment, Faculty of Mechanical Engineering, Technical University of Liberec, Studentska 2, 46117, Liberec, the Czech Republic
[2]Department of Heating and Cooling Technics, the Faculty of Mechanical Engineering, Bialystok University of Technology, ul. Wiejska 45C, 15-351 Bialystok, Poland

Abstract. These days, much effort is being put into lowering the consumption of electric energy and involving renewable energy sources. Many engineers and designers are trying to develop environment-friendly technologies worldwide. It is related to incorporating appropriate devices into such technologies. The object of this paper is to investigate these devices in connection with refrigeration systems. Ejectors can be considered such as these devices. The primary interest of this paper is to investigate the suitability of a numerical model for an ejector, which is incorporated into a refrigeration system. In the present paper, there have been investigated seven different test runs of working of the ejector with a working fluid R-1234ze(E). Some of the investigated cases seem to have a good agreement and there are no significant discrepancies between them, however, there are also cases that do not correspond to the experimental data at all. The ejector has been investigated in both on-design and off-design working modes. A comparison between the experimental and numerical data (CFD) performed by Ansys Fluent software is presented and discussed for both an ideal and a real gas model. In addition, an enhanced analytical model has been introduced for all runs of the ejector.

1 Introduction

Ejectors are devices which are able to utilize low-grade thermal energy to drive refrigeration systems and therefore reduce consumption of electric energy which is otherwise needed for working of a refrigeration system with a compressor. Refrigeration system has been probably one of the most common applications of ejectors in recent years. In addition, ejectors are relatively easy to manufacture and compared to, for example, sorption systems require lower investment costs. Ejectors are commonly used for their simple design, reliability and lack of moving parts. Probably the main problem that prevents ejectors from being used more is their low efficiency which does not exceed 30 percent. There has been done much work on investigation of ejectors so far. However, there are still many unexplored areas concerning ejectors that require more research efforts and a rigorous investigation. It should be noted that the flow and especially the mixing of flows in ejector are very difficult to analyse.

As previously mentioned, an ejector is a substitution of a compressor in a classic vapour compression system. The compression of the secondary fluid flowing from the evaporator, which would be otherwise performed by a compressor, is achieved by compressing the fluid in the ejector and then it is further discharged to the condenser [1].

There are two major types of mixing in ejector according to work by Keenan and Neumann [2] distinguished: the constant-area mixing and the constant-pressure mixing. The design of constant-pressure ejectors is rather complex and causes lots of problems for researchers due to its very strict requirements on a precise determination of mixing chamber wall shape to ensure that the static pressure remains constant throughout the whole mixing chamber. Ejectors designed based on constant-pressure mixing should give higher performance in most cases. However, there has not been established a definite link between the performance of constant area and constant pressure ejectors [3]. The flow in constant-pressure ejectors was for example investigated by Kracik and Dvorak in work [4].

A commonly used ejector refrigeration system can be seen in Fig. 1. In this case, the motive vapour is generated in generator and the heat for its heating is usually a low-grade heat, i.e. the heat is extracted from a low-temperature heat source. The possibility of using a source with a relatively low temperature seems to make ejectors so interesting and remarkable.

The refrigeration system has already been investigated in more detail by Smierciew et al. in work [1]. However, in that work the system was investigated

* Corresponding author: jan.kracik@tul.cz

 https://doi.org/10.1051/epjconf/201818002047

for isobutane as a working fluid. Isobutane is classified as explosive, therefore, it is not suitable for some applications. The same author also investigated an ejector system with R-1234ze(E) as a refrigerant [5]; however, in this case, the ejector geometry was slightly different. Fig. 2 illustrates the ejector geometry used in the present paper.

This paper presents an ejector operating with R-1234ze(E), which is considered as a safe alternative to isobutane (R-600a). According to Smierciew et al. [5] there are no research studies dealing with either experimental or numerical investigation on ejectors operating with R-1234ze(E) at this scale. Furthermore, this paper attempts to present both experimental and numerical results accompanied by a theoretical method of ejector performance calculation and show dissimilarities and discrepancies between the different numerical modelling approaches.

Each ejector works under specific conditions and three main modes of ejector performance are distinguished. These modes are obvious from Fig. 3. It can be seen that they divide the performance line of a supersonic ejector into three sections depending on the value of the condenser pressure or back-pressure as shown in the figure. It is highly desirable to design supersonic ejectors for working in the on-design region. This region is recognised as a part with a maximal entrainment ratio and it is also a sign that both the primary and secondary flow are choked, i.e. double choking region. There is no dependency of entrainment ratio or secondary mass flow rate on the condenser pressure in this region. The region should be as wide as possible so that an ejector could sustain its maximal entrainment ratio within a wide range of condenser pressures. The second region is often denoted as the off-design region. In this case, there is a strong dependency of entrainment ratio on the condenser pressure. The secondary mass flow rate and the back-pressure are closely coupled, therefore, the ejector performance is highly sensitive to the adjusted parameters and the secondary flow is not choked anymore. The pressure which separates these two regions is called the critical pressure and the point corresponding to this pressure is the critical point. The location of the critical point on the performance line depends on a given ejector geometry and given parameters on the inlets of ejector. The last region is the region of reverse flow, i.e. an ejector is not able to entrain any fluid from the evaporator and stops working properly. It should be noted that it is always desirable to avoid working ejectors under these conditions.

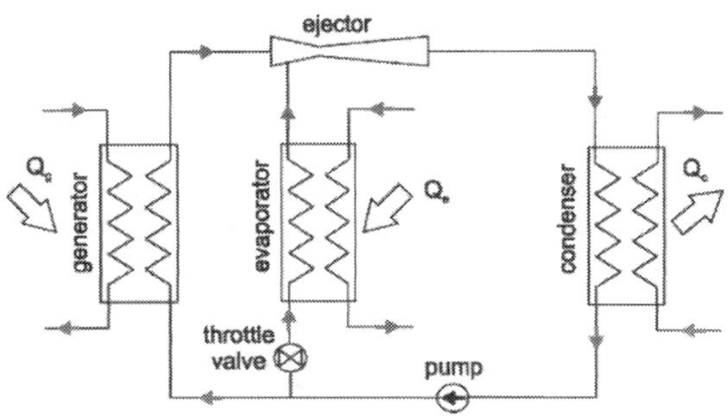

Fig. 1. Schematic diagram of a common ejector refrigeration system.

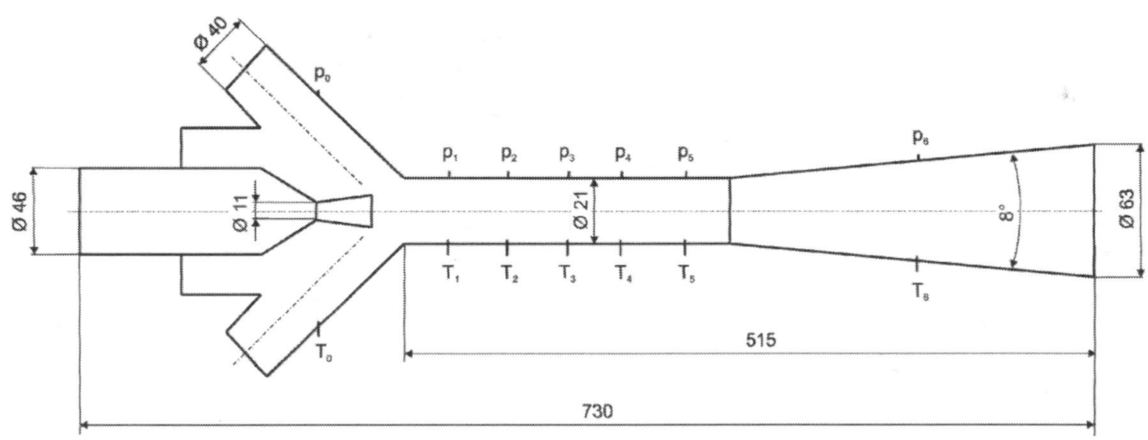

Fig. 2. Ejector geometry with main dimensions in millimetres.

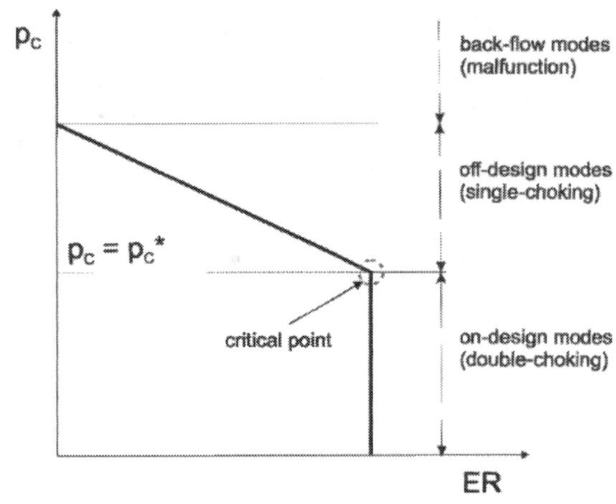

Fig. 3. Typical performance line of supersonic ejectors.

2 Methods and experimental installation

Performance of any ejector can be described by two major parameters. The first parameter is the entrainment ratio that is defined as

$$ER = \frac{\dot{m}_e}{\dot{m}_g}, \quad (1)$$

where $\dot{m}$ is mass flow rate and indices e and g stand for evaporator and generator, respectively. The other parameter is the compression ratio and can be calculated as

$$CR = \frac{p_c - p_e}{p_g - p_e}, \quad (2)$$

where p is the absolute pressure and index c stands for condenser.

The experimental apparatus is presented in Fig. 4 and 5. The apparatus was built at the Bialystok University of Technology in Bialystok, Poland. The main components of the apparatus are listed in the figure legend. The ejector itself was equipped with a pair of temperature and pressure sensors in the suction chamber, five pairs of sensors in the mixing chamber and a pair of sensors in the diffuser (see Fig. 2). The throat diameter of the motive nozzle was 11 mm. The location of the temperature sensors (RTD) and pressure sensors (P) on the experimental apparatus is obvious from Fig. 4. Accuracy of the sensors were as follows: 0.25 % for the pressure sensors, 0.20 % for the temperature sensors, and 0.15 % for the Coriolis mass flow meters. The measurement error was calculated by means of the total derivative approach. All of the measurement sensors and the entire measurement system were calibrated before each experimental run.

Three auxiliary loops were applied in the experimental apparatus that serve for thermal load of the evaporator, for condenser cooling, and as a heat source for the vapour generator. The Coriolis mass flow meters and temperature sensors at the heat exchangers inlets and outlets were installed on these loops. The auxiliary loops allowed to adjust the operation parameters in a wide range of thermal capacity. Water was used as a working fluid in the generator heat source system and ethylene glycol was used as a heat transfer fluid in the evaporator heat load system and in the condenser cooling system. A dry cooler is used in the condenser cooling system.

The presented system is proposed for operation with a district heating system, which is considered as the low-grade motive source. During experiments, operation of the district heating system was simulated by the electric heater system of thermal capacity approximately 100 kW. The electric heater operated at its full capacity. The mass flow rate of the circulating water in the heating loop was set to maintain its temperature at the vapour generator inlet below 70 °C. The pressure and the temperature of the motive vapour at the ejector inlet were controlled by control valves. Moreover, a regenerative heat exchanger was applied for improvement of the effectiveness of the ejection device.

A diaphragm pump was used for the circulation of the refrigerant. The pump is driven by motor power of 3 kW achieved at 700 RPM. In general, the pump is used for circulation of part of the refrigerant. In other words, the pump feeds only the vapour generator. However, as it can be seen in Fig. 4, the system pump is located right behind the condenser before the mixed fluid is divided into the motive and the entrained fluid. It means that the pump draws all the amount of the refrigerant, and consequently the pump consumes more energy. However, location of the pump in the presented system enables more precise and easier regulation and control of the system operation. In refrigeration system operating with commonly used refrigerants like R-404A or R-134a the throttle valve that feeds the evaporator operates at a pressure difference between the condenser and the evaporator of 500 kPa or more. For low-pressure refrigerants this pressure difference is between 200 kPa and 300 kPa. Without the initial increase in pressure in the throttle valve, the operation would be less stable and unpredictable. This is the reason for the location of the pump. However, since the liquid pump consumes almost negligible amount of energy in the total motive power consumption of the system, the current system configuration does not lead to significant deterioration of COP and provides easier control of the operation.

The measurement system was based on two National Instrument systems. The computer used LabVIEW software which is capable to receive on-line data from the refrigeration properties software REFPROP [6, 7]. The sampling frequency during measurement was 1 Hz. The measurement time was between 90 - 120 seconds. All readings were averaged to form one experimental point. The operating conditions were considered as stable when the pressure and temperature profiles did not show significant changes. Table 1 illustrates the investigated runs with the operating conditions.

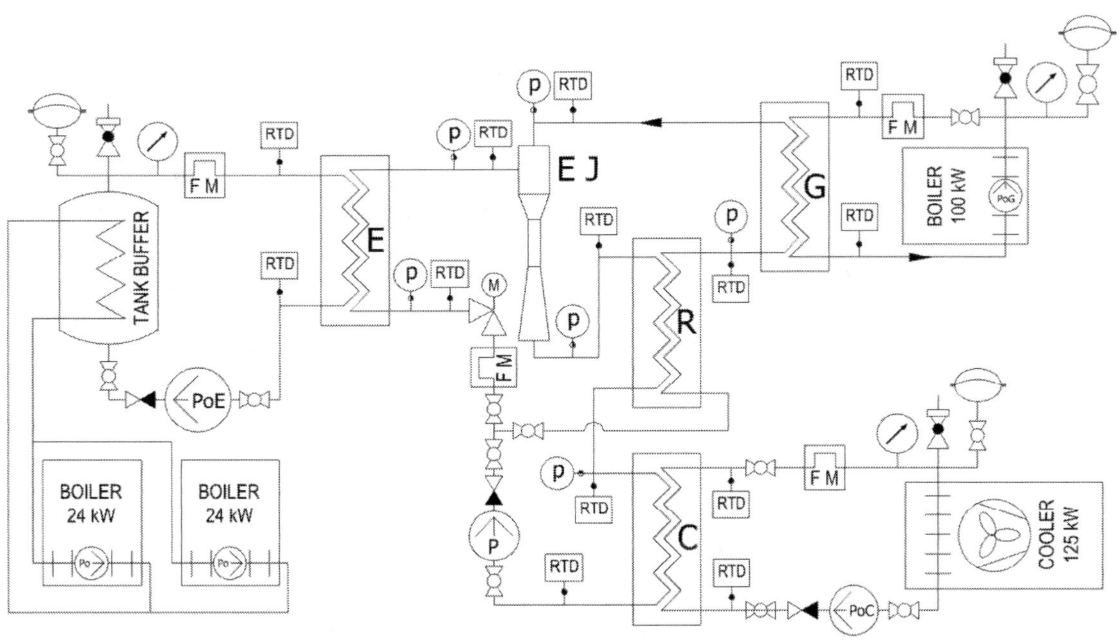

Fig. 4. Schematic diagram of the experimental apparatus: vapour generator (G), condenser (C), evaporator (E), ejector (EJ), pump (P), internal heat exchanger (R), M - throttle valve, RTD - temperature sensor, p - pressure sensor, FM - flow meter.

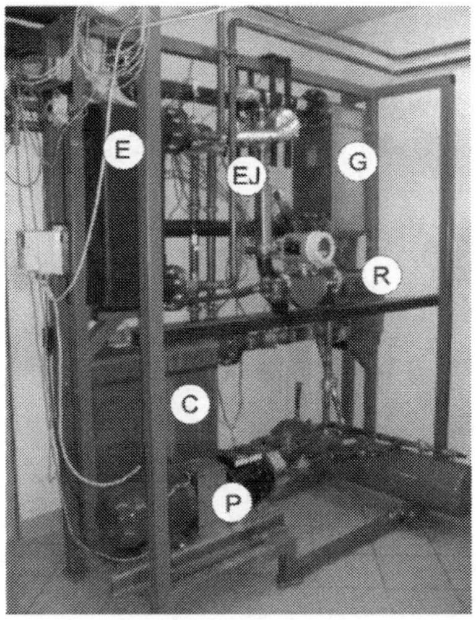

Fig. 5. Photo of the experimental apparatus.

Table 1. Ejector operating conditions.

	p_g (MPa)	p_e (MPa)	p_c (MPa)	T_g (K)	T_e (K)
run no. 1	1.174	0.216	0.420	338.3	278.9
run no. 2	1.180	0.210	0.479	335.1	284.0
run no. 3	1.170	0.203	0.503	335.5	281.8
run no. 4	1.165	0.207	0.509	337.2	279.5
run no. 5	1.179	0.256	0.541	338.1	283.4
run no. 6	1.182	0.265	0.552	338.2	284.1
run no. 7	1.205	0.320	0.612	341.7	284.9

3 Numerical modelling

Since this paper aims at assessment the ability of numerical methods (CFD) for calculation of the flow in supersonic ejectors, especially in refrigeration applications, this part is crucial. The most important issue is probably to be able to obtain plausible results in a wide range of operation of the ejector. In other words, it is highly desirable to obtain such a numerical solution which is very close to the experimental results in the different modes of supersonic ejector operation, from the on-design mode through the off-design mode to the back-flow mode (malfunction).

It should be noted that there have been done many studies for air as a working fluid comparing CFD with experiments. The results seem to be in a good agreement in the on-design mode, however, the further from the critical point towards the back-flow regimes an ejector working regime is the larger discrepancy between CFD and experiment is often recognized.

In order to reduce the complexity of the numerical model, one quarter of a three-dimensional model of the ejector was used for all calculations (Fig. 6).

The flow in the ejector is described by the compressible form of the conservation equations since they have proved to be more suitable for describing compressible flows with significant density changes. The influence of thermodynamic and transport properties of the used refrigerant are not assumed to be significant, therefore, these properties are considered as constants during the simulations. The working fluid was modelled according to both the ideal and the real gas model.

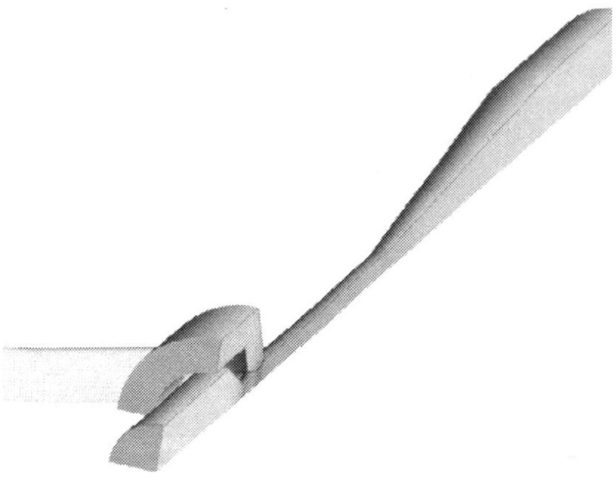

Fig. 6. The numerical model of the ejector.

The ideal gas was modelled with the following properties: specific heat $c_p = 291.74\,J \cdot kg^{-1} \cdot K^{-1}$, thermal conductivity $k = 1.36 \cdot 10^{-2}\,W \cdot m^{-1} \cdot K^{-1}$, viscosity $\mu = 1.224 \cdot 10^{-5}\,Pa \cdot s$ and molecular weight $M = 114\,kg \cdot kmol^{-1}$.

As regards the other modelling approach, the real gas was modelled using the NIST REFPROP database provided by Ansys Fluent Inc. The main limitation of this approach is that it assumes that the fluid used in the ANSYS Fluent calculation is superheated vapour, supercritical fluid, or liquid. It should be noted that subcritical flow conditions, where vapour coexists with liquid in two-phase flow, are not supported.

The flow governing equations are written in the form as follows:

$$\frac{\partial \rho}{\partial t} + \frac{\partial}{\partial x_i}(\rho u_i) = 0\,, \tag{3}$$

$$\frac{\partial}{\partial t}(\rho u_i) + \frac{\partial}{\partial x_j}(\rho u_i u_j) = -\frac{\partial}{\partial x_j}(p\delta_{ij}) + \frac{\partial \tau_{ij}}{\partial x_j}\,, \tag{4}$$

$$\frac{\partial}{\partial t}(\rho e_0) + \frac{\partial}{\partial x_j}(\rho u_j e_0 + u_j p + q_j - u_i \tau_{ij}) = 0\,, \tag{5}$$

$$p = \rho r T \tag{6}$$

and for a Newtonian fluid the viscous stress term can be written in form

$$\tau_{ij} = \mu_{eff}\left(\frac{\partial u_i}{\partial x_j} + \frac{\partial u_j}{\partial x_i}\right) - \frac{2}{3}\mu_{eff}\frac{\partial u_k}{\partial x_k}\delta_{ij}\,. \tag{7}$$

The governing equations are solved using the commercial CFD package Fluent provided by Ansys, Inc. Three-dimensional hexahedral mesh with nearly 1,700,000 elements was created for calculations. The ejector mesh is illustrated in more detail in Fig. 7. The y+ was in most of the cases below value 5. Double precision option was used to get more precise solution. For all variables the second-order upwind scheme was utilized. As already mentioned, all the thermodynamics and transport properties as well as boundary conditions are held constant during the simulations.

Although the use of the transient term in the governing equations seems to be beneficial since this term, in some cases, is able to stabilize the solution during the calculation, the steady state governing equations were used for the calculations with the real gas, and in these cases the transient term was omitted. The reason for doing it in this way was that it appeared almost impossible to achieve convergence in a transient way when using the real gas model. On the other hand, all calculations with the ideal gas model were performed in a transient manner.

To take turbulence effects into account the k-ω SST turbulence model, which has been proved to be the most suitable for ejectors calculations [8], was used. However, most of the calculations for validating suitability of a proper turbulence model have been done for air as a working fluid. The proper choice of the most suitable turbulence model for calculating flows with a different fluid than the air could be an issue and may require more attention in future works as each refrigerant may require its own CFD study on the most suitable turbulence model. Nevertheless, it is believed that for the current calculations the most suitable turbulence model for air ejectors is also the most suitable turbulence model for any other working fluid. Furthermore, it is worth to mention that unlike the air under a wide range of conditions, most refrigerants behave like a real gas. For more information on this topic, see Ansys Fluent documentation [9].

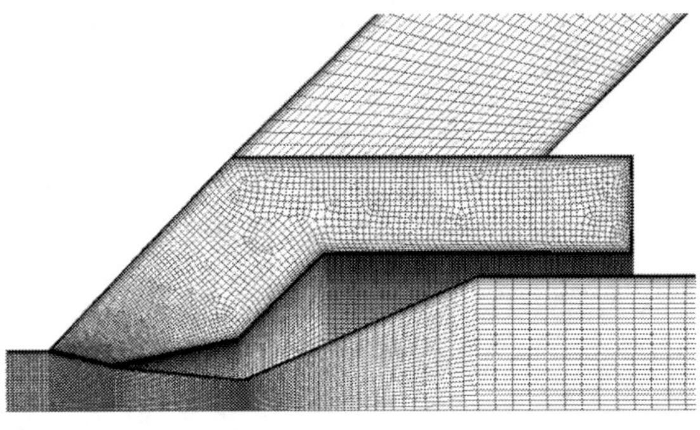

Fig. 7. Detail of the ejector mesh in the vicinity of the primary nozzle.

4 Analytical approach

There was used an analytical model for theoretical investigations. The model is currently under development at the Technical University of Liberec, and is based on the 1D model by Chen et al. [10] and the model by Abramovich [11]. The authors of this paper modified both models and combined them together as well as added some new features. The model will be investigated in more details, and introduced in future works.

5 Results and discussion

Fig. 8 provides an approximate illustration all the test runs obtained from Table 1. As can be seen, two test runs are in the on-design modes with double choking, while the rest of them belong to the off-design modes with single choking.

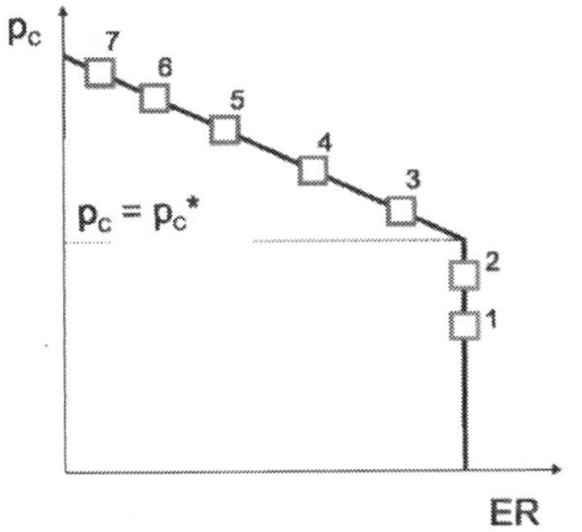

Fig. 8. Approximate representation of the investigated test runs on the performance line of the ejector (boundary conditions on inlets are not constant on this curve).

The courses of the static pressure distribution on the ejector wall for all seven test runs are plotted in figures 9-15. A comparison between both CFD approaches the ideal gas and the real gas with experimental data is obvious from these figures. It can be seen that discrepancies in the static pressure distribution on the ejector wall are low. There are negligible differences for test run no. 1 (Fig. 9). There is obvious a higher discrepancy between the real gas model and the ideal gas model for test run no. 2 (Fig. 10). However, in this case there is a reasonable agreement between the real gas model and the experimental data. As regards the test run no. 3, there is an agreement between the real gas model and the ideal gas model, however, the experimental data is inconsistent with both of these numerical approaches (Fig. 11). The main discrepancy is further expansion of the flow in the mixing chamber for the experiment.

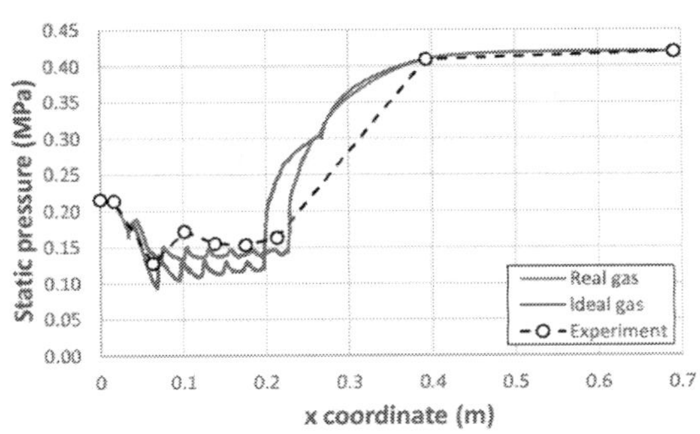

Fig. 9. Static pressure distribution on the ejector wall for the test run no. 1.

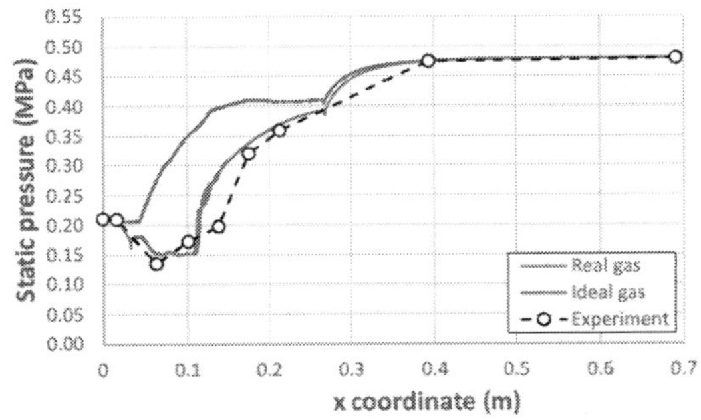

Fig. 10. Static pressure distribution on the ejector wall for the test run no. 2.

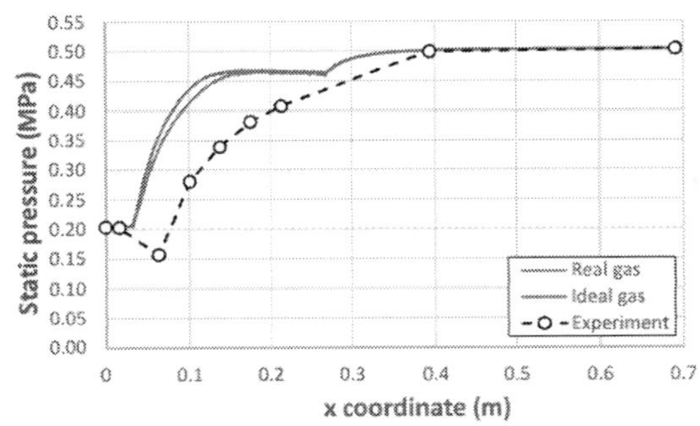

Fig. 11. Static pressure distribution on the ejector wall for the test run no. 3.

A similar trend can be seen in Fig. 12 for test run no. 4, apart from the expansion of the flow in the mixing chamber. The test runs no. 5 and 6 are also similar to each other as Fig. 13 and Fig. 14 illustrate, respectively. In both cases, there are only insignificant discrepancies. Obviously, the lowest discrepancies in static pressure courses can be seen for test run no. 7 (Fig. 15). Based on the experimental static pressure courses it can be stated that the ejector is working in the on-design mode for the test run no. 1 and test run no. 2 (Fig. 3 and Fig. 8).

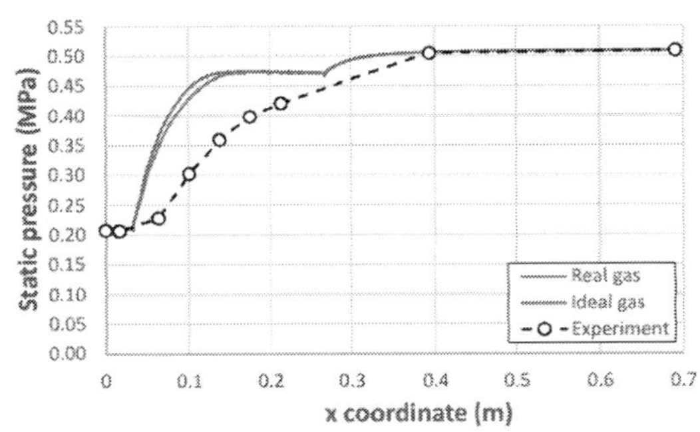

Fig. 12. Static pressure distribution on the ejector wall for the test run no. 4.

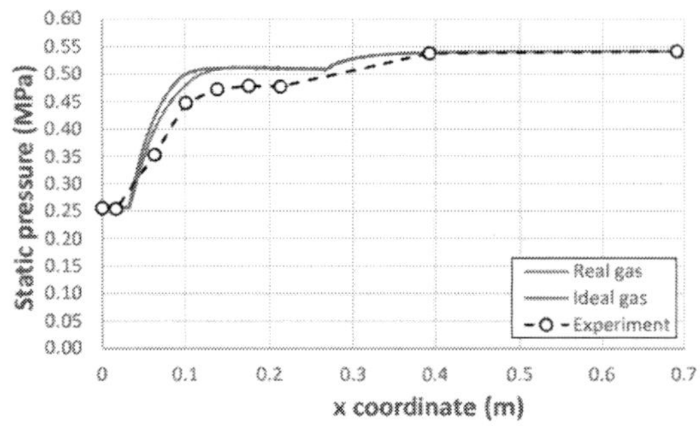

Fig. 13. Static pressure distribution on the ejector wall for the test run no. 5.

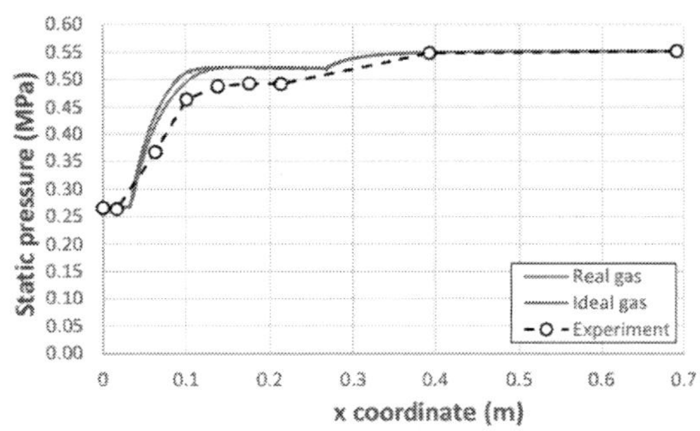

Fig. 14. Static pressure distribution on the ejector wall for the test run no. 6.

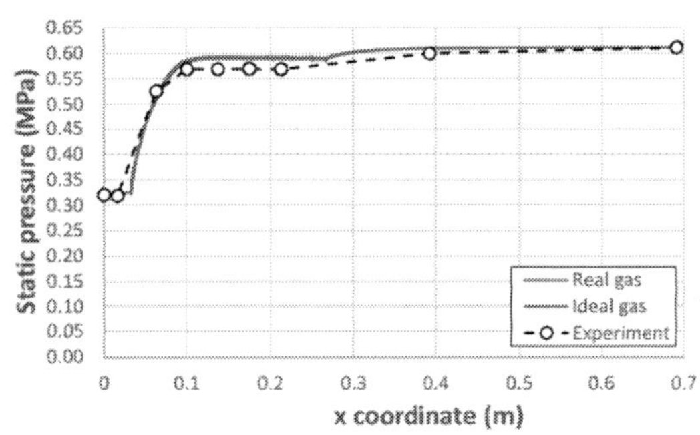

Fig. 15. Static pressure distribution on the ejector wall for the test run no. 7.

The entrainment ratio comparison for all tested runs is shown in Fig. 16. Significant discrepancies in the entrainment ratio between the experiment and the CFD data based on the real gas model can be observed. Similarly, there are also discrepancies between the experiment and the CFD data based on the ideal gas model.

These discrepancies are lowest for test run no. 1 and test run no. 2. These runs were earlier in the paper identified as the on-design runs, i.e. cases where the secondary flow is choked. Other cases lie beyond ±20 % relative error limit and are not worth mentioning in more detail. It should be noted that there were not obtained any negative values of entrainment ratio from the experiment.

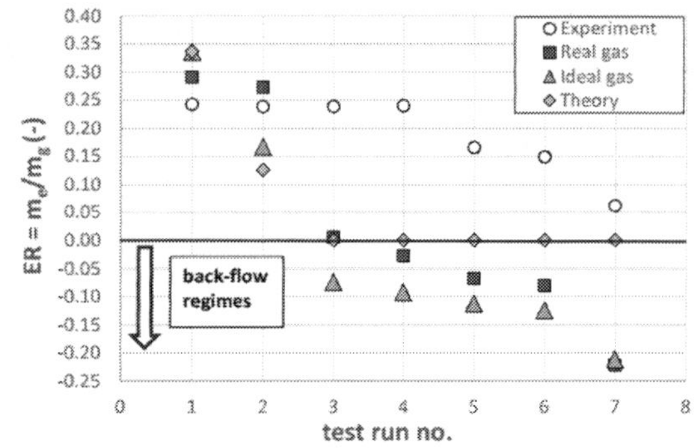

Fig. 16. Entrainment ratio comparison for experiment, real gas (CFD), ideal gas (CFD) and theoretical model.

The highest entrainment ratio obtained from experiment was 0.242 with a corresponding compression ratio of 0.213 (for test run no. 1). On the contrary, the lowest entrainment ratio was found for test run no. 7, ER = 0.062 and CR = 0.329. It can also be seen in Table 2 that the entrainment ratio is inversely proportional to the compression ratio, which is consistent with literature.

Table 2. Values of the entrainment ratio for all test runs.

	CR (-)	Experiment ER (-)	Real gas ER (-)	Ideal gas ER (-)	Theory ER (-)
run no. 1	0.213	0.242	0.291	0.336	0.336
run no. 2	0.277	0.238	0.273	0.167	0.126
run no. 3	0.310	0.239	0.006	-0.075	0.000
run no. 4	0.315	0.240	-0.028	-0.093	0.000
run no. 5	0.309	0.166	-0.068	-0.113	0.000
run no. 6	0.312	0.149	-0.080	-0.125	0.000
run no. 7	0.329	0.062	-0.222	-0.212	0.000

Apart from the runs no. 1 and 2, the ideal gas model gives negative values of entrainment ratio. On the other hand, the real gas model gives negative values of entrainment ratio for test runs no. 4 to 7. The entrainment ratio for test run no. 3 is slightly positive, however, it could be argued that the numerical precision plays here a significant role as the value is very close to zero.

As can be seen in Table 3, there are reasonable agreements in the primary mass flow rate for all investigated cases. This can be caused by the fact that the primary flow was choked in all calculated cases. As a matter of fact, it is obvious that both CFD approaches and theory tend to slightly underestimate the mass flow rate of the primary stream. The highest value of the primary mass flow rate was for each test run obtained from the experiment.

Table 3. Values of the primary (generator) mass flow rate.

m_g	Experiment (kg/s)	Real gas (kg/s)	Ideal gas (kg/s)	Theory (kg/s)
run no. 1	0.486	0.461	0.460	0.461
run no. 2	0.493	0.478	0.465	0.463
run no. 3	0.488	0.462	0.461	0.459
run no. 4	0.483	0.458	0.458	0.457
run no. 5	0.488	0.460	0.462	0.463
run no. 6	0.489	0.464	0.464	0.464
run no. 7	0.495	0.470	0.470	0.473

Table 4. Values of the secondary (evaporator) mass flow rate.

m_e	Experiment (kg/s)	Real gas (kg/s)	Ideal gas (kg/s)	Theory (kg/s)
run no. 1	0.118	0.134	0.154	0.155
run no. 2	0.118	0.130	0.078	0.058
run no. 3	0.116	0.003	-0.034	0.000
run no. 4	0.116	-0.013	-0.042	0.000
run no. 5	0.081	-0.031	-0.052	0.000
run no. 6	0.073	-0.037	-0.058	0.000
run no. 7	0.031	-0.104	-0.100	0.000

Table 4 illustrates results for the secondary mass flow rate. It is obvious that the discrepancies in the entrainment ratio are mainly caused by an improperly computed secondary mass flow rate. In this regard, all used methods for predicting flow in an ejector fail. Moreover, it should be pointed out that the presented analytical model is unable to predict back-flow regimes, i.e. regimes where the secondary mass flow rate has got a negative value. In such cases the analytical model gives a zero secondary mass flow rate.

Probably the key parameter for assessing the quality of the mixing the motive and the entrained flow is the mixing length and the area where the secondary flow reaches its maximal velocity and in some cases (on-design modes) becomes choked. CFD can help to evaluate that, but it needs to be in a reasonable agreement with experimental data first.

Figures 17 to 20 illustrate the contours of Mach number in the motive nozzle and the mixing chamber for test run no. 1 and 2.

The flow in the ejector during the test run no. 1 is for the ideal gas model and the real gas model obvious from Fig. 17 and Fig. 18, respectively. Evidently, the primary nozzle operates in the over-expanded regime in both cases the ideal and the real gas model, because the jet of the primary stream is slightly converged toward the ejector centre and a complex pattern of shocks and reflections is evolved in the mixing chamber. As a result, the choking of the secondary flow occurs further downstream in comparison to the under-expanded jet. Additionally, the ideal gas model tends to predict higher Mach numbers in the entire ejector. As regards the position of the closing shock wave, the real gas model predicts the shock wave further downstream compared to the ideal gas model. Nevertheless, the choking of the secondary stream occurs approximately at the same distance downstream the motive nozzle, and forms almost the same fictive throat for both the ideal and real gas model. This could be the reason for the low discrepancy in the secondary mass flow rates between the ideal and real gas model.

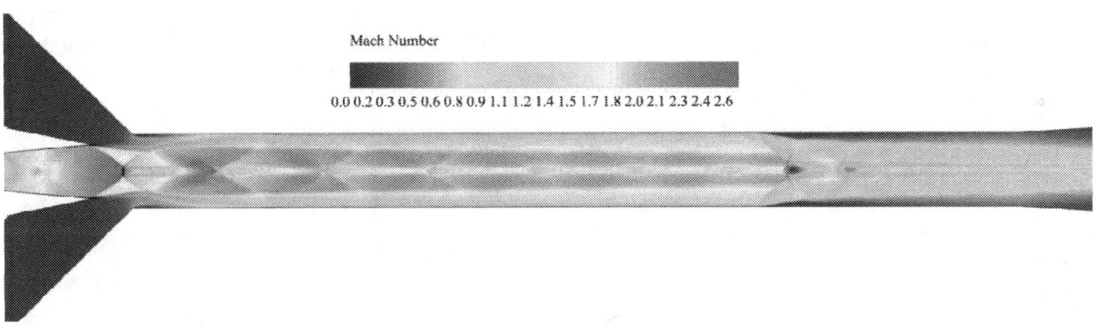

Fig. 17. Mach number contours for test run no. 1 and the ideal gas model; ER = 0.336, CR = 0.213.

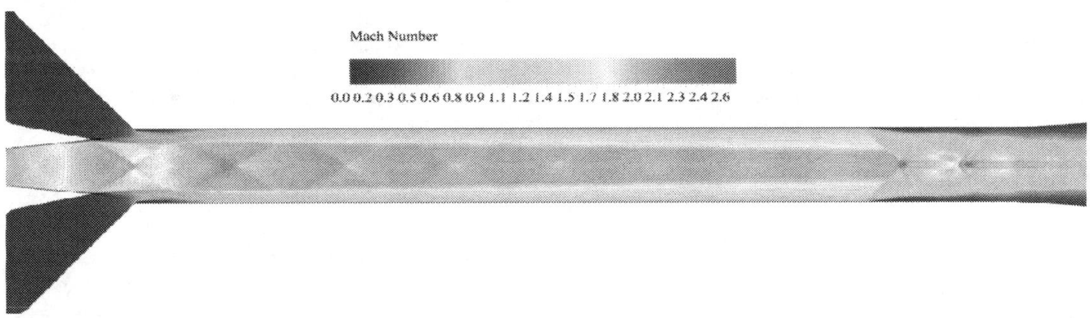

Fig. 18. Mach number contours for test run no. 1 and the real gas model; ER = 0.291, CR = 0.213.

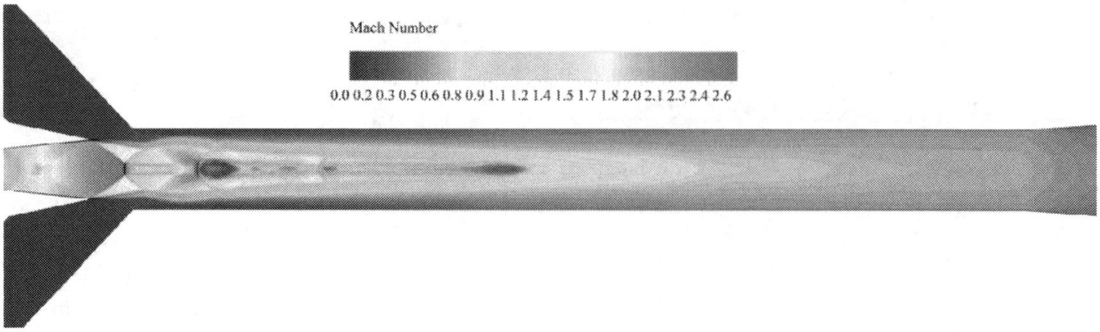

Fig. 19. Mach number contours for test run no. 2 and the ideal gas model; ER = 0.167, CR = 0.277.

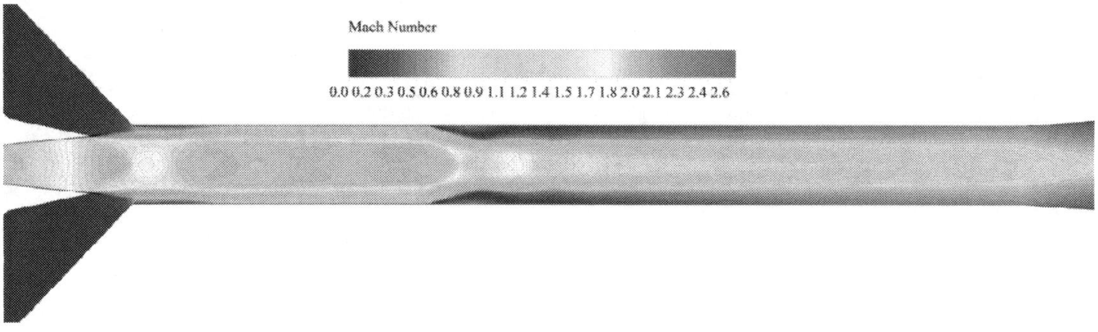

Fig. 20. Mach number contours for test run no. 2 and the real gas model; ER = 0.273, CR = 0.277.

Nevertheless, in both cases is observed the growth of boundary layer until the closing shock wave appears. Moreover, a boundary layer separation is provoked right behind the closing shock wave.

Fig. 19 and Fig. 20 give an illustration of the flow in the ejector during the test run no. 2 for the ideal gas model and the real gas model, respectively. It is observed that the flow patterns are completely different. Since the calculated secondary mass flow rate for the real gas model is closer to the experimental data, it can be assumed that for the given conditions the ideal gas model fails and probably cannot be used for calculations. The secondary mass flow rate for the ideal gas is smaller than for the real gas (Table 4). This is probably due to the fact that the secondary flow is not choked in the former case.

Obviously, in the case with the ideal gas, the ejector is able to entrain only negligible amount of the secondary fluid. For the case with the real gas, the secondary flow becomes choked slightly further compared to the test run no. 1. On the contrary, the closing shock wave is weaker and occurs more upstream for this case. As a result, a strong boundary layer separation can be seen in the mixing chamber section (Fig. 20). In conclusion, it is obvious that further tests are required to obtain more accurate results.

6 Summary and conclusion

This paper concentrates on the validation of a CFD tool for a supersonic ejector working with R-1234ze(E) as a low global warming potential refrigerant. The selection of this fluid was on the basis of the analysis of the working fluids for ejector system with respect to Regulation of the European Parliament and the EU Council No. 517/2014 [5].

The CFD results obtained on simplified 3D models have been presented for both the ideal and real gas. The comparison with the experimental data shows significant discrepancies in mass flow rates, especially in the off-design region. This could be caused by simplified numerical models or more likely by inability of the physical models of the solver to capture all the relevant phenomena that occur within the mixing chamber and diffuser. On the other hand, the comparison of static pressure distributions shows reasonable discrepancies. Furthermore, the contours of Mach number have shown several common phenomena in ejectors: a motive nozzle working regime, a region of the choking of the secondary flow, a closing shock wave, and a boundary layer separation.

The highest entrainment ratio during experiments was 0.242 (test run no. 1). Nevertheless, this test run gives the lowest compression ratio of 0.213. On the other hand, the lowest entrainment ratio was 0.062 (test run no. 7) and corresponding compression ratio was 0.329. Based on the experimental data, any ejector test run has not been proved to be working in the back-flow mode.

On the contrary, the calculation of the flow by means of the real gas approach gives the back-flow mode for four test runs; and the ideal gas approach even five out of seven test runs.

The primary mass flow rates were in a good agreement for all tested runs. This was probably caused by the fact that the primary flow was choked in all calculated cases. However, it is worth mentioning that both CFD approaches and theory tend to slightly underestimate the mass flow rate of the primary flow. The highest value of the primary mass flow rate was for each test run obtained from the experiment.

It seems that most of the difficulties during the calculation of the flow in ejectors are caused by the incorrectly computed secondary mass flow rate. It is generally well-known fact that prediction of the secondary mass flow rate is a challenging task in many branches.

It appears that there is still need for a deeper investigation on ejectors in conjunction with refrigeration applications as there are significant discrepancies in mass flow rates between CFD and experimental data, especially the secondary mass flow rate in the off-design mode. Only two tested runs out of seven seem to be in a sufficient agreement with the experimental data. It is believed that it cannot be a coincidence, however, there needs to be an extensive validation performed in order to be able to predict the flow in ejectors working with common refrigerants more correctly.

Validation test runs over the entire range of operating conditions are still required. It should be a motivation for further research to prove the ability of CFD to model different operating regimes. Furthermore, a more comprehensive analysis of the experimental data and CFD is desired for better understanding the flow processes that occur within the ejector mixing chamber and diffuser. In addition, it has been also shown that the presented analytical model is unable to predict back-flow modes, i.e. modes with a negative secondary mass flow rate, and in such cases gives a zero secondary mass flow rate regardless the magnitude of the negative flow rate. It is worth mentioning that the model is still in development since a deeper understanding of the processes in ejectors is highly desirable. Finally, further tests will be required to obtain more accurate results.

Acknowledgement

This publication was written at the Technical University of Liberec as part of the Czech-Polish project "Research of processes in supersonic ejectors with isobutane" no. 7AMB16PL011 and as part of the project "Experimental and numerical investigation in applied fluid mechanics and energy devices, no. 21124" with the support of the Specific University Research Grant, as provided by the Ministry of Education, Youth and Sports of the Czech Republic in the year 2017

References

1. K. Smierciew, D. Butrymowicz, J. Gagan, V. Dvorak. *Int. Conf. Compressors Cool.* **9**, (2017).
2. J.H. Keenan, E.P. Neumann. *J. Appl. Mech.* **64,** 75–81 (1942).
3. D. W. Sun, I. W. Eames. *J. Ins. Energy*, **68**, 65–79, (1995)
4. J. Kracik, V. Dvorak. *Exp. Fluid Mech.* **92**, (2015).
5. Smierciew, K., Gagan, J., Butrymowicz, D., Łukaszuk, M. & Kubiczek. *Appl. Therm. Eng.* **110**, 115–125 (2017).
6. E.W. Lemmon , M.L. Huber, M.O. McLinden. *Natl Std. Ref. Data Series.* **9.1**, (2013)
7. M. Thol, E.W. Lemmon, *Int. J. of Thermophys.* **37**, (2016)
8. J. Kolář, V. Dvořák. *Int. J. Mech. Mechatronics Eng.* **5**, 1715–1719 (2011).
9. ANSYS. ANSYS FLUENT User's Guide. **15317**, 2498 (2011).
10. W. Chen, M. Liu, D. Chong, J. Yan, A. B. Little. *Int. J. Refrig.* **36**, 1750–1761 (2013).
11. G. N. Abramovich. *Applied Gas Dynamics. Third Edition.* (Defense Technical Information Center, 1973). at https://books.google.cz/books?id=-LISOAAACAAJ

Validation of laser-penetration- and electrical-conductivity-based methods of concentration measurement in flow with intense transport of coarse sediment

Jan Krupička[1,*], *Václav* Matoušek[1], *Tomáš* Picek[1] and *Štěpán* Zrostlík[1]

[1]Faculty of Civil Engineering CTU in Prague, Department of Hydraulics and Hydrology, Thákurova 7, 16000 Prague, Czech Republic

Abstract. Lack of experimentally determined information on inner structure of flows carrying large amount of coarse sediment is an important limitation for development and validation of appropriate mathematical models. Measurement of local flow properties is particularly difficult in case of coarse sediment due to specific features associated with high ratio of the sediment size to the flow depth. This paper focusses on two methods of evaluation of vertical concentration profiles in the flow of a mixture of plastic grains and water in a laboratory flume. The first one was proposed by B. Spinewine et al. in 2011 and it is based on the relation between concentration of solids and the depth of penetration of a laser stripe to the flow, which is evaluated from a high speed camera record. The second method is based on well-known relation between the concentration and electrical conductivity of a mixture sensed by immersed probe. A sensitivity analysis of both the methods is performed to show the most crucial parameters affecting accuracy of the results. The methods are validated on data measured in fluidization cell with controlled particle concentration. Presented results on flow show limitations and potential of the methods for laboratory studies on liquid-granular flow.

1 Introduction

High amount of solid material can be carried by flows at high discharges in steep channels with erodible bed. Using laboratory flume to experimentally investigate inner structure of such flows can provide essential information for development and validation of mathematical models based on relations between local concentration, velocity and fluid/solids stresses (e.g. Berzi 2011, Capart & Fraccarollo 2011, Berzi & Fraccarollo 2013). Flow conditions in merit of interest can be characterised by presence of layered structure with majority of the solid material transported within a transport layer adjacent to the bed. High solids concentration and predominating effect of interparticle interactions are typical of the transport layer.

To model such conditions physically in the scale of laboratory flumes, relatively light and large particles are often used (Sumer et al., 1996; Capart and Fraccarollo, 2011; Revil-Baudard et al., 2015). A need to measure with a spatial resolution comparable with the particle size, high concentration (implying opacity of mixture) and a wide range of flow velocities make measurement of local concentration difficult. From a variety of measuring methods developed in past (see e.g. Nasr-el-din et al., 1984; Graham et al., 2002; Shook and Daniel, 1965; Revil-Baudard et al., 2015; Capart et al., 2002), we selected two for testing. The first one was recently proposed by Spinewine et al. (2011) for measurement of vertical concentration profiles in laboratory flumes. It is a non-invasive technique based on observation of depth of penetration of laser sheet to the mixture. The second method, on the other hand, implements well known principle based on measurement of electrical conductivity in vicinity of electrodes submerged into the flow.

2 Principle of selected methods

2.1 Laser stripe measurement

Principle of the method is described in detail in Spinewine et al. (2011). Flow is illuminated at selected cross-section of the flume by a laser sheet penetrating the flow through a transparent side-wall of the flume. The principle is novel and differs from earlier proposed methods based on image processing (Capart et al., 2002; Spinewine et al., 2003). Laser sheet is oriented normal to the flume wall and to the flow direction. At the each vertical position, the particle nearest to the flume wall is hit by the laser.

Depth of laser penetration can be evaluated from a photo of illuminated cross-section taken in direction deviated from the laser sheet (through the side wall) by an angle α. Position of a pixel in the photo can be related to a certain position in the illuminated cross-section via a holomorphic transformation. Using the transformation, each photo is ortho-rectified to the coordinate system of

* Corresponding author: jan.krupicka@fsv.cvut.cz

flume cross-section. Then a pixel of the maximal brightness, B_{max}, is identified at the each vertical position and its distance from the flume side wall, y, is recorded. If the brightness is higher than a certain threshold value B_{th}, this pixel is interpreted as a particle hit by laser and distance y is identified as the depth of laser penetration. The threshold brightness was introduced to filter out unwanted optical effects such as reflections and shadows.

Parameters of the transformation between position of a pixel in the photo and a position in the illuminated cross-section must be calibrated before or after measurement of concentration profile. This is done by taking a calibration photo of a target of known geometry placed at known position in the flume.

As particles enter and leave illuminated cross-section, depth of laser penetration fluctuates in time, but the mean value can be deduced from recorded sequence of photos if observation angle α is known. This mean is related to the concentration: the higher concentration at given vertical position, the higher probability that some particle will be hit by the laser near the wall and the smaller mean depth of laser penetration is observed. Spinewine et al. (2011) derived, under some simplifying assumptions, a mathematical form of this relationship. Hence, a vertical concentration profile can be evaluated from recorded photos of flume cross-section illuminated by a laser sheet.

2.2 Measurement of local electrical conductivity

In the simplest configuration, electrical conductivity was evaluated from measurement of voltage and current on pair of electrodes (Ong a Beck, 1975). In configuration proposed by Nasr-El-Din et al. (1987) two pairs of electrodes mounted on vertically shifted gauge were used to evaluate vertical concentration profile in pressurized pipe. Electrical current was set to a pair of field electrodes generating electrical filed and voltage was sensed by a pair of measuring electrodes. This configuration enabled minimization of polarization effects caused by current at surface of field electrodes. Field electrodes were at top and bottom side of cylindrical probe of diameter of 5 mm. Measuring electrodes were at the side of probe in mutual distance of 1 mm. Size of particles in mixture ranged from 0.5 to 5.5 mm. Nasr-El-Din et al used alternating current 1 mA of high frequency (1 kHz) to further minimize polarization. MacTaggart et al. (1993) used the same frequency and amplitude but slightly different geometrical configuration of electrodes to measure concentration of particles up to 0.5 mm. In CCM probe (Ribberink and Al-Salem, 1995) measuring electrodes are placed in one row between field electrodes in mutual distance of 0.6 mm. The probe was developed to measure concentration of fine sand in oscillatory flow. In all of these applications, particles were relatively small in comparison with characteristic dimensions of flow. The only exception was measurement with 5.5-mm particles, for which Nasr-El-Din et al. (1987) reported considerable distortion of calibration curve for calculation of concentration from measured voltage. Moreover, MacTaggart et al. (1993) and Nasr-El-Din et al. (1987) pointed out that measurement is affected close to flow boundary.

Principe of measurement is based on relation between electrical conductivity of mixture and concentration of particles. Particles are assumed to be non-conductive. Directly measured quantities are voltage U and current I between electrodes. The ratio $U / I = R_{eq}$ characterises electrical resistivity of mixture in measuring volume of electrodes. If homogeneous distribution of concentration is assumed in this volume, R_{eq} depends only on the concentration, conductivity of fluid and geometry of probe. Conductivity of fluid σ_f must be measured independently by conductometer to separate its effect $R_n = R_{eq} \cdot \sigma_f$. Effect of probe geometry is constant for given probe and can be quantified by $R_{n,pf}$, i.e. value of R_n calculated from measurement with pure fluid. Then relative conductivity of mixture $\sigma_{rel} = R_{n,pf} / R_n$ depends only on local concentration c: $\sigma_{rel} = F(c)$. Several forms of function F were suggested in literature (for overview see e.g. MacTaggart et al., 1993). The relation of Rue & Tobias (1959) has a very simple form of

$$\sigma_{rel} = (1\text{-}c)^{m}. \qquad (1)$$

3 Experiment in fluidization cell

3.1 Fluidization cell

Experiment in fluidization cell is suitable way to test method of concentration measurement under control conditions (Spinewine et al., 2011). We performed set of experiments to evaluated potential of selected methods for measurement of concentration in laboratory flume. Fluidization cell 105 cm high with a rectangular cross-section 22 cm x 20 cm (Fig. 1).

System of meshes in the bottom section of cell ensured homogeneous inflow of water into the cell. After filling the cell by white plastic particles (rounded, $d_{50} = 3.14$ mm, $\rho = 2.36$ kg/m^3), flow rate thru the cell was gradually increased as long as homogeneous mixture occupying almost entire height of the cell established. This height was measured visually using a scale fixed to the cell wall. Mean concentration in mixture c_{fc} was calculated as a ratio of volume of solid fraction (calculated from known weight and density of particles) and volume occupied by the mixture. Interface between the mixture and particle-free water above the mixture was most distinct and stable at highest concentration of 46%. At lowest concentration, uncertainty in determination of interface position was approximately 6 cm, i.e. approximately 6.5% of the measured value. This is a small value in comparison with expected errors of laser stripe measurement and measurement of electrical conductivity. Experiment was performed for seven distinct concentrations between 7 and 46%. Measurement above 46% was not possible, because homogenous fluidization suddenly switched to regime with convective cells.

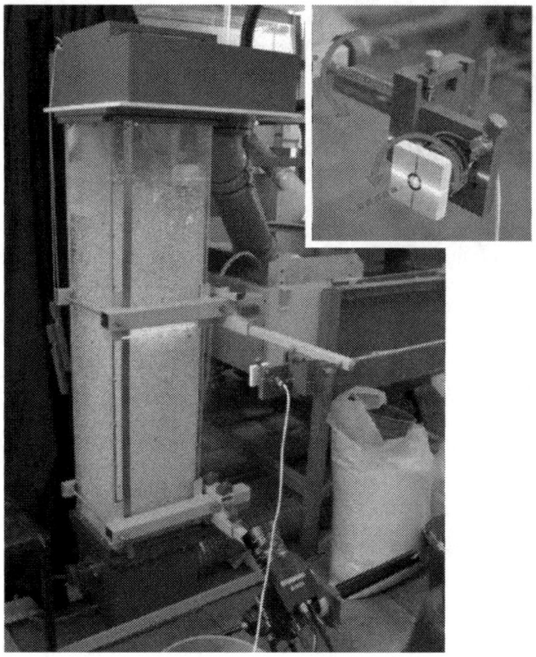

Fig. 1. Fluidization cell with camera and laser mounting. In detail: laser mounting and laser illuminated by reflected laser sheet when adjusting perpendicularity.

3.2 Laser stripe measurement

Arrangement of laser and camera enabled evaluation of horizontal concentration profile across the fluidization cell (Fig. 1). We used diode source of green laser sheet (Blau Optoelektronik FP-Mvnano-520-30-30-F, 30 mW, 520 nm). Mounting of the laser enabled adjustment of laser vertical position and inclination. Distance of the laser from the fluidization cell was 34 cm. Perpendicularity of laser sheet and fluidization cell wall was adjusted by employing a mirror attached to the wall – laser inclination was gradually varied as long as reflected laser sheet hit the laser alone (see detail in Fig. 1). High speed camera (Dantec Dynamics FlowSense EO 4M-41 with objective Carl Zeiss Planar 1.4/50 ZF) was used to acquire sequence of photos. Position of an objective of the camera was 39 cm below the laser and the camera was inclined from horizontal by an angle of about 36°.

Calibration photo was taken when the cell was filled with water but before particles were inserted. Spinewine et all (2011) used a target with the face attached to the wall. Hence, calibration points lied in the plane of wall instead of in the plane of illuminated cross-section, in which depth of laser penetration is evaluated. Then the transformation must be extrapolated to this cross-section. To reduce possible errors resulting from the extrapolation we used different arrangement with a target face normal to the wall as is shown in the figure 2. When the target was attached to the inner surface of wall, position of laser was carefully adjusted to illuminate edge of the target (the edge divided laser sheet thickness to halves). Then the laser was turned off and the calibration photo was taken. Using described arrangement, transformation is calibrated in the same plane as is that of measurement.

Maximum possible exposition time of 1.6 ms was chosen for measurement of concentration to achieve high aperture (f/11 and f/16 according actual conditions), and hence high depth of sharpness. Frame rate was set to 58 fps and sequence of 1575 photos (resolution 1168x2352) was acquired at the each concentration.

Fig. 2. Calibration target placed in fluidization cell and illuminated by laser sheet.

3.3 Measurement of local electrical conductivity

Employing numerical simulation of electrical filed generated by electrodes, configuration with pair of measuring electrodes between pair of field electrodes in one line was selected to achieve best spatial resolution (see Fig. 3). Electrodes were oriented against the flow direction to minimize disturbances introduced to the flow in front of the probe. Electrodes were made from stainless steel; the measuring pair has 1.0 mm and the field pair 1.6 mm in diameter. Characteristic dimension of measuring volume is given by distance of electrodes, which was chosen to be several particle diameters to minimize effects near particle-probe contacts and influence of particle size on relation between measured voltage and concentration.

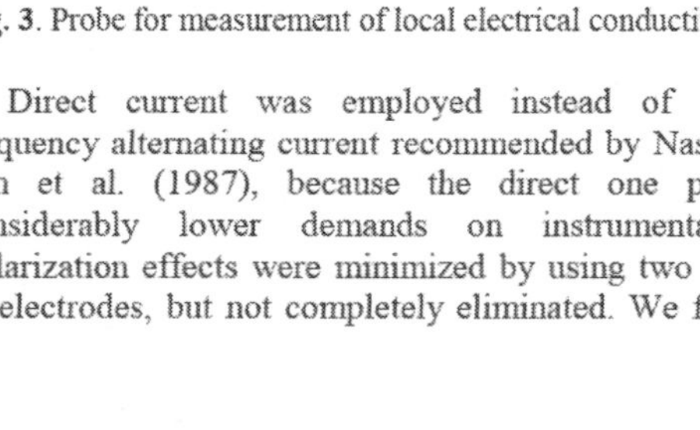

Fig. 3. Probe for measurement of local electrical conductivity.

Direct current was employed instead of high-frequency alternating current recommended by Nasr-El-Din et al. (1987), because the direct one places considerably lower demands on instrumentation. Polarization effects were minimized by using two pairs of electrodes, but not completely eliminated. We found that sufficiently accurate measurement with direct current is possible when the voltage at measuring electrodes is measured immediately after the current is supplied to field electrodes. Then the current must be turned off again. Hence, the measurement cannot be continuous. Current of 2 mA was applied on field electrodes by adjustable current source. Voltage at measuring electrodes was sensed by Analog Devices AD694 transmitter with a voltage input range of 0-2 V. Transmitter output 4-20 mA current signal was led to an A/D convertor via a shielded line. A galvanic separator was inserted to the current loop to prevent closing ground loops.

First, photos for laser stripe measurement were acquired, and then electrical conductivity has been measured at the each concentration. The probe with electrodes has been placed into the fluidization cell to the horizontal position of laser sheet and to the centre of cell cross-section. Numerical simulation of electrical filed generated by electrodes showed that distance from the wall is sufficient to prevent possible effect of wall vicinity reported by MacTaggart et al. (1993). After current was applied to electrodes and voltage was sensed in the mixture, flow through the fluidization cell was stopped and the procedure was repeated in pure fluid to obtain value of $R_{n,pf}$. Water conductivity was regularly measured by a conductometer (ExStick EC500).

4 Results and sensitivity analysis

4.1 Laser stripe measurement

Transformation of position in photo to physical coordinates was calibrated and an observation angle of $\alpha = 33.6°$ was evaluated. Spinewine at all (2011) didn't proposed any guideline how to determine threshold value B_{th} of brightness needed to filter out unwanted optical effects. We employed an optimization procedure to find the value of B_{th} producing minimal mean squared deviation between concentrations evaluated from laser stripe measurement, c_{lsm}, and concentrations in fluidization cell, c_{fc}. Spinewine at al. (2011) found out that calibrated transformation between position in photo and real position in the illuminated cross-section is usually not sufficiently accurate in the near wall region. This causes inaccurate transformation of wall position, and hence evaluated depth of laser penetration to deviates from proper value by a systematic error which has to be corrected. When we included the correction y_0 to the optimisation as an additional parameter, the mean squared deviation dropped significantly. Figure 4 shows results for the optimised values of $B_{th,opt} = 0.78$ (value of grey at grayscale between 0=black and 1=white) and $y_{0,opt} = 0.4$ mm.

To investigate effect of B_{th} on evaluated c_{lsm}, we performed sensitivity analysis, results of which are plotted in the figure 5. Each colour indicates experiment at distinct concentration. Values of c_{fc} are indicated by thin lines and the thick lines show variation of c_{lsm} with B_{th}. Correction y_0 is kept on the optimised value. Discrepancies $c_{lsm} - c_{fc}$ are indicated by numerical values and by vertical dashed lines at $B_{th} = B_{th,opt}$. All calculations were made at $y_0 = y_{0,opt}$. The three curves at lower concentrations show somewhat different course then the other. This is caused by smaller objective aperture used for photos acquired at higher concentrations. Using aperture of f/11, most of photo pixels representing illuminated practices fell close to the maximal brightness of 1 and smaller brightness can be attributed to other optical effects such as shadows and reflections. Using aperture of f/16, lower value of B_{th} is necessary to filter out unwanted optical effects. When higher than necessary B_{th} is applied, evaluated concentration changes only slightly. Finally, it starts to oscillate at very high B_{th} becase insufficient amount of relevant data passes through the filter.

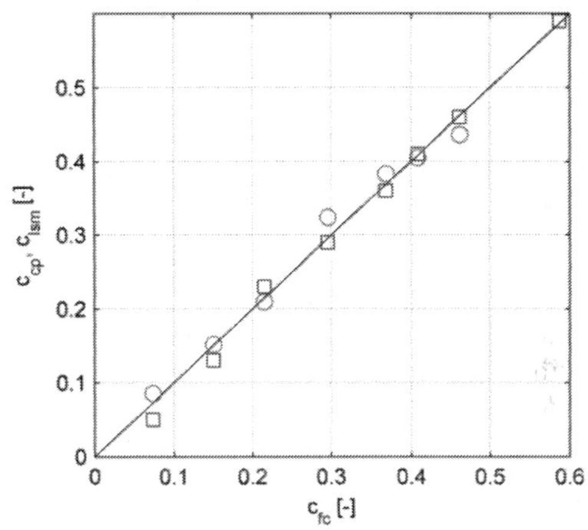

Fig. 4. Evaluated concentrations plotted against the concentrations in fluidization cell: circles – laser stripe measurements, squares – measurement of local conductivity.

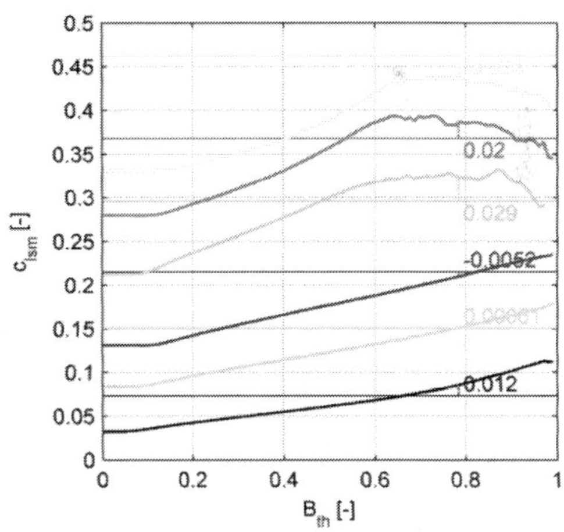

Fig. 5. Effect of threshold brightness on concentration evaluated from laser stripe measurement. Thick line – evaluated concentration, thin line – concentration in fluidization cell.

Sensitivity of results to wall position correction y_0 is shown in the figure 6. The effect is insignificant at low concentrations, but it becomes crucial as depth of laser

https://doi.org/10.1051/epjconf/201818002048

penetration decreases at high concentration. The correction y_0 is subtracted from the identified depth of laser penetration y. Hence, initial increase of c_{lsm} with y_0 is observed. However, when y_0 exceeds real position of the wall, data at position $y<y_0$ are excluded from analysis. This affects mean value of y and the trend in c_{lsm} changes.

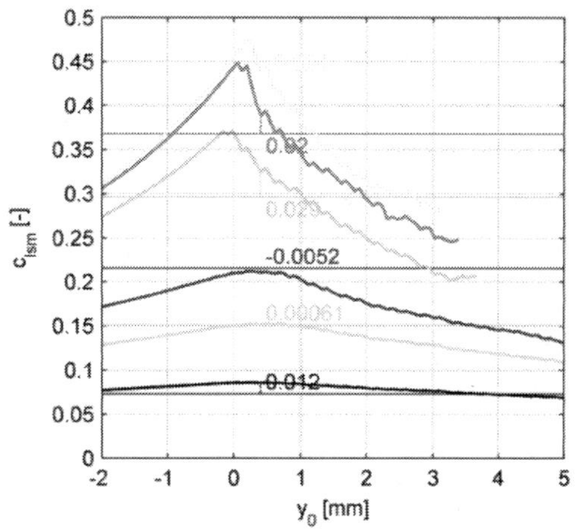

Fig. 6. Effect of wall position correction on concentration evaluated from laser stripe measurement. Thick line – evaluated concentration, thin line – concentration in fluidization cell.

Observation angle α is determined from calibrated parameters of holomorphic transformation. This procedure cases that its value is uncertain. Fortunately, figure 7 shows, that evaluated concentration is not very sensitive to errors in α around the value of 30°.

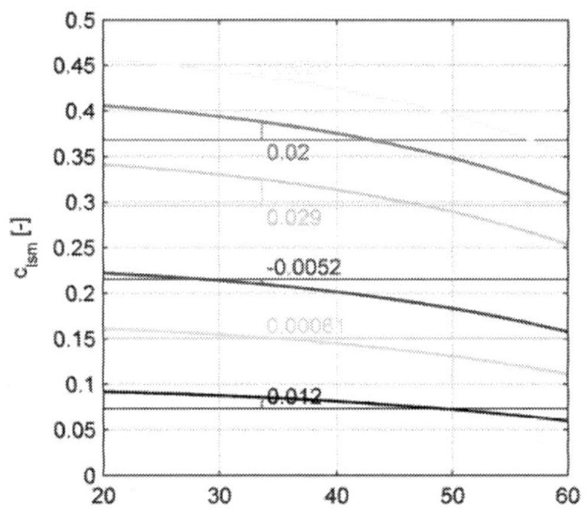

Fig. 7. Effect of observation angle on concentration evaluated from laser stripe measurement. Thick line – evaluated concentration, thin line – concentration in fluidization cell.

4.2 Measurement of local electrical conductivity

Exponent m in eq. (1) can differ from the value of 1.5 reported by Rue & Tobias (1959) for spherical particles. It can be easily evaluated from measurement under well-defined conditions of loosely poured particles – we found the value of 1.33±0.02 for rounded particles used in our experiments. Comparison of concentrations evaluated from conductivity measurement using $m = 1.33$ and mean concentration in fluidization cell is given in figure 4. The best fit is achieved when the value of $m = 1.32$ is instead of 1.33 indicating that the calibration of m is correct. Sensitivity of evaluated concentrations to m varying in the range of ±10% from the best fitting value is shown in the figure 8. It is seen that error of 10% causes

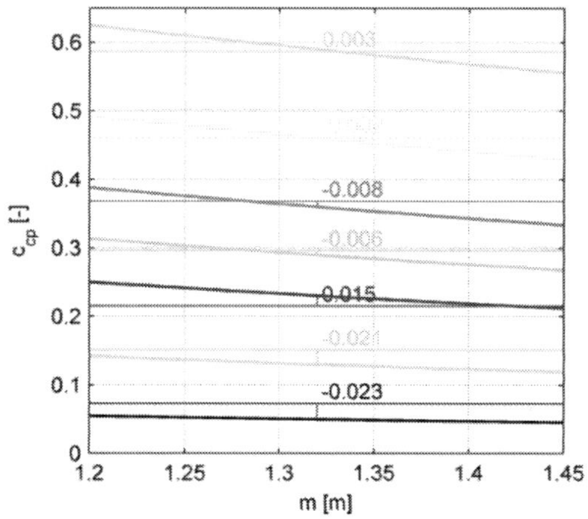

Fig. 8. Effect of exponent in equation (1) on concentration evaluated measurement of electrical conductivity. Thick line – evaluated concentration, thin line – concentration in fluidization cell.

A very good match is seen in the figure 8 between c_{cp} and c_{fc}. However, it should be noted at this point that the method was applied under the ideal conditions of homogeneous mixture and at low velocity up to 13 cm/s – the settling velocity of used particles. When employing simple procedure of data evaluation described in section 2.2, spatial resolution of the probe is expected to be insufficient for measurement of local concentrations in transport layer only several grains thick.

5 Conclusions

Both tested methods, when properly calibrated, provide accurate values of concentration of particles in homogeneous mixture prepared in fluidization cell.

The method based on measurement of mixture conductivity seems to be more accurate at higher concentrations. It is easy to use from point of view of instrumentation as well as data evaluation. However, in the form tested in this study, its application is limited only to flows with small concentrations gradients.

The method based on laser stripe measurement provides high spatial resolution, and hence it is suitable for application in laboratory flume experiments focused on flow with high concentration gradients. Sensitivity analysis showed high sensitivity of results to parameter B_{th}, for determination of which no general guideline is available. Hence, a lack of the method robustness can be considered as its disadvantage.

Acknowledgement

The research has been supported by the Czech Science Foundation through the grant project No. 16-21421S.

References

1. D. Berzi, *J. Hydraul. Eng.*, **137**(10), 1200-1207, (2011)
2. D. Berzi, L. Fraccarollo, *Phys. Fluids*, **25**, 106601, (2013)
3. H. Capart, L. Fraccarollo, *Geophys. Res. Lett.*, **38**, L20402 (2011)
4. B.M. Sumer, A. Kozakiewicz, J. Fredsøe, R. Deigaard, *J. Hydraul. Eng.* **122**, 549–558, (1996)
5. T. Revil-Baudard, J. Chauchat, D. Hurther, P.A. Barraud, *J. Fluid Mech.*, **767** 1-30, (2015)
6. H. Nasr-el-din, C.A. Shook, M.N. Esmail, Can. J. *Chem. Eng.*, **62**, 179–185, (1984)
7. L. Graham, R. Hamilton, R. Rudman, P. Strode, L. Pullum, *Proceedings Hydrotransport 15. Banff, Canada*, 149–158, (2002)
8. R.G. Gillies, *PhD thesis, University of Saskatchewan, Saskatoon, Canada,* (1993)
9. R.E. De La Rue, , C.W. Tobias, , *J. Electrochem. Soc.*, **106**, 827-833, (1959)
10. R. S. MacTaggart, H. A. Nasr-El-Din, J. H. Masliyah, *Sep. Tech.*, **3**(3), 151-160, (1993)
11. H. Nasr-El-Din, C. A. Shook, J. Colwell,. *Int. J. Multiphase Flow*, **13**(3), 365-378, (1987)
12. K. H. Ong, M. S. Beck, *Measurement and Control*, **8**(11), 453-463, (1975)
13. J. S. Ribberink, A. A. Al-Salem, *Coastal Engineering*, **25**(3-4), 205-225, (1995)
14. B. Spinewine, H. Capart, L. Fraccarollo, M. Larcer, *Exp. Fluids* **50**, 6 (2011)
15. H. Capart, D.L. Young, Y. Zech, *Exp. Fluids* **32**, 2 (2002)

Bulk viscosity of water in acoustic modal analysis and experiment

Jan Kůrečka[1,⋆], *Vladimír* Habán[1,⋆⋆], and *Daniel* Himr[1,⋆⋆⋆]

[1] *Victor Kaplan Department of Fluid Engineering, Brno University of Technology, Technická 2896/2, Brno, 616 69, Czech Republic*

Abstract. Bulk viscosity is an important factor in the damping properties of fluid systems and exhibits frequency dependent behaviour. A comparison between modal analysis in ANSYS Acoustics, custom code and experimental data is presented in this paper. The measured system consists of closed ended water-filled steel pipes of different lengths. The influence of a pipe wall, flanges on both ends and longitudinal waves in the structural part were included in measurement evaluation. Therefore, the obtained values of sound speed and bulk viscosity are parameters of the fluid. A numerical simulation was carried out only using fluid volume in a range of bulk viscosity. Damping characteristics in this range were compared to measured values. The results show a significant influence of sound speed and subsequently, the use of sound speed value regressed from experimental data yields a better fit between the measurement and the computation.

1 Introduction

Bulk viscosity is a parameter describing dissipation of energy in a fluid caused by compression and dilatation and can be defined as a resistance of fluid against change of volume. It is different from shear viscosity describing energy loss of flowing liquid. More information from studies focused on determination of bulk viscosity of different fluids is summarized in [2]. Previous studies show that bulk viscosity is frequency dependent[3].

Finite element method (FEM) acoustic analysis can be used to predict sound propagation, vibroacoustic problems, response of fluid system to excitation and other applications related to fluid acoustic [4], [5], [6]. Acoustic simulation applied to a design of a muffler, design of a damper of pressure pulsations and modification of industrial fan after failure are described in mentioned references. ANSYS acoustic simulation software is solving Navier-Stokes and continuity equations linearized and modified into wave equation. Analogically to degrees of freedom in mechanical solver the acoustic simulation solves degrees of freedom of acoustic pressure and velocity [7]. Aim of this contribution is to discuss how commonly used acoustic analysis software treats bulk viscosity and compare simulation to experiment and theory used in evaluation process.

2 Experiment description

The experimental system consists of pipe equipped with four sensors, two accelerometers and two pressure sensors on each end of pipe as depicted in Fig. 1. The pipe is at one end connected to a pump and there is a ball valve in between that is used to close the pipe after pressurization. At the opposite end of pipe the pipe is connected to a valve used to remove trapped air from system.

The pipe was excited by hitting it with a hammer and pressure oscillations and accelerations were measured. Real and imaginary part of eigen value was determined by FFT. Real part was taken as width of resonance range at half amplitude.

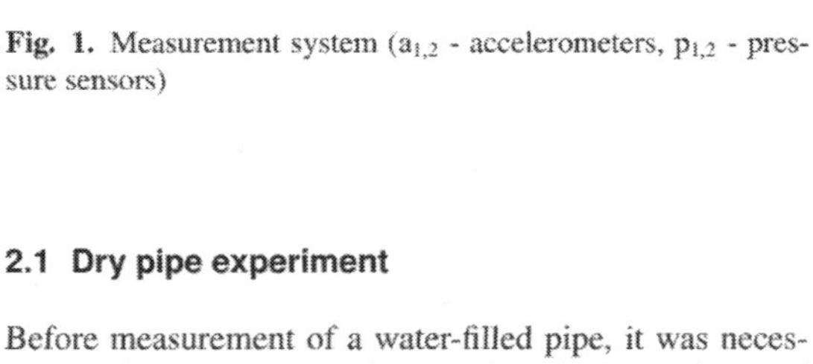

Fig. 1. Measurement system ($a_{1,2}$ - accelerometers, $p_{1,2}$ - pressure sensors)

2.1 Dry pipe experiment

Before measurement of a water-filled pipe, it was necessary to determine material properties of structural part of the pipe. For this reason, experiment with an empty tube was performed. In the following part, theory behind evaluation of a measured data is described.

⋆e-mail: 133644@vutbr.cz
⋆⋆e-mail: haban@fme.vutbr.cz
⋆⋆⋆e-mail: himr@fme.vutbr.cz

2.2 Transfer matrix and material properties of dry pipe

There is a rod (assuming the situation is same for a pipe) from elastic material. Its length dimension is significantly bigger than its diameter, its cross-section is constant, cross-sections perpendicular to the rod axis remain planar even when the rod is deformed and the stress is equally distributed in a cross-section. Moment equation of a rod element can be written as:

$$S_p \cdot \rho \frac{\partial^2 x_1}{\partial t^2} \cdot dx = S_p \cdot \frac{\partial \sigma}{\partial x} \cdot dx \tag{1}$$

It can be adapted into form describing a longitudal rod displacement (index 1 means direction along the length of the pipe or axial direction).

$$\rho \frac{\partial^2 x_1}{\partial t^2} = \frac{\partial \sigma}{\partial x} \tag{2}$$

Subsequently the stress σ can be expressed using Voigt-Kelvin material model.

$$\sigma = E_1 \cdot \frac{\partial x_1}{\partial x} + b_1 \cdot \frac{\partial^2 x_1}{\partial x \cdot \partial t} \tag{3}$$

Combining Eq. (3) and Eq. (2) and minor modifications:

$$\rho \cdot \frac{\partial^2 x_1}{\partial t^2} = E_1 \cdot \frac{\partial^2 x_1}{\partial x^2} + b_1 \cdot \frac{\partial^3 x_1}{\partial x^2 \cdot \partial t} \tag{4}$$

Laplace transform of this equation and separation of differential on the right hand side:

$$s^2 \cdot \tilde{x}_1 \cdot \frac{\rho}{(E_1 + b_1 \cdot s)} = \frac{\partial^2 \tilde{x}_1}{\partial x^2} \tag{5}$$

Using speed of sound c_0 formulated in Eq. (6) the moment equation can be further modified.

$$\frac{1}{c_0^2} = \sqrt{\frac{\rho}{E_1} + b_1 \cdot s} \tag{6}$$

$$\frac{\partial^2 \tilde{x}_1}{\partial x^2} = \frac{1}{c_0^2} \cdot s^2 \cdot \tilde{x}_1 \tag{7}$$

Steady vibrations in axial direction formulated as a function:

$$\tilde{x}_1 = U(x) \cdot e^{i \cdot s \cdot t} \tag{8}$$

Voigt-Kelvin model after Laplace transform:

$$\tilde{F} = \tilde{S}_p \cdot (E_1 + b_1 \cdot s) \cdot \frac{\partial \tilde{x}_1}{\partial x} \tag{9}$$

Solution of momentum equation with respect to steady vibrations formulation in Eq. (8) is:

$$\tilde{x}_1(x) = C_1 \cdot \cosh\left(\frac{s}{c_0} \cdot x\right) - C_2 \cdot \sinh\left(\frac{s}{c_0} \cdot x\right) \tag{10}$$

Derivation of this equation gives following formulation:

$$\frac{\partial \tilde{x}_1}{\partial x} = C_2 \frac{s}{c_0} \cosh\left(\frac{s}{c_0} \cdot x\right) - C_1 \frac{s}{c_0} \sinh\left(\frac{s}{c_0} \cdot x\right) \tag{11}$$

Constants C_1 and C_2 can be determined from the boundary condition at x = 0 along with known Laplace images of force $\tilde{F}$ and displacement $\tilde{x}_1$:

$$\tilde{x}_1(x = 0) = C_1 \cdot \cosh(0) + C_2 \cdot \sinh(0) \tag{12}$$

Therefore, a Laplace image of displacement at start of pipe equals constant C_1:

$$\tilde{x}_1(x = 0) = C_1 \tag{13}$$

To obtain constant C_2 equations describing V-K model after Laplace transformation Eq. (9) and derivation of momentum equation Eq. (11) can be used.

$$\tilde{F}(x = 0) = \tilde{S}_p \cdot (E_1 + b_1 \cdot s) \cdot \frac{s}{c_0} \cdot C_2 \tag{14}$$

Then second constant can be expressed in a following formulation:

$$C_2 = \frac{\tilde{F}}{\tilde{S}_p \cdot (E_1 + b_1 \cdot s) \cdot \frac{s}{c_0}} \tag{15}$$

Finally, the matrix formulation can be written using the transform matrix $\mathbf{P}_T$, describing vector $\mathbf{u}$ at any length of pipe.

$$\mathbf{u}(x, s) = \mathbf{P}_T(x, s) \cdot \mathbf{u}(0, s) \tag{16}$$

$$\mathbf{u} = \begin{pmatrix} \tilde{x}_1 \\ \tilde{F} \end{pmatrix} \tag{17}$$

$$\mathbf{P}_T = \begin{pmatrix} \cosh\left(\frac{s}{c_0} \cdot x\right) & P_{T12} \\ P_{T21} & \cosh\left(\frac{s}{c_0} \cdot x\right) \end{pmatrix} \tag{18}$$

Elements P_{T12} and P_{T21} are formulated in following equations.

$$P_{T12} = \frac{1}{\tilde{S}_p \cdot (E_1 + b_1 \cdot s) \cdot \frac{s}{c_0}} \cdot \sinh\left(\frac{s}{c_0} \cdot x\right) \tag{19}$$

$$P_{T21} = \tilde{S}_p \cdot (E_1 + b_1 \cdot s) \cdot \frac{s}{c_0} \cdot \sinh\left(\frac{s}{c_0} \cdot x\right) \tag{20}$$

Transfer matrices $\mathbf{P}_{10}$ and $\mathbf{P}_{23}$ were used to account for weight of flanges, connected hoses and valves on both ends (index 0 means start of pipe, index 3 end of pipe, 1 is position right after start of pipe, 2 analogical to 1).

$$\mathbf{P}_{10} = \mathbf{P}_{23} = \begin{pmatrix} 1 & 0 \\ M \cdot s^2 & 1 \end{pmatrix} \tag{21}$$

Equation in matrix form using above mentioned transfer matrices:

$$\begin{pmatrix} \tilde{x}_{1_3} \\ \tilde{F}_3 \end{pmatrix} = \mathbf{P}_{10} \cdot \mathbf{P}_T \cdot \mathbf{P}_{23} \begin{pmatrix} \tilde{x}_{1_0} \\ \tilde{F}_0 \end{pmatrix} \tag{22}$$

With boundary conditions, Eq.(23), (24) and a few modifications, the final equation Eq.(25) used to obtain material properties of the tube (E_l and b_l) is:

$$S_0 = 0 \tag{23}$$

$$S_3 = 0 \tag{24}$$

$$\begin{pmatrix} 0 \\ 0 \\ 0 \\ 0 \end{pmatrix} = \begin{pmatrix} 0 & 1 & 0 & 0 \\ 0 & 0 & 0 & 1 \\ P_{Tf31} & P_{T12} & -1 & 0 \\ P_{Tf41} & P_{Tf42} & 0 & -1 \end{pmatrix} \cdot \begin{pmatrix} \tilde{x}_{1_0} \\ \tilde{F}_0 \\ \tilde{x}_{1_3} \\ \tilde{F}_3 \end{pmatrix} \tag{25}$$

And elements of this matrix are written in Eq. 26, 27 and 28.

$$P_{Tf31} = P_{T11} + P_{T12} \cdot M \cdot s^2 \tag{26}$$

$$P_{Tf41} = P_{T21} + M \cdot s^2 \cdot (P_{T12} \cdot M \cdot s^2 + 2 \cdot P_{T11}) \tag{27}$$

$$P_{Tf42} = P_{T22} + P_{T12} \cdot M \cdot s^2 \tag{28}$$

Values of Young modulus (E_l) and material damping (b_l) of the pipe can be obtained by making determinant of transfer matrix equal to zero, with s (eigen number) chosen from measurement data.

2.3 Water-filled pipe experiment

Transfer matrix was formulated in a same way as in previous chapter. The process of formulation exceeds the scope of this contribution and as such is not covered. Navier-Stokes and continuity equations were applied so the final form allows us to enumerate values of bulk viscosity μ^B and speed of sound c_0 for water inside pipe. Material properties obtained by a manner covered in previous chapter were used so the Young modulus and damping of pipe are separate properties of structural part and the evaluated bulk viscosity and speed of sound are properties of fluid for a given simulated instance.

$$\begin{pmatrix} 0 \\ 0 \\ 0 \\ 0 \end{pmatrix} = \mathbf{P}_{Twater} \cdot \begin{pmatrix} \tilde{Q}_0 \\ \tilde{p}_0 \\ \tilde{Q}_3 \\ \tilde{p}_3 \end{pmatrix} \tag{29}$$

Table 1. Mesh size independence test

No. of elements	Frequency [Hz]	Damping parameter [Hz]
1274	691,45	-5,2292
200	691,45	-5,2292
80	691,45	-5,2292
30	691,45	-5,2293
12	691,62	-5,2319
6	694,04	-5,2686

3 Simulation

The computed model was only water-filled part of a pipe. Simulation was performed using ANSYS Mechanical program with Acoustics extension. This extension makes acoustic commands accessible in GUI of ANSYS FEM software. According to [7], there should be at least twelve linear elements of mesh per wavelength or six quadratic ones. Number of elements in a cross section of modelled cylinder is not as important as number of elements along the length of pipe, since the modal solution is longitudinal half wave.

The mesh size dependence analysis shows that few hundred quadratic elements is more than sufficient for accurate results. Table 1 contains values of simulation one-meter-long pipe meshed with quadratic elements HEX20.

Boundary conditions on the cylindrical surface as well as on the circular top and bottom is rigid wall. The rigid wall is default boundary condition in ANSYS Acoustics and means that acoustic velocity on a selected geometry is zero. In other words, acoustic velocity does not fluctuate and stays constant. Properties of fluid domain were density and viscosity (shear viscosity or volume viscosity), these parameters were same for all simulations. Speed of sound was regressed from an experimental data. The bulk viscosity varied in a range of measured bulk viscosities to encompass needed amount of values and compare them to the experiment.

4 Custom code

Software based on TMM (transfer matrix method) was used to determine if there is any difference in implementation of the bulk viscosity in ANSYS Acoustics and in custom code named FAchar. In Introduction to Acoustic [7], ANSYS states momentum equation with bulk viscosity μ^B Eq. (30).

$$\rho \frac{\partial v_{a_i}}{\partial t} = -\nabla p_a + (\frac{4}{3}\mu + \mu^B) \cdot \nabla(\nabla \cdot v_{a_i}) \tag{30}$$

And FAchar's bulk viscosity implementation in momentum equation is written as Eq. (31):

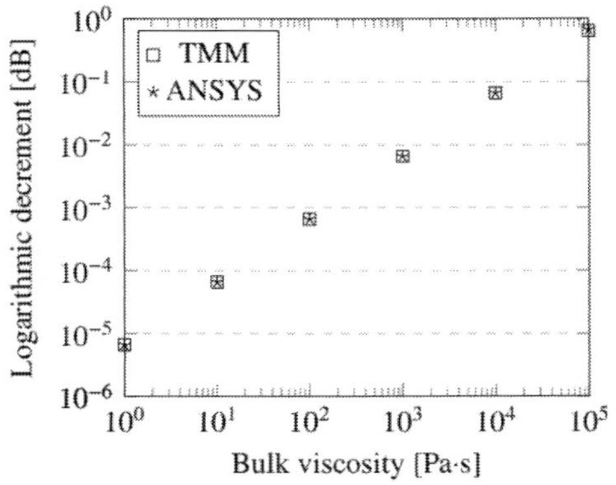

Fig. 2. ANSYS and custom code comparison

Table 2. ANSYS and custom code percental differences (ANSYS minus FAchar and divided by custom code value as a base)

bulk v. [Pa·s]	freq. diff. [%]	log. dec. diff. [%]
1	0,00000017	0,138667
10	0,00000017	0,013868
100	0,00000017	0,001387
1000	0,00000017	0,000085
10000	0,00000018	-0,00547
100000	0,006218615	-0,56218

$$\rho\frac{\partial v_i}{\partial t} = -\nabla p + \nabla\Pi_{ij} \quad (31)$$

$$\Pi_{ij} = \mu\frac{1}{2}(\frac{\partial v_i}{\partial x_j} + \frac{\partial v_j}{\partial x_i}) + \mu^B\frac{\partial x_i \cdot \partial v_k}{\partial x_j \cdot \partial x_k} \quad (32)$$

The computation was performed for both cases on a model of one-meter-long pipe with diameter of 50 mm. Speed of sound was 1500 Pa·s, density 1000 kg·m^{-3} and bulk viscosities were six values from 1 to 100 000 m^2·s^{-1}.

It can be concluded from Figure 2 and Table 2 that with an exception at highest and lowest bulk viscosities the differences in data are very small so author concludes the implementations are sufficiently even and variations come from numeric methods used, discretization and precision.

5 Results

The results compared between the experiment and the simulation are damped frequency and logarithmic decrement of damping. Damped frequency values differ by a few percent, frequency obtained from the experiment and numerical model is close to the theoretical frequency value. Analytical frequency can be calculated for a wave twice as long as a pipe. In Table 3 damped frequencies from experiment and simulation are compared to analytical value calculated for given speed of sound. Bigger difference of experimental values is expected result because there are more factors involved and those are omitted in simulation.

Table 3. Damped frequency comparison; pipe length 1 m, diameter 23 mm

Sound speed [m·s^{-1}]	Frequency [Hz]	Analytic freq. [Hz]
Experiment		
1381,381578	695,418	690,691
1383,582975	696,510	691,791
1383,567324	696,510	691,784
1383,572992	696,510	691,786
1383,581693	696,510	691,791
1383,58942	696,510	691,795
Acoustic simulation		
1387,84	693,905	693,920
1386,085	693,024	693,043
1384,33	692,143	692,165
1382,575	691,261	691,288
1380,82	690,378	690,410

Logarithmic decrements show very similar trends in relation to bulk viscosity for all pipes but the differences are significantly bigger and up to ten percent. This is documented in Fig. 3, 4 and 5.

6 Discussion

Discrepancy in measurement and simulation data could be attributed to various causes. As already stated, in computation there is omission of pipe structure, so simulation has no influence of wall thickness, mass of flanges and measuring apparatus or damping in pipe material. In experiment a bending of pipe occurs and that is not implemented in simulation as well. Also, even small amount of air trapped in pipe can alter measured data [1]. Dissipation of energy caused by vibrations of pipe in air is another neglected effect.

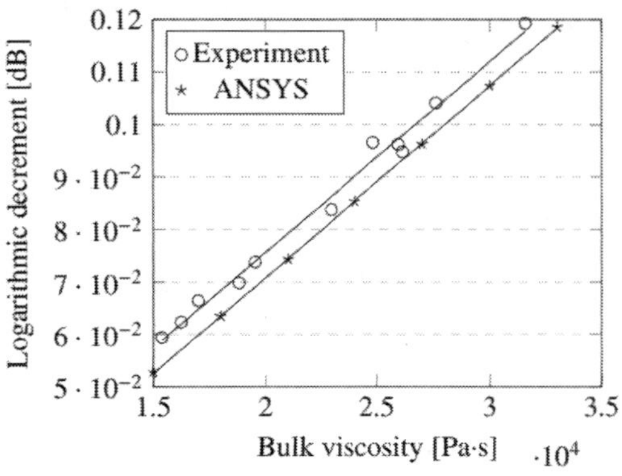

Fig. 3. ANSYS and experiment; pipe length 2 m; diameter 32 mm

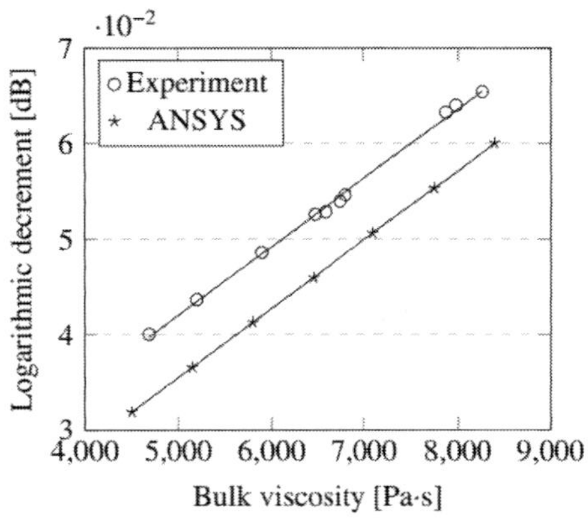

Fig. 4. ANSYS and experiment; pipe length 1 m; diameter 32 mm

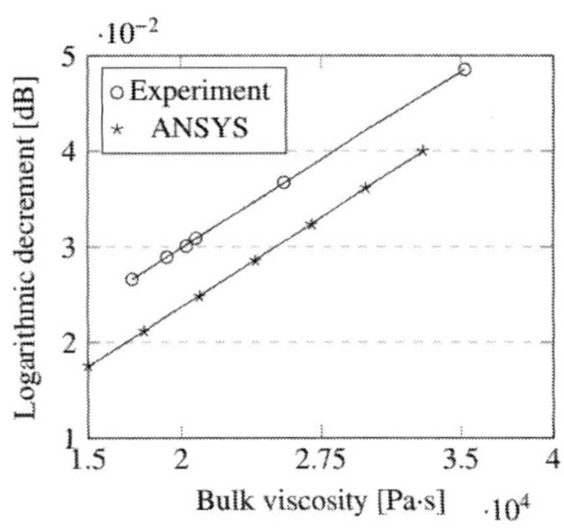

Fig. 5. ANSYS and experiment; pipe length 6 m; diameter 42,5 mm

7 Conclusion

Paper has described measurement of bulk viscosity and part of the mathematical model needed to evaluate data from obtained values of accelerations and pressure pulsations. Acquired bulk viscosities and eigen numbers were used with simulation. Ability to simulate model with bulk viscosity in ANSYS Acoustic and custom code is presented and results show conformity to measured values. Experiment and simulation vary, but given the simplification of computed model, it can still be considered good fit. Overall conclusion to be made is that acoustic simulation with bulk viscosity can yield results with acceptable precision.

Further work could focus on Fluid Structure interaction to take structural part into account and improve fidelity of modeled hydraulic system.

Acknowledgements

This work was supported by NETME Centre, regional R&D centre built with the financial support from the Operational Programme Research and Development for Innovations within the project NETME Centre (New Technologies for Mechanical Engineering), Reg.no. CZ.1.05/2.1.00/ 01.0002 and, in the follow-up sustainability stage, supported through NETME CENTRE PLUS (LO1202) by financial means from the Ministry of Education, Youth and Sports under the "National Sustainability Programme I" and from the 2017 Science fund of Faculty of Mechanical engineering, Brno University of Technology project id.: FV 17-25.

Nomenclature

b_1	[N s m^{-1}]	damping of a pipe
c_0	[m s^{-1}]	speed of sound
E_1	[Pa]	stiffness constant of a pipe
F	[N]	force
M	[kg]	mass
p	[Pa]	pressure
p_a	[Pa]	acoustic pressure, amplitude of fluctuation
$\mathbf{P}$		transfer matrix
Q	[m^3 s^{-1}]	discharge
s	[s^{-1}]	parameter of Laplace transform
S	[m^2]	area
S_p	[m^2]	area of cross-section
$\mathbf{u}$		state vector
v	[m s^{-1}]	velocity
v_a	[m s^{-1}]	acoustic velocity
x	[m]	dimension coordinate
ρ	[kg m^{-3}]	density
σ	[Pa]	stress
μ	[Pa s]	viscosity
μ^B	[Pa s]	bulk viscosity
$\tilde{}$		variable after Laplace transform

References

1. HIMR D., HABÁN V.. Simulation of low pressure water hammer. IOP Conference Series: Earth and Environmental Science. 2010, 12, 012087-. DOI: 10.1088/1755-1315/12/1/012087. ISSN 1755-1315.
2. GRAVES R. E., ARGROW. B. M.. Bulk Viscosity: Past to Present. *Journal of Thermophysics and Heat Transfer*. 1999, 13(3), 337-342. DOI: 10.2514/2.6443. ISSN 0887-8722.
3. POCHYLÝ F., HABÁN V., FIALOVÁ S.. Bulk viscosity – constitutive equations. *International Review of Mechanical Engineering*, September 2011, Vol. 5 n. 6, pp. 1043-1051. ISSN: 1970-8734.
4. KORECK J., MAESS M., GAUL L.. Acoustic-Structure Simulation of Passive Fluid Pulsation Dampers, Fortschritte der Akustik: Plenarvorträge und Fachbeiträge der 33. Deutschen Jahrestagung für Akustik, DAGA 2007, Stuttgart. Berlin: DEGA, 2007. ISBN: 978-3-9808659-3-7.
5. EISINGER F. L., SULLIVAN R. E.. Vibration Fatigue of Centrifugal Fan Impeller Due to Structural-Acoustic Coupling and Its Prevention: A Case Study. *Journal of Pressure Vessel Technology.* 2007, 129(4), 771-. DOI: 10.1115/1.2767371. ISSN 00949930.
6. JONES P., KESSISSOGLOU N.. (2009). An evaluation of current commercial acoustic FEA software for modelling small complex muffler geometries: Prediction vs experiment. Annual Conference of the Australian Acoustical Society 2009 - Acoustics 2009: Research to Consulting.
7. ANSYS Inc., Introduction to acoustics, ANSYS Acoustics Lectures, 2016

Main rotor-body action for virtual blades model

Alexander Kusyumov[1,a], Sergey Kusyumov[2], Sergey Mikhailov[3], Elena Romanova[4], Konstantin Phayzullin[5], Evgeny Lopatin[6] and G. Barakos[7]

[1-6]Kazan National Research Technical University n.a. A. N. Tupolev (KNRTU-KAI), Kazan, Russia
[7]University of Glasgow, Glasgow, UK

Abstract. This research aims to investigate a virtual blade model and assess rotor influence on helicopter fuselage aerodynamics. The rotor disk is discretized in the azimuthal direction, and a time-varied pressure jump is applied in regions occupied by the blades. To obtain the pressure jump, an actuator disk is employed using uniform and non-uniform blade load distribution, based on momentum theory.

1 Introduction

Rotor Computational Fluid Dynamics (CFD) calculations are challenging. The complexity of the problem is due to the flow unsteadiness, the coupled aerodynamics and aeroelasticity of rotor blades, and the presence of vortical effects and wakes characterized by a range of flow scales, both laminar and turbulent. One can also include the large variety of flight conditions. Despite these difficulties, CFD methods now are becoming mature enough to simulate the flow characteristics around helicopters [1, 2].

Computations with detailed resolution of the helicopter blades are still expensive. For this reason, actuator models of rotors and propellers have their place in the tool-chain of engineering methods available to helicopter designers. An advantage of the actuator disk model is its simplicity, efficiency and relatively good accuracy. Classic actuator disk models allow the use of steady-state CFD computation, significantly reducing the required computer time and memory. The actuator disk approximates the rotor using an infinitely thin source of momentum. This is modeled as a pressure jump across the disk.

The momentum source is modeled independently of the fuselage without any coupling between the two. For this reason, the effect of the rotor disk is simplified and the method, although computationally efficient, can only be used for initial estimates of the fuselage loads [3, 4]. The other peculiarity of this formulation is the steady-state formulation that allows for time averaged estimates of rotor body interactions. More advanced concepts of the actuator-disk methods are the regionalized actuator disk models considered in [5] for flows around propellers.

In reality, the effect of blades on the fuselage is unsteady. For this reason the vortical structure of the flow around a fuselage in rotor-body interaction is different than what is predicted by steady-state actuator-disk methods. This problem can be partly solved using the unsteady actuator disk model considered in [5] for flows around propellers. The propeller disk surface is divided along azimuthally direction, and time-varied pressure jump and tangential velocity in some regions are specified. The flow parameters are updated with respect to the azimuth angle at each time step to simulate the unsteady flow effect induced by the rotating blades.

Corresponding author: postbox7@mail.ru

The aim of the present paper is to simulate the flow around a helicopter rotor/fuselage case in forward flight mode using the Virtual Blade Actuator (VBA) model instead of resolved blades. To obtain the pressure jump for the mean rotor thrust the uniform actuator disk was modeled using uniform and non-uniform blade load distribution, based on momentum theory.

Rotor action changes the aerodynamic performance of fuselage and the properties of the flow (including turbulence modeling parameters) around the helicopter. The parameters of turbulence determine the broadband noise of the flow [6].

2 Computational Case

A CAD model of a helicopter fuselage corresponding to an early design of the ANSAT-P model of the modern, light-medium ANSAT helicopter (produced by the JSC Kazan Helicopters) is used. The shape of the ANSAT-P model is similar to the prototype of the ANSAT helicopter (Figure 1) used for preliminary studies in the T-1K wind tunnel of Kazan National Research Technical University named after A.N. Tupolev (KNRTU-KAI). The aerodynamic performance for this model was first obtained for a "blind" CFD modeling exercise [7].

Some geometric parameters of the model are presented in Table 1.

Figure 2 shows the CAD model of the fuselage with the actuator disk and the virtual blades. In this paper a sector of the virtual blades is used to simplify the computational procedure (the area of the virtual blades corresponds to the planform area of the resolved blades).

https://doi.org/10.1051/epjconf/201818002050

The location of virtual blades varies in time according to the location of resolved blades.

Table 1. CAD model parameters

Parameter	Value
Number of blades N	4
Radius of rotor R, m	0.575
Chord c, m	0.032
Root radius R_0, m	0.144
Length of fuselage L_F, m	1.1

Fig. 1. ANSAT-P fuselage model in the test section

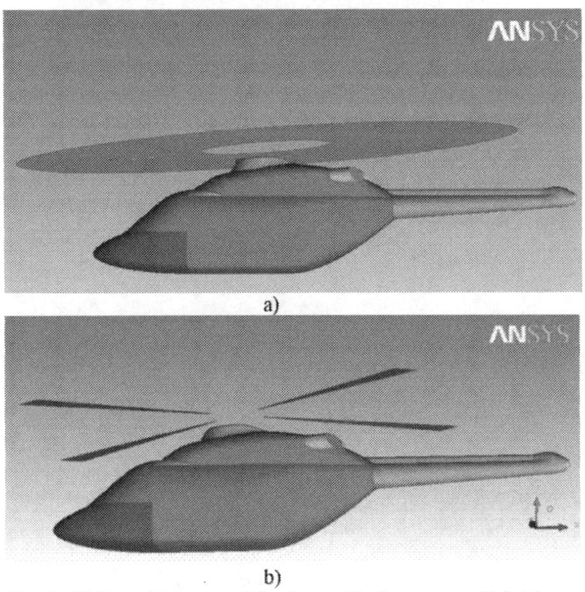

a)

b)

Fig. 2. CAD model image of fuselage with the actuator disk (a) and the virtual blades (b)

3 Numerical approach

The simulation of the flow around the isolated fuselage ANSAT-P was conducted using the FLUENT CFD code. For modeling an unstructured tetra-grid with 4.800.000 elements was used that was built using the ANSYS ICEM program. The grid in the immediate vicinity of the body was designed to resolve the flow boundary layer and was constructed using about 20 prismatic layers.

The value of $y+$ for all turbulence models was up to 1. At least 15 elements were kept inside the boundary layer for sufficient resolution. The mesh had the first cell height of 0.005 mm above the wall. Other parameters of grid and the grid sensitivity question were considered in reference [7].

Results using the tetra-mesh, were compared against the the HMB code of Liverpool University that employed a multi-block hexa-mesh (11.000.000 elements).

Table 2 presents the drag (C_D) and lift (C_L) obtained using CFD approach for pith angle α=0 degrees in comparison to experimental data. For the numerical simulation, the k-ω and the Spalart-Allmaras (SA) models of turbulence were used with Fluent and k-ω model was used with HMB. The conditions of the wind tunnel experiment and CFD modelling corresponded to the free stream Mach number and the Reynolds numbers were of 0.1 and $4.4\cdot 10^6$, respectively.

Table 2. CFD results

Computation mode	CFD		Experiment	
	C_D	C_L	C_D	C_L
Fluent, SA	0.101	-0.07	0.09	-0.07
Fluent, k-ω	0.102	-0.076		
HMB, k-ω	0.095	-0.075		

As the VBA model of actuator requires a transient formulation the Spalart-Allmaras turbulence model was chosen for computations to minimize time of simulation.

4 Flow conditions

The actuator disk approximates the rotor using an infinitely thin source of momentum. This is modeled as a pressure jump across the disk. The momentum source is modeled independently of the fuselage without any feedback between the two.

To obtain the pressure jump for the mean rotor thrust the uniform actuator disk was modeled proceeding from the thrust force T (corresponding to the weight of helicopter). The constant pressure jump was applied to the grid points of the actuator disk, inside a radius R of 0.574 m (Figure 2), with a disk tilt of α_a=4.5 degrees. Computations were performed for free stream velocity V_∞ of 53 m/s, angular frequency of rotation $\omega=2\pi f$ (f=50 Hz). For this condition the C_T/μ^2 was 0.1744 (Δp=300 Pa). Here $C_T=2T/(\rho\pi R^2 V_t^2)$ is the trust coefficient, $\mu=V_\infty/V_t$ is the advanced ratio, $V_t=\omega R$ is the rotor tip velocity, ρ is the air density.

For a non-uniform disk a "typical" distribution of the pressure was estimated based on references [8, 9]. The distribution of the pressure jump on the disk surface is given by:

$$\Delta p = \rho\gamma[\gamma\,\mathrm{sign}(\delta)/2 + V_\infty\cos(\alpha-\alpha_a+\delta)]$$

(the same approach was used in [10]). Here δ is the angle of vortex cylinder slope (a small value for cruise velocity conditions), α_a is the angle of actuator disk tilt. A general expression for the function γ of circulation distribution on the disk surface can be written as

$$\gamma(r,\psi)=\gamma_r(r)+\gamma_s(r)\sin(\psi),$$

where ψ is azimuth angle of considered point on disk surface. One can note here that rotor thrust value is determined basically by the $\gamma_r(r)$ function. The $\gamma_s(r)$ determines inequality of circulation distribution for advancing and retreating blades and for the present research is accepted $\gamma_s(r)\equiv 0$ (according to the recommendation of reference [8]).

The distribution of the averaged blade load γ_r can be written in the form

$$\gamma_r = Cf_r(r_n),$$

where $f_r(r_n)=r_n^2\left(2-r_n^2-r_n^4\right)$, $r_n=r/R$ is normalized radius. The value of C is determined by the formula

$$C=1.989\,V_\infty\left[-\cos(\alpha-\alpha_a+\delta)+\right.$$

$$\left.+\sqrt{\cos^2(\alpha-\alpha_a+\delta)+1.207\,C_T/\mu^2}\right].$$

Eventually in this work it is assumed that the pressure distribution depends on the local disk radius r only:

$$\Delta p_0=\rho\left[\gamma_r^2/2+\gamma_r V_\infty\cos(\alpha-\alpha_a+\delta)\right].$$

The uniform and non-uniform disk was implemented in the HMB solver and in Fluent using user-defined functions. Figure 3 presents the uniform and non-uniform actuator disk pressure distributions vs normalized rotor disk radius for the considered conditions.

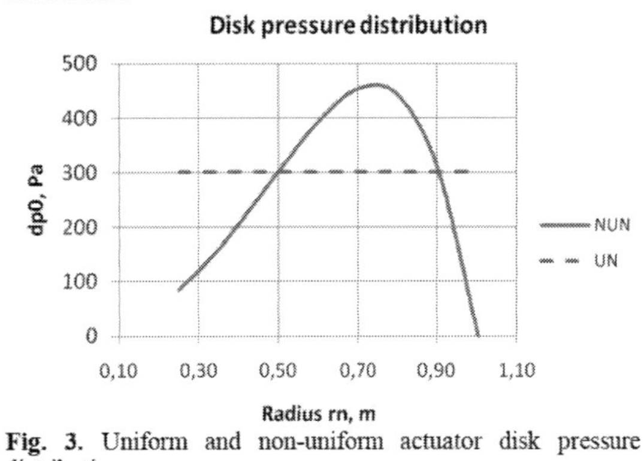

Fig. 3. Uniform and non-uniform actuator disk pressure distributions

5 Results for classical actuator disk

The pressure coefficient $Cp=2(p-p_\infty)/(\rho V_\infty^2)$ distribution on the fuselage surface (p is a local pressure, p_∞ is the free stream pressure) without actuator is shown in Figure 4a (for visualization the ANSYS CFD-Post code is used). Figures 4b and 4c show the difference $dCp=Cp-Cp_a$ for the uniform and nonuniform models (C_{pa} is the pressure coefficient with the actuator).

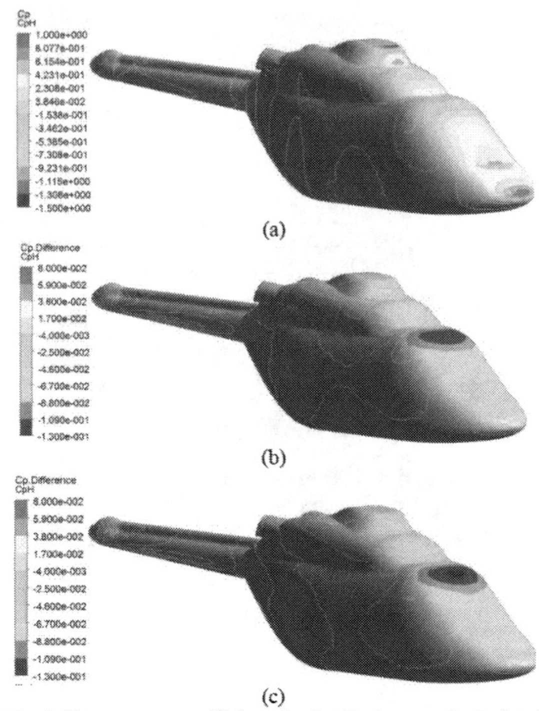

Fig 4. The pressure coefficient *Cp* distributions on the isolated fuselage surface (a), difference *dCp* with uniform (b) and nonuniform (c) actuators

From Figure 4 it follows that the presence of the actuator disk does not lead to strong redistribution of the pressure coefficient for the considered conditions (relatively high value of the advanced ratio coefficient μ=0.29). Some discrepancy can be observed at the front part of fuselage.

The steady-state formulation of the CFD computations allows for a simple approach (based on Proudman's formula [6]) to estimate acoustic emissions of the turbulent flow, assuming that the emitted noise is continuously distributed over a broad range of frequencies. In this case the broadband noise power, can be estimated from the RANS equations using the mean flow field, turbulent kinetic energy k and the dissipation rate ε.

According to Proudman the total radiated acoustic power per unit volume of turbulence is:

$$P_A=\alpha_\varepsilon\rho\varepsilon M_t^5,$$

where

$$M_t=(2k)^{1/2}/a.$$

Here a is the speed of sound, ρ is the air density, M_t is the turbulent Mach number. The re-scaled coefficient, α_ε, is set to 0.1.

Fluent has already implemented Proudman's formula for acoustic power per unit of volume, generated by isotropic turbulence without mean flow. As the Proudman's approach requires determination of the turbulent kinetic energy k in the present paper the broadband noise is considered for k-ω model.

A 2D distribution of the Sound Pressure Level (SPL, dB) is given in a across the flow, located close to the junction of the tail boom and fuselage (Figure 5).

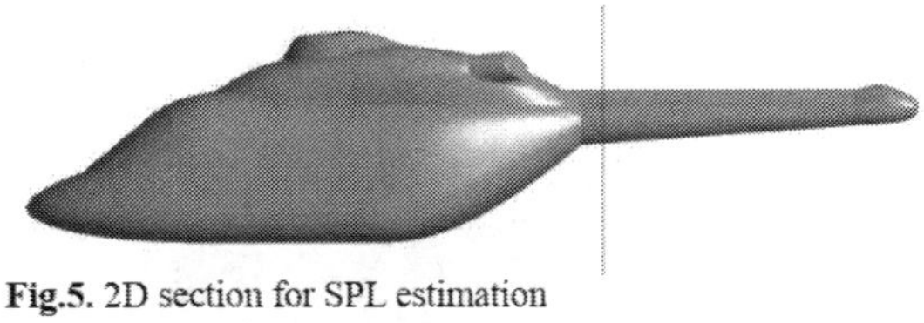

Fig.5. 2D section for SPL estimation

Figure 6a shows the 2D SPL_0 distribution for the isolated fuselage with and without actuator at 0 degrees of pitch angle. Figures 6b and 6c show the difference $dSPL=SPL_0-SPL_a$ for the uniform and nonuniform classical actuator models (SPL_a is the result with the actuator).

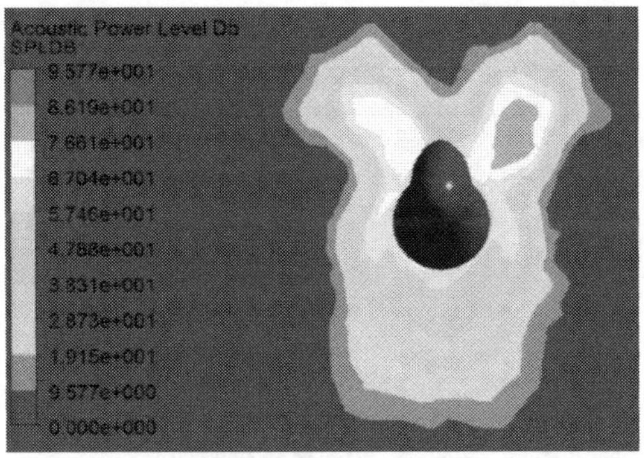

(a)

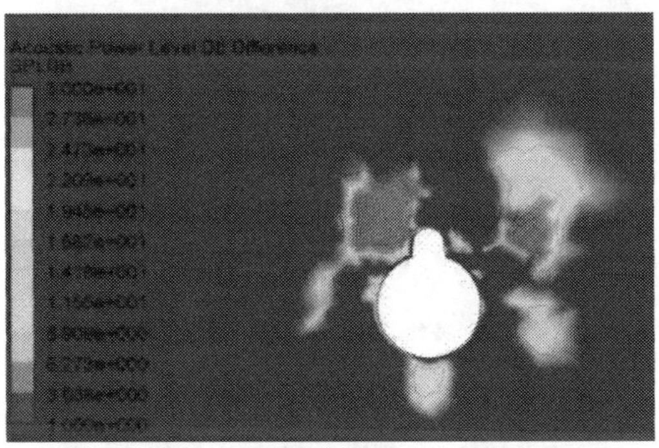

(b)

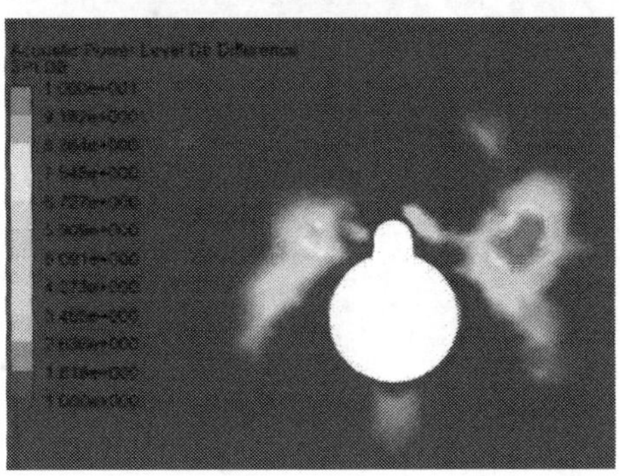

(c)

Fig.6. 2D SPL_0 distribution for the isolated fuselage (a) and dSPL for the fuselage with uniform (b) and nonuniform (c) actuator

The uniform load distribution suggests a strong influence of the actuator on the turbulent flow properties in comparison to the nonuniform load distribution.

6 Results for VBA model

For the uniform and nonuniform VBA disk models simulations were conducted in unsteady mode. The simulations are described in section 4 and time a step of Δt=0.0001 sec was used corresponding to an azimuth angle step of 1.8 degrees.

Figures 7, 8 present the pressure coefficient distribution at the actuator disc area and on the fuselage surface for two VBA models: uniform and nonuniform. Figure 8 shows $dCp=Cp_{VBA}-Cp_a$ between the surface pressure coefficient values. Here Cp_a is the pressure coefficient for the classical actuator disk model and Cp_{VBA} is the pressure coefficient for the VBA actuator disk models (uniform and nonuniform). The Lift and Drag coefficients time behavior for the uniform and nonuniform actuator disk models are presented in Figure 9. The location of the virtual blades in Figures 7, 8 corresponds approximately to time t_1 (Figures 7a, 7b, 8a, 8b) and t_2 (Figures 7c, 7d, 8c, 8d), so that t_2-t_1=0.002 s. Points 1 and 2 in Figure 9 also correspond to t_1 and t_2, respectively.

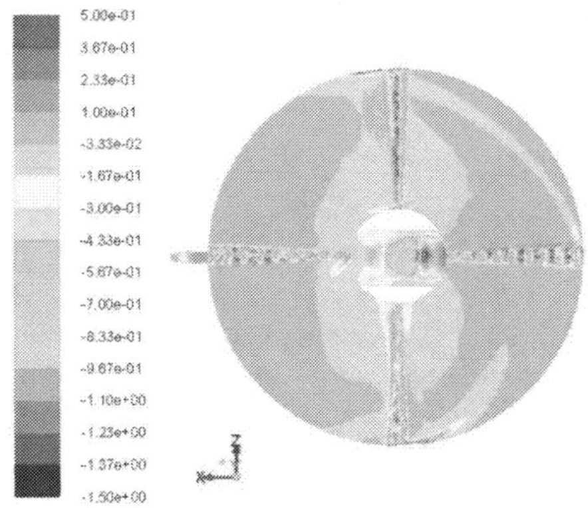

(a)

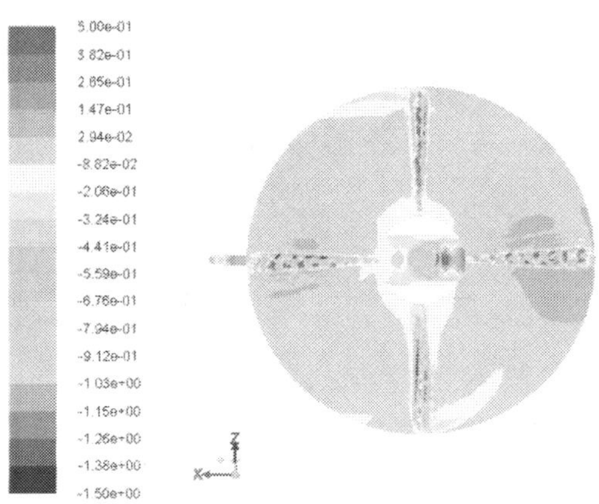

(b)

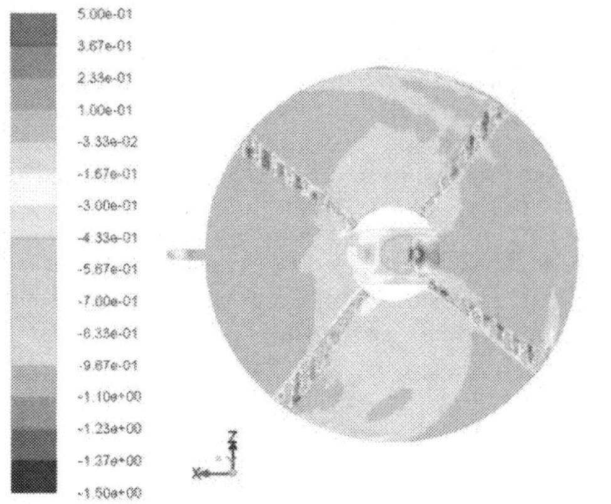

(c)

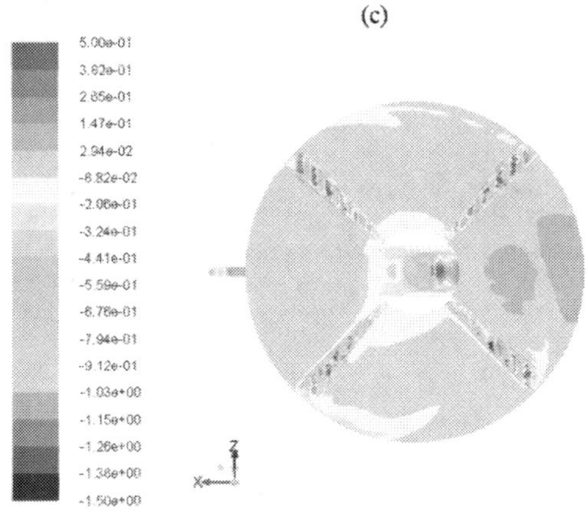

(d)

Fig.7. Distribution of the disk pressure for the uniform (a, c) and nonuniform (b, d) VBA models

For both locations of the rotor blades (presented in Figure 7) it can be seen that there is some discrepancy of the pressure coefficient near the tips of blades. Figure 8 shows minor difference between the pressure coefficients on the top front part of the fuselage surface. These discrepancies suggest differences between the values of the lift (C_L) and drag (C_D) fuselage coefficients for the uniform and nonuniform VBA model (Figure 9).

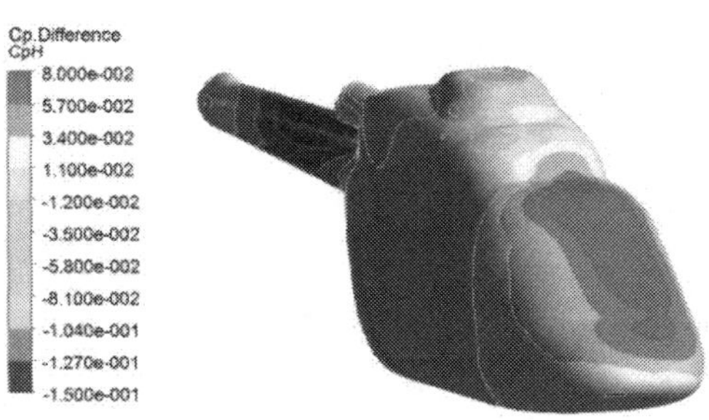

(a)

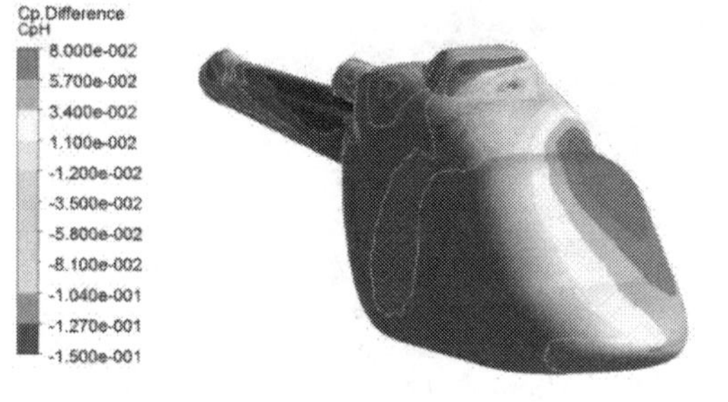

(b)

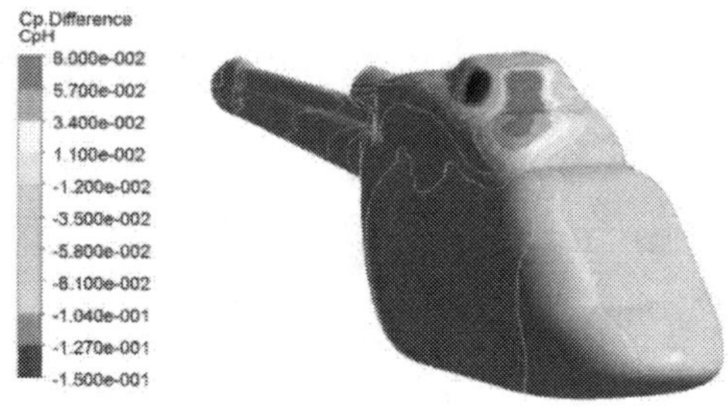

(c)

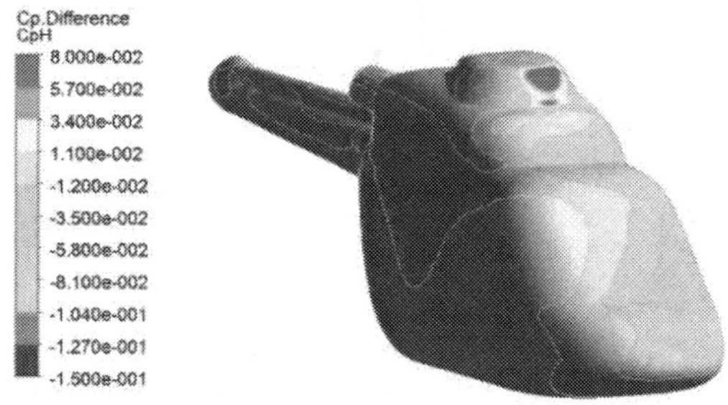

(d)

Fig.8. Distribution of the fuselage pressure for the uniform (a, c) and nonuniform (b, d) VBA models

Enhanced positive ΔCp area on the top front part of the fuselage surface for the nonuniform VBA disk model (in comparison to uniform VBA model) leads to lower C_L peak values (t_1 instant). For the same reason (different pressure distribution) C_D values at t_2 instant for the uniform VBA disk model are higher in comparison to the nonuniform VBA model.

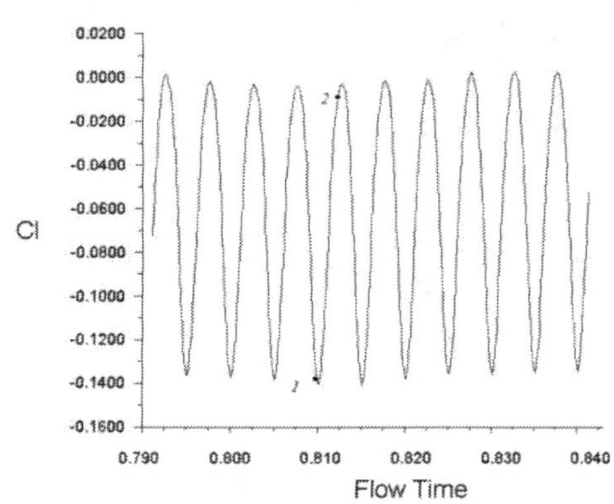

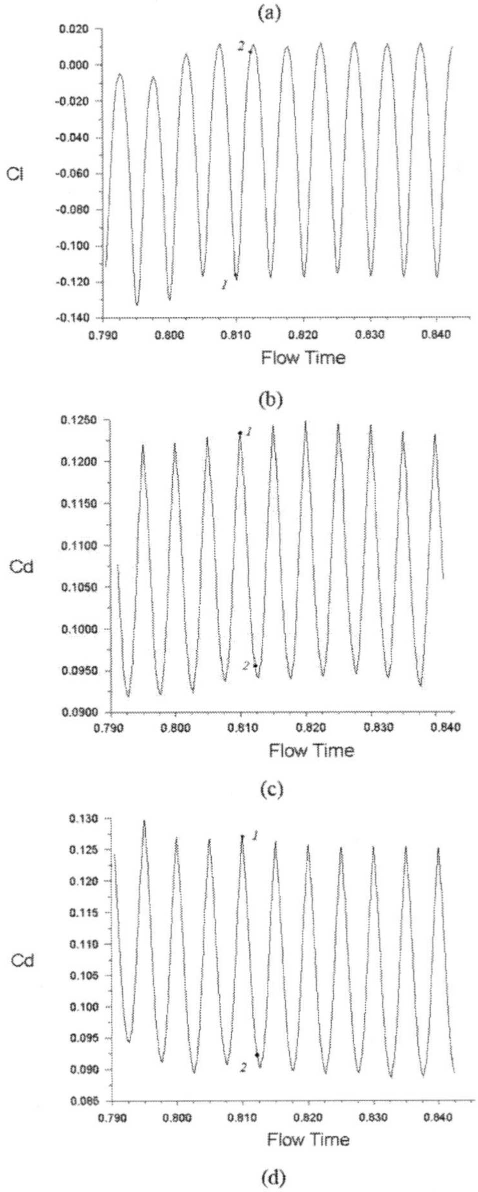

Fig.9. Lift and Drag coefficients for the uniform (a, c) and nonuniform (b, d) VBA models

For both uniform and nonuniform VBA disk models averaged values of lift and drag coefficients correspond to the aerodynamical coefficients for the considered steady simulation mode: C_L=-0.07, C_D=0.101.

7 Conclusion

The structure and the properties of the flow around an idealized fuselage of the ANSAT helicopter prototype were examined. The influence of the rotor on the fuselage aerodynamics was modeled with the classical disk model and the virtual blades. Uniform and nonuniform actuator load distributions were used to reproduce the main rotor flow properties. The acoustic properties (broadband noise) of the flow and the frequency range of the vortical structures were estimated using Proudman's approach and the RANS equations, in terms of mean flow field, turbulent kinetic energy and the dissipation rate.

In the future, the results of virtual blades will be compared to resolved blade data.

Acknowledgments

The support of this work via the "State tasks of the Education and Science Ministry of Russian Federation" grant (No 9.1577.2017/P CH) is gratefully acknowledged.

References

1. R. Steijl , and G. N. Barakos, *Aerospace Science and Technology*, **19**, 1 (2012).
2. A. F. Antoniadis et al., *Aerospace Science and Technology*, **19**, 1 (2012).
3. Ph. Beaumier, J.-M. Bousquet, *24-th ICAS congress* (2004).
4. T. Renaud, M. D. O'Brien, M. Smith, M. Potsdam, *American Helicopter Society 60th Annual Forum,* (2004).
5. Q. Wang, Z. Jiang and Q. Zhang, *Engineering Applications of Computational Fluid Mechanics*, **8**, 1 (2014).
6. I. Proudman, *Proc. Roy. Soc*., **214** (1952).
7. A. Kusyumov, S. Mikhailov, A. Garipov, E. Nikolaev, G. Barakos, *Transactions on Control and Mechanical Systems*, **1**, 7 (2012).
8. H. Heyson, S. Katzoff, *NACA TR*, **1319** (1957).
9. V.I. Shaidakov, *Proceedings of Moscow Aviation Institute*, **31**. (1978). [in Russian]
10. A. Kusuymov, S. Mikhailov, E. Nikolaev, N. Shilova, A. Garipov, *Proceedings of the ASME 2011 International Mechanical Engineering Congress & Exposition IMECE 2011*, (2011)

The high hall ventilation with the simplified simulation of the fan

Martin Kyncl[1,a] and *Jaroslav* Pelant[1]

[1] VZLÚ - Czech Aerospace Research Centre, Beranových 130, 199 05 Praha - Letňany, Czech Republic

Abstract. Here we work with the system of equations describing the non-stationary compressible turbulent multi-component flow in the gravitational field. We focus on the numerical simulation of the fan situated inside the high hall. The RANS equations are discretized with the use of the finite volume method. The original modification of the Riemann problem and its solution is used at the boundaries. The combination of specific boundary conditions is used for the simulation of the fan. The presented computational results are computed with own-developed code (C, FORTRAN, multiprocessor, unstructured meshes in general).

1 Introduction

In this contribution we consider the Reynolds-Averaged Navier-Stokes equations (RANS) with the k-omega model of turbulence, shown in [1,2]. We focus on the real gas flow in the gravitational field. The special method for the simulation of the simple fan is presented, using two compatible boundary conditions, used previously in [3]. We focus on the numerical solution of these equations, using own-developed software.

2 Methods

For the discretization of the system we use either explicit or implicit finite volume method in order to discretize the analytical problem, represented by the system of equations in generalized (integral) form. In order to apply this method we split the area of the interest into the elements, and we construct a piecewise constant solution in time, as described in [4]. The crucial problem of this method lies in the evaluation of the so-called fluxes (and its Jacobians) through the edges/faces of the particular elements. We use the exact Riemann solver for the solution of the local problem at each face, with the theory shown in [5]. At the boundary faces it is necessary to solve the incomplete Riemann problem, where the right-hand side initial condition is not known. It can be shown, that this right-hand side initial condition for the local problem can be partially replaced by the suitable complementary condition. Various original modifications of the Riemann problem (and exact solutions of these modifications) were shown and analyzed in [6–10]. These boundary conditions were also used in [11, 14, 13]. Here we present the simulation of the fan, using two connected boundaries. At these boundaries we solve the conservation laws, using the modification of the Riemann problem by the preference of the total quantities at the inlet, and the modification by the preference of the mass flow at the outlet. Using such boundary conditions it is necessary to compute the solution of the resulting non-linear problems. The combination of these two boundary conditions was used previously for the simulation of the propeller disk in [3]. The partial boundary condition with the preference of mass flow is sometimes being implemented with the use of some iterative process, guessing the correct values (for the pressure, density, velocity) in order to match the given mass flow through the boundary. In our approach we try to be as exact as possible, using our own original procedures. We follow the exact solution of the initial-value problem for the system of hyperbolic partial differential equations, the original analysis was shown in [14]. The own-developed software (C, FORTRAN) is based on the finite volume method with the implicit or explicit time discretization, solution is computed on unstructured 3D meshes in general. Multiple types of parallelization are used OpenMPI (for the computers with shared memory), MPICH (for the computers with the distributed memory), CUDA (graphic cards computations) are used, see also [15–18]. The large linear systems within the implicit method are solved with the implemented preconditioned generalized minimal residual (GMRES) method.

[a] e-mail: kyncl@vzlu.cz

3 The Riemann problem for the split Euler equations

For many numerical methods dealing with the two or three dimensional equations, describing the compressible flow, it is useful to solve the Riemann problem for the *3D split* Euler equations. We search the solution of the system of partial differential equations in time t and space (x, y, z)

$$\frac{\partial \varrho}{\partial t} + \frac{\partial \varrho u}{\partial x} = 0 \tag{1}$$

$$\frac{\partial \varrho u}{\partial t} + \frac{\partial (p + \varrho u^2)}{\partial x} = 0 \tag{2}$$

$$\frac{\partial \varrho v}{\partial t} + \frac{\partial \varrho u v}{\partial x} = 0 \tag{3}$$

https://doi.org/10.1051/epjconf/201818002051

$$\frac{\partial \varrho w}{\partial t} + \frac{\partial \varrho u w}{\partial x} = 0 \tag{4}$$

$$\frac{\partial E}{\partial t} + \frac{\partial u(E+p)}{\partial x} = 0, \tag{5}$$

equipped with the initial conditions

$$\varrho(x,t) = \varrho_L, \boldsymbol{v}(x,t) = \boldsymbol{v}_L, p(x,t) = p_L, \quad x < 0, \tag{6}$$

$$\varrho(x,t) = \varrho_R, \boldsymbol{v}(x,t) = \boldsymbol{v}_R, p(x,t) = p_R, \quad x > 0. \tag{7}$$

Vector $\boldsymbol{v} = (u, v, w)$ denotes the velocity, ϱ density, p pressure, $E = \varrho e + \varrho|\boldsymbol{v}|^2$ is the total energy, with e denoting the specific internal energy. We assume the equation of state for the calorically ideal gas

$$e = \frac{p}{\varrho(\gamma-1)}.$$

'Split' means here that we still have 5 equations in 3D, but the dependence on y, z (space coordinates) is neglected, and we deal with the system for one space variable x. The system (1)-(5) is considered in the set $Q_\infty = (-\infty, \infty) \times (0, +\infty)$.

The solution of this problem is fundamentally the same as the solution of the Riemann problem for the 1D Euler equations, see [4, page 138]. In fact, the solution for the pressure, the first component of the velocity, and the density is exactly the same as in one-dimensional case. It is a characteristic feature of the hyperbolic equations, that there is a possible raise of discontinuities in solutions, even in the case when the initial conditions are smooth, see [19, page 390]. Here the concept of the classical solution is too restrictive, and therefore we seek a weak solution of this problem. To distinguish physically admissible solutions from nonphysical ones, entropy condition must be introduced, see [19, page 396]. By the solution of the problem (1)-(5) ,(6),(7) we mean the *weak entropy solution* of this problem in Q_∞. The analysis to the solution of this problem can be found in many books, i.e. [4], [19], [5]. The general theorem on the solvability of the Riemann problem can be found in [4, page 88].

The solution is piecewise **smooth** and its general form can be seen in Fig. 1, where the system of half lines is drawn.

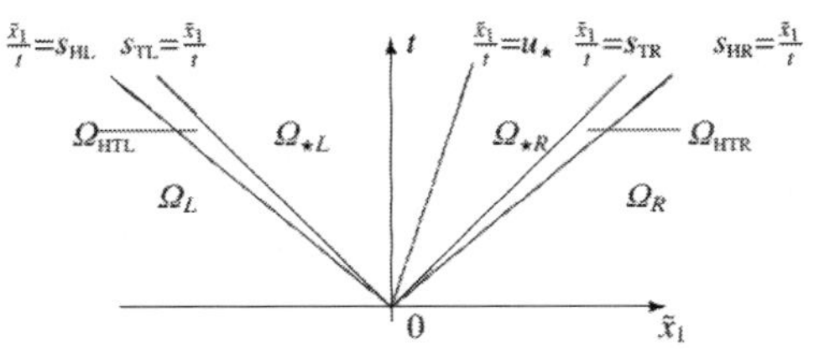

Fig. 1: Structure of the solution of the Riemann problem (1)-(5) ,(6),(7)

These half lines define regions, where solution is either constant or given by a smooth function. Let us define the open sets called **wedges** (see Fig. 1):

$$\begin{aligned} \Omega_L &= \text{wedge}(L_{-\infty}, s_{\text{HL}}), & \Omega_{\star R} &= \text{wedge}(u_\star, s_{\text{TR}}), \\ \Omega_{\text{HTL}} &= \text{wedge}(s_{\text{HL}}, s_{\text{TL}}), & \Omega_{\text{HTR}} &= \text{wedge}(s_{\text{TR}}, s_{\text{HR}}), \\ \Omega_{\star L} &= \text{wedge}(s_{\text{TL}}, u_*), & \Omega_R &= \text{wedge}(s_{\text{HR}}, L_{+\infty}). \end{aligned}$$

Using the theory in [4], [19], [5], we can write the solution for the primitive variables as

$$\begin{aligned} (\varrho, u, v, w, p)|_{\Omega_L} &= (\varrho_L, u_L, v_L, w_L, p_L), \\ (\varrho, u, v, w, p)|_{\Omega_{\star L}} &= (\varrho_{\star L}, u_\star, v_L, w_L, p_\star), \\ (\varrho, u, v, w, p)|_{\Omega_{\star R}} &= (\varrho_{\star R}, u_\star, v_R, w_R, p_\star), \\ (\varrho, u, v, w, p)|_{\Omega_R} &= (\varrho_R, u_R, v_R, w_R, p_R). \end{aligned}$$

The folowing relations for these variables hold:

$$u_\star = u_L + \begin{cases} -(p_\star - p_L)\left(\dfrac{\frac{2}{(\gamma+1)\varrho_L}}{p_\star + \frac{\gamma-1}{\gamma+1}p_L}\right)^{\frac{1}{2}}, & p_\star > p_L \\ \dfrac{2}{\gamma-1}a_L\left[1 - \left(\dfrac{p_\star}{p_L}\right)^{(\gamma-1)/2\gamma}\right], & p_\star \le p_L \end{cases} \tag{8}$$

$$\varrho_{\star L} = \begin{cases} \varrho_L \dfrac{\frac{\gamma-1}{\gamma+1}\frac{p_L}{p_\star} + 1}{\frac{p_L}{p_\star} + \frac{\gamma-1}{\gamma+1}}, & p_\star > p_L \\ \varrho_L \left(\dfrac{p_\star}{p_L}\right)^{\frac{1}{\gamma}}, & p_\star \le p_L \end{cases} \tag{9}$$

$$s^1_{TL} = \begin{cases} u_L - a_L\sqrt{\dfrac{\gamma+1}{2\gamma}\dfrac{p_\star}{p_L} + \dfrac{\gamma-1}{2\gamma}}, & p_\star > p_L \\ u_\star - a_L\left(\dfrac{p_\star}{p_L}\right)^{(\gamma-1)/2\gamma}, & p_\star \le p_L \end{cases} \tag{10}$$

$$u_\star = u_R + \begin{cases} (p_\star - p_R)\left(\dfrac{\frac{2}{(\gamma+1)\varrho_R}}{p_\star + \frac{\gamma-1}{\gamma+1}p_R}\right)^{\frac{1}{2}}, & p_\star > p_R, \\ -\dfrac{2}{\gamma-1}a_R\left[1 - \left(\dfrac{p_\star}{p_R}\right)^{(\gamma-1)/2\gamma}\right], & p_\star \le p_R \end{cases} \tag{11}$$

$$\varrho_{\star R} = \begin{cases} \varrho_R \dfrac{\frac{p_\star}{p_R} + \frac{\gamma-1}{\gamma+1}}{\frac{\gamma-1}{\gamma+1}\frac{p_\star}{p_R} + 1}, & p_\star > p_R \\ \varrho_R \left(\dfrac{p_\star}{p_R}\right)^{\frac{1}{\gamma}}, & p_\star \le p_R \end{cases} \tag{12}$$

$$s^3_{TR} = \begin{cases} u_R + a_R\sqrt{\dfrac{\gamma+1}{2\gamma}\dfrac{p_\star}{p_R} + \dfrac{\gamma-1}{2\gamma}}, & p_\star > p_R \\ u_\star + a_R\left(\dfrac{p_\star}{p_R}\right)^{(\gamma-1)/2\gamma}, & p_\star \le p_R \end{cases} \tag{13}$$

Here $a_L = \sqrt{\gamma p_L/\varrho_L}$, $a_R = \sqrt{\gamma p_R/\varrho_R}$, and γ denotes the adiabatic constant. Further s^1_{TL} denotes "unknown left wave speed", s^3_{TR} "unknown right wave speed". Note, that the system (8) - (13) is the system of 6 equations for 6 unknowns $p_\star, u_\star, \varrho_{\star L}\ \varrho_{\star R}, s^1_{TL}, s^3_{TR}$.

Solution for the Pressure $p_\star$
The combination of the equations (8), (11) gives the implicit equation for the unknown pressure $p_\star$

$$u_L + F_{1L}(p_\star) = u_R + F_{3R}(p_\star),$$

with

$$F_{1L}(p) = \begin{cases} -(p - p_L)\left(\dfrac{\frac{2}{(\gamma+1)\varrho_L}}{p + \frac{\gamma-1}{\gamma+1}p_L}\right)^{\frac{1}{2}}, & p > p_L \\ \dfrac{2}{\gamma-1}a_L\left[1 - \left(\dfrac{p}{p_L}\right)^{(\gamma-1)/2\gamma}\right], & p \le p_L. \end{cases},$$

$$F_{3R}(p) = \begin{cases} (p - p_R)\left(\frac{\frac{2}{(\gamma+1)\varrho_R}}{p + \frac{\gamma-1}{\gamma+1}p_R}\right)^{\frac{1}{2}}, & p > p_R, \\ -\frac{2}{\gamma-1}a_R\left[1 - \left(\frac{p}{p_R}\right)^{(\gamma-1)/2\gamma}\right], & p \leq p_R. \end{cases}$$

This is a nonlinear algebraic equation, and one cannot express the analytical solution of this problem in a closed form. The solution $p_\star$ can be found as the root of the function $F(p)$ defined as

$$F(p) = F_{3R}(p) - F_{1L}(p) + u_R - u_L. \tag{14}$$

The analysis of this function can be found in [4]. Here we state, that $F(p)$ is monotone and concave. Also the analytic expression for its derivative is available. For a positive solution for the pressure $F(0) < 0$ is required. This gives the *pressure positivity condition*

$$u_R - u_L < \frac{2}{\gamma - 1}(a_L + a_R). \tag{15}$$

The Newton method can be used to find the root of $F(p) = 0$. With the pressure $p_\star$ known, we use the equations (8)-(13) to obtain the whole solution.

Remarks

- Once the pressure $p_\star$ is known, the solution on the left-hand side of the contact discontinuity depends only on the left-hand side initial condition (6). And similarily, with $p_\star$ known, only the right-hand side initial condition (7) is used to compute the solution on the right-hand side of the contact discontinuity.
- The solution in $\Omega_L \cup \Omega_{HTL} \cup \Omega_{\star L}$ (across 1 wave) (solvability in general case)
  The solution components in $\Omega_L \cup \Omega_{HTL} \cup \Omega_{\star L}$ region must satisfy the system of equations (8)-(10). It is a system of three equations for four unknowns ($\varrho_{\star L}$, $p_\star$, $u_\star$, s^1_{TL}). We have to add another equation in order to get the uniquely solvable system in $\Omega_L \cup \Omega_{HTL} \cup \Omega_{\star L}$. This is the key problem for the outlet boundary condition.
- The equation (8) can be written as the *equation for pressure*, see [7]

$$p_\star = P(u_\star), \tag{16}$$

$$P(u) = \begin{cases} P_S(u), & u < u_L, \\ P_R(u), & u_L \leq u < u_L + \frac{2}{\gamma-1}a_L, \end{cases}$$

with

$$P_S(u) = p_L + \frac{\gamma+1}{4}\varrho_L(u_L - u)^2 + \frac{(u_L - u)\sqrt{4\varrho_L\gamma p_L + \varrho_L^2(\frac{\gamma+1}{2})^2(u_L - u)^2}}{2},$$

$$P_R(u) = p_L\left(\frac{-u + u_L + \frac{2}{\gamma-1}a_L}{\frac{2}{\gamma-1}a_L}\right)^{\frac{2\gamma}{\gamma-1}}.$$

In our work we use the analysis of this problem also for the solution of the initial-boundary value at the boundary faces.

4 Boundary condition by preference of total quantities and the tangential component of velocity

Using this boundary condition, we try to prescribe given total quantities at the face, comlete analysis was shown in [7], [20]. The conservation laws must be satisfied in close vicnity to the boundary face. We use the analysis of the incomplete Riemann problem to construct the values for the density, pressure and velocity. The following notation is used (some values are used only for an INLET case)

ϱ_B	static density at the boundary (unknown)
p_B	static pressure at the boundary (unknown)
$\boldsymbol{v}_B$	velocity vector at the boundary (in global coordinates) (unknown)
u_B	normal component of velocity at the boundary (local coordinates) (unknown)
$\boldsymbol{n}$	unit outer normal to face $\boldsymbol{n} = (n_1, n_2, n_3)$
ϱ_L	static density near the wall, time 0
p_L	static pressure near the wall, time 0
$\boldsymbol{v}_L$	velocity vector, state near the wall at time 0
u_L	normal component of velocity, state near the wall at time 0, $u_L = \boldsymbol{v}_L \cdot \boldsymbol{n}$
p_o	given value for the total pressure (=desired total pressure at the face)
θ_o	given value for the total temperature (=desired total temperature at the face)
$\boldsymbol{d}$	given INLET velocity direction $\boldsymbol{d} = (d_1, d_2, d_3)$

At the vicinity of the boundary we solve the modified Riemann problem for the split Euler equations (1)-(5), introduced in Section 3, with the left-hand side initial condition (6) and the **complementary conditions** (prescribed direction of velocity and given total quantities).

$$\boldsymbol{v}_B = \boldsymbol{d}|\boldsymbol{v}_B|, \text{ with } \boldsymbol{v}_B \cdot \boldsymbol{n} = u_\star. \tag{17}$$

$$\theta_{\star R} = \theta_o\left(1 - \frac{\gamma - 1}{2a_o^2}u_\star^2\right), \text{ with } a_0^2 = \gamma R\theta_0(\boldsymbol{d} \cdot \boldsymbol{n})^2, \tag{18}$$

$$p_\star = p_o\left(\frac{\theta_{\star R}}{\theta_0}\right)^{\gamma/(\gamma-1)} \tag{19}$$

$$u_\star < 0. \tag{20}$$

Here $\theta_o > 0, p_o > 0$ are given constants, R denotes the characteristic gas constant, and γ is the adiabatic constant. The equations (18),(19) are considered for the velocity $u_\star \in \left(-\sqrt{\frac{2a_o^2}{(\gamma-1)}}, \sqrt{\frac{2a_o^2}{(\gamma-1)}}\right)$. The equations (18),(19) come from the idea, that the *total pressure* p_o and the *total temperature* θ_o are known (prescribed) in the region $\Omega_{\star R}$.

According to analysis shown in [20], there are two possibillities for the 1-wave, either there is a **1-shock wave** and the sought velocity $u_\star$ can be found as a root of the function

$$F(u) = p_o\left(1 - \frac{\gamma - 1}{2a_o^2}u^2\right)^{\gamma/(\gamma-1)} - P_S(u), \tag{21}$$

considered for $u < u_L, -\sqrt{\frac{2a_o^2}{(\gamma-1)}} < u < 0$, and with $P_S(u)$ defined in (16).

Or there is a **1-rarefaction wave**, and the velocity $u_\star$ can be computed combining (18),(19),(16) as

$$u_\star = \frac{2\left(u_L + \frac{2}{\gamma-1}a_L\right) - \sqrt{DIS}}{2\left[1 + \frac{2a_L^2}{(\gamma-1)a_o^2}\left(\frac{p_o}{p_L}\right)^{(\gamma-1)/\gamma}\right]}, \qquad (22)$$

$$DIS = 4\left(\frac{p_o}{p_L}\right)^{(\gamma-1)/\gamma}\frac{2a_L^2}{(\gamma-1)a_o^2}\left[\frac{2a_o^2}{(\gamma-1)} + \frac{4a_L^2}{(\gamma-1)^2}\left(\frac{p_o}{p_L}\right)^{(\gamma-1)/\gamma} - \left(u_L + \frac{2a_L}{\gamma-1}\right)^2\right].$$

Summarizing the above possibilities of the 1-shock and the 1-rarefaction wave we can construct the algorithm in figure 2. for the computation of the velocity $u_\star$ in the *star region*.

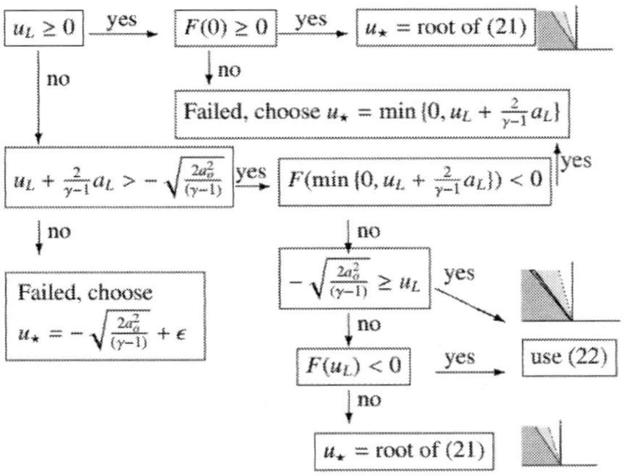

Fig. 2: Algorithm for the velocity solution $u_\star < 0$.In the case of failure, the problem doesn't have a solution, and the values are chosen.

Knowing the velocity $u_\star$ we compute the pressure $p_\star$ using (19), and the density $\varrho_{\star R} = \frac{p_\star}{R\theta_{\star R}}$. At last we set the boundary values as

$$p_B = p_\star, \quad \varrho_B = \varrho_{\star R}, \quad \boldsymbol{v}_B = \frac{\boldsymbol{d}u_\star}{\boldsymbol{d}\cdot\boldsymbol{n}}.$$

The problem (1)-(5) ,(6), (19), (18) has a unique solution for the initial data satisfying $u_L + \frac{2}{\gamma-1}a_L > -\sqrt{\frac{2a_o^2}{(\gamma-1)}}$, and $F(\min\{0, u_L + \frac{2}{\gamma-1}a_L\}) > 0$. In the case of $u_L + \frac{2}{\gamma-1}a_L \leq -\sqrt{\frac{2a_o^2}{(\gamma-1)}}$ there is no solution. In this case we prescribe the velocity $u_\star = -\sqrt{\frac{2a_o^2}{(\gamma-1)}} + \epsilon$, with $\epsilon > 0$ being a small positive constant. If $F(\min\{0, u_L + \frac{2}{\gamma-1}a_L\}) \leq 0$ then the problem does not have a negative solution, and for the practical applications we choose the velocity $u_\star = \min\{0, u_L + \frac{2}{\gamma-1}a_L\}$, or we use the boundary condition preferring the static pressure $p_\star = p_o$, see [10,7], in this case.

Let $\tilde{p_0}, \tilde{\theta_0}$ be the given total quantities together with the given **tangential velocity components** : v_t and w_t are the velocity components in the direction of $\boldsymbol{o}$ and $\boldsymbol{p}$. Here $\boldsymbol{o}\cdot\boldsymbol{n} = 0$, $|\boldsymbol{o}| = 1$, and $\boldsymbol{p} = \boldsymbol{n}\times\boldsymbol{o}$. Then we set

$$\boldsymbol{d} = -\boldsymbol{n}, \quad \theta_0 = \tilde{\theta_o} - \frac{\gamma-1}{2\gamma R}(v_t^2 + w_t^2), \quad p_0 = \tilde{p_0}\left(\frac{\theta_0}{\tilde{\theta_0}}\right)^{\frac{\gamma}{\gamma-1}},$$

and we use the same algorithm (figure 2.) for the computation of the velocity $u_\star$. Then it is $p_B = p_\star$, $\varrho_B = \varrho_{\star R}$, and

$$\boldsymbol{v}_B = (n_1u_\star + o_1v_t + p_1w_t, n_2u_\star + o_2v_t + p_2w_t, n_3u_\star + o_3v_t + p_3w_t,).$$

5 Outlet boundary condition by preference of massflow

Using this boundary condition, we try to prescribe given total quantities at the face, comlete analysis was shown in [8, 14]. The conservation laws must be satisfied in close vicnity to the boundary face. We use the analysis of the incomplete Riemann problem to construct the values for the density, pressure and velocity. The following notation is used (some values are used only for an INLET case)

- ϱ_B static density at the boundary (unknown)
- p_B static pressure at the boundary (unknown)
- $\boldsymbol{v}_B$ velocity vector at the boundary (in global coordinates) (unknown)
- u_B normal component of velocity at the boundary (local coordinates) (unknown)
- $\boldsymbol{n}$ unit outer normal to face $\boldsymbol{n} = (n_1, n_2, n_3)$
- ϱ_L static density near the wall, time 0
- p_L static pressure near the wall, time 0
- $\boldsymbol{v}_L$ velocity vector, state near the wall at time 0
- u_L normal component of velocity, state near the wall at time 0, $u_L = \boldsymbol{v}_L\cdot\boldsymbol{n}$
- $G_\star$ given inlet massflow (=desired massflow at the face)

At the vicinity of the boundary we solve the modified Riemann problem for the split Euler equations (1)-(5), introduced in Section 3, with the left-hand side initial condition (6) and the **complementary conditions** prescribing the mass flow

$$u_\star\varrho_{\star R} = G_\star, \qquad (23)$$

where $G_\star \geq 0$ is given constant ($G_\star = \frac{massflow}{face\ area}$). We are interested in the solution with $u_\star > 0$, and $s_{L?} < 0$, which guarantees the possibility of the values to be prescribed at the boundary. In general, there are two possibilities of the wave pattern, which may interest us.

- **1-shock wave**
  Here, we are interested in the solution with $s_1 < 0, u_\star > 0$. Let us construct the function $F_S(u)$, using the relations (8),(9)

$$F_S(u) = u\varrho_{\star L}(u) - G_\star, \qquad (24)$$

where

$$\varrho_{\star L}(u) = \varrho_L\frac{(\gamma-1)p_L + (\gamma+1)P_S(u)}{(\gamma+1)p_L + (\gamma-1)P_S(u)},$$

and $P_S(u)$ defined in (16). The sought velocity $u_\star$ is the root of this function $F_S(u)$. It is $s_1 < 0$ for $u_\star < u_X = u_L\frac{2+(\gamma-1)M_L^2}{(\gamma+1)M_L^2}$, with $M_L^2 = u_L^2\varrho/\gamma p$, see [7]. The first derivative is $F_S' > 0$ for $u < min\{u_L, u_X\}$. The problem has a solution with the 1-shock wave if $F_S(min\{u_L, u_X\}) > 0$.

- **1-rarefaction wave**
Let us assume the solution with the 1-rarefaction wave. Using (8),(9),(23), the velocity $u_\star$ is the root of the function

$$F_R(u) = u\varrho_L\left[1 - \frac{\gamma-1}{2a_L}(u-u_L)\right]^{\frac{2}{\gamma-1}} - G_\star, \qquad (25)$$

defined in the interval $u_L < u < u_L + \frac{2a_L}{\gamma-1}$. It is $F_R(u_L) = \varrho_L u_L - G_\star$. The first derivative is zero at the points $u_1 = u_L + \frac{2a_L}{\gamma-1}$, $u_0 = \frac{\gamma-1}{\gamma+1}\left(u_L + \frac{2a_L}{\gamma-1}\right)$. Derivative $F_R'(u) > 0$ in the interval $(0, u_0)$ and $F'(u) < 0$ for $u \in (u_0, u_1)$. The maximum of the function F_R is at the point u_0. We are interested in the solution with $s_{TL} < 0$. This is satisfied only fo $u_\star < u_0$.

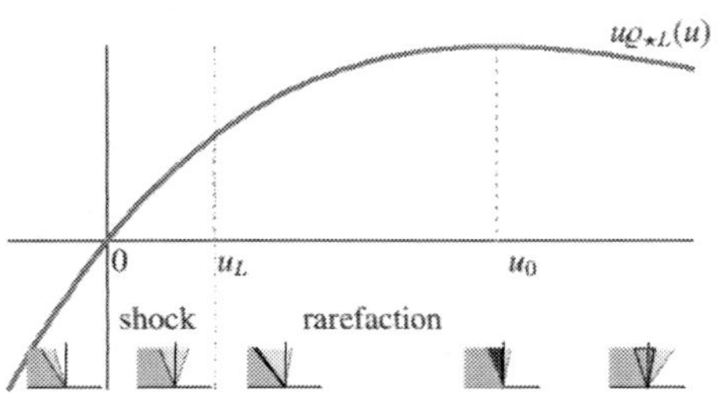

Fig. 3: Example of the function $u\varrho_{\star L}(u)$, here $u_L = 20$, $\gamma = 1.4$, $\varrho_L = 1.25$, $p_L = 100000$.

Algorithm
Here we present the possible algorithm for the solution of the sought values at the boundary.

```
F(u_L) = u_L ϱ_L − G⋆
if F(u_L) > 0 then
  u⋆ = root of F_S(u), u ∈ (0, min{u_X, u_L}), (24).
else
  if u_L > u_0 then
    u⋆ = u_L
  else
    if (u_1 < u_0)
      u⋆ = u_1 − ε
    else
      if F(u_0) < 0 then
        u⋆ = u_0
      else
        u⋆ = root of F_R(u), u ∈ (u_L, u_0), (25).
      end
    endif
  end
end
Compute p_B = p⋆, ϱ_B = ϱ⋆L using (16),(9).
```

The resulting boundary velocity in the global coordinates is $\boldsymbol{v}_B = \boldsymbol{n}(u_B - u_L) + \boldsymbol{v}_L$.

6 Simple fan simulation

The simple fan simulation is based on the boundary conditions shown above in sections 4, 5. The inlet boundary condition with the given total temperature and the total pressure is connected with the outlet boundary condition with the preference of the mass flow. This procedure was succesfully used by authors in [3].

Here we suppose, that the total quantities θ_{0FAN}, p_{0FAN} and the tangential velocity components at the fan outlet are known. Then we may use the boundary condition in section 4 to compute the boundary state (at the fan outlet). The conservation law of mass is valid across the fan, and therefore we may use the known computed massflow (at the fan outlet) to complement the incomplete the Riemann problem at the fan inlet. The solution to this problem gives us the boundary state at the fan inlet, analysis and the algorithm is shown in section 5.

The approach is based on the estimation of the total quantities θ_{0FAN}, p_{0FAN} and the tangential velocity at the fan, wich may be rather complicated. Let us for example suppose, that the distribution of the velocity $\boldsymbol{v}_{FAN}$ is known at the fan outlet (at some time instant), together with the total quantities p_{0X}, θ_{0X} before the fan rotor. Then we may estimate

$$\theta_{0FAN} = \theta_{0X}\left(1 + \frac{\gamma-1}{2\gamma R\theta_{0X}}\boldsymbol{v}^2_{FAN}\right),$$

$$p_{0FAN} = p_{0X}\left(1 + \frac{\gamma-1}{2\gamma R p_{0X}}\boldsymbol{v}^2_{FAN}\right)^{\gamma/(\gamma-1)}.$$

The figures 4-5 show the test simulation of the fan $|\boldsymbol{v}_{FAN}| = 20$ [m s^{-1}] with the regime $p_{0X} = 101325$[Pa], $p_{0X} = 293.15$[K].

Table 1: Outlet boundary condition by the preference of massflow. The input data u_L, $G_\star$ and the solution (rounded) ϱ_B, u_B, p_B, $G_B = \varrho_B u_B$ for the test problems with $\varrho_L = 1.25$, $p_L = 100000$.

input data		solution			
u_L	$G_\star$	G_B	ϱ_B	u_B	p_B
-20	1	1.00	1.1740	0.8517646	91596.49
-20	50	50.00	1.0113	49.44231	74326.85
-20	150	130.35	0.47304	275.5533	25655.28
20	1	1.00	1.3235	0.7555568	108333.1
20	100	100.00	0.96848	103.2548	69960.27
20	200	150.45	0.53309	282.2200	30328.54
400	10.0	10.00	3.2188	3.106711	421919.0
400	100	100.00	3.0591	32.68887	385170.7
400	600	500	1.25	400	100000

7 Examples

Here we show the computational results for the flow around and inside the high hall in the isothermal atmosphere. The initial condition was constant in the whole domain with the total temperature $\theta_0 = 293.15$[K], zero velocity, and the total pressure distribution modified in the gravitational field as $p_0 = p_{00}e^{9.81z/(R\theta_0)}$, where $p_{00} = 101325$[Pa], z denotes the vertical coordinate, and R is the gas constant $R = 287.04$[J kg^{-1} K^{-1}]. The fan is placed horizontally in the high hall, both direction of the flow through the fan were simulated. The magnitude of the fan velocity (parameter) $|\boldsymbol{v}_{FAN}| = 5$[m s^{-1}] was chosen. Another simulations with two fans were computed. Results are shown in figures 6-8.

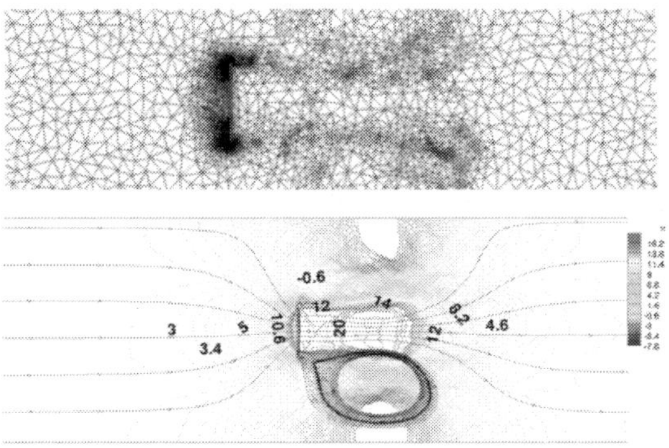

Fig. 4: The fan flow simulation example (fan depicted by the red vertical line), inviscid flow. Isolines of horizontal velocity, streamlines at time $t = 0.15$ [s].

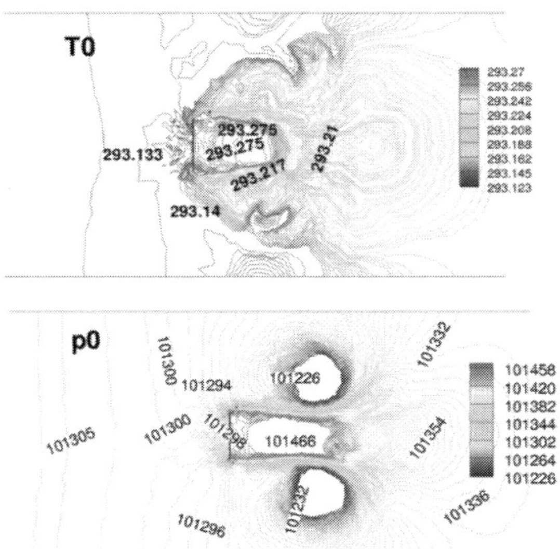

Fig. 5: The fan flow simulation example (fan depicted by the red vertical line), inviscid flow. Isolines of the total temperature and the total pressure, streamlines at time $t = 0.15$ [s].

Conclusion

This paper shows the numerical simulation of the fan situated inside the high hall. The partial differential equations describing the conservation laws are solved numerically with the use of the finite volume method. Own software was programmed. The special modification of the Riemann problem and its solution was used at the boundaries. The combination of these boundary conditions was used for the simulation of the fan, which is the original result of this work.

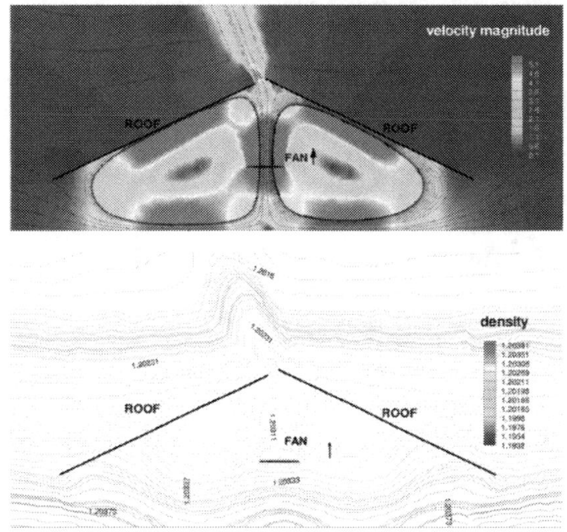

Fig. 6: The flow inside the high hall induced by the fan. One horizontally placed fan, facing upward. Density and velocity magnitude isolines and streamtraces shown.

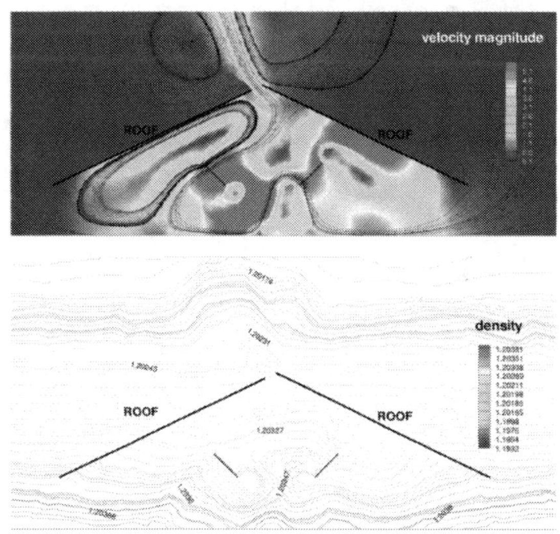

Fig. 7: The flow inside the high hall induced by two fans. Density and velocity magnitude isolines and streamtraces shown.

Acknowledgment

The results originated with the support of Ministry of the Interior of the Czech Republic, project SCENT, Grant MSM 0001066902 of the Ministry of Education of the Czech Republic, and Ministry of Industry and Trade of the Czech Republic for the long-term strategic development of the research organization. The authors acknowledge this support.

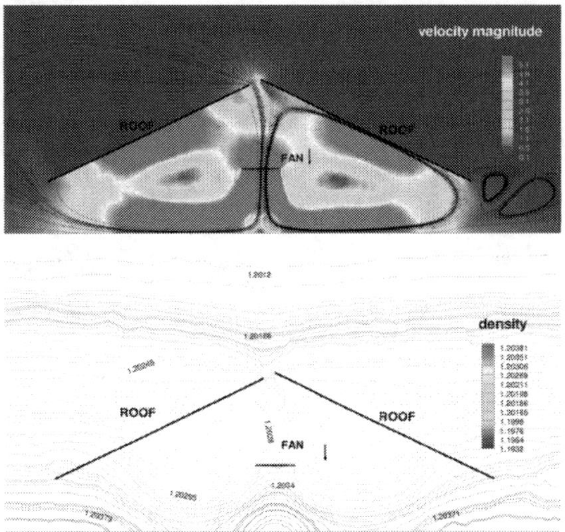

Fig. 8: The flow inside the high hall induced by the horizontally placed fan, facing downwards. Density and velocity magnitude isolines and streamtraces shown.

References

1. D. C. Wilcox, *Turbulence Modeling for CFD*. KNI, Inc., Anaheim, California, (July 1998).
2. C. J. Kok, AIAA Journal, Vol. **38.**, No. 7., (2000).
3. M. Kyncl and J. Pelant, *EPJ WoC*, **67**, 02064 (2014)
4. M. Feistauer, J. Felcman, and I. Straškraba, *Mathematical and Computational Methods for Compressible Flow*, Oxford University Press, Oxford (2003)
5. E. F. Toro, *Riemann Solvers and Numerical Methods for Fluid Dynamics*, Springer, Berlin, (1997)
6. J. Pelant, *ARTI Reports VZLÚ, Z-65, Z-67 to Z-73*,VZLÚ, Prague (1996-2000)
7. M. Kyncl, *Numerical solution of the three-dimensional compressible flow.* Doctoral Thesis, Prague (2011)
8. M. Kyncl and J. Pelant, *Appl. Mech. and Mat.*, **821**, 70-78, (2016)
9. M. Kyncl and J. Pelant, *EPJ WoC*, **143**, 02061 (2017)
10. M. Kyncl and J. Pelant, *Applications of the Navier-Stokes equations for 3d viscous laminar flow for symmetric inlet and outlet parts of turbine engines with the use of various boundary conditions,* Tech. rep. R3998, VZLÚ (2006)
11. J. Michálek, P. Straka, *Journal of Thermal Science*, **22** (5), pp. 413-423, (2013)
12. P. Straka, *Proc. ECCOMAS 2016, 5.-10.6.2016*, pp. 7532-7543 , Greece (2016).
13. P. Straka, *The European Physical Journal of Conferences*, **114**, 02115 (2016)
14. M. Kyncl and J. Pelant, *Proc. ECCOMAS 2016, 5.-10.6.2016*, Crete Island, Greece (2016).
15. M. Kyncl and J. Pelant, *Implicit method for the 3d RANS equations with the k-ω (Kok) Turbulent Model.* Tech. rep. R5453, Prague (2012)
16. M. Kyncl and J. Pelant, *Implicit method for the 3D Euler equations.* Tech. rep. R5375, VZLÚ (2012)
17. M. Kyncl and J. Pelant, *EPJ WoC*, **92**, 02044 (2015)
18. M. Kyncl and J. Pelant, *EPJ WoC*, **45**, 01097 (2016)
19. M. Feistauer, *Mathematical Methods in Fluid Dynamics*, Longman Scientific & Technical, Harlow, (1993)
20. M. Kyncl and J. Pelant, *Proc. ECCOMAS 2014, ECFD VI, 20.-25.7.2014*, Barcelona, Spain (2014)

The simulation of the flow through the porous media and diffusible barriers

Martin Kyncl[1,a] and *Jaroslav* Pelant[1]

[1] VZLÚ - Czech Aerospace Research Centre, Beranových 130, 199 05 Praha - Letňany, Czech Republic

Abstract. Here we work with the RANS equations describing the non-stationary viscous compressible fluid flow. We focus on the numerical simulation of the flow through the porous media, characterized by the loss of momentum. Further we simulate the flow through the set of diffusible barriers. Here we analyze the modification of the Riemann problem with one-side initial condition, complemented with the Darcy's law and added inertial loss. We show the computational results obtained with the own-developed code for the solution of the compressible gas flow.

1 Introduction

The physical theory of the compressible fluid motion is based on the principles of conservation laws of mass, momentum, and energy. The mathematical equations describing these fundamental conservation laws form a system of partial differential equations (the Euler equations, the Navier-Stokes equations, the Navier Stokes equations with turbulent models). We focus on the flow through the porous media and throug the diffusible barriers (fences). We choose the well-known finite volume method to discretize the analytical problem, represented by the system of the equations in generalized (integral) form. To apply this method we split the area of the interest into the elements, and we construct a piecewise constant solution in time. The crucial problem of this method lies in the evaluation of the so-called fluxes through the edges/faces of the particular elements. In order to compute these fluxes, various methods can be used. We decided to use the analysis of the exact solution also for the discretization of the fluxes through the boundary edges/faces and on the edges/faces simulating the diffusible barrier, as was shown in [1]. We construct own algorithm for the solution of the boundary problem at the diffusible barrier, and we use it in the numerical examples.

2 Formulation of the Equations

We consider the conservation laws for viscous compressible turbulent flow of ideal gas with the zero heat sources in a domain $\Omega \in \boldsymbol{R}^N$, and time interval $(0, T)$, with $T > 0$. The system of the Reynolds-Averaged Navier-Stokes equations in 3D has the form

$$\frac{\partial \boldsymbol{w}}{\partial t} + \sum_{s=1}^{3} \frac{\partial \boldsymbol{f}_s(\boldsymbol{w})}{\partial x_s} = \sum_{s=1}^{3} \frac{\partial \boldsymbol{R}_s(\boldsymbol{w}, \nabla \boldsymbol{w})}{\partial x_s} + \boldsymbol{S} \quad \text{in } Q_T = \Omega \times (0, T). \tag{1}$$

[a] e-mail: kyncl@vzlu.cz

Here x_1, x_2, x_3 are the space coordinates, t the time, $\boldsymbol{w}$ is the state vector, $\boldsymbol{f}_s$ are the inviscid fluxes, $\boldsymbol{R}_s$ are the viscous fluxes, $\boldsymbol{S}$ are additional sources. $\boldsymbol{v} = (v_1, v_2, v_3)^T$ denotes the velocity vector, ϱ is the density, p the pressure, θ the absolute temperature, $E = \varrho e + \frac{1}{2}\varrho \boldsymbol{v}^2 + \varrho k$ the total energy.

$$\boldsymbol{w}(x,t) = (\varrho,\ \varrho v_1,\ \varrho v_2,\ \varrho v_3,\ E)^T$$
$$\boldsymbol{f}_s = (\varrho v_s,\ \varrho v_s v_1 + \delta_{s1} p, \ldots, \varrho v_s v_3 + \delta_{s3} p,\ (E + p) v_s)^T$$
$$\boldsymbol{R}_s = (0,\ \tau_{s1}, \tau_{s2}, \tau_{s3},\ \textstyle\sum_{r=1}^{3} \tau_{sr} v_r + C_k \partial\theta/\partial x_s)^T$$

Further $\tau_{ij} = \begin{cases} (\mu + \mu_T) S_{ij}, & i \neq j \\ (\mu + \mu_T) S_{ij} - \frac{2}{3}\varrho k, & i = j \end{cases}$, with $S_{11} = \frac{2}{3}\left(2\frac{\partial v_1}{\partial x_1} - \frac{\partial v_2}{\partial x_2} - \frac{\partial v_3}{\partial x_3}\right), S_{12} = \frac{\partial v_1}{\partial x_2} + \frac{\partial v_2}{\partial x_1}, S_{13} = \frac{\partial v_1}{\partial x_3} + \frac{\partial v_3}{\partial x_1}, S_{21} = S_{12}, S_{22} = \frac{2}{3}\left(-\frac{\partial v_1}{\partial x_1} + 2\frac{\partial v_2}{\partial x_2} - \frac{\partial v_3}{\partial x_3}\right), S_{23} = \frac{\partial v_2}{\partial x_3} + \frac{\partial v_3}{\partial x_2}$,
$S_{31} = S_{13}, S_{32} = S_{23}, S_{33} = \frac{2}{3}\left(-\frac{\partial v_1}{\partial x_1} - \frac{\partial v_2}{\partial x_2} + 2\frac{\partial v_3}{\partial x_3}\right)$, where μ is the dynamic viscosity coefficient dependent on temperature, μ_T is the eddy-viscosity coefficient. For the specific internal energy $e = c_v\theta$ we assume the caloric equation of state $e = p/\varrho(\gamma - 1)$, c_v is the specific heat at constant volume, $\gamma > 1$ is called the *Poisson adiabatic constant*. The constant C_k denotes the heat conduction coefficient $C_k = (\frac{\mu}{P_r} + \frac{\mu_T}{P_{r_T}})c_v\gamma$, and P_r is laminar and P_{r_T} is turbulent Prandtl constant number. In our application of flow in the gravitational field we set the source terms to $\boldsymbol{S} = (0, \varrho g_1, \varrho g_2, \varrho g_3, \varrho \boldsymbol{g} \cdot \boldsymbol{v})$, where $\boldsymbol{g} = (g_1, g_2, g_3)$ is the gravitational acceleration vector.

Here we assume the system (1) equipped with the two-equation turbulent model $k-\omega$ (Kok), described in [2]. The effective turbulent viscosity is $\mu_T = \varrho k/\omega$.

$$\frac{\partial \varrho k}{\partial t} + \frac{\partial \varrho k v_1}{\partial x_1} + \frac{\partial \varrho k v_2}{\partial x_2} + \frac{\partial \varrho k v_3}{\partial x_3} = P_k - \beta^* \varrho \omega k + \\ + \frac{\partial}{\partial x_1}\left((\mu + \sigma_k \mu_T)\frac{\partial k}{\partial x_1}\right) + \\ + \frac{\partial}{\partial x_2}\left((\mu + \sigma_k \mu_T)\frac{\partial k}{\partial x_2}\right) + \frac{\partial}{\partial x_3}\left((\mu + \sigma_k \mu_T)\frac{\partial k}{\partial x_3}\right), \tag{2}$$

$$\frac{\partial \varrho\omega}{\partial t} + \frac{\partial \varrho\omega v_1}{\partial x_1} + \frac{\partial \varrho\omega v_2}{\partial x_2} + \frac{\partial \varrho\omega v_3}{\partial x_3} = P_\omega - \beta\varrho\omega\omega + C_D + \frac{\partial}{\partial x_1}\left((\mu + \sigma_\omega\mu_T)\frac{\partial\omega}{\partial x_1}\right) + \\ + \frac{\partial}{\partial x_2}\left((\mu + \sigma_\omega\mu_T)\frac{\partial\omega}{\partial x_2}\right) + \frac{\partial}{\partial x_3}\left((\mu + \sigma_\omega\mu_T)\frac{\partial\omega}{\partial x_3}\right), \quad (3)$$

where k the specific turbulent kinetic energy and ω the turbulent dissipation are functions of time t and space coordinates x_1, x_2, x_3. The production terms P_k and P_ω are given by formulas

$$P_k = \tau_{11}\frac{\partial v_1}{\partial x_1} + \tau_{12}\frac{\partial v_1}{\partial x_2} + \tau_{21}\frac{\partial v_2}{\partial x_1} + \tau_{22}\frac{\partial v_2}{\partial x_2} + \tau_{13}\frac{\partial v_1}{\partial x_3} + \\ +\tau_{31}\frac{\partial v_3}{\partial x_1} + \tau_{23}\frac{\partial v_2}{\partial x_3} + \tau_{32}\frac{\partial v_3}{\partial x_2} + \tau_{33}\frac{\partial v_3}{\partial x_3},$$

$$P_\omega = \frac{\alpha_\omega\omega P_k}{k},$$

where functions τ are defined in (1) with $\mu = 0$, $\alpha_\omega = \frac{\beta}{\beta^*} - \frac{\sigma_\omega\kappa^2}{\sqrt{\beta^*}}$ and $\sigma_k = \frac{2}{3}$, $\beta^* = 0.09$, $\beta = \frac{5}{6}\beta^*$, $\sigma_\omega = 0.5$, $\kappa = 0.41$. The cross-diffusion term C_D is defined as

$$C_D = \sigma_d \frac{\varrho}{\omega} \max\left\{\frac{\partial k}{\partial x_1}\frac{\partial\omega}{\partial x_1} + \frac{\partial k}{\partial x_2}\frac{\partial\omega}{\partial x_2} + \frac{\partial k}{\partial x_3}\frac{\partial\omega}{\partial x_3}, 0\right\},$$

where $\sigma_d = 0.5$ is constant.

3 Porous media simulation

The porous media is simulated using the modification of the system of equations (1). The simple porous media can be simulated via the new source term, written as

$$\boldsymbol{S} = \begin{pmatrix} 0 \\ -\frac{\mu}{\alpha}v_1 - C_0\frac{\varrho}{2}|\boldsymbol{v}|v_1 + \varrho g_1 \\ -\frac{\mu}{\alpha}v_2 - C_0\frac{\varrho}{2}|\boldsymbol{v}|v_2 + \varrho g_2 \\ -\frac{\mu}{\alpha}v_3 - C_0\frac{\varrho}{2}|\boldsymbol{v}|v_3 + \varrho g_3 \\ -\frac{\mu}{\alpha}\boldsymbol{v}^2 - C_0\frac{\varrho}{2}|\boldsymbol{v}|\boldsymbol{v}^2 + \varrho\boldsymbol{g}\cdot\boldsymbol{v} \end{pmatrix} \quad (4)$$

Here α is the permeability coefficient, C_o is the pressure gradient coefficient.

4 Numerical method

For the discretization of the system we proceed as described in [3,4]. We use either explicit or implicit finite volume method to solve the systems sequentionally. Here we present the discretization of the system (1). By Ω_h let us the denote the polyhedral approximation of Ω. The system of the closed polyhedrons with mutually disjoint interiors $\mathcal{D}_h = \{D_i\}_{i\in J}$, where $J \subset Z^+ = \{0, 1, \ldots\}$ is an index set and $h > 0$, will be called a *finite volume mesh*. This system $\mathcal{D}_h$ approximates the domain Ω, we write $\overline{\Omega}_h = \bigcup_{i\in J} D_i$. The elements $D_i \in \mathcal{D}_h$ are called the *finite volumes*. For two neighboring elements D_i, D_j we set $\Gamma_{ij} = \partial D_i \cap \partial D_j = \Gamma_{ji}$. Similarly, using the negative index j we may denote the boundary faces. Here we will work with the so-called *regular* meshes, i.e. the intersection of two arbitrary (different) elements is either empty or it consists of a common vertex or a common edge or a common face (in 3D). The boundary ∂D_i of each element D_i is

$$\partial D_i = \bigcup_{\Gamma_{ij}\in\Gamma_{D_i}} \Gamma_{ij}. \quad (5)$$

Here the set $\Gamma_{D_i} = \{\Gamma_{ij}; \Gamma_{ij} \subset \partial D_i\}$ forms the boundary ∂D_i. By $\boldsymbol{n}_{ij}$ let us denote the unit outer normal to ∂D_i on Γ_{ij}. Let us construct a partition $0 = t_0 < t_1 < \ldots$ of the time interval $[0, T]$ and denote the time steps $\tau_k = t_{k+1} - t_k$. We integrate the system (1) over the set $D_i \times (t_k, t_{k+1})$. With the integral form of the equations we can study a flow with discontinuities, such as shock waves, too.

$$\int_{D_i}\int_{t_k}^{t_{k+1}} \frac{\partial\boldsymbol{w}}{\partial t}dx\,dt + \int_{t_k}^{t_{k+1}}\int_{D_i}\sum_{s=1}^{3}\frac{\partial\boldsymbol{f}_s(\boldsymbol{w})}{\partial x_s}dx\,dt = \\ = \int_{t_k}^{t_{k+1}}\int_{D_i}\sum_{s=1}^{3}\frac{\partial\boldsymbol{R}_s(\boldsymbol{w},\nabla\boldsymbol{w})}{\partial x_s}dx\,dt \quad (6)$$

Using the Green's theorem on D_i it is

$$\int_{D_i}\sum_{s=1}^{3}\frac{\partial\boldsymbol{f}_s(\boldsymbol{w})}{\partial x_s}dx = \int_{\partial D_i}\sum_{s=1}^{3}\boldsymbol{f}_s(\boldsymbol{w})n_s\,dS, \quad (7)$$

$$\int_{D_i}\sum_{s=1}^{3}\frac{\partial\boldsymbol{R}_s(\boldsymbol{w},\nabla\boldsymbol{w})}{\partial x_s}dx = \int_{\partial D_i}\sum_{s=1}^{3}\boldsymbol{R}_s(\boldsymbol{w},\nabla\boldsymbol{w})n_s\,dS.$$

Here $\boldsymbol{n} = (n_1, n_2, n_3)$ is the unit outer normal to ∂D_i. Further we use (5), and we rewrite (6)

$$\int_{D_i}(\boldsymbol{w}(x, t_{k+1}) - \boldsymbol{w}(x, t_k))\,dx + \quad (8)$$

$$\int_{t_k}^{t_{k+1}}\sum_{\Gamma_{ij}\in\Gamma_{D_i}}\int_{\Gamma_{ij}}\sum_{s=1}^{3}(\boldsymbol{f}_s(\boldsymbol{w}) - \boldsymbol{R}_s(\boldsymbol{w},\nabla\boldsymbol{w}))(n_{ij})_s\,dS\,dt = 0$$

We define a finite volume approximate solution of the system studied (1) as a piecewise constant vector-valued functions $\mathbf{w}_h^k$, $k = 0, 1, \ldots$, where $\boldsymbol{w}_h^k$ is constant on each element D_i, and t_k is the time instant. By $\boldsymbol{w}_i^k$ we denote the value of the approximate solution on D_i at time t_k. We approximate the integral over the element D_i

$$\int_{D_i}\boldsymbol{w}(x, t_k)\,dx \approx |D_i|\boldsymbol{w}_i^k. \quad (9)$$

Further we proceed with the approximation of the fluxes. Usually the flux $\sum_{s=1}^{3}\boldsymbol{f}_s(\boldsymbol{w})(n_{ij})_s|_{\Gamma_{ij}}$ is being approximated by a *numerical flux* at suitable time instant t_l

$$\sum_{s=1}^{3}\boldsymbol{f}_s(\boldsymbol{w})(n_{ij})_s|_{\Gamma_{ij}} \approx \boldsymbol{H}(\boldsymbol{w}_i^l, \boldsymbol{w}_j^l, \boldsymbol{n}_{ij}),$$

with $\boldsymbol{w}_i^l$, $\boldsymbol{w}_j^l$ denoting the approximate solution on the elements adjacent to the edge Γ_{ij} at the time instant t_l. In the case of a boundary face the vector $\boldsymbol{w}_j^l$ has to be specified. Here we show the numerical flux based on the solution of the Riemann problem for the split Euler equations. By $\boldsymbol{w}_{\Gamma_{ij}}^l$ let us denote the state vector $\boldsymbol{w}$ at the center of the edge Γ_{ij} at the time instant t_l, and let us suppose $\boldsymbol{w}_{\Gamma_{ij}}^l$ is known. Evaluation of these values will be a question of the further analysis, here we use them to approximate the integrals with the one-point rule

$$\int_{\Gamma_{ij}}\sum_{s=1}^{3}\boldsymbol{f}_s(\boldsymbol{w}(x, t_l))(n_{ij})_s\,dS \approx |\Gamma_{ij}|\sum_{s=1}^{3}\boldsymbol{f}_s(\boldsymbol{w}_{\Gamma_{ij}}^l)(n_{ij})_s. \quad (10)$$

$$\int_{\Gamma_{ij}} \sum_{s=1}^{3} \boldsymbol{R}_s(\boldsymbol{w}(x,t_l), \nabla \boldsymbol{w}(x,t_l))(n_{ij})_s \, dS \approx |\Gamma_{ij}| \sum_{s=1}^{3} \boldsymbol{R}_s(\boldsymbol{w}^l_{\Gamma_{ij}}, \nabla \boldsymbol{w}^l_{\Gamma_{ij}})(n_{ij})_s.$$

Here $\nabla \boldsymbol{w}^l_{\Gamma_{ij}}$ denotes the $\nabla \boldsymbol{w}$ at the center of the edge Γ_{ij} at time instant t_l. Now it is possible to approximate the system (8) by the following explicit finite volume scheme

$$(\boldsymbol{w}_i^{k+1} - \boldsymbol{w}_i^k) + \Phi(\boldsymbol{w}_i^k) = 0 \qquad (11)$$

with

$$\Phi(\boldsymbol{w}_i^k) = \frac{\tau_k}{|D_i|} \sum_{\Gamma_{ij} \in \Gamma_{D_i}} |\Gamma_{ij}| \left(\boldsymbol{H}(\boldsymbol{w}_i^k, \boldsymbol{w}_j^k, \boldsymbol{n}_{ij}) - \sum_{s=1}^{3} \boldsymbol{R}_s(\boldsymbol{w}^k_{\Gamma_{ij}}, \nabla \boldsymbol{w}^k_{\Gamma_{ij}}) n_s \right).$$

With this finite volume formula one computes the values of the approximate solution at the time instant t_{k+1}, using the values from the time instant t_k, and by evaluating the values $\boldsymbol{w}^k_{\Gamma_{ij}}$ at the faces Γ_{ij}. In order to achieve the stability of the used method, the time step τ_k must be restricted by the so-called CFL condition, see [5]. The crucial problem of this discretization lies with the evaluation of the edge values $\boldsymbol{w}^k_{\Gamma_{ij}}$. Or one deals with the problem of finding the face fluxes $\boldsymbol{H}(\boldsymbol{w}_i^k, \boldsymbol{w}_j^k, \boldsymbol{n}_{ij})$. It is also possible to use the **implicit scheme**

$$(\boldsymbol{w}_i^{k+1} - \boldsymbol{w}_i^k) + \Phi(\boldsymbol{w}_i^{k+1}) = 0 \qquad (12)$$

The Newton method.
We may proceed as shown in [5][page 216]. At each time level t_{k+1} the following iterative scheme is applied:

$$\boldsymbol{w}_i^{k+1,0} = \boldsymbol{w}_i^k \qquad (13)$$

$$\left(\mathbb{I} + \frac{D\Phi(\boldsymbol{w}_i^{k+1,r})}{D\boldsymbol{w}} \right) (\boldsymbol{w}_i^{k+1,r+1} - \boldsymbol{w}_i^{k+1,r}) = \qquad (14)$$
$$-\boldsymbol{w}_i^{k+1,r} + \boldsymbol{w}_i^k - \Phi(\boldsymbol{w}_i^{k+1,r}), \quad r = 0, 1, \dots$$

Here $\frac{D\Phi(\boldsymbol{W})}{D\boldsymbol{w}}$ is the Jacobian matrix of the mapping Φ. Using such process we obtain the sequence $\left\{\boldsymbol{w}_i^{k+1,r}\right\}$ converging to $\boldsymbol{w}_i^{k+1}$ as $r \to \infty$.

The crucial problem of this discretization lies with the evaluation of the face fluxes $\boldsymbol{H}(\boldsymbol{w}_i^{k+1}, \boldsymbol{w}_j^{k+1}, \boldsymbol{n}_{ij})$. One possibility is to use the linearization via the Taylor expansion of the vector function $\boldsymbol{H}(\boldsymbol{wI}, \boldsymbol{wJ}, \boldsymbol{n})$, this was shown in [3].

5 The Riemann problem for the split Euler equations

For many numerical methods dealing with the two or three dimensional equations, describing the compressible flow, it is useful to solve the Riemann problem for the *3D split* Euler equations. We search the solution of the system of partial differential equations in time t and space (x, y, z)

$$\begin{aligned}
\frac{\partial \varrho}{\partial t} + \frac{\partial \varrho u}{\partial x} &= 0 \\
\frac{\partial \varrho u}{\partial t} + \frac{\partial (p + \varrho u^2)}{\partial x} &= 0 \\
\frac{\partial \varrho v}{\partial t} + \frac{\partial \varrho u v}{\partial x} &= 0 \\
\frac{\partial \varrho w}{\partial t} + \frac{\partial \varrho u w}{\partial x} &= 0 \\
\frac{\partial E}{\partial t} + \frac{\partial u(E + p)}{\partial x} &= 0,
\end{aligned} \qquad (15)$$

equipped with the initial conditions

$$\varrho(x,t) = \varrho_L, \boldsymbol{v}(x,t) = \boldsymbol{v}_L, p(x,t) = p_L, \quad x < 0, \qquad (16)$$

$$\varrho(x,t) = \varrho_R, \boldsymbol{v}(x,t) = \boldsymbol{v}_R, p(x,t) = p_R, \quad x > 0. \qquad (17)$$

Vector $\boldsymbol{v} = (u, v, w)$ denotes the velocity, ϱ density, p pressure, $E = \varrho e + \varrho |\boldsymbol{v}|^2$ is the total energy, with e denoting the specific internal energy. We assume the equation of state for the calorically ideal gas

$$e = \frac{p}{\varrho(\gamma - 1)}.$$

'Split' means here that we still have 5 equations in 3D, but the dependence on y, z (space coordinates) is neglected, and we deal with the system for one space variable x. The system (15) is considered in the set $Q_\infty = (-\infty, \infty) \times (0, +\infty)$.

The solution of this problem is fundamentally the same as the solution of the Riemann problem for the 1D Euler equations, see [5, page 138]. In fact, the solution for the pressure, the first component of the velocity, and the density is exactly the same as in one-dimensional case. It is a characteristic feature of the hyperbolic equations, that there is a possible raise of discontinuities in solutions, even in the case when the initial conditions are smooth, see [6, page 390]. Here the concept of the classical solution is too restrictive, and therefore we seek a weak solution of this problem. To distinguish physically admissible solutions from nonphysical ones, entropy condition must be introduced, see [6, page 396]. By the solution of the problem (15),(16),(17) we mean the *weak entropy solution* of this problem in Q_∞. The analysis to the solution of this problem can be found in many books, i.e. [5], [6], [7]. The general theorem on the solvability of the Riemann problem can be found in [5, page 88].

The solution is piecewise **smooth** and its general form can be seen in figure 1, where the system of half lines is drawn.

These half lines define regions, where solution is either constant or given by a smooth function. Let us define the open sets called **wedges** (see figure 1):

$$\begin{aligned}
\Omega_L &= \text{wedge}(L_{-\infty}, s_{\text{HL}}), & \Omega_{\star R} &= \text{wedge}(u_\star, s_{\text{TR}}), \\
\Omega_{\text{HTL}} &= \text{wedge}(s_{\text{HL}}, s_{\text{TL}}), & \Omega_{\text{HTR}} &= \text{wedge}(s_{\text{TR}}, s_{\text{HR}}), \\
\Omega_{\star L} &= \text{wedge}(s_{\text{TL}}, u_\star), & \Omega_R &= \text{wedge}(s_{\text{HR}}, L_{+\infty}).
\end{aligned}$$

Using the theory in [5], [6], [7], we can write the solution for the primitive variables as

$$\begin{aligned}
(\varrho, u, v, w, p)|_{\Omega_L} &= (\varrho_L, u_L, v_L, w_L, p_L), \\
(\varrho, u, v, w, p)|_{\Omega_{\star L}} &= (\varrho_{\star L}, u_\star, v_L, w_L, p_\star), \\
(\varrho, u, v, w, p)|_{\Omega_{\star R}} &= (\varrho_{\star R}, u_\star, v_R, w_R, p_\star), \\
(\varrho, u, v, w, p)|_{\Omega_R} &= (\varrho_R, u_R, v_R, w_R, p_R).
\end{aligned}$$

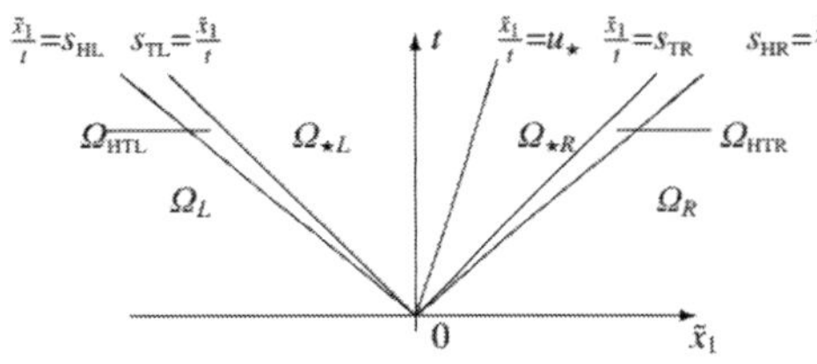

Fig. 1. Structure of the solution of the Riemann problem (15),(16),(17)

The folowing relations for these variables hold:

$$u_\star = u_L + \begin{cases} -(p_\star - p_L)\left(\frac{\frac{2}{(\gamma+1)\varrho_L}}{p_\star + \frac{\gamma-1}{\gamma+1}p_L}\right)^{\frac{1}{2}}, p_\star > p_L \\ \frac{2}{\gamma-1}a_L\left[1-\left(\frac{p_\star}{p_L}\right)^{(\gamma-1)/2\gamma}\right], p_\star \le p_L \end{cases} \tag{18}$$

$$\varrho_{\star L} = \begin{cases} \varrho_L \frac{\frac{\gamma-1}{\gamma+1}\frac{p_L}{p_\star}+1}{\frac{p_L}{p_\star}+\frac{\gamma-1}{\gamma+1}}, \; p_\star > p_L \\ \varrho_L\left(\frac{p_\star}{p_L}\right)^{\frac{1}{\gamma}}, \quad p_\star \le p_L \end{cases} \tag{19}$$

$$s^1_{TL} = \begin{cases} u_L - a_L\sqrt{\frac{\gamma+1}{2\gamma}\frac{p_\star}{p_L}+\frac{\gamma-1}{2\gamma}}, \; p_\star > p_L \\ u_\star - a_L\left(\frac{p_\star}{p_L}\right)^{(\gamma-1)/2\gamma}, \quad p_\star \le p_L \end{cases} \tag{20}$$

$$u_\star = u_R + \begin{cases} (p_\star - p_R)\left(\frac{\frac{2}{(\gamma+1)\varrho_R}}{p_\star + \frac{\gamma-1}{\gamma+1}p_R}\right)^{\frac{1}{2}}, p_\star > p_R, \\ -\frac{2}{\gamma-1}a_R\left[1-\left(\frac{p_\star}{p_R}\right)^{(\gamma-1)/2\gamma}\right], p_\star \le p_R \end{cases} \tag{21}$$

$$\varrho_{\star R} = \begin{cases} \varrho_R \frac{\frac{p_\star}{p_R}+\frac{\gamma-1}{\gamma+1}}{\frac{\gamma-1}{\gamma+1}\frac{p_\star}{p_R}+1}, \; p_\star > p_R \\ \varrho_R\left(\frac{p_\star}{p_R}\right)^{\frac{1}{\gamma}}, \quad p_\star \le p_R \end{cases} \tag{22}$$

$$s^3_{TR} = \begin{cases} u_R + a_R\sqrt{\frac{\gamma+1}{2\gamma}\frac{p_\star}{p_R}+\frac{\gamma-1}{2\gamma}}, \; p_\star > p_R \\ u_\star + a_R\left(\frac{p_\star}{p_R}\right)^{(\gamma-1)/2\gamma}, \quad p_\star \le p_R \end{cases} \tag{23}$$

Here $a_L = \sqrt{\gamma p_L/\varrho_L}$, $a_R = \sqrt{\gamma p_R/\varrho_R}$, and γ denotes the adiabatic constant. Further s^1_{TL} denotes "unknown left wave speed", s^3_{TR} "unknown right wave speed". Note, that the system (18) - (23) is the system of 6 equations for 6 unknowns $p_\star, u_\star, \varrho_{\star L}\, \varrho_{\star R}, s^1_{TL}, s^3_{TR}$.

Solution for the Pressure $p_\star$
The combination of the equations (18), (21) gives the implicit equation for the unknown pressure $p_\star$

$$u_L + F_{1L}(p_\star) = u_R + F_{3R}(p_\star),$$

with

$$F_{1L}(p) = \begin{cases} -(p - p_L)\left(\frac{\frac{2}{(\gamma+1)\varrho_L}}{p + \frac{\gamma-1}{\gamma+1}p_L}\right)^{\frac{1}{2}}, \quad p > p_L \\ \frac{2}{\gamma-1}a_L\left[1-\left(\frac{p}{p_L}\right)^{(\gamma-1)/2\gamma}\right], \; p \le p_L. \end{cases},$$

$$F_{3R}(p) = \begin{cases} (p - p_R)\left(\frac{\frac{2}{(\gamma+1)\varrho_R}}{p + \frac{\gamma-1}{\gamma+1}p_R}\right)^{\frac{1}{2}}, \quad p > p_R, \\ -\frac{2}{\gamma-1}a_R\left[1-\left(\frac{p}{p_R}\right)^{(\gamma-1)/2\gamma}\right], \; p \le p_R. \end{cases}$$

This is a nonlinear algebraic equation, and one cannot express the analytical solution of this problem in a closed form. The solution $p_\star$ can be found as the root of the function $F(p)$ defined as

$$F(p) = F_{3R}(p) - F_{1L}(p) + u_R - u_L. \tag{24}$$

The analysis of this function can be found in [5]. Here we state, that $F(p)$ is monotone and concave. Also the analytic expression for its derivative is available. For a positive solution for the pressure $F(0) < 0$ is required. This gives the *pressure positivity condition*

$$u_R - u_L < \frac{2}{\gamma - 1}(a_L + a_R). \tag{25}$$

The Newton method can be used to find the root of $F(p) = 0$. With the pressure $p_\star$ known, we use the equations (18)-(23) to obtain the whole solution.
In our work we use the analysis of this problem also for the solution of the initial-boundary value at the boundary faces, see also [8–11].

6 Boundary condition for the diffusible barrier

Here we present the diffusible barrier condition as a combination of Darcy's law with the additional innertial losses.

$$\frac{\Delta p}{\Delta m} = -\left(\frac{\mu}{\alpha}U + C_o\frac{1}{2}RU^2\right), \text{where} \tag{26}$$

$\Delta p = p_2 - p_1$ ... pressure difference across the barrier
Δm ... thickness of the barrier, example $\Delta m \in (0.01, 0.05)$
μ ... dynamic viscosity
α ... permeability coefficient, example $\alpha = 10^{-8}$
C_o ... pressure gradient coefficient, example $C_o = 10^2$
R ... density at the barrier
U ... velocity at the barrier
The equation (26) can be rewritten in the form

$$RU^2 + CU = d(p_1 - p_2), \tag{27}$$

here $C = \frac{2\mu}{\alpha C_o} > 0$, $d = \frac{2}{\Delta m C_o} > 0$, $(p_1 - p_2) \ge 0$. We are interested in the boundary values R, U, P at the barrier. Further we require that the conservation laws (1D Euler equations) are satisfied in the close vicinity of the barrier

$$\frac{\partial}{\partial t}\begin{pmatrix}\varrho \\ \varrho u \\ E\end{pmatrix} + \frac{\partial}{\partial \tilde{x}_1}\begin{pmatrix}\varrho u \\ \varrho u^2 + p \\ (E+p)u\end{pmatrix} = 0, \quad t \in (0,\infty),\ \tilde{x}_1 \in (-\infty, \infty). \tag{28}$$

Here the axis $\tilde{x}_1$ is perpendicular to the barrier, $\varrho(\tilde{x}_1, t)$ denotes the density, $p(\tilde{x}_1, t)$ is the pressure, $u(\tilde{x}_1, t)$ is the velocity (with the direction perpendicular to the barrier), $E(\tilde{x}_1, t)$ denotes the total energy: $E = \varrho u^2/2 + p/(\gamma - 1)$. The initial condition is formed by the two states near the barrier. Let us denote these states ϱ_1, u_1, p_1 and ϱ_2, u_2, p_2:

$$\begin{aligned} &\varrho(\tilde{x}_1, 0) = \varrho_1 \text{ for } \tilde{x}_1 < 0, \quad \varrho(\tilde{x}_1, 0) = \varrho_2 \text{ for } \tilde{x}_1 > 0, \\ &u(\tilde{x}_1, 0) = u_1 \text{ for } \tilde{x}_1 < 0, \quad u(\tilde{x}_1, 0) = u_2 \text{ for } \tilde{x}_1 > 0, \\ &p(\tilde{x}_1, 0) = p_1 \text{ for } \tilde{x}_1 < 0, \quad p(\tilde{x}_1, 0) = p_2 \text{ for } \tilde{x}_1 > 0. \end{aligned} \tag{29}$$

In Section 5 we have shown the unique solution (entropy weak) of the Riemann problem (28),(29). Adding another condition (27) into this complete system doesn't make sense. Therefore we think of the barrier problem as of two boundary problems with the two particular solutions (different in general).

- Problem 1. Find the solution R_1, U_1, P_1 of the system (28), equipped with the one-side initial condition ϱ_1, u_1, p_1 and the boundary condition (27).
- Problem 2. Find the solution R_2, U_2, P_2 of the system (28), equipped with the one-side initial condition ϱ_2, u_2, p_2 and the boundary condition (27).

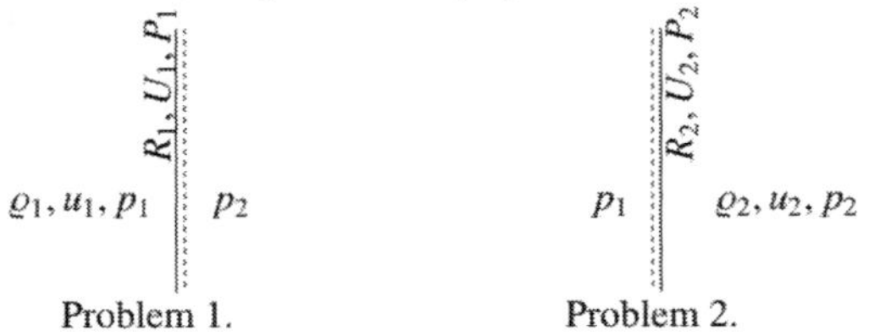

Further we will show the solution of these two problems. The analysis of these problems was shown in [1], here we write the resulting algorithm. We suppose the initial velocity $u_1 \geq 0$, and we seek the solution with $U > 0$.

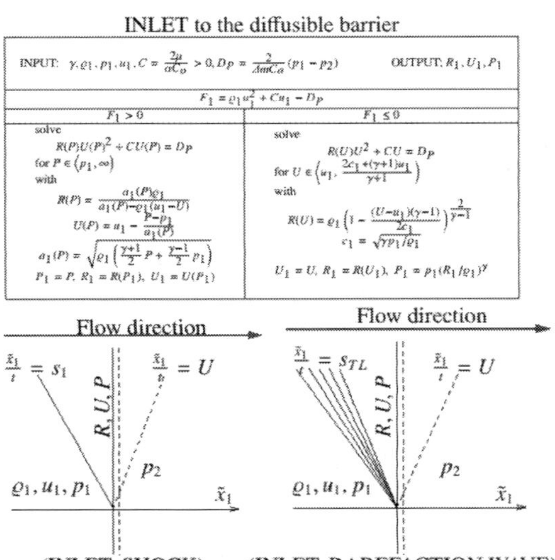

Fig. 2. Algorithm for the solution of the Problem 1 (INLET to the barrier). Both posibilities are taken into account.

7 Examples

Here we show the computational results of the compressible gas flow, obtained with the own-developed codes (the RANS with the k-ω model of turbulence). The gas constants for the air are $\gamma = 1.4$ and $R = 287.04$ [J kg^{-1} K^{-1}]. We use the viscosity coefficient (depending on the local temperature) computed with the use of the Sutherland's law $\mu = \mu_{ref}\left(\frac{T}{T_{ref}}\right)^{3/2}\frac{T_{ref}+S}{T+S}$, with $S = 110.4$[K],

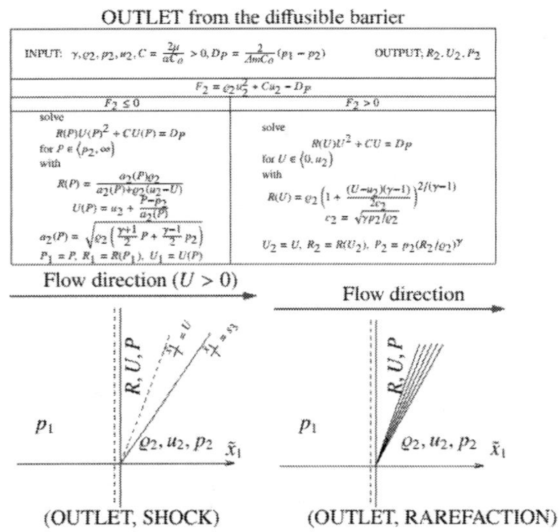

Fig. 3. Algorithm for the solution of the Problem 2 (OUTLET from the barrier). Both posibilities are taken into account.

$T_{ref} = 273.15$[K], $\mu_{ref} = 1.716e-05$[Pa s], and the heat conduction coefficient $k = 0.0211$. The total temperature $\theta_0 = 273.15$ [K], and the total pressure $p_0 = 101325$ [Pa] was chosen.

The figure 4. shows the computed result of the flow through the porous media, with the velocity regime 15 [m s^{-1}]. Simple porous area was chosen, and various near-ground wind blast intensities were tested.

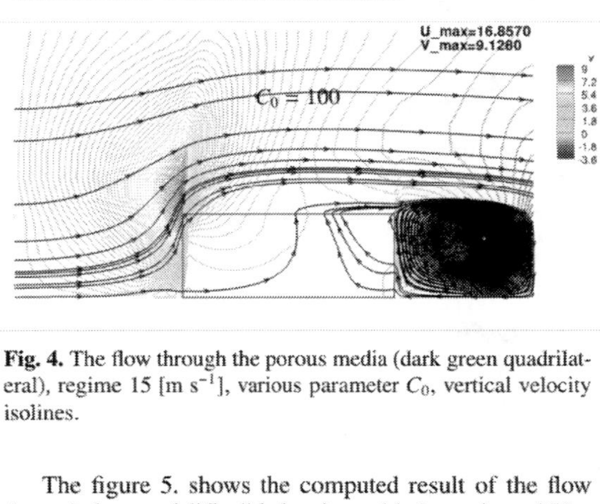

Fig. 4. The flow through the porous media (dark green quadrilateral), regime 15 [m s^{-1}], various parameter C_0, vertical velocity isolines.

The figure 5. shows the computed result of the flow through the set of diffusible barriers with the regimes 15 [m s^{-1}] and 25 [m s^{-1}]. The diffusible barrier was simulated using the boundary condition described in section 6. with constants $C_0 = 25, C_0 = 100$, and $\Delta m = 0.01$, $\alpha = 10^{-8}$. Further, the inlet blast with the total pressure increased by 12882.888[Pa] was tested.

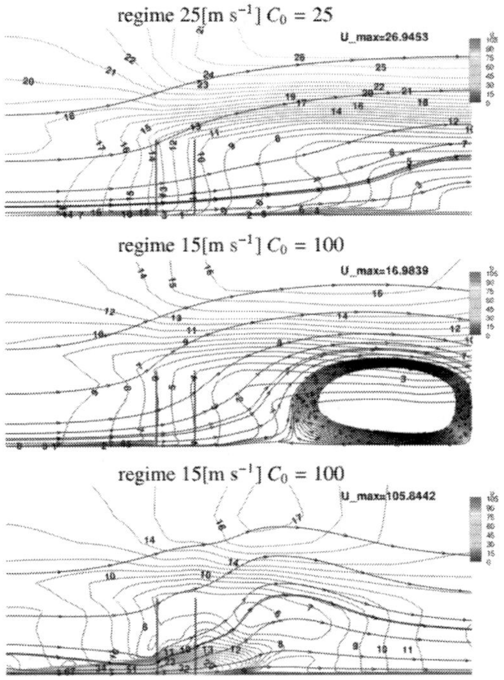

Fig. 5. The flow through the diffusible barrier (two vertical red lines represent the barriers), various regimes and barrier parameters, near-ground wind blast at the bottom picture. Horizontal component of velocity isolines and streamtraces shown.

Conclusion

This paper is focused on the viscous compressible flow, with the focus on the porous media and the diffusible barrier. The boundary condition for the diffusible barrier (analyzed by authors in [1]) is shown together with the computational algorithm. The work is based on the analysis of the Riemann Problem for the split Euler equations and the modifications of this problem. Here the left hand side initial condition is replaced by various complementary conditions. All codes were implemented into the own-developed software. The numerical examples were presented.

Acknowledgment

The results originated with the support of Ministry of the Interior of the Czech Republic, project SCENT, Grant MSM 0001066902 of the Ministry of Education of the Czech Republic, and Ministry of Industry and Trade of the Czech Republic for the long-term strategic development of the research organization. The authors acknowledge this support.

References

1. M. Kyncl and J. Pelant, *EPJ WoC*, **45**, 01054, (2013)
2. C. J. Kok, *AIAA Journal*, **38**, 7, (2000)
3. M. Kyncl and J. Pelant, *Implicit method for the 3d RANS equations with the k-ω (Kok) Turbulent Model.* Technical report R5453, (2012)
4. M. Kyncl and J. Pelant, *Implicit method for the 3d Euler equations.* Technical report R5375, VZLÚ, (2012)
5. M. Feistauer, J. Felcman, and I. Straškraba, *Mathematical and Computational Methods for Compressible Flow.* Oxford University Press, Oxford, (2003)
6. M. Feistauer, *Mathematical Methods in Fluid Dynamics.* Longman Scientific & Technical, Harlow, (1993)
7. E. F. Toro, *Riemann Solvers and Numerical Methods for Fluid Dynamics.* Springer, Berlin, (1997)
8. M. Kyncl. *Numerical solution of the three-dimensional compressible flow.* Doctoral Thesis, Prague, (2011)
9. M. Kyncl and J. Pelant, *Proc. ECCOMAS 2014, ECFD VI, 20.-25.7.2014*, Spain (2014)
10. M. Kyncl and J. Pelant, *Appl. Mech. and Mat.*, **821**, 70-78, (2016)
11. M. Kyncl and J. Pelant, *Proc. ECCOMAS 2016, 5.-10.6.2016*, Greece (2016)

https://doi.org/10.1051/epjconf/201818002053

Droplets size evolution of dispersion in a stirred tank

Bohus Kysela[1a], Jiri Konfrst[1], Zdenek Chara[1], Radek Sulc[2], and Darina Jasikova[3]

[1] The Czech Academy of Sciences, Institute of Hydrodynamics
Pod Patankou 30/5, 166 12 Prague 6, Czech Republic
[2] Czech Technical University in Prague, Faculty of Mechanical Enginnering, Department of Process Engineering, Technicka 4, 166 07 Prague 6, Czech Republic
[3] Technical University of Liberec, Institute for Nanomaterials, Advanced Technology and Innovation, Studentska 1402/2, 461 17 Liberec, Czech Republic

Abstract. Dispersion of two immiscible liquids is commonly used in chemical industry as wall as in metallurgical industry e. g. extraction process. The governing property is droplet size distribution. The droplet sizes are given by the physical properties of both liquids and flow properties inside a stirred tank. The first investigation stage is focused on in-situ droplet size measurement using image analysis and optimizing of the evaluation method to achieve maximal result reproducibility. The obtained experimental results are compared with multiphase flow simulation based on Euler-Euler approach combined with PBM (Population Balance Modelling). The population balance model was, in that specific case, simplified with assumption of pure breakage of droplets.

1 Introduction

Particle size distribution is a crucial property in particulate processes and it affects a lot of quantities - flow of suspension, dispersion, eventuallly separation processes, mass transfer etc. This contribution is focused on in-situ measurements and evaluation of particle size distribution of suspended solid spherical particles and spherical droplets of oil in water-oil dispersion. The dispersions were processed in an agitated vessel. These investigated processes are commonly used in chemical or metallurgical industry (e.g. extraction, catalytic reactions, polymerization etc.). The dynamic evolution of droplet size distribution in time inside an agitated vessel has been investigated by many researchers [1–5]. They tried to develop a simple model of droplet size distribution developing in time and depending on parameters of the mixing system. The agitated vessel geometry or impeller types were also examined [5,4]. Several approaches applying of population balance modelling [6–9] were used. The generalized equation of population balance can be used to solve grow of crystals (molecular transport, nucleation, breakage, aggregation), aggregation process of solid particles in fluid, liquid-liquid dispersion process and many others. The method is based on the solution of integro-differential equations with probability density functions. There are many methods how to solve population balance, one of the commonly used is the QMOM (Quadrature method of moments) based on quadrature approximation [10,11] which can be very easily implemented to existing solvers based on finite volume method. Hence it is used in coupling with CFD (Computational fluid dynamics). The coupling is based on media transport, application of turbulence diffusion as well as application of dissipation rate of turbulent kinetic energy in breakage kernel functions. The main aim of our investigation is development of very precise measurement of the droplet size distribution in a pseudo steady state, when the system is in balance and particle size distribution is stable (balance for one impeller speed in agitated vessel system), and assumed breakage process only. Then the initial conditions and calculation setting could by adjusted to the numerical simulation. The next step is validation of the time dependent evolution droplet size modelling in future.

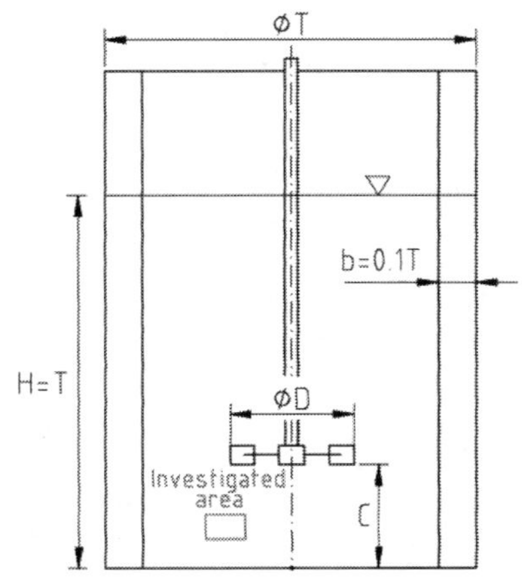

Fig. 1. Geometry of the mixing tank (T = 300 mm, H/T = 1, T/D = 3, b/T = 1/10, four baffles).

2 Droplet size measurement

Measurement was performed using two immiscible liquids - water and silicone oil in an agitated system schematically described in Fig.1. The particle sizes were evaluated from the raw image records by the image analysis.

[a] e-mail: kysela@ih.cas.cz

2.1 Experimental set-up

The experiments were performed in the agitated vessel with standard cylindrical geometry and with the flat bottom. The vessel was equipped with four radial baffles (width b = 0.1 T). The diameter of the vessel was T = 300 *mm*. The vessel was filled with water and the liquid level was H = T. The standard Rushton impeller with ratio T/D = 3 and off-bottom clearance C = 0.75 D (see Fig.1) was used. The impeller speed was 150; 200; 250; 300 *rpm*. The model liquids were distilled water and silicone oil AP 200 (density = 1070 $kg \cdot m^3$, kinematic viscosity = $2 \cdot 10^{-4}\ m^2 \cdot s^{-1}$ at 25°*C*). Used dispersed oil phase fraction was 0.0047, where the oil was added in before measurement by syringe injection. The size of the investigated area (Fig.1) was approximately 15 × 12 *mm*. The area was located 40 *mm* from impeller axis and off-bottom clearance was 30 *mm*. The focused plane was approximately 50 *mm* from the vessel wall. This position was chosen due to the circulation loop generated by the radial impeller. The dispersion in that place is moving uniformly.

2.2 Image recording

The image recording was realised by a high speed camera system SpeedSence MK III with frame rate lowered to 30 *fps*, shutter time 0.1 *ms*, full resolution 1280 × 1024 *pixels*. The frames were captured at the region under the impeller (see Fig. 1), where the particle velocities were below 1 $m \cdot s^{-1}$. The image resolution was 0.0118 *mm/pixel* in this configuration. Each record was 1000 images (total time 33.3 *s*) and time of 10 *min*. was taken between impeller speed changes to stabilise the droplet distribution.

2.3 Image analysis

The detected objects evaluation was investigated by several approaches. At first the illumination of the captured area was tested. The light sheets as well as various light sources were used. The light scattering and refraction trough the transparent droplets disallowed precise evaluation. Moreover, it was also shown that the evaluation by a standard threshold was not quite accurate. The back lighting combined with enhanced approach of image evaluation based on finding edges of spherical particles was found as the best final configuration (see Fig. 2. The three basic conditions had to be fulfilled:

i) Only different particles should be detected on the subsequent frames (that is why frame rate was only 30 fps).
ii) The moving particles should have sharp boundary (very short shutter time is necessary as well as high-power light source)
iii) Using precise spherical particles or providing droplets with minimal deformation.

The final approach for the image analysis was: 1) edges finding (high gradients), 2) threshold, 3) evaluation of particle area using fill-holes filter and enhanced by circularity evaluation.

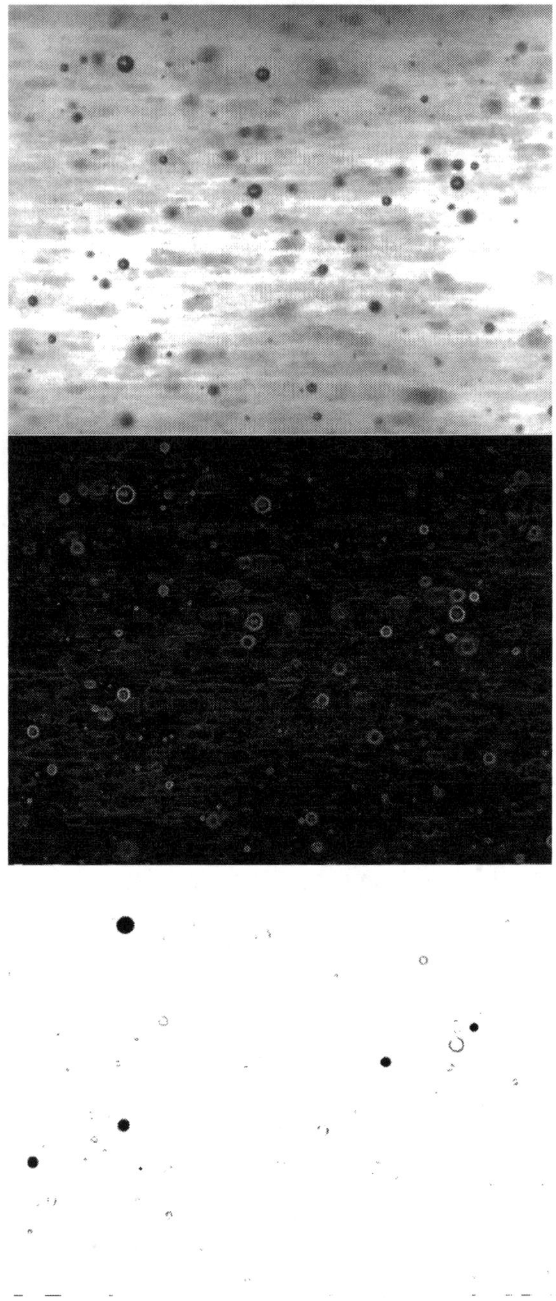

Fig. 2. Example of obtained raw image (top), finding edges - high gradients (middle), and evaluated result (bottom).

3 Population balance modelling (PBM)

The dispersion process of the oil droplets was also modelled using CFD-PBM approach. The special add-on of ANSYS Fluent software package was used. The numerical simulations were performed using combination of fluid flow solution in the modelled agitated vessel and implemented PBM solution.

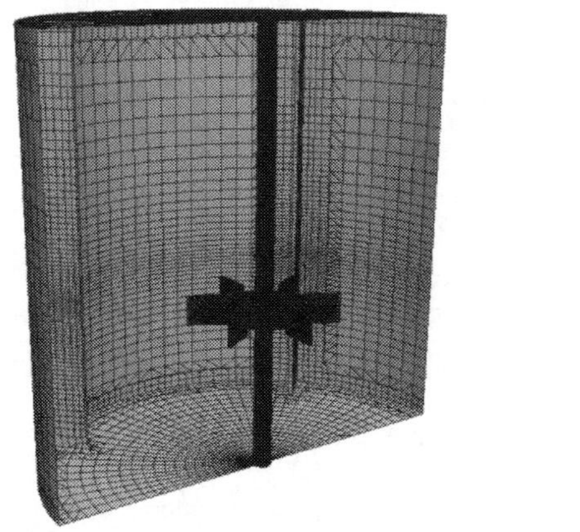

Fig. 3. Mesh of the agitated vessel and Rushton turbine impeller.

The flow was solved with transient RANS model k-ε Realizable combined with MRF(moving reference frame) method for simulation of the impeller movement. The MRF region has the cylindrical shape with distance D/10 under and below impeller cylindrical envelope and with diameter 2D. The mesh size was approximately 0.5 mil. mainly tetrahedral cells more dense in impeller region (Fig. 3). Two types of boundary conditions were defined: no slip walls (impeller, tank) and moving wall (impeller shaft outside of MRF zone). Standard wall functions were used for the walls. The population balance is given in that specific case only by the breakage of droplets without nucleation and molecular growth and also without aggregation because the droplets are in very low concentration. Thus the population balance [9] can be simplified:

$$\frac{\partial n(L;\mathbf{x},t)}{\partial t} + \langle u_i \rangle \frac{\partial n(L;\mathbf{x},t)}{\partial x_i} - \frac{\partial}{\partial x_i}\left[\Gamma_t \frac{\partial n(L;\mathbf{x},t)}{\partial x_i}\right] = B(L;\mathbf{x},t) - D(L;\mathbf{x},t), \quad (1)$$

where $n(L;\mathbf{x},t)$ is particle size distribution function, and L represents particle size, $\mathbf{x}$ is position vector, t is time. This equation is specified for combination with CFD calculations and $\langle u_i \rangle$ is Reynolds-averaged velocity in the i-th direction, and Γ_t represents turbulent diffusivity. The terms on the left side $B(L;\mathbf{x},t)$ and $D(L;\mathbf{x},t)$ are, respectively, the bird and death rates due to breakage. The moments of the PSD are defined as follows:

$$m_k(\mathbf{x},t) = \int_0^{\infty} n(L;\mathbf{x},t) L^k \mathrm{d}L \quad (2)$$

and thus the transport equation for the k-th moment is

$$\frac{\partial m_k(\mathbf{x},t)}{\partial t} + \langle u_i \rangle \frac{\partial m_k(\mathbf{x},t)}{\partial x_i} - \frac{\partial}{\partial x_i}\left[\Gamma_t \frac{\partial m_k(\mathbf{x},t)}{\partial x_i}\right] = \overline{B_k}(\mathbf{x},t) - \overline{D_k}(\mathbf{x},t), \quad (3)$$

where the moments of the bird and death are

$$\overline{B_k}(\mathbf{x},t) = \int_0^{+\infty} L_k \int_0^{+\infty} a(\lambda)\, b(L|\lambda)\, n(\lambda;\mathbf{x},t)\, \mathrm{d}\lambda\, \mathrm{d}L, \quad (4)$$

$$\overline{D_k}(\mathbf{x},t) = \int_0^{+\infty} L^k\, a(L)\, n(L;\mathbf{x},t)\, \mathrm{d}L. \quad (5)$$

The breakage kernel $a(L)$ defines the rate coefficient for breakage particle of the length L. The fragment distribution function for the breakage of a particle of size L is given by $b(L|\lambda)$, where L and λ are lengths of two particles. The breakage kernel was used Luo-Svensen [12] for liquid-liquid system:

$$a(L) = c_4 \left(\frac{\varepsilon}{L^2}\right)^{1/3} \int_{\xi_{min}}^{1} \frac{(1+\xi)^2}{\xi^{11/3}} \exp\left[-\frac{12 c_f \sigma}{\beta \rho_d \varepsilon^{2/3} L^{5/3} \xi^{11/3}}\right] \mathrm{d}\xi \quad (6)$$

where the $\xi = \eta/L$, η is the eddy length-scale, c_4 is a constant of order unity ($c_4 = 0.92$), c_f is the increase coefficient in the surface area $[c_f = f^{2/3} + (1-f)^{2/3} - 1]$ and f is the volume fraction of one of two fragments.

The quadrature method of moments (QMOM) is based on the following quadrature approximation:

$$m_k(\mathbf{x},t) = \int_0^{\infty} n(L) L^k \mathrm{d}L \approx \sum_{i=1}^{N_d} w_i L_i^k \quad (7)$$

where weights (w_i) and abscissas (L_i) are determined through the product-difference (PD) algorithm from the low-order moments [9]. By using the PD algorithm, a quadrature approximation with N_d weights and N_d abscissas can be constructed using the first $2N_d$ moments of PSD. For example, if $N_d = 3$ only the first six moments ($m_0, ..., m_5$) are tracked, and the quadrature approximation is given by

$$m_k = \sum_{i=1}^{N_d=3} w_i L_i^k = w_1 L_1^k + w_2 L_2^k + w_3 L_3^k. \quad (8)$$

Knowledge of w_i and L_i suffices to close the transport equations for the moments, and in our case with the presumption of pure breakage the moment equation become

$$\frac{\partial m_k(\mathbf{x},t)}{\partial t} + \langle u_i \rangle \frac{\partial m_k(\mathbf{x},t)}{\partial x_i} - \frac{\partial}{\partial x_i}\left[\Gamma_t \frac{\partial m_k(\mathbf{x},t)}{\partial x_i}\right] = \sum_i a_i \overline{b}_i^{(k)} w_i - \sum_i L_i^k a_i w_i, \quad (9)$$

where $a_i = a(L_i)$, and

$$\overline{b}_i^{(k)} = \int_0^{+\infty} L^k b(L|L_i) \mathrm{d}L. \quad (10)$$

It has been shown [9,10] that by using this quadrature approximation it is possible to track the moments of the PSD with very high accuracy. For the reconstruction of distribution from the finite number of its moments can be used several techniques [13,14].

4 Results

The measurement based on image analysis is limited by a number of evaluated particles. Moreover, the proposed more accurate method this number even decreases. The

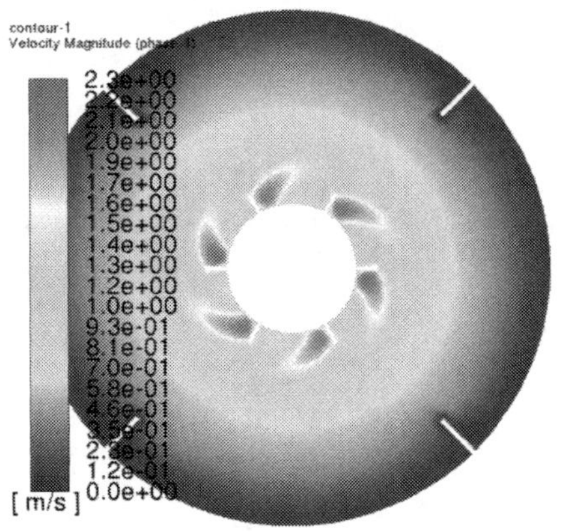

Fig. 4. Distribution of the velocity magnitude in the impeller plane (300rpm).

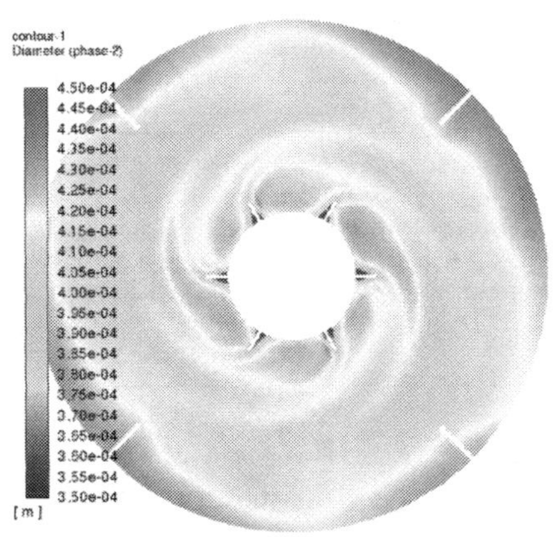

Fig. 5. Distribution of the particle mean Sauter (d_{32}) diameter in the impeller plane (300rpm).

dependency of mean Sauter droplet diameter on number of droplets was evaluated. The results corresponding to all impeller speeds are depicted in Fig. 6. There are visible changes approximately up to 1000 of evaluated droplets in the case of all impeller speeds. The decrease of mean diameter with number of evaluated droplets at 150 *rpm* was evidently caused by not absolutely stable state at the beginning of the measurement. The big jump in this dependency was induced by random appearing of several big droplets. It confirms an opinion of unstable beginning state. The real accuracy of image analysis evaluation was investigated using the precise spherical stainless steel particles with diameter 0.635 ± 0.001 *mm*. The evaluation process was set in accordance with the best obtained results for the spherical particles. The variance was under 3 %. The evolution of droplet volume cumulative distribution with impeller speed is shown in Fig. 7. Depicted results indicate a similar tendency at the increasing impeller speed as transient drop size distribution change obtained by Bak and Podgorska [2] for droplets in liquid-liquid system in an agitated vessel. The real experimental system of agitated vessel was subsequently simulated using CFD-PBM approach with identical geometry and identical conditions (impeller speeds, densities, viscosities etc.). The inter-facial tension between oil and water was determined $\sigma = 0.0035\ N \cdot m^{-1}$ [15] as an input value for the calculations. The breakage kernel was used Luo-Svensen [12] described in eq. 6. Input distribution of droplets was mono-disperse with 6*mm* in diameter (similar to the syringe dosage in real case). The simulation time was 10 *min* as well as in experiments. An example of both results flow field and mean Sauter ($d_{32} = m_3/m_2$) droplets diameter are depicted in Fig. 4 and Fig. 5, respectively. The results are shown in impeller plane, where regions with the maximal shear rates and high turbulent kinetic dissipation rates are usually found. The comparison of the droplet distribution from experiments and calculations are shown in Fig. 7. The distribution at 300 *rpm* is slightly underestimated by calculations. It could be caused by calculations of the turbulent kinetic energy dissipation rate (that are usually underestimated too) and also by the discrepancy in initial conditions and PBM inputs. The results at the 150 *rpm* indicate the discrepancy of distribution shape that is probably caused by the non-stable conditions in experimental measurements. Nevertheless, the trend of results for both measurements and simulations is consistent.

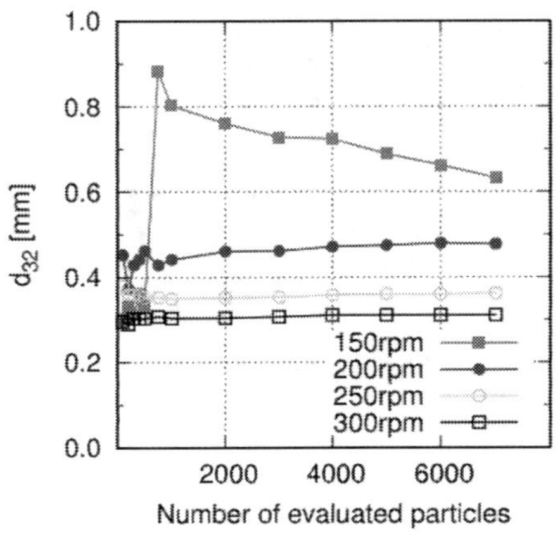

Fig. 6. Mean Sauter (d_{32}) droplet diameter changes with the number of evaluated particles.

5 Conclusions

The method of in-situ measurement of particle size distribution based on image analysis was improved with finding edges (high gradients in images) evaluation enhanced by circularity analysis. The high accuracy and good reproducibility allows to apply this method to precise measurement of the drop size distribution. The results were compared with calculations based on the combination of the CFD with the PBM. The results verified that only the breakage process is involved in our case (in very diluted dispersions). But it was found that the final results of CFD-PBM calculations are not only dependent on the model settings, but also on correctly defined conditions in comparison with experiments.

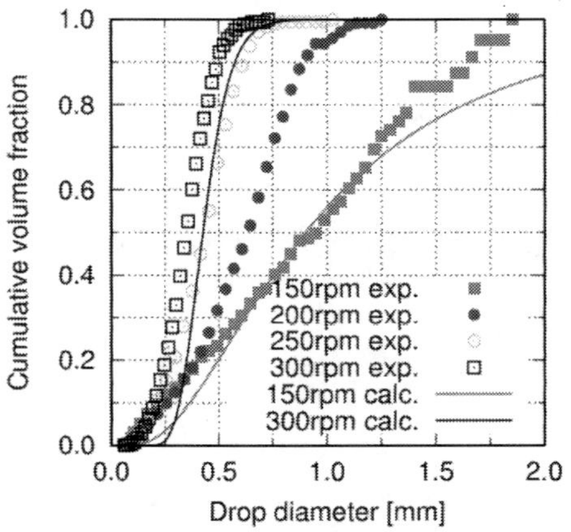

Fig. 7. Relation of the cumulative volume fraction on drop diameter at 150; 200; 250; 300 *rpm* (points - experiments, continuous line - simulations)

Acknowledgments

This research has been subsidized by the research project No. GA ČR 16-20175S and RVO: 67985874.

References

1. J. Baldyga, J. Bourne, A. Pacek, A. Amanullah, A. Nienow, Chemical Engineering Science **56**, 3377 (2001)
2. A. Bak, W. Podgarska, Chemical Engineering Research and Design **91**, 2142 (2013), mixing
3. P. Ditl, R. Sulc, Technical Transactions. **M-1**, 83 (2016)
4. G. Zhou, S.M. Kresta, Chemical Engineering Science **53**, 2099 (1998)
5. H. Tang, L. Wrobel, International Journal of Engineering Science **43**, 1234 (2005)
6. N.B. Raikar, S.R. Bhatia, M.F. Malone, D.J. McClements, C. Almeida-Rivera, P. Bongers, M.A. Henson, Colloids and Surfaces A: Physicochemical and Engineering Aspects **361**, 96 (2010)
7. J. Baldyga, W. Orciuch, Chemical Engineering Science **56**, 2435 (2001), industrial Crystallisation
8. A. Sathyagal, D. Ramkrishna, G. Narsimhan, Computers and Chemical Engineering **19**, 437 (1995)
9. D.L. Marchisio, R.D. Vigil, R.O. Fox, Chemical Engineering Science **58**, 3337 (2003)
10. D.L. Marchisio, R. Vigil, R.O. Fox, Journal of Colloid and Interface Science **258**, 322 (2003)
11. T. Nguyen, F. Laurent, R. Fox, M. Massot, Journal of Computational Physics **325**, 129 (2016)
12. H. Luo, H.F. Svendsen, AIChE Journal **42**, 1225 (1996)
13. V. John, I. Angelov, A. Oncul, D. Thevenin, Chemical Engineering Science **62**, 2890 (2007)
14. L. de Souza, G. Janiga, V. John, D. Thevenin, Chemical Engineering Science **65**, 2741 (2010)
15. F. Peters, D. Arabali, Colloids and Surfaces A: Physicochemical and Engineering Aspects **426**, 1 (2013)

Large-Eddy Simulation of Internal Flow through Human Vocal Folds

Martin Lasota[1] and *Petr* Šidlof[1,2]

[1] *Institute of New Technologies and Applied Informatics, Technical University of Liberec 1, Studentska 2, 461 17 Liberec, Czech Republic*
[2] *Institute of Thermodynamics, Academy of Sciences of the Czech Republic, Dolejskova 5, 182 00 Prague 8, Czech Republic*

Abstract. The phonatory process occurs when air is expelled from the lungs through the glottis and the pressure drop causes flow-induced oscillations of the vocal folds. The flow fields created in phonation are highly unsteady and the coherent vortex structures are also generated. For accuracy it is essential to compute on human-like computational domain and appropriate mathematical model. The work deals with numerical simulation of air flow within the space between plicae vocales and plicae vestibulares. In addition to the dynamic width of the rima glottidis, where the sound is generated, there are lateral ventriculus laryngis and sacculus laryngis included in the computational domain as well. The paper presents the results from OpenFOAM which are obtained with a large-eddy simulation using second-order finite volume discretization of incompressible Navier-Stokes equations. Large-eddy simulations with different subgrid scale models are executed on structured mesh. In these cases are used only the subgrid scale models which model turbulence via turbulent viscosity and Boussinesq approximation in subglottal and supraglottal area in larynx.

1 Introduction

Large-eddy simulation was proposed in as early as 1963 by Smagorinsky [7]. In LES the large scale motions of turbulent flow are computed directly and small scale motions (sub-grid scale = SGS) are modelled. LES has significant reduction in computational cost compared to DNS and is more accurate than the RANS approach. The large eddies contain most of the turbulent energy and these eddies are responsible for most of the momentum transfer and turbulent mixing. The small eddies incline to be isotropic and homogenous (and therefore the SGS motions should be easier to model than model all scales within a domain). Nowadays, LES is promising numerical tool for simulating realistic turbulent flows. For instance inside the complicated measurable place in human larynx.

2 Mathematical formulation

The Navier-Stokes equations are derived from the conservation laws for mass, momentum and energy. In LES a low-pass spatial filter is applied to the instantaneous conservation equations to formulate equations for large scale motions (explicit filtering). If the finite volume method (FVM) is used for solving the instantaneous governing equations numerically, then there is no need to apply a filter to the equations explicitly (it is called implicit filtering, equivalent to convolution with a top-hat filter). The filtered equations in a Newtonian incompressible flow can be written as

$$\partial_i \bar{u}_i = 0, \tag{1}$$

$$\partial_t(\rho \bar{u}_i) + \partial_j(\rho \bar{u}_i \bar{u}_j) = -\partial_i \bar{p} + 2\partial_j(\mu \overline{S_{ij}}) - \partial_j(\tau_{ij}), \tag{2}$$

$$\overline{S_{ij}} = \frac{1}{2}(\partial_i \bar{u}_j) + \partial_j \bar{u}_i), \tag{3}$$

$$\tau_{ij} = \rho(\overline{u_i u_j} - \bar{u}_i \bar{u}_j), \tag{4}$$

where $\bar{u}_i$ is filtered velocity, ρ is density, μ is molecular viscosity, $\overline{S_{ij}}$ is resolved scale strain rate tensor. [10]

In the context of LES τ_{ij} is called the subgrid scale (SGS) Reynolds stress. It plays a role in LES similar to the Reynolds stress in RANS, but physics that it models is different. The SGS energy is usually a much smaller part of the total energy at the turbulent velocity field than the RANS turbulent energy. Thus a model accuracy may be less crucial in LES in comparison with RANS. [1] To model SGS stress tensor we can use eddy-viscosity assumption (Boussinesq's hypothesis) as follows

$$\tau_{ij} = -2\mu_t \overline{S_{ij}} + \frac{1}{3}\delta_{ij}\tau_{ll}, \tag{5}$$

where μ_t is SGS eddy viscosity. If μ_t is substitute into the equation (2) then becomes

$$\partial_t(\rho \bar{u}_i) + \partial_j(\rho \bar{u}_i \bar{u}_j) = -\partial_i \bar{p} + 2\partial_j[(\mu + \mu_t)\overline{S_{ij}}], \tag{6}$$

and the remaining problem now is how to determine μ_t. The Smagorinsky model contains

$$\mu_t = \rho(C_s \bar{\Delta})^2 S, \tag{7}$$

Corresponding author: martin.lasota@tul.cz

$$S = (2\bar{S}_{ij}\bar{S}_{ij})^{\frac{1}{2}}, \quad (8)$$

$$\Delta = (\Delta_x\Delta_y\Delta_z)^{\frac{1}{3}}, \quad (9)$$

where C_S is the Smagorinsky constant (0.18). [4] The value gives reasonable results for isotropic turbulence and for flows near a solid wall it should be 0.1. [10] Nevertheless, this model has shortcomings as for example being too dissipative. Improvement of this model was suggested by Germano [3]. In OpenFOAM (OF) the μ_t is determined as follows

$$\mu_t = C_k\sqrt{k}\Delta, \quad (10)$$

$$k = \sqrt{-b + \frac{\sqrt{b^2 + 4ac}}{2a}}, \quad (11)$$

$$a = \frac{C_E}{\Delta},\ b = \frac{2}{3}\mathrm{tr}(\bar{\mathbf{S}}),\ c = 2C_k\Delta(\mathrm{dev}(\bar{\mathbf{S}} : \bar{\mathbf{S}}), \quad (12)$$

$$\Delta = c'V_C^{\frac{1}{3}}, \quad (13)$$

where C_k and C_E are model constants (C_k = 0.094, C_E = 1.048). Symbol Δ represents a filter width, c' is deltaCoeff (in OF=1) and V_C is the volume of the cell.

3 Geometry and mesh

The geometry of the vocal folds was specified by Scherer [6]. For LES computations, a structural mesh from 400k volume cells was created and it included 5 cells in the z-axis (5x 0.2 mm). The one of the important parameter for testing dynamic mesh (where the vocal folds have two-degrees of freedom) is maximal non-orthogonality factor. The analysis of this parametes make sense in the time, when the vocal folds are in the nearest position to each other, i.e. t = 0.003 s and the maximal non-orthogonality factor is 55.14).

Mesh spacings in wall units are commonly used to indicate LES adequacy. Especially the theoretical limits for LES and DNS are

$$50 \leq \Delta x^+ \leq 150,\ \Delta y^+_{wall} < 1,\ 15 \leq \Delta z^+ \leq 40, \quad (14)$$

$$10 \leq \Delta x^+ \leq 20,\ \Delta y^+_{wall} < 1,\ 5 \leq \Delta z^+ \leq 10, \quad (15)$$

where the x is streamwise direction, the y direction is wall normal and the z is spanwise and homogeneous. [2] In this study the y^+ for the critical time when the mesh is maximally deformed is: y^+_{avg} = 1.17, y^+_{max} = 17.29 and in different time is usually about y^+_{avg} = 0.31, y^+_{max} = 3.29.

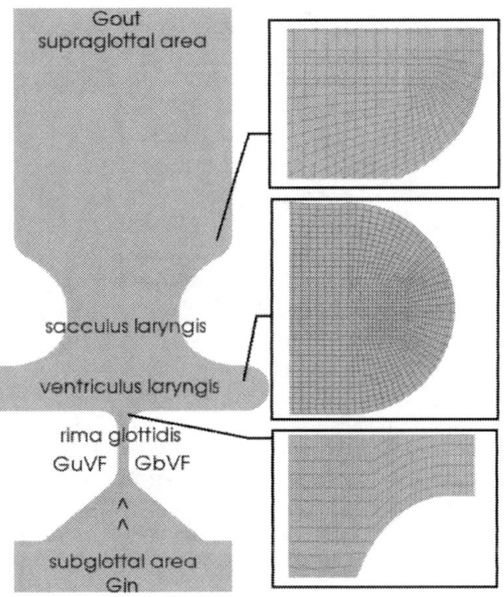

Fig. 1: Geometry and mesh

Table 1: Boundary conditions for the velocity $\boldsymbol{u}$, kinematic pressure $P = (p/\rho)$ and turbulent kinetic energy k

region	$\boldsymbol{u}$ [=] m.s^{-1}	P [=] m^2.s^{-2}	k [=] m^2.s^{-2}
Γ_{in}	from flux	300	2.10^{-5}
Γ_{out}	$\frac{\partial \mathbf{u}}{\partial \mathbf{n}} = 0$	0	$\frac{\partial k}{\partial \mathbf{n}} = 0$
Γ_{bVF}	$\boldsymbol{u} = \frac{\partial h(x,y)}{\partial t}$	$\frac{\partial P}{\partial \mathbf{n}} = 0$	$k = 0$
Γ_{uVF}	$\boldsymbol{u} = \frac{\partial h(x,y)}{\partial t}$	$\frac{\partial P}{\partial \mathbf{n}} = 0$	$k = 0$
Γ_{wall}	$\boldsymbol{u} = 0$	$\frac{\partial P}{\partial \mathbf{n}} = 0$	$k = 0$

Table 1 species the boundary condition of this presented case. Inlet velocity is computed from the flux if it is applied the condition $\boldsymbol{u}.\boldsymbol{n} < 0$ and if the $\boldsymbol{u}.\boldsymbol{n} > 0$ then $\boldsymbol{u} = 0$. The function $h(x, y)$ is calculated from (16), where f = 100 Hz, $A_1 = A_2$ = 0.3 mm, $\eta = \pi/2$

$$w_1 = A_1 sin(2\pi f + \eta),\ w_2 = A_2 sin(2\pi f) \quad (16)$$

4 Numerical solution

In this LES study, central differencing scheme (CDS) is used for spatial discretization of the diffusive term. The CDS can be obtained by a Taylor series expansion where the terms involve derivatives of the second order (higher are neglected). The second order CDS is non-dissipative and conservative which is essential for LES. Upwind-based schemes are not used in LES, because of the production of high numerical dissipation. For spatial discretization of convective term, a total variation diminishing scheme (TVD) is used. A numerical method is said to be TVD, if the total variation (TV) is not increased in time, see (17), where ϕ is a variable, i is an index of the node. A monotone scheme is TVD and a TVD scheme is monotonicity preserving. It does not create any new local extrema within solution domain. In our case we have used high resolution scheme, see (18), where the term called the anti-diffusive flux creates 2nd order of accuracy (which it decreases numerical diffusion, but also it leads to unphysical oscillations). To preserve the good properties, i.e. the stability from the 1st order scheme and the accuracy from the 2nd order scheme is to multiple the flux limiter ψ to the anti-diffusive term, see (19), where indexes are D-downwind (left), C-center, U-upwind (right), f-face (between C and U) and r_f is taken as the ratio of two consecutive gradients. [5] In this study was used the flux limiter of the MUSCL TVD scheme as the discretization scheme of the convective term, see (20) and CD TVD scheme with $\psi(r_f) = 1$ as discretization scheme of the diffusive term.

$$TV^{t+\Delta t} = \sum_i |\phi_{i+1}^{t+\Delta t} - \phi_i^{t+\Delta t}| \leq \sum_i |\phi_{i+1}^{t} - \phi_i^{t}| = TV^t \quad (17)$$

$$\phi_f = \underbrace{\frac{1}{2}(\phi_D - \phi_C)}_{CD} = \underbrace{\phi_C}_{upwind} + \underbrace{\frac{1}{2}(\phi_D - \phi_C)}_{anti-diffusive flux} \quad (18)$$

$$\phi_f = \phi_C + \frac{1}{2}\psi(r_f)(\phi_D - \phi_C),\ r_f = \frac{\phi_C - \phi_U}{\phi_D - \phi_C} \quad (19)$$

$$\psi(r_f) = max(0, min(2r_f, (r_f + 1)/2, 2)) \quad (20)$$

For the temporal discretization a second-order backward implicit Euler was used. Since the time steps are small in LES, it is not essential to use different temporal scheme than second-order.

5 Results

After one period which is composed from one convergent and one divergent motion of the vocal folds, this state is captured on Fig. 2. The velocity field shows the airflow from subglottal area to supraglottal area (vocal folds are in divergent position). An air jet in glottal area is formed on the surface of the right vocal fold margin. The velocity of an airflow is much lower (circa 50 % lower) than in the similar study [9] with no turbulent model, because the effective viscosity ($\nu_{eff} = \nu + \nu_{SGS}$) is computed higher in the high shear regions. The velocity is also slightly lower with 2D Smagorinsky and 2D One-Equation compared with [9]. Detailed studies deal with vocal folds are also [11] and [8]. On the Fig. 3 the high shear regions are shown: 1) The shear layer within rima glottidis, 2) The upper parts of the sacculus laryngis walls, 3) The peripherals of large eddies in supraglottal area. In the ideal case the influence of the mesh the mesh should not be seen.

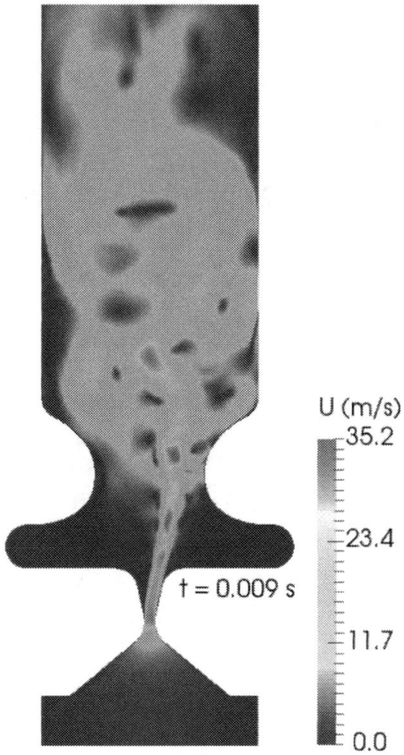

Fig. 2: Velocity field

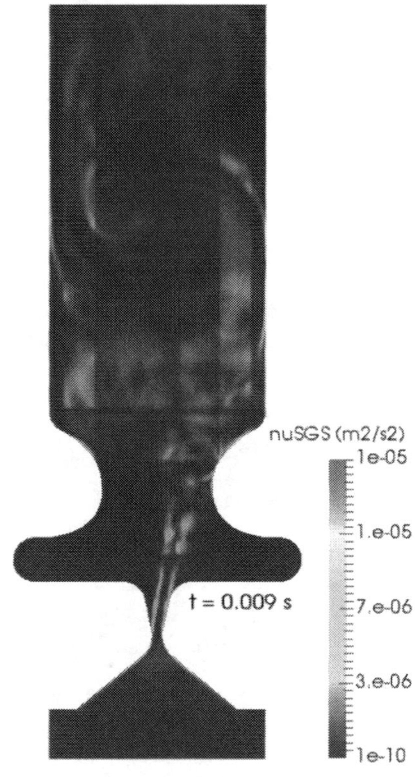

Fig. 3: Subgrid scale viscosity field

On the Fig. 4 are shown isosurfaces of the Q-criterion which detect the regions where coherent structures can exist. If the laplacian of the pressure is positive then the condition about generating the vortex tubes is satisfied. In the rima glottidis we can recognize the so-called hairpin horseshoe vortex which is generally formed near the wall as a result of the Tollmien-Schlichting instability. The leg of the vortex is in a buffer layer in the longitudinal direction, the neck and the head of the vortex are also seen. A regular shape of this kind of vortex is usually an exception, because the transition of the turbulence regime is stormy dynamic process in which the shear flow created from the main flow interact with the neighbouring structures. This kind of vortices are significant for generating and self-sustaining the turbulent structures.

The phenomenon in the supraglottal section is the so-called merging of vortices where these vortices have the same orientation of rotation and create cirles. The spatial effect of this mechanism is related to pseudoperiodic disturbances and the energy within the vocal fold domain is transfered from small eddies via merging to create one larger eddy (reverse energy cascade). The interactions of vortex tubes are also captured and it looks like „8". Two vortices are connected in the one point and subsequently disconnected vice versa than how they were connected before (the so-called bridging).

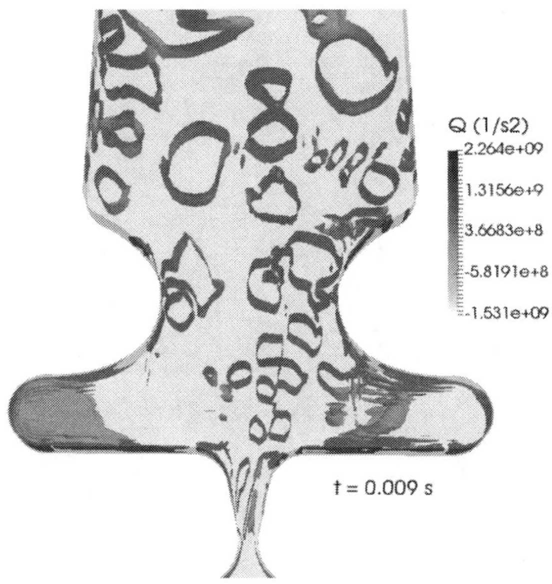

Fig. 4: Iso-surface Q-criterium

6 Conclusion

In this study, 3D LES case was simulated with Smagorinsky SGS model on structured mesh in OpenFOAM 4.1. The weaknesses of this study are following. The Smagorinsky model overpredicts the dissipation rate and for further work is highly recommend to use one-equation SGS model with van Driest damping function and more requirement because of the additional transport equation for the SGS kinetic energy. Another point for correct 3D modeling is a guarantee fine mesh refinement for wall-resolved LES in all three directions. Last but not least important point is initialization of turbulence at inlet. Current inflow boundary condition generation methods in LES are generally two basic categories: the precursor method (an addition simulation) and synthesis method (random fluctuations at the inlet). This is the main issue of the further work, because the vocal folds domain is computed from the pressure difference (not by prescribed velocity).

Acknowledgemnt

This work was supported by the Ministry of Education of the Czech Republic through the SGS project no. 21176/115 of the Technical University of Liberec.

References

[1] Ferziger, J. H. (1998). Direct and large eddy simulation of turbulence. *Numerical methods in fluid mechanics*, 16, 53-73.

[2] Georgiadis, N. J., Rizzetta, D. P., & Fureby, C. (2010). Large-eddy simulation: current capabilities, recommended practices, and future research. *AIAA journal*, 48(8), 1772-1784.

[3] Germano P., Piomelli U, Moin P, Cabot WH A dynamic subgrid scale eddy viscosity model. *Pdys Fluids* 1991;3(7):1760-5.

[4] Lesieur, M.; Métais, O.; Comte, P. *Large-eddy simulations of turbulence*. Cambridge University Press, 2005.

[5] Moukalled, F., Mangani, L., Darwish, M. *The Finite Volume Method in Computational Fluid Dynamics: An Advanced Introduction with OpenFOAM and Matlab.* American University of Beirut, Lebanon: Springer International Publishing, 2015. ISBN 978-3-319-16873-9.

[6] Scherer R., Shinwari D., Witt K. D., Zhang C., Kucinschi B., and Afjeh A., Intraglottal pressure profiles for a symmetric and oblique glottis with a divergence angle of 10 degrees, *Journal of the Acoustical Society of America*, vol. 109, pp. 1616-1630, Apr. 2001.

[7] Smagorinsky J. General circulation experiments with the prim- itive equations I. the basic experiment. Mon. Weather Rev. 1963;91(3):99–164.

[8] Šidlof P., Zörner S., Hüppe A. (2015), A hybrid approach to computational aeroacoustics of human voice production, *Biomechanics and Modeling in Mechanobiology 14*, pp. 473-488, DOI 10.1007/s10237-014-0617-1

[9] Šidlof P., Large-Eddy Simulation of Airflow in Human Vocal Folds, *Topical problems of fluid mechanics*, 2015

[10] Zhiyin, Y. (2015). Large-eddy simulation: Past, present and the future. Chinese journal of Aeronautics, 28(1), 11-24.

[11] Zörner S., Šidlof P., Hüppe A., Kaltenbacher M. (2016), Flow and Acoustic Effects in the Larynx for Varying Geometries, *Acta Acustica united with Acustica* 102(2), pp. 257-267, DOI 10.3813/AAA.918942

Variations of flow in human airways as a consequence of lung diseases

Frantisek Lizal[1,*], *David* Stejskal[1], *Miloslav* Belka[1], *Jan* Jedelsky[1], *Miroslav* Jicha[1], *Kristian* Brat[2], *Vladimir* Herout[2] and *Elena* Lizalova Sujanska[2]

[1]Brno University of Technology, Faculty of Mechanical Engineering, Energy Institute, Technicka 2896/2, Brno 616 69, Czech Republic

[2]The University Hospital Brno, Clinic of Pulmonary Diseases and Tuberculosis, Jihlavska 20, Brno 625 00, Czech Republic

Abstract. The efficiency of drug delivery administered by inhalation depends, among other factors, such as size and shape of aerosol particles, significantly also on the flow in the airways. As many lung diseases change both the breathing pattern and the shape of airways, we focus in this study on the influence of several selected diseases on the distribution of flow between the lung lobes and on changes the diseases induce on the course of flowrate. First, we present results of a literature survey focused on the published records of pathological breathing patterns. In the second part, we describe the newly designed breathing simulator and the implementation of the patterns into it. The last part is focused on the experimental verification of fidelity of the simulated breathing patterns.

1 Introduction

A proper delivery of a prescribed medication to afflicted lung regions is a crucial point of the therapy of respiratory diseases. The efficiency of the drug delivery administered by inhalation depends, among other factors, such as size and shape of aerosol particles, also on the flow in the airways. However, our ability to predict the distribution of inhaled particles in diseased airways is insufficient [1].

Human airways consist of the oral (and nasal) cavity, pharynx, larynx, trachea and 23 generations of bronchial branching. The lungs are divided into five lobes, while two lobes are formed in the left lung and three lobes are formed in the right lung. The respiratory cycle comprises the active action of inspiration, which is driven by the diaphragm and intercostal muscles, and the passive phase of expiration, which uses natural elasticity of the lungs and thorax.

1.1 Lung volumes and capacities

The lung function monitoring and diagnostics of certain diseases can be performed using lung volumes and lung capacities. The volume of air that is inhaled and exhaled during one breath cycle under resting conditions is called tidal volume (*TV*). Its value is around 0.5 l for healthy adults. Residual volume (*RV*) is the volume of air remaining in the lungs after a maximal exhalation. Inspiratory reserve volume (*IRV*) is the maximal volume that can be inhaled from the end-inspiratory level. Inspiratory capacity (*IC*) is the volume of air that can be inhaled after normal exhalation. It is the sum of *IRV* and *TV*. Expiratory reserve volume (*ERV*) is the maximal volume of air that can be exhaled from the normal end-expiratory position. Vital capacity (*VC*) is the volume of air which can be exhaled after the deepest inhalation. It is the sum of *IRV*, *ERV* and *TV*. Total lung capacity (*TLC*) is the volume in the lungs at maximal inflation, the sum of *VC* and *RV*.

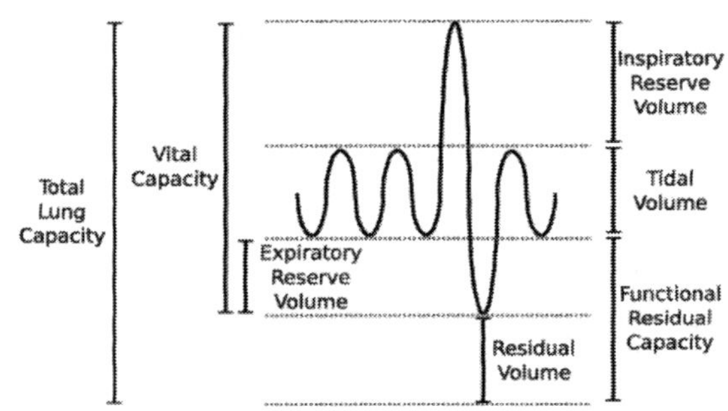

Fig. 1. An illustration of the lung volumes and capacities [2].

Functional residual capacity (FRC) is the volume in the lungs at the end-expiratory position. Forced vital capacity (FVC) is the maximal volume of air that can be exhaled after a deep inspiration. Unlike the VC, which is measured during the quiet breathing, the FVC is measured during vigorous forced expiration. FVC is a dynamic value. Forced expiratory volume (FEV1) is the volume of exhaled air during the first second of forced expiration. FEV1 is a dynamic value and is usually given as a percentage of FVC.

* Corresponding author: lizal@fme.vutbr.cz

1.2 Measurement of the lung volumes and capacities

The data on the flow and inhaled air volume distribution among the lobes presented in this text originate from several measuring methods. The most common method is spirometry (see Fig. 2), which serves as a main pulmonary function test in the clinical practice.

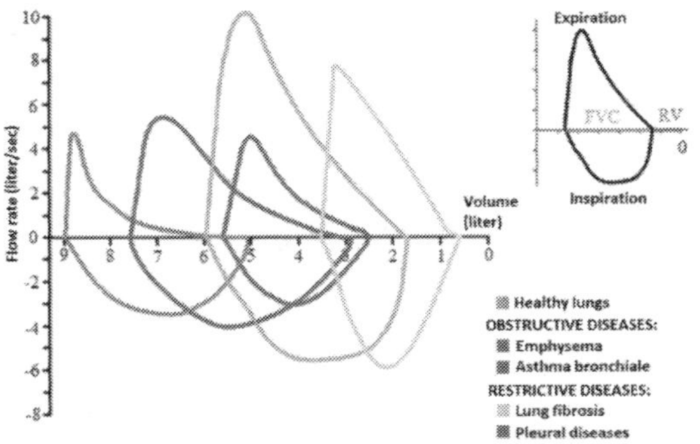

Fig. 2. Flow-Volume loop (output of the spirometry) showing *FVC* manoeuvre for healthy and diseased lungs [3].

All the major lung volumes and capacities except RV and FRC can be measured using spirometry, however, only the data for whole lungs are obtained, and hence the volume distribution among the lobes must be acquired using different methods.

Such a suitable method is Computed Tomography (CT), which can be performed even for dynamic characteristics, in the case that the CT scanner allows for fast volumetric imaging. As supplementary methods also Positron Emission Tomography (PET) and Single Photon Emission Computed Tomography (SPECT) can be used [4].

1.3 Respiratory diseases and their influence on the lung functions

Major inter-subject variability in particle deposition in the airways (up to 30%) exists among healthy individuals [5]. This variability is highlighted during pathological conditions represented by different respiratory diseases including bronchial asthma, chronic obstructive pulmonary disease (COPD), interstitial lung diseases (ILDs) or cystic fibrosis (CF) [5]. Both bronchial asthma and COPD patients suffer with bronchial obstruction, i.e. limitation of airflow in the lower airways. Differences in breathing cycles of healthy peers and asthmatic patients were studied by Choi et al. [6]. The pathological changes of airflow and lung mechanics in COPD and asthma patients increase rapidly during periods of disease progression (i.e. exacerbation of asthma/COPD). Moreover, in many COPD patients, multiple areas of emphysema develop during the course of the disease, causing dramatic changes in the lung architecture. Abnormal patterns of particle deposition are an obvious consequence of this process in patients with this group of pulmonary diseases.

In contrast, the group of ILDs and CF are diseases characterized by a progressive accumulation of collagen fibres in the pulmonary intercellular space. This process results in dynamic loss of lung elasticity and compliance as well as in the development of bronchiectases, i.e. aneurysmatic remodelling of the bronchial lumina. In consequence, the patterns of particle deposition may change very dramatically in course of the disease. Importantly, inhalation of hypothetic new agents might be one of the potential ways to target mechanisms of the fibrotic process effectively. Thus, more precise data are warranted to better understand the progressive and long-term impacts of ILDs and CF on lung mechanics and bronchial airflow.

For study purposes (focusing on ILDs and obstructive pulmonary diseases), computed tomography (CT scanning) and magnetic resonance imaging (MRI) of the lungs are most widely used to follow-up the patterns of pathological changes in bronchial or lung architecture [7]. For illustration, CT scans of normal lungs and of lungs of a patient with CF are presented as Figs. 3 and 4, respectively. However, obtaining data on ILDs and CF lobar flow distribution is difficult, therefore we decided to begin with diseases for which the data were already published.

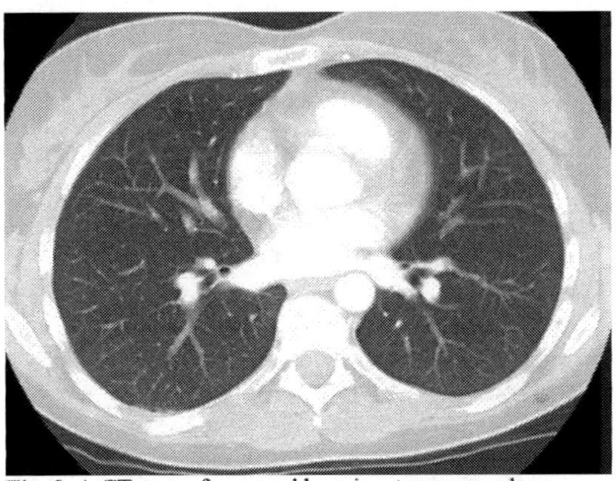

Fig. 3. A CT scan of a normal lung in a transverse plane.

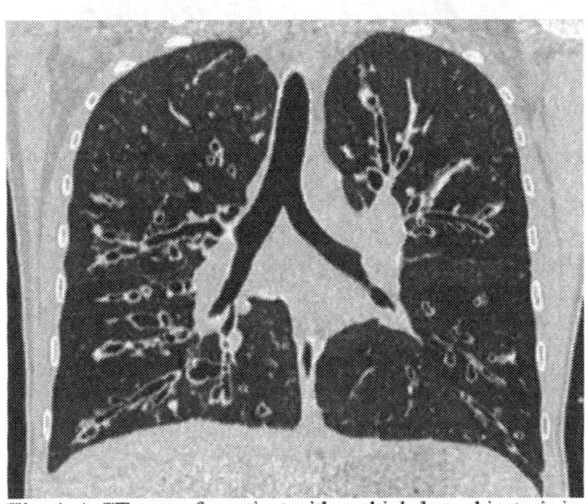

Fig. 4. A CT scan of a patient with multiple bronchiectasis in a coronal plane (clinical diagnosis: cystic fibrosis).

The main aim of this work is to acquire data on flow and inhaled air volume distribution among the lung lobes and implement these into a newly designed breathing simulator. The simulator will be employed during experiments studying regional deposition of aerosols in human airway replicas [8]. As the differences among the healthy and diseased lungs are less apparent during normal breathing, and also due to the fact, that inhalation of medical aerosols follows a pattern significantly different than normal breathing, we decided to set up and simulate the IC manoeuvre. After implementation of the obtained breathing patterns into the simulator it was necessary to verify the ability of the simulator to follow the selected regime and to close the electro-pneumatic valves to allow separation of inspiration and expiration phase for aerosol and clean air path.

2 Methods

The whole process of obtaining, processing and implementation of the breathing patterns was described in the bachelor thesis of David Stejskal [3]. Briefly, as a first step the distribution of volumes among the lobes for healthy lungs during normal breathing and *IC* manoeuvre were obtained from the literature [6,9]. Then the data for several diseases were acquired. The data for asthmatics were obtained also from the study of Choi et al. [6], the data on emphysema were from [10], on cancer from [11] and on lobectomy from [12].

2.1 Breathing simulator setup

The newly designed breathing simulator (custom-built, based on FESTO technology, Esslingen am Neckar, Germany) comprises five computer-controlled motors which drive pistons in five pairs of cylinders. One cylinder is always for the clean air, the second serves for the aerosol path. This division is necessary, as separate measurement of inspiration and expiration (and hence separate dosing of aerosol into inspiration and expiration phase) is required.

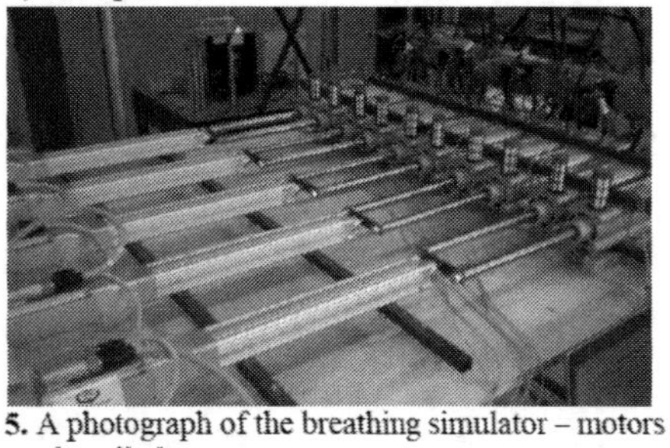

Fig. 5. A photograph of the breathing simulator – motors and pneumatic cylinders.

The obtained patterns were normalized in their duration and distribution of flows and volumes among the lobes were prescribed. The resulting courses of flowrates were transformed into prescription of piston positions for each lobe (see examples in Figures 6 to 8, where LUL means left upper lobe, LLL – left lower lobe, RUL – right upper lobe, RML – right middle lobe and RLL – right lower lobe).

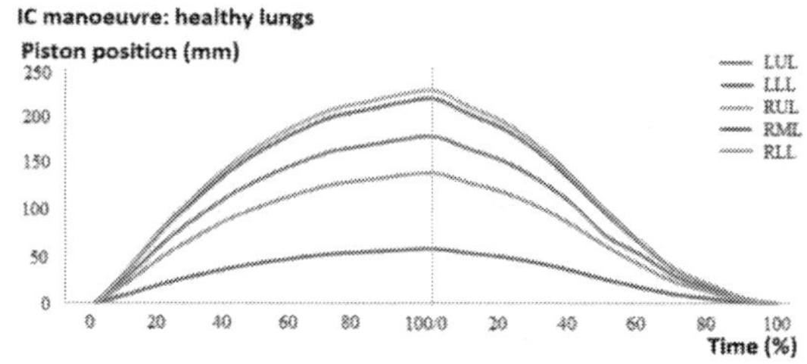

Fig. 6. *IC* manoeuvre – transformation of the flow distribution among the lobes to the prescription of the piston position for healthy lungs.

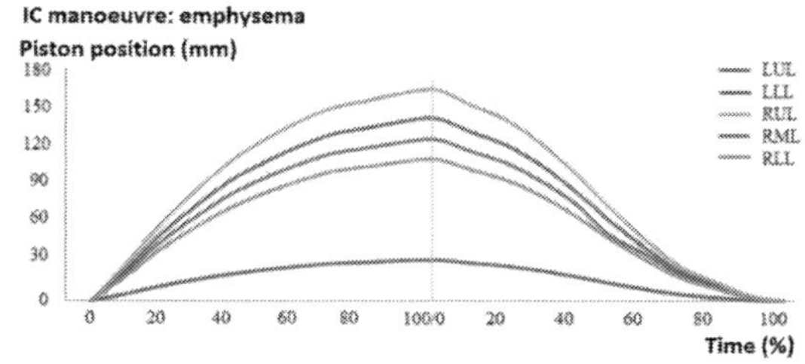

Fig. 7. *IC* manoeuvre – transformation of the flow distribution among the lobes to the prescription of the piston position for lungs with emphysema.

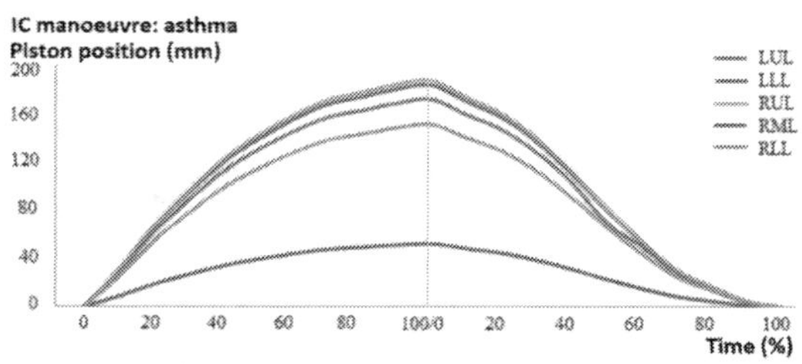

Fig. 8. *IC* manoeuvre – transformation of the flow distribution among the lobes to the prescription of the piston position for severe asthma.

2.2 Measurement setup

The duration of switching of the closing valves on the pneumatic cylinders, which is important for accurate separation of the inspiratory and expiratory phase during experiments with aerosol deposition, was measured by Constant Temperature Anemometry (CTA).

The StreamLine hot-wire anemometer (Dantec Dynamics A/S, Skovlunde, Denmark) was used for measurement of the velocity in the clean air and aerosol path tubes. Two single-wire probes (Dantec 55P13) were used. The calibration was performed by Dantec StreamLine® Pro Automatic calibrator in the range of 0.02 to 30 m/s.

The following four cases were measured: 1) the clean air path with disconnected valve (i.e. no closing), 2) the clean air path with connected and activated valve, 3) the aerosol path with disconnected valve, and 4) the aerosol path with connected and activated valve. The measurements were made three times for each case, while nine breath cycles were recorded during each run.

3 Results and discussion

The resulting records of velocity were averaged and interpolated by cubic B spline and compared. The measurements showed that there were no significant differences between the clean air and aerosol path for the cases with disconnected valves. It means that any differences in the courses of velocities in the following cases can be attributed to the role of the valves.

Figs. 9 and 10 show results with connected and activated valves for aerosol and clean air path, respectively. The red curve represents the velocity in the tube from the pneumatic cylinder to the electro-pneumatic valve, the blue curve represents the velocity in the tube behind the valve branch which is being closed during expiration. The velocities differ before and after the valve, however, it is just a consequence of misalignment of the probes with the centre of the tube and different velocity profiles in the tubes. Both the curves basically overlap during the inspiration for both aerosol and clean air path. However, the delay in the valve closing at the beginning of the expiration can be clearly identified in both graphs. The reaction time is about 0.5 s and the flow terminates in additional approximately 0.5 s.

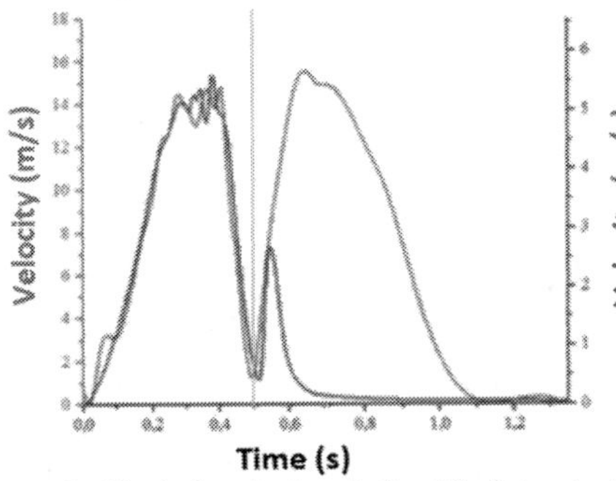

Fig. 9. Velocities before (red) and after (blue) the closing valve on the aerosol path

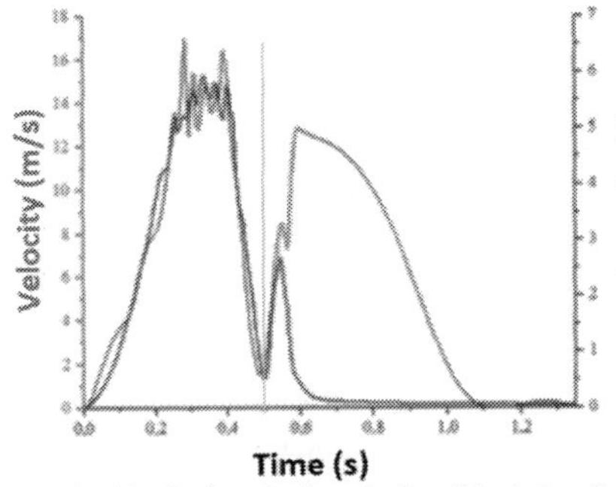

Fig. 10. Velocities before (red) and after (blue) the closing valve on the clean air path

4 Conclusions

The results showed that it is necessary to decrease the delay in closing time of the valve. The valves were not originally set to the minimal reaction time, therefore the minimal time of 3.7 ms was set. The verification of this setup is in progress. Further decrease of the closing delay can be achieved by opening of the valves in advance, however, this setup requires profound changes in the software of the simulator.

The tests also proved that the simulator accurately follows the prescribed breathing pattern and that there were no significant differences in the velocity profile on both aerosol and clean air paths when the electro-pneumatic valves were open.

Acknowledgement: This work was supported by the Czech Science Foundation under the grant GA16-23675S, by the project FSI-S-17-4444, and by the project LTC17087 of the programme INTER-EXCELLENCE, sub-programme INTER-COST.

References

1 A. V. Kolanjiyil, C. Kleinstreuer, and R. T. Sadikot, Computers in biology and medicine **84**, 247 (2017); B. Hajian, J. De Backer, W. Vos, J. Aerts, J. Cluckers, and W. De Backer, Expert Opinion on Drug Delivery **13** (12), 1719 (2016).
2 Pathway medicine, (2017), Vol. http://www.pathwaymedicine.org/lung-volumes.
3 D. Stejskal, Bachelor thesis, Brno University of Technology, 2017.
4 F. Lizal, J. Jedelsky, K. Morgan, K. Bauer, J. Llop, U. Cossio, S. Kassinos, S. Verbanck, J. Ruiz-Cabello, A. Santos, E. Koch, and C. Schnabel, European Journal of Pharmaceutical Sciences (2017).
5 M. Hussain, W.-H. Renate, and H. Werner, Journal of Thoracic Disease **3** (3), 156 (2011).
6 S. Choi, E. A. Hoffman, S. E. Wenzel, M. H. Tawhai, Y. B. Yin, M. Castro, and C. L. Lin, Journal of Applied Physiology **115** (5), 730 (2013).
7 P. Bokov, B. Mauroy, M. P. Revel, P. A. Brun, C. Peiffer, C. Daniel, M. M. Nay, B. Mahut, and C. Delclaux, Respiratory Physiology & Neurobiology **173** (1), 1 (2010); C. P. Heussel, S. Ley, A. Biedermann, A. Rist, K. K. Gast, W. G. Schreiber, and H. U. Kauczor, European radiology **14** (12), 2188 (2004).
8 F. Lizal, J. Elcner, P. K. Hopke, J. Jedelsky, and M. Jicha, Proceedings of the Institution of Mechanical Engineers Part H-Journal of Engineering in Medicine **226** (H3), 197 (2012); F. Lizal, M. Belka, J. Adam, J. Jedelsky, and M. Jicha, Proceedings of the Institution of Mechanical Engineers Part H-Journal of Engineering in Medicine **229** (10), 750 (2015).
9 N. Jahani, S. Choi, J. Choi, K. Iyer, E. A. Hoffman, and C. L. Lin, Journal of Applied Physiology **119** (10), 1064 (2015).
10 H. N. E. Bastos, I. Neves, M. Redondo, R. Cunha, J. M. Pereira, A. Magalhaes, and G. Fernandes, J Bras Pneumol **41** (6), 489 (2015).
11 S. G. Yuan, K. A. Frey, M. D. Gross, J. A. Hayman, D. Arenberg, X. W. Cai, N. Ramnath, K. Hassan, J. Moran, A. Eisbruch, R. K. Ten Haken, and F. M. Kong, Int J Radiat Oncol **82** (4), E631 (2012).
12 N. Berend, A. J. Woolcock, and G. E. Marlin, Thorax **35** (2), 145 (1980).

Simulation of transonic flows through a turbine blade cascade with various prescription of outlet boundary conditions

Petr Louda[1,a], *Petr* Straka[2], *Jaromír* Příhoda[3]

[1,a] *Faculty of Mechanical Engineering, Czech Technical University in Prague, Czech Republic*
[2] *VZLU Czech Aerospace Research Centre, Plc, Prague, Czech Republic*
[2] *Institute of Thermomechanics AS CR, v.v.i., Prague, Czech Republic*

Abstract. The contribution deals with the numerical simulation of transonic flows through a linear turbine blade cascade. Numerical simulations were carried partly for the standard computational domain with various outlet boundary conditions by the algebraic transition model of Straka and Příhoda [1] connected with the EARSM turbulence model of Hellsten [2] and partly for the computational domain corresponding to the geometrical arrangement in the wind tunnel by the γ-ζ transition model of Dick et al. [3] with the SST turbulence model. Numerical results were compared with experimental data. The agreement of numerical results with experimental results is acceptable through a complicated experimental configuration.

1 Introduction

The problem of the shock-wave/boundary-layer interaction is studied in the long term nevertheless the majority of results are focused on the interaction with the turbulent boundary layer. A review of results related to the shock-wave/boundary layer interaction is given e.g. by Gaitonde [4] and/or Arnal and Délery [5].

A relatively significant attention is recently given to the problem of the separation-induced transition due to the interaction of the shock wave with the laminar boundary layer, as e.g. experimental and numerical investigations of the interaction in the framework of the TFAST project "Transition Location Effect on Shock Wave Boundary Layer Interaction", see Flaszynski et al. [6] and Piotrowicz et al. [7].

The structure of the transonic flow through a turbine blade cascade considerably depends on the character of the shock-wave/boundary-layer interaction. In case of the laminar boundary layer, the shock-wave/boundary-layer interaction is usually connected with the flow separation followed by the transition in separated flow.

Existing simulations of transonic flows through a linear turbine blade cascade have shown a rather significant dependence of results on the prescription of the outlet boundary condition, i.e. outlet static pressure. Due to the relatively complicated geometrical arrangement of the test section with the immediately connecting settling chamber, the outlet of the computational domain is usually prescribed in the traversing plane behind the blade cascade where the static pressure is given on the basis of mean values of measured data.

Therefore there were realized numerical simulations partly for the computational domain corresponding to the geometrical arrangement of the wind tunnel by the γ-ζ transition model of Dick et al. [3] connected with the SST turbulence model and partly for the standard computational domain with various outlet boundary conditions by the algebraic transition model of Straka and Příhoda [1] connected with the EARSM turbulence model. Numerical simulations were carried out for transonic flow through a linear turbine blade cascade corresponding to a rotor mid-section of the last stage of a large output steam turbine.

2 Mathematical model

The numerical simulation of compressible flows is based on the solution of the conditionally-averaged Navier-Stokes equations completed by the constitutional relations and by a turbulence model, a model of turbulent heat transfer and by a model of the laminar/turbulent transition. The basic mathematical model consists of the EARSM turbulence model of Hellsten [2], the algebraic transition model proposed by Straka and Příhoda [1] and the turbulent heat transfer model based on the generalized gradient hypothesis according to Launder [8].

The EARSM turbulence model was used in the form corresponding to models with the turbulent viscosity. Turbulent scales are expressed by the turbulent energy k and by the specific dissipation rate ω by means of the SST turbulence model of Menter [9]. Turbulent time scale $\tau \sim 1/\omega$ near the wall is given by the Kolmogorov viscous time scale. The production term in the turbulent energy equation is modified for the reduction of the undesirable overproduction of the turbulent energy in the stagnation region. Production and destruction terms in the equation for the turbulent energy are multiplied by the intermittency coefficient γ. Similarly, the effective viscosity is given by $\mu_{ef} = \mu + \gamma\mu_t$ in the transition region.

[a] Corresponding author: petr.louda@fs.cvut.cz

The transition model is based on the concept of different values of the intermittency coefficient in the boundary layer γ_i and in the free stream γ_e. The intermittency coefficient in the boundary layer γ_i is described by the relation

$$\gamma_i = 1 - \exp\left[-\hat{n}\sigma\left(Re_x - Re_{xt}\right)^2\right] \quad (1)$$

according to Narasimha [10]. The transition onset is given by the empirical correlation for the momentum Reynolds number $Re_{\theta t} = f(Tu, \lambda_t)$ where Tu (%) is the free-stream turbulence level and λ_t is the pressure-gradient parameter. The transition length is expressed using the parameter $N = \hat{n}\sigma Re_{\theta t}^3$ where $\hat{n}$ is the spot generation rate and σ is the spot propagation rate introduced by Narasimha [10]. The parameter N for the attached flow is similarly given by the empirical relation $N = g(Tu, \lambda_t)$ proposed by Solomon et al. [11]. The onset of transition in separated flow is given by the correlation according to Mayle [12] in the form

$$Re_{xt} = 300\, Re_{\theta s}^{0.7} + Re_{xs} \quad (2)$$

where $Re_{\theta s}$ is the momentum Reynolds number at the separation and Re_{xs} is the Reynolds number related to the distance of the separation from the leading edge. The transition length of the separation is given according to Walker [13] by the relation

$$\hat{n}\sigma = \frac{Re_{xt}^{-1.34}}{40} \quad (3)$$

and so the same approach can be applied as in the attached flow.

For prediction of transitional flows in complex geometries, the application of local variables is necessary. The momentum Reynolds number is replaced according to Langtry and Menter [14] by the maximum of the vorticity Reynolds number in the form

$$Re_{\Omega} = y^2 |\Omega| / \nu \quad (4)$$

where y is the distance from the wall and Ω is the absolute value of the vorticity tensor. The link between both Reynolds numbers is expressed by the relation $Re_\theta = Re_{\Omega max}/C$ where the parameter C depends on the pressure gradient. The variation of the parameter C with the pressure gradient parameter was estimated by means of similar solutions of Falkner-Skan velocity profiles.

The mathematical model was implemented into the in-house numerical code based on the finite volume method. The multi-block quadrilateral structured grid with a block overlapping implementation refined near walls was applied for numerical simulations.

The γ-ζ transition model according to Dick et al. [3] (see also Lodefier and Dick [15]) is based on two transport equations for the near-wall part γ and free-stream part of the intermittency ζ used in the form

$$\frac{\partial(\rho\gamma)}{\partial t} + \frac{\partial(\rho U_i \gamma)}{\partial x_i} = P_\gamma + \frac{\partial}{\partial x_i}\left[\left(\mu + \sigma_\gamma \mu_t\right)\frac{\partial \gamma}{\partial x_i}\right] \quad (5)$$

$$\frac{\partial(\rho\zeta)}{\partial t} + \frac{\partial(\rho U_i \zeta)}{\partial x_i} = -E_\zeta + \frac{\partial}{\partial x_i}\left[\left(\mu + \sigma_\zeta \mu_t\right)\frac{\partial \zeta}{\partial x_i}\right] \quad (6)$$

with the production term in the eq. (5)

$$P_\gamma = 2\beta_\gamma\left(1-\gamma\right)\sqrt{-\ln\left(1-\gamma\right)}\rho U_\gamma F_s \quad (7)$$

where F_s is a starting function set to unity at the start of transition. The dissipation term in the eq. (6) is given by the relation

$$E_\zeta = C_2 \mu_\zeta \frac{U}{U_e^2}\frac{\partial U}{\partial n}\frac{\partial \zeta}{\partial n} \quad (8)$$

where U is the local velocity and U_e is the mean velocity at the boundary layer edge. The term E_ζ guarantees that ζ vanishes in the boundary layer, so that both intermittencies are complementary in the turbulent flow. The relation between the growth parameter β_γ and the spot growth rate is

$$\beta_\gamma = \sqrt{\frac{N\sigma}{U_e}} \quad (9)$$

The boundary conditions are $\gamma = 0$, $\zeta = 1$ in the inlet, $\partial\zeta/\partial n = \partial\gamma/\partial n = 0$ in the outlet and $\partial\gamma/\partial n = 0$, $\zeta = 0$ on the wall. The criteria of Abu-Ghannam and Shaw [16] and Mayle [12] are used for the transition onset in the attached flow. For transition in separated flow the Mayle criterion modified by Dick et al. [3] was used in the form

$$Re_{st} = 700\, Re_{\theta s}^{0.7} \exp\left(-0.3 Tu_e\right) \quad (10)$$

with parameters related to the separation point. The used criteria are functions of boundary layer parameters as the momentum thickness, free-stream velocity and turbulence intensity. The used transition model and its implementtation into the in-house numerical code are described in detail by Louda et al. [17].

3 Results

The both transition models i.e. the EARSM turbulence model connected with the algebraic transition model and the SST turbulence model with the γ-ζ transition model were applied for the numerical simulation of 2D transonic flow through a linear turbine blade cascade corresponding to a rotor mid-section of the last stage of a large output steam turbine. Numerical results were compared with experimental results covering optical and pressure measurements, see Luxa et al. [18].

The predictions were carried out for the relative spacing $t/c = 0.718$, inlet angle $\alpha_1 = 29.1°$ and for various outlet isentropic Mach number M_{2is} corresponding to the static pressure p_2 determined by the data reduction method in the traversing plane behind the blade cascade.

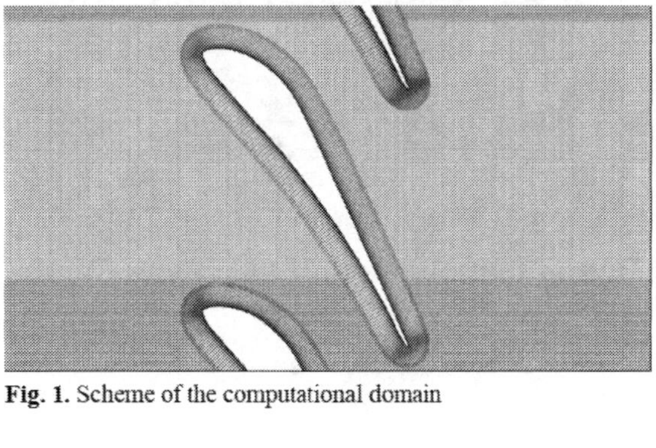

Fig. 1. Scheme of the computational domain

The multi-block quadrilateral structured computational mesh refined near walls was used for numerical simulations by means the algebraic transition model implemented into the in-house numerical code. The computational domain is demonstrated in Fig. 1. This approach allows the application of structured grids for complex geometries as well.

The inlet of the domain was in the distance of 0.55*c* from the blade leading edge and the outlet in the distance 0.7*c* behind trailing edges. The inlet boundary conditions were prescribed by the constant total pressure, total temperature and inlet flow angle. The outlet boundary condition was defined by the constant static pressure determined according to the outlet isentropic Mach number. Periodicity conditions were applied on corresponding free side boundaries of the computation domain. The inlet free-stream turbulence parameters corresponding to turbulence intensity Tu = 1.5% and the ratio of the turbulent and molecular viscosity μ_t/μ = 50 were chosen.

Predictions were focused on the adequate simulation of the shock-wave/boundary-layer interaction. The flow separation on the blade suction side is caused by the interaction of the reflected shock wave with the laminar boundary layer.

Fig. 2. Interferometric picture for M_{2is} = 1.515

Fig. 3. Schlieren picture for M_{2is} = 1.515

The interferometric and schlieren pictures obtained for M_{2is} = 1.515 are shown in Fig. 2 and 3. The interaction of the inner branch of the exit shock wave with the boundary layer on the suction side is connected with the separation on the laminar boundary layer.

Simulations by means of the algebraic transition model were carried out for various outlet static pressures p_2/p_{01} i.e. for various Mach numbers M_{2is} with the aim to reach a good agreement of the predicted flow field in the blade cascade expressed by the Mach number isolines. This agreement was tested by the position of the inner branch of the reflected shock wave especially by the angle ψ between the shock wave and the linear part of the blade suction side, see Fig. 2.

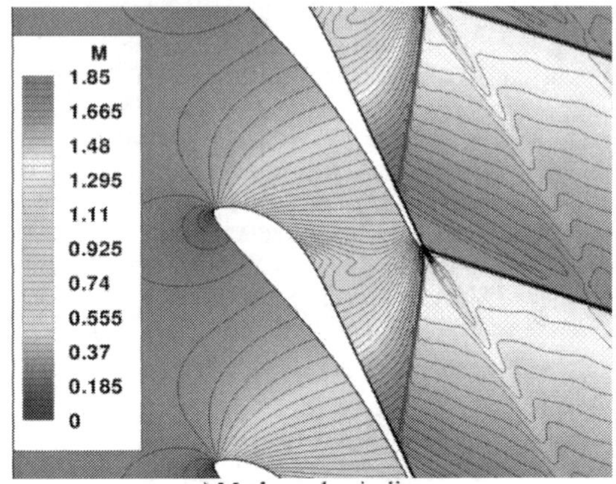

a) Mach number isolines

b) Numerical schlieren

Fig. 4. Flow field for M_{2is} = 1.515 (algebraic transition model)

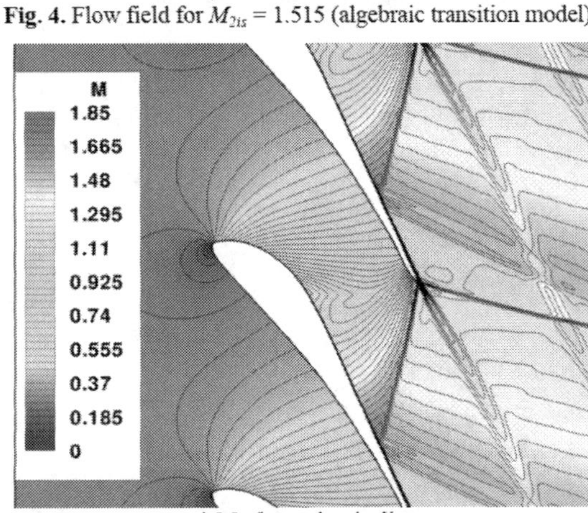

a) Mach number isolines

b) Numerical schlieren

Fig. 5. Flow field for M_{2is} = 1.422 (algebraic transition model)

The best agreement was achieved for the pressure ratio p_2/p_{01} = 0.3048 corresponding to M_{2is} = 1.422. The Mach number isolines and numerical schlieren pictures obtained by the algebraic transition model are shown for M_{2is} = 1.422 and 1.515 in Fig. 4 and 5. The variation of the shock wave angle with the pressure ratio p_2/p_{01} is presented in Fig. 6.

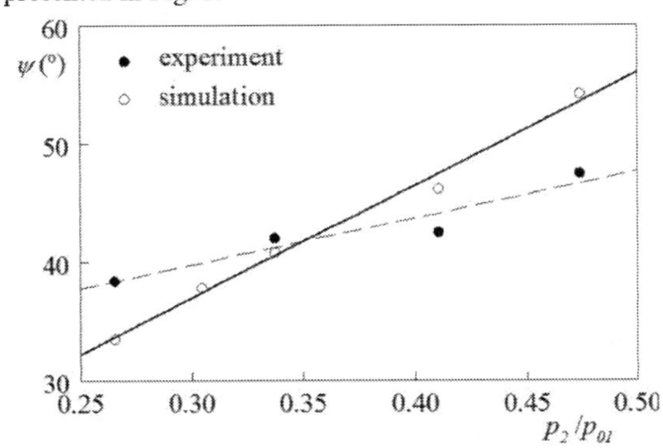

Fig. 6. Variation of the shock wave angle with the outlet static pressure

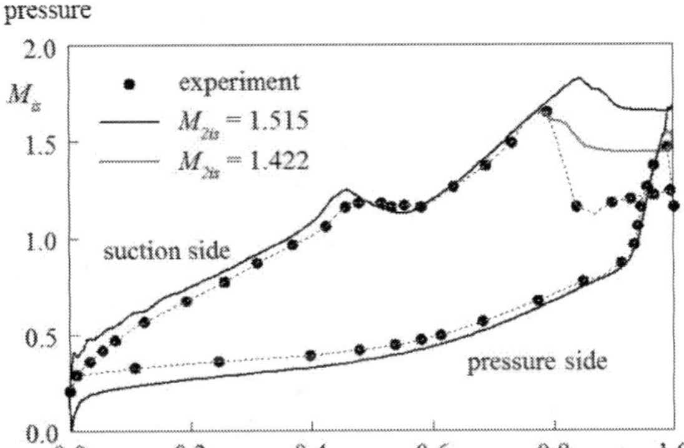

Fig. 7. Distribution of the isentropic Mach number along the blade (algebraic transition model)

The distribution of the isentropic Mach number along the blade surface is compared with experimental data in Fig. 7. The distance b represents the axial chord length. The only difference between the both Mach numbers M_{2is} is the position of the shock-wave impact on the suction side where the prediction for M_{2is} = 1.422 agrees well with experiments.

The distribution of the skin friction on the blade surface is presented in Fig. 8. The skin friction coefficient is defined by the relation $C_f = \tau_w/(p_{01}-p_1)$. The skin friction distribution reflects the Mach number distribution as the only difference is downstream of the predicted shock-wave impact.

The detail of skin friction distribution with separation zones is given in Fig. 9. The separation zone estimated from experimental results for the M_{2is} = 1.515 is marked by the rectangle. The extent of the separation zone is approximately from $x/b \approx 0.771$ to 0.832. The interaction of the inner branch of the exit shock wave with the boundary layer on the suction side and the separation zone is acceptably predicted even though the predicted separation length is rather shorter.

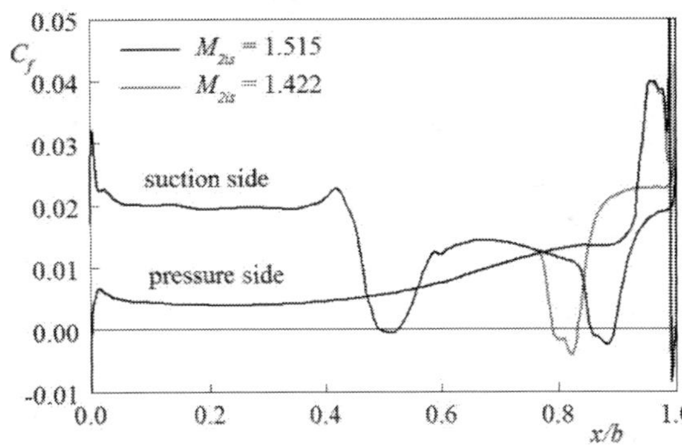

Fig. 8. Skin friction distribution (algebraic transition model)

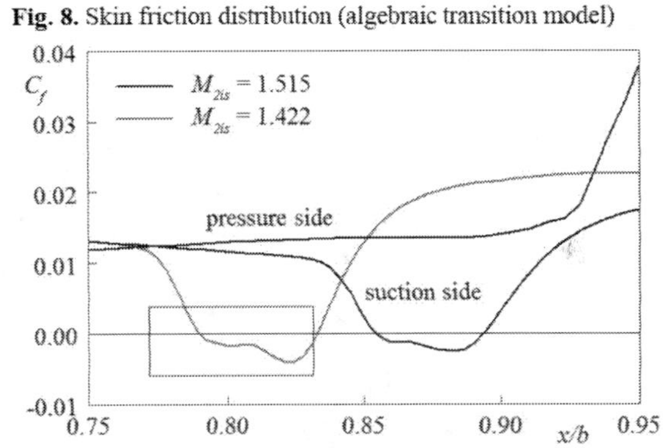

Fig. 9. Detail of the skin friction distribution (algebraic transition model)

Further, it can be seen from the distribution of the Mach number and the skin friction that the sudden change of the surface curvature of the suction side at the distance $x/b \approx 0.55$ leads to a short laminar separation bubble. Besides, a very short separation can be seen as well on the blade pressure side just upstream of the trailing edge, see Fig. 8.

The computational domain used for application of the γ-ζ transition model covers the inlet part, the test section with six blades and a part of the settling chamber. The upper blade is extended by the perforated tailbox. The chosen computational domain corresponds as much as possible to the experimental arrangement in the wind tunnel.

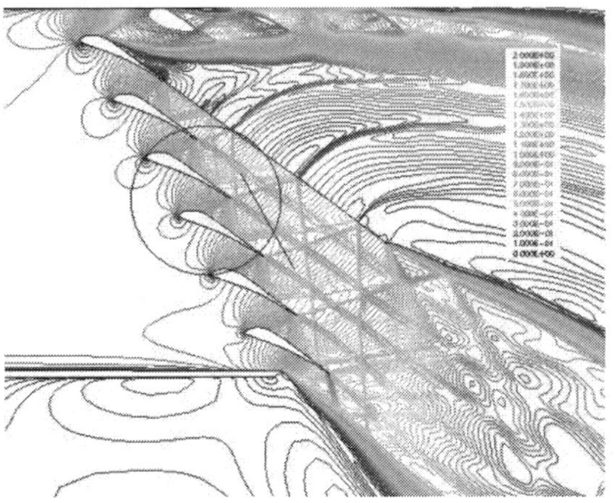

Fig. 10. Mach number isolines (γ-ζ transition model)

A cell-centered finite volume method with quadrilateral (in 2D) or hexahedral (in 3D) finite volumes composing a structured grid was used for the spatial discretization of the γ-ζ transition model. The inlet boundary conditions were prescribed in the same way as for the algebraic transition model. Due to the used configuration, the prescription of the outlet boundary condition is somewhat difficult and so the trial-and-error method was used.

The results obtained for the ratio p_2/p_{01} = 0.300 corresponds well with experimental data for the M_{2is} = 1.356 related to the static pressure in the traversing plane. The whole configuration is apparent from Fig. 10 where the Mach number isolines are shown. The circle marks the optical window in the test section and the line is the traversing plane.

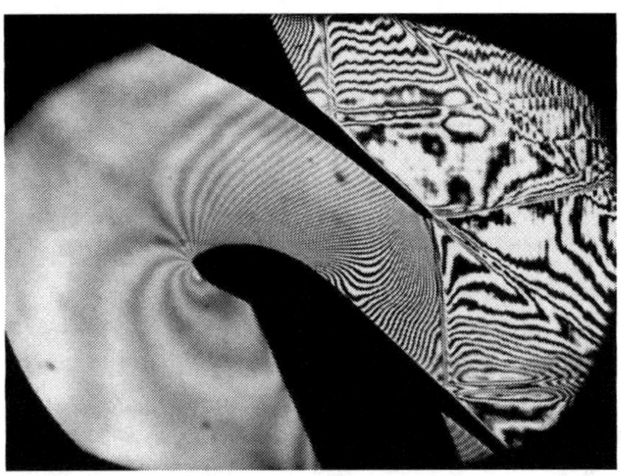

Fig. 11. Interferometric picture for M_{2is} = 1.356

The interferometric picture for the isentropic outlet Mach number M_{2is} = 1.356 is shown in Fig. 11. The detail of the Mach number isolines corresponding to the measured area is shown in Fig. 12. The agreement of the both flow fields is acceptable. The distribution of the isentropic Mach number along the blade is presented for the two middle blades in Fig. 13. It can be seen that the interaction of the inner branch of the outlet shock wave with the boundary layer is on the both blades somewhat different including the shock wave angle.

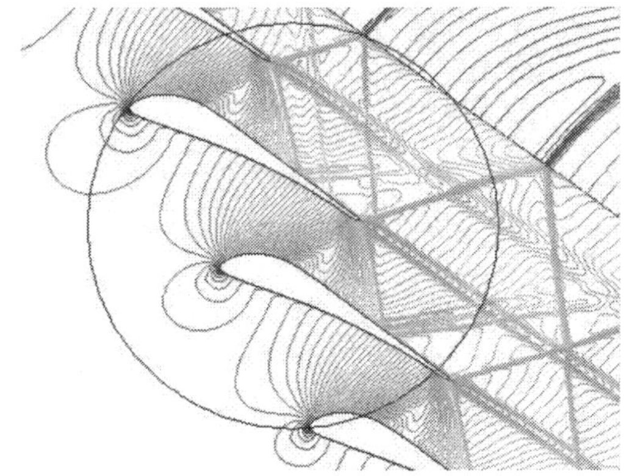

Fig. 12. Detail of the Mach number isolines (γ-ζ transition model)

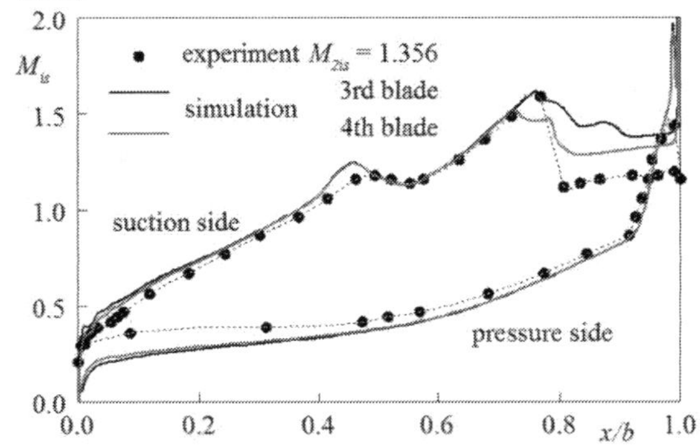

Fig. 13. Distribution of the isentropic Mach number (γ-ζ transition model)

The corresponding distribution of the skin friction coefficient is shown in Fig. 14 where the separation position agrees with the shock wave impact. The detail of the skin friction distribution is shown in Fig. 15. The experimentally determined separation zone covering the suction side form from $x/b \approx 0.75$ to 0.817 is marked by the rectangle.

The agreement of the prediction of the separation location with experimental data is quite acceptable. Similarly as the algebraic transition model, the γ-ζ transition model predicts a nearly separation in the region of the sudden change of the surface curvature of the suction side and a very short separation near the trailing edge on the blade pressure side.

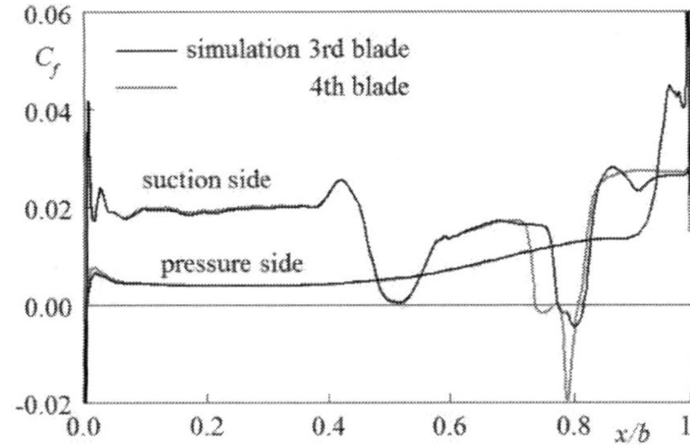

Fig. 14 Skin friction distribution (γ-ζ transition model)

https://doi.org/10.1051/epjconf/201818002056

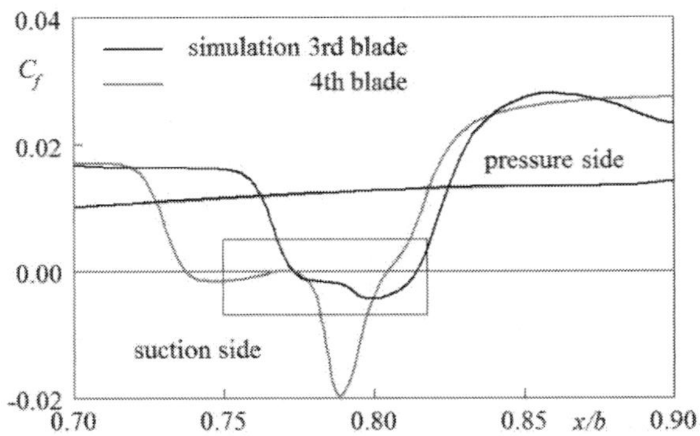

Fig. 15 Detail of the skin friction distribution (γ-ζ transition model)

Conclusions

Numerical simulations of 2D transonic flow through a linear mid-span turbine blade cascade were carried out for various outlet boundary conditions. Predictions were accomplished partly for the standard computational domain with various outlet boundary conditions by the algebraic transition model of Straka and Příhoda [1] connected with the EARSM turbulence model of Hellsten [2] and partly for the computational domain corresponding to the geometrical arrangement in the wind tunnel by the γ-ζ transition model of Dick et al. [3] with the SST turbulence model.

The accomplished simulations have shown the significant dependence of results on the prescription of the outlet boundary condition. Simulations carried out for the standard computational domain with the output at the traversing plane behind the blade cascade and/or for the computational domain corresponding to the experimental arrangement have shown that numerical results obtained on the basis of the numerical optimization correspond quite well to experimental data. The agreement of predicted flow fields and main parameters including energy losses is acceptable.

The work was supported by the Technology Agency of the Czech Republic under the grants TH02020057 and TA04020129. Institutional support RVO 61388998 is also gratefully acknowledged.

References

1. P. Straka, J. Příhoda, Proc. Conf. Topical Problems of Fluid Mechanics, Prague, 131-134 (2014)
2. A. Hellsten, *New two-equation turbulence model for aerodynamics applications*, PhD. Thesis, Helsinki University of Technology (2004)
3. E. Dick, S. Kubacki, K. Lodefier, W. Elsner, Intermittency modelling of transitional boundary layer flows on steam and gas turbine blades, *Engineering Applications of Computational Fluid Dynamics*, Vol. 2, International Energy and Environment Foundation. 173-216 (2012)
4. D. V. Gaitonde, Progress in Aerospace Sciences, **72**, 80-99 (2015)
5. D. Arnal, J. Délery, *Laminar-turbulent transition and shock wave/boundary layer interaction*, RTO-EN-AVT-116, 46 p. (2005)
6. P. Flaszynski, P. Doerffer, R. Szwaba, P. Kaczynski, M. Piotrowicz, Jour. of Thermal Science, **24**, 510-515 (2015)
7. M. Piotrowicz, P. Flaszyński, Jour. of Physics: Conference Series, **760,** 012023 (2016)
8. B. E. Launder, Jour. of Heat Transfer, **110**, 1112-1128, (1988)
9. F. R. Menter, AIAA Jour., **32**, 1598-1605 (1994)
10. R. Narasimha, Prog. Aerospace Science, **22**, 29-80, (1985)
11. W. J. Solomon, G. J. Walker, J. P. Gostelow, Jour. of Turbomachinery, **118**, 744-751 (1996)
12. R. E. Mayle, Jour. of Turbomachinery, **113**, 509-537 (1991)
13. G. J. Walker, AIAA Jour., **27**, 595-607 (1989)
14. R. Langtry, F. R. Menter, AIAA Jour., **47**, 2894-2906 (2009)
15. K. Lodefier, E. Dick, Flow, Turbulence and Combustion, **76**, 103-132 (2006)
16. B. J. Abu-Ghannam, R. Shaw, Jour. of Mech. Eng. Science, **22**, 213-228 (1980)
17. P. Louda, J. Příhoda, K. Kozel, Proc. Conf. Topical Problems of Fluid Mechanics, Prague, 211-220 (2017)
18. M. Luxa, D. Šimurda, P. Šafařík, J. Synáč, B. Rudas, Proc. 10th European Conference on Turbomachinery, Fluid Dynamics and Thermo-Dynamics, Lappeenranta, 360-369 (2013)

Numerical modelling of turbulent transition in complex geometries

Petr Louda[1,2a], *Jaromír* Příhoda[1], and *Karel* Kozel[2]

[1] Institute of Thermomechanics AS CR, Prague, Czech Republic
[2] Czech Technical University in Prague, Czech Republic

Abstract. The work deals with numerical simulation of laminar-turbulent transition in transonic flows in turbine cascades. The 3D cascade geometry as well as 2D model cascade in a wind tunnel is simulated. The γ-ζ transition model is based on empirical criteria for start of the transition. The implementation of the model is discussed including reformulation of the criterion for transition on separation bubble.

Key words. turbine cascade, transition modelling, shock wave-boundary layer interaction, wind tunnel

1 Introduction

Mathematical models based on the Favre-averaged Navier-Stokes equations are well accepted tools in predicting turbulent flows, however, the accuracy is influenced by turbulence model and often by its capability to predict transition to turbulence. Common two-equation eddy-viscosity turbulence models usually predict too early start of transition and then the transition is too fast. This problem is further emphasized by over-prediction of the turbulent energy production on the leading edge of the blade which has its origin in the eddy-viscosity assumption. Some ad hoc remedies of the later problem has been proposed e.g. limiter by Medic, Durbin [1]. However the transition still requires explicit triggering. Considering transition models based on transport equations which seem more general than algebraic ones, the models contain equation for an intermittency variable and also for other auxiliary variable or variables. More recent examples are 3-equation model by Walters and Cokljat [2] or 2-equation model by Menter, Langtry [3]. Also 1-equation model is proposed by Durbin [4]. These models have "local" form enabling easy implementation especially on parallel computers. However they also share disadvantage of containing transition criteria implicitly, or as Menter, Langtry model require evaluating criteria in every point of the flow-field which is time consuming and ambiguous especially when moving walls are present. Although there are new attempts to improve upon these models, their reliability remains to be confirmed. In this work we apply a more conservative γ-ζ model of Lodefier, Dick [5] and Kubacki et al [6] instead. The model contains transition criteria explicitly and any new criterion can be added easily. The empirical transition criteria are reliable. The downside is that the model distinguishes free-stream and boundary layer and thus is not local. Nevertheless we show that also for non-trivial geometry when using multi-block grids the model can be block-local and does not require case-specific input under assumption that the whole thickness of boundary layer is contained in one block, at least in region where transition occurs. The model is applied here in domains with non-trivial boundary as 2D cascade in wind tunnel or 3D periodic cascade The implementation does not require any case-specific user input.

[a] Corresponding autor: petr.louda@fs.cvut.cz

2 Mathematical model

The mathematical model of turbulent flow is based on Favre-averaged Navier-Stokes (NS) equations, see e.g. Wilcox [7]. The system consisting of continuity, 3 momentum and energy equations can be written in 3D in Cartesian coordinates as

$$\int_V \frac{\partial W}{\partial t} + \oint_{\partial V} F^I dS = \oint_{\partial V} F^V dS, \tag{1}$$

$$W = \begin{bmatrix} \rho \\ \rho u_1 \\ \rho u_2 \\ \rho u_3 \\ \rho E \end{bmatrix}, \quad F^I = u_c \begin{bmatrix} \rho \\ \rho u_1 \\ \rho u_2 \\ \rho u_3 \\ \rho H \end{bmatrix} + \begin{bmatrix} 0 \\ pn_1 \\ pn_2 \\ pn_3 \\ 0 \end{bmatrix}, \tag{2}$$

$$F^V = \begin{bmatrix} 0 \\ t_{i1} + \tau_{i1} \\ t_{i2} + \tau_{i2} \\ t_{i3} + \tau_{i3} \\ (t_{ij} + \tau_{ij})u_j - q_i - q_i^t \end{bmatrix} n_i, \tag{3}$$

where V is control volume, n_i outer unit normal vector components of its surface, t time, ρ density, u_i velocity vector components, E total energy per unit volume, $H = E + p/\rho$ is total enthalpy and p static pressure. The normal velocity $u_c = u_i n_i$. Summation convention is used for repeated indices. Equation of state for perfect gas is prescribed in the form

$$E = \frac{1}{\gamma - 1}\frac{p}{\rho} + \frac{1}{2}(u_1^2 + u_2^2 + u_3^2) + k, \tag{4}$$

with the ratio of specific heats $\gamma = 1.4$ and k being turbulent energy. The molecular stress tensor and heat flux vector respectively are assumed in the form

$$t_{ij} = \mu 2S_{ij}, \quad S_{ij} = \frac{1}{2}\left(\frac{\partial u_i}{\partial x_j} + \frac{\partial u_j}{\partial x_i}\right) - \frac{2}{3}\delta_{ij}\frac{\partial u_k}{\partial x_k}, \quad (5)$$

$$q_i = -\frac{\gamma}{\gamma - 1}\frac{\mu}{Pr}\frac{\partial(p/\rho)}{\partial x_i}, \quad (6)$$

where δ_{ij} is the Kronecker delta. The dynamic viscosity μ and Prandtl number satisfy

$$\mu = const, \quad Pr = const. \quad (7)$$

The effect of turbulent fluctuations is present by the Reynolds stress tensor τ_{ij} and turbulent heat flux q_i^t, which need to be modeled. An eddy viscosity model assumes

$$\tau_{ij} = \mu_t 2S_{ij} - \frac{2}{3}\delta_{ij}\rho k, \quad (8)$$

$$q_i^t = q_i \frac{Pr}{\mu}\frac{\mu_t}{Pr_t}, \quad (9)$$

where μ_t is eddy viscosity and the turbulent Prandtl number is set $Pr_t = 0.91$. In a k-ω model, the eddy viscosity $\mu_t \sim \rho k/\omega$. The k-ω system can be written

$$\int_V \frac{\partial}{\partial t}\begin{bmatrix}\rho k\\ \rho\omega\end{bmatrix} d\mathcal{V} + \oint_{\partial V} u_c \begin{bmatrix}\rho k\\ \rho\omega\end{bmatrix} dS =$$
$$\oint_{\partial V}\begin{bmatrix}(\mu + \sigma_k\mu_t)\frac{\partial k}{\partial x_i}\\ (\mu + \sigma_\omega\mu_t)\frac{\partial\omega}{\partial x_i}\end{bmatrix} n_i dS + \int_V \begin{bmatrix}P_k - \beta^*\rho\omega k\\ \alpha\frac{\omega}{k}P_k - \beta\rho\omega^2 + CD\end{bmatrix} d\mathcal{V}, \quad (10)$$

where the turbulent production $P_k = \tau_{ij}\frac{\partial u_i}{\partial x_j}$, the α, β, β^*, σ_k, σ_ω are model coefficients and CD a cross-diffusion term, $CD \sim (\partial k/\partial x_i)(\partial\omega/\partial x_i)$. In this work the SST variant of k-ω model is use, see Menter [8].

3 Transition modelling

The two-equation turbulence models for RANS alone do not predict the transition to turbulence with satisfactory precision. Therefore some intermittency function multiplying turbulent stresses or merely turbulence production is introduced. In this work the γ-*zeta* intermittency model of Dick et al. [9] is chosen. It has been developed with emphasis on transition phenomena occurring in turbomachinery as is the bypass transition, transition on separation bubble or wake induced transition. For better prediction of the later case the intermittency is split into near-wall part γ and free-stream part ζ each governed by its transport equation

$$\frac{\partial\rho\gamma}{\partial t} + \frac{\partial\rho u_i\gamma}{\partial x_i} = \frac{\partial}{\partial x_i}\left[(\mu + \sigma_\gamma\mu_t)\frac{\partial\gamma}{\partial x_i}\right] + S_\gamma \quad (11)$$

$$\frac{\partial\rho\zeta}{\partial t} + \frac{\partial\rho u_i\zeta}{\partial x_i} = \frac{\partial}{\partial x_i}\left[(\mu + \sigma_\zeta\mu_\zeta)\frac{\partial\zeta}{\partial x_i}\right] + S_\zeta \quad (12)$$

with source terms

$$S_\gamma = 2\beta_\gamma(1-\gamma)\sqrt{-\ln(1-\gamma)}\rho U_\gamma F_s \quad (13)$$

$$S_\zeta = -C_2\mu_\zeta\frac{U}{U_e^2}\frac{\partial U}{\partial n}\frac{\partial\zeta}{\partial n} \quad (14)$$

where μ_t is eddy viscosity, μ_ζ viscosity for ζ (see [9]), β_γ growth parameter, U_γ mean velocity scale, n wall-normal direction. The Prandtl numbers $\sigma_\gamma = \sigma_\zeta = 1$. The F_s is switching function equal either 0 in laminar boundary layer or 1 if a criterion for turbulent flow is satisfied. The transition criteria can be chosen according to expected transition mechanism. The source term S_ζ guarantees that ζ vanishes in the boundary layer, so that both intermittencies are complementary in the turbulent flow. Since the source terms are not directly linked, the total intermittency $(\gamma + \zeta)$ has to be limited to $< 0, 1 >$ explicitly. The boundary conditions are $\gamma = 0$, $\zeta = 1$ in the inlet, $\partial\gamma/\partial n = \partial\zeta/\partial n = 0$ in the outlet and $\partial\gamma/\partial n = 0$, $\zeta = 0$ on the wall.

The γ-ζ transition model relies on meeting various transition criteria in switching source term for near-wall intermittency γ. For bypass transition the Mayle and Abu-Ghannam, Shaw criteria are used, see [10,6]. The criteria are functions of boundary layer parameters as momentum thickness,free-stream velocity and turbulence intensity. These have to be computed preferably in an automatic manner. The boundary layer edge detection algorhitm is taken from the original article [9]. The boundary layer edge is defined as distance from wall where the vorticity reaches 1% of its maximum, found on the wall in the attached flow. In separated flow the local maximum farthest from the wall is taken. This thickness is further increased by 30 % to reach the free-stream.

The criteria for transition on separation bubble applicable in internal aerodynamics moreover depend explicitly on the upstream history of the flow. Here the Mayle criterion in the form

$$Re_{st} = 700 Re_{\theta s}^{0.7}\exp(-0.3Tu_e) \quad (15)$$

is used, where Re_{st} is the Reynolds number from the distance from the point of separation x_s, $Re_{\theta s}$ is the Reynolds number from momentum thickness at the point of separation and Tu_e free-stream turbulence intensity. Obviously, the tracking of separation is very cumbersome in 2-dimensional case (although it has also been done by the authors) and plain ambiguous and thus impossible in 3D case. Therefore the following re-formulation is proposed, based on the observation of the simulated separation bubbles. To an acceptable degree of precision, one can assume that the separation thickness δ_0 starts growing linearly with th distance from separation x_s. Then

$$x_s = \delta_0/(A\frac{\partial\delta_0}{\partial s}) \quad (16)$$

where δ_0 is the distance of zero velocity streamline from the wall and the slope in streamwise direction $\frac{\partial\delta_0}{\partial s}$ is evaluated on standard stencil parallel to wall. The non-local formulation persists only in the wall-normal direction which is acceptable. The empirical coefficient $A = 0.9$ is not very critical. If the slope $\frac{\partial\delta_0}{\partial s}$ is negative, the maximum thickness of the bubble

https://doi.org/10.1051/epjconf/201818002057

has been reached. It is assumed that in this part the transition started at latest. In the attached state the criterion for attached boundary layer takes over.

Further step to application of the transition model in separated flow is necessary. Since the source term in γ-equation is always non-negative, once the intermittency is captured in the recirculation it can not diminish except by diffusion which, however, has mostly opposite effect. The separation is then predicted as too short. Therefore a numerical sink term is switched-on inside the recirculation bubble.

In 3D cases a flow parallel to a corner can be present. The vorticity magnitude here has no more maxima on the wall. Therefore the boundary layers on both walls are observed and if they are no more well defined, the average of free-stream parameters from both walls are used for the whole corner area. The intersection of 3 walls is excluded, not least because of the lack of physically valid transition criteria. Such corner in principle can be dealt with by placing arbitrarily small grid block into the corner and setting the intermittency production there ad hoc, e.g. to zero.

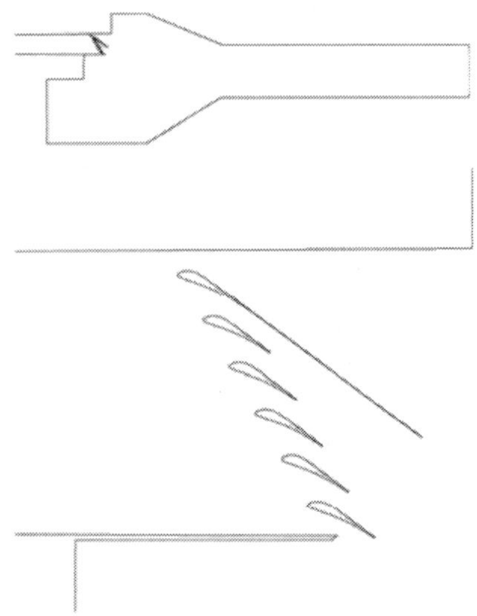

Fig. 1. Domain boundary for the cascade in wind tunnel

4 Numerical method

The above mathematical model is solved numerically by an in-house Fortran 90 code implementing implicit finite volume upwind solver. For spatial discretization we use a cell centered finite volume method with quadrilateral (in 2D) or hexahedral (in 3D) finite volumes composing multiple blocks of structured grid. The numerical inviscid flux is computed by the AUSMPW+ splitting [11]. The higher order of accuracy is achieved by linear interpolation in the direction of grid lines with e.g. van Leer limiter. The discretization of diffusive flux is central. The approximation of cell face derivatives needed in diffusive terms uses octahedral dual finite volumes constructed over each face of primary volume – the vertices are located in vertices of primary face and in centers of adjacent primary volumes. For time discretization, the implicit backward Euler scheme is employed where the steady residual at new time level is approximated by linear extrapolation. The Jacobi matrices of the flux are obtained as derivatives of discrete expressions for flux with respect to nodal values from the stencil of implicit operator. We chose 7-point stencil, which leads to block 7-diagonal system of linear equations (not considering boundary conditions). The size of a block equals to the number of coupled equations. For more details see [12]. The turbulence model equations are solved in the same manner, however, decoupled from the RANS equations. The decoupling saves memory and computational time since there is no need for matrix-vector multiplication if the source terms Jacobi matrix is chosen diagonal.

Some remarks concerning transition model have already been presented. In parallel implementation it is natural to distribute the work block-wise, although it cannot be optimal in all cases. Therefore we require the model be at least block-local. This is satisfied if the whole thickness of a boundary layer is contained in same block. It can be satisfied by sufficiently thick O-grid around the blade.

5 Computational results

In this part we show application of the model to 2D and 3D flow through turbine cascades. In 2D we consider flow through model cascade consisting of six mid-section blades in a wind tunnel. The boundary approximates wind tunnel facility of the Institute of Thermomechanics AS CR and is shown in Fig. 1. To improve periodicity of the flow around cascade, internal wall is located behind top blade. In measurements, carried out by Luxa et al. [13,14] the internal wall is perforated plate with staggered rows of circular holes. Obviously, in 2D a model is needed. In present simulations it is simplified to non-permeable line of zero thickness. The isentropic outlet Mach number of the tunnel is 1.120, the isentropic Reynolds number for the blade $1.2 \cdot 10^6$.

The computational grid consist of approx. 310 000 finite volumes divided into 192 blocks. Each blade is discretised by 260 grid points. We note that the model of transition does not require any case specific user input other than merely activating the model.

The Fig. 2 shows simulations of the experimental cascade in the wind tunnel, the Fig. 3 same cases with internal wall. The internal wall greatly improves the periodicity of the flow through cascade, although only impermeable internal wall has been applied. It can be expected that with perforated wall the flow pattern will be similar but with weaker shock wave reflexions. The transition causes maximum Mach numbers little bit higher than in fully turbulent case. Both cases with internal wall are very similar except that with transition model the boundary layer on the suction sides is laminar up to the interaction with the incident shock wave. In more detail, the middle channel is shown in Figs. 4 and 5. Fully turbulent simulation shows simple reflection of the shock wave from turbulent boundary layer. With the model of transition the boundary layer in front of the interaction is still laminar and small separation exists under the shock wave. The deformation of boundary layer causes formation of separation shock at beginning of the recirculation bubble and reattachment shock at its end. In between expansion waves are present. A better resolution of the interaction is possible on finer grid but for the purpose of the whole tunnel study we stop here. The corresponding near wall intermittency γ is shown in Fig. 6, where red color denotes turbulent boundary layer ($\gamma = 1$).

Considering comparison with experiment finding corresponding regimes appears difficult. The reference is isentropic Mach number in a certain point in the settling chamber. This is different for each tunnel simulation although the outlet isentropic Mach number is same ($M_{2is} = 1.120$) and also the mass flow is practically same (44 kg/s/m). The reference Mach number varied between 1.244 and 1.320 in mentioned 4 simulations. Therefore we show two measured regimes in Fig. 7 for two Mach numbers. The configuration is without internal wall and should correspond to simulation from Fig. 4. The computed shock wave-boundary layer interaction is correct but the angle of the shock wave seems to correspond to higher reference Mach number.

As an example of 3D simulation, the SE1050 mid-section turbine cascade is chosen. The outlet Mach number $M_{2is} = 1.2$ and $Re_{2is} = 1.5 \cdot 10^6$. One period of the cascade is solved and the blade is prismatic with symmetry condition in the midplane. The blade has 230×95 grid points on its surface. The inlet angle is design angle. A small recirculation bubble exists where incident shock reaches suction side. This is shown on streamwise velocity component in blue in Fig. 8 whereas the interaction with incident shock is turbulent as shown on the isolines of Mach number in the same figure. Indeed the boundary layer on the largest part of the suction side is turbulent as shown on the near wall intermittency in Fig. 9. The boundary layer on the pressure side is laminar.

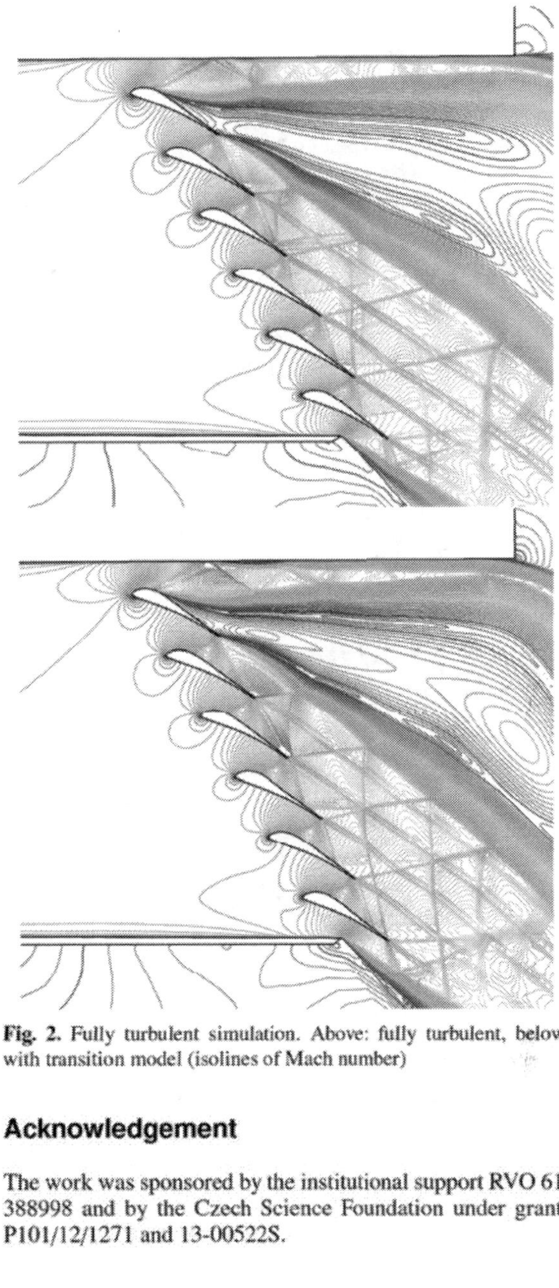

Fig. 2. Fully turbulent simulation. Above: fully turbulent, below: with transition model (isolines of Mach number)

6 Conclusions

The work presented simulations of 2D and 3D turbulent flow through turbine cascades with eddy-viscosity turbulence model complemented with the γ-ζ model of transition to turbulence. The mathematical model is solved by implicit AUSM finite volume method on multi-block structured grids. The implementation of transition model does not rely on explicit prescription of boundary layer edge and is adaptive as long as the whole thickness of boundary layer is contained in one block, which is typically O-grid around the blade consisting of several blocks in tangential direction. Also the treatment of corners is automatic. The transition mechanisms include bypass transition and transition on the separation bubble. The transition criterion for the later mechanism has been re-formulated in order to be local in streamwise direction. The physical correctness of transition prediction in the flow in convex corner however still needs to be confirmed by measurement. The results are shown for 2D turbine cascade in wind tunnel where the transition model enables to capture measured interaction of shock wave with laminar boundary layer. Further the model has been applied to 3D periodic turbine cascade.

Acknowledgement

The work was sponsored by the institutional support RVO 61-388998 and by the Czech Science Foundation under grants P101/12/1271 and 13-00522S.

References

1. G. Medic, P. Durbin, Journal of Turbomachinery **124**, 187 (2002)
2. D.K. Walters, D. Cokljat, J. of Fluids Engineering **130**, 121401 (2008)
3. R.B. Langtry, F.R. Menter, AIAA journal **47**, 2894 (2009)
4. P. Durbin, Int. J. of Heat and Fluid Flow **36**, 1 (2012)
5. K. Lodefier, E. Dick, Flow, Turbulence and Combustion **76**, 103 (2006)

 https://doi.org/10.1051/epjconf/201818002057

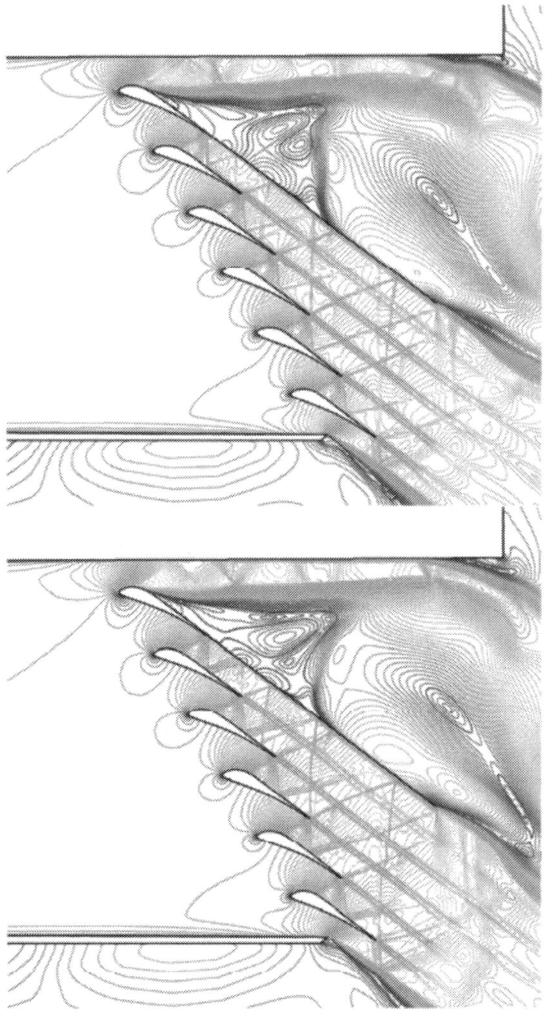

Fig. 3. Simulation with transition model. Above: fully turbulent, below: with transition model (isolines of Mach number)

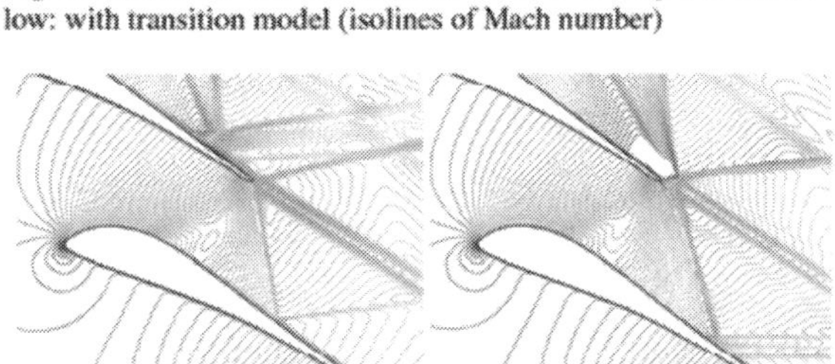

Fig. 4. Interaction of incident shock wave with boundary layer, cases without internal wall. Left: fully turbulent, right: with transition

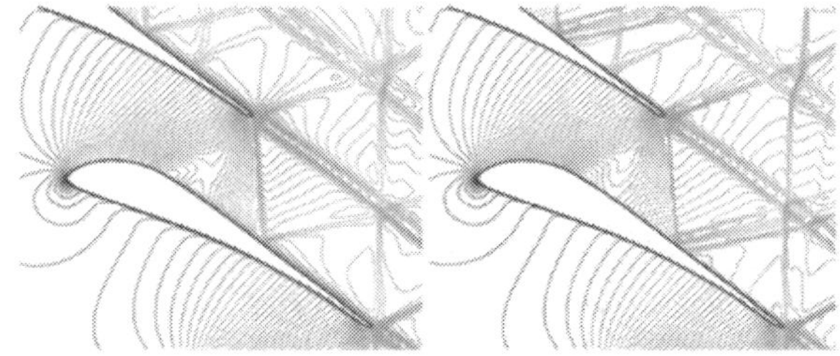

Fig. 5. Interaction of incident shock wave with boundary layer, cases with internal wall. Left: fully turbulent, right: with transition

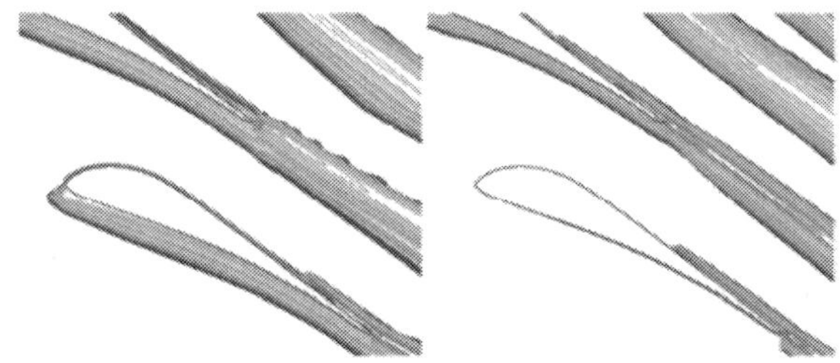

Fig. 6. Isolines of near wall intermittency γ. Left: case without internal wall, right: with internal wall

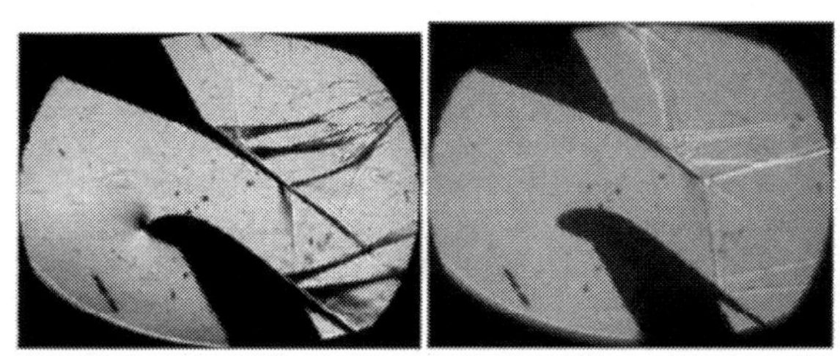

Fig. 7. Schlieren images from measurements by Luxa et al. (Institute of Thermomechanics AS CR). Letf: M_{ref} = 1.328, right: M_{ref} = 1.407

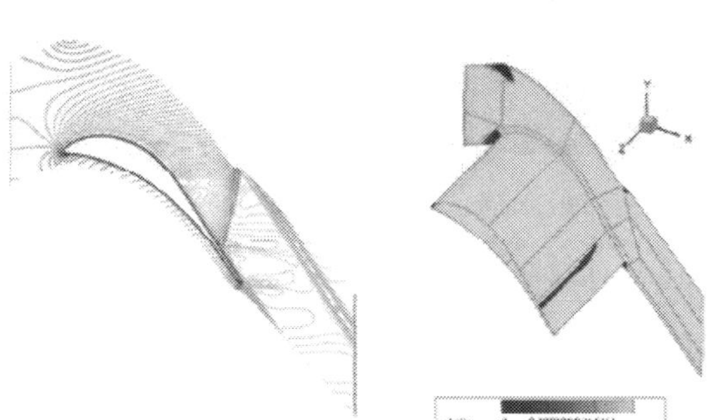

Fig. 8. Left: isolines of Mach number in the center-plane of 3D cascade, right: streamwise velocity near the wall - negative values in blue

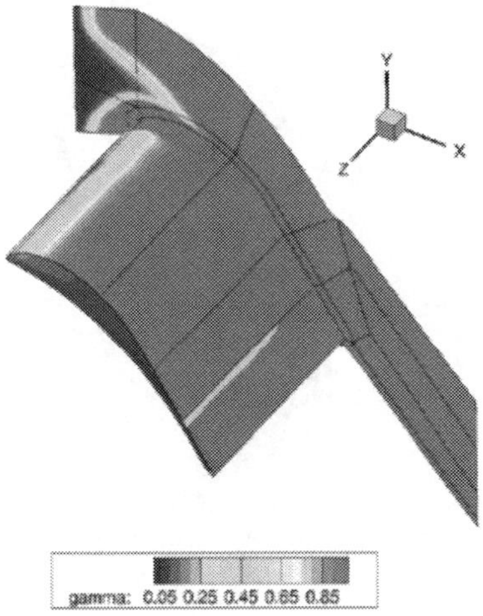

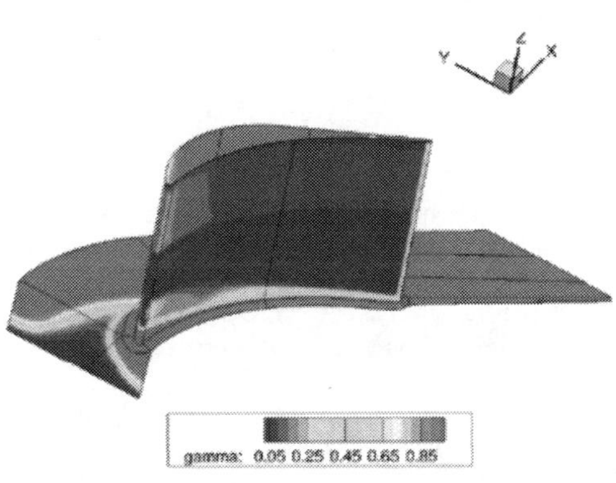

Fig. 9. Near wall intermittency on the 3D cascade

6. S. Kubacki, K. Lodefier, R. Zarzycki, W. Elsner, E. Dick, Flow, Turbulence and Combustion **83**, 539 (2009)
7. D.C. Wilcox, *Turbulence modeling for CFD* (DCW Industries, Inc., California, 1998), 2nd edition
8. F.R. Menter, AIAA Journal **32**, 1598 (1994)
9. E. Dick, S. Kubacki, K. Lodefier, W. Elsner, *Enginnering Applications of Computational Fluid Dynamics* (International Energy and Environment Foundation, 2012), Vol. 2, chap. Intermittency modelling of transitional boundary layer flows on steam and gas turbine blades, pp. 173–216
10. R.E. Mayle, J. Turbomachinery **113**, 509 (1991), trans. ASME
11. K.H. Kim, C. Kim, O.H. Rho, J. of Computational Physics **174**, 38 (2001)
12. P. Louda, K. Kozel, J. Příhoda, Int. J. for Numerical Methods in Fluids **56**, 1399 (2008)
13. M. Luxa, J. Synáč, P. Šafařík, D. Šimurda, Komunikacie **14**, 23 (2012)
14. M. Luxa, D. Šimurda, P. Šafařík, J. Synáč, B. Rudas, *High-speed Aerodynamic Investigation of the Midsection of a 48" Rotor Blade for the Last Stage of Steam Turbine*, in *Proceedings 10th European Conference on Turbomachinery Fluid Dynamics and Thermodynamics* (2013), pp. 360–369

Rainbow schlieren measurements in underexpanded jets from square supersonic micro nozzles

Hiroaki Maeda[1,*], *Hikaru* Fukuda[1], *Shinichiro* Nakao[1], *Yoshiaki* Miyazato[1], and *Yojiro* Ishino[2]

[1]The University of Kitakyushu, 1-1, Hibikino, Wakamatsu-ku, Kitakyushu, 808-0135, Japan
[2]Nagoya Institute of Technology, Gokiso-cho, Showa-ku, Nagoya, 466-8555, Japan

Abstract. The slightly underexpanded free jet issued from a supersonic micro nozzle with a design Mach number of 1.5 and a square shape of 1 mm × 1 mm at the nozzle exit has been experimentally investigated using the rainbow schlieren deflectometry combined with the computed tomography where the nozzle operating pressure ratio is held constant at 4.0. Density fields in the free jet obtained from the rainbow schlieren deflectometry are the vortex sheet theory proposed by Tam. It is shown that there is good quantitative agreement between experiment and theory. It is shown that the rainbow schlieren deflectometry is a useful tool for studying the three-dimensional structure of shock containing free jets from micro-scale supersonic nozzles.

1 Introduction

In recent years, the design and fabrication of a micro-electro mechanical system (MEMS) has increased the need for an understanding of fluid flow in microscale devices including inkjet printers, small satellite thrusters, critical nozzles at low Reynolds numbers and flowmeters of next-generation fuel cell vehicles (FCV). Especially, the systems will need miniaturization in the space engineering. A detailed and deep comprehension of the flow characteristics through such a device requires the information of vector and scalar quantities in the flow field.

The Reynolds-averaged Navier-Stokes simulation or large eddy simulation often becomes effective tools to investigate the physical properties in various areas such as aerospace and mechanical engineering because of cost curtailment and time saving. However, such numerical simulations include turbulent models to close the governing equations. The capability of these models in simulating a complex physical phenomenon must be ascertained before the numerical results can be accepted. Numerical results are usually validated by comparing them with reliable experimental data. Qualitative flow visualization including schlieren and shadowgraph techniques [1] is routinely used to evaluate the computed flowfield. Nevertheless, in these techniques, quantitative assessment of physical properties such as the temperature, density, pressure, and velocity in a simulated flowfield are not feasible.

Many researchers often compare simulated schlieren or shadowgraph images directly with those from experiments having the same flow conditions in order to validate the simulation. However, even though a computed schlieren picture is in good agreement with an experimental one, genuine scalar or vector fields cannot always be captured correctly by the computation because conventional schlieren images can only provide integrated information about the density gradient. Therefore, a comparison between experiment and simulation is only meaningful if the numerical results are spatially averaged along the view direction.

In the present study, the three-dimensional density field of the slightly underexpanded supersonic jet issued from a square micro Laval nozzle with a design Mach number of 1.5 and a square shape of 1 mm × 1 mm at the nozzle exit plane is for the first time clarified by the rainbow schlieren deflectometry [2~4] combined with the computed tomography. Also, the three-dimensional free jet flow fields under the same conditions as the experiment are solved using the vortex sheet theory [5]. A quantitative comparison between experiment and theory is carried out to investigate three-dimensional spatial variations of the shock-containing free jet. It is found that the rainbow schlieren deflectometry combined with the computed tomography is a very efficient tool to measure three-dimensional density fields of shock-containing free jets from micro nozzles and to validate numerical simulation codes.

2 Vortex Sheet Model

In general, the vortex sheet model [5] proposed by Tam can quantitatively predict the gross features of a moderately underexpanded jet from an axisymmetric or rectangular nozzle, including the shock-cell spacing, screech frequencies except for higher levels of underexpanded or overexpanded conditions with the formation of a Mach disk or Mach shock inside the jet. A general solution of the vortex sheet shock cell boundary value problem can be found by the method of eigenfunction expansion.

* Hiroaki Maeda: x6mba020@eng.kitakyu-u.ac.jp

Figure 1 shows a linear shock cell model for the underexpanded free jet from a square supersonic nozzle with a width h_e at the exit and a design Mach number M_d. In the case of the underexpanded jet, the Prandtl-Meyer expansion waves occur from the nozzle lip because the nozzle exit pressure p_e is larger than the back pressure p_b. The expansion waves are reflected as the compression waves on the opposite jet boundary and one shock cell is formed.

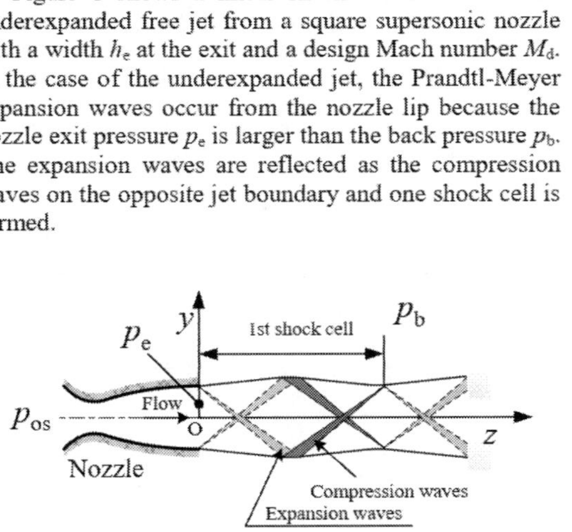

Fig. 1. Vortex sheet model

It is assumed that inside the jet the flow is inviscid and the mixing layer is thin vortex sheet, and outside the jet there is no disturbance. By combining the equation of continuity, the linearized equation of motion and the isentropic equation, it is straightforward to obtain the analytical solution for the pressure p and density ρ inside the jet as follows:

$$p(r,z) = \sum_{m=1}^{\infty}\sum_{n=1}^{\infty} A_{mn}\phi_{mn}\cos\left(\lambda_{mn}z\right) + p_b \tag{1}$$

$$\rho(r,z) = \rho_j + \frac{p(r,z) - p_b}{a_j^2} \tag{2}$$

with

$$A_{mn} = \frac{4\Delta p}{nm\pi^2}\left[1 - \cos\left(\frac{m\pi h_e}{h_j}\right)\right]\left[1 - \cos\left(\frac{n\pi h_e}{h_j}\right)\right] \tag{3}$$

$$\phi_{mn} = \sin\left[\frac{m\pi}{h_j}\left(x + \frac{h_j}{2}\right)\right]\sin\left[\frac{n\pi}{h_j}\left(y + \frac{h_j}{2}\right)\right] \tag{4}$$

$$\lambda_{mn} = \frac{\pi}{\left(M_j^2 - 1\right)^{\frac{1}{2}}}\left(\frac{m^2 + n^2}{h_j^2}\right)^{\frac{1}{2}} \quad (m, n = 1, 2, \text{[redacted]} \tag{5}$$

$$\Delta p = p_e - p_b \tag{6}$$

$$M_j = \sqrt{\frac{2}{\gamma - 1}\left\{\left(\frac{p_{os}}{p_b}\right)^{\frac{\gamma-1}{\gamma}} - 1\right\}} \tag{7}$$

$$\frac{h_e}{h_j} = \left(\frac{M_j}{M_d}\right)^{\frac{1}{2}}\left[\frac{2 + (\gamma - 1)M_d^2}{2 + (\gamma - 1)M_j^2}\right]^{\frac{\gamma+1}{4(\gamma-1)}} \tag{8}$$

where x is the horizontal axis, y the perpendicular axis, and z the longitudinal axis, γ the specific heat ratio, Δp the pressure difference between p_e and p_b, n and m are the natural numbers, ϕ_{nm} the eigenfunctions, and λ_{nm} the eigenvalues. Also, h_j, ρ_j, M_j, and a_j are the width, density, Mach number, and the speed of sound of the fully expanded jet, which are calculated based on the assumption of the isentropic flow from the nozzle inlet to the back pressure in the downstream of the nozzle exit. This shock-cell model can apply over a range of fully expanded Mach number M_j given by

$$\left|M_j^2 - M_d^2\right| < 1 \tag{9}$$

for underexpanded and overexpanded jet conditions. In addition, Suda and Kaji [6] investigated unsteady shock behavior in a rectangular underexpanded supersonic jet using a laser schlieren system and found that although the first and second shock cells slightly oscillate across the time-mean location, the third shock cell flaps dynamically with a travelling shock wave in it. Therefore, the shock cell theory appears to be reasonable over the region from the nozzle exit to the end of the 2nd shock cell.

3 Rainbow Schlieren Tomography

The problem of obtaining reliable quantitative information about the flow structure of three-dimensional supersonic jets arises in many industrial fields. Although direct imaging such as the conventional schlieren cannot provide quantitative information about the structure of three dimensional jets, the schlieren CT which combines the rainbow schlieren deflectometry with so-called computed tomography, often abbreviated as CT must be utilized. Computed tomography was originally developed for use in medical diagnostics with X-rays, but was later adapted for other applications in many scientific fields including physics, chemistry, astronomy, and geophysics, using in many cases other kinds of radiation and even elementary particles. A brief outline of computed tomography for density fields in the three-dimensional jet flows is given by Faris et al. [7] and Agrawal et al. [8].

Figure 2 shows a light ray traveling in a cross-section (x, y plane) of an asymmetric jet issued from a nozzle where the x, y, z rectangular Cartesian coordinate system is used and the z axis is perpendicular to the x, y plane which includes the vector along the optical axis direction of the schlieren system. Also, the n and n_a in Fig. 2 indicate the refractive index in the jet cross-section and that of the surrounding air, respectively.

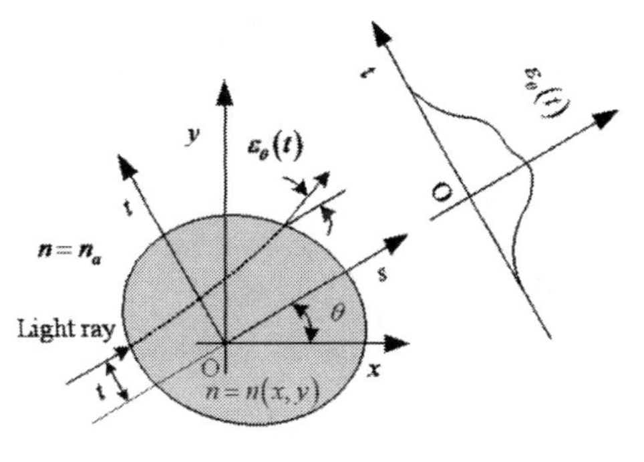

Fig. 2. Projection for tomographic reconstruction

The refractive index or density fields will now be investigated for a cross section of z = constant. As shown in Fig. 2, let us consider the rotated coordinates (s, t) inclined at an angle θ away from the fixed-original coordinates (x, y). Then, a light ray traveling in the s direction with an offset of t from the axis s is bent by the interaction with the jet flow and has an angle of inclination $\varepsilon_\theta(t)$ with respect to its original path. As the incoming ray is initially parallel to the s axis, the deflection angle $\varepsilon_\theta(t)$ after passing through the refractive index field is given by the line integral

$$\varepsilon_\theta(t) = \int_{-\infty}^{\infty} \frac{\partial \eta(s,t)}{\partial t} ds \quad \textbf{(10)}$$

along the s direction of the partial derivative $\partial\eta/\partial t$ with respect to the t variable of the normalized refractive index difference

$$\eta(s,t) \equiv \frac{n - n_a}{n_a} \quad \textbf{(11)}$$

for small ray deflections. The deflection angle $\varepsilon_\theta(t)$ is taken for a range of various angles from $\theta = 0$ deg to 180 deg. The task of tomographic reconstruction in the present investigation is to find $\eta(x,y)$ based upon a given knowledge of $\varepsilon_\theta(t)$ and then density fields can be obtained through a well-known linear relation between refractive index and density, as described later.

After the one-dimensional Fourier transform of Eq. (10) with respect to the t variable and using the Leibniz rule for differentiation under the integral sign, it reduces to a relation,

$$\int_{-\infty}^{\infty} \varepsilon_\theta(t) \exp(-i2\pi t\zeta) dt = 2\pi i\zeta \int_{-\infty}^{\infty}\int_{-\infty}^{\infty} \eta(s,t) \exp(-i2\pi t\zeta) ds dt \quad \textbf{(12)}$$

The transformation from the (s,t)-coordinate to (x,y)-coordinate for the integral of the right-hand side of Eq. (12) yields the following form with $u = -\zeta\sin\theta$ and $v = \zeta\cos\theta$

$$\int_{-\infty}^{\infty}\int_{-\infty}^{\infty} \eta(s,t) \exp(-i2\pi t\zeta) ds dt = \int_{-\infty}^{\infty}\int_{-\infty}^{\infty} \eta(x,y) \exp[-i2\pi(ux+vy)] dx dy \quad \textbf{(13)}$$

This equation expresses the two-dimensional Fourier transform of $\eta(x,y)$. Hence, the inverse Fourier transform of Eq. (12) using the result of Eq. (13) with the convolution theorem leads to

$$\eta(x,y) = \int_0^{\pi} \left[\varepsilon_\theta(t) * k(t)\right]_{t=-x\sin\theta+y\cos\theta} d\theta \quad \textbf{(14)}$$

where $k(t)$ is given by [8]

$$k(t) = \frac{\sin^2(\pi t f_{\max})}{\pi^2 t} \quad \textbf{(15)}$$

with the Nyquist frequency $f_{\max}$ and the symbol * denotes convolution between $\varepsilon_\theta(t)$ and $k(t)$.

When deflection data are sampled at a spacing of Δt, only frequencies below the Nyquist frequency

$$f_{\max} = \frac{1}{2\Delta t} \quad \textbf{(16)}$$

are adequately sampled.

The ray deflection angle $\varepsilon_\theta(t)$ is correlated with the focal length f_d of a decollimating lens and the ray transverse displacement $d_\theta(t)$ at the cut-off plane of the schlieren system and given by

$$\varepsilon_\theta(t) = \frac{d_\theta(t)}{f_d} \quad \textbf{(17)}$$

For air there is a simple linear relation between the refractive index n and the gas density ρ:

$$\rho = \frac{n-1}{K} \quad \textbf{(18)}$$

where K is the Gradstone-Dale constant. As the K hardly depends on the wavelength of light, the same constant is used for all wavelengths.

Combination of Eqs. (11), (17) and (18) yields three-dimensional density fields of shock-containing jets. The more detailed descriptions for the reconstruction process to obtain three-dimensional density fields are reported in a paper of Agrawal et al. [8].

4 Experimental Apparatus

Experiments have been performed in a blow-down supersonic wind tunnel with the jet issued in the quiescent laboratory air. A schematic diagram of the experimental apparatus with a schlieren optical system is shown in Fig. 3. The air supplied by a compressor that pressurizes the ambient air up to 1 MPa is filtered, dried and stored in a high-pressure reservoir consisting of two tanks with a total capacity of 2 m^3. The high-pressure dry air from the reservoir is stagnated in a plenum chamber and then discharged into the atmosphere through a test nozzle. The total temperature in the plenum chamber is equal to the room temperature, and the plenum pressure is controlled and maintained constant during the testing by a valve.

A Laval nozzle with a design Mach number of 1.5 was used as the test nozzle and it is schematically shown in Fig. 4. It has a two-dimensional wall contour with a 850 μm height at the throat and 1000 μm height at the exit.

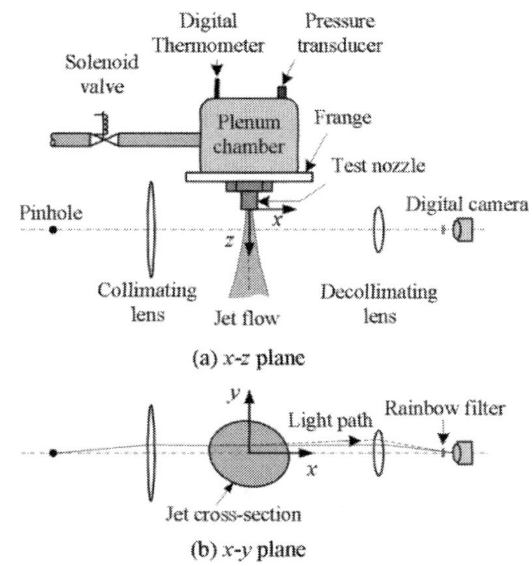

Fig. 3. Schematic diagram of experimental apparatus

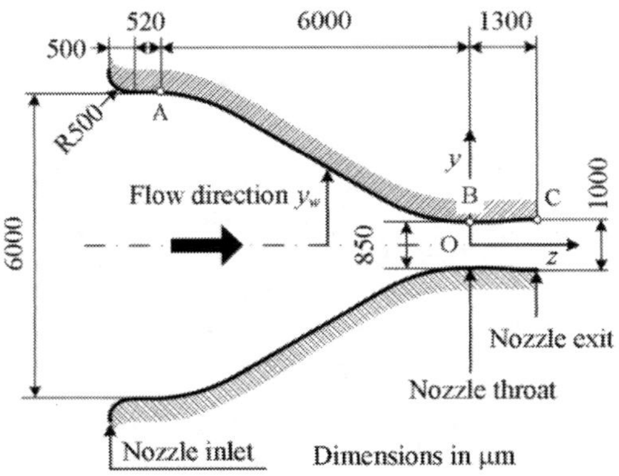

Part A to B: $y_w = 1288 \sin\left[\frac{\pi}{6000}(z - 3000)\right] + 1713$

Part B to C: Method of characteristics

Fig. 4. Schematic drawing of test nozzle

The nozzle wall contour over a range of parts A to B is designed to be a sinusoidal curve and that over a range of parts B to C is designed by the method of characteristic [9] to provide uniform and parallel flow in the nozzle exit plane at the design condition.

The jet issued from the nozzle is visualized by the rainbow schlieren deflectometry with a field of view of 30 mm diameter for a nozzle pressure ratio of $p_{os}/p_b = 4.0$ where p_{os} is the plenum pressure and p_b is the ambient pressure or back pressure. Also, the ambient pressure, temperature, and density for the present experiment are p_b = 100.5 kPa, T_b = 298.7 K, and ρ_b = 1.17 kg/m^3, respectively. Since the theoretical nozzle pressure ratio for the present nozzle is 3.67, the nozzle is operated at slightly underexpanded condition.

The rainbow schlieren system consists of rail-mounted optical components including a 50 μm diameter pinhole, a collimating achromatic lens with 100 mm diameter of f_c = 500 mm focal length, a decollimating achromatic lens with 30 mm diameter of f_d = 250 mm focal length, a computer generated 35 mm wide slide with color gradation in a 1.4 mm wide strip, and a digital camera (Nikon D7100) with a 30 mm diameter focusing lens of 600 mm focal length. The source image magnification factor defined as the ratio of the source image size to the original source size at the pinhole plane for the present rainbow system can be readily estimated from a relationship of the geometrical optics as $f_d/f_c = 0.5$. Therefore, the source image with a 25 μm diameter is formed at the focal point of the decollimating lens, wherein a rainbow filter is placed. It should be noted that the resolution and sensitivity of the rainbow schlieren system depend primarily on the size of the source image, the focal length of the decollimating lens, and resolution of color gradients on the rainbow filter [2].

A continuous 250 W metal halide light source connected to a 50 μm diameter fiber optic cable provides the light input at the pinhole through a 16.56 mm focal length objective lens. The camera output in the RGB format is digitized by a personal computer with 24 bit color frame grabber. The rainbow filter used in the present experiments is shown in Fig. 5. The filter was fabricated in computer software and then printed digitally on a high resolution 23 mm color film recorder. It has continuous hue variation from *hue* = 0 to 360 deg in a 1.4 mm wide strip and the origin $y = 0$ corresponds to *Hue* = 200 deg. The characteristics of the rainbow filter were performed by traversing the filter automatically in intervals of 0.05 mm in the y direction at the schlieren cut-off plane before starting experiments. The calibration result is shown as open symbols in Fig. 6. The abscissa is *Hue* and the ordinate is the transverse ray displacement d from the x axis at the cut-off plane. The solid line indicates a least-squares regression line of the experimental data using fifth degree polynomials.

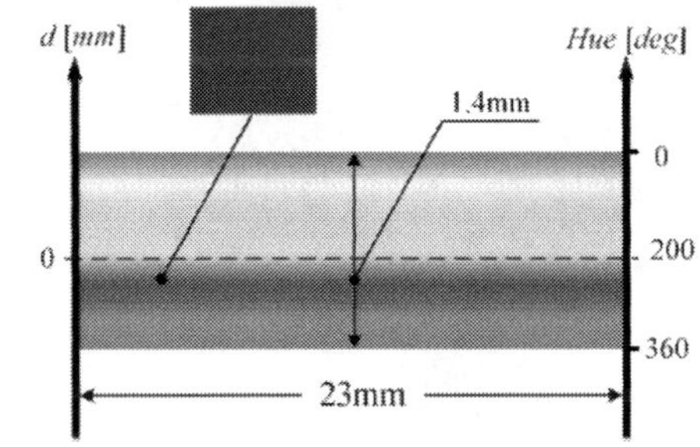

Fig. 5 Rainbow filter image

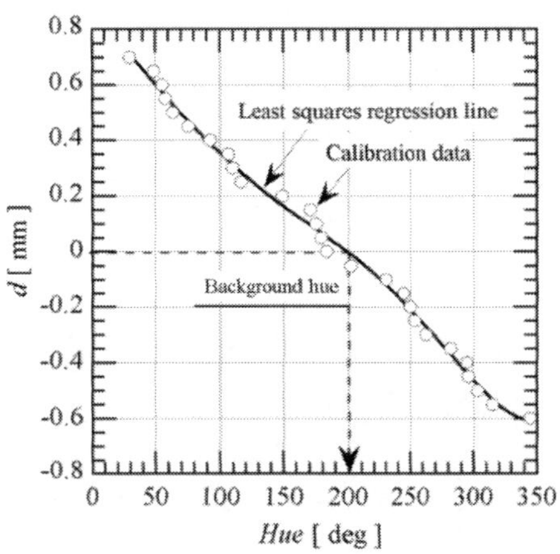

Fig. 6. Calibration curve of rainbow filter

As shown in Fig. 7, the test nozzle is installed inside a pulley with a gear ratio of 1:4 and it can be rotated about the center axis of the nozzle by a stepping motor with a gear ratio of 1:36 connected through a timing belt. In the present experiment, rainbow schlieren pictures are acquired over a range of nozzle angular angles from $\theta = 0$ deg to −180 deg by rotating the nozzle about its longitudinal axis (z axis) in equal angular intervals of −5

 https://doi.org/10.1051/epjconf/201818002058

deg. These 37 schlieren pictures are used for reconstruction of the jet density field by the schlieren CT method. The jet density profiles obtained from the schlieren CT method are compared with the vortex sheet model.

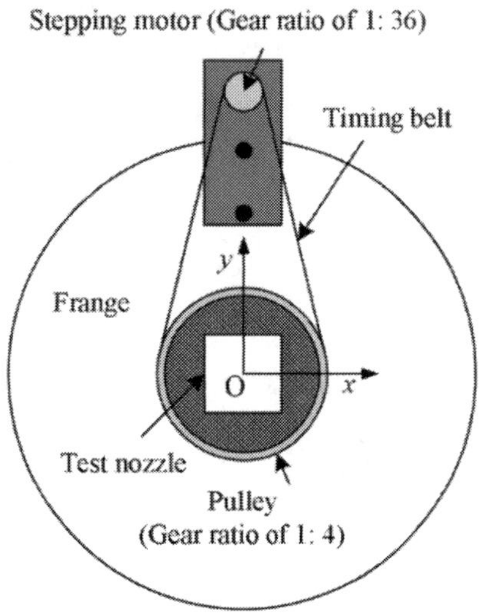

Fig. 7. Nozzle rotating device

5 Results and Discussion

A typical rainbow schlieren picture of slightly underexpanded supersonic jet issued from a square micro Laval nozzle is shown in Fig. 8 with the flow from left to right. The ratio p_{os}/p_b of the plenum pressure p_{os} to back pressure p_b is held constant at 4.0. As shown on the upper right corner in the picture of Fig. 8, the rainbow filter is placed horizontal with respect to the z axis at the schlieren cut-off plane and the background hue corresponds to the color of the location indicated as the dashed line on the filter image.

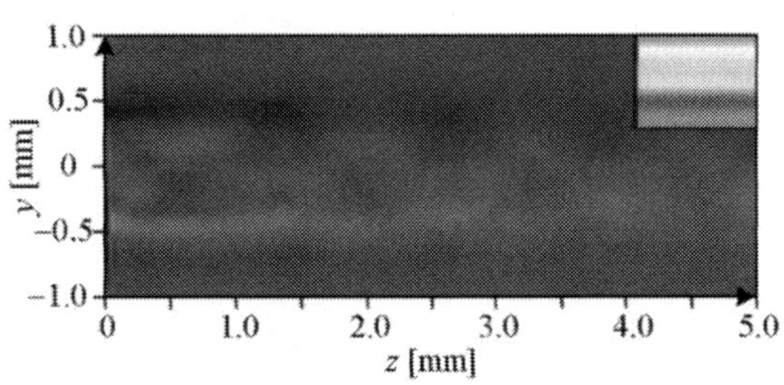

Fig. 8. Rainbow schlieren picture of underexpanded supersonic jet from micro Laval nozzle for $p_{os}/p_b = 4.0$

The schlieren picture has a high resolution of 426 pixels/mm and it corresponds to a spatial resolution of around 2.35 μm for the present schlieren set up. The schlieren picture of Fig. 8 shows the archetypal structure of underexpanded jet, which is the quasi-periodic shock cell structure, but it is blurred over the full range of the shock cell structure and the jet free boundaries are ambiguous. Therefore, with rainbow schlieren pictures only, it is very difficult to detect significant flow features of micro supersonic jets.

The three-dimensional density field corresponding to the schlieren picture of Fig. 8 can be inferred using the comvolution-backprojection (CBP) algorithm, which is valid even for reconstruction of an asymmetric jet. A density contour plot of the underexpanded micro jet , which includes the jet centerline (z axis), is given in Fig. 9 with a color bar representation showing density values and the solid lines with a numeral in the contour indicate isopycnic where experimental data near the nozzle exit plane are excluded in advance for the reconstruction because they include unrealistic density values.

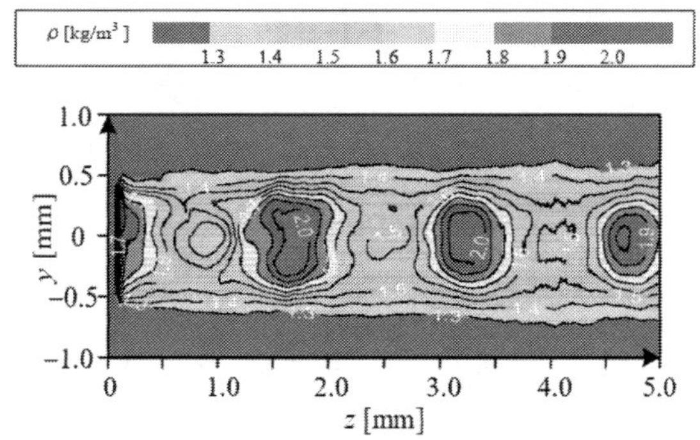

Fig. 9. Density contour plot of underexpanded supersonic jet at y-z plane

Unlike the schlieren picture of Fig. 8, the density contour plot illustrates the various flow features of the shock cell structure quantitatively such as the shape and size of expansion and compression regions, shock cell intervals, jet boundaries, and mixing layers. The density contour of Fig. 9 exhibits nearly symmetry about the jet centerline.

A comparison between experiment and theory for the jet density profile in the streamwise direction at a vertical location of $y = 0$ mm is given in Fig. 10 where the blue line shows the density reconstructed from the rainbow schlieren and the red line is calculated from the theory with only the fundamental Fourier mode of $n = m = 1$ for Fig. 10(a) and the theory for Fig. 10(b) consists of the Fourier modes added sequentially from $n = m = 1$ to $n = m = 3$. Also, the left pointing arrow in Fig. 10 shows the density at the nozzle exit plane which is calculated based on the assumption of the isentropic flow from the nozzle inlet to the exit. The density distribution of the theory shows a typical sinusoidal wave form consisting of a smooth repetitive oscillation with a constant wavelength and amplitude. The streamwise locations of the local maxima and minima and the wavelength for the experiment are in good agreement with those for the theory, but there is some difference quantitatively for the density amplitudes. As can be seen in Fig. 10(b), there is a very favorable overall agreement between experiment and theory over the full length of the region measured except for some deviation at the second shock cell where the theoretical density profile is shifted downstream

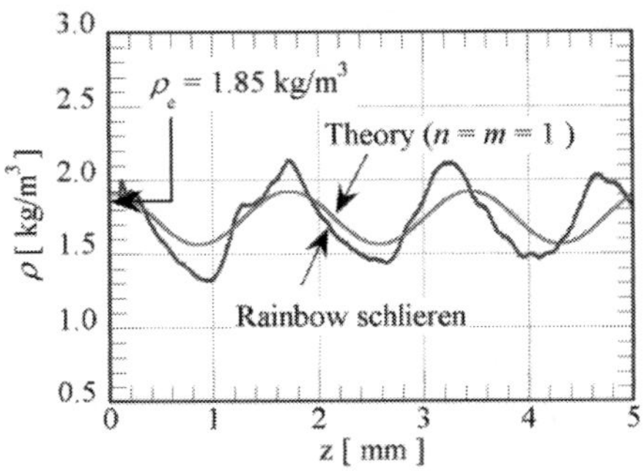

(a) Comparison between experiment and theory with fundamental Fourier mode alone

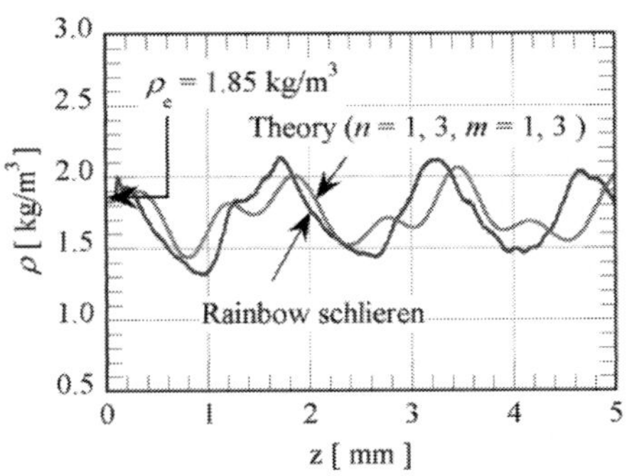

(b) Comparison between experiment and theory with Fourier modes added sequentially from $n = m = 1$ to $n = m = 3$

Fig. 10. Streamwise density distributions at $y = 0$

CONCLUDING REMARKS

Three-dimensional density fields of the shock containing free jet issued from a square Laval nozzle with a design Mach number of 1.5 were quantitatively obtained by the rainbow schlieren deflectometry combined with the computed tomography based on the convolution backprojection algorithm. As a result, it was found that a density contour plot at any cross-section of the jet can be efficiently inferred from the schlieren CT. Density fields reconstructed from the schlieren CT are in good quantitative agreement with those from vortex sheet model.

It was demonstrated by continuous color gradation that the density field including the jet axis shows the representative shock-cell structure of a repeated compression and expansion waves. Also, the density profile along the jet centerline exhibits a significant structure consisting of fine-scale spatial oscillations with the wavelengths of the higher-order modes in addition to the fundamental mode.

NOMENCLATURE

a_j : fully expanded speed of sound, m/s
f_c :focal length of collimating lens, mm
f_d :focal length of decollimating lens, mm
h_e : width at nozzle exit plane, mm
h_j :fully expanded jet width, mm
Hue : color on rainbow filter, deg
M_d : design Mach number, -
M_j : fully expanded jet Mach number, -
n_a :refractive index at surrounding air, -
p_e :static pressure at nozzle exit, Pa
p_{os} : plenum pressure, Pa
p_b : back pressure, Pa
T_{os} : plenum temperature, K
T_b :ambient temperature, K
x,y,z : Cartesian coordinate system
γ : specific heat ratio, -
θ: angle at polar coordinate, deg
ρ_b :ambient density, kg/m^3
ρ_j :fully expanded jet density, kg/m^3

REFERENCES

1. Settles, G. S., Schlieren and Shadowgraph Techniques, Springer, (2001).
2. Al-Ammar, K., Agrawal, A.K., Gollahalli,S.R., and Griffin, D., Application of Rainbow Schlieren Deflectometry for Concentration Measurements in an Axisymmetric Helium Jet, *Applied Optics*, 25-2, (1998), pp. 89-95.
3. Kolhe,P.S. and Agrawal, A.K., Density Measurements in a Supersonic Microjet Using Miniture Rainbow Schlieren Deflectometry, *AIAA J.*, 47-4, (2009), pp. 830-838.
4. Takano, H., Kamikihara, D., Ono, D., Nakao, S., Yamamoto, H., and Miyazato, Y., Three-Dimensional Rainbow Schlieren Measurements in Underexpanded Sonic Jets from Axisymmetric Convergent Nozzles, *J. Thermal Science*, 25-1, (2016), pp. 78-83.
5. Tam, C.K.W., The Shock-Cell Structures and Screech Tone Frequencies of Rectangular and Non-Axisymmetric Supersonic Jets, *J. Sound and Vibration*, 121-1, (1988), pp. 135-147.
6. Kaji, S., and Nishijima, N., Pressure Field Around a Rectangular Supersonic Jet in Screech, *AIAA J.*, 34-10, (1996), pp. 1990-1996.
7. Faris, G.W., and Byer, R. L., Three-Dimensional Beam-Deflection Optical Tomography of a Supersonic Jet, *Applied Optics*, 27-24, (1988), pp. 5202-5212.
8. Agrawal, A.K., Butuk, N.K., Gollahalli, S.R., and Griffin, D., Three-Dimensional Rainbow Schlieren Tomography of a Temperature Field in Gas Flows, *Applied Optics*, 37- 3, (1998), pp. 479-485.
9. Foelsch, K., The Analytical Design of an Axially Symmetric Laval Nozzle for a Parallel and Uniform Jet, *J. Aeronautical Sciences*, 16-3, (1949), pp. 161-166, and p. 188.

Internal flow characteristics in scaled pressure-swirl atomizer

Milan Malý[1,*], *Marcel* Sapík [1], *Jan* Jedelský[1], *Lada* Janáčková[1], *Miroslav* Jícha[1], *Jaroslav* Sláma[2] and *Graham* Wigley[3]

[1]Brno University of Technology, Faculty of Mechanical Engineering, Czech Republic
[2]Provyko s.r.o. Czech Republic
[3]Loughborough University, United Kingdom

Abstract. Pressure-swirl atomizers are used in a wide range of industrial applications, e.g.: combustion, cooling, painting, food processing etc. Their spray characteristics are closely linked to the internal flow which predetermines the parameters of the liquid sheet formed at the discharge orifice. To achieve a better understanding of the spray formation process, the internal flow was characterised using Laser Doppler Anemometry (LDA) and high-speed imaging in a transparent model made of cast PMMA (Poly(methyl methacrylate)). The design of the transparent atomizer was derived from a pressure-swirl atomizer as used in a small gas turbine. Due to the small dimensions, it was manufactured in a scale of 10:1. It has modular concept and consists of three parts which were ground, polished and bolted together. The original kerosene-type jet A-1 fuel had to be replaced due to the necessity of a refractive index match. The new working liquid should also be colourless, non-aggressive to the PMMA and have the appropriate viscosity to achieve the same Reynolds number as in the original atomizer. Several liquids were chosen and tested to satisfy these requirements. P-Cymene was chosen as the suitable working liquid. The internal flow characteristics were consequently examined by LDA and high-speed camera using p-Cymene and Kerosene-type jet A-1 in comparative manner.

1 Introduction

Pressure-swirl (PS) atomizers are easy to manufacture and provide good atomization. They are frequently used in various applications where a large surface area of droplets is needed or a surface must be coated by a liquid e.g. combustion, fire suspension or air conditioning. In principle, the pumped liquid is injected via tangential ports into a swirl chamber where it gains a swirl motion, under which it leaves the exit orifice as a conical liquid sheet which consequently breaks up into small droplets due to aerodynamic forces. The centrifugal motion of the swirling liquid creates a low-pressure zone in the centre of the swirl chamber and generates an air core along the centreline. The flow pattern inside the atomizer is rather complex; it is two-phase flow with secondary flow effects. There is a strong link between the internal flow conditions and the resulting spray characteristics; however, not all aspects of the internal flow are well understood.

A number of authors have studied and visualized the internal flow and air core characteristics using transparent models of the PS atomizers. One of the first were Horvay and Leuckel [1, 2] who used scaled transparent modular atomizers with various internal geometries. They investigated different sized inlet ports and geometries of the swirl chamber. They used a mixture of castor (ricinus) oil, tetraline (1,2,3,4- Tetrahydronaphthalene) and turpentine oil as working liquid to match the refraction index. De Keukelare [3] also used scaled PMMA model to measure the flow rate, pressure distribution and the velocity profiles inside the swirl chamber together with the air core dimensions. He used water and presented a correlation of experimental data with theoretical predictions. Chinn and Yule [4] followed, using LDA and high-speed imaging with water as the working liquid to acquire a velocity field inside the atomizers. They used three different swirl chamber geometries (conical, curved and square). Their experimental data of velocity fields and air-core growth are in accordance with CFD predictions.

Due to the complicated geometry of the swirl chamber with its curved surfaces, it is advisable to have the same refractive index for the liquid and atomizer body in order to make optical measurements.

In addition to Horvay and Leuckel [1, 2], Liu [5] and Thomson [6] also used the same liquid mixture (castor oil, tetraline and turpentine oil) which, however, causes chemical degradation of the PMMA parts. Hassan and Dominguez-Ontiveros [7] investigated matching the PMMA refractive index with p-Cymene and a sodium iodide mixture. Due to difficulty of preparation of sodium iodide, its high cost and occurrence of small particles, they preferred p-Cymene. Muguercia [8] matched the refractive index of the glass tube with a mixture of water and sugar and also with a mixture of water and glycerine.

In this paper, the scaled transparent model of a small pressure-swirl atomizer was manufactured from PMAA in which high-speed imaging and LDA measurement were performed. Several liquids with similar refractive indices were tested and discussed in terms of corrosiveness to PMAA. The air-core shape and velocity profiles were measured for two different liquids and the necessity of refractive index matching is discussed.

* Corresponding author: milan.maly@vutbr.cz

2 Experimental setup

The experiments were performed on a specially designed test bench at the Spray laboratory in the Brno University of Technology. All tests were performed at room temperature (25 °C).

The pressurized liquid was supplied by a fuel supply system – see fig. 1. The liquid was pumped by centrifugal pump (3) from the main tank (1) via a filter (2) to the Coriolis mass flow meter (4). Temperature (5) and pressure sensors (7) were mounted before the atomizer (8). The mass flow rate was set by changing pump speed. The spray freely falls into a collector and returns into the main tank.

2.1 The atomizer

The small-sized simplex atomizer as used in a combustion chamber of turbojet aircraft engine, was investigated. Its main dimension are listed in fig. 2. Due to the small dimensions, the optical measurements inside the transparent model would be extremely difficult. For this reason, a scaled transparent copy, with a modular construction was manufactured. It consisted of three pieces, one with the swirl chamber, the second part with the tangential inlet ports and the last one a cap. This construction allowed changing each piece for another with a different geometry thus providing easy modification of the atomizer geometry.

A thin transparent gasket material was used to prevent leakage of the liquid between the interfaces. The inner surfaces of every piece were grounded and polished to achieve transparency for the optical measurements.

2.2 Flow match and tested liquids

The scaled model is ten times larger than the original atomizer and its construction is shown in fig. 3. To maintain the flow behaviour, dimensionless numbers such as Reynolds number, swirl number and Froude number must remain the same as in the original atomizer. Thus a new inlet condition had to be proposed to fulfil these requirements.

The Reynolds number is defined as the ratio of inertial force to the viscous force and it is a crucial parameter that must be maintained to simulate the behaviour of liquid in the original atomizer [9]:

$$\mathrm{Re} = \frac{w_i D_i}{\nu} \tag{1}$$

where w_i is the velocity inside the inlet port, D_i is the hydraulic diameter of the inlet port and ν is the liquid kinematic viscosity.

Another important dimensionless number is the swirl number, S_0. This number is basically the ratio of the radius of the swirl chamber, the discharge orifice and cross-section of the inlet ports [3]. It has to be the same for the scaled model and the original atomizer. If the each part of the atomizer is scaled equally then the swirl number remains the same:

$$S_0 = \frac{\pi R r_0}{A_i} \tag{2}$$

where R is radius of the swirl chamber, r_0 is radius of the exit orifice and A_i is total cross-section of inlet ports.

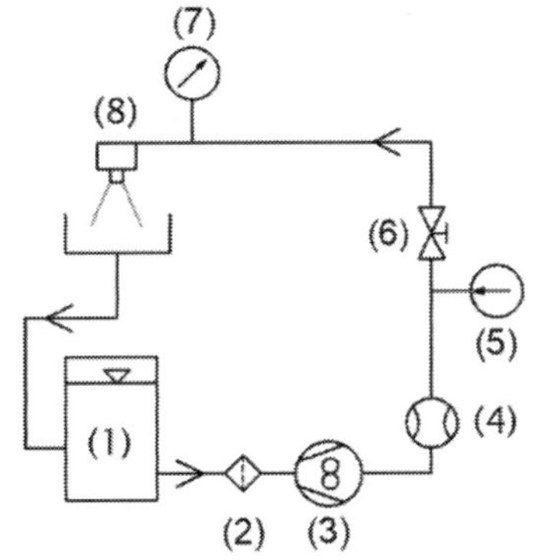

Fig. 1. Test bench

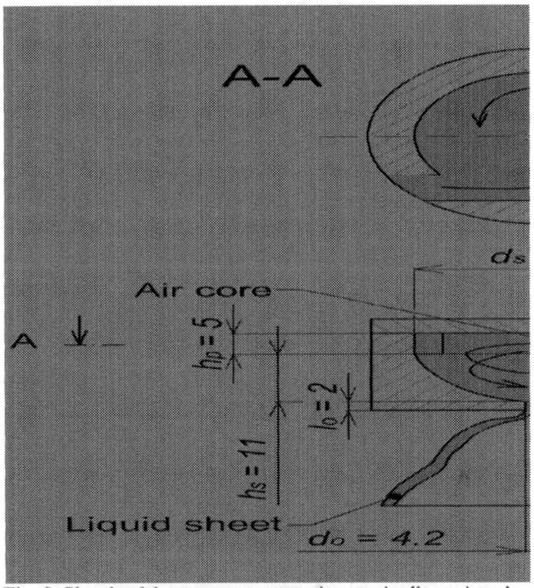

Fig. 2. Sketch of the transparent atomizer, main dimensions in millimetres. The original atomizer is ten times smaller.

Fig. 3. Model of the scaled transparent modular atomizer

The Freude number shows the effect of gravity in comparison to the energy of the bulk flow [9]. This number must be high enough to minimalize the influence of gravity on the liquid flow:

$$Fr = \sqrt{\frac{u^2}{4r_0 g}} \qquad (3)$$

Where u is axial velocity, r_0 is diameter of the orifice and g is gravitational acceleration.

Moreover, it is necessary to match the refractive index of the fluid with the one of the PMMA to allow optical measurement and visualizations inside the swirl chamber. The refractive index of original JET A-1 kerosene is 1.44 whereas the refractive index for PMMA is 1.49. Therefore, a number of different liquids and mixtures were tested to find the optimum liquid for the model

Table 1. Liquids properties at 20°C, refractive indices based on wavelength 660 nm

Liquid	ρ [kg/m³]	n [-]	μ [mPa·s]	Damage after 10 days
Castor oil	960	1.479	986	None
Paraffin oil	850	1.48	21	None
Anise oil	981	1.557	2.37	Heavy
Turpentine	857	1.473	1.49	Slight
Tetraline	973	1.541	2.1	Partial
p-Cymene	860	1.49	0.876	Slight
Glycerine	1262	1.47	1480	None
Kerosene	785	1.44	1.6	None

The liquid corrosiveness to PMMA was tested using several liquids in undiluted form as well as in mixtures to achieve the exact refraction index of PMMA – see Table 1 and 2. Each mixture in table 2 has approximately the same refraction index as PMMA.

Table 2. Mixtures of liquids, refractive index matched with PMAA

Mixtures of liquids		Damage after 10 days
Anise oil +	Paraffin oil	Slight
	Turpentine	Partial
	Castor oil	Slight
	p-Cymene	Slight
	Glycerine	Not soluble
Tetraline +	Paraffin oil	Slight
	Turpentine	Slight
	Castor oil	Slight
	p-Cymene	Slight
	Glycerine	Not soluble

It can be concluded that the castor (ricinus) oil, paraffin oil and glycerol do not damage the PMAA at all. However, it would be difficult to use them due to their high viscosity and low refractive index. The liquids with a close refractive index to PMAA, e.g. p-Cymene and mixtures based on Anise oil or Tetraline always corroded the PMAA. For the sake of simplicity p-Cymene was chosen as a one-component operating liquid. Moreover, the refractive index of clean P-Cymene can be tuned also by temperature where an increase in temperature of about 1°C results in a decrease in the refractive index of approximately 0.0004 [10]. It is also colourless, has a low frothiness and damage to the PMMA was slight but acceptable.

After the first trial of p-Cymene as the working liquid, the atomizer was damaged, especially the parts with inlets and threads. The problem was linked to the internal stresses, which accelerates deterioration of the PMMA. This effect was not considered during the static tests.

Table 3. List of operating regimes

Original atomizer			
Re [–]	Δp [MPa]	m_l [kg/h]	Fr [–]
1560	**0.5**	5.41	137.4
2560	**1.5**	8.97	359.8

Transparent atomizer				
	Kerosene		p-cymene	
Re [–]	m_l [kg/h]	Fr [–]	m_l [kg/h]	Fr [–]
1560	54.1	6.9	29.0	3.4
2560	89.7	11.4	48.1	5.7

2.3 High speed imaging and LDA system

A Photron SA-Z high-speed camera was used to characterise the spatial and temporal behaviour of the air core. The atomizer was illuminated by a background light using an LED panel. The camera frame rate was 20,000 fps and image resolution was 1024 × 1024 px. The shutter speed was set to 20 μs.

The laser Doppler anemometer (LDA), a 2D FlowExplorer (Dantec Dynamics A/S), was employed for the point-wise measurement of the swirl and radial velocity in three cross-sections across the swirl chamber, see fig 5. The LDA used diode-pumped solid-state lasers and generated beams with 660 nm and 785 nm wave lengths. The beam was split into two pairs of parallel beams. Frequency of one beam from each pair was shifted by 80 MHz. A converging transmitting lens with 300 mm focal length was used to form an ellipsoidal probe volume with size approx. 0.1×0.1×0.7 mm at the beam crossing point. The LDA was configured in a backscatter mode.

The flow was seeded by silica particles e-spheres SL75 with mean diameter of 45 μm. The Stokes number, based on the swirl velocity and diameter of the swirl chamber, was less than 0.01 for each regime, which should ensure a sufficiently small flow traceability error.

During LDA measurements, several possible issues must be considered. The differences in refractive indices between the surrounding air, atomizer body and operating liquid cause displacements and distortion of the measuring volume due to different optical paths of each beam, see fig 4.

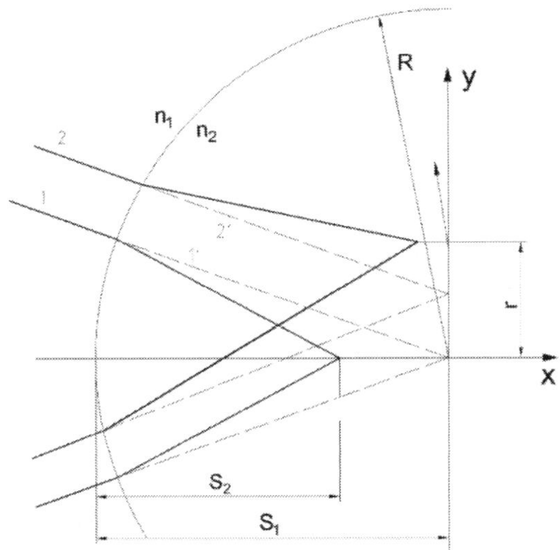

Fig. 4. Schematic drawing of beam path through the transparent model

Different issues arise in measurements of the swirl, radial and axial velocity component and must be taken in to account. Considering the matched refractive indices of liquid and atomizer body, the problem can be simplified to measurement in a planar window and thus only the position of the measurement volume must be corrected for as:

$$S_2 = \frac{n_2}{n_1} S_1 \tag{4}$$

where n_1 and n_2 are refractive indexes of surrounding air and atomizer body, S_1 is the position of the measuring volume in the air, S_2 is the real position of the measuring volume inside the atomizer.

However, when the refractive index is not matched i.e. a difference greater than 0.01 then the circular geometry of the swirl chamber must be considered – see fig. 4. In the case of the swirl velocity component measurement, the optical plane is perpendicular to the atomizer axis, therefore, the properties of the measurement volume are dependent on the local position of the measurement point into flow [14]. Both, the position and velocity must be corrected.

Measurement of radial velocity component is even more complex and in the positions further from the atomizer centre, the optical aberration and measurement volume dislocation make the measurement almost impossible [14].

3 Results and discussion

The results are divided into two main parts: The air core shape based on the high-speed visualization and the velocity profiles based on the LDA measurements. Emphasis is placed on evaluation of refractive index effect in both cases.

Fig. 5. Measurement positions

3.1 High speed visualization

In a comparative manner, the high-speed records with kerosene and p-cymene as the working liquid is shown in fig. 6. For the kerosene image there are darker region towards the edge of the swirl chamber. This is caused by light refraction at the swirl chamber wall. It is not evident in the atomizer centre due to the small relative curvature. This problem is overcome using the liquid with the same refractive index as the atomizer body as can be seen for the results from p-cymene.

A fully developed air core was found in all tested cases. It was cylindrically shaped with an extended diameter inside the exit orifice. Similarly shaped air cores were described by several authors [11, 12, 13]. The air core diameter (d_{ac}) inside the exit orifice was 3.09 ± 0.05 mm and it was equal for both liquids and inlet conditions. In the swirl chamber, the d_{ac} was 2.1 mm for both liquids at Re = 1560 and it decreased by 5 % with increasing Re.

Fig. 6. High-speed image, from top: Re = 1560 and 2560, left: Kerosene, right: p-cymene

3.2 Velocity profiles

The swirl velocity profiles show a sharp maximum near the interface which is typical for Rankine vortex – see fig 7. This swirl velocity is almost independent of the axial distance from the top of the swirl chamber. For both liquids, the profiles had the same slope, however, in case of kerosene, the values are higher due to the higher flowrate used in order to keep the same Re. In positions close to the air-core, the raw velocity data were filtered since strong noise was generated by reflection from the air core surface.

The turbulence intensity (TI) was calculated as the ratio of the standard mean deviation to an average swirl velocity. Its dependence on radial distance is shown in fig. 8 for both liquids and pressure regimes in axial distance of 2.5 mm.

Increasing the inlet velocity (Re) in the inlet ports slightly decreased the turbulence intensity. Kerosene featured slightly higher turbulence level compared to p-cymene at the same Re. Strong increases in TI was found in positions near the wall and close to the air core.

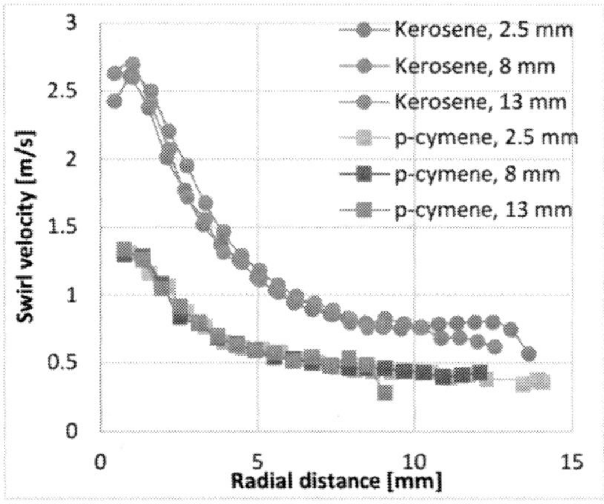

Fig. 7. Swirl velocity profiles for Re = 1560

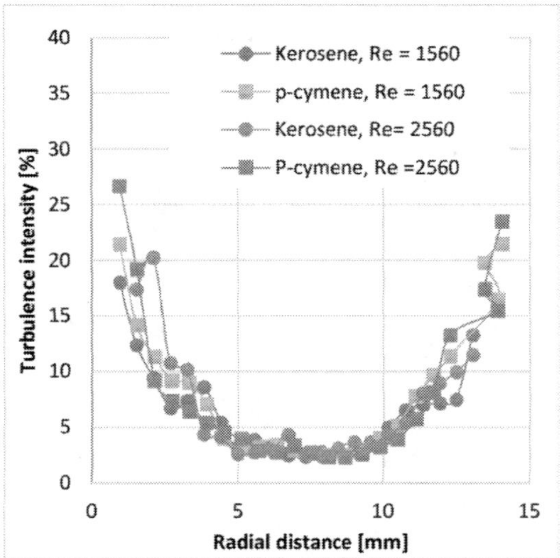

Fig. 8. Turbulence intensity, Z = 2.5 mm

However, the raw velocity data for TI were not filtered to remove effect of optical noise. This is evident in a sharp increase in TI for position very close to the atomizer centre. However, the impact for both liquid is very similar. The presence of large gradients in the swirl velocity may also result in a change of the velocity magnitude along the measuring volume itself which consequently leads to an apparent increase in the turbulence results. In the centre of the measurement plane across the swirl chamber, the TI was approximately 3 % which correspond to low turbulent flow. However, due to low Re in the inlet ports and the fact that the swirling flow tends to laminarise the flow itself [9], this near laminar nature of the flow is to be expected. Measured TI data may also be overestimated due to time-resolved flow fluctuations and general measurement inaccuracies.

To evaluate the effect of different liquid refractive indices, the LDA data-rate and validation rates were compared – see fig 9 and fig 10. In the case of swirl velocity (fig. 9), both liquids behaved similarly which was not expected. P-cymene had higher validation rate in positions closer to the atomizer centre. However, in positions close to the wall, both liquids had similar validation rates of over 70 %. The data-rate was found to be even less dependent on the liquid used. The difference is noticeable only close to the atomizer centre where the kerosene had higher data-rates compared to p-cymene. This behaviour may be caused by different concentrations of tracking particles which have a strong impact on data-rate. It can be concluded, that the difference in refractive index between kerosene and the atomizer body ($\Delta n = 0.05$) caused very small distortions of the LDA measuring volume in the case of the swirl velocity measurement.

The situation is different when the radial velocity is measured. The data-rate in the case of kerosene was about one order of magnitude lower compared to p-cymene. The difference is smaller in the positions near the atomizer centre as the relative curvature is also smaller.

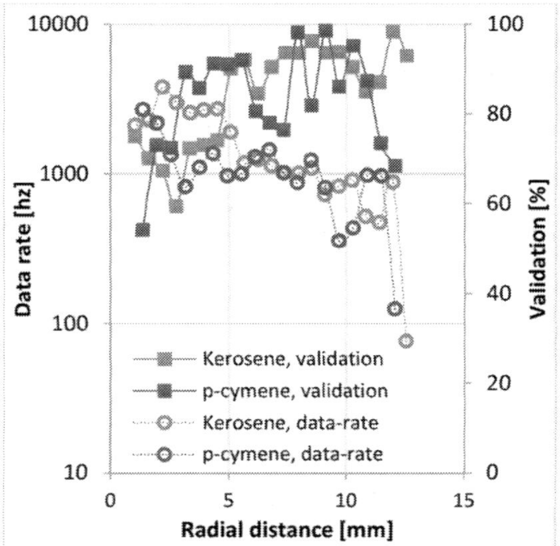

Fig. 9. Swirl velocity, data-rate and validation rate, Re = 1560, Z= 8 mm

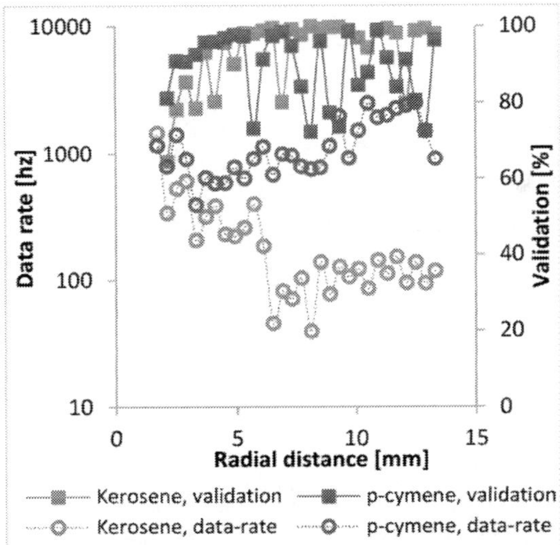

Fig. 10. Radial velocity, data-rate and validation rate, Re = 1560, Z = 8 mm

For positions higher then R/2 the laser beam deviations and optical aberrations corrupt the measurement volume and make measurements almost impossible. This is in agreement with [14]. The measured velocity data in positions with a greater diameter than R/2 are caused probably by noise, as the velocity magnitude is close to zero.

4 Conclusion

The internal liquid flow of a transparent pressure-swirl atomizer was investigated by high-speed imaging and LDA measurements using two liquids with different refractive indices.

The internal flow generates a stable, cylindrically shaped air core for both liquids under the pressure regimes studied. Swirl velocity profiles were found to be independent of axial location in the swirl chamber and had a sharp maximum near the air-core boundary, typical for a Rankine vortex.

Refractive index matching is necessary for the LDA measurement of radial and axial velocity components. However, there were only small difference in the case of swirl velocity measurements. Results from the high-speed imaging were independent of the liquid used.

This work has been supported by the project No. GA15-09040S funded by the Czech Science Foundation, the project LO1202 NETME CENTRE PLUS with the financial support from the Ministry of Education, Youth and Sports of the Czech Republic under the "National Sustainability Programme I" and project Reg. No. FSI-S-17-4444 funded by the Brno University of Technology.

References

1. M. Horvay, W. Leuckel, ALAFM, **2**, 18, (1985)
2. M. Horvay, W. Leuckel, *Experimental and Theoretical Investigation of Swirl Nozzles for Pressure-Jet Atomization.* (1986)
3. H.J.K De Keukelaere, MSc Dissertation, Dept. Mech. Eng., UMIST, (1995)
4. D. Cooper, J.J. Chinn, A.J. Yule, ILASS, **15**, (1999)
5. C.H. Liu, C. Vafidis, J.H. Whitelaw, EF, **10**, pp. 6, (1990)
6. B.E. Thomson, J. Senaldi, C. Vafidis, J.H. Whitelaw, H. McDonald, JSR, **29**, pp. 247-252, (1992)
7. Y.A. Hassan, E.E. Dominguez-Onziveros, NED, **238**, 6, (2008)
8. I. Muguercia, B. Cazanas, W. Li, M.A. Ebadian, SPIE, **2052**, 8 (1993)
9. J.J. Chinn, *The Numerics of the Swirl Atomizer.* Dept. Mech. Eng., UMIST, (2008)
10. M.P. Bardet, D.C. Fu, E.C. Sickel, A.N. Weichselbaum, ISALTFM, **17**, (2014)
11. M. Halder, S. Dash, S. Som, ETFS, **26(8)**, 8, (2002)
12. E.J. Lee, S.Y. Oh, H.Y. Kim, S.C. James, S.S. Yoon, ETFS, **34(8)**, 8, (2010)
13. G. Amini, IJMF, **79**, 10, (2016)
14. Z. Zhang, *LDA Application Methods: Laser Doppler Anemometry for Fluid Dynamics*, Springer, (2010)

Experimental investigation and numerical modelling of 3D radial compressor stage and influence of the technological holes on the working characteristics

Richard Matas[1,*], *Tomáš* Syka[1] and *Lukáš* Hurda[1]

[1]University of West Bohemia in Pilsen, New Technologies Research Centre, Univerzitní 2732/8, 301 00 Plzeň, Czech Republic

Abstract. The article deals with a description of results from research and development of a radial compressor stage with 3D rotor blades. The experimental facility and the measurement and evaluation process is described briefly in the first part. The comparison of measured and computed characteristics can be found in the second part. The last part of this contribution is the evaluation of the rotor blades technological holes influence on the compressor stage characteristics.

1 Introduction

Centrifugal compressors are widely used in the industry. A traditional producer of multi-stage centrifugal compressors, the company Howden ČKD Compressors s.r.o. (HCKD) has developed a new family of compressor stages. The development project was realized in the cooperation among PCA Engineers Limited and University of West Bohemia in Pilsen (UWB). The main goal of the project is to achieve higher stage efficiency and better operational properties.

The description of processes, operations and performance of the centrifugal compressors are described in many publications, for example in [1]. Performances of the centrifugal compressors stages are analysed through two of the main properties: the pressure ratio and the efficiency. These two parameters vary with volume flow and with Mach number.

The verification of the new stages properties requires detailed measurement of performance characteristics to obtain reliable data for compressor design. Measured data are also important for validation of CFD numerical models. The validated numerical models can give also the complex information about the flow in the stage and help to design the profile of the channel.

2 Experimental facilities

2.1 Experimental rig

The current development of the centrifugal compressor stages follows the systematic research and development program in the former ČKD company in the past; see for example the paper [2]. The experimental compressor stages verification is being realized in the development test facility HCKD built in Pilsen in 2013, where the single-stage test compressor DARINA IV with the exchangeable flow part is placed. The compressor consists of an inlet cone, an impeller, a diffuser, a return bend, a de-swirl channel and an exit part.

This quite unique compressor is driven by the electromotor with maximum input wattage 1.2 MW, which is driven by the frequency converter for the continuous revolution change. The test rig is shown in the upper part of Fig. 1.

The test rig is designed for impellers with diameter of 440 mm. It is possible to change shaft revolutions from 4 000 to 17 500 rpm. A set of throttle-valves for the aerodynamic resistance adjusting is placed in the discharge piping. Mass-flow rates of the air through the suction and discharge piping are measured by orifices.

Fig. 1. Photo of the centrifugal compressor stage together with the cross-section with red colour highlighted evaluated planes

* Corresponding author: mata@ntc.zcu.cz

https://doi.org/10.1051/epjconf/201818002060

2.2 Experimental equipment

The parameters of the measured stage are obtained by complex measurement and data acquiring system.

The flow parameters are measured in the evaluation planes. The description of the planes in the stage is long-term used, see [3]. The main evaluation planes are the planes 0 and 7 depicted in the bottom part of Fig. 1.

There is a number of static and total pressure and total temperature probes (multiple sensors are placed in the planes 0 and 7). The torquemeter gives the value of torque and rotating speed of the compressor rotor.

Physical models of the new designed stages were prescribed in HCKD. For the evaluation the stage RTK01 was selected in two modifications with various blade lengths, which is the stage for high flow rates. The impellers and stator parts of the stages RTK01W and RTK02N are depicted in Fig. 2.

Fig. 2. Models of the stage RTK01W and RTK01N for the experimental compressor.

2.3 Evaluation of measurement

The measurement campaigns are established to produce curves of operational characteristics – branches of compressor map. The number of points on the curve marked as proper for curve reconstruction is kept minimal, although there are some parts of the curve which the measurement should never omit. The campaign starts at the maximum flow rate and minimum pressure ratio and then the flow rate is slowly reduced by discharge control shutter.

The operators find several points to determine carefully the shape of curve during the diminishing choking of flow. Then a maximum discharge pressure point is searched carefully. After that, an approximate position of surge initiation is found.

The aim is to measure one operational curve keeping the reference (blade) Mach number (1) constant. This number is based on speed of sound in gas at stage intake state and the velocity of blades at outer diameter of the impeller:

$$M2u = \frac{2 \cdot \pi \cdot \Omega \cdot D_2}{a_0} = \frac{u_2}{a_0} \qquad (1)$$

This means the rotational speed is being adjusted slightly based on the actual intake conditions.

The measurement evaluations has two main branches based on the way of determination of the stage required power.

A simple evaluation based on the torquemeter direct measurement of the power throughput is supplied by the bearings losses determined by the measurement of heat-up of the lubricating oil continuous stream. These losses are subtracted from the measured power.

The more complex set of calculations is needed to express the required power based on the flow temperature measurements. The polytrophic change of state of gas is accounted for after the exact total and static parameters in evaluation planes 0 and 7 are calculated. Iterative procedures are employed several times.

The reason for the implementation of iterative algorithms (if it is not required even for a perfect gas) is the required option of performing the calculations with real gas properties based on pressure and temperature explicit equation of state.

At first, the restitution factor of the total temperature probes is evaluated based on the mean flow speed in the intake and discharge plane respectively, while the static states i.e. the effective densities are not known a priori.

A theoretical isentropic discharge temperature and specific work is the main output of the polytrophic compression analysis as other parameters can be readily obtained from already analysed states at the endpoints of the real polytrophic curve. The theoretical isentropic discharge temperature is searched to assure zero entropy change while assuming the same total pressure ratio as the real compression measured by total pressure probes.

The key parts when thinking of data comparison (current measured with older measured or even CFD results) are the procedures of corrections for reference conditions. The test rig is prone to various effects of the environment, the intake air state being the strongest. Not only the state of air but also its composition change when the humidity differences are involved.

One correction procedure carries out the effects of the intake gas state and composition on compressor operation employing the classic similitude relations in its most general form to readily utilize real gas equation of state. Five non-dimensionless numbers are kept constant for the actual and reference intake compressions: blade Mach number, flow coefficient, power coefficient, total pressure ratio and polytrophic efficiency. This assumption allows the absolute performance characteristics to be transformed using algebraic equations. The polytrophic efficiency is then used to back-check the evaluation of corrected polytrophic discharge temperature based on the predicted enthalpy rise.

The second correction procedure is used to account for the inequality of the blade Mach numbers in measurement and the desired reference. Efforts are put to minimize this difference during measurement, but the weather conditions and a non-sensitive rotational speed control system cannot be overcome perfectly.

Idealized relations are utilized to cancel the above mentioned effect. Those are functional relations of the compressor performance characteristics with the impeller rotational speed being the function argument. The result of this procedure is an approximation, but when the actual revolutions are close to the desired, it should be very accurate while the magnitude of corrections is very low.

As well as for the CFD results, dimensionless numbers can be expressed to be used in compressor design tasks.

Besides the whole stage evaluation, a part-by-part analysis can be carried out using the data acquired from the internal planes visible in lower part of Fig. 1. Mean velocity vectors, performance characteristics of diffuser or the losses of the de-swirl channel could be addressed.

2.4 Parameters of stages

The evaluation of various compressor stage parameters was done according to the formulas below. Values of the evaluated parameters were averaged by mass flow in postprocessing. The parameters are: specific total work (2), specific isentropic work (3), pressure ratio (4), isentropic efficiency (5) and flow coefficient (6).

$$\Delta h_t = c_p \cdot (T_{t7} - T_{t0}) \tag{2}$$

$$\Delta h_{is} = z \cdot r \cdot T_{t0} \cdot \frac{\kappa}{\kappa - 1}\left[\left(\frac{p_{t7}}{p_{t0}}\right)^{\frac{\kappa-1}{\kappa}} - 1\right] \tag{3}$$

$$PI = \frac{p_{t7}}{p_{t0}} \tag{4}$$

$$ETA = \frac{\Delta h_{is}}{\Delta h_t} \tag{5}$$

$$FI0 = \frac{4 \cdot Q_0}{\pi \cdot D_2^2 \cdot u_2} \tag{6}$$

The obtained characteristics include the confidential data of HCKD and only relative values are presented in the following sections. The values PI_N, ETA_N and m_N are normalized by the design point values in the measured characteristic.

The results from stages CFD simulations were compared to the measured data converted to the standard inlet conditions.

3 Numerical simulations of the measured stage

3.1. Software

CFD simulations are now an essential tool for the design of all types of devices in turbomachinery. They enable to gain much time between the concept and the test. The verification of simulations is carried out virtually in all development turbomachinery groups and widely published; for example in [4] and many others. For the purposes of this work many CFD simulations of various compressor stages were solved in the NUMECA FINE/Turbo (NFT) software environment, where the main goal was to determine the suitable turbulence model and computational grid parameters.

3.2 Simulations of the stages

The methodology of simulation and properties of the model has also been evaluated and improved [6]. The mentioned stages were modelled without and with leakage. The geometry of the numerical models, computational domains and an example of the computational mesh for the stage with leakages can be seen in Fig. 3.

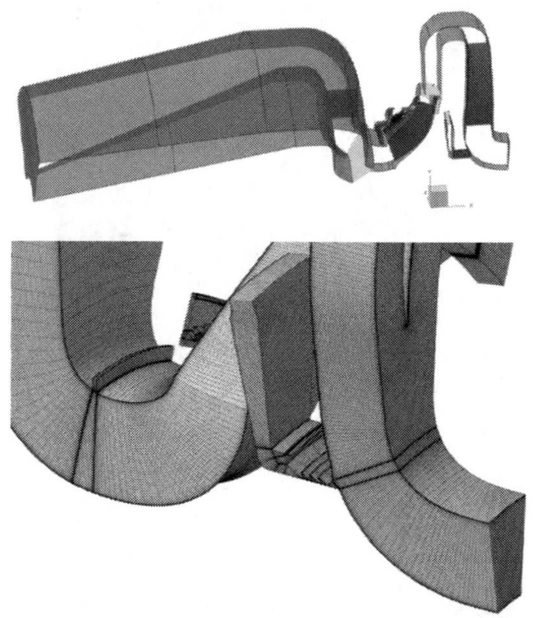

Fig. 3. Stages geometries of the computational domains and an example of the computational mesh.

The air, modelled as an ideal gas, was used as the flow medium, SST k-ω and EARSM turbulence models were used for the described simulations.

3.3 Comparison of results - stage RTK01W

The values for this stage were already published in [5]. In Fig. 4 you can see the performance characteristics for M2u = 0.9. It is possible to compare the results of described CFD simulations with measured data. The measured pressure ratio of the stage is lower than the ratio obtained from the CFD simulations. The reason can be the shape of the impeller – the fillets or other inaccuracies [6]. The efficiency values for the EARSM model with leakages are predicted well. The question is why all the CFD simulations (the authors have other results from ANSYS CFX available) predict higher flow rate than the measurement shows. The shift between flow rates in the measured and computed characteristics is typical for this stage and despite the repetition of measurements and calculations with different models this remains unclear.

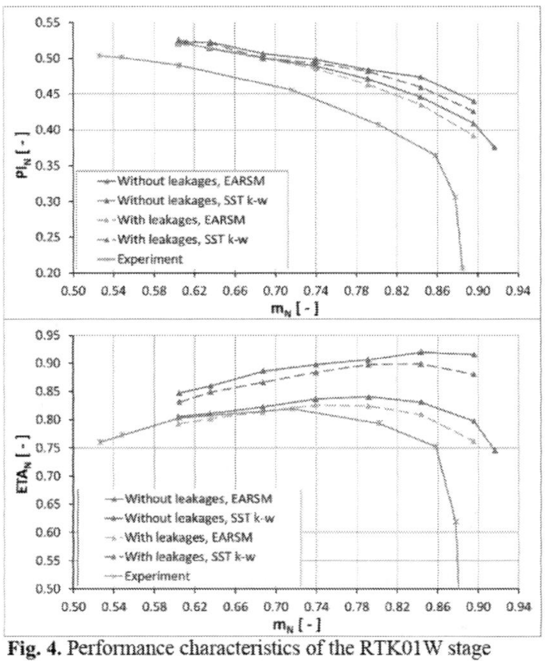

Fig. 4. Performance characteristics of the RTK01W stage

On the basis of the above, it can be concluded that the efficiencies and pressure ratios are kept in the used CFD models and those can be regarded as credible.

3.4 Comparison of results - stage RTK01N

The same process as for the stage RTK01W was implemented for the stage RTK01N.

In Fig. 5, the performance characteristics are presented. The results of numerical simulations can be compared sufficiently to the measured values. The situation here is very similar as for the stage with the longer blades in the impeller shown in the Fig. 4.

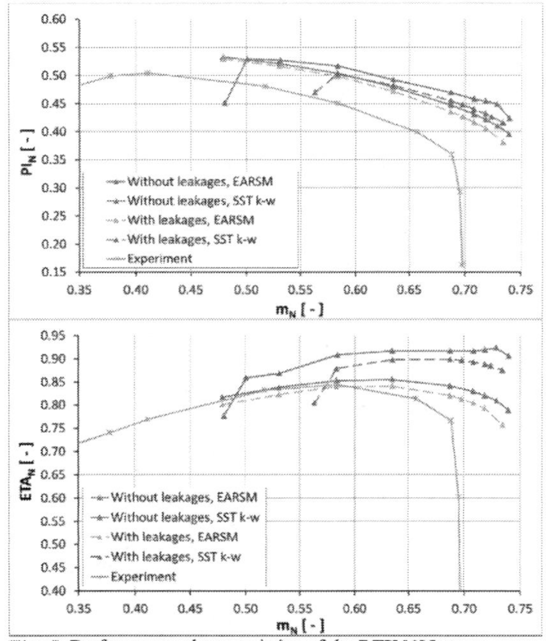

Fig. 5. Performance characteristics of the RTK01N stage

4 Influence of the technological holes to the performance characteristics

4.1. Technological holes

For the most part the impellers with 3D blades are manufactured by milling on a 5 axes CNC machine. In some cases the technology combining the classical machining and welding can be preferable. The welding of impellers requires the technological holes in the blade; you can see an example of such hole in Fig. 6.

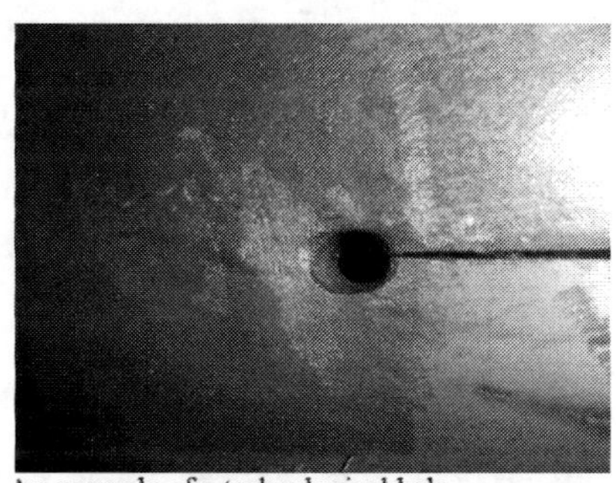

Fig. 6. An example of a technological hole

The goal is to find out the suitable holes location depending on the blade type and achieved stage efficiency.

Due to the complexity of shape and with regard to the demanded precision of the blades it is necessary to place the weld and also the holes near the tip of the blades, so only low ribs remain on the cover disc to form the shape of blades. The holes are then formed from two semi-circular notches in both welded parts.

Important factors are the diameter of the hole and its orientation. In all cases the perpendicular direction of the axis of the hole to the central equidistant surfaces of the blade is handled. Since this is the first analysis of this type and the parameters of the hole depending on its specific position, the central cross section was modelled as constant with the diameter of 5 mm with regard to the technology.

The current position on the edge of the blade is defined by the distance from the leading edge. Hole in the geometrical variants were placed in positions 24, 41, 59, 65, 71 and 87 % of blade length from the leading edge. The Fig. 7 describes the positions of the holes. The basic position of the hole in the 65 % is indicated in green ink, this should be closest to the real positioning. Other possible locations are in range between 59 and 71 % from the leading edge.

The remaining positions probably will not be usable because of the manufacturing technology or the high stress in the compressor impeller and they are only theoretical.

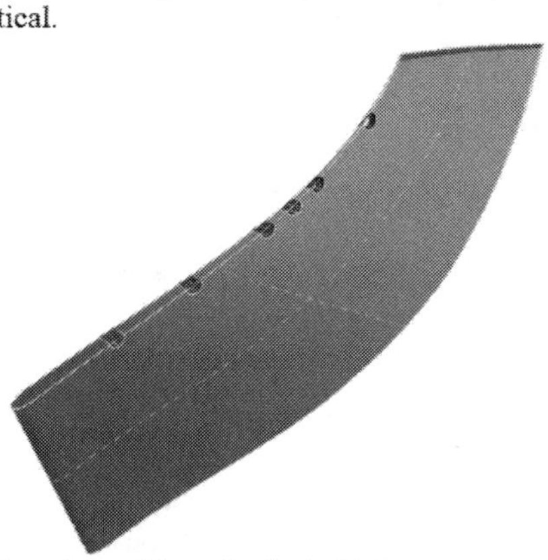

Fig. 7. Locations of the technological holes

4.2 Computational domains and meshes

For the CFD computations the quality a topology of the computational mesh is often very important. It was not easy to include the holes into the mesh of the blade. Two types of meshes were tested.

The first one is relatively simple and it is based on using non-conformal grid interfaces applied directly on the volume representing the hole, see. Fig. 8.

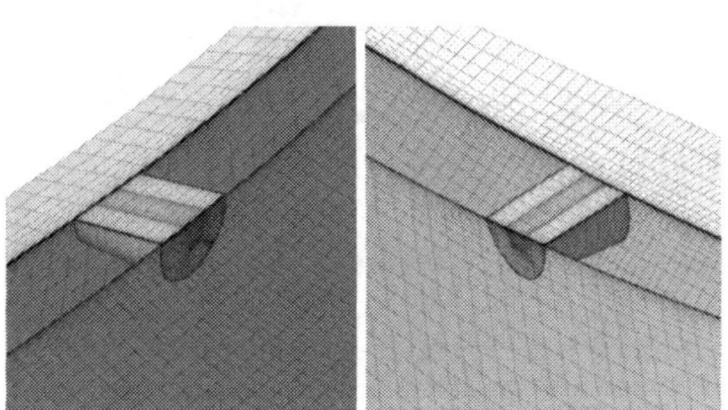

Fig. 8. Domains of the technological hole connected using non-conformal interfaces

The second method also uses the non-conformal grid interfaces, but the connection of the domain with the hole is performed away from the blade surface and it is possible to keep the geometry of the hole including edges. New blocks of the mesh and the connection to the channel mesh is depicted in Fig. 9.

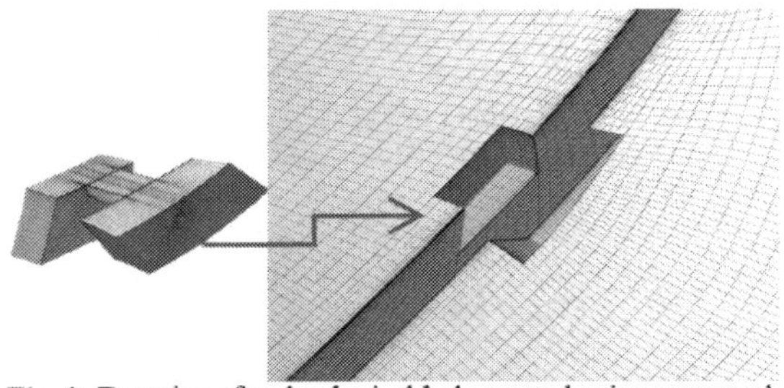

Fig. 9. Domains of technological hole created using new mesh blocks

The comparison of both methods shows that the results of the first method are worse. The different density of the mesh on the non-conformal interface is followed by the distortion of the flow field. That was the reason why all of the described simulations used the second, more complicated method.

All numerical simulations in this chapter were carried out using simplified models without leakages and with the same settings as the models mentioned in chapter 3.

4.3 Results of simulations - stage RTK01W

The results were evaluated as for the stage without holes. The Fig. 10 illustrates the flow in the hole.

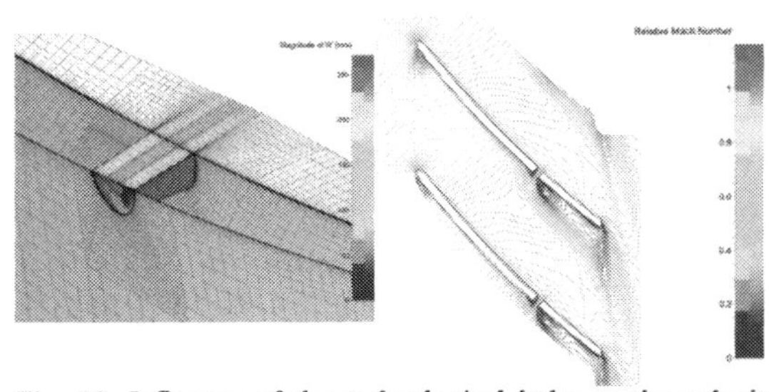

Fig. 10. Influence of the technological holes to the velocity field

The shapes of the performance characteristics are similar to the above depicted ones and therefore it is more appropriate to use the relative influence on the efficiency characteristics. The influence on the pressure ratio is inconsiderable and the pressure ratios obtained by CFD models with and without holes are almost the same.

The Fig. 11 shows the relative influence of the holes to the efficiency of the stage. The influence was tested for three basic values of the M2u (i. e. three values of the rotor speed). The normalised mass flow rate m_N is encident to the computed point for the stage without hole in the characteristics in the Fig 4.

The influence of the hole is different for various M2u, but usually lies between 0 and 0.5 %. The comparison shows the small decreases in efficiency depending on the compression point. The effect is the greatest in case of higher flow rates above 0.5 and does not exceed 1.8 %. These values show that the usage of the welded impellers with small technological holes is possible without a significant impact on efficiency.

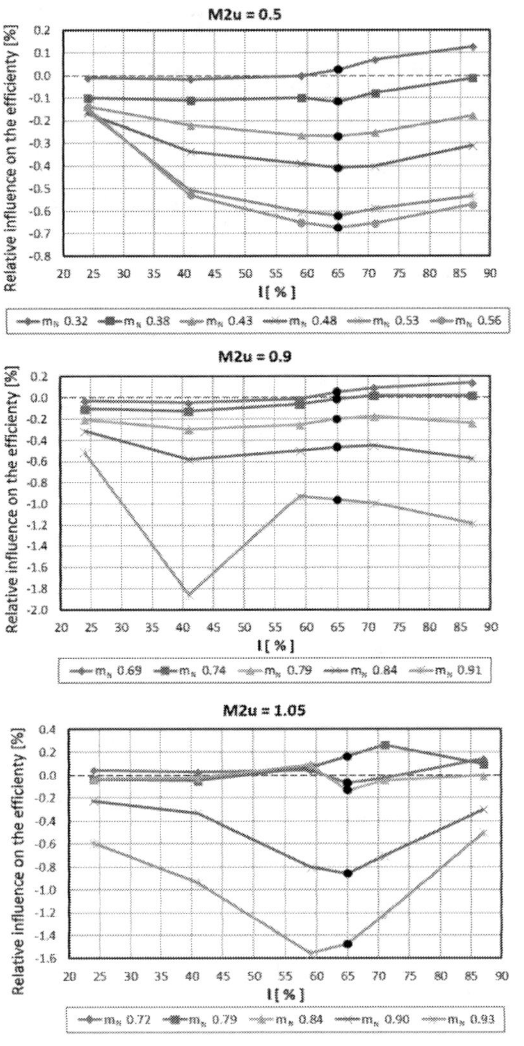

Fig. 11. CFD simulations of the RTK01W stage – influence of the technological holes on the efficiency

4.4 Results of simulations - stage RTK01N

The same set of simulations and evaluation was made for the stage RTK01N with the technological holes. You can see the influence on the efficiency in Fig 12. It is evident that the results are similar to the results of the impeller with the wide channels but not the same. Obviously the results are about 0 and 0.5 % again, but the impact is different and the maximum value is only 1 %.

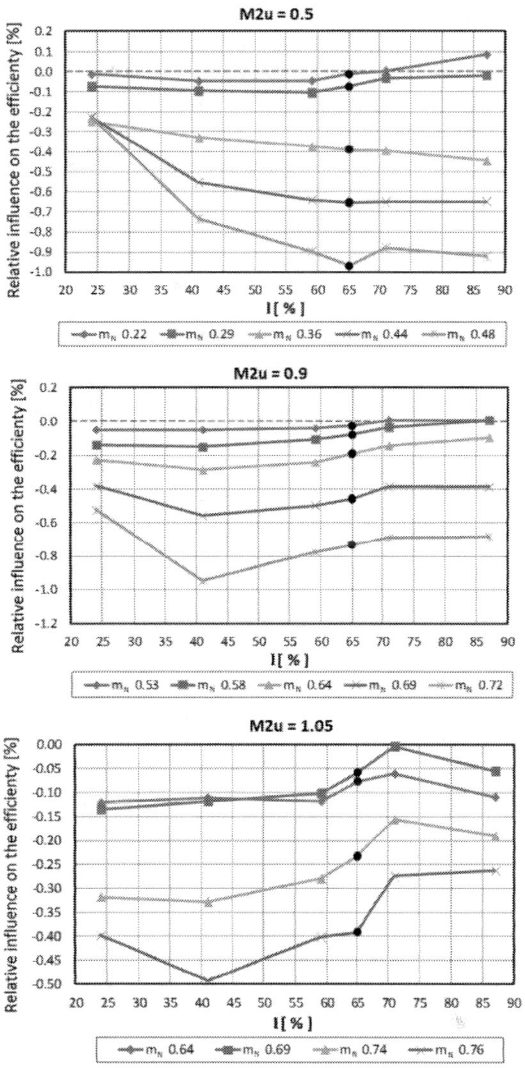

Fig. 12. CFD simulations of the RTK01N stage – influence of the technological holes on the efficiency

5 Conclusions

The verification of the compressor stages with the relatively wide range of parameters is possible to perform on the HCKD DARINA IV compressor with good repeatability. The test facility is able to evaluate complete integral and some local quantities in order to prepare appropriate data for the thermodynamic design.

It is possible to verify CFD models and their applicability by mutual comparison of experimental data and CFD simulations results. The methodology for the centrifugal compressor stages CFD simulations was verified in the NUMECA FINE/Turbo software for the stage with 3D blades in the impeller. The verification for stages with 2D blades was published in [7].

The presented results of two variations the stage RTK01 shows good agreement in efficiency, relatively good agreement in the pressure ratio but yet unexplained slight differences in the flow rates. These simulations and comparisons are the basis for the last part of the contribution, for the influence of the technological holes to the performance characteristics.

The appropriate model of the stages with the holes was developed. The results of the simulations with the technological holes show very small influence on the pressure characteristic and to the flow rate. Influence to the efficiency varies with the position of the hole from zero to 1.8 % in specific conditions.

The result show that in case of a welded impeller it is necessary to assess working conditions due to the relative size of the hole and where appropriate, reduce the values of the efficiency expected for compressor design.

Acknowledgement

The results were developed within the CENTEM PLUS (LO1402) by financial means from the Ministry of Education, Youth and Sports under the "National Sustainability Programme I." and specific research. The authors would like to thank the Howden ČKD Compressors Ltd. for permission to publish this paper and all colleagues who participated on this work.

Nomenclature

a	Speed of sound
c_p	Isobaric specific heat capacity
D	Impeller outer diameter
ETA	Isentropic efficiency
$FI0$	Flow coefficient for intake volumetric flowrate
l	Hole position measured from blade tip
m	Mass flow rate
$M2u$	Blade Mach number
PI	Total pressure ratio
p_t	Total pressure
r	Specific gas constant
T_t	Total temperature
u_2	Blade speed at impeller exit
z	Real gas compressibility factor
Δh	Specific enthalpy difference
κ	Ratio of specific heats
Ω	Rotational speed of the impeller

Common indices:

0	Plane 0 (stage intake)
2	Impeller exit
7	Plane 7 (stage discharge)
is	Isentropic
N	Normalized

References

1. M. P. Boyce, "Principles of Operation and Performance Estimation of Centrifugal Compressors," Proceedings of the twenty-second turbomachinery symposium, Dallas, TX. (1993)
2. J. Paroubek and J. Kynčl, "The Influence of Impeller Flow Channel on Aerodynamic Performance of a Centrifugal Compressor Stage," ASME Paper **98-GT-40** (1998).
3. J. Paroubek and V. Cyrus and J. Kynčl, „The Influence of Outlet Stator Part Surface Roughness on the Performance of a Very Low Flow Coefficient Centrifugal Compressor Stage", ASME Paper **96-OT-179** (1996)
4. P. Le Sausse and P. Fabrie and D. Arnou and F. Clunet, "CFD comparison with centrifugal compressor measurements on a wide operating range," EPJ Web of Conferences, Volume **45** (2013)
5. Numerical and Experimental Modelling of the Radial Compressor Stage Syka, Tomáš and Matas, Richard and Luňáček, Ondřej, AIP Conference Proceedings, **1745, 020059** (2016).
6. T. Syka, "Geometry structure influence on the radial compressor stage efficiency," Ph.D. thesis, in Czech, ZČU in Pilsen (2015)
7. Matas R., Syka T., Luňáček O., "Numerical and experimental modelling of the centrifugal compressor stage – setting the model of impellers with 2D blades", EPJ Web of Conferences, Volume **143**, 02073 (2017)

EPJ Web of Conferences **180**, 02061 (2018) https://doi.org/10.1051/epjconf/201818002061
EFM 2017

Experimentally determined distribution of granular-flow characteristics in collisional bed load transport

Václav Matoušek[1,*], *Štěpán* Zrostlík[1], *Luigi* Fraccarollo[2], *Anna* Prati[2], and *Michele* Larcher[3]

[1]Czech Technical University in Prague, Faculty of Civil Engineering, 166 29 Prague 6, Czech Republic
[2]University of Trento, Department of Civil, Environmental and Mechanical Engineering, Trento, Italy
[3]Free University of Bozen-Bolzano, Department of Science and Technology, Bolzano, Italy

Abstract. A series of laboratory experiments on turbulent open-channel two-phase flow in a form of intense bed load transport is reported. Measurements in a laboratory tilting flume included camera based imaging techniques to identify the structure of the flow at the local level. Obtained experimental distributions of two-phase flow related parameters - granular velocity, concentration, and temperature - across a collisional transport layer are discussed. The results are analysed together with additional measured quantities (discharges of mixture and grains, flow depth, bed slope etc). Our major goal is to evaluate the distribution of granular stresses across the transport layer with a special attention paid to the interface between the transport layer and the bed. Furthermore, comparisons are discussed between the experimental results and predictions produced by suitable kinetic-theory based models.

1 Introduction

Intense bed load transport is associated with flood discharges in steep-slope mountain torrents and rivers. It can be also generated by transient flows, e.g. dam-break flows, in a channel with a mobile bed. The liquid- and solid phases interact in a complex way within a bed load layer developed above a mobile bed. Transported sediment grains are supported by mutual contacts, primarily collisions, and by residual turbulent uplift. It is poorly understood how to describe mechanisms that interrelate parameters like the sediment discharge and the flow depth. The parameters are dependent on distributions of velocity and concentration of solid grains in the flow and hence also on distributions of normal- and shear stresses by solids.

Principles of granular rheology as expressed in the granular kinetic theory [1-2] should serve to model such flows. Kinetic theory based models are available for various granular flow conditions and geometries [3-5]. However, for open channel bed load flows with typically large gradients of local concentrations and velocities the models have not been sufficiently verified by experiments yet. The goal of our work is to produce reliable experimental data suitable to evaluate predictive abilities of the kinetic theory based models. It means that granular flow must be observed on the local level enabling to identify vertical distributions of grain concentration and velocity in a shear layer through which bed load is transported above the bed.

* Corresponding author: v.matousek@fsv.cvut.cz

2 Experimental work

We present experiments carried out in the Water Engineering Laboratory of Czech Technical University in Prague in June 2016.

2.1 Flow conditions

We study turbulent two-phase flow under the following conditions:

- gravity-driven open-channel flow, steady-state uniform flow;
- flow over mobile bed at upper-stage plane bed regime (high bed shear);
- transported sediment grains supported by mutual contacts, no turbulent suspension;
- broad range of bed slopes, flows depths, sediment flow rates, and total flow rates.

For the experiment presented in this paper, we used one fraction of plastic grains (mass-median diameter 3.18 mm, density 1358 kg/m^3, terminal settling velocity 0.131 m/s, round shape of grains) as model sediment in order to ensure intense collisional transport and create suitable conditions for particle tracking by cameras.

2.2 Experimental set up and measurement techniques

2.2.1 Tilting flume

We used our recirculating flume with a rectanular cross section. It is 8 m long, 0.20 m wide, 0.27 m deep. It is equipped for continuous measurement of flow rate of carrying liquid and transported solids along the closed circuit. In the flume, a position of flow surface is sensed in 5 measuring cross sections along the length of the flume. The flume and its permanent measuring equipment (magnetic flowmeter, differential pressure transducers, water surface acoustic sensors, flume inclination meter etc.) are described in detail elsewhere [6].

2.2.2 Camera based measuring techniques

We used camera based measuring techniques which were developed and used (in cooperation with other researchers) in the laboratory of University in Trento.

Laser stripe technique [7] using a camera capturing a deformation of a laser strip (Figure 1) is used to determine a vertical distribution of concentration of grains across a collisional transport layer. It covers local solids concentration within a broad range of values (0 to 0.6 by volume) in the near-wall region of channel flow and compensates for the effect of a presence of the transparent wall on the local distribution of the grains.

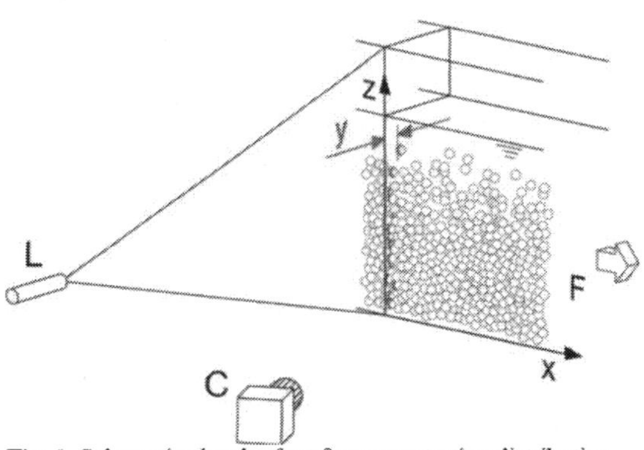

Fig. 1. Schematic sketch of set for concentration distribution measurement using laser stripe method [7].

To measure local velocities of colliding grains in time and space we adopted a Voronoi imaging technique based on Particle Tracking Velocimetry [8] using two synchronized high-speed cameras (resolution 1024 × 1024 pixels, scale ≈3 pixels/mm, frame rate 500 Hz over duration 3.07 s) (Figure 2). The method captures three components of local velocity of colliding grains and enables to conduct statistics of velocity fluctuations.

Fig. 2. Stereo camera set for velocity distribution measurement using imaging technique.

2.3 Experimental results

2.2.1 Global quantities

We carried out 4 test runs at approximately constant flow rate of mixture and 4 different bed slopes (the flume was tilted to 4 different angles starting at low angle for Run 1 and reaching the steepest slope for Run 4). As a result we obtained different solids flow rates (expressed here as the transport concentration defined as the ratio of the volumetric flow rate of grains and of the total volumetric flow rate) and corresponding bed shear stresses (expressed by the dimensionless Shields parameter). Table 1 summarizes values of the global quantities of flow and shows that both the transport concentration and the Shields parameters increased with the steeper installed slope.

Table 1. Major global quantities of flow by test runs.

Quantity	Run 1	Run 2	Run 3	Run 4
Flow velocity (m/s)	0.59	0.71	0.77	0.73
Flow depth (mm)	54	49	45	46
Bed slope (-)	0.011	0.017	0.025	0.036
Transport concentration (-)	0.013	0.033	0.062	0.118
Shields parameter (-)	0.48	0.68	0.92	1.43
Froude number (-)	0.80	1.03	1.16	1.09

A detailed analysis of flows carrying the plastic grains including the fraction tested here was published recently [9]. The analysis evaluates also an interesting change in a trend in a relation between the flow depth and the bed slope (the depth no longer decreases with the increasing slope if the Shields parameter exceeds a threshold value of about 1). Now we focus on measured local flow quantities and their relationship to the global quantities.

2.2.2 Distribution of concentration of grains

The measured distribution of the local volumetric concentration of grains is plotted in Figure 3. The local concentration, c, is normalized by the maximum concentration, c_{max}, which is the bed concentration and is considered 0.65 here). The concentration is plotted against the relative vertical position within the collisional transport layer, y, normalized by the thickness of the collisional layer, h_{coll}.

The distribution seems to be approximately linear with a very similar slope for the test runs 1-3, i.e. for runs at the Shields parameters below a value of one, although there might be a slight difference in local concentrations just above the bottom of the collisional layer. The distribution of the test run 4 exhibits a considerably lower slope and reaches higher local values of the concentration near the bottom of the collisional layer. This observation is consistent with our analysis of the earlier experiments [10].

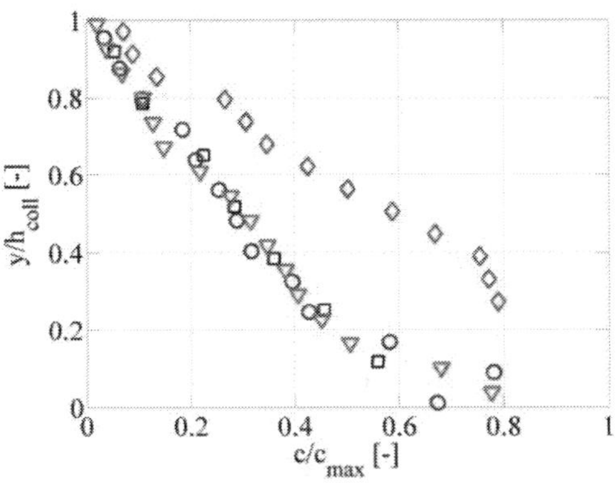

Fig. 3. Measured vertical distribution of volumetric concentration of grains across collisional layer (expressed as relative value to maximum concentration c_{max} = 0.65). Legend: black square – Run 1, blue circle – Run 2, magenta triangle – Run 3, red diamond – Run 4.

2.2.3 Distribution of velocity of grains

The measured distribution of the local velocity of grains is plotted in Figure 3. The local solids velocity, u_s, is normalized in the way used in the kinetic theory (e.g. [1]).

The distribution seems to be approximately linear for all test runs. The slope tends to increase with the increasing Shields parameter provided that the Shields does not exceed a value of one. For the Shields above one, the slope does not further increase and remains approximately constant, i.e. independent of the Shields parameter.

This observation is consistent with our experimental results obtained for a distribution of velocities in the collisional transport layer by acoustic methods [10].

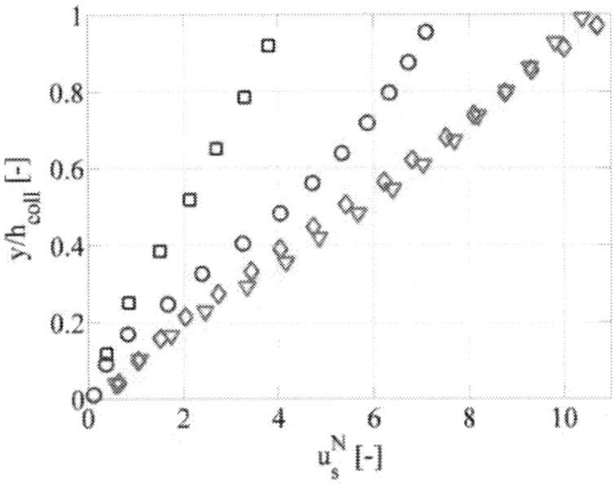

Fig. 4. Measured vertical distribution of velocity of grains across collisional layer (expressed in dimensionless form). Legend: black square – Run 1, blue circle – Run 2, magenta triangle – Run 3, red diamond – Run 4.

2.2.4 Distribution of granular temperature

The granular temperature is a quantity expressing a kinetic energy of intergranular collisions [1]. It is obtained from the stereo camera measurements of local granular velocities by applying a statistical method which determines the fluctuating component of the local velocity of a colliding grain in all three directions.

In Figure 5, the dimensionless granular temperature (it is made dimensionless in the way the kinetic theory applies, e.g. [1]) is plotted as a function of the relative vertical position within the collisional layer.

The measurements reveal an interesting trend in a development of the distribution of the granular temperature across the collisional transport layer. While the temperature gradually increases from the approximately zero at the bottom of the colliding layer to a certain maximum at the top of the layer in flows at low Shields parameter, in flows of high Shields parameter the maximum value is reached within the layer (Figure 5).

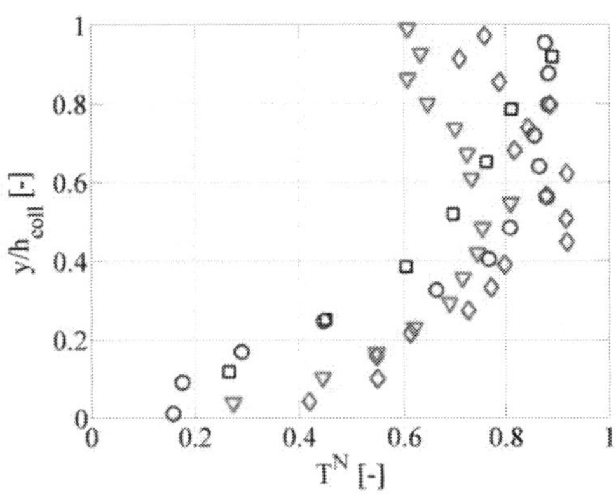

Fig. 5. Measured vertical distribution of granular temperature across collisional layer (expressed in dimensionless form). Legend: black square – Run 1, blue circle – Run 2, magenta triangle – Run 3, red diamond – Run 4.

3 Analysis and discussion

3.1. Distribution of granular stresses

Kinetic theory for granular flows offers constitutive relations which relate depth distributions of granular stresses with depth distributions of concentration and velocity in solid-fluid flows dominated by granular collisions. Distributions produced by the constitutive relations have not yet been verified experimentally for bed-load transport.

We use our experimental data (measured profiles estimated for convenience as linear profiles) in order to produce granular stress distributions using kinetic-theory based relations. A procedure for calculating the stresses from the test based vertical profiles of granular velocity and concentration is described in [11].

To summarize the procedure, momentum balance equations are employed to get the distributions of the total shear stress and of the solids normal stress across the collisional layer. The force balance between the driving force and the resisting force assumes that the total shear stress $s+S$ (composed of the granular component, s, and the liquid component, S) at each vertical position, y, balances the longitudinal component of the weight of overlaying burden of liquid and solids. The granular normal stress is supposed to balance the normal component of the submerged weight of grains above the vertical position y. However, the momentum balance does not directly produce the granular shear stress corresponding with the granular normal shear stress. The kinetic theory must be employed to find this relationship. In the kinetic theory, the local granular stresses (normal and shear components) depend on grain density, velocity gradient (strain rate), and granular temperature. Relationships based on the classical kinetic theory contain no empirical constants. Measured distributions of granular velocity and concentration are used to calculate the distribution of the shear stresses by grains using kinetic theory based equations. An assumption of a negligible contribution of the fluid component of the total shear stress at the bottom of the collisional layer enables to check the equality of the local total shear stress by the momentum balance with the granular shear stress by a kinetic theory model at this boundary. Hence, predicted value of solids shear stress at the bottom of the collisional transport layer is an appropriate indicator of quality of a prediction of the solids shear stress by a kinetic theory based model.

Different kinetic theory based models used in the evaluation [1, 12, 13] give a slightly different distributions of the solids shear stress, but there is generally a good agreement between the predictions of the granular shear stress and the measurement derived total shear stress at the bottom of the collisional layer provided that the Shields parameter of the flow is lower than the threshold value of about one (Figure 6). No reasonable agreement is found if the Shields parameter exceeds the threshold value (Figure 7). This seems to confirm that classical kinetic theory works well for collision-dominated flows of low to moderate local concentrations but fails at higher concentrations where granular collisions are no longer uncorrelated. In flows exhibiting the dimensionless bed shear stress higher than one, the local concentration in the vicinity of the bottom of the transport layer approaches c_{max} and the develops the dense sliding layer meaning that flow conditions are unsuitable for applying principles of the classical kinetic theory.

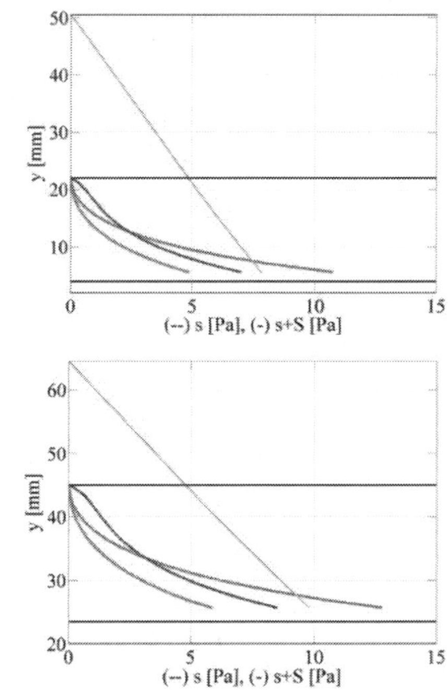

Fig. 6. Distribution of total shear stress (s+S) and solids shear stress (s), solids stress in calculated by kinetic-theory based methods for flows of low Shields parameter (up = Run 2, down = Run 3). Legend: green - distribution of total shear stress; blue - distribution of solids shear stress by [1]; magenta - distribution of solids shear stress by [12]; red - distribution of solids shear stress using the method by [13]; black = positions of visually observed interfaces (bottom and top of transport layer, water surface).

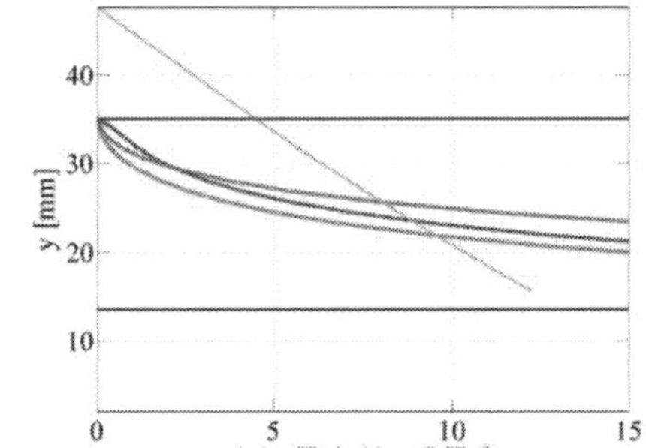

Fig. 7. Distribution of total shear stress (s+S) and solids shear stress (s), solids stress in calculated by kinetic-theory based methods for flow of high Shields parameter (Run 4). Legend: as in Fig. 6.

https://doi.org/10.1051/epjconf/201818002061

3.2 Comparison with other experiments

Similar experiments with intense bed load transport of plastic grains was carried out in a laboratory flume at University of Trento in 2009 [13]. The same optical techniques were adopted to sense and analyse the granular distributions in an observed collisional layer.

Plots in Figures 8 and 9 compare the Prague experimental results with the Trento results for similar flow conditions (values of Shields parameter). The plots show results in the entire sensed range of vertical positions including positions below and above the collisional layer where the results are actually not valid.

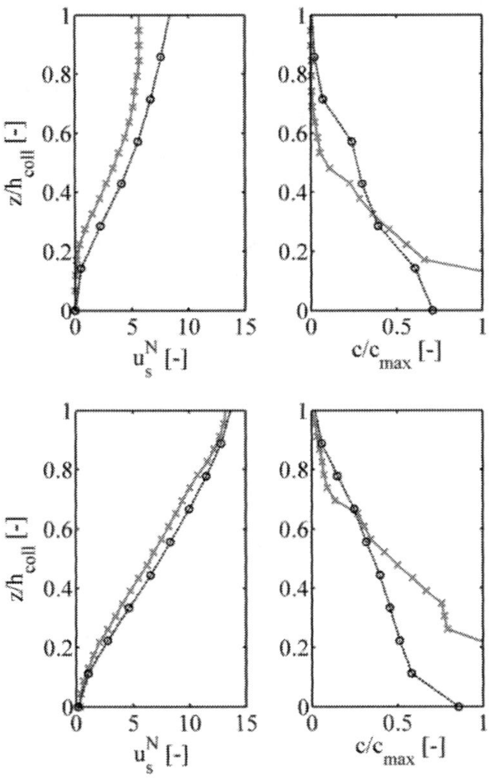

Fig. 8. Distribution of granular velocity and concentration across flow depth (up = Run 1, down = Run 4). Legend: red cross – 2016 Prague tests; black circle – 2009 Trento tests.

In the collisional layer, the agreement between the results of the two tests is satisfactory. Higher quality of camera images was reached in the Prague tests which enabled to more accurate and reliable processing of the velocity fluctuations and hence the granular temperatures than in the Trento tests (the temperature results of the Trento tests are published here for the first time).

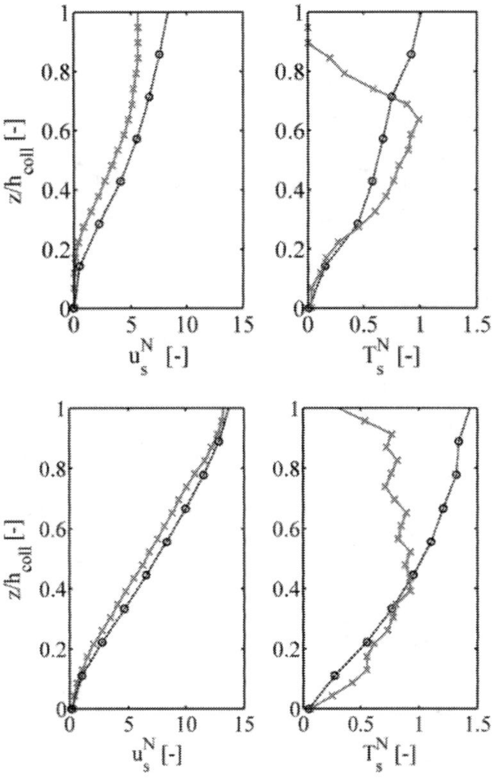

Fig. 9. Distribution of granular temperature across flow depth (up = Run 1, down = Run 4). Legend: as in Figure 8.

4 Conclusions

Kinetic theory based predictions of depth distribution of granular shear stress are evaluated using results of our contact-load experiments and assuming linear profiles of grain concentration and velocity across the collisional layer.

Solids stress at the bottom of the collisional layer resulting from the predicted distribution agreed well with the experimentally determined total shear stress at the same interface (provided that the bed Shields parameter was lower than say 1).

The classical kinetic theory works well for collision-dominated flows of low to moderate local concentrations where granular collisions are uncorrelated.

The granular temperature tends to reach a maximum value near the centre of the collisional layer thickness provided that the thickness (and hence the Shields parameter) is high enough to develop sufficiently high local concentration at the bottom of the collisional layer. Otherwise, the maximum of the temperature occurs near the top of the collisional layer.

Acknowledgement

The research has been supported by the Czech Science Foundation through the grant project No. 16-21421S and by Faculty of Civil Engineering of the Czech Technical University in Prague through the student grant project No. SGS17/065/OHK1/1T/11. An assistance of V. Bareš, Z. Chára, J. Hlom, J. Krupička, and T. Picek with the Prague experiments is highly acknowledged.

References

1. J.T. Jenkins, D.M. Hanes, J. Fluid Mech. **370**, 29-52 (1998)
2. V. Garzó, J.W. Dufty, Phys. Rev. E **59**(5), 5895-5911 (1999)
3. A. Armanini, H. Capart, L. Fraccarollo, M. Larcher, J. Fluid Mech. **532**, 269–319 (2005)
4. D. Berzi, L. Fraccarollo, Geophys. Phys. Fluids **25**, 106601 (2013)
5. B. Spinewine, H. Capart, J. Fluid Mech. **731**, 579–613 (2013)
6. V. Matoušek, V. Bareš, J. Krupička, T. Picek, Š. Zrostlík, J. Hydrol. Hydromech. **63**(4), 318-326 (2015)
7. B. Spinewine, H. Capart, L. Fraccarollo, M. Larcher, Exp. Fluids **50**(6), 1507–1525 (2011)
8. H. Capart, D. L. Young, Y. Zech, Exp. Fluids **32**, 121–135 (2002)
9. V. Matoušek, V. Bareš, J. Krupička, T. Picek, Š. Zrostlík, Can. J. Chem. Eng. **94**, 1076-1083 (2016)
10. V. Matoušek, V. Bareš, J. Krupička, T. Picek, Š. Zrostlík, *River Flow 2016* (IAHR, Saint Louis, 2016)
11. V. Matoušek, *Hydrotransport 2017* (BHRG, Melbourne, 2017)
12. D. Berzi, ASCE J. Hydraul. Eng. **137**(10), 1200-1207 (2011)
13. H. Capart, L. Fraccarollo, Geophys. Res. Lett. 38. L20402 (2011)

Modelling approaches for pipe inclination effect on deposition limit velocity of settling slurry flow

Václav Matoušek[1,*], *Mikoláš* Kesely[1], *Pavel* Vlasák[1]

[1]Institute of Hydrodynamics of Czech Academy of Sciences, Pod Patankou 30/5, 166 12 Prague 6, Czech Republic

Abstract. The deposition velocity is an important operation parameter in hydraulic transport of solid particles in pipelines. It represents flow velocity at which transported particles start to settle out at the bottom of the pipe and are no longer transported. A number of predictive models has been developed to determine this threshold velocity for slurry flows of different solids fractions (fractions of different grain size and density). Most of the models consider flow in a horizontal pipe only, modelling approaches for inclined flows are extremely scarce due partially to a lack of experimental information about the effect of pipe inclination on the slurry flow pattern and behaviour. We survey different approaches to modelling of particle deposition in flowing slurry and discuss mechanisms on which deposition-limit models are based. Furthermore, we analyse possibilities to incorporate the effect of flow inclination into the predictive models and select the most appropriate ones based on their ability to modify the modelled deposition mechanisms to conditions associated with the flow inclination. A usefulness of the selected modelling approaches and their modifications are demonstrated by comparing model predictions with experimental results for inclined slurry flows from our own laboratory and from the literature.

1 Introduction

One of the most important and yet one of the most difficult problems associated with the design and operation of a pipeline system transporting settling slurries is a prediction of the deposition limit velocity of the flowing slurry. Coarse settling slurries are transported at very different conditions (pipe size and geometry, solids contents in slurry etc) and exhibit very different properties (slurry density, degree of stratification etc). Moreover, they are often transported in pipelines containing sections that are inclined from the horizontal. Examples include in-plant transfer lines, tailings disposal operations, dredging, deep sea mining, and long-distance pipelining [1].

2 Definition and identification of threshold velocity

2.1 Issue of definition

In slurry pipeline practice, the term "critical velocity" is often used to determine a threshold below which the flow of slurry should not drop in a pipe in order to avoid operational problems. In the literature, many names and definitions can be found for the so called critical velocity. Some authors refer to it as to velocity at which a stationary bed deposit starts sliding, other authors as to velocity at which the stationary bed or sliding bed disappears. Others consider the critical velocity as the velocity at which the hydraulic gradient is at a minimum on a pipe resistance curve for mixture at certain delivered concentration in a pipe of a certain size.

We consider the deposition limit velocity, V_{dl}, as an average flow velocity at which grains first stop moving and start to develop a deposit at the bottom of a pipe. If coarse grains are transported at high delivered concentrations C_{vd} (say $C_{vd} > 0.2$, see e.g. [2]), they tend to form a sliding bed at velocities higher than V_{dl} and hence V_{dl} is a threshold at which the bed stops sliding. In flow of low concentration of coarse grains, no compact sliding bed is formed if the flow decelerates towards V_{dl}. Instead the grains tend to form temporal clusters at the bottom of the pipe if the flow velocity gets close to V_{dl}, and the clusters interrupt movement before they come to a final stop at V_{dl}. If fine grains are transported instead of coarse grains in settling slurry flow, they do not form a sliding bed or any clusters before the deposit develops. In such a flow, a development of the deposit is a much more stable process than in coarse settling slurry flow, where it is often associated with a development of density waves, particularly in closed loops [3].

2.2 Issue of experimental identification

A measurement of V_{dl}, or its determination by visual observation in a clear pipe, is associated with

* Corresponding author: matousek@ih.cas.cz

considerable uncertainty caused both by the unstable process of bed forming and by the individual character of each observer's decision making. Hence, a considerable scatter must be expected to occur in experimental results and an expected deviation of predictions from experimental data must be considered generally larger than the deviation for a frictional pressure drop at standard operating velocity in slurry flow. While the deviation up to say 10% is considered good for frictional pressure drop predictions, for V_{dl} the deviation may be considered satisfactory if it does not exceed 20% [4].

3 Observations in literature

The deposition limit velocity is sensitive to pipe diameter, to properties of both solids (size, density) and carrying liquid (density, viscosity) and to the concentration of transported solids. Experimental data from large pipes of industrial size are extremely scarce, and a majority of available data are from pipes of diameters up to 100 or 150 mm.

3.1 Effect of grain size

Much more experimental experience than for an effect of the pipe size is available for an effect of the particle size on V_{dl}, although data for particles smaller than say 100 micron are not as common. Experimental observations reveal that there is a certain critical size at which V_{dl} is maximal in a given pipe. This size is of about 0.5 mm (medium sand in water), with coarser and finer grains of the same density forming granular deposits at lower flow velocities. In Figure 1, the effect of a variation of V_{dl} (expressed using the dimensionless number $F_L = V_{dl} / \sqrt{2 \cdot g \cdot (S_s - S_f) \cdot D}$, in which g is gravitational acceleration, S_s and S_f relative densities of solids and liquid, D pipe inner diameter) with the grain size (expressed using Archimedes number Ar) is demonstrated on results of a recent experiment.

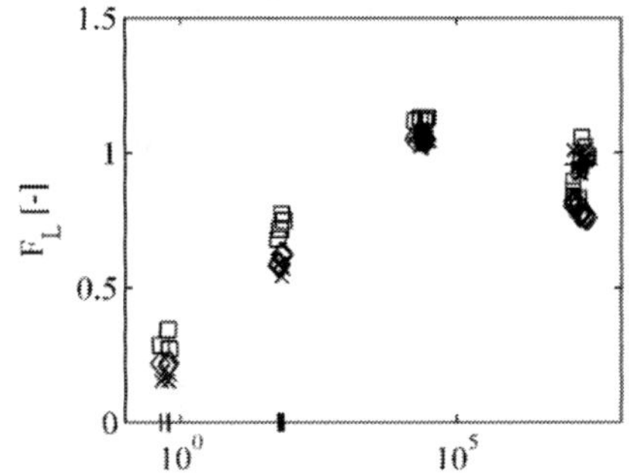

Fig. 1. Relationship between dimensionless deposit velocity F_L and Archimedes number Ar for four individual fractions in 203-mm pipe [4]. Legend: square = experiment in GIW Hydraulic Laboratory, diamond = VSCALC-V_{sm} model, x = 1952-Durand model.

A comparison of the experimental results with predictive models also shown in Figure 1 is discussed in Chapter 4 below.

3.2 Effect of broad grain size distribution

Industrial settling slurries often consist of particles of very different sizes; the particle size distribution may cover sizes which differ with two orders of magnitude. A broad particle size distribution affects parameters of slurry flow including deposition limit velocity. Recently [4], experimental results were analysed of a comprehensive experimental campaign testing slurry flows composed of solids of different fractions in the GIW Hydraulic Laboratory in 2016.

The experimental results showed that interactions among fractions affected the resulting deposition limit velocity in flows of broadly graded settling slurries. The effect of particle size distribution on the deposition limit velocity was not benign. The deposit velocity was not necessarily lower in a flow of slurry composed of four fractions of different sizes than in slurry flow of the individual fraction with the highest deposit velocity from the four fractions.

3.3 Effect of pipe inclination

Compared to the horizontal flow, inclined flow produces an additional force (a component of the submerged weight of grains) which acts either against the direction of flow (an ascending pipe) or in the flow direction (a descending pipe). In the ascending pipe (more often applied in the industry), the upwards inclined flow tends to require higher throughput velocity than the horizontal flow in order to avoid deposition. This is particularly significant for coarse-particle flows.

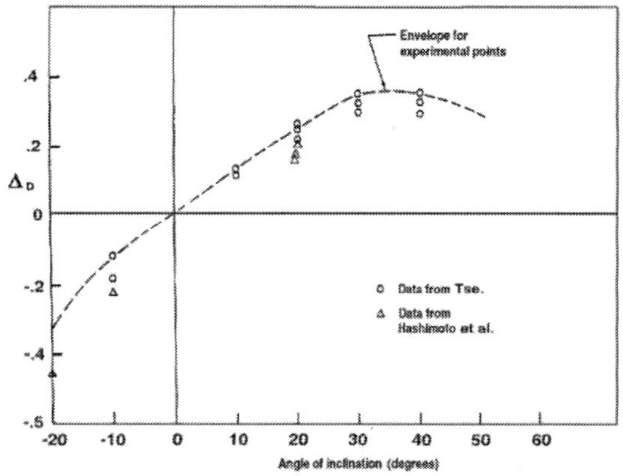

Fig. 2. Effect of angle of inclination on Durand deposition parameter [5].

Wilson and Tse [5] conducted experiments with very coarse grains (1 to 6 mm) in a 76-mm pipe that could be inclined up to 40 deg from the horizontal. The effect of pipe inclination on V_{dl} was expressed as the change in the dimensionless Durand parameter ($\Delta_D = F_{L,\theta} - F_{L,0}$, where $F_{L,\theta}$ is for V_{dl} in a pipe inclined to angle θ and $F_{L,0}$ for V_{dl} in a horizontal pipe) with the pipe inclination angle θ (Fig. 2).

Matoušek [6] demonstrated that the deposition limit velocity is associated with a degree of flow stratification. His experimental work with a 150 mm diameter pipe and inclinations angles between -35 degrees and +35 degrees showed a strong influence of pipe inclination on slurry flow stratification. The observed concentration profiles were less stratified for rising inclinations and more stratified for descending inclinations in flows with coarse sands and gravel while they remained very similar in flows with fine to medium sands. This effect of different flow velocity relative to the sliding bed developed in the ascending flow (slow sliding) and in the descending flow (fast sliding) is a result of the mechanics of partially-stratified inclined flow.

Spelay et al. [1] proposed a criterion (turbulent suspension efficiency parameter) to evaluate the extent and impact pipe inclination will have on the deposition limit velocity.

De Hoog et al. [7] verified a usefulness of the Wilson-Tse diagram (Fig. 2) for very coarse grains by experiments with gravel fractions (4.6, 6.3 and 12 mm) in a 100-mm pipe inclined up to 52 degree. They found the maximum V_{dl} at of about 30 degrees.

4 Approaches for predictive modelling

4.1 Semi-empirical approach

An estimation of deposition limit velocity has been subject of numerous studies over last 60 years with pioneer works dating to early 1950s. The two most often used semi-empirical models are the Durand model [8,9] and the Wilson-GIW model (described as the VSCALC model in [10]).

The Durand model uses an empirical nomograph which relates the dimensionless V_{dl} (F_L) with the particle size and solids concentration (C_v up to 0.15). There are two versions of the nomograph in the literature [11].

The VSCALC model combines the Wilson formulae for medium and coarse fractions with the Thomas formula for fractions finer than say 100 microns. The model gives a maximum value of deposit velocity for given sizes of particle and pipe (V_{sm}) and a value modified for different delivered concentrations (V_s).

A modification was proposed for a deposit velocity predictive model in order to take effects of a disproportionate contribution of individual fractions to the deposition limit velocity in flow of broadly graded slurry into account [4].

In the VSCALC model, the effect of pipe inclination on V_{dl} is calculated using the Wilson-Tse method.

4.2 Mechanistic approach: two-layer model

The horizontal-flow Wilson-GIW correlation is calibrated by predictions of an early version of a two-layer model for fully stratified flow. A more refined and complete two-layer model for partially stratified flow can predict the effect provided that proper transport- and friction formulae are used for the top of the bed in the model.

5 Experimental modelling for testing predictive models

In order to evaluate predictive modelling approaches, we produce experimental results by testing flow behaviour of heterogeneous mixtures using solids fractions of very different grain sizes over the wide range of volumetric solids concentrations in a 100-mm pipe loop with and inclinable section at Institute of Hydrodynamics (IH) of Czech Academy of Sciences. The test loop is described in details elsewhere [12].

5.1 Horizontal flow

Tests in a horizontal pipe were carried out for four narrow-graded fractions of glass beads (B134 of d_{50} = 0.18 mm, B9 of d_{50} = 0.44 mm, B8 of d_{50} = 0.53 mm, TK0810 of d_{50} = 0.9 mm).

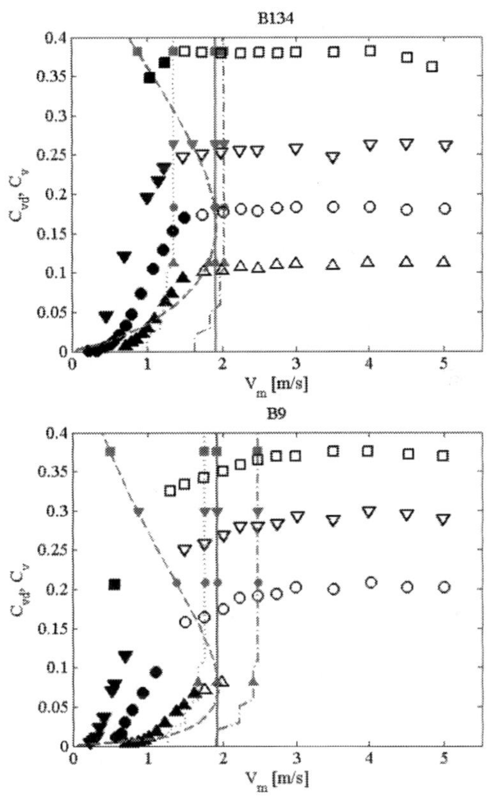

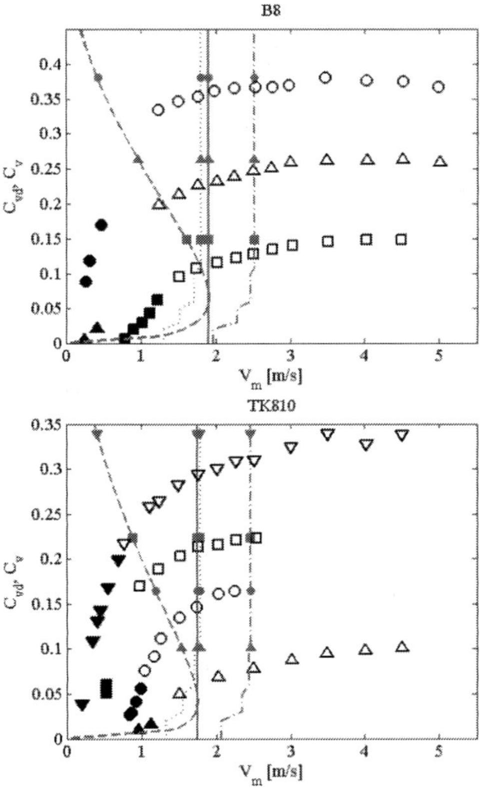

Fig. 3. Deposition limit velocity depicted in plot of delivered concentration of solids (C_{vd}) against average velocity of slurry flow (V_m) in flow of slurries of different solids contents. Plots are for four different tested solids. [13] Legend: black marker = V_m below deposition limit velocity, blank marker = V_m above deposition limit velocity, lines = predictions using different models, best prediction by VSCALC-Vsm (dash line).

Flows of different fractions exhibited rather different behaviour near the deposition limit velocity, where the stationary deposit had started to be formed. While slurry flow of the fine fraction (B134) was quite stable and exhibited a smooth transition between deposit-free flow and flow with stationary deposit, the coarser fractions exhibited transition associated with instabilities and formation of density waves. Under such unstable conditions, measurement of V_{dl} was impossible; hence missing markers around the depoisition limit in the plots for B9, B8 and TK0810 in Figure 3. The experimental results from the horizontal pipe and their comparison with predictive models are described in [13].

5.2 Inclined flow

Our new inclined-pipe experiments include measurements of local concentration at the bottom of the pipe in flow at each installed slurry velocity (above and below V_{dl}). Also, profiles of local concentration are measured across a pipe cross section at (or very near to) the deposition limit velocity. The local concentration is measured by radiometric instruments with which our test loop is equipped and which are described e.g. in [12].

Our goals are twofold. First, we want to make the experimental identification of V_{dl} less subjective by including the new measuring technique: an establishment of the deposit is sensed through the measurement of local concentration at the bottom of the pipe (at the vertical position 10 mm above the pipe wall). This enables to compare V_{dl} obtained by the traditional (and subjective) visual observations with V_{dl} by the radiometric technique. Second, we aim to evaluate a relation between the distribution of solids across the flow and the establishment of the deposit at the bottom of the flow. We hypothesize that mechanisms responsible for grain support in the flow, and hence for the solids distribution (thus for the shape of a concentration profile), affect the way the deposit is established.

5.2.1 Experimental identification of deposition limit velocity

The new tests in an inclined pipe were carried out for one narrow-graded fraction of glass beads (B134 of d_{50} = 0.18 mm) and one narrow-graded fraction of sand (P0612 of d_{50} = 0.87 mm). Experimental data for medium-to-coarse sands as P0612 are not available in the literature database.

The use of a measurement of local concentration at the bottom of the pipe, c_b, for a determination of the deposition limit velocity is demonstrated in Figure 4. During a test run the flow velocity V_m is gradually decreased and c_b remains approximately constant until the deposition limit is reached. At the deposition limit, c_b jumps to a higher value typical for deposit formed at the bottom of the pipe. In Figure 4, c_b starts to increase at V_m about 2.4 m/s and the deposit starts to be formed (i.e. the deposition limit velocity is reached) at approximately 2.2 m/s.

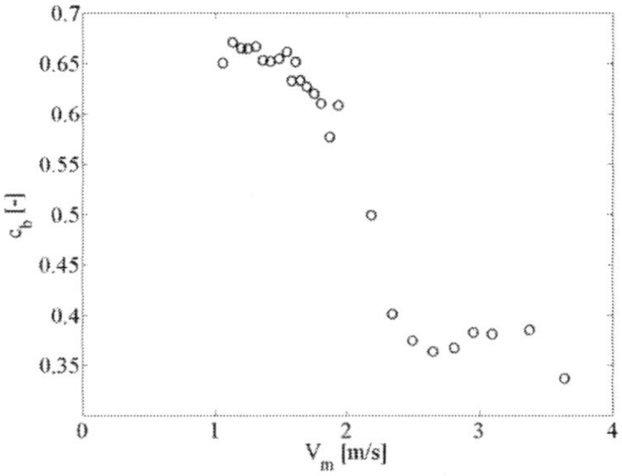

Fig. 4. Development of local concentration at bottom of pipe with velocity of P0612-slurry flow (C_{vd} = 0.11) in pipe inclined to +25 degree.

Values of V_{dl} determined using the radiometric method are compared with results of the traditional visual observation in Figure 5. In general, the agreement is good. The visually observed values tend to be slightly lower than the values identified as the velocity at which c_b starts to increase. Typically, a visually observed value corresponds with velocity at which c_b reaches

https://doi.org/10.1051/epjconf/201818002062

approximately 0.6, which is a value typical for granular deposit (it is bigger than c_b for the sliding bed).

The identified values of V_{dl} are less sensitive to a variation of the pipe inclination angle than the VSCALC model (including the Wilson-Tse method) predicts for the P0612-slurry (medium to coarse sand slurry) at C_{vd} = 0.11 in the 100-mm pipe inclined between 0 degree (horizontal pipe) and +45 degree (ascending pipe) (Figure 5). The Wilson-Tse method was constructed [5] and verified [1,7] for considerably coarser slurries (gravel slurries). We hypothesize that the over-prediction by the method for the medium to coarse sand is caused by the fact that the sand flow is less stratified than the gravel flow. The degree of stratification (expressed by a shape of a concentration profile) varies significantly with a grain size and also with a slope to which flow is inclined. This is demonstrated on our measured profiles below.

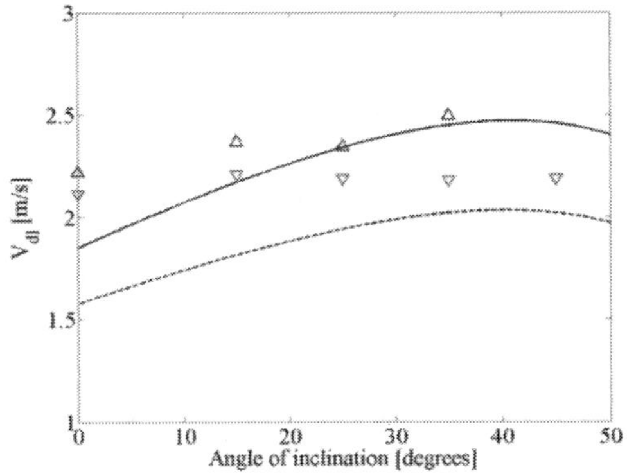

Fig. 5. Deposition limit velocity of P0612-slurry flow (C_{vd} = 0.11) in pipe at different inclination angles. Legend: red marker = experiment - visual observation, blue marker = experiment – c_b-method, full line = prediction VSCALC-Vsm, dash line = prediction VSCALC-Vs.

5.2.2 Distribution of solids concentration at deposition limit velocity

Solids distributions were measured simultaneously in the ascending leg and in the descending leg of the inclinable invert U-tube, which is a part of our test loop. The measurements were carried out for the fractions of very different grain size and very similar grain density (finer B134 and coarser P0612). This enables to demonstrate the effect of both the grain size and the pipe inclination angle on the shape of the concentration profile.

In Figures 6-9, concentration profiles are plotted for the two fractions at the same delivered concentration (C_{vd} = 0.11) and flow velocity slightly above the deposition limit velocity. It is typical for the finer fraction (Figs. 6-7) that the concentration gradient is mild across the flow suggesting that the degree of stratification is very weak and c_b does not reach a high value even if the flow velocity is only slightly above V_{dl}. Moreover, the effect of pipe inclination on the shape of concentration profiles is weak too as the flow exhibit very similar shapes in the ascending pipe (Fig. 6) and descending pipe (Fig. 7).

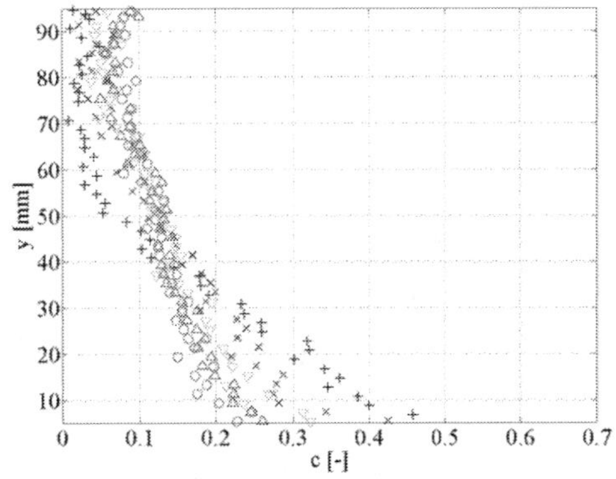

Fig. 6. Distribution of granular concentration at velocity near deposition limit for flow of B134-slurry (C_{vd} = 0.11) at different pipe inclinations (ascending pipe). Legend: black = 0 deg, blue = +15 deg, green = +25 deg, red = +35 deg, magenta = +45 deg.

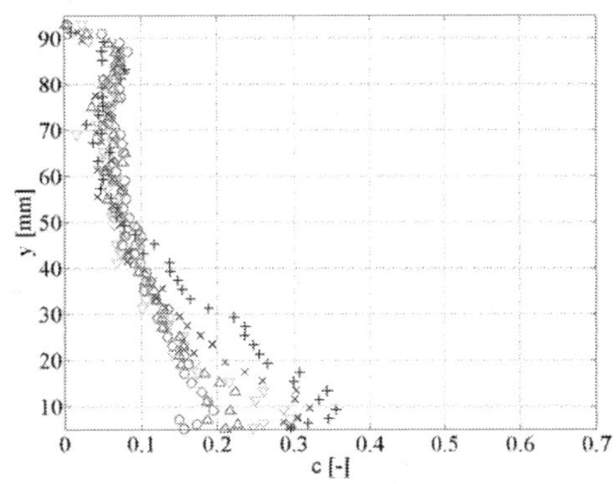

Fig. 7. Distribution of granular concentration at velocity near deposition limit for flow of B134-slurry (C_{vd} = 0.11) at different pipe inclinations (descending pipe). Legend: black = 0 deg, blue = -15 deg, green = -25 deg, red = -35 deg, magenta = -45 deg.

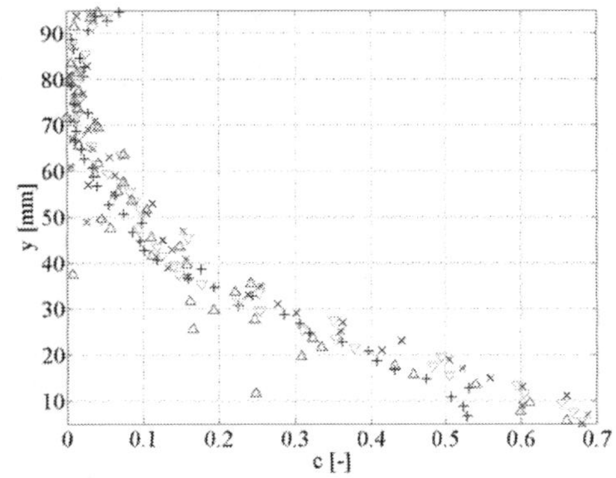

Fig. 8. Distribution of granular concentration at velocity near deposition limit for flow of P0612-slurry (C_{vd} = 0.11) at different pipe inclinations (ascending pipe). Legend: black = 0 deg, blue = +15 deg, green = +25 deg, red = +35 deg.

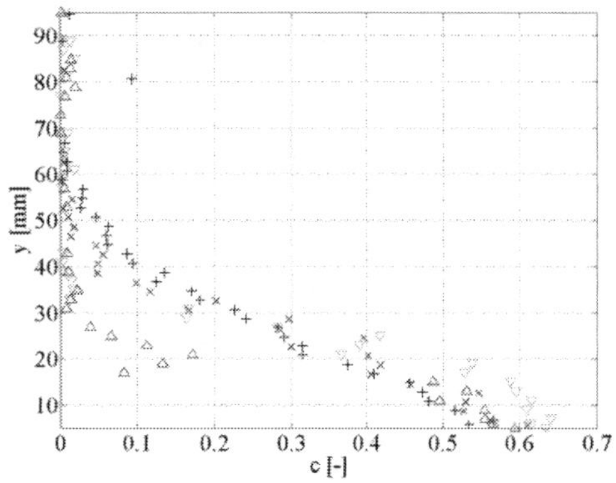

Fig. 9. Distribution of granular concentration at velocity near deposition limit for flow of P0612-slurry (C_{vd} = 0.11) at different pipe inclinations (descending pipe). Legend: black = 0 deg, blue = -15 deg, green = -25 deg, red = -35 deg.

A situation is very different flow carrying the medium-to-coarse sand (Fig. 8-9). The flow is partially stratified with a steep concentration gradient and with high values of the local c_b at the velocity slightly above the deposition limit. Furthermore, the effect of pipe inclination is strong as the flow tends to be less stratified in the ascending pipe (Fig. 8) that in the descending pipe (Fig. 9) at the same slope. This is in an agreement with our earlier observation at velocities higher above the deposition limit in other pipe loop [6]. Mechanisms leading to this effect are described in [6] as well, their incorporation to an appropriate model for the effect of pipe inclination on the deposition limit velocity in partially stratified settling slurry flows in work currently in progress.

6 Conclusions

Existing modelling approaches for pipe inclination effect on the deposition limit velocity seem to be inappropriate for settling slurry flow which are partially stratified.

Experimental results for a medium-to-coarse sand are new to the literature database. They indicate that the effect of the angle of flow inclination on the deposition limit velocity is weaker than predicted by the Wilson-Tse method calibrated and validated for coarser (gravel) fractions but not negligible as expected for finer fractions (fine to medium sands). A new modelling approach must be found for the medium to coarse sand slurries.

A new experimental technique of an identification of the deposition limit velocity by sensing a change in the local concentration at the bottom of the pipe using a collimated gamma-ray gives reliable results provided that the solids concentration in a pipe is high enough to constitute a deposit. The results agree well with the visual observations.

Solids distribution in a pipe cross section tends to vary with the angle of pipe inclination. Modelling of the process of deposit formation using mechanisms responsible for solids distribution in settling slurry flow in work currently in progress.

The research has been supported by the Czech Science Foundation through the grant project No. 17-14271S.

References

1. R.B. Spelay, R.G. Gillies, S.A. Hashemi, R.S. Sanders, Can. J. Chem. Eng. **94**, 1032-1039 (2016)
2. V. Matoušek, J. Krupička, *ASME 2013 Fluids Engineering Division Summer Meeting* (ASME, Incline Village, USA, 2013)
3. V. Matoušek, J. Krupička, *Transport and Sedimentation of Solids Particles* (Rostock, Germany, 2013)
4. V. Matoušek, R. Visintainer, J. Furlan, G. McCall II, A. Sellgren, *Transport and Sedimentation of Solid Particles* (Prague, Czech Republic, 2017)
5. K.C. Wilson, J.K.P. Tse, *Hydrotransport 9* (BHRA, Rome, 1984)
6. V. Matoušek, *Hydrotransport 13* (BHRG, Johannesburg, 1996)
7. E. de Hoog, M. in 't Veld, J. van Wijk, A. Talmon, *Transport and Sedimentation of Solids Particles* (Prague, Czech Republic, 2017)
8. R. Durand, E. Condolios, *2émes Journées de l'Hydralique* (SHF, Grenoble, 1952)
9. R. Durand, *IAHR Minnesota International Hydraulic Convention* (IAHR, Minneapolis, Minnesota, 1953)
10. Wilson, K.C., Addie, G.R., Sellgren, A., Clift, R., 1997. Slurry Transport Using Centrifugal Pumps, 2nd Edition, Blackie Academic & Professional, London
11. Miedema, S.A., 2016. Slurry Transport: Fundamentals, A Historical Overview & The Delft Head Loss & Limit Deposit Velocity Framework, Delft University of Technology, Delft
12. V. Matoušek, J. Krupička, V. Pěník, Particul. Sci. Technol. **32**(2), 186-196 (2014)
13. M. Kesely, V. Matoušek, P. Vlasák, *RELPOWFLO V* (Skien, Norway, 2017)

Two-dimensional quantitative visualization of isolator shock trains by rainbow schlieren deflectometry

Atsushi Matsuyama[1,*], *Shinichiro* Nakao[1], *Daisuke* Ono[1], *Yoshiaki* Miyazato[1], *Masashi* Kashitani[2]

[1]Department of Environmental Engineering, The University of Kitakyushu, 1-1, Hibikino, Wakamatsu-ku, Kitakyushu, 808-0135, Japan
[2]Department of Aerospace Engineering, National Defence Academy, 1-10-20, Hashirimizu, Yokosuka, Kanagawa, 239-8686, Japan

Abstract. The rainbow schlieren deflectometry is effective in studying quantitatively the density fields in shock-containing free jets at high precision and high spatial resolution. However, there has been no practical application of rainbow schlieren deflectometry for shock trains in a confined duct. Therefore, in the present study, the rainbow schlieren deflectometry is applied to the flow field including a shock train in a constant-area duct where just upstream of the shock train the freestream Mach number is 1.34, the unit Reynolds number is 5.39×10^7 m^{-1}, and the boundary layer displacement thickness is 0.149 mm. As a result, a two-dimensional density field of the shock train is for the first time quantitatively displayed and the fine structure of the shock train is illustrated as a color gradation representation.

1 Introduction

Over the past few decades, the development of dual-mode scramjet engines for realizing next-generation propulsion systems has been aggressively conducted. Dual-mode scramjet engines provide a practical solution to supersonic flight by operating under a wide range of Mach numbers. In the ramjet mode, the incoming air is compressed through a sequence of shocks known as precombustion shocks before entering the combustion chamber. To isolate the effect of flight conditions on precombustion shocks, a nearly constant area duct referred to as an isolator is placed between the inlet and the combustor. The isolator is effective in holding the entire shocks. However, the isolator length is typically determined from empirical correlations to withstand the maximum pressure rise expected between the inlet and the exit. Being able to predict the length of a precombustion shock structure is the key component of isolator design for dual-mode scramjet engines. Also such a repeated shock structure appears in a variety of flow-devices including supersonic wind tunnels, supersonic ejectors, supersonic inlets of aircraft engines, etc and has been called in many ways by many researches, multiple shocks, shock system, for example. We refer to such a series of shocks as a "shock train". The shock train is usually followed by an adverse pressure gradient region, if the duct is long enough. The interaction region including both the shock train and the subsequent static pressure recovery region is referred to as a "pseudo-shock". However, the technical term "shock train" as well as "pseudo-shock" has been used in confusing way in most of previous articles. A review of the past experimental, theoretical, and numerical investigations on the subject of shock train and pseudo-shock phenomena in internal gas flows is given by Matsuo et al [1].

As a result from the past investigations, it is found that a shock train occurs as a result of the interaction between a normal shock wave and a boundary layer developing along the wall surface in a confined duct. In almost all the previous experimental investigations, shock trains were studied by wall pressure measurements and schlieren optical observation. Global properties of shock trains including the pressure rise due to the shock train, the distances between each shock in a shock train, the number of shocks constituting a shock train, and the longitudinal length of a shock train are significantly influenced by three main parameters just upstream of the shock train, i.e., the freestream Mach number, the Reynolds number based upon the boundary layer characteristic thickness such as the boundary layer thickness or boundary layer momentum thickness, and the blockage ratio or the confinement levels as characterized by the ratio of the boundary layer thickness to the duct half height in case of two-dimensional duct or duct radius in case of circular duct.

However, regardless of innovatory developments such as experimental apparatuses, flow measuring systems, powerful supercomputer with high speed and enormous memory, little is investigated about the detail of the density field within a shock train except for the study of Sugiyama et al [2], who investigated the internal structure of a shock-train with a freestream Mach number of 1.77 in a straight-square duct using a two-component dual-beam laser Doppler velocimeter (LDV) and the double exposure laser holographic interferograms. They showed that the central flow in the shock-train is considered to be the isentropic flow, while to be the adiabatic flow near the wall.

* Atsushi Matsuyama: x6mba021@eng.kitakyu-u.ac.jp

Another possible measuring technique to obtain the density field of shock train is considered to be the rainbow schlieren method [3]. Howes et al. [4] devised conventional schlieren methods using the filter with continuous color gradation instead of an usual tricolour filter used in colour schlieren techniques. Kolhe et al. [5] applied the rainbow schlieren deflectometry for supersonic micro jet flow issued from a orifice, and measured the flow characteristics of the shock-containing jet qualitatively and quantitatively. Yamamoto et al. [6] measured the density fields in the correctly expanded supersonic jets issued from an axisymmetric convergent-divergent nozzle with a design Mach number of 1.6. Takano et al. [7] measured the density field of an underexpanded sonic jet by the rainbow schlieren deflectometry combined with the computed tomography. However, almost all the studies by rainbow schlieren deflectometry have been applied for free jet issued from nozzles. To the authors' knowledge, Chaganti et al. [8] was for the first time applied a rainbow schlieren deflectometry for internal shock containing flows to understand the physics of the shock-wave boundary-layer interaction. However, their rainbow schlieren technique does not have accurate enough to describe the density field of the shock wave boundary layer interaction.

Many researchers often compare simulated schlieren or shadowgraph images directly with those from experiments having the same flow conditions in order to validate the simulation. In some cases, the validation of the simulation has been performed by comparing the simulated result with the experimental wall pressure distribution in the region where the shock train occurs. However, even though simulated schlieren picture or wall pressure is in good agreement with an experimental one, a whole internal flow characteristics including the shock train cannot always be captured correctly by the computation. Until recently, very little quantitative assessments of simulations with scalar or vector properties such as the temperature, density, pressure, and velocity have been available in the flow field of shock trains. Therefore, the aims of the present study are to provide the reliable experimental data of density fields for validating a variety of numerical simulation codes on shock trains and to clarify the details of shock train structure.

2 Experimental apparatus

Experiments were carried out in a blow down facility with a rainbow schlieren optical system of the High-Speed Gasdynamics Research Laboratory at the University of Kitakyushu. A schematic diagram of the experimental apparatus is shown in Fig 1. The air supplied by a compressor is filtered, dried and stored in a plenum chamber and it is discharged into the atmosphere through a test section. The pressure in the plenum chamber is automatically controlled and maintained constant during the testing by a solenoid valve. The operating pressure ratio of plenum to back pressures varies from 1.5 to 2.5.

Figure 2 shows a schematic diagram of a test section. The test section consists of a supersonic nozzle followed by a constant area two-dimensional duct. The nozzle is designed by the method of characteristics to provide uniform and parallel flow at the nozzle exit plane and its design Mach number is 1.5. The nozzle has heights of 4.24 mm and 5 mm at the throat and exit, respectively, and it is a constant width of 13 mm over the full length from the inlet to exit. A two-dimensional constant area duct with a height of 5 mm and a width of 13 mm is connected to the exit of the supersonic nozzle. The dashed line in Fig. 2 indicates the range of an optical window for flow visualization by the rainbow schlieren deflectometry.

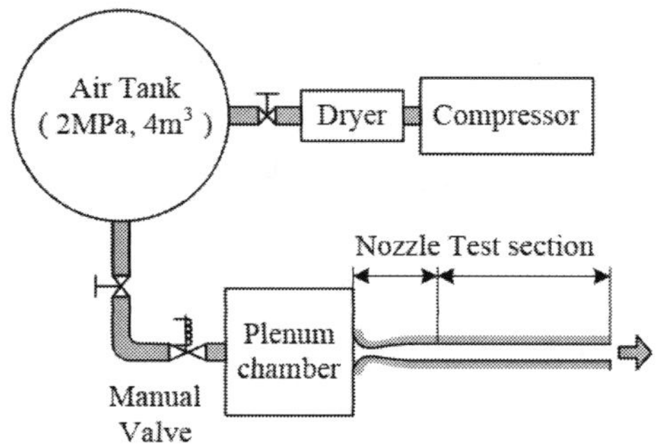

Fig 1. Schematic experimental apparatus

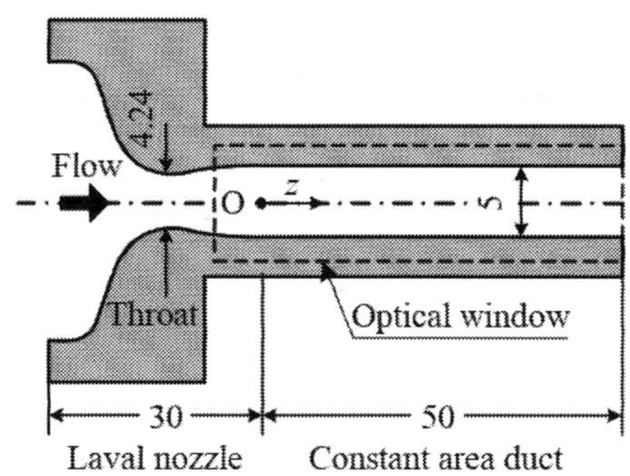

Fig 2. Schematic drawing of test section

3 Rainbow schlieren optical system

The present rainbow schlieren system consists of rail-mounted optical components including a 3 mm × 50 μm rectangular pinhole, two 100 mm diameter, 500 mm focal length achromatic lenses, a computer generated 23 mm wide slide with color gradation in a 1.4 mm wide strip, and a digital camera. A continuous LED light source is connected to a 50 μm diameter optical fiber.

Figure 3 shows a schematic drawing of a part of the present rainbow schlieren system including the test section, the decollimating lens, rainbow filter and digital camera. The parallel light rays coming from the collimating lens are passing through the test section, refracted at the decollimating lens, and then passing through the location of the background hue on the rainbow filter at the focal plane of the decollimating lens before reaching the digital camera. Rainbow schlieren methods are eliminate problems associated with inhomogeneous absorption of light by the test medium and nonlinearities in recording the image.

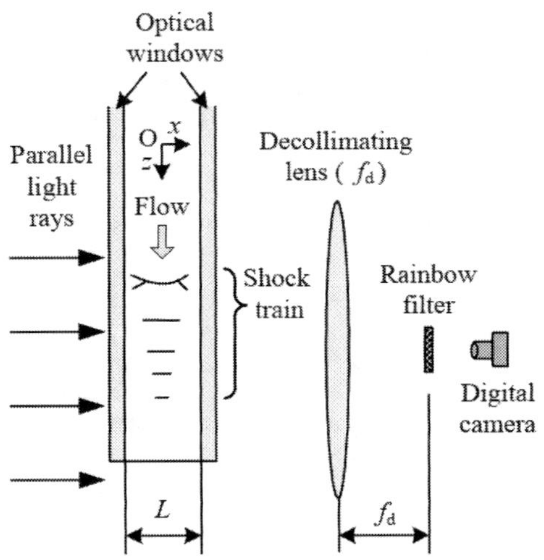

Fig 3. Sketch of rainbow schlieren system

The rainbow filter used in the present experiments is shown in Fig 4. It was fabricated in computer software and then printed digitally on a high resolution 23 mm color film recorder. It has continuous hue variation from *Hue* = 0 to 360 deg in a 1.4 mm wide strip. The characteristics of the rainbow filter were performed by traversing the filter in intervals of 0.05 mm in the z direction at the schlieren cut-off plane before starting experiments.

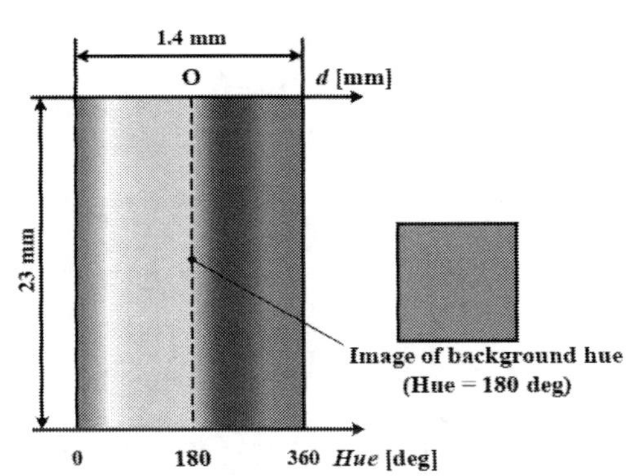

Fig 4. Rainbow filter

The calibration curve of the rainbow filter used in the present study is shown in Fig. 5 with the experimental data expressed as open symbols. The solid line indicates a least squares regression of the experimental data. As shown in Fig. 4, the background hue in the rainbow filter for the present experiment is *Hue* = 180 deg.

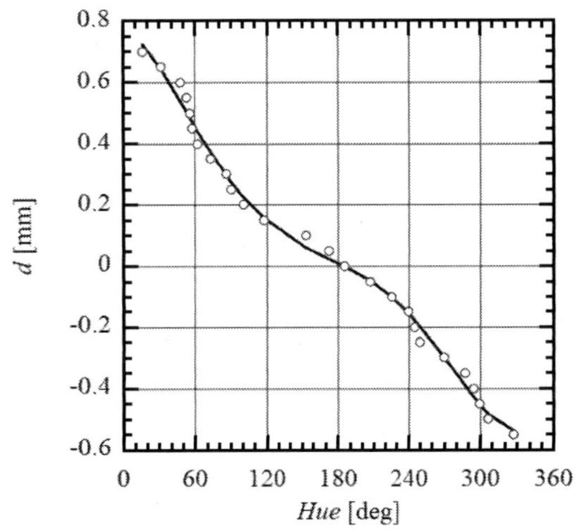

Fig 5. Calibration curve of rainbow filter

4 Method for obtaining density fields

If there is no air flow in the test section, the light rays from the collimating lens pass through the location of the background hue on the rainbow filter. On the other hand, if there is air flow with density gradients in the test section, some part of the light rays from the collimating lens is deflected when passing through the test section. The deflection angle ε of a light ray after passing through the test section is given by the following relation.

$$\varepsilon = \frac{d}{f_d} \tag{1}$$

where f_d is the focal length of the decollimating lens and d is the ray transverse displacement from the location of the background hue. The direction at which the light ray after passing through a refracted index field in the test section is travelling are not influenced by the existence of optical windows and it remains the same regardless of whether or not there are optical windows if the windows have uniform refractive index and both the surfaces of each window are parallel to each other. In the case the light ray from the test section is passing through the same location on the rainbow filter regardless of the existence of the optical windows because the rainbow filter is positioned at the focal plane of the decollimating lens. For the schlieren picture with the vertical rainbow filter corresponding to the vertical knife edge for the conventional gray scale schlieren, density gradients in the flow direction (z axis) after passing through the test section with a width L are given by the following relation.

$$\frac{\partial \rho}{\partial z} = \frac{\varepsilon}{KL} \tag{2}$$

$K = 2.26 \times 10^{-4}$ m^3/kg is Gladstone-Dale constant, L = 13 mm is the duct width over the span direction as shown in Fig. 3. As K hardly depends on the wavelength of light, the same constant is used for all wavelengths. The integration of Eq. (2) with respect to the z yields the density distribution within the shock trains.

$$\rho(z,y)=\frac{1}{KL}\int^{z}\varepsilon dz+C(y) \tag{3}$$

where $C(y)$ is the integral constant and it can be obtained from boundary conditions as follows. There are two boundary conditions to specify the $C(y)$ of Eq. (3). One is the density just upstream of the shock train and the other is that at the duct exit when the exit flow is subsonic.

For the former boundary condition, the density upstream of the shock train is specified under the assumption of isentropic flow from the plenum chamber to the nozzle exit. In case of the latter boundary condition, the Mach number at the duct exit can be derived from the equation of continuity between the nozzle throat and the duct exit under the assumption that the flow is isentropic from the plenum chamber to the throat and adiabatic from the throat to the duct exit.

From the assumption described above, a relation on the exit Mach number can be derived

$$\frac{p_{os}}{p_b}\frac{A^*}{A_e}=\left(\frac{\gamma+1}{2}\right)^{\frac{\gamma+1}{2(\gamma-1)}}M_e\left(1+\frac{\gamma-1}{2}M_e^2\right)^{\frac{1}{2}} \tag{4}$$

The equation of state at the duct exit gives

$$\rho_e=\frac{p_e}{RT_e} \tag{5}$$

where p_e is equal to the back pressure p_b because of subsonic flows at the duct exit and the exit temperature T_e is obtained from the condition of adiabatic flow $T_{os}=T_b$ through the whole flow field including the shock train and M_e from Eq. (4), i.e.,

$$\frac{T_{os}}{T_e}=1+\frac{\gamma-1}{2}M_e^2 \tag{6}$$

Combination of Eqs. (3) ~ (6) yields the density field within the shock train. In the present study, the density upstream of the shock train was specified as the boundary condition.

5 Results and discussion

5.1 Rainbow schlieren visualization

A typical rainbow schlieren picture of a shock train is presented in Fig. 6 with the flow from left to right. The picture was taken with a vertical rainbow filter to visualize the density variation in the flow direction where the plenum pressure p_{os} is 185.6 kPa, the back pressure p_b = 103.1 kPa is equal to the atmospheric pressure, and the atmospheric temperature T_b is 291.9 K.

The boundary layer thickness δ_1 just upstream of the shock train can be estimated from relations shown below. For compressible boundary layers on a flat-plate,

$$\frac{\delta_1^*}{\delta_1}\equiv\int_0^1\left(1-\frac{\rho_1 u_1}{\rho_{1\infty}u_{1\infty}}\right)d\eta \tag{7}$$

where η denotes the normal distance from the duct wall normalized by δ_1.

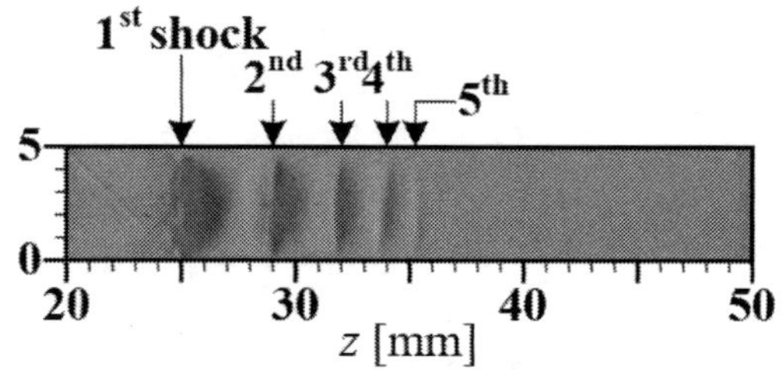

Fig 6. Rainbow schlieren picture

From the assumptions of constant total temperature over the whole flow field and constant static pressure in the lateral direction at just ahead of the shock train, the density ratio inside and at the edge of the boundary layer is obtained as follows:

$$\frac{\rho_1}{\rho_{1\infty}}=\frac{1}{1+\frac{\gamma-1}{2}M_{1\infty}^2\left[1-\left(\frac{u_1}{u_{1\infty}}\right)^2\right]} \tag{8}$$

For the boundary layer just ahead of the shock train, the one-nth power velocity profile is assumed by

$$\frac{u_1}{u_{1\infty}}=\begin{cases}\eta^{1/n} & for\ \ 0\le\eta\le1\\ 1 & for\ \ 1\le\eta\le H/2\end{cases} \tag{9}$$

where H is the duct height.

Although the power law index n generally depends on the freestream Mach number and Reynolds number, a value of 7 is chosen for n.

Substitution of Eqs. (5), (8) and (9) into Eq. (7) leads to [9]

$$\frac{\delta_1^*}{\delta_1}=1-7(D-1)\times\left[\frac{D^3}{2}\ln\left(\frac{D}{D-1}\right)-\frac{D^2}{2}-\frac{D}{4}-\frac{1}{6}\right] \tag{10}$$

where

$$D\equiv1+\frac{2}{(\gamma-1)M_{1\infty}^2} \tag{11}$$

A value of δ_1 = 0.80 mm for the present experiment is obtained.

Just upstream of the shock train, the freestream Mach number $M_{1\infty}$ is 1.34, the unit Reynolds number Re/m is 5.39 x 10^7, the boundary layer displacement thickness δ^*_1 is 0.149 mm where the δ^*_1 was obtained from the freestream and the equation of continuity between the nozzle throat and the location just upstream of the shock train under the assumption of isentropic flow from the plenum chamber to the location, and zero pressure gradient over the duct section at the location. The picture of Fig. 6 shows the shock train positioned near the nozzle exit where the boundary layer is thinnest. However, a shock train appears as a result of the interaction between a normal shock wave and boundary layer. The positions of each shock constituting the shock train are indicated in Fig. 6 as the down-pointing arrows. The first shock wave is bifurcated at the near the wall surface, while following secondary shock waves are not. The five shock waves can be observed in the rainbow schlieren picture.

5.2 Density distribution

A density gradient contour map of a two-dimensional flow field including a shock train was obtained from the schlieren picture of Fig. 6 with the calibration curve of Fig. 5 and the results is presented in Fig. 7. This picture quantitatively captures several features of a shock train. A shock train consisting of around five shocks is clearly seen where the first shock wave is bifurcated at the near the wall surface, while following secondary shock waves are not. The interval between successive downstream shocks decreases.

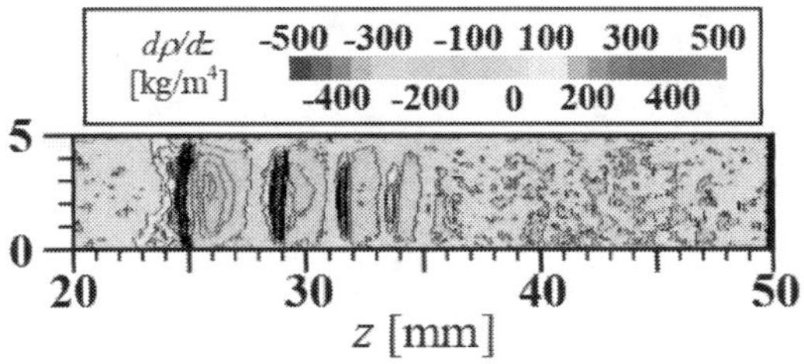

Fig 7. Density gradient contour map

The centerline density distribution in the shock train is indicated in Fig. 8 where the downward pointing arrows show the locations of each shock in the shock train. Figure 8 shows that a successive sudden increase in the density is due to the presence of the successive shocks constituting the shock-train. The steepest slope in the density is observed at the location of the first shock and the slopes of the following shocks gradually decrease in the downstream direction because the strength of the shock decreases toward the downstream.

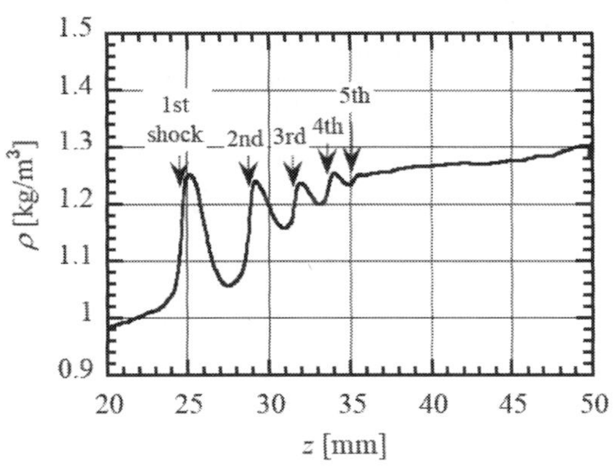

Fig. 8. Streamwise density distribution

The density abruptly increases with a density spike at the location just behind of the first shock and then decreases soon before increase again by the second shock and a similar density variation is repeated between the successive downstream shocks. It is found that the rainbow schlieren deflectometry is an implement pertinent to obtain the fine structure of the shock train.

6 Concluding remarks

In the present study, we applied a rainbow schlieren deflectometry for the internal flow including a shock train. As a result, the centerline density distribution along the streamwise direction exhibits a wavy form with several peaks, corresponding to each shock wave constituting the shock train. The two-dimensional density field of a shock train is quantitatively shown as a density gradient contour plot. Also, the rainbow schlieren deflectometry is effective to observe flow-fields of two-dimensional shock wave boundary layer interaction quantitatively. The rainbow schlieren deflectometry appears to be a very promising approach for two-dimensional quantitative density measurements in a shock containing flowfield.

Acknowledgements

The authors would like to gratefully acknowledge the assistance of graduate students Hirofumi TAKANO and Taishi TAKESHITA of the University of Kitakyushu for their invaluable support to the experimental work and exceptional skills in the fabrication of the facility.

NOMENCLATURE

A cross-sectional area, mm^2
C integral constant
d displacement of light ray, mm
f_d focal length of decollimating lens, mm
K Gladstone-Dale constant, m^3/kg
L spanwise width of duct, mm
M Mach number
p static pressure, Pa
R gas constant, J/kg·K
Re/m unit Reynolds number
T temperature, K
γ specific heat ratio of gas
δ^* boundary layer displacement thickness, mm
δ boundary layer thickness, mm
ε deflection angle of light ray, deg
ρ density, kg/m^3

Subscripts
b value at surrounding air
e duct exit
os total or stagnation state
1 just upstream of shock train
∞ freestream outside boundary layer

Superscripts
* sonic state or throat state

References

1. Matsuo, K., Miyazato, Y., and Kim, H. D., (1999): "*Shock Train and Pseudo-Shock Phenomena in Internal Gas Flows*", Progress in Aerospace Sciences, Vol. **35**, No. 1, pp. 33-100.
2. Sugiyama, H., Arai, T., Abe, H., Takahashi, T., and Takayama, K., (1990): "*Flow Mechanism of λ-type Pseudoshock Waves in a Straight-Square Duct*", Bull, JSME **56**(522), B, 330-335.
3. Settles, G., S., (2001): "*Schlieren and Shadowgraph Techniques*",Springer.
4. Howes, L. W., (1984): "*Rainbow Schlieren and Its Applications*", Applied Optics, 23-14, pp. 2449-2460.
5. Kolhe, P. S., and Agrawal, A. K., (2009): "*Density Measurements in a Supersonic Microjet Using Miniture Rainbow Schlieren Deflectometry*", AIAA J, Vol. **47**, No. 4, pp. 830-838.
6. Yamamoto, H., Irie, M., Miyazato, Y., and Matsuo, K., (2010): "*Application of Rainbow Schlieren Deflectometry for Axisymmetric Supersonic Jets (Comparison of Experiments with Numerical Analysis)*", J. Thermal Sciences, Vol. **19**, No. 3, pp. 218-221.
7. Takano, H., Kamikihara, D., Ono, D., Nakao, S., Yamamoto, H., and Miyazato, Y., (2016): "*Three-Dimensional Rainbow Schlieren Measurements in Underexpanded Sonic Jets from Axisymmetric Convergent Nozzles*", J. Thermal Science, Vol. **25**, No. 1, pp. 78-83.
8. Chaganti, N., Kurup ,A.L., and Ölçmen, S., (2013): "*Study of Unsteadiness of Shock Wave Boundary Layer Interaction Using Rainbow Schlieren Deflectometry and Proper Orthogonal Decomposition*", AIAA Paper, No. 2013-0536.
9. Tamaki, T., Tomita, Y., and Yamane, R., (1970) : "*A Study of Pseudo-Shock (1st Report, l-type Pseudo-Shock)* ", Bulletin of the JSME, Vol. **13**, No. 55, pp. 51-58.

Experimental Characterization of the Jet Wiping Process

Miguel Alfonso Mendez[1,⋆], *Adriana* Enache[1], *Anne* Gosset[2], and *Jean-Marie* Buchlin[1]

[1] *EA Dep., von Karman Institute for Fluid Dynamics, Sint-Genesius-Rode, Belgium*
[2] *Naval and Oceanic Engineering Department, University of A Coruña, Ferrol, Spain*

Abstract. This paper presents an experimental characterization of the jet wiping process, used in continuous coating applications to control the thickness of a liquid coat using an impinging gas jet. Time Resolved Particle Image Velocimetry (TR-PIV) is used to characterize the impinging gas flow, while an automatic interface detection algorithm is developed to track the liquid interface at the impact. The study of the flow interaction is combined with time resolved 3D thickness measurements of the liquid film remaining after the wiping, via Time Resolved Light Absorption (TR-LAbs). The simultaneous frequency analysis of liquid and gas flows allows to correlate their respective instability, provide an experimental data set for the validation of numerical studies and allows for formulating a working hypothesis on the origin of the coat non-uniformity encountered in many jet wiping processes.

1 Introduction

In the jet wiping process, an impinging gas jet is used to control the thickness of a liquid coating dragged by a moving substrate. The substrate is typically a large sheet withdrawn vertically from a bath at a constant speed so that the removed liquid runs back by gravity. This process is widely used in photographic film manufacturing, paper industry and in the hot-dip galvanizing process [1–4].

Because of its industrial relevance, the development of jet wiping lines has been supported by a large number of analytical and experimental investigations, which led to the formulation of several simplified physical models [5–9]. These models provide the engineering tools to predict the final coating thickness as a function of all the operating parameters and therefore provide the framework for the automatic control of these lines [10–12].

The drawback of the method, on the other hand, is the limited surface quality achievable, which constitute a major concern for applications demanding high surface finish quality such as, for example, automotive industry. Understanding the mechanisms responsible for the coating nonuniformities requires a time-resolved characterization of the interaction between the gas jet flow and the liquid film, up to day only achieved by means of numerical simulations [13–15] or simplified laboratory models of the wiping process [16–18]. Most of the experimental works on the jet wiping, in fact, have been focused on average quantities to validate physical models [8, 19, 20] or in the definition of operational maps to avoid the liquid film break up, a phenomenon known as splashing [20, 21].

This work provides a time resolved experimental characterization of the jet wiping process combining Time Resolved Particle Image Velocimetry (TR-PIV) on the gas jet with an automatic tracking of the liquid interface and Time Resolved Light Absorption (LAbs) thickness measurements of the final coat.

The parameters and the dimensionless quantities governing the process are described in Section 2, recalling the operative range of the method in several industrial applications. The experiments were conducted in the Ondule laboratory, developed at the von Karman Institute (VKI) to simulate the jet wiping process. The facility and its instrumentation are presented in Section 3.

The data processing of each measurement technique is presented in Section 4. These include the image preprocessing by means of an automatic masking algorithm to track the gas liquid interface from the TR-PIV images (Section 4.1, [18, 22–24]), the gas flow field modal analysis via Proper Orthogonal Decomposition (POD, Section 4.2, [25, 26]) and the 3D film thickness reconstruction of the final coat (Section 4.3, [23, 27–29]). Each of these techniques is used to construct a characteristic frequency spectrum for the gas jet flow, the final coat and the liquid layer removed in the wiping process, referred to as runback flow. These spectra are presented and discussed in Section 5; the conclusions are drawn in Section 6.

2 Problem Formulation and Scaling Laws

A schematic of the jet wiping process in its basic configuration is proposed in Figure 1, recalling the process parameters: the substrate speed U_s, the nozzle stagnation pressure ΔP_n, the nozzle opening d and the nozzle stand-off distance Z. The relevant liquid properties are density ρ_l, dynamic viscosity μ_l and surface tension σ_l.

This section aims at presenting the dimensionless numbers and the reference quantities used to scale the

⋆e-mail: mendez@vki.ac.be

measurements presented in this work and to discuss the similarity between the investigated configuration and the ones encountered in different industrial processes. These scaling parameters are inferred from the formulation of the simplified models used in the literature of the jet wiping process.

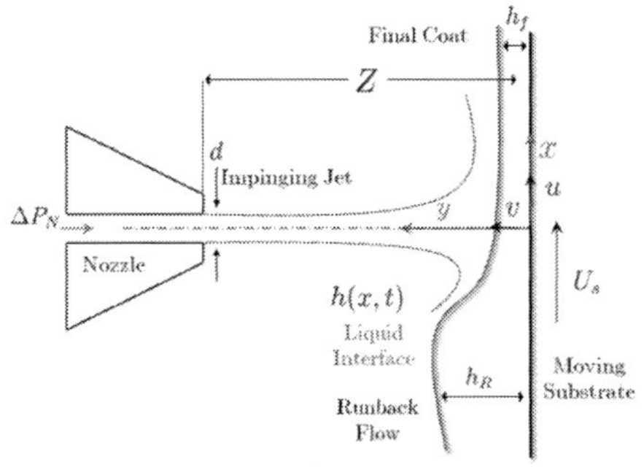

Fig. 1. Schematic of the jet wiping process: a nozzle at a pressure ΔP_N releases a jet flow at a distance Z from a substrate moving at a speed U_s. The impingement produces a run-back liquid layer of thickness h_R and a thinner final coat of thickness h_f.

Simplified models build from the governing equations of the thin liquid film flow, considered an incompressible flow with constant properties, under the assumption of lubrication theory ($\partial_x h \to 0$, $v \to 0$, $\partial_x \to 0$) and steady state ($\partial_t \to 0$) conditions. It is thus assumed that the liquid film has a boundary layer behavior ($\partial_y p \to 0$) and that the coupling between the two phases is one-way: the pressure and the shear stress distribution produced at the wall by the impinging gas jet are not influenced by the liquid film, and are therefore inferred from well established experimental correlations obtained for jet impinging on a dry wall. Under these conditions, and neglecting the effects of surface tension, the film thickness at each stream-wise coordinate must satisfy the following dimensionless cubic equation [6–9]:

$$\left(\partial_{\hat{x}}\hat{P}_g + 1\right)\hat{h}^3 - \frac{3}{2}\hat{T}_g\hat{h}^2 - \hat{h} + 2\hat{q} = 0\ , \qquad (1)$$

where $\hat{h} = h/h_0$ is the dimensionless film thickness, with $h_0 = \sqrt{\nu U_s/g}$ the reference thickness; $\partial_{\hat{x}}\hat{P}_g = \partial_x \hat{P}_g/(\rho_l g)$ and $\hat{T}_g = T_g/\sqrt{\mu_l \rho_l U_s g}$ are the dimensionless pressure gradient and shear stress distributions produced by the impinging gas jet; $\hat{q} = 2q/3\sqrt{\nu U_s^3/g}$ is the net dimensionless flow rate remaining in the final coat. Simplified physical models differ in the degree of simplification involved in the solution of eq.(1), which can include the surface tension [8, 30], neglect the shear stress contribution [5], or force the simplification by assuming that the maximum pressure gradient ($\partial_{\hat{x}}\hat{\mathcal{P}}_g$) and the maximum shear stress ($\hat{\mathcal{T}}_g$) act in the same location [8]. In this case, assuming that the system operates in optimal conditions ($\partial_h q = 0$), the final thickness $\hat{h}_f$ obeys a simple equation:

$$\hat{h}_f = \frac{\hat{\mathcal{T}}_g + \sqrt{\hat{\mathcal{T}}_g^2 + 4(\partial_{\hat{x}}\hat{\mathcal{P}}_g + 1)}}{2(\partial_{\hat{x}}\hat{\mathcal{P}}_g + 1)}\ . \qquad (2)$$

The maximum dimensionless pressure gradient, from the correlations in [31, 32], reads

$$\partial_x \mathcal{P}_g = \left(\frac{0.7141}{b}P_M\right)\frac{1}{\rho_l g}\ , \qquad (3)$$

where

$$P_M = 6.5\Delta P_n \frac{d}{Z} \qquad (4a)$$

$$b = \left[0.0019\left(\frac{Z}{d}\right)^2 + 0.0551\left(\frac{Z}{d}\right) + 0.4035\right]d\ . \qquad (4b)$$

Similarly, the maximum dimensionless shear stress is:

$$\hat{\mathcal{T}}_g = -\left(P_n C_\tau \frac{d}{Z}\right)\frac{1}{\sqrt{\mu_l \rho_l U_s g}}\ , \qquad (5)$$

where the constant C_τ is a function of the Reynolds number $Re_J = U_J d/\nu_g$ of the gas jet:

$$C_\tau = \begin{cases} 0.1287 - 10^{-5}Re_j & \text{if } Re_J \leq 6170 \\ 0.067 & \text{if } Re_J > 6170\ . \end{cases} \qquad (6)$$

The validity of this approximated model for predicting the final thickness is limited to the case of strong wiping ($\partial_x \hat{P}_g \gg 1$), as discussed in [8] and [30]. In the regimes of weak wiping ($\partial_x \hat{P}_g \sim 1$), where surface tension becomes important, this model is known to significantly over-predict the wiping capabilities of the jet flow. Moreover, the extreme simplifications involved in the derivation of eq.(2) do not allow for investigating the jet wiping process in the dynamic conditions, i.e. including both splashing occurrence [20] and film instability [36]. Yet, this physical model is indicative of the impact of the jet parameters on the liquid flow. In this work it is used as a benchmark and to derive the scaling quantities for the non-dimensionalization of the measurements. These dimensionless numbers, moreover, allows for investigating the flow similarity between the laboratory conditions and the conditions of several industrial processes.

It is interesting to observe, for example, that under the assumption leading to eq.(2), the thickness of the run-back flow $\hat{h}_R$ is linked to the final thickness $\hat{h}_f$ by mass conservation. In the far field conditions ($x \to \pm\infty$), in fact, where $P_g \to 0$ and $T_g \to 0$, eq.(1) reduces to a universal cubic equation, shown in Fig.2 for an exemplary case. This cubic equation results from the balance of viscous and gravitational forces and has two admissible solutions for a given flow rate $\hat{q}$. Fixing the first, corresponding to the final thickness $\hat{h}_f$ in eq.(2), the second is easily computed numerically. This thicker solution corresponds to the run-back flow thickness $\hat{h}_R$.

In absence of jet wiping, then, as only one solution is possible, the coating thickness withdrawn by the plate

Table 1. Indicative operation range of the jet wiping processes in several industrial configuration, including the VKI jet wiping laboratory *Ondule* used in this work. The references from which the data are taken (or inferred from) are indicated in each entry. The final coating thickness and the run-back flow thickness obtained using the 0D wiping model described in Section 2.

	Dimensional Quantities (Indicative values)							
	ρ_l $[g/cm^3]$	μ $[cP]$	U_s $[m/min]$	U_J $[m/s]$	d $[mm]$	Z $[mm]$	h_f $[mm]$	h_R $[mm]$
Ondule Laboratory	1.02	50 − 100	6 − 30	20 − 40	1.3	13 − 20	0.2 − 0.5	1.5 − 2.5
Photographic Film Coater [1, 3, 33]	1 − 1.1	1 − 40	10 − 100	15 − 30	0.5 − 1	2 − 10	0.05 − 0.2	0.4 − 3
Paper Coater [2, 34, 35]	0.8 − 1.5	5 − 300	400 − 700	50 − 150	3 − 10	20 − 70	0.2 − 1	4 − 15
Galvanizing Line [4–6]	6.54	2.5 − 3.5	50 − 150	100 − 250	1 − 1.5	6 − 20	0.01 − 0.07	0.3 − 0.5

Table 2. Reference dimensionless quantities for the ranges presented in Table.1. The dimensionless pressure gradient and shear stress are computed using the correlations in (3) and (5). The final coating thickness $\hat{h}_f$ and the run-back $\hat{h}_R$ are scaled with respect to the optimal thickness $h_0 = \sqrt{\nu_l U_s/g}$ and the film thickness ratio is defined in eq.(7). The Reynolds number in the jet flow and the liquid film are defined as $Re_J = U_J d/\nu_g$ and $Re_L = U_s h_0/\nu_l$.

	Dimensionless Quantities						
	$\partial_x \mathcal{P}^*$	$\mathcal{T}^*$	$\hat{h}_f$	$\hat{h}_R$	ζ	$Re_l\,(\times 10^3)$	$Re_J\,(\times 10^3)$
Ondule Laboratory	10 − 30	0.4 − 1	0.2 − 0.3	1.5 − 1.6	1.2 − 1.4	0.001 − 0.01	2.5 − 3.5
Photographic Film Coater [1, 3, 33]	35 − 40	0.2 − 3	0.15 − 0.2	1.65 − 1.7	1.4 − 1.5	0.01 − 0.04	1 − 1.5
Paper Coater [2, 34, 35]	6 − 400	0.2 − 2.5	0.05 − 0.4	1.5 − 1.7	1.1 − 1.6	0.4 − 5	10 − 300
Galvanizing Line [4–6]	30 − 600	2 − 40	0.08 − 0.23	1.6 − 1.65	1.3 − 1.6	0.3 − 0.9	10 − 25

is[1] the one corresponding to the maximum flux. Since this point is considered to build the reference quantities, in such conditions one recovers $\hat{q} = 1$ and $\hat{h} = 1$.

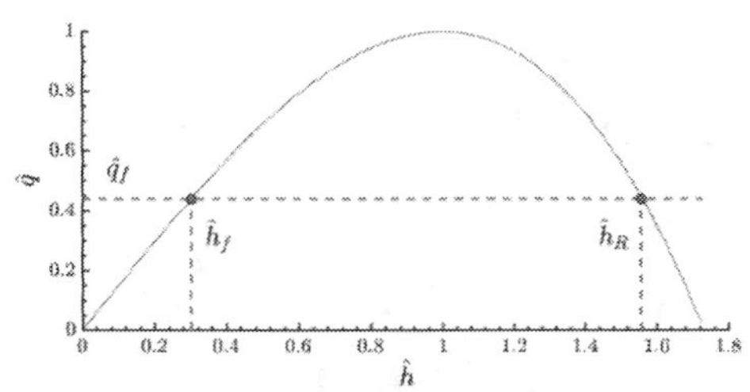

Fig. 2. Cubic relation between coating thickness $\hat{h}$ and flow rate $\hat{q}$ in the far field conditions for eq.(1). For a given flow rate, two possible solutions are possible. The first corresponds to the final thickness $\hat{h}_f$ (estimated from eq.2); the second corresponds to the run-back flow thickness $\hat{h}_R$.

The relevant dimensional and dimensionless parameters for the experimental set up used in this work are collected in Tables 1 and 2, together with those encountered in typical industrial configurations such as photographic film manufacturing [1, 3, 33], paper industry [2, 34, 35] and galvanizing lines [4–6]. In addition to the aforementioned dimensionless quantities, the table includes the Reynolds number in the liquid film $Re_L = U_s h_0/\nu_l$ and the wiping thickness aspect ratio ζ, defined as:

$$\zeta = \frac{h_R - h_f}{h_0} \tag{7}$$

Although only indicative, these ranges give an overview of the variety of operating conditions in which the jet wiping process is used: from the extremely high jet velocity used in the galvanizing lines, to the large viscosities encountered in a paper coater and the thinnest coatings used in photographic film manufacturing. Yet, there exists a wide range of dimensionless film thickness which overlaps among all the configurations and leads to similar average thickness profiles. The Ondule facility, presented in Section 3, was designed at the von Karman Institute (VKI) to cover this range while operating in conditions for which the nonintrusive experimental techniques presented in Section 3 and Section 4 give adequate accuracy and resolution.

3 The VKI Ondule Facility

The jet wiping facility Ondule is described in Fig.3a. In this simplified laboratory model, the rising substrate is simulated by a cylinder (1) in acrylic glass with 450 mm diameter and 350 mm width. This cylinder is dipped in a liquid bath (2) for about 5 cm and it is driven by an electrical motor rotating at speed in the range 5 − 20 rpm ($U_s \approx 0.1 - 0.45\,m/s$). A doctoring blade (3) removes the coating layer before immersion.

[1] This results, independent of surface tension, is valid for sufficiently high Reynolds and Capillary numbers. For a discussion on the free-drag out problem the reader is referred to [6, 37–39].

https://doi.org/10.1051/epjconf/201818002064

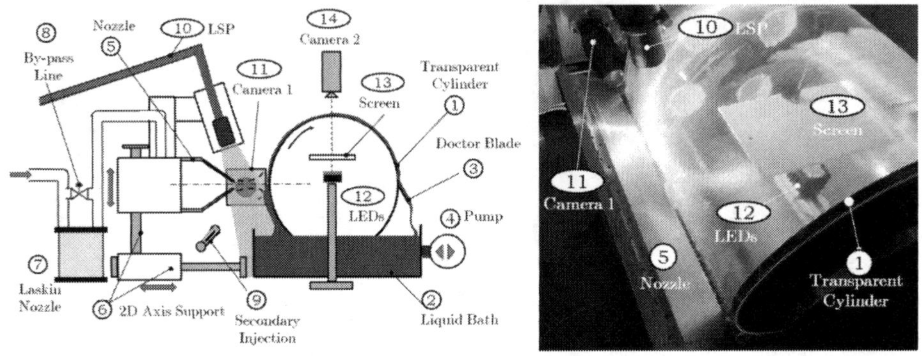

a) b)

Fig. 3. a) sketch of the VKI Ondule facility used to reproduce the jet wiping process in Fig.1, along with the required hardware for the TR-PIV and TR-LAbs measurements. Figure b: photograph of the test section showing some of the key components listed in Figure a: the nozzle (5), the PIV camera (11) and the optical arm (10), the transparent cylinder (1) hosting the LEDs (12) and the screen (13) for the liquid film back-lightning.

The working liquid is Dipropylene Glycol (ρ_l = 1026± 1% Kg/m^3, μ = 50 ± 2% cP, σ = 32 ± 1% N/mm), introduced in the bath by a volumetric pump Pollard PR40 (4) at the beginning of the test. The nozzle (5) releasing the gas jet is installed on a two axis mechanism (6) to adjust the impingement position and the stand-off distance.

To perform TR-PIV and flow visualization, the nozzle is supplied by a Laskin Nozzle PIVTEC45-M (7 in Fig.3a) operating with mineral oil Ondina Shell 917 to provide seeding particles. An additional bypass line (8) allows to control the seeding density in the jet flow and to supply in parallel a secondary injection (9) positioned below the jet flow to maximize the seeding uniformity. The light source is a dual diode-pumped ND:YLF laser (Quantronix Darwin Duo 527), equipped with a light sheet probe LSO from ILA (10) to produce the laser sheet. The interaction between the jet flow and the liquid film is recorded by a CMOS Photron FASTCAM SA1 (11) positioned on one side of the facility at a distance of about 16cm from the investigated plane. This camera is equipped with a zoom objective Tamron CZ-735 70 – 300mm, providing a scaling factor of 20 ± 0.1$pixel/mm$. The video size is 5.5GB, corresponding to n_t = 21000 images (10500 image pairs) with a resolution of 512 × 512 pixels, taken at a frequency of f_s = 3 kHz. In order to enhance the interface visibility via laser induced fluorescence, Rhodamine B is diluted in the liquid with a concentration of ≈ 10 mg/Kg. An exemplary image of the TR-PIV acquisition is shown in Fig.4.

To perform TR-LAbs, a diffuse light source is installed inside the transparent cylinder. This consists of two arrays of 4 × 4 LEDs of ≈ 0.7 W each, with emittance in the range of 636 ± 26 μm (12 in Fig.3a). These LEDs are installed behind a 3 mm thick sheet of Opaline acrylic glass (13). The receiver (14) is a rolling shutter 16 $bits$ CMOS camera (Hamamatsu ORCE-Flash 4.0) synchronized with the LEDs, to acquire 2000 back-lighting images at 300 Hz with a resolution of 660 × 1800 pixels. This camera is installed perpendicularly to the test section, at about 40cm from the cylinder. It is equipped with a 35 mm Nikkor-s objective, yielding a scaling factor of 13.4 ± 0.3 $pixel/mm$, and a band-pass filter (band pass range 640 ± 40 μm) to limit the camera sensitivity in the wavelength range where the light absorption takes places. To maximize and match the absorption spectra of the liquid with the emission spectra of the light, a methylene blue colorant is diluted in the liquid, with a concentration of 21 mg/Kg. The definition of the optimal colorant concentration for minimizing the measurement uncertainties is presented in [29, 40]. An exemplary image of the TR-Labs acquisition is shown in Fig.5 (top), together with the corresponding reference image (bottom) of the dry test section.

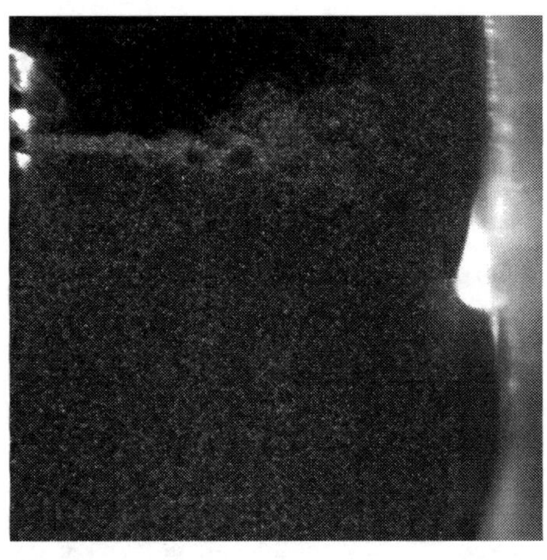

Fig. 4. Exemplary TR-PIV image of the gas jet liquid film interaction. The liquid film visibility is enhanced by the laser induced fluorescence. Both seeding injections (4,5 in Fig.3 a) are active, producing a uniform seeding.

A picture of the test section with some of the key components for both measurement techniques is shown in Fig.3b.

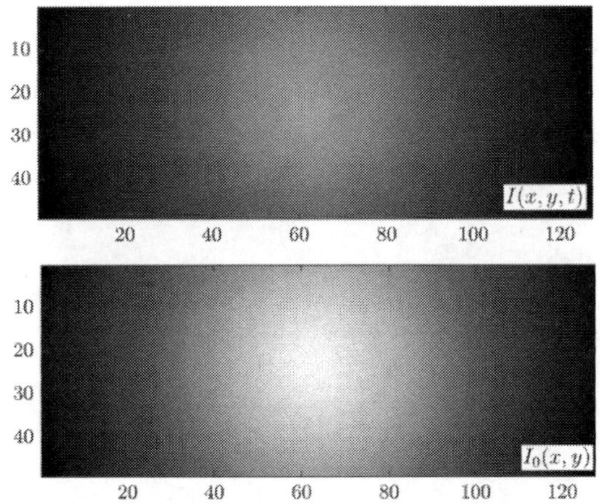

Fig. 5. Exemplary TR-LAbs image of the final coat (top) and corresponding reference image (bottom).

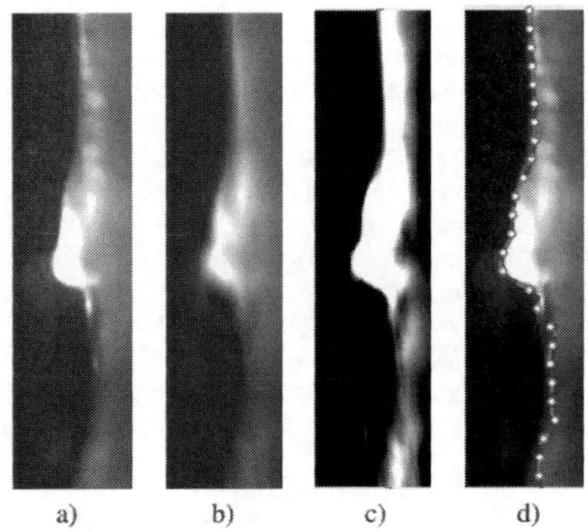

Fig. 6. Image processing steps for the automatic liquid interface tracking. Each image is cropped in the portion displaying the liquid (a), and filtered using a POD background removal algorithm (b) to remove the PIV particles. The result is sharpened via Gaussian Kernel (c) and the interface retrieved by gray scale gradient analysis (d).

4 Data Processing

4.1 Runback Flow Analysis and Masking

The liquid interface is monitored from the PIV images using an image processing routine developed in *Matlab*. This routine consists of four steps. First, the PIV images are cropped around the liquid interface, as shown in Fig.6, for the example image presented in Fig.4. Second, the set of images is filtered by means of the POD-based masking procedure described in [23, 24]. This filter consists of decomposing the video sequence via Proper Orthogonal Decomposition (POD) and keeping only a small fraction of the dominant modes, in order to capture the motion of large objects while removing the uncorrelated contribution of the particle pattern. The result of this step is the complete removal of the PIV particles from the images, as shown in Fig.6b. This operation significantly simplifies the interface detection by means of standard edge enhancement methods, performed by using a Gaussian sharpening kernel (Fig.6-c) and followed by the peak detection of the gray scale profiles along the image horizontal lines. The result is finally filtered by means of a Savitzky–Golay smoothing filter [41]. For the exemplary image sample proposed, the interface is shown in Fig.6-d.

The interface detection in each time step is used to assembly the characteristic map $h(x, t)$, describing the propagation of interface disturbances in space and time. An example of such map is shown in Fig.7a. This map displays the characteristic lines along which the waves in the impinged liquid film propagates. The automatic detection of these lines and their slope, related to the propagating wave speed, is carried out by means of Hough Transform as described in [29]. The characteristic spectrum of the run back liquid film is constructed by choosing a location in space x^* (indicated in Fig.7a) and computing its spectra $\mathcal{H}_r(f)$ using the Welch's spectral density estimation [42, 43] with 5 windows and 30% overlap. A typical result is shown in Fig.7b.

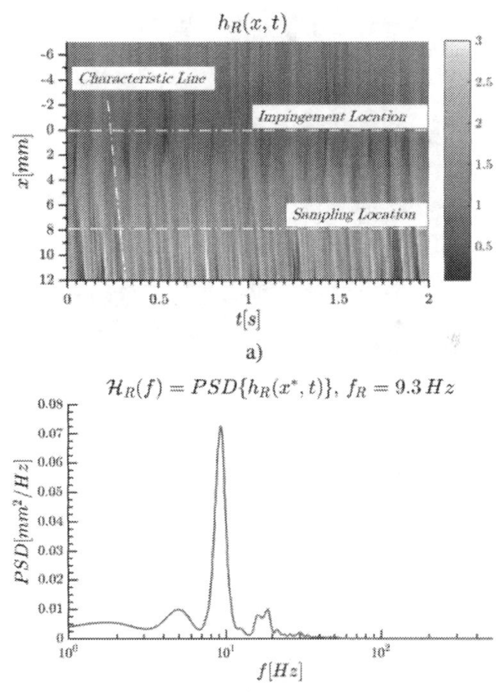

Fig. 7. Exemplary characteristic map of the liquid interface retrieved from the TR-PIV images. This map shows the characteristic lines of the wave propagation and it is used to compute the characteristic spectrum of the run back liquid film, shown in b).

4.2 Jet Flow Analysis via Modal Decomposition

The PIV images are first masked using the interface detection described in the previous section and then pre-processed to increase the particle-background contrast. This is achieved combining the POD-based background removal in [24] with the intensity capping [44] and the CLAHE histogram equalization [45]. The PIV interrogation is carried out with the open-source *Matlab* routine PIVlab [22], using a standard iterative multi-step interrogation (three passes from 64×64 to 8×8) with spline window deformation and 2D Gaussian sub-pixel interpolation. To increase the flow coherence, limit the random noise and the size of the dataset, the original vector field $\mathbf{u}_o = (u_o, v_o)$ is low pass filtered and down-sampled with a moving average kernel of size 3:

$$\mathbf{u}[i,j,k] = \frac{1}{3} \sum_{p=3k-2}^{p=3k} \mathbf{u}_o[i,j,p] \; . \tag{8}$$

The vector validation is carried out both before and after the down-sampling by setting a threshold on the histogram standard deviation and using the median test [46]. The resulting dataset consists of $n_t = 3500$ velocity fields having $n_x \times n_y = 63 \times 63$ vectors, with a spatial resolution of about $\Delta x = 0.35mm$, and a time resolution of $\Delta t = 1ms$. An examplary velocity field is shown in Fig.8.

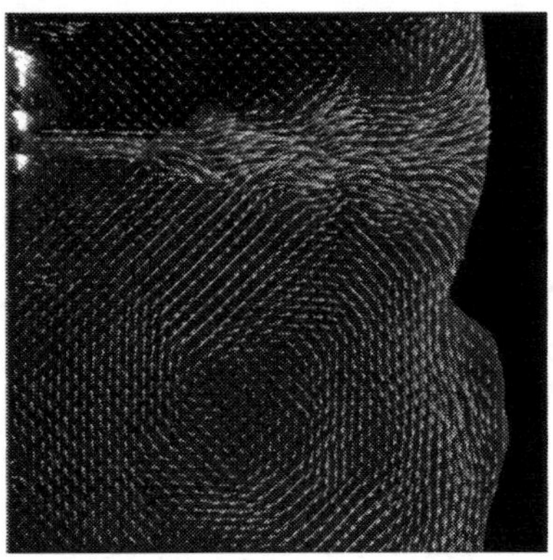

Fig. 8. Down-sampled velocity field with automatic masking of the liquid film interface. The spatial and temporal resolutions are 0.35 *mm* and 1 *ms* respectively.

The frequency response of the gas flow field is analyzed from the spectral content of the dominant POD modes (see also [25, 26]) of the down-sampled velocity field.

That is, the zero mean shifted data set $\mathbf{u}' = \mathbf{u} - 1/n_t \sum_{k=1}^{n_t} \mathbf{u}_k$ is decomposed into a linear combination of modes composed of a spatial structure $\phi_r[i,j]$, a temporal evolution $\psi_r[k]$ and an energy content σ_r:

$$\mathbf{u}'[i,j,k] = \sum_{r=1}^{n_t} \sigma_r \phi_r[i,j] \psi_r[k] \; . \tag{9}$$

By definition, the spatial structures ϕ_r are eigenvectors of the spatial correlation C matrix; the temporal evolutions ψ_r are eigenvectors of the temporal correlation matrix K; the energy contents σ_r are the square roots of the common eigenvalues of C and K. These matrices are defined as

$$K[i,j] = \frac{1}{2n_s} \sum_{m=1}^{n_x} \sum_{n=1}^{n_y} \mathbf{u}'[m,n,i]\mathbf{u}'[m,n,j] \tag{10a}$$

$$C[\mathbf{i},\mathbf{j}] = \frac{1}{n_t} \sum_{m=1}^{n_t} \mathbf{u}'[\mathbf{i},m]\mathbf{u}'[\mathbf{j},m] \tag{10b}$$

where $n_s = n_x \times n_y$ is the number of spatial points, $\mathbf{i},\mathbf{j} = 1, \ldots n_s$ are the matrix linear indices. A typical POD spectrum is shown in Fig.9.

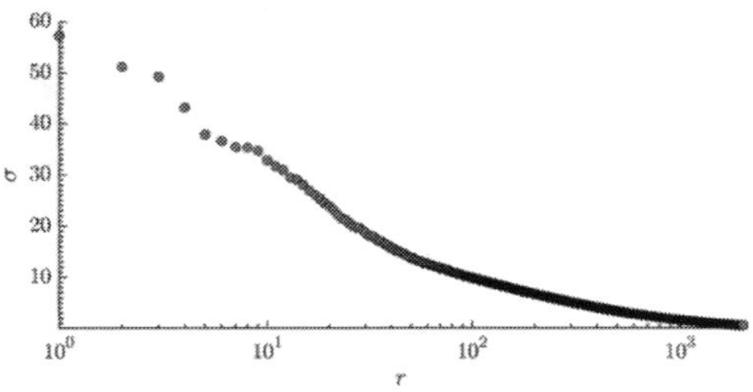

Fig. 9. Energy content in each POD mode for an exemplary test case. Typically, the first five modes capture a large amount of the total energy in the dataset, while following gentle decay is due to the highly turbulent nature of the flow field.

While the number of modes required to capture the large scale features of the flow is of the order of ten, the energy decay in the following modes remains rather smooth, due to the highly turbulent nature of the dataset. In all the experiments performed, however, the dominant modes are characterized by a highly harmonic behavior, which accounts for a large scale and coherent oscillation of the impinging jet flow.

The frequency content of the first six modes for the exemplary test case presented in Fig.8 is shown in Fig.10. The harmonic response of the jet oscillation is present in the dominant mode related to such an instability. This response is linked to the run-back film pulsation, characterized by the spectrum in Fig.7b, as discussed in Section 5. The characteristic spectra of the jet flow, therefore, is taken from the power spectral density of the temporal evolution of the first POD modes, i.e. $\mathcal{Y}_J = |\hat{\psi}_1|(f)$.

4.3 Final Coat Characterization

The TR-LAbs measures the liquid film thickness contour map from the amount of light absorbed from a backlighting source. The absorption is increased by means of a Methylene blue colorant, which absorption spectrum match the emission spectrum of the light source. For sufficiently diluted compounds, the light absorption and the film thickness are linked by the Beer-Lambert's law:

$$h(x,y,t) = \frac{1}{\gamma} \ln\left(\frac{I_0(x,y)}{I(x,y,t)}\right) = \frac{1}{\gamma} A(x,y,t) \; , \tag{11}$$

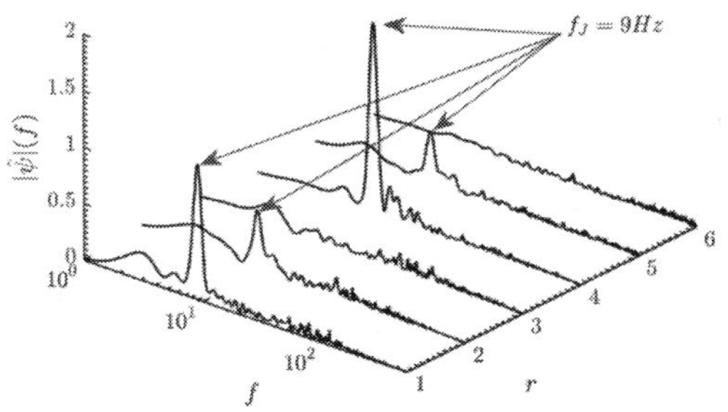

Fig. 10. Exemplary spectra of the first five dominant POD temporal modes $\psi(t)$ of the TR-PIV velocity field of the jet flow. The pulsation of the run back flow (spectra shown in Fig.7b) introduce an harmonic response on the jet flow. Test case: $\hat{Z} = 16$, $U_J = 29m/s$, $U_s = 0.35m/s$.

where $I_0(x, y)$ is the reference light intensity, taken from the dry test section (see Fig.5a), $I(x, y, t)$ is the intensity of the light transmitted through the liquid layer (cf. Fig.5b) and γ is the absorption coefficient, which mostly depends on the colorant concentration. Both light intensities I_0 and I are measured in terms of image gray levels by the camera. For a colorant concentration of 21 mg/Kg, the absorbance coefficient is $\gamma = 0.75 \pm 1\%mm^{-1}$.

The major difficulty encountered in this work is the motion of the cylinder which makes the time independence of the reference image questionable. While this image $I_0(x, y)$ is taken as a time average of a video recording with the dry test section, defects in the acrylic glass and the entrainment of air bubbles results in moving regions where the thickness measurement is compromised. An example of thickness reconstruction obtained by using eq.(11) is shown in Fig.11a.

An image processing routine was therefore developed to remove these defects in the post-processing step. The range of acceptable thickness measurements is first set by a threshold in the standard deviation of the thickness map. The outliers regions are converted to a binary image and the connected pixel clusters are labeled using connectivity analysis ([47], Fig.11b). Connected areas smaller than a user defined threshold are removed and re-interpolated by means of the penalized least squares method proposed in [48]. A typical liquid film interface reconstruction, after the outliers removal and the smoothing step, is shown in Fig.11-c.

For the purpose of this work, it was of interest to evaluate the film thickness average, the interface smoothness and a characteristic liquid film spectra. The thickness average is computed by averaging the film reconstruction both in space and in time:

$$h_f = \frac{1}{n_s n_t} \sum_{\mathbf{i}=1}^{n_s} \sum_{k=1}^{n_t} h[\mathbf{i}, k] \tag{12}$$

where $\mathbf{i} \in [1, n_s]$ are spatial points (pixels) and $k = [1, n_t]$ is the image index number within a video. The interface smoothness is evaluated by considering the peak to peak amplitude within a typical interface wave. This is evaluated by averaging in space the difference between the envelope of the wave peak $h_M[\mathbf{i}]$ and the envelope of the wave minima $h_m[\mathbf{i}]$:

$$h_{pp} = \frac{1}{n_s} \sum_{\mathbf{i}=1}^{n_s} \Big(h_M[\mathbf{i}] - h_m[\mathbf{i}] \Big) \tag{13}$$

where the envelopes are defined as:

$$h_M[\mathbf{i}] = \max_k^{n_t} \Big\{ h[\mathbf{i}, k] \Big\} \tag{14a}$$

$$h_m[\mathbf{i}] = \min_k^{n_t} \Big\{ h[\mathbf{i}, k] \Big\} \tag{14b}$$

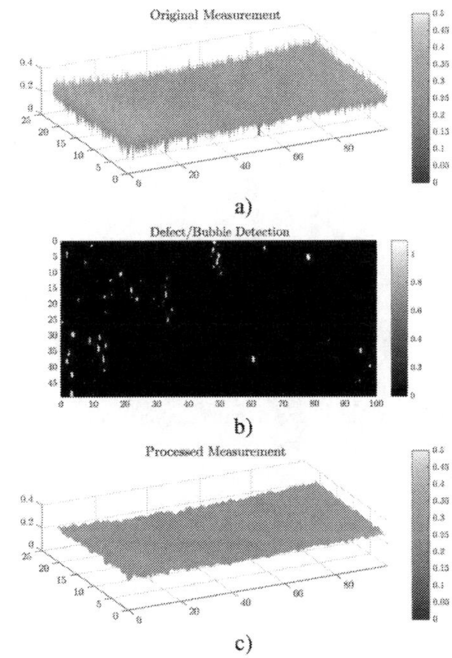

Fig. 11. a) Example of TR-LAbs measurement with no pre-processing; b) bubble/defect detection using morphological labeling and histogram threshold; c) processed measurement with outliers interpolation and smoothing.

Finally, the characteristic spectra of the liquid is computed as the spatial and the temporal ensemble average of the stream wise (thus along the image rows) thickness film profiles, that is:

$$\mathcal{H}_f[n] = \frac{1}{n_t n_y} \sum_{i=1}^{n_y} \sum_{k=1}^{n_t} H_i^k[n] \tag{15}$$

where $H_i^k[n]$ is thickness spectrum computed along the columns $j \in [1, n_x]$, for each $i \in [1, n_y]$ raw of the $k \in [1, n_t]$ image:

$$H_i^k[n] = \frac{2}{n_x} \sum_{j=1}^{n_x} h[i, j, k] \exp\left(\frac{-2\pi \mathrm{i}(j-1)(n-1)}{n_x} \right). \tag{16}$$

5 Results and Discussion

Keeping the cylinder velocity of $U_s = 0.35m/s$, the experimental campain consists of six jet velocities (from $U_J = 23m/s$ to $U_J = 39m/s$) at two stand-off distances, namely $\hat{Z} = 12$ and $\hat{Z} = 16$. The analysis is herein divided in terms of statistical quantities (Section 5.1) and frequency analysis (Section 5.2).

5.1 Averaged Quantities of the Final Coat

The measured average film thickness is shown in Fig.12, as a function of the dimensionless pressure gradient and compared with the 0D wiping model in eq.(2). The pressure gradient for the experimental points is computed from eq.(3) and the measurements are scaled with respect to the optimal thickness $h_0 = \sqrt{(\nu U_s/g)}$.

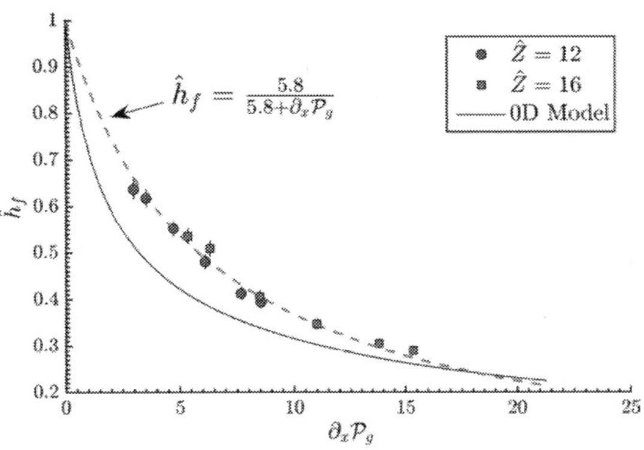

Fig. 12. Comparison of the measured average film thickness of the final coat $\hat{h}_f$ in dimensionless form with the 0D wiping model in eq.(2), as function of the dimensionless pressure gradient in eq.(3).

As expected, the 0D model overestimates the wiping capabilities at the lowest pressure gradients, while the agreement with the experimental points improves for stronger wiping performances. A simple rational correlation is proposed to fit the experimental points. This correlation, shown in Fig.12 is constructed such that $\hat{h}_f = 1$ for $\partial_x \mathcal{P}_g \to 0$ and can be considered as an empirical correction to the eq.(2) for $\partial_x \mathcal{P}_g < 18$.

The peak to peak amplitudes of the waves in the final coat are shown in Fig.13, scaled with respect to the optimal thickness and correlated to the final coating thickness regardless of the stand off distance or the jet velocity. Such correlation suggests that the characteristic of the final coat solely depends on the coating thickness, and therefore on the jet pressure gradient and shear stress at the impact, regardless of the conditions at which the wiping takes place ($\Delta P_N, Z, d, U_s$). Concerning the order of magnitudes, in the range $5-10\%$, one should take into account that the film thickness is analyzed at a distance of about $L \approx 1/2\pi R \approx 330mm$, with R the cylinder radius, from the wiping point. In all these cases then, the waves reach the measurement test section $\approx 1s$ after the wiping has taken place and during which they are subject to the damping action of surface tension. Concerning the trend, the monotonically increasing behavior implies that stronger wiping conditions ($h_f \to 0$) lead to smaller waves in absolute values. As the proportionality of the amplitude is $\hat{A}_f \propto \hat{h}_f^{0.6}$ these results suggest that in relative terms the amplitude of the coating non-uniformities increases ($\hat{A}_f/\hat{h}_f \propto \hat{h}_f^{-0.4}$) for stronger wiping.

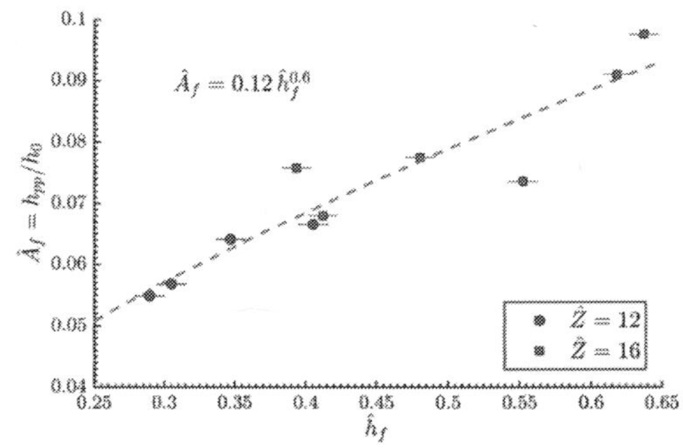

Fig. 13. Peak to peak amplitude of the waves in the final coat.

Whether such trends can be justified on the basis of a liquid film instability requires further investigation, although several authors have shown, using standard linear stability analysis, that the flow in the final coat is neutrally stable [49, 50]. On the other hand, on the basis of the average thickness curve in Fig.12, such trend is in line with the hypothesis that undulation in the final coat results from an instability of the impinging jet flow, which produces time variations of the pressure gradient [15–18]. The wiping curve in Fig.12 predicts, in fact, a saturation of the final thickness at the largest pressure gradients, which implies that thinner coats are less sensitive to variations in the wiping conditions than thicker ones.

5.2 Frequency Analysis of the Wiping Process

The characteristic spectra for the run back liquid film ($\mathcal{H}_R$), the gas jet flow ($\mathcal{Y}_J$) and the final coating thickness ($\mathcal{H}_f$), computed as described in Section 4, are shown for the six gas jet velocities in Fig.14 for the stand-off distance $\hat{Z} = 12$ (left) and $\hat{Z} = 16$ (right). Each spectrum is divided by its maximum for plotting purposes.

Interestingly, the dominant peak in spectra of the gas jet always match the one in the spectra of the run back liquid flow and the corresponding frequency increases monotonically with the velocity of the impinging jet. To investigate the role of the jet flow and that of the liquid film in this spectral matching, these results are scaled using a different characteristic time for the two flows in Fig.15. For the liquid film, the characteristic time considered is based on the liquid viscosity and the withdrawal speed $[t] = h_0/U_s = \sqrt{\nu/(U_s g)}$ and the resulting dimensionless frequency is shown in Fig.15a) as a function of the final coating thickness $\hat{h}_f$ and the run back film thickness $\hat{h}_R$. The latter is computed by introducing the measured

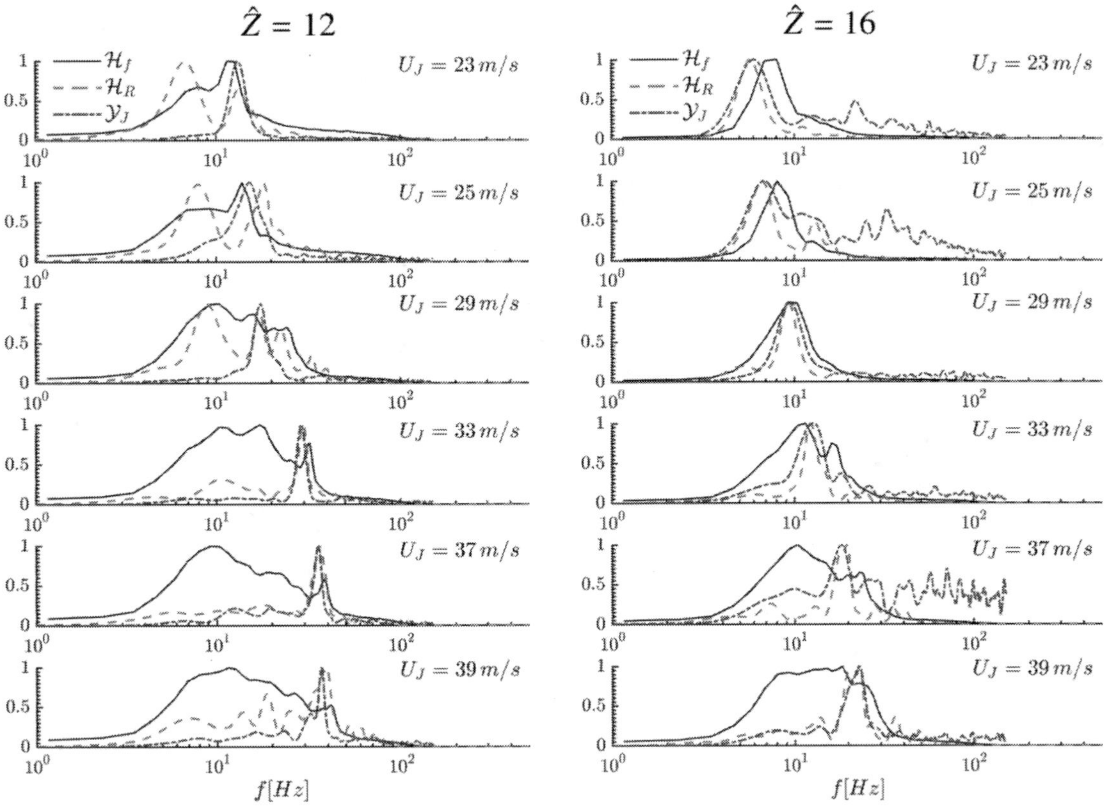

Fig. 14. Spectra of the final coating layer ($\mathcal{H}_f$, black continuous lines), the run back liquid flow ($\mathcal{H}_R$, red dashed lines) and the gas jet flow ($\mathcal{Y}_J$, blue dotted lines) for $\hat{Z}$ = 12 (left) and $\hat{Z}$ = 16 (right) for six velocities of the impinging jet flow. These spectra are normalized with respect to their peak; the amplitude of the coating defect is discussed in Fig.13.

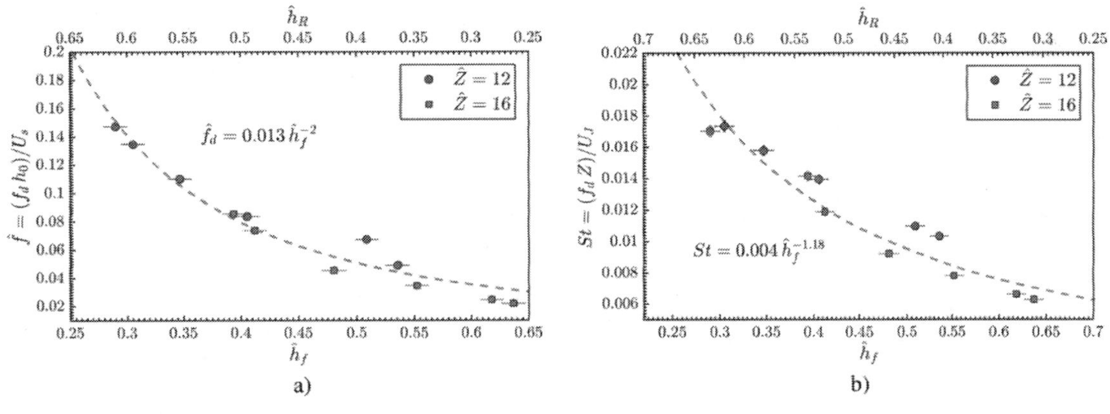

Fig. 15. Dimensionless correlation for the dominant frequency in the run back liquid film and in the impinging jet flow scaled with respect to the liquid film parameters ($\hat{f} = fh_0/U_s$, Fig. a) and with respect to the gas jet parameters ($St = fZ/U_J$, Fig. b).

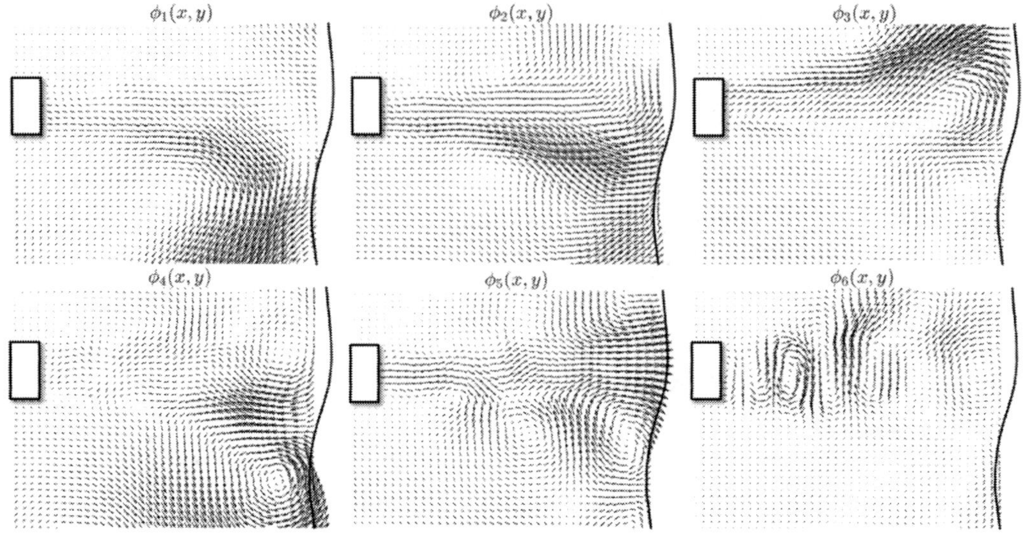

Fig. 16. Spatial structures $\phi[i,j]$ of the six POD modes of the TR-PIV dataset, associated with the spectra shown in Fig.10. The combination of the first five modes describes a sustained oscillation of the impinging jet, linked to the run back flow pulsation. Modes above the 5th are linked to shear layer instabilities and characterized by much higher frequencies, disregarded in this work.

average thickness in the final coat in the universal cubic equation in Fig.2 for the far field conditions.

The collapsing of the measurements with the final thickness is once again showing that the interaction between the two flows is solely governed by the wiping strength, regardless of other jet parameters. In particular, the frequency of the run back flow pulsation increases for stronger wiping, and thus increases with the run back flow thickness. This result is expected, both on a qualitative level and in terms of order of magnitude, for the most amplified waves in the linear stability analysis of a liquid film falling freely over a vertical fixed wall [36]. A quantitative agreement is on the other hand not expected since the run-back flow is in a highly nonlinear regime and strongly sheared by the impinging gas jet. It is therefore reasonable to argue that the run back flow pulsation is the result of a liquid film instability, that would exist even for an ideally stable impinging gas jet.

Concerning the gas jet oscillation, the dominant frequency is scaled with respect to the jet flow parameters, i.e. taking a reference time scale as $[t] = Z/U_J$, in Fig.15b. The dimensionless frequency of the jet oscillation is about one order of magnitude lower than what is observed for a gas jet impinging on a solid interface which models the shape of the run back flow interface, for which it is found in [18, 23, 51] a constant Strouhal number $St = fZ/U_J \approx 0.07$. Moreover, while the Strouhal number of the oscillation for a gas jet impinging on these solid models is independent of the stand-off distance and the maximum interface height [18, 23, 51], the Strouhal number in the wiping condition is here found to vary with the stand off distance and to be solely a function of the wiping performance. This behavior is unexpected for a gas jet impinging on a solid wall, and it is therefore argued that the jet oscillations are driven by the pulsations of the run back flow.

The hydrodynamic mechanism which is responsible for the link between the run-back flow pulsations and the impinging gas jet oscillation is well captured by the first five POD modes of the TR-PIV data of the jet flow, an exemplary set of which is shown in Fig.16. These modes have in fact a similar temporal evolution, dominated by the frequency of the run-back flow pulsation (in this example, shown in Fig.7b) and visible from their spectra, shown in Fig.10. The combination of these modes describes a sustained oscillation of the impinging jet flow (mostly captured by the modes 1-2-3), driven by the unsteady flow recirculation produced in the run back flow (captured by the modes 1-4-5). Higher modes appear to be uncorrelated (typically starting from the mode 6) with the run back flow pulsation and are linked to shear layer instabilities of the jet flow, occurring at much higher frequencies.

The mechanics of the low-frequency oscillation of an impinging gas jet, driven by the unsteadiness of the impinged interface were reported in a similar configuration in [16–18]. Regardless of the geometrical configuration, several common features can be identified: the periodic deflection of one of the wall jets due to the interface pulsation and the formation of unsteady recirculation produced by the entrainment flow, the impinging jet, and the deflected wall jet.

Finally, while it is clear the pulsation in the run back flow drives the oscillation of the jet, it not clear whether the undulation in the final coat results from the stream-wise propagation of the run back flow pulsation or the oscillations of the impinging gas jet. A perfect match of the peaks of $\mathcal{Y}_J$ and $\mathcal{H}_f$ appears at the lowest velocities for both the stand-off distances, but vanishes at the largest velocities, for which the frequency spectra in the final coat broadens considerably. Particularly interesting are the cases at low velocity and close stand-off distance, where the run back flow features two harmonic responses, both observed also in the final liquid film. It is therefore reasonable to conclude that the oscillation of the impinging gas jet, produced by the coupled instability between the gas jet flow and the run back liquid film, is one of the key sources for the final coating non-uniformity. However, the discrepancy in the frequency spectra at the fastest jet speed implies that this is not the only source and that the final coat could in principle assume a wavy pattern regardless of the impinging jet flow stability.

6 Conclusions

An experimental characterization of a jet wiping process, in which a 2D gas jet impinges on a liquid film to control the coating thickness over a moving substrate, has been presented. After discussing the relevant dimensionless numbers and the scaling laws of this two-phase flow configuration, the jet wiping facility Ondule, developed at the von Karman Institute, was presented together with the measurement test sections used for the temporal and spatially resolved investigation.

These include Time Resolved Particle Image Velocimetry (TR-PIV) and liquid interface tracking via image processing to study the wiping process in the impact region, and 3D Time-Resolved Light Absorption (LAbs) to characterize the interface dynamics of the coating film flow after the wiping.

The statistical properties of the final liquid film confirm the prediction of simplified engineering models of the wiping process, while the frequency analysis of the interface dynamics at the impact, combined with the modal analysis of the jet flow via Proper Orthogonal Decomposition (POD), revealed an unstable and periodic mechanism. In particular, it is shown that over the entire range of experimental conditions, the liquid film wiped from the plate and running back to the bath is unstable and develops a periodic train of large amplitude waves. These waves unsteadily confine the impinging gas jet, periodically producing a large scale recirculation which destabilizes the gas flow and promotes its oscillation. Simple correlations for the characteristic frequency of this coupling phenomenon are given in terms of liquid film and gas jet characteristic scales as a function of the wiping condition, to provide a solid experimental support for validating numerical or theoretical approaches.

As this unsteady phenomenon results in time dependent wiping capabilities of the jet, it is argued that its occurrence can be printed in the final coat and therefore be responsible for the limited coating quality achieved by jet wiped products. However, since the spectral matching between jet oscillations and the waves in the final coat is lost in some of the presented test cases it is concluded that other source of instability of the jet wiping process must exist.

Acknowledgments

The authors gratefully acknowledge the Belgian FNRS for supporting M. A. Mendez with a FRIA grant, and Arcelor-Mittal, for the financial support and for the permission to publish the material presented.

List Of Symbols

Acronyms

TR-PIV	Time Resolved Particle Image Velocimetry
TR-LAbs	Time Resolved Light absorption
VKI	von Karman Institute
POD	Proper Orthogonal Decomposition
PSD	Power Spectral Density
CLAHE	Contrast-Limited Adaptive Histogram Equalization

Notation

$[x]$	Characteristic scale for x
$\hat{x}$	Nondimensionalized x
x_M	Maximum component of x
x_m	Minimum component of x
x_f	x property of the final coat
x_R	x property of the run-back flow
x_s	x property of the solid substrate
x_j	x property of the jet flow
x_l	x liquid property
x_g	x gas property
x_0	x reference quantity
$\mathbf{x}'$	mean subtracted quantity
h_{pp}	peak to peak amplitude
u	velocity vector field
$\mathcal{H}$	characteristic spectrum
C	spatial correlation matrix
K	temporal correlation matrix
ϕ_r	modes spatial structure
ψ_r	modes temporal evolution
σ_r	modes energy content
n_t	number of samples
n_s	number of spatial points
n_x	number of columns
n_y	number of rows

Dimensionless Quantitites

Re_L	Reynolds number of the liquid film
Re_j	Reynolds number of the gas jet flow
St	Strouhal number of the gas jet flow
$\mathcal{Y}_J$	characteristic spectrum of the gas jet flow
C_τ	empirical coefficient
$\hat{A}_f$	dimensionless peak to peak amplitude
ζ	wiping thickness aspect ratio
$\partial_x \mathcal{P}_g$	pressure gradient

Dimensional Quantities

U	velocity	m/s
ΔP_n	nozzle stagnation pressure	Pa
P	pressure	Pa
Z	nozzle stand-off distance	mm
d	nozzle opening	mm
ρ	density	kg/m^3
μ	dynamic viscosity	$mPa.s$
ν	kinematic viscosity	m^2/s
σ	surface tension	N/mm
h	thickness	mm
h_0	optimal thickness	mm
g	gravitational acceleration	m/s^2
q	final coat flow rate	m^3/s
T_g	shear stress	Pa
t	time	s
R	Cylinder radius	mm
b	x position where $P = P_M/2$	mm
Δx	spatial resolution	mm
Δt	temporal resolution	ms
f	frequency	Hz
γ	light absorption coefficient	mm^{-1}
I	light intensity	$grayscale$

References

1. G.E. Cohen, E., *Modern Coating and Drying Technology* (New York: VCH, 1992)
2. TAPPI, Tech. rep., Atlanta: TAPPI Press (1981)
3. Deryagin, B.M, Levi, S.,M., *Film Coating Theory* (New York: Focal Press, 1964)
4. Butler, JJ., Beam, D.J, Hawkins, J.C., Iron and Steel Engineer **8**, 77 (1970)
5. Thornton, J. A., Graff, H.F., Metallurgical Transactions B **7**, 607 (1976)
6. Ellen, C H, Tu, C V, J. Fluids Eng. **106**, 399 (1984)
7. Tuck, E.O., Phys.Fluids **9**, pp. 2352 (Sept,1983)
8. Buchlin, J.-M., *Modelling of Gas Jet Wiping*, in *Thin Liquid Films and Coating Processes,VKI Lecture Series* (von Karman Institute for Fluid Dynamics, 1640 Rhode-Saint-Genese, Belgium, 1997)
9. Elsaadawy, E.A., Hanumanth, G.S., Balthazaar, A.K.S., McDermid, J.R., HryMak, A.N., Forbes, J.F., Metallurgical and Materials Transactions B **38**, pp. 413 (2007)
10. Bertin, M.C., Brown, L., Whitehead, R.L., Yrisarri, E., IFAC Proceedings Volumes **11**, pp 199 (1978)
11. Fenot, C. Rolland, F., Vigneron, G., Landau, E.D., IFAC Proceedings Volumes **25**, pp. 347 (1992)
12. Guelton, N., Lopes, C., Sordini, H., Metallurgical and Materials Transactions B **47**, pp 2666 (2016)
13. Myrillas, K., Gosset, A., Rambaud, P., Buchlin, J.-M, Eur. Phys. J. Spec. Top **166**, pp. 93 (2009)
14. Myrillas, K., Rambaud, P., Mataigne, J.-M., Gardin, P., Vincent, S., Buchlin, J.-M, Chem. Eng. Process. **68**, pp. 26 (2013)
15. Pfeiler, C., Ebi, W., Reiss, G. Riener, C.K., Angeli. G. Kharicha, A., Steel Res. Int. **87**, pp 1 (2017)
16. Mendez, M.A. , *Dynamics of Gas Jet Impingement on Vertical Falling Liquid Films*, in *Proceedings of the 6th Symposium of VKI PhD Research* (Rhode Saint Genese, BE, 2015)
17. Mendez, M.A., Buchlin, J.-M. , *A research methodology to study jet wiping processes*, in *Proceedings of the 11th European Coating Symposium ECS2015* (Eindhoven, NL, 2015)
18. Mendez, M.A. , *Oscillation of a Planar Gas Jet Impinging on a Deformable Wall*, in *Proceedings of the 7th Symposium of VKI PhD Research* (Rhode Saint Genese, BE, 2016)
19. Lacanette, D, Gosset, A., Vincent S., Buchlin, J-M, Arquis E., Phys. Fluids **18**, 042103 (2006)
20. Gosset, A., Buchlin, J.-M., Journal of Fluids Engineering **129**, 466 (2007)
21. Dubois, M., Reithmuller, M.L., Buchlin, J.-M., Arnalsteen, M., *The Gas-Jet Wiping Limit: The Splashing Phenomenon*, in *GalvatechConference* (Chicago, Iron and SteelInstitute, 1995), pp. 667–673
22. Thielicke, W., Stamhuis, E.J., Journal of Open Research Software **2**, 30 (2014)
23. Mendez, M.A., Buchlin, J.-M., *Quantitative flow visualization of confinement-driven instabilities of an impinging slot jet*, in *11th International Symposium on Particle Image Velocimetry (PIV2015)* (Santa Barbara, California, US, 2015)
24. Mendez, M.A., Raiola, M., Masullo, A., Discetti,S., Ianiro,A., Theunissen, R., Buchlin, J.-M., Exp Therm Fluid Sci **80**, 181 (2017)
25. P. Holmes, J. Lumley, G. Berkooz, Cambridge University Press (1996)
26. L. Sirovich, Quart. Appl. Math. **45**, 561 (1987)
27. Gosset, A., *Measurement techniques for unstable film flows*, in *VKI Lecture Series on Thermo-hydraulic Instabilities* (2006), pp. 1–25
28. Ozgen, S., Carbonaro, M, Sarma, G.S.R., Phys. Fluids **14**, pp. 3391 (2002)
29. Mendez, M.A., Scheid, B, Buchlin, J-M, Chemical Engineering Science **170**, 122 (2017), 13th International Conference on Gas-Liquid and Gas-Liquid-Solid Reactor Engineering
30. Yoneda H., Ph.D. thesis, University of Minnesota, Minneapolis, MN (1993)
31. Beltaos,S.,Rajaratnam,N., J. Hydraul. Res. **11**, 29 (1973)
32. Tu, C.V., Wood, D.H., Exp Therm Fluid Sci **13**, 364 (1996)
33. Saydlowski, B. K., *Air knife* (1966)
34. Brezinski, J.P., *Method and apparatus for coating paper* (1965)
35. Bicknell, J.A., *Method and apparatus for coating paper* (1964)
36. Kalliadasis, S., Ruyer-Quil, C. , Scheid, B., Velarde, M.G., *Falling Liquid Films*, Vol. 176 (Springer, Applied Mathematical Sciences, 2012)
37. Groenveld, P., Chemical Engineering Science **25**, 1267 (1970)

38. De Ryck, A., Quéré, D., J. Colloid Interface Sci. **203**, pp. 278 (1998)
39. Spiers, R.P., Subbaraman, C.V., Wilkinson, W. L., Chem. Eng. Sci. **29**, pp. 389 (1974)
40. Mendez, M.A. , Nemeth, L., Buchlin, J.-M, EPJ Web of Conferences **114**, pp. 1 (2016)
41. Savitzky, A., Golay, M. J. E., Anal. Chem. **36**, 1627 (1964)
42. Welch, P. D., IEEE Transactions on Audio and Electroacoustics **AU-15**, 70 (1967)
43. Oppenheim, A. V.; Schafer, R. W. , *Digital signal processing* (Prentice-Hall, 1975)
44. Shavit, U., Lowe, R., Steinbuck, J., Exp. Fluids **42**, 225 (2007)
45. Pizer, S.M., Amburn, E.P., Austin, J.D., Cromartie, R., Geselowitz, A., Greer, T. Romeny, B.T.H., Zimmerman, J.B., Computer vision, Graphics, and Image Processing **39**, 355 (1987)
46. Westerweel, J., Scarano, F., Exp. Fluids **39**, 1096 (2005)
47. Gonzalez, R. C., Woods, R. E., *Digital Image Processing* (Prentice-Hall, Inc., 2014)
48. Garcia, D., Computational Statistics and Data Analysis **54**, 1167 (2010)
49. Tu, CV, Ellen, C H, *Stability of Liquid Coating in the Jet Stripping Process*, in *9th Australasian Fluid Mechanics Conference* (1986)
50. Hocking, G. C., Sweatman, W. L., Fitt, A. D., Breward, C., Journal of Engineering Mathematics **70**, 297 (2011)
51. Mendez, M.A., Scelzo, M.T., Buchlin, J.-M., Exp Therm Fluid Sci **91**, 256 (2018)

Fluidic Vectoring of a Planar Incompressible Jet Flow

Miguel Alfonso Mendez[1,*], *Maria Teresa* Scelzo[1], *Adriana* Enache[1], and *Jean-Marie* Buchlin[1]

[1] *EA Dep., von Karman Institute for Fluid Dynamics, Sint-Genesius-Rode, Belgium*

Abstract. This paper presents an experimental, a numerical and a theoretical analysis of the performances of a fluidic vectoring device for controlling the direction of a turbulent, bi-dimensional and low Mach number (incompressible) jet flow. The investigated design is the co-flow secondary injection with Coanda surface, which allows for vectoring angles up to 25° with no need of moving mechanical parts. A simple empirical model of the vectoring process is presented and validated via experimental and numerical data. The experiments consist of flow visualization and image processing for the automatic detection of the jet centerline; the numerical simulations are carried out solving the Unsteady Reynolds Average Navier- Stokes (URANS) closed with the $k-\omega$ SST turbulence model, using the *PisoFoam* solver from OpenFOAM. The experimental validation on three different geometrical configurations has shown that the model is capable of providing a fast and reliable evaluation of the device performance as a function of the operating conditions.

1 Introduction

Controlling the direction of a jet flow is a crucial operation in several engineering applications. Examples range from thrust control in propulsion systems [1–4] to particle sorting in powder-based industrial processes [5, 6].

Such control is known as jet vectoring and can be mechanic, that is by means of movable surfaces, or aerodynamic, that is by means of a secondary injection or suction. While the mechanic vectoring is always effective and offers large vectoring performances, the presence of moving mechanical parts leads to control devices that are complex, heavy, aerodynamically inefficient and hard to integrate on existing nozzles [7]. By contrast, aerodynamic systems are light and simple to integrate, but their vectoring performance and controllability are limited to a relatively narrower range of operating conditions, which depends on the particular scheme adopted [8–10].

These systems, referred to as fluidic vectoring devices, exploit the interaction of the jet flow with adjacent secondary fluid streams and can be classified depending on how such interaction is generated.

Fluidic vectoring devices can operate in steady or unsteady conditions. Devices that operate in steady conditions can be further classified based on the direction of the secondary flow, which can be in co-flow (blowing, [4, 11]), counter-flow (suctioning, [12–14]) or a suitable combination of both (blowing and suctioning, [6, 15]). These designs usually combine the secondary flow with a smooth extension of the nozzle to exploit the well known Coanda effect, that is the adherence of a fluid stream to a wall, as a result of an unbalance in the entrainment of surrounding fluid [16–18]. Devices that operate in unsteady conditions can be further classified based on the net mass flux introduced, which can be zero if the secondary stream is produced by synthetic jets [9, 19] or not if it is produced by pulsing jets [20, 21].

Regardless of the fluidic scheme adopted, the design and the control of any aerodynamic vectoring system require a precise evaluation of its vectoring performance as a function of the operating conditions. The scope of this work is to present a complete characterization of an aerodynamic vectoring system combining experimental, numerical and theoretical analysis.

In particular, the investigated design is the Coanda co-flow fluidic scheme, to be implemented in the vectoring of a turbulent bi-dimensional jet flow at low speed, namely at a Mach number sufficiently low to neglect compressibility effects. This design allows for avoiding the use of high-frequency actuators as in synthetic jets vectoring, or the installation of injection and suction lines at a short distance from the jet. On the other hand, its use for low Mach number flows is not documented in the literature. Although several experimental works have been presented [4, 22], a definition of the role of its geometrical and operational parameters is still lacking and most of the numerical and analytical modeling attempts are focused on the high-speed regimes typical of aircraft propulsion [23–28].

To support design engineers in the development of fluidic vectoring schemes, this work proposes an empirical model of the jet deflection, developed on the basis of simple physical considerations and validated via numerical and experimental data. Such model does not aim to exhaustively describe all the physical mechanisms involved in the vectoring process but to provide an engineering tool to analyze the impact of the operating parameters on the

*e-mail: mendez@vki.ac.be

system performance and to provide a physical insight on the vectoring phenomenon.

The investigated co-flow fluidic design is described in Section 2, along with the proposed empirical model. A set of 2D simulations were used to support the simplifying assumptions in the model derivation, using Unsteady Reynolds Average Navier Stokes (RANS) modeling. The numerical set up is presented in Section 3. Finally, the empirical model is validated experimentally, by combining flow visualization and image processing, as described in Section 4. The results are presented and discussed in Section 5; the conclusions are drawn in Section 6.

2 Problem Formulation

The co-flow fluidic vectoring design under consideration is shown in the schematic in Fig.1, which recalls the main control variable: the opening for the primary and the secondary jets h_p and h_s, their respective average velocities u_p and u_s and the radius of the Coanda surface R, which is tangent to the secondary channel. The distance L between the primary and the secondary channels should be minimized, compatibly with mechanical constraints, to promote the interaction between the two jets.

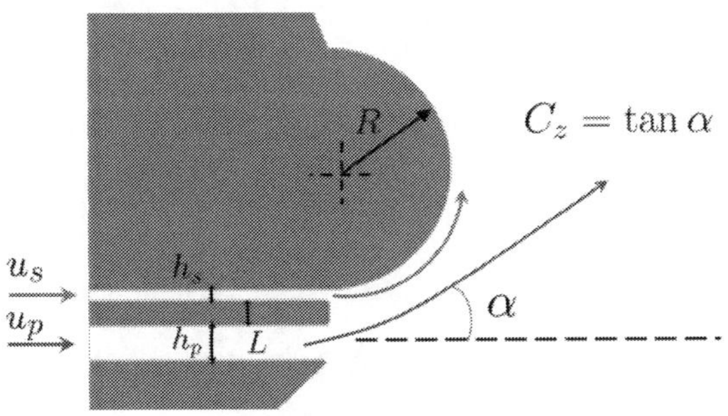

Fig. 1. Geometrical configuration and operating parameters for the co-flow fluidic vectoring considered in this work.

For a given fluidic nozzle, that is fixing h_s, h_p, L, R, the operating condition of the device is typically given in terms of mass flux ratio β or momentum ratio β_M. For an incompressible flow configuration, with the density constant and equal in both the injections, these are

$$\beta = \frac{h_s u_s}{h_p u_p} \;;\quad \beta_M = \frac{h_s u_s^2}{h_p u_p^2} = \beta\,\frac{u_s}{u_p} = \beta^2\,\frac{h_p}{h_s}\;. \quad (1)$$

Although these quantities are interchangeable, the momentum ratio β_M is considered to better represent the physics of the vectoring process and it is hereinafter chosen as the dimensionless parameter describing the operation condition.

The performance of the device, i.e its operational map, is typically defined in terms of thrust vectoring coefficient, that is the slope $C_Z = \tan(\alpha)$ of the achieved vectoring angle α, as function of the momentum ratio β_M. This operational map is qualitatively divided in three zones [4, 28], referred to as dead zone, control zone and saturation zone.

The dead zone is the region where the momentum ratio β_M is below a critical value β_M^C, and is characterized by no jet vectoring or even negative vectoring ($\alpha < 0$ in Fig.1). The saturation zone is the region above a critical value β_M^S, where the vectoring of the jet no longer increases with increased β_M. The co-flow device, therefore, can operate within a range $\beta_M \in [\beta_M^C, \beta_M^S]$ which depends on its geometry, in the so called control zone.

The empirical model proposed in this work hinges upon the formulation of three physical principles, each described in a dedicated subsection.

First, a general vectoring law is formulated in Section 2.1 by modeling the jet deflection as the result of a transversal pressure gradient on the primary jet flow. This pressure gradient originates from the Coanda effect acting on the secondary jet flow, as formulated in Section 2.2. Depending on both the nozzle geometry and the operating condition, the interaction between the two jets strongly influences the vectoring physics, as formulated in Section 2.3. Section 2.4 finally collects all the theoretical considerations into a simple equation to describe the operational map of the device.

The flow of both jets is assumed to be incompressible, isothermal and mono-dimensional, that is concentrated along their respective centerlines. Turbulent and viscous effects are introduced empirically.

2.1 Flow deflection of the Primary Jet

In a Lagrangian framework, neglecting the effects of viscosity, the forces acting on a control fluid volume Ω of density ρ, traveling at the flow velocity $u_M(\zeta)$, only includes inertia and pressure gradient as shown in the schematic of Fig.2.

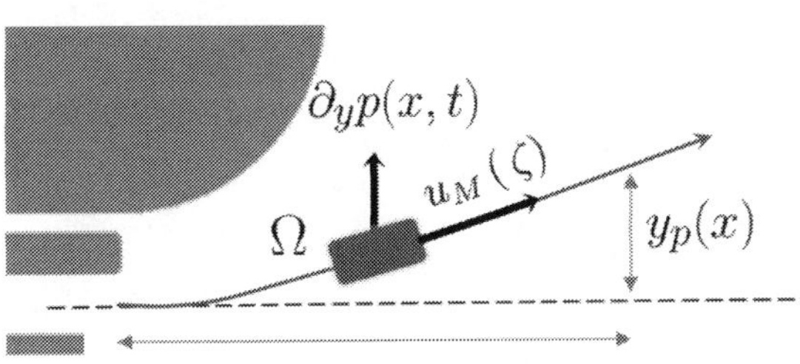

Fig. 2. Force balance of a control volume in the primary jet, advected at the averaged velocity u_M and traveling under the action of a transversal pressure gradient $\partial_y p$. The trajectory of jet becomes linear after a short distance.

Here the flow velocity $u_M(\zeta)$ is an average of the velocity of the two injections, described in Section 2.3, and ζ is the curvilinear coordinate following the primary jet centerline. In a first approximation, we neglect the curvature effect on the primary jet, which is on the other hand crucial for the secondary jet as described in Section2.2. Assuming that the jet travels along a line inclined by the vectoring angle α, the link between the curvilinear coordinate and the horizontal axis is simply $x \approx \zeta/\cos(\alpha)$.

As the flow is released in atmosphere (i.e. is sufficiently far from the walls) the stream-wise pressure gradient is negligible and the force balance only includes a transversal pressure gradient $\partial_y p(x)$:

$$\rho\Omega\frac{D^2 y_p(x)}{Dt^2} = \Omega\,\partial_y p(x,t)\ , \tag{2}$$

where $y_p(x)$ is the vertical displacement of the control volume with respect to the origin, located at the nozzle geometrical centerline. Introducing the material derivative $D(\bullet)/Dt = \partial_t(\bullet) + u_p(x)\partial_x(\bullet)$, eq. (2) becomes:

$$\partial_{tt}y_p + 2u_M\partial_{xt}y_p + u_M^2\partial_{xx}y_p = -\frac{1}{\rho}\partial_y p(x,t)\ . \tag{3}$$

The complete solution of eq.(3) by means of Laplace transform is proposed in [29]. In this work, considering only the steady solution reached at the end of the transitory, the time derivatives can be dropped and eq.(3) is easily integrated to retrieve the trajectory of the jet centerline:

$$\partial_{xx}y_p(x) = \frac{1}{\rho}\frac{\partial_y p(x)}{u_M^2(x)} \tag{4a}$$

$$\partial_x y_p(x) = C_Z = \int_0^x \frac{1}{\rho}\frac{\partial_y p(s)}{u_M^2(s)}ds + C_1 \tag{4b}$$

$$y_p(x) = \int_0^x\int_0^x \frac{1}{\rho}\frac{\partial_y p(s)}{u_M^2(s)}ds + C_1 x + C_2 \tag{4c}$$

The integration constants C_1 and C_2 can be set to zero by assuming that the primary jet flow is infinitely 'rigid' at the nozzle outlet, and thus $\partial_x y_p = y_p = 0$ at $x = 0$.

Equations (4) allows for several important considerations on the thrust vectoring phenomenon. First, no thrust vectoring is produced if the pressure gradient $\partial_y p$ is identically null over the entire domain. Second, provided that the pressure gradient only acts within a certain region, say $\partial_y p(x) = 0\ \forall x > x^*$, eq.(4c) leads to a constant value, that is the trajectory of the jet flow $y_p(x)$ becomes a straight line with constant slope:

$$C_Z(\beta_M) = \int_0^{x^*} \frac{1}{\rho}\frac{\partial_y p(s,\beta_M)}{u_M^2(s,\beta_M)}ds\ . \tag{5}$$

This equation can be considered as a general vectoring law, which shows that the vectoring performance depends on the strength $\partial_y p(x,\beta_M)$ and the extension x^* of the pressure gradient, and on the averaged advection velocity of the jet $u_M(x,\beta_M)$. In principle, then, a faster jet requires a stronger (or more extended) pressure gradient to achieve the same vectoring angle. Moreover, the characteristic curve $C_Z(\beta_M)$ of the device can be inferred from the trends of these two terms as function of the momentum ratio β_M: the control zone can be seen as the region in which $\partial_y p(\beta_M)$ grows faster than $u_M^2(\beta_M)$, while saturation occurs then these two quantities have the same functional dependency on β_M, so that their ratio remains constant. These two terms are analyzed in Section 2.2 and 2.3 respectively.

2.2 Vectoring performance of the Secondary Jet

As long as the secondary jet flow is attached on the cylindrical surface, a pressure drop is produced at the wall. This phenomenon, known as the Coanda effect, is usually described from the force balance of an inviscid fluid [16–18], which is shown in the schematic of Figure 3.

As the velocity is uniform along the radial direction, the force balance of a fluid element at a distance r from the Coanda center reads:

$$\partial_r p = \frac{\rho u_s^2}{r} \tag{6}$$

Integrating eq.(6) over the thickness of the secondary jet, assumed to remain $\sim h_s$, yields the pressure drop over its cross section:

$$\Delta P_c = \rho u_s^2(\theta)\int_R^{R+h_s}\frac{dr}{r} = \rho u_s^2\ \ln\!\left(\frac{R+h_s}{R}\right) \tag{7}$$

This result was used in a first attempt to define the pressure gradient driving the vectoring as the average value $\partial_y p \sim \Delta P_c/h_s$. In the range of geometrical parameters, however, a better reference value was found to be the maximum pressure gradient produced at the wall, i.e. taking $r = R$ in eq.(6):

$$\partial_r p\Big|_{r=R} = \frac{\rho u_s^2}{R}\ . \tag{8}$$

Any result provided by inviscid theory provides nevertheless an upper and ideal limit for the achievable pressure gradient. Viscosity and turbulence diffusion, in fact, introduce a non-uniform velocity profile, a decay of its maximum value along the polar coordinate and a thickening of the jet boundary h_s. An analytical, self-similar solution

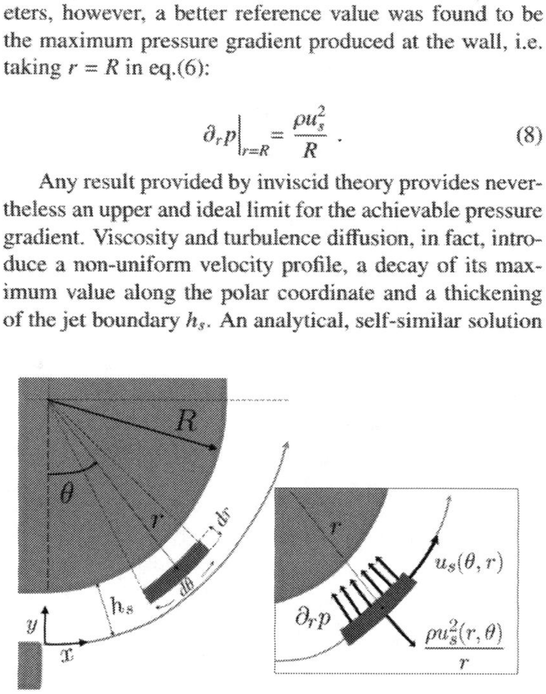

Fig. 3. Radial equilibrium of a control volume in the secondary jet flow, attached on the cylindrical surface by the Coanda effect. The flow is assumed to be inviscid, so that the velocity profile is constant and the jet thickness remains $\sim h_s$.

which describes the entire evolution of a Coanda wall jet is proposed in [30]. For the purpose of this work, we postulate that eq.(8) requires at least two empirical corrections, written in variable separated form as

$$\partial_y p \approx \left(\frac{\rho u_s^2}{R}\right) c_1\left(\frac{h_s}{R}, h_p, Re_s, \beta_M\right) P(x) . \qquad (9)$$

The first correction c_1 takes into account the reduction due to the curvature ratio h_s/R, a term often referred to as curvature parameter [30], turbulence and viscous diffusion, and the momentum exchange with the primary jet flow. This term depends on the geometrical parameters $h_s/R, h_p$, on the Reynolds number of the secondary injection $Re_s = h_s u_s/\nu$, with ν the kinematic viscosity, and the blowing ratio β_M. In a first simplification, however, we assume that this correction is solely a function of the geometrical parameters.

The second correction $P(x)$ takes into account the decay law of the net pressure gradient, due to both the dissipation and the momentum exchange between the two jets. This function should vary in the range $P \in [0, 1]$ and lead to a vanishing of the suctioning effect above a certain value x^* from the inlet. An attempt to define it from URANS simulations is presented in Section 5.1. Here, we limit the analysis of the pressure gradient to the introduction of the momentum ratio in eq.(9):

$$\frac{\partial_y p(x,\beta)}{\rho} \approx \beta_M u_p^2 \frac{h_p}{h_s} c_1\left(\frac{h_s}{R}, h_p\right) \frac{1}{R} P(x) . \qquad (10)$$

That is, if the dependency of the correction c_1 on the blowing ratio is neglected, the pressure gradient grows linearly with β_M for a given velocity in the primary jet flow.

2.3 Momentum exchange between Jet Flows

The average velocity in eq.(5) takes into account the momentum exchange between the two jets. An approach solely based on the momentum conservation appeared nevertheless particularly cumbersome since it requires the evaluation of the viscous dissipation on the Coanda surface and a clear definition of the jet flow boundaries. This approach is therefore dropped in favor of a more heuristic average, which takes the form

$$u_M^2 = \xi(\beta_M)\, u_s^2 + [1 - \xi(\beta_M)] u_p^2 \qquad (11)$$

The mixing coefficient $\xi(\beta_M)$ takes into account that at low momentum ratios ($\beta_M \ll 1$), when $u_s \ll u_p$, the average velocity must be close to that of the primary (i.e. $\xi \approx 0$) while at large momentum ratios ($\beta_M > 1$), when $u_s \gg u_p$, the average velocity must be close to that of the secondary (i.e. $\xi \approx 1$).

Observing that the pressure gradient driving the jet vectoring in eq.(9) is approximately $\partial_y p \propto u_s^2$, one could see that the condition $\xi = 1$ defines the saturation of the device, since it yields $u_M^2 = u_s^2$ and the velocity of the secondary jet –that is the operating conditions– cancels out from eq.(5). Introducing the momentum ratio in eq.(12), and introducing a shape function $G(x)$ for the stream-wise velocity evolution yields:

$$u_M^2(x,\beta) = u_p^2\left[\xi(\beta_M)\frac{h_p}{h_s}\beta_M + 1 - \xi(\beta_M)\right] G^2(x) . \qquad (12)$$

The shape function $G(x)$ ranges from 1, (for $x \to 0$) to 0 (for $x \to \infty$). Eq.(12) can be seen as a variable separated form of the average velocity, of the kind often used in the study of self-similar flows.

For a turbulent free jet released at a velocity u_j, for example, this shape function $G_j(x) = u_j(x)/u_j$ is available in the framework of boundary layer theory, using a mixing length turbulence model [29, 31–33]:

$$G_j(x) = \begin{cases} G_j = 1 & \text{if } x \leq x_c \\ G_j(x) = (x_c/x)^{0.5} & \text{if } x > x_c \end{cases}$$

where x_c is the length of the potential core, typically of the order of $2 - 3$ times the nozzle opening.

A similar approach could in principle be followed for both the primary and the secondary jet flows, and use the averaging process point-wise. Here, however, we make no analytical attempt in defining a general function $G(x)$, nor in seeking for a self-similar form of both jets during their interaction. This shape function is presented for several numerical test cases in Section 5.1. It is nevertheless worth observing that the influence of the average decay $G(x)$ becomes important in eq.(5) only as long as the length of the 'potential core' of the two flows is much shorter than the critical position x^* within which the pressure gradient acts, appearing in the upper limit of the integral.

2.4 Model Formulation

Introducing eq. (12) and (10) into in eq.(5) leads to the final form of the empirical model proposed for the thrust vectoring effect:

$$C_Z\big(\beta_M\big) = C(h_s, h_p, R)\frac{\beta_M}{\xi(\beta_M)\frac{h_p}{h_s}\beta_M + 1 - \xi(\beta_M)} , \qquad (13)$$

with the model coefficient

$$C = \frac{h_s}{h_p} c_1(h_s, h_p, R)\frac{1}{R}\int_0^{x^*} \frac{F(s)}{G^2(s)} ds \qquad (14)$$

This coefficient is assumed to be independent from the operating condition and thus defined by a given nozzle design. An evaluation of such assumption is provided in Section 5.1. Here, the model closure requires the definition of the functional $\xi(\beta_M)$ and the constraints required for its identification. The procedure proposed is based on the observation that eq.(16), or more generally (5), allows for a saturation of the vectoring angle provided that the average velocity becomes $u_M \sim u_s$, that is $\xi \approx 1$. To close the model, therefore, we write the functional $\xi(\beta_M)$ as a polynomial expansion in β_M of order $O(P)$:

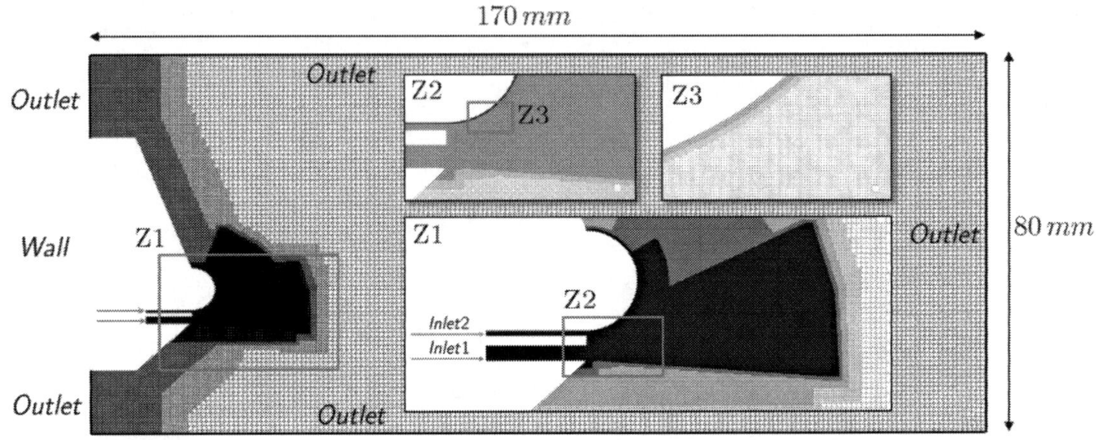

Fig. 4. Computational domain and discretization mesh, with the boundary conditions in each patch. The different level of mesh refinements are shown in three zooms (Z1,Z2 and Z3). The mesh consists of about 2×10^5 cells, and the domain is sufficiently large to simulate the entire evolution of the jets before they reach the numerical boundaries.

$$\xi(\beta_M) = \sum_{j=1}^{j=P} a_j \beta_M^j \,, \tag{15}$$

defined so that $\xi(\beta_M = 0) = 0$, and constrained so that $\xi(\beta_S) = 1$. As the order of the expansion is increased above one, a set of $n-1$ constraints on its derivatives must be defined. A plausible hypothesis is that the growth of ξ saturates as it approaches critical value β_M^S for which $\partial_{\beta_M^S} C_Z = 0$, and so $\partial^n_{\beta_M}\xi(\beta_M^S) = 0$ for any derivative $n > 1$. The opposite approach is to postulate an initially gentle growth of $\xi(\beta_M)$ followed by a strong growth towards β_M^S, that is $\partial^n_{\beta_M}\xi(\beta_M^C) = 0$ for $n > 1$.

In what follows, we illustrate the procedure only for the first order expansion, which provided a sufficiently accurate comparison with the presented experimental data. Assuming that $\xi = a_1\beta_M$, eq.(16) becomes:

$$C_Z\big(\beta_M\big) = C(h_s, h_p, R)\frac{\beta_M}{a_1 \frac{h_p}{h_s}\beta_M^2 - a_1\beta_M + 1} \,. \tag{16}$$

The maximum deflection ($\partial_{\beta_M} C_Z = 0$) is therefore obtained when

$$\beta_M^S = \sqrt{\frac{h_s}{h_p a_1}} \tag{17}$$

Imposing that this corresponds to $\xi = 1$, the coefficient becomes $a_1 = h_p/h_s$. The model in eq.(16) reduces therefore only to one unknown, assumed to be independent from the operating conditions. Interestingly, the momentum ratio at which the maximum deflection occurs becomes $\beta_S = h_s/h_p$, and the corresponding maximum vectoring coefficient is $C_Z\big(\beta_S\big) = C\, h_s/h_p$. The model coefficient C is deduced by fitting eq.(16), with $a_1 = h_p/h_s$, to the experimental points in Section 5.2; the terms in eq.(14) and their dependency on the operating conditions is investigated in Section 5.1. Remarkably, such simple model predicts a negligible impact of the Coanda radius, and a much stronger influence of the opening ratio h_p/h_s.

3 Numerical Methods

The 2D numeral simulations supporting the development of the proposed engineering model are performed using a classical Unsteady Reynolds Averaged Navier Stokes (URANS) approach, implemented in the *Piso-FOAM* solver in *OpenFOAM* with the well-established pressure-based finite-volume methodology.

The computational domain and the mesh discretization are shown in Figure 4, together with the definition of the boundary conditions. The domain is created using the open-source *FreeCAD* software, and consists of a rectangular box of $170 \times 80 mm^2$ from which the nozzle geometry is removed via Boolean subtraction. The size of the domain is large enough to allow for the complete development of the two jets and their interaction, as well as for simulating the entrainment flow from the external boundaries without undesired boundary effects.

The discretization mesh is unstructured, prepared using the *HEXPRESS™* package from Numeca and consisting of about with an 2×10^5 polyhedral cells, with different levels of refinement in the entrainment regions, the zone of jet mixing and development (cf. Fig.4, zoomed view Z1), and the Coanda wall (cf. Fig.4, zoomed views Z2 and Z3). Here, in particular, the mesh allows for an average $y^+ \approx 1$ and no wall function are employed. The URANS equations are therefore solved up to the wall, using the well-known $k-\omega SST$ turbulence model [34].

All the wall patches are set to no-slip condition, with a zero gradient condition for the pressure, a large value of specific dissipation ω, and a zero turbulence kinetic energy k. In the outlet patches, a zero gradient condition is imposed to both the turbulence quantities and the

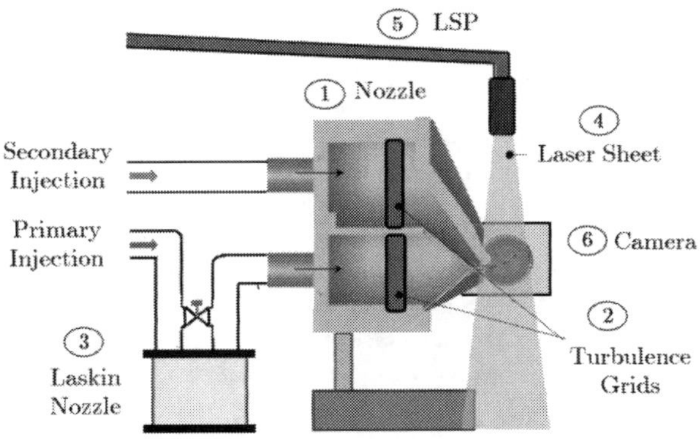

a)

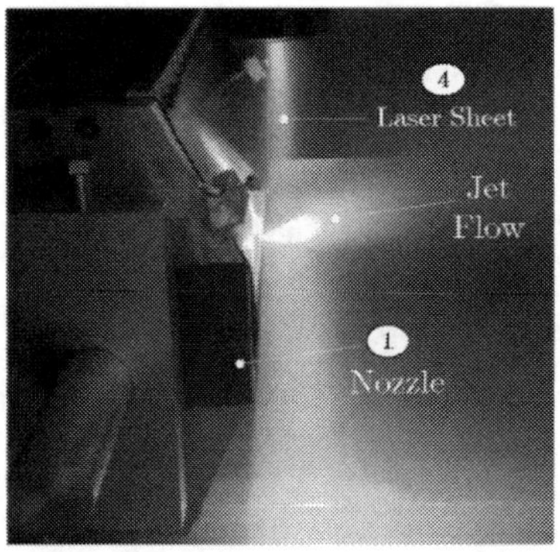

b)

Fig. 5. a) Schematic of the VKI facility used to study the Coanda co-flow fluidic vectoring scheme, with the main components for the flow visualization. b): picture of the set-up during a jet vectoring experiment, showing the nozzle (1), the laser sheet (3) and the deflected jet flow.

pressure, while the velocity field is set to the OpenFOAM *pressureInletOutletVelocity* condition. This boundary condition allows for correctly modeling both the flow outlet and the entrainment flows by automatically imposing a zero gradient in regions where fluid flows out of the domain and an inlet value (set to 10^{-3}) in the regions where fluid flows enter the domain. The inlet condition for the two injections is a uniform velocity, with the fixed value calculated from the pressure measurement in the experiments. The flow in the channel develops for about $6h_p$ before the release in open pressure, similarly to what happens in the real nozzle. The turbulent kinetic energy is calculated as $k = 1.5(U_r\,T.I.)^2$ where the reference velocity U_r is taken from the inlet value, and $T.I.$ is the turbulence intensity, set to $T.I. = 10\%$ in agreement with the maximum value obtained from a preliminary experimental characterization of the nozzle via hot wire anemometry. The turbulence dissipation is computed as $\varepsilon = C_\mu k^{3/2}/l$, with $C_\mu = 0.09$ and the mixing length l is set to 0.007 times the nozzle exit dimension as suggested in [35].

Concerning the numerical schemes, both the advection and the diffusion terms are discretized with a second order Gauss linear schemes (upwind for the advection terms; corrected for the diffusion terms). The time discretization is the implicit Euler scheme, running with a time-step of about $2 \times 10^{-7}s$ to ensure a CFL number of approximately 0.8. The physical simulation time to overcome the transitory and reach a steady state solution is generally of the order $0.02sec$.

4 Experimental Methods

4.1 Experimental set-up

The experimental set-up for fluidic control testing developed at the von Karman Institute is shown in the schematic of Figure 5a. The nozzle (1) is composed of two independent chambers, equipped with turbulence grids (2) to uniform turbulence intensity and the span-wise air distribution. These chambers are supplied by two independent air pressure lines, controlled by upstream pressure regulators. A pressure tap is flash mounted on the back of each chamber and connected to two piezoresistive pressure transducers AMS 5812 to acquire the stagnation pressure $P_{p,s}$ of the two injections. Given the chamber discharge coefficient $D_{s,p}$, available from a pressure-velocity calibration via hot-wire anemometry, the velocity of the two injections is computed as $u_{p,s} = D_{s,p}\sqrt{2P_{p,s}/\rho}$. These pressure transducers have a full scale of about $0 - 2000Pa$, corresponding to a maximum of $58\,m/s$, and an absolute uncertainty of the order of $10Pa$.

A Lasking Nozzle PIVTEC45-M (3), operating with mineral oil (Ondina Shell 917), is used to produce seeding particles in the primary jet and thus to allow for high-speed flow visualization. At the scope, the middle plane of the nozzle is illuminated by a laser sheet (4) of approximately $2mm$ thickness, produced by a light sheet probe (5) LSO, from ILA. The light source is a dual diode-pumped Nd:YLF laser system (Quantronix Darwin Duo 527) running in continuous mode. The images are acquired by a CMOS Photron FASTCAM SA1 (6), placed at about 16 *cm* from the test section and equipped with a Tamron CZ-735 70-300mm zoom objective, leading to a scaling factor of about $21 pixels/mm$. The videos have a maximum size of 5.5 GB, consisting of $n_t = 12473$ images with a resolution of $n_p = 512x896\ pixels^2$, sampled at a frequency of $f_s = 1kHz$, corresponding to a acquisition time of about $t \approx 12s$. The exposure time is set to the maximum possible ($\approx 1/f_s$). This, together with the continuous light source and the high seeding concentration, allows for displaying the jet flow as a uniform and bright region, clearly distinguishable from the background. A picture of the facility during an experiment is shown in Fig.5b.

The nozzle facility allows for varying the Coanda radius between $R = 2\,mm$ and $R = 5\,mm$, while the secondary opening can be changed between $h_s = 0.5\,mm$ and $h_s = 1\,mm$. The primary opening is kept to $h_p = 1.5mm$ with a gap of $L = 1mm$ between the two channels.

https://doi.org/10.1051/epjconf/201818002065

4.2 Data processing

As the jet flow appears rather stable in the high-speed flow visualization, each video sequence is down-sampled by a factor of 100 and the centerline of the jet flow is reconstructed from each of the resulting $n_t = 124$ images.

The centerline detection is carried out with the image processing algorithm proposed in [36, 37]. This algorithm consists of three steps. First, each image is horizontally re-contrasted to uniform the light intensity within the jet flow and low-pass filtered with a Gaussian kernel to remove high-frequency noise. An exemplary result of this step is shown in Figure 6a. Second, the grayscale profile in each column is further smoothed via a Savizty-Sgolay filter and the peak of each profile detected. Finally, the jet centerline is constructed as the loci of the grayscale peak in each column and the result from the down-sampled images is time averaged. Three exemplary profiles corresponding to the vertical lines denoted with x_1, x_2, x_3 in Figure 6a are shown in Figure 6b as a function of the vertical coordinate and together with the peak detection, marked in black.

The time averaged result is interpolated with a linear fit, from which the thrust vectoring angle is computed. A systematic uncertainty in the fitting process is estimated to be of the order of $\pm 1^o$. A random relative uncertainty of approximately $\pm 2\%$, linked to the averaging process, is computed from the standard deviation of the vectoring angle retreived in each of the down-sampled images.

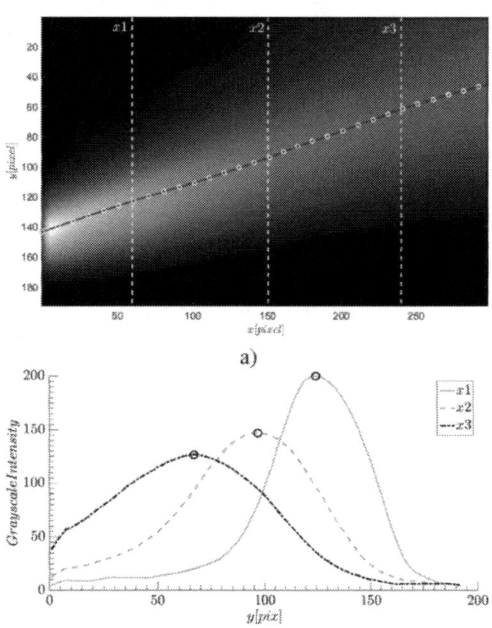

Fig. 6. a) Exemplary image horizontal re-contrasted to uniform the average gray scale value in each column. In each column, the location of the jet centerline is identified as the peak of gray scale intensity. b) Gray scale profiles at three sample locations $x1 = 60$, $x2 = 150$ and $x3 = 240$ *pixels* are chosen from the image in a).

5 Results and Discussion

5.1 Numerical Results

The contour plot of the velocity magnitude and several representative velocity profiles along the radial coordinate are shown in Figure.5.1 for four numerical simulations. These test cases include a reference test with $u_s = 57\,m/s$ and $u_p = 0\,m/s$ (5.1a) and three active control cases with blowing ratios of $\beta_M = 1.47, 0.48, 0.12$ (Figure 5.1b,c,d).

The geometrical parameters are $h_s = 0.5\,mm$, $h_p = 1.5\,mm$ and $R = 5\,mm$, with $L = 1\,mm$, while the velocity of the primary jet is kept at $u_p = 38\,m/s$. The velocity profiles extracted along $\theta = 10^o, 30^o, 50^o$ and 70° and scaled with respect to the average velocity in eq.(11), with $\xi = (h_p/h_s)\beta_M$ as described in Section 2.4. The radial coordinate ranges from $r = R$ (at the Coanda wall) to the half width of the primary jet, namely $r = R + h_s + L + h_p/2$ and it is shifted and scaled to the range $\hat{r} \in [0, 1]$.

The evolution of the pressure gradient at the wall $P(\theta) = \partial_r p(\theta)/p(0)$ and the average velocity $G(\theta) = u_A(\theta)/u_A(0)$ is shown in Figure 8a and 8b respectively. Here, the average velocity is taken as the integral average in the radial coordinates, that is

$$u_A(\theta) = \int_{\hat{r}} u(\hat{r}, \theta) d\hat{r} \tag{18}$$

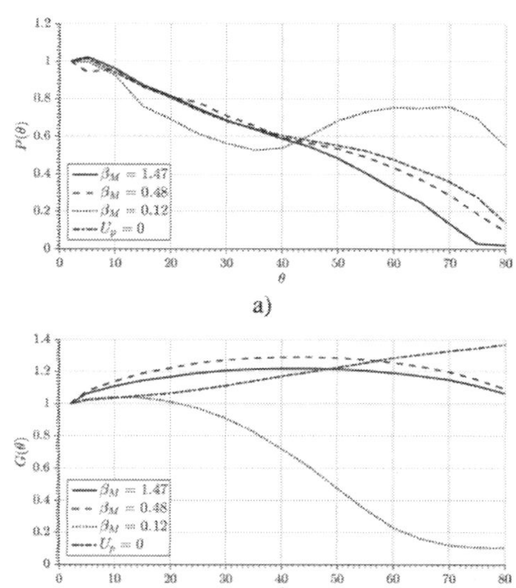

Fig. 8. Evolution of the radial pressure gradient $P(\theta)$ along the Coanda surface (a) and the averaged velocity $G(\theta)$ as function of the angular coordinate θ for the four numerical test cases shown in Figure 5.1.

Concerning the reference test case in Figure 5.1a with no primary injection ($u_p = 0\,m/s$), the velocity profiles display the classical shape of a wall jet, that is a combination of a boundary layer profile in the bottom part, close to the wall, and a free jet profile in the upper part.

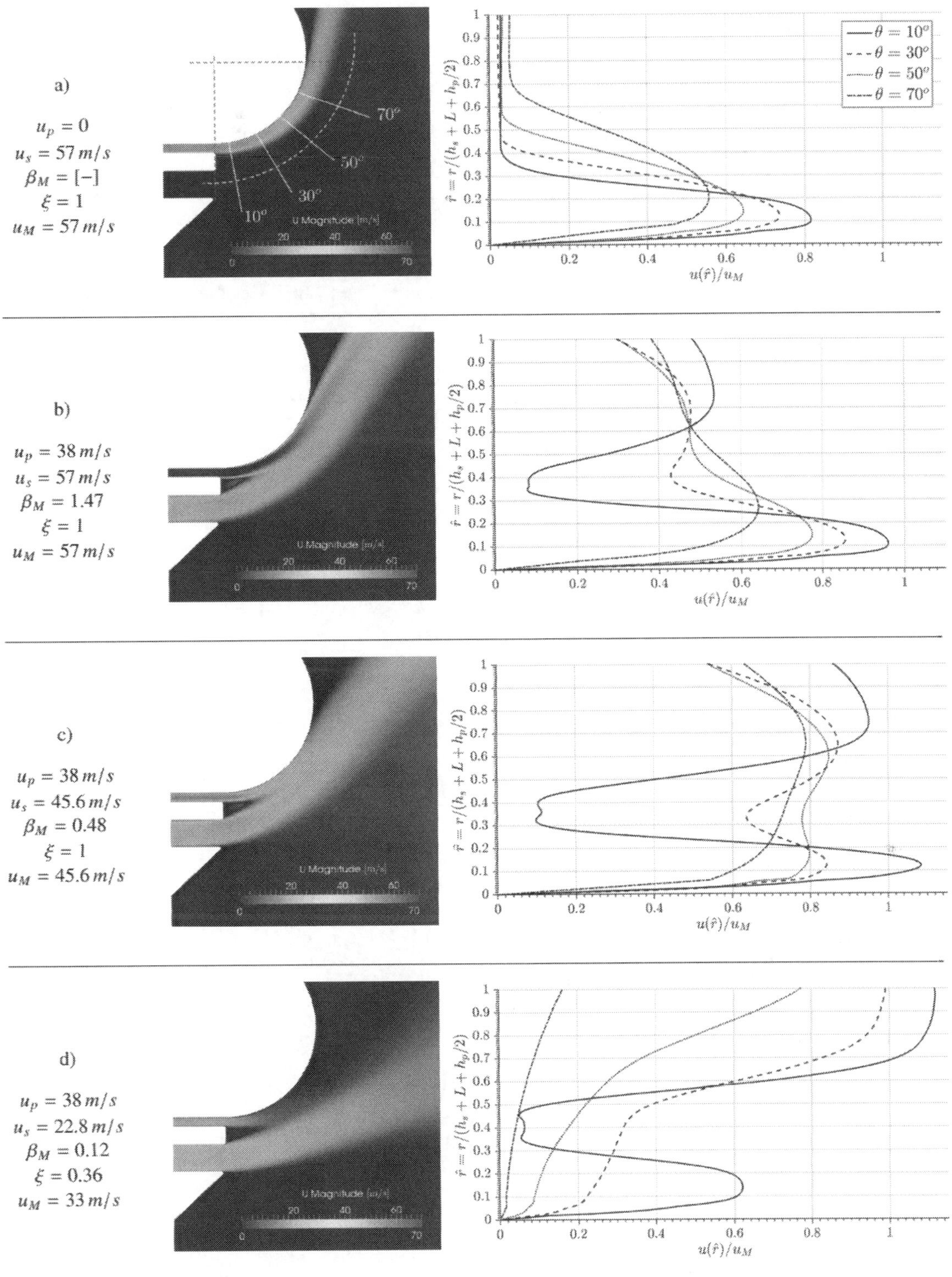

Fig. 7. Contour of the velocity magnitude at the nozzle outlet (left) and evolution of the velocity profiles (right) for four numerical test cases. The geometrical parameters are $h_s = 0.5\,mm$, $h_p = 1.5\,mm$, $R = 5\,mm$ and $L = 1mm$; the operating conditions are indicated on the left, including the average velocity u_M from eq.(11) that is used to scale the velocity profiles on the right.

The effects of turbulent and viscous diffusion is that of decreasing the maximum velocity and to increase the width of the jet. As a result, the pressure gradient at the wall (dash-dotted line in Figure 8a) drops almost linearly with the angular coordinate, while the average velocity (dash-dotted line in Figure 8b) increases. That is also visible from the velocity profiles: the decay of the maximum velocity is largely compensated by the thickening of the jet flow due to the entrainment of surrounding fluid.

When both injections are active, the trend of these macroscopic quantities remains rather unchanged for a wide range of momentum ratios. Figure5.1b reproduces the same results of Figure5.1a for a momentum ratio as large as $\beta_M = 1.47$, that is well beyond the saturation condition. This implies $\xi = 1$ and thus $u_M = u_s$ from equation (11). The velocity peak is only slightly influenced by the momentum ratio, despite the considerable broadening of the velocity profiles due to the momentum exchange. The pressure gradient (continuous line in Figure 8a) at the wall remains close to the reference case, departing slightly only for $\theta > 40^o$. The evolution of the average velocity (continuous line in Figure 8b) displays a slight parabolic shape, but remains of the order of $G(x) \approx 1$. Decreasing the momentum ratio to $\beta_M = 0.48$ does not significantly alter these behavior, as visible from Figure 5.1c and from the dashed line in Figures 8a,b.

These results prove that the model coefficient in eq.(14) remains independent from the momentum ratio β_M over a large range of operating parameters.

Such simplification, however, reveals its limit at very low momentum ratio, as can be seen from Fig.5.1c and the dotted line in Figure8a-b, regarding the numerical test case with $\beta_M = 0.12$. In this case, the evolution of the average velocity is mostly governed by the velocity of primary jet flow, as correctly predicted by eq (12). However, the separation of the secondary jet occurs much sooner than in the previous cases and this reduces the range $x \in [0, x^*]$ over which the pressure gradient driving the vectoring is acting. The limits of the integral in eq.(14), therefore the model coefficient (14), depend on the momentum ratio at the limit $\beta_M \ll \beta_M^S$. At larger momentum ratios, on the other hand the simplification of a constant model coefficient in eq.(14) appears largely justified.

5.2 Experimental Validation

Three geometrical configurations have been analyzed over a wide range of momentum ratios. The first two investigate the impact of the secondary opening h_s, changing from $h_s = 0.5\,mm$ to $h_s = 1\,mm$ while keeping the primary opening at $h_p = 1.5\,mm$ and the Coanda radius to $R = 5\,mm$. The third case investigates the impact of the Coanda radius, reduced to $R = 2\,mm$, while keeping $h_s = 0.5\,mm$ and $h_p = 1\,mm$. All the cases are performed with two velocities in the primary jet flow, namely $u_p = 28m/s$ and $u_p = 38m/s$. Four exemplary flow visualizations for the cases with $u_p = 28\,m/s$ at different momentum ratios are shown in Figure 9, together with the automatic centerline detection.

The measured vectoring coefficient as a function of the momentum ratio is shown in Figures 10a,b,c, and compared to the prediction of the empirical model in eq.(16) taking the model coefficient in eq.(14) as $C = 0.39(h_p/h_s)$. For the three investigated geometrical configurations, the model prediction are in close agreement with the experimental measurements. In particular, a comparison between Figure 10a and Figure 10b reveals the impact of the ratio of the two openings h_s/h_p, which, with remarkable agreement with the model prediction, sets the momentum ratio at the saturation at $\beta_M^S \approx h_s/h_p$. A comparison between Figure 10-b with Figure 10-c, on the other hand, shows the limited influence of the Coanda radius, in agreement with the formulation of the empirical model.

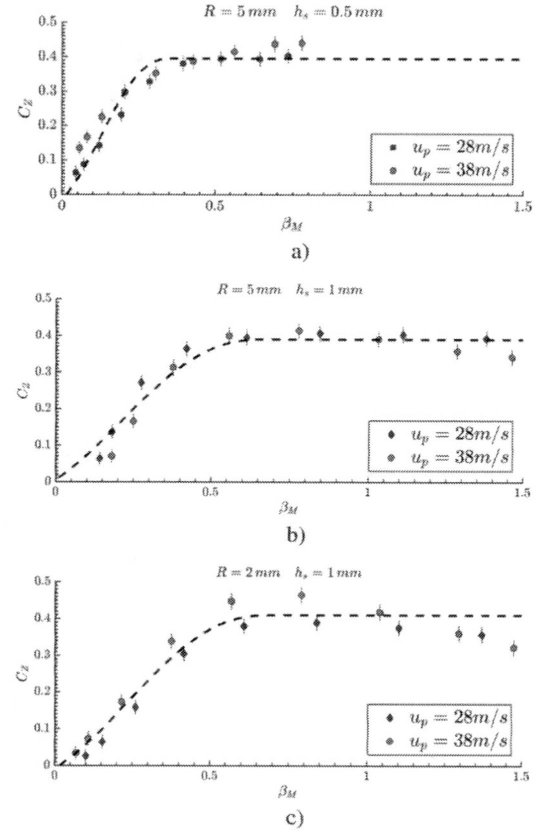

Fig. 10. Thrust vectoring coefficient C_Z as a function of the momentum ratio β_M for three geometrical configurations, comparing the prediction of the proposed empirical model with two sets of experimental points with $u_P = 28\,m/s$ and $u_P = 38\,m/s$. The primary opening is fixed at $h_p = 1.5\,mm$, the Coanda radius and the secondary opening are indicated in the plot title.

Finally, the influence of the primary velocity appears reasonably negligible for the large Coanda radius (10a,b), once again in agreement with the empirical model, but not at the lowest radius (10c). In this case a proper saturation is not reached at the highest velocity, but the vectoring performances appear to decrease after having reached

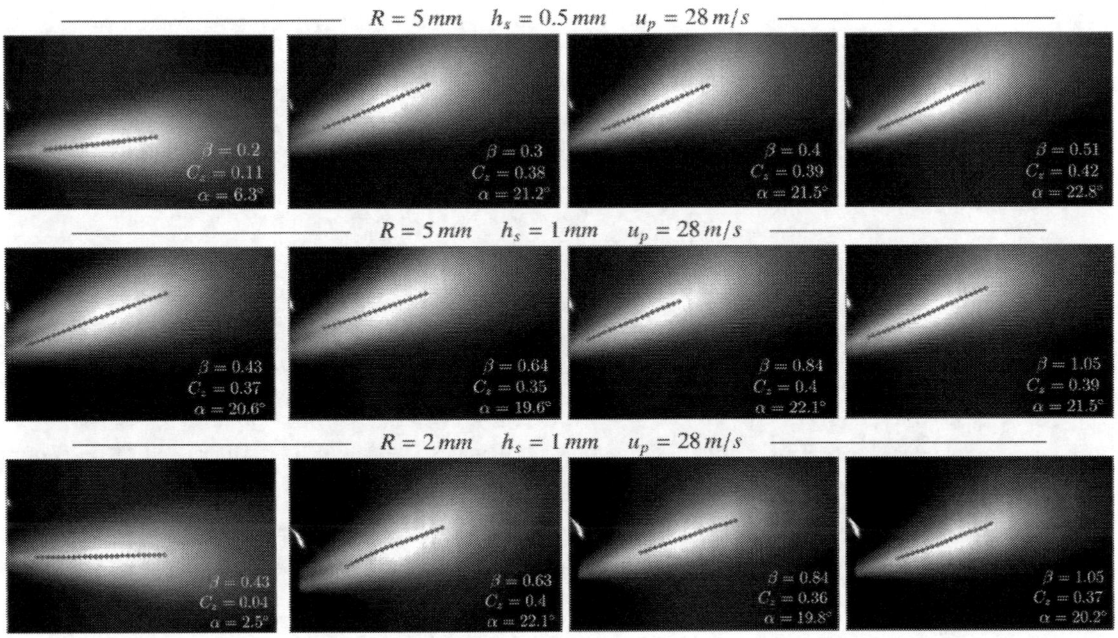

Fig. 9. Set of four down-sampled images and jet centerline detection for the three geometrical configurations investigated experimentally and for the primary jet flow at $u_p = 28\,m/s$. The measured vectoring angle and coefficient are indicated in each picture, together with the mass ratio β in equation.(1).

the maximum value. Such behavior, not predictable by the proposed model, could be due to the separation of the secondary jet flow, which might become more strongly linked to the momentum ratio for higher curvature parameters h_s/R. In this case, the assumption of constant C in eq.(14)– therefore the assumption of linear dependency of the transversal pressure gradient versus the momentum ratio– is clearly too restrictive.

6 Conclusions

An experimental, numerical and theoretical investigation of a co-flow Coanda fluidic design for the vectoring of a slot, incompressible jet has been presented. This design consists of a secondary injection, parallel to the primary, and a cylindrical surface to exploit the Coanda effect. An empirical model for predicting the vectoring performance as a function of the geometrical and the operating parameters has been formulated. This model describes the vectoring process as the result of a transversal pressure gradient, governed by the Coanda effect, and inertia, governed by the momentum exchange between the two jet flows.

The simplifying hypothesis postulated in the model derivation are validated numerically by means of 2D URANS simulations and the model predictions are compared to a large set of experimental data from flow visualization and image processing.

Within the investigated range of geometrical parameters, the results have shown that the jet vectoring is mostly governed by the ratio of the secondary to the primary opening of the jet flows, rather than the Coanda radius. This parameter defines in fact the momentum ratio above which the achievable vectoring angle saturates. For three different geometrical configuration and two primary velocities, the prediction of the model are found to be in close agreement with the experimental data over a large range of momentum ratios.

Acknowledgments

The authors gratefully acknowledge the Belgian FNRS for supporting M. A. Mendez with a FRIA grant and ArcelorMittal, for the financial support and for the permission to publish the material presented. We finally thank D. Resende and L. Paoli for their contribution in a preliminary experimental and the numerical work, conducted in the framework of a Short Training Program (STP) at the von Karman Institute for Fluid Dynamics.

List Of Symbols

Notation and Subscripts

$\partial_y(\bullet)$	partial derivative along y
$\partial_r(\bullet)$	partial derivative along r
$\partial_t(\bullet)$	partial derivative in time
p	primary injection
s	secondary injection

Symbols

C_z	thrust vectoring coefficient	–
D	nozzle discharge coefficient	–
f_s	image acquisition frequency	Hz
G	averaged velocity decay	–
h	nozzle aperture	m
L	distance between the two jet apertures	m
n_p	image resolution	$pixel^2$
n_t	number of images	–
p	pressure	Pa
P	wall pressure gradient decay	–
r	radial coordinate	m
R	Coanda radius	m
u	jet velocity	m/s
u_A	averaged profile velocity along r	m/s
u_M	averaged velocity from eq.(12)	m/s
x	horizontal coordinate	m
x_c	jet potential core length	m
y	vertical coordinate	m
α	vectoring angle	deg
β	mass flux ratio	–
β_M	momentum ratio	–
β^c	critical β	–
β^s	saturation β	–
ζ	curvilinear coordinate	m
θ	angular coordinate	deg
ξ	mixing coefficient	–
ρ	air density	kg/m^3

References

1. Deere, K., *Summary of Fluidic Thrust Vectoring Research Conducted at NASA Langley Research Center*, in *21st AIAA Applied Aerodynamics Conference* (2003), Vol. 2003-3800
2. Deere, K., *Computational Investigation of the Aerodynamic Effects on Fluidic thrust Vectoring*, in *36th AIAA/ASME/SAE/ASEE Joint Propulsion Conference and Exhibit* (2000), Vol. 2000-3598
3. Deere, K., *A Computational Study of a New Dual Throat Fluidic Thrust Vectoring*, in *41st AIAA/ASME/SAE/ASEE Joint Propulsion Conference and Exhibit* (2005), Vol. 2005-3502
4. Mason, M., S., Crowther, W.J., *Fluidic Thrust Vectoring of Low Observable Aircraft*, in *CEAS Aerospace Aerodynamic Research Conference* (2002)
5. Jackson, D. N., Smith, B.L., J. Fluid Eng. **129**, 902 (2007)
6. Bettridge, M.W., Smith B.L., Spall, R.E., Exp. Fluids **40**, 776 (2006)
7. Walker, S., *Lessons learned in the development of a national cooperative program*, in *33rd Joint Propulsion Conference and Exhibit, Joint Propulsion Conferences* (1997)
8. Hammond, D.A., Redekopp, L.G., J. Fluid Mech **338**, 231 (1997)
9. Smith, B.L., Glezer, A., Bull. Am. Phys. Soc. **39**, 1894 (1994)
10. Smith, B.L., Glezer, A., AIA Journal **43**, 2117 (2005)
11. Heo, J.-Y., Yoo, K.-H., Lee, Y., Sung, H.-G., Cho, S.-H., Jeon, Y.-J., Journal of Propulsion and Power **28**, 858 (2012)
12. Strykowski, P. J., Krothapalli, A., Forliti, D. J., AIA Journal **34**, 2306 (1996)
13. Flamm, J. D., *Experimental study of a nozzle using fluidic counterflow for thrust vectoring*, in *34th Joint Propulsion Conferences* (1998), Vol. 98-3255
14. Hunter, C.A., Deere, K.A., *Computational Investigation of Fluidic Counterflow Thrust Vectoring*, in *35th Joint Propulsion Conference and Exhibit, Joint Propulsion Conferences* (1999), Vol. 99-2669
15. Humes, Z., Smith, B.L., Minichiello, A.L., *Particle sosorting by aerodyamic vectoring*, in *Bull. Am. Phys. Soc.* (2005)
16. B. Newman, *Boundary Layer and Flow Control Principles and Applications* (Pergamon Press Inc., 1961), Vol. 1, chap. The deflection of Plane Jets by Adjacent Boundaries- Coanda Effect, pp. 232–264
17. Bradshaw, P., Tech. rep., AGARDograph n 169 (1973)
18. Carpenter, P.W., Green, P.N., Journal of Sound and Vibration **208**, 777 (1997)
19. Smith, B.L., Glezer, A., J. Fluid Mech **458**, 1 (2002)
20. Bremhorst, K., Hollis, P.G., AIAA Journal **AIAA 97-3348**, 2043 (1990)
21. Pack, L.G., Seifer, A., Journal of Aircraft **38**, 486 (2001)
22. Al-Asady, A.A.A., Abdullah,A.M., NJES **20**, 5 (2017)
23. Saghafi, F., Banazadeh, A., The Aeronautical Journal **3117**, 17 (2008)
24. Saghafi, F., Banazadeh, A., *Coanda Surface Geometry Optimization for Multi-Directional Co-flow Fluidic Thrust Vectoring*, in *Proceedings of ASME Turbo Expo 2009* (2009)
25. Song, M.J., Yoon, S.H., Chang, H.B., Cho, Y.H., Lee, Y, *A Study on the Performance of the Fluidic Thrust Vector Control Utilizing Supersonic Coanda Effects*, in *5th Symposium on Integrating CFD and Experiments in Aerodynamics* (2012), pp. 253–261
26. Zmijanovic, V., Lago, V., Leger, L, Depussay, E., Sellam, M., Chpoun,A., Progress in Propulsion Physics **4**, 227 (2013)
27. Bharathwaj, R, Giridharan, P, Karthick,K., Prasath, C.H., Prakash, K.,M., IOP Conf. Ser.: Mater. Sci. Eng. **149**, 012210 (2016)
28. Trancossi, M., Dumas, A., SAE Technical Papers (2011)
29. Kirshner, J.M., Katz, S., *Design Theory of FlFluidic Components* (Academic Press, 1975)
30. Dumitrache, A., Frunzulica,F., Ionescu,T.C. , in *Nonlinearity, Bifurcation and Chaos - Theory and Applications*, edited by Awrejcewicz, J., Hagedorn, P. (InTech, 2012)
31. Reichardt, H., Z. Angew. Math. Mech. **21** (1941)

32. Schlichting, H., *Boundary-Layer Theory* (McGraw-Hill, 1960)
33. S.L. Abramovich, G.N., *The Theory of Turbulent Jets* (The MIT Press, 1963)
34. Menter, F.R., AIAA Journal **32**, 1598 (1994)
35. Versteeg, H.K., Malalasekera, W., *An Introduction to Computational Fluid Dynamics* (Longman Scientific & Technical, 1995)
36. Mendez, M.A., Buchlin, J.-M., *Quantitative flow visualization of confinement-driven instabilities of an impinging slot jet*, in *11th International Symposium on Particle Image Velocimetry (PIV2015)* (Santa Barbara, California, US, 2015)
37. Mendez, M.A., Scelzo, M.T., Buchlin, J.-M., Exp Therm Fluid Sci **91**, 256 (2018)

Gas dispersion modelling in a critical infrastructure area

Petr Michálek[1,*], *David* Zacho[1] and *Nikola* Žižkovský[1]

[1]VZLU – Czech Aerospace Research Centre, Department of Aerodynamics, Beranových 130, 199 05Praha 9, Czech Republic

Abstract. Experimental investigation of gas dispersion in a critical infrastructure area was performed using a boundary layer wind tunnel. The dispersion of gases is evaluated from the point of view on required function of security systems in case of critical infrastructure accident involving airplane attack. Physical model of the critical infrastructure installed in the wind tunnel is subjected to measurement of gas concentration with flame ionisation detectors.

1 Introduction

Critical infrastructures, such as power plants, are fundamental part of state infrastructure and must provide power for the other necessary parts to function properly. The power plants might be vulnerable to terrorist attacks or accidents using large aircrafts, i. e. Boeing 737 or Airbus A330, which contains thousands litres of jet fuel. The impact of the aircraft and subsequent fire of the fuel might damage the critical infrastructure and thus cause its malfunction. This paper describes modelling of gas dispersion in an area of the critical infrastructure facility in case of a large aircraft impact.

The fire and smoke dispersion of burning jet fuel in the vicinity of critical infrastructure buildings will be evaluated in a computational model. The fire and smoke dispersion will be evaluated from the point of view on required function of security systems in case of critical infrastructure accident involving intentional airplane attack. The physical modelling of the dispersion inside critical infrastructure area will serve for validation and comparison with the computational model.

Recent and not so recent work in VZLU in the area of gas dispersion included dispersion around train station [1] and dispersion in large urban area [2] and detailed model from the same area [3]. Dispersion over an coal mine [4] and effects of wind fences porosity were studied in Czech Republic as well [5]. The safety of nuclear power plants against aircraft attack was studied e.g. in [6].

2 Gas dispersion modeling

2.1 The boundary layer wind tunnel

The experimental work was done in the boundary layer wind tunnel (BLWT) in VZLU, a. s. The BLWT is an open-circuit wind tunnel with cross section 1.8 m wide and 1.5 m high with adjustable ceiling. The test section for boundary layer development is 13.6 m long. The tunnel is powered by 55 kW fan. Maximal velocity above the boundary layer is ca. 25 $m \cdot s^{-1}$. Reference velocity above the boundary layer is measured with Pitot-static probe and hot-wire probe. The model space is equipped with 1.75 m dia. turntable and traversing device for moving the probe in all three directions x, y, z. The simulation means can create agricultural, suburban or urban boundary layer according to Eurocode 1 [7]. The boundary layer developed over the entire test section with plastic sheet with 7 mm high truncated cones and a rectangular barrier 140 mm high, which served as turbulence generator. This boundary layer corresponds with terrain type III (suburban).

2.2 The critical infrastructure model

The physical model of the critical infrastructure was manufactured in scale 1:660 on the basis of supplied CAD model. The model includes buildings larger than 10 m in full scale and connecting above-ground pipelines. Railway tracks and power lines were not modelled. Buildings were cut out of synthetic wood and cooling towers were 3D printed. The terrain inside the studied area is nearly flat. The model is depicted in Figure 1; main flow direction is from right to the left.

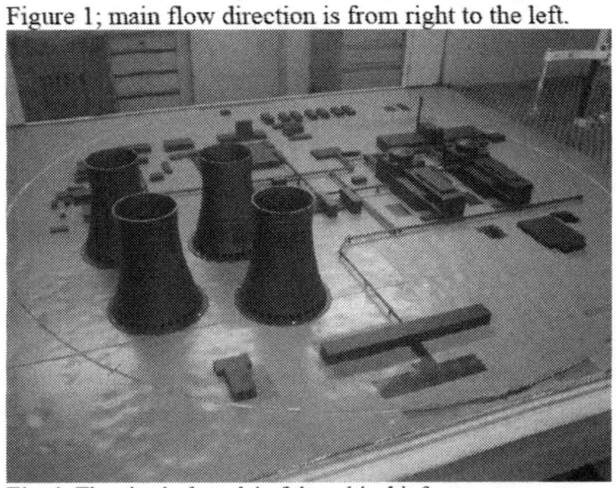

Fig. 1. The physical model of the critical infrastructure

* Corresponding author: michalek@vzlu.cz

The attack scenario describes worst possible impact area between the two main block buildings, so the emission source was installed there near the second main block building. The emission source is a small chamber sized 25 × 25 × 10 mm covered with filtration fabric. The source is visible as white square in Figure 1 between the main block buildings at left side. The main block buildings are especially vulnerable through the ventilation openings on their facades, so the concentration measurements were focused on the positions of the openings. The model was installed on a turntable, so measurements of different wind directions were made possible. Scheme of ventilation openings are presented in Figure 2, where the openings are marked on the buildings with numbers. The southwest facade (not depicted) also contains one ventilation opening No. 23. The scheme of entire critical infrastructure area is shown in Figure 3, where north direction is at the top of the picture. The impact point and the source of emission in marked with red dot.

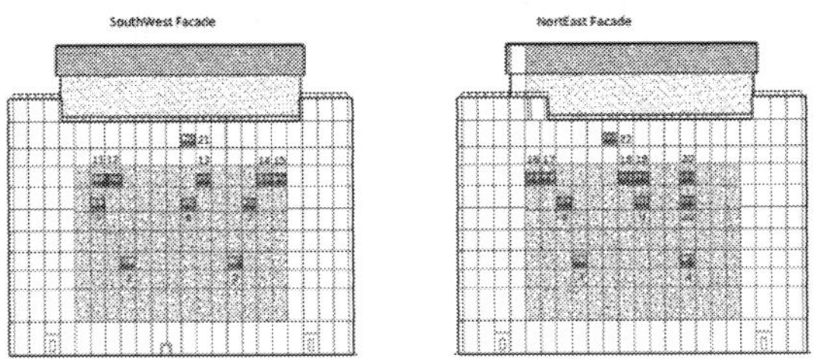

Fig. 2. Ventilation openings on the main block building

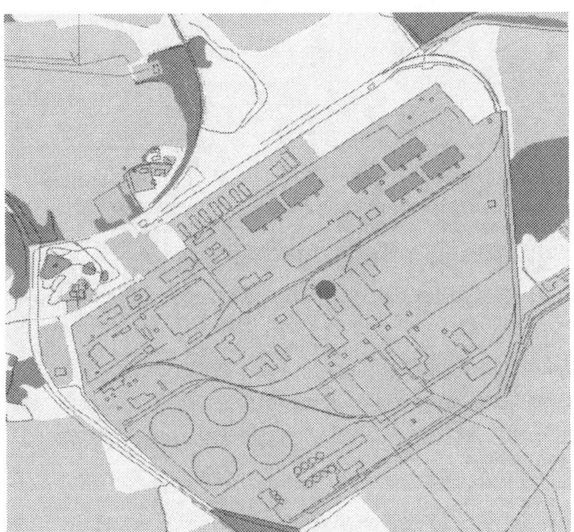

Fig. 3. Scheme of the entire area

2.3 Flow and dispersion measurements

The flow field of the incoming boundary layer was measured with single-sensor hot-wire probe and the concentration at desired points was measured with flame ionization detector (FID). The vertical mean velocity profile is shown in Figure 4, where U4 ($m \cdot s^{-1}$) is longitudinal component of velocity vector, I_{u4} (%) is turbulence intensity of the component U and z (mm) is vertical height above the ground. The velocity profile corresponds with its typical logarithmic shape.

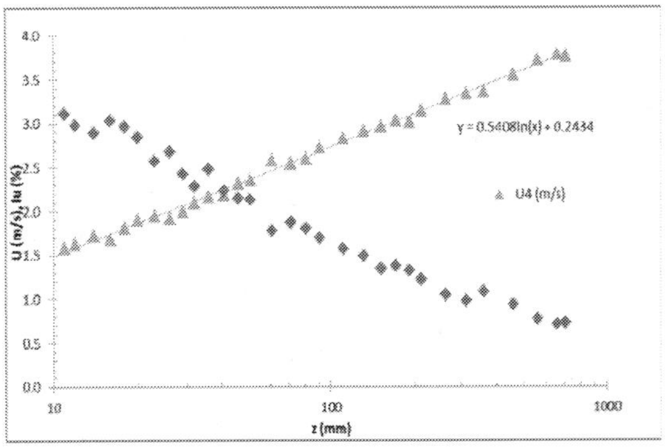

Fig. 4. The boundary layer velocity profile

The wind tunnel velocity was set to ca. 4.4 $m \cdot s^{-1}$ above the boundary layer and this reference velocity was used in the dispersion experiments as well. For this size of the modeled area, the small-scale similarity criteria applies, which means that the minimal Reynolds roughness number Re* has to be satisfied. This ensures invariance of the flow pattern around the structure on the main flow velocity. The Reynolds roughness number is defined as

$$Re^* = z_0 \times u^* / \nu = 8.7 \geq 1 \tag{1}$$

where ν is kinematic viscosity of air ($m^2 \cdot s^{-1}$), u* is frictional velocity ($m \cdot s^{-1}$) and z_0 is aerodynamic roughness (m). Values of $u^* = 0.22\ m \cdot s^{-1}$ and $z_0 = 0.64$ mm were derived from the logarithmic fit of the mean velocity profile.

A mixture of tracer gas (ethane) and compressed air was released from the source using flow controllers. The volume flow of ethane was 0.3 $l \cdot min^{-1}$ and air flow was 5.1 $l \cdot min^{-1}$. Concentration field was measured using suction tube mounted on a traversing device. The suction tube was connected via stainless steel tubing to a peristaltic pump and a flame ionization detector (FID). FIDs use ionized hydrogen flame to burn the incoming sample and then measure the electric current through the flame. The FID signal is measured with the electrometer and then sampled and saved in computer. Sampling frequency was 100 Hz and sampling time was 1 min for each measuring point and then 30 s waiting to move to the next point so that the previous sample clears the tubing. FID calibration was made with calibration gas mixture with precise concentration 1000 ppm (parts per million) of ethane in air.

3 Results

The concentration measurements were made at the ventilation points for all of the main wind directions and selected other wind directions.

The FID signals were converted into concentration with LabView software and then mean values were calculated for each point. The response of the measuring chain does not allow fluctuation measurements. Figure 5 presents concentration near ventilation openings at SE wall of block 1, where Sv1 etc. corresponds with numbers of ventilation openings in Figure 2. Wind direction at 0 ° corresponds to SE wind. Maximal concentration values appear at wind direction 180 °, when the SE wall of block 1 lies in the wake of block 2 and thus accumulation of emissions from the source can occur in the wake. Other values near zero can be considered as zero.

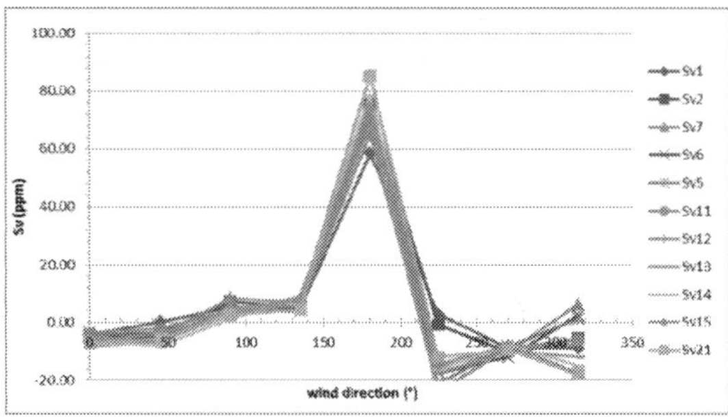

Fig. 5. Concentration near ventilation openings, SE wall of block 1

Figure 6 presents concentration at ventilation openings at NW wall of block 1, except of opening Sv23, which lies at SW wall. High concentration was measured at wind direction 0 °, which equals to SE wind here. In this wind direction the NW wall lies in the wake of block 1, which supports accumulation of the emission. Lower values were measured also at 180 degrees (NW wind), where NW wall lies in the wake of block 2.

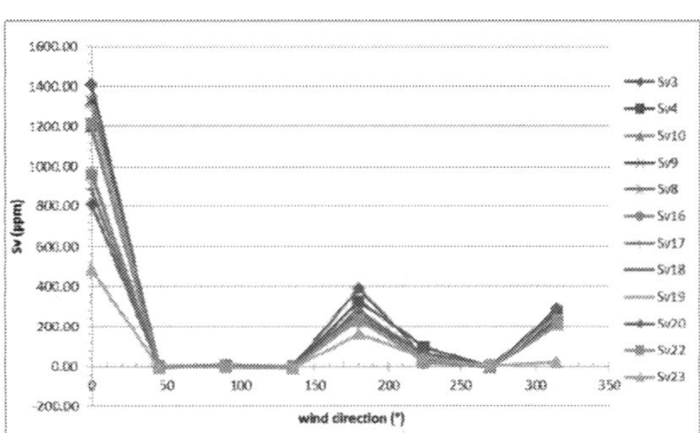

Fig. 6. Concentration near ventilation openings, NW wall of block 1

Figure 7 presents concentration at ventilation openings at SE wall of block 2. These concentration exhibit high values again at wind direction 180 ° (NW) and 225 ° (N), where SE wall lies in the wake of block 2. The values reach the highest level, because this wall is the nearest from the source.

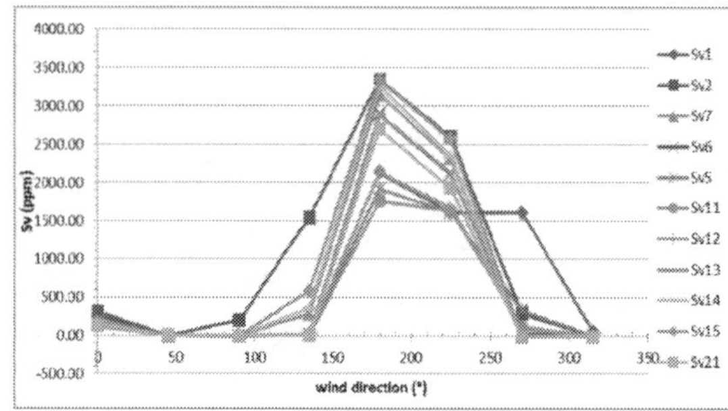

Fig. 7. Concentration near ventilation openings, SE wall of block 2

Figure 8 presents concentration at ventilation openings at NW wall of block 2. The values rise at wind direction 45 ° (S), where the NW wall lies in the wake of block 2. Line representing opening Sv23 reaches its highest value at 225 ° (NE), where the opening No. 23 lies in the wake of block 2 and at 0 ° (SE), where it lies near the wake of block 1.

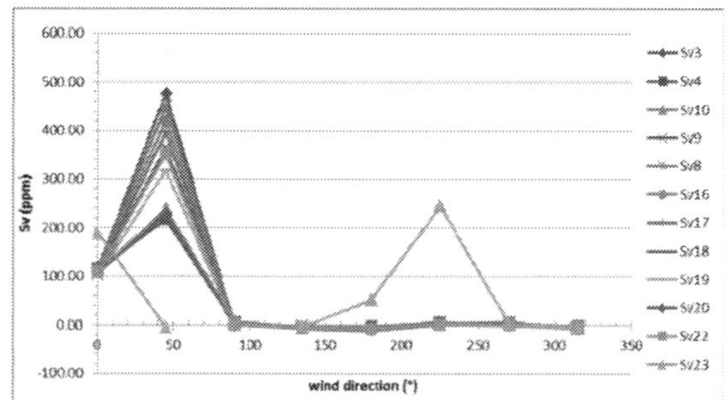

Fig. 8. Concentration near ventilation openings, NW wall of block 2

4 Conclusions

Gas dispersion was studied on a physical model of critical infrastructure area. Concentration was measured at the ventilation openings of the main block buildings. Highest values were found in the case where the walls with openings lie in the wake of the main block building. The experimental data will be used for verification and validation of the computational model. The results and routines of the computational model could be used for other critical infrastructure objects in the future, for increase of reliability and safety of buildings and technology.

Acknowledgements

Authors thank for the support from the Ministry of Interior of Czech Republic under project VI20172020092 – Simulation of fire and smoke dispersion in critical infrastructure area in case of accident or aircraft attack.

References

1. R. Ulman, Z. Janour, H. Sedenkova, J. Drbohlav, *Proceedings of EACWE4*, Prague (2005), paper #189
2. P. Michálek, D. Zacho, *Engineering Mechanics 2015, p. 85* (2015)
3. P. Michálek, D. Zacho, *EPJ Web of Conferences* **143,** 02074 (2017)
4. S. Nosek, Z. Janour, L. Kukacka, K. Jurcakova, R. Kellnerova, E. Gulikova, *EPJ Web of Conferences* **67,** 02084 (2014)
5. S. Nosek, R. Kellnerova, K. Jurcakova, Z. Janour, H. Chaloupecka, M. Jakubcova, *EPJ Web of Conferences* **114,** 02084 (2016)
6. J. Králik, *Applied Mechanics and Materials* Vol. **617** (2014), pp 76-80
7. Eurocode 1: Actions on structures – Part 1-4: General actions – Wind loads (EN 1991-1-4), European Committee for Standardization, Brussels (2005)

Influence of boundary conditions on the hydrodynamic forces of an oscillating sphere

Domenica Mirauda[1,*], *Marco* Negri[2], *Luca* Martinelli[2], and *Stefano* Malavasi[2]

[1]School of Engineering, Basilicata University, 85100 Potenza, Italy
[2]Department of Civil and Environmental Engineering, Politecnico di Milano, 20133 Milano, Italy

Abstract. The design of submerged structures in sea currents presents certain problems that are not only connected to the shape of the obstacle but also to the number of acting forces as well as the correct modelling of the structures dynamic response. Currently, the common approach is that of integrated numerical modelling, which considers the contribution of both current and structure. The reliability of such an approach is better verified with experimental tests performed on models of simple geometry. On the basis of these considerations, the present work analyses the hydrodynamic forces acting on a sphere, which is characterised by a low mass ratio and damping. The sphere is immersed in a free surface flow and can oscillate along the streamwise and transverse flow direction. It is located at three different positions inside the current: close to the channel bottom, near the free surface and in the middle, and equally distant from both the bottom and free surface. The obtained results for different boundaries and flow kinematic conditions show a relevant influence of the free surface on the hydrodynamic forces along both the streamwise and transverse flow directions.

1 Introduction

Knowledge of the dynamic response of submerged and not-completely submerged structures in sea currents is fundamental for the correct design of a large number of off-shore structures (e.g., ocean risers, wind farms, marine turbines, oil conduits, platforms, buoys, power wave converters, and submerged floating tunnels). Although this topic has been considered in several previous studies, many questions remain unanswered. For example, in these types of structures, the modelling of the hydrodynamic response is usually subject to a linearization of the hydrodynamic force components considering: the inviscid flow, body subject only to conservative forces due to gravity and irrotational motion. Such linearization implies that the fluid oscillations are smaller than the wave length. As a consequence, the modelling of the hydrodynamic response of the oscillating structure does not allow us to consider the effects of refracted or steep waves or secondary, or of superior order, hydrodynamic effects. In addition, modelling situations in which the displacements of both submerged and floating bodies are larger than their typical dimensions is not allowed. Moreover, in most cases, these structures are modelled without considering the presence of free surface (water-air interface) effects, which therefore overlooks an important interference factor for body movements. In fact, this hypothesis can hide the effects of vortex-induced vibrations (VIV), which are not negligible whenever resonance or lock-in phenomena (where the frequency of VIV is close to the natural frequency of the submerged or floating system) occur. In fact, such phenomena are two of the most critical mechanical conditions for these structures.

Considering the complexity of the research topics and numerous different forces stressing the structure, integrated numerical modelling seems to be the most suitable method to consider the contributions of the most significant forces [1-3]. Additionally, the presence of a free surface influences hydrodynamic actions in a peculiar way, both in the case of a steady flow, such as a sea current, and unsteady ones, such as waves [4].

To assess their reliability, the resolving methods derived from integrated numerical modelling need to be compared with the results of the experimental analyses. Given the complexity of the phenomena, these involve models of simplified geometries (e.g., spheres, cylinders, and rectangles). In this complicated scenario, most studies in the literature focus on two-dimensional structures, such as rectangles and cylinders [5-7].

Few studies report the analysis of hydrodynamic forces acting on a three-dimensional body, such as a sphere [8-10]. In particular, Govardhan and Williamson [8] investigated the fluctuating transverse forces acting on an elastically mounted sphere that was constrained to move transversely to the flow direction. These forces were measured by force balance and utilising linear variable displacement transducers. The other two papers [9,10] instead show a trend of the mean drag coefficient of tethered buoyant spheres that are free to oscillate

* Corresponding author: domenica.mirauda@unibas.it

 https://doi.org/10.1051/epjconf/201818002067

along both the streamwise and transverse flow directions.

All of the above mentioned studies refer to spheres that are completely immersed in the current and distant from the free surface and channel bottom. In contrast to previous studies, this paper analyses the influence of both the free surface and channel bottom on the hydrodynamic forces acting on an elastically mounted sphere that is free to move in the streamwise (x) and transverse (y) flow directions.

2 Experimental set-up

The experimental set-up consisted of a deformable rod-sphere system immersed in a steady flow and in a horizontal, non-tilted, Plexiglas open water channel with a rectangular cross section of 0.5 x 0.6 m and a length of approximately 7 m (Fig. 1a). The sphere was composed of PVC, water filled, and with a diameter $D= 0.087\ m$ [11,14]. The surface of the sphere was covered with paint to reduce the surface roughness. A flexible cylindrical rod was used to retain the sphere and connect it to a fixed support structure.

The rod had a circular cross-section. The longer upper part was composed of Polyoxymethylene (Delrin) with a diameter $D_1= 0.012\ m$; the shorter lower part was composed of stainless steel with a much smaller cross-section and a diameter $D_2= 0.003\ m$ to avoid perturbing the flow in the proximity of the sphere.

A clutch linked the rod to a fixed support structure that was located above the channel (Fig. 1) and allowed the deformable length of the rod and the immersion of the sphere to change.

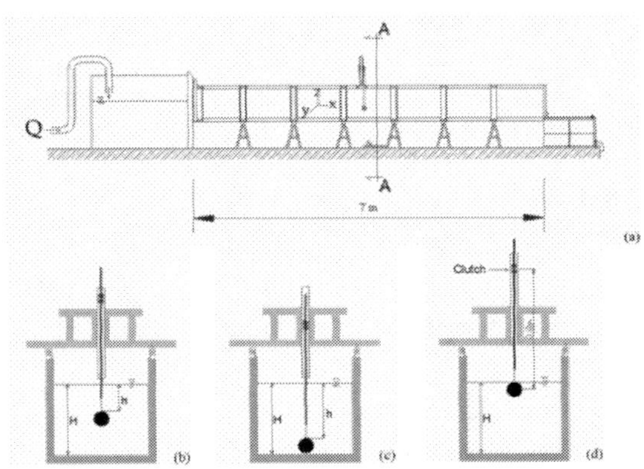

Fig. 1. Experimental channel; the transverse section of the channel and sphere positions at (b) h*=2, (c) h*=3.97, and (d) h*=0.

The rod flexibility allowed the sphere two degrees of freedom (in the streamwise, x, and transverse, y, directions).

The spherical body was immersed in a free surface flow with a water depth equal to 5 times the sphere diameter ($H= 0.435\ m$) to maintain the blockage coefficient ($\gamma_b= 0.03$, the ratio between the body frontal area and transverse section of the channel) constant. The flow had a constant turbulence level of approximately 10 %, whereas the mean flow velocity, U, changed between 0.2 and 0.5 m/s.

Three different boundary conditions of the flow around the sphere were tested and were denoted by value h^* of the ratio between h (distance of the sphere upper surface from the free surface) and the sphere diameter, D. We considered a quasi-symmetric condition ($h^*= 2$) with the sphere equally distant from the free surface and channel bottom (Fig. 1b) as well as two other conditions of asymmetric bounded flow (Figs. 1c and 1d): one with the sphere close to the channel bottom ($h^*= 3.97$) and the other with the sphere close to the free surface ($h^*= 0$).

The sphere displacements for the different flow velocities, U, were measured by image analysis through a Charge Coupled Device (CCD) camera, which allowed reconstruction of the sphere trajectories in the horizontal plane (x-y). The camera, which was located below the water channel (Fig. 1a), permitted us to memorise the body *2D* displacements due to the presence of a marker placed below the sphere in-line with its barycentre [11-14]. The time history of the *2D* movement of the sphere was provided by using a blob analysis algorithm to recognise the position, dimension and shape of the marker on each image. The sequence was pre-processed by a filter procedure. We set the level of the threshold filter at 100/255, which is a value that assured the negligible influence of the filtering on the calculated displacement. The high quality of the acquisitions and blob detection procedure used for the identification of the marker minimised the measurement uncertainty of the sphere displacements. This uncertainty was assumed to be less than the image's pixel dimension, which ranged between 0.4mm and 0.5 mm (with an image resolution between 18.5 and 23 pixel/cm). The position of the sphere on each frame was provided considering the barycentre of the marker, which had a mean area of approximately 100 pixel^2. Moreover, during the procedure of the displacement calculation, we controlled the dimension of the marker on the image so that it did not change more than 5 % along the sequence [11-14].

3 Dynamic characterisation of the rod-sphere system and determination of the hydrodynamic forces

From a mechanical point of view, the composite rod acts as an elastic restraint for a sphere in the x and y directions. Furthermore, due to its small mass, its dynamics can be disregarded, and the fluid forces on the sphere can be simply computed by use of the dynamic equilibrium equation of a sphere that is considered to be a rigid body as well as by knowledge of its velocity, acceleration and forces exerted by the rod.

We assumed that the oscillations were of a small amplitude so that: (a) the forces exerted by the rod could be linearly related to the sphere displacement through a stiffness coefficient, k, and (b) the rod could also be linearly related to the velocity through a damping coefficient, c. The values of the spring stiffness, k, and damping coefficient, c, were determined from the modal properties estimated through dynamic experimentation in air as well as a parametric technique with reference to the geometry and structural boundary conditions of the experimental set-up described in Section 2. In particular, they were evaluated by disregarding the mass of the rod and by examining the motion of the sphere in the y direction, assuming that the rod-sphere system behaves like a linear single degree-of-freedom (SDOF) oscillator of mass equal to that of a sphere filled with water ($m= 0.46\ kg$). A comparison of the results with the theory of the SDOF oscillators permitted an assessment of the validity of the assumed linearity hypothesis for the rod forces at the displacement level involved by the testing in the water channel described in Section 2.

When using dynamic testing, it is easy and reliable to estimate the stiffness, k, if the mass is known, but damping is one of the most difficult modal structural properties to estimate. This is due to several reasons. One is the representation through viscous damping of different forms of dissipation, which can be due to several mechanisms of an internal or external nature, such as hysteretic damping, Coulomb-friction damping, velocity squared damping (hydrodynamic force), or other cinematic variables that are different from the velocity. Another reason is the measurement noise, which requires some form of averaging in the estimation process to eliminate its effects.

The classical method of the logarithmic decrement (Log-Dec) was adopted here to estimate the modal parameters of the rod-sphere system and was supplemented with averaging to mitigate the noise effects. The Log-Dec method can strictly only deal with SDOF systems, whereas in the case study, the length of the rod, its mass (albeit small) and the spatial distribution of the mass on the sphere only lead to an approximation of a SDOF oscillator, and therefore, more than one mode is possible. However, in practice, whenever the natural frequencies of the system are sufficiently separated, as in the case analysed here, the response of a SDOF oscillator can be obtained by isolating the single mode of interest utilising a band-pass filter. Care must be exerted in choosing the type of filter since it distort the signal in the time domain and cause bias error.

The adopted procedure is as follows:

a) recording of several free decay oscillations (Fig. 2) of the sphere were set free to oscillate from an initial position that is displaced in the flow-transverse direction, y, by 5 cm with respect to the static equilibrium position. A Brüel & Kjær Type 4384 piezoelectric accelerometer with a sensitivity of $1.004\ pC/ms^{-2}$ at $159.2\ Hz$ was used along with a *16 bits* ADC at the sampling frequency of 500 Hz;

b) estimate the power spectral density (PSD) function from the recorded signal using the Welch's method [15] and divide the input signal into eight overlapping segments with a 50 % overlap (a Hamming window is used on each segment before estimating the PSD);

c) identify the PSD of the spectral bell and the damped natural frequency ($f_d=1.422\ Hz$) of the first mode in the air (Figs. 3 and 4). Due to the small value of damping, this frequency can be considered as that of the un-damped structure, f_n;

d) filter the free decay signal (Figs. 6 and 7) with a band-pass filter centred on f_n to obtain the response of an SDOF (an order two band-pass Butterworth filter, Fig. 5, with lower and upper cut-off frequencies of 0.5 Hz and 2 Hz, respectively, is applied to the time series first in the forward direction and then backwards to avoid phase distortion of the filtered signal);

e) separate the recorded signal into single episodes of free decay oscillations;

f) transform to the absolute value;

g) represent the results in a semi-log scale of each free decay oscillation (Figs. 8 and 9) and linearly interpolate the peaks to obtain the angular coefficient of the line, which when divided by $2\pi f_n$, directly provides the value of the critic damping $\zeta = c/[2(km)^{0.5}]$.

Some tests were prematurely stopped (e.g., at $t=340\ s$ in Fig. 2) due to the large coupling of the longitudinal (x) and lateral (y) vibration components. The non-perfect linear shape of the elastic rod induces this type of coupling and is also partially responsible for the deviation from linearity in the system response at low amplitudes, which is clearly shown by the scattering of the peaks in Figures 8 and 9 from 55 s onward. Also note that for the small values of acceleration at approximately 100 s shown in Figure 9, the accelerometer is close to its operational limits, and the effect of the signal to noise ratio may have contributed to the scattering in spite of the band-pass filtering.

This scattering can adversely influence the estimation of the damping factor value. Therefore, this was calculated for each single free decay episode; first, by using all of the peaks, and then, by subsequent confirmation using only the larger amplitude peaks corresponding to the first 20 % of each signal, which are less affected by the effects of the rod geometric non-linearity. The results are listed in Table 1. Nearly negligible differences were obtained from the two different estimates. A mean value of $\zeta= 0.00443$ was finally selected from the results of the second column in Table 1.

Assuming a mass $m= 0.46\ kg$ from the natural frequency $f_n = 1.422\ Hz$ of the first mode, the stiffness and the damping coefficient (with $\zeta= 0.00443$) in the air of the SDOF system can be computed as:

$$k = (2\pi f_n)2m=36.72\ kg/s^2 \quad (1)$$

$$c = 2\zeta(km)^{0.5}=0.036\ kg/s \quad (2)$$

The values of f_{nw} and ζ_w in water, which were equal to 1.213 Hz and 0.0038, respectively, were obtained by multiplying the values of the same parameters in the air by the coefficient $[m^*/(m^*+C_A)]^{1/2}$ in which m^* is the ratio between the system mass, m, and displaced fluid mass, m_d ($m^*=m/\pi\rho D^3/6= 1.34$) and C_A is the potential added mass coefficient ($C_A= 0.5$ for a sphere) when taking into account the water presence according to the paper by [8].

https://doi.org/10.1051/epjconf/201818002067

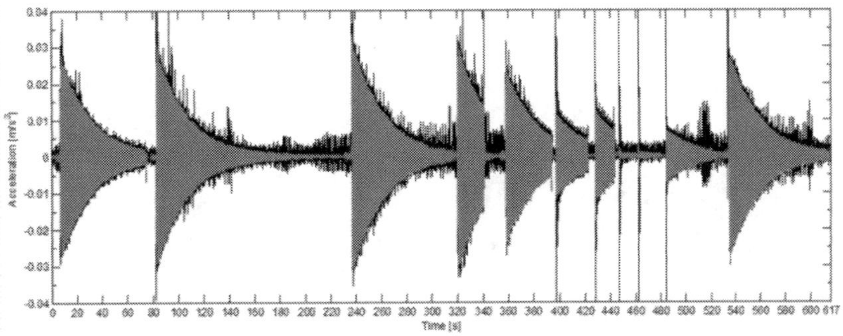

Fig. 2. The recorded acceleration signal showing several free decay episodes: black = unfiltered, grey = after filtering.

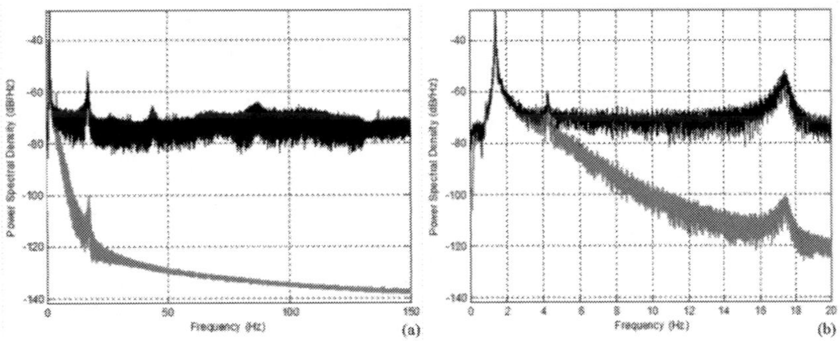

Fig. 3. Estimated power spectral density functions of accelerations: black = unfiltered acceleration signal, grey = after filtering. Overall view (a) and zoom-in (b).

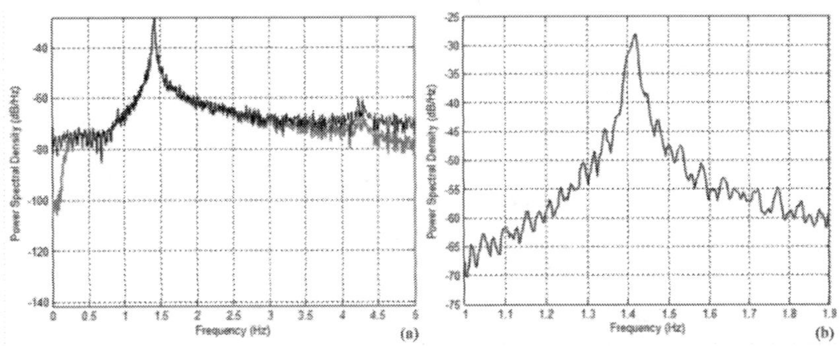

Fig. 4. Estimated power spectral density functions of accelerations: black = unfiltered acceleration signal, grey = after filtering. The larger zoom-in level (a and b) showing no attenuation of the spectral bell due to band-pass filtering.

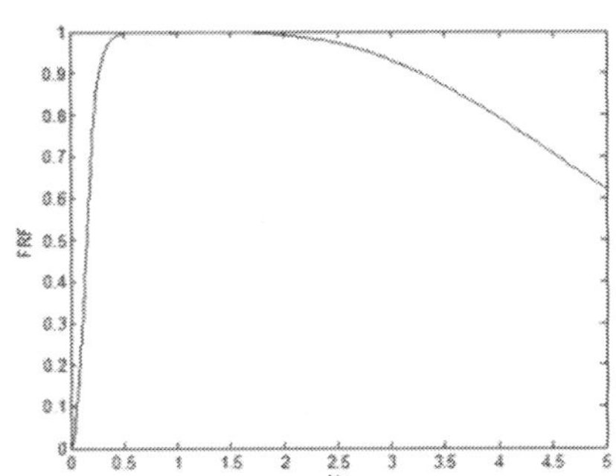

Fig. 5. The Frequency Response Function (FRF) of the adopted Butterworth 2nd- order filter.

* Corresponding author: domenica.mirauda@unibas.it

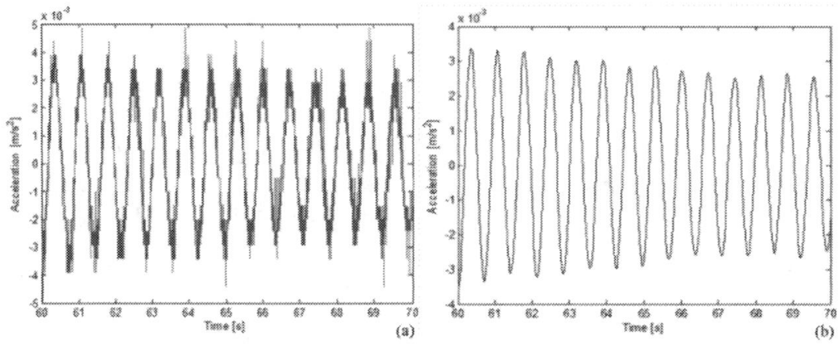

Fig. 6. The detail of the acceleration signal (a) before and (b) after filtering for the free decay oscillations episode at a time of 7-73 s.

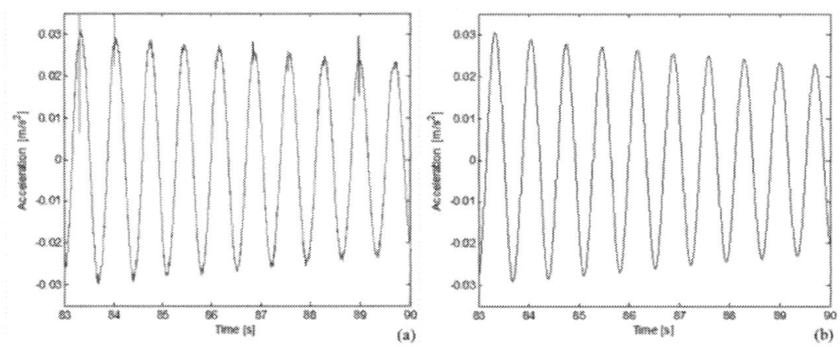

Fig. 7. Detail of acceleration signal (a) before and (b) after filtering for the free decay oscillations episode at a time of 83-200 s.

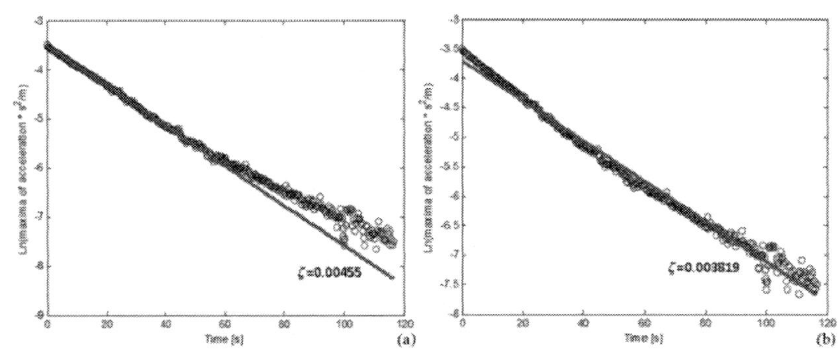

Fig. 8. Linear fit for the free decay oscillations episodes at a time of 83-200 s: (a) using the peaks of the first 20 % of the signal and (b) using all of the peaks.

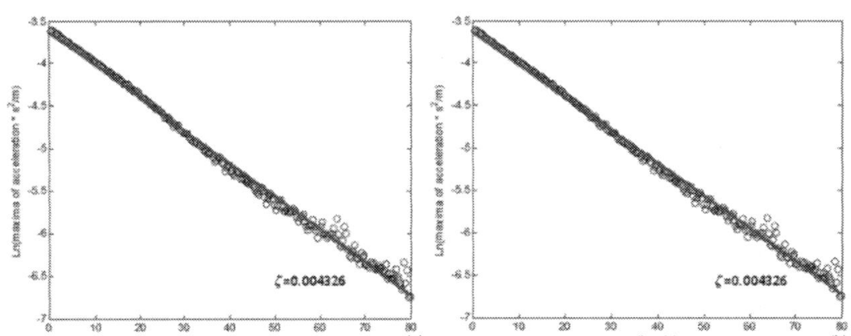

Fig. 9. Linear fit for the free decay oscillation episodes at a time of 537-616 s: (a) using the peaks of the first 20 % of the signal and (b) using all of the peaks.

Table 1. Damping ratio values from episodes of different free decay oscillations.

Time [s]	ζ [%] First 20% of the peaks	ζ [%] 100% of the peaks
7.7 – 73	0.4561	0.4430
83 – 200	0.4550	0.3819
273 – 316	0.4360	0.4782
486 – 520	0.4360	0.4782
537 – 616	0.4326	0.4326

Under the linearity and decoupling hypotheses assumed herein, the decoupled equations of motion of the sphere along the streamwise, x and transverse, y directions are:

$$F_x = m\ddot{x}+c\dot{x}+kx \quad (3)$$

$$F_y = m\ddot{y}+c\dot{y}+ky \quad (4)$$

where F_x and F_y are the total fluid forces in the streamwise and transverse direction, respectively; m is the total oscillating structural mass (i.e., not including fluid added mass); and k and c are the stiffness and the damping coefficients, respectively, that were previously estimated.

The instantaneous velocities $(\dot{x},\dot{y})$ and accelerations $(\ddot{x},\ddot{y})$ in (3,4) were obtained by the Centred Finite Difference Method (CFDM) starting from the streamwise and transverse displacements (x,y) as measured within the experimental campaign described in Section 2.

4 Discussion of the results

In this study, we investigated fluid-dynamic loading on an oscillating sphere located at three different relative submergence conditions $h^*= 2$, $h^*= 3.97$ and $h^*= 0$ in steady free surface flows (Table 2) to analyse how the presence of both the flow free surface and channel bottom influences the fluid-dynamic loading acting on the sphere.

First, we report the dynamic response in terms of amplitude, A^*_y, and frequency, f^*, as related to the three submergence conditions already considered by Mirauda et al. [11-14]. In particular, the experimental condition of $h^*= 2$ (a completely submerged sphere) is compared with those observed in literature. The dynamic response in the function of the reduced velocity U^* ($U^*= U/f_{nw}D$) is reported in Figure 10. This figure highlights how the trend of oscillation amplitudes is crescent with U^*, which is similar to the first branch of the literature curve (Fig. 10a) and tends to reach Mode I. Moreover, from the frequency response in Figure 10b, it is possible to observe how the system tends to reach a resonance condition that is typical of the periodic Mode I, in which the oscillation frequency is close to the natural frequency for the higher flow velocity ($U^*= 4.74$).

Table 2. Experimental tests [11].

Test	h^*	U^*
1	2	1.90
2	3.97	1.90
3	0	1.90
4	2	2.84
5	3.97	2.84
6	0	2.84
7	2	3.79
8	3.97	3.79
9	0	3.79
10	2	4.74
11	3.97	4.74
12	0	4.74

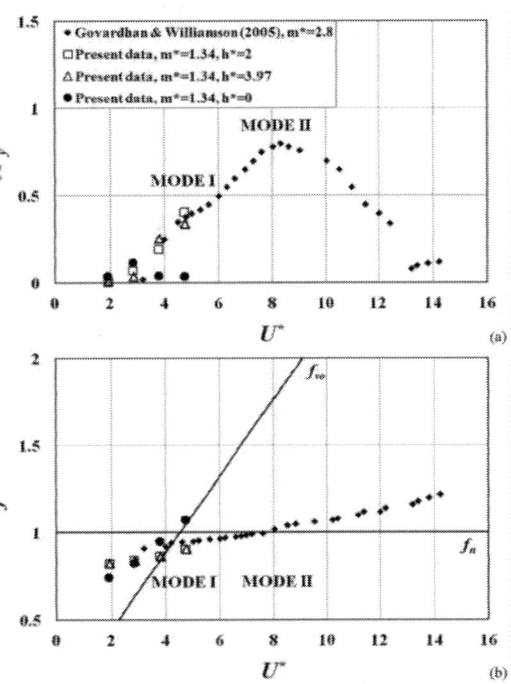

Fig. 10. (a) Transverse normalised amplitudes (A^*_y) and (b) frequency ratios (f^*) at different relative submergences (h^*) in the function of U^*.

The other two conditions, $h^*= 3.97$ and $h^*= 0$, were never analysed in the literature and show different behaviours (Fig.10). In detail, for $h^*= 3.97$, the trend of the oscillation amplitudes and frequencies are similar to those of $h^*= 2$, which show how the presence of the bottom does not influence the dynamic response of the system. On the contrary and in the case of $h^*= 0$, a first increase and subsequent decrease of A^*_y for $U^*> 2.84$ are observed, and the latter was induced by the distortion of the free surface, which inhibits the development of a dynamic response typical of Mode I. The frequencies trend in the case of $h^*= 0$ is always a crescent but seems to be distant from Mode I (Fig. 10b). Within the experimental range shown in Figure 10, the fluid-

* Corresponding author: domenica.mirauda@unibas.it

dynamic loading that was acting on the sphere was calculated with (3,4); moreover, it was also split into a "potential force" component (F_p) and a "vortex force" component (F_v) due to the dynamics of vortices as performed by [8,16,17] and several other studies:

$$F_{(x,y)}=F_{p(x,y)}+F_{v(x,y)} \quad (5)$$

where the potential force can be calculated as:

$$F_{p(x,y)} = -C_A m_d(\ddot{x}, \ddot{y})t=-m_d(\ddot{x}, \ddot{y})t \quad (6)$$

in which m_a is the added mass ($C_A m_d$) and $\ddot{x}(t)$ and $\ddot{y}(t)$ are the instantaneous accelerations of the sphere in the streamwise and transverse directions, respectively. Normalising the forces by $1/2\rho U^2\pi D^2/4$, the force coefficients (C_x and C_y) and potential components (C_p) can be computed, whereas the vortex force coefficients (C_v) were computed as the difference between the total and potential components.

When the experimental results evidence the resonance condition of Mode I, we also considered the hypothesis of sinusoidal behaviour to solve the equation of motion as reported by [8]. In most cases of vortex-induced vibrations, the frequency of body oscillations are synchronised with the periodic vortex wake mode downstream of the body, and therefore, the transverse amplitude response $y(t)$ and the fluid force $F_y(t)$ are well represented by:

$$y(t) = - A \sin(2\pi f t) \quad (7)$$

$$F_y(t) = F_0 \sin(2\pi f t+\phi) \quad (8)$$

where f is the transverse oscillation frequency (f_y) and φ is the phase angle between the total fluid force and the transverse body displacement. Therefore, the solution of Eq. (4) is:

$$A^*_y = y/D= 3/32\pi^2 \; C_y \sin\phi/(m^*+C_A)\zeta_w \; (U^*/f^*)^2 f^* \quad (9)$$

where:

$$f^*= f/f_{nw}= [(m^*+C_A)/(m^*+C_{EA})]^{0.5} \quad (10)$$

and C_{EA} is an effective added mass coefficient that includes an apparent effect due to the transverse total fluid force in phase with the body acceleration ($C_y \cos\varphi$):

$$C_{EA} = 3/16\pi^2 A^*_y \; (U^*/f^*)^2 \; C_y \cos\phi \quad (11)$$

By combining these three equations, we computed the transverse force coefficient (C_y) acting on the sphere when knowing the transverse displacements, y, and oscillation frequency, f_y.

Figures 11 and 12 report the transverse displacements (y/D) normalised with respect to the sphere diameter and coefficient (C_y) compared for each different h^*; moreover, we report the transverse coefficients split into the potential ($C_{p,y}$) and vortex ($C_{v,y}$) components, as discussed before. The figures refer to Tests 1, 2, 3 (Figure 11), 10, 11 and 12 (Fig. 12) of Table 2. The mean component of the displacements and force coefficients remain constant around the zero value due to the symmetry of the set-up. The comparison between Figures 11 and 12 highlight the increase of the displacement and force fluctuating components with the increasing velocity when the sphere is far from the free surface ($h^*= 2$ and $h^*= 3.97$). Instead, and different from these two submergence conditions, a reduction of the displacement and force fluctuating components occurs at $h^*= 0$ with the increase of the reduced velocity (Figs. 11a, 11b, 12a and 12b). In the case of the forces, the splitting of the transverse coefficient into its potential ($C_{p,y}$) and vortex ($C_{v,y}$) components highlights that this behaviour is imputable to the vortex component of the transverse force coefficient.

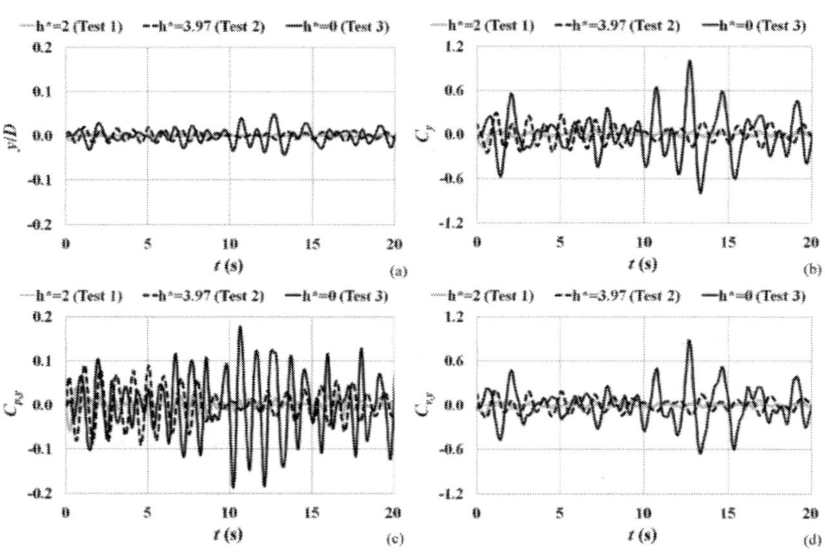

Fig. 11. (a) Trend of normalised transverse oscillations (y/D); (b) transverse total force C_y; and (c) and (d) are the potential and vortex component of the transverse total force for low flow velocity ($U^*= 1.90$).

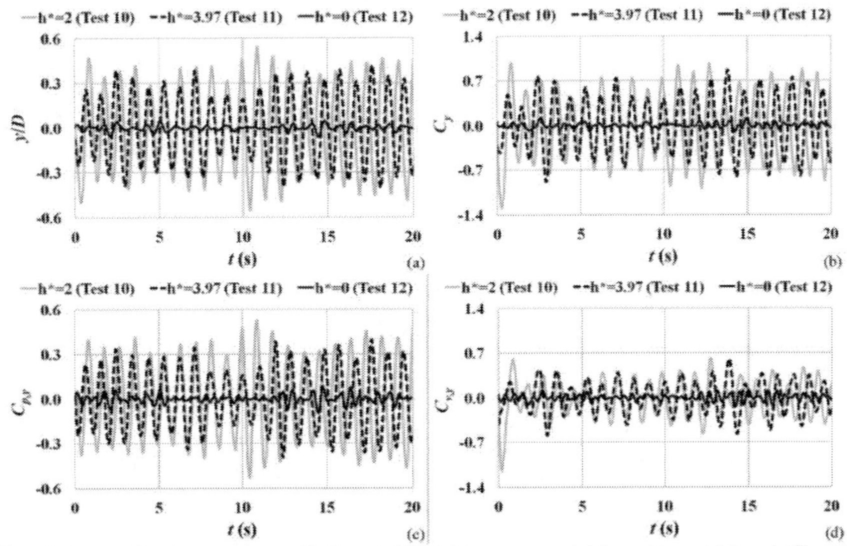

Fig. 12. (a) Trend of normalised transverse oscillations (y/D); (b) transverse total force C_y; and (c) and (d) are the potential and vortex component of the transverse total force for high flow velocity ($U^*= 4.74$).

The experimental data lead us to consider that the strong distortion of the free surface, which occurs when $h^*= 0$ and the flow velocity is larger than $U^*= 2.84$, noticeably reduces the fluctuating component of both the displacements and transverse force coefficient acting on the sphere. By contrast, for values of $U^*<2.84$, we observe an increase of the transverse displacement fluctuating component and a significant influence of the vortex component on the transverse force coefficient. As discussed before, when Mode I is achieved, the transverse force coefficient (C_y) acting on the sphere could be computed by knowing the transverse displacements, y, and oscillation frequency, f_y, under the hypothesis of sinusoidal behaviour. With the aim of verifying the force coefficient evaluation, we compared the methods discussed before. We considered the case of $U^*= 4.74$ with $h^*= 2$ and $h^*= 3.97$; which are conditions when the sinusoidal behaviour is prevalent in the transverse displacements (Figs. 13a-f) and Mode I is achieved (Fig. 10). Instead, we could not consider the condition of $h^*= 0$ for the low periodicity of the signal as observed in Figures 13g-i, probably due to the strong deformation of the free surface above the sphere that influences body movement and induces random motion.

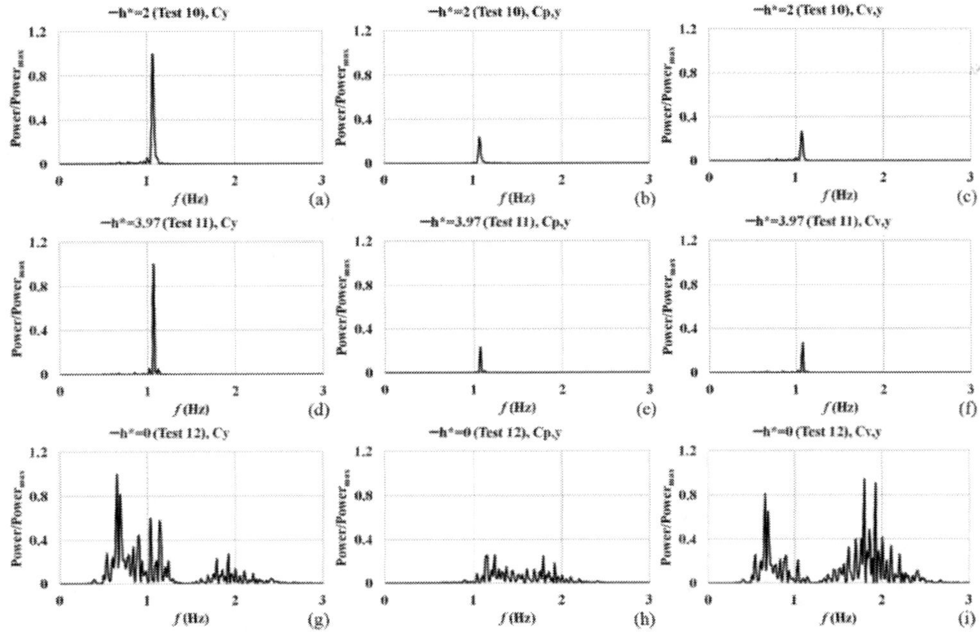

Fig. 13. Power spectra of the transverse force coefficients at (a), (b), and (c) $h^*= 2$; (d), (e), and (f) $h^*= 3.97$; and (g), (h), and (i) $h^*= 0$ for high flow velocity ($U^*= 4.74$).

Figures 14 and 15 report the trends of the force coefficients derived from the computation of the body instantaneous velocities and accelerations by means of the CFDM (black line). These trends also come from the solution of the motion equation according to [8] (grey line) for the conditions of relative submergence $h^*= 2$ and $h^*= 3.97$ and for the higher flow velocity ($U^*= 4.74$, Test 10 and 11). For this velocity of the flow, the sphere is close to the resonance condition of Mode I. The analysis of the curves shows how the results are, in the resonance situation, very similar.

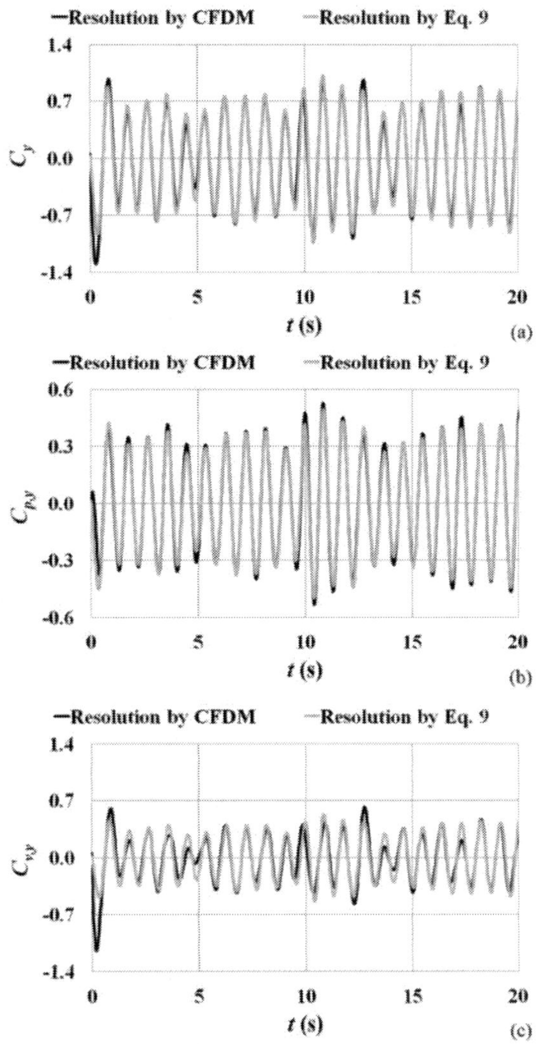

Fig. 14. (a) Transverse total force C_y and (b) and (c) are the potential and vortex component, respectively, of the transverse total force for high flow velocity ($U^*= 4.74$) at $h^*= 2$ (Test 10).

Instead, Figures 16a and 16b report that the behaviour of the streamwise movement was normalised with respect to the sphere diameter and the drag coefficients for the three relative submergences h^* at a low flow velocity ($U^*=1.90$; Test 1, 2 and 3). It is important to note how the mean and fluctuating components of the drag coefficient are significantly greater in the asymmetrical boundary conditions ($h^*= 0$ and $h^*= 3.97$).

In addition, Figures 16c and 16d show that the drag coefficients split into the potential ($C_{p,x}$) and vortex ($C_{v,x}$) components, as previously discussed. As expected, the potential component of drag (Fig. 16c) moves around zero, whereas the vortex component (Fig. 16d) contains the mean and part of the fluctuating components of the drag coefficient.

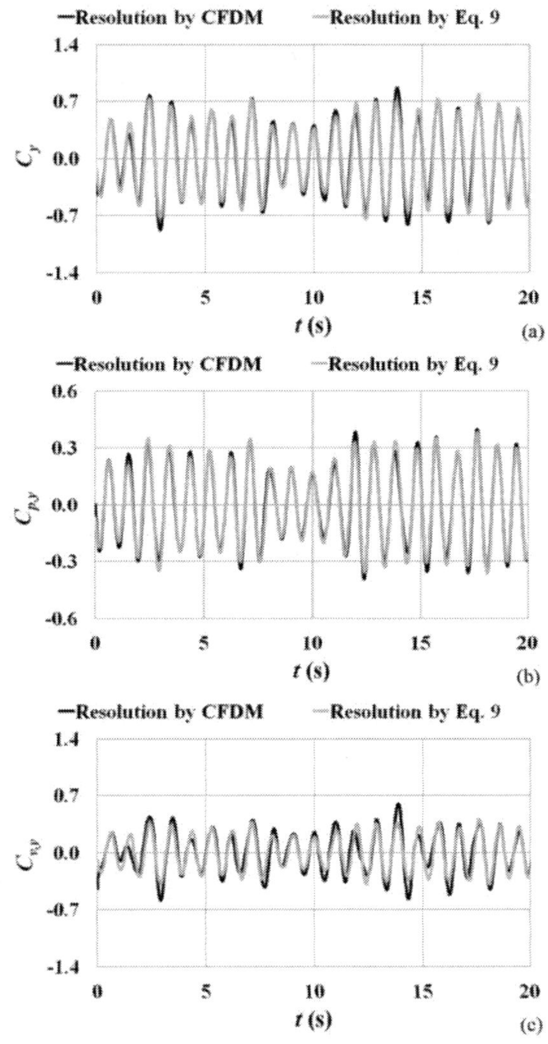

Fig. 15. (a) Transverse total force C_y and (b) and (c) are the potential and vortex component, respectively, of the transverse total force for high flow velocity ($U^*= 4.74$) at $h^*= 3.97$ (Test 11).

In addition to, as it is possible to note through the power spectra of Figure 17, in the cases of $h^*= 3.97$ and $h^*= 0$, the signal is periodic with one predominant peak of approximately *1Hz*, whereas under the condition of $h^*=2$, an overlapping of non-periodic signals is observed

* Corresponding author: domenica.mirauda@unibas.it

https://doi.org/10.1051/epjconf/201818002067

with a range of frequencies between *0.7* and *1.5 Hz*. By increasing the velocity from $U^*= 1.90$ to $U^*= 4.74$ (Fig. 18a), the mean displacements of the sphere in the x direction, with respect to its static position (zero value), are amplified. However, for this higher flow velocity (Fig. 18), interestingly enough, the results are different in terms of the mean and fluctuating components of the force coefficients along the x direction. As Figure 18b depicts, the mean values of C_x follow the ordering displayed in Figure 18a for the displacements. For $h^*= 2$, an increase of the mean value of C_x is noted as well as of the fluctuating component, whereas the case of $h^*= 3.97$ also leads to a larger mean value associated with a larger loss of periodicity. Moreover, $h^*= 0$ leads to the lowest value of C_x (the mean value of C_x is 1.25 while the standard deviation is $stdC_x= 0.05$ for $U^*= 4.74$), which is a value lower than that found for $U^*= 1.90$ ($C_x=1.47$ and $stdC_x= 0.43$). In this case, the amplitude of the fluctuating part is also reduced, which suggests a radical change in the mechanism of the vortex detachment.

Such behaviour could be due to the presence of a jet-like flow between the upper surface of the sphere and free surface, which reduces both the mean and fluctuating component of the forces.

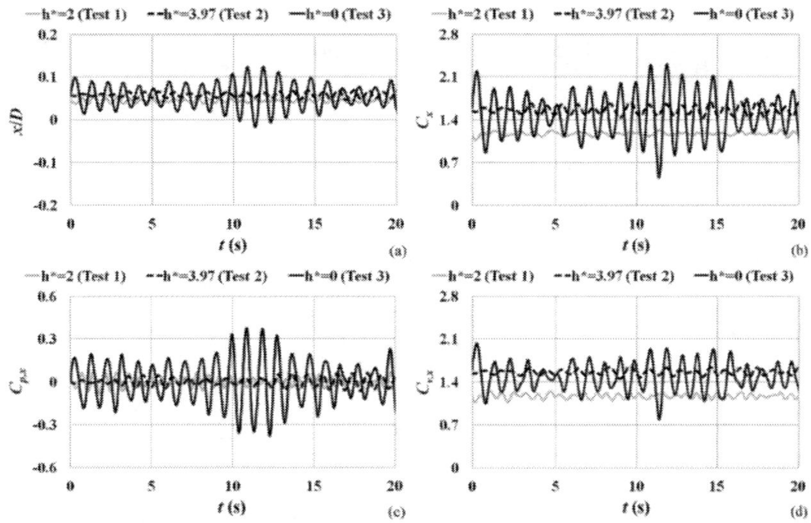

Fig. 16. (a) Trend of normalised streamwise oscillations (x/D); (b) streamwise total force C_x; and (c) and (d) are the potential and vortex component, respectively, of the streamwise total force for low flow velocity ($U^*= 1.90$).

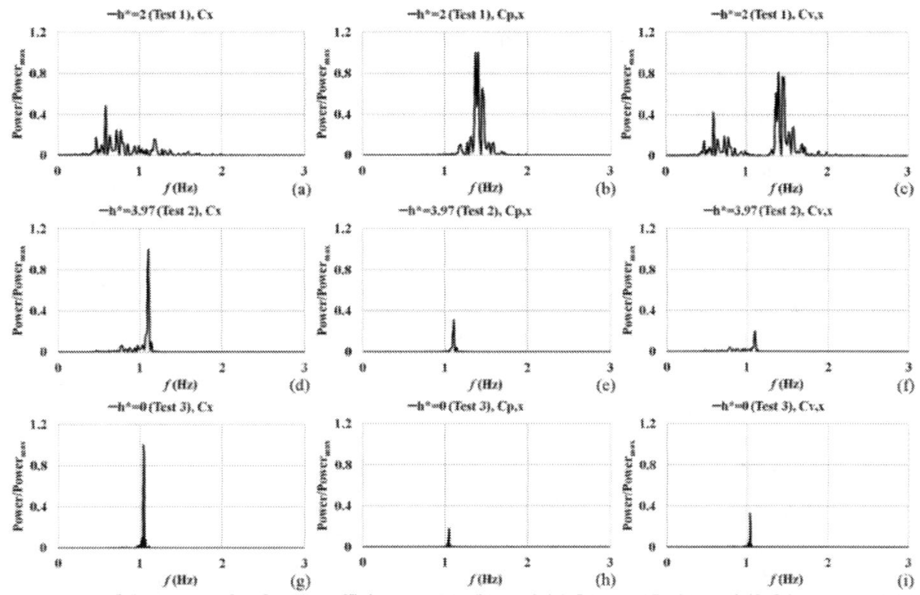

Fig. 17. Power spectra of the streamwise force coefficients at (a), (b), and (c) $h^*= 2$; (d), (e), and (f) $h^*= 3.97$ and (g), (h), and (i) $h^*= 0$ for low flow velocity ($U^*= 1.90$).

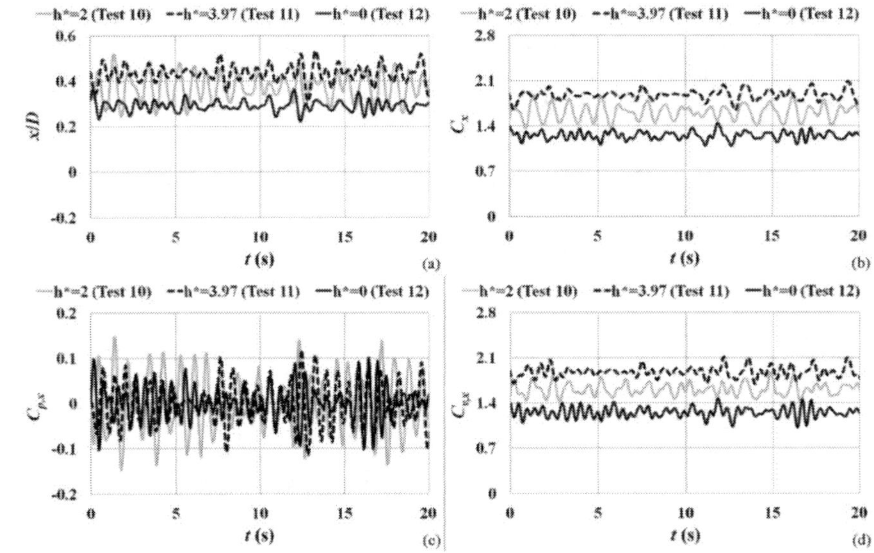

Fig. 18. (a) Trend of normalised streamwise oscillations (*x/D*); (b) streamwise total force C_x; and (c) and (d) are the potential and vortex component, respectively, of the streamwise total force for high flow velocity ($U^*= 4.74$).

4 Conclusions

The measurements of the forces acting on the structure with simplified geometry described in this paper represent a necessary step for the numerical resolution of coupled mechanical-fluid dynamic models. Knowledge of the internal forces in the mechanical model permits computation of the fluid dynamics from its motion. The internal forces were reconstructed after establishing a correlation between the internal forces and motion of the mechanical model.

The results underlined an increase of the mean components of the force coefficients along the streamwise direction in the asymmetrical conditions of flow for low velocity with respect to the condition of $h^*= 2$. Instead, a reduction of the mean component is observed for the higher velocity in the case of $h^*= 0$. Such behaviour is explained by the presence of a free surface distortion, which reduces the mean component of the force. Instead, the fluctuating components show similar values for all of the conditions. Only in the case of $h^*= 0$, at the investigated low velocity, is it possible to observe larger values of displacement and force coefficients due to the presence of a jet-like flow between the upper surface of the sphere and free surface, which increases the movements and forces acting on the sphere.

Along the transverse direction and for a low velocity, it is interesting to note a behaviour of the system similar to the case of the streamwise direction with higher fluctuations of displacements and forces for $h^*= 0$, whereas for high velocities, it evolves in the opposite manner. In fact, for $h^*=0$, the strong distortion of the free surface above the sphere reduces the displacements and force fluctuating components and influences the body to induce a random movement. In the case of $h^*= 3.97$ and $h^*= 2$, we instead note a considerable increase of fluctuating components and a repeatable trend with sinusoidal functions, which is typical of the resonance condition already observed in the literature by several authors.

Finally, an additional interesting result is the significant values assumed by the vortex fluid force, which as discussed, could provide an important contribution to the analysis of this type of phenomenon.

References

1. J.M. Jonkman, P.D. Sclavounos, Development of fully coupled aeroelastic and hydrodynamic models for offshore wind turbines, *44th AIAA Conference: Aerospace Sciences Meeting and Exhibit Online Proceedings*, Reno, Nevada, 10-12 January (2006)
2. A. Amicarelli, R. Albano, D. Mirauda, G. Agate, A. Sole, R. Guandalini, A smoothed particle hydrodynamics model 3D solid body transport in free surface flows, Computers & Fluids, **116**, 205-228 (2015)
3. R. Albano, A. Sole, D. Mirauda, J. Adamowski, Modelling large floating bodies in urban area flash-floods via a smoothed particle hydrodynamics model, J. of Hydrology, **541**, 344-358 (2016)
4. S.K. Chakrabarti, Hydrodynamic of offshore structures, Springer-Verlag, New York (1987)
5. M. Brankovic, P.W. Bearman, Measurements of transverse forces on circular cylinders undergoing vortex-induced vibration, J. Fluids and Struct. **22**, 829-836 (2006)
6. T.L. Morse, C.H.K. Williamson, Employing controlled vibrations to predict fluid forces on a

cylinder undergoing vortex-induced vibration, J. Fluids and Struct. **22**, 877-884 (2006)
7. C.H.K. Williamson, R. Govardhan, A brief review of recent results in vortex-induced vibrations, J. Wind Eng. and Ind. Aerodyn. **96**, 713-735 (2008)
8. R. Govardhan, C.H.K. Williamson, Vortex induced vibration of a sphere, J. Fluid Mech. **531**, 11-47 (2005)
9. R. Govardhan, C.H.K. Williamson, Vortex induced motions of a tethered sphere, J. Wind Eng. and Ind. Aerodyn. **69-71**, 375-385 (1997)
10. C.H.K. Williamson, R. Govardhan, Dynamics and forcing of a tethered sphere in a fluid flow, J. Fluids and Struct. **11**, 293-305 (1997)
11. D. Mirauda, A. Volpe Plantamura, S. Malavasi, Hydrodynamic forces acting on an oscillating structure, *Proc. 2nd International Conference on Physical Coastal Processes, Management and Engineering, Coastal Processes 2011, WIT Transactions on Ecology and Environment, WIT Press,* Naples, Italy **149**, 321-331 (2011)
12. D. Mirauda, A. Volpe Plantamura, S. Malavasi, Boundaries effects on the movements of a sphere immersed in a free surface flow, J. Offshore Mech. and Arct. Eng. **133**, 041301-1-041301-5 (2011)
13. D. Mirauda, A. Volpe Plantamura, S. Malavasi, Boundaries influence on the flow field around an oscillating sphere, *Proceedings of the ASME 2013 32nd International Conference on Ocean, Offshore and Arctic Engineering OMAE2013*, June 9-14, Nantes, France (2013)
14. D. Mirauda, A. Volpe Plantamura, S. Malavasi, Dynamic response of a sphere immersed in a shallow water flow, J. Offshore Mech. and Arct. Eng. **136**, 021101-1-021101-6 (2014)
15. P.D. Welch, The Use of Fast Fourier Transform for the Estimation of Power Spectra: A Method Based on Time Averaging Over Short, Modified Periodograms, IEEE Trans. Audio Electroacust. AU-**15**, 70-73 (1967)
16. R. Govardhan, C.H.K. Williamson, Modes of vortex formation and frequency response of a freely vibrating cylinder, J. Fluid Mech. **420**, 85-130 (2000)
17. J. Lightill, Wave loading on offshore structures, J. Fluid Mech. **173**, 667-681 (1986)

Inhalers and nebulizers: basic principles and preliminary measurements

Ondrej Misik[1*], *Frantisek* Lizal[1], *Vahid* Farhikhteh Asl[2], *Miloslav* Belka[1], *Jan* Jedelsky[1], *Jakub* Elcner[1] and *Miroslav* Jicha[1]

[1]Brno University of Technology, Faculty of Mechanical Engineering, Energy Institute, Technicka 2896/2, Brno 616 69, Czech Republic
[2]Dresden University of Technology, Faculty of Environmental Science, Urban and Industrial Water Management Institute, Bergstraße 66, 01069 Dresden, Germany

Abstract. Inhalers are hand-held devices which are used for administration of therapeutic aerosols via inhalation. Nebulizers are larger devices serving for home and hospital care using inhaled medication. This contribution describes the basic principles of dispersion of aerosol particles used in various types of inhalers and nebulizers, and lists the basic physical mechanisms contributing to the deposition of inhaled particles in the human airways. The second part of this article presents experimental setup, methodology and preliminary results of particle size distributions produced by several selected inhalers and nebulizers.

1 Introduction

The inhalation treatment becomes very popular nowadays. It is one of the most effective non-invasive way of the treatment among the large range of diseases from cancer to migraines [1]. The effect of the inhalation therapy depends on rate and place of particle deposition in the airway. There are three essential mechanisms of particle deposition in lungs: inertial impaction, gravitational sedimentation and Brownian diffusion. The largest particles, mostly larger than 5 μm, are deposited by inertial impaction particularly in upper airways. The Stokes number plays an essential role in this mechanism of deposition [2; 3]. Then, in the tracheobronchial tree, the airstream is decelerated as the total area of the cross-section increases. It causes that particles reaching this part of the lungs are settled down by gravitational sedimentation. Particles with size between 1 μm and 5 μm are usually deposited by this mechanism. Particles in sizes of 0.1–1 μm are mostly exhaled from lungs, but particles smaller than 100 nm are deposited well with the same probability in all parts of the lung due to the diffusion [1; 2; 4]. Generally, it is not necessary to deposit the particles in the alveolar regions, where absorption of drug to the blood circulation is highest. Particularly in the case of lung diseases, the demanded deposition is in the regions, where the receptors for wanted effect are located. The targeted deposition of drugs is a task for inhalation systems [5].

Generally, the current inhalation systems can be divided to three main categories: nebulizers, metered dose inhalers (MDI) and dry powder inhalers (DPI). The background of the inhalation treatment originates from antiquity. First reference of inhalation therapy was the papyrus from ancient Egypt in 1554 BC. This papyrus

Fig. 1. The Sales-Girons nebulizer. [8]

* Corresponding author: misik.ondro@gmail.com

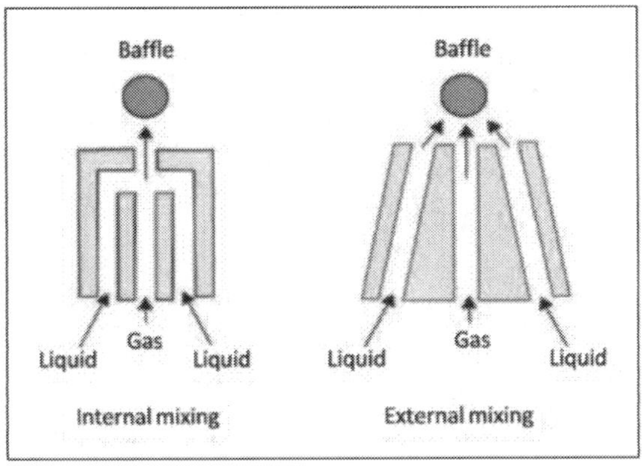

Fig. 2. Typical designs of the jet nebulizers. [8]

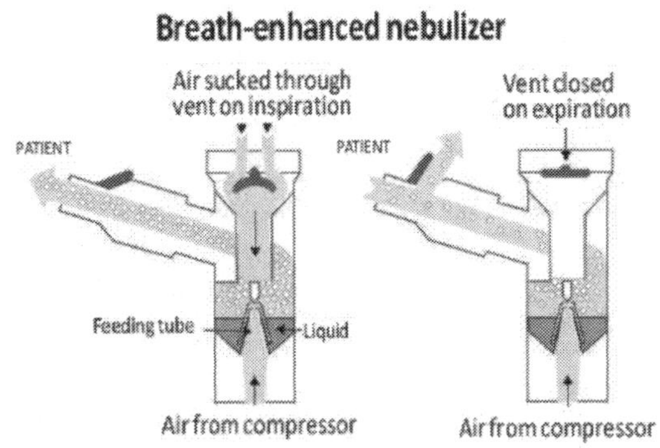

Fig. 3. The principle of the BEN. [8]

describes inhaling of the vapours created from the black henbane *(Hyoscyamus niger)* put onto hot bricks. Over the many years, there were a lot of improvised ways and devices for inhalation therapy. After the industrial revolution, the inhalation devices became mass produced. In the middle of 19th century, inhalation was used for the first time to the anaesthesia. The first nebulizer (Sales-Girons nebuliser) was invented in 1858 as a replacement of spas for people who cannot afford it *(Fig. 1)*. Smoking the asthma cigarettes and inhaling the smoke of burned powders was also very popular at that time. Inhalation devices in the current form had been developing during the 20th century. The first DPI, Aerohaler, was designed in 1948 for inhaling the benzylpenicillin. MDIs are produced since 1956, when the Riker laboratories brought Medihaler to the market. MDIs were firstly manufactured with CFCs (Chlorofluorocarbons) formulation. However, the usage of CFCs was banned by Montreal Protocol in 1987 because of the ozone layer depletion and CFC was replaced by HFA (hydrofluoroalkane) [6; 7].

The nebulizers are non-portable devices which nebulize a liquid drug solution. They require connection to the electric network during the usage. They are classified based on the technology into three categories: jet, ultrasound and mesh nebulizers. The principle of jet nebulizers is formed on the Venturi tube, where air from the compressor is accelerated through the jet and goes near the small capillary tube in which a negative pressure is created. In view of the fact that liquid is sucked up from a reservoir in contact with airstream, it is atomized because of shear forces. A baffle is placed in front of the airflow in order to catch the larger aerosol particles and return them back to the reservoir. The common designs of the jet nebulizers are shown on the *Fig. 2* [5; 8]. For more economical usage were invented so called Breath-enhanced inhalers (BEN) and Breath-actuated inhalers (BAN). Whereas the standard jet nebulizers generate aerosol constantly BEN and BAN try to reduce its aerosol production during the exhalation. The principle of the aerosol generation in BEN is the same as in the case of the standard jet nebulizer, but a one-way valve is located above the nebulizing chamber. Another one-way valve is located on the mouthpiece and the aerosol is generated constantly in the nebulizing chamber. The valve above the nebulizing chamber is open during the inspiration which makes the aerosol to come out from the nebulizer. During the expiration, the first valve is closed and the valve on the mouthpiece is opened. This causes that the aerosol remains in the device during expiration and exhaled air is blown out through the valve on the mouthpiece *(Fig. 3)*. BANs are often the nebulizers with electronic control which switch on during inspiration in order to decrease a wastage. For instance, the AeroEclipse nebulizer has a green button on the top of the device. This button moves down during the inhalation due to the negative pressure. The piston connected to the button is placed on the spring and the movement activates the aerosol production. During the exhalation, the piston moves back and switches off the device. The green button also signalizes to the patient that they are inhaling correctly. This nebuliser is suitable for old patients and children because of the high device sensitivity [8; 9; 10]. Principally, ultrasound nebulizers create a standing wave on a liquid, which produce an aerosol on the liquid surface. The vibrations heat up the solution, so this device is suitable for thermostable drugs only [5; 10]. In the mesh nebulizers, the liquid drug solution is forced through a vibrating mesh with a number of little orifices producing an aerosol. Although

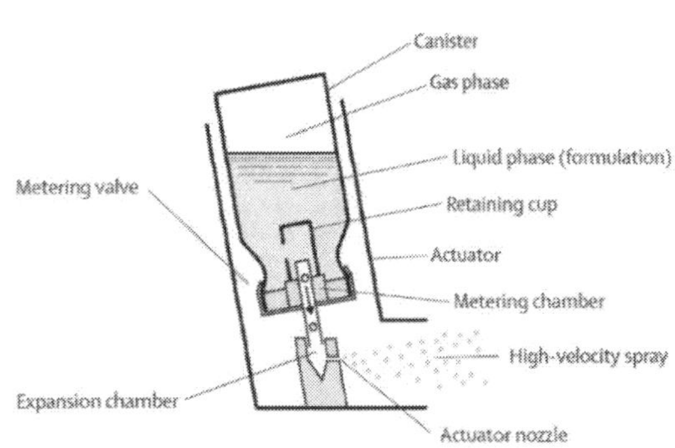

Fig. 4. The scheme of the usual MDI. [21]

the mesh nebulizers are the most effective, they are also the most expensive and the most complicated [5; 10].

Another big group of inhalers are metered dose inhalers (MDI) which have been popular for a long time because they are small, portable and easy to use. General components of the MDI are a canister, metering valves including inner and outer one, and an actuator mouthpiece as you can see on the *Fig. 4* [11; 5]. The canister is filled with solution of the propellant, drug, surfactant and some other constituents. The initial state before usage of the MDI occurs when the inner valve is open and the outer one is closed. During the usage, the inner valve is closed to meter the precise dose of the formulation, subsequently the outer one is allowed to open and emits the aerosol, and the aforementioned cycle repeats in the same way [11]. A coordination between actuation of the device and inspiration causes many complications. This was solved by an invention of a breath-actuated MDI, such as Autohaler. This inhaler contains a compressed spring, which causes actuation of the device when the jet is opened by the inspiration [5]. Owing to the propellant usage, velocity of the emitted aerosol is high which causes significant oropharyngeal deposition. That is the reason to use additional devices, such as spacers or chambers. These components prolong the distance between the mouthpiece and the mouth cavity which slows down the aerosol flow. Furthermore, it eliminates the problem with coordination of the inspiration and actuation [12].

The usage of propellants is non-ecological. Moreover, high velocity causes complications with coordination. Avoiding that was the aim of invention of a Soft Mist Inhalers (SMI), such as Respimat. These devices are light-weight portable devices which nebulize the liquid drug solution using mechanical energy of spring and emit the metered dose of soft mist. When the patient rotates the inhaler body by 180°, the cartridge from the reservoir is loaded to a metering chamber. The gear transforms a rotation into a linear movement and compresses the spring. By clicking the dose-release button, the spring is released which forces the dose through a tiny twin-nozzle in a uniblock part, where nozzles are at the controlled angle. It makes the two streams collide and produces a soft mist with low velocity [13; 5; 14].

The last category of inhalers is dry powder inhalers (DPI) which have three divisions: Single-dose, Multiunit-dose and Multi-dose inhalers [15]. Single-dose inhalers have metered dose of drug powder packed into a capsule. Prior to an inhalation, the capsule is set into a device and pierced to release the dose. The Multiunit-dose inhalers encompass plenty doses in one blister, so there are many doses available. The Multi-dose inhaler possesses a reservoir with powder formulation and it meters a precise dose of powder before the inhalation [15]. The powder is influenced by three types of forces: the interparticle force, the dispersion force created by a device and the force of the inspiration. The aim is to keep these forces in balance, which means the balance between resistance of the device and airflow of inspiration. The comparison of the selected DPI is shown in *Tab 1*. Strong airflow results in the deposition in the oropharyngeal region, whereas the higher device resistance makes it difficult to inhale for children, old people or people with advanced-stage diseases. So, we can sort devices also by their resistance as low resistance, medium resistance and high resistance devices [16; 17; 5]. Another important point is deagglomeration of particle clusters. As was mentioned above, the interparticle forces act between the particles creating particle agglomerates. Particles should be deagglomerated during inhalation. For this purpose, a mesh can be set into the airflow or the mouthpiece geometry can be shaped to create whirls and turbulent flows. A humidity causes huge problems for deagglomeration, hence the DPI should be stored on a dry place and patient must not exhale into the device [5].

Table 1. The comparison of the resistance and the airflow during the inspiration in selected devices

Inhalation devices	Resistance of the device [kPa^{0,5} l/min]	Flow during inspiration [l/min]
Breezhaler®	0,017	111
Aerolizer®	0,019	102
Ellipta®	0,027	74
Novolizer®	0,027	72
Accuhaler/Diskus®	0,027	72
Genuair®	0,031	64
Nexthaler®	0,036	54
Turbohaler®	0,039	54
Handihaler®	0,058	37

Experimental measurements of particle size distributions coming from various kinds of inhalers were performed. The obtained data were used to estimate lung deposition.

2 Experimental measurement

The experimental measurements which are described in this part are based on Misik's bachelor thesis *Inhalers and nebulizers for medical use* [5]. Particle size distributions of selected devices were measured and their FPF (Fine Particle Fraction) were evaluated. FPF represents the percentage of particles which deposit well in the lung, i.e.

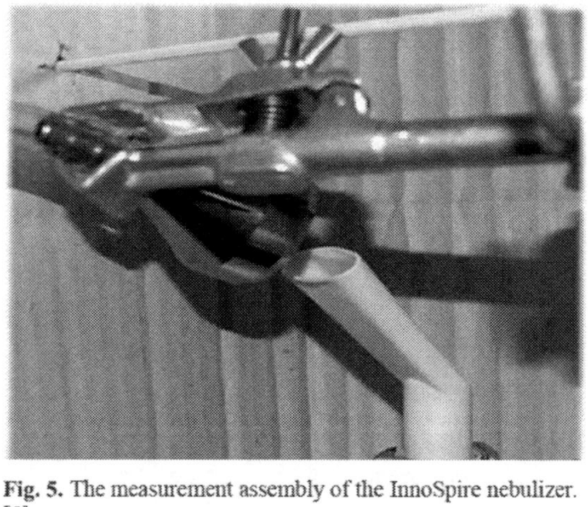

Fig. 5. The measurement assembly of the InnoSpire nebulizer. [5]

particles in the range 1–5 μm. Measurements were done using the Aerodynamic Particle Sizer (APS, TSI 3321, TSI Inc., Shoreview, MN, USA), which measures an aerodynamic particle diameter. The inhaler was connected to the APS via a conductive tube to eliminate deposition on the walls of the tube because of the electrostatic force. An isokinetic probe was connected at the inlet of the tube to eliminate the effect of a stream angle and velocity differences. Nebulizers and MDI were sampled in the open air because sampling into the closed box results in coagulation of particles. In addition, the nebulizer particles were coagulating in open space due to constant generation of aerosol for quite a long time [5]. In terms of DPI, it was necessary to simulate an inspiration, therefore the DPIs were sampled into a plastic canister connected to a simulator of breathing. Since the canister was not air-tight, it was necessary to induce an inspiration flow of 1.5 l/s on the simulator. In general, measurement was more precise when the distance between the mouthpiece and isokinetic probe was short [5]. Following devices were measured: nebulizer InnoSpire® Elegance (Philips), metered dose inhaler Atrovent® (Boehringer Ingelheim), soft mist inhaler Respimat® (Boehringer Ingelheim) and dry powder inhalers Ellipta®(Glaxosmithkline) and Breezhaler® (Novartis) [5].

Fig. 6. The measuring assembly of the Atrovent MDI. [5]

2.1 InnoSpire® Elegance

During the measurement, the nebulization chamber was connected to a compressor with maximum pressure 317 kPa and airflow of 6 l/min. The chamber was filled with distilled water and oriented in the vertical direction. The hose connected to the APS had to be bended, what might have caused deposition in the hose by impaction. Since the aerosol flow from nebulizers is quite low, the direction of flow is susceptible to external influences like airstreams in the laboratory. To prevent these effects, it was required to make a partial cover which had to be open to avoid coagulation. The photo of the measuring assembly of InnoSpire nebulizer can be seen on the *Fig. 5* [5].

2.2 Atrovent®

During the measurement of the Atrovent inhaler, an effort was made to reach a constant flow of aerosol, so several doses were emitted during one sample. Similar procedure was done in the measurement of particle size distribution of the Respimat® SMI. Each time, the device had to be actuated manually which might have caused small movements of the device during the sampling. The MDI emits the aerosol with high velocity so small movements could have caused variability in results. The photo of the experiment with Atrovent is on the *Fig. 6* [5].

2.3 Dry Powder Inhalers

As was mentioned previously, it was necessary to use the canister and the simulator of breathing for sampling during the measurement of dry powder inhalers. Another complication was that any effort to provide the continual sampling was unsuccessful. Therefore, the sampling time was shortened to 4 s. The mean diameter decreased during the sample time, as only particles with small diameter remained in the canister. To sum up, these measurements were least accurate [5].

3 Results and discussion

Table 2. Fine particle fractions and CMADs of measured inhalers. [5]

Inhalation device	FPF [%]	CMAD [μm]
InnoSpire® Elegance	75,05 ± 5,09	3,66 ± 0,22
Atrovent® 1.	59,37 ± 6,93	1,08 ± 0.06
Atrovent® 2.	69,30 ± 4,26	1,20 ± 0,05
Atrovent® 3.	78,59 ± 0,89	1,29 ± 0,01
Respimat® SMI Placebo	58,74 ± 3,65	3,94 ± 0,25
Respimat® SMI Spiolto	56,67 ± 4.23	4,30 ± 0,27
Breezehaler®	83,16 ± 2,51	2,80 ± 0,18
Brimica® Genuair®	66,21 ± 5,66	1,31 ± 0,131
ANORO® Ellipta®	80,09 ± 0,95	1,85 ± 0,10

In data analyses FPF and count median of aerodynamic diameter (CMAD) were calculated. In our case the

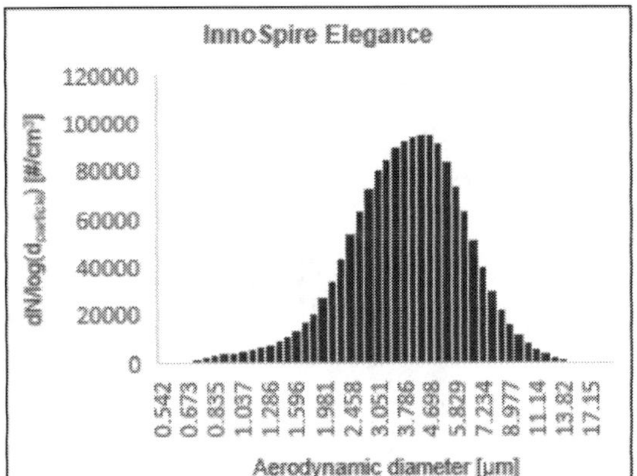

Fig. 7. The measured particle size distribution of the InnoSpire nebulizer. [5]

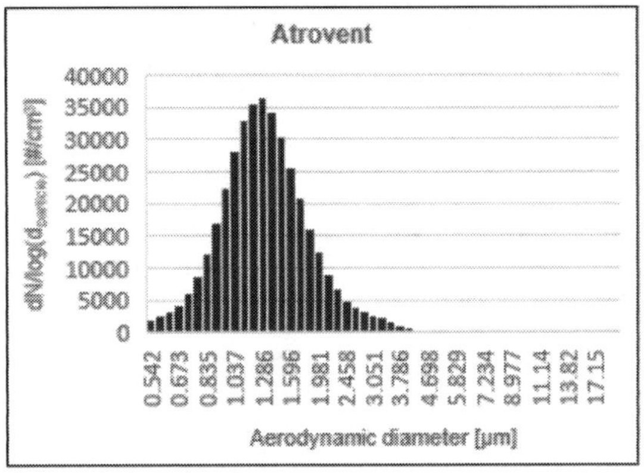

Fig. 8. The measured particle size distribution of the Atrovent MDI [5].

calculated FPF was represented by particles in the range of 1 – 4.7 µm [5]. The obtained data are shown in the *Tab 2.*

The technical description of the InnoSpire device declares that its FPF calculated as a fraction of particles < 5 µm is 77 % [18]. In our measurements, the average FPF of this nebulizer was about 75%. Therefore, very good agreement was found. However, from obtained data we can see that particles coagulated considerably. The nebulizer was turned on for approximately 2.5 minutes and the measurements were done in a sequence without long pauses. Based on this data, it can be seen that during the experiment, on one hand, the number of small particles was decreasing and on the other hand, the amount of the large particles was increasing. At the beginning, the FPF of this device was about 65 % and it grew till the end of the experiment, when it reached about 80 %. So, in this case the coagulation had a positive impact. However, the inhalation therapy with this device should last 6 – 8 minutes and the results can be different. For future measurement, it would be suitable to simulate a breathing pattern during the experiment to check up the coagulation in the case of real flow in the mouthpiece. The measured size distribution of InnoSpire nebulizer is shown on the *Fig. 7* [5].

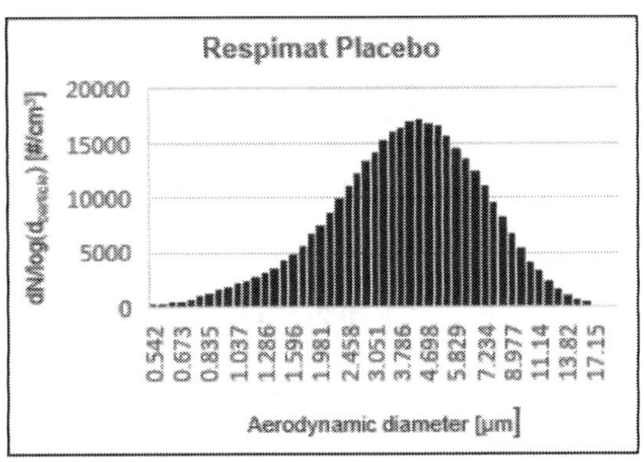

Fig. 9. The measured particle size distribution of the Respimat Placebo SMI [5].

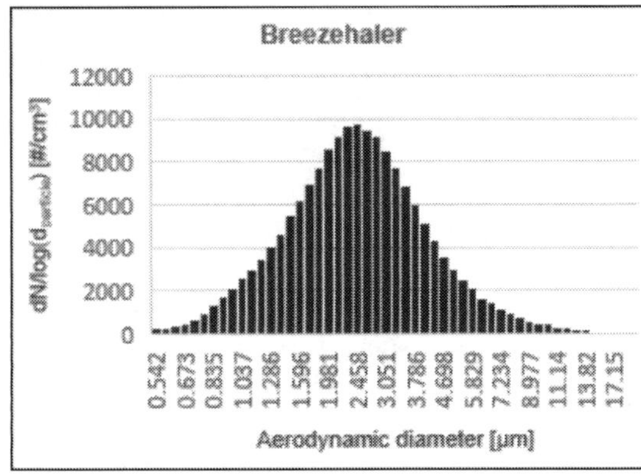

Fig. 10. The measured particle size distribution of Breezehaler DPI [5].

In the data from measurement of the Atrovent particle size distribution, a big variability was observed between each set of data. Because of that, three different sets of the measurements were done (Atrovent 1., Atrovent 2. Atrovent 3.). There were small changes in the assembly setting (like distance and angle between isokinetic probe and the mouthpiece of the device) between each set; and each set consists at least from 4 measurements. As was already mentioned, the manual actuation of the device was performed which could cause small movements of the device. This probably caused the variability of aerosol velocity, the movements and variable direction of mouthpiece could have a considerable impact on the results. There was also a different distance between the isokinetic probe and the mouthpiece in each measurement. *Mitchel et al.* [19] measured the particle size distribution with Aerosizer. In their experiment, the FPF was about 97 % and representing the particles < 4.6 µm, while in the case that particles smaller than 1 µm were added to calculation, the obtained values were about 99 %, which is in agreement with the data from Aerosizer. The measured particle size distribution can be seen on the *Fig. 8* [5].

Results of both Respimat SMI formulations were nearly identical. Respimat SMI is developed to create particles smaller than 5.8 µm and FPF should range between 65 – 80 %. In this case, particle fraction calculated with range of particles < 5.8 µm was 77 % for Placebo and 73 % for Spiolto formulation. In this case the different formulation had no influence on the particle size distributions (*Fig. 9*) [5].

As mentioned before, there was a lot of inaccuracies in the measurements of the DPI. The canister was not air tight, because of the imperfect sealing in the places of the hose connection. Another problem of the sampling was the inability of the simulator to stop the cycle after the inspiration. Inspiration sucked the powder from the device and then the flow induced by simulator during expiration disturbed the flow inside canister and blowed out the particle from the isokinetic probe. Lastly, the measured distributions gave us an overview of DPI particle sizes. The measured particle size distribution is shown on the *Fig. 10* [5].

4 Conclusions

In this paper, the preliminary results of the particle size distributions from several selected inhalers were presented and practical issues about accuracy of the measurements were discussed.

In the InnoSpire Elegance nebulizer, the coagulation of the particles was detected. In the future measurement, it would be appropriate to investigate if the mouthpiece is the reason of the coagulation and if the particles are coagulating also in case of a cyclic breathing regime [5].

MDI Atrovent showed significant variability of the results. It was attributed to the high velocity of the aerosol stream and instability of the device during the measurement. This could be solved by using the spacer to enlarge the distance between the mouthpiece and the isokinetic probe and slow down the aerosol before entering the airways. To reduce the complications with the non-continual dosing during the sampling time, it could be useful to perform the measurements with the Cascade Impactor where the sampling time is not relevant. The ipratropium bromide is the effective component of the Atrovent formulation which makes the maximal bronchodilatation effect when MMAD (mass median of aerodynamic diameter) of the aerosol is about 3 μm [20]. MMADs calculated from CMADs specified in the table of the results were between 1.6 – 2.0 μm. It is probable that after the application of the spacer and after solving the problem of non-continual dosing, MMAD will increase and approach values about 3 μm [5].

The measurements of DPI bring information about particle size distribution, but the results were influenced by a few deficiencies, where the most important one was probably that the device was able to release just one dose during the sampling time. Because of this, the future measurement should include also the Cascade Impactor to avoid the influence of the sampling time [5].

Acknowledgement

This work was supported by the Czech Science Foundation under the grant GA16-23675S, by the project FSI-S-17-4444, and by the project LTC17087 of the programme INTER-EXCELLENCE, sub-programme INTER-COST.

References

1. J. Patton and P. Byron, Nature Reviews Drug Discovery **vol. 6,** 67-74 , (2007).
2. W. Hinds, *Aerosol technology: properties, behavior, and measurement of airborne particles*, 2nd ed. (1999).
3. C. Loira-Pastoriza, J. Todoroff and R. Vanbever, Advanced Drug Delivery Reviews **vol. 75**, 81-91, (2014).
4. A. Fernández Tena and P. Casan Clarà, Archivos de Bronconeumología (English Edition) **vol. 48**, 240-246, (2012).
5. O. Mišík, Bachelor thesis, Brno University of Technology, (2017).
6. S. Stein, Journal of Aerosol Medicine and Pulmonary Drug Delivery **vol. 30**, 20-41, (2016).
7. P. J Anderson MD, Respiratory Care, **vol. 50**, 1139-1150, (2005).
8. O. Nerbrink, Inhalation, **vol. 10**, 9-14, (2016).
9. K. Leung, E. Louca and A. Coates, Chest, **vol. 126**, 1619-1629, (2004).
10. A. Ari, Eurasian Journal of Pulmonology, **vol. 16**, 1-7, (2014).
11. S. Stein, P. Sheth, P. Hodson and P. Myrdal, AAPS PharmSciTech **vol. 15**, 326-338 (2014).
12. S. Newman, Respiratory Care **vol. 50**, 1177-1190 (2005).
13. R. Dalby, J. Eicher and B. Zierenberg, Medical Devices: Evidence and Research **2011,** 4, 145-155 (2011).
14. V. Kašák and E. Feketeová, Farmakoterapie **vol. 6** 4, (2010).
15. L. Garcia-Contreras, M. Ibrahim and R. Verma, Medical Devices: Evidence and Research **2015,** 8, (2015).
16. R. Dal Negro, Multidisciplinary Respiratory Medicine **vol. 10,** 1, 1-4 (2015).
17. M. Hoppentocht, P. Hagedoorn, H. Frijlink and A. de Boer, Advanced Drug Delivery Reviews **vol. 75**, 18-31 (2014).
18. R. Hatley, R. Potter and Y. Dagtyareva, (n.d.), (2016)
19. J. Mitchell, M. Nagel and Y. Chengà, J. Aerosol Sci. **30**, (1999).
20. P. Zanen, L. Go and J. Lammers, Thorax **vol. 51**, 977-980 (1996).
21. M. Dolovich and R. Dhand, The Lancet **vol. 377**, 1032-1045 (2011).

On the spray pulsations of the effervescent atomizers

Marek Mlkvik[1], *Branislav Knizat*[1]

[1]Faculty of mechanical engineering, STU in Bratislava, Namestie slobody 17,81231, Bratislava, Slovak Republic

Abstract. The presented paper focuses on the comparison of the two effervescent atomizer configurations—the outside-in-gas (OIG) and the outside-in-liquid (OIL). The comparison was based on the spray pulsation assessment by different methods. The atomizers were tested under the same operating conditions given by the constant injection pressure (0.14 MPa) and the gas to the liquid mass ratio (GLR) varying from 2.5 to 5%. The aqueous maltodextrin solution was used as the working liquid (μ = 60 and 146 mPa·s). We found that the time-averaging method does not provide sufficient spray quality description. Based on the cumulative distribution function (CDF) we found that the OIG atomizer generated the spray with non-uniform droplet size distribution at all investigated GLRs. Exceptionally large droplets were present even in the spray which appeared stable when was analyzed by the time-averaging method.

1 Introduction

Twin-fluid atomizers are widely used, as an alternative to pressure nozzles, for spraying of viscous liquids. These spraying devices are specific by using a pressurized gas to enhance the liquid breakup. The need to reduce atomizing gas consumption and working pressure led to the development of internal-mixing twin-fluid atomizers that are more energy saving than their external mixing counterparts. A number of internal mixing atomizer designs have been proposed and studied in detail in past decades [6], [9], [3]. Though these works provide sufficient amount of information about particular atomizer designs and their spray properties, there is a lack of their systematic comparison and applicability analysis [1]. In our previous works, we provided a comparison of the OIG, OIL, Y-jet and CFT atomizers based on the internal flow regime and a primary breakup conditions [6].

As we analyzed the time-averaged Sauter mean diameter ($D32_a$) of the droplet sizes, we found significant differences between the spray quality of the OIG and OIL atomizers. Therefore, we will focus our interest on the deeper investigation of the spray temporal behavior in this paper. We will analyze the measured time-dependent droplet sizes to judge the spray quality. This analysis will be further used to compare the OIL and OIG atomizers and their applicability potential.

2 OIG and OIL atomizers

The main feature of the investigated atomizers is the mixing chamber, where the internal flow is developed. This part of both the investigated atomizers has the same design (Fig. 1).

It is designed as the tube with upper and side injection ports. These are used to introduce the liquid and the atomizing gas.

The OIG and the OIL configurations differ in the way of the working fluids injection. At the OIG atomizer is the liquid introduced through the top inlet port (port A, Fig. 1) and the gas is injected by the number of the side ports (ports B, Fig. 1). The only difference of the OIG and the OIL atomizers is that the liquid and the gas injection ports are switched. Even when this slight change may seem unimportant. It causes a dramatic change of the internal flow. As we previously observed [6], the internal flow of the OIG atomizer tends to develop as the plug or slug flow, while the liquid in the OIL atomizer tends to attach on the wall of the mixing chamber, which leads to the annular structure of the internal flow.

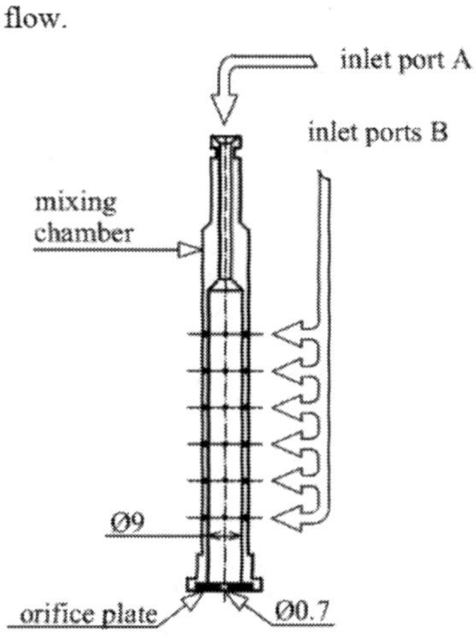

Fig. 1 Schematic drawing of the effervescent atomizer mixing chamber (both the configurations)

* Corresponding author: marek.mlkvik@stuba.sk

3 Measurement of droplet sizes

The experiments were conducted on the test bench schematically drawn in Fig.2.
An eccentric screw pump at constant rotation speed was used to drive the liquid and the liquid mass-flow was controlled by a bypass line. The constant pressure of the atomizing gas (air) was maintained by a pressure relief valve. The air pressure regulation was thus independent of its mass-flow. All the experiments were performed at the room temperature. The detailed test bench parameters can be found in [6].
The atomizer working regimes were defined by the Δp and the GLR. We have chosen the inlet air pressure as an independent parameter because its contribution to the total energy balance dominates over the liquid inlet pressure [4]. Therefore, the inlet pressure can be related to the potential energy of the atomization process. The GLR has been chosen because it is a parameter which can be simply related to the effectiveness of the atomizers work.

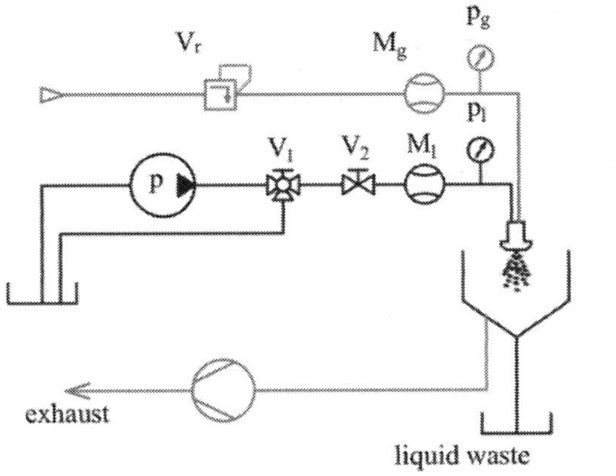

Fig. 2 Test rig

The droplet sizes were measured by the laser diffraction system (Malvern Spraytec) at the distance 100 mm downstream to the discharge orifice. The measurement frequency was 500 Hz and the measurement period was 25 s.

4 Spray pulsation analysis

The one way of the spray pulsation analysis is to use the time-averaged Sauter mean diameter (averaging was done over 12 500 samples for each working regime). The estimated standard deviation of the $D32_a$ can be therefore related to the spray pulsation. This method, however previously used and accepted, provides an only perfunctory view into the investigated problem. It can only differ between the regimes with and without pulsations and it cannot provide any other information about its nature (range of droplet sizes, periodicity....)

In our effort to investigate the spray pulsations more deeply, we investigated the time-dependent data. This was done by applying the cumulative distribution function, as proposed by [5]. This method provided us the more-relevant spray quality assessment tool.

5 Results

The simplest, but only qualitative, insight into the temporal spray behavior provide the records of the time-dependent droplet sizes in Fig. 3, 4. These figures show the comparison of the two atomizers at the same GLR and inlet pressure (Fig. 3) and the GLR atomizer working at two different gas consumption regimes (Fig. 4). It is obvious that both, the atomizer configuration as well as the GLR, strongly influences the spray quality. It is well-known that the droplet sizes generated by the twin-fluid atomizers decrease when the gas consumption rise. It is caused by the higher energy potential given by the compressed gas [4]. More surprising discover is that the same atomizer can produce spray with different quality when the gas and the liquid ports are switched, especially at the low GLRs. The Fig. 3 shows the difference.

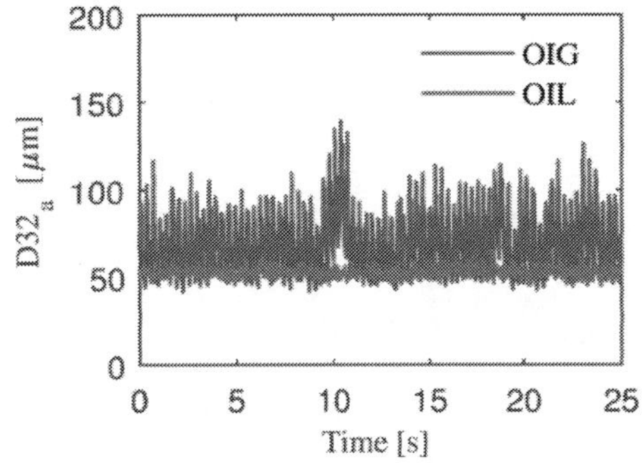

Fig. 3 Comparison of the time-dependent droplet sizes generated by the different atomizers for the same working regime ($\Delta p = 0.14$ MPa, GLR = 5%)

The droplet sizes time-record indicates that both, the D32 and its fluctuations are higher for the OIG atomizer configuration. Method of the spray comparison presented in Fig. 3, 4, however, simple to read, provides poor quantitative insight. The more suitable analysis is provided by time-averaged droplet sizes (Fig. 5). The diagrams show the dependence of time-averaged Sauter mean diameter in relation to the GLR for the different liquid viscosity. We can observe the decreasing trend of the droplet sizes as well as the pulsation intensity related to the standard deviation of the $D32_a$ (error bars in Fig. 5) for both the atomizers. As was outlined by the Fig. 3, the pulsations were more intense when the OIG atomizer was used, especially at the low GLRs. Now focus on the data measured for the less viscous liquid to simplify further analysis.

The error bars in Fig. 5 indicate, that for GLR > 5% was the spray pulsations intensity comparable. Now, the question is if this analysis provided all necessary information about the spray quality. For example, some applications, like spray-drying or oil burning require highly uniform droplet size distribution, which corresponds to the narrow error bars in Fig. 5.

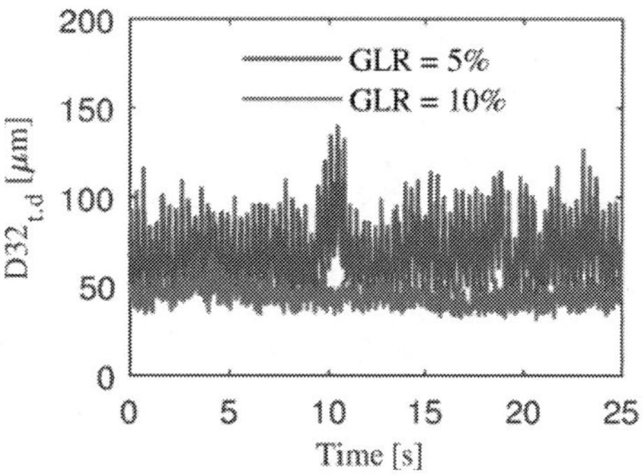

Fig. 4 Comparison of the time-dependent droplet sizes generated by the OIG atomizer at different GLRs

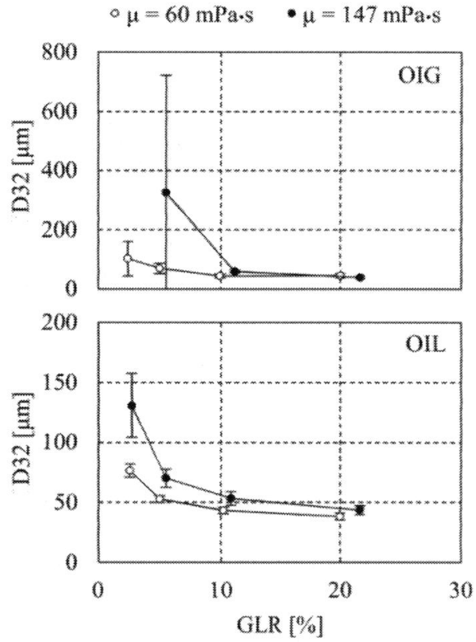

Fig. 5 Comparison of the cumulative distribution function for the two atomizers at GLR = 2.5% and inlet pressure 0.14 MPa

Even the small amount of the very large drops can cause serious problems in such applications. Problem is that a few the droplets highly excessing the average size do not affect the standard deviation of $D32_a$ and, therefore, is the presence of such droplets in Fig. 5 neglected.

A method for more-detailed droplet size analysis as proposed by [5] is based on the application of cumulative distribution function to the time-record of the measured droplet sizes. The results of this analysis can be seen in Fig. 6 and 7. The first diagram (Fig. 6) compares both the atomizers at the regime with the very low gas consumption. The results of this analysis show that the OIG atomizer produced generally larger droplets (~100 μm) than the OIL device (~75 μm), which is in accordance with the previous analysis. The additional information about spray pulsations (droplet sizes variations) is encoded in the shape of the CDF curve. The cumulative distribution function for the OIL is narrow and its slope is steep, which indicates that the droplet sizes variations were low (which also corresponds to the error bar in Fig. 5). For the OIG device is the CDF curve wider, which indicates more intense spray pulsations. In advance, we can analyze the composition of the droplets in the spray. The spray produced by this atomizer contained a number of the very small (50 μm), but also the very large (>150 μm) droplets. This can be seen in the bottom (CDF < 0.05) and top (CDF > 0.95) parts of the curve. However interesting are these findings they do not provide new essential knowledge about the spray, for the regime where the pulsations were obvious from Fig. 5 (GLR = 2.5%).

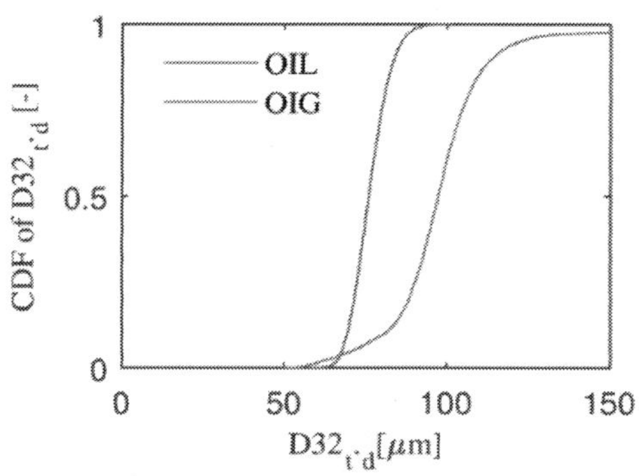

Fig. 6 Comparison of the cumulative distribution function for the two atomizers at GLR = 2.5%

The more important is this analysis for the regimes where the error bars did not signalize any significant droplet size variations (GLR > 5%). In Fig. 7, are the CDF curves for the regime characterized by the GLR = 5%. The OIL atomizer produced a very fine spray with the average droplet size about 55 microns. Also, the steep CDF curve indicates the uniform droplet size distribution. The measured data for the OIG configuration suggest that the spray contained a wide range of droplet sizes (from 45 to 115 microns). Even more important is that there were present a number of droplets larger than 100 μm. The CDF, therefore, revealed that even at relatively stable regimes (according to the Fig. 5), the spray quality might be insufficient for some applications where the high droplet sizes uniformity is required.

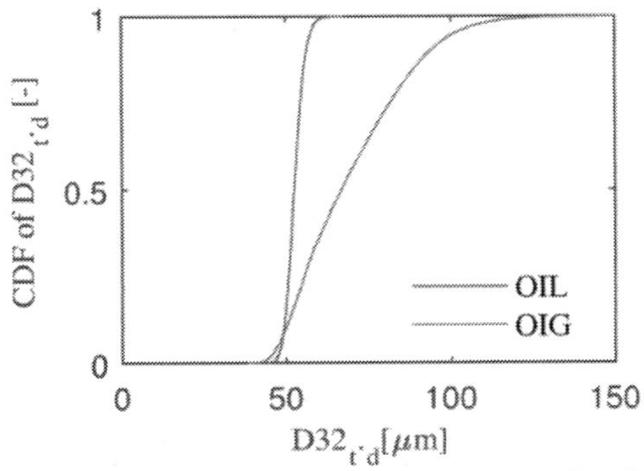

Fig. 7 Comparison of the cumulative distribution function for the two atomizers at GLR = 5%

From the previous analysis is clear that the spray quality (judged by the droplet size variations or pulsations) is highly sensitive not only to the GLR but also on the configuration of the effervescent atomizer injector ports. The main difference between the OIL and OIG atomizers lies in the internal flows. It was estimated [7] that the internal flows of these atomizers were of a different pattern. The flow in the OIG atomizer was the plug or slug flow. Such a flow is typical by varying of the large gas and liquid structures. This led to the temporal GLR variations in the discharge orifice of the OIG atomizer and caused spray pulsations. On the other hand, the flow in the mixing chamber of the OIL atomizer was wall attached with separated gas and the liquid streams. This caused relatively constant gas to the liquid mass ratio in the discharge orifice and therefore stable conditions for the liquid breakup.

6 Conclusion

The presented paper deals with the comparison of the two effervescent atomizer configurations. The comparison was based on the spray pulsation assessment by different methods. The main findings can be shortly concluded in the following points:

- The important factor influencing the spray quality is, except the working regime, also the configuration of the effervescent atomizer (OIL, OIG).
- The OIL atomizer not only produced the smaller droplets for all the investigated cases but also generated uniform spray with the low temporal droplet sizes variations. It was caused by the stable annular flow in the discharge orifice of the atomizer.
- The OIG atomizer generated the spray with the larger droplets. Its work was characteristic by the pulsations of the spray which led to the large standard deviation of the time-averaged sauter mean diameter of the droplets at the lowest GLR. At the GLR = 5% were the pulsations considerably lower, but the further analysis by the CDF showed, that the spray contained a considerable number of the very large (>100 microns) droplets. It was caused by the plug internal two-phase flow.
- When testing the atomizers dedicated for the spray quality sensitive applications (such the spray-drying), the detailed statistical analysis of the droplet sizes must be done. The time-averaging is not suitable as it neglects occasional presence of the very large (or small) droplets. These droplets can further cause non-optimal operation of the device.

Acknowledgement

The authors also appreciate the support from Prof. Dr.-Ing. Heike P. Schuchmann, Dr. rer. nat. Dipl.-Ing. Volker Gaukel and Dipl.-Ing Philipp Stähle from Karlsruhe Institute of Technology for providing us their test rig for experiments and other support and cooperation.
We want to also appreciate the help from doc. Ing. Jan Jedelsky, PhD from Brno University of Technology for providing his atomizers geometry details.

References

1. G.Ferreira, J.A.García, F.Barreras, A.Lozano, E.Lincheta, Fuel Process. Technol. **90** (2009), 270–278
2. G.Ferreira, J. C. F. Teixeira, C. J. Bates, P. J. Bowen, Atomization Sprays **11** (2001),
3. J. S. Chin, A. H. Lefebvre, Journal of Engineering for Gas Turbines and Power **117** (1995), 266-271
4. J. Jedelsky, M. Jicha, J, Fuel 1 **11** (2013)
5. A. Kleinhans, K. Georgieva, M. Wagner, V. Gaukel, chemical engineering and processing **104** (2016), 212-218
6. M.Mlkvik, , P.Stähle, ,H. P.Schuchmann, , V.Gaukel, J. Jedelsky, M.Jicha, International Journal of Multiphase Flow 77, (2015), 19-31
7. M.Mlkvik, R. Olsiak, *In: Experimental Fluid Mechanics 2016, (2016)*
8. J.Schröder, S.Kraus, B.B.Rocha, V.Gaukel, H.P.Schuchmann, J. Food Eng. 105 (2011), 656–662.
9. M. R. O. Panao, A. L. N. Morreira, Experimental, Thermal and Fluid Science **30** (2005), 79-96

The computational study in the unloading slot of the control valve

Lukáš Mrózek[1,2*], *Ladislav* Tajč[2], *Jan* Vimmr[3], *Ondřej* Bublík[4], *Václav* Sláma[1] and *Robert* Kalista[1,2]

[1]Doosan Skoda Power, Experimental Research of Flow, Tylova 1/57, Pilsen, Czech Republic
[2]University of West Bohemia, Department of Power System Engineering, Univerzitni 8, Pilsen, 306 14, Czech Republic
[3]University of West Bohemia, Department of Mechanics, Univerzitni 8, Pilsen, 306 14, Czech Republic
[4]University of West Bohemia, NTIS – New Technologies for the Information Society, Univerzitni 8, Pilsen, 306 14, Czech Republic

Abstract. The flow in the unloading system of the control valve is modelled using the numerical simulation. The proportional flow values for various widths of the slot and openings are evaluated. The changes of the pressure and Mach number in the space of the valve are recorded for the selected pressure ratios. The influence of the length of the slot on the mass flow changes is assessed. The results are compared with the experimental findings.

1 Introduction

Control valves enable supply to the steam turbine with such a quantity of steam which is necessary to achieve the required performance. Their dimensions are chosen with respect to the admission pressure of the steam and the steam flow rate. The principle of the regulation is based on the change of the flow area using the controlled lift of the valve cone. Abilities of actuators are limited in terms of force effects. If these limits are exceeded, it is necessary to divide the steam supply into more valves or to use unloading valves. In these types of valves the pressure acting on the certain area of the cone is reduced artificially, which facilitates the manipulation with the cone. The pressure decrease is achieved by throttling the steam in the bypass slot. In order to be able to dimension the unloading system correctly, it is necessary to have verified materials for the design of slot dimensions and to know their impact on the values of the flow coefficients. It is necessary to be able to design a suitable width of the slot with respect to the basic dimensions of the valve. For this reason the experiment, which models the flow in the bypass system of the control valve, is prepared [1]. The aim of the experiment is to obtain, for its various widths of the slot, the data about mass flows, flow coefficients of individual parts of the bypass system and their impacts on the pressure force acting on the valve cone. The use of computations of the media flow through the slot is also examined using the modern discontinuous Galerkin method in the numerical simulation.

2 Model of the unloading system of the valve

One example of possible designs of the unloading valve is shown in Figure 1. This is the valve designed by Professor A.E. Zarjankin from the Moscow Energy Institute. Doosan Skoda Power uses own specific design where the slot function is the same.

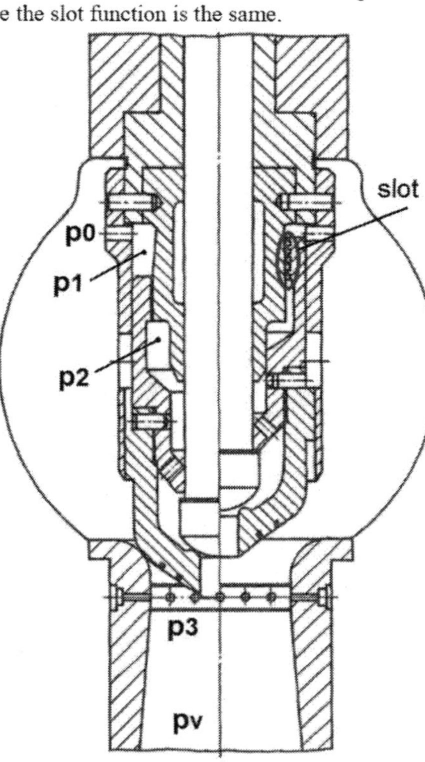

Fig. 1. Design of the unloading valve

* Corresponding author: lukas.mrozek@doosan.com

The steam of unloading system is led through a number of openings in the guide sleeve to the slot where it is throttled from the pressure p_1 to the pressure p_2. Then the steam passing through the area under the bypass cone gets to the diffuser, where it expands to the pressure p_3. The lift of valve cone and the pressure p_3 is determined by the turbine operation. The length of the slot changes with the lift of the cone as well. From a certain lift of the cone h the supply of the bypass steam from the slot is closed and the valve begins to work in the non-unloading mode. This causes the increase of the stabilization power acting on the valve cone again.

On the model of the bypass system of the valve shown in Figure 2, the aerodynamic parameters of three parts, i.e. the set of input openings, the slots themselves and the output section, can be examined. For the initial phase of experiments, eight openings with the diameter $d_1 = 9\ mm$ were chosen in order to model the input section. The width of the slot $\delta = 0.3, 0.5$ and $1\ mm$ is considered. The slot with $\delta = 0.15\ mm$ has also been tested. In this case the problems with the correct evaluation of the mass flow already occur. The experiments are also completed with the variant of the slot with $\delta = 0.3\ mm$. The length of the slot is $l = 50\ mm$ in all cases. The output section is created by six openings $d_3 = 9.5\ mm$. The model itself is connected to the compressed air distribution, which enables its expansion from the pressure at the level about *7.5 bar (a)* to the barometric pressure. In addition to the experiments the flow on the model of the valve was verified using numerical simulations.

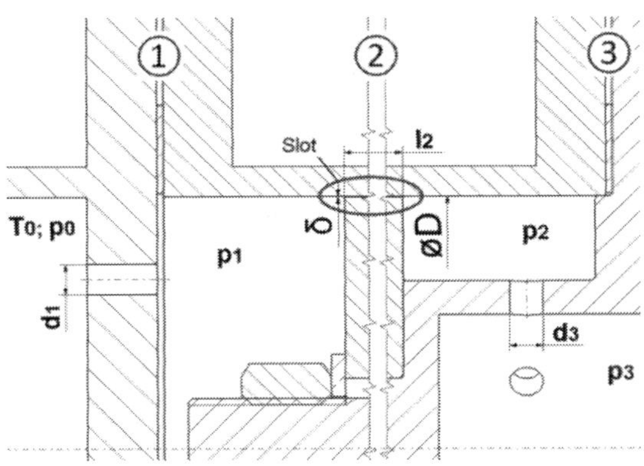

Fig. 2. Model of the bypass system of the valve

3 Modelling the flow in the slot using the numerical simulation

A part of the experimental test rig containing the throttling slot is considered for the computations themselves. This is the model of the bypass system of the valve which is presented in Figure 2. The total computational mesh is presented in Figure 3.

For the numerical simulation of the air flow through the slot, the nonlinear system of Favre-averaged Navier-Stokes equations is considered. The system is completed by the Spalart-Allmaras turbulence model [2] for transport equation of eddy viscosity. The discontinuous Galerkin finite element method (DGFEM) is used for the spatial discretization of the system of equations [3, 4]. The advantages of the method are high order of spatial accuracy and low artificial dissipation. As the basis functions, the linear polynomials were considered. This choice ensured second order of accuracy in space. For the approximation of inviscid fluxes on the element boundaries, the numerical Lax-Friedrichs flux was used [5]. Similarly, the viscous fluxes are evaluated using the appropriate numerical fluxes. In present paper, the approximation known as interior penalty (IP) method was chosen. The developed software is applied for numerical simulations of the flow in the slot having the width *0.15*, *0.3* and *1 mm* for different pressure gradients.

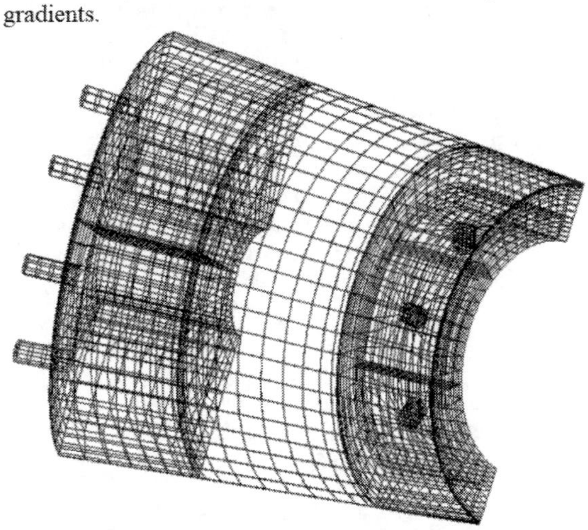

Fig. 3. Computational mesh of the model of the bypass system

The computations were compared with the experiments. The measurement was carried out for two widths of the slot $\delta = 0.3\ mm$ and $\delta = 1\ mm$ with the same length $l = 50\ mm$. Comparison of computational mass flow rate with the experimentally measured data is presented in Figure 4 and 5. A good agreement for the tested examples is verified.

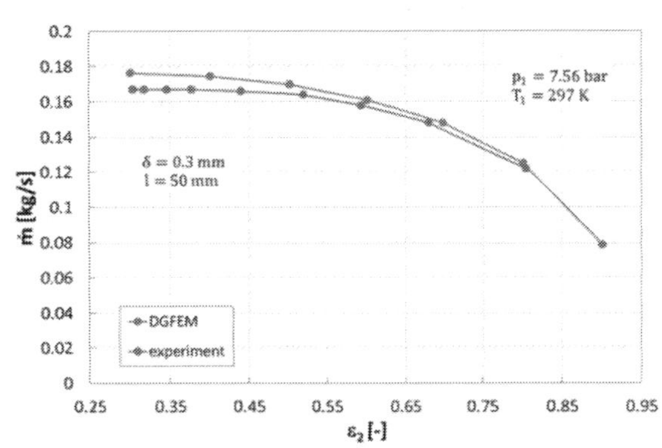

Fig. 4. Computational and experimentally evaluated mass flow through the slot having the width $\delta = 0.3\ mm$ and $l/\delta = 167$

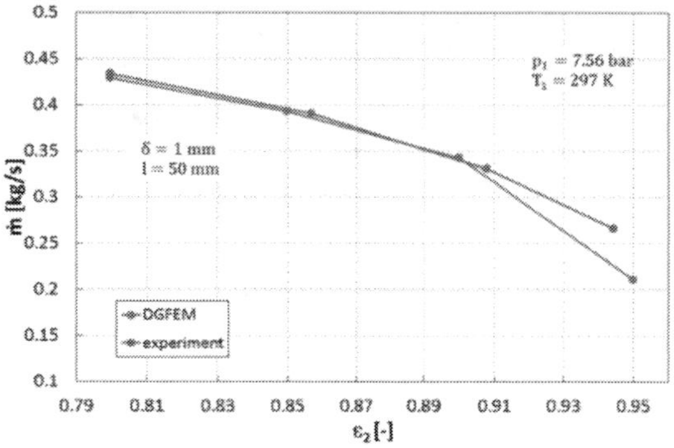

Fig. 5. Computational and experimentally evaluated mass flow through the slot having the width $\delta = 1\ mm$ and $l/\delta = 50$

By using computational studies it is possible to verify whether modelling is applied for slots when the geometric similarity is maintained. The slot having the width $\delta = 0.3\ mm$ was used as the base. The comparison with the slots having the widths $\delta = 0.5\ mm$ and $\delta = 1\ mm$ was carried out. Considered dimensions and air input parameters are shown in Table 1. Computed mass flows are shown in Figure 6.

Table 1. Dimensions and input parameters

Width	δ	[mm]	0.3	0.5	1
Diameter	D	[mm]	150	150	150
Length	l	[mm]	50	83.3	167
Pressure	P_1	[bar]	7.56	7.56	7.56
Temperature	T_1	[K]	297.15	297.15	297.15
Critical flow	$\dot{m}_*$	[kg/s]	0.2505	0.4175	0.8349

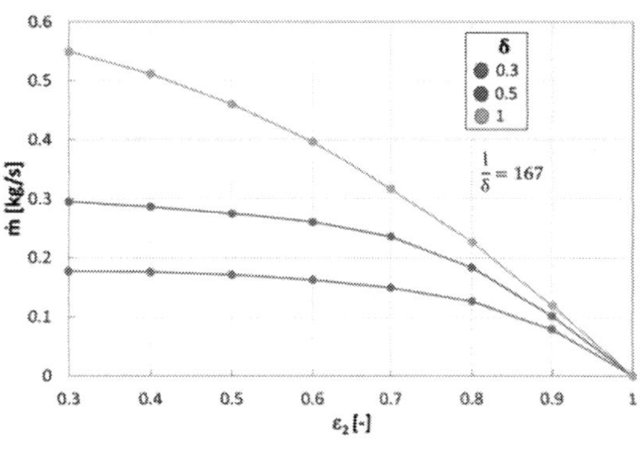

Fig. 6. Mass flows through the slots having different widths and the same ratio l/δ

The same diameter, on which the slot is located, was considered. However, the length of the slot was used to keep the ratio l/δ in all cases. Comparison of flows in dimensionless parameters is presented in Figure 7.

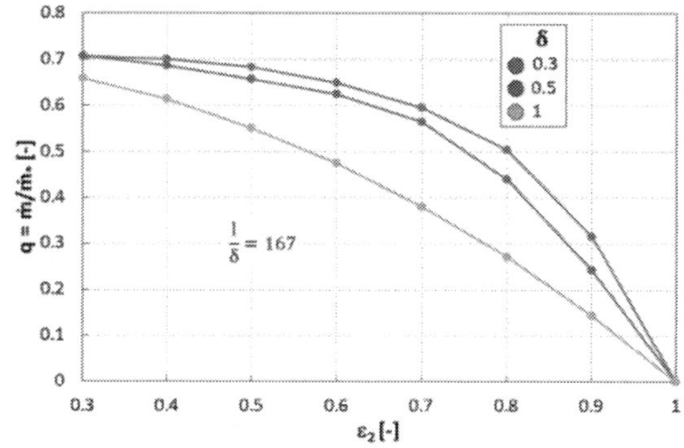

Fig. 7. Mass flows through the slots for $l/\delta = 167$

Assuming that modelling is kept, the dimensionless flow for all three slots should be the same. However, with the increasing width of the slot at the same pressure ratios the dimensionless mass flow decreases. There is only a small difference between the variants with $\delta = 0.3\ mm$ and $0.5\ mm$. However, there is a significant difference in the variant with the slot $\delta = 1\ mm$. It might be expected that with the increasing width of the slot the influence of the friction in the boundary layer on the total flow will be reduced and therefore the flow will increase. According to the computations carried out the flow is influenced by the length of the slot and the relevant influence on l/δ is not linear. For comparison, the correction of the flow from the variant $\delta = 0.3\ mm$ and $l = 50\ mm$ for the arrangement with $\delta = 0.5\ mm$ and $\delta = 1\ mm$ for the same slot $l = 50\ mm$ is presented in Figure. 8. The same average velocity for all slots is assumed. In this case with the increasing width of the slot, the mass flow increases. Never the less, the parameter l/δ changes. The flow decreases with the increase of the value l/δ. The length of the slot has a major influence on the resulting mass flow.

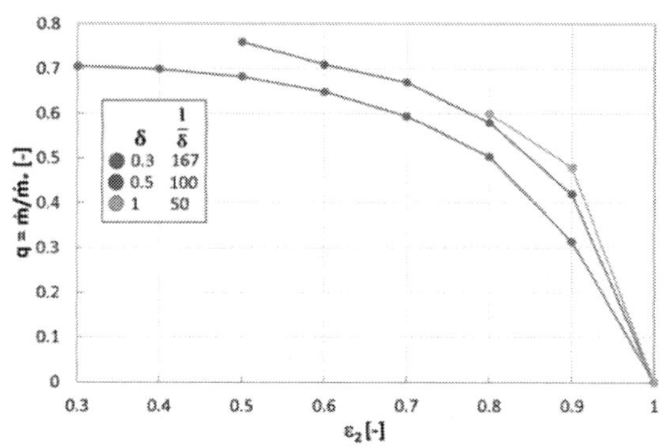

Fig. 8. Correction of the proportional flow q for $\delta = 0.3\ m$ for other widths of the slot

A series of computations of the flow through the slot having $\delta = 0.3\ mm$ and length $l = 1 \div 80\ mm$ was carried out. When evaluating the dimensionless parameter $q = \dot{m}/\dot{m}_*$, the superheated steam and the air in particular cases was considered as the working medium. The experimental findings were also used. The resulting mass

 https://doi.org/10.1051/epjconf/201818002070

flows are shown in Figure 9. With increasing length of the slot the relative mass flow decreases.

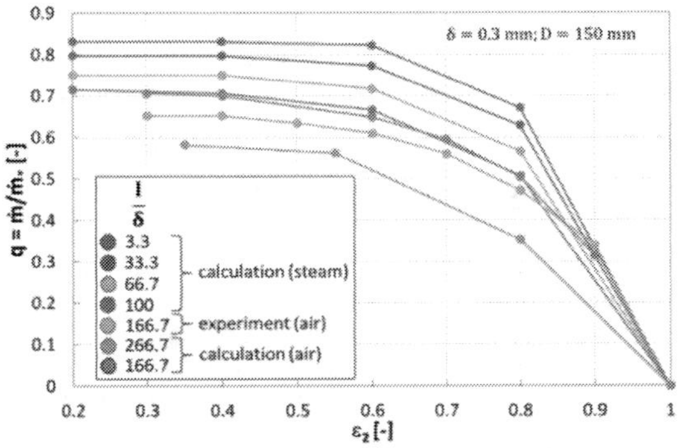

Fig. 9. Influence of the relative length l/δ on the relative flow q for the slot having $\delta = 0.3\ mm$

As its course for the constant pressure ratio $\varepsilon_2 = 0.4$ and $\varepsilon_2 = 0.8$ in Figure 10 shows, this is not linear dependence q on relative length l/δ.

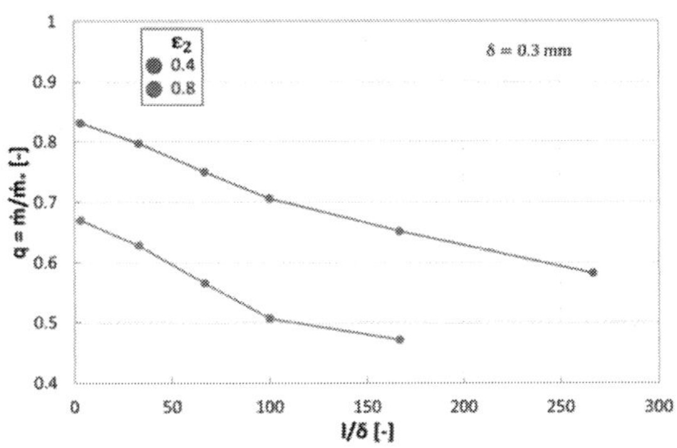

Fig. 10. Dependence of the relative mass flow q on the relative length of the slot

Similar findings concerning the slot having the width $\delta = 0.3\ mm$ are applied for the slot with $\delta = 0.5\ mm$ as well. The results from computations and experiments are shown in Figure 11 and 12.

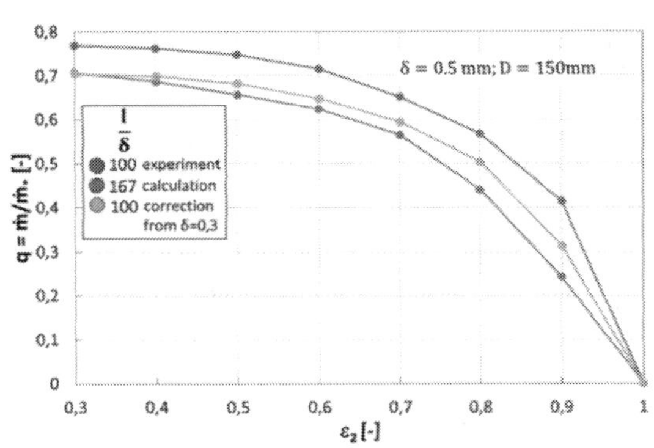

Fig. 11. Influence of the relative length l/δ on the relative flow q for the slot having $\delta = 0.5\ mm$

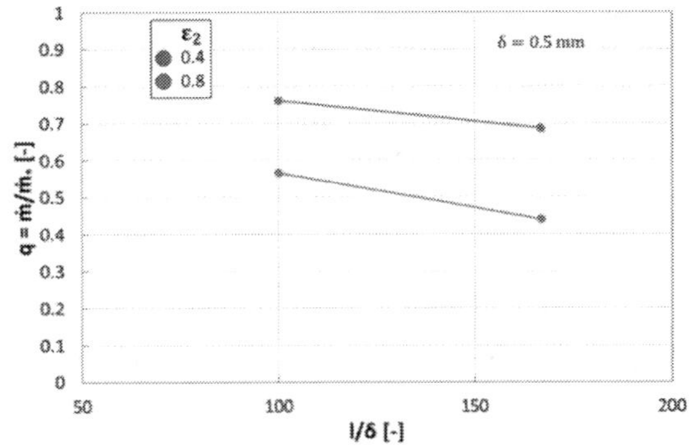

Fig. 12. Dependence of the relative mass flow q on the relative length of the slot

The data for the slot with the width $\delta = 1\ mm$ are also available from the computations and experiments. Dependence of the relative flow q on the pressure ratio ε_2 and on the relative length of the slot is presented in Figure 13 and Figure 14.

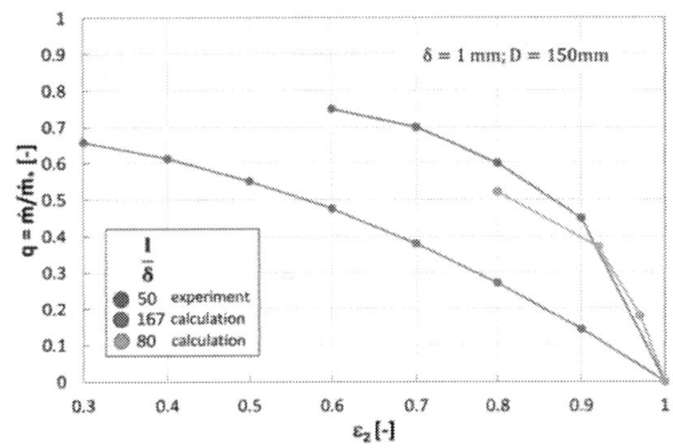

Fig. 13. Influence of the relative length l/δ on the relative flow q for the slot having $\delta = 1\ mm$

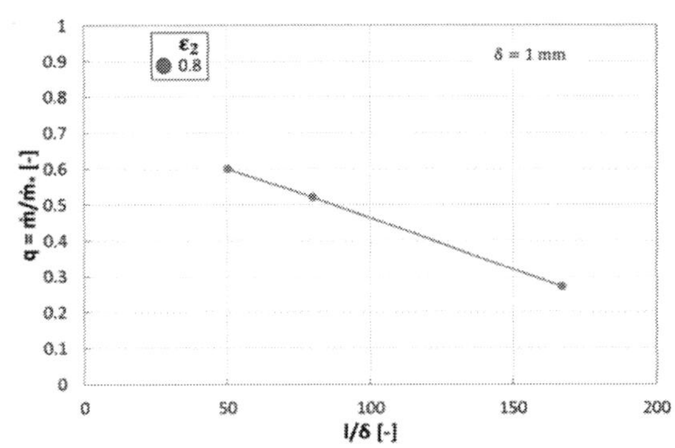

Fig. 14. Dependence of the relative mass flow q on the relative length of the slot

The CFD study shows a rather significant decrease of the flow for the slot having $\delta = 1\ mm$ at a longer length and the pressure ratio $\varepsilon_2 > 0.5$. It is not clear what causes this phenomenon. In practical applications on the valves maximum lengths of the slots s $l/\delta \approx 50 - 80$ can occur.

This is the area out of the course $q = f(\varepsilon_2)$ with more significant uncertainty of the dependence.

The numerical simulation enables evaluation of courses of aerodynamic parameters in all parts of the tested model. It particularly concerns the distribution of the pressure, velocity, Mach number and Reynolds number

Three pressure ratios ε for all model arrangement are considered. How the pressure ratios on the slot are different from the pressure ratio across all model is presented in Table 2.

Immediately at the inlet of the slot there is a sudden decrease of the pressure and a sudden increase in velocity. For the slot with the width $\delta = 0.15\ mm$ the aerodynamic blockage was achieved at the exit from the slot.

Table 2. Pressure ratios

δ	$\varepsilon = p_3/p_0$	$\dot{m}$	$\varepsilon = p_2/p_1$	q
[mm]	[-]	[kg/s]	[-]	[-]
0.15	0.35	0.062	0.43	0.611
	0.55	0.058	0.6	0.572
	0.85	0.02	0.92	0.197
0.3	0.35	0.135	0.52	0.665
	0.55	0.122	0.67	0.601
	0.85	0.065	0.92	0.32

Two pressure ratios are considered on the slot. The flow separation at the inlet of the slot and reduction of the flow area is characteristic. At the outlet from the slot, the flow clings to the outer limiting wall. At the pressure ratio $\varepsilon_2 = 0.3$, the maximum velocity occurs behind the slot, but at the higher pressure ratio the maximum velocity already occurs inside the slot.

4 Conclusion

It was investigated and validated that the discontinuous Galerkin method is an appropriate tool to describe the flows in the slots of different dimensional configurations. The acceptable agreement between experiments and computations was verified. The linear dependence of the mass flow on the relative length of the slow is not verified. However, for the area of real values l/δ on the control valves the linear dependence is acceptable. The numerical simulation enables mapping velocity and the pressure field across the entire flow area of the tested model. Pressure ratios on the slot itself differ from the pressure ratios on the whole model.

5 Acknowledgements

This study was supported by the project L01506 of the Ministry of Education, Youth and Sports of the Czech Republic and by the internal grant project SGS-2016-038 of the University of West Bohemia.

References

1. L. Mrózek, L. Tajč, *Modelling of flow in the unloading slot of the control valve* (AIP Conference Proceedings 1889, 020024 (2017), http://dx.doi.org/10.1063/1.5004358)
2. P. R. Spalart, S. R. Allmaras, *A One-Equation Turbulence model for Aerodynamic Flows* (AIAA Paper 92-439 (1992)).
3. B. Cockburn, C.-W. Shu, *Runge-Kutta Discontinuous Galerkin Methods for Convection- Dominated Problems* (Journal of Scientific Computing 16 (3), 173-261, 2001).
4. J. Vimmr, O. Bublík, A. Pecka, *A parallel implementation of an implicit discontinuous Galerkin finite element scheme for fluid flow problems* (Advances in Engineering Software, article in press, http://dx.doi.org/10.1016/j.advengsoft.2016.11.007)
5. E. F. Toro: *Riemann Solvers and Numerical Methods for Fluid Dynamics*, (Springer, Heidelberg, pp. 325, 1999)

Clocking of stators in one and half stage of axial steam turbine

Martin Němec[1,*], *Tomáš* Jelínek[1], and *Petr* Milčák[2]

[1]VZLÚ, Beranových 130, Prague 9, Czech Republic
[2]Doosan Škoda Power, Tylova 1/57, 301 57, Pilsen, Czech Republic

Abstract. An investigation of one and half axial turbine stage configuration was carried out in a closed-loop wind tunnel. The investigation was addressed to that impact how the previous stage outlet flow field influences the flow structures in the next stator in steam multistage turbines. The stage - stator interaction has been studied in this work. The detailed measurement with a pneumatic probes and fast response pressure probes behind the rotor and the second stator were performed to gain the useful data to analyze the impact. The detailed flow field measurement was carried out in the nominal stage regime (given by the stage isentropic Mach number 0.3 and velocity ratio *u/c* 0.68). The clocking effect of the stators is discussed and detailed unsteady flow analysis is shown.

1 Introduction

This work is a continuation of a long-term cooperation in turbine aerodynamics research with the leading producer of steam turbines Doosan Škoda Power (DSPW). DSPW has recently expanded a spectrum of stages offered by a reaction stages used for high-pressure and medium-pressure steam turbines. A multistage arrangement design has to be done with the most accurate models. These models must include kinetic energy losses caused by a profile and secondary structures and also include enthalpy changes due to a mutual interaction of stages and overflow through a shroud seals.

As a support of improvement of such design processes a research of a flow filed in a multistage turbine is realized in cooperation of VZLU and DSPW. Effects of the inlet flow parameters on a flow field structures in the axial turbine stage were studied as an initiative activity within this work. Various inlet turbulence intensity and inlet end-wall boundary layer thicknesses were studied in refs. [1, 2]. The results show that the typical steam turbine reaction stage is nearly insensitive on such inlet flow parameters changes.

At present, the main effort is directed to investigating the influence of the multi stage flow structures interaction.

The stator preceded by the full stage operated under various operational conditions and with different rotor shroud geometry configurations was experimentally studied in this work. Overall performances and time averaged results are discussed in ref. [3], investigation by an in-house CFD codes is presented in refs. [4, 5]. This contribution is focused on a specific field in multistage turbines - the effect of stators clocking. Usually, the clocking is studied with the same number of vanes in both stators. In this specific case, vane numbers are different and indivisible. The configuration was studied in order to find possible specific behaviours caused by interactions of this case.

2 Facility configuration

The test rig is a part of the closed-loop aerodynamic wind tunnel located in VZLU turbomachinery laboratory. A level of pressure in the wind tunnel is maintained by a system of vacuum pumps and a desired air mass flow rate is then provided by means of a twelve-stage radial compressor driven by a 1.3 MW electric motor. The air temperature and humidity are controlled by inter-stage heat exchangers and a condenser dryer, respectively.

The one and half axial turbine stage arrangement is depicted in Fig. 1. It is consisted of the stator and the shrouded rotor of the first stage which was followed by the stator of the second stage (second rotor was not present). Blade rows axial gaps were kept similar to real steam turbine geometry. The facility configuration allowed measuring the flow field structures in two planes: at the inlet and at the outlet of the second vane. The facility description is introduced in refs. [3, 6] more precisely.

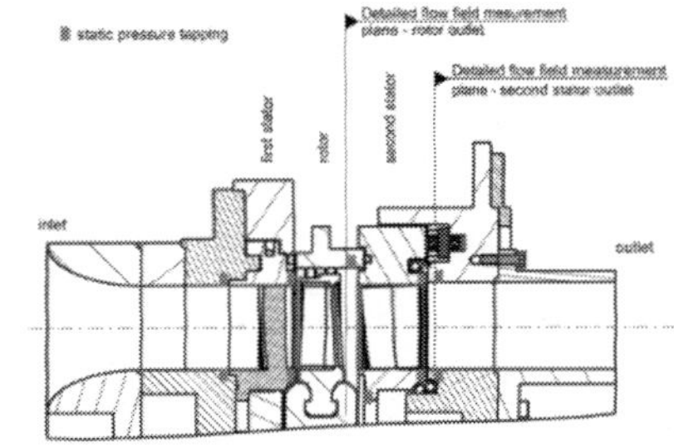

Fig. 1. The one and half stage meridian cut

* Corresponding author: nemec@vzlu.cz

https://doi.org/10.1051/epjconf/201818002071

Detailed flow field investigation was performed by a miniature five-hole pressure probes (tip diameter 1.2 mm) and with a fast response pressure probes. All probes were designed and manufactured in VZLU. The design of the fast response pressure probe was focused on the measurement of unsteady total pressure. The probe combines time averaged measurements of yaw angle and static pressure with unsteady measurement of the total pressure. Dynamic calibration was performed in shock tube obtaining probe own frequency (approximately 20kHz) and damping factor. Methodologies used for such fast response pressure probes development are presented in ref. [7].

3 Turbine stage specification

The one and half turbine stage model is consisted of the axial turbine stage REAC_v2-1 in configuration with nominal stator - rotor axial gap. The second stator geometry is REAC_v2-2. The rotor - second stator axial gap corresponds to the largest inter-stage axial gap typically used by DSPW. The blades geometry represents shaping of an experimental reaction turbine stage with increased outlet angle, which is commonly used for higher volumetric flow stages. The stage geometry is given in Tab. 1. The stage regime is specified in Tab. 2.

Table 1. Stage geometry

	1st stator	Rotor	2nd stator
Designation	v2-1	v2	v2-2
Vane/blade number	100	92	56
Aspect ratio	1.9	1.7	1.1
Pitch to chord ratio	0.75	0.68	0.75
Inlet pitch angle	90	70	90

Table 2. Stage regime

Stage isentropic Mach number	0.3
Velocity ratio u/c	0.68
The first stator isentropic Mach number	0.23
The second stator isentropic Mach number	0.24
The first stator Reynolds number	215 000
The second stator Reynolds number	210 000

4 Stators clocking

As mentioned before, stators with the same vanes numbers are usually studied. In this case, numbers of vanes are different and moreover indivisible. Such geometry is the clocking insensitive in the overall performance point of view. It is caused by a variation of the first and the second stator vanes relative circumferential positions. It is depicted in Fig. 2 - circumferential angle between the second vane and the closest first vane are shown here for 16 vanes of the second stator. Nevertheless, the stators interaction and influence of the relative position of vanes on the flow field is possible to study also in this case. As an advantage, it wasn't necessary to change relative position of the stators, because of the vanes relative position variation mentioned before. As a common divisor of vanes numbers is 4, many vanes relative positions are contained in one quarter of the wheel. The quarter is consisted of three periods and each period contains five relative vanes positions. These positions are shifted by 0.26° between periods.

The facility enables to measure 82° sector. So, the quarter is not fully covered (and also three periods are not captured completely), but, as it is shown further, this sector measurement gives reliable results.

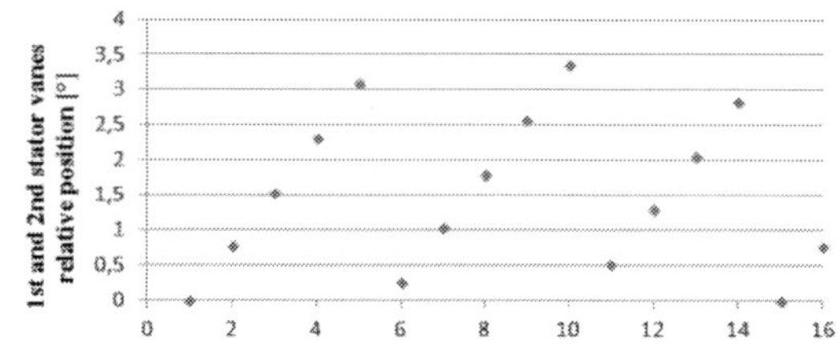

Fig. 2. The first and second stator relative vanes angles

4.1 Time averaged characteristics

Figure 3 shows a distribution of kinetic energy loss coefficient at the outlet of the second stator. No significant differences are possible to identify, except of vane number 4. The same conclusion is possible to derive from radial distributions of kinetic energy loss coefficient and yaw angle (Fig. 4). Vanes numbers from 6 to 10 are drawn here. These vanes cover one period in the quarter of the stator and measurement of this sector was performed at one time, thus there is no influence of the facility regime setup. Peculiarity of the vane number 4 is also shown in the radial distributions. A damage on the vane trailing edge (on a pressure side) was found after dismounting the facility. Interesting thing is, that the kinetic energy loss coefficient radial distribution (and also integral value) was not affected a lot by the damage. On the other hand, the outlet angle was changed quite significantly. The change of the outlet angle one can derive already from the distribution of the kinetic energy loss in the Fig. 3 (the deformation of the wake in the hub region).

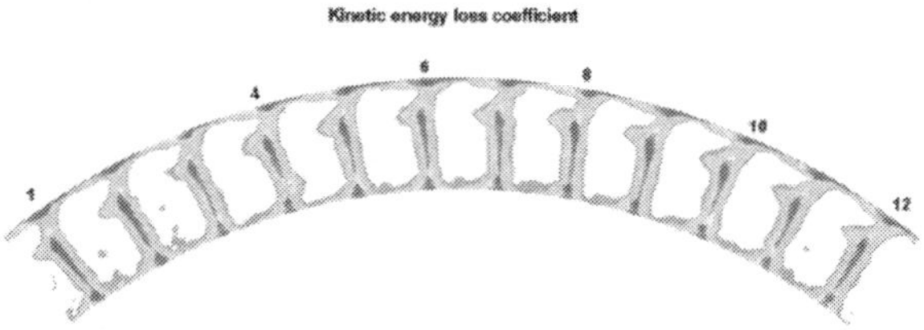

Fig. 3. The second stator kinetic energy loss coefficient distribution

As a conclusion, the time averaged characteristics did not showed an effect of clocking on the second stator behaviours.

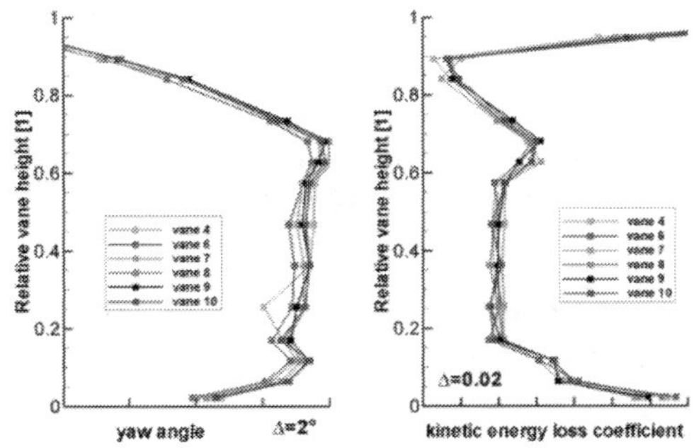

Fig. 4. The second stator outlet angle and kinetic energy loss coefficient radial distribution

(RMS) value computed from a complete signal acquired (Fig. 5). The second is RMS of a phase lock averaged signal (averaging respecting the rotor position - Fig. 6) and the third is blade passing frequency amplitude extracted with the Fast Fourier transform (FFT) (Fig. 7). Short explanation, what these results show, follows.

The simple RMS computation highlights fluctuations referenced to the second stator flow field development together with fluctuations caused by the rotor blade passing. Nevertheless, the RMS caused by the second stator flow field development is orderly higher in comparison with RMS caused by the passing of the rotor blades. So, it is possible to say, that the basic RMS shows us fluctuations mainly related to the second stator flow field development. The distribution of the basic RMS is shown in Fig. 5. Regions of a secondary flow together with wakes are clearly visible.

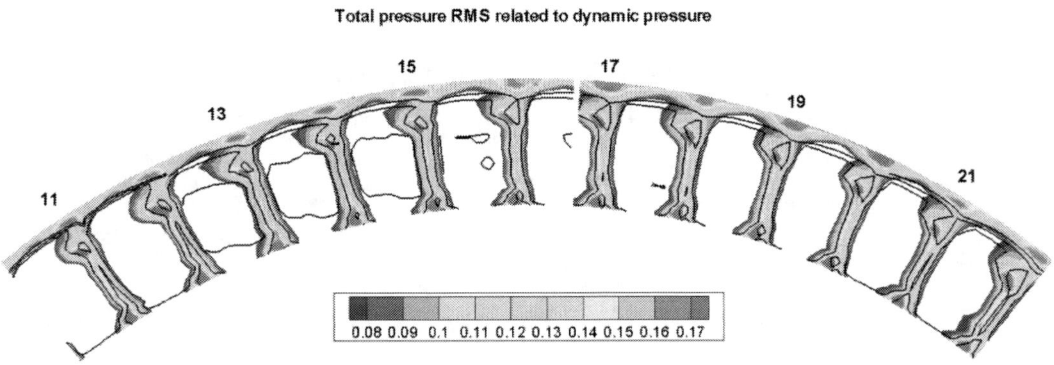

Fig. 5. RMS distribution at the second stator outlet

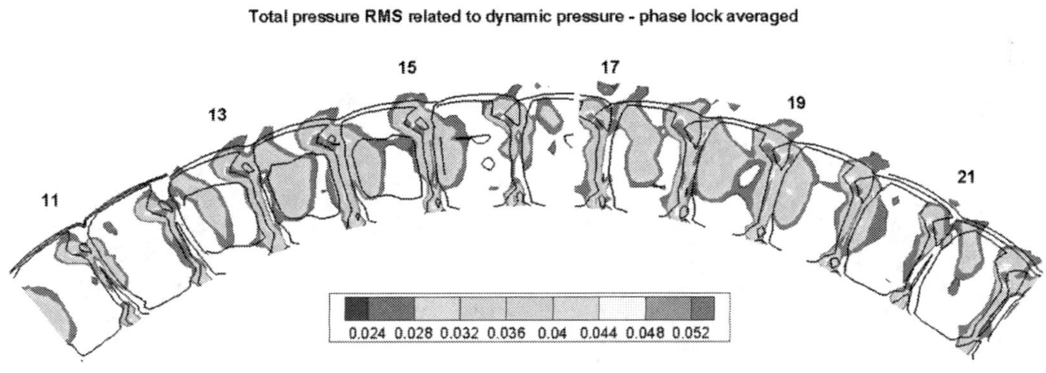

Fig. 6. Phase lock averaged RMS distribution at the second stator outlet

4.2 Unsteady pressure measurement results

The experiment was focused on the total pressure fluctuations capturing to analyze unsteady flow field evolution. Three main characteristics are obtained by the postprocessing. The first one is the root mean square

The RMS distribution is not distinctly affected by the stators clocking, what confirms the previous conclusion, that the second stator is not affected by the previous stator.

Phase lock averaging highlights phenomena related to the rotor flow field. An average rotor revolution is computed, then RMS value of the signal is determined.

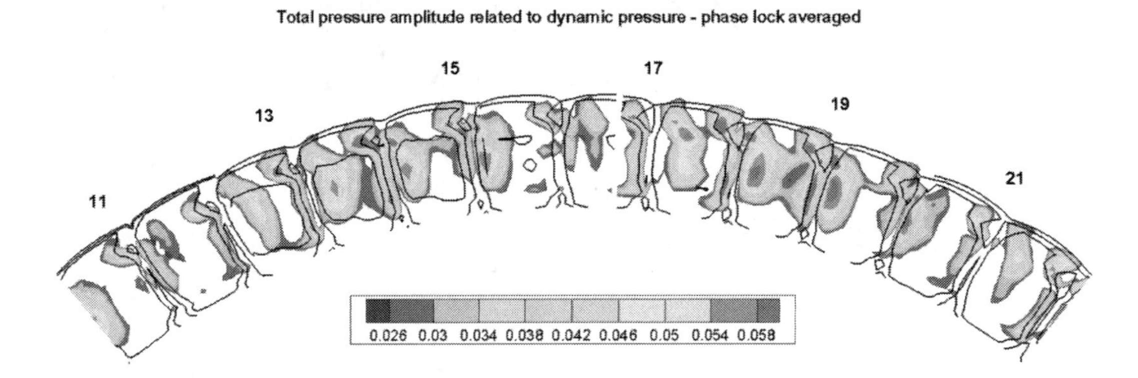

Fig. 7. Amplitude of the blade passing frequency at the second stator outlet

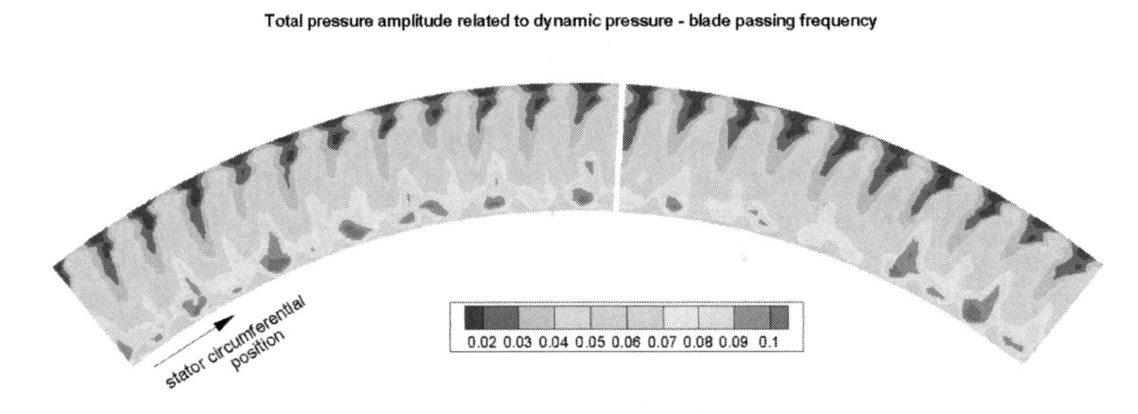

Fig. 8. Amplitude of the blade passing frequency at the rotor outlet

This RMS distribution is affected by the stators clocking, thus the clocking has an effect on the rotor flow field transformation through the second stator. Periodic changes in the flow field activity are shown in Fig. 6 and also in Fig. 7 with total pressure amplitude of blade passing frequency. The period is approximately five vanes, what is consistent with the vanes relative position variation shown in Fig. 2. With respect to the aforementioned displacement, the positions with almost identical distributions are not present exactly after five vanes, but the closest similar distributions are seen in a rhythm 5 5 4 5 5 4 vanes. A zones characterised by low total pressure RMS and amplitude (vanes 11(and 12)-16-21) are alternated by zones with high RMS and amplitude - vanes 14-18(and 19).

Finally, let's highlight the fact, that the most important exciting forces acting rotor behind the second stator are caused by the second stator flow field (compare RMS values in Fig. 5 and Fig. 6). Nevertheless, the unsteadiness caused by the rotor and the clocking effect on the rotor flow field transformation can bring a specific exciting frequencies due to circumferential periodic changes.

4.3 Rotor flow field and its transformation in the second stator

Firstly, we will focus on the flow field in the rotor outlet plane. Flow parameters distribution in this plane is given by the single stage flow field development. The rotor outlet is therefore affected only by the stator-rotor interaction and will be a reference for the transformation analysis. Figure 8 shows a distribution of total pressure amplitude (blade passing frequency) at the rotor outlet. The first stator wakes regions are clearly visible in this distribution.

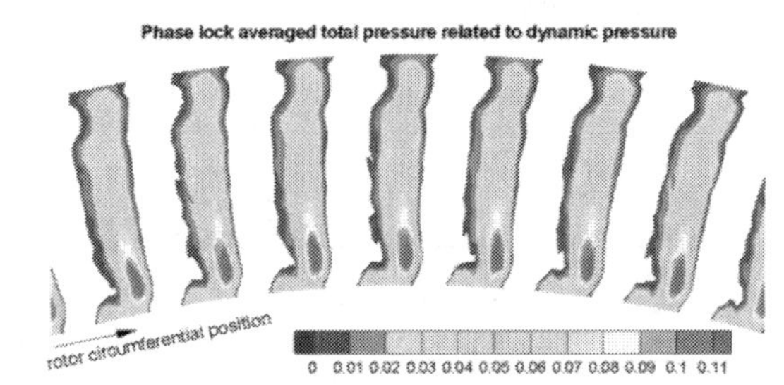

Fig. 9. Phase lock averaged total pressure difference distribution at the rotor outlet

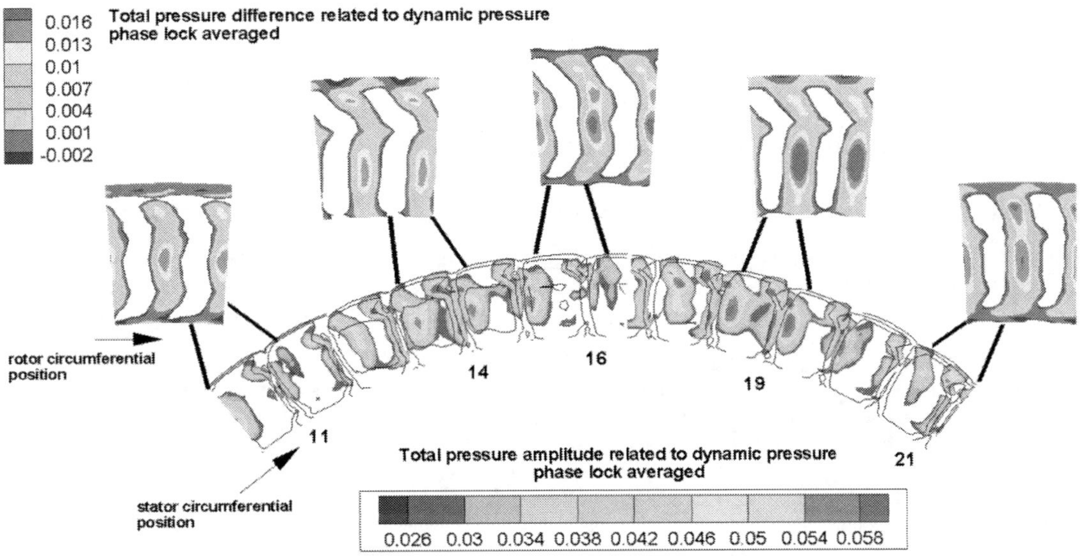

Fig. 10. Phase lock averaged total pressure difference distributions at the second stator outlet

To show the transformation process in the second stator, an averaged rotor outlet distribution is used. The averaged distribution is got by averaging across traversed area and is shown in Fig. 9. The influence of the first stator is averaged in this distribution. Let's note, that absolute values (total pressure is used here) do not provide a sufficient resolution for the rotor structures identification. Therefore, a difference between a local total pressure and total pressure averaged for a relevant diameter is used in order to highlight rotor structures. The highest total pressure amplitude is in the hub region, where the rotor hub passage vortex is developed. It is clearly visible in Fig. 9. The rotor blade wakes are clearly visible in this figure as well. One can also expect a region of high amplitude in the tip area, which should be related to the rotor tip passage vortex. In the case of the shrouded rotor, there is a mixing process with the rotor shroud leakage flow, what leads to a suppression of the tip passage vortex intensity.

Contrary to the rotor outlet plane, the second stator outlet plane is characterized by the position of the highest amplitude in the midspan and in the tip regions. Moreover, the regions are influenced also by the clocking of the stators as mentioned before. The variation of the rotor-related flow field is shown in Fig. 10. Distributions of phase lock averaged total pressure difference are shown here for a few representative areas.

5 Conclusion

The specific case of stators clocking was studied with unsteady pressure measurement instrumentation. It is shown, that the case with indivisible numbers of stators vanes is clocking insensitive in time averaged characteristics point of view. The clocking effect is possible to observe in unsteady evolution of the flow field. Specific frequencies of exciting forces appear due to an interaction of stators and rotor flow fields.

It is shown, that the second stator moves the unsteadiness caused by the rotor hub passage vortex in the tip direction, what may be interesting namely for a following rotor excitation. The forces caused by these unsteady behaviours are likely to be orderly lower in comparison with exciting by the second stator. Thus, wakes and secondary flow regions of the second stator stays the most important for the following rotor excitation.

Acknowledgment

The work was done with financial support of the Technology Agency of the Czech Republic on the project TA04020129 "Investigation of flows in a multi-stage axial turbine stages arrangement".
Authors thanks to Doosan Škoda Power for permission to publish these results.

References

1. T. Jelínek, P. Straka, V. Uruba, AIP Conf. Proc. 1745, 020019 (2016)AIP, 2016
2. T. Jelínek, M. Němec, V. Uruba, *22nd International Conference Engineering Mechanics 2016*, pp. 254-257, 2016
3. M. Němec, T. Jelínek, P. Milčák, *16th conference on Power System Engineering, Thermodynamics & Fluid Flow - PSE 2017*, 2017
4. P. Straka, J. Pelant, *Engineering Mechanics 2016*, pp. 522-525, 2016
5. P. Straka, J. Pelant, M. Němec, T. Jelínek, P. Milčák, *EPJ Web of conferences*, Vol. 143, 2016
6. T. Jelínek, M. Němec, *Turbostroje 2010*, 2010
7. M. Němec, *PSE-ES 2009*, 2009

Dustproof cooling of the electrical box

Patrik Nemec[1,*], *Katarína* Kaduchová[1], and *Milan* Malcho[1]

[1]University of Zilina, Faculty of Mechanical Engineering, Department of Power Engineering, Univerzitna 8215/1, 010 26 Zilina, Slovak republic

Abstract. In present are electrical boxes cooled by air through the intake hole on the bottom electrical box to the box space with electrotechnical elements and exhaust through the hole at the top to the surrounding by natural convection. This cooling method is effective but operate with the risk of contamination electrotechnical elements by dust sucking from surrounding air. The goal of this work is solution of the dustproof cooling of the electrical box by natural convection. The work deal with design of the device with the heat transfer by the phase change of the working fluid and experimental measuring its thermal performance at the cooling electrotechnical elements loaded by heat 1 200 W in the dustproof electrical box.

1 Introduction

The durability and reliability of electronic devices and electronic components are decreasing with high temperature. When the temperature increases, the parameters of the electronic components change and component can get damaged. This paper deals with heat dissipation problem for the multi-system static converter for a two-storey personal wagon. The static converter is a power electronic device with no moving parts that alter the parameters of electrical energy of one type to other parameters of the desired electrical power of another kind (voltage and current size and time, frequency and number of phases). In the experiment, the Joule heat generated by the static power converter has been simulated by resistive heating elements. Resistive element converts electrical energy into heat through resistance also known as Joule heat. Heating conductor can be explained by forwarding part of the kinetic energy of the particles causing electrical current (typically electrons) to particles, which are not participating in the electric current (usually positive ions in a fixed position). This increases the thermal motion of the particles, and the wire is heated [1, 2, 3].

The aim of this work is to design and construct a cooling device operating as a thermosyphon loop heat pipe, which would be able to dissipate the heat generated from electronic components, so that their efficiency is to be the highest and the lifespan is to be the longest.

For proper operation of the cooling device, it is important to correctly design evaporating and condensing section. Additional conditions for proper operation of the device are hermetic closures, which contain the working medium (in our case, distilled water) at a defined pressure. The principle of operation is that the evaporator absorbs heat from the environment which causes the working medium inside the evaporator to begin to evaporate at a certain temperature. The steam passes through the steam line into the condensing portion of the loop heat pipe. The condensation part is cooled by the cooling medium (water, air, etc.), which causes cooling of the vapor in the condensing part and condensation of the working medium. Subsequently, the working medium is gravitated to the liquid pipeline and back to the vaporized portion of the loop heat pipe, where it evaporates again. This results in circulation of the working medium connected to the heat transfer. Important factor of cooling device design is the working temperature (max. temperature at which the cooling device will work) of the work medium. Other important factors are location of the heat source from cooling device and method of removal transferred heat by the cooling device (natural or forced convection) [4, 5, 6].

2 Design of cooling device

One of the important parameters for manufacturing and subsequent proper functioning of the cooling device is design of the evaporation and condensing part. The most important parameter in the design is going out of the existing dimensional parameters available to placing the proposed cooling device. The new design is supposed to replace the old method of cooling the power transducer, where electronic devices inside the cabinet have been dusting. The Fig. 1 shows the place of storage of a new cooling device. By changing the phase of the working fluid from the evaporation heat exchanger, which will take waste heat generated by static power converters to the condensing heat exchanger, which will be by forced convection remove transferred heat [7].

* Corresponding author: patrik.nemec@fstroj.uniza.sk

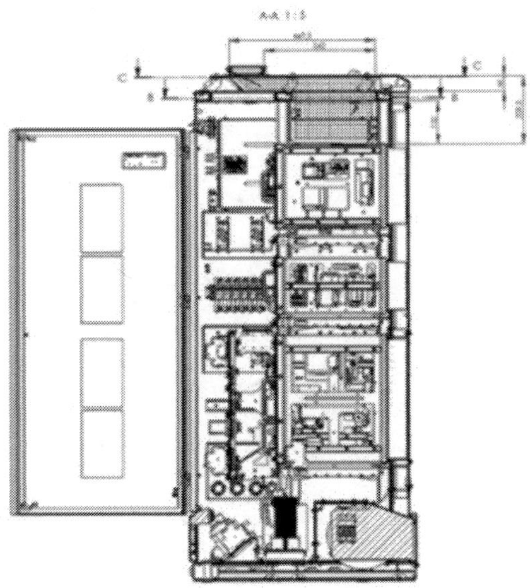

Fig. 1. Cooling device placing in static power converter.

2.1 Evaporator

The evaporation heat exchanger is one of two heat exchangers of cooling device. Main focus in designing of evaporator is to achieve the best possible absorption of heat dissipation. The evaporating part in the experimental device consists of two reservoirs connected by a copper pipes. For better absorption of generated heat, the lamellas were designed on the copper pipes. At the bottom of the evaporator is a drainage pipe that connects the two reservoirs. Parameters of the evaporating part of the cooling device are shown in Table 1. In the Fig. 2 and Fig. 3 are views of the evaporating part of the cooling device.

Table 1. Structural composition of cooling device evaporation part.

Lamels	Number	408
	Dimension	240 x 120 mm
	Thickness	0.25 mm
	Pitch	2 mm
	Material	aluminum
Pipe	Number	32
	Diameter	Φ16 x 0.8 mm
	Length	944 mm
	Material	copper
Reservoir	Dimensions (t x v x l)	120 x 240 x 30 mm
	Material	copper

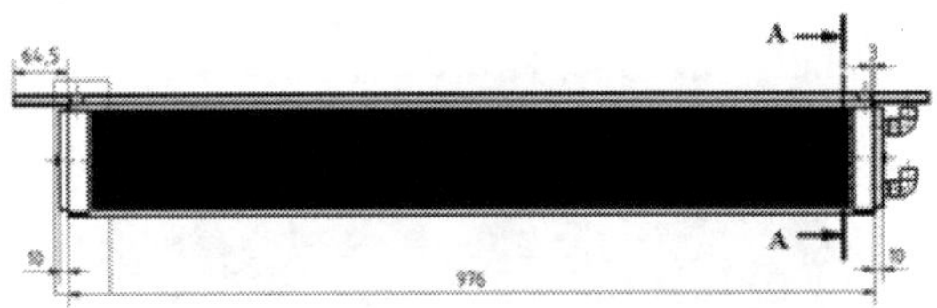

Fig. 2. Evaporating part of cooling device – front view.

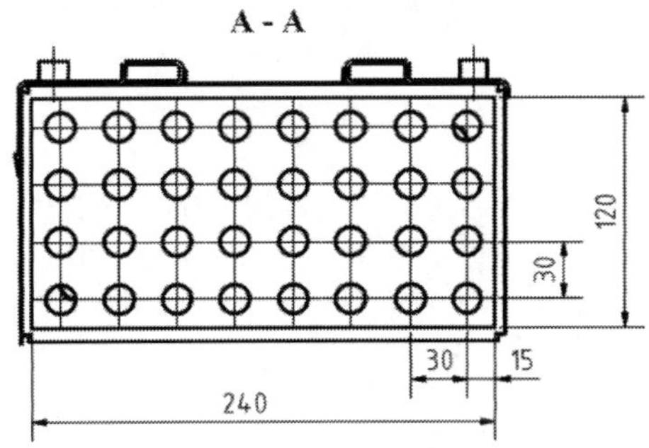

Fig. 3. Evaporating part of the cooling device – side view.

2.2 Condenser

The condenser of the cooling device is made of a meander-lined copper pipe with lamellas to increase the heat exchange surface with the surroundings. The condenser is located above the evaporation part. A T-piece fitting is provided on the top of the condensing pipeline for filling working fluid and measuring the pressure inside the cooling device. Parameters of the condensing part of the cooling device are shown in Table 2. In Fig. 4 is a condenser of a cooling device.

Table 2. Structural composition of cooling device condenser part.

Lamels	Number	332
	Dimension	420 x 30 mm
	Thickness	0.25 mm
	Pitch	2 mm
	Material	aluminum
Pipe	Number	14
	Diameter	Φ16 x 0.8 mm
	Length	785 mm
	Material	copper
Fittings	Elbow	R15
	Material	copper

Fig. 4. Condensing part of cooling device – upper view.

3 Experimental device for verifying functionality and determining the operating parameters of the cooling device

Verification of functionality and determinating the performance parameters of the cooling device were carried out on an experimental measuring device made from real model of multisystem static power transducer model according to the actual state and is shown in the Fig. 5 and Fig. 6. The model consisted of steel construction of cube profile with dimensions 800 x 593 x 1 146 mm (height x thickness x length). In the steel frame are attached wooden plates of thickness 2 cm. On the back side of the construction is formed profile for connecting air duct to supply air to cool the condenser. Inside cube construction is a wall for the installation of heating devices and to create a space for air flow with the possibility for measuring the temperature at front and behind the evaporator. To reduce heat losses to the environment, the entire experimental device was insulated with 5 cm thick polystyrene.

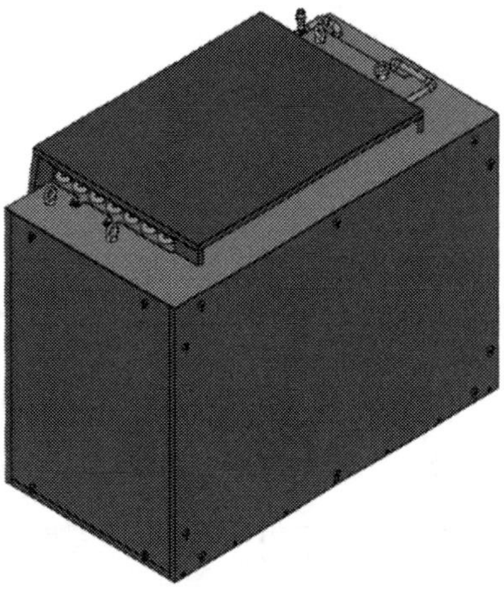

Fig. 5. Experimental device – outside view.

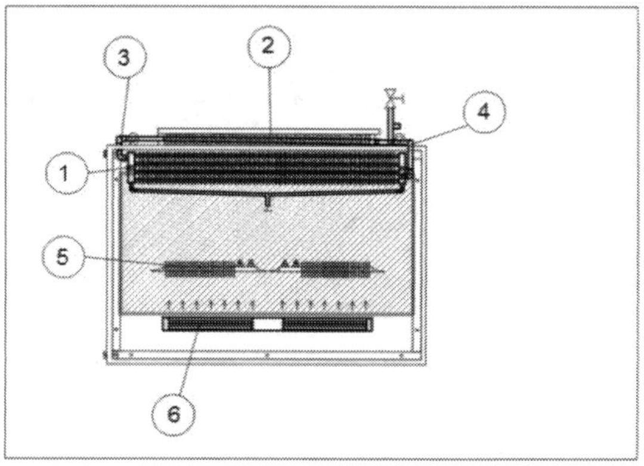

Fig. 6. Experimental device – inside view: 1 – evaporator, 2 – condenser, 3 – vapor pipeline, 4 – liquid pipeline, 5 – electrical heating element, 6 – tangential fan.

In the Fig. 7 is a schematic connection of the measuring devices to simulation operating condition of electronics elements and to measure operating parameters of the cooling device (amount of the working substance and the cooling air flow rate).

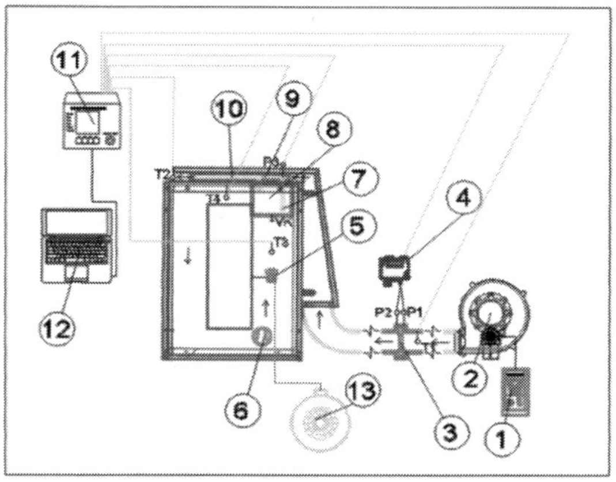

Fig. 7. Schema of entire measuring device: 1 – frequency inverter, 2 – medium pressure fan, 3 – normalized aperture, 4 – differential pressure sensor, 5 – electrical heating element, 6 – tangential fan, 7 – liquid pipeline, 8 – evaporation part of cooling device, 9 – condensation part of cooling device, 10 – vapor pipeline, 11 – data center, 12 – PC, 13 – regulated autotransformer.

4 Measurement

Experimental measurement of the operating parameters of the cooling device was performed at a 1 200 W heat load for a various volume of working media. Heat was generated by two 1 000 W electric heating elements connected in parallel to an electrical power supply (regulated autotransformer) via an active power transducer to define strict value of the electrical power. The tangential fan located inside the multisystem inverter model was designed to better dissipate heat from the heating elements to the evaporating heat exchanger. Cooling of the condensing part was made by the air flow controlled by operating speeds of fan using a frequency converter. Measurements were performed at the cooling air flow rate corresponding to operating speed of fan 10, 20, 30, 40 and 50 Hz. The cooling air flow rate was measured using a standard aperture in the pipeline where a difference in static pressure before and after the diaphragm was scanned by the DPT sensor. For temperatures determining the operating parameters of the cooling device were measured using the Pt100 temperature sensors.

Setting the optimum operating parameters of the cooling device consists of determining the volume of working substance and the rate of cooling air at which the cooling device will most efficiently dissipate heat. The quantity of the working substance is the main parameter for the proper functionality of the cooling device. At small volume of the working substance inside cooling device may entire quantity evaporate and the heat transfer will stop. At high volume of the working substance may liquid phase of working substance overflow entire device and the device will not operate in a two-phase heat transfer mode.

An ideal amount of the working substance in the cooling device is, when it always will have just so much liquid phase in the evaporation part, that consumes all supplied energy at the given conditions to the phase change of its entire volume during its operation.

To determine amount of working fluid in the cooling device, it was assumed that the maximum amount of working fluid corresponded to the entire volume of the evaporator part. The test quantities of the working fluid were determined by the percentage ratios of the evaporator part of the cooling device. The filling of the cooling device was carried out by the method of vacuuming whole volume by the vacuum pump and then filling with a quantity proportional to 20 %, 40 %, 60 %, 80 % and 100 % of the evaporator volume.

The total volume of the evaporator portion of the cooling device was determined by the sum of the volumes of the individual parts constituting the evaporation part.

$$V_V = (\pi . r^2 . l) . n + (a . b . c) \quad (1)$$
$$V_V = 6.07\ l$$

Where r (mm) is radius of pipes which connect collecting containers, l (mm) is the length of pipes, n (-) number of pipes, a, b, c (mm) are the dimensions of collecting containers.

Distilled water was used as the working fluid in the cooling device. The use of distilled water is in order to obtain a clean working substance free of impurities and minerals which are removed during the distillation process from the water. The change of the liquid phase to vapor phase is determined by operating pressure and temperature evaporating part of the cooling device [8, 9].

5 Evaluation of measured operating parameters

The calorimetric method was used to determine the amount of power transferred by the cooling device:

$$Q = m .\ c .\ (T_2 - T_1) \quad (2)$$

Where Q (W) is the heat flux, m (kg.s^{-1}) is the mass flow, T_2 (K) is output temperature, T_1 (K) is input temperature and c (J.kg^{-1}.K^{-1}) is specific thermal capacity of substance [10, 11].

In Table 3 are calculated mass flow rates and cooling air flow rates corresponding to the fan operating speed.

Table 3. Calculation of cooling air velocity.

Frequency	10 Hz	20 Hz	30 Hz	40 Hz	50 Hz
q_m (kg.s^{-1})	0.0698	0.1436	0.2140	0.2817	0.3470
q_v (m^3.s^{-1})	0.0601	0.1239	0.1848	0.2429	0.2987
w (m.s^{-1})	1.9192	3.9569	5.9028	7.7581	9.5402

In Fig. 8 are shown the performance values transferred by the cooling device at different quantities of the working fluid, cooling air velocity at 1 200 W thermal load. According results in the graph are not clear what the amount of the working substance is most suitable for the cooling device, because with change of the cooling air flow rate measured performances are changed too. Based on these results, it is possible to say that at the lower airflow rates, the highest thermal performance was measured.

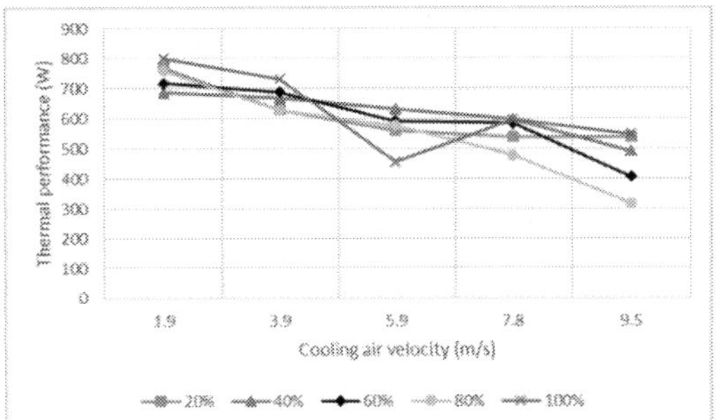

Fig. 8. Thermal performance of the cooling device at different quantities of the working fluid and the cooling air velocity.

Another important indicator during cooling device operation is the ambient temperature measured in front and behind evaporator inside the box with the power electronic components. These temperatures affect the life of electronic components. Lower indoor temperature mean the better operating conditions of the electronic components. When the cooling air velocity was changed, there were no significant differences in the temperatures inside the box. In Fig. 9 are average values of temperatures measured inside the box at individual volumes of the working substance in the cooling device. Fig. 9 shows that increasing volume of the working substance in cooling device increase temperature in the box.

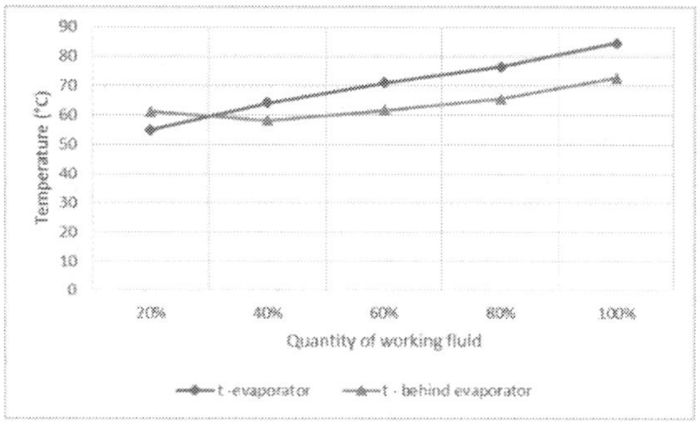

Fig. 9. Temperatures inside the electrical box.

6 Conclusion

The aim of this work was to design and construct a cooling device operating as a thermosyphon loop heat pipe, which would be able to remove generated waste heat from the electronic components so that their efficiency was as high as possible and their lifetime was as long as possible. Experimental measurement of the operating parameters of a cooling device with different quantities of the working fluid in the range of 20 to 100 % of the total evaporator volume and cooling air flow rates ranging from 2 to 10 $m.s^{-1}$ at a 1 200 W heat load showed that the ideal amount of the working substance in the cooling device is 20 to 40 % and the cooling air flow rate is 2 $m.s^{-1}$. At higher volumes of working substance and higher cooling air rates there was a decreasing in heat removal from the surroundings by the evaporating heat exchanger and insufficient heat transfer through the cooling device, resulting in overheating of the evaporator and raising the temperature inside the static power converter model.

This article was created with financial support of the project APVV -15-0778 "Limits of Radiative and Convective Cooling through the Phase Changes of Working Fluid in Loop Thermosyphon" and 042ŽU - 4/2016 "Cooling on the basis of physical and chemical processes.

References

1. P. Hrabovský, Š. Papučík, R. Lenhard, EPJ Web of Conference, **114** (2016)
2. D. V. Kozak, Yu. E. Nikolaenko, *International Conference on Electronics and Information Technology, EIT 2016 - Conference Proceedings*, (2016)
3. J. Jandačka J. Mičieta, M. Holubčík, R. Nosek, Energy & fuels, **31**, 3 (2017)
4. M. Brůna, Manufacturing technology, **15** (2015)
5. R. Nosek, S. Gavlas, R. Lenhard, V. Sedlak, H. Mollerhagen Arvesen, Communications, **16** (2014)
6. P. Pilát, M. Patsch, Š. Papučík, M. Vantúch, AIP Conf. Proc., 1608 (2014)
7. L. J. Orman, R. Chatys, Transport Means - Proceedings of the International Conference (2011)
8. R. Lenhard, J. Jandačka, AIP Conf. Proc., **1558** (2013)
9. J. Cernecky, J. Koniar, Z. Brodnianska, J. of Heat Transfere, **136**, 4 (2014)
10. M. Mahdavi, S. Qiu, S. Tiari, Apl. Therm. Eng., **81** (2015)
11. M. Holubčík, P. Ďurčanský, J. Jandačka, R. Nosek, Scient. World J., **2014** (2014)

Influence of the ambient temperature on the cooling efficiency of the high performance cooling device with thermosiphon effect

Patrik Nemec[1,*], *Milan* Malcho[1]

[1]University of Zilina, Faculty of Mechanical Engineering, Department of Power Engineering, Univerzitna 8215/1, 010 26 Zilina, Slovak republic

Abstract. This work deal with experimental measurement and calculation cooling efficiency of the cooling device working with a heat pipe technology. The referred device in the article is cooling device capable transfer high heat fluxes from electric elements to the surrounding. The work contain description, working principle and construction of cooling device. The main factor affected the dissipation of high heat flux from electronic elements through the cooling device to the surrounding is condenser construction, its capacity and option of heat removal. Experimental part describe the measuring method cooling efficiency of the cooling device depending on ambient temperature in range -20 to 40°C and at heat load of electronic components 750 W. Measured results are compared with results calculation based on physical phenomena of boiling, condensation and natural convection heat transfer.

1 Introduction

The cooling device with heat pipe technology may be visualised as a long hollow pipe, bent and the ends joined to form a continuous loop, filled with working fluid and orientated in a vertical plane. If the one side of the loop is heated and the other side cooled, the average density of the fluid in the heated side is less than in the cooled side. An essentially hydrostatic pressure difference, as a result of the thermally induced temperature gradient between the hot and the cold sides, rives the fluid flows around the loop. The 'buoyancy' force, as it is often termed, driving the fluid is in turn counteracted by an opposing frictional force that tends to retard the flow [1].

Cooling device can transfer heat from the interior of a microelectronic system to a central location where space limitations are less stringent. The advantages that a cooling device with heat pipe technology enjoys over a conventional refrigeration system include: (1) absence of moving parts leading to a more reliable system operation, (2) increased choices for selecting a working fluid compatible with microelectronics chips since it does not have to go through a refrigeration cycle, (3) reducing the decomposition rate of the working fluid as the higher temperatures at the compressor discharge in a vapour compression refrigeration system are not encountered, (4) clean operation as no oil is circulated through the system. In comparison to pool boiling systems employing vapour space condensation, a heat pipe technology offers more flexibility in terms of providing a centralized condenser with different feed lines to individual evaporator stations. Further, with the addition of a liquid circulating pump in a cooling device, higher heat transfer coefficients associated with flow boiling systems could be realized [2].

2 Design of cooling device

The main parts of cooling device are:

- evaporator,
- condenser,
- pipeline,
- filling and closing valve,
- working fluid

The model of cooling device is shown on Fig. 1.The evaporator enables on the base of a phase change (boiling) of the working fluid an intensive heat removal from its surface. It has to be constructed so that it will prevent the leakage of the working fluid, maintain pressure differences in all the walls and enable heat transfer from the electronic component into the working fluid as well as a suitable distribution of liquid and vapour phases of the working fluid. When choosing the material suitable for the construction of evaporator, it is necessary to pay attention to its thermokinetic characteristics. To provide a minimal temperature drop between a heat source and evaporator, the evaporator material must feature high thermal conductivity. To prevent escape of vapour, it should not be porous. The material should have high strength but, at the same time, it should be easily machineable and compatible with the working fluid [3].

The evaporator body is a plate with dimensions 116 x 206 x 30 mm. To provide the working fluid circulation there are two 12 mm openings drilled horizontally on the plate and connected with twenty one 6 mm vertical

* Corresponding author: patrik.nemec@fstroj.uniza.sk

https://doi.org/10.1051/epjconf/201818002073

connecting channels. They provide the transport of heated fluid vapour from the bottom to the top section of the evaporator. On the outer contact surface of the evaporator and electronic component there are three grooves with a mounted temperature sensor.

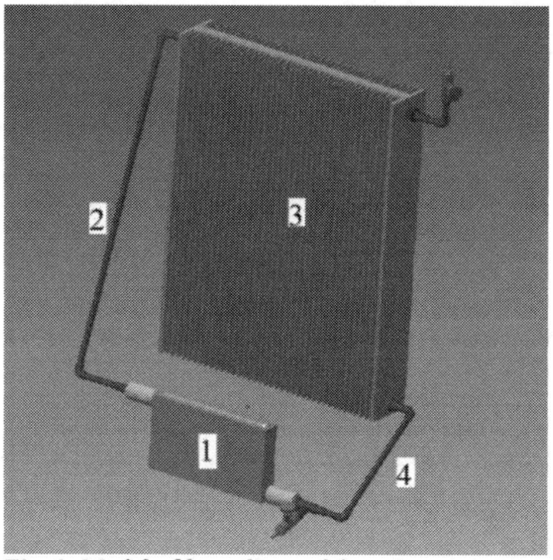

Fig. 1. Model of loop thermosiphon: 1 – evaporator, 2 – vapour pipeline, 3 – condenser, 4 – liquid pipeline.

The condenser is proposed so that it was possible removed heat in to the surrounding by natural convection. It is created from two alumina fin coolers with dimensions 400 x 480 mm and fin height 45 mm. The each cooler has on the back side along the length ninety 7.5 mm deep and wide vertical grooves, and on the top and bottom two 10 mm deep and wide horizontal grooves. The cooler are soldered together on the back side and through the grooves can the working fluid flow down to the evaporator. This robust condenser construction allow remove high heat fluxes loaded to the evaporator [4].

The vapour pipeline and liquid pipeline connect evaporator with condenser and are transportation parts through the working fluid circulate in closed loop. The pipelines are made from copper material, due to its good installation properties and compatibility with almost all mediums used in heat pipe [5].

The working fluid in cooling device is Fluorinert FC 72, due to its compatibility with most metals, low boiling temperature (56 °C) and solidification (-90 °C) and, first of all, due to its excellent dielectric characteristics [6].

3 Experiment

Fig. 2 show scheme to measure cooling efficiency of cooling device. Measurement consist from heat load of evaporator part and temperature detection on the evaporator part. On the evaporator part of the cooling device are fixed electric components connected to the DC power supply source and load heat flux to cooling device. Between surface of electronic element and evaporator surface are inserted three thermocouples (on the left side, on the right side and in the centre) to scanning electronic component temperature because the highest admissible temperature on the contact surface is 100 °C. The temperature of the condenser is measured by two pair's temperature sensors. One thermocouple pairs is placed in the centre of condenser and second thermocouple pairs is placed on the top of condenser. On the condenser is connected pressure sensor to measure pressure in cooling device. All sensors are connected on the measuring unit where measured parameters convert and transmit signal to the PC. The measurement of the cooling efficiency of the cooling device was performed at input electric power 750 W. Value 750 W was limited by the maximal electric power capacity of the DC power supply source. The measurement take place in the thermostatic chamber BINDER with regulation of ambient temperature in range -20 to 40 °C.

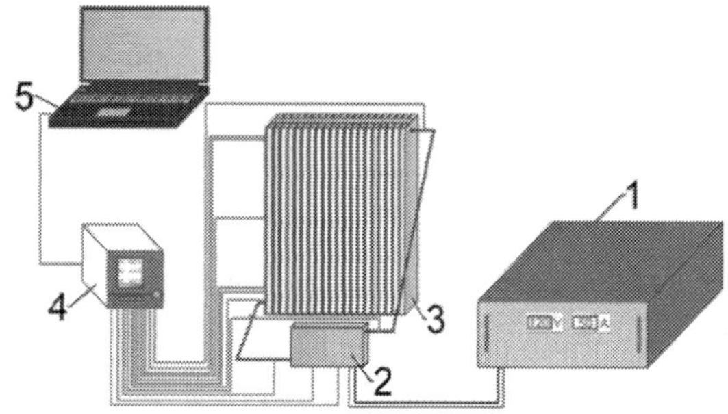

Fig. 2. Measuring scheme of loop thermosiphon cooling efficiency: 1 – power supply, 2 – evaporator, 3 – condenser, 4 – measuring centre, 5 – PC.

4 Mathematic calculation

Basic of the mathematical calculation are heat transfer equations and criterial equations. The heat flux equates to the electrical power of the electronic components, reduced by the heat transfer into the air is transferred through the evaporator wall into the working fluid.

$$Q = Pel = UI\ [W] \tag{1}$$

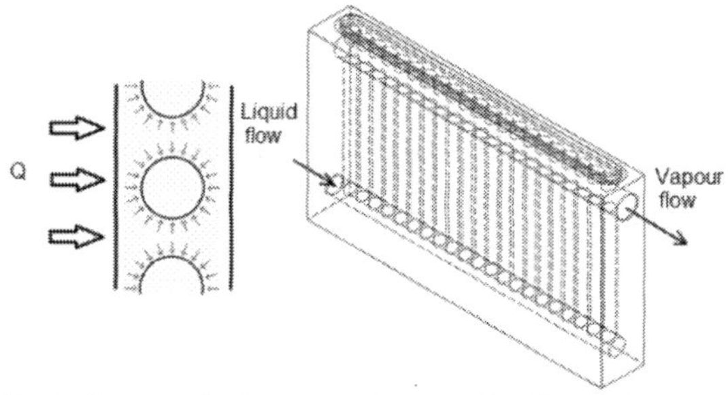

Fig. 3. Heat transfer in evaporation part of cooling device

The heat flux through the wall of the evaporator shown in Fig. 3 is determined by Furierer equation [7].

$$Q = q_{sp}S\ [W] \tag{2}$$

where q_{sp} is heat flux density on the contact surface of the electronic components expressed by relationship 3.

$$q_{sp} = (\lambda_{sv}/\delta_s)(t_{sp}-t_{sv})[W/m^2.K] \quad (3)$$

where λ_{sv} is thermal conductivity of the evaporator wall, t_{sp} is temperature of the outside surface of the evaporator wall, t_{sv} is temperature of the inside surface of the evaporator wall and δ_s is evaporator wall thickness expressed by relationship 4.

$$\delta_s = \delta_v - (\pi d/8)\ [m] \quad (4)$$

The liquid is heated to the boiling point due to the heat flux, when phase changes from liquid to vapour occur. Heat flux is determined by Newton's relationship expressed by relationship 5 [8].

$$Q = \alpha_v S_v(t_{sv} - t_{2,3})\ [W] \quad (5)$$

where α_v heat transfer coefficient at liquid boiling, S_v is inner surface of the evaporator wall and $t_{2,3}$ is temperature change of the state from liquid to vapour.

The heat transfer coefficient is then determined by Nusselt criterion according to the relationship 6.

$$Nu = \alpha_v L/\lambda_l\ [-] \quad (6)$$

where λ_l is thermal conductivity of the liquid phase and L is characteristic dimension expressed by relationship 7.

$$L = \sigma\rho_l c_p T_{2,3}/(\rho_v l_v)^2\ [m] \quad (7)$$

where σ is surface tension of the liquid, ρ_l is density of the liquid, c_p is specific thermal capacity, $T_{2,3}$ is thermodynamic temperature change of the state from liquid to vapour, ρ_v is density of vapour and l_v is latent heat of vaporization.

Relations 8 and 9 express Nusellt criterion in pressure range of $1.10^5 < p < 170.10^5$ Pa for bubble boiling at free flow convection [9].

$$Nu = 0.0625Re_B0.5Pr_K^{0.333}\ 10^{-5} \leq Re_B \leq 10^{-2} \quad (8)$$

$$Nu = 0.125Re_B0.65Pr_K^{0.333}\ 10^{-2} \leq Re_B \leq 10^4 \quad (9)$$

where Re_B is Reynolds criterion, expressed by relationship 10 and Pr_K is Prandtl criterion expressed by relationship 11.

$$Re_B = \omega_v L/\nu_l\ [-] \quad (10)$$

$$Pr = \nu_l/a_l\ [-] \quad (11)$$

where ν_l is viscosity of the liquid and ω_v is velocity generation of vapour affected by the heat flux density expressed by relation 12. The vapour is then continuously fed through the pipeline from the evaporator to the condenser. In condenser vapour condenses and returns back in the liquid phase of condensate [10].

$$\omega_v = q_v/\rho_v l_v\ [m.s^{-1}] \quad (12)$$

Then is temperature of the evaporator inner wall t_{sv} determined by the relationship 13 resulting from the relationship 5.

$$t_{sv} = (q_v/\alpha_v) + t_{2,3}\ [°C] \quad (13)$$

Temperature of contact area is determined by modification of the relationship 3 and expressed by relationship 14 [11].

$$t_{sv} = q_{sp}/\lambda_{sv} + t_{sv}\ [°C] \quad (14)$$

5 Results and discussion

Under the term cooling efficiency is meaning cooling effectivity of the cooling device. The cooling effectivity is expressed by ability of the cooling device to remove heat from the electronic element connected on the evaporator depending on the ambient temperature. In Fig. 4 is shown measured and calculated evaporator temperature of cooling device depending on ambient temperature. There is seen that with increase of ambient temperature increase evaporator temperature. Increasing evaporator temperature mean that he cooling effectivity of the cooling device decrease. Even though the cooling device has high cooling effectivity, because is able to remove high heat flux from the electronic element and keep its temperature under admissible temperature of 100 °C.

Comparison calculated and measured temperature of cooling device evaporator proves that the formulas chosen for evaporator temperature calculation in this case was correct and can be applied in other similar cases of heat transfer calculations.

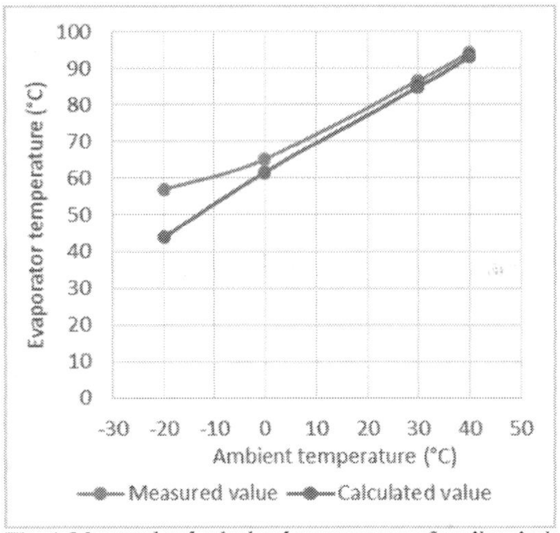

Fig. 4. Measured and calculated temperatures of cooling device evaporator depending on ambient temperature.

6 Conclusion

Experiment was aim at evaluation of evaporator temperature by measurement and calculation of cooling device working with a heat pipe technology. Conclusion of results comparison obtained by calculation and measurement is that cooling device is able operate in temperature range from –20 to 40 °C at high heat load and chosen formulas in the calculation can be applied in other similar cases of heat transfer calculations.

Acknowledgment

This work is supported by the financial assistance of the projects APVV – 15-0778 "Limits of Radiative and Convective Cooling through the Phase Changes of Working Fluid in Loop Thermosyphon" and 042ŽU - 4/2016 "Cooling on the basis of physical and chemical processes.

References

1. C. Ruppersberg, R.T. Dobson, J. of Energy in South Africa, **18**, 3 (2007)
2. J. Hartenstine, R. Boner III, J. Montgomery, T. Semenic, Proc. of ASME Inter PAC conf., (2007)
3. P. Pilat, M. Patsch, S. Papucik, M. Vantuch, AIP Conf. Proc., **1608** (2014)
4. F. Urban, M. Kubín, L. Kučák, AIP Conference Proceedings, **1608** (2014)
5. R. Lenhard, J. Jandačka, AIP Conf. Proc., **1558** (2013)
6. M. Holubčik, P. Ďurčanský, J. Jandačka, R. Nosek, Scient. World J., **2014** (2014)
7. D. V. Kozak, Yu. E. Nikolaenko, *International Conference on Electronics and Information Technology, EIT 2016 - Conference Proceedings*, (2016).
8. M. Mahdavi, S. Qiu, S. Tiari, Apl. Therm. Eng., **81** (2015)
9. Ľ. Šooš, P. Križan, M. Matúš, J. Beniak, Int. J. of Mechanics, **9** (2015)
10. T. Orzechowski, L. J. Orman, Polution Engineering, **38**, 7 (2006)
11. J. Jandačka J. Mičieta, M. Holubčík, R. Nosek, Energy & fuels, **31**, 3 (2017)

 https://doi.org/10.1051/epjconf/201818002074

Influence of yaw on propeller aerodynamic characteristics

Van Bang Nguyen[1,*], *Dalibor* Rozehnal[1], *Jakub* Hnidka[1], and *Vu Uy* Pham[2]

[1]University of Defence, Kounicova 65, 66210, Brno, Czech Republic
[2]Military Technical Academy, Hoang Quoc Viet 236, Hanoi, Vietnam

Abstract. Between the propeller axis and free stream direction, it can still be a non-zero yaw angle. This paper introduces some propeller experiments, in which the propeller aerodynamic characteristics have been determined in various yaw angle and different rotational speeds. The experimental aerodynamic characteristics are acquired dynamic values, from which the influence of yaw conditions on the frequency and the amplitude of propeller thrust and torque can be obtained.

1 Introduction

The research of air propeller assumes that the axis of the propeller disc is aligned with inflow direction [1]. However, between the propeller axis and airflow direction can still be a non-zero angle, when the propeller aircraft is in some special flight regimes as climbing, flying horizontally at low speed, fighter maneuvers or flying after an engine failure in multi-engine propeller aircraft [2], [3]. The small multicopters (rotorcraft with more than one main rotor) operate usually in the similar conditions (Fig. 1.), because all their rotors spend most of its flight time in yaw flow at the same yaw angle θ.

Fig. 1. The kinetic scheme of the propeller in yaw [4].

This report aims to determine the influence of yaw conditions (yaw angle and rotational speed) on the frequency and the amplitude of dynamic component of propeller thrust and torque using experimental results from simple wind tunnel experiment. The propeller has been measured at the various yaw angle from 0° to 90° (with constant rotational speed) and various rotational speeds (with constant yaw angle 45°).

2 Experiment setup

Propeller and wind tunnel

The two-blade fixed pitch propeller Bambula 18x8 with a diameter of 18 inches was used in this experiment. The test bench is optimized to measure the dynamic components of both force (thrust) and moment (torque) of propeller shaft, at various yaw angles and rotational speeds.

Fig. 2. Propeller setup in yaw.

The placement of test bench in the wind tunnel, which is located in a wind tunnel of the University of Defence in Brno (Fig. 2). It is the low-speed return-flow wind tunnel with an open test section. At the core of the wind tunnel is a 500 mm diameter wooden propeller fan driven by an 18000 W electric motor [4]. The free stream speed V in the test section is constant at V = 12 m.s^{-1}. The yaw mechanism allows changing the yaw angle from 0° to 90° with step of yaw angle 5°.

* Corresponding author: vanbang.nguyen@unob.cz

Sensor and data acquisition

The test bench implements a piezoelectric 2-component transducer Kistler 9365B, which reacts very quickly to all changes of measured parameters. That means that the results obtain thrust or torque pulsation, if they occur. The NI DAQ hardware and an interface programmed in LabVIEW software have been prepared for the data acquisition

The thrust of the propeller can be measured in range from -20kN to 20 kN, and the range of torque is from -200 Nm to 200 Nm. It is important to note that despite the significant range of the sensor, it can measure forces with precision and it has a resolution of 3,6 pC/N and of 140 pC/Nm. .

Fig. 3. The 2-component piezo sensor Kistler 9365B

The experimental data are compared only in a constant time 0.2 s. In this time section, only the change of measuring parameters is acquired and the average value of their parameters in this time is always setup to zero.

Repeat of measurements

In the first series of measurements, the rotational speed n of the propeller in each repeating cycle of measurement remained nearly constant at about n = 5400 rpm and the yaw angle changes from 0° to 90° with step of 15°. The aim of this first experiment is to identify the influence of yaw angle on thrust and torque characteristics and its frequency analysis performed via fast Fourier transform (FFT).

In the second series of measurements, the rotational speed n changes from 2200 rpm to 5380 rpm, while the yaw angle is constant at 45°. The aim of this experiment is to identify the influence of rotational speed n on thrust characteristics of propeller.

3 Results and discussion

The theoretical frequency f_o can be estimated by using equation (1). During one revolution of the propeller rotor in yaw condition, every blade section operate an oscillation [5], [6]. The angle of attack (AoA) of blade section is changed from α_{min} to α_{max} and later returns to the minimum value with a frequency [5]. This periodical change of every blade section also results to characteristics of propeller periodical oscillation. The theoretical frequency f_o of oscillation generated by the change of AoA of blade section can be calculated by the expression:

$$f_0 = k\frac{n}{60} \tag{1}$$

The value of frequency f_o of the first experiments (k= 2, n= 5400 rpm) according to (1) must be around 180 Hz.

4.1. Influence of yaw on thrust

Fig. 4 shows the thrust oscillation in time domain and s-domain. The thrust generated by propeller in case of non-zero yaw angle alters around average value and appears to perform nearly periodical oscillation.

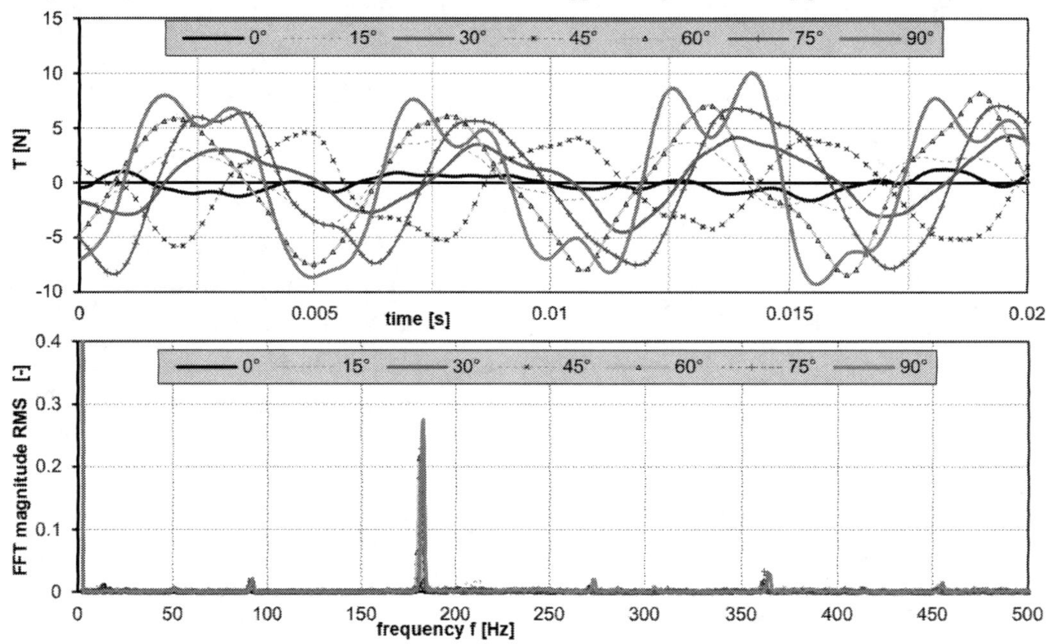

Fig. 4. Thrust - time curves and its FFT for different yaw angles

The frequency analysis is coincided with the estimation, because the main frequency occurred in every yaw angle is always near with the calculated frequency f_o =180 Hz. Smaller peaks occur around values 90Hz (1/2 f_o), 270Hz (3/2 f_o), and 360 Hz (2 f_o).

4.2. Influence of yaw on torque

The main frequency of torque in every yaw angle is also close to the calculated frequency f_o =180 Hz. The next most common frequency is 360 Hz (2 f_o).

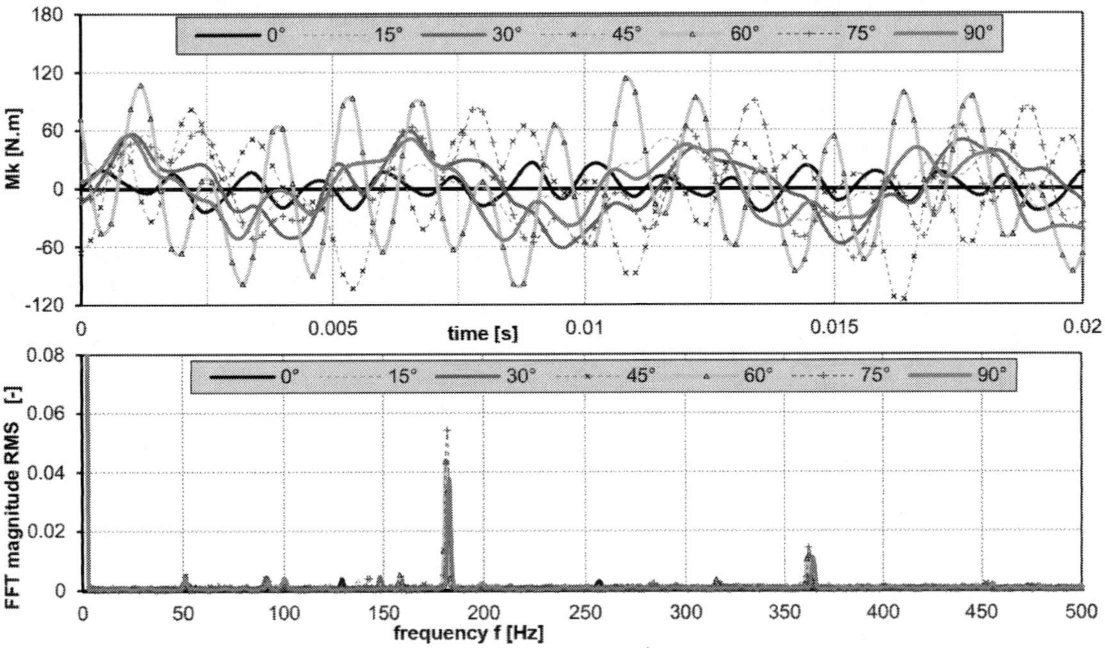

Fig. 5. Torque - time curves and its FFT for different yaw angles, V=12 m.s^{-1}, n = 5400 rpm

There are some frequencies, which occur without any rules. They can be caused by the construction of electric motor, in which the movement of rotor depends on specific number of slots and number of poles.

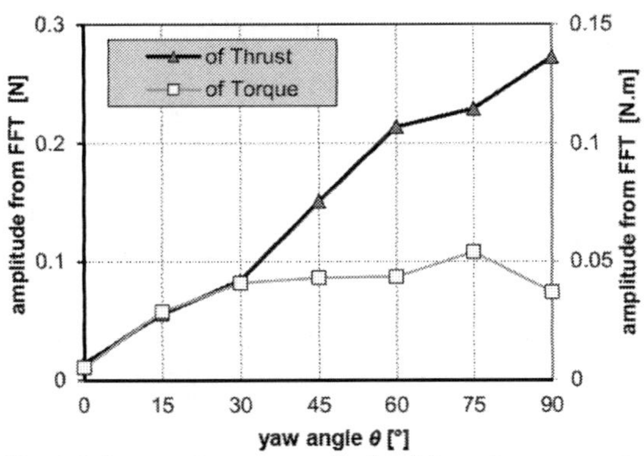

Fig. 6. Influence of yaw on amplitude of dynamic components

Fig. 6. shows the influence of yaw angle on magnitude of amplitude in frequency analysis. It also points out that the magnitude of amplitude of thrust oscillation is significantly depended on yaw angle while it is not for the torque. The magnitude of torque oscillation increases with yaw angle only in range from 0° to 75°, and then it conversely decreases according to yaw angle. The torque achieves maximum values for yaw angle around 75°.

4.3 The influence of rotation speed

This next experiment identifies the influence of rotational speed to thrust characteristics and its frequency analysis. The freestream speed V in the test section in this experiment is 12 m.s^{-1} and the rotational speed n changes from 2200 rpm to 5380 rpm, while the yaw angle is constant at 45°.

* Corresponding author: vanbang.nguyen@unob.cz

 https://doi.org/10.1051/epjconf/201818002074

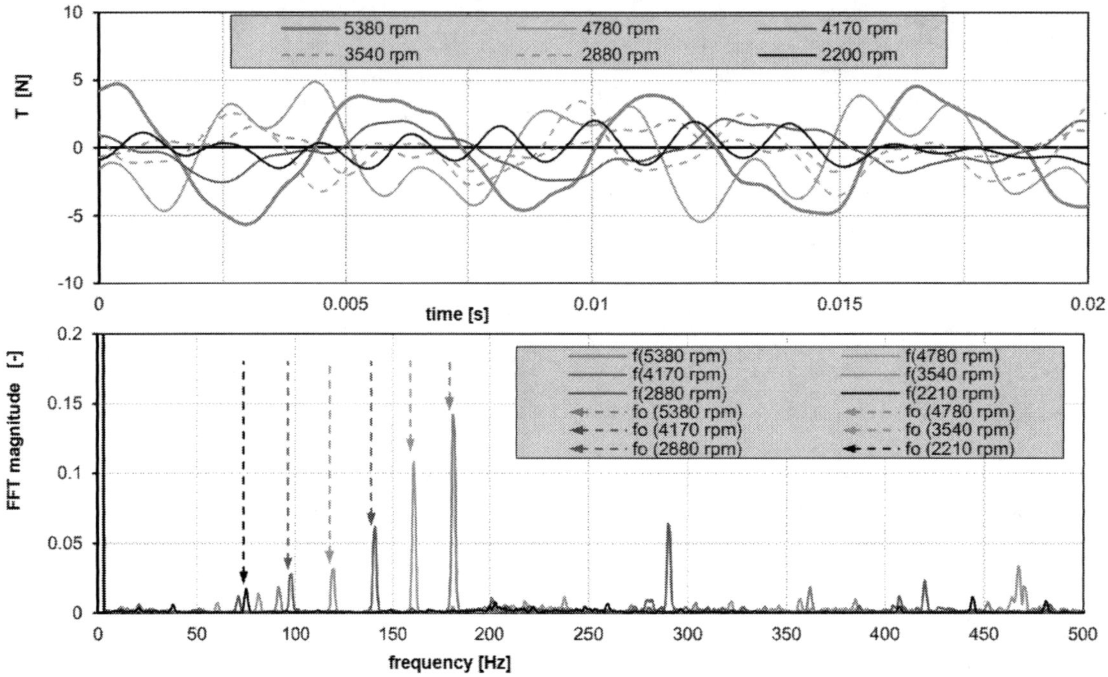

Fig. 7. Thrust - time curves and its FFT for different rotational velocities

The magnitude of thrust oscillation increases proportionally with increasing of rotational speed. The frequency analysis in Fig. 7. shows all peaks and the calculated frequency f_o for every yaw condition. The FFT of measured thrust in every rotational speed n always shows one peak coincided with the frequency f_o calculated for the corresponding n by expression (1).

4 Conclusion

The experimental results show in which frequencies the thrust or torque pulsation occur and how its amplitude change according to yaw condition.

- The main frequency find from FFT were always near with the calculated frequency f_o (Fig. 4., Fig. 5., Fig. 7.).

- The magnitude of thrust oscillation increased clearly proportionally with yaw angle (Fig. 6.).

- The magnitude of torque oscillation increased with yaw angle in range from 0° to 75°, and then it conversely decreases (Fig. 6.).

- The magnitude of thrust oscillation increases nearly parabolic with increasing of rotational speed (Fig. 7.).

The information about dynamic component of propeller aerodynamic characteristic in yaw can be used to verify the theoretical calculation [7].

References

1. T. Theodorsen, *The theory of propellers*. Washington: U. S. Govt. Print. Off (1945).
2. H. S. Ribner, Propellers in yaw. NACA report No. 820 (1945).
3. H. B. Freeman, *The effect of small angles of Yaw and Pitch on the characteristics of airplane propellers*. NACA report No. 389, (1932). A.
4. D. Rozehnal, *Rozvoj experimentální aerodynamiky na UO*. Habilitation thesis, University of defense in Brno (2009).
5. N. V. Bang, D. Rozehnal, *Aerodynamic hysteresis of oscillating airfoil*. In: 23rd International Conference Engineering Mechanics 2017. Brno: Brno University of Technology, Faculty of Mechanical Engineering, Institute of Solid Mechanics, Mechatronics and Biomechanics, Brno, 2017, s. 106-109. ISSN 1805-8248. ISBN 978-80-214-5497-2, (2017).
6. N. V. Bang, D. Rozehnal, *Measurement of Aircraft Propeller in Yaw Flow*. In: International Conference on Military Technologies. ICMT: University of defence in Brno, 2017, p. 566-569. ISBN 978-1-5386-1988-9, (2017).
7. J. Sveda, J. Salga, B. Horeni, *Výpočet aerodynamických charakteristik šikmo ofukované vrtule*. In: Report VAAZ/635, Brno (1977).

* Corresponding author: vanbang.nguyen@unob.cz

CFD simulation of ejector: is it worth to use real gas models?

Nguyen Van Vu [a] and Jan Kracik

Department of Power Engineering Equipment, Faculty of Mechanical engineering, Technical University of Liberec, Czech Republic

Abstract. This paper presents progress of evaluation existing ejector of previous work with different gases, included real gas. This work was done by the Ansys simulation tool. The paper also discussed about the two solvers used for simulating real gas difluoromethane (R32). The results showed that entrainment ratios are almost not influenced by different gases but by the working conditions. Secondly, the solver pressure-based can obtain comparable results with the density-based one, while it is much more stable and solution converges faster than the density-based.

Keywords: difluoromethane, real gas, ejector, entrainment ratio, pressure-based, density-based

1 Introduction

Low-grade energy ejector refrigeration system (ERS) drew great attention because of its simplicity and the ability to use low energy sources that other while wasted, such as solar, geothermal, waste energy from plants. The most essential part of this system is the ejector; design of ejector decides the performance of the system.

Thanks to many advantages of CFD, it has been used widely in many areas and has been proved as a power tool for research and development [1][2][3]. CFD is very handy for analysis ejector because CFD simulation can give detail picture about almost every property the supersonic flow, which neither mathematical calculation, nor experimental measurement can do. Moreover, the experimental setups are quite expensive and can analyse only a limited number of flow conditions. CFD simulation offers all characteristic of simulated fluid at once. We can obtain detail, yet complete picture of the subject we study with CFD simulation.

Come to study ejector, most researchers concern about one working fluid, it can be either water vapor [4], air as ideal gas [5], or R134a [6], etc. Apparently, there is no paper mention about the performance of an ejector with different working fluid. Therefore, with this paper, I would like to do a quick comparison between fluids.

Likewise, as far as I know, there is no paper talked about the flow simulation within ejector of real gas, which is high compressible, supersonic flow with shock waves. Hence I try to do so and give a comparison with ideal gas model to see if it worth simulating real gas model. Some engineering problems involve fluids that do not behave as ideal gases. For example, at very high-pressure or very low-temperature conditions (for example, the flow of a refrigerant through a compressor) the flow cannot typically be modelled accurately using the ideal-gas assumption.

There are very few papers on difluoromethane (R32) as refrigerant for ERS. R32 has very good thermo-physical properties as well as good thermal conductivity and low viscosity [7]. It is a perfect substitute for R410a, which is very common in refrigeration applicants but is being phased out because of it high global warming potential (GWP). GWP index of R410a is 2088 while R32 is 675, more than three times higher. In near future, R32 is very attracting refrigerant because of its properties.

2 Model approach

2.1 Ejector model

This is a continuation of the work [8]. J. Kracik made a detail numerical and experimental study about relationship of primary pressure and entrainment ratio (ER) of ejector. This is a great ejector model, since it can obtain the ER as high as 0.82, one of the highest values of ER that have been reported. Kracik used air – ideal gas as working fluid. Different values of nozzle exiting position (NXP) were tested. And at the NXP = 3 mm and the following working conditions: $p_p = 3$ bars, p_s and $p_b = 0$ (pressure of primary, secondary and total flow, respectively) delivered the best agreement between CFD simulation and experimental measurement. Accordingly, I used this setup to verify my numerical model.

The model was created in a two dimension domain. However, the axisymmetric solver was applied, hence the three dimensional effect was taken into account in the simulation. The mesh was made of 266000 structured quadrilateral elements, as shown in figure 2. The concentration of grid density was focused on the area from the primary throat to the diffusor of the ejector.

[a] Corresponding author: nguyen.van.vu@tul.cz

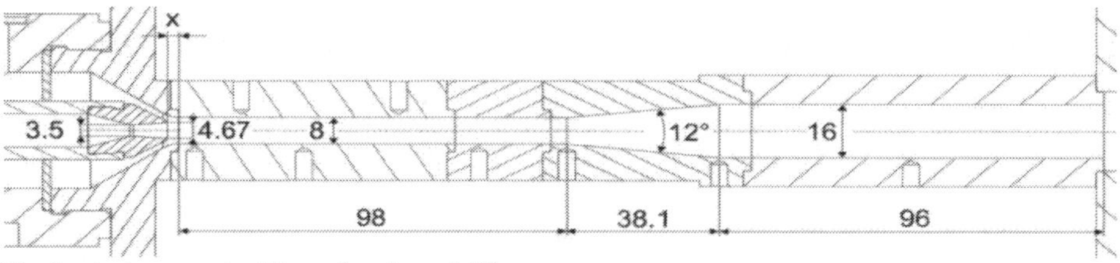

Fig. 1.: ejector geometry (dimensions in mm) [8]

2.2 Solver setup

Because the flow inside the ejector is high velocity, high compressible, the density-based solver was chosen (the pressure-based solver also used for comparison). Steady flow was selected. Second order upwind scheme was used for all calculations.

In a research on ejector for refrigeration system by CFD simulation and confirmed by experiment, Ruangtrakoon et al use CFD to study the effect of primary throat nozzle [10], they found that k-ω SST provided more accurate results than two other turbulence models. Therefore, k-ω SST was chosen as default model. Mathematical model of two-equations k-ω SST is already emphasized on flow close to wall [10], it is not necessary to use wall enhance function.

The density-based implicit solver was selected to solve the governing equations which have been proven to be suitable for a supersonic flow field [1], [10] and the flow was set based on steady-state. The pressure-based coupled solver was used for comparing with the other solver in the work of real gas simulation

Least Square Cell-based gradient is less expensive to compute than node-based gradient and it is also much more superior than cell-based gradient [9]. Therefore, Least Square Cell-based gradient was selected as default for all calculation.

The primary inlet and the secondary inlet were set as pressure-inlet type, the outlet was set as pressure-outlet.

The computational results are compared with available experimental data to assess the accuracy of the numerical results obtained with different turbulence models.

2.3 Working fluid properties

Air, carbon dioxin were used as working fluids. They are ideal gases, called directly from Fluent database. Pseudo difluoromethane R32 was also used as working fluid of the model. The properties of pseudo R32 at room temperature is shown in table 1.

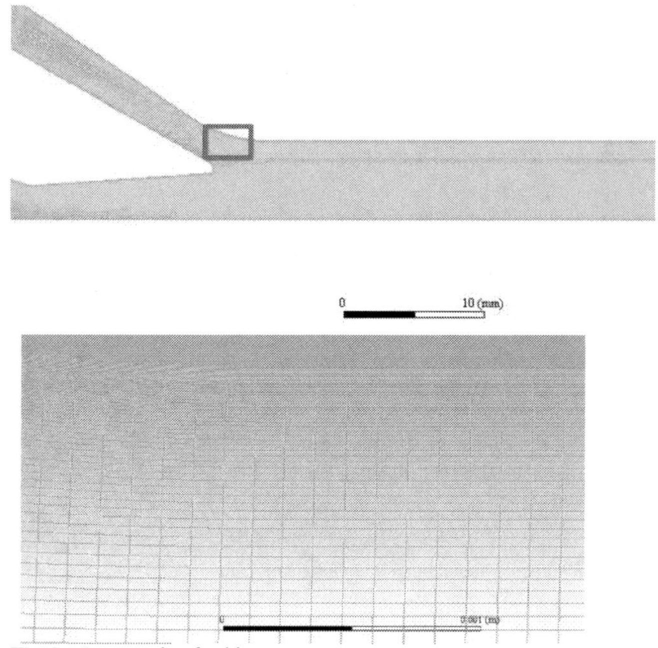

Fig. 2.: Computational grid

Table 1. Properties *of pseudo R32 at room conditions (REFPROP 9.1)*

Properties	Values
Thermal conductivity, k	12.508 $(\frac{mW}{m \cdot K})$
Viscosity, μ	12.607 $(\mu\ Pa\ s)$
Specific heat capacity, C_p	0.84716 $(\frac{kJ}{kg \cdot K})$
Molecular weight, M	52.024 $\frac{kg}{kmol}$

2.4 Boundary conditions

Boundary conditions of primary and secondary inlet were set as pressure-inlet, the outlet of the ejector was set as pressure-outlet. Temperatures were set to 300 K in all cases. If not mentioned, the default primary pressure was set to 3 bars, secondary pressure (overpressure) was zero and backpressure varied. Ambient pressure was set according to local average value in Liberec, 97000 Pa.

2.5 Simulations of real gas

The work used NIST real gas database integrated directly to Ansys Fluent for the calculation. The limitation of this real gas model that matter to this study is: the model assumes that the fluid is either superheated vapor, or supercritical fluid, or liquid. Vapor coexists with liquid (two phase flow) is not supported, which is the most disadvantage of this model.

Pressure-inlet, mass flow-inlet, and pressure-outlet are the only inflow and outflow boundaries available for use with the real gas models.

Note that once the NIST real gas model is activated, control of relevant property evaluations is relinquished to the REFPROP database.

Simulation of real gas is challenging, especially for supersonic flows. As far as I know, there are very few, if any, real gas CFD simulations of ejector explicitly reported. Researchers assumed that ideal gas simulations are able to give sufficient results but ironically, there is no comparison by actual results. There are many limitations of real gas models in Ansys Fluent [9], causing difficulties for simulation of compressible gas with supersonic flow. In previous Fluent versions, it was impossible to use pressure-based solver for NIST real gas models [9]. Recently, Fluent supports both pressure-based and density-based solver. A comparison of two solvers was made.

The real gas simulations were more time-consuming than the one with ideal gases because the flow modelling of NIST real-gas flow is much more complex than ideal-gas flow. As what I observed from the calculations, the pressure-based solver was converged at least three times quicker and residuals were more stable when compared to the other solver.

3 Results and discussion

The calculations were considered as converged when these criteria were met: (1) residuals got lower than 10^{-5} (10^{-4} for real gas model simulation), due to difficulties of convergence), (2) mass fluxes of inlet and outlet flows were stable and they had to satisfy the principle of conservation of mass, the net must lower than 10^{-6} kg/s. Most of calculations in this study converged when the iterations reached the value of 100 000.

3.1 Verification

A model was created to verify with the work of Kracik [8] at different backpressures (or pressure ratio between backpressure and secondary flow (μ). The numerical results obtained showed some discrepancies, mainly in the off-design region. As we can see in the table 2, the entrainment ratio between experimental and numerical method within the on-design region are good, less than 10% of discrepancies [8]. The results obtained from current work were a little bit higher than the simulation results from Kracik. This can be explained by the difference of meshes between the two. However, it is safe to claim that these two works gave consistent results.

Table 2. Comparison results from Kracik [8] with current work

μ	1	1.1	1.2	1.3	1.4
Experiment [8]	0.774	0.774	0.763	0.714	0.627
CFD [8]	0.839	0.839	0.839	0.839	0.776
Current work	0.838	0.838	0.838	0.838	0.774

3.2 Relation between entrainment ratio and turbulence kinetic energy

The shockwave at the beginning of diffusor section when the backpressure is low, and it headed to primary nozzle as the backpressure increased. Figure 3 showed the contours of ejector's Mach number at on-design regime. Many researchers claimed that the length of mixing chamber does not significantly affect to performance of ejector [2] [13]. However, it was observed that the length of mixing chamber help stabilize ejector's entrainment ratio, means the performance remains stable while working pressures change in a limited values.

The figure 3 shows the contours of Mach number in mixing chamber area. At the pressure ratio μ of 1, there is a strong shock wave at beginning of diffusor section. Which is corresponding to the highest kinetic energy k with the peak is around 2800 $\mathrm{m^2 \cdot s^{-2}}$ (Fig. 4).

Then with the increase of μ to 1.1, 1.2, 1.3, the flow becomes less intense and no shockwave is seen at ending section of mixing chamber (Fig. 3). This phenomenon is related to the decrease of turbulence kinetic energy. Peak values k also gradually move to the left as μ increases. In spite of those, the ER does not seem to be affected by μ. The flow at the end of mixing chamber at these working pressures is still supersonic with Mach numbers slightly greater than 1.

At value μ of 1.4, the area that has Mach number greater than 1.8 was broken down to pieces, the flow at the end of mixing chamber is subsonic (around Mach 0.4) as seen in last contour of figure 3, and the curve of turbulence kinetic energy significantly move away from the others forward the primary nozzle. The peak value of k at this working condition is greater than the one at μ = 1.3 (Fig. **4**). These are signs of the critical pressure, at which ejector begins off-design regime, the ER decreases from 0.836 to 0.784 (Table 2) and quick reaches to zero as the μ increases further.

https://doi.org/10.1051/epjconf/201818002075

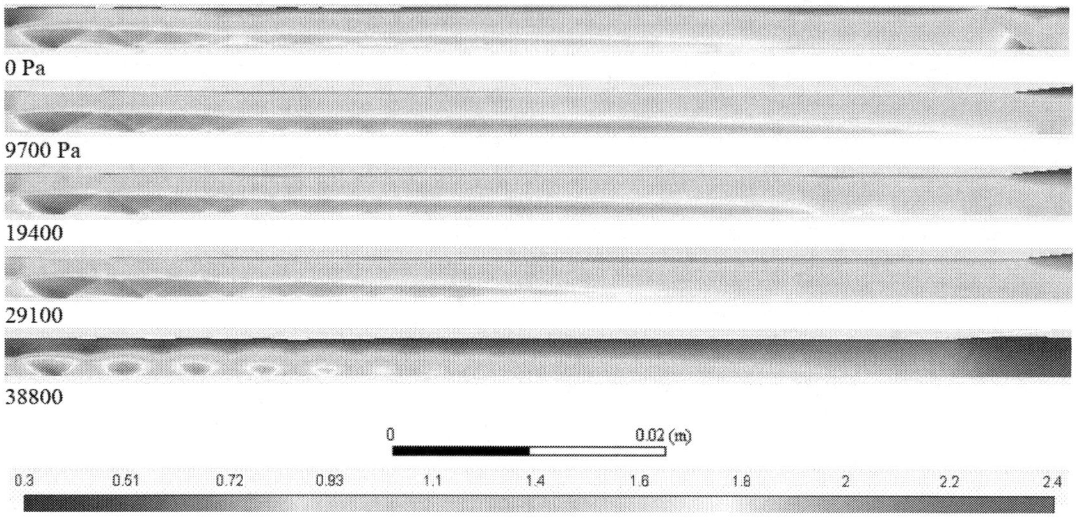

Fig. 3.:Contours of Mach number at different pressure ratio

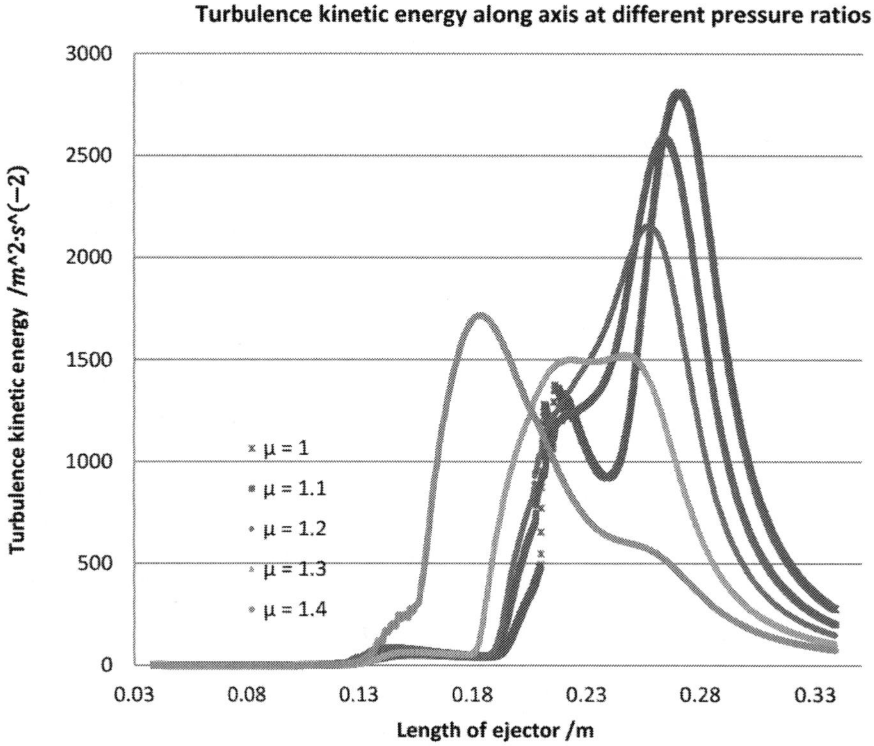

Fig. 4.: Turbulence kinetic energy along axis at different pressure ratios

https://doi.org/10.1051/epjconf/201818002075

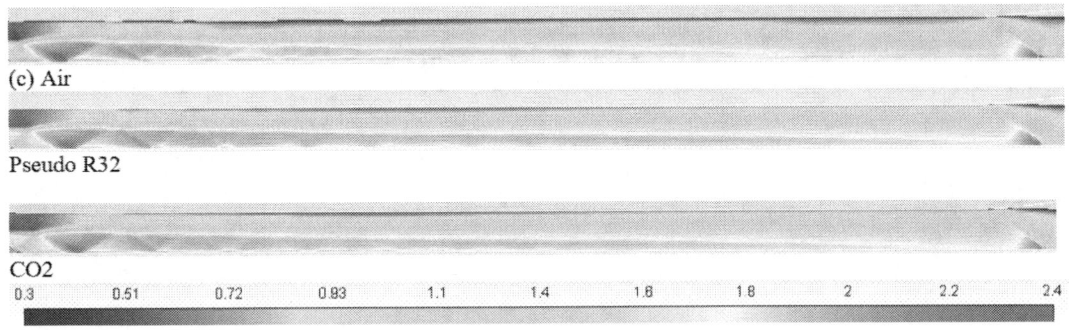

Fig. 5. Mach contours of different ideal gases

There is certainly a relation between critical pressure ratio and positions of peak values k, if μ exceeds critical value, the kinetic energy will transform to turbulence immediately after the primary nozzle. Therefore, the kinetic energy of flow will diminish and entrains fewer medium from suction chamber as consequence. This observation gives good explanation why ER decreases suddenly when μ reaches critical value.

3.3 Entrainment ratio of different gases

In contrast to other characteristics, mass fluxes varied related to the molecular weight but not in linear behaviour, though (Table 3). The entrainment ratios of all gases were remained the same despite of the changes of mass flow rates. From this we can go to a conclusion: the ER is not affected by working gases (at least for ideal gases). It means that changing working fluid does not significantly affect performance of ejector. This is an important point, because basically, we can apply high performance ejector from other researches for our desired working fluids without the needs of geometry reinvestigation as long as they share same working conditions.

Table 3. *Mass fluxes of ideal gases*

	R32	Air	CO_2
Primary-inlet (kg/s)	0.0113	0.00861	0.0103
Secondary-inlet (kg/s)	0.0095	0.00722	0.0086
Outlet (kg/s)	-0.0208	-0.01584	-0.0190
Net (kg/s)	2.59E-07	-1.73E-06	-4.49E-06
ER	0.841	0.838	0.835

3.4 Density-based and pressure-based, a comparison of real gas simulations

Historical speaking, the pressure-based solver was developed for low speeds incompressible flows. However, recently it has been reformulated and extended to solve wider range from incompressible to high compressible. This part gives comparison between a well-known solver for shock wave, density-based and pressure-based, which was believed as unsuitable for high compressible flow.

We would like have the answer for the question: can pressure-based solver give proper results in simulation high compressible with shockwave flow?

Again, all working conditions were remained but working fluid. NIST real gas model of R32 was called in Ansys Fluent. Then two solvers were applied. These calculations were very time consuming.

Figure 6 shows contours of Mach number of pressure-based and density-based solver of real gas model. Both gave very similar results, in graphic and in number. As seen in the figure 6, there are differences in intensiveness of shockwaves at the beginning and at the end mixing chamber but it hard to tell any difference at other aspects. The calculation of the pressure-based solver took significant less time than the other. In addition, density-based solver was very sensitive to working pressure, for example the problem went to divergence just by slightly increase primary pressure, while the pressure based still got convergence. In the figure 7 we can notice that the mass flux of secondary flow in case of pressure-based is slightly greater, leading a higher ER by 1%. From these two points, pressure-based solver is obvious the winner. The pressure-based solver in the latest Ansys Fluent version is mature to solve complex problem such as high compressible with shockwave real gas flow.

Another question: is it worth to trade-off the preciousness of real gas model over the simplicity of ideal gas model? First let's look at the Mach contours of these models (Fig. 7). We can see that the flow of ideal gas model show higher gradient, the subsonic layer at the wall is thicker than in real gas model. In other hand, the shockwave at the end of real gas models are much more intensive. This might caused by viscosity of working fluids are different. The real gas models took a many factors into account while the ideal gas model simply ignores, such as the change of specific heat capacity, viscosity along with the change of fluid temperature and pressure, etc. It is not easy to tell exactly the reason for this and it is not scope of this paper.

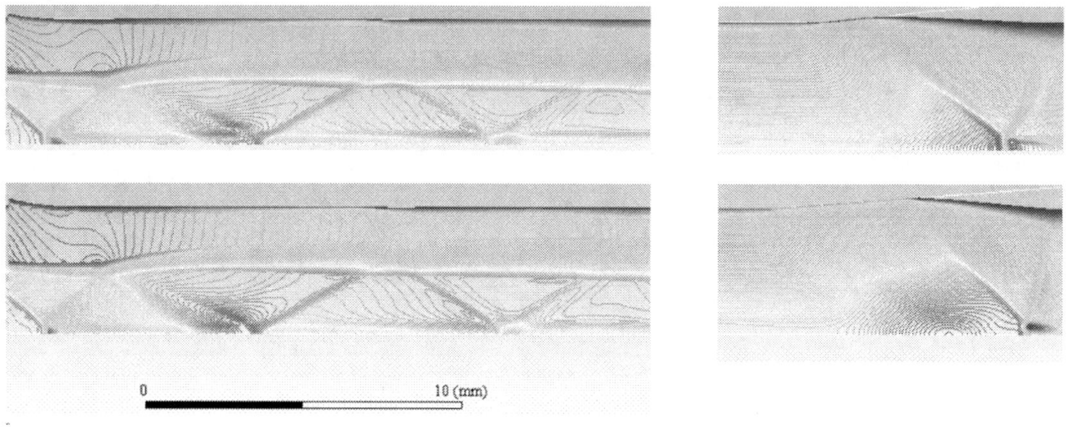

Fig. 6. Mach contours of pressure-based (upper) and density-based solver

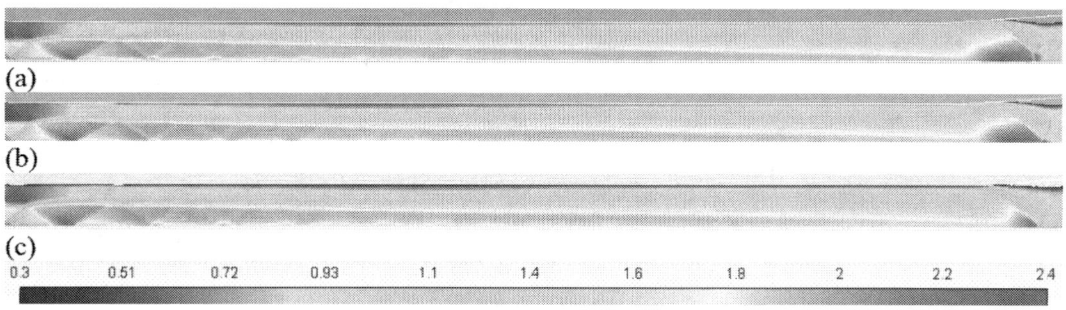

Fig. 7. Mach contours of pressure-based, real gas (a) and density-based, real gas (b) solver and ideal gas Pseudo R32 density-based solver

4 Conclusions

The entrainment ratios of ejector are certainly has connection with the turbulence kinetic energy changes inside the ejector, as point out in the last section. It was observed that if there is a jump of the turbulence value right after primary nozzle, means the primary flow at the exiting nozzle is highly turbulent, then the ER is low, off-design regime. In contrast, ejector is on-design regime when that primary flow are has low turbulence kinetic energy after exiting primary nozzle for a certain distance. This will give a sense for further detail investigation of relationship of ER and μ, as well as with the length of mixing chamber.

Secondly, the entrainment ratios of ejector were remained constantly for different working fluids (as least for ideal gases). This means that the performance of ejector is not varied significant with different refrigerants in certain conditions.

Lastly, real gas model simulations seemed to be unnecessary because too much effort was put into it with just a little benefit paid back. And the pressure-based solver (with couple solver) could handle the supersonic flow of real gas model well enough, unlike traditional belief. Therefore, researchers should consider pressure-based as prior solver even for high compressible, supersonic flow with complex thermo-physical changing such as shock waves to take advantages of this solver (less time-consuming, more stable). Obviously, these observations are needed to verify by further works.

Acknowledgment

We gratefully acknowledge the support of the Student Grant Agency of the Technical University of Liberec (project no. 21124).

References

1. A. Hemidi, F. Henry, S. Leclaire, J. M. Seynhaeve, and Y. Bartosiewicz, *Appl. Therm. Eng.*, vol. **29**, no. 8–9, pp. 1523–1531, (2009).
2. L. Su, "Cfd Simulation and Shape Optimization of Supersonic Ejectors for Refrigeration" (2015).
3. Y. Allouche, C. Bouden, and S. Varga, *Int. J. Refrig.*, vol. **39**, pp. 186–195, (2014).
4. R. Yapıcı, *Energy Convers. Manag.*, vol. **49**, no. 5, pp. 953–961, (2008).
5. A. Hemidi, F. Henry, S. Leclaire, J.-M. Seynhaeve, and Y. Bartosiewicz, *Appl. Therm. Eng.*, vol. **29**, no. 14, pp. 2990–2998, (2009).
6. J. Yan, W. Cai, and Y. Li, *Renew. Energy*, vol. **46**, pp. 155–163, (2012).
7. "Service manual for products using R32." web access at "*http://www.icglimited.co.uk/library/web/1. Daikin/4. Emura/Emura - Wall Mounted Units (with R32 Refrigerant)/RXJ M2V1B/RXJ M2V1B - Service Manual.pdf*"
8. J. Kracik and V. Dvorak, *EPJ Web Conf.*, vol. **143**, p. 2056, (2017).
9. ANSYS, "ANSYS FLUENT User's Guide," vol. **15317**, no. November, p. 2498, (2011).
10 N. Ruangtrakoon, T. Thongtip, S. Aphornratana, and T. Sriveerakul, *Int. J. Therm. Sci.*, vol. **63**, pp. 133–145, (2013).
11. S. He, Y. Li, and R. Z. Wang, vol. **13**, no. 8, pp. 1760–1780, (2009).
12. N. Van VU, "Study on ejector using CFD simulation." (to be published)
13. T. Thongtip and S. Aphornratana, *Appl. Therm. Eng.*, vol. **110**, pp. 89–101, (2017).

https://doi.org/10.1051/epjconf/201818002076

Development of fire shutters based on numerical optimizations

Ondrej Novak[1,a], Petr Kulhavy [1], Tomas Martinec[1], Michal Petru[1] and Pavel Srb[1]

[1] Institute for Nanomaterials, Advanced Technologies and Innovation, Technical University of Liberec, Studentská 2, 461 17, Liberec 1, Czech Republic

Abstract. This article deals with a prototype concept, real experiment and numerical simulation of a layered industrial fire shutter, based on some new insulating composite materials. The real fire shutter has been developed and optimized in laboratory and subsequently tested in the certified test room. A simulation of whole concept has been carried out as the non-premixed combustion process in the commercial final volume sw Pyrosim. Model of the combustion based on a stoichiometric defined mixture of gas and the tested layered samples showed good conformity with experimental results – i.e. thermal distribution inside and heat release rate that has gone through the sample.

1 Introduction

Fire safety is an inseparable aspect during the buildings design. One of the basic requirements especially for industrial buildings is dividing the building into fire zones, each other protected by fire partition structures, to prevent spreading of a potential fire and minimization looses. In our case some light fireproof materials that are used in fast fire barriers (e.g. movable gate, blinds) have been tested. Their function is to create a thermal shield between place which should stay safe and cool even if they are affected by high temperature radiation and convection from a near heat sources at the other side [1] for a by declared standard given time.

This article deals with numerical and experimental study of thermal transfer between flame of a propan butan jet and measured layered samples. Solving of a similar problem requires simultaneously application of several disciplines. Fire safety comprises a significant area of materials, design and technology research that bring the safety of buildings and another constructions. Fire shutters are designed to protect not only a property but especially health and lives. Fulfilment of the demanding requirements of fire shutters will provide sufficient time for evacuation after a fire outbreak, but also increase the time required for to reach of the place of fire by fire fighters [2]. The issue of designing and testing of heat-resistant materials and fire shutters is very complex. The reason for that is the complexity of the physical nature of the flame construction process. The reason can be found in the complicated physical nature of the burning process and its impact on the fire construction. The problem lies in the combination of multiple heat transfer mechanisms from heat source to construction and also in fire-resistant construction. While in conventional structures at normal temperatures, the heat transfer is predominantly realised by the conduction, at high temperatures there is also a significant influence of radiation transmission. Another problem is the unsteady behavior of materials, because thermal insulation properties vary with temperature, but very often manufacturers specify only one value. Therefore, in the modelling process, it is necessary to take this fact into account and to set the coefficients of the material model based on the results of real measurements carried out on reduced scale models of fire doors. Only after tuning material model and obtaining the corresponding FE models the virtual prototyping using modelling can begin.

2 Materials and methods

Standardized test [3,4] is conducted in a special chamber where the process of fire simulating two LPG burners while the mass flow of fuel and oxidant could be regulated.

A homogenous distribution of the temperature in testing chamber is required for corresponding data of tests that are used for construction optimization of fire doors. Temperature distribution in the testing chamber near of sample was tested during measurement cycle. The temperature curve is prescribed with equation (1).

$$T = 345 \log[8\,(t-20)+1]+20 \qquad (1)$$

Where t is time in minutes and T is required temperature in the chamber in Celsius degree.

2.1 Fireproof layered sample

The fire door is composed of five individual layers. Applied materials are put without air gape. The

[a] Corresponding author: ondrej.novak@tul.cz

composition of this sandwich is put in Table 1. and the real segment of the tested shutter could be seen in Fig.1.

Table 1. Composition of sandwich

Material	Thickness [mm]	Density [kg.m^{-3}]	Specific heat capacity in the dry state [J.kg^{-1}.K^{-1}]	Thermal conductivity coefficient λ [W.m^{-1}.K^{-1}]
Steel (zinc coating)	1	7850	490	50
Gypsum	12,5	750	1060	0,22
Mineral wool	60	200	880	0,064
Gypsum	12,5	750	1060	0,22
Steel (zinc coating)	1	7850	490	50

Fig. 1. Cross section of the lamella

2.2 Experimental device

The experiment was executed in chamber with dimension 1 x 1 x 1 meter. The sample with dimension of 1x1 m created one vertical wall of the testing chamber. On outer side of the sample four thermocouples were placed into the square shape in the centre of sample. For heating a propane /butane gas was applied. The one burner was placed on vertical side longitudinally to the sample. The temperature was regulated manually depending on the required heating curve in accordance with average temperature measured in chamber by six thermocouples.

3 Experiments

The combustion process in the chamber is accompanied by a flow of gas, air and combustion products. This process will affect the distribution of heat in the chamber and around the sample. The temperature field on the surface of the sample facing towards the fire should not show significant variations with regard to position.

For heat loading a standardized heating curve was applied. Its course is done by the equation (eg. 1).

3.1 Temperature on a sample

Fig. 2 shows temperature on outer side of the sample. The thermocouple signal was averaged and obtained data was put in the graph. There is also added the standardized curve.

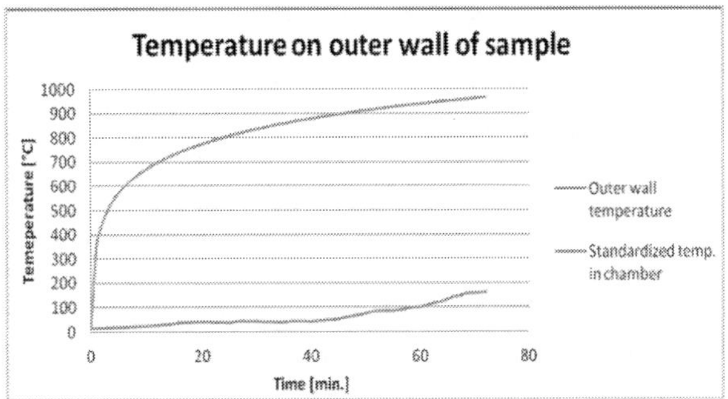

Fig. 2. Experimental results of temperature taken on outer sample surface

4 Numerical simulation

Analytical or numerical solution of turbulent flow problems can be achieved using various levels of approximation. The easiest approach to modeling combustion is according to [5] use semi-empirical correlations, Moody's diagram or a more sophisticated method involves the use of Reynolds' averaging (2):

$$\langle f \rangle = \frac{1}{T}\int_{t}^{t+T} f(\tau)d\tau \qquad (2)$$

Where f is the long time average of quantity and T the time interval - longer than all the time scales of the turbulent flow.

If the averaging is applied to the equations of motion, we obtain the Reynolds-averaged Navier-Stokes equations (RANS) that describe the evolution of the mean quantities [6].

The effect of turbulent fluctuations appears in a Reynolds stress term must be modeled to close the system [6, 7].

Used software PyroSim (Fire Dynamics Simulator developed at NIST) solves the Navier-Stokes equations using the Large Eddy Simulation method. This solution is transient and will vary with time at any point [7, 8]. In contrast, the Reynolds Averaged Navier-Stokes approach provides a single time averaged solution. Large-eddy simulations (LE) are a technique between the direct simulation of turbulent flows and the solution of the Reynolds-averaged equations.

If a sufficiently refined mesh is used, FDS can accurately simulate a free jet. If the mesh refinement is too coarse, the solution will not observe turbulence in the shear layer that entertains the surrounding fluid and the predicted centerline velocity will not decay [8]. Generally, it is of course better to assess the quality of the mesh size in terms of nondimensional parameter, rather than an absolute size.

In FDS is the cell ratio used to describe relative mesh size. Due to the accuracy the cell ratio should not been bigger than 1/5 of the characteristic fire diameter. According to [6, 8] a mesh of even 1/16 the jet side is needed to obtain a reasonably accurate solution.

4.1 Model definition

The non-premixed combustion model with a mixture of propane-butane (C2H6, CH4) has been created. The LE theorem and the solution with considered baroclinic torque and the non ideal energy approach, that consider arising CO, H_2 and soot, has been used.

Three-dimensional flow was modelled by large Eddy model. Particle of defined gas, presented a moveable fragmented droplets (5000 per second), have been through nozzle sprayed into a model domain with defined volume flow rate (L/min), velocity (m/s) through a 12/10°conical jet. The value of the volume flow was necessary to adjust during the process of combustion, in order to hold the normalized temperature. That was done in ten time steps, and as could be seen in the Fig. 3, there is possible to recognize a small changing of the tangent vector. In specific processes such as atomization or spraying a wide variety of mathematical and empirical distribution functions, such as the normal, log-normal or Rosin-Rammler distribution of particles is used [9, 10]. The particles with median diameter 50 μm via Rosin-Rammler log normal function have been distributed. This statistical model is suitable for description of materials composed of huge number of small particles like powder, emulsion and in our case small gas particles. More about using of this technique could be read there [11, 12].

As had been already mentioned, the model of combustion are really sensitive to amount of small details, where the vortexes and nonlinearities arising. Because of that, it was necessary to create several model domains, with the best quality of the mesh in the place, where the jet is located and where the fire process is happen. This could be seen in the Fig. 3, where also the thermocouples used for measuring the in/out temperature and the tested layered sample are.

4.2 Results and discussion of numerical simulation

At the carried simulation we were primarily interested in the most accurate definition of the fire and subsequent in studying its effects on the tested layered sample. The most important results for us is how the heat is transferred through the sample and especially the temperature measured on the outside sheet.

First results showed different course of sample temperatures in initial state of the heating. In the graph (Fig. 4) a less steep rise in the temperature of the real sample can be observed compared to a model that shows a rapid increase in temperature range 0 – 10 minutes. The experimental curve is fitted by the polynomial function. The cause has been investigated and an explanation could be the fact that gypsum contains big amount of water, especially under high relative humidity.

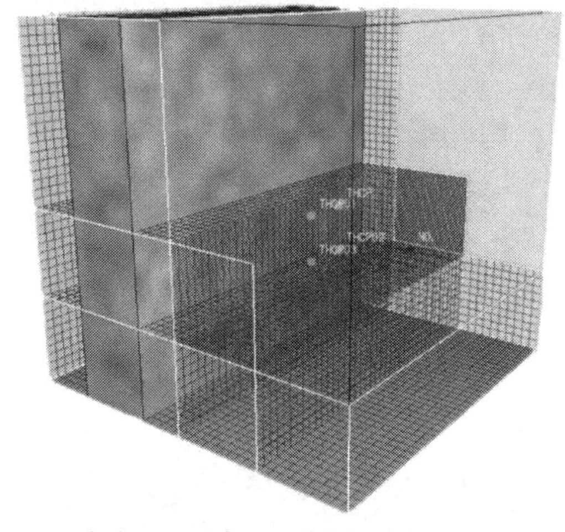

Fig. 3. The inner mesh bases and devices of the model

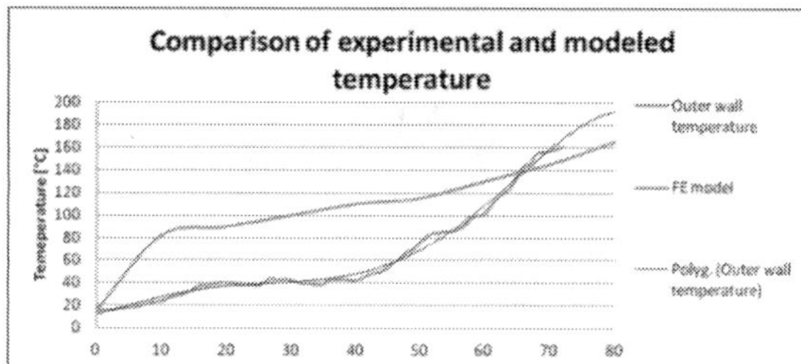

Fig. 4. Experimental and FE model temperatures

In the experiment calcium sulphate dihydrate (CaSO4·2H2O) was applied. Thus, the water is chemically bonded. Also it can be underlined that this type of the gypsum is significantly porous and water is physically bonded in pore channels. Under higher relative humidity these channels are full filled with the water. If higher temperature above 100°C is applied, water receives an energy required for its warming, then the energy required for state change (from liquid to steam) and the energy required for the heating of steam to temperatures above 100°C. If these energies are compared with tabled energy of gypsum (Tab. 2), then it can be concluded that for the model the change of the energy depending on temperature is required. This could be done by setting a table dependency of the material conductivity on temperature either directly in the material, or at any of chosen thermocouple used in the model.

Table 2. Energy for heating of wet gypsum

Specific heat capacity [J.kg^{-1}.K^{-1}]		Specific heat of boiling [kJ.kg^{-1}]	Enthalpy of saturated water [kJ.kg^{-1}]
Gypsum	Water		
490	4180	2 257	2670

It is evident that higher influence has water content and the coefficient has to be depending on the temperature, especially in range of 0 to 100°C. Also thermal conductivity coefficient λ will be in this range higher.

The results after the refinement of this coefficient are more close to reality how it is seen in Fig. 5.

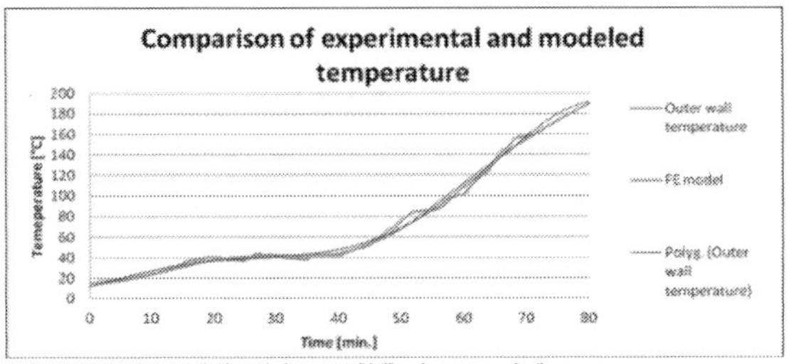

Fig. 5. FE model with modified material parameters

A graphical representation of results is put in Fig. 6-13. They show time course of temperatures each ten minutes from 0 to 70 min.

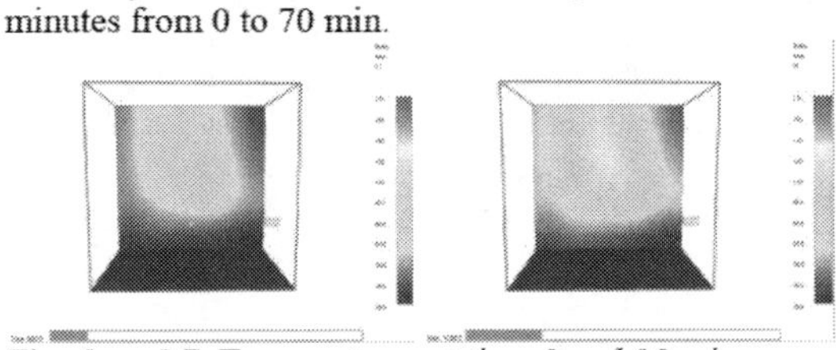

Fig. 6. and 7. Temperatures at time 0 and 10 min.

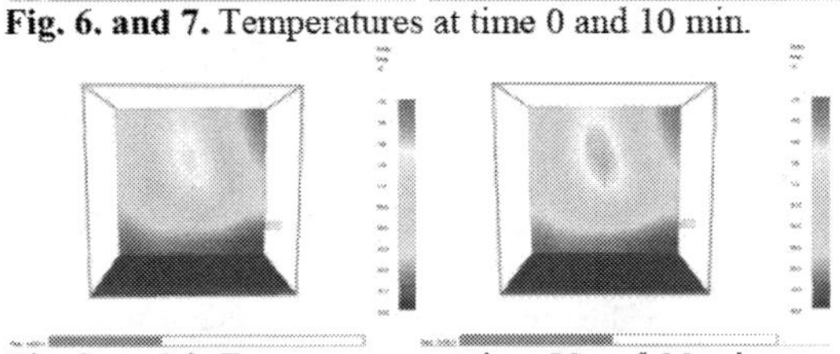

Fig. 8. and 9. Temperatures at time 20 and 30 min.

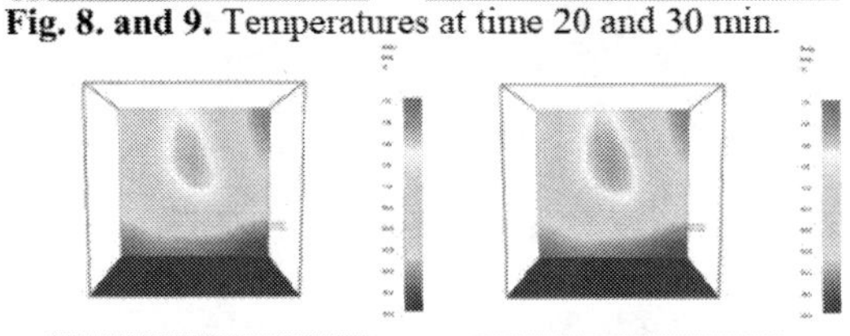

Fig. 10. and 11. Temperatures at time 40 and 50 min.

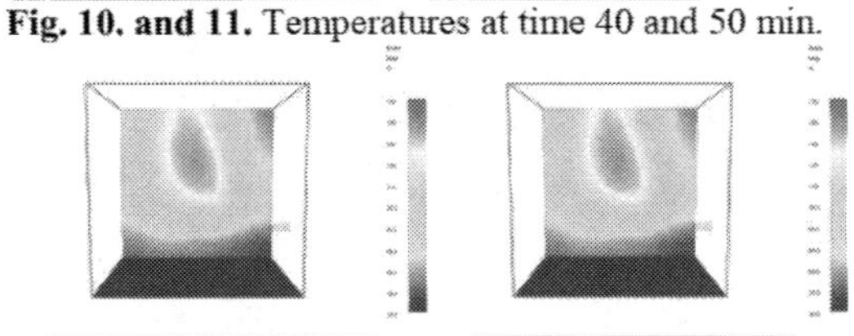

Fig. 12. and 13. Temperatures at time 60 and 70 min.

The experimental and numerical determination of the temperatures arising on the sample are in quite a good conformity. From the ordinary identifiable physical properties, which can be accurately determined experimentally it was in the fuel definition necessary to gone through the description of quantum chemical particles. At this moment, is due to the complexity of solved task necessary to rely on literature, which is based on empirically verified and tested data [1, 5, 10, 11, 12]. After a thorough study of this issue is using and adjusting of these parameters advisable to tweak the simulation nearer the experiment.

5 Conclusion

In the case of a fire inside building it is necessary to limit the spreading by the safe isolation fire zones, ie. the closure of fire doors. In the presented work the new fire shutter consisted on combination of heat-resistant materials was shown. Using the software Pyrosim from the manufacturer Nist, the numerical model based on Large Eddy approach has been created.

In comparison of the simulation and experiment, we can talk about a relatively good agreement in the measured temperatures inside and outside the reference sample (inner and outer sheets). Use of simulation is therefore very suitable for direct determination of the moment and duration critical temperature values which must be coordinated with the knowledge obtained during the experiment. It should be emphasized that some parameters in material model has to be changed based on measurements of samples under high temperatures and also on experience.

Acknowledgments

The results of this project VI20152018046 were obtained with co-funding from the Ministry of the Interior of the Czech Republic as part of targeted support from the project "Využití moderních metod modelování při vývoji a zkoušení protipožárních uzávěrů".

References

1. P. Kulhavy, T. Martinec, O. Novak, M. Petru and P. Srb, Testing fireproof materials in a combustion chamber, EPJ Web of Conf. **143**, 02058 (2017)
2. R. L. Webb, N. KIM, *Principles of enhanced heat transfer*, Taylor & Francis, (2005)
3. ČSN EN 1363-1 - *Testing of fire resistance* - Part 1, ÚNMZ, Praha, (2000)
4. ISO 834-10:2014 *Fire resistance tests* - Elements of building construction, (2014)
5. B. J. M. Giesen, et. al., *Modelling and simulation of a jet fan*, Env. Modelling and Software, **26**, (2011)
6. U. Piomelli, Large-eddy: Achievements and challenges, Progress in Aerospace Sc., **35**, (1999)
7. K. McGrattan, et al. *Fire Dynamics Simulator Technical Reference Guide,* **6**, *(2016)*
8. FDS-SMV. *Fire Dynamics Simulator.*, USA: National Ins. of Standards and Technology.
9. D. Frenkel, B. Smit, *Understanding Molecular Simulation,* San Diego: Academic Press, **2**, (2002)
10. E. Jurado, et al, Estimation of the distribution of droplet size, Colloids and Surfaces A: Physicochem. Eng. Aspects, **295** (2007)
11. K. Akselvoll, P. Moin, Large-eddy simulation of turbulent conned jets, J. Fluid Mech., (1996)
12. A. H. Lefebvre, *Atomization and Spray*, Hemisphere Publishing, Washington, (1989)

Natural convection heat transfer around a horizontal circular cylinder near an isothermal vertical wall

Marcel Novomestský[1], *Richard* Lenhard[1], and *Ján* Siažik[1]

[1]University of Žilina, Faculty of Mechanical Engineering, Department of Power Engineering, Univerzitná 1, 010 26 Žilina, Slovakia

Abstract. Natural convection heat transfer from a single horizontal cylinder near an isothermal wall is experimentally investigated in the environmental chamber. When cylinder is heated the plume rising from the heated cylinder interacts with the wall and significantly affects the surface heat transfer distribution. Thus, the distance between horizontal cylinder and side wall can influence the heat transfer. The heat transfer from heated cylinder with different distances was compared. It was found that the influence of the wall on the convection heat transfer from the circular cylinder began when the distance between the wall and cylinder is less than 80 mm.

1 Introduction

Natural convection heat transfer is an inexpensive, reliable and noise-free method of heat transfer. For instance, horizontal cylinder has many engineering applications such as radiators, heat exchangers, thermal storage systems, and cooling of electronic components.

Past decade has witnessed extensive work on convective heat transfer from cylindrical surface. Morgan [1], Churchill and Chu [2] have determined empirical correlation equations which focus mainly on Nusselt number. Kuehn and Goldstein [3] investigated natural convection around an isothermal circular cylinder for Rayleigh numbers of $< 10^7$ and have solved the full Navier –Stokes and energy equations. Kuehn and Goldstien have shown that the heat transfer is at its maximum at the bottom of the cylinder and decreases toward the top of the cylinder. The decrease in heat transfer is attributed to an increase of the thermal boundary layer thickness.

Later, Fujii *et al.* [4] presented experimental results on natural convection heat transfer from horizontal wire. They used two computational domains, one is a domain near the cylinder and the other is one outside this domain. For order to check the validity of the boundary conditions they used an open and a solid boundary condition.

In particular, natural convection from horizontal cylinder has received considerable attention in the literature because of the fundamental nature of the geometry. The heat transfer from a cylinder near the side wall is quite different with respect to the one from a single cylinder due to the interaction of the side wall. This interaction is owing to the buoyant plume that is generated by cylinder.

The current research investigates the distribution of the heat transfer from horizontal heated cylinder near the side wall under natural convection conditions in air. This experimental study considers the influence of distance between wall and cylinder. The range of cylinder and wall spacing was taking from no inflation distance to distance closed to the wall.

In this article, we consider natural convection, alternatively called free convection, where any fluid motion occurs by natural means such as buoyancy. The buoyancy force is caused by the density difference between the heated fluid adjacent to the surface and the fluid surrounding it, and is proportional to this density difference and the volume occupied by the warmer fluid.

It was assumed that the flow is steady and no slip condition occurs. Thermo-physical properties of the air (Tab. 1) are assumed to be constant except for the density variation, which is approximated by the Boussinesq model. The cylinder surface is maintained at a constant temperature T_s which is higher than the outer temperature T_∞. The side wall temperature T_w is equal to the outer temperature T_∞. The momentum equation, for laminar natural convection flow over cylinder may be written under the Boussinesq approximation as follows [5]:

$$(\rho - \rho_\infty)\, g \approx - \rho_\infty\, \beta\, (T - T_\infty)\, g \qquad (1)$$

where ρ_∞ is the (constant) density of the flow, T_∞ is the operating temperature, and β is the thermal expansion coefficient. Equation 1 is obtained by using the Boussinesq approximation $\rho = \rho_\infty\ (1 - \beta\Delta T)$ to eliminate ρ from the buoyancy term. This approximation is accurate as long as changes in actual density are small, specifically, the Boussinesq approximation is valid when $\beta\ (T - T_\infty) < 1$. The Boussinesq module should not be used if the temperature differences in domain are large.

* Corresponding author: marcel.novomestsky@fstroj.uniza.sk

Table 1. Thermo-physical properties of working fluid

Fluid	*Pr*	ρ (kg/m^3)	*k* (W/m K)	βx10^{-5} (1/K)
Air	0.7	1.2	0.026	340

2 Experimental apparatus

The experimental apparatus consists essentially of a heated circular cylinder mounted horizontally on holding construction, and placed in the environmental chamber.

2.1 Experimental cylinder and installation of thermocouples

The experimental device is circular cylinder (1) Fig. 1 made of stainless steel with length of 1 m. The air flows due to the buoyancy effect. The cylinder is heated electrically by using an electrical heater as shown in Fig. 1. It consists of nickel-chrome wire (3) electrically isolated by ceramic beads and wound uniformly along the cylinder as a coil in order to give uniform heat flux. For measuring the cylinder surface temperature were used NiCr-Ni (type K) thermocouples (2) mounted inside of the cylinder close to the surface in the middle of the cylinder. The outlet air temperature was measured by one resistance thermometer placed in the middle of the environmental chamber [6].

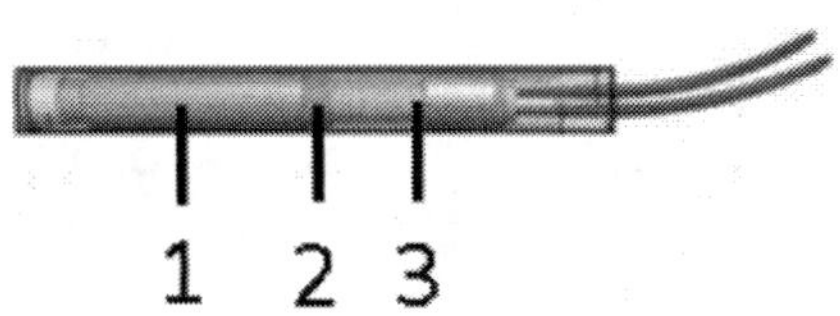

Fig. 1. Longitudinal section of experimental cylinder

2.2 Environmental chamber

Experimental investigation was obtained in the environmental chamber Fig.2, where the surround temperature T_∞ was set to constant 20°C.

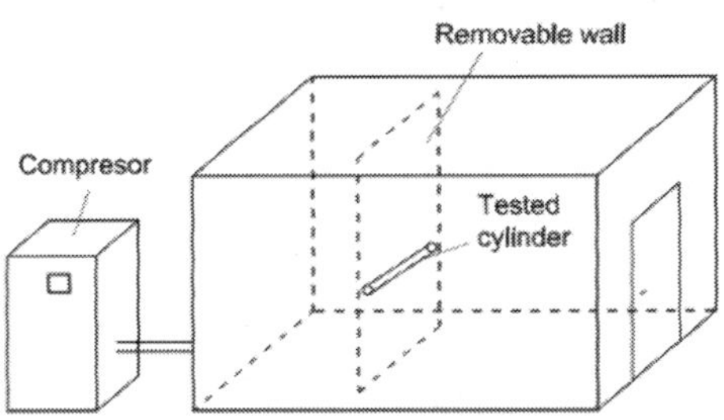

Fig. 2. An environmental chamber for performance tests

The environmental chamber has almost constant conditions tab. 2. The relative humidity was between 52 – 67 % which has only small influence on the air density.

Table 2. Environmental chamber conditions

T_∞ [°C]	*P* [kPa]	*rH* [%]
20±0,05	96,8	52 - 67

2.2.1 Test area

Thermal output measurement is based on the norm STN EN 442: 2000. According to this norm, the test area must be equipped with:

- closed chamber without ventilation with space for installation of the test specimen and a water-cooled surface in order to achieve a certain temperature conditions independent on the external environment,
- device for cooling the water which run in the wall of the chamber,
- primary heating circuit constructed to supply test specimen,
- measuring instruments and control equipment meeting the requirements of norm [7].

2.2 Experimental procedure

A voltage regulator (matrix), digital ammeter and digital voltmeter were used to control and measure the input power to the working cylinder. The apparatus was allowed to run till pseudo-steady state conditions were achieved. The reading of thermometer and resistance thermometer were recorded every 2 min by digital electronic device until the reading became constant, and then, after half hour, the final reading was recorded [8].

2.3 Data analysis method

The confinement ratio *C* is the first control parameter of the experiments ($C = X/r$), where *X* is distance between the wall and the cylinder as shown Fig. 3, and *r* is a cylinder radius). Nine different confinement ratios *C* will be investigated in this study: *C* = 1,5; 2; 2,5; 3; 4; 5; 7,5 and 10. The second control parameter (ΔT) is the difference between the cylinder surface temperature (T_s) and the air bulk (T_∞). The air bulk temperature is measured in the middle of the environmental chamber. This second parameter dictates the natural convection flow regime and is generally expressed in the form of the Rayleigh number define as:

$$Ra = \beta g D^3 \Delta T Pr / \upsilon^2 \quad (2)$$

where β is the volumetric thermal expansion coefficient of the fluid, *g* is the gravitational constant, *Pr* is the Prandtl number and υ is the kinematic viscosity of air. All parameters are evaluated at the fluid temperature define as:

$$T_f = (Ts + T_\infty) / 2 \quad (3)$$

In the work reported here, the maximum Rayleigh number was of the order of 10^5, meaning that the flow is nominally still laminar.

The convection heat transferred from the heated cylindrical surface can be calculated as:

$$Q_{conv.} = Q_{IP} - Q_{rad.} \quad (4)$$

The total input power supplied to the heated cylinder can be calculated as

$$Q_{IP} = V \times I \quad (5)$$

The radiation heat transfer from the tube can be calculated from:

$$Q_{rad.} = A_s \varepsilon \sigma \,(T_s^4 - T_\infty^4) \quad (6)$$

where A_s is the surface area of the tube = $\pi \times d \times L$; ε is emissivity of given material; and σ is Stefan-Boltzmann constant = 5.66×10^{-8} W/m^2 K^4.

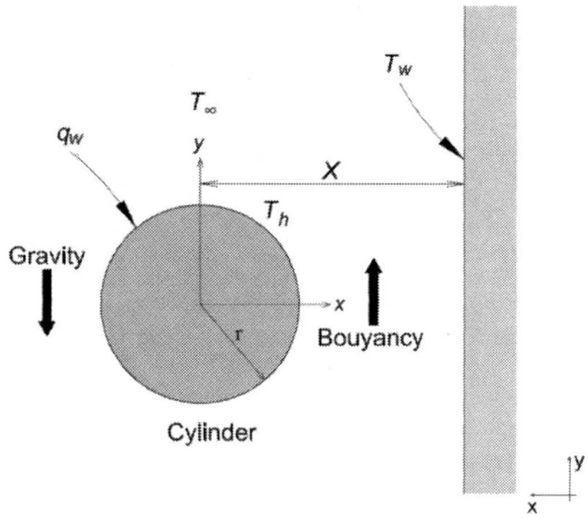

Fig. 3. Heated circular cylinder near an isothermal wall

After calculating the radiative heat transfer from the cylinder, it was noticed that it is very small compared with the total heat transfer, and consequently, it can be ignored.

The convection heat transfer can be obtained from:

$$q_{conv.} = Q_{conv.} / A_s \quad (7)$$

where $A_s = \pi \times d \times L$.

3 Experimental results

A total of 45 test runs were conducted to cover 9 different distances for heated circular cylinder with length of 1 m (L / D = 5). The range of heat flux was varied from 159 W/m^2 to 796 W/m^2.

3.1 Heat flux

For analysis of the natural convection heat transfer around a horizontal cylinder with the constant heat flux the temperature parameter was analysed for 10^4 Rayleigh number. The surface temperature of the cylinder is taken with one thermocouple placed in the middle of the cylinder close to the surface. Fig. 4 – 8 shows that the surface temperature is decreased with distance X with direction to the wall.

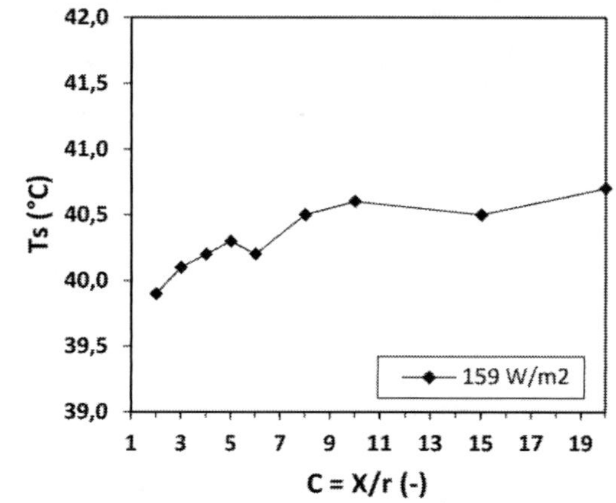

Fig 4. Surface temperature for q = 159 W/m^2

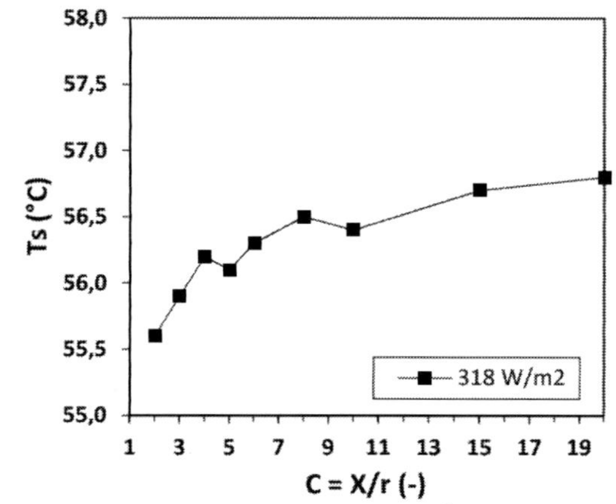

Fig. 5. Surface temperature for q = 318 W/m^2

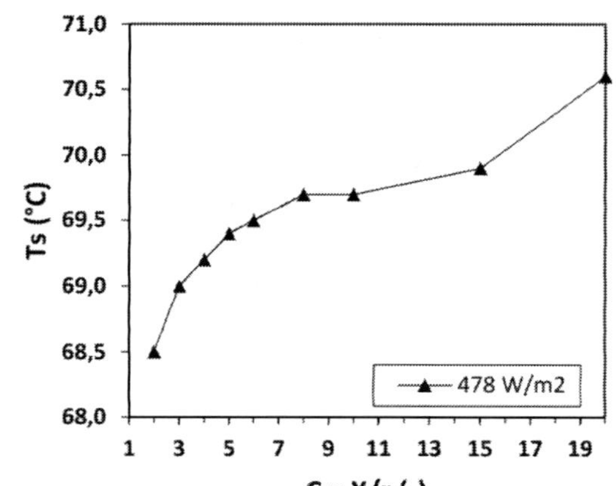

Fig. 6. Surface temperature for q = 478 W/m^2

Fig. 9 shown for all tests that temperatures exhibit a larger change when the distance between the cylinder and the wall is less than 80 for all tests.

https://doi.org/10.1051/epjconf/201818002077

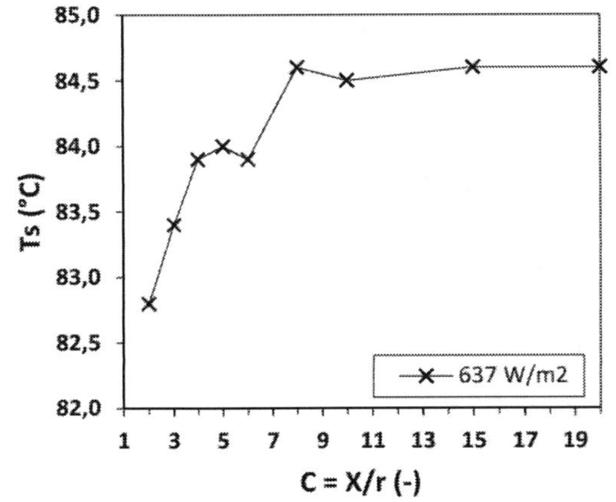

Fig. 7. Surface temperature for $q = 637$ W/m^2

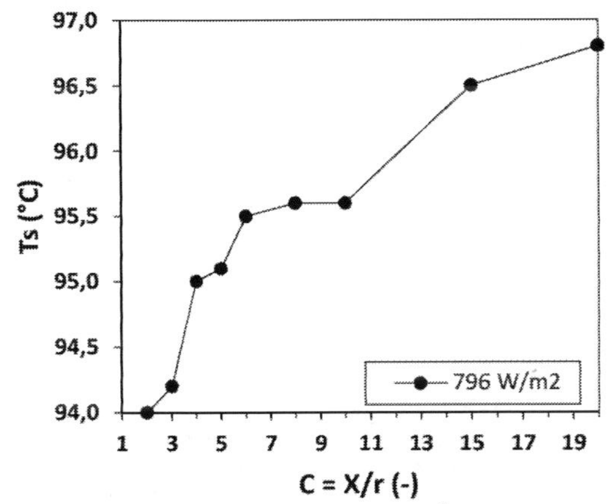

Fig. 8. Surface temperature for $q = 796$ W/m^2

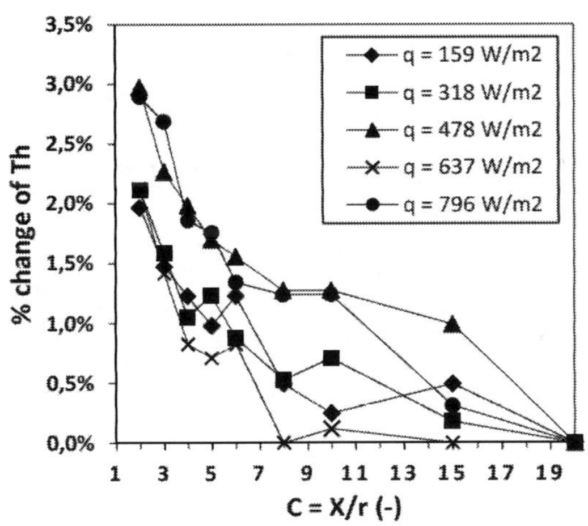

Fig. 9. Percentage change of temperature with distance from wall to centre of the cylinder

3.2 Effect of wall confinement

The experimental data obtained from this investigation analyses the effects of the confining wall on the rate of heat rejection from surface of the heated cylinder. For this study, the maximum average Nusselt number occurs at a confinement ratio 2.

3.3 Average heat transfer correlation

The general correlation obtained by dimensional analysis from Incropera and DeWitt [10] for heat transfer by natural convection is:

$$Nu_D = f_1 (Gr, Pr)^n \quad (8)$$

In the case of heat transfer from cylinder to vertical wall, one expects that there is effect of distance between the wall and the cylinder. For similarity with a horizontal cylinder, the characteristic linear dimension in the *Nu* and *Gr* may be taken as the cylinder diameter (*D*) and the effect of the distance between wall and cylinder may be taken as $C = X/r$. Then, Eq. (8) becomes:

$$Nu_D = f_2 (Gr, Pr)^n + f_3 C^m \quad (9)$$

Therefore, the following correlation was obtained from the present study for smallest and biggest heat flux and a general correlation was obtained for all distances shown in Fig. 10 – 11.

$$Nu_D = 0,48\,(Gr, Pr)^{0,25} + 1,05C^{-0,15} \quad (10)$$

As a comparison between the present study and normal case, which included a horizontal cylinder in the free air without any restrictions, the correlation obtained for the normal case from Incropera [10] is,

$$Nu_D = 0,48\,(Gr, Pr)^{0,25} \quad (11)$$

Comparing Eq. 10 with the Eq. 11 for this study shown in the Fig. 10-11, is proven that the distance between the wall and the cylinder has an effect on the heat transfer results.

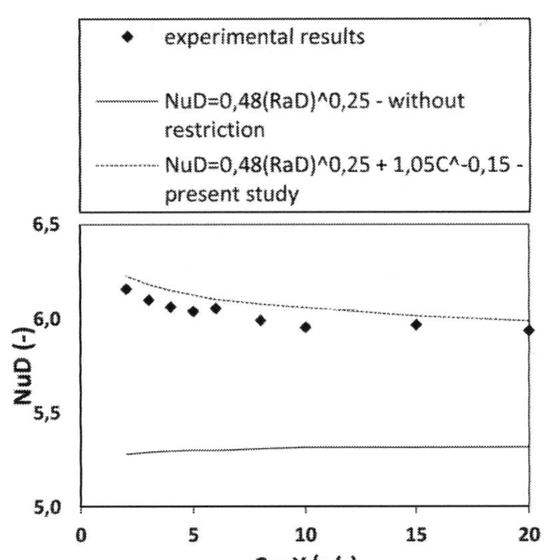

Fig. 10. Correlation between average Nusselt number versus confinement ratio for smallest $q = 159$ W/m^2

The difference of present correlation with Incropera work is about 15% which is not counting with wall confinement.

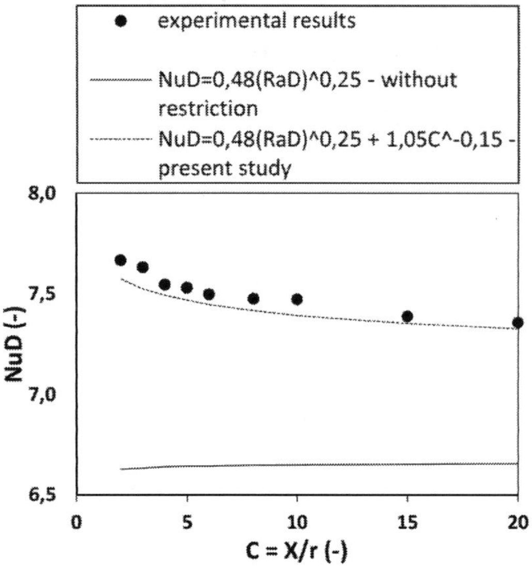

Fig. 11. Correlation between average Nusselt number versus confinement ratio for highest $q = 796$ W/m^2

4 Conclusions

An experimental study of a laminar natural convection heat transfer from a single horizontal cylinder near an isothermal wall was conducted. Laminar convection heat transfer causes flow because the density differences. The flow moved up around the hot cylinder and continued upward when the cylinder was not interacted with the wall. The flow moved across the wall when the upward flow is interacted with cold wall and began mainly when the distance between the wall and the cylinder of 20 mm diameter was less than 80 mm. The highest temperature change occurs at confinement ratio 2. The improvement of correlation in this study is up to 15%.

Nomenclature

A_s = Cylinder surface area, (m^2)
C = Confinement ratio, $C = X/r$ (-)
D = Diameter of circular cylinder, (m)
g = Gravitational acceleration, (m/s^2)
Gr = Grashoff number, (-)
I = Heater current, (A)
k = Thermal conductivity, (W/m.k)
Nu = Average Nusselt number, (-)
Pr = Prandtl number, (-)
q = Heat flux, (W/m^2)
$Q_{conv.}$ = Convection heat transfer, (W)
Q_{IP} = Total input power, (W)
Q_{rad} = Radiant heat transfer, (W)
r = Radius of circular cylinder, (m)
Ra = Rayleigh number, (-)
T = Temperature, (°C)
T_s = Temperature of the surface, (°C)
T_W = Temperature of the wall, (°C)
T_∞ = Temperature of the cold surface, (°C)
ΔT = Temperature difference, (-)
k = Thermal conductivity, (W/m.k)
V = Heater voltage, (V)
X = Distance from centre of square cylinder to wall
x = Cartesian coordinate in horizontal direction, (m)
y = Cartesian coordinate in vertical direction, (m)

Greek Symbols

α = Thermal diffusivity, (m^2/s)
β = Volumetric coefficient of thermal expansion, (1/K)
μ = dynamic viscosity of the fluid, (kg/m.s)
ε = Emissivity, (-)
σ = Stefan-Boltzmann constant, (W/m^2 K^4)
ρ = Density of the fluid, (kg/m^3)
υ = Kinematic viscosity of the fluid, (m^2/s)

Acknowledgments

This work was supported by Department of Power Engineering under APVV-15-0778 Limits of Radiative and Convective Cooling through the Phase Changes of Working Fluid in Loop Thermosyphon.

References

1. V. T. Morgan, Adv. Heat Transfer, **11**, 199 (1975)
2. S. W. Churchill, H. H. S. Chu, International Journal of Heat and Mass Transfer, **18**, 1323 (1975)
3. T. H. Kuehn, R. J. Goldstein, International Journal of Heat and Mass Transfer, **23,** 971 (1982)
4. T. Fujii, M. Fujii, T. Honda, Proceedings of the 7th international heat transfer conference, **2,** 311 (1982)
5. Ansys fluent, User's Guide, **15**, 765 (2009)
6. A. Kapjor, *Prenos tepla z orientovaných teplovýmenných plôch,* (2012)
7. STN EN 442-2:2000, (1996)
8. R. Lenhard, J. Jandačka, M. Malcho, Vetranie a klimatizácia, *Energetická hospodárnosť vetrania a klimatizácie pri tvorbe vnútornej klímy*, 11, (2008).
9. M. Novomestský, H. Smatanova, A. Kapjor, The application of experimental and numerical methods in fluid mechanics and energy, **1745,** (2016)
10. F. P. Incropera, D. P. DeWitt, T.L. Bergman, A.S. Lavine, *Fundamentals of heat and mass transfer*, **7,** (2011)

How much can be saved by changing the regulation of the apartment house boiler room

Petr Novotný, *Jan* Hujer

TU of Liberec, faculty of Mechanical Engineering, Department Power Engineering Equipment, Czech Republic

Abstract. Boiler room for apartment building is in operation since 2009 and data are collected for cost evaluation. Question arises after a minor reconstruction. How much energy has been saved? The article selects and processes data that was available. It shows that the importance of the heating circuit pump control cannot be underestimated or overestimated. The article derives savings from gathered data.

1 Introduction

An older apartment building disconnected from the central heat supply in 2009 and built its own gas boiler room. It operated according to the original project until 2016. During this year, a change of regulation of the house heating was made by means of a mixing valve in the heating circuit.

There are two logs on boiler room operation, one of which contains information on energy consumption. This information is written manually twice during the month. By evaluating these data, we obtain information about the energy consumption before and after reconstruction.

2 Project of 2009

Three condensing boilers operate in cascade and prefer hot water. Boilers are controlled by equithermal regulation, which uses historical data of approximately three days to calculate the heating water temperature.The boilers are connected via a torus to the heating system with a separate heating pump. The individual risers of the house are controlled by differential pressure regulators, the radiators have thermostatic valves. The DHW heating in two water tanks is controlled by the control unit and a separate charging pump, the heating circuit pump is switched off in the case of DHW heating.The water for the heating of the containers is taken behind the toroid Fig. 1. According to the project, when the hot water supply is depleted, the heating system pump is switched off and the hot water heating pump switched on. Water heating is performed by all three boiler units. Their power is greater than that of the hot water heat exchanger. Heating time varies depending on the current consumption. Water tanks have a capacity of 500 liters, average daily consumption of hot water is about 4 m^3. Heating does not start after the hot water supply has been depleted but sooner. It is due to the position of the thermometer in the tank and the setting of the minimum temperature. In reality, due to losses in hot water circulation, the amount of heated water is doubled (corresponding to gas consumption for water heating).

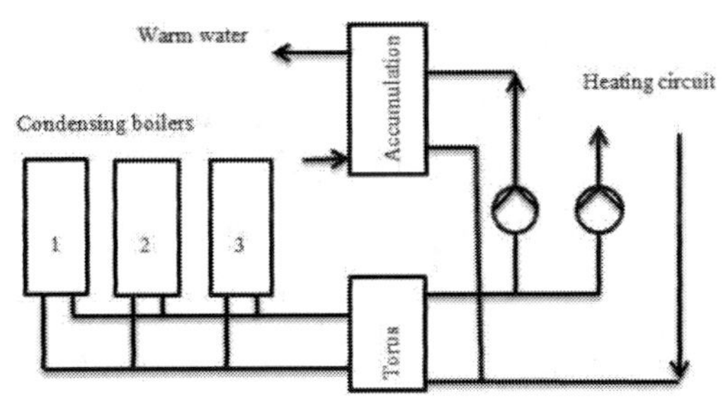

Fig.1. Simplified scheme of engagemet from the year 2009

Hot water supply system has a separate circulating pump that interrupts circulating during the night hours. During the day circulation occurs every 15 minutes for a duration of about 5 minutes. The duration of hot water heating throughout the day can be estimated in a length of 15 extremely up to 25 minutes. Heating system starts gradually after completion of hot water heating. First, the heating circuit pump is started and when boiler water is cooled below required heating water temperature, boiler units are gradually started up. This extends the time of reduced heat supply to the radiators and, in addition, opens the thermostatic heads and increases the flow of heating water. If we consider a minimum of 12 heating cycles per day (in fact, it is more) for 15 minutes, it is a three-hour interruption of the heating water supply over a day, which is logically replaced by high temperature in the heating system and a larger amount of heating water. Additionally, the pump (in Adaptive Auto mode) looks for the optimum setting, but restarts every time it is turned off [1].

The control unit has enabled mixing valve control and simultaneous DHW and heating. In 2009, however, the designer described this possibility of installing a mixing

* Corresponding author: petr.novotny@tul.cz

valve as something a expensive and unnecessary when the pump runs in autoadaptive mode. Grundfos Magna 50-120 pump operated in this setup often at 60% of the delivered quantity, exceptionally at 40%. At the start of the heating circuit, it increased the output to 100% in the short term. The pump has a min. 35 W and max. 800 W. It does not measure the amount of supplied water but calculates from the pump operating parameters.

3 Reconstruction

After the heating season in 2016, the boiler room was reconstructed. Mixing valve has been mounted in the heating circuit in Fig. 2. The mixing valve is at the inlet of the heating circuit pump. The pump has been set to Constant Output Pressure at a discharge height of 3 meters. The pump has not been set to auto-adaptive mode. The setpoint temperature of the heating water has been lowered by two degrees, because the heating takes place without interruption throughout the day.

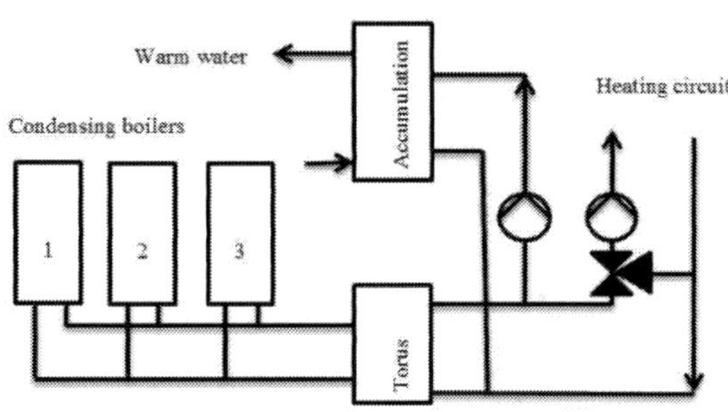

Fig. 2. Simplified diagram of 2016

After reconstruction, when water temperature in the storage tank drops below the set value, the DHW charging pump is triggered. At the same time, if all the boiler units are not working, they start and gradually increase the power. Boiler output temperature is 80°C. The three-way valve adjusts the outlet temperature to the heating circuit according to the outside temperature. It responds with a delay, so that the heating water temperature increases by several degrees during DHW heating. The control unit gradually eliminates this temperature increase, but only after the hot water heating has finished. In the heating season 2016/17, during the boiler inspections, the pump operated according to display when it froze at 40% of the supplied amount of heating water. Most of the time pump displayed 20%.

4 Measured operation data

Data recorded in the log is written manually. Data records are in some cases longer than 15 or 30 days. This makes evaluation difficult, but on the other hand it was important to provide this information for economical comparison of energy consumption and the amount of finance for the boilers room operation.
Data from 2010 to 2014 were selected from the log for months from October to April Fig. 3. Data on average monthly temperatures were obtained from the Czech portal TZB for calculating degree days (on link « vytápění, tabulky a výpočty, výpočet denostupňů ») [2]. Here are the meteorological data from Liberec airport. The airport is situated on the outskirts of the city at an altitude of 400 m above sea level.
It should also be noted that there are differences between the temperatures of different districts of the city and the airport. The city is located in hilly terrain with various buildings and vegetation. Differences in temperatures are not constant but are not critical to our measurement and analysis. These temperature differences are not critical for boiler room data and are attenuated by the house and system dynamics. Monthly data on energy consumption are mainly exposed to errors in the uneven recording of operational data in the boiler room.

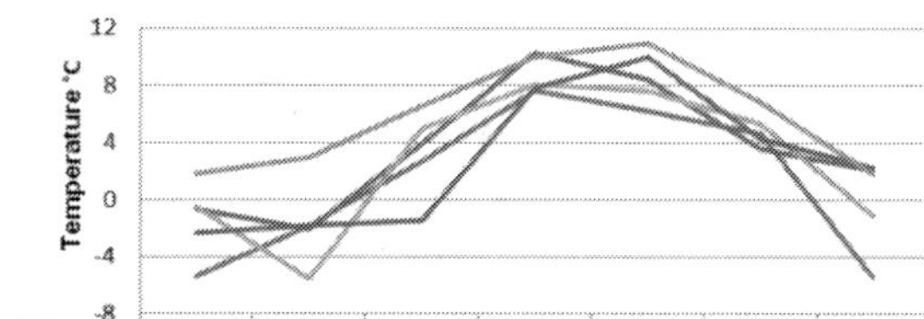

2010	-5,4	-1,8	2,7	7,6	6,2	4,6	-5,4
2011	-0,6	-2	3,9	10,3	8,4	3,6	2,1
2012	-0,5	-5,6	5	8,1	7,6	5,3	-1,1
2013	-2,3	-1,8	-1,4	7,8	10	4,3	2,3
2014	1,8	3	6,6	10	11	6,8	1,8

Fig. 3. Monthly temperature data

4.1 Electricity consumption

4.1.1 All monthly data before reconstruction

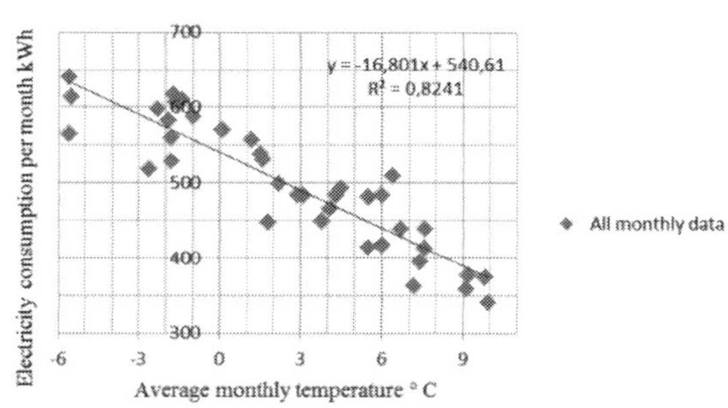

Fig.4. All monthly data

The uneven write period ranges from 28 days to 33 days in Figure 4. This is naturally reflected in the reliability of the approximation result using the formula equation.

4.1.2 Selected data for 30 and 31 days

All data that did not meet the condition was removed from the file. The confidence factor has improved, but the impact of one day's tolerance is high. For example, where the average monthly temperature is -5.4°C for two months of 30 and 31 days of Figure 5. Other inaccuracies are, for example, that the month, which has 31 days, has recorded data for only 30 days and vice versa. We can improve this by using a longer time span. During the year we have 7 months. We divide them into 3 autumn and 4 spring.

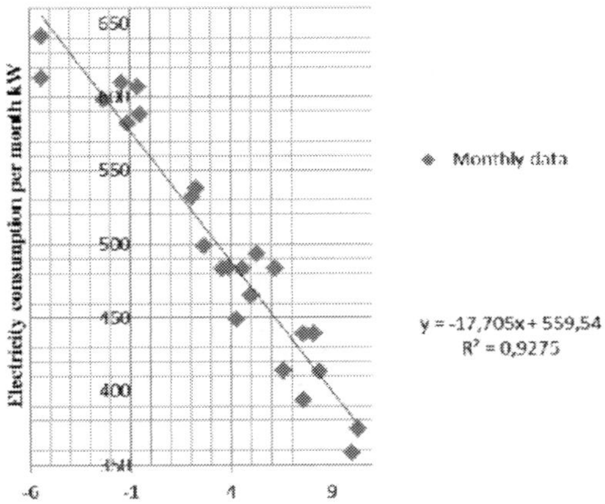

Fig.5. Monthly electricity consumption in kWh besed on average monthly temperature

4.1.3 Editing data

Table 1. Autumn

Year/autumn	2010	2011	2012	2013	2014
Temperature °C	1,566	4,033	3,1	5,533	6,533
Electricity kWh	1607	1428	1516	1363	1313

Table 2. Spring

Year/spring	2010	2011	2012	2013	2014
Temperature °C	0,775	2,9	1,75	0,575	5
Electricity kWh	2066	1983	2057	2106	1756

The sum of the months will be ninety and one hundred and twenty days where the error one day has little effect. Table 1 shows the results for October, November and December (Autumn) and Table 2 for January, February, March and April (Spring).

According to the results in Fig. 6, the dependence of the energy consumption on the average temperature can be considered as linear in the range of the measured data.

4.2 Gas consumption

Similar inaccurate data is also on gas consumption. Another error that affects measured data is warm water. The annual consumption is in Table. 3. The average consumption of an apartment building per year is

Table 3. Warm water

Year	2010	2011	2012	2013	2014	2015	2016
m^3	1595	1522	1518	1539	1594	1543	1533

1549 m^3 and a deviation range from +3% to -2% in different years.

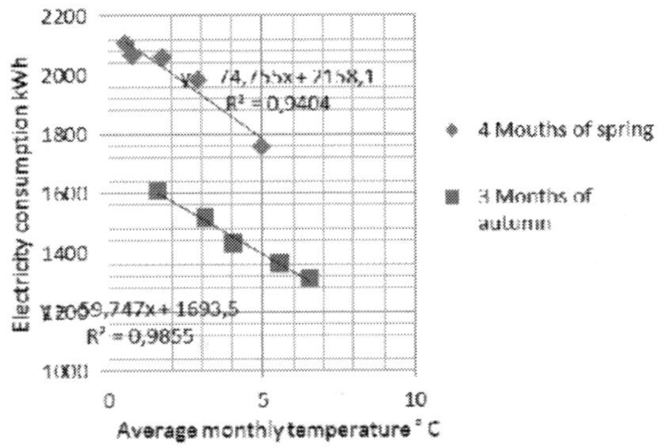

Fig.6. Dependence of consumption on temperature

Monthly differences are similar (%) in water consumption. In winter, the water is colder.

The temperature of the water at the inlet before heating is not measured and the consumption of gas for heating the water is not measured. The consumption of gas for DHW heating can be determined outside the heating season. Water heating will be viewed as a disorder that affects some result.

4.2.1 Selected data for 30 and 31 days

In Fig. 7 the data is evenly loaded with error. There is a relatively high confidence coefficient of 0.98 at line approximation. Improving the reliability factor by further editing due to data acquisition inaccuracies can not be expected.

4.2.2 Editing data

We will perform the sum of months October, November and December (autumn) and the remaining 4 months as "Spring". It is clear from Figure 8 that further improvement of results has not occurred even though we have been working with more days (90 and 120). We will use the monthly gas consumption data to evaluate the reconstruction.

5 Reducing energy consumption

The reconstructed boiler room was in operation from autumn 2016 until spring 2017. However, we can't use data for all months, because some records are missing or are deducted at the wrong time interval.

It has been confirmed that the electricity consumption for this house is changing linearly proportional to the average temperature. This is a dependence on the measured temperature in the evaluated period. Whether this dependency applies to temperatures up to -15 °C is not confirmed. Before the reconstruction and after the reconstruction, the boiler room worked without damping. The house has small apartments and is inhabited by a wide range of people from old citizens to families with small children (not the exception of a four-person family in a 1+1 apartment). Every apartment occupant adjusted the temperature according to his own needs by adjusting the thermostatic valve. Uninterrupted heating increased satisfaction of the residents and removed noise caused by dilatations, which could not be eliminated for years.

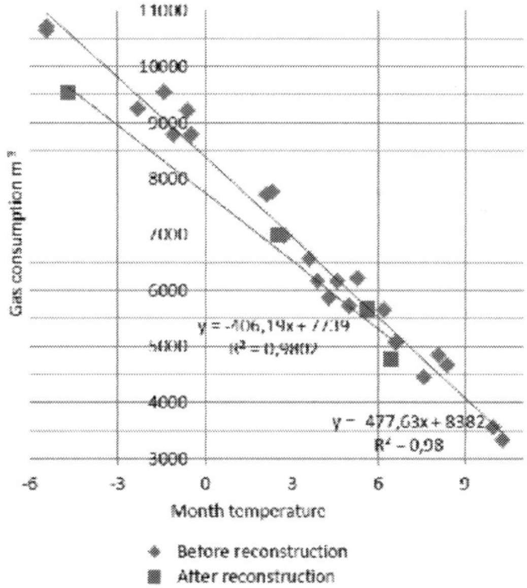

Fig.7. Gas consumption data for 30 and 31 days

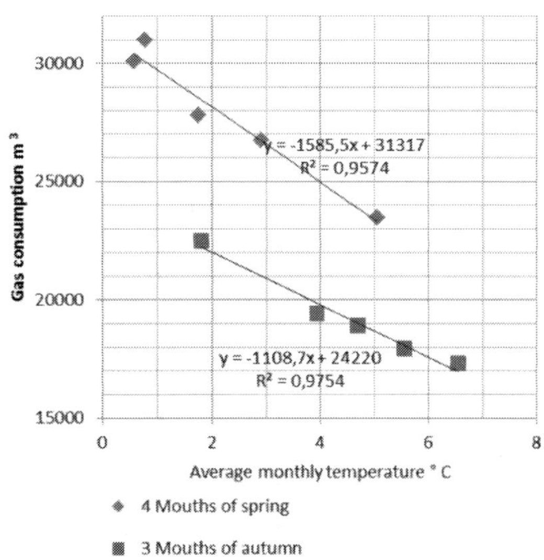

Fig.8. Dependence of consumption on temperature

In the heating season 2016/17, the adjustments made were reflected in electricity consumption. Although both pumps were operated at the same time and the mixing valve actuator was operating, there are savings in electricity consumption. Linear equations obtained by simple regression can be considered to be sufficiently accurate to estimate the savings in electricity consumption. In autumn 2016 after reconstruction, when the average temperature in three months (autumn) was 2.9°C, consumption was 1342 kWh boiler room. If the reconstruction was not carried out, the linear model consumption would be approximately 1514 kWh. The saving is about 11% of the electricity. In the spring of 2017, at an average temperature of 1.5°C, and a consumption of 1739 kWh, savings of approximately 274,7 kWh were calculated, is 13% of electricity.

Total annual electricity consumption averages around 4 MWh. Total consumption decreased by about 446 kWh, which is 10%.

Figure 7 shows five months of gas consumption in the boiler room after reconstruction. Statistically, this is an insignificant amount dat that can't be used even if the reliability is high. Using the results shown in Figure 8 is worse coefficient reliability. Usable data is for a period of four months (spring) in 2017, when the average temperature was 2.15°C. The gas consumption was 26880.1 m^3 which according to the linear model corresponds to 27908 m^3 of gas for this average temperature. The difference is therefore 1027.9 m^3 and corresponds to the inaccuracy of the linear model. The gas saving can't be determined from the measured data because it is a small quantity ranging from 2 to 4%. Savings can be achieved by modifying the boiler room. These savings are not significant but ultimately reduce the energy requirements to ensure thermal comfort in a residential building.

Conclusion

- The data obtained was not accurate to detect gas savings
- Significant savings are in electricity consumption
- We can only estimate gas savings of around 2%, which at an annual consumption of 50 thousand m^3 is 1000 m^3 of gas

Acknowledgment

This publication was written at the Technical University of Liberec as part of the project "Experimental and numerical investigation in applied fluid mechanics and energy devices, no. 21124" with the support of the Specific University Research Grant, as provided by the Ministry of Education, Youth and Sports of the Czech Republic in the year 2017.

References

1. Adaptive Selection of Control-Curves for Domestic Circulators; Carsten Skovmose Kallesøe*, Niels Bidstrup*, Manfred Bayer; **Grundfos Management A/S, Poul Due Jensens Vej 7, DK-8850 Bjerringbro, Denmark (e-mail: {ckallesoe,nbidstrup}@grundfos.com)*
2. Data Available May 31, 2017 at http://vytapeni.tzb-info.cz/tabulky-a-vypocty/103-vypocet-denostupnu
3. *P. Novotný*, workbook boiler rooms, unavailable.

https://doi.org/10.1051/epjconf/201818002079

Simulation and analysis of collapsing vapor-bubble clusters with special emphasis on potentially erosive impact loads at walls

Daria Ogloblina*, *Steffen J.* Schmidt, and *Nikolaus A.* Adams

Technische Universität München, Institute of Aerodynamics and Fluid Mechanics, 85748 Garching bei München, Germany

Abstract. Cavitation is a process where a liquid evaporates due to a pressure drop and re-condenses violently. Noise, material erosion and altered system dynamics characterize for such a process for which shock waves, rarefaction waves and vapor generation are typical phenomena. The current paper presents novel results for collapsing vapour-bubble clusters in a liquid environment close to a wall obtained by computational fluid mechanics (CFD) simulations. The driving pressure initially is 10 MPa in the liquid. Computations are carried out by using a fully compressible single-fluid flow model in combination with a conservative finite volume method (FVM). The investigated bubble clusters (referred to as "clouds") differ by their initial vapor volume fractions, initial stand-off distances to the wall and by initial bubble radii. The effects of collapse focusing due to bubble-bubble interaction are analysed by investigating the intensities and positions of individual bubble collapses, as well as by the resulting shock-induced pressure field at the wall. Stronger interaction of the bubbles leads to an intensification of the collapse strength for individual bubbles, collapse focusing towards the center of the cloud and enhanced re-evaporation. The obtained results reveal collapse features which are common for all cases, as well as case-specific differences during collapse-rebound cycles. Simultaneous measurements of maximum pressures at the wall and within the flow field and of the vapor volume evolution show that not only the primary collapse but also subsequent collapses are potentially relevant for erosion.

1 Introduction

Collapsing single bubbles have been extensively investigated in the past and results obtained by experiments, theory and simulations are broadly represented in the literature [1], [2]. Significantly fewer publications can be found where larger bubble clusters are considered [3-5]. Although there is a simple theory provided by Brennen et al [6], clusters of bubbles are difficult to investigate experimentally. One reason is that a repeatable generation of a cluster of well-defined bubbles (radii, position) is hardly achievable. Therefore, numerical simulation of collapsing bubble clusters is a suitable way to enhance physical insight into collapse processes.

To estimate the interaction among bubbles a *cloud interaction parameter* β was introduced [4]. This parameter is defined as $\alpha_0 \cdot (1- \alpha_0) \cdot A_0^2/R_0^2$, where α_0 is initial cloud void fraction and A_0 and R_0 are the cloud and the bubble radius, respectively.

It was shown in previous studies [7], [8] that cloud collapse mechanisms are significantly depending on the cloud interaction parameter. For high values ($\beta >>1$) the shock develops and propagates from the outside towards the center of a cloud, therefore producing a focused collapse. At low values ($\beta < 1$) the interaction of the bubbles is weak and one might rather observe individual collapses. In this investigation the interaction parameter is chosen to be 4, 8 and 11, indicating that bubble interaction is relevant.

There are several common approaches to model liquid-vapor interfaces, for example: two-fluid approaches [9] with sharp interface treatment [1], the volume of fluid technique [10], and so called single-fluid approaches with implicit interface capturing techniques. Two-fluid methods are capable of representing the phase interphase with high precision and allow for inclusion of surface tension. However, they require the solution of two sets of governing equations and an explicit interface-interaction model.

In the present work a single-fluid approach coupled with a thermodynamic equilibrium model is adopted. When the grid resolution is sufficiently fine, the thermodynamic equilibrium model implicitly resolves the interface and becomes comparable to a sharp interface technique without surface tension [3]. It enables the prediction of re-evaporation processes, such as bubble rebound. However, the inclusion of surface tension is difficult and thus, the method is suitable for processes where surface tension is negligible, such as the investigated inertia dominated collapses of bubble clusters under high ambient pressure.

In this investigation, two major physical aspects, namely the cloud interaction parameter and the standoff-distance, are investigated.

First, a series of three bubble clusters at a fixed stand-off distance from a wall are generated. The radii of the approximately spherical clusters are kept constant while the common radii of the involved bubbles are varied. This leads to three different cloud interaction parameters.

In a second series of computations, the cloud interaction parameter is kept constant while four different stand-off distances are investigated.

For all configurations, the dominant collapse mechanism, rebound structures and maximum impact loads are analysed. A major finding is that the secondary collapse after the first rebound may lead to the highest loads at the solid wall. This indicates that the secondary collapse can considerably contribute to cavitation induced material erosion.

2 Methodology

2.1 Numerical approach

The main focus of this investigation is on inertia-driven effects and wave dynamics arising during the collapse of a bubble cloud. Surface tension and viscous effects are neglected and the governing equations are the compressible Euler equations. For this investigation the flow simulation code CaTUM is used [11]. All numerical procedures have been extensively validated and the results have been published in the literature [12-16]. CaTUM is a density based finite volume method employing a Low-Mach-number consistent flux function and an explicit time marching procedure. The compressibility of the fluid is taken into account in order to capture shock formation and wave propagation. Hence the numerical time step is proportional to the ratio of the minimum grid size and the fastest signal speed (speed of sound in the liquid).

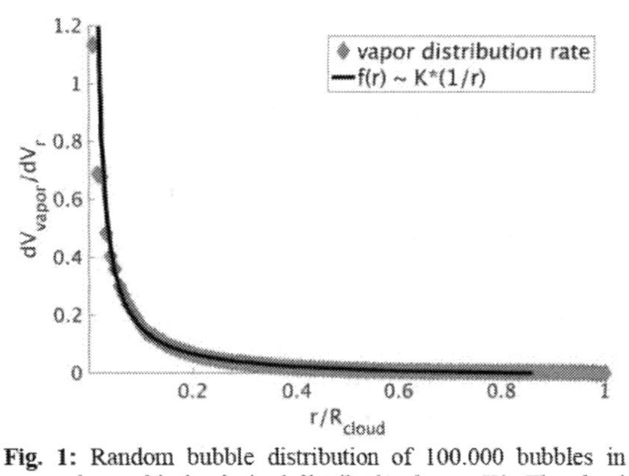

Fig. 1: Random bubble distribution of 100.000 bubbles in comparison with the desired distribution law ~ K/r. The cloud interaction parameter is $\beta = 11$

Phase change is modelled with equilibrium phase transition, i.e. both phases are in thermal, mechanical and phase equilibrium. In this study, the thermodynamic properties of water (in its liquid and vapor phases) are described by efficient barotropic equations of state. Pure liquid can be modelled by a Tait equation while saturated mixtures are modelled by a barotropic relation $p = p\ (\rho)$, which is obtained by integration of the equilibrium speed of sound along an isentrope. The vapor volume fraction of a saturated mixture of liquid and vapor can be derived from the saturation densities of both phases.

2.2 Cloud specification and computational domain

As there is insufficient experimental data of "reference clouds" available, a random procedure proposed by *Schmidt et al* [3] is utilized to generate bubble clusters. The following criteria are assumed:

The bubble cluster covers a spherical domain with a diameter of 30 mm. Within this domain, 150 spherical bubbles of equal radii are randomly generated. The minimum distance between neighbouring bubbles is one bubble radius. It is further assumed that the bubble number density reaches its maximum at the center of the cloud and follows a K/r rule. Here, K is a positive constant dependent on the selected bubble radius and r is the radial coordinate from the center to the outside of the bubble cluster.

Numerical tests with 100.000 bubbles showed that the random bubble generator produces the desired discrete bubble distribution consistently with the analytic model. **Figure 1** shows a comparison of the distribution achieved by the random generator in comparison with the analytic model.

The generated cloud is embedded within a small computational domain of 35x35x40 mm³ (**see Fig. 2**) which itself is embedded in a larger domain of 4x4x2 m³. In the small domain a fine Cartesian grid is applied in order to minimize numerical errors.

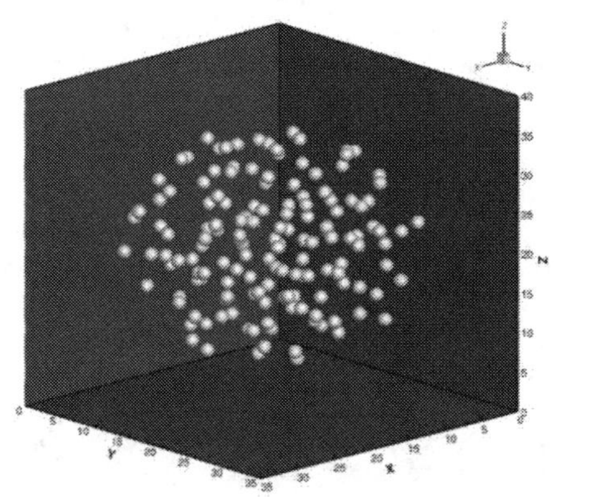

Fig. 2: Random distribution of spherical bubbles within the small computational domain

The large outer domain contains a hexahedral grid where grid cells are stretched towards the boundary of the domain in order to reduce computational costs. The boundaries of the large outer domain are inviscid adiabatic walls.

All bubbles are filled with water vapor while the surrounding domain contains liquid water at an initial pressure of p_∞ = 100 bar. The initial pressure inside the bubbles is equal to the vapor pressure p_{sat} = 2340 Pa at T = 293K. The velocity field is initially at rest.

The effects of surface tension, viscosity, gravity and non-condensable gas content are neglected.

2.3 Recorded simulation data

A numerically defined "pressure transducer" is used to record the static pressure on an area of 15x15 mm^2 at each time step. The numerical transducer is located directly below the bubble cluster at the center of the bottom wall. The numerical sampling frequency of the transducer is $49.3 \cdot 10^6$ Hz. Aside from the transducer at the wall, the maximum pressure within the entire flow field is monitored.

By monitoring the total vapor volume V_{vap} within the entire domain the temporal evolution of evaporation and condensation can be analysed. Instead of showing vapor volume versus time directly, it is more intuitive to present the vapor volume through an equivalent radius R_{equiv}. This is achieved by defining $V_{vap}: = 4/3 \cdot \pi \cdot R^3_{equiv}$.

Collapsing bubbles or vapor patterns arising from rebounds are tracked by application of a "Collapse detector approach" developed by *Mihatsch et al* [13]. This algorithm detects isolated collapses during the simulation and characterizes the strength of a collapse by recording the maximum pressure at the focal point. Additionally, the positions and time instants where collapses have been detected are stored. One application of this "Collapse detector" is to estimate the potential of cavitation erosion by analysing strong collapses close to a material surface. [13-15].

3 Problem setup

The aim of this investigation is to examine the influence of two parameters (cloud interaction parameter β and stand-off distance d) with respect to the collapse process and on maximum loads at the wall. Therefore, two setups are generated and simulated. For both setups and all computations conducted, the radius of the generated cloud is A_0 = 15mm.

The first setup contains three random clouds with different initial bubble radii of R_0 = 0.5 mm, 1.0 mm and 1.5 mm, respectively. As the number of bubbles equals 150 for all cases, the initial vapor volume fraction α_0 and the interaction parameter β depend on the initial bubble radius only. The initial vapor volume fraction is computed by $\alpha_0 = 150 \cdot V_{bubble}/V_{cloud}$ and the interaction parameter is β = 4, 8 and 11 respectively. As the distribution of bubbles is randomly generated for each configuration, the positions of individual bubbles are case dependent. The stand-off distance d = 3.2 mm of the cloud from the wall is kept constant.

The second setup is designed to investigate the effects of different stand-off distances to the wall. The clouds are identical in terms of bubble distribution and initial bubble radius, which is R_0 = 1 mm. The stand-off distance is defined as the minimum distance between the virtual surface of the cloud and the bottom wall. In the following, stand-off distances of d = 1.2 mm, 3.2 mm, 6.0 mm and 10.0 mm are applied.

Setup details are shown in the **Table 1** and visualized in **Figure 3**.

Table 1: Properties of the designed setups

Parameters	Setup I	Setup II
Bubble radius, R_0 [mm]	R_0 = 0.5, 1.0, 1.5	R_0 = 1.0
Vapor ratio, α_0 % [-]	α_0 = 0.4%, 3%, 8%	α_0 = 3%
Cloud interaction parameter, β [-]	β = 4, 8, 11	β = 8
Standoff distance, d [mm]	d = 3.2	d = 1.2, 3.2, 6.0, 10.0

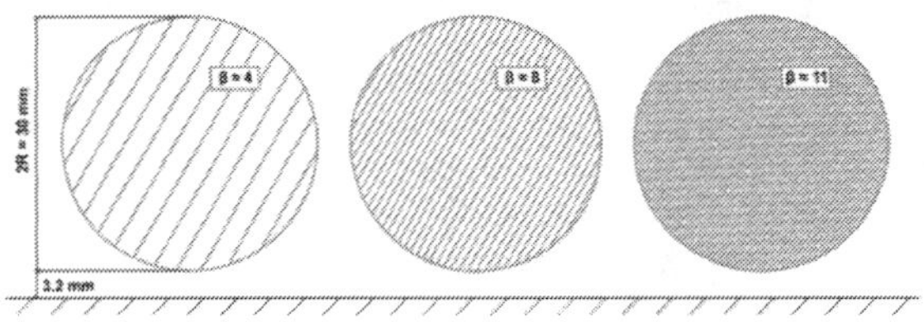

Fig. 3a: Schematic visualization of Setup I: different β

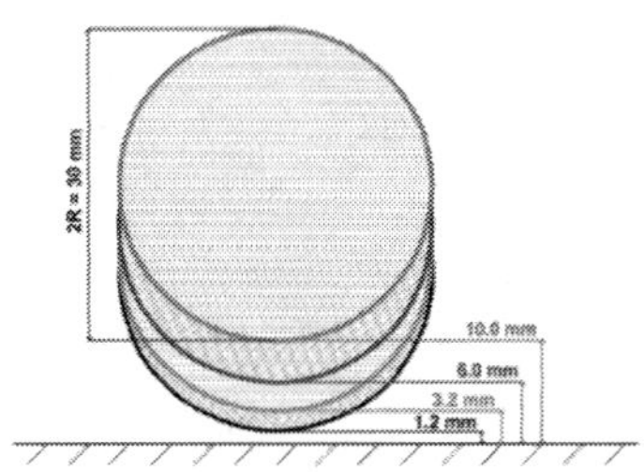

Fig. 3b: Schematic visualization of Setup II: different stand-off distances d

4 Results

This section describes major results where the prior interest is to analyse collapse mechanisms depending on the investigated configuration. The analysis contains the pressure field within the domain, the maximum load at the wall, and the temporal evolution of the amount of vapor. In particular, attention is paid to secondary clouds caused by rebounds and the location and intensity of their collapses.

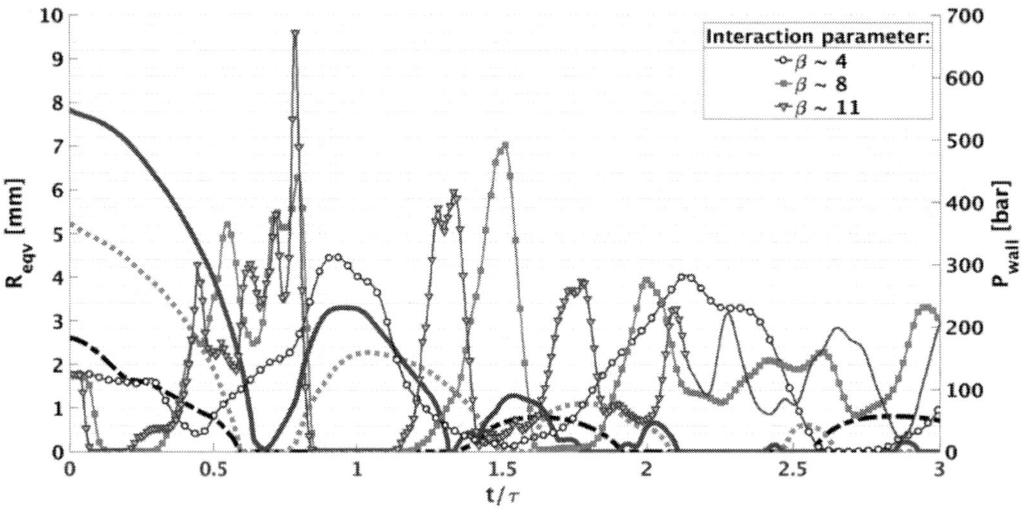

Fig. 4: Setup I: Thick line – equivalent bubble radius evolution; thin line - pressure recorded by the wall transducer; both against dimensionless time scaled with Rayleigh time for equivalent bubble

4.1 Setup I: Variation of interaction parameter β

An increase of the bubble radius leads to an increase of the cloud interaction parameter at constant cloud radius and a constant number of bubbles. Therefore, the variation of this quantity will affect the cloud collapse behaviour, e.g. the strength of bubble-bubble interaction, focussing effects towards the center, shielding effects [4], as well as the intensities of collapses of individual bubbles and of the focused collapse at the center of the bubble cluster.

Figure 4 shows the dimensionless temporal evolution of the equivalent radius R_{equiv} as well as of the output from the pressure transducer at the wall for all three interaction parameters. Note that the Rayleigh time τ [17] used to nondimensionalize the time axis has to be computed individually for each configuration based on the case-specific vapor volume in order to obtain a similarity of the temporal evolution of the equivalent radius. It can be seen that the relative duration of the cloud collapse is about 60% of the case-specific Rayleigh time for all three configurations. However, the subsequent re-evaporation (rebound) is quite different between the three cases. The smaller the interaction parameter, the more pronounced is the delay between collapse and rebound. This tendency seems to hold for subsequent collapses of re-evaporated structures as well. As a result, the number of rebounds during unit time (nondimensionalized with the Rayleigh time) increases with increased interaction parameter.

With respect to the evolution of the pressure recorded at the wall, the following trends are found. Weak interaction leads to a low frequency increase of the pressure. Higher interaction causes high frequency pressure oscillations and larger amplitudes (see **Fig. 4**). However, a comparison of the maximum pressure recorded by the numerical transducers subsequent to the collapse phases show an unexpected behaviour, as depicted in **Figure 5**. While the load at the wall decays more or less monotonically in case of strong interaction, it stays approximately constant in case of very weak interaction.

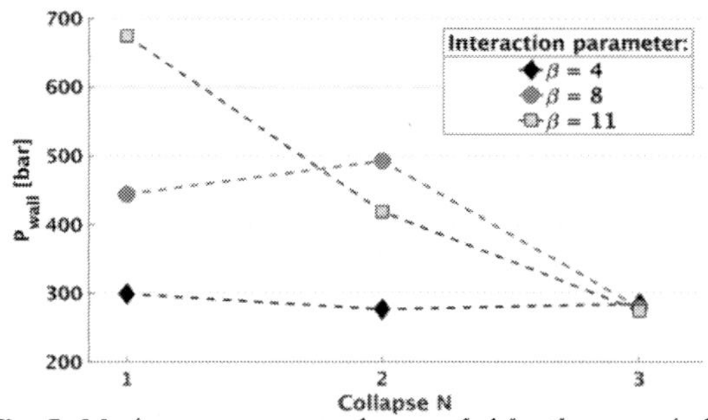

Fig. 5: Maximum pressure peaks recorded by the numerical wall pressure transducer subsequent to consecutive collapses

In case of intermediate interaction (here for β = 8) a nonlinear trend is predicted. In this case, the most intense load at the wall occurs after the secondary collapse

In the following, the physical details during the collapse and rebound phases are discussed for each configuration individually.

4.1.1 Weak interaction at β = 4

The first cloud contains the least amount of vapor of $\alpha_0 = 0.004$ and the initial bubble radius is $R_0 = 0.5\ mm$.

Figure 6 shows the output of the collapse detector. Size and color of a diamond indicate maximum pressure of a collapse and its position. The cloud starts to collapse from the outer layer where individual bubble collapses with intensities of 10^4 bar. Bubbles continue to collapse independently in the whole cloud volume without pronounced interaction. All collapses are of comparable strength and there is no focussing or shielding effect visible.

The first rebound is rather weak and does not result in a pronounced secondary cloud (**Fig. 7a**). The collapse of the original cloud initiates a velocity field towards the center of the cloud. As the lower part of the cloud is confined by the adjacent wall, the upward motion is slightly slower than the downward motion. Thereby, the center of the rebound structure is shifted towards the wall. Consequently, the collapse of this structure takes place closer to the wall as compared to the original cloud (**Fig. 7d**). The pressure peak reaches $1.1 \cdot 10^4$ bar. The previously described processes reappear, leading to a rebound structure that finally collapses directly at the wall (**Fig. 7g**). The intensity of this collapse reaches $0.4 \cdot 10^4$ bar.

In despite of the decrease in pressure at the focal points of the collapses, the respond at the wall remains more or less the same. This explains the nonlinear behavior of the recorded maximum pressure at the wall as already shown in **Figure 5**. The shift of the focal point compensates the weakening of the collapse intensity, leading to almost constant loads at the wall.

4.1.2 Intermediate interaction at β = 8

The second cloud contains an intermediate amount of vapor of α_0 = 0.03. The initial bubble radius is R_0 = 1.0 mm.

Compared to the previous case with weak interaction, the intensities of individual collapses increase up to $1.5 \cdot 10^4$ bar. Individual collapses appear mainly in the outer region of the cloud and a pronounced focusing of the collapse towards the center is detected. The output of the collapse detector (**Fig. 6b**) shows two strong collapses in the center region with intensities of 1.9 and $2.1 \cdot 10^4$ bar.

After the primary collapse the secondary cloud appears as a result of rarefaction waves which are reflected from the wall. Thereby, the pressure reduction is amplified by the wall and re-evaporation takes place. Most of the resulting vapour structures are in contact to the wall (**Fig. 7b**). The collapse of these vapour structures leads to a pressure peak of $2.2 \cdot 10^4$ bar (**Fig. 7e**). This kind of impact might be highly erosive.

The following rebound appears again at the wall (**Fig. 7h**) and the subsequent collapse exhibits a relatively moderate magnitude of $0.8 \cdot 10^4$ bar.

4.1.3 Strong interaction at β = 11

The cloud featuring the largest amount of vapour among the considered clouds exhibits an initial vapor volume α_0 = 0.08 and the initial bubble radius is R_0 = 1.5mm.

Strong focussing is observed where the intensities of individual collapses are successively increasing towards the center of the cloud (**Fig. 6c**). The pressure peak at the focal point reaches $2.6 \cdot 10^4$ bar. In this case bubble-bubble interaction and shielding is significant. A snapshot of rebounded single bubbles and a pronounced

a: $\beta = 4$

Most intense collapses in outer layer of a cloud (~ 10^4 bar). No intense collapses at the center. Very weak rebound close to the wall

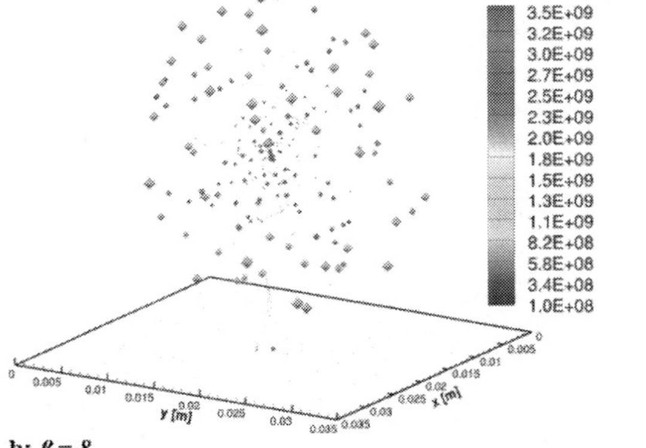

b: $\beta = 8$

Intensity of individual collapses increases (~ $1.5 \cdot 10^4$ bar). Focusing towards the center observed (~ 1.9 - $2.1 \cdot 10^4$ bar). Noticeable rebound at the wall with peak pressure ~ $2.2 \cdot 10^4$ bar. Second rebound collapses again at the wall (~ $0.8 \cdot 10^4$ bar)

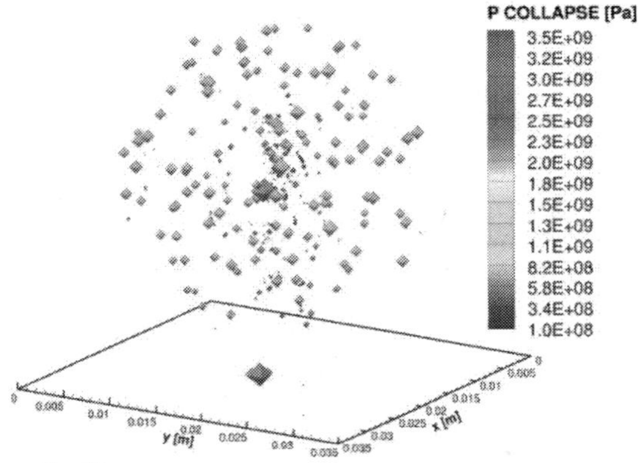

c: $\beta = 11$

Further increase of collapses intensities of individual bubbles. Intense focusing and high collapse pressure at the center (~ $2.6 \cdot 10^4$ bar). First rebound causes most severe collapse close to a wall (~ $3.5 \cdot 10^4$ bar). Further collapses occur rather individually above the wall

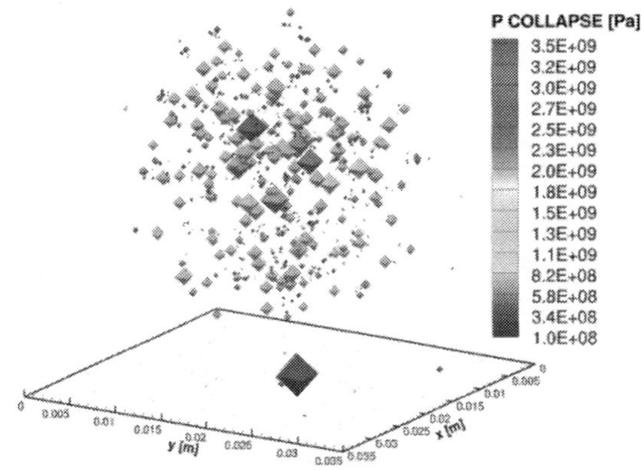

Fig. 6: Setup I, d = 3.2 mm. Collapse detector: collapses below1000 bar are not shown, size and color of diamonds represent collapse locations and predicted pressure magnitudes

 https://doi.org/10.1051/epjconf/201818002079

a = 0.4%, β = 4　　*a = 3%, β = 8*　　*a = 8%, β = 11*

a – c: Pressure field and vapor iso-surface after the prior collapse before the second. In the first case no pronounced cloud appears during rebound. In the other two cases vapor rebound occurs right at the wall, a secondary cloud in the center of the domain. Clouds collapse from top to bottom in both latter cases.

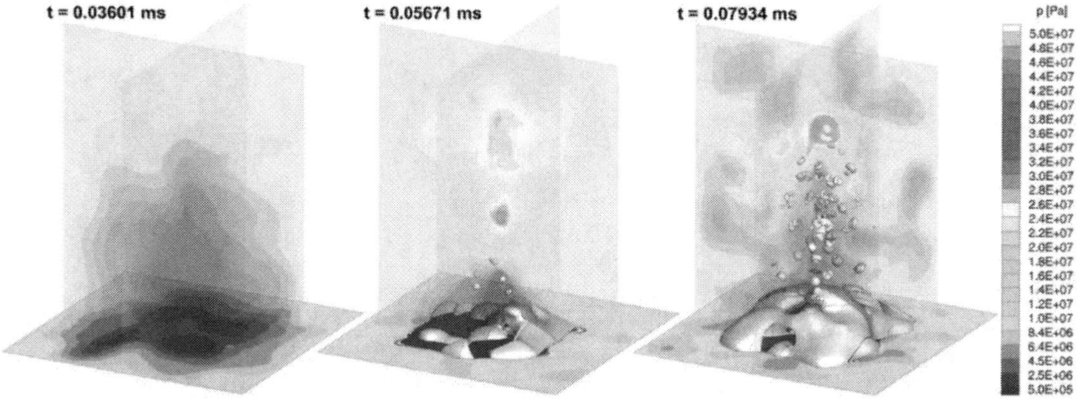

d – f: Pressure field and vapor iso-surface at the second collapse. In the first case relatively small collapse of $1.1 \cdot 10^4$ bar appears above the wall. In the middle case vapor at the wall collapses producing pressure peak of $1.78 \cdot 10^4$ bar. The last case shows the most violent collapse at the wall of $2.26 \cdot 10^4$ bar. This type of collapse is supposed to be highly relevant for cavitation erosion.

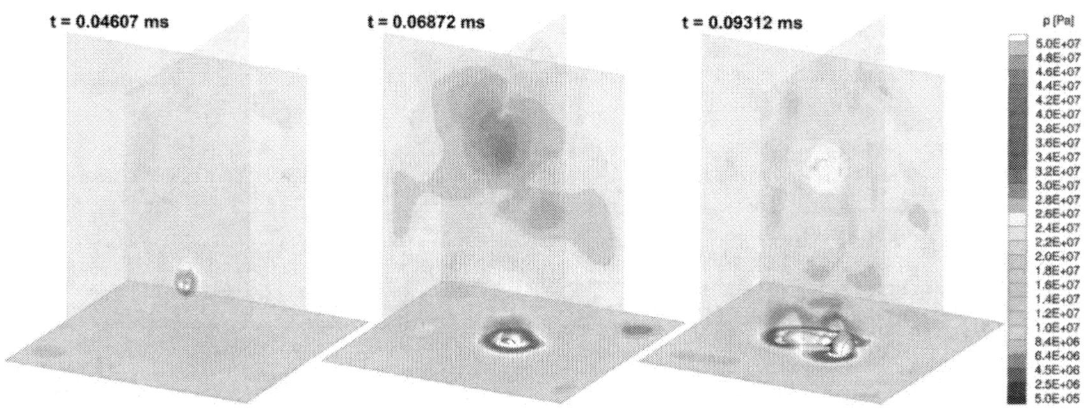

g – i: Pressure field and vapor iso-surface at the third collapse. In the first case the position of a collapse shifts further towards the wall and it occurs right at the wall with pressure magnitude $0.4 \cdot 10^4$ bar. In the middle case the third collapse of $0.8 \cdot 10^4$ bar is also at the wall. In opposite, the last case shows no significant impact on the wall since vapor structures are driven far above the wall and then collapse with $0.07 \cdot 10^4 \cdot$bar. Two latter clouds collapse from bottom to top.

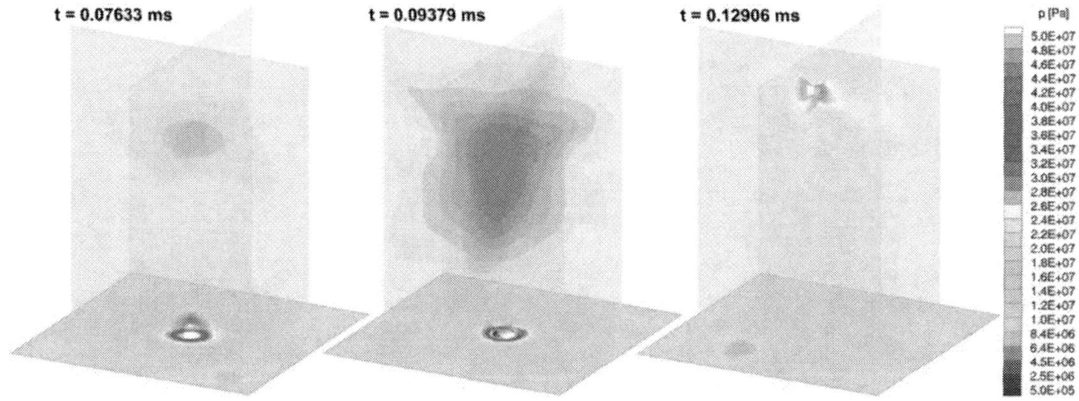

Fig. 7. Setup I: Pressure field depicted on two intersecting slices of the inner domain. White structures are iso-surfaces of vapor ratio equal to 5%. The initial vapor parameter and therefore interaction parameter increase from left to right .

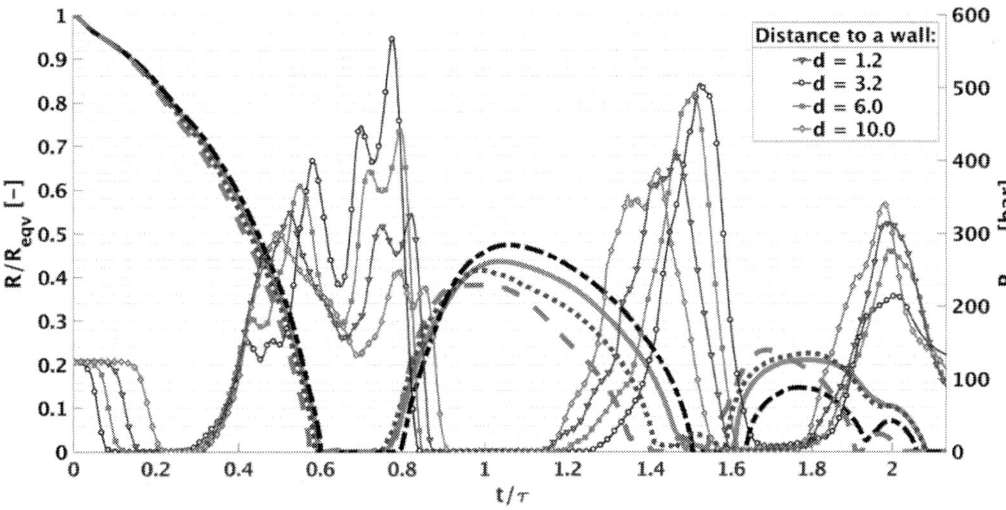

Fig. 9: Setup II: Thick line – equivalent bubble radius evolution; thin line - pressure recorded by the wall transducer; both against dimensionless time scaled with Rayleigh time for equivalent bubble

secondary cloud is presented in **Figure 8.**

It can be seen that bubbles exhibit a toroidal shape due to liquid jet impingement during the collapse. The collapse of the secondary structure is focussed towards the wall **(Fig. 7c).** Thereafter, the most severe collapse occurs at the wall and reaches an intensity of $3.5 \cdot 10^4$ bar **(Fig. 7f)**. The third cloud mainly consists of very weak vapor toroids. In this case, the onset of the collapse is located at the wall. The collapse front travels upwards without intensification until it terminates with a perceptible pressure peak of $0.07 \cdot 10^4 \cdot$bar **(Fig. 7i).** .

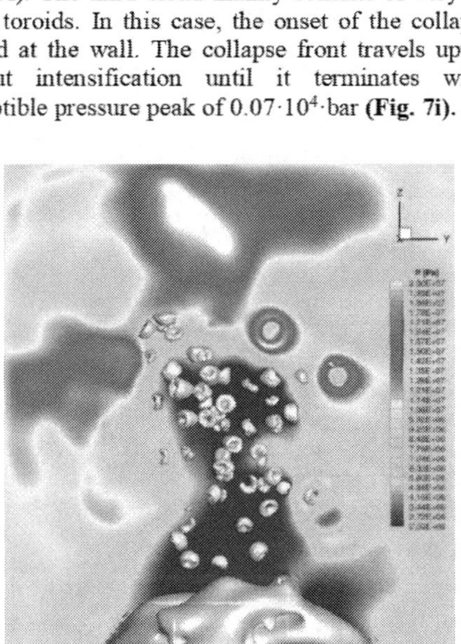

Fig. 8: Case β = 11, d = 3.2: Pressure and vapor structures between the primal and first rebound collapses, middle slice in y-z plane. Non-spherical focused collapse of individual bubbles and large vapor region at the wall. The bubble deformation is characterized by induced jets pointing towards the cloud center

4.2 Setup II: Variation of stand-off distance d

Setup II consists of four identical clouds regarding the initial size and position of bubbles, which are located at four different distances from the wall: d = 1.2, 3.2, 6.0 and 10.0 mm. The distance d is defined as the minimum space between the wall and the virtual surface of the cloud. The cloud interaction parameter is β = 8 for all cases.

4.2.1 Vapor volume evolution

Figure 9 shows the dimensionless temporal evolution of the equivalent radius R_{equiv} as well as of the output from the pressure transducer at the wall for all four stand-off distances. In contrast to the corresponding analysis for setup I, the Rayleigh time τ used to nondimensionalize the time axis is in this case identical for all four configurations. Again, the duration of the collapse of the primary cloud is approximately 60% of the Rayleigh time of an equivalent bubble for all cases. It can be seen that the duration of the primary collapse is slightly shorter when the stand-off distance is increased. The wall partially confines the flow field for small stand-off distances, while it has minor effects when the cloud is located sufficiently far away from it. As β = 8, moderate interaction intensity and collapse focussing occurs for all cases.

An unexpected behaviour is seen for the second and the third rebound of vapour structures. The maximum amount of vapour is inversely proportional to the initial stand-off distance during the first rebound. This is, however, not the case for the second rebound, where the tendency seems to be reversed.

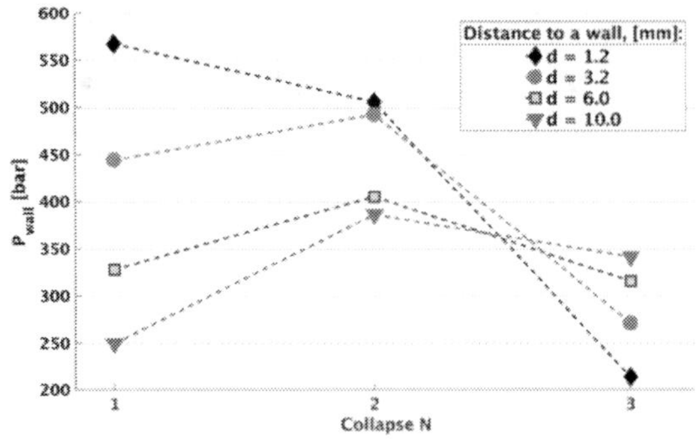

Fig. 10: Pressure peaks caused by primary collapse and two rebounds. Pressure of primary collapse is proportional to cloud distance to the wall. Intensity of second rebound collapse reverses intensity of primary collapse

4.2.2 Pressure on the wall

The output of the numerical pressure transducer at the wall shows that the pressure increases during the end of a collapse phase and features several maxima during collapse and rebound (**Fig. 9**). This observation holds for all four stand-off distances and all computed collapse-rebound-cycles.

Figure 10 shows maximum pressures recorded at the wall after three collapse phases for all investigated stand-off distances. As expected, the intensity of the primary collapse is inversely proportional to the initial stand-off distance. This trend can be explained by the location of the focal point of the collapse, which locates more closely to the wall if the stand-off distance is smaller. Interestingly, the pressures recorded after the secondary collapses and especially subsequent to the third collapses show a nonlinear behaviour: for a stand-off distance of d = 1.2 mm a decrease in pressure is predicted for subsequent collapses. In case of d = 3.2 mm to d = 10 mm the pressures recorded after the secondary collapses exceed those recorded after the primary collapses. Although the collapse intensities decrease when comparing collapses 2 and 3, an unexpected trend is found: the pressures recorded after the third collapse are now proportional to the initial stand-off distance.

In order to understand this behavior, the maximum vapor volume during rebound and the position of the focal point of the subsequent collapse are investigated. The maximum vapour volume (or the radius of the equivalent bubble) is reversed in order when comparing rebound 1 to rebound 2. Additionally, the focal points are shifted towards the wall in all cases. Both contributions affect the pressure recorded at the wall. However, the maximum pressure at the focal point of a (focussed) cloud collapse is not necessarily related to the pressure recorded at the wall. For example, in case of d = 10 mm, the pressure detected at the focal point is much higher during the primary collapse compared to the subsequent collapses.

5 Discussion

5.1 Impact on the wall

The "numerical pressure transducer" is used in order to obtain the force acting on the sensor area of 15 x 15 mm^2. Since this area is relatively large compared to the size of individual bubbles, pressure signals caused by single bubble collapses are blurred over the sensor area. Although this averaging effect of a large sensor area is insufficient to identify the intensity of strong collapses that occur locally, it allows for analysis of the overall temporal evolution of the pressure at the wall. As an example, **Figure 6c** shows a strong collapse right at the wall, while the pressure transducer (**Fig. 5, blue line**) records just a medium peak, which is orders in magnitude lower than the value at the focal point. Thus, it is important to analyse simultaneously outputs of various detectors to obtain sufficient insight into the collapse dynamics.

However, the data obtained by the collapse detector can be directly used to investigate collapse focussing.

For setup II the identical initial bubble configuration is applied for all four cases, which allows to compare directly the output of the numerical transducer at the wall. For the same reason the collapse detector would show nearly identical results among different stand-off distances and is not used for comparison. This is, however, strictly true for the primary collapses, only.

5.2 Pressure magnitude comparison

The pressure peaks captured by the collapse detector are grid dependent as long as the spatial resolution is not sufficient to capture the smallest relevant physical length scales [3]. Therefore, the output of the collapse detector can be used for a relative comparison of different cases. *Mihatsch et al* [13] showed that a scaling procedure can be applied to compensate resolution dependencies but in order to perform this scaling, experimental data is required. Due to the lack of experimental validation possibilities for the cases investigated in this study, a scaling of the data obtained by the collapse detector is not yet possible. Therefore, the pressure magnitudes reported in this paper, when obtained with the collapse detector, are a measure for relative comparison between cases and setups but not necessarily absolute values.

5.3 Pressure initialization process

For this work, the initial pressure distribution was simplified in the following way. The pressure inside the bubbles is the saturation pressure, while the pressure within the whole domain filled with liquid was initialized with the value of 100 bar. This setup enables a straight forward reproducibility for other research groups who might be interested in reproducing the cases presented in this paper. However, from a physical point of view, it would be more realistic to assume that the Laplacian of the initial pressure field is zero but without featuring a jump in pressure at the surfaces of the bubbles Such initialization is described by *Schmidt et al* [3] and will be used for upcoming investigations as well.

6 Summary and conclusions

This study shows the effects of the cloud interaction parameter and of the stand-off distance on the collapse of a cloud of vapour bubbles in a liquid ambient. Maximum loads at the surface are recorded by application of a numerical force/pressure transducer. Collapse focussing is analysed by application of a procedure which detects isolated collapses (collapse detection algorithm by *Mihatsch at al* [13]). Two series of numerical simulations with in total seven cases were performed and analysed.

The major findings are:

1) The cloud interaction parameter proposed by Brennen [4] is a suitable measure for bubble interactions and shock focussing. From this investigation it can be concluded that an interaction parameter of O(1) or smaller results in weak interaction without shock/collapse focussing. Once the interaction parameter is O(10) or higher, strong interaction and pronounced focussing are to be expected. For partial cavities [17] a large interaction parameter can be estimated by visual inspection of clouds observed in experiments. Thus, collapse focussing might be a very relevant physical process in cavitating flows occurring in hydraulic machinery. Increasing the interaction parameter β leads to an enhancement of geometrical focusing and of the collapse pressure of individual bubbles. Additionally, the intensification of vapor production close to the wall during rebound is observed and, consequently, an increase of maximum loads at the wall is detected.

2) The stand-off distance is well suited to quantify the intensity of the primary cloud collapse with respect to its erosive potential at an adjacent material surface. However, this investigation shows that secondary and even third collapses subsequent to rebound processes can cause intense loads at material surfaces, too. Therefore, primary collapses and collapses of rebounded vapor structures are both potentially relevant for erosion. The underlying physical mechanism is purely inertia controlled. The primary collapse induces a velocity field which targets towards the wall and rebounding vapour patters are advected with this velocity field. Secondary and tertiary collapses occur closer and closer to the wall and the resulting load at the wall increases as well.

Further studies of this work will include non-homogeneous initial bubble radii in order to assess effects of micro-scale details aside from macro-scale quantities such as the interaction parameter. Further numerical developments will be the inclusions of viscous effects and non-condensable gas content. The investigation of collapse processes using non-Newtonian liquids is currently under development.

7 Acknowledgment

This study is part of a project that has received funding from the European Union's Horizon 2020 research and innovation programme under grant agreement № 642536.

References

1. Lauer E., Hu X. Y., Hickel S., Adams N. A., Comput. Fluids **69**, 1–19 (2012)
2. Fujikawa S. and Akamatsu T., J. Fluid Mech. **97**, 481–512 (1980).
3. Schmidt S.J., Mihatsch M., Thalhamer M., Adams N.A., *WIMRC Cavitation Forum*, Warwick, UK, (2011).
4. F Wang Y.-C., Brennen C. E., J. Fluids Eng **121**(4), (1999)
5. Rossinelli D., Hejazialhosseini B., Hadjidoukas P., Bekas C., Curioni A., Bertsch A., Futral S., Schmidt S. J., Adams N. A. and Koumoutsakos P, *International Conference on High Performance Computing, Networking, Storage and Analysis* (SC '13. ACM, NY, USA, 2013).
6. Brennen C. E., *Cavitation and Bubble Dynamics*, Cambridge University Press, (1995).
7. d'Agostino L., Brennen C. E., *AESME Cavitation and Multiphase Flow Forum*, 72-75, (1983)
8. d'Agostino L., Brennen C. E., J. Acous. Soc. Am. **84**, 2126-2134, (1999).
9. Rudman M., Int. J. Num. Meth. Fluids, **24**, 671-691, (1997).
10. Hirt C. W., Nichols B. D., J. Comp Phys **39**, 201-225, (1981).
11. Sezal I. H., *Compressible Dynamics of Cavitating 3-D Multi-Phase Flows*, PhD Thesis, TUM (2009).
12. Egerer C. P., Schmidt S. J., Hickel S., Adams N. A., J. Comp Phys, (2016).
13. Mihatsch M., Schmidt S.J., Thalhamer M., Adams N.A., *8th International Symposium on Cavitation*, CAV2012 (2012).
14. Budich B., Schmidt S. J., Adams N. A., *31st Symposium on Naval Hydrodynamics*, 12 Sept (2016).
15. Budich B., Schmidt S. J., Adams N. A., *31st Symposium on Naval Hydrodynamics*, 15 Sept (2016).
16. Beban B., Legat S., Schmidt S.J., Adams N. A., *9th International Symposium on Cavitation*, CAV2015, J. Phys **656** (2015)
17. Franc J.-P., Michel J.-M, *Fundamentals of Cavitation*, Kluwer Academic Publishers, (2004).

Flow Behavior in Side-View Plane of Pitching Delta Wing

Mehmet Can Pektas[1], *Mehmet Oguz* Tasci[1], *Ilyas* Karasu[2], *Besir* Sahin[1,*] and *Huseyin* Akilli[1]

[1] Department of Mechanical Engineering, Faculty of Engineering and Architecture, Cukurova University, 01330, Adana/ Turkey

[2] Departments of Aerospace Engineering, Faculty of Aeronautics and Aerospace, Gaziantep University, 27310, Gaziantep/Turkey

Abstract. In the present investigation, a delta wing which has 70° sweep angle, Λ was oscillated on its mid-cord according to the equation of $\alpha(t)=\alpha_m+\alpha_o\sin(\omega_e t)$. This study focused on understanding the effect of pitching and characterizing the interaction of vortex breakdown with oscillating leading edges under different yaw angles, β over a slender delta wing. The value of mean angle of attack, α_m was taken as 25°. The yaw angle, β was varied with an interval of 4° over the range of $0°\leq\beta\leq 16°$. The delta wing was sinusoidally pitched within the range of period of time $5s\leq T_e\leq 60s$ and reduced frequency was set as K=0.16, 0.25, 0.49, 1.96 and lastly amplitude of pitching motion was arranged as $\alpha_o=\pm5°$.Formations and locations of vortex breakdown were investigated by using the dye visualization technique in side view plane.

1 Introduction

As a result of past researches, delta wings are main element of combat aircrafts. The control of vortex dynamics, unsteady flow characteristic and physics are also considerably important to improve the maneuverability of air vehicles [1]. Vortex breakdown can be described as a vortex with a bursting point located between apex and trailing edge of the delta wing [2].

Vortex breakdown is one of the most important parameter on the delta wing surface. [3] showed that bubble and spiral type vortex breakdown and [4] also stated that double helix vortex breakdown can be visualized in a diverging cylindrical tube as shown in Figure 1.

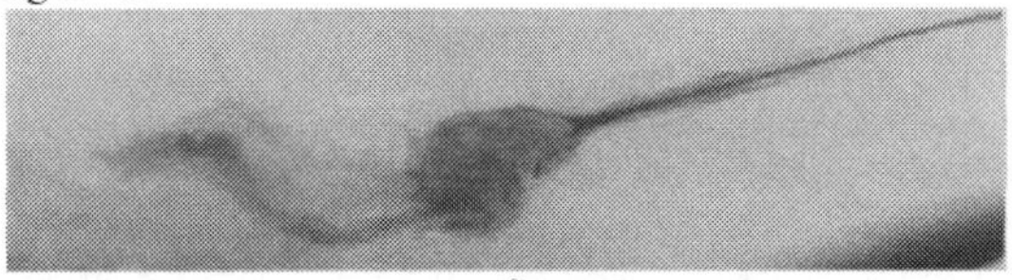

a)

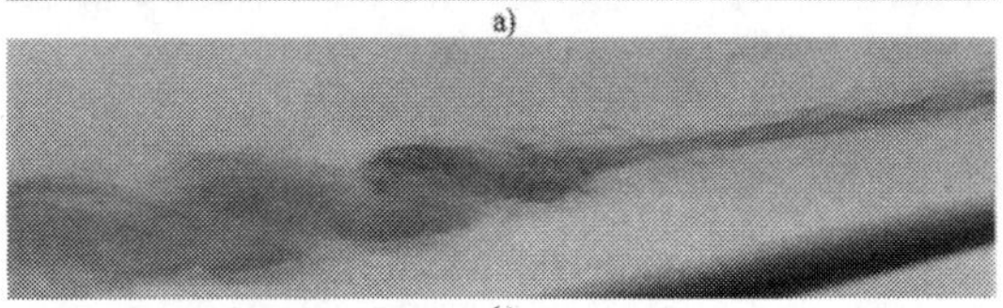

b)

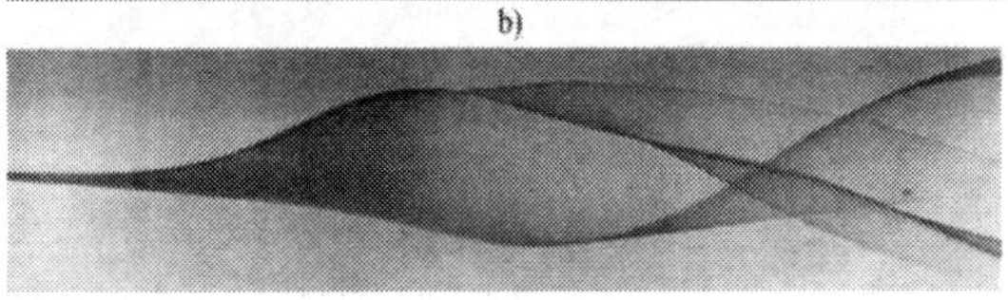

c)

Fig. 1. Types of vortex breakdown a) Bubble type vortex breakdown b) Spiral type vortex breakdown [3] c) Double helix vortex breakdown [4]

[5] examined flow characteristics over the lambda wing under variation of angle of attack within the range of $7°\leq\alpha\leq17°$ using dye visualization and the stereo PIV technique. They proved that flow structure and development of vortex breakdown profoundly are affected by angle of attack, α.

[6] demonstrated that structures of vortex bursting and vortical flow over a non-slender diamond wing are sensitively effected by yaw angle, β using dye visualization and the PIV technique and time averaged flow data over the wing surface changes profoundly increasing over the yaw angle, β=6°.

[10] investigated vortical flow structures over the pitching delta wing having 75° sweep angle, Λ using the PIV technique and comparing to vortical flow development of stationary and oscillating wing with and without the impingement plate as shown in Figure 2.

[7] studied structures of leading edge vortices in side view planes as shown in Figure2. He showed that yaw angle, β influences flow structures in side view plane substantially. When yaw angle, β is increased to the value of 20°, well-defined Kelvin-Helmholtz (K-H) instabilities appear on both sides of leading edge vortex before and after onset of vortex breakdown clearly. [8] defined flow separation and integration on the surface of wing via numerical methods for the delta wings having sweep angles, Λ of 65° and 76°/40°.

[9] experimentally investigated the flow structure and turbulence statistics of non-slender delta and diamond wings having sweep angle, of Λ=40°. As shown in Figure 3, wings are subjected to perturbation with the amplitude of $\alpha_o=\pm0.5°$ and oscillating period of T_e=0.5s. They performed experiments using dye and the stereo PIV technique to observe vertical flow structures in plan-view plane by varying angles of attack, α at within the range of $7°\leq\alpha\leq17°$. Patterns of time averaged vorticity and contours of transverse velocity components of the

* Corresponding author: bsahin@cu.edu.tr

perturbed delta wing are compared with the case of a stationary delta wing for angle of attack, α of 10°. It was revealed that perturbation alters the vertical flow structures even close to the surface of the delta wing. Similar works are also available in the literature such as [11, 12, 13, 14, 15, 16, and 17].

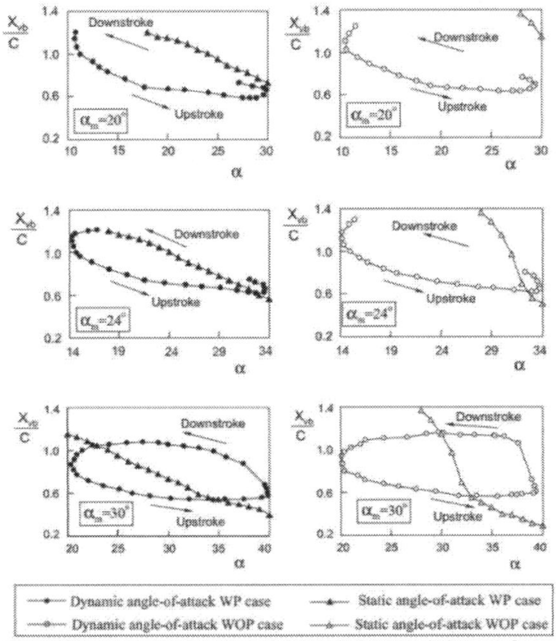

Fig. 2. Comparison of static and dynamic hysteresis loops of onset of vortex breakdown versus attack angle, α. Oscillating period, T_e=22.54 s and reduced frequency, K=0.74 and amplitude of pitching motion, α_0=10° [10]

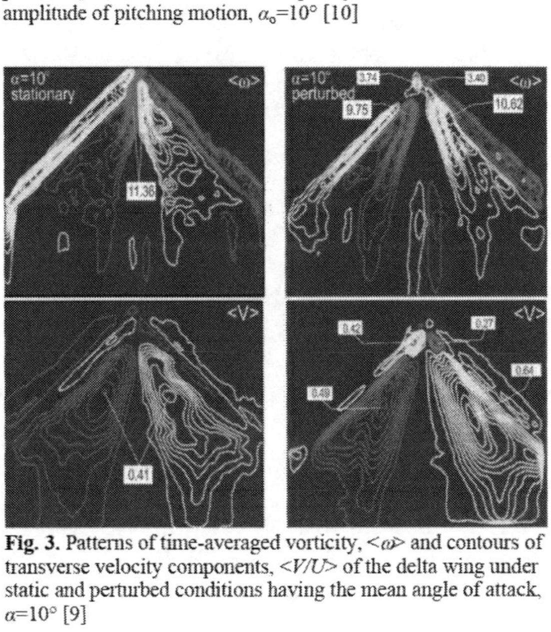

Fig. 3. Patterns of time-averaged vorticity, <ω> and contours of transverse velocity components, <V/U> of the delta wing under static and perturbed conditions having the mean angle of attack, α=10° [9]

2 Material and Method

Experiments on the hydrodynamics delta wings were performed in the free surface water channel having length of 8000 mm, width of 1000 mm and height of 750 mm. The channel was made from 15 mm thick Plexiglas. Water channel system has 2 tanks; one of them is situated at inlet and the other one at outlet of the channel as shown in Figure 4. Two sets of honeycombs were placed in water channel before the test region where contraction of 2:1 is present to regulate the upstream flow. This design of channel with a double honeycombs provide turbulence intensity lower 0.5 %. Circulation of water was provided with a water pump having frequency control unit and power of 15 kW.

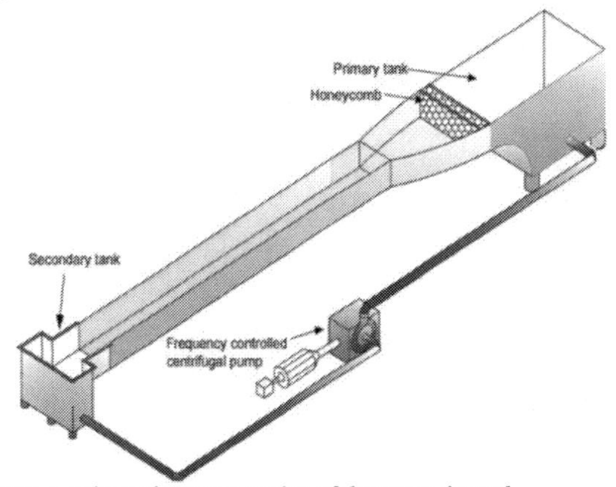

Fig. 4. Schematic representation of the water channel

The dye visualization technique is defined as the method which gives preliminary or qualitative information about flow structures over the delta wing. In the present study, dye experiments were performed by using fluorescent dye which changes color and shines under the laser sheet. Fluorescent dye is arranged by mixing powder of Rodamine 6G with water. Dye was injected in the test region by taking the advantage of fluorescent dye reservoir height. Furthermore, continuous laser source was used during the dye observations. Dye experiments were carried out in side view plane as seen in Figure 5. Laser sheet was set by passing through the center of leading edge vortex and laser sheet is perpendicular to the plane of the delta wing. Flow structure was observed by ejecting Rodamine 6G dye which was illuminated by laser sheet in the test region. A small size needle and plastic tube were used to eject dye. Images of hydrodynamic of flow structures were record with a SONY HD-SR1 video camera and Sony Play Memories Software was used to capture the picture of instantaneous flow data.

Figure 6 shows pitch up, pitch down motion and location of laser sheet in side and plan-view planes schematically.

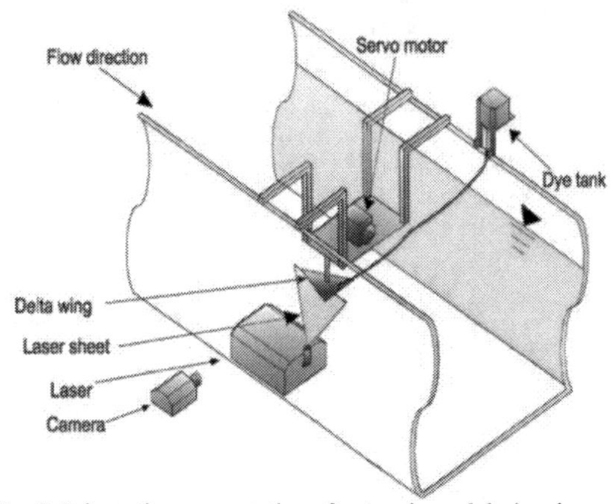

Fig. 5. Schematic representation of water channel during dye experiment in side view plane

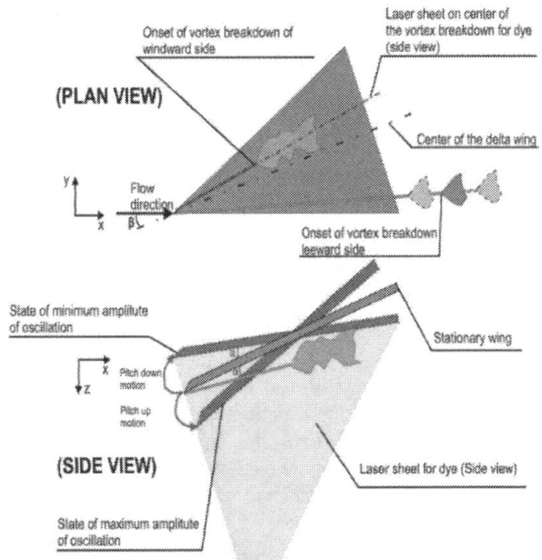

Fig. 6. Schematic representation of pitching motion of delta wing and experimental set up for dye experiment

3 Result and Discussion

In dye visualization, the effect of yaw angle, β, period of oscillation, T_e, mean angle of attack, α_m amplitude of the pitching motion, α_o and maximum minimum positions of vortex breakdowns are clearly seen. Vortex breakdown locations with maximum and minimum lengths are defined as maximum-minimum distances which mean that the location of vortex breakdown changes arbitrarily when delta wing is fixed at any angles of attack, α. These minimum and maximum lengths of pitching delta wing were also defined during the experiment. Figure 7 shows that vortex breakdown location in pitching delta wing motion for the maximum value of amplitude of the pitching motion, α_o=+5° at a mean angle of attack, α_m=25° does not proceed in accordance with pitching motion. Location of vortex breakdown moves towards the trailing edge of delta wing at the state of maximum dynamic angle of attack, α(t)=30°.

On the other hand, dimensionless length between vortex breakdown location and leading edge of delta wing, X_{vb}/C relatively increases in the case of pitching delta wing comparing to the stationary delta wing case. Here, X_{vb} is the length between vortex breakdown location leading edge of delta wing and C is the chord length. Onset of vortex break down occurs at about X_{vb}=0,75C at a static delta wing case, for angle of attack of α=25+5=30° and yaw angle of β=0° depending on the oscillation period, T_e. It is clearly seen that onset of vortex breakdown moves towards apex of delta wing by providing yaw angle, β and increasing mean angle of attack, α_m=25°, dimensionless number of cord length, X_{vb}/C regularly are decreased.

Also, effect of oscillation period on the vortex breakdown location, X_{vb}/C is seen in figures. Thus, it is understood that varying yaw angle, β and period of oscillation, T_e, effects the variation of vortex breakdown locations.

As specified in the above Figure 7, location of vortex breakdown is shown for maximum value of amplitude of the pitching motion, α_o=+5° or the end of the portion of cycle in which the angle of attack, α increases which is stated as upstroke. In addition, it is seen that location of vortex breakdown, X_{vb}/C moves outside of the delta wing in minimum value of amplitude of the pitching motion, α_o=-5° or the end of the portion of cycle in which the angle of attack, α decreases which is denoted as downstroke.

To correctly compare stationary and dynamic cases of delta wing, the value of angle of attack, α for stationary delta wing cases in dye images is similar with the minimum value of dynamic angle of attack of pitching delta wing such as α(t)=25-5=20°. Formation of onset of vortex breakdown in the cases of stationary and oscillating delta wings under variation of yaw angle, β within the range of 0°≤β≤16°, period of oscillation, T_e ranging from 5s to 60s at mean angle of attack, α_m of 25°, amplitude of delta wing pitching motion, α_o of ±5° are presented in Figure 8. Images of onset of vortex breakdown are presented for stationary delta wing for angle of attack, α of 25° under variation of yaw angle, β within the range of 0°≤β≤16°. But the rest of images presented in Figure 8 are visualized at a dynamic angle of attack, α(t) of 20° for periods of oscillation, T_e ranging from 5s to 60s. Images for oscillating delta wing at a dynamic angle of attack, α(t) of 20° (down-stroke) shown in the first column of Figure 8 indicate that excluding the results of zero yaw angle, β and 5s period onset of vortex breakdown occurs over the surface of delta wing for 4°≤β≤16° and 20s≤T_e≤60s. Examining all images for pitching delta wing reveals that the time lags between angles of attack, α of delta wing and onset of vortex breakdown is higher. But this period of phase shift get smaller for a high time period, for example for the cases of 60s time period, T_e the time lag between angles of attack, α of delta wing and onset of vortex breakdown becomes zero. This conclusion is derived from the comparisons of images presented in the first and last line of Figure 8. Considering mean angle of attack, α_m and amplitude of the pitching motion, α_o for dynamic case vortex breakdown location moved away from the rear of the delta wing to the location of X_{vb}=1.2*C*. As dye experiments demonstrate in the cases of amplitude of the pitching motion, α_o=+5° (upstroke) and α_o=-5° (downstroke), it was determined that effect of yaw angle, β in the case of low period of oscillation, T_e in the motion of pitching delta wing in down stroke direction. Figure 8 shows that the lower period of oscillation, T_e causes a high rate of phase differences. But the level of phase shift gets smaller for a higher time period, T_e for example, in the case of 60s time period, the time lag between angles of attack, α of delta wing and onset of vortex breakdown becomes zero for α_m=25°.

Hysteresis presented in Figure 9 to 11 demonstrates the record of locations of vortex breakdown, X_{vb}/C, during pitching motion of delta wing in a closed loop. When the delta wing under perturbation with a high amplitude, α_m and time period, T_e a time delay takes place between the position of delta wing and locations of onset of vortex breakdown comparing to the stationary delta wing. As it is shown in Figure 9, when the mean angle of attack is, α_m=25°, amplitude of pitching motion is, α_o=±5° and yaw angle is, β=0°, at the period of oscillation, T_e =5s, there is a discrepancy between static and oscillating wing cases caused by the dimensionless reduced frequency, *K*. When period of oscillation, T_e, increases so the dimensionless reduced frequency, *K*, decreases, the large discrepancies between the locations of vortex breakdown, X_{vb}/C and dynamics angle of attack, α(t) disappear. The minimum dimensionless length of X_{vb}/C can be seen at a dynamic attack angle of α(t)=20°. At T_e=60s, the minimum dimensionless length of X_{vb}/C occurs at the dynamic angle of attack of α(t)=27°. However, we can conclude that the time delay decreases as the period of oscillation, T_e increases. As seen, comparison of varied yaw angles, β is clearly shown in figures. During this pitching of delta wing dynamics of angles of attack changes from higher angles of attack, α and lower angles of attack, α. In both extreme cases there is leading edge vortex bursting with time delay. If we think of this unstable effect as an advantage, high performance aircraft can make certain maneuvers faster and more efficiently [11].

Whereas red line represents the stationary delta wing, blue line represents the oscillating delta wing.

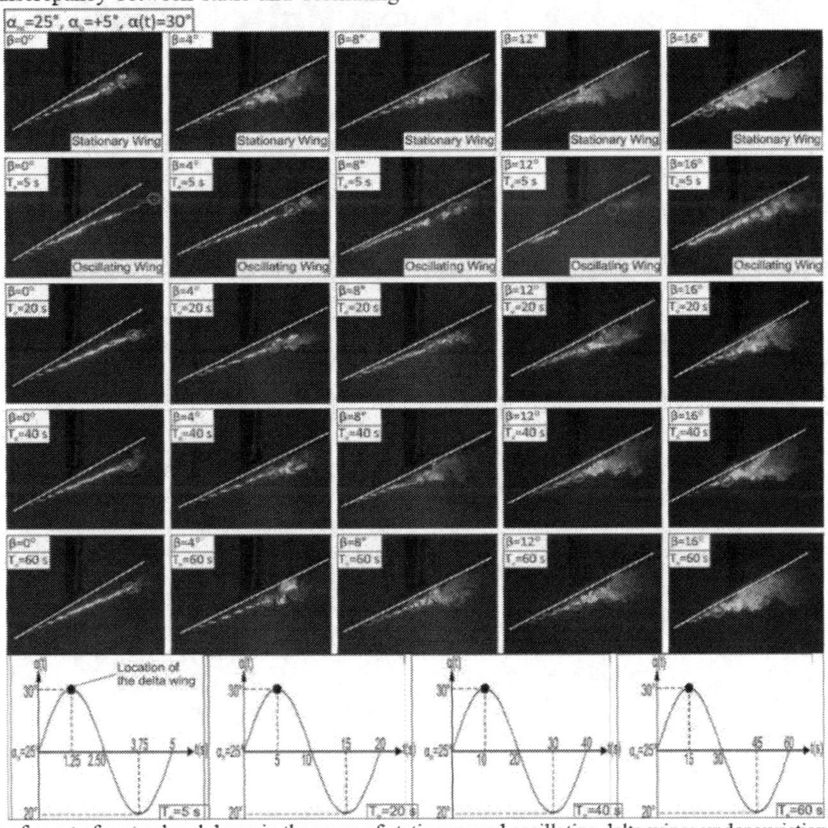

Fig. 7. Formation of onset of vortex breakdown in the cases of stationary and oscillating delta wings under variation of yaw angle, β within the range of 0°≤β≤16°, period of oscillation, T_e ranging from 5s to 60s at mean angle of attack, α_m of 25°, amplitude of delta wing pitching motion, α_o of +5° (upstroke) and dynamic angles of attack, α(t) vary between 20° and 30°. All images are taken at angle of attack, α=30°

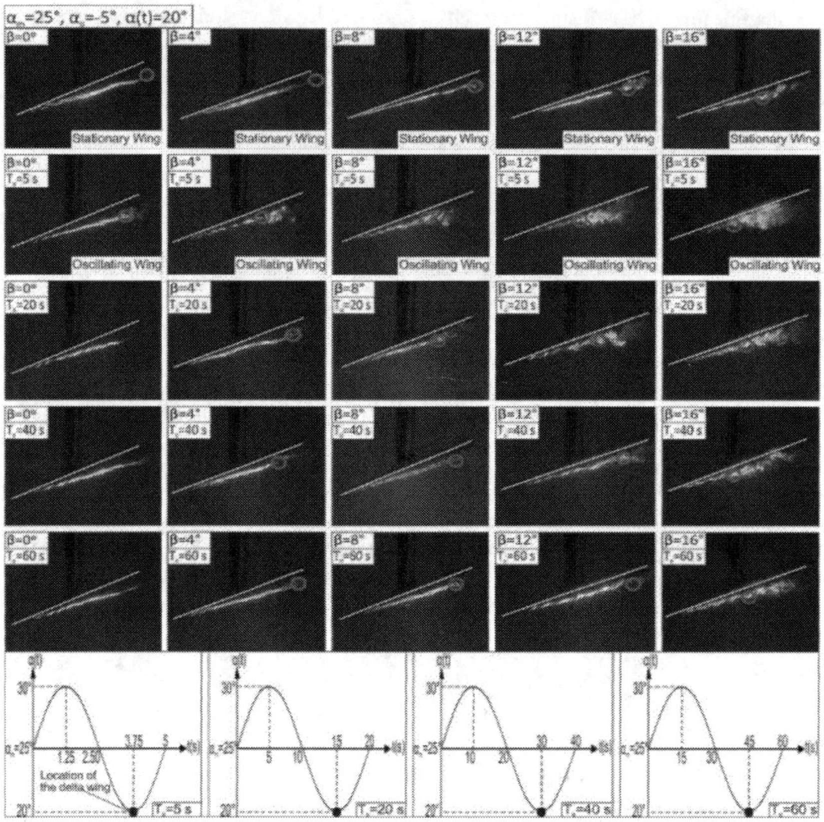

Fig. 8. Formation of onset of vortex breakdown in the cases of stationary and oscillating delta wings under variation of yaw angle, β within the range of $0° \leq \beta \leq 16°$, period of oscillation, T_e ranging from 5s to 60s at mean angle of attack, α_m of 25°, amplitude of delta wing pitching motion, α_o of -5° (downstroke) and dynamic angles of attack, $\alpha(t)$ vary between 20° and 30°. All images are taken at angle of attack, α=20°

4 Conclusions

Breakdown position of leading edge vortex on the windward side moves towards to the apex of the delta wing and vorticity concentrations spreads over majority of the delta wing surface. But the other side of leading edge vortex moves in forward direction without bursting underneath the delta wing for a higher yaw angle, β. Dye visualizations show that with increasing angle of attack, α and yaw angle, β, a strong Kelvin-Helmholtz instabilities form and leading edge vortices breakdown earlier causing a large scale vorticity concentrations and these vorticity concentrations interact with the surface of the delta wing leading to unsteady loading or buffeting the delta wing surface on the windward side of the wing. Increasing the amplitude of the delta wing pitching motion, α_o causes alterations in locations of vortex breakdowns for angles of attack, α. The locations of vortex breakdown travels towards the apex of the delta wing more closer compared with the amplitude of α_o=+5°. When the delta wing is under pitching motion in upstroke or downstroke directions, a time delay between dynamic angles of attack, $\alpha(t)$ and dimensionless length of X_{vb}/C, that is to say, a delay of vortex breakdown formation occurs in both high and low dynamic angles of attack, $\alpha(t)$. In summary, small yaw angle, β is more effective in altering the location of vortex breakdown in both cases either the stationary or oscillating delta wing. Leading edge vortex breakdown can be controlled partially using active or passive control techniques in order to delay onset of vortex breakdown to improve hydrodynamic performance and decline unsteady loading effects on the surface of delta wing.

In conclusions, an experimental work should be performed to control the windward side of leading edge vortex breakdown either to delay leading edge vortex bursting process or to induce onset of leading edge vortex breakdown at an earlier stage.

Acknowledgments

This study was supported by The Scientific and Technological Research Council of Turkey (TÜBİTAK) under contract number of 114M497 and Cukurova University Scientific Research Unit under contract number of FLY-2016-6283.

References

1. Heron, I., & Myose, R. Y. On the impingement of a von kàrmàn vortex street on a delta wing. *Collection of Technical Papers - AIAA 22nd Applied Aerodynamics Conference*, 1 178-188 (2004).

2. Su, W., Liu, M., & Liu, Z. Topological structures of separated flows about a series of sharp-edged delta wings at angles of attack up to 90°. *Topological Fluid Mechanics, edited by H. K. Moffatt, and A. Tsinober, Cambridge Univ. Press*, New York (1990).
3. NASA, (2016). Jet Exit Facility-Bubble Type Vortex Breakdown and Spiral Type Vortex Breakdown. https://crgis.ndc.nasa.gov/historic/Jet_Exit_Facility
4. Sarpkaya, T. On stationary and travelling vortex breakdowns. *Journal of Fluid Mechanics*, 45(3), 545-559. doi:10.1017/S0022112071000181 (1971).
5. Yayla, S., Canpolat, C., Sahin, B., & Akilli, H. The effect of angle of attack on the flow structure over the nonslender lambda wing. *Aerospace Science and Technology*, 28 (1), 417-430. doi:10.1016/j.ast.2012.12.007 (2013).
6. Sahin, B., Yayla, S., Canpolat, C., & Akilli, H. Flow structure over the yawed nonslender diamond wing. *Aerospace Science and Technology*, 23(1), 108-119. doi:10.1016/j.ast.2011.06.008 (2012).
7. Karasu, I. Effect of yaw angle on vortex formation over a slender delta wing. Çukurova University Institute of Natural and Applied Sciences, Ph. D. Thesis, Adana, 159p (2015).
8. Woodiga, S., Liu, T., Ramasamy, R. S. V., & Kode, S. K. Effects of pitch, yaw, and roll on delta wing skin friction topology. Proceedings of the Institution of Mechanical Engineers, Part G: *Journal of Aerospace Engineering*, 230(4), 639-652. doi:10.1177/0954410015594823 (2016).
9. Canpolat, C., Sahin, B., Yayla, S., & Akilli, H. Effects of perturbation on the flow over nonslender delta wings. *45th AIAA Fluid Dynamics Conference*, 1-10 (2015)
10. Ozgoren, M., & Sahin, B. (2002). Effect of pitching delta wing on vortex structures with and without impingement plate. *Turkish J. Eng. Env. Sci.*, 26, 325- 343.
11. LeMay, S. P., Batill, S. M., & Nelson, R. C. (1990). Vortex dynamics on a pitching delta wing. *Journal of Aircraft*, 27(2), 131-138. doi:10.2514/3.45908
12. Canpolat C., Yayla S., Sahin B., Akilli H., “Effects of Trailing-Edge Attachment on the Flow Structure over a Generic Delta Wing” Journal of Aerospace Engineering 30, 06017003, 2017
13. Yayla S., Canpolat C., Sahin B., Akilli H.,” Yaw Angle Effect on Flow Structure over the Nonslender Diamond Wing” AIAA JOURNAL, Vol.48, No.10,pp2457-2461,2010
14. Canpolat C., Yayla S., Sahin B., “Observation of the Vortical Flow over a Yawed Delta Wing”, Journal of Aerospace Engineering, Vol. 25, Issue. 4, pp. 613-626, 2012
15. Canpolat C., Yayla S., Sahin B., Akıllı H., "Dye Visualization of the Flow Structure over a Yawed Nonslender Delta Wing" Journal of Aircraft, Vol. 46, No.5, September-October 2009.
16. Karasu, I, Tasci M., O., Sahin, B., and Akilli H., Stereo PIV investigation on windward side leading edge vortex of a yawed slender delta wing, Conferences on Advances in Mechanical Engineering, 11-13 May 2016, Yildiz Technical University, Istanbul, Turkey
17. Sahin, B., Taşcı., M., O., Karasu, I., Akıllı H., Flow structures in end-view plane of slender delta wing, Experimental Fşuid Mechanics 2016, November 15-18, 2016 Czech Republic

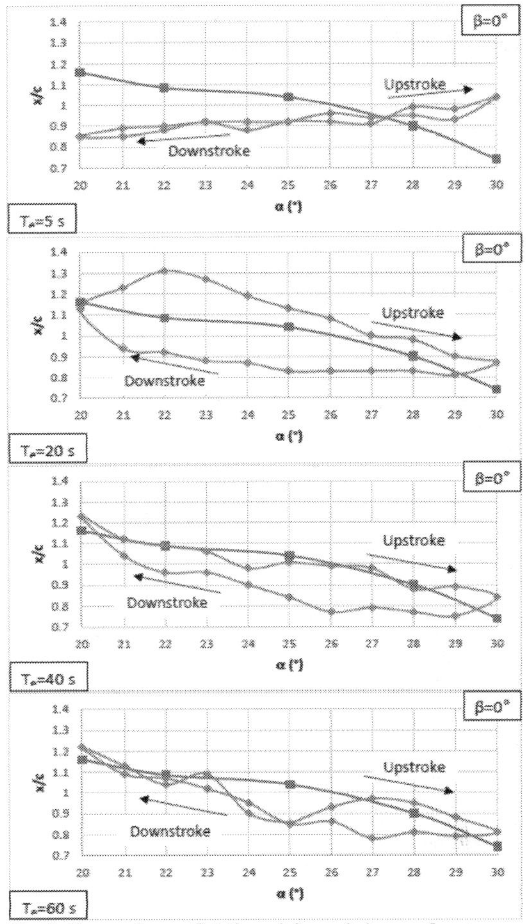

Fig. 9. Comparisons of static and dynamic loops of vortex breakdown as a function of angle of attack, α. Mean angle of attack is α_m=25°, amplitude of pitching motion is α_o=±5°, yaw angle is β=0°, period of oscillation is T_e=5s, 20s, 40s, 60s and reduced frequency is K= 0.16, 0.25, 0.49, 1.96

 https://doi.org/10.1051/epjconf/201818002080

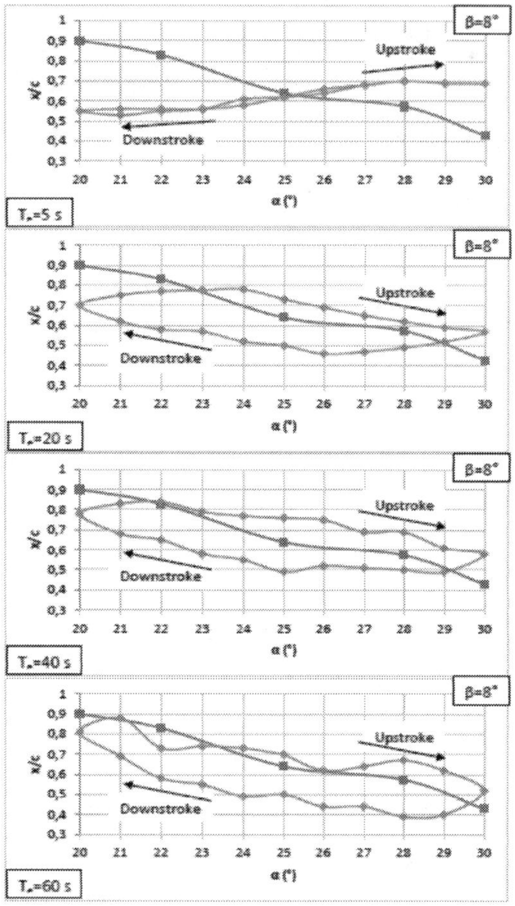

Fig. 10. Comparisons of static and dynamic loops of vortex breakdown as a function of angle of attack, α. Mean angle of attack is α_m=25°, amplitude of pitching motion is α_o=±5°, yaw angle is β=8°, period of oscillation is T_e=5s, 20s, 40s, 60s and reduced frequency is K= 0.16, 0.25, 0.49, 1.96

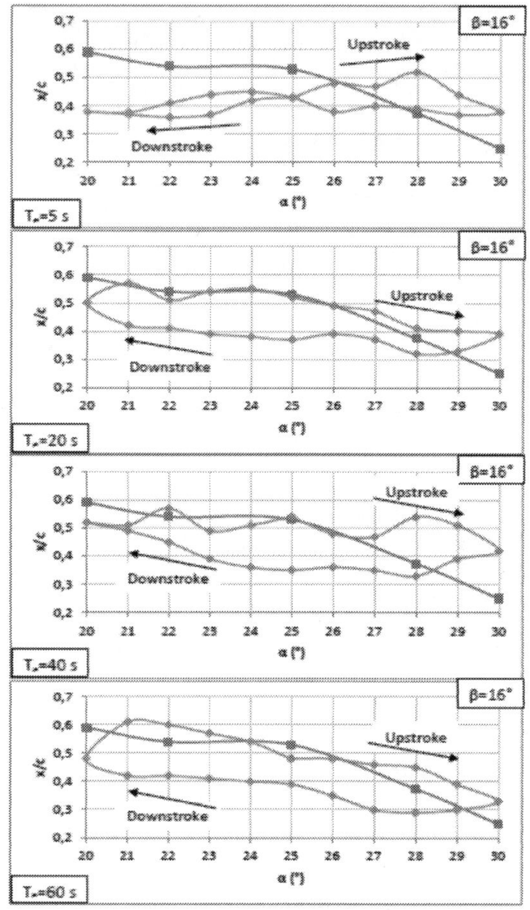

Fig. 11. Comparisons of static and dynamic loops of vortex breakdown as a function of angle of attack, α. Mean angle of attack is α_m=25°, amplitude of pitching motion is α_o=±5°, yaw angle is β=16°, period of oscillation is T_e=5s, 20s, 40s, 60s and reduced frequency is K= 0.16, 0.25, 0.49, 1.96

Analysis of the velocity distribution in different types of ventilation system ducts

Kazimierz Peszyński[1,*], *Lukasz* Olszewski[1], *Emil* Smyk[1], and *Daniel* Perczyński[1]

[1]UTP University of Science and Technology in Bydgoszcz, Faculty of Mechanical Engineering, Control Group, Poland

Abstract. The paper presents the results obtained during the preliminary studies of circular and rectangular ducts before testing the properties elements (elbows, tees, etc.)of rectangular with rounded corners ducts. The fundamental problem of the studies was to determine the flow rate in the ventilation duct. Due to the size of the channel it was decided to determine the flow rate based on the integration of flow velocity over the considered cross-section. This method requires knowledge of the velocity distribution in the cross section. Approximation of the measured actual profile by the classic and modified Prandtl power-law velocity profile was analysed.

1 Preliminary remarks

The testing of ventilation systems components and their design can be greatly simplified by finding analytical solutions which determine the velocity distribution of air flow in a duct. Control Group of Mechanical Engineering Faculty UTP in Bydgoszcz in cooperation with Nucair Technologies Sp. z o.o., Solec Kujawski, Poland conducts research of new type ventilation ducts of rectangular with rounded corners cross-section. This duct type has significant operational advantages, but the aerodynamic phenomena occurring in these ducts have not been studied yet.

2 Circular cross-section of duct

In the state of art literature, the ventilation ducts with a circular cross-section are well examined in the terms of analysis. The ducts with rectangular cross-section are also well described. This paper presents the results of own research related to these two types of channels.

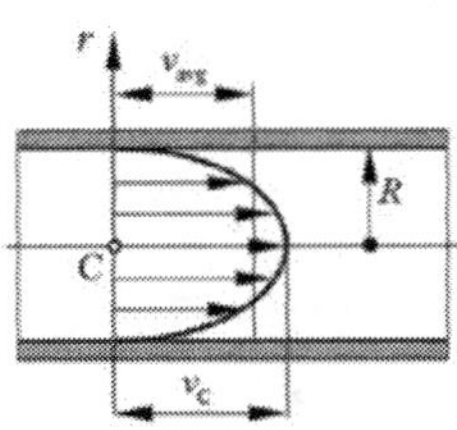

Fig. 1. Average velocity v_{avg} for circular cross-section

During the study of pressure drop (hydraulic resistance), one of the main problems has been to determine the flow rate of air. These tests were aimed to determine the flow rate based on velocity distribution in a duct. In fluid flow, it is convenient to work with an average velocity v_{avg} which remains constant in incompressible flow when the cross-section of the duct is constant (**Fig. 1**). A semi-empirical dependence was used due to the lack of accurate analytical solutions. It is known as Prandtl power-law velocity profile, which is expressed as [1, 2] for a duct of circular cross-section

$$v = v_C\left(1-\frac{r}{R}\right)^{\frac{1}{n}} \tag{1}$$

where: v, $v_C = v_{max}$, r, R are the quantities set out in **Fig. 1**, and $1/n = 1/6 \div 1/10$ is the exponent of the power for a range of Reynolds numbers $Re = 4\cdot10^3 \div 3,2\cdot10^6$ [1]. These values $1/n$ refer to the smooth walls, and for the rough walls are $1/n = 1/4 \div 1/5$. Equation (1) expires near the wall.

2.1 Analysis of the Prandtl power-law velocity profile

The first doubts have risen during the numerical analysis of the problem in the ANSYS-FLUENT environment. The curve illustrating the velocity distribution along a diameter in Fig. 2 suggests that the tangent to velocity profile in the axis point of the circular cross-section duct is perpendicular to the axis. This can never be achieved when using the Prandtl approximation model, because after its differentiation with respect to r we obtain

$$\frac{dv}{dr} = \left(v_C\left(1-\frac{r}{R}\right)^{\frac{1}{n}}\right)' = -\frac{1}{nR}v_C\left(1-\frac{r}{R}\right)^{\frac{1}{n}-1} \tag{2}$$

hence substituting into (2) $r = 0$, we get

$$\frac{dv}{dr} = -\frac{v_C}{nR} \neq 0 \tag{3}$$

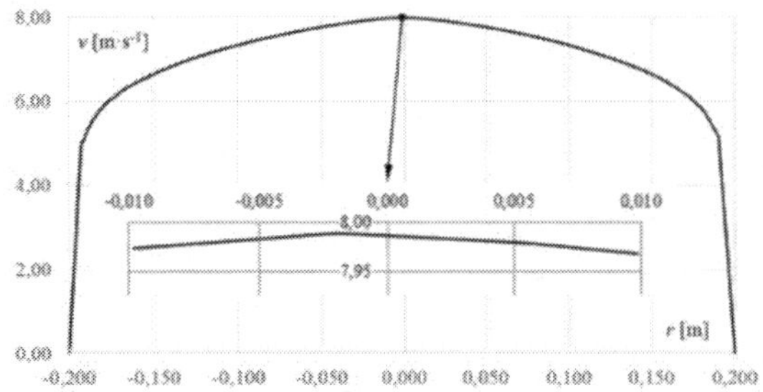

Fig. 2. Numerical solution in ANSYS-FLUENT

Because previously were studied rectangular ducts and rectangular ducts with rounded corners, where well they fits the modified Prandtl power-law velocity profile in the form

$$v = v_C\left(1-\left(\frac{r}{R}\right)^2\right)^{\frac{1}{n}} \tag{4}$$

for the comparative analysis of the ducts shapes was adopted precisely this equation. Earlier, a comparison of velocity distributions of the both equations was done. The results of this analysis are shown in **Fig. 3**

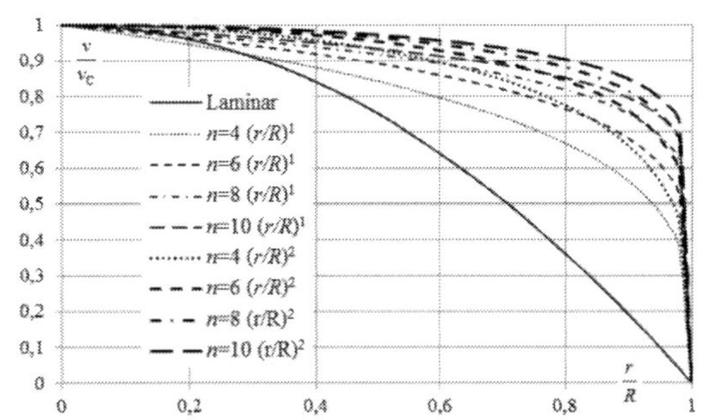

Fig. 3. Velocity distribution for fully developed flow

The figure **Fig. 3** shows that the difference of distribution decreases with the increase of n . We can use smaller parameter n by using distribution equation which includes $(r/R)^2$. Thanks to this operation, the function of distribution $v(r)$ will be the same for all types of ducts.

2.2 Analysis of the tests results

The study involved a linear section of the duct composed of 12 same segments with a radius $R = 0{,}2\,\text{m}$ and length 1 m. The measurements were performed along the horizontal diameter. **Fig. 4** presents the results of measurements and the approximation of this course by the two methods mentioned above. The approximation in this case turned out to be better by classic Prandtl equation. This conclusion has been drawn from the analysis of the sum of squares of deviations $\Delta = (v_a - v_m)^2$ for all measuring point.

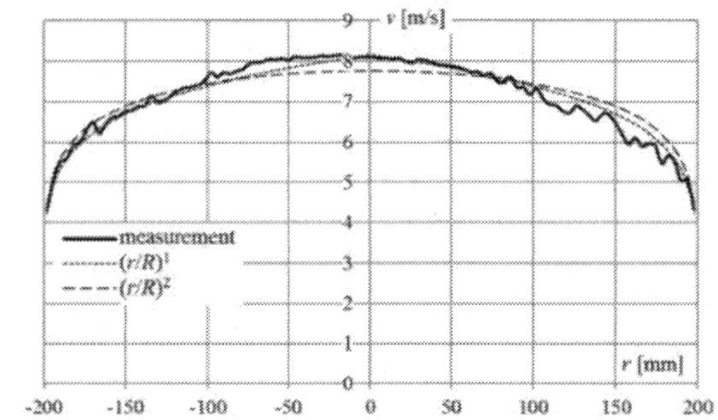

Fig. 4. Velocity distribution in circular cross-section

The maximum velocity on the duct axis determined on the basis of 20 measurements made with an interval of 1 s was $v_{Cm} = 8{,}11\,\text{m}\cdot\text{s}^{-1}$. The velocity resulting from the approximation by the classic Prandtl equation was $v_{C1} = 8{,}16\,\text{m}\cdot\text{s}^{-1}$ while the velocity after the approximation by modified Prandtl equation was $v_{C2} = 7{,}76\,\text{m}\cdot\text{s}^{-1}$. The difference appears to be significant; therefore we examined how this affects the average velocity. Following the numerically calculated flow rates $Q_m = 0{,}833\,\text{m}^3\cdot\text{s}^{-1}$, $Q_1 = 0{,}831\,\text{m}^3\cdot\text{s}^{-1}$, and $Q_2 = 0{,}825\,\text{m}^3\cdot\text{s}^{-1}$ divided by cross-section area $A_{cir} = 0{,}126\,\text{m}^2$, we obtain average values of velocities which are necessary to determine the Reynolds number: measured $v_{avg\,m} = 6{,}63\,\text{m}\cdot\text{s}^{-1}$, approximated by the classic Prandtl equation $v_{avg1} = 6{,}61\,\text{m}\cdot\text{s}^{-1}$, and approximated by the modified Prandtl equation $v_{avg2} = 6{,}57\,\text{m}\cdot\text{s}^{-1}$. Dispersion of these velocities related to the value of v_m does not exceed 1%.

2.3 Analytical calculations

Later in the research, the average velocity for the modified Prandtl distribution was attempted to be determined. The following equation (5) was used for this purpose

$$v_{avg} = \frac{4\int_0^R v_C\frac{\pi}{2}\left(1-\left(\frac{r}{R}\right)^2\right)^{\frac{1}{n}} r\,dr}{A_{cir}} = \frac{\frac{2\pi v_C}{R^{\frac{2}{n}}}\int_0^R \left(R^2-r^2\right)^{\frac{1}{n}} r\,dr}{\pi R^2} \tag{5}$$

after substituting $z^2 = R^2 - r^2 \rightarrow r\text{d}r = -z\text{d}z$ we obtained

$$v_{avg} = \frac{-\frac{2\pi v_C}{R^{\frac{2}{n}}}\int_R^0 z^{2\frac{1}{n}} z\,\text{d}z}{\pi R^2} = \frac{-\frac{2\pi v_C}{R^{\frac{2}{n}}}\int_R^0 z^{2\frac{1}{n}+1}\,\text{d}z}{\pi R^2} \tag{6}$$

and ultimately after the appropriate transformations

$$v_{avg} = \frac{n}{n+1}v_C \tag{7}$$

The above formula shows the fact that the study should determine the dependence of $n = f(Re)$, because the exponent $1/n$ has a decisive influence on the value of the average velocity in the cross-section. It happens very often that it is uncritically accepted as $1/n = 1/7$.

Therefore, the average velocity in tested cross-section is $v_{avg2'} = 6,79\,\mathrm{m\cdot s^{-1}}$ on the basis of a formula (7) for velocity $v_{C2} = 7,76\,\mathrm{m\cdot s^{-1}}$ and $n = 7$, which is 102.4% of experimentally determined value. Excess of average velocity may prove that n was too high.

3 Rectangular cross-section

Rectangular cross-section with rounded corners is a combination of parts of the rectangular cross-section and circular cross-section. Therefore, this type of ducts was part of compare analysis.

3.1 The axial velocity distributions

Experimental study of velocity distribution is made for a rectangular cross-section with width $W = 0,5\,\mathrm{m}$ and height $H = 0,25\,\mathrm{m}$. The study involved a linear section of the duct composed of 12 same segments of 1 m length. The profiles were measured on symmetry axes. The traversing step of the thermo-anemometer probe was $\Delta w = \Delta h = 4\,\mathrm{mm}$. During the measurements, the same velocity v_C of the air flow at the point of intersection of the two axes should be provided. The test results are shown in the figures **Fig. 5** and **Fig. 6**.

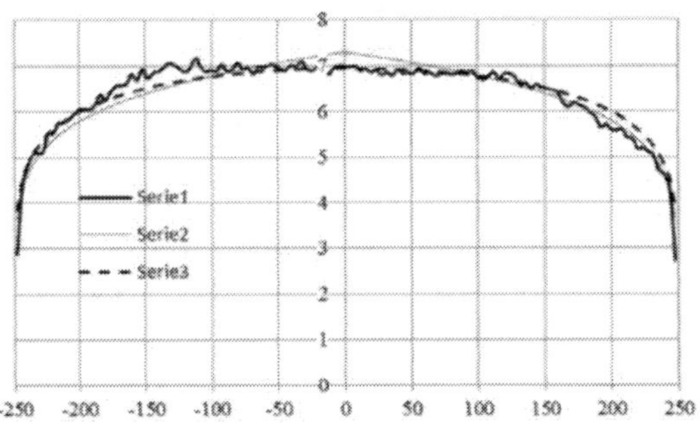

Fig. 5. Velocity distribution along the axis w

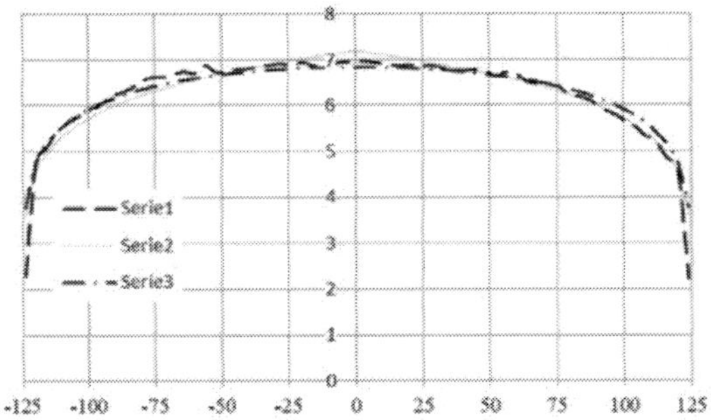

Fig. 6. Velocity distribution along the axis h

The figures **Fig. 5** and **Fig. 6** also present the approximating curves of the measured results by the formulas:

$$v_w = v_C\left(1-\left(\frac{w}{0,5W}\right)\right)^{\frac{1}{n}} \text{ and } v_w = v_C\left(1-\left(\frac{w}{0,5W}\right)^2\right)^{\frac{1}{n}} \quad (8)$$

$$v_h = v_C\left(1-\left(\frac{h}{0,5H}\right)\right)^{\frac{1}{n}} \text{ and } v_h = v_C\left(1-\left(\frac{h}{0,5H}\right)^2\right)^{\frac{1}{n}} \quad (9)$$

Approximating formulas were adopted in the same form as for a circular cross-section. Radius r was replaced by w and h respectively.

Consequently, flow velocity at any point in cross section is given by

$$v(w,h) = v_C\left(1-\left(\frac{w}{0,5W}\right)\right)^{\frac{1}{n}}\left(1-\left(\frac{h}{0,5H}\right)\right)^{\frac{1}{n}} \quad (10)$$

or

$$v(w,h) = v_C\left(1-\left(\frac{w}{0,5W}\right)^2\right)^{\frac{1}{n}}\left(1-\left(\frac{h}{0,5H}\right)^2\right)^{\frac{1}{n}} \quad (11)$$

On the basis of the formula (11), the average velocity $v_{avg} = 5,88\,\mathrm{m\cdot s^{-1}}$ was calculated by numerical integration and it is necessary to determine the Reynolds number. This result was verified by the method described in Section 3.2.

3.2 Determining the average velocity using *log-Chebyshev method*

The duct is divided into rectangular areas (**Fig. 7**) by using the log-Chebycheff method. Velocity is measured at corners of these rectangulars.

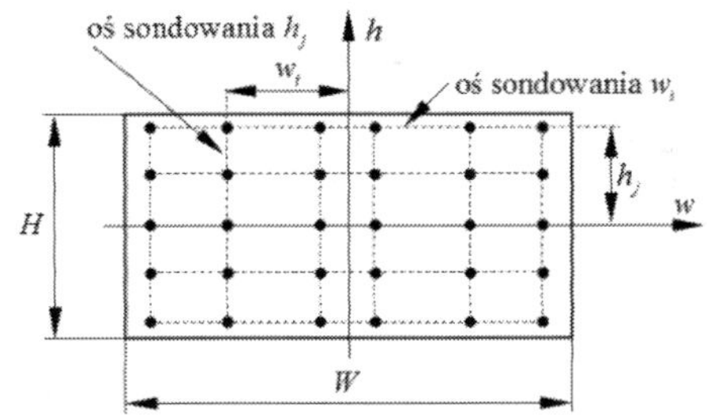

Fig. 7. Example of the measurement points determined in a rectangular cross-section

Table 1. Location of measurement points in a rectangular cross-section [3]

Number of traversing axes i or j	Values $\frac{w_i}{W}$ or $\frac{h_j}{H}$
5	0, ±0,212, ±0,426
6	±0,063, ±0,265, ±0,439
7	0, ±0,134, ±0,297, ±0,447

The points are detrmined by standard ISO 5221 Air distribution and air diffusion – Rules to methods of measuring air flow rate in an air handling duct [3]. The minimum of 25 $(i = 5; j = 5)$ points must be measured in order to get a good average, while the maximum is 49 $(i=7;\ j=7)$ (Table 1). Any intermediate combination is permissible. Velocity measurements were made at various points by using Prandtl tube. For $i=5$ and $j=5$, air velocities obtained at particular points are presented in Table 2.

Table 2. Coordinates and measurement results for $i = j = 5$

v $\mathrm{m \cdot s^{-1}}$		w_i mm				
		-213	-106	0	106	213
h_j mm	-106,5	5,03	5,15	5,07	5,02	4,8
	-53	5,69	6,34	6,36	6,27	5,55
	0	5,75	6,87	6,95	6,68	5,96
	53	5,81	6,78	6,55	6,05	5,58
	106,5	5,04	5,17	5,21	5,15	4,95

On the basis of numerical values contained in Table 2, we obtain $v_{\mathrm{avg}} = 5{,}7512\,\mathrm{m \cdot s^{-1}}$, which is 97.8% of value obtained by numerical integration of curves approximated by formula (10).

3.3 Analytical method

Velocity distribution defined by the formula (10) can be the basis for analytical determination of the volumetric flow rate

$$Q = 4v_C \int_0^{0,5W} \int_0^{0,5H} \left(1-\left(\frac{w}{0,5W}\right)^2\right)^{\frac{1}{n}} \left(1-\left(\frac{h}{0,5H}\right)^2\right)^{\frac{1}{n}} \mathrm{d}h\,\mathrm{d}w \quad (11)$$

The velocity at the duct symmetry axis was adopted $v_C = 6{,}95\,\mathrm{m \cdot s^{-1}}$, the same as for the numerical integration.

Unfortunately, the integral defined by the formula (11) is not expressed in explicit form. Numerical calculation of this integral can be achieved by using the integral of the special Γ function. Using Γ function we obtain the following solution of equation (10)

$$Q = 0{,}25WHv_C\left(\frac{\Gamma(1/2)\Gamma(1+1/n)}{\Gamma(3/2+1/n)}\right)^2 \quad (12)$$

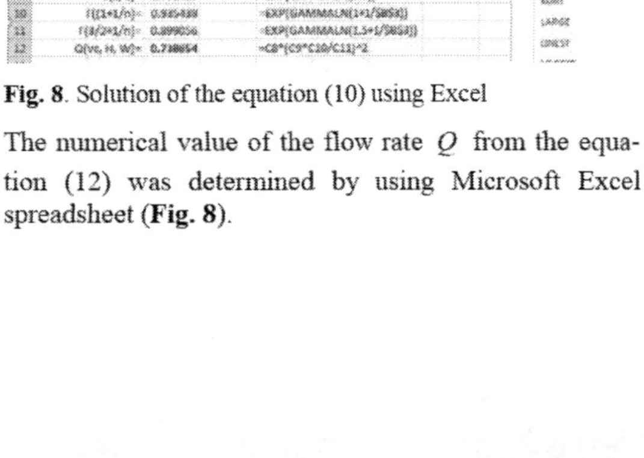

Fig. 8. Solution of the equation (10) using Excel

The numerical value of the flow rate Q from the equation (12) was determined by using Microsoft Excel spreadsheet (**Fig. 8**).

Due to the GAMMALN command counts $ln(\Gamma(x))$, it was necessary to enter this function as an internal function of the EXP command. Function $\Gamma(x)$ is determined from the formula

$$\Gamma(x) = \int_0^{\infty} e^{-u}u^{x-1}\mathrm{d}u \quad (13)$$

The individual values of the function for the arguments defined by the equation (12) are: $\Gamma(1/2) \approx 1{,}77454$, $\Gamma(1+1/7) \approx 0{,}93438$, $\Gamma(3/2+1/7) \approx 0{,}899056$. where $n=7$. After substituting these values and $W=0{,}5\,\mathrm{m}$, $H=0{,}25\,\mathrm{m}$, $v_C = 6{,}95\,\mathrm{m \cdot s^{-1}}$ into the equation (12), we received $Q = 0{,}739\,\mathrm{m \cdot s^{-1}}$, which corresponds to the average velocity $v_{\mathrm{avg}} = 5{,}909\,\mathrm{m \cdot s^{-1}}$. This is 100.5% of value obtained during the numerical integration profiles approximated by the equation (10).

4 Summary

The analysis shows that the basic form of a power Prandtl formula (1) as well as it modified version (4) can be used for the approximation of the actual velocity distributions.

The classic form of power-law Prandtl is better for circular cross-section, but the modified one is more accurate for rectangular cross-section.

The modified Prandtl formula allows the derivation of a simple relation, which defines the value of the average velocity v_{avg} as a function of maximum velocity v_C on the symmetry axis of the cross-section.

It is accepted that the modified power-law form Prandtl will be used during testing of rectangular with rounded corners ventilation ducts, because the rectangular cross-section plays a dominant role in this cross-section.

Acknowledgements

The authors would like to kindly thank the company Nucair Technologies Sp. z o.o., Solec Kujawski, Poland for the performance of tested ducts and the delivery of air supply system.

References

1. Z., Orzechowski, J. Prywer, R. Zarzycki: *Mechanika płynów w inżynierii środowiska* (Eng. Fluid Mechanics in Environmental Engineering), WNT, Warszawa, (1997)
2. Y.A Cengel, J.M Cimbala: *Fluid Mechanics. Fundamentals and Applications*, McGraw-Hill, (2006)
3. Standard ISO 5221: Air distribution and air diffusion – Rules to methods of measuring air flow rate in an handling duct) (1994)

https://doi.org/10.1051/epjconf/201818002082

Modelling of air flow rate in significantly flattened rounded rectangular ventilation ducts

Kazimierz Peszynski[1,*], *Jan* Novosád[2], *Emil* Smyk[1], *Lukasz* Olszewski[1], *Petra* Dančová[2]

[1]UTP University of Science and Technology in Bydgoszcz, Faculty of Mechanical Engineering, Kaliskiego 7, Poland
[2]Department of Power Engineering Equipment, Faculty of Mechanical Engineering, Technical University of Liberec, Studentska 2, 46117 Liberec, Czech Republic

Abstract. Paper presents new mathematical model for air flow velocity distribution in rounded rectangular ducts and its experimental verification. In papers [1, 2] an mathematical model based on modified Prandtl equation for power power-law velocity profile was determined. It works very well for smaller cross sections. During the study of larger cross sections new phenomena in flowing air have been observed, it forced the search for a new model. The new model is based on a rounded rectangular division into two parts: slot and rounded square.

1 Introduction

An important element in the design of ventilation systems (first of all connected with air conditioning) is the selection of appropriate ducts. So far ducts of circular or rectangular shape are commonly used, sometimes of oval cross section area. For some time a new ventilation system has been developed, in which channels have a rounded rectangular cross section area.

This new system combines the advantages of circular and rectangular channels. It provides as well as rectangular ducts good use of space in rooms, allows good cleansing of them (no sharp corners), and has good flow properties.

Λ		W [mm]										
		300	400	500	600	800	1000	1200	1400	1600	1800	2000
H [mm]	200	0.667	0.500	0.400	0.333	0.250						
	250	0.833	0.625	0.500	0.417	0.313	0.250					
	300	1.000	0.750	0.600	0.500	0.375	0.300	0.250				
	400	1.333	1.000	0.800	0.667	0.500	0.400	0.333	0.286			
	500	1.667	1.250	1.000	0.833	0.625	0.500	0.417	0.357	0.313		
	600	2.000	1.500	1.200	1.000	0.750	0.600	0.500	0.429	0.375	0.333	0.300
	800	2.667	2.000	1.600	1.333	1.000	0.800	0.667	0.571	0.500	0.444	0.400
	1000	3.333	2.500	2.000	1.667	1.250	1.000	0.833	0.714	0.625	0.556	0.500
	1200	4.000	3.000	2.400	2.000	1.500	1.200	1.000	0.857	0.750	0.667	0.600

P		W [mm]										
		300	400	500	600	800	1000	1200	1400	1600	1800	2000
R [mm]	100	0.333	0.250	0.200	0.167	0.125	0.100	0.083	0.071	0.063	0.056	0.050

Fig. 1 Determination of dimensionless parameters.

Although the theory of air flow in ducts is reasonably well understood, theoretical solutions are obtained only for a few simple cases such as fully developed laminar flow in a circular pipe [3]. Therefore, we must rely on experimental results and empirical relations for most air flows in ducts rather than closed-form analytical solutions. Noting that the experimental results of ventilation ducts are obtained under carefully controlled laboratory conditions and that no two systems are exactly alike, we must not be so naive as to view the results obtained as "exact." An error of 10 percent (or more) in friction factors calculated using values presented in different design manuals is the "norm" rather than the "exception."

On the other hand, mathematical modelling of the air flow distribution in cross-sectional area of the duct is necessary. In Fig. 1, all dimension series of the developed new system ducts are shown, 79 cells are filled in this table. For each dimension there are about 40 different elements, this means that more than 3,000 items should be tested. This is not possible from a time point of view.

For this reason, it is necessary to develop a certain empirical model of air flow in cross section based on experimental research. The first step in the development of this model was the adoption of dimensionless parameters for characterization of the duct cross-section shape Λ and P, see Fig.1, Fig. 2.

2 Geometrical similarity

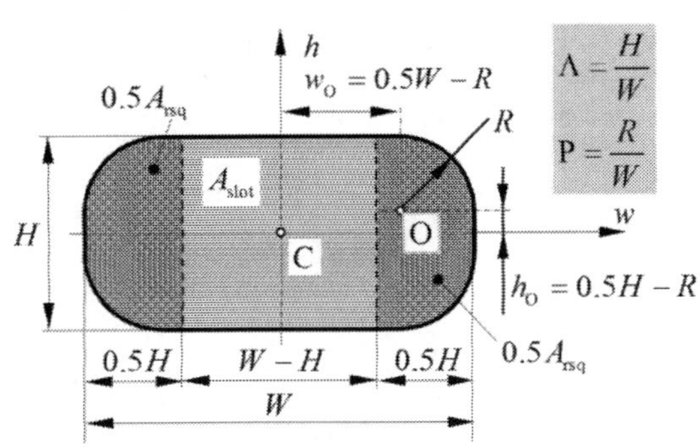

Fig. 2 Basic parameters of cross sectional area and idea it division into two basic parts.

* Corresponding author: kazimierz.peszynski@utp.edu.pl

Cross sectional area is characterized by three parameters: width W, height H, and rounding radius R (Fig. 2). These parameters, during the analysis, are used in two non-dimensional forms: flatness of rectangle Λ and relative rounding P, where width W is the basic parameter.

The cross-sectional area of ducts varies from $A_{rr\_\min} = 0.051\,\mathrm{m}^2$ to $A_{rr\_\max} = 2.323\,\mathrm{m}^2$. This wide scope of areas imposes high demands relating to the measurement stand. Rounding radius R is substantial for the degree of the rounded rectangular reduction area with respect to the rectangular cross section.

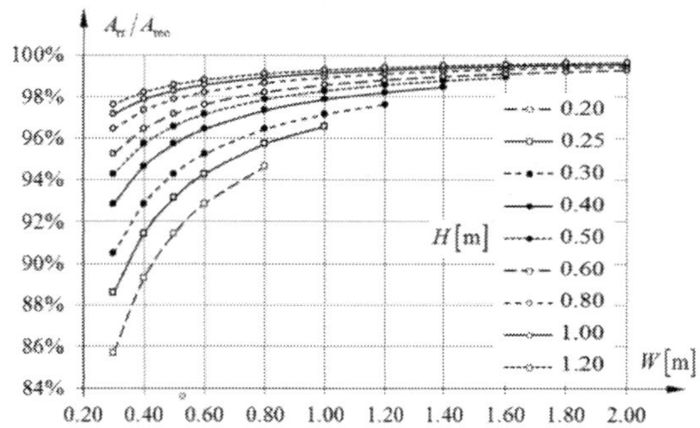

Fig. 3 Degree of use of full rectangular cross section in ducts with rounded rectangulars

In Fig. 3 impact of the rounding radius on the size of rounded rectangular the cross sectional area related to full rectangular cross section. This effect was determined by the formula:

$$\frac{A_{rr}}{A_{rec}} = \frac{W \cdot H - (4-\pi)R^2}{W \cdot H} = 1 - \frac{(4-\pi)\mathrm{P}^2}{\Lambda} \tag{1}$$

where: A_{rr} – rounded rectangular area, A_{rec} – full rectangular area (without rounded corners)

For technological reasons only one rectangle rounding radius $R = 0.1\,\mathrm{m}$ was tested.

3 Modelling of volume flow rate

Modelling of volume flow rate is concerned mainly to rectangular cross sections for $\Lambda < 0.5$ or $\Lambda > 2.0$. Therefore, a new model of velocity distribution in the cross section was to be found. The analysis below is only case-sensitive $\Lambda < 0.5$, because the cross–sections for $\Lambda > 2.0$ are only rotate the section by angle 90°, what is irrelevant from the mathematical point of view. It was noted that within a wide range of coordinates around the central point C, the local flow velocity was not changed. This suggested a presumption that velocity distribution in this area is similar to the velocity distribution in a slot. This observation confirms the well-known fact that the Reynolds number based on the hydraulic diameter D_h for cross sections different from circular better fit the similarity of flows, if the cross section is closer to the circle., e.g. square, hexagon and so on. Therefore, the new model was taken over that the flow in the channel is twofold: one–dimensional as in slot and two–dimensional as in circular cross-section (Fig 2).

Cross–section marked in the figure Fig. 2 as A_{slot} is a rectangle of dimensions $w \times h = (W-H)H$. The flow in this section is similar to the flow in an infinitely broad slot of height h. This means that the flow velocity depends only on the dimension (high) h and is independent of w. Across the slot axis the velocity is maximum and is equal $\left[v(h,w)_{h=0}\right]_{w=-\frac{W-H}{2}}^{w=\frac{W-H}{2}} = v_C$. Two side parts of the duct cross section together they form a square with rounded corners of side H, marked in the figure Fig. 2 as A_{rsq}. For a parameter $R = 0(\mathrm{P}=0)$ this cross section becomes a square, while for $R = 0.5H$ the cross section becomes a circle.

Volume flow rate in the slot

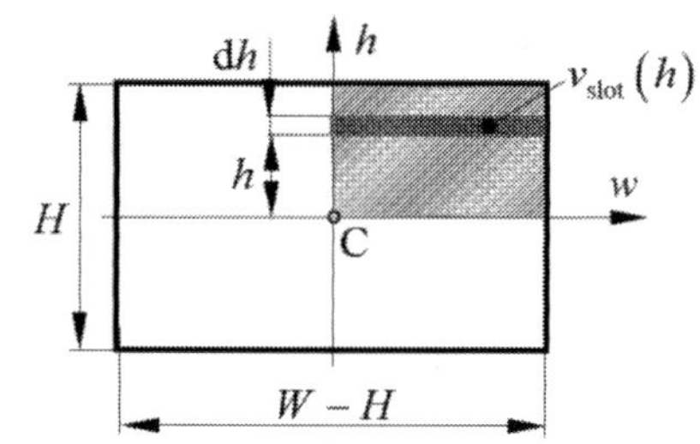

Fig. 4 Idea of the rectangular slot

The volume flow rate through the elementary cross section is defined by the formula:

$$\mathrm{d}\dot{V}_{slot} = 0.5(W-H)v_{slot}(h)\mathrm{d}h \tag{2}$$

When the integration of flow rate over the entire surface we obtain:

$$\begin{aligned}\dot{V}_{slot} &= 4\int_0^{0.5H} 0.5(W-H)v_{slot}\mathrm{d}h \\ &= 2(W-H)\int_0^{0.5H} v_{slot}\mathrm{d}h\end{aligned} \tag{3}$$

Based on empirical studies, it can be assumed that for large sections the model resulting from the modified Prandtl formula is superior to that of its original model. According to the modified Prandtl formula (power-law velocity profile), the flow velocity distribution along the coordinate is defined by the formula [1]

$$v_{slot}(h) = v_C\left(1 - \left(\frac{h}{0.5H}\right)^2\right)^{\frac{1}{n}} \tag{4}$$

After substituting (4) to (3) we obtain

$$\dot{V}_{slot} = 2(W-H)v_C\int_0^{0.5H}\left(1-\left(\frac{h}{0.5H}\right)^2\right)^{\frac{1}{n}}\mathrm{d}h \tag{5}$$

The integral defined by the formula (5) can be determined after the substitution $h^2 = 0.25H^2t$, or $h = 0.5H\sqrt{t}$, so we get:

$$\mathrm{d}h = 0.5H\frac{1}{2\sqrt{t}} = 0.25Ht^{-\frac{1}{2}}\mathrm{d}t \tag{6}$$

After substituting (6) to (5) we obtain

$$Q_{slot} = 0.5\left(W-H\right)Hv_C\int_0^1 t^{\frac{1}{2}-1}\left(1-t\right)^{\frac{1}{n}+1-1}\,dh \qquad (7)$$

The integral expression in formula (7) is a beta function $B\left(\frac{1}{2},\frac{1}{n}+1\right)$, also called the Euler integral of the first kind. A key property of the Beta function is its relationship to the Gamma function Γ

$$B\left(\frac{1}{2},\frac{1}{n}+1\right) = \frac{\Gamma\left(\frac{1}{2}\right)\Gamma\left(\frac{1}{n}+1\right)}{\Gamma\left(\frac{3}{2}+\frac{1}{n}\right)} \qquad (8)$$

Finally, we can write

$$\dot{V}_{slot} = 0.5\left(W-H\right)H\frac{\Gamma\left(\frac{1}{2}\right)\Gamma\left(\frac{1}{n}+1\right)}{\Gamma\left(\frac{3}{2}+\frac{1}{n}\right)}v_C \qquad (9)$$

or using non-dimensional parameter Λ of cross sectional area

$$\dot{V}_{slot} = 0.5\left(1-\Lambda\right)\Lambda\frac{\Gamma\left(\frac{1}{2}\right)\Gamma\left(\frac{1}{n}+1\right)}{\Gamma\left(\frac{3}{2}+\frac{1}{n}\right)}W^2v_C \qquad (10)$$

2.2 Volume flow rate through the rounded square

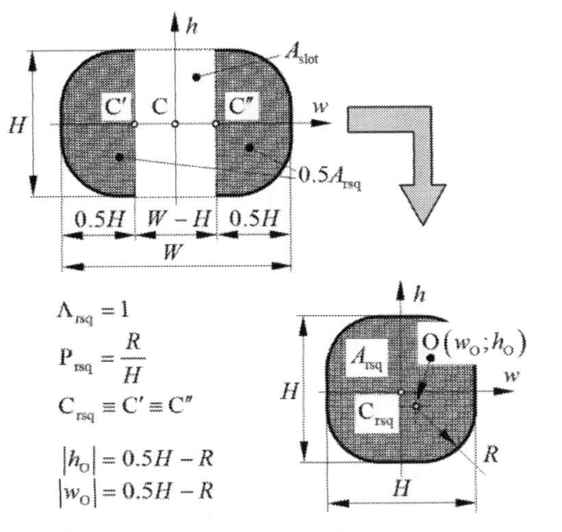

Fig. 5 Idea of virtual rounded square duct

Two outer parts of duct cross section together form a square of side H, because of the symmetry and the same velocities on the perpendicular to the axis w dividing lines, passing through points C' and C''. Virtual both parts can be connected to a rounded square duct with centre of symmetry C_{rsq}. Note that

$$v_{C_{rsq}} = v_{C'} = v_{C''} = v_C \qquad (11)$$

This duct of cross section A_{rsq} perimeter is created by side walls where the air velocity is zero (Fig. 5). Instead of deriving formulas for $\dot{V}_{rsq}$ as for slot, may be used equation (6) presented in [2]:

$$v_{avg} = \frac{\Lambda-\left(4-\pi\right)P^2v_C}{\Lambda\frac{n}{n+1}\left(\frac{n}{n+1}\left(1-\left(\frac{4P^2}{\Lambda}\right)^{\frac{n+1}{n}}\right)+\pi 4^{\frac{1}{n}}\left(\frac{P^2}{\Lambda}\right)^{\frac{n+1}{n}}\right)} \qquad (12)$$

Above mentioned equation (12) was is derived for a rounded rectangle, to get the equation for rounded square, just we can substitute $\Lambda = 1$ (note that $\Lambda = 1$ is only for virtual square) in this case, because $W = H$:

$$v_{avg} = \frac{1-\left(4-\pi\right)P^2v_C}{\frac{n}{n+1}\left(\frac{n}{n+1}\left(1-\left(4P^2\right)^{\frac{n+1}{n}}\right)+\pi 4^{\frac{1}{n}}P^{\frac{2(n+1)}{n}}\right)} \qquad (13)$$

and for volume flow rate $\dot{V}_{rsq}$ we can write

$$\dot{V}_{rsq} = \frac{\left(1-\left(4-\pi\right)P^2\right)\Lambda^2\left(1-\left(4-\pi\right)\right)W^2v_C}{\frac{n}{n+1}\left(\frac{n}{n+1}\left(1-\left(4P^2\right)^{\frac{n+1}{n}}\right)+\pi 4^{\frac{1}{n}}P^{\frac{2(n+1)}{n}}\right)} \qquad (14)$$

2.3 Volume flow rate through the rounded rectangular

Volume flow rate through the rounded rectangular is sum of flow rates through slot and rounded square $\dot{V}_{rr} = \dot{V}_{slot} + \dot{V}_{rsq}$, than

$$\dot{V}_{rr} = \left(0.5\left(1-\Lambda\right)\Lambda\frac{\Gamma\left(\frac{1}{2}\right)\Gamma\left(\frac{1}{n}+1\right)}{\Gamma\left(\frac{3}{2}+\frac{1}{n}\right)}+\frac{\left(1-\left(4-\pi\right)P^2\right)\Lambda^2\left(1-\left(4-\pi\right)\right)}{\frac{n}{n+1}\left(\frac{n}{n+1}\left(1-\left(4P^2\right)^{\frac{n+1}{n}}\right)+\pi 4^{\frac{1}{n}}P^{\frac{2(n+1)}{n}}\right)}\right)W^2v_C \qquad (15)$$

4 Measurement results

Received model that allows to determine volume flow rate through the rounded rectangular $\dot{V}_{rr}$ verified by measurements on a test bench. Test bench consist of flowmeter "testo", fan, tubular air deflector [4], and ventilation duct of $20\times W$ meters long. Initial part of test bench for measurement volume flow rate $\dot{V}_{rr}$ is presented in Fig. 6. Flowmeter "testo" was mounted on the fan inlet. This solution allows to use this measurement device for many measurements. Of course, it is assumed that the amount of incoming air is equal to the amount of exhaust air from duct line.

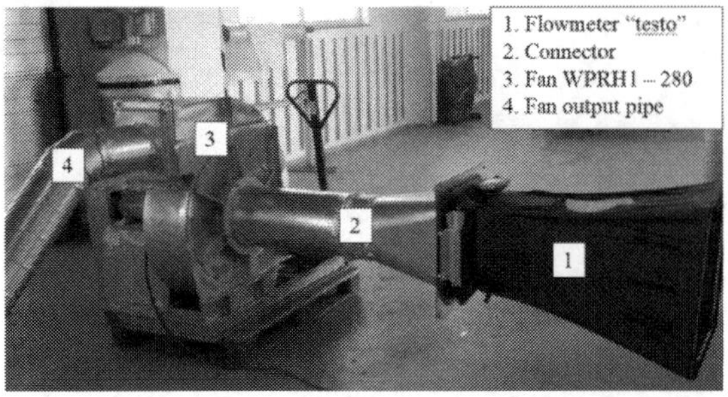

Fig. 6 Initial part of test bench for measurement volume flow rate

Flowmeter "testo" allows to measure the volume flow rate of air in the range $\dot{V} \in \langle 0.12 \div 1.2 \rangle \, \mathrm{m^3 s^{-1}}$. It has own tubular defector to calm the input flow. The fan rotational speed was controlled by the inverter.

All measurements were provided near the end of duct. In Fig.7 is presented traversing system (left) and HWA probe placed 6 mm before the end of duct.

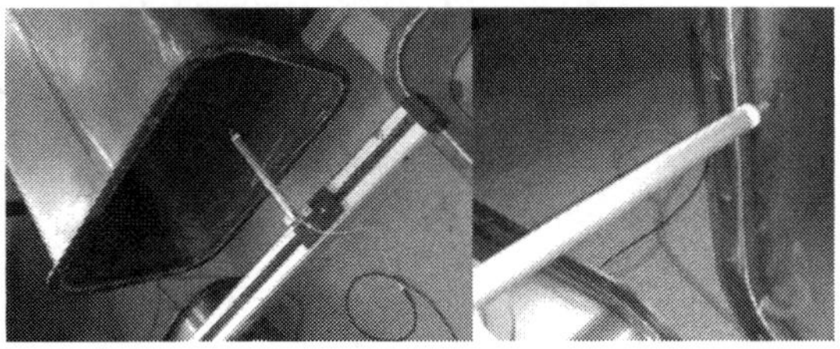

Fig. 7 Measurement of velocity distribution in duct

Usefulness (verification) of the new model is presented in Fig. 8. These are the results of measurements and calculations for the duct of $W = 0.6\,\mathrm{m}$, $H = 0.2\,\mathrm{m}$, and $R = 0{,}1\,\mathrm{m}$, therefore $\Lambda = 0.33$ and $\mathrm{P} = 0.167$.

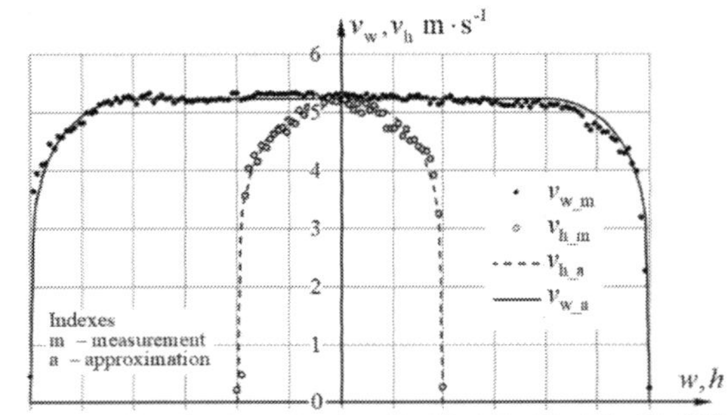

Fig. 8 An approximation of the measured velocity distribution using the new model

Presented in Fig. 8 approximation was made for $n = 5$ and velocity $v_C = 5.24\,\mathrm{m \cdot s^{-1}}$. Velocity v_C was determined as the mean value of the all measurements between $w = \pm 200\,\mathrm{mm}$. Value of measured average velocity, calculated as $v_{avg\_m} = \dot{V}_{tr} / A_{tr}$, was $4.65\,\mathrm{m \cdot s^{-1}}$. Value of counted average velocity was $4.54\,\mathrm{m \cdot s^{-1}}$ which is 97,6% of v_{avg_m}. Cross sectional area of duct was $A_{tr} = 0.111\,\mathrm{m^2}$.

5 Conclusions

The new model is very useful especially when modelling large significantly flattened rounded rectangular ventilation ducts. In this case the models based on Prandtl equation for power power-law velocity fail. They needs big number n, e.g. $n > 20$. Unfortunately this model is more complicated. The main disadvantage is the gamma function, which requires application of advanced computational technique. However, in some applications it can be very useful, for example, developing a new type of flow meter, based on flow velocity measurement at the central point C.

This new model can also be used for smaller cross sectional, it is more universal.

The required measuring range of volume flow rate $\dot{V} \in \langle 0.257 \div 11.957 \rangle \, \mathrm{m^3 s^{-1}}$ exceeded the measuring range of the flowmeter. Therefore, the new model developed was used to determine the value of volume flow rate $\dot{V}$. This was necessary for the 57 ducts examined. Only for 22 cross sections it was possible use flowmeter "testo", therefore the new developed model has greatly helped in carrying out research.

Acknowledgement

The authors would like to kindly thank the company Nucair Technologies Sp. z o.o., Solec Kujawski, Poland for the performance of tested ducts and delivery of the air supply system. Authors also received institutional support BS 16/2013granted by Faculty of Mechanical Engineering of UTP University.

References

1. K. Peszyński, L. Olszewski, E. Smyk E, and D. Perczyński, (2016): Analysis of the Velocity Distribution in Different Types of Ventilation System Ducts, International Conference Experimental Fluid Mechanics 2016, Nov. 15th – 18th, 2016, Mariánské Lázně, Czech Republic, p. 577 – 580
2. K. Peszynski, L. Olszewski, E. Smyk, T. Kasprowicz (2017): Development of New Type Ventilation Ducts System, Proceedings of 23rd International Conference Engineering Mechanics 2017, Svratka, Czech Republic, p. 50 – 53
3. Z. Orzechowski, J. Prywer, R. Zarzycki: Mechanika płynów w inżynierii środowiska (1997), (Eng. Fluid Mechanics in Environmental Engineering), WNT, Warszawa
4. E. Smyk, P. Mrozik, S. Wawrzyniak, K. Peszynski (2017): Tabular Air Deflector in Ventilation Ducts, Proceedings of 23rd International Conference Engineering Mechanics 2017, Svratka, Czech Republic, p. 882 – 885

Development of an anti-flood board to protect the interiors and exteriors of the infrastructure

Michal Petru[1a], Pavel Srb[1], Ladislav Sevcik [1], Tomas Martinec[1], and Petr Kulhavy [1]

[1] Institute for Nanomaterials, Advanced Technologies and Innovation, Technical University of Liberec, Studentská 2, 461 17, Liberec 1, Czech Republic

Abstract. This article deals with the development of an anti-flood board to protect the interior and exterior of various infrastructures, such a houses, cottages or industrial buildings. It was designed prototypes and assembled numerical simulations. In Central Europe and in particular in the Czech Republic, floods are an integral part of the natural water cycle and cause great loss of life and great property damage. The development of new types of mobile anti-flood boards is very important as the design solution is developed for flood protection with regard to minimizing weight, cost of production, easy manipulation, simplicity and speed of installation.

1 Introduction

Floods are a natural part of the water cycle. The Czech Republic was hit by devastating floods in past years. These floods have triggered an effort in the development of flood protection measures. For larger cities, it seems advantageous to use flood barriers with the substructure. Barriers with substructure are usually too expensive for small towns and villages. Mobile barriers are here better because they can be easily placed to the affected area. The whole system can be easily moved to great distance, it is appropriate to shared system with towns in various flood areas, where floods are typically at different times. For protection of personal property in areas with low building density are these mobile barriers expensive. Here are the best solution secure individual buildings against water ingress. Hence began the development of lightweight anti-flood board for installation to building bores such as doors and windows. Flood barriers have to meet a number of tests [1, 2], such as the impact of a log, resistance to waves etc.

2 Experimental measurements of mobile flood barriers

As part of the development of anti-flood systems without substructure, functional prototypes have been assembled and subjected to tests for interior and exterior installation and also in water testing channel in Jap-Jacina company fig.1. Testing conditions based on the prepared certification methodology for testing flood protection systems have been used. An interior and exterior protection layout was prepared. Various case studies of hydrostatic and hydrodynamic loads were tested for a new design solution. A new design has been prepared with regard to minimize weight, cost of production, and ease of handling, simplicity and speed of installation, as shown in fig. 2.

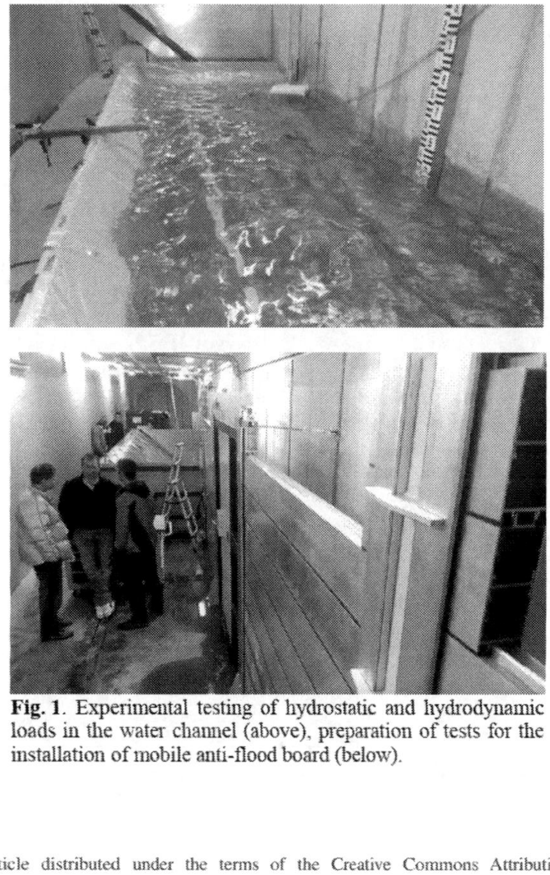

Fig. 1. Experimental testing of hydrostatic and hydrodynamic loads in the water channel (above), preparation of tests for the installation of mobile anti-flood board (below).

[a] Corresponding author: michal.petru@tul.cz

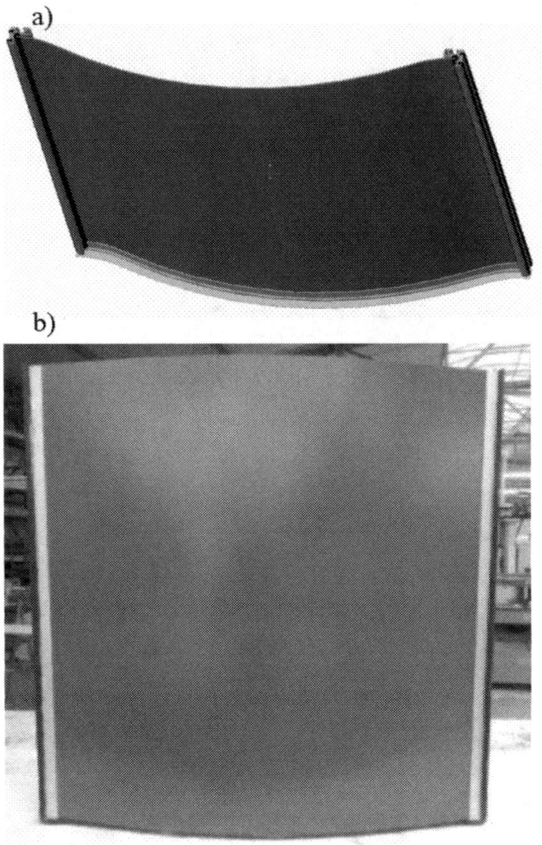

Fig. 2. Development of the prototype of mobile anti-flood board: a) CAD model, b) real prototype

3 Numerical simulations

For virtual study of material properties, structural optimization and critical regions of mobile anti-flood board was used coupled Euler and Lagrange technique (CEL), which combine both approaches in one analysis. [manual abaqus] This method was originally developed to simulate of airbag inflation, later to solve problems such as bird – plane collision, aquaplaning, water tank impact load etc. The method consist mesh adjustment, surface reconstruction and solution of liquids and structures equations. The purpose of this technique is to avoid mesh problems in simulations where an extreme deformation occurs. In the simulations is commonly used Lagrangian approach, which is not suitable for cases of excessive deformation. For those cases Eulerian meshing approach is more suitable. The CEL technique captures the strengths of both Lagrange and Euler method. The CEL method allows the user to selectively mesh individual analysis components. Bodies reaching large deformations (liquids, gases, etc.) can be meshed using Eulerian technique, and conventional Lagrange technique can be used for other bodies. Interaction between bodies is determined by the definition of contact. In Lagrange simulations each element represents the volume of material. Movement of material is possible only by mesh deformation. The Lagrange approach is suitable for most structural problems, but in simulations where large deformations are achieved, the elements are distorted and the simulation cannot continue. In Eulerian simulations the mesh remains stationary during the simulation and the material flows through elements that do not deform. The Euler mesh is typically a simple rectangular grid of elements designed to extend well beyond the Eulerian material boundaries, giving the material space to move and deform. For material assignment in Eulerian analysis, it is necessary to use a predefined field of material that uses the principle of material volume fractions. During Eulerian analysis, the presence of material in each element is monitored in terms of material volume fractions. Volume fraction is the percentage of the volume of the element that occupies the material. For elements which are only partially filled with material, the exact geometric composition of the material inside of the element is not known. Abaqus interpolates volume fractions in adjacent elements to estimate material boundaries within an element. Volume fractions are expressed as a number between zero and one, this number means the percentage of filling, e.g. 0.23 means that 23 % of element is filled by assigned material.

3.1 Material model of water

Formulation of Equation of state Mie Grüneisen $U_S - U_P$ form was used as a material model for water in the Abaqus simulation [3]. The most common form of Mie-Grüneisen is:

$$p - p_H = \Gamma \rho (E_m - E_H) \quad (1)$$

where p_H and E_H are the Hugoniot pressure and specific energy (per unit mass) and they are functions of density only. Γ is the Grüneisen ratio, which describes the alteration in a crystal lattice's vibration frequency, based on the lattice's increase or decrease in volume as a result of temperature change. It is used in quantum mechanics and it is defined as:

$$\Gamma = \Gamma_0 \frac{\rho_0}{\rho} \quad (2)$$

Where Γ_0 is the initial Grüneisen ratio (material constant) and ρ_0 is the reference density. The Hugoniot energy E_H is related to the Hugoniot pressure p_H by:

$$E_H = \frac{p_H \eta}{2 \rho_0} \quad (3)$$

Where $\eta = 1 - \rho_0/\rho$ is the nominal volumetric compressive strain. Elimination of Γ and E_H from the above equations yields:

$$p = p_H \left(1 - \frac{\Gamma_0 \eta}{2}\right) + \Gamma_0 \rho_0 E_m \quad (4)$$

The state and the energy equations represent coupled equations for pressure and internal energy. Abaqus solves these equations simultaneously at each material point.

A common form of the Hugoniot data is given by

$$p_H = \frac{\rho_0 c_0^2 \eta}{(1 - s\eta)^2} \quad (5)$$

Where c_0 and s define the linear relationship between the shock velocity U_S and the particle velocity U_P as follows:

$$U_S = c_0 + sU_p \tag{6}$$

With the above assumptions the linear $U_S - U_P$ Hugoniot form is written as

$$p = \frac{\rho_0 c_0^2 \eta}{(1-s\eta)^2}\left(1 - \frac{\Gamma_0 \eta}{2}\right) + \Gamma_0 \rho_0 E_m \tag{7}$$

Where

$$\rho_0 c_0^2 > 0 \tag{8}$$

is equivalent to the elastic bulk modulus at small nominal strains. There is a limiting compression given by the denominator of this form of the equation of state

$$\eta_{lim} = \frac{1}{s} \tag{9}$$

$$\rho_{lim} = \frac{s\rho_0}{s-1} \tag{10}$$

3.2 Material model of board

Numerical simulations of mobile anti-flood board have been developed to find the appropriate dimensions, thickness and material to meet cost, production possibilities and assembly requirements. The simulations were designed that the both edges of board were fixed to replace the mounting board in the building bore, and then the hydrostatic pressure was applied to different heights depending on the type of board. A number of simulations have been carried out with different materials and thicknesses for several types of building bores. Composite materials have been selected for their unique properties at low weight compared to steel. Corresponding mechanical properties, depending on weight optimization as well as price, give the glass fiber reinforced composites. For simulations a material model based on the basic equation for describing the mechanical properties of the composite (11) was used. The mechanical properties of this composite structure are influenced by volume of the fibers and the matrix, the thickness of the composite, the number and orientation of layers, etc. [4]. For the optimal price and mechanical properties ratio is the range of fibers volume between 35 % - 50 % (Tab.1.).

$$\begin{Bmatrix} \varepsilon_{11} \\ \varepsilon_{22} \\ \varepsilon_{33} \\ \gamma_{12} \\ \gamma_{23} \\ \gamma_{13} \end{Bmatrix} = \begin{bmatrix} 1/E_{11} & -\nu_{12}/E_{11} & -\nu_{12}/E_{11} & 0 & 0 & 0 \\ -\nu_{12}/E_{11} & 1/E_{22} & -\nu_{23}/E_{22} & 0 & 0 & 0 \\ -\nu_{12}/E_{11} & -\nu_{23}/E_{22} & 1/E_{22} & 0 & 0 & 0 \\ 0 & 0 & 0 & 1/G_{23} & 0 & 0 \\ 0 & 0 & 0 & 0 & 1/G_{12} & 0 \\ 0 & 0 & 0 & 0 & 0 & 1/G_{12} \end{bmatrix} \cdot \begin{Bmatrix} \sigma_{11} \\ \sigma_{22} \\ \sigma_{33} \\ \tau_{12} \\ \tau_{23} \\ \tau_{13} \end{Bmatrix} \tag{11}$$

Tab. 1. Mechanical properties of composites with epoxy matrix reinforced with glass fibers

V_f [%]	E_{11} [GPa]	E_{22} [GPa]	G_{12} [GPa]	G_{23} [GPa]	ν_{12} [-]	ν_{23} [-]
10	8,8	4,8	1,8	1,74	0,329	0,48
20	16,1	5,4	2,2	2,21	0,319	0,53
30	22,7	6,3	2,6	2,67	0,31	0,54
40	29,2	8,4	3,1	3,23	0,289	0,52
50	34,3	9,9	4	4,02	0,279	0,54

4 Results and discussion

Figure 3 shows the results of numerical simulations of anti-flood board. Numerical simulation of board for flexural stress analysis at hydrodynamic load was carried out. Lightened glass fibre reinforced composite boards of thickness of 1.5, 2 and 3 mm, was tested. Simulated boards were designed for building bores from 400 mm to 1000 mm wide. The simulation of hydrostatic pressure show that a 1.5 mm thick board is significantly deformed compared to 3 mm. A composite 3 mm thick board is suitable for hydrodynamic loading. It is necessary to have a fibres volume at least 35 %. This gives us the construction material with high toughness at impact load. With increasing volumetric filling decreasing the tension in board, thus also decreasing load on the expansion mechanism.

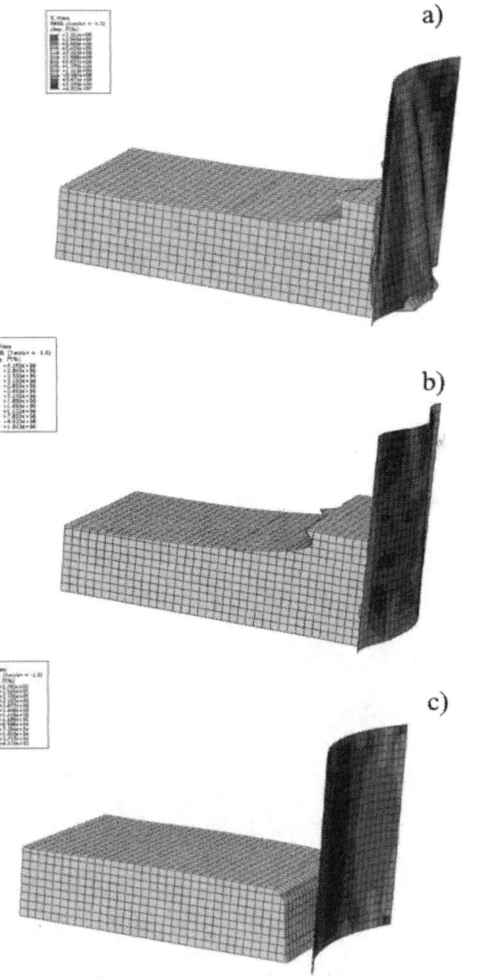

Fig. 3. FEM model of anti-flood board: a) thick 1,5 mm, b) thick 2 mm, c) thick 3 mm.

Figure 4 shows results of the experiments. The horizontal axis of the plot shows the change in the building bore (longitudinal deformation of board). The vertical axis shows the force of board buckling for the given deformation and the force from the water pressure.

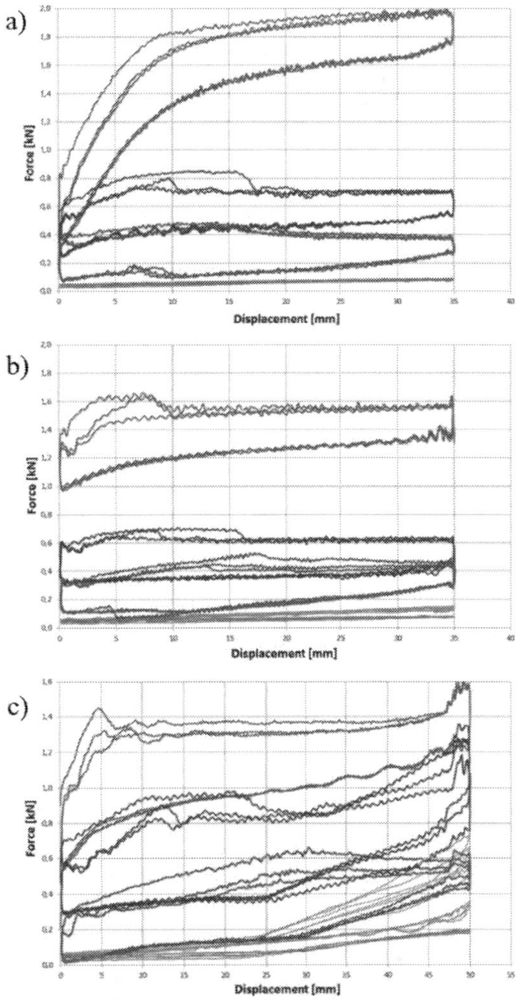

Fig. 4. Experimental measurement of the forces of loaded prototype: bore width a) 550 mm, b) 650 mm, c) 750 mm.

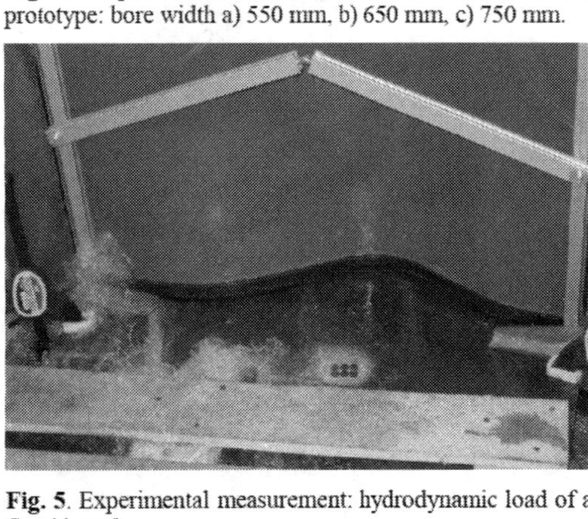

Fig. 5. Experimental measurement: hydrodynamic load of anti-flood board

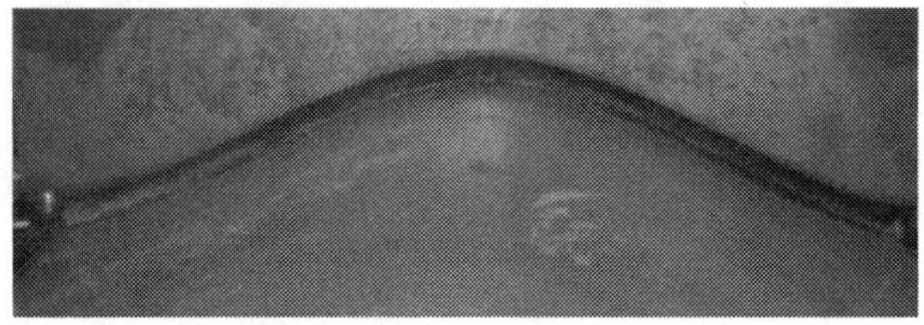

Fig. 6. Experimental measurement: detail of sealing.

Composites boards of 3 different thicknesses were measured (in plots 3 mm – green, 2 mm – blue, 1.5 mm – red). The dark tint of colour displays the values of the vertical force, the light tint of the colour shows value of the side force. The board was place to testing frame first preload of height 15 mm was applied then were performed 3 cycles of height 35 mm loading. Measurements were carried out for three bores width 500 mm, 650 mm and 750 mm. The tests show that the design is satisfactory with minimal fictional coefficient 0.15. Due to the fact that the rubber seal (fig. 5, 6) increases this coefficient, the solution is applicable.

5 Conclusions

In case of a flood it is necessary to divert water out of the inhabited area as soon as possible. Fixed or mobile barriers are usually used in high urban density areas. To protect personal property in a low urban density areas are these barriers to expensive. Here are the best solution secure individual buildings against water ingress. This can be achieved by the using anti-flood board installed in building bores such as doors and windows. Board must meet many requirements, especially low weight and easy installation. Based on many mechanical tests and numerical simulations have been developed anti-flood board made of composite material. This new design meets all requirements and is characterized by its low weight and excellent sealing capability.

Acknowledgements

The results of this project VI20152018005 were obtained with co-funding from the Ministry of the Interior of the Czech Republic as part of targeted support from the project "Vývoj protipovodňových systémů pro zvýšení ochrany obyvatelstva a infrastruktury".

References

1. Approval Standard for Flood Abatement Equipment, Class Number 2510, FM Approvals LLC. (2013)
2. Muttray, M.O., Oumeraci, H. Theoretical and experimental study on wave damping inside a rubble mound breakwater (2005)
3. P. Srb, M. Petru and P. Kulhavy, Numerical simulation of flood barriers, EPJ Web of Conf. **143**, 02115 (2017)
4. Syrovátková, M., Kulhavý, P., Srb, P., Petrů, M., Testing of tensile properties of carbon prepreg composite rods with adding of a non-composite part, (2016), Defect and Diffusion Forum

New boundary conditions for fluid interaction with hydrophobic surface

František Pochylý [1], *Simona* Fialová [1*], and *Michal* Havlásek [1]

[1]Brno University of Technology, Faculty of Mechanical Engineering, Victor Kaplan Department of Fluids Engineering, Technická 2896/2, 61669 Brno, Czech Republic

Abstract: Solution of both laminar and turbulent flow with consideration of hydrophobic surface is based on the original Navier assumption that the shear stress on the hydrophobic surface is directly proportional to the slipping velocity. In the previous work a laminar flow analysis with different boundary conditions was performed. The shear stress value on the tube walls directly depends on the pressure gradient. In the solution of the turbulent flow by the k-ε model, the occurrence of the fluctuation components of velocity on the hydrophobic surface is considered. The fluctuation components of the velocity affect the size of the adhesive forces. We assume that the boundary condition for ε depending on the velocity gradients will not need to be changed. When the liquid slips over the surface, non-zero fluctuation velocity components occur in the turbulent flow. These determine the non-zero value of the turbulent kinetic energy *K*. In addition, the fluctuation velocity components also influence the value of the adhesive forces, so it is necessary to include these in the formulation of new boundary conditions for turbulent flow on the hydrophobic surface.

1 Introduction

Solution of both laminar and turbulent flow with consideration of hydrophobic surface is based on the original Navier assumption that the shear stress on the hydrophobic surface is directly proportional to the slipping velocity 1. This condition is supplemented by Tesk's assumption that the slip will occur after a certain shear stress threshold has been exceeded. To solve more complicated liquid flow tasks over a curved surface 2, the mentioned boundary condition is supplemented by other terms. Generally on the surface S for the shear stress vector $\boldsymbol{\tau}$ applies:

$$c = 0: \ |\boldsymbol{\tau}| \leq |\boldsymbol{\tau}|_K \qquad c \neq 0: \ \boldsymbol{\tau} = kc \tag{1}$$

This boundary condition is also advantageous because it determines the zero normal component of the velocity to the surface that must be met if the boundary is not porous.

In previous work, a laminar flow analysis with different boundary conditions was performed, depending on the shear stress characterizing the slip of the liquid (water in this case) on the hydrophobic surface with the following conclusions:

- The fluid slip on the surface is subject of a low surface energy.
- The surface energy can be judged by the value of the contact angle of the liquid droplet on the surface 8.
  $\Theta < 80°$ - hydrophilic surface (wettable surface).
  $\Theta > 80°$ - hydrophobic surface (surface is non wettable).
- Surface energy fundamentally affects the shear stress value on the solid-liquid interphase 3. The shear stress value on the tube walls directly depends on the pressure gradient. At low surface energy the fluid slips on the surface. The hydrophobic surface is characterized by adhesive forces dependent on adhesion coefficient α. For $\alpha = 100$, the surface already behaves as hydrophilic.

Based on experiments, it has been found that the hydrophobic surface has the ability to bind air molecules 4, 5, 6. It retains this property even during the laminar flow of the water in pipes. The air layer adjacent to the surface changes the viscosity in this area 7.

* Corresponding author: fialova@fme.vutbr.cz

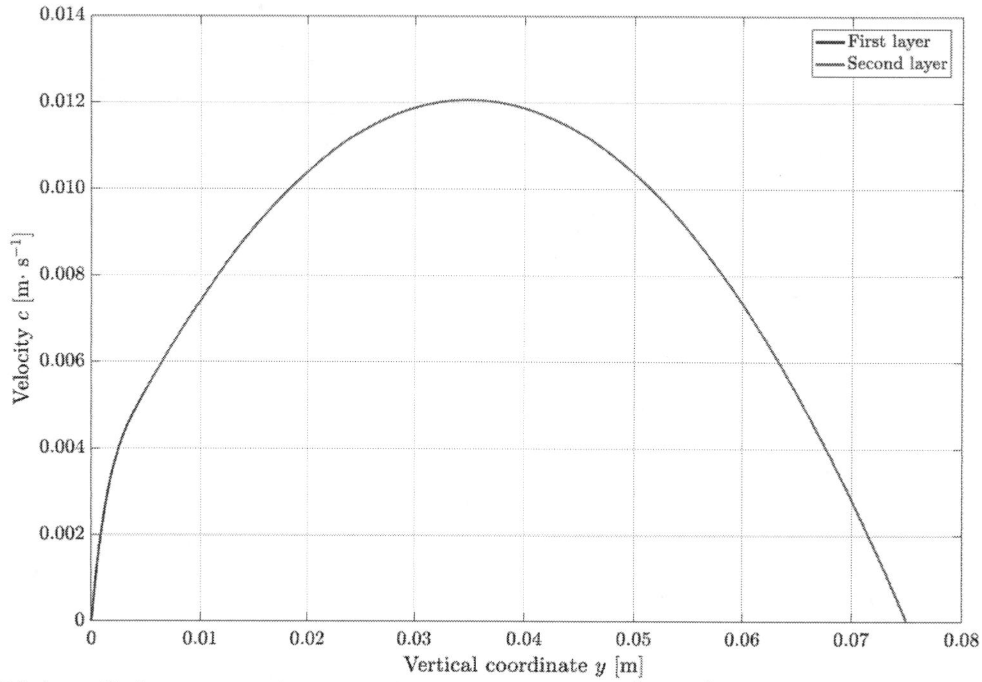

Fig. 1. Velocity profile shape $\Delta p = 0.015\text{Pa}, \tau_0 = 0{,}05379\tau_{KR}, \chi = 3{,}752457.10^9 1/\text{Pas}m^2$

2 Turbulent Flow

It is supposed that the viscosity in the interval varies according to:

$$\eta_0 = \frac{\eta}{1 + \chi\eta(h-y)^2} \tag{2}$$

The viscosity value depends on the constant χ that represents the degree of saturation by gas:

For $\chi > 0$, η_0 on the surface decreases
For $\chi < 0$, η_0 on the surface increases

this value also depends on the shear stresses on the surface. In terms of qualitative analysis, it is particularly interesting to see Fig.1, which shows the laminar velocity profile of the flow in two layers of different viscosity (on the left side there is η_0 and on the right side $\eta > \eta_0$).

From Fig. 1, a steep increase of the velocity profile at low viscosity is evident. The velocity profile is characteristic especially for the air layer adhering to the surface of the tube ($c = 0$). When evaluating experiments using the PIV method (which cannot measure the velocity profile near the wall), it could be incorrectly judged from the figure, that the liquid slips along the wall of the tube. This is because we would have the last measured point where the first layer ends (the end of the blue color). That is why certain caution should be applied when evaluating results of the PIV experiments 6.

The experiment also shows that the thickness of the air layer decreases as the Reynolds number increases and in turbulent flow, at higher Reynolds number values it is completely flushed.

When the air plastron is flushed, the adhesion forces act on the liquid at the surface. Adhesion forces are characterized by the adhesion coefficient k or its non-dimensional variant γ. Coefficient k and γ correlate as follows:

$$k = \gamma\rho c_t \tag{3}$$

The coefficient k was determined based on surface energy, when a drop of liquid slides over an inclined plane 8 However, when the fluid flows in the tube near the surface, the adhesive forces will be influenced by the pressure gradient, so it can be assumed that they will change depending on the fluid velocity in the vicinity of the wall.

To explain the effect of the flowing liquid on the value of the adhesion forces in the turbulent flow, let us proceed from a simple example of liquid flow between two parallel walls, as shown in Fig. 2.

By the integration of the centered Navier-Stokes equations for steady flow between two parallel plates a relationship can be derived for shear stresses at a selected point of the surface induced by the effects of adhesive forces:

$$\tau(0) + \tau(H) = -Hgradp \tag{4}$$

It is clear from expression (4) that the shear stress at one point on the boundary will depend on the pressure gradient, so it can be assumed that the pressure gradient value here will affect the value of adhesive forces. We suppose that the surface $y = 0$ is hydrophobic and the surface $y = H$ is hydrophilic.

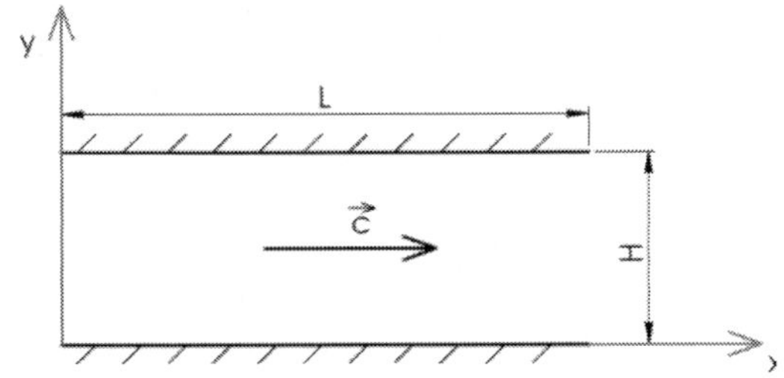

Fig. 2. Coordination system

Therefore, the stresses on the surfaces $y = 0$ and $y = H$ will qualitatively vary depending on the turbulent velocity fluctuations. Turbulent fluctuations can be described by different models. Let us speak about the *K-ε* model 9,4, where *K* and *ε* are defined by the relations:

$$K = \frac{1}{T}\int_{t-\frac{T}{2}}^{t+\frac{T}{2}} c_i' c_i' d\tau ; \tag{5}$$

$$\mathcal{E} = \nu\frac{1}{T}\int_{t-\frac{T}{2}}^{t+\frac{T}{2}} \frac{\partial c_i'}{\partial x_j}\frac{\partial c_i'}{\partial x_j} d\tau \tag{6}$$

Based on K and ε, the turbulent viscosity η_T is determined, for which holds:

$$\eta_T = \rho C\mu \frac{K^2}{\varepsilon} \tag{7}$$

From the above mentioned definitions, it is understandable that at the hydrophilic surface to which the fluid adheres the fluctuation components of the velocity c_i' will be zero. Therefore, for $y = H$:

$$y = H: \quad c_i' = 0; \quad \frac{\partial c_i'}{\partial x_j} \neq 0 \qquad K = 0; \quad \varepsilon \neq 0; \quad \eta_T = 0 \tag{8}$$

$$\tau(H) = \eta \frac{\partial c(H)}{\partial y} \tag{9}$$

Another situation occurs in the case of the hydrophobic surface $y = 0$. If we consider a liquid slip, the following holds:

$$y = 0: \quad c_i' \neq 0; \quad \frac{\partial c_i'}{\partial x_j} \neq 0 \qquad K \neq 0; \quad \varepsilon \neq 0; \quad \eta_T \neq 0 \tag{10}$$

$$\tau(0) = (\eta + \eta_T)\frac{\partial c(0)}{\partial y} \tag{11}$$

If both surfaces are hydrophobic, equation (4) can be written in the form:

$$\tau(0) = -\frac{H}{2} gradp \tag{12}$$

Hence, the adhesive stress is a linear function of the pressure gradient. After fitting from (10) and considering the assumption that the adhesive force is a linear function of the slip velocity c_s, it can also be written:

$$\tau(0) = (\eta + \eta_T)\frac{\partial c(0)}{\partial y} = -\frac{H}{2} gradp = kc_s \tag{13}$$

From the foregoing idea it is important to note that in the boundary condition on the hydrophobic surface in the turbulent flow, it is no longer possible to only consider the molecular viscosity , but the turbulent viscosity η_T as well.

In order to perform a qualitative analysis of laminar and turbulent flow for hydrophilic and hydrophobic surfaces, it is necessary to express the stress tensor by a single, resilient simplified model. For this case we will unfold from the following power model of turbulence.

For stress tensor:

$$\sigma_{ij} = -\delta_{ij}\, p + \tau_{ij} \qquad \tau_{ij} = 2\alpha_n 2^{\frac{n-1}{2}} \left(c_{ij}c_{ij}\right)^{\frac{n-1}{2}} c_{ij} \tag{14}$$

from (14) for the turbulent viscosity applies:

$$\eta + \eta_T = 2\alpha_n 2^{\frac{n-1}{2}} \left(c_{ij}c_{ij}\right)^{\frac{n-1}{2}} = \eta_s \tag{15}$$

η_s represents the sum of molecular and turbulent viscosities. Based on this model, we are now dealing with the problem of stationary flow of incompressible liquid in a circular cross section tube as is shown in Fig. 3.

In the solution of the turbulent flow by the k-epsilon model, the occurrence of the fluctuation components of velocity on the hydrophobic surface is considered. The fluctuation components of the velocity affect the size of the adhesive forces. This fact must also be taken into account when defining the coefficient of adhesion. We assume that the boundary condition for ε depending on the velocity gradients will not need to be changed.

By analyzing the simplified model of turbulence, it is possible to assess differences in laminar and turbulent flow with consideration of surface hydrophobicity. When the liquid slips over the surface, non-zero fluctuation velocity components occur in the turbulent flow. These determine the non-zero value of the turbulent kinetic energy *K*. This results in a non-zero value of the turbulent viscosity of the liquid on the surface. In addition, the fluctuation velocity components also influence the value of the adhesive forces, so it is necessary to include these in the formulation of new boundary conditions for turbulent flow on the hydrophobic surface 11.

3 Velocity profiles for laminar and turbulent flow

Considering the coordinate system of Fig. 2 and the following assumptions,

$$\boldsymbol{c} = (0, c, 0), \qquad c = c(r) \tag{16}$$

on the basis of this simplification, considering (14) and by the integration of the Navier-Stokes equation, we obtain the relationship for the velocity gradient:

$$\frac{\partial c}{\partial r} = -\left(\frac{p_1 - p_2}{2L\alpha_n}\right)^{\frac{1}{n}} r^{\frac{1}{n}} \tag{17}$$

$$\Delta p = p_1 - P_2$$

When we mark

$$Z_n = \left(\frac{\Delta p}{2L\alpha_n}\right)^{\frac{1}{n}} \tag{18}$$

By the solution of equations (17) we obtain the relation for velocity component c:

$$c = A - Z_n \frac{n}{n+1} r^{\frac{n+1}{n}} \tag{19}$$

We define the integration constant A from the condition on the boundary of the hydrophobic surface, for which, considering (13), holds:

$$r = R: \ \eta_s \frac{\partial c}{\partial r} = -kc \tag{20}$$

$$\frac{\partial c}{\partial r} = -Zr^{\frac{n+1}{n}}$$
$$r = R: \ \frac{\partial c(R)}{\partial r} = -ZR^{\frac{1}{n}} \tag{21}$$

From the combination of equations (19) - (21) for A holds:

$$A = \left(\frac{\eta_s}{k} Z_n R^{\frac{1}{n}}\right) + Z\frac{n}{n+1} R^{\frac{n+1}{n}} \tag{22}$$

When the surface is hydrophilic

$$\alpha_n = \eta \tag{23}$$

If the surface is hydrophobic

$$\alpha_n = \eta_s \left|\frac{\partial c_2(R)}{\partial r}\right|^{1-n} \tag{24}$$

Considering these assumptions, it holds:

$$c = \Delta p \frac{R}{2L}\left\{\frac{1}{k} + \frac{n}{n+1}\frac{R}{\eta_s}\left[1 - \left(\frac{r}{R}\right)^{\frac{n+1}{n}}\right]\right\} \tag{25}$$

For the average velocity, it can be written:

$$c_{ST} = \frac{\Delta p}{2}\frac{R}{L}\left[\frac{1}{k} + \frac{n}{3n+1}\frac{R}{\eta_s}\right] \tag{26}$$

The velocity on the hydrophobic surface:

$$c_s = \Delta p \frac{R}{2L}\frac{1}{k} \tag{27}$$

The following conclusions follow from this solution. The relationship (25) is general and applies to both laminar and turbulent flows. For laminar flow $n = 1$, so it will be for:

Hydrophobic surface - laminar flow

$$c = \Delta p \frac{R}{2L}\left\{\frac{1}{k} + \frac{1}{2}\frac{R}{\eta}\left[1 - \left(\frac{r}{R}\right)^2\right]\right\} \tag{28}$$

$$c_s = \Delta p \frac{R}{2L}\frac{1}{k} \tag{29}$$

Hydrophilic surface - laminar flow

$$c = \Delta p \frac{R^2}{4L}\frac{1}{\eta}\left[1 - \left(\frac{r}{R}\right)^2\right] \tag{30}$$

Hydrophobic surface - turbulent flow

The n value depends on the Reynolds number, $n < 1$

$$c = \Delta p \frac{R}{2L}\left\{\frac{1}{k} + \frac{n}{n+1}\frac{R}{\eta_s}\left[1 - \left(\frac{r}{R}\right)^{\frac{n+1}{n}}\right]\right\} \tag{31}$$

$$\Delta p = 2\frac{L}{R^2}\frac{(3n+1)}{\frac{1}{kR}\frac{3n+1}{n} + \frac{1}{\eta_s}}\frac{c_{ST}}{n} \tag{32}$$

Hydrophilic surface - turbulent flow

$$c = \Delta p \frac{R}{2L}\frac{n}{n+1}\frac{R}{\eta}\left[1 - \left(\frac{r}{R}\right)^{\frac{n+1}{n}}\right] \tag{33}$$

$$\Delta p = 2\frac{L}{R^2}\eta(3n+1)\frac{c_{ST}}{n} \tag{34}$$

$$c_s = 0 \tag{35}$$

From the above mentioned results, the following statements arise:

- The sliding velocity in both laminar and turbulent flows is formally the same. Compare (28) and (33).

- From relation (32), it is obvious that the coefficient of adhesion k and the viscosity on the surface of the pipe $\eta_s = \eta + \eta_T$ have qualitatively the same effect on the differential pressure. For constant k an increase of η_s will lead to an increase of the differential pressure.
- It follows from relation (31) that at low values k there will be high values of steady velocity on and near the surface. However, these values generate high fluctuation velocities c_i', and increase K and η_T. This, however, causes the velocity to decrease both near the surface and on the surface itself, so that also k will be inversely affected. However, this does not affect the mathematical model because constant k and η_T are considered.
  Therefore, the model boundary condition (20) must be modified so that k and also η_T is dependent on n. For the K-ε model, it will be necessary to adjust the boundary condition for K at the hydrophobic surface. Since now the boundary condition for the K relation is in the form of

$$K = 0 \tag{36}$$

4 The specification of boundary conditions

The value of the kinetic turbulent energy density K (36) on the hydrophobic surface will depend on the velocity slip rate. In this sense, it is necessary to formulate a new boundary condition.

4.1 $K-\varepsilon$ turbulence model

Due to the occurrence of the fluctuation components of velocity c_i' on the hydrophobic surface and assuming that the fluctuation components of velocity also affect the size of the adhesive forces, it can be assumed:

$$K = \beta c_t c \tag{37}$$

where β is a dimensionless constant, and c is the value of velocity slip of the fluid on the hydrophobic surface, in the tangent direction to the surface. The turbulent viscosity can be determined from definition (3).

The fluctuation velocity components of the liquid on the hydrophobic surface will affect the size of the adhesive forces. Assuming that, it must be taken into account when defining the coefficient of adhesion:

$$k = \rho c_t(\alpha + \beta) \tag{38}$$

We assume that the boundary condition for ε, depending on the velocity gradients, will not need to be changed.

4.2 Simplified turbulence model for a tube of circular cross section

The effect of the fluctuation components of velocity will depend on the Reynolds number and will influence the value of the adhesive forces characterized by the adhesion coefficient k 12. The value of the exponent n is dependent on the Reynolds number. Therefore, for k the following relationship can be proposed, expressing the interaction of the adhesive forces and the force effect of the fluctuation velocity components represented by the kinetic energy of turbulence K.

$$k = \rho c_t \left(\alpha + \beta \frac{1-n}{n}\right) \tag{38a}$$

According to the simplified model (11) applied to the liquid flow in a circular cross-section tube, the relationship for turbulent viscosity can be derived. On the basis of the definition of the mixing length l, we can say:

$$l = \psi(1-n)r, \eta_T = -\rho l^2 \frac{\partial c}{\partial r}$$

From here and from (31) we get an expression for the resulting viscosity of the liquid on the hydrophobic surface:

$$\eta_s = \frac{\eta}{2} + \frac{1}{2}\sqrt{\eta^2 + 4\rho\left(\frac{\Delta p}{2L}\right)\psi^2(1-n)^2 R^3} \tag{39}$$

The new boundary condition expressing the slip of the velocity on the hydrophobic surface will take the form:

$$\eta_s \frac{\partial c}{\partial r} = \rho c_t \left(\alpha + \beta \frac{1-n}{n}\right) c \tag{40}$$

α, β, Υ are dimensionless constants. The relationship for the velocity profile (31) will be as follows:

$$c = \frac{\Delta p}{2} \frac{R}{L} \left\{ \begin{matrix} \dfrac{n}{\rho c_t[\alpha n + \beta(1-n)]} + \\ + \dfrac{nR}{\eta_s(n+1)}\left[1 - \left(\dfrac{r}{R}\right)^{\frac{n+1}{n}}\right] \end{matrix} \right\} \tag{41}$$

$$c_{ST} = \frac{\Delta p}{2} \frac{R}{L} \left[\begin{matrix} \dfrac{n}{\rho c_t[\alpha n + \beta(1-n)]} + \\ + \dfrac{n}{3n+1} \dfrac{R}{\eta_s} \end{matrix} \right] \tag{42}$$

$$c_S = \frac{\Delta p}{2} \frac{R}{L} \left[\frac{n}{\rho c_t[\alpha n + \beta(1-n)]}\right] \tag{43}$$

From expression (43), which represents the liquid slip on the hydrophobic surface, the influence of turbulent

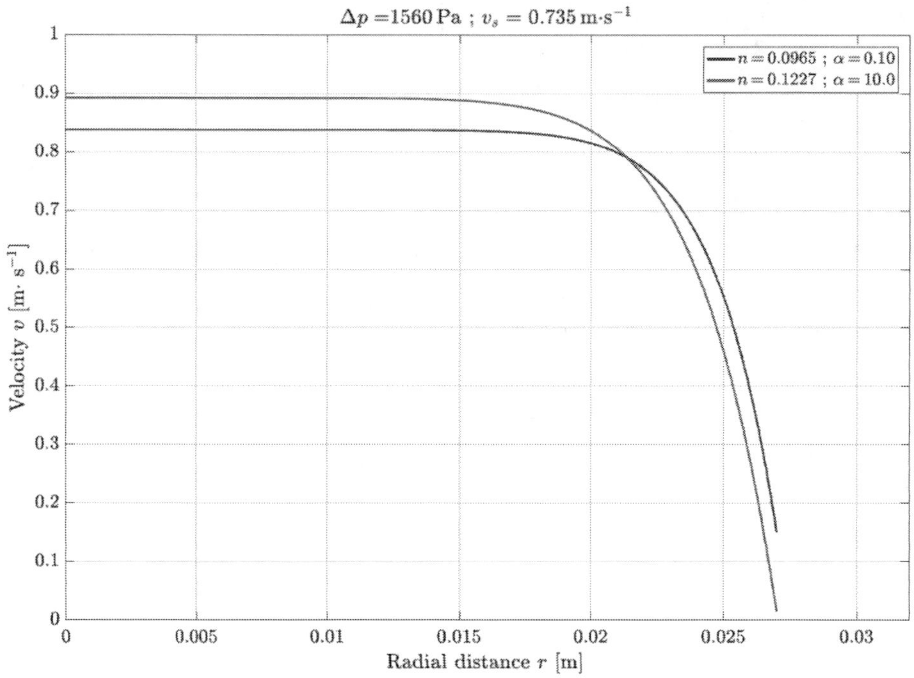

Fig. 3. Turbulent velocity profiles for a constant pressure drop and a constant mean velocity. $\beta = 0{,}1$

fluctuations on the adhesive forces effect is already evident. Figure 3 shows velocity profiles for different adhesion coefficient values. From this, it is clear that the simplified turbulence model for a circular cross-section tube is a good illustration of the shape of the turbulent velocity profile. Note that $\Delta p, c_{ST}$ corresponds to an experiment carried out on a 4m length tube.

The new boundary conditions for turbulent flow on the hydrophobic surface can be for the turbulence model (K - ε) written in the form:

$$c = 0: \quad |\tau| \leq |\tau|_K \qquad c \neq 0: \tau = \rho c_t(\alpha + \beta)c \tag{44}$$

$$K = \beta c_t c \tag{45}$$

5 Conclusion

By analyzing the simplified model of turbulence, it was possible to assess differences in laminar and turbulent flow with consideration of surface hydrophobicity. If there is a fluid slip on the surface, a non-zero fluctuation component of the velocity occurs in the turbulent flow. These are determining the non-zero value of the turbulent kinetic energy K. This results in a non-zero value of the turbulent viscosity of the liquid on the surface. In addition, the fluctuation components of velocity influence the value of the adhesive forces, so it is necessary to include these facts in the formulation of new boundary conditions (44), (45) for turbulent flow on the hydrophobic surface.

Grant Agency of the Czech Republic, within the projects GA101/15-06621S, GA101/17-19444S is gratefully acknowledged for support of this work.

References

1. Navier C. L. M. H., "Memoire Surles du Movement des." *Mem Acad. Sci. Inst. France*, vol. 1, No. 6, pp. 414-416. (1823)
2. Grunceli BRK, Sandham ND and McHale G. Simulation of laminar flow past a superhydrophobic sqhere with drag reduction and separation delay. *Phys Fluids 2013*; 25: 043601.
3. Thomas Young: An Essay on the Cohesion of Fluids. *Philosophical Transactions of the Royal Society of London*. Vol. 95 (1805), 65-87.
4. Vakarelski I. U., Chan D. Y. C., Marston J. O., Thoroddsen S. T.: Dynamic Air Layer on Textured Superhydrophobic Surfaces, *Langmuir*, 2013, *29* (35), pp 11074–11081
5. Jašíková, D., Němcová, L., Kopecký V., The Methodic for Study of Smart Surfaces Using PIV Technique. *AIP Conf. Proc. vol. 80, nr. 8*. pp. 80 – 87 (2014)

6. Fialová, S.; Pochylý, F.; Kotek, M; Jašíková, D.: Velocity profiles of fluid flow close to a hydrophobic surface, *EPJ Web of Conferences 143*, 02023 (2017) DOI: 10.1051/epjconf/201714302023
7. Fialová, S., Pochylý, F., Havlásek, M., Malík, J.: Influence of boundary conditions on fluid flow on hydrophobic surfaces, *AIP Conf. Proc - 36th Meeting of Departments of Fluid Mechanics and Thermodynamics* (2017)
8. Fialová, S.; Pochylý, F.: Identification and Experimental verification of the adhesive coefficient of hydrophobic materials, *Wasserwirtschaft Extra, 1/2015*, ISSN 0043 0978, pp. 125-129
9. Mohammadi, B., & Pironneau, O. (1993). *Analysis of the K-epsilon turbulence model*. France: Editions MASSON.
10. Mansour, N. N.; Kim, J.; Moin, P. Near-wall k-epsilon turbulence modeling, *6th Symposium on Turbulent Shear Flows, France, Proceedings (A88-38951 15-34). University Park, PA, Pennsylvania* (1987), p. 17-4-1 to 17-4-6.
11. Kučera, R.; Šátek, V.; Haslinger, J.; Fialová, S.; Pochylý, F. Modelling of Hydrophobic Surfaces by the Stokes Problem with the Stick- Slip Boundary Conditions. *Journal of Fluids Engineering-Transactions of the ASME*, 2017, vol. 139, no. 1, p. 0112021-0112029. ISSN: 0098-2202.
12. Volkov AV, Parygin AG, Lukin MV, et al. Analysis of the effect of hydrophobic properties of surfaces in the flow part of centrifugal pumps on their operational performance. *Therm Eng 2015*; 62: 817–824.

Simulator with integrated HW and SW for prediction of thermal comfort to provide feedback to the climate control system

Jan Pokorný[1,*], *Barbora* Kopečková[1], *Jan* Fišer[1] and *Miroslav* Jícha[1]

[1]Brno University of Technology, Faculty of Mechanical Engineering, Energy Institute, Technická 2896/2, Brno, Czech Republic

Abstract. The aim of the paper is to assemble a simulator for evaluation of thermal comfort in car cabins in order to give a feedback to the HVAC (heating, ventilation and air conditioning) system. The HW (hardware) part of simulator is formed by thermal manikin Newton and RH (relative humidity), velocity and temperature probes. The SW (software) part consists of the Thermal Comfort Analyser (using ISO 14505-2) and Virtual Testing Stand of Car Cabin defining the heat loads of car cabin. Simulator can provide recommendation for the climate control how to improve thermal comfort in cabin by distribution and directing of air flow, and also by amount of ventilation power to keep optimal temperature inside a cabin. The methods of evaluation of thermal comfort were verified by tests with 10 test subjects for summer (summer clothing, ambient air temperature 30 °C, HVAC setup: +24 °C auto) and winter conditions (winter clothing, ambient air temperature -5 °C, HVAC setup: +18 °C auto). The tests confirmed the validity of the thermal comfort evaluation using the thermal manikin and ISO 14505-2.

1 Introduction

The issue of thermal comfort in cars is still relevant especially in the context of energy savings. To achieve thermal comfort with less energy demands is possible by improving the used materials (glazing, car painting, etc.) HVAC (heating, ventilation and air conditioning) technologies and also car climate control systems.

This paper refers to assembly a simulator with integrated hardware (thermal manikin Newton with probes) and software (TCA, VTSCC) for prediction of thermal comfort to provide feedback to the climate control system.

The main feature of simulator is evaluation of effect of clothing on thermal comfort using thermal manikin [1] or at least sensors of equivalent temperature [2]. Both work on the principle of EHT (equivalent homogenous temperature) and comfort zones [3]. The advantage of thermal manikin compared to the sensors of EHT is that it follows the shape of the human body, that influences the air flow around the measured point and also the heated surfaces generate similar plumes as real human. Thermal manikins can be used to evaluate new strategies of cabin ventilation, which is currently widely studied, not only for electric vehicles, where air-conditioning (A/C) system can significantly impact their performance, but also for self-driving vehicles, etc.

The related question is how much energy will be needed to obtain optimal conditions. To answer this question, the car cabin heat load models can be used as a simple pre-developmental design tool. In this context, several car cabins and HVAC models have been developed up to now. Kiss [4] developed automotive A/C system simulation tool on the MATLAB/Simulink platform. The model consists of a detailed cooling circuit model and relatively simple cabin model. In the study of Fayazbakhsh [5], a comprehensive heat balance model was developed for usage in mobile air-conditioning design. The model takes the vehicle geometry, material properties, and driving conditions characteristics as input. Macros [6] proposed a thermal model of a vehicle cabin. The model was verified by considering the cabin of a real car outfitted with a temperature sensor network. Torregrosa-Jaime [7] developed a lumped-parameter thermal model of the passenger's compartment of a vehicle. This transient thermal model was also validated with experimental data under varies ambient conditions. In the study of Lee [8], the transient thermal model of the passenger car's cabin was developed to investigate the dynamic behaviour of cabin thermal conditions. Brèque [9] developed a transient mono-zonal thermal model of a vehicle cabin in order to predict the thermal needs. The model considered conduction, convection, radiation heat transfers as well as the HVAC and moisture air impacts.

The aim of this work is to verify these methods and demonstrate that the simulator is able to help evaluate climate control efficiency.

2 Methods

As mentioned above, the simulator consists of hardware and software part. The hardware part is formed by thermal manikin Newton and additional probes of air velocity, RH, and temperature sensors. The software part consists of TCA (Thermal Comfort Analyzer) [10] and

* Corresponding author: pokorny.j@fme.vutbr.cz

VTSCC (Virtual Testing Stand of Car Cabin) [11, 12]. TCA is empirical based model that evaluates thermal comfort by using Newton manikin and comfort zones method [3]. VTSCC is heat balance model able to predict temperatures and heat loads of the car cabin based on real operational data. Both software models were developed in Matlab and compiled into the executable applications.

The first step was to verify comfort zones method against human subject's tests by using thermal manikin and TCA (chapter 2.1 and 2.2). When this method was verified, the comfort zones were used to evaluate local thermal comfort to provide recommendations for distribution and directing of air flow to the human surface in order to improve thermal comfort.

The second step was to apply VTSCC to calculate car cabin heat loads to provide recommendation how to change the HVAC power (feeding air temperature and total mass flow) to improve thermal comfort inside a car cabin (chapter 2.3).

2.1 Thermal manikin and Comfort zones

Thermal manikin (Figure 1) is the human body shape device for measuring heat fluxes from its surface for given temperatures. In this study, we prefer this method to evaluate EHT. The second method is to keep constant heat fluxes and monitor the surface temperatures. EHT is defined as temperature of a uniform space, in which all surface temperatures are equal to air temperature, there is no air movement other than the self-convection of the manikin and the rate of heat loss would be the same as was actually measured. EHT is useful as a basis for comparison for complex and highly non-uniform thermal environment, such as automobiles [13].

The diagram of comfort zones (Figure 3 and 8) imagines the subjects' votes of larger group of people to 18 body parts. The black line in these diagrams represents the objective measurement of EHT, more specifically heat losses from thermal manikin body when is tempered on 34 °C. Based on the positon of the line in the diagram the local thermal sensation/comfort can be evaluated and thus the recommendation to climate control can be done.

Comfort zones are defined by the range of EHT which is associated with the range of mean thermal vote (MTV). MTV uses the 7-point Bedford scale comfort, the diagram has 5 zones and only 3 zones represent comfort, outside these comfort zones it is discomfort anyway, see Table 1.

Table. 1 MTV scale and comfort zones

MTV	Scale	Comfort zones
+3	Much too hot	MTV>1.5
+2	Too hot	
+1	Hot but comfortable	MTV>0.5
0	Neutral	-0.5<MTV<0.5
-1	Cold but comfortable	MTV<-0.5
-2	Too cold	MTV<-1.5
-3	Much too cold	

The diagram shape i.e. range of specific comfort zone depends on the clothing (more insulation means wider zones and higher tolerance to lower EHT). To identify shape of the diagram the calibration for the specific clothing ensemble is necessary.

In this paper, the calibration data of indoor summer and winter are used. These data were measured separately before the own tests with human subjects in climate chamber. For summer clothing, the total insulation was 1.13 clo and for winter clothing 1.46 clo. The intrinsic insulation was 0.49 clo (summer) and 0.82 clo (winter), where 1 clo = 0.155 m^2K/W.

We applied these "clo" values to diagram and evaluated thermal comfort by the comfort zones method and thermal manikin Newton. It provides together with inherited probes of RH, velocity and temperatures information about local heat fluxes and cabin environment (e.g. temperatures at the feet and head level of manikin *T1_Feet*, *T2_Head)*. TCA calculated from these heat fluxes Newton's local EHTs and MTVs (see black line in comfort zones diagrams) and also diagrams of comfort zones were plotted in the Figure 3 and 8.

2.2 Climate chamber tests with human subjects

The aim of the test procedure was to answer the question: "Does the manikin report the same thermal comfort as the test subject?" To be able to compare results of thermal manikin Newton and test subjects, they had to be exposure to the same well defined environmental conditions. For this reason, the experiment was carried out in the climate chamber (NETME centre, Brno), where a test car was prepared for the experiment. Thermal manikin was sitting on the co-driver's seat, while test subject was sitting on the driver's place.

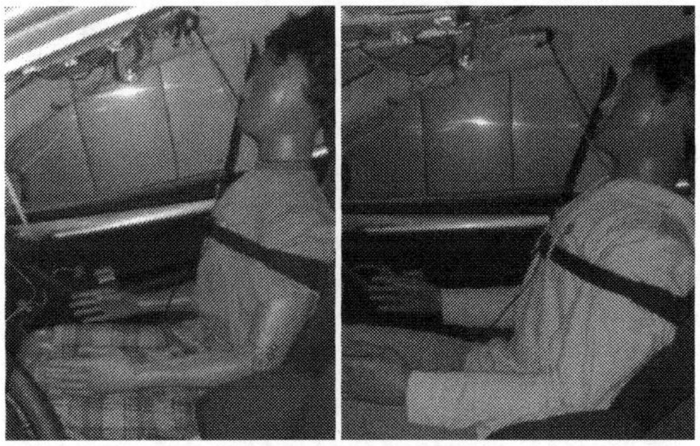

Fig. 1. Thermal manikin with summer and winter clothing.

The test procedure included summer and winter driving conditions. We investigated a group of 10 people (young males 23-35 years old) for both conditions, altogether we collected 20 questionnaires. We were able to measure four volunteers (test subjects) per day, and the whole experimental procedure took six days. The time schedule of typical test day was analogical to test day 1 with test subjects N.1-4.

The test car was parked into the climate chamber and preconditioned for at least two hours in given conditions, including switching off thermal manikin that was dressed in summer/winter clothing (Figure 1).

Then the HVAC was setup to expected comfort level (summer: +24 °C auto, winter: +18 °C auto) and the cabin environment was preconditioned for next 30 minutes. Meanwhile test subjects were stabilized in the control room outside the climate chamber.

The first test subject N.1 entered the cabin with the supervisor of the experiment. Thus, two people were in the cabin during filling questionnaires and one thermal manikin. The questionnaires were filled immediately after they got into the car, and then after 20 and 30 minutes.

When subject N.1 ended procedure the HVAC was still in operation and no people were inside for 30 min to obtain similar starting environmental condition for the second test subject N.2.

After the subject N.2, there was at least one hour break for the lunch. The HVAC system was off during this period and thus the temperature in the cabin was influenced only by conditions in the climate chamber. In the afternoon, tests were performed with subjects N.3 and N.4.

During the test procedure, the conditions in climate chamber were monitored (temperature, humidity, solar intensity). Also, the test car was equipped by PT100 sensors to monitor temperature inside a car cabin, including feeding air form vents. The amount of feeding ventilation air was calculated from the voltage on fan terminals. In addition, some data from inbuilt sensors were possible to read using the CAN (Controller Area Network) of the car.

2.3 Calculation of cabin heat loads

The measured data of ambient environment and HVAC setup were used as input data to the VTSCC model. More specifically, following boundary conditions together with description of geometry and material compositions of car cabin define VTSSC input interface.

T_AMB (°C) – ambient temperature outside the car was controlled by climate chamber: summer test ~ 30-34 °C, winter test ~ -5 °C (~ means approximate value).

T_VENT (°C) – average temperature of feeding air measured in all opened vent (summer test ~ 25 °C, winter test ~ 23 °C).

U_VENT (V) – for summer conditions the voltage was considered 3.7 V and for winter 4.3 V. Volumetric flow rate of feeding air was then 143 m^3/h and 169 m^3/h.

N_OCC (-) – number of occupants defines internal sources. In the simulation, the heating power of human was considered 100 W and the power of manikin was directly measured by himself and it was generally in range of 50–150 W (only at start of the test day manikin needed more power to heat up itself).

T_OPT (°C) – optimal EHT inside a car cabin. Defined by position of Wholebody neutral comfort zone (MTV=0). It is different for summer and winter clothing (25.7 °C and 23 °C).

I_SOL (W/m^2) – during all tests solar intensity was zero.

SPEED (km/h) - during all tests car was idling and the relative air speed was ~ 5 km/h.

RH (%) – the effect of relative humidity was not considered in this study, because the RH of feeding air was not measured (it was assumed 50 %).

These boundary conditions were used to calculate the air temperature inside a cabin *T_IN* (°C), and also the heat loads of cabin: from vents *Q_VENT* (W), by heat generation of all passengers including thermal manikin *Q_MET* (W), by conduction through cabin envelope *Q_COND* (W) and by radiation, especially solar radiation *Q_SOL* (W). Also, the heat transfer by convection and radiation was considered to calculate temperatures inside the cabin.

The validity of heat loads calculation was proved by the comparison of predicted *T_IN* and measured air temperatures *T1_Feet* and *T2_Head* (Figure 5 and 10). In addition, optimal ventilation power *Q_VENT_OPT* (W) was calculated to express the necessary power to reach optimal temperature *T_OPT*. The difference between the actual *Q_VENT* and optimal *Q_VENT_OPT*, is then possible to use as recommendation for the change of the feeding air parameters such as temperature and volumetric flow rate.

3 Results

In this chapter, the results of human subject tests are presented for both conditions – summer and winter. We verified comfort zones method by average value of Wholebody MTV over all test subjects (N.1-10) in times after getting in car, after 20 min and after 30 min. To demonstrate also evaluation of local thermal comfort, comfort zones diagram is plotted for the case: test subject N.2 after 30 min. Finally, the VTSCC prediction of cabin air temperature and cabin heat loads is demonstrated for test day with test subjects N.1-4.

3.1 Summer conditions

For the summer conditions the thermal manikin Newton and comfort zones evaluate thermal comfort very similar to human subjects at observed moments. The both approaches detect neutral sensation, tending to warm but comfortable. Figure 2 indicates that Newton's data best match the Global AVG data. It means that although the overall comfort can be evaluated from local thermal votes (Parts AVG), it is still more relevant to ask test subjects also for the overall comfort.

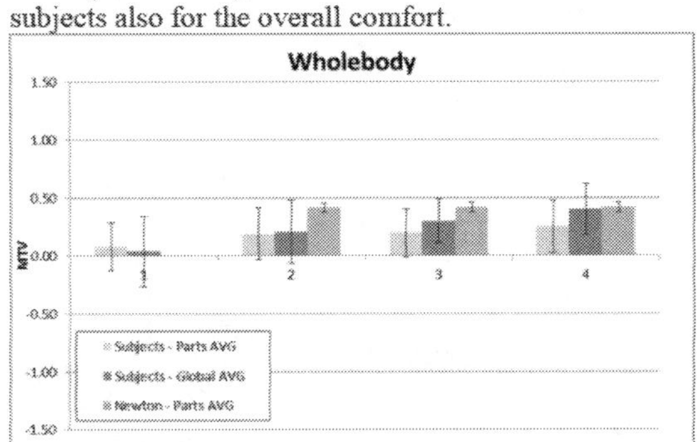

Fig. 2. Summer conditions evaluation of human subjects' tests – Wholebody MTV for all subjects N.1-10.

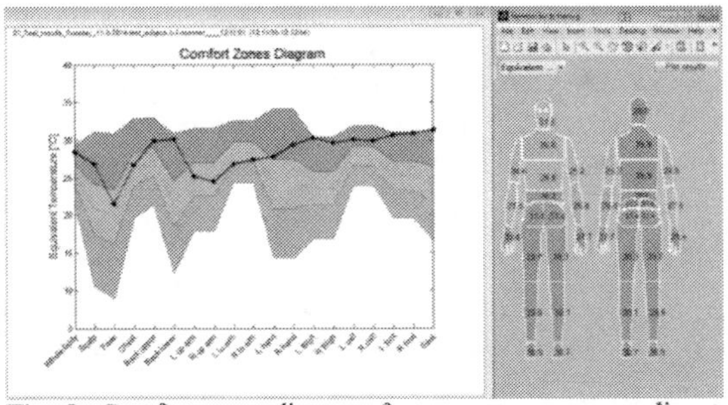

Fig. 3. Comfort zone diagram for summer; corresponding to the conditions of test subject N.2 after 30 min.

A more detailed view on the conditions of test subject N.2 after 30 minutes (time 12:12) is shown in the Figure 3: Face, Chest, Arms were in neutral (except cold Left_Lower_Arm), whilst other parts were warm but comfortable. The average cabin air temperature was then 28.3 °C (*T1_Feet*=30.6 °C, *T2_Head*=25.9 °C), which was higher than the recommended EHT for optimal thermal comfort. Optimal EHT value for the summer clothing was determined based on the comfort zones diagram as 25.7 °C.

Mean radiant temperature does not differ significantly from cabin air temperatures. Because *T_Surf_AVG* including interior temperatures such as dashboard, roof, floor, doors was in the range of air temperatures *T1_Feet* and *T2_Head*. Only *T_Windshield* representing glazing parts was a little bit higher, see Figure 5. This implies that in this case EHT can be estimated by cabin air temperatures (no solar radiation, low air speed around manikin).

Figure 4 shows specific boundary conditions such as ambient temperature inside a climate chamber, ventilation air temperature, the voltage of fan, which is proportional to the amount of feeding air, and also the heat generated by passengers including thermal manikin.

Figure 5 presents prediction of cabin mean air temperature *T_IN*, which was mostly between measured air temperatures *T1_Feet* and *T2_Head*, measured by probes near Newton's feet and head. These results support the validity of VTSCC model, which was then used for calculation of car cabin heat loads (Figure 6).

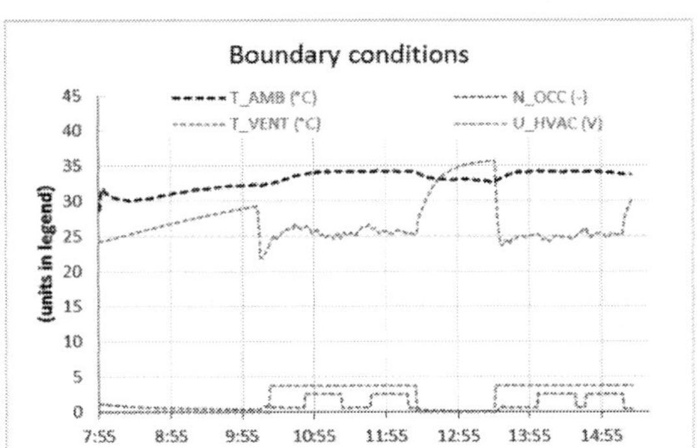

Fig. 4. Summer test for test subject. Boundary conditions.

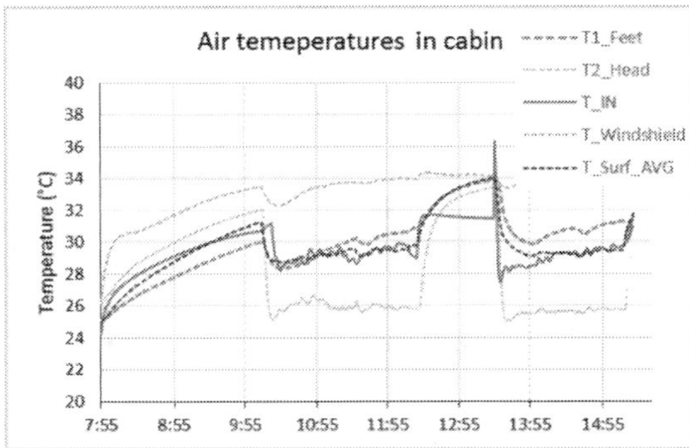

Fig. 5. Summer test for subject N.1-4. VTSCC prediction.

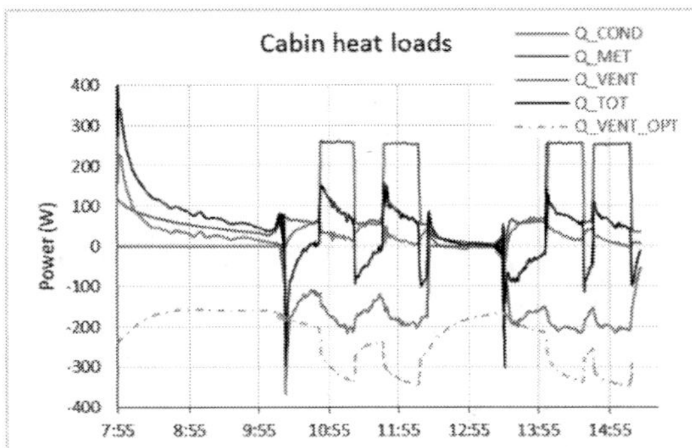

Fig. 6. Summer test for subject N.1-4. Cabin heat loads.

Q_MET was up to 250 W when there were people inside.
Q_COND was below 50 W and tended to the 0 W, when the inside air temperature approached to the outside temperature.
Q_TOT expresses a total heat load of car cabin, which is responsible for the change of temperature inside a cabin. The value was predominantly positive during the test, which implies slightly gradual increase of temperature inside a cabin. This was also observed during the test.
Q_VENT was always higher when people were into the car because the automatic A/C system regulates the air temperature inside cabin.
Q_VENT_OPT was calculated for optimal cabin temperature 23 °C. To achieve comfort level, it is recommended to cool down cabin by ~300 W from vents instead of ~200 W. More details can be seen in Figure 6 as the difference between green dashed line and blue solid line.

These results can be used for the recommendation to the HVAC system, what should be changed. In this case, there is recommended to cool down lower parts of human body, and to achieve the optimal comfort it is recommended to cool down a feeding air by ~100 W.

3.2 Winter conditions

In the winter conditions, the thermal manikin Newton tends to be slightly more sensitive to cold than human (Figure 7). The thermal manikin detects already cold but comfortable, whilst the human subject refers just neutral state. But the difference is up to ~0.5 of MTV scale, which is the minimal distinction of the comfort scale (Table 1). As in summer case, Newton's data better

https://doi.org/10.1051/epjconf/201818002085

match the experimental data considering also the global vote of test subjects, not only local votes.

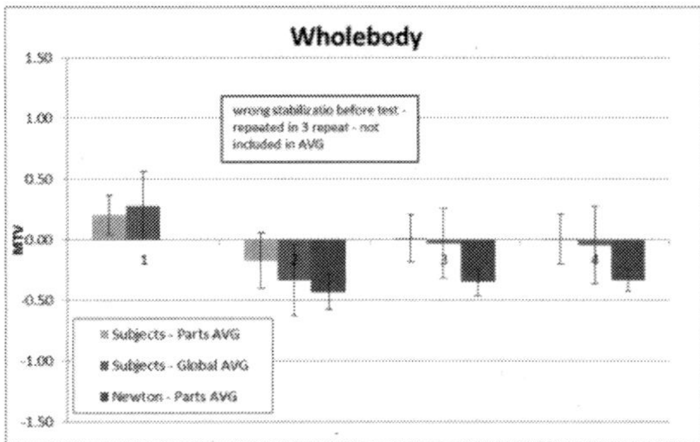

Fig. 7. Winter conditions evaluation of human subjects' tests – Wholebody MTV for all subjects N.1-10.

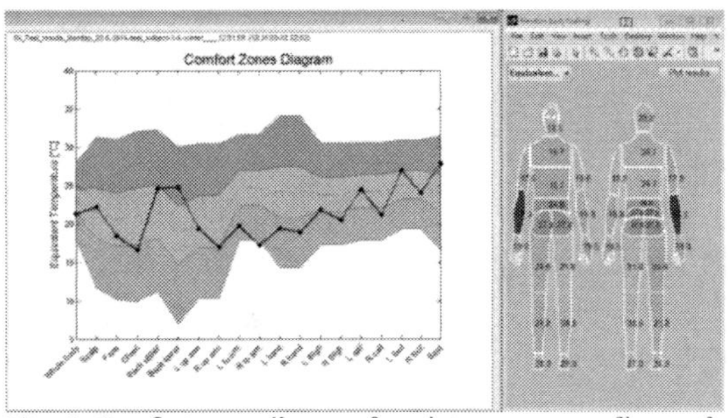

Fig. 8. Comfort zone diagram for winter; corresponding to the conditions of test subject N. 2 after 30 min.

A more detailed view on the conditions of test subject N.2 after 30 minutes (time 12:32) is shown in the Figure 8: Scalp, Face, Back Upper, Left Upper Arm, Left Calf and Feet were in neutral, whilst other parts were cold but comfortable, or too cold (Right Lower Arm). Back Lower and Seat (Pelvis) were the only warm parts, because of their contact with the seat. The cabin air temperature was then 23.7 °C (*T1_Feet*=23.2 °C, *T2_Head*=24.2 °C), which was even higher than the optimal EHT. For the winter clothing it was determined as 23 °C.

However, the mean radiant temperature differs from the air temperatures substantially, which is documented in the Figure 10. Interior surface temperature *T_Surf_AVG* including dashboard, roof, floor, doors was only ~20 °C and *T_Windshield* was even below 10 °C. This implies mean radiant temperature MRT ~18.3 °C and EHT ~ 21 °C, which explains why negative MTV was reported by manikin and human subjects even the air temperature was acceptable.

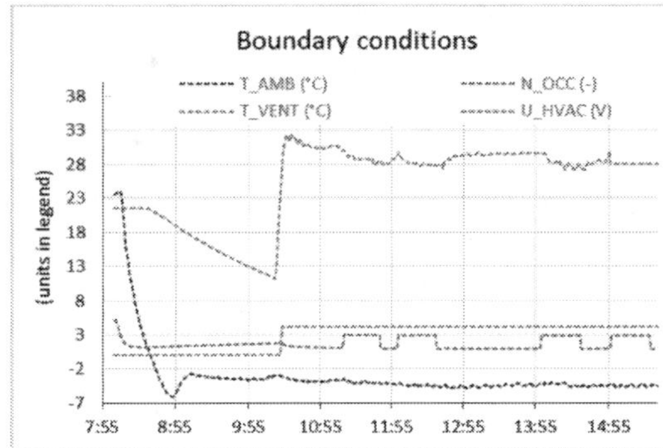

Fig. 9. Winter test for subject N.1-4. Boundary conditions.

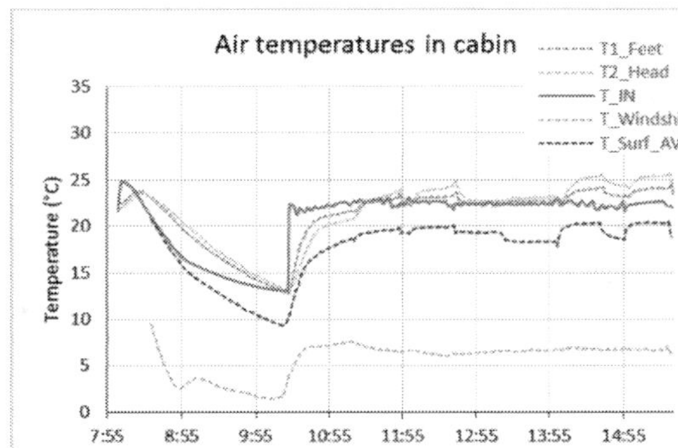

Fig. 10. Winter test for subject N.1-4. VSTCC prediction.

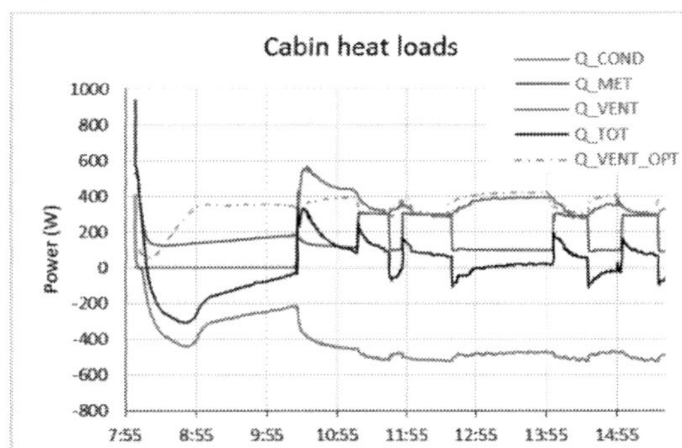

Fig. 11. Winter test for subjects N.1-4. Cabin heat loads.

Figure 9 shows specific boundary conditions of winter tests for test subjects N. 1-4. The predicted cabin air temperature *T_IN* was in accordance to the measured temperatures at the feet and head level (Figure 10). Calculated car cabin heat loads are then shown in the Figure 11.

Q_MET was up to 350 W except the start of preconditioning when the thermal manikin needed even more power to heat up itself.

Q_COND was below -500 W during the climate control system keeping temperature on the optimal temperature.

Q_TOT was positive just in case when test subjects were inside a car.

Q_VENT was ~300 W. Only at the beginning of the heating, before the subject N.1 entered into cabin, the ventilation heat load by feeding air was almost 600 W. This illustrates why test subjects waited 30 minutes after

HVAC was turned on: stabilization of conditions inside a cabin.

Q_VENT_OPT was also ~300 W to achieve optimal cabin temperature 23 °C see green dotted line and solid blue in the Figure 11, when the test subjects were inside a cabin. In case, when test subjects were not inside and HVAC was on (e.g. 13:55) it rises to ~400 W (similarly as real *Q_VENT*). It means that amount of heating power was adequate, but the distribution of heating power should be improved.

The most demanding parts for increase of heating were right legs and arms, especially right lower arm. The slightly cool sensations on these parts were caused by the effect of radiation from side windows, which were very cold. This effect was minimalized by climate control which distributed substantial part of energy to the feet and side vents, but still remained apparent and measureable by thermal manikin.

4 Discussion

The comfort zones method was adopted to evaluate thermal comfort by using thermal manikin Newton. By this method is possible to analyse local thermal comfort, which allows to make recommendation about distribution and directioning of air flow from vents. Results of human subjects proved that the method is reliable in both summer and winter conditions

Based on the VTSCC simulation is possible to recommend how many power is necessary to deliver into the cabin to reach an optimal comfort temperature inside a cabin. The VTSCC model was designed as 3-airzone model. However, only 1-airzone configuration was used for calculations in this study to simplify input data file. We incline to an opinion that for such kind of models as VTSCC, it is better to keep versatility of the model. It means to use simple model for different type of cars instead of more complex and detailed model valid only for one specific car. Which would require additional input data to define advection scheme of multi-airzone model and other localized parameters. The more detailed studies are better to carry out using the CFD (Computational Fluid Dynamics) simulations [14].

Despite the simplicity of the model we can obtain interesting information about the HVAC power needs to improve thermal comfort. The model also allows to considered effect of RH, however in this study the RH of feeding air was not measured.

In this moment only the total energy needed to provide thermal comfort temperature was considered. The next step would be the algorithm solving the problem how to prepare the ventilation air to provide optimal conditions.

5 Conclusion

The evaluation of thermal comfort using thermal manikin was verified. The information about the local thermal comfort by the simulator was relevant and helpful to investigate car cabin environment. In addition, the simulation of heat loads was performed to reveal energy demand to improve thermal comfort in car cabin

In this moment, the testing of real-time conditions and the real-time recommendation based on the CAN data provided by the car is under development.

Acknowledgement

The research was supported by the project LO1202 Netme Centre Plus with the financial support from the Ministry of Education, Youth and Sports of the Czech Republic under the "National Sustainability Programme I" and the project of the Josef Božek Competence Centre for Automotive Industry TE01020020.

References

1. E. Foda, *Evaluating local and overall thermal comfort in buildings using thermal manikins* (Phd Thesis, Aalto University, 2012)
2. M. Fojtlín, J. Fišer, J. Pokorný, A. Povalač, T. Urbanec, M. Jícha, Journal of Thermal Biology, **69** (2017)
3. ISO 14505-2, *Ergonomics of the thermal environment: Evaluation of thermal environments in vehicles: Part 2: Determination of equivalent temperature* (ISO standard, 2006)
4. T. Kiss, L. Chaney, J. Meyer, SAE Int. J. Passeng. Cars - Mech. Syst., **6**, 15 (2013)
5. M. Fayazbakhsh, M. Bahrami, SAE Technical Paper, 14 (2013)
6. D. Marcos, F. J. Pino, C. Bordons, J. Guerra, Appl. Therm. Eng., **66**, 11 (2014)
7. B. Torregrosa-Jaime, F. Bjurling, J. M. Corberán, F. Di Sciullo, J. Payá, Appl. Therm. Eng., **75**, 9 (2015)
8. H. Lee, Y. Hwang, I. Song, K. Jang, Energy, **90**, 10 (2015)
9. F. Brèque, M. Nemer, *12th international Modelica Conference* (2017)
10. J. Pokorný, J. Fišer, M. Jícha. *In Proceedings of Ambience14& 10i3m*, 6 (Tampere, 2014)
11. J. Pokorný, J. Fišer, M. Jícha, Advances In Engineering Software, **76**, 8 (2014)
12. J. Pokorný, J. Fišer, M. Fojtlín, M. Jícha, *Simulace budov a techniky prostředí* (2016)
13. D.P. Wyon, S. Larsson, B. Forsgren, and I. Lundgren, *SAE-Technical Paper 890049*, 11 (1989)
14. R. Ambs, *SAE Technical Paper,* 10 (2002)

On the effect of moving blade grid on the flow field characteristics

Pavel Procházka[1,*], *Václav* Uruba[1], *Luděk* Pešek[1] and *Vítězslav* Bula[1]

[1]Institute of Thermomechanics, Czech Academy of Sciences, Czech Republic

Abstract. The motivation of this paper is the continual development of the blades for the last stage of the steam turbine. The biggest problem is the slenderness of such blades and the extreme sensitivity to aeroelastic vibrations (flutter) caused by the instabilities present in the flow. This experimental research is dealing with the aeroelastic binding of the moving blades located in the blade grid with the flow field and vice versa. A parallelogram is used to ensure one order of freedom of the blade. The grid has five blades in total, three of them are driven by force control using three shakers. The deviation as well as force response is measured by strain gauges and dynamometers. The flow field statistical as well as dynamical characteristics are measured by optical method Particle Image Velocimetry. The grid is connected to the blow-down wind tunnel with velocity range up to 40 m/s. The aeroelastic binding is investigated in dependency on used actuation frequency and maximal amplitude (the intensity of force actuation) and on different Reynolds numbers. The flow field and the wake behind each individual blade are studied and the maximal interaction is examined for individual inter-blade phase angle of the grid.

1 Introduction

This work was established as a responce to a necessary need to further improve the efficiency of the steam energy conversion in the last stages of the steam turbine. The blade-tip region, whose airfoils have very sharp leading edge and low camber, is high-loaded by low supersonic relative velocities and flutter phenomenon or even high cycle fatigue damage are present. The experimental or numerical investigation of the flutter in low pressure turbine (LPT) blades is then in the center of interest for many research groups which have been still dealing with turbomachinery blade dynamic stability. The aeroelastic binding is very often studied by employing of self-excited blades in the cascade.

This research is related to modern turbofan engines [1] as well as to steam turbines. The numerical studies are validated by experimental investigation where the aerodynamical forces and momentum [2] and unsteady pressures [3] are measured, or the velocity components are determined directly. Nowinski's work [3] is based on unsteady pressure measurement made along the test section outer wall with the goal to eliminate flutter in LPT blades. This was conducted in annular cascade. The investigation of planar blade cascade is more common case. Tsymbalyuk et al. [2] has investigated the blade dynamic stability. His experiment is based on self-excited vibrations of the airfoil cascade using the airfoil suspension system which has two degrees of freedom – the displacement and the rotation. The measurement of flutter conducted in high subsonic aerodynamic tunnel as well as PIV measurement of single vibrating wing is described in [4]. Results of ref. [4] revealed that the flow is mainly affected by the changing wing angle and the effect of displacement is rather small.

This experimental work is closely linked to the project dealing with aeroelastic couplings and dynamic behaviour of rotational periodic bodies [5, 6]. Vimmr et al. [5] carried out a numerical study which uses a simplified model of blade cascade (3 airfoils) which perform harmonic motion and with individual inter-blade phase angle. This angle is chosen in order to set a travelling wave mode of vibration which is present in the rotational solid bodies. The aim of this study is to determine the aerodynamic damping coefficients caused by the aerodynamic forces acting on the flat plates. Půst et al. [6] has described analytically the dynamic properties of the five-blade-cascade excited by the running harmonic external forces and aerodynamic self-excited forces were investigated.

The aim of this study is to examine the flow properties in the vicinity of moving blade cascade. The mean flow statistics are specified in dependency on the various phase angles. The fast dynamical interaction between adjacent profiles is examined with the help of the time-resolved Particle Image Velocimetry (TR-PIV). In this article, the effect of different Reynolds numbers, various amplitudes and frequency values are analyzed with respect to fluid-structure interaction problems.

* Corresponding author: prochap@it.cas.cz

https://doi.org/10.1051/epjconf/201818002086

2 Experimental setup

The blade cascade was designed and fabricated in a geometrical arrangement as it can be seen from figures 1 and 2. The blade profile is simple flat plate with semi-cylindrical leading and trailing edges. The chord length is 52.8 mm and each blade takes the angle of attack of about 11.4°. The distance of two adjacent blades in perpendicular sense to the chord is 16 mm. Each blade is attached using parallelogram (two beams linked by the four spring-steel belts) to the supporting sidewall. This mechanism enables to each blade only the displacement in the perpendicular direction. No torsion and tilting are available. The profile length is finite, it was designed to spread across the wind tunnel outlet nozzle and its value is 80 mm. The leading and the trailing blades are auxiliary. The second, third and fourth blades (blade A, B, C) were self-excitated by LDS electrodynamic exciters. The exciter was linked with the blade by a rod. To measure inertial, elastic and aerodynamical forces, a force transducer was placed at the end of this rod. The movement of the blades is measured by the foil strain-gauges. The blade excitation is established independently of others and it is given by prescribed amplitude, frequency and phase angle. The harmonic movement of blade C and blade A were phase shifted $\Delta\phi$ and $-\Delta\phi$ to middle blade. Five distinct regimes of the excitation were tested. The parameters are available in table 1.

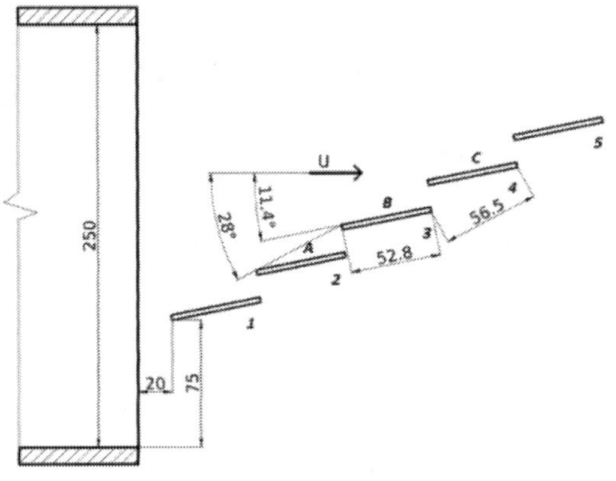

Fig. 1. Blade cascade, plane of symmetry.

Table 1. Tested regimes.

Regime number	Frequency (Hz)	Force (N)	Peak-to-peak ampl. (mm)
I.	20	15	2.233
II.	20	19	2.946
III.	30	13.5	2.729
IV.	40	5	1.884
V.	40	7.5	2.828

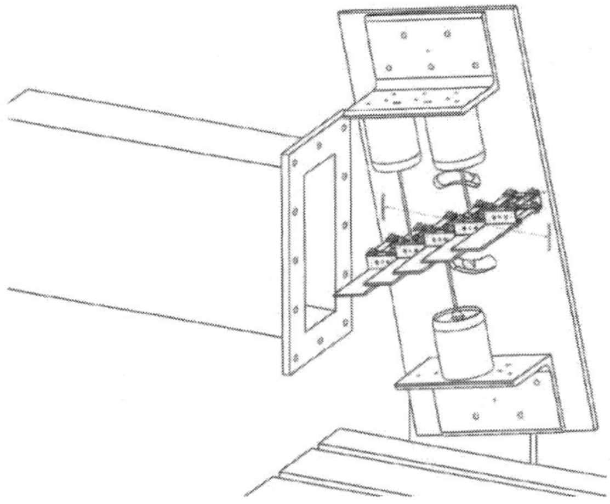

Fig. 2. Experiment arrangement.

The blade cascade was connected to the outlet nozzle of the blow-down wind tunnel with cross section dimensions of 100x250mm. The leading edge of the first blade was 20 mm from the outlet nozzle. The aeroelastic behaviour was studied also in dependency on various Reynolds numbers which was achieved by different outlet velocity (10, 20, 30 and 40 m/s). The turbulence intensity level is about 0.1% measured at distance of 0.26 m from the nozzle [7]. The outcoming flow is perfect steady in time and the flow enters the blades at the angle -11.4° (the suction side is below one). The preferred velocity was 40 m/s and the main task was to study the effect of the changing inter-blade phase angle. This angle corresponds to different travelling wave mode and the strongest interaction between two adjacent profiles is searched. The phase angle equal to zero corresponds to the situation when all three blades are performing the harmonic motion synchronously. The phase angle equal to π describes the state when the blade A and C make the motion synchronously and the blade B goes in opposite sense. The PIV measurements were conducted for following values of phase angle $\phi=0$, $\pi/4$, $\pi/2$, $3\pi/4$, π, $-3\pi/4$, $-\pi/2$, $-\pi/4$.

To study the flow field, an optical anemometry method with good time and space resolution is desired. Since the interactions of the wakes of adjacent airfoils should be examined, TR-PIV is very convenient as it is not single-point measurement method. There is possible to acquire data with sufficient high acquisition frequency and make dynamical analysis of the flow. The measurement apparatus is from Dantec Dynamics company. To illuminate the tracing particles (safex generator), laser New Wave Pegasus Nd:YLF with double-head is used (frequency 10 kHz, wavelength 527nm, energy in one pulse 10mJ for 1 kHz). Fast PIV camera Phantom V611 (up to 6000 double snaps per second, spatial resolution 1280x800 pix, memory 8 GB) is used to capture the particle movement. The system synchronization is performed via timer box. Dynamic Studio ver. 3.4 is utilized to capture the data and post-process data. Adaptive correlation with embedded validation is used to calculate velocity vector fields.

Interrogation area of 16x16 pixels results in 159x99 vectors (maximal 5% invalid vectors). Further data analysis and plotting is performed in Matlab or in Tecplot software.

Since the flow field in the vicinity of this cascade with moving profiles has pseudo-periodical nature, it is very convenient to sample the data not only for statistical and dynamical properties calculation but it is also desirable to do phase-locked measurement. The flow is acquired with the frequency of 100 Hz during at least ten seconds to be able to calculate mean flow field and all higher statistical moments. As it will be apparent later in the following chapter, these results will not be so helpful in aeroelastic binding description. The main focus of this article is devoted to the dynamical analysis. The acquisition frequency should be as high as possible. For flutter phenomenon investigation, the data were acquired with the frequency of 2 kHz during two seconds. The phase-locked measurement is very convenient to suppress the random (turbulent) component of the flow and to highlight the periodical component which is given by harmonic periodic motion of the blades. Eight phases per period (which was given by the excitation frequency) were acquired during several seconds (there is limitation due to camera memory). The dataset of each phase is averaged over all measured periods and consequently the standard deviation (STD) is calculated for each phase averaging.

3 Results

Although the blade cascade was tested during four different tunnel outlet velocities, the main focus is set to 40 m/s in this article. The low-speed incoming flow is considered as incompressible. Base case (the regime without self-excitation) was investigated in detail using PIV as well as several regimes for different excitation parameters given table 1 and different inter-blade phase angles. In additional, several regimes were tested when only one or two blades were performing harmonic motion. The topology of the mean flow field can be clearly seen from figure 3. Black vectors represent the velocity vector orientation and the scalar value is plotted as vector module. The origin of the coordinate system is placed in the leading edge of the third profile (blade B). The first part is plotted for base case, second one depicts one of the regimes from table 1. The middle profile is not excited in the third part. Although all flow fields were taken during very different conditions, the resulting flow topology is exactly the same. There is a massive flow separation at the first blade (which is not present at these pictures) and the flow is deflected toward the other blades 2, 3, 4 and 5. The angle of attack is significantly reduced for other blades in cascade. There is clear separation at the suction side of each blade right behind the leading edge. The flow is re-attached at approximately 20% of the chord length again. The flow is accelerated in this region about 10-20% compared with the incoming flow velocity. The blade wake is slender and it contains a sequence of vortices separated periodically from the blade surface at relative high frequency. Note that flow field is incomplete in front of the cascade as the laser illumination was located from the back side. Gentle increase of the velocity in the inter-blade channel is caused by the blade thickness and of by course the existence of boundary layers. The figure 4 is plotted for velocity variances (a sum of variances of both components). There is also no visible effect if we compare different regimes (the comparison is made here for regime I and several phase angles).

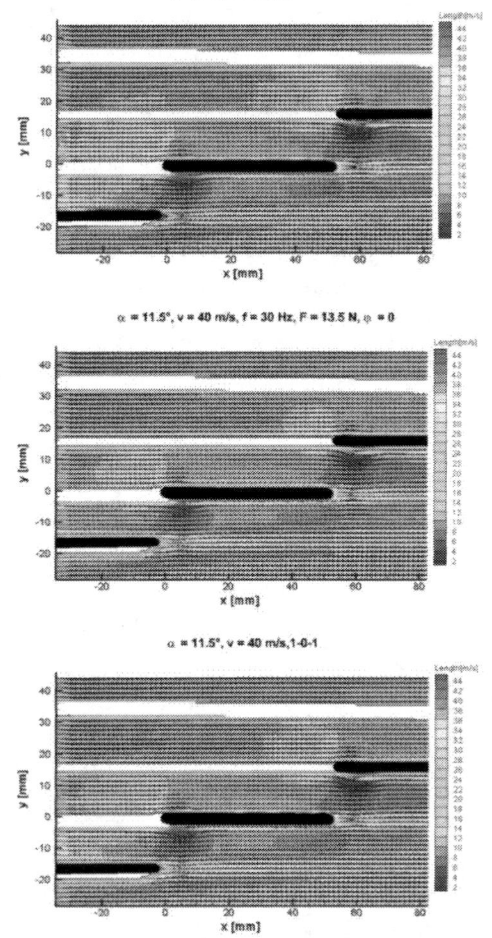

Fig. 3a, b, c. Comparison of mean flow field topology.

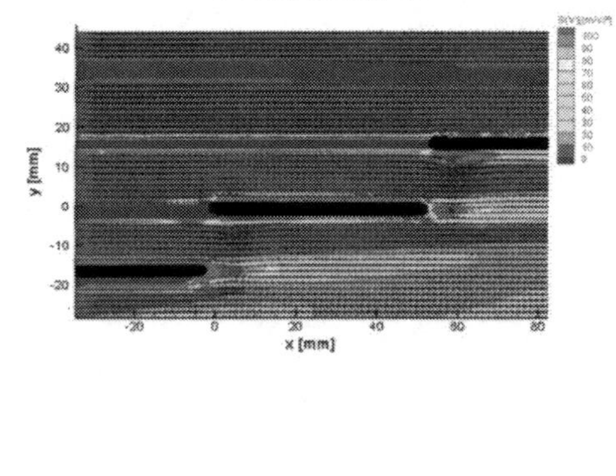

https://doi.org/10.1051/epjconf/201818002086

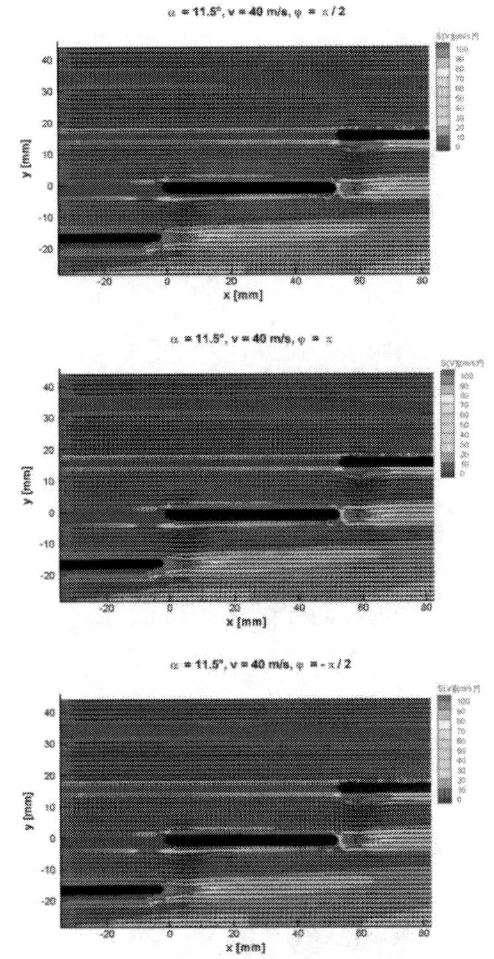

Fig. 4a, b, c, d. Velocity variances. Comparison of four distinct phase angles ϕ=0, $\pi/2$, π, - $\pi/2$.

The analysis of statistical quantities has not brought any clarification. There is a need to study the time-dependent flow quantities and especially frequencies of the velocity fluctuation. Figure 5 is plotted for time-dependent record of the velocity magnitude at distinct place far from the blade surface for base case (5a) and for regime I (5b). The records are plotted for four periods (given by excitation frequency). The excitation caused clear periodical changes of the velocity whose frequency is related to the excitation frequency. Turbulent noise has been superimposed on both waveforms.

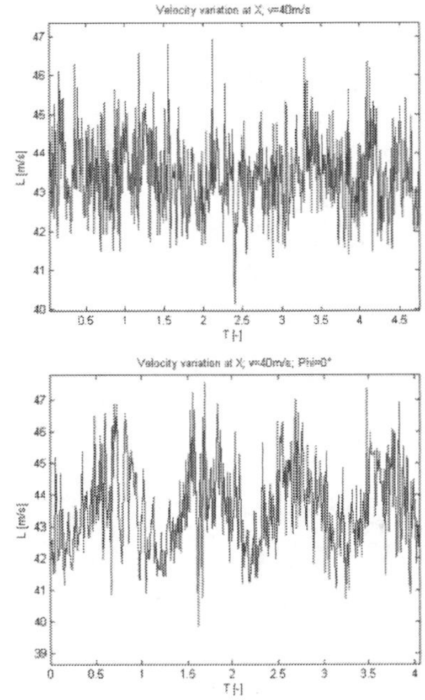

Fig. 5a, b. Velocity variation during 4 periods at point X. Base case versus regime I.

Fast Fourier transform (FFT) has been computed in each point of the flow domain. Power spectral density (PSD) of related frequency (excitation frequency) was consequently plotted as a scalar value in the flow domain. The areas of maximal intensity correspond with the position of maximal aeroelastic binding between the flow and blade cascade. These positions can be seen in figure 6 where the diagram is divided for u and v velocity component and the effect of Reynolds number is also visible. The stronger binding is present for velocity of 40 m/s and u velocity component. The most interesting spots are present in the vicinity of the leading edges and also in the wakes. A shadow in front of blades is caused due to missing data. At upper and lower side from the leading edge, there are two spot of augmented activity and there is also remarkable formation at the location of the separation bubble. This formation is very close to the wake of the preceding blade and the binding would increase rapidly if these two areas crossed. Nevertheless, all of these structures relate to motion of the blades as there are no such structures at all for the base case. Figure 7 shows the dependency on phase angle. This angle varies to 0, $\pi/2$, π and $-\pi/2$. The binding is the weakest paradoxically for the angle π when the blades move again each other. On the other hand, the angle equal to zero gives much stronger binding. Similar result is possible to see from the vertical velocity component. The increasing of the excitation frequency has led to a growth of the binding area in the volume and also in the maximal intensity. This effect proves at most for 40 Hz (regime V), the increment is not so significant for 30 Hz (regime III). Figure 8 shows the comparison of regimes I, III and V for phase angle equal to zero.

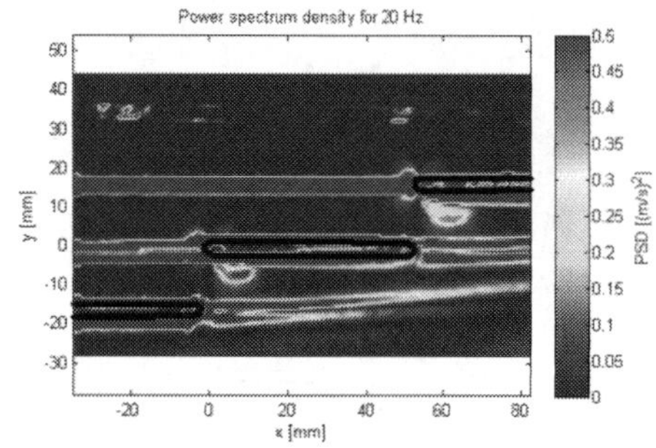

Fig. 6a. PSD scalar map for 20Hz, 20m/s for u-component.

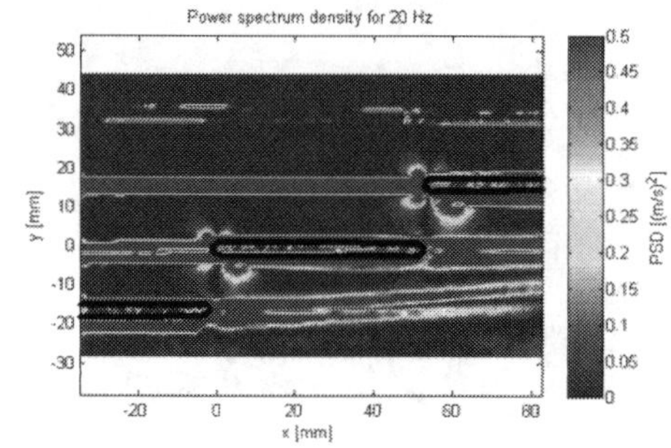

Fig. 7a. PSD scalar map for 20Hz, phase angle ϕ=0.

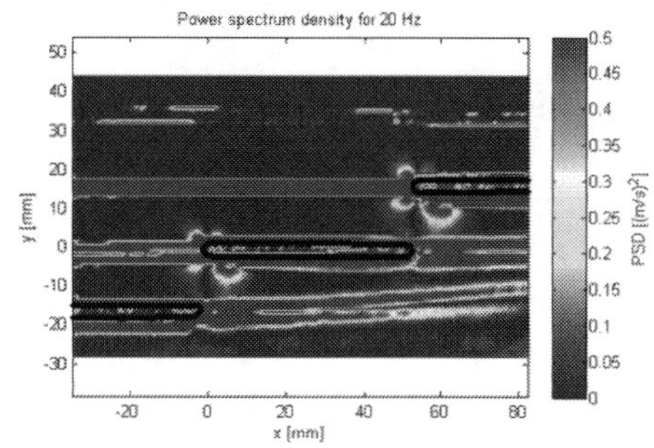

Fig. 6b. PSD scalar map for 20Hz, 20m/s for v-component.

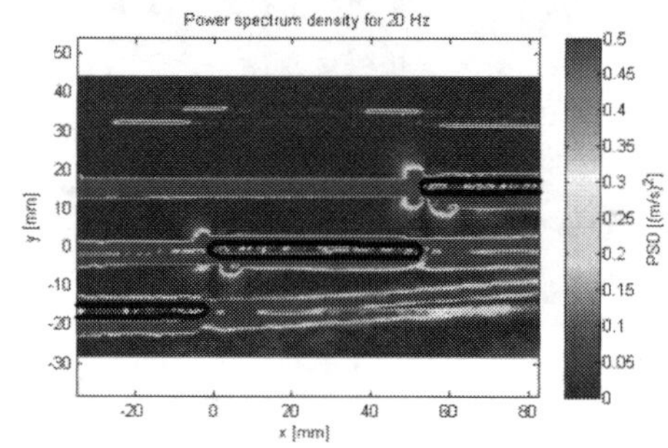

Fig. 7b. PSD scalar map for 20Hz, phase angle $\phi = \pi/2$.

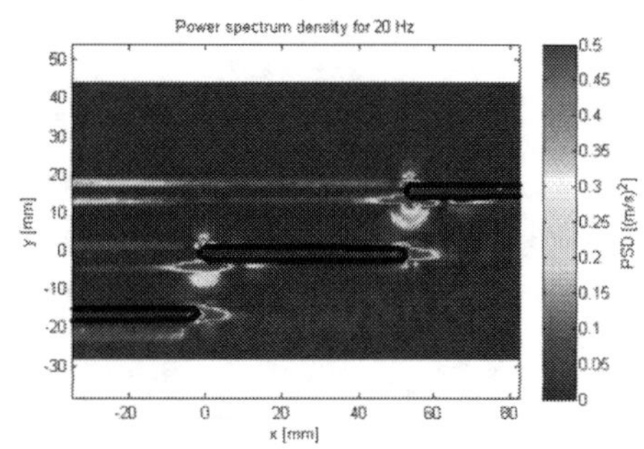

Fig. 6c. PSD scalar map for 20Hz, 40m/s for u-component.

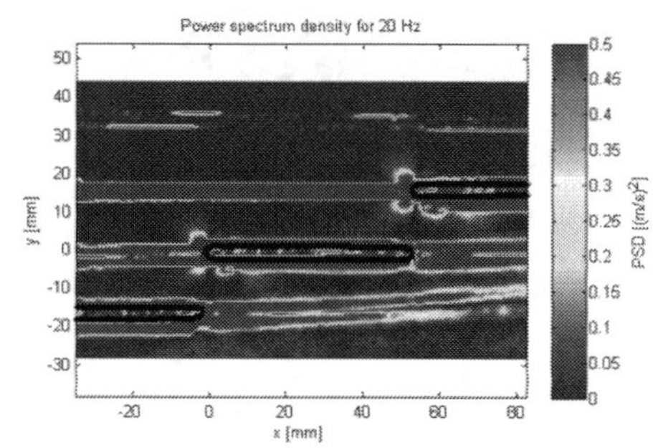

Fig. 7c. PSD scalar map for 20Hz, phase angle $\phi = \pi$.

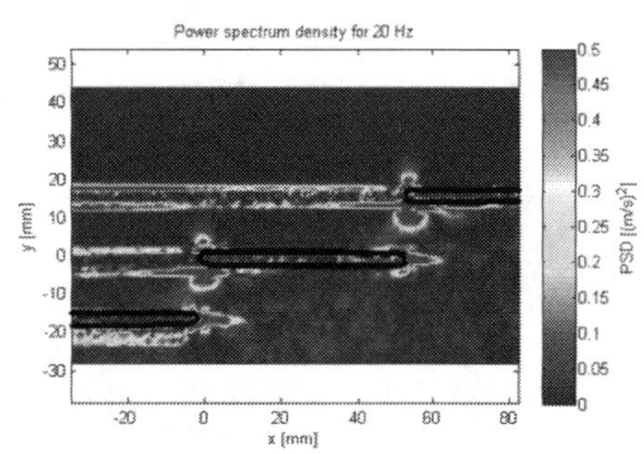

Fig. 6d. PSD scalar map for 20Hz, 40m/s for v-component.

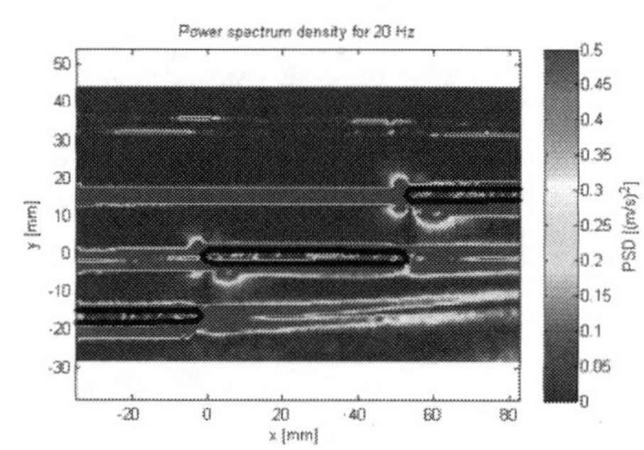

Fig. 7d. PSD scalar map for 20Hz, phase angle $\phi = -\pi/2$.

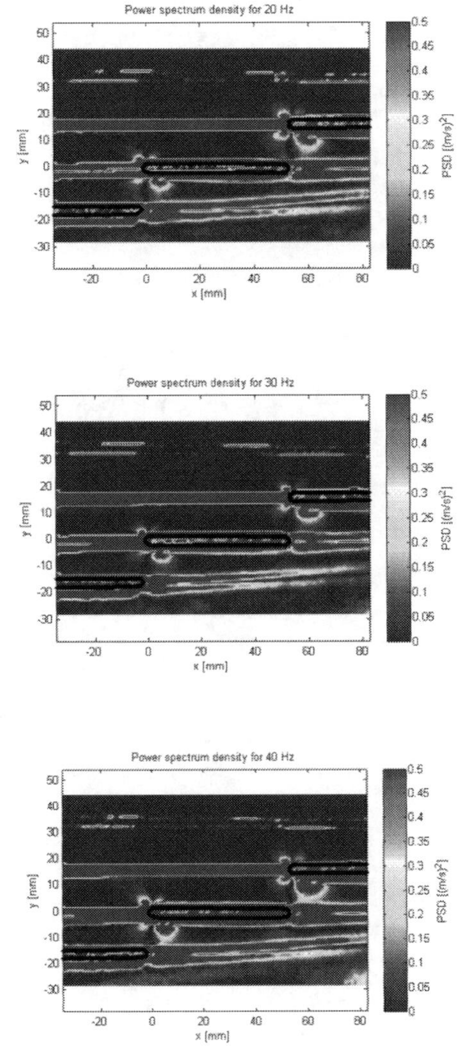

Fig. 8a, b, c. Phase angle ϕ=0, regimes I, III and V.

Detailed analysis of spectral density was performed in specific points (figure 9). The emphasis was given to points 2, 3 and X. X-point was calculated from all experimental data as an average coordinate of the maximal intensity locations. There is power spectral density for frequencies up to 50 Hz in figure 10– the calculation is performed separately for u and v velocity component (as previously). Maximal peak value of PSD was identified for vertical velocity component at points 2 and 3. However, the situation at the point X is completely different. There is a huge peak value for horizontal component and its value is two orders bigger compared to previous points. Also peak value of vertical component is orderly higher. The aeroelastic binding is also the strongest at point X. Next comparison was evaluated for the spectral density at point X for different phase angles. This influence corresponds to figure 7. The biggest peak value occurs for the angle equal to zero and on the contrary the binding is rather weak for angle π. The regimes generated at π/2 and – π/2 are similar. The peak value of PSD was analyzed also in dependency on excitation regime. Figure 11 shows spectral density for angle ϕ= π for regimes I, II, III and V. The maximal peak value occurs for the second regime. This is given due to the fact that this regime disposes of maximal deviation amplitude. The regime V as the only one has the significant peak value also for vertical velocity component making it the most binding regime.

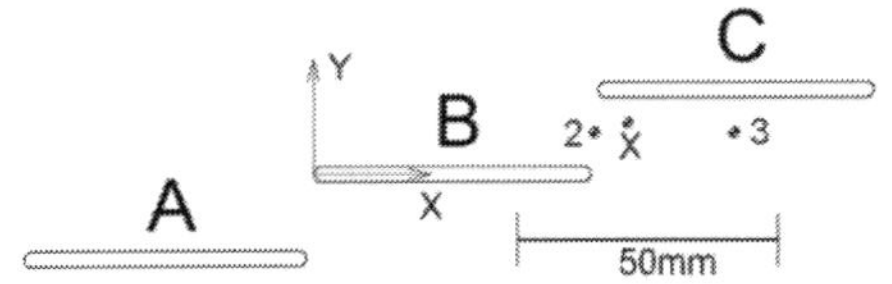

Fig. 9. Location of points under investigation.

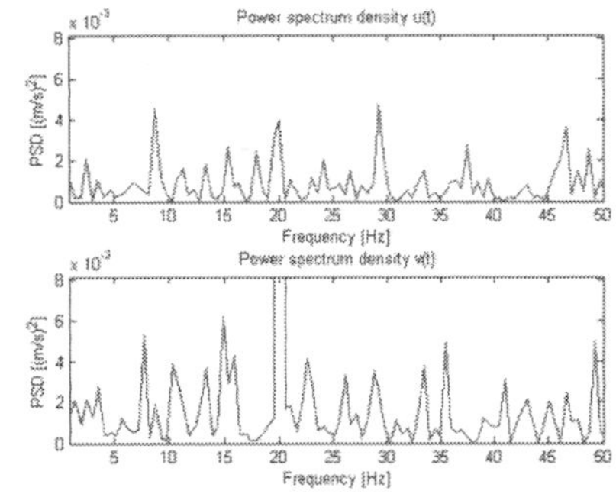

Fig. 10a. Phase angle ϕ=π. Frequency spectrum at point 2 for regime I.

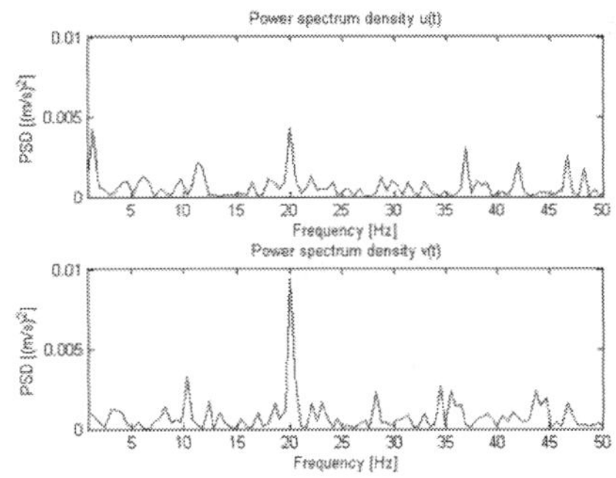

Fig. 10b. Phase angle ϕ=π. Frequency spectrum at point 3 for regime I.

https://doi.org/10.1051/epjconf/201818002086

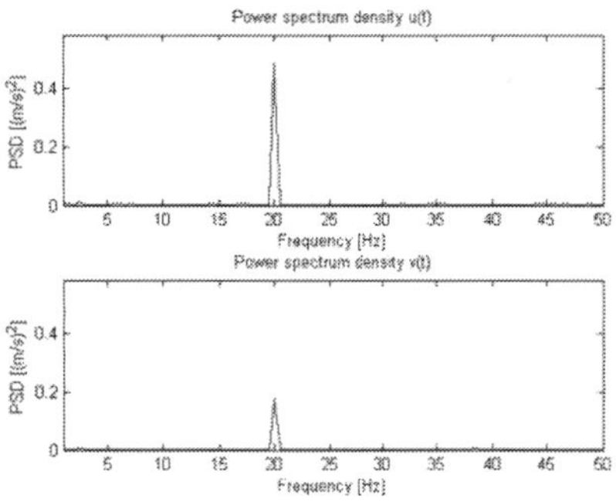

Fig. 10c. Phase angle $\phi=\pi$. Frequency spectrum at point X for regime I.

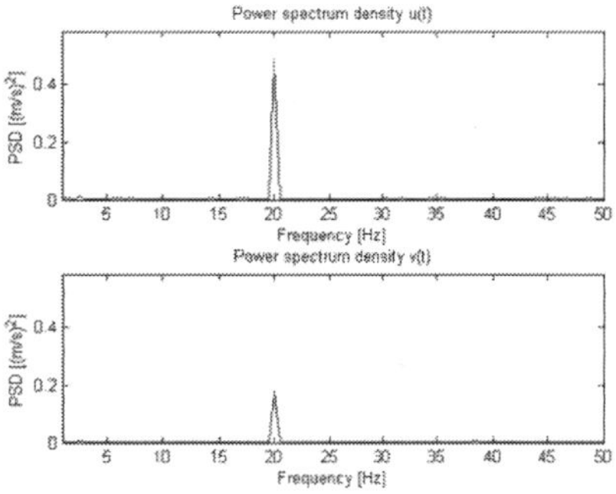

Fig. 11a. $\phi=\pi$. Freq. spectrum at point X for regime I.

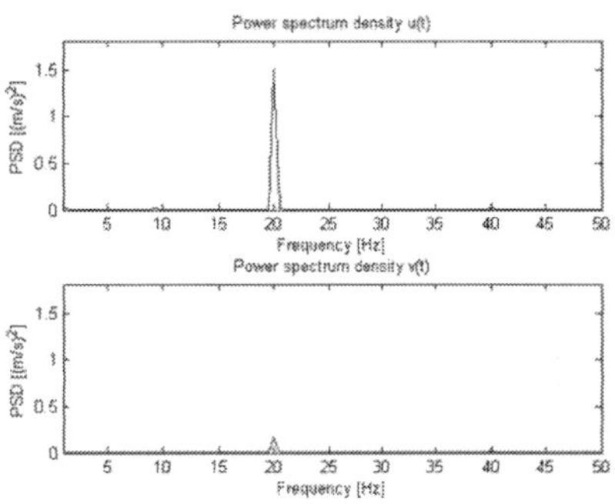

Fig. 11b. $\phi=\pi$. Freq. spectrum at point X for regime II.

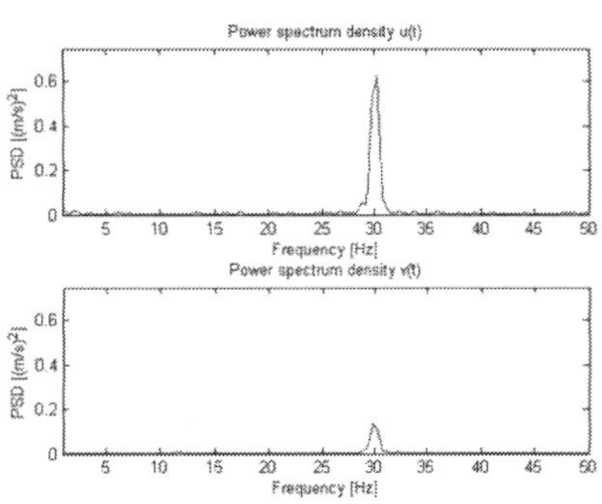

Fig. 11c. $\phi=\pi$. Freq. spectrum at point X for regime III.

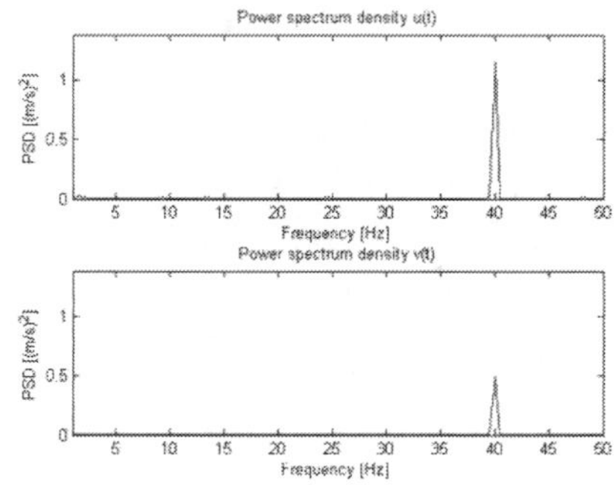

Fig. 11d. $\phi=\pi$. Freq. spectrum at point X for regime V.

The fact, that aeroelastic binding between flow field and moving blades are rather weak, is best characterized by data gained using phase-locked measurements. Figure 12 is depicted for 2 periods of harmonic motion of the blades. Each period is divided into eight phases. The phase-mean flow velocity approximately corresponds to harmonic motion. The lower diagram is plotted for standard deviation of phase-mean velocity. There is a comparison between points 2 and X for inter-blade phase angle of 0 in figure 12. The mean-phase velocity varies in much smaller range (one order lower) than the mean value of STD for point 2 (and also 3). The peak-to-peak amplitude is slightly higher than mean STD only at point X. It is obvious that turbulent random noise predominates over periodic velocity changes given by moving blades for absolute majority points located in inter-blade channel. It is necessary to emphasize that actual location of the point X differs slightly for every regime thus it is very sensitive task. The similar situation can be observed for inter-blade phase angle of π. No dominant periodic velocities are present.

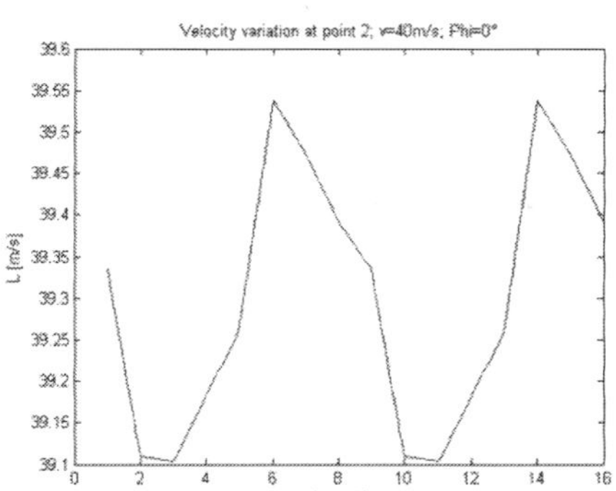

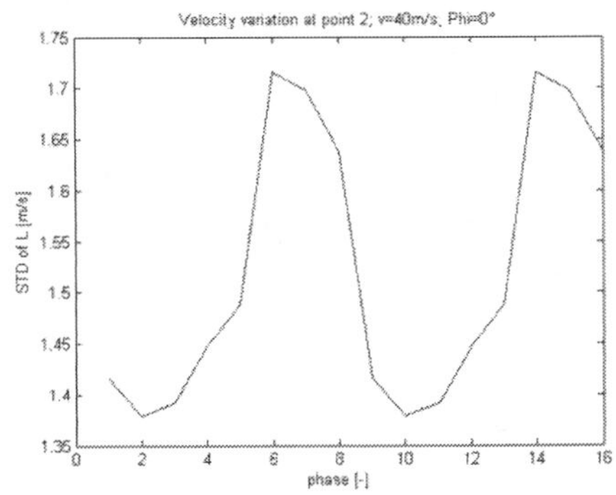

Fig. 12a, b. Phase-locked measurement – mean-phase velocity and STD value - at point 2.

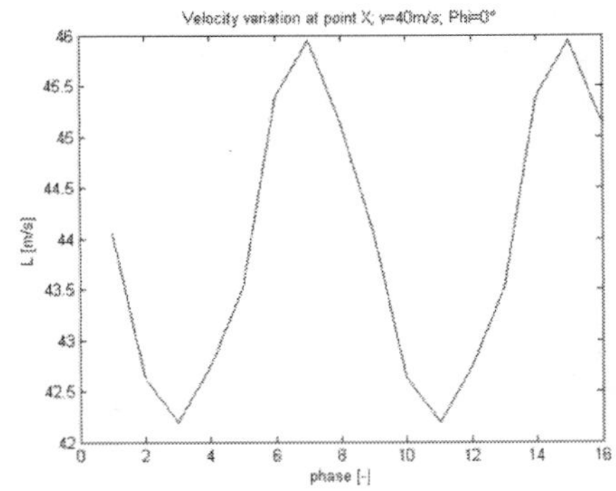

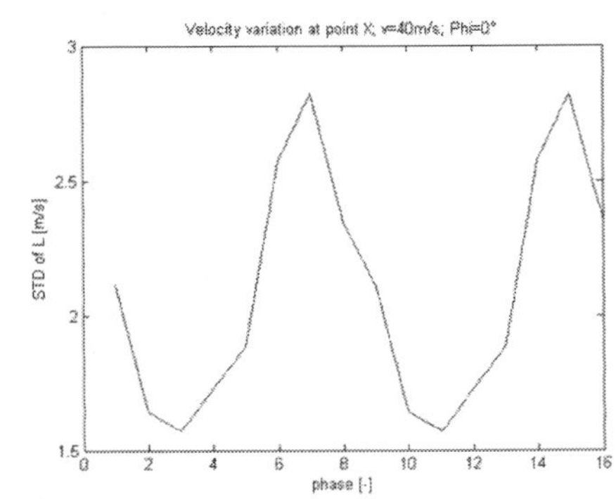

Fig. 12c, d. Phase-locked measurement – mean-phase velocity and STD value - at point X.

The effect of phase angle choice is apparent also from phase-locked measurements. Again it is valid that for angle equal to zero the binding is more powerful and the weakest binding is observed for $\phi = \pi$. While the peak-to-peak amplitude of mean-phase velocity is 3,9 m/s for $\phi = 0$, moving blades again each other gives only 2,2 m/s. Travelling wave defined by angle $\phi = -\pi/2$ or $\pi/2$ gives approximately 2,5 m/s. The average STD value is 2,25 m/s for all mentioned cases.

4 Conclusion

The experimental investigation of the flow field in the vicinity of the blade cascade with self-excitation blades was performed using TR-PIV. The outlet tunnel velocity varied from 10 to 40 m/s. The aeroelastic binding between the flow and moving blades was observed and quantitatively evaluation was done from spectral analysis and phase-locked velocities. It seems that the biggest interaction occurs for inter-blade angle equal to zero and for high magnitudes of peak-to-peak blade deviations.

The binding would increase further if the areas were interdigitated. These could arise after that the excitation frequency or blade deviation is increased, optimally both. Now the movement of the blade in one degree of freedom is too low compared to the incoming flow velocity. The velocity of the blade is two orders of magnitude smaller compared to the flow speed. Hence, it is not enough to create the stronger binding. The flow structure has only tiny information about the blade movement during its flow through the inter-blade channel. To increase the velocity of the blade deviation is limited due to parameters of the excitation actuators. Further option is add the second degree of freedom which is rotational movement of each individual blade which should result in much stronger binding due to changing value of angle of attack.

Acknowledgement

This work was supported by the research project of the Czech Science Foundation No. 16-04546S "Aero-elastic couplings and dynamic behaviour of rotational periodic bodies".

References

1. J. Lepičovský, E. R. McFarland, V. R. Capece, T. A. Jett, R. G. Senyitko, NASA/TM-211894 (2002)
2. V. Tsymbalyuk, J. Linhart, XVII IMEKO World Congress, TC3 (2003)
3. M. Nowinski, J. Panovsky, J. Eng. Gas Turbines Power 122(1), 82-88 (1999)
4. V. Uruba, M. Knob, V. Vlček, M. Luxa, J. Lukáč, R. Dvořák, Interaction and Feedbacks, 109-114 (2008)
5. J. Vimmr, O. Bublík, A. Pecka, L. Pešek, The 13th ISAIF in Okinawa, ISAIF13-S-0089 (2017)
6. L. Půst, L. Pešek, International Journal of Bifurcation and Chaos, IJBC-D-16-00353R2 (2017) – to be published
7. J. Jonáš, O. Mazur, V. Uruba, IV. Watermanagement conference, Prague (2004)

https://doi.org/10.1051/epjconf/201818002087

Dynamic interferometric measurement with extended unambiguity range in flow measurement

Pavel Psota[1*], *Roman* Doleček[1], *Vít* Lédl[1], and *Tomáš* Vít[2]

[1]Institute of New Technologies and Applied Informatics (NTI), Faculty of Mechatronics, Informatics and Interdisciplinary Studies, Technical University of Liberec, Studentská 2, 461 17 Liberec, Czech Republic
[2]Institute of Energetic Devices (KEZ), Faculty of Mechanical Engineering, Technical University of Liberec, Studentská 2, 461 17 Liberec, Czech Republic

Abstract. This paper reports on a new approach to measure dynamic processes in fluid mechanics using interferometry with extended dynamic range. A key factor is the use of two wavelengths and the recording of interferograms from both wavelengths in one frame. Phase map evaluation is based on the Fourier transform. The difference between the obtained phase fields creates a synthetic phase whose dynamic range also covers large changes of the measured quantity.

1 Introduction

Interferometric methods such as digital holographic interferometry and interferometry are generally understood to belong to the group of most accurate measuring devices [1]. Their fundamental measure is the wavelength of light and the todays interferometer can resolve even a fraction of the wavelength, so their sensitivity is outstanding. Due to their great accuracy interferometers in different variations are used in many branches of measurement including the fluid mechanics. However, due to the bounded nature of a light wave, the measurement range given by the unambiguity is squeezed within the range of one cycle of the wave. Measured phase maps are therefore wrapped within the range of 2π radians.

Unless the measured temperature field shows a small gradient (there is no violation of the Nyquist conditions when sampling the interferogram by the camera), the phase field can be expanded using standard 2D algorithms to obtain a smooth phase field without jumps. The ambiguity of interferometric measurement requires knowledge of phase change at least at one point. If we consider the phenomenon taking place only in a small area in the middle part of the field of view (e.g. jets), it can be assumed that the change at the edge of the field of view will be minimal and does not cause a phase change of more than 2π. This area can be considered as a reference wave and can be used as a starting position for unwrapping procedure. The measurement is then absolute in terms of temperature change.

However, many tasks in the flow area do not have these properties. Often there is a significant change in the measured quantity across the whole field of view. As an example, you can measure heat transfer inside the prime mover resonator or flow around the wing at high speeds. If we want to quantify the observed phenomenon correctly, we must measure this quantity at least at one point by another method. For instance, if we measure the temperature gradient inside the cavity, a single-point temperature probe can be placed here. The value at this point will be the reference value at which the measured values can be related to the interferometric value. This approach is naturally in contradiction with the big advantage of contactlessness of interferometric measurement.

Multiple wavelengths [2] interferometry can be used to increase the dynamic range of the measurement. Two or more interferograms at different wavelength are successively captured and the phase information is combined in order to get higher dynamic range of the measurement. These measurement methods are suitable for stationary phenomena that are quite rare in the fluid mechanics measurement. In this paper we propose a method based on carrier frequency interferometry with simultaneous use of two wavelengths. This makes it possible to vary the dynamic range of measurement almost arbitrarily and it is therefore possible to measure in real time phenomena causing large phase changes without the need for a reference gauge.

2 Principle of the method

Digital interferometry is a very accurate measurement method that is sensitive to the change of optical wave phase. By recombining two coherent waves, an interference pattern I is created in Cartesian coordinates (x,y) described by the equation:

$$I(x,y) = A(x,y) + B(x,y)\cos(\varphi(x,y)) \quad (1)$$

* Corresponding author: pavel.psota@tul.cz

where A represents additional component, B is multiplicative component and φ stands for phase. Phase-shifting interferometry (PSI) is usually used for quantitative phase evaluation of the phase map $\varphi(x, y)$. In PSI the phase shift is repeatedly added to the cosine function argument of equation (1). There are many ways how to generate the phase shift in interferometers like a slight movement of mirror in reference arm or wavelength change. From (1) follows that at least 3 interferograms must be recorded (three equations for three variables A, B, and φ), and therefore PSI is not suitable for measuring of dynamic events.

In contrast, the spatial carrier frequency interferometry (SCI) introduces carrier frequencies υ_x, υ_y into a well-known interference equation (1):

$$I(x,y) = A(x,y) + B(x,y)\cos\left(2\pi\upsilon_x x + 2\pi\upsilon_y y + \varphi(x,y)\right). \quad (2)$$

This generates a high density interference pattern. Due to the carrier frequencies, we are able to separate the wavefront, its conjugate part and the DC term, see Fig.1. After performing the FT, we get the phase map directly as an angle of the complex field. SCI is therefore suitable for measurement of fast phenomena since the phase is evaluated from single hologram. For proper working of SCI it is necessary to set the carrier frequencies correctly so that all the spectral parts can be separated and in the same time the sampling criterion (Nyquist frequency υ_N), which is given by the pixel extension of the digital image sensor, is not violated. In practice one can observe the Fourier spectra of the interferogram in real time and shift the wavefront in e.g. $\upsilon_x = \upsilon_y = 0.5\upsilon_N$ by adjusting the tip/tilt of a mirror in the reference arm of the interferometer, see Fig. 1c.

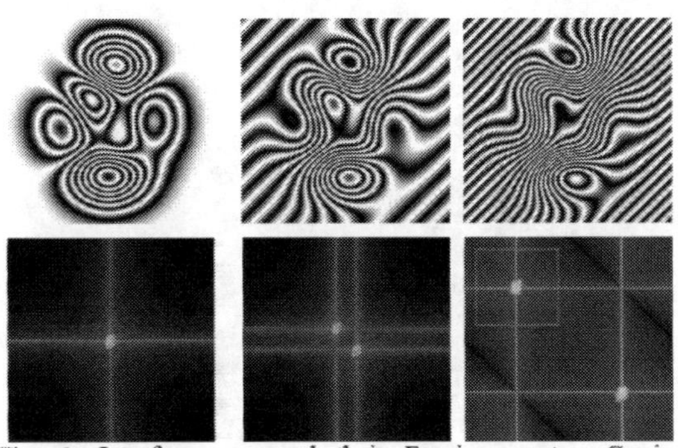

Fig. 1. Interferograms and their Fourier spectra. Carrier frequencies are set to a) $\upsilon_x = \upsilon_y = 0$; b) $\upsilon_x = \upsilon_y = 0.1\upsilon_N$; c) $\upsilon_x = \upsilon_y = 0.5\upsilon_N$. The red rectangle denotes spectral part of the wavefront used for the phase retrieval.

Graphically the process of processing is described in Fig. 2. The output is a wrapped phase map that is in the interval [-π, π]. This limitation causes ambiguity in interferometric measurement as described in the introduction chapter. Particularly, the phenomenon can generate maximal phase change to be π.

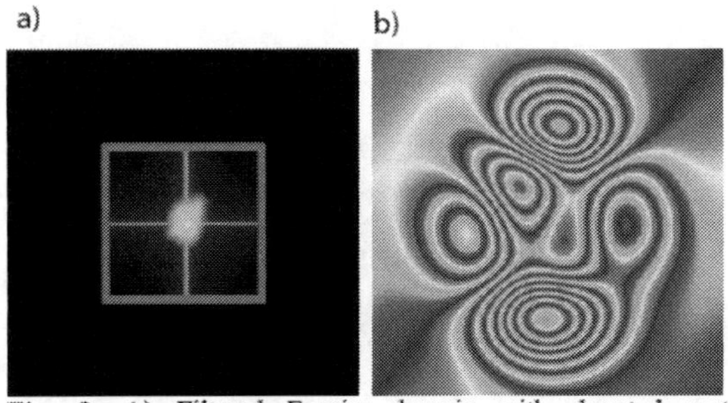

Fig. 2. A) Filtered Fourier domain with denoted part corresponding to Fig. 1c and b) wrapped phase obtained as an angle of the complex field computed by inverse Fourier transform of spectrum shown in a).

Multi-wavelength interferometry allows for extension of the measurement range. If the interferometer employs two wavelengths λ_1, λ_2, the difference between the phases φ_1, φ_2 is called synthetic phase Φ :

$$\Phi = \varphi_2 - \varphi_1 = \frac{4\pi}{\Lambda} z \quad (3)$$

that corresponds to the phase with the synthetic wavelength:

$$\Lambda = \frac{1}{1/\lambda_1 - 1/\lambda_2}. \quad (4)$$

The synthetic wavelength is larger than the optical wavelengths λ_1, λ_2 and determines the range of unambiguousness to $\Lambda/2$. The synthetic phase therefore provides greater measurement range. On the other hand, the uncertainty in interferometric measurement is in fraction of the used wavelength. Therefore the synthetic phase Φ suffers from noise more than the individual phases φ_1, φ_2.

In the case of dynamic measurement, the condition of recording and evaluation from one interferogram remains. It is therefore not possible to first record one interferogram with use of the first wavelength and then repeat the measurement with the second wavelength as the measured variable changes over time.

However, formula (2) and Fig. 1 show that the magnitude carrier frequencies and their directions determine the position of the wavefront in the spectral domain. Taking advantage of this property and considering interferometry with two different wavelengths at the same time on the camera chip, we can achieve their spectral separation and reconstruct the phase fields without affecting them. The interference equation (2) is then extended by another interference member yielding in:

$$I(x,y) = A(x,y) + \\ + B_1(x,y)\cos\left(\omega_{x1}x + \omega_{y1}y + \varphi_1(x,y)\right) + \\ + B_2(x,y)\cos\left(\omega_{x2}x + \omega_{y2}y + \varphi_2(x,y)\right) \quad (5)$$

Same subscripts correspond to the same wavelengths. By adjusting the carrier frequencies for the different wavelengths in the perpendicular directions, spectral separation of the phase information can be achieved.

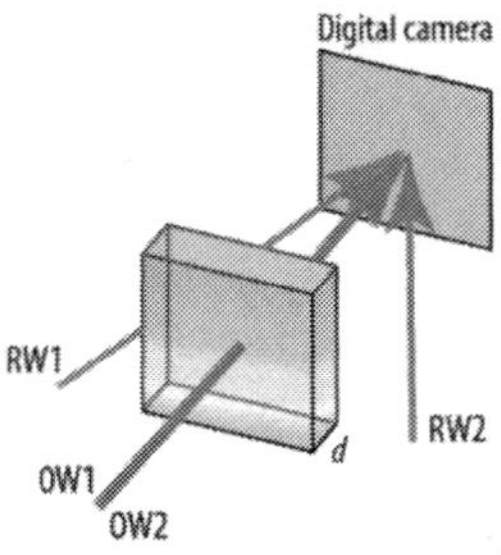

Fig. 3. Principle scheme of DWSCI. Objet waves (OW1, OW2) integrating the measured phase change imping perpendicularly on the CCD or CMOS while the reference waves for different wavelengths (RW1, RW2) imping tha sensor at different angles in order to generate different spatial carrier frequencies.

The phases at the both wavelengths are then retrieved in the same manner as described in the previous section and the synthetic phase can be computed (3). The synthetic phase has an increased range but for the price of lower accuracy. Therefore, the synthetic phase is suitable as a good estimate and one can use its values in order to correct the phase map obtained from single wavelength measurement. This approach yields in dual wavelength spatial carrier interferometry (DWSCI). This enables high dynamic range measurement and high accuracy in real-time measurement. For the sake of clarity, it is necessary to note that the recording depends on the frame rate of the camera (can be high speed camera with frame rate in kHz), and the evaluation is then given by the computing time on the computer used (in order of few seconds for common PC). However, the evaluation can also be done off-line after the experiment ends.

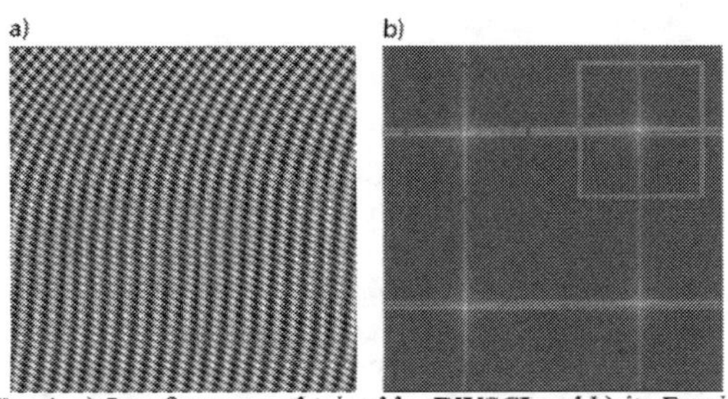

Fig. 4. a) Interferogram obtained by DWSCI and b) its Fourier sepctrum. Red and green rectangles represent wavefronts of generated by the measured phenomena at different wavelengths.

3 DWSCI - case study

This section will introduce simulations result showing basic features and benefits of DWSCI.

Let us consider a 2D temperature field (the temperature values along z-axis are constant e.g. temperature field in a cavity) [3] having length *d=10e⁻²m* and the initial temperature *25°C*. In the measurement time the temperature field has changed and the temperature change map is introduced in Fig.5a. The maximum temperature change is cca 290° C and the minimum is 30° C. The temperature field is related to the refractive index n by formula [4]:

$$T = \frac{\frac{A}{n-1} - 1}{B}, \quad (6)$$

where A=0.000294036 and B=0.00368184. And finally the interferometers are sensitive to phase φ change caused by the variation in refractive index Δn:

$$\varphi = \frac{2\pi}{\lambda}\Delta n d\,. \quad (7)$$

The typically used coherent source in interferometry is a double-frequency Nd:YAG laser having wavelength of 532 nm. The phase change caused by the heating of the space in the field of view (Fig. 5 a) is shown in Fig. 5 b.

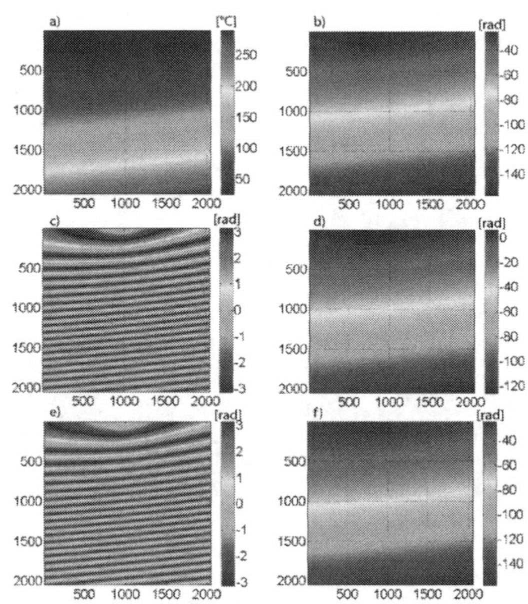

Fig. 5. a) Simulated temperature change; b) absolute phase corresponding to temperature change in a); c) wrapped phase as measured by single wavelength interferometry at 532 nm; d) unwrapped phase map from c) with wrong scale due to ambiguity of interferometric measurement; e) wrapped phase as measured by single wavelength interferometry at 514.5 nm; f) resulting phase map computed from synthetic phase and providing scale in agreement with b).

https://doi.org/10.1051/epjconf/201818002087

From the computed (ideal) phase field we generated the interference pattern and processed the data in order to get phase information again. Due to the ambiguity of the intererometric measurement, we obtained measured wrapped phase field regardless of the way of evaluation (PSI, SCI, ...). The wrapped phase map is shown in Fig. 5c. After application of phase unwrapping algorithm, we obtain a smooth (without 2π jumps) phase field, see Fig. 5d. Qualitatively the two phase fields (ideal and measured) are in agreement; however, their absolute value varies and therefore the absolute temperature distribution cannot be measured accurately.

Therefore, another approach, such as DWSCI, must be applied. Let us add a second light source – e.g. Ar laser - with wavelength 514.5nm to the experiment. By setting the correct carrier frequencies (the incident angle of the reference waves on the sensor), we get the interferogram as shown in Fig. 4a described by

$$\begin{aligned} I(x,y) &= A(x,y)+ \\ &+B_1(x,y)\cos\left(0.5\omega_N x+0.5\omega_N y+\varphi_1(x,y)\right)+ \\ &+B_2(x,y)\cos\left(-0.5\omega_N x+0.5\omega_N y+\varphi_2(x,y)\right) \end{aligned} \quad (8)$$

The frequency spectrum of the interferogram is shown in Fig. 4b. It is clearly seen the spectral separation of the both spectral components of the captured wavefronts marked with rectangles. Different colors denote different wavelengths. The phase fields are retrieved as mentioned above by filtering out the carrier frequency in Fourier domain followed by inverse Fourier transform. Reconstructed wrapped phase fields are shown in Fig, 5e, respectively. From the both phase maps, a synthetic phase can be computed (3). Using formula (4) synthetic wavelength yields in 15.6 μm and provides sufficient measurement range.

As the noise in the phase field is directly proportional to the wavelength, the synthetic phase map suffers much more with noise than the phase field computed from single wavelength measurement. Several procedures can be used to refine the synthetic phase map in order to gain advantage from the both approaches: large measurement unambiguity of synthetic phase and accurate measurement of single wavelength phase.

The easiest and most straightforward procedure is to use the synthetic phase to detect the average phase shift (in terms of fringe order N) across the field of view by:

$$N = round\left(\frac{\left\langle \Phi(x,y)\dfrac{\Lambda}{\lambda_2}\right\rangle - \left\langle \varphi_2(x,y)\right\rangle}{2\pi}\right), \quad (9)$$

where $\langle\ \rangle$ operator stands for average. In described example the fringe order is N=-5. The unwrapped phase φ_2 providing relative information can be adjusted in order to have absolute measurement by

$$\varphi_2^{ABS}(x,y) = \varphi_2(x,y) + 2\pi N . \quad (10)$$

Fig. 5f shows the phase field obtained by (10) that is in agreement with ideal phase map and thus the method provides absolute interferometric measurement.

Other procedures, which can be used for DWSCI are performed independently in each pixel. This is for example a hierarchical phase unwrapping [5] or a robust retrieval introduced by authors in [6]. All approaches will theoretically finish with the same results; however, in the case of disturbing signals, it is more appropriate to use a robust evaluation that gives reliable results even in the presence of, for example, dispersion caused by a large wavelength difference.

4 Conclusions

This article introduces a new approach to measure dynamic processes in fluid mechanics using interferometry with extended dynamic range. A key factor is the use of two wavelengths and the recording of interferograms from both wavelengths in one frame. Phase map evaluation is based on the Fourier transform, and therefore the spectral separation of the phase information from both wavelengths can be achieved when the hardware of the experimental arrangement (angles of the reference waves) is correctly set. The difference between the obtained phase fields creates a synthetic phase whose dynamic range also covers large changes of the measured quantity. However, the synthetic phase is more influenced by noise. By combining the synthetic phase and the phase obtained from single wavelength, measurement can be made more precise to the value of standard interferometric technique. In conclusion, DWSCI has the same accuracy as SCI but with a significantly higher range of measurement.

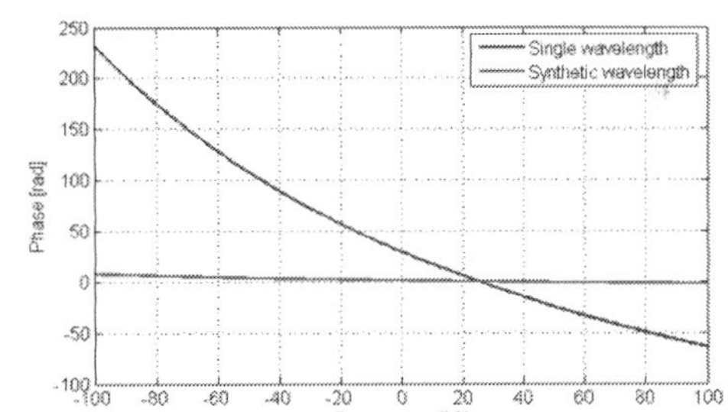

Fig. 6. Dependence of phase change on temperature change. As the temperature varies from its initial value T0=25°C the phase of the object wave changes. Phase ambiguity range [-π,π] in case of single wavelength is reached for small temperature change while the synthetic phase allows for much larger temperature changes.

Fig. 6 shows how the measurement range can be extended. Considering the same parameters as in previous section, the maximal change in temperature of SCI (being inside ambiguity interval [-π, π]) is approximately 25°C ± 3°C. In case of DWSCI (green line) the interval is approximately [-41, 146] °C.

Acknowledgments

We gratefully acknowledge the support of the Grant Agency of the Czech Republic - Czech Science Foundation (projects no. 16-16596S).

References

1. D. Malacara, *Optical Shop Testing. 2007.*
2. K. Meiners-Hagen, R. Schödel, F. Pollinger, A. Abou-Zeid, *MEASUREMENT SCIENCE REVIEW,* **9** (2009).
3. R. Doleček, P. Psota, V. Lédl, T. Vít, J. Václavík, and V. Kopecký, *Applied Optics*, **52**, 1, (2013).
4. C. M. Vest. *Holographic Interferometry*. 1979.
5. Ch. Wagner, W. Osten, S. Seebacher, *Opt. Eng.*, **39**, 1, (2000).
6. P. Psota, V. Lédl, P. Vojtíšek, O. Matoušek, *Proc. SPIE 10151*, **1015113,** (2016).

Numerical simulation of airflow around the evaporator in the closed space

Tomáš Puchor[1,*], *Roman* Banovčan[1], and *Richard* Lenhard[1]

[1]University of Žilina, Faculty of Mechanical Engineering, Department of Power Engineering, Univerzitna 1, 010 26 Žilina, Slovakia

Abstract. The article deals with a numerical simulation of the forced airflow around a evaporator with the finned tubes in the electrotechnical box, by finite volume method in the program ANSYS Workbench. The work contains an analysis of the impact of forced airflow on the evaporator with the various seated the electrical components. The aim of the work is to find out the most effective way of heat dissipation by forced convection from the electrical components in the closed space with lowest pressure loss.

1 Introduction

One of the biggest constraints on the electronics industry is the efficient cooling of the system, which ensures continuous production of high heat. Because the heat generated by the devices needs to be efficiently taken off. All electronic components generate heat during their operation. This heat needs to be dissipated to avoid partial or total damage to components and thus avoid making the entire device stopped working. One way to remove heat is to use a thermosiphon loop heat pipe. Cooling systems used in electrical cabinets have been dealt with generated heat by the supply of ambient air at a lower temperature and the extraction of heated air from the inside of the cabinet through the gratings on the cabinet walls. The new solution focuses on separate enclosed air circulation inside the electrical cabinet to prevent contamination of the surface of the electrical components with dust from the surrounding environment. The electrotechnical box with heat exchanger and whole loop thermosyphon can be seen in figure 1 [1]. In this way, maintaining the clean surface of the components and achieving a non-changeable thermal bridge caused by dust is achieved.

The method for removing generated heat inside the electrical cabinet is solved by separated circuit which consists of a heat source and an exchanger. The heat is supplied from source to exchanger by fan. The source of heat transfer mediation between the heat source and the exchanger is air. The air flow between the ribs of the heat exchanger. The space between fins acts like pressure barrier that the air flowing from the fan must overcome. When the air flows through the space, the part of mechanical energy is converted to heat by the influence of viscosity. If we do not consider air compressibility, the loss of mechanical energy will result in a loss of pressure. We are talking about pressure losses [2].

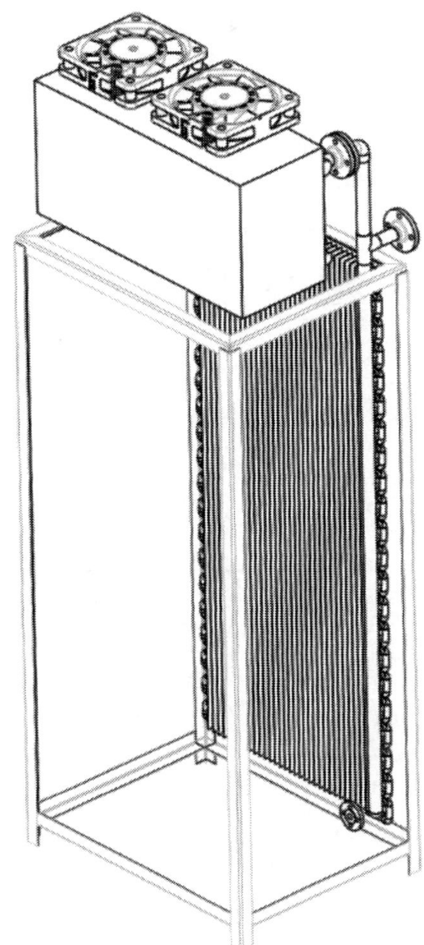

Fig. 1. Electrotechnical box with heat exchanger.

* Corresponding author: tomas.puchor@fstroj.uniza.sk

2 Heat exchanger

The heat exchanger consists of vertical fins placed on the horizontal tube of figure 2. The dimensions of the fins are shown in Table 1. Tubes with a diameter of 15 mm are passing through the blades at a distance of 50 mm from each other. The front and the back wall of the exchanger is closed to form rectangular profile ducts in space between the fins. The space of the exchanger fins is examined for the pressure loss and changes gradually by 0.5 mm from 1.5 mm to 3.5 mm. The thickness of the ribbons is 0.25 mm. To simulate the flow, the thickness of the lamellae is negligible [3].

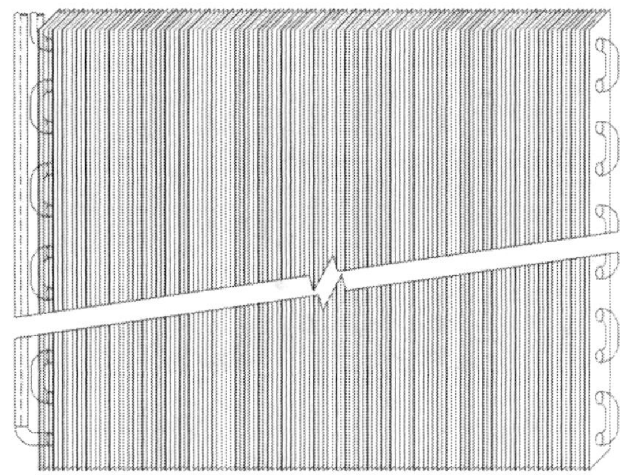

Fig. 2. Heat exchanger (detail).

Table 1. Dimensions of exchanger.

Fins	
Quantity	240
Diemsnions (mm)	50 x 1600
Thickness (mm)	0.25
Distance (mm)	from 1.5 to 3.5
Material	aluminum
Pipes	
Quantity	32
Diameter (mm)	15
Lenght (mm)	660
Material	cupper

2.1 Simplified model

Due to the complexity and size of the model, therefore, a simpler model was made to fit only necessary part to perform simulation and to obtain the best results. One space could be used as a model due to the repeated fins spacing. However, the model which consisted of three fins and thus two spaces was chosen. Marginal spaces were modelled with half thickness to match symmetry on both sides of the model. Part of this modified model can be seen in figure 3.

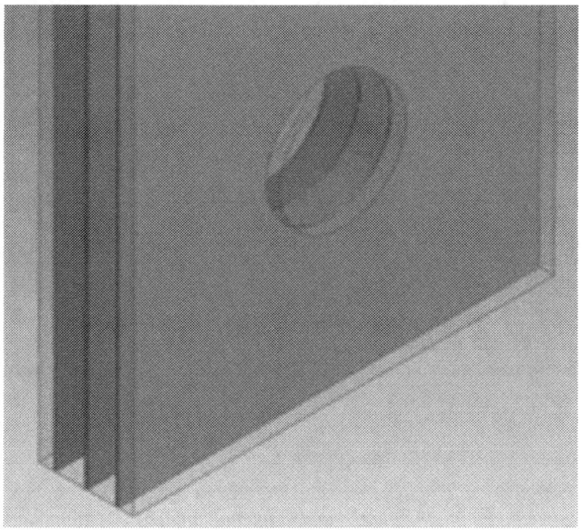

Fig. 3. Model part.

3 Mesh quality

To make simulations, it was necessary to design a sufficient model of the part of the exchanger and the mesh network to suit the correctness of the calculation. Model consists from seven bodies in one part. On the first body, the symmetry part was created edge sizing including top and bottom edge and all 32 holes. Edges sizing was chosen as definition for Number of Divisions with total of 80. The next step was to apply sweep method on this first body with manual source on one face. Number of divisions was chosen with 5 sweep number of divisions. The next sweep method was applied on second body with same settings. Only the sweep number of divisions was setted up to 3. On the third body, represented as the full air space, the sweep method was with 10 sweep number of divisions. The steps were applied similar for all fins bodies, air spaces and symmetries. Each second body was suppressed because it was considered as not necessary for current simulation. This suppressed bodies matched fins. These hidden bodies will be unsuppressed in another simulation with porous media. The mesh was generated with no problems as seen on figure 4. In Table 2. is the number of cells for each rib spacing along with orthogonal quality and orthogonal skewness [4].

Table 2. Mesh information's.

	Space (mm)				
	1.5	2	2.5	3	3.5
Number of cells	1 561 370	1 374 330	1 287 960	1 148 260	1 109 990
Orthogonal quality	0.12	0.16	0.19	0.23	0.21
Ortho skewness	0.82	0.79	0.74	0.71	0.67

Fig. 4. Meshed model.

4 Setting up the simulation

The general simulation setting for each rib spacing was as follows. Energy model and Viscous standard k–e model with enhanced wall function was activated. Medium flowing in the inner-space, was air of a density of 1.225 kg.m^{-3} and a viscosity of 1.7894.10^{-5} [5]. In boundary conditions all air inlets were set as a velocity-inlet with 2.5 m.s^{-1} speed. Turbulence specification method with intensity and hydraulic diameter was chosen with hydraulic diameter for each inlets shown in Table 3. First and last inlet is symmetrical. All outlets was chosen as pressure-outlet. The last step in boundary conditions was to select symmetry on both symmetrical body faces. Solution methods was chosen as coupled and pseudo transient with least squares cell based gradient and momentum as second order upwind. All other spatial discretization was left do default. Solution controls was left to default. Solution initialization was setted up for standard with reference frame as relative to cell zone. Initial values was left do default as well. Solution was initialized. Number of iteration was setted up to 400 and later to more until the simulation was considered as convergent.

Table 3. Hydraulic diameter for distance.

	Distance (mm)				
	1.5	2	2.5	3	3.5
Hydraulic diameter of inlets (mm)	2.9	3.9	4.8	5.6	6.5

The speed 2.5 m.s^{-1} was considered based on the chosen fan with characteristic as sin in figure 5. The fan was chosen as smallest one as possible based on space inside the electrotechnical box and dimensions of evaporator exchanger. Fan is from company HPM therm s.r.o., from where we also took its characteristics.

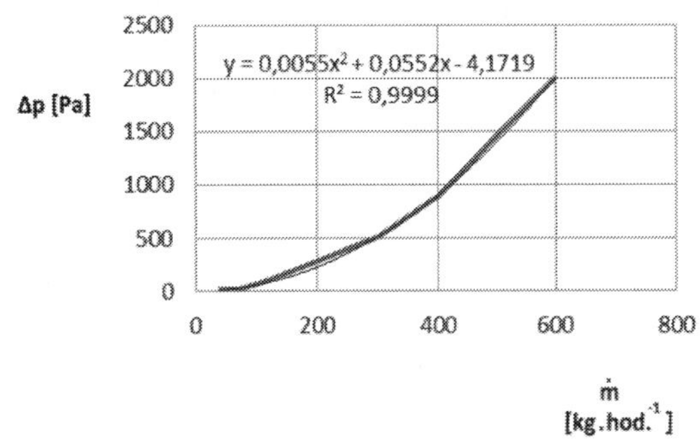

Fig. 5. Characteristics of fan.

5 Calculations and results

According to the set parameters, the pressure loss simulations were performed and the results are written in Table 4, and in figure 6 the pressure loss dependence of the rib spacing is seen. From the values we can see that the pressure loss due to rib spacing has an exponential characteristic. The figure 7 shows the pressure loss over the entire length of the gap.

Table 4. Pressure loss.

	Distance (mm)				
	1.5	2	2.5	3	3.5
Pressure (Pa)	271.7	174.9	126.8	98.82	81.4

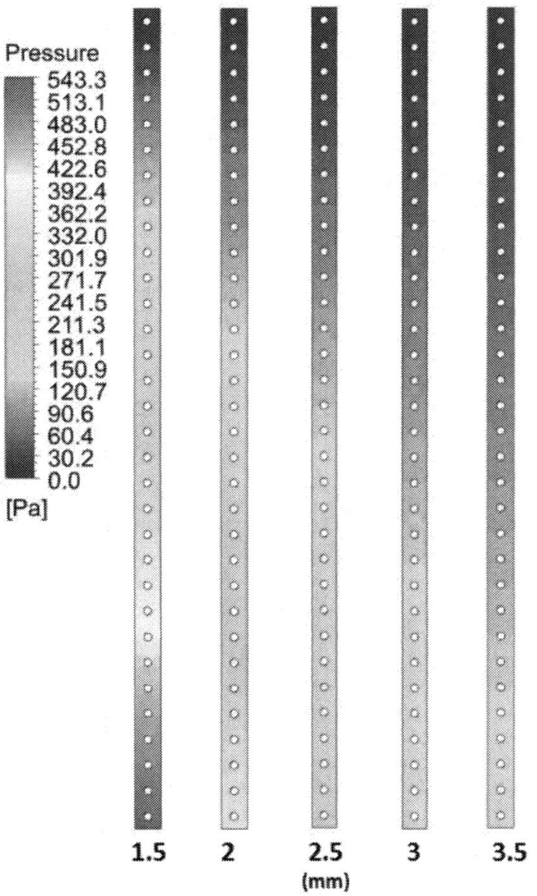

Fig. 6. Pressure loss.

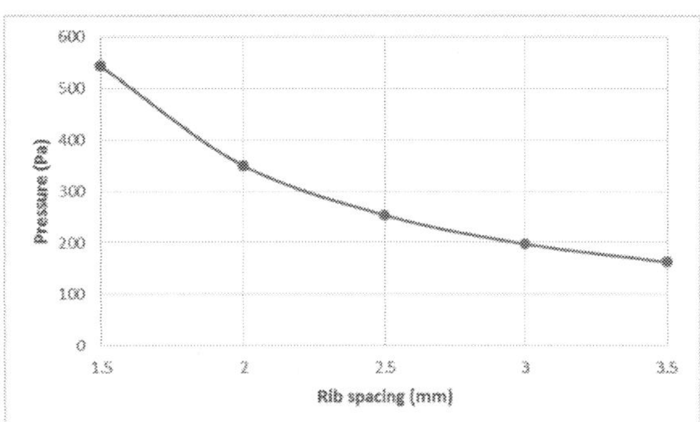

Fig. 7. Pressure loss drop dependance on distance of ribs.

6 Conclusion

The aim of the article was to compare the pressure loss in the air flow between the ribs of the loop thermosiphon heat exchanger. By simulations, I found optimal assembly to be a 2.5 mm gap with a pressure loss of 126.8 Pa. As the ribs approach, the pressure increase sharply, and the rib distance from the 2.5 mm cause the pressure to decreases gradually negligibly.

These simulations was performed for detecting pressure loss for each space between the fins. The next step will be replacing the ribs with a porous material and therefore it was necessary to determine these pressure drops for comparison with the pressure drops in the porous material. Based on the findings I'll be able to set the correct porosity to achieve the same pressure drops.

This paper was written within the framework of the project APVV-15-0778. Limits of Radiative and Convective Cooling through the Phase Changes of Working Fluid in Loop Thermosiphon.

References

1. J. Hartenstine, R. Bonner III, J. Motgomery, T. Semenic, *Loop thermosyphone design for cooling of large area high flux sources* (Proc. Of IPAC, Vancouver, 207)
2. P. Nemec, A. Čaja, M. Malcho, Matematical and computer modelling **57**, 126 (2013)
3. R. Lenhard, M. Malcho, Matematical and computer modelling **57**, 112 (2013)
4. D. Reay, P. Kew, R. McGlen, Heat pipes – Theory, design and applications **6** (2014)
5. R. Lenhard, J. Jandačka, AIP Conference Proceedings **1558**, 2138 (2013)
6. P. Ďurčanský, R. Lenhard, J. Jandačka, Acta Polytechnica **55**, 4 (2015)
7. M. Brůna, Manufacturing technology **15** (2015)
8. T. Kruczek, Energy **91** (2015)
9. B. Skočilasová, J. Skočilas, AIP Conf. Proc. **1608** (2014)
10. P. Nemec, M. Malcho, K. Kaduchová, EPJ Web of Conferences **143** (2017)
11. J. Kňourek, R. Matas, O. Prokeš, D. Tenkrát, EPJ Web of Conferences (2013).

CFD study of the thermal transfer of a non-Newtonian fluid within a tank mechanically stirred by an anchor-shaped impeller

L. Rahmani [1], O. Seghier [1], A. Benmoussa [1] and B. Draoui [1]

[1]Arids zones energetic laboratory (ENERGARID), TAHRI Mohamed University of Béchar, BP 417, 08000 Béchar, Algeria.

Abstract. The most of operations of chemical, biochemical or petrochemical industries are carried out in tanks or in reactors which are mechanically-controlled. The optimum mode of operation of these devices requires a finalized knowledge of the thermo-hydrodynamic behavior induced by the agitator. In the present work, the characterization of the incompressible hydrodynamic and thermal fields of a non-Newtonian fluid (Bingham) in a flat, non-baffled cylindrical vessel fitted with anchor agitator was undertaken by numerical simulation, using the CFD code Fluent (6.3.26) based on the finite volume discretization method of the energy equation and the Navier-Stokes equations which are formulated in (U.V.P) variables. We have summarized this simulated system by comparing of the consumed power and the Nusselt number for this type of mobile (Anchor agitator).

1 Introduction

Many industries (chemical, pharmaceutical and food) use, produce or transform complex fluids. These are high viscosity fluids, often with non-Newtonian behaviour. Their preparation generally requires a heating step followed by a cooling step. Heating is essential to facilitate manufacture (chemical reaction and / or physical-chemical transformation).

The cooling or the finishing step consists in bringing the fluids to a temperature close to ambient temperature in order to approach the conditioning step. The cooling is most often carried out in the same stirred tank.

There are two strong demands in the industry: firstly, ensure the quality of finished products. Secondly, work in a more and more concentrated environment for pollution and cost problems. In order to know and check all the involved aspects in the process, in particular hydrodynamics and heat transfer.

The objective of this work is the thermal study of incompressible laminar flows in a tank mechanically agitated by an anchor agitator and the fluids used are viscoplastic with Bingham model. The modeling difficulties of the problem lie in the viscoplastic nature of the fluid. We perform a numerical simulation of hydrodynamics and heat transfer using the industrial code (Fluent ®). The results obtained give us the possibility to make a comparative study between our results and previous work from the point of view of energy consumption (Power Number), and heat transfer (Nusselt number).

Corresponding author: lakhdar.rahmani08@gmail.com

2 Numerical models

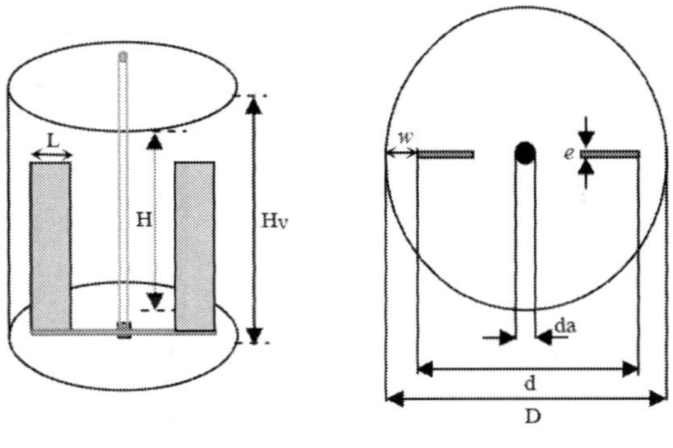

Fig. 1. The geometry of mixing system.

The mixing systems are presented in Fig1. The tank is a flat-bottomed vessel (inside diameter: D) equipped with an anchor agitator (diameter d). H is the fluid height. Values for the agitator width 'L', agitator-to-wall clearance 'w', impairs thickness 'e' and geometrical characteristics are reported in Table 1.

Table 1. Geometrical characteristics of the agitator.

D	**1**
d	**0.96**
da	**0.023**
e	**0.027**
L	**0.067**
w	**0.02**

3 Comportment law of fluid

To study the yield stress fluids which constitute the purpose of this paper, mixing systems are dealt with a numerical simulation using Viscoplastic fluids modeled by Bingham law:

$$\overline{\overline{D}} = 0 \quad \text{For} \quad \left\|\tau\right\| < \tau_0 \tag{1}$$

$$\overline{\overline{\tau}} = \left(\frac{\tau_0}{\dot{\gamma}} + \eta_\infty\right)\overline{\overline{D}} \quad \text{For} \quad \left\|\overline{\overline{\tau}}\right\| > \tau_0 \tag{2}$$

Where $\overline{\overline{D}}$ and $\overline{\overline{\tau}}$, are respectively the rate of strain tensor and the stress tensor. τ_0 is the yield stress.

The shear rate is defined as $\dot{\gamma} = \sqrt{2\, tr\, \overline{\overline{D}}^2}$ where *tr* stands for the trace.

$\overline{\overline{D}}$ And $\overline{\overline{\tau}}$ are defined as:

$$\overline{\overline{D}} = 1/2\left(\nabla V + \nabla V^T\right)$$

$$\left|\overline{\overline{\tau}}\right| = \sqrt{\frac{1}{2}\overline{\overline{\tau}} : \overline{\overline{\tau}}^T}$$

According to equations (1) and (2), the flow domain for Bingham fluid is characterized by two distinct regions. In the regions where $\left|\overline{\overline{\tau}}\right| \prec \tau_0$ the material behaves as a solid (rigid zone), and in the regions where $\left|\overline{\overline{\tau}}\right| \succ \tau_0$ the material flow with an apparent viscosity η_{ap}.

$$\eta_{ap} = \eta_\infty + \frac{\tau_0}{\dot{\gamma}} \tag{3}$$

The major difficulty with the constitutive equation (1) and (2) when used for numerical simulation is the discontinuity associated with infinite value of the viscosity when $|\tau|$ approaches τ_0. This is reached closed to the yield surface delimiting plug regions. Some numerical works have been published on methods adapted to the initial set of equations [(1), (2)] completed by continuity equation and momentum equation in stress formulation. But most of the published works on this topic use modified versions which consist in replacing equations. (1) + (2) By a unique and continuous equation written with a variable viscosity η_{ap} [9]:

$$\tau = 2\eta_{ap}\overline{\overline{D}} \tag{4}$$

4 Model equations and computational procedure

The equations describing fluid flow are derived from the conservation of mass and momentum which are also known as Navier–Stokes equations for fluids of constant physical properties:

$$\nabla \cdot V = 0 \tag{5}$$

$$\rho\frac{\partial V}{\partial t} + \rho(V.\nabla) - \eta_{ap}\Delta V - 2\overline{\overline{D}}\cdot\nabla\eta_{ap} + \nabla P = 0 \tag{6}$$

$$\frac{\partial\theta}{\partial\tau} + U\frac{\partial\theta}{\partial X} + V\frac{\partial\theta}{\partial Y} = \frac{1}{\mathrm{Re.Pr}}\left(\frac{\partial^2\theta}{\partial X^2} + \frac{\partial^2\theta}{\partial Y^2}\right) \tag{7}$$

With Prandtl number :

$$\mathrm{Pr} = \frac{v}{a}$$

4.1 Boundary conditions

In the CFD code used two reference frames can be treated. One is stationary, the impeller is in rotation and the tank is fixed. The other frame, which is rotating, the impeller is kept stationary while the outer wall of the vessel is given an angular velocity equal and opposite to the velocity of the rotating frame. Constant boundary conditions have been set respecting a fixed reference frame (RRF) approach. The boundary conditions for velocity are fixed on the gate impeller and the vessel. On the gate impeller: Moving reference frame implemented in the CFD code.

4.2 Validation

To verify the model of the fluid and to validate our results we compared our power number with previous works on an anchor agitator. Figure 2 gives an example of comparison between our results and a numerical work of Marouche [14]. The results show a satisfactory agreement.

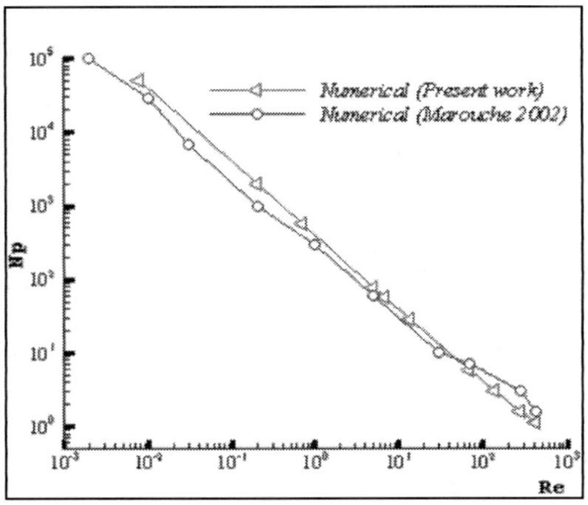

Fig. 2. Variation of power number as function of Reynolds number.

5 Results and interpretations

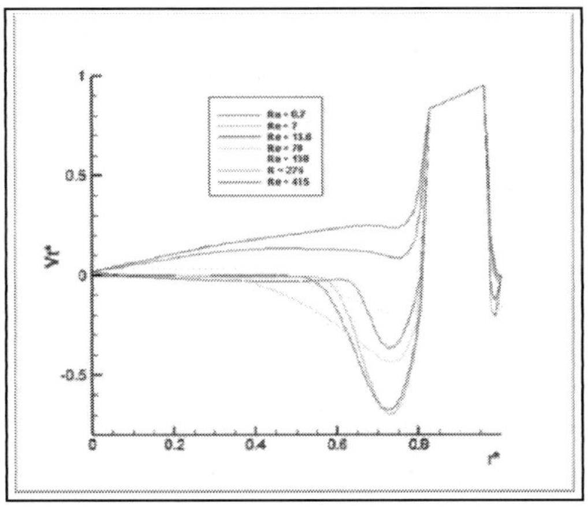

Fig. 3. Influence of inertia on the tangential velocity at the blade and its extension.

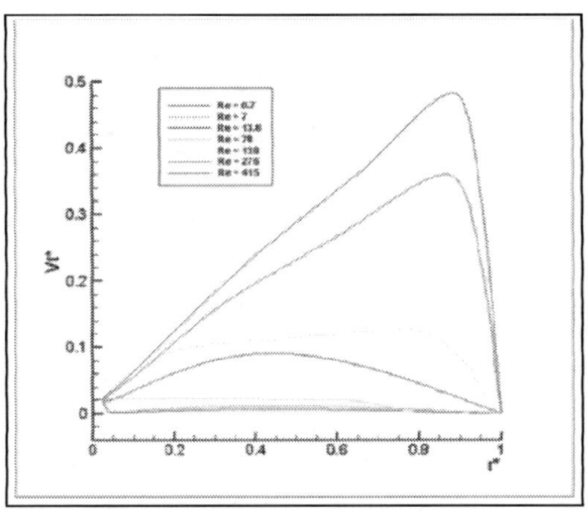

Fig. 4. Influence of inertia on the radial velocity at the blade and its extension.

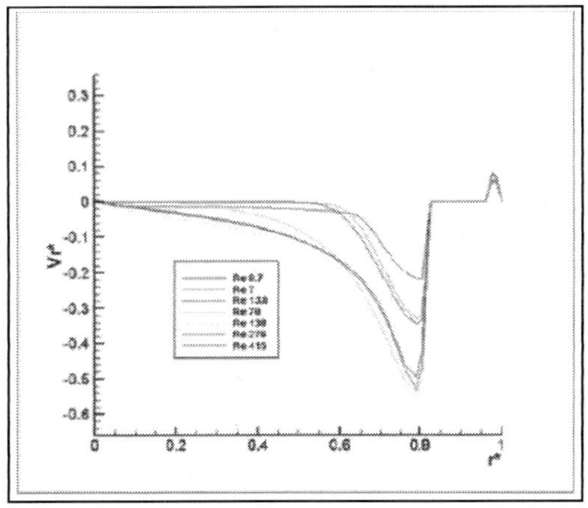

Fig. 5. Influence of inertia on the tangential velocity at the mediator.

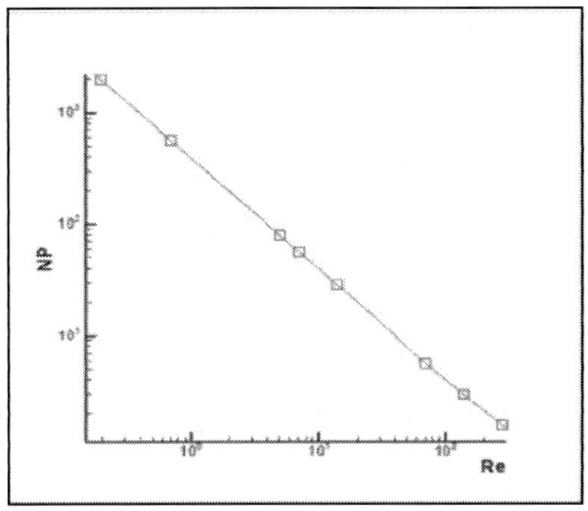

Fig. 6. Influence of the anchor on the power consumption.

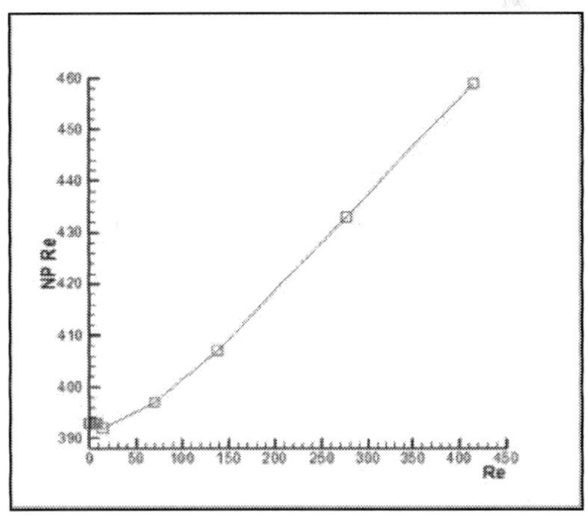

Fig. 7. Influence of the anchor on the Product (Np * Re).

https://doi.org/10.1051/epjconf/201818002089

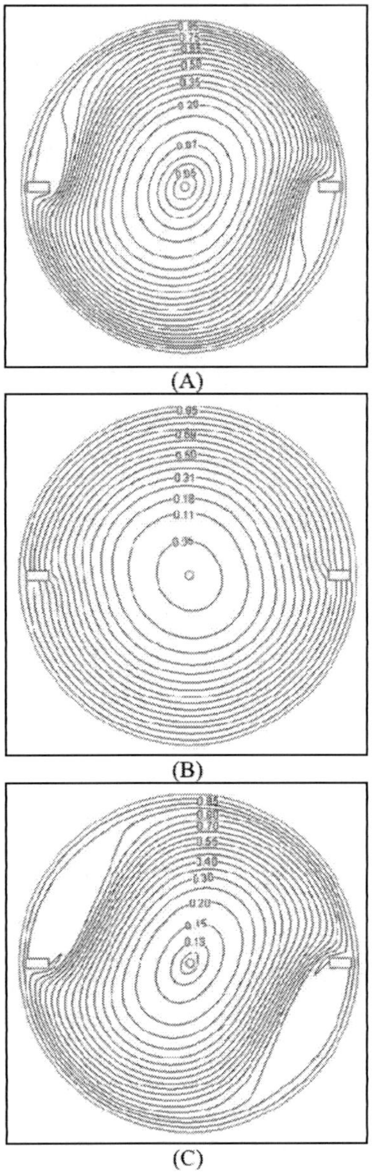

Fig. 8. Isotherms lines (a): Re=0.7, (b): Re=13.8, (c): Re=415 for Pr=7

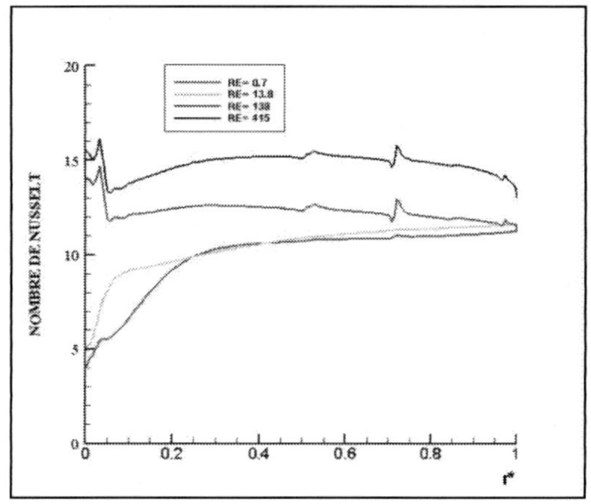

Fig. 9. Inertia effect on the Nusselt number as a function of Reynonlds number.

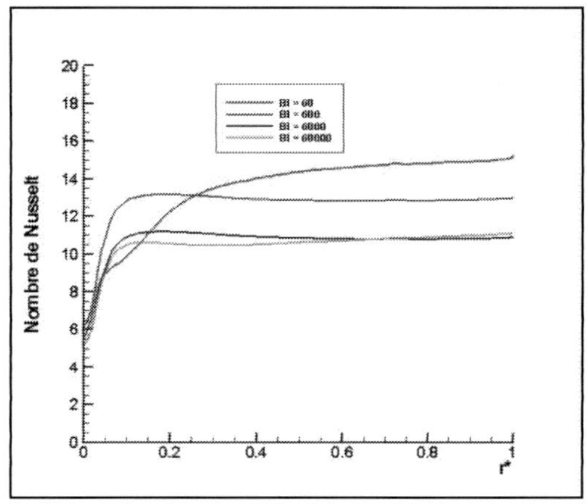

Fig. 10. Plasticity effect on the Nusselt number as a function of Bingham number.

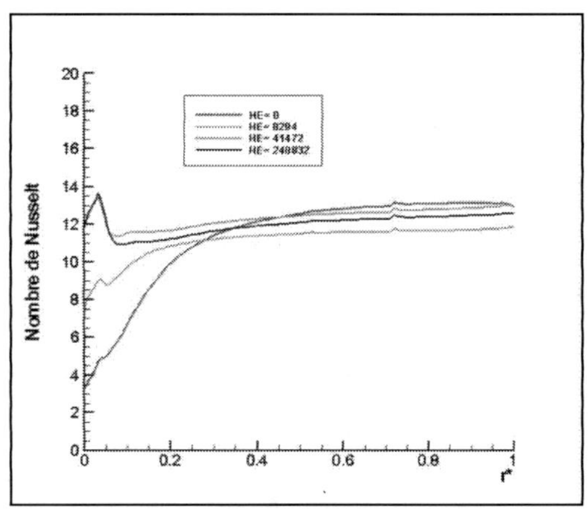

Fig. 11. Effect of Hedstrom number on the Nusselt number for different radious position.

Figures 3, 4, 5, 6 and 7 show the variation of the tangential and radial velocity of the agitator and its extension and on the mediator according to the radial position. We studied the influence of Reynolds number on the nature of the flow by varying this parameter from 0.7 to 415 for a Bingham fluid threshold τ 0= 1 Pa.

We note that the isotherms are in perfect harmony with the stream function lines, i.e. they take the same shape as the current stream lines. It is clear that these movements in the radial direction contribute to supporting heat exchange within the vessel. The trajectory of the heated fluid tends to be extended and deflected only by the action of the tangential velocities. We also find that in the immediate vicinity of the heating surface, the fluid adhering to the wall of the tank shows a large temperature gradient. On the one hand, the lowest temperatures reside within the recirculation zones. Between adjacent threads of fluid, the heat is probably conductive (Figure 8).

The thermal performance is intimately related to the hydrodynamics of the state of agitated tank. Two different processes combine to ensure homogenization: the first is a mechanical process which allows the penetration of the hot flow of the fluid inside the tank. The second is probable in the course of which the microscopic molecular diffusion due to its increase in the flow threshold is important and the differences between the lines of the current. At this stage, on the one hand, the zone or the stream lines would be very different, the temperature should be higher and on the other hand, local temperature gradients would be very large which leads to a very important exchange for the heat.

Figure (8) show the isotherms for different flow regimes. For small numbers of Reynolds they are symmetrical by contribution to the mediator, on the other hand when increasing number of Reynolds this symmetric will be disappearing.

Figure (9) illustrates the evolution of the Nusselt number as a function of the position by varying the flow regime. Note that this parameter increases by increasing the Reynolds number by 10 to 15 at most. Also notes that this parameter is higher in the wall than the center of the vessel for low numbers of Reynolds, also observes maximum local values near the blades and in the center of the tank from the recirculation zone which generated in these areas.

Figure (10) represents the evolution of the Nusselt number as a function of the position by varying the viscoplasticity, it is noted that this parameter decreases when increasing the plasticity of the fluid. The Nusselt number varies from 6 to 16 in maximum, we remark as well as from the radial position r = 0.3 this parameter remains constant for a determined Bingham number.

The figure (11) shows the evolution of the Nusselt number as a function of the number of Hedstrom (He), observes an increase of this parameter to the radial position r = 0.3 with the increase of the Hedstrom number from this position, this number will be decreased slightly with the increasing of the Hedstrom number.

6 Conclusions

This study was motivated by the need to introduce fluid mechanics in the process of design and optimization of thermal transfer in process engineering. Our work has concentrated on viscoplastic fluids whose behaviour is still poorly controlled and which are involved in many applications.

During this process, the behaviour evolves over time: practically Newtonian at the beginning of the reaction, it becomes rheofluidifier and then viscoplastic when the concentrations increase. The non-Newtonian behaviours of the rheofluidifying type have been widely studied in the literature, and their influence studied both on the global quantities, such as the power consumed, as well as on the local variables such as the velocity field.

On the other hand, the hydrodynamics of viscoplastic fluids in complex geometries, such as agitation systems, have been little studied. However, it plays an important role in the conduct and efficiency of a process since it conditions transfers, so this work is a step in the study of these concentrated systems. We have thus studied the hydrodynamics of a model fluid for viscoplastic behaviour and heat transfer within a classical agitation system for highly viscous fluids, namely a Bingham fluid in a mechanically agitated system.

We have analyzed this problem numerically. A first work was to validate this approach with previous work. This work was coupled with an analytical study on the revolution of the power consumed and heat transfer within such a system. On this simple flow on which the inertia does not intervene, we know that the coefficient Ks of Metzner and Otto is constant when the flow is totally sheared and the fluid then has a pseudo-plastic behaviour. It tends towards a constant value for the very large numbers of Bingham.

Numerical simulations, carried out in 2D geometries, allowed to show the important unifications induced by the viscoplastic behaviour and in particular the quasi mobilization of the fluids in a large region when the threshold effects are dominant. This configuration is obviously detrimental to a heat transfer operation for the tested anchor agitator.

It has been observed in particular that the plasticity considerably reduces the axial flow. We have shown that this extension is not a simple calibration of geometry but depends on the crossed effects of the Bingham number of and Reynolds. This led us to note that for this fluid type the choice of an anchor is appropriate and consumes less power.

References

[1] Kerim Yapici, Bulent Karasozen, Michael Schafer, Yusuf Uludag. Numerical investigation of the effect of the Rushton type turbine design factors on agitated tank flow characteristics. Chemical Engineering and Processing. (2007).

[2] F. Ein-Mozaffari, C.P.J. Bennington, G.A. Dumont, Suspension yield stress and the dynamic response of agitated pulp chests, Chem. Eng. Sci. **60** (2005).

[3] T.P. Elson, The growth of caverns formed around rotating impellers during the mixing of a yield stress fluid, Chem. Eng. **96** 303–319 (1990).

[4] D. Anne-Archard, M. Marouche, H.C. Boisson. Hydrodynamics and Metzner–Otto correlation in stirred vessels for yield stress fluids. Chemical Engineering Journal, **125**, 15–24 (2006).

[5] J.N. Haque, T. Mahmud, K. Roberts, Modeling turbulent flows with freesurface in unbaffled agitated vessels, Ind. Eng. Chem. Res. **45**, 2881–2891 (2006).

[6] V.V. Ranade, S.M.S. Dommeti, Computational snapshot of flow generated by axial impellers in baffled stirred vessels, Trans. Inst. Chem. Eng. **74**, 476–484 (1996).

[7] J.J. Derksen, Numerical simulation of solid suspension in a stirred tank, AIChE J. **49**, 2700–2714 (2003).

[8] M. Marouche, Hydrodynamique d'un système d'agitation en fluide viscoplastique et en régime laminaire inertiel, Thèse de doctorat, Institut Nationale Polytechnique de Toulouse. (2002).

[9] L. Rahmani1 et al. Heat transfer to Bingham fluid during laminar flow in agitated tank. *International Review of Mechanical Engineering (I.RE.M.E.), Vol. 09, (2013).*

[10] O'Donovan EJ, Tanner RI, Numerical analysis of the Bingham squeeze film problem, journal of Non-New. Flu. Mec. **15** ,75-83, (1984).

[11] PATANKAR S. V. *Numerical heat transfer and fluid flow*. Mc-Graw Hill. (1980).

[12] G.C. Vradis, M.V. Otugen, The axisymmetric sudden expansion flow of a non-Newtonian viscoplastic fluid, J. Fluid Eng. **119**, 193–200 (1997).

[13] M. Marouche, D. Anne-Archard, H.C. Boisson, A numerical model of yield stress fluid dynamics in a mixing vessel, Appl. Rheol. **12**, 182–191 (2002).

[14] M. Marouche, Hydrodynamique d'un système d'agitation en fluide viscoplastique et en régime laminaire inertiel, Thèse de Doctorat de l'INP Toulouse, France, (2002).

[15] A.B. Metzner, R.E. Otto, *Agitation of non-Newtonian fluids*, AIChE J. **3**, 3–10 (1957).

Manipulation of the swirling flow instability in hydraulic turbine diffuser by different methods of water injection

Pavel Rudolf [1,*], *Jiří* Litera[1], *Germán* Alejandro Ibarra Bolanos[2] and *David* Štefan[1]

[1]V. Kaplan Dept. of Fluid Engineering, Faculty of Mechanical Engineering, Brno University of Technology, Technická 2896/2, Brno 61669, Czech Republic
[2]Research Group in Fluid Dynamics, School of Mechanical Engineering, Universidad del Valle, Calle 13 No. 100-00, Cali, Colombia

Abstract. Vortex rope, which induces substantial pressure pulsations, arises in the draft tube (diffuser) of Francis turbine for off-design operating conditions. Present paper focuses on mitigation of those pulsations using active water jet injection control. Several modifications of the original Susan-Resiga's idea were proposed. All modifications are driven by manipulation of the shear layer region, which is believed to play important role in swirling flow instability. While some of the methods provide results close to the original one, none of them works in such a wide range. Series of numerical experiments support the idea that the necessary condition for vortex rope pulsation mitigation is increasing the fluid momentum along the draft tube axis.

1 Introduction

Both financial subsidies and technological advancement enabled wide spread of renewable energy sources, especially wind and photovoltaic power plants. Increasing share of these sources, which are highly dependent on meteorology conditions or daylight raises concern about stability of the electrical grid. Grid control can be realized by such devices that have both sufficient power and short response time. These criteria are met especially in hydro power plants. Their role in the grid control increases and providing of so called ancillary services becomes even more important nowadays. Hydro power plants offer all three types of grid control (primary, secondary, tertiary), but also voltage support or compensation of active power loss with response time within tens of seconds. Another advantage is black start and island operation capability.

Most (more than 60%) of hydro power plants with installed power bigger than 10 MW is equipped with Francis turbines, moreover pump turbines (reversible Francis turbines) installed in pump storage power plants make up almost another 10% of all hydraulic turbines worldwide.

2 Vortex rope in hydraulic turbine diffuser

Requirements on the control ability of hydro power plants enforce their operation in off-design conditions (deep part load, part load, upper part load, overload). However single controlled hydraulic turbines (e.g. Francis turbine) impose highly swirling flow on the turbine runner exit. The swirl can be explained by mismatch of moment of momentum produced by guide vanes and moment of momentum consumed by the runner. The residual swirl leaves from the runner to turbine draft tube in form of circumferential velocity v_u (see Fig.1). Draft tube is a diffuser, which converts dynamic pressure into static pressure and thereby increases turbine efficiency. Swirling flow is very susceptible to instabilities. For case of the part load the instability is in form of spiral vortex, which is a result of so called spiral vortex breakdown (see Fig.2). This strong coherent vortical structure is well known also in swirl burners or flow over the airplanes. Vortex breakdown is usually defined as sudden change of the flow structure at critical swirl level and is characterized by flow deceleration in vicinity of the axis, formation of stagnation point, onset of recirculation bubble and eventually vortices are rolled-up (KH instability) along the shear layer in-between the two opposing streams. It means that the resulting vortex rope is wrapped around the low velocity zone, Fig.3.

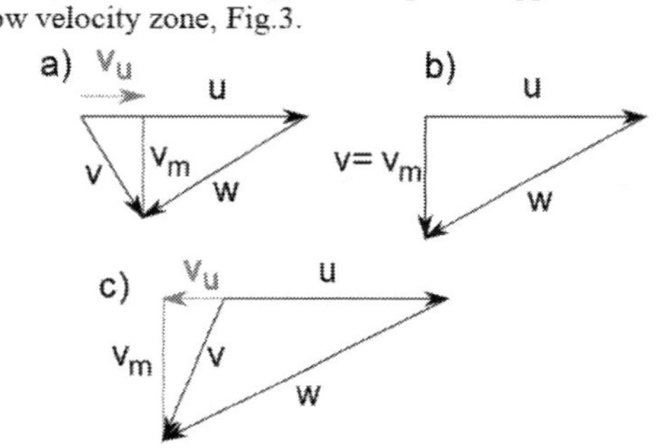

Fig. 1. Connection between the velocity triangle on runner exit and the residual swirl for a) partload, b) best efficiency point, c) overload

* Corresponding author: rudolf@fme.vutbr.cz

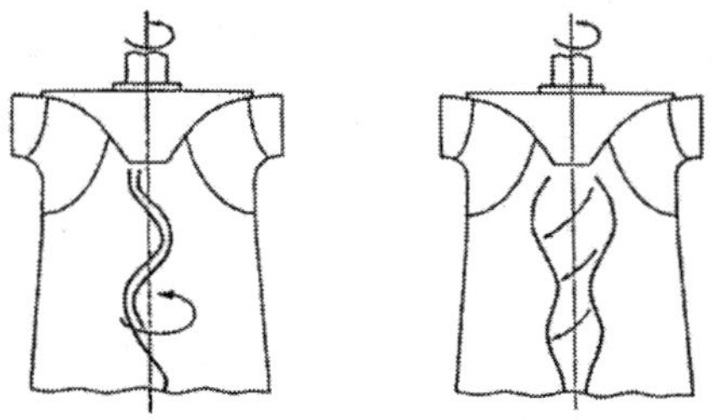

Fig. 2. Vortex rope in case of partload (left) and overload (right) turbine operation

The corkscrew vortex in turbine diffuser is also called a precessing vortex core (PVC) due to its kinematics [1, 2].

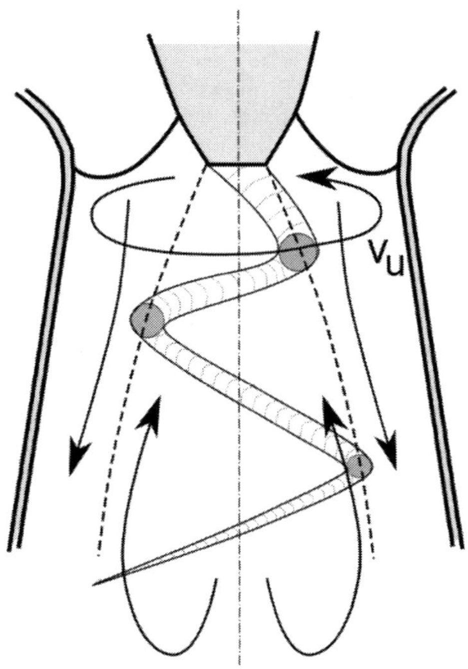

Fig. 3. Spiral vortex breakdown (dashed line marks the shear layer)

Precessing motion of the vortex rope is a source of pressure pulsations both in radial and axial direction. These pulsations can reach up to 5-10% of the turbine head and are manifested by vibrations and noise. Pulsations can also be transmitted upstream and cause runner vibrations. From point of dynamical description this type of pulsations belongs to low frequency ones, because their typical frequency is about one fifth to one fourth of the runner rotational frequency.

3 Vortex rope control

3.1 Active water jet control designed by Politehnica Timisoara

Since vortex rope is very dominant and strong swirling structure its mitigation is rather difficult. In practice usually air injection through runner cone is applied. However admission of air into the diffuser may cause additional dynamical problems related to increased compliance. Another way for vortex rope suppression is mounting the fins on draft tube walls, which decelerates the rotating flow, but also increases hydraulic losses.

Susan-Resiga et al put forward the idea of continuous water injection through runner cone in 1996 [3]. A complex test apparatus was built at Politehnica Timisoara to test the water injection in terms of pressure amplitudes suppression and the necessary water discharge [4], see Fig.4. Test apparatus uses a swirl generator with free runner to imitate flow field corresponding to partial load Francis turbine operation. Smart and elaborate idea of collecting the water for injection from downstream end of the draft tube was employed (so called flow feedback). This method takes advantage of the adverse pressure gradient in turbine diffuser.

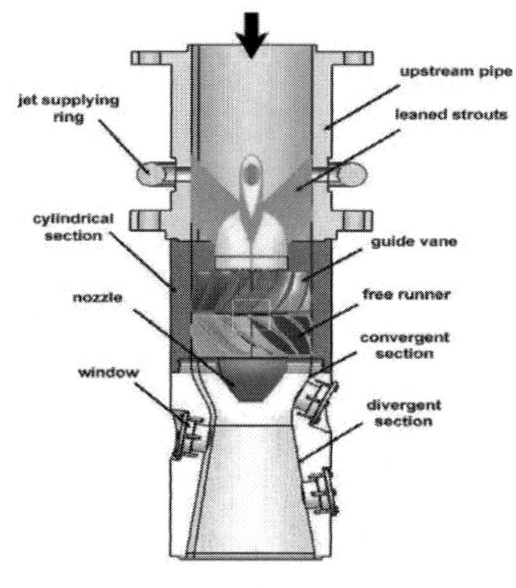

Fig. 4. Test apparatus at Politehnica Timisoara with free runner, plexiglass draft tube and set of pressure probes [4]

Static pressure probes were mounted in four locations on the draft tube wall (Fig. 5). Extensive testing proved viability of the proposed method at least in the laboratory conditions. However it was found that the discharge supplied by the adverse pressure gradient is not sufficient to provide full vortex rope elimination and additional discharge by external pump has to be employed. It means that the overall discharge for water injection is relatively high and potentially there might be more suitable locations were to apply this flow control. In terms static

 https://doi.org/10.1051/epjconf/201818002090

pressure amplitude mitigation, energy and discharge consumption.

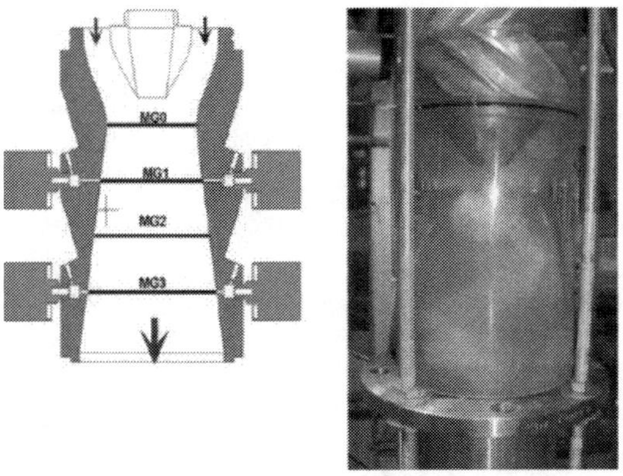

Fig. 5. Locations of pressure probes (left) and visualization of the vortex rope (right) [5]

3.2 Computational modelling of the water injection and shear layer instability

Commercial CFD code ANSYS Fluent was used for computational simulations of the flow downstream of the swirl generator. Boundary conditions for velocity and turbulence were adopted from [5]. Computational domain consisted of about half a million hexagonal elements with finer regions in location of the swirling flow (Figs. 6, 7). Maximum wall y^+ values were around 130, maximum aspect ratio 12. Turbulence was modelled using Reynolds stress model to include the effects of turbulence anisotropy in strongly swirling flows.

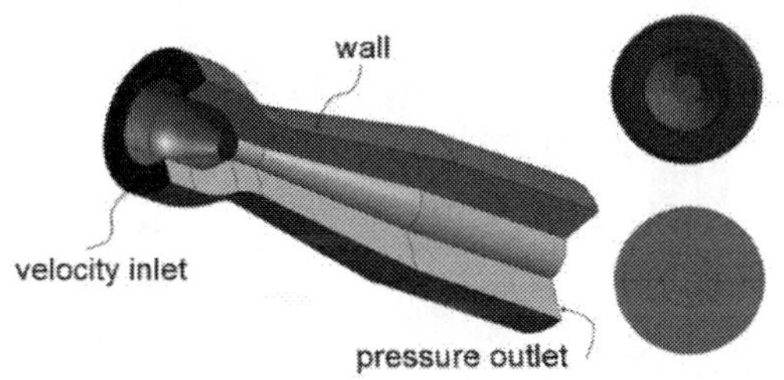

Fig. 6. Boundary conditions and decomposition of the flow domain

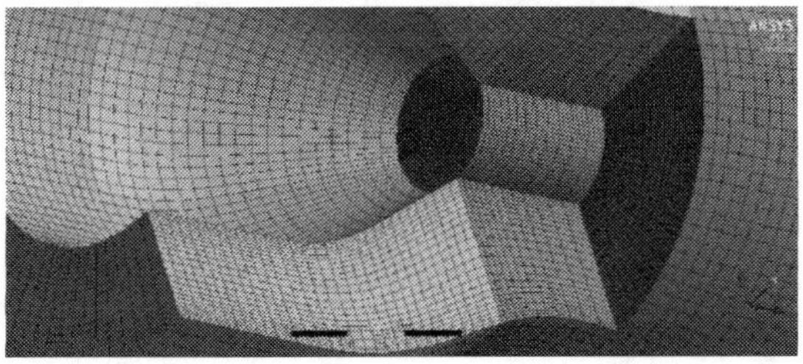

Fig. 7. Computational mesh close to the swirl generator cone

Resulting vortex rope has shape similar to the one observed during experimental testing (compare Fig. 5 and Fig.8). The dominant frequency connected with vortex rope rotation is 14.57 Hz, which is very close to the experimental one (14.95 Hz). Amplitude of the first harmonic in position MG1 is underestimated by 20%. This fact is quite common for transient flow simulations. Modeling also captured the plunging mode on lower frequency (about one third of the first harmonics) and with about 80% of the amplitude of the first harmonics (overestimation compared to experimental results).

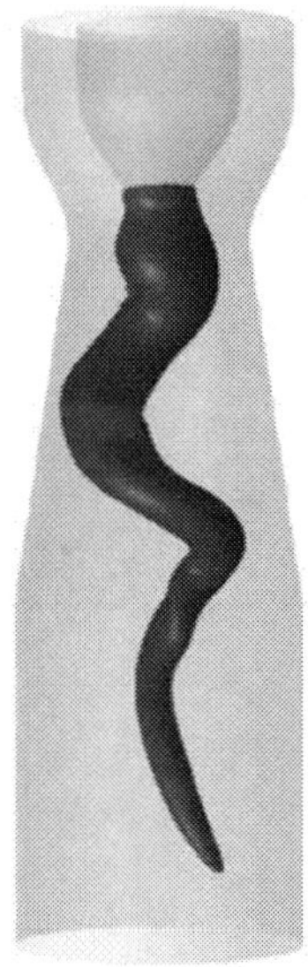

Fig. 8. Snapshot of the instantaneous pressure field visualizing the pressure contour of the vortex rope

Vortex rope onset is connected with instability of the shear layer in between two opposing streams under conditions of adverse pressure gradient (similar to Kelvin-Helmholtz instability). Therefore it was assumed that manipulation of the shear layer in location of highest strain rate would be most sensitive and provide best results. Magnitude of the strain rate tensor was computed in region according to Fig. 9. Strain rate tensor is defined by equation (1).

$$v_{ij} = \frac{1}{2}\left(\frac{\partial v_i}{\partial x_j} + \frac{\partial v_j}{\partial x_i}\right) \tag{1}$$

And its magnitude according to equation (2).

$$\|v_{ij}\| = \sqrt{v_{ij}v_{ij}} \tag{2}$$

Distribution of the strain rate tensor magnitude over the investigated region is plotted in figure 10.

https://doi.org/10.1051/epjconf/201818002090

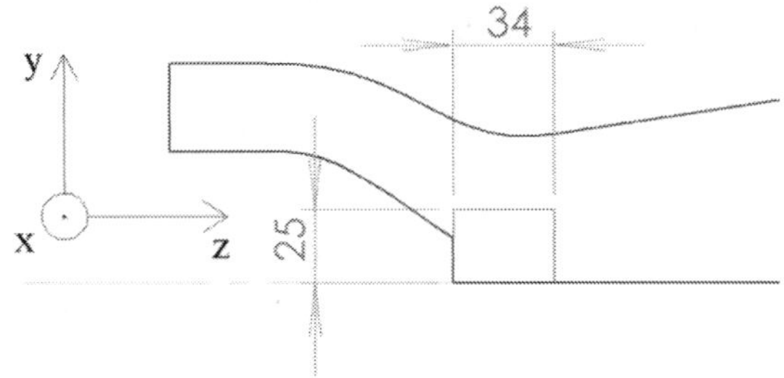

Fig. 9. Region for strain rate tensor computation

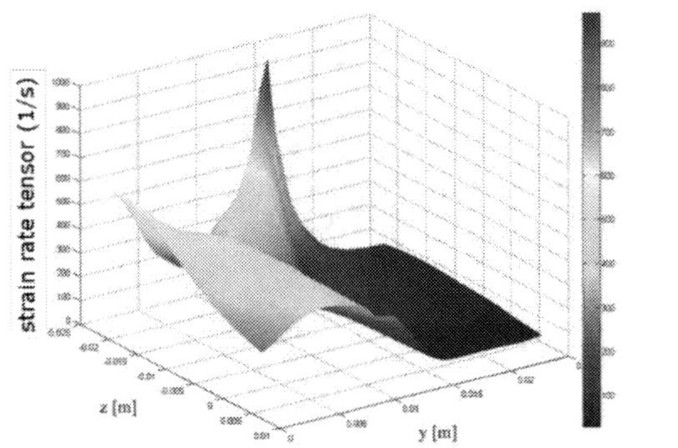

Fig. 10. Strain rate tensor magnitude

The highest value and hence the most sensitive location is placed just at the end of the conical region on the outer diameter of the circular hole, which was used by Susan-Resiga for water injection. This conclusion was later used to investigate different prospective types of water injection control. Their summary is provided in indicative figure 11.

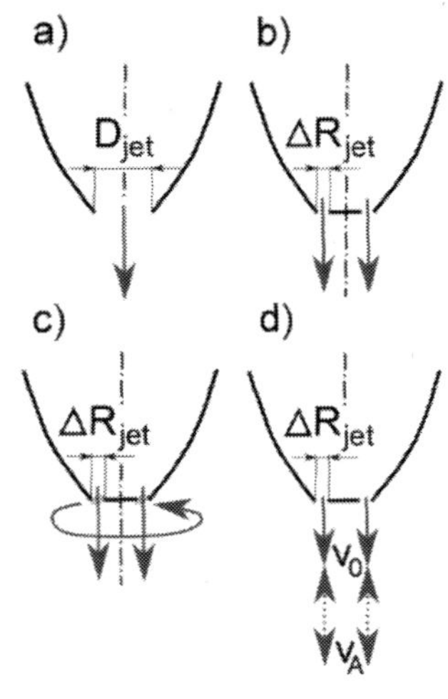

Fig. 11. Different types of water injection: a)full circular injection (Susan-Resiga [3, 4]), b) annular injection, c) annular injection with opposite swirl, d) annular injection with pulsating jet

3.2 Different methods of water injection control [6]

3.2.1 Full circular jet control with D_{jet} = 30.6 mm

Water injection through full circular cross-section (see Fig.11a), which corresponds to the one applied by Susan-Resiga. It was used as a benchmark for comparison with other methods.

3.2.2 Full circular jet control with D_{jet} = 20.6 mm

Similar to previous case, injection along the diffuser axis (see Fig.11a). Diameter for the injection is reduced to 67.3%, flow area reduction is to 45.6%. The jet is continuous in time.

3.2.3 Annular jet with gap (ΔR_{jet} = 2 mm)

Annular gap is designed at the edge of the conical region in the location, where highest magnitude of strain rate tensor was identified (see Fig.11b). Width of the gap is only 2 mm. jet is continuous in time.

3.2.4 Annular jet with gap (ΔR_{jet} = 5 mm)

Similar to previous case, but annular gap is 5 mm (see Fig.11b). Jet is continuous in time.

3.2.5 Annular jet (ΔR_{jet} = 2 mm) with gap producing opposite swirl

Stator blade cascade producing swirl with velocity angle of 45 degrees with axis and opposite sense of rotation than the vortex rope is placed in the annular gap of 2 mm, see Fig. 11c. Hydraulic loss in the blade cascade is superimposed to the energy consumption.

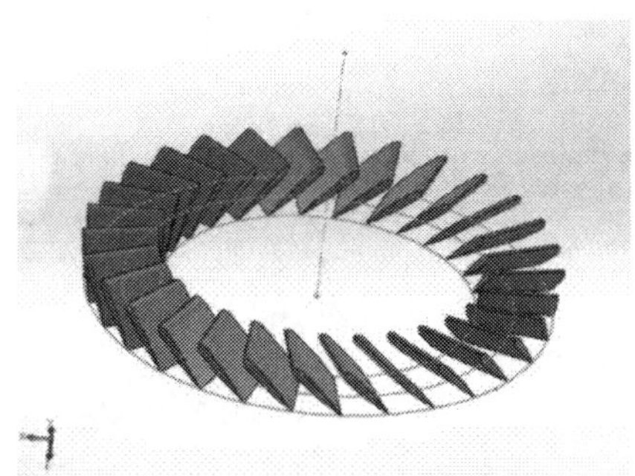

Fig. 12. Stator blade cascade in the annular gap

3.2.5 Annular gap (ΔR_{jet} = 2 mm) with pulsating jet

To reduce the discharge consumption a pulsating jet was designed with constant stationary velocity component and with a pulsating one, whose frequency was modulated. Jet was emitting from annular gap of 2 mm, see Fig. 11d.

$$v(t) = v_0 + v_A \sin(\omega t) \tag{3}$$

Constant stationary value v_0 corresponding to discharge 3 l/s ($v_0 = v_A = 14.5$ m.s^{-1}) was selected and pulsation frequency ranged from 0 to 29 Hz.

4 Results

First, results were evaluated in terms of reduction of static pressure pulsations in location MG1 (Fig. 13). From the plot over the range of relative discharges Q_{jet}/Q from 0 to 0.5 it is clear that the best results are obtained with the original full circular jet. Acceptable results are also achieved using annular jet with opposite swirl and annular jet with 5 mm gap.

Best results in terms of energy consumption are again provided by the original full circular gap jet. Energy consumption increases with square of velocity.

It should be noted that pulsating jet did not disrupt the original flow field, but pulsations were superimposed to the pressure pulsations induced by vortex rope. Therefore pressure amplitudes have always been higher than the pulsations of the rope itself over whole range of the modulating frequencies. Results with pulsating jet are not plotted in Fig. 13, 14.

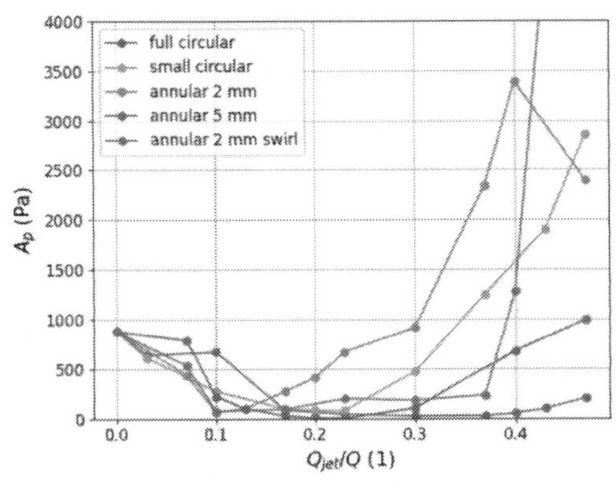

Fig. 13. Amplitudes of the static pressure in position MG1 for different methods of vortex rope mitigation

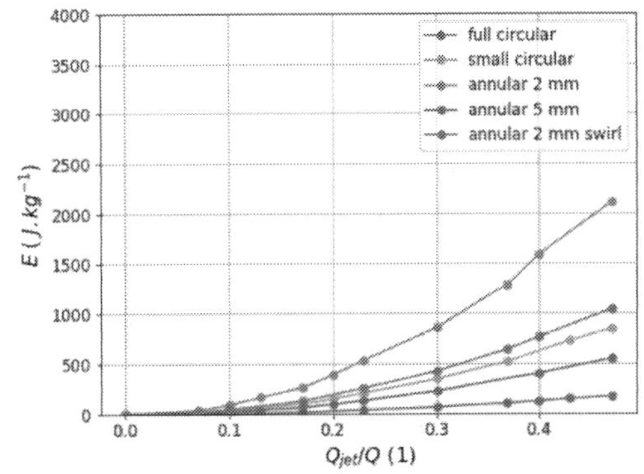

Fig. 14. Energy required for water jet supply for different methods of vortex rope mitigation

5 Conclusion

Numerical experiments with water jet injection for vortex rope pulsations mitigation were carried out. Altogether seven design modifications were tested. It was shown that none of them was better in pressure amplitudes suppression than the original one proposed by Susan-Resiga [3].

Manipulating the shear layer in location of highest strain rate did not prove to be a successful approach. Presented results indicate that vortex rope behaviour is especially influenced by flow along the diffuser axis in region of momentum deficit, where high amounts of the injected water provide best results. This finding is in agreement with recent observations of Goyal et al [7], who found out that the velocity profile in the decelerated region has to be influenced in appropriate way by bringing higher momentum of the fluid.

While the current design modifications did not generate better method than the original one, they provided important results for our future effort in the active control of the diffuser flow based on reduced order models built upon proper orthogonal decomposition of the swirling flow field [8], [9].

Acknowledgements

This research was supported by Czech Science Agency under project 17-01088S "3D instability of a shear layer in adverse pressure gradient" and Faculty of Mechanical Engineering, Brno University of Technology project FSI-S-17-4615.

References

1. A.T. Favrel, Dynamics of cavitation precessing vortex core for Francis turbines at part load operating conditions, PhD thesis, EPFL Lausanne, 2016
2. D. Štefan, P. Rudolf, Proper Orthogonal Decomposition of Pressure Fields in a Draft Tube Cone of the Francis (Tokke) Turbine Model. In Journal of Physics Conference Series. Journal of

https://doi.org/10.1051/epjconf/201818002090

Physics: Conference Series, 2015. pp. 1-14, ISSN: 1742-6596.
3. R. Susan-Resiga, T.C. Vu, S. Muntean, G.D. Ciocan, B. Nennemann, Jet control of the draft tube vortex rope in Francis turbines at partial discharge, 23rd IAHR Symposium, Yokohama, 2006
4. A.I. Bosioc, C. Tanasa, S. Muntean, R. Susan-Resiga, Unsteady pressure measurements and numerical investigation of jet control method in a conical diffuser with swirling flow, IOP Conference Series: Earth and Environmental Science, 12 (1), 2010
5. O. Petit, A.I. Bosioc, H. Nilsson, S. Muntean, R. Susan-Resiga, Unsteady simulations of the flow in a swirl generator using OpenFOAM, Int. Journal of Fluid Machinery and Systems, 4 (1), 2011, pp. 199-208
6. J. Litera, Flow control in hydraulic turbine draft tube, diploma thesis, Brno University of Technology, 2017
7. R. Goyal, B.K. Gandhi, M.J. Cervantes, Experimental study of mitigation of a spiral vortex at high Reynolds number under an adverse pressure gradient, Physics of fluids, 29 (10), 2017
8. D. Štefan, P. Rudolf, S. Muntean, R. Susan-Resiga, Proper Orthogonal Decomposition of Self- Induced Instabilities in Decelerated Swirling Flows and Their Mitigation Through Axial Water Injection. Journal of Fluids Engineering - Transictions of the ASME, 2017, 139 (8), pp. 1-25. ISSN: 0098-2202
9. P. Rudolf, D. Štefan, R. Klas, Spatio- Temporal Description of the Swirling Flow in Hydraulic Turbine, Wasserwirtschaft, 1, 2015, pp. 18-22, ISSN: 0043- 0978

Influence of Eccentricity and Angular Velocity on Force Effects on Rotor of Magnetorheological Damper

Dominik Šedivý[1,*], *Petr* Ferfecki[2], *Simona* Fialová[1]

[1]Brno University of Technology, Victor Kaplan Department of Fluid Engineering, 61669 Brno, Czech Republic
[2]VŠB - Technical University of Ostrava, IT4Innovations, 70833 Ostrava, Czech Republic

Abstract. This article presents the evaluation of force effects on squeeze film damper rotor. The rotor is placed eccentrically and its motion is translate-circular. The amplitude of rotor motion is smaller than its initial eccentricity. The force effects are calculated from pressure and viscous forces which were measured by using computational modeling. Damper was filled with magnetorheological fluid. Viscosity of this non-Newtonian fluid is given using Bingham rheology model. Yield stress is not constant and it is a function of magnetic induction which is described by many variables. The most important variables of magnetic induction are electric current and gap width between rotor and stator. The simulations were made in finite volume method based solver. The motion of the inner ring of squeeze film damper was carried out by dynamic mesh. Numerical solution was solved for five different initial eccentricities and angular velocities of rotor motion.

1 Introduction

Change of fluid properties in working damper is one of possible ways to improve damping ability of damper. However this change has to be very fast. Magnetorheological (MR) fluids are liquids which show fast transition from a liquid to a nearly solid state under the presence of external magnetic fields. Generally, their viscosity depends on magnetic induction and therefor MR fluids belong to non-Newtonian fluids. This special property makes them very good candidates for applications in mechanical systems which require active control of vibrations. Classical examples are shock absorbers or vibration dampers. Shaft vibrations are generated from the imbalance rotor motion which can be caused by assembling or manufacturing inaccuracies. These negative effects can be noticeably minimized if active or semi-active damping elements are installed between the shaft and the stationary part of the rotating machine. Positive parameters of MR fluids in connection with conventional squeeze film dampers provide a possibilities to design a semi-active damping elements. Nowadays the MR fluids are a subject of an intensive experimental and theoretical research. Mathematical models of squeeze film damper were published in works [2, 4]. A computational simulation was done in article [1]. An interaction between surface wettability and MR fluids was solved in works [5, 6]. Force effects on squeeze film damper were studied in paper [7].This article continues in research, which was started in work [7].

2 Methods

Application of squeeze film damper is shown in Fig. 1. The rotor shaft is attached in bearings which allow its rotation. There is a fixed connection between bearings and housing of bearings by a squirrel spring. Squeeze film damper is consisting of an inner and an outer ring and magnetorheological oil. The primary aim of damper is to eliminate unwanted vibrations. Magnitude of viscosity is controlled by the electric current in electric coil which generates magnetic induction. The domain for numerical simulation was just the place between inner and outer ring.

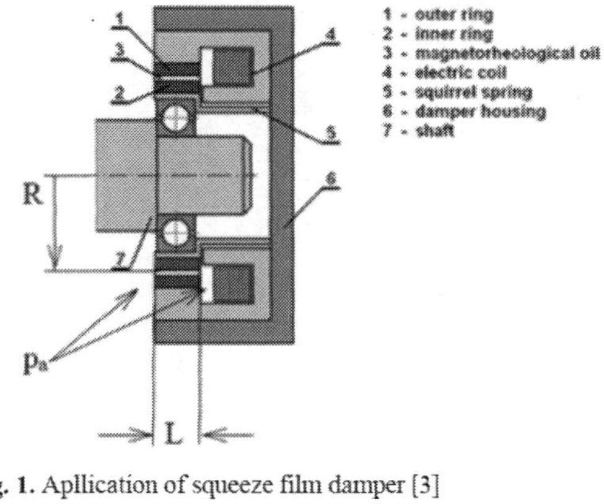

Fig. 1. Apllication of squeeze film damper [3]

The outer ring is a stationary wall and its diameter is 152 mm. The inner ring is a moving wall with 150 mm diameter and is eccentric placed in relations to outer ring

* Corresponding author: Dominik.Sedivy@vutbr.cz

Motion of inner ring is translate-circular and its amplitude of deviation is variable. Mean value of eccentricity (center of movement) in direction of x axis was solved for five different values. For every case was maximal eccentricity in direction of x axis 0.55 mm. The motion for case that mean value of eccentricity is 0.5 mm and amplitude of deviation is 0.05 mm is shown in Fig. 2.

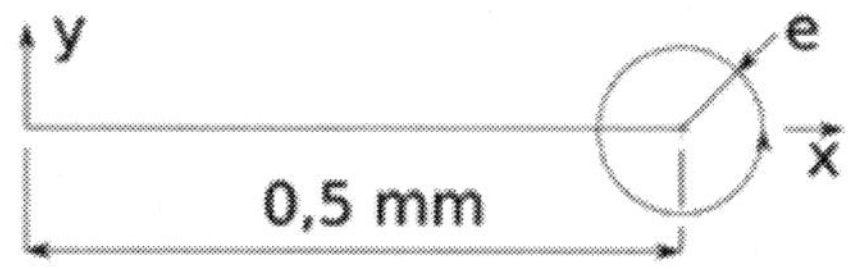

Fig. 2. Eccentric motion

Combinations of deviation amplitudes and mean values of eccentricities are shown in Table 1.

Table 1. Combination of eccentricity and deviation

Mean value of eccentricity [mm]	Amplitude of deviation [mm]
0.52	0.03
0.51	0.04
0.5	0.05
0.49	0.06
0.48	0.07

The real length of the damper is 44 mm. The numerical solution was solved by commercial CFD (computational fluid dynamics) software (ANSYS Fluent). Following boundary conditions were set up for numerical simulation. To obtain lower simulation time, the geometry was cut in the middle of damper length and symmetry condition was used. The outer ring was determined as a stationary wall. At the outlet from damper, a pressure outlet with magnitude of relative pressure 0 Pa was set. The inner ring of squeeze film damper was defined as a moving wall. Its motion was prescribed by velocities functions for every axis:

$$v_x = -e \cdot \omega \cdot \sin(\omega \cdot t) \quad (1)$$

$$v_y = e \cdot \omega \cdot \cos(\omega \cdot t) \quad (2)$$

where v_x and v_y are components of velocity in direction of x and y axis,
e is amplitude of deviation.

Frequency of inner ring motion is prescribed by its angular velocity $\omega = (60; 70; 80; 90; 100) \cdot \pi\, s^{-1}$. The velocity in direction of z axis was equal to $0\ m \cdot s^{-1}$.

Numerical simulations were calculated by using the dynamic mesh. Displacements were small in these solutions, therefore the meshes of computational domains did not have to be remeshed (mesh is redesigned - number of elements are changing during the simulation). Mesh of computational domain has 320 000 hexahedral elements (minimum orthogonal quality = 0,888; maximum ortho skew = 0,112; maximum aspect ratio = 18,928). Because the mesh was just deformed (stretched and compressed), Smoothnig method was applied, which is included in ANSYS Fluent. Diffusion parameter was defined as Cell-volume and its value was set equal to 0. The motion of inner ring was prescribed by user defined function (UDF) and inner ring was set as rigid body.

3 Mathematical model of fluid

Density of magnetorheological fluid has value $970\ kg \cdot m^{-3}$. For case there is no electric current in the coil, magnetorheological fluids have constant viscosities and the fluid behavior is Newtonian. Electric current in coil generates a magnetic field and the magnetorheological fluid will change its properties. For the mixed viscosity the Bingham rheology model applies which was defined in work [1]. This model was also used in paper [7]. The model is described by equation:

$$\tau = \tau_y + \mu \cdot \dot{\gamma} \quad (3)$$

where τ [Pa] is shear stress which acts upon the element of fluid,
τ_y [Pa] is shear yield stress,
μ [Pa·s] is dynamic viscosity,
$\dot{\gamma}$ [s^{-1}] expresses shear rate.

The shear yield stress is considered as constant commonly. However, in this case, it is a function of many variables and it is defined by the following equation:

$$\tau_y = k_y \cdot B^{n_y} \quad (4)$$

Where B is magnetic induction,
k_y and n_y are material constant of magnetorheological fluid.

For this simulation following values were chosen $k_y = 10000\ Pa \cdot T^{-2}$; $n_y = 2$. Magnetic induction is also dependent on the thickness of the lubricating film in radial direction. Definition of magnetic induction in squeeze film damper is described by the following relation:

$$B = k_B \cdot \mu_0 \cdot \mu_r \cdot \frac{I}{h(t)} \quad (5)$$

where k_B expresses construction parameter of damper (100 threads and efficiency 60 % means its value is $k_B = 0{,}6$),
μ_0 is a permeability of vacuum ($4\pi \cdot 10^{-7}$ Hm^{-1}),
μ_r is relative permeability of magneto-rheological fluid which for our case has value 5,
I is electric current which flows through coil, which was set on value 0.2 A in this paper.

Time dependent size of the thickness of the lubricating film is defined by $h(t)$:

$$h = c - e \cdot \cos(\varphi - \gamma) \quad (6)$$

where h - thickness of the oil film,
c - the width of the gap between the inner and outer rings of the damper,
e - the eccentricity,
φ - the circumferential coordinate,
γ - the position angle of the line of centre (Fig. 3).

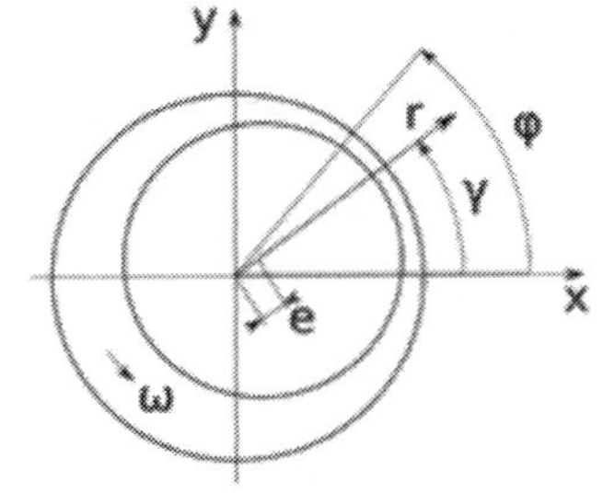

Fig. 3. Inner ring motion

4 Solution and results

Motion of squeeze film damper inner ring is time dependent. Therefore, the solution was simulated as transient. The size of time step has values 10^{-4} s. The viscosity of filling fluid in damper was relatively high, velocities and the size of gap width are small. Therefore, also the Reynolds numbers come out small. And for this reason no turbulence model was used and laminar model was chosen.

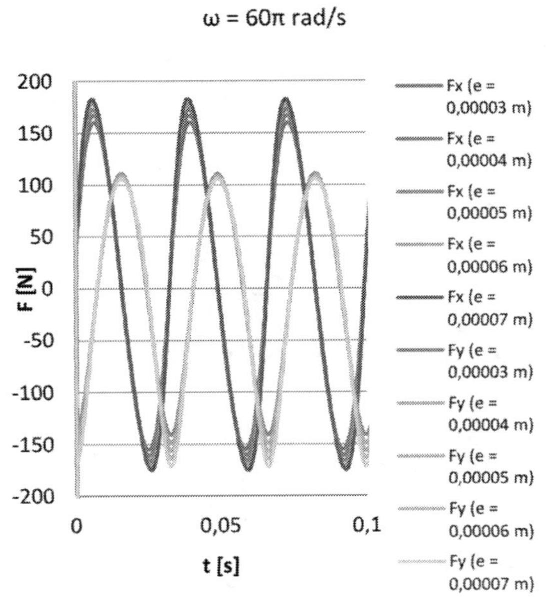

Fig. 4. Force components for ω = 60π rad/s

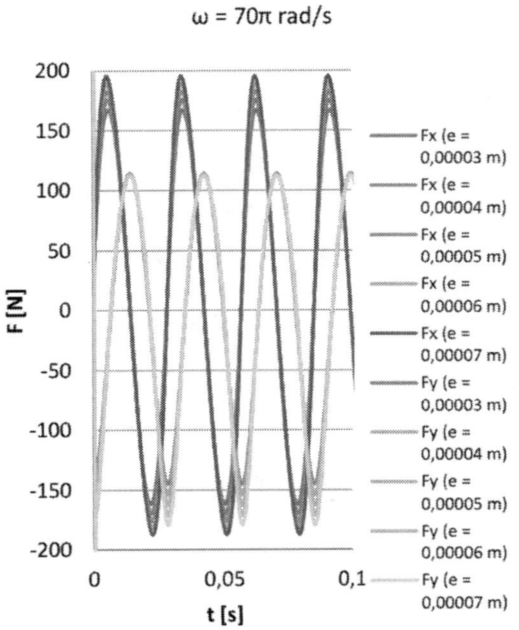

Fig. 5. Force components for ω = 70π rad/s

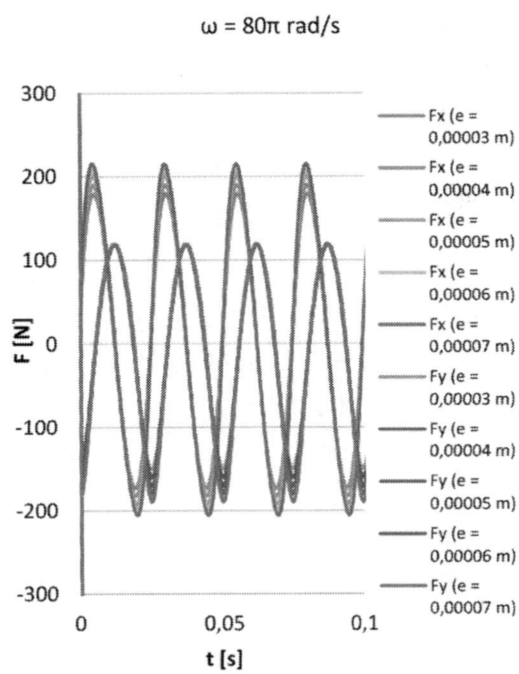

Fig. 6. Force components for ω = 80π rad/s

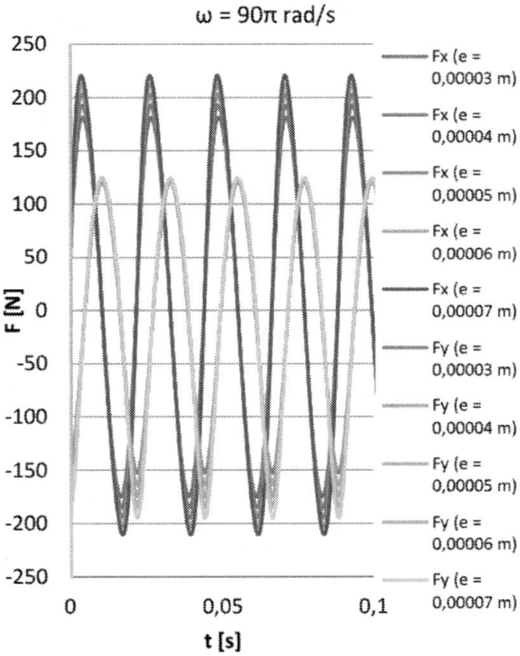

Fig. 7. Force components for $\omega = 90\pi$ rad/s

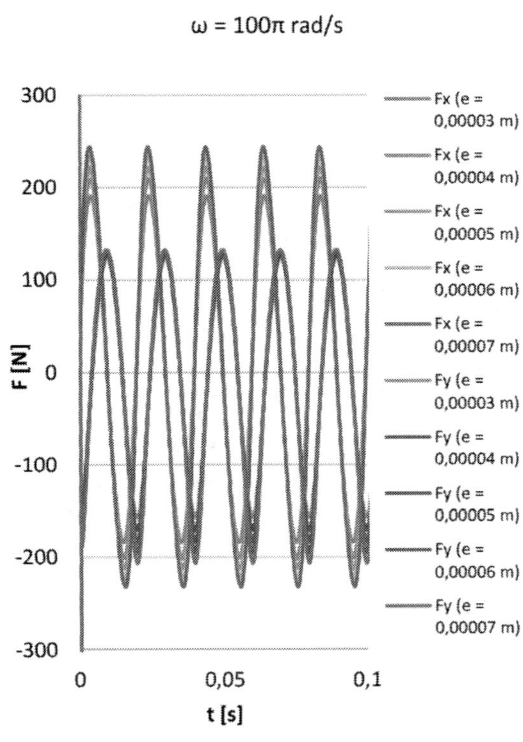

Fig. 8. Force components for $\omega = 100\pi$ rad/s

In Fig. 4.-8. force components are displayed for different eccentricities and different angular velocities. Values are periodically changing. Values of positive and negative amplitudes of forces in x axis direction are almost same. That does not apply to amplitudes of forces in y axis direction as well. Positive force amplitudes in y axis direction are almost constant for one value of angular velocity, but negative values are different.

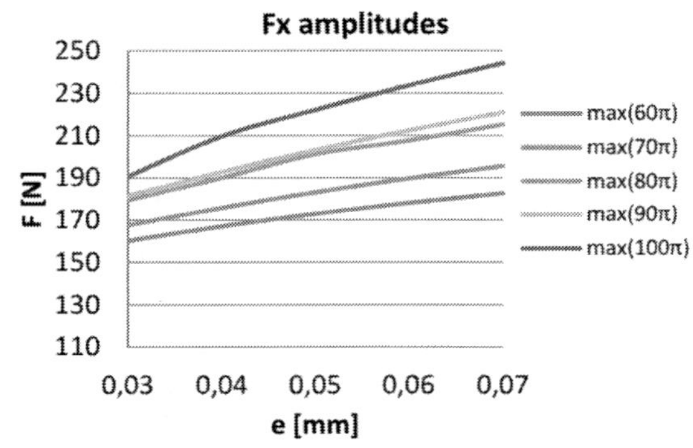

Fig. 9. Fx positive amplitudes

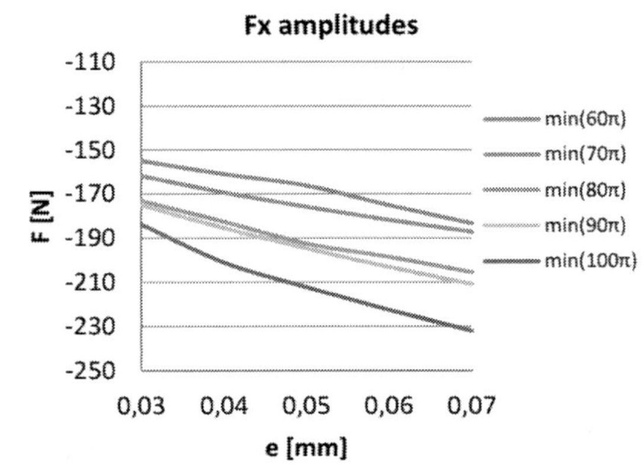

Fig. 10. Fx negative amplitudes

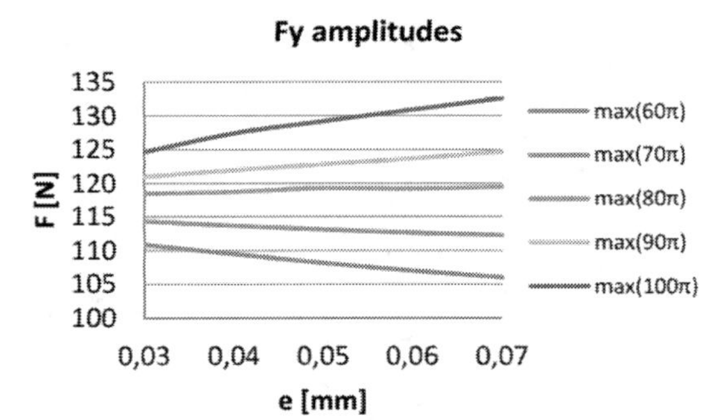

Fig. 11. Fy positive amplitudes

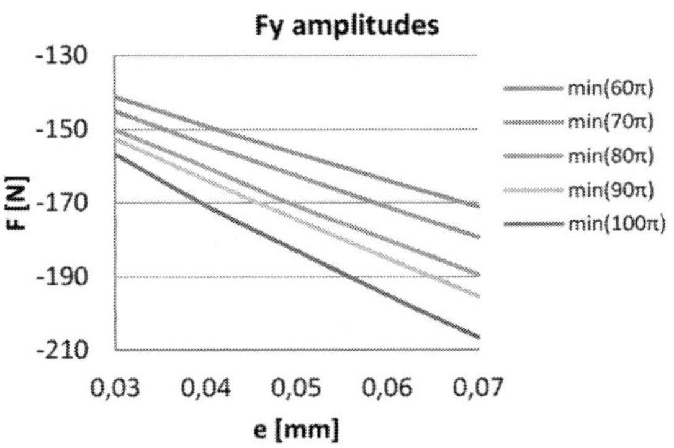

Fig. 12. Fy negative amplitudes

Values of force amplitudes as function of amplitude of deviation are displayed in Fig. 9. – 12. Values of force amplitudes in direction of x axis are increasing with increasing values of angular velocities and with increasing values of amplitude of deviation. The same applies to Fy negative amplitudes as well. However positive values of force amplitudes in direction of y axis do not have these trends. Fy positive amplitudes are almost constant in case, when angular velocity was equal to 80π rad/s. Force effects are decreasing for smaller angular velocities and increasing for higher angular velocities. Erroneous assumption was to expected that the force effects will be increasing with increasing values of angular velocities. Small differences can be caused by inaccuracy of numerical solutions. This phenomenon has to be studied in detail.

5 Conclusions and future work

Force effects are dependent on a size of an eccentricity of inner ring motion and its angular velocity. Force amplitudes in both direction increases with increasing motion eccentricity. However results show that description of force amplitudes on inner ring as function of angular velocity is more complicated than it was expected. This phenomenon will be studied in future work more depth. Next plan is to extend the present numerical model by new boundary condition of different wettability of surface.

Acknowledgment

Grant Agency of Czech Republic, within the project GA15-06621S, and Faculty of mechanical engineering BUT, within the project FSI-S-17-4615, are gratefully acknowledged for support of this work.

References

1. Zapoměl, J.; P. Ferfecki a P. Forte. A computational investigation of the transient response of an unbalanced rigid rotor flexibly supported and damped by short magnetorheological squeeze film dampers. Smart Materials and Structures. 2012, 21(10), 105011-. DOI: 10.1088/0964-1726/21/10/105011. ISSN 0964-1726.
2. Zapomněl, J.; Ferfecki, P.; Kozánek, J. Mathematical modelling of a damping element working on the principle of squeezing two layers of normal and magnetorheological oils arranged in series and its application for vibration attenuation of a rigid rotor. In Náprstek, J.; Fischer, C. (ed.). Engineering Mechanics 2012. Prague: ITAM AS CR, v. v. i., 2012, s. 1581-1588. ISBN 978-80-86246-40-6. [Engineering Mechanics 2012 /18./, Svratka, 14.05.2012-17.05.2012, CZ].second literature
3. Beran, J.; Bílek, M.; Žabka, P., ed. Advances in mechanism design II: Proceedings of the XII International Conference on the Theory of Machines and Mechanisms. Springer International Publishing. ISBN 978-331-9440-866. ISSN 2211-0984.third literature
4. Krutil, J.; Pochylý, F.; Fialová, S. (2015) Nonlinear Mathematical Model of the Rotor Motion in a Thin Hydrodynamic gap. International Science Index Conference, Amsterdam, Netherlands, 2015 - proceedings, pp 305-309
5. Pochylý, F.; Fialová, S.; Malenovský, E. (2012) Bearing with magnetic fluid and hydrophobic surface of the lining. IOP Conference Series: Earth and Environmental Science. 2012. 15 (2). p. 1 - 9. ISSN 1755-1315
6. Pochylý, F.; Fialová, S.; Kozubková, M. (2010) Assessment of cavitation creation depending on the surface wettability, IOP Conference Series: Earth and Environmental Science, WOS: 000325657000106Luigi T. De Luca, *Propulsion physics* (EDP Sciences, Les Ulis, 2009)
7. Šedivý, D. ; Ferfecki, P. ; Fialová, S. Force Effects on Rotor of Squeeze Film Damper Using Newtonian and Non-Newtonian Fluid. Proceedings of the 36th Meeting of Departments of Fluid Mechanics and Thermomechanics. Pilsen, Czech Republic: AIP Publishing, 2017. http://aip.scitation.org/doi/abs/10.1063/1.5004368

Processing of Cells' Trajectories Data for Blood Flow Simulation Model*

Martin Slavík [+], *Kristína* Kovalčíková, *Hynek* Bachratý [+], *Katarína* Bachratá, and *Monika* Smiešková

Cell-in-fluid Research Group, http://cell-in-fluid.fri.uniza.sk; Department of Software Technologies; Faculty of Management Science and Informatics, Žilina, Slovakia

Abstract. Simulations of the red blood cells (RBCs) flow as a movement of elastic objects in a fluid, are developed to optimize microfluidic devices used for a blood sample analysis for diagnostic purposes in the medicine. Tracking cell behaviour during simulation helps to improve the model and adjust its parameters. For the optimization of the microfluidic devices, it is also necessary to analyse cell trajectories as well as likelihood and frequency of their occurrence in a particular device area, especially in the parts, where they can affect circulating tumour cells capture. In this article, we propose and verify several ways of processing and analysing the typology and trajectory stability in simulations with single or with a large number of red blood cells (RBCs) in devices with different topologies containing cylindrical obstacles.

1 Introduction

Microfluidic devices with periodic obstacle arrays inside them are often used for diagnosis or even for a treatment of various diseases related to the blood. Such devices are designed for capturing, detecting and sorting rare blood cells or microorganisms. Shapes and topologies of the devices vary depending on their purpose. Many applications are in the research of circulating tumour cells (CTCs). These cells get to the blood circulation from primary tumours and can create metastasises in other parts of the human body. In [1], a device consisting of a microchannel with periodic obstacle array is described. The obstacles are covered with biologically active substance, adhesive for chosen type of cells. Obstacles without biologically active substance are used in [2]. The authors describe experiments using different angles of fluid flow inside the channel. Sometimes an array of obstacles is used, in which every row is shifted compared to the previous one [3]. This device is designed based on a diagnostics device used for extraction and enrichment of parasites from human blood. Obstacle do not have to be necessarily cylindrical, as we can see in [4], where the periodic array of obstacles with triangle cross-section is used.

Recently, computational modelling has been successfully used for investigation of rare cell behaviour in a channel with periodic obstacle arrays. One of the tools, which allows such approach is Object-in-fluid framework implemented in open source software ESPResSo [5], [6], [7]. In [6] author studies a rare cell capture rate according to the cell-fluid ratio in a channel with shifted obstacles. The cells were represented only by spherical objects, what is a typical shape of CTC. In [7] the trajectories of single CTC were investigated in a channel with regular periodic obstacle array, with different cell densities. In this case, only CTC had a spherical shape and other cells were represented by RBCs.

The key ability of the microfluidic devices considered in this study, is their capacity to capture CTC in the blood, respectively in blood plasma. To achieve high capture efficiency, it is necessary to optimize CTC trajectories to maximize contact with obstacles. A particular problem is their very rare occurrence in blood plasma, and the small ratio of CTC compared to RBC. In the literature, it ranges from $1:5x10^4$ to $1:10^9$. CTC will therefore be significantly affected by the RBC trajectories. In this article, we propose basic methods of investigation and statistical evaluation of the properties of RBC trajectories, mainly in terms of their stability, specifically changes in their position during their way through the obstacle array. For this purpose, we have also used Object-in-fluid framework. Due to the large computational difficulty of simulations, we examined the problem on simplified types of channels with cylindrical obstacles.

2 Simulation technical parameters

All simulation experiments described in this work were carried out in a cuboid-shaped channel with the length of 100 μm (x-axis), the width of 50 μm (y-axis), and the height of 30 μm (z-axis). The channel was closed from four sides, i.e. it had its bottom, top and side walls. It was open in the X-direction to let the fluid flow. This

* Supported by the Slovak Research and Development Agency under the contract No. APVV-15-0751, by the Ministry of Education, Science, Research and Sport of the Slovak Republic under the contract No. VEGA 1/0643/17 and by FVG 2017 grant of Faculty of Management Science and Informatics, University of Žilina.

+ Corresponding authors: martin.slavik@fri.uniza.sk, hynek.bachraty@fri.uniza.sk was also the direction of the periodic conditions applied which made the experiment less demanding on calculations also for models of larger channels. Cylindrical obstacles (pillars) with the radius of 5 μm were placed inside the channel. For a precise description of the channel shape, see Figure 1.

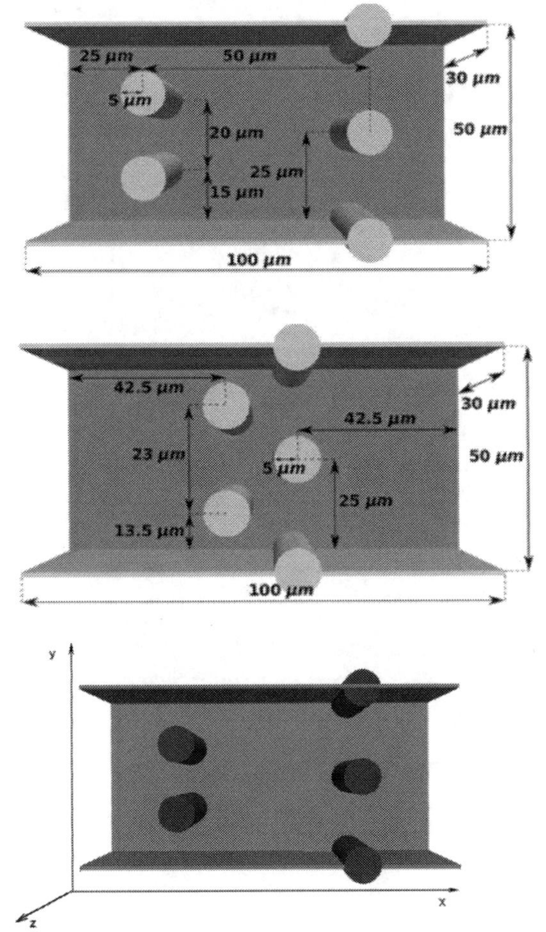

Fig. 1. Topology and proportions for channel A, channel C and orientation of coordinate axis

Each of the experiments ran for 380,000 simulation steps, which, with a time step of 0.2 μs, corresponds to the real time of the experiment of 0.0762 s. In the time, the fastest cells passed the simulated channel section many times, which means they did hundreds of micrometres in total.

All experiments were performed in the simulation software ESPResSo with usage of Object-in-fluid framework for modelling cells and LB for modelling fluid. As a model of each of the RBCs we had chosen a surface mesh with 141 nodes and the dimensions of 7.82μm × 7.82μm × 2.56μm. The elastic coefficients of the mesh were ks = 0.0044, kb = 0.0715, kal = 0.005, kag = 1, kv =1.25. The simulation mass of every cell was 0.05957 pg. The interactions among the cells were modelled by membrane collision interaction with the parameters of mc K = 0.005, mc n = 2.0, mc cut = 0.5. The interactions between the cells and surface of the channel were ensured by soft sphere interaction with the parameters of sof t K = 0.00035, sof t n = 1.0, sof t cut = 0.5. The fluid was discretized by a three-dimensional regular lattice with the space step 1 μm. The fluid viscosity was 1.5 mPa.s and the density was 1000 kg/m^3, representing blood plasma at 20 ◦C. The interaction parameter between fluid and object called friction was 0:025865531914893616. Different external volume forces were used to induce the flow in the X-direction.

3 Stability of isolated RBC trajectories

To obtain basic information about the properties of RBC trajectories in Type A and Type C channels, we first performed 15 calibration simulation experiments with a single RBC in the channel. This single RBC gradually passed through the channel from five different starting positions (with y-coordinate, i.e. the distance from channel edge, of 5, 10, 15, 20 and 25 μm) and from 3 starting orientations (called flat, cross and straight), Figure 2.

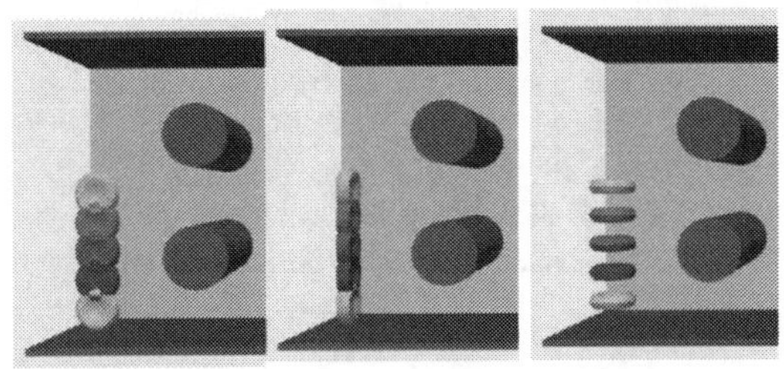

Fig. 2. Conjoint picture of RBCs in each starting position (from 5 to 25 μm) and each orientation (flat, cross and straight).

In addition to these initial conditions, the trajectories of each RBC were only influenced by the velocity field (stream lines) of the blood plasma flow and collisions with the canal surface. Under these simple conditions, it was our goal to examine the stability of the trajectories of the cells. In the first place, we will examine the y-coordinate of the RBC trajectory when it repeatedly (periodically) passes through the same channel regions. We would like to determine if the trajectory has a stable, random, increasing or decreasing tendency.

Despite the relative simplicity of simulations, a complete data description of the movements of individual RBCs is intricate and to obtain answers, statistical processing and reduction of the data is required. For each of the 15 trajectories we investigated, the y-coordinate was measured. Specifically, the y-coordinate of the cell transitioning through the first and the second row of obstacles (measured in direction of x-axis, for channel A positions 25 + k.100 μm and 75 + k.100 μm, for channel C the positions 42.5 + k .100 μm and C 57.5 + k.100 μm), and through the center and end of the channel (positions 50 + k.100 μm and 100 + k.100 μm). According to the cell velocity in the individual simulations, we have obtained sequence of y-coordinates for each of these cutting positions for 6 to 8 cell transitions. (To obtain this data, we created a custom sw tool in which the number and location of the cut positions are an optional parameter.) We also calculated the slope coefficient of the linear regression line to describe the trend of coordinate changes. If the absolute value of the slope exceeded 0.1 we regarded transitions through these cutting positions as unstable with the trends of its change. Its value at 0.3 corresponds approximately to the relative change of the coordinate by 5% of the total channel width, which is 20% of the size of a typical slot in the obstacle array.

To further verify the obtained results, we performed 10 experiments in which we add other RBCs to original one in the channel A. In five simulations by adding 10 and in five by adding 20 randomly seeded RBCs.

We illustrate the results of the simulations analysis in several pictures. At this point, we summarize conclusions that were, to a certain extent, predictable. We have verified and confirmed them with simulations and statistics.

• The trajectory stability (the y-coordinate progress of the positions of the cells at important cutting positions) depends strongly on the starting position of the RBC. For some positions, the sequence is almost constant, for others significantly increasing or decreasing. For these positions, the RBC trajectory is "typically" shifted toward a certain optimum given only by the topology of the device.

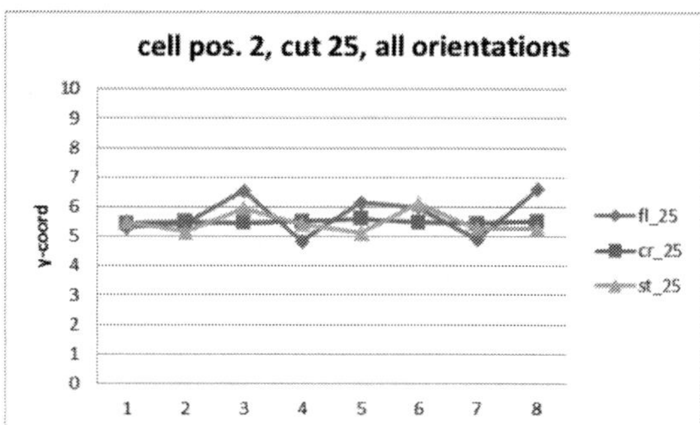

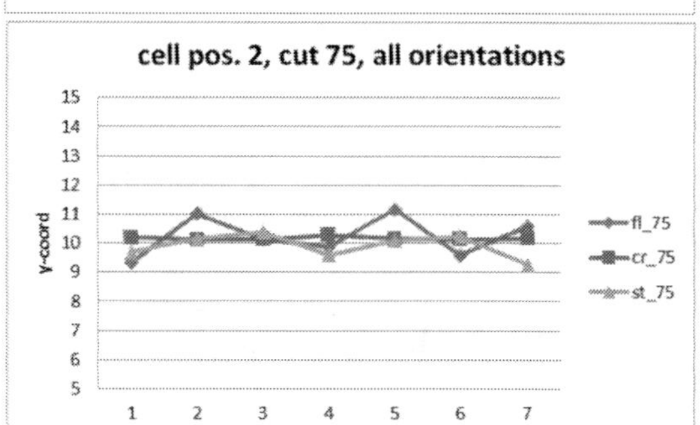

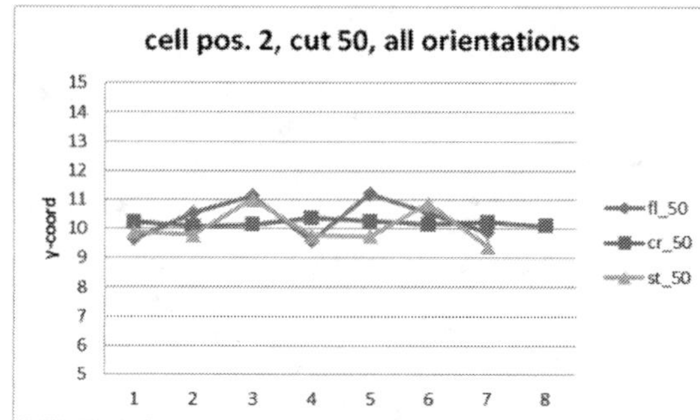

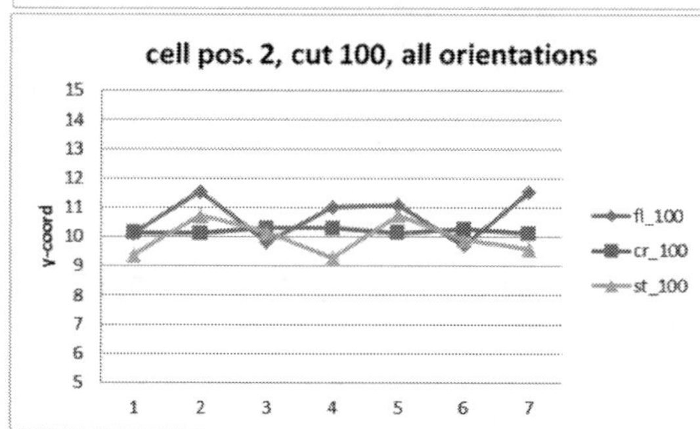

Fig.3. For cell in starting position 2 (10 μm from edge) the stability of trajectory in all cutting positions and all cell orientations is visible. The slope coefficients (in order of graphs and its lines) are 0.07207, 0.00484, -0.00748, 0.06780, -0.00037, -0.04995, 0.02895, -0.00346, -0.01948, 0.06583, 0.00273, -0.01419

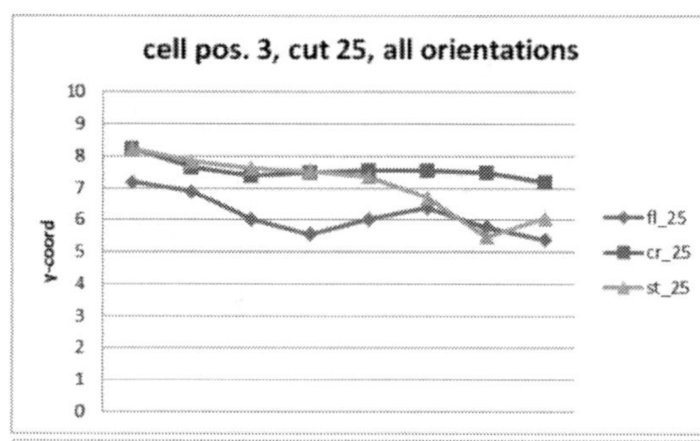

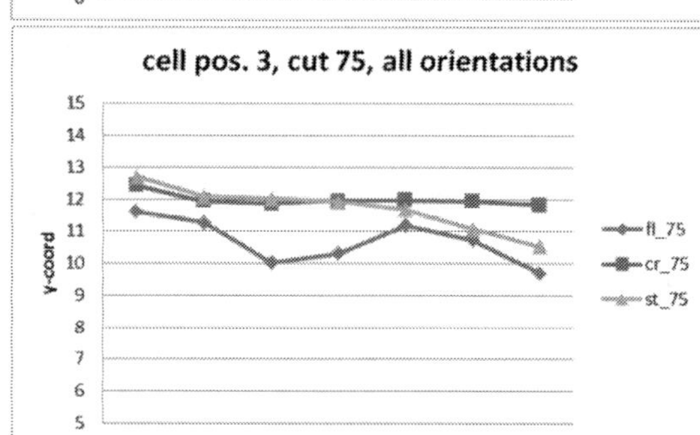

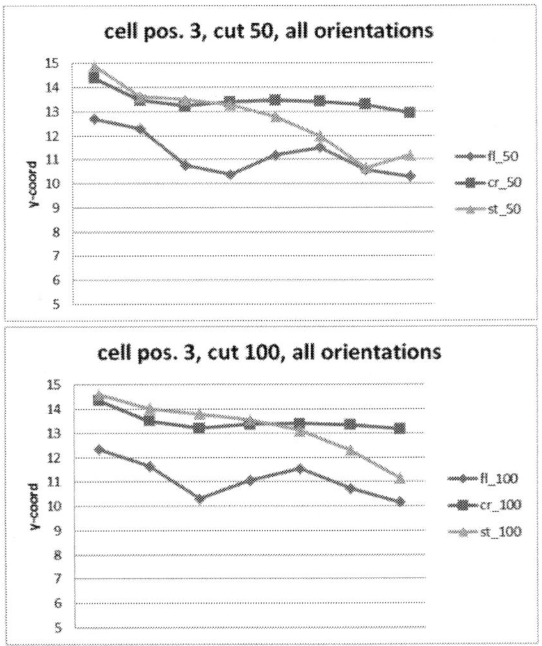

Fig.4. For cell in starting position 3 (15 μm from edge), for all orientations and cutting positions the tendency to change the trajectory is clearly visible. In this case the slope coefficients are -0.19813, -0.08924, -0.35882, -0.19956, -0.05978, -0.31496, -0.26794, -0.12479, -0.54526, -0.25306, -0.13050, -0.51749

• Trajectory stability is also dependent on the initial RBC orientation. All the simulations we performed showed that the "flat" orientation is highly problematic and corresponds primarily to random and chaotic (changing but not monotone) changes in the y-coordinate sequence. However, it is valid (and can be statistically substantiated) that this orientation is quite rare in simulations with a greater number of RBCs, as RBC collisions (both with cells and with the surface of the device) lead to the "increase" of RBC orientation that are much more stable.

• Behaviour of isolated cells changes significantly after addition of 10 or 20 randomized RBCs (approximately corresponding to low hematocrit of approximately 2-5%). Significant y-coordinate sequences characterized by high absolute values of the slope of the linear regression lines are lost and their randomness is increased.

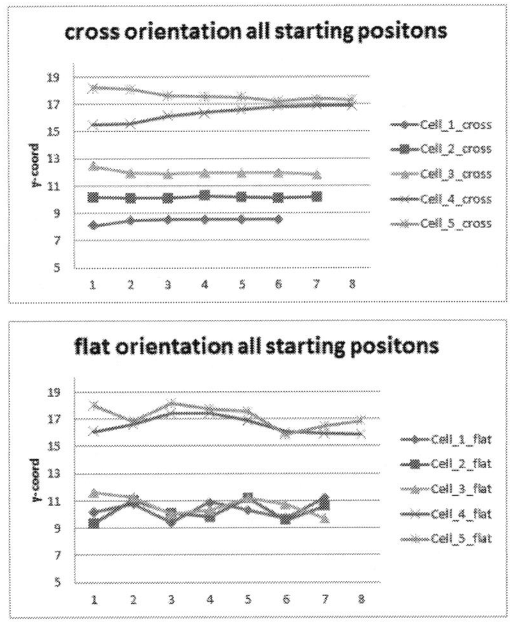

Fig.5. For cutting position 75 + k.100 μm and each cell starting positions we illustrate the different behaviour of y-coordinate changes for cross and flat starting orientation.

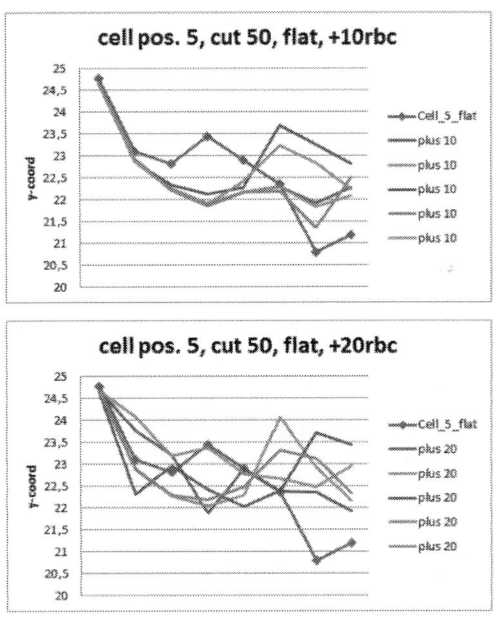

Fig.6. For cutting position 50 + k.100 μm and cell with starting position 5 (25 μm from edge), flat orientation, we can see clear differences between y-coordinate sequences for "solo" experiment and experiments with 10 and 20 randomly added cells

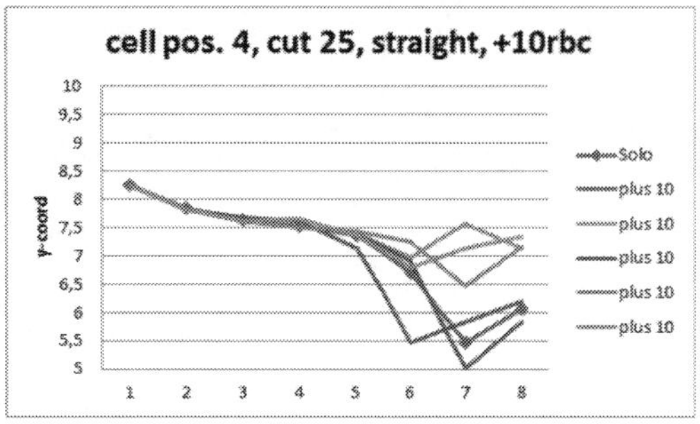

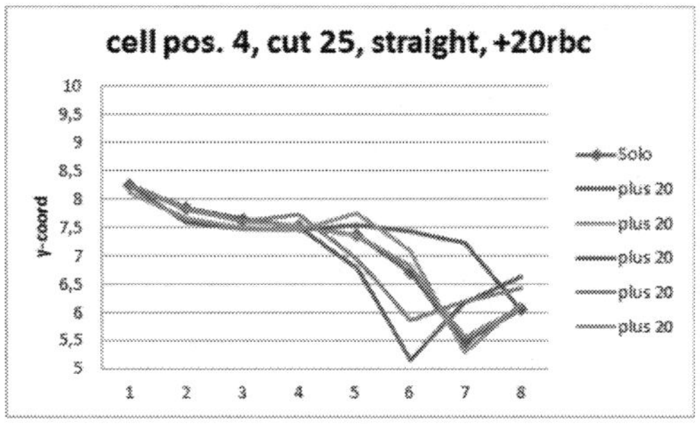

Fig.7. For cutting position 25 + k.100 μm and cell with starting position 4 (20 μm from edge), straight orientation, the differences after adding 10 and 20 cells are not so distinct

• Based on these results, it is clear that the characteristics of the RBC trajectories and the special (un) stability of the trajectories need to be examined in simulations with a larger number of RBCs whose initial positions and orientations are statistically "colorful" and possibly "fine-tuned" by some specific start of the simulation. To examine the sensitive trajectory properties, it is possible to exclude RBC in some "insensitive" positions and orientations.

4 Stability analysis of trajectories for large numbers of RBC

In the context of these conclusions, we have proposed a methodology for statistical monitoring and evaluation of the stability of trajectories for a larger number of RBC. Our main intention was to perform the necessary processing and reduction of trajectory data and to decide the issue of trajectory stability using standard statistical methods.

We have verified the method with simulations in type A channel for 20, 50, and 100 RBC, and in channel C for 50 RBC. Our analysis originated from the cell transitions at the 4 cut positions mentioned in Section 3. For each of the cells, we captured its position at the first and at the last transition trough each of the cutting positions. Thus, we obtained a set of 20 (50, 100) positions of the first and the last cell transition positions at this particular channel location. Assuming that the RBC trajectories are stable, these sets should be statistically identical, respectively very similar. On the contrary, if trajectories tend to change, for example, shift to "typical" trajectories; we should monitor some statistical "simplification" of data and the like.

With the perspective of using the methodology for other types of channels, we have used the distance (in the y-axis direction) of the cell's centre from the border of the closest obstacle, as the analysed value of the cell position, when passing through a series of obstacles (cutting positions 25 + k.100 μm and 75 + k.100 μm). We used the distance (in the y-axis direction) of the cell centre from the channel boundaries to measure cell positions in the channel sections between the obstacle arrays (cutting positions 50 + k.100 μm and 100 + k.100 μm).

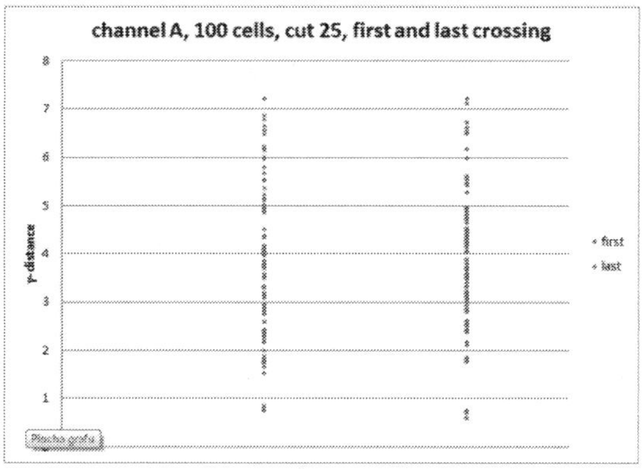

Fig. 8. For 100 cell in channel A and cutting position 25 + k.100 μm two datasets of distances for first and last crossing are depicted. The difference in their distribution is not clear visible

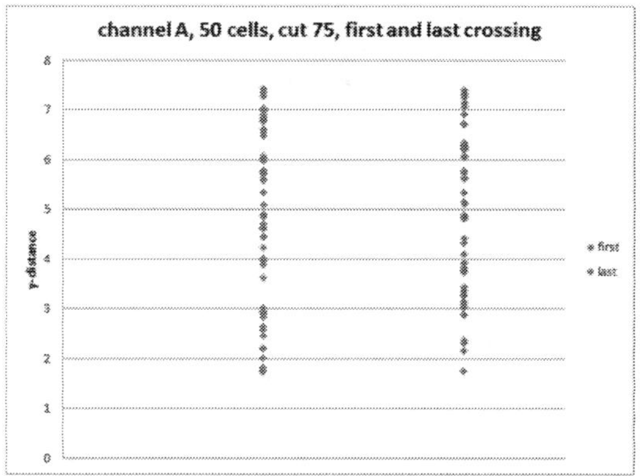

Fig. 9. For 50 cell in channel A and cutting position 75 + k.100 μm the distributions of distances for first and last crossing are much more similar but not visible without statistical analysis

To evaluate the statistical similarity of these data for the first and last cell transition trough the cutting position, we used the standard Two-Sample Kolmogor-Smirnov test. It tests the hypothesis that both sets of the data come from the same (unspecified) distribution.

Overall, we compared pairs of data sets for 5 simulations and 4 cutting positions in each of them. In 12 cases, based on the KS-test statistical value, we could accept the hypothesis that the data is from the same distribution. In 2 cases, however, this hypothesis could be rejected at a 5% level of significance, once at 10% level and 5 times on a not very typical but used for the analysis at 15% level. Based on these results, we consider the methodology to be sufficiently sensitive and ready to process and assess the stability of RBC trajectories in more complex channels designed for real use.

Tab. 1. Table with results of K-S testing. Logical value 0 indicates acceptance of hypothesis that both datasets are from the same distribution. In other cases we depict the significance level, on which the hypothesis can be refused.

		A100	A50a	A50b	A20		C50
cut 25	logical	5%	0	15%	15%	cut 42,5	0
	p-value	0,26	0,9541	0,5077	0,4973		0,6779
	ks2stat	0,14	0,1	0,16	0,25		0,14
cut 50	logical	0	0	0	15%	cut 50	0
	p-value	0,556	0,6779	0,6779	0,4973		0,8409
	ks2stat	0,11	0,14	0,14	0,25		0,12
cut 75	logical	0	15%	0	0	cut 57,5	10%
	p-value	0,961	0,5077	0,6779	0,9655		0,3584
	ks2stat	0,07	0,16	0,14	0,15		0,18
cut 100	logical	0	15%	0	5%	cut 100	0
	p-value	0,961	0,5077	0,8409	0,2753		0,8409
	ks2stat	0,07	0,16	0,12	0,3		0,12

5 RBC's Trajectories Data Processing for Various Geometries

Our task was to analyse the blood flow in the microfluidic devices regarding the behaviour of individual cells and the capture of cancer cells on the device columns. In the case where only the fluid is moving in the microfluidic device, we can illustrate its flow through the stream lines. These are different for the various layouts of columns that capture the CTCs. In the case of a homogeneous fluid that does not contain blood cells, we could predict the transition of the cancer cell better than in the case of non-homogenous fluid, such as blood that in addition to the blood plasma contains also red blood cells, white blood cells, blood platelets and other blood components. The movement of cancer cells in the fluid (without other blood cells), in the microfluidic devices, is used to find suitable chemical bonds for the capture of cancer cells, for example in the article [14]. However, this model does not consider the influence of other objects in the blood. The role of diagnostics is to detect the presence of CTC cells between the healthy cells in the blood. In our model, we use red blood cells that are modelled as elastic bodies with properties similar to real objects, not only in shape, size and weight, but also in their elasticity.

The elastic properties of the blood cells cause that their movement in a microfluidic device is not predictable with only the knowledge of the fluid stream lines. For the further development of the device, we are interested in the movement of the CTC cell in a microfluidic device and the characteristics that will describe the behaviour of the blood cells in terms of their movement in the device. At this stage in the development of the simulation model, it is our role to identify such statistical characteristics of the model that will allow us to predict the trajectory and the nature of the cell movement. We would like to obtain different values of these characteristics for different device geometries and different hematocrit levels. On the contrary, if the simulations differ only by the initial blood cells seeding, but their hematocrit levels and the geometry of the device (as well as other parameters such as the total flow rate) will be the same, we would like to obtain same values of the characteristics.

In this part of the paper we will look at the characteristics based on the mass properties of the whole set of blood cells in two different geometries, which are marked A and C, see Figure 10 and Figure 11, and by the number of used cells as A20, A50, A100, C50. The first experiments with these geometries are described in articles [11], [12], [10]. For our experiment, we used geometries with the same number of obstacles that were the same size, the only difference between these geometries was the distance of the obstacles. In each of the geometries, we did several simulations with different seedings.

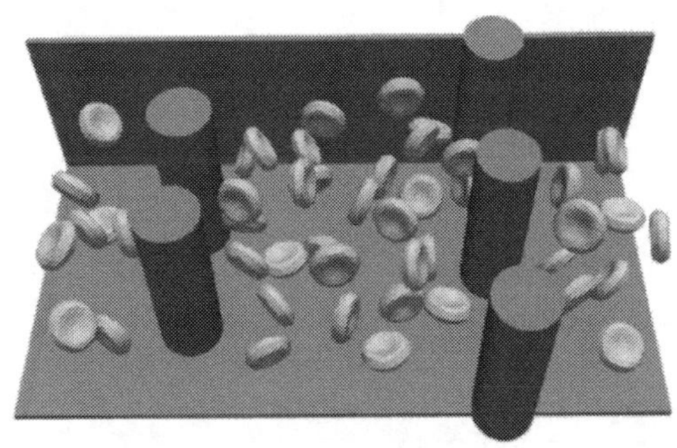

Fig. 10. Geometry A with 50 randomly distributed red blood cells.

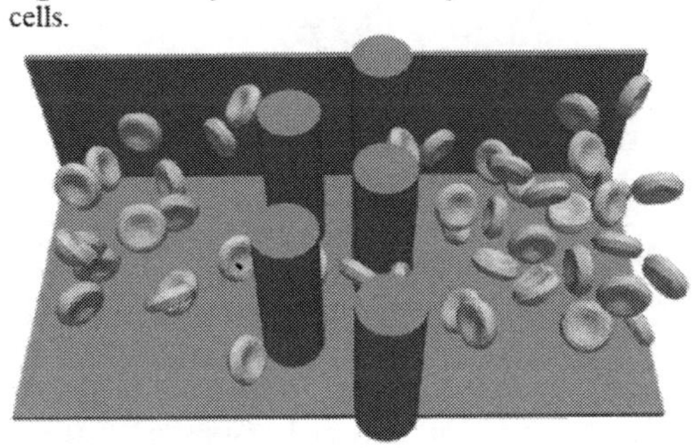

Fig. 11. Geometry C with 50 randomly distributed red blood cells.

The fluid flows in the direction of the x-axis, the channel width is in the y-axis direction and its height is in the direction of the z-axis as indicated in Figure 1.

5.1 Analysis of trajectories and output preparation for further statistical processing

While examining a video from simulation, as well as from our previous analyses [13], it is clear that in each simulation steps, some cells move faster and others slower. Therefore, we cannot analyse only one of the coordinates when plotting and subsequently processing the trajectories of the blood cells. Figure 12 shows the simulation progress of X-coordinates of 50 blood cells centres, where the simulation step was used as an independent variable. The trajectories of moving blood cells, however, can be seen much better when the X-coordinate is chosen as the independent variable, as we can see in Figure 13.

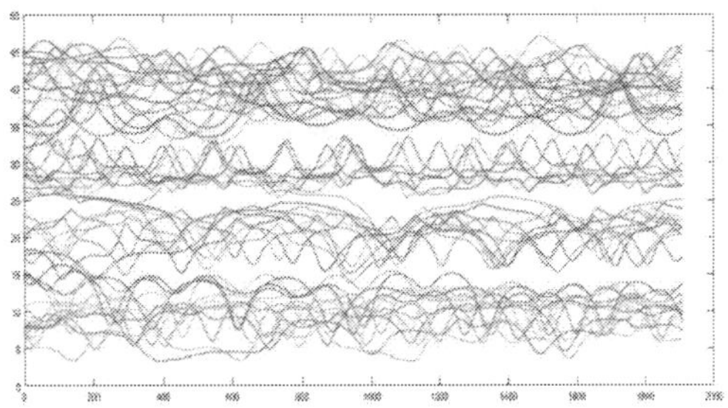

Fig. 12. Geometry A50, the position of the Y-coordinates of the cells' centre at each step of the simulation.

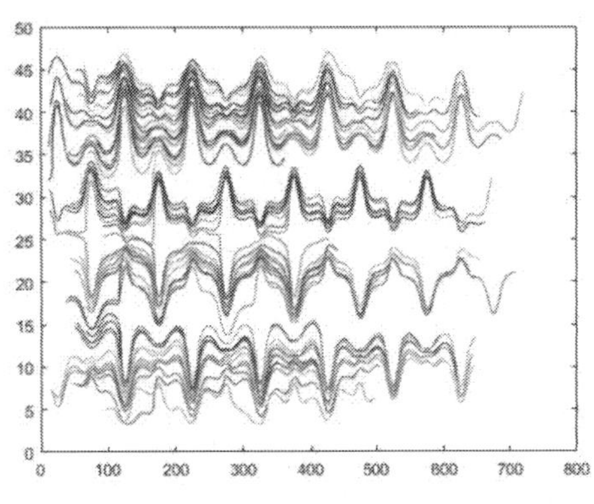

Fig. 13. Geometry A50, the position of the Y-coordinates of the cells' centre depending on X-coordinate.

In Figure 14 we see the trajectories of moving cells in the direction of the Z-axis in relation to the X-coordinate. Since the obstacles in the microfluidic device do not change in this direction, the number of contacts with the obstacle that catches CTC cells, will not be dependent on the position in the Z-axis direction. Therefore, we do not include analysis of the Z-coordinate of the blood cells centres.

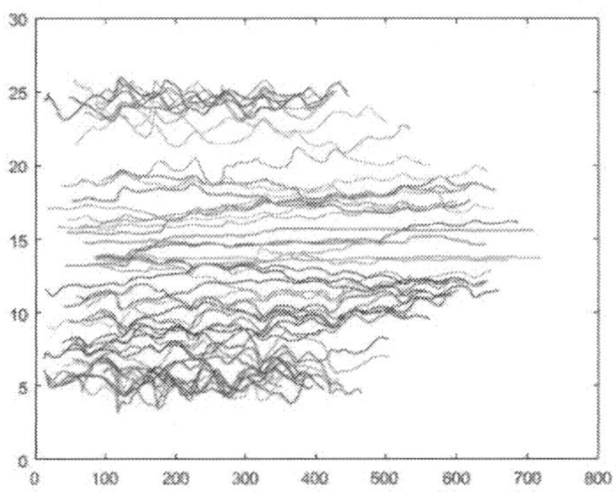

Fig. 14. Geometry A50, the location of the Z-coordinate of the cell centre in relation to X-coordinate.

Figure 15 shows the trajectories of the moving cells in the Y-direction in relation to the X-coordinate for the geometry A for 50 cells. This is also done for the trajectories of 50 cells for geometry C in Figure 16.

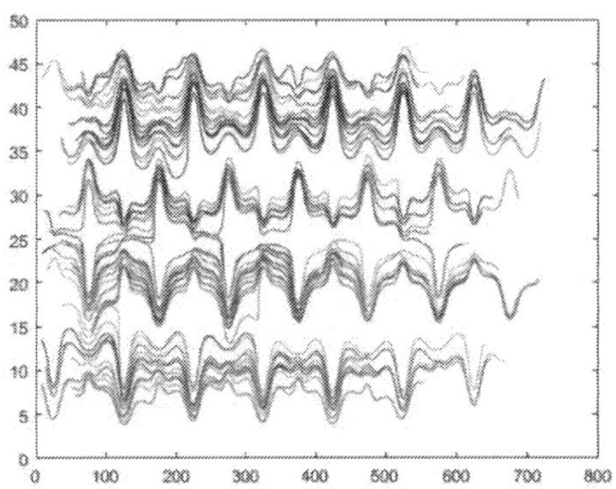

Fig. 15. Geometry A50, two different seedings, Y-coordinate position in relation to X-coordinate

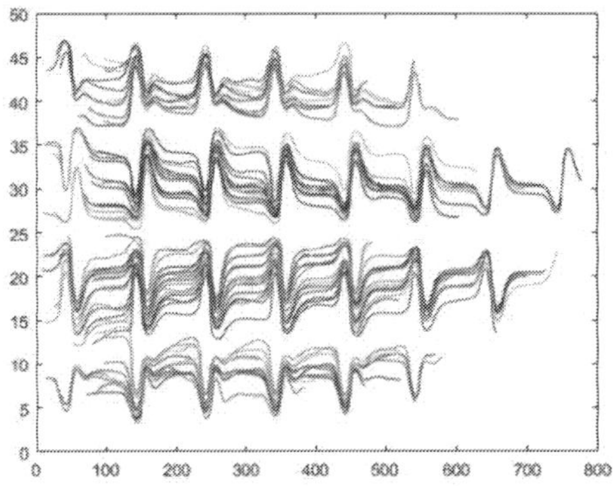

Fig. 16. Geometry C50, two different seedings, Y-coordinate position in relation to X-coordinate.

At various starting positions of the cells and their random rotation, we can find, for each geometry, typical trajectories as well as clusters of cells with the same typical patterns. In Figure 17, three seedings for geometry A are shown, and two seedings for geometry C are shown in Figure 18. In both cases, we synchronized the start of the trajectory only from the simulation step in which the cells reached the same value of the X-coordinate, corresponding to the end of the first crossing of the simulation channel.

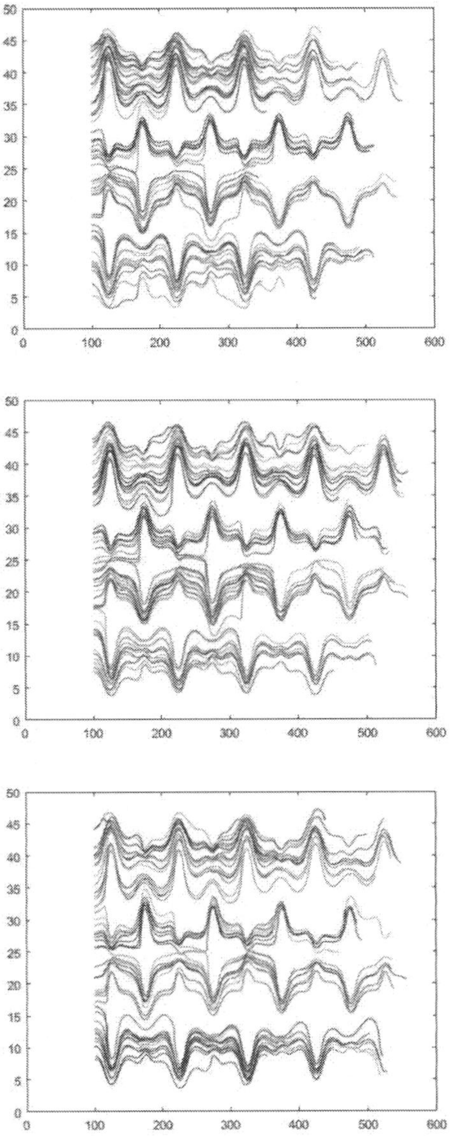

Fig. 17. Geometry A50, three different seedings, the Y-coordinate position in relation to X-coordinate. Values are shifted in the X-direction, to obtain the same starting position for all the cells.

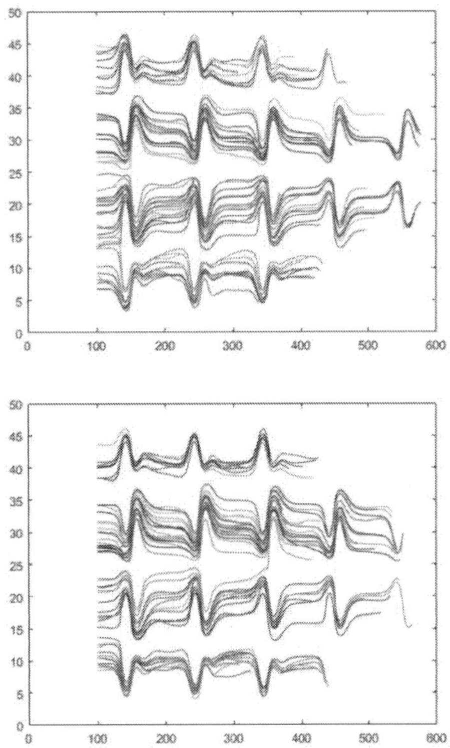

Fig. 18. Geometry C50, two different seedings, the Y-coordinate position in relation to X-coordinate. Values are shifted in the X-direction, to obtain the same starting position for all the cells.

For a smaller number of blood cells in the fluid, i.e., for a smaller hematocrit, there is a small number of interactions, whether between blood cells or between blood cells and obstacles in the microfluidic device. But even when simulating the flow of twenty blood cells in geometry A, in addition to the trend of copying, the study showed in the trajectory analysis of the site a slight change in the movement of the blood cells as can be seen in Figure 19.

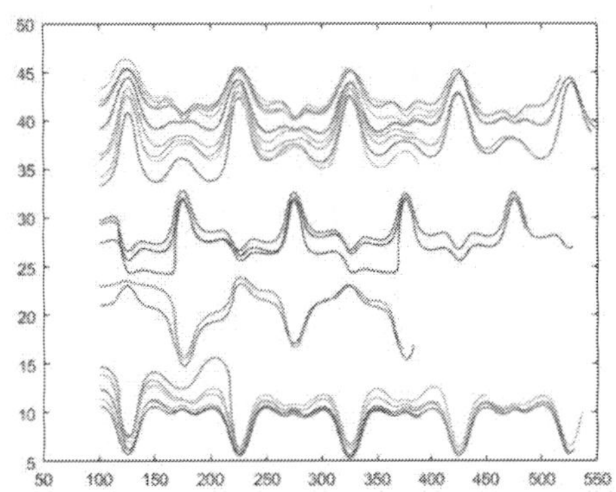

Fig. 19. Geometry A, 20 blood cells.

Similarly, when analysing the flow of 100 cells in geometry A, on the one hand, we can see a strong trend to copy the trajectory. On the other hand, it is evident in Figure 20 that a lot of the cells move also in the Z-axis direction, or significantly slow down, which leads to smaller distance travelled in the X-axis direction (with the same number of time steps).

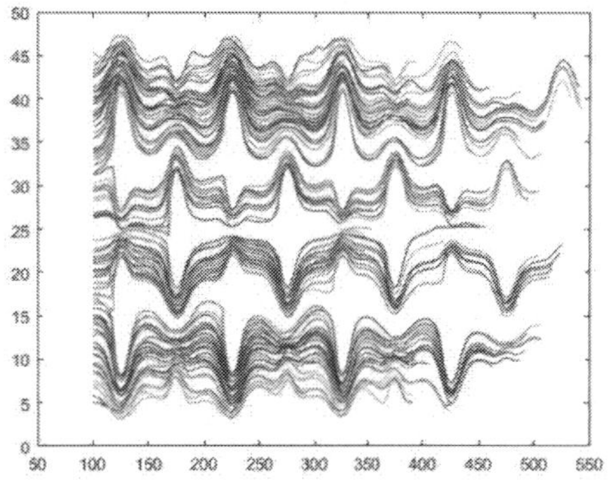

Fig. 20. Geometry A, 100 blood cells.

Distance profiles for each geometry and different amounts of cells are shown in the following charts. The distances are in ascending order. The values for the device with geometry A for 20, 50 and 100 cells are plotted on Figure 21, with two different initial seedings for 50 blood cells. For devices with geometry C, there are also two different initial seedings.

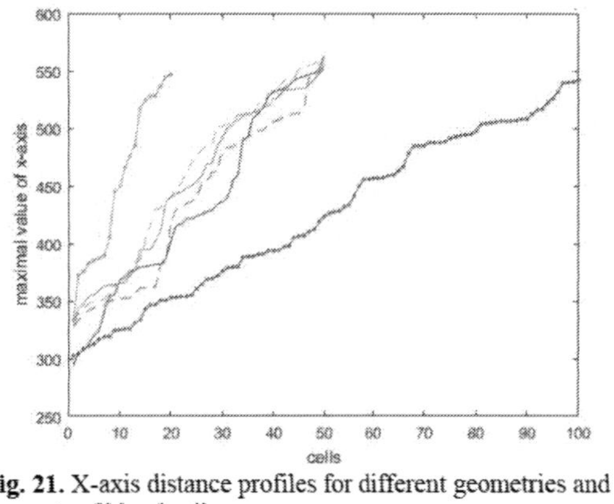

Fig. 21. X-axis distance profiles for different geometries and amounts of blood cells.

5.2. Principal component analysis and clustering of RBC centre trajectories

The processing of trajectories of the ensemble of the cells allow us to predict typical behaviour of blood cells and to estimate their trajectories based on the cluster in which the cell originates. The number of clusters and the variability of their patterns for the different microfluidic device geometries and for the different numbers of blood cells present in the device will be a characteristic used to optimize microfluidic devices. It is desirable that this characteristic will be the same for different initial seedings in the same geometry and with the same number of blood cells, and differences will occur at different geometries and at different number of cells. Clusters for RBC centre trajectories are easily indefinable, in simulations that we have analysed so far, thus will not be further analyzed in this article. We will now focus on the Principal component analysis (PCA) method, which can help us not only distinguish between devices, but mainly can help us to determine the possible rate of success of the device capturing blood cells. In Figure 22, we can see differences in PCA results for the same geometry and different numbers of blood cells.

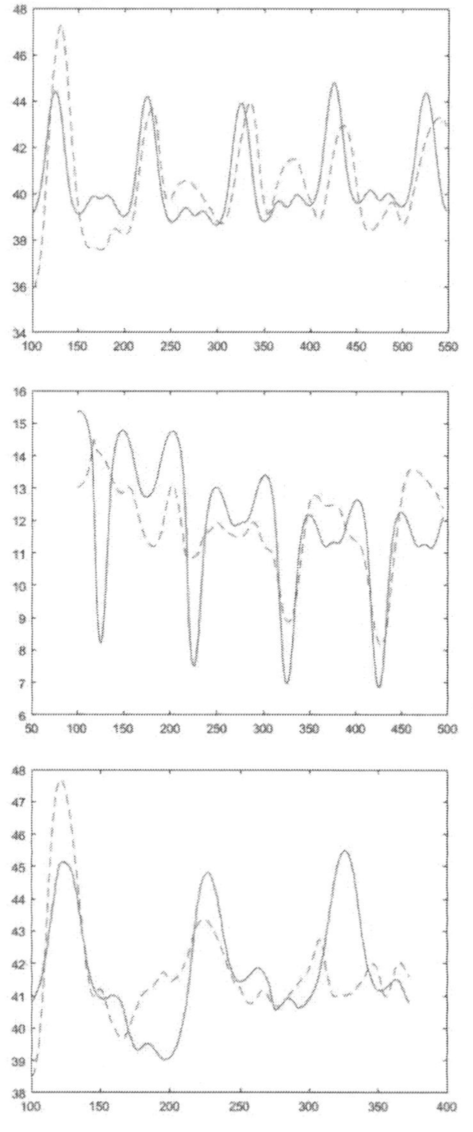

Fig. 22. Approximation of one of the trajectories using the first 4 principal components in geometry A. Gradually for 20, 50 and 100 cells.

According to this result, the behaviour of the blood cells is more predictable and closer to the stream lines when there is less blood cells in the plasma. When there are more blood cells, it is necessary to take more principal components for a better reconstruction of the trajectory, as seen in Figure 23 and Figure 24.

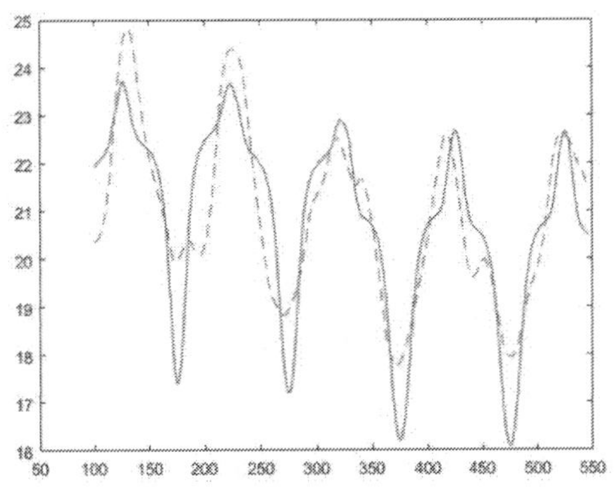

Fig. 23. Trajectory approximation of one of 50 cells using the first 9principal components in geometry A, seeding a.

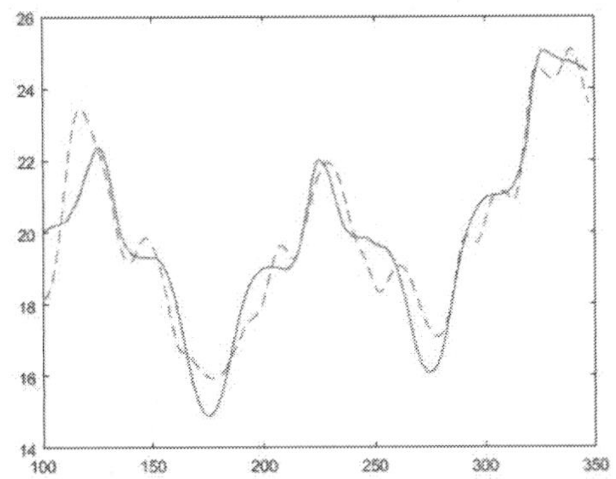

Fig. 24. Trajectory approximation of one of 100 cells using the first 10 principal components in geometry A.

For each geometry and different number of RBC in them, in Figures 25, 26, 27, 28 and 29 we have always drawn three principal components. These determine the course of the most common trajectory and thus are a specific characteristic of the channel.

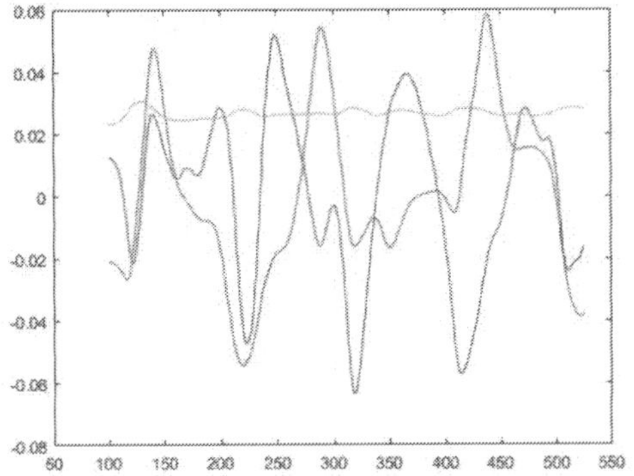

Fig. 25. The first three principal components for 20 cells in geometry A.

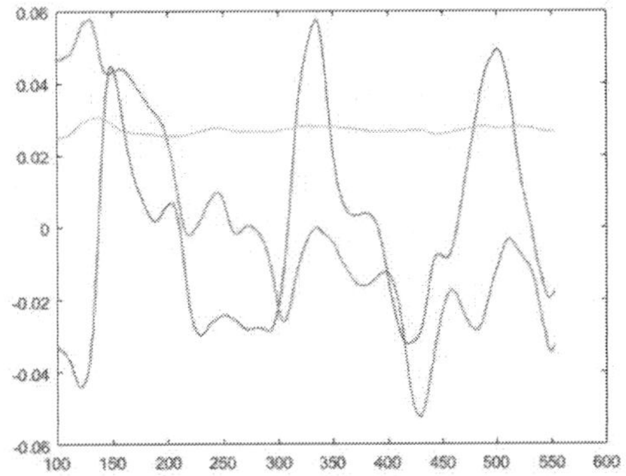

Fig. 26. The first three principal components for 50 cells in geometry A., seeding a.

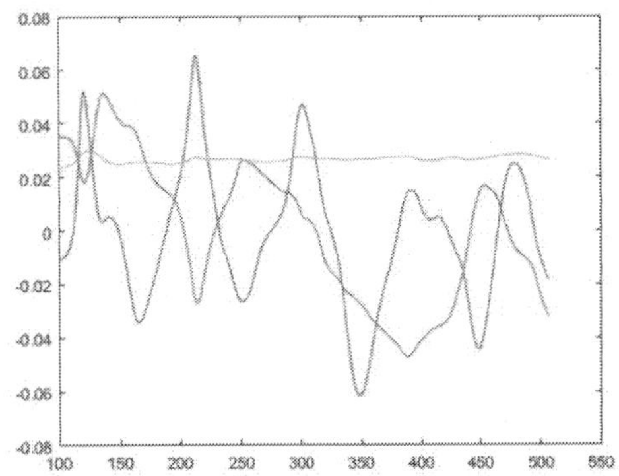

Fig. 27. The first three principal components for 50 cells in geometry A, seeding b.

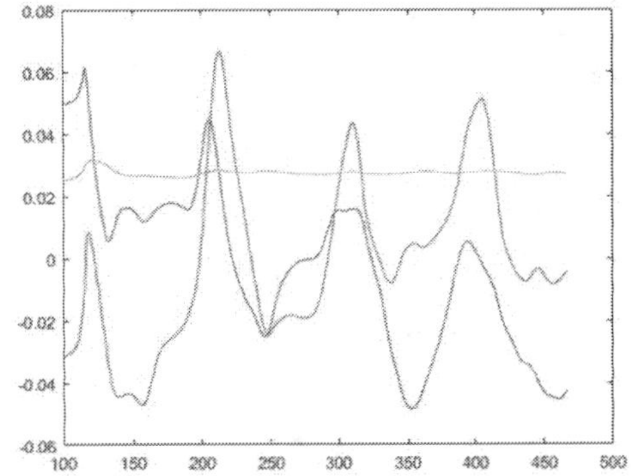

Fig. 28. The first three principal components for 100 cells in geometry A.

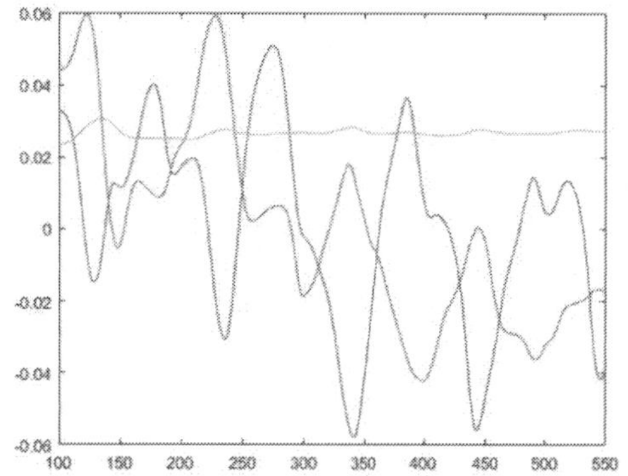

Fig. 29. The first three principal components for 50 cells in geometry C, seeding a.

In addition to the channel trajectory analysis, using the first three principal components, it is possible to interpret the PCA through trajectory reconstructions using the first principal components. Although the reconstructed trajectory differs from the actual trajectory, the error that will arise is in favour of the typical behaviour of all the blood cells. In the future, it will be useful to look mainly to the components of individual clusters that are visible from the trajectory even without a computer analysis. These PCA will allow us to predict the movement of the blood cell and later the cancer cell in the channel.

6 Applying of the investigated characteristics to development of the simulation model

In this article, we have presented a methodology to investigate the trajectories of the cells moving in microfluidic channels. We have verified that proposed methodology help us to recognize different modes of cell-movement in the channels. The proposed methodology will be than applied to several models of existing microfluidic channels. The next step of our study will be to evaluate the measure of collisions between the cells and the obstacles in the micro channel. Subsequently, the measures should be done with models containing a cancer cell(s). After that, we can compare the efficiency of tested microfluidic channels, eventually to propose an improvement of those channels.

References

1. S. Nagrath et al., Nature, **450**, 1235–1239 (2007)
2. S. Hanasoge, R. Devendra, et al., Microfluidics and Nanofluidics, **18**, 1195-1200 (2015)
3. S. Holm et al., Analytical Methods, **8.16**, 3291-3300 (2016)
4. A. F. Sarioglu et al., Nature methods, **12.7**, 685-691 (2015)
5. I. Cimrák, M. Gusenbauer, I. Jančigová, Computer Physics Communications, **185**, 900-907 (2014)
6. I. Cimrák, I. Jančigová, R. Tóthová, M. Gusenbauer, Applications of Computational Intelligence in Biomedical Technology, **606**, 1-28 (2015)
7. I. Cimrák, K. Bachratá, H. Bachratý, I. Jančigová, R. Tóthová, M. Bušík, M. Slavík and M. Gusenbauer, Communications, Scientific Letters of the University of Zilina, **18/1a**, 13-20 (2016)
8. I. Cimrák, Computer Methods in Biomechanics and Biomedical Engineering, **1-6**, 1525-1530 (2016)
9. M. Bušík, I. Jančigová, R. Tóthová, I. Cimrák, Journal of Computational Science, **17**, 370-376 (2016)
10. K. Kovalčíková, M. Slavík, Central European researchers journal, **3**, 21-27 (2017)
11. M. Slavík, K. Bachratá, H. Bachratý and K. Kovalčíková, International Conference on Information and Digital Technologies 2017 (IDT) International Conference on. IEEE, 344-349 (2017)
12. H. Bachratý, K. Kovalčíková, K. Bachratá and M. Slavík, International Conference on Information and Digital Technologies 2017 (IDT) International Conference on. IEEE, 36-46 (2017)
13. K. Bachratá, H. Bachratý, M. Slavík, EPJ Web of Conferences, **143**, (2016)
14. D. Horák et al., Biomed Mater Res, **I. 101A**, 23–32 (2013)

Numerical simulation of minor losses coefficient on the example of elbows

Smyk Emil*, *Mrozik* Dariusz, *Olszewski* Łukasz, *Peszyński* Kazimierz

UTP University of Science and Technology, Faculty of Mechanical Engineering, Poland

Abstract. Determining of minor losses coefficient is very complicated problem. Analytical methods are often very difficult and experimental methods are very expensive and time-consuming. Consequently, the use of numerical methods seems to be a good solution, but there are no publications describing this issue. Therefore, the paper is describing the numerical method of determining the minor loss coefficient ξ on the example of elbows with circular cross-section.

1 Introduction

Pneumatic, hydraulic or ventilating systems are just a few examples of systems in which various types of fittings are used. Examples of fittings used in different systems are: elbow, bow, diffuser, confusor, tee or valve [1]. One of the most important parameters of all type of fittings is minor loss coefficient ξ.

Minor (ξ) and major/head (λ) loss coefficients are used to calculation of pressure losses in different kind of systems. Major loss coefficient λ refers to energy losses caused by friction, while minor loss coefficient ξ to energy losses caused by flow disorder. However, during the major loss coefficient λ is designate for practically every case (experimental models, practically independent of channel shape), the minor loss coefficient ξ only for specific cases (tables with specific fittings).

Minor loss coefficient ξ can be defined in an analytical, experimental [2, 3] or numerical [4-6] way. However, it should be noted that analytical way is very complicated and can't be used in any case. The most universal and generally used is the experimental way. It is time-consuming and expensive process, because the minor loss coefficient ξ is dependent on Reynolds number, duct dimensions and other [7]. For these reason design and use in real object of new type of ducts (with another than circular or rectangular cross-section) is unprofitable. Amount of work and money necessary to designate a minor loss coefficient ξ for all new fittings is disproportionate to the potential benefits. Taking into consideration, it should be found numerical way to determination of minor loss coefficient ξ for different kind of fittings. It would reduce the amount of experimental research.

In this investigation is presented numerical way of minor loss coefficient ξ designation for elbow with circular cross-section. This kind of fitting is commonly used in different systems and its minor loss coefficients ξ, can be found in literature [7, 8] and also in standards [9]. It must be noted that the authors failed to find articles dedicated the issue presented in this paper.

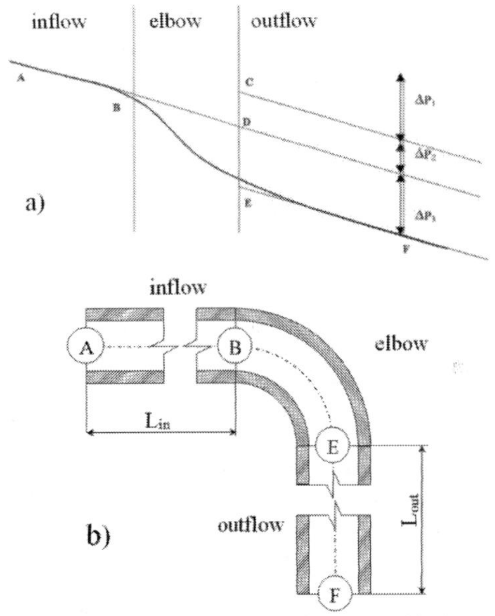

Fig.1 a) Schematic variation of centerline pressure in duct with bend with identification of various pressure drop components [6], b) model of measurement system

1.1 Minor loss coefficient for elbow–methodology of determination

The determination of minor loss coefficient ξ for elbow is performed by measuring the static pressure on the duct section with elbow (fig. 1b). As can be seen the pressure different at the points A and F is caused by (fig. 1a):

- major losses in the inflow and outflow Δp_1;
- major losses in elbow Δp_2;
- minor losses caused by elbow Δp_3.

* Corresponding author: emil.smyk@utp.edu.pl

Because the major losses in the inflow and outflow Δp_1 are known and can be calculate, based on the static pressure measurement in point A and F the pressure loss coefficient K_b for elbow can be calculated [6]:

$$K_b = \frac{2(\Delta p_2 + \Delta p_3)}{\rho v^2} \quad (1)$$

where ρ is density of working fluid and v is mean velocity of working fluid. Eq 1 can be also expressed as:

$$K_b = \frac{p_A - p_F}{\rho v^2} - \lambda \frac{l_{in} + l_{out}}{D_h} \quad (2)$$

where l_{in} is length of inlet, l_{out} is length outlet, and D_h is hydraulic diameter. The hydraulic diameter D_h can be calculated from:

$$D_h = \frac{4 \cdot A}{P} \quad (3)$$

where A is cross-sectional area and P is wetted perimeter of the cross-section. In case of ventilation systems P is equal duct perimeter. For a circular duct the hydraulic diameter D_h is simply the diameter of the duct.

The minor loss coefficient ξ can be calculated by reduction of pressure loss coefficient K_b by major losses in elbow Δp_2. Minor loss coefficient ξ can be calculated from [6]:

$$\xi = K_b - \lambda C \theta \quad (4)$$

where C is the curvature ratio and θ is bend angle in radian.

It must be noted that the length of inlet and outlet must be equal or longer than 10 hydraulic diameter D_h. This length allows for the disappearance of the disorder caused by fitting.

During investigation Reynolds number Re was define as:

$$Re = \frac{v D_h}{\mu} \quad (4)$$

where μ is kinematic viscosity of working fluid.

2 Numerical simulation

Numerical simulation was made in Ansys Fluent software. The first step of calculation was choice of simulation model. Calculated minor loss coefficient ξ, according to eq. 2 and 4, is depended on the adopted major loss coefficient λ. Consequently, it was decided to choose the turbulent model gives major loss coefficient λ results consistent with the theory. The next stage was designation of the minor loss coefficient ξ for elbows with selected parameters. All simulation was made with air as working fluid without compressibility.

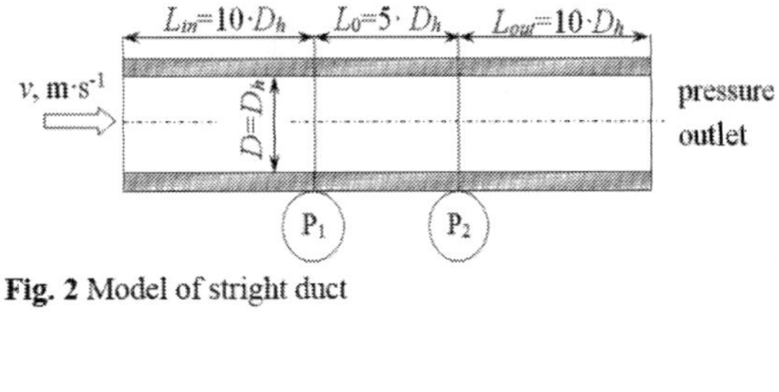

Fig. 2 Model of stright duct

2.1 Preliminary research

In this part of research was used three generally used turbulence models (Spalart– Allmaras, k-ε and k-ω) to determine how they simulate major losses. For this purpose, a straight duct model was created. The model was consist from inlet L_{in}, measuring section L_0 and outlet L_{out}, lengths as shown in fig. 2. The plane of symmetry was used in the model, and during meshing was used only hexadron elements.

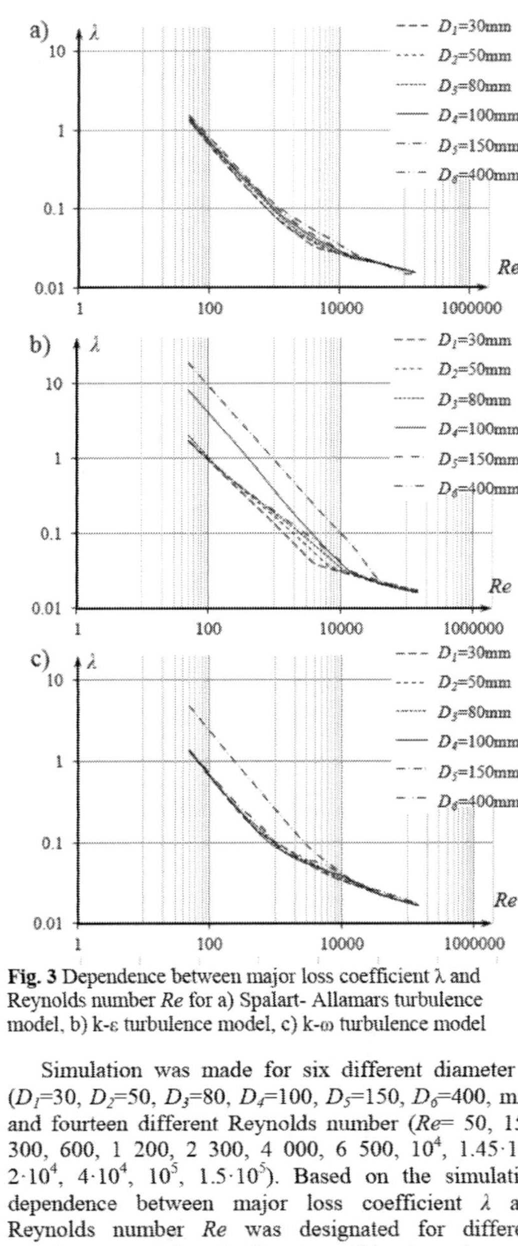

Fig. 3 Dependence between major loss coefficient λ and Reynolds number Re for a) Spalart- Allamars turbulence model, b) k-ε turbulence model, c) k-ω turbulence model

Simulation was made for six different diameter D (D_1=30, D_2=50, D_3=80, D_4=100, D_5=150, D_6=400, mm) and fourteen different Reynolds number (Re= 50, 150, 300, 600, 1 200, 2 300, 4 000, 6 500, 10^4, $1.45 \cdot 10^4$, $2 \cdot 10^4$, $4 \cdot 10^4$, 10^5, $1.5 \cdot 10^5$). Based on the simulation dependence between major loss coefficient λ and Reynolds number Re was designated for different turbulence model and diameter. This dependence is shown in fig. 3, logarithmic scale was used. Additionally comparison of averaged major loss coefficients λ from simulations and major loss coefficient λ calculated on the basis of theory [1, 7-9].As can be seen in the fig. 3 the results are very different. The first difference is dependence of results from diameter. In case of Spalart- Allmaras turbulence model the calculated major loss coefficients λ are similar and only Reynolds number *Re* influences their value (fig. 3a). It is consistent with the theory [1, 7-9]. However, in the case of k-ε turbulence model (fig. 3b) the minor loss coefficient λ for diameter D_1 and D_5 stand out from the rest of results, and they values are bigger than other. In case D_1 for $Re<2\cdot10^4$ and in case D_5 for $Re<4\cdot10^3$.

For k-ω turbulence model results for only one diameter stand out from the rest. The major loss coefficients λ for diameter D_5 are bigger than other in case $Re<10^4$.

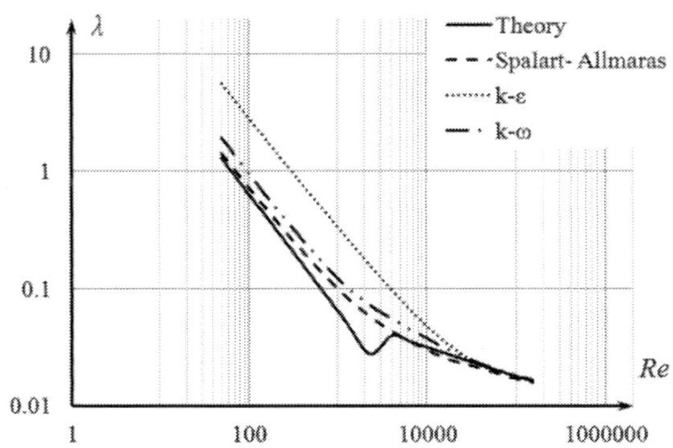

Fig. 4 Comparision of avereged major loss coefficients λ from simulation with theoretical [1, 7-9]

As can be seen on the fig. 4 the averages major loss coefficients λ for Spalart- Allamars turbulence model are the most similar to the theoretical values. However in case of big Reynolds number $Re>2\cdot10^4$ all used turbulence models give similar convergence of results with theory.

Based on the above results Spalart- Allmaras turbulence model was chosen to simulations which the determination of minor loss coefficient ξ is purpose. The authors are aware that this model may not be the best model to simulation of turbulence, which arise in the elbow. However, due to local velocity changes, at the wall in elbow and behind them, it seems that the use of model which simulates the most accurate major losses will help to improve the accuracy of the results.

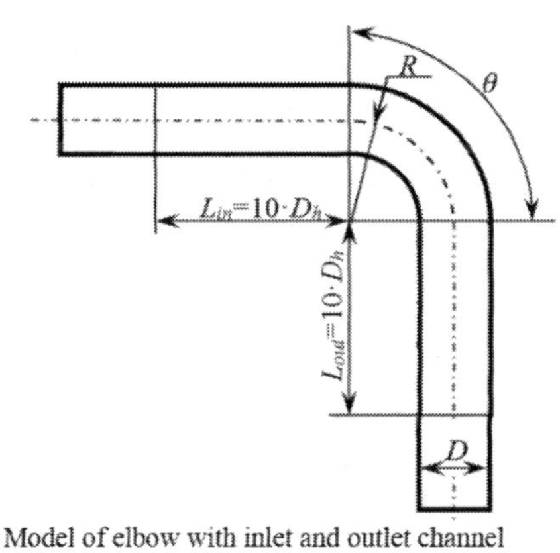

Fig. 5 Model of elbow with inlet and outlet channel

2.1 Main simulation

2.1.1 Simulation model and parameters

Simulation model is shown in fig. 5. Before and behind the elbow was modelled inlet and outlet channel of length $L_{in}=L_{out}=10\cdot D_h$, where hydraulic diameter is equal normal duct diameter $D_h=D$. The diameter of duct D=400mm was chose in such a way as to have the least effect on the designated minor loss coefficient ξ (in [7] has been shown that for duct diameter $D<300$ the minor loss coefficient ξ isn't constant and it is strongly dependent on diameter D). The models were made in different version for θ=20, 30, 45, 60, 75 and 90° and for ratio of elbow ratio to duct diameter R/D=0.5, 0.75, 1, 1.5, 2 and 2.5. On the end of duct was used boundary layer *pressure outlet*, and on the inlet of duct *velocity inlet* was selected as to Reynolds number was equal $Re=10^4$, $2\cdot10^4$, $4\cdot10^4$, 10^5 or $1.5\cdot10^5$.

As above mentioned, 36 different models were made and each was examine for 5 different velocity. This allows you to create a table with minor loss coefficients ξ for a elbow with circular cross-section, which is similar to tables found in the literature [1, 7-9].

Minor loss coefficients ξ were calculated from formulas (2) and (4). During calculation major loss coefficient λ was adopted from preliminary research.

2.1.2 Simulation results

As mentioned simulation was made with five different velocity for each from 36 models. Exemplary results of calculated minor loss coefficients ξ for elbow θ= 75° is presented in table 1.

Table 1. Calculated minor loss coefficients ξ for elbow θ= 75° for different Reynolds number *Re* and their average and standard deviation

R/D	Reynolds number *Re*					Average ± standard deviation
	10^4	$2\cdot10^4$	$4\cdot10^4$	10^5	$1.5\cdot10^5$	
0.5	0.51	0.50	0.52	0.51	0.51	0.51±0.0071
0.75	0.29	0.24	0.24	0.22	0.21	0.24±0.0308
1	0.25	0.19	0.18	0.16	0.15	0.186±0.0391
1.5	0.19	0.16	0.15	0.13	0.12	0.15±0.0274
2	0.16	0.13	0.14	0.12	0.11	0.132±0.0192
2.5	0.14	0.10	0.10	0.08	0.08	0.10±0.0245

In table 1 are presented minor loss coefficients ξ calculated for different Reynolds number *Re*. The results are rounded to the second decimal place. Additionally it was calculated their average and standard deviation. As can be seen the minor loss coefficient ξ is dependent on *R/D* ratio (this is an obvious and well-known fact) and Reynolds number *Re*. Generally the bigger the Reynolds number *Re*, the smaller the minor loss coefficient ξ. The results for the same *R/D* ratio are very different and they can vary by as much as 40% from each other for different Reynolds number *Re* e.g. for *R/D*=1 if $Re=10^4$ then ξ=0.25 and if $Re=1.5\cdot10^5$ then ξ=0.15

It has big influence on presented in table 1 average of results and them standard deviation. In case of mentioned earlier elbow *θ*= 75°, *R/D*=1 the standard deviation is about 21% of the average value of designated minor loss coefficients *ξ*. To highlight how the results differ from each other in table 2 are showed what percentage of the average value is the standard deviation.

Table 2. Ratio of absolute value of standard deviation to the average value expressed in percent

R/D	Angle *θ*					
	20°	**30°**	**45°**	**60°**	**75°**	**90°**
0.5	29.88%	42.70%	4.32%	13.36%	1.39%	0.96%
0.75	30.62%	40.16%	23.16%	33.72%	12.84%	9.48%
1	36.38%	39.46%	32.71%	33.72%	21.03%	16.39%
1.5	30.99%	34.58%	28.04%	33.72%	18.26%	12.65%
2	30.99%	31.11%	25.25%	31.11%	14.57%	12.04%
2.5	67.71%	59.43%	36.36%	35.94%	24.49%	20.79%

As can be seen in the table 2 the results was significantly different from each other. The large value in table 2 tells us that the standard deviation was considerable in relation to average value and therefore it mean that during simulation the Reynolds number *Re* has huge influence on the designed minor loss coefficient *ξ*. It can be seen that this dependence is the stronger the lower the angle *θ* and the bigger the ratio *R/D* is. This is a negative phenomenon which influence which negatively affects the quality of results. However, despite this it was decided compare the results obtained from the simulations with the results from the literature [8]. The average values of results are presented in table 3 and minor loss coefficients *ξ* values from literature are presented in table 4.

It can be seen that the value of calculated and literary minor loss coefficients *ξ* are close to each other especially for elbow with *θ*=75° and *θ*=90° and for elbow with ratio *R/D*≥1.5. To better illustrate in table 5 are presented modulus of relative errors of coefficients from simulation (table 3) relative to coefficients from [8] (table 4). Relative error was calculated from:

$$e = \left|\frac{\xi_c - \xi_l}{\xi_l}\right| \cdot 100\% \qquad (5)$$

where ξ_c is minor loss coefficients from calculation and ξ_l is minor loss coefficients from [8] for specific elbow.

In table 3 are presented rounded values. However during table 5 creation was used not rounded values, which was calculated on base of simulation

Additionally in the table 5 are marked with a gray background the cases where the relative error is smaller than 30%. Authors took this error threshold as being sufficient to recognize the results as correct.

As mentioned earlier results for elbow with *θ*=75° and *θ*=90° and for elbow with ratio *R/D*≥1.5 are very good and except elbow *θ*=20° *R/D*=2 and elbow *θ*=45° *R/D*=2.5 minor loss coefficients ξ for all elbows in this area are very close to literary value (relative error smaller than 30%). In case of elbow with small *R/D* ratio and small angle the calculations results are completely different for example for elbow *θ*=20° *R/D*=0.5 or elbow *θ*=20° *R/D*=2.

On this basis it can be stated that the proposed method of simulation, of minor loss coefficients ξ in elbow, gives correct results only for elbow with a large angle or with a large ratio *R/D*. However high correlation of the results to the Reynolds number *Re* is a big inconvenience. In addition, it requires an additional check for other fittings.

Table 3. Designated from the simulations minor loss coefficients *ξ* for elbow with circular cross section

R/D	Angle *θ*					
	20°	**30°**	**45°**	**60°**	**75°**	**90°**
0.5	0.03	0.05	0.14	0.30	0.51	0.74
0.75	0.04	0.06	0.10	0.17	0.24	0.30
1	0.05	0.07	0.10	0.15	0.19	0.21
1.5	0.05	0.07	0.11	0.13	0.15	0.15
2	0.05	0.07	0.07	0.12	0.13	0.13
2.5	0.03	0.05	0.11	0.10	0.10	0.09

Table 4. The minor loss coefficients *ξ* for elbow with circular cross section from [8]

R/D	Angle *θ*					
	20°	**30°**	**45°**	**60°**	**75°**	**90°**
0.5	0.22	0.32	0.43	0.55	0.64	0.71
0.75	0.10	0.15	0.20	0.26	0.30	0.33
1	0.07	0.10	0.13	0.17	0.20	0.22
1.5	0.05	0.07	0.09	0.12	0.14	0.15
2	0.04	0.06	0.08	0.10	0.12	0.13
2.5	0.04	0.05	0.07	0.09	0.11	0.12

Table 5. Modulus of relative error of coefficients from simulation relative to coefficients from [8]

R/D	Angle *θ*					
	20°	**30°**	**45°**	**60°**	**75°**	**90°**
0.5	86.7%	85.1%	67.3%	44.7%	20.2%	4.2%
0.75	59.9%	58.6%	48.4%	34.4%	20.2%	7.7%
1	34.0%	32.6%	19.4%	13.3%	6.8%	3.1%
1.5	3.3%	6.3%	17.1%	12.2%	6.8%	0.3%
2	33.3%	24.6%	8.7%	24.4%	9.8%	3.4%
2.5	13.4%	9.4%	58.0%	7.6%	10.0%	24.8%

It must be noted that obtained results was compare to data included in [8]. This may seem dubious in terms of scientific value. However, in the world's publications there are missing articles dealing with minor losses (omitting few articles like [2-6]). Research group, to which the authors of this article belong, deals with the issue of minor losses in ventilation ducts. Initial results of research are presented in [10-12].

In [6] are presented methods of minor losses ξ calculation in elbow for different Reynolds number. However, in work [6] the results of presented calculations aren't compared to experimental data and refer to elbow with very small value of hydraulic diameter D_h (paper [6] refers to channels in fuel cell stacks). For this reason presented in this investigations results can't be compared to data from [6].

5 Conclusion

In this article is presented method of simulation of minor loss coefficients ξ in elbow with circular cross section. It has been shown that this method gives satisfactory results in case of some tested models. Presented results needs additionally research that will help make a results independent from Reynolds number *Re*, what may increase quality of results.

Additionally during the simulation the different types of turbulence models were tested how they simulate the head losses.

Please note that the presented method is intended to simplify the testing process of the fittings and to assist the experimental methods

Acknowledgment

Scientific work financed from the budget for science in the years2015/2018 as a research project within the program under the name "Diamond Grant "

Authors also received institutional support BSM – 60/2017 granted by Faculty of Mechanical Engineering of UTP University of Science and Technology.

References

1 A. Pełcha, *Wentylacja i klimatyzacja - podstawy (Polish: Wentilation and air conditioning– base)* (Oficyna Wydawnicza Politechniki Wrocławskiej, Wrocław, 2008)

2 K. Strzelecka, K. Jeżowiecka–Kabsch, Ochrona środowiska **30**, 2 (2008)

3 Ł. Mika, Inżyn. i Apara. Chemi. **49**, 1 (2010)

4 A.S. Kheireddine, M.H. Sanda, S.H. Chaturvedi, T.O. Mohieldin, Energy **22**, 4 (1997)

5 G. Gan, S.B. Riffat, Applied Energy. **54**, 2 (1996)

6 S. Maharudrayya, S. Jayanti, A.P. Deshpande, J. of Power Sour. **138**, 1-2 (2004)

7 T. Klinke, *Wentylacja, tablice do obliczeń strat ciśnienia (Polish: Ventilation, pressure loss calculation boards)* (Oficyna Wydawnicza Politechniki Warszawskiej, Warszawa, 2007)

8 J. Hendiger, P. Ziętek, M. Chlidzinska, *Wentylacja i klimatyzacja. Materiały pomocnicze do projektowania (Polish: Wentilation and air conditioning. Design aids)* (Venture Industries Sp. z.o.o., Warszawa, 2013)

9 Polish standard PN-76 M-34034

10 E. Smyk, D. Mrozik, S. Wawrzyniak, K. Peszyński, Eng. Mech. 2017 (post-conference materials), 882-885 (2017)

11 K. Peszyński, Ł. Olszewski, E Smyk, T. Kasprowicz, Eng. Mech. 2017 (post-conference materials), 50-53 (2017)

12 K. Peszyński, D. Perczyński, E Smyk, P. Kolber, Eng. Mech. 2017 (post-conference materials), 770-773 (2017)

Experimental study of the flow field disturbance in the vicinity of single sensor hot-wire anemometer

Jacek Sobczyk[1,*]

[1]Strata Mechanics Research Institute of the Polish Academy of Sciences, Reymonta 27, 30-059 Krakow, Poland

Abstract. Preliminary experimental study of the flow field disturbance in the vicinity of single sensor normal hot-wire anemometer (SN) probe was carried out. Regular 2D particle image velocimetry (PIV) setup equipped with micro lens and distance rings was applied to measurements of macroscopic flow around microscopic elements. Experimental results revealed complexity of the flow around the wire and its strong dependence on both – the velocity magnitude and the probe orientation in relation to freestream direction. Examination of the velocity fields in the vicinity of SN probe suggests that it may not be such a "point" measurement method as it is commonly assumed to be.

1 Introduction

Metrology of flows (MOF) is a well-developed field of science in the present day [1-3]. What is more, continuous technological progress and growing industry demands are the main stimuli forcing this field to develop accordingly [4]. Nowadays measurements of flows make use of more and more sophisticated equipment. Regardless of the main working principle modern sensors, transducers as well as complex measurement systems are able to supply us with more and more accurate measurement results in a blink of an eye. New frontiers are being reached especially in the field of non-intrusive, optical measurements [5-10].

It is really true, that we are able to measure so complex phenomena in a so challenging conditions that a few decades ago it would simply seem a science-fiction. On the other hand, frequently, when using an expensive and sophisticated equipment or, just the opposite, "well-known" and "easy-to-use" one we are losing healthy criticism, believing in almost every result we get. This approach seems natural, as we want to believe that such equipment should serve us with only good results that we could rely on. Especially when we are sure that it is working properly and the phenomena we are measuring lays within limits covered by the device measuring range.

Implications of such an approach may be quite significant especially when high accuracy is indispensable and ambient conditions are variable in a range much wider than the device was designed for [1-3]. A simple example is the use of hot-wire anemometer (HWA) in the flow with variable temperature and/or absolute humidity, pressure and especially variable gas mixture composition. Even though velocity of the flow lays within HWA calibration limits, relatively small parallel change of one or more of the mentioned parameters adds unexpected shifts to the measurement results, which are not always easy to compensate. What is more, single sensor normal HWA (SN) probe is usually regarded as a point measurement device/method, that virtually does not disturb the flow under investigation.

To avoid the former issue knowledge of the device constrains and the details of measurement conditions is indispensible. To understand the simplification of the latter one non-intrusive, optical measurement method such as particle image velocimetry (PIV) [2-6, 8-10] can be of a help. In fact, it is relatively easy to demonstrate by means of PIV that this common believe is not correct, especially in the case of low velocities.

There is a significant amount of comprehensive studies of various types of HWA probes aerodynamics. They were conducted with use of various techniques, predominantly pointwise, such as another HWA probe or laser Doppler anemometry [11-12]. There is only a few attempts of using PIV technique to analyze flow around HWA probe of any type present in the literature [13-14]. Those known to the author of the present study are very incomprehensive. Therefore the subject under investigation is original and – in the light of what was written above – important.

Considerations mentioned above as well as measurement results presented below possess many various aspects, that will not be raised. In the present paper only the disturbance of flow in the vicinity of the SN wire and tips of prongs will be emphasized. For a chosen type of SN probe the measured disturbances result only from its geometry as well as applied velocities of air flow and chosen orientations in space.

* Corresponding author: sobczyk@img-pan.krakow.pl

2 Measurements

2.1 Measurement stand

All measurements were carried out in the new, closed-circuit, low speed wind tunnel (fig. 1) belonged to the Laboratory of Optical Measurements Methods (LOMM) of the Strata Mechanics Research Institute of the Polish Academy of sciences (SMRI PAS) [15]. Basic parameters of the air stream potential core in the 0.5 m x 0.5 m x 1.5 m measurement chamber are: velocity in the range of 0.1÷60.0 m/s, level of turbulences (square root of the mean variance of the velocity components referred to the mean velocity) not higher than 0.5%, controlled temperature and relative humidity, full optical access to enable measurements by means of all optical techniques available on the market.

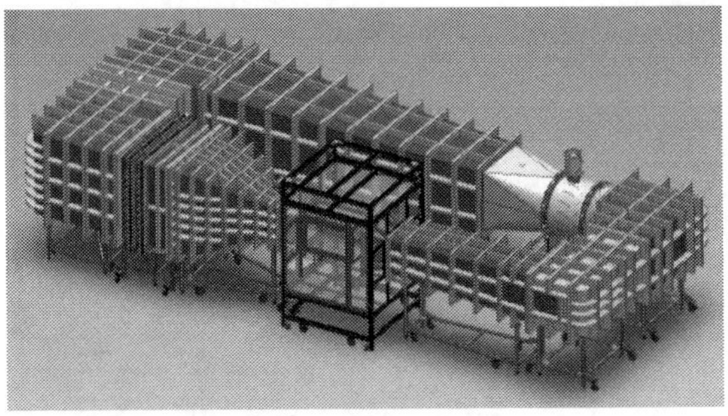

Fig. 1. Visualization of the wind tunnel.

The wind tunnel measurement stand is equipped with the set for stereo particle image velocimetry (SPIV) method. This measuring set is based on two sCMOS, 5.5 Mpx cameras, Nd:YaG, double impulse, 200 mJ per impulse, green laser and seeding generator filled with Di-Ethyl-Hexyl-Sebacate (DEHS). For the purpose of carrying out the planned measurements only one camera (PIV not SPIV configuration) was used together with 180 mm micro 1:1 lens and 2 sets of distance rings (fig. 2). This allowed to achieve projection of area c.a. 6.1 mm x 7.2 mm onto 2160 px x 2560 px sCMOS sensor. 150 pairs of images were taken at the maximum rate of the system i.e. 15 Hz in every single measurement. More than 3000 in total. This amount was suitable for the described research problem, however it is usually not sufficient for statistical considerations.

Fig. 2. The sCMOS camera used in measurements connected to two sets of distance rings and 180 mm 1:1 micro lens.

For the research a regular type of SN probe was chosen (made in the SMRI PAS): 5 microns in diameter, 1.35 mm long, tungsten wire welded to two straight, conical, 20 mm long prongs. Stem short, 5 mm in diameter, fixing to the prongs covered by means of a convex droplet of glue (fig. 3). The probe was located close to the measuring chamber inlet and close to its axis.

Fig. 3. The SWHWA under investigation. Picture taken during the calibration procedure. The head of the Schmidt type anemometer visible in the bottom right corner.

2.2 Measurement conditions

Four different orientations of the SN probe with regard to the air freestream direction (fig. 4) and four different freestream velocities (0.16, 0.56, 2.0 and 8.5 m/s) were chosen for analysis. The orientations do not exhaust all the possibilities, but give a good introduction to the problem itself. Reynolds numbers calculated for the wire and prongs are gathered in table 1.

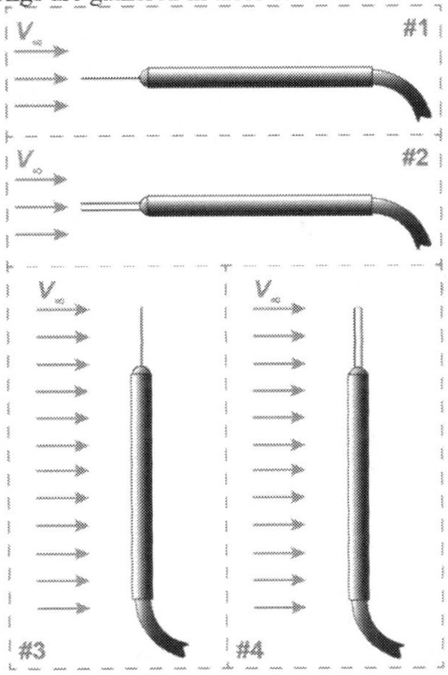

Fig. 4. Sketches presenting four different orientations of the SWHWA probe applied during measurements.

Measurements were carried out at the following microclimatic conditions: temperature $20.0 \pm 0.5°C$, relative humidity $58 \pm 1\%$, barometric pressure 993 ± 1 hPa. Due to wind tunnel construction short- and medium-term airflow velocity changes and turbulence intensity in the measurement chamber were stabilized at the level of 1.5% [16] and 0.5% [17], respectively.

 https://doi.org/10.1051/epjconf/201818002094

Table 1. Re values calculated for the wire and prongs.

	Diameter* [m]	Velocity [m/s]	Re**
Wire	5.0E-06	0.16	0.05
		0.56	0.18
		2.00	0.66
		8.55	2.82
Prongs	2.0E-04	0.16	2.11
		0.56	7.38
		2.00	26.4
		8.55	112.8

*) The diameter of the wire is 5 um. However due to presence of the submicronic seeding particles which condensate on its surface, the diameter may increase in an uncontrolled way depending on the wire orientation. Diameter of the visible parts of prongs on PIV images varies in 0.1-0.3 range. Mean value of 0.2 was chosen as a representative. DEHS condensation on the prongs leads to appearing of droplets which slowly grow and flow down additionally disturbing the flow.

**) Re values were calculated for air properties resulting from microclimatic conditions. Presence of the DEHS suspension, slightly increasing the apparent density of air, was not taken into account.

2.3 Results

Full comparison of the achieved results in the form of velocity magnitude scalar maps is presented in fig. 5. The first 8 maps has a constant 0.0÷1.0 relative velocity scale to enable comparison of the type of flow. Whereas on another 8 maps velocity scales were adjusted so as to expose complexity of the flow field in every single case.

Due to strong reflections, which obstruct estimation of vectors, contour maps were cleared in the close vicinity of prongs and wire. The laser light sheet was directed perpendicularly to the camera and set in the plane going:

a) through the middle of the wire in case of orientations #1 and #3 (see fig. 4),

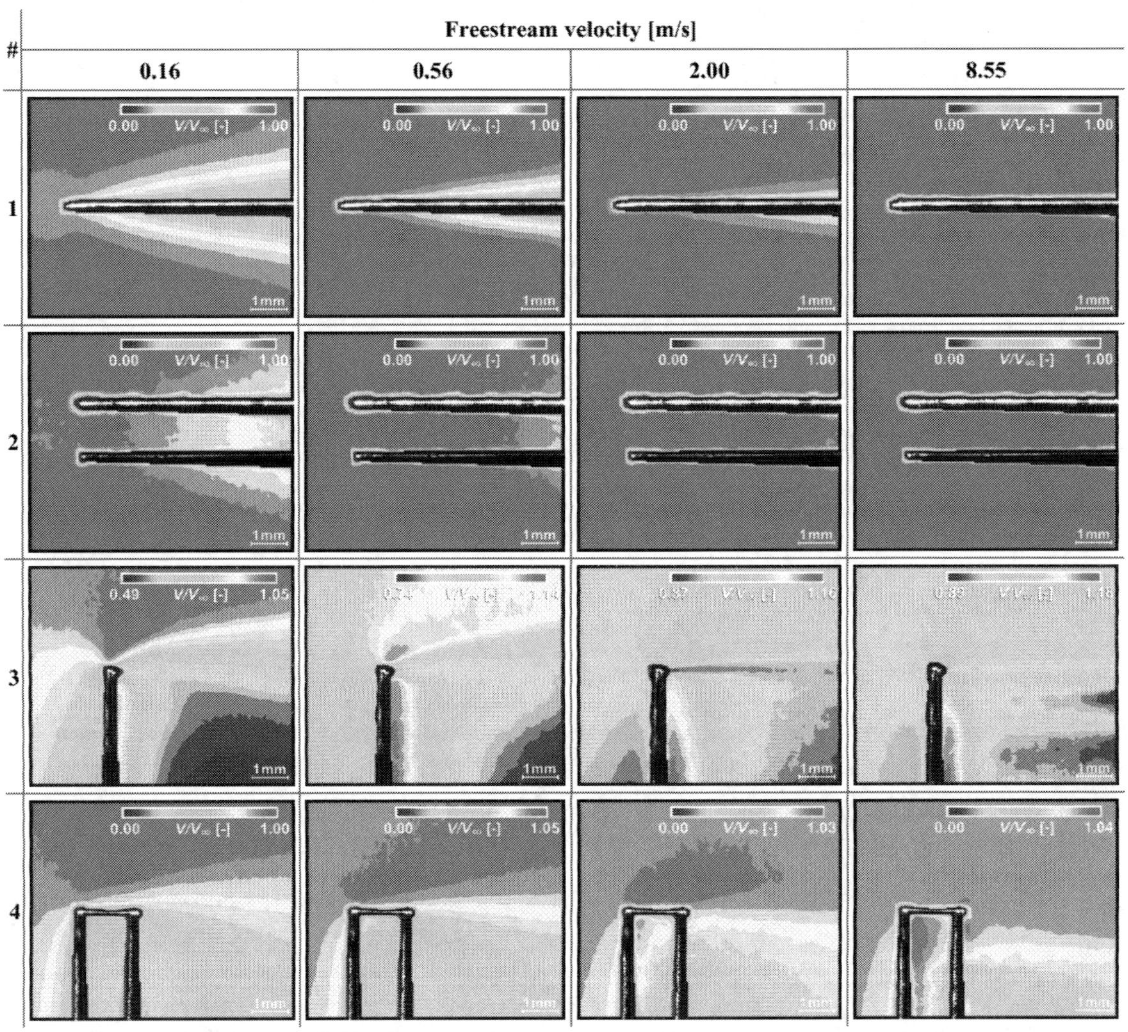

Fig. 5. Contour plots of the averaged (over time) fields of the velocity magnitude. Plots in columns relate to this same freestream velocity, whereas in rows relate to this same orientation in space.

b) immediately behind the prongs in case of orientation #2 and
c) along the prongs in case of orientation #4.

Contour plots presented in fig. 5 depicture the SN probe influence on air stream in close vicinity of the wire and the tips of prongs. This influence strongly depends on both – velocity magnitude and the probe orientation in relation to freestream direction.

For orientation #1 one can easily see increasing area of the field of lowered velocity progressing in the upstream direction as the velocity decreases. This phenomena may explain for example the voltage-velocity dependence deviation from the King's equation for low velocities [18]. In case of orientation #2 this effect is slightly less visible only due to measurement plane shift off the plane in which the wire and the prongs lie.

The SN probe orientations perpendicular to the stream introduce much more disturbance and therefore observed velocity fields are more diversified. Again, complex flow patterns created in low velocities are being deformed and then almost blown off in higher velocities.

An interesting observation is a relatively strong wake behind the wire in orientation #3. This phenomena was raised for example in paper [14], where also the difference between heated and cold wire wakes by means of PIV measurements was qualitatively expressed. More detailed and quantitative analysis one may find in the monograph [19], author of which measured velocity profiles in the wakes behind 30 mm long and 5÷10 um thick tungsten wires (heated and unheated) by means of another SN. It was found, that minimal velocity in these Gaussian-shaped wake profiles exhibit lower values and their full widths at half maxima (FWHM) are bigger behind heated wire than behind a cold one. No significant differences for wires of various diameters were observed.

All the PIV measurements described here were conducted for heated wire only, with the overheat ratio $a = 0.8$. At this stage of the research it is impossible to univocally state that there was no DEHS film covering the wire as it is the case for a cold one. Heating the wire may prevent DEHS from condensing and/or promote its efficient evaporation as the DEHS boiling point, depending on its chemical purity, lays within the range of 232-256°C [20-21]. Further detailed research is required to ascertain what is the most suitable temperature of the wire for PIV measurements, as too low may allow DEHS film growing and too high may provoke violent vaporisation. The influence of orientation in respect to the gravity direction also needs to be examined.

On the other hand calibration procedure repeated during and after PIV measurements revealed the deviation of calibration curve of less than 5% and 1%, respectively, in comparison to one made prior to experiments. This suggests that the wire at $a = 0.8$ was relatively resistant to DEHS condensation and not too hot at the same time. Therefore the results of the present study (fig. 6a-c) were compared to those from [19] (fig. 6d-f) in a quantitative manner.

During analysis of the fig. 6 it is important to bear in mind, that the SN probe used in the present study had 22 times shorter wire comparing to one used in [19]. Small

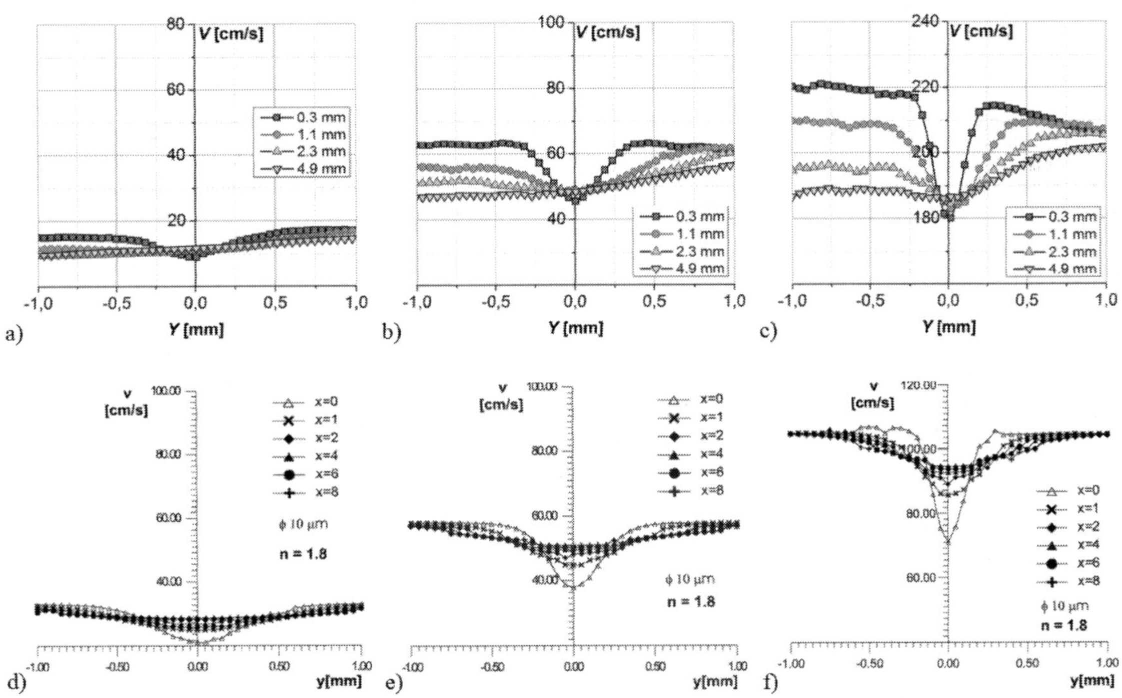

Fig. 6. Transverse velocity profiles of the wake behind the heated wire ($a = 0.8$). Results achieved by the author of the present paper – upper row, and by the author of [19] – bottom row. In both cases values in legends define the distances in millimetres between the wire and the profiles.

space between prongs resulted in the velocity field alternation in this area and above (*vide* fig. 5 #3). This in turn was responsible for braking the symmetry of velocity profiles in fig. 6a-c.

Graphs presented in fig. 6a-c were prepared in coordinate systems corresponding to those from [19]. The origin of abscissa overlap with the centre of the wire and spreads 1 mm above and 1 mm below it. Ordinate, regardless of the velocity mean value, covers the range of 80 cm/s.

The first two couples under comparison (fig. 6a,d and fig. 6b,e) have very similar mean velocities. The third couple (fig. 6c,f) differs in terms of the mean velocity by the factor 2, as the author of [19] conducted all measurements at low velocities only. Still, this couple can be easily compared qualitatively.

Simple, visual comparison of graphs presented in fig. 6 leads to three basic observations:

- it is clear that both series of experiments recorded very similar wakes, also in terms of quantity,
- discrepancies result from the difference in distance between prongs,
- the disturbing behaviour of prongs expresses stronger and closer to the wire as the velocity increases.

3 Conclusions

A series of PIV measurements made in the new facility belonging to SMRI PAN revealed interesting properties of flows around wire of the SN probe. Four different flow patterns related to probe orientations in space were recorded and presented. They depict very complex 3D velocity fields, also in the close vicinity of the wire, especially for the transverse orientation of probe. This leads to the conclusion, that even simple SN probe is by no means a real pointwise measuring device. This also explains why measurements in the close proximity of any other body such as surface or even another probe require probes of a special design and dedicated corrections of the results.

Another conclusion is that flow pattern change in correlation with the probe orientation is responsible for its directional response. It means, that even if during measurements the probe is oriented exactly this same way as it was during calibration the results cannot be accurate if the flow is highly disturbed/turbulent or its main direction significantly deviates in time. This is one of a well-known restriction of the HWA technique and SN in particular [11, 13, 18, 22-25].

Flow pattern change in correlation with the free-stream velocity was also revealed. Most of all, it illustrates change of the disturbing character of prongs. This phenomena contributes i.a. to the calibration curve parameters – unnoticeable in higher velocities (small or similar to King's equation type of dependence) and more explicitly in the lower ones (as it was already mentioned above).

Knowledge about velocity field in the vicinity of SN probe, and especially close to its wire may be very useful as it allows for quick estimation of the probe interaction with the flow itself as well as with adjacent objects. Similar research conducted for all other types of HWA probes would create a valuable database.

Acknowledgment

This publication presents results of the work supported by the Statutory Research of IMG-PAN.

References

1. M. Kochsiek, M. Gläser, *Handbook of Metrology* (John Wiley & Sons, 2010)
2. R.W. Miller, *Flow Measurement Engineering Handbook* (McGraw-Hill Inc., New York, 1996).
3. C. Tropea, A.L. Yarin, *Springer Handbook of Experimental Fluid Mechanics* (Springer Science & Business Media, 2007)
4. S. Discetti, A. Ianiro, *Experimental Aerodynamics* (CRC Press, 2017)
5. T. Yoshizawa, *Handbook of Optical Metrology: Principles and Applications* (CRC Press, 2017)
6. C. Mercer, *Optical Metrology for Fluids, Combustion and Solids* (Springer Science & Business Media, 2013)
7. A. Boutier, *Laser Metrology in Fluid Mechanics: Granulometry, Temperature and Concentration Measurements* (John Wiley & Sons, 2013)
8. M. Raffel, Ch.E. Willert, S. Wereley, J. Kompenhans, *Particle Image Velocimetry: A Practical Guide* (Springer Science & Business Media, 2007)
9. A. Schröder, Ch.E. Willert, *Particle Image Velocimetry: New Developments and Recent Applications* (Springer Science & Business Media, 2008)
10. R.J. Adrian, J. Westerweel, *Particle Image Velocimetry* (Cambridge University Press, 2011)
11. R.J. Adrian, R.E. Johnson, B.G. Jones, P. Merati, A.T.C. Tung, J. Phys. E **17**, 1 (1984)
12. G. Comte-Bellot, A. Strohl, E. Alcaraz, J. Appl. Mech **38,** 4 (1971)
13. L.J.W., Graham, J. Soria, K. Bremhorst, Exp. Fluids **19**, 379-382 (1995)
14. J. Sobczyk, M. Gawor, Przegląd Górniczy, **11/2015**, 117-123 (2015), (*in Polish*)
15. M. Bujalski, M. Gawor, J. Sobczyk, Transactions Of The Strata Mechanics Research Institute, **15**, 1-2 (2013), (*in Polish*)
16. J. Sobczyk, M. Gawor, Intensity of turbulence in the potential core of the SMRI PAS new wind tunnel, SMRI PAS Internal Reporting Conference, 2016, (*in Polish*)
17. J. Sobczyk, M. Gawor, Transactions Of The Strata Mechanics Research Institute, **18**, 4 (2016), (*in Polish*)
18. H.H. Bruun, *Hot-Wire Anemometry Principles, Principles and Signal Analysis* (Oxford University Press, 1995)

19. E. Poleszczyk, Transactions of the Strata Mechanics Research Institute, Series: Dissertations, Monographs, **1** (Krakow, 2002), (*in Polish*)
20. https://pubchem.ncbi.nlm.nih.gov/compound/bis_2-ethylhexyl__sebacate, (*retrieved 19.09.2017*)
21. DEHS exemplary datasheets: https://www.topas-gmbh.de/dateien/prospekt/dehs_prspe.pdf; https://www.ila.de/fileadmin/documents/en/datasheets/piv/seeds/ILA_data_DEHS.pdf; https://www.palas.de/en/product/download/dehs/datasheet/pdf, (*retrieved 19.09.2017*)
22. J.W. Elsner, *Regeneracja sond Termoanemometrycznych*, Cieplne maszyny przepływowe, 68-69, (1968), (*in Polish*)
23. F.E. Jorgensen, DISA Information, **11**, 31-37, (1971)
24. O.O. Mojola, DISA Information, **23**, 24-27 (1978)
25. L. di Mare, T.O. Jelly, I.J. Day, Meas. Sci. Technol. **28**, (2017)

Computational domain discretization in numerical analysis of flow within granular materials

Marcin Sosnowski[1,*]

[1]Jan Dlugosz University in Czestochowa, Faculty of Mathematics and Natural Sciences, 13/15 Armii Krajowej Av, 42-200 Czestochowa, Poland

Abstract. The discretization of computational domain is a crucial step in Computational Fluid Dynamics (CFD) because it influences not only the numerical stability of the analysed model but also the agreement of obtained results and real data. Modelling flow in packed beds of granular materials is a very challenging task in terms of discretization due to the existence of narrow spaces between spherical granules contacting tangentially in a single point. Standard approach to this issue results in a low quality mesh and unreliable results in consequence. Therefore the common method is to reduce the diameter of the modelled granules in order to eliminate the single-point contact between the individual granules. The drawback of such method is the adulteration of flow and contact heat resistance among others. Therefore an innovative method is proposed in the paper: single-point contact is extended to a cylinder-shaped volume contact. Such approach eliminates the low quality mesh elements and simultaneously introduces only slight distortion to the flow as well as contact heat transfer. The performed analysis of numerous test cases prove the great potential of the proposed method of meshing the packed beds of granular materials.

1 Introduction

The widespread prevalence of granular materials in nature and their common use in many industries, make packed beds of granular materials the focus of numerous experimental and theoretical studies [1-4]. Such research are often performed using advanced computational techniques [5-7], including Computational Fluid Dynamics (CFD), which applies numerical methods to solve fluid flow problems including Fluid-Solid Interactions (FSI).

Numerical methods of solving the differential equations are implemented in commercial CFD codes but the user has been left with detailed control over the discretization of the computational domain. This process is commonly called meshing and consists in dividing the analysed geometry into numerous small control volumes commonly called cells or elements. This step of CFD influences the numerical stability of the model and consequently the accuracy of the obtained results [8, 9].

Moreover the common challenge for CFD is the expensive computational cost in terms of time required to reach a converged solution [10]. The very first issue influencing the convergence as well as the accuracy of the results is the applied method of computational domain discretization [8, 11, 12].

Therefore the carried out research focus on the innovative approach to discretization method of the analysed domain representing the packed bed of granular material.

2 Methods

The performed research concern the development of the optimal method of discretization of the computational domain which represents solid granules submerged in the fluid domain - packed bed of granular material. The solid bodies of spherical shape (homogeneous granules) are in direct contact between each other in a single point between two adjacent granules.

The standard approach to computational domain discretization where two spherical bodies contact tangentially in one point (Fig. 1a) leads to extremely low-quality mesh cells, resulting in numerical instability of the flow calculations.

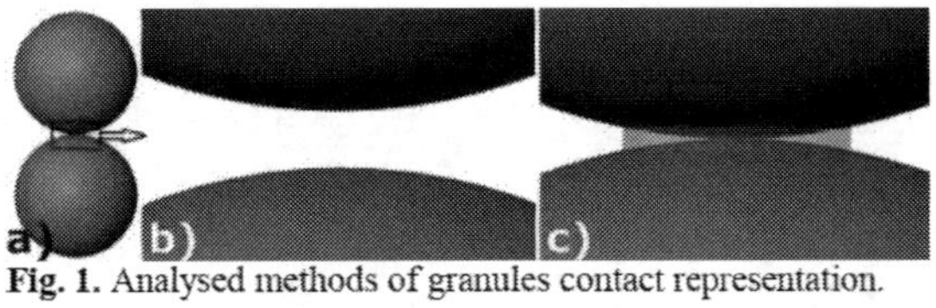

Fig. 1. Analysed methods of granules contact representation.

The reduction of radius of individual granules (Fig. 1b) can be applied in order to avoid the above mentioned effect of low quality mesh. It allows to obtain the acceptable mesh quality but entails the necessity of significant size reduction of mesh cells. In consequence the total number of cells increases and it directly affects the overall calculation time.

* Corresponding author: m.sosnowski@ajd.czest.pl

Unfortunately this approach disturbs the flow and heat transfer between individual granules which no longer have direct contact with each other.

Therefore an innovative approach to discretization of packed beds of granular materials is proposed. It consists in extending the contact between the individual granules from single point to the cylindrical region as shown in Fig. 1c. Such approach allows to achieve high quality mesh and simultaneously guarantees the contact of the adjacent granules. Therefore the flow disturbance between the granules is insignificant.

2.1 Geometry and analysed cases

The geometry representing the packed bed of granular material (Fig. 2) was generated using SolidWorks with ANSYS 18.1 plugin, which allows the bi-directional associativity between the CAD geometry (SolidWorks) and mesh generator (ANSYS Meshing).

Fig. 2. Axonometric view of the analysed packed bed of granular material.

Two global variables defining the relative radius of individual granule (eq. 1) as well as relative radius of the cylindrical volume (eq. 2) were introduced in order to speed up the process of preparing the geometry for different cases.

$$R_{g\%} = \frac{R_g}{R} \cdot 100\% \quad (1)$$

$$R_{cyl\%} = \frac{R_{cyl}}{R} \cdot 100\% \quad (2)$$

where :

- R radius of the granules contacting in a single point;
- R_g reduced radius of individual granules;
- R_{cyl} radius of the additional cylindrical volume between individual granules.

The geometry was prepared in two configurations in order to generate the mesh using two methods (Fig. 3):

- with additional cylindrical volumes between the individual granules,
- with reduction of granules radius.

Five cases (a-e) were analysed for each method according to Table 1.

Table 1. Analysed configurations and cases.

Case	Method of computational domain discretization			
	additional cylindrical volume		reduction of granule radius	
	$R_{cyl\%}$	mesh elements	$R_{g\%}$	mesh elements
a	5.00%	37325	99.75%	37731
b	6.25%	37612	99.50%	38035
c	7.50%	37886	99.25%	38247
d	8.75%	37516	99.00%	38851
e	20.00%	36590	95.00%	47300

The CAD model of the packed bed was prepared as a sector of the whole bed consisting of one spherical granule (marked red in Fig. 3) surrounded by eight sections of granules. Each section represents one-eight of a granule (marked green in Fig. 3). Such procedure allows to minimize the volume of the computational domain which directly results in CPU time reduction. Simultaneously the possibility to generate the entire packed bed by simple multiplication of the sector CAD model is maintained.

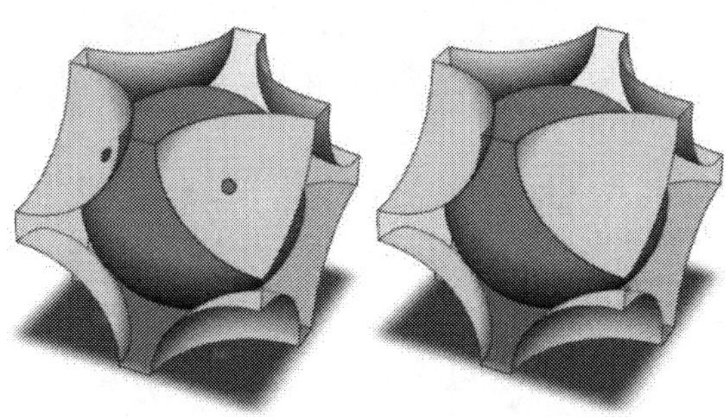

Fig. 3. Axonometric view of the analysed geometry with (left) and without (right) additional cylindrical volumes between the individual granules.

2.2 Meshing

The meshing process of all ten configurations of the analysed geometries (five cases for each of the two methods) was performed using ANSYS Meshing 18.1. Tetrahedral (TET) elements were used to discretize the computational domain. The size function was set to UNIFORM in order to obtain the most regular mesh. Moreover the minimal size, maximal face size as well as maximal TET size constrains were set to the value of $0.1R$. The above meshing settings resulted in the final mesh consisting of 36k to 47k elements (depending on the method and case). The final mesh for selected case of each of the two methods is depicted in Fig. 4.

The output mesh quality was analysed. The quality analysis provides a composite metric that ranges between

0 and 1 according to formula (3). A value of 1 indicates a perfect element while a value of 0 indicates that the element has a zero or negative volume.

The element of the worst quality generated throughout the entire research (ten meshes) was characterized by the quality index of 0.17 and therefore it was assumed that the mesh did not affect the numerical stability of the computational model.

$$Q = C \cdot \frac{V}{\sqrt{(\sum l^2)^3}} \quad (3)$$

where :

- V volume of the computational domain;
- l length of the element edge;
- C constant dependent on the element type (eg. 124.708 for TETs).

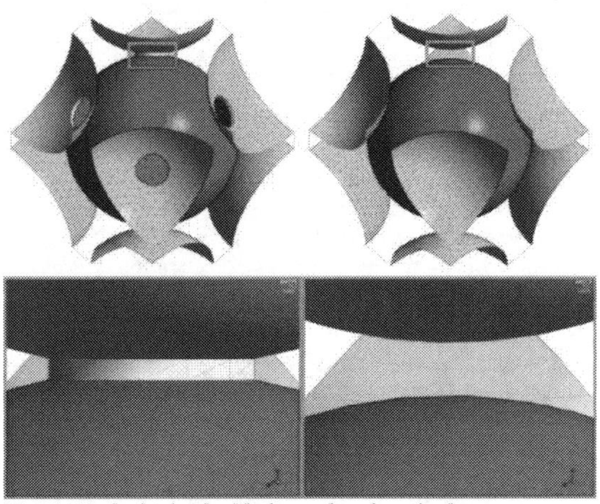

Fig. 4. Mesh obtained with the analysed methods:
left – with additional cylinderical volumes ($R_{cyl\%}$=20%);
right – with reduction of granules radius ($R_{g\%}$=95%).

2.3 Computational model

The exactly identical computational model (except for the domain geometry) was used for all analysed cases and both discretization methods. The solver was configured as pressure-based and the analysis were performed for steady state.

Pressure-velocity coupling by SIMPLE algorithm was used as a solution method. This algorithm uses a relationship between velocity and pressure corrections to enforce mass conservation and to obtain the pressure field.

Air at normal conditions was the fluid medium. Velocity-inlet boundary condition type was assigned to the inlet (magenta surface in Fig. 5) with velocity magnitude normal to the boundary equal 1 m/s. The outlet of the computational domain (yellow surface in Fig. 5) was defined as outflow. Wall with no slip shear condition was assigned to the surfaces of granules (red and green surfaces in Fig. 5) whereas wall with specified shear condition of shear stress X, Y and Z components equal 0 was assigned to the surfaces of module interfaces (cyan surfaces in Fig. 5) in order to assure the periodicity of the flow in each of the two directions.

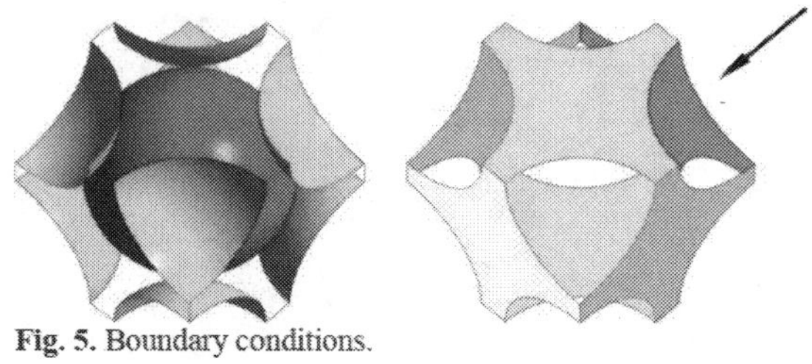

Fig. 5. Boundary conditions.

The calculations were performed for only 100 iterations in order to compare the rate of convergence for all analysed cases.

3 Results

The obtained results concerning the velocity magnitude averaged along outlet line and averaged on outlet surface are depicted in Fig. 6 and Fig. 7 respectively. Cases a-e for both methods are parametrized according to Table 1. The average velocity magnitude is referenced to the inlet velocity.

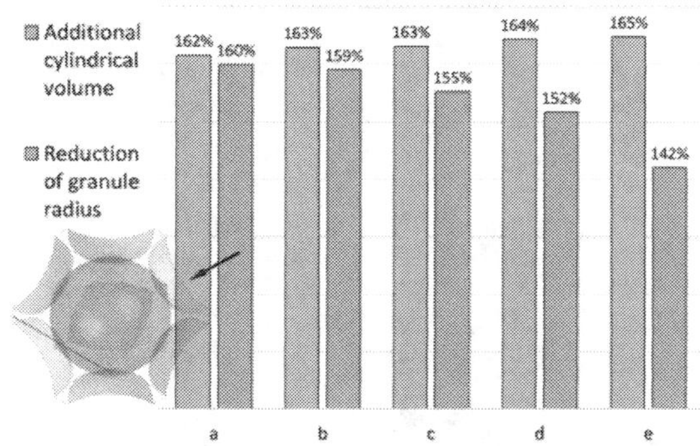

Fig. 6. Average relative velocity magnitude along outlet line

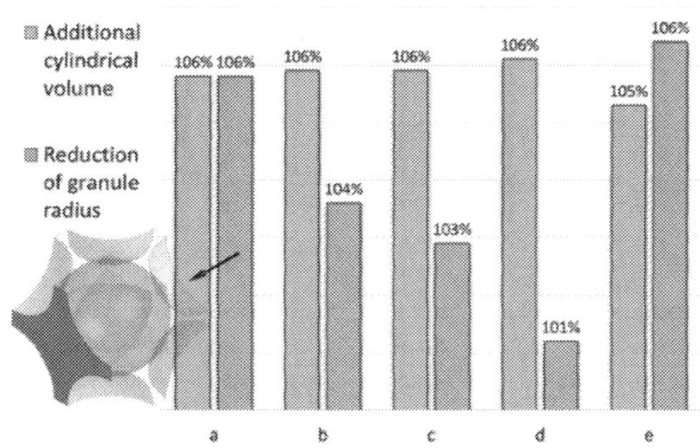

Fig. 7. Average relative velocity magnitude on outlet surface

The discretization method with additional cylindrical volume is characterized by relatively stable average velocity along outlet line (variation of 3%) and on outlet surface (variation of 1%) for each of the analysed radius of the cylindrical volume ($R_{cyl\%}$). The variations of average velocity are more significant (18% and 5% respectively) for the method of granule radius reduction.

Contour plots of velocity magnitude on outlet surface, XY and XZ planes for both methods and all configurations are presented in Fig. 8-Fig. 13.

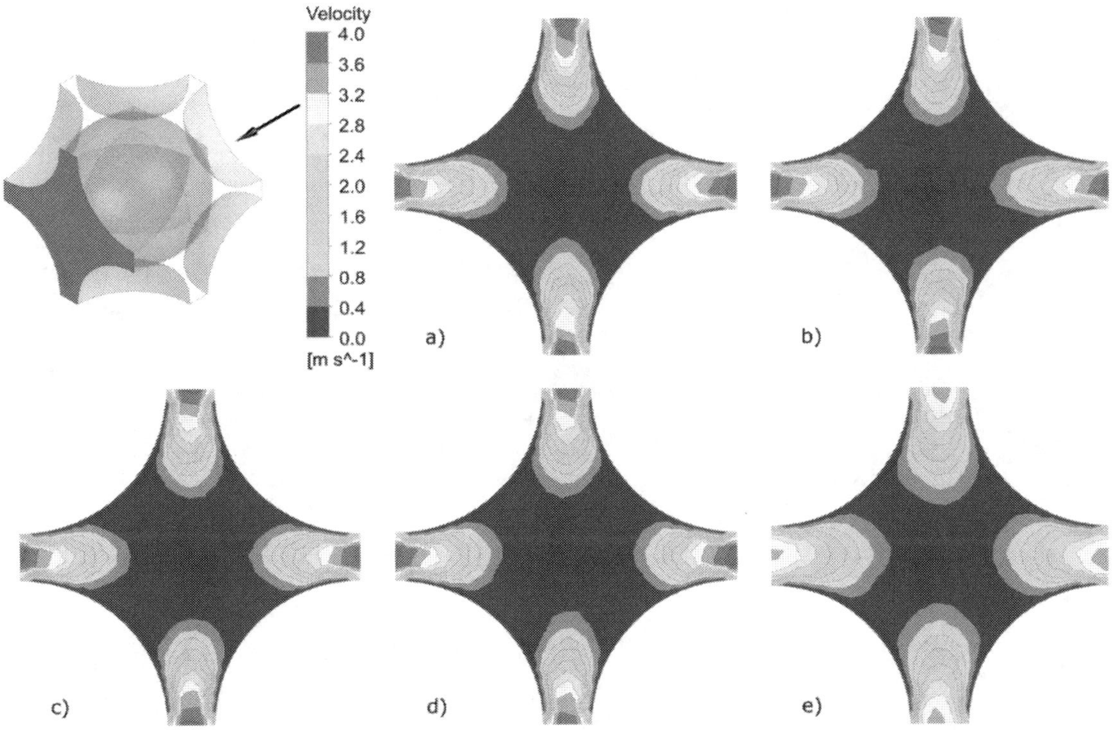

Fig. 8. Velocity magnitude on outlet surface obtained with the discretization method of granule radius reduction $R_{g\%}$ equal: a - 99.75%. b - 99.50%. c - 99.25%. d - 99.00%. e - 95.00%.

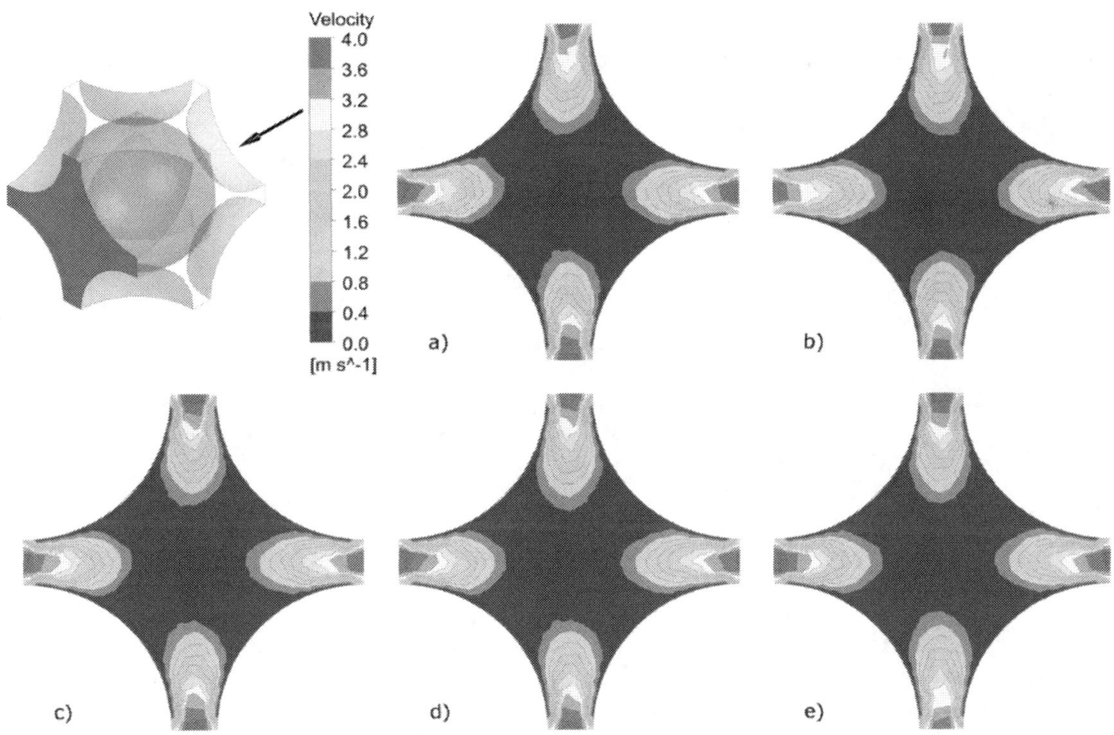

Fig. 9. Velocity magnitude on outlet surface obtained with the discretization method with additional cylindrical volumes $R_{cyl\%}$ equal: a – 5.00%. b – 6.25%. c – 7.50%. d – 8.75%. e - 20.00%.

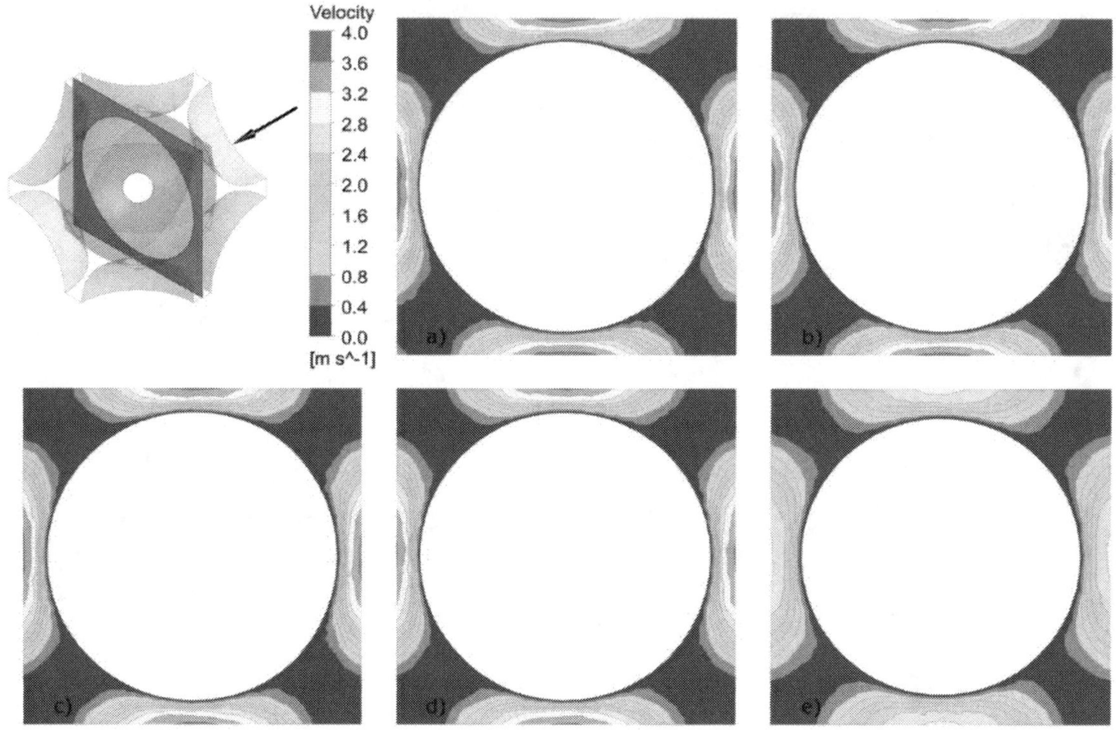

Fig. 10. Velocity magnitude on XY plane obtained with the discretization method of granule radius reduction $R_{g\%}$ equal: a - 99.75%. b - 99.50%. c - 99.25%. d - 99.00%. e - 95.00%.

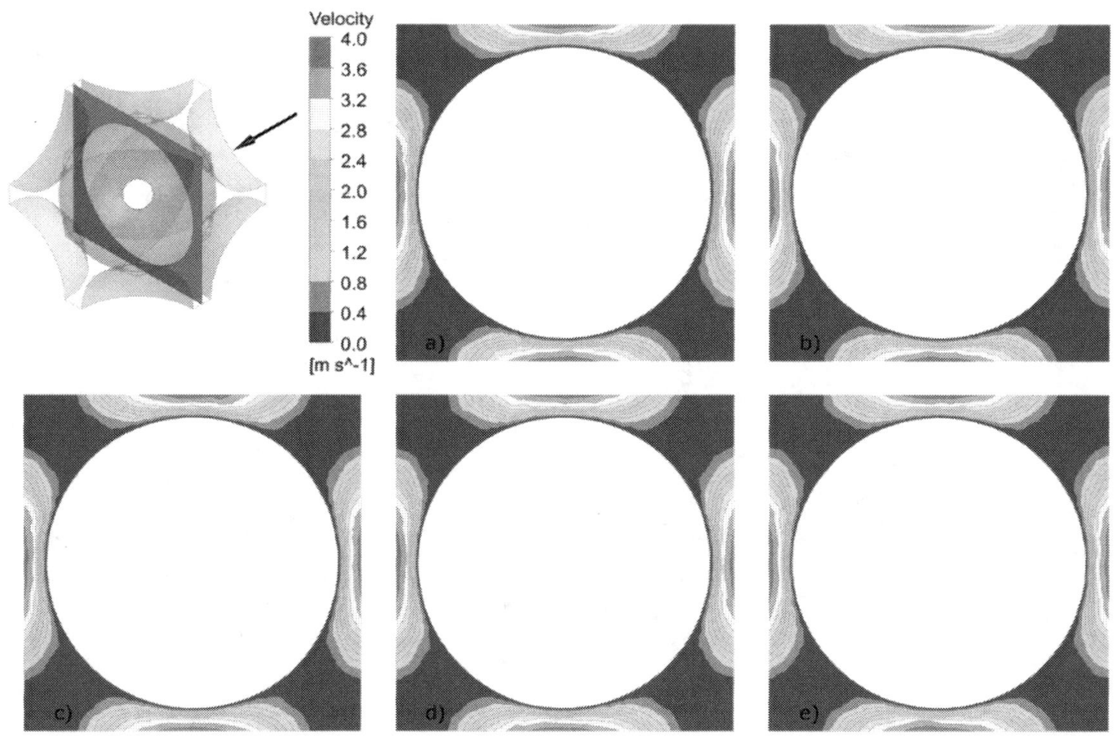

Fig. 11. Velocity magnitude on XY plane obtained with the discretization method with additional cylindrical volumes $R_{cyl\%}$ equal: a – 5.00%. b – 6.25%. c – 7.50%. d – 8.75%. e - 20.00%.

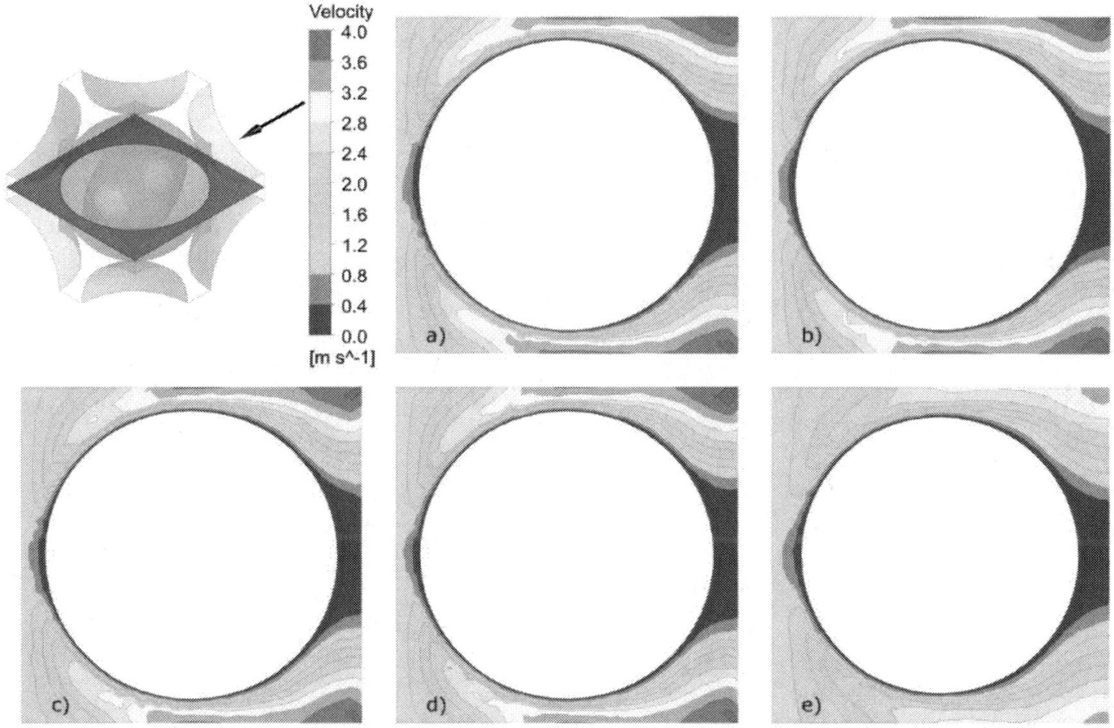

Fig. 12. Velocity magnitude on XZ plane obtained with the discretization method of granule radius reduction $R_{g\%}$ equal: a - 99.75%. b - 99.50%. c - 99.25%. d - 99.00%. e - 95.00%.

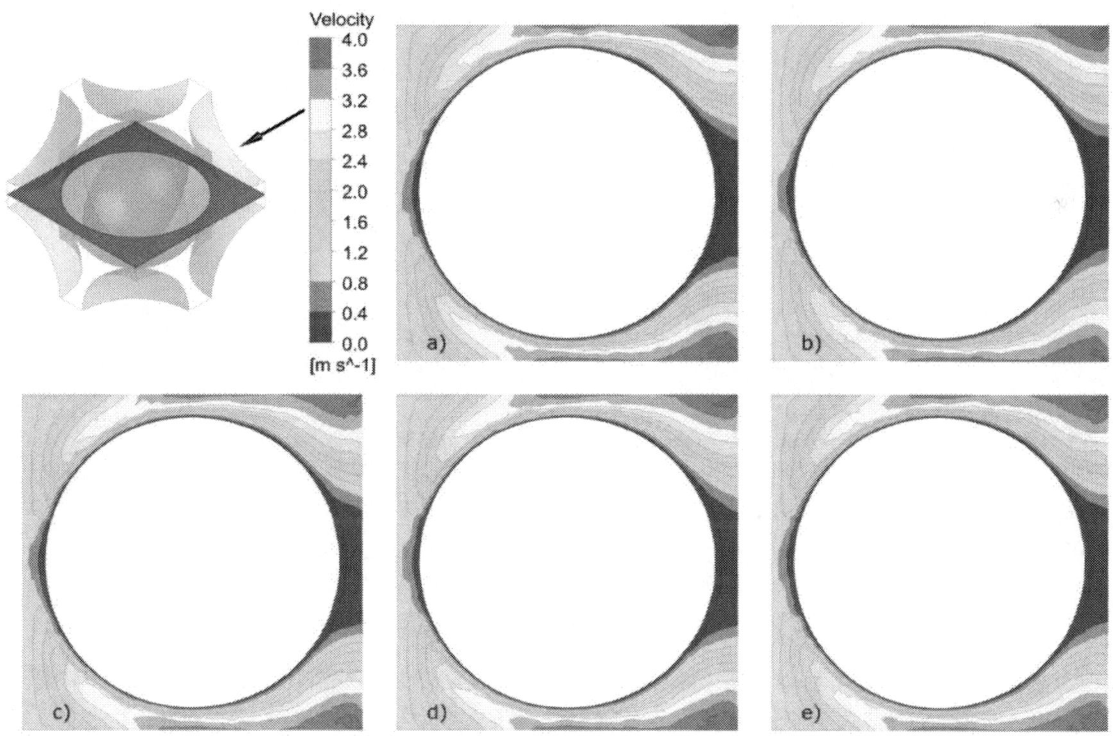

Fig. 13. Velocity magnitude on XZ plane obtained with the discretization method with additional cylindrical volumes $R_{cyl\%}$ equal: a – 5.00%. b – 6.25%. c – 7.50%. d – 8.75%. e - 20.00%.

It is clearly seen that the results obtained with the mesh prepared according to the proposed innovative method of computational domain discretization are similar irrespective of the radius of the additional cylindrical volume ($R_{cyl\%}$) for all post-processing surfaces.

The method of granule radius reduction is sensitive to the level of reduction $R_{g\%}$ - the velocity field becomes more uniform along with the decrease of granule radius.

4 Discussion

The proposed innovative method of computational domain discretisation in CFD analysis of packed beds of granular materials is characterized by lower variation of mesh elements count (37 325 – 36 590) in comparison to the commonly used method of granules radius reduction (37 731 – 47 300).

The variations of average velocity for the analysed cases are significant for the granules radius reduction method and negligible for the innovative method proposed in the paper.

The same issue occurs in terms of the velocity field obtained using the mesh generated with the analysed methods: the proposed method is independent to the radius of the additional cylindrical volume ($R_{cyl\%}$) while the level of reduction of granule radius ($R_{g\%}$) in the standard method strongly affects the velocity field.

Experimental validation of the numerical results is necessary in order to properly and fully define the applicability of the presented method. The experimental test stand has already been designed and it is under construction.

5 Conclusions

It can be concluded that the proposed innovative method of computational domain discretization is characterized by more stable behaviour of the CFD results in comparison to the commonly used method.

The carried out research indicate the purposefulness of further investigation in order to evaluate the optimal $R_{cyl\%}$ value as well as the correlation of mesh element size and the radius of the additional cylindrical volume.

Moreover the Grid Convergence Index (GCI) [13] has to be checked to assure the mesh-independent solution and fully evaluate the proposed method.

Finally the experimental research is necessary to validate the method against real data.

The work is partialy suported by Narodowe Centrum Nauki (National Science Centre) grant number 2017/01/X/ST8/00019

References

1. Gan, J.Q., Z.Y. Zhou, and A.B. Yu, *Effect of particle shape and size on effective thermal conductivity of packed beds.* Powder Technology, 2017. **311**: p. 157-166.
2. Lee, C.H., et al., *Evaluation of effective thermal conductivity of unsaturated granular materials using random network model.* Geothermics, 2017. **67**: p. 76-85.
3. Koekemoer, A. and A. Luckos, *Effect of material type and particle size distribution on pressure drop in packed beds of large particles: Extending the Ergun equation.* Fuel, 2015. **158**: p. 232-238.
4. Hurley, R.C., et al., *Dynamic Inter-Particle Force Inference in Granular Materials: Method and Application.* Experimental Mechanics, 2016. **56**(2): p. 217-229.
5. Dayal, R. and T. Gambaryan-Roisman, *Heat transfer in granular medium for application to selective laser melting: A numerical study.* International Journal of Thermal Sciences, 2017. **113**: p. 38-50.
6. Gui, N., et al., *TDEM simulation and analysis of thermal conduction through packed granular beds.* Canadian Journal of Physics, 2016. **94**(9): p. 826-833.
7. Wang, X.L., M.Y. Ye, and H.L. Chen, *Computational study on the behaviors of granular materials under mechanical cycling.* Journal of Applied Physics, 2015. **118**(17).
8. Bianchi, G., et al., *Deforming grid generation for numerical simulations of fluid dynamics in sliding vane rotary machines.* Advances in Engineering Software, 2017. **112**: p. 180-191.
9. Li, H., L. Rong, and G.Q. Zhang, *Reliability of turbulence models and mesh types for CFD simulations of a mechanically ventilated pig house containing animals.* Biosystems Engineering, 2017. **161**: p. 37-52.
10. Haddadi, B., C. Jordan, and M. Harasek, *Cost efficient CFD simulations: Proper selection of domain partitioning strategies.* Computer Physics Communications, 2017. **219**: p. 121-134.
11. Gnatowska, R., M. Sosnowski, and V. Uruba, *CFD modelling and PIV experimental validation of flow fields in urban environments.* E3S Web of Conferences, 2017. **14**.
12. Sosnowski, M., J. Krzywanski, and R. Gnatowska, *Polyhedral meshing as an innovative approach to computational domain discretization of a cyclone in a fluidized bed CLC unit.* E3S Web of Conferences, 2017. **14**: p. 01027.
13. Celik, I.B., et al., *Procedure for estimation and reporting of uncertainty due to discretization in CFD applications.* Journal of Fluids Engineering-Transactions of the Asme, 2008. **130**(7).

Polyhedral meshing in numerical analysis of conjugate heat transfer

Marcin Sosnowski[1,*], *Jaroslaw* Krzywanski[1], *Karolina* Grabowska[1], and *Renata* Gnatowska[2]

[1]Jan Dlugosz University in Czestochowa, Faculty of Mathematics and Natural Sciences, 13/15 Armii Krajowej Av, 42-200 Czestochowa, Poland
[2]Czestochowa University of Technology, Faculty of Mechanical Engineering and Computer Science, 21 Armii Krajowej Av, 42-200 Czestochowa, Poland

Abstract. Computational methods have been widely applied in conjugate heat transfer analysis. The very first and crucial step in such research is the meshing process which consists in dividing the analysed geometry into numerous small control volumes (cells). In Computational Fluid Dynamics (CFD) applications it is desirable to use the hexahedral cells as the resulting mesh is characterized by low numerical diffusion. Unfortunately generating such mesh can be a very time-consuming task and in case of complicated geometry - it may not be possible to generate cells of good quality. Therefore tetrahedral cells have been implemented into commercial pre-processors. Their advantage is the ease of its generation even in case of very complex geometry. On the other hand tetrahedrons cannot be stretched excessively without decreasing the mesh quality factor, so significantly larger number of cells has to be used in comparison to hexahedral mesh in order to achieve a reasonable accuracy. Moreover the numerical diffusion of tetrahedral elements is significantly higher. Therefore the polyhedral cells are proposed within the paper in order to combine the advantages of hexahedrons (low numerical diffusion resulting in accurate solution) and tetrahedrons (rapid semi-automatic generation) as well as to overcome the disadvantages of both the above mentioned mesh types. The major benefit of polyhedral mesh is that each individual cell has many neighbours, so gradients can be well approximated. Polyhedrons are also less sensitive to stretching than tetrahedrons which results in better mesh quality leading to improved numerical stability of the model. In addition, numerical diffusion is reduced due to mass exchange over numerous faces. This leads to a more accurate solution achieved with a lower cell count. Therefore detailed comparison of numerical modelling results concerning conjugate heat transfer using tetrahedral and polyhedral meshes is presented in the paper.

1 Introduction

The application of Computational Fluid Dynamics (CFD) as an advanced and valuable research tool has recently been common in various branches of industry such as aerospace [1], automotive [2, 3], civil engineering [4-8], power engineering [9], medicine [10] and others. It allows to achieve flawless product designing at relatively low cost by combining the prediction of fluid flow, heat and mass transfer, chemical reactions and related phenomena.

CFD is a numerical approach for solving nonlinear differential equations describing the analysed fluid flow. Majority of these equations do not have analytical solution hence approximate numerical methods are applied in order to solve them. The level of approximation directly influences the compatibility of modelling results with real data. The round-off errors due to finite word size will always interfere the solution but the truncation errors due to approximations in numerical models can be minimized by improved pre-processing (meshing), which is the first step in CFD analysis. This stage of CFD consists in discretizing the computational domain into numerous finite volumes (cells). Refining the computational mesh strongly contributes to the numerical stability of the model and consequently the accuracy of the obtained results [11, 12] because it drastically reduces the truncation error. Moreover the type of mesh elements strongly influences the numerical diffusion as well as the quality and time of convergence. Two mesh types are commonly applied in CFD: hexahedral HEX elements (Fig. 1 red) and tetrahedral TETRA elements (Fig. 1 blue)

HEX mesh is characterized by low numerical diffusion, hence in most cases it results in the lowest discretization error. But on the other hand the numerical diffusion increases in case of flow not perpendicular to the cell faces and it is not always possible to construct structured HEX mesh for complex geometries. Therefore this mesh type is frequently ineffective in industrial applications.

For that reason algorithms of generation TETRA mesh have been implemented into commercial pre-processors. The generation of this type of mesh can be automated at a large degree, which is the biggest advantage of TETRA mesh. But such cells cannot be stretched excessively, so

* Corresponding author: m.sosnowski@ajd.czest.pl

significantly larger number of elements has to be used in comparison to structured HEX mesh in order to represent specific geometry issues. Moreover the neighbouring nodes of a TETRA cell may lie in nearly one plane and therefore computing gradients can be difficult, which may result in poor convergence [10].

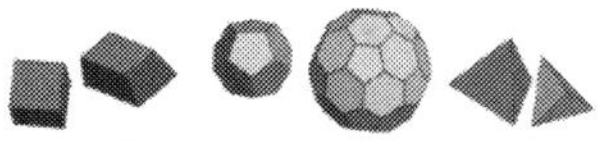

Fig. 1. Types of mesh elements:
HEX – red, TETRA – blue, POLY – green.

POLY mesh has recently been introduced in order to combine the advantages of HEX and TETRA meshes: low numerical diffusion of HEX and rapid semi-automatic generation of TETRA [5, 13, 14]. The individual POLY neighbours with more cells in comparison to HEX and TETRA, which improves the calculation of gradients and allows to interchange the mass over numerous faces, reducing numerical diffusion effects caused by flows not perpendicular to any of the cell face. It is advantageous in situations where no prevailing flow direction can be identified and leads to more accurate solution achieved with a lower cell count. Moreover POLY cell is not oversensitive to stretching and therefore assures improved numerical stability of the model.

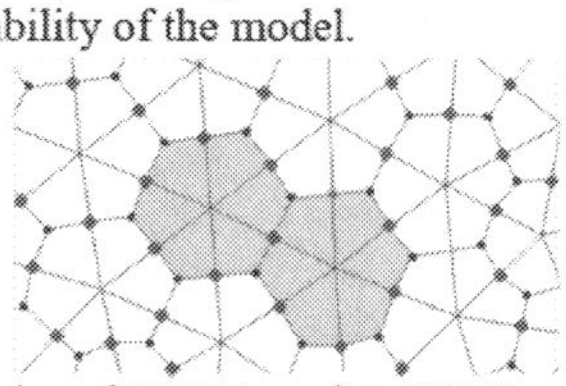

Fig. 2. Conversion of TETRA mesh to POLY mesh

POLY mesh generation is a simple conversion of TETRA mesh to POLY mesh by decomposition of cell into multiple sub-volumes depicted in Fig. 2. In order to do so, new edges are created on each face between the face centroid (dots in Fig. 2) and the centroids of the edges of that face (hexagons in Fig. 2). Subsequently new faces are created within the cell by connecting the cell centroid to the new edges on each face. The newly-created faces may be adjusted and merged with the neighbouring ones during the agglomeration process in order to minimize the number of faces of the resultant polyhedral cell.

2 Methods

2.1 The research object

Air-water heat exchanger was the research object. It consisted of the insulated cuboid (180 mm x 120 mm x 60 mm) with air inlet/outlet and aluminium heat pipe of 8 mm internal radius and 1 mm wall thickness (Fig. 3). The CAD model of the heat exchanger was prepared in SolidWorks as an assembly of three parts: cuboid-shaped air domain, aluminium pipe and domain of water in the heat pipe. ANSYS 18.1 plugin was used in order to parametrize the CAD model for subsequent meshing and analysis by defining the named selections for boundary conditions assignment. The plugin allows also the bi-directional associativity between the CAD geometry and mesh generator. Moreover the model was split into two symmetric parts in order to minimize the volume of the computational domain which directly influences the CPU time.

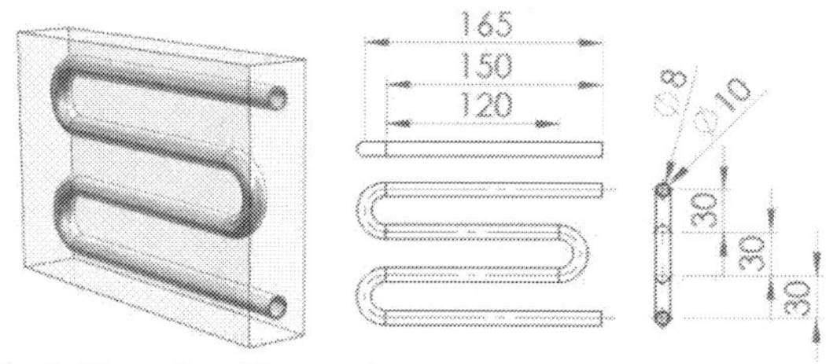

Fig. 3. The analysed heat exchanger

2.2 Pre-processing

ANSYS Meshing 18.1 was used as a pre-processor. The patch conforming algorithm was applied in order to generate fully conformal tetrahedral mesh. Element size representing the aluminium heat pipe as well as maximal face and TETRA sizes allowed in the computational domain were selected as mesh size constrains. Four configurations for both mesh types depicted in Fig. 4 (TETRA and POLY) were generated in order to perform mesh dependency studies. The mesh configurations differed in the total number of elements but the relative mesh sizes were maintained. The POLY meshes were generated using ANSYS Fluent 18.1 by conversion of the TETRA mesh. The number of elements of the final meshes of both analysed types was approx. 60k, 150k, 230k and 340k.

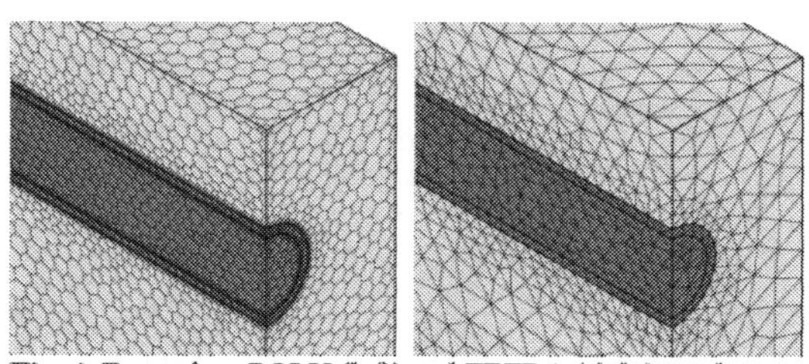

Fig. 4. Exemplary POLY (left) and TETRA (right) meshes

The output mesh quality analysis provided a composite metric that ranges between 0 and 1 according to formula (1). A value of 1 indicates a perfect element while a value of 0 indicates that the element has a zero or negative volume. The element of the worst quality generated throughout the entire research (eight meshes) was characterized by the quality index of 0.11 and therefore it was assumed that the mesh did not affect the numerical stability of the computational model.

$$Q = C \cdot \frac{V}{\sqrt{(\sum l^2)^3}} \qquad (1)$$

where :

- V volume of the computational domain;
- l length of the element edge;
- C constant dependent on the element type.

https://doi.org/10.1051/epjconf/201818002096

2.3 Numerical accuracy and discretization error

The numerical accuracy and its estimation is crucial for CFD. One issue connected with the numerical accuracy estimation is the iterative convergence. In most cases it can be assumed that it is achieved with at least three orders of magnitude decrease in the normalized residuals for each equation solved . This guideline was applied within the carried out analysis.

The mesh resolution is also a crucial factor influencing the numerical accuracy [15]. One of the methods of discretization error estimation is the Richardson Extrapolation (RE) [16]. The local RE values of the predicted variables may not exhibit a monotonic dependence on mesh resolution. Moreover in transient calculations, RE will also be a function of time and space. Despite its shortcomings and generalizations, it is currently the most reliable method available for the prediction of numerical uncertainty.

The mesh dependency analysis applied within this paper to the estimation of discretization error is based on the RE. The representative mesh size h was defined according to equation (2) [16].

$$h = \sqrt[3]{\frac{1}{N}\sum_{i=1}^{N}(\Delta V_i)} \quad (2)$$

where:
ΔV_i - the volume of the ith cell,
N - the total number of cells.

The mesh refinement factor r was calculated as quotient of representative size of coarse and fine mesh (3) with assumption (4).

$$r = \frac{h_{coarse}}{h_{fine}} \quad (3)$$

$$h_1 < h_2 < h_3; \quad r_{21} = \frac{h_2}{h_1}; \; r_{32} = \frac{h_3}{h_2} \quad (4)$$

The order of convergence p was calculated based on equation (5) where $\varepsilon_{32}=\phi_3-\phi_2$ and $\varepsilon_{21}=\phi_2-\phi_1$. It was solved using fixed-point iteration with the initial guess equal to the first term.

$$p = \frac{\left| ln\left|\frac{\varepsilon_{32}}{\varepsilon_{21}}\right| + ln\left(\frac{r_{21}^{p}-1\cdot sgn\left(\frac{\varepsilon_{32}}{\varepsilon_{21}}\right)}{r_{32}^{p}-1\cdot sgn\left(\frac{\varepsilon_{32}}{\varepsilon_{21}}\right)}\right)\right|}{ln(r_{21})} \quad (5)$$

ϕ_k denotes the value of the variable important to the objective of the simulation study for solution obtained with the kth mesh. The average temperature on outlet of the domain was chosen as the most representative variable.

The extrapolated values were calculated on the basis of equation (6), approximate relative error on the basis of eq. (7), extrapolated relative error on the basis of eq. (8) and mesh convergence index C on the basis of eq. (9).

$$\phi_{ext}^{21} = \frac{r_{21}^{p}\phi_1-\phi_2}{r_{21}^{p}-1}; \quad \phi_{ext}^{32} = \frac{r_{32}^{p}\phi_2-\phi_1}{r_{32}^{p}-1} \quad (6)$$

$$e_a^{21} = \left|\frac{\phi_1-\phi_2}{\phi_1}\right|; \quad e_a^{32} = \left|\frac{\phi_2-\phi_3}{\phi_2}\right| \quad (7)$$

$$e_{ext}^{21} = \left|\frac{\phi_{ext}^{12}-\phi_1}{\phi_{ext}^{12}}\right|; \quad e_{ext}^{32} = \left|\frac{\phi_{ext}^{23}-\phi_2}{\phi_{ext}^{23}}\right| \quad (8)$$

$$C_{21} = \frac{1.25\cdot e_a^{21}}{r_{21}^{p}-1}; \quad C_{32} = \frac{1.25\cdot e_a^{32}}{r_{32}^{p}-1} \quad (9)$$

The type of convergence can be evaluated using formula (10).

$$\frac{\varepsilon_{coarse}}{\varepsilon_{fine}}\begin{cases} < 0 & converged\ (oscilatory) \\ > 1 & not\ converged \\ [0,1] & converged \end{cases} \quad (10)$$

In order to evaluate the discretization error, four TETRA and four POLY meshes were generated of approx. 60k, 150k, 230k and 340k cells. As can be seen in Table 1, the meshes of 340k cells can be assumed as converged.

Table 1. Discretization errors.

	Ave. temp. on centerline ϕ	Convergence $\varepsilon_{coarse}/\varepsilon_{fine}$	ϕextrapolated ϕ_{ext}	Approximated rel. error e_a	Extrapolated rel. error e_{ext}	Mesh convergence index C
POLY_340k	312,73	converged	312,73	0,01%	0,00%	0,00%
POLY_230k	312,76	not conv.	312,76	0,00%	0,00%	0,00%
POLY_150k	312,77	-	-	0,03%	-	-
POLY_60k	312,87	-				
TETRA_340k	312,76	converged	312,71	0,02%	0,02%	0,02%
TETRA_230k	312,82	osc. conv.	312,63	0,02%	0,06%	0,01%
TETRA_150k	312,88	-	-	0,03%	-	-
TETRA_60k	312,79	-				

2.4 Computational model

The computational models used for all eight cases (four configurations of TETRA mesh and four configurations of POLY mesh) were parametrized identically. The only difference was the computational mesh. The solver was configured as pressure-based and the analysis were performed for steady state. Standard k-epsilon viscous model was used along with standard wall functions. The turbulence model constants are listed in Table 2.

Table 2. Turbulence model parameters.

Model Constant	Value
Cmu	0.09
C1-Epsilon	1.44
C2-Epsilon	1.92
TKE Prandtl Number	1
TDR Prandtl Number	1.3

Velocity-inlet boundary condition type was assigned to the inlet to the air domain with velocity magnitude normal to the boundary equal 0.1 m/s (Fig. 5). The turbulence parameters were specified as intensity and hydraulic diameter equal 5% and 0.09 m respectively. The temperature of the inlet air was 293 K. The water inlet was also configured as velocity-inlet with velocity magnitude normal to the boundary equal 0.01 m/s. The turbulence parameters were specified as intensity and hydraulic diameter equal 5% and 0.008 m respectively. The temperature of the inlet air was 323 K.

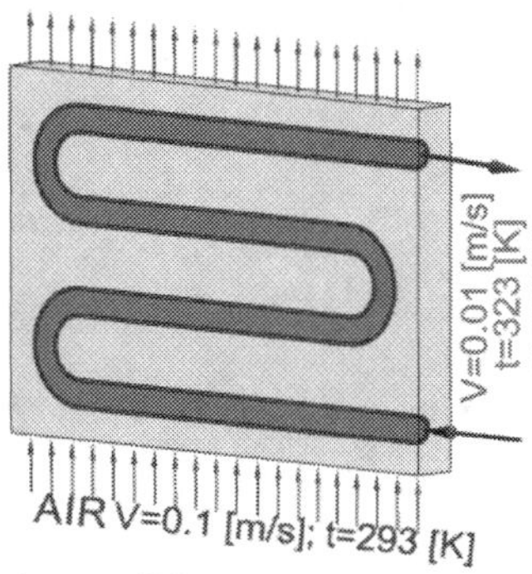

Fig. 5. Boundary conditions

Two fluids (liquid-water, air) and one solid (aluminium) of parameters given in Table 3 were chosen as cell zone conditions.

The outlet of the computational domain was defined as pressure outlet. Symmetry boundary condition was defined on the symmetry plane. The air-aluminium and water-aluminium interface surfaces were defined as thermally coupled in order to assure the conjugate heat transfer. Wall type condition was assigned to the remaining surfaces.

Table 3. Parameters of cell zones.

material	**density [kg/m³]**	**specific heat [J/kg·K]**	**thermal conductivity [W/m·K]**	**viscosity [kg/m·s]**
water-liquid	998.2	4182	0.6	1003e-6
air	1.225	1006,43	0,0242	17.89e-6
aluminium	2719	871	202.4	-

Pressure-velocity coupling by SIMPLE algorithm was used as a solution method. This algorithm uses a relationship between velocity and pressure corrections to enforce mass conservation and obtain the pressure field.

3 Results

3.1 Convergence

The numerical calculations were performed with ANSYS Fluent 18.1 on Windows 10 workstation equipped with Intel i7 4510U 2.00GHz CPU and 16GB of RAM. The solver was operating in a single mode.

The analysis were assumed as converged when all normalized residuals decreased below the level of 1e-3.

The number of iterations to convergence varied dependent on the mesh element type and the number of mesh elements as depicted in Fig. 6. The converged solution was obtained with fewer iterations in case of POLY mesh regardless of the number of mesh elements. The reduction of iterations varied between 20% (150k cells) and 32% (340k cells).

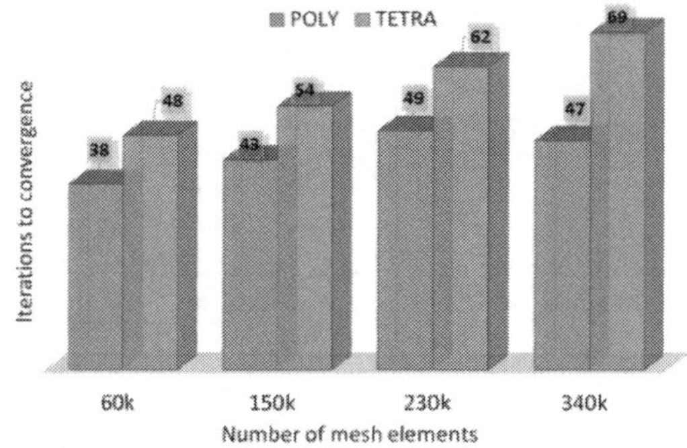

Fig. 6. Number of iterations to convergence of the analysed cases

Moreover the calculation time to convergence was measured with a stopwatch in order to evaluate the CPU time necessary to obtain the solution. The comparison of the calculation time for all analysed meshes is depicted in Fig. 7. The reduction of calculation time for POLY mesh was observed in case of the 340k cells (18% reduction) while calculations with coarser meshes were approx. 15% slower in comparison to TETRA.

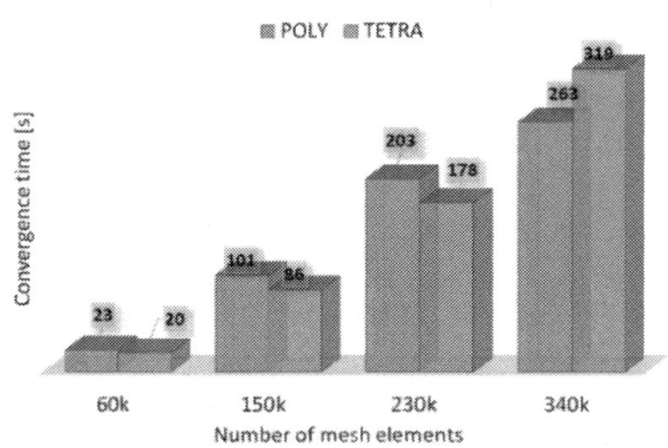

Fig. 7. Time to convergence of the analysed cases

3.2 Temperature field

The temperature distribution in the computational domain and in the air outlet from the heat exchanger is depicted for both mesh types and all mesh resolutions in Fig. 8 and Fig. 9. The figures are created with exactly the same temperature scale in order to easily distinguish the differences in the temperature fields resulting from the mesh type as well as the number of mesh elements.

Both figures reveal evident influence of the number of mesh elements on the results in case of TETRA while the results for POLYs are almost identical except for the mesh of the lowest element count. Moreover the temperature field for TETRAs is characterised by the artefacts in the form of jagged edges of the colour bands which directly indicates the discretization error.

https://doi.org/10.1051/epjconf/201818002096

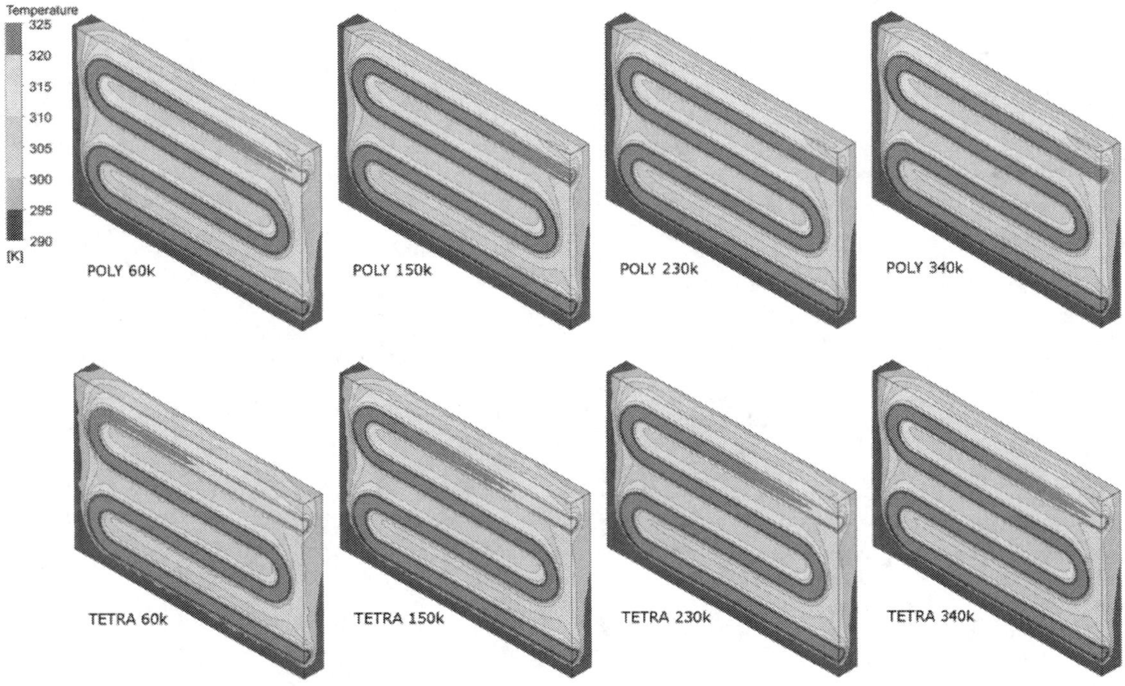

Fig. 8. Temperature distribution in the computational domain

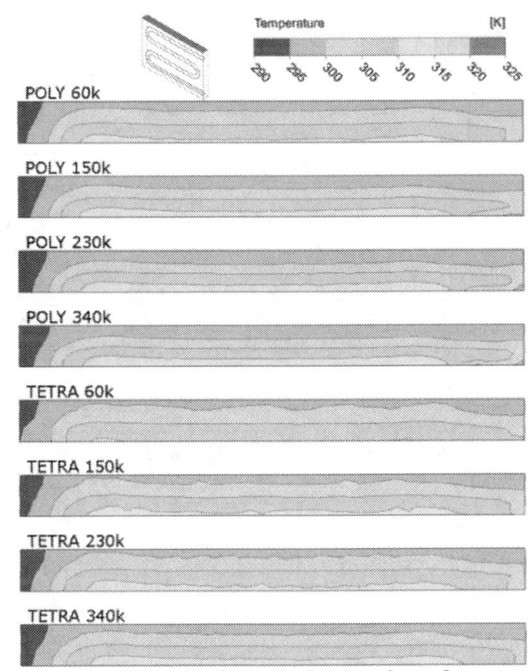

Fig. 9. Temperature distribution in the air outlet surface

3.3 Temperature profile

The variation of air temperature along the centerline (Fig. 10) of the analysed heat exchanger depicts the efficiency of conjugate heat transfer between the fluid/aluminium/air domains.

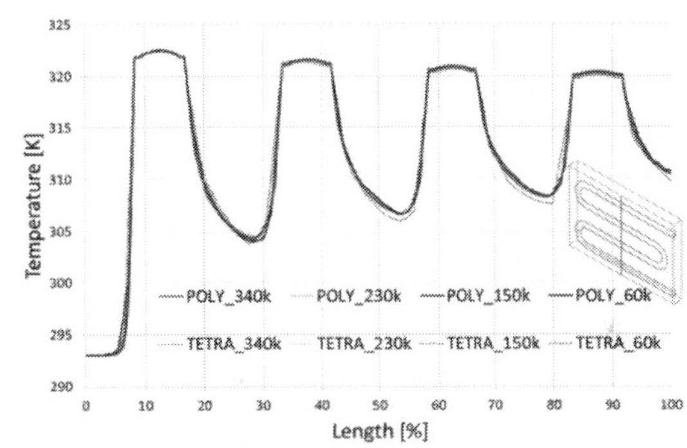

Fig. 10. Temperature along the centerline of the heat exchanger

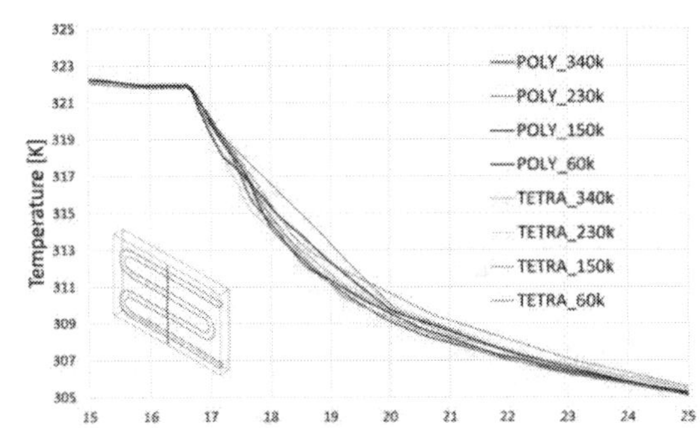

Fig. 11. Temperature along the segment of the centerline of the heat exchanger

The temperature profile obtained with all POLY meshes is similar but the results for TETRA mesh of 60k elements is noticeably different. The segment of the centerline concerning the 15%-25% of the heat exchanger length depicted in Fig. 11 allows to clearly visualize the difference in the efficiency of conjugate heat transfer due to mesh type and number of mesh elements. It confirms the rough estimation that the dispersion between the TETRA meshes of different number of elements is significantly greater in comparison to POLY meshes.

4 Discussion

The presented results of conjugate heat transfer analysis with the use of computational methods proves the importance of the very first step in CFD analysis which is the pre-processing. The obtained temperature field and temperature profile differed not only due to mesh resolution (number of cells) but also due to the mesh type (TETRA/POLY).

The carried out research confirmed that the discretization of the computational domain with POLY elements is beneficial not only in terms of the number of iterations necessary to obtain the converged solution but also the homogeneity of the results obtained with different mesh resolutions. Therefore the performed research prove that the application of POLY mesh leads to a more accurate solution achieved with a lower cell count.

Experimental validation of the numerical results is necessary in order to properly and fully define the applicability of polyhedral meshing in conjugate heat transfer numerical analysis.

5 Conclusions

The carried out research indicate the purposefulness of further investigation concerning the application of polyhedral elements in CFD.

Detailed experimental research is necessary in order to validate the computational results against real data. The experimental test stand allowing to perform the above mentioned experimental tests is under construction.

The work is partialy suported by Narodowe Centrum Nauki (National Science Centre) grant number 2017/01/X/ST8/00019

References

1. Kroll, N., *The ADIGMA Project*, in *Adigma-a European Initiative on the Development of Adaptive Higher-Order Variational Methods for Aerospace Applications: Results of a Collaborative Research Project Funded by the European Union, 2006-2009*, N. Kroll, et al., Editors. 2010, Springer-Verlag Berlin: Berlin. p. 1-9.
2. Jamrozik, A., et al., *Numerical simulation of two-stage combustion in SI engine with prechamber*. Applied Mathematical Modelling, 2013. **37**(5): p. 2961-2982.
3. Jamrozik, A., et al., *Modeling of Thermal Cycle CI Engine with Multi-Stage Fuel Injection*. Advances in Science and Technology Research Journal, 2017. **37**(5): p. 179-186.
4. Gnatowska, R., *A Study of Downwash Effects on Flow and Dispersion Processes around Buildings in Tandem Arrangement*. Polish Journal of Environmental Studies, 2015. **24**(4): p. 1571-1577.
5. Lanzafame, R., S. Mauro, and M. Messina, *Wind turbine CFD modeling using a correlation-based transitional model*. Renewable Energy, 2013. **52**: p. 31-39.
6. Gnatowska, R., M. Sosnowski, and V. Uruba, *CFD modelling and PIV experimental validation of flow fields in urban environments*. E3S Web of Conferences, 2017. **14**.
7. Gnatowska, R. and T. Rybak, *Numerical Analysis of Heat Transfer around 2D Circular Cylinder in Pulsation Inflow*, in *Proceedings of the International Conference of Numerical Analysis and Applied Mathematics 2014*, T.E. Simos and C. Tsitouras, Editors. 2015.
8. Gnatowska, R. and S. Sikora, *Numerical Analysis of Wind Flow and Erosion in Flow around the Bump Terrain*, in *Proceedings of the International Conference of Numerical Analysis and Applied Mathematics 2014*, T.E. Simos and C. Tsitouras, Editors. 2015.
9. Zylka, A., et al. *Numerical simulations of fluidization dynamics in a hot model of a CLC process*. in *4th Scientific and Technical Conference on Modern Technologies and Energy Systems (WTiUE)*. 2016. Cracow, POLAND.
10. Spiegel, M., et al., *Tetrahedral vs. polyhedral mesh size evaluation on flow velocity and wall shear stress for cerebral hemodynamic simulation*. Computer Methods in Biomechanics and Biomedical Engineering, 2011. **14**(1): p. 9-22.
11. Bianchi, G., et al., *Deforming grid generation for numerical simulations of fluid dynamics in sliding vane rotary machines*. Advances in Engineering Software, 2017. **112**: p. 180-191.
12. Li, H., L. Rong, and G.Q. Zhang, *Reliability of turbulence models and mesh types for CFD simulations of a mechanically ventilated pig house containing animals*. Biosystems Engineering, 2017. **161**: p. 37-52.
13. Su, J.W., et al., *A two-layer mesh method for discrete element simulation of gas-particle systems with arbitrarily polyhedral mesh*. International Journal for Numerical Methods in Engineering, 2015. **103**(10): p. 759-780.
14. Sosnowski, M., J. Krzywanski, and R. Gnatowska, *Polyhedral meshing as an innovative approach to computational domain discretization of a cyclone in a fluidized bed CLC unit*. E3S Web of Conferences, 2017. **14**: p. 01027.
15. Eca, L. and M. Hoekstra, *A procedure for the estimation of the numerical uncertainty of CFD calculations based on grid refinement studies*. Journal of Computational Physics, 2014. **262**: p. 104-130.
16. Celik, I.B., et al., *Procedure for estimation and reporting of uncertainty due to discretization in CFD applications*. Journal of Fluids Engineering-Transactions of the Asme, 2008. **130**(7).

Analysis of mean and fluctuating helicity measured by TomoPIV in swirling jet

Rodion Stepanov[1,*], *Peter* Frick[1], *Vladimir* Dulin[2,3], and *Dmitriy* Markovich[2,3]

[1] *Institute of Continuous Media Mechanics, Korolyov 1, Perm, 614013, Russia*
[2] *Kutateladze Institute of Thermophysics, Lavrentyev av. 1, 630090 Novosibirsk, Russia*
[3] *Novosibirsk State University, Pirogova 2, 630090 Novosibirsk, Russia*

Abstract. Important role of helicity was theoretically predicted for the generation of large-scale magnetic fields and atmospheric vortices. Helicity can lead to a reduction of turbulent dissipation in the atmosphere or in a specific constrained flow, e.g. in pipe. We use the TomoPIV data (42 cube of grid points, resolution 0.84 mm) to measure 3D velocity field of turbulent swirling flows. We study spatial distribution of the mean and fluctuating components of energy and helicity. We find that helical turbulence excitation and decay along stream of the jet strongly depend on the inflow swirl. We observe spatial separation of turbulent flow with different sign of helicity while integrated values are conserves. It is shown that large scale swirling flow induces helicity at the small scales. Our results bring valuable materials for benchmark the modern numerical simulations with turbulent closure technique.

1 Introduction

Helicity is a pseudoscalar characteristic of the flow which is defined by the correlation of the velocity and vorticity fields. It is known to be a topological characteristic of linkage of vortex lines [1]. In absence of viscosity the helicity is a conserved quantity as well as energy. Important role of helicity was theoretically predicted for the generation of large-scale magnetic fields (MHD dynamo) [2] and atmospheric vortices (hydrodynamic alpha effect) [3]. Helical features of atmospheric flows can be important to understated the tropical cyclone genesis [4, 5]. Helicity can lead to a change of turbulent transport coefficients, e.g. reduction of turbulent dissipation [6, 7]. Different helical effects were revealed by theoretical studies or numerical simulations of 3D turbulence under idealized conditions: inverse energy cascade by helical triads [8], sustainment of large-scale vorticity field in a rotating turbulence with nonuniform helicity [9], reduction of direct energy cascade in highly helical turbulent flow [10, 11]. However a comparison of experimental and theoretical results of helical phenomena remains a challenging problem.

Analysis of helicity generation, advection and dissipation mechanisms seems as an affordable task in frame of numerical simulations [12]. Experimental investigation of helicity of the mean flow is also possible. Mean velocity components can be measured with accuracy, enabling their differentiation to obtain the vorticity. However experimental study of helicity fluctuations in space and time is extremely difficult since they require measurements of instantaneous distributions of both velocity and vorticity vector fields. Helicity in MHD turbulence can be detected indirectly by measurements of magnetic field induction [13]. As far as we know there is only one successful attempt to measure the helicity power spectrum [14]. Modern optical techniques such as dual-plane or tomographic PIV (particle image velocimetry) make possible the reconstruction of helicity density distribution in a plane [15] or even in a volume [16]. The reconnection of linked vortex loops through helicity transformation demonstrated in recent successful experiments with high-speed scanning tomography apparatus [17]. In this work we use data of the tomographic PIV measurements of 3D velocity field in turbulent swirling jet for analysis of distribution of mean and fluctuating helicity. We suggest a processing of measurements that allow us to study the relation between the mean and fluctuating components of helicity and compare it with theoretical predictions.

2 Experimental setup and flow configuration

Swirling jets are generated in a transparent box by a convergent axisymmetric nozzle with mounted inside vane swirler (highly lighted part in figure 1). The detailed description of the experiment can be found in [16], where coherent structures in low-swirl (LS) and high-swirl (HS) turbulent jets are discussed. The Reynolds number Re $= U_b d/\nu$ is equal to 8900 (here $U_b = 0.48$ m/s is the bulk velocity of the jet, $d = 15$ mm is the outlet diameter of the nozzle and ν is the viscosity). 3D velocity measure-

*e-mail: rodion@icmm.ru

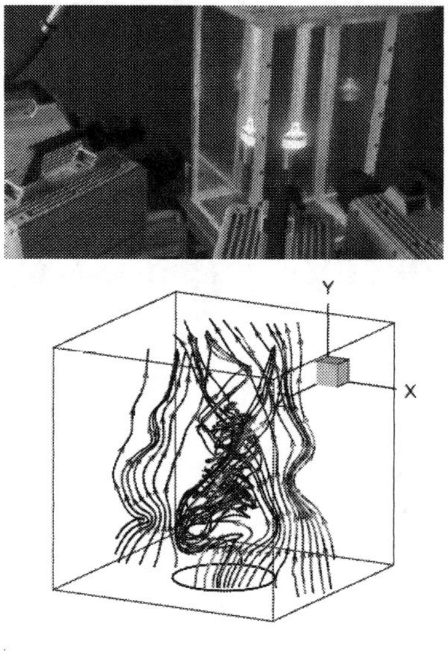

Fig. 1. Experimental setup (top) and structure of the swirling jet flow (bottom).

ments by tomographic PIV are performed for a cubic domain 40 × 40 × 40 mm downstream the nozzle rim. 2000 snapshots of 3D velocity are captured within one second, whereas the characteristic timescale is approximately forty times smaller [18]. Flow precession period for the HS jet is fifteen times smaller than the measurements interval. Reliable agreement between the mean velocity and Reynolds stresses obtained during the current high-repetition-rate experiments and previous low-frequency measurements is reported in [19].

In both, LS and HS jets, the superimposed swirl promotes development of helical instability modes in the jet mixing layer. The swirl is characterized by the swirl rate S, defined by the ratio of the angular jet momentum to the axial jet momentum, multiplied by the nozzle radius. $S = 0.47$ and $S = 1.0$ for LS and HS, respectively. These swirl intensities are correspondingly well below and above the critical swirl rate $S_{cr} = 0.6$ for the swirling jet's vortex breakdown and formation of the recirculation zone at the jet axis [20]. Thus, in the latter case the flow is featured by a pronounced bubble-type vortex breakdown with flow precession. Direct 3D PIV measurements confirm the scenario, in which the flow dynamics is dominated by a global azimuthal instability mode, corresponding to a rotating coherent structure composed of spiraling vortex core and outer large-scale helical vortex [21].

3 Results

Applying decomposition of total fields in the mean and fluctuating parts to the velocity $\mathbf{U} = \mathbf{V} + \mathbf{v}$ and vorticity $\nabla \times \mathbf{U} = \mathbf{W} + \mathbf{w}$ (by definition, $\langle \mathbf{U} \rangle = \mathbf{V}$, $\langle \mathbf{v} \rangle = 0$ and time averaging is supposed for statistically stationary flow), one gets the energy of the mean flow $E = |\mathbf{V}|^2$ and energy

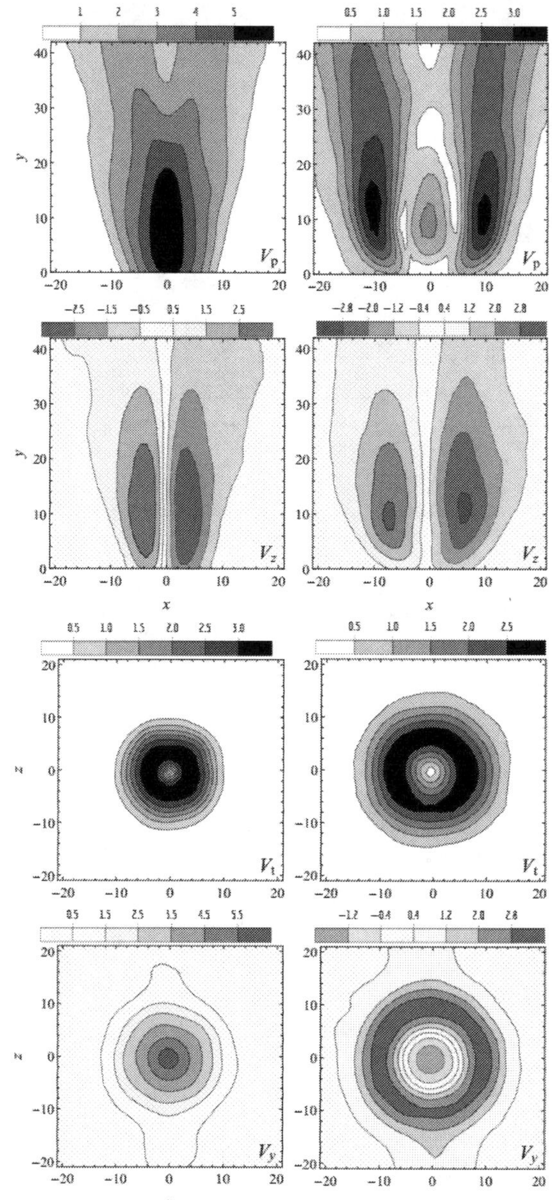

Fig. 2. Mean velocity components in the plane $z = 0$ (two top rows) and in the plane $y = 10$ (two bottom rows): LS (left column), HS (right column) jets.

of fluctuations $e = \langle |\mathbf{v}|^2 \rangle$ and the helicity of mean flow $H = \mathbf{V} \cdot \mathbf{W}$ and helicity of fluctuations $h = \langle \mathbf{v} \cdot \mathbf{w} \rangle$. We analyze E, e, H and h separately to understand the relative dynamics of energy and helicity in the turbulent flow.

In figure 2 we show the distributions of velocity components in the central vertical cross-section ($z = 0$) and in one selected horizontal cross-section at $y = 10$ of both LS and HS jets. Velocity components perpendicular to the presented plains are demonstrated by its values with sign (

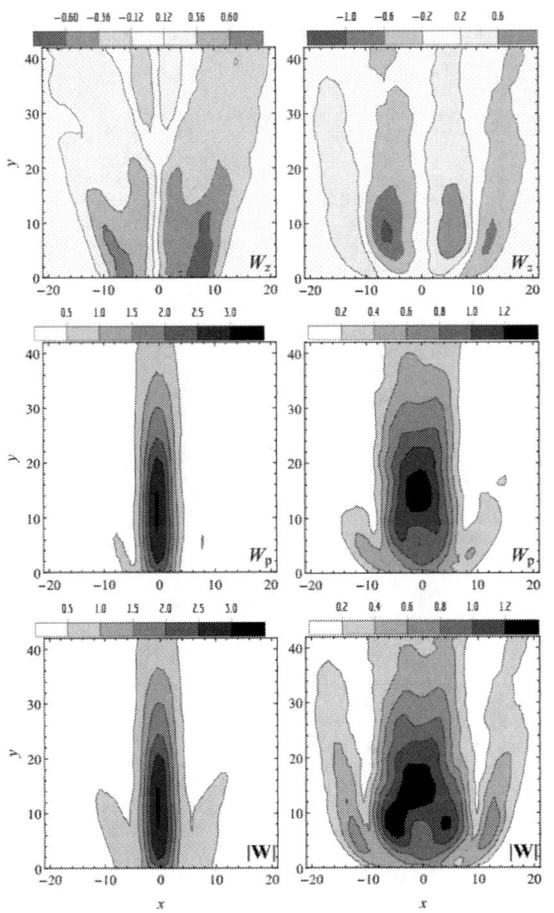

Fig. 3. Mean flow vorticity in the plane $z = 0$ for LS (left column) and HS (right column) jets: W_z (top row), intensity W_p (middle row), total intensity $|\mathbf{W}|$ (bottom row).

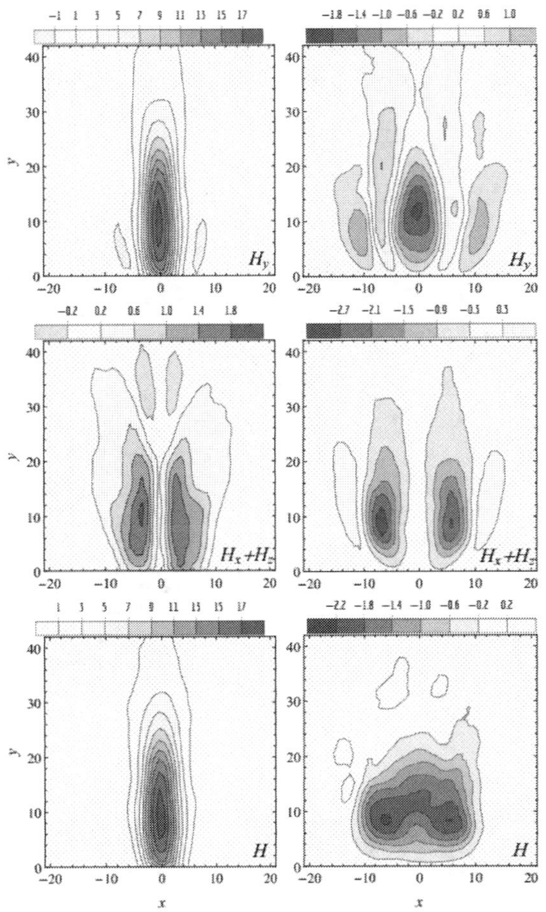

Fig. 4. Helicity of mean flow in the plane $z = 0$ for LS (left column) and HS (right column) jets: H_y (top row), $H_x + H_z$ (middle row) and H (bottom row).

V_z and V_y) while velocity components parallel to the presented plains are depicted by its absolute values (intensity of poloidal velocity component $V_p = (V_x^2 + V_y^2)^{1/2}$ and intensity of toroidal velocity component $V_t = (V_x^2 + V_z^2)^{1/2}$). The horizontal cross-section at $y = 10$ is selected because the HS jet has inner region with reversed flow at this hight. Dimensional quantities here and below are given in units of length 0.07 mm and units of time 0.0005 s so unit of velocity is 0.14 m/s.

The difference is more pronounced in the structure of mean vorticity $\mathbf{W}$ presented in figure 3. In the LS jet the map of the W_z component reveals a strong toroidal vortex nearby the nozzle and a weak contrarotating toroidal vortex at the downstream. In the HS jet this contrarotating vortex goes down to the nozzle pushing aside the weakened lower vortex. The distribution of total vorticity intensity becomes also more complicate in HS jet.

Resulting helicity of the mean flow is shown in figure 4. We split the total helicity in two parts: the parallel and perpendicular to jet flow, $H_y = V_y W_y$ and $H_t = V_x W_x + V_z W_z$. The structure of helicity field is completely different in two jets. In both jets a strong anisotropy is seen, the H_y component dominates in LS jet, while H_t is stronger in HS jet. Moreover, the sign of dominating helicity is opposite in LS and HS. Thus, the vortex breakdown in HS jet and formation of the central recirculation zone locally change the sign of the mean flow helicity.

Characteristics of fluctuating fields are shown in figure 5 and confirm the particular difference between LS and HS flows. Note that the similar structure of energy and enstrophy distributions (correspondingly, e and $e_w = \langle|\mathbf{w}|^2\rangle$) in each jet (LS and HS) mean a moderate isotropy of fluctuations. Sign defined helicity of fluctuations indicates a correlation of $\mathbf{v}$ and $\mathbf{w}$ fields (see figure 6). Again one can see strong difference in sign and distribution of small-scale helicity between LS and HS. The accordance of the dominating sign of h and H in each flow supports the transfer of helicity from large scales to small scales.

It is interesting to follow the evolution of helicity, mainly generated at the onset of jet flow, along the jet. We

 https://doi.org/10.1051/epjconf/201818002097

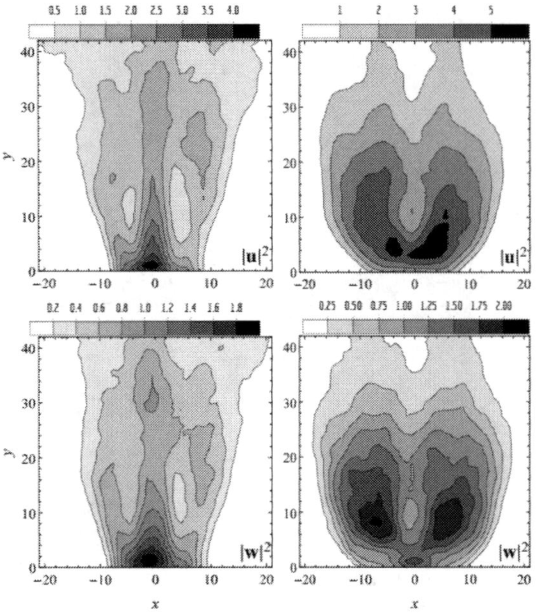

Fig. 5. Characteristics of fluctuations in the plane $z = 0$ for LS (left column) and HS (right column) jets: energy of velocity fluctuations e (top row), enstrophy of fluctuations $e_\omega = \langle|\mathbf{w}|^2\rangle$ (bottom row).

plot in figure 7 the y-dependency of flow characteristics integrated over xOz plane, which illustrate the generation and decay of helical turbulence. The large-scale helicity is generated in the beginning of the jet up to $y \approx 10$. Note again that the helicity input rate depends on the degree of swirling in a very complicate way. Moderate swirling produce positive helicity, i.e. the sign of helicity corresponds to the hand of helix in the nozzle. The strong swirling increases the opening angle and generates an inverse axial flow, producing negative mean-flow (large-scale) helicity. In both jets the mean-flow helicity reaches the maximum at $y \approx 10$ (although HS evolves faster) and monotonically decays in what follows. In HS the decay is essentially faster – the mean-flow helicity vanishes at $y \approx 30$. The energy of the mean flow at $y > 30$ becomes very similar in both jets. More surprisingly seams the equality of energy of fluctuations at large y, because at the maximum (at $y \approx 12$) the energy of fluctuations in HS was about 5 times larger as in LS. Note, that despite different spatial distributions of helicity components H_y and H_t the integrated values are getting in balance. Similar character of evolution (isotropization) was found by a helicity analysis in the swirling pipe flow [22].

The mean helicity of fluctuations in LS jet is very weak at low y. It is slightly negative and changes the sign in the cross-section, where H reaches the maximum. Downstream, the small-scale helicity monotonically increases up to $y \approx 30$. In this part of the jet ($10 < y < 30$) the energy of small-scale pulsations also increases, supporting the concept of energy and helicity and helicity transfer

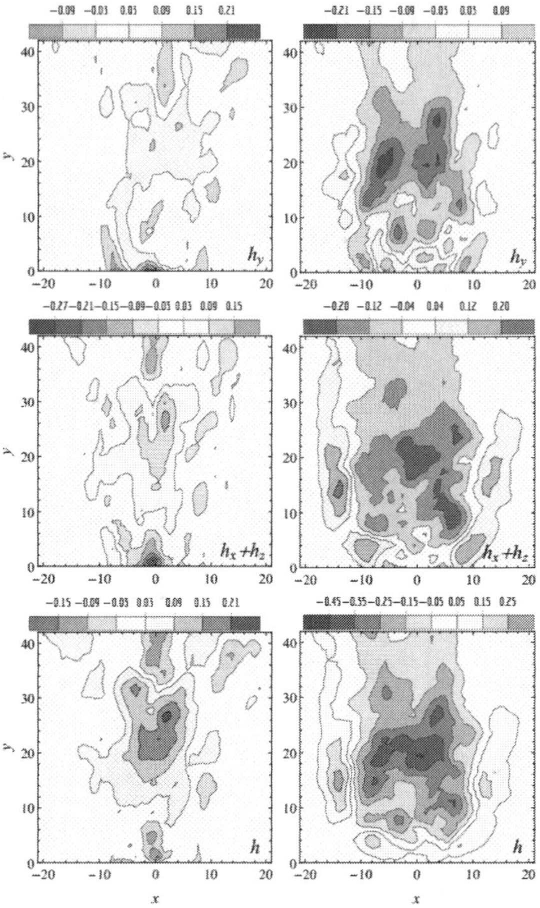

Fig. 6. Helicity of fluctuations in the plane $z = 0$ for LS (left column) and HS (right column) jets: h_y (top row), $h_t = h_x + h_z$ (middle row) and total h (bottom row).

from large scales to small scales. Helicity of fluctuations in two horizontal cross-sections at $y = 16$ and $y = 32$ are shown in figure 8.

4 Conclusions

Tomographic particle image velocimetry measurements of turbulent swirling jets allows us to study the mean and pulsation components of the flow. Spatial distribution of energy and helicity reveals the anisotropic structure of the jet flow. Scenario of helicity generation and decay along the stream of the jet depends dramatically on the inflow swirl. It may correspond to bifurcation character of filament formation of helical vortex which depends on swirl intensity [16, 21, 23]. We observe spatial separation of turbulent flow with different sign of helicity while integrated values are conserved. Applying time and space averaging we demonstrated energy and helicity transfers from large scales to small scales. Our results bring valuable materials to benchmark the modern numerical simulations with turbulent closure technique, i.e. LES. Smogarinski's model

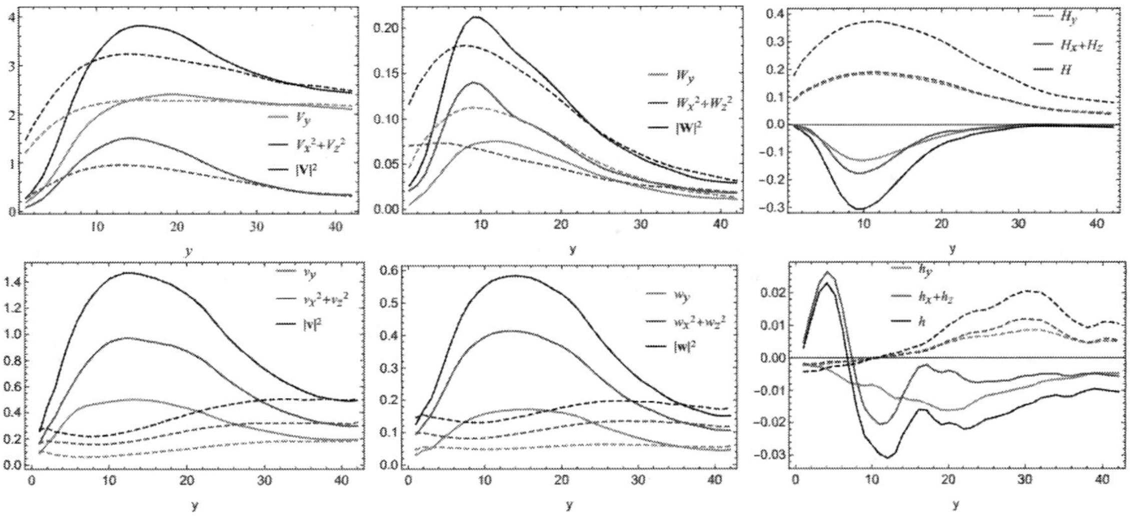

Fig. 7. Dependencies of flow characteristics integrated over xOz plain on y coordinate (along jet stream) for LS (dashed lines) and HS (solid lines) jets.

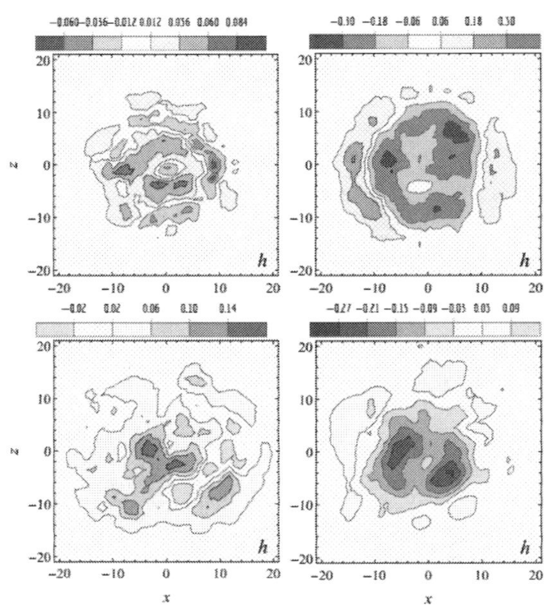

Fig. 8. Helicity of fluctuations in the planes $y = 16$ (top row) and $y = 32$ (bottom row) for LS (left column) and HS (right column) jets.

including helicity and inhomogeneity of turbulence might be revised and improved.

Acknowledgements. This work was supported by the Russian Science Foundation (grand RSF-16-41-02012).

References

[1] H.K. Moffatt, J. Fluid Mech. **35**, 117–129 (1969)
[2] M. Steenbeck, F. Krause, K.H. Rädler, Z. Naturforsch., Part A **21**, 369 (1966)
[3] S.S. Moiseev, R.Z. Sagdeev, A.V. Tur, G.A. Khomenko, A.M. Shukurov, Sov. Phys. Dokl. **28**, 926 (1983)
[4] G. Levina, M. Montgomery, Doklady Earth Sciences **434**, 1285 (2010)
[5] G.V. Levina, M.T. Montgomery, Doklady Earth Sciences **458**, 1143 (2014)
[6] D.K. Lilly, J. Atmos. Sci. **43**, 126 (1986)
[7] N. Yokoi, A. Yoshizawa, Phys. Fluids **5**, 464 (1993)
[8] L. Biferale, S. Musacchio, F. Toschi, Phys. Rev. Lett. **108**, 164501 (2012)
[9] N. Yokoi, A. Brandenburg, Phys. Rev. E **93**, 033125 (2016), 1511.08983
[10] R. Stepanov, E. Golbraikh, P. Frick, A. Shestakov, Phys. Rev. Lett.**115**, 234501 (2015), 1508.07236
[11] M. Kessar, F. Plunian, R. Stepanov, G. Balarac, Phys. Rev. E **92**, 031004 (2015)
[12] A. Teimurazov, A. Sukhanovskii, A. Evgrafova, R. Stepanov, Journal of Physics: Conference Series **899**, 022017 (2017)
[13] R. Stepanov, R. Volk, S. Denisov, P. Frick, V. Noskov, J.F. Pinton, Phys. Rev. E**73**, 046310 (2006)
[14] B.M. Koprov, V.M. Koprov, V.M. Ponomarev, O.G. Chkhetiani, Dokl. Phys. **50**, 419 (2005)
[15] G.S. Regunath, W.B. Zimmerman, V. Tesař, B.N. Hewakandamby, Exp. Fluids **45**, 973 (2008)
[16] D.M. Markovich, V.M. Dulin, S.S. Abdurakipov, L.A. Kozinkin, M.P. Tokarev, K. Hanjalić, J. Turbulence **17**, 678 (2016)
[17] D. Kleckner, W.T.M. Irvine, Nat. Phys. **9**, 253 (2013)

[18] S.V. Alekseenko, V.M. Dulin, Y.S. Kozorezov, D.M. Markovich, International Journal of Heat and Fluid Flow **29**, 1699 (2008)

[19] V. Dulin, S. Abdurakipov, D. Markovich, K. Hanjalic, *Dynamics of large-scale vortices in non-swirling and swirling turbulent jets. Time-resolved tomographic PIV measurements*, in *Proceedings of the 10th International Symposium on Turbulence and Shear Flow Phenomena*, edited by L. Smits, A. Johansson, M. Oberlack, H.J. Sung, S. Tavoularis, H. Nagib (2017), Vol. 3, p. 233

[20] H. Liang, T. Maxworthy, Journal of Fluid Mechanics **525**, 115 (2005)

[21] K. Oberleithner, C. Paschereit, R. Seele, I. Wygnanski, AIAA journal **50**, 1437 (2012)

[22] A. Vasiliev, A. Sukhanovskii, R. Stepanov, IOP Conference Series: Materials Science and Engineering **208** (2017)

[23] L.G. Sarasua, A.C. Sicardi Schifino, R. González, Physics of Fluids **17**, 044104 (2005)

Impact of different thickness of the smooth heated surface on flow boiling heat transfer

Kinga Strąk*, *Magdalena* Piasecka

Faculty of Mechatronics and Mechanical Engineering
Kielce University of Technology, Al. 1000-lecia P.P. 7, 25-314 Kielce, Poland

Abstract. This paper presents a comparison of the performance of three smooth heated surfaces with different thicknesses. Analysis was carried out on an experimental setup for flow boiling heat transfer. The most important element of the setup was the test section with a rectangular minichannel, 1.7 mm deep, 16 mm wide and 180 mm long, oriented vertically. The heated element for the FC-72 Fluorinert flowing in the minichannel was designated as a Haynes-230 alloy plate (0.10 mm and 0.45 mm thick) or a Hastelloy X alloy plate (0.65 mm thick). Infrared thermography was used to measure the temperature of the outer plate surface. The local values of the heat transfer coefficient for stationary state conditions were calculated using a simple one-dimensional method. The experimental results were presented as the relationship between the heat transfer coefficients in the subcooled boiling region and the distance along the minichannel length and boiling curves. The highest local heat transfer coefficients were recorded for the surface of 0.10 mm thick heated plate at the outlet and 0.45 mm thick plate at the minichannel inlet. All boiling curves were typical in shape.

1 Introduction

The micro and macro-scale devices such as compact heat exchangers, cooling devices, high-powered lasers, and micro-electronics manufacturing systems use flow boiling with phase change on enhanced surfaces to provide the highest possible efficiency of heat transfer. Compared to single-phase flow, larger amounts of thermal energy can be removed when micro- and mini-scale channels are used in flow boiling. The flow along minichannels varies considerably from that along macrochannels [1-2].

The first classification of small channels, proposed by Shah and based on the hydraulic diameter, divides them into two groups [3]:

- conventional channels: d_h >6 mm,
- minichannels: d_h <6 mm.

Based on the same criterion, hydraulic diameter (d_h), Kandlikar and Grande introduce new groups [4]:

- conventional channels - $d_h > 3$mm
- minichannels - 3mm $\geq d_h >$ 200 μm
- microchannels - 200 μm $\geq d_h >$ 10 μm
- transitional channels- 10 μm $\geq d_h >$ 0.1 μm
- transitional microchannels -10 μm $\geq d_h >$ 1 μm
- transitional nanochannels - 1 μm $\geq d_h >$ 0.1 μm
- molecular nanochannels - 0.1 μm $\geq d_h$.

Kew and Cornwell [5] suggest basing the division on the confinement number, *Co*:

- microchannels, for $Co < 0.5$,
- macrochannels, for $Co > 0.5$.

The Kielce University of Technology has conducted a series of studies devoted to flow boiling heat transfer with refrigerants flowing along minichannels with enhanced heated plates [6-12] at different minichannel orientations [10-12]. The use of enhanced surfaces produced in thermal processes is known to intensify the heat transfer process as shown in [13-15]). Such surfaces are applied also in pool boiling heat transfer [8, 16-20]. Studies on heat transfer in confined spaces as face seals were described in [21-23].

2 Experiment

2.1 Experimental stand

The study was carried out for a flow boiling process by using the experimental setup shown in Fig. 1. The major part of the setup was the test section with a minichannel, described in detail in [7, 12, 24]

Temperature changes on one side of the plate were measured with an E60 FLIR infrared camera (2) of accuracy: ± 1°C or ± 1%, temperature range: 0 ÷ 120°C [25]. To obtain emissivity of 0.83 [24], the plate surface was coated with black paint. Narrow micanite panels placed on both sides of the plate prevented its deformation. The other side of the plate, in contact with the fluid, was observed through a glass panel. K-type

* Corresponding author:kzietala@tu.kielce.pl

 https://doi.org/10.1051/epjconf/201818002098

thermocouples and pressure sensors (8) were installed at the inlet and outlet of the minichannel.

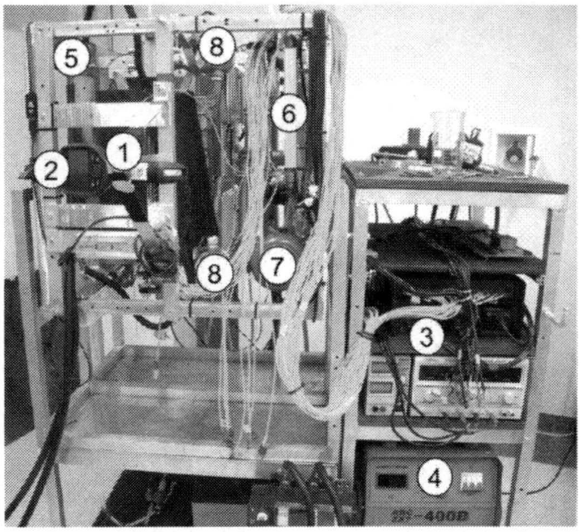

Fig. 1. The image of the main systems of the experimental setup, 1-test section; 2-infrared camera, 3-data acquisition station, 4-inverter welder, 5-shunt, 6-mass flowmeter, 7- gear pump, 8- pressure sensor.

2.2 Test section and heated plates

The test section with a rectangular, vertically oriented, 1.7 mm deep, 16 mm wide and 180 mm long minichannel was the most important element of the experimental setup. Figure 2 shows the photograph of the test section. The channel depth is provided by Teflon plate of adequate thickness (2). One of the three smooth surfaces below acted as the heating element (7) for FC-72 flowing in the minichannel:

- Hastelloy X alloy plate, about 0.65 mm thick,
- Haynes-230 alloy plate, about 0.10 mm thick,
- Haynes-230 alloy plate, about 0.45 mm thick.

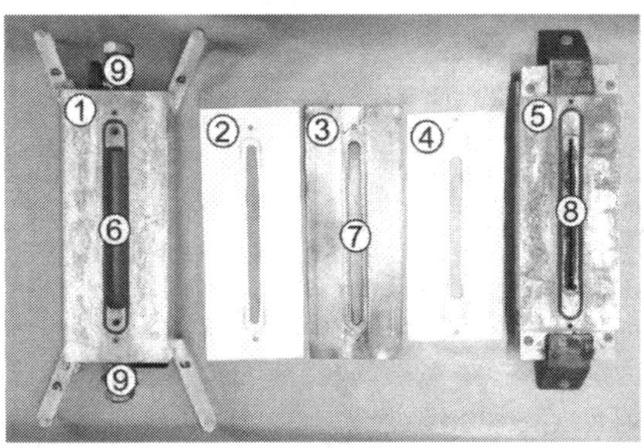

Fig. 2. The image of the test section with the minichannel: 1-front cover, 2, 4-Teflon plate, 3-graphite plate, 5-channel body, 6- glass panel, 7-heated plate, 8-micanite plate, 9-thermocouple.

All the heated plates were manufactured by Haynes Int. Inc. (USA) and chosen for their electrical resistivity, assuming minor resistivity changes with temperature. The Hastelloy X and Haynes-230 alloy plates are made mainly of Ni-Cr-Fe-Mo and Ni-Cr-W-Mo alloys, respectively. In addition to excellent high-temperature strength and oxidation resistance, the alloys exhibit superior long-term stability and fabricability [7, 26].

Figures 3-5 refer to the three types of smooth heated surfaces, 0.10 mm, 0.45 mm and 0.65 mm thick, respectively, showing the topography of a fragment of the melts (a), 3D topography (b) and a profilogram of a fragment of the single path surface (c).

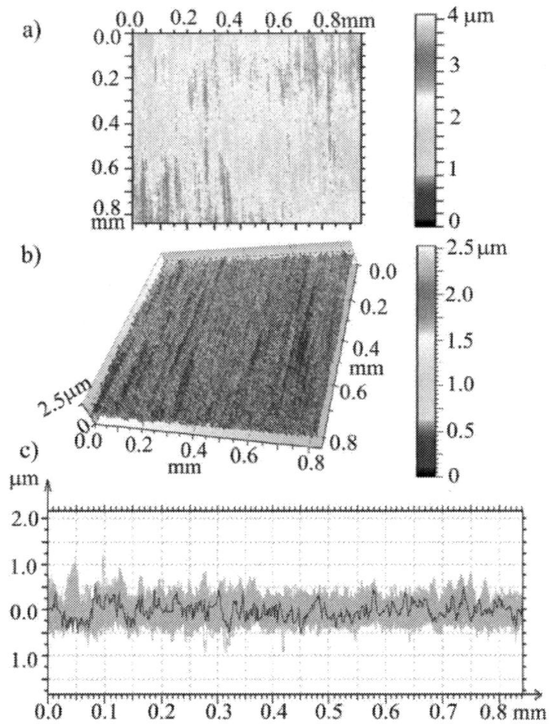

Fig. 3. Fragment of the plate's smooth surface 0.10 mm thick: a) topography of a fragment of the melts, b) 3D topography, c) roughness profile.

Table 1 compiles main roughness parameters of the surfaces, measured using the optical profilometer Talysurf CCI to ISO 25178-2:2012 manufactured by Taylor Hobson.

Table 1. Main roughness parameters.

Parameter of the surfaces	Smooth surface thickness:		
	0.10 mm	0.45 mm	0.65 mm
Ra	0.1432	0.1752	0.1327
Sa	0.1491	0.3455	0.1444

Ra - arithmetic mean deviation of the roughness profile, μm
Sa - arithmetic mean height, μm.

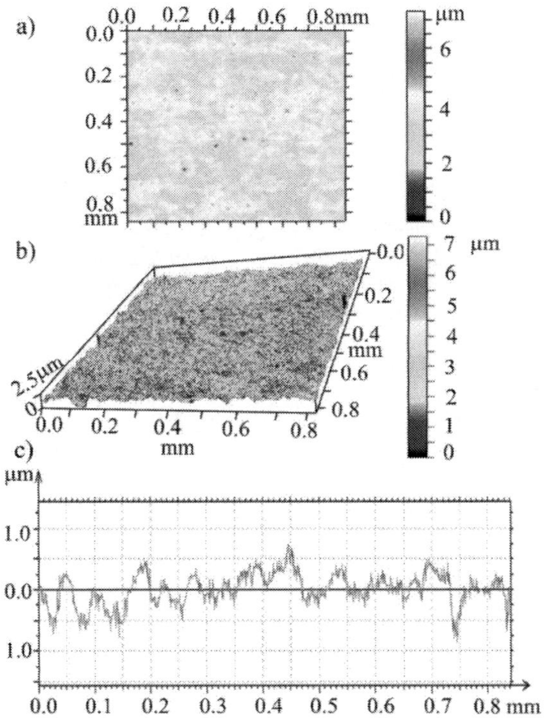

Fig. 4. Fragment of the plate's smooth surface 0.45 mm thick: a) topography of a fragment of the melts, b) 3D topography, c) roughness profile.

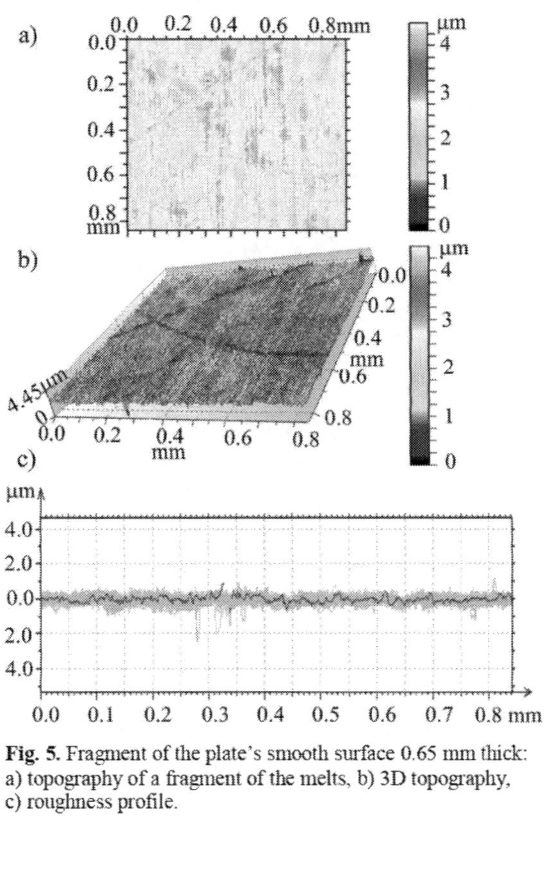

Fig. 5. Fragment of the plate's smooth surface 0.65 mm thick: a) topography of a fragment of the melts, b) 3D topography, c) roughness profile.

The study of roughness parameters was conducted at the Geometric Computer Measurement Laboratory at the Kielce University of Technology. The profilometer enables non-contact measurement and analysis of the geometric structure of the tested surface [7].

The arithmetic mean deviation of the roughness profile for various surface thicknesses was similar. The values ranged from 0.13 µm to 0.18 µm. The arithmetic mean height for 0.10 mm and 0.65 mm surface thicknesses was similar, with *Sa* twice as large for the surface 0.45 mm thick.

3 Experiment

During the experimental series, electric power supplied to the heated plate was gradually increased, as was the heat flux transferred to the fluid in the minichannel, thus initiating nucleate boiling and enhancing heat transfer. The first region after the boiling onset was the subcooled boiling region - the liquid became superheated only in the vicinity of the heated surface; it was highly subcooled in the core of the flow.

The experimental parameters of tested series are presented in Table 2.

Table 2. Experimental parameters of tested series.

Thickness of the heated plate δ_P (mm)	Heat flux density q_w (kW·m^{-2})	Mass flux G (kg·m^{-2}s^{-1})	Inlet pressure p_{in} (kPa)	Inlet liquid sub-cooling ΔT_{sub} (K)
0.10	26	243	198	55
	40			
0.45	26	423	102	39
	40			
0.65	26	421	106	38
	40			

4 Calculations

The authors of [6, 11, 12, 24, 27-30] present various methods of determining the local heat transfer coefficient in the two-dimensional approaches employing Trefftz or Beck methods obtained from solving the inverse problem. In this article, the one-dimensional approach is used, taking into account the heat flow direction, which is perpendicular to the direction of the fluid flow in the minichannel.

Local heat transfer coefficients were determined using the one-dimensional method, described also in [7, 24, 31]. In the subcooled boiling region local coefficients are determined from the following equation:

$$\alpha(x) = q_w / (T_P(x,\delta) - T_f(x) - q_w \cdot \frac{\delta_P}{\lambda_P}) \quad (1)$$

x - distance from the minichannel inlet along the direction of the flow, q_w - density of the heat flux transferred from the heated plate to the fluid, $T_p(x)$ - plate temperature, measured by using infrared camera, $T_f(x)$ - fluid temperature calculated from the assumption of the linear distribution of the fluid temperature along the minichannel, λ_P - thermal conductivity of the heated plate, δ_P - thickness of the heated plate.

5 Results and discussion

The experiments were recorded for the adopted heat flux supplied to the heated plate under similar stable thermal and flow conditions. The experimental results pertain to the steady state (stationary condition). The analysis includes the data from the experiments performed for three thicknesses of the heated plate: 0.10 mm, 0.45 mm and 0.65 mm. The results are presented graphically as the heat transfer coefficient versus the distance from the minichannel inlet at the the subcooled boiling region for two heat fluxes: 26 kW·m^{-2} and 40 kW·m^{-2}, see Figs. 6 and 7, respectively.

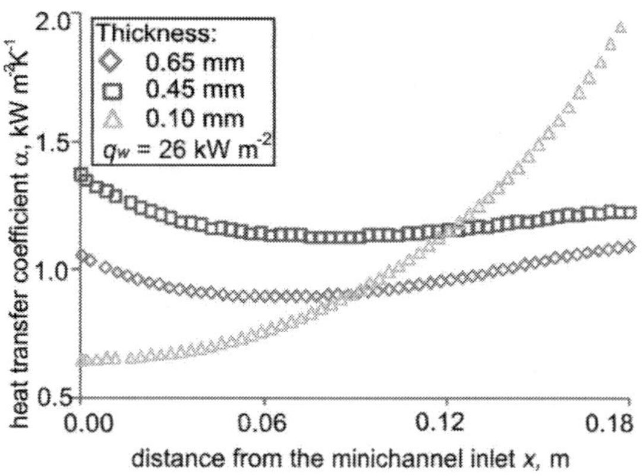

Fig. 6. Heat transfer coefficient vs. distance from the minichannel length obtained for the subcooled boiling region for three tested thicknesses of the heated plate, q_w = 26 kW·m^{-2}.

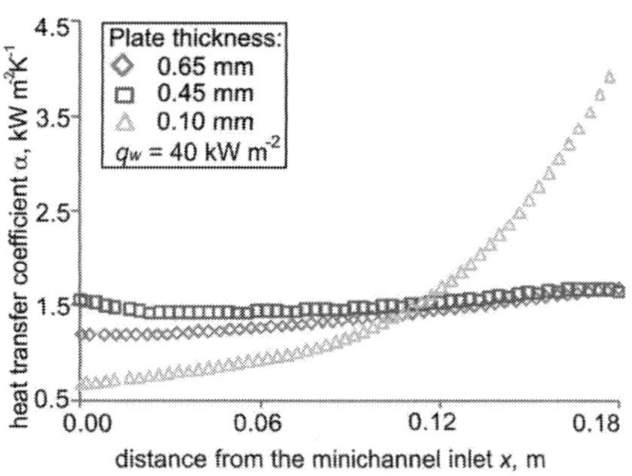

Fig. 7. Heat transfer coefficient vs. distance from the minichannel length obtained for the subcooled boiling region for three tested thicknesses of the heated plate, q_w = 40 kW·m^{-2}.

In the subcooled boiling region (Figs. 6-7), heat transfer coefficients were relatively low in comparison to those at the saturated boiling region [10-12, 27, 29]. The local heat transfer coefficient values increased with the distance from the minichannel inlet to the outlet and achieved maximum of about 4 kW·m^{-2}·K^{-1} for 0.10 mm thickness of the surface at the 40 kW·m^{-2} heat flux (Fig. 7). The highest local heat transfer coefficients were recorded for the surface of 0.10 mm thick heated plate (from 0.12 m to 0.18 m distance from the minichannel inlet) and 0.45 mm thick plate (from 0 m to 0.12 m distance from the minichannel inlet) at all heat fluxes analysed. The surfaces of 0.65 mm (from 0.12 m to 0.18 m distance from the minichannel inlet) and 0.10 mm (from 0 m to 0.09 m distance from the minichannel inlet) thick plates provided the lowest coefficients.

To determine the effect of the heated plate thickness on the value of the heat transfer coefficient, the following quantities were calculated:

1. the value of heat transfer through the heated plate

$$z(x) = \alpha(x) - q_w / (T_P(x,\delta) - T_f(x)) \tag{2}$$

2. the value of heat transfer coefficient from eq. (1) for each heated plate thickness under analysis,
3. the percentage value of heat transfer through the heated plate for various thicknesses of the plate, according to the following relation

$$y = z / \alpha(x) \cdot 100\% \tag{3}$$

Results of calculations are presented in Table 3.

Table 3. Effect of the heated plate thickness on the heat transfer coefficient.

	Percentage value of heat transfer through the heated plate, y (%)		
Heat flux density, q_w (kW·m^{-2})	Smooth surface thickness (mm)		
	0.10	0.45	0.65
26	1.2	6.4	7.6
40	1.8	8.2	10.8

As the thickness of the plate and the heat flux increased, the percentage of heat transfer through the surface of the plate increased. The smallest percentage was observed for the thinnest plate and the lowest heat flux, while the largest was for the thickest plate and the highest heat flux. The results meet the expectations resulting from the description of heat transfer according to the Fourier law.

The boiling curves shown in Fig. 8 were constructed for the three plate thickness. The curves were plotted as the heat flux density against the temperature difference T_p-T_f for three selected distances: 0.06 m - 1/3 of the channel length from the inlet (a), 0.09 m - 1/2 of the channel length from the inlet (b), and - 0.12 m - 1/3 of the channel length from the outlet (c).

All curves were typical in shape. Spontaneous nucleation at boiling incipience caused a drop in the temperature of the heated surface, the phenomenon being known as "nucleation hysteresis" characteristic of highly wetting dielectric fluids (e.g. refrigerants). The bubbles act as internal heat sinks, absorbing a significant amount of energy transferred to the liquid [9, 11, 12]. The arrangement of points on the boiling curves shows the highest scatter and variability of the data obtained from experiments with the plate of 0.10 mm thickness. The highest temperature drops characteristic of boiling initiation are observed in boiling curves plotted for 0.10 mm and 0.65 mm thick plates. The boiling curves are close to parallel for the 0.45 mm and 0.65 mm thick plates. The relationships are shifted in the graph by the heat flux value of 20-30 kW·m^{-2} and temperature difference drops about 10 K.

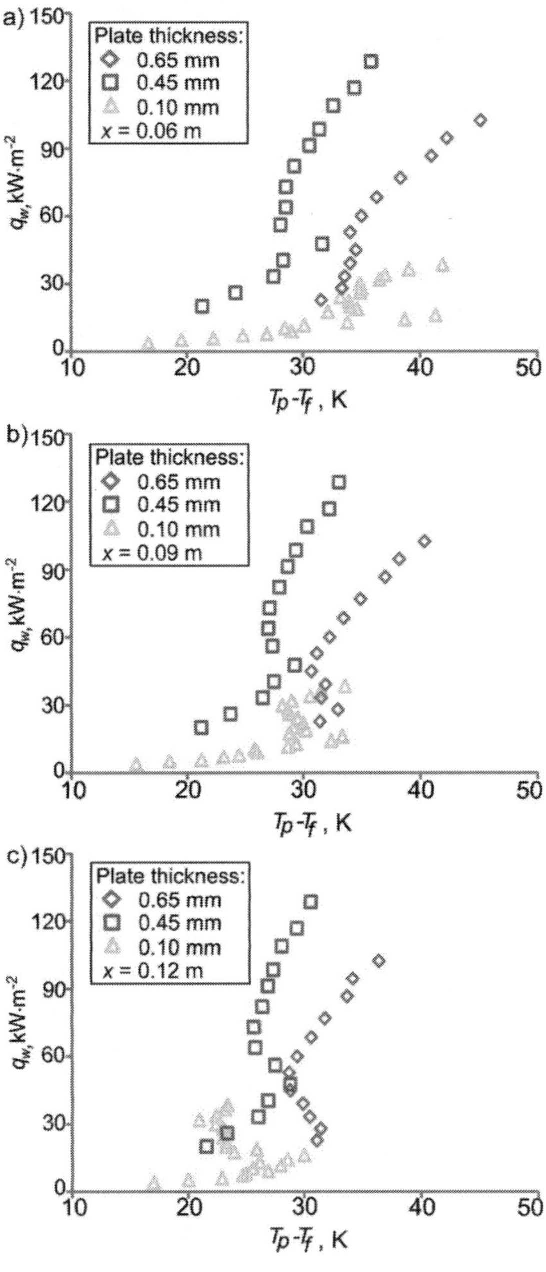

Fig. 8. Boiling curves generated for the three distances from the minichannel inlet: 0.06 m (a), 0.09 m (b) and 0.12 m (c) constructed as: q_w vs. T_p-T_f, T_p - heated plate temperature, T_f - bulk fluid temperature.

6 Conclusions

This paper presents a comparison of the performance of three smooth heated surfaces with different thicknesses. Analysis was carried out on an experimental stand for flow boiling heat transfer. The test section with a rectangular minichannel, 1.7 mm deep, 16 mm wide and 180 mm long, oriented vertically. The heated element for the working fluid (FC-72) was designated as a Haynes-230 alloy plate: 0.10 mm and 0.45 mm thick or Hastelloy X alloy plate about 0.65 mm thick. Infrared thermography was used to measure the temperature of the outer plate surface.

The local values of the heat transfer coefficient were calculated using one-dimensional method in the subcooled boiling region at two heat fluxes supplied to the plates. The experimental results were presented as the relationship between the heat transfer coefficient vs the distance along the minichannel length and boiling curves.

In the tested boiling region was observed that the heat transfer coefficients were relatively low in comparison to those at the saturated boiling region. The highest local heat transfer coefficients were recorded for the surface of 0.10 mm thick heated plate at the outlet and 0.45 mm thick plate at the minichannel inlet.

The boiling curves were plotted as the heat flux density against the temperature difference T_p-T_f for three selected distances from the minichannel inlet: 0.06 m, 0.09 m and 0.12 m. The highest temperature drops characteristic of boiling initiation were observed in boiling curves plotted for 0.10 mm and 0.65 mm thick plates.

Acknowledgments

The research reported herein was supported by a grant from the National Science Centre, Poland (No. DEC-2016/23/N/ST8/01247).

Nomenclature

A	surface area, m^2
Co	confinement number, -
d_h	hydraulic diameter, m
G	mass flux, kg·m^{-2}s^{-1}
p_{in}	inlet pressure, Pa
q_w	heat flux density, W·m^{-2}
Ra	arithmetic mean deviation of the roughness profile, μm
Sa	arithmetic mean height, μm
T	temperature, K
x	distance from the minichannel inlet, m
y	the percentage value of heat transfer through the heated plate, %
z	the value of heat transfer through the heated plate, W·m^{-2}·K^{-1}

Greek symbols

α	heat transfer coefficient, W·m^{-2}·K^{-1}

ΔT_{sub} inlet liquid subcooling, K
δ thickness, depth, m
λ thermal conductivity, W·m^{-1} K^{-1}

Subscripts
f fluid
in at the inlet
p plate

References

1. R. Ali, B. Palm, M. H. Maqbool, J. Heat Transfer **133**(8), 81501 (2011)
2. M.R. Sohel, R. Saidur, N.H. Hassan, M.M. Elias, S.S. Khaleduzzaman, I.M. Mahbubul, Int. Commun. Heat Mass Transf. **46**, 85–91 (2013)
3. Shah, R. K., *Classification of Heat Exchangers* (Eds., Hemisphere Publishing Washington DC, 9–46, 1981)
4. S. G., Kandlikar, W. J. Grande, Heat Transf. Eng. **24**(1), 3–17 (2003)
5. P. A. Kew, K. Cornwell, Appl. Therm. Eng. **17**(8–10), 705–715 (1997).
6. B. Maciejewska, K. Strąk, M. Piasecka, Procedia Eng. **157**, 82–88 (2016)
7. M. Piasecka, K. Strąk, B.Grabas, Arch. Metall. Mater. **62**(4), 1983–1990 (2017)
8. M. Piasecka, Ann. Nucl. Energy **73**, 282–293 (2014)
9. M. Piasecka, Int. J. Refrig. **56**, 198-212 (2015)
10. M. Piasecka, Int. J. Heat Mass Transf. **81**, 114–121 (2014)
11. M. Piasecka, B. Maciejewska, Exp. Therm. Fluid Sci. **68**, 459–467 (2015)
12. K. Strąk, M. Piasecka, B. Maciejewska, Int. J. Heat Mass Transf. **117**, 375–387 (2017)
13. W. Depczyński, J. Achiev. Mater. Manuf. Eng. **66**(2), 67–72 (2014)
14. B. Grabas, Exp. Therm. Fluid Sci. **68**, 499–508 (2015)
15. N. Radek, J. Konstanty, M. Scendo, Arch. Metall. Mater. **60**, 2579–2584 (2015)
16. R. Pastuszko, R Kaniowski, EPJ Web Conf. **25**, paper No. 02019 (2012)
17. R. Pastuszko, R Kaniowski, J. Phys. Conf. Ser. **745**(3), paper No. 32137 (2016).
18. R. Pastuszko, EPJ Web Conf. **45**, paper No. 01020 (2013)
19. R. Pastuszko, M. Piasecka, J. Phys. Conf. Ser. **395**, paper No. 12137 (2012)
20. J. T. Cieśliński, Rev. Chem. Eng. **27**(3–4), 179–190 (2011)
21. S. Błasiak, A. Zahorulko, Tribol. Int. **94**, 126–137 (2016)
22. S. Błasiak, Int. J. Heat Mass Transf. **100**, 78–88 (2016)
23. S. Błasiak, 2015, Int. J. Heat Mass Transf. **81**, 90–102 (2015)
24. M. Piasecka, K. Strąk, B. Maciejewska, Heat Transf. Eng. **38**(3), 332–346 (2017)
25. Calibration Certificate No. K1501035, Calibration Laboratory No. 2372, Accredited by Czech Accreditation Institute under ČSN EN ISO/IEC 17025:2005
26. www.haynesintl.com/pdf/h3009.pdf
27. B. Maciejewska, M. Piasecka, Int. J. Heat Mass Transf. **107**, 925–933 (2017)
28. S. Hozejowska, M. Piasecka, Heat Mass Transf. **50**(8), 1053–1063 (2014)
29. B. Maciejewska, M. Piasecka, Heat Mass Transf. **53**(4), 1211–1224 (2017)
30. S. Hożejowska, R. Kaniowski, M. E. Poniewski, Exp. Therm. Fluid Sci. **78**, 18–29 (2016)
31. M. Piasecka, K. Strąk, Trans. Inst. Fuid-Flow Mach. **128**, 97–118 (2015)

 https://doi.org/10.1051/epjconf/201818002099

The heat transfer coefficient determination with the use of the Beck-Trefftz method in flow boiling in a minichannel

Kinga Strąk[1], *Beata* Maciejewska[2*], *Magdalena* Piasecka[3]

[1,3] Faculty of Mechatronics and Mechanical Engineering
[2] Faculty of Management and Computer Modelling
Kielce University of Technology, Al. 1000-lecia P.P. 7, 25-314 Kielce, Poland

Abstract. In this paper, the solution of the two-dimensional inverse heat transfer problem with the use of the Beck method coupled with the Trefftz method is proposed. This method was applied for solving an inverse heat conduction problem. The aim of the calculation was to determine the boiling heat transfer coefficient on the basis of temperature measurements taken by infrared thermography. The experimental data of flow boiling heat transfer in a single vertical minichannel of 1.7 mm depth, heated asymmetrically, were used in calculations. The heating element for two refrigerants (FC-72 and HFE-7100, 3M) flowing in the minichannel was the plate enhanced on the side contacting with the fluid. The analysis of the results was performed on the basis of experimental series obtained for the same heat flux and two different mass flow velocities. The results were presented as infrared thermographs, heated wall temperature and heat transfer coefficient as a function of the distance from the minichannel inlet. The results was discussed for the subcooled and saturated boiling regions separately.

1 Introduction

Boiling is an effective heat transfer process, which due to the phase change provides high heat transfer capacity. Flow boiling heat transfer is used in many applications utilizing the heat-removal processes. Variations in the process intensity may be, however, a limitation to the efficiency of boiling heat transfer, characterized by heat transfer coefficient values and can be estimated due to the heat transfer coefficient values.

Inverse heat conduction problems [1, 2], such as heat transfer coefficient identification, are solved with a number of numerical methods. One of these methods is the Beck method [3 ,4], which introduces sensitivity coefficients as derivatives of the measured quantity with respect to the identified quantity and transforms an inverse problem into several direct problems. In this paper, the Trefftz method [5] was used to solve direct problems by approximating the unknown solution to a differential equation with a linear combination of functions strictly satisfying the governing differential equation. The Trefftz method was explored in more detail in [6-14].

The presented method is used to solve the heat transfer problem in the flow in channels. Similar issues are discussed in articles [15-22].

2 Experimental setup and methodology

The major elements of the experimental stand are presented in Fig. 1a. The essential part of the stand is the test section (Fig. 1b) with a single rectangular minichannel, 1.7 mm in depth, 16 mm in width and 180 mm in length. The heated plate (12) made of Haynes-230 alloy, 0.45 mm thick, was enhanced by vibration-assisted laser surface on the side contacting refrigerant.

It was possible to measure temperature on the outer side of the plate due to infrared camera (Fig. 1, 2). The E60 FLIR infrared camera (2) had an accuracy of ± 1 °C or ± 1% in the temperature range 0 ÷ 120 °C [23]. This smooth plate surface was coated with a black paint of known emissivity (of 0.83) [24]. The enhanced surface of the heated plate was observed through a glass panel (14). The flow patterns were recorded using a quick shot camera (3). K-type thermocouples (17) and pressure sensors measured temperature and pressure at the inlet and outlet of the minichannel, respectively.

During experiment, there was a laminar flow of refrigerant along a minichannel and a gradual increase in the electric power supplied to the plate followed by an increase in the heat flux transferred to the fluid. Experimental parameters for several settings of the heat flux supplied to the heated plate during subcooled and saturated boiling region were collected. The experimental parameters are presented in Table 1 and physical properties of refrigerants in Table 2.

The results were analysed for:

- two refrigerants, FC-72 and HFE-7100 (3M), used as working fluids in the flow loop,
- two mass flow velocities: $Q_m = 12\cdot10^{-3}$ kg·s^{-1} and $Q_m = 18\cdot10^{-3}$ kg·s^{-1},

* Corresponding author:beatam@tu.kielce.pl

- four heat fluxes: 40 kW·m^{-2}, 52 kW·m^{-2}, 59 kW·m^{-2} and 68 kW·m^{-2} set at the subcooled boiling region,
- two heat fluxes: 120 kW·m^{-2} and 125 kW·m^{-2}, set at the saturated boiling region.

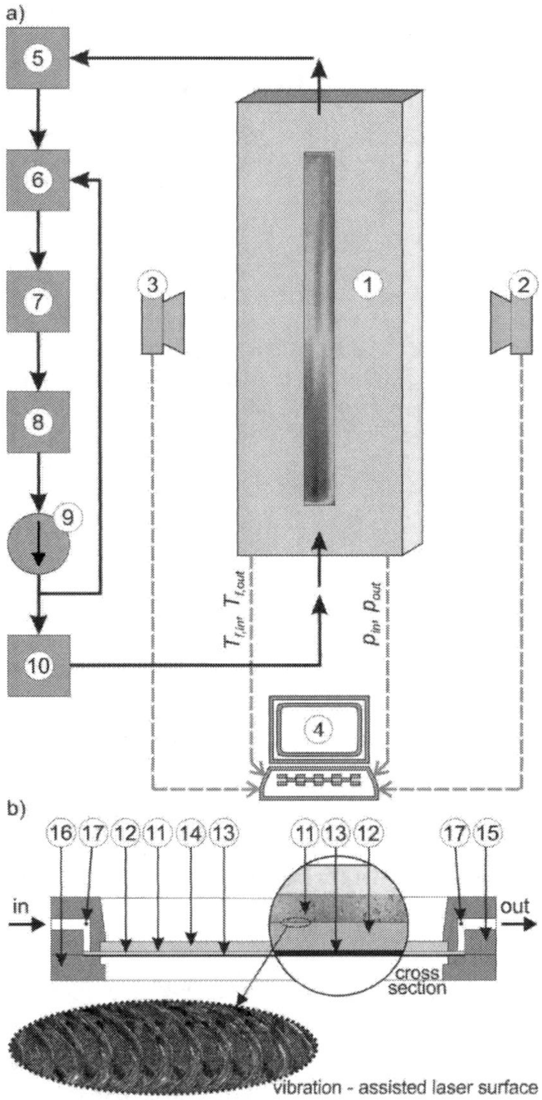

Fig. 1. a) The schematic diagram of the main elements of the experimental stand: 1-test section, 2-infrared camera, 3-quick shot camera, 4-data acquisition station with pc computer, 5-tube-type heat exchanger, 6-compensating tank/pressure regulator, 7-deaerator, 8-filter, 9- gear pump, 10-mass flowmeter; b) the schematic diagram of the test section: 11- minichannel, 12-heated plate, 13-black paint layer, 14-glass panel, 15-channel body, 16-front cover, 17-thermocouple.

3 Heat transfer coefficient determination

3.1 Main assumptions

It was assumed that the heat flow in the minichannel was stationary and two-dimensional. The variation in temperature along the channel width was neglected.

Local values of the heat transfer coefficient were calculated from Newton's law:

$$\alpha(x) = \frac{q(x)}{T(x,\delta) - T_f(x)} \tag{1}$$

where: x – the flow direction, y – the direction perpendicular to the flow direction, related to the thickness of the plate, q – density of the heat flux transferred from the plate to the fluid, T – plate temperature, δ – thickness of the plate, T_f – temperature of the fluid, in the subcooled boiling region it equals to $T_l(x)$ and in the saturated nucleate boiling region it equals to $T_{sat}(x)$, $T_l(x)$ – liquid temperature calculated from the assumption of the linear distribution of the liquid temperature along the channel and T_{sat} – saturation temperature determined on the basis of the linear distribution of the fluid pressure along the channel. The heat flux density q was calculated from Fourier's law.

Table 1 Experimental parameters of experimental series for the subcooled and saturated boiling regions.

Set of heat flux (No.)	Range of mass flux G (kg m^{-2}s^{-1})	Range of inlet pressure p_{in} (kPa)	Range of inlet liquid subcooling ΔT_{sub} (K)	Heat flux density q_w (kWm^{-2})
Subcooled boiling region				
1	423-650	111-210	43-70	40
2				52
3				59
4				68
Set of heat flux (No.)	Range of mass flux G (kg m^{-2}s^{-1})	Range of inlet pressure p_{in} (kPa)	Range of inlet liquid subcooling ΔT_{sub} (K)	Heat flux density q_w (kWm^{-2})
Saturated boiling region				
1	440-638	147-229	44-66	120
2				125

Table 2 The basic thermal and flow properties of and Novec™ HFE-7100 and Fluorinert FC-72 [25].

Properties		Fluids	
	Unit	**HFE-7100**	**FC-72**
Boiling point	K	334	329
Molecular weight	kg·mol^{-1}	0.250	0.338
Vapor pressure	Pa	27·10^{3}	30·10^{3}
Heat of vaporization	J·kg^{-1}	112·10^{3}	88·10^{3}
Liquid density	kg·m^{-3}	1510	1680
Coefficient of expansion	K^{-1}	0.0018	0.0016
Kinematic viscosity	m^{2}·s^{-1}	0.38·10^{3}	0.38·10^{3}
Specific heat	J·kg^{-1}·K^{-1}	1183	1100
Thermal conductivity	W·m^{-1}·K^{-1}	0.069	0.057
Surface tension	N·m^{-1}	0.0136	0.01

3.2 Methodology of calculations, boundary conditions

The plate temperature T was determined by solving the inverse heat conduction problem in the heated plate:

$$\frac{\partial^2 T}{\partial x^2}+\frac{\partial^2 T}{\partial y^2}=-\frac{I\,\Delta U}{A\,\delta\,\lambda}=-\frac{q_w}{\delta\,\lambda}\ \text{ for } (x,y)\in\Omega \qquad (2)$$

where $\Omega=\{(x,y)\in R^2 : x_1<x<x_P,\ \ 0<y<\delta\}$,
and the boundary conditions (see Fig. 2):

$$T(x_1,y)=T_1 \text{ for } 0<y<\delta \qquad (3)$$

$$T(x_P,y)=T_P \text{ for } 0<y<\delta \qquad (4)$$

$$\frac{\partial T}{\partial y}(0,y)=0 \text{ for } x_1<x<x_P \qquad (5)$$

$$T(x_p,0)=T_p \text{ for } p=1,2,\dots P \qquad (6)$$

where: I – current supplied to the heated plate, ΔU – drop in voltage along the length of the plate, A – surface area of the heated plate in contact with the fluid, δ – thickness of the heated plate, q_w – heat flux density, x_1 – location of the first measurement point at the boundary $y=0$, x_P – location of the last measurement point, P – number of measurements, T_p – measured temperature of the heated plate, λ – thermal conductivity of the heated plate.

The inverse problem (Eqs. 2-6) was solved using the Beck method.

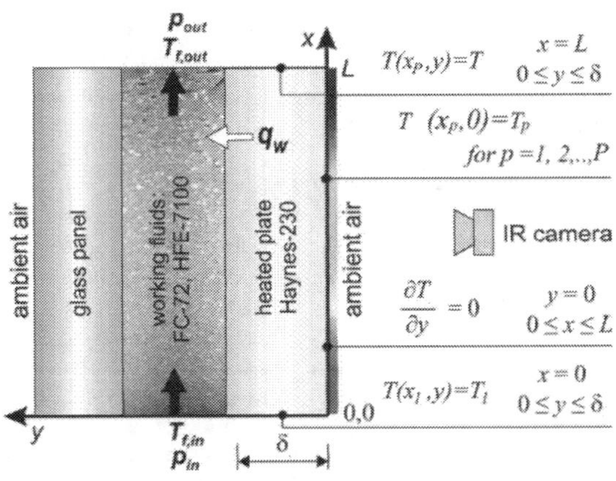

Fig. 2. Boundary conditions.

3.3 The Beck method

The plate temperature T dependent on the boundary temperature T_m for $m=1,2,\dots,K$ at the boundary $y=\delta$ for $x_1\le x\le x_P$, like in [4], is expanded into a Taylor series around fixed point $(T_{01},\dots,T_{0K})$. Since the problem being considered is a linear problem, higher order derivatives disappear and the Taylor series has the form:

$$T(x,y,T_1,\dots,T_K)=\Theta(x,y)+\sum_{m=1}^{K}Z_m(x,y)(T_m-T_{0m}) \qquad (7)$$

where $\Theta(x,y)=T(x,y,T_{01},\dots,T_{0K})$ and $Z_m(x,y)$, for $m=1,2,\dots,K$ are the so-called sensitivity coefficients defined according to the formula:

$$Z_m(x,y)=\left.\frac{\partial T}{\partial T_m}\right|_{T_m=T_{0m}} \qquad (8)$$

Substituting Eq. (7) into Eq. (2) leads to 1+K direct problems which were solved using the Trefftz method [8].

The values of T_m for $m=1,2,\dots K$ in expression Eq. (7) were calculated by minimizing the following functional:

$$J_P=\sum_{p=1}^{P}\left(T(x_p,y_p,T_1,\dots,T_K)-T_p\right)^2 \qquad (9)$$

4 Results and discussion

The study focused on heat transfer coefficient identification in the regions of subcooled and saturated boiling. The coefficient values were determined by solving the inverse heat conduction problem using the Beck method in combination with the Trefftz method.

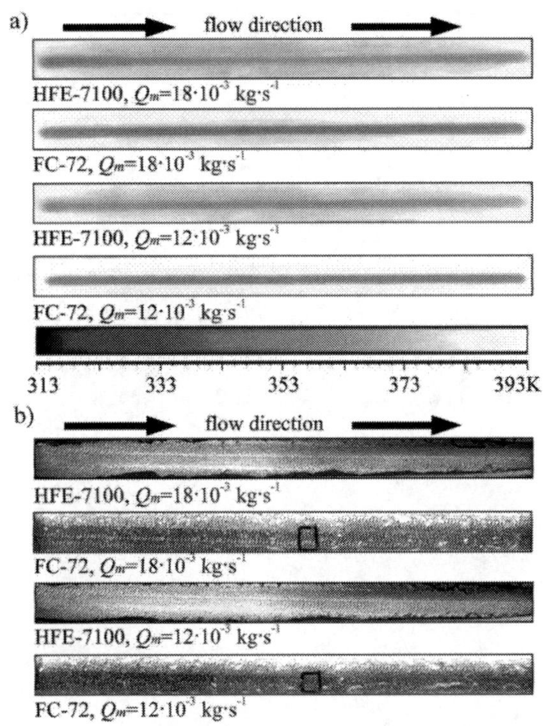

Fig. 3. a) Infrared thermograms, b) two-phase flow sctuctures images, for q_w = 120 kW·m^{-2}, two mass flow velocities, working fluids: HFE-7100 and FC-72, the saturated boiling region, experimental parameters shown in Table 1.

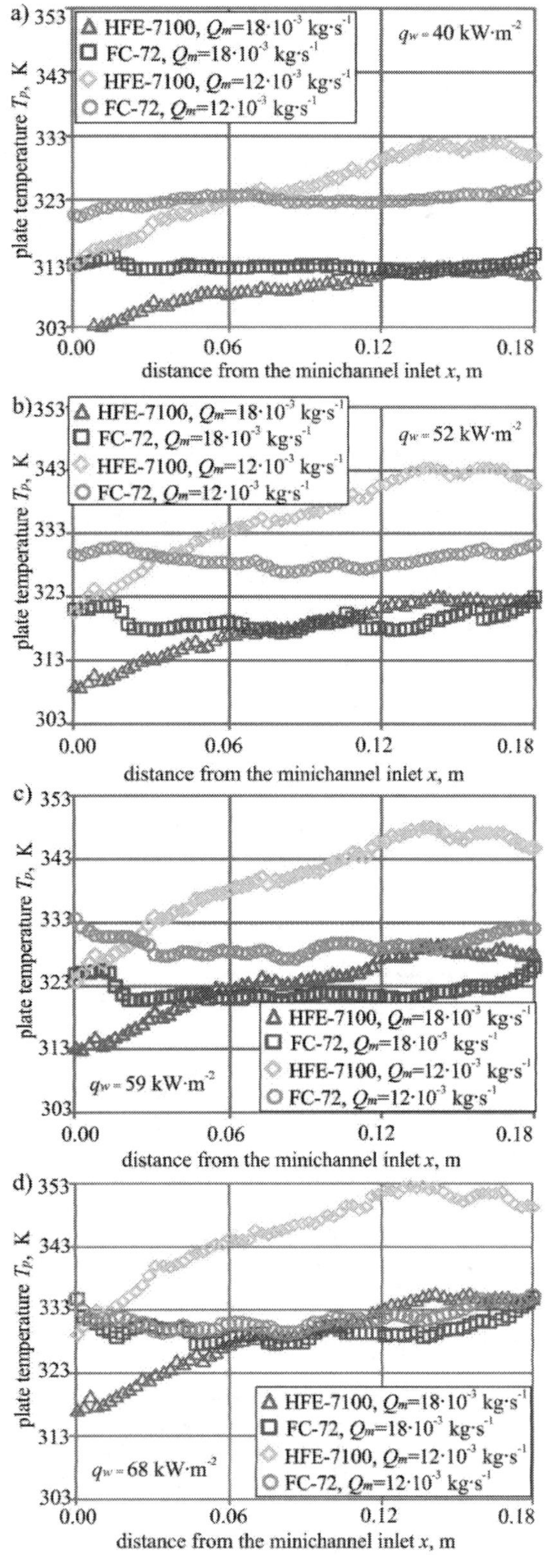

Fig. 4. Plate temperature vs. distance from the minichannel length obtained for the subcooled boiling region, experimental parameters shown in Table 1.

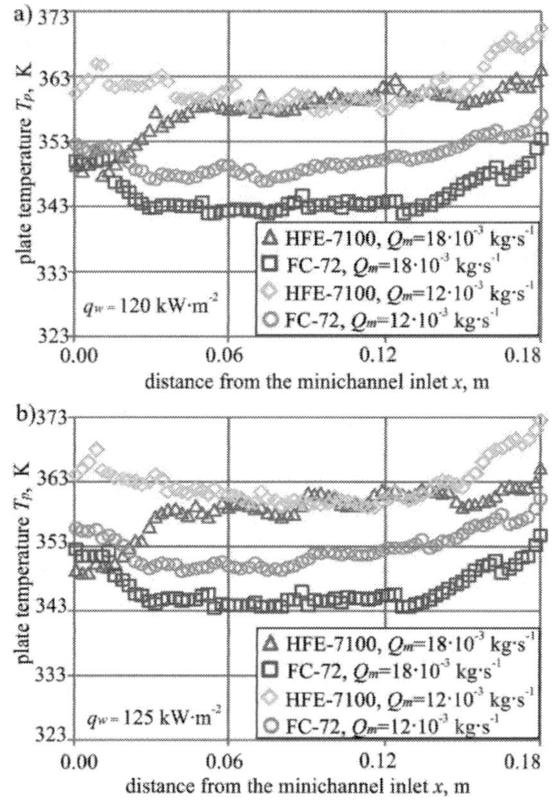

Fig. 5. Plate temperature vs. distance from the minichannel length obtained for the saturated boiling region, experimental parameters shown in Table 1.

The results are presented graphically as :

- infrared thermograms (Fig. 3a) and two-phase flow structures images (Fig. 3b), for q_w = 120 kW·m^{-2}, two mass flow velocities, working fluids: HFE-7100 and FC-72, the saturated boiling region;
- the temperature of the outer heated plate surface determined through infrared camera measurements, for all anayzed settings of the heat flux versus the distance from the minichannel inlet, obtained for the subcooled boiling region (Fig. 4a-d) and the saturated boiling region (Fig. 5a,b);
- the heat transfer coefficient for all anayzed settings of the heat flux versus the distance from the minichannel inlet obtained for the subcooled boiling region (Fig. 6a-d) and the saturated boiling region, Fig.(7a,b).

The highest temperatures of the heated plate surface were recorded for HFE-7100 at mass flow velocity of Q_m = 12·10^{-3} kg·s^{-1} in the subcooled and saturated boiling regions (except for the inlet part of the channel at subcooled boiling region). Lower temperatures were recorded for FC-72 at mass flow velocity of Q_m = 18·10^{-3} kg·s^{-1} (except for the inlet part of the channel).

For HFE-7100, at the subcooled boiling region, the temperature rose uniformly with increasing distance from the minichannel inlet, whereas for FC-72, the temperature changes were minor.

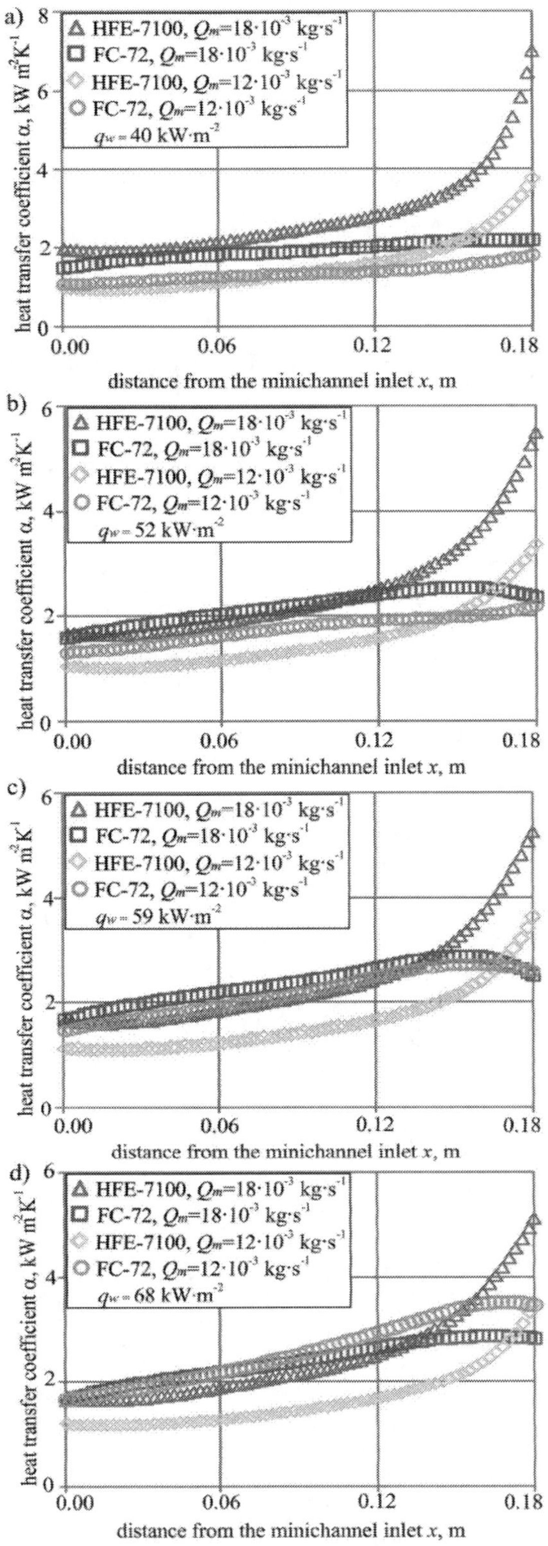

Fig. 6. Heat transfer coefficient vs. distance from the minichannel length obtained for the subcooled boiling region, experimental parameters shown in Table 1.

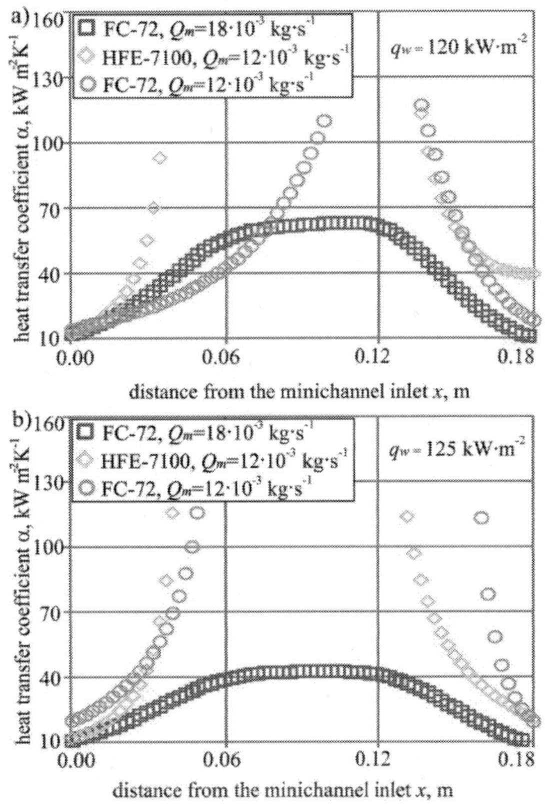

Fig. 7. Heat transfer coefficient vs. distance from the minichannel length obtained for the saturated boiling region, experimental parameters shown in Table 1.

The heated plate temperatures recorded for HFE-7100 at the saturated boiling region were higher (irrespective of the mass flow velocity) than those provided by FC-72. At this region, no substantial plate temperature increase was observed.

In the subcooled boiling region, heat transfer coefficients were in the range 1.5 to 7.0 $kW{\cdot}m^{-2}{\cdot}K^{-1}$, that is, much lower than those obtained for the saturated boiling region. The lowest coefficients for both liquids were at the inlet to the minichannel, and the highest - at the outlet. The heat transfer coefficient calculated when HFE-7100 was used, started rising with the increasing distance from the minichannel inlet, reaching the highest values at a 2/3 length of the channel from the inlet. Heat transfer coefficients obtained with HFE-7100 at mass flow velocity of $Q_m = 18{\cdot}10^{-3}$ $kg{\cdot}s^{-1}$ at all the heat fluxes were recorded along the same outlet section of the minichannel. The lowest values of the heat transfer coefficient when this liquid were recorded at $Q_m = 12{\cdot}10^{-3}$ $kg{\cdot}s^{-1}$ along the whole length of the channel, except for the section near the minichannel outlet. For FC-72, the coefficients had similar values in the range of 1.5-3.5 $kW{\cdot}m^{-2}{\cdot}K^{-1}$ at both mass flow velocities. The graph of the function showed a slight coefficient drop at the very outlet of the channel. At higher heat fluxes (59 $kW{\cdot}m^{-2}$ and 68 $kW{\cdot}m^{-2}$), FC-72 provided higher heat transfer coefficients for both mass flow velocities compared to HFE-7100, except for the near-outlet section of the minichannel.

In the saturated boiling region, the heat transfer coefficient reached substantially higher values compared to the subcooled boiling region (similarly as observed in [26-28]) with the highest being from 10 kW·m^{-2} K^{-1} to 120 kW·m^{-2} K^{-1}. No data were collected for HFE-7100 for Q_m = 18·10^{-3} kg·s^{-1} at heat fluxes of q_w = 120 kW·m^{-2} and q_w = 125 kW·m^{-2} for both fluids for Q_m = 18·10^{-3} kg·s^{-1} in the central part of the minichannel (Fig. 7a-b). It was assumed that the measurement error, T_p-T_{sat}, was about 0.5 K, and the values T_p-T_{sat} < 1 K were not considered in the calculations. At the saturated boiling of FC-72 and at higher mass flow velocity, the heat transfer coefficient reached the lowest values of all tested liquid and mass flow velocities up to 65 kW·m^{-2}·K^{-1}. At lower mass flow velocity for both liquids, the coefficient had much higher values (2-3 times higher) up to about 120 kW·m^{-2}·K^{-1}. At the given mass flow velocity, heat transfer coefficients provided by HFE-7100 near the minichannel inlet were higher compared to FC-72.

5 Conclusion

In this work, the solution of the two-dimensional inverse heat transfer problem with the use of the Beck method coupled with the Trefftz method was proposed. The experimental data of flow boiling heat transfer in a single vertical minichannel of 1.7 mm depth, heated asymmetrically, were used in calculations. The heating element for two refrigerants (FC-72 and HFE-7100) flowing in the minichannel was the plate enhanced on the side contacting with the fluid. The results were presented as infrared thermographs, heated wall temperature and heat transfer coefficient as a function of the distance from the channel inlet.

The highest temperatures of the heated plate surface were recorded for HFE-7100 at mass flow velocity of Q_m = 12·10^{-3} kg·s^{-1} in the subcooled and saturated boiling regions. Lower temperatures were recorded for FC-72 at mass flow velocity of Q_m = 18·10^{-3} kg·s^{-1}.

In the subcooled boiling region, local heat transfer coefficients were in the range 1.5 to 7.0 kW·m^{-2} K^{-1}. The lowest values of the coefficients were obtained with HFE-7100 at mass flow velocity of Q_m = 18·10^{-3} kg·s^{-1} at all the heat fluxes were recorded along the same outlet section of the minichannel.

In the saturated boiling region, the heat transfer coefficient reached substantially higher values compared to the subcooled boiling region, with the highest being from 10 to 120 kW·m^{-2} K^{-1}. For FC-72 and at higher mass flow velocity, the heat transfer coefficient reached the lowest values of all tested refrigerants and mass flow velocities.

Acknowledgments

The research reported herein was supported by a grant from the National Science Centre, Poland (No. DEC-2016/23/N/ST8/01247).

References

1. A. Tikhonov, V. Arsenin, *Solution of ill-posed problems* (Wiley, New York, 1977)
2. M. N. Özisįk., H. R. B. Orlande, *Inverse heat transfer: fundamentals and applications. Fundamentals and Applications* (Taylor & Francis, New York, 2000)
3. J. V. Beck, B. Blackwell, C. R. St. Clair, *Inverse heat conduction. Ill-posed Problems* (Wiley - Interscience Publ., New York, 1985)
4. B. Kruk, M Sokała, J. Appl. Math. Mech. **3**, 693–694 (1999)
5. E. Trefftz, Proc. Int. Kongress für Technische Mechanik, Zürich, 131-137 (1926)
6. I. Herrera, Meth. Partial Diff. Eqs. **16**, 561-580 (2000)
7. V. Kompis, F. Konkol, M. Vasko, Comp. Assis. Mech. Eng. Sci. **8**, 385-395 (2001)
8. M. J. Ciałkowski, J. Therm. Sci. **11**, 148–162 (2002)
9. K. Grysa, A. Maciąg, A. Pawińska, Int. J. Heat Mass Transf. **55**, 7336–7340 (2012)
10. K. Grysa, B. Maciejewska, J. Theor. Appl. Mech. **51**, 251-264 (2013)
11. S. Hożejowska, R. Kaniowski, M.E. Poniewski, Exp. Therm. Fluid Sci. **78**, 18–29 (2016)
12. B. Maciejewska, M. Piasecka, Heat Mass Transf. **53**, 1211-1224 (2017)
13. S. Blasiak, A. Pawinska, Int. J. Heat Mass Transf. **90**, 710-718 (2015)
14. B. Maciejewska, M. Piasecka, Int. J. Heat Mass Transf. **107**, 925-933 (2016)
15. J. R. Thome, V. Dupont, A. M. Jacobi, Int. J. Heat Mass Transf. **47**, 3375-3385 (2004)
16. V. Dupont J. R. Thome, A. M. Jacobi, Int. J. Heat Mass Transf. **47**, 3387-3401 (2004)
17. P. Duda, J. Taler, Int. J. Heat Mass Transf. **52**, 1517-1524, (2009)
18. P. Ocłoń, S. Łopata, M. Nowak, Heat Mass Transf. **51**, 553-566 (2015)
19. D. Mikielewicz, J. Wajs, M. Gliński, A.B.R.S Zrooga., Exp. Therm. Fluid Sci. **44**, 556–564 (2013)
20. R.T. Lahey, D.A. Drew, Nuclear Eng. Design **204**, 29-44 (2001)
21. M. Jaszczur, M Zych, R. Hanus, EPJ Web Conf. **143**, paper No. 02046 (2017)
22. W. Kuczyński, H. Charun, T. Bohdal, Int. J. Heat Mass Transf. **111**, 83-93 (2017)
23. Calibration certificate No. K1501035, Calibration laboratory No. 2372, accredited by Czech accreditation Institute under ČSN EN ISO/IEC 17025:2005
24. M. Piasecka, K. Strąk, B. Maciejewska, Heat Transf. Eng. **38**, 332–346 (2017)
25. www.3M.cm/electronics
26. M. Piasecka, B. Maciejewska, Exp. Therm. Fluid Sci. **68**, 459-467 (2015)
27. M. Piasecka, Int. J. Refrig. **56**, 198-212 (2015)
28. M. Piasecka, Int. J. Heat Mass Transf. **81**, 114-121 (2015)

Numerical Study of Shaft-Seal Parameters for Various Geometry Configurations and Operation Regimes

Petr Straka[1,*]

[1]VZLU – Czech Aerospace Research Centre, Beranových 130, 19905 Prague 9, Czech Republic

Abstract. The contribution deals with numerical modelling of flow through the shaft labyrinth seal for various geometric configurations and operating states. The objective is to obtain dependency of the mass flow rate through the seal and the outlet flow angle from the seal on the pressure ratio and the rotation speed for various seal clearances and other geometrical parameters. The results will be used as a background for modification of the test-rig for axial turbine stage.

1 Introduction

In order to reduce the problem complexity the aerodynamic and energy properties of the axial turbines are usually investigated (theoretically, experimentally, numerically) using simplified models. One of the simplest among these models is the linear (or straight) blade cascade, which represents an unrolled circumferential section of the axial blade wheel (see e.g. [1, 2]). A little more complex is model of one separated blade wheel without any effect of surrounding blade wheels or other flow-parts of the axial blade machines. The model of single axial turbine stage, composed from stator and rotor blade wheels, represents a natural extension the complexity of previous model. The experimental test rig was built for measurement of the aerodynamic and energy properties of the axial turbine stages ([4, 5], fig. 1).

Fig. 1. Turbine test rig.

This test rig was then used for investigation of several axial turbine stages in single-stage configuration (details can be found in [6, 7]).

In case of multistage configuration, the flow field in the axial turbine stage is affected by flow through the foregoing stages. This effect is particularly apparent in peripheral parts of the blade-span as a consequence of the secondary flows. Therefore, the existing test rig has been modified to be able to set the inlet flow parameters (the turbulent intensity and the span-wise velocity profile) in order to simulate the effect of foregoing turbine stages [9]. Subsequently the ability of the test rig has been extended for measuring of so called *one and half stage* configuration. The *one and half stage* represents further complexity extension of previous single stage model. This model consists of one axial turbine stage (stator and rotor wheel) followed with next stator wheel. That configuration allows catching the effect of the foregoing turbine stage on the flow field in subsequent stator blade wheel [10, 11].

One of the significant effects, that influences the flow in peripheral parts of the blade-span, is the outlet stream from the labyrinth-seal (shaft seal at the root part or shroud seal at the tip part). For this reason, the effort is to extend the test rig capability to include the effect of the shaft- and shroud-labyrinth-seal. The construction of mentioned test rig is such that the rotor blades are carried on the rotating disc. In case of the axial turbine stage in the drum-type rotor configuration, the rotating drum-type rotor is replaced by the stationary hub wall (see e.g. [7]), which allows to simplify the design and remove some problems with the rotor dynamics. Similarly, the shaft-seal can be designed as rotary with rotating seal-fins and the bottom wall connected with the rotor disc (fig. 2 left) or as stationary with neglecting of the seal-fins and the bottom wall rotation (fig. 2 right). Design with rotary shaft-seal may lead to problems with the rotor dynamics because of the additional mass of the rotor disc, whereas design with stationary shaft-seal is much simpler for realization. On the other hand the loss of the rotary effect may have a significant impact on the mass flow rate through the seal and on the outlet angle from the seal.

* Corresponding author: straka@vzlu.cz

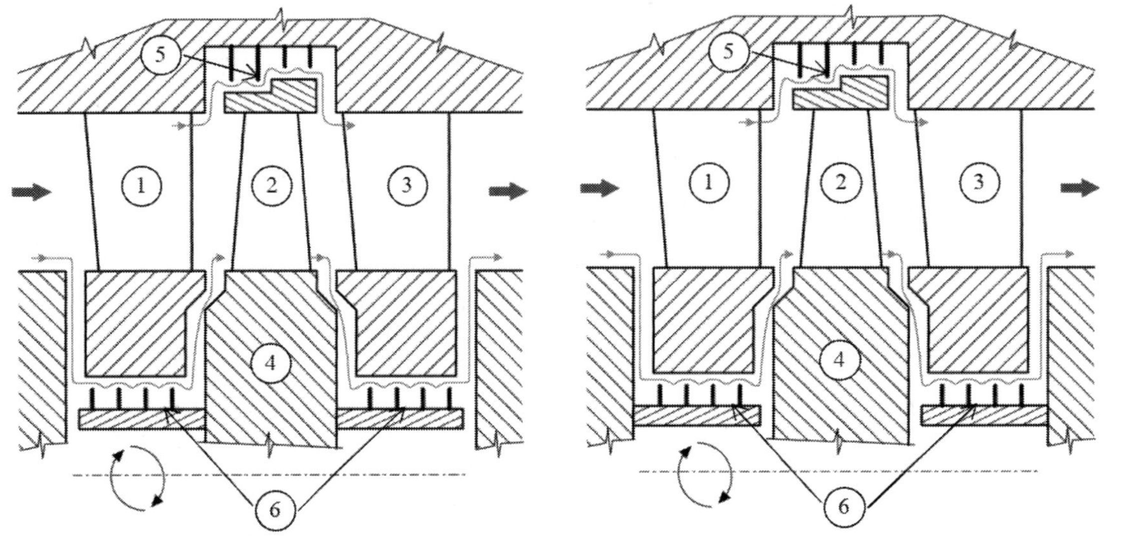

Fig. 2. Scheme of the *one and half stage* with rotary shaft-seal (left) and stationary shaft-seal (right); 1 – first stator blade, 2 – rotor blade, 3 – second stator blade, 4 – rotor disc, 5 – shroud-seal, 6 – shaft-seal.

Numerical simulation of flow through simplified model of the shaft-seal was carried out as a basic background for selection of variant with rotary or stationary shaft-seal. The numerical simulation was performed for various clearances, pressure ratios, rotation speeds and number of seal-fins in configurations with rotary and stationary seal. In fig. 2, the flow-paths through the labyrinth seals are depicted. Let us look on the shaft-seal now. After passing the area of seal-fins the stream flows in radial direction along the rotor disc. Length of this radial part is given by structural limitations of the test rig and is much longer than in real configuration with the drum-type rotor (fig. 3). Therefore, in order to describe the effect of the rotor disc on conditions inside the shaft-seal, the numerical simulation was carried out also for geometrical configuration which is close to the real configuration with the drum-type rotor.

Figure 4 shows scheme of the simplified shaft-seal model with description of the boundaries. The outer shaft diameter is for all configurations $\varnothing d_1$ = 220 mm, diameter of the bottom wall is $\varnothing d_2$ = 177.5 mm for the test rig configuration and $\varnothing d_2$ = 206.5 mm for real configuration, the seal-clearance is c = 0.4, 0.8, 1.2 mm, pitch of the seal-fins is b = 6 mm for the test rig configuration and b = 7.5 mm for real configuration, the height of the seal area is h = 4.5 mm for all configurations, number of seal-fins is 3, 4 and 5 for the test rig configuration, in case of real configuration only 4 seal-fins was studied. The thickness of the seal-fins is 0.3 mm.

2 Numerical results

Flow through the shaft-seal is modelled as steady, compressible, turbulent flow of the ideal gas. The flow field in presented configurations can be supposed as axisymmetric. This means that it could be modelled using the quasi two-dimensional model of the axisymmetric flow in the cylindrical coordinate system. However, it is planned in following work to investigate

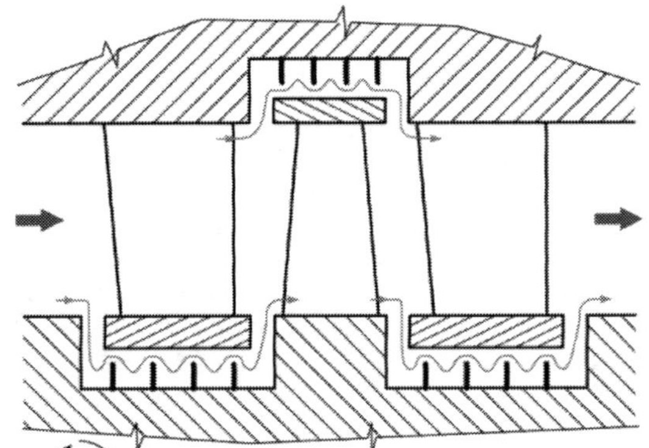

Fig. 3. Scheme of real configuration with the drum-type rotor.

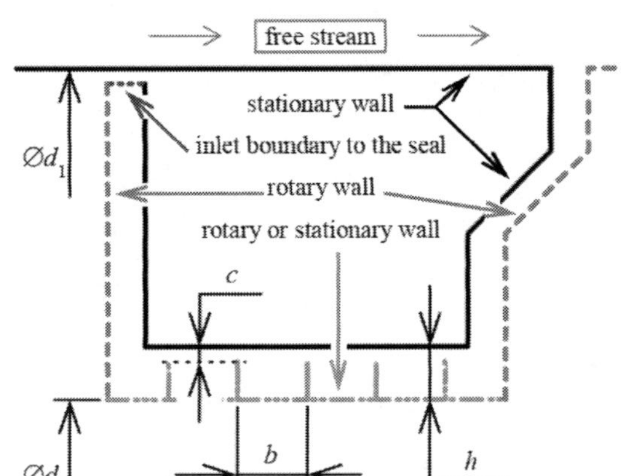

Fig. 4. Scheme of simplified model of the shaft-seal.

general 3D configurations – interaction with the rotor blades, spatial deformations of the seal-fins, etc. Therefore the flow is modelled as 3D in the Cartesian coordinate system from the beginning. In fig. 4 there is shown the computational domain meridian. 3D computational domain takes angular section related to one pitch of the first stator blades.

At the inlet boundary to the shaft-seal there are prescribed the total pressure $p_T = 102$ kPa, the total temperature $T_T = 300$ K, turbulence intensity $Tu = 2.2\%$, ratio of the turbulent to the molecular viscosity $\mu_t / \mu = 100$ and radial flow direction. The total pressure and temperature of the free stream are the same as at the inlet boundary to the shaft-seal. Flow direction of the free stream is axial. The free stream velocity corresponds to the static to total pressure ratio. The computations were carried out for the static to total pressure ratio in range $0.778 \leq p_S / p_T \leq 0.975$, rotation speed of the rotating walls in range $900 \leq \text{RPM} \leq 5100$.

Calculations of flow were performed using the in-house numerical software, which is based on solution of RANS equations closed with two-equations turbulence model (for more details see [12, 13, 14]). The nonlinear explicit algebraic model of the Reynolds stress is used in this work [15].

The computational domain is discretized by multi-block structured computational mesh of hexahedral cells (detail in fig. 5c). The computational mesh is refined close to the walls to ensure that the viscous sub-layer is covered at least by five cells.

Fields of the mass flux density along the meridian plane for the pressure ratio $p_S / p_T = 0.778$ and 900 RPM are shown in fig. 5a, 5b and 5d. The mass flux density along the meridian plane is defined as

$$\dot{m} = \sqrt{(\rho u_{ax})^2 + (\rho u_{rad})^2}\ , \tag{1}$$

where ρ is the density, u_{ax} and u_{rad} are axial and radial velocity vector components respectively. Parts a) and b) of fig. 5 shows configuration of the test rig whereas the real configuration is shown in parts d), e) and f). Figure 5e shows projection of the stream lines to the meridian

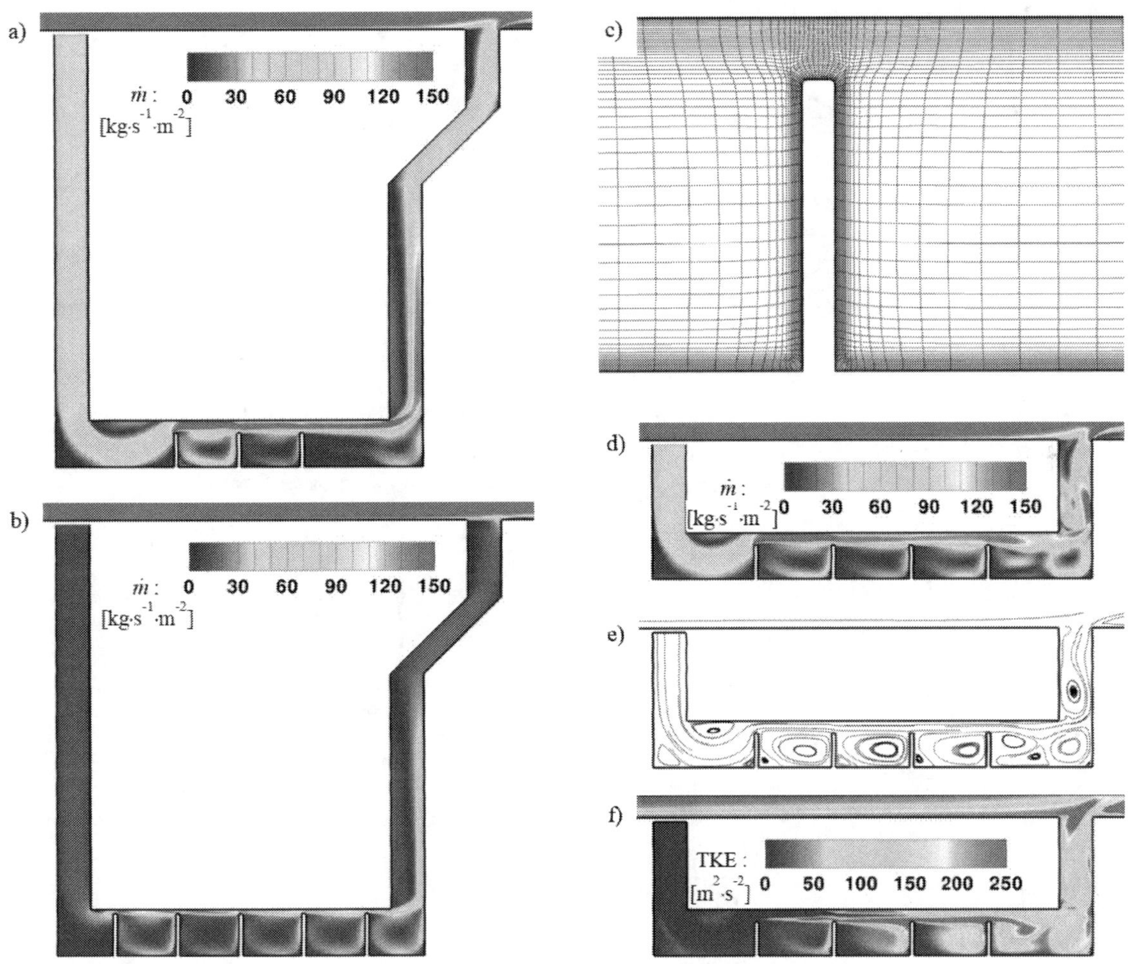

Fig. 5. a) mass flux density, test rig configuration, 3 seal-fins, seal clearance 1.2 mm; b) mass flux density, test rig configuration, 5 seal-fins, seal clearance 0.4 mm; c) detail of the computational mesh around the seal-fin tip; d) mass flux density, real configuration, 4 seal-fins, seal clearance 1.2 mm; e) meridian stream lines projection, real configuration, 4 seal-fins, seal clearance 1.2 mm; f) turbulent kinetic energy, real configuration, 4 seal-fins, seal clearance 1.2 mm. All flow fields: 900 RPM, $p_S / p_T = 0.778$.

plane. Field of the turbulent kinetic energy (TKE) is shown in fig. 5f. A comparison of figs. 5a and 5b illustrates the difference of flow rate through the shaft-seal in configuration with the lowest number of the seal-fins and with the widest seal clearance and in configuration with highest number of the seal-fins and with the narrowest seal clearance. Note that all flow fields presented in fig. 5 corresponds to the case with rotating seal-fins and bottom wall.

The dependency of the mass flux Q through the shaft-seal on various geometric and operating parameters is shown in fig. 6 for several investigated cases. It is clear that the mass flux increases with increasing size of the seal clearance, with the decreasing number of the seal-fins, with the decreasing pressure ratio and with the decreasing speed of rotation. Further it can be found that the dependency of the mass flux on the seal clearance, on the number of the seal-fins and on the rotation speed is linear. The effect of rotation of the seal-fins and the bottom wall is such that the mass flux through the stationary shaft-seal is higher than the mass-flux through the rotary shaft-seal. The difference between the mass flux through the stationary and rotary shaft-seal increases with increasing of the rotation speed.

From chart in fig. 6a it is evident that the mass flux through the stationary shaft-seal for given seal clearance corresponds with the rotary shaft-seal of enlarged seal clearance of δc. It has been found that this imaginary enlargement of the seal clearance in case of the stationary shaft-seal does not exceed 9% in investigated range of the rotation speed and the pressure ratio. It may be noted that the higher mass flux through the shaft-seal in the real configuration compared to the case of the test rig configuration is given by the fact that the seal

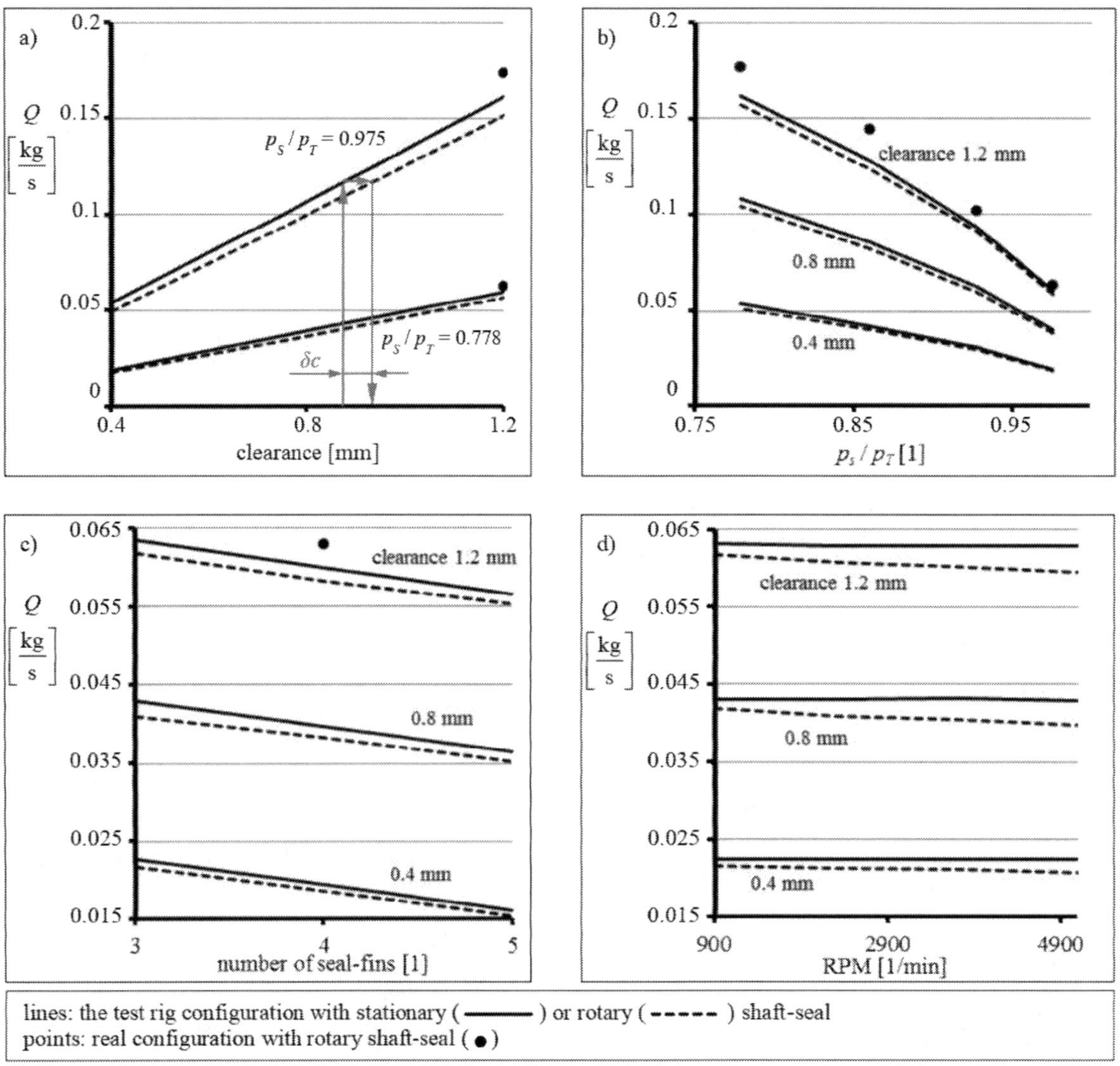

Fig. 6. a) mass flux dependency on the seal clearance for variant of 4 seal-fins, 5100 RPM and two values of the pressure ratio; b) mass flux dependency on the pressure ratio for variant of 3 seal-fins, 900 RPM and three values of the seal clearance; c) mass flux dependency on number of the seal-fins for variant of 900 RPM, $p_S/p_T = 0.975$ and three values of the seal clearance; d) mass flux dependency on the rotation speed for variant of three seal-fins, $p_S/p_T = 0.975$ and three values of the seal clearance.

https://doi.org/10.1051/epjconf/201818002100

clearance is on higher diameter in the real configuration, therefore the through-flow area is larger.

Another observed parameter besides the mass flux is the flow angle ψ at the outlet from the shaft-seal. This angle is defined as deflection from the radial direction as illustrates fig. 7. The dependency of the deflection angle ψ on various geometric and operating parameters is shown in fig. 8 for some chosen variants. As expected, the deflection angle ψ grows with rising the rotation speed, decreases with decreasing of the pressure ratio, with increasing of the seal clearance and with decreasing the number of the seal-fins. In other words, the deflection angle ψ decreases with decreasing of the mass flux through the shaft-seal. This trend is shown in fig. 9 and fig. 10. Note that the points in fig. 9 are drawn for

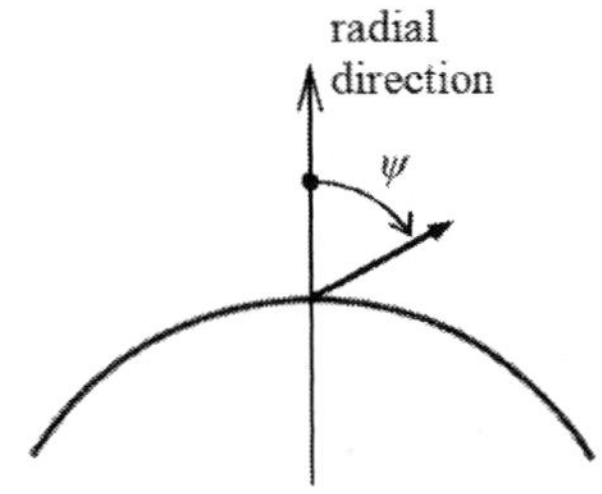

Fig. 7. Deflection angle definition.

various pressure ratios, various seal clearances and various number of the seal-fins. Similarly as in case of the mass flux also the deflection angle ψ from the stationary shaft-seal corresponds with the deflection

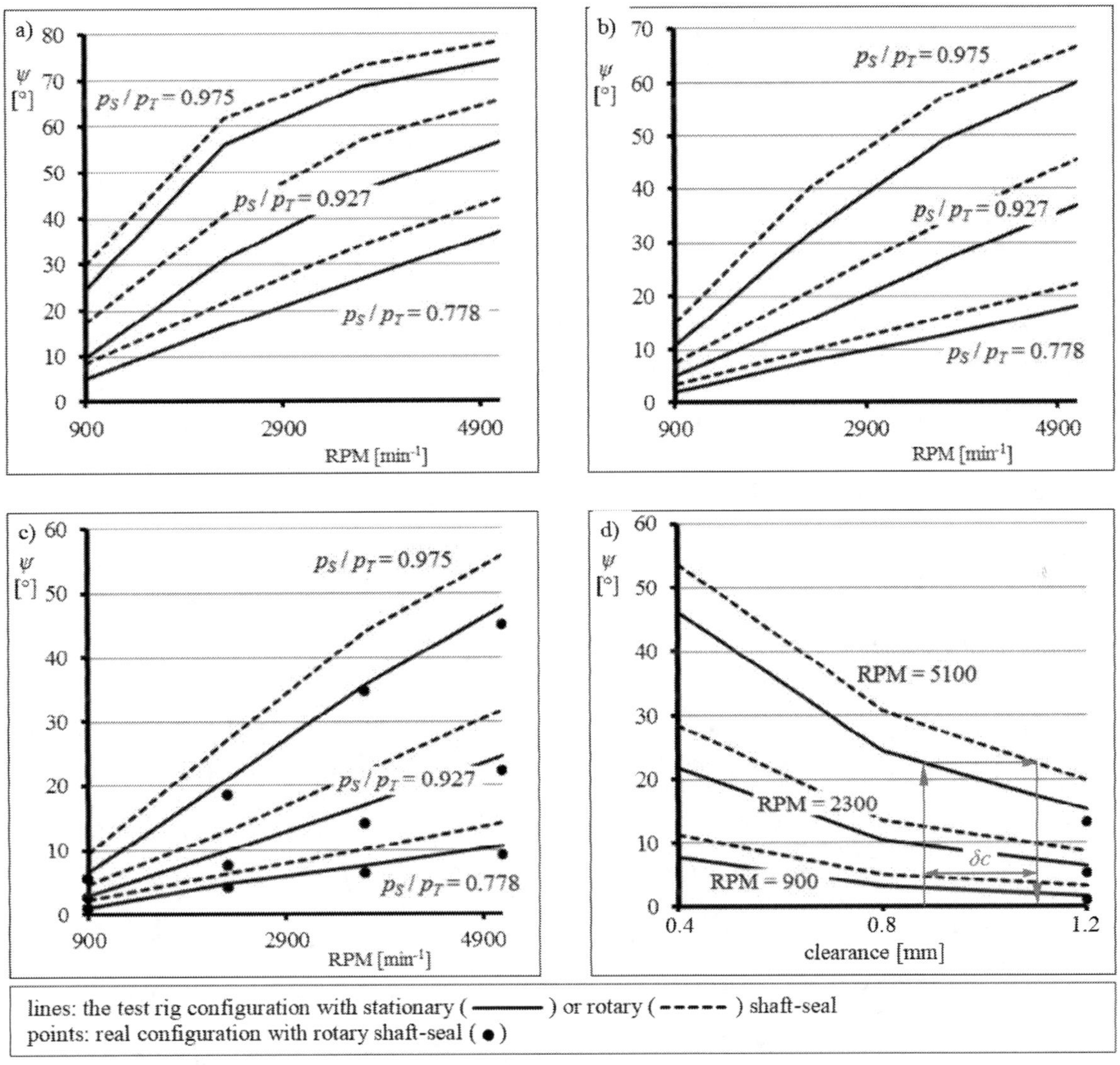

Fig. 8. a) deflection angle dependency on the rotation speed for variant of 4 seal-fins, seal clearance 0.4 mm and three values of the pressure ratio; b) deflection angle dependency on the rotation speed for variant of 4 seal-fins, seal clearance 0.8 mm and three values of the pressure ratio; c) deflection angle dependency on the rotation speed for variant of 4 seal-fins, seal clearance 1.2 mm and three values of the pressure ratio; d) deflection angle dependency on the seal clearance for variant of 4 seal-fins, $p_S / p_T = 0.860$ and three values of the rotation speed.

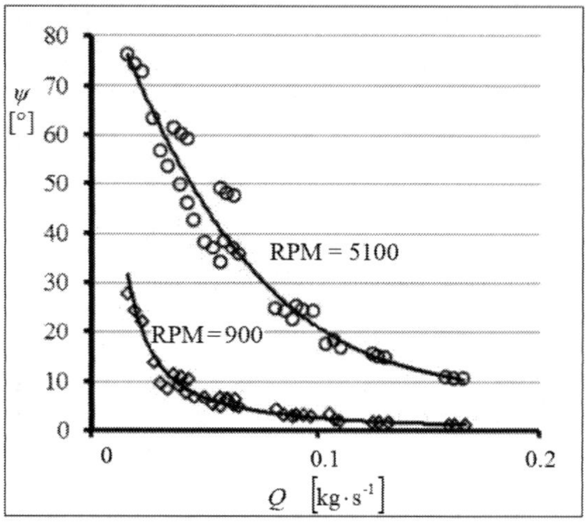

Fig. 9. Dependency of the deflection angle on the mass flux – test rig configuration, stationary shaft-seal, two rotation speed (points are for various pressure ratios, various seal clearance and various number of the seal-fins).

angle from the rotary shaft-seal with enlarged seal clearance about δc as shown in fig. 8d. Furthermore, it can be stated that the deflection angle from the rotary shaft-seal is significantly larger (up to ten degrees) compared to the stationary shaft-seal, which is an effect of the rotating rotor disc. The deflection angle from the stationary shaft-seal is much closer to the case of the real configuration as shown in figures 8c and 8d.

3 Conclusions

The results of numerical calculation of flow through the shaft-seal in the test rig configuration show that:

- the mass flux through the rotary shaft-seal is lower than the mass flux through the stationary shaft-seal at the same geometric and operating conditions,
- the deflection angle at the outlet from the rotary shaft-seal is larger than in case of the stationary shaft-seal at the same geometric and operating conditions,
- the flow properties of the stationary shaft-seal correspond to the rotary shaft-seal with enlarged seal clearance at the same operating conditions,
- the flow properties of the stationary shaft-seal correspond better to properties of the real configuration compared to properties of the rotary shaft-seal.

Based on obtained results it will be possible to propose the stationary shaft-seal with enlarged seal clearance to preserve the same mass flux as in case of the real configuration where the influence of the shaft-seal rotation on the deflection angle at the outlet from the seal will be substituted by the effect of the rotor disc. The configuration of the stationary shaft-seal is useful with respect to prevent of some problems with the rotor dynamics, also with respect to much simpler construction of the test rig and with respect to much more accurate concentric seating of the seal-fins.

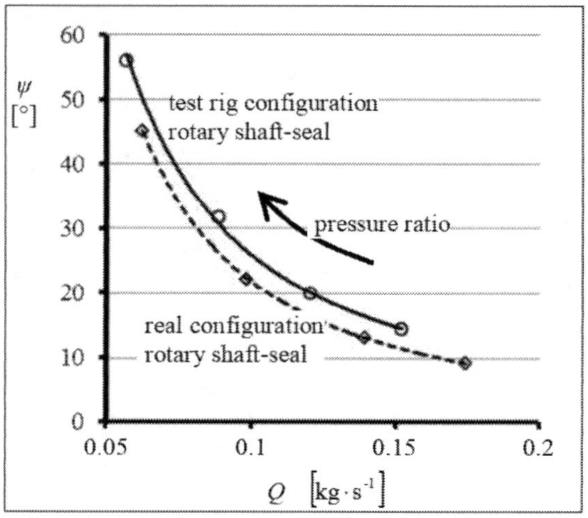

Fig. 10. Dependency of the deflection angle on the mass flux – four seal-fins, 5100 RPM, seal clearance 1.2 mm.

Acknowledgement

This work was supported by the project TH02020086 of the Technology Agency of the Czech Republic and by the Long-term Framework Advancement Plan provided by the Ministry of Industry and Trade of the Czech Republic.

References

1. T. Jelínek, P. Straka, M. Kladrubský, Applied Mechanics and Materials, **821**, pp. 48-56 (2016)
2. M. Němec, P. Straka, T. Jelínek, M. Babák, ASME paper GT2010-22490 (2010)
3. M. Němec, J. Michálek, V. Strach, The XVIII Symposium on Measuring Techniques in Turbomachinery, Thessaloniki (2006)
4. T. Jelínek, M. Němec, Proc. Conf. EFM 2011, pp. 706-716, Jičín (2011)
5. M. Němec, T. Jelínek, M. Babák, ASME paper GT2012-69693 (2012)
6. P. Straka, M. Němec, T. Jelínek, EPJ, **92**, 02088 (2015)
7. T. Jelínek, M. Němec, V. Uruba, Proc. Conf. Eng. Mechanics 2016, pp. 254-257, Svratka (2016)
8. M. Němec, T. Jelínek, P. Milčák, Proc. Conf. PSE 2017, Pilsen (2017)
9. M. Němec, T. Jelínek, P. Milčák, Proc. Conf. EFM 2017, Mikulov, (2017, in-press)
10. P. Straka, EPJ Web of conferences, **25**, 01090 (2012)
11. P. Straka, Report VZLÚ R-5904, Prague (2013), (in Czech)
12. P. Straka, Report VZLÚ R-4910, Prague (2010), (in Czech)
13. C.L. Rumsay, T.B. Gatski, Jour. Of Aircraft, **83**, pp. 904-910 (2001)

Beer foam decay: effect of glass surface quality and CO_2 content

Radek Šulc[1,*], and *Jiří* Bojas[1]

[1]Czech Technical University in Prague, Faculty of Mechanical Engineering, Department of Process Engineering, Technická 4, 166 07 Prague, Czech Republic

Abstract. The effect of beer glass surface quality and CO_2 content in beer on foam decay was investigated using our experimental method. The effect of beer glass surface quality on foam decay was experimentally investigated for: i) cold & clean glass surface, ii) warm & clean glass surface, iii) cold & greasy glass surface, and iv) cold & dusty glass surface. The fastest foam decay was observed for greasy glass surface. It was found that increasing CO_2 content in beer: i) the liquid content in the foam decreases, and ii) the foam breaks down faster. The foam decay and growth kinetics of foam-liquid interface were statistically treated using own models.

1 Introduction

The quality of beer foam (volume, structure, stability, colour etc.) is a determining factor in the appearance of a good beer, especially for pilsner style beers. The pilsner style beers (i,e. beers produced by bottom-fermenting yeast) form large volume of dense and stable foam which clings on glass surface [1]. The foam of pilsner style beer should be stable approx. 200 seconds [2].

The perfectly draught beer is presented in Fig. 1. The formed foam has creamy appearance and protects beer against oxidation. The formed foam bubbles are small as possible. The absence of bubbles on glass surface signalizes the glass cleanliness. On the perfectly clean and cold glass the foam clings and forms circles on the glass surface (Fig. 2).

Fig. 1. Perfectly draught beer (Pilsner Urquell tank beer, restaurant Konvikt, Prague).

Fig. 2. Cold & clean glass surface – cling after drinking.

* Corresponding author: Radek.Sulc@fs.cvut.cz

The beer foam stability is usually determined using device of NIBEM [3]. The beer is filled straight from the bottle or can into the measuring vessel which imitates the pouring of beer. The method measures the time period of drop of foam- beer interface at three different heights after 10, 20 and 30 mm. The measurement starts once the foam-beer interface drops for 10 mm under the edge of the cuvette.

Šrogl and Klasova [4] tested own method for foam stability testing. They foamed beer in laboratory mixer and then they measured time required for returning foam back into liquid. Šavel and Brož [5] summarized the principles and methods of measurement of beer foaming power including kinetic equations describing foam disintegration. Novak et al. [6] foamed beer in bubbled column by defined gas flowrate. The steady foam height in column was determined. In this state rate of foam forming equals to rate of foam decomposition.

Šavel [7] tested two models of the beer foam decomposition based on kinetics of 1st order of isolated (foam → beer) and serial reaction (foam → beer in foam → beer). The foam decomposition was investigated in a cylindrical vessel with the diameter of 55 mm and the height of 130 mm. The beer presented in the cylinder was foamed using the beer outflow from bottle or by the injection of defoamed beer. The second model was found to be better comparing with the first model that described only a part of the foam decomposition curve.

The experimental error of foam decay measurement using standard methods can reach up to 20 % depending on the experimenter´s experience. In our point of view the short decay time due to lower initial foam height is responsible for higher error rate of measuring methods.

The aim of this contribution is: i) to test own simple method for stability testing of beer foam without expensive measuring device, ii) to investigate the effect of beer glass surface quality and CO_2 content in beer on foam decay and foam-liquid interface. The foam decay and growth kinetics of foam-liquid interface were statistically treated using own models.

2 Theoretical background

2.1 Foam decay model

The time dependence of dimensionless foam height h_F^* was described by S-curve as follows [8]:

$$h_F^*(t_F) = \frac{1}{1+\exp(a_0 + a_1 \cdot t_F)} \quad (1)$$

where: a_0 (-), a_1 (min^{-1}) are parameters, t_F (min) is the time and dimensionless foam height h_F^* at time t_F is defined as follows:

$$h_F^*(t_F) = \frac{h_F(t_F) - h_{Ffin}}{h_{F0}(t_{F0}) - h_{Ffin}} \quad (2)$$

where h_{F0} (t_{F0}) is the foam height in the initial state, h_{Ffin} is the foam height in the final state, and h_F (t_F) is the foam height at time t_F. The S-curve form given by Eq. (1) can be transformed as follows:

$$h_F^*(t_F) = \frac{1}{1+\exp\left(\ln(45^2/55^2)\cdot\dfrac{(t_F - t_{F0.5})}{(t_{F0.55} - t_{F0.45})}\right)} \quad (3)$$

where $t_{F\,0.5}$ (min) is the time when the dimensionless foam height reaches 50%, ($t_{F0.55} - t_{F0.45}$) (min) is time difference from 45 % to 55% of dimensionless foam height. Near the 50% of dimensionless foam height the curve is linear, thus we can estimate foam decay rate $v_{F\,0.50}$ in this point as follows:

$$v_{F0.5} \approx \frac{\Delta h_F^*}{\Delta t_F} = \frac{0.55-0.45}{t_{F0.55} - t_{F0.45}} \quad (4)$$

where $t_{F0.55}$ and $t_{F0.45}$ are the times when the dimensionless foam height reaches 55 % and 45 %, respectively. Combining Eqs. (3) and (4) the final formula describing the foam decay is obtained as follows:

$$h_F^*(t_F) = \frac{1}{1+\exp(10\cdot\ln(45^2/55^2)\cdot v_{F0.5}\cdot(t_F - t_{F0.5}))} \quad (5)$$

2.2 Growth model of foam-liquid interface

The time dependence of liquid height is usually described [7] by kinetics of 1st order in final form as follows:

$$h_L^*(t_L) = 1 - \exp(-t/t_{Lalt}) \quad (6)$$

where: t_{Lalt} (min) is the average lifetime, t_L (min) is the time and dimensionless liquid height h_L^* at time t_L is defined as follows:

$$h_L^*(t_L) = \frac{h_L(t_L) - h_{L0}(t_{L0})}{h_{Lfin} - h_{L0}(t_{L0})} \quad (7)$$

where h_{L0} (t_{L0}) is the liquid height in the initial state, h_{Lfin} is the liquid height in the final state, and h_L (t_L) is the liquid height at time t_L. The parameter $t_{L\,alt}$ represents the time in which 63.2 % of dimensionless liquid height is reached.

We found that model given by Eq. (6) is unsatisfactory at the beginning of liquid level formation. The delay of liquid level formation was observed in this period. Therefore, the model given by Eq. (6) was modified taking this effect into account as follows:

$$h_L^*(t_L) = 1 - (1-\tau_L/t_{Lalt})^{-1} \times$$
$$\times(\exp(-t/t_{Lalt}) - (\tau_L/t_{Lalt})\cdot\exp(-t/\tau_L)) \quad (8)$$

where τ_L is the time delay.

The measured properties are shown in Fig. 3.

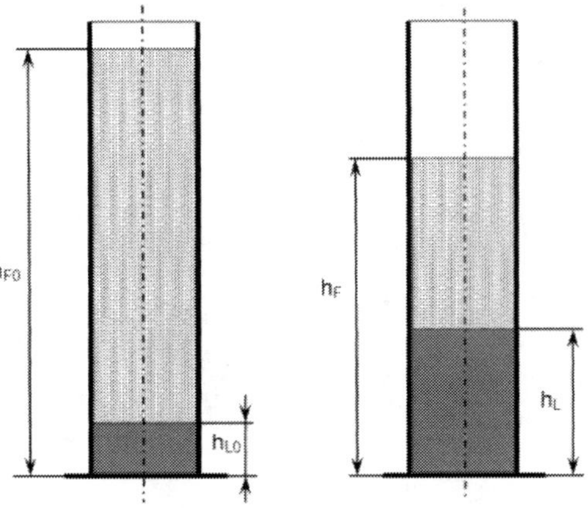

Fig. 3. Measured properties.

3 Experimental

The experimental procedure of our method occurs in following steps: i) Filling of glass cylinder of height of 190 mm by beer foam, ii) Visual observation of foam decay in time. We assume that the uncertainty should be decreased increasing the initial foam height.

The beer foam was formed by special way of beer serving. In this case, the beer foam outflows only from beer tap (in Czech slang this serving style is called "mliko").

3.1 Effect of beer glass surface quality

The cold & clean glass surface was prepared by the following procedure: i) washing with water and degreasing agent, ii) rinsing with tap water, and iii) cooling in the refrigerator to temperature of 7°C. The warm & clean glass surface was prepared by heating of clean glass to temperature of 37°C. This temperature corresponds to the typical temperature of glass from dishwasher. The cold & greasy glass surface was prepared by laying of a thin layer of pork lard on cold & clean glass surface. The greasy glass corresponds to wrongly washed glass. The cold & dusty glass surface was prepared by laying of a dust on cold & clean glass surface. The 10° Gambrinus beer was used for experiments.

3.2 Effect of beer CO_2 content

The 12° unfiltered Hubertus lager produced by brewery Kácov was used for experiments. The beer was stored in beer barrel at 6°C. The CO_2 content was increased gradually from 4.6 to 9.7 g/L in the barrel by means of gaseous CO_2 atmosphere under liquid. The initial value of CO_2 content of 4.6 g/L lies in the range of CO_2 content (4-5 g/L) typical for pilsner style beer.

4 Data analysis

4.1 Effect of beer glass surface quality

4.1.1 Foam decay

The evaluated model parameters for various quality of glass surface are presented in Table 1. The comparison of experimental data and predicted foam decomposition is presented in Fig. 4 for tested surfaces. The agreement of model curve and experimental data is relatively very well for all tested surfaces. Some discrepancies can be found at the initial and final state of foam decomposition.

As it follows from Fig. 4, the foam decay depends on glass surface quality. As expected, the longest time for the total foam decomposition was observed for clean & cold glass (12 minutes). The shortest time was observed for cold & greasy glass (8 minutes). As it follows from Fig. 4, the foam height in the final state is the same for all tested types of glass surface. This fact signalizes that the liquid content present in foam was the same for all tests. For illustration, the foam quality is presented in Fig. 5 for cold & dusty glass surface and cold & clean glass surface.

Table 1. Foam decay - effect of beer glass surface quality: S-curve parameters.

Beer glass surface quality	$t_{F0.5}$ **(min)**	$v_{F0.5}$ **(min^{-1})**	h_{F0} **(mm)**	$h_{F fin}$ **(mm)**	**R (-)**
cold & clean glass surface	6.3	- 0.143	190	87.5	0.994
warm & clean glass surface	5.76	- 0.202	190	87	0.996
cold & greasy glass surface	5.67	- 0.184	190	88	0.993
cold & dusty glass surface	4.84	- 0.200	190	88	0.975

https://doi.org/10.1051/epjconf/201818002101

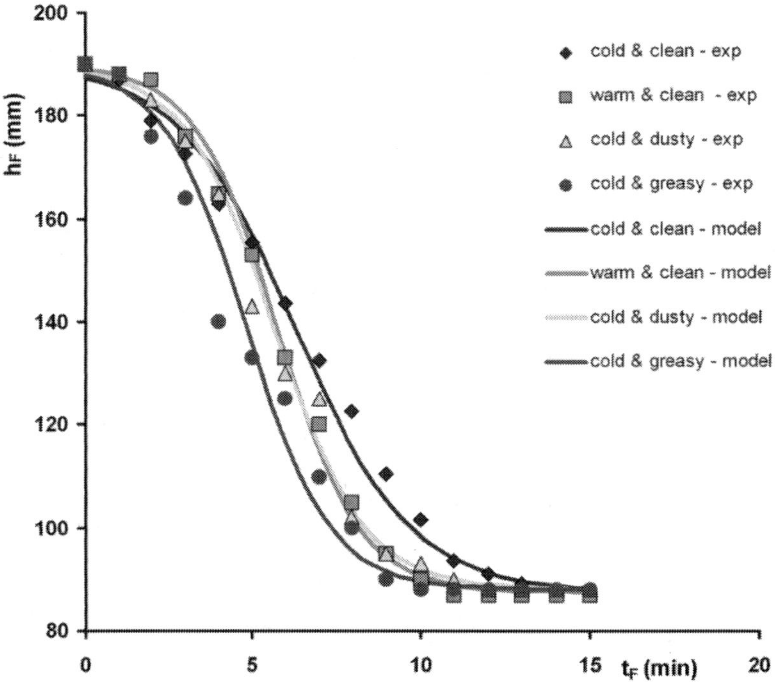

Fig. 4. Foam decay – effect of beer glass surface quality.

a) b)

a) cold & dusty glass surface
The thin foam contains large bubbles.
The glass surface is covered by small bubbles.

b) cold & clean glass surface
The dense foam contains small bubbles.
The glass surface clean without captured bubbles.

Fig. 5. Foam quality – effect of glass surface quality

Table 2. Liquid formation - effect of beer glass surface quality.

Beer glass surface quality	τ_L (min)	$t_{L\,alt}$ (min)	τ_L / $t_{L\,alt}$ (-)	h_{L0} (mm)	$h_{L\,fin}$ (mm)	$h_{L\,fin}$ - h_{L0} (mm)	R (-)
cold & clean glass surface	0.09	3.35	0.03	16	93	77	0.999
warm & clean glass surface	1.19	2.13	0.56	14	81	67	0.999
cold & greasy glass surface	1.06	1.44	0.74	18	82	64	0.997
cold & dusty glass surface	1.2	1.29	0.93	15	81.5	66.5	0.997

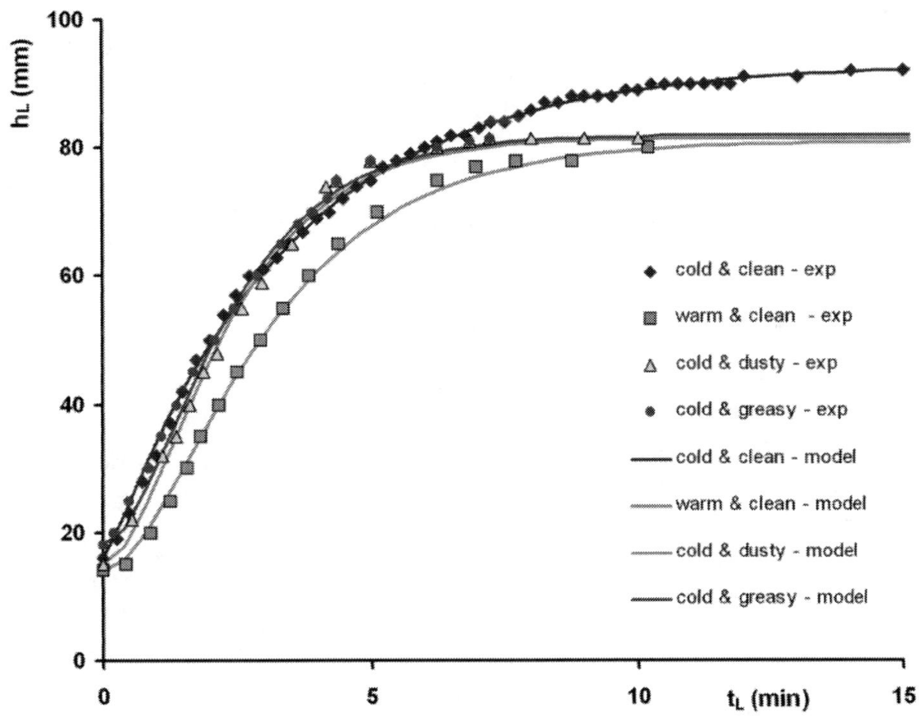

Fig. 6. Liquid formation – effect of beer glass surface quality.

4.1.2 Liquid formation

The evaluated model parameters for various quality of glass surface are presented in Table 2. The comparison of experimental data and liquid height growth predicted by Eq. (8) is presented for tested surfaces in Fig. 6. The agreement of model and experimental data is very well for all tested surfaces.

As it follows from Fig. 6, the liquid formation depends on glass surface quality. As expected, the longest average lifetime and the highest liquid height in the final state were observed for clean & cold glass (3.35 minutes and 77 milimeters). As it follows from Fig. 6, the liquid heights in the final state are different for tested types of glass surface. This fact signalizes that the liquid content present in foam was not the same. The lowest liquid content was observed for cold & greasy glass. The significant effect of glass surface was observed for time delay τ_L. The value close to zero was found for cold & clean glass surface. For other surfaces the values greater than 1 minute were evaluated.

 https://doi.org/10.1051/epjconf/201818002101

4.2 Effect of beer CO_2 content

4.2.1 Foam decay

The evaluated model parameters for various beer CO_2 content are presented in Table 3. The comparison of experimental data and predicted foam decomposition is presented in Fig. 7. The agreement of model curve and experimental data is relatively very well for all tested CO_2 contents. Some discrepancies can be found at the initial and final state of foam decomposition.

Table 3. Foam decay - effect of beer CO_2 content: S-curve parameters.

c_{CO2} (g/L)	$t_{F\,0.5}$ (min)	$v_{F0.5}$ (min^{-1})	h_{F0} (mm)	$h_{F\,fin}$ (mm)	$h_{F\,fin}$ / h_{F0} (-)	R (-)
4.6	7.20	- 0.111	190	80	0.421	0.996
6.2	7.80	- 0.114	190	66	0.347	0.997
7.9	7.83	- 0.086	190	55	0.289	0.994
8.7	6.91	- 0.119	190	52	0.274	0.995
9.7	6.77	- 0.108	190	46	0.242	0.994

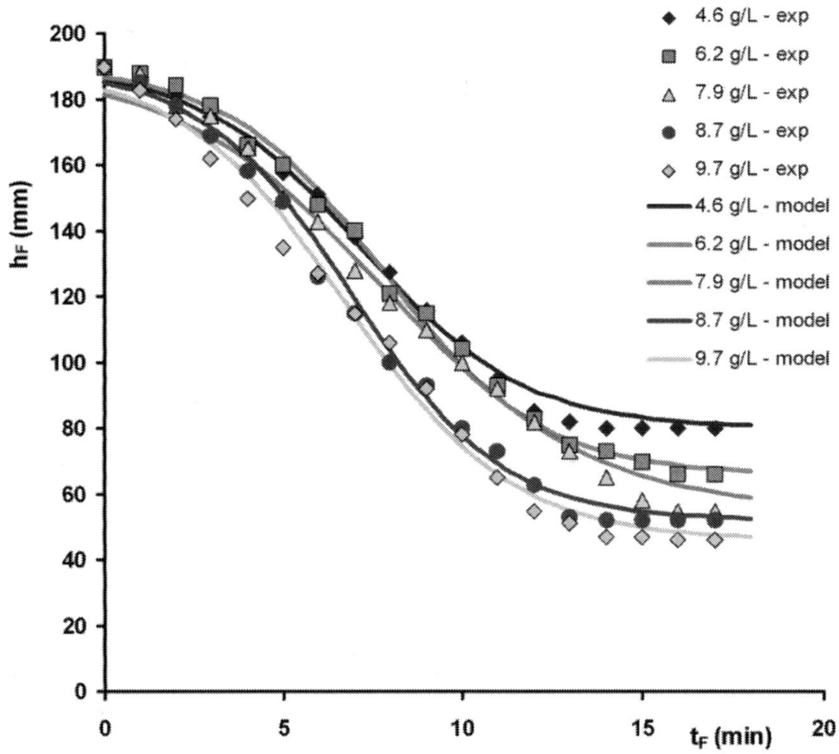

Fig. 7. Foam decay – effect of beer CO_2 content.

Increasing CO_2 content the foam is decomposed faster. The decreasing foam height in the final state for higher CO_2 content signalizes that the higher CO_2 content the lower liquid content is present in beer foam. The following dimensional relations were found for parameters $t_{F0.5}$, $v_{F0.5}$ and h_{Ffin} / h_{F0} :

$$t_{F0.5} = -0.1433 \cdot c_{CO2}^2 + 1.9711 \cdot c_{CO2} + 1.1457 \quad (R = 0.995) \qquad (9)$$

$$v_{F0.5} \approx -0.109 \qquad (10)$$

$$h_{Ffin} / h_{F0} = 1.276 \cdot c_{CO2}^{-0.72} \quad (R = 0.997) \qquad (11)$$

where $t_{F0.5}$ (min), $v_{F0.5}$ (min-1) and c_{CO2} (g/L). The effect of CO_2 content on foam decay rate was found to be negligible. For illustration, the obtained relations are graphically shown in Figs. 8, 9 and 10.

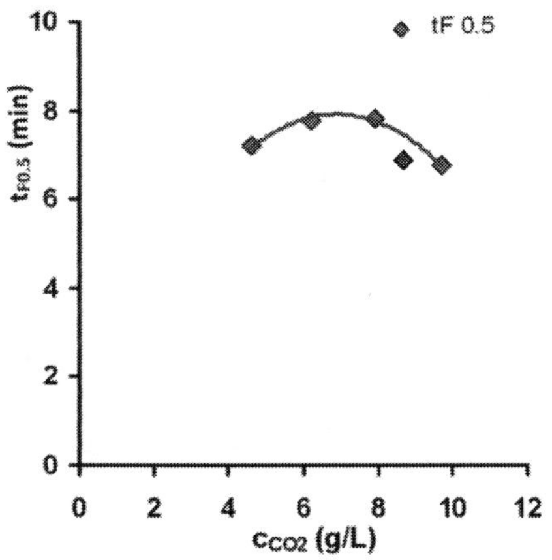

Fig. 8. Foam decay – effect of CO_2 content: parameter $t_{F0.5}$.

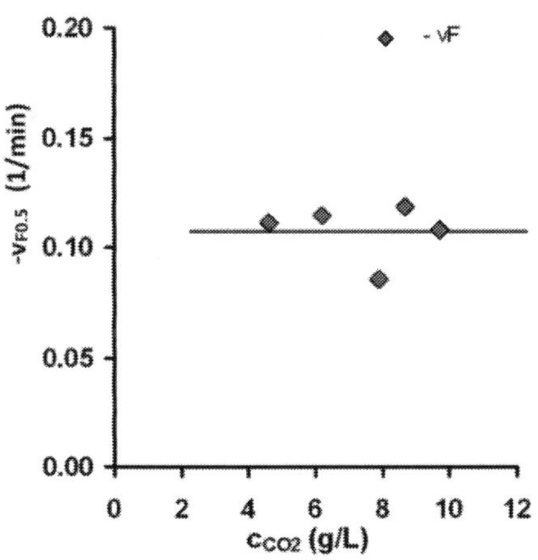

Fig. 9. Foam decay – effect of CO_2 content: parameter $v_{F0.5}$.

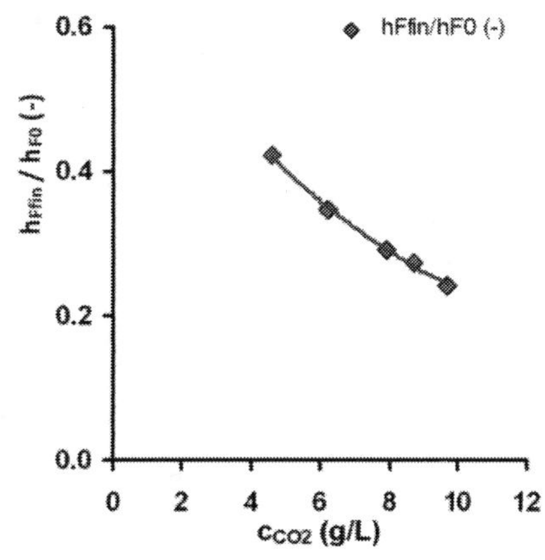

Fig. 10. Foam decay – effect of CO_2 content: h_{Ffin} / h_{F0} ratio.

4.2.2 Liquid formation

The evaluated model parameters for various beer CO_2 content are presented in Table 4. The comparison of experimental data and liquid height growth predicted by Eq. (8) is presented for tested CO_2 contents in Fig. 11. The agreement of model and experimental data is relatively very well for all tested CO_2 contents.

The time delay τ_L and average lifetime $t_{L\,alt}$ decrease with increasing CO_2 content in the beer. The effect of CO_2 content is clearly visible for parameters τ_L, $\tau_L / t_{L\,alt}$ ratio and h_{Lfin}. For CO_2 content greater then 6.2 g/L the liquid is released from foam immediately after foaming.

The following dimensional relations were found for parameters τ_L, $t_{L\,alt}$ and h_{Lfin}:

$$\tau_L = 12.047 \cdot c_{CO2}^{-1.2} \quad (R = 0.996) \qquad (12)$$

$$t_{Lalt} = 4.318 \cdot c_{CO2}^{-0.25} \quad (R = 0.84) \qquad (13)$$

$$\tau_L / t_{Lalt} = 2.791 \cdot c_{CO2}^{-0.95} \quad (R = 0.969) \qquad (14)$$

$$h_{Lfin} = 195 \cdot c_{CO2}^{-0.6} \quad (R = 0.982) \qquad (15)$$

where τ_L (min), $t_{L\,alt}$ (min), h_{Lfin} (mm) and c_{CO2} (g/L). For illustration, the obtained relations are graphically shown in Figs. 12, 13 and 14.

 https://doi.org/10.1051/epjconf/201818002101

Table 4. Liquid formation - effect of beer CO_2 content.

c_{CO2} (g/L)	τ_L (min)	$t_{L\ alt}$ (min)	$\tau_L / t_{L\ alt}$ (-)	h_{L0} (mm)	$h_{L\ fin}$ (mm)	$h_{L\ fin} - h_{L0}$ (mm)	R (-)
4.6	1.97	2.92	0.675	0	75	75	0.998
6.2	1.29	2.90	0.445	5	61.5	56.5	0.999
7.9	1.02	2.37	0.430	4	54	50	0.997
8.7	0.88	2.60	0.338	6	52	46	0.999
9.7	0.80	2.44	0.330	5	46	41	0.997

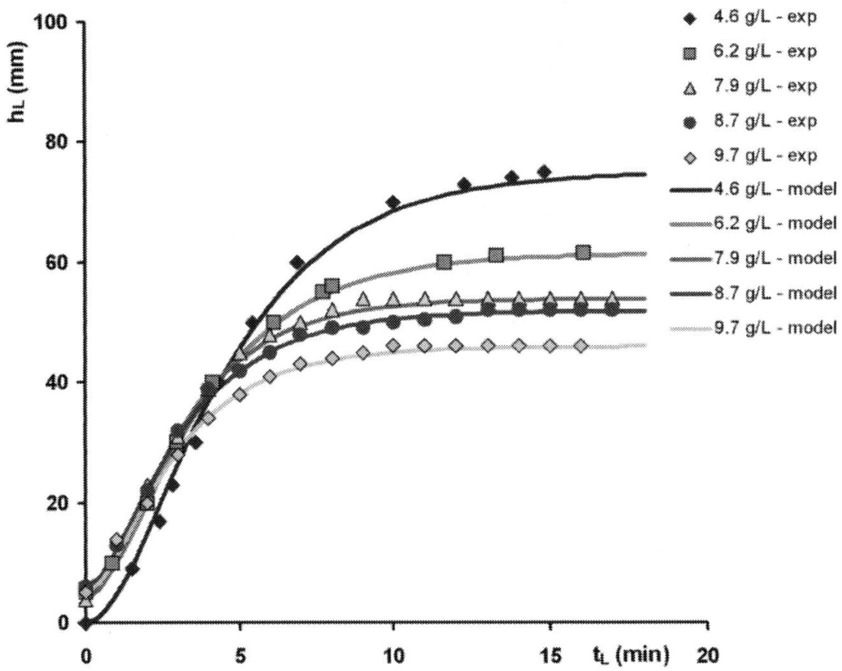

Fig. 11. Liquid formation – effect of CO_2 content.

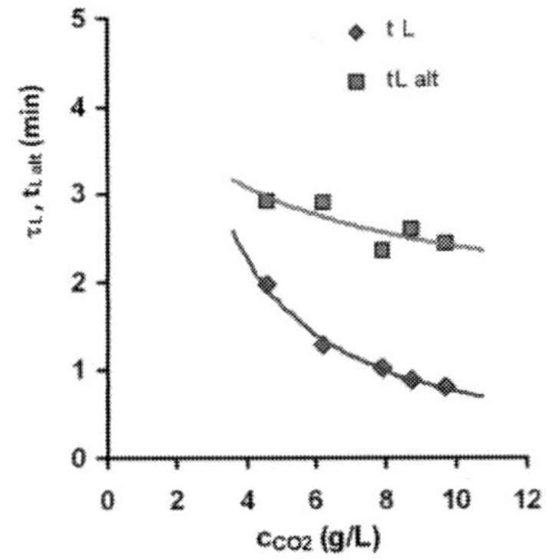

Fig. 12. Liquid formation – effect of CO_2 content: parameters τ_L and $t_{L\ alt}$.

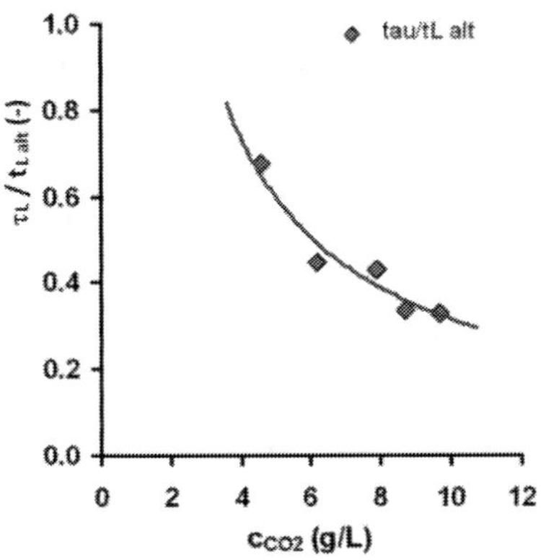

Fig. 13. Liquid formation – effect of CO_2 content: $\tau_L/t_{L\ alt}$ ratio.

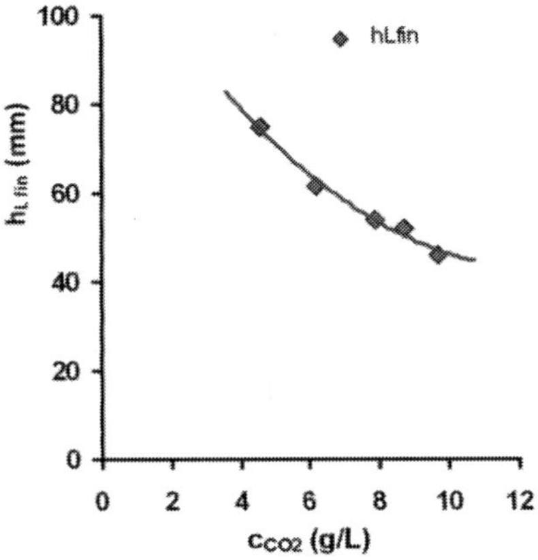

Fig. 14. Liquid formation – effect of CO_2 content: parameter $h_{L\ fin}$.

5 Conclusions

The effect of beer glass surface quality and CO_2 content in beer on foam decay was investigated using our experimental method.

The effect of beer glass surface quality on foam decay was experimentally investigated for: i) cold & clean glass surface, ii) warm & clean glass surface, iii) cold & greasy glass surface, and iv) cold & dusty glass surface. The fastest foam decay was observed for greasy glass surface.

Increasing beer CO_2 content the foam decomposition is faster. It was found that increasing CO_2 content in beer: i) the liquid content in the foam decreases, and ii) the foam breaks down faster. The foam decay and growth kinetics of foam-liquid interface were statistically treated using own models.

This work was supported by grant No. SGS16/149/OHK2/2T/12 of the Grant Agency of the Czech Technical University in Prague and RVO: 68407700 MSMT ČR.

Symbols

c_{CO2} mass concentration of CO_2 dissolved in beer, g/L
h_F foam height, mm
h_F^* dimensionless foam height, -
h_{F0} foam height in the initial state, mm
h_{Ffin} foam height in the final state, mm
h_L liquid height, mm
h_L^* dimensionless liquid height,-
h_{L0} liquid height in the initial state, mm
h_{Lfin} liquid height in the final state, mm
R correlation index, -
t_F time of foam decomposition, min
$t_{F\,0.5}$ time when the dimensionless foam height reaches 50%, min
t_L time of liquid formation, min
t_{Lalt} average lifetime, min
$v_{F\,0.50}$ - foam decay rate, min^{-1}

Greek symbols
τ_L time delay, min

References

1. G. Basařova, *Czech beer* (in Czech) (NUGA, Prague, 2009)
2. M. Baszcyňski, T. Brányik, Struktura pivní pěny, In: *Proceedings of Conference "Kvasná chemie a bioinženýrství 2010"* (Prague, 2010)
3. W.J. Klopper, H.A. Vermeire, Brauwissenschaft **30**, 276 (1977)
4. J. Šrogl, V. Klasová, Kvasný průmysl **22**, 28 (1976)
5. J. Šavel, A. Brož, Kvasný průmysl **52**, 314 (2006)
6. P. Novák, M. Baszcyňski, T. Brányik, M.C. Růžička, Vliv vybraných fyzikálně-chemických vlastností piva na stabilitu pěny, In: *Proceedings of Conference "Kvasná chemie a bioinženýrství 2010"* (Prague, 2010)
7. J. Šavel, Kvasný průmysl **32**, 76 (1986)
8. L.J. Reed, J. Berkson, J. Phys. Chem. **33**, 760 (1929)

Local velocity scaling in an impeller discharge flow in T400 vessel agitated by tooth impeller in a fully turbulent region

Radek Šulc[1,*], *Pavel* Ditl[1], *Ivan* Fořt[1], *Darina* Jašíkova[2], *Michal* Kotek[2], *Václav* Kopecký[2], and *Bohuš* Kysela[3]

[1]Czech Technical University in Prague, Faculty of Mechanical Engineering, Department of Process Engineering, Technická 4, 166 07 Prague, Czech Republic
[2]Technical University of Liberec, Institute for Nanomaterials, Advanced Technology and Innovation, Studentská 1402/2, 461 17 Liberec, Czech Republic
[3]Czech Academy of Sciences, Institute of Hydrodynamics v.v.i, Pod Paťankou 30/5, 166 12 Prague, Czech Republic

Abstract. Hydrodynamics and flow field were measured in an agitated vessel using 2-D Time Resolved Particle Image Velocimetry (2-D TR PIV). The experiments were carried out in a fully baffled cylindrical flat bottom vessel 400 mm in inner diameter agitated by a tooth impeller 133 mm in diameter. The velocity fields were measured in the impeller discharge flow for impeller rotation speeds from 300 rpm to 700 rpm and three liquids of different viscosities (i.e. (i) distilled water, ii) a 28% vol. aqueous solution of glycol, and iii) a 43% vol. aqueous solution of glycol), corresponding to the impeller Reynolds number in the range 68 000 < Re < 221 000. This Re range secures the fully-developed turbulent flow of agitated liquid. In accordance with the theory of mixing, the dimensionless mean and fluctuation velocities in the measured directions were found to be constant and independent of the impeller Reynolds number. On the basis of the test results the spatial distributions of dimensionless velocities were calculated. The radial turbulence intensity was found to be in the majority in the range from 0.3 to 0.9, which corresponds to the high level of this quantity.

1 Introduction

It is important to know the flow and the flow pattern in an agitated vessel in order to determine many impeller and turbulence characteristics, e.g. impeller pumping capacity, intensity of turbulence, turbulent kinetic energy, convective velocity, and the turbulent energy dissipation rate. The information and data that are obtained can also be used for CFD verification.

In PIV studies, the spatial distributions of various properties are often presented. However, the inspection analysis shows that the validity of the spatial distribution of any dimensionless property for arbitrary process conditions in geometrically similar agitated vessels requires independence of a given dimensionless property from the impeller Reynolds number. The spatial distribution that is obtained is in general valid only when this independence is both theoretically predicted and experimentally verified. Unfortunately, in many studies the results are presented only for one impeller speed and, in addition, the experiments are often carried out under the transient flow regime.

Kysela et al. [1] investigated the flow discharging from a Ruhston turbine, Šulc et al. [2] presented the local velocity profiles in the zone in upward flow to the impeller for three impeller rotation speeds in a vessel 300 mm in the inner diameter filled by a water and agitated by a Rushton turbine, as well. In our previous work [3] the scaling of the velocity field in the zone in upward flow to the impeller for three liquids of different viscosities in a vessel 400 mm in the inner diameter agitated by a Rushton turbine were presented.

Agitation of two immiscible liquids or solid-liquid suspension is a frequent operation in chemical and metallurgical industries [4]. In this case the tooth impeller has been used usually for dispersion process [5].

The aim of this work is to study scaling of the velocity field in the impeller discharge flow in a vessel 400 mm in inner diameter mechanically agitated by a tooth impeller (CVS 691038.1) in a fully turbulent region for three liquids of different viscosities. The independency of dimensionless spatial velocity distribution on the impeller Reynolds number was tested statistically. The hydrodynamics and the flow field were measured in an agitated vessel using Time Resolved Particle Image Velocimetry (TR PIV).

2 Theoretical background

2.1 Inspection analysis of flow in an agitated vessel

The flow of a Newtonian fluid in an agitated vessel has been described by the Navier – Stokes equation:

* Corresponding author: radek.sulc@fs.cvut.cz

$$\rho \cdot \left(\frac{\partial \vec{U}}{\partial t} + \vec{U} \bullet \nabla \vec{U} \right) = -\nabla p + \mu \cdot \nabla^2 \vec{U} + \rho \cdot \vec{g} \quad (1)$$

This equation can be rewritten into dimensionless form, as follows (e.g. [6]):

$$\left(\frac{\partial \vec{U}^*}{\partial t^*} + \vec{U}^* \bullet \nabla^* \vec{U}^* \right) = -\nabla^* p^* + \frac{1}{\mathrm{Re}} \cdot \nabla^{*2} \vec{U}^* + \frac{1}{Fr} \cdot \vec{n}^* \quad (2)$$

where the dimensionless properties are defined as follows:

- dimensionless instantaneous velocity:

$$\vec{U}^* = \frac{\vec{U}}{N \cdot D}, \quad (3a)$$

- dimensionless instantaneous pressure:

$$p^* = \frac{p}{\rho \cdot N^2 \cdot D^2}, \quad (3b)$$

- dimensionless space gradient: $\nabla^* = \frac{\nabla}{D}$, (3c)

- impeller Reynolds number $\mathrm{Re} = \frac{N \cdot D^2}{\nu}$, (3d)

- impeller Froude number: $Fr = \frac{N^2 \cdot D}{g}$, (3e)

and where $\vec{n}$ is a unity vector.

Similarly, the equation of continuity for stationary flow of a non-compressible fluid is given

$$\nabla \bullet \vec{U} = 0 \quad (4)$$

and can be rewritten into the dimensionless form as follows:

$$\nabla^* \bullet \vec{U}^* = 0 \quad (5)$$

The following relations can be obtained for the dimensionless velocity components and the dimensionless pressure, respectively, by inspection analysis of Eqs. (2) and (5), as follows:

$$\vec{U}^* = f_1(\vec{x}^*, t^*, \mathrm{Re}, Fr), \quad (6)$$

$$p^* = f_2(\vec{x}^*, t^*, \mathrm{Re}, Fr), \quad (7)$$

where $\vec{x}^*$ is a dimensionless location vector.

For stationary flow with a periodic character the velocity time dependence can be eliminated by substituting the velocity and pressure by time-averaged properties. For highly turbulent flow in an agitated baffled vessel, the viscous and gravitational forces can be neglected and finally the time-averaged dimensionless velocity components and the pressure are independent of the impeller Reynolds and the Froude numbers, and depend on location, only:

$$\overline{U_i}^* = f_1(\vec{x}^*), \quad (8)$$

$$p^* = f_2(\vec{x}^*). \quad (9)$$

Reynolds decomposition of instantaneous velocity components has been applied for the velocity profiles studied in this work.

2.2 Mean and fluctuation velocity

Using PIV, the instantaneous velocity data set $U_i\ (t_j)$ in the ith direction for $j = 1, 2, \ldots N_R$ at observation times t_j with equidistant time step Δt_S (i.e. $\Delta t_S = t_{j+1} - t_j$) was obtained in a given location. Assuming the ergodic hypothesis the time-averaged mean velocity $\overline{U_i}$ was determined as the average value of velocity data set $U_i\ (t_j)$:

$$\overline{U_i} = \frac{1}{N_R} \cdot \sum_j U(t_j) \quad \text{for } j = 1, 2, \ldots, N_R, \quad (10)$$

where $\overline{U_i}$ is mean velocity in the ith direction, $U_i\ (t_j)$ is instantaneous velocity in the ith direction at the observation time t_j, and N_R is the number of data items in the velocity data set.

Consequently, the fluctuation velocity in the ith direction $u_i\ (t_j)$ at observation time t_j is obtained by decomposition of the instantaneous velocity:

$$u_i(t_j) = U_i(t_j) - \overline{U_i} \quad \text{for } j = 1, 2, \ldots, N_R, \quad (11)$$

where $u_i\ (t_j)$ is the fluctuation velocity in the ith direction at observation time t_i, $\overline{U_i}$ is mean velocity in the ith direction, $U_i\ (t_i)$ is instantaneous velocity in the ith direction at the observation time t_i.

The root mean squared fluctuation velocity is determined as follows:

$$\overline{u_i} = \left(\frac{1}{N_R} \cdot \sum_i u_i^2(t_j) \right)^{1/2} \quad \text{for } j = 1, 2, \ldots, N_R, \quad (12)$$

where $\overline{u_i}$ is the root mean squared fluctuation velocity, and $u\ (t_i)$ is the fluctuation velocity at observation time t_i.

3 Experimental

The hydrodynamics and the flow field were investigated in an agitated vessel using Time Resolved Particle Image Velocimetry (TR PIV). The experiments were carried out in a fully baffled cylindrical flat bottom vessel 400 mm in the inner diameter. The tank was agitated by a tooth impeller 133 mm in diameter, i.e. the dimensionless impeller diameter D/T was 1/3.

 https://doi.org/10.1051/epjconf/201818002102

Table 1. Operational parameters.

Run #	liquid	N (rpm)	ν (m^2/s)	Re (-)	Record time T_R (s)
w-N300	water	300	9.35×10^{-7}	94 641	8
w-N500	water	500	9.35×10^{-7}	157 735	8
w-N700	water	700	9.35×10^{-7}	220 829	8
g28-N700	28 % glycol	700	2×10^{-6}	103 186	8
g43-N700	43 % glycol	700	3×10^{-6}	68 791	8

The dimensionless impeller clearance *C/D* taken from the lower impeller edge was 0.75. The tank was filled with degassed liquid, and the liquid height was 400 mm, i.e. the dimensionless liquid height *H/T* was 1. The dimensionless baffle width *B/T* was 1/10. To prevent air suction the vessel was covered by a lid.

Three liquids of different viscosities were used as the agitated liquid: i) distilled water ($\nu = 9.35 \times 10^{-7}$ m^2/s), ii) a 28% vol. aqueous solution of glycol ($\nu = 2 \times 10^{-6}$ m^2/s), and iii) a 43% vol. aqueous solution of glycol ($\nu = 3 \times 10^{-6}$ m^2/s). The velocity fields were detected at an impeller rotation speed in the range from 300 rpm to 700 rpm, which covers the impeller Reynolds number range from 68 000 to 221 000. This Re range secures the fully-developed turbulent flow regime. The operational conditions are summarized in Table 1.

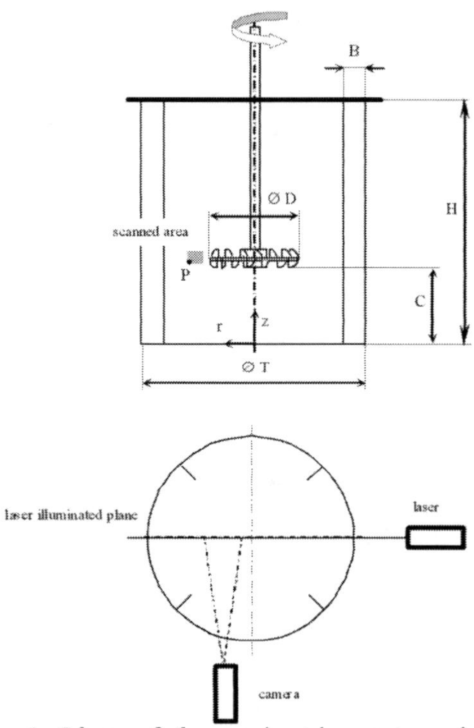

Fig. 1. Scheme of the experimental apparatus and the investigated area.

The time-resolved LITRON LDY 302 2D-PIV system (Dantec Dynamics (Denmark)) consists of a Neodyme-YLF laser (light wave length 532 nm, impulse energy 2x15 mJ), a SpeedSence 611 high speed PIV-regime camera (working on frequency 1kHz with resolution 1280 x 800 pixels) with a Nikon Macro 200 optical lens system equipped with an optical filter of wave length 570 nm. Rhodamine B fluorescent particles 11.95 ± 0.25 μm in mean diameter were used as seeding particles. The fluorescent particles lit by 532 nm light emit 570 nm light. In this way, non-seeding particles such as impurities and bubbles are separated, and are not recorded. The operating frame rate was 1 kHz (1000 vector fields per second), i.e. the sampling time Δt_s was 1 ms. The measured vertical plane was located in the center of the vessel and in the middle of the baffles. The plane was illuminated by a laser sheet. The investigated area was 25.6 x 16 mm, and the PIV spatial resolution $\Delta = 0.64$ mm was maintained the same during the experiment. The length of the record time was selected so that relevant results were obtained [7].

The position of left bottom apex P [r_P; z_P] was [106; 105], i.e. dimensionless radius $2r/T = 0.53$ and dimensionless height $z/T = 0.2625$, i.e. the left bottom edge was located 8 mm below the impeller paddle axis and 106 mm from the impeller axis. The scheme of the experimental apparatus and the investigated area is depicted in Fig. 1. The camera was positioned orthogonally to the laser sheet. The investigated area corresponds to region O according to the classification given by Fořt et al. [8].

4 Experimental data evaluation

According to the inspection analysis, the dimensionless velocities normalized by the product of impeller speed *N* and impeller diameter *D* should be independent of the impeller Reynolds number.

The effect of impeller Reynolds number on dimensionless velocities was tested by hypothesis testing ([9]). The statistical method of hypothesis testing can estimate whether the differences between the predicted parameter values (e.g. according to some proposed theory) and then the parameter values evaluated from the measured data are negligible. In this case, we assumed dependence of the tested parameter on the impeller Reynolds number, described by the simple power law correlation $parameter = B.Re^{\beta}$, and the difference between predicted exponent β_{pred} and evaluated exponent β_{calc} was tested. The hypothesis test characteristics are given as $t = (\beta_{calc} - \beta_{pred}) / s_{\beta}$ where s_{β} is the standard error of parameter β_{calc}.

Table 2. Dimensionless velocities – effect of the impeller Reynolds number

	Hypothesis: parameter = B.(Re)^0 percentage / t-characteristics |t|	
Parameter	acceptable	not acceptable
$\overline{U}_r/(ND)$ (-)	86.9 % / 1.37	13.1 % / 5.6
$\overline{U}_{ax}/(ND)$ (-)	90.9 % / 1.0	9.1 % / 5.41
$\overline{u}_r/(ND)$ (-)	86.3 % / 1.36	13.7 % / 4.94
$\overline{u}_{ax}/(ND)$	77.3 % / 1.1	22.7 % / 6.97
$\overline{u}_{ax}/\overline{u}_r$ (-)	100 % / 0.33	0 % / 0
T.I.$_r$ (-)	100 % / 0.37	0 % / 0

If the calculated $|t|$ value is less than the critical value of the t-distribution for (m-2) degrees of freedom and significance level α, the difference between β_{calc} and β_{pred} is statistically negligible (statisticians state: "the hypothesis cannot be rejected"). In our case, the independence of dimensionless velocities of the impeller Reynolds number was tested as the hypothesis, i.e. $parameter = B.(Re)^0$ = const., i.e. $\beta_{pred} = 0$. The t-distribution coefficient $t_{(m-2),\ \alpha}$ for five impeller Reynolds numbers and significance level $\alpha = 0.05$ is 3.1825. The hypothesis testing was done for each point in the investigated area. The hypothesis test results are presented for investigated area in Table 2 by the percentage of points in which the above-formulated hypothesis *parameter = const.* is satisfied and by the percentage of points in which the hypothesis *parameter = const.* can not be accepted. For illustration, the average values of calculated $|t|$ value are presented here also.

The dimensionless radial mean velocities were found in the range from 0.1 to 1.3. The dimensionless axial mean velocities were found to be close to zero, in the range from 0.004 to 0.079. These findings correspond to characteristics of the given zone, according to Fořt et al. [7]. This region contains predominantly radial stream from the impeller. Due to lower impeller clearance C/D the axis of the discharge flow is diverted from the impeller paddle axis. The dimensionless axial mean velocities are lower than the dimensionless values for radial mean velocity as expected for this region. The tested hypothesis can be accepted in the majority of profile points of the investigated area.

The dimensionless radial r.m.s. fluctuation velocities were found to be in the range from 0.184 to 0.536. The tested hypothesis can be accepted in the majority of profile points, as is signalized by the low calculated $|t|$ values. The dimensionless axial r.m.s. fluctuation velocities were found to be in the range from 0.2 to 0.552. The tested hypothesis can be accepted in the majority of profile points as is again signalized by the low calculated $|t|$ values.

On the basis of the results of this hypothesis test, we assume that all dimensionless velocities can be statistically taken as constant and independent of the impeller Reynolds number. The spatial distributions of dimensionless velocities for seven operational conditions were averaged and are presented in Figures 2 – 5. For illustration, the axial profiles of velocity components are presented in Figures 6 - 9 for selected radial positions r^* = 0.4165, 0.4485 and 0.4837. The axial profiles exhibit typical bell shape with the maximum, especially for U_r^* and u_{ax}^*. The investigated area is shown in Fig. 1 by shadow form .The profiles obtained are in accordance with velocity profiles typical for region O as reported by Fořt et al. [8].

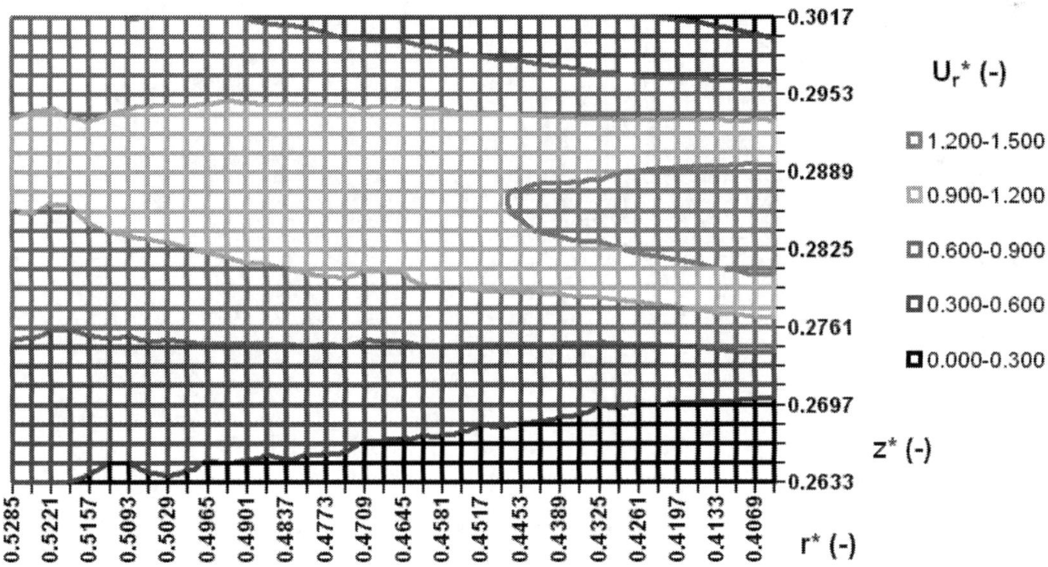

Fig. 2. Dimensionless radial mean velocity distribution $U_r^* = \overline{U}_r/(ND) = f(r^*, z^*)$ in the impeller discharge stream.

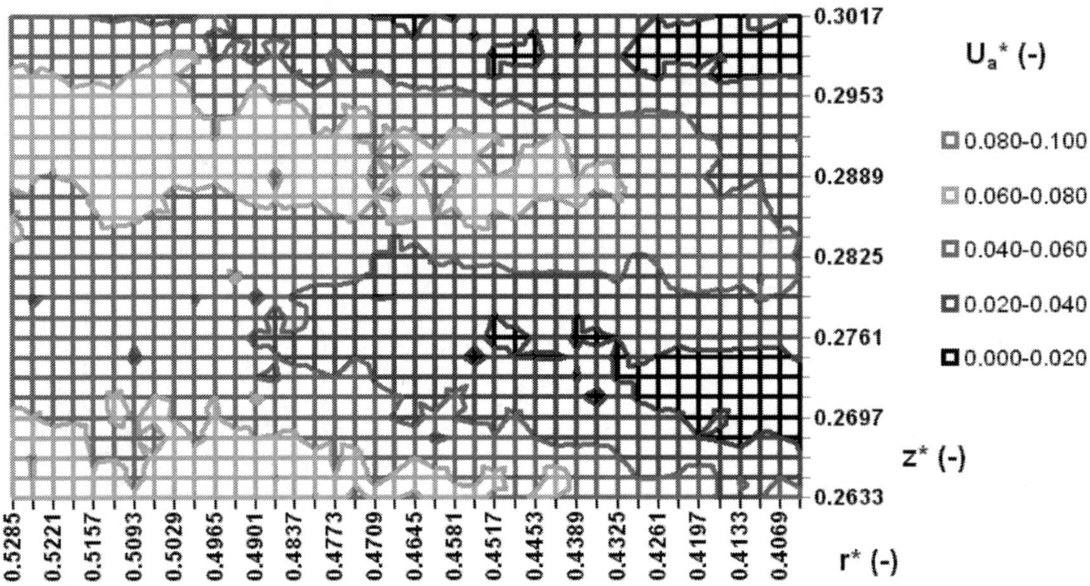

Fig. 3. Dimensionless axial mean velocity distribution $U_{ax}^* = \overline{U}_{ax}/(ND) = f(r^*, z^*)$ in the impeller discharge stream.

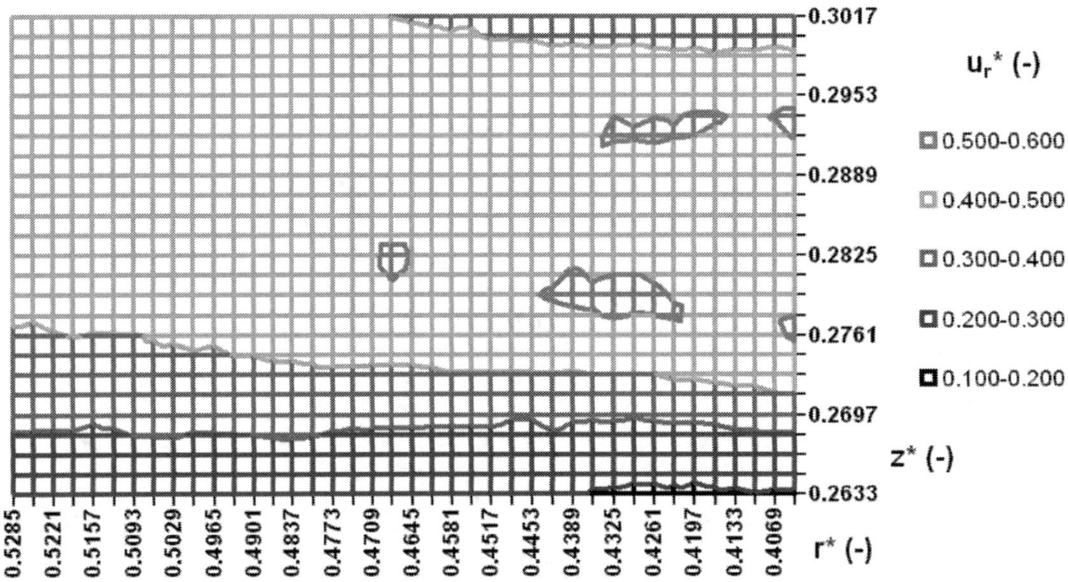

Fig. 4. Dimensionless radial fluctuation velocity distribution $u_r^* = \overline{u}_r/(ND) = f(r^*, z^*)$ in the impeller discharge stream.

https://doi.org/10.1051/epjconf/201818002102

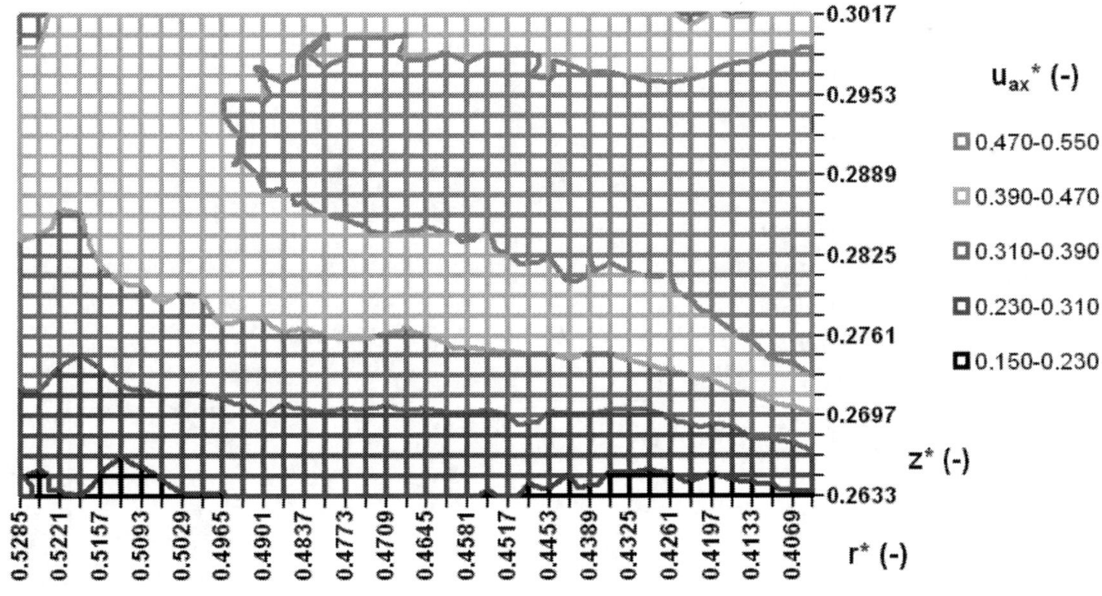

Fig. 5. Dimensionless axial fluctuation velocity distribution $u_{ax}^* = \overline{u}_{ax}/(ND) = \mathrm{f}(r^*, z^*)$ in the impeller discharge stream.

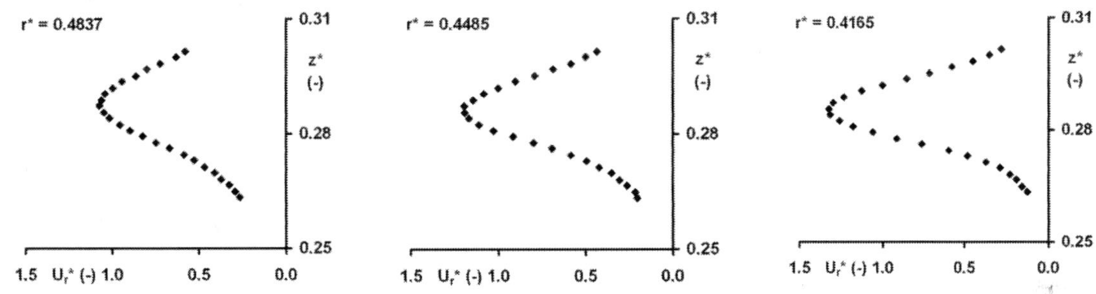

Fig. 6. Dimensionless radial mean velocity profiles $\overline{U}_r/(ND) = \mathrm{f}(z^*)$ in the impeller discharge stream.

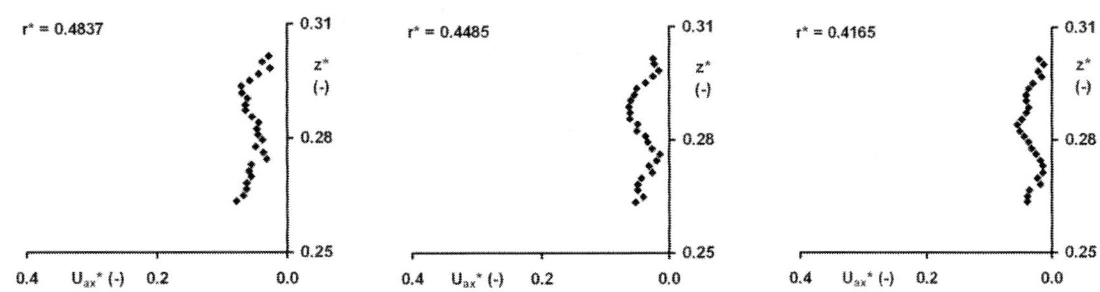

Fig. 7. Dimensionless axial mean velocity profiles $\overline{U}_{ax}/(ND) = \mathrm{f}(z^*)$ in the impeller discharge stream.

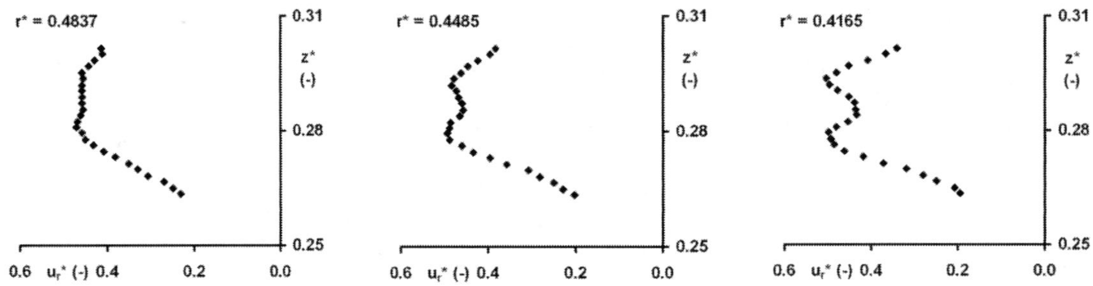

Fig. 8. Dimensionless radial fluctuation velocity profiles $\overline{u}_r/(ND) = \mathrm{f}(z^*)$ in the impeller discharge stream.

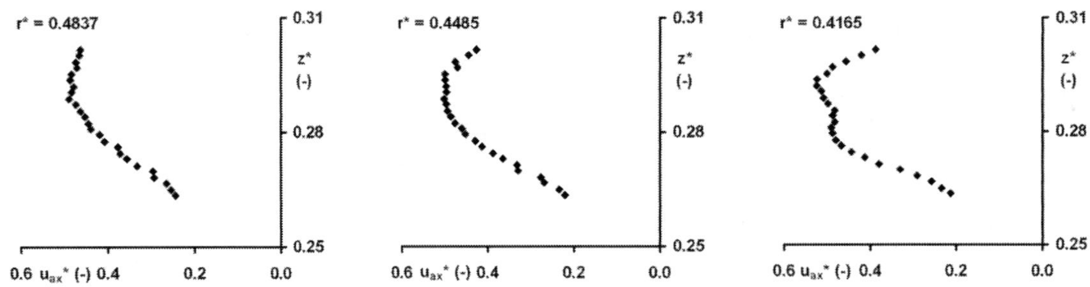

Fig. 9. Dimensionless axial fluctuation velocity profiles $\overline{u}_{ax}/(ND) = \mathrm{f}(z^*)$ in the impeller discharge stream.

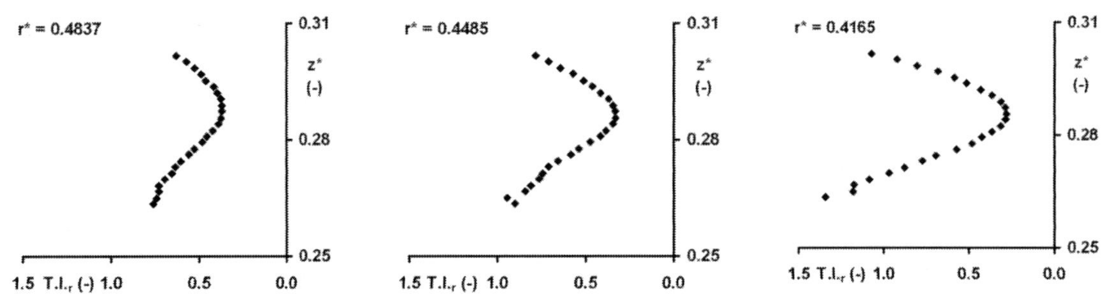

Fig. 10. Radial turbulence intensity profiles $T.I._r = \mathrm{f}(z^*)$ in the impeller discharge stream.

Intensity of turbulence

The radial turbulence intensity was calculated for each point in the investigated area as follows:

$$T.I._{,r} = \frac{\overline{u}_r}{\overline{U}_r} \tag{13}$$

where $\overline{u}_r$ is radial r.m.s. fluctuation velocity, $\overline{U}_r$ is radial mean velocity. For dimensionless velocities independent of the impeller Reynolds number, the turbulence intensity should also be independent of this quantity. The independence of dimensionless velocities of the impeller Reynolds number was again tested by hypothesis testing. The t-distribution coefficient $t_{(m-2),\ \alpha}$ for five impeller Reynolds numbers and significance level α = 0.05 is 3.1825. Hypothesis testing was done for each point in the investigated area. The hypothesis test results are presented in Table 2, as well.

 https://doi.org/10.1051/epjconf/201818002102

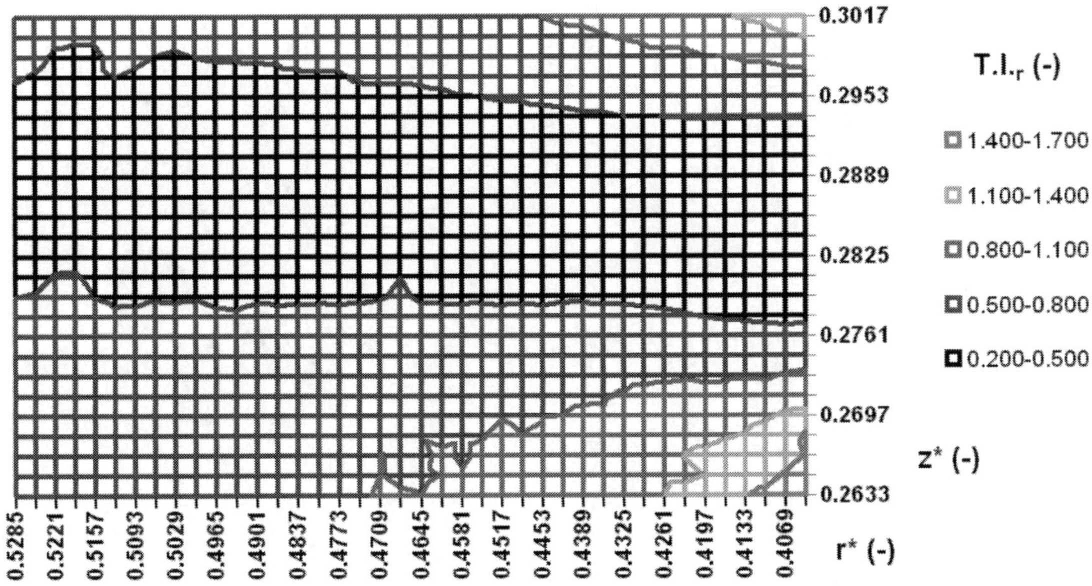

Fig. 11. Radial turbulence intensity distribution $T.I._r = f(r^*, z^*)$ in the impeller discharge stream.

The tested hypothesis can be accepted in the majority of profile points. On the basis of these hypothesis test results, we assume that the radial turbulence intensity can be statistically taken as constant and independent of the impeller Reynolds number, as expected. As it was shown, the calculated values are in the majority of points in the range from 0.27 to 0.9. These values correspond to high turbulence intensity. The radial profiles of radial turbulence intensity are presented in Figure 10 for selected radial positions $r^* = 0.4165$, 0.4485 and 0.4837. The spatial distribution of quantity $T.I._r$ in the impeller discharge stream is shown in Figure 11.

5 Conclusions

The following results have been obtained in this study:

The hydrodynamics and flow field were measured in a vessel 400 mm in the inner diameter agitated by a tooth impeller using the 2-D Time Resolved Particle Image Velocimetry (2-D TR PIV). The velocity fields were detected in the impeller discharge stream for impeller rotation speeds from 300 rpm to 700 rpm and three liquids of different viscosities, corresponding to the impeller Reynolds number in the range 68 000 < Re < 221 000.

The dimensionless axial mean velocities were found to be close to zero. These findings correspond to the characteristics of the given zone according to Fořt et al. [5]. This region contains predominantly radial flow from the rotating impeller.

In accordance with the theory of mixing, the dimensionless mean and fluctuation velocities in measured directions were found to be constant and independent of the impeller Reynolds number. On the basis of the test results the spatial distributions of dimensionless velocities were calculated.

The radial turbulence intensity was calculated and was found to be in the majority of points in the range from 0.27 to 0.9, which corresponds to high level of turbulence intensity. It was found that the radial turbulence intensity can be statistically taken as constant and independent of the impeller Reynolds number.

Acknowledgement

This research has been supported by Grant Agency of the Czech Republic project No. 16-20175S and by Ministry of Education, Youth and Sports of the Czech Republic project No. LO1201 (National Programme for Sustainability I) and RVO:67985874.

Symbols

B baffle width, m
C impeller clearance, m
D impeller diameter, m
H liquid height, m
m number of experimental points, -
N impeller rotational speed, 1/s
N_R number of data in velocity data set, -
r radius (radial coordinate), m
r^* dimensionless radius; $r^* = 2r/T$, -
Re impeller Reynolds number; $Re = N.D^2/\nu$, -
t time, s
t_i observation time, s
t_R record time, s
$t_{(m-2),\,a=0.05}$ t-distribution for (m-2) degrees of freedom and significance level α, -
T vessel diameter, m

T.I. turbulence intensity, -
u_i fluctuation velocity in i^{th} direction, m/s
$\overline{u}_i$ root mean squared fluctuation velocity in i^{th} direction, m/s
U_i instantaneous velocity in i^{th} direction, m/s
$\overline{U}_i$ mean velocity in i^{th} direction, m/s
z actual height (axial coordinate), m
z^* dimensionless height; $z^* = z/T$, -

References

1. B. Kysela, J. Konfršt, I. Fořt, M. Kotek, Z. Chára, Chem. and process Eng. **35**, 137-147 (2014)
2. R. Šulc, V. Pešava, P. Ditl, Acta Polytechnica **54**, 430-438 (2014)
3. R. Šulc, P. Ditl, I. Fořt, D. Jašíková, M. Kotek, V. Kopecký, B. Kysela, Local velocity scaling in T400 vessel agitated by Rushton turbine in a fully turbulent region. EPJ Web Conf., **143**, Article No. 02120 (2017)
4. E.L. Paul, V.A. Atiemo-Obeng, S.M. Kresta (eds), *Handbook of Industrial Mixing* (Wiley & Sons, USA, 2004)
5. P. Ditl, R. Šulc, Technical Transactions **113**, 83-94 (2016)
6. V. Novák, F. Rieger, K. Vavro, *Hydraulické pochody v chemickém a potravinářském průmyslu* (SNTL, Prague, 1989)
7. B. Kysela, J. Konfršt, Z. Chára, M. Kotek, EPJ Web of Conferences **67**, 02065 (2014)
8. I. Fořt, A. Obeid, V. Březina, Coll. Czechoslov. Chem. Comm. **47**, 226-239 (1982)
9. B.L. Bowerman, R.T.O´Connell, *Applied statistics: improving business processes* (Richard D. Irwin, USA, 1997)

CFD study on the effects of boundary conditions on air flow through an air-cooled condenser

Zdeněk Sumara[1*], *Michal* Šochman[1]

[1]*CTU in Prague, Faculty of Mechanical Engineering, Department of Fluid Mechanics and Thermodynamics, Czech Republic*

Abstract. This study focuses on the effects of boundary conditions on effectiveness of an air-cooled condenser (ACC). Heat duty of ACC is very often calculated for ideal uniform velocity field which does not correspond to reality. Therefore, this study studies the effect of wind and different landscapes on air flow through ACC. For this study software OpenFOAM was used and the flow was simulated with the use of RANS equations. For verification of numerical setup a model of one ACC cell with dimensions of platform 1.5x1.5 [m] was used. In this experiment static pressures behind fan and air flows through a model of surface of condenser for different rpm of fan were measured. In OpenFOAM software a virtual clone of this experiment was built and different meshes, turbulent models and numerical schemes were tested. After tuning up numerical setup virtual model of real ACC system was built. Influence of wind, landscape and height of ACC on air flow through ACC has been investigated.

1 Introduction

In standard design of air-cooled condenser (ACC) the influence of wind is not considered, as well as the surrounding terrain or other aspects that certainly affects the effectiveness of ACC [1]. This paper presents results based on CFD model of real smaller 2 cells of ACC. In CFD model the influence of wind and surrounding buildings is considered. From resulting data, the change of flow rate with the change of boundary conditions is evaluated as well as the influence of uniformity of velocity field on condenser on the effectiveness of ACC.

The study was provided with the software OpenFOAM and CFD model was validated on a smaller model of one cell ACC [2-3].

This study shows that the change of air volume flow with the change of boundary conditions can be very significant. The change of air volume flow through condenser can reach 25 [%] with the change of wind velocity. The influence of surrounding building can change the volume flow by about 10 [%]

2 Validation of CFD model

Validation of used CFD model is the step that cannot be omitted in any CFD study. For this study a model of one ACC cell was built with dimensions 1.5 x 1.5 [m]. The scheme of this model can be seen in Fig. 1.

Single row condenser that is used in modern ACC was replaced by screens and honeycombs in the model. Pressure drop of screen and honeycombs was measured in a wind tunnel.

In CFD a virtual image of this model was made. For simulation of pressure drop of screens and honeycombs a porous medium approach was used. Fan was simulated as a pressure jump. In Figure 2 below there is a computational domain.

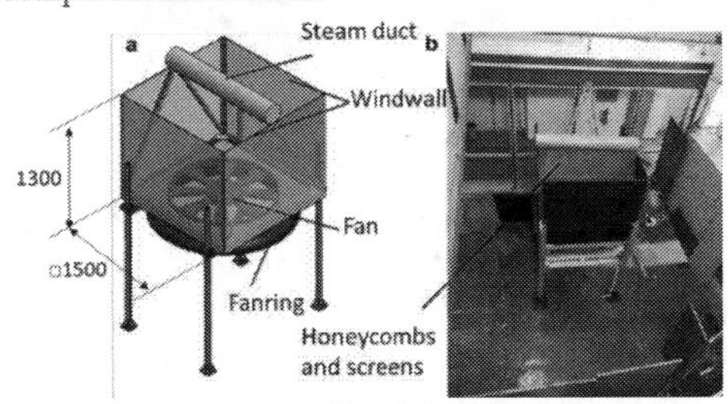

Fig. 1. Model of one cell ACC

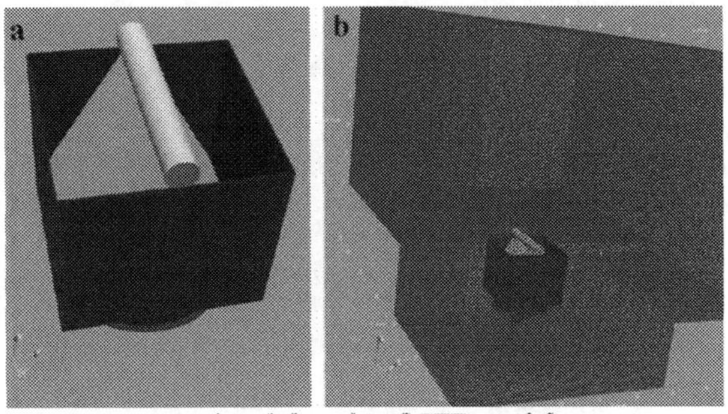

Fig. 2. Computational domain of CFD model

Static pressure behind the fan and air volume flow through screen for several fan rpm were measured on the experimental model. For measurement of air flow a multi constant anemometer (MCCA) was used. In the following figure there is a picture from measurement of

* Corresponding author: zdenek.sumara@fs.cvut.cz

air volume flow for one position. Air flow and static pressure behind the fan were measured for 237, 444, 591 and 884 [r/min] of the fan.

Fig. 3. One position of MCCA

Every position in which air flow rate was measured is in the scheme in Figure 4.

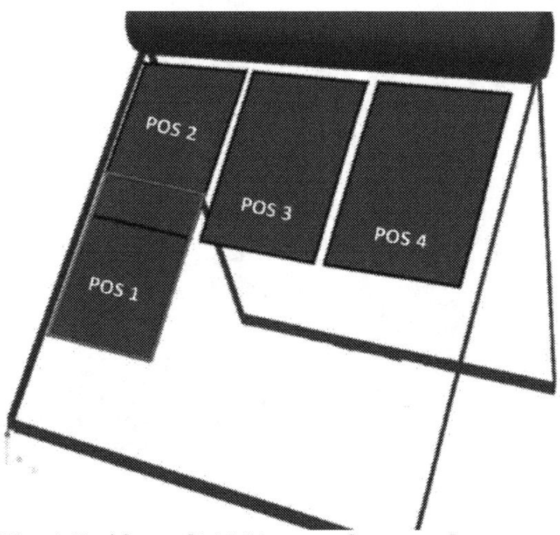

Fig. 4. Positions of MCCA on condenser surface

Experimental data were used for the validation of computational mesh as well as numerical schemes and model of turbulence. Sensitivity of simulation on mesh was tested in the first place. Meshes with 4, 6, and 13 millions of cells were tested. Results of this first testing can be found in Figure 5.

In this figure the comparison of experiment with the simulation for air flow through position 1 and 3 can be found. The biggest difference between CFD and experiment is for *Mesh 1*. The difference between *Mesh 2* and *Mesh 3* is only 4 % on average. Therefore, structure of *Mesh 2* was chosen for CFD study.

The results of simulation depend to a great degree on chosen turbulent model, that is why 3 models were tested – *standard k-ε*, *realizable k-ε* and *k-ω SST*. Deviation of simulation with *standard k-ε* model is about 20% on average. For model *realizable k-ε* and *k-ω SST* is deviation about 10% on average. The simulation with model *k-ω SST* ends very often with divergence. Model *realizable k-ε* was very stable during testing. For reasons given above the turbulence model *realizable k-ε* was chosen for CFD study [4].

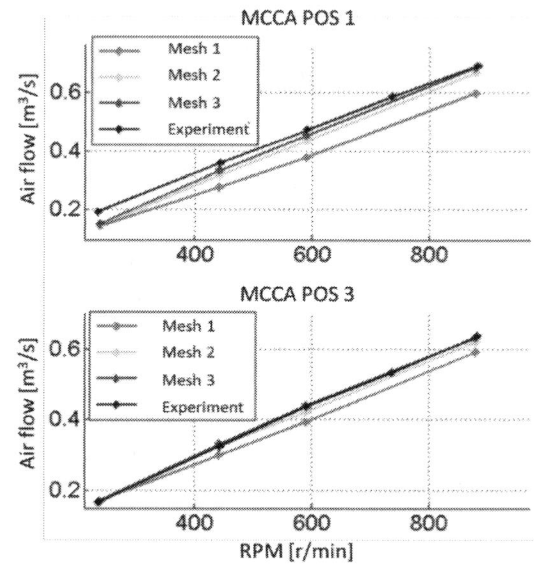

Fig. 5. CFD simulation sensitivity on mesh

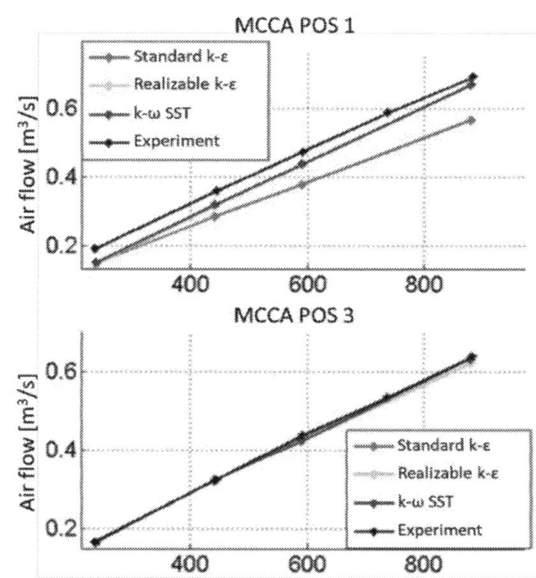

Fig. 6. Simulation results for different models of turbulence

3 CFD study

Geometry and surrounding buildings for study was inspired by real ACC system located in Slovakia.

Dimensions of two ACC cells in CFD model are equal to real ACC. In CFD model buildings with similar positions and dimensions were also placed as in surrounding of real ACC system. It is fact, that the building placed directly under ACC certainly influences an air flow inlet towards a condenser. The same can be said about the building on side of ACC that blocks an air inlet to ACC.

In figure 8 there is a scheme of model of ACC system. In red building under ACC there is a steam turbine, a generator, a control room etc. A warehouse (green building) is also the building that can influence significantly the air flow through ACC. Last building in model is a 48 [m] high chimney.

Fig. 7. Picture of real ACC system

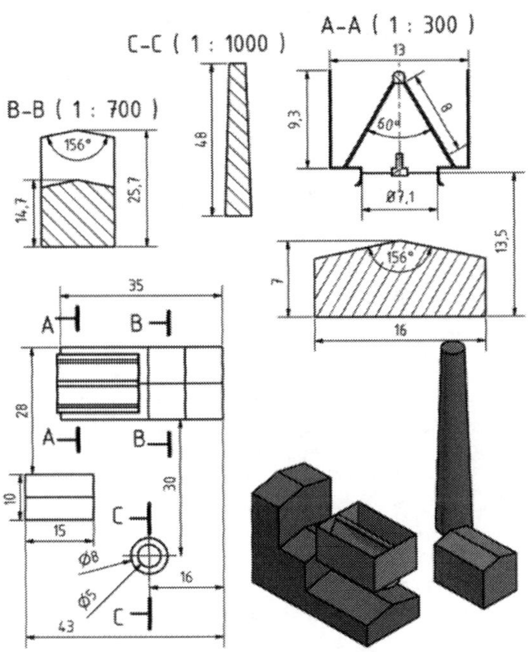

Fig. 8. Scheme of ACC model with dimensions in meters

3.1 Computational domain

Dimensions of computational domain is 600 x 600 [m] and the height is 180 [m]. For every simulation one of domain patch is the inlet and the other one is the outlet. Boundary conditions on ground are set as slip for course mesh region and as noslip for fine and finer mesh region. Boundary conditions for other walls of domain are set to slip condition.

For every calculation mesh has around 21 M cells. The scheme of domain with one of 4 inlets positon can be found in Figure 9.

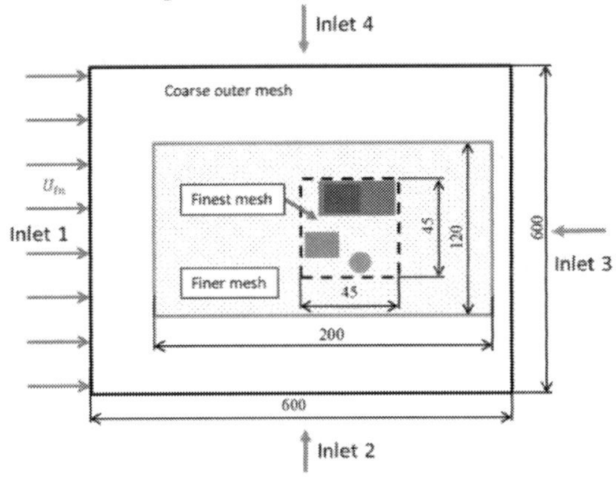

Fig. 9. Computational domain

3.2 Definition of simulation

Simulation is designed as a steady and incompressible flow without heat transfer. This is a reasonable simplification because of low velocities and quite low temperature of the stem in duct – heat exchange doesn't change velocity field much.

Simulation was provided for 3 velocities of wind – 0, 2 and 4 [m/s] and for 4 inlets. Mesh was especially adjusted for every inlet. In Figure 9 mesh adjust for inlet 1 can be found. Simulation was also made with and without buildings for every wind velocity and inlet.

4 Results

There is almost no difference between volume air flow through ACC from simulation with and without buildings for wind velocity 0 [m/s]. The building under ACC was probably considered in the design. Therefore, suction height is sufficient even with building under it and higher suction height cannot increase air flow through a condenser. The difference between these two main settings is found for wind velocity higher than 0 [m/s]. In Figure 10 there are graphs with the change of air volume flow through a condenser with the change of wind velocity. Base volume flow is the volume flow for wind velocity 0 [m/s]. For both settings - with/without buildings the base flow is almost the same. For wind velocity 2 [m/s] volume flow is higher for both cases. Highest increasing of volume flow is caused by wind velocity 2 [m/s].

Simulation without building does not depend on wind inlet direction which was expected and the increase of volume flow is about 24 [%]. In case with buildings there is the highest increase of volume flow for about 29 [%] for wind inlet 4 which makes sense.

The lowest effectiveness is shown in simulation with the wind inlet 2. That is because of chimney and green warehouse in Figure 8. Inlet air must flow around or above these buildings and the wake forms behind these buildings. Wake causes the decrease of pressure in inlet of ACC. The whole system has then bigger pressure drop.

Wind velocity 4 [m/s] causes the decrease of volume flow in the case without building. Volume flow is about 5 [%] lower than for conditions without wind.

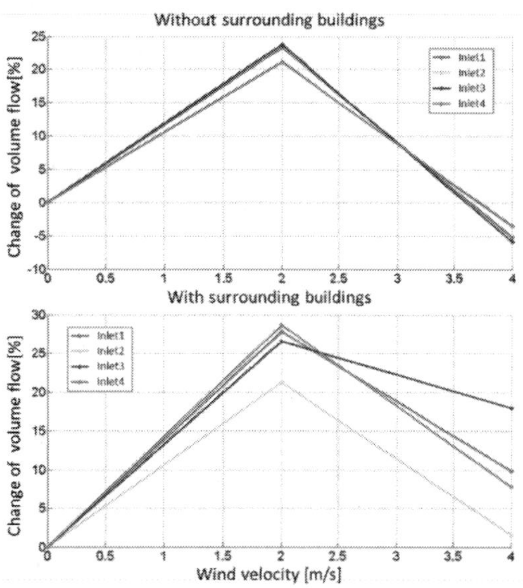

Fig. 10. Change of air volume flow with wind velocity

Reason for decreasing the volume is the decrease of a pressure at ACC inlet.

On other hand, volume flow in case with buildings is still bigger than for conditions with zero wind velocity. Buildings have positive impact on effectiveness of ACC. Significant effect is seen for wind inlet 3 – between inlet and ACC there is a red building as can be seen in Figures 9 and 8. The wake behind this building decreases the pressure at ACC outlet and at the same time ACC inlet is shielded so that decrease of inlet pressure is not so huge as without shielding.

Influence of non-uniform velocity field on condenser has been investigated which is not considered in standard design of ACC system. Therefore, with known equation for calculation of heat transfer coefficient heat duty of ACC was calculated for uniform velocity field and for the same volume flow but non-uniform velocity field known from CFD simulation. Fraction of these two heat duties is plot in graphs in Figure 11 for different wind velocities and wind inlets. Sensitivity of this fraction on wind velocity, wind inlet or surrounding buildings is small. On average heat duty for non-uniform velocity field is about 5 [%] smaller which is for sure not negligible.

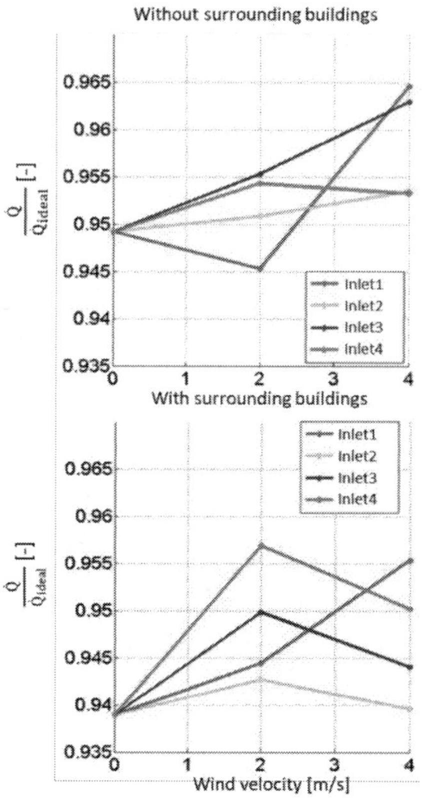

Fig. 11. Heat duty for uniform and real velocity field

5 Conclusion

In this study the impact of boundary conditions on effectiveness of ACC has been investigated. The results show significant influence of wind and surrounding buildings on the effectiveness of ACC. Wind can increase volume of air flow through condenser about 28 [%] or decrease volume of air flow about 5 [%]. Whether wind has negative or positive impact, it depends on wind velocity and on surrounding buildings. Next, the influence of non-uniformity of velocity field on surface condenser cannot be negligible. It causes about 5 [%] decrease of heat duty.

This study shows benefit of CFD simulation in design of ACC system. It would take into account an influence of surrounding conditions in design of ACC. Cooling systems based on ACC could be then smaller or their effectiveness could be higher and with that efficiency of steam turbine would be also higher.

In next studies a real ACC system with all important details included terrain and exact wind characteristics could be simulated. Results of this simulation should be compared with the design data of ACC and with measurement of air volume flow and heat duty direct on real ACC.

https://doi.org/10.1051/epjconf/201818002103

References

1. Kroeger, Detlev G. Air-cooled heat exchangers and cooling towers. Tulsa, Okl.: Penwell, c2004. ISBN 08-781-4896-5
2. Greenshields, Christopher J. OpenFOAM - Programmer's Guide. London, UK, 2017. OpenFOAM Foundation Ltd.
3. Greenshields, Christopher J. OpenFOAM - User Guide. London, UK, 2017.
4. Ferziger, Joel H. a M. Peric. Computational methods for fluid dynamics. 3rd, rev. ed. New York: Springer, c2002. ISBN 978-3-540-42074-3.

On mathematical modelling of aeroelastic problems with finite element method

Petr Sváček[1,*]

[1] *Dep. of Technical Mathematics, Faculty of Mechanical Engineering, Center of Advanced Aerospace Technology, Czech Technical University in Prague, Technická 4, 166 01 Praha 6, Czech Republic*

Abstract. This paper is interested in solution of two-dimensional aeroelastic problems. Two mathematical models are compared for a benchmark problem. First, the classical approach of linearized aerodynamical forces is described to determine the aeroelastic instability and the aeroelastic response in terms of frequency and damping coefficient. This approach is compared to the coupled fluid-structure model solved with the aid of finite element method used for approximation of the incompressible Navier-Stokes equations. The finite element approximations are coupled to the non-linear motion equations of a flexibly supported airfoil. Both methods are first compared for the case of small displacement, where the linearized approach can be well adopted. The influence of nonlinearities for the case of post-critical regime is discussed.

1 Introduction

There are many technical or scientific applications in a wide range of technical disciplines, where the mathematical modelling of fluid-structure interaction (FSI) problems is important. Let us mention for instance a wide range of aeroelastic tools used in aerospace engineering to investigate the aircraft safety, see e.g. [1], aeroelastic stability of bridges in civil engineering, see e.g. [2], or scientific applications in biomechanics, e.g. [3] or [4]. The classical aeroelastic approach based on the asymptotic aeroelasticity is fast, efficient and reliable for thin airfoils, see e.g. [5], and that is why it is still very popular in the technical practice.

The aeroelasticians in practice use the linearized methods to find the solution of aeroelastic problems usually in terms of the prediction of the aeroelastic instability, see [6]. The use of linearized aerodynamics allows to determine the aeroelastic asymptotic stability in terms of the critical velocity, damping and frequency graphs, see [1], [7], or [8]. Although similar approaches are well studied it needs to be mentioned that the asymptotic stability determines the critical velocity, but does not guarantee the safety, see [9, 10]. The external excitation can lead to transient growth and consequently to the structural failure, although the system is aeroelastically stable. Thus the use of an alternative accurate aeroelastic simulations is very attractive because it reduces the development risk, the number of experiments and the possibile future design modifications. which could provide additional information to that one provided from the linearized approach.

However, the mathematical modelling of FSI problems in general is much more complicated, because one needs to consider the viscous possibly turbulent flow which is moreover in the interaction with the nonlinear behavior of the elastic structure. Moreover, the time changes of the flow domain due to the vibrating structure cannot be neglected and the methods applicable on moving domain or meshes must be employed, see e.g. [11]. Last, the coupled fluid-structure system for the fluid flow and for the oscillating structure needs to be solved simultaneously using a coupled strategy at any time instant, which possibly can create an artificial instability of such a scheme, see e.g. [12].

In applications the FSI problems are usually solved with the aid of the so-called partitioned or de-coupled approach, which approximates the fluid flow and structure motion by different solvers. The coupling is then enforced by using suitable conditions prescribed at the fluid-structure interface. These approaches can be characterized further either as strongly coupled scheme (if subiterations are used to enforce the interface conditions) or as loosely coupled one. Such a loosely coupled scheme was used e.g. in [13] for FSI problems tested on a proposed benchmark problem and the unconditional stability was derived based on an energy estimate. Nevertheless the loosely coupled algorithms were observed in many cases to be instable due to the poor fluid-solid coupling, see [12]. In order to get a stable scheme either the partitioned approach with relaxed strongly coupled algorithm or the monolithic approaches can be used, cf. [14, 15] or [16]. Nevertheless, the high fidelity aeroelastic models are still very demanding to be solved and thus its direct use in industry is rather rare, see [17].

* e-mail: petr.svacek@fs.cvut.cz

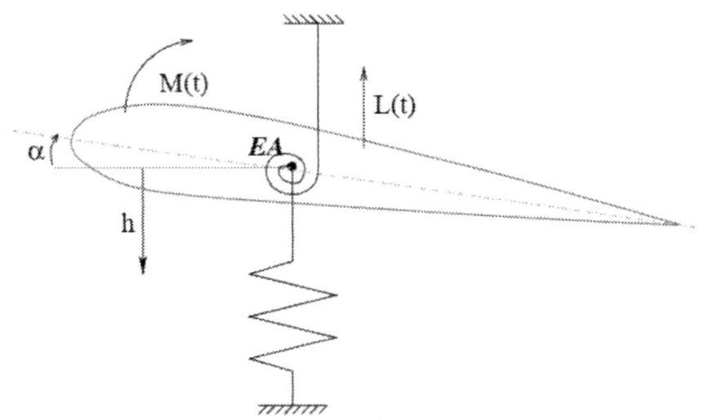

Fig. 1. Elastically supported airfoil with two degrees of freedom bending h and pitching α (on the left). Airfoil in undeformed position showing the elastic axis (EA) and center of gravity (CG) (on the right).

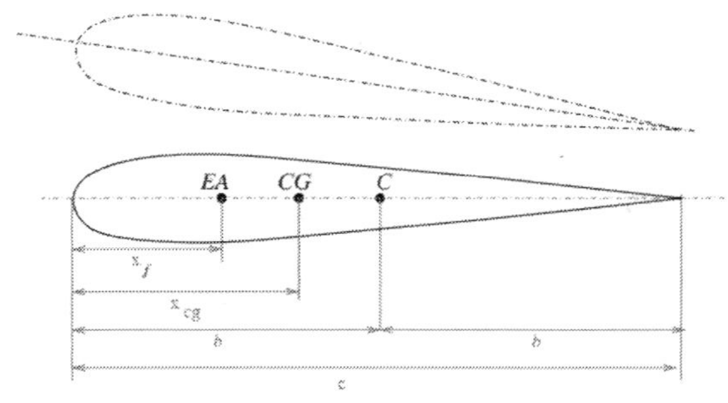

Fig. 2. Elastically supported airfoil with two degrees of freedom bending h and pitching α (on the left). Airfoil in undeformed position showing the elastic axis (EA) and center of gravity (CG) (on the right).

For the solution of each part in FSI problem a suitable numerical method needs to be applied. One of the most popular methods is the finite element method, but in its application for numerical simulation of incompressible flow problem one must overcome several sources of instabilities. Particularly, the possible instability caused by the incompatibility of the pressure and velocity pairs of finite elements needs to be taken into account, see e.g. [18], [19]. Furthermore, the other instability is caused by the dominating convection terms, see [20]. Let us mention that there is also another not so well-studied instability source in the Galerkin discretization method which is related to the possible poor resolution of pressure, see [21], [22], [23]. For the case of high Reynolds numbers the turbulence effects should be also taken into account using a wide variate of the mathematical models, see e.g. [24], [25], [26], [27] or [28] as few examples how to include a turbulence model in the numerical simulation.

The other thing is that for approximation of FSI problem the applied numerical method should be able to treat the moving domain/meshes. The most popular method is the arbitrary Lagrangian-Eulerian(ALE) method, see [29]. There are several possibilities how to apply the finite element method, see [30]. Usually in order to provide higher order accuracy or stability some additional assumptions on the ALE are required, see [31], [32]. Particularly the so-called geometrical conservation law is important, see [33].

In this paper the numerical analysis of an interaction of the incompressible viscous flow with a vibrating airfoil is considered, see e.g. [30], [34] or [35]. The attention is paid to an approximation on moving meshes with the aid of suitable arbitrary Lagrangian-Eulerian(ALE) method. For the case of high Reynolds numbers anisotropically refined mesh is used and the choice of the stabilizing parameters is discussed. Two finite element pairs are considered for the numerical approximation and the stabilization procedure for both of them is presented based on the so-called residual based stabilization, see e.g. [23]. A benchmark problem of mutual interaction of the airflow and a flexibly supported airfoil with two degrees of freedom (2DOF) (the bending and the torsion modes) is considered. The problem is solved using a partitioned scheme with strongly coupled/loosely-coupled scheme. The problem of the solution of the non-linear flow problem per-each time step is discussed. The paper is organized as follows: first the mathematical model is presented, then the weak form suitable for application of the finite element method is introduced, then the numerical approximation is presented and the numerical results are shown.

2 Structure vibrations

In this paper an airfoil section that can be vertically displaced by h (downwards positive) and rotated by angle α (clockwise positive) is considered, see Fig. 1. These two degrees of freedom h and α are also called the bending and the pitching, respectively. The nonlinear equations of motion then read (see [30])

$$\begin{aligned} m\ddot{h} + S_\alpha \ddot{\alpha} \cos\alpha - S_\alpha \dot{\alpha}^2 \sin\alpha + k_h h &= -L(t), \\ S_\alpha \ddot{h} \cos\alpha + I_\alpha \ddot{\alpha} + k_\alpha \alpha &= M(t). \end{aligned} \quad (1)$$

where m is the mass of the airfoil, S_α is the static moment around the elastic axis (EA), and I_α is the inertia moment around EA. The parameters k_h and k_α denote the stiffness coefficients. On the right-hand side the aerodynamical lift force $L(t)$ and aerodynamical torsional moment $M(t)$ are involved.

Under the assumption of small displacements (i.e. $\alpha \approx 0$, $\sin\alpha \approx 0$ and $\cos\alpha \approx 1$) the system of equations can be linearized as

$$\begin{aligned} m\ddot{h} + S_\alpha \ddot{\alpha} + k_h h &= -L(t), \\ S_\alpha \ddot{h} + I_\alpha \ddot{\alpha} + k_\alpha \alpha &= M(t). \end{aligned} \quad (2)$$

In order to solve system of ordinary differential equations (1) or (2) the aerodynamical lift force $L(t)$ and the aerodynamical moment $M(t)$ needs to be evaluated for any time instant t. These aerodynamical quantities can be expressed with the aid of flow quantitiesas pressure p, flow velocity $\boldsymbol{u}$, fluid density ρ and fluid kinematic viscosity ν, or using a suitable aerodynamical model.

The system of equations (2) is also completed by the prescription of suitable initial conditions at $t_0 = 0$ for $\alpha(t_0), h(t_0), \dot{\alpha}(t_0)$ and $\dot{h}(t_0)$.

3 Aerodynamical model

Two aerodynamical models are considered in this paper. First, the aerodynamical forces are determined with for the case of a thin oscillating airfoil with focus on the determination of the aeroelastic stability of a vibrating airfoil. Using the asymptotic stability approach only the critical velocity needs to be determined, which is characterized as the flow velocity for which the airfoil starts to harmonically oscillate. This means that the applied aerodynamical model does not have to describe the aerodynamical forces in general but only for the case of harmonically vibrating airfoil.

Second, the aerodynamical forces are evaluated using the surface integral of the density of the aerodynamical forces obtained from the flow velocity and pressure. The flow velocity and pressure are then assumed to be solution of the incompressible Navier-Stokes equations. The solution of the Navier-Stokes equations is approximated using a stabilized finite element method.

3.1 Linearized model.

The first model of the linearized aerodynamics is based on the assumption of thin airfoil, for which the aerodynamical forces acting on the harmonically oscillating airfoil can be evaluated. Using the potential function approach, conformal transformation and writting the solution in terms of known analytical solutions, the aerodynamical lift force and moment can be expressed, [7]. Moreover, these aerodynamical forces depend only on the reduced frequency $k = b\omega/U_\infty$ and the far field velocity U_∞ (see [8]) as

$$\begin{aligned} \frac{L}{\pi\rho b} &= b\left(\ddot{h} - d_m\ddot{\alpha} + U_\infty\dot{\alpha}\right) \\ &+ 2U_\infty C(k)\left(\dot{h} + U_\infty\alpha + d_t\dot{\alpha}\right) \\ \frac{M}{\pi\rho b^2} &= \left(d_m\ddot{h} - (\frac{b^2}{8} + d_m^2)\ddot{\alpha} - U_\infty d_t\dot{\alpha}\right) \\ &+ \frac{2U_\infty}{b} d_e C(k)\left(\dot{h} + U_\infty\alpha + d_t\dot{\alpha}\right), \end{aligned} \quad (3)$$

where x_f denotes the distance of the elastic axis from the leading edge, $b = \frac{c}{2}$ denotes the midchord of the airfoil (see Fig. 2, $d_e = x_f - \frac{c}{4}$ is the distance of the elastic axis from the aerodynamical center, $d_m = x_f - b$ is the distance of the elastic axis from the midchord position, $d_t = \frac{3}{4}c - x_f$ is the distance of the elastic axis from the 75 % of the airfoil and $C(k)$ is the Theodorsen function.

3.2 Incompressible Navier-Stokes equations

The flow velocity $\boldsymbol{u}$ and the pressure p needs to be determined for instance using a flow model. Here, the flow in the bounded time-dependent domain $\Omega_t \subset R^2$ is governed by the system of incompressible Navier-Stokes equations

$$\rho\frac{\partial \boldsymbol{u}}{\partial t} + \rho\nabla\cdot(\boldsymbol{u}\otimes\boldsymbol{u}) + \nabla p - \nabla\cdot\left(2\mu\left(\nabla\boldsymbol{u} + (\nabla\boldsymbol{u})^T\right)\right) = 0, \quad (4)$$

coupled with the continuity equation

$$\nabla\cdot\boldsymbol{u} = 0, \quad (5)$$

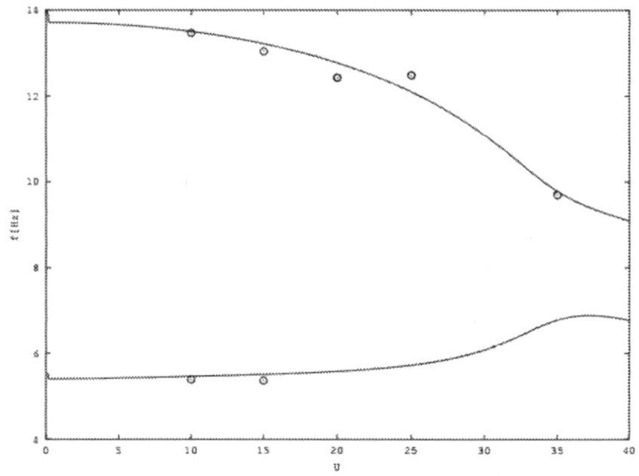

Fig. 3. The frequency of vibrating airfoil in dependence on the far field velocity U_∞ for the first considered aeroelastic problem. The results obtained by the linearized aerodynamical theory (solid line) and by the finite element method (symbols) are shown.

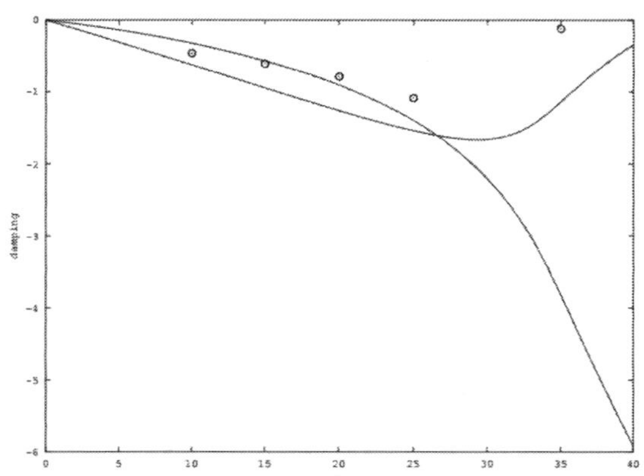

Fig. 4. The aerodynamical damping in dependence on the far field velocity U_∞ for the first considered aeroelastic problem. The results obtained by the linearized aerodynamical theory (solid line) and by the finite element method (symbols) are shown.

where $\boldsymbol{u} = \boldsymbol{u}(x,t)$ is the flow velocity, $p = p(x,t)$ is the pressure, ρ is the constant fluid density ρ_∞, and μ is the fluid viscosity. In order to treat the flow in time dependent computational domain the arbitrary Lagrangian Eulerian method is used, for the sake of brevity the description of the ALE method is omitted here. The boundary of the computational domain $\partial\Omega_t$ consists of mutually disjoint parts Γ_I (inlet), Γ_O (outlet), Γ_{Wt} (the moving part, vibrating airfoil). The following boundary conditions are prescribed

$$\begin{aligned} &a) \quad \boldsymbol{u} = \boldsymbol{w}_D \quad \text{on } \Gamma_{Wt}, \\ &b) \quad \boldsymbol{u} = \mathbf{v}_D \text{ on } \Gamma_I, \\ &c) \quad -p\boldsymbol{n} + \frac{1}{2}(\boldsymbol{u}\cdot\boldsymbol{n})^-\boldsymbol{u} + \nu\frac{\partial \boldsymbol{u}}{\partial \boldsymbol{n}} = 0, \text{ on } \Gamma_O, \end{aligned} \quad (6)$$

where the domain velocity $\boldsymbol{w}_D$ is on Γ_{Wt} equal to the local boundary velocity, $\boldsymbol{u}$ is the flow velocity, p is the kinematic pressure, $\boldsymbol{n}$ denotes the unit outward normal vector to Ω_t and $\alpha^- = \min(\alpha, 0)$ denotes the negative part of a real number α. The computations start from the initial condition $\boldsymbol{u}(x, 0) = \boldsymbol{u}^0(x)$ for $x \in \Omega_0$.

The aeroelastic lift force and aerodynamical moment can be also determined using the flow velocity and pressure, i.e. the aerodynamical force read

$$L = -l \int_{\Gamma_{Wt}} \sigma_{2j} n_j \, dS, \tag{7}$$

and the aerodynamical moment is given by

$$M = l \int_{\Gamma_{Wt}} \sigma_{ij} n_j r_i^{\text{ort}} \, dS, \tag{8}$$

where $\sigma_{ij} = \rho\left[-p\delta_{ij} + \nu(\partial_{x_j} u_i + \partial_{x_i} u_j)\right]$, $r_1^{\text{ort}} = -(x_2 - x_2^{\text{EA}})$, $r_2^{\text{ort}} = x_1 - x_1^{\text{EA}}$, l denotes the considered depth of the airfoil section, and $x^{\text{EA}} = (x_1^{\text{EA}}, x_2^{\text{EA}})$ is the position of EA of the airfoil at the time instant t.

4 Numerical simulations

Here, two different approaches were used for numerical simulation using the two aerodynamical models mentioned above. First, the model of linearized aerodynamics is used. In this case however two types of instability needs to be considered, the static instability called the divergence type of instability the dynamic type if instability known as flutter. For the finite element method application we focus here only on approximation of the fluid flow.

4.1 Linearized aerodynamics

4.1.1 *Static instability.*

First, we consider the problem of static instability (called divergence type of instability). To this end we introduce the non-dimensional lift coefficient c_L defined as

$$c_L = \frac{L}{\frac{1}{2}\rho U_\infty^2 c l},$$

where U_∞ denotes the far-field flow velocity, c denotes the airfoil chord, ρ denotes the constant fluid density and l denotes the depth of the considered airfoil section (the dimension of the airfoil section in the z-direction). Furthermore, for the thin airfoil the lift coefficient (approximately) depends on the angle α linearly as

$$c_L = 2\pi(\alpha - \alpha_0),$$

where α_0 denotes the angle of attack of zero lift force, which for the sake of simplicity is asssumed to equal zero in what follows, i.e. $\alpha_0 = 0$. Similarly, the non-dimensional aerodynamical moment around the aerodynamical center x_{AC} (aerodynamical center is the point of the airfoil for which the aerodynamical moment is independent of the angle of attack α, for subsonic flow it is 1/4 of the airfoil) reads

$$c_{Mac} = \frac{M_{ac}}{\frac{1}{2}\rho U_\infty^2 c^2 l}.$$

The aerodynamical lift force then reads

$$L = \pi\rho U_\infty^2 c l \alpha$$

and the aerodynamical moment with respect to the elastic axis can be expressed as

$$M = \frac{1}{2}\rho U_\infty^2 c^2 l c_{Mac} + L(x_f - x_{ac}).$$

Assuming the static case (i.e. $\dot\alpha = \dot h = 0$ similarly as the second derivatives) and using the expressions for L and M in (2) we get the homogenous system of linear equations

$$k_h h + \pi\rho U_\infty^2 c l \alpha = 0, \tag{9}$$
$$(k_\alpha - (x_f - x_{ac})\pi\rho U_\infty^2 c l)\alpha = \frac{1}{2}\rho U_\infty^2 c^2 l c_{Mac},$$

whose solution depends on the value

$$k_\alpha - (x_f - x_{ac})\pi\rho U_\infty^2 c l.$$

If this value is non-zero, there exists a unique solution. The interesting case is the zero value, which means that there is no (unique) solution of the system and consequently no static equilibrium exists. This means that the system is aeroelastically unstable with the divergence type of instability and the velocity U_{div} for which it occurs is called the divergence type of instability. This velocity can be expressed as

$$U_{div} = \sqrt{\frac{k_\alpha}{(x_f - x_{ac})\pi\rho c l}}.$$

4.1.2 *Flutter determinant and pk-method*

Using the linearization (3) in the linearized system of equations (1) leads to the system of two nonlinear equations with two parameters, the reduced frequency k and the far field velocity U_∞. Based on this nonlinear system the critical flutter velocity can be determined and using the so-called $p - k$ method the vibration frequency and damping for a given flow velocity U_∞ can be computed.

4.2 Finite element method for flow problem

In order to discretize the problem (4) in time, the equidistant partition $0 = t_0 < t_1 < \cdots < T$, $t_k = k\Delta t$ of the time interval $[0, T]$ is considered with the constant time step $\Delta t > 0$. The time derivative in (4) is approximated by the second order backward difference formula

$$\left.\frac{\partial \boldsymbol{u}}{\partial t}\right|_{t=t_{n+1}} \approx \frac{3\boldsymbol{u}^{n+1} - 4\boldsymbol{u}^n + \boldsymbol{u}^{n-1}}{2\Delta t}. \tag{10}$$

The time discretized problem is than spatially discretized by the finite element method. To this end the approximations $\boldsymbol{u}^k$ and p^k are sought in the finite element spaces $\boldsymbol{W}_h$

and Q_h, respectively. In the practical computations these spaces are defined with the aid an admissible triangulation $\mathcal{T}_h$ of Ω and the well known Taylor-Hood finite element is used.

The discrete formulation contains the Galerkin terms defined for any $U = (\boldsymbol{u}, p)$ and $V = (\boldsymbol{\varphi}, q)$ by

$$\begin{aligned} a(U,V) &= \left(\frac{3\boldsymbol{u}}{2\Delta t} + \boldsymbol{u}\cdot\nabla\boldsymbol{u}, \boldsymbol{\varphi}\right)_\Omega \\ &+ (-p\mathbb{I} + \nu\nabla\boldsymbol{u}, \nabla\boldsymbol{\varphi})_\Omega + (\nabla\cdot\boldsymbol{u}, q)_\Omega, \quad (11) \\ f(V) &= \left(\frac{4\boldsymbol{u}^n - \boldsymbol{u}^{n-1}}{2\Delta t}, \boldsymbol{\varphi}\right)_\Omega . \end{aligned}$$

The stabilization terms containing the SUPG/PSPG stabilization are defined by

$$\begin{aligned} \mathcal{L}(U,V) &= \sum_{K\in\mathcal{T}_h} \delta_K \times \\ \Big(\frac{3\boldsymbol{u}}{2\Delta t} &- \nu\Delta\boldsymbol{u} + (\boldsymbol{u}\cdot\nabla)\boldsymbol{u} + \nabla p, (\boldsymbol{u}\cdot\nabla)\boldsymbol{\varphi} + \nabla q\Big)_K, \\ \mathcal{F}(V) &= \sum_{K\in\mathcal{T}_h} \delta_K\Big(\frac{4\boldsymbol{u}^n - \boldsymbol{u}^{n-1}}{2\Delta t}, (\boldsymbol{u}\cdot\nabla)\boldsymbol{\varphi} + \nabla q\Big)_K, \end{aligned}$$

and the stabilization term for div-div stabilization reads $\mathcal{P}(U,V) = \sum_{K\in\mathcal{T}_h} \tau_K(\nabla\cdot\boldsymbol{u}, \nabla\cdot\boldsymbol{\varphi})_K$ with the choice of δ_K and τ_K according to [30]. This approach was previously numerically verified on a number of benchmark problems, see [30], and also it satisfy the optimal error estimates and thus it should be more precise compared e.g. to the classical finite volume methods.

Then the non-linear *stabilized discrete problem* at a time instant $t = t^{n+1}$ reads: Find $p^{n+1} := p$, $\boldsymbol{u}^{n+1} := \boldsymbol{u}$, $U = (\boldsymbol{u}, p) \in \boldsymbol{W}_h \times Q_h$, such that $\boldsymbol{u}$ satisfies approximately the conditions (6 a-b) and

$$a(U,V) + \mathcal{L}(U,V) + \mathcal{P}(U,V) = f(V) + \mathcal{F}(V), \qquad (12)$$

holds for all $V = (\boldsymbol{\varphi}, q) \in \boldsymbol{X}_h \times Q_h$. The solution of the non-linear problem is realized by the Oseen linearization. The nonlinear problem (12) is then coupled with the system (1) and solved simultaneously.

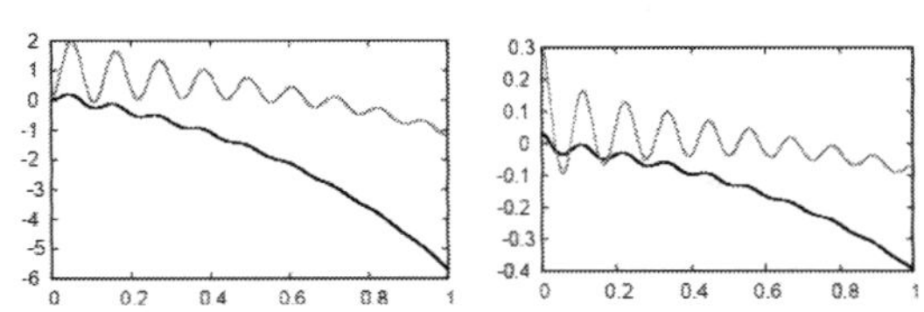

Fig. 5. Aeroelastic response in terms $h(t)$ (on the left) and $\alpha(t)$ (on the right) for $U = 40m/s$ in dependence on the value of the initial displacement of $\alpha(0)$.

5 Numerical results

In this paper two aeroelastic problems were considered, see [30] and [7]. The first one is an example of the divergence type of instability and with the further increase of the far field velocity the flutter type of instability occurs. The numerical approximation of such an example can be a little bit difficult particularly for the post-critical velocities, where the behaviour of the aeroelastic is non-linear and for instance its behaviour can be dependent on the initial displacement. The second example is the classical flutter type of instability, where the instability occurs due to the coupling between the two modes of bending and torsion.

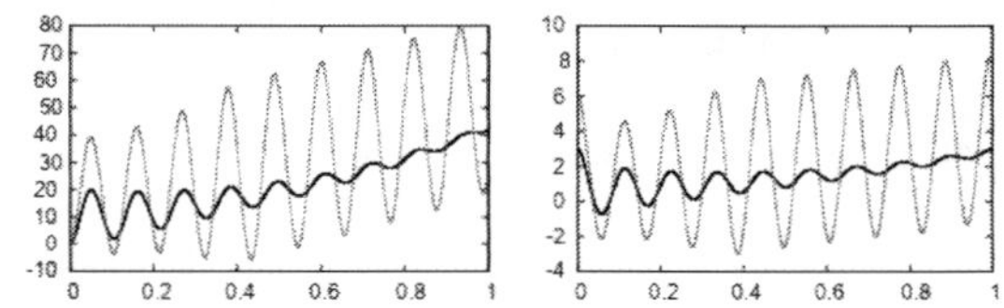

Fig. 6. Aeroelastic response in terms $h(t)$ (on the left) and $\alpha(t)$ (on the right) for $U = 40m/s$ in dependence on the value of the initial displacement of $\alpha(0)$.

5.1 Divergence type of instability case

A model problem originally studied in [30] is considered here. In this case the airfoil NACA 0012 was considered in interaction with incompressible fluid with the constant density$\rho = 1.225\text{kg/m}^3$, the fluid kinematic viscosity $\nu = \mu/\rho = 1.5 \times 10^{-5}\,\text{m}^2\,\text{s}$, the airfoil chord $c = 0.3\,\text{m}$, the airfoil airfoil mass $m = 1.7324\,\text{kg}$ (related to the depth 1 m), the airfoil inertia moment $I_a = 0.0097458\,\text{kg m}^2$, the airfoil static moment $S_a = -0.015592\,\text{kg m}$, the bending stiffness was $k_h = 2102.2$ N / m and the torsional stiffness $k_a = 73.912$ N m / rad. The elastic axis was located at 40 percent of the airfoil chord measured from the leading edge of the airfoil, $x_f = 0.4c = 0.12\,\text{m}$ and the center of gravity of the airfoil was located at 37 percent of the airfoil, i.e. $x_{cg} = 0.37c$.

The flutter velocity was determined approximately as 42.5 m/s, whereas the system becomes unstable due to divergence type of instability at lower far field velocity $U_\infty = 37.7$ m/s. The linearized approach prediction (realized with the aid of MATLAB/Octave script) in terms of aerodynamical damping and vibrations frequency is shown in Fig. 3 in terms of frequency and in Fig. 4 in terms of damping. In the graphs the flutter velocity 42.5 can be identified from the graph as the lowest positive velocity with the zero damping coefficient. Further, the finite element method was applied on the same aeroelastic problem and the similar results in terms of frequency and damping was obtained for the sub-critical far field velocities in the range 5 - 35 m/s, see Fig. 4. However, with the increase of the far field velocity a difference can be observed namely in the damping coefficient. This is particularly due to the fact, that the applied numerical approach is able to capture the divergence type of instability, which is not captured by the linearized approach. This also corresponds well with the results for the post-critical velocity $U_\infty = 40$ m/s and different initial displacement $\alpha(0) = 0.03, 0.3, 3$ and 6 degrees. The aeroelastic response shown in Figs. 5, 6 have the typical divergence type of

https://doi.org/10.1051/epjconf/201818002104

instability behaviour, which especially for the initial displacement of 6 degrees start to resemble rather the flutter type of instability. The flutter velocity determined by the linearized approach is however determined approximately as 42.5 m/s, see Fig. 3-4. This decrease of the flutter velocity due to the high initial displacement can not be captured by the linearized approach, but it well agrees with the real physical phenomena.

5.2 Flutter instability case

For the second considered case the airfoil NACA 0012 was considered in interaction with incompressible fluid with the constant densityρ = 1.225kg/m^3, the fluid kinematic viscosity $\nu = \mu/\rho = 1.5 \times 10^{-5}$ m^2 s, the airfoil chord $c = 0.254$ m, the airfoil airfoil mass $m = 4.7174$ kg, the airfoil inertia moment $I_a = 0.029522$ kg m^2, the airfoil static moment $S_a = 0.14978$ kg m, the bending stiffness was k_h = 14741 N / m and the torsional stiffness k_a = 121.30 N m / rad. The elastic axis was located at 42.5 percent of the airfoil chord measured from the leading edge of the airfoil, $x_f = 0.425c = 0.10795$ m and similarly the center of gravity of the airfoil was located at 55 percent of the airfoil, i.e. $x_{cg} = 0.55c = 0.13970$ m.

The aeroelastic system was investigated using the flutter determinant and the flutter velocity was determined as U_{crit} = 27.387 m/s with the flutter frequency f = 9.3181 Hz. The application of the $p - k$ method lead to the determined flutter velocity U_{crit} = 27.8 m/s with the found flutter frequency f = 9.4 Hz, which agrees well with the flutter determinant method. Let us also mention that for this case the divergence critical velocity is much higher (U_{div} = 52.8 m/s) then the flutter velocity as the position of the elastic is before the centre of gravity.

The finite element method was applied for solution of this aeroealstic system for the prescribed far field velocities U_∞ = 5, 10, 15, 20, 22, 24, 26 and 28 m/s. Results are shown in Figures 9-16. The aeroelastic response agrees well for lower flow velocities 5 – 20 m/s both in the frequency and also the increase of the aerodynamicl damping can be clearly observed, check Figures 9, 10, 11, 12.

6 Conclusion

This paper described two possible strategies of solution of an classical aeroelastic problem. The first on based on the classical approach of linearized aerodynamics was susccessfully applied for solution of two aeroelastic problems from references. Its results were also compared to the other method described in this paper, which is the finite element approximations of the FSI problem. Both approaches were found to be very similar in terms of the predicted frequency and damping for sub-critical velocities. However for far field velocities close to the critical one it was found that the latter approach based on the finite element method can be influenced by the nonlinearities presented in the system as large displacements, high initial displacement etc.

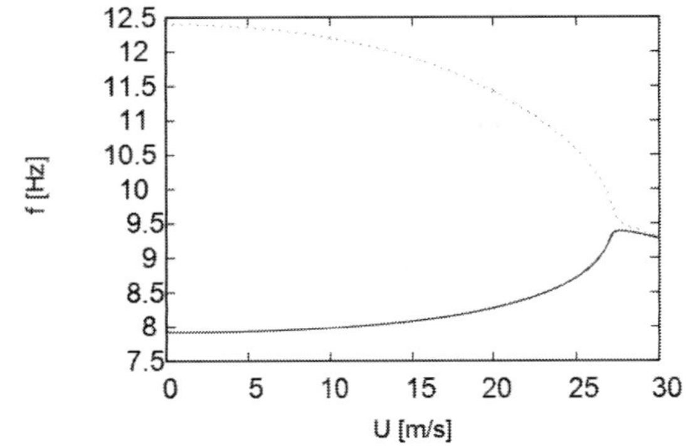

Fig. 7. The frequency of vibrating airfoil in dependence on the far field velocity U_∞ for the second considered aeroelastic problem. The results obtained by the linearized aerodynamical theory (solid line) and by the finite element method (symbols) are shown.

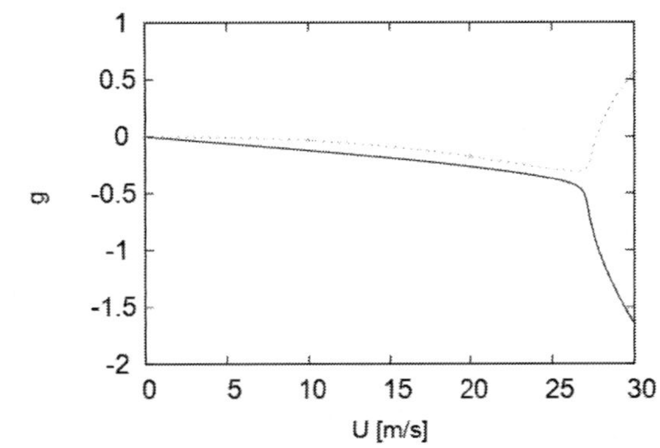

Fig. 8. The aerodynamic damping in dependence on the far field velocity U_∞ for the second considered aeroelastic problem. The results obtained by the linearized aerodynamical theory (solid line) and by the finite element method (symbols) are shown.

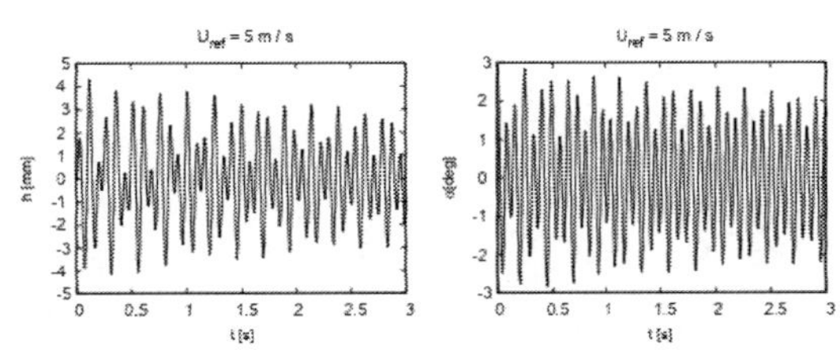

Fig. 9. Aeroelastic response for the seconde considered case in terms of h and α in time domain for far field velocity U_∞ = 5 m/s

 https://doi.org/10.1051/epjconf/201818002104

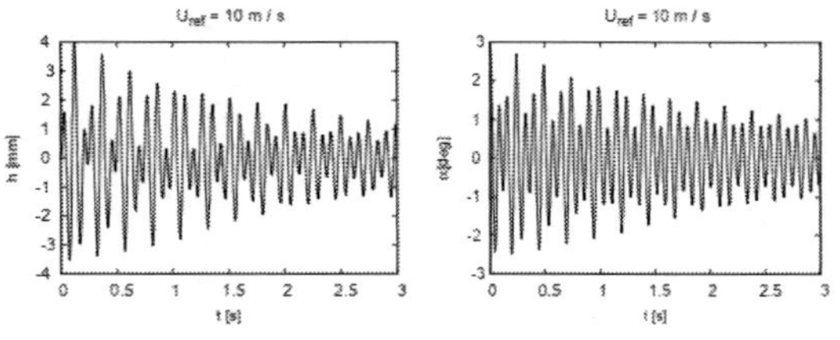

Fig. 10. Aeroelastic response for the seconde considered case in terms of h and α in time domain for far field velocity $U_\infty = 10$ m/s

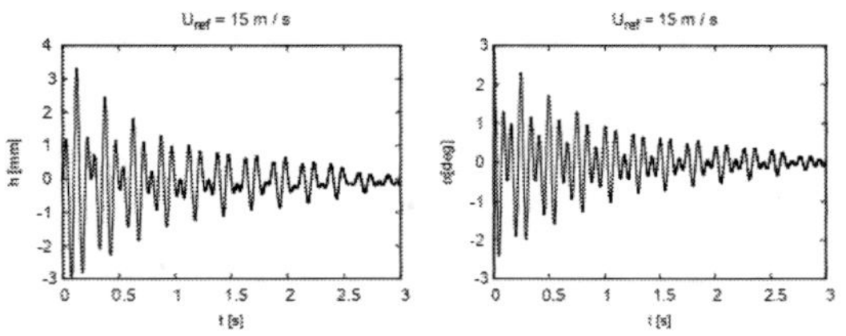

Fig. 11. Aeroelastic response for the seconde considered case in terms of h and α in time domain for far field velocity $U_\infty = 15$ m/s

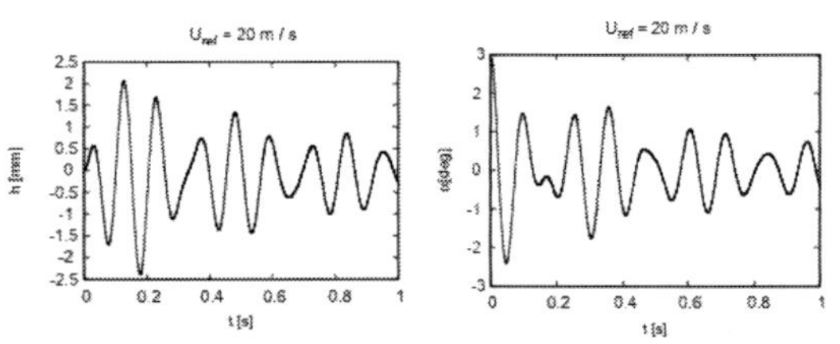

Fig. 12. Aeroelastic response for the seconde considered case in terms of h and α in time domain for far field velocity $U_\infty = 20$ m/s

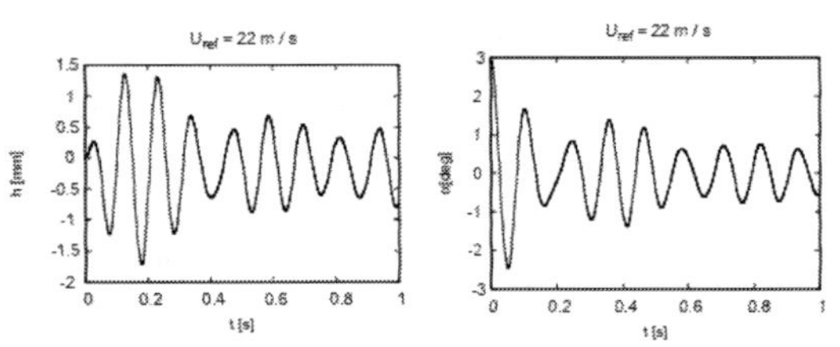

Fig. 13. Aeroelastic response for the seconde considered case in terms of h and α in time domain for far field velocity $U_\infty = 22$ m/s

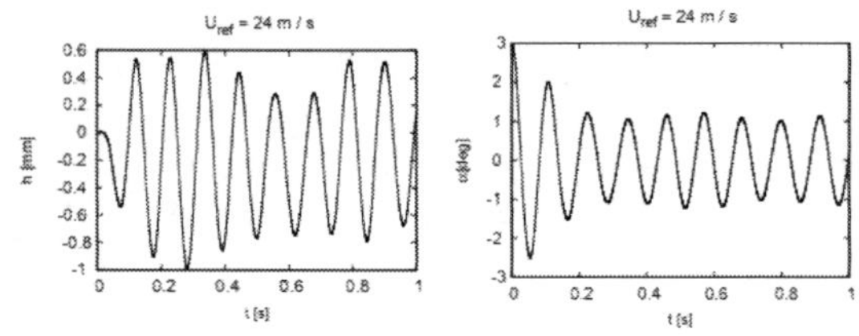

Fig. 14. Aeroelastic response for the seconde considered case in terms of h and α in time domain for far field velocity $U_\infty = 24$ m/s

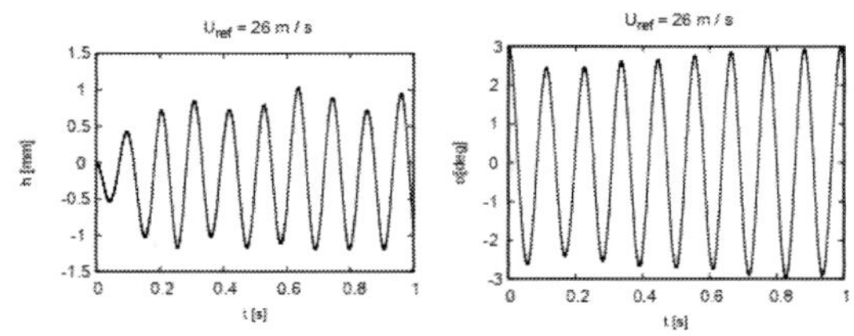

Fig. 15. Aeroelastic response for the seconde considered case in terms of h and α in time domain for far field velocity $U_\infty = 26$ m/s

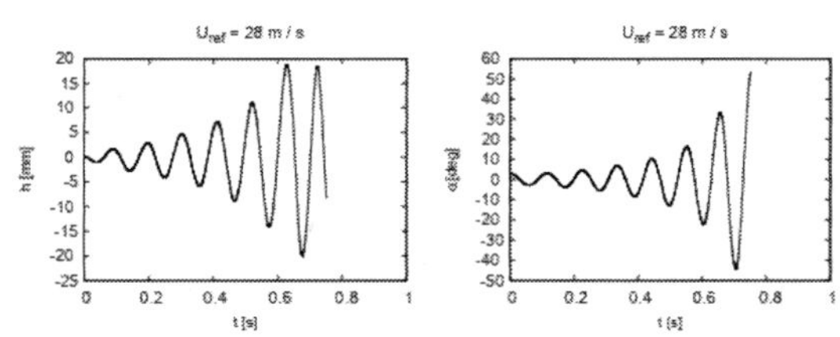

Fig. 16. Aeroelastic response for the seconde considered case in terms of h and α in time domain for far field velocity $U_\infty = 28$ m/s

Acknowledgements. Authors acknowledge support from the EU Operational Programme Research, Development and Education, and from the Center of Advanced Aerospace Technology (CZ.02.1.01/0.0/0.0/16_019/0000826), Faculty of Mechanical Engineering, Czech Technical University in Prague.

References

[1] E.H. Dowell, R.N. Clark, *A modern course in aeroelasticity*, Solid mechanics and its applications (Kluwer Academic Publishers, Dordrecht, Boston, 2004)

[2] E. Simiu, R.H. Scanlan, *Wind effects on structures : fundamentals and applications to design*, 3rd edn. (New York : John Wiley, 1996)

[3] S. Kaneko, T. Nakamura, F. Inada, M. Kato, *Flow-Induced Vibrations. Classifications and Lessons from*

Practical Experiences. (Technical Section on Flow-Induced Vibrations, JSME DMC Division, Elsevier, 2008)

[4] T. Bodnár, G.P. Galdi, v. Nečasová, eds., *Fluid-Structure Interaction and Biomedical Applications* (Birkhäuser Basel, 2014)

[5] T. Theodorsen, W. Mutchler, U.S.N.A.C. for Aeronautics, *General Theory of Aerodynamic Instability and the Mechanism of Flutter*, National Advisory Committee for Aeronautics: Report (National Advisory Committee for Aeronautics, 1935)

[6] E. Naudasher, D. Rockwell, *Flow-Induced Vibrations* (A.A. Balkema, Rotterdam, 1994)

[7] Y. Fung, *An Introduction to the Theory of Aeroelasticity*, Dover Phoenix Edition: Engineering (Dover Publications, 2002)

[8] R. Bisplinghoff, H. Ashley, R. Halfman, *Aeroelasticity*, Dover Books on Aeronautical Engineering Series (Dover Publications, 1996), `https://books.google.cz/books?id=jtqDQ2nTvvcC`

[9] C. Verhoosel, T. Scholcz, S. Hulshoff, M. Gutierrez, *Transient stability analysis of an aeroelastic problem with random fluid and structure properties*, in *Proc of the 10th International Conference on Structural Safety and Reliability, ICOSSAR10* (Osaka, Japan, 2009), pp. 377–384

[10] D. Jones, A. Gaitonde, in *Innovation for Sustainable Aviation in a Global Environment*, edited by D. Knorzer, J. Szodruch (IOS Press BV, 2012), pp. 110–115

[11] P. Le Tallec, J. Mouro, Computer Methods in Applied Mechanics and Engineering **190**, 3039 (2001)

[12] C. Förster, W.A. Wall, E. Ramm, Computer Methods in Applied Mechanics and Engineering **196**, 1278 (2007)

[13] M. Bukač, S. Čanič, B. Muha, Journal of Computational Physics **281**, 493 (2015)

[14] E. Aulisa, G. Bornia, S. Manservisi, Communications in Computational Physics **18**, 621 (2015)

[15] S. Manservisi, SIAM Journal on Numerical Analysis **44**, 2025 (2006)

[16] F. Nobile, C. Vergara, SIAM Journal on Scientific Computing **30**, 731 (2008)

[17] J. Vos, A. Rizzi, D. Darracq, E. Hirschel, Progress in Aerospace Sciences **38**, 601 (2002)

[18] P.M. Gresho, R.L. Sani, *Incompressible Flow and the Finite Element Method* (Wiley, Chichester, 2000)

[19] V. Girault, P.A. Raviart, *Finite Element Methods for the Navier-Stokes Equations* (Springer, Berlin, 1986)

[20] H.G. Roos, M. Stynes, L. Tobiska, *Robust Numerical Methods for Singulary Perturbed Differential Equations: Convection Diffusion and Flow Problems*, Vol. 24 (Springer, Berlin, Heidelberg, 2008)

[21] G. Lube, G. Rapin, Computer Methods in Applied Mechanics and Engineering **195**, 4124 (2006)

[22] A. Linke, L.G. Rebholz, N.E. Wilson, Journal of Mathematical Analysis and Applications **381**, 612 (2011)

[23] T. Gelhard, G. Lube, M.A. Olshanskii, J.H. Starcke, Journal of Computational and Applied Mathematics **177**, 243 (2005)

[24] P. Sagaut, *Large Eddy Simulations for Incompressible Flows* (Springer, Berlin, 2000)

[25] M. Avila, R. Codina, J. Principe, Computers & Fluids **99**, 44 (2014)

[26] D.C. Wilcox, *Turbulence Modeling for CFD* (DCW Industries, 1993)

[27] G. De Nayer, M. Breuer, International Journal of Heat and Fluid Flow **50**, 300 (2014)

[28] M. Feistauer, J. Horáček, P. Sváček, Communications in Computational Physics **17**, 146 (2015)

[29] T. Nomura, T.J.R. Hughes, Computer Methods in Applied Mechanics and Engineering **95**, 115 (1992)

[30] P. Sváček, M. Feistauer, J. Horáček, Journal of Fluids and Structures **23**, 391 (2007)

[31] D. Boffi, L. Gastaldi, Computational Methods in Applied Mechanical Engineering **193**, 4717 (2004)

[32] L. Formaggia, F. Nobile, Computational Methods in Applied Mechanical Engineering **193**, 4097 (2004)

[33] C. Farhat, P. Geuzaine, Computational Methods in Applied Mechanical Engineering **193**, 4073 (2004)

[34] P. Sváček, Mathematics and Computers in Simulation **80**, 1713 (2010)

[35] M. Feistauer, J. Horáček, M. Ružička, P. Sváček, Computers & Fluids **49**, 110 (2011)

Using the adsorption chillers for utilisation of waste heat from rotary kilns

Karol Sztekler[1*], *Wojciech* Kalawa[1], *Wojciech* Nowak[1], *Sebastian* Stefański[1], *Jarosław* Krzywański[2] and *Karolina* Grabowska[2]

[1]AGH University of Science and Technology, A. Mickiewicza Av. 30, 30-059 Krakow, Poland
[2]Jan Dlugosz University, Armii Krajowej Av. 13/15, 42-200 Czestochowa, Poland

Abstract. Waste heat utilisation and its reclamation is important for increasing the efficiency of the electric power production and for decreased consumption of primary energy. Waste heat can be utilised for the electric power production or for manufacturing and processing purposes. According to long-term forecasts, the consumption of electric power with current consumption of about 150 TWh, can be increased to about 230 TWh in around 2040, what may lead to the increased energy consumption from fuels and the increased emission of harmful contaminants to the atmosphere. The cement subsector, next to the glass industry, is among the most power-consuming industries, and it consumes about 12-15% of total energy consumed by industry in total, whereas the CO_2 emission from such industry exceeds about 5% in worldwide scale. In the cement sector, there is an increasing need for useful reclamation of waste energy, in order to obtain high energy-saving factors, and hence the financial and environmental benefits. In this paper the chances for waste energy reclamation from a rotary kiln are presented, with means of the used adsorption chillers.

1 Introduction

The efficiency of energy utilised in the business is the important factor influencing on the level of manufacturing costs, company's profits, competitiveness of products, on social upkeep and on standards of living of citizens. As a result, when the energy is used ineffectively, high consumption of energy resources and problems with natural environment contamination are observed. Strategic goal in the ecological and energy policy in Poland is currently the increased efficiency of energy economy. The energy efficiency is related with the increased competitiveness of Polish companies and with many other elements, when compared to other EU countries. The problem of energy efficiency is treated as priority, because the advancement in such field is very important for the execution of all energy policy goals and for the majority of ecological and climatic policy goals. A basic goal in the area of effectiveness, next to the goals stipulated in the directive, is currently energy consumption decreased with 20%, in comparison to forecasts for 2020, from the energy efficiency increase [1,2]. As it was mentioned, the cement sub-sector is among the most power-consuming industries, that consumes about 12–15% of total energy consumed by whole industry. A cement plant consumes in average about 4 GJ of energy in order to produce 1 tonne of cement (up to 5.4 GJ/t in China [3]), therefore it is among the most power-consuming manufacturing processes. The production of cement is related with the emission of many harmful substances, whereas carbon dioxide emission from the cement industry comprises about 5% of total worldwide emission originating from the activity of human being. Emission from the cement industry in Poland in 2002 amounted to about 7824 thousand tonnes, what can be interpreted as the unit emission of about 0.69 kg/tonne for the produced cement [1]. The yearly cement sales in Poland in 2014 amounted to 15.2 million tonnes, whereas in 2015 it was about 16.4 million tonnes and in 2016 it should be on the similar level [4]. According to long-term forecasts, the amount of cement sold may increase, what obviously generates the increased electric power consumption and the emission of gases, therefore all works in search for the decreased cement plant's impact or the increased exploitation level for primary energy are advisable.

2 Methods for waste heat utilisation from the cement plant

In the cement sector, there is an increasing need for useful reclamation of waste energy, in order to obtain high energy-saving factors, and hence the financial and environmental benefits. The waste heat was used for the first time in the cement industry more than 100 years ago. It was waste heat utilised from flue gases with means of the system of heating tubes. In the cement manufacturing process, much energy is irretrievably lost; decrease in such loses in a global scale with at least fraction of a percent generates visible savings from the combusted fuel.

* Corresponding author: sztekler@agh.edu.pl

The increased thermal efficiency for rotary kilns is available with means of retrofit actions, such as: implementation and improvement in the heat exchanger construction, clinker coolers or other devices cooperating with the cement kiln. The most beneficial retrofit solution will be searched in this papers, that is related with the recovery of waste heat possible for implementation in the Rudniki Cement Plant nearby Czestochowa. Utilisation of the waste heat depends on its temperature in large extent, as well as on the speed of heat flow, its availability in a day and in a year. Possible and the most often used heat sources for reclamation purposes in Cement Plants are flue gases, radiation from the internal surface of the rotary kiln, a heat transferred by a hot clinker leaving the kiln, as well as a cooling air for a burned material (Fig. 1).

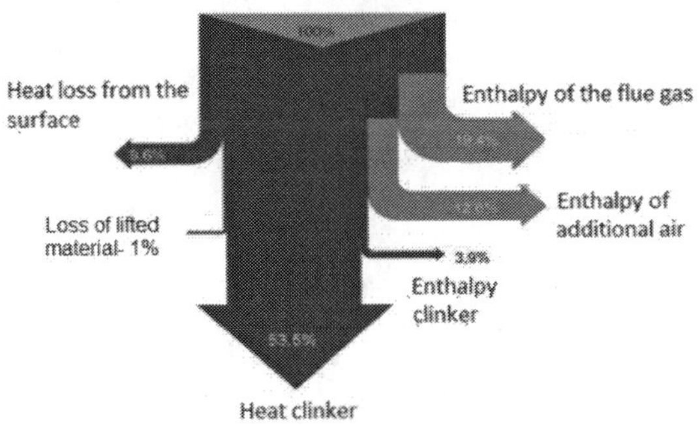

Fig. 1. Heat balance of cement kiln [5].

We can discern the following methods for waste heat utilisation in the cement plant: Utilisation of hot air from clinker coolers - it is used as the primary or secondary air for the combustion of coal dust or as a drying agent for the coal and input materials in the mills; The heat reclaimed from the rotary kiln's surface - the hot kiln's surface that can reach temperature exceeding 400°C enables the reception of heat with means of external heat exchangers located around the kiln; Heat utilisation from waste gases and clinker coolers air - the effective reclamation of waste heat from the stream of gases and air is possible with means of Clausius-Rankine water-vapour circuits and Clausius-Rankine organic cycle (ORC), as well as with means of air turbine [6,7,8,9].

In this papers the utilisation of waste heat coming from the rotary kiln's surface is contemplated, with means of the radiation exchanger that is used for the production of hot water t = 65°C used as the energy source powering the adsorption chiller. The adsorption chiller is the device, in which the water adsorption phenomenon in the silica-gel deposits or the zeolite deposits is used, that can be regenerated with the heat with temperature of about 55°C. Such a low powering temperature for the chiller opens a variety of possible applications for this device. The chiller may be powered by a geothermal heat, a thermal energy from solar collectors, a waste energy from some processes or a heat from district heating network. Description for operation of the adsorption chiller is presented in the literature [10–14]. Adsorption chillers can produce the chilled water in the range of 7–12°C, that can be used for air-conditioning purposes. Adsorption chillers are characterised with low chilling performance coefficient (COP) ranging from 0.2 to 0.6, thus such systems are mainly dedicated in the fields where waste heat is at disposal [11,12,15]. Application of such type of installation enables the use of the waste heat from the rotary kiln.

3 Utilising waste heat from the Rudniki Cement Plant

Rudniki Cement Plant nearby Czestochowa manufactures: cements (slag, mixed, namely multi-component, lime, metallurgical), road adhesive (since 2009) and limestone dust. Rudniki and Chelm Cement Plants in the Cemex group are dominating in Poland in terms of waste utilisation as the alternative fuel. Rudniki Cement Plant has been in the ownership of the Cemex group since 2005. In the Rudniki Cement Plant two waste heat reclamation methods exist. The first one is using the heated air in the clinker cooling process as the secondary heat guided to the rotary kiln. The second waste heat utilisation method from the industrial installation is using flue gases received from the kiln, in order to dry raw materials fed to the limestone dust and to dry the coal in the ball mill [5,16].

3.1. Radiating exchanger

Rudniki Cement Plant is equipped with three rotary kilns with capacity of 650 t/day and length of 140 m. Such rotary kilns are made from steel in the shape of elongated, cylindrical tube. They are inclined at the level of about 3–5%, what enables the movement of input material to the front of the kiln in a natural way. Maximum temperatures in the cement kiln are the flame temperature and temperature of gases in the sintering zone, where they reach 2000°C. Such high temperatures are reached in the kiln with means of the rotary kiln operating in counter current. Final product leaving the kiln reaches the temperatures of 900–1300°C. The kiln surface may reach a temperature exceeding 400°C and is another important heat loss source. Heat losses by convection and radiation may comprise about 8–15% of input energy [6,7,17,18]. Rotary kilns emit significant amounts of heat on the radiation route, that may be usefully exploited, for i.e. domestic hot water warming purposes for satisfaction of the Plant's needs, or for the heated water used when powering the adsorption chillers. The external radiator exchangers may be found in the literature, that most often exist in the form of longitudinal series of tubes located coaxially on the cylindrical surface of the external layer of the rotary kiln or it has the semi-circular shape in profile, in the form of bent tubes (Fig. 2).

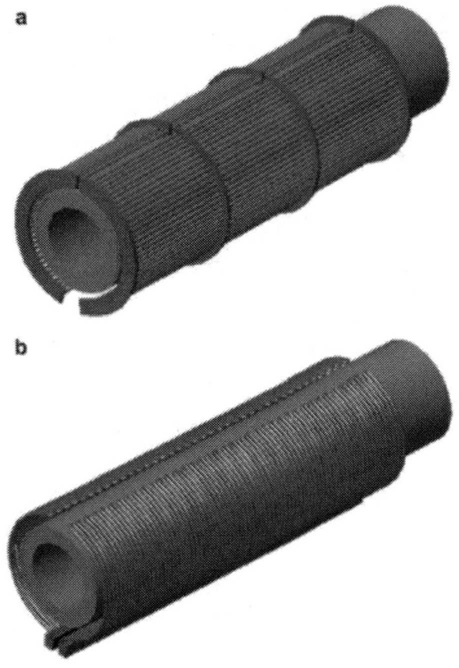

Fig. 2. The heat exchanger: a) in tube arrangement located coaxially on the external layer of the rotary kiln, b) in the arrangement of bent tubes [18].

The exchangers with oval or cylindrical tubes with ribs are also used in order to increase the heat exchange surface, and such exchangers are used in the places where media participating in the heat exchange are gases (air, flue gases) and any fluids (e.g. water, oils). An interesting solution for the heat recovery from the external surface of the rotary kiln is the axial superposition of fixed housing, that may cover up to about 80% of total kiln length. It forms the heat recovery chamber, with which an exhausting fan is connected, where the heated air from the surface of the kiln may be used in order to dry the coal.

3.2. The selection of the exchanger's operation parameters

The proposed exchanger will have two functions in the cement plant, in the winter period it can be used for domestic hot water production purposes, and in the summer period the generated hot water will be used for powering the adsorption chiller and for production of the chilled water. Thermal-imaging measurement was used for the evaluation of temperature distribution on the kiln's mantel, (Fig. 3), on the 16–28 m section of initial kiln's length, where temperature was 348.7°C.

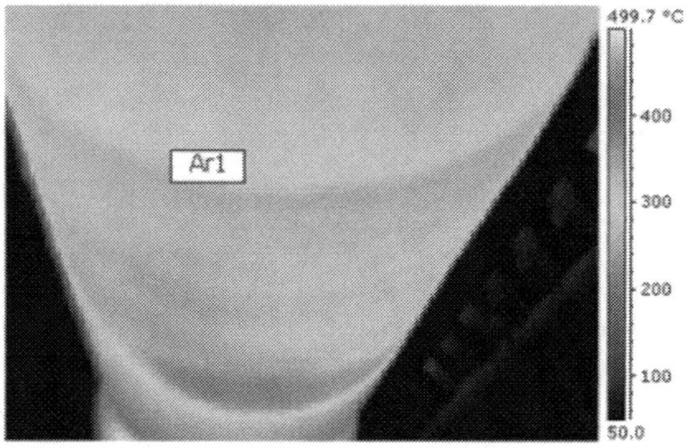

Fig. 3. Temperature distribution on the no. 4 kiln's mantel.

It was assumed in the heat-flow evaluations, that the exchanger: will be made as a single-row screen made from seamless steel tubes; the screen will be made in the form of cylinder spur covering 1/6 of the rotary kiln circumference; minimum distance from the kiln 1m; the boiler seamless steel tubes are used for the construction, with dimensions of: $\frac{d_z}{d_w}=\frac{51\text{ mm}}{44.6\text{ mm}}$, water temperature supplied to the radiator exchanger t_1 = 10°C; distance between the tubes 0.15m; the heat exchanger length 8-12 m, maximum daily heat consumption for the plant 0.12 l/s; average temperature value on the kiln's mantle at the section of 16-28 m t_1 = 316°C ; energy supply is guided into the adsorption chiller in the form of water with temperature of t = 65°C, The assumed COP for the adsorption chiller COP = 0.42, when producing the chilled water with temperature of t = 12/7°C [12,13,14,15]. Thermal-flow evaluations are made according to the literature [19,20,21,22].

4 Evaluation results

Thermal-flow evaluations were made for the exchanger with length of 12 m and 8 m, according to the literature [19–22]. Where *s* is distance of the exchanger from the rotary kiln surface, *m*; t_2 – the exchanger surface temperature, *°C*; $Q_{1\text{-}2}$ – amount of the received heat from the kiln's surface with means of the radiator exchanger, kW_c; *v* – water volume stream heated to temperature of 65°C; *Qch* – cold production in the adsorption chillers, kW_{ch} (Table 1).

Table 1. Evaluation results for the exchanger with length of *l = 12 m* and l = 8 m.

Exchanger length 12 m					Exchanger length 8 m				
s	t_2	$Q_{1\text{-}2}$	v	Q_{ch}	s	t_2	$Q_{1\text{-}2}$	v	Q_{ch}
m	°C	kW_c	dm^3/s	kW_{ch}	m	°C	kW_c	dm^3/s	kW_{ch}
1.0	39.4	163.34	0.716	68.60	1.0	40.2	64.80	0.285	27.22
1.1	39.4	139.68	0.613	58.67	1.1	39.8	55.45	0.243	23.29
1.2	39.3	121.33	0.532	50.96	1.2	39.8	48.16	0.211	20.23
1.3	39.3	106.75	0.468	44.83	1.3	39.8	42.37	0.186	17.79
1.4	39.2	94.95	0.416	39.88	1.4	39.8	37.68	0.165	15.83

Exchanger length 12 m					Exchanger length 8 m				
s	t_2	Q_{1-2}	v	Q_{ch}	s	t_2	Q_{1-2}	v	Q_{ch}
m	°C	kW_c	dm^3/s	kW_{ch}	m	°C	kW_c	dm^3/s	kW_{ch}
1.5	39.2	85.24	0.374	35.80	1.5	39.8	33.83	0.148	14.21
1.6	39.1	77.14	0.338	32.40	1.6	39.8	30.61	0.134	12.86
1.7	39.1	70.30	0.308	29.53	1.7	39.9	27.89	0.122	11.71
1.8	39.1	64.46	0.283	27.07	1.8	39.9	25.58	0.112	10.74
1.9	39.1	59.43	0.261	24.96	1.9	40.0	23.58	0.104	9.90
2.0	39.1	55.06	0.241	23.12	2.0	40.1	21.84	0.096	9.17
3.0	39.1	30.78	0.135	12.93	3.0	42.9	12.17	0.054	5.11
3.1	39.1	29.42	0.129	12.36	3.1	42.5	11.61	0.051	4.87
3.2	39.1	28.17	0.124	11.83	3.2	41.7	11.10	0.049	4.66
4.0	39.3	20.86	0.092	8.76	4.0	41.7	8.26	0.0367	3.47

4 Conclusion

The assumed 0.12 l/s domestic hot water demand in the Plant can be satisfied with both proposed exchanger versions. The exchanger with length of 12 m may reach the values of volume yield in the range of 0.716–0.092 l/s, whereas in the exchanger with length of 8 m: 0.285–0.036 l/s. In case of the first exchanger, it should be placed at least at distance of 3m, in order to obtain the yield of about 0.122 l/s. In case of the second exchanger, when it is placed at distance of 1.7 m, the 0.122 l/s yield can be obtained. Maximum thermal output possible in case of the installed 12m and 8m exchanger at distance of 1m, with simultaneously obtained about 163 kW_c and 64.80 kW_c, respectively, similarly to the case of the installed adsorption pump, it enables the production of about 68 kW_{ch} and 27.2 kW_{ch}. Whereas, when simultaneous production of domestic hot water for the plant's needs and cold is assumed, we can produce with means of the 12-meter-long exchanger about 56 kW_{ch}; and in case of the 8-meter-long exchanger about 15 kW_{ch}. Such solutions provide complete satisfaction of domestic hot water and cold needs in the Plant, with relatively small investment costs, what also enables primary energy savings, that would be otherwise used in domestic hot water and cold production.

Acknowledgment

The paper was funded from government money Faculty of Energy and Fuels number 11.11.210.216.

References

1. B. Vanderborght, B. Brodmann, *The Cement CO_2 Protocol: CO_2 Emissions Monitoring and Reporting Protocol for the Cement Industry* (Guide to the Protocol 1.6, 2001)
2. K. Pikoń, M. Kajda-Szcześniak, M. Bogacka, *Chemical Industry* **94** 1548 (2015) (in Polish)
3. E. Mokrzycki, A. Uliasz-Bocheńczyk, *Energy Policy* **9** (2006)
4. http://www.polskicement.pl/aktualnosci/W_2015_roku_wzrost_sprzedazy_cementu_w_Polsce_do_16_mln_ton-308 (2017)
5. K. Sztekler, M. Komorowski, M. Tarnowska, Ł. Posak *Proc. of the Int. Conf. on the Sustainable Energy and Environment Development, E3S Web of Conferences* **10** (2016)
6. A. Atmaca, R. Yumrutas, *Appl. Thermal Eng.* **66** 435 (2014)
7. J. Duda, *The Scientific Works of the Institute of Ceramics And Building Materials* **32** (2012) (in Polish)
8. S. Karellas, A. Leontaritis, G. Panousis, E. Kakaras, *Energy* **58** 147 (2013)
9. F. Campana, M. Bianchi, L. Branchini, A. De Pascale, A. Peretto, M. Baresi, *Energy Conv. and Management* **76** 244 (2013)
10. R. Wang, L. Wang, Wu J, *Adsorption Refrigeration Technology: Theory and Application* (2014)
11. K. C. Ng, B. B. Saha, *Advances in Adsorption Technology* (2010)
12. X. Wang, H.T. Chua, *Int. J. of Refrig.* **30** 1417 (2007)
13. Q. W. Pan, R. Z. Wang, L. W. Wang, D. Liu, *Int. J. of Refrig.* **67** 336 (2016)
14. M. Gwardera, K. Kupiec, *Chemical Engineerign and Equipment* **50** 38 (2011) (in Polish)
15. M. Chorowski, *Nowa Energia* **4** (2014)
16. K. Sztekler, K. Wojciechowski, M. Komorowski, M. Tarnowska, *Proc. of Energy and Fuels Conf., E3S Web of Conferences* **14** (2016)
17. S. Żurakowki, S. Hojarczyk, *Rotary kilns – desing and contruction* (WNT, Warsaw 1969) (in Polish)
18. A. Caputo, P. Pelagagge, P.Salini, *Appl. Thermal Eng.* **31** 2578 (2011)
19. J. P. Holman, *Heat Transfer* 10th Edition (McGraw-Hill, New York, 2010)
20. T. Hobler, *Heat transfer and heat exchangers* (PWT, Warsaw, 1959) (in Polish)
21. U. Lubczyńska, *Applied hydraulics* (Kielce 1996) (in Polish)
22. A. Sala, *Radiaton heat transfer* (WNT, Warsaw, 1982) (in Polish)

Using the adsorption chillers for waste heat utilisation from the CCS installation

Karol Sztekler[1*], *Wojciech* Kalawa[1] , *Wojciech* Nowak[1], *Sebastian* Stefański[1], *Jarosław* Krzywański[2] and *Karolina* Grabowska[2]

[1]AGH University of Science and Technology, A. Mickiewicza Av. 30, 30-059 Krakow, Poland
[2]Jan Dlugosz University, Armii Krajowej Av. 13/15, 42-200 Czestochowa, Poland

Abstract. Worldwide tendencies in the scope of environmental protection demonstrate the requirement for the limited carbon dioxide emission, that influences on the development of greenhouse effect. As a result of coal as a basic fuel used in the professional power industry, this industry sector is the greatest CO_2 polluter and it means that works on the reduction of carbon dioxide in such industry are completely justified. In the IPSEpro programming environment, a reference block model for a conventional coal power station was elaborated, including the CO_2 separation unit basing on the adsorption methods with the CO_2 preparation installation to liquid state. Simulation researches were conducted with means of numeric techniques, that enabled the system analysis for the CO_2 separation unit with the CO_2 preparation system to the liquid state, as well as analysis was made for the use of chiller systems, basing on the adsorption technology for waste heat use originating from the compression of CO_2 in a cascade system, as well as for potential opportunities for further exploitation of the produced chilled water in the CCS cycle. We analysed in these papers the opportunities for chiller systems application, based on the adsorption chillers in the CCS installation used for the reduction of CO_2 emission in the coal power station and its influence on the operation of a power station cycle.

1 Introduction

Climatic changes have been more and more visible in last years and have been related with the problem of global warming. CO_2 is the greatest component of the greenhouse gas emission in percentage value (about 72% of total emission), then it is methane CH_4 (18%) and nitrogen monoxide N_2O (9%). According to scientists, carbon dioxide is most harmful for environment and has the greatest influence on the greenhouse effect, thus development of technology limiting the CO_2 emission is purposeful [1]. Main mechanisms contributing to the generation of CO_2 are processes related with the production of electric power, that are responsible for even 40% of worldwide carbon dioxide emission. Electric power industry is generally based on the combustion of coal, thus this industry is the largest emitter of CO_2. It is expected, that in developed countries the consumption of electric power will increase threefold until 2020, when compared to 1970. Whereas, in developing countries, where the number of citizens may be doubled until 2020, in relation to 1970, consumption of electric power may increase even twentyfold. It is forecasted in other resources, that until 2030 the yearly increase in demand for electric power will amount to about 2–3% [2,3].

The problem of CO_2 emission is not the problem of explicitly one country, but it pertains to the whole world. Worldwide economy is developing very rapidly and such development leads to the increased demand for electric power. In order to meet such demand, electric power should be obtained from fossil fuels, what enables its stable and uninterrupted production process. The coal may be still such fuel, with very large resources enabling constant deliveries of electric power. In order to make the energy from the coal the least burdensome for natural environment, the CO_2 separation and sequestration technologies should be implemented, that certainly enable the reduction in the emission of such gas. The CCS Technology (Carbon Capture Storage) is based on the separation and sequestration of gas, subsequently its transportation and storage. Limitation of carbon dioxide emission can be pursued with means of the following methods: post-combustion, pre-combustion, oxy-fuel combustion. The CO_2 separation before the combustion process is related with initial preparation of the fuel. This fuel is gassed, and after this the synthesis gas is obtained (CO and water vapour). In order to obtain the CO_2 and H_2 streams, the synthesis gas can be treated in the systems based on: processes combined with the conversion of CO into CO_2, processes without the conversion of CO into CO_2, coal gasification in pure oxygen, Integrated Gasification Combined Cycle – IGCC. CO_2 separation processes after the combustion of fuel. Such technologies are characterised with the separation of carbon dioxide after the combustion

* Corresponding author: sztekler@agh.edu.pl

process from flue gases. Systems for carbon dioxide separation after the combustion process are separate constructional units, that can be implemented in the existing power units, without interference with theirs construction. The CO_2 separation from flue gases can be provided with following methods [4–15]: absorption, adsorption, membrane, cryogenic. The separation system based on the adsorption method is contemplated in this papers. The separation process is not the final step in the CO_2 treatment, and the subsequent step is the preparation of separated gas for transportation in liquid form. This process is performed with means of the multi-step compression system, up to the pressure in which the CO_2 is changed from gaseous state to liquid state. Between particular compression levels, the waste heat can be utilised and the cold can be produced. When producing the cold, highly efficient and reliable sorption heat recovery technologies can be utilised, such as Adsorption Chillers with desalination function, or Absorption Chillers [16]. The adsorption chiller unit is the industry technical device designed for the production of chilled water and desalinated water with means of low-parameter heat used as a powering energy (60°C–90°C). Principle of operation is detailed in the literature in the broader extent [17–21]. The COP value depends on water temperature leading to the deposit regeneration, as well as on the chilling water, and it is in the range of 0.49-0.56. Such a low COP value is related with low level of thermal conductivity for an adsorbent forming such deposit, as well as its sorption properties [21]. Whereas, the COP is the ratio of heat amount received from evaporation, to the heat amount delivered in the deposit heating step and desorption step, and it is the factor of chilling efficiency:

$$COP = \frac{Q_p}{Q_o + Q_d} \tag{1}$$

Such device provides simultaneous production of chilled water in temperature range: 7-12°C. The chilled water produced in such devices can be used for air-conditioning purposes, or for other technological processes [20].

2 Developing simulation models

Installations capturing carbon dioxide on the basis of adsorption processes are not currently operating in full scale in real systems with a power unit. Therefore, we do not have completely documented data, that enables the assessment of its influence on the operation of a power unit. When using the IPSEpro software from SimTech company, we can make models for: a power unit cycle, a carbon dioxide capture process, basing on the mixed PTSA method; a complete technological system for CO_2 separation and sequestration, as well as we can make analysis for the chilling installation operation. When basing on data provided in [22], the heat cycle model for a power unit, also called the reference cycle, was developed. Basic parameters for the reference heat cycle are presented in table below.

Table 1. Basic parameters of the reference power cycle [22].

Parameter	Value	Unit
Electric power	833	MW
Fresh vapour pressure	266	bar
Fresh vapour temperature	554	°C
Fresh vapour mass stream	625	kg/s
Efficiency of reference cycle	45.1	%
Mass flow for furnace flue gases	1016	kg/s
Mass flow for CO_2	189.6	kg/s
Fresh vapour pressure	266	bar

The CO_2 separation system was introduced in the reference cycle, then it was integrated with the whole power unit's cycle. Standard IPSEpro library does not contain the model for the CO_2 separation system from flue gases based on the adsorption processes, therefore a numerical model for this process had to be developed, as well as all required balance equations, required from the scope of occurring phenomena. It was assumed, that complete separation process will be described in the form of a single module, that will be available in the form of a single object from the level of IPSEpro software library. The separation module based on the PTSA (*Pressure Temperature Swing* Adsorption) was accepted for analyses. Detailed description and principle of operation for such module was presented in the following literature [2,3,7,23]. It was assumed in the model, that after the separation process, clean carbon dioxide is obtained. The reference cycle diagram including the integrated unit for carbon dioxide capture, altogether with its compression installation, is presented on fig. 1.

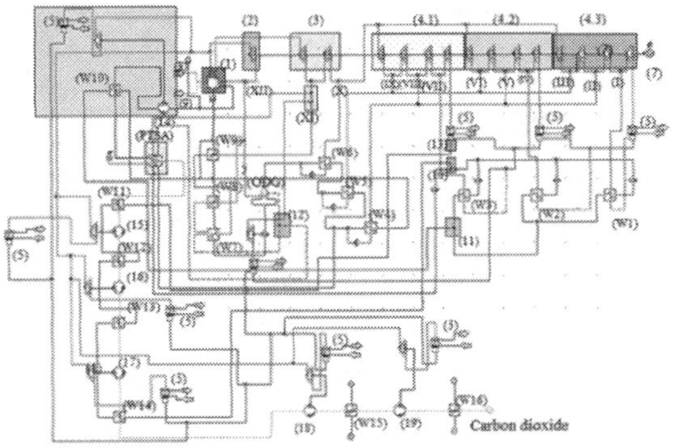

Fig. 1. The reference cycle model with integrated PTSA unit including CO_2 preparation system to liquid form.

The released carbon dioxide is guided outside the PTSA system in the amount of about 190 kg/s, what amounts to 684 t/h (with $\dot{Q}_{flue\_gases}$ = 1016 kg/s for 100%

CO_2 separation level). However, in order to make it usable in other industry branches or for storing purposes, it must be transported to destination place in liquid form. The separated carbon dioxide can be prepared for transportation with means of the multi-step compression from 0.15 bar to 100 bar, with means of five used compressors driven by a turbine powered on fresh vapour (15), (16), (17), (18), (19). It is assumed, that the compression level at particular compressor steps will be greater than $\Pi = 7$. Before the first compression level (15), carbon dioxide with temperature of 207°C (CO_2 temperature leaving the capturing unit) is cooled down in the exchanger (W11) to temperature of 35°C, then after each CO_2 compression level it is cooled in the exchangers (W12), (W13), (W14) to temperature of 35°C, and in final step in the exchangers (W15), (W16) to temperature of 25°C, to liquid form. The exchangers in which the CO_2 is cooled down, may be used for the production of hot water guided to the adsorption chillers, after the deposit regeneration process.

3 Simulation results

In this section analysis is made for waste heat utilisation from the CO_2 preparation installation, until the liquid form. For the purpose of such simulation researches, the followings assumptions are made:

1. Before introducing the flue gases to the CO_2 capture system, they must be dehydrated, as well as gaseous contaminants (SO_2, NO_X) and solid contaminants must be removed.
2. The CO_2 separation unit operates in the fixed state.
3. Adsorption process parameters in the unit PTSA p = 2.0 bar, temperature t = 110°C for desorption p = 0.15 bar
4. The carbon dioxide separation efficiency from the stream of flue gases amounts to 100%.
5. Powering energy in the form of hot water with temperature of t = 90°C is delivered to the adsorption chiller, with the assumed temperature decrease in the regenerated deposit $\Delta T = 7°C$ [16,24,25].
6. The COP for the adsorption chiller is assumed, COP = 0.56, when producing the chilled water with temperature of t = 12/7°C.

Simulation results demonstrate that altogether with the increased stream of carbon dioxide separated from the flue gases, the amount of generated heat increases (Fig. 2). The amount of generated heat also increases during the CO_2 compression and received by a chilling agent from the carbon dioxide [19,20].

For total stream of flue gases and with 100% separation level (CO_2 stream 189.6 kg/s), the heat output transferred to the chilling agent amounts to about 180 MW_t, whereas for 10% of flue gas stream amount of heat output transferred to the water is about 18 MW_t. In table 2, amount of heat received at particular carbon dioxide chilling steps is presented. Table 2 includes the evaluation for amount of the produced chilled water t = 12/7°C and the chillers chilling output at particular CO2 chilling steps, as well as the production of hot water 90/83°C used in the regeneration process.

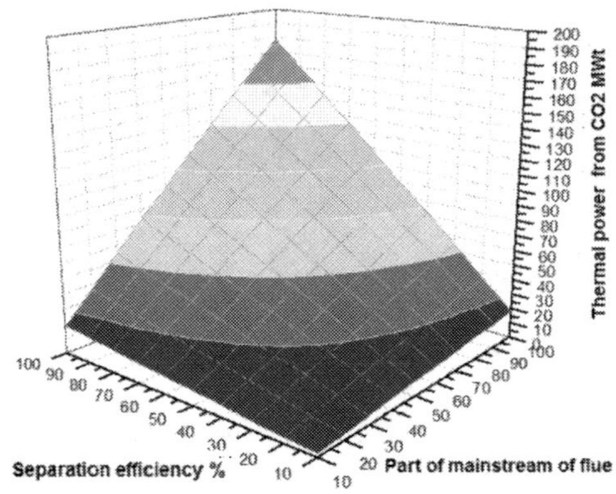

Fig. 2. Total heat output received from the CO_2 during its chilling.

Table 2. Heat recovery from particular CO_2 chilling steps.

Heat received from CO_2 after the first chilling step (W11)	Heat received from CO_2 $Q_{re.} = 28\ MW_t$ Hot water stream $m_h = 918.67$ kg/s Chilled water stream $m_{ch} = 747$ kg/s Chiller's chilling output $Q_{ch} = 15.68\ MW_{ch}$
Heat received from CO_2 after the second chilling step (W12)	Heat received from CO_2 $Q_{re.} = 29\ MW_t$ Hot water stream $m_h = 986.72$ kg/s Chilled water stream $m_{ch} = 774.2$ kg/s Chiller's chilling output $Q_{ch} = 15.68\ MW_{ch}$ $Q_{ch} = 16.24\ MW_{ch}$
Heat received from CO_2 after the third chilling step (W13)	Heat received from CO_2 $Q_{re.} = 29\ MW_t$ Hot water stream $m_h = 986.72$ kg/s Chilled water stream $m_{ch} = 774.2$ kg/s Chiller's chilling output $Q_{ch} = 16.24\ MW_{ch}$
Heat received from CO_2 after the fourth chilling step (W14)	Heat received from CO_2 $Q_{re.} = 21\ MW_t$ Hot water stream $m_h = 714.52$ kg/s Chilled water stream $m_{ch} = 560.6$ kg/s Chiller's chilling output $Q_{ch} = 11.76\ MW_{ch}$
Heat received from CO_2 after the fifth chilling step (W15)	Heat received from CO_2 $Q_{re.} = 40\ MW_t$ Hot water stream $m_h = 1361$ kg/s Chilled water stream $m_{ch} = 1068$ kg/s Chiller's chilling output $Q_{ch} = 22.4\ MW_{ch}$
Heat received from CO_2 after the sixth chilling step (W16)	Heat received from CO_2 $Q_{re.} = 28\ MW_t$ Hot water stream $m_h = 918.67$ kg/s Chilled water stream $m_{ch} = 1068$ kg/s Chiller chilling power $Q_{ch} = 15.68\ MW_{ch}$

4 Summary

Following the collected data it is determined, that heat recovery at particular carbon dioxide chilling steps and the production of chilled water in amount of m = 4.67 t/s is possible, what conforms with 100 MW_{ch} total chilling output. According to data [22], currently the largest manufactured adsorption chillers have 2 MW_{ch} output, what means that 50 such units should be installed, in order to capture the heat coming from the CO_2

compression. In the analysed case, we cope with polygeneration, because the production of electric power, heat, cold and CO_2 takes simultaneously place, and the CO_2 may be treated as useful product. The capture system in which 100% of carbon dioxide separation is provided from total stream of flue gases produced by a power unit with nominal power of 833 MW_e (the reference unit modelled according to the Belchatow Power Plant power unit) [26], including devices for technical execution of the adsorption, desorption, compression and CO_2 condensation process, generates about 16% decrease in the heat cycle efficiency and about 350 MW_e [3,23] demand for the energy necessary for complete carbon dioxide sequestration. Additional product in form of the chilled water is produced in the adsorption chillers and can be sold or reused in other industrial processes, and the utilised waste heat should contribute to the increased efficiency of chemical energy coming from fuel in the production of electric power in such power plant. In the EU-15 countries in 2000, 27% of office spaces and 5% of houses were cooled. Following data analysis for such 15 EU countries, it is assumed that target share of cooled space („the European saturation rate") in the residential objects will amount to 40%, whereas in office objects it will be 60%. It is evaluated, that this level will be obtained in 2012. In Poland, such conditions can be achieved in a significantly later period, namely about 2020. According to forecasts presented in the work [27], demand for cold will amount to 5000 MW_{ch} in 2020 in Poland. One power plant with CCS, in which the heat recovery systems from carbon dioxide chilling installations are installed, can satisfy 2% demand for total cold production in Poland in 2020. Another step in such analysis is evaluation for the influence of waste heat utilisation installations on the operation of CCS installation, and on the operation of power unit.

Acknowledgment

The paper was funded from government money Faculty of Energy and Fuels number 11.11.210.216.

References

1. K. Pikoń, M. Kajda-Szcześniak, M. Bogacka, *Przemysł Chemiczny* **94** 1548 (2015)
2. K. Sztekler, W. Kalawa, M. Panowski, *Energy Policy J.* **17** 137 (2014)
3. K. Sztekler, W. Kalawa, M. Panowski, R. Klajny, *Instal* **12** 17 (2013)
4. J. Kotowicz, K. Janusz, *Rynek Energii* **1** (2007)
5. I. Majchrzak-Kucęba, W. Nowak, *Proc. of the 2nd Int. Conf. on Contemporary Problems of Thermal Enginnering* 641 (2004)
6. L. Szczygieł, *Biuletyn Urzędu Regulacji Energetyki* **5** (2007)
7. K. Sztekler, M. Panowski, R. Klajny, *Prace Naukowe Politechniki Warszawskiej* **26** 317 (2009)
8. M. Mazurkiewicz, A. Uliasz-Boheńczyk, E. Mokrzycki, Z. Piotrowski, R. Pomykała, *Polityka Energetyczna* **8** 527 (2005)
9. J. Rakowski, *Energetyka* **4** (2010)
10. R. Lensen, *The promoter effect of piperazine on the removal of carbon dioxide* (2004)
11. S. Chakravarti, A. Gupta, B. Hunek, *Proc. of First National Conf. on Carbon Sequestration* (Washington DC, 2001)
12. T. Czakiert, K. Sztekler, S. Karski, *Fuel Processing Technology* **91** 1617 (2010)
13. J. Rolker, M. Seiler, *Adv. in Chemical Eng. and Science* **1** 280 (2011)
14. F. Colin, *CO_2 Capture With MEA: Integrating the Absorption Process and Steam Cycle of an Existing Coal-Fired Power Plant*, (Waterloo, Ontario, Canada, 2004)
15. M. Wanga, A. Lawala, P. Stephensonb, J. Siddersb, C. Ramshawa, H. Yeunga, *Chem. Engineering Research and Design* **89** 1609 (2011)
16. R. Wang, L. Wang, Wu J, *Adsorption Refrigeration Technology: Theory and Application* (2014)
17. K. Sztekler, M. Komorowski, M. Tarnowska, Ł. Posak *Proc. of the Int. Conf. on the Sustainable Energy and Environment Development, E3S Web of Conferences* **10** (2016)
18. K. Sztekler, K. Wojciechowski, M. Komorowski, M. Tarnowska, *Proc. of Energy and Fuels Conf., E3S Web of Conferences* **14** (2016)
19. K. C. Ng, B. B. Saha, *Advances in Adsorption Technology* (2010)
20. M. Chorowski, *Nowa Energia* **4** (2014)
21. M. Gwardera, K. Kupiec, *Inż. Ap. Chem.*, **50** 38 (2011)
22. http://www.net-sa.eu/ (2017)
23. K. Sztekler, T. Siwek, M. Komorowski, *Problemy Energetyki i Ochrony Środowiska* **62** 111 (2013)
24. X. Wang, H.T. Chua, *Int. J. of Refrig.* **30** 1417 (2007)
25. Q. W. Pan, R. Z. Wang, L. W. Wang, D. Liu, *Int. J. of Refrig.* **67** 336 (2016)
26. EPConstruction Sp. z o.o.: *Building project of 833MW power unit for BOT Elektrownia Bełchatów SA, Z-398, 3, The environmental impact assessment report, Technical description 01* (2005)
27. UCB Energ. I Ochr. Środ., ITC Politechniki Śląskiej, ARE S.A., *Raport z II etapu pracy: Analiza krajowego potencjału wysokosprawnej kogeneracji* (2007)

https://doi.org/10.1051/epjconf/201818002107

Recent advances in characterisation of subsonic axisymmetric nozzles

Václav Tesař [1,a]

[1] *Institute of Thermomechanics v.v.i., Czech Academy of Sciences, Dolejškova 5, 182 00 Praha-Kobylisy, Czech Republic*

Abstract. Nozzles are devices generating jets. They are widely used in fluidics and also in active control of flows past bodies. Being practically always a component of larger system, design and optimisation of the system needs characterisation of nozzle properties by an invariant quantity. Perhaps surprisingly, no suitable invariant has been so far introduced. This article surveys approaches to characterisation quantities and presents several examples of their typical use in systems such as parallel operation of two nozzles, matching a nozzle to its fluid supply source, apparent resistance increase in flows with pulsation, and the secondary invariants of a family of quasi-similar nozzles.

1 Introduction

Nozzles generate jets by allowing fluid to leave closed cavities and issue into a more or less open space. Some uses of nozzles are long established, in fact since prehistoric times, shown by example Fig. 1. Quite old are also various uses in heating and drying by impinging jets [2]. Recently rapidly increasing in importance are nozzles for jet control of flow past bodies (airplane wings or turbine blades [7]), where the jets replace earlier mechanical components (flaps, flow spoilers, etc.) inserted into the flow. Substantially dependent on nozzles is also fluidics, the technique of generation and control of fluid flows. It recently became widely used especially in the small scale microfluidics [3] and also - with large potential in energy engineering – in fuel producing bioreactors [8] or the new approaches to combustion [6].

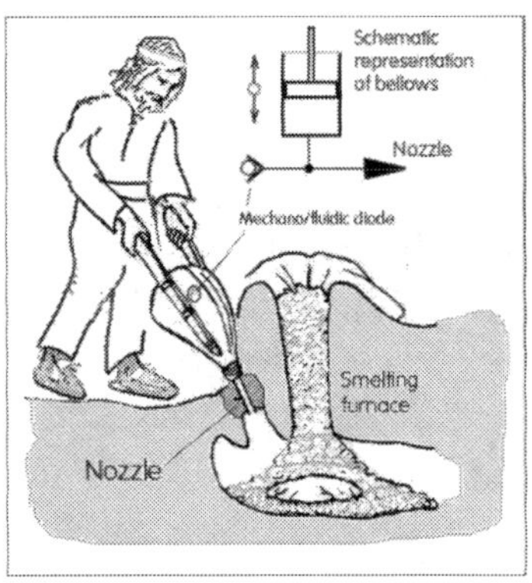

Fig. 1. An example of earliest use of nozzles. Since the Early Bronze Age (~ 4 000 B.C.) nozzles were applied to increasing the temperature in producing metals from ore [4]. Air is here seen being blown into the smelting furnace by bellows through a ceramic nozzle. Reminders of such nozzles are found in large quantities in archaeological excavations [5].

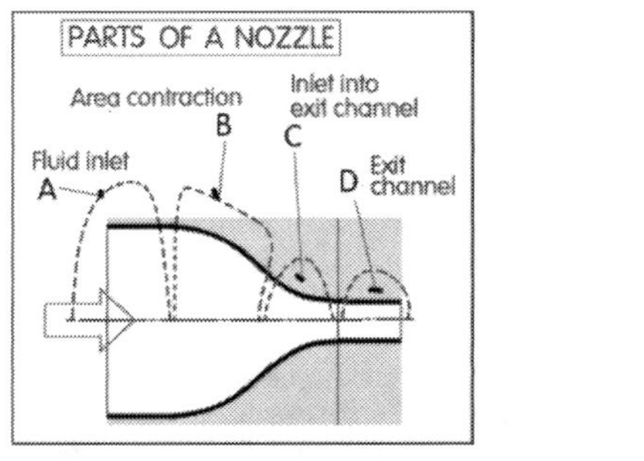

Fig. 2. Component parts of typical low-loss (i.e. smoothly contoured) axisymmetric nozzle. Fluid is accelerated in the area contraction B to counter the slowing down of the generated jet by viscous and vortex momentum transport to surroundings.

2 Nozzles

Nozzles - even if limited in this article to only subsonic and axisymmetric versions - exist in vast numbers of various shapes. In contrast to diffusers (which perform the opposite conversion of deceleration, Fig. 3), design of which for good performance may be difficult, nozzles pose less problems. Almost any nozzle geometry is likely to perform acceptably well. This is reflected in literature, vast one about diffusers and much less extensive about nozzles. In fact the nozzle literature does not very much address the question of pressure difference across nozzles discussed in this paper. Instead, it mostly concentrates on ensuring a particular shape of velocity profile at the nozzle exit. Especially wind tunnel nozzles are required to generate a special, rectangular profile. To exhibit low loss, nozzles require rounding all internal edges, as shown in the shape example in Fig. 2. The rounding can be made relatively easily if the nozzle is large (as in the wind tunnel case). At small scales the rounding may be a problem since it may be difficult to get with the machine

[a] Corresponding author: tesar@it.cas.cz

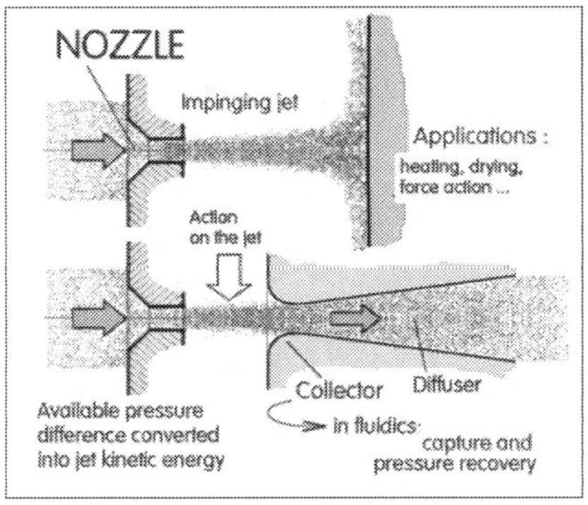

Fig. 3. There are applications with complicating effects caused by a device exposed to the jet flow. Fortunately, the impingement (top) and jet-capturing-collector flows (bottom) do not contribute significantly to pressure difference across the nozzle — the problem discussed here.

tool to the internal surfaces. Making small nozzles thus may require a special manufacturing process (e.g., the "3D printing"), which tend to be slow and expensive. As a result, unless there is requirement of very low loss, small size nozzles are quite often made with manufacturing economy as the chief factor. Nozzles are thus not infrequently made with the sharp edges as left by drilling – or, as a compromise, with just the entrance edge removed by chamfering [11] as seen in Fig. 4.

Present paper discusses axisymmetric nozzles with round exit channel of diameter d (Fig. 5). Typical feature of most nozzles is the decrease of area cross section in the flow direction in the locations B and C in Fig. 2. The reason is to counter the gradual momentum loss of generated jet especially if - as is often the case - the jet is required to reach with a reasonable velocity to a certain axial distance from the nozzle exit. The momentum loss is particularly remarkable if the jet is of submerged type, i.e. if the jet fluid is gas issuing into a gas-filled space or a liquid jet is issuing into space filled with liquid).

The problem caused by the sharp edges of the convex type "a" in Fig. 4, is they give rise to separation of the flow from the walls downstream from the edge - forming there an annular vortex ring. This vortex tends to

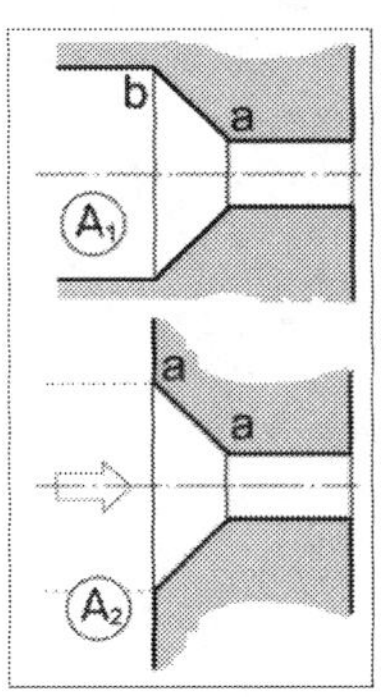

Fig. 4. Fluid is supplied into a nozzle using one of these two inlet configurations: either by the pipe entry A_1 or from the half-space A_2. The effect of these inlets on nozzle pressure drop is fortunately insignificant. More important are vortices generated at the convex corners "a" and concave "b".

be shed, usually in an irregular manner. The result is unwelcome pulsation in the generated jet flow. Also, the vortex shedding extracts from the main flow the energy tneeded for generating a new ring that replaces the shed one. Similar re-circulating vortex rings are also formed in the concave corner "b" shown in Fig. 4. This is energetically less demanding (velocity of motion inside the corner "b" is lower), but the unpleasant effect of vortex shedding may be found even there.

An important feature of good nozzles is their exit channel (the part D in Fig. 2). Its task is to stabilise the generated jet and ensure its proper flow direction. In an attempt to decrease the friction losses some nozzle designers disregard the need of this channel and make it short or absent. It is not a good idea. To be effective, the exit channel should not be shorter than ~ 2d. A shorter exit channel or missing immediately downstream from the contraction generates at the exit a phenomenon known under the historic term "*vena contracta*", causing significant energetic losses.

Only in a few cases (room heating, ventilation) the jets are left to dissipate their kinetic energy completely in the downstream space. Much more often they are used to perform there various actions like impingement (top of Fig. 3), or are captured into a collector (bottom of the picture). A literature on these phenomena exists, like [2] or [3]. From the point of view of the present article these phenomena are generally of little relevance. Perhaps more important may be the pressure decrease downstream from the nozzle exit due to the jet-pumping effect (entrainment of the outer fluid and dragging it along with the jet). Also this effect on nozzle pressure drop is relatively negligible unless the objects on which the jet acts are at very short distance from the nozzle exit.

3 Characterisation of nozzles in their role of circuit components

Prior to development of no-moving-part fluidics, the hydraulic and pneumatic systems almost always operated with some mechanical movable components. From the systems point of view these arrangements were usually quite simple. Their design could be thus handled by classical fluid mechanics, evaluating local velocity and local pressure – and then integrating these values to compute the acting forces. It was fluidics [3] that came with circuit solutions ideas that follow the well developed analogous theory of circuits in electric engineering. Unfortunately, in this approach fluidics encountered several significant problems. Behaviour of the components from which electric circuits are built is in steady states fully described by resistance R. This characterises the dependence between two parameters of state, the current I [A] and voltage difference ΔU [V], mutually related by the simple linear Ohm's law,

$$\Delta U = R \; I \qquad \text{... (1)}$$

In the case of fluidic devices – and nozzles in particular - the situation is very much different.

(a) While the electric state variables are directly measurable by instruments, in fluidics the values measured by manometer and flowmeter are insufficient.

They serve just as inputs into a more complex computation procedure [21]. Pressure difference ΔP [Pa] measured by manometers may be considered an analogy to ΔU in eq. (1) — while the role of the through variable, corresponding to the electric current, may be assumed by the volume flow rate $\dot{V}$ [m^3/s] measured by flowmeters. In an electric system, the solution then progresses by input of the measured quantities into the Kirchoff laws, basically conservation relations for circuit solutions. Unfortunately, there are no conservation laws for the quantities ΔP and $\dot{V}$. Obviously, in a real fluid of non-negligible compressibility its volume is not constant and increases along the pressure decrease. Thus $\dot{V}$ typically grows in the flow direction. The problems with ΔP as opposed to decrease of ΔU in the direction of electric current is particularly apparent in a diffuser — the pressure in which may actually increase in the direction of fluid flow. Yet ΔP and $\dot{V}$ were used traditionally and it took considerable time and effort to break this tradition to accept the introduction of the not directly measurable conservation quantities, the specific energy difference Δe [m^2/s^2] together with the mass flow rate $\dot{M}$ [kg/s].

b) In contrast to the simple across variable ΔU in eq. (1), the specific energy difference Δe, which plays in fluidics the analogous role, is a multi-component vari-able. The drop in specific energy across a nozzle (or another fluidic device) is due to a part of it being spent on driving the fluid through. It is a sum of differences is several (usually three) components – as discussed in eq. (1) in [11]:

$$\Delta e = (e_K + e_P + e_{pos})_X - (e_K + e_P + e_{pos})_Y \quad \dots (2)$$

- between an inlet X and exit Y. In eq. (2) e_K is specific kinetic energy of fluid while specific sum of pressure and thermal energy is e_P. If the fluid is not gas, it is necessary to consider also the difference in specific vertical position energy of the liquid), which is e_{pos}.

c) In contrast to the Ohm's law eq. (1), fluidic devices mostly behave in nonlinear manner, complicating the solution mathematics.

Since the dissipated energy in eq.(2) is mostly the kinetic component Δe_K (organised flow motion is converted into the chaotic energy of thermal motions) and this is proportional to the square of the velocity, eqs. (3, 4, 5) in [11], the equation of characteristic for most nozzles is near to quadratic parabola.

$$\Delta e \sim \dot{M}^2 \quad \dots(3).$$

An important tool for solutions of problems of fluid mechanics in general is the dimensional analysis and the associated Buckingham's π theorem. It simplifies solved problems by reducing the number of variables (by conversion of primary simplexes into complex quantities). A governing equation that involves n simple variables can be re-written as an equation of $n - m$ dimensionless parameters, where m is the number of independent dimensions. This analysis may even suggest the characters of the initially unknown governing equation. Among the dimensionless complex parameters, of essential importance for nozzles is Reynolds number

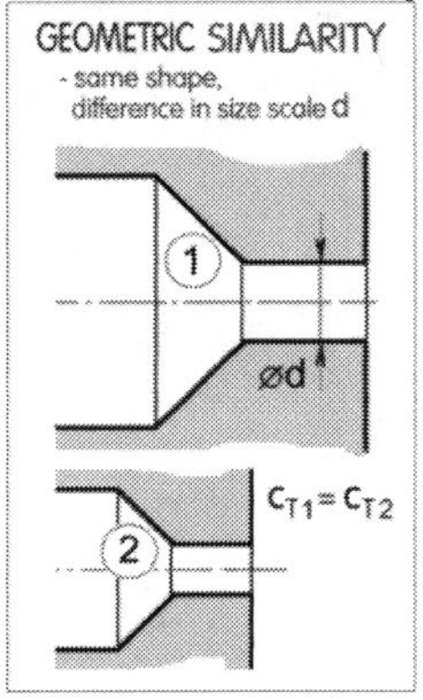

Fig. 5. Two nozzles are fully similar if they differ only in magnitude of the exit diameter d. Such full similarity has been always considered a necessary condition for application of the dimensional analysis.

Re of the nozzle exit flow, eq. (1) in [12].

$$Re = \frac{wd}{\nu} \quad \dots (4)$$

Apart from the size scale expressed by the diameter d and the scale of velocities of motion, expressed by the nozzle exit velocity w [m/s], Re involves also the variable that expresses the properties of the fluid, viscosity ν [m^2/s] characterising the magnitude of internal friction – energy dissipation in the fluid

It has been always considered without questions that the basic condition necessary for the similarity solutions is the geometric similarity, as shown in Fig. 5. It means the devices for which the solution is valid must be of the same geometry and differ only in the size scale.

Dynamics of flowfields – the problem of momentum transport in a flow with dependence on more than a single spatial co-ordinate - is described by partial differential equations. Very important consequence of what the similarity approach can do is reducing the number of variables. This is essential simplification mathematics: the partial differential equation with two or even three independent variables is converted into the much easier to solve ordinary differential equation — with derivatives only by the single similarity-transformed independent dimensionless distance. This similarity-based simplification is described, e.g., in [19]. Later, in [14], it was attempted to apply this similarity-transform approach to compute the flowfield of synthetic jets, generated by alternating inflow and outflow from a nozzle. In this case, however, the complicated conditions made the same full similarity solution not possible. Nevertheless, the situation was not completely hopeless. What was possible, is applying the concept of "quasi-similarity". It means the transformation of variables and of the governing partial differential equation was not made at once to be valid over the whole investigated flowfield. Instead, the key parameter in the similarity transformation was gradually varied as the solution progressed along the jet axis.

There is a fundamental principle of physical sciences in claiming that solutions of physical problems in general

 https://doi.org/10.1051/epjconf/201818002107

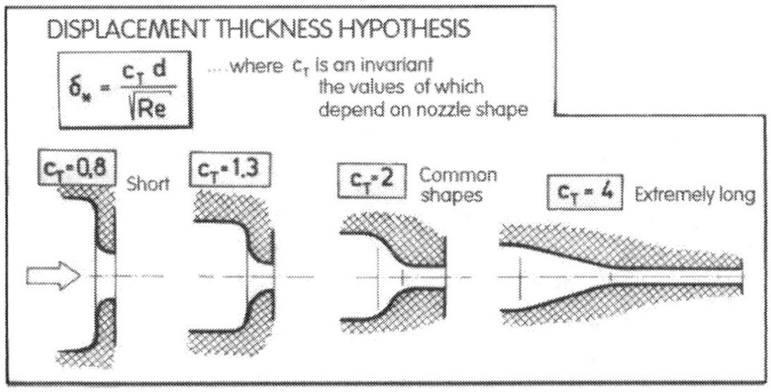

Fig. 6. Characterisation invariant c_T of nozzles was introduced initially in [1, 9] for nozzles with smooth, rounded internal shape – but it could be later extended to characterise the internal geometry of nozzles in general. Increasing c_T value means a longer friction path of the internal flow, the consequence being a thicker boundary layer in the nozzle exit plane.

are actually identification of problem ***invariants***. In the present case of nozzles, the characterisation invariant was found by this author to be the parameter c_T, introduced already in 1985 [1] to specify the properties of nozzle geometry. Typical meaning of this characterisation quantity is shown in the example of nozzles of various axial length shown in Fig. 6. Introduction of the parameter c_T was based in [1] on two hypotheses. Both were valid for smoothly shaped nozzles (no sharp internal edges). An easier to follow explanation of the approach, than [1] is available the later paper [9].

The idea of the invariant characterising a nozzle was later extended to be applicable also to nozzles with the sharp internal edges [10]. This is useful because of the importance of such nozzles in the practical applications requiring inexpensive, simply made nozzles. Analysis of the two ideas [9] and [10] later found that the approach in [10] is actually convertible to c_T in [9].

The basic idea of this characterisation is the fact that parameter c_T fully specifies nozzle behaviour as presented by the characteristic in Fig. 8. In other words, two nozzles with the same magnitude of the shape parameter c_T exhibit the same specific energy loss Δe at any given flow rate $\dot{M}$ — even though they may be geometrically very much different. This means here is not valid the standard definition of similarity according to which similar nozzles differ only in their absolute size expressed by their exit diameter d.

Interestingly, this new idea of similarity may be extended also to quasi-similarity: geometric variability of the numerical value of a shape parameter. In [12] it was applied to a family of nozzles, geometrically otherwise similar apart from having different exit channel lengths, as presented in Fig. 7.

Before advancing to presenting examples of prac-tical use of the nozzle characterisation, it may be useful to discuss the simpler approach to the characterisation problem. It may be seen as member of the following list:

0-th order characterisation: neglects frictional losses inside the nozzle.

1st order characterisation: takes friction loss into account but considers it constant, not Re-dependent.

2nd order characterisation: full description of be-haviour in the most important range of Reynolds numbers.

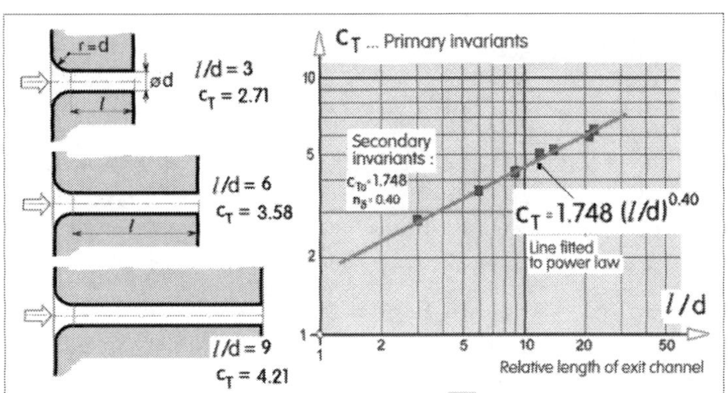

Fig. 7. An example of nozzle family with geometric quasi-similarity of having different exit-channel lengths which makes the standard π theorem not applicable. Since in the pressure-difference analysis the interest is only in a single quantity of each nozzle — its characterisation invariant c_T — it became possible to introduce [12] the idea of the ***secondary*** invariants, c_{T0} and n_δ.

 https://doi.org/10.1051/epjconf/201818002107

4 Fist-order characterisation by dissipance Q

Obvious idea considered useful at the time when were designed and tested first fluidic circuits was to follow the analogy with the circuits theory in electrical engineering. In the steady-state regimes, behaviour is there described by the value of R, the multiplicative parameter in the Ohm's law, eq. (1). As already mentioned above, in the fluidic devices the analogy of eq. (1) is nonlinear, nearest perhaps to quadratic dependence.

$$\Delta e = Q\,\dot{M}^2 \qquad \text{...(5)}$$

— this is eq. (3) with introduced multiplication factor Q [m²/kg²] known as ***quadratic dissipance***. In many practical situations the deviations of real behaviour from eq. (5) may be small and negligible. Then the numerical value of dissipance **Q** may be considered constant — and may be then considered as characterising the first-order behaviour of a nozzle, Fig. 8.

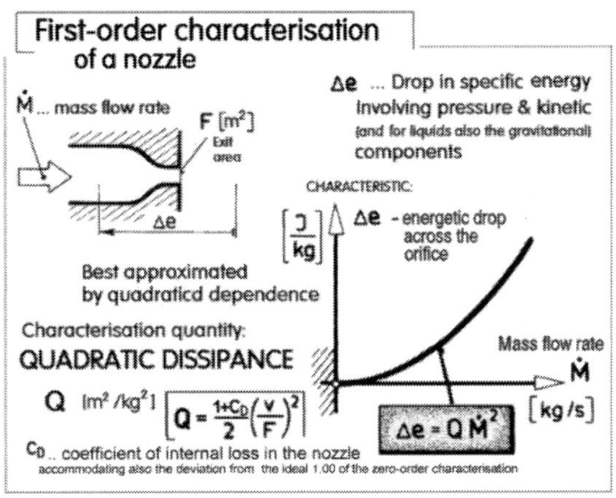

Fig. 8. Nozzles may be approximately characterised by the single parameter Q in the simple quadratic dependence eq. (5) between the two state parameters. Strictly speaking Q is not an invariant – but deviations are small and handling it as if it were simplifies solution of many typical problems.

5 Practical uses of characterisation

In experimental investigation of a nozzle the mass flow rate at the right-hand side of eq. (5) is usually obtained by readings of a flowmeter – while in a numerical flowfield computation its quantity is input by simple summation of inlet velocity boundary condition over the inlet area. The specific energy drop Δe across the nozzle on the left-hand side of eq. (5) may be an object of complicated computations [3, 21] if the conditions make it necessary to take into account compressibility of the fluid. If, however, the compres- sibility can be neglected, then Δe is simply proportional to pressure difference ΔP measured by a manometer

$$\Delta e = v\,\Delta P \qquad \text{... (6)}$$

- where the proportionality constant v [m³/kg], is the specific volume of the fluid. For flows of a gas this quantity v is computed from temperature, pressure, and gas constant r:

$$v = \frac{r\,T}{P} \qquad \text{... (7)}$$

The value of the dissipance **Q** is then evaluated as

$$Q = (1 + c_D)\,\frac{8\,v^2}{\pi^2 d^4} \qquad \text{...(8)}$$

-where c_D is the loss (or drag) coefficient of the internal flow in the nozzle

Energy dissipation
in the jet
inside nozzle

$$Eu = 1 + c_D$$

while **Eu** is the loss Euler number, mostly involving the dissipation of the kinetic energy of the jet. In the zeroeth-order approach, useful for initial assessments of nozzle behaviour, c_D is neglected and only the jet dissipation is considered

$$Eu = 1 \qquad \text{...(9)}$$

Second-order characterisation also operates with Euler number. Important property of this approach, however, is variation of Eu with Reynolds number [9,10]

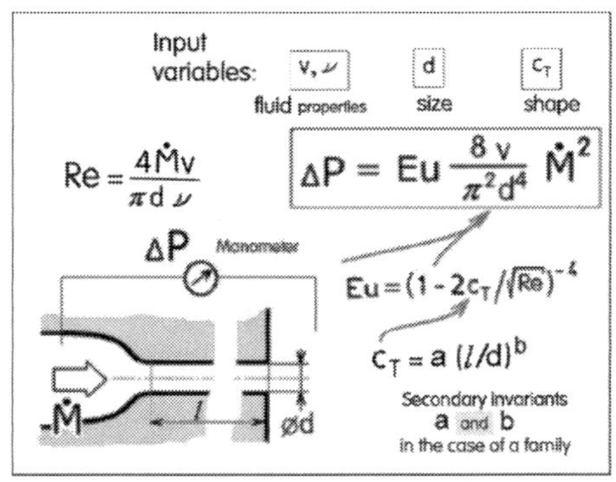

Fig. 9. Schematic representation of the steps in computation of pressure drop across a nozzle from known c_T (or known secondary invariants, Fig. 7). It is assumed that Reynolds number of the nozzle exit flow is Re > 2 000 (otherwise, in the creeping flow regime, it would be necessary to apply higher characterisation approach [12]).

After identification of the numerical value of the invariant coefficient c_T, the Eu is calculated using the second-order universal characteristic

$$Eu = \frac{Re^2}{c_T^4} \Big/ \left(\frac{\sqrt{Re}}{c_T} - 2\right)^4 \qquad \text{... (10)}$$

Validity is demonstrated using experimental data for two extreme nozzle cases in Fig. 10. On one hand is there he case of extremely long exit channel nozzle l = 13.9 d, (nozzles so long are not used in standard engineering practice) for which was evaluated c_T = 4.771. The other case in Fig. 10 is the extremely short orifice (rather than a nozzle). Its c_T is 1.288 — again an extreme geometry not used in usual engineering applications. Despite these geometries being so extremely different, in Fig. 10 they are seen to agree with the second-order universal characteristic eq. (10).

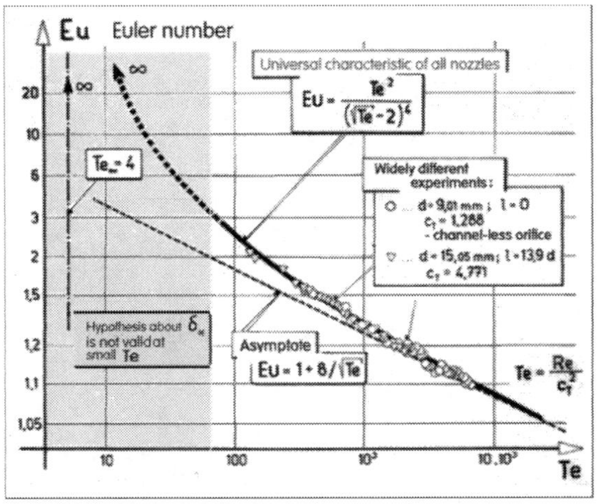

Fig. 10. Author's universal characteristic of all nozzles compared with two extreme cases, the shortest and the longest nozzle for which were by the author accumulated experimental pressure loss data.

6 Application examples

6.1 Two nozzles in parallel, characterised by Q

This is a rather simple case of a fluidic circuit solution problem. Classical approach is laborious step-by-step evaluation of velocities and pressures, usually complicated by initial absence of knowledge of the flow division into the two paths (note that the nozzles need not be identical or even of similar properties). Thus the computation has to progress in iteration cycles. Introduction of the dissipances Q_1 and Q_2 charac-terising the nozzles simplifies the procedure substantially. Presented at left in Fig. 11 is schematic representation of the two nozzles both connected to the same source of fluid – and hence applying at both nozzles the same constant specific energy drop Δe. The solution steps in Fig. 11 shows the solution to be quite simple. It may be useful to introduce the concept of a unit consisting of the two nozzles and evaluate effective dissipance Q_{ef} of this unit. In the bottom left part of Fig. 11 is the expression for this unit's effective dissipance.

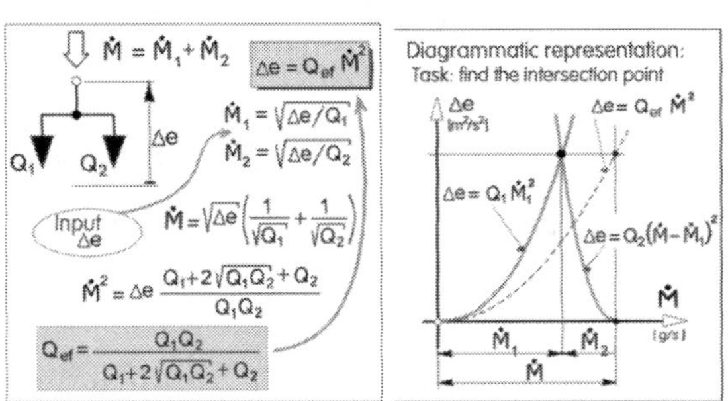

Fig. 11. An example of usefulness of nozzle characterisation by quadratic dissipances Q. For a given specific energy drop Δe and dissipances Q_1 and Q_2 of the two nozzles the approach evaluates their mass flow rates. A useful step is to evaluate the effective dissipance Q_{ef} of the unit consisting of the two nozzles.

It is a result of the sum of reciprocal square roots of the two dissipances. It is easy to evaluate the total mass flow rate through the unit as the sum of the two mass flow rates. The first step in the solution is evaluation of the two flows by inversions of eq. (5). Their sum is then squared - and this immediately leads to computing the value Q_{ef}. The procedure may be guided by the graphical solution shown in the right-hand side of Fig. 11. Two quadratic-parabola characteristics are drawn, with their origins on the horizontal mass flow rate axis — one of them is drawn on a transparent paper, which is then inverted and shifted along the horizontal axis until the intersection of the conditions (vertical positions of the interaction points) is fulfilled. Of course, real drawing the two parabolic curves and search for their intersection point is nowadays unlikely, but a mental picture of the two curves may serve as a useful support for the actual computation on a computer. The procedures of Fig. 11 may be easily generalised to the case of a larger number of parallel nozzles — note that coolant nozzles are often arranged in such clusters.

6.2 Application example #2: Nozzle characterised by dissipance Q matched to its fluid supply source

Solutions like the result in previous #1 may assume a constant specific energy drop at the exit of fluid flow source. This may be done, but complicates the circuit by added pressure regulator – making the system less efficient, due to the loss in the source, and more expensive. Without regulator, the available specific energy (and of course also pressure in a simple incompressible case) decreases with increasing delivered flow rate. A realistic model of a fluid source with exit loading, shown in the bottom left part of Fig. 12, has in the output a fluidic resistor representative of the decrease. Its presence models the typical losses that the fluid must overcome on its way out to the load.

The task is to find the optimum delivery of fluid power from the supply to the nozzle — as shown schematically at left in Fig. 12. Behaviour of the nozzle is described by its dissipance Q_N, dependent upon the nozzle exit diameter d. If this diameter is small (the case of high nozzle dissipance) the delivered power is small because of the small nozzle exit flow rate. On the other hand, with a large-diameter, low dissipance nozzle the fluid flow issuing from the nozzle may be large, but the available pressure drop on it would be so low that the delivered power would be also uselessly small. Optimum behaviour of this simple circuit is obviously somewhere between these two extremes. It is defined by the maximum product of the two exit parameters, nozzle output mass flow rate and specific energy drop. The derivation in [17] is as follows:

The power $\dot{A}$ [W] delivered into the nozzle is

$$\dot{A} = \Delta e_Y \, \dot{M}_Y \qquad (11)$$

The characteristic of the nozzle is

$$\Delta e_Y = Q_N \, \dot{M}_Y^2 \qquad (12)$$

 https://doi.org/10.1051/epjconf/201818002107

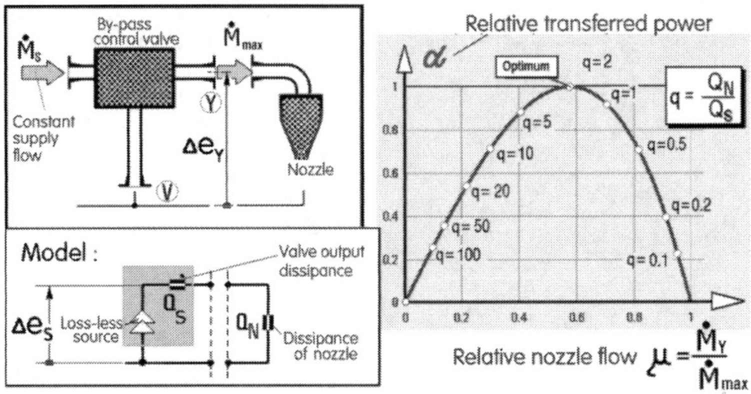

Fig. 12. An example of matching the nozzle, characterised by its dissipance Q_N, and its fluid supplying valve with output dissipance Q_S. The ratio of the two dissipances q is related in the unique manner to the ratio of supplied and by-passed flow rates, the dependence exhibiting an optimum for transferring maximum fluid power (product of mass flow rate and specific energy drop).

while the loading characteristic of the source is

$$\Delta e_Y = \Delta e_S - Q_S \dot{M}_Y^2 \tag{13}$$

(from the constant specific energy Δe_S of an ideal loss-less source is subtracted the internal loss inside the source). Combination of eqs. (12) and (13) is

$$Q_N \dot{M}_Y^2 = Q_S \dot{M}_{max}^2 - Q_S \dot{M}_Y^2 \tag{14}$$

and the power delivered into the nozzle is

$$\dot{A} = \Delta e_S \dot{M}_Y - Q_S \dot{M}_Y^3 \tag{15}$$

The optimum is defined by

$$\frac{d\dot{A}}{d\dot{M}_Y} = 0 \tag{16}$$

Applying this condition to eq. (15) results in

$$\Delta e_S = 3 Q_S \dot{M}_Y^2 \tag{17}$$

$$\dot{M}_Y = \sqrt{\frac{\Delta e_S}{3 Q_S}} \tag{18}$$

It is useful to operate with relative, universally valid dimensionless vaslues.

$$\mu = \frac{\dot{M}_Y}{\dot{M}_{max}} \qquad \alpha = \dot{A}/\dot{A}_{max} \tag{19}$$

- where theoretical maximum flow rate ideally obtainable from the source (at zero specific energy drop at the output terminal) is

$$\dot{M}_{max} = \sqrt{\frac{\Delta e_S}{Q_S}} \tag{20}$$

- which, compared with eq. (18) means the optimum is at

$$\mu_{opt} = \frac{1}{\sqrt{3}} \tag{21}$$

while the optimum value of the loading factor q – the ratio of nozzle and source dissipances is

$$q = \frac{Q_N}{Q_S} \tag{22}$$

in re-written eq.(14) is

$$q \mu^2 = 1 - \mu^2 \tag{23}$$

which means (Fig. 12)

$$q_{opt} = 2 \qquad Q_N = 2\, Q_S \tag{24}$$

6.3 Application example #3: Request for no spillover regime

Difficulty with no-moving-part fluidic devices and circuits is their generally more laborious design than in of corresponding hydraulic and pneumatic devices with moving parts. In the latter cases the active mechanical component is placed to a position in which it dictates where the fluid must and must not go The undesirable other flowpaths are simply mechanically blocked.. In the pure, no-moving-part fluidics such blockage does not exist and the choice of desirable flow path among the existing ones depends on the proper distribution of pressure. At the same time, there are now increasingly often situations in which precise determination of the fluid flowpath is absolutely necessary. Typical for present-day fluidics and microfluidics are biomedical applications where the handled fluids are small samples of biological material – sometimes difficult to obtain (perhaps at a cost of pain to the patient) and at any rate the samples should be small. This means any spilling over of the sample into the wrong path - and even a small partial loss by this spillover, as suggested in Fig. 13 - is definitely unwelcome and sometimes forbidden. Another area of increasing popularity of fluidics is microchemistry. typical tasks there are discoveries of new substances (a case is discovery of new catalysts [3, 22]). For the reliability of the obtained results it is absolutely necessary absolutely necessary to prevent any cross-contamination between the very small handled amounts .

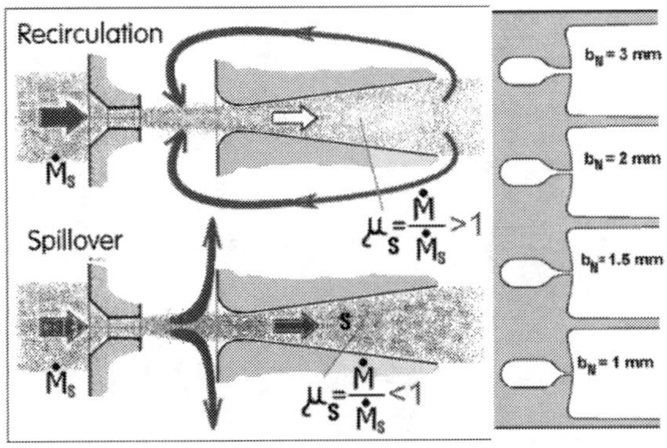

Fig. 13. In the absence of solid mechanical blockage in pure fluidics, it depends on the loading whether there will be jet pumping or spillover – or, mostly desired, neither of these two = the no-spillover regime. The study in [17] used a family of nozzles (right) with equal depth and different exit widths.

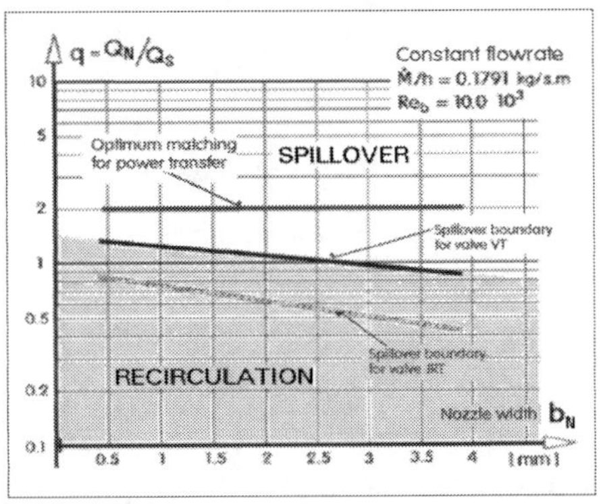

Fig. 14. Optimality criterion quite often imposed is request for the no-spillover regime. In this example from [17], various nozzles were connected to fluidic valve VT and JRT designs. Note that achieving the no-spillover $\mu_S = 1$ was in both cases incompatible with the power transfer optimum q= 2 of Fig. 12.

of the reactants. This is another typical task requiring precise adjustments of pressure distributions that can ensure proper resultant flow directions, either the recirculation or spilling the fluid over (Fig. 13) or, the most often demanded, the exact no-spillover regime. The circuit design had to develop relations for the characterisation quantities Q of the valve as well of the connected fluidic loads. As demonstrated on the example in Fig. 14 reproduced from [17], quite often the performance criteria are, unfortunately, mutually incompatible: For example, the no-spillover conditions, different for the two valves discussed in [17], cannot be achieved simultaneously with the power transfer optimum $q_{opt} = 2$.

6.4 Example #4: Apparent Q increase in pulsatile-flow

Nonlinearity of nozzle characteristic causes phenomena that may surprise unprepared, especially those who intent to measure the time-mean pressure drop across a nozzle. If the flow contains, apart from the steady component, also a periodic pulsation component — such as the harmonic one in Fig. 15 — the measured time-mean pressure difference increases above the expected value. The nozzle apparently behaves as if its exit were becoming smaller: the pressure drop across it has increased. The effect is, in fact, a special case of flow rectification. In practice it is complicated by influence of capacitance of connected accumulation devices. The time-mean value of the pressure (or, in Fig. 16, the specific energy difference) increases because the second-power component $\sin^2(2\pi f t)$ is everywhere

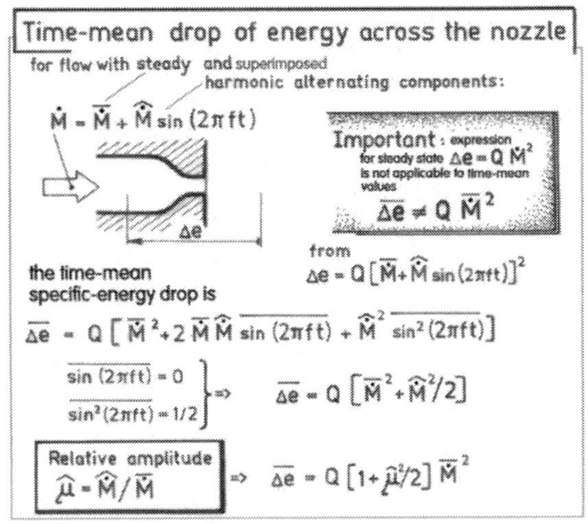

Fig. 15. Because of non-linearity of the nozzle characteristic, superposition of periodic pulsation (here the simple harmonic alternating flow) results in the nozzle behaving as if its time-mean dissipance increases.

positive and adds an amount to the result – an amount proportional to the square of the relative amplitude of the alternating component.

Author performed simple experimental verifications of these results. A nozzle of exit diameter d = 6.03 mm, with quadrant-shaped-shaped contraction corresponding to the nozzles presented in Fig. 7, but with zero channel length l /d = 0 — was connected to a positive-displacement blower. This type of blowers supplies air at a constant adjusted flow rate in a flow rate not dependent on the loading connected to it. The other, pulsatory flow component was supplied by piston-type generator with the piston driven by the Scotch-yoke mechanism ensuring a simple harmonic flow oscillation. The time-mean pressure difference across the nozzle was measured by highly damped liquid micromanometer. Evaluated from the time-mean pressure drop was the effective dissipance Q_{ef} compared with the steady-state flow (no pulsation) nozzle dissipance Q. Each of these four experiments, the results of which are presented in Fig. 16, was run starting from the non-moving piston generator so that it was possible to check the initial value Q. Then the piston generator was switched on and the amplitude of piston motion was gradually increased. The stroke s of the piston was kept constant s/d = 1.16 10^3. As expected on the basis of the derivation shown in Fig. 15, the time-

https://doi.org/10.1051/epjconf/201818002107

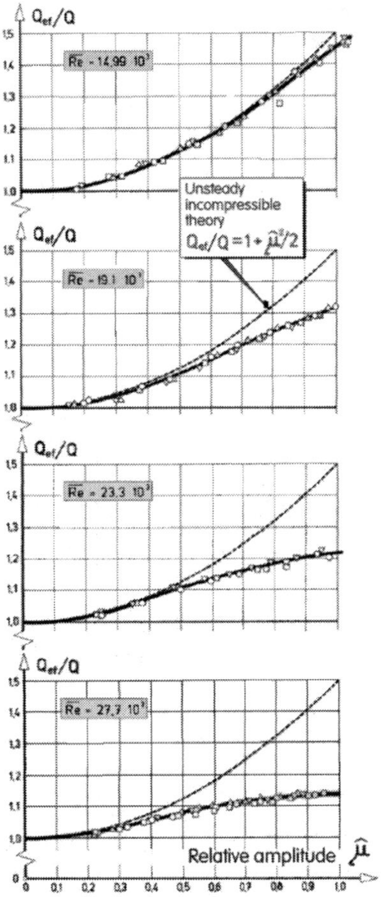

Fig. 16. Examples of time-mean pressure drops across the nozzle measured by the author. The values are smaller than the theory presented in Fig. 15. This pressure-decreasing effect was stronger with increased magnitude of the DC flow component (in the diagrams characterised by the time-mean Reynolds number — although later analysis has shown that this is not a Re effect).

mean pressure across the nozzle has gradually increased. Each of the four experiments finished when the maximum amplitude reached the magnitude of AC component. Were the piston amplitude increased further, the air flow at some segment of the oscillation would reverse sign — obviously undesirable in this experiment because the used quadrant-shaped nozzle would exhibit different properties in the reverse flow direction. In the four experiments with results shown in Fig. 16 the agreement with the theory from Fig. 15 was satisfactory in the case with smallest flow rate. The time-man flow rate is in Fig. 16 characterised by the value of Reynolds number $\overline{\mathrm{Re}}$ — because the deviations from the theory were at that time thought to be a Reynolds-number effect (it should be emphasised that the disadvantage of the used first-order characterisation of the nozzle is the variation of Q with varying Re).

The correct explanation of the pressure decreasing effect as seen in Fig. 16 was later provided in ref. [13]. It

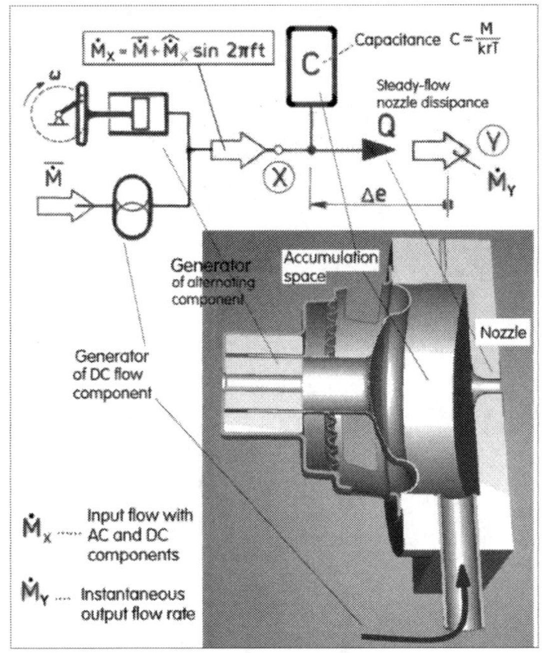

Fig. 17. An example and schematic representation of the test configurations used for investigating the effect of attached accumulation chamber. Instead of the original piston-type devices, with inertial-limited frequency range, these later devices used loudspeaker driver generating the AC flow component.

mulations space upstream from the nozzle for two purposes. One of them, is the usual flow recovery after the disturbance caused by the asymmetric inflow (note that in Fig. 17 the steady component comes through the tube from below). The other purpose is providing a space for the moving piston — or, in the present case the membrane of the speaker. The compression capability of this space is characterised by its capacitance C(Fig. 17) .

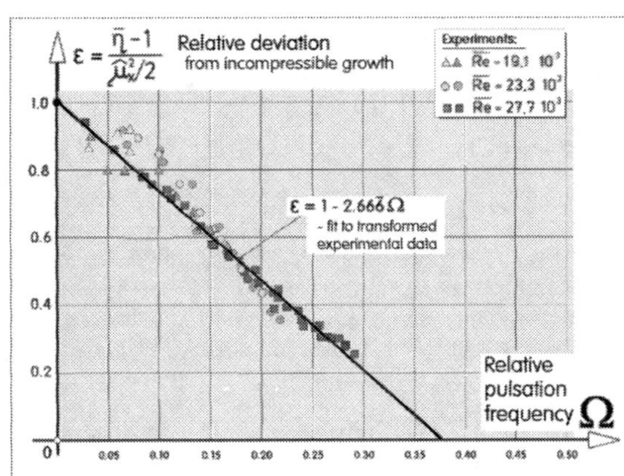

Fig. 18. Demonstration of the correctness of the explanation of the pressure decrease in Fig. 15 by the damping effect of the inevitable air-accommodating chamber C in Fig. 15. There is a universal linear law for the compressibility effect as a function of the relative pulsation frequency.

It determines, together with the nozzle dissipance Q, the characteristic time of the accumulation

$$t_c = C\,Q\,\overline{M_X} \qquad (27)$$

- where $\overline{M_X}$ is the constant time-mean input flow delivered by the DC flow generator.

The discussion in [13] has introduced several dimensionless variables by means of which it is possible to express the results. in general terms. One of them is the dimensionless frequency

$$\Omega = \pi f t_c \qquad (28)$$

The diagram in Fig. 18 shows the universal dependence of the relative deviations from the incompressible expression (from Fig. 15) on this dimensionless frequency.

9 Conclusions

This paper surveys the parameters useful for characterising hydrodynamic properties of nozzles. There is actually a hierarchy of characterisation approaches. Known and so fat widely used is the simplest **0-th** order characterisation eq. (9), which neglects frictional losses inside the nozzle. Later was introduced the **1**st order characterisation by quadratic dissipance Q, which takes into account the internal (mainly friction) losses and considers them a constant. not Re-dependent eq. (8). Applicable within much wider range of Reynolds number values is the recently introduced **2**nd order approach eq. (1, 9, 10, 12), which characterises a nozzle by its invariant coefficient c_T, in the expression eq. (10). Recently this second-order approach eq. (10) was extended for use in family of quasi-similar nozzles (in fact of dissimilar geometry) with properties described by secondary invariants, Fig. 7. Paper provides several examples of the practical applications nozzle computations using the characterisation.

Acknowledgements

Author obtained institutional support RVO:61388998 and also Czech Republic Grant Agency research grant Nr. 17-08218S

References

1. V. Tesař, *Application PO 17-85 for Certificate of Discovery*, Czechoslovak Patent and Discovery Office, Prague April 1985
2. V. Tesař, chapter „*Impinging Jets*" pages. 191 – 231 in: *Vortex Rings and Jets, Recent development in Near-Field Dynamics*. Springer Science + Business Media, 2015
3. V. Tesař, *Pressure-driven microfluidics*, Artech House Publishers, Boston USA 2007
4. A. Hauptmann, I. Wagner, Metals and Mines, *Studies in Archeology, British Museum Pub.*, London, 67 (2007)
5. Z. Váňa, Artia Praha 1983
6. B. C. Bobusch, *Dissertation TU Berlin*, 2015
7 H. Choi, W.-P. Jeon, J. Kim, Annular review of fluid mechanics, Vol. 49, 2008
8. W. B. J. Zimmerman, V. Tesař, *European Patent.* EP2081666, filed Oct. 2007
9. V.Tesař, *Chemical Engineering Research and Design* **86** 1253 (2008)
10. V. Tesař, *Chemical Engineering Research and Design* **88** 1433 (2010)
11. V. Tesař, *Sensors and Actuators A*, **263**, 147 (2017)
12. V. Tesař, *Sensors and Actuators A*, **248** 246 (2016)
13. V. Tesař, *Sensors and Actuators A*, **179** 211 (2012)
14. V. Tesař, J. Kordík, *Sensors and Actuators A*, **149** 255 (2009)
15. V. Tesař, J. Kordík, *Sensors and Actuators A*, **199** 379 (2013)
16. V. Tesař, J. Kordík, *Sensors and Actuators A*, **199** 391 (2013)
17. V. Tesař, *Chemical Engineering Research and Design*, **87** 817 (2009)
18. E. Jarkovský, *Foundations of practical computation of orifices, nozzles, and Venturi tubes*, SNTL Prague (1958)
19. V. Tesař, *Zeitschrift für Angewandte Mathematik und Mechanik*, 77, 333 (1997)
20. H. Blasius, *Grenzschichen in Flussigkeiten mit kleiner Reibung*, Teubner 1907,
21. V. Tesař, Chapter in *Microfluidics: History, Theory, and Applications*, Springer-Verlag, Wien-New York 2006
22. V. Tesař, *Role of microfluidics in discovering new marketable substances – a survey*, Proc. of 19th Internat. Conf., p. 614, Svratka, Czech Republic, May 2013
23. V. Tesař, J. Kordík, *Sensors and Actuators A*, **191** 34 (2013)

Wind tunnel for studies of latent heat storage

Václav Tesař [1,a]

[1] *Institute of Thermomechanics v.v.i., Czech Academy of Sciences, Dolejškova 5, 182 00 Praha-Kobylisy, Czech Republic*

Abstract. When a heated solid body temperature reaches the melting point, temperature stops increasing and remains constant until the whole body is completely molten. The heat input during this melting is spent on freeing the body molecules. This ***latent*** heat of melting remains inside the body and may be released when the body is cooled and solidifies. This heat was suggested, already several decades ago, for storing thermal energy. The advantage it offers is avoiding high temperature differences - which otherwise decrease effectiveness of storage (by inevitable heat escape by conduction). Also the mass of the body needed to store a given amount of heat is much smaller. For investigations of the melting and solidification processes a special wind tunnel has been designed and is being built in this study. The tested sample of phase change material, encapsulated in a spherical shell, will be exposed in the tunnel to recirculating hot air flow in a 140 mm x 140 mm test section. Sudden decrease in air flow temperature is made by shifting away the whole closed-circuit part of the tunnel and exposing the test section to flow of cold (room temperature) air.

1 Introduction

One of the main problems with energy – perhaps the crucial one - is mismatch between availability and demand. This creates a strong interest in efficient energy storage. Of the existing principles listed in Fig. 1, this paper is concerned with temporary storing of heat between the times when it is abundant and when demanded. While some other storage principles need a mechanical machinery or another complex system, the heating and cooling of a body is extremely simple and fail-safe. Intensity of the heating and cooling processes in the simplest case depends on available temperature differences – while total storage capacity depends on the mass of the charged body. High temperature differences suffice with a small storage body but may cause problems – such as too much heat escaping through inevitably imperfect insulation by conduction.

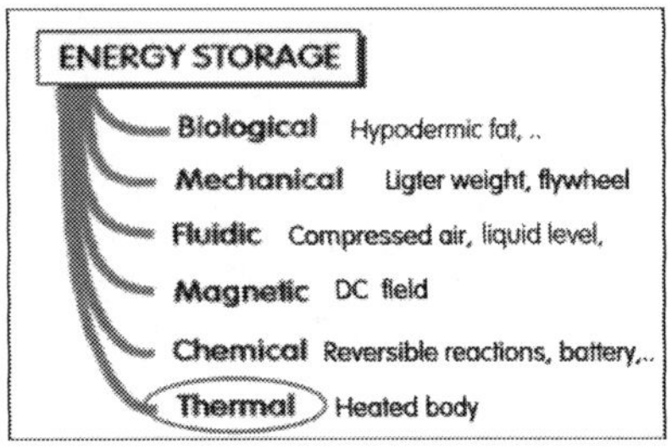

Fig. 1. Among energy storage methods, the thermal storage handles lower quality energy, not directly convertible into other forms – particularly not into the most often demanded electric energy. It is thus particularly useful for simple applications such as keeping a more or less constant temperature of living bodies (humans) and objects with which humans come in contact.

An interesting solution avoiding the high temperatures and yet needing only rather small heated body is offered by using phase changes. In principle, the charging and discharging can take place in the region of constant temperature seen on the equilibrium states diagram (valid for paraffins) in Fig. 2.

The variability of parameters of a substance is governed by the Gibb's Phase Rule [1]

$$\mathbf{F} = 3 - \mathbf{P} \quad (1)$$

- where **F** is the number of degrees of freedom while **P** is the number of phases in mutual contact. With a single-phase material considered here it is possible to vary pressure and temperature independently (F = 2). Upon reaching the phase change surface (Fig. 2) the degrees of freedom reduce to F = 1. This means the temperature of an isobarically heated material remains constant until all the investigated contents is changed from the state **a** into **b.** The heat input without corresponding temperature increase is described as "*latent heat*", the term introduced in 1762 by J. Black, [2] (professor of "Medicine and Chemistry" at the University of Edinburgh) in experiments supported by Scottish whisky manufacturers.

In principle it is possible to use for the storage the melting as well as vapourisation phase changes. It was the latter that was studied by Black - but practical aspects in the storage problem generally give preference to the former. Magnitudes of the stored heat is presented in Fig. 3. The same amount of heat is released from the melt during its solidification. The advantages obtainable from the (very nearly – note the differences needed at the process beginning and end) isothermal energy storage were soon recognised as attractive possibility – apparent by comparing in Fig. 3 the difference between the final state 2* without the phase change and 2 with the latent heat. Attractiveness of the latent heat storage idea is reflected in the large and steady rising numbers of annually published research papers on this subject,

[a]Corresponding author: tesar@it.cas.cz

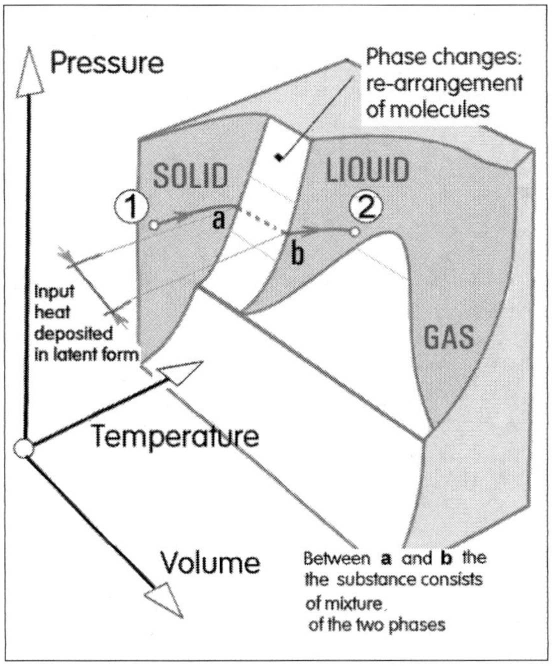

Fig. 2. When a solid substance is heated, the limit **a** is reached at which all additional thermal input is used to liberate the molecules into free motion in liquid phase. Temperature is not increased until all molecules undergo the phase change to **b**. The diagram applies to paraffins the volume of which increases during melting (solid paraffin body does not float on the liquid phase surface and sinks to the bottom).

presented in Fig. 4. Many storage processes among those listed in Fig. 1 are proposed and currently developed aiming at the problem of mismatch in electric energy, nowadays increasing in importance as the result of poor controllability of nuclear reactors on one hand and unreliability of renewable energy sources on the other one. Thermal energy storage, however, is not really suited for this particular purpose. Converting heat into electricity at industrial scale requires engines operating in a thermal cycle to evade the fact that thermal energy – energy of chaotic motions – is of lower "quality" that the organised motion of electrons in electric conductors.

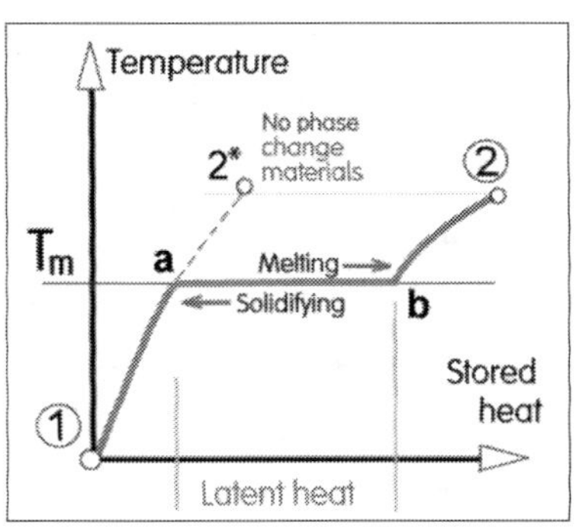

Fig. 3. Dependence between variations of temperature and internal thermal energy of a substance with and without phase change between the input state 1 and final states 2 or 2*.

Nevertheless, there are many cases where the electric form is not necessary. There are energy consumers needing it in the thermal form. In particular, there is a significant activity in development of "smart" thermal insulators accumulating heat for a certain, usually not long period at quite low temperature levels. Especially promising is the current development of "smart" textiles, containing tiny encapsulated volumes of the phase change material among the fibres of the woven yarn. With this textile material, typically for use in battle-dress uniforms in desert combat areas, the latent-heat beads can cool the soldier during the hot desert day (while being charged by the solar heat) – and heating him in the cold desert night (discharging the accumulated heat) [3]. Another very typical application is cooling the heat generated in very compact, high-density electronic components. In particular, there are now active research projects investigating cooling of hand-held smart phones [4] that in the most compact versions can become heated to temperature levels above 45 °C — the limit at which plastic objects in experiments [5] could be held for a long time in hand.

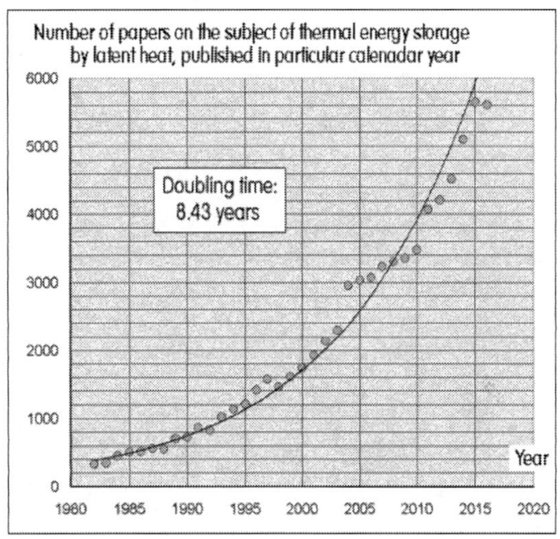

Fig. 4. Interest in storage of thermal energy in latent heat of a phase change has been steadily increasing as documented by the annual number of published papers – data from Scopus.

2 Phase change materials

The earliest observations of the processes associated with melting and solidification phase changes used to involve water and its change into ice. These, however, take place at inconveniently low temperature from the point of above mentioned human-related applications. Interest therefore shifted to studies of other materials, especially those that could be used in the temperature levels typical for living bodies – indicated by the horizontal red line in Fig. 5. The material developments resulted in four substance families

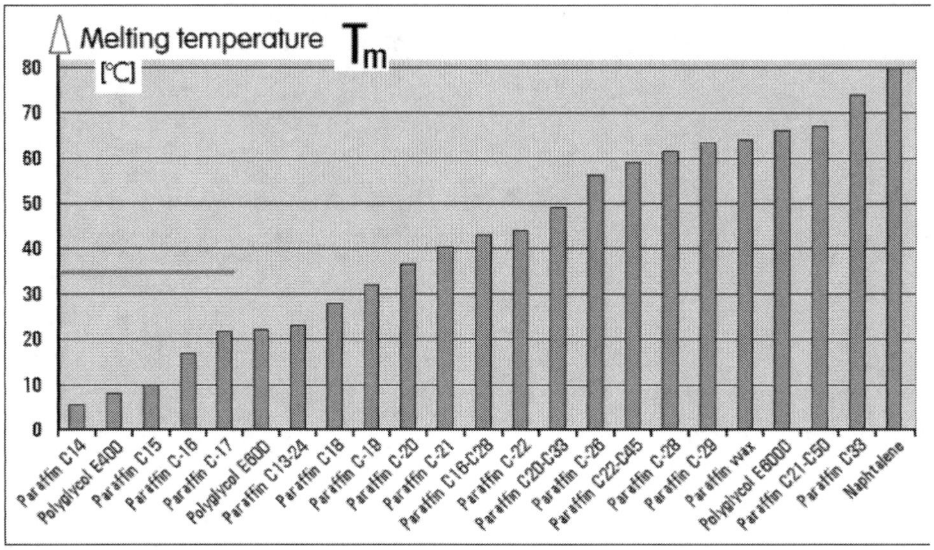

Fig. 5. Present prevailing development interests in application of latent heat storage tend to concentrate on keeping comfortable thermal conditions for humans (red line) and also of devices and instruments that should be comfortably touched even if they generate heat. As shown here, existing range of commercially available paraffins offers a suitable choice of the melting temperature level.

considered to be most promising – not only from the point of view of the melting temperatures, but considering also several other criteria. The total list of requirements is as follows:

(1) Suitable level of the melting temperature T_m
(2) Large enthalpy of phase change e_L (Fig. 8)
(3) High thermal conductivity
(4) Little or no superheating and subcooling
(5) Availability
(6) Low cost
(7) Inflammable as well as chemically inert
(8) Non-toxic and convenient for manual handling

The importance of the proper melting temperature has finally led to present-day commercial availability of phase change materials with quite finely adjustable T_m values, an example of which is presented in Fig. 5. This now provides an opportunity to concentrate on other properties from the above list.

The importance of large enthalpy of fusion – i.e. the horizontal width e_L of the phase change zone between **a** and **b** in Figs. 3 and 8 is also obvious. Typical values for the four material families are presented in Fig. 6.

It may be perhaps less apparent why it is also important for the phase change material to have high thermal conductivity – both in the solid and liquid phases. The lower is the heat flow resistance, the smaller are the deviations from temperature constancies at both ends of the heating curve shown in Figs. 3 and 8, i.e. the difference between the points **1** and **a** in the discharging (solidification) regime and between the points **b** and **2** in the charging (melting) process.

Some materials exhibiting a proper melting temperature value T_m and thus seemingly suitable for the thermal energy storage may be actually not practically useful for this role because of their large hysteresis of the heating curve. Instead of the single horizontal line between **a** and **b** in Figs. 3, 7 and 8, there may be two horizontal lines at different temperatures – one for heating and the other for cooling. The subcooling means the molten phase does not solidify at the nominal T_m - and the superheating means the material stays solid at temperatures above the nominal T_m value. The four families of materials most promising for meeting the criteria (1) to (8) in the previous page are:

(a) Low melting temperature metals,
(b) Paraffins
(c) Linear alkyl hydrocarbons
(d) Salt hydrates

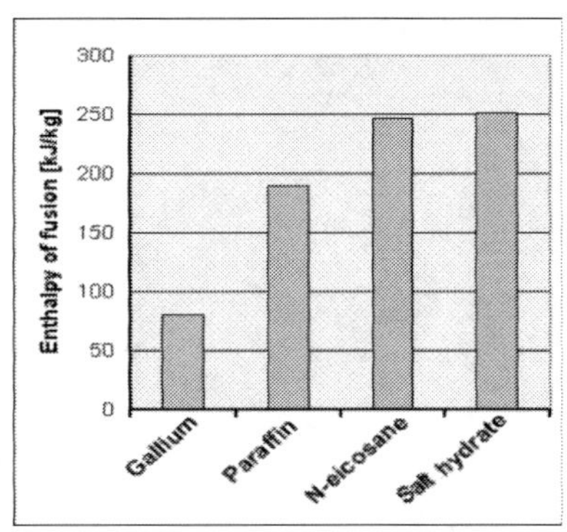

Fig. 6. Selected examples of heat stored per unit of mass – one from each of the four discussed latent-heat storage material families: low-melting-point metals, paraffins, alkyls, and salt hydrates. Of course, high value of the fusion heat e_L shown here is, together with the temperature presented in Fig. 5, the prime factor in choice of the used material. There are, however, other properties that may be of importance in particular cases.

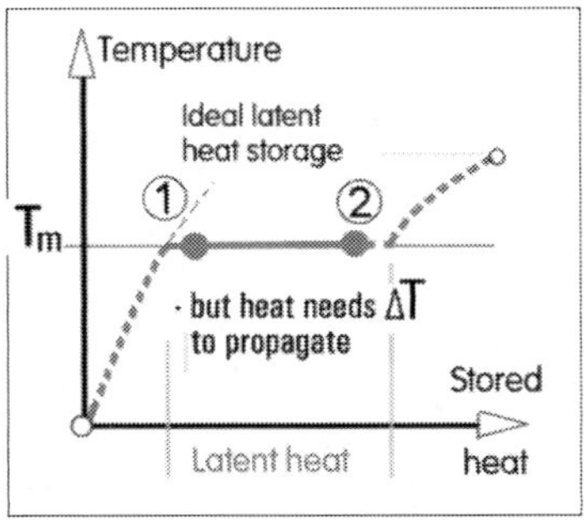

Fig. 7. Ideal loss-less heat storage would follow the constant temperature level as shown here between **1** and **2.** Unfortunately, transport of heat to or from the storage body is not possible without temperature differences necessary – Fig. 8 – for charging as well as discharging.

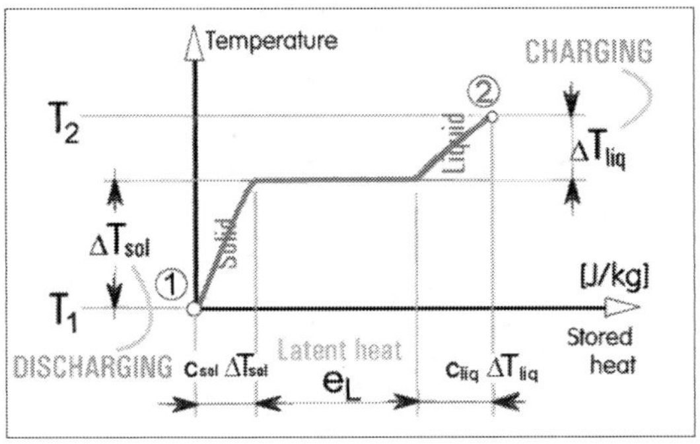

Fig. 8. Actual latent-heat storage contains the inevitable conductivity-governed transport into or from the storage material. The inevitable conduction parts of charging and discharging, between **1** and T_m as well as between T_m and **2**, decrease storage efficiency.

Among the metals remaining liquid at low temperatures, the most attractive from the here discussed point of view seems to be **gallium**, melting nominally at T_m = 29,76 °C. The temperature T_m may be adjusted to a desirable value by alloying gallium with other, higher melting point metals. Advantage of gallium is its availability (it is in production at quite large scale for use in the plutonium-gallium alloy in nuclear weapons, e.g. [16]).

Gallium latent heat of melting is 80.3 kJ/kg, Fig. 6. This is not much, but would be acceptable if there were not other problems. One of them is quite large subcooling [6, 7]. Also unpleasant is the fact that gallium attacks most metals by diffusing into their lattice thus making the other metal brittle. Many low-melting-point metals, such as caesium and rubidium are even less suitable, being highly reactive and toxic. Commercially available is the alloy called *"Cerrolow"* , with melting temperature T_m = 47.2 °C. Another commercially available alloy is the six-component eutectic consisting of Bi-Pb-Sn-Cd-In-Tl exhibiting Tm = 41.5 °C.

Organic substances considered or used in the phase-change material heat storage role due to having reasonably low melting temperature T_m are of two types.

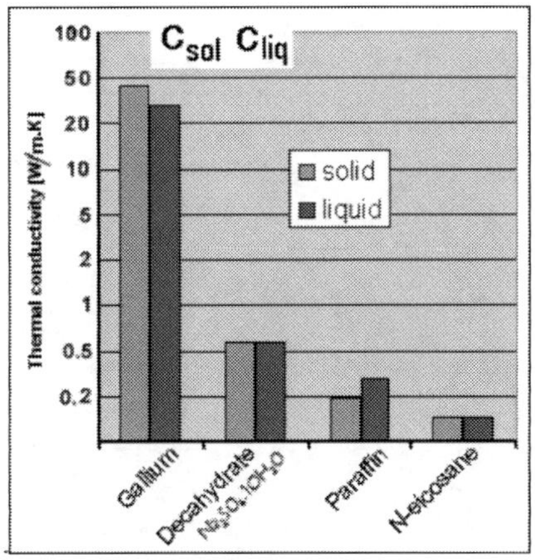

Fig. 9. Characteristic property of the discussed four families of phase change materials are their very different thermal conductivities — note that the values are here plotted using logarithmic scale. The different conductivity coefficients can significantly influence the overall efficiency of the storage.

One of them are **paraffins** with various T_m levels listed in Fig. 5. Their latent heat of fusion (Fig. 6) is higher than that of liquid metals, and their other advantages is the fact that they are chemically inert and friendly in manual handling. They are also commercially available at low cost. Their weak point is typically very low thermal conductivity, cf. Fig. 9, and - perhaps less importantly - also their mild flammability.

Paraffins are at present attractive objects for experimental investigations of latent heat - and are currently often discussed in heat storage literature - due to the useful fact that their liquid and solid phases are easily discriminated visually: molten paraffins are transparent and allow observation of the decreasing or increasing volume of the solid phase, which is opaque.

Other family of interesting organic compounds are **alkyls** – linear hydrocarbons. Typical, quite widely considered because of some excellent properties, is particularly N-eicosane — listed in Figs. 6 and 9. Alkyls in general are chemically stable, non-corrosive, exhibit no subcooling and posses very high latent heat per unit of weight – Fig. 6. On the negative side they are flammable (which is not so important property when used inside an encapsulation) and possess very low thermal conductivity, as is presented in Fig. 9 (note the logarithmic scale).

The fourth family of materials currently considered are **salt hydrates,** inorganic compounds. They were and still are objects of several investigations because of their absolutely highest known values of enthalpy of fusion, cf. Fig. 6. In particular considered as quite attractive is sodium sulfate decahydrate $Na_2SO_4.10H_2O$. Its problem is separation effect in solidification: the solid salt generated in the discharging process settles down at the bottom of the vessel in which this material is kept. As a result of this spatial differentiation the salt ceases to be easily available for recombination with water, necessary during the subsequent charging part of the cycle.

The question of low thermal conductivity, which is a typical property of the otherwise quite attractive organic heat storage materials, is not in general given the adequate attention it deserves. It is not uncommon that comparisons of the effectiveness of alternative processes in literature is evaluated based on the assumption presented in Fig. 7 — that both beginning **1** and end **2**

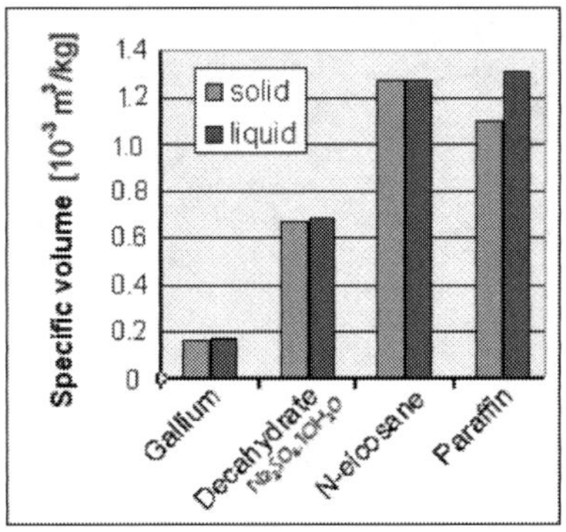

Fig. 10. Also of considerable importance in designing the latent heat storage system are the specific volumes of the materials – and the changes that the volume undergoes in the melting and solidification. The typical larger value of the liquid phase means the solid body, instead of floating (like ice) on the liquid surface, would be kept at the bottom of the storage vessel. Values in the diagram are valid for T ~20 °C.

of the heating process are at their common isotherm. There are actually two facts that should be kept in mind. The first of them is the unavoidable presence of the conduction for charging and discharging, at the ends of the line as shown in Fig. 8. The inlet and outlet temperatures must be above or below the T_m isotherm. With no temperature differences the heat would not be transported so that the storage would not work at all. The second fact are wide differences between the conductivity values in the four material families, as seen in Fig. 9 (less apparent there because of the logarithmic scale). The problem is obviously more serious with the organic materials.

There is yet another property of the phase change materials that may play a non-negligible role. It is their specific volume and its variation during the phase change. Of particularly complicating influence may be the large change of volume in the case of paraffins (note the larger difference of the values between the solid and liquid phase in Fig. 10).

4 Encapsulation

This volume difference may be so large that the storage material may significantly stress the walls of the vessel in which it is encapsulated – and perhaps even cause them to buckle or collapse. To eliminate this danger, the top of the storage vessel may be left open into atmosphere — but then it is necessary to take into account the existence of a quite large part of the capsule occupied in solidification uselessly by the air in the cavity. Perhaps more importantly, the storage system must ensure that the opening for the communication of the vessel with the atmosphere stays always on capsule or vessel top. Keeping the hole at the required positioning is not so easy task as it might appear.

For his decision on the overall layout of thermal energy storage system with phase change materials, the designer must consider the alternative configurations ensuring that the material in its liquid phase remains where he want it to be – and not spilled elsewhere. This means there has to be for this purpose an encapsulating vessel or cavity.

As shown in Fig. 11, there are three available basic configurations used for containing the phase changing material and keeping it immobilised:

(a) Storage in a tank – with external or internal heaters
(b) Encapsulation - the material is kept in many smaller capsules to decrease the heat transport distance
(c) Microencapsulation: large number of very fine capsules, usually of size less than 1 mm

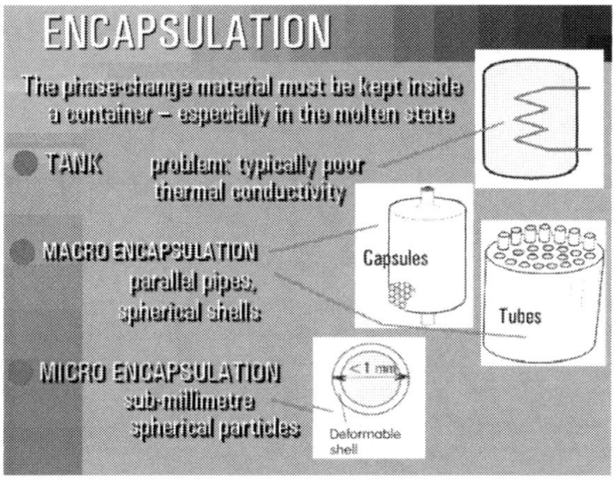

Fig. 11. Especially during the melting process – and, of course, also later when the phase changing material is in the molten state – it is necessary to keep this material within a suitable container preventing the escape. There are three basic configurations.

(a) Tank (or a large cavity inside the solid body): It is the simplest and also least expensive approach. The total volume of the phase change material is kept inside a single space (or perhaps in a few smaller-size spaces). Transfer of heat into the molten material in this case takes place across a quite large internal distances. This may be suitable for slowly operating storage systems, with time scales typically of the order of hours. It may be nevertheless necessary to select the phase change material exhibiting a high thermal conductivity (Fig. 9). The thermal transport may be influenced by the natural convection in the molten material – the convection intensity may be significant due to the larger size of the tank. The heat transport distances may be decreased (thus decreasing the time scale) by complex shaping of the heater, perhaps providing it with some fins or ribs.

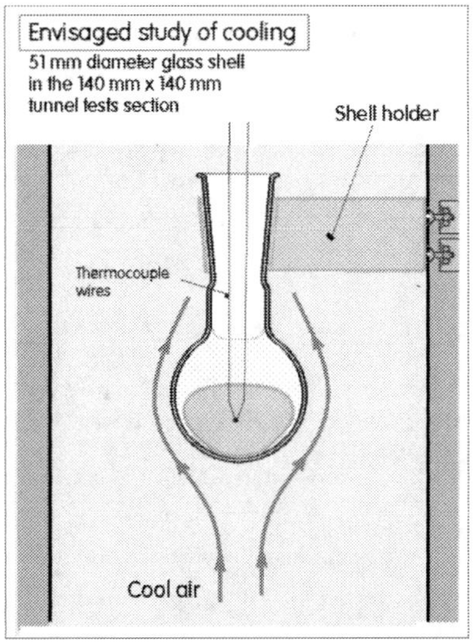

Fig. 12. Proposed configuration of the planned experiments. The heating/cooling fluid (air) flows vertically past the scaled-up model of microcapsule. The spherical shell is actually a commercially available glass vessel (flask) attached to the holder at the vessel neck.

(b) Storage material encapsulated: In this case the total volume of the phase change material is divided into a considerable number - or even large number - of solid containment shells. The disadvantage of this approach is increased size and, more important, higher cost of the storage system. On the other hand, the encapsulation improves storage dynamics by decreasing significantly the distance over which the heat is transported. Typical shell configurations are pipes or spheres, Fig. 11. It is also necessary to provide a suitable mechanism for carrying the heat up to each shell. Typical solution is use of another, auxiliary heat carrying liquid which fills the space between the shells on their outer side.

(c) Microencapsulation. Essentially similar to (b) but with substantially smaller shells, typically with their outer size less than 1 mm. This can decrease even more the response time of heating or cooling. Of course, making the large number of the small shells as separate objects would be laborious and therefore expensive. This is circumvented by the shells grown or deposited on the surface of a phase change material droplets – usually by a chemical (or physical) deposition process. Typically of spherical shape, the shell material must be deformable to account for the volume changes during the phase transition. The microencapsulation is a rather new idea now increasing in popularity – especially for the above-mentioned use in the "smart" textiles where the sub-millimetre capsules are fixed between the textile fibres.

5 Planned experiments

Because of the expected later use of the information gained in experiments for applications similar to the thermal-insulation hot climate field service uniforms, planned configuration in the tests is with the phase

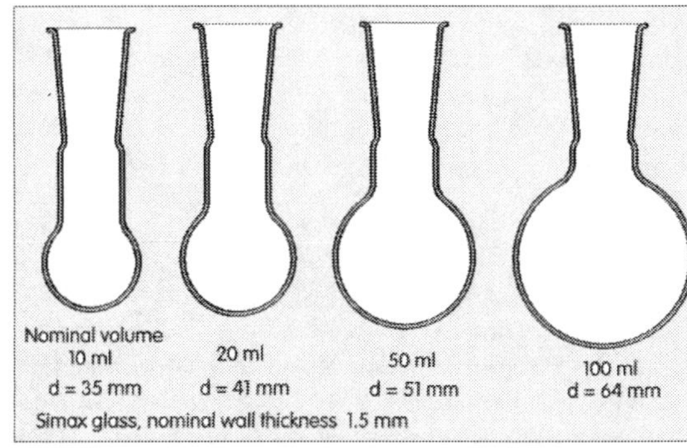

Fig. 13. Vessels obtainable from the supplier to be used in the test runs as the scaled-up shell model. The four available sizes can be usefully applied for investigation of Reynolds number effects at the same air flow velocity. The inlet necks on the top are, of course, a deviation from the ideal spherical geometry, but the downstream location is hopefully in the wake of the flow.

change material filling an internal space of spherical shells. Since the parameter characterising the progress of the phase change processes is going to be the volume fraction of the material in its solid phase, the experiments are planned to use paraffins — due to their transparency in molten state. The fraction will be computed from timed camera-record images of the boundaries of visibly not yet molten solid-state material (as it is approximately shown near the shell bottom in Fig. 12). Precision of the evaluated boundaries is, of course, the better the larger is the investigated object. This is the reason for selecting for observation and recording only a single microcapsule body model, scaled up substantially.

The tests are going to be set up as shown in Fig. 12, with the spherical shell containing the paraffin exposed to flow of air that may be alternatively hot (for melting) or cold (for studies of solidification). The scaling up will also make possible placing into various location inside the paraffin tiny thermocouple probes for local temperature measurement. Of course, also the temperature of the air flowing past the capsule model will be measured by thermocouples, mainly for computing from the air temperature its viscosity. This is necessary for identification of the flow Reynolds number.

$$\mathrm{Re} = \frac{wd}{\nu} \qquad (2)$$

- where w [m/s] is the characteristic velocity of the air flow, d [m] is the characteristic length, which in the flows past spherical objects is the sphere diameter (Fig. 13), and υ [m^2/s] is air viscosity (dependent on temperature). Because of the kinematic viscosity increasing with air temperature and the expected flow velocities being not large, Reynolds number of the flow past the test model are likely to be very low, corresponding to the not very high flow velocities in the microcapsules applications. The walls of the vertical air flow channels will be provided with transparent windows for illumination of the scene and for the lens of externally positioned camera. Spherical shells of constant wall thickness would be difficult to manufacture in the workshop of the Institute of thermomechanics, author's employer, but fortunately suitable objects for use in test models were found in small glass flasks – as shown in Fig. 13 – manufactured and offered for sale by local supplier of laboratory glass products. On their offer are four flask diameters (all with the same glass wall thicknesses). These size alternatives with geometric similarity make possible testing the transformation of the phase-change variables and equation into the similarity-transformed co-ordinates. Certain disadvantage of the use of the flasks is the presence of the neck, which the manufacturer makes on the flask top for insertion into it the tested materials. Of course, some access into the internal volume is necessary anyway, but here the necks appear in relation to the flask body diameters somewhat too large, as may be judged by inspection of Fig. 13. It need not be an unsurmountable problem. Wind tunnel aerodynamic experiments with spherical objects, especially at low Reynolds numbers, indicate that the boundary layer on object surface separates and forms downstream from it a wake inside which presence of material objects like the flask neck does not necessarily substantially influence the flow. The wake is filled with more or less chaotic and fast varying vortices. At the extremely low expected Reynolds numbers, from Re ~ 10 to Re ~ 100, there is a certain regularity of the flow with stationary vortex ring inside the wake, which may interact with the flask neck, but preliminary estimates indicate higher Re values.

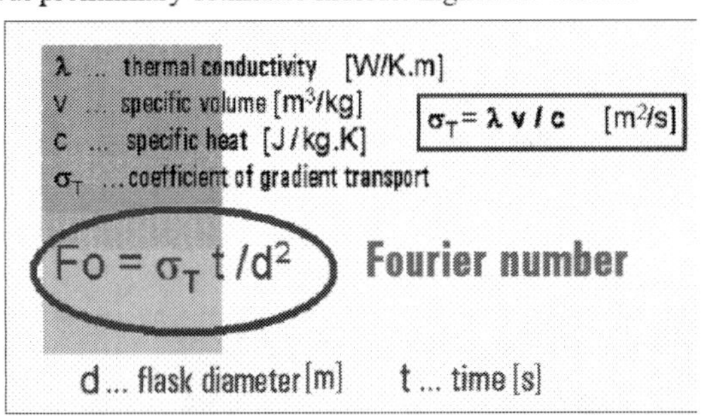

Fig. 14. Fourier number Fo is the governing parameter of unsteady heat transfer processes. It may be expected quite safely that in the discussed case it will be the dimensionless measure of melting and solidification time.

The wake past a sphere begins at the location where the boundary layer separates from the body surface. In laminar flows the typical angular distance from the front stagnation point to the separation point is ~80 deg – i.e. already upstream the position where, at ~90 deg, the sphere has its largest diameter. It is unlikely, especially in the hot air flow, that the conditions in the here discussed wind tunnel will reach fully developed turbulent flow. Nevertheless, even if it happens, the location of the turbulent flow separation is roughly at ~130 deg. This means in both cases separation locations will be sufficiently far upstream from the position where there is the neck of the flask. The neck is likely to be inside the wake where the flowfield condition are not suitable anyway for detailed investigations.

6 Governing dimensionless parameters

While the flow of heated air past the test object is governed by the Reynolds number, eq. (2), the processes of heat transfer are expected to depend upon the product of two dimensionless parameters, Fourier number Fo, defined in Fig. 14, and the Stefan number Sf, defined in Fig. 15.

$$t \sim Fo.Sf \qquad (3)$$

Fo characterises time dependent, unsteady processes without phase change, in which it represents the non-dimensionalised time **t**. The multiplication by Sf results in the expressions which characterise the processes involving the latent heat.

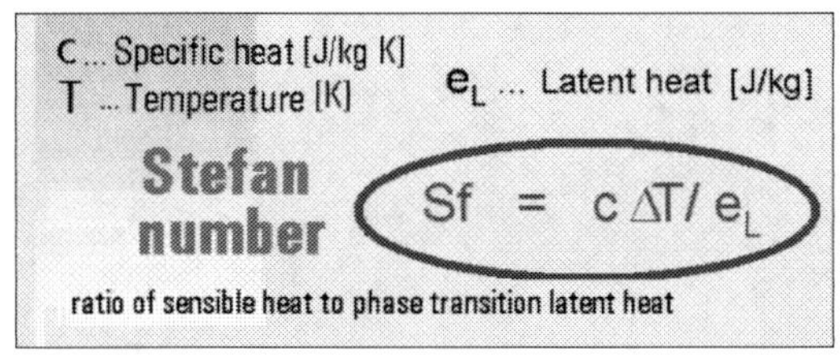

Fig. 15. Stefan number Sf is expected to be the governing dimensionless parameter of prime importance in processing the experimental data obtained for the melting and solidification studies in the capsule models.

It is difficult to estimate to what degree the heat propagation inside the encapsulation will be limited only to heat transport by conduction. It cannot be excluded that at the larger geometric scales there may appear effects of internal natural convection. In that case a certain role may be also played by Grasshoff number, defined

$$Gr = (g\, \boldsymbol{\beta}_v (T - T_m)\, d^3)/\upsilon^2 \; ... (4)$$

- where $g = 9.81\ m\ s^{-2}$ is gravitational acceleration, β_V [-] is coefficient of volumetric expansion of the sample in liquid state, T [K] is the temperature inside the sample in solid state, and the rest are parameters that were already introduced above.

It may be expected that plotting the time dependence of the solidification processes after conversion into the dimensionless variables will be similar to what is presented in Fig. 16. On the horizontal co-ordinate is the dimensionless time eq. (3) while on the vertical co-ordinate is plotted the relative percentage of the volume of tested material that remains during solidification in molten state (related to the initial 100 % volume). Typical heat transfer processes, because of the character of the governing Fourier equation, usually exhibit exponential dependence. In the semi-logarithmic co-ordinates used to plot the data in Fig. 16, such dependence would be fitted by a straight line. This is what is likely to be found in the settled regime at a longer time after the start of cooling. Of as yet not well understood character will be quite probably the initial

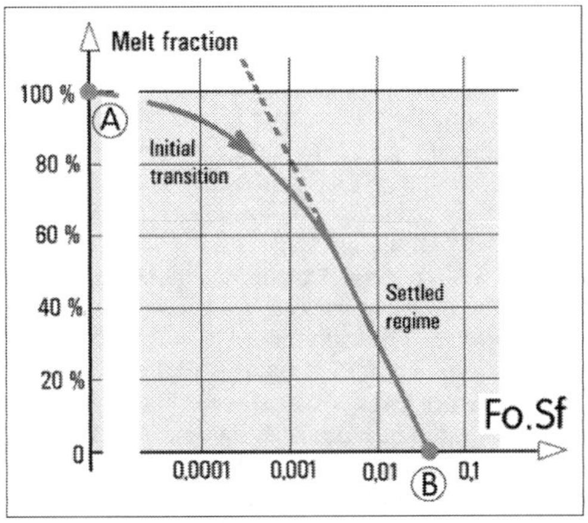

Fig. 16. Expected character of measured data on the solidification inside the spherical capsule model after their conversion into dimensionless dependence in semi-logarithmic co-ordinates . Essential part of the plotted curve is exponential (and therefore a shown here as straight line). Very little is so far known about the very short time of initial transition.

regime of small Fo.Sf values. To answer this question, it seems to be necessary to concentrate the attention in the planned experiments on the initial regime and its transition into the exponential dependence — i.e. in the time range largely so far out of interest.

7 Wind tunnel layout

The experiments aimed at answering the above mentioned questions (and some other ones related to it) are to be performed in the configuration as shown in Fig. 12 — where the initially hot air flow past the tested spherical shell, securing a complete melting, is to be suddenly and very rapidly switched to a cold air flow. The suddenness is important for obtaining data in the initial transition regime seen in Fig. 16. This is a requirement that was so far never posed in wind tunnel designs.

The requirements placed on general tunnel configuration included the starting point heating to (and, if desirable, also above) approximately T ~ 60 °C followed by the sudden switching to cold air flow. If such temperature changes were made in a direct manner it would be necessary to apply very high heating an cooling power input. The obvious problem is the thermal lag: heated components of the tunnel are likely to keep considerable stored heat after the switching. Considering these aspects — and also the availability of reasonable power input in the laboratory — it became obvious that the molten-material regime necessitates re-circulating the hot air flow. Thus the wind tunnel design ideas had to concentrate on the closed-circuit tunnel layouts. The vertical flow in the test section as presented in Fig. 12 – which, of course, is a part of the recirculatory airflow - necessitated orientation of the circuit in the vertical plane. The resultant configuration is as presented in Fig. 17. To avoid a large structure that would be necessary with

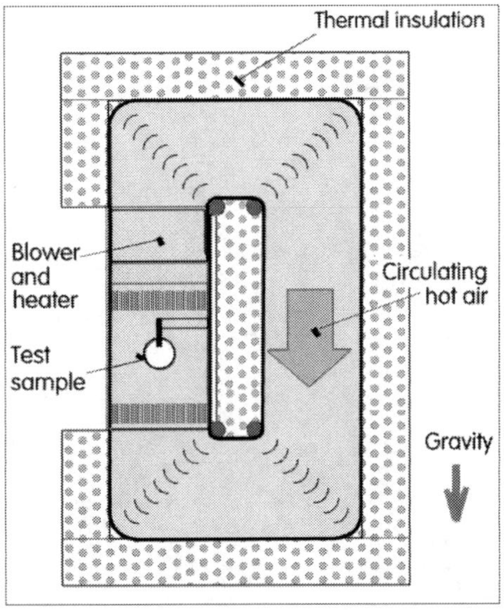

Fig. 17. Layout of the currently built wind tunnel for the latent heat tests. Hot air flow circulates in a vertically oriented loop. In the "up" arm at left is the test section and above it the combined unit for driving and heating the flow.

simple large turning-radius elbows (to avoid flow separations from the walls) in the return part of the tunnel, the design was focused on the historic approach of Prandtl's Göttingen tunnel. The elbows chosen were of sudden 90 deg turning type the flow separation eliminated by a rather dense turning blades. As shown in

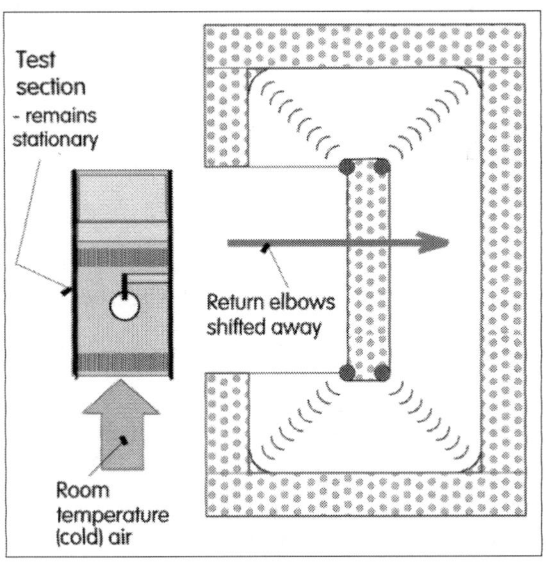

Fig. 18. Operation of the wind tunnel in the sample solidifying test. At the start of the cooling run the whole return-flow part of the tunnel is shifted away while the test section remains in place and is fed by cold air from the room. The stationary position of the test section is necessary for retaining properly focused cameras and lights (also not shifted).

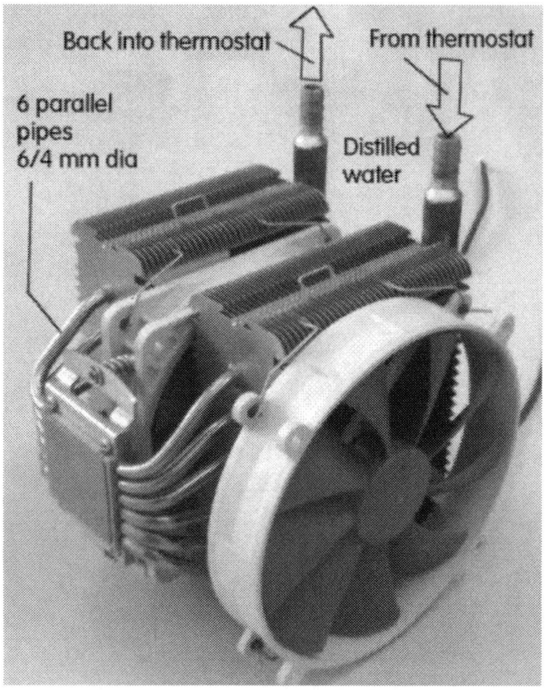

Fig. 19. Photograph of the heating and driving unit. Note the two series of the sheet-metal fins and the Ni coated copper pipes supplying them with heat from the thermostat.

Fig. 17, each of the four elbows are provided with a row of 8 turning blades. For further improvement of the velocity profile inside the test section, below and above the test section the whole tunnel cross section is filled with shortened packed drinking straws.

The blower unit inevitably destroys somewhat the character of the flow and to minimise its effect it is positioned above the test section. Swirl flow is suppressed by the heat-transfer find in the unit, as described below.

Heating of the air flow is secured by heat exchanger forming a part of the unit together with the two fans. Heat is transferred into the air through a grid of rather densely packed thin metal fins. Transfer into the fins is by tubes connected into a thermostat operating with recirculating distilled water. Thermostat allows precise adjustment of the distilled water temperature. The thermostat is positioned immediately below the wind tunnel.

For the sudden switching from the hot air flow above the melting temperature into a cold air flow regime, the solution was found in placing the whole return-flow part of the wind tunnel on rails. As shown in Fig. 18, the return duct may be suddenly shifted away so that the blower starts immediately delivering into the test section the cool room temperature air.

8 Driving and heating unit

The wind tunnel design discussed above was based on the idea of commercial availability the suitable unit generating the hot air flow. In this unit the air flow is to be driven and simultaneously heated to and above the paraffin melting temperature T_m. The suitable unit, originally intended for a different task, was selected in manufacturers' catalogue on the Internet. It is presented in the photograph Fig. 19. It was designed for use in cooling a computer CPU.– and here it is adapted for a slightly different role. It is specified as Noctua NH-D14 6-heatpipe with dual NF-A15 140 fans — the last number 140 indicates the diameter of the fans in millimetres. In its original use unit is provided with the baseplate (seen at left in Fig. 19) for fixing to this location the CPU. The heat from the baseplate was to be removed by 12 (= 2 x 6) heat pipes and transferred to two series-connected grids of thin aluminium fins exposed to the air flow.

For the use in the developed wind tunnel the internal components of the heat pipes were removed and the ends of the pipes were provided with two manifolds. Each manifold was tin-soldered to six pipes and contained a ferrule to serve as inlet or (on the opposite side) an outlet of heated distilled water from the thermostat. The baseplate for the CPU is retained, though it now does not serve for any particular purpose. The positioning of the unit inside the test section is shown in Fig. 20. Each of the two fans, operating in series, has an airscrew of 180 mm diameter with 9 blades. Driving power for each airscrew is 0.6 W.

The unit is to be positioned into the 140 mm x 140 mm test section of the tunnel as presented in Fig. 12. The rotational speed of the fans is adjustable in the PWM (pulse width mode) regime. In the manufacturer's catalogue is stated that the fans are fully controllable from 600 RPM minimum speed to maximum 1300 RPM. The flow rate at this maximum rotation speed is

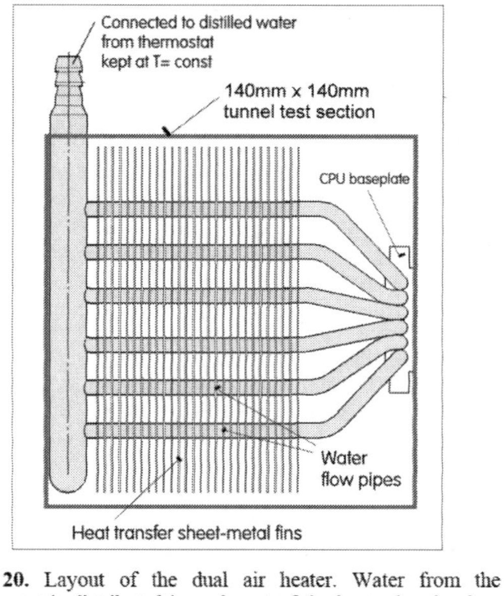

Fig. 20. Layout of the dual air heater. Water from the thermostat is distributed in each part of the heater by six pipes to a series of parallel thin aluminium fins. The original baseplate for CPU was retained despite its now having no useful role.

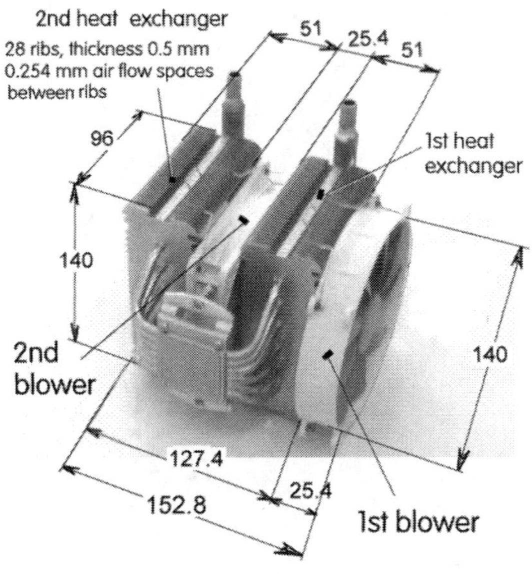

Fig. 21. The unit used to drive the circulating heated air in the wind tunnel by the two blowers - and also maintaining its temperature by means of the two-component heat exchangers. Liquid with precisely adjusted temperature is supplied from an external thermostat unit.

64.9 CFM (cubic feet per minute) which represents 0.0332 m^3/s. The cross-section area of the empty wind tunnel test section is 0.0196 m^2. If the wind tunnel is run with the largest, d = 64 mm diameter spherical shell (cf. Figs. 12 and 13), then the available cross-sectional area for the air flow is reduced to 0.0164 m^2 . Neglecting the losses in the return duct of the wind tunnel, at the maximum rotational speed of the fans the airflow velocity is approximately 2.0 m/s. With cold air of viscosity 15.7 10^{-6} m^2/s the Reynolds number of the flow past the tested sample under these conditions is roughly Re ~ 8.25 10^3. When run with 60 °C hot air of viscosity 18.8 10^{-6} m^2/s, the Reynolds number will be approximately 7.1 10^3.

The lowest Reynolds number of the hot air flow obtained by continuous electronic PWM control, i.e. when run at the value 600 RPM minimum fan speed indicated in the manufacturer's catalogue with the smallest, d = 35 mm diameter spherical shell (cf. Fig. 13), the same calculation shows Reynolds number Re = $1 \cdot 10^3$. It is, of course, easily possible to switch off one of the two fans to obtain approximately Re = 0.5 10^3 — and even much lower values may be obtained by mechanical area blockage in the return flow.

9 Wind tunnel body and progress report

At the time of writing this paper (June 2017) the wind tunnel is still being built. The associated photographs Figs. 22 and 26 show it in its not yet finished state.

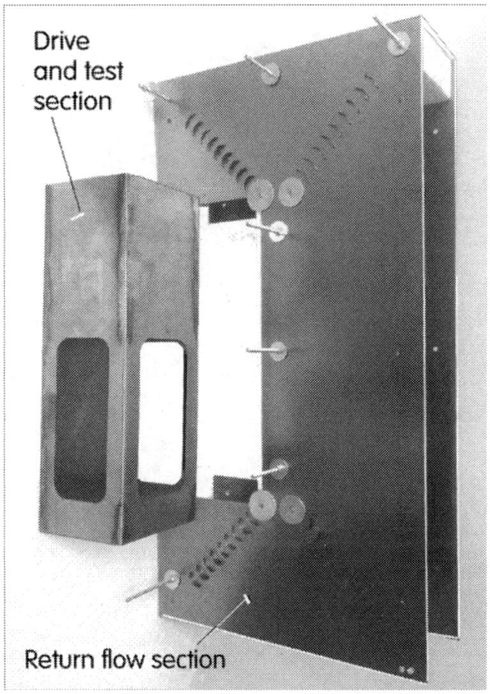

Fig. 22. Photograph of the two main components of the wind tunnel in construction. Both are made of steel sheets as the shape-carrying structure – at left is the test section with the large rectangular windows, at right is the flow-return part. The latter will be covered with thermal insulation cladding of thermal insulation thick polystyrene foam plates.

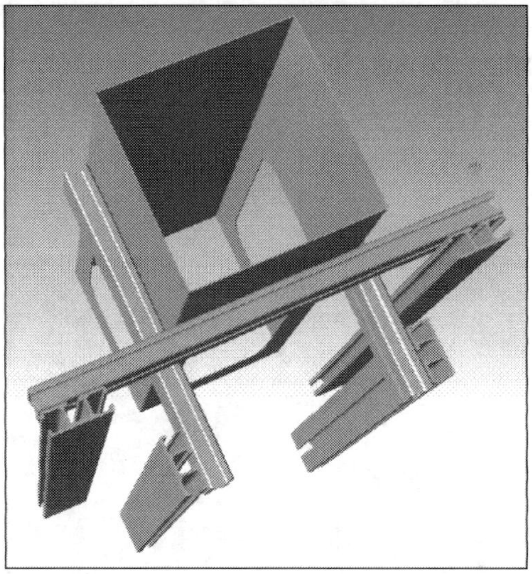

Fig. 23. Rather complicated design of fixating the test section (while the rest of the tunnels is translatable on rails) was necessitated by the request all three large access windows being simultaneously accessible for cameras. The components of the fixation structure are standard aluminium-alloy profile rods, commercially available. Triangular holders as well as the screws are here not shown.

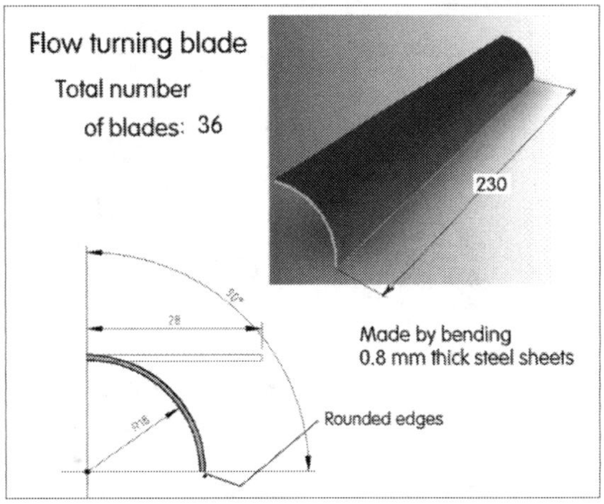

Fig. 24. Simple turning blades made by bending a 0.8 mm sheet metal rectangles, of constant thickness with rounded leading edge.

There are two basic parts: at left side of both photographs it is the test section housing the scaled-up model under test as well as the unit from Fig. 21. On the right-hand side in these photographs is the body of the return-flow tunnel part. It is on wheels running in rails so that this whole large part may be shifted away from the test section (at the time of taking the photographs the wheels and rails were still missing).

The test section is made from steel sheets bent and welded into a rectangular object. Well visible feature are the large rectangular holes that serve as access and observation windows. One is in the front wall and the other two are two on both sides. The holes are going to be covered by polymethylmethacrylate (transparent plexiglass) plates — held on screws so that they be removed for access to the test model. The front window may be replaced by thermocamera observation port is now already provided in the wind-tunnel test section as an alternative to one of the plexiglass windows. It is a round, 30 mm dia disk of ZnSe, supplied by Eksma

Fig. 25. The close-up photograph of the flow-turning blades (their detail is in Fig. 24) already fixed to their position in the return-flow duct.

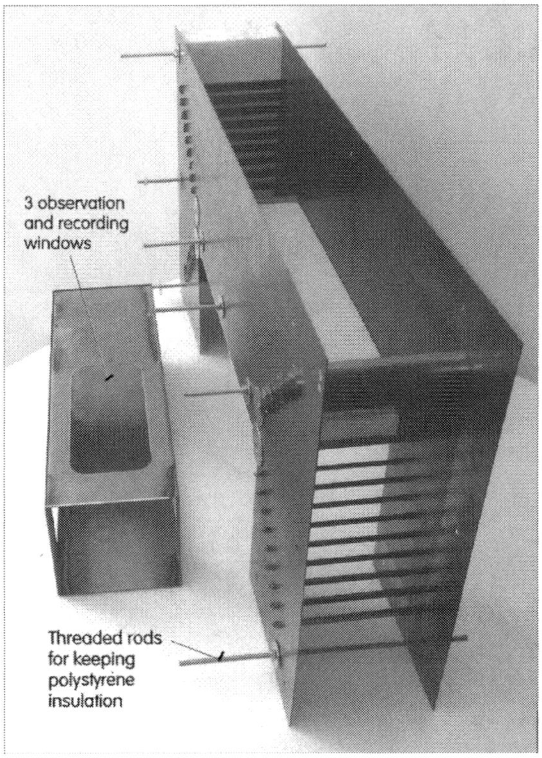

Fig. 26. Photograph of the two components of the tunnel. The polystyrene foam thermal insulation cladding will be held on the threaded rods seen in this photograph. The cladding will leave for the air flow the flowpath as seen in Fig. 27.

Optics in Lithuania. It is necessary because plexiglass blocks infrared light of wavelengths longer than 2 800 nm [17]. The test section is fixed to the stand by the structure shown in Fig. 23. Somewhat complex design is due to the request of no obstruction of any of three windows – which are quite wide so that the no-obstruction requirement resulted in placing the vertical members of the structure outside the test section corners, as is visible in the picture Fig. 23.

On the right-hand aside of the photographs Figs. 22 and 26 is the return duct part of the wind tunnel. Its basic components are the quite large sheet-metal plates — the left and right ones. Both are held together by the total of 32 corner flow-turning blades with the whole welded together. On the bottom part of this whole unit (not seen the photographs here) are standard commercially available metal rails – and wheels running in them. Workshop drawing of the detail of flow-turning blades is presented in Fig. 24. The blades are very simply made by bending from a 0.8 mm thick steel sheet rectangles – and are therefore of the everywhere constant thickness. Only the leading edges are rounded, as may be recognised from the close-up photograph of the blades in their positions in Fig. 25. As was mentioned above, the blades, apart from their flow-turning role, also hold the two steel sheets (as seen in Fig. 26) together at the 140 mm mutual separation distance.

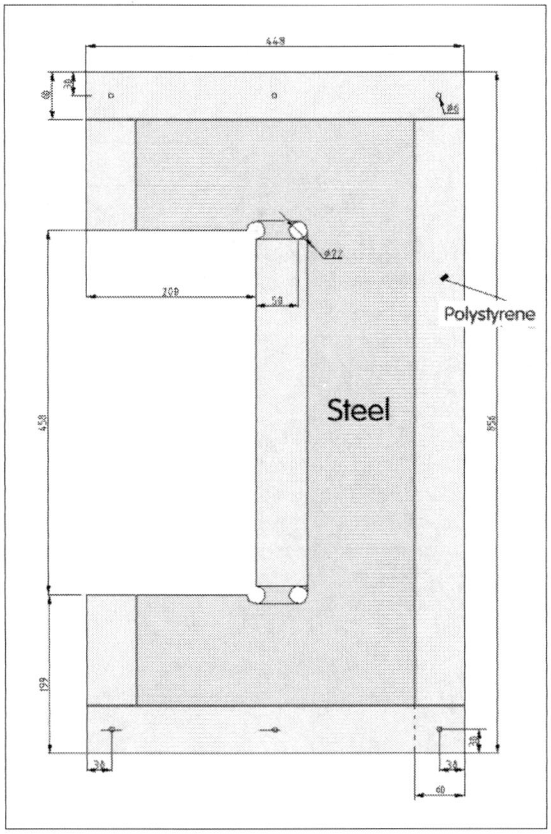

Fig. 27. Position of the 60 mm thick polystyrene block inside the sheet-metal walls. Available for the airflow is the grey colour area.

Still missing in the attached photographs is the covering by 60 mm thick thermal protection polystyrene foam plates. They will be held in position on the outer side of the sheet-metal plates using the threaded rods particularly well seen as projecting from the steel plate in Fig. 26. Made of polystyrene foam plates, 60 mm thick and 140 mm wide, will be also the components placed between both steel plates in the configuration presented in Fig. 26. They will delimit the return flowpath of the hot air. Since the melting and solidification tests are planned to be performed with paraffin melted at temperatures only slightly above the temperature of human body, there is no problem foreseen with the direct contact of the hot air and the plates of polystyrene which melts in the temperature range 210 - 249 ° C (depending on the precise sort).

10 Conclusions

This paper provides essential information about progress in the grant project started in January 2017. It aim is obtaining basic information and its use in development of thermal energy storage system using the energy stored in the form of latent heat — i.e. by using it to melting a solid object. The advantage of this approach is the possibility of storing quite large amount of heat in a relatively small melted body, without operating with high temperatures. The first part of the paper provides collected information and some useful data about the phase-change materials. Described in the second part of the paper is a very special small wind tunnel. It should serve for studies of melting and solidifying paraffins, encapsulated in a scaled up spherical shell. The extraordinary feature of the tunnel, which is now built, is the capability of performing a sudden step change of the temperature. .

Acknowledgements

Author was financially supported by the Czech Republic Grant Agency research grant Nr. 17-08218S and also by institutional support RVO: 61388998.

References

1. J. W. Gibbs, Scientific Papers, Dover, New York, 1961
2. H. Chisholm, "Black Joseph", Encyclopaedia Britannica **4**, Cambridge Univ. Press, 1911
3. K. Karthik et al., Journal of Textile and Apparel, Technology and Management **7**, 3, 2012
4. G. Haoshan, L. Jing, Journal of Heat Transfer **135**, 054503, 2013
5 N. Leoni, C.H. Amon, C. H., *ASME HTD* **343**, pp. 49–56, 1997
6. I. Harary et al., International Journal of Thermodynamics **19**, 36-41, 2016
7 K.G. Steenbergen, Modelling the Melting of Gallium Clusters, PhD Thesis, Univ. of Wellington, New Zealand, 2013
8. Ł. Wierbicki, Journal of Achievements in Materials and Manufacturing Engineering, p. 26, 2014
9. Anonymous, "Praffin hydrocarbon", Encyclopaedia Britannica, 2015
10. Anonymous, "Alkyl Groups", IUPAC, Compendium of Chemical technology, 2nd. ed., 2015
11. H. Plessen, "Sodium Sulfates" in Ullmann's Encyclopedia of Industrial Chemistry Wiley-VCH, 2000
12. E. Assis, G. Ziskind, Letan R., Journal of Heat Transfer **131**, 024502-1, 2009
13. A. Felix Regin, S.C. Solanki, J.S. Saini, Renewable and Sustainable Energy Reviews **12**, 2438-2458, 2008
14. F.A.P.S. Scacchetti, E. Pinto, G.M.B. Soares, Prograss in Oreganic Coatings, 107, 64-74, 2017
15. S. M. Zhang, T. Noll, D. Houtman, Thermal and Thermomechanical Phenomena in Electronic Systems, ITHERM 2002. Conf. San Diego, 2002
16. W.G.Wolfer, B. Oudot, N. Baclet, Journal of Nuclear Materials, 359, 185-191, 2006
17. N. L. Chiromawa, K. Ibrahim, Internat. Conf. of Global Network for Innovative Technology IGNITE 2014

HYSTERESIS OF PLANAR DOUBLE SLOT IMPINGING AIR JETS

Tomáš Tisovský[a,1], Tomáš Vít[1]

[1]*Department of Power Engineering Equipment, Faculty of Mechanical Engineering, Technical University of Liberec, Czech Republic*

Abstract. In present study, the bistability and hysteresis of the non-isothermal flow behind two slot nozzle is numerically investigated. Theoretical review of bistability of isothermal stagnation flow is presented and both flow field patterns that exist in the region of bistability are commented on. Hysteresis is found in number of conducted simulations and effects of various parameter changes are discussed. Moreover, the mechanism of transition from one flow field pattern to another is discussed for both transition directions. In conclusion, validity of 2D simulation results against real 3D problems is of concern and general contribution of this research is discussed. Throughout this work, emphasis is on applications regarding heat transfer.

1 Introduction

In many engineering applications, heat and mass transfer are of great importance. Due to the perpendicular orientation of the flow towards the wall, impinging jets achieve the highest heat transfer rates $\dot{q}\ [\frac{W}{m^2}]$ among all the convective processes. [1]

The heat transfer rate is given by Newton law of cooling which separates the contribution of temperature difference of the heated surface and the fluid (at sufficiently large distance from the surface) and all other factors represented by heat transfer coefficient $\alpha\ [\frac{W}{m^2 K}]$. The equation therefore is

$$\dot{q} = \alpha\, \Delta T = \alpha(T_w - T_\infty). \quad (1)$$

Compared to flow parallel to the cooled surface, jet impingement produces heat transfer coefficients that are up to three times higher at a given maximum flow speed. [2]

In the case of cooling of electronic devices, it is common requirement to keep the temperature of cooled component below given value. Since the temperature of cooling substance is, in case of an air jet, usually given by external conditions and can be therefore quite high, it is desirable to employ cooling process that produces as high heat transfer coefficient as possible.

An impinging jet is configured by high momentum flow discharged through a nozzle (or series of nozzles) and its subsequent impingement on the target surface.

The greatest drawback in the endeavor to achieve highest possible heat transfer coefficient is the existence of so-called viscous sublayer, which forms on the impingement wall. Fluid in this sublayer is effectively stationary and through its thickness, the heat is therefore transferred only by conduction governed by Fourier law

$$\dot{q} = -\lambda\, \nabla T, \quad (2)$$

where $\lambda\ [\frac{W}{mK}]$ is the coefficient of thermal conductivity. Due to favourable orientation of the flow relative to the target surface, perpendicular impinging jets achieve lowest possible viscous sublayer thickness.

Impinging jets have therefore large potential in heat and mass transfer applications and are subject of extensive research. Over past years, many reviews were published, [1, 3, 4, 5, 6] to name a few.

This paper deals with double slot impinging jets and aims to shed some light on hysteresis phenomenon which this configuration exhibits. Hysteresis describes the response to a quasi-steady change of one of the parameters that influence the flow field. If the response of a system depends on the direction of parameter change, hysteresis is present and the relationship between specific parameters forms the hysteresis loop. Bistability, which is closely related to hysteresis, occurs when boundary conditions alone can not determine the flow field uniquely. This leads to the existence of two stable steady states.

Hysteresis in double slot isothermal impinging jets was already observed by Travnicek and Krizek [7] and some theoretical background was given by Marsik et al. [8]. Shariatmadar et al. presented a study concerning heat transfer characteristics of laminar slot jet arrays impinging on a constant target surface temperature. [9] In their paper, they also investigated double slot impinging jet configuration and they obtained results corresponding to both flow patterns of concern in this paper. However, they did not comment on hysteresis nor bistability.

Experimental research of hysteresis in impinging jets is recently focused on annular jets, which exhibit similar flow patterns as the double slot jets (main difference being

[a] Corresponding author: tomas.tisovsky@tul.cz

that the vortices in double slot impinging jets are in fact vortex rings in annular impinging jets).

Bistability and hysteresis on annular impinging jets was theoretically predicted in 1991 by Kokoshima et al. [10].

Travnicek et al. [11] used fluidic control (the term fluidics means a technology of flow handling based on the fluid flow interaction without action of moving components) and achieved two stable flow field patterns of annular impinging jet both experimentally and numericaly. Moreover, hysteresis was observed in their experiments.

Furthermore, Travnicek and Tesar found two different regimes of the time-mean flow field, differing in the size of the central recirculation zone, with either the single stagnation point or the stagnation circle. They used acoustic excitation to switch between the two regimes. [12]

In another publication, Travnicek and Tesar experimentaly investigated annular impinging air jet [13]. In their work, the bistability and hysteresis were found, but only at small Reynolds numbers, and at small annular nozzle slot widths. Also, they concluded that an extension of the nozzle center-body promotes the tendency to the hysteretic behaviour.

On the other, hysteresis in annular impinging jets is not universal phenomenon as is suggestested by work of Tesar et al. [19]. They reported on five flow regimes of annular jet, however, no hysteresis was observed.

While there is noticeble effort to study hysteresis in impinging jets experimentaly, there is significant lack of data from numerical simulations. Similar approach as in this paper was used by one of the authors to model hysteresis in annular impinging isothermal flow [15]. Other than that, no contribution employing computational fluid dynamics to investigate hysteresis of impinging jets was found in available literature.

In present study, hysteresis is evaluated via nozzle moving with relatively low velocity relative to the target surface. Therefore the hysteresis was evaluated only to some extent. That is because hysteresis is defined as response to the directional change of some parameter alone. Here, the movement of the nozzle has also some influence.

We attempted to control the flow structure be means of motion of the nozzle. This approach is, off course, not suitable for use in applications in heat transfer where some hypothetical device would need to function efficiently continuously for relatively long period of time. However, it gives some insight to the flow field pattern change.

Effects connected with hysteresis might become highly important in microelectronics to cool thermally highly loaded components. Length scales of impinging jets suitable for applications in microelectronics are generally too low for turbulence to take place. Jets are therefore effectively laminar, heat transfer rates are than not enhanced by turbulence and need rises for other mechanisms to compensate for it.

2 Double Slot Impinging Jets

2.1 Single slot impinging jet

Single slot impinging jet is, in this case, composed of single high aspect ratio rectangular nozzle and a target plate below it. Here, the flow is modelled as 2D with the gradients along longer side of the nozzle profile being omitted.

2.1.1 Transition to turbulence

It was found by Lemanov et al. [16] that submerged air jet exiting from plane channel of width 600 μm exhibits transition to turbulence roughly at the distance of 30 mm from the exit. Since the slot width in our simulations was 0.5 mm and the target surface was not further than 14 mm from nozzle exit, it is justified to assume that the flow is laminar.

2.2 Double slot impinging jets

Double slot impinging jets are formed by two parallel single impinging jets in proximity to each other.

2.2.1 Two stable flow field patterns

The existence of two flow patterns in double slot impinging jet was already investigated by Marsik et al. [8]. In their paper, theoretical guidelines for the bistability of double slot nozzle jet were given. They even provided some experimental results, however they did not succeed in reproducing the results by CFD simulation.

Double slot impinging jets exhibit two stable flow field patterns. Existence of specific pattern is influenced mostly by the nozzle to wall distance. Other parameters, such as distance between the two slots, width of the slot, flow rate through the nozzles and others play significant role in influencing the hysteresis itself.

The two main flow field patterns are illustrated in figure 1.

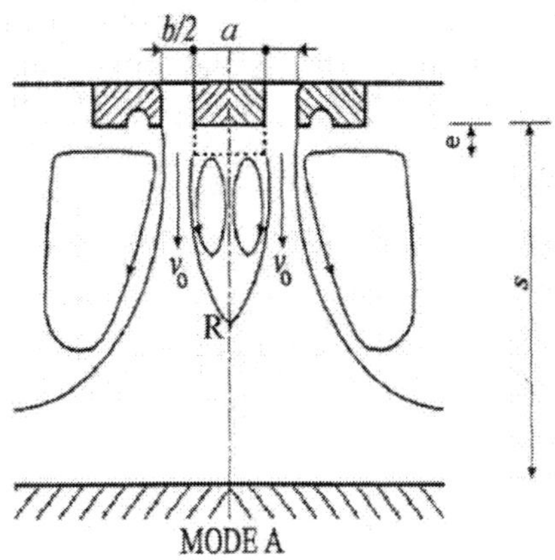

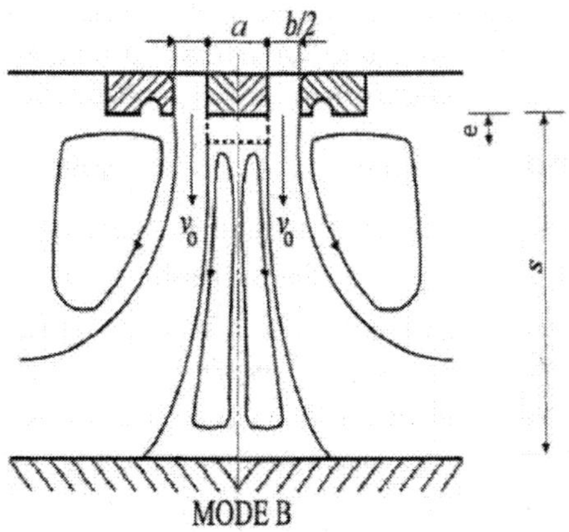

Fig. 1. Flow field patterns of type A and B [8]

Here, the two main flow field patterns of the annular impinging jets will be referred to as A and B. The main characteristic of the pattern A is the existence of the reattachment point R. Below the point R, the flow field has a similar character to the flow field behind a single slot nozzle. If flow pattern A occurs, the region upstream of point R will be called the intermediate merging zone and the region downstream of point R will be called the fully merged zone. Flow pattern A is generally found for larger nozzle-to-wall distances.

Flow field pattern B is, on the other hand, characterized by the tendency of the flow to bend away from the nozzle axis and is generally found for lower nozzle-to-wall distances.

3 Hysteresis Evaluation

In present paper, the hysteresis of double slot impinging jet is evaluated using the computational fluid dynamics (CFD) simulation. This approach was chosen mainly because it provides information about the entire flow field. Moreover, numerical simulations have the advantage over experiment in investigating wide range of geometry modifications and other input parameters easily.

3.1 Mathematical model

Hysteretic behavior of annular impinging jets was investigated numerically in Fluent using the finite volume approach for solving appropriate conservation equations of fluid mechanics. Hysteretic behavior was confirmed through the dynamic mesh simulation where the nozzle moved relatively to the target surface with the velocity magnitude significantly smaller than the magnitude of the flow nozzle exit velocity.

3.1.1 Governing equations and boundary conditions

Mass, momentum and energy conservation equations together with the equation of state of ideal gas form a closed set of 6 coupled scalar equations for 6 unknown variables. The equations are [17]

$$\frac{\partial \rho}{\partial t} + \nabla.(\rho\vec{v}) = 0. \tag{3}$$

$$\frac{\partial}{\partial t}(\rho\vec{v}) + \nabla.(\rho\vec{v}\vec{v}) = -\nabla p + \nabla.\vec{\vec{\tau}} + \rho\vec{g}. \tag{4}$$

$$\frac{\partial}{\partial t}(\rho e) + \nabla.[\vec{v}(\rho e + p)] = \nabla.(k\,\nabla T) + \nabla.\left(\vec{\vec{\tau}}\vec{v}\right) \tag{5}$$

$$\rho = \frac{p}{r_{air} T} \tag{6}$$

$$e = h - \frac{p}{\rho} + \frac{v^2}{2} \tag{7}$$

$$\vec{\vec{\tau}} = \mu[(\nabla\vec{v} + \nabla\vec{v}^T) - \frac{2}{3}\nabla.\vec{v}\vec{\vec{I}}] \tag{8}$$

Although the dimensions and air density are small in value, gravity term in momentum balance must be included in order to account for buoyancy forces acting on air at different temperatures.

Equations (3) to (8) together with appropriate boundary conditions give a solvable problem.

The velocity inlet profile is given by the analytical solution to a steady 2D flow between two infinite planes driven by pressure difference in one direction [18], i.e.

$$u_z = \frac{1}{\mu}\frac{\Delta p}{L}\left(\frac{x^2}{2} - \frac{(2a+b)}{4}x + \frac{ab}{8} + \frac{a^2}{8}\right) \tag{9}$$

where x is distance from the jet axis. The ratio of pressure difference to the slot length is the source term in equation (9).

Although it is custom to give fluid inlet velocity constant profile [1], a fully developed channel profile was used here. This is because long channel of the nozzle can help the flow achieve steady state without any disturbances which would influence the flow downstream from the nozzle and which would be therefore an unknown effect in future experimental study.

3.1.1 Computing

As was already mentioned, the flow was modelled as two dimensional. Also, in respect of low velocities and low dimensions of the investigated flow, simple model of laminar flow was used. Furthermore, the fluid thermophysical properties (i.e. viscosity and thermal conductivity) were assumed constant. This assumption is justified if there the temperature differences in the domain are kept small. Pressure-based solver was used to calculate the fields of variables that were solved for. Coupling between continuity and momentum equations were solved for iteratively by means of the PISO algorithm [19]. Second order discretization schemes are used for pressure and second order upwind schemes are used for solution of momentum and energy equations. Whether convergence has been reached was evaluated based on scaled residual evolutions. Limits for scaled residuals were set to 10^{-3} for continuity and momentum equations and to 10^{-6} in case of energy equation.

3.1.2 Hysteresis evaluation method

Several configurations were investigated. For future reference, those will be denoted Jet1 to Jet8. Table 1 gives details of individual configurations in terms of dimension and table 2 gives main characteristics of the flow.

For the purpose of characterization of the flow, Reynolds number is defined with b as the characteristic dimension and with the average nozzle exit velocity.

$$\mathrm{Re} = \frac{b v_{exit\_average}}{\nu} \tag{10}$$

Table 1. Double slot jet configurations - dimensions

Configuration name	a [mm]	b/2 [mm]	e [mm]
Jet1	4	0.5	0
Jet2	4	0.5	0
Jet3	4	0.5	0
Jet4	4	0.5	+0.5
Jet5	4	0.5	-0.5
Jet6	3.4	0.5	0
Jet7	4.6	0.5	0

Table 2. Double slot jet configurations – parameters of the flow

Configuration name	v_{exit_max} [m/s]	$v_{exit_{av}}$ [m/s]	v_{nozzle} [m/s]
Jet1	1.05	0.7	0.007
Jet2	1.05	0.7	0.0035
Jet3	1.05	0.7	0.00175
Jet4	1.05	0.7	0.007
Jet5	1.05	0.7	0.007
Jet6	1.05	0.7	0.007
Jet7	1.05	0.7	0.007

From equation (10), the Reynolds number for all the cases is 50.

The dynamic mesh case with layering was simulated, there have been sporadic step-wise deviations from the smooth trend of investigated variables. To produce a nice plots, data was smoothed by fitting successive sub-sets of adjacent data points with low-degree polynomial. This method is known as the Savitzky-Golay filter. [20]

3.2 Results

In all following graphs, vertical lines of corresponding colour denote the point of greatest heat transfer coefficient for given configuration. Furthermore, all the graphs are non-dimensionalized.

It should be noted that a in the non-dimensionalized plots is different for Jet6 and Jet7 as compared to the rest of the cases.

3.2.1 Streamlines of flow field patterns A and B

In this section, streamlines for both patterns as well as for chosen state of transition between individual patterns is shown.

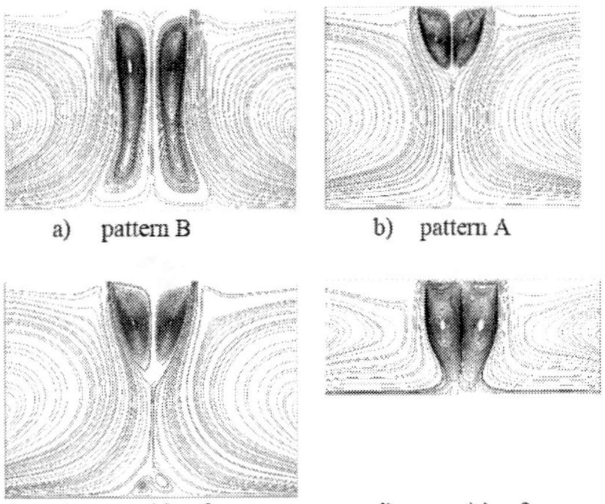

a) pattern B b) pattern A

c) transition from pattern B to pattern A d) transition from pattern A to pattern B

Fig. 2. Hysteresis loops of stagnation pressure at the intersection of jet axis with target surface

Figure 2a and 2b shows patterns B and A respectively in order as they emerged when the nozzle moved away from the target surface. Figure 2c) then show some state of transition between pattern B and pattern A (meaning that the nozzle now moves away from the surface). The transition occurs by splitting the two main vorticle structures of pattern B into four. Transition to pattern A is complete once the vertical structures next to target surface vanish completely. Similarly, the transition from pattern A to pattern B is again accompanied by splitting the two existing vortices into four.

The difference between the two transitions is that in case of transition from pattern B to pattern A, the two vortices are stretched in the direction perpendicular to the wall until they become instable and are then split into four roughly separate structures, two of which then decrease in size. On the other hand, in the case of transition from pattern A to pattern B, the two main vortical structures reach close enough to the target surface, the main flow does not reach to the intersection of double slot nozzle axis with the target surface and the pattern B is formed. No new vortex pair next to the target surface is formed in this case.

Above reasoning suggests that the bistability region is in fact occupied by patterns similar in nature to those shown in figure 2c and 2d rather than the two basic pattern A and pattern B.

3.2.2 Stagnation pressure at the center of target surface

Hysteresis is best identified by comparison of a pressure at a stagnation point at the intersection of jet axis with the target surface (refered to as central stagnation point in this paper) for a case when the domain height was diminishing and for the case when the domain was stretching.

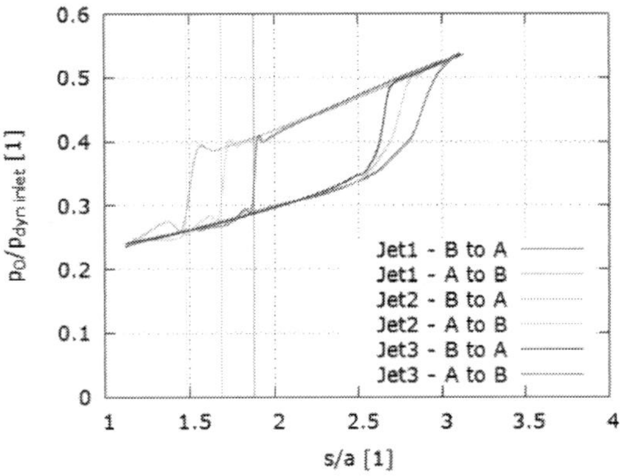

a) different relative nozzle velocities

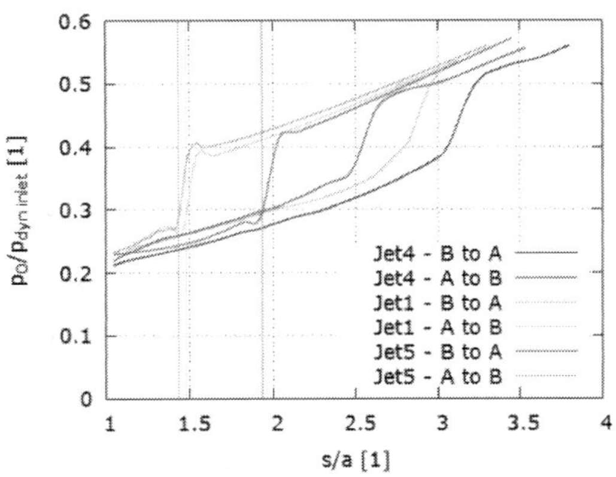

b) nozzle center-body extention/no extention/insertion

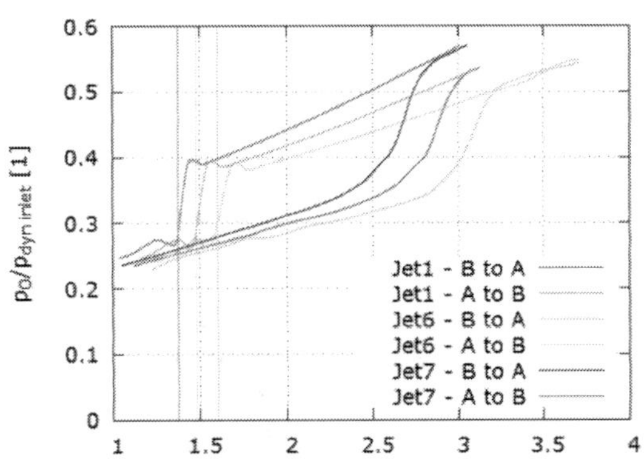

c) different double slot spacing

Fig. 3. Hysteresis loops of central stagnation pressure at the intersection of jet axis with target surface

Figure 3 shows stagnation pressure at the intersection of jet axis with target surface and provides a nice hysteresis loops for all investigated configurations.

Figure 2a shows configurations that only differ in relative velocity between the two slot nozzle and the target surface, therefore providing means to determine the effects of change in relative velocity. As the nozzle velocity gets lower, the flow field has more time to adapt to new circumstances. As a result, the hysteresis loop gets narrower.

Figure 2b shows the effect on nozzle center-body extention and insertion on the hysteresis. It is clear that nozzle center-body extention shifts the hysteresis loop towards higher nozzle-to-wall distances. This is because the two main vortex structures are left with less space than in the case of Jet1. Interestingly, the hysteresis loop is shifted of its original position by a distance that is approximately equal to four times the extention value. The insertion makes the transition from pattern B to pattern A occur earlier. This is because the two vortical structures of pattern B are elongated thanks to the space made free by the insertion.

Finally, figure 2c demonstrates the effect of the double slot nozzle spacing. Naturaly, hysteresis region shifts towards lower nozzle-to-wall distances for lower double slot nozzle spacing. The central stagnation pressure for case of pattern A is greater as the slots are closer to each other. This is because the vortical structure below the center-body are smaller and they therefore require less energy from the main flow.

Whether any of the concidered configurations of the planar double slit jet exhibits bistability remains open question. However, some conclusions can be made for configuration Jet1. In this case, as the velocity of the nozzle approaches zero (the relative nozzle velocity is reduced by half and by quarter in configurations Jet2 and Jet3 respectively), the region of hysteresis loop that is shown in figure 3a should decrease to the region of bistability. It is believed by the authors that the region of bistability would be achieved when the slope of the two transition lines would become $\frac{\pi}{2}$. Based on the character of the slope increase, it can be concluded that the region of bistability does in fact exist for this particular configuration.

Courtesy of the famous Courant-Friedrichs-Lewy condition [21], simulation takes longer and longer for lower nozzle velocities and investigating flow bistability through this approach is therefore not possible.

3.2.3 Maximum stagnation pressure

From the point of view of practical applications, much more interesting parameter is the maximum stagnation pressure. Figure 3 shows maximum stagnation pressure as a function of nozzle-to-wall distance. Since these maxima are not at constant distance from the jet axis, hysteresis loops such as the ones in figure 4 do not not emerge.

Here, discussion will be focused on local maxima obtained in transition from pattern A to pattern B.

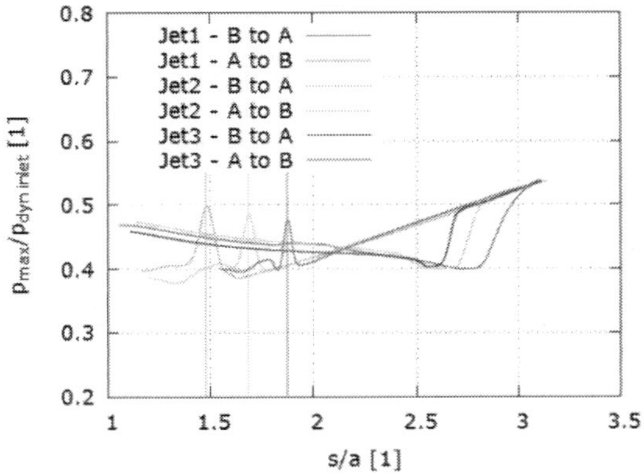

a) different relative nozzle velocities

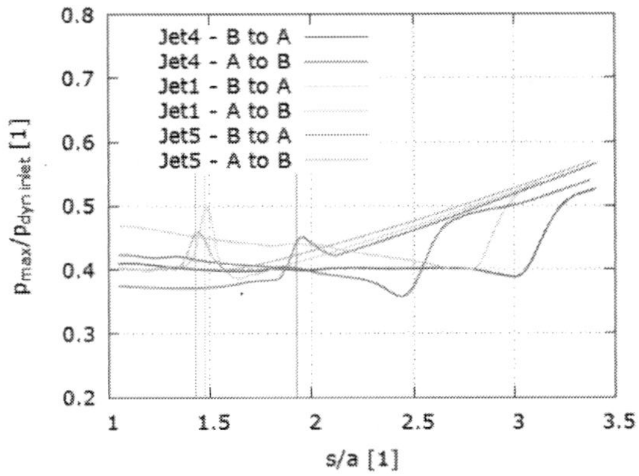

b) nozzle center-body extention/no extention/insertion

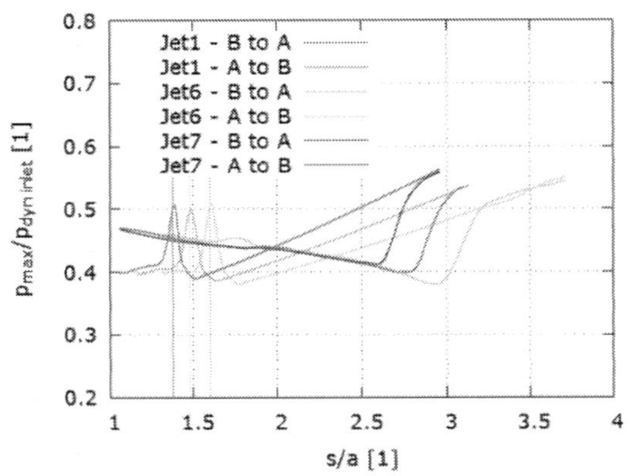

c) different double slot spacing

Fig. 4. Maximum stagnation pressure

Figure 4 shows that regions of maximum stagnation pressure are also regions of maximum heat transfer coefficient (as was previously mentioned, vertical lines indicate corresponding maxima in heat transfer coefficient).

Figure 3a shows that, for larger nozzle velocities, higher maximum stagnation pressure is achieved. To illustrate the significance of this increase, table was set to calculate relative changes of important parameters between successive configurations. Since configuration Jet3 has no predecessor, nothing will be computed. About 1.7% increase of maximum stagnation pressure was found from case Jet1 to case Jet2 and about 2.5% increase of maximum stagnation pressure was found from case Jet2 to case Jet3.

Effect of nozzle center-body extention or insertion is not beneficial with respect to maximum stagnation pressure. But still, nozzle center-body extentions might be useful in cases where there is need to shift the maximum stagnation pressure position towards higher nozzle-to-wall distances.

Figure 3c shows that the maximum of stagnation pressure increases with decreasing double slot spacing and, at the same time, also shifts towards lower nozzle-to-wall distances.

3.2.4 Heat transfer

Finally, maximum heat transfer coefficient, non-dimensionalized to the form of Nusselt number, is plotted as a function of the nozzle-to-wall distance.

In order to determine the heat transfer coefficient, it had to be calculated from the total surface heat flux.

Goldstein et al. [22] carried out experimental tests and found that the convective heat transfer coefficient is independent of the temperature difference between jet and ambient as long as it is defined with the difference between the heated wall temperature and the adiabatic wall temperature. This gives the heat transfer coefficient as

$$\alpha = \frac{\dot{q}}{T_{\text{target surface}} - T_{\infty}}, \qquad (6)$$

where $\dot{q}$ is the total surface heat flux from the target surface.

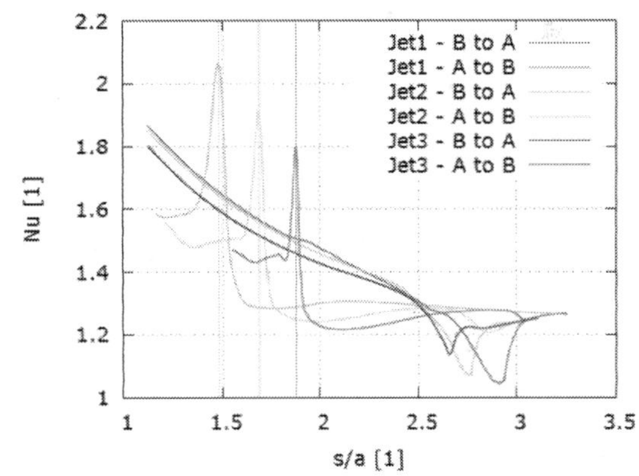

a) different relative nozzle velocities

https://doi.org/10.1051/epjconf/201818002109

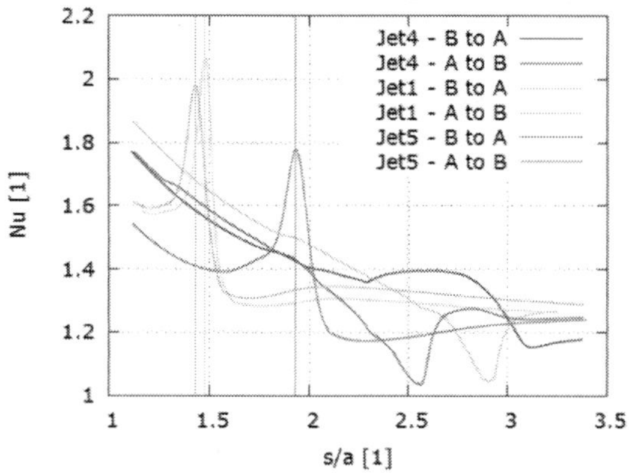

b) nozzle center-body extention/no extention/insertion

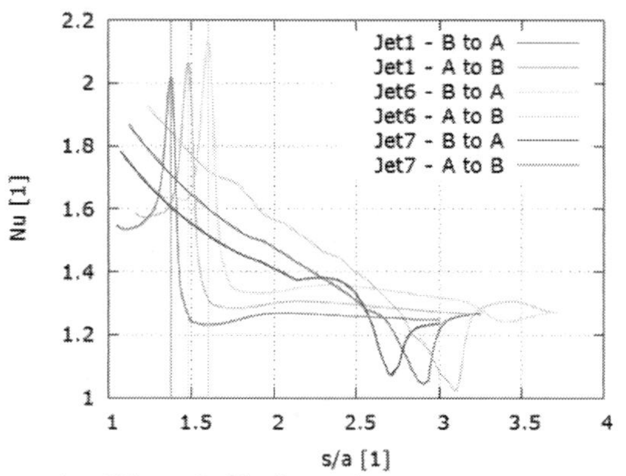

c) different double slot spacing

Fig. 5. Maximum heat transfer coefficient

Figure 5 shows the maximum local Nusselt number as a function of non-dimensionalized distance of the target surface from the nozzle exit. In respect of investigated configurations, overall best heat transfer rate is obtained for larger relative velocity of the nozzle, for the case where no insertion or extention of nozzle center-body is employed and for the case where the two slots are as close as possible. For all the cases, the increase is quite significant. Both with respect of comparison between individual cases and with respect to the heat transfer coefficient for other surface to nozzle distances.

3.2.5 Positions of maximum stagnation pressure and maximum heat transfer coefficient as a function of nozzle-to-wall distance for transition region from pattern A to pattern B

From the viewpoint of cooling applications, the most promising seems to be the region transition from pattern A to pattern B. This region has, as is shown in figure 4, local maximum of the stagnation pressure and consequently also local maximum of heat transfer coefficient.

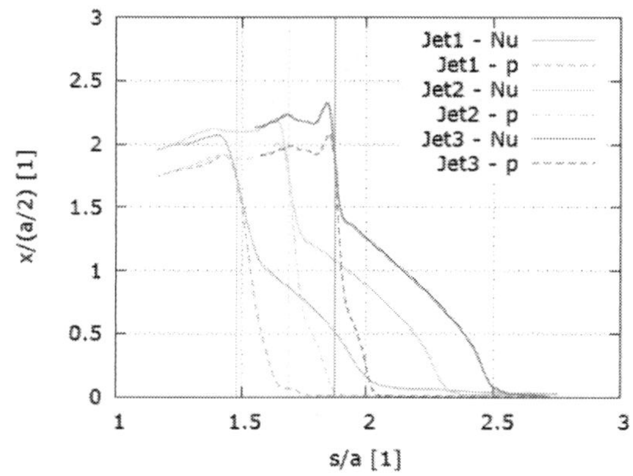

a) different relative nozzle velocities

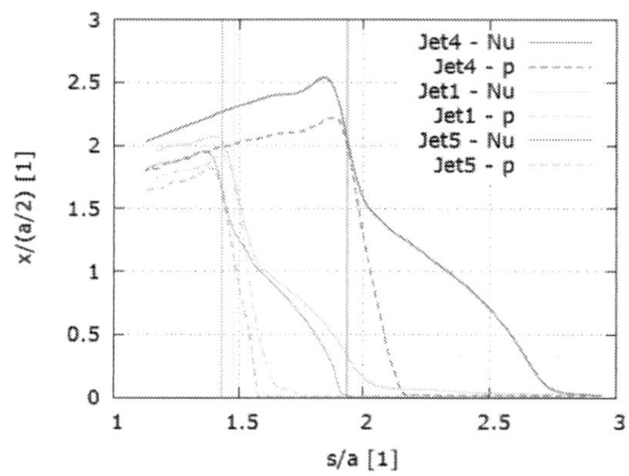

b) nozzle center-body extention/no extention/insertion

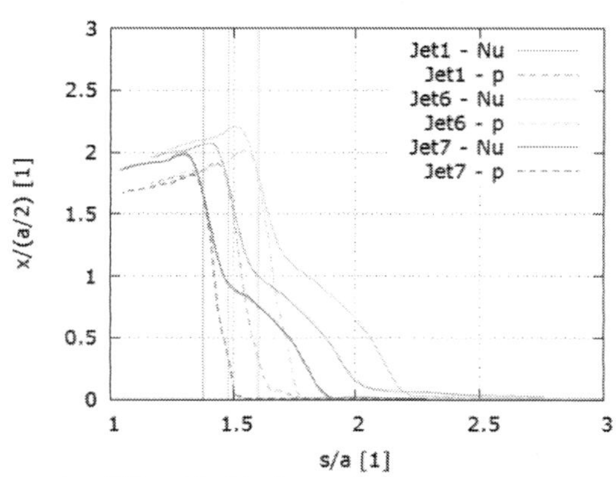

c) different double slot spacing

Fig. 6. Distances of local maxima from the jet axis

Figure 6 deals only with the pattern A to pattern B transitions and shows positions of stagnation pressure (dashed line) and heat transfer coefficient (solid line) maxima against the normalized nozzle-to-wall distance. Vertical lines show that the overall maxima in heat transfer coefficient coincide with the overall maxima for stagnation pressure. Interesting is that generally, in this transition regime, maxima of heat transfer coefficient for other nozzle-to-wall distances are not located at the same place as the stagnation pressure maxima. Maxima in heat transfer coefficient, with the exception of the overall maximum, tend to be further away from the jet axis than the stagnation pressure maxima. This suggests that the effect of overall maxima in heat transfer coefficient is achieved by some combination of impinging jet and wall jet.

As discussed above, the overall maxima in heat transfer coefficient can be shifted further away from the jet axis either by addition of nozzle center-body extention or by increasing the nozzle slot spacing. It is interesting to note that in case of lower nozzle velocities, the overall maximum is also shifted away from the axis. However, this effect is not as significant as the other two mentioned above.

3.3 Stability and Validity of 2D model

For lower nozzle exit velocities, investigated flows were all stable and results from corresponding simulations were provided in previous text. However, for larger nozzle exit velocities, flow became unstable.

Generally, flow might be instable when there are destabilizing factors present. Those factors are namely the presence of shear region within viscous fluid and adverse pressure gradient. [18]

Instable flow was obtained for the case when the domain height was diminishing, while, for the same nozzle exit velocity, the flow was stable when the domain height was stretching. This is because the relative movement of the nozzle toward the target surface is a source of the adverse pressure gradient, which is, as mentioned, one of the destabilizing factors.

The other destabilizing factor is the shear region within viscous fluid and it is present in all flows of real fluids, double slot jets not excluded. Therefore there is also a Reynolds number threshold at which, flow with nozzle moving away from the target surface becomes instable.

3.3.1 Assumption of 2D flow

Real 3D double slot jet with slot lengths much larger than slot widths might be mathematically modeled as 2D flow, taking into account only single section, supposing the same flow picture for any value of the invariant parameter (i.e. the position along the length of the slot). Such assumption is justified everywhere except at the ends of the slots.

Uruba shows [15] that even for low Reynolds numbers, omitting 3D structure of the flow could lead to misinterpretation of the flow field and wrong indentification of physical mechanisms involved in flow process. This was demonstrated for the case of a wake behind a cylinder which is geometrically similar to the double slot jet case presented here (it can be considered a nozzle centerbody in a cross-flow). However, the results might be correct on average.

4 Conclusions

Present paper deals with the hysteresis of 2D double slot jets. Hysteresis was confirmed for range of configurations and some quantitative understanding, applicable to real word applications, is obtained. Moreover, the process of transition of one flow field to another was describes and some interesting observations were made. To the best knowledge of the authors, no article on numerical evaluation of hysteresis in double slot impinging jets was yet published in available literature.

Provided results do not aim to give a complex view on the problem of hysteresis of planar double slot jets. However, some insight was provided to the effect of nozzle extention/insertion and slot spacing and possible means to identify the region of bistability were proposed.

Motivation for research concerned with bistability and hysteresis of jets lies in the development of more efficient systems for heat and mass transfer technologies. Presented results suggest a way to significantly increase local heat transfer coefficient. Such result is extremely important for the field of microfluidics where turbulence is generally not accessible to improve heat transfer.

The aim to achieve highest possible heat transfer rate is generally not the only consideration. From construction point of view, location of the heat transfer maximum as well as the effect of nozzle to wall distances are of concern. Insight for both these concerns and more are given by results provided in this paper.

Greatest drawback of present work is the fact that presented results are not numerical grid resolution independent. In the hysteresis region, one configuration is generally stable while the other is instable. The transition from one configuration to the other can therefore be caused by both the physical effect as well as numerical effect relevant to the actual solution of governing equations. The main focus regarding future work will therefore be in arriving at the numerical grid independent solution. In light of this paragraph, it is further emphasized that the value of this article lies mainly at the application of method of dynamic mesh for investigating hysteresis of double slot impinging jets and in comparison of various simulation results which were obtained under similar numerical conditions.

As for 2D flow, the instability thresholds with respect to at least some parameters are still to be investigated for the case of stretching domain as well as for the case of diminishing domain. Moreover, effect of different nozzle exit velocities is still to be examined. This is made difficult by the strong dependence of the resulting hysteresis loop on the nozzle velocity. In this paper, nozzle velocity was referred to as some small fraction of the average nozzle exit velocity. However, there is no guarantee that the effect, even approximately and for low nozzle velocities, depends linearly on the average nozzle exit velocity. Therefore, in order to really understand the effect of nozzle exit velocity on hysteresis, it is appropriate to investigate the bistability, which corresponds to the limit of velocity of the nozzle approaching zero. One of the topics of future research therefore should be to propose a method to investigate the bistability of double slot impinging jets and consequently also the annular impinging jets.

Acknowledgement

We gratefully acknowledge the support of the Grant Agency of the Czech Republic (Project No. 16-16596S).

References

1. V. Tesar, Impinging jets. In Vortex Rings and Jets (ed. D. T. H.New & S. C. M.Yu), pp. 191–233. Springer Singapore.
2. N. Zuckerman, N. Lior, Jet Impingement Heat Transfer: Physics, Correlations, and Numerical Modeling, Advances in heat transfer, vol. 39, 2006, ISBN-13: 978-0120200399.
3. Martin H., Heat and mass transfer between impinging gas jets and solid surfaces, Advances in Heat Transfer, Vol. 13, 1977, pp. 1–60.
4. S. J. Downs, E. H. James, Jet impingement heat transfer – a literature survey, Proceedings of the National Heat Transfer Conference, ASME, Pennsylvania, 1987.
5. R. Viskanta, Heat transfer to impinging isothermal gas and flame jets, Experimental Thermal and Fluid Science, Vol. 6, 1993, pp. 111–134.
6. S. V. Garimella, Heat transfer and flow fields in confined jet impingement, Annual Review of Heat Transfer, Vol. XI, 2000, pp. 413–494.
7. Z. Travnicek, F. Krizek, Impinging jet and combined slot nozzle, Heat Mass Transfer 35 (5) (1999) 351–356.
8. F. Marsik, P. Safarik, Z. Travnicek, Instability and Bistability of Stagnation Flow, International Symposium on Experimental and Computational Aerothermodynamics of Internal Flows, Gdanks, Poland, Sept 4-7, 2001.
9. H. Shariatmadar, A. Momeni, A. Karimi, M. Ashjaee, Heat transfer characteristics of laminar slot jet arrays impinging on a constant target surface temperature, Applied Thermal Engineering 76 (2015) 252-260.
10. Y. Kokoshima, A. Shimizu, T. Murao, Numerical analysis of annular turbulent jet impinging on a flat plate, in: Proceedings of the 3rd Triennial International Symposium on Fluid Control, Measurement, and Visualization, FLU-COME 91, ASME, San Francisco, 1991, pp. 205–210.
11. Z. Travnicek, K. Peszynski, J. Hosek, S. Wawrzyniak, Aerodynamic and mass transfer characteristics of an annular bistable impinging jet with a fluidic flip–flop control, Int. J. Heat Mass Trans. 46 (7) (2003) 1265–1278.
12. Z. Travnicek, V. Tesar, An annular impinging jet with recirculation zone expanded by acoustic excitation, Int. J. Heat Mass Trans. 47 (10) (2004) 2329–2341.
13. Z. Travnicek, V. Tesar, Hysteresis in annular impinging jets. Experimental Thermal and Fluid Science 44 (2013) 565-570.
14. V. Tesar, M. Jilek, Z. Randa, Topology changes in an annular impinging jet flow, in: Proceedings of the Topical Problems in Fluid Mechanics 2001, Institute of Thermomechanics CAS, Prague, 2001, pp. 121–124.
15. T. Tisovsky, Bistability and Hysteresis of Annular Impinging Jet, The application of experimental and numerical methods in fluid mechanics and energy, Terchova – Vratna dolina, Slovakia, 2016.
16. V. V. Lemanov, V. I. Terekhov, K. A. Sharov, A. A. Shumeiko, An Experimental Study of Submerged Jets at Low Reynolds Numbers, Technical Physics Letters, 2013, Vol. 39, No. 5, pp. 421–423. © Pleiades Publishing, Ltd., 2013.
17. FLUENT 17.1 Users Guide. Fluent Inc., © 2016 SAS IP, Inc.
18. L. D. Landau, E. M. Lifshitz, Fluid Mechanics, Second Edition: Volume 6 (Course of Theoretical Physics), 1987.
19. R. I. Issa, Solution of implicitly discretized fluid flow equations by operator/splitting. J. Comput. Phys., 62, 40-65, 1986.

A. Savitzky, M. J. E. Golay, Smoothing and Differentiation of Data by Simplified Least Squares Procedures. Analytical Chemistry 36 (8): 1627–39, 1964. doi:10.1021/ac60214a047

20. J. H. Ferziger, M. Peric, Computational Methods for Fluid Dynamics, 2001.
21. R. J. Goldstein, K. A. Sobolik, W. S. Seol, Effect of entrainment on the heat transfer to a heated circular air jet impinging on a flat surface, Trans. ASME, Vol. 112, pp. 608-611, 1990.
22. V. Uruba, On 3D Instability of Wake behind a Cylinder, The application of experimental and numerical methods in fluid mechanics and energy, Terchova – Vratna dolina, Slovakia, 2016.

Instantaneous fluid force acting on a circular cylinder control by a small rod

Takayuki Tsutsui[1,*]

[1] Department of Mechanical Engineering, The National Defense Academy, 1-10-20 Hashirimizu, Yokosuka , Kanagawa 238-8686 Japan

Abstract. Two unique bluff body flow control methods using a small rod have been proposed in previous studies. The first is the forced reattachment method, which is a type of separated shear layer control. This method reduces drag and generates a lift. The second is the front rod method, which involves the placement of a small rod upstream of the bluff body to control the incoming flow and reduce drag. This paper describes the features of the instantaneous fluid force achieved using these flow control methods. These methods were experimentally applied to a cylinder, and the instantaneous pressure field and the flow visualizations of these methods are presented. When the forced reattachment method was applied, a lift force was generated, and the vibration was less than that in the case of the front rod method. When the front rod method was applied, the drag force was reduced by over 50% in comparison with those in the uncontrolled case and the case with the forced reattachment method.

1 Introduction

The ability to manipulate a flow field is of immense technological importance. For example, if drag of vehicles and buildings can be reduced, much fuel cost and construction materials for buildings would be saved. Flow control around bluff bodies is of importance and of interest for engineering field.

To investigate flow control around a circular cylinder, two previously proposed flow control methods using a small rod were applied [1 – 9]. The first is the forced reattachment method [1 – 4]. Figure 1 shows a visualization of the flow and the distribution of the pressure coefficient C_p around a circular cylinder with forced reattachment flow control. This method achieves shear layer separation and reattachment using a small rod. Using this method reduced the drag of the circular cylinder by 20 – 30 % and generated a lift coefficient C_L of 1.0 [1]. In this method, the separated shear layer from the upper side of the cylinder adheres to the rear of the cylinder as a result of the control by the small rod, causing a stagnant region to form behind the cylinder. The C_p distribution on the lower side is higher than that without a small rod, thus generating a lift.

The second method is the front rod method. In this method, a small rod is placed upstream of a bluff body. When this method is applied to a circular cylinder, it has been reported that the total drag coefficient is reduced by 63% relative to that in the uncontrolled case [5, 6]. Figure 2 shows the flow visualization and the C_p distribution around the circular cylinder for this method. This method has been applied to bluff bodies of several shapes: a flat plate [7], a square prism [8], and an inclined square prism [9]. In these cases, the total drag coefficients were decreased by 34 – 75 % relative to that without the rod.

The general features, such as the static and time-averaged field fluid forces, of the two flow control methods have been previously reported. However, these studies did not report the instantaneous fluid force acting on cylinders. The oscillation of the fluid force can sometimes cause vibrations in civil engineering structures. It is therefore important to understand the instantaneous features of these flow control methods.

The aim of this study was to describe the instantaneous behaviors of a flow around a circular cylinder under these flow control methods. This paper presents the instantaneous pressure field and flow visualizations obtained using these two methods. From these results, the features of the oscillating drag and lift were clarified.

2 Experimental apparatus and methods

The coordinate systems and notation used in the present study for the forced reattachment and front rod methods are shown in Figure 3(a) and (b), respectively. For the forced reattachment method, the diameters of the circular cylinder and small rod were D = 40 mm and d = 10 mm, respectively. The small rod was placed in the shear layer of the cylinder. The gap between the cylinder and small rod had a width of G = 4 mm, and the angular position of the rod was α = 125 °. The forced reattachment phenomenon can be induced by a rod as small as 2 mm in diameter [1 – 4]; however, in this study, to ensure the steady occurrence of the forced reattachment phenomenon, the diameter of the small rod was 10 mm.

For the front rod method, the diameters of the

* tsutsui@nda.ac.jp

https://doi.org/10.1051/epjconf/201818002110

circular cylinder and small rod were D = 40 mm and d = 10 mm, respectively. The small rod was placed upstream of the cylinder. The longitudinal distance between the axes of the cylinder and small rod was L = 70 mm.

Figure 4 shows a schematic of the wind tunnel and pressure measurement equipment. The experiments were performed in a low-speed wind tunnel, the working section of which has a height of 1,000 mm, a width of 150 mm, and a length of 1,200 mm. The free-stream velocity U was varied from 6 to 20 m / s, and the turbulent intensity was approximately 0.4 % in this velocity range. The Reynolds number based on D ranged from 3.8×10^4 to 1.3×10^5. The velocity fluctuation in the wake and the time-averaged and instantaneous pressures on the surfaces of the cylinders were measured using this setup. Flow visualization was performed using the smoke–wire method.

A detailed view of the circular cylinder is shown in Figure 5(a). The cylinder is made of acrylic resin and has two pressure taps of 1.6 mm in diameter connected to a manometer and a semi-conductor pressure sensor. The pressure distribution over the entire surface of the cylinder can be determined by rotating it about its center axis.

Figure 5(b) shows the measurement technique used to measure the instantaneous pressure around the circular cylinder. The velocity fluctuating signal was used as a standard signal to align the phase. A hot-wire probe was placed behind the cylinder to measure the vortex shedding frequency. The hot wire was placed at a radial distance of D and a downstream distance of 3D from the center of the cylinder. The pressure signals at ϕ were recorded simultaneously while rotating the cylinders by increments of 10°. The oscillating pressure distribution was obtained from the instantaneous pressure coefficients measured at the same phase for a given cylinder rotation angle ϕ.

Flow visualization was conducted using the smoke–wire method. Figure 6 shows a schematic diagram of the smoke–wire system. The bottom of the fluctuating velocity signal is the trigger position for flashing the stroboscope. Images of the flow around the cylinders can be captured at any phase by setting the delay time.

(a) flow visualization (Re = 1×10^4) (b) time averaged pressure distribution (Re = 4.2×10^4)

Fig. 1. Flow controlled by a small rod (forced reattachment method, D = 40 mm, d = 10 mm, G = 4 mm, α = 125°)

(a) flow visualization (Re = 1×10^4) **(b) time averaged pressure distribution (Re = 4.2×10^4)**

Fig.2. Flow controlled by a small rod (front rod method, D = 40 mm, d = 10 mm, L = 70 mm)

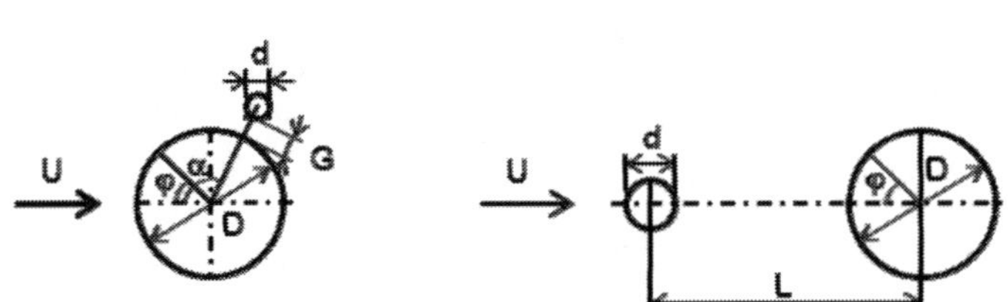

(a) forced reattachment method **(b) front rod method**

Fig. 3. Corrdinate system and notation

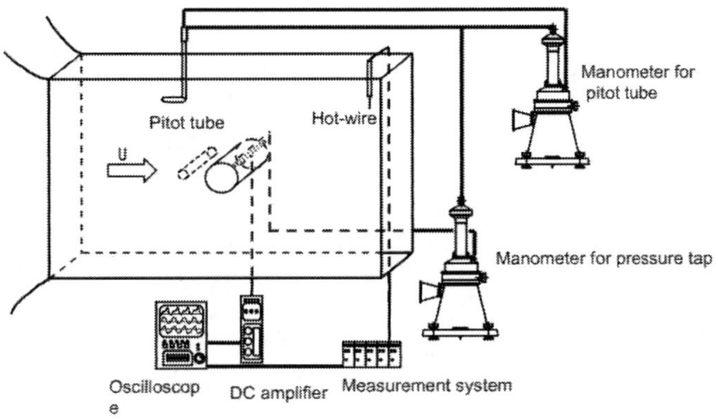

Fig. 4. Pressure measurement system

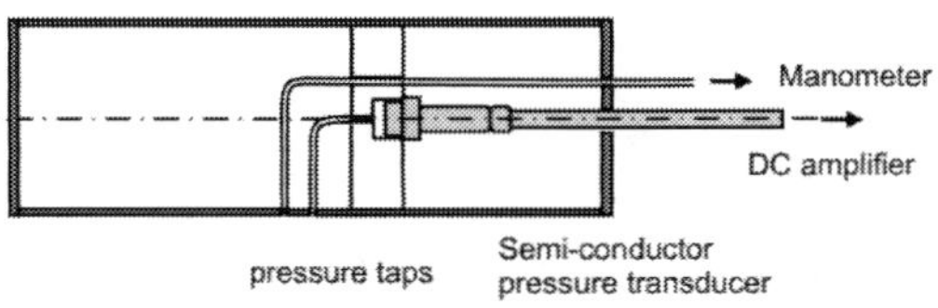

(a) pressure measurement model

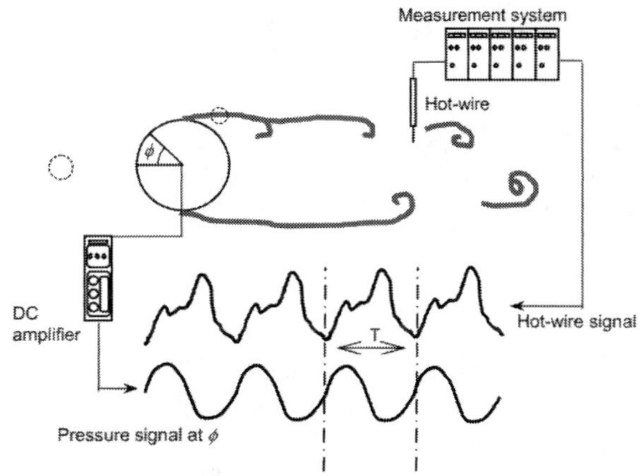

(b) instantaneous pressure measurement technique

Fig. 5. Pressure measurement technique

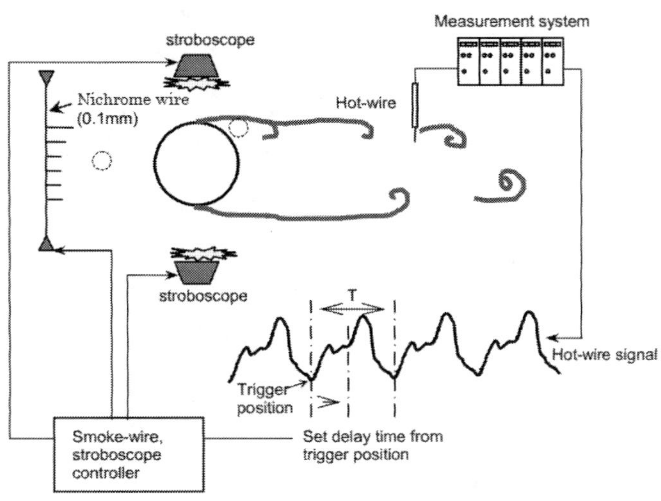

Fig. 6. Flow visualization technique

3 Results and discussion

3.1 Single cylinder

To confirm the validity of the measurement methods, the flow was visualized, and the time-averaged and instantaneous pressures over a circular cylinder by a small rod were measured. Figure 7 shows flow visualization photographs of the circular cylinder without a small rod from t / T = 0 to 7 / 8 at intervals of 1 / 8, where T and t are the period of vortex shedding and the time, respectively. The Strouhal number St is defined as f D / U, where f is the vortex shedding frequency; in this experiment, a Strouhal number was St = 0.20.

The instantaneous pressure distributions at different stages of the vortex shedding cycle are shown in Fig. 8. Negative and positive pressures are plotted inside and outside of the circle, respectively. The motion of the flow around a bluff body, such as a cylinder, is a periodic motion that regularly returns to a given position. These pressure distributions correspond to the flow visualization photographs in Figure 7.

The oscillating drag and lift coefficients C_D and C_L of the circular cylinder obtained in the present study and are shown in Figure 9 along with those obtained in a previous study by Nishimura and Taniike [10]. The drag and lift coefficients were obtained by integrating the pressure distribution. The present results show the same tendencies as the previous results.

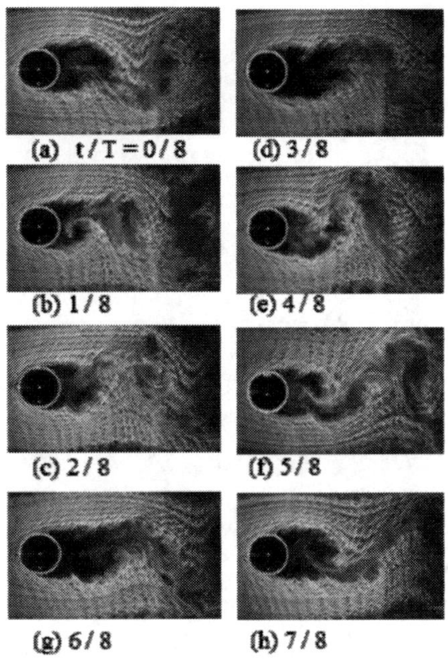

Fig. 7. Sequential photographs of a single circular cylinder (D = 40 mm, Re = 4.2 × 10⁴, St = 0.20, f = 80 Hz, T = 12.5 ms)

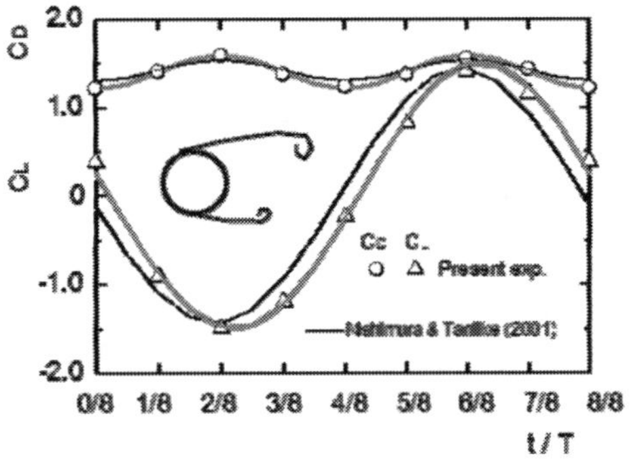

Fig. 9. Oscillating drag and lift coefficient of a single circular cylinder

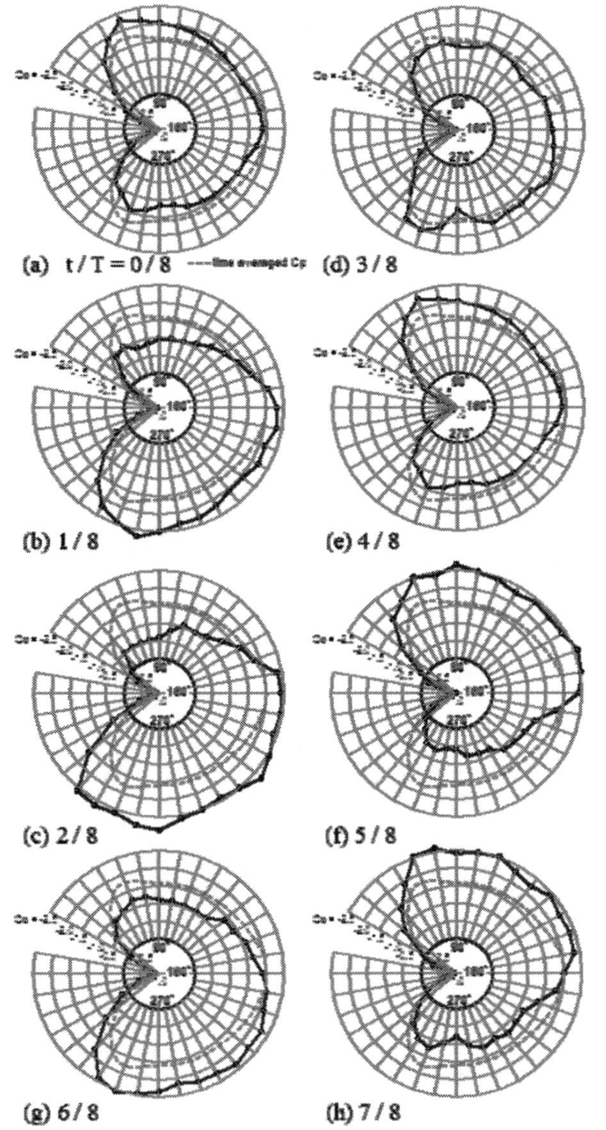

Fig. 8. Instantaneous pressure distribution of a single circular cylinder at different stages of a vortex shedding cycle

3.2 Forced reattachment method

First, the forced reattachment phenomenon was considered. Figure 10 shows a schematic of the forced reattachment phenomenon. The small rod was positioned just inside of the shear layer separated from the upper side of the circular cylinder so that it divided the shear layer. In this setup, the lower shear layer reattaches and adheres to the rear side of the cylinder, and the upper shear layer is supported and elongated by the small rod. A large stagnant region forms behind the cylinder. This forced reattachment phenomenon is not an intermittent phenomenon, and thus the wake motion is not more active than that of single circular cylinder. Under this control method, the mean pressure is asymmetrically distributed between the upper and lower sides of the cylinder. As a result, a lift force is generated.

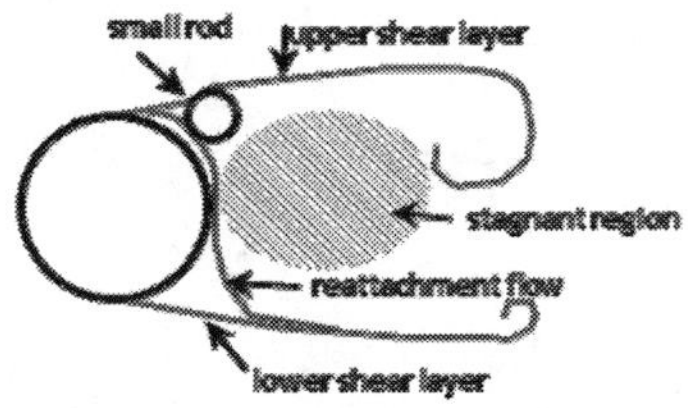

Fig. 10. Description of the forced reattachment phenomenon

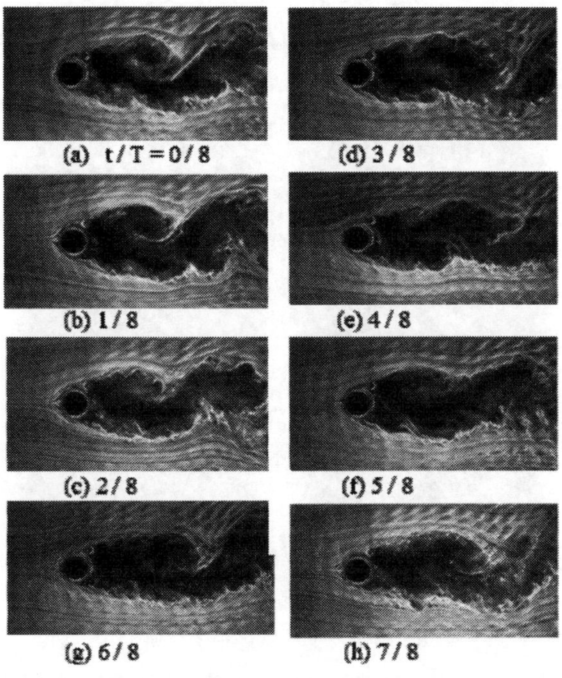

Fig. 11. Sequential photographs of the controlled circular cylinder by the forced reattachment method (D = 40 mm, d = 10 mm, G = 4 mm, α = 125° , Re = 4.2 × 10⁴, St = 0.19, f = 74 Hz, T = 13.5 ms)

Figure 11 shows flow visualization photographs at eight different stages of the vortex shedding cycle (t / T = 0 to 7 / 8 at intervals of 1 / 8) of the circular cylinder flow controlled by the forced reattachment method. The flow visualization photographs show that the upper and lower shear layers were elongated and the vortex street was not observed. This means that the wake motion is not more active than that of single circular cylinder, because the wake motion was constrained by the upper shear layer and the reattachment flow.

Figure 12 shows the instantaneous pressure distribution at eight different stages of the vortex shedding cycle (t / T = 0 to 7 / 8 at intervals of 1 / 8) along with the time-averaged pressure distributions. At every value of t / T, the C_p distribution was asymmetric about the horizontal center line, therefore, a lift acting in the upward direction was generated.

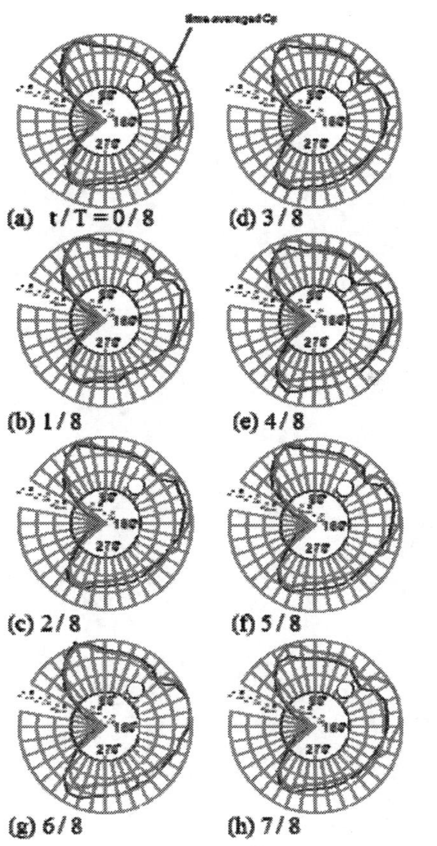

Fig. 12. Instantaneous pressure distribution of the controlled circular cylinder at different stages of a vortex shedding cycle

3.3 Front rod method

Figure 13 shows a brief schematic of the front rod method. In this setup, the shear layer separated from the small rod reattaches to the front side of the cylinder and a quasi-static vortex is formed between the rod and cylinder. A separation bubble forms, causing the separation point to move backward on the cylinder and the wake to decrease relative to that in the case with a single cylinder. Figure 14 shows flow visualization photographs of the cylinder flow controlled using the front rod method. A vortex street was clearly observed behind the cylinder.

The instantaneous pressure distributions for this case at eight different stages of the vortex shedding cycle (t / T = 0 to 7 / 8 at intervals of 1 / 8) are shown in Figure 15. The pressure is always negative in the front region, and C_p has two maximum values at $\phi = \pm 30$ °, which are the reattachment points of the shear layer separated from the small rod. The C_p value at the front side of the cylinder has negative value because of the quasi-static vortex. At every value of t / T, the back pressure is higher than the corresponding pressures obtained using the forced reattachment method and those of the single cylinder. This indicates that a reduction in the drag reduction can be expected.

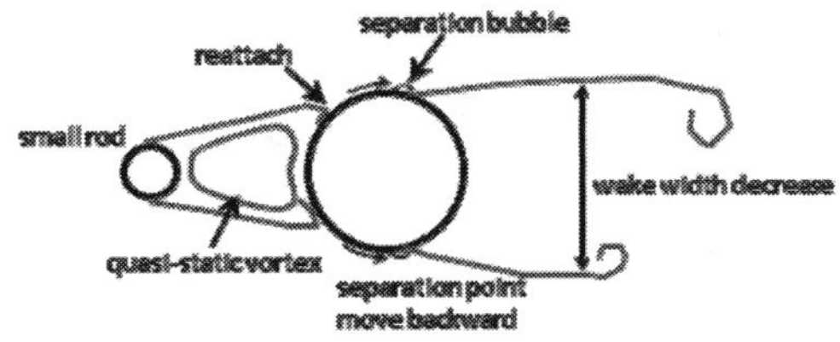

Fig. 13. Description of the flow control by a rod

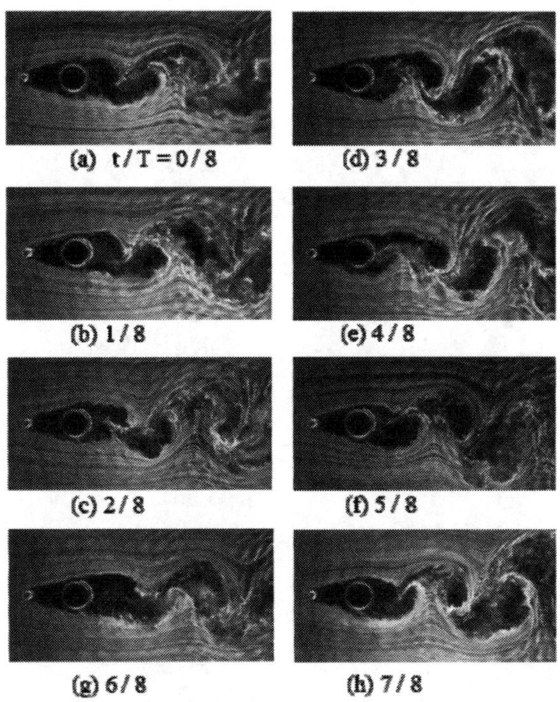

Fig. 14. Sequential photographs of the controlled circular cylinder by the front rod method
(D = 40 mm, d = 10 mm, L = 70 mm, Re = 4.2 × 10^4, St = 0.22, f = 90 Hz, T = 11.1 ms)

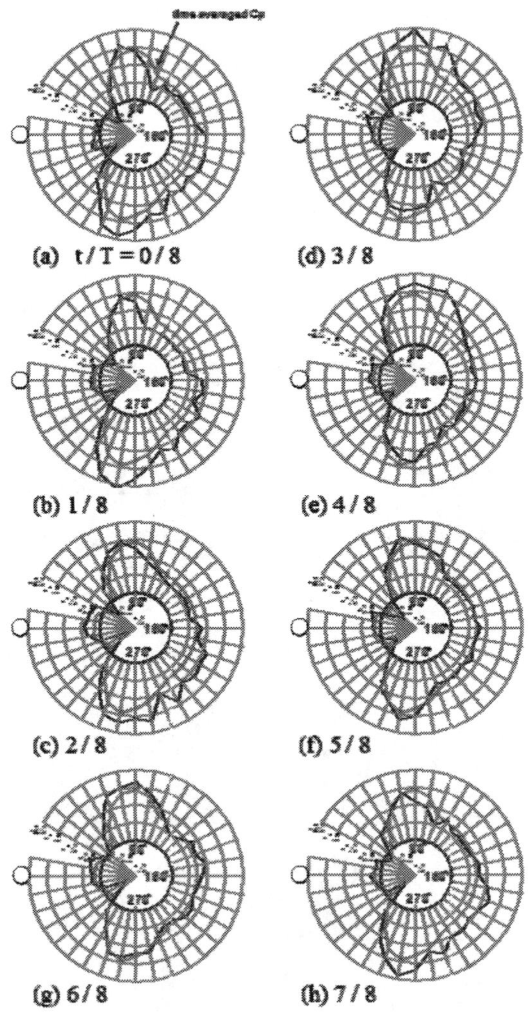

Fig. 15. Instantaneous pressure distribution of the controlled circular cylinder at different stages of a vortex shedding cycle

https://doi.org/10.1051/epjconf/201818002110

3.4 Oscillating fluid force of the controlled cylinders

The effectiveness of applying this technique to a circular cylinder was then evaluated. Figure 16 shows the Lissajous figures of the oscillating drag and lift coefficients C_D and C_L of the cylinders with their flows controlled by the two different methods and that for the single cylinder.

For the forced reattachment method, the center point of the figure has coordinates of $C_D = 1.1$ and $C_L = 0.27$, and the amplitude is 0.39. In this case, a lift force is generated, and the vibration is less than that when using the front rod method.

For the front rod method, the center point of the figure has coordinates of $C_D = 0.58$ and $C_L = 0.0$, and the amplitude is 0.50. In comparison with the single cylinder and the forced reattachment method, the drag force in this case is reduced by over 50 %.

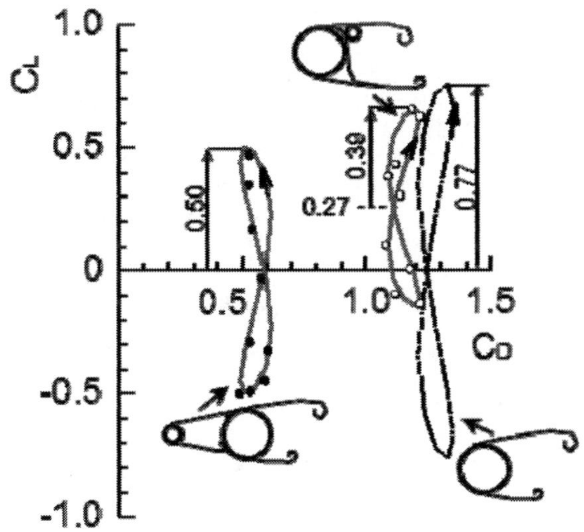

Fig. 16. Lissajous figure

4 Conclusions

Two flow control methods, the forced reattachment and front rod methods, were considered in this study. The characteristics of the flow and the oscillating fluid force of a cylinder with the flow controlled by a small rod were investigated experimentally.

When the forced reattachment method was applied, a lift force was generated, and the vibration was less than that in case of the front rod method. In comparison with the single cylinder and the forced reattachment method, the drag force was reduced by over 50 % when the front rod method was used.

References

1. T. Tsutsui, T. Igarashi, *Flow Control around a Circular cylinder by a Small Cylinder*, Proc. of the 11th Australasian Fluid Mechanics Conference. Hobart, Australia. pp. 14-18, (1992)
2. T. Igarashi ,T. Tsutsui, *Flow Force Acting on a Circular cylinder Controlled by a Small Rod.* Proc. of the 3rd JSME-KSME Fluids Engineering Conference. Sendai, Japan. pp. 571-576, (1994)
3. T. Tsutsui, T. Igarashi and K. Kamemoto, Interactive Flow around Two Circular cylinders of Different Diameters at Close Proximity Experiment and Numerical Analysis by Vortex Method, J. Wind Eng. Ind. Aerodyn., 69-71, pp. 279-291, (1997)
4. T. Tsutsui, T. Igarashi, *Flow Control around a Circular cylinder by a Small Cylinder (Properties of Reattachment Jet).* Proc. of the 3rd ASME-JSME Joint Fluids Eng. Conf., San Francisco, California, FEDSM99-6943, (1999)
5. T. Tsutsui, T. Igarashi, Drag Reduction of a Circular cylinder in an Air-stream, J. Wind Eng. Ind. Aerodyn., 90, pp. 527-541, (2002)
6. T. Tsutsui, T. Igarashi, Heat Transfer Enhancement of a Circular cylinder, Trans. ASME J. Heat Transfer. 128, pp. 226-233 (2006)
7. T. Igarashi, N. Terachi, Heat transfer enhancement and drag reduction of a flat plate normal to the airstream (flow control using a rod). Heat transfer – Asian Res. 27 (2), 99 – 113, (1998)
8. T. Tsutsui, T. Igarashi, H. Nakamura, Drag Reduction and Heat Transfer Enhancement of a Square Prism, JSME Int. Ind. J. B, 44, No. 4, pp. 575 – 583, (2001)
9. T. Igarashi, T. Tsutsui, *Enhancement of heat transfer and reduction of drag of a square prism arranged diamond – shape in an air stream*, Proc. of the 11th IHTC, Kyongiu, Korea, Vol. 5, pp261 – 266, (1998)
10. H. Nishimura and Y. Taniike, Aerodynamic characteristics of fluctuating forces on a circular cylinder, J. Wind Eng. Ind. Aerodyn, **89**, pp. 713-723, (2001)

Flow Structure behind a Wing at High Reynolds Numbers

Václav Uruba[1,2,*], *Zdeněk* Pátek[3], *Pavel* Procházka[1], *Vladislav* Skála[1], *David* Zacho[3], *Robert* Kulhánek[3]

[1]Institute of Thermomechanics, ASCR, v.v.i., Dolejškova 5, Praha 8, CR
[2]UWB, Faculty of Mechanical Engineering, Department of Power System Engineering, Universitní 8, Plzeň, CR
[3]VZLU, a. s., Beranových 130, Praha – Letňany, CR

Abstract. The stereo PIV measurement were performed behind a wing in the plane perpendicular to the flow to study the vortical structures oriented in the streamwise direction, which take place both in suction and pressure sides of the wing. The Reynolds numbers during the experiments in the 3 m wind tunnel range from 0.5 million up to 1.5 million.

1 Introduction

Despite of the fact, that the aeronautics is during the last century one of the most successful and developing industrial fields, not all physical aspects of the phenomena involved have not been fully clarified yet. Generation of the lift force is still explained using non-physical potential flow theory in combination with Kutta-Joukovski theorem, which is nonphysical as well. The estimated lift force value corresponds to the real value satisfactory only for a very specific situation – small angle of attack and regular airfoil. To control successfully the process of lift generation in any situation we should have detailed information on the real physical mechanisms behind this phenomenon. This fact was clearly pointed out by Hoffmann and Johnson, who formulated the new theory of flight [1].

The new theory of flight relies on presence of streamwise vortices located close to the suction surface of the airfoil, which are dynamical in nature. The authors supported the new theory by mathematical simulations, however relevant experiments are still missing. An attempt of experimental prove of the new flight mechanism hypothesis was carried out in [2], showing clearly the existence of the streamwise vortices for the flat plate with angle of attack 5 and 10 degrees, next paper [3] studies the flow close to the plate surface. Both experiments suffer from a very low Reynolds number flow, of about 25 000.

The presented paper describes experiments on prismatic wing performed in the large wind tunnel allowing for relatively high Reynolds numbers from half up to one and half million.

2 Experimental setup

Experiments have been carried out in the large close circuit wind tunnel VZLU diameter of the test-section 3 m, velocity up to 75 m/s.

The airfoil with chord 600 mm and span 1200 mm was designed for a high-performance glider. The section with maximal thickness 14.5% in 43% chord, asymmetrical design, maximal curvature 4.4% in 45% chord. The airfoil aerodynamic properties were thoroughly studied beforehand. Zero lift was detected at AOA -4.5 deg, the AOA 5.5 deg used in the presented experiments results in lift coefficient of about 1.1 and drag coefficient 0.025. The airfoil was equipped by circular endplates.

The Litron Nano L 200-15 PIV laser was used with a pulse energy 200 mJ. For the stereo PIV the 2 cameras NanoSense 4MKII with resolution 4M pixels 2048 x 2048, lenses focal length 60 mm. The cameras were located outside the measuring chamber of diameter 3 m. The standard chequered target was used for calibration. The maximal acquisition frequency of the PIV system was 7 Hz.

As a seeding material the olive oil atomized using system of Laskin nozzle aerosol generator was used producing fog consisting of droplets a few microns in diameter. The PIV technique relies on agglomeration of those droplets, as the resolution of the cameras do not allow for detection of individual droplets. Unfortunately the fog in the close circuit wind tunnel was homogenized very quickly lowering the PIV correlation quality rapidly. However in the most interesting region on the airfoil suction side the quality of seeding was relatively good, because of strong turbulence and vortices presented. After all we cannot completely avoid presence of error vectors in the field of interest, situation gets worse for higher velocities.

The stereo PIV technique has been used with measuring plane located 100 mm behind the airfoil trailing edge, perpendicular to the incoming flow direction. Each measurement consist of 380 double-snaps resulting in the same number of vector-fields acquired with the frequency 7 Hz.

The experiment schematic setup is depicted in Fig. 1.

* Corresponding author: uruba@it.cas.cz

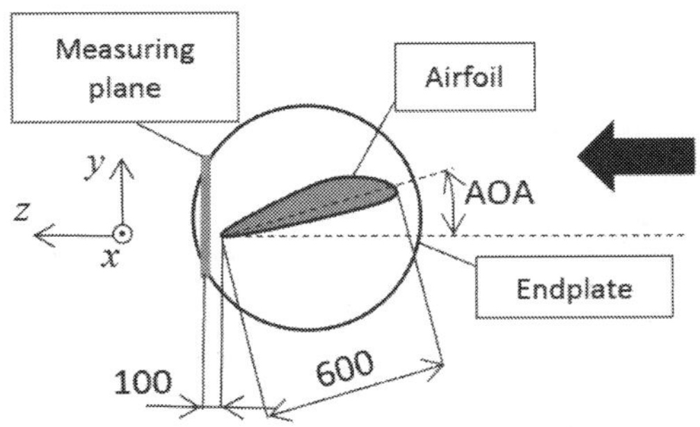

Fig. 1. Experimental setup schematics, side view.

The Cartesian coordinate is introduced in Fig. 1, the measuring plane is (*xy*).

3 Results

The velocity vector-fields were evaluated for 3 Reynolds numbers: 0.5, 1 and 1.5 million. The following pages contain figures depicting central part of the airfoil. The *x* direction is horizontal spanwise, while *y* direction is vertical, both *x* and *y* axis are perpendicular to the flow direction in the test section inlet. The broken line in all figures in $y = 0$ corresponds to the airfoil trailing edge position. Only part of the measuring plane close to the airfoil mid-span will be presented, showing reasonable number of error vectors, however still present. The values of coordinates are nondimensional, reference dimension is the airfoil chord as usual.

The 3 following subchapters contain the set figures of the same type, each for one particular Reynolds number.

3.1 Reynolds number 0.5 million

The experiment results obtained in the Reynolds number half million are to be presented first. The corresponding inlet velocity was about 12.5 m/s.

As the very first result the example of instantaneous velocity field is shown in Fig. 2.

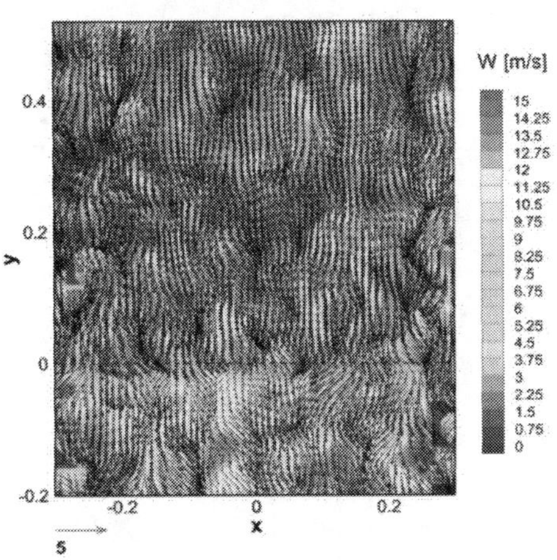

Fig. 2. Example of the instantaneous velocity field. Re = 0.5M.

The vectors in Fig. 2 represent in-plane instantaneous velocity, components in x and y directions, colour represent the third velocity component in z direction. In the left lower corner the reference vector is added, magnitude 5 m/s. The streamwise velocity component ranges from low values (green – about 5 m/s) up to highest values (red – 15 m/s). The blue regions on sides are not well evaluated and consist of error data.

Low streamwise velocity is located around the trailing edge, higher values are detected in spots in upper suction region. The velocity field suggests cellular structure of the flow-field, however the topology is changing from snap to snap.

In the measuring plane there is predominant downward direction of mean in-plane velocity. This fact is evident from Fig. 3, where time-mean velocity components are shown in the same system as in preceding figure.

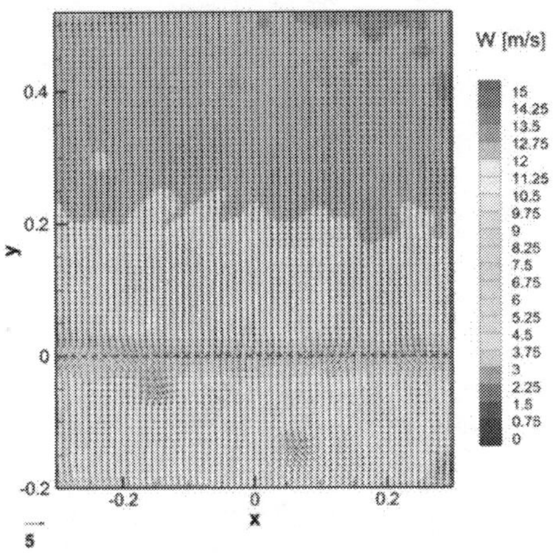

Fig. 3. Mean of-plane velocity component with in-plane velocity vector components. Re = 0.5M.

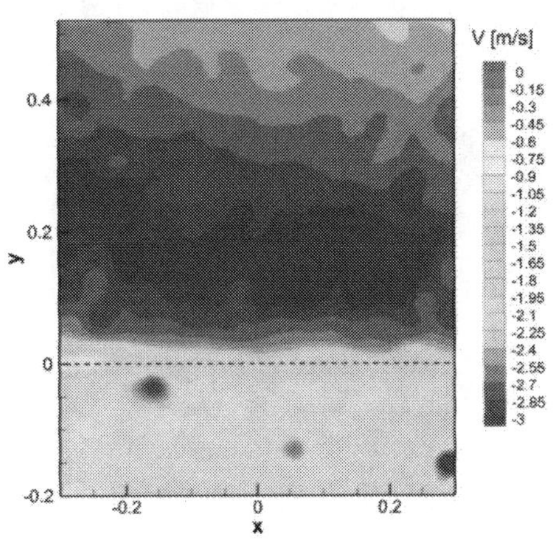

Fig. 4. Mean vertical velocity component distribution. Re = 0.5M.

In Fig. 3 the higher streamwise velocity in the upper suction region is evident as well as the downward in-plane velocity indicating the downwash flow. However the downwash is more intensive in the upper suction part in comparison with the lower pressure region, as shown in Fig. 4, where distribution of vertical mean velocity component is depicted.

The turbulent activity could be characterized by sum of the three velocity components variances, as shown in Fig. 5.

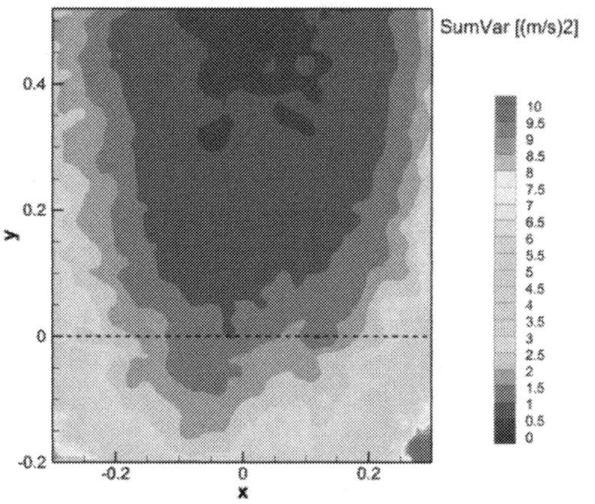

Fig. 5. Sum of velocity components variances. Re = 0.5M.

Fig. 5 suggests, that the turbulence activity is more intense in the pressure (lower) region. The maxima of the activity close to the sides of the region of interest are connected rather with relatively frequent appearance of error data here.

The last portion of results related to the lowest Reynolds number is oriented to the vorticity streamwise component. Fig. 6 shows the instantaneous distribution of the vorticity for one randomly chosen moment.

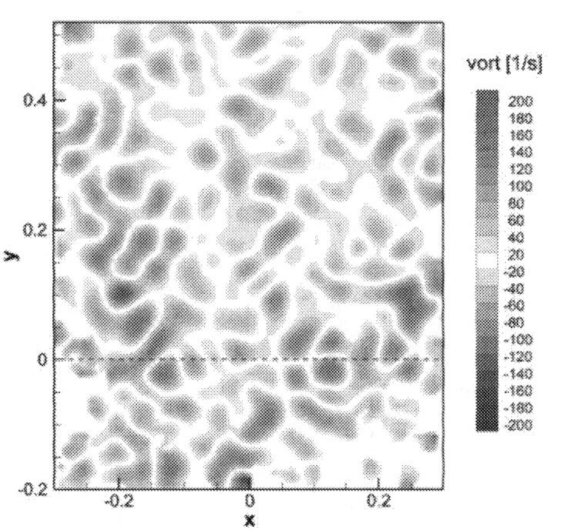

Fig. 6. Example of the instantaneous vorticity field. Re = 0.5M.

Blue colour corresponds to negative value (clockwise orientation), while red is positive. Concentration of the vorticity could be interpreted as presence of a vortex crossing the plane of measurement. However we have no information on the vortex centre-line shape in space.

To test the randomness of the vorticity field, the time-mean vorticity field has been evaluated, result is presented in Fig. 7.

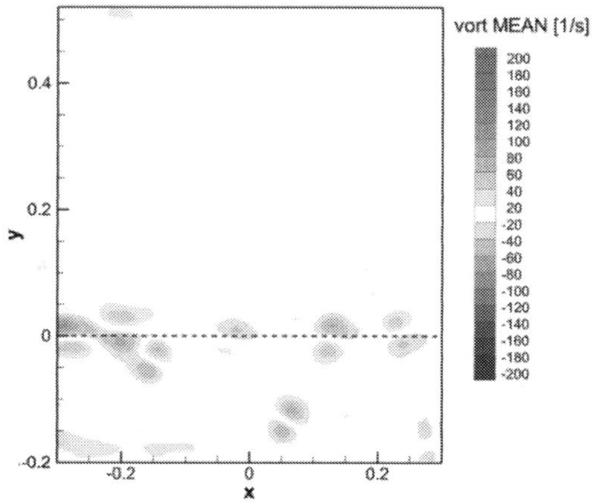

Fig. 7. Mean vorticity distribution. Re = 0.5M.

Almost all vorticity is thus eliminated apart from the region close to the trailing edge. However, our statistics of 380 realisations could not be sufficient.

The intensity if vortex dynamics could be quantified using the local root mean square (RMS) value shown in Fig. 8.

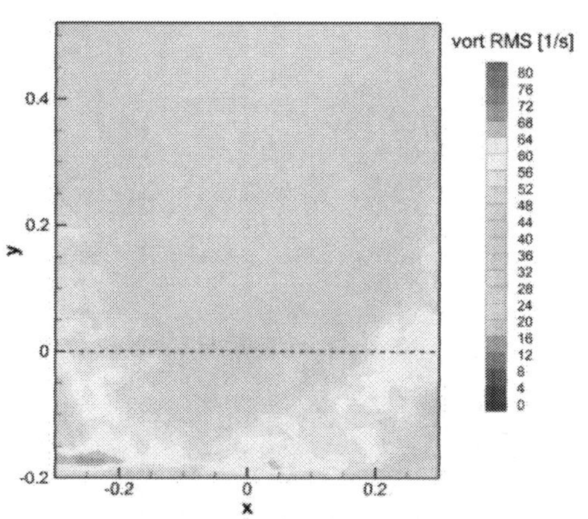

Fig. 8. RMS vorticity distribution. Re = 0.5M.

Maximum of vorticity dynamical activity is located in the pressure region below the trailing edge position.

3.2 Reynolds number 1 million

The experiments in Reynolds number of 1 million were performed in the same way as for the lower Re. Velocity of the incoming flow was about 25 m/s in this case.

The same type of analysis producing the same set of graphical outputs was carried out.

https://doi.org/10.1051/epjconf/201818002111

First, an example of instantaneous velocity field is shown in Fig. 9.

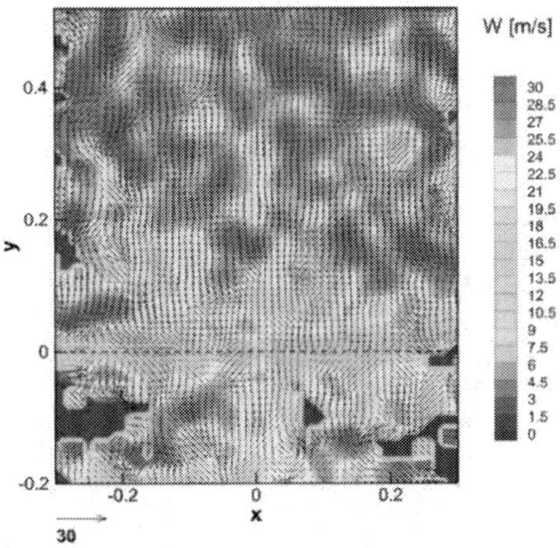

Fig. 9. Example of the instantaneous velocity field. Re = 1M.

The character of the instantaneous velocity field is similar, only values of magnitudes are roughly double as high.

Distribution of the mean velocity shows acceleration well above the airfoil, 0.5 chord above the trailing edge the velocity approaches 30 m/s while pressure region exhibits barely 20 m/s.

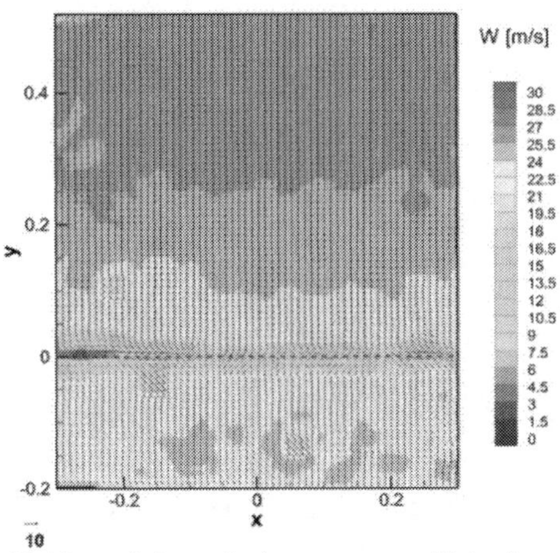

Fig. 10. Mean of-plane velocity component with in-plane velocity vector components. Re = 1M.

The downwash effect is demonstrated in Fig. 11, where distribution of the vertical mean velocity component indicate twice as high values in the suction region in comparison with the pressure region.

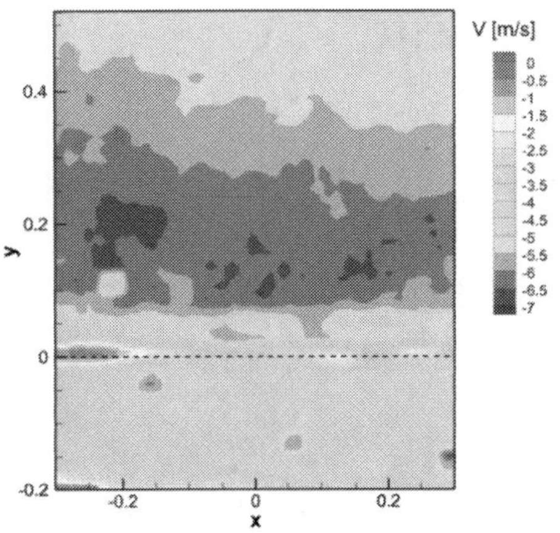

Fig. 11. Mean vertical velocity component distribution. Re = 1M.

The dynamical activity is higher in the pressure region, as shows Fig. 12.

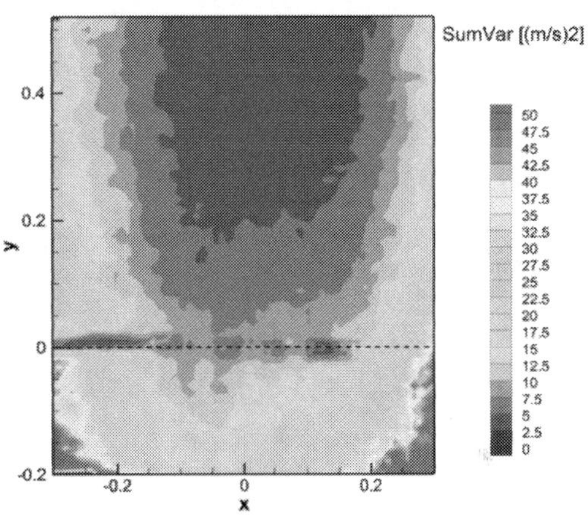

Fig. 12. Sum of velocity components variances. Re = 1M.

Distribution of instantaneous streamwise vorticity component in Fig. 13 confirms similar vortical structure in both regions, pressure and suction.

Mean streamwise vorticity component distribution in Fig. 14 suggests relatively stationary streamwise vortices located close to the trailing edge both in pressure and suction regions. The orientation of vortices is complementary, the vortex couples are detected with opposite orientation in suction and pressure regions.

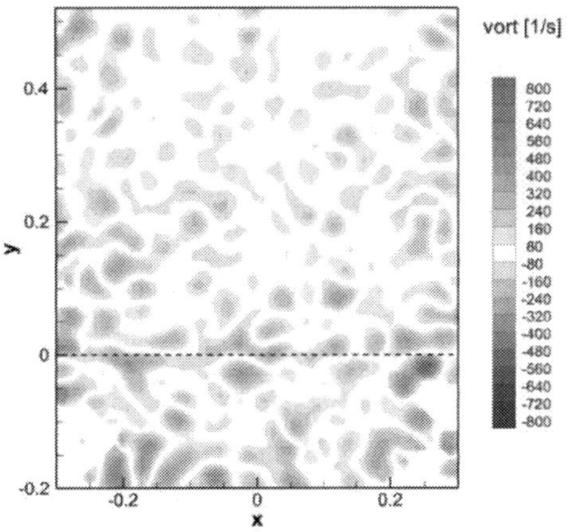

Fig. 13. Example of the instantaneous vorticity field. Re = 1M.

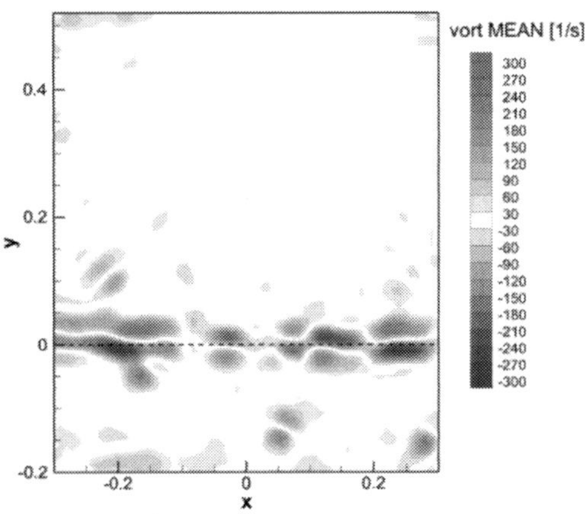

Fig. 14. Mean vorticity distribution. Re = 1M.

The RMS streamwise vorticity component presented in Fig. 15 shows higher dynamical activity in the pressure region.

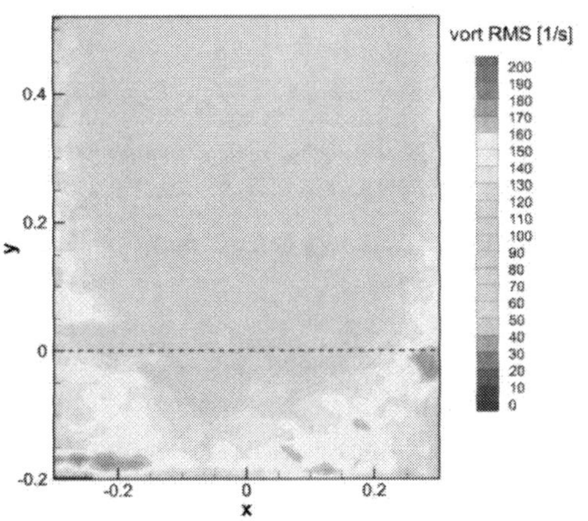

Fig. 15. RMS vorticity distribution. Re = 1M.

3.3 Reynolds number 1.5 million

The similar analysis is to be shown for the Reynolds number 1 and half million. Velocity in the test section inlet was about 38 m/s in this last case.

The same set of results are to be presented. The results are qualitatively very similar to the cases of 0.5 and 1 million Reynolds number, however proportionally higher velocity and vorticity magnitudes. More error regions are detected than before.

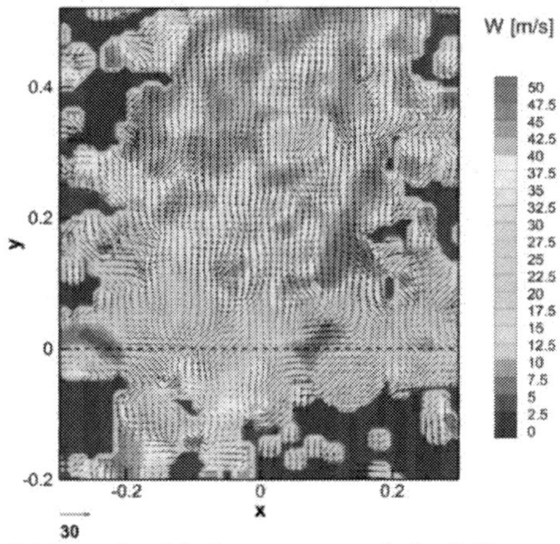

Fig. 16. Example of the instantaneous velocity field. Re = 1.5M.

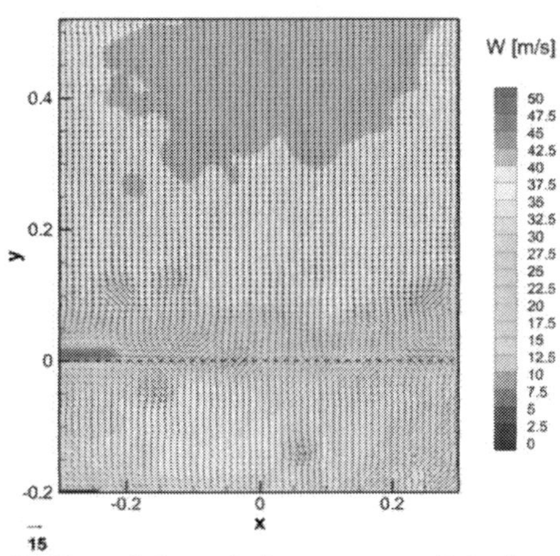

Fig. 17. Mean of-plane velocity component with in-plane velocity vector components. Re = 1.5M.

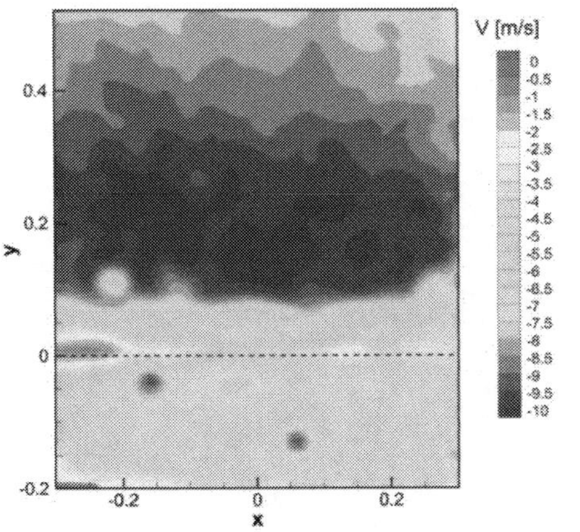

Fig. 18. Mean vertical velocity component distribution. Re = 1.5M.

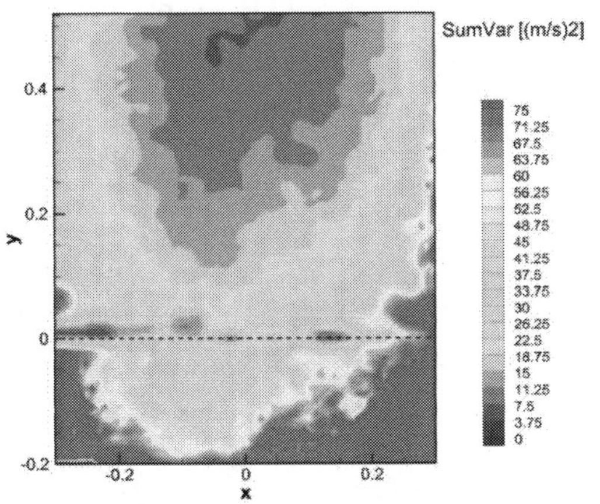

Fig. 19. Sum of velocity components variances. Re = 1.5M.

The observations on velocity distributions comply with those given above for lower Reynolds number.

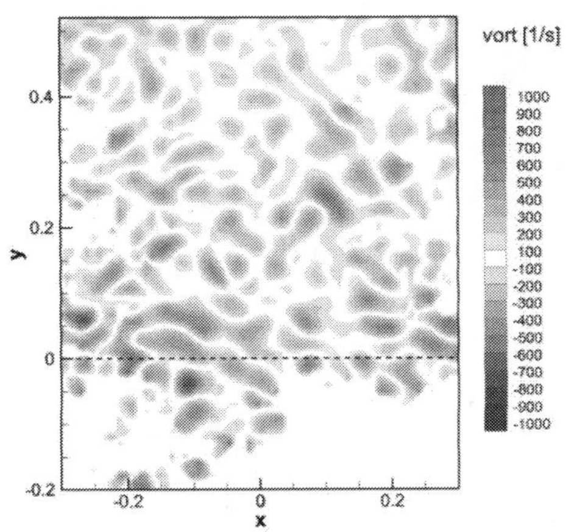

Fig. 20. Example of the instantaneous vorticity field. Re = 1.5M.

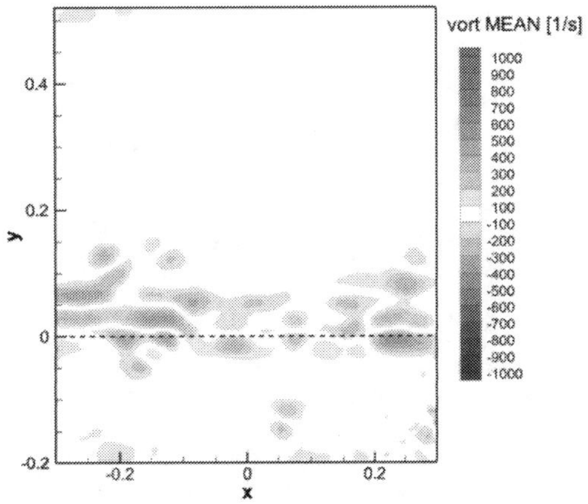

Fig. 21. Mean vorticity distribution. Re = 1.5M.

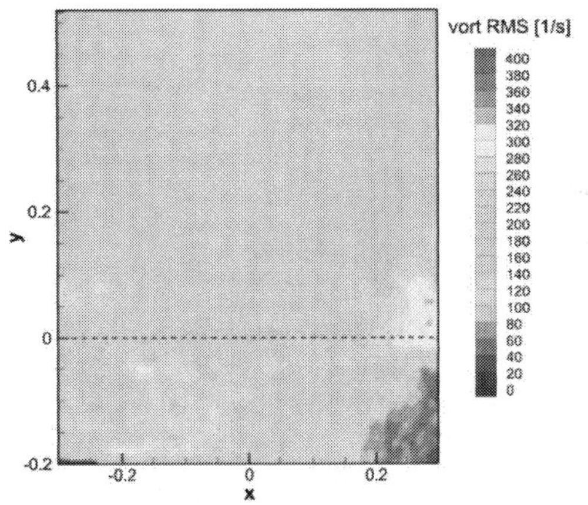

Fig. 22. RMS vorticity distribution. Re = 1.5M.

Results on streamwise vorticity components confirm statements concerning lower Reynolds numbers as well.

4 Conclusions

Experiments on flow structure behind an airfoil with Reynolds number of the order of million and moderate angle of attack have been carried out.

The results show downwash effect with higher both streamwise and vertical velocity components in the suction region (above the trailing edge) than in the pressure region (bellow the trailing edge).

Presence of streamwise vortical structures in the instantaneous fields is proved.

The dynamical activity seems to be higher in pressure region, however in this part more problems with the data quality were detected, this fact could influence the dynamical results.

No qualitative differences in the presented experimental results for Reynolds numbers 0.5, 1 and 1.5 million were found out.

Acknowledgment

This work was supported by the Czech Aerospace Research Centre VZLU Praha, University of West Bohemia and the Grant Agency of the Czech Republic, project No. 17-01088S.

References

1. J. Hoffman, C. Johnson, Resolution of d'Alembert's paradox, J. Math. Fluid Mech., **12** (3), pp.321-334, (2010)
2. V. Uruba, D. Pavlík, P. Procházka, V. Skála, V. Kopecký, On 3D flow-structures behind an inclined plate, EPJ Web of Conferences, **143**, Article number 02137 (2017)
3. V. Uruba, P. Procházka, V. Skála, On 3D Flow Structure of the Boundary Layer on the Suction Side of a Blade, to be published

https://doi.org/10.1051/epjconf/201818002112

On 3D Flow Structure of the Boundary Layer on the Suction Side of a Plate

Václav Uruba[1,2,*], *Pavel* Procházka[1], *Vladislav* Skála[1]

[1]Institute of Thermomechanics, ASCR, v.v.i., Dolejškova 5, Praha 8, CR
[2]UWB, Faculty of Mechanical Engineering, Department of Power System Engineering, Universitní 8, Plzeň, CR

Abstract. Experimental study of the structure of the boundary layer on the suction side of a plate is to be presented. The flat plate with rounded edges and angle of attack of 7° is used. The boundary layer flow will be explored using the time-resolved PIV technique, measuring plane was located very close to the wall. Analysis of the flow dynamics is to be presented using the POD technique applied on both velocity and vorticity fields.

1 Introduction

The presented study is motivated by new ideas about principle of flight by Hoffman and Johnson from KTH Stockholm, see [1,2]. They formulated the new hypothesis of physical mechanism of flight which relies on existence of streamwise vortical structures on the suction side of the airfoil. The vortices origin is supposed to be the instability of the boundary layer subjected to adverse pressure gradient on the airfoil pressure side (i.e. upper).

The study [3] was focused on structure of the wake behind the plate, the presented study concentrates on structure of the flow close to the suction surface of the plate.

2 Experimental setup

Flat plate inclined with angle of attack (AOA) 7 degrees has been placed in a uniform low turbulence stream. The blow-down facility produces a jet with uniform velocity distribution, mean velocity about 10 m/s, intensity of turbulence less than 0.2 %. The plate of thickness 2 mm had rounded edges, chord c = 100 mm and span 300 mm.

The flat plate chord was 100 mm resulting in Reynolds number about 50 thousands.

2.1 Instrumentation

The time-resolved PIV method was used for the experiments. The measuring system DANTEC consists of the double-pulse laser with cylindrical optics and CCD camera. The software Dynamics Studio 3.4 was used for velocity-fields evaluation. Laser New Wave Pegasus Nd:YLF, double head, wavelength 527 nm, maximal frequency 10 kHz, a shot energy is 10 mJ for 1 kHz (corresponding power 10 W per head). Camera Phantom V711 has maximal resolution 1280 x 800 pixels and corresponding maximal frequency 3000 double-snaps per second. The presented experiments are intended to cover low-frequency dynamics offering good statistics. Thus, for the measurements, the frequency 1 kHz and 4000 double-snaps in sequence corresponding to 4 s of record for mean evaluation was acquired. Details are given in references [5,6].

The plane of measurement has been located parallel to the plate suction surface in the distance 1 mm above the surface – see Fig. 1.

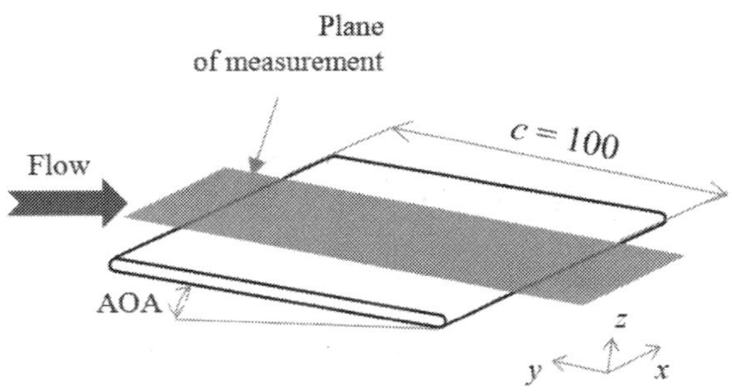

Fig. 1. Schematics of the experiment.

However the laser sheet thickness was of the same value as the distance (approx. 1 mm) representing uncertainty of the measurement location. If the measuring plane is well defined in a distinct distance from the surface within the boundary layer viscous sublayer, then the value of the measured velocity is proportional to the local skin friction. Unfortunately, there is relatively important uncertainty in the distance value and the structure of the boundary layer is not precisely known, the evaluated skin friction is rather of qualitative nature.

The results should be considered as indicators only of the flow topology close to the surface. The interpretation is similar as results of surface flow visualisation using

* Corresponding author: uruba@it.cas.cz

 https://doi.org/10.1051/epjconf/201818002112

surface paintings. The vectorlines could be supposed to follow the skin friction on the surface.

In Fig. 1 the Cartesian coordinate system is introduced, defined as follows: x axis is parallel to the leading and trailing edges, z is perpendicular to the plate and y is parallel to the plate in the flow direction.

2.2 Analysis methods

Instantaneous velocity fields in the measuring plane have been evaluated using "Adaptive PIV" method. The mean velocity field has been evaluated first to obtain the flow statistics and remove the mean flow. The dynamical behaviour is characterized by the turbulent kinetic energy defined in usual way as a half of sum of individual velocity components variances.

The Proper Orthogonal Decomposition (POD) method has been used to assess the flow dynamics. The POD was introduced in the context of turbulence by Lumley [3] as an objective definition of what was previously called big eddies and which is now widely known as coherent structures. Extraction of deterministic features from a random, fine grained turbulent flow is a challenging problem. The POD provides an unbiased technique for identifying such structures. The method extracts the candidate which is best correlated, in a statistical sense, with the background velocity field. The different structures are identified with the orthogonal eigenfunctions of the decomposition theorem of probability theory. This is thus a systematic way to find organized motions in a given set of realizations of a random field.

3 Results

The results are divided into two parts. The first part is oriented on time-mean results, while the second part characterizes the flow dynamics.

3.1 Time mean

The presented results cover the upper surface of the plate from leading edge to trailing edge. Width of the evaluated area was about 0.3 c, located in the centre of the plate. Position $y = 0$ corresponds to the leading edge while $y = 1$ indicates position of the trailing edge.

To have some preliminary idea about the flow structure in the measuring plane, an example of the instantaneous velocity field is presented in Fig. 2. To help with the vectors interpretation, the vector lines are added arbitrarily.

The flow-field could be divided into several parts along the y axis:

- $y \in (0;0.08)$: presence of the free stream, however above the surface. In this region the measuring plane is above the boundary layer, which is very thin here, less than 1 mm,
- $y \in (0.08;0.25)$: separation bubble with backward oriented flow,
- $y \in (0.25;1)$: normal streamwise flow, however perturbed.

Separation of the flow is located very close to the leading edge, reattachment point is located at $y = 0.25$. This result is supported by the mean velocity vectors distribution in Fig. 3, colours represent velocity modulus.

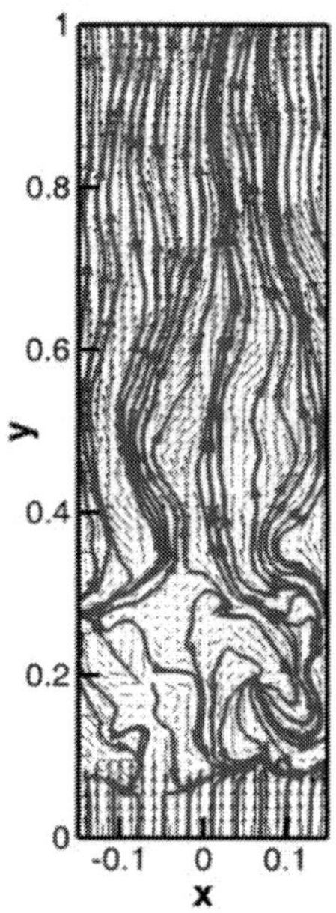

Fig. 2. Instantaneous velocity vectors.

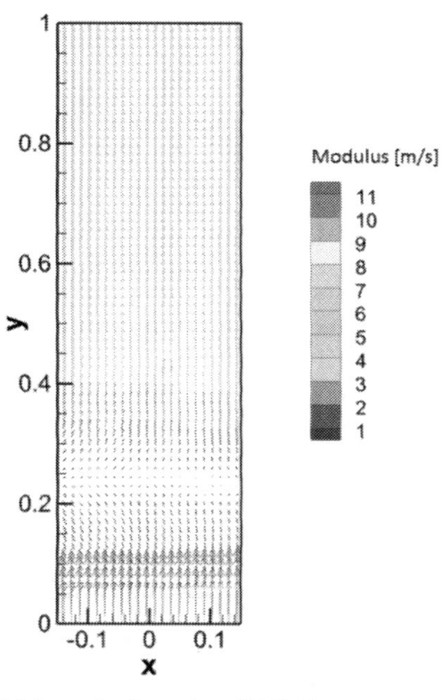

Fig. 3. Mean velocity vectors distribution.

 https://doi.org/10.1051/epjconf/201818002112

In Fig. 3 the vectors are coloured according to the local mean velocity magnitude ("Length"). Additional confirmation shows Fig. 4, where distribution of the streamwise mean velocity component V is shown. The backward flow region is in blue.

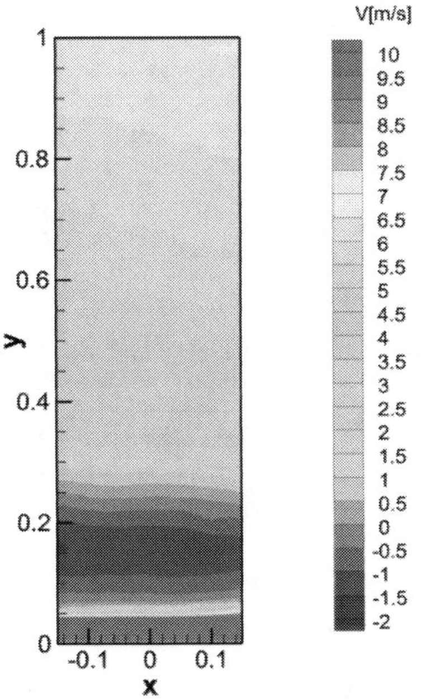

Fig. 4. Streamwise mean velocity component distribution.

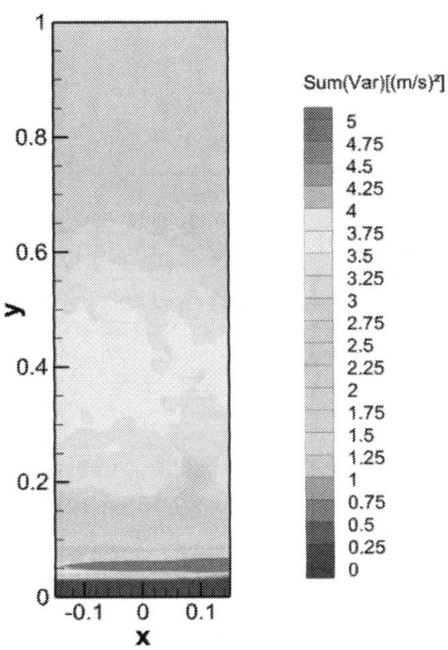

Fig. 5. Sum of in-plane velocity components variances distribution.

The dynamic activity is quantified using the sum of velocity components variances, shown in Fig. 5.

The results indicates maximal dynamical activity within the separation bubble, especially the bubble leading edge in the plane of measurement is intensively changing its position.

To visualise possible vortical structures oriented in the z-direction perpendicular to the plate, the corresponding vorticity component has been evaluated. In Fig. 6 there is an example of distribution of the instantaneous vorticity component in z-direction.

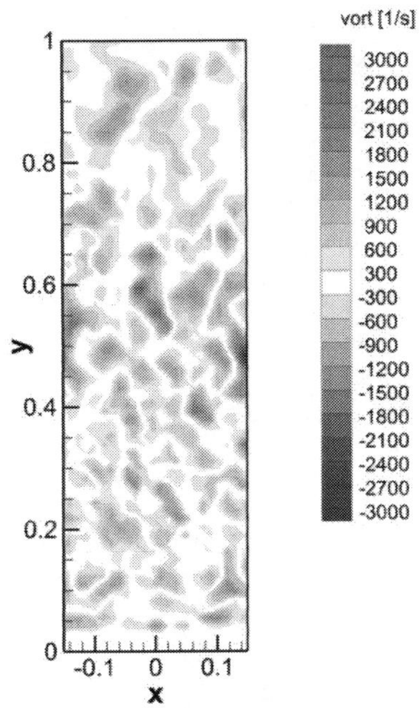

Fig. 6. Instantaneous vorticity distribution.

Both positive and negative vorticity is present in the instantaneous field, structure is rather random, however the maximum magnitude indicates presence of waves in the velocity field – compare with Fig. 2.

Statistical characteristics of the vorticity have been evaluated as well. In Figs. 7 and 8 distributions of the vorticity mean value and RMS are depicted respectively. The maximal vorticity activity is detected within the separation bubble and in close to the $y = 0.4$ the vortex couple of opposite sign has been detected. It will be subjected to further analysis in future.

In statistical quantities distributions the spanwise direction x is supposed to be irrelevant, any changes in this direction should be attributed to insufficient number of statistical data input into averaging process.

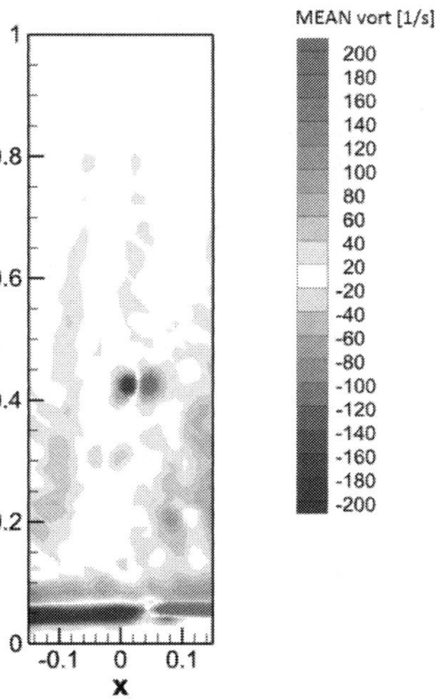

Fig. 7. Mean vorticity distribution.

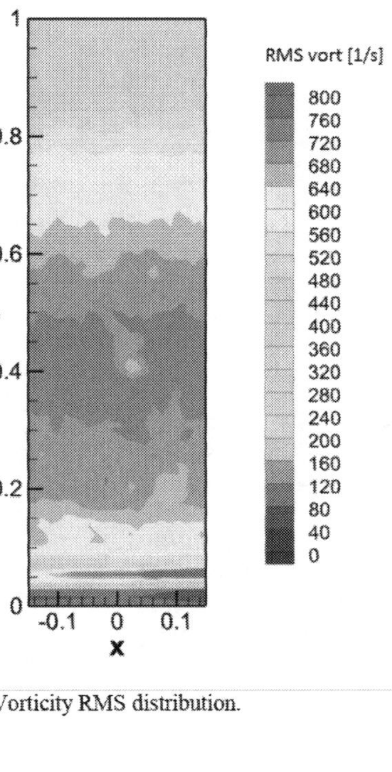

Fig. 8. Vorticity RMS distribution.

3.2 Dynamics

Dynamical analysis has been carried out using the POD method. As vortical structures are to be studied, the POD analysis has been applied both on velocity and vorticity fields. The POD method maximizes the analysed quantity variance, in the case of velocity this corresponds to kinetic energy (double), while for vorticity the half variance equals to enstrophy.

The region close to the leading edge has been excluded from the space domain for the analysis, as dynamics of separation bubble has not been in the scope of our interest. The area of interest is reduced for the study of dynamics, the range for dynamical analysis is defined as follows: $x \in (-1.5;1.5)$, $y \in (0.15;1)$.

3.2.1 Dynamics of velocity field

The dynamical analysis is applied on the velocity fields, the velocity POD modes nave been evaluated. Each mode is characterised by its energy. Distribution of the energy across the velocity POD modes is shown in Fig. 9 for the lowest 1000 modes, note that the vertical axis is log.

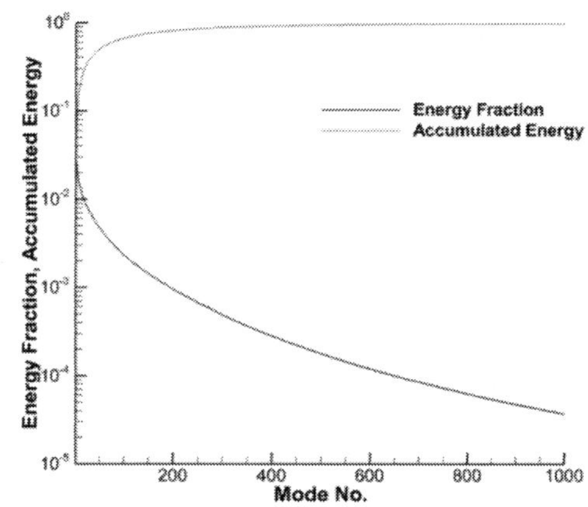

Fig. 9. Energy distribution across the velocity POD modes.

In Fig. 9 there is the "Energy Fraction" indicating relative energy of a given POD mode and the "Accumulated Energy" which represents the energy contained in all modes with number lower and equal to the given mode. The Total Kinetic Energy (TKE) is considered to be 1. For example the 1st velocity POD mode contains 2.8% of the TKE, 10 first POD modes cover 20% and 100 modes about 67% of the TKE.

The low order modes are the most energetic ones. The velocity POD modes topology will be shown in following Figures. The first 3 modes are depicted in Figs. 10-12, the modes are represented by vector fields with arbitrarily added vector lines in blue. In Figs. 13 and 14 an example of higher order vorticity POD modes is given, modes Nos. 10 and 100.

https://doi.org/10.1051/epjconf/201818002112

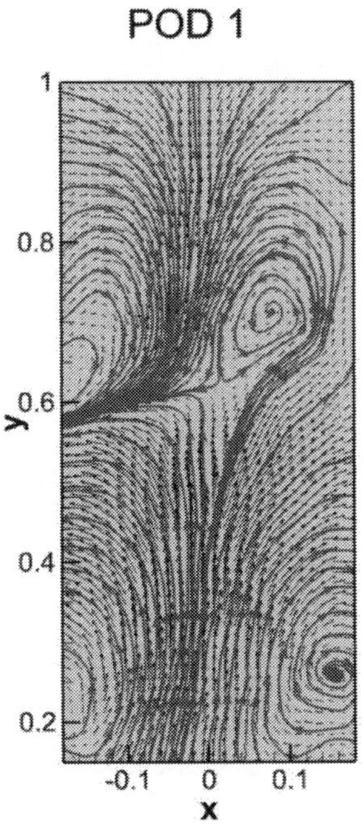

Fig. 10. Velocity POD mode No. 1, 2.8% TKE.

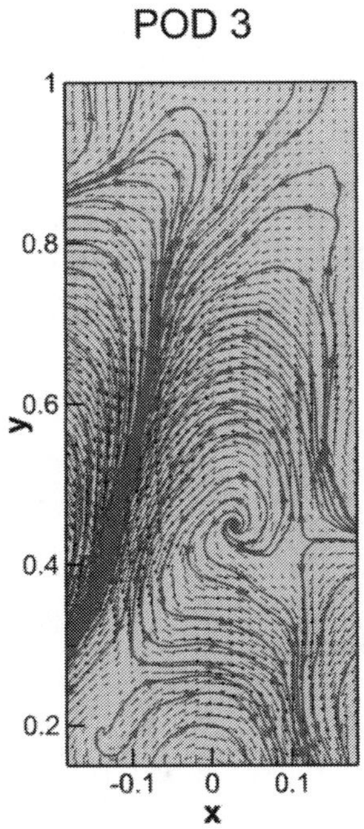

Fig. 12. Velocity POD mode No. 3, 2.4% TKE.

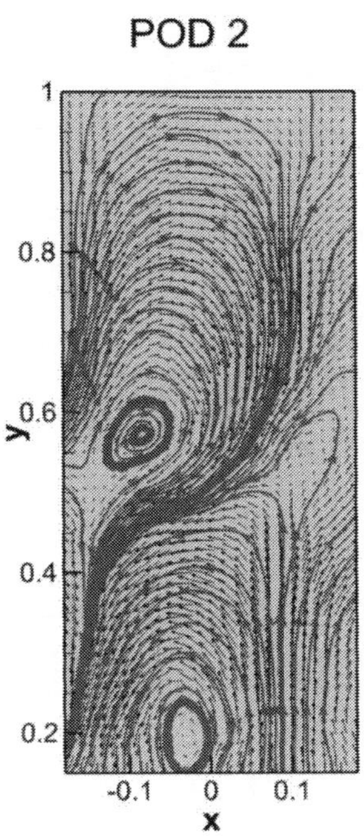

Fig. 11. Velocity POD mode No. 2, 2.6% TKE.

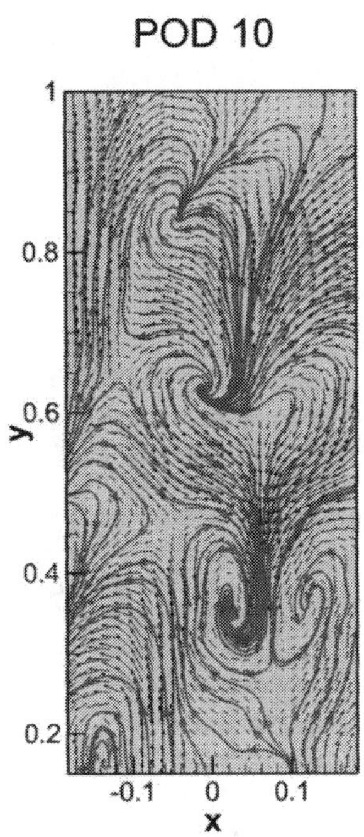

Fig. 13. Velocity POD mode No. 10, 1.4% TKE.

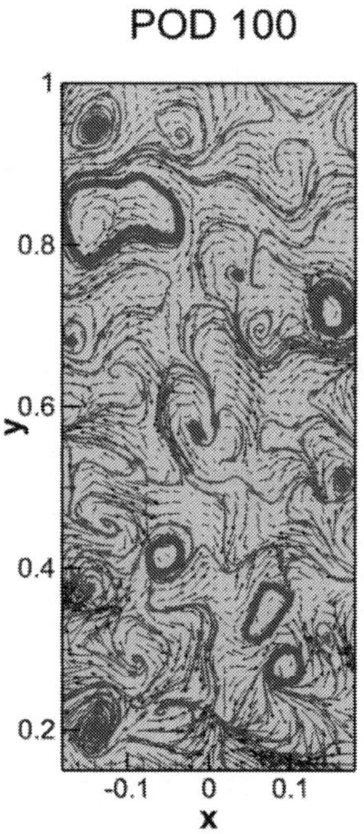

Fig. 14. Velocity POD mode No. 100, 0.23% TKE.

The typical dynamics of the velocity field is characterized by low-order modes, they consist of a small number of big vortices distributed in the velocity field. The vortices are located in the middle part of the plate chord.

The higher order velocity POD modes contain relatively small amount of TKE and they consist of a big number of small vortices distributed regularly in the whole domain.

3.2.2 Dynamics of vorticity field

The similar dynamical analysis is applied on the vorticity fields, the vorticity POD modes have been evaluated. However, there is a difference in the procedure, as now we work with scalar fields instead of vectors. Now, each mode is characterised by its enstrophy. Distribution of the enstrophy across the vorticity POD modes is shown in Fig. 15 for 1000 modes, the vertical axis is log. The figures on both "Enstrophy Fraction" and "Accumulated Enstrophy" are relative to the Total Enstrophy (TE).

The enstrophy of the 1st vorticity POD mode is about 1.05% of the TE, while 10 lowest order modes cover 9% and 100 modes 53% of the TE.

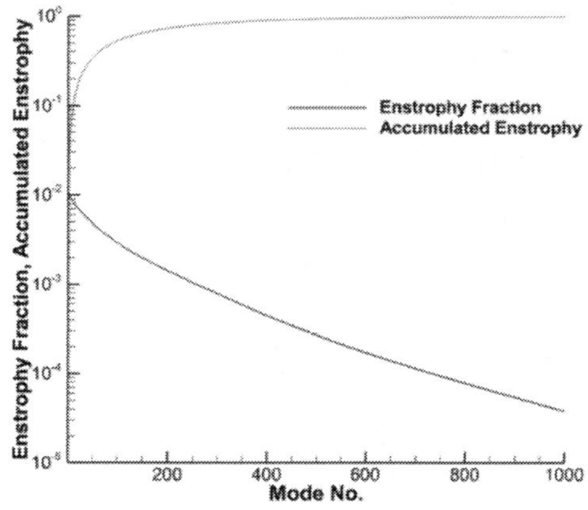

Fig. 15. Enstrophy distribution across vorticity POD modes.

Examples of the vorticity POD modes topologies are to be shown in Figs. 16-20 representing the modes Nos. 1, 2, 3, 20 and 100 respectively. Distribution of the vorticity component in z direction, red – positive, blue – negative (clockwise).

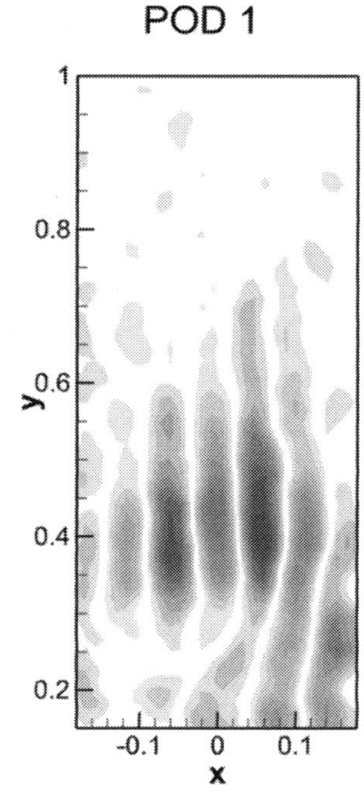

Fig. 16. Vorticity POD mode No. 1, 1.05% TE.

https://doi.org/10.1051/epjconf/201818002112

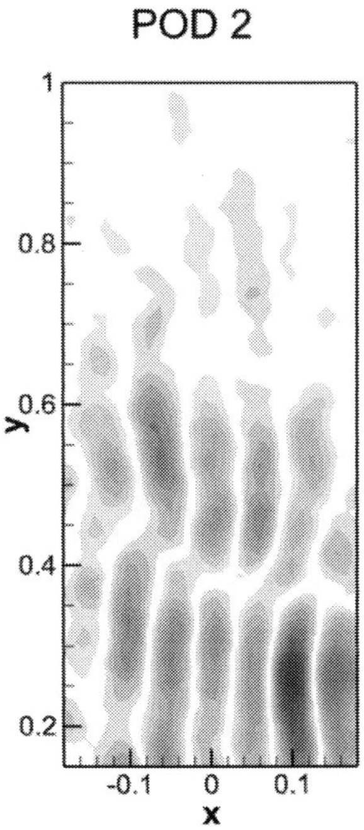

Fig. 17. Vorticity POD mode No. 2, 1.03 TE.

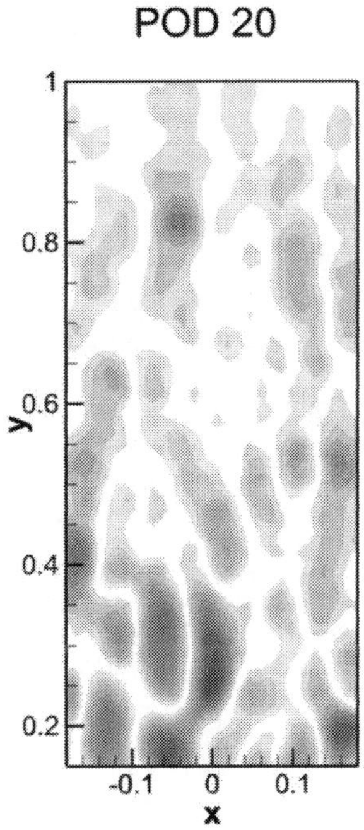

Fig. 19. Vorticity POD mode No. 20, 0.71% TE.

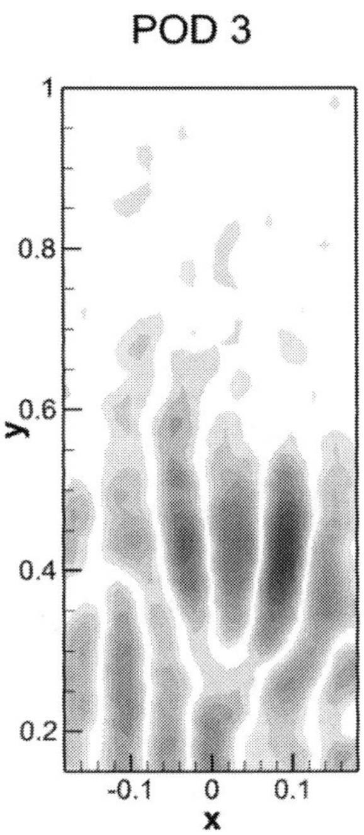

Fig. 18. Vorticity POD mode No. 3, 1% TE.

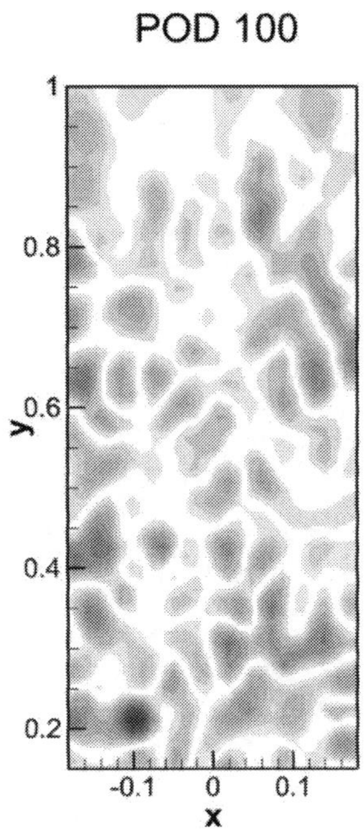

Fig. 20. Vorticity POD mode No. 100, 0.29% TE.

The lower order vorticity POD modes (1, 2, 3) are typical with vortical activity organized in strips with streamwise direction. The strips are located close to the middle of the chord and are of alternating orientation.

The high order vorticity POD modes are characterized by more or less regularly distributed vorticity in spots of positive and negative vorticity.

4 Conclusions

Topology of the flow in the vicinity of suction surface of the flat plate in regular stream with angle of attack 7° was studied experimentally. Time resolved PIV method has been applied with measuring plane located close to the plate surface parallel to it. Detected velocity could be linked to skin friction on the suction surface of the plate.

Both velocity and vorticity fields have been explored using classical statistical methods and POD method. Typical vortical structures have been detected.

Acknowledgments

This work was supported by the Grant Agency of the Czech Republic, project No. 17-01088S.

References

1. J. Hoffman, C. Johnson, The Mathematical Secret of Flight, *Normat* **57**, 4, pp.1–25, (2009)
2. J. Hoffman, C. Johnson, Resolution of d'Alembert's paradox, J. Math. Fluid Mech., **12** (3), pp.321-334, (2010)
3. J.L. Lumley, The structure of inhomogeneous turbulent flows, Atm.Turb. and Radio Wave Prop., Yaglom and Tatarsky eds., Nauka, Moskva, pp. 166-178 (1967)
4. L. Sirovich, Turbulence and the dynamics of coherent structures. Quart. Appl. Math., **45**, pp. 561-590 (1987)
5. V. Uruba, D. Pavlík, P. Procházka, V. Skála, V. Kopecký, On 3D flow-structures behind an inclined plate, EPJ Web of Conferences, **143**, Article number 02137 (2017)
6. V. Uruba, Near Wake Dynamics around a Vibrating Airfoil by Means of PIV and Oscillation Pattern Decomposition at Reynolds Number of 65 000, *Journal of Fluids and Structures*, **55**, pp. 372-383 (2015)
7. V. Uruba, Decomposition methods in turbulent research, EFM11, *EPJ Web of Conferences*, **25** 01095 (2012)

EPJ Web of Conferences 180, 02113 (2018) https://doi.org/10.1051/epjconf/201818002113
EFM 2017

Aeroacoustic computation of fluid-structure interaction problems with low Mach numbers

Jan Valášek[1,2,*] and *Petr* Sváček[1,2]

[1] Department of Technical Mathematics, Faculty of Mechanical Engineering, Czech Technical University in Prague, Karlovo nám. 13, Praha 2, 121 35, Czech Republic
[2] Dept. of Instrumentation and Control Engineering, Faculty of Mechanical Engineering, Center of Advanced Aerospace Technology, Czech Technical University in Prague, Technicka Street 4, 16607, Prague 6, Czech Republic

Abstract. This contribution deals with the acoustic simulation of aerodynamical noise generated by a flow over an airfoil or by flow in a flexible channel. Since the considered flow has low Mach number the hybrid approach of acoustic analogies can be applied here with benefits. The fluid-structure-acoustic interaction problem is generally described as a quite complicated problem comprising of three different physical fields - the vibration of the elastic body, the unsteady fluid flow and the acoustics together with mutual couplings. The fluid flow in time dependent domain is governed by the incompressible Navier-Stokes equations in arbitrary Langrangian-Eulerian formulation and the elastic structure is modelled by the means of linear elasticity theory. The Lighthill analogy and acoustic perturbation equation (APE) is considered to describe the sound propagation. The simulation of fluid-structure (FSI) interaction and acoustic field is implemented using the FEM in an in-house solver. The sound sources computed from FSI results are analyzed and within sound propagation simulation the perfectly matched layer technique is used. In the end the results of Lighthill and APE analogy are compared.

1 INTRODUCTION

The simulation of noise creation and propagation is of a quite big importance today due to increasing number and power of noise producing devices. Good examples of external and internal problems with aerodynamically generated sound are air flow around airfoil, air-conditioning or sound produced by human vocal fold model. These problems are studied by experimental, analytical and numerical methods in many applications, see for example [1] or [2], nevertheless the acoustic simulation within fluid-structure interaction (FSI) problem is used only rarely, see for example [3].

The acoustic quantities can be possibly obtained from solution of compressible Navier-Stokes equations and by later filtering of acoustic parts but such an approach is extremely computational demanding if not infeasible in many cases. That is why usually simplified approaches are used. The development of these approaches, called acoustic analogies, started by Lighthill in 1952, see [4]. Then the new idea of acoustic/viscous splitting were introduced in 1994 by Hardin and Pope [5] and the cornerstone of modern computational aeroacoustics was laid by Ewert and Schröder in 2003, see [6]. They derived new system of equations from Navier-Stokes equations by filtering based on splitting flow variables in three parts. Here, we focus on the acoustic perturbation equation (APE) analogy and also the well-known Lighthill analogy is explained and used to verify the results of APE analogy.

In general the fluid-structure-acoustic interaction (FSAI) problem, see e.g. [2], is very complex phenomenon composed of three different physical fields - the deformation of the elastic structure, the unsteady fluid flow and the acoustics together with mutual couplings. The hybrid approach of acoustic analogies allows to decouple fluid-structure interaction (FSI) problem and acoustics, i.e. the effects of acoustic coupling to FSI problem is neglected due to the small intensity of resulting acoustic field. It enables us to solve acoustic problem in parallel or at the end of FSI computation, where from the stored fluid velocities and pressures the sound source is calculated. Moreover the interpolation onto (usually) coarser acoustic grid, [7], can be performed to save the computational time. During solution of sound propagation the perfectly matched layer (PML) technique is used due to the simulation in the open domains. This additional layer of elements introduces effective damping of outgoing waves without reflection at the interface between PML and acoustic propagation domain, [8].

The fluid flow is governed by the incompressible Navier-Stokes equations with the aid of the Arbitrary Lagrangian-Eulerian (ALE) method, which allows the computational domain of the fluid to be deformed. The

* e-mail: jan.valasek1@fs.cvut.cz

elastic structure is modelled by the means of linear elasticity theory, where material with more layers is used. All the three subproblems are numerically approximated by the finite element method. Specially, the fluid flow simulation is stabilized by the modified Streamline-Upwind/Petrov-Galerkin stabilization, see [9].

In the end the numerical simulation of sound propagation is given and the results of the Lighthill analogy and APE analogy is presented.

2 MATHEMATICAL DESCRIPTION

For the sake of simplicity only a two-dimensional problem is described here. The FSI problem is solved on a small subdomain of the whole domain, where the sound propagation problem is considered. The scheme of FSI domain is shown in Figure 1 and the whole computational domain for acoustic problem is depicted in Figure 2. The FSI domain consists of an elastic body domain Ω^s and a fluid domain Ω^f, where the sound sources are computed from the flow quantities. The elastic structure is described in Lagrange coordinate system, therefore Ω^s_{ref} denotes the reference state, which remains the same for any time t. For the description of the fluid flow on moving domain Ω^f_t the ALE method is used and we distinguish between the reference domain Ω^f_{ref} at time instant $t = 0$ and the deformed domain Ω^f_t at arbitrary time t. Similarly, by $\Gamma_{W_{\text{ref}}}$ and Γ_{W_t} we denote the reference state and deformed state at arbitrary time t of the interface, respectively.

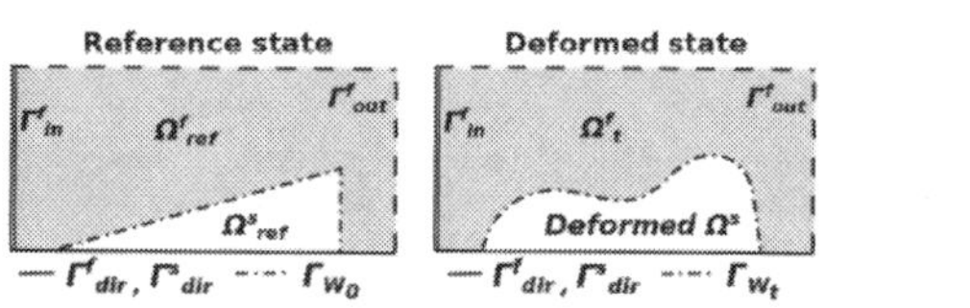

Fig. 1. Scheme of FSI domain consisting of elastic structure domain and fluid domain marked before (left) and after (right) a deformation.

2.1 ALE method

The ALE method enables to describe the fluid flow in a time-dependent domain. The basis of this method lies in the construction of a diffeomorphism A_t which maps the reference (undistorted) domain Ω^f_{ref} on to the domain Ω^f_t at any instant time $t \in (0, \mathrm{T})$. Under assumption of continuity of the time derivative $\frac{\partial A_t}{\partial t} \in C(\overline{\Omega^f_{\text{ref}}})$ the ALE domain velocity $\mathbf{w}_D$ can be defined as

$$\mathbf{w}_D(A_t(X), t) = \frac{\partial}{\partial t} A_t(X), \quad t \in (0, \mathrm{T}),\ X \in \Omega^f_{\text{ref}}. \tag{1}$$

Finally, the ALE derivative is introduced as the time derivative with respect to a fixed point $X \in \Omega^f_{ref}$. The ALE derivative can be expressed as

$$\frac{D^A}{Dt} f(x, t) = \frac{\partial f}{\partial t}(x, t) + \mathbf{w}_D(x, t) \cdot \nabla f(x, t). \tag{2}$$

More details and practical construction of ALE mapping is described e.g. in [10] or [11].

2.2 Fluid flow

The flow of the viscous incompressible fluid in time variable domain Ω^f_t obeys the Navier-Stokes equations in the ALE form

$$\frac{D^A \mathbf{v}}{Dt} + ((\mathbf{v} - \mathbf{w}_D) \cdot \nabla)\mathbf{v} - \nu^f \Delta \mathbf{v} + \nabla p = \mathbf{0}, \quad \operatorname{div} \mathbf{v} = 0 \text{ in } \Omega^f_t, \tag{3}$$

where $\mathbf{v}(x, t)$ is the fluid velocity, p denotes the kinematic pressure and ν^f is the kinematic fluid viscosity. Equations (3) are supplied by the zero initial conditions and following boundary conditions. For $x \in \Gamma^f_{\text{Dir}} \cup \Gamma_{W_t}$ and $t \in (0, \mathrm{T})$ holds

$$\mathbf{v}(x, t) = \mathbf{w}_D(x, t), \tag{4}$$

and the pressure fulfills for $x \in \Gamma^f_{\text{In}} \cup \Gamma^f_{\text{Out}},\ t \in (0, \mathrm{T})$

$$p(x, t)\, \mathbf{n}^f - \nu^f \frac{\partial \mathbf{v}}{\partial \mathbf{n}^f}(x, t) = -\frac{1}{2} \mathbf{v}(\mathbf{v} \cdot \mathbf{n}^f)^- + p_{\text{ref}}\, \mathbf{n}^f, \tag{5}$$

where $\mathbf{n}^f$ is unit outer normal to boundary $\partial\Omega^f_t$ and p_{ref} is a reference pressure, which has possibly different values at the inlet Γ^f_{In} and at the outlet Γ^f_{Out}. The condition (5) is the modified do-nothing boundary condition, which suppresses the possible backward inlet through the outlet part of the boundary, see [12].

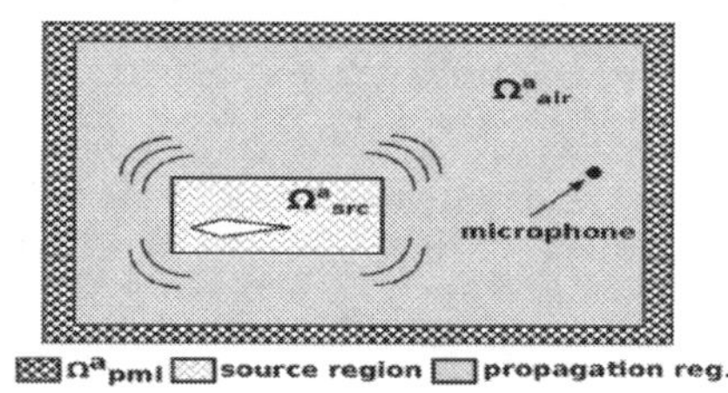

Fig. 2. Scheme of acoustic domain. The domain, where the sound propagation is computed, is composed of the sound source region Ω^a_{src} (equals FSI domain without structure domain, i.e. it equals reference fluid domain Ω^f_{ref}) and the propagation region. It is closed by the PML blocks.

2.3 Elastic body

The deformation $\mathbf{u}(X, t) = (u_1, u_2)$ of the elastic body Ω^s is modelled by the partial differential equations

$$\rho^s \frac{\partial^2 u_i}{\partial t^2} - \frac{\partial \tau^s_{ij}(\mathbf{u})}{\partial X_j} = f^s_i \quad \text{in } \Omega^s \times (0, \mathrm{T}), \tag{6}$$

where ρ^s denotes the structure density, the tensor τ^s_{ij} is the Cauchy stress tensor, the vector $\mathbf{f}^s = (f^s_1, f^s_2)$ is the volume density of an acting force and $X = (X_1, X_2)$ are the reference coordinates. In the case of isotropic material the Hook law reads

$$\tau^s_{ij} = \lambda^s (\operatorname{div} \mathbf{u})\, \delta_{ij} + 2\mu^s e^s_{ij}(\mathbf{u}), \tag{7}$$

where e^s_{ij} are components of the strain tensor of small displacements $e^s(\mathbf{u}) = \frac{1}{2}\left(\frac{\partial u_j}{\partial X_k} + \frac{\partial u_k}{\partial X_j}\right)$, $\mathbb{I} = (\delta_{ij})$ is the Kronecker's delta and λ^s, μ^s are Lamé coefficients. They possibly can have different values for different parts of the body. The equations (6) are supplied by the initial and boundary conditions

$$\begin{aligned} &\text{a)} \quad \mathbf{u}(X,0) = \mathbf{u}_0(X), \ \frac{\partial \mathbf{u}}{\partial t}(X,0) = \mathbf{u}_1(X), \quad \text{for } X \in \Omega^s, \\ &\text{b)} \quad \mathbf{u}(X,t) = \mathbf{u}_{\mathrm{Dir}}(X,t), \quad \text{for } X \in \Gamma^s_{\mathrm{Dir}},\ t \in (0,\mathrm{T}), \quad (8) \\ &\text{c)} \quad \tau^s_{ij}(X,t)\, n^s_j(X) = q^s_i(X,t), \ \text{for } X \in \Gamma^s_{\mathrm{W_{ref}}},\ t \in (0,\mathrm{T}), \end{aligned}$$

where the $\Gamma_{\mathrm{W_{ref}}}, \Gamma^s_{\mathrm{Dir}}$ are mutually disjoint parts of the boundary $\partial\Omega^s = \Gamma_{\mathrm{W_{ref}}} \cup \Gamma^s_{\mathrm{Dir}}$ (see Figure 1) and $n^s_j(X)$ are the components of the unit outer normal to $\Gamma_{\mathrm{W_{ref}}}$.

2.4 Aeroacoustics

The acoustic domain Ω^a is composed of three subdomains $\Omega^a_{\mathrm{src}}, \Omega^a_{\mathrm{air}}$ and Ω^a_{pml}, see Figure 2. The domain Ω^a_{src} of acoustic sources is the same as the reference fluid domain and sound sources are here evaluated. The domain Ω^a_{air} represents an air region without sound sources, where acoustic waves propagate freely. The both domains $\Omega^a_{\mathrm{src}}, \Omega^a_{\mathrm{air}}$ of free sound propagation are enclosed by PML domain Ω^a_{pml} to damp the outgoing sound waves.

The first considered analogy is the Lighthill acoustic analogy proposed in 1952, which has the form of the inhomogeneous wave equation

$$\frac{1}{c_0^2}\frac{\partial^2 p'}{\partial t^2} - \frac{\partial^2 p'}{\partial x_i^2} = \frac{\partial^2 T_{ij}}{\partial x_i \partial x_j}, \tag{9}$$

for unknown pressure fluctuation $p' = p - p_0$ and with a given speed of sound c_0. It is derived from supposition of sound propagating from a small source region with fluid flow through a voluminous medium in the rest state, see [4]. Hence the Lighthill tensor T_{ij} can be expressed as (hypothetical) stress difference between the flow and the rest state of the fluid

$$T_{ij} = \rho^f v_i v_j + \left((p - p_0) - c_0^2(\rho^f - \rho^f_0)\right)\delta_{ij} - \tau^f_{ij} \approx \rho^f v_i v_j, \tag{10}$$

where p_0, ρ^f_0 denotes the pressure and the density of the fluid at rest state, respectively. The used approximation in equation (10) for small Mach numbers was proposed already in [4]. The wave equation (9) is equipped with the zero initial condition and the boundary of acoustic domain $\partial\Omega^a$ with the outer unit normal $\mathbf{n}^a$ is considered as sound hard

$$\frac{\partial p'}{\partial \mathbf{n}^a}(x,t) = 0 \qquad \text{for } x \in \partial\Omega^a,\ t \in (0,\mathrm{T}), \tag{11}$$

which means that sound waves are there reflected back to the acoustic domain Ω^a. The Lighthill analogy is quite easy to understand and it constitutes very general approach. It's disadvantage is that the pressure fluctuations p' can be considered as purely acoustic pressure p_a only outside the fluid flow region, where no additional pressure is present.

The second presented acoustic analogy is the APE. It is analogy based on the splitting of physical quantities into the mean part $\bar{\mathbf{v}}, \bar{p}$, the fluctuating non-acoustic part $\mathbf{v}_{ic}, p_{ic}$ and the acoustic part $\mathbf{v}_a, p_a$

$$\mathbf{v} = \bar{\mathbf{v}} + \mathbf{v}_{ic} + \mathbf{v}_a, \quad p = \bar{p} + p_{ic} + p_a. \tag{12}$$

Inserting these expressions into the Navier-Stokes equations of incompressible fluid together with assumptions of the irrotational acoustic field ($\nabla \times \mathbf{v}_a = \mathbf{0}$), isentropic flow, incompressible fluid ($\nabla \times \mathbf{v}_{ic} = 0$) and neglecting nonlinear terms results into system of equations

$$\frac{\partial p^a}{\partial t} + \bar{\mathbf{v}} \cdot \nabla p^a + \rho^f_0 c_0^2 \nabla \cdot \mathbf{v}^a = -\frac{Dp_{ic}}{Dt}, \tag{13a}$$

$$\frac{\partial \mathbf{v}^a}{\partial t} + \nabla(\bar{\mathbf{v}} \cdot \mathbf{v}^a) + \frac{1}{\rho^f_0}\nabla p^a = \mathbf{0}. \tag{13b}$$

For the precise derivation and the discussion see [13]. This system can be further simplified into one scalar equation by introduction of the acoustic potential ψ^a related to the acoustic particle velocity by $\mathbf{v}^a = -\nabla\psi^a$.

$$\frac{1}{c_0^2}\frac{D^2\psi^a}{Dt^2} - \Delta\psi^a = -\frac{1}{c_0^2}\frac{Dp_{ic}}{Dt}, \tag{14}$$

where p_{ic} denotes the pressure connected with incompressible flow and the substantial derivative equals $\frac{D}{Dt} = \frac{\partial}{\partial t} + \bar{\mathbf{v}} \cdot \nabla$. For small velocities the mean velocity can be neglected ($\bar{\mathbf{v}} = 0$) and formulation (14) get again the form of inhomogenous wave equation

$$\frac{1}{c_0^2}\frac{\partial^2\psi^a}{\partial t^2} - \Delta\psi^a = -\frac{1}{\rho^f_0 c_0^2}\frac{\partial p_{ic}}{\partial t}. \tag{15}$$

Equation (15) is called acoustic perturbation equation (APE).

Equation (14) or (15) is considered again with sound hard boundaries $\frac{\partial\psi^a}{\partial \mathbf{n}^a}(x,t) = 0,\ x \in \partial\Omega^a,\ t \in (0,\mathrm{T})$. The analogies based on the quantity splitting approach is physically more complicated than Lighthill analogy, on the other hand it delivers directly the acoustic pressure (through relation $p^a = \rho^f_0 \frac{\partial\psi^a}{\partial t}$) in the whole acoustic domain.

2.5 Coupling conditions

The FSI problem is a coupled problem, where the location of the interface Γ_{W_t} at any time t depends on the solution of the FSI problem itself. It corresponds to the establishing force equilibrium between the aerodynamic and the elastic forces and it is implicitly given by deformation of the interface $\Gamma_{W_{\mathrm{ref}}}$.

Moreover the requirement of stress continuity across the interface leads to the Neumann boundary condition (8 c) for the elastic problem in the form

$$q^s_i(X,t) = -\sum_{j=1}^{2} \sigma^f_{ij}(x)\, n^f_j(x), \tag{16}$$

where $x = X + \mathbf{u}(X,t)$ for any $X \in \Gamma_{\mathrm{W_{ref}}}$. The components of the fluid stress tensor are $\sigma^f_{ij} = -\rho^f_0 p \delta_{ij} + \rho^f_0 \nu^f (\frac{\partial v_i}{\partial x_j} + \frac{\partial v_j}{\partial x_i})$ for $i, j \in \{1, 2\}$.

The Dirichlet boundary condition postulated by equation (4) for the fluid flow problem is originated by the enforcement of velocity continuity across the interface.

Since the acoustic field has substantially less energy than the flow field, the backward coupling of acoustics to the flow field is not considered.

https://doi.org/10.1051/epjconf/201818002113

3 NUMERICAL SCHEME

All three subproblems of FSAI problem is solved by finite element method. Therefore partial differential equations (6), (3) and (9) or (15) are reformulated in a weak sense in space and then this semi-discrete system is discretized in time by the finite difference scheme with the equidistant time step $\Delta t = \frac{T}{N}, N >> 1$.

3.1 Elastic body

First, equation (6) is multiplied by a test function ψ and integrated over the whole domain Ω^s. The application of the Green theorem and Hooke law (7) leads to the classical weak form of elastic problem

$$\left(\rho^s \frac{\partial^2 \mathbf{u}}{\partial t^2}, \psi\right)_{\Omega^s} + (\lambda^s (\mathrm{div}\, \mathbf{u}) \mathbb{I} + 2\mu^s \mathbf{e}^s(\mathbf{u}), \mathbf{e}^s(\psi))_{\Omega^s} = = (\mathbf{f}^s, \psi)_{\Omega^s} + (\mathbf{q}^s, \psi)_{\Gamma^s_{\mathrm{Neu}}}, \tag{17}$$

which needs to be fulfilled by all $\psi \in \mathbf{V}$. We use the notation $(\cdot,\cdot)_{\mathcal{D}}$ for the scalar product of the Lebesque spaces $L^2(\mathcal{D})$ or $\mathbf{L}^2(\mathcal{D})$. The space $\mathbf{V} = V \times V$, where

$$V = \{\phi \in W^{1,2}(\Omega^s) | \phi = 0 \text{ on } \Gamma^s_{\mathrm{Dir}}\}, \tag{18}$$

and $W^{1,2}(\Omega)$ is the Sobolev space, see [14]. The approximative solution $\mathbf{u}_h$ is sought in a finite element (FE) space, where it can be expressed as

$$\mathbf{u}_h(x,t) = \sum_{j=1}^{N_h} \alpha_j(t) \psi_j(x), \tag{19}$$

where $\psi_1, \ldots, \psi_{N_h}$ are basis functions of FE space. Substituting (19) into (17) and setting $\psi = \psi_i$ gives the system of ordinary differential equations of second order for the unknown coeffients $\alpha_j(t)$

$$\mathbb{M}^T \ddot{\alpha} + \mathbb{C}^T \dot{\alpha} + \mathbb{K}^T \alpha = \mathbf{b}(t), \tag{20}$$

where the matrix $\mathbb{C} = \epsilon_1 \mathbb{M} + \epsilon_2 \mathbb{K}$ were added as the proportional damping with a proper choice of parameters ϵ_1, ϵ_2. The vector $\mathbf{b}(t)$ has components $b_i(t) = (\mathbf{f}^s, \psi_i)_{\Omega^s} + (\mathbf{q}^s, \psi_i)_{\Gamma^s_{\mathrm{Neu}}}$ and the elements of matrices $\mathbb{M} = (m_{ij}), \mathbb{K} = (k_{ij})$ are given by

$$m_{ij} = \left(\rho^s \psi_j, \psi_i\right)_{\Omega^s}, k_{ij} = \left(\lambda^s (\mathrm{div}\, \psi_j) \mathbb{I} + 2\mu^s \mathbf{e}^s(\psi_j), \mathbf{e}^s(\psi_i)\right)_{\Omega^s}. \tag{21}$$

System (18) is numerically discretized in time by the Newmark method, see for example [15] or [10].

3.2 Fluid flow

The discretization procedure of equation (3) is reverse – it is firstly discretized in time and after then in space. The ALE derivative is approximated by the BDF2 scheme

$$\frac{D^A \mathbf{v}}{Dt}(t_{n+1}) \approx \frac{3\mathbf{v}^{n+1} - 4\overline{\mathbf{v}}^n + \overline{\mathbf{v}}^{n-1}}{2\Delta t}, \tag{22}$$

where the notation $\overline{\mathbf{v}}^i(x) = \mathbf{v}^i(\tilde{x})$ for $\tilde{x} = A_{t_i}(A^{-1}_{t_{n+1}}(x))$, $i \in \{n-1, n\}$, $x \in \Omega^f_{n+1}$ for a fixed time t_{n+1} was used. In the rest of paper we omit the upper time index $^{n+1}$ and set $\Omega^f := \Omega^f_{t_{n+1}}$.

The similar steps as in the previous paragraph leads to the weak spatial formulation of equation (3) in time t_{n+1}. It can written as the abstract problem of seeking a function pair $V = (\mathbf{v}, p)$ such, that

$$a(V, \mathbf{\Phi}) = f(\mathbf{\Phi}), \tag{23}$$

is satisfied for any test function $\mathbf{\Phi} = (\varphi, q)$ from space $\mathbf{X} \times M$. The space $M = L^2(\Omega^f)$ and $\mathbf{X} = X \times X$ is given as

$$X = \{f \in W^{1,2}(\Omega^f) | f = 0 \text{ on } \Gamma^f_{\mathrm{Dir}} \cup \Gamma^f_{\mathrm{In}} \cup \Gamma^f_{\mathrm{W}_{t_{n+1}}}\} \subset W^{1,2}(\Omega^f) \tag{24}$$

The form $a(\cdot,\cdot)$ is defined by

$$a(V, \mathbf{\Phi}) = \left(\frac{3\mathbf{v}}{2\Delta t}, \varphi\right)_{\Omega^f} + \frac{1}{2}(((\mathbf{v} - 2\mathbf{w}^{n+1}_D) \cdot \nabla)\mathbf{v}, \varphi)_{\Omega^f} - \\ - \frac{1}{2}((\mathbf{v} \cdot \nabla)\varphi, \mathbf{v})_{\Omega^f} + \frac{1}{2}((\mathbf{v} \cdot \mathbf{n})^+ \mathbf{v}, \varphi)_{\Gamma^f_{\mathrm{Out}}} \\ + \nu^f (\nabla \mathbf{v}, \nabla \varphi)_{\Omega^f} - (p, \mathrm{div}\, \varphi)_{\Omega^f} + (q, \mathrm{div}\, \mathbf{v})_{\Omega^f} \tag{25}$$

and functional $f(\mathbf{\Phi}) = \left(\frac{4\overline{\mathbf{v}}^n - \overline{\mathbf{v}}^{n-1}}{2\Delta t}, \varphi\right)_{\Omega^f}$. The nonlinear system of equations (23) is solved by fixed point iteration.

This formulation was further stabilized by the streamline-upwind/Petrov-Galerkin (SUPG) together with 'div-div' stabilization and the pressure-stabilization/Petrov-Galerkin (PSPG) in order to effectively decrease/reduce numerical instabilities associated with high Reynolds numbers. For details see [16] or [10].

3.3 Coupling algorithm of FSI problem

The advantage of hybrid approach is that FSI problem can be computed independently on the later acoustic simulation. Only the fluid pressure and velocities are needed to be saved for subsequent computing of acoustic sources given by (29) or (30). Then the relevant wave equation is solved.

The algorithm solving FSI problem is implemented in the form of strong coupling, see e.g. [11].

3.4 Aeroacoustics

In the case of Lighthill analogy, equation (10) is after multiplication by test function η integrated over the whole acoustic domain Ω^a and the application of the Green theorem together with boundary condition (14) yields

$$\left(\frac{1}{c_0^2} \frac{\partial^2 p'}{\partial t^2}, \eta\right)_{\Omega^a} + (\nabla p', \nabla \eta)_{\Omega^a} = -((\mathrm{div}\, \mathbf{T}), \nabla \eta)_{\Omega^a}, \tag{26}$$

Finally, if we restrict us to to finite element space and the solution is sought in form of $p'_h(t,x) = \sum_{j=1}^{N_{h2}} \gamma_j(t) \eta_j(x) \in V_h \subset W^{1,2}(\Omega^a)$, it leads to

$$\mathbb{M}^a \ddot{\gamma} + \mathbb{K}^a \gamma = \mathbf{b}(t), \tag{27}$$

where elements of matrices $\mathbb{M}^a = (m_{ij}^a), \mathbb{K}^a = (k_{ij}^a)$ are given as

$$m_{ij}^a = \left(\frac{1}{c_0^2}\eta_i, \eta_j\right)_{\Omega^a}, \quad k_{ij}^a = \left(\frac{\partial \eta_i}{\partial x_m}, \frac{\partial \eta_j}{\partial x_m}\right)_{\Omega^a}, \tag{28}$$

and components of vector $\mathbf{b}(t)$ are

$$b_i^{LH} = -\left(\frac{\partial T_{jm}}{\partial x_m}, \frac{\partial \eta_i}{\partial x_j}\right)_{\Omega^a}, \tag{29}$$

with upper index ^{LH} notifying the Lighthill case.

Repeating the same steps in the case of APE (15) ends with the form as (27), only the right hand side vector $\mathbf{b}(t) := \mathbf{b}^{APE}(t) = (b_i^{APE})$ is different

$$b_i^{APE} = -\left(\frac{1}{\rho_0^f c_0^2}\left(\frac{\partial p_{ic}}{\partial t} + \bar{\mathbf{v}} \cdot \nabla p_{ic}\right), \eta_i\right)_{\Omega^a}. \tag{30}$$

In both cases system (27) is discretized in time by the Newmark method.

4 NUMERICAL RESULTS

In this section are listed the numerical results of acoustic field generated by flow around an airfoil or in a channel.

4.1 Numerical results of the flow around an airfoil

For the flow over an airfoil only the fluid-acoustic coupling was considered and the motion of the structural part was omitted in order to simplify the model. The results are shown for NACA0012 profile under angle of attack $\alpha = 5°$ with inlet velocity $\mathbf{u}_{\text{In}} = 10.0\,\text{m/s}$ and related Reynold number $\text{Re} = 5000$. Time step was chosen as $\Delta t = 10^{-6}\,\text{s}$. This setting excites the regular separation of flow at the trailing edge as in the case of the von Kármán vortex street as it was expected. The flow separation produces vorticity eddies in the wake behind profile, see Fig. 3.

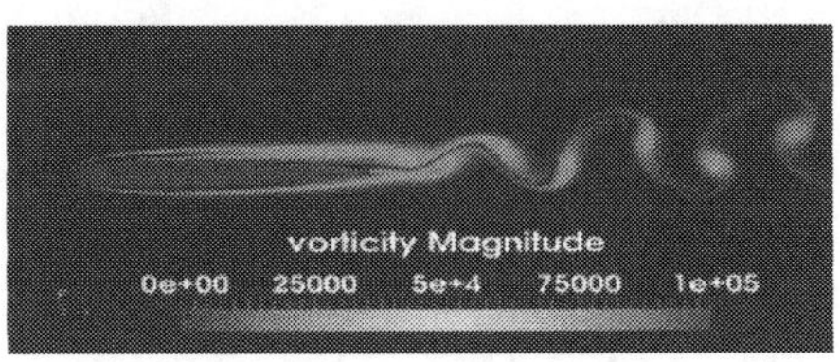

Fig. 3. The characteristic distribution of vorticity.

The sound sources given by (10) and (15) are plotted in Fig. 4. The sound sources in the Lighthill case correspond well with the plotted vorticity, meanwhile the sound sources of APE analogy are closely connected with the flow of pressure depression in the center of each vortex rather than with high velocity gradient at the vinicity of the leading edge.

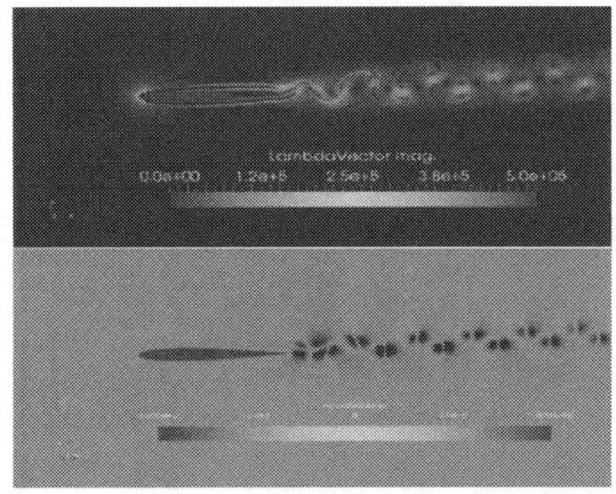

Fig. 4. Top: The distribution of the Lighthill tensor magnitude. **Down:** The sound source density in the case of APE analogy.

4.2 Numerical results of FSAI

The FSI results were achieved by in-house FEM program with implemented P1-bubble/P1 elements, acoustics were computed by academic solver CFS++ from TU Vienna. For the FSI as well as for the transient acoustic simulation the time step was chosen as $2.5 \cdot 10^{-5}\,\text{s}$. The four layered vocal fold (VF) model was the same as in [16], but the geometry of channel behind glottis was changed, see schematic figure 5. The fluid flow was driven by the kinematic pressure difference between inlet and outlet $\Delta p = 300\,\text{Pa} \cdot \text{m}^3/\text{kg}$. The airflow excited model to vibrate two dominant frequencies equal to first two eigenfrequencies obtained by the modal analysis.

Then the sound sources were computed by Lighthill analogy or by APE, see figure 6. The structure of the sources are very similar, however the resulting magnitudes of sources are quite different (figures are normalized).

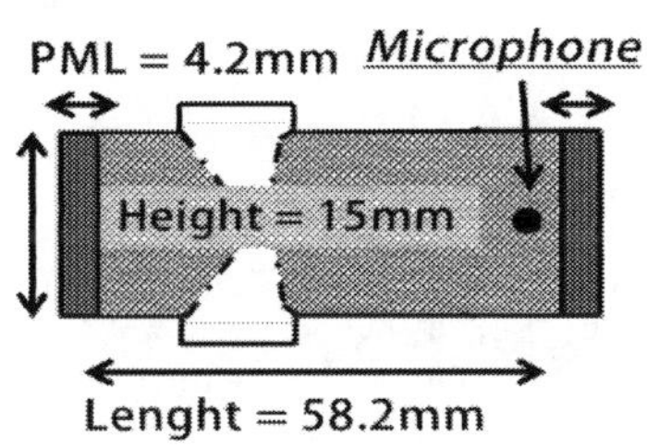

Fig. 5. Dimensions of the vocal fold model with PML blocks at inlet and outlet.

The magnitude difference can be seen on the results of acoustic simulation. The results of Lighthill analogy have overestimated the low frequencies under 600 Hz, the higher frequencies already correspond relatively well to APE, see Fig. 7. The sound at the microphone position could be expected to have the distinct peaks equal to the dominant frequencies of VF motion and higher harmonics, what does not have to coincide with the basic frequencies of a specific vowel created by filtering in a vocal tract, see [3]. However our results do not confirm such hypothesis, but the reason can be caused by 2D FSI simulation with overestimation of induced vorticity behind the glottis.

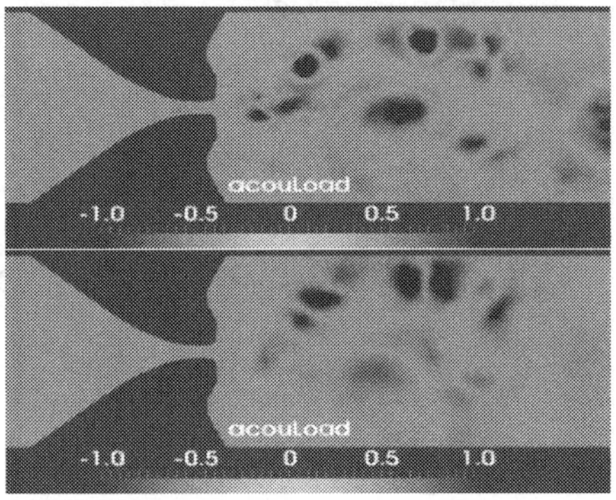

Fig. 6. The sound sources projected onto acoustic grid and normalized to 1 at time 0.24 s. **Top:** Lighthill analogy. **Bottom:** APE analogy.

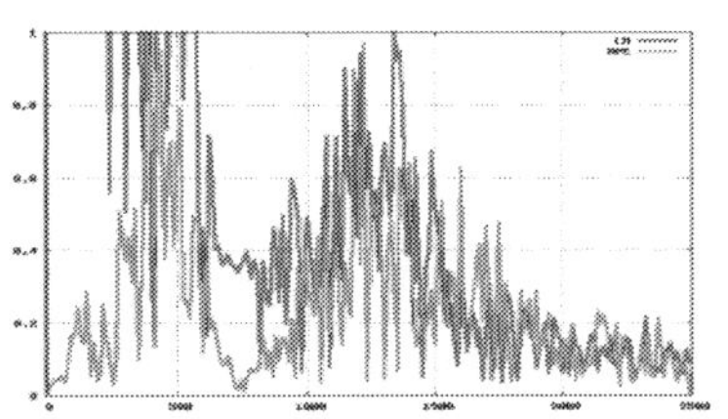

Fig. 7. The frequency spectra of sound pressure level at the microphone position. Lighthill results were normalized to correspond with APE at frequency 1258 Hz.

5 CONCLUSION

This contribution presents the mathematical description of the FSAI, where the hybrid partitioned approach was applied. The Lighthill acoustic analogy and APE analogy was used for the determination of the acoustic sources and computation of their propagation through surrounding acoustic domain.

The all considered physical problems are numerically solved by the FEM. The fluid solver was stabilized by SUPG, PSPG and 'div-div' method. The sound sources were computed and interpolated on coarser acoustic grid. PML layer was added to the acoustic domain to a account for free radiation. In the end induced VF vibration by fluid flow was computed with the help of in-house developed program and then the results of sound propagation were obtain by CFS++ solver. The results showed the general applicability of presented methodology for problems of low Mach number flows.

Acknowledgements. Authors acknowledge support from the EU Operational Programme Research, Development and Education, and from the Center of Advanced Aerospace Technology (CZ.02.1.01/0.0/0.0/16_019/0000826), Faculty of Mechanical Engineering, Czech Technical University in Prague.

References

1. M. De Gennaro, H. Kühnelt, M. Kaltenbacher, in *18th AIAA/CEAS Aeroacoustics Conference (33rd AIAA Aeroacoustics Conference)* (2012), p. 2135
2. M. Kaltenbacher, *Numerical simulation of mechatronic sensors and actuators: finite elements for computational multiphysics* (Springer, 2015)
3. G. Link, M. Kaltenbacher, M. Breuer, M. Döllinger, Computer Methods in Applied Mechanics and Engineering **198**, 3321 (2009)
4. M.J. Lighthill, in *Proceedings of the Royal Society of London*, vol. 211 (The Royal Society, 1952), vol. 211, pp. 564–587
5. J. Hardin, D. Pope, Theoretical and Computational Fluid Dynamics **6**(5), 323 (1994)
6. R. Ewert, W. Schröder, Journal of Computational Physics **188**(2), 365 (2003)
7. M. Kaltenbacher, M. Escobar, S. Becker, I. Ali, International Journal for Numerical Methods in Fluids **63**(9), 1103 (2010)
8. B. Kaltenbacher, M. Kaltenbacher, I. Sim, Journal of Computational Physics **235**, 407 (2013)
9. T. Gelhard, G. Lube, M.A. Olshanskii, J.H. Starcke, Journal of Computational and Applied Mathematics **177**(2), 243 (2005).
10. M. Feistauer, P. Sváček, J. Horáček, in *Fluid-structure Interaction and Biomedical Applications*, ed. by T. Bodnár, G.P. Galdi, S. Nečasová (Birkhauser, 2014), pp. 312–393
11. J. Valášek, P. Sváček, J. Horáček, in *EFM15 – Experimental Fluid Mechanics 2015*, vol. 114, ed. by V. M., D. P. (EPJ Web of Conferences, 2016), vol. 114.
12. M. Braack, P.B. Mucha, Journal of Computational Mathematics **32**, 507 (2014)
13. A. Hüppe, Spectral finite elements for acoustic field computation. Ph.D. thesis, Alpen-Adria-Universität Klagenfurt (2012)
14. R.A. Adams, *Sobolev Spaces* (Academic Press, New York, 1975)
15. A. Curnier, *Computational methods in solid mechanics* (Springer Netherlands, 1994)
16. J. Valášek, M. Kaltenbacher, P. Sváček, in *Topical problems of fluid mechanics 2017*, ed. by D. Šimurda, T. Bodnár (Institute of Thermomechanics, AS CR, v.v.i., 2017), pp. 303–312.

Description of saturation curves and boiling process of dry air

Magda Vestfálová[1,*], *Markéta* Petříková[1], *Martin* Šimko[1]

[1]KEZ, TU Liberec, Studentská 1402/2, 461 17 Liberec 1, Czech Republic

Abstract. Air is a mixture of gases forming the gas wrap of Earth. It is formed by dry air, moisture and other pollutants. Dry air is a substance whose thermodynamic properties in gaseous state, as well as the thermodynamic properties of its main constituents in gaseous state, are generally known and described in detail in the literature. The liquid air is a bluish liquid and is industrially used to produce oxygen, nitrogen, argon and helium by distillation. The transition between the gaseous and liquid state (the condensation process, resp. boiling process), is usually displayed in the basic thermodynamic diagrams using the saturation curves. The saturation curves of all pure substances are of a similar shape. However, since the dry air is a mixture, the shapes of its saturation curves are modified relative to the shapes corresponding to the pure substances. This paper deals with the description of the dry air saturation curves as a mixture, i.e. with a description of the process of phase change of dry air (boiling process). The dry air saturation curves are constructed in the basic thermodynamic charts based on the values obtained from the literature. On the basis of diagrams, data appearing in various publications are interpreted and put into context with boiling process of dry air.

1 Dry air composition

Dry air is a mixture of gases, most of which are nitrogen N_2 and oxygen O_2. The composition of clean dry air does not change up to about $95\,km$ above sea level and corresponds to the values in Table 1 according to the International Standard Atmosphere [1].

Table 1. Composition of clean dry air [1].

gas	content of volume [%]
Nitrogen N_2	78.084
Oxygen O_2	20.9476
Argon Ar	0.934
Carbon dioxide CO_2	0.0314 *
Neon Ne	0.001818
Helium He	0.000524
Krypton Kr	0.000114
Xenon Xe	0.0000087
Hydrogen H_2	0.0005
Nitrogen monoxide N_2O	0.00005 *
Methane CH_4	0.0002
Ozone O_3 in summer in winter	up to 0.000007 * up to 0.000002 *
Sulphur dioxide SO_2	up to 0.0001 *
Nitrogen Dioxide NO_2	up to 0.000002 *
Iodine I_2 I2	up to 0.000001 *
Air	100

(The content of the gas which is marked by the asterisk may undergo significant variations from time to time or from place to place.)

Table 2. Composition of clean dry air [2].

gas	content of volume [%]
Nitrogen N_2	78.12
Oxygen O_2	20.96
Argon Ar	0.92
Air	100

The dry air considered in [2] contains no carbon dioxide CO_2 or other trace elements. The composition of clean dry air considered in [2] corresponds to the values in Table 2.

According to [2], errors due to measurement are greater than differences due to deviations in the air composition under consideration.

2 Boiling of the components of dry air

If we focus on the phase change between the liquid and gas phases, the parameters (i.e. pressure, temperature and volume or density) of the critical point cr and triple point tr are given for O_2, N_2 and Argon Ar in [2] – see Table 3 and Table 4. The critical point is the state with the highest possible pressure and at the same time the highest possible temperature at which boiling of the pure substance may occur. The triple point is conversely the state with the lowest possible pressure and at the

* Corresponding author: magda.vestfalova@tul.cz

same time the lowest possible temperature at which boiling of the pure substance may occur.

Table 3. Parameters of the critical point cr of dry air components [2].

gas	critical point temperature t_{cr} [°C]	critical point pressure p_{cr} [MPa]	specific volume of critical point v_{cr} [$m^3 \cdot kg^{-1}$]
Nitrogen N_2	−146.958	3.3958	0.0031918
Oxygen O_2	−118.555 °C	5.046	0.0022928
Argon Ar	−122.463	4.863	0.0018671

Table 4. Parameters of the triple point tr of dry air components [2].

gas	triple point temperature t_{tr} [°C]	triple point pressure p_{tr} [Pa]	specific volume of triple point v_{tr} [$m^3 \cdot kg^{-1}$]
Nitrogen N_2	−209.999	12523	1.4830536
Oxygen O_2	−218.789	146.33	96.543627
Argon Ar	−189.3442	68891	0.246626

The liquid-vapor saturation curve in the equilibrium $p-t$ diagram, which passes through the boiling points, begins in a triple point tr and ends at a critical point cr. For nitrogen N_2 and oxygen O_2, it is possible to find boiling parameters in [3], i.e. for each temperature it is possible to find the pressure at which boiling occurs. For argon Ar, the required data was searched in [4]. From the above data, a saturation curve in the equilibrium phase $p-t$ diagram can be constructed for each of the dry air components – see Figure 1.

During the change phase at constant pressure of the pure substance, i.e. nitrogen N_2, oxygen O_2 and argon Ar, the temperature does not change, but the volume increases. In [3], for nitrogen N_2 and O_2 for each temperature, it is possible to find not only pressure but also density (resp. the specific volume) at the beginning and at the end of boiling process. From these data, for each of the components, a curve connecting the saturated liquid states (liquid saturation curve) and a curve connecting the saturated vapor state (vapor saturation curve) can be constructed in the $p-v$ diagram, see Figure 2. The critical point for each of these curves lies at the "highest" point of the corresponding saturation curve, i.e. at a point corresponding to the maximum pressure at which the substance can boil. The diagram also shows the boiling process for both gases for several different temperatures (from $-210\,°C$ to $-120\,°C$).

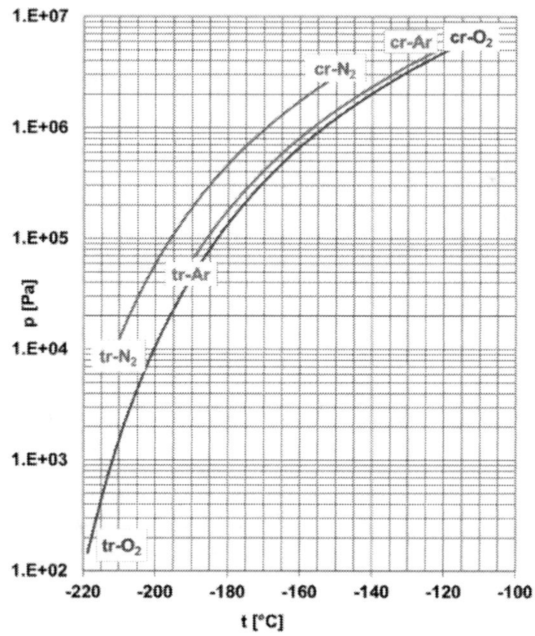

Fig. 1. Liquid-vapor saturation curve in the phase $p-t$ diagram for the three most important components of dry air: nitrogen N_2 - green, oxygen O_2 - blue, argon Ar - red; cr - critical point, tr - triple point.

3 Phase change of dry air

Boiling of mixture has a different course from the boiling of the pure component [5], [6]. Let us assume the binary mixture only (we only consider nitrogen N_2 and oxygen O_2). By isobaric heating of the binary liquid mixture at a certain temperature we reach the state in which the mixture begins to boil (point of starting the boiling). The point of starting the boiling at the given pressure is uniquely determined by the boiling point of the components at the corresponding pressure and by the proportional composition of the individual components in the mixture. During the boiling of the liquid phase, the more volatile component of the mixture – in our case nitrogen N_2 – evaporates more intensively. This means that the emerging gas phase will be "richer" about the nitrogen N_2, on the other hand the remaining liquid phase will be "poorer" about nitrogen N_2. Thus, the composition of the liquid phase varies – in the liquid phase there is a higher oxygen ratio than it was in the beginning of boiling. If the boiling should continue, the temperature must gradually increase – i.e. the boiling (phase change) of the mixture at constant pressure does not proceed under a constant temperature but with a gradual increase of temperature! At a certain temperature, all the liquid phase is converted to the gas phase – the boiling is finished. The boiling end temperature is higher than the starting boiling temperature at a given pressure, and is again uniquely determined by the boiling point of the individual components at the corresponding pressure and at the relative proportion of the individual components in the mixture.

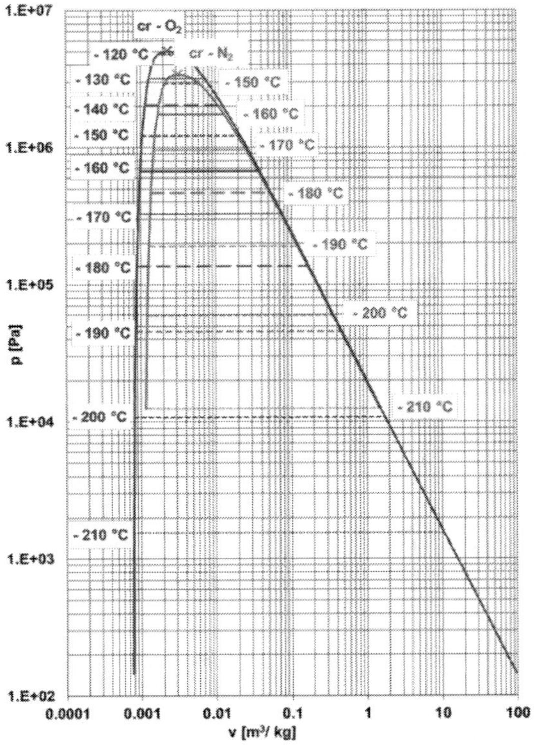

Fig. 2. Liquid and vapor saturation curves in $p-v$ diagram for dry air components: nitrogen N_2 - green, oxygen O_2 - blue; cr -critical point.

In the equilibrium phase $p-t$ diagram, the boiling of the mixture of given composition is not represented by a single curve (in which points the boiling would run at constant pressure and temperature) but by a loop, on the left branch of which are the starting boiling points and on the right branch of which are the boiling end points for the mixture of this composition. For dry air, it is possible to find the start and end melting states parameters in [2]. Figure 3 shows an equilibrium phase $p-t$ diagram with a plotted saturation curve of oxygen O_2, nitrogen N_2 and dry air, i.e. a mixture of oxygen O_2 and nitrogen N_2 containing $23\,\%$ by volume of oxygen O_2. For mixtures of oxygen O_2 and nitrogen N_2 of another composition, the loops in the equilibrium phase $p-t$ diagram would have a different shape.

If we construct an envelope of saturation loops for different compositions, we would obtain a curve of so-called pseudocritical states, originating from the oxygen O_2 critical point and ending at critical point nitrogen N_2 (the dotted line in Figure 3). The pseudocritical state, unlike the critical state, is not defined by the highest possible pressure of boiling and the highest possible temperature of boiling (see Table 5). The pseudocritical state is defined as envelope curve of the saturation loops.

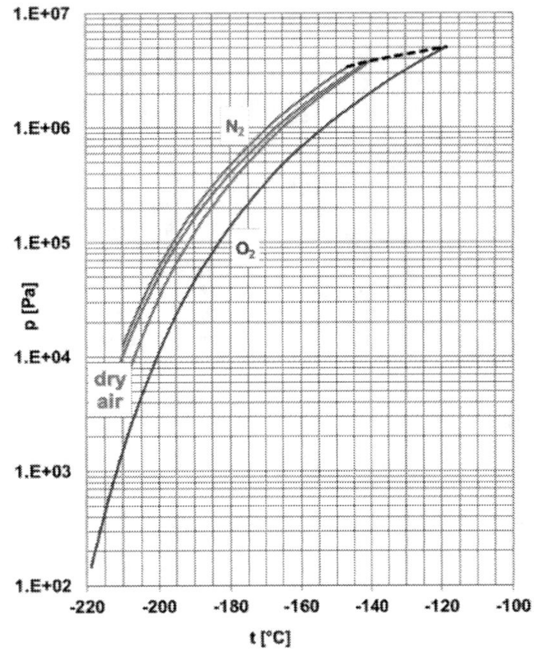

Fig. 3. Liquid-vapor saturation curve in the phase $p-t$ diagram for the dry air (red loop), nitrogen N_2 (green curve) and oxygen O_2 (blue curve); pseudocritical states of the mixture of nitrogen N_2 and oxygen O_2 - black dotted curve.

Looking at the top of the dry air curve (see Figure 4), we can see that in pseudocritical condition the pressure is lower than the maximum pressure at the saturation loop, i.e. if the boiling of air at the pseudocritical pressure $p = 3.7860\,MPa$ would run, then at the beginning of boiling, the air temperature (at that point is the air in the whole volume in liquid phase) would correspond to the pseudocritical temperature $t = -140.6194\,°C$, but the temperature would increase during boiling. The boiling could however take place at a higher pressure: the maximum pressure, at which the elementary boiling process (phase change) could take place in the air, is $p = 3.7891\,MPa$. The corresponding temperature would be $t = -140.5465\,°C$ (i.e. higher than pseudocritical). This state is called maxcondenbar state.

Looking at the top of the dry air curve (see Figure 4), we can also see that in pseudocritical condition the temperature is lower than the maximum temperature at the saturation loop. When we observe a process of phase change during isothermal expansion, so the maximum temperature, at which the elementary phase change could take place in the air, is $t = -140.5188\,°C$. The corresponding pressure would be $p = 3.78502\,MPa$

https://doi.org/10.1051/epjconf/201818002114

(i.e. lower than pseudocritical). This state is called maxcondentherm state.

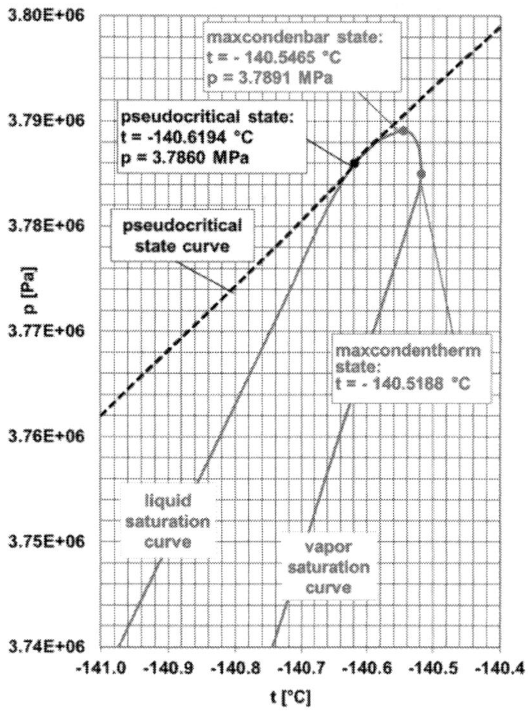

Fig. 4. Liquid and vapor saturation curves in p-v diagram for dry air components: nitrogen N_2 - green, oxygen O_2 - blue.

Table 5. Parameters of the pseudocritical point of air [2].

point	temperature $[°C]$	pressure $[MPa]$	specific volume $v_m\,[m^3 \cdot kg^{-1}]$
"Critical"	−140.6194	3.7860	0.0029188
Maxcondentherm	−140.5188	3.78502	0.0033052
Maxcondenbar	−140.5465	3.7891	0.0031125

The difference in the course of the phase change of the mixture from the phase change of the pure substance can also be seen in the $p-v$ diagram: during the phase change (for example during boiling) at constant pressure, volume of mixture increases but temperature of the mixture increases too. Conversely, during the phase change (for example from a liquid phase to a gaseous phase) under constant temperature conditions, volume of mixture increases but pressure of the mixture decreases. In Figure 5, the saturation curves of the beginning and end of the boiling process are plotted in the $p-v$ diagram for dry air (and for comparison for nitrogen N_2 and oxygen O_2 too), including the network of isotherms.

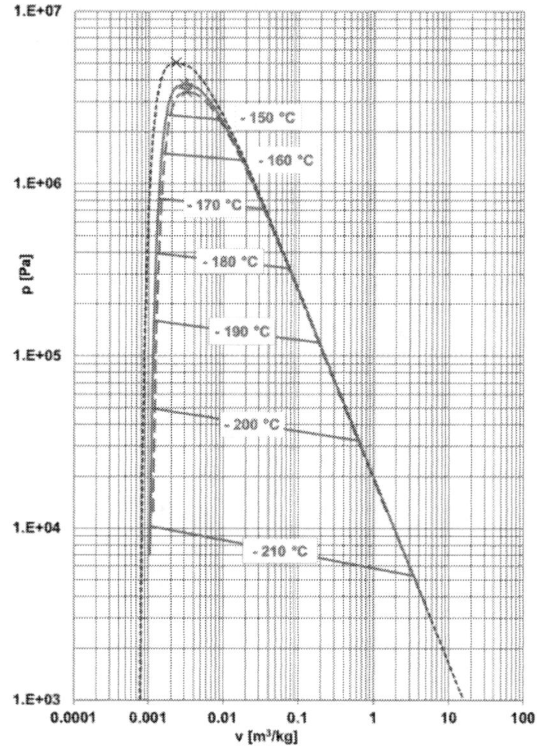

Fig. 5. Liquid and vapor saturation curves in the phase $p-t$ diagram for the dry air.

4 Conclusion

Dry air is a mixture of many gases. Except for nitrogen N_2 and oxygen O_2, gases are present in the air in a small amount and their effect on the thermodynamic behavior of the air can be neglected. Thus, from a thermodynamic point of view, dry air can be considered as a binary mixture. Changing the phase of the mixture takes place (at a constant pressure) while changing the temperature. Therefore, in the equilibrium phase $p-t$ diagram of dry air, the saturation curve passes in the saturation surface bounded by saturation loop. On its lower branch are the points of starting of phase change and on the upper branch there are the points of the ending of phase change. At the liquid-vapor saturation curve of dry air, there is an interesting area near "critical" point with three important points: a pseudocritical point, a point corresponding to the maximum possible condensing temperature – maxcondentherm point, and a point corresponding to the maximum possible condensing pressure – maxcondenbar point.

Acknowledgement

This publication was written at the Technical University of Liberec as part of the project SGS 21135 with the support of the Specific University Research Grant, as provided by the Ministry of Education, Youth and Sports of the Czech Republic in the year 2017.

References

1. *International Standard Atmosphere,* ISO 2533:1975, International Organization for Standardization (1975)
2. E, W. Lemmon, R. T. Jacobsen, S. G. Penoncello, D. G. Friend, Journal of Physical and Chemical Reference Data, **29**, 3: 331-385 (2000)
3. *VDI Wärmeatlas,* Springer (2006)
4. J. Šesták, J. Bukovský, M. Houška, *Thermal processes - transport and thermodynamic data,* CTU Publishing House (in Czech) (1993)
5. E. Hála, J. Pick, V. Fried, O. Vilím, *Equilibrium between liquid and vapour,* Foreign Literature Publishing House (in Russian) (1962)
6. W. J. Moore, *Physical Chemistry,* State publishing house of technical literature (in Czech) (1972)

Determination of the applicability limits of the ideal gas model for the calculation of moist air properties

Magda Vestfálová[1,*], and *Pavel* Šafařík[2]

[1]KEZ, TU Liberec, Studentská 1402/2, 461 17 Liberec 1, Czech Republic
[2]CTU in Prague, FME, Department of Fluid Dynamics and Thermodynamics, Technická 4, 166 07 Praha 6, Czech Republic

Abstract. The submitted paper deals with the finding of such moist air states in which the components of moist air and hence the humid air itself can be described by the ideal gas model while maintaining a predefined accuracy. Both components of moist air (dry air and water vapor) can be described by a model of ideal gas at sufficiently low pressures and sufficiently high temperatures. In the paper, we are looking for such combinations of pressures and temperatures for both components, where the relative deviation in the density calculation using the ideal gas model does not exceed the desired value. In addition, on the basis of the mixture theory, such moist air conditions (characterized by pressure, temperature and specific humidity) are searched, on which the accuracy of the calculation meets the required conditions. Subsequently, diagrams are constructed that can be used to help identify the interface between a moist air area that can be described by a simple ideal gas model, and areas where it is necessary to use a more accurate model for one of the components.

1 Introduction

Moist air is a mixture of dry air and water. To simplify the physical description of moist air that occurs in nature or participates in technical processes, only these components can be considered. The presence of other substances in the air (such as various gaseous pollutants, liquid components of different technologies, solids dust) will not be considered.

If all the humidity in the moist air is in gaseous state, the moist air is a homogeneous mixture. The moist air as a heterogeneous mixture contains water partly in the gas state and partly in the liquid state or partly in the gas state and partly in the solid state or partly in the gas state, partly in the liquid state and partly in the solid state.

Humidity is present in relatively small quantities in atmospheric air but is of considerable importance both in the environment and in technological processes too.

In the case of homogeneous moist air, the properties of its components are very often described by the ideal gas model. Therefore, the aim of the present paper is to determine the applicability range of this model when the ideal gas is defined [1] as a theoretical model of compressible fluid, which is characterized by the relationship between the quantities of pressure $p\,[Pa]$, temperature $T\,[K]$ and density $\rho\,[kg \cdot m^{3}]$

$$p = \rho \cdot r \cdot T \,, \tag{1}$$

* Corresponding author: magda.vestfalova@tul.cz

where $r\,[J \cdot kg^{-1} \cdot K]$ is the specific gas constant, and by constant values of specific heat capacities at constant pressure $c_p\,[J \cdot kg^{-1} \cdot K] = const$. and at constant volume $c_v\,[J \cdot kg^{-1} \cdot K] = const$ ..

2 Components of moist air

Dry air

The basic physical and thermodynamic properties of dry air as an ideal gas are, according to NIST (National Institute for Standards and Technology, USA) [2]:

middle molar mass:

$$M = 28.958\,3 \frac{kg}{kmol}$$

specific gas constant:

$$r = 287.12 \frac{J}{kg \cdot K}$$

specific heat capacity at constant pressure:

$$c_p = 1\,005.9 \frac{J}{kg \cdot K}$$

specific heat capacity at constant volume:

$$c_v = 718.8 \frac{J}{kg \cdot K}$$

Poisson constant:

$$\kappa = 1.399$$

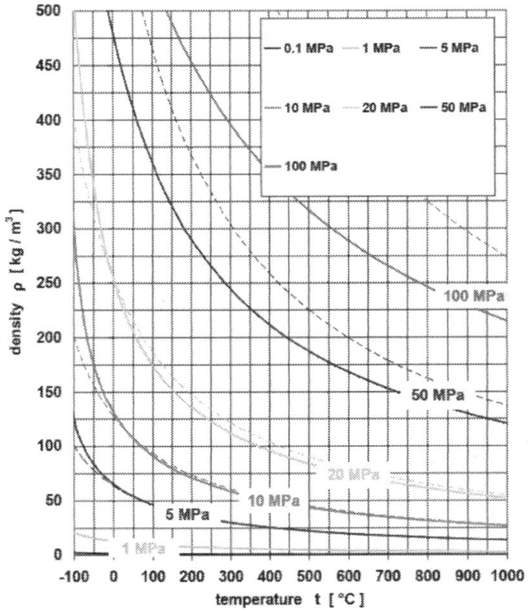

Fig. 1. Dependence of dry air density on pressures of 0.1 MPa (black), 1 MPa (yellow), 5 MPa (blue), 10 MPa (purple), 20 MPa (green), 50 MPa (brown) and 100 MPa (red) according to [2] (solid lines) and dependence of dry air density obtained with the ideal gas model (dashed lines).

The theoretical model of ideal gas is very advantageous for its simplicity, and is usually used to calculate the dry air properties near the atmospheric conditions. For conditions significantly different from standard atmospheric, the ideal gas model exhibits large deviations from the actual values. Figure 1 shows the actual temperature-dependence of density for selected constant pressure values (solid lines) and for a comparison of the same dependences obtained with the ideal gas model (dashed lines).

It can be seen that the ideal gas model is useful for calculating the density of dry air at low pressures and high temperatures. With increasing pressure and decreasing temperature, the differences between actual density and density calculated using the ideal gas model increase.

Figure 2 shows relative dry air density deviations calculated using the ideal gas model for the same pressures as were used in Figure 1. The area of possible use of the ideal gas model depends on the required accuracy: if we would consider the accuracy of 10 % to be sufficient, it is possible to use the ideal gas model to calculate the dry air density up to pressures of 20 MPa (for temperatures higher than − 50 °C) (for pressures of 50 MPa and 100 MPa , the ideal gas model in the entire temperature range of − 150 °C to 1 000 °C gives deviations greater than 10 %). However, if we require an accuracy of up to 1 % , it would be possible to use the ideal gas model to calculate the dry air density only within a certain range of pressures and temperatures. For example, for a pressure of 0.1 MPa , the ideal gas model is usable for temperatures above about − 140 °C , for a pressure of 1 MPa is usable for temperatures above about − 25 °C , for a pressure of 5 MPa in a temperature range of 25 °C to 105 °C , for a pressure of 10 MPa only in a temperature range of 25 °C to 75 °C .

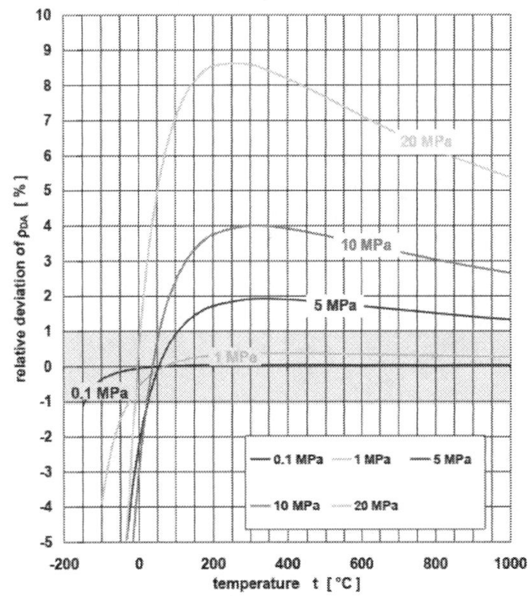

Fig. 2. The course of relative deviations of dry air densities as an ideal gas in dependence on temperature for pressures of 0.1 MPa (black), 1 MPa (yellow), 5 MPa (blue), 10 MPa (purple), 20 MPa (green), 50 MPa (brown) a 100 MPa (red).

Water is a relatively complex substance from a thermodynamic point of view. It is the most widely used and best available energy transfer medium. Therefore, extensive research has focused on it. For the work of experts, documents are gradually approved by the International Association for Properties of Water Steam (IAPWS) as recommendations for determination of thermodynamic and thermophysical properties of water and water vapour. Important documents are:

- 1995 (revision 2014) – Formulation of the thermodynamic properties of a common water substance for general and scientific use (in short IAPWS-95),
- 1997 – New international formulation of thermodynamic properties of water and steam for industrial purposes (in short IAPWS IF-97).

The state equations of water and water vapour according to these two documents are substantially distant from the state equation of the ideal gas.

The basic physical properties of water and water vapour are according to [3]:

molar mass: $M = 18.015 \ \frac{kg}{kmol}$

specific gas constant: $r = 461.526 \ \frac{J}{kg \cdot K}$

triple point:

temperature of triple point:

$$t_{tr} = 0.01 \ °C = 273.16 \ K$$

pressure of triple point:

$$p_{tr} = 611.657 \ Pa$$

critical point:

critical temperature:

$$t_{cr} = 373.946 \ °C = 647.096 \ K$$

critical pressure:

$$p_{cr} = 22.064 \ MPa$$

The thermodynamic behaviour of water occurring in homogeneous moist air (i.e. water vapour) can be described under normal atmospheric conditions by the ideal gas model.

The basic thermodynamic properties of water vapour as an ideal gas:

specific heat capacity at constant pressure:

$$c_p = 1898.56 \ \frac{J}{kg \cdot K}$$

specific heat capacity at constant volume:

$$c_V = 1398.56 \ \frac{J}{kg \cdot K}$$

Poisson constant:

$$\kappa = 1.33$$

For thermodynamics of humid air, it is important to determine the thermodynamic properties of water vapour within the limits of saturation.

On the liquid-steam saturation limit from the triple point to the critical point, the scientific formulation (IAPWS 95) [4] determines the dependence of the saturated water vapour pressure on the temperature function

$$p'' = p_{cr} \cdot \exp\left[\frac{1}{1-\tau} \cdot \left(a_1 \cdot \tau + a_2 \cdot \tau^{1.5} + a_3 \cdot \tau^3 + a_4 \cdot \tau^{3.5} + a_5 \cdot \tau^4 + a_6 \cdot \tau^{7.5}\right)\right], \quad (2)$$

where:

$$\tau = 1 - \frac{T}{T_{cr}} \quad (3)$$

is the temperature function,

$T \ [K]$ is the temperature,

the constants in equation (2) are given in Table 1.

Table 1. Constants in equation (2).

a_1	−7.859 517 83	a_4	22.680 741 1
a_2	1.844 082 59	a_5	−15.961 871 9
a_3	−11.786 649 7	a_6	1.801 225 02

On the saturation limit between the solid and gaseous phases H_2O, i.e. for sublimation and desublimation, the pressure in relation to the temperature from $50 \ K$ to the triple point is solved according to equation [5]

$$p'' = p_{tr} \cdot \exp\left(\theta^{-1} \cdot \sum_{i=1}^{3} a_i \cdot \theta^{b_i}\right), \quad (4)$$

where:

$$\theta = \frac{T}{T_{tr}}, \quad (5)$$

is the temperature function,

$T \ [K]$ is the temperature,

the constants in equation (4) are given in Table 2.

Table 2. Constants in equation (4).

a_1	−21.214 400 6	b_1	0.003 333 333 33
a_2	27.320 381 9	b_2	1.206 666 67
a_3	−6.105 981 3	b_3	1.703 333 33

Similarly to dry air, it is advantageous to describe the properties of water steam be the ideal gas model. The scope of application of the ideal gas model is limited by the beginning of condensation, it is by achieving a saturation state (a condition where moist air becomes a heterogeneous mixture, it is when water begins excreted from moist air in the form of a liquid or solid phase), and next by the range of pressures and temperatures where the steam is a homogeneous mixture, but the ideal gas model has large deviations from actual data. Figure 3 shows the actual dependence of density of water vapor at the temperature for selected constant pressure values (full line) and, for comparison, the same dependence obtained using the ideal gas model (dashed lines).

Similarly to dry air, at lower pressures, the ideal gas model produces results comparable to real values, even up to the saturation curve. At higher pressures, the densities calculated using the ideal gas model deviate from the actual properties near the saturation curve, and for very high pressures, the ideal gas model does not produce the right results even at high temperatures.

Figure 4 shows relative water vapor density deviations calculated using the ideal gas model for the pressures used in Figure 3. As with dry air, the area of possible use of the ideal gas model depends on the required accuracy: if we would consider the accuracy of $10 \ \%$ to be sufficient, it is possible to use the ideal gas model to calculate the water vapor density up to pressure of $2 \ MPa$ and it up to the saturation curve, or even up to pressure of $10 \ MPa$ for temperatures over $500 \ °C$. However, if we require an accuracy of up to 1%, it would be possible to use the ideal gas model to calculate the water vapor density only up to pressure about $50 \ kPa$ (an accuracy of up to 1% would be met up to the saturation vaporous state), or to pressure about $0.1 \ MPa$, but only for temperatures higher than about $135 \ °C$, or to pressure about $0.2 \ MPa$, but only for temperatures higher than about $205 \ °C$; at a pressure of $1 \ MPa$ the temperature would have to be above $450 \ °C$.

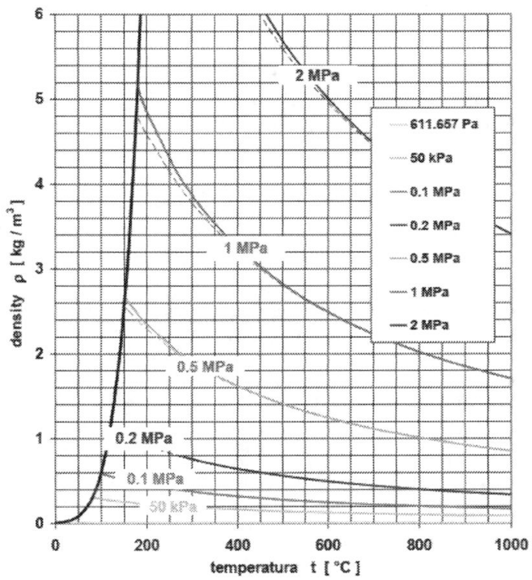

Fig. 3. The dependence of the density of water vapor at the temperature for pressures $611.657\ Pa$ (light blue), $50\ kPa$ (yellow), $0.1\ MPa$ (purple), $0.2\ MPa$ (blue), $0.5\ MPa$ (green), $1\ MPa$ (red) and $2\ MPa$ (brown) according to [3] (solid lines) and the dependence of water vapor density obtained with the ideal gas model (dashed lines).

3 Moist air

Moist air as a thermodynamic system is a mixture of dry air and water substances. The water contained in the moist air is always in the form of steam, but in addition (in the case of a heterogeneous mixture), part of the water in moist air may also be in a liquid or solid state.

Temperature of moist air $t_{MA}\,[°C]$ is equal to the temperature of its components

$$t_{MA} = t_{DA} = t_V = t\,. \tag{6}$$

Pressure of moist air $p_{MA}\,[Pa]$ is equal according to Dalton's law to the sum of the partial pressures of dry air and water vapor

$$p_{MA} = p_{DA} + p_V\,. \tag{7}$$

Mass of moist air $m_{MA}\,[kg]$ is equal

$$m_{MA} = m_{DA} + m_{H_2O}\,. \tag{8}$$

Density of moist air $\rho_{MA}\,[kg\cdot m^{-3}]$ is the ratio of the mass of moist air to its volume, it is provided that the water in the humid air is only as vapor

$$\rho_{MA} = \rho_{DA} + \rho_V\,. \tag{9}$$

Pressure of saturated water vapor $p_V''\,[Pa]$ is a significant variable for moist air, which determines the water content limit in the air for the existence of a homogeneous mixture:

- if $p_V < p_V''$, the moist air is unsaturated and the mixture homogeneous;
- if $p_V = p_V''$, the moist air is saturated and the mixture homogeneous;
- if $p_V > p_V''$, the moist air is supersaturated and the mixture is heterogeneous.

Pressure of saturated water vapor is a function of temperature

$$p_V'' = p_V''(t)\,, \tag{10}$$

(see equation (2), (4)).

Dew point temperature $t_{DP}\,[°C]$ is the moist air temperature at which the moist air reaches the saturation state under isobaric cooling

$$p_V(t) = p_V''(t_{DP})\,, \tag{11}$$

it is the water vapor contained therein is in the state of saturated vapor.

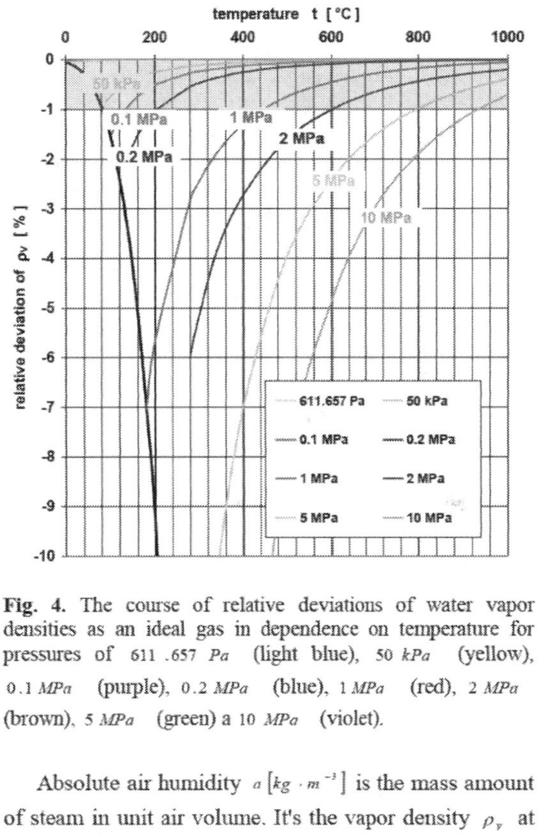

Fig. 4. The course of relative deviations of water vapor densities as an ideal gas in dependence on temperature for pressures of $611.657\ Pa$ (light blue), $50\ kPa$ (yellow), $0.1\ MPa$ (purple), $0.2\ MPa$ (blue), $1\ MPa$ (red), $2\ MPa$ (brown), $5\ MPa$ (green) a $10\ MPa$ (violet).

Absolute air humidity $a\,[kg\cdot m^{-3}]$ is the mass amount of steam in unit air volume. It's the vapor density ρ_V at vapor pressure p_V and temperature t. Absolute humidity is defined only for a homogeneous mixture and for given temperature can be within a range $\langle 0;\, a'' = \rho_V''(t)\rangle$.

Relative air humidity $\varphi\,[-]$ is the ratio of absolute humidity to absolute humidity of saturated humid air at a given temperature

$$\varphi = \frac{a}{a''}\,. \tag{12}$$

Relative humidity is defined only for a homogeneous mixture and may be within the range $\langle 0;1 \rangle$. Assuming water vapor contained in moist air can be described by the ideal gas model

$$\varphi = \frac{p_V}{p_V''} . \quad (13)$$

Specific air humidity $x\left[kg_V \cdot kg_{DA}^{-1}\right]$ is the ratio of the mass of water to the dry air mass

$$x = \frac{m_{H_2O}}{m_{DA}} . \quad (14)$$

In the case of a homogeneous mixture in which water is contained in the air only in the vapor form

$$x = \frac{m_V}{m_{DA}} = \frac{\rho_V}{\rho_{DA}} . \quad (15)$$

Specific gas constant of moist air $r_{MA}\left[J \cdot kg^{-1} \cdot K^{-1}\right]$ is equal

$$r_{MA} = \frac{1}{1+x} r_{DA} + \frac{x}{1+x} r_V \quad (16)$$

and specific heat capacity at constant pressure of moist air $c_{p\,MA}\left[J \cdot kg^{-1} \cdot K^{-1}\right]$ is equal

$$c_{p\,MA} = \frac{1}{1+x} c_{p\,DA} + \frac{x}{1+x} c_{p\,V} . \quad (17)$$

So that thermodynamic constant of moist air as and ideal gas depend on specific humidity x; at $x = const$. they are constant.

If the moist air forms a homogeneous mixture and its components (dry air and steam) can be described by the ideal gas model, it is possible to derive the relationship between specific and relative humidity

$$x = 0.622 \cdot \frac{p_V}{p_{MA} - p_V} , \quad (18)$$

resp.

$$x = 0.622 \cdot \frac{\varphi\, p_V''}{p_{MA} - \varphi\, p_V''} . \quad (19)$$

For the specific humidity of the saturated homogeneous mixture $(\varphi = 1)$

$$x'' = 0.622 \cdot \frac{p_V''(t)}{p_{MA} - p_V''(t)} = x''(t) . \quad (20)$$

Our task is to find a boundaries in moist air (combination of pressure, temperature and humidity), where there is possible to use the ideal gas model to calculate the properties of components of the moist air with sufficient accuracy.

One limit is the boundary on which ends the existence of water in the form of steam only. The paper [6] deals by searching for this boundary: For a given value of specific humidity of moist air, the dependence of saturated pressure of moist air on the temperature was solved from equations (20) and (2), respectively at the case of sublimation limit may be (4). In the equilibrium $p-T$ diagram, a curve (with a parameter of the specific humidity x) was constructed, on which water would be excreted from moist air (the curves of the saturated moist air for different water contents, it is for different values of specific humidity x), see Figure 5.

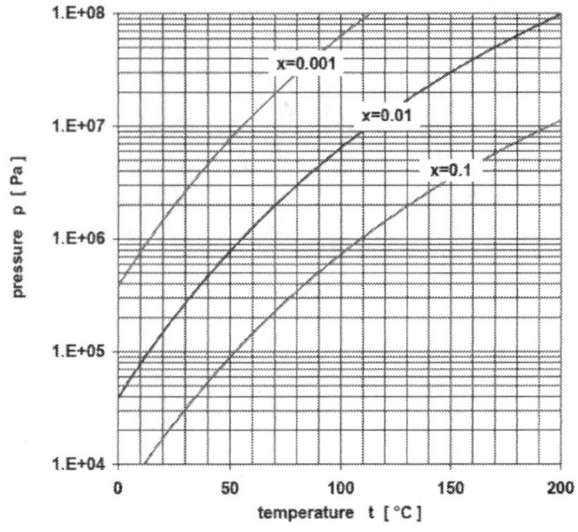

Fig. 5. Curves of saturated moist air for different values of specific humidity x [6].

The aim of this paper is to find additional limits (it means the areas of states of the homogeneous moist air), where both components can be described by the ideal gas model with the required accuracy.

The curves in Figure 5 show the states of saturated (it means homogeneous) moist air for three different values of specific humidity x. The vapor contained in the saturated moist air is in the form of saturated vapor. However, according to Figure 4, the ideal gas model used for vapor exhibits increasing deviations, when the pressure increases. So, for higher pressures, on the curves in the diagram in Figure 5 there are although the homogeneous states of moist air, but if we want to keep the necessary accuracy, it is no longer possible to describe the water vapor contained in the moist air by the ideal gas model!

For the desired maximum deviation, it is possible to find in the diagram in Figure 4 the dependence of partial pressure of vapor on temperature. Then, for the selected specific humidity x, we calculate from the relation (18) the maximum pressure of the humid air, at which the density of the water vapor contained in the moist air at a given temperature and at given specific humidity can be calculated using the ideal gas model (when the required accuracy is achieved). The obtained curve ends in the $p-T$ diagram on the saturated moist air curve corresponding to the given specific humidity x. In Figures 6, 7 and 8, these curves are plotted with a dashed line of the same color as the saturated moist air curves for the corresponding value of specific humidity x. Thus, if the state of moist air of a given specific humidity in the $p-T$ diagram is to the right of the saturated moist air curve and at the same time to the right of the corresponding dashed curve, (the boundary of the area of the given accuracy), it is possible to calculate the density of vapor contained in the moist air according to the ideal gas model, guaranteeing the required accuracy; between the two curves is the area where the moist air is still homogeneous, but for calculating the vapor properties it is necessary to use one of the more precise models; to the left of the saturated moist air curves, the humid air is already heterogeneous.

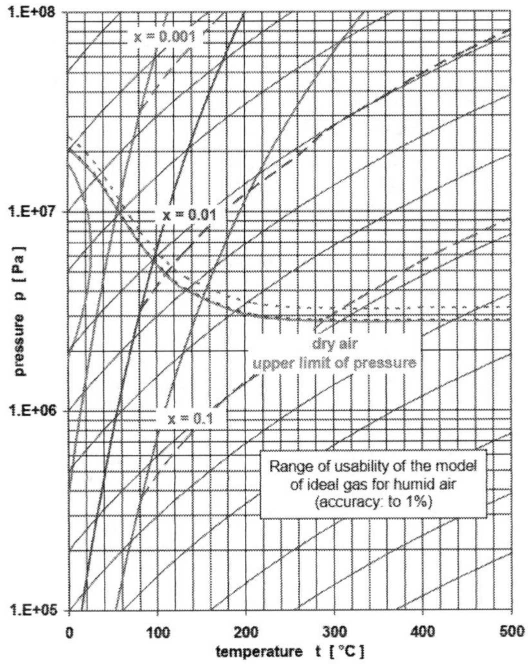

Fig. 6. Area of usability of the ideal gas model for the calculation of properties of components of moist air at the required accuracy of the calculation to 1% for specific humidity x of 0.1, 0.01 and 0.001.

Other limit, which have to be found, is the field of application of the ideal gas model for dry air contained in the moist air. As in the case of water vapor, we can find in the diagram in Figure 2 the dependence of dry air partial pressure on the temperature for the desired maximum relative deviation. By adjusting the equation (18) the relationship between the pressure of moist air, specific humidity and partial pressure of dry air can be obtained

$$p_{MA} = p_{DA}\left(1 + \frac{x}{0.622}\right), \tag{21}$$

from which we can calculate the maximum moist air pressure for the selected specific humidity x in which the density of the dry air contained in the moist air at a given temperature and specific humidity can be calculated using the ideal gas model to meet the required accuracy.

In Figure 6, for the selected values of specific humidity x is plotted line (dotted curve) of maximum pressure of moist air at which it is possible to calculate the partial dry air density using the ideal gas model, while meeting the required accuracy. It can be seen from the Figure 6 that the upper limit of pressure of moist air depends very little on the specific humidity x and the curves almost coincide with the dry air partial pressure curve (for the maximum humidity used by us: $x = 0.1$, the total pressure is equal to 1.16 times the dry air partial pressure; for humidity $x = 0.01$ is equal to 1.016 times and for humidity $x = 0.001$ it's just 1.0016 times). Thus, in the diagram in Figure 6, the dependence of the maximum dry air partial pressure on the temperature to fulfill the required accuracy is plotted. This curve is only a little different from the curves describing the total pressure of moist air for the selected values of specific humidity x. In the other figures, therefore, only the dependence of dry air partial pressure is plotted for expression of boundary of usability of model of ideal gas for dry air.

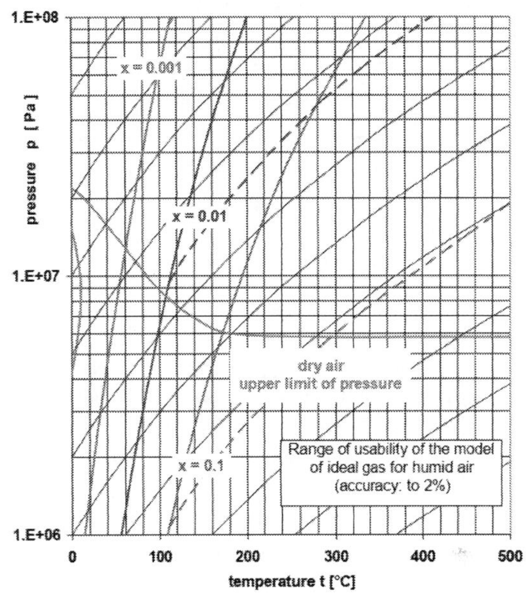

Fig. 7. Area of usability of the ideal gas model for the calculation of properties of components of moist air at the required accuracy of the calculation to 2% for specific humidity x of 0.1, 0.01 and 0.001.

Thus, if the state of moist air is in the $p-T$ diagram shown by a point below the maximum dry air partial pressure curve, the partial density of the dry air (contained in the moist air) may be calculated with sufficient precision using the ideal gas model.

The area where the densities of the both components of moist air can be calculated using the equation state of the ideal gas lies in the $p-T$ diagrams "bottom right": it is bounded by a saturated moist air curve for the corresponding specific humidity x, by a curve showing the maximum value of the air pressure (at given temperature and specific humidity x), when it is possible to model water vapor with the ideal gas model, and by a curve showing the maximum pressure value (at given temperature) when it is possible to model dry air with the ideal gas model.

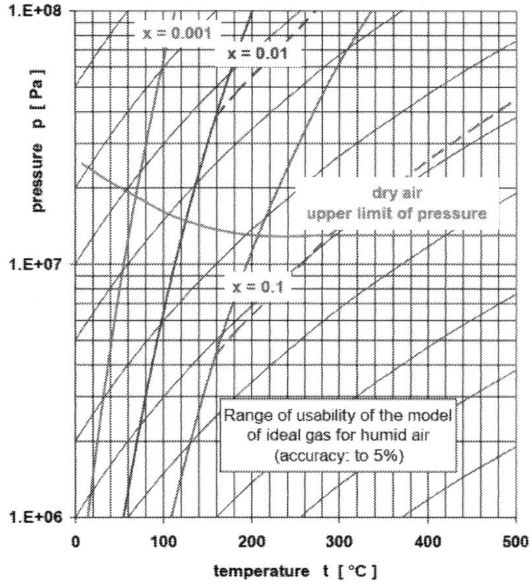

Fig. 8. Area of usability of the ideal gas model for the calculation of properties of components of moist air at the required accuracy of the calculation to $5\,\%$ for specific humidity of 0.1, 0.01 and 0.001.

Figure 6 shows an area in which the ideal gas model achieves an accuracy of $1\,\%$, in Figure 7 an area with accuracy of $2\,\%$, and in Figure 8 an area with accuracy of $5\,\%$.

In all of the figures, an isentropy network (a set of brown curves) is plotted for better idea:

1) If the reversible adiabatic compression occurs in homogeneous moist air, the moist air will "move away" from the saturation curve, it means: it is not possible to get from homogeneous moist air to heterogeneous moist air during isentropic compression, on the other hand, in a reversible adiabatic expansion, it is possible to reach a state of saturated moist air.

2) If the reversible adiabatic compression occurs in homogeneous moist air, the accuracy of the description of the water vapor properties by the ideal gas model will be increased ("we move away" from boundaries limiting the usability of the ideal gas model for steam water), but the accuracy of the calculation of the dry air properties by the ideal gas model will be reduced; in reversible adiabatic compression it is possible to cross the boundary where the ideal gas model is still usable for calculating the dry air properties.

3) In reversible adiabatic expansion, the accuracy of the ideal gas model for the dry air contained in moist air increases; the accuracy of the ideal gas model for the water vapor contained in moist air reduces, but not too significantly - the curve limiting the pressure of moist air for using the ideal gas model for the water vapor has only a slightly greater slope than the isentropic curves; The main "danger" in reversible adiabatic expansion in moist air is the possibility of reaching the condensation boundary.

Conclusion

In the equilibrium $p - T$ diagram, the boundary of range of usability of the ideal gas model for the calculation of properties of components of moist air were constructed for different relative accuracy values ($1\,\%$, $2\,\%$, $5\,\%$). Once the boundary is reached, a more accurate model is required to use (for calculation of properties of the components of moist air) for preserve the accuracy of calculation.

Acknowledgment

This publication was written at the Technical University of Liberec, Faculty of Mechanical Engineering with the support of the Institutional Endowment for the Long Term Conceptual Development of Research Institutes, as provided by the Ministry of Education, Youth and Sports of the Czech Republic in the year 2017.

The second author, P.Šafařík, expresses thanks for support from the Project No. CZ.2.16/3.100/21569 Centre 3D Volumetric Anemometry.

References

1. P. Šafařík, M. Vestfálová, *Thermodynamics of Moist Air*, CTU Publishing House (in Czech) (2016)
2. E.W. Lemmon, R.T. Jacobsen, S.G. Penoncello, D.G. Friend, Journal of Physical and Chemical Reference Data, **29**, 3: 331-385 (2000)
3. O. Šifner, J. Klomfar, *International Standards of Thermophysical Properties of Water and Steam*, Academia (in Czech) (1996)
4. IAPWS: Revised Release on the IAPWS Formulation 1995 for the Thermodynamic Properties of Ordinary Water Substance for General and Scientific Use (2014)
5. IAPWS: Revised Release on the Pressure along the Melting and Sublimation Curves of Ordinary Water Substance (2011)
6. M. Vestfálová, P. Šafařík, EPJ Web of Conferences **143**, 02139 (2016)

Numerical and experimental study of fluid flow in simplified blade cascade with prescribed harmonic motion

Jan Vimmr[1,*], *Ondřej* Bublík[1], *Aleš* Pecka[1], *Luděk* Pešek[2,**], and *Pavel* Procházka[2]

[1] *NTIS, Faculty of Applied Sciences, University of West Bohemia, Pilsen, Czech Republic*
[2] *Institute of Thermomechanics AS CR, v.v.i., Academy of Sciences of the Czech Republic, Prague, Czech Republic*

Abstract. In this paper, a numerical simulation and experimental measurements of compressible viscous fluid flow in simplified blade cascade are compared. The cascade consists of five flat plates three of which perform prescribed harmonic motion. The computed unsteady velocity field is compared with experimental measurements at selected points. Moreover, the power spectral density corresponding to the frequency of harmonic motion is computed for both the numerical and experimental data and compared. The numerical simulation was performed using the developed in-house CFD software based on the discontinuous Galerkin method, which offers high order of accuracy.

1 Introduction

The first papers focusing on the aero-elasticity of blades in axial turbines appeared in the forties of the 20th century [1], [2]. Here an empirical criterion for flutter occurrence was introduced. It is based on a reduced blade frequency for bending blade vibration. This criterion however neglects other influences, such as geometric configuration of blade cascade or angle of attack, which may also lead to flutter. Nowadays the problem of aero-elasticity is still a subject of topical interest because modern turbines are designed for higher efficiencies and higher power under higher operational temperatures and flow rates. Higher operational, safety and economical demands force the designers to be more precise during the process of determination of safe operational condition laying out of the area with loss of stability.

In the work [3] it has been shown that the critical flutter point, i.e. the highest aero-dynamical excitation, arises for a particular inter-blade phase angle. Most of aerodynamic studies are based on the so called travelling wave mode of vibration, where one assumes that all blades vibrate at the same frequency, amplitude and inter-blade phase angle. Other aerodynamic studies use the influence coefficient technique, which is based on the calculation of aerodynamic forces of vibrating blade acting on non-vibrating adjacent blades.

In our previous work [4] we used the travelling wave mode approach for the numerical modelling of the unsteady flow in the simplified blade cascade formed by three flat plates. We prescribed a harmonic motion for all three flat plates in the cascade such that their mutual movement produces a travelling wave. The flutter instability occurred for specific phase angles as expected.

In this study, we use the same concept of the travelling wave mode approach. In order for the configuration to correspond to a real blade disc better, we have added two fixed plates on the top and the bottom. The experiments have also been carried out for the same configuration. The numerical and experimental data were compared in both the steady case, where all of the plates are fixed in time, and for the transient case, where the three middle plates perform a harmonic motion and the two outermost blades are still.

The numerical simulations are based on the solution of the two-dimensional system of Favre-averaged Navier-Stokes equations, in dimensionless arbitrary Lagrangian-Eulerian (ALE) form. The computations were performed using the developed in-house CFD software Flow-Pro based on the discontinuous Galerkin finite element method.

*e-mail: jvimmr@kme.zcu.cz
**e-mail: pesek@it.cas.cz

2 Mathematical model

The two-dimensional problem of compressible viscous fluid flow through simplified blade cascade with prescribed harmonic motion is described by Favre-averaged system of Navier-Stokes equations, in dimensionless arbitrary Lagrangian-Eulerian (ALE) formulation written as

$$\frac{D^A\bar{\varrho}}{Dt}+\frac{\partial(\bar{\varrho}\tilde{u}_j)}{\partial x_j}-U_j\frac{\partial\bar{\varrho}}{\partial x_j}=0,$$

$$\frac{D^A}{Dt}(\bar{\varrho}\tilde{u}_i)+\frac{\partial}{\partial x_j}(\bar{\varrho}\tilde{u}_i\tilde{u}_j+\bar{p}\delta_{ij})-U_j\frac{\partial(\bar{\varrho}\tilde{u}_i)}{\partial x_j}=\frac{1}{Re}\frac{\partial}{\partial x_j}(\tilde{t}_{ij}+\tau_{ij}),$$

$$\frac{D^A}{Dt}(\bar{\varrho}\tilde{e})+\frac{\partial}{\partial x_j}(\bar{\varrho}\tilde{e}\tilde{u}_j+\bar{p}\tilde{u}_j)-U_j\frac{\partial(\bar{\varrho}\tilde{e})}{\partial x_j}=\frac{1}{Re}\frac{\partial}{\partial x_j}\left[(\tilde{t}_{ij}+\tau_{ij})\tilde{u}_i+\frac{\kappa}{\kappa-1}\left(\frac{\mu}{Pr}+\frac{\mu_t}{Pr_t}\right)\frac{\partial}{\partial x_j}\left(\frac{\bar{p}}{\bar{\varrho}}\right)\right],$$

where $i, j = 1, 2$, $\bar{\varrho}$ and $\bar{p}$ is the dimensionless time-averaged density and pressure, $\tilde{u}_i$ and $\tilde{e}$ is the dimensionless mass-averaged velocity component and energy, Re is the Reynolds number, Pr and Pr_t is the Prandtl number and the turbulent Prandtl number, μ and μ_t is the dynamic and turbulent viscosity and U_j is the j-th component of mesh velocity. For a flow around flat plates, the value of the turbulent Prandtl number is $Pr_t = 0.89$. Mass-averaged stress tensor and Reynolds stress tensor are given by

$$\tilde{t}_{ij}=2\mu\tilde{S}_{ij},\quad \tau_{ij}=2\mu_t\tilde{S}_{ij},$$

where

$$\tilde{S}_{ij}=\frac{1}{2}\left(\frac{\partial\tilde{u}_i}{\partial x_j}+\frac{\partial\tilde{u}_j}{\partial x_i}\right)-\frac{1}{3}\delta_{ij}\frac{\partial\tilde{u}_k}{\partial x_k}.$$

The equation of state for ideal gas is considered, i.e.

$$\bar{p}=(\kappa-1)\left[\bar{\varrho}\tilde{e}-\frac{1}{2}\bar{\varrho}\sum_{i=1}^{2}\tilde{u}_i^2\right].$$

Finally the symbol $\frac{D^A}{Dt}$ denotes the ALE derivative [5].

To include the influence of turbulent fluctuations on the mean flow a one-equation turbulence model of Spalart and Allmaras [6] is used. It is given by the transport equation for eddy viscosity $\tilde{\nu}$ written as

$$\frac{D^A(\bar{\varrho}\tilde{\nu})}{Dt}+\frac{\partial(\bar{\varrho}\tilde{\nu}\tilde{u}_j)}{\partial x_j}-U_j\frac{(\partial\bar{\varrho}\tilde{\nu})}{\partial x_j}=\frac{1}{Re}\frac{1}{\sigma}\left[\frac{\partial}{\partial x_j}\left((\mu+\bar{\varrho}\tilde{\nu})\frac{\partial\tilde{\nu}}{\partial x_j}\right)+\bar{\varrho}c_{b2}\frac{\partial\tilde{\nu}}{\partial x_j}\frac{\partial\tilde{\nu}}{\partial x_j}\right]+c_{b1}\bar{\varrho}\tilde{S}\tilde{\nu}-\frac{1}{Re}c_{w1}f_w\frac{1}{\bar{\varrho}}\left(\frac{\bar{\varrho}\tilde{\nu}}{D}\right)^2$$

and completed by the following relations

$$\mu_t=\bar{\varrho}\tilde{\nu}f_{v1},$$

$$\tilde{S}=\bar{\Omega}+\frac{1}{Re}\frac{\tilde{\nu}}{\kappa^2D^2}f_{v2},$$

$$f_w=g\left(\frac{1+c_{w3}^6}{g^6+c_{w3}^6}\right)^{\frac{1}{6}},$$

$$g=r+c_{w2}\left(r^6-r\right),$$

$$r=\min\left(\frac{\tilde{\nu}}{Re\,\tilde{S}\kappa^2D^2},10\right),$$

$$f_{v1}=\frac{\chi^3}{\chi^3+c_{v1}^3},\quad \chi=\frac{\tilde{\nu}}{\nu},$$

$$f_{v2}=1-\frac{\chi}{1+\chi f_{v1}},$$

where D is the distance to the nearest wall and $\bar{\Omega}$ is the vorticity magnitude. The constants for the turbulence model are $c_{b1} = 0.1355$, $c_{b2} = 0.622$, $\sigma = \frac{2}{3}$, $\kappa = 0.41$, $c_{w1} = \frac{c_{b1}}{\kappa^2} + \frac{1+c_{b2}}{\sigma}$, $c_{w2} = 0.3$, $c_{w3} = 2.0$, $c_{v1} = 7.1$.

The discretization of the mathematical model in ALE formulation using the discontinuous Galerkin method [7], [8] is described in previous works [4] and [9] in detail. Here, a simple algorithm for mesh deformation is also described. The mentioned mathematical model is incorporated in the developed inhouse numerical software called FlowPro. The software is validated on an inviscid compressible unsteady fluid flow around oscillating NACA0012 airfoil for a free stream Mach number $M_\infty = 0.755$ [4]. The instantaneous motion of the airfoil is given as a function of time

$$\alpha(t)=\alpha_m+\alpha_o\sin(\omega t),$$

which defines the angle of tilt of the airfoil. Here the mean angle $\alpha_m = 0.016°$ and the amplitude of oscillation $\alpha_o = 2.51°$. The angular velocity is defined as $\omega = 2ku_\infty/c$, where reduced frequency $k = 0.0814$, profile chord length $c = 1$ and u_∞ is the free-stream velocity. The coordinates of the reference point of rotation are $x_{ref} = [0.25, 0]$. Figure 1 shows a comparison of obtained numerical results (red) by the developed code with the results published in [10] and [11] (black), which are computed by finite volume solvers. A reasonably good agreement is apparent from both of the test problems, which confirms a correct implementation of the implicit DGFEM solver along with the ALE approach.

3 Experimental setup

The simplified blade cascade was chosen for the study of flutter in order to achieve a simple experimental setup. The experimental setup that was already constructed at the Institute of Thermomechanics AS CR is described in this section. The experimental stand of the simplified blade cascade is shown in Figure 2. The cascade consists of 5 flat plates numbered 2, 1, 0, -1 and -2 from top to bottom. Each plate is on one of its ends attached to the base by four spring steel belts. The outermost plates -2 and 2 are steady, thus have zero degrees of freedom and are present in order to simulate the periodic boundary condition. The three middle plates -1, 0 and 1 can move only in the horizontal

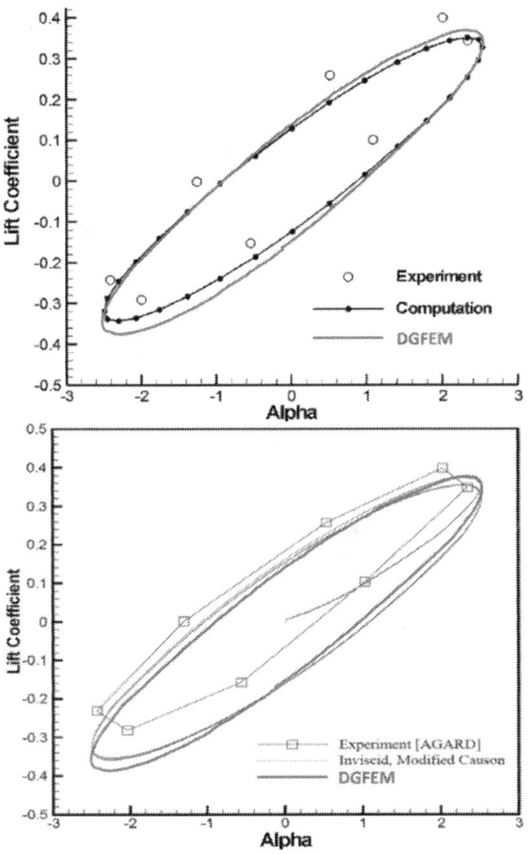

Fig. 1. Dependence of the lift coefficient c_L on the angle of the airfoil. Comparison with the numerical and experimental results published in [10] (top) and [11] (bottom).

direction, hence the simplified blade cascade has three degrees of freedom. The harmonic motion with prescribed amplitude, frequency and phase of the three middle plates is accomplished by electrodynamic exciters, which are attached to the base through two L-shaped beams and linked to the plates by a rod. A force transducer, which records inertial and elastic forces of the vibrating plate and aerodynamic forces of the fluid, is placed at the end of each rod. The movement of the plate is recorded by the foil strain gauges, which are calibrated for the displacement of the plates. Two strain gauges are glued onto both sides of each spring steel belt and they are connected to the semi-bridge circuit. The incidental angle of the flow and the cascade can be changed by rotating the whole base.

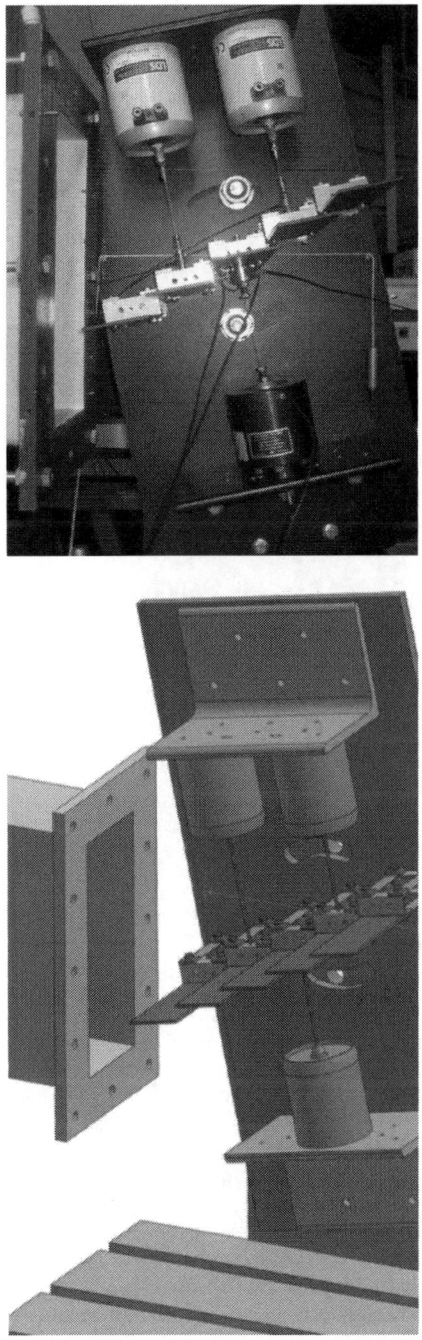

Fig. 2. Experimental stand of the simplified blade cascade (top) and its model created with CAD software (bottom).

4 Results

In this section, the experimental data are compared with the numerical results computed by the developed in-house CFD software FlowPro based on the discontinuous Galerkin method. Two following cases are considered. In the first case, the plates are fixed in time and the steady state is reached. The resulting velocity fields are compared. In the second case, the three middle flat plates perform kinematic harmonic motion. In both cases the same boundary condition are applied. The computation domain is shown in Figures 3 and 4. In order to obtain correct results, it turns out that it is necessary to include the inlet part of stand in the geometry as shown in Figure 3. The inlet velocity magnitude is set to $u = 40\,\mathrm{m\,s^{-1}}$, the angle of attack to $\alpha = 11.4°$ and inlet pressure to $p_i = 101325\,\mathrm{Pa}$. The computed and measured velocity fields for steady plates are shown in Figures 5 and 6, respectively. One can see that the velocity fields have similar character. The values of the velocity magnitude are approximately the same in both cases.

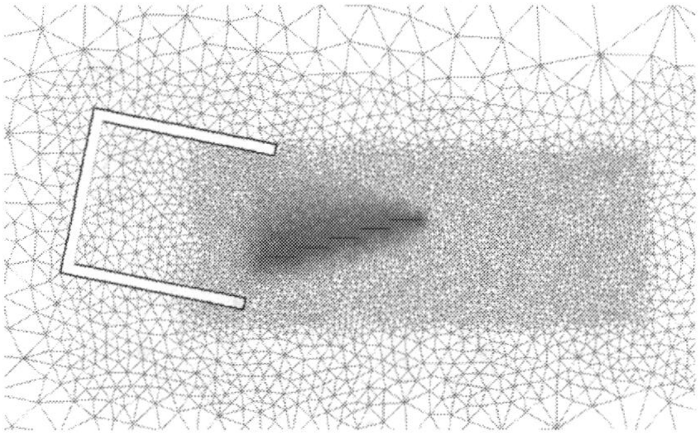

Fig. 3. Computational domain and triangular mesh.

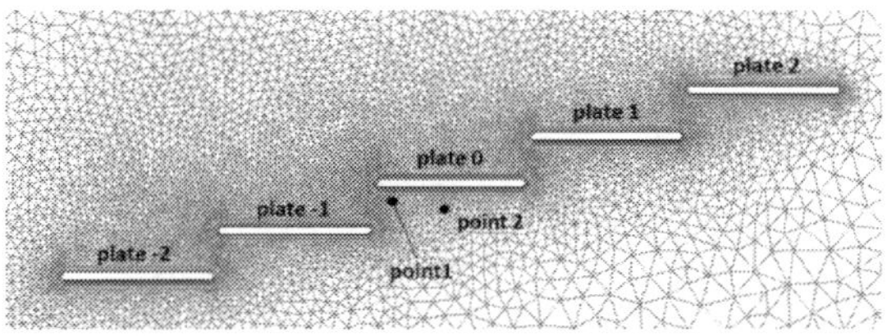

Fig. 4. Detail of computational mesh in the vicinity of simplified cascade.

In the second test case, the three middle plates perform kinematic harmonic motion in the y-direction $y = A\sin(2\pi ft + \phi)$ with an inter-blade phase angle $\phi = \pi/2$ for the upper plate (plate 1), $\phi = 0$ for the middle plate (plate 0) and $\phi = -\pi/2$ for the bottom plate (plate -1), see Figure 4. The amplitude and frequency of the harmonic motion is $A = 0.00283\,\mathrm{m}$ and $f = 40\,\mathrm{Hz}$, respectively. In order to compare the unsteady velocity fields two points between plates 0 and -1 were chosen as shown in Figure 4. One period of the oscillating velocity at point 1 and 2 is plotted in Figure 7 and Figure 8, respectively. A reasonable agreement between numerical and experimental data is apparent at point 1, on the other hand the difference in amplitude is significant at point 2. Finally, Figures 9 and 10 show power spectral density of y and x component of velocity corresponding to the frequency of 40 Hz obtained by the numerical simulation and Figures 12 and 13 also show power spectral density of y and x component of velocity this time obtained from the experiment. Note that the experimental data are not defined in front of the flat plates because of their shade with respect to the laser beams. In addition, the computed power spectral density of pressure is shown in Figure 11. A qualitative comparison of results reveal a good agreement.

Fig. 5. Computed steady velocity field

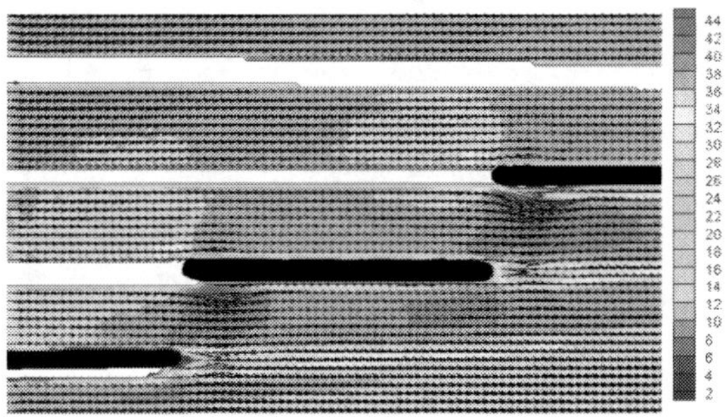

Fig. 6. PIV measured steady velocity field

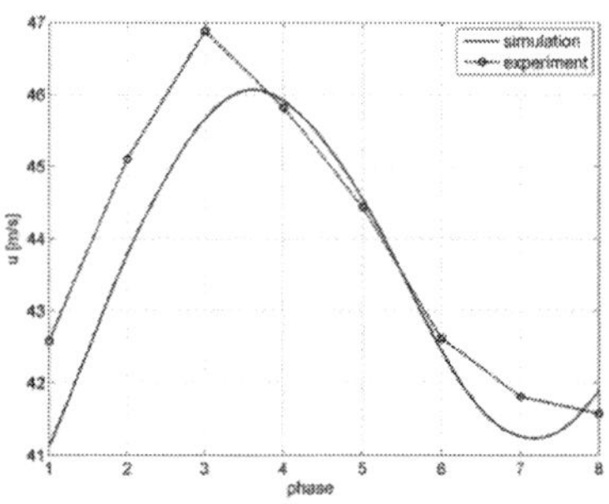

Fig. 7. Comparison of x component of velocity at point 1

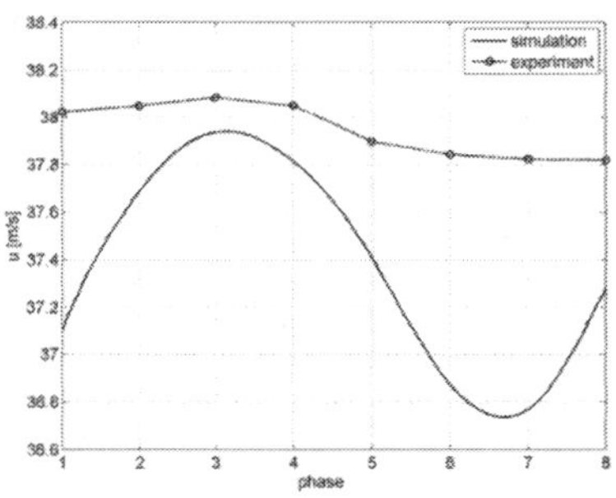

Fig. 8. Comparison of x component of velocity at point 2

5 Conclusion

The comparison between numerical and experimental data show a reasonably good agreement. Hence, the in-house

https://doi.org/10.1051/epjconf/201818002116

CFD solver based on the discontinuous Galerkin method is applicable for solution of blade cascade problems. The results of power spectral densities shows the higher values between the tip of the plate and the end of the bottom plate, see Figures 9, 11, 12, 13. This higher values represents aerodynamic coupling between plates related to oscillation frequency 40Hz. The use of higher oscillation frequency will probably lead to higher aerodynamic coupling between the plates. This assumption will be studied in future work.

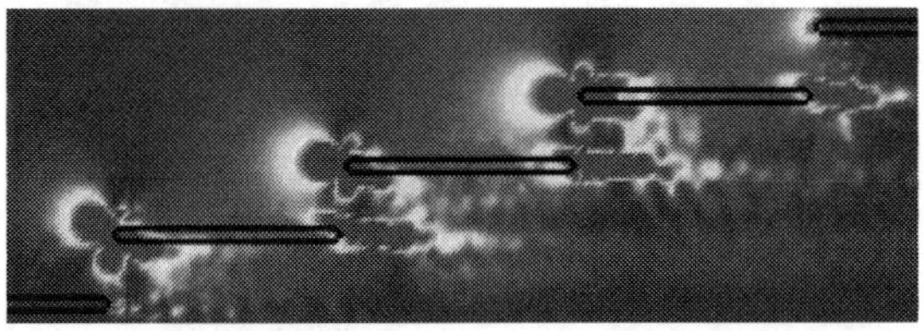

Fig. 9. Computed power spectral density of y-component of velocity

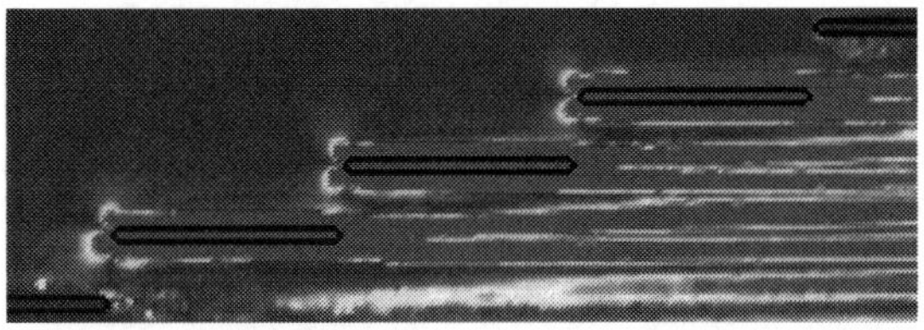

Fig. 10. Computed power spectral density of x-component of velocity

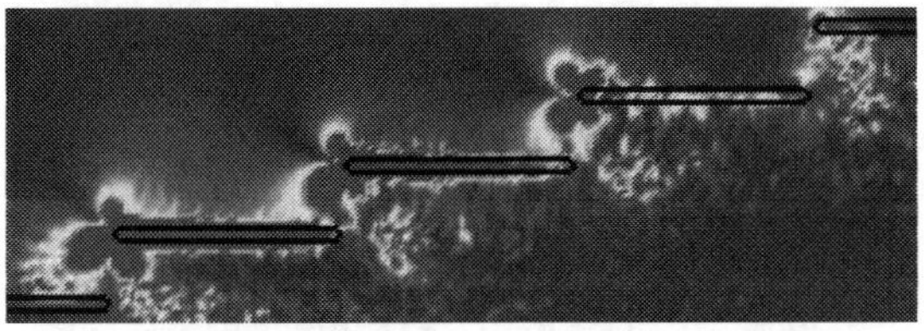

Fig. 11. Computed power spectral density of pressure

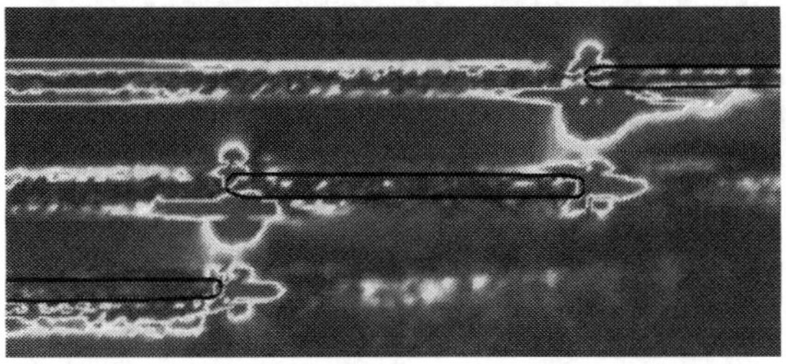

Fig. 12. Measured power spectral density of y-component of velocity

Acknowledgements

The authors appreciate the kind support by the grant GA 16-04546S "Aero-elastic couplings and dynamic behaviour of rotational periodic bodies" of the Czech Science Foundation and by the internal student grant projects SGS-2016-038 of the University of West Bohemia.

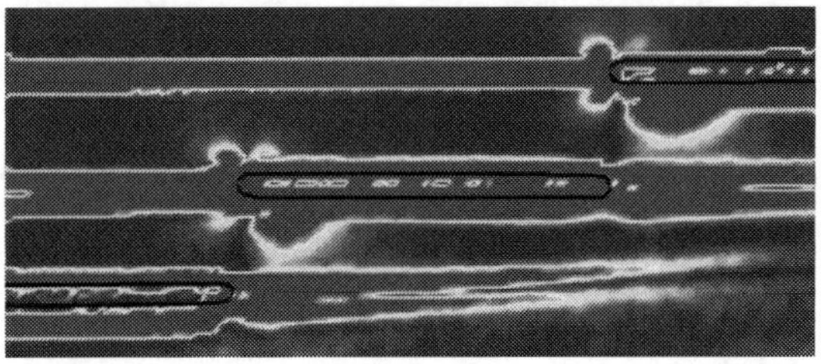

Fig. 13. Measured power spectral density of x-component of velocity

References

[1] H. M. Owens, W. E. Trumpler, ASME Paper, 49-A-64 (1949)

[2] E. K. Armstrong, R. E. Stevenson, J. R. Aero. Soc., **64** (591), 117-130 (1960)

[3] D. Schlafli, Experimentelle untersuchung der instationaren stromung in oscillierenden ringgittern. PhD Thesis No. 800, Lausanne (1989)

[4] J. Vimmr, O. Bublík, A. Pecka, L. Pešek, In Proceedings 6th ECCOMAS Thematic Conference on Computational Methods in Structural Dynamics and Earthquake Engineering (COMPDYN 2017), Rhodes Island, Greece, (2017)

[5] M. Feistauer, J. Česenek, NM&A, **60** (46), 1-13 (2011)

[6] P. R. Spalart, S. R. Allmaras, AIAA Paper, 92-439 (1992)

[7] W. H. Reed, T. R. Hill, Los Alamos Scientific Laboratory, Report No. LA-UR-73-479 (1973)

[8] B. Cockburn, C.-W. Shu, J. Sci. Comput., **16** (3), 173-261 (2001)

[9] O. Bublík, J. Vimmr, A. Jonášová, Appl. Math. Comput., **267**, 329-340 (2015)

[10] D. J. Kirshman, F. Liu, Comput. Fluids, **35** (6), 571-586, (2006)

[11] J. Fürst, K. Kozel, P. Furmánek, Kybernetika, **45** (4), 567-579 (2009)

Numerical simulation of fluid flow through simplified blade cascade with prescribed harmonic motion using discontinuous Galerkin method

Jan Vimmr[1,★], *Ondřej* Bublík[1], *Helena* Prausová[1], *Jindřich* Hála[2], and *Luděk* Pešek[2,★★]

[1] *NTIS, Faculty of Applied Sciences, University of West Bohemia, Pilsen, Czech Republic*
[2] *Institute of Thermomechanics AS CR, v.v.i., Academy of Sciences of the Czech Republic, Prague, Czech Republic*

Abstract. This paper deals with a numerical simulation of compressible viscous fluid flow around three flat plates with prescribed harmonic motion. This arrangement presents a simplified blade cascade with forward wave motion. The aim of this simulation is to determine the aerodynamic forces acting on the flat plates. The mathematical model describing this problem is formed by Favre-averaged system of Navier-Stokes equations in arbitrary Lagrangian-Eulerian (ALE) formulation completed by one-equation Spalart-Allmaras turbulence model. The simulation was performed using the developed in-house CFD software based on discontinuous Galerkin method, which offers high order of accuracy.

1 Introduction

In the forties of the 20th century, first papers focused on the aero-elasticity of blades in axial turbines appeared [1]. Empirical criterion on arising of flutter is introduced in the paper [2]. This criterion is based on a reduced blade frequency for bending blade vibration. If the reduced blade frequency is less than 0.33, the flutter is not excited. Nevertheless, other influences such as geometric configuration of blade cascade and angle of attack can also lead to flutter.

The problem of aero-elasticity is still subject of topical interest because modern turbines are designed for higher efficiencies and higher power under higher operational temperatures and flow rates. Higher operational, safety and economical demands force the designers to be more precise during the process of determination of safe operational condition laying out of the area with loss of stability. Originally, the term flutter was established for flow attached to the vibration of blade when phase displacement between blade motion and induced aerodynamic forces occurs. In dependence on time shift between the moving structure and unsteady aerodynamic forces acting on the structure, following cases can occur:

- the energy of the flow is absorbed into the blade,
- kinetic energy of the blade is transmitted to the flow,
- the energy of the flow stays unchanged during particular cycles of blade vibration.

If the blades add the energy to the flow, we talk about damped vibration from the aero-elasticity point of view. If the flow adds the energy to blades and their amplitudes increase, we talk about unstable excited vibration. In the work [3] it has been shown that the critical flutter point, i.e. the highest aero-dynamical excitation, arises by a particular inter-blade phase angle. Most of aerodynamic studies are based on the so called travelling wave mode of vibration, where one assumes that all blades vibrate with the same frequency, amplitude and inter-blade phase angle. Other aerodynamic studies use the influence coefficient technique which is based on the calculation of aerodynamic forces of vibrating blade acting on non-vibrating adjacent blades.

For purpose of this study, we choose the travelling wave mode approach for the numerical modelling of the unsteady flow and flutter analysis in the simplified blade cascade formed by three flat plates. Thus, all flat plates in the cascade vibrate and their mutual movement prescribes the travelling waves in the cascade. The aim of this numerical study is to determine aerodynamic forces acting on the flat plates in cascade, when the harmonic motion with phase delay is prescribed. The numerical simulation is performed using developed in-house CFD software based on the discontinuous Galerkin finite element method. In section 4 we present numerical results of unsteady fluid flow for the selected configuration of the simplified blade cascade described in [3]. For the numerical simulation one flow speed and forward travelling wave mode in the cascade were considered.

★e-mail: jvimmr@kme.zcu.cz
★★e-mail: pesek@it.cas.cz

2 Mathematical model

The two-dimensional problem of compressible viscous fluid flow through simplified blade cascade with prescribed harmonic motion is described by Favre-averaged system of Navier-Stokes equations, in dimensionless arbitrary Lagrangian-Eulerian (ALE) formulation written as

$$\frac{D^A\bar{\varrho}}{Dt}+\frac{\partial(\bar{\varrho}\tilde{u}_j)}{\partial x_j}-U_j\frac{\partial\bar{\varrho}}{\partial x_j}=0, \tag{1}$$

$$\frac{D^A}{Dt}(\bar{\varrho}\tilde{u}_i)+\frac{\partial}{\partial x_j}(\bar{\varrho}\tilde{u}_i\tilde{u}_j+\bar{p}\delta_{ij})-U_j\frac{\partial(\bar{\varrho}\tilde{u}_i)}{\partial x_j}=\frac{1}{Re}\frac{\partial}{\partial x_j}(\tilde{t}_{ij}+\tau_{ij}),$$

$$\frac{D^A}{Dt}(\bar{\varrho}\tilde{e})+\frac{\partial}{\partial x_j}(\bar{\varrho}\tilde{e}\tilde{u}_j+\bar{p}\tilde{u}_j)-U_j\frac{\partial(\bar{\varrho}\tilde{e})}{\partial x_j}=\frac{1}{Re}\frac{\partial}{\partial x_j}\left[(\tilde{t}_{ij}+\tau_{ij})\tilde{u}_i+\frac{\kappa}{\kappa-1}\left(\frac{\mu}{Pr}+\frac{\mu_t}{Pr_t}\right)\frac{\partial}{\partial x_j}\left(\frac{\bar{p}}{\bar{\varrho}}\right)\right],$$

where $i,j=1,2$, $\bar{\varrho}$ and $\bar{p}$ are dimensionless time-averaged values of density and pressure, $\tilde{u}_i$ and $\tilde{e}$ are dimensionless mass-averaged velocity components and energy, Re is Reynolds number, Pr is Prandtl number, $Pr_t=0.89$ is turbulent Prandtl number for flow around flat plate, μ and μ_t are dynamic and turbulent viscosity. U_j are components of mesh velocity. Mass-averaged stress tensor and Reynolds stress tensor are given by

$$\tilde{t}_{ij}=2\mu\tilde{S}_{ij},\quad \tau_{ij}=2\mu_t\tilde{S}_{ij},$$

where

$$\tilde{S}_{ij}=\frac{1}{2}\left(\frac{\partial\tilde{u}_i}{\partial x_j}+\frac{\partial\tilde{u}_j}{\partial x_i}\right)-\frac{1}{3}\delta_{ij}\frac{\partial\tilde{u}_k}{\partial x_k}.$$

The equation of state for ideal gas is considered as

$$\bar{p}=(\kappa-1)\left[\bar{\varrho}\tilde{e}-\frac{1}{2}\bar{\varrho}\sum_{i=1}^{2}\tilde{u}_i^2\right].$$

The symbol $\frac{D^A}{Dt}$ denotes the ALE derivative [5].

To include the influence of turbulent fluctuations on the mean flow a one-equation turbulence model of Spalart and Allmaras [4] is used. It is given by transport equation for eddy viscosity $\tilde{\nu}$ written as

$$\frac{D^A(\bar{\varrho}\tilde{\nu})}{Dt}+\frac{\partial(\bar{\varrho}\tilde{\nu}\tilde{u}_j)}{\partial x_j}-U_j\frac{(\partial\bar{\varrho}\tilde{\nu})}{\partial x_j}=\frac{1}{Re}\frac{1}{\sigma}\left[\frac{\partial}{\partial x_j}\left((\mu+\bar{\varrho}\tilde{\nu})\frac{\partial\tilde{\nu}}{\partial x_j}\right)+\bar{\varrho}c_{b2}\frac{\partial\tilde{\nu}}{\partial x_j}\frac{\partial\tilde{\nu}}{\partial x_j}\right]+c_{b1}\bar{\varrho}\tilde{S}\tilde{\nu}-\frac{1}{Re}c_{w1}f_w\frac{1}{\bar{\varrho}}\left(\frac{\bar{\varrho}\tilde{\nu}}{D}\right)^2 \tag{2}$$

and completed by the following relations

$$\mu_t=\bar{\varrho}\tilde{\nu}f_{v1},$$

$$\tilde{S}=\bar{\Omega}+\frac{1}{Re}\frac{\tilde{\nu}}{\kappa^2D^2}f_{v2},$$

$$f_w=g\left(\frac{1+c_{w3}^6}{g^6+c_{w3}^6}\right)^{\frac{1}{6}},$$

$$g=r+c_{w2}\left(r^6-r\right),$$

$$r=\min\left(\frac{\tilde{\nu}}{Re\,\tilde{S}\,\kappa^2D^2},10\right),$$

$$f_{v1}=\frac{\chi^3}{\chi^3+c_{v1}^3},\quad \chi=\frac{\tilde{\nu}}{\nu},$$

$$f_{v2}=1-\frac{\chi}{1+\chi f_{v1}}$$

and constants $c_{b1}=0.1355$, $c_{b2}=0.622$, $\sigma=\frac{2}{3}$, $\kappa=0.41$, $c_{w1}=\frac{c_{b1}}{\kappa^2}+\frac{1+c_{b2}}{\sigma}$, $c_{w2}=0.3$, $c_{w3}=2.0$, $c_{v1}=7.1$, where D is the distance to the nearest wall and $\bar{\Omega}$ is the vorticity magnitude.

The system of Favre-averaged Navier-Stokes equations (1) and the transport equation for eddy viscosity (2) are numerically solved together, the whole mathematical model can be written in a dimensionless compact flux vector form as

$$\frac{D^A\boldsymbol{w}}{Dt}+\sum_{j=1}^{2}\left(\frac{\partial}{\partial x_j}\boldsymbol{f}_j(\boldsymbol{w})-U_j\frac{\partial\boldsymbol{w}}{\partial x_j}\right)=\frac{1}{Re}\sum_{j=1}^{2}\frac{\partial}{\partial x_j}\boldsymbol{f}_j^v(\boldsymbol{w},\nabla\boldsymbol{w})+\boldsymbol{p}(\boldsymbol{w},\nabla\boldsymbol{w}), \tag{3}$$

where $\boldsymbol{w}$ is a vector of flow variables, $\boldsymbol{f}_j(\boldsymbol{w})$ are the inviscid fluxes, $\boldsymbol{f}_j^v(\boldsymbol{w},\nabla\boldsymbol{w})$ are the viscid fluxes and $\boldsymbol{p}(\boldsymbol{w},\nabla\boldsymbol{w})$ is a production term. The vectors are then written as

$$\boldsymbol{w}=\begin{bmatrix}\bar{\varrho}\\ \bar{\varrho}\tilde{u}_1\\ \bar{\varrho}\tilde{u}_2\\ \bar{\varrho}\tilde{e}\\ \bar{\varrho}\tilde{\nu}\end{bmatrix},\quad \boldsymbol{f}_1(\boldsymbol{w})=\begin{bmatrix}\bar{\varrho}\tilde{u}_1\\ \bar{\varrho}\tilde{u}_1^2+\bar{p}\\ \bar{\varrho}\tilde{u}_1\tilde{u}_2\\ \tilde{u}_1(\bar{\varrho}\tilde{e}+\bar{p})\\ \bar{\varrho}\tilde{\nu}\tilde{u}_1\end{bmatrix},\quad \boldsymbol{f}_2(\boldsymbol{w})=\begin{bmatrix}\bar{\varrho}\tilde{u}_2\\ \bar{\varrho}\tilde{u}_1\tilde{u}_2\\ \bar{\varrho}\tilde{u}_2^2+\bar{p}\\ \tilde{u}_2(\bar{\varrho}\tilde{e}+\bar{p})\\ \bar{\varrho}\tilde{\nu}\tilde{u}_2\end{bmatrix},$$

$$\boldsymbol{f}_1^v(\boldsymbol{w},\nabla\boldsymbol{w})=\begin{bmatrix}0\\ \tilde{t}_{11}+\tau_{11}\\ \tilde{t}_{12}+\tau_{12}\\ \sum_{i=1}^{2}\tilde{u}_i(\tilde{t}_{i1}+\tau_{i1})+\frac{\kappa}{\kappa-1}\left(\frac{\mu}{Pr}+\frac{\mu_t}{Pr_t}\right)\frac{\partial}{\partial x_1}\left(\frac{\bar{p}}{\bar{\varrho}}\right)\\ \frac{1}{\sigma}(\mu+\bar{\varrho}\tilde{\nu})\frac{\partial\tilde{\nu}}{\partial x_1}\end{bmatrix},$$

$$\boldsymbol{f}_2^v(\boldsymbol{w},\nabla\boldsymbol{w})=\begin{bmatrix}0\\ \tilde{t}_{12}+\tau_{12}\\ \tilde{t}_{22}+\tau_{22}\\ \sum_{i=1}^{2}\tilde{u}_i(\tilde{t}_{i2}+\tau_{i2})+\frac{\kappa}{\kappa-1}\left(\frac{\mu}{Pr}+\frac{\mu_t}{Pr_t}\right)\frac{\partial}{\partial x_2}\left(\frac{\bar{p}}{\bar{\varrho}}\right)\\ \frac{1}{\sigma}(\mu+\bar{\varrho}\tilde{\nu})\frac{\partial\tilde{\nu}}{\partial x_2}\end{bmatrix},$$

 https://doi.org/10.1051/epjconf/201818002117

$$\boldsymbol{p}(\boldsymbol{w}, \nabla \boldsymbol{w}) = \begin{bmatrix} 0 \\ 0 \\ 0 \\ 0 \\ \frac{1}{Re}\frac{1}{\sigma}\bar{\varrho}c_{b2}\left(\frac{\partial\tilde{\nu}}{\partial x_1}\frac{\partial\tilde{\nu}}{\partial x_1} + \frac{\partial\tilde{\nu}}{\partial x_2}\frac{\partial\tilde{\nu}}{\partial x_2}\right) + \\ +c_{b1}\bar{\varrho}\tilde{S}\,\tilde{\nu} - \frac{1}{Re}c_{w1}f_w\frac{1}{\bar{\varrho}}\left(\frac{\bar{\varrho}\tilde{\nu}}{D}\right)^2 \end{bmatrix}.$$

3 Discretization

The spatial discretization of equations (3) is performed using the discontinuous Galerkin finite element method (DGFEM), [6], [7], [8].

Let $\mathcal{T}_h = \{\Omega_k : \Omega_k \in \Omega, k = 1, \ldots, K\}$ be a triangulation of the time varying computational domain $\Omega(t) \in \mathcal{R}^2$, with boundary $\partial\Omega$. Let Γ^{int} and $\Gamma^b \subset \partial\Omega$ are sets of inner and boundary edges of elements Ω_k. The solution of nonlinear system of equations (3) is considered in the space of vector functions $\mathbf{S}_h = S_h \times S_h \times S_h \times S_h$, where

$$\begin{aligned} S_h = \{&v(\mathbf{x}) : v(\mathbf{x}) \in \mathcal{P}^q(\Omega_k) \text{ if } \mathbf{x} \in \Omega_k, \\ &v(\mathbf{x}) = 0 \text{ if } \mathbf{x} \notin \Omega_k, \forall \Omega_k \in \mathcal{T}_h\} \end{aligned}$$

and $\mathcal{P}^q(\Omega_k)$ is the space of polynomials of q-th order at the most.

Multiplying the system of equations (3) by a test function $\mathbf{v} \in \mathbf{S}_h$, integrating it over the element Ω_k and using the Green's theorem, the following integral identity is obtained

$$\begin{aligned} &\frac{d}{dt}\int_{\Omega_k} \boldsymbol{w}\cdot\boldsymbol{v}\,d\Omega - \int_{\Omega_k}\sum_{j=1}^{2}(\boldsymbol{f}_j(\boldsymbol{w}) - U_j\boldsymbol{w})\cdot\frac{\partial\boldsymbol{v}}{\partial x_j}\,d\Omega + \\ &\oint_{\partial\Omega_k}\sum_{j=1}^{2}(\boldsymbol{f}_j(\boldsymbol{w}) - U_j\boldsymbol{w})n_j\cdot\boldsymbol{v}\,dl = \\ &\frac{1}{Re}\oint_{\partial\Omega_k}\sum_{j=1}^{2}\boldsymbol{f}_j^v(\boldsymbol{w}, \nabla\boldsymbol{w})n_j\cdot\boldsymbol{v}\,dl - \\ &\frac{1}{Re}\int_{\Omega_k}\sum_{j=1}^{2}\boldsymbol{f}_j^v(\boldsymbol{w}, \nabla\boldsymbol{w})\cdot\frac{\partial\boldsymbol{v}}{\partial x_j}\,d\Omega + \\ &\int_{\Omega_k}\boldsymbol{p}(\boldsymbol{w}, \nabla\boldsymbol{w})\cdot\boldsymbol{v}\,d\Omega. \end{aligned}$$

The key part of the method is an approximation of curve integrals. They are evaluated in the same way as it is done in the case of finite volume method. The inviscid fluxes are evaluated as

$$\sum_{j=1}^{2}\boldsymbol{f}_j(\boldsymbol{w})n_j \approx \mathcal{H}(\boldsymbol{w}^+, \boldsymbol{w}^-), \quad \text{on } \Gamma^{int},$$

$$\sum_{j=1}^{2}\boldsymbol{f}_j(\boldsymbol{w})n_j \approx \mathcal{H}(\boldsymbol{w}^b, \boldsymbol{w}^-), \quad \text{on } \Gamma^b,$$

where $\mathcal{H}$ is Lax-Friedrichs flux, [9], $\boldsymbol{n} = [n_1, n_2]^T$ is vector of outer normal, values $\boldsymbol{w}^-$ resp. $\boldsymbol{w}^+$ are the left resp. right limits of the vector $\boldsymbol{w}$ on the edge $\Gamma^{int} \cup \Gamma^b$ and $\boldsymbol{w}^b$ is a boundary value of $\boldsymbol{w}$ computed from boundary conditions.

Similarly, the value of viscous fluxes are evaluated using the appropriate numerical fluxes. In present paper, the approximation known as interior penalty (IP) method was chosen, [8].

Let $\boldsymbol{w}_k$ be a part of solution $\boldsymbol{w}$ corresponding to control element Ω_k and $\varphi_{k,i}$ be i-th basis function of the space $\mathcal{P}^q(\Omega_k)$. The m components w_k^m, $m = 1, 2, ..., 4$, of vector $\boldsymbol{w}_k$ can be expressed as a linear combination

$$w_k^m = \sum_{i=1}^{q} w_{k,i}^m(t)\varphi_{k,i}(\boldsymbol{x}). \tag{4}$$

Inserting it into the integral identity and evaluating the volume and curve integrals using the Gauss integration rules of appropriate order, the following system of ordinary differential equations is obtained

$$\frac{d(\mathbf{MW})}{dt} = \mathbf{R}(\mathbf{W}), \tag{5}$$

where vector $\mathbf{W}$ consists of coefficients of linear combination (4) and $\mathbf{M}$ is a mass matrix. In this study, linear Lagrange polynomials are chosen as the basis functions of the space $\mathcal{P}^q(\Omega_k)$.

Time integration of equation (5) is performed using the implicit backward Euler method, which is solved by the Newton's method

$$\begin{aligned} \boldsymbol{W}^{s=0} &= \boldsymbol{W}^n, \\ \left[\frac{\boldsymbol{M}^{n+1}}{\Delta t} - \frac{\partial\boldsymbol{R}}{\partial\boldsymbol{W}}(\boldsymbol{W}^s)\right]\Delta\boldsymbol{W} &= \\ \boldsymbol{R}(\boldsymbol{W}^s) - &\frac{\boldsymbol{M}^{n+1}\boldsymbol{W}^s - \boldsymbol{M}^n\boldsymbol{W}^n}{\Delta t}, \\ \boldsymbol{W}^{s+1} &= \boldsymbol{W}^s + \Delta\boldsymbol{W}, \\ \boldsymbol{W}^{s+1} &\longrightarrow \boldsymbol{W}^{n+1}. \end{aligned} \tag{6}$$

The system of linear equations (6) is solved iteratively using the GMRES method with block diagonal Jacobi preconditioner.

For the computation of deformed mesh points coordinates, the blending function approach was used [10], [11]. For our purpose, the blending function was generalized for three independently moving blades. Let $\boldsymbol{x}_0$ denotes the vector of initial coordinates of each mesh node. For each blade ($i = 1, 2, ..., n$, in our study $n = 3$), the vector of initial coordinates $\boldsymbol{x}_0$ is transformed into a new vector of coordinates $\boldsymbol{y}_i$ using the following formula:

$$\boldsymbol{y}_i = \boldsymbol{p}_i + \mathbf{T}(\alpha)_i\boldsymbol{x}_0,$$

where $\boldsymbol{p}_i = \boldsymbol{p}_i(t)$ is the time dependent translation vector and $\mathbf{T}(\alpha)_i = \mathbf{T}(\alpha(t))_i$ is the time dependent rotation matrix of i-th blade. The coordinates $\boldsymbol{x} = \boldsymbol{x}(t)$ of the new mesh points are computed by applying the blending functions as

$$\boldsymbol{x} = \frac{1}{n}\sum_{i=1}^{n}(b_i(\boldsymbol{x}_0)\boldsymbol{x}_0 + (1 - b_i(\boldsymbol{x}_0))\boldsymbol{y}_i).$$

The key part of this algorithm is to determine the blending functions b_i. The blending function b_i have to satisfy two basic properties:

1. the value of i-th blending function b_i on the boundary of i-th blade have to be equal to 0,

2. the value of i-th blending function b_i on the boundary of j-th blade or on boundary of computational domain have to be equal to 1.

The blending function b_i fulfilling these properties can be obtained using the solutions of boundary value problems

$$\frac{d^2 b_i}{dx^2} + \frac{d^2 b_i}{dy^2} = 0,$$

together with boundary condition $b_i = 0$ on the i-th blade and $b_i = 1$ on the rest of boundaries.

4 Numerical results

The developed in-house CFD software based on the discontinuous Galerkin method was applied to the numerical simulation of compressible viscous fluid flow through a simplified blade cascade with prescribed harmonic motion. The geometric configuration of the simplified blade cascade and the angle of attack of the flow were adopted from the test case no. 3, which was experimentally studied in [3]. The only difference is that the thin airfoils of test cascade in [3] were replaced by 3 millimetres thick flat plates. The geometric configuration of the simplified test cascade is shown in Figure 1.

The considered two-dimensional computational domain and corresponding unstructured triangular mesh with 11932 elements is shown in Figure 2. The detailed view of the mesh in the vicinity of the three flat plates is shown in Figure 3. Each plate performs kinematic harmonic motion in vertical direction $y = A\sin(2\pi f t + \phi)$ with an inter-blade phase angle $\phi = \pi/2$ for the upper plate, $\phi = 0$ for the middle plate and $\phi = -\pi/2$ for the bottom plate. The amplitude and the frequency of the harmonic motion are $A = 0.003$ m and $f = 20$ Hz, respectively. At the inlet of the computational domain, see Figure 2, the stagnation pressure $p_{01} = 101325$ Pa, the stagnation temperature $T_{01} = 293.15$ K and angle of attack $\alpha = 11.4$ were prescribed. At the outlet, the static pressure $p_2 = 95520$ Pa was prescribed. The value of p_2 was calculated out of the free stream velocity $100\,\mathrm{ms}^{-1}$ under the consideration of standard sea level air conditions.

The pressure contours are shown at three selected times $t_1 = 0.0433$ s, $t_2 = 0.1109$ s and $t_3 = 0.1345$ s in Figures 4-6. The times t_1, t_2 and t_3 correspond respectively to the zero, maximal and minimal vertical displacement of the middle flat plate. The time evolution of y-component of aerodynamic forces acting on upper, middle and bottom plates is shown in Figure 7. The results produced by the developed in-house solver were compared to the results obtained by the commercial CFD software package ANSYS Fluent. The y-components of aerodynamic forces acting on upper, middle and bottom plates obtained by both the in-house and commercial solvers are shown in Figure 8. Finally, vertical displacement of the middle plate along with y-component of the aerodynamic force acting on the middle plate are plotted in Figure 9.

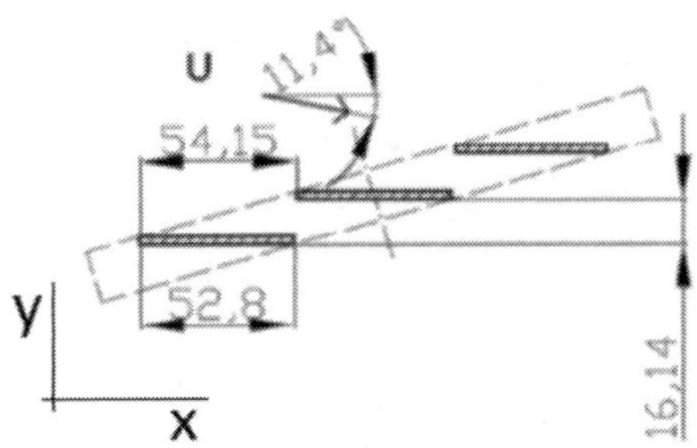

Fig. 1. Test geometric configuration of the simplified cascade which consists of three oscillating flat plates. The dimensions are in millimetres. The width of the plates is 0.08 m.

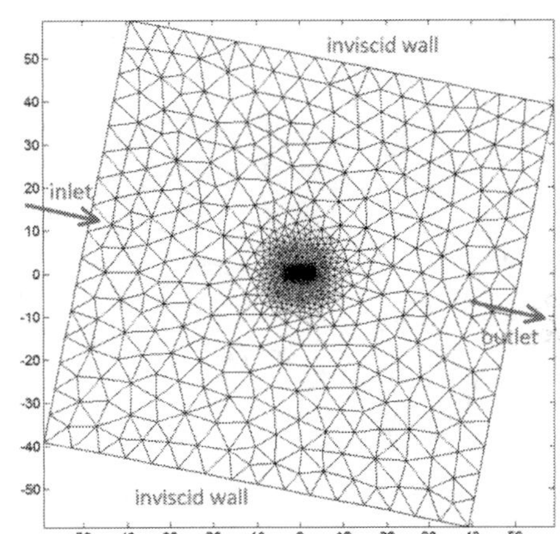

Fig. 2. Computational domain and triangular mesh.

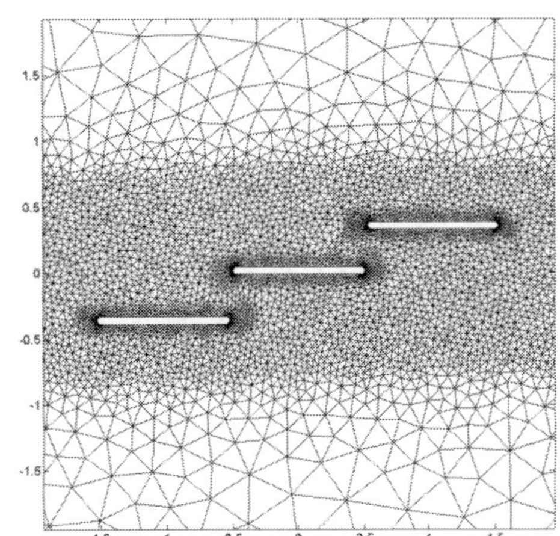

Fig. 3. Detail of computational mesh in the vicinity of simplified cascade.

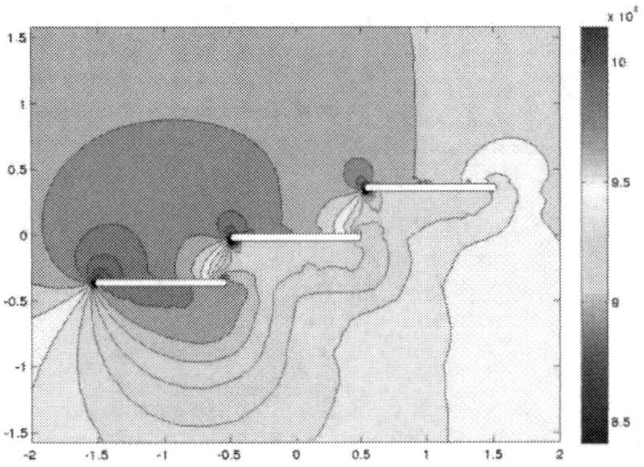

Fig. 4. Pressure contours at time t = 0.0433 s, where the middle plate has a zero displacement.

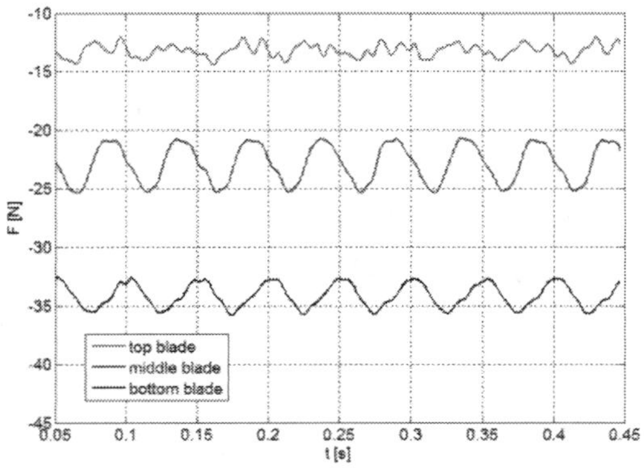

Fig. 7. Time development of y-components of aerodynamic forces acting on the upper (red), middle (blue) and bottom (black) plates.

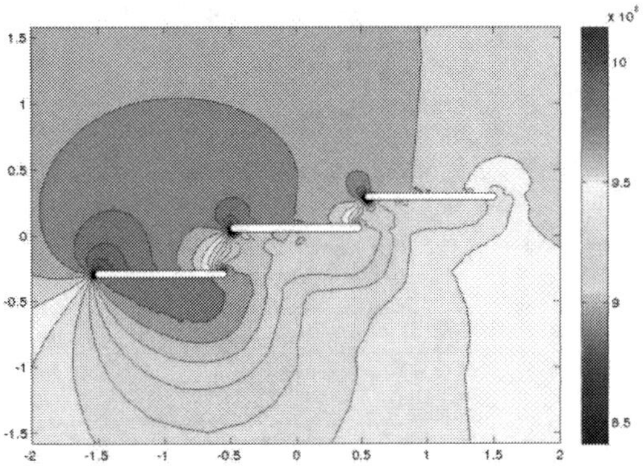

Fig. 5. Pressure contours at time t = 0.1109 s, where the middle plate reaches a highest displacement.

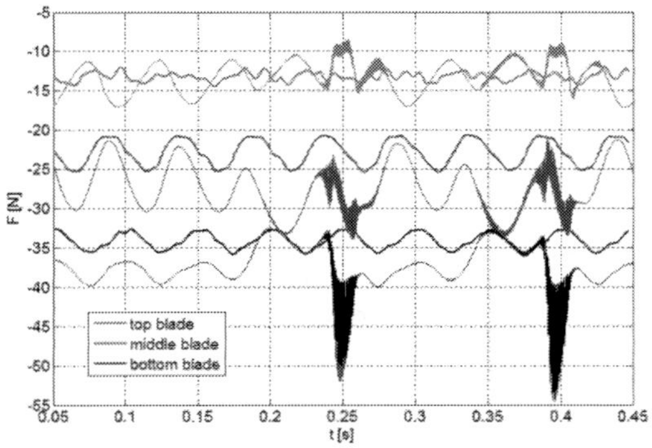

Fig. 8. Time development of y-components of aerodynamic forces produced by the developed solver in comparison to the results obtained be the commercial CFD software package ANSYS Fluent.

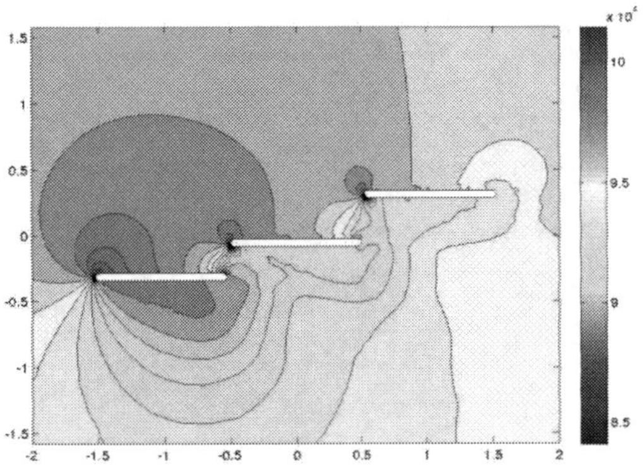

Fig. 6. Pressure contours at time t = 0.1345 s, where the middle plate reaches a lowest displacement.

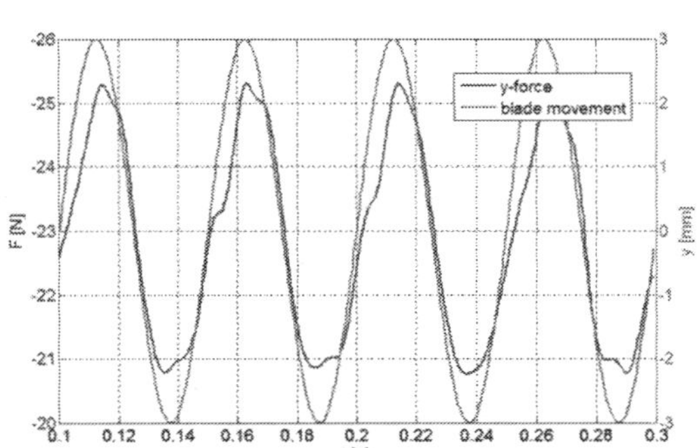

Fig. 9. Time development of vertical displacement (green) and y-component of aerodynamic force (blue) for the middle plate.

5 Conclusion

The in-house CFD solver based on the discontinuous Galerkin method was developed for the numerical solution of compressible viscous flow problems in domains with moving boundaries. This solver was applied to the numerical simulation of compressible viscous fluid flow through the simplified blade cascade formed by three flat plates with prescribed harmonic motion.

The paper presents numerical results in the form of pressure distribution and time evolution of aerodynamic forces acting on each flat plate for the free stream velocity $100\,\mathrm{ms}^{-1}$ and for the forward travelling wave mode, i.e. the inter-blade phase angle $\phi = \pi/2$. The y-components of aerodynamic forces acting on the flat plates obtained by the developed in-house CFD solver show qualitative agreement with those from the commercial CFD software package ANSYS Fluent. Comparing the time evolution of the middle plate vertical displacement with the time evolution of the y-component of aerodynamic force acting on the middle plate, one can observe a small time delay of the force. Such delay is typical for the case of forward wave motion through cascade when kinetic energy of the plate is transmitted to the flow.

The objective of our future study will be aimed at numerical computation of the y-components of aerodynamic forces acting on blades and their time delays to the blade's vertical displacements for different travelling waves through the blade cascade in order to assess the flutter stability of the cascade in the whole range of the inter-blade phase angle ϕ from 0 to 2π. Obtained numerical results will be experimentally validated in the Institute of Thermomechanics of the Czech Academy of Sciences.

Acknowledgements

The authors appreciate the kind support by the grant GA 16-04546S "Aero-elastic couplings and dynamic behaviour of rotational periodic bodies" of the Czech Science Foundation.

References

[1] H. M. Owens, W. E. Trumpler, ASME Paper, 49-A-64 (1949)

[2] E. K. Armstrong, R. E. Stevenson, J. R. Aero. Soc., **64** (591), 117-130 (1960)

[3] D. Schlafli, Experimentelle untersuchung der instationaren stromung in oscillierenden ringgittern. PhD Thesis No. 800, Lausanne (1989)

[4] P. R. Spalart, S. R. Allmaras, AIAA Paper, 92-439 (1992)

[5] M. Feistauer, J. Česenek, NM&A, **60** (46), 1-13 (2011)

[6] W. H. Reed, T. R. Hill, Los Alamos Scientific Laboratory, Report No. LA-UR-73-479 (1973)

[7] B. Cockburn, C.-W. Shu, J. Sci. Comput., **16** (3), 173-261 (2001)

[8] V. Dolejší, M. Holík, J. Hozman, J. Comput. Phys., **230** (11), 4176-4200 (2011)

[9] E. F. Toro, Riemann Solvers and Numerical Methods for Fluid Dynamics. Springer, Heidelberg (1999)

[10] P.-O. Persson, J. Bonet, J. Peraire, Comput. Methods Appl. Mech. Eng., **198** (17-20), 1585-1595 (2009)

[11] O. Bublík, J. Vimmr, A. Jonášová, Appl. Math. Comput., **267**, 329-340 (2015)

Surface tension of aqueous binary mixtures under the supercooled conditions – Development of the measuring technique and preliminary data for water + lower alcohols

Václav Vinš[1,2*], Jiří Hykl[1], Zbyněk Nikl[1], Miroslav Čenský[1], and Jan Hrubý[1]

[1]Institute of Thermomechanics of the CAS, Dolejškova 1402/5, Prague 8, 182 00, Czech Republic
[2]Škoda Auto, tř. Václava Klementa 869, Mladá Boleslav, 293 01, Czech Republic

Abstract. An experimental apparatus originally developed for the measurement of surface tension of supercooled water was modified such that it allows for measurement of binary aqueous mixtures. The measuring principle based on the capillary rise technique is similar to that employed in the previous measurements with pure water [J. Hrubý et al., *J. Phys. Chem. Lett.* **5** (2014) 425 and V. Vinš et al., *J. Phys. Chem. B* **119** (2015) 5567]. The temperature dependence of the surface tension is determined from the measured height of the liquid column elevated in a vertical capillary tube with an inner diameter of 0.32 mm. The aqueous liquid rises to a height of 60 mm to 95 mm depending on the solute concentration and the temperature. Preliminary tests performed with the binary mixtures of water with methanol, ethanol, and *n*-propanol at various concentrations and temperatures down to –28 °C approved functionality of the measuring technique. Some difficulties, such as influence of impurities on the liquid column elevation or formation of bubbles in the liquid column, were observed. The experimental apparatus is further being modified in order to obtain more accurate data for various aqueous mixtures.

1 Introduction

This study is part of a long-term research of our group focussing on the experimental investigation of thermophysical properties of water and aqueous systems under the supercooled conditions, i.e. on the properties of the metastable liquid aqueous phase at temperatures below the equilibrium freezing temperature. Water shows several anomalies at low temperatures, e.g., the commonly known maximum in density at +4 °C or a sharp increase of isobaric heat capacity and isothermal compressibility at temperatures approaching –38 °C [1,2], i.e. a macroscopic supercooling limit of homogeneous freezing. The surface tension was expected to show another anomaly referred as the SIP (second inflection point) anomaly in its temperature trend at temperatures close to –8 °C. However, the existence of SIP was recently refuted down to –26 °C by our experiments [3-5].

In this work, the experimental apparatus and the measuring technique used for the measurement of the surface tension of pure supercooled water [3-7] was modified such that it allows for investigation of aqueous binary mixtures. The new accurate data for the surface tension of aqueous systems, such as mixtures with alcohols or salts, under the supercooled conditions are important for scientific communities focused on the atmospheric and oceanographic research [8,9] and on the investigation of anomalous behaviour of water systems at low temperatures [1,2]. Functionality of the new measuring technique was successfully tested by initial measurements of water mixed with several short-chain alcohols; namely with methanol, ethanol, and *n*-propanol.

2 Experimental Apparatus

The experimental apparatus for the measurement of surface tension of pure water under the supercooled conditions was described in previous studies [3,4,7] in detail. The measuring principle is based on a modified capillary-rise technique, when the upper part of a vertical capillary tube is positioned in a temperature-controlled chamber. The capillary tube is located in the chamber such that the temperature of the liquid meniscus inside the capillary tube could be varied. The surface tension σ and eventually its temperature trend can be determined from the measured height h of the liquid column elevated in the capillary tube of a known inner diameter d.

$$\sigma = \frac{h\rho g d}{4\cos\theta} \tag{1}$$

In equation (1), ρ stands for the density of liquid, g is the gravity, and θ denotes the contact angle between the liquid and the capillary tube wall. Basic principle of the measuring technique is depicted in Figure 1. Height h, i.e. the distance between the liquid meniscus inside the capillary tube and the liquid level inside the lower container is measured by using a digital camera installed on a high-precision height gauge. The standard uncertainty of the measured height was below 0.012 mm.

* Corresponding author: vins.vaclav@seznam.cz

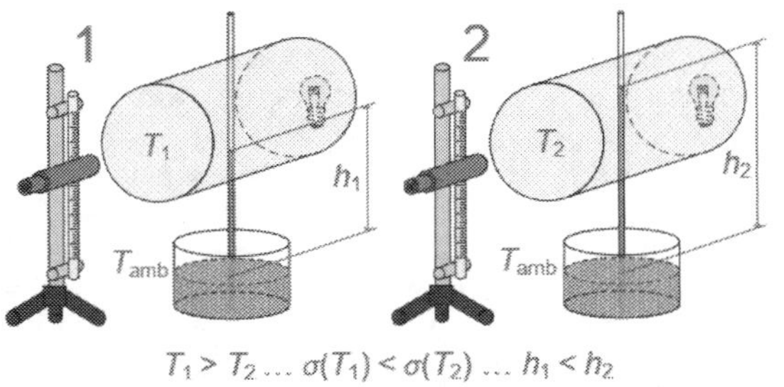

Figure 1. Scheme of the experimental apparatus with the vertical capillary tube partially placed inside a temperature-controlled chamber. Case 1 corresponds to the reference measurement, e.g., at +15 °C, and case 2 is performed at the target temperature.

3 Description of Measurement

The mixture samples were prepared by the weighing method using analytical scale *Mettler Toledo XSE 205 DU* with accuracy better than 0.1 mg. The samples were mixed using an ultrasonic homogenizer in order to assure homogeneous distribution of the composition. The uncertainty of the mass fraction of the binary samples was better than 0.0001. Ultrapure water, prepared from the serially connected purification devices reverse osmosis unit *Watrex Rowapur* and *Neptune Analytical*, was used both for the capillary tube cleaning and for the preparation of measured samples. The ultrapure water had a constant resistivity of 18.2 MΩ, total organic carbon content below 5 ppb and was free of particles larger than 0.2 μm. A sample of methanol was purchased from Lach-Ner and had declared purity of 99.99 %. Sample of absolute ethanol was purchased from Penta Chemicals and had declared purity of 99.99 %. Sample of *n*-propanol was purchased from Penta Chemicals and had declared purity of 99.94 %.

Temperature inside the temperature-controlled chamber was measured with two in-house calibrated resistive thermometers Pt100 located close to the capillary tube. The standard uncertainty of the measured temperature was better than 0.040 °C. The capillary tube was repeatedly cleaned with chromosulfuric acid and ultrapure water. The upper end of the capillary tube was connected to the helium distribution setup [6], which allowed flushing of the capillary tube with pure helium. All measurements were carried out with descending meniscus by applying a small underpressure above the liquid meniscus using a syringe installed at the helium distribution setup.

The surface tension was measured with a technique similar to that employed in the previous studies with pure water [3,4] referred as the *height* measurement. The main difference was that two independent measurements with pure water were conducted before and after the measurements with an aqueous mixture. The reference measurements with pure water at a constant temperature of +15 °C were used for the evaluation of the inner diameter of the capillary tube. The inner diameter was determined from the reference measurements and the international standard for the surface tension of ordinary water [10] approved by IAPWS [11]. Consequently, all measurements presented in this work are relative and related to the surface tension of pure water at +15 °C. The magnitude of the inner diameter of all capillary tubes employed in this study lied in the range from 0.3220 mm to 0.3230 mm with a standard uncertainty better than 0.0020 mm.

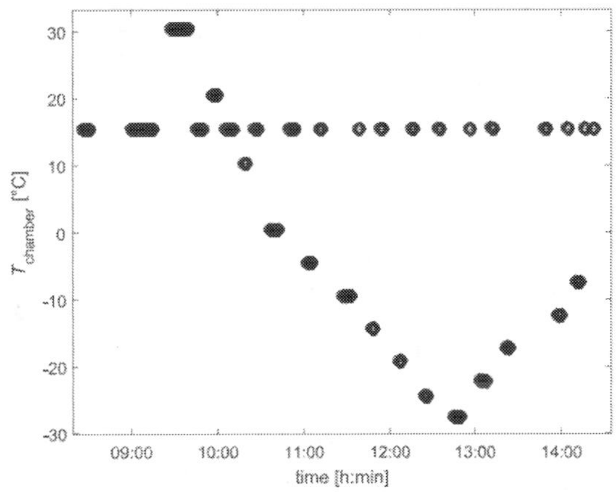

Figure 2. Temperature variation during one experimental run. Measurement with 10 % mass fraction of methanol.

Figure 2 shows variation of temperature inside the temperature-controlled chamber over time. It shall be noted that only the stabilized conditions are shown. The measurement at the desired temperature was every time followed by the measurement at the reference temperature of +15 °C. The reference measurements helped to detect possible negative effects during the experiments, such as presence of impurities on the capillary tube wall, degradation of the sample, i.e. variation of its composition, etc. We note that no variation of the sample composition over the time of measurement was detected during our experiments.

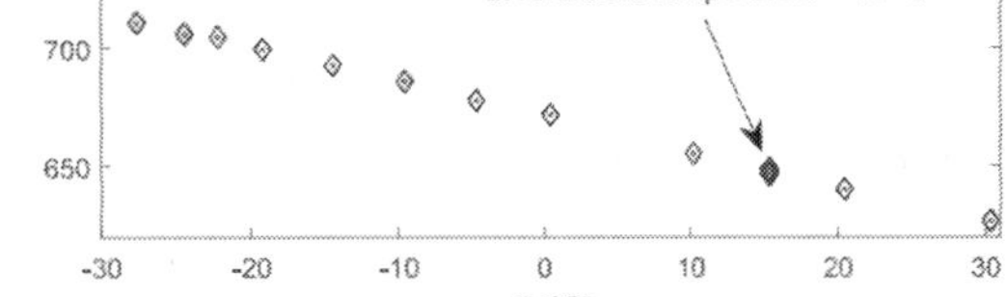

Figure 3. Hydrostatic pressure of the liquid column inside the capillary tube measured for the aqueous binary mixture with 15 % mass fraction of methanol. Comparison of the linear temperature change over the transient region of 3 mm height and the sudden temperature jump.

Hydrostatic pressure of the liquid column elevated in the capillary tube given as follows

$$p_{\mathrm{hyd}} = h\rho g \quad (2)$$

is depicted in Figure 3 for a mixture of water with a 15 % mass fraction of methanol. Due to a significant decrease in the surface tension caused by the presence of methanol, the hydrostatic pressure at +15 °C dropped from 909 Pa corresponding to the pure water sample down to 648 Pa in case of the mixed sample.

Both the measurements with mixed sample at the reference temperature of +15 °C performed throughout the experimental run and the reference measurements with pure water conducted before and after the mixture measurement show considerably good internal consistency as can be seen in Figure 3. No negative effects during the measurement were observed in this case.

Nevertheless, not every experimental test could be treated as successful. Measurements under the metastable conditions are rather difficult as the purity of the sample, the cleanness of the experimental apparatus and above all the time scale of the measurement matter a lot. Figure 4 shows an interesting phenomenon of an unsuccessful measurement, when small bubbles appeared in the liquid column elevated in the capillary tube. Three stages of the gradual bubble formation can be seen in the figure. Possible reason for the bubble formation could be a presence of impurities or small cracks on the capillary tube inner surface avoiding continuous movement of the liquid column together with an inhomogeneous composition of the mixed sample along the capillary tube length.

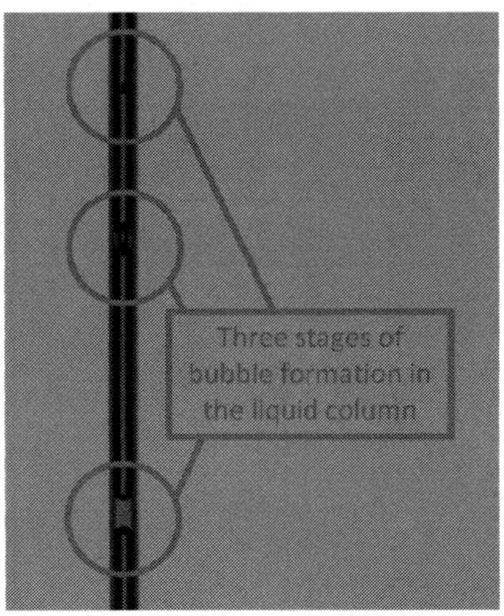

Figure 4. Photo of the capillary tube inside the temperature-controlled chamber with three stages of the bubble formation within the liquid column of water and lower alcohol mixture.

4 Results

Functionality of the new measuring technique was investigated on the measurements of aqueous binary mixtures with methanol, ethanol, and *n*-propanol. The data were collected in the temperature range from −28 °C to +31 °C, i.e. both in the stable region and under the supercooled metastable conditions. Figure 5 shows preliminary testing measurements for two mixtures with ethanol and *n*-propanol with mass fraction of alcohol of 5 % in both cases. The experimental data are compared with the IAPWS correlation [11] for the surface tension of pure water and the water–lower alcohol experimental parametrization suggested by Němec [12]. At the reference temperature of +15 °C, adding 5 % of ethanol resulted in a decrease of the surface tension by 16.5 mN m^{-1} compared to pure water. In case of *n*-propanol the decrease became even more significant of around 26.5 mN m^{-1}. For both mixtures, the experimental data show quite good agreement with the theoretical parametrization at temperatures above +15 °C. However at sub-zero temperatures, the experimental data deviate from the correlation. The deviation is particularly significant in case of *n*-propanol at temperatures below −5 °C. The continuous increase of the surface tension with decreasing temperature is qualitatively similar to the SIP anomaly of pure water detected at temperature around −8 °C [3,13]. Nevertheless, the presented data are only preliminary and further experimental investigation of this phenomenon is needed. Unfortunately this represents an uneasy task due to some difficulties by the measurements with ethanol and *n*-propanol, such as formation of bubbles within the liquid column shown in Figure 4.

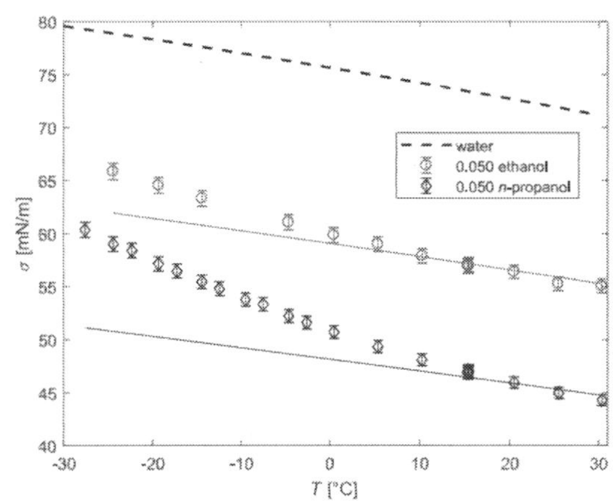

Figure 5. Results of preliminary testing measurements of the surface tension of water + ethanol and water + *n*-propanol. The mass fraction of alcohol was 5 % in both cases. Black dashed line represents IAPWS correlation [11] for pure water extrapolated below 0.01 °C. Red and blue solid lines are predictions of the water–short chain alcohol correlation [12].

The water–lower alcohol model combines the surface tension of pure water obtained from the IAPWS correlation [11] and the surface tension of pure alcohols predicted using correlations by Mulero et al. [16]. The values for pure components are combined using Connors-Wright mixing formula [17]. In case of water + ethanol mixture, the correlation by Wilemski [18] was used.

In the data analysis, density of water + lower alcohol systems required in equations (1) and (2) was predicted with the Redlich-Kister type of parametrization, in which the densities of pure water and pure alcohols were calculated using the reference equation of state IAPWS-95 [19] and the parametrizations from the Landolt-Börnstein series [20], respectively. For more details see ref. [12]. A constant contact angle $\theta = 0.5°$ was considered for all measurements with mixed samples. In the evaluation of the capillary tube inner diameter, a constant contact angle of $3 \pm 1°$ was considered for pure water and fused silica in case of the descending meniscus [14,15]. We note that the influence of the contact angle is minor compared to other quantities in equation (1).

Figure 6 shows data for the surface tension of water + methanol binary systems at five different mass fractions of methanol ranging from 0.03 to 0.20. The data are compared to the IAPWS correlation [11] for pure water and the water + lower-alcohol mixture correlation [12]. As can be seen the data agree quite well with the mixture model over the entire temperature range from –28 °C to +31 °C. The only considerable difference can be seen in case of the lowest concentration of 0.03. Unlike in case of mixtures with ethanol and *n*-propanol depicted in Figure 5, no significant growth in the surface tension at low temperatures was observed for the water + methanol system. The data have almost constant slope in the experimental temperature range.

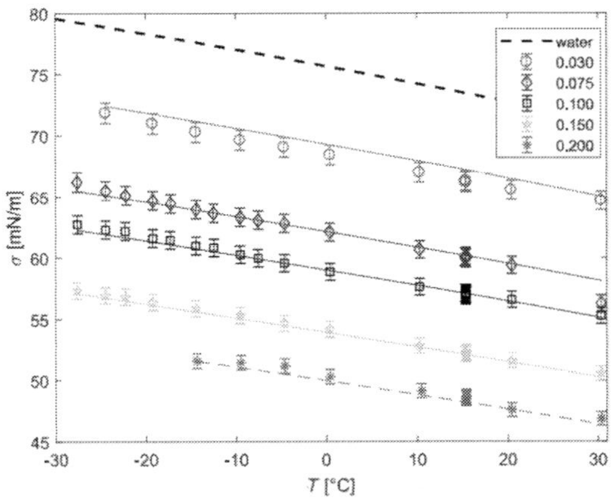

Figure 6. Surface tension of water + methanol including supercooled region. Values in the legend correspond to mass fraction of methanol. Black dashed line represents IAPWS [11] correlation for the surface tension of ordinary water extrapolated below 0.01 °C, other lines are predictions of experimental correlation [12].

As the presented data have preliminary character, the standard uncertainty of the resulting surface tension is estimated relatively large $u(\sigma) \leq 0.80$ mN m^{-1} compared to pure water measurements with the maximum uncertainty of around 0.40 mN m^{-1} [3-5]. Main sources of the uncertainty are coming from the uncertainty of the IAPWS correlation [11] for the surface tension of pure water at the reference temperature of +15 °C $u(\sigma_{IAPWS}) = 0.37$ mN m^{-1}, which is used in the determination of the capillary tube inner diameter, and from the density of the binary mixture, which is considered quite large $u_r(\rho) = 1.0$ % as it contains also the uncertainty of the mixture composition.

5 Conclusions

The experimental apparatus for the measurement of surface tension of water under the metastable supercooled conditions was successfully modified in order to investigate aqueous binary mixtures. The preliminary measurements with water + short-chain alcohols proved functionality of the new measuring technique. Data for the binary system water + methanol at mass fractions of methanol varying from 0.03 to 0.20 show good agreement with the theoretical parametrization [12] over the entire experimental temperature range from –28 °C to +31 °C. On the other hand, preliminary measurements of binary aqueous systems with ethanol and *n*-propanol deviate from the model, as a significant increase in the surface tension was detected at sub-zero temperatures. The experimental temperature trend similar to the SIP anomaly of the pure water [3] is particularly pronounced in case of *n*-propanol. Further experimental investigation is required in order to clarify this phenomenon.

We note that the apparatus is further being modified in order to improve the accuracy of the measured data. The experimental setup is currently being used for the measurements of surface tension of water + sodium chloride and the supercooled seawater.

Acknowledgement

The study was supported by the Czech Science Foundation grant GJ15-07129Y and by the institutional support RVO:61388998 from IT CAS.

References

1. P.H. Poole, F. Sciortino, U. Essmann, H.E. Stanley, Nature **360**, 324 (1992)
2. O. Mishima, H.E. Stanley, Nature **396**, 329 (1998)
3. J. Hrubý, V. Vinš, R. Mareš, J. Hykl, J. Kalová, J. Phys. Chem. Lett. **5**, 425 (2014)
4. V. Vinš, M. Fransen, J. Hykl, J. Hrubý, J. Phys. Chem. B **119**, 5567 (2015)
5. V. Vinš, J. Hošek, J. Hykl, J. Hrubý, J. Chem. Eng. Data **62**, 3823 (2017)
6. V. Vinš, J. Hrubý, J. Hykl, J. Blaha, B. Šmíd, EPJ Web Conf. **45**, 01094 (2013)
7. V. Vinš, J. Hošek, J. Hykl, J. Hrubý, EPJ Web Conf. **92**, 02108 (2015)
8. Y. Viisanen, R. Strey, H. Reiss, J. Chem. Phys. **99**, 4680 (1993)
9. M.A.L.J. Fransen, E. Sachteleben, J. Hrubý, D.M.J. Smeulders, Exp. Fluids **55**, 1780-1 (2014)
10. N.B. Vargaftik, B.N. Volkov, L.D. Voljak. J. Phys. Chem. Ref. Data **12**, 817 (1983)
11. IAPWS revised release on surface tension of ordinary water substance. 2014; http://www.iapws.org
12. T. Němec, Institute of Thermomechanics of the CAS, Prague, Czech Republic, private communication
13. P.T. Hacker, NACA TN **2510** (1951)

14. V.G. Stepanov, L.D. Volyak, Y.V. Tarlakov, J. Eng. Phys. Thermophys. **32**, 646 (1977). Translated from Inzhenerno-Fizicheskii Zhurnal **32**, 1000 (1977)
15. L.D. Volyak, V.G. Stepanov, Y.V. Tarlakov, Zh. Fiz. Khim. **49**, 2931 (1975)
16. A. Mulero, I. Cachadina, and E. L. Sanjuán, J. Phys. Chem. Ref. Data **44**, 033104 (2015)
17. C.-W. Chang, T.-L. Hsiung, C.-P. Lui, C.-H. Tu, Fluid Phase Equilib. **389**, 28 (2015)
18. G. Wilemski, J. Chem. Phys. **91**, 2492 (1987).
19. W. Wagner, A. Pruß, J. Phys. Chem. Ref. Data 31, 387 (2002).
20. M. Frenkel, X. Hong, R. C. Wilhoit, K. R. Hall, Densities of Alcohols, edited by K. R. Hall and K. N. Marsh, Landolt-Börnstein, New Series, Vol. IV/8G (Springer, 1998).

Control of droplet size in rain-zone in wet cooling tower

Rut Vitkovicova [1*], *Pavol* Vitkovic[1]

[1]Czech Technical University in Prague, Department of Fluid Dynamics and Thermodynamics, 16607 Prague, Czech Republic

Abstract. The performance of the wet cooling tower is significantly affected by the droplet size occurring in the rain zone. In order to effectively manage the size of these droplets, it was necessary to experimentally determine the effect of the fills of the cooling towers on droplets. Five types of cooling fillers were used for experimental measurements: 3 film fills and 2 splash fills - trickle and grid. Drop size measurements were performed using the LIF method. Histograms of droplets size were obtained from measured droplet sizes under each fill, and for each fill, the Sauter droplet diameter was then calculated. According to a theoretical analysis of a breakdown of droplets, the combinations of some fills and the effect of their surface treatment on the droplet diameter were then measured for comparison.

1 Introduction

The task of the cooling tower is to dissipate away waste heat to the atmosphere [1]. In the wet cooling towers, cooled water is piped to nozzles that spray water on the cooling fills [2]. Significant heat dissipation occurs in three areas: in the area between the nozzles and the fill zone - the spray zone, the cooling fill itself, and the area under the cooling fills - the rain zone. The most significant heat transfer takes place directly on the cooling fills. The type of cooling fill determines how significant the heat transfer will be directly on the fill but also in the rain zone. Its mathematical model is described in [3] and [4].

There are many types of cooling fills, but the most common fills are film fills and splash fills (Figures 1-3). In the first type, as suggested by the name, water after impact on the surface of the fill creates a thin film of water, which allows to intensify the transfer of heat. The droplets under the fill are consequently mainly formed by dripping - drip effect. The second type of fill is based on breaking big water structures into smaller droplets - splash effect. With the splash mechanism, a large number of small droplets can be obtained, thereby increasing the heat transfer surface. Film fills achieve greater efficiency, but the advantage of splash fills lies in their resistance to the influence of dirty water.

In order to effectively control the droplet size in the rain zone and hence the efficiency of cooling, it is necessary to know the range of droplet sizes formed under the individual fills. Consequently, experiments were carried out to determine droplet size in the rain zone for several types of fills. Based on these measurements and the theoretical analysis of the principle of droplet formation under cooling fills, two variants of fills were proposed to improve the heat transfer efficiency in the rain zone. The effect of these proposals was subsequently verified experimentally.

2 Measurements

2.1 Measured fills

The measurement of droplet size under the fill was performed for 5 commonly used fills. Three fills were the film fill type (Fig.1): a straight film fill, a slope film fill and an offset film fill. These fills are made up of blocks and the water runs down through the channel on the surface of the fill. The three measured film fills varied in the shape of this channel and its orientation in the fill. The straight film fill has a channel axis parallel to the direction of flow, the axis of the channel of the slope film fill is inclined to the direction of flow and the offset film fill has a curved channel.

The splash fills that were measured were a trickle and a grid type. The trickle fill consists of a dense network block (Fig 2).

The grid fill (Fig. 3) is also made up of a mesh, but not as dense as the trickle fill has it. The grid fill does not form a compact block, but it is composed of layers of individual panels. One layer was measured in this experiment.

* Corresponding author: rut.vitkovicova@fs.cvut.cz

https://doi.org/10.1051/epjconf/201818002119

Fig. 1. Film fill

Fig. 2. Trickle fill

Fig. 3. Grid fill

2.2 Experimental method and measurement facilities

The Laser-Induced Fluorescence (LIF) method is used for obtaining a size of droplets below the fills. LIF has excellent detection sensitivity because a signal is observed against a dark background [5]. Main components of LIF system are a laser, camera with an optical filter and some fluorescent dye. Fluorescent solution (in this case Rhodamine B) in water can absorb light (green light) at one frequency and subsequently reemit light at a different frequency. The laser beam is expanded into a sheet using two cylindrical lenses. The cylindrical lens produces a light sheet of few millimeter thick. Rhodamine B, as a tracer dye, emits light of wavelength between 500 and 700 nm after absorbing light with the wavelength of 532 nm. Excitation light was filtered by a red filter with the cut-on wavelength about 600 nm. All the reflections and other droplets that are in front and behind the light sheet are eliminated through the filter. The red filter was installed at the lens of a camera. In Fig. 4 is shown scheme of this used method.

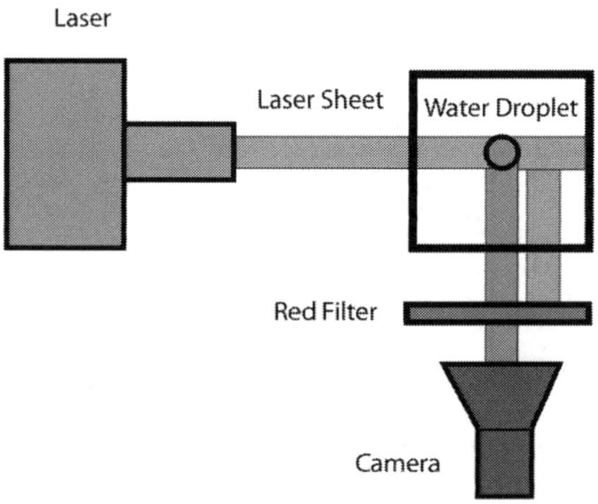

Fig. 4. LIF measurement

The experimental facility is consisting of a basin, a test section, an extending part, a water pump, a flow meter and a droplet generator. The test section is made from a plexiglass (Fig. 5). The dimension of the test section is 0.6 x 0.6 m and a total height of this section is 1 m. The extending part is consisting of several modules with a height of 1 m and 0.5 m.

https://doi.org/10.1051/epjconf/201818002119

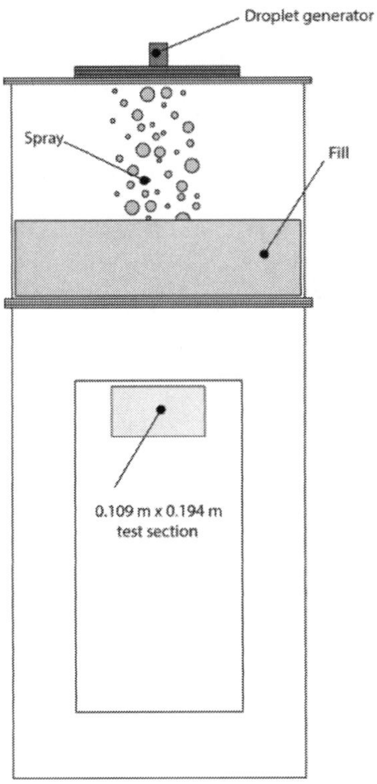

Fig. 5. Experimental facility

The modules are made from a galvanized metal sheet. And it can be arbitrarily combined with each other. The water droplets generator is supplied by the water feed pump connected to the water basin below the test section. The water flow rate is measured by the electromagnetic flow meter. The droplets generator was made from plexiglass sheets [6]. It is consisting of a three layers. The last layer has holes with 1 mm diameter arranged into the grid 0.01 x 0.01 m. The grid of the holes is creating an active zone in the test section. It dimension is 0.2 x 0.2 m. The droplets generator was placed on the top of the experimental facility. Droplets size distribution of droplet generator is based on [6].

2.3 Set – up of experiments

In this case, it was used measuring system which consist of the industry camera SONY XCG/H280E with 85 mm lens. The exposure time was set to 1/50 sec. The aperture was set at 1.8. The Coherent Verdi G Series semiconductor laser was used. The laser power was set at 1 W.

The water flow rate was 11 l/s. It was measured 4 distances (1 m, 2 m, 3, m, 4 m) between the bottom of the fill and test section for 5 types of fills. Another series of measurement was finding out an influence of a combination of two type of fills and two type fills, where one of them was covered by a hydrophobic surface, on the size of droplets in rain zone.

2.4 Data evaluation

The capture images from the camera were evaluated in Matlab. A first step was increased contrast to staying black background and white blobs. This process is simply because a signal from the measurement is great due to use the Rhodamine G and the green laser light. In the Fig. (6) is showed this process.

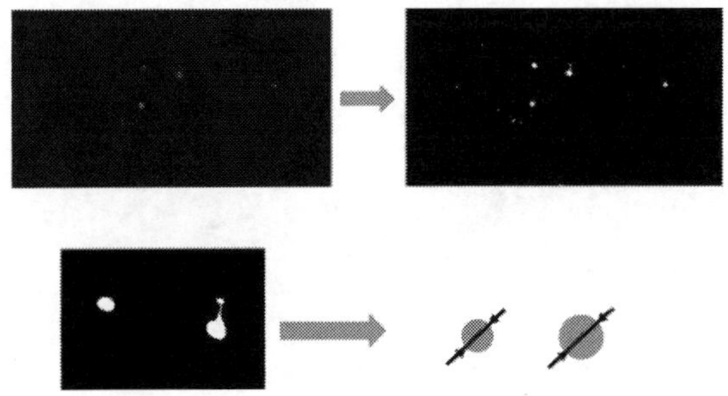

Fig.6. Process of droplets diameters evaluating

A second step is to calculate from the white blobs equivalent spherical droplet diameters d_i. Third step was a calculation of Sauter mean diameter (SMD) of all (N) calculated droplets from the equation:

$$SMD = \frac{\sum_{i=1}^{N} d_i^3}{\sum_{i=1}^{N} d_i^2} \tag{1}$$

The Sauter mean diameter allows a comparison of measured fills. When the Sauter mean diameter smaller, the efficiency of heat transfer is better in rain zone.

3 Results

Histograms of normalized droplets size were obtained from the evaluated data. In Fig. 7-11 are shown this normalized histograms. It is obvious that a change of droplets diameter is not very significant for different distances.

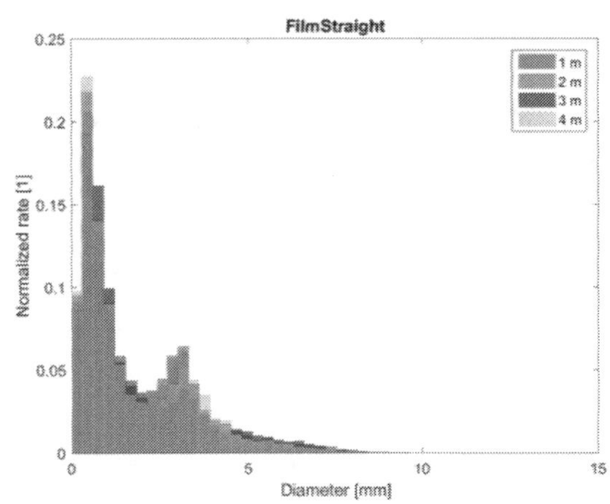

Fig. 7. Histogram of droplets diameters under film fill straight.

 https://doi.org/10.1051/epjconf/201818002119

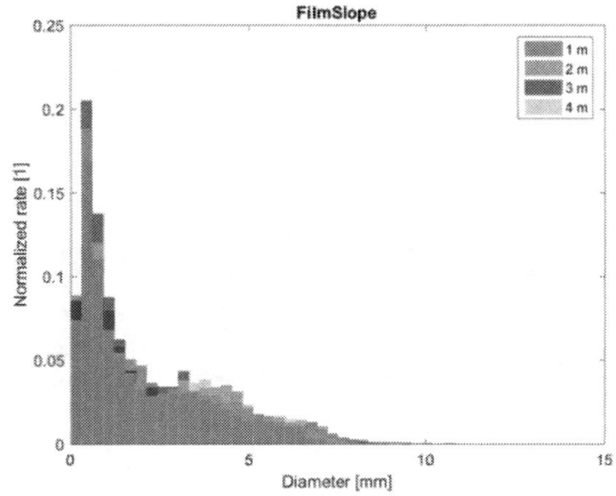

Fig. 8. Histogram of droplets diameters under film fill slope.

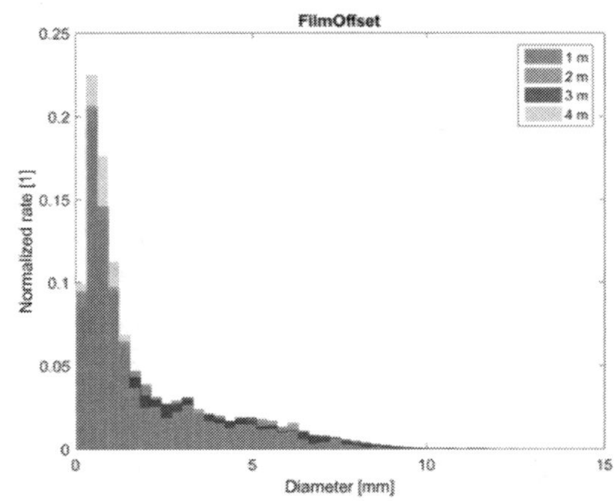

Fig. 9. Histogram of droplets diameters under film fill offset.

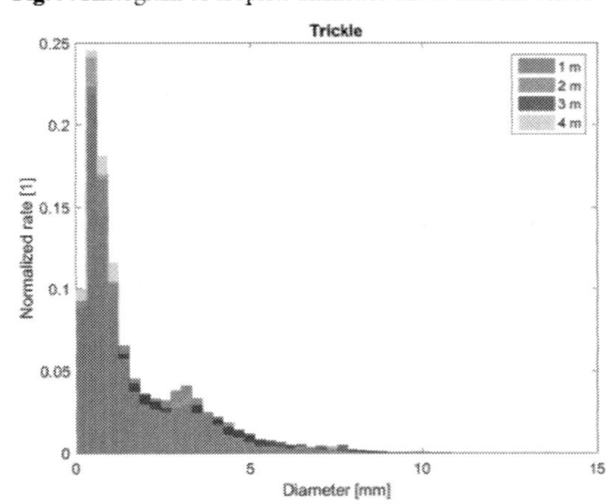

Fig. 10. Histogram of droplets diameters under trickle fill.

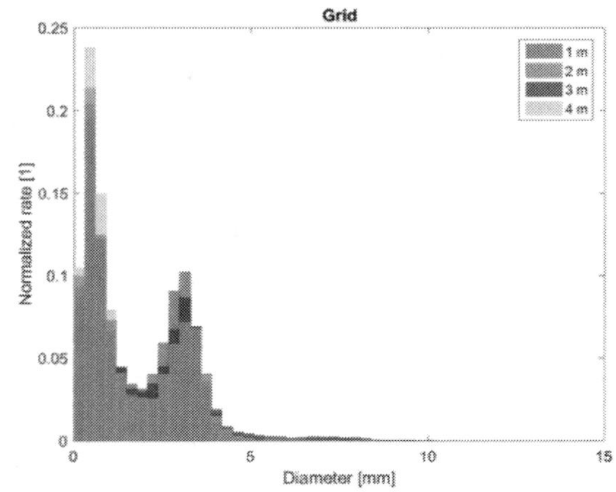

Fig. 11. Histogram of droplets diameters under grid fill.

It was calculated Sauter mean diameter for each fill. A comparison of Sauter mean diameter of this measured fills is in Tab. 1. The values of Sauter mean diameter show that the grid fill is the best fill for making a small droplets.

Table 1. Sauter mean diameter of each fill.

Type of fill	Sauter mean diameter [mm]
Straight film fill	4.23
Slope film fill	4.87
Offset film fill	5.26
Trickle fill	4.37
Grid fill	3.52

Based on previous experiments, theoretical analysis and analysis of a formation and a breakup of droplets [7, 8, 9, 10, 11], a combination of two fills was chosen to improve cooling performance in the rain zone while maintaining a high heat transfer efficiency directly on the fill. This option uses the splash mechanism (breaking a droplet on satellite droplets). The film fill was selected as the main fill, and a grid fill placed under film fill was selected as a complement fill to the splash mechanism. Two variants of this combination were measured. The first variant was the combination of these fills without any modification, the second combination was a grid fill with a hydrophobic layer. Hydrophobic layer prevents the formation of water film on the surface and allows smaller droplets to form. For experimental work, the hydrophobic layer was made of wax (Fig.12).

Fig. 12. Grid fill with hydrophobic surface.

Figures 13 and 14 show histograms comparing the rate of droplets between a combination of film fill and grid fill with wax with film fill itself (Fig.13) and film fill and grid fill with wax and fill and grid fill without surface treatment (Fig.14).

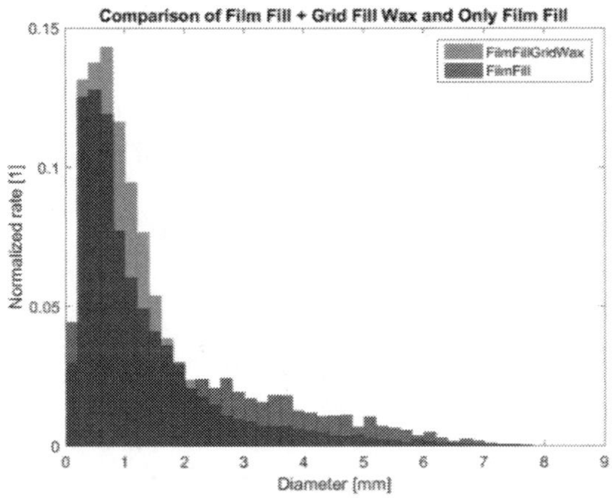

Fig. 13. Histogram of comparison combinations of film fill and grid fill with wax with film fill itself.

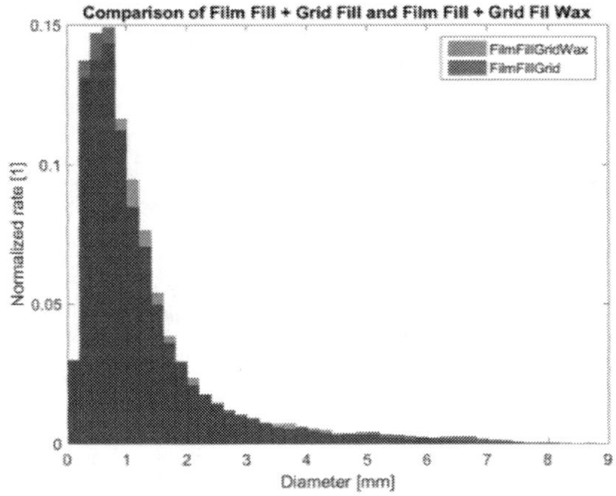

Fig. 14. Histogram of comparison combinations of film fill and grid fill with wax and fill and grid fill without surface treatment.

As is clear from the histograms, the film fill and grid fill combination with wax greatly reduces droplets in the rain zone to the film fill itself. The difference between the combination of film fill and grid fill without surface layer and the film fill and grid fill with wax is not very significant (Sauter mean diameter difference is about 0.5 mm), but it can be noticed that the grid fill with wax does not make nearly any droplets larger than 6.5 mm, but at the grid fill the maximum diameter is 7.5 mm. This is due to the limitation of drip mechanism of grid fill with the hydrophobic layer. The Sauter mean diameters comparison is in Tab. 2, where the Sauter mean diameter for the droplet generator itself is mentioned for a comparison.

Table 2. Sauter mean diameter of combinations.

Type of fill	Sauter mean diameter [mm]
Droplet generator (DG)	2.79
DG + film fill	4.87
DG + film fill + grid fill	3.90
DG + film fill + grid fill with wax	3.37

4 Conclusion

From the droplet size measurements at four distances below the fill, it appeared that the droplet size range under the fill doesn't change much with a distance, but only a small increase in the number of smaller droplets with a greater distance from the fill can be observed. If we compare the Sauter mean diameters, it is clear, as already mentioned, that the grid fill makes the smallest droplets. The reason is the predominant splash effect over dripping. The film fill straight and the trickle fill have the Sauter mean diameter of approximately the same value. This may be due to the fact that some droplets pass the fill straight through the channel without touching the walls of the fill. The trickle fill has a larger Sauter mean diameter against to the grid fill, because on the surface of the trickle fill the water film is formed to a greater extent and therefore it occurs more the drip mechanism. The largest drops are made by the film fill offset. The reason may be the dominating drip mechanism.

For a conservation of a high efficiency of heat transfer on the fill but also to increase efficiency in the rain zone, based on these previous measurements and theoretical analysis, a solution was chosen where a grid fill for small droplets was added under highly effective film fills. It was also found out to what extent the improvements will occur if this fill is covered with a hydrophobic layer. It is clear from the presented results that the grid fill placed under the film fill will greatly contribute to the reduction of droplets diameter in the rain zone. The difference between the standard and grid fill with the hydrophobic surface is not so obvious, however, it can be seen that the hydrophobic surface prevents more water film formation and reduces the drip mechanism and supports the spalsh mechanism. The results show that in cooling towers where film fill can be used, the efficiency of cooling can be improved by adding an additional build in order to minimize pressure loss.

Acknowledgments

The author acknowledges the support received from Advanced technologies for heat and electricity production TE01020036 and Centre of 3D volumetric velocimetry – COLA supported by the European Union (CZ.2.16/3.1.00/21569) Operating Program Prague competitiveness.

References

1. D.G. Kröger, *Air-Cooled Heat Exchangers and Cooling Tower*, (2004)
2. H.C. Reuter, D.G. Kröger ASME IHTC, **4**, 619 – 628, (2010)
3. T. Hyhlík, Engineering Mechanics, **21**, 240-243, (2014)
4. T. Hyhlík, ICCT 2014 International Conference, 219-226, (2014)
5. R. N. Zare, Annual Review of Analytical Chemistry, **5**, 1-14, (2012)
6. P. Vitkovic, EPJ Web of Conferences, **114**, 02137, (2016)
7. A. L. Yarin, Annu. Rev. Fluid Mech. **38**, 159-92,(2006)
8. A. A. Dreyer, P.J. Erens, J. Heat Mass Transfer, 39, 109-123, (1996)
9. C. Josserand, S. T. Thoroddsen, Annu. Rev. Fluid Mech., **48**, 365-91, (2015)
10. P. Vitkovic, EPJ Web of Conferences, **143**, 02142, (2017)
11. M. Rein, Fluid Dyn. Research, 12, 61-93, (1993)

Flow behaviour and structure of heterogeneous particles-water mixture in horizontal and inclined pipes

Pavel Vlasák[1*], *Zdeněk* Chára[1], *Jiří* Konfršt[1]

Institute of Hydrodynamics CAS, v. v. i., 166 12 Prague 6, Czech Republic

Abstract. The effect of slurry velocity and mean concentration of heterogeneous particle-water mixture on flow behaviour and structure in the turbulent regime was studied in horizontal and inclined pipe sections of inner diameter D = 100 mm. The stratified flow pattern of heterogeneous particle-water mixture in the inclined pipe sections was revealed. The particles moved mostly near to the pipe invert. Concentration distribution in ascending and descending vertical pipe sections confirmed the effect of fall velocity on particle-carrier liquid slip velocity and increase of in situ concentration in the ascending pipe section. Slip velocity in two-phase flow, which is defined as the velocity difference between the solid and liquid phase, is one of mechanism of particle movement in two-phase flow. Due to the slip velocity, there is difference between transport and in situ concentrations, and the slip velocity can be determined from comparison of the in situ and transport concentration. For heterogeneous particle-water mixture flow the slip velocity depends on the flow structure.

1 Introduction

Hydraulic pipeline transport is commonly used for transport of different loose (granular, powdery) materials, e.g. in dredging, building, mining, or land reclamation [1]. A lot of theoretical or experimental studies have been carried out on transport of sand or fine particles in horizontal pipes [2, 3]. However, a relatively little research has been done on hydraulic conveying in vertical and inclined pipes, especially of coarse particles as gravel or pebbles.

From operational point of view transport concentration, operational velocity and pressure drops are the most important parameters for pipeline transport design and operation. The particles in the turbulent flow are supported by turbulent diffusion, and near the pipe wall a lift force, associated with slip velocity and concentration profile, contributed to the particle lift-off, too. For the particles with size larger than the thickness of viscous sub-layer, Saffman force, induced due to the shear of the fluid, supports particle movement and together with Magnus force (due to the particle rotation) [4] could reach a significant fraction of the total weight of particles [5].

Based on macroscopic modelling, where the conservation equations are formulated using averaged quantities over the control volume, Wilson [6] proposed a two-layer model for heterogeneous slurries with a fully stratified flow pattern. Heterogeneous mixture flow in a pipe may be defined as the flow with an asymmetrical velocity and concentration distribution. When the Reynolds number, friction factor and Coulomb type friction are defined for each layer as well as the interfacial friction factor, the flow parameters could be determined [7]. Because the layers differ in the local solids concentration and velocity, there is a difference in the mean velocities of the particles and the liquid. Slip between the particles and the liquid results in a continuous transfer of energy from the fluid to the particle and from the particle to the pipe wall [8].

Friction losses in pipeline flow of heterogeneous solid particle-water mixtures are strongly dependent on the flow pattern [9]. If the operational velocity of the mixture is close to deposition limit, a granular bed forms at the pipe invert. The bed slide over the pipe invert at velocities above the deposition limit and it is stationary below the deposition limit. The contact bed is important contributor to solid friction in mixture flow.

2 Experimental equipment and material

The experimental investigation was carried out on the pipe loop of inner diameter D = 100 mm with horizontal (A) and inclinable (B) pipe sections, see Figure 1. The total length of the loop is 93 m, the loop is suitable for studying of mixture velocity and concentration effect on the mixture flow behaviour, concentration distribution and pressure drop.

The measured solid-liquid mixture was prepared in a mixing tank (1) and pumped by a centrifugal slurry pump GIW LCC-M 80-300 (2) with variable speed drive Siemens 1LG4283-2AB60-Z A11 (3) to a measurement section of transport pipe. Transparent viewing pipe sections (7) for visual observation were situated on the end of the horizontal pipe section (A) and in the inclinable section (B). The transparent viewing pipe sections (7) for visual observation were situated on the end of the horizontal pipe section (A).

* Corresponding author: vlasak@ih.cas.cz

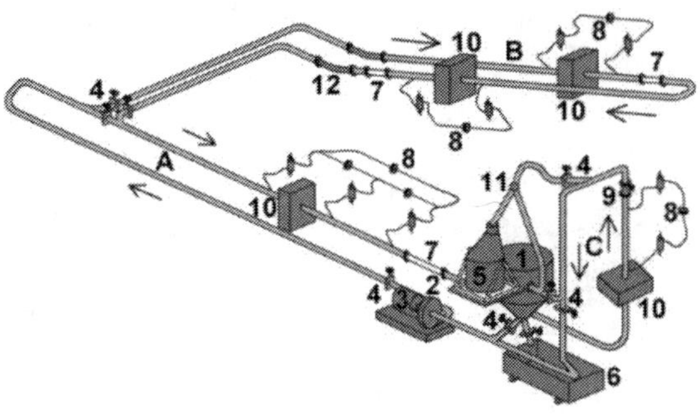

Fig. 1. Experimental test loop D = 100 mm (IH AS CR)

A – horizontal section, B – inclinable section, C – short vertical section; 1 – mixing tank, 2 – centrifugal dredge pump, 3 – electric motor, 4 – slide valves, 5 – pycnometer, 6 – slurry output tank, 7 – transparent viewing pipe section , 8 – differential pressure transducer, 9 – electromagnetic flow meter, 10 – radiometric devices, 11 – flow divider, 12 – flexible hose.

The pressure drops were measured by the Rosemount DP transmitters 1151DP (8) in the horizontal and inclined measuring sections within a broad range of mean mixture velocities, from values close to the deposition limit velocity ($V_{dl} \approx 1$ m/s), to maximum values $V_{max} \approx 4.0$/s.

Slurry mean velocities were measured by a Krohne OPTIFLUX 5000 magnetic flow meter (9), mounted in the short vertical section (C) at the end of the circuit. The flow divider (11) allows collection of slurry samples in the calibrated sampling tank (5) and measuring of the delivered concentration, c_v, and the flow rate Q. The vertical U-tube (B) enables evaluating the delivered concentration of solid phase.

To measure local concentration and concentration profiles the loop is equipped with radiometric density meters (10) placed on a special support controlled by the computer [10]. The support serves for vertical linear positioning of both the source and the detector to measure chord averaged vertical concentration profiles.

To determine local concentration distribution through the pipe cross-section (the concentration maps) the parallel projections of γ-ray beam were provided at several angles around the pipe axis (i.e. step of 15° from zero to 165°), and the collected data were processed by computer tomography method.

The radiometric density meters consist of a γ-ray source (Caesium^{137}Cs, activity 740 MBq) and of a detector (a scintillating crystal of NaI(Tl)). A multi-channel digital analyser enables an evaluation of the energy spectrum of the detected signal. The measuring time period of 16 seconds was used to sense the local concentration at each position [11].

The radiometric device was mounted upstream of the measuring and viewing sections in the horizontal or inclinable sections of the loop.

The studied solid-liquid mixtures consist of basalt pebbles (particle diameter, d, ranging from 8 to 16 mm, the mean diameter $d_{50} \approx 11.5$ mm, particle density ρ_p = 2 895 kg m^{-3}) and water, the overall volumetric concentration, c_v, ranged from 3 to 15%.

3 Slip velocity

Slip velocity in two-phase flow is defined as the velocity difference between the solid and liquid phase. It is one of mechanism of particle movement in two-phase flow. In the homogeneous model of two-phase flow the slip velocity V_{sl} is by definition assumed to be close to zero (no slip). For heterogeneous slurry flow it was experimentally observed that the slip velocity depends on the flow pattern (e.g. suspended flow, stratified flow, slug flow). Due to slip velocity, there is difference between delivery (transport c_d) and spatial (in situ c_v) concentrations, and the slip ratio ($c_d / c_v = V_{sl} / V_s$) is the parameter describing flow stratification in a pipe. In vertical two-phase flow the slip velocity can be approximated by hindered settling velocity. The slip-velocity will be determined from comparison of the mean in situ and mean transport concentration.

Effect of mean mixture velocity V_s and pipe inclination angle α on values of slip ratio in the horizontal, vertical, and inclined ascending ($V_{sl,a} / V_s$) and descending ($V_{sl,d} / V_s$) pipe sections, determined from the measured mixture transport and in-situ concentrations is indicated in Table 1 for mean mixture velocity about V_s = 2.85 m/s and in Table 2 for mean mixture velocity about V_s = 3.85 m/s and transport concentration $c_d \approx 0.06$. From Figure 2 it is evident that value of the slip ratio increases with the inclination angle α and with decreasing flow velocity V_s. The slip ratio is practically constant for inclination angle in range 0 -15° and 45-90°.

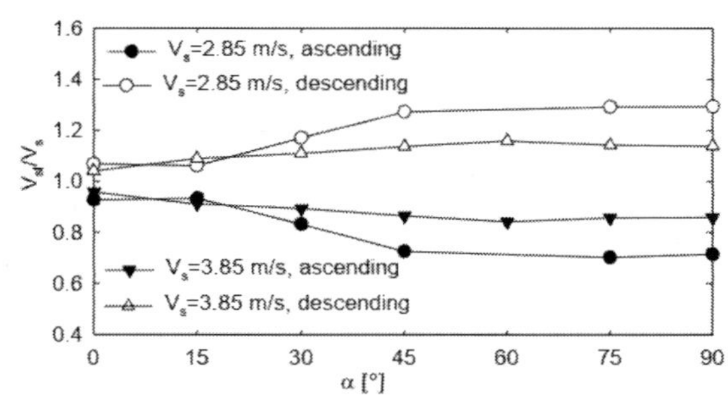

Fig. 2. Effect of pipe inclination angle α on slip ratio V_{sl} / V_s.

Table 1. Effect of the pipe inclination angle α on slip ratio in ascending $V_{sl,a} / V_s$, and descending $V_{sl,d} / V_s$ pipe sections.

V_s = **2.85** [m/s]					
α [°]	c_d	$c_{v,a}$	$c_{v,d}$	$V_{sl,a}/V_s$	$V_{sl,d}/V_s$
0°	0,0385	0,0360	0,0415	1,06944	0,92771
15°	0,0397	0,0425	0,0374	0,93412	1,06150
30°	0,039	0,0469	0,0333	0,83156	1,17117
45°	0,0392	0,0541	0,0308	0,72458	1,27273
75°	0,0372	0,0530	0,0288	0,70189	1,29167
90°	0,0370	0,0518	0,0286	0,71429	1,29371

Effect of the mixture velocity on value of the slip ratio in horizontal and vertical ascending or descending pipe sections is illustrated on Figure 3. It is evident that in vertical flow the differences between slip ratios V_{slip} / V_s in ascending and in descending pipe sections decreases with increasing flow velocity V_s. These slip ratios are nearly symmetrical to line $V_{slip} / V_s = 1$. The values of the slip ratio in horizontal and ascending vertical flow are very similar.

Table 2. Effect of pipe inclination angle α on slip ratio in ascending $V_{sl,a}/V_s$, and descending $V_{sl,d}/V_s$ pipe sections.

V_s = **3.85** [m/s]					
α [°]	c_d	$c_{v,a}$	$c_{v,d}$	$V_{sl,a}/V_s$	$V_{sl,d}/V_s$
0	0,06110	0,0637	0,0587	0,9592	1,0409
15	0,06410	0,0703	0,0588	0,9118	1,0901
30	0,06050	0,0677	0,0545	0,8936	1,1101
45	0,05470	0,0633	0,0481	0,8641	1,1372
60	0,05380	0,0639	0,0464	0,8419	1,1595
75	0,05065	0,0591	0,0443	0,8570	1,1433
90	0,05330	0,0621	0,0468	0,8580	1,1390

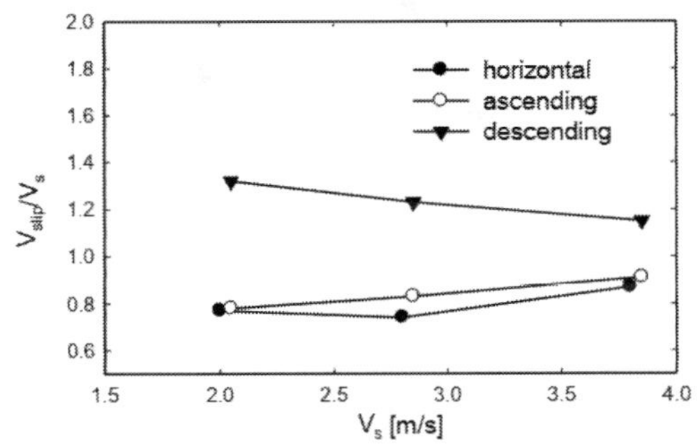

Fig. 3. Slip ratio V_{slip}/V_s in horizontal and vertical pipe sections.

4 Local concentration distribution

Distribution of the local concentration in the pipe cross-section has a great effect on both the mixture's flow behaviour and pressure drop and it is important for understanding the mechanism of the heterogeneous mixture flow. Various methods have been used for measurement of the local concentration, e.g. isokinetic sampling, different visualization techniques, electrical resistance and capacity, or radiometric methods [5, 6, 12-19].

The effect of different flow parameters on the local concentration distribution in chord-averaged vertical concentration profiles in pipe sections, e.g. the pipe inclination α and the mixture velocity V_s was studied for three values of the transport concentration and three mixture flow velocities. The measured concentration profiles confirmed the stratified flow pattern of the coarse particle-water mixture in the inclined pipe sections, see Figure 4. The chord-averaged concentration profiles can be divided into three parts similarly as in the chord-averged horizontal pipe sections [20-21].

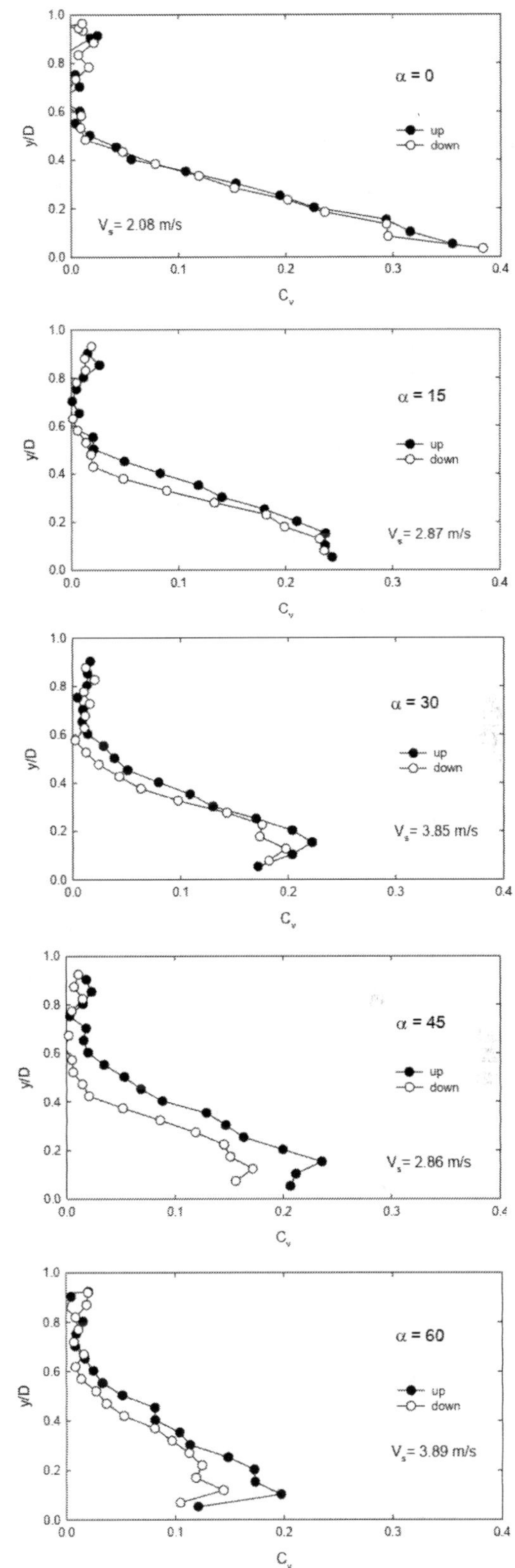

https://doi.org/10.1051/epjconf/201818002120

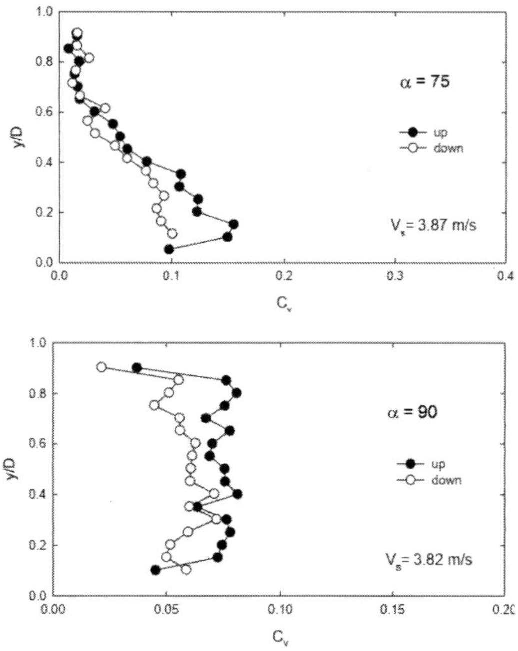

Fig. 4. Effect of the inclination angle α and the mean mixture velocity V_s on profiles of local concentration c_v (transport concentration $c_d \approx 0.06$).

The local concentration c_v approaches practically zero in the upper portion of the pipe, this region increases for the descending flow with decreasing mixture velocity V_s and mean transport concentration c_d. A nearly linear concentration distribution was determined in the central portion of the concentration profiles.

Solids concentration reached maximum near the pipe invert, and for the higher mixture velocity a thin layer with nearly constant local concentration formed. However, for studied conditions the bed concentration in inclined pipe sections never reached values close to the loose-packed value. It demonstrates the permanent movement of bed particles along the bed of the pipe (sliding, rolling or saltation mode) for given flow condition (i.e. for velocities higher than the deposition limit).

Similarly as it was confirmed for pressure drop, the effect of pipe inclination for low values of angle α (up to about 30°) is not significant, see Figure 5, especially in the upper and central portions of the pipe [8, 10]. Local concentration c_v at the pipe invert slightly decreased with the increasing pipe inclination. For higher values of angle α, a decrease in concentration close to the pipe invert was observed; the local concentration in the central portion of the pipe increased and for the vertical pipe a nearly constant concentration distribution was observed. The local concentration in the bed layer decreased with increasing velocity and with increasing inclination angle α. No maximum of local concentration was observed for descending flow direction and the inclination angle α = 15°, concentration profiles were nearly linear in the lower portion of the pipe. The effect of inclination was similar for both higher mixture velocities, for low mixture velocity (V_s = 2.05 m/s) a bed layer was observed.

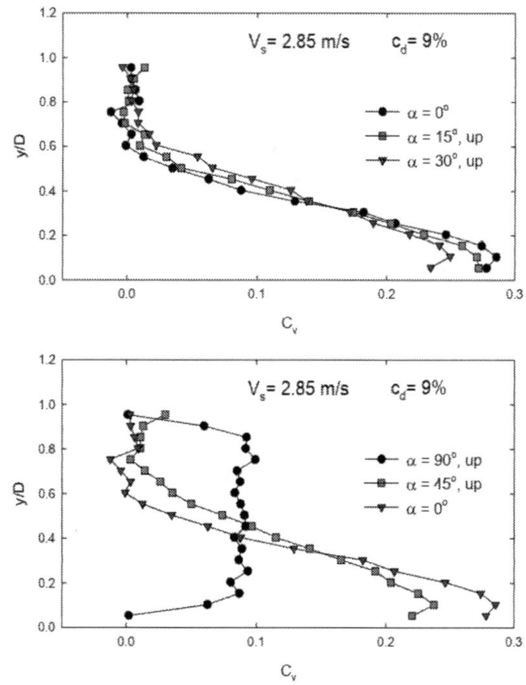

Fig. 5. Effect of the inclination angle on local concentration c_v.

The effect of the mean transport concentration c_d for the inclination angle α = 30° and the mixture velocity V_s = 2.85 m s^{-1}, is illustrated on Figure 6, both for the ascending and descending flow directions.

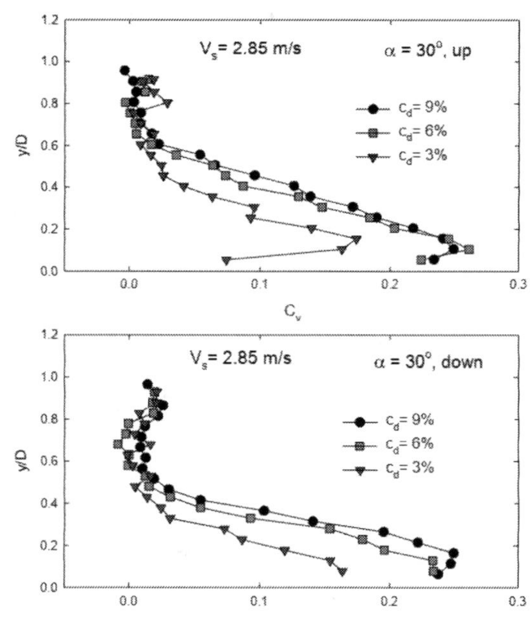

Fig. 6. Effect of the transport concentration c_d on local concentration c_v.

With the increasing mean transport concentration the local concentration in the bed layer, as well as in central proportion of the pipe increased, and in ascending flow direction the bed layer was distinct even for low concentration value. For the descending flow direction concentration profiles are nearly linear in the lower portion of the pipe, and the zero concentration part of the pipe cross-section is significantly bigger than that for the ascending flow direction due to the braking effect of gravity force on ascending flow and accelerating effect of gravity force on descending flow [22]. The positive and negative effect of gravity decreases and increases, respectively, the particle-liquid slip velocity.

The same was observed from Figure 7, which illustrated the effect of mean mixture velocity V_s for low values of the pipe inclination angle α =15° and 30°. For moderate and higher mixture velocities bed layers with thickness of about 20% of the pipe diameter were formed. The local concentration in the bed layer decreased with increasing velocity; probably due to increasing saltation intensity more particles reach higher portion of the pipe. This effect increased with increasing inclination angle.

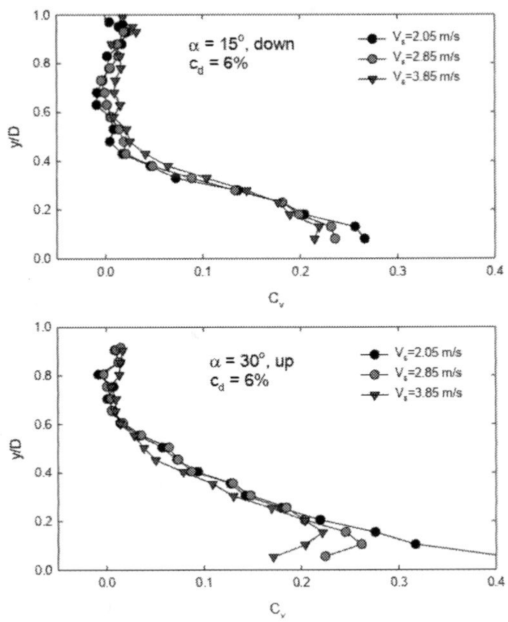

Fig. 7. Effect of the mean mixture velocity V_s on local concentration c_v.

The effect of the ascending and descending flow is clearly illustrated in Figure 8, for the low and high mixture velocities (V_s = 2.05 and 3.85 m/s), and wide range of the inclination angle α, from 15° to 90°. The local in situ concentration in ascending pipe sections is due to the effect of particle fall velocity always higher than that in descending pipe sections. Solid particles move in the ascending section with lower velocity than in the descending section, of course the transport concentration is constant over the whole pipe loop. It is valid also for vertical up-ward and down-ward flow, where difference between the concentration values corresponds to particle slip velocity.

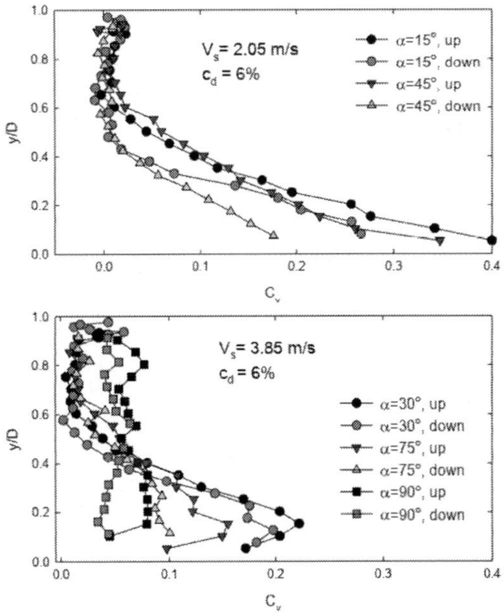

Fig. 8. Effect of the ascending and descending flow on local concentration c_v

4 Conclusions

The coarse particles (basalt pebbles) – water mixtures flow behaviour was studied in horizontal and inclined smooth pipe sections of inner diameter D = 100 mm. The visualization and local concentration measurements revealed the stratified flow pattern of the coarse particle-water mixture in the inclined pipe sections.

Similarly to horizontal flow the chord averaged concentration profiles in the inclined pipe sections can be divided in three parts: a zero-concentration region in the upper portion of the pipe, a region with nearly linear concentration distribution in the central portion of the pipe, and a region with a maximum concentration near the pipe invert.

For velocities close to deposition limit dune formations or sliding bed were formed. For moderate and higher mixture velocities, particle saltation became the dominant mode of sediment transport.

The in situ concentration reached higher values in the ascending section than in the descending section, the particles moved mostly near to the pipe invert.

Value of slip ratio increases with the inclination angle and with decreasing flow velocity; the slip ratio is practically constant for the inclination angle in range 0 - 15° and 45-90°.

For the inclination angle α up to about 30°, the effect of the pipe inclination on local concentration distribution is not significant. The zero concentration region increases for descending flow direction with decreasing mixture velocity and mean transport concentration and it is bigger than that for the ascending flow direction. With increasing mean mixture velocity the local concentration in the bed layer slightly decreased; this effect increased with increasing inclination angle.

Acknowledgements

Supports under the project 17-14271S of the Grant Agency of the Czech Republic, and RVO: 67985874 of the Czech Academy of Sciences are gratefully acknowledged.

References

1. P. Vlasak, B. Kysela, Z. Chara, J. Hydrol. Hydromech., **60**, 115 (2012).
2. P. Vlasak, Z. Chara, Powder Technol, **104**, 200 (1999).
3. P. Vlasak, Z. Chara, Part. Sci.Technol,, **27**, 428 (2009).
4. N. Lukerchenko, Yu. Kvurt, I. Keita, Z. Chara, P. Vlasak, Part. Sci.Technol,, **30**, 55 (2012).
5. K.C. Wilson, R.S. Sanders, R.G. Gillies, C.A. Shook, Powder Technol, **197**, 247 (2010).
6. K.C. Wilson, *Proc. HYDROTRANSPORT 4*, Banff (Canada), Paper A (1976).
7. K.C. Wilson, R.G. Addie, A. Sellgren, R. Clift, *Slurry Transport Using Centrifugal Pumps* (Springer,US., 2006)
8. P. Vlasak, Z. Chara, J. Konfrst, J. Hydrol. Hydromech. **65**, 183 (2017).
9. V. Matousek, V., Experimental Thermal and Fluid Science, **26**, 693 (2002).
10. P. Vlasak, Z. Chara, J. Krupicka, J. Konfrst, J. Hydrol. Hydromech. **62**, 241 (2014).
11. J. Krupicka, V. Matousek, J. Hydrol. Hydromech., **62**, 126 (2014).
12. J. Dolanský, Z. Chara, P. Vlasak, B. Kysela, J. Hydrol. Hydromech. **65**, 105 (2017).
13. V. Matousek, P. Vlasak, Z. Chara, J. Konfrst, Maritime Engineering, **168**, 93 (2015).
14. K. Przewlocki, A. Michalik, K. Wolski, K. Korbel, *Proc. HYDROTRANSPORT 6*, Canterbury (U.K.), **1**, 105 (1979).
15. F.J. Pugh, K.C. Wilson, Powder Technol, **104**, 221 (1999).
16. J. Sobota, F. Plewa, EJPAU 3(2), #01, http://www.ejpau.media.pl/volume3/issue2/environment/art-01.html (2000)
17. J. Sobota, P. Vlasak, G. Strozik, F. Plewa, *Proc. Eighth ISOPE Ocean Mining (& Gas Hydrates) Symp.*, Chennai (India), 220 (2009).
18. P. Vlasak, B. Kysela, Z. Chara, Part. Sci.Technol, **32**, 179 (2014).
19. M. Zych, R. Hanus, P. Vlasak, M. Jaszczurd, L. Petryka,*Powder Technol*, **318**, 491 (2017).
20. P. Vlasak, Z. Chara, J. Konfrst, J. Krupicka, *EPJ Web of Conferences* **92**, 02111(2015)
21. P. Vlasak, Z. Chara, J. Konfrst, J. Krupicka, Can. J. Chem Engng, **94**, 1040 (2016).
22. P. Vlasak, Z. Chara, J. Konfrst, Proc. 23rd Int. Conf. Eng. Mech. 2017, Svratka (CR), 1034 (2017).

PIV measurements of airflow past multiple cylinders

Waldemar Wodziak[1], *Jacek* Sobczyk [1]

The Strata Mechanics Research Institute of The Polish Academy of Sciences, Reymonta 27, 30-059 Krakow, Poland

Abstract.. Flow characteristics in vicinity of six circular cylinders aligned inline was investigated experimentally by means of PIV method. Experiments were conducted in a low speed closed circuit wind tunnel. Inflow velocity was 1.2 m/s which corresponds to *Re*=1600 based on the cylinder diameter. Spacing ratio between cylinders *L/D* was 1.5. Instantaneous and averaged velocity fields were presented. Experiments were designed in order to use their results as a test case for future numerical calculations.

1 Introduction

Fluid flow around cylinders is subject of interest theoretical and applied investigations. Knowledge about flow characteristic is essential for example to determine heat transfer intensity in rod structure of nuclear reactors or heat exchangers. Fluid flow over one and two cylinders has been studied extensively and is well known [1], [2], [3]. Depending on Reynolds number and distance between cylinders different flow patterns are observed.

Airflows in mining excavations are disturbed by presence of various devices. Velocity field changes and separation and recirculation zones may appear. Support roof system is widely used in mining longwalls for ceiling protection. One of the components of support roof system are legs which are approximately cylinder-shaped and distributed inline. This configuration may be considered as flow through multiple cylinders.

Configuration of multiple array of cylinders has been discussed in several works. Effect of gap spacing on vortex shedding of laminar flow past six cylinders arranged in-line are studied by [4] and [5]. They found a flow pattern transition between spacing ratios 3.6 and 4 when drag and lift coefficients change significantly. Studies of flow around six cylinders are mostly based on CFD calculations and only for low Reynolds numbers. However, the main limitation of computer modelling is the need to verify the applied numerical models, typically by experimental testing [6],[7].

Experimental results of turbulent airflow through six cylinders presented in this paper can be used as a benchmark for future CFD validation.

2 Experimental setup

Measurements were performed a in low speed closed-circuit wind tunnel. Total dimensions of wind tunnel are 9,79 x 4,08 x 2,34 m. A 30 kW fan is used to create a desired flow in the test section. Velocity range is from 0.1 to 60 m/s and longitudinal free stream turbulent intensity is less than 0.5%. Wind tunnel is equipped with cooling system to control temperature and humidity. Test section have dimensions: a length of 1.5 m, height and width 0.5m and is adapted to Stereo PIV measurements [8].

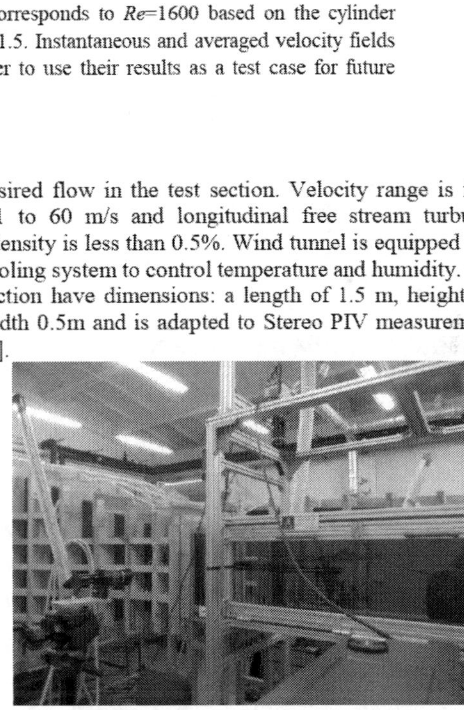

Fig. 1. Experimental setup

Particle image velocimetry is a non-intrusive method used to obtain velocity fields. Particle markers are introduced to measured area and illuminated twice by laser sheet. Camera placed commonly perpendicularly to illuminated plane records two images. Particles on the first frame are distributed stochastically. Next, this pattern is found on second frame. Particle displacement is calculated by means of statistical methods. Time between laser pulses is known therefore velocity vector can be obtained [9].

The seeding for the PIV method was DEHS oil sprayed through a seeding generator. The tested planes were illuminated by a two-impulse Nd:YAG laser and particle images were registered by two sCMOS cameras with a 5,5 MPx resolution. Each time 300 images were recorded with the frequency 15Hz. Calibration was carried out before measurements. Adaptive correlation with 32x32 px interrogation window was used to

* Corresponding author: wodziak@img-pan.krakow.pl

calculate velocity fields. The free-stream velocity was measured using a Schmidt anemometer.

Velocity range for airflow through legs in mining excavations is from 0.1 to 20 m/s which corresponds to *Re* number (based on leg diameter D=0.2m) from 1300 to 270000. In wind tunnel experiments inflow velocity was V_0= 1.2 m/s which gives a Reynolds number about 1600 (based on the cylinder diameter). 2D measurements were conducted for constant temperature (21^{0}C), humidity (φ=64,9%) and atmospheric pressure (980 hPa). Spacing between cylinder was L/D=1.5. Measured areas were located in lateral region and second on the wake region of airflow (Fig.2). Coordinate system origin is placed at the center of the upstream cylinder.

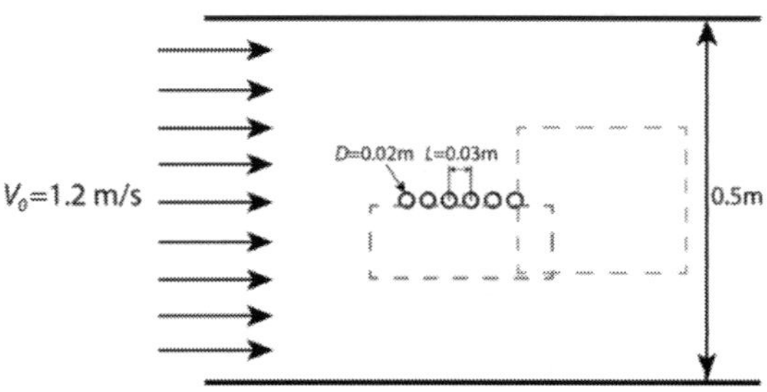

Fig. 2. Measured areas at test section

3 Results of PIV measurements

Normalized mean velocity components and their standard deviations contours calculated for lateral region are presented in Fig.3-6. Mean values are averaged over 300 images. There is an aerodynamic shadow between first two cylinders with virtually no mass exchange in the y direction. Following cylinders are more and more infected by vortices shed from the first one.

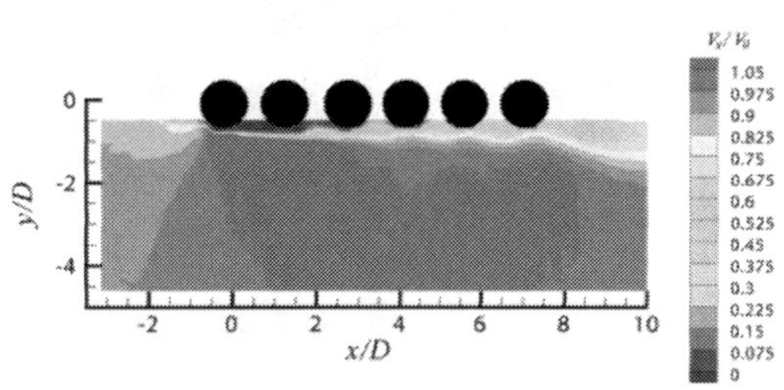

Fig. 3. Normalized mean streamwise velocity component contour plot obtained for lateral region

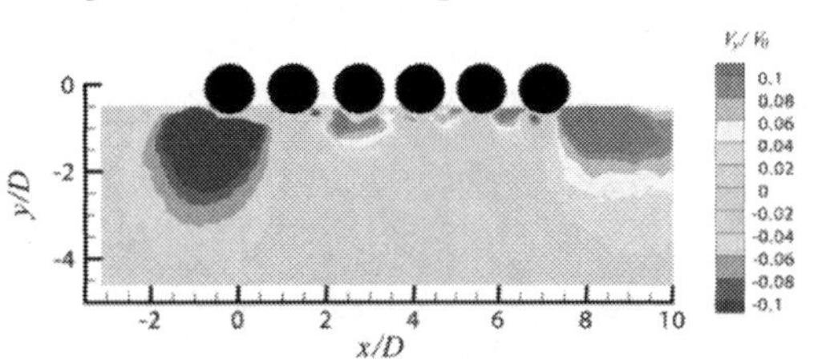

Fig. 4. Normalized mean transverse velocity component contour plot for obtained lateral region

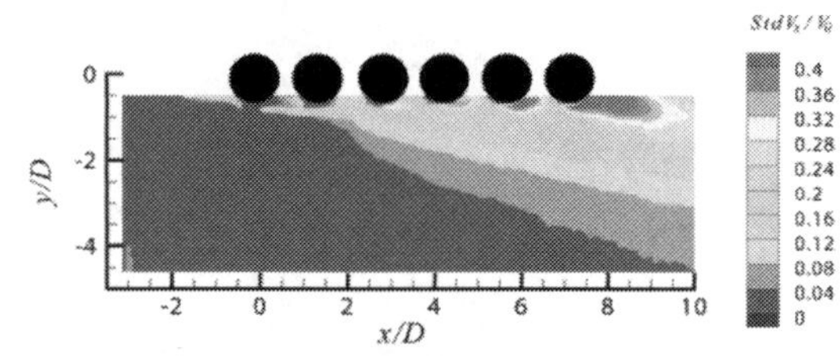

Fig. 5. Normalized mean standard deviation of streamwise velocity component contour plot for lateral region

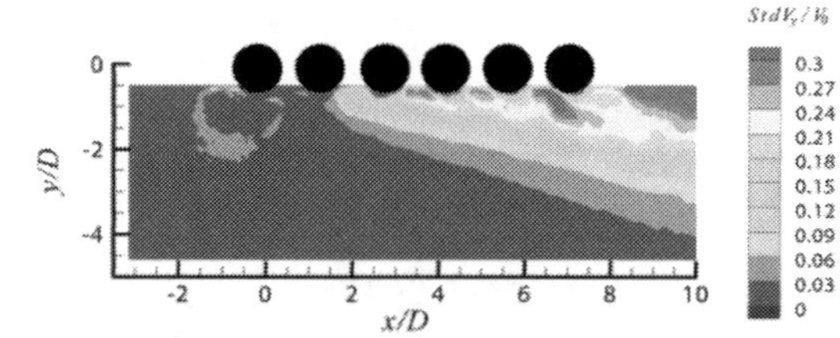

Fig. 6. Normalized mean standard deviation of transverse velocity component contour plot obtained for lateral region

PIV method provides instantaneous measurements of velocity vectors. Normalized velocity magnitude V contours at four time steps obtained for lateral region are presented in Fig 7-10. For better visualization of vortices from instantaneous velocity fields free stream velocity V_0 was subtracted and streamlines on the basis of remaining information were plotted. Contour plots presented on Fig. 7-10 and 15-18 contain complete information about V.

Time step between frames was ~0.06 s. Vortex shed from the first cylinder moved along following cylinders trying to enter in between them to create wakes. Remaining cylinders interfered with the wake dynamics and vortex formation region of the first cylinder. Vortices slid over them to finally form von Karman street in the wake region.

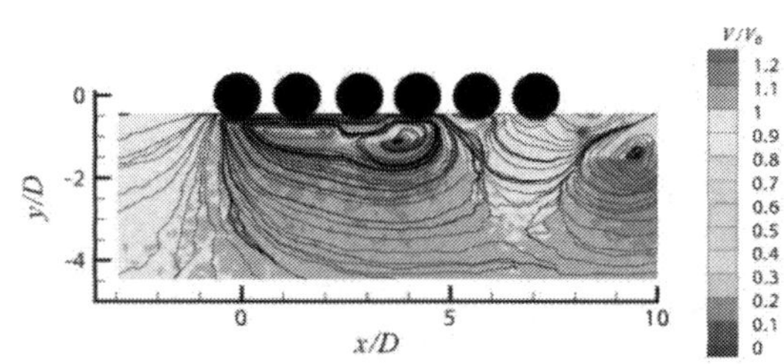

Fig. 7. Normalized velocity magnitude contour plot at first time step obtained for lateral region

 https://doi.org/10.1051/epjconf/201818002121

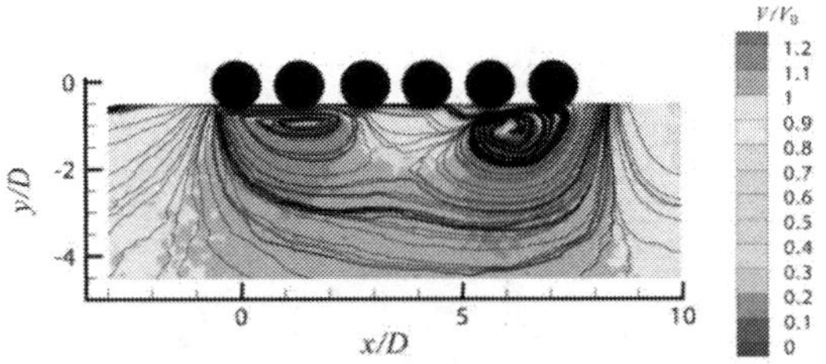

Fig. 8. Normalized velocity magnitude contour plot at second time step obtained for lateral region

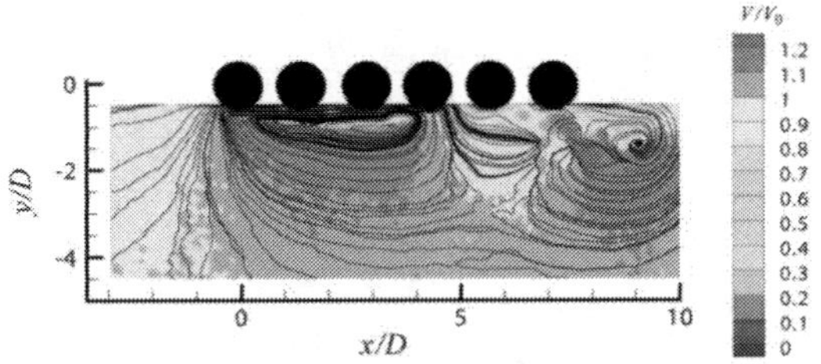

Fig. 9. Normalized velocity magnitude contours and third time step obtained for lateral region

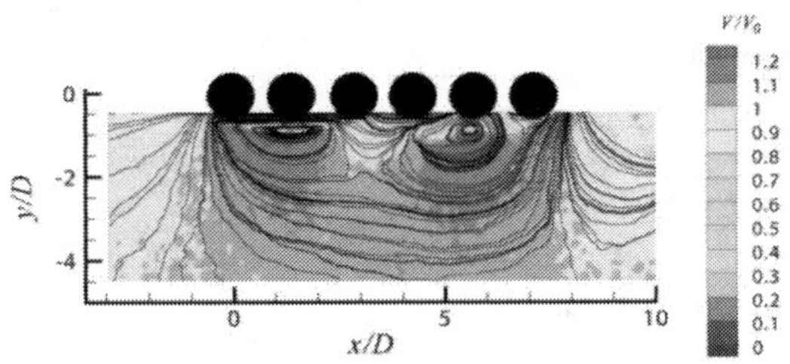

Fig. 10 Normalized velocity magnitude contours and fourth time step obtained for lateral region

Mean velocity components and their standard deviations are showed in Fig. 11-14.

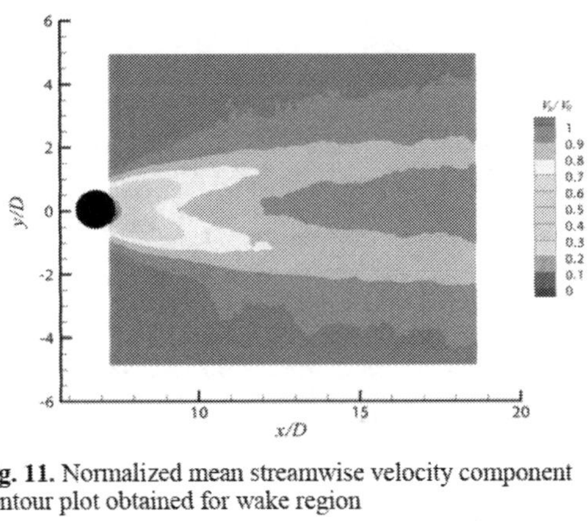

Fig. 11. Normalized mean streamwise velocity component contour plot obtained for wake region

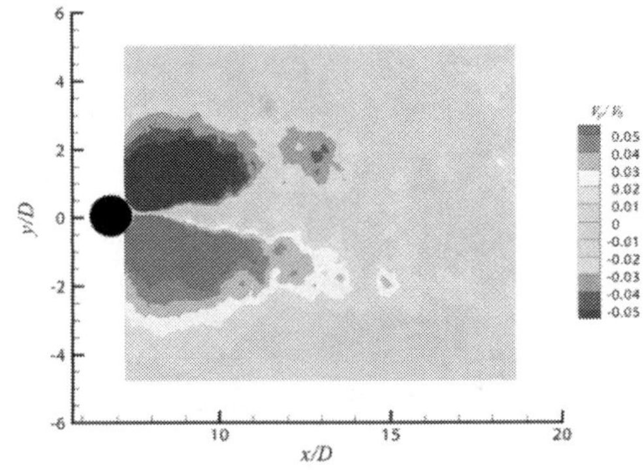

Fig. 12. Normalized mean transverse velocity component contour plot obtained for wake region

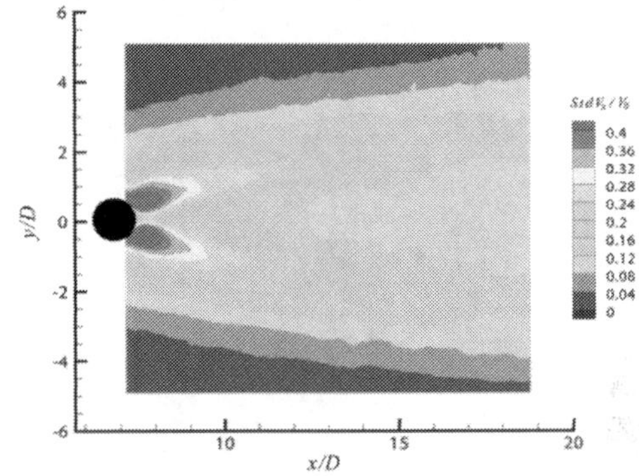

Fig. 13. Normalized mean standard deviation of streamwise velocity component contour plot obtained for wake region

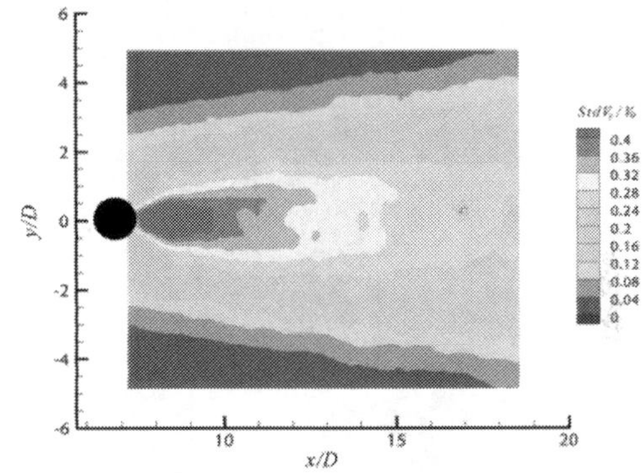

Fig. 14. Normalized mean standard deviation of transverse velocity component contour plot obtained for wake region

Normalized velocity value vector contours and vortex visualization at four time steps obtained for wake region are presented on Figures 15-18. Time difference between frames is 0.06 s. Vortex shedding from the outer side of cylinders forms a single vortex street.

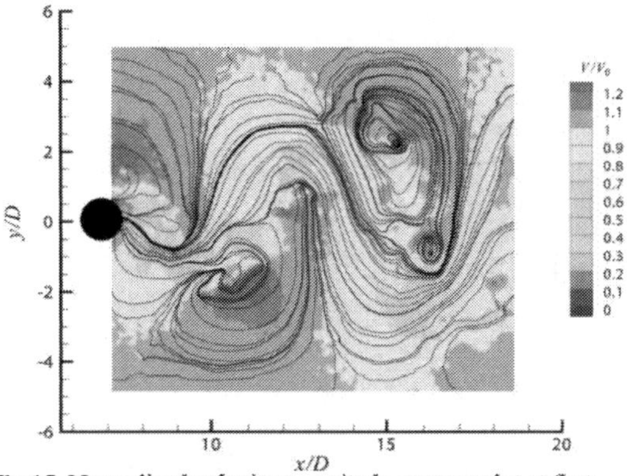

Fig.15. Normalized velocity magnitude contour plot at first time step obtained for wake region

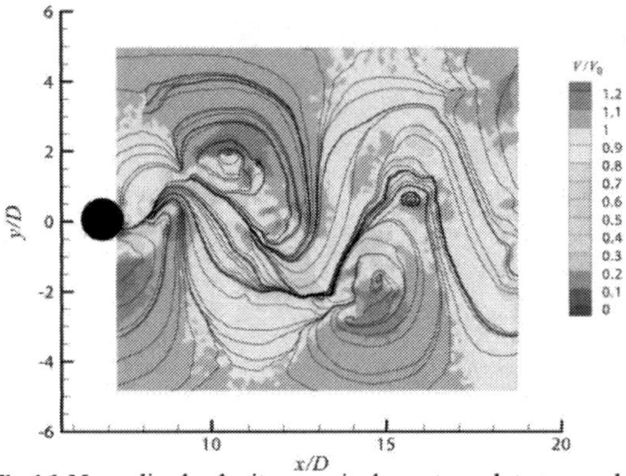

Fig.16. Normalized velocity magnitude contour plot at second time step obtained for wake region

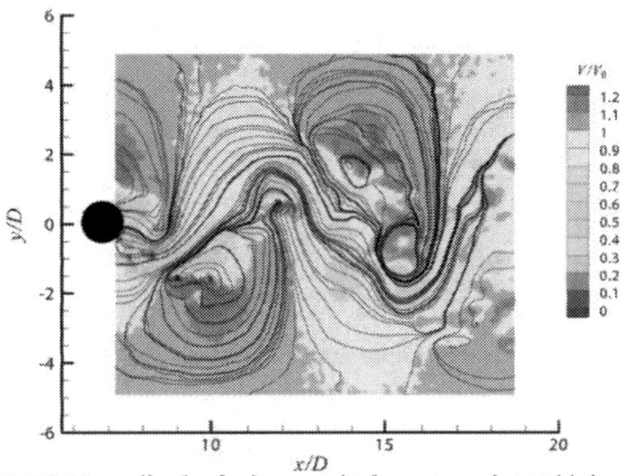

Fig.17. Normalized velocity magnitude contour plot at third time step obtained for wake region

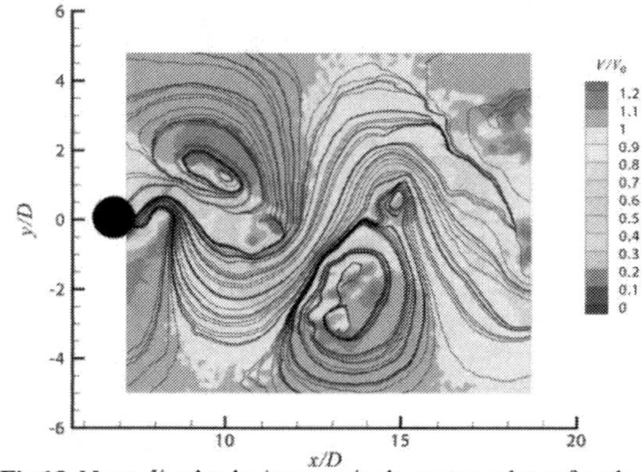

Fig.18. Normalized velocity magnitude contour plot at fourth time step obtained for wake region

4. Conclusions

Preliminary experimental results of airflow past six cylinders including mean and instantaneous velocity vector fields are presented. Flow past multiple cylinders is frequently seen in engineering. Knowledge about velocity fields in mining excavations is important in terms of proper ventilation. Cylinder-shaped legs modify velocity fields in the longwall area. In future works PIV results can be compared to CFD calculations.

Acknowledgment

This publication presents results of the work supported by the Grant for the Young Researchers the Polish Minstry of Science and Higher Education (7443/E-43/M/2017) and the Statutory Research of IMG-PAN

References

1. D. Sumner, J. Fluid Struct, **26**, 849-899 (2010)
2. Y.Zhou, Md. Mahbub Alam, J. Heat and Fluid Flow, **62**, 510-537, (2016)
3. C. Demartino, F. Ricciardelli, Eng. Struct, **137**, 76-114, (2017)
4. C. Liang, X. Luo, Comput. Fluids, **38,** 950-964, (2009)
5. F. Fornarelii, P. Oresta, A. Lippolis, J. Heat Mass Trans., **11**, 1-28, (2015)
6. J. Szmyd, M. Branny, M. Karch, W. Wodziak, M. Jaszczur, R. Nowak, Arch. Min. Sci., **58**, 333-348, (2013)
7. M. Branny, M. Karch, W. Wodziak, M. Jaszczur, R. Nowak, J. Szmyd, J. Phys. Conf.. Ser., **530,** 012029-1–012029-8, (2014)
8. M. Bujalski, M. Gawor, J. Sobczyk, Transactions Of The Strata Mechanics Research Institute (in Polish), **15** (2013)
9. M. Raffel, C.E Willert, S. Wereley, J.Kompenhans, *Particle Image Velocimetry. A Practical Guide,* Springer, 2007

Pilot study about the micro hydropower generation by use of flow induced vibration phenomenon

Yoshifumi Yokoi[1,*]

[1]National Defence Academy of Japan, Department of Mechanical Engineering, 1-10-20 Hashirimizu, Yokosuka 239-8686, Japan

Abstract. In order to construct a micro power generation system using a piezo-electric element, power generation was tried using excitation oscillation of the bluff cylinder by the vortex shedding from the bluff cylinder. The bluff cylinder consists of a board spring section in which the piezo-electric element was attached, and a body section. The bluff cylinder was inserted into the water flow, the shape and the submersion depth of the bluff cylinder, and the flow velocity were varied, and the power generation characteristic was investigated. As a result, it was found that it can generate electricity by vortex excitation. It was found that the length and the submersion depth of the body section influence power generation. It was shown that the power generation characteristic changes with cross-sectional shape of the bluff cylinder. The most suitable state was the case where the submersion depth was set to 140 mm with a circular cylinder with a span length of 250 mm. It is important to choose the power generation object which suited the use purpose.

1 Introduction

Considering an energy problem in recent years, it is interesting to exploit effectively the hydraulic power energy of the small scale and medium scale which is not yet used. As the example, the development and utilization in a river, an irrigation canal, etc. of low drop micro hydropower generation are performed actively. Those trends were reported by Turbomachinery Society of Japan [1, 2]. In realization of the plan, there are many matters which cannot be suitable by extension of the technique developed in the large-scale water-power generation of the concentration of all aspect of a society in one place. Therefore, development of new power generation technique or some method is needed. In response to such background, the power generation method using the self oscillation (vortex excitation) by the vortex shedding from the bluff cylinder placed into the flow is proposed by Bemitsas et al. [3], Uno & Kawashima [4], Koide et al. [5] and Hiejima et al. [6]. In order not to have a rotation mechanism the merit of vortex excitation type power generation is a maintenance-free. And the assembly and installation are also easy. Furthermore, since there is no impeller, power generation is possible, without damaging the living thing which lives in water by rotation of the impeller.

In this study, the piezo-electric element was used for the power generation element, and power generation was tried using excitation oscillation of the bluff cylinder by the vortex shedding from the bluff cylinder. Design manufacture of the power generation bluff cylinder was performed. And some tests were performed by the power generation bluff cylinder which was inserted into the water flow. In such tests, the shape and the submersion depth of the submersion section, and the flow velocity were varied, and the power generation characteristic was investigated.

2 Experimental apparatus and method

2.1 Outline of experimental apparatus

Experimental apparatus consists of a closed circuit water channel, a test bluff cylinder-traverse apparatus, and an oscillating frequency measuring device. The closed circuit water channel is the apparatus (water capacity 4 m^3) used by the previous report [7]. The sizes of the test section were 2 m in length, 800 mm in width, and 430 mm in depth, the flow velocity range was 0 - 1.2 m/s, and the turbulence to the average flow velocity was less than ±1.5 %. The test bluff cylinder-traverse apparatus is 2 axis dial slight movement type, and those moving ranges are 72 mm and 64 mm, respectively. Those directions of moving are the depth direction and the width direction. As for the oscillation frequency measuring device, the storage oscilloscope (KENWOOD, DCS-7020) was used.

2.2 Test cylinders and generators

The schematic drawing of the test bluff cylinder (oscillating pendulum) is shown in Fig.1. The test bluff cylinder consists of a spring section (upper part) and a body section (lower part). The spring is a board with a length of 70 mm, a width of 16 mm, and a thickness of 2

* Corresponding author: yokoi@nda.ac.jp

https://doi.org/10.1051/epjconf/201818002122

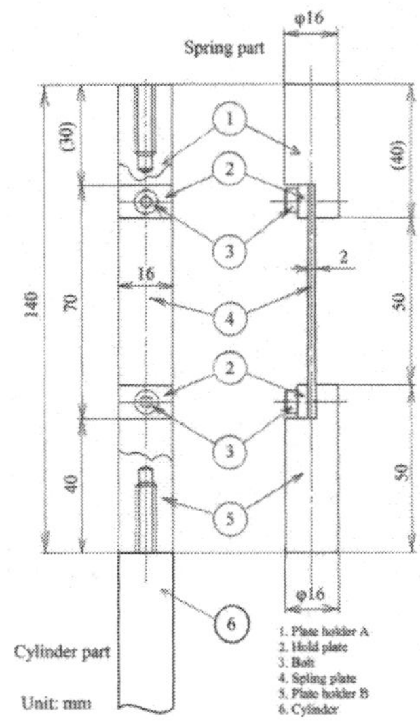

Fig. 1. The schematic drawing of a test cylinder spring section

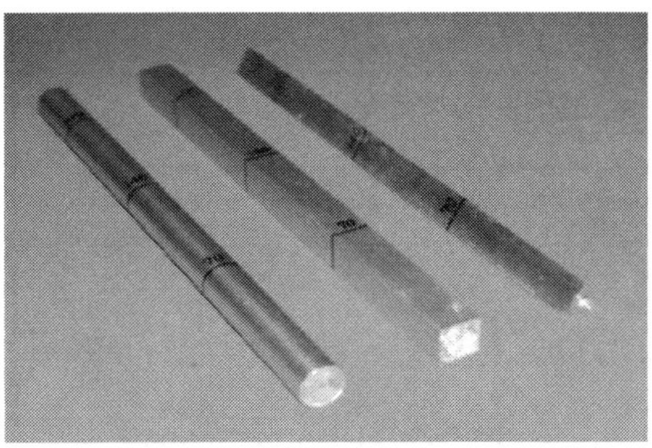

Fig. 2. Body section shape of the test cylinder (circular cylinder, square cylinder, and triangular cylinder)

mm made from vinyl chloride. As for the body, the circular cylinder, the triangular cylinder, and the square cylinder were prepared. Figure 2 shows the shape of those cylinders. The length of circular cylinder is two kinds, 250 mm and 500 mm, and the diameter is 16 mm. The length of a triangular cylinder and a square cylinder is 250 mm, and those side lengths are 15 mm.

The power generation object was set in the section of the board spring. The used power generation objects were shown in Fig. 3. One of which is a piezo-electric sensor (Tokyo sensor company, DT1-028 K/L) and the other is a piezo-electric buzzer (SPL company, PT08-Z185). These two kinds of power generation objects are made by different material and the manufacture method.

2.3 Experimental parameters

Experiment parameters are the length of bluff cylinder, the submersion depth of the body (draft), its setting

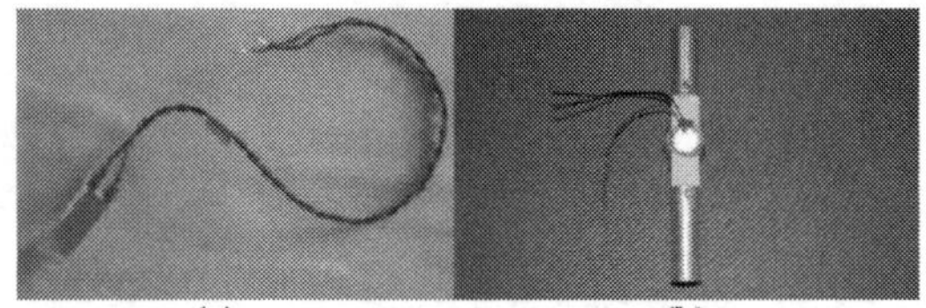

(a) (b)

Fig. 3. The aspect of the power generation object, (a) piezo-electric sensor, (b) piezo-electric buzzer

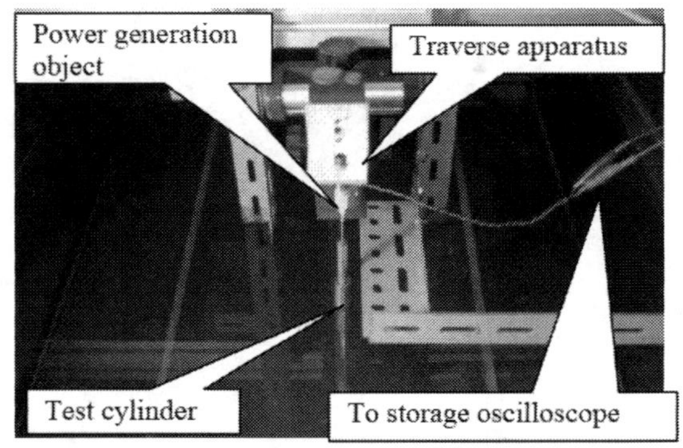

Fig. 4. The photograph of the situation where the test cylinder was set in the traverse apparatus

posture, and water flow velocity. The submersion depth of the body is three kinds (70, 140 and 210 mm). The posture of a triangular cylinder and a square cylinder is in the state where the angle was directed to the flow, and the state which directed the plane. The water flow velocity was varied to 44 steps in the range of 0 - 1.2 m/s.

2.4 Experimental procedure

The experiment procedure is the following. The power generation object is set in the board spring section of the test bluff cylinder, and the test bluff cylinder is attached in the traverse apparatus. The direction is decided to oscillate right-angled to the flow in the case of the attachment. The body is sunk in the target submersion depth. An aspect that it was set based on the procedure is shown in Fig. 4. The target flow velocity is caused by inverter operation of the pump of the closed circuit water channel. The voltage variation waveform from the power generation object sensor is observed with the storage oscilloscope, and output voltage and frequency are measured. Here, the frequency measured the signal of voltage waveform, observed the time jitter with the storage oscilloscope, and determined for fluctuation frequency.

https://doi.org/10.1051/epjconf/201818002122

Tab. 1. Synthetic character frequency of the test cylinder

Situation (mm)		Cylinder shape (Hz)				
span	draft	C.C.	S.S.	F.S.	S.T.	F.T.
	70	2.5	2.0	1.9	2	2.2
250	140	2.2	1.7	1.7	1.9	2
	210	2.1	1.6	1.6	1.9	
500	205	0.9				
	380	0.8				

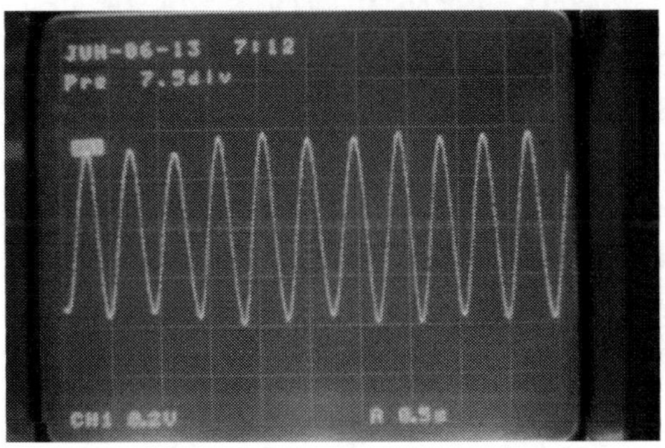

Fig. 5. An example of the output waveform from the power generation object

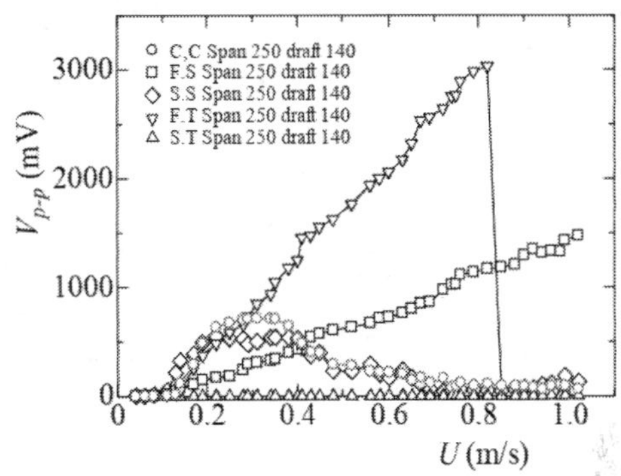

Fig. 6. Relationship between flow velocity and output voltage, in the case of circular cylinder

3 Experimental result and discussion

3.1 Measurement of natural frequency

The result of having measured the characteristic vibration of the overall system of the test bluff cylinder with a board spring is shown in Tab. 1. The meaning of the symbols used in the table are the following. The symbol "C.C." is a circular cylinder, the symbol "S.S." is the square cylinder which directed the angle to the flow, the symbol "F.S." is the square cylinder which directed the plane to the flow, the symbol "S.T." is the triangular cylinder which directed the angle to the flow, and the symbol "F.T." is the triangular cylinder which directed the plane to the flow. When the submersion depth of the body (draft) became shallow, it was confirmed that the character frequency as the overall system tends to become high.

3.2 Power generation characteristic test

Since the piezo-electric element reacts to unsteady stress, it can generate electricity by the lift alternately produced by the Karman vortex. An example of the power generation voltage waveform is shown in Fig. 5. In the beam which generally has a uniform section, if amplitude becomes large near critical oscillation frequency and it passes over it, it is known that amplitude will become small. If the vortex excitation based on the Karman vortex shedding also makes the flow velocity increase from a halt condition, vortex shedding frequency also increases, and the bluff cylinder begins to oscillate based on critical oscillation frequency, and obtains a soon large amplitude point. Moreover, amplitude will become small if it passes over this point, and it is predicted that oscillation ceases. Figure 6 shows the relationship between the flow velocity and the produced voltage. An abscissa is the average flow velocity in the test section of the closed circuit water channel, and the ordinate is the average value of the peak-to-peak voltage observed by the oscilloscope waveform. In the case of the circular cylinder (C.C.) and the sharp square cylinder (S.S.), the number of peaks is one and after the peak with the increase in the flow velocity, the voltage shows a gently-sloping reduction. On the other hand, in the case of a flat triangular cylinder (F.T.) and a flat square cylinder (F.S.), there is no particular peak and the tendency which the voltage produced to the increase in the flow velocity is increasing is seen. In the case of a sharp triangular cylinder (S.T.), there was no response to the increase in the flow velocity. In the case of a circular cylinder and a sharp square cylinder, one peak with wide width can see. Since voltage is obtained even if the flow velocity changes, the peak with wide width is a thing desirable in use. Since the voltage of 500 mV or more is obtained in

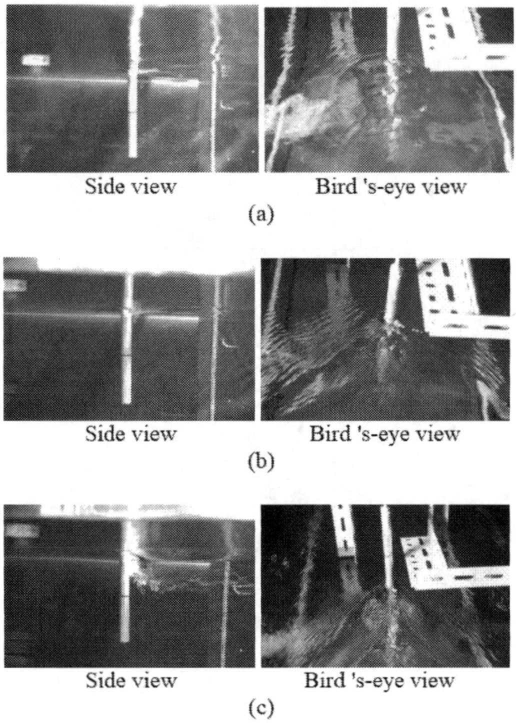

Fig. 7. Flow feature around circular cylinder, (a) flow velocity; 0.13 m/s, (b) flow velocity; 0.38 m/s, (c) flow velocity; 0.79 m/s

the range of the flow velocity 0.22 - 0.38 m/s, considering this matter, as for the circular cylinder body and the sharp square body, it turns out that it is in the setting situation of having been most suitable for the use.

3.3 The aspect of the flow around the bluff cylinder

Figures 7 - 9 show the aspect of the flow of the circular cylinder, the square cylinder, and the triangular cylinder. The length of those bluff cylinders used here was 250mm. Figure 7 shows the aspect of the flow when sinking the circular cylinder 140 mm and varying the flow velocity. Where a right-hand side photograph is seen from the width side (underwater), a left-hand side photograph is in the state which looked at behind from slant. In Fig. 7(a), it is shown that the dimple of the water surface has produced alternately behind the circular cylinder. That shows that the Karman vortex is formed. The difference of a water level has not occurred between the upstream side of the circular cylinder, and the down stream side. In Fig. 7(b), the length which forms the dimple of the water surface of the circular cylinder behind becomes short, and an aspect that the wave is pulled from the circular cylinder can be seen. The difference of the water level can be seen have occurred between the upstream side of the circular cylinder, and the down stream side. As for this state, the value of the produced voltage becomes the maximum so that may be seen in Fig. 5. In Fig. 7(c), the dimple of the water surface produced alternately is not seen, but the state where the towing wave has occurred from the circular cylinder is seen. The difference of the water level having occurred between the upstream side of the circular cylinder and the down stream side, and a cavity producing behind the circular cylinder, and involving in air bubbles is seen. This state is in the state in the middle of carrying out reduction with a gently-sloping value of the voltage produced in Fig. 5. Figure 8 shows the aspect of the flow when sinking the square cylinder 140 mm and varying the flow velocity. Figures 8(a) and 8(b) are in the state which directed the angle to the flow (S.S.), and those flow velocities are 0.29 m/s and 0.6 m/s, respectively. In Fig. 8(a), generating of the Karman vortex is seen behind the square cylinder. And there are a few differences of the water level between the

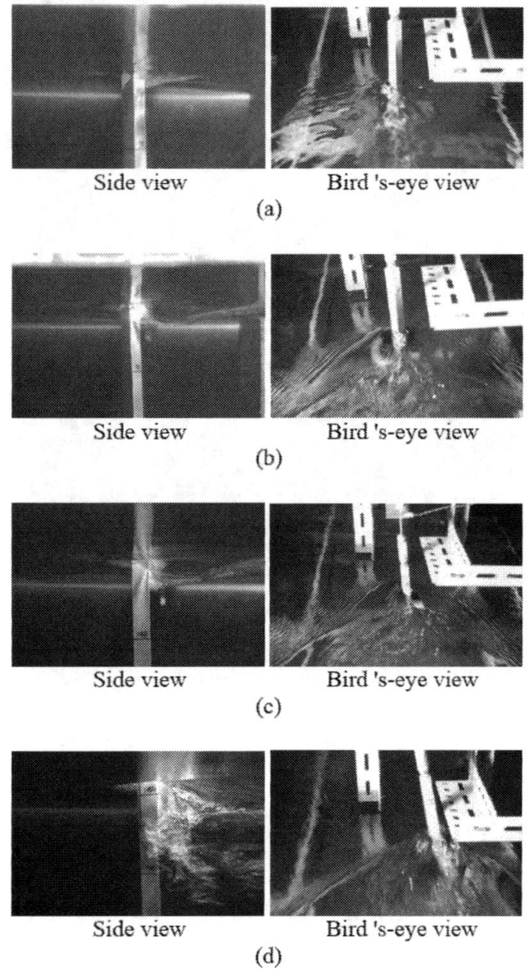

Fig. 8. Flow feature around square cylinder, (a) sharp square cylinder (S. S.), flow velocity; 0.29 m/s, (b) sharp square cylinder (S. S.), flow velocity; 0.6 m/s, (c) Flat square cylinder (F. S.), flow velocity; 0.6 m/s, (d) Flat square cylinder (F. S.), flow velocity; 1.02 m/s

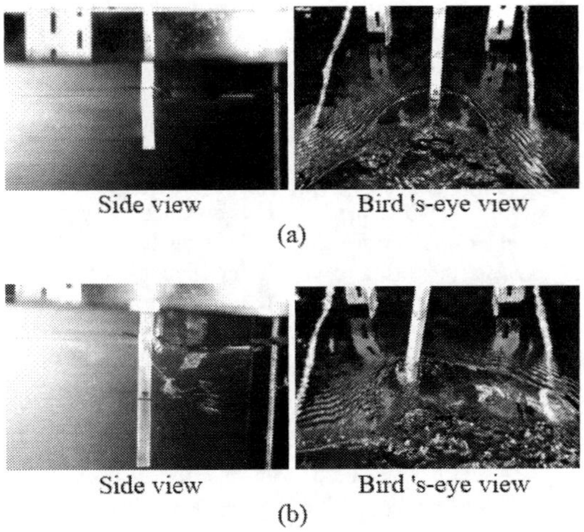

Fig. 9. Flow feature around triangular cylinder, (a) sharp triangular cylinder (S. T.), draft; 70 mm, flow velocity; 0.6 m/s, (b) Flat triangular cylinder (F. T.), draft; 140 mm, flow velocity; 0.6 m/s

upstream side of the square cylinder, and the down stream side. At this time, the value of the produced voltage was in the state which was a little less than the peak. On the other hand, generating of the Karman vortex is not seen by Fig. 8(b). An aspect that the towing wave has arisen from the square cylinder is shown. At this time, the difference of the water level occurs around the square cylinder, and the state where it became depressed from the water surface is remarkable. Figures 8(c) and 8(d) are in the state which directed the face (side) to the flow (F.S.), and those flow velocities are 0.6 m/s and 1.02 m/s, respectively. Formation of the Karman vortex is not seen behind of the square cylinder, but it is shown that the towing wave has arisen from the square cylinder. The difference of the water level between the upstream side of the square cylinder and the down stream side is also large, and involving in air bubbles is shown. Moreover, it is shown that the cellular involvement becomes intense with the increase in the flow velocity. This state does not exist at the self oscillation by the Karman vortex shedding, and the square cylinder is made for movement slowly carried out in a lift component direction to be carried out by the disproportionate flow which flows into the dimple produced behind square cylinder, and is considered that the voltage value produced with the increase in the flow velocity as shown in Fig. 6 increases in monotone. Figure 9 shows the aspect of the flow at the time of using the triangular cylinder. Figure 9(a) shows the state when making it sink 70 mm, where the angle of the triangular cylinder is directed to the flow (S.T.), and setting the flow velocity as 0.6 m/s. The Karman vortex which should be formed behind the triangular cylinder is negated by interference of the towing wave, and is not seen. The difference of the water level exists between the upstream side of the triangular cylinder, and the down stream side. In the difference of this water level, it seems that it does not become what moves the triangular

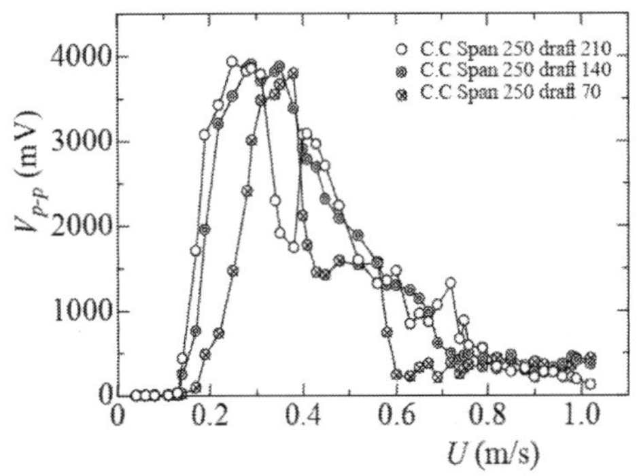

Fig. 10. Relationship between flow velocity and output voltage, in the case of circular cylinder, the piezo-electric buzzer was used

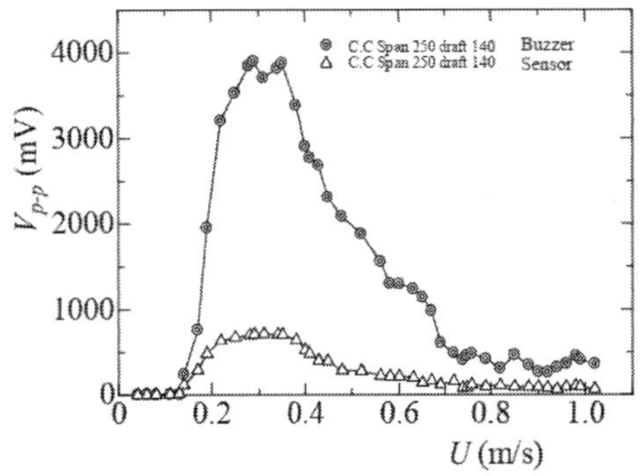

Fig. 11. Relationship between flow velocity and output voltage, in the case of circular cylinder, comparison of the piezo-electric sensor and the piezo-electric buzzer

cylinder since it has equal balance on both sides of the triangular cylinder in the case of the inflow to the down stream side. On the other hand, Fig. 9(b) shows the state when making it sink 70 mm, where the side of the triangular cylinder is directed to the flow (S.T.), and setting the flow velocity as 0.6 m/s. Although formation of the Karman vortex is not seen behind the triangular cylinder, an aspect that the flat surface triangular cylinder is moving slowly greatly like the time of the flat surface square cylinder is seen. The difference of the water level in front and in rear of the triangular cylinder is also large, and an aspect that air bubbles are entangled in is shown. In the bluff cylinder which directed the "side" to the flow, although the Karman vortex was not produced behind the bluff cylinder, the bluff cylinder performed slowly large movement. The cause of that is because the flow which turns behind the bluff cylinder carries out unbalance on both sides. Therefore, since amplitude also becomes large by the increase in the flow velocity, the value of the voltage produced as shown in Fig. 6 also becomes large.

3.4 Comparison of power generation elements

The experimental result when using a piezo-electric buzzer as the power generation object is shown in Fig. 10. In this experiment, using the circular cylinder with length of 250 mm, the submersion depth of the body was set to 70, 140, 210 mm, and water flow velocity was varied to 44 steps in the range of 0 - 1.2 m/s. Even when it changes into the piezo-electric buzzer, the power generation characteristic does not change, but since a case with a submersion depth of 140 mm of the body has wide peak width, it is thought that it is most suitable for the use. Figure 11 shows and compares the power generation voltage characteristic of two kinds of power generation objects (piezo-electric sensor and piezo-electric buzzer). Here, the circular cylinder of 250 mm length with submersion depth of 140 mm is used. In any case, the peak of power generation voltage is shown at the flow velocity range of 0.22 - 0.38 m/s. However, the clear difference is shown among both. It is shown that the voltage which a piezo-electric buzzer produces is about 4 times the voltage which a piezo-electric sensor produces. Since there is a difference on manufacture in the used product of two piezo-electric elements, it is guessed that such a result was obtained. It was found that it is important to choose the piezo-electric element corresponding to the purpose.

4 Conclusions

The following conclusions were obtained as a result of performing the characteristic test of the bluff cylinder pendulum aiming at generating electricity by vortex excitation.

(1) Generating of voltage was accepted by vortex excitation through the piezo-electric element.

(2) The length and the submersion depth of bluff cylinder influence the movement of the oscillating pendulum.

(3) The magnitude and its peak width of the voltage produced by vortex excitation corresponding to the flow velocity are changed by cross-sectional shape of the body section.

(4) The most suitable state was the case where the submersion depth was set to 140 mm with the circular cylinder with span length of 250 mm.

(5) It is producing pendulum movement of the bluff cylinder, even when the balance of the inflow produced according to the difference of the water level collapses. In this case, the value of voltage continues increasing in monotone with the increase in the flow velocity.

(6) It is important to choose the power generation object which suited the use purpose.

Author is thankful to Mr. Jalernsk Kamon, Mr. Srisuntornsiri Rungsak and Mr. Kouichirou Ooue who helped this work and who are the student staff of my laboratory.

References

1. Turbomachinery Society of Japan ed., Turbomachinery, **30**(3), 129 (2002)
2. Turbomachinery Society of Japan ed., Turbomachinery, **30**(4), 193 (2002)
3. M. M. Bemitsas,K. Raghavan,Y. Ben-Simon,E. M. H. Garcia, Journal of Offshore Mechanics and Arctic Engineering, **130**, 041101(2008)
4. M. Uno, R. Kawashima, Turbomachinery, **38**(2), 121 (2010)
5. M. Koide, T. Sekizaki, S. Yamada, T. Takahashi, M. Shirakashi, Transactions of the Japan Society of Mechanical Engineers, **77**(775), 702 (2011)
6. S. Hiejima, K. Oka, K. Hayashi, H. Inoue, Annual Journal of Civil Engineering in the Ocean, **69**(1), 12 (2013)
7. Y. Yokoi, *57th International Conference of Machine Design Departments* (ICMD2016), 175 (2016)

Evaluation of the structures size in the liquid-gas flow by gamma-ray absorption

Marcin Zych[1,*], *Robert* Hanus[2], *Marek* Jaszczur[3], *Dariusz* Świsulski[4], *Leszek* Petryka[5], *Paweł* Jodłowski[5] and *Piotr* Zych[6]

[1]AGH University Science and Technology, Faculty of Geology, Geophysics and Environmental Protection, 30-059 Kraków, Poland
[2]Rzeszów University of Technology, Faculty of Electrical and Computer Engineering, 35-959 Rzeszów, Poland
[3]AGH University Science and Technology, Faculty of Energy and Fuels, 30-059 Kraków, Poland
[4] Gdańsk University of Technology, Faculty of Electrical and Control Engineering, 80-233 Gdańsk, Poland
[5]AGH University of Science and Technology, Faculty of Physics and Applied Computer Science, 30-059 Kraków, Poland
[6] Warsaw University of Technology, The Institute of Telecommunications, 00-661 Warszawa, Poland

Abstract. The rapid development of tomography methods particularly electrical, X and gamma rays allows for a wide range of the information about flow structure. However, all of such methods are quite complicated. At the same time much simpler systems as the measuring system of gamma rays absorption, allows to obtain a all key flow information which describe the two-phase flow. In the article the results of analyzes of radiometric signal that not only allow to recognize the type of flow, but also the assessment of forming structures are presented. Calculation and interpretation of the data were based on the cross-correlation and cross-spectral density function. In order to verify the calculations the photographic documentation made during the measurements was used.

1 Introduction

Depending on the velocity of the two-phase mixture of gas-liquid in a pipe and the void fraction α, one can obtain different flow structures. The recognition of transport method of the gas phase and the assessment of the size of the resulting structures are being increasingly undertaken as research topics. The most often used in such studies are Tomographic methods: electric, X-rays and gamma rays and optical methods [1-5]. Tomography systems require the installation of sensors around the selected sections of the pipeline and from the obtained signals to reconstruct the image showing the distribution of individual gas bubbles, and also outline the basic flow parameters. However, the authors believe that in the case of small diameter pipelines one can use a simpler measurement systems that also allow the assessment of the flow characteristics, the type of identification and estimation of the size of the resulting structures.

With a small diameter pipe (less than 50 mm), the flow system as a pipeline is often reduced to the two-dimensional problem, where it is sufficient to know the length and depth of the resulting structures.

The measuring system proposed in this article is a set of measurement devices based on the absorption of gamma radiation, which is being perfected by the authors for many years [6-8].

2 Laboratory stand

The hydraulic system built in the Sedimentological Laboratory of the Faculty of Geology, Geophysics and Environmental Protection, AGH in Krakow was used for measurements. A fragment of the experimental system is shown in Fig. 1.

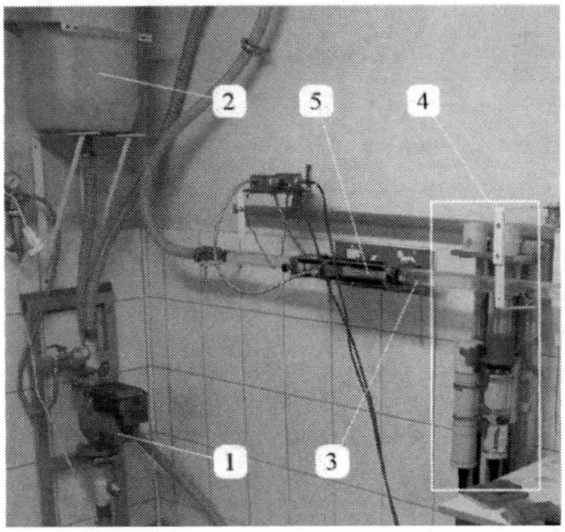

Fig. 1. Fragment of laboratory stand for the measurement of two-phase liquid-gas flow: 1 – pump, 2 - expansion tank, 3 – pipeline, 4 – gamma absorption measurement system, 5 – ultrasonic flowmeter.

* Corresponding author: zych@geol.agh.edu.pl

A set for measuring the absorption of gamma radiation consists of two scintillation probes and a gamma-ray Am-241 sources of about 100 mCi activity. The pipe through which flows the water-air mixture is formed with a horizontal pipe made of metapleks with a length of 4.50 m and an internal diameter of 30 mm. The flow velocity of the liquid is contained in the range of 0.17 – 3.60 m/s, allowing to obtain the following types of movement: stratified (wavy), slug, plug and bubble flow.

The measuring stand, in addition to the absorption of gamma radiation and ultrasonic flowmeter, is equipped with a camera for independent tracking of the movements of the gas phase.

The laboratory stand along with a description of the measurement system for the absorption of gamma radiation has been widely described in articles [9,10].

3 Analysis of time and frequency signals

In the experiments carried out, stochastic signals were obtained, which are characterized by a large proportion of noise. Depending on the amount of the gas and the flow rate of the mixture the signal changes as can be observed in the frequency and time domains [10]. For the presentation of the possibilities of the proposed method, an experiment was chosen for analysis in which the structure of slug flow was achieved. This type of movement arising from the large amount of air supply is characterized by very large blisters of gas filling almost the entire cross-section of the pipe (Fig. 2), and smaller bubbles flowing in the spaces between large structures (Fig. 3). These structures clearly exhibit the time course of signal, a portion of which is shown in Fig. 4.

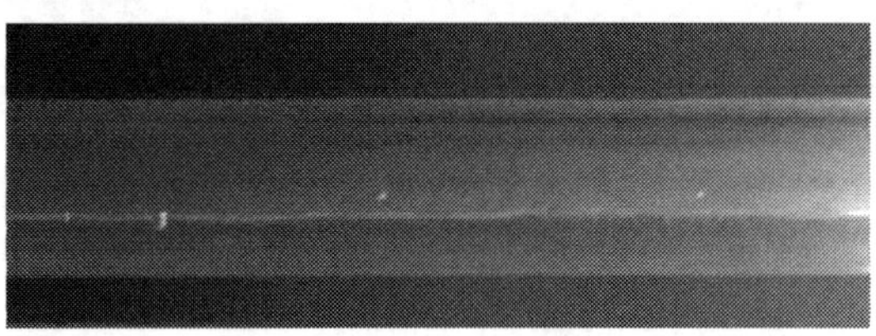

Fig. 2. Fragment of a large bubble entrusts resulting from the slug flow.

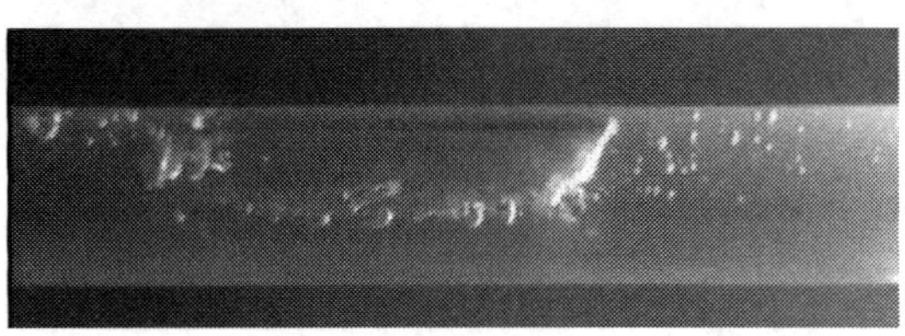

Fig. 3. Small bubble formed in the slug flow.

The experiment analyzed in the article was conducted at an average velocity of water $\upsilon_W = 2.08$ m/s (measurement done by ultrasonic flowmeter), the average velocity of the air $\upsilon_A = (1.325 \pm 0.053)$ m/s and an average gas void fraction $\alpha = 0.434$ [11].

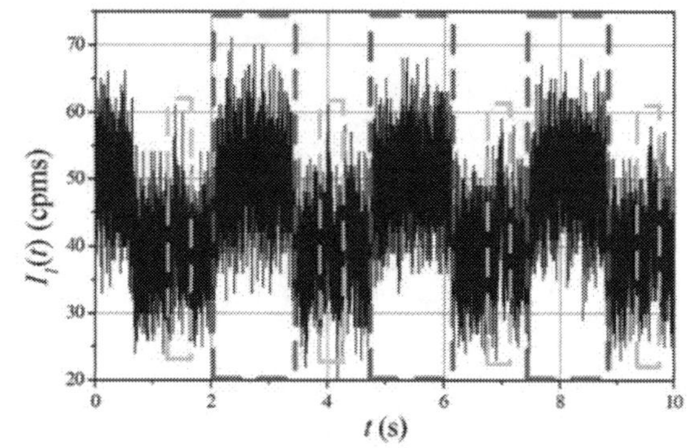

Fig. 4. The first 10 s of signal registered for the described experiment. Red frame indicates the signal coming from the flow of large air bubbles, and an orange frame from smaller structures.

Using filtering of signal from Fig. 4 one receives smoothed data as shown in Fig. 5.

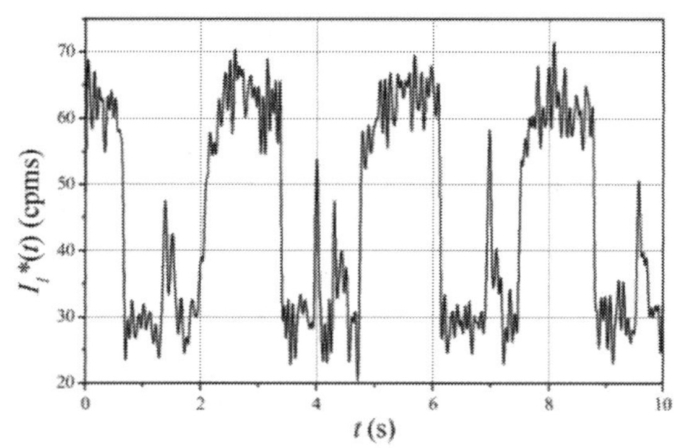

Fig. 5. The first 10 s of filtered signal registered for the described experiment.

The shown representation of the signal in the time domain shows that it is a superposition of a square wave function associated with the movement of large bubbles of gas, and features related to the movement of small bubbles. Such a course of time should include a frequency characteristic in the frequency domain. For large structures, it will be a distinctive set of frequencies that appear with fallowing ones being odd multiples of the fundamental frequency, according to:

$$f_n = (2 \cdot n + 1) \cdot f_0 \tag{1}$$

where: f_0 – fundamental harmonic with the highest amplitude, f_n – another characteristic harmonic in the spectrum $n \in \{1, 2, \ldots\}$.

For the analysis in the frequency domain a cross spectral density function (CSDF) was used, obtained by the Fourier transform of the cross-correlation function (CCF) [10,12]. For reason of clarity, Fig. 6 shows the frequency spectrum with marked characteristic frequencies. The red lines indicate the first six harmonics associated with the movement of large bubbles described by equation (1). In turn, the first 5 harmonics f_k marked

https://doi.org/10.1051/epjconf/201818002123

with orange lines are even multiples of dominant frequency f_0 described by the relationship:

$$f_k = 2 \cdot k \cdot f_0 \tag{2}$$

for $k \in \{1, 2, 3, \ldots\}$.
On the basis of the signal waveforms and photographic documentation it has been identified that these are frequencies related to the movement of smaller structures, located in the spaces between large blisters. The red dashed line, forming a frame indicates a useful range of signal frequencies on the outskirts of the determined cut-off frequency in order to reduce noise [10].

a) b)

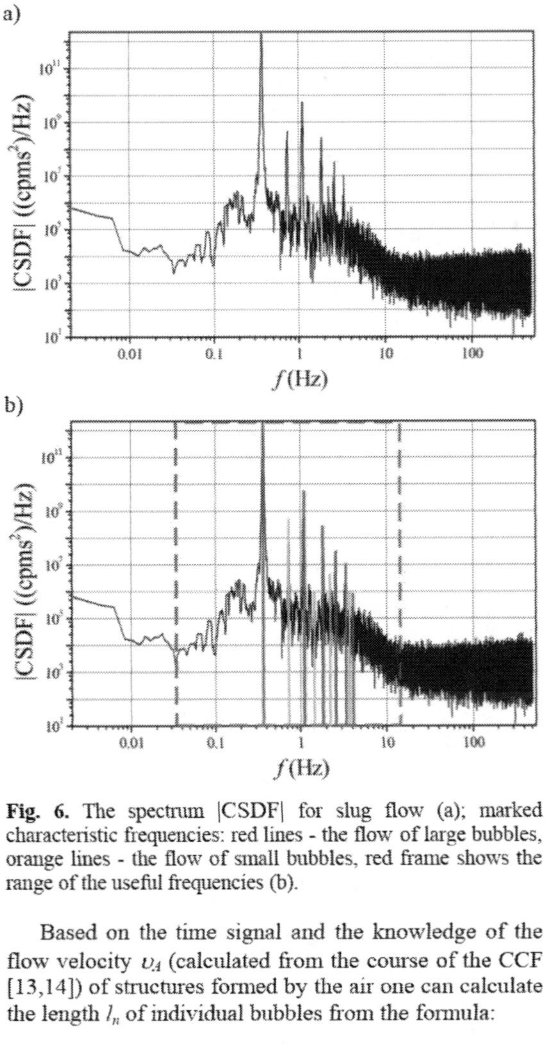

Fig. 6. The spectrum |CSDF| for slug flow (a); marked characteristic frequencies: red lines - the flow of large bubbles, orange lines - the flow of small bubbles, red frame shows the range of the useful frequencies (b).

Based on the time signal and the knowledge of the flow velocity υ_A (calculated from the course of the CCF [13,14]) of structures formed by the air one can calculate the length l_n of individual bubbles from the formula:

$$l_n = t_n \cdot \upsilon_A \tag{3}$$

where, t_n – the flow time of a large bubble of air registered by a scintillation probe. This time is reflected in the data by an important increase in the number of counts.

The second method for determining the length of the largest structures created by the air, is the analysis of the characteristic frequencies of the spectrum |CSDF|. The harmonic which is associated with the movement of the large bubbles is the domain frequency f_0, which allows one to calculate the average length l of the largest structures in the entire duration of the measurement (480 s), from the relationship:

$$l = \frac{\upsilon_A}{2 \cdot f_0} \tag{4}$$

The average size of smaller structures in the flow are associated with even-numbered multiples of f_0. For the medium bubbles it is for example $f_{10} = 10 \cdot f_0$.

Another important issue is the depth of the resulting structures, which can be determined based on the absorption of gamma radiation. However, this requires the prior calibration of the measuring system, which is described in the article [11]. The calibration procedure is the same as when determining the void fraction α. The method makes it possible to assess the depth of single large bubbles d_n from the relationship:

$$d_n = ln\left(\frac{I_n}{I_0}\right) \Big/ 0.0111 \tag{5}$$

where: I_n – intensity of radiation in the interval, where a big bubble flows, I_0 – intensity of radiation for the pipeline completely filled with water (reference point).

The method also allows the determination of the mean depth d of single large bubbles. However, when calculating the average value one should calculate the coefficient C which is the logarithm of the ration of the number of counts for passages where there were large parts of the structure to the signal, which formed small bubbles. After performing the calculations for a number of cycles of the flow of large and smaller structures, it was found that the value of $C = 2$. So finally the average depth d of a big bubble for the entire experiment (i.e. 480 s) is given by:

$$d = 2 \cdot ln\left(\frac{I}{I_0}\right) \Big/ 0.0111 \tag{6}$$

where: I – the average radiation intensity for the entire measurement.

The uncertainty of the size of the individual values u_c, which are determined in an indirect way, is based on the law of propagation of uncertainty as recommended by IMEKO [15].

4 The results of the experiment

Table 1 shows the current and average length and depth of large bubbles for the first 10 seconds of the flow and the average values for the whole experiment, lasting 480 seconds.

Table 1. The values of instantaneous and average length l_n and depth d_n of large bubbles.

No. of large structure	l_n (m)	$u_c(l_n)$ (m)	d_n (mm)	$u_c(d_n)$ (mm)
1	1.78	0.07	23	1.0
2	1.84	0.07	23	1.0
3	1.77	0.07	22	1.0
4	1.83	0.07	23	1.0
5	1.84	0.07	23	1.0
6	1.92	0.08	24	1.0
7	1.90	0.08	22	1.0
8	1.95	0.08	23	1.0
9	1.96	0.08	22	1.0
10	1.99	0.08	22	1.0
average 10s	1.88	0.08	23	1.2
average 480s	1.82	0.07	25	0.7

Shown in Table 1, the length l_n of large structures was calculated according to the formula (3) and the depth d_n from the formula (5). In contrast, the average length l of a large bubble for the whole measurement time of 480 s was calculated from the formula (4), and the depth d from equation (6).

Analyzing values summarized in Table 1 it can be seen that the length of the large structures has changed during the first 10 seconds in the range from 1.77 m to 1.99 m, oscillating around the average value of 1.88 m. This is the value lying in the range of uncertainty for the average for the whole measurement (480 s) of 1.82 m. The noticeable difference is probably related to the greater variability in length l_n of large bubbles in the longer duration of the experiment. Also, analysis of the depth d_n and d of various large structures gives similar values. The accuracy of the proposed methods is not high because of uncertainty does not exceed 5% of the measured values.

5 Conclusions

The article presents two ways of determining the size of structures of two-phase gas-liquid flow of slug flow type based on signal analysis of radiometric probes. Both analyses in the time domain and frequency domain show a high similarity of the results within the measurement uncertainty. The results characterize the variability of the length l_n and depth d_n of large bubbles very well, what was shown by the detailed analysis within the first 10 seconds of the experiment. Additional comparison of the absorption measurement with the photographic documentation has enabled the identification of a part of the characteristic frequencies in the spectrum of |CSDF|, allowing them to bind with the flow and the size of the smaller structures produced for this type of flow.

The conducted experiments showed that for the flow in horizontal pipelines with diameters comparable to the size of the crystals of scintillation probes presented the discussed system can be competitive with tomographic methods. The use of mathematical methods for signal analysis in both time and frequency domains allows to determine a larger number of flow parameters than which was measured before with gamma ray absorption.

In order to confirm the suitability of the considerations described in the article, further tests of the method of absorption of gamma radiation to other types of flow and conduits positioned at different angles will be carried out. In addition, the authors plan to conduct simulation experiments, which would allow to optimize the geometry of the measurement and calibration system without the need to conduct the experiment.

Funding sources

This publication is funded by AGH University of Science and Technology (No 11.11.140.645).

References

1. M. Bottin, et al, Int. J. Multiphas. Flow **60**, 161-179 (2014)
2. T. Xue, L. Qu, Z. Cao, T. Zhang, Flow Meas. Instrum. **27**, 29–36 (2012)
3. M.R. Rzasa, Nucl. Eng. Des. **239**, 699-707 (2009)
4. Z. Yu, B. Qincheng, H. Richa, Appl. Therm. Eng. **60,** 398-410 (2013)
5. V. Mosorov, Meas. Sci. Technol. **17**, 753-760 (2006)
6. M. Zych, et al, Flow Meas. Instrum. **35**, 11–15 (2014)
7. R. Hanus, L. Petryka, M. Zych, Flow Meas. Instrum. **40**, 58–63 (2014)
8. L. Petryka, M. Zych, R. Murzyn, Nukleonika **50**, 43–46 (2005)
9. R. Hanus, Int. J. Multiphas. Flow **72**, 210–217 (2015)
10. M. Zych, R. Hanus, L. Petryka, *MATEC WoC* **18**, 02004 (2014)
11. M. Zych, et al., *JPCS* (to be published)
12. J.S. Bendat, A.G. Piersol, *Random data - analysis and measurement procedures* (John Wiley, New York, 2010)
13. W.Q. Yang, M.S. Beck, Flow Meas. Instrum. **8,** 77-84 (1998)
14. M.S. Beck, A. Plaskowski, *Cross correlation flowmeters - their design and application* (Adam Hilger, Bristol, 1987)
15. *Guide to the expression of uncertainly in measurement.* (International Organisation for Standardisation, 1995)

https://doi.org/10.1051/epjconf/201818002124

Radioisotope measurement of selected parameters of liquid-gas flow using single detector system

Marcin Zych[1,*], *Robert* Hanus[2], *Marek* Jaszczur[3], *Volodymyr* Mosorov[4], and *Dariusz* Świsulski[5]

[1] AGH University Science and Technology, Faculty of Geology, Geophysics and Environmental Protection, 30-059 Kraków, Poland
[2] Rzeszów University of Technology, Faculty of Electrical and Computer Engineering, 35-959 Rzeszów, Poland
[3]AGH University Science and Technology, Faculty of Energy and Fuels, 30-059 Kraków, Poland
[4]Łódź University of Technology, Institute of Applied Computer Science, 90-924 Łódź, Poland
[5]Gdańsk University of Technology, Faculty of Electrical and Control Engineering, 80-233 Gdańsk, Poland

Abstract. To determine the parameters of two-phase flows using radioisotopes, usually two detectors are used. Knowing the distance between them, the velocity of the dispersed phase is calculated based on time delay estimation. Such a measurement system requires the use of two gamma-ray sealed sources. But in some situations it is also possible to determine velocity of dispersed phase using only one scintillation probe and one gamma-ray source. However, this requires proper signal analysis and prior calibration. This may also cause larger measurement errors. On the other hand, it allows measurements in hard to reach areas where there is often no place for the second detector. Additionally, by performing a previous calibration, it is possible to determine the void fraction or concentration of the selected phase. In this work an autocorrelation function was used to analyze the signal from the scintillation detector, which allowed for the determination of air velocities in slug and plug flows with an accuracy of 8.5%. Based on the analysis of the same signal, a void fraction with error of 15% was determined.

1 Introduction

Radioisotope meters in flow mixtures studies of two and multiple phases are most commonly used as single-mode absorption spectrometric systems [1-5]. With them, after the appropriate calibration, one can measure the average density of the flowing mixture, void fraction or concentration of one of the phases. In addition, the flow type can also be recognized by analyzing the signal from the detector [5].

Up-to-date single-detector solutions do not allow to determine another important flow parameter, which is the velocity of the dispersed phase. Only systems extended by two or more detectors allow to measure this quantity [6-10].

The article presents the method of determining two parameters in a single detector system for liquid-gas flow: average velocity and void fraction. The air velocity was determined by the analysis of the autocorrelation function (ACF) of the detector signal. This method requires prior calibration for standard air velocity values. Based on analysis of the same signal, the void fraction was also determined. Also, setting this parameter required a prior calibration.

2 Experimental set-up

The results of the measurements obtained at the research facility located in the Sediment Laboratory of the Faculty of Geology, Geophysics and Environmental Protection of AGH - UST in Krakow [9] were used. The installation diagram is shown in Fig. 1. Using this installation, liquid-gas flows can be tested using single- and dual-detectors. The configuration included two radiometric sets composed of gamma radiation Am-241 (1) with 3.7 GBq activity in lead collimators and two scintillation probes (2) with 2" × 2" NaI(Tl) crystals. Moving the system along the pipeline allows for the shift system (9). Water flow control is possible through an inverter mounted on a flow pump (5), and an ultrasonic meter (3). As a result, the flow velocity of the water could be adjusted in the range (0.9 ÷ 3.0) m/s. A two-phase water-air mixture is obtained by dispensing air from the compressor (6) through an injector (8) to a 4.5 m horizontal pipeline. The amount of air injected was determined using a Brooks 4800 (4) flow controller. Measurements were made at an airflow of 10.73 l/min. The venting of the installation takes place through the expansion tank (7). Photographic documentation was also performed during the measurements, the analysis of which allows to determine the type of flow.

On the station presented, air velocity measurements in a two detector system were done which were later used to benchmark the single-detector method. Calibration measurements for void fraction were also performed. The idea of this type of measurement is presented in articles [4,9].

* Corresponding author: zych@geol.agh.edu.pl

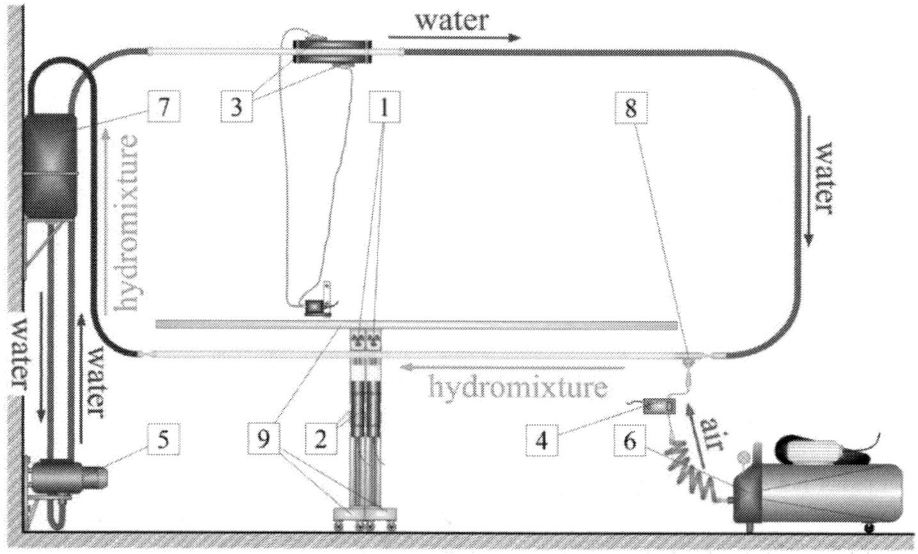

Fig. 1. Schema of laboratory experimental set-up: 1 –sealed Am-241 sources with collimators, 2 – scintillation probes, 3 – ultrasonic flow meter of water, 4 – mass flow meter of air, 5 – pump with power inverter, 6 – compressor 50l, 7 – expansion tank, 8 – injector, 9 – shift system.

3 Measurement of selected parameters of liquid-gas flow

3.1 Air velocity measurement

Typically, two sets of source-probe located at a known distance from each other are used in radioisotope liquid-gas flow measurements. The average velocity of the dispersed phase is calculated based on the time delay of the obtained signals, and the known distance of the probes [8-10,13]. In this work the above method was used to determine the reference air velocity υ_A.

The main purpose of the research was to investigate the possibility of using the autocorrelation function to measure the velocity of the dispersed phase in a single-detector system. The ACF for the discrete values of the signal $x(n)$ is defined by the relation [12,13]:

$$R_{xx}(\tau) = \frac{1}{N}\sum_{n=0}^{N-1} x(n)\cdot x(n+\tau) \tag{1}$$

where: $R_{xx}(\tau)$ – autocorrelation function, τ – time delay, N – number of samples, $x(n)$ – digital signal from the probe $n = t/\Delta t$, t – time of measurement, Δt – sampling time. The most frequently used is the normalized ACF:

$$\rho_{xx}(\tau) = \frac{R_{xx}(\tau)}{\sigma_x^2} \tag{2}$$

where: $\rho_{xx}(\tau)$ – normalized ACF, σ_x – standard deviation of analyzed signal.

Such prepared experiment enabled the ACF sensitivity to be checked for changes to the velocity of the gas phase in the pipeline.

Fig. 2 shows an example of an ACF waveform for a measurement in which the mean water velocity was $\upsilon_W = 2.05$ m/s and the average air velocity $\upsilon_A = 1.732$ m/s. The following parameter values were adopted: N=180000, Δt = 1 ms.

The studies show that a change in velocity of the air passage changes the ACF indicated in Fig. 2 by a red frame. This fragment is shown enlarged in Fig. 3. As the curve resembles half of a normal distribution, it has been found that a suitable parameter for determining the velocity of the gas phase may be a standard deviation σ_{ACF}.

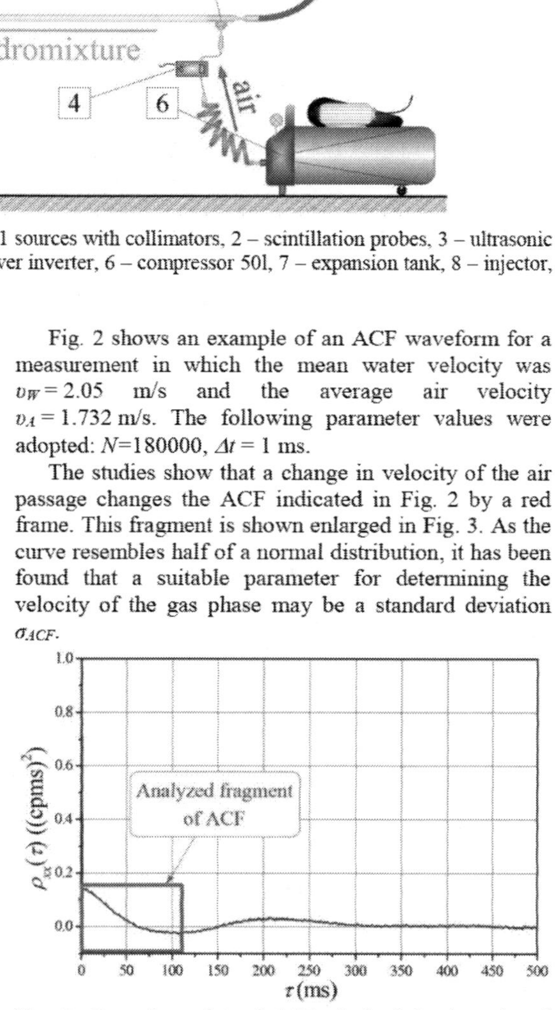

Fig. 2. Exemplary plot of ACF obtained in the selected experiment. Red frame indicated the fragment which is related to the change of air velocity.

ACF runs for different types of flows were performed in experiments. Changes in the ACF distribution for intermittent flows are shown in Fig. 4a and 4b. What is important, for the bubble flow such changes have not been observed. This suggests that the ACF method is only suitable for determining gas phase velocities in slug and plug flows.

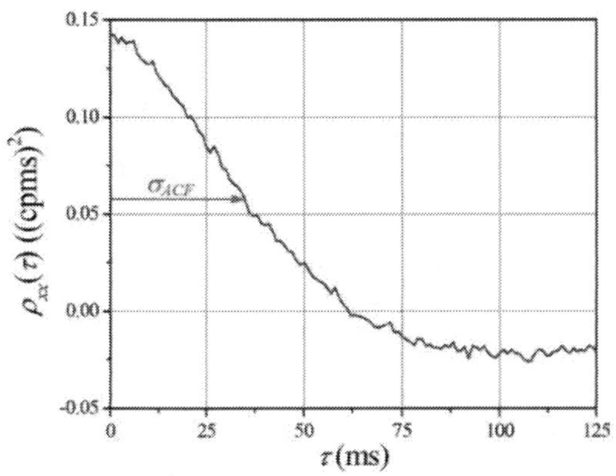

Fig. 3. Analyzed fragment of ACF from Fig. 2.

a)

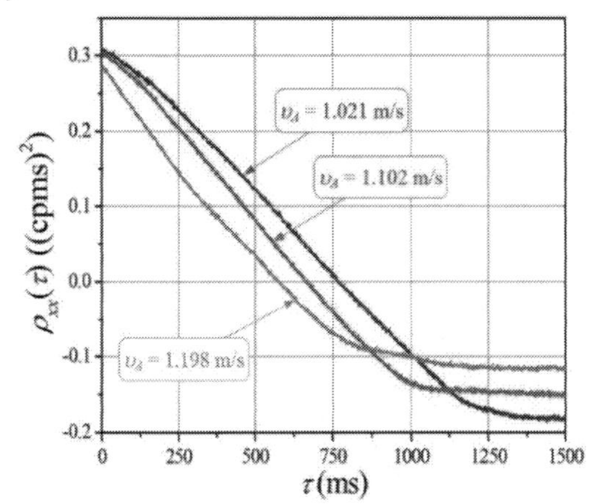

b)

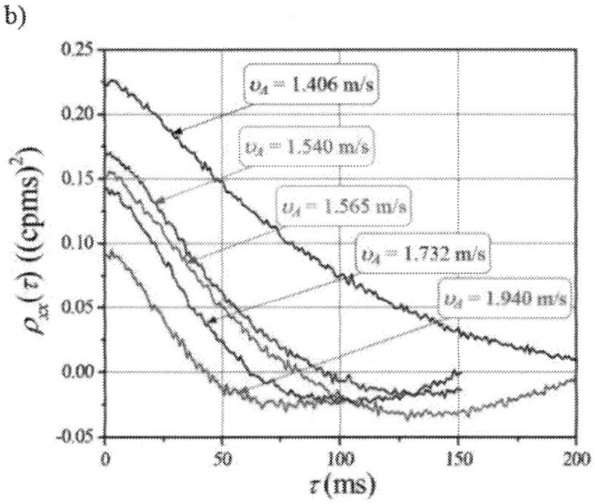

Fig. 4. Changing the ACF shape for a) slug flows ; b) plug flows.

Table 1 summarizes the results. Water velocity υ_W was measured by ultrasonic flowmeter. Determined values of σ_{ACF} and corresponding air velocity υ_A allow to make a calibration graph, which is shown in Fig. 5.

Table 1. Obtained results of experiments: υ_W - water velocity. $u(\upsilon_A)$ - uncertainty of air velocity

υ_W (m/s)	σ_{ACF} (ms)	υ_A (m/s)	$u(\upsilon_A)$ (m/s)	**Type of flow**	
1.24	548	1.021	0.036	slug	intermittent
1.33	446	1.102	0.042		
1.41	378	1.198	0.046		
1.59	71	1.406	0.054	plug	
1.78	44	1.540	0.061		
1.82	41	1.565	0.064		
2.05	32	1.732	0.070		
2.19	26	1.940	0.079		

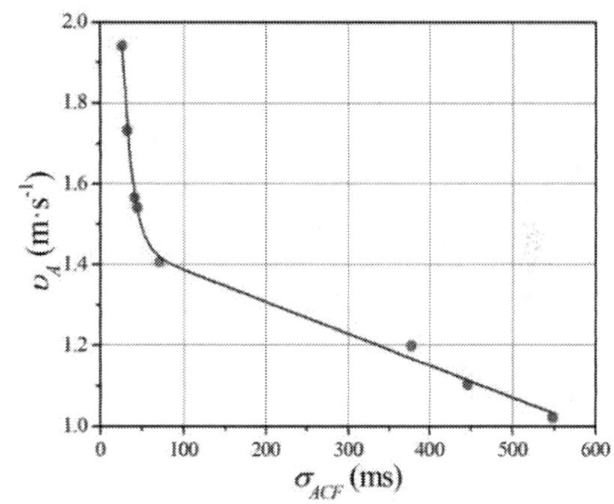

Fig. 5. Calibration curve $\upsilon_A = f(\sigma_{ACF})$

Based on the distribution of the points in Fig. 5, the relationship between air velocity and standard deviation σ_{ACF} is obtained:

$$\upsilon_A(\sigma_{ACF}) = A_1 \cdot exp\left(-\frac{\sigma_{ACF}}{a_1}\right) + A_2 \cdot exp\left(-\frac{\sigma_{ACF}}{a_2}\right) + B \quad (3)$$

where: A_1 = (4.7 ± 1.5) m/s; a_1 = (11.5 ± 1.7) ms; A_2 = (1260.443 ± 0.034) m/s; a_2 = (1.600 ± 0.043)·10^6 ms; B = (-1258.9795 ± 0.0068) m/s.

The calculated coefficient of determination R^2 = 0.998 indicates the good fit of the curve to the measuring points. Analysis with the use of uncertainty propagation method [14] shows that the air velocity can be measured to an accuracy of 8.5% of its value. The main influence on measurement uncertainty is σ_{ACF} value which is the smaller the larger the velocity determination error is. It should be noted that the presented calibration curve applies only to the described flow conditions. In case of different air flow, the measuring system should be calibrated again.

https://doi.org/10.1051/epjconf/201818002124

3.2 Determination of void fraction

The publication [4] describes exactly the method of calibration of the gamma-absorption system for liquid-gas two-phase flows to determine average void fraction α for air. Because the experiments used in this article have changed the geometry of the measurement section for the radiometric set, the calibration was re-done. As a result, the calibration relationship shown in Fig. 6 was obtained.

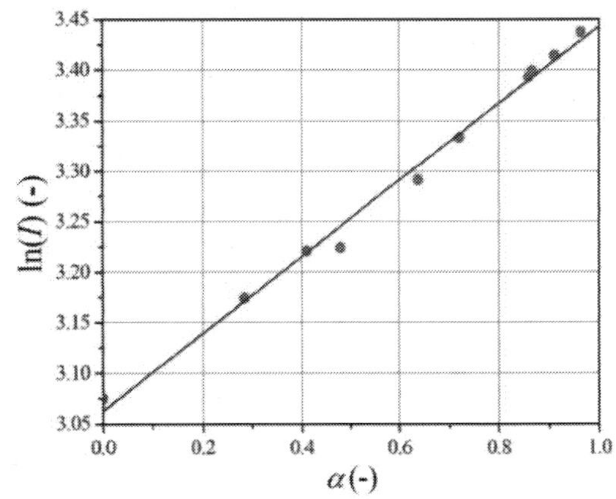

Fig. 6. The relationship between the natural logarithm of the gamma radiation intensity I (recorded by scintillation probe) and the void fraction α.

The dependency $\ln(I) = f(\alpha)$, after fitting the least squares method, can be described by the equation:

$$\ln(I) = c \cdot \alpha + \ln(I_0) \tag{4}$$

where: I – intensity of gamma radiation recorded by the probe, $c = 0.381 \pm 0.012$; $\ln(I_0) = 3.0630 \pm 0.0080$; I_0 – the radiation intensity at the pipeline completely filled with water. The determination factor is: $R^2 = 0.993$.

Table 2. Valuation of void fraction α and uncertainty values $u(\alpha)$ for the analyzed experiments.

v_W (m/s)	α (-)	$u(\alpha)$ (-)
1.24	0.276	0.023
1.33	0.250	0.022
1.41	0.228	0.022
1.59	0.202	0.022
1.78	0.199	0.022
1.82	0.180	0.022
2.05	0.164	0.022
2.19	0.155	0.021

The dependence (4) is more universal than Equation (3) because after measuring the intensity of radiation I_0 it can be used to determine the average value of α irrespective of the flow type of water-air mixture for pipelines of the same diameter and of the same material as the test installation. The measurement system used allows the void fraction to be evaluated to an accuracy of approximately 15%. Table 2 summarizes the mean α values for the analyzed experiments. Uncertainty values $u(\alpha)$ were determined on the basis of the law of uncertainty propagation based on formula (4) [14].

4 Conclusions

The paper presents the method for determining the velocity of the dispersed phase and the void fraction for liquid-gas flow using a single detector gamma radiation absorption system.

On the basis of studies it was found that a single set of probe scintillation - sealed source of gamma radiation allows the determination of the above-mentioned size.

However, determining the velocity of the gas phase in a single-detector system requires prior calibration, which is troublesome. In addition, the results are more error-prone than with the two-detector system. An autocorrelation function was used to analyze the signal from the scintillation detector, which allowed the determination of air velocity to an accuracy of 8.5%. The presented studies show that the gas velocity can only be determined to a limited extent, i.e. for intermittent flows. In addition, based on the analysis of the same signal from the detector, a void fraction with accuracy of 15% was determined.

It should be emphasized that the studies of the one detector system should continue and the possibility of linking other statistical parameters to the airflow signal should be explored. This would increase the accuracy and extend the measurement capabilities. Such an operation is economically justified because one-detector systems are cheaper than multi-detectors. Additionally, while reducing costs, the radiation level is reduced, which is extremely important for the operation of this type of equipment.

Acknowledgment

The authors would like to thank dr Leszek Petryka for his cooperation during the measurements.

This publication is funded by AGH University of Science and Technology (No 11.11.140.645) and by Gdańsk University of Technology.

References

1. V. Matoušek, Exp. Therm. Fluid Sci. **26**, 693–702 (2002)
2. G.A. Johansen, P. Jackson, *Radioisotope gauges for industrial process measurements* (Wiley, New York, 2004)
3. P. Vlasak et al., J. Hydrol. Hydrom. **62**, 241–247 (2014)
4. M. Zych et al., JPCS **745**, 032124 (2016)
5. G.H. Roshani et al., Flow Meas. Instrum. **54**, 39–45 (2017)
6. P.R. Tortora et al., Int. J. Multiphas. Flow **32**, 972–995 (2006)
7. H.J. Pant, et al., Appl. Radiat. Isotop. **67,** 1609–1615 (2009)
8. R. Hanus et al., Flow Meas. Instrum. **40**, 58–63 (2014)
9. R. Hanus, Int. J. Multiphas. Flow **72,** 210–217 (2015)
10. M. Zych et al., Powder Technol. **318**, 491–500 (2017)
11. R. Hanus et al., EPJ WoC **114**, 02035 (2016)
12. J.S. Bendat, A.G. Piersol, *Random data - analysis and measurement procedures* (John Wiley, New York, 2010)
13. M.S. Beck, A. Plaskowski, *Cross correlation flowmeters - their design and application* (Adam Hilger, Bristol, 1987
14. *Guide to the expression of uncertainly in measurement.* (International Organisation for Standardisation, 1995)

AUTHOR INDEX

AUTHOR INDEX

AUTHOR INDEX

EDP Sciences – Web of Conferences
17, Avenue du Hoggar
Parc d'Activité de Courtabouf
BP 112
F-91944 Les Ulis Cedex A
France

ISBN 978-1-5108-6391-0

9 781510 863910

CURRAN ASSOCIATES INC.
proceedings.com